2011年柯珊（左）、李曙光（中）、滕方振（右）参与MC-ICP-MS仪器的安装

2011年何永胜（左）记录MC-ICP-MS的安装步骤和注意事项

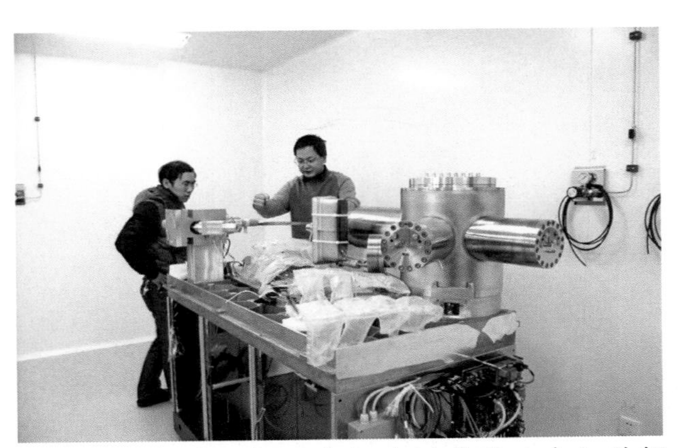

2011 年滕方振（右）为刘盛遨（左）讲解MC-ICP-MS仪器的工作原理

2011年李王晔（左）和鲁颖淮（右）帮助工程师把仪器部件运入屋内

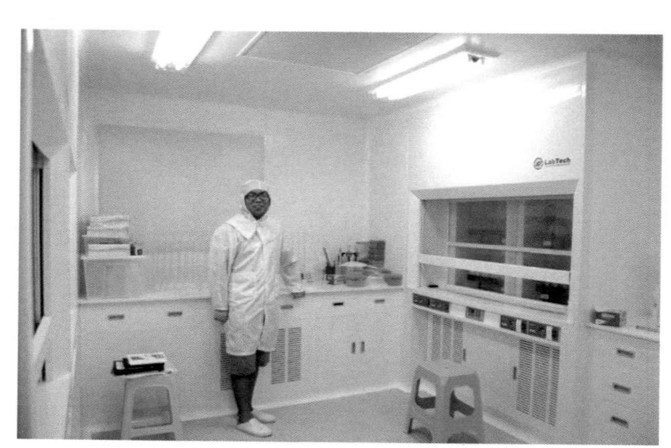

超净化学实验室的超净操作柜，样品的加热酸分解、被测元素的离子交换柱分离、纯化和溶液蒸干实验均在此操作柜内进行

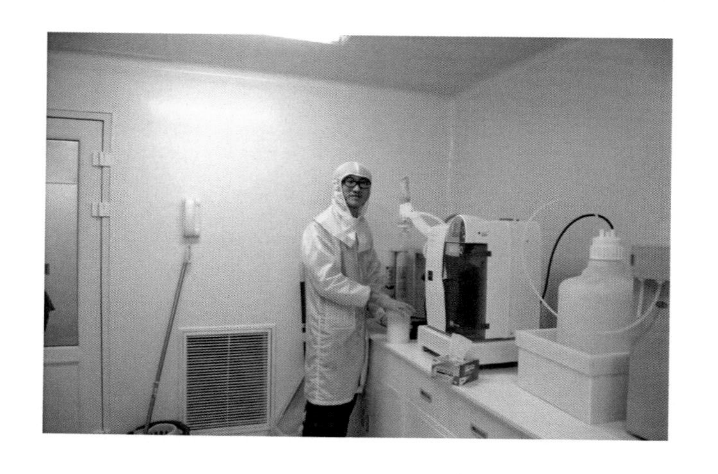

超净实验室的超纯去离子水净化设备

左上图，2014 年由国家自然科学基金委员会资助，中国地质大学（北京）举办的首届非传统稳定同位素暑期学校在中国地质大学(北京)综合楼 101 教室开课，所有师生热情洋溢，座无虚席。

左下图，李曙光给学生们分享他对非传统稳定同位素地球化学的认识过程。

右下图，中国科学技术大学黄方教授给学生讲授 V 同位素地球化学，李曙光主持讨论。

右上图，2014 年中国科学院广州地球化学研究所张兆峰在暑期学校作 Ca 同位素地球化学报告。

左下图，2014 年美国华盛顿大学滕方振在暑期学校作 Mg 同位素地球化学报告。

右下图，2014 年中国科学院广州地球化学研究所韦刚建在暑期学校作稳定 Sr-Nd 同位素体系报告。

2014 年首届非传统稳定同位素暑期学校在中国地质大学（北京）的国际会议中心闭幕，众师生合影留念

左图，2019 年李曙光在地球深部氧循环研讨会上总结发言。

下图，2019 年地球深部氧循环与大气氧升高研讨会合影。

2019 年中国地质科学院朱祥坤研究员在深部氧循环研讨会上作报告

2019 年何永胜在深部氧循环研讨会上作报告

2019 年王水炯在深部氧循环研讨会上作报告

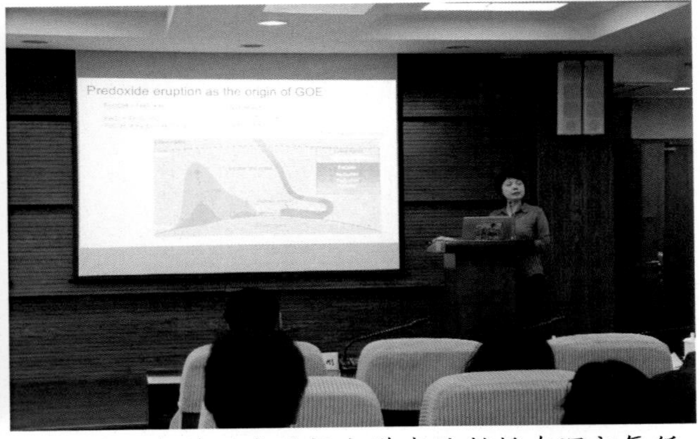

2019 年美国密歇根大学李洁教授在深部氧循环研讨会上作报告

左上图，2018 李曙光与来访美国 McDonough 教授合影。

左下图，2018 李曙光与来访美国 Lundstrom 教授合影。

右上图，2017 年滕方振（中）与他大学本科指导老师李曙光（右）和博士指导老师 Rudnick 教授（左）在中国地质大学（北京）国际会议中心合影。

左图，李曙光与其在中国地质大学（北京）指导的第一个博士生王水炯在办公室合影。王水炯博士毕业后在美国做了4年博士后，2018年作为高级人才引进回国，入职中国地质大学（北京）同位素地球化学实验室。

右图，李曙光和同事朱建明教授在办公室门口合影。

左上图，李曙光和课题组众师生开组会，报告人为何永胜。

右上图，李曙光介绍一楼的仪器间，包括 Neptune MC-ICP-MS 和 Element ICP-MS。

左下图，李曙光在办公室审阅英文文章。

2017 年李曙光与同位素地球化学实验室老师和来访美国 Rudnick 教授在中国地质大学（北京）逸夫楼前合影。

2017 年中国地质大学（北京）同位素地球化学实验室师生在逸夫楼前留下全家福合影。

RESEARCH HIGHLIGHTS

GEOSCIENCES

A store of subducted carbon beneath Eastern China

Albrecht W. Hofmann

Geochemists' toolkit for tracing the global deep carbon cycle has recently been augmented by the analysis of magnesium isotopes. Light Mg-isotopic compositions normally occur uniquely in carbonates precipitated from seawater, either in the form of sedimentary limestones/dolomites or as sea-water-precipitated alteration products impregnating the oceanic crust. But Chinese geochemists have found that large numbers of volcanic rocks from the eastern part of China also contain the tell-tale signature of marine carbonates, indicating that these have been recycled through the upper mantle and into the sources of the volcanic rocks. Li and co-authors have reviewed the available data, adding many new analyses, in order to map the regional distribution of these anomalous Mg-isotopic compositions [1]. What emerges is a remarkable isotopic map of young volcanic rocks from eastern China, showing light Mg-isotopic compositions in a swath of land extending from the east coast approximately 500 km inland and from Hainan Island in the south to the Russian border in the north. Even more remarkably, this largely matches the extent of a slab of flat-lying but deeply subducted lithosphere, which is located at a depth of 400–600 km in the so-called mantle transition zone, near the base of the upper mantle beneath eastern China [2]. This so-called 'stagnant slab' has been identified and imaged by the techniques of seismic tomography, which works in ways analogous to medical X-ray tomography but uses seismic (earthquake) waves, which traverse the Earth's interior.

The authors argue that the two phenomena are related. Subduction of altered oceanic crust introduced (originally) marine carbonate into the deep upper mantle. The carbonates that survive subduction to this depth are predominantly Mg carbonates, and their melting point is several hundred degrees lower than the melting point of silicates. They will therefore form carbonate melt, which infiltrates the overlying more ordinary mantle rocks, the so-called asthenosphere. During this process, the carbonate melt is likely to be reduced to diamond, and consequently the infiltrated rock assemblage solidifies. Ultimately, though, this material is transported to shallower levels by solid-state mantle convection until it undergoes a second stage of melting at depths between 300 and 100 km. These melts eventually reach the surface to generate the volcanoes that still remember the subducted magnesium isotopes.

A potentially important ramification is that, if it is possible to identify large mantle regions containing subducted carbonates, this will give us new information about the down-going part of the global carbon cycle. Previous estimates of carbon flux entering the convecting mantle via subduction range from essentially zero to 52 Mt carbon per year [3], and it has therefore not been clear whether the carbon exchange between mantle and exosphere is actually a cycle or a one-way process of cumulative CO_2 outgassing. Future investigations will show whether or not the present story of extensive carbon subduction will hold up elsewhere. Thus, 'stagnant slabs' have also been found in other parts of the world [4]. If similar signatures with subducted carbonate can be identified in those regions as well, we may finally get a better grip on the down-going limb of the deep-mantle carbon cycle.

Albrecht W. Hofmann
Max Planck Institute for Chemistry,
Hahn-Meitnerweg 1, Mainz, Germany
E-mail: albrecht.hofmann@mpic.de

REFERENCES

1. Li SG, Yang W and Ke S *et al. Natl Sci Rev* 2017; **4**: 111–20.
2. Huang J and Zhao D. *J Geophys Res: Solid Earth* 2006; **111**: B09305.
3. Kelemen PB and Manning CE. *Proc Natl Acad Sci USA* 2015; **112**: E3997–4006.
4. Fukao Y, Obayashi M and Nakakuki T. *Annu Rev Earth Planet Sci* 2009; **37**: 19–46.

National Science Review
4: 2, 2017
doi: 10.1093/nsr/nwx002

NSR National Science Review

2019 Best Paper

Awarded to:

Shu-Guang Li, Wei Yang, Shan Ke, Xunan Meng, Hengci Tian, Lijuan Xu, Yongsheng He, Jian Huang, Xuan-Ce Wang, Qunke Xia, Weidong Sun, Xiaoyong Yang, Zhong-Yuan Ren, Haiquan Wei, Yongsheng Liu, Fancong Meng and Jun Yan

For the paper entitled:

Deep carbon cycles constrained by a large-scale mantle Mg isotope anomaly in eastern China

This paper was published in

Natl Sci Rev (2017) 4 (1): 111-120

Chunli Bai, Editor-in-Chief
Mu-ming Poo, Executive Editor-in-Chief
Zhonghe Zhou, Associate Editor-in-Chief

15 April 2020

RESEARCH ARTICLE

National Science Review
0: 1–10, 2016
doi: 10.1093/nsr/nww070
Advance access publication 13 November 2016

GEOSCIENCES

Deep carbon cycles constrained by a large-scale mantle Mg isotope anomaly in eastern China

Shu-Guang Li[1,2,*], Wei Yang[3,*], Shan Ke[1], Xunan Meng[1], Hengci Tian[3], Lijuan Xu[1], Yongsheng He[1], Jian Huang[2], Xuan-Ce Wang[4], Qunke Xia[5], Weidong Sun[6], Xiaoyong Yang[2], Zhong-Yuan Ren[6], Haiquan Wei[7], Yongsheng Liu[8], Fancong Meng[9] and Jun Yan[10]

[1]State Key Laboratory of Geological Processes and Mineral Resources, China University of Geosciences, Beijing 100083, China; [2]CAS Key Laboratory of Crust-Mantle Materials and Environments, School of Earth and Space Sciences, University of Science and Technology of China, Hefei 230026, China; [3]Key Laboratory of Earth and Planetary Physics, Institute of Geology and Geophysics, Chinese Academy of

应用Mg同位素示踪发现中东部大地幔楔是一个巨大的俯冲碳酸盐储库。该文被世界著名地球化学家A.W.Hofmann给予亮点评论，并被NSR评为2019最佳论文

李曙光院士论文选集

(卷二)

金属稳定同位素示踪深部碳循环研究

李曙光　刘盛遨　杨　蔚 等 编

科 学 出 版 社

北　京

内 容 简 介

本卷有选择地汇集了李曙光研究团队2012～2022年在金属稳定同位素示踪深部碳循环方面的主要学术论文。本卷汇集的论文涉及金属同位素分析测试方法的建立，标定地质储库的同位素组成和地质过程的同位素分馏，以及应用Mg-Zn-Fe等金属同位素示踪深部碳-氧循环的实例研究等，探讨了俯冲和岩浆过程中金属同位素行为，不同俯冲带和不同俯冲深度的深部碳循环过程以及再循环碳酸盐种属及通量等许多重要科学问题。这些论文大部分发表在高水平期刊上，引用率高。其中论文（Li et al.，2017，NSR）被NSR评为2019最佳论文。

本文集所收集论文源自多种学术期刊，各源刊格式标准可能不统一，本着尊重历史、忠于原文的精神，所用物理量单位、符号、图例、参考文献等尽量保留了原文风貌。

本书可供地质学领域科技工作者和相关领域教师、研究生与本科生参考。

审图号：GS京（2024）2233号

图书在版编目（CIP）数据

李曙光院士论文选集. 卷二, 金属稳定同位素示踪深部碳循环研究 / 李曙光等编. -- 北京 : 科学出版社, 2024.11

ISBN 978-7-03-077563-4

Ⅰ. ①李… Ⅱ. ①李… Ⅲ. ①地球物理学-文集②同位素年代学-文集③碳循环-文集 Ⅳ. ①P3-53②P597-53③X511-53

中国国家版本馆CIP数据核字（2024）第013770号

责任编辑：焦 健 / 责任校对：何艳萍
责任印制：肖 兴 / 封面设计：北京图阅盛世

科学出版社 出版
北京东黄城根北街16号
邮政编码：100717
http://www.sciencep.com
北京建宏印刷有限公司印刷
科学出版社发行 各地新华书店经销
*
2024年11月第 一 版 开本：889×1194 1/16
2024年11月第一次印刷 印张：59 3/4 插页：6
字数：1920000

定价：698.00元

（如有印装质量问题，我社负责调换）

《李曙光院士论文选集》（卷二）

编委会名单

前　　言

我自 2008 年开始受聘中国地质大学（北京）兼职教授，并在 2012 年正式聘为中国地质大学（北京）全职教授，学校领导要求我在中国地质大学（北京）负责建设一个一流的同位素地球化学实验室。20 世纪末，由于多接收电感耦合等离子体质谱仪的成功研制，过去难于精确测量的金属同位素组成得以精确测量。这促使非传统金属稳定同位素地球化学研究得以快速发展，成为同位素地球化学新的前沿研究领域。当时我国在该领域研究刚刚起步。鉴于此，我决定该新建同位素地球化学实验室以开展非传统金属稳定同位素地球化学研究为主要目标。我们与美国华盛顿大学滕方振教授开展深度合作，并于 2011 年起，该实验室陆续建立了 Mg、Fe、Ca、Zn、Cu、Cr、Ni 等同位素分析方法。随后，深部碳循环及其对地球宜居环境的形成和演化的影响成为 21 世纪地球科学重要的前沿研究领域，如何应用同位素示踪深部碳循环引起我的兴趣。受美国哈佛大学黄士春博士等在 *Geochimica et Cosmochimica Acta* 发表的首次应用 Ca 同位素示踪夏威夷玄武岩源区有再循环的俯冲碳酸盐论文的启发，我在中国地质大学（北京）新建的同位素实验室和研究团队自 2012 年起在国际上率先开展 Mg、Zn、Fe 稳定同位素示踪深部碳循环研究。近十年，在国家自然科学基金两个重点项目和一个国家科技部变革性技术重大专项的资助下，本研究团队已在金属稳定同位素示踪深部碳循环领域取得一系列重要成果。本学术论文选集第二卷汇集了近十年（2012～2022）李曙光院士团队在示踪深部碳循环相关的金属稳定同位素分析方法、地质储库的同位素组成、地质过程的同位素分馏和应用 Mg、Zn、Fe 示踪深部碳循环等四个领域发表的论文 49 篇。这些工作开拓并完善了 Mg、Zn 同位素用作示踪深部碳循环的工具，并且至少在以下八个方面增加和改进了我们对深部碳循环的认知。

（1）板块俯冲过程中俯冲洋壳和碳酸盐的 Mg、Ca 同位素地球化学行为。通过对来自同一俯冲板片、不同俯冲深度和变质程度的基性变质岩（绿片岩、斜长角闪岩和榴辉岩）的 Mg 同位素调查，发现这一进变质和脱水过程不能改变玄武岩的 Mg 同位素组成。而俯冲变质的碳酸盐（大理岩）与它裹挟的变玄武岩（榴辉岩）之间可发生 Ca-Mg 元素及 Mg 同位素交换。该交换反应可导致纯方解石碳酸盐变成含镁方解石或白云质碳酸盐，并提升其 δ^{26}Mg 值（从–4.0‰升到>–2.5‰，但<–0.5‰）；而碳酸盐化榴辉岩的 δ^{26}Mg 值可降低至–1.9‰。这一板块俯冲过程中碳酸盐岩和变玄武岩的 Mg 同位素地球化学行为调查为应用 Mg 同位素示踪深部碳循环奠定了理论基础。其重要意义有两个方面：①由于俯冲碳酸盐与周围的镁铁或超镁铁质岩石的 Ca-Mg 交换，部分富 Ca 碳酸盐转变为含 Mg 或富 Mg 碳酸盐，降低了其在俯冲板片析出水流体中的溶解度，从而得以随俯冲板片进入深部地幔；②碳酸盐化榴辉岩的轻 Mg 同位素组成是应用 Mg 同位素示踪深部碳循环的干扰因素，需要加以甄别。

（2）发现东亚大地幔楔是一个巨大的俯冲碳酸盐储库及两个深部碳循环圈。对中国东部以及俄罗斯远东年龄小于 106Ma 的晚白垩世和新生代来自对流上地幔的大陆玄武岩的 Mg、Zn 同位素填图，发现东亚北起俄罗斯远东，南至中国福建的对流上地幔存在大尺度轻 Mg 同位素和重 Zn 同位素异常。该 Mg-Zn 同位素异常区与西太平洋俯冲板片在地幔过渡带滞留分布区在空间上高度吻合，显示这一对流上地幔大尺度轻 Mg-重 Zn 同位素异常的成因与地幔过渡带滞留的西太平洋俯冲板片有关。结合它们的主-微量元素及 Mg-Sr 同位素特征，证明该 Mg-Zn 同位素异常是富镁碳酸盐（菱镁矿+白云石）随俯冲洋壳进入地幔过渡带并发生熔融，其熔体交代上覆的对流上地幔形成的。因此，由西太平洋俯冲板片在东亚大陆下面的地幔过渡带滞留形成的大地幔楔是一个巨大的俯冲碳酸盐储库。对巴尔干半岛源自地幔过渡带滞留俯冲板片的富钠碱性玄武岩的最新研究表明它们同样具有轻镁重锌的同位素特征，证明俯冲板片形成的大地幔楔是俯冲碳酸盐储库是普遍现象。

（3）发现俯冲 Ca、Mg 碳酸盐在板片析出水流体和超临界流体的溶解度差异。岛弧玄武岩的 Mg 同位素组成与地幔值类似或略重，没有轻 Mg 同位素异常。根据不同碳酸盐种属在高压水热流体中溶解度的差异，提出俯冲板片在岛弧区较浅部位（70～120km）析出的水流体优先选择溶解 Ca 质碳酸盐，因此不能改造岛弧区地幔楔的 Mg 同位素组成。此外还发现俯冲板片在>150km 深度析出的超临界流体可溶解白云石类含镁碳酸盐，它交代的深部地幔楔（如大别山毛屋超镁铁岩块）及其部分熔融产生的玄武质岩浆（如腾冲火山岩）具有轻 Mg 同位素特征。据此，提出岛弧系统和大地幔楔系统三个深部碳循环圈及各自 Mg、Zn 同位素特征模型。

（4）发现深俯冲碳酸盐部分还原成金刚石并氧化地幔硅酸盐的歧化反应。通过铁同位素研究揭示了具有轻镁同位素组成的中国东部霞石岩、碧玄岩、碱性玄武岩系列有异常重的 Fe 同位素组成和异常高的 $Fe^{3+}/\sum Fe$，且与 $\delta^{26}Mg$ 值呈负相关，论证了该高氧逸度硅酸盐岩浆是由于深俯冲（>250km）碳酸盐还原成金刚石，并氧化了金属铁和硅酸盐的二价铁离子。此外，通过强亲铁元素研究还揭示了这一歧化反应还氧化了地幔硫化物。这些发现揭示了深俯冲碳酸盐驱动的氧循环，这对维持显生宙大气氧含量由重要作用。

（5）发现华北克拉通岩石圈地幔部分碳酸盐化，并在其减薄过程中释放碳。山东省胶东地区 125Ma 的高镁煌斑岩具有轻 Mg 同位素组成，其熔体包体有很高的 CO_2 含量。据此提出华北克拉通破坏及岩石圈减薄可导致克拉通岩石圈地幔储存的碳高通量释放，促进了早白垩世全球温室效应。

（6）发现新特提斯造山带的高钾碱性玄武岩具有 Mg-Zn 同位素解耦特征。发现云南-缅甸西部，青藏高原雅江缝合带和环地中海的高钾碱性玄武岩具有 Mg-Zn 同位素解耦特征，即 Mg 同位素轻，但 Zn 同位素组成与地幔类似。该特征证明这种高钾碱性玄武岩的源区是被俯冲的含碳酸盐泥质沉积物交代或混杂的岩石圈地幔。

（7）发现大地幔楔碱性玄武岩是霞石岩超碱性熔体与岩石圈相互作用的产物。对华北克拉通东部的新生代陆内超碱-碱性玄武岩系列的 Mg-Zn 同位素和痕量元素地球化学研究揭示了大地幔楔的碳酸盐化地幔熔融产生的富 CO_2 超碱性霞石岩岩浆，上升到岩石圈底部与岩石圈相互作用可导致霞石岩岩浆演化为碱性玄武岩岩浆，并导致岩石圈地幔的进一步减薄。

（8）发现内蒙古大兴安岭重力梯度带以西的新生代玄武岩也具有轻 Mg 同位素特征，其成因与古亚洲洋板片南东向俯冲输送的俯冲碳酸盐有关。上述已获得的对深部碳循环的认知主要还是定性的。未来需要做更深入的定量研究以解决深部碳循环进出地幔的通量问题。鉴于目前国内外已有越来越多的同位素地球化学实验室和研究团队开展了应用金属稳定同位素示踪深部碳循环的研究，期望这本文集有助于推动该领域研究工作的深入和发展。

目　　录

第四部分　金属稳定同位素示踪深部碳循环

第一部分　非传统稳定同位素分析方法

High-precision copper and iron isotope analysis of igneous rock standards by MC-ICP-MS*

Sheng-Ao Liu[1], Dandan Li[1], Shuguang Li[1], Fang-Zhen Teng[1,2], Shan Ke[1], Yongsheng He[1] and Yinghuai Lu[1]

1. State Key Laboratory of Geological Processes and Mineral Resources, China University of Geosciences, Beijing 100083, China
2. Isotope Laboratory, Department of Earth and Space Sciences, University of Washington, Seattle, WA 98195, USA

亮点介绍：文中介绍了一种新的实验方法，使用强阴离子树脂 AG-MP-1M，在同一流程内分离火成岩Fe和Cu同位素，获得长期外精度δ^{65}Cu的±0.05‰(2SD)和δ^{56}Fe的±0.049‰(2SD)。应用此方法测量了目前火成岩的Fe、Cu同位素组成，结果显示火成岩δ^{65}Cu值在−0.01‰至+0.39‰（n=11），表明火成岩Cu同位素组成不均匀。Fe、Cu同位素同一流程分离和高精度Fe、Cu同位素比值测定对于在地质和生物领域中Fe、Cu同位素的研究具有巨大优势。

Abstract Stable isotopic systematics of Cu and Fe are two important tracers for geological and biological processes. Generally, separation of Cu and Fe from a matrix was achieved by two independent, completely different methods. In this study, we report a method for one-step anion-exchange separation of Cu and Fe from a matrix for igneous rocks using strong anion resin AG-MP-1M. Cu and Fe isotopic ratios were measured by multi-collector inductively coupled plasma mass-spectrometry (Neptune plus) using a sample–standard bracketing method. External normalization using Zn to correct for instrumental bias was also adopted for Cu isotopic measurement of some samples. In addition, all parameters that could affect the accuracy and precision of isotopic measurements were examined. Long-term external reproducibility better than ±0.05‰ (2SD) for δ^{65}Cu and ±0.049‰ (2SD) for δ^{56}Fe was routinely obtained. Cu and Fe isotopic compositions of commercially accessible igneous rock standards including basalt, diabase, amphibolite, andesite and granodiorite were measured using this method. δ^{65}Cu values of igneous rock standards vary from −0.01‰ to +0.39‰ (n = 11) with an overall range (0.40‰) that exceeds about 8 times that of the current analytical precision. The improved precisions of stable Cu isotopic analysis thus demonstrate that igneous rocks are not homogeneous in Cu isotopic composition. The procedure for one-step separation of Cu and Fe and high-precision analysis of Cu and Fe isotopic ratios have an important advantage for economical and efficient study of stable Cu and Fe isotopic systematics in geological and biological fields.

* 本文发表在：Journal of Analytical Atomic Spectrometry, 2014, 29: 122-123

1 Introduction

Copper is a transition metal and has two stable isotopes of mass 63 and 65, whose average abundances are 69.17% and 30.83%, respectively.[1] Natural variations in stable Cu isotope abundances have been observed since the 1950s.[2] The search by Walker et al. using thermal ionization mass spectrometry (TIMS) had identified a total range of ~9‰ in $^{65}Cu/^{63}Cu$ ratios in Cu-rich minerals, sediments and organic samples, although the precision (1–1.5‰) they achieved was relatively poor compared with the modern standard. In the past decade, lots of high-precision Cu isotopic data obtained using MC-ICP-MS have been reported.[3-12] These new results have provided profound insights into processes that link planetary accretion in high-temperature environments as well as inorganic and biological chemistry in low temperature supergene environments. However, Cu isotopic ratios of most igneous rocks span a narrow range of <±0.4‰.[12] Application of Cu isotopes to high-temperature geochemistry thus requires high-precision and accurate analyses.

In the recent decade, Fe isotope geochemistry has gained particular interest due to its relatively high planetary abundance, multiple redox states and biological utilization. For instance, over 5 ‰ $\delta^{56}Fe$ variation has been observed during low temperature geological and biological processes.[13-15] High-temperature equilibrium Fe isotope fractionation is, however, limited as well. Recently, high-precision Fe isotope data better than ±0.03‰ have been obtained using high-resolution (HR) MC-ICP-MS.[16]

With the development of analytical precision, the combined utilization of Cu and Fe isotopes as geochemical and biological tracers has been recently undertaken on both experimental and field work.[6,17,18] The combined use of Cu and Fe isotopes has an important advantage because Cu and Fe behave distinctly in several aspects. For example, Cu(II) is more fluid-mobile than Cu(I) and Fe(III) is less mobile than Fe(II). This difference may result in contrasting behaviors of Cu and Fe isotopes during mineral dissolution. Typically, Cu and Fe in rocks or aqueous solutions were purified by two independent, completely different methods, using anion resin AG-MP-1 and AG-X4 or X8 respectively. In an original paper, Maréchal et al.[4] managed one-step anion-exchange separation of Cu and Fe using strong anion resin AG-MP-1 by involving stepwise decreases in concentrations of hydrochloric acid. Regrettably, they did not measure Fe isotopes along with Cu isotopes after chemical purification, and thus the quality of Fe isotope data obtained using this method was unknown. In a recent paper, Borrok et al.[6] outlined a method to separate Cu and Fe through a single anion-exchange column and measure Cu and Fe isotopic ratios in complex aqueous solutions. They obtained 2σ precisions better than ~±0.1‰ for Cu and Fe isotopic analysis. Because most rocks have remarkably different chemical compositions from aqueous solutions, the procedures for isolation of Cu and Fe from a matrix could be different for rocks and aqueous systems. To date, no systematic study has been carried out to separate Cu and Fe in a single column for rocks and to measure their isotopic compositions with high precision. In addition, Cu cannot be completely separated from Fe using the AG-X4 or X8 resin, but high Cu/Fe (>20) can cause significant offset on $\delta^{56}Fe$ ratio analysis of >0.2‰.[16] Therefore, accurate analysis of Fe isotopes on samples with high Cu/Fe (e.g., Cu-rich sulfides) is impossible using the general procedure.

In this paper, we report a method for one-step anion-exchange separation of Cu and Fe from matrix elements for igneous rocks using strong anion resin AG-MP-1M. We measured both Cu and Fe isotopic compositions of eleven commercially accessible igneous rock standards (e.g., BHVO-2, BIR-1 and BCR-2, etc.) using this method. All parameters that could potentially affect the quality of isotopic analysis were well evaluated. A long-term external reproducibility of better than ±0.05‰ (2SD) for $\delta^{65}Cu$ and $\delta^{56}Fe$ measurements has been obtained.

2 Analytical methods

The detailed procedures for sample dissolution, column chemistry and instrumental analysis are presented in the following three separate sections.

2.1 Sample dissolution

Copper concentration in the analyzed igneous rock standards mostly varies from ~20 to ~200 μg g^{-1}, and thus 10–20 mg samples were weighed to contain at least ~0.4 μg Cu for isotopic analysis. The amounts of iron in the weighed samples are typically <1 mg. The samples were dissolved in a 1 : 1 (v/v) mixture of double-distilled HF and HNO_3 in Savillex screw-top beakers, followed by heating at 160 ℃ on a hotplate in an exhaust hood (Class 100). The solutions were dried down at 150 ℃ to expel the fluorine. The dried residues were refluxed with a 1 : 3 (v/v) mixed HNO_3 and HCl, followed by heating and then evaporating to dryness at 80 ℃. The samples were refluxed with concentrated HNO_3 until complete dissolution was achieved, and subsequently dried down at 80 ℃. 1 ml of 8 N HCl + 0.001% H_2O_2 was added to the beaker and the sample was heated to dryness. This process was repeated three times to ensure that all cations were converted to chloride species. The final material was dissolved in 1 ml of 8 N HCl + 0.001% H_2O_2 in preparation for ion-exchange separation.

2.2 Ion-exchange chromatography

The chemical purification method in this study is modified from Maréchal et al. [4] The major difference between this study and Maréchal et al.[4] is that we used 8 N HCl for Cu separation from matrix elements instead of 7 N HCl used by Maréchal et al.[4] Bio-Rad AG-MP-1M strong anion exchange resin (100–200 mesh; chloride form) was used for separation of Cu and Fe from matrix elements. The resin was pre-cleaned with 0.5 N HNO_3 and 8 N HCl alternating with MQ H_2O (18.2 MΩ) 12 times. The pre-cleaned column (4 mm in diameter and 9 cm long; Poly-Prep Chromatography) was filled with pre-cleaned AG-MP-1M resin, and washed with 7 ml 0.5 N HNO_3 and 5 ml 8 N HCl alternating with MQ H_2O three times. The volume of resin was adjusted to 2 ml in 8 N HCl. 7 ml of 8 N HCl was added to the column for conditioning and then samples dissolved in 1 ml 8 N HCl (+0.001% H_2O_2) were loaded onto the column. Matrix elements (e.g. Na, Mg, Al, K, Ca, Ti, Cr, Ni and Mn) were eluted in the first 10 ml 8 N HCl, leaving Fe, Co, Cu and Zn retained on the resin. The one-step anion-exchange chromatographic method, involving stepwise decreases in concentrations of hydrochloric acid, can separate Cu and Fe from other ions.[4] Copper was collected in the following 24 ml of 8 N HCl. Iron fraction was collected in the following 18 ml of 2 N HCl. Analysis of basalt and granodiorite samples yields consistent elution curves (Fig. 1), suggesting that this method is suitable to samples with various matrix compositions. Both the Cu and Fe fractions were evaporated to dryness, dissolved in 3% (m/m) HNO_3, and then re-evaporated to dryness and re-dissolved in 3% HNO_3 to remove all chlorine prior to isotopic ratio analysis.

Total procedural blanks (from sample dissolution to mass spectrometry) were routinely measured and had a long-term average of ~1.5 ng (1–2 ng, n=10) for Cu and ~6 ng (2–10 ng; n=10) for Fe, which are considered neglected during mass spectrometry. The contribution from blank is still insignificant when the amount of Cu loaded is as low as ~0.4 μg (see Section 3.4).

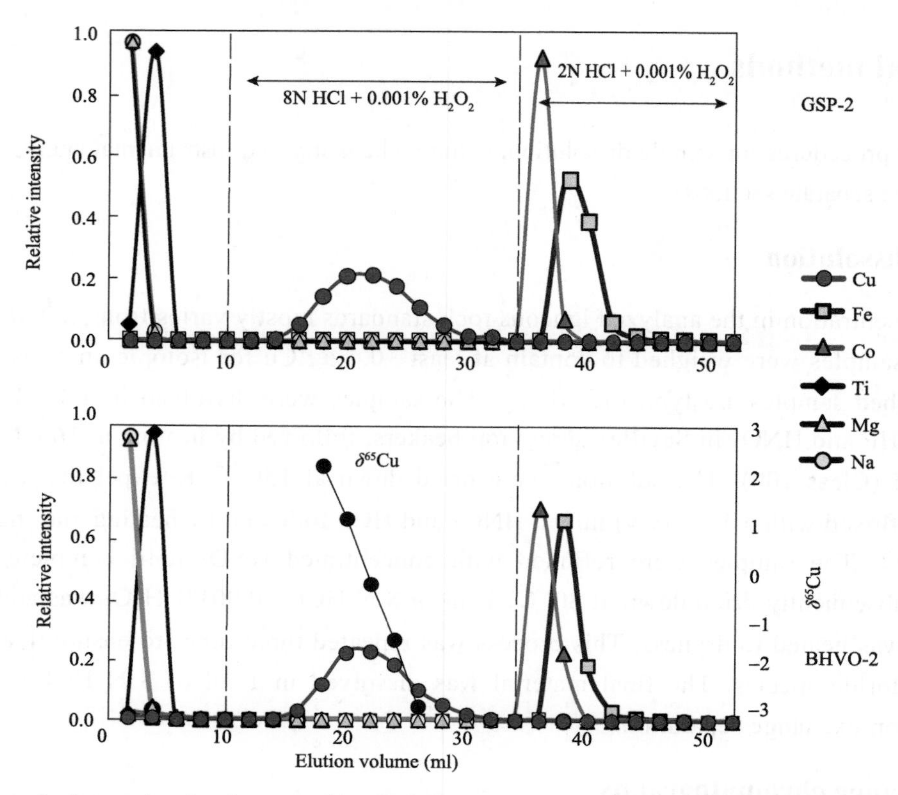

Fig. 1 Elution curves for international rock standard basalt (BHVO-2; 127 μg g^{-1} Cu and 8.36 wt.% Fe) and granodiorite (GSP-2; with 43 μg g^{-1} Cu and 3.43 wt.% Fe) on a 2 ml resin bed of AG-MP-1M. Cu and Fe yields are 99.7±0.8% (2SD, *n*=5) and 99.9±0.6% (2SD, *n*=5) respectively. The Cu cuts eluted from BHVO-2 were separately analyzed for Cu isotopic ratios to evaluate whether or not there is isotope fractionation during chemical purification. Note that cobalt was completely separated from Cu using the 8 N HCl medium.

2.3 Mass spectrometry

Copper and iron isotopic ratios were measured by a sample-standard bracketing method using a Thermo-Finnigan Neptune plus MC-ICP-MS at the Isotope Geochemistry Laboratory of the China University of Geosciences, Beijing. The sample-standard bracketing (SSB) method has been successfully used for Cu and Fe isotopic analysis.[6,9,16,19] Element-doping using Zn or Ni as an external standard was also commonly adopted to correct for instrumental mass bias during Cu isotopic analysis.[1,3,4] We performed external normalizing using Zn-doping with an aim to evaluate whether there is significant difference in precisions between the two methods. The results will be discussed in Section 3.5.

The instrument is equipped with a Cetac ASX-110 automatic sampler and a PFA Teflon self-aspirating micronebulizer system. The mass spectrometry parameters are outlined in Table 1. Prior to sample introduction, samples and standards were diluted to produce ~100 ppb Cu solution and ~1 ppm Fe solution in 3% (m/m) HNO_3 respectively. The uptake rate was ~50 or 100 μl min^{-1}, and no difference in accuracy and precision was found at different uptake rates. The take-up time was 80 s. Prior to each analysis sequential rinses of two separate 3% HNO_3 of 100 s were used to reduce baselines to <1 mV on the ^{63}Cu and ^{56}Fe channels.

The sampler and skimmer cones are made of Ni, and the high-sensitivity (X) cones are used to increase transmission by a factor of 2−3 relative to the routine H-cones. For example, the ^{63}Cu signal was typically ~6 V/ 100 ppb when we used the X-cone. The high sensitivity allows samples containing ~0.2 μg Cu to be measured for at least four blocks of 40 cycles each (100 ppb in 2 ml solution). This is particularly important for measurement of samples with a small amount of Cu but a large amount of Fe, which is true for most silicates and Fe-sulfides. Otherwise, this needs considerable amounts of digested rocks, which may exceed the loading capacity of the column. Cu isotopic ratios were analyzed in low-resolution mode with ^{63}Cu in the Central cup and ^{65}Cu in the H2 Faraday cup. A measurement consists of at least four blocks of 40 cycles of ~10 s each, and thus each value reported is the average of at least 160 ratios. Cu isotopic data are reported in standard δ-notation in per mil relative to standard reference material (SRM) NIST 976:

$$\delta^{65}Cu = ((^{65}Cu/^{63}Cu)_{sample}/(^{65}Cu/^{63}Cu)_{NIST\ 976} - 1) \times 1000$$

Table 1 Neptune plus operating conditions for Cu isotopic ratio measurements[a]

Instrument parameters	
Rf power	1250 W
Cooling Ar	~16 l min^{-1}
Auxiliary Ar	~1.0 l min^{-1}
Nebuliser Ar	~1.0 l min^{-1}
Extraction voltage (hard)	−2000 V
Vacuum	4–8×10^{-9} Pa
Cu sensitivity	~60 V ppm^{-1} (LR)
Fe sensitivity	~10 V ppm^{-1}(HR)
Cones	Ni (X)
Sample uptake	~50 μl min^{-1}

[a] LR: low-resolution; HR: high-resolution.

Iron isotopic ratios were measured in high-resolution mode ($M/\Delta M$ = ~10 000). ^{53}Cr, ^{54}(Fe, Cr), ^{56}Fe, ^{57}Fe, ^{58}(Fe, Ni) and ^{60}Ni isotopes were measured in the static mode by Faraday cups at Low 3, Low 1, Central, High 1, High 2 and High 4 positions, respectively. The measured ^{53}Cr was used to correct any ^{54}Cr interference on ^{54}Fe. The ^{56}Fe signal was ~10 V for the analyzed 1 ppm solution using the X-cone. The ^{54}Fe signal is typically >500 mV which is important to obtain high-precision iron isotopic measurement.[16] A measurement consists of four blocks of 40 cycles of ~8 s each. Fe isotope data are reported in standard δ-notation in per mil relative to the reference material IRMM-014, as follows:

$$\delta^{x}Fe = ((^{x}Fe/^{54}Fe)_{sample}/(^{x}Fe/^{54}Fe)_{IRMM-014} - 1) \times 1000$$

where x refers to mass 56 or 57.

3 Accuracy and precision check

In the following sections, we address several important parameters that can lessen the quality of Cu and Fe isotopic analysis. These parameters include incomplete recovery, the effects of matrix elements (Co, Na, Ti and Fe) on the instrumental mass bias, amounts of loaded Cu and Fe, storage of Cu and Fe standard solution, acid molarity, and Cu–Fe concentration of samples.

3.1 Incomplete recovery

Significant Cu and Fe isotope fractionations can occur during ion-exchange chromatography due to incomplete recovery of Cu (ref. 9 and 20) or Fe.[21] Similar ion-exchange fractionation has also been found for other metal isotopes, e.g., Mg.[22,23] Cu and Fe that were eluted earlier were always isotopically heavier than those eluted later, probably reflecting isotope fractionation between the resin bound and the free Cu or Fe species. We obtained similar results by analyzing the Cu cuts at a 2 ml interval eluted from USGS basalt standard BHVO-2 (Fig. 1). For example, Cu that occurs at the 17–18 ml cut is >5‰ heavier in δ^{65}Cu than that eluted at the 25–26 ml cut (Fig. 1). Calculation with the fractionation factor shows that ~90% recovery will produce up to 0.4‰ shift of measured δ^{65}Cu values relative to the true value. Similar isotope fractionation was observed during Fe elution. Therefore, to reduce the impact from chemical purification and achieve accurate Cu and Fe isotope data, complete recovery must be achieved.

The Cu and Fe recovery in this study has been estimated in two ways. The first was to collect the Cu cut (total 24 ml) or Fe (18 ml) eluted from natural samples (BHVO-2 and GSP-2) and then compare them with the total Cu or Fe signal in all cuts (52 ml; Fig. 1). This way yielded a recovery of Cu = 99.7±0.8% (2SD, n = 5) and Fe = 99.9±0.6% (2SD, n = 5). The second was to purify a given amount of pure Cu and Fe solutions and check the yields. This yielded a recovery of Cu = 99.9±0.5% (2SD, n = 9) and 100.4±0.8% (2SD, n = 9). Clearly, both methods yielded complete recovery for Cu and Fe during chemical purification.

3.2 Acid molarity and concentration mismatch

The possible influence of acid molarity and Cu or Fe concentration of samples and standards on isotopic analysis must be evaluated when using the sample–standard bracketing method. The effect of acid molarity on Cu isotopic analysis was evaluated by changing the acid molarity of samples (NIST 976 was used here) at certain acid molarity of bracketing standards (3% HNO_3; 0.325 N) and the same Cu concentration (100 ppb). The results indicate that for high acid molarity the observed effect is a shift towards heavy isotopic composition (Table 2). For example, a 10% difference of acid molarity between samples and standards caused a shift of δ^{65}Cu values by larger than 0.3‰. To eliminate the effect of acid molarity, the same batch of newly made 3% HNO_3 was always used for samples and bracketing standards in this study.

Table 2 Test of influences of matrix elements and concentration mismatch on Cu and Fe isotopic analysis

Name	Ti/Cu	δ^{65}Cu	2SD	Name	Co/Cu	δ^{65}Cu/δ^{56}Fe	2SD/2SD
Ti doping test				**Co doping test**			
Ti1	0.001	−0.01	0.05	Co1	0.001	0.00/−0.01	0.04/0.05
Ti2	0.01	−0.02	0.05	Co2	0.01	0.01/0.01	0.04/0.04
Ti3	0.1	0.00	0.03	Co3	0.1	0.03/0.00	0.04/0.04
Ti4	0.3	0.07	0.04	Co4	0.5	−0.01/−0.02	0.06/0.02
Ti5	0.5	0.14	0.05	Co5	1	0.02/−0.03	0.02/0.03
Ti6	1.0	0.28	0.05	Co6	2	0.04/−0.03	0.05/0.04
Ti7	10	3.20	0.07	Co7	5	0.05/0.01	0.02/0.04
				Co8	10	0.05	0.04
Na doping test	**Na/Cu**			**Fe doping test**	**Fe/Cu**		
Na1	0.1	0.00	0.04	Fe1	0.1	−0.02	0.07
Na2	0.5	−0.01	0.06	Fe2	0.5	−0.02	0.04
Na3	1	−0.02	0.03	Fe3	1	−0.05	0.03
Na4	1.2	−0.04	0.03	Fe4	2	−0.01	0.02
Na5	1.5	−0.05	0.04	Fe5	4	−0.23	0.03
Na6	2	−0.08	0.04				
Na7	5	−0.18	0.06				

Continued

Name	Ti/Cu	δ^{65}Cu	2SD	Name	Co/Cu	δ^{65}Cu/δ^{56}Fe	2SD/2SD
Concentration match test	**Without on-peak zero correction**			**Concentration match test**	**On-peak zero correction**		
Test	$C_{sample}/C_{standard}$			Test	$C_{sample}/C_{standard}$		
Cm-1	0.1	−4.39	0.06	Cm-1	0.1	−0.01	0.04
Cm-2	0.2	−2.03	0.06	Cm-2	0.2	0.05	0.07
Cm-3	0.5	−0.55	0.04	Cm-3	0.5	0.01	0.04
Cm-4	0.7	−0/31	0.03	Cm-4	0.7	0.03	0.05
Cm-5	0.8	−0.15	0.02	Cm-5	0.8	−0.02	0.04
Cm-6	0.9	−0.01	0.06	Cm-6	0.9	0.01	0.05
Cm-7	1.01.1	0.02	0.05	Cm-7	1	0.01	0.06
Cm-8	1.2	0.08	0.02	Cm-8	1.1	0.00	0.06
Cm-9	1.3	0.10	0.05	Cm-9	1.2	0.01	0.06
Cm-10	1.5	0.16	0.03	Cm-10	1.5	0.00	0.04
Cm-11	2	0.20	0.02	Cm-11	1.6	0.03	0.06
Cm-12	5	0.32	0.06	Cm-12	2	0.01	0.06
Cm-13		0.47	0.06	Cm-13	5	0.00	0.06
Acid match	**Acid molarity (sample/std.)**			**Acid match**	**Acid molarity (sample/std.)**		
AM-1	0.3	−2.04	0.07	AM-6	1.1	0.33	0.05
AM-2	0.5	−1.16	0.06	AM-7	1.2	0.44	0.06
AM-3	0.7	−0.72	0.07	AM-8	1.3	1.07	0.07
AM-4	0.8	−0.41	0.05	AM-9	1.7	1.1	0.07
AM-5	1.0	0.04	0.05				

The effect of the Cu concentration on Cu isotopic analysis was evaluated by changing the Cu concentration of samples (NIST 976 was used here) at certain Cu concentrations of bracketing standards (100 ppb). The results demonstrate that imperfect concentration match (>10%) can largely affect the accuracy of Cu isotopic measurements (Fig. 2a). The positive correlation between δ^{65}Cu and the concentration ratio of sample to standard ($C_{sample}/C_{standard}$) suggests a small interference on ^{63}Cu relative to ^{65}Cu when concentrations of the sample and standard are inconsistent. We modeled the effect by assuming no interference on ^{65}Cu. Consequently, δ^{65}Cu values of NIST 976 Cu standard solutions relative to the standard itself can be calculated as follows:[24]

$$\delta^{65}\text{Cu} = 1000 \times f \times (R - 1)/(R + f)$$

where R is the ratio of Cu concentration in "sample" relative to "standard" and f is fractional contribution of interference signals on ^{63}Cu. The modeling results obtain a best fitting with measured data with $f = 0.00055$ (Fig. 2a), indicating that even a small interference (<3 mV) on ^{63}Cu (see Section 3.3 for possible molecular spectral interferents) can cause large Cu isotopic offset.

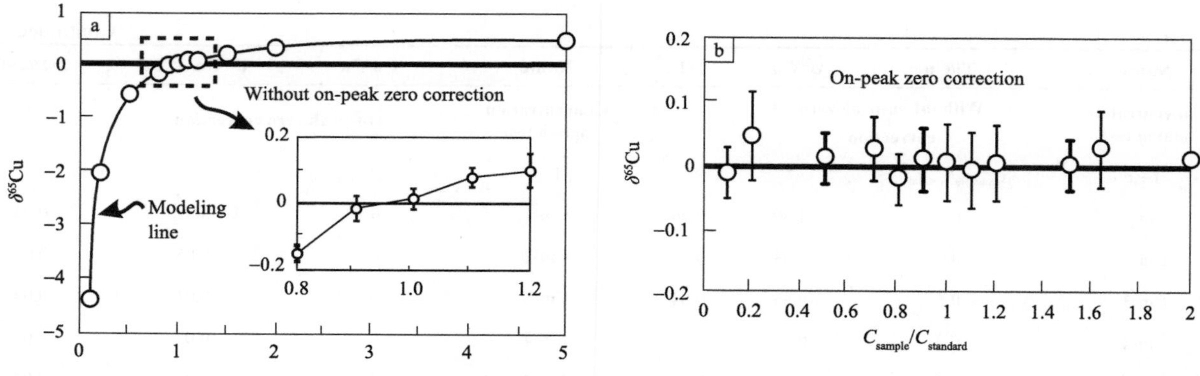

Fig. 2　Cu isotopic ratio variation of pure Cu standard solutions (NIST 976) with changing Cu concentrations relative to the bracketing standard (NIST 976) with certain Cu concentration (100 ppb). Results obtained without on-peak zero correction (upper figure) and those obtained with on-peak zero correction (lower figure) are presented for comparison. The bold line in the upper figure indicates the modeling results by assuming that there is only interference on ^{63}Cu. See text for details. The errors (2SD) were calculated on the basis of four times replicate measurements in an analytical session. Data are reported in Table 2.

Nevertheless, when on-peak zero (OPZ) correction was applied, up to 90% concentration difference between samples and standards yields results which are still close to zero within analytical uncertainty (Fig. 2b). This suggests that limited interference on ^{63}Cu can be effectively offset when blank contribution from acid was reasonably corrected. However, given that the composition of blanks may greatly vary with time, this correction may not be prevalent under different working conditions. For Fe isotopic analysis, up to 80% concentration difference between samples and standards also yields consistent results within analytical uncertainty when OPZ correction was applied. During the course of sample analysis, the concentration of Cu or Fe in samples is strictly set within ±10% of the standards and OPZ correction was always used.

3.3　Matrix effect

The potential molecular spectral interferents for Cu isotopic analysis include ^{47}Ti^{16}O, ^{23}Na^{40}Ar on ^{63}Cu, ^{25}Mg^{40}Ar and ^{47}Ti^{18}O on ^{65}Cu, etc.[8] We found that Cu isotopic analysis in our working conditions is sensitive to the presence of matrix Na, Fe and Ti (Fig. 3a–c). The influence of Na on Cu isotopic analysis towards a light isotopic composition is likely a result of (^{23}Na^{40}Ar)$^+$ interference on the lighter isotope of Cu (^{63}Cu). The significant effect of Fe on Cu (and Zn) isotopic measurement was previously found by Archer and Vance.[25] By contrast, Zhu et al.[5] showed that Fe/Cu molar ratios up to 15 cause insignificant influence on Cu isotopic analysis using the Nu Instruments. The reason for these differences is unclear. There is no known molecular spectral interference from iron on mass 63 or 65. The significant influence of Fe on Cu isotopic analysis found in this study, however, points out that chemical purification may be necessary for Cu isotopic analysis of Fe-bearing Cu sulfides (e.g., chalcopyrite).

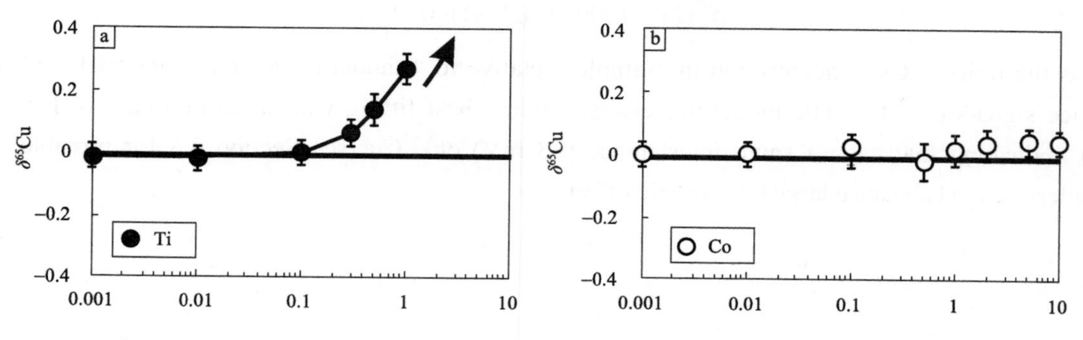

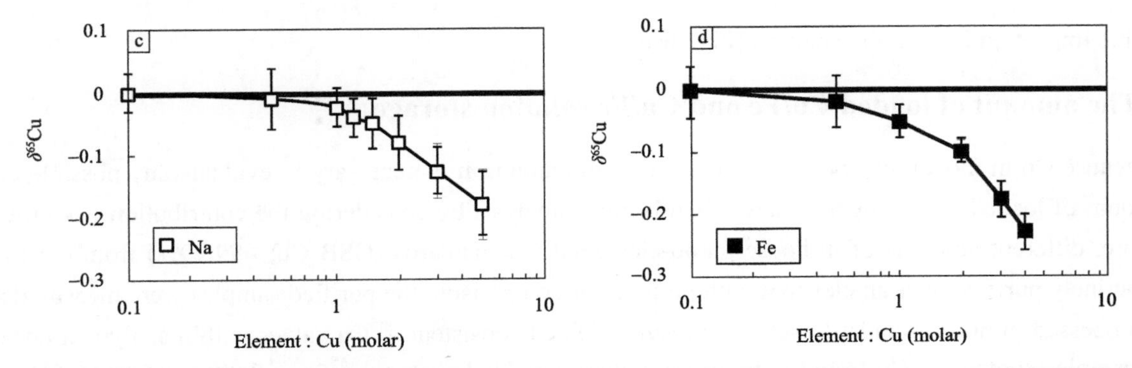

Fig. 3 Cu isotopic variations of NIST Cu standard solutions spiked with different amounts of Ti, Co, Fe and Na relative to the unspiked Cu solution. The Cu concentration for samples and bracketing standards (NIST 976) is the same (100 ppb). The variations of Fe isotopic composition with Co/Fe are also plotted in this figure. The errors (2SD) were calculated based on four times replicate measurements. Data are reported in Table 2.

The polyatomic interference from Ti on Cu isotopic analysis may be attributed to the oxides of Ti ($^{16}O^{47}Ti$ and $^{16}O^{49}Ti$) on mass 63 and 65 respectively. In addition, the hydroxides ($^{16}O^{1}H$) of ^{48}Ti and ^{46}Ti also have the same mass number with the two isotopes of Cu.[12] Because ^{47}Ti (7.44%) has higher natural abundance than ^{49}Ti (5.41%), the contributions from polyatomic ions of Ti-oxides would lower the mass 65/mass 63 ratio. By contrast, ^{48}Ti (73.72%) is more abundant than ^{46}Ti (8.25%), and thus the contribution from polyatomic ions of Ti-hydroxides would cause the measured results towards heavy δ^{65}Cu values when Ti is present.[12] Li et al.[12] found significant influence of Ti on Cu isotopic analysis towards heavy isotopic compositions using a *Nu Plasma* MC-ICP-MS. By contrast, Bigalke et al.[7] reported a remarkable influence of Ti on Cu isotopic ratio analysis towards a light value, using the Neptune MC-ICP-MS. We measured a set of Cu-free Ti solutions with concentrations varying from 100 ppb to 1 ppm in low-resolution mode. The results showed that signals of both mass 63 and 65 increased significantly compared with the blank baseline (3% HNO_3) but the mass 65/mass 53 ratios also increased from ~0.47 to ~0.76. This clearly demonstrates a major interference of Ti on mass 65 over 63. We also performed high-resolution (M/DM = ~10 000) measurement for a mixed Ti and Cu solution (each 1 ppm) on the mass 63 and 65. There was no clearly visible shoulder, particularly at mass 65. The reasons remain unresolved.

To overcome the matrix interference, the only way would be sufficient purification. Analysis of the Cu cuts eluted from basaltic and granitic rocks shows that the ratios of major ions (Mg, Ca, Fe, Na, Mn, etc.) to Cu were less than 0.01 after one time purification. The ratios can be markedly reduced (<0.001) after double column chemistry. The low signal of matrix elements yielded neglected influence on Cu isotopic measurement (Fig. 3). Titanium, however, was commonly found in the Cu cuts eluted from these rocks, with Ti/Cu up to ~0.3 (e.g., basalt BHVO-2) after one purification due to high Ti/Cu in the rocks (>100). The modest Ti/Cu ratio of 0.3 would cause an offset of ~0.15‰ relative to the true δ^{65}Cu value (Fig. 3a). A second purification is thus needed. After double column chemistry, Ti/Cu can be reduced to less than 0.03 for all analyzed samples which contributed neglectful influence on Cu isotope ratio analysis.

Different from Cu, only one column chemistry has been undertaken for Fe in all analyzed rock samples. After single column chemistry, the ratios of all ions to Fe were found to be less than 0.01. The signal ratio of $^{53}Cr/^{54}Fe$ is commonly at or below the level of 10^{-5}. Such low signals of interferents did not generate any detectable influence on Fe isotopic measurement. It is noted that Co was completely separated from Cu in the 8 N HCl medium but it was shifted into the fraction of Fe in 2 N HCl. Experiments were thus designed to evaluate the possible interference of Co on Fe isotopic analysis. The results show that Co/Fe ratios up to 5 did not produce detectable impact on Fe isotopic analysis (Fig. 3b).

3.4 The amount of loaded Cu/Fe and Cu/Fe solution storage

Because Cu in most rocks only constitutes a small amount, it is necessary to evaluate any possible effect of the amount of loaded Cu on the accuracy of Cu isotopic analysis by considering the contribution from the blank. Therefore, different amounts of in-house mono-elemental Cu standards (GSB Cu; >99.99%) from 0.4 to 20 μg were routinely purified over an eleven month period. For comparison, the purified samples were measured against the unprocessed in-house standard itself. The results yielded consistent δ^{65}Cu values within analytical uncertainty for all samples containing Cu from 0.4 to 20 μg, with a weighted average δ^{65}Cu = 0.006±0.05‰ (2SD; n = 21) relative to the unprocessed standard (Fig. 4). This suggests that the contribution from blank (averaged ~1.5 ng) is still insignificant when the amount of Cu loaded to the column is as low as 0.4 mg (blank: sample = ~0.3%). A calculation should indicate that contamination of the sample with 0.3% blank, for which an extreme positive δ^{65}Cu value of +10‰ (ref. 26) is assumed, would cause an undetectable shift (+0.03‰) on Cu isotopic composition. If the extreme negative value (~−17‰)[26] is assumed, the shift would be ca. +0.05‰. Furthermore, the average δ^{65}Cu values of all purified samples are close to zero relative to the unpurified standards, suggesting that no Cu isotope fractionation has been generated during chemical purification. The loaded amounts of Fe in all samples analyzed in this study range from ~0.6 to 0.9 mg, which are one hundred thousand times larger than the total procedural blank (~6 ng). The blank contributions on Fe isotopic analysis are thus considered neglectful.

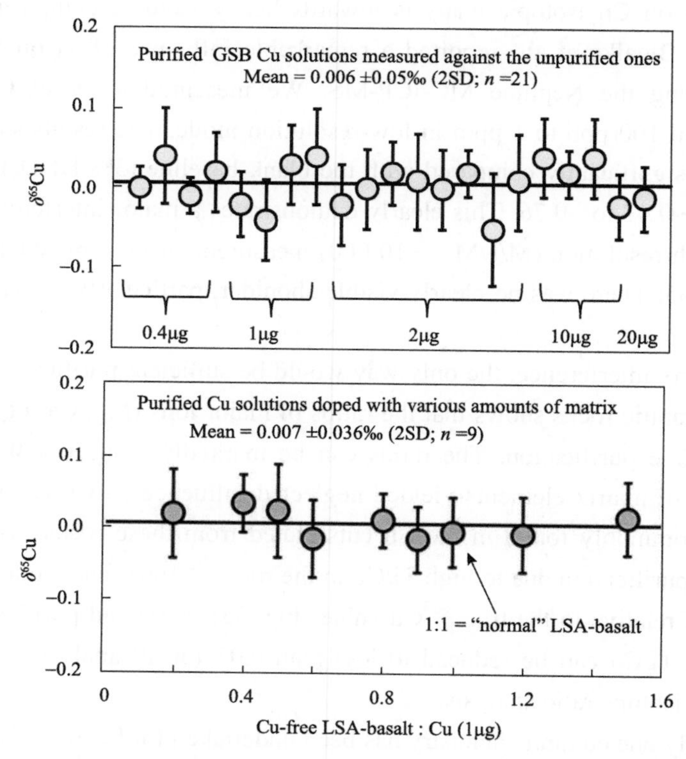

Fig. 4 Test of the effect of the amount of loaded Cu on the accuracy of Cu isotopic analysis (upper diagram). In-house mono-element standard solutions (GSB Cu) were prepared to contain different amounts of Cu (0.4–20 μg) and were purified through column chemistry. Cu isotopic ratios were measured relative to the in-house standard itself. A set of solutions with 1 mg pure Cu (NIST 976), mixed with various amounts of synthetic Cu-free LSA-basalt, were processed through column chemistry (twice) and measured against NIST 976 (lower diagram). The 1:1 ratio of Cu-free LSA-basalt to Cu is equal to that in the "normal" Cu-containing LSA-basalt. The errors (2SD) were based on four times replicate measurements. Data are reported in Table 3.

To avoid an important systematic bias, it is critical to ensure that no isotopic changes occur in the bracketing standard. One primary concern is the effect of long-term storage of working standards in plastic bottles. Significant deviation of isotopic ratios of standards with time has been observed for Mg.[27] Storage of the pure, concentrated GSB Cu and IRMM-014 Fe standards (100 ppm) in 50 ml clean fluorinated plastic (Teflon®) bottles for one year has not caused any detectable deviations in Cu and Fe isotopic ratios. This indicates that no any systematic bias occurred in the bracketing standards, and thus, the samples analyzed.

3.5 Cu isotopic analysis with Zn-doping

Apart from the SSB method, external (inter-element) normalization using Zn or Ni was also commonly pursued to correct for instrumental mass bias during Cu isotopic measurement.[1,3,4] Compared with the external normalization, sample–standard bracketing does not require either introduction of a known $^{66/68}Zn/^{64}Zn$ ratio or removal of natural Zn in a sample. By contrast, Zn must be completely removed and an external standard with known $^{66/68}Zn/^{64}Zn$ ratio must be introduced before isotopic analysis. Compared with external normalization, however, sample–standard bracketing may not account for machine drift such as variations in the plasma, temperature, etc. We measured a processed (twice) natural basalt sample (BHVO-2) by using Zn (SRM 3168a standard solution) as an external standard. Mass fractionation was first corrected with the exponential mass bias function.[4] The delta values were then calculated by calibrating the mass bias-corrected Cu isotopic ratios against the mean of two adjacent standards (NIST 976). The slopes (S) on the plot of $\ln^{64}Zn/^{66}Zn$ versus $\ln^{63}Cu/^{65}Cu$, measured in mixed Cu + Zn (each 100 ppb) standard solutions, slightly vary on different days (from 0.99 to 1.10) but are almost constant within one session (24 h). The inconstant variation of S with time has been previously observed.[4] The $^{65}Cu/^{63}Cu$ ratios were therefore calculated by reference to the standard regression line measured on each day. The measured results for BHVO-2 are plotted in Fig. 5. The average $\delta^{65}Cu$ value is +0.132±0.042‰ (2SD; n = 16). This value is in agreement within uncertainty with the result measured by the SSB method (+0.150± 0.050‰). Compared with the SSB method, it seems that there is no significant improvement of the analytical precision. This may be in part due to the stability of the machines that have either constant or negligible drift.

3.6 Precision and accuracy check

3.6.1 Cu isotopic analysis

The $\delta^{65}Cu$ variations of bracketing standards relative to the mean of two neighbouring standards were also calculated during the course of sample analysis. The values are commonly in the range of ±0.06‰ (2SD). Prior to analysis of samples, at least one in-house standard was repeatedly run. Only after the precision obtained for standards was better than ±0.05‰ (2SD), samples were measured. This makes sure high-precision analysis of Cu (and Fe) isotopic ratios.

Repeat analyses of the in-house Cu standards (GSB Cu) and well-studied igneous rock standards allow evaluation of our long-term analytical precision and accuracy. Long-term analysis of the GSB Cu solutions over a six month period gave an average $\delta^{65}Cu$ of +0.44±0.04‰ (2SD; n = 32) relative to NIST 976 (Fig. 6a). The precision reflects the long-term external reproducibility of pure Cu solution measurement. Compared with the precision obtained from processed GSB Cu solution (±0.05‰; 2SD), the results indicate that the purification processes do not result in a significant shift in analytical precision. A synthetic "basalt" (LSA-basalt) was made to have a chemical composition similar to the average LCC[28] by spiking the Cu standard (NIST 976) with Cu : Fe : Zn : Cr : Ni : Ti : Na : Mg : K : Al : Ca : Mn = 1 : 2500 : 3 : 8 : 3.4 : 190 : 750 : 1700 : 200 : 4500 : 2600 : 30. The "basalt" sample was processed through column chemistry (two times) as the same as done for the natural rock samples. The long-term analysis (over ten months) yielded a mean δ^{65}Cu=–0.004±0.048‰ (2SD; n=9; Table 3). This value is identical within uncertainty to zero, indicating accurate and precise Cu isotopic analysis.

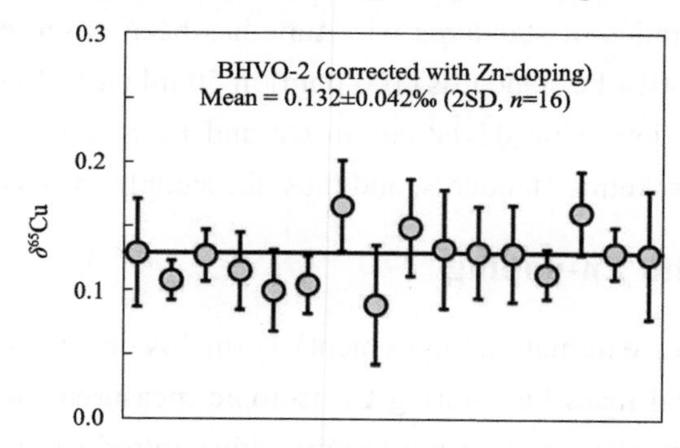

Fig. 5　Measurement of the purified basalt sample (BHVO-2) using Zn as an external standard (SRM 3168a). The corrected ^{65}Cu/^{63}Cu ratios were calculated relative to the mean of two neighboring standards (NIST 976). See text for details.

In addition, we separated a set of synthetic solutions with a fixed amount of Cu contained in solution of variable ionic strength. The aim was to test the effect of the amount of matrix on Cu purification. NIST 976 Cu standard (1 mg) was mixed with the remade Cu-free LSA-basalt, with the ratios of Cu-free LSA-basalt to Cu varying from 0.2 to 1.5 (note that the ratio of 1:1 is equal to that in the original Cu-containing LSA-basalt). All mixed samples were processed through column and all values were close to zero with an average δ^{65}Cu= 0.007±0.038‰ (2SD; n=9) (Fig. 4). The results indicate that Cu can be well separated from the matrix for considerably high ions/Cu samples with complete recovery.

Table 3　Cu isotopic composition of igneous rock standards reported in this study

Sample type	Cu (mg/g)	Session	δ^{65}Cu	2SD	n^a	Ti/Cu after purification	Comments[b]
BHVO-2, Basalt, Hawaiian,	127	1	0.14	0.05	4	0.20	1–3 used the same bulk raw
USA		2	0.11	0.06	4	0.01	solution; 1 was processed only one
		3	0.19	0.05	4	0.01	time, each 10 mg
		4	0.17	0.03	4	0.01	Merged from Cu cuts, 20 mg
		5	0.16	0.05	4	<0.01	New digestion; 20 mg
		6	0.12	0.04	4	0.01	New digestion, 10 mg
		7	0.15	0.05	4	0.02	New digestion, 10 mg
		8	0.13	0.06	4	0.01	New digestion, 10 mg
		9	0.17	0.05	4	0.01	9–13 used the same bulk raw
		10	0.18	0.06	4	<0.01	solution, each 10 mg
		11	0.13	0.03	4	<0.01	
		12	0.18	0.05	4	0.01	
		13	0.17	0.05	4	0.01	
		14	0.14	0.03	4	0.01	13–15 used the sample purified
		15	0.18	0.06	4	0.01	solution measured in different
							days (over 3 months)
		16	0.15	0.05	6	<0.01	16 and 17 used the same bulk raw
		17	0.12	0.04	6	<0.01	solution, each 10 mg
		18	0.15	0.05	6	0.01	New digestion, 10 mg

Continued

Sample type	Cu (mg/g)	Session	δ^{65}Cu	2SD	n[a]	Ti/Cu after purification	Comments[b]
Average (n=18)			0.15	0.05			This study
			0.10	0.10			Weinstein et al. (2011)
BIR-1a, Basalt, Iceland	125	1	–0.02	0.05	2	<0.01	1 and 2 used the same bulk raw
		2	0.01	0.05	4	<0.01	solution, each 10 mg
		3	0.02	0.04	4	<0.01	New digestion, 20 mg
		4	0.01	0.05	6	<0.01	4 and 5 used the sample purified
		5	–0.03	0.05	4	<0.01	solution measured on different days (over 2 months).
		6	0.03	0.06	6	<0.01	New digestion, 10mg
Average (n=6)			0.00	0.05			This study
BIR-1, Basalt, Iceland	125	1	–0.02	0.05	4	<0.01	1 and 2 used the same bulk raw
		2	–0.01	0.04	6	<0.01	solution, each 10 mg
		3	–0.03	0.04	4	<0.01	New digestion, 10 mg
		4	0.01	0.05	4	0.01	New digestion, 20 mg
		5	0.02	0.04	4	<0.01	New digestion, 10 mg
Average (n=5)			–0.01	0.04			This study
			–0.02	0.10			Li et al. (2009)
JB-3, Basalt, Japan	199	1	0.18	0.07	4	0.01	1 and 2 used the same bulk raw
		2	0.16	0.06	4	0.01	solution, each 10 mg
		3	0.15	0.03	4	<0.01	New digestion, 20 mg
Average (n=3)			0.16	0.06			This study
BCR-2	19	1	0.22	0.05	4	0.02	New digestion, 20 mg
Basalt, USGS		2	0.22	0.04	4	0.03	New digestion, 20 mg
Average (n=2)			0.22	0.04			
			0.22	0.06			Bigalke et al. (2010a)
			0.18	0.09			Bigalke et al. (2011)
GSP-2, Granodiorite, USGS	43	1	0.32	0.05	4	0.01	1 and 2 used the same bulk raw
		2	0.31	0.05	4	0.01	solution, each 20 mg
		3	0.28	0.03	4	<0.01	New digestion, 20 mg
Average (n=3)			0.30	0.04			This study
			0.35	0.06			Bigalke et al. (2010b)
			0.25	0.03			Bigalke et al. (2010a)
AGV-2, Andesite, USGS		1	0.06	0.04	4	0.01	New digestion, 20 mg
		2	0.05	0.04		0.01	New digestion, 20 mg
Average (n=2)			0.05	0.04			This study
			0.10	0.10			Weinstein et al. (2011)
GBW07105, Basalt, China	49	1	0.09	0.06	4	0.03	New digestion, 20 mg
		2	0.11	0.07	4	0.02	New digestion, 10 mg
		3	0.08	0.04	4	0.02	3 and 4 used the same bulk raw
		4	0.09	0.06	4	0.02	solution, each 20 mg

Continued

Sample type	Cu (mg/g)	Session	$\delta^{65}Cu$	2SD	n[a]	Ti/Cu after purification	Comments[b]
Average (n=4)			0.09	0.03			This study
GBW07122, Amphibolite,	84	1	0.38	0.04	4	<0.01	1 and 2 used the same bulk raw
China		2	0.43	0.06	4	0.01	solution, each 20 mg
		3	0.37	0.07	4	0.01	New digestion, 10 mg
Average (n=3)			0.39	0.06			This study
W-2a, Diabase, Virginia	110	1	0.10	0.08	4	<0.01	1 and 2 used the same bulk raw
		2	0.11	0.05	4	<0.01	solution, each 10 mg
		3	0.11	0.04	4	<0.01	New digestion, 10 mg
Average (n=2)			0.11	0.02			This study
JA-1	42	1	0.31	0.04	4	<0.01	1 and 2 used the same bulk raw
		2	0.28	0.07	4	<0.01	solution, each 10 mg
		3	0.29	0.04	4	<0.01	New digestion, 10 mg
Average (n=2)			0.29	0.03			This study
	Cu-free LSA-basalt : Cu						
Mixed Cu + LSA-basalt[c]	0.2		0.02	0.06	4	—	
	0.4		0.03	0.04	4	—	
	0.5		0.02	0.06	4	—	
	0.6		−0.01	0.05	4	—	
	0.8		0.01	0.04	4	—	
	0.9		−0.01	0.05	4	—	
	1.0		0.00	0.05	4	—	Mean of 9 repeat analyses
	1.2		−0.01	0.05	4	—	
	1.5		0.02	0.05	4	—	
Average (n=9)			0.007	0.036			

[a] The times of repeat measurements of the same purification solution by MC-ICP-MS. 2SD 2 times the standard deviation of the population of n repeat measurements of a sample solution. [b] All samples were processed two times through column chemistry except the one of the standard BHVO- 2 as indicated. 10 or 20 mg denotes the weight of primary sample powder which was dissolved and loaded into the column. [c] 1 mg Cu (NIST 976 standard) was spiked with various amounts of Cu-free LSA-basalt (Fe : Zn : Cr : Ni : Ti : Na : Mg : K : Al : Ca : Mn = 2500 : 3 : 8 : 3.4 : 190 : 750 : 1700 : 200 : 4500 : 2600 : 30). The "mixed" sample was processed through column chemistry as the same as done for the samples. If no isotope fractionation occurs during column chemistry the value should be close to zero.

At least two repeat measurements were performed over a ten month period for all igneous rock geostandards in this study. These analyses include independent digestion of the same rock powder, duplicate column chemistry using aliquots of the same bulk raw solution, different amounts of loaded Cu, duplicate measurements of purified Cu solutions on different days, as well as combination of Cu cuts (Table 2). Hawaiian basalt BHVO-2 was most frequently analyzed, which has an average $\delta^{65}Cu$=+0.15‰±0.05‰ (2SD; n=18). The consistent values among samples with independent digestion suggest homogeneous Cu isotopic composition of the rock powers of basalt standard BHVO-2 (Fig. 6b). A purified solution was measured on different days (over 3 months) and yielded consistent results (Table 3), again suggesting that the Cu isotopic composition of the Cu solution did not deviate with time. The $\delta^{65}Cu$ value of BHVO-2 obtained here is lightly heavier than but similar within uncertainty to the value (+0.10±0.10‰; 2SD) reported by Weinstein et al.[29] Given the most frequent analyses, we recommend a reference δ^{65}Cu value of +0.15‰ for the international basalt standard material BHVO-2.

The Columbia River basalt standard (BCR-2) has an average δ^{65}Cu=+0.22±0.04‰ (2SD). This δ^{65}Cu value is in agreement within uncertainty with that (+0.22±0.06‰) reported by Bigalke et al.[30] and +0.18±0.09‰ reported by Bigalke et al.[7] The USGS granodiorite standard GSP-2 has an average δ^{65}Cu=+0.30±0.04‰ (2SD). Bigalke et al.[7,30] reported two values (+0.25±0.05‰ and +0.35‰) for GSP-2, with a difference of 0.10‰. The value obtained in this study is slightly different but agrees within uncertainty with their results. The Icelandic basalt standard BIR-1a has an average Cu isotopic composition equivalent to the NIST 976 Cu standard, with δ^{65}Cu=0.00±0.05‰ (2SD; n=6). Another set (BIR-1) of the Icelandic basalt standard has an average δ^{65}Cu=–0.01±0.04‰ (2SD; n=5) identical to the value of BIR-1a. The value for BIR-1 reported here is in agreement with the value (–0.02±0.10‰) reported by Li et al.[12]

3.6.2 Fe isotopic analysis

High-precision Fe isotope data of the commercially accessible geostandards have been widely reported in the literature. In this study, only the international standard BHVO-2 was repeatedly analyzed over a period of ten months. The Fe isotope data for BHVO-2 and other geo-standards are reported in Table 4. The data define a linear trend in three-isotope space with a slope of 1.460 indicative of mass-dependent isotope fractionation. The long-term analysis of BHVO-2 yielded an average δ^{56}Fe=+0.121±0.049‰ (2SD; n=12) (Fig. 7). This value agrees within ±0.01‰ with the high-precision values of BHVO-2 reported in the literature (including those obtained with ^{57}Fe–^{58}Fe double spike), e.g., +0.128±0.019‰ (ref. 31) and +0.114±0.011‰. The long-term analysis of synthetic LSA-basalt obtained an average δ^{56}Fe = –0.008±0.041‰ (2SD) relative to the original Fe solution.

Table 4 Fe isotopic composition of geostandards reported in this study

Sample	Session[a]	δ^{56}Fe	2SD	δ^{57}Fe	2SD	n	Comments
BHVO-2	1	0.085	0.059	0.140	0.038	4	1–3 used the same bulk raw solution
	2	0.137	0.025	0.190	0.074	4	
	3	0.143	0.020	0.197	0.061	4	
	4	0.148	0.050	0.246	0.042	4	New digestion
	5	0.149	0.043	0.239	0.063	4	New digestion
	6	0.132	0.043	0.223	0.029	4	New digestion
	7	0.111	0.040	0.191	0.038	4	New digestion
	8	0.090	0.047	0.167	0.037	4	New digestion
	9	0.116	0.042	0.192	0.057	4	9 and 10 used the same bulk
	10	0.109	0.048	0.189	0.054	4	raw solution
	11	0.124	0.038	0.199	0.052	4	New digestion
	12	0.114	0.041	0.175	0.049	4	New digestion
Average (n=12)		0.121	0.049	0.175	0.064		
BIR-1a		0.060	0.042	0.085	0.072	9	
BIR-1		0.078	0.027	0.130	0.069	4	
JB-3	1	0.099	0.033	0.149	0.046	4	
	2	0.103	0.050	0.171	0.059	4	
BCR-2		0.107	0.025	0.170	0.013	3	

Continued

Sample	Session[a]	δ^{56}Fe	2SD	δ^{57}Fe	2SD	n	Comments
GSP-2	1	0.173	0.031	0.250	0.067	4	New digestion
	2	0.164	0.060	0.246	0.089	4	New digestion
AGV-2		0.106	0.036	0.179	0.025	4	
GBW07105		0.146	0.035	0.221	0.056	4	
GBW07122		0.069	0.020	0.096	0.069	4	
W-2a		0.036	0.053	0.054	0.016	3	
JA-1		0.057	0.019	0.100	0.048	3	
LSA-basalt[b]		–0.008	0.041	–0.010	0.059		40

[a] All samples were processed through only one column chemistry. The session numbers correspond to the same numbers as in Cu isotopic analysis (Table 3), during which Cu and Fe were eluted through a single column. [b] Iron in the LSA-basalt was made from the GSB Fe solution and the purified (one time) samples were measured against the original GSB Fe solution.

The Cu and Fe isotopic results obtained in this study are plotted against the literature data in Fig. 8a and b. Our data are generally consistent with literature data for all analyzed standards. In summary, accurate and precise analysis of Cu and Fe isotopic ratios can be achieved using the established procedure. A long-term external reproducibility better than ±0.05‰ (2SD) of both δ^{65}Cu and δ^{56}Fe measurement for silicate rocks can be routinely obtained. This allows for economical and efficient study of stable Cu and Fe isotopic systematics in geological and biological fields.

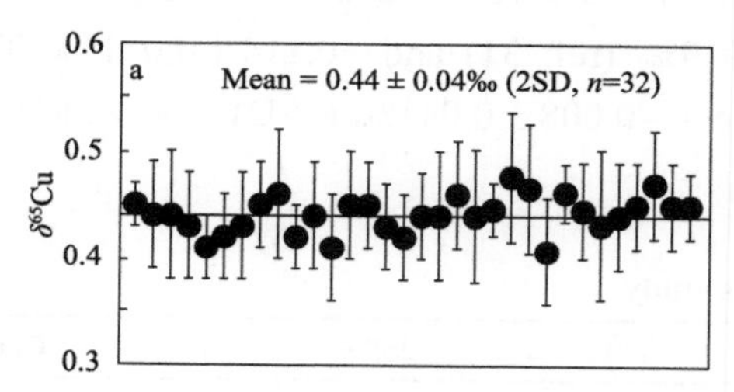

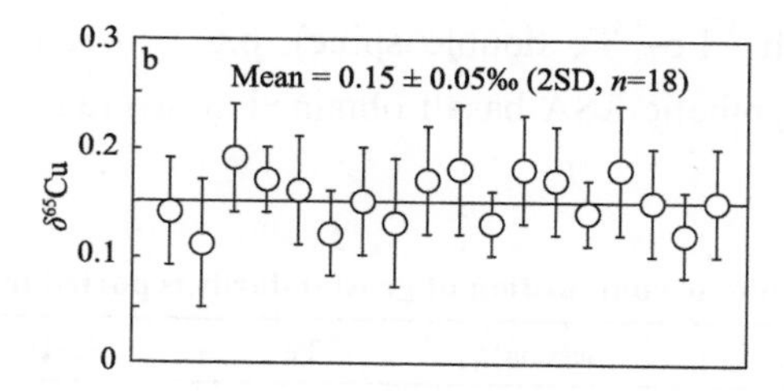

Fig. 6 Long-term analysis of in-house mono-element standard solutions (GSB Cu) (upper diagram) and USGS basalt standard BHVO-2 (lower diagram) relative to NIST 976. The in-house standard has an average δ^{65}Cu=+0.44±0.04‰ (2SD; n=32); the precision (±0.04‰) represents long-term reproducibility of analysis of pure Cu solution. BHVO-2 has an average δ^{65}Cu=0.15±0.05‰ (2SD; n=18).

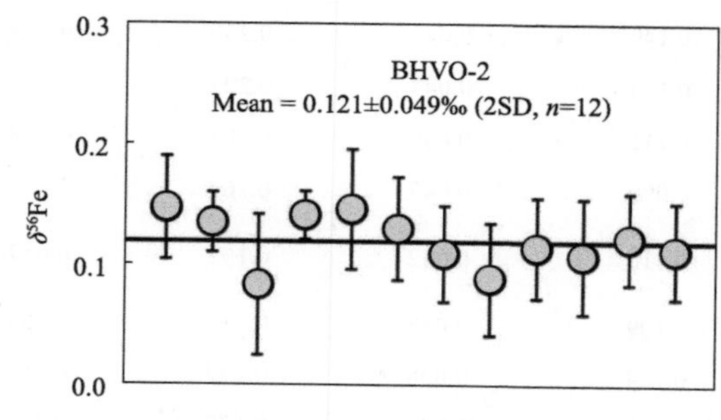

Fig. 7 Long-term analyses of Fe isotopic compositions of international basalt standard BHVO-2. The data are reported in Table 4. The mean δ^{56}Fe value is 0.121±0.049‰ (2SD; n = 12).

4 Copper and iron isotopic composition of igneous rock standards

Totally eleven international igneous rock standards were analyzed in this study. The δ^{56}Fe values of all standards are in the range of ±0.10‰. However, the overall range of δ^{65}Cu is up to 0.40‰ (–0.01 to +0.39‰). Basalt standards BIR-1 and BIR-1a have the lightest Cu isotopic composition of zero among all geostandards analyzed, and the amphibolite standard from China (GBW07122) has the heaviest Cu isotopic composition (+0.39‰). Given the long-term reproducibility of Cu isotopic analysis (±0.05‰; 2SD) routinely obtained in this study, we conclude that the range can be significantly discriminated, which represents about 8 times that of the analytical precision.

Previous studies have suggested a similar Cu isotopic composition among mid-oceanic ridge basalt (MORB),[31] oceanic island basalt (OIB),[23] continental basalt,[25] peridotite[33] and granite.[12] A mean value of zero relative to NIST 976 has been recommended for the Cu isotopic composition of these silicate reservoirs in the Earth. The bulk silicate Earth (BSE) is thus believed to have δ^{65}Cu close to zero. However, the results from some natural rock standards obtained in this study have shown that Cu isotopic compositions of basalts or diabases (BCR-2, BIR-1, JB-3, W-2a and GBW07105) are significantly different (Fig. 9). Although only two andesite geostandards (AGV-2 and JA-1) have been analyzed, they also have different Cu isotopic compositions (Fig. 9). This suggests that the Cu isotopic composition of intermediate-felsic rocks is also not homogeneous.

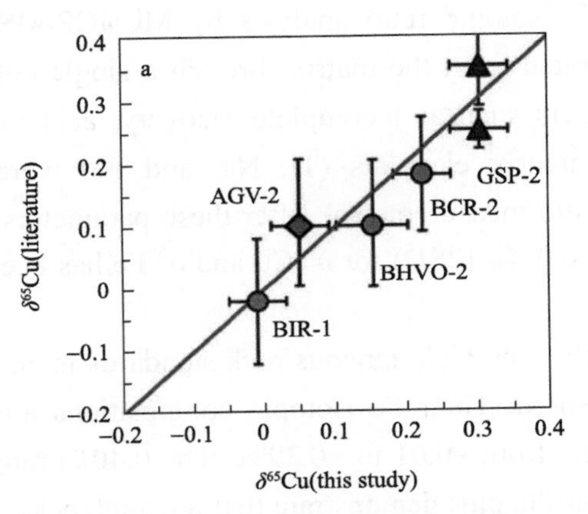

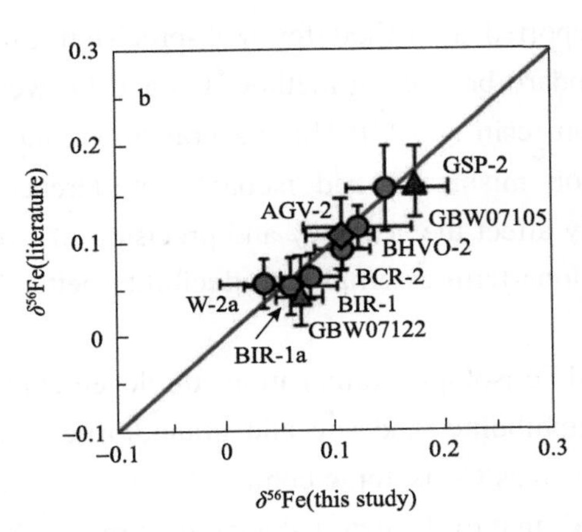

Fig. 8 Comparison of Cu and Fe isotopic compositions of igneous rock standards reported in this study and those reported in the literature. The data from this study and literature are listed in Tables 3 and 4. Iron isotopic data of igneous rock standards are widely available in the literature and only the data from Carddock and Dauphas[32] are plotted here for comparison.

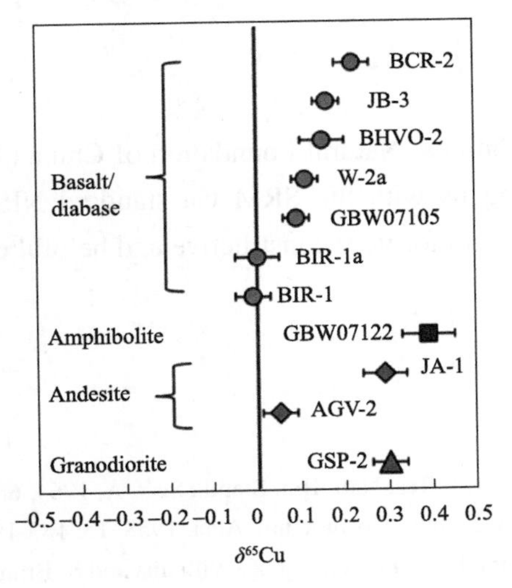

Fig. 9 Cu isotopic composition of silicate rock standards reported in this study. The results clearly demonstrate that the Cu isotopic composition of igneous rocks, including basaltic and felsic rocks, is not homogeneous. The overall δ^{65}Cu variation can be distinguished by the current analytical precision. Data are reported in Table 3.

If one assumes that these rock standards were significantly free to surface alteration after intrusion or eruption, the detectable Cu isotopic variation among igneous rocks should reflect either high-temperature magmatic processes or isotopic heterogeneity in the source regions. Cu isotope fractionation during crystal-melt differentiation of granitic magmas may be small as revealed by a granite study,[12] although they reported an overall δ^{65}Cu variation of >0.4‰. It is currently unclear that to what extents these variations reflect magmatic differentiation. A detailed evaluation of these mechanisms is beyond the scope of the present study. Nevertheless, the results indicate that Cu isotopic variations should not be confined to the realm of biology or low temperature aqueous geochemistry but may also occur at high temperature magmatic processes. This makes the Cu isotope a potential tracer for high-temperature magmatic processes in addition to its wide application to low-temperature geochemistry. Further studies are needed to better address (i) to what extents Cu isotopic compositions of igneous rocks may vary and (ii) how these variations were caused.

5 Conclusion

We reported a method for high-precision Cu and Fe isotopic ratio analysis by MC-ICP-MS using the sample-standard bracketing method. Cu and Fe were separated from the matrix through a single column using strong anion resin AG-MP-1M. Several important parameters such as incomplete recovery, acid molarity and concentration mismatch, and isobaric interference from matrix elements (Ti, Na, and Fe) were found to significantly affect the accuracy and precision of isotopic ratio measurements. After these parameters were fully addressed, long-term external reproducibility better than ± 0.05‰ (2SD) for δ^{65}Cu and δ^{56}Fe has been routinely obtained.

Cu and Fe isotopic compositions of eleven commercially accessible igneous rock standards including basalt, diabase, amphibolite, andesite and granodiorite were measured. Their Fe isotopic compositions are relatively uniform, whereas Cu isotopic compositions vary significantly from −0.01 to +0.39‰. The 0.40‰ range exceeds about 8 times that of the external analytical precision. The results thus demonstrate that igneous rocks may not be homogeneous in Cu isotopic composition, and the Cu isotope could be used to trace high-temperature magma processes in addition to its wide application to low-temperature geochemistry.

Acknowledgements

This work is supported by the National Natural Foundation of China (41203013) to SAL. We are grateful to Prof. X.-K. Zhu for kindly providing us with the SRM Cu standard NIST 976. We thank the editor Harriet Brewerton and two anonymous reviewers for very constructive and helpful comments which largely improved the manuscript.

References

1 W. R. Shields, T. J. Murphy and E. L. Garner, J. Res. Natl. Bur. Stand., Sect. A, 1964, 68, 589-592.
2 E. C. Walker, F. Cuttitta and F. E. Senftle, Geochim. Cosmochim. Acta, 1958, 15, 183-194.
3 R. Mathur, L. Jin, V. Prush, J. Paul, C. Ebersole, A. Fornadel, J. Z. Williams and S. Brantley, Chem. Geol., 2012, 304-305, 175-184.
4 C. N. Mar′echal, P. T′elouk and F. Albarède, Chem. Geol., 1999, 156, 251-273.
5 X. K. Zhu, R. K. O'Nions, Y. Guo, N. S. Belshaw and D. Rickard, Chem. Geol., 2000, 163, 139-149.
6 D. M. Borrok, R. B. Wanty, W. I. Ridley, R. Wolf, P. J. Lamothe and M. Adams, Chem. Geol., 2007, 242, 400-414.

7 M. Bigalke, S. Weyer and W. Wilcke, Geochim. Cosmochim. Acta, 2011, 75, 3119-3134.

8 T. F. D. Mason, D. J. Weiss, M. S. A. Horstwood, R. R. Parrish, S. S. Russell, E. Mullane and B. J. Coles, J. Anal. At. Spectrom., 2004, 19, 209-217.

9 X. K. Zhu, Y. Guo, R. J. P. Williams, R. K. O'Nions, A. Matthews, N. S. Belshaw, G. W. Canters, E. C. de Waal, U. Weser, B. K. Burgess and B. Salvato, Earth Planet. Sci. Lett., 2002, 200, 47-62.

10 J.-M. Luck, D. B. Othman and F. Albarède, Geochim. Cosmochim. Acta, 2005, 69, 5351-5363.

11 J. Bermin, D. Vance, C. Archer and P. J. Statham, Chem. Geol., 2006, 226, 280-297.

12 W. Q. Li, S. E. Jackson, N. J. Pearson, O. Alard and B. W. Chappell, Chem. Geol., 2009, 258, 38-49.

13 B. L. Beard, C. M. Johnson, L. Cox, H. Sun, K. H. Nealson and C. Aguilar, Science, 1999, 285, 1889-1892.

14 C. M. Johnson, B. L. Beard, C. Klein, N. J. Beukes and E. E. Roden, Geochim. Cosmochim. Acta, 2008, 72, 151-169.

15 M. S. Fantle and D. J. DePaolo, Earth Planet. Sci. Lett., 2004, 228, 547-562.

16 N. Dauphas, A. Pourmand and F.-Z. Teng, Chem. Geol., 2009, 267, 175-184.

17 A. Fernandez and D. M. Borrok, Chem. Geol., 2009, 264, 1-12.

18 S. Graham, N. Pearson, S. Jackson, W. Griffin and S. Y. O'Reilly, Chem. Geol., 2004, 207, 147-169.

19 P. B. Larson, K. Maher, F. C. Ramos, Z. Chang, M. Gaspar and L. D. Meinert, Chem. Geol., 2003, 201, 337-350.

20 C. Maréchal and F. Albarède, Geochim. Cosmochim. Acta, 2002, 66, 1499-1509.

21 A. Anbar, J. Roe, J. Barling and K. Nealson, Science, 2000, 288, 126-128.

22 F.-Z. Teng, M. Wadhwa and R. T. Helz, Earth Planet. Sci. Lett., 2007, 261, 84-92.

23 J. A. Baker, M. Schiller and M. Bizzarro, Geochim. Cosmochim. Acta, 2012, 77, 415-431.

24 X. K. Zhu, A. Makishima, Y. Guo, N. S. Belshaw and R. K. O'Nions, Int. J. Mass Spectrom., 2002, 220, 21-29.

25 C. Archer and D. Vance, J. Anal. At. Spectrom., 2004, 19, 656-665.

26 R. Mathur, S. Titley, F. Barra, S. Brantley, M. Wilson, A. Phillips, F. Munizaga, V. Maksaev, J. Vervoort and G. Hart, J. Geochem. Explor., 2009, 102, 1-6.

27 F. Huang, J. Glessner, A. Ianno, C. Lundstrom and Z. Zhang, Chem. Geol., 2009, 268, 15-23.

28 R. L. Rudnick and S. Gao, Treatise Geochem., 2003, 1-64.

29 C. Weinstein, F. Moynier, K. Wang, R. Paniello, J. Foriel, J. Catalano and S. Pichat, Chem. Geol., 2011, 286, 266-271.

30 M. Bigalke, S. Weyer and W. Wilcke, Soil Sci. Soc. Am. J., 2010, 74, 60-73.

31 M.-A. Millet, J. A. Baker and C. E. Payne, Chem. Geol., 2012, 304-305, 18-25.

32 P. R. Carddock and N. Dauphas, Geostand. Geoanal. Res., 2010, 35, 101-123.

33 B. Othman, D. Luck, J. M. Bodinier, N. T. Arndt and F. Albarede, Geochim. Cosmochim. Acta, 2006, 70, A46.

High-precision iron isotope analysis of geological reference materials by high-resolution MC-ICP-MS*

Yongsheng He[1], Shan Ke[1], Fang-Zhen Teng[1,2], Tiantian Wang[1], Hongjie Wu[1], Yinhuai Lu[1] and Shuguang Li[1]

1. State Key Laboratory of Geological Processes and Mineral Resources, China University of Geosciences, Beijing, 100083, China
2. Isotope Laboratory, Department of Earth and Space Sciences, University of Washington, Seattle, WA 98195, USA

亮点介绍：开发了高精度Fe同位素的MC-ICP-MS测试方法，并报道了22个硅酸岩、火成碳酸岩、页岩、碳酸盐和黏土标样的Fe同位素组成。

Abstract We report high-precision iron isotopic data for twenty-two commercially available geological reference materials, including silicates, carbonatite, shale, carbonate and clay. Accuracy was checked by analyses of synthetic solutions with known Fe isotopic compositions but different matrices ranging from felsic to ultramafic igneous rocks, high Ca and low Fe limestone, to samples enriched in transition group elements (e.g., Cu, Co, and Ni). Analyses over a 2-year period of these synthetic samples and pure Fe solutions that were processed through the whole chemistry procedure yielded an average δ^{56}Fe value of -0.001 ± 0.025‰ (2*s*, *n*=74), identical to the expected true value of 0. This demonstrates a long-term reproducibility and accuracy of <0.03‰ for determination of $^{56}Fe/^{54}Fe$ ratios. Reproducibility and accuracy were further confirmed by replicate measurements of the twenty-two RMs, which yielded results that perfectly match the mean values of published data within quoted uncertainties. New recommended values and associated uncertainties are presented for interlaboratory calibration in the future.

Keywords Fe isotopes, geological reference materials, sedimentary RMs, recommended values, calibration.

1 Introduction

Iron (Fe) is a major element in the Earth's crust, mantle and core, with multiple redox states, and has four stable isotopes: ^{54}Fe, ^{56}Fe, ^{57}Fe, and ^{58}Fe. Significant Fe isotope fractionation has been predicted theoretically and determined experimentally between dissolved Fe^{2+} and Fe^{3+} during redox changes (Schauble et al. 2001, Johnson et al. 2002, Balci et al. 2006); thus, Fe isotope geochemistry has been used as an important tool to trace oxygen

* 本文发表在：Geostandards and Geoanalytical Research, 2015, 39(3): 341-356

fugacity of the hydrosphere, and the rise of atmospheric oxygen (Rouxel 2005, Li et al. 2013). Microbial utilisation may also promote the change of oxygen fugacity and/or the mobilisation of insoluble Fe^{3+} species, which leads to Fe isotope fractionations termed 'biosignatures' (Beard et al. 1999, Johnson et al. 2008, Heimann et al. 2010). In addition, significant Fe isotope fractionation also occurs during partial melting (Williams et al. 2004, Weyer and Ionov 2007, Dauphas et al. 2009a) and magma differentiation (Teng et al. 2008, 2011, 2013, Schuessler et al. 2009, Sossi et al. 2012, Telus et al. 2012) with the melt generally enriched in Fe^{3+} and thus 'heavier' Fe. Iron isotope geochemistry thus may also record the oxygen fugacity of magmatic systems. For the reasons mentioned above, Fe isotope geochemistry has gained particular interest in the past decade.

High precision and accuracy are required for Fe isotope determination, especially for high temperature processes, during which isotope fractionation is generally small (e.g., Beard et al. 2003a, Heimann et al. 2008, Teng et al. 2008, 2011, 2013, Dauphas et al. 2009a, Schuessler et al. 2009, Telus et al. 2012, Wang et al. 2012, Weyer and Seitz 2012, Craddock et al. 2013). To date, most Fe isotope data have been determined by multi-collector inductively coupled plasma-mass spectrometry (MC-ICP-MS) (e.g., Anbar et al. 2000, Belshaw et al. 2000, Beard et al. 2003a, Kehm et al. 2003, Malinovsky et al. 2003, Weyer and Schwieters 2003, Dauphas et al. 2004, 2009b, Poitrasson and Freydier 2005, Schoenberg and Blanckenburg 2005, Dideriksen et al. 2006, Millet et al. 2012). Argide interferences (e.g., $^{40}Ar^{14}N^{+}$, $^{40}Ar^{16}O^{+}$, and $^{40}Ar^{16}O^{1}H^{+}$ on $^{54}Fe^{+}$, $^{56}Fe^{+}$, and $^{57}Fe^{+}$, respectively) inherent to the ICP source have been resolved by a number of methods, such as use of desolvating nebulisers (Anbar et al. 2000, Belshaw et al. 2000), or hexapole collision cell (Beard et al. 2003a, Dauphas et al. 2004), by operating the MC-ICP-MS at reduced power (cold plasma) (Kehm et al. 2003), or by high mass resolution (Malinovsky et al. 2003, Weyer and Schwieters 2003). Another challenge for accurate Fe isotope determination is how to correct the >3% mass bias per atom mass unit during the MC-ICP-MS measurement. Mass bias has been fundamentally corrected by calibrator-sample bracketing by assuming that the mass bias is the same for the sample and the calibrator (Zhu et al. 2002, Malinovsky et al. 2003, Weyer and Schwieters 2003, Schoenberg and von Blanckenburg 2005, Dauphas et al. 2009b). Furthermore, elemental spike (termed 'Ni-doping' and 'Cu-doping', and referred to here as 'Cu-spiking') (Arnold et al. 2004, Poitrasson and Freydier 2005, Schoenberg and von Blanckenburg 2005) or double spike techniques (Dideriksen et al. 2006, Millet et al. 2012) have been additionally adopted to improve the long-term reproducibility.

Precision better than <0.03‰ (two standard deviations) can be routinely achieved for $^{56}Fe/^{54}Fe$ ratio determination (Dauphas et al. 2009b, Schuessler et al. 2009, Craddock and Dauphas 2010, Millet et al. 2012). Assurance of accuracy for such high precision is thus very important and usually is verified by analyses of well-characterised geological reference materials. For this purpose, Craddock and Dauphas (2010) reported recommended values for thirty-seven geological RMs, based on their own duplicated analyses, with an accuracy for $^{56}Fe/^{54}Fe$ of <0.03‰. Millet et al. (2012) proposed reference values for five commonly used igneous rock RMs based on a compilation from five independent laboratories. Even though Fe isotope geochemistry has been intensively used to trace Fe cycling in Earth surface systems, few high-precision data have been reported for sedimentary RMs(Craddock and Dauphas 2010). Better characterisation of Fe isotopic compositions of geological reference materials becomes urgent for interlaboratory calibration purposes, given that the supply of the commonly used CRM, IRMM-014 Fe is exhausted.

In this study, we demonstrate that $^{56}Fe/^{54}Fe$ could be precisely and accurately measured within 0.03‰ (2*s*) using high-mass resolution MC-ICP-MS by a calibrator-sample bracketing technique. We then report Fe isotopic data for twenty-two commonly used and commercially available international geological reference materials, including sixteen igneous rocks, five sedimentary rocks, and one metamorphic rock. After compilation of >200 data obtained in this study and those published by other laboratories, we present reference values for these geological reference materials for future interlaboratory calibration and quality control.

2 Experimental procedure

2.1 Sample dissolution and column chemistry

Chemical separation procedures for iron rely on its affinity to anion-exchange resins in HCl media (Strelow 1980, Anbar et al. 2000, Zhu et al. 2002, Beard et al. 2003b, Kehm et al. 2003, Arnold et al. 2004, Dauphas et al. 2004, Poitrasson et al. 2004, Schoenberg and von Blanckenburg 2006, Millet et al. 2012, Sossi et al. 2014). Here, we adopt the sample dissolution and chemical separation procedures described by Dauphas et al.(2009b). A brief description is given below.

Typically, sample powders (3–50 mg) were dissolved in a mixture of HF-HNO_3-HCl-$HClO_4$. Iron was purified using the AG1X-8 (200–400 mesh chloride form, Bio-Rad, Hercules, CA, USA) resin in HCl media. Matrix elements were removed by 8 ml 6 mol l^{-1} HCl, and Fe was then collected by 9 ml 0.4 mol l^{-1} HCl. The same column procedure was repeated twice to ensure complete elimination of the matrices. The final Fe eluate was acidified with 100 μl of concentrated HNO_3. The dried sample was then dissolved in 3% HNO_3 for isotopic determination. Double-distilled acids and Optima^{TM} ultra pure $HClO_4$ were used. The whole procedure Fe blank was < 10 ng with an average of 6.9±7.0 ng (2*s*, *n* = 15), which is <0.01% of the processed samples and therefore considered negligible.

The elution sequence was checked by USGS reference materials BHVO-2 (basalt), AGV-2 (andesite), GSP-2 (granodiorite) and COQ-1 (carbonatite) (Fig. 1). Ten millilitres of 6 mol l^{-1} HCl and 20 ml 0.4 mol l^{-1} HCl were added to examine tails of matrix and interfering elements and Fe, respectively. All major matrices and isobaric elemental Cr were completely removed in the first 8 ml 6 mol l^{-1} HCl (Fig. 1). Copper and Co cannot be quantitatively separated from Fe by this procedure. However, this does not affect the accuracy and precision of Fe isotopic determinations of most geological samples that have Cu/Fe and Co/Fe << 1 (see 'Optimisation of isotopic determinations'), except for samples such as chalcopyrite. Further tests showed that 98.4% Cu and 99.9% Co could be eluted by 16 ml 6 mol l^{-1} HCl without loss of Fe. Therefore, 8 ml 6 mol l^{-1} HCl is adequate to remove matrices in routine analyses, whilst 16 ml 6 mol l^{-1} HCl is necessary for samples with high Cu/Fe ratios. Iron was collected by 9 ml 0.4 mol l^{-1} HCl (0.5 ml + 0.5 ml + 1 ml + 3 ml + 4 ml increments). Iron isotope compositions for the second (0.5 ml), third (1.0 ml) and fourth (3.0 ml) fractions of eluate 0.4 mol l^{-1} HCl for BHVO-2 were measured and yielded $\delta^{56}Fe_{IRMM}$ as 0.526±0.050‰, –0.476±0.050‰ and –1.268±0.068‰, respectively (Fig. 1a, Table S2). This agrees with previous observations that anion-exchange resins can strongly fractionate Fe isotopes with heavier Fe being preferentially eluted (Anbar et al. 2000). Quantitative recovery of Fe, thus, is important for accurate isotopic determination. Yields of Fe in this study were always >99.8%, with an average value of 99.95±0.13% (2*s*, *n* = 5).

Separation of Fe from Zn has been a problem in previous studies when HNO_3 or 0.05 mol l^{-1} HCl was used as the eluent solution to collect Fe (Poitrasson and Freydier 2005, Schoenberg and von Blanckenburg 2005). The elution sequence here shows that quantitative Fe recovery was achieved using 0.4 mol l^{-1} HCl with all Zn retained in the resin.

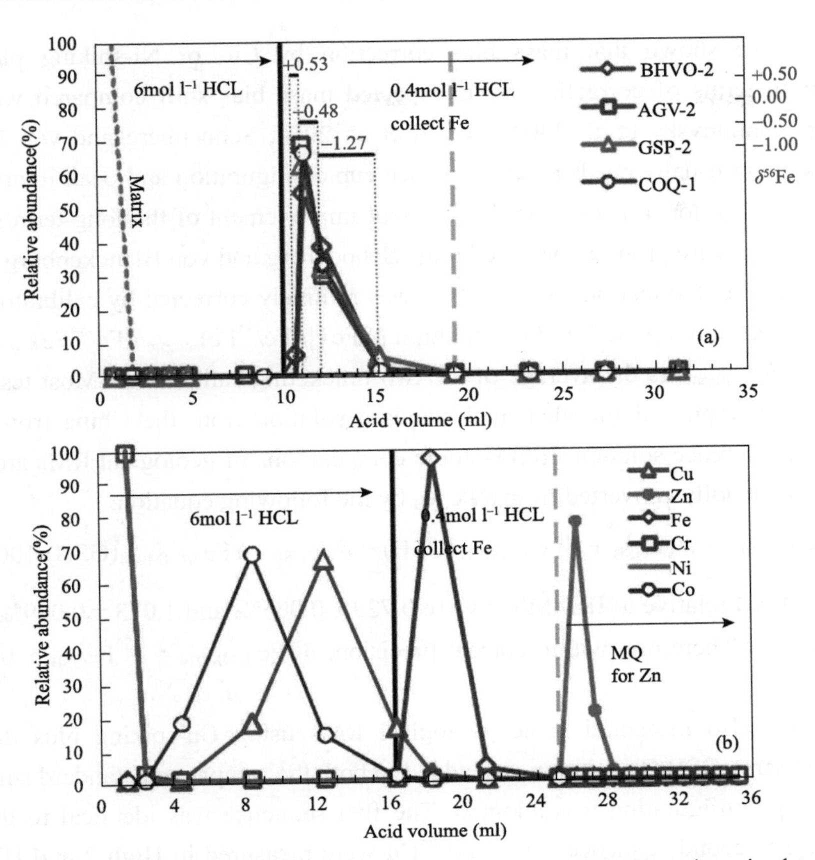

Fig. 1 Elution curves of Fe and matrix elements for geological reference materials (a) and a mixed transition element standard solution, A-Mix2 (b). The Fe isotopic compositions of some fractions of the Fe cut for BHVO-2 are also plotted. Matrix is the concentration weighted mean of Na, Al, Ca, Mg, Ti and Mn.

2.2 Mass spectrometry

Iron isotopic measurements were conducted on a Thermo-Finnigan Neptune Plus MC-ICP-MS at the Isotope Geochemistry Laboratory, China University of Geosciences, Beijing (CUGB). The MC-ICP-MS provided one low- and two high-mass resolution modes (termed 'medium' and 'high' mass resolution, MR and HR). The MR and HR modes provided mass resolutions defined as $m/(m_{0.95}-m_{0.05})$ equal to ~8000 and ~10000, both of which allowed the Fe^{+} peaks be partially resolved from the argide interferences as flat-topped shoulders on the low mass side of mixed peaks (Malinovsky et al. 2003, Weyer and Schwieters 2003, Dauphas et al. 2009b). Both modes were tested in this study.

^{53}Cr, ^{54}(Fe + Cr), ^{56}Fe, ^{57}Fe, ^{58}(Fe + Ni) and ^{60}Ni isotopes were collected in static mode by Faraday cups at Low 3, Low 1, Central, High 1, High 2 and High 4 positions, respectively. A routine analysis consisted of a 30-s baseline correction and twenty-five cycles of ~8.389 s integrations and 3 s idle. The measurement repeatability (RSE) of the $^{56}Fe/^{54}Fe$ ratio was around 10 ppm and rarely exceeded 20 ppm.

Samples were introduced into the Ar plasma using a Cetac ASX-100 auto sampler through a Thermo Scientific Stable Introduction System (SIS), which consisted of a quartz cyclonic and Scott-type spray chambers and a set of PFA Teflon self-aspirating micro-nebulisers with different uptake rates (45, 50, 100 μl min^{-1}). No difference was found in signal sensitivity, precision or accuracy of isotopic results for different uptake rates. The sample and standard solutions were prepared in 3% HNO_3 (ca. 0.35 mol l^{-1}) with Fe concentration ranging between 1 and 3 μg ml^{-1} for MR analysis and between 3 and 10 μg ml^{-1} for HR analysis to obtain a stable signal of 9–20 V on ^{56}Fe. The 3% HNO_3 was measured as blank only at the beginning of a sequence, whilst on-peak zero correction was carried out for every analysis on both RMs and samples.

Previous studies have shown that mass bias correction by Cu- or Ni-spiking plus calibrator-sample bracketing is superior in terms of correcting for unexpected mass bias shift compared with calibrator-sample bracketing alone (e.g., Malinovsky et al. 2003, Arnold et al. 2004, Schoenberg and von Blanckenburg 2005). However, Cu- or Ni-spiking usually needs a two-sequence cup configuration and peak jumping, which increases the measurement time by a factor of about two, but without improvement of the long-term reproducibility when sample and standard solutions are properly prepared (e.g., Schoenberg and von Blanckenburg 2005, Dauphas et al. 2009b, see below). Therefore, the instrument mass bias was routinely corrected by calibrator-sample bracketing. Data were processed offline and reported in the δ notation ($\delta^{i}Fe=[(^{i}Fe/^{54}Fe)_{sample}/(^{i}Fe/^{54}Fe)_{calibrator}-1]*1000$, where i is 56 or 57 and $(^{i}Fe/^{54}Fe)_{calibrator}$ is the average of the two bracketing calibrators). Most tests carried out in this study used GSB Fe (an ultrapure single elemental standard solution from the China Iron and Steel Research Institute) as an in-house reference solution. Iron isotopic compositions of geological RMs are reported relative to IRMM-014, which can be readily converted from $\delta^{i}Fe_{GSB}$ by the following equation:

$$\delta^{i}Fe_{IRMM014} = \delta^{i}Fe_{GSB} + \delta^{i}Fe_{IRMM014}(GSB) + \delta^{i}Fe_{GSB} \times \delta^{i}Fe_{IRMM014}(GSB)/1000$$

$\delta^{56}Fe$ and $\delta^{57}Fe$ of GSB relative to IRMM014 were $0.729\pm0.005‰$ and $1.073\pm0.009‰$ (2SE, 95% CI, N = 33) (Table 1), respectively. Therefore, within current precision, $\delta^{56}Fe_{IRMM014} = \delta^{56}Fe_{GSB} + 0.729$; $\delta^{57}Fe_{IRMM014} = \delta^{57}Fe_{GSB} + 1.073$.

For comparison, we also measured some geological RMs using Cu-spiking plus the calibrator-sample bracketing method. The same GSB Cu solution was added to both the sample and standard solutions with Fe:Cu = 1:1. A two-sequence cup configuration was adopted. The first sequence was identical to the calibrator-sample bracketing method. In the second sequence, ^{63}Cu and ^{65}Cu were measured in High 2 and High 4 Faraday cups. The dispersion and focus lens were used to correct the relative mass dispersion between $^{63}Cu/^{65}Cu$ and $^{58}Fe/^{60}Ni$ to achieve square peak shapes for Cu isotopes. Iron isotope ratios were corrected by $^{65}Cu/^{63}Cu$ to the first measured value by assuming the same instrument fractionation between Fe and Cu and an exponential law; then, sample δ values were calculated against the bracketing calibrators.

To achieve better measurement reproducibility, the calibrator-sample sequences were repeated four or nine times; the reported isotopic compositions are the averages of repeat analyses. The instrument error of each sample was reported after Dauphas et al. (2009b) assuming $s_{sample} = s_{calibrator}$. Briefly, a delta value was calculated for each bracketing calibrator according to the calibrator before and after, and the two standard deviation ($2s$) of this large number of δ values provided a robust estimate of the instrumental reproducibility. The instrument error for each sample was calculated as $2s_{bracketing}/\sqrt{n}$ (n represents the number of measurements for each sample). Where a sample eluate was measured nine times, the mean and error for the first four values were also calculated to test the reproducibility when samples were measured four times each. As discussed by Dauphas et al. (2009b) and Millet et al. (2012), the instrumental uncertainty alone does not account for all sources of analytical uncertainty (see section 'Reproducibility, accuracy and error report').

Table 1　Summary of duplicated measurement of GSB Fe and A-samples passed through chemistry

Sample	chemistry	$\delta^{56}Fe^{*}$	$2s$	$\delta^{57}Fe^{*}$	$2s$	2SE	N	$\delta^{56}Fe^{**}$	$2s$	$\delta^{57}Fe^{**}$	$2s$	2SE	N
GSB Fe solutions													
GSB[a]		0.001	0.020	0.002	0.040	0.022	27	0.001	0.028	0.005	0.054	0.032	27
GSB[b]		0.724	0.030	1.066	0.064	0.035	18						
GSB[c]		0.733	0.024	1.080	0.043	0.027	9	0.736	0.024	1.081	0.062	0.036	15
Weighted mean and 2SE, 95% CI		0.729	0.005	1.073	0.009		33						

Continued

Sample	chemistry	$\delta^{56}Fe^{*}$	2s	$\delta^{57}Fe^{*}$	2s	2SE	N	$\delta^{56}Fe^{**}$	2s	$\delta^{57}Fe^{**}$	2s	2SE	N
Standard solutions and artificial samples solutions through the chemical procedures[a]													
GSB	once column	−0.002	0.025	−0.005	0.043	0.028	13	−0.006	0.028	0.000	0.052	0.039	13
	twice column	0.001	0.023	0.000	0.059	0.031	6	0.014	0.053	0.019	0.086	0.046	6
	whole procedure	0.000	0.018	0.003	0.042	0.016	7	−0.005	0.023	0.010	0.053	0.024	7
A-DM	once column	−0.001	0.017	0.010	0.056	0.020	7	0.007	0.024	0.023	0.049	0.026	8
	twice column	0.010	0.048	0.021	0.069	0.019	4	0.003	0.037	0.008	0.051	0.024	6
A-MORB	once column	−0.004	0.011	0.000	0.037	0.024	8	−0.005	0.016	0.006	0.066	0.032	9
	twice column	0.006	0.016	0.001	0.055	0.018	5	0.004	0.029	0.009	0.043	0.025	6
A-UCC	once column	−0.003	0.005	−0.010	0.053	0.017	3	−0.003	0.028	−0.015	0.081	0.025	4
	twice column	0.027	0.017	0.016	0.031	0.015	1	−0.004	0.030	−0.022	0.021	0.026	3
A-LCC	once column	−0.004	0.034	−0.004	0.092	0.025	6	0.003	0.042	0.010	0.122	0.027	5
	twice column	0.000	0.034	0.001	0.055	0.021	2	−0.001	0.033	0.005	0.071	0.025	4
A-HFG	once column	−0.018	0.014	−0.035	0.038	0.012	1	−0.006	0.030	0.009	0.132	0.023	2
	twice column							0.000	0.062	−0.024	0.077	0.023	1
A-LS	once column	0.014	0.020	0.033	0.045	0.014	1	0.003	0.034	−0.007	0.095	0.031	3
A-Mix1	once column	−0.003	0.023	−0.014	0.013	0.017	3	−0.009	0.026	−0.010	0.020	0.025	3
	twice column	−0.009	0.029	−0.013	0.077	0.025	3	−0.005	0.039	−0.008	0.060	0.029	4
A-Mix2	once column	−0.012	0.010	−0.015	0.003	0.017	3	0.000	0.029	−0.018	0.038	0.025	4
	twice column	−0.020	0.024	−0.023	0.051	0.012	1	−0.002	0.027	−0.016	0.025	0.027	2
Total		−0.001	0.025	−0.001	0.052	0.024	74	−0.001	0.031	0.003	0.064	0.031	90

Compositions for A-sample solutions and raw data are referred to Table S1 and Table 2. [a] Denotes the result against GSB Fe itself and is reported relative to GSB Fe. [b] Denotes values calculated from reference material data measured both against IRMM014 and GSB separately, while [c] denotes data measured directly against IRMM-014. 2SE is the mean of $\delta^{56}Fe$ 2SE for individual replicate measurements on each RM. Measurement repeatability is expressed when a single measurement was conducted. *N* denotes the times of duplicated measurements, each of which includes nine or four replicate analyses on an aliquot of Fe solution (marked as * and **, respectively). Few of these aliquots were from the same eluate Fe, while most of them are independent from column chemistry. For details refer to the text and Table S2.

2.3 Optimisation of isotopic determinations

Faraday cup alignment and argide interferences: A proper Faraday cup alignment and a good resolution of argide interferences are necessary for high-precision Fe isotopic determination (Malinovsky et al. 2003, Weyer and Schwieters 2003, Dauphas et al. 2009b). Faraday cup alignment was tested in MR mode. The virtual mass increased by a step of 0.001 atomic mass unit (amu) after each eleven measurements of $^{56}Fe/^{54}Fe$ and $^{57}Fe/^{54}Fe$ ratios on a 1.5 μg ml^{-1} GSB Fe solution, bracketed by measurements of the blank 3% HNO_3 for on-peak zero correction. Five groups of $\delta^{56}Fe$ and $\delta^{57}Fe$ thus could be calculated for each virtual mass. Highly reproducible and accurate data (2*s* ranges from 0.024 to 0.075, and mean $\delta^{56}Fe$ and $\delta^{57}Fe$ values within 0.02‰ and 0.03‰, respectively) could be achieved over a mass range of 0.005 amu (Fig. 2). This supports a previous observation that an optimum cup alignment can provide low mass shoulders over a mass range of 0.005 amu for high-precision Fe isotopic determination by a Thermo Neptune instrument in MR mode (Dauphas et al. 2009b). The most precise and accurate data were produced at the visual centre of the shoulder, where the Ar argide tails were insignificant. All data during the course of this study yielded a slope of 1.476±0.010 (2SE, 95% confidence intervals; calculated by Isoplot, Ludwig 2001) in a $\delta^{57}Fe$ versus $\delta^{56}Fe$ diagram (Fig. 3), indistinguishable within error from the equilibrium (1.475) mass-dependent fractionation law (Young et al. 2002). This suggests insignificant contribution of interfering polyatomic ions to the Fe shoulders.

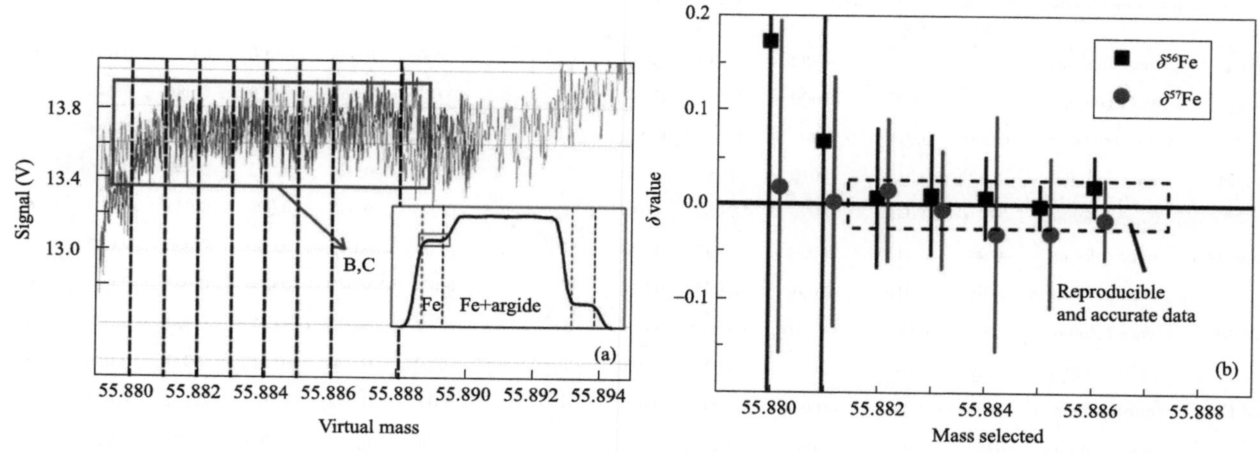

Fig. 2　Test of the peak flatness and interference tailing on Fe isotope measurement in MR using the SIS and 1.5 μg ml^{-1} GSB Fe solutions in 3% HNO_3. The top panel (a) shows peak scans at 54 (green) and 56 (red) of the left shoulder plateau. The signal intensity of 54 was multiplied by 16.7. Eleven runs of isotopic measurements were continuously made at each virtual mass marked by the vertical lines, which provides five delta value for $^{56}Fe/^{54}Fe$ and $^{57}Fe/^{54}Fe$. The mean values and 2*s* of delta values measured are shown in (b), and those for $\delta^{57}Fe$ were slightly and arbitrarily moved to the right for the purpose of clarity.

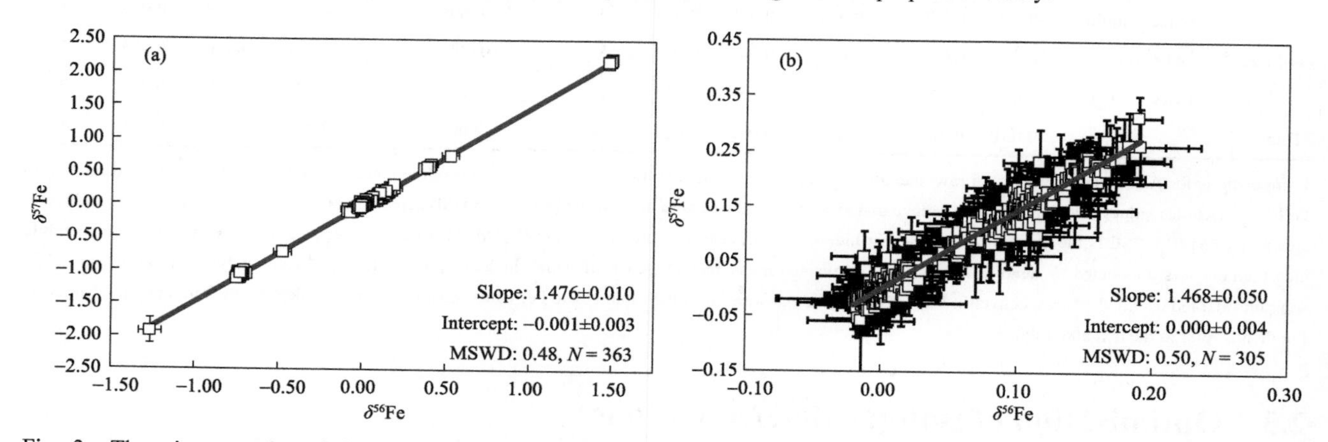

Fig. 3　Three-isotope plot of δ-values measured for standard solutions, elution cuts and geological samples (N = 363) (a). Three-isotope plot of the data with $\delta^{56}Fe$ ranging from −0.1 to +0.2‰ are detailed in (b), suggesting that the observed slope is not randomly determined by the two or three most fractionated samples, but reflects the mass-dependent fractionation law.

Concentration matching: Early studies show that matching of Fe concentration between samples and bracketing calibrators can be important for accurate Fe isotopic determinations (Belshaw et al. 2000, Zhu et al. 2002, Malinovsky et al. 2003). Sample GSB Fe solutions with concentrations ranging from 0.75 to 3.0 μg ml^{-1} were measured against a 1.5 μg ml^{-1} GSB Fe solution in MR mode three to four times, yielding $\delta^{56}Fe$ within 0.000 ± 0.020‰ (Fig. 4). No correlation between $\delta^{56}Fe$ and the degree of mismatch between samples and standard solutions was observed. This supports the observation of recent studies that the effect of mismatch of concentration can be reduced for the Neptune Plus instrument by using 'wet' plasma with the SIS uptake system (Schoenberg and von Blanckenburg 2005, Dauphas et al. 2009b).

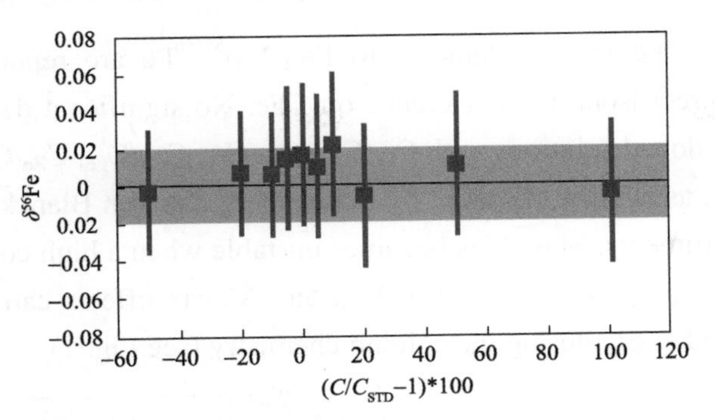

Fig. 4 Effects of concentration matching between sample and bracketing standard solutions on isotopic measurement. The horizontal solid line and grey bar represent the long-term reproducibility for measuring δ^{56}Fe of pure Fe solutions (0 ± 0.020‰).

Acid matching: Concentration of the acid used for iron isotope measurement can significantly affect the instrument mass bias and thus the accuracy of data (Malinovsky et al. 2003, Schoenberg and von Blanckenburg 2005, Dauphas et al. 2009b). This effect was tested by measuring 1.5 μg ml^{-1} GSB Fe solutions with HNO_3 concentrations from 1 to 9% against that in 3% HNO_3. The test sequence was performed twice. Measurement of GSB Fe, with variable acidity, yielded a very large isotopic variation up to 1.00‰ on δ^{56}Fe. The stronger the acid used, the heavier apparent Fe isotopic composition was measured (Fig. 5a). These results agree with observations reported by previous studies. Mass-independent fractionation was also produced when the difference in acidity was large (Fig. 5b). This is not a consequence of variation in ArN^+, because (a) argide interferences were insignificant due to high mass resolution, and (b) if ArN^+ tails are present and increase with increasing acidity, a larger ArN^+ interference and thus an apparently lighter isotope composition would be expected at higher acidity. Stronger space charge effects, which may occur at higher acidity, can yield apparent heavier isotope compositions by preferential loss of light isotopes and produce a slope steeper than the theoretical mass fractionation law (ca. 1.5). Therefore, it is crucial to match the acidity between samples and standard solutions for high accuracy. For each routine analytical sequence, the same batch of 3% HNO_3 was used to prepare blank, sample and standard solutions.

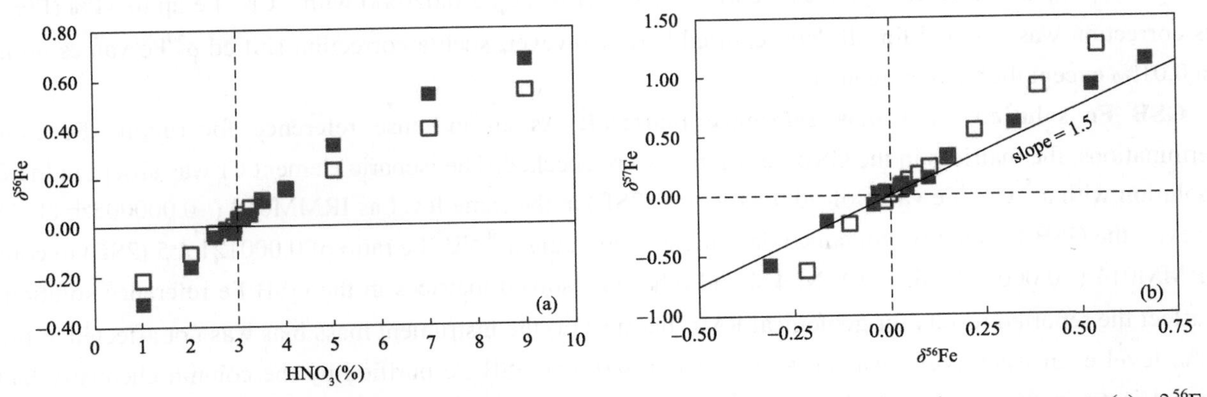

Fig. 5 Effects of acidity matching between sample and bracketing standard solutions on isotopic measurement: (a) δ^{56}Fe vs. HNO_3 (%) (a) and (b)δ^{57}Fe vs. δ^{56}Fe. The 1.5 μg ml^{-1} GSB Fe with HNO_3 concentration from 1 to 9% was measured against that in 3% HNO_3. The testing sequence was repeated twice, and the filled and open symbols show the replicate results.

Matrix effects: Matrix elements present in sample solutions may significantly reduce the precision and accuracy of Fe isotopic determinations (Arnold et al. 2004). Thereofore, the influence of Mg, Cr, Ni, Cu, Co and Zn on Fe isotopic determinations was investigated. Solutions of 1.5 μg ml^{-1} Fe mixed with up to 15 μg ml^{-1} single matrix elements were measured by bracketing with pure Fe solutions. The resulting δ^{56}Fe data are plotted in Fig. 6

against concentration ratios of the matrix elements to Fe. $2\times\delta^{57/56}Fe$ are reported rather than $\delta^{56}Fe$ for the Cr-doping tests, as ^{54}Cr is a direct isobaric interference on ^{54}Fe. No significant deviation on $\delta^{56}Fe$ was observed within the quoted error for the doped solutions with C_{Cr}/C_{Fe}, C_{Ni}/C_{Fe}, C_{Co}/C_{Fe}, C_{Zn}/C_{Fe} up to 1, C_{Cu}/C_{Fe} up to 2 and C_{Mg}/C_{Fe} up to 10, respectively, as already observed by Schoenberg and von Blanckenburg (2005) and Dauphas et al. (2009b) using Neptune instruments. Mass bias becomes unstable when a high concentration of matrix elements (e.g., Cu, Co, Zn) exists (see $C_{Cu, Co, Zn}/C_{Fe}$ at 10 in Fig. 6a). Matrix effects can be routinely eliminated since matrix elements can be removed easily during the column chemistry (see Fig. 1).

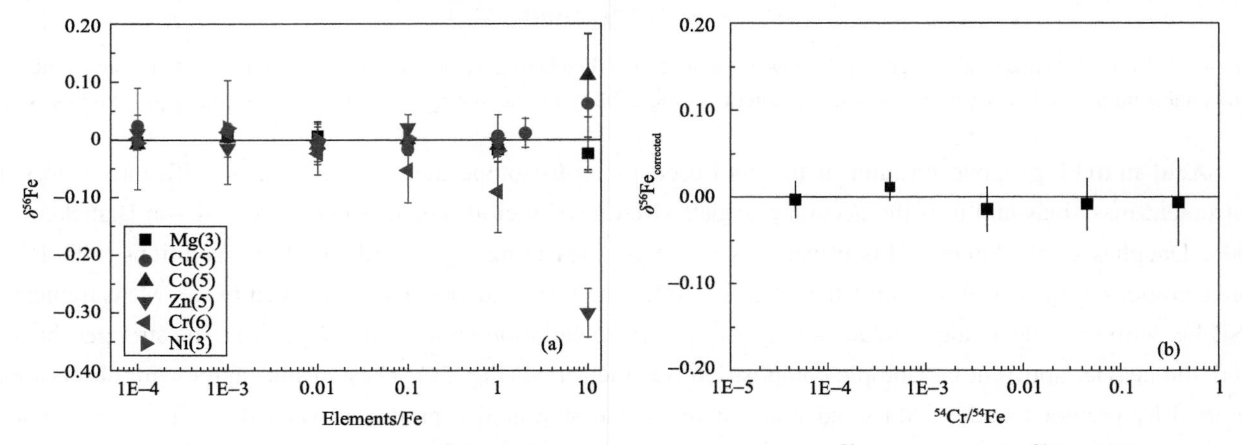

Fig. 6 Effects of matrix elements on Fe isotopic measurement (a) and the isobaric ^{54}Cr correction on ^{54}Fe (b). Because of an isobaric interference of ^{54}Cr on ^{54}Fe, the isotopic compositions for Cr-doping experiments are plotted as $2\times\delta^{57/56}Fe$ in (a). Chromium interference was monitored by ^{53}Cr and corrected by assuming $^{54}Cr/^{53}Cr$ as ~0.2571 (see text for details).

Correction of Cr interferences: Quantitative removal of Cr was straightforward with the chemistry procedure presented above. However, trace amounts of remaining Cr may substantially affect the ^{54}Fe measurement. Chromium interference was monitored by ^{53}Cr and corrected by assuming a $^{54}Cr/^{53}Cr$ ratio of ~0.2571, which is the measured ratio on a 100 ng ml^{-1} GSB Cr solution during the Cr-doping measurement. Tests on the Cr-doping measurements yielded a correct value for $\delta^{56}Fe$ (±0.020‰) with $^{54}Cr/^{54}Fe$ up to 41% (Fig. 6b). This correction was adopted for all data reported here. However, such a correction shifted $\delta^{56}Fe$ values no more than 0.01‰ except the Cr-doping data.

GSB Fe solution as a new reference material: As an in-house reference for routine Fe isotopic determinations, the matrices in the GSB Fe solution were checked. The isobaric element Cr was absent in the GSB Fe solution with a $^{53}Cr/^{54}Fe$ value of −0.000003±7 (2SE) at the same level as IRMM014 (−0.000005±11, 2SE). However, the GSB Fe solution contained detectable Ni yielding a $^{60}Ni/^{54}Fe$ ratio of 0.000121±5 (2SE) in contrast to IRMM014 (−0.000004±6, 2SE). Nickel and other unresolved matrices in the GSB Fe reference solution did not affect the reported results for geological RMs, because (a) the instrument mass bias was not affected at the ca. 0.02‰ level even when Ni/Fe was up to 1 (see Fig. 6a), (b) GSB Fe purified by the column chemistry had an identical isotopic composition to that unprocessed (Table 1), and (c) the geological samples, measured against GSB and IRMM014 Fe separately, yielded consistent $\delta^{56}Fe_{IRMM014}$ values within 0.02‰. Given that the commonly used IRMM-014 Fe is no longer available, it is suggested that GSB Fe be used as a new RM for precise cross-laboratory calibration of Fe isotopes. Aliquots of this Fe solution are available upon request from the authors.

2.4 Reproducibility, accuracy and error report

Reproducibility of the instrument: Multiple analyses for each sample can significantly enhance long-term reproducibility (Poitrasson and Freydier 2005, Schoenberg and von Blanckenburg 2005, Weyer et al. 2005, Dauphas et al. 2009b, Millet et al. 2012). The effect on the reproducibility of replicate analyses was evaluated following the methods of Poitrasson and Freydier (2005) and Millet et al. (2012), and plotted in Fig. 7a. The 1783 data of the bracketing calibrators measured from June 2012 to June 2013 were randomly subdivided into subgroups with N analyses (N=2–19). Two standard deviations of the mean values of these subgroups were calculated, and the calculation was repeated 1000 times to give a reliable estimate. The results show that the $2s$ values of multiple analyses decreases dramatically with increasing N when $N \leqslant 5$, decreases gently when $5 < N < 10$ and very slowly when $N \geqslant 10$ (Fig. 7a). The expected $2s$ values for N=4 and 9 are 0.033‰ and 0.022‰ for δ^{56}Fe, respectively, which are consistent with the reproducibility obtained by repeated analyses of the GSB Fe reference solution (0.020‰, Table 1). It is important to note that the calculated $2s$ for measurements including N replicate analysis each approximately equals that for a single analysis divided by $\sqrt{N}$, possibly reflecting the intrinsic nature of a normal distribution. This suggests consistency between $2s$ and the average error (2SE) for a large data set (Fig. 7a), because the errors reported here were calculated as $2s_{\text{bracketing calibrator}} / \sqrt{N}$.

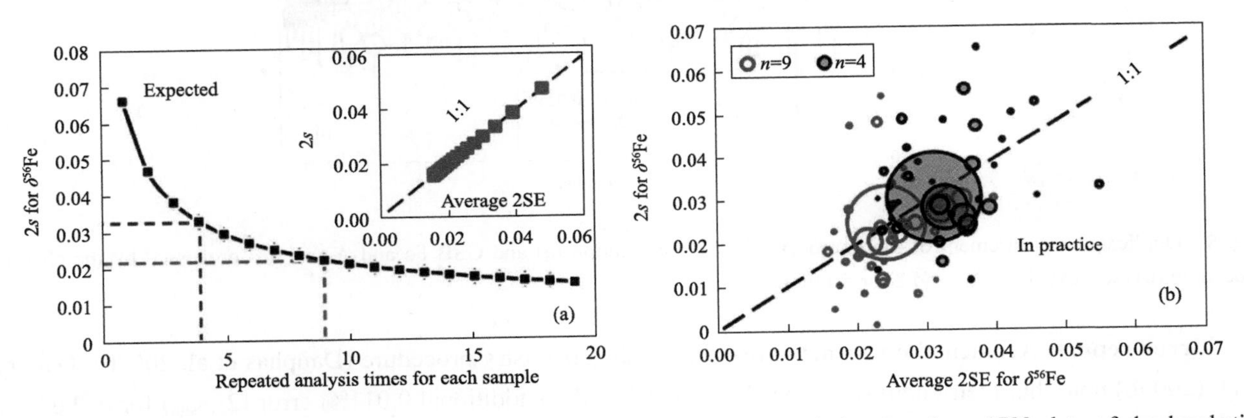

Fig. 7 The expected $2s$ versus number of replicate analyses on each sample solution based on 1783 data of the bracketing calabrators from June 2012 to June 2013 (a), and $2s$ versus average 2SE for replicate analyses of pure Fe solutions, A-samples and geological RMs (b). The inset in (a) shows the consistency between the expected $2s$ and 2SE, which is verified by the practice data. The larger bubble size in (b) reflects more replicate measurements on individual samples (N). A-samples are a series of synthetic solutions that mimic the compositions of representative geological samples (see the text and supplementary Table S1 for details).

Reproducibility of the chemistry and accuracy: Given the potential heterogeneity of natural geological reference materials, a series of synthetic solutions (referred as 'A-samples' in the following discussion) was prepared to evaluate the errors from sample processing by using GSB single element solutions from the China Iron and Steel Research Institute (Table S1). The compositions of these synthetic samples mimic those of the depleted mantle, average mid ocean ridge basalt (Salters and Stracke 2004), the upper and lower continental crust (Rudnick and Gao 2003), a highly fractionated granite (07SHU-5, He et al. 2011), a Fe barren limestone (GBW-07120 commercially available from the IGGE) and samples enriched in transition group elements with Fe : Cr, Ni, Cu, Co, Zn as 100 : 1 and 10 : 1, respectively.

These synthetic samples (A-samples) and GSB Fe were processed through the column chemistry (1–2 passes) as well as the whole chemistry procedure and yielded an overall average $\delta^{56}\text{Fe}_{\text{GSB}}$ of -0.001 ± 0.025‰ ($2s$, n=74) relative to the unprocessed ones (expected to be zero) (Fig. 8,Table 1). This demonstrates that no systematic isotopic fractionation was associated with column chemistry within 0.025‰ ($2s$, n=74) for δ^{56}Fe. Geological RMs processed through the column chemistry once or twice gave δ^{56}Fe values identical within the quoted errors (Table S3).

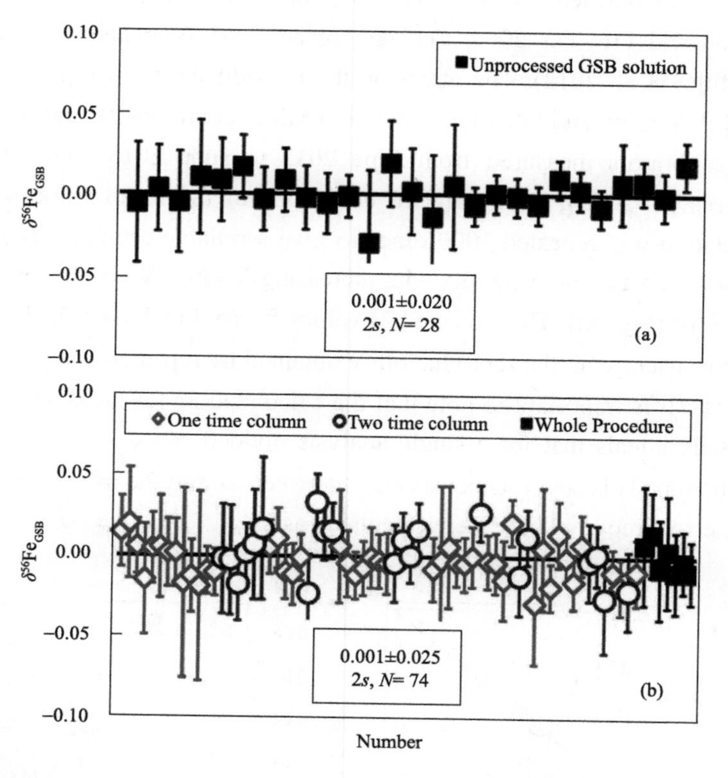

Fig. 8　Duplicated measurement of the unprocessed GSB Fe solution (a) and GSB Fe and A-samples processed by the chemical procedure (b) (see text, Table S1 and 2 for details).

Error report: Accidental errors may arise during the chemistry procedure (Dauphas et al. 2009b). Dauphas et al. (2009b) note that their chemistry procedure contributed an additional 0.011‰ error ($2s_{\text{Chem}}$) for δ^{56}Fe based on repeated analysis of BHVO-1 over 1 year. Millet et al. (2012) also estimated a 0.011‰ error from their chemical separation protocol by the difference between practical and theoretical reproducibility of δ^{56}Fe.

Both GSB Fe without chemistry and GSB Fe and A-samples with chemistry were repeatedly measured in this study (Fig. 8, Table 1 and Table S2). From this, we estimate $2s_{\text{Chem}}$ to be 0.005‰ and 0.012‰ for δ^{56}Fe and δ^{57}Fe, respectively, from the $2s$-differences of these two groups of measurements (Millet et al. 2012). Accidental error from the chemistry was also evaluated following Dauphas et al. (2009b) based on distribution of seventy-four duplicate analyses of GSB Fe and A-samples through chemistry. In brief, a constant s_{Chem} was incorporated into s_{Data} ($s_{\text{Data}}^2 = s_{\text{MS}}^2 + s_{\text{Chem}}^2$) to make the MSWD (mean square of the weighted deviates, or reduced χ^2) of the replicate results equal to 1.0. Consistent $2s_{\text{Chem}}$ values were estimated to be 0.005‰ and 0.013‰ for δ^{56}Fe and δ^{57}Fe, respectively, suggesting that our estimates are robust. These estimated $2s_{\text{Chem}}$ values are insignificant and negligible, and were not integrated into the final δ-values for natural samples, because calibration from $\delta^{56}\text{Fe}_{\text{GSB}}$ to $\delta^{56}\text{Fe}_{\text{IRMM014}}$, for example, had already included an extra uncertainty of 0.005‰ (see 'Optimisation of isotopic determinations' and Table 1, $(2s)_{\text{IRMM014}}^2 = (2s)_{\text{GSB}}^2 + 0.005^2$). This calibration may cause a systematic bias rather than an increase in the dispersion of the data. The average of errors (2SE, 95% confidence interval) for individual replicate measurements and $2s$ of δ^{56}Fe were compiled for GSB Fe without chemistry, GSB and eight A-samples with chemistry and twenty-two geological RMs (Tables 1 and 2, Fig. 7b). Most samples, especially those measured frequently, plot along the 1 : 1 line, indicating a very good consistency between the measurement error reported and the measurement reproducibility.

2.5 Iron isotope compositions of geological reference materials

The iron isotopic composition of twenty-two geological RMs, including sixteen igneous rocks, one metamorphic rock, two carbonates, one sandstone, one shale and one clay, were measured at least three times. The replicate results are presented in Table S3 and summarised in Table 2. Few 'Cu-spiking' data yielded results consistent with those by a solely calibrator-sample bracketing technique within the quoted errors and thus will not be discussed separately.

Sixteen igneous RMs, including one peridotite, six basalts, one diabase, five andesites, two granitoids, and one carbonatite, yielded δ^{56}Fe values ranging from −0.07 to 0.16‰. One amphibolite, GBW-07122 (GSR-15) from IGGE, yielded a δ^{56}Fe value of 0.06‰. Fourteen of these seventeen igneous and metamorphic RMs have been measured previously (Beard et al. 2003b, Craddock and Dauphas 2010, Craddock et al. 2013, Dauphas et al. 2009a,b, Huang et al. 2011, Millet et al. 2012, Poitrasson and Freydier 2005, Poitrasson et al. 2004, Schoenberg and von Blanckenburg 2006, Schuessler et al. 2009, Sossi et al. 2012, Telus et al. 2012, Teng et al. 2008, Wang et al. 2012, Weyer et al. 2005, Zhao et al. 2012), and only the data with reported measurement reproducibility better than 0.1‰ are compiled. As shown in Table S3, different sets of powder for USGS geological RMs, such as BHVO-1 vs. BHVO-2, BCR-1 vs. BCR-2, etc., gave identical Fe isotopic compositions and will be considered to be identical in the following. Our values perfectly match with the mean values derived from nine laboratories, excluding data from the University of Copenhagen and except for COQ-1 (Table S3, Fig. 9), despite the reported reproducibility varying from 0.02 to 0.08‰ for δ^{56}Fe. The two standard deviation value of the mean δ^{56}Fe values from our measurements and the other nine laboratories was better than 0.03‰, indicating insignificant systematic bias. This consistency indicates two observations: (a) that complete recovery of Fe and elimination of matrices can be easily achieved, and (b) that all data reported have been measured or once calibrated against IRMM-014 Fe. With respect to the data from the University of Copenhagen, Fe isotope determination was not conducted on a high-mass resolution MC-ICP-MS (Dideriksen et al. 2006), and thus, it is not surprising that remaining polyatomic interferences may affect the data reduction of double-spiked measurements leading to erroneous δ^{56}Fe. The mean δ^{56}Fe value obtained for COQ-1 in this study is larger than that from the University of Chicago by 0.050‰ (Craddock and Dauphas 2010). Given the perfect match of data for the other geological RMs from these two laboratories, the relatively large discrepancy on COQ-1 is probably due to greater uncertainties, for example 0.042 to 0.067‰ on δ^{56}Fe, for this sample in Craddock and Dauphas (2010).

Table 2 Summary of duplicate measurements of geological reference materials

Sample	Description	$\delta^{56}Fe^{*}$	$2s$	$\delta^{57}Fe^{*}$	$2s$	2SE	N	$\delta^{56}Fe^{**}$	$2s$	$\delta^{57}Fe^{**}$	$2s$	2SE	N
Igneous and metamorphic rock RMs													
BHVO-2	Basalt	0.112	0.021	0.163	0.040	0.024	27	0.109	0.028	0.160	0.058	0.032	47
BCR-2	Basalt	0.080	0.024	0.123	0.036	0.023	10	0.084	0.029	0.130	0.048	0.032	17
W-2a	Diabase	0.053	0.025	0.074	0.054	0.026	8	0.054	0.038	0.071	0.070	0.037	13
AGV-2	Andesite	0.102	0.022	0.151	0.036	0.026	16	0.096	0.027	0.148	0.039	0.035	25

Continued

Sample	Description	$\delta^{56}Fe^{*}$	2*s*	$\delta^{57}Fe^{*}$	2*s*	2SE	*N*	$\delta^{56}Fe^{**}$	2*s*	$\delta^{57}Fe^{**}$	2*s*	2SE	*N*
GSP-2	Granodiorite	0.154	0.012	0.230	0.032	0.024	11	0.157	0.025	0.222	0.038	0.036	16
COQ-1	Carbonatite	−0.065	0.028	−0.106	0.026	0.019	6	−0.065	0.036	−0.094	0.055	0.027	7
JP-1	Peridotite	0.006	0.011	0.012	0.016	0.031	2	−0.006	0.031	0.015	0.068	0.046	3
JB-1b	Basalt	0.095	0.054	0.125	0.053	0.024	3	0.099	0.066	0.119	0.046	0.037	4
JB-2	Basalt	0.060	0.035	0.091	0.058	0.028	3	0.077	0.050	0.096	0.060	0.042	4
JA-1	Andesite	0.056	0.008	0.087	0.030	0.029	5	0.051	0.044	0.066	0.052	0.043	6
GBW-07103, GSR-1	Granite	0.156	0.021	0.213	0.010	0.025	8	0.148	0.047	0.214	0.084	0.037	10
GBW-07104, SDR-2	Andesite	0.091	0.016	0.116	0.037	0.024	3	0.089	0.030	0.113	0.072	0.038	4
GBW-07105, GSR-3	Basalt	0.159	0.017	0.230	0.015	0.020	3	0.171	0.027	0.244	0.041	0.034	4
GBW-07122, GSR-15	Amphibolite	0.057	0.009	0.095	0.048	0.021	4	0.054	0.014	0.087	0.059	0.023	4
JB-3	Basalt	0.080	0.036	0.117	0.057	0.027	3	0.087	0.044	0.114	0.092	0.041	4
JA-2	Andesite	0.105	0.037	0.142	0.032	0.027	3	0.097	0.038	0.136	0.082	0.040	3
JA-3	Andesite	0.086	0.001	0.120	0.034	0.023	3	0.081	0.011	0.116	0.042	0.036	4
Sedimentary rock and mineral RMs													
GBW-07106, GSR-4	Sandstone	0.384	0.033	0.576	0.029	0.025	3	0.383	0.049	0.573	0.068	0.032	4
GBW-07107, GSR-5	Shale	0.136	0.015	0.190	0.028	0.022	6	0.142	0.020	0.209	0.035	0.032	9
GBW-07108, GSR-6	Limestone	0.096	0.030	0.116	0.039	0.040	5	0.099	0.033	0.100	0.049	0.055	6
GBW-07120, GSR-13	Limestone	0.139	0.048	0.212	0.093	0.023	8	0.141	0.049	0.213	0.083	0.027	8
NAu-2	Clay	1.482	0.018	2.193	0.047	0.020	6	1.482	0.034	2.194	0.073	0.020	6

Data are reported relative to IRMM-014. 2SE is the mean of δ^{56}Fe 2SE for individual replicate measurements on each RM. *N* denotes the number of duplicate measurements, each of which included nine or four replicate analyses on an aliquot of Fe solution (marked as * and **, respectively). Few of these aliquots were from the same eluate Fe, while most were independent from chemistry to isotope determination. For details and raw data refer to the text and Table S3.

Five sedimentary rock reference materials were measured in this study with δ^{56}Fe ranging from 0.09 to 1.47‰. The shale GBW-07107 (GSR-5) from IGGE gave a δ^{56}Fe of 0.14‰, similar to that of igneous rocks and in the range of global shales (Beard et al. 2003b). Two carbonates (GSR-6 and GSR-13) yielded δ^{56}Fe ranging from 0.10 to 0.14‰, also close to igneous rocks and shales. This may suggest that Fe is dominantly present as silica particles in the carbonates, given the very low dissolved Fe content in oxidised Phanerozoic Ocean water (Radic et al. 2011) and the fact that these carbonates have SiO_2 contents from 6.65 to 15.60% m/m(IGGE certificates). Sandstone GSR-4 had a δ^{56}Fe value of 0.384±0.033‰ (2*s*, *n*=3), significantly higher than igneous rocks. Clay NAu-2 from the Clay Mineral Society yielded δ^{56}Fe as high as 1.482±0.018‰ (2*s*, *n*=6). This coincides with the enrichments of Fe^{3+} in the sandstone and nontronite (IGGE certificate, Keeling et al. 2000). Fe^{3+} species tend to enrich heavier isotopes (Polyakov and Mineev 2000).

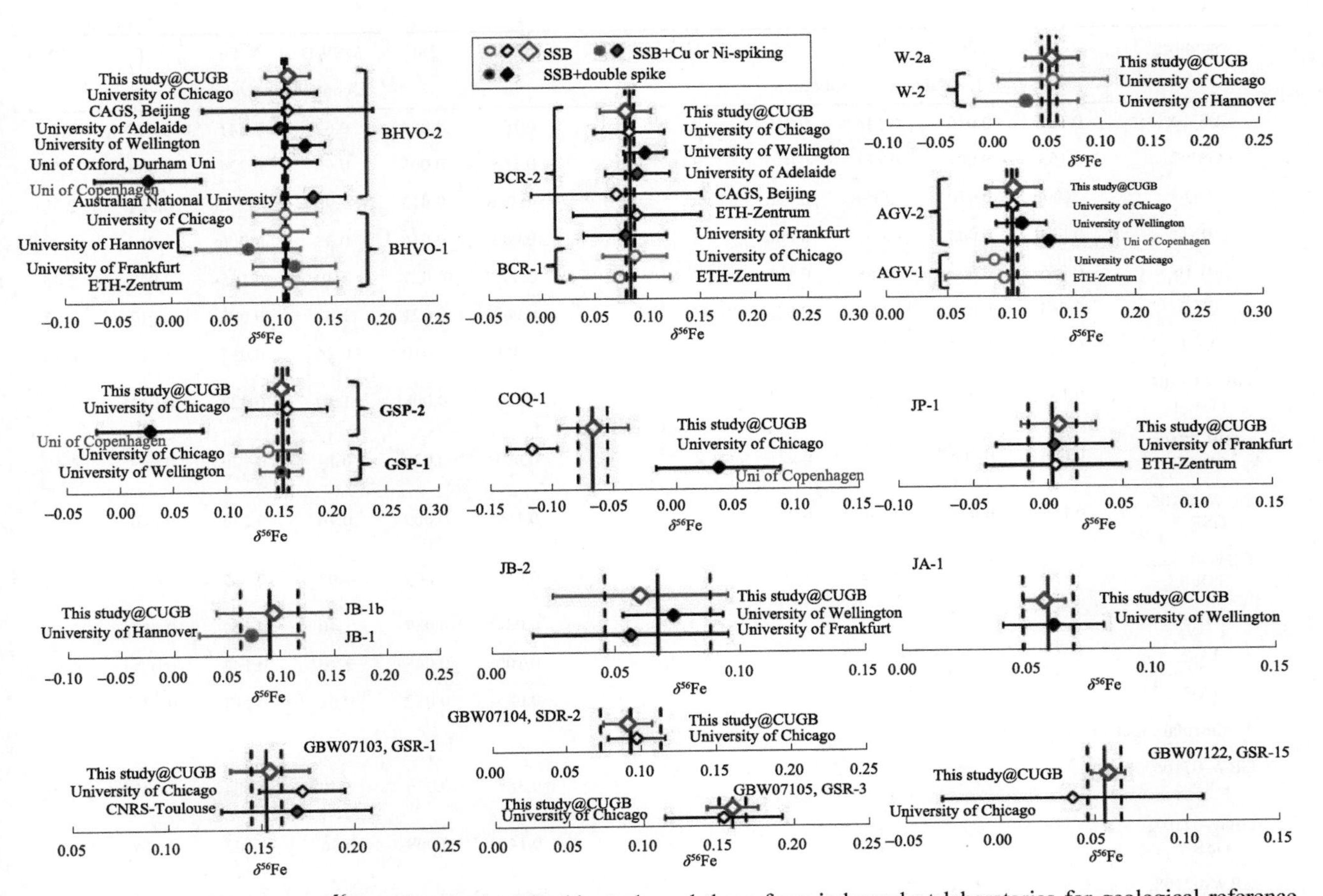

Fig. 9 Comparison of mean δ^{56}Fe value obtained in this study and those from independent laboratories for geological reference materials. Error bars represent two standard deviations. Reported long-term reproducibilities are adopted when numbers of duplicate measurements are lower than three. Only the data with $N = 9$ multiple analysis each are shown for this study. The vertical solid and dotted lines represent the new recommended values and corresponding 2SE (95% CI).

Based on the available high-precision data from ten independent laboratories, new recommended values are presented for twenty-two geological RMs as a weighted averages, weighting by assigned errors only, and were calculated by using the program Isoplot (Ludwig 2001) (Table 3). The reported uncertainties are within the 95% confidence interval and reflect the actual scatter of the data with application of Student's t-distribution. The new recommended values do not vary much from recent studies (Craddock and Dauphas 2010, Millet et al. 2012), but are more robust due to the large, high-precision data set and the compilation from ten independent laboratories. They may help in the interlaboratory calibration of Fe isotope measurements in the future, and the more so since the commonly used CRM, IRMM-014 is no longer available.

Table 3 Comparison of mean values of ten independent laboratories and new recommended values

Sample	δ^{56}Fe	$2s$	δ^{57}Fe	$2s$	N	δ^{56}Fe	2SE	MSWD	δ^{57}Fe	2SE	MSWD
	Interlaboratory calibration					Recommended values					
Igneous and metamorphic rock RMs											
BHVO-2	0.110	0.023	0.168	0.049	9	0.109	0.002	0.86	0.169	0.005	1.60
BCR-2	0.084	0.017	0.126	0.023	8	0.084	0.004	0.82	0.123	0.007	0.91
W-2a	0.045	0.029	0.053	0.079	3	0.051	0.007	1.70	0.070	0.014	1.70

Continued

Sample	δ^{56}Fe	2*s*	δ^{57}Fe	2*s*	*N*	δ^{56}Fe	2SE	MSWD	δ^{57}Fe	2SE	MSWD
	Interlaboratory calibration					Recommended values					
AGV-2	0.101	0.012	0.148	0.019	4	0.101	0.004	0.62	0.141	0.005	0.84
GSP-2	0.153	0.001	0.223	0.016	3	0.155	0.005	0.68	0.224	0.009	1.02
COQ-1	−0.090	0.070	−0.136	0.085	2	−0.066	0.013	2.90	−0.113	0.024	2.30
JP-1	0.004	0.003	0.007	0.008	3	0.002	0.016	0.38	0.006	0.019	0.28
JB-1b	0.085	0.030	0.088	0.104	2	0.091	0.027	3.30	0.115	0.042	2.40
JB-2	0.063	0.018	0.094	0.026	3	0.067	0.021	1.80	0.095	0.017	1.07
JA-1	0.058	0.005	0.088	0.002	2	0.058	0.010	0.24	0.087	0.014	0.43
GBW-07103, GSR-1	0.166	0.018	0.240	0.051	3	0.154	0.008	1.50	0.227	0.014	1.20
GBW-07104, SDR-2	0.094	0.008	0.123	0.021	2	0.093	0.020	0.14	0.120	0.049	1.40
GBW-07105, GSR-3	0.156	0.010	0.230	0.002	2	0.159	0.009	0.76	0.232	0.016	0.37
GBW-07122, GSR-15	0.048	0.027	0.072	0.066	2	0.055	0.009	1.02	0.085	0.028	2.30
JB-3						0.084	0.027	1.40	0.109	0.052	2.10
JA-2						0.100	0.045	2.20	0.142	0.024	0.55
JA-3						0.085	0.012	0.02	0.117	0.023	0.59
Sedimentary rock and mineral RMs											
GBW-07106, GSR-4						0.387	0.026	1.50	0.584	0.025	0.70
GBW-07107, GSR-5						0.140	0.008	0.36	0.202	0.016	0.37
GBW-07108, GSR-6						0.093	0.015	0.43	0.113	0.027	0.22
GBW-07120, GSR-13						0.143	0.024	6.80	0.213	0.041	5.50
NAu-2						1.482	0.009	0.61	2.193	0.025	1.40

Inter-laboratory comparison is provided as mean and 2*s* of mean values from each laboratory. *N* denotes the number of independent laboratories from which these data were derived. New recommended values were calculated by Isoplot (Ludwig 2001) according to the consistent results from ten independent laboratories, and uncertainties are reported as 2SE, 95% c.i. with the actual distribution of individual data considered.

3 Concluding remarks

This study presents a thorough set of tests for precision and accuracy of a Fe isotope determination procedure using sample-calibrator bracketing techniques on a Neptune Plus MC-ICP-MS, which demonstrates reproducibility and accuracy better than 0.03‰ for δ^{56}Fe. We provide a comprehensive set of Fe isotope data for twenty-two commonly used and commercially available geological reference materials of highly variable matrices. Our data contribute to a better characterisation of Fe isotopic compositions of igneous and metamorphic reference materials. In addition, we for the first time report high-quality Fe isotope data for five sedimentary RMs. This should help interlaboratory calibration of Fe isotope measurement in the future.

Acknowledgements

Constructive comments from four anonymous reviewers and efficient editorial handling of Bill McDonough are highly appreciated. We thank Jochen Hoefs for helpful suggestions and English polishing. Dong Hailiang and Zeng Qiang kindly provided the NAu-2 powder. This work is supported by State Key Laboratory of Geological Processes and Mineral Resources, the Fundamental Research Funds for the Central Universities (2652014056) and the National Natural Science Foundation of China (41103028) to YSH.

References

Anbar A., Roe J.E., Barling J. and Nealson K.H. (2000) Non-biological fractionation of iron isotopes. Science, 288, 126-128.

Arnold G.L., Weyer S. and Anbar A.D. (2004) Fe isotope variations in natural materials measured using high mass resolution multiple collector ICP-MS. Analytical Chemistry, 76, 322-327.

Balci N., Bullen T.D., Witte-Lien K., Shanks W.C., Motelica M. and Mandernack K.W. (2006) Iron isotope fractionation during microbially stimulated Fe(II) oxidation and Fe(III) precipitation. Geochimica et Cosmochimica Acta, 70, 622-639.

Beard B.L., Johnson C.M., Cox L., Sun H., Nealson K.H.and Aguilar C. (1999) Iron isotope biosignatures. Science, 285, 1889-1892.

Beard B.L., Johnson C.M., Skulan J.L., Nealson K.H., Cox L. and Sun H. (2003a) Application of Fe isotopes to tracing the geochemical and biological cycling of Fe. Chemical Geology, 195, 87-117.

Beard B.L., Johnson C.M., von Damm K.L. and Poulson R.L. (2003b) Iron isotope constraints on Fe cycling and mass balance in oxygenated Earth oceans. Geology, 31, 629-632.

Belshaw N.S., Zhu X.K., Guo Y. and O'Nions R.K. (2000) High precision measurement of iron isotopes by plasma source mass spectrometry. International Journal of Mass Spectrometry, 197, 191-195.

Craddock P.R. and Dauphas N. (2010) Iron isotopic compositions of geological reference materials and chondrites. Geostandards and Geoanalytical Research, 35, 101-123.

Craddock P.R., Warren J.M. and Dauphas N. (2013) Abyssal peridotites reveal the near-chondritic Fe isotopic composition of the Earth. Earth and Planetary Science Letters, 365, 63-76.

Dauphas N., Janney P.E., Mendybaev R.A., Wadhwa M., Richter F.M., Davis A., van Zuilen M., Hines R. and Foley C.N. (2004) Chromatographic separation and multicollection-ICP-MS analysis of iron: Investigating mass-dependent and - independent isotope effects. Analytical Chemistry, 76, 5855-5863.

Dauphas N., Craddock P.R., Asimow P.D., Bennett V.C., Nutman A.P. and Ohnenstetter D. (2009a) Iron isotopes may reveal the redox conditions of mantle melting from Archaean to Present. Earth and Planetary Science Letters, 288, 255-267.

Dauphas N., Pourmand A. and Teng F.-Z. (2009b) Routine isotopic analysis of iron by HR-MC-ICP-MS: How precise and how accurate? Chemical Geology, 267, 175-184.

Dideriksen K., Baker J.A. and Stipp S.L.S. (2006) Iron isotopes in natural carbonate minerals determined by MC-ICP-MS with a ^{58}Fe-^{54}Fe double spike. Geochimica et Cosmochimica Acta, 70, 118-132.

He Y.S., Li S.G., Hoefs J., Huang F., Liu S.A. and Hou Z.H. (2011) Post-collisional granitoids from the Dabie orogen: New evidence for partial melting of a thickened continental crust. Geochimica et Cosmochimica Acta, 75, 3815-3838.

Heimann A., Beard B.L. and Johnson C.M. (2008) The role of volatile exsolution and sub-solidus fluid/rock interactions in producing high $^{56}Fe/^{54}Fe$ ratios in siliceous igneous rocks. Geochimica et Cosmochimica Acta, 72, 4379-4396.

Heimann A., Johnson C.M., Beard B.L., Valley J.W., Roden E.E., Spicuzza M.J. and Beukes N.J. (2010) Fe, C, and O isotope compositions of banded iron formation carbonates demonstrate a major role for dissimilatory iron reduction in ~ 2.5 Ga marine environments. Earth and Planetary Science Letters, 294, 8-18.

Huang F., Zhang Z., Lundstrom C.C. and Zhi X. (2011) Iron and magnesium isotopic compositions of peridotite xenoliths from Eastern China. Geochimica et Cosmochimica Acta, 75, 3318-3334.

Johnson C.M., Skulan J.L., Beard B.L., Sun H., Nealson K.H. and Braterman P.S. (2002) Isotopic fractionation between Fe(III) and Fe(II) in aqueous solutions. Earth and Planetary Science Letters, 195, 141-153.

Johnson C.M., Beard B.L. and Roden E.E. (2008) The iron isotope fingerprints of redox and biogeochemical cycling in modern and ancient Earth. Annual Review of Earth and Planetary Sciences, 36, 457-493.

Keeling J.L., Raven M.D. and Gates W.P. (2000) Geology and characterization of two hydrothermal nontronites from weathered metamorphic rocks at the Uley graphite mine, South Australia. Clays and Clay Minerals, 48, 537-548.

Kehm K., Hauri E.H., Alexander C.M.O.D. and Carlson R.W. (2003) High precision iron isotope measurements of meteoritic material by cold plasma ICP-MS. Geochimica et Cosmochimica Acta, 67, 2879-2891.

Li W.Q., Czaja A.D., Kranendonk M.J.V., Beard B.L., Roden E.E. and Johnson C.M. (2013) An anoxic, Fe(II)-rich, U-poor ocean 3.46 billion years ago. Geochimica et Cosmochimica Acta, 120, 65-79.

Ludwig K.R. (2001) User manual for Isoplot/Ex (rev. 2.4.9): A geochronological toolkit for Microsoft Excel. Berkeley Geochronology Center, Special Publication, No. 1a, 55 pp.

Malinovsky D., Stenberg A., Rodushkin I., Andren H., Ingri J., Ohlander B. and Baxter D.C. (2003) Performance of high resolution MC-ICP-MS for Fe isotope ratio measurements in sedimentary geological materials. Journal of Analytical Atomic Spectrometry, 18, 687-695.

Millet M.-A., Baker J.A. and Payne C.E. (2012) Ultra-precise stable Fe isotope measurements by high resolution multiple-collector inductively coupled plasma-mass spectrometry with a ^{57}Fe-^{58}Fe double spike. Chemical Geology, 304-305, 18-25.

Poitrasson F. and Freydier R. (2005) Heavy iron isotope composition of granites determined by high resolution MC-ICP-MS. Chemical Geology, 222, 132-147.

Poitrasson F., Halliday A.N., Lee D.-C., Levasseur S. and Teutsch N. (2004) Iron isotope differences between Earth, Moon, Mars and Vesta as possible records of contrasted accretion mechanisms. Earth and Planetary Science Letters, 223, 253-266.

Polyakov V.B. and Mineev S.D. (2000) The use of Mössbauer spectroscopy in stable isotope geochemistry. Geochimica et Cosmochimica Acta, 64, 849-865.

Radic A., Lacan F. and Murray J.W. (2011) Iron isotopes in the seawater of the equatorial Pacific Ocean: New constraints for the oceanic iron cycle. Earth and Planetary Science Letters, 306, 1-10.

Rouxel O.J. (2005) Iron isotope constraints on the Archean and Paleoproterozoic ocean redox state. Science, 307(5712), 1088-1091.

Rudnick R.L. and Gao S. (2003) Composition of the continental crust. In: Holland, H.D. and Turekian, K.K. (eds), Treatise on geochemistry. Elsevier-Pergamon (Oxford), 1-64.

Salters V. and Stracke A. (2004) Composition of the depleted mantle. Geochemistry, Geophysics, Geosystems, 5(5), 1525-2027.

Schauble E.A., Rossman G.R. and Taylor H.P. (2001) Theoretical estimates of equilibrium Fe-isotope fractionations from vibrational spectroscopy. Geochimica et Cosmochimica Acta, 65, 2487-2497.

Schoenberg R. and von Blanckenburg F. (2005) An assessment of the accuracy of stable Fe isotope ratio measurements on samples with organic and inorganic matrices by high-resolution multicollector ICP-MS. International Journal of Mass Spectrometry, 242, 257-272.

Schoenberg R. and von Blanckenburg F. (2006) Modes of planetary-scale Fe isotope fractionation. Earth and Planetary Science Letters, 252, 342-359.

Schuessler J.A., Schoenberg R. and Sigmarsson O. (2009) Iron and lithium isotope systematics of the Hekla volcano, Iceland - Evidence for Fe isotope fractionation during magma differentiation. Chemical Geology, 258, 78-91.

Sossi P.A., Foden J.D. and Halverson G.P. (2012) Redox-controlled iron isotope fractionation during magmatic differentiation: An example from the Red Hill intrusion, S. Tasmania. Contributions to Mineralogy and Petrology, 164, 757-772.

Sossi P.A., Halverson G.P., Nebel O. and Eggins S.M. (2014) Combined separation of Cu, Fe and Zn from rock matrices and improved analytical protocols for stable isotope determination. Geostandards and Geoanalytical Research, Doi: 10.1111/j.1751-908X.2014.00298.x.

Strelow F.W.E. (1980) Improved separation of iron from copper and other elements by anion-exchange chromatography on a 4% cross-linked resin with high concentrations of hydrochloric acid. Talanta, 27, 727-732.

Telus M., Dauphas N., Moynier F., Tissot F.L.H., Teng F.Z., Nabelek P.I., Craddock P.R. and Groat L.A. (2012) Iron, zinc, magnesium and uranium isotopic fractionation during continental crust differentiation: The tale from migmatites, granitoids, and pegmatites. Geochimica et Cosmochimica Acta, 97, 247-265.

Teng F.-Z., Dauphas N. and Helz R.T. (2008) Iron isotope fractionation during magmatic differentiation in Kilauea Iki lava lake. Science, 320(5883), 1620-1622.

Teng F.-Z., Dauphas N., Helz R.T., Gao S. and Huang S. (2011) Diffusion-driven magnesium and iron isotope fractionation in Hawaiian olivine. Earth and Planetary Science Letters, 308, 317-324.

Teng F.-Z., Dauphas N., Huang S. and Marty B. (2013) Iron isotopic systematics of oceanic basalts. Geochimica et Cosmochimica Acta, 107, 12-26.

Wang K., Moynier F., Dauphas N., Barrat J.A., Craddock P.R. and Sio C.K. (2012) Iron isotope fractionation in planetary crusts. Geochimica et Cosmochimica Acta, 89, 31-45.

Weyer S. and Ionov D.A. (2007) Partial melting and melt percolation in the mantle: The message from Fe isotopes. Earth and Planetary Science Letters, 259, 119-133.

Weyer S. and Schwieters J.B. (2003) High precision Fe isotope measurements with high mass resolution MC-ICP-MS. International Journal of Mass Spectrometry, 226, 355-368.

Weyer S. and Seitz H.M. (2012) Coupled lithium- and iron isotope fractionation during magmatic differentiation. Chemical Geology, 294-295, 42-50.

Weyer S., Anbar A.D., Brey G.P., Münker C., Mezger K. and Woodland A.B. (2005) Iron isotope fractionation during planetary differentiation. Earth and Planetary Science Letters, 240, 251-264.

Williams H.M., McCammon A.H., Peslier A.H., Halliday A.N., Teutsch N., Levasseu S. and Burg J.P. (2004) Iron isotope fractionation and the oxygen fugacity of the mantle. Science, 304, 1656-1659.

Young E.D., Galy A. and Nagahara H. (2002) Kinetic and equilibrium mass-dependent isotope fractionation laws in nature and their geochemical and cosmochemical significance. Geochimica et Cosmochimica Acta, 66, 1095-1104.

Zhao X., Zhang H., Zhu X., Tang S. and Yan B. (2012) Iron isotope evidence for multistage melt-peridotite interactions in the lithospheric mantle of eastern China. Chemical Geology, 292-293, 127-139.

Zhu X.K., Guo Y., Williams R.J.P., O'Nions R.K., Matthews A., Belshaw N.S., Canters G.W., de Waal E.C., Weser U., Burgess B.K. and Salvato B. (2002) Mass fractionation processes of transition metal isotopes. Earth and Planetary Science Letters, 200, 47-62.

Supporting information

The following supporting information may be found in the online version of this article:

Table S1. Composition of synthetic samples prepared for accuracy checking.

Table S2. Iron isotope composition of repeated measurement of GSB Fe and A-samples through chemistry.

Table S3. Repeated analyses of geological reference materials and compilation of literature values.

This material is available as part of the online article from: http://onlinelibrary.wiley.com/doi/10.1111/j.1751-908X.2014.00304.x/abstract (This link will take you to the article abstract).

Mass-independent and mass-dependent Ca isotopic compositions of thirteen geological reference materials measured by thermal ionization mass spectrometry*

Yongsheng He[1], Yang Wang[1], Chuanwei Zhu[1], Shichun Huang[2] and Shuguang Li[1]

1. State Key Laboratory of Geological Processes and Mineral Resources, School of Earth Sciences and Mineral Resources, China University of Geosciences, Beijing 100083, China
2. Department of Geoscience, University of Nevada, 4505. Maryland Parkway, Las Vegas, NV 89154, USA

亮点介绍： 本文报道了一种利用热电离质谱（TIMS）配合双稀释剂技术测定样品高精度 Ca 同位素组成的分析方法。本方法可同时获得样品的放射性成因 Ca 同位素组成（$\varepsilon^{40/44}$Ca）和稳定 Ca 同位素组成（$\delta^{44/42}$Ca），分析精度分别为$\pm$1.0 ε 和$\pm$0.03‰（2SE, $n\geqslant8$），达国际先进水平。利用该方法对一系列地质样品进行了高精度 Ca 同位素组成的测定，研究发现（1）地幔岩石相对常用国际标准样品 NIST SRM 915a 亏损 $\varepsilon^{40/44}$Ca 约 0.79$\pm$0.60 ε。（2）火成岩标样的 $\delta^{44/42}$Ca 存在显著变化（0.27‰～0.54‰）。其中，基性–酸性岩标样 Ca 同位素组成系统低于上地幔，且与 MgO、CaO 等分异指标没有相关性，揭示岩浆分异过程中 Ca 同位素分馏不显著。$(Dy/Yb)_N$较高的标样 Ca 同位素组成系统偏轻，这可能与石榴子石相对富集重 Ca 同位素有关。

Abstract We report mass-independent and mass-dependent Ca isotopic compositions for thirteen geological reference materials, including carbonates (NIST SRM 915a and 915b), Atlantic seawater as well as ten rock reference materials ranging from peridotite to sandstone, using traditional ε and δ values relative to NIST SRM 915a, respectively. Isotope ratio determinations were conducted by independent unspiked and ^{43}Ca-^{48}Ca double-spiked measurements using a customised Triton Plus TIMS. The mean of twelve measurement results gave $\varepsilon^{40/44}$Ca values within $\pm$1.1, except for GSP-2 that had $\varepsilon^{40/44}$Ca= 4.04$\pm$0.15 (2SE). Significant radiogenic ^{40}Ca enrichment was evident in some high K/Ca samples. At an uncertainty level of $\pm$0.6, all reference materials had the same $\varepsilon^{43/44}$Ca and $\varepsilon^{48/44}$Ca values. We suggest the use of $\delta^{44/42}$Ca to report mass-dependent Ca isotopic compositions. The precision under intermediate measurement conditions for $\delta^{44/42}$Ca over eight months in our laboratory was $\pm$0.03‰ (with $n\geqslant8$ repeat measurements). Measured igneous reference materials gave $\delta^{44/42}$Ca values ranging from 0.27‰ to 0.54‰. Significant Ca isotope fractionation may occur during magmatic and metasomatism processes. Studied reference materials with higher (Dy_n/Yb_n) tend to have lower $\delta^{44/42}$Ca, implying a potential role of garnet in producing magmas with low $\delta^{44/42}$Ca. Sandstone GBW07106 had a $\delta^{44/42}$Ca value of 0.22‰, lower than all igneous rocks studied so far.

* 本文发表在：Geostandards and Geoanalytical Research, 2017, 41: 283-302

1 Introduction

Calcium is a major element on the Earth and plays an essential role in geological and biological processes. It has six stable isotopes: ^{40}Ca, ^{42}Ca, ^{43}Ca, ^{44}Ca, ^{46}Ca and ^{48}Ca. Variations in Ca isotopic ratios can be caused by nucleosynthetic anomaly, the decay of ^{40}K to ^{40}Ca and mass-dependent isotope fractionation. Calcium isotopic variations thus may provide important constraints on the origin and evolution of planets (Simon et al. 2009, Huang et al. 2012), age information by the ^{40}K-^{40}Ca isotopic system (Marshall and DePaolo 1982, 1989, Caro et al. 2010) and the global Ca cycle from the surface to deep mantle (De La Rocha and DePaolo 2000, Farkaš et al. 2007, 2011, Huang et al. 2011, Zhu et al. 2015). The reported $\delta^{44/42}Ca$ variation in terrestrial samples is about 2.0‰ (Fantle and Tipper 2014), and thus, precise and accurate Ca isotopic measurement is essential for the use of calcium isotopes as geochemical tools.

Difficulties in obtaining high-precision Ca isotopic data include the large natural abundance difference, the large relative mass range, asymmetrical peak shapes, strong isobaric interferences in multi-collector-inductively coupled plasma-mass spectrometry (MC-ICP-MS), nonideal evaporation and ionisation processes in thermal ionisation mass spectrometry (TIMS) and degradation of Faraday cups (Heuser et al. 2002, Fantle and Bullen 2009, Boulyga 2010, Holmden and Bélanger 2010, Morgan et al. 2011, Lehn et al. 2013, Fantle and Tipper 2014, Zhu et al. 2016). Typical intermediate measurement precision for $\delta^{44/40}Ca$ is ~0.1‰ (Tipper et al. 2006, Huang et al. 2010, Lehn et al. 2013, Valdes et al. 2014, Zhu et al. 2016).

Here, we report (a) new analytical developments for Ca isotopic measurement at the Isotope Geochemistry Laboratory (State Key Laboratory of Geological Processes and Mineral Resources, China University of Geosciences, Beijing) and (b) the characterisation of Ca isotopic compositions of thirteen geological reference materials. These include commonly used carbonate reference materials (NIST SRM 915a and 915b), Atlantic seawater, as well as ten rock reference materials ranging in type from peridotite to sandstone.

1.1 Two types of Ca isotopic effects

Natural Ca isotopic effects can be divided into two types, namely mass-independent and mass-dependent isotopic effects (e.g., Niederer and Papanastassiou 1984, Huang et al. 2012). During unspiked measurements, natural and instrumental mass-dependent fractionation is internally corrected to $^{42}Ca/^{44}Ca = 0.31221$ using the exponential law (Russell et al. 1978, Hart and Zindler 1989); consequently, mass-independent Ca isotopic data describe radiogenic ^{40}Ca excess due to ^{40}K decay and addition of nucleosynthetic components with nonsolar isotopic compositions. Mass-independent Ca isotopic data are commonly reported in traditional ε values relative to a certain geological reservoir, for example, carbonate NIST SRM 915a (Huang et al. 2012, Naumenko-Dèzes et al. 2015).

$$\varepsilon^{i/44}\mathrm{Ca}(\mathrm{SAM})=\left[\frac{\left({}^{i}\mathrm{Ca}/{}^{44}\mathrm{Ca}_{\mathrm{SAM}}\right)_{\mathrm{N}}}{\left({}^{i}\mathrm{Ca}/{}^{44}\mathrm{Ca}_{\mathrm{SRM915a}}\right)_{\mathrm{N}}}-1\right]\times 1000 \quad (1)$$

where i can be 40, 43, 46 and 48, and subscript N represents internal normalization to $^{42}Ca/^{44}Ca = 0.31221$.

Mass-dependent (stable) isotopic data describe the natural isotopic variations intrinsic to the samples that occur due to atomic mass difference among isotopes. Therefore, only instrumental fractionation is corrected using the double-spike technique with TIMS (Heuser et al. 2002, Huang et al. 2010, Lehn et al. 2013, Zhu et al. 2016) and the calibrator-sample bracketing method with MC-ICP-MS (Tipper et al. 2006, Valdes et al. 2014). Mass-dependent isotopic data are reported as $\delta^{44/40}Ca$ or $\delta^{44/42}Ca$:

$$\delta^{44/i}\mathrm{Ca}(\mathrm{SAM})=\left(\frac{^{44}\mathrm{Ca}/^{i}\mathrm{Ca}_{\mathrm{SAM}}}{^{44}\mathrm{Ca}/^{i}\mathrm{Ca}_{\mathrm{SRM915a}}}-1\right)\times 1000 \quad (2)$$

where i can be 40 or 42. It should be noted that $\delta^{44/40}$Ca includes contributions both from mass-dependent isotope fractionation and from the decay of ^{40}K. The radiogenic excess of ^{40}Ca relative to NIST SRM 915a can be calculated from the measured $\delta^{44/40}$Ca and $\delta^{44/42}$Ca based on the following expression (Farkaš et al. 2011):

$$\varepsilon^{40/44}\mathrm{Ca}=\left[\left(\delta^{44/42}\mathrm{Ca}\times 2.0483\right)-\delta^{44/40}\mathrm{Ca}\right]\times 10 \quad (3)$$

where the coefficient of 2.0483 represents the slope of the mass-dependent fractionation between $\delta^{44/40}$Ca and $\delta^{44/42}$Ca following the exponential law (Farkaš et al. 2011, Huang et al. 2012).

1.2 Published studies and associated problems

The entire $\delta^{44/42}$Ca range of terrestrial samples is about 2.0‰, and thus, the use of Ca isotopes as a proxy requires precise and accurate measurement (Fantle and Tipper 2014). Currently, stable Ca isotopic ratios can be measured with an intermediate precision ('external precision') of $\pm$0.10‰ on the $\delta^{44/40}$Ca scale by TIMS and by MC-ICP-MS (Tipper et al. 2006, Huang et al. 2010, Lehn et al. 2013, Valdes et al. 2014, Zhu et al. 2015, 2016). Comparison of stable Ca isotopic data from different laboratories is complicated, because the commonly used $\delta^{44/40}$Ca reflects the combined effect of radiogenic ^{40}Ca enrichment and mass-dependent fractionation (Fantle and Tipper 2014). Discrimination of these two types of effects can be obtained using combined mass-independent and mass-dependent isotopic measurements (Huang et al. 2012, Schiller et al. 2012) or based on simultaneously reported $\delta^{44/40}$Ca and $\delta^{44/42}$Ca using a ^{43}Ca-^{48}Ca double-spike (Huang et al. 2010, Farkaš et al. 2011). However, these combined data have been rarely reported in the literature. For interlaboratory comparison purposes, it is necessary to better characterise both mass-independent and mass-dependent Ca isotopic compositions of geological reference materials.

It remains difficult to compare directly mass-independent Ca isotopic ratios obtained by different laboratories. Published $(^{40}\mathrm{Ca}/^{44}\mathrm{Ca})_N$ of NIST SRM 915a ranges from 46.106 to 47.1649 (Naumenko-Dèzes et al. 2015 and references therein). Naumenko-Dèzes et al. (2015) report internally consistent Ca isotopic ratios for NIST SRM 915a and 915b using a customised Triton Plus TIMS instrument that allowed simultaneous measurement of ^{40}Ca to ^{48}Ca. An improved exponential law using two stable isotopic ratio pairs (e.g., ^{42}Ca/^{44}Ca and ^{48}Ca/^{44}Ca) has been shown to properly correct the instrumental mass bias, and argued to work for natural samples (Naumenko-Dèzes et al. 2015). It is necessary to check whether these absolute values can be reproduced in other laboratories.

2 Measurement procedure

2.1 Chemistry procedures

Typically, sample powders weighing 3–50 mg were dissolved in a mixture of 1.5 ml concentrated HF and 0.5 ml concentrated HNO_3 in 6-ml Teflon^{TM} beakers at 100 ℃ for three days. The sample solutions were then dried down and evaporated with 0.5 ml concentrated HCl at 120 ℃ three times. The products were finally dissolved in 0.5 ml 2.5 mol l^{-1} HCl. For mass-independent isotopic measurement, an aliquot of each sample solution containing ca. 50 μg Ca was evaporated to dryness, and re-dissolved in 50 μl 2.5 mol l^{-1} HCl for column chemistry. For mass-dependent isotopic measurement, an aliquot of each sample solution containing ca. 50 μg Ca was mixed with an appropriate amount of ^{43}Ca-^{48}Ca double-spike solution (Table 1) so that the ^{48}Ca/^{40}Ca ratio of the mixture was ca. 0.1145. The spiked samples were then homogenised by refluxing at 120 ℃ overnight, evaporated to dryness again and dissolved in 50 μl 2.5 mol l^{-1} HCl. Calcium was purified using quartz columns (3 mm in diameter and ca. 11 cm in length) filled with AG50W-X12 (200–400 mesh; Bio-Rad, Hercules, CA, USA) resin. Columns were precleaned and conditioned with 10 ml 6 mol l^{-1} HCl, 5 ml high-purity water and 2 ml 2.5 mol l^{-1} HCl in a stepwise fashion. Sample solutions with ~50 μg Ca were loaded using 50 μl 2.5 mol l^{-1} HCl. Matrix elements (e.g., Fe, Mg, Ti, K and Al) were removed by 5 ml 2.5 mol l^{-1} HCl, and Ca was collected using 3.5 ml 2.5 mol l^{-1} HCl without collecting significant amount of Sr (Fig. 1). Each sample was passed through the column twice. The Ca yield after two-column chemistry was about 90%. The purified Ca was then evaporated to dryness with several drops of concentrated HNO_3 for isotopic measurement. Double-distilled acids were used, and the whole procedure Ca blank typically ranged from 13 to 58 ng, which is <0.2% of the processed samples.

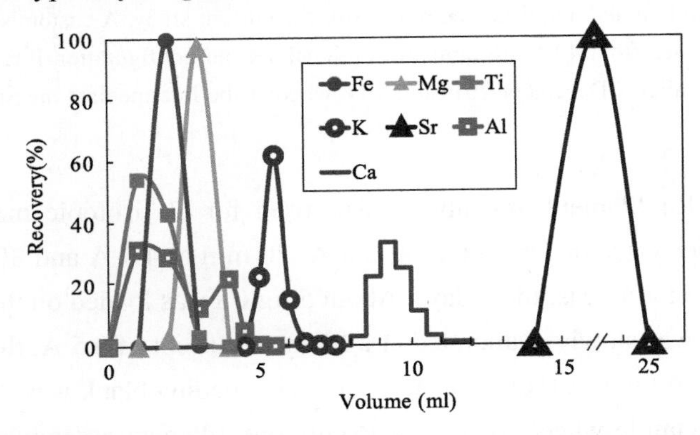

Fig. 1 The Ca elution curve of BHVO-2 passed through columns (0.3 cm × 11 cm) filled with AG50W-X12 (200 to 400 mesh). [Colour figure can be viewed at wileyonlinelibrary.com]

Table 1 Internally normalised isotopic composition for NIST SRM 915a and ^{43}Ca-^{48}Ca double-spike (DS)

Samples	SRM 915a	DS (meas.)	DS (calib.)
$^{40}Ca/^{44}Ca$	47.1516(6)	3.57458(2)	3.57534(4)
$^{42}Ca/^{44}Ca$	0.31221	0.172803(2)	0.172821(1)
$^{43}Ca/^{44}Ca$	0.0648681(11)	15.9830(2)	15.9838(1)
$^{46}Ca/^{44}Ca$	0.0015125(19)	0.00045859	0.000446(1)
$^{48}Ca/^{44}Ca$	0.0886516(28)	32.0444(3)	32.0382(4)

Isotopic composition of NIST SRM 915a is given by a long-term mean of 237 times measurement over nine months (Table S1). Double-spike composition was first measured independently for four times (5 μg per load) and then calibrated for the remaining unresolved mass bias using double-spiked NIST SRM 915a data. The details are referred to the main text and Appendix S1. The errors are two times the standard error of the mean and correspond to the last figures shown.

2.2 Mass spectrometry

The Triton Plus TIMS at the Isotope Geochemistry Lab was equipped with a customised Faraday cup (L5) for ^{40}Ca and a Faraday cup (H4) for ^{48}Ca, allowing static measurement of masses from 40 to 48 amu. All Faraday cups were connected to 10^{11} Ω resistors. All Ca isotopes and ^{41}K were routinely collected in one single cup configuration line as shown in Fig. 2. Calcium isotopes were typically measured with a total intensity of 100–300 pA, which yielded a ^{40}Ca signal of ca. 8–30 V. An optional second configuration line for ^{42}Ca and ^{44}Ca at the L3 and central Faraday cups, respectively, was added after 7 December 2015 to check whether the intermediate measurement precision was dominated by the low ^{42}Ca signal (typically ~ 100 mV). The isobaric interference of ^{40}K on ^{40}Ca was generally < 6 ppm and did not require correction.

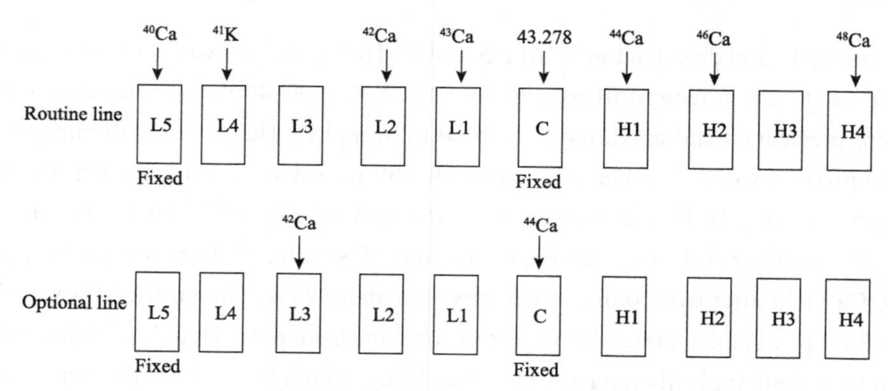

Fig. 2　The Faraday cup configuration and amplifiers employed for the current study. A customised Faraday cup alignment allows simultaneous measurement of masses from 40 to 48 amu. An optional second configuration line for ^{42}Ca and ^{44}Ca at the L3 and central Faraday cups was adopted after 7 December 2015 to check whether the intermediate measurement precision was dominated by the low ^{42}Ca signal.

A double-zone-refined Re filament assemblage was used for Ca isotopic measurements. After welding, filaments were degassed under vacuum for 10 min at 3.5 A, 20 min at 4.5 A and 30 s at 3.0 A with subsequent oxidising at room temperature for at least three days. About 5 μg Ca was loaded on the evaporation filament using 1 μl 3% HNO_3. The solution was slowly evaporated at a filament current of 0.5 A, then heated at 1.5 A for 1 min and finally heated up to dull red for 30 s (Huang et al. 2010). The loading blank was < 100 pg.

The Triton Plus TIMS sample wheel could host twenty-one filament assemblages. A typical measurement session involved up to twenty-one filament assemblages hosted in a single wheel, which typically contained 2–3 NIST SRM 915a, 1–2 Atlantic Seawater or NIST SRM 915b and 16–18 unknown samples. During a measurement routine, the ion and evaporation filaments were heated up automatically. The ion filament was rapidly heated to 2500 mA in 3 min, and then, the evaporation filament was heated to 50 mA in 1 min. The ion and evaporation filaments were then heated up by increments of 50 mA and 30 mA alternately with a slope of 30 mA min^{-1} until a stable ^{40}Ca signal ranging from 10 to 30 V was obtained. Evaporation filament current was kept below 800 mA, which was important to obtain a stable instrumental mass bias. During the heating routine, ^{44}Ca was monitored for ion intensity. Source lenses were focused when ^{44}Ca reached 5 mV and then at the end of the heating routine. A routine measurement consisted of a peak centre on ^{44}Ca at H1 at the beginning and every eight blocks of data acquisition. Each block included a 60-s baseline measurement and fifteen cycles with an integration time of 16.776 s. Amplifier rotation for every block was enabled. Total analytical time was ca. 90 min for one filament assemblage, including a 30-min heating routine. Typically, each sample was loaded on three to four filaments, and each filament assemblage was measured two to three times, independently, from heating up to data acquisition. Mass-independent Ca isotopic data were internally corrected to $^{42}Ca/^{44}Ca$ = 0.31221 and reported in conventional ε values. For the mass-dependent isotopic effect, isotope fractionation in the laboratory, including low column chemistry yield and isotopic fractionation on TIMS, was corrected by a ^{43}Ca-^{48}Ca double-spike technique (Table 1). $^{42}Ca/^{44}Ca$, $^{43}Ca/^{44}Ca$ and $^{48}Ca/^{44}Ca$ were used for double-spike data deduction with a procedure modified from Johnson and Beard (1999), Rudge et al. (2009) and Li et al. (2011), which is detailed in Appendix S1. Mass-dependent isotopic compositions of geological reference materials are reported in $\delta^{44/42}Ca$ values after Farkaš et al. (2011).

2.3　Double-spike calibration

The ^{43}Ca-^{48}Ca double-spike was prepared by mixing ^{43}Ca and ^{48}Ca single spikes obtained from the Oak Ridge National Laboratory (USA) according to the optimised proportion predicted by Rudge et al. (2009). Its isotopic composition was then measured by four runs independently prepared from loading (5 μg of each filament) to data acquisition. The instrumental mass bias was preliminarily corrected to $^{43}Ca/^{48}Ca$=0.49878, obtained by a gravimetric method using the certificate values for single spikes, using the exponential law. As shown in Fig. 3, linear correlations were found among the measured $(^{43}Ca/^{40}Ca)_N$, $(^{48}Ca/^{40}Ca)_N$ and $(^{44}Ca/^{40}Ca)_N$ values, after normalisation to $^{43}Ca/^{48}Ca$=0.49878, which reflect mixtures of double-spike and blank Ca during the sample loading procedure. The loading blank was then estimated to be < 100 pg. The run with the highest $(^{43}Ca/^{40}Ca)_N$ and $(^{48}Ca/^{40}Ca)_N$ represents the least contaminated double-spike, which was used to calculate the double-spike isotopic composition (Table 1). These measured isotopic ratios differ from the 'true' values because the instrumental mass bias was corrected to $^{43}Ca/^{48}Ca$=0.49878 with a large uncertainty. This $^{43}Ca/^{48}Ca$ normalisation value was obtained using a gravimetric method by weighing several milligrams of ^{43}Ca and ^{48}Ca single spikes during double-spike solution preparation. The double-spike composition was further calibrated based on 186 spiked NIST SRM 915a runs during May 2015 to September 2015 using a procedure detailed in Appendix S1, which makes a correction on $^{43}Ca/^{48}Ca$ by ~0.25‰. The NIST SRM 915a-calibrated ^{43}Ca-^{48}Ca double-spike composition is reported in Table 1. Using these calibrated values, 100 independent runs of spiked NIST SRM 915a from October 2015 to December 2015 yielded mean $\delta^{44/40}Ca$ and $\delta^{44/42}Ca$ values of 0.00±0.12‰ (2*s*) and –0.03±0.08‰ (2*s*), respectively. The slight shift in mean $\delta^{44/42}Ca$ may be attributed to drift in the H4 Faraday cup efficiency when aging, and thus, all $\delta^{44/42}Ca$ data obtained after 25 September 2015 were re-normalised by subtracting –0.03 accordingly.

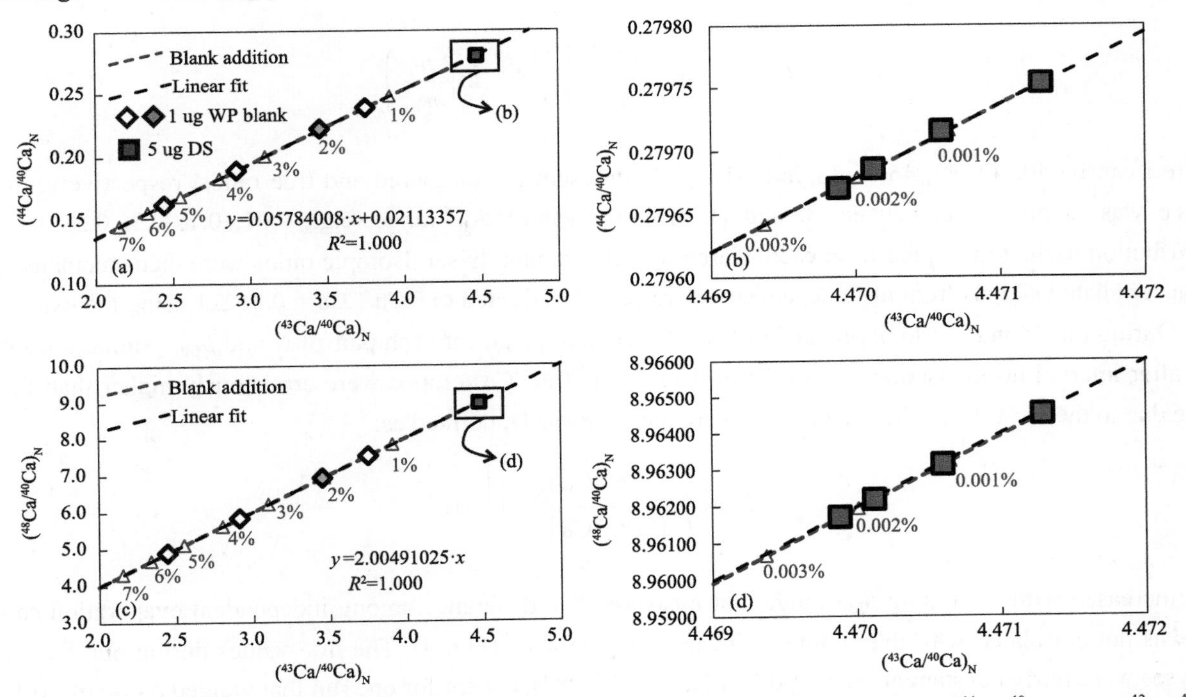

Fig. 3 Determination of the isotopic composition of the double-spike by blank subtraction: (a) $(^{44}Ca/^{40}Ca)_N$ vs. $(^{43}Ca/^{40}Ca)_N$ diagram, (b) inset to Fig. 3a, (c) $(^{48}Ca/^{40}Ca)_N$ vs. $(^{43}Ca/^{40}Ca)_N$ diagram and (d) inset to Fig. 3c. Mass bias was normalised against $^{43}Ca/^{48}Ca$=0.49878. Linear fitting was conducted for four runs on the double-spike independent from loading (5 μg each, green filled squares) to data acquisition and three runs of whole procedure blank (traced by 1 μg double-spike each, unfilled diamonds) prior to September 2015. The whole procedure blank measured on 29 November 2015 (grey filled diamond) was not used for linear fitting, but is also plotted here for comparison. Measurement repeatability uncertainties are below the size of symbols. [Colour figure can be viewed at wileyonlinelibrary.com]

3 Effect of heterogeneous evaporation: A Monte Carlo approach

During Ca isotopic measurement, the intermediate precision is always worse than the measurement repeatability ('internal precision') (Heuser et al. 2002, Fantle and Bullen 2009, Lehn et al. 2013, Lehn and Jacobson 2015, Naumenko-Dèzes et al. 2015). This is also true for our measurements of NIST SRM 915a: the intermediate precision was two to eight times of that of measurement repeatability for single mass-independent and mass-dependent Ca isotopic measurements. For example, the two standard deviation value of $\delta^{44/40}$Ca was 0.16‰, compared with the mean measurement repeatability of 0.02‰ (Tables S1 and S3). During TIMS measurement, non-ideal evaporation and ionisation processes have been proposed as a major reason for the worse intermediate precisions observed (Fantle and Bullen 2009, Fantle and Tipper 2014, Lehn and Jacobson 2015). Sample Ca loaded on the Re or Ta filament is not an ideal point source (Fantle and Bullen 2009, Lehn and Jacobson 2015, Naumenko-Dèzes et al. 2015), and mixing of signals from independent evaporation sources can lead to the violation of the commonly adopted exponential law, which leads to erroneous data (Fig. 4a; Fantle and Bullen 2009, Lehn and Jacobson 2015). The phenomenon is referred to as heterogeneous evaporation (HE) effect after Fantle and Bullen (2009). We used a Monte Carlo approach to evaluate this effect and to test how it can be corrected for. In this approach, Ca with an isotopic composition identical to NIST SRM 915a (Table 1) was to evaporate from n (2 to 20) independent sources on a filament, and each followed the exponential law. The fractionation factor

$$\beta_{\mathrm{MC}} = \ln\left(\frac{\left({}^{i}\mathrm{Ca}/{}^{44}\mathrm{Ca}\right)_{\mathrm{M}}}{\left({}^{i}\mathrm{Ca}/{}^{44}\mathrm{Ca}\right)_{\mathrm{T}}}\right) / \ln\left(\frac{m_i}{m_{44}}\right) \tag{4}$$

where i can be 40, 42, 43, 46 or 48, and M and T represent the measured and true ratios, respectively, of each source was randomly set between 0 and an arbitrary value (β_{MC} range, e.g., 0.01, 0.1, 0.3, 0.5, 0.7, 1.0). Contribution to the total signal from each source was also randomly set. Isotopic ratios were then calculated using the accumulated signals from all independent sources, and corrected to ^{42}Ca/^{44}Ca = 0.31221 using the exponential law. During our Monte Carlo approach, 100 cases were simulated for each pair of N and β_{MC}. Simulations reveal that after internal normalisation to ^{42}Ca/^{44}Ca=0.31221, (^{40}Ca/^{44}Ca)$_{\mathrm{N}}$ ratios were erroneously higher than the true value due to the HE effect. Their deviations were measured as D, defined as:

$$D = \left[\frac{\left({}^{i}\mathrm{Ca}/{}^{44}\mathrm{Ca}\right)_{\mathrm{N}}}{\left({}^{i}\mathrm{Ca}/{}^{44}\mathrm{Ca}\right)_{\mathrm{T}}} - 1\right] \times 10^4 \tag{5}$$

This increased with increasing β_{MC} range that measures β_{MC} difference among independent evaporation sources, and was not correlated with the number of evaporating sources (Fig. 4). The β_{MC} values during our Ca isotopic analyses were fairly constant at -0.17 ± 0.02 ($2s$, $n = 987/988$), except for one run that yielded a β_{MC} of -0.11 and was treated as an outlier. In addition, the β_{MC} value in each individual run did not change significantly with time. Therefore, the error introduced by the HE effect for Ca isotopic analyses was estimated to be <10 ppm, which is negligible. Additional sources of uncertainty must account for the larger intermediate precision and will be discussed in the next section.

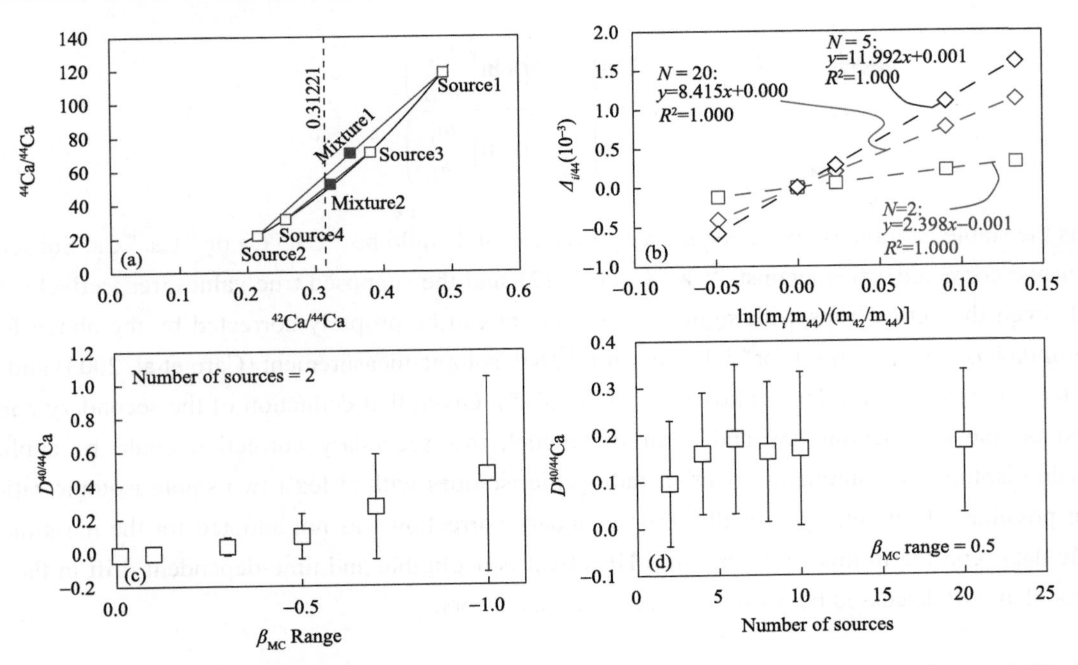

Fig. 4 Simulation of the heterogeneous evaporation effect using a Monte Carlo approach. A cartoon shown mixing processes of multiple sources and regression of data to ^{42}Ca/^{44}Ca=0.31221 based on the exponential law (a). The simulation approach was detailed in the main text. $\Delta_{i/44}$, where i can be 40, 42, 43, 46 and 48, was calculated by the difference between the theoretically measured ratios and the true values following a formula of the exponential law (Equations 13 and 14 in Appendix S1). As shown in (b), $\Delta_{i/44}$ varies with the isotopic ratio pair mass dependently. Difference from the measured (^{40}Ca/^{44}Ca)$_N$ and the true value was expressed as a D value ($D=\left[\frac{\left({}^{40}\mathrm{Ca}/{}^{44}\mathrm{Ca}\right)_N}{\left({}^{40}\mathrm{Ca}/{}^{44}\mathrm{Ca}\right)_T}-1\right]\times10^4$). The mean D values were plotted against the β_{MC} range and number of independent evaporation sources (c and d), with error bars representing two standard deviation. [Colour figure can be viewed at wileyonlinelibrary.com]

The Monte Carlo simulations also reveal that the extent of deviation from the exponential law due to the HE effect is mass-dependent, which can be expressed following a formula of the exponential law:

$$\Delta_{\frac{i}{44}}=\ln\left(\frac{\left({}^{i}\mathrm{Ca}/{}^{44}\mathrm{Ca}\right)_N}{\left({}^{i}\mathrm{Ca}/{}^{44}\mathrm{Ca}\right)_T}\right)/\ln\left(\frac{m_x}{m_{44}}\right) \tag{6}$$

where i can be 40, 42, 43, 46 and 48 and m refers to the atomic masses. Our simulations thus provide a theoretical explanation for the concept that the HE effect during mass-dependent Ca isotopic measurement can be optimised by choosing double-spike pairs based on an 'average mass rule' (Lehn and Jacobson 2015). The 'average mass rule' predicts a minimised HE effect when the average mass of a double-spike is close to that of the normalising ratio (e.g., using ^{42}Ca-^{43}Ca double-spike to monitor ^{44}Ca/^{40}Ca). It is interesting to note that $\Delta_{i/44}$ and $\ln[(m_i/m_{44})/(m_{42}/m_{44})]$ rigorously follow a linear trend crossing the origin (Fig. 4b). This validates a secondary correction using a second stable isotopic ratio other than ^{42}Ca/^{44}Ca=0.31221 to correct the HE effect for mass-independent isotopic measurement (see Appendix S1 for details):

$$\frac{\left({}^{x}\mathrm{Ca}/{}^{44}\mathrm{Ca}\right)_N}{\left({}^{x}\mathrm{Ca}/{}^{44}\mathrm{Ca}\right)_T}=\left[\frac{\left({}^{y}\mathrm{Ca}/{}^{44}\mathrm{Ca}\right)_N}{\left({}^{y}\mathrm{Ca}/{}^{44}\mathrm{Ca}\right)_T}\right]^{A} \tag{7}$$

where

$$A=\frac{\ln\left(\frac{m_x}{m_{42}}\right)\times\ln\left(\frac{m_x}{m_{44}}\right)}{\ln\left(\frac{m_y}{m_{42}}\right)\times\ln\left(\frac{m_y}{m_{44}}\right)} \quad (8)$$

$^{y}Ca/^{44}Ca$ is the stable isotopic ratio for secondary correction and could be $^{43}Ca/^{44}Ca$ or $^{48}Ca/^{44}Ca$. Subscript N and T represent the corrected ratios against $^{42}Ca/^{44}Ca$=0.31221 and the supposed true values, respectively. As shown in Fig. S1, even theoretical runs with significant HE effect can be properly corrected by the above formula. A similar formula has been proposed for high-precision ^{142}Nd isotopic measurement (Caro et al. 2003) and also used for Ca isotopic measurement (Naumenko-Dèzes et al. 2015). Given that deduction of the secondary correction is only based on the exponential law and a mixing model, this secondary correction could be applicable for high-precision isotopic measurement for other radiogenic isotopes with at least two stable isotopic ratio pairs to correct for possible HE effect. Specifically, this secondary correction was not adopted for the mass-independent Ca isotopic data reported in this study, because HE effect is negligible and time-dependent drift in the measured $(^{48}Ca/^{44}Ca)_N$ that will discussed below may cause additional errors.

4 Mass-independent isotopic compositions of geological materials

4.1 Time-dependent drift of Faraday cup efficiency

During seventeen analytical sessions, 256 unspiked NIST SRM 915a measurements were carried out to monitor possible Faraday cup efficiency drift and the intermediate measurement precision (over 9 months) of mass-independent Ca isotopic determinations (Fig. 5 and Table S1). Nineteen out of 256 runs, identified by $(^{40}Ca/^{44}Ca)_N$ < median−(maximum−median), yield $(^{40}Ca/^{44}Ca)_N$ substantially lower than average and were treated as outliers (Fig. 5a). Contamination of spikes from the ion sources, possibly due to spiked analytical sessions that were measured alternately with unspiked ones, can be ruled out by the measured $(^{43}Ca/^{44}Ca)_N$ and $(^{48}Ca/^{44}Ca)_N$ ratios of these outliers comparable to the majority data (Fig. 5a and Table S1). Several parameters have been proposed that affect the accuracy of Ca isotopic determinations, including Faraday cup efficiency drift (Simon et al. 2009, Caro et al. 2010, Holmden and Bélanger 2010, Lehn et al. 2013), heterogeneous evaporation effect (Hart and Zindler 1989, Fantle and Bullen 2009) and ion optical aberrations (Fletcher et al. 1997, Holmden and Bélanger 2010). No systematic drift on the measured $(^{40}Ca/^{44}Ca)_N$ was evident between March 2015 and December 2015. The width of the graphite insert for the Faraday cup L5 was doubled and customised for the most intense signal (^{40}Ca). This study thus confirms that the customised L5 cup is immune to efficiency drift (Naumenko-Dèzes et al. 2015). In addition, the heterogeneous evaporation effect leads to measured $(^{40}Ca/^{44}Ca)_N$ higher than the true value (Fig. 4a and section 'Effect of heterogeneous evaporation'), and thus, it cannot account for the low $(^{40}Ca/^{44}Ca)_N$ values (Fig. 5a). The low $(^{40}Ca/^{44}Ca)_N$ of the outlier runs can be explained by selective loss of the ^{40}Ca ion collected in the low mass edge Faraday cup, but without any effect on other isotopic ratios. The only explanation of this phenomenon is clipping on ^{40}Ca after ion-beam dispersion, for example, interaction with the metal mask in front of the L5 cup, which is used to keep the peak shape of ^{40}Ca same to other masses.

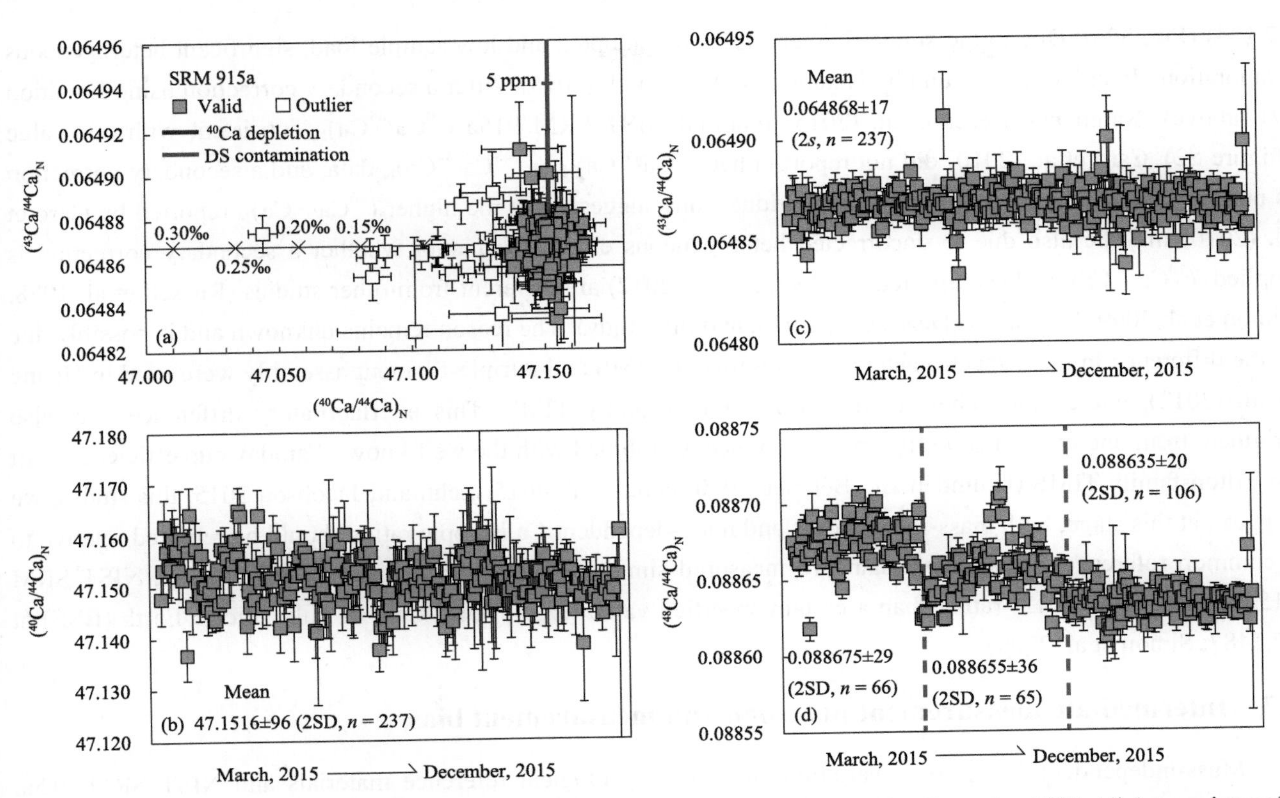

Fig. 5 Mass-independent isotopic ratios for NIST SRM 915a measured from March 2015 to December 2015. All data are internally normalised to $^{42}Ca/^{44}Ca$ = 0.31221 (Russell et al. 1978) assuming the exponential law. Nineteen out of 256 runs, identified by $(^{40}Ca/^{44}Ca)_N$ < median − (maximum − median), yield $(^{40}Ca/^{44}Ca)_N$ substantially lower than average and were treated as outliers (unfilled squares in (a)). The outlier data are not shown in (b)–(d). Theoretical double-spike contamination curve (vertical line) was calculated based on the isotopic composition of NIST SRM 915a and ^{43}Ca-^{48}Ca double-spike listed in Table 1. Selective depletion of ^{40}Ca (the horizontal line in (a)) was modelled by decreasing $(^{40}Ca/^{44}Ca)_N$ arbitrarily. Range bars represent two standard error of the mean. [Colour figure can be viewed at wileyonlinelibrary.com]

No $(^{43}Ca/^{44}Ca)_N$ drift was observed for NIST SRM 915a, but a systematic drift was evident for $(^{48}Ca/^{44}Ca)_N$ (Fig. 5c, d). In detail, measured $(^{48}Ca/^{44}Ca)_N$ gradually decreased from 0.088675±29 (2*s*, *n*=66) between March 7 and March 16, 0.088655±36 (2*s*, *n*=65) between April 16 and July 7, to 0.088635±20 (2*s*, *n*=106) between July 9 and December 30. This may reflect a drift in the H4 Faraday cup efficiency, possibly due to the fact that ^{48}Ca is the second brightest beam during the spiked runs and collected by the edge H4 Faraday cup. The ^{48}Ca ion beam with a large incident angle may hit the sides of the H4 collector and smooth its surface over time, which may cause the change in collector efficiency (Holmden and Bélanger 2010).

4.2 Mass-independent isotopic compositions of NIST SRM 915a

It is important to check whether internally corrected Ca isotopic ratios of NIST SRM 915a from different laboratories are consistent. Two hundred and thirty-seven runs of unspiked NIST SRM 915a yielded mean $(^{40}Ca/^{44}Ca)_N$ and $(^{43}Ca/^{44}Ca)_N$ values of 47.1516 ± 0.0096 (±2.0 ε, 2*s*) and 0.064868 ± 0.000017 (±2.6 ε, 2*s*), respectively (Table 1 and Fig. 5b, c). Despite the time-dependent drift mentioned above, a long-term mean $(^{48}Ca/^{44}Ca)_N$ value for NIST SRM 915a was 0.0886516±0.0000029 (2SE). This value is used for double-spike calibration and interlaboratory comparison. Our mean NIST SRM 915a $(^{40}Ca/^{44}Ca)_N$ is consistent with those (47.1519–47.1531) reported by Russell et al. (1978) and Simon et al. (2009), but significantly different from those reported by Caro et al. (2010), Naumenko-Dèzes et al. (2015) (47.1622 to 47.1649) and Huang et al. (2012) (ca.

47.134) (Fig. S2a). Due to the single filament assembly adopted and low sample load, significant heterogeneous evaporation effect has been found by Naumenko-Dèzes et al. (2015). After a secondary correction using Equation (7) (above), Naumenko-Dèzes et al. (2015) reported NIST SRM 915a $(^{40}Ca/^{44}Ca)_N$ consistent with our value (Figure S2). Caro et al. (2010) did not report either $(^{43}Ca/^{44}Ca)_N$ or $(^{48}Ca/^{44}Ca)_N$ data, and a secondary correction is not allowed. Changing in β_{MC} during individual runs suggests that the higher $(^{40}Ca/^{44}Ca)_N$ reported by Caro et al. (2010) may be also due to uncorrected heterogeneous effect. No matter whether a secondary correction is applied, $(^{40}Ca/^{44}Ca)_N$ values reported by Huang et al. (2012) are different from other studies (Russell et al. 1978, Simon et al. 2009, Naumenko-Dèzes et al. 2015, and this study). The reason remains unknown and is possibly due to the difference in mass spectrometers. An Isoprobe-T TIMS and a triple-filament assembly were used in Huang et al. (2012), while most other studies used a Triton-family TIMS. This interlaboratory difference may also originate from uncorrected cup efficiency difference. Combined with the well-known Faraday cup efficiency drift on Triton-family TIMS (Holmden and Bélange 2010, Feng et al. 2015, Lehn and Jacobson 2015, this study), we suggest, at this stage, both mass-independent and mass-dependent Ca isotopic ratios should be reported relative to a common reference measurement standard measured simultaneously with unknown samples (e.g., NIST SRM 915a; Huang et al. 2012) rather than a certain specified value (e.g., $(^{40}Ca/^{44}Ca)_N$ of bulk silicate Earth (BSE) at 47.1487; Simon et al. 2009).

4.3 Intermediate measurement precision and measurement bias

Mass-independent Ca isotopic variations of twelve geological reference materials and NIST SRM 915a, including carbonate, seawater and rock reference materials ranging from peridotite to rhyolite, were measured and reported relative to the mean NIST SRM 915a value of each session as conventional ε values (Table 2). The intermediate precision of $\varepsilon^{40/44}Ca$, $\varepsilon^{43/44}Ca$ and $\varepsilon^{48/44}Ca$ for single mass-independent Ca isotopic measurement was ±1.5, ±3.0 and ±2.8, respectively. Where each sample was measured more than eleven times, 2SE values for $\varepsilon^{40/44}Ca$, $\varepsilon^{43/44}Ca$ and $\varepsilon^{48/44}Ca$ were better than 0.30, 0.57 and 0.76, respectively. $\varepsilon^{40/44}Ca$ and $\varepsilon^{43/44}Ca$ for NIST SRM 915b from this study were -0.58 ± 0.08 (2SE) and 0.36 ± 0.16 (2SE), respectively, consistent with values (-1.2 ± 1.4 and -0.25 ± 0.91) recently reported by Naumenko-Dèzes et al. (2015) after a secondary correction to $^{48}Ca/^{44}Ca = 0.0886516$.

Table 2　Mass-independent isotopic compositions of reference materials relative to NIST SRM 915a

Material	$\varepsilon^{40/44}Ca$	$2s$	2SE	$\varepsilon^{43/44}Ca$	$2s$	2SE	$\varepsilon^{46/44}Ca$	$2s$	2SE	$\varepsilon^{48/44}Ca$	$2s$	2SE	n
NIST SRM 915b	−0.58	1.26	0.08	0.36	1.99	0.16	2.3	101	4.8	−0.34	2.16	0.21	36
NIST SRM 915b (literature)[a]	−1.22		1.35	−0.25		0.91							
Atlantic seawater	−0.49	0.91	0.11	0.36	2.3	0.20	2.6	138	6.0	−0.11	2.08	0.27	20
Seawater1 (literature)[b]	−0.02		0.15										
Seawater2 (literature)[b]	−0.06		0.16										
Seawater3 (literature)[c]	−0.41		0.50	−0.04		1.34							
JP-1, peridotite	−0.47	1.93	0.14	0.27	3.23	0.27	−6.3	119	8.5	−0.47	2.38	0.33	34
BHVO-2, basalt	−0.60	1.68	0.14	0.27	3.27	0.29	2.7	117	9.1	−0.48	2.85	0.37	35
BHVO-2 (literature)[d]	1.1		1.6										
BCR-2, basalt	−1.09	1.75	0.15	−0.18	3.06	0.31	−4.7	122	9.9	−0.48	2.36	0.38	28
BCR-1 (literature)[c]	0.1		0.6	0.3		1.2				0.7		0.9	
BCR-1 (literature)[e]	−0.56		0.26	0.37		0.39							

Continued

Material	$\varepsilon^{40/44}$Ca	2*s*	2SE	$\varepsilon^{43/44}$Ca	2*s*	2SE	$\varepsilon^{46/44}$Ca	2*s*	2SE	$\varepsilon^{48/44}$Ca	2*s*	2SE	*n*
GBW07105, basalt	−0.98	2.02	0.30	−0.27	4.55	0.57	3.7	222	20	−0.46	4.42	0.76	11
AGV-2, andesite	−0.82	1.91	0.24	−0.54	1.58	0.44	−20	131	13	0.09	2.55	0.58	12
GBW07104, andesite	−0.24	1.31	0.24	−0.43	5.17	0.47	−8.9	213	15	0.01	4.16	0.62	13
GSP-2, granodiorite	4.04	1.24	0.15	0.07	3.02	0.30	−16	120	9	−0.36	2.73	0.37	25
GBW07103, granite	−0.35	1.20	0.28	−0.57	3.81	0.55	−0.9	196	18	−0.08	4.52	0.66	12
COQ-1, carbonatite	0.15	1.90	0.16	0.07	2.37	0.32	−4.9	118	9.6	0.08	2.68	0.38	24
GBW07106, Quartz sandstone	0.49	1.14	0.26	−0.04	2.86	0.5	13	160	17	−0.28	3.88	0.60	11

Raw data are listed in Table S2, and ε values are calculated relative to the mean values of NIST SRM 915a for each session. Literature data sources: a, Naumenko-Dèzes et al. (2015); b, Caro et al. (2010); c, Huang et al. (2012); d, Amini et al. (2009); e, Simon et al. (2009).

4.4 Mass-independent Ca isotopic compositions of geological reference materials

Twelve geological reference materials, including NIST SRM 915b, Atlantic seawater, one peridotite, three basalts, four intermediate to felsic igneous rocks, one carbonatite, and one sandstone yielded identical $\varepsilon^{43/44}$Ca and $\varepsilon^{48/44}$Ca values within ±0.6. This provides a fundamental basis that terrestrial samples are fractionated from a single Ca isotopic composition (Appendix S1). Accordingly, different $\varepsilon^{40/44}$Ca observed in these reference materials should reflect variable radiogenic ^{40}Ca enrichment.

Eleven of twelve geological reference materials showed $\varepsilon^{40/44}$Ca within ±1.1 relative to NIST SRM 915a. The granodiorite reference material GSP-2 with K/Ca ~ 3.05 (atomic ratio) showed a measurable ^{40}Ca enrichment with $\varepsilon^{40/44}$Ca up to 4.04±0.15 (2SE). Despite a high K/Ca ratio (up to 3.85), granite reference material GBW07103 (GSR-1) yielded a $\varepsilon^{40/44}$Ca value of −0.35±0.28 (2SE) and no detectable radiogenic ^{40}Ca ingrowth compared with peridotite and basalt reference materials. The emplacement age of GBW07103 and GSP-2 are unknown, but the former must be substantially younger than that of the latter. Previous studies have shown that only ancient samples with high K/Ca ratios may have significant ^{40}Ca enrichment, since ^{40}K has a very low relative abundance (ca. 0.12‰), while ^{40}Ca is the most abundant Ca isotope (Marshall and DePaolo 1989, Simon et al. 2009, Caro et al. 2010). Whether the commonly used reference material NIST SRM 915a has measurable radiogenic ^{40}Ca excess is still under debate (Simon et al. 2009, Caro et al. 2010, Fantle and Tipper 2014, Lehn and Jacobson 2015). For example, Simon et al. (2009) reported that BSE is slightly depleted in radiogenic ^{40}Ca with $\varepsilon^{40/44}$Ca of −0.68 to −0.93, while Caro et al. (2010) measured eight ultramafic to mafic rocks that have $\varepsilon^{40/44}$Ca around 0 relative to NIST SRM 915a within quoted errors. Peridotite JP-1 and other three basalts (BHVO-2, BCR-2 and GBW07105) yield $\varepsilon^{40/44}$Ca ranging from −1.09 to −0.47, with an average of −0.79±0.60 (2*s*), consistent with Simon et al. (2009). It is noted that these three studies rely on a totally different set of samples. However, at ±1 ε level, all studies found no measurable ^{40}Ca excess for NIST SRM 915a compared with most silicate rocks.

5 Stable isotopic compositions of geological materials

5.1 Optimization of double-spike to sample Ca proportion

Measurement repeatability can be improved by optimising the spike-to-sample ratio in the spike-sample mixture (q) (Rudge et al. 2009, Lehn et al. 2013, Feng et al. 2015). The measurement repeatability of single double-spiked runs in this study was modelled after Lehn et al. (2013), assuming a total ion beam of 20 V, 120 cycles of static data acquisition in a single Faraday cup line and an integration time of 16.776 s. The modelled results are shown in Fig. 6, which suggest an optimised q ranging from 0.05 to 0.20. This leads to a theoretical measurement repeatability on $\delta^{44/42}$Ca of ~0.025‰. q was practically adopted to be 0.15 in this study. Twenty-eight runs of double-spiked NIST SRM 915a with a total ion beam within 20±1 V yielded a mean measurement repeatability of 0.030±0.014‰ (2s) for $\delta^{44/42}$Ca, comparable to the theoretical prediction. One over-spiked run with q=0.275 and another under-spiked run with q = 0.039 yielded $\delta^{44/42}$Ca values of 0.00 ± 0.02‰ and 0.00±0.03‰, respectively. This confirms the wide range of optimised q for high precision and accuracy predicted by the theoretical calculations (Fig. 6).

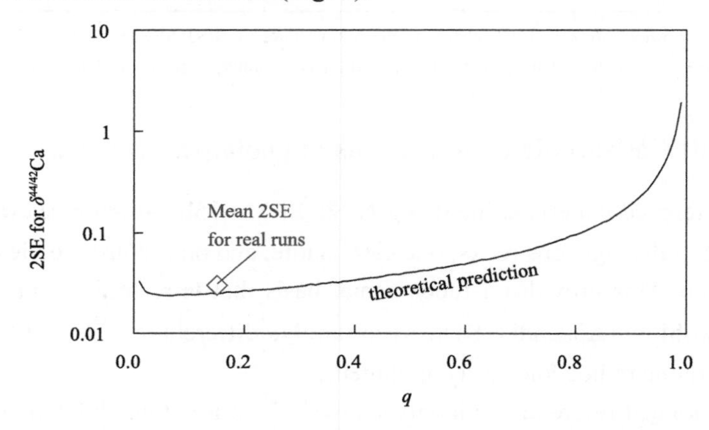

Fig. 6 Theoretical $\delta^{44/42}$Ca measurement repeatability (2SE) using the ^{43}Ca-^{48}Ca double-spike from this study. The model reflects a 20-V total beam and 120 cycles with an integration time of 16.776 s. q, representative of the mole fraction of double-spike in the mixture, was adopted to be 0.15 in this study. Twenty-eight runs of double-spiked NIST SRM 915a with a total beam within 20±1 V yielded a mean measurement repeatability of 0.030±0.014‰ (2s), comparable to the theoretical prediction.

5.2 Intermediate measurement precision and measurement bias

Intermediate measurement precision and accuracy for stable Ca isotopic determinations were estimated based on duplicate results of geological reference materials and interlaboratory comparison. Most previous studies use $\delta^{44/40}$Ca to report mass-dependent Ca isotopic variation. Because of possible radiogenic ^{40}Ca contribution (see Mass-independent isotopic compositions of geological materials), we used $\delta^{44/42}$Ca. Published $\delta^{44/40}$Ca data were re-calculated to $\delta^{44/42}$Ca using the equation $\delta^{44/42}\text{Ca} = (\delta^{44/40}\text{Ca} + \varepsilon^{40/44}\text{Ca}/10)/2.0483$, and mass-independent isotopic compositions are reported in Table 2. Mean $\delta^{44/42}$Ca values for NIST SRM 915a, 915b and Atlantic seawater from each measurement session during the period May 2015 to December 2015 are listed in Table S3 and plotted in Fig. 7. No long-term drift was observed for $\delta^{44/42}$Ca at the ±0.09 level. Based on 286 double-spiked runs of NIST SRM 915a, the intermediate precision of $\delta^{44/42}$Ca was 0.09‰ for single measurement, slightly better than those reported for stable Ca measurement using a ^{43}Ca-^{48}Ca double-spike recently (0.09‰ to 0.17‰ for $\delta^{44/42}$Ca) (Huang et al. 2010, Lehn and Jacobson 2015). Intermediate precision ('external') can be improved by multiple analyses of the same sample (Poitrasson and Freydier 2005, He et al. 2015). Each dissolved rock powder reference material was therefore measured on average eight times, and the mean values with 2SE are reported. This led to an average $\delta^{44/42}$Ca intermediate precision of 0.03‰, based on the duplicate measurements on ten rock reference materials (Table 3). Comparison of results with literature data was performed on the same basis (Table 3 and Fig. 8). For example, the mean $\delta^{44/42}$Ca value for Atlantic seawater was 0.907±0.007‰ (2SE, n=147), consistent with the median value, 0.904‰, compiled by Fantle and Tipper (2014).

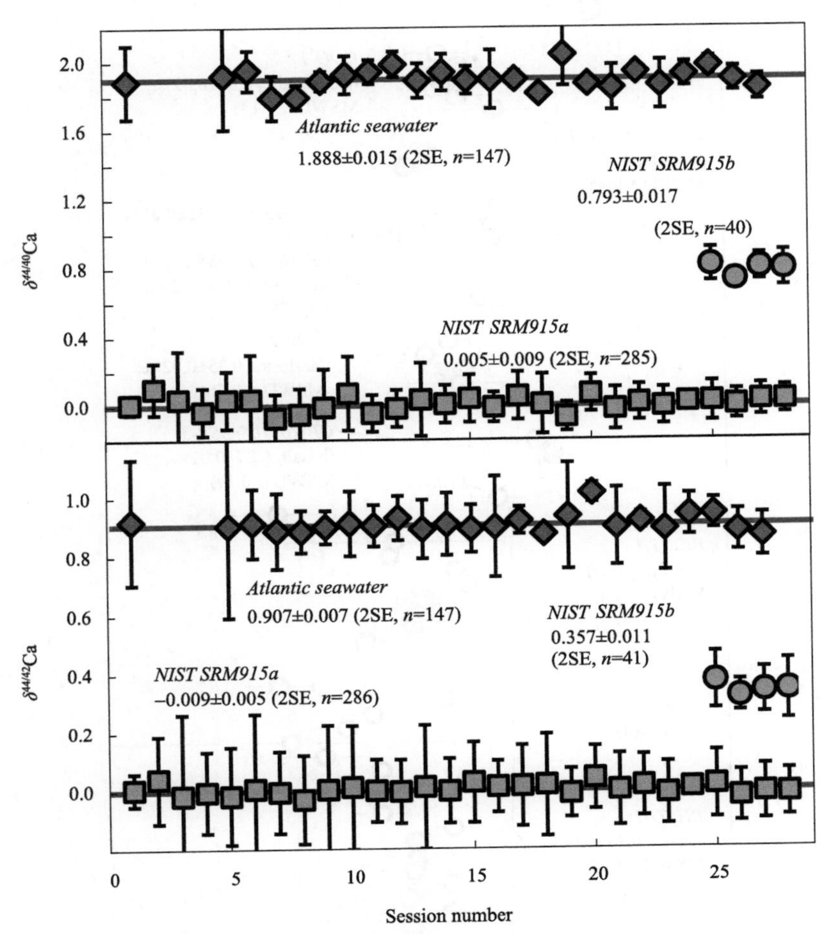

Fig. 7 Intermediate measurement precision of stable Ca isotopic determinations exhibited by the mean values of NIST SRM 915a, 915b and Atlantic seawater from each session. Range bars represent two standard deviations. Horizontal lines are the true and recommended values for NIST SRM 915a and Atlantic seawater, respectively. Recommended values for Atlantic seawater are the median of the compilation from Fantle and Tipper (2014). [Colour figure can be viewed at wileyonlinelibrary.com]

$\delta^{44/40}$Ca was also obtained by our ^{43}Ca-^{48}Ca double-spike measurements, which allowed $\varepsilon^{44/40}$Ca to be calculated using the double-spiked runs (Farkaš et al. 2011). It is important to assess the measurement precision and bias of $\varepsilon^{40/44}$Ca obtained using this approach. Based on duplicated measurements of double-spiked NIST SRM 915a, 915b and Atlantic seawater, the intermediate precision of $\varepsilon^{40/44}$Ca was between 1.0 and 1.9 for single measurements (Table S3), comparable to that of unspiked runs (see Time-dependent drift of Faraday cup efficiency). Duplicate analyses of ten rock reference material powders indicate that intermediate precision for $\varepsilon^{40/44}$Ca could be better than 1.0, with an average of 0.68, provided each sample was analysed eight times (Table 3). $\varepsilon^{40/44}$Ca values calculated using the double-spiked runs were consistent within ± 1.0 with those measured in independent unspiked runs (Fig. 9). In summary, $\delta^{44/42}$Ca and $\varepsilon^{44/40}$Ca could be measured simultaneously within ± 0.03‰ and ±1.0, respectively, if each sample was analysed eight times using a ^{43}Ca-^{48}Ca double-spike technique and the data processing procedures reported here.

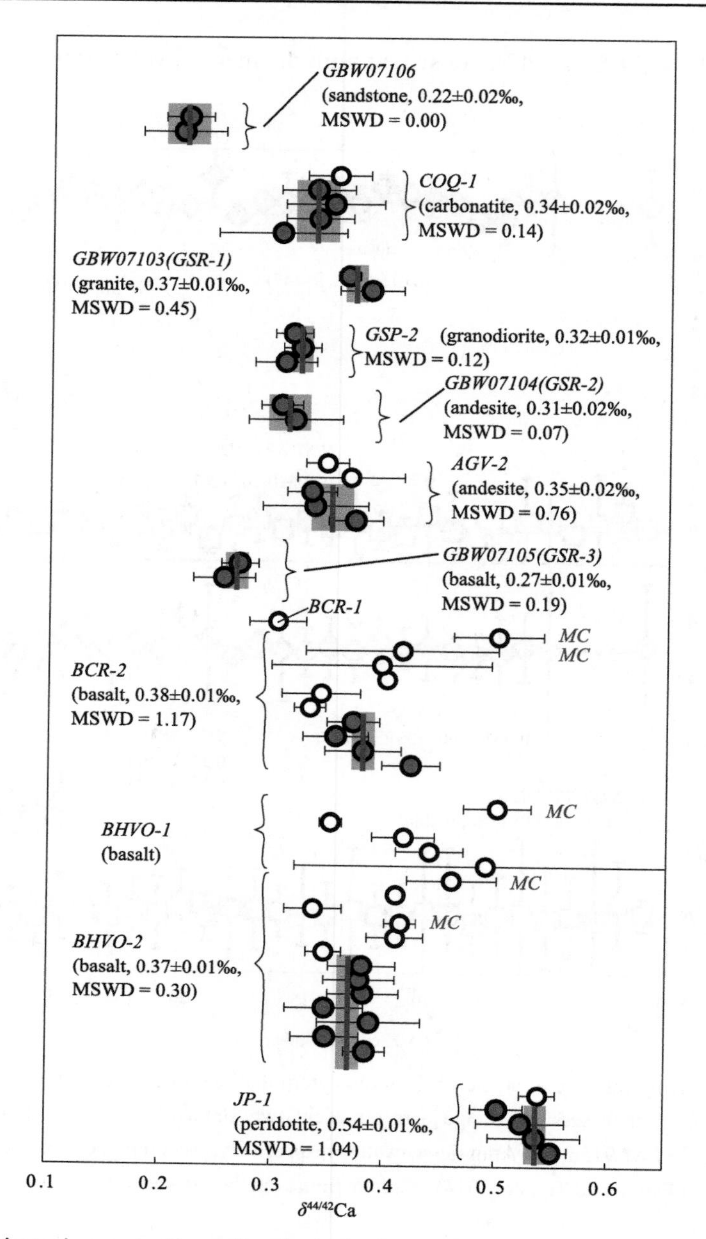

Fig. 8　Summary of stable Ca isotopic compositions of geological reference materials. Ca isotopic data obtained in this study (filled symbols) are compared against literature data (open symbols). The weighted mean values and their relevant errors of our own data are plotted as (vertical) red lines and grey bars. All the literature data were converted to the $\delta^{44/42}$Ca scale using $\delta^{44/42}\text{Ca} = (\delta^{44/42}\text{Ca} + \varepsilon^{40/44}\text{Ca}/10)/2.0483$ if only $\delta^{44/40}$Ca has been reported. 'MC' indicates data obtained by MC-ICP-MS. Data sources refer to Table 3, and range bars represent two standard error of the mean. [Colour figure can be viewed at wileyonlinelibrary.com]

Table 3　Stable isotopic compositions of geological reference materials relative to NIST SRM 915a

Sample/Laboratory ID	$\delta^{44/40}$Ca	2s	2SE	$\delta^{44/42}$Ca	2s	2SE	$\varepsilon^{40/44}$Ca	2s	2SE	n	Data sources
The Atlantic SW	1.888	0.183	0.015	0.907	0.082	0.007	−0.31	1.89	0.16	147	This study
Seawater (literature)	1.900		0.025	**0.904** [a]		**0.013**					Compilation from Fantle and Tipper (2014)
SRM915b	0.793	0.105	0.017	0.357	0.069	0.011	−0.60	0.98	0.16	41	This study
SRM915b (literature)	0.72	0.11	0.02	**0.323** [a]		**0.010**					TIMS, ^{42}Ca-^{48}Ca; Feng et al. (2016)
SRM915b (literature)	0.695	0.234	0.054	0.345	0.136	0.031				19	TIMS, ^{43}Ca-^{48}Ca; Lehn and Jacobson (2015)

Continued

Sample/Laboratory ID	$\delta^{44/40}$Ca	2s	2SE	$\delta^{44/42}$Ca	2s	2SE	$\varepsilon^{40/44}$Ca	2s	2SE	n	Data sources
SRM915b (literature)	0.731		0.006	**0.329** [a]		**0.005**				37	TIMS, ^{42}Ca-^{43}Ca; Lehn et al. (2013)
SRM915b (literature)	0.72	0.22	0.04	**0.323** [a]		**0.020**				56	TIMS, ^{43}Ca-^{48}Ca; Heuser and Eisenhauer (2008)
SRM915b (literature)				0.35		0.01				74	SSB, MC-ICP-MS; Valdes et al. (2014)
SRM915b (literature)				0.340	0.06	0.015				15	SSB, MC-ICP-MS; Harouaka et al. (2016)
SRM915b (literature)				0.42		0.02					SSB, MC-ICP-MS; Schiller et al. (2012)
SRM915b (RV)				**0.334**		**0.004**					
JP-1, peridotite											
He89	1.174	0.065	0.025	0.551	0.040	0.015	−0.46	1.12	0.42	7	This study
He139	1.093	0.079	0.026	0.536	0.123	0.041	0.05	2.25	0.75	9	This study
He101	1.219	0.128	0.043	0.524	0.118	0.039	−1.46	1.88	0.63	9	This study
He122	1.083	0.087	0.031	0.503	0.066	0.023	−0.54	1.67	0.59	8	This study
Weighted Mean	**1.134**			**0.535**			**−0.61**				
2s	**0.131**			**0.041**			**1.26**				
2SE	0.015			0.012			0.28				
MSWD	**3.94**			**1.04**							
JP-1 (literature)	1.15	0.07	0.03	**0.538** [a]		**0.016**				4	TIMS, ^{42}Ca-^{43}Ca; Magna et al. (2015)
JP-1 (RV)				**0.536**		**0.009**					
BHVO-2, basalt											
He52	0.759	0.044	0.015	0.384	0.055	0.018	0.27	0.95	0.32	9	This study
He115	0.777	0.133	0.054	0.349	0.074	0.030	−0.63	1.04	0.42	6	This study
He102	0.826	0.210	0.074	0.388	0.130	0.046	−0.32	1.52	0.54	8	This study
He108	0.723	0.116	0.039	0.348	0.105	0.035	−0.11	1.54	0.51	9	This study
He140	0.828	0.171	0.057	0.382	0.095	0.032	−0.46	1.43	0.48	9	This study
He141	0.821	0.056	0.019	0.378	0.095	0.032	−0.46	1.88	0.63	9	This study
He142	0.850	0.091	0.032	0.381	0.084	0.030	−0.70	1.54	0.54	8	This study
Weighted Mean	**0.787**			**0.375**			**−0.24**				
2s	**0.091**			**0.034**			**0.67**				
2SE	0.010			0.011			0.17				
MSWD	**2.39**			**0.30**							
BHVO-1 (literature)	1.00		0.12	0.49		0.17					TIMS, ^{43}Ca-^{48}Ca; Huang et al. (2010)
BHVO-1 (literature)	0.96		0.05	0.44		0.03					TIMS, ^{43}Ca-^{48}Ca; Huang et al. (2011)
BHVO-1 (literature)				0.50		0.03					SSB, MC-ICP-MS; Schiller et al. (2012)
BHVO-1 (literature)	0.804	0.203	0.083	0.417	0.115	0.028				6	TIMS, ^{43}Ca-^{48}Ca; Lehn and Jacobson (2015)
BHVO-1 (literature)	0.780	0.033	0.014	**0.352** [a]		**0.010**				6	TIMS, ^{42}Ca-^{43}Ca; Lehn and Jacobson (2015)
BHVO-1 (literature)	0.77	0.10	0.03	**0.347** [a]		**0.029**				3	TIMS, ^{42}Ca-^{48}Ca; Feng et al. (2016)
BHVO-2 (literature)	0.77	0.10	0.03	**0.347** [a]		**0.016**				12	TIMS, ^{42}Ca-^{48}Ca; Feng et al. (2016)
BHVO-2 (literature)	0.90	0.11	0.05	**0.41** [a]		**0.03**				5	TIMS, ^{42}Ca-^{43}Ca; Magna et al. (2015)
BHVO-2 (literature)				0.41		0.01					SSB, MC-ICP-MS; Valdes et al. (2014)
BHVO-2 (literature)	0.75	0.22	0.05	**0.34** [a]		**0.03**				20	TIMS, ^{42}Ca-^{43}Ca; Amini et al. (2009)
BHVO-2 (literature)	0.90			**0.41** [a]							TIMS, ^{42}Ca-^{43}Ca; Amini et al. (2009)
BHVO-2 (literature)				0.46		0.04					SSB, MC-ICP-MS; Schiller et al. (2012)
BHVO-1(RV)				**0.364**		**0.008**					
BHVO-2(RV)				**0.381**		**0.007**					

Continued

Sample/Laboratory ID	$\delta^{44/40}$Ca	2s	2SE	$\delta^{44/42}$Ca	2s	2SE	$\varepsilon^{40/44}$Ca	2s	2SE	n	Data sources
BCR-2, basalts											
He90	0.839	0.145	0.048	0.423	0.078	0.026	0.27	1.47	0.49	9	This study
He116	0.816	0.173	0.087	0.380	0.068	0.034	−0.37	1.29	0.65	4	This study
He103	0.792	0.157	0.052	0.356	0.087	0.029	−0.64	2.15	0.72	9	This study
He109	0.785	0.077	0.027	0.371	0.066	0.023	−0.26	1.34	0.47	8	This study
Weighted Mean	**0.798**			**0.383**			−0.17				
2s	**0.049**			**0.058**			0.76				
2SE	0.021			0.014			0.28				
MSWD	**0.33**			**1.17**							
BCR-1 (literature)	0.73		0.05	**0.30** [a]		**0.03**					TIMS, ^{42}Ca-^{48}Ca; Simon and DePaolo (2010)
BCR-2 (literature)				0.41		0.09					SSB, MC-ICP-MS; Valdes et al. (2014)
BCR-2 (literature)	0.79	0.12	0.02	**0.332** [a]		**0.014**				24	TIMS, ^{42}Ca-^{48}Ca; Feng et al. (2016)
BCR-2 (literature)	0.81	0.17	0.07	**0.34** [a]		**0.03**				6	TIMS, ^{42}Ca-^{43}Ca; Amini et al. (2009)
BCR-2 (literature)	0.93			**0.40** [a]							TIMS, ^{42}Ca-^{43}Ca; Amini et al. (2009)
BCR-2 (literature)	0.92	0.40	0.20	**0.40** [a]		**0.10**				4	TIMS, ^{43}Ca-^{48}Ca; Wombacher et al. (2009)
BCR-2 (literature)				0.50		0.04					SSB, MC-ICP-MS; Schiller et al. (2012)
BCR-2(RV)				**0.359**		**0.009**					
GBW07105 (GSR-3), basalt											
He144	0.566	0.065	0.022	0.255	0.082	0.027	−0.43	1.61	0.54	9	This study
He226	0.633	0.088	0.031	0.269	0.047	0.017	−0.82	0.70	0.25	8	This study
Weighted Mean	0.587			**0.265**			**−0.76**				
2s	0.095			**0.020**			**0.55**				
2SE	0.018			0.014			0.22				
MSWD	**3.14**			**0.19**							
AGV-2, andesite											
He54	0.785	0.06	0.02	0.371	0.07	0.02	−0.37	0.72	0.29	9	This study
He111	0.679	0.12	0.05	0.335	0.12	0.05	0.07	2.01	0.82	6	This study
He104	0.676	0.14	0.05	0.332	0.07	0.02	0.05	1.39	0.46	9	This study
Weighted Mean	**0.751**			**0.349**			**−0.23**				
2s	**0.124**			**0.044**			**0.50**				
2SE	0.019			0.015			0.24				
MSWD	**3.57**			**0.76**							
AGV-2 (literature)				0.37		0.05					SSB, MC-ICP-MS; Valdes et al. (2014)
AGV-2 (literature)	0.79	0.09	0.03	**0.346** [a]		**0.019**				9	TIMS, ^{42}Ca-^{48}Ca; Feng et al. (2016)
AGV-2(RV)				**0.349**		**0.012**					
GBW07104 (GSR-2), andesite											
He144	0.664	0.124	0.041	0.317	0.126	0.042	−0.15	2.06	0.69	9	This study
He225	0.691	0.118	0.042	0.305	0.053	0.019	−0.66	0.83	0.29	8	This study
Weighted Mean	**0.677**			**0.307**			**−0.58**				
2s	**0.037**			**0.017**			**0.72**				
2SE	0.029			0.017			0.27				
MSWD	**0.20**			**0.07**							

Continued

Sample/Laboratory ID	$\delta^{44/40}$Ca	2s	2SE	$\delta^{44/42}$Ca	2s	2SE	$\varepsilon^{40/44}$Ca	2s	2SE	n	Data sources
GSP-2, granodiorite											
He105 GSP-2	0.294	0.169	0.056	0.308	0.082	0.027	3.37	1.44	0.48	9	This study
He229 GSP-2	0.299	0.036	0.013	0.322	0.047	0.017	3.62	1.05	0.37	8	This study
He230 GSP-2	0.270	0.091	0.032	0.315	0.047	0.017	3.75	0.52	0.19	8	This study
Weighted Mean	**0.295**			**0.317**			**3.69**				
2s	**0.031**			**0.014**			**0.39**				
2SE	0.012			0.011			0.16				
MSWD	**0.35**			**0.12**							
GBW07103 (GSR-1), granite											
He143	0.727	0.135	0.045	0.384	0.085	0.028	0.59	1.40	0.47	9	This study
He224	0.750	0.045	0.016	0.364	0.028	0.010	−0.05	0.65	0.23	8	This study
Weighted Mean	**0.747**			**0.366**			**0.07**				
2s	**0.032**			**0.029**			**0.91**				
2SE	0.015			0.009			0.21				
MSWD	**0.23**			**0.45**							
COQ-1, carbonatite											
He90	0.672	0.039	0.022	0.304	0.098	0.057	−0.49	1.76	1.02	3	This study
He113	0.731	0.093	0.042	0.337	0.073	0.030	−0.28	0.92	0.41	6	This study
He119[b]	0.614		0.16	0.350		0.09	1.04			1	This study
He106	0.746	0.141	0.050	0.335	0.092	0.033	−0.60	2.13	0.75	8	This study
Weighted Mean	**0.659**			**0.335**			**−0.37**				
2s	**0.121**			**0.039**			**0.33**				
2SE	0.014			0.019			0.34				
MSWD	**3.68**			**0.14**							
COQ-1 (literature)	0.71	0.11	0.06	**0.354**[a]		**0.028**				4	TIMS, ^{42}Ca-^{48}Ca; Feng et al. (2016)
COQ-1(RV)				**0.341**		**0.015**					
GBW07106, sandstone											
He146	0.379	0.092	0.031	0.216	0.111	0.037	0.64	1.90	0.63	9	This study
He228	0.439	0.124	0.047	0.221	0.056	0.021	0.13	0.85	0.32	7	This study
Weighted Mean	**0.397**			**0.220**			**0.24**				
2s	**0.085**			**0.007**			**0.71**				
2SE	0.026			0.018			0.29				
MSWD	**0.38**			**0.00**							

Superscript a denotes $\delta^{44/42}$Ca values calculated by ($\delta^{44/40}$Ca + $\varepsilon^{40/44}$Ca/10)/2.0483 with correction for the radiogenic ^{40}Ca anomaly obtained here, where only $\delta^{44/40}$Ca has been previously reported. Literature data for seawater represent median values of the data compilation from Fantle and Tipper (2014). He119 (COQ-1) was only measured on TIMS one time, and intermediate precision for single measurement is given for the errors. These data were not used to calculate the mean value and standard deviation for $\varepsilon^{40/44}$Ca. Recommended values (RV) are given, if consistent data from at least two independent laboratories were integrated.

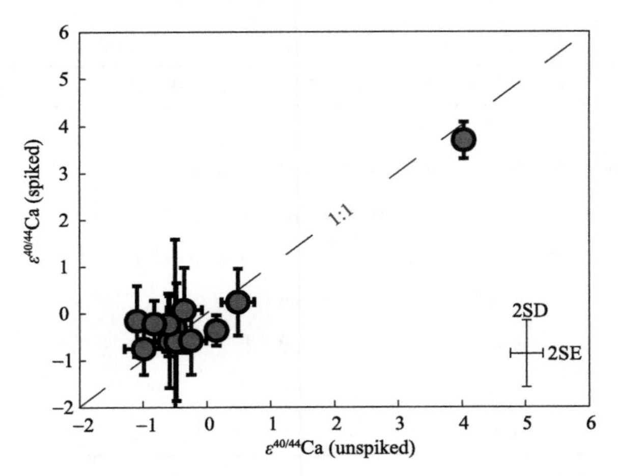

Fig. 9　A comparison of $\varepsilon^{40/44}$Ca obtained for geological reference materials from the results of ^{43}Ca-^{48}Ca double-spiked and unspiked runs.

5.3　Stable Ca isotopic compositions of geological reference materials

We report stable Ca isotopic compositions of twelve commonly used geological reference materials relative to NIST SRM 915a (Table 3). As noted above, these data are in general consistent with the literature data within quoted errors after correction of radiogenic ^{40}Ca contribution. Combined with literature data, new values have been calculated for eleven of twelve reference materials as error-weighted means after Craddock and Dauphas (2010) and listed in Table 3. Data from Schiller et al. (2012) were not integrated, because they are systematically higher than other published values. No further update for recommended values of the seawater is necessary since the data reported here are very close to the median of the literature data (Table 3; Fantle and Tipper 2014).

Nine igneous rocks yielded $\delta^{44/42}$Ca values ranging from 0.265±0.014‰ (2SE) to 0.536±0.009‰ (2SE), demonstrating a detectable stable Ca isotope fractionation during high-temperature processes. Carbonatite COQ-1 has a $\delta^{44/42}$Ca of 0.335±0.019‰ (2SE), similar to the value (~0.35‰) previously reported by Feng et al. (2016). Three basaltic reference materials, BHVO-2, BCR-2 and GBW07105, have $\delta^{44/42}$Ca from 0.265±0.014‰ (2SE) to 0.381±0.007‰ (2SE), systematically lower than peridotite JP-1 (0.536±0.009‰; 2SE). This is consistent with the previous observation that basalts have stable Ca isotopic compositions lighter than the mean upper mantle (Fig. 10) (Huang et al. 2010, 2011, Valdes et al. 2014). Significant lower and variable $\delta^{44/42}$Ca observed in Hawaii basalts have been attributed to incorporating surface carbonates enriched in isotopically light Ca into their mantle sources (Huang et al. 2011). Recycled carbonates can be identified by the elevated Sr/Nb in basalts (Huang et al. 2011). GBW07105, a Cenozoic basalt from Zhangjiakou, eastern China, with the lowest $\delta^{44/42}$Ca, however, does not have high Sr/Nb (Fig. 10c), indicating that low $\delta^{44/42}$Ca magmas may be produced by other processes.

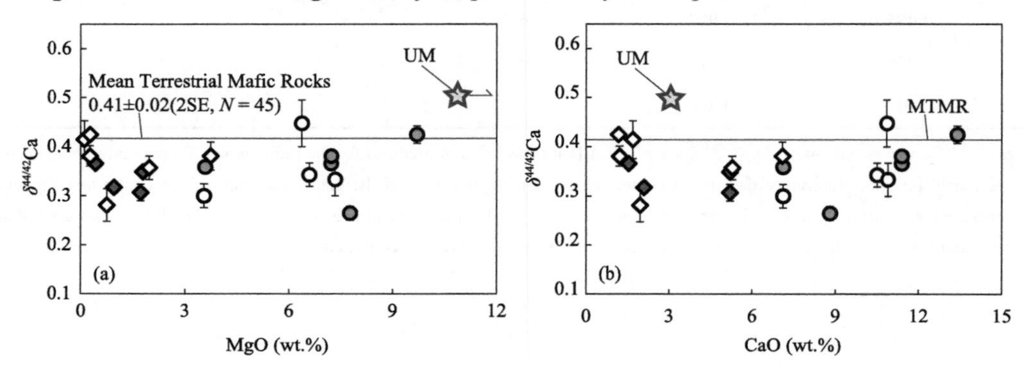

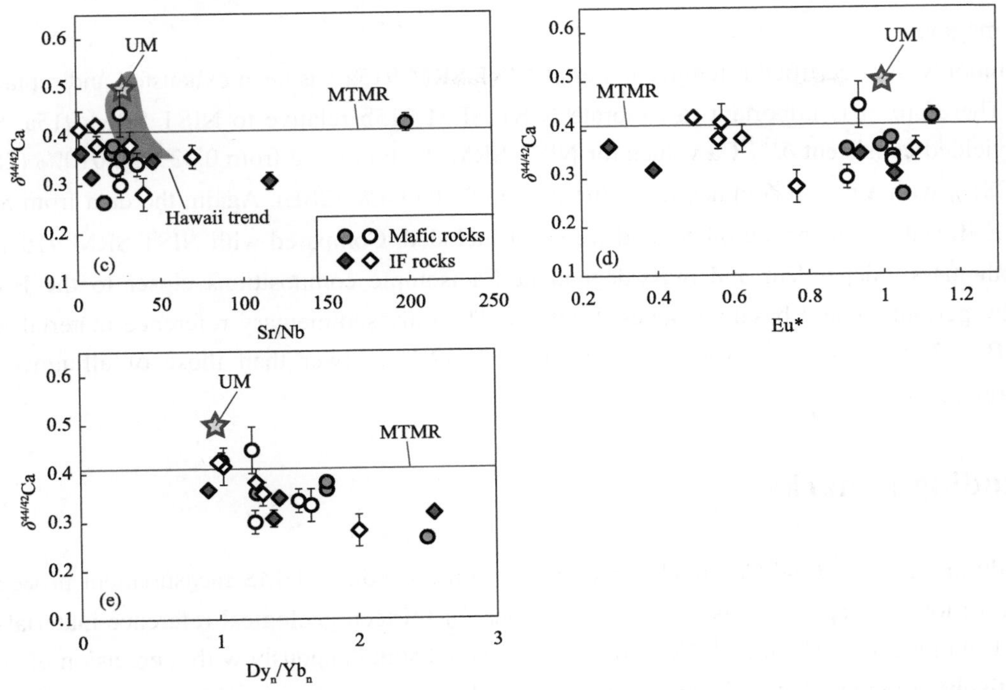

Fig. 10 $\delta^{44/42}$Ca vs. MgO, CaO, Sr/Nb, Eu* and (Dy_n/Yb_n) diagrams for mafic and intermediate to felsic (IF) igneous reference materials. Newly reported data and those compiled from at least two independent laboratories (filled) as well as the literature data from only one laboratory (open) are plotted. Mean terrestrial basaltic rock was calculated from forty-five mafic rocks recently reported (Amini et al. 2009, Huang et al. 2010, 2011, Simon and DePaolo 2010, Valdes et al. 2014, Jacobson et al. 2015). The upper mantle estimate is from Huang et al. (2010) and Sun and McDonough (1989). Eu* = $Eu_N/\sqrt{(Sm_N * Gd_N)}$ where subscript N represents normalisation to the chondrite values (Sun and McDonough 1989). The Hawaii trend is after Huang et al. (2011). Elemental data for geological reference materials are from the GeoReM database (http://georem.mpch-mainz.gwdg.de) and the certificates from IGGE and USGS. [Colour figure can be viewed at wileyonlinelibrary.com]

It is still under debate whether Ca isotopes fractionate during magma differentiation due to limited high-precision Ca isotopic data of differentiated magmas with known elemental compositions. Data for intermediate to felsic geological reference materials measured in this and previous studies (Amini et al. 2009, Valdes et al. 2014, Feng et al. 2016) allow a first-order constraint on this issue. Four intermediate to felsic igneous reference materials (AGV-2, GSP-2, GBW07103 and GBW07104) have highly variable CaO from 5.20% m/m to 1.55% m/m, but only reveal a limited $\delta^{44/42}$Ca variation from 0.307±0.017‰ (2SE) to 0.366±0.009‰ (2SE). Combined with literature data, similar to the compilation of Schiller et al. (2016), no trend can be identified between $\delta^{44/42}$Ca and MgO, CaO for mafic to felsic geological reference materials (Fig. 10a, b), suggesting that Ca isotope fractionation is either insignificant or not systematic (e.g., crystallisation of different minerals has opposite isotopic effect) during magma differentiation. Plagioclase is a major Ca-rich mineral in additional to pyroxene and amphibole. Substantial plagioclase crystallisation decreases the Eu*, defined as $Eu/\sqrt{(Sm \times Gd)}$ normalised to CI chondrite values of magmas. The lack of correlation between $\delta^{44/42}$Ca and Eu* (Fig. 10d) may further suggest insignificant Ca isotope fractionation during fractional crystallisation of plagioclase. All mafic to felsic igneous samples with light Ca isotopic compositions tend to have high (Dy_n/Yb_n) (Fig. 10e), an indicator of residual garnet (He et al. 2011) or garnet crystallisation during differentiation (Macpherson et al. 2006). Garnet has a stronger Ca-O bond than clinopyroxene and thus likely has substantially heavier Ca isotopes than clinopyroxene and other Ca-rich minerals (e.g., amphibole and plagioclase) (Magna et al. 2015). Partial melting with abundant residual garnet and/or magma differentiation with substantial crystalline garnet may be another way to produce low $\delta^{44/42}$Ca magmas.

The commonly used certified reference material NIST SRM 915a has been exhausted and replaced by NIST SRM 915b. Therefore, it is important to calibrate NIST SRM 915b relative to NIST SRM 915a. Six of seven laboratories yielded consistent $\delta^{44/42}$Ca values for NIST SRM 915b ranging from 0.323±0.010‰ (2SE) to 0.357 ±0.011‰ (2SE), with a new recommended value of 0.334±0.004‰ (2SE). Again, the data from Schiller et al. (2012) are significantly higher than others and were not included. Compared with NIST SRM 915a, NIST SRM 915b has both mass-independent and mass-dependent Ca isotopic compositions closer to the BSE estimates represented by peridotites and basalts (Tables 2 and 3). The only sedimentary reference material in our study, sandstone GBW07106, has a $\delta^{44/42}$Ca of 0.220±0.018‰ (2SE), lower than those of all igneous reference materials reported.

6 Concluding remarks

This study presents a ^{43}Ca-^{48}Ca double-spike calibration procedure, TIMS measurement procedures for Ca isotopic composition and reports Ca isotopic compositions for thirteen geological reference materials. Using our double-spike technique, $\delta^{44/42}$Ca and $\varepsilon^{40/44}$Ca could be measured simultaneously with a precision of ±0.03‰ and ±1.0, respectively, provided that each sample was analysed eight times. We provide a comprehensive data set of both mass-independent and mass-dependent Ca isotopic compositions of thirteen geological reference materials. This will allow future interlaboratory calibration of Ca isotopic measurement.

Measurement of igneous reference materials yielded $\delta^{44/42}$Ca values between 0.27‰ and 0.54‰ relative to NIST SRM 915a, well outside the intermediate precision of the measurements. Basaltic to felsic rocks generally have 'lighter' Ca isotopic compositions than the upper mantle to different extents. Therefore, either significant Ca isotope fractionation occurs during magmatic processes, or the variation can be attributed to recycling of surface materials. No correlation between $\delta^{44/42}$Ca and MgO or CaO was observed for mafic to felsic igneous reference materials, suggesting that Ca isotope fractionation is either insignificant or not systematic during magma differentiation. Given the rather low $\delta^{44/42}$Ca for a sandstone and that the lower $\delta^{44/42}$Ca reference materials tend to have higher (Dy_n/Yb_n), we suggest a potential role for residual or crystalline garnet and incorporation of silicate sediments into sources to produce magmas with low $\delta^{44/42}$Ca.

Acknowledgements

Constructive comments from two anonymous reviewers and efficient editoral handling of Mary Horan are highly appreciated. This work is supported by the National Natural Science Foundation of China (41230209 to LSG, 41473016 and 41673012 to YSH), and the State Key Laboratory of Geological Processes and Mineral Resources(Open Research Program GPMR201510). SCH acknowledges support from NSF award EAR-1524387. This is CUGB petrogeochemical contribution No. PGC-201514.

References

Amini M., Eisenhauer A., Bohm F., Holmden C., Kreissig K., Hauff F. and Jochum K.P. (2009) Calcium isotopes ($\delta^{44/40}$Ca) in MPI-DING reference glasses, USGS rock powders and various rocks: Evidence for Ca isotope fractionation in terrestrial silicates. Geostandards and Geoanalytical Research, 33, 231-247.

Boulyga S.F. (2010) Calcium isotope analysis by mass spectrometry. Mass Spectrometry Reviews, 29, 685-716.

Caro G., Bourdon B., Birck J.-L. and Moorbath S. (2003) ^{146}Sm-^{142}Nd evidence from Isua metamorphosed sediments for early differentiation of the Earth's mantle. Nature, 423, 428-432.

Caro G., Papanastassiou D.A. and Wasserburg G.J. (2010) ^{40}K–^{40}Ca isotopic constraints on the oceanic calcium cycle. Earth and Planetary Science Letters, 296, 124-132.

Craddock P.R. and Dauphas N. (2010) Iron isotopic compositions of geological reference materials and chondrites. Geostandards and Geoanalytical Research, 35, 101-123.

De La Rocha C.L. and DePaolo D.J. (2000) Isotopic evidence for variations in the marine calcium cycle over the Cenozoic. Science, 289, 1176-1178.

Fantle M.S. and Bullen T.D. (2009) Essentials of iron, chromium, and calcium isotope analysis of natural materials by thermal ionization mass spectrometry. Chemical Geology, 258, 50-64.

Fantle M.S. and Tipper E.T. (2014) Calcium isotopes in the global biogeochemical Ca cycle: Implications for development of a Ca isotope proxy. Earth-Science Reviews, 129, 148-177.

Farkaš J., Böhm F., Wallmann K., Blenkinsop J., Eisenhauer A., van Geldern R., Munnecke A., Voigt S. and Veizer J. (2007) Calcium isotope record of Phanerozoic oceans: Implications for chemical evolution of seawater and its causative mechanisms. Geochimica et Cosmochimica Acta, 71, 5117-5134.

Farkaš J., Déjeant A., Novák M. and Jacobsen S.B. (2011) Calcium isotope constraints on the uptake and sources of Ca^{2+} in a base-poor forest: A new concept of combining stable ($\delta^{44/42}Ca$) and radiogenic (εCa) signals. Geochimica et Cosmochimica Acta, 75, 7031-7046.

Feng L.P., Zhou L., Yang L., Tong S.Y., Hu Z.C. and Gao S. (2015) Optimization of the double spike technique using peak jump collection by a Monte Carlo method: An example for the determination of Ca isotope ratios. Journal of Analytical Atomic Spectrometry, 30, 2403-2411.

Feng L.P., Zhou L., Yang L., DePaolo D.J., Tong S.Y., Liu Y.S., Owens T.L. and Gao S. (2016) Calcium isotopic compositions of sixteen USGS reference materials. Geostandards and Geoanalytical Research. doi:10.1111/ggr.12131.

Fletcher I.R., Maggi A.L., Rosman K.J.R. and McNaughton N. (1997) Isotopic abundance measurements of K and Ca using a wide-dispersion multi-collector mass spectrometer and low-fractionation ionisation techniques. International Journal of Mass Spectrometry and Ion Processes, 163, 1-17.

Harouaka K., Mansor M., Macalady J.L. and Fantle M.S. (2016) Calcium isotopic fractionation in microbially mediated gypsum precipitates. Geochimica et Cosmochimica Acta, 184, 114-131.

Hart S.R. and Zindler A. (1989) Isotope fractionation laws: A test using calcium. International Journal of Mass Spectrometry and Ion Processes, 89, 287-301.

He Y.S., Li S.G., Hoefs J., Huang F., Liu S.A. and Hou Z.H. (2011) Post-collisional granitoids from the Dabie orogen: New evidence for partial melting of a thickened continental crust. Geochimica et Cosmochimica Acta, 75, 3815-3838.

He Y.S., Ke S., Teng F.Z., Wang T.T., Wu H.J., Lu Y.H. and Li S.G. (2015) High-precision iron isotope analysis of geological reference materials by high-resolution MC-ICP-MS. Geostandards and Geoanalytical Research, 39, 341-356.

Heuser A. and Eisenhauer A. (2008) The calcium isotope composition ($\delta^{44/40}Ca$) of NIST SRM 915b and NIST SRM 1486. Geostandards and Geoanalytical Research, 32, 311-315.

Heuser A., Eisenhauer A., Gussone N., Bock B., Hansen B.T. and Nägler T.F. (2002) Measurement of calcium isotopes ($\delta^{44}Ca$) using a multi-collector TIMS technique. International Journal of Mass Spectrometry, 220, 385-397.

Holmden C. and Bélanger N. (2010) Ca isotope cycling in a forested ecosystem. Geochimica et Cosmochimica Acta, 74, 995-1015.

Huang S.C., Farkaš J. and Jacobsen S.B. (2010) Calcium isotopic fractionation between clinopyroxene and orthopyroxene from mantle peridotites. Earth and Planetary Science Letters, 292, 337-344.

Huang S.C., Farkas J. and Jacobsen S.B. (2011) Stable calcium isotopic compositions of Hawaiian shield lavas: Evidence for recycling of ancient marine carbonates into the mantle. Geochimica et Cosmochimica Acta, 75, 4987-4997.

Huang S.C., Farkaš J., Yu G., Petaev M.I. and Jacobsen S.B. (2012) Calcium isotopic ratios and rare earth element abundances in refractory inclusions from the Allende CV3 chondrite. Geochimica et Cosmochimica Acta, 77, 252-265.

Jacobson A.D., Grace A.M., Lehn G.O. and Holmden C. (2015) Silicate versus carbonate weathering in Iceland: New insights from Ca isotopes. Earth and Planetary Science Letters, 416, 132-142.

Johnson C.M. and Beard B.L. (1999) Correction of instrumentally produced mass fractionation during isotopic analysis of Fe by thermal ionization mass spectrometry. International Journal of Mass Spectrometry, 193, 87-99.

Lehn G.O. and Jacobson A.D. (2015) Optimization of a ^{48}Ca-^{43}Ca double-spike MC-TIMS method for measuring Ca isotope ratios ($\delta^{44/40}$Ca and $\delta^{44/42}$Ca): Limitations from filament reservoir mixing. Journal of Analytical Atomic Spectrometry, 30, 1571-1581.

Lehn G.O., Jacobson A.D. and Holmden C. (2013) Precise analysis of Ca isotope ratios ($\delta^{44/40}$Ca) using an optimized ^{43}Ca–^{42}Ca double-spike MC-TIMS method. International Journal of Mass Spectrometry, 351, 69-75.

Li J., Zhu X.K. and Tang S.H. (2011) The application of double spike in non-traditional stable isotopes: A case study on Mo isotopes. Rock and Mineral Analysis, 30, 138-143.

Macpherson C.G., Dreher S.T. and Thirlwall M.F. (2006) Adakites without slab melting: High pressure differentiation of island arc magma, Mindanao, the Philippines. Earth and Planetary Science Letters, 243, 581-593.

Magna T., Gussone N. and Mezger K. (2015) The calcium isotope systematics of Mars. Earth and Planetary Science Letters, 430, 86-94.

Marshall B.D. and DePaolo D.J. (1982) Precise age determinations and petrogenetic studies using the K-Ca method. Geochimica et Cosmochimica Acta, 46, 2537-2545.

Marshall B.D. and DePaolo D.J. (1989) Calcium isotopes in igneous rocks and the origin of granite. Geochimica et Cosmochimica Acta, 53, 917-922.

Morgan J.L.L., Gordon G.W., Arrua R.C., Skulan J.L., Anbar A.D. and Bullen T.D. (2011) High-precision measurement of variations in calcium isotope ratios in urine by multiple collector inductively coupled plasma-mass spectrometry. Analytical Chemistry, 83, 6956-6962.

Naumenko-Dèzes M.O., Bouman C., Nägler T.F., Mezger K. and Villa I.M. (2015) TIMS measurements of full range of natural Ca isotopes with internally consistent fractionation correction. International Journal of Mass Spectrometry, 387, 60-68.

Niederer F.R. and Papanastassiou D.A. (1984) Ca isotopes in refractory inclusions. Geochemica et Cosmochimica Acta, 48, 1279-1293.

Poitrasson F. and Freydier R. (2005) Heavy iron isotope composition of granites determined by high resolution MC-ICP-MS. Chemical Geology, 222, 132-147.

Rudge J.F., Reynolds B.C. and Bourdon B. (2009) The double spike toolbox. Chemical Geology, 265, 420-431.

Russell W.A., Papanastassiou D.A. and Tombrello T.A. (1978) Ca isotope fractionation on the Earth and other solar system materials. Geochimica et Cosmochimica Acta, 42, 1075-1090.

Schiller M., Paton C. and Bizzarro M. (2012) Calcium isotope measurement by combined HR-MC-ICP-MS and TIMS. Journal of Analytical Atomic Spectrometry, 27, 38-49.

Schiller M., Gussone N. and Wombacher F. (2016) High temperature Ca isotope geochemistry of terrestrial silicate rocks, minerals and melts. In: Gussone N., Schmitt A.D., Heuser A., Wombacher F., Dietzel M., Tipper E. and Schiller M. (eds), Calcium stable geochemistry. Springer (Berlin), 223-229.

Simon J.I. and DePaolo D.J. (2010) Stable calcium isotopic composition of meteorites and rocky planets. Earth and Planetary Science Letters, 289, 457-466.

Simon J.I., DePaolo D.J. and Moynier F. (2009) Calcium isotope composition of meteorites, Earth, and Mars. The Astrophysical Journal, 702, 707-715.

Sun S.S. and McDonough W.F. (1989) Chemical and isotopic systematics of oceanic basalts: Implication for mantle composition and process. In: Saunders A.D. and Norry M.J. (eds), Magmatism in the ocean basins. Geological Society of London, Special Publication, 42, 313-345.

Tipper E., Galy A. and Bickle M. (2006) Riverine evidence for a fractionated reservoir of Ca and Mg on the continents: Implications for the oceanic Ca cycle. Earth and Planetary Science Letters, 247, 267-279.

Valdes M.C., Moreira M., Foriel J. and Moynier F. (2014) The nature of Earth's building blocks as revealed by calcium isotopes. Earth and Planetary Science Letters, 394, 135-145.

Wombacher F., Eisenhauer A., Heuser A. and Weyer S. (2009) Separation of Mg, Ca and Fe from geological reference materials for stable isotope ratio analyses by MC-ICP-MS and double-spike TIMS. Journal of Analytical Atomic Spectrometry, 24, 627-636.

Zhu H.L., Zhang Z.F., Liu Y.F., Liu F. and Kang J.T. (2015) Calcium isotope geochemistry review. Earth Science Frontiers, 22, 44-53.

Zhu H.L., Zhang Z.F., Liu Y.F., Fang L., Xin L. and Sun W.D. (2016) Calcium isotopic fractionation during ion-exchange column chemistry and thermal ionisation mass spectrometry (TIMS) determination. Geostandards and Geoanalytical Research, 40, 185-194.

Supporting information

The following information may be found in the online version of this article:

Appendix S1. Data deduction procedures.

Figure S1. Measured $(^{40}Ca/^{44}Ca)_N$ from 100 theoretical cases with significant heterogeneous effect using a Monte Carlo approach.

Figure S2. Comparison of the absolute $^{40}Ca/^{44}Ca$ ratios observed in this study with the literature data.

Table S1. Raw data for unspiked NIST SRM 915a runs.

Table S2. Raw data of unspiked runs for other twelve geological reference materials relative to NIST SRM 915a.

Table S3. Raw data of double-spiked runs for NIST SRM 915a, NIST SRM 915b and Atlantic Seawater.

This material is available as part of the online article from: http://onlinelibrary.wiley.com/doi/10.1111/ggr.12153/ abstract (This link will take you to the article abstract).

High-precision measurement of stable Cr isotopic in geological reference materials by double-spike TIMS method*

Chun-Yang Liu[1], Li-Juan Xu[1*], Chun-Tao Liu[1], Jia Liu[2*], Li-Ping Qin[2], Zi-Da Zhang[1], Sheng-Ao Liu[1] and Shu-Guang Li[1,2]

1. State Key Laboratory of Geological Processes and Mineral Resources, School of Scientific Research, China University of Geosciences, Beijing 100083, China
2. CAS Key Laboratory of Crust-Mantle Materials and Environments, School of Earth and Space Sciences, University of Science and Technology of China, Hefei 230026, China

亮点介绍： 针对目前 Cr 同位素分析方法存在化学流程复杂，TIMS 测试时有机质干扰多等问题，本方法通过优化化学纯化流程，使用高氯酸与浓硝酸有效去除有机质，从而显著提高样品分析时的信号强度和信号稳定性，有效降低了 TIMS 分析所需的样品量，可对低 Cr 含量地质样品进行高精度同位素分析。本方法对 BHVO-2 测试的长期精度优于 0.031‰。对七个国际标准物质进行同位素分析，结果与国际上其他实验室结果一致，精度达到国际一流水平。

Abstract Chromium (Cr) isotopes have been widely used in various fields of earth and planetary sciences. However, high-precision measurements of Cr stable isotope ratios is still challenged by difficulties in purifying Cr and organic matter interference from resin using double-spike thermal ionization mass spectrometry (DS-TIMS). In this study, an improved and easily operated (without oxidizing reagent) two-column chemical separation procedure using AG50W-X12 (200–400 m) resin is introduced. This resin has a higher cross-linking density than AG50W-X8, and this higher density generates better separation efficiency and higher saturation. Organic matter from the resin is a common cause of inhibition of the emission of Cr during analysis by TIMS. Here, perchloric and nitric acids are utilized to eliminate organic matter interference. The Cr isotope ratios of samples with lower Cr contents can be measured precisely by TIMS. The long-term measurement intermediate precision of $\delta^{53/52}Cr_{NIST\ SRM\ 979}$ for BHVO-2 is better than ±0.031‰ (2s) over one year. Replicated digestions and measurements of geological reference materials (OKUM, MUH-1, JP-1, BHVO-1, BHVO-2, AGV-2 and, GSP-2) yield $\delta^{53/52}Cr_{NIST\ SRM\ 979}$ results raging from –0.129‰ to –0.032‰, which are distinguishable within our intermediate precision. The Cr isotope ratios of geological reference materials are consistent with the $\delta^{53/52}Cr_{NIST\ SRM\ 979}$ values reported by previous studies, and the measurement uncertainty (±0.031‰, 2s) is significantly improved.

Keywords Cr isotope, TIMS, double spike, reference materials, high-precision

* 本文发表在：Geostandards and Geoanalytical Research, 2019, 43: 647-661

1 Introduction

Chromium (Cr) is one of the transition metals, compatible and moderately siderophile, and widely distributed in the crust, mantle and core with a large content range (i.e., <1 to 7600 μg g^{-1}, Allègre et al. 1995, Rudnick and Gao 2003). Chromium has four stable isotopes, ^{50}Cr, ^{52}Cr, ^{53}Cr, and ^{54}Cr, with relative abundances of 4.345%, 83.789%, 9.501%, and 2.365%, respectively (Meija et al. 2016). Based on the short-lived ^{53}Mn-^{53}Cr extinctive radionuclides (3.7 ± 0.4 Myr), previous studies on Cr isotopes have focused on dating meteorites or identifying the formation process occurring in the solar system during the first 20 Myr by mass-independent Cr isotope variation (Birck and Allègre 1988, Lugmair and Shukolyukov 1998, Trinquier et al. 2008, Yamakawa et al. 2009, Qin et al. 2010a). In addition, $^{54}Cr/^{52}Cr$ ratios in extraterrestrial samples that display planetary scale isotopic anomalies have been used to fingerprint different nucleosynthetic provenances (Trinquier et al. 2007, Qin et al. 2010b, Qin and Carlson 2016).

Recently, the chromium isotope have been applied as a paleoredox proxy in low-temperature environments and especially exploited as a tracer of Earth's atmospheric oxygenation (Frei et al. 2009, Crowe et al. 2013, Planavsky et al. 2014, Cole et al. 2016). Chromium is typically present in nature in three valence states of Cr(II), Cr(III) and Cr(VI). Hexavalent chromium Cr(VI), as either a chromate (alkaline pH) (CrO_4^{2-}) or a hydrogen chromate (acidic pH) ion ($HCrO_4^-$), is more soluble and mobile than trivalent chromium Cr(III) (Rai et al. 1989, Ellis and Bullen 2002). To date, up to ~6–7‰ variation of $\delta^{53/52}Cr_{NIST\ SRM\ 979}$ has been found during reduction and oxidation reactions (Schauble et al. 2004, Zink et al. 2010), and a large range from −0.27‰ to 1.23‰ was also found during non-redox-dependent processes (Saad et al. 2017). Accordingly, Cr isotopes have been utilised to trace the natural attenuation of Cr(VI) in groundwater (Blowes 2002) or to constrain the redox state of modern and ancient seawater (Frei et al. 2011, Bonnand et al. 2013, Scheiderich et al. 2015, Gueguen et al. 2016). A few studies have been devoted to the inventory of the solid Earth and planetary processes at high-temperature, and these studies improve our understanding of the Cr cycle (Schoenberg et al. 2008, Rudge et al. 2009, Moynier et al. 2011, Farkaš et al. 2013, Li et al. 2016, Xia et al. 2017). About the inventory of the solid Earth, Schoenberg et al. (2008) suggested that the $\delta^{53/52}Cr_{NIST\ SRM\ 979}$ value of the igneous silicate Earth is -0.124 ± 0.101‰ by analysing oceanic and continental basalts, mantle xenoliths, ultramafic rocks and cumulates. Farkaš et al. (2013) suggested that the $\delta^{53/52}Cr_{NIST\ SRM\ 979}$ value of the bulk silicate Earth (BSE) is -0.079 ± 0.129‰ by analysing mantle-derived chromites. Recently, Xia et al. (2017) suggested that the $\delta^{53/52}Cr_{NIST\ SRM\ 979}$ values of fresh fertile peridotites are -0.14 ± 0.12‰ on average and the variation of $\delta^{53/52}Cr_{NIST\ SRM\ 979}$ values of mantle peridotites is caused by partial melting. The $\delta^{53/52}Cr_{NIST\ SRM\ 979}$ value of BSE was given as -0.11 ± 0.11‰ by analysing komatiites (Sossi et al. 2018). For the planetary processes, Moynier et al. (2011) suggested that there could be Cr isotopic fractionation during core segregation, as the $\delta^{53/52}Cr_{NIST\ SRM\ 979}$ values of meteorites (−0.2‰ to ~−0.4‰) are lighter than that of the BSE. Schiller et al. (2014) also found light $\delta^{53/52}Cr_{NIST\ SRM\ 979}$ of meteorites (~−0.30‰). However, Bonnand et al. (2016b) found that meteorites have a BSE-like stable Cr isotopic ratio and suggested that chromium isotopes are fractionated during the magmatic process. Recently, Bonnand and Halliday (2018) suggested that equilibrium fractionation between iron liquid and sulfides or kinetic fractionation during oxidation of Cr may be responsible for Cr isotopic fractionation during fractional crystallisation of meteorite rocks.

Compared with the very large Cr isotopic fractionation at low temperatures, Cr isotopic fractionation in high-temperature environments is much smaller. Thus, small measurement uncertainties during measurement of natural samples are a prerequisite to improve the application of Cr stable isotopic fractionation in high-temperature geochemistry. High-precision Cr isotopic measurement have been obtained for samples with extremely high chromium contents such as meteorite and extraterrestrial samples (Moynier et al. 2011, Schiller et al. 2014, Bonnand et al. 2016a, 2016b). An improved Cr isotope ratio measurement procedure for terrestrial samples that contain less chromium content is also established in this study.

The small measurement uncertainty of stable Cr isotope ratio measurement results requires not only a high-quality chemical separation process but also appropriate methods for mass bias correction, which occurs during measurement and chemical purification. High quality Cr purification enables a reduction in the isobaric interference from ^{54}Fe, ^{50}V, and ^{50}Ti, and matrix element effects that interfere with the production of Cr^+ thermal ions (Ball and Bassett 2000). Chromium mainly has two valence states in solution, and the behaviours of Cr(III) and Cr(VI) during ion exchange separation are very different (Schoenberg et al. 2008, Trinquier et al. 2008, Yamakawa et al. 2009, Li et al. 2016, Li et al. 2017). To convert all chromium ions to the same valence state, various oxidising or reducing agents ($(NH_4)_2S_2O_8$, $K_2S_2O_8$, $KMnO_4$, and HCl) were used in previous studies in the anion or cation resin stage, and correspondingly, two, three, or even four columns were used (Table 2, Lugmair and Shukolyukov 1998, Ball and Bassett 2000, Ellis and Bullen 2002, Johnson and Bullen 2004, Halicz et al. 2008, Trinquier et al. 2008, Schoenberg et al. 2008, Døssing et al. 2011, Rodler et al. 2015, Li et al. 2016, Li et al. 2017, Yamakawa et al. 2009, Schoenberg et al. 2016, Schiller et al. 2014, Zhu et al. 2018). These methods can obtain satisfactory pure Cr cuts from multiple kinds of samples. Generally, 0.5–1 μg of Cr must be loaded onto the filament for precise measurement results with the TIMS. Thus, the chemical purification process must be guaranteed to be applicable for larger test portions sizes. The residual of the oxidising reagents (SO_4^{2-}) can severely inhibit the Cr^+ signal during TIMS measurements, but this residual is difficult to remove (Ball and Bassett 2000). In addition, the high ionisation potential of Cr and the inhibition effect of the Cr^+ signal from organic matter during instrument measurements are the main reasons impeding the acquisition of high-precision isotope ratio results in compositionally complex geological materials during TIMS measurements (Johnson and Bullen 2004, Yamakawa et al. 2009, et al. 2013, Li et al. 2016).

Previous studies achieved Cr purification from meteorite and terrestrial samples by a two-step column method with AGW50-X8 (Birck and Allègre 1988, Lugmair and Shukolyukov 1998, Trinquier et al. 2008, Qin et al. 2010a, Bonnand et al. 2016b). Bonnand et al. (2011) also obtained the precise Cr isotope ratio of carbonates based on a one-step method, which has the same Cr speciation and resin types. In this study, we report an improved two-step column purification scheme without additional oxidizing/reducing agent for high-precision measurement of stable Cr isotope ratios based on the method mentioned above. The resin is modified to Bio-Rad AG50W-X12 (200–400 mesh) cation exchange resins. The inhibition effects of the isobaric interference, matrix and organic matter are tested and assessed. $HClO_4$ and HNO_3 acids were utilized to eliminate organic matter interference from the resin and were removed without Cr loss. High-precision Cr isotope ratios of geological reference materials ranging in lithologies from ultramafic rock to granitoid are presented in this study.

2 Measurement procedure

2.1 Reagents

In our experiments, all optima-grade HF, HCl, and HNO_3 acids were further purified twice by a Savillex™ DST-1000 sub-boiling distillation system (Minnetonka, USA). High purity water (HPW) with a resistivity of 18.2 MΩ cm^{-1} was obtained from a Milli-Q (MQ) Element system (Millipore, USA) and used in the whole procedure. PFA Savillex™ beakers were cleaned with HNO_3 (14 mol l^{-1}) and/or HCl (11 mol l^{-1}), diluted to 50% (v/v) with water for 24 h on a hotplate at 130 ℃, followed by cleaning with purified HCl (11 mol l^{-1}) and were finally rinsed with HPW. Optima-grade perchloric acid was obtained from Thermo Fisher and purified by Bio-Rad AG50W-X12 resin (200–400 mesh) with the chromatographic method. The GSB-Ti, Fe, V, K, Na, Ca, Mg, and Mn ultrapure standard solutions used in this study were from the China Iron and Steel Research Institute.

2.2 Sample digestion

Sample powders containing 0.5–1 μg Cr were digested in a combination of concentrated HF (29 mol l^{-1}) and HNO_3 (14 mol l^{-1}, 2:1 in volume) in Teflon™ PFA beakers at 130 ℃ for two or three days until the solid particles disappeared. Next, the solutions were heated to evaporation. Then aqua regia (HCl: HNO_3 = 3:1) was added until the steamed sample completely dissolved. The sample solutions were evaporated to dryness on a hot plate at ~130 ℃. The products were finally dissolved in 0.5 ml of concentrated HCl (11 mol l^{-1}). The dissolved sample solutions were mixed with 0.4 ml ^{50}Cr–^{54}Cr double-spike solution for every ~1 μg Cr, and then the mixtures were heated on the hotplate at 130 ℃ overnight to achieve homogenization. After evaporation to dryness, the mixtures were re-dissolved in 0.2 ml 6 mol l^{-1} HCl.

2.3 Column chemistry

A modified method for Cr purification was developed by a two-step column procedure (Fig. 1b, c, Table 1). In previous studies, cation and anion exchange resins (100–200 mesh/ 200–400 mesh) were usually used in the Cr column separation procedure (Table 2, Lugmair and Shukolyukov 1998, Ball and Bassett 2000, Ellis and Bullen 2002, Johnson and Bullen 2004, Halicz et al. 2008, Trinquier et al. 2008, Schoenberg et al. 2008, Døssing et al. 2011, Rodler et al. 2015, Li et al. 2016, Li et al. 2017, Yamakawa et al. 2009, Schoenberg et al. 2016, Schiller et al. 2014). AG50W-X8 was applied as a conventional cation exchange resin during Cr purification (Table 2, Bonnand et al. 2011, Bonnand et al. 2016a, 2016b, Qin et al. 2010a, Trinquier et al. 2008). The purification methods applicable to silicate and meteorite materials have a limit on the minimum Cr mass fraction of the samples, and this limit is greater than approximately 45 μg g^{-1} (Table 2). Additionally, a yield of more than 0.5 μg Cr is normally required after separation for high-precision measurement results by MS. Here, analytical-grade Bio-Rad AG50W-X12 (200–400 mesh) cation exchange resins were used, and these resins have higher crosslinkage (12%), smaller wet bead size (53–106 μm) and smaller molecular weight limits (~400) than the Bio-Rad AG50W-X8 (200–400 mesh; crosslinkage: 8%; wet bead size: 63–150 μm; molecular weight limits: ~1000) cation exchange resins (http://www.bio-rad.com/en-hk/category/ion-exchange-resins/). Thus, the AG50W-X12 (200–400 mesh) resin has a high sample load capacity so that this scheme is suitable for geological samples with variable of Cr content. Silicate samples with Cr mass fractions greater than 17 μg g^{-1} are suitable for our column chemistry which mean the the maximum mass for sample powders is about 50 mg. The yield of this method is as good as published purification schemes (Table 2). The adoption of the AG50W-X12 resin also improves the separation efficiency and selectivity, which result in better separation efficiency than the Bio-Rad AG50W-X8 resin (Fig. 1a, b). The retardation of Ti and V ions avoids the overlap of the elution of Cr ions and isobaric interference cations (Fig. 1a, b). A better purification effect for Cr is also obtained from matrix cations such as Mg^+ and Ca^+ (Fig. 1a, b).

Table 1　Chromium chemistry purification procedure and the function of each step

Separation stage	Reagent	Volume (ml)
Column 1: Cation exchange resin (AG50W-X12, 200–400 mesh, 1 ml)		
Condition	1 mol l^{-1} HCl	5
Load sample and collect Cr	1 mol l^{-1} HCl	1.2
Collect Cr	1 mol l^{-1} HCl	1
Collect Cr	1 mol l^{-1} HCl	2.5
Elute Fe, V, and matrix	6 mol l^{-1} HCl	7
Column 2: Cation exchange resin (AG50W-X12, 200–400 mesh, 0.33 ml)		
Condition	0.16 mol l^{-1} HNO_3	4
Load sample	0.16 mol l^{-1} HNO_3	2
Elute Ti and V	0.5 mol l^{-1} HF	3
Elute Mn and matrix	1 mol l^{-1} HCl	1
Elute Mn and matrix	1 mol l^{-1} HCl	3
Elute Mn and matrix	1 mol l^{-1} HCl	3
Elute Mn and matrix	1 mol l^{-1} HCl	3
Elute Mn and matrix	1 mol l^{-1} HCl	1
Collect Cr	2 mol l^{-1} HCl	2

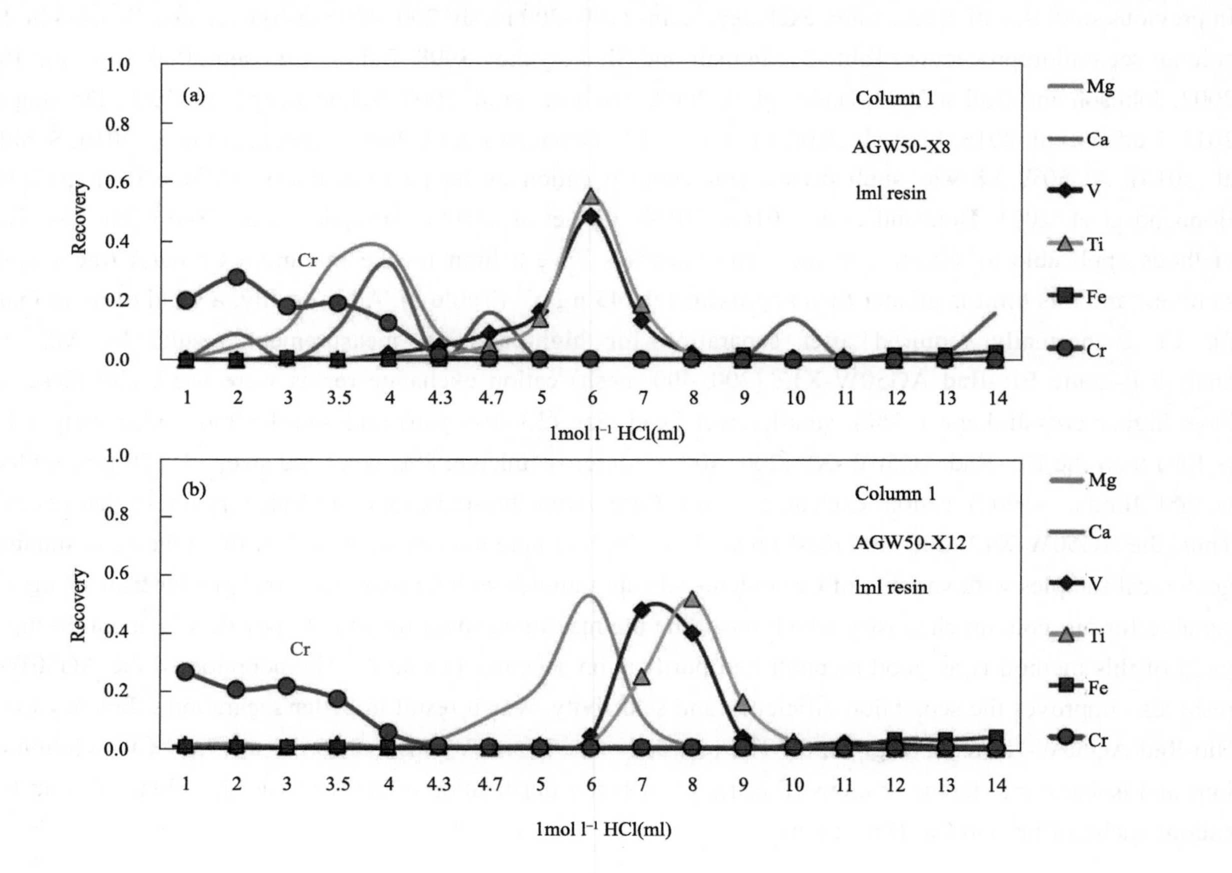

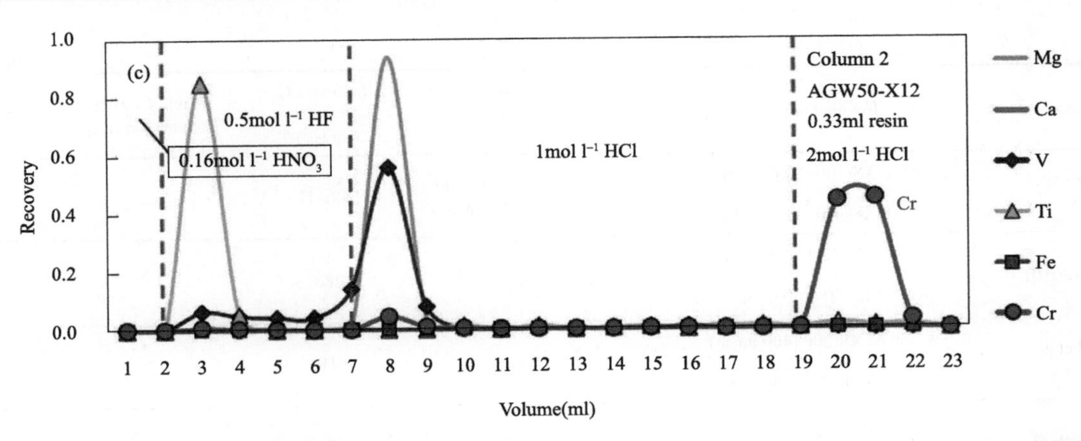

Fig. 1 The Cr elution curve of BHVO-2 for column 1 (4 mm in diameter and 10 cm in length) and column 2 (4 mm in diameter and 3 cm in length) using AG50W-X12 resin (200–400 mesh).

In detail, 1 ml and 0.33 ml of Bio-Rad AG50W-X12 (200–400 mesh) cation exchange resins were loaded in columns (4 mm in diameter) with different lengths named column 1 and 2, respectively. For column 1 (Fig. 1b, Table 1), the resin was washed with 5 ml 6 mol l^{-1} HCl and 3 ml HPW, and was pre-conditioned with 5 ml 1 mol l^{-1} HCl. Prior to launching the Cr column, the samples were diluted with HPW to a total volume of 1.2 ml 1 mol l^{-1} HCl solution, and then loaded carefully onto the resin beds. Chromium in the samples formed as Cr(III)-Cl complexes in concentrated HCl at 130 ℃ and was further eluted with 3.5 ml 1 mol l^{-1} HCl (Larsen et al. 2016). Then sub-purified Cr samples were re-digested with 20 μl concentrated HNO_3 and diluted with 2 ml HPW (~0.16 mol l^{-1} HNO_3) before loading on column 2. The Fe, V, and matrix elements were almost completely removed by column 1 (Fig. 1b, Table 1).

For column 2 (Fig. 1c, Table 1), the resin was washed with 4 ml 6 mol l^{-1} HCl and 3 ml HPW. The chromium aliquots from column 1 were loaded onto column 2, which was pre-conditioned with 5 ml 0.16 mol l^{-1} HNO_3. After sample loading, Ti and Al ions were removed by 3 ml 0.5 mol l^{-1} HF, and V and other matrix elements were eluted again by following 10 ml 1 mol l^{-1} HCl. Chromium was eluted with 4 ml 2 mol l^{-1} HCl. After column separation, the final Cr cuts were died down in concentrated HNO_3 (14 mol l^{-1}) several times and then dissolved with concentrated HNO_3 and $HClO_4$ acid to eliminate organic matter before isotopic measurement.

By implementing a two-step separation procedure, we obtained a highly purified Cr fraction (Fig. 1b, c). The Cr yields after the two-column chemistry were higher than 85%, and the whole-procedure Cr blank was less than 3–4 ng, which was considered to be negligible relative to total amount of loaded Cr (1000 ng). The Cr yields were tested using BHVO-2 by ICP-MS, and the whole-procedure Cr blank was also measured by ICP-MS.

Table 2 Reported methods for Cr separation

Reference	Ion exchanger	Sample type	The lowest Cr mass fraction (μg g^{-1})	Yield Cr after separation (μg)	Yield (%)
This study	AG50W-X12(200–400 mesh) AG50W-X12(200–400 mesh)	Silicate	17	1	85
Frei et al. (2009)	AG1-X12 AG1-X8	BIF	0.5	2–5	80–90
Schoenberg et al. (2016)	AG1-X8(100–200 mesh) AG50W-X8(200–400 mesh) AG50W-X8(200–400 mesh)	Chondrite and silicate	45.4	4–20	n.d.

Continued

Reference	Ion exchanger	Sample type	The lowest Cr mass fraction (μg g^{-1})	Yield Cr after separation (μg)	Yield (%)
Schoenberg et al. (2008)	AG1-X8(100–200 mesh) AG1-X8(100–200 mesh)	Basalt, shale,ultramafic rocks and cumulates	42.51	n.d.	70–85
Qin et al. (2010)	AG50W-X8(200–400 mesh) AG1-X8(100–200 mesh)	Terrestrial rock and chondrite	280	1–2	80
Bonnand et al. (2016)	AG50W-X8(200–400 mesh) AG50W-X8(200–400 mesh)	Mare basalts	1410	1	80
Li et al. (2017)	Ln Spec resin AG1-X8(200–400 mesh)	Carbonate and silicate	5	> 0.05	94.7–97.5
Scheiderich et al. (2015)	AG1-X8(100–200 mesh) AG50W-X8(100–200 mesh)	Seawater	n.d.	0.08–0.3	n.d.
Trinquier et al. (2008)	AG50W-X8(200–400 mesh) AG50W-X8(200–400 mesh)	Meteorites	n.d.	n.d.	80
Gilleaudeau et al. (2016)	AG1-X8(100–200 mesh) AG1-X8(100–200 mesh) AG50W-X8(200–400 mesh)	Carbonates	0.19	0.5–6	n.d.
Bonnand et al. (2011)	AG50W-X8(200–400 mesh)	Carbonates	1.05	0.25	70–80
Gueguen et al.(2016)	AG1-X8(100–200 mesh) AG50W-X8(200–400 mesh)	Marine sediments	17	n.d.	75–85
Yamakawa et al. (2009)	AG1-X8(200–400 mesh) AG50W-X8(200–400 mesh) AG50W-X8 (200–400 mesh)	Silicate	n.d.	0.5–3	n.d.
Dossing et al. (2010)	AG1-X8(100–200 mesh) AG1-X8(100–200 mesh)	Cr (VI) stock solution	0.08	43230	> 95
Ball et al. (2010)	AG1-X8 AG50W-X8	Water	n.d.	0.665	90
Farkaš et al. (2013)	AG1-X8(100–200 mesh)	Chromites, ultra-mafic rocks	976	2	90–95
Johnson and Bullen (2004)	AG 1-X8 AG 1-X8		n.d.	0.25	n.d.
Halicz et al. (2008)	AG1-X8 AG50W-X12	Silicate	320	n.d.	> 95
Schiller et al. (2014)	AG1-X4(200–400 mesh) AG50W-X8(200–400 mesh) TODGA	Dunite, meteorites	n.d.	3–6	n.d.
Zhu et al. (2018)	AG50W-X8 AG1-X8 AG1-X8 AG1-X8	Carbonate, silicate, shale, soil, plants and animals.	0.41	0.3–0.6	80–98

n.d.: no data was reported.

2.4 Cr double-spike technique

The ^{50}Cr-^{54}Cr double spike is usually used to correct the mass dependence fractionation generated during sample processing as well as by instrumental mass bias effects for the Cr isotope ratio of samples (Ellis and Bullen 2002, Zink et al. 2010, Schoenberg et al. 2008, Frei et al. 2009, Qin et al. 2010a, Bonnand et al. 2011, Farkaš et al. 2013, Planavsky et al. 2014, Wang et al. 2016, Cole et al. 2016, Li et al. 2016, Li et al. 2017, Xia et al. 2017, Zhu et al. 2018). The mass-dependent isotopic fractionation in the laboratory, from the column chemistry and from isotopic fractionation by TIMS, was corrected by a ^{50}Cr-^{54}Cr double-spike technique. Two single spikes of ^{50}Cr and ^{54}Cr were purchased from Oak Ridge National Laboratory. The double spike was prepared by mixing the single spikes in an appropriate proportion (^{50}Cr :$^{54}Cr \approx 1:1$) and was calibrated by TIMS at the China University of Geosciences. To achieve high-precision isotopic determinations, the proportions of two single spikes and between double spike and sample must be optimised. Our double-spike correction method assumed an exponential mass fractionation law and calculated by iterative Newton-Raphson procedure (Albarède and Beard 2004) to solve the double-spike equations described in Rudge et al. (2009).

The Monte Carlo method was used to predict optimized the spike/ (spike + sample) ratios (Q) and ratios of double spike (P). Lehn et al. (2013) further proposed an improved method to simulate TIMS analysis by considering Faraday collector damage and achieved excellent repeatability precision of ~0.024‰ (2SE) for Calcium isotope ratio measurement results. Here, according to the mathematical equations presented by Lehn et al. (2013), the Monte Carlo simulation, mainly, takes counting statistics (s_{cs}) and Johnson noise (s_{jn}) into consideration to optimize spike/ (spike+sample) ratios (Q) and the ratios of ^{50}Cr spike and ^{54}Cr spike (P) of Cr isotopes (Fig. 2a). The simulation was implemented by MATLAB and the results are presented in Figure 2a. The ion-beam voltage of each isotope, integration time, cycles of each analysis, temperature of the collectors (298 K) and amplifier resistance are all involved in calculation of the theoretical uncertainty curve, which was consistent with that in the experiment. We mixed two single spikes as a $^{54}Cr/Cr_{total}$ ratio of 0.484. The $\delta^{53/52}Cr_{NIST\ SRM\ 979}$ of NIST SRM 979 mixed with different double-spike proportions was determined by TIMS, and the experimental error matched the theoretical simulation uncertainty curve of $\delta^{53/52}Cr_{NIST\ SRM\ 979}$ (Fig. 2b, Table S1). Given that the ratio of $^{54}Cr/Cr_{total}$ in double-spike is 0.484, the optimal spike proportion should be 10% to 40% of the total Cr ($Cr_{spike}+Cr_{sample}$), hence, we selected a Cr_{spike}/Cr_{total} ratio of 0.25 in our method.

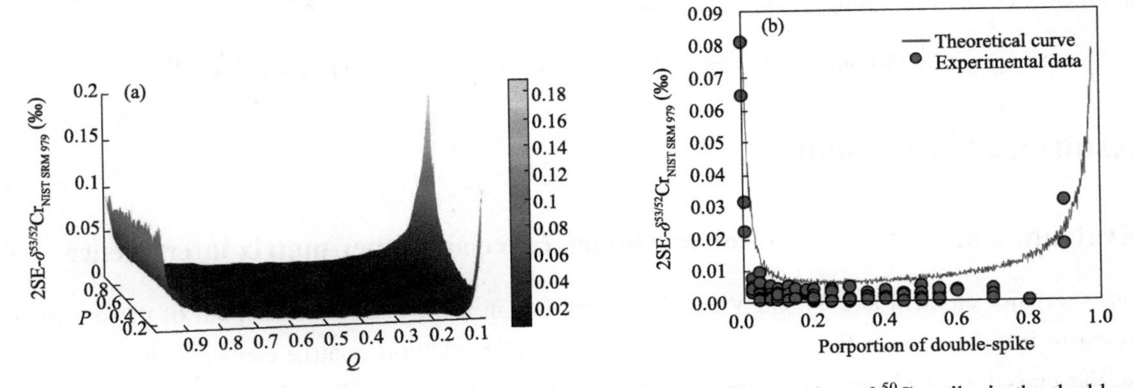

Fig. 2 (a) 3D contour plot of doubling the standard error (2SE), as a function of proportion of ^{50}Cr spike in the double-spike and spike/sample ratios. The horizontal axis gives the proportion Q of double spike in the double spike-sample mixture, and the vertical axis gives the proportion P of ^{54}Cr in the double spike. (b) Comparison between theoretically simulated and measured error under the ^{54}Cr spike proportion is 0.484. The theoretical doubling the standard error of $\delta^{53}Cr$ was calculated by the double-spike equation described in Lehn et al. (2013). Noted that there is a broad region around the optimum ratios.

2.5 Mass spectrometric measurements

The measurements of Cr isotope ratios were performed on a ThermoFisher Scientific *TRITON Plus* TIMS in the Isotope Geochemistry Lab of the China University of Geosciences, Beijing. This instrument was equipped with nine Faraday cups linked to 10^{11} Ω amplifiers. Chromium was loaded in 2 μl of 3% m/v HNO_3 onto outgassed Re single filaments under a binocular microscope using a 0.5–10 μl-range digital pipette. Narrow parafilm dams were placed on the filaments to facilitate the core formation of sample aliquots and the remaining organics were removed by a brief heating procedure. Then, a mixture solution of 1.3 μl high purity (99.99%) silica gel (<100 nm) and 1.3 μl saturated H_3BO_3 was added to the sample drops. The mixture was first dried by slow heating under low current conditions (~0.5 A) through the filament to form a glass, and then the current was increased slowly until the filament was dull red (~2.2 A) for 2 seconds. Every filament load standard or sample contained 0.5–1 μg of Cr. One sample was loaded on 1–2 filaments. All Cr isotopic data were acquired in static multi-collection mode by the collector array summarised in Table 3. Ion beam measurements at m/z $^{49}(Ti^+)$, $^{51}(V^+)$, and $^{56}(Fe^+)$ allowed us to correct isobaric interferences from Ti^+ and V^+ on Cr^+ and Fe^+ on Cr^+. During a measurement routine, the filament was heated to 1500 mA in 5 min and then heated to 1800 mA in 3 min. Generally, data acquisition started when the current beam of $^{52}(Cr^+)$ was stable and higher than 4×10^{-11} A and the temperature of the filaments was ca. 1330 ℃. For each analysis, fourteen blocks of twelve cycles at integration time of 8 s were obtained using amplifier rotation. Peak Centre and auto-focus were applied at the beginning and after every seven blocks of data acquisition. Each block included a 60 s baseline measurement. Amplifier gains were calibrated at the start of each day to eliminate all gain calibration errors. The total acquisition time was approximately 60 min for one filament, including a 15 min heating routine. Each filament was monitored three or four times.

Table 3　Cup configuration for Cr isotope measurement

Element	L4	L2	L1	C	H1	H2	H3
Cr	^{49}Ti	^{50}Cr	^{51}V	^{52}Cr	^{53}Cr	^{54}Cr	^{56}Fe

The $^{53}Cr/^{52}Cr$ ratios of the samples are expressed as the per mil deviation from the Cr isotope reference material NIST SRM 979 measured in the same barrel:

$$\delta^{53/52}Cr_{NIST\ SRM\ 979}\ (‰) = R(^{53}Cr/^{52}Cr)_{sample}/R(^{53}Cr/^{52}Cr)_{NIST\ SRM\ 979} - 1$$

3 Results and discussion

3.1 Evaluation of Ti, V, and Fe isobaric interference and other matrix interference

Iron is a major element and Ti and V are trace elements in geological materials. All of these elements have direct isobaric interference on the ion beam of Cr isotopes. In addition, matrix elements (K, Na, Ca and Mg) suppress the emission of Cr on the filament, and this suppression may affect the uncertainty of measurement results of Cr isotopes. Hence, it is important to evaluate the potential effects of isobaric interference from Ti, V, Fe and other matrix elements, even though our Cr separation method provides good separation of Cr from terrestrial samples. To evaluate the isobaric interferences from Ti, V, and Fe, 1 μg of unspiked GSB-Cr calibrator was doped with different amounts of GSB- Ti, V, and Fe calibration standard solutions (Fig. 3, Table S2). The isobaric elements/Cr concentration ratios ranged from 0.001 to 10. We also doped a mixture of K, Na, Ca and Mg to evaluate the inhibition effects from these matrix elements on the Cr^+ signal intensity during determination (the ratio of mixture to Cr ranged from 0.01 to 0.1).

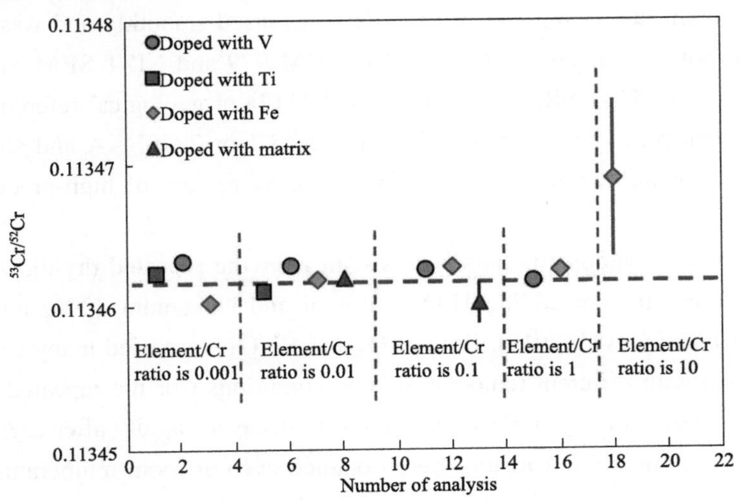

Fig. 3 Measured Cr isotope ratios of unspiked GSB Cr (containing 1 μg of Cr) doped with different masses of GSB Ti, V and Fe. The red dashed line represents the measured isotope ratio of pure GSB Cr.

As a result of high ionization potential, Ti and V are very difficult to ionize with TIMS, a finding also proposed by Li et al. (2016) (Fig. 2). Nevertheless, we could not obtain a stable Cr^+ single when the ratio of Ti to Cr is greater than 0.01. The presence of Ti suppresses the signal intensity of Cr^+ and results in poor precision (Li et al. 2016). We also noticed that the signal of Cr^+ attenuates rapidly during determination. No signals of $^{51}(V^+)$ and $^{49}(Ti^+)$ were detected during instrument measurements, as the $^{49}Ti/^{52}Cr$ and $^{51}V/^{52}Cr$ ratios of samples doped with GSB-Ti and V were indistinguishable from those of the pure GSB-Cr calibration material. This finding suggests that Ti and V only inhibit the emission of Cr but show no isobaric interferences. The mixture of K, Na, Ca and Mg provides a similar effect as that of Ti and V for Cr isotope measurement (Fig. 3).

However, the presence of both Cr^+ and Fe^+ beams can be observed in pure GSB-Cr doped with Fe during measurement, and a weak positive relationship was observed between the $^{53}Cr/^{52}Cr$ ratios and the amount of added Fe (Fig. 3). The samples doped with less than 1 μg Fe could show bias free $^{53}Cr/^{52}Cr$ ratios depending our analysis (Fig.3). Iron can be ionised with Cr due to the lower ionisation potential than Ti and V at slightly higher temperature. Excess Fe caused strong $^{56}(Fe^+)$ signal interference and inhibited the emission of Cr^+ for a sample doped with 10 μg Fe (Fig. 3). It is noteworthy that the current beam of $^{56}(Fe^+)$ increases with temperature, especially when the temperature is higher than 1350 ℃. Therefore, in our experience, data acquisition should be treated carefully when the temperature of the filaments reaches 1350 ℃.

In contrast to analysis by MC-ICP-MS, Cr isotopic measurements by TIMS do not have strong isobaric and matrix interferences, as the Ti, V and most of the Fe have low ionisation efficiency or even no ionization due to the high ionisation potential and different ionisation temperature relative to those of the Cr isotope (Li et al., 2016). The addition of Ti, V and Fe inhibits the ionisation of Cr, and only Fe can be ionised with Cr (Fig. 3).

3.2 Elimination of organic matter in the sample after chemical separation

During the measurement of Fe, Cr, Pb, and Cu isotopes using TIMS, organic matter is considered a significant interference (Kuritani and Nakamura 2002, Johnson and Bullen 2004, Yamakawa et al. 2009, Chrastný et al. 2013). The organic matter in samples is most likely derived from the exchange resin or colloidal silica.

Previous work used ultraviolet radiation, H_2O_2 or $HClO_4$ treatment to eliminate organic residues originating from the column resin (Kuritani and Nakamura 2002, Johnson and Bullen 2004, Yamakawa et al. 2009, Chrastný et al. 2013). During our experiment, the beam current of $^{52}(Cr^+)$ increased smoothly and was stable over 10×10^{-11} A when 0.5 μg of pure Cr isotope standard materials NIST SRM 979 and NIST SRM 3112a were measured. By contrast, the beam of $^{52}(Cr^+)$ in NIST SRM 979, NIST SRM 3112a or geological reference materials (~0.5–1 μg Cr) that underwent chemical procedures, usually did not exceed 1 or 2×10^{-11} A and attenuated to zero rapidly. Thus, the elimination of organic matter is required for the measurement of high-precision Cr isotope ratio in complex rock samples.

The organic component in samples is highly persistent after the repeated drying of samples with drops of nitric acid at elevated temperature or adding H_2O_2 (Kuritani and Nakamura 2002, Johnson and Bullen 2004, Yamakawa et al. 2009, Chrastný et al. 2013). Both H_2O_2 and HNO_3 were tried many times to eliminate organic matter after Cr purification with different temperature/time conditions (for the repeated digestion of a BHVO-2 sample with 1 μg Cr). As shown in Fig. 4, the signal intensity decayed rapidly after drying 4–8 times repeatedly samples with drops of H_2O_2, and similar results were obtained even at room temperature / heating the samples with H_2O_2 at 70 ℃ for 1–14 days. The signal intensity also decayed after repeatedly drying samples 5–8 times with drops of HNO_3.

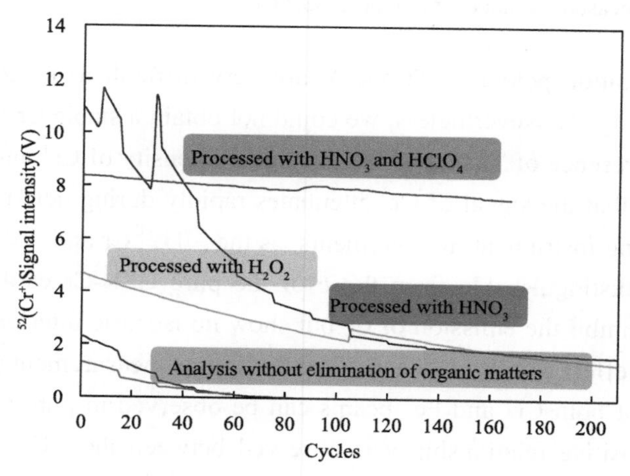

Fig. 4　The beam intensity curve of $^{52}(Cr^+)$ of samples processed with different methods for eliminating organic matter. Compared with H_2O_2 and HNO_3, the mixture solution of $HClO_4$ and HNO_3 achieved the best effect. A stable and high intensity current beam of Cr^+ is obtained.

$HClO_4$ has not been commonly used recently, as perchloric results in Cr loss by evaporation of Cr (Makishima et al. 2002). However, in this study, an optimised process performed with $HClO_4$, which is the most effective medium to eliminate organic interference after trying different processes (HNO_3/H_2O_2 with different temperature/time conditions). All samples were heated on a 120 ℃ hotplate with 500μl concentrated HNO_3 and 30 μl $HClO_4$ for two weeks to obtain a high and stable beam intensity of $^{52}(Cr^+)$ (Fig. 4, Table S3). The reaction rate does not clearly increase with increasing temperature but is coupled with the reaction time. Attempts to shorten the sample processing time at 120 ℃ or higher did not yield the satisfactory effect of the eliminating organic interference. In addition, the degree of elimination of the organic matters with pure $HClO_4$ is limited without cooperation of HNO_3. The effect of procedural blank from $HClO_4$ is negligible, as $HClO_4$ is purified. As shown in Fig. 5b, the $\delta^{53/52}Cr_{NIST\ SRM\ 979}$ values at every point are obtained during repeated measurements of one sample on one loaded filament with repeated sample digestion. High-precision Cr isotope measurement results with a typical high and stable beam of $^{52}(Cr^+)$ ($>4\times10^{-11}$ A) is achieved for every repeated geological sample after elimination of the organic component by HNO_3 and $HClO_4$ (Fig. 5b). The long-term intermediate precision for both USGS BHVO-2 and JP-1 was less than 0.035‰ for the two-year measurement (Fig. 5b). As suggested by Makishima et al. (2002), the addition of perchloric acid results in Cr loss by evaporation of Cr in the form of CrO_2Cl_2 in the process of drying (Makishima et al. 2002, Li et al. 2017). The boiling point temperature of CrO_2Cl_2 is 116.85 ℃ (390 K). In this study, because of the low boiling temperature of concentrated HNO_3 (83 ℃), $HClO_4$ can evaporate the solution to dryness with HNO_3 at only 105 ℃ to avoid the loss of Cr.

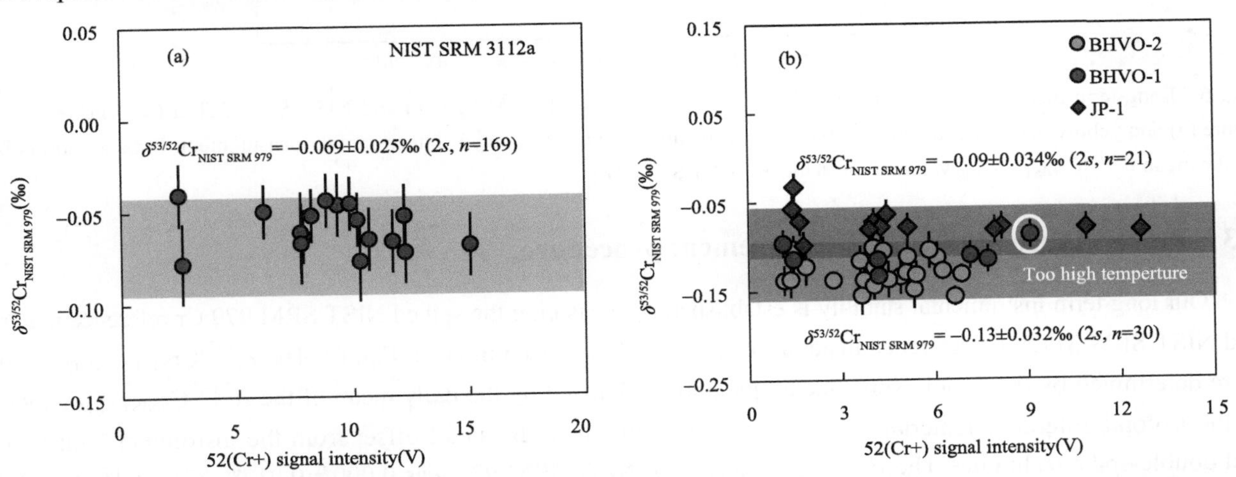

Fig. 5 The $\delta^{53/52}Cr_{NIST\ SRM\ 979}$ value for NIST SRM 3112a (a) and three geological reference materials (BHVO-1, BHVO-2, and JP-1) (b) obtained at different beam intensities of $^{52}Cr^{+}$. The $\delta^{53/52}Cr_{NIST\ SRM\ 979}$ values of JP-1 and BHVO-1 are acquired by duplicate measurements on one loaded filament. The beam intensity could easily reach 10 V or higher with a 10^{11} amplifier after eliminating organic matter. Shaded areas of different colours represent the 2 standard deviation for the mean $\delta^{53/52}Cr_{NIST\ SRM\ 979}$ values of NIST SRM 3112a and the other reference materials in our lab (BHVO-1, BHVO-2, and JP-1).

Geological reference materials such as BHVO-1, BHVO-2, and JP-1 all produced a stable (Fig. 5b) and high beam intensity of $^{52}Cr^{+}$ (i.e., 4–12 V) after eliminating organic matter. One measurement result of the BHVO-1 was slightly higher than the mean value of BHVO-1 in our lab, as the repeated measurements on the filament resulted in an excessive filament temperature (1380 ℃). The $^{56}Fe/^{52}Cr$ ion beam ratio of this BHVO-1 sample (2.74×10^{-4}) is two orders of magnitude higher than that of the pure Cr isotope reference material, and these this data are consistent with mean value of $\delta^{53/52}Cr_{NIST\ SRM\ 979}$ of BHVO-1 in our laboratory within the given precision. Neverless, a small mass of Fe is ionised at this temperature, perhaps resulting of a shift of the $\delta^{53/52}Cr_{NIST\ SRM\ 979}$ of BHVO-1. This observation is consistent with our previous discussion. To achieve high-precision Cr isotope determination, a steady filament temperature should be maintained at approximately 1350℃ during measurement.

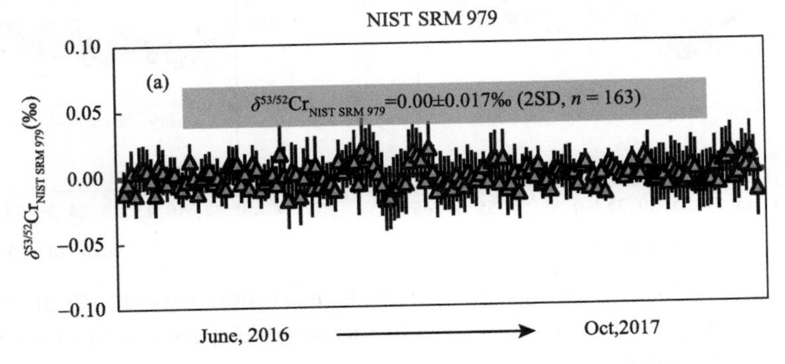

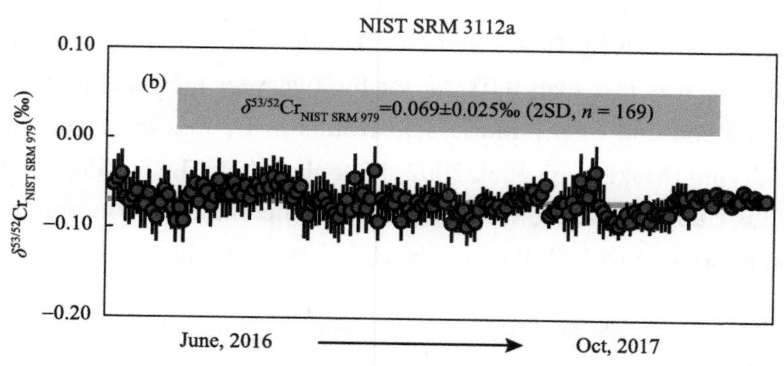

Fig. 6　Long-term measurement results of $\delta^{53/52}Cr_{NIST\ SRM\ 979}$ for NIST SRM 979 (a) and NIST SRM 3112a (b). The red line in figure (a) and yellow line in figure (b) represent the mean values obtained during a sixteen-month measurement. Data are normalized by the mean $\delta^{53/52}Cr_{NIST\ SRM\ 979}$ value of the standard NIST SRM 979.

3.3　Performance of the new measurement procedure

Our long-term instrumental stability is established by measuring the spiked NIST SRM 979 Cr reference material and NIST SRM 3112a Cr reference material over a period of sixteen months (Fig. 6). The $\delta^{53/52}Cr_{NIST\ SRM\ 979}$ values were determined by the double-spike technique and normalised by the daily mean of the $\delta^{53/52}Cr_{NIST\ SRM\ 979}$ value of the isotopic reference material NIST SRM 979 to eliminate the small offset from the instrument's mass bias and double-spike technique. The $\delta^{53/52}Cr_{NIST\ SRM\ 979}$ of NIST SRM 979 was 0.000±0.017‰ (2s, n=163), and the $\delta^{53/52}Cr_{NIST\ SRM\ 979}$ of NIST SRM 3112a is −0.069±0.025‰ (2s, n=169) relative to that of NIST SRM 979. Our NIST SRM 3112a $\delta^{53/52}Cr_{NIST\ SRM\ 979}$ value are in excellent agreement with those in previous studies (i.e., −0.067±0.024 ‰, Schoenberg et al. 2008, −0.09±0.04 ‰, Xia et al. 2017, −0.07±0.04 ‰, Shen et al. 2018). To check the stability of the measurement results, we also analysed the mass-independent Cr isotope ratios (after internal normalisation using the $^{50}Cr/^{52}Cr$ ratio) of these two isotope reference materials without double-spike addition over a period of six months. The $^{53}Cr/^{52}Cr$ ratios and $^{54}Cr/^{52}Cr$ ratios of NIST SRM 979 and NIST SRM 3112a were stable and uniform, with $^{53}Cr/^{52}Cr$ ratios of NIST SRM 979 and NIST SRM 3112a of 0.1134616±0.00002 (2s, n=22) and 0.1134617±0.00002 (2s, n=63) and $^{54}Cr/^{52}Cr$ ratios of NIST SRM 979 and NIST SRM 3112a of 0.028214±0.00001 (2s, n=22) and 0.028214±0.00001 (2s, n=63), respectively (Fig. 7).

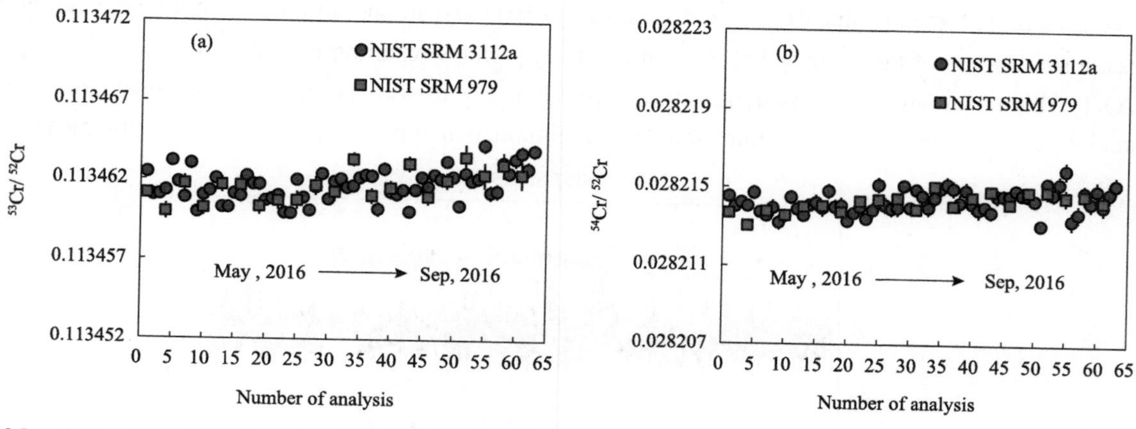

Fig. 7　Mass-independent isotopic ratios ($^{53}Cr/^{52}Cr$ (a), $^{54}Cr/^{52}Cr$ (b)) for NIST SRM 979 and NIST SRM 3112a measured from May 2016 to September 2016. The instrumental mass fractionation of Cr was corrected by assuming a constant ratio of 0.051859 (Shields et al. 1966) for $^{50}Cr/^{52}Cr$ and an exponential mass fractionation law.

3.4 $^{53/52}Cr_{NIST\ SRM\ 979}$ of geological reference materials

We report stable Cr isotope ratio of seven commercially available geological reference materials relative to NIST SRM 979 (Table 4 and Fig. 8). All the $\delta^{53/52}Cr$ of isotope reference materials are in good agreement with the published data, and have the same or smaller measurement uncertainty (Table 4). These analyses include independent digestion and duplicate measurement results of geological reference materials.

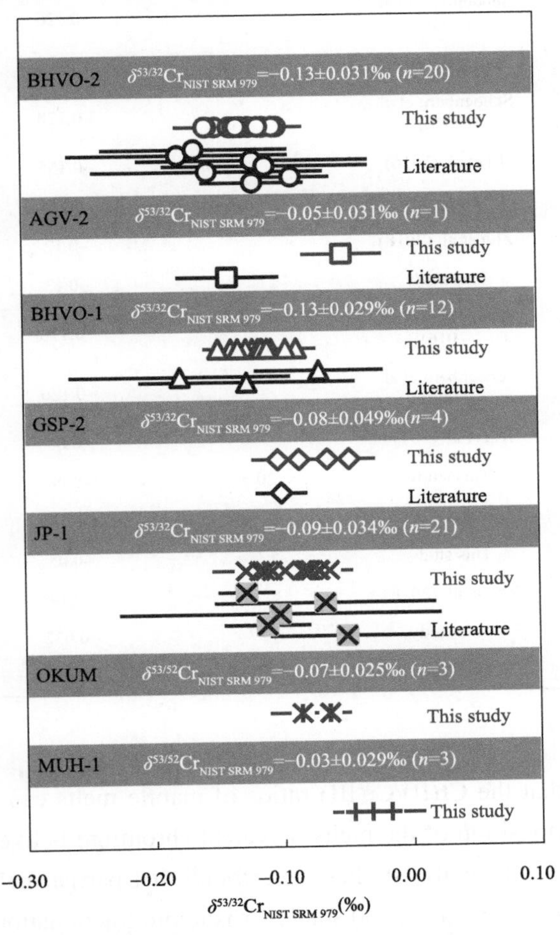

Fig. 8 $\delta^{53/52}Cr_{NIST\ SRM\ 979}$ values relative to NIST SRM 979 for geological reference materials in this study. Comparison of geological reference materials data analysed in this study (red symbols) with literature values (blue symbols) (Schoenbetg et al. 2008, Schoenbetg et al. 2016, Bonnand et al. 2016a, 2016b, Gueguen et al. 2016, Wang et al. 2016, Li et al. 2016, Li et al. 2017, Xia et al. 2017, Wu et al. 2017, Zhu et al. 2018).

Table 4 $\delta^{53/52}Cr_{NIST\ SRM\ 979}$ value of geological reference materials in this study and in the literature

Sample name	Sample type	Reference	Cr (μg g^{-1})	$\delta^{53/52}Cr_{NIST\ SRM\ 979}$(‰)	2s	N
JP-1	Peridotite	This study	2807	−0.09	0.034	21
		Schoenberg et al. (2016)		−0.067	n.d.	n.d.
		Bonnand et al. (2016a)		−0.128	0.022	14
		Bonnand et al. (2016b)		−0.102	0.012	5
		Li et al. (2016)		−0.112	n.d.	n.d.
		Zhu et al. (2018)		−0.05	0.06	3

Continued

Sample name	Sample type	Reference	Cr (μg g^{-1})	$\delta^{53/52}Cr_{NIST\ SRM\ 979}$(‰)	2s	N
BHVO-2	Basalt	This study	280	−0.13	0.032	18
		Wang et al.(2016)		0.11	0.08	7
		Gueguen et al. (2016)		−0.12	0.09	25
		Schoenberg et al. (2008)		−0.126	0.084	3
		Wu et al. (2017)		−0.09	0.03	8
		Schoenberg et al. (2016)		−0.178	n.d.	n.d.
		Li et al. (2016)		−0.155	n.d.	n.d.
		Li et al. (2017)		−0.165	n.d.	n.d.
		Zhu et al. (2018)		−0.12	0.04	15
BHVO-1	Basalt	This study	271	−0.13	0.029	12
		Schoenberg et al. (2016)		−0.178	n.d.	n.d.
		Schoenberg et al. (2008)		−0.126	0.084	3
		Xia et al. (2017)		−0.07	0.05	
GSP-2	Granodiorite	This study	20	−0.08	0.049	4
		Zhu et al. (2018)		−0.10	0.02	3
AGV-2	Andesite	This study	17	−0.05	0.031	1
		Zhu et al. (2018)		−0.14	0.04	4
MUH-1	Ultramafic rock	This study	2727	−0.032	0.029	3
OKUM	Komatiite	This study	2385	−0.072	0.025	3

n.d.: no data was reported.

Previous studies revealed that the Cr(II)/Cr(III) ratios of mantle melts can significantly vary depending on the oxygen fugacity and the composition of the melts. Divalent chromium is likely to be the dominant species in basaltic melts (Berry et al. 2006). Recently, studies have found that partial melting of the mantle may cause a small but detectable (~0.4‰ of $\delta^{53/52}Cr_{NIST\ SRM\ 979}$) stable Cr isotopic fractionation (Schoenberg et al. 2016, Xia et al. 2017). The oxygen fugacity variation may dominate the mass-dependent chromium stable isotope fractionation in high-temperature processes (Bonnand and Halliday 2018, Shen et al. 2018, Sossi et al. 2018). Thus, stable Cr isotopic fractionation may be observed between mantle and basalts, given the different Cr(II)/Cr(III) ratios of mantle melts and basaltic melts (Schoenberg et al. 2008). In this study, two basalt reference materials from the USGS, BHVO-1 (−0.120±0.029‰, 2s) and BHVO-2 (−0.129±0.032‰, 2s), gave slightly lower $\delta^{53/52}Cr_{NIST\ SRM\ 979}$ values than that of peridotite reference material JP-1 from JSG (−0.088±0.034‰; 2s) with the mean results over a period of two years. The data obtained in this study for these basalt and peridotite reference materials are consistent with the $\delta^{53/52}Cr_{NIST\ SRM\ 979}$ values reported by previous studies within the precision (Schoenberg et al. 2008, Schoenberg et al. 2016, Bonnand et al. 2016a, 2016b, Wang et al. 2016, Li et al. 2016, Li et al. 2017, Gueguen et al. 2016, Wu et al. 2017, Xia et al. 2017, Zhu et al. 2018) (Fig. 5b).

We also report the $\delta^{53/52}Cr_{NIST\ SRM\ 979}$ values for komatiite reference material (OKUM), ultramafic rocks reference material (MUH-1), and two intermediate and felsic igneous reference materials (AGV-2 and GSP-2). The $\delta^{53/52}Cr_{NIST\ SRM\ 979}$ values of OKUM, MUH-1, AGV-2, and GSP-2 are −0.072±0.025‰ (2s), −0.032±0.029‰ (2s), −0.051±0.031‰ (2s) and −0.075±0.049‰ (2s), respectively. Each sample is loaded onto one filament and measured twice with a beam of ^{52}Cr over 2×10^{-11} A. Thus, the Cr isotope ratio of geological samples with a lower chromium mass fraction (20 μg g^{-1}) can also be determined precisely without an inhibition effect from organic matter. Overall, seven igneous rock reference materials analysed in this study yielded $\delta^{53/52}Cr_{NIST\ SRM\ 979}$ values ranging from −0.129±0.032‰ (2*s*) to −0.032±0.029‰ (2*s*), which is distinguishable within our measurement uncertainty, demonstrating detectable stable Cr isotopic fractionation during high-temperature processes.

4 Summary

High-precision Cr isotope measurement results for natural samples were obtained with double-spike TIMS. A two-step separation of Cr from geological materials without using additional oxidisation reagent was presented in this study. A stable and strong Cr^{+} beam was obtained with the elimination of organic matter in the sample after the described chemistry procedure involving perchloric acid and concentrated HNO_3. The measurement uncertainty in this study was assessed by multiple analyses of the NIST SRM 979 and NIST SRM 3112a reference materials. The $\delta^{53/52}Cr_{NIST\ SRM\ 979}$ values obtained are 0.000±0.017‰ (2*s*, *n*=163) and −0.069±0.025‰ (2*s*, *n*=169), respectively, relative to that of NIST SRM 979. We also reported Cr isotopic data for seven geological reference materials relative to NIST SRM 979. Basalt reference materials ($\delta^{53/52}Cr_{NIST\ SRM\ 979}$ of BHVO-1=−0.120±0.029‰, 2SD; BHVO-2=—0.129±0.032‰, 2*s*) have lower lighter $\delta^{53/52}Cr_{NIST\ SRM\ 979}$ values than peridotite reference material (JP-1=—0.088±0.034‰, 2*s*). These reference materials ranging in composition from ultramafic to felsic yielded $\delta^{53/52}Cr_{NIST\ SRM\ 979}$ values ranging from −0.129±0.032‰ (2*s*) to −0.032±0.029‰ (2*s*), which are distinguishable within our measurement uncertainty.

Acknowledgements

We greatly appreciate two anonymous reviewers for their constructive comments and the efficient editorial handling of chief editor Prof. Thomas Meisel. We appreciate J.-M. Zhu for providing the NIST SRM 979 standard. This work is supported by MOST Special Fund from the State Key Laboratory of Geological Processes and Mineral Resources, China University of Geosciences (MSFGPMR201811) to XLJ, the National Natural Foundation of China (No. 41403010 and No. 41622303) and the Fok Ying Tung Education Foundation (2-2-2016-01) to LSA.

Reference

Albarède F. and Beard B. (2004) Analytical methods for non-traditional isotopes. Reviews in Mineralogy and Geochemistry, 55, 113-152.

Allègre C.J., Poirier J.P., Humler E. and Hofmann A.W.(1995)The chemical composition of the Earth. Earth and Planetary Science Letters, 134, 515-526.

Ball J.W. and Bassett R.L. (2000) Ion exchange separation of chromium from natural water matrix for stable isotope mass spectrometric analysis. Chemical Geology, 168, 123-134.

Berry A.J., O'Neill H.S.C., Scott D.R., Foran G.J. and Shelley J.M.G. (2006) The effect of composition on Cr^{2+}/Cr^{3+} in silicate melts. American Mineralogist, 91, 1901-1908.

Birck J.-L. and Allègre C.J. (1988) Manganese-chromium isotope systematics and the development of the early Solar System. Nature, 331, 579-584.

Blowes D. (2002) Environmental chemistry: Tracking hexavalent Cr in groundwater. Science, 295, 2024-2025.

Bonnand P. and Halliday A.N. (2018) Oxidized conditions in iron meteorite parent bodies. Nature Geoscience, 11, 401-404.

Bonnand P., Parkinson I.J., James R.H., Karjalainen A.M. and Fehr M.A. (2011) Accurate and precise determination of stable Cr isotope compositions in carbonates by double spike MC-ICP-MS. Journal of Analytical Atomic Spectrometry, 26, 528-535.

Bonnand P., James R.H., Parkinson I.J., Connelly D.P. and Fairchild I.J. (2013) The chromium isotopic composition of seawater and marine carbonates. Earth and Planetary Science Letters, 382, 10-20.

Bonnand P., Parkinson I.J. and Anand M. (2016a) Mass dependent fractionation of stable chromium isotopes in mare basalts: Implications for the formation and the differentiation of the Moon. Geochimica et Cosmochimica Acta, 175, 208-221.

Bonnand P., Williams H.M., Parkinson I.J., Wood B.J. and Halliday A.N. (2016b) Stable chromium isotopic composition of meteorites and metal-silicate experiments: Implications for fractionation during core formation. Earth and Planetary Science Letters, 435, 14-21.

Chrastný V., Rohovec J., Čadková E., Pašava J., Farkaš J. and Novak M. (2013) A new method for low-temperature decomposition of chromites and dichromium trioxide using bromic acid evaluated by chromium isotope measurements. Geostandards and Geoanalytical Research, 38, 103-110.

Cole D.B., Reinhard C.T., Wang X., Gueguen B., Halverson G.P., Gibson T., Planavsky N.J. (2016) A shale-hosted Cr isotope record of low atmospheric oxygen during the Proterozoic. Geology, 44, 555-558.

Crowe S.A., Døssing L.N., Beukes N.J., Bau M., Kruger S.J., Frei R. and Canfield D.E. (2013) Atmospheric oxygenation three billion years ago. Nature, 501, 535.

Døssing L.N., Dideriksen K., Stipp S.L.S. and Frei R. (2011) Reduction of hexavalent chromium by ferrous iron: A process of chromium isotope fractionation and its relevance to natural environments. Chemical Geology, 285, 157-166.

Ellis A.S. and Bullen T.D. (2002) Chromium isotopes and the fate of hexavalent chromium in the environment. Science, 295, 2060-2062.

FarkašJ., ChrastnýV., Novák M., Čadkova E., Pašava J., Chakrabarti R., Jacobsen S.B., Ackerman L. and Bullen T.D. (2013) Chromium isotope variations ($\delta^{53}/^{52}Cr$) in mantle-derived sources and their weathering products: Implications for environmental studies and the evolution of $\delta^{53}/^{52}Cr$ in the Earth's mantle over geologic time. Geochimica et Cosmochimica Acta, 123, 74-92.

Frei R., Gaucher C., Poulton S.W. and Canfield D.E. (2009) Fluctuations in Precambrian atmospheric oxygenation recorded by chromium isotopes. Nature, 461, 250-253.

Frei R., Gaucher C., Døssing L.N. and Sial A.N. (2011) Chromium isotopes in carbonates-A tracer for climate change and for reconstructing the redox state of ancient seawater. Earth and Planetary Science Letters, 312, 114-125.

Gilleaudeau G.J., Frei R., Kaufman A.J., Kah L.C., Azmy K., Bartley J.K. and Knoll A.H. (2016) Oxygenation of the mid-Proterozoic atmosphere: Clues from chromium isotopes in carbonates. Geochemical Perspectives Letters, 2, 178-187.

Gueguen B., Reinhard C.T., Algeo T.J., Peterson L.C., Nielsen S.G., Wang X., Rowe H. and Planavsky N.J. (2016) The chromium isotope composition of reducing and oxic marine sediments. Geochimica et Cosmochimica Acta, 184, 1-19.

Halicz L., Yang L., Teplyakov N., Burg A., Sturgeon R. and Kolodny Y. (2008) High precision determination of chromium isotope ratios in geological samples by MC-ICP-MS. Journal of Analytical Atomic Spectrometry, 23, 1622-1627.

Johnson T.M. and Bullen T.D. (2004). Selenium, iron and chromium stable isotope ratio measurements by the double isotope spike TIMS method. In: DeGroot P. (ed.), Handbook of stable isotope analytical techniques. Elsevier BV (Amsterdam, The Netherlands), 632-651.

Kuritani T. and Nakamura E. (2002) Precise isotope analysis of nanogram-level Pb for natural rock samples without use of double spikes. Chemical Geology, 186, 31-43.

Larsen K.K., Wielandt D., Schiller M. and Bizzarro M. (2016) Chromatographic speciation of Cr(III)-species, inter-species equilibrium isotope fractionation and improved chemical purification strategies for high-precision isotope analysis. Journal of Chromatography A, 1443, 162-174.

Lehn G.O., Jacobson A.D. and Holmden C. (2013) Precise analysis of Ca isotope ratios ($\delta^{44}/^{40}Ca$) using an optimized ^{43}Ca–^{42}Ca double-spike MC-TIMS method. International Journal of Mass Spectrometry, 351, 69-75.

Li C.F., Feng L.J., Wang X.C., Chu Z.Y., Guo J.H. and Wilde S.A. (2016) Precise measurement of Cr isotope ratios using a highly sensitive Nb_2O_5 emitter by thermal ionization mass spectrometry and an improved procedure for separating Cr from geological materials. Journal of Analytical Atomic Spectrometry, 31, 2375-2383.

Li C.F., Feng L.J., Wang X.C., Wilde S.A., Chu Z.Y. and Guo J.H. (2017) A low-blank two-column chromatography separation strategy based on a $KMnO_4$ oxidizing reagent for Cr isotope determination in micro-silicate samples by thermal ionization mass spectrometry. Journal of Analytical Atomic Spectrometry, 32, 1938-1945.

Lugmair G.W. and Shukolyukov A. (1998) Early solar system timescales according to ^{53}Mn-^{53}Cr systematics. Geochimica et Cosmochimica Acta, 62, 2863-2886.

Makishima A., Kobayashi K. and Nakamura E. (2002) Determination of chromium, nickel, copper and zinc in milligram samples of geological materials using isotope dilution high resolution inductively coupled plasma-mass spectrometry. Geostandards and Geoanalytical Research, 26, 41-51.

Meija J., Coplen T.B., Berglund M., Brand W.A., De Biévre P., Gröning M. and Prohaska T. (2016) Isotopic compositions of the elements2013(IUPAC technical Report). Pure and Applied Chemistry, 88, 293-306.

Moynier F., Yin Q.Z. and Schauble E. (2011) Isotopic evidence of Cr partitioning into Earth's core. Science, 331, 1417.

Planavsky N.J., Reinhard C.T., Wang X., Thomson D., McGoldrick P., Rainbird R.H. and Lyons T.W. (2014) Low Mid-Proterozoic atmospheric oxygen levels and the delayed rise of animals. Science, 346, 635-638.

Qin L. and Carlson R.W. (2016) Nucleosynthetic isotope anomalies and their cosmochemical significance. Geochemical Journal, 50, 43-65.

Qin L., Alexander C.M.O., Carlson R.W., Horan M.F. and Yokoyama T. (2010a) Contributors to chromium isotope variation of meteorites. Geochimica et Cosmochimica Acta, 74, 1122-1145.

Qin L., Rumble D., Alexander C.M.O'.D., Carlson R.W., Jenniskens P. and Shaddad M.H. (2010b) The chromium isotopic composition of Almahata Sitta. Meteoritics and Planetary Science, 45, 1771-1777.

Rai D., Eary L.E. and Zachara J.M. (1989) Environmental chemistry of chromium. Science of the Total Environment, 86, 15-23.

Rodler A., Sánchez-Pastor N., Fernández-Díaz L. and FreiR. (2015) Fractionation behavior of chromium isotopes during coprecipitation with calcium carbonate: Implications for their use as paleoclimatic proxy. Geochimica et Cosmochimica Acta, 164, 221-235.

Rudge J.F., Reynolds B.C. and Bourdon B. (2009) The double spiketoolbox.Chemical Geology, 265, 420-431.

Rudnick R.L. and Gao S. (2003) Composition of the continental crust. In: Turekian K.K. and Holland H.D. (eds), Treatise on geochemistry. Pergamon (Oxford), vol. 3, 1-64.

Saad E.M., Wang X., Planavsky N.J., Reinhard C.T. and Tang Y. (2017) Redox-independent chromium isotope fractionation induced by ligand-promoted dissolution. Nature Communications, 8, 1590.

Schauble E., Rossman G.R. and Taylor H.P. (2004) Theoretical estimates of equilibrium chromium-isotope fractionations. Chemical Geology, 205, 99-114.

Scheiderich K., Amini M., Holmden C. and Francois R. (2015) Global variability of chromium isotopes in seawater demonstrated by Pacific, Atlantic, and Arctic Ocean samples. Earth and Planetary Science Letters, 423, 87-97.

Schiller M., Van Kooten E., Holst J.C., Olsen M.B. and Bizzarro M. (2014) Precise measurement of chromium isotopes by MC-ICP-MS. Journal of Analytical Atomic Spectrometry,29, 1406-1416.

Schoenberg R., Zink S., Staubwasser M. and von Blanckenburg F. (2008) The stable Cr isotope inventory of solid Earth reservoirs determined by double spike MC-ICP-MS. Chemical Geology, 249, 294-306.

Schoenberg R., Merdian A., Holmden C., Kleinhanns I.C., Haßler K., Wille M. and Reitter E. (2016) The stable Cr isotopic compositions of chondrites and silicate planetary reservoirs. Geochimica et Cosmochimica Acta, 183, 14-30.

Shen J., Qin L., Fang Z., Zhang Y., Liu J., Liu W., Wang F., Xiao Y., Yu H. and Wei S. (2018) High-temperature inter-mineral Cr isotope fractionation: A comparison of ionic model predictions and experimental investigations of mantle xenoliths from the North China Craton. Earth and Planetary Science Letters, 499, 278-290.

Shields W.R., Murphy T.J., Garner E.L. and Dibeler V.H. (1963) Absolute isotopic abundance ratio and the atomic weight of chlorine. Advances in Mass Spectrometry, 84, 163-173.

Sossi P.A., Moynier F. and van Zuilen K. (2018) Volatile loss following cooling and accretion of the Moon revealed by chromium isotopes. Proceedings of the National Academy of Sciences of the United States of America, 115, 10920-10925.

Trinquier A., Birck J.-L. and Allègre C.J. (2007) Widespread ^{54}Cr heterogeneity in the inner solar system. Astrophysical Journal, 655, 1179-1185.

Trinquier A., Birck J.-L. and Allègre C.J. (2008) High-precision analysis of chromium isotopes in terrestrial and meteorite samples by thermal ionization mass spectrometry. Journal of Analytical Atomic Spectrometry, 23, 1565-1574.

Wang X., Planavsky N.J., Reinhard C.T., Zou H., Ague J.J., Wu Y., Gill B.C., Schwarzenbach E.M. and Peucker-Ehrenbrink B. (2016) Chromium isotope fractionation during subduction-related metamorphism, black shale weathering, and hydrothermal alteration. Chemical Geology, 423, 19-33.

Wu W., Wang X., Reinhard C.T. and Planavsky N.J. (2017) Chromium isotope systematics in the Connecticut River. Chemical Geology, 456, 98-111.

Xia J., Qin L., Shen J., Carlson R.W., Ionov D.A. and Mock T.D. (2017) Chromium isotope heterogeneity in the mantle. Earth and Planetary Science Letters, 464, 103-115.

Yamakawa A., Yamashita K., Makishima A. and Nakamura E. (2009) Chemical separation and mass spectrometry of Cr, Fe, Ni, Zn, and Cu in terrestrial and extraterrestrial materials using thermal ionization mass spectrometry. Analytical Chemistry, 81, 9787-9794.

Zhu J.M., Wu G., Wang X., Han G. and Zhang L. (2018) An improved method of Cr purification for high precision measurement of Cr isotopes by double spike MC-ICP-MS. Journal of Analytical Atomic Spectrometry, 33, 809-821.

Zink S., Schoenberg R. and Staubwasser M. (2010) Isotopic fractionation and reaction kinetics between Cr(III) and Cr(VI) in aqueous media. Geochimica et Cosmochimica Acta, 74, 5729-5745.

Supporting information

The following supporting information may be found in the online version of this article:

Table S1. Raw data for NIST SRM 979 with different spike ratios.

Table S2. Raw data for unspiked GSB-Cr doped with different proportion of Ti, V, Fe and matrix elements.

Table S3. Raw data for samples after column chemistry with different procedures.

This material is available from: http://onlinelibrary.wiley.com/doi/10.1111/ggr.12283/abstract (This link will take you to the article abstract).

第二部分　地质储库的金属稳定同位素组成

Magnesium isotopic compositions of altered oceanic basalts and gabbros from IODP Site 1256 at the East Pacific Rise*

Jian Huang[1], Shan Ke[2], Yongjun Gao[3], Yilin Xiao[1] and Shuguang Li[2, 1]

1. CAS Key Laboratory of Crust-Mantle Materials and Environments, School of Earth and Space Sciences, University of Science and Technology of China, Hefei 230026, China
2. State Key Laboratory of Geological Processes and Mineral Resources, China University of Geosciences, Beijing 100083, China
3. Department of Earth and Atmospheric Sciences, University of Houston, Houston, TX 77204, USA

亮点介绍：本文首次报道了近洋脊蚀变洋壳的镁同位素组成，发现贫碳酸盐的蚀变洋壳具有类似于地幔的镁同位素组成，说明再循环蚀变洋壳不会导致地幔镁同位素不均一。

Abstract To investigate the behaviour of Mg isotopes during alteration of oceanic crust and constrain the Mg isotopic compositions of the altered oceanic crust (AOC), high-precision Mg isotope analyses have been conducted on forty-four altered basalts and gabbros recovered from IODP Site 1256, which represent the carbonate-barren AOC formed at the East Pacific Rise (EPR). These samples were altered by interaction of seawater-derived fluids with oceanic crust at different temperatures and water/rock ratios. With the exception of one sample that has a slightly heavier Mg isotopic composition (δ^{26}Mg = 0.01±0.08‰), all the other samples have relatively homogenous and mantle-like Mg isotopic compositions, with δ^{26}Mg ranging from –0.36 to –0.14‰ (an average value of –0.25±0.11‰, 2SD, n = 43). This suggests that limited Mg isotope fractionation occurred during alteration of oceanic crust at the EPR at bulk rock scale, irrespective of highly variable alteration temperatures and variable water/rock ratios. Thus, our study suggests that the offset of δ^{26}Mg values between seawater and global runoff dominantly results from the formation of marine dolomite as a sink for Mg. The mantle-like Mg isotopic composition further indicates that recycling of carbonate-barren AOC would not result in Mg isotope heterogeneity in the mantle at global scale. Consequently, the light Mg isotopic compositions of the mantle at local scale must result from incorporation of recycled Mg isotopically light carbonates.

Keywords Magnesium isotopes, basalt and gabbro, oceanic crust, fluid-rock interaction, isotope fractionation

1 Introduction

Magnesium is a fluid-mobile, major element in Earth's hydrosphere and lithosphere. It has three isotopes

* 本文发表在：Lithos, 2015, 231: 53-61

(^{24}Mg, ^{25}Mg, and ^{26}Mg) with a >8% mass difference between ^{26}Mg and ^{24}Mg that is large enough to result in detectable Mg isotope fractionation in geological and biological processes (e.g., Young and Galy, 2004). Previous studies demonstrated that during continental weathering, the residual products (e.g., soil or saprolite) are enriched in heavy Mg isotopes compared to their precursors (e.g., Brenot et al., 2008; Li et al., 2010; Liu et al., 2014; Pogge von Strandmann et al., 2008; Teng et al., 2010a; Tipper et al., 2006a,b, 2010, 2012a,b; Wimpenny et al., 2010). These studies have attributed this enrichment to a preferential uptake of ^{26}Mg by secondary silicate minerals (e.g., smectite) in the residues, leading to isotopically light weathering fluids. Such a mechanism is confirmed by the light Mg isotopic compositions observed for pore waters and river waters draining from silicate rocks (e.g., Tipper et al., 2006a, 2010).

Based on the Mg isotopic compositions of the world's largest rivers, Tipper et al. (2006a) estimated the flux weighted average δ^{26}Mg value of global runoff at -1.09 ± 0.05‰, which is much lighter than that of seawater (-0.83 ± 0.09‰, Foster et al., 2010; Ling et al., 2011). This difference indicates that the oceanic Mg budget is not at steady state and/or that the Mg isotope ratios are fractionated in the ocean (Tipper et al., 2006a). The former hypothesis is in consistent with variations of the seawater Mg concentrations and Mg/Ca ratios over the past 30 Ma (e.g., Holland, 2005; Raush et al., 2013), while the latter is in agreement with Mg uptake from the ocean by marine dolomite precipitation and hydrothermal sink (Tipper et al., 2006a). At steady state, where riverine input and hydrothermal and dolomite output sustain the oceanic Mg mass balance (Tipper et al., 2006a), estimates of the hydrothermal sink of Mg at mid-ocean ridges vary from 61 to 100% (Elderfield et al., 1996; Holland, 2005; Tipper et al., 2006a; Beinlich et al., 2014). Such large amounts of hydrothermal sink of Mg may modify Mg isotopic compositions of the AOC, as seawater and fresh MORBs have distinct Mg isotopic compositions.

Teng et al. (2010a) speculated that the bulk AOC should have lighter Mg isotopic compositions relative to fresh MORBs, however, they did not consider dolomite output for the ocean. Another study on cratonic eclogites with affinity to oceanic crust from South Africa attributed the observed light Mg isotopic compositions (δ^{26}Mg as low as -0.80‰) to the AOC precursor (Wang et al., 2012). However, these studies only indirectly constrained the Mg isotopic compositions of the AOC. This significantly hampers our understanding of Mg isotope geochemistry with respect to several issues that need to be constrained, including (1) Are the Mg isotopic compositions variable in a profile through the AOC as is observed for Li and O isotopic compositions (e.g., Gao et al., 2012)? (2) What is the average Mg isotopic composition of the AOC? (3) Does the bulk AOC have a lighter Mg isotopic composition relative to fresh MORBs, as speculated by previous studies (Teng et al., 2010a)? Knowledge of these issues is of great importance to understand the behaviour of Mg isotopes during hydrothermal alteration and to evaluate whether Mg isotopes can be used to trace the recycled AOC and distinct mantle reservoirs as defined by Li-O-Sr-Nd-Pb isotope systematics (e.g., Zindler and Hart, 1986; Hofmann, 1997; Eiler, 2001; Elliot et al., 2004).

To better constrain these issues, we report for the first time Mg isotopic compositions of altered basalts and gabbros recovered from the IODP Site 1256 on the Cocos plate in the eastern equatorial Pacific Ocean (Wilson, 1996). Previous studies showed that although both low temperature seawater alteration and high temperature hydrothermal alteration occurred in these mafic rocks (e.g., Alt et al., 2010; Gao et al., 2012), the abundance of carbonates is low, as indicated by petrological studies and their relatively low loss on ignition (LOI <4.71wt.%, Alt et al., 2006; Neo et al., 2009). Although these altered rocks display large variations in δ^{18}O and δ^{7}Li values (Gao et al., 2012), our results show limited Mg isotopic variations that are similar to global oceanic basalts and mantle peridotites. Therefore, fractionation of Mg isotopes is limited during hydrothermal alteration of oceanic crust and the light Mg isotopic compositions of the mantle must result from the contamination of the low δ^{26}Mg carbonates rather than the recycled carbonate-barren AOC.

2 Geological settings and sample descriptions

IODP Site 1256 (6°44.2′N, 91°56.1′W) is located in the Guatemala Basin on the Cocos plate in the eastern equatorial Pacific Ocean (Fig. 1). The oceanic crust at the drill site formed ca. 15 Ma ago during a sustained episode of superfast ocean ridge spreading with rates up to 200 mm/yr (Wilson, 1996; Wilson et al., 2003). Basement rocks were recovered from Holes 1256C and 1256D, but the uppermost lavas were sampled only in Hole 1256C (Wilson et al., 2003). Pilot Hole 1256C penetrated a 250.7 m sediment section and extends 88.5 m into basement that consists of a 32 m thick lava pond with thin sheet flow above and below (Fig. 2). The main Hole 1256D is located ~30 m to the south, starts coring at 276 m below seafloor (mbsf), and extends to a depth of 1507.1 mbsf (Fig. 2).

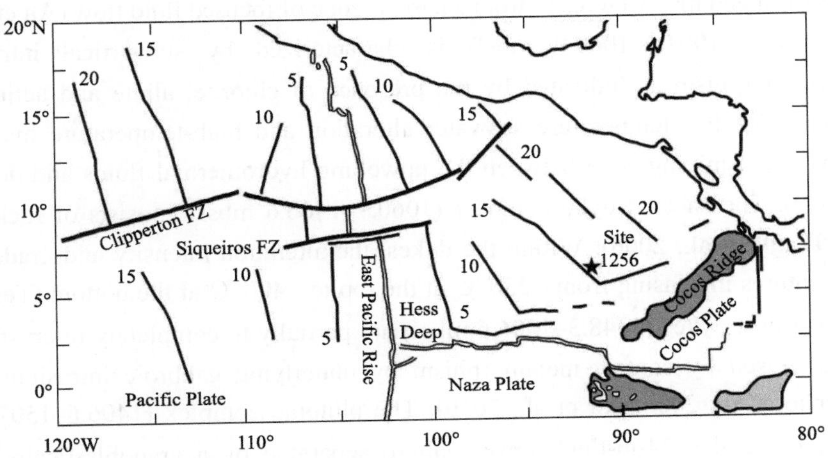

Fig. 1 Simplified geological map showing the location of IODP Site 1256 (modified after Gao et al., 2012). Isochrones with 5Ma intervals are also shown.

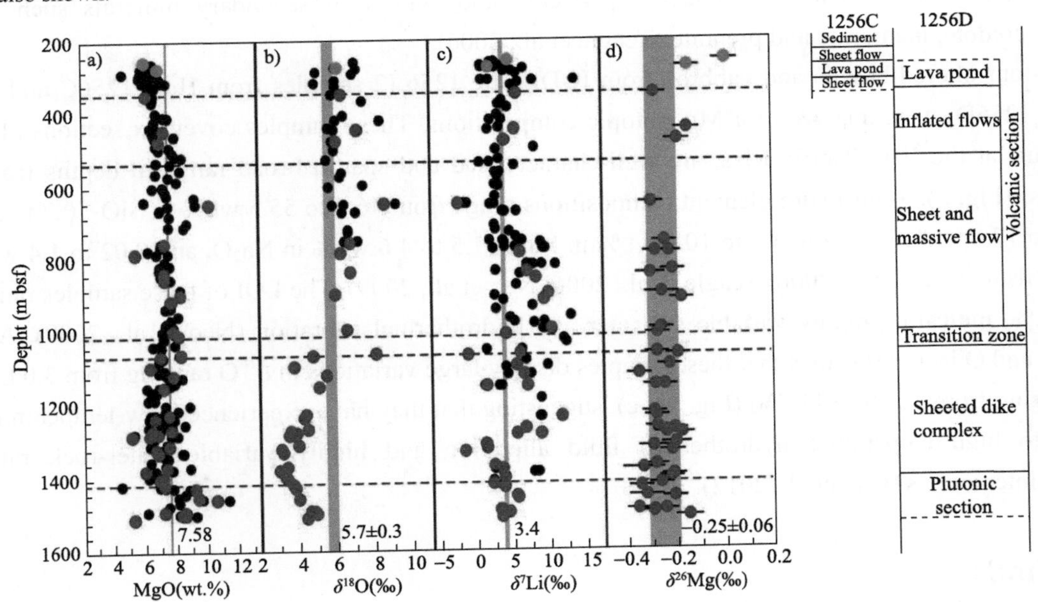

Fig. 2 Downhole variations of MgO contents and O-Li-Mg isotopic compositions of the altered basalts and gabbros from the IODP Site 1256. The MgO contents are taken from references (Neo et al., 2009; Teagle et al., 2006; Wilson et al., 2003, 2006), and $\delta^{18}O$ and $\delta^{7}Li$ values are taken from Gao et al. (2012). The average MgO contents and isotopic compositions of fresh MORBs (denoted by gray lines or bars) are taken from references (Hamon and Hoefs, 1995; Hofmann, 1988; Teng et al., 2010b; Tomasack et al., 2008). Green and red solid circles denote the studied samples from the Hole 1256C and D, respectively.

The oceanic crust sampled at the Hole 1256D was subdivided into six sections based on the igneous stratigraphy, including, from top to bottom, the lava pond, inflated flows, sheet and massive flows, transition zone, sheeted dyke complex and plutonic complex (Fig. 2) (Wilson et al., 2003, 2006; Teagle et al., 2006). The lava pond (~276–350.3 mbsf), inflated flows (350.3–533.9 mbsf) and sheet and massive flows (533.9–1004.2 mbsf) can be further classified on the basis of the processes involved in their alteration (Alt et al., 2010). The upper volcanic section (down to 946 mbsf) has experienced low-temperature (<100 ℃) seawater alteration in reducing conditions as evidenced by the presence of saponite ± pyrite and the low abundance of oxyhydrioxide enriched alteration halos. The lower volcanic section below 964 mbsf has suffered elevated alteration temperatures as indicated by the presence of pyrite alteration halos, mixed-layer chlorite/smectite and anhydrite (Teagle et al., 2006; Alt et al., 2010). Overall, the volcanic section is slightly to moderately altered, but a highly altered 41 cm interval occurs at 648 mbsf and may well result from a narrow zone of focused fluid flow (Alt et al., 2010).

The transition zone (1004.2–1060.9 mbsf) is characterized by subvertical intrusive contacts and greenschist-facies metamorphism as indicated by the presence of chlorite, albite and actinolite (Teagle et al., 2006). In this section, both low-temperature seawater alteration and high-temperature hydrothermal alteration occur, indicating that it is a mixing zone between the upwelling hydrothermal fluids and downwelling seawater (Alt et al., 1996, 2010). The sheeted dyke complex (1060.9–1406.6 mbsf) consists of rocks that are highly to completely altered (Teagle et al., 2006). Within the dykes, the alteration intensity and grade increase downhole with alteration temperatures increasing from ~250 ℃ at the top to ~400 ℃ at the bottom (Teagle et al., 2006; Alt et al., 2010). The lowermost dykes (1348.3–1406.6 mbsf) are partially to completely re-crystallized to distinctive granoblastic textures caused by contact metamorphism by underlying gabbroic intrusions at temperatures of ~900–1050 ℃ (Koepke et al., 2008; Alt et al., 2010). The plutonic complex (1406.6–1507.1 mbsf) contains a 52-m-thick upper gabbro and a 24-m-thick lower gabbro separated by a granoblastically recrystallized dyke screen. Metamorphic conditions in this section are similar to those in the lower dykes (Alt et al., 2010), but the margins of the gabbro bodies are moderately altered and consist of secondary minerals such as chlorite, amphibole, epidote, laumonite and prehnite (Teagle et al., 2006).

Forty-four altered basalts and gabbros from IODP Site 1256 (2 samples from Hole 1256C and 42 samples from Hole 1256D) were analyzed for Mg isotopic compositions. These samples cover six sections of the altered oceanic crust at the East Pacific Rise, are well-characterised and span a broad range in depths from 258.5 to 1502.7 mbsf (Fig. 2). Their major element compositions range from 46.4 to 55.5 wt.% in SiO_2, 0.7 to 2.3 wt.% in TiO_2, 8.1 to 17.0 wt.% in Fe_2O_3, 5.2 to 10.6 wt.% in MgO, 1.5 to 4.6 wt.% in Na_2O, and 0.02 to 1.4 wt.% in K_2O (Fig. 2a, Wilson et al., 2003, 2006; Teagle et al., 2006; Neo et al., 2009). The LOI of these samples range from ~0 to 4.71 wt.%, indicating highly variable seawater and hydrothermal alteration (Neo et al., 2009). A combined study of Li and O isotopes shows that these samples display large variations in $\delta^{18}O$ ranging from 3.0 to 8.6‰ and in $\delta^{7}Li$ ranging from –2.29 to 11.7‰ (Fig. 2b, c), suggesting that they have experienced low-temperature seawater alteration to high-temperature hydrothermal fluid alteration and highly variable water-rock ratios during water-rock interactions (Gao et al., 2012).

3 Methods

Measurements of Mg isotope ratios were conducted at the State Key Laboratory of Geological Processes and Mineral Resources, China University of Geosciences (Beijing) (CUGB) following the procedures similar to those established by Yang et al. (2009), Li et al. (2010) and Teng et al. (2010b). All chemical procedures were conducted in a clean laboratory environment. Sample powders were precisely weighted into and dissolved in

Savillex screw-up beakers in a mixture of concentrated ultrapure HF-HNO_3 (~3:1). The capped beakers were heated at a temperature of 160 ℃ on a hot plate in a laminar flow exhaust for 24 hours. In order to achieve 100% dissolution, the solutions were evaporated to dryness, dissolved in aqua regia and heated overnight at a temperature of 130 ℃. The solutions were then evaporated to dryness in the following day, refluxed with concentrated ultrapure HNO_3 until completely dissolved and then evaporated again to dryness. The dried residue was finally dissolved in 1 N HNO_3 for chromatographic separation.

Chemical separation of Mg was achieved by cation exchange chromatography. The column was loaded with 2 ml of Bio-Rad 200-400 mesh AG50W-X8 pre-cleaned resin (rinsed with >14 times column volume of 6 N HCl and 18.2 $M\Omega.cm^{-1}$ Milli-Q® water). The resin was further cleaned with 5ml Milli-Q® water and conditioned with 6 ml 1 N HNO_3. Samples containing ~10 μg Mg were loaded on the resin and eluted with 1 N HNO_3. The same column procedure was processed twice for all samples in order to obtain a pure Mg solution. The solutions were first evaporated to dryness and then re-dissolved in 3% HNO_3, ready for final dilution immediately prior to analysis. The concentrations of cations (such as Ti, Al, Fe, Ca, Na, and K) in the solutions were detected by ICP-MS, and the results show that cation/Mg (mass/mass) ratios are <0.02. The total procedural blanks were negligible and ⩽10 ng, which is comparable with that of Teng et al. (2010a).

Magnesium isotope ratios were measured by the sample-standard bracketing method on a Neptune Plasma MC-ICPMS at the CUGB. Magnesium isotopes were analyzed in a low-resolution mode, with ^{26}Mg, ^{25}Mg and ^{24}Mg measured simultaneously in separate Faraday cups (H3, IC and L3). The uptake rate was set to 100μl/min, and a 100 ppb solution typically yields a 3.5–4V ^{24}Mg signal (with a $10^{11}\Omega$ resistor for the Faraday cups). The in-run precision on the $^{26}Mg/^{24}Mg$ ratio for a single block of 40 ratios is <±0.02‰ (2SD). The internal precision on the measured $^{26}Mg/^{24}Mg$ ratio based on 3 or 4 repeated analyses of the same sample solution during a single analytical session is ⩽0.05‰ (2SD) for most of samples. Repeated analyses of synthetic pure Mg solution and natural materials are indicative of a long-term precision of ⩽±0.05‰ (Ke et al. in preparation). The results are reported in the conventional δ notation that is defined as $\delta^{X}Mg = [(^{X}Mg/^{24}Mg)_{sample}/(^{X}Mg/^{24}Mg)_{DSM3} - 1] \times 1000$, where X = 25 or 26, and DSM3 is an international reference of solution made from pure Mg metal (Galy et al., 2003). Analyses of the well-characterized USGS standard BHVO-2 in this study yielded $\delta^{26}Mg = -0.28 \pm 0.04$‰ (2SD, n=6, Table 1), in agreement with previously reported values (Teng et al., 2007, 2010b; Huang et al., 2009; Pogge von Strandmann et al., 2011; An et al., 2014).

Table 1 Mg, O and Li isotopic compositions together with SiO_2 and MgO contents of crustal rocks from IODP site 1256

Cor/Sc[a]	T-B cm	Depth mbsf	Sample description	SiO_2[b] wt.%	MgO[b] wt.%	Li/Yb[b]	LOI[b] wt.%	CIA[b]	$\delta^{7}Li$[b] ‰	$\delta^{18}O$[b] ‰	$\delta^{26}Mg$ ‰	2SD	$\delta^{25}Mg$ ‰	2SD	$\Delta^{25}Mg'$[c]	N[d]
6R/2	3–11	258.46	coarse-grained altered basalt	55.5	6.18	1.75	NA	23.9	4.75	NA	0.01	0.08	0.00	0.05	0.00	4
Repli.[e]											0.03	0.02	0.03	0.01	0.01	4
8R/1	101–108	276.51	fine-grained basalt	53.6	7.05	1.38	NA	15.6	2.06	6.2	−0.14	0.09	−0.06	0.03	0.01	4
Repli.											−0.11	0.05	−0.06	0.03	0.00	4
12R/8	71–79	351.21	microcrystalline basalt	50.1	6.28	1.26	NA	36.5	5.86	6.4	−0.28	0.09	−0.17	0.01	−0.02	4
27R/1	130–137	446.7	microcrystalline basalt	51.3	7.38	1.62	NA	33.1	5.57	7.2	−0.15	0.06	−0.05	0.05	0.02	4
Repli.											−0.15	0.09	−0.07	0.07	0.01	4
32R/1	114–120	476.34	microcrystalline basalt	50.9	7.07	1.47	NA	36.8	4.14	6.1	−0.19	0.08	−0.09	0.05	0.01	4
57R/2	117–127	648.02	red + green altered basalt	46.4	10.31	5.93	NA	52.6	−2.29	8.6	−0.30	0.03	−0.16	0.05	0.00	4
75R/1	131–133	753.05	patchy basalt	49.6	7.36	1.17	1.69	37.3	NA	6.8	−0.24	0.07	−0.13	0.04	0.00	4
80R/1	103–107	781.47	aphyric cryptocrystalline basalt	51.9	5.49	0.99	1.27	37.8	NA	6.3	−0.23	0.04	−0.12	0.05	0.00	4
80R/2	92–102	781.47	fine-grained basalt with mixed halo	51.9	5.49	1.28	1.27	37.8	6.13	NA	−0.28	0.03	−0.13	0.04	0.02	4

Continued

Cor/Sc[a]	T-B cm	Depth mbsf	Sample description	SiO_2[b] wt.%	MgO[b] wt.%	Li/Yb[b]	LOI[b] wt.%	CIA[b]	δ^7Li[b] ‰	$\delta^{18}O$[b] ‰	$\delta^{26}Mg$ ‰	2SD	$\delta^{25}Mg$ ‰	2SD	$\Delta^{25}Mg'$[c]	N[d]
87R/2	66–68	831.06	phyric fine-grained basalt	51.1	7.33	1.50	0.63	35.7	7.08	6.8	−0.21	0.05	−0.12	0.03	−0.01	4
89R/1	70–73	840.68	phyric cryptocrystalline basalt	50.7	7.35	1.32	0.66	36.4	8.3	NA	−0.30	0.04	−0.18	0.06	−0.02	4
96R/1	29–31	893.29	aphyric microcrystalline basalt	50.2	8.39	0.98	NA	36.1	10.02	6.0	−0.26	0.04	−0.14	0.07	0.00	4
99R/2	101–120	909.63	fine-grained background alteration	50.9	7.74	1.45	0.46	36.1	9.42	NA	−0.18	0.05	−0.08	0.02	0.01	4
Repli.											−0.23	0.04	−0.10	0.11	0.02	4
114R/2	54–56	990.04	altered phyric microcrystalline basalt	50.8	7.88	1.03	0.96	37.1	10.37	6.6	−0.26	0.04	−0.12	0.05	0.01	4
117R/1	97–107	1004.17	fine-grained cataclastic basalt	51.1	8.23	1.00	2.45	38.9	4.15	NA	−0.28	0.06	−0.14	0.04	0.00	4
128R/1	58–63	1056.77	fine-grained background basalt	53.2	8.39	0.69	1.38	35.2	6.1	8.1	−0.28	0.08	−0.16	0.04	−0.01	3
129R/1	34–51	1061.51	fine-grained background basalt	50.8	7.06	0.69	1.01	35.4	−1.07	4.8	−0.19	0.06	−0.10	0.05	0.00	3
135R/1	54–64	1090.86	patchy microcrystalline basalt	55.5	6.38	0.81	4.71	42.5	5.58	3.8	−0.23	0.04	−0.12	0.04	0.00	3
140R/1	76–79	1114.72	volcanic breccia	51.6	7.87	3.69	4.09	43.0	6.6	5.5	−0.35	0.07	−0.19	0.01	−0.01	3
Repli.											−0.38	0.03	−0.17	0.03	0.02	4
147R/1	40–48	1145.98	doleritic basalt	50.9	7.22	0.96	2.13	36.1	1.47	NA	−0.25	0.03	−0.12	0.04	0.00	3
147R/1	75–77	1145.98	aphyric fine-grained dolerite	50.9	7.22	0.69	2.13	36.1	7.05	5.1	−0.29	0.03	−0.15	0.05	0.00	3
163R/3	59–62	1219.91	phyric cryptocrystalline basalt	50.8	7.56	0.67	1.86	36.4	NA	4.4	−0.29	0.05	−0.15	0.03	0.01	3
165R/3	101–103	1230.19	aphyric microcrystalline basalt	51.9	6.65	0.38	1.64	35.4	11.7	4.3	−0.28	0.04	−0.13	0.13	0.01	3
173R/2	6–10	1260.6	phyric medium-to fine-grained basalt	51.3	7.71	0.78	1.09	36.3	6.76	4.5	−0.28	0.03	−0.13	0.02	0.02	3
174R/1	130–134	1266.13	aphyric fine-grained basalt	51.5	6.44	0.35	1.33	37.2	NA	3.6	−0.21	0.06	−0.09	0.04	0.02	3
175R/1	58–62	1272.05	aphyric cryptocrystalline basalt	51.2	6.82	0.40	1.04	36.3	NA	3.9	−0.25	0.03	−0.13	0.01	0.00	3
176R/1	133–136	1277.27	aphyric fine-grained basalt	50.5	6.3	0.35	1.42	36.5	5.56	3.5	−0.24	0.03	−0.13	0.01	0.00	3
Repli.											−0.20	0.05	−0.11	0.04	0.00	4
176R/2	22–25	1278.03	ahyric fine-grained basalt	50.2	5.16	0.26	1.00	37.1	8.86	3.8	−0.18	0.02	−0.09	0.04	0.01	3
182R/1	25–28	1305.09	aphyric fine-grained basalt	50.6	7.55	0.42	0.28	36.4	1.31	4.0	−0.20	0.07	−0.11	0.06	−0.01	3
184R/1	98–104	1314.5	aphyric fine-grained basalt	51.0	6.71	0.38	0.45	36.2	1.45	3.1	−0.30	0.04	−0.13	0.06	0.02	3
187R/1	15–17	1325.88	aphyric microcrystallinebasalt	50.8	6.84	0.29	0.39	36.5	−1.6	3.2	−0.22	0.07	−0.10	0.05	0.01	3
196R/1	30–32	1363.86	aphyric fine-grained basalt	50.8	7.41	0.40	0.22	36.1	2.42	3.4	−0.29	0.03	−0.16	0.04	−0.01	3
202R/1	37–42	1373.05	aphyric fine-gained basalt	50.3	6.01	0.28	0.00	37.3	3.54	3.0	−0.34	0.03	−0.16	0.09	0.02	3
Repli.											−0.27	0.04	−0.15	0.02	−0.01	4
207R/1	10–15	1390.8	aphyric cryptocrystlalline basalt	52.0	7.13	0.29	NA	36.2	NA	3.3	−0.21	0.07	−0.11	0.04	0.00	4
209R/1	15–19	1396.65	aphyric microcrystalline basalt	52.2	6.91	0.32	NA	36.1	3.82	3.4	−0.22	0.02	−0.10	0.01	0.02	4
214R/2	50–55	1413.55	disseminated oxide gabbro	49.5	7.22	0.63	0.46	36.0	2.26	3.7	−0.27	0.03	−0.12	0.01	0.02	4
217R/1	4–9	1421.77	disseminated oxide gabbro	49.4	9.23	0.88	0.71	35.3	3.76	3.8	−0.35	0.05	−0.16	0.04	0.02	4
222R/2	25–35	1446.37	green altered gabbro	49.0	9.45	0.52	0.61	35.4	5.46	NA	−0.33	0.04	−0.16	0.07	0.01	4
223R/2	41–48	1450.68	olivine gabbronorite	48.7	10.65	0.68	1.66	35.7	5.59	4.0	−0.21	0.01	−0.10	0.05	0.01	4
230R/1	68–72	1483.72	fine-grained opx bearing oxide gabbro	47.5	7.23	0.50	0.15	35.1	3.22	4.6	−0.36	0.04	−0.17	0.03	0.01	4
Repli.											−0.34	0.06	−0.18	0.11	−0.01	4
230R/2	36–40	1484.99	disseminated oxide gabbro-gabbronorite	49.5	8.52	0.61	0.68	35.0	3.3	4.8	−0.25	0.05	−0.11	0.01	0.02	4
231R/3	21–27	1488.8	oxide gabbronorite	49.1	8.36	0.68	0.86	34.4	4.39	4.9	−0.30	0.10	−0.14	0.03	0.02	4
231R/3	80–98	1491.36	opx bearing gabbro	49.4	8.59	0.88	0.51	34.4	3.56	NA	−0.30	0.08	−0.16	0.04	0.00	4

Continued

Cor/Sc[a]	T-B cm	Depth mbsf	Sample description	SiO_2[b] wt.%	MgO[b] wt.%	Li/Yb[b]	LOI[b] wt.%	CIA[b]	δ^7Li[b] ‰	δ^{18}O[b] ‰	δ^{26}Mg ‰	2SD	δ^{25}Mg ‰	2SD	Δ^{25}Mg'[c]	N[d]
234R/1	19–22	1502.76	fine-grained basalt	49.5	5.26	0.57	2.19	34.2	3.4	4.4	–0.15	0.03	–0.08	0.06	0.00	4
BHVO-2											–0.27	0.05	–0.15	0.05	–0.01	3
Repli.											–0.28	0.03	–0.14	0.03	0.00	3
Repli.											–0.27	0.02	–0.14	0.05	0.00	3
Repli.											–0.27	0.07	–0.13	0.03	0.01	4
Repli.											–0.29	0.03	–0.15	0.02	0.00	4
Repli.											–0.32	0.07	–0.16	0.03	0.00	4
Avg.											–0.28	0.04	–0.14	0.02		

[a] Two samples, Cor/Sc 6R/2 (258.46 mbsf) and 8R/1 (276.51 mbsf) were collected from Hole 1256C, and the others from Hole 1256D.

[b] The data for contents of SiO_2, MgO, LOI, and CIA are taken from Wilson et al. (2003, 2006), Teagle et al. (2006) and Neo et al. (2009), and for Li/Yb ratios and O-Li isotopic compositions are taken from Gao et al. (2012).

[c] $\Delta^{25}Mg'$ ($= \delta^{25}Mg' - 0.521\delta^{26}Mg'$) was defined by Young and Galy (2004), and $\delta^{X}Mg' = 1000\ln(\delta^{X}Mg/1000 + 1)$, where X refers to mass 25 or 26.

[d] N represents the numbers of mass spectrometry analyses for one sample.

[d] Repli. = repeated column chemistry and instrumental measurement of different aliquots of the same stock solution.

4 Results

The Mg isotopic compositions of the altered basalts and gabbros investigated in this study are listed in Table 1. In the plot of δ^{25}Mg vs. δ^{26}Mg, our samples define a linear line with a slope of 0.522 (Fig. 3), similar to that of the predicted equilibrium mass-dependent fractionation line (0.521, Young and Galy, 2004). The $\Delta^{25}Mg'$ ($\Delta^{25}Mg' = \delta^{25}Mg' - 0.521\delta^{26}Mg'$, where $\delta^{X}Mg' = 1000\ln(\delta^{X}Mg/1000+1)$, and X refers to mass 25 or 26, Young and Galy, 2004) values range between –0.02 and 0.02, very close to zero. These results indicate that Mg isotopes obey the mass-dependent fractionation law (Young and Galy, 2004).

Except for one sample from Hole 1256C (258.46 mbsf) that has a slightly heavier Mg isotopic composition ($\delta^{26}Mg = 0.01 \pm 0.08$‰), the δ^{26}Mg values of the other samples range from –0.36 to –0.14‰ (Figs. 3-5), with an average value of -0.25 ± 0.11‰ (n=43, 2SD). This average value is identical to that of the global MORBs (-0.25 ± 0.06‰, Teng et al., 2010b) and the upper mantle (-0.25 ± 0.07‰, Teng et al., 2010b).

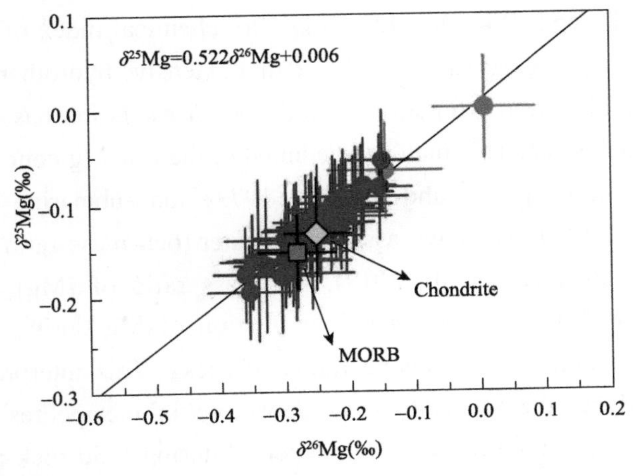

Fig. 3 δ^{26}Mg vs. δ^{25}Mg of the altered basalts and gabbros from the IODP Site 1256. Also shown for comparison are the average δ^{26}Mg values of global MORBs and Chondrites (Teng et al., 2010b). The symbols are the same as in Fig. 2.

5 Discussion

5.1 Behaviour of Mg isotopes during seawater hydrothermal alteration

Previous oxygen isotope studies of ophiolite complexes, generally recognized to represent fragments of ancient oceanic crust and mantle, indicate both ^{18}O enrichments (e.g., pillow lavas) and ^{18}O depletions (e.g., gabbros) relative to the primary mantle-derived magmatic $\delta^{18}O$ values (e.g., Muehlenbacks and Clayton, 1971; Gregory and Taylor, 1981). This is consistent with direct measurements of in situ profile of O isotopes for core samples through the oceanic crust from IODP site 1256, which show ^{18}O enrichments in the upper extrusive basalts and ^{18}O depletion in basalts and gabbros from the sheeted dyke complex and plutonic section (e.g., Alt et al., 1996; Teagle et al., 1996; Alt and Teagle, 2000; Gao et al., 2012). The variations in $\delta^{18}O$ values in both ophiolite complexes and in situ altered oceanic crust have been attributed to hydrothermal alteration and weathering processes at different temperatures, as low temperature (<250 ℃) alteration tends to enrich mafic rocks (e.g., basalts and gabbros) in ^{18}O, while high temperature (>250 ℃) alteration tends to enrich rocks in ^{16}O (e.g., Muehlenbachs and Clayton, 1972; Gregory and Taylor, 1981; Alt and Teagle, 2000; Gao et al., 2012).

The altered basalts and gabbros investigated in this study have highly variable $\delta^{18}O$ and $\delta^{7}Li$ values, ranging from 3.0 to 8.6‰ and –2.3 to 11.7‰, respectively (Fig. 2b, c). Specifically, altered basalts from the volcanic section and the transition zone (>1060.9 mbsf) display relatively heavier O isotopic compositions with $\delta^{18}O$ values ranging between 6.0‰ and 8.6‰ (Fig. 2b), much higher than those of fresh MORBs (5.7±0.3‰, Harmon and Hoefs, 1995). This indicates that these basalts have experienced pronounced low-temperature hydrothermal alteration (<250 ℃, Gao et al., 2012). By contrast, altered basalts from the sheeted dyke complex and altered basalts/gabbros from the plutonic section, have relatively lower $\delta^{18}O$ values (3.0 to 5.5‰), reflecting interaction between seawater-derived fluids and oceanic crust at relatively high temperatures (>250 ℃, Gao et al., 2012). In addition, quantitative modelling has demonstrated that enrichment of ^{7}Li in the altered basalts and gabbros reflects high water/rock ratios (i.e., >10 for hydrothermal fluids and >20 for seawater) during fluid-rock interaction, whereas depletion of ^{7}Li in altered rocks corresponds to low water/rock ratios (<10) (Gao et al., 2012).

In contrast to the highly variable $\delta^{18}O$ and $\delta^{7}Li$ values, 43 out of 44 altered basalts and gabbros at this ODP site have Mg isotopic compositions similar to those of fresh MORBs (−0.25±0.06‰, Teng et al., 2010b). Their $\delta^{26}Mg$ values are invariant with depth (Fig. 2d), LOI (Fig. 4a), chemical index of alteration (CIA, Fig. 4b), and $\delta^{18}O$ and $\delta^{7}Li$ values (Fig. 4c). These observations indicate that extensive hydrothermal alteration of oceanic crust at the EPR at highly variable alteration temperatures and water/rock ratios has caused only very limited Mg isotope fractionation at bulk-rock scale. This may be attributed to the low Mg concentrations of seawater ([Mg] = 53 mmol/L or MgO=0.21 wt.%, Carpenter and Manella, 1973) and submarine hydrothermal vent fluids (<1 mmol/L, Wetzel and Shock, 2000). Even if we assume seawater/rock mass ratio is up to 1, the highest value obtained from Li isotope results (Gao et al., 2012), the mass ratio of $[Mg]_{seawater}/[Mg]_{rock}$ for our altered basalts/gabbros (MgO≥5.16 wt.%) is <0.04. Such low fraction of Mg during fluid-rock interaction is thus unlikely to cause a significant shift in Mg isotope ratios of rocks. This interpretation is consistent with the conclusion obtained from studying the Mg isotopic compositions of metapelites from Onawa contact aureole (Maine, USA), where limited Mg isotope fractionation occurred during fluid-rock interaction due to much lower Mg concentrations in fluids relative to rocks (Li et al., 2014). It is unclear why one sample (258.46 mbsf) from the top of the Hole 1256C has a heavier Mg isotopic composition than fresh MORBs (Fig. 2d). We speculate that this heavy Mg isotopic signature may come from the contamination of marine silicate sediments, because ① this sample is the closest to a 250.7 m sediment cap (Fig. 2d); ② its high SiO_2 contents (55.5 wt.%) (Table 1) may result from contamination of silica-rich marine sediments; and ③ marine silicate sediments have heavy Mg isotopic compositions (δ^{26}Mg = up to 0.08‰, Mavromatis et al., 2014).

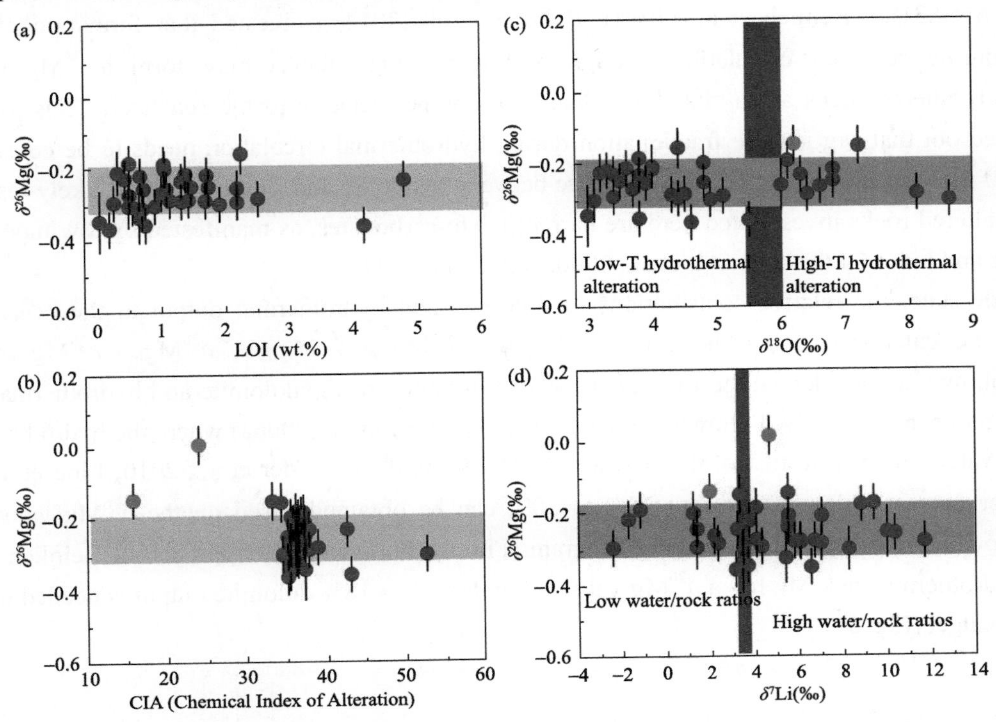

Fig. 4 Plots of δ^{26}Mg vs. LOI (a), CIA (b), δ^{18}O (c), and δ^{7}Li (d) of the altered basalts and gabbros from the IODP Site 1256. The parallel and vertical grey bars represent the average δ^{26}Mg and δ^{7}Li values of fresh MORBs (Tomascak et al., 2008; Teng et al., 2010b). The symbols are the same as in Fig. 2.

5.2 Implications for the oceanic Mg budget

The remarkable difference in Mg isotopic compositions between the global runoff and seawater (−1.09± 0.05‰ vs. −0.83±0.09‰, Tipper et al., 2006a; Foster et al., 2010; Ling et al., 2011) indicates that Mg isotope ratios of seawater is currently not at steady state and/or that the fractionation of Mg isotopes occurs in the ocean (Tipper et al., 2006a). Variations in the Mg concentrations and Mg/Ca ratios of seawater over the past 30 Ma support the former possibility (e.g., Holland, 2005; Raush et al., 2013). The latter possibility is in accordance with Mg uptake by the formation of marine dolomite and hydrothermal sink (Tipper et al., 2006a).

Teng et al. (2010a) speculated that the bulk AOC may preferentially take up ^{24}Mg from seawater and thus should be light in Mg isotopic compositions. However, Teng et al. (2010a) admitted that this speculation is oversimplified without consideration of Mg removal from the ocean by dolomite formation. In this study, direct measurements of core samples through a section of the altered oceanic basalts and gabbros from the IODP Site 1256 show that with the exception of one sample (δ^{26}Mg = 0.01±0.08‰), the AOC has a relatively homogeneous Mg isotopic composition, with an average δ^{26}Mg of −0.25±0.11‰ (2SD, n=43). This value is identical to that of global fresh MORBs (−0.25±0.06‰, Teng et al., 2010b). A possible implication is that the difference in Mg isotope ratios between seawater (−0.83‰) and global runoff (−1.09‰) is not due to hydrothermal circulation at mid-ocean ridges. This is in agreement with the prediction (Tipper et al., 2006a) that Mg uptake into the AOC by hydrothermal circulation has a δ^{26}Mg that is equal to that of seawater. In this case, the offset of Mg isotopic compositions between global runoff and seawater most likely results from precipitation of marine dolomite that has extremely light Mg isotopic compositions (δ^{26}Mg = −3.5 to −1.7‰, Higgins and Schrag, 2010; Fantle and Higgins, 2014; Mavromatis et al., 2014). However, a more recent study found that magnesite generated via carbonation of antigorite in hydrothermally altered ultramafic rocks in Norway has light Mg isotopic compositions (δ^{26}Mg=−0.96±0.31‰, Belinch et al., 2014). Belinch et al. (2014) suggested that formation of Mg-bearing carbonates during seawater circulation through mid-ocean ridge flanks may form a ^{24}Mg sink and the hydrothermally altered rocks are a sink for lighter Mg isotopes relative to the seawater. Thus, Beinlich et al. (2014) pointed out that Mg isotope fractionation during hydrothermal circulation needs to be considered in any assessment of the oceanic Mg cycle. The difference between our study and Beinlich's most likely results from the fact that the altered rocks investigated here are barren in Mg-carbonates, as manifested by low modal abundance of carbonates and low LOI (Table 1, Alt et al., 2006; Neo et al., 2009).

At steady state, the relative contributions of dolomite and hydrothermal output to the oceanic Mg mass balance can be calculated using the equation $J_{dol}/J_{riv}=(\delta^{26}Mg_{hyd}-\delta^{26}Mg_{riv})/(\delta^{26}Mg_{hyd}-\delta^{26}Mg_{dol})$, where the subscripts dol, hyd and riv denote the fluxes (J) and δ^{26}Mg values of the dolomite and hydrothermal output, and the riverine source, respectively. Following the approach of Tipper et al. (2006a) where the hydrothermal sink Mg has a δ^{26}Mg value identical to that of the seawater (−0.83‰±0.09‰, Foster et al., 2010; Ling et al., 2011), the relative importance of dolomite output of 9.7 to 30% can be obtained based on the δ^{26}Mg values of marine dolomite recovered from the Ocean Drilling Program (Fig. 5). Following the approach of Beinlich et al. (2014) where the hydrothermal sink Mg has a δ^{26}Mg value of −0.96, 5.1 to 18% dolomite output is needed to balance the oceanic Mg budget (Fig. 5).

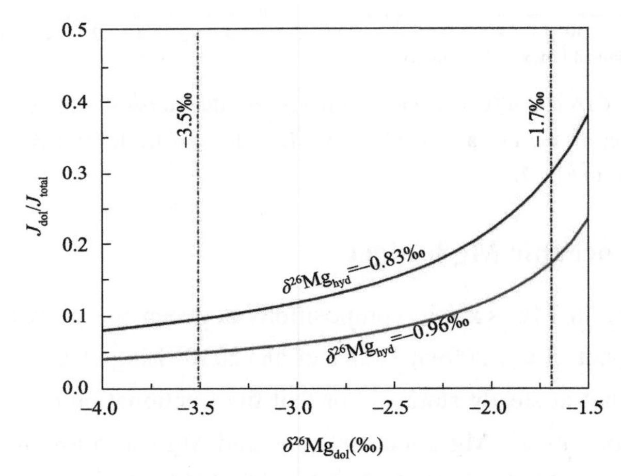

Fig. 5 The relative importance of dolomite output in controlling the oceanic Mg budget. The results were modelled using $J_{dol}/J_{tot} = (\delta_{hyd}-\delta_{riv})/(\delta_{hyd}-\delta_{dol})$, where the subscripts riv, hyd and dol represent the fluxes (J) and δ^{26}Mg values (δ) of the river source and hydrothermal and dolomite output, respectively. In this calculation, the δ^{26}Mg values of marine dolomite recovered from Ocean Drilling Program were applied (−3.5 to −1.7‰, Higgins and Schrag, 2010; Fantle and Higgins, 2014; Mavromatis et al., 2014), and the δ^{26}Mg values of hydrothermal sink Mg were set as −0.83‰ (Tipper et al., 2006a) and −0.96‰ (Beinlich et al., 2014), respectively.

5.3 Implications for tracing recycled crustal materials using Mg isotopes

Crustal recycling is an important process for the chemical evolution of the Earth's mantle and can be traced by stable and radiogenic isotopes, such as Li, O, C, Sr, Nd, and Pb isotopes (e.g., Hofmann, 1997; Eiler, 2001; Deines, 2002; Elloit et al., 2004). The Mg isotopic composition is relatively homogeneous in the mantle (δ^{26}Mg = −0.25±0.07‰) (Teng et al., 2007, 2010b; Handler et al., 2009; Yang et al., 2009; Bourdon et al., 2010; Huang et al., 2011; Liu et al., 2011; Pogge von Strandmann et al., 2011; Xiao et al., 2013), but is highly heterogeneous in crustal rocks and in sedimentary rocks in particular (δ^{26}Mg = –5.57 to 1.8‰) (e.g., Galy et al., 2002; Young and Galy, 2004; Tipper et al., 2006b; Pogge von Strandmann et al., 2008; Li et al., 2010; Higgins and Schrag, 2010; Teng et al., 2010a, Jacobson et al., 2010; Wombacher et al., 2011; Huang et al., 2012, 2013; Geske et al., 2014; Liu et al., 2014). Therefore, Mg isotopes have also been used in attempts to trace recycled crustal materials in the mantle (Shen et al., 2009; Yang et al., 2012; Xiao et al., 2013; Huang et al., 2015). Since metamorphic dehydration and high-temperature partial melting only result in limited Mg isotope fractionation (e.g., Teng et al., 2007, 2010b; Li et al., 2010, 2014; Liu et al., 2010; Telus et al., 2012; Wang et al., 2014a), it is expected that the recycled materials will preserve their original Mg isotopic signatures through subduction zone processes. Thus, the similarity in Mg isotopic compositions of the carbonate-barren AOC and that of the mantle that we demonstrate in this study implies that recycling of such AOC is unlikely to result in heterogeneous Mg isotopic compositions of the mantle at a global scale. This conclusion is consistent with the result obtained from first-order modelling (Teng et al., 2010a). Such a suggestion is also consistent with the mantle-like Mg isotopic compositions of global oceanic island basalts (OIBs) (–0.26±0.08‰, Teng et al., 2010b), whose mantle source may contain a few percent of recycled AOC (Hofmann, 1997 and references therein).

As the carbonate-barren AOC have a mantle-like Mg isotopic composition, the observed light Mg isotopic compositions in some cratonic eclogites (Fig. 6), representing the residues of the AOC after subduction-zone dehydration/melting, may reflect local enrichments of carbonates that are extremely enriched in light Mg isotopes (Wang et al., 2014b). During alteration of oceanic crust, a significant proportion of CO_2 in seawater together with

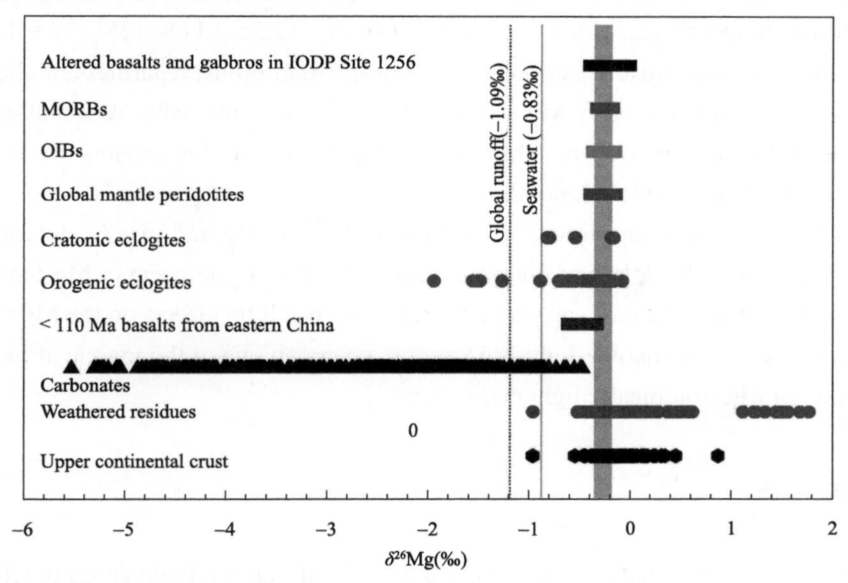

Fig. 6 δ^{26}Mg values of terrestrial materials, including altered basalts and gabbros from IODP Site 1256 (this study), mid-ocean ridge basalts (MORBs, Bourdon et al., 2010; Teng et al., 2010b), oceanic island basalts (OIBs, Bourdon et al., 2010; Teng et al., 2010b), global mantle peridotites (Yang et al., 2009; Young et al., 2009; Bourdon et al., 2010; Teng et al., 2010b; Huang et al., 2011; Liu et al., 2011; Pogge von Strandmann et al., 2011; Xiao et al., 2013), cratonic and orogenic eclogites (Li et al., 2011; Wang et al., 2014a, b), <110 Ma basalts from eastern China (Yang et al., 2012; Huang et al., 2015), carbonates (Galy et al., 2001, 2002; Chang et al., 2003; Tipper et al., 2006b; Buhl et al., 2007; Brenot et al., 2008; Pogge von Strandmann et al., 2008; Hippler et al., 2009; Wombacher et al., 2009, 2011; Higgins and Schrag, 2010; Immenhauser et al., 2010; Jacobson et al., 2010; Pokrovsky et al., 2011; Geske et al., 2012, 2014; Riechelmann et al., 2012; Azmy et al., 2013; Fantle and Higgins, 2014; Kasemann et al., 2014; Wang et al., 2014b), weathered residues developed on silicate rocks (Tipper et al., 2006b, 2010, 2012a,b; Brenot et al., 2008; Pogge von Strandmann et al., 2008; Teng et al., 2010a; Bolou-Bi et al., 2012; Huang et al., 2012; Liu et al., 2014), and upper continental crust (Li et al., 2010).

the Ca and Mg cations originally released from oceanic basalts (Wilkinson and Algeo, 1989), may form carbonates (Staudigel et al., 1989; Alt and Teagle, 1999). The carbonates usually occur as carbonate veins in the AOC and can be locally enriched with modal abundance as high as 75 vol.% (Staudigel et al., 1996). Considering the extremely light Mg isotopic compositions of such carbonates (Fig. 6), part of the altered oceanic crust enriched in carbonates may have a distinct, light Mg isotopic composition. Apart from cratonic eclogites, carbonate sediments and some orogenic eclogites also display extremely light Mg isotopic compositions (Fig. 6). Recycling of such isotopically light crustal materials would shift some parts of the mantle towards lighter Mg isotopic compositions. Such a mechanism may account for the light Mg isotopic compositions of the <110 Ma continental basalts and some mantle peridotites from eastern China (δ^{26}Mg = –0.60 to 0.30‰, Yang et al., 2009, 2012; Xiao et al., 2013; Huang et al., 2015). These light Mg isotopic compositions, however, require rather large proportions (5–20%) of dolomite with δ^{26}Mg = –2.6‰ and 22 wt.% MgO (Higgins and Schrage, 2010; Fantle and Higgins, 2014; Mavromatis et al., 2014) into a depleted mantle (MgO = 37.8 wt.%, McDonough and Sun, 1995; δ^{26}Mg = –0.25‰, Teng et al., 2010b).

6 Conclusions

High-precision Mg isotope analyses on the altered basalts and gabbros recovered from IODP 1256 Site on the Cocos plate in the eastern equatorial Pacific Ocean, demonstrate that the carbonate-barren AOC has mantle-like Mg isotopic compositions, with an average δ^{26}Mg of –0.25±0.11‰ (2SD, n = 43). This observation suggests that alteration of oceanic crust by seawater and hydrothermal fluids, regardless of alteration temperatures and water-rock ratios, is unlikely to modify Mg isotope ratios of basalts and gabbros at bulk-rock scale. This thus indicates that marine dolomite enriched in ^{24}Mg is the dominant sink for causing the heavier Mg isotopic compositions of seawater relative to global runoff.

Using different Mg isotopic compositions of hydrothermal sink Mg and marine dolomite, a simple model calculation suggests that 5 to 30% dolomite output is needed to sustain the oceanic Mg balance at steady state. Our results also suggests that recycling of carbonate-barren AOC has little effects on the Mg isotopic composition of the mantle at a global scale and that the light Mg isotopic compositions of the mantle at local scale must result from the contamination of Mg isotopically light carbonates.

Acknowledgements

This work is financially supported by grants from the National Science Foundation of China (NSFC) to S.G. Li (41230209), the Chinese Ministry of Science and Technology to J. Huang (2015CB856102), and the NSFC to J. Huang (41303015) and Y.L. Xiao (41273037). Thanks are due to Drs. G. Wörner and W.Y. Li for polishing English and discussions on early version of this manuscript. Help from T. Gao, Z.J. Li, and Z.Z. Wang during the course of Mg isotope analyses is thanked. We also gratefully acknowledge two anonymous reviewers for thorough and constructive reviews of the manuscript, and Dr. Andrew Kerr for the editorial handling.

References

Alt, J.C., Teagle, D.A.H., 1999. The uptake of carbon during alteration of ocean crust. Geochimica et Cosmochimica Acta 63, 1527-1535.

Alt, J.C., Teagle, D.A.H., 2000. Hydrothermal alteration and fluid fluxes in ophiolites and oceanic crust. Geological Society of America Special Papers 349, 273-282.

Alt, J.C., Teagle, D.A.H., Laverne, C., Vanko, D., Bach, W., Honnorez, J., Becker, K., Ayadi, M., Pezard, P.A., 1996. Ridge flank alteration of upper ocean crust in the eastern Pacific: A synthesis of results for volcanic rocks of holes 504B and 896A. Proceedings of the Ocean Drilling Program, Scientific Results 148, 435-452.

Alt, J.C., Laverne, C., Coggon, R.M., Teagle, D.A.H., Banerjee, N.R., Morgan, S., Smith-Duque, C.E., Harris, M., Galli, L., 2010. Subsurface structure of a submarine hydrothermal system in ocean crust formed at the East Pacific Rise, ODP/IODP Site 1256. Geochemistry, Geophysics, Geosystems 11, Q10010.

An, Y., Wu, F., Xiang, Y., Nan, X., Yu, X., Yang, J., Yu, H., Xie, L., Huang, F., 2014. High-Precision Mg Isotope Analyses of Low-Mg rocks by MC-ICP-MS. Chemical Geology 390, 9-21.

Azmy, K., Lavoie, D., Wang, Z., Brand, U., Al-Aasm, I., Jackson, S., Girard, I., 2013. Magnesium-isotope and REE compositions of Lower Ordovician carbonates from eastern Laurentia: implications for the origin of dolomites and limestones. Chemical Geology 356, 64-79.

Beinlich, A., Mavromatis, V., Austrheim, H., Oelkers, E.H., 2014. Inter-mineral Mg isotope fractionation during hydrothermal ultramafic rock alteration – Implications for the global Mg-cycle. Earth and Planetary Science Letters 392, 166-176.

Bolou-Bi, E.B., Vigier, N., Poszwa, A., Boudot, J.-P., Dambrine, E., 2012. Effects of biogeochemical processes on magnesium isotope variations in a forested catchment in the Vosges Mountains (France). Geochimica et Cosmochimica Acta 87, 341-355.

Bourdon, B., Tipper, E.T., Fitoussi, C., Stracke, A., 2010. Chondritic Mg isotope composition of the Earth. Geochimica et Cosmochimica Acta 74, 5069-5083.

Brenot, A., Cloquet, C., Vigier, N., Carignan, J., France-Lanord, C., 2008. Magnesium isotope systematics of the lithologically varied Moselle river basin, France. Geochimica et Cosmochimica Acta 72, 5070-5089.

Buhl, D., Immenhauser, A., Smeulders, G., Kabiri, L., Richter, D.K., 2007. Time series δ^{26}Mg analysis in speleothem calcite: Kinetic versus equilibrium fractionation, comparison with other proxies and implications for palaeoclimate research. Chemical Geology 244, 715-729.

Carpenter, J.H., Manella, M.E., 1973. Magnesium to chlorinity ratios in seawater. Journal of Geophysical Research 78, 3621-3626.

Chang, V.T.C., Makishima, A., Belshaw, N.S., O'Nions, R.K., 2003. Purification of Mg from low-Mg biogenic carbonates for isotope ratio determination using multiple collector ICP-MS. Journal of Analytical Atomic Spectrometry 18, 296-301.

Deines, P., 2002. The carbon isotope geochemistry of mantle xenoliths. Earth-Science Reviews 58, 247-278.

Eiler, J.M., 2001. Oxygen Isotope Variations of Basaltic Lavas and Upper Mantle Rocks. Reviews in Mineralogy and Geochemistry 43, 319-364.

Elderfield, H., Schultz, A., 1996. Mid-ocean ridge hydrothermal fluxes and the chemical composition of the ocean. Annual Review of Earth and Planetary Sciences 24, 191-224.

Elliott, T., Jeffcoate, A., Bouman, C., 2004. The terrestrial Li isotope cycle: light-weight constraints on mantle convection. Earth and Planetary Science Letters 220, 231-245.

Fantle, M.S., Higgins, J., 2014. The effects of diagenesis and dolomitization on Ca and Mg isotopes in marine platform carbonates: Implications for the geochemical cycles of Ca and Mg. Geochimica et Cosmochimica Acta.

Foster, G.L., Pogge von Strandmann, P.A.E., Rae, J.W.B., 2010. Boron and magnesium isotopic composition of seawater. Geochemistry, Geophysics, Geosystems 11, Q08015.

Galy, A., Belshaw, N.S., Halicz, L., O'Nions, R.K., 2001. High-precision measurement of magnesium isotopes by multiple-collector inductively coupled plasma mass spectrometry. International Journal of Mass Spectrometry 208, 89-98.

Galy, A., Bar-Matthews, M., Halicz, L., O'Nions, R.K., 2002. Mg isotopic composition of carbonate: insight from speleothem formation. Earth and Planetary Science Letters 201, 105-115.

Galy, A., Yoffe, O., Janney, P.E., Williams, R.W., Cloquet, C., Alard, O., Halicz, L., Wadhwa, M., Hutcheon, I.D., Ramon, E., Carignan, J., 2003. Magnesium isotope heterogeneity of the isotopic standard SRM980 and new reference materials for magnesium-isotope-ratio measurements. Journal of Analytical Atomic Spectrometry 18, 1352-1356.

Gao, Y.J., Vils, F., Cooper, K.M., Banerjee, N., Harris, M., Hoefs, J., Teagle, D.A.H., Casey, J.F., Elliott, T., Laverne, C., Alt, J.C., Muehlenbachs, K., 2012. Downhole variation of lithium and oxygen isotopic compositions of oceanic crust at East Pacific Rise, ODP Site 1256. Geochemistry, Geophysics, Geosystems 13.

Geske, A., Zorlu, J., Richter, D.K., Buhl, D., Niedermayr, A., Immenhauser, A., 2012. Impact of diagenesis and low grade metamorphosis on isotope (δ^{26}Mg, δ^{13}C, δ^{18}O and ^{87}Sr/^{86}Sr) and elemental (Ca, Mg, Mn, Fe and Sr) signatures of Triassic sabkha

dolomites. Chemical Geology 332-333, 45-64.
Geske, A., Goldstein, R.H., Mavromatis, V., Richter, D.K., Buhl, D., Kluge, T., John, C.M., Immenhauser, A., 2014. The magnesium isotope (δ^{26}Mg) signature of dolomites. Geochimica et Cosmochimica Acta.
Gregory, R.T., Taylor, H.P., 1981. An oxygen isotope profile in a section of Cretaceous oceanic crust, Samail Ophiolite, Oman: Evidence for δ18O buffering of the oceans by deep (>5 km) seawater-hydrothermal circulation at mid-ocean ridges. Journal of Geophysical Research: Solid Earth 86, 2737-2755.
Handler, M.R., Baker, J.A., Schiller, M., Bennett, V.C., Yaxley, G.M., 2009. Magnesium stable isotope composition of Earth's upper mantle. Earth and Planetary Science Letters 282, 306-313.
Harmon, R., Hoefs, J., 1995. Oxygen isotope heterogeneity of the mantle deduced from global ^{18}O systematics of basalts from different geotectonic settings. Contributions to Mineralogy and Petrology 120, 95-114.
Higgins, J.A., Schrag, D.P., 2010. Constraining magnesium cycling in marine sediments using magnesium isotopes. Geochimica et Cosmochimica Acta 74, 5039-5053.
Hippler, D., Buhl, D., Witbaard, R., Richter, D.K., Immenhauser, A., 2009. Towards a better understanding of magnesium-isotope ratios from marine skeletal carbonates. Geochimica et Cosmochimica Acta 73, 6134-6146.
Hofmann, A.W., 1988. Chemical differentiation of the Earth:the relationship between mantle, continental crust, and oceanic crust. Earth and Planetary Science Letters 90, 297-314.
Hofmann, A.W., 1997. Mantle geochemistry: the message from oceanic volcanism. Nature 385, 219-229.
Holland, H.D., 2005. Sea level, sediments and the composition of seawater. American Journal of Science 305, 220-239.
Huang, F., Glessner, J., Ianno, A., Lundstrom, C., Zhang, Z., 2009. Magnesium isotopic composition of igneous rock standards measured by MC-ICP-MS. Chemical Geology 268, 15-23.
Huang, F., Zhang, Z., Lundstrom, C.C., Zhi, X., 2011. Iron and magnesium isotopic compositions of peridotite xenoliths from Eastern China. Geochimica et Cosmochimica Acta 75, 3318-3334.
Huang, J., Li, S.-G., Xiao, Y., Ke, S., Li, W.-Y., Tian, Y., 2015. Origin of low δ^{26}Mg Cenozoic basalts from South China Block and their geodynamic implications. Geochimica et Cosmochimica Acta 164, 298-317.
Huang, K.-J., Teng, F.-Z., Wei, G.-J., Ma, J.-L., Bao, Z.-Y., 2012. Adsorption- and desorption-controlled magnesium isotope fractionation during extreme weathering of basalt in Hainan Island, China. Earth and Planetary Science Letters 359-360, 73-83.
Huang, K.-J., Teng, F.-Z., Elsenouy, A., Li, W.-Y., Bao, Z.-Y., 2013. Magnesium isotopic variations in loess: Origins and implications. Earth and Planetary Science Letters 374, 60-70.
Immenhauser, A., Buhl, D., Richter, D., Niedermayr, A., Riechelmann, D., Dietzel, M., Schulte, U., 2010. Magnesium-isotope fractionation during low-Mg calcite precipitation in a limestone cave-Field study and experiments. Geochimica et Cosmochimica Acta 74, 4346-4364.
Jacobson, A.D., Zhang, Z., Lundstrom, C., Huang, F., 2010. Behavior of Mg isotopes during dedolomitization in the Madison Aquifer, South Dakota. Earth and Planetary Science Letters 297, 446-452.
Kasemann, S.A., Pogge von Strandmann, P.A.E., Prave, A.R., Fallick, A.E., Elliott, T., Hoffmann, K.-H., 2014. Continental weathering following a Cryogenian glaciation: Evidence from calcium and magnesium isotopes. Earth and Planetary Science Letters 396, 66-77.
Koepke, J., Christie, D.M., Dziony, W., Holtz, F., Lattard, D., Maclennan, J., Park, S., Scheibner, B., Yamasaki, T., Yamazaki, S., 2008. Petrography of the dike-gabbro transition at IODP Site 1256 (equatorial Pacific): The evolution of the granoblastic dikes. Geochemistry, Geophysics, Geosystems 9, Q07O09.
Li, W.-Y., Teng, F.-Z., Ke, S., Rudnick, R.L., Gao, S., Wu, F.-Y., Chappell, B.W., 2010. Heterogeneous magnesium isotopic composition of the upper continental crust. Geochimica et Cosmochimica Acta 74, 6867-6884.
Li, W.-Y., Teng, F.-Z., Xiao, Y., Huang, J., 2011. High-temperature inter-mineral magnesium isotope fractionation in eclogite from the Dabie orogen, China. Earth and Planetary Science Letters 304, 224-230.
Li, W.-Y., Teng, F.-Z., Wing, B.A., Xiao, Y., 2014. Limited magnesium isotope fractionation during metamorphic dehydration in metapelites from the Onawa contact aureole, Maine. Geochemistry, Geophysics, Geosystems 15, 408-415.
Ling, M.-X., Sedaghatpour, F., Teng, F.-Z., Hays, P.D., Strauss, J., Sun, W., 2011. Homogeneous magnesium isotopic composition of seawater: an excellent geostandard for Mg isotope analysis. Rapid Communications in Mass Spectrometry 25, 2828-2836.
Liu, S.-A., Teng, F.-Z., He, Y., Ke, S., Li, S., 2010. Investigation of magnesium isotope fractionation during granite differentiation: Implication for Mg isotopic composition of the continental crust. Earth and Planetary Science Letters 297, 646-654.
Liu, S.-A., Teng, F.-Z., Yang, W., Wu, F.-Y., 2011. High-temperature inter-mineral magnesium isotope fractionation in mantle

xenoliths from the North China craton. Earth and Planetary Science Letters 308, 131-140.

Liu, X.-M., Teng, F.-Z., Rudnick, R.L., McDonough, W.F., Cummings, M.L., 2014. Massive magnesium depletion and isotope fractionation in weathered basalts. Geochimica et Cosmochimica Acta 135, 336-349.

Mavromatis, V., Meister, P., Oelkers, E.H., 2014. Using stable Mg isotopes to distinguish dolomite formation mechanisms: A case study from the Peru Margin. Chemical Geology 385, 84-91.

Muehlenbachs, K., Clayton, R.N., 1972. Oxygen Isotope Studies of Fresh and Weathered Submarine Basalts. Canadian Journal of Earth Sciences 9, 172-184.

Neo, N., Yamazaki, S., Miyashita, S., 2009. Data report: Bulk rock compositions of samples from the IODP Expedition 309/312 sample pool, ODP Hole 1256D. Superfast Spreading Rate Crust, Proceedings of the Integrated Ocean Drilling Program,309/312.

Pogge von Strandmann, P.A.E., Burton, K.W., James, R.H., van Calsteren, P., Gislason, S.R., Sigfússon, B., 2008. The influence of weathering processes on riverine magnesium isotopes in a basaltic terrain. Earth and Planetary Science Letters 276, 187-197.

Pogge von Strandmann, P.A.E., Elliott, T., Marschall, H.R., Coath, C., Lai, Y.-J., Jeffcoate, A.B., Ionov, D.A., 2011. Variations of Li and Mg isotope ratios in bulk chondrites and mantle xenoliths. Geochimica et Cosmochimica Acta 75, 5247-5268.

Pokrovsky, B.G., Mavromatis, V., Pokrovsky, O.S., 2011. Co-variation of Mg and C isotopes in late Precambrian carbonates of the Siberian Platform: A new tool for tracing the change in weathering regime? Chemical Geology 290, 67-74.

Rausch, S., Böhm, F., Bach, W., Klügel, A., Eisenhauer, A., 2013. Calcium carbonate veins in ocean crust record a threefold increase of seawater Mg/Ca in the past 30 million years. Earth and Planetary Science Letters 362, 215-224.

Riechelmann, S., Buhl, D., Schröder-Ritzrau, A., Spötl, C., Riechelmann, D.F.C., Richter, D.K., Kluge, T., Marx, T., Immenhauser, A., 2012. Hydrogeochemistry and fractionation pathways of Mg isotopes in a continental weathering system: Lessons from field experiments. Chemical Geology 300-301, 109-122.

Shen, B., Jacobsen, B., Lee, C.-T.A., Yin, Q.-Z., Morton, D.M., 2009. The Mg isotopic systematics of granitoids in continental arcs and implications for the role of chemical weathering in crust formation. Proceedings of the National Academy of Sciences 106, 20652-20657.

Staudigel, H., Hart, S.R., Schmincke, H.-U., Smith, B.M., 1989. Cretaceous ocean crust at DSDP Sites 417 and 418: Carbon uptake from weathering versus loss by magmatic outgassing. Geochimica et Cosmochimica Acta 53, 3091-3094.

Staudigel, H., Plank, T., White, B., Schmincke, H.-U., 1996, Geochemical Fluxes During Seafloor Alteration of the Basaltic Upper Oceanic Crust: DSDP Sites 417 and 418, Subduction Top to Bottom, American Geophysical Union, p. 19-38.

Teagle, D.A.H., Alt, J.C., Bach, W., Halliday, A.N., Erzinger, J., 1996. Alteration of upper ocean crust in a ridgeflank hydrothermal upflow zone: Mineral, chemical and isotopic constraints from ODP Hole 896A. Proceedings of the Ocean Drilling Program Scientific Results 148, 119-150.

Teagle, D.A.H., Alt, J.C., Umino, S., Miyashita, S., Banerjee, N.R., Wilson, D.S., Scientists, t.E., 2006. Superfast Spreading Rate Crust, Proc. Integr. Ocean Drill. Program, 309/312.

Telus, M., Dauphas, N., Moynier, F., Tissot, F.L.H., Teng, F.-Z., Nabelek, P.I., Craddock, P.R., Groat, L.A., 2012. Iron, zinc, magnesium and uranium isotopic fractionation during continental crust differentiation: The tale from migmatites, granitoids, and pegmatites. Geochimica et Cosmochimica Acta 97, 247-265.

Teng, F.-Z., Wadhwa, M., Helz, R.T., 2007. Investigation of magnesium isotope fractionation during basalt differentiation: Implications for a chondritic composition of the terrestrial mantle. Earth and Planetary Science Letters 261, 84-92.

Teng, F.-Z., Li, W.-Y., Rudnick, R.L., Gardner, L.R., 2010a. Contrasting lithium and magnesium isotope fractionation during continental weathering. Earth and Planetary Science Letters 300, 63-71.

Teng, F.-Z., Li, W.-Y., Ke, S., Marty, B., Dauphas, N., Huang, S., Wu, F.-Y., Pourmand, A., 2010b. Magnesium isotopic composition of the Earth and chondrites. Geochimica et Cosmochimica Acta 74, 4150-4166.

Tipper, E.T., Galy, A., Gaillardet, J., Bickle, M.J., Elderfield, H., Carder, E.A., 2006a. The magnesium isotope budget of the modern ocean: Constraints from riverine magnesium isotope ratios. Earth and Planetary Science Letters 250, 241-253.

Tipper, E.T., Bickle, M.J., Galy, A., West, A.J., Pomiès, C., Chapman, H.J., 2006b. The short term climatic sensitivity of carbonate and silicate weathering fluxes: Insight from seasonal variations in river chemistry. Geochimica et Cosmochimica Acta 70, 2737-2754.

Tipper, E.T., Gaillardet, J., Louvat, P., Capmas, F., White, A.F., 2010. Mg isotope constraints on soil pore-fluid chemistry: Evidence from Santa Cruz, California. Geochimica et Cosmochimica Acta 74, 3883-3896.

Tipper, E.T., Calmels, D., Gaillardet, J., Louvat, P., Capmas, F., Dubacq, B., 2012a. Positive correlation between Li and Mg isotope ratios in the river waters of the Mackenzie Basin challenges the interpretation of apparent isotopic fractionation during

weathering. Earth and Planetary Science Letters 333-334, 35-45.
Tipper, E.T., Lemarchand, E., Hindshaw, R.S., Reynolds, B.C., Bourdon, B., 2012b. Seasonal sensitivity of weathering processes: Hints from magnesium isotopes in a glacial stream. Chemical Geology 312-313, 80-92.
Tomascak, P.B., Langmuir, C.H., le Roux, P.J., Shirey, S.B., 2008. Lithium isotopes in global mid-ocean ridge basalts. Geochimica et Cosmochimica Acta 72, 1626-1637.
Wang, S.-J., Teng, F.-Z., Williams, H.M., Li, S.-G., 2012. Magnesium isotopic variations in cratonic eclogites: Origins and implications. Earth and Planetary Science Letters 359-360, 219-226.
Wang, S.-J., Teng, F.-Z., Li, S.-G., Hong, J.-A., 2014a. Magnesium isotopic systematics of mafic rocks during continental subduction. Geochimica et Cosmochimica Acta 143, 34-48.
Wang, S.-J., Teng, F.-Z., Li, S.-G., 2014b. Tracing carbonate-silicate interaction during subduction using magnesium and oxygen isotopes. Nature Communications (DOI: 10.1038/ncomms6328).
Wetzel, L.R., Shock, E.L., 2000. Distinguishing ultramafic-from basalt-hosted submarine hydrothermal systems by comparing calculated vent fluid compositions. Journal of Geophysical Research: Solid Earth 105, 8319-8340.
Wilkinson, B.H., Algeo, T.J., 1989. Sedimentary carbonate record of calcium-magnesium cycling. American Journal of Science 289, 1158-1194.
Wilson, D.S., 1996. Fastest known spreading on the Miocene Cocos-Pacific Plate Boundary. Geophysical Research Letters 23, 3003-3006.
Wilson, D.S., Teagle, D.A.H., Acton, G.D., 2003. Proceedings of the Ocean Drilling Program, Initial Reports, vol. 206, Ocean Drill. Program, College Station, Tex.
Wilson, D.S., Teagle, D.A.H., Alt, J.C., et al., 2006. Drilling to Gabbro in Intact Ocean Crust. Science 312, 1016-1020.
Wimpenny, J., Gíslason, S.R., James, R.H., Gannoun, A., Pogge Von Strandmann, P.A.E., Burton, K.W., 2010. The behaviour of Li and Mg isotopes during primary phase dissolution and secondary mineral formation in basalt. Geochimica et Cosmochimica Acta 74, 5259-5279.
Wombacher, F., Eisenhauer, A., Heuser, A., Weyer, S., 2009. Separation of Mg, Ca and Fe from geological reference materials for stable isotope ratio analyses by MC-ICP-MS and double-spike TIMS. Journal of Analytical Atomic Spectrometry 24, 627-636.
Wombacher, F., Eisenhauer, A., Böhm, F., Gussone, N., Regenberg, M., Dullo, W.C., Rüggeberg, A., 2011. Magnesium stable isotope fractionation in marine biogenic calcite and aragonite. Geochimica et Cosmochimica Acta 75, 5797-5818.
Xiao, Y., Teng, F.-Z., Zhang, H.-F., Yang, W., 2013. Large magnesium isotope fractionation in peridotite xenoliths from eastern North China craton: Product of melt-rock interaction. Geochimica et Cosmochimica Acta 115, 241-261.
Yang, W., Teng, F.-Z., Zhang, H.-F., 2009. Chondritic magnesium isotopic composition of the terrestrial mantle: A case study of peridotite xenoliths from the North China craton. Earth and Planetary Science Letters 288, 475-482.
Yang, W., Teng, F.-Z., Zhang, H.-F., Li, S.-G., 2012. Magnesium isotopic systematics of continental basalts from the North China craton: Implications for tracing subducted carbonate in the mantle. Chemical Geology 328, 185-194.
Young, E.D., Galy, A., 2004. The Isotope Geochemistry and Cosmochemistry of Magnesium. Reviews in Mineralogy and Geochemistry 55, 197-230.
Young, E.D., Tonui, E., Manning, C.E., Schauble, E., Macris, C.A., 2009. Spinel-olivine magnesium isotope thermometry in the mantle and implications for the Mg isotopic composition of Earth. Earth and Planetary Science Letters 288, 524-533.
Zindler, A., Hart, S., 1986. Chemical geodynamics. Annual Review of Earth and Planetary Sciences 14, 493-571.

The behavior of magnesium isotopes in low-grade metamorphosed mudrocks*

Shui-Jiong Wang[1, 2], Fang-Zhen Teng[2], Roberta L. Rudnick[3] and Shu-Guang Li[1, 4]

1. State Key Laboratory of Geological Processes and Mineral Resources, China University of Geosciences, Beijing 100083, China
2. Isotope Laboratory, Department of Earth and Space Sciences, University of Washington, Seattle, WA 98195-1310, USA
3. Geochemical Laboratory, Department of Geology, University of Maryland, College Park, MD 20742, USA
4. CAS Key Laboratory of Crust-Mantle Materials and Environments, School of Earth and Space Sciences, University of Science and Technology of China, Hefei 230026, Anhui, China

亮点介绍： 本文测定了英国加里东期三个下古生界盆地（北部湖区、南部湖区和南部高地）亚绿片岩相变质泥岩的 Mg 同位素组成，发现成岩作用及低变质作用过程不会改变泥岩 Mg 同位素组成及其碳酸盐成分与硅酸盐成分比例。通过对比分析含碳酸盐及不含碳酸盐的泥岩 Mg 同位素组成差异，结合稀酸淋洗实验，证明了碳酸盐加入能够降低泥岩的 Mg 同位素组成，而泥岩的硅酸盐组分的 Mg 同位素组成受伊利石/白云母和绿泥石的比例控制，并能够指示其原岩的 Mg 同位素组成。综上，Mg 同位素具有研究沉积成岩作用和示踪沉积物再循环的潜力。

Abstract Magnesium isotopic compositions of mudrocks metamorphosed at sub-greenschist facies from three lower Paleozoic basins (northern Lake District, southern Lake District, and Southern Uplands) in the British Caledonides were measured in order to understand the behavior of Mg isotopes during diagenesis and low-grade metamorphism. Carbonate-free mudrocks from the northern Lake District have heavy δ^{26}Mg values varying from –0.17 to +0.25. By contrast, Mg isotopic compositions of carbonate-bearing mudrocks from the southern Lake District and Southern Uplands vary more widely, with δ^{26}Mg ranging from –0.74 to –0.08. Acid leaching experiments on the latter show that the leachates have higher Ca/Al and Ca/K ratios than the residues due to the dissolution of leachable carbonates. The δ^{26}Mg values of leachates (–1.54 to –0.21) are always lower than the corresponding residues (δ^{26}Mg = –0.39 to +0.09), consistent with isotopically light Mg in carbonates. A rough, negative correlation between δ^{26}Mg and Mg/Al for the residual silicate fraction of mudrocks suggests that their Mg isotopic compositions are controlled by the relative proportion of illite/muscovite and chlorite. Global clastic sediments display highly variable Mg isotopic compositions that are negatively correlated with CaO/Al_2O_3 and CaO/TiO_2, implying that carbonates introduce light Mg isotopes to sediments, although the silicate end member itself has a wide range of δ^{26}Mg, depending on its mineralogy. Magnesium

* 本文发表在：Geochimica et Cosmochimica Acta, 2015, 165: 435-448

isotopic compositions of mudrocks, as well as their silicate and carbonate fractions, do not vary systemically as metamorphism proceeds from diagenesis to low-grade metamorphism, suggesting limited Mg isotope fractionation during low-temperature metamorphic dehydration (<300 ℃). The general decrease of Mg fraction (by mass) contributed by carbonate with increasing metamorphic grade suggests that dissolution or decomposition of carbonates during metamorphism expelled light Mg isotopes. Thus, the Mg isotopic compositions of the silicate fractions in clastic sediments more faithfully reflect their provenance signatures. Our study shows that Mg isotopes can be used to study sedimentary diagenesis, and Mg isotopes may prove a useful tracer of sediments recycled into the mantle given their heterogeneous δ^{26}Mg values.

Keywords Magnesium isotopes, metamorphic dehydration, mudrock, carbonate, leaching

1 Introduction

Magnesium (Mg), with its three isotopes of ^{24}Mg, ^{25}Mg and ^{26}Mg, is a soluble major element in Earth's mantle and crust. The significant mass difference among the three isotopes (e.g., >8% between ^{24}Mg and ^{26}Mg) leads to potentially large isotope fractionations associated with geological processes. The terrestrial mantle, as represented by peridotite xenoliths and oceanic basalts, displays a restricted range of Mg isotopic composition (Teng et al., 2007b, 2010a; Handler et al., 2009; Yang et al., 2009; Bourdon et al., 2010; Dauphas et al., 2010; Bizzarro et al., 2011; Huang et al., 2011; Pogge von Strandmann et al., 2011; Xiao et al., 2013; Lai et al., 2015), with the average δ^{26}Mg of −0.25±0.07 (2SD; $\delta^{26}\text{Mg} = [(^{26}\text{Mg}/^{24}\text{Mg})_{\text{sample}}/(^{26}\text{Mg}/^{24}\text{Mg})_{\text{DSM-3}}-1]\times1000$; Teng et al., 2010a). Although the average δ^{26}Mg value of bulk upper continental crust (−0.22) is estimated to be similar to the normal mantle value (Li et al., 2010), significant heterogeneity (varying by up to 7‰) has been documented in sedimentary rocks and soils (e.g., Galy et al., 2002; Tipper et al., 2006b; Pogge von Strandmann et al., 2008a; Immenhauser et al., 2010; Huang et al., 2013; Liu et al., 2014). For example, carbonate minerals have the lowest δ^{26}Mg of terrestrial rocks (e.g., Higgins and Schrag, 2010), whereas weathered regoliths have heavy δ^{26}Mg values up to +1.81 (e.g., Liu et al., 2014). With respect to the hydrosphere, seawater has a homogenous δ^{26}Mg around −0.83 (Foster et al., 2010; Ling et al., 2011), while global rivers have variable δ^{26}Mg values, with the average (−1.09) generally lighter than that of seawater (Tipper et al., 2006b, 2008; Brenot et al., 2008; Pogge von Strandmann et al., 2008b; Wimpenny et al., 2011). The large Mg isotopic variations observed in sedimentary rocks and the systematic Mg isotopic difference between lithosphere and hydrosphere are thought to result from low-temperature chemical weathering and sedimentation (Tipper et al., 2006a, b, 2008, 2010; Brenot et al., 2008; Teng et al., 2010b; Wimpenny et al., 2010, 2011, 2014a; Huang et al., 2012; Opfergelt et al., 2012, 2014; Pogge von Strandmann et al., 2008b, 2012; Liu et al., 2014; Ma et al., 2015).

Chemical weathering transports isotopically light Mg from silicate rocks to the hydrosphere, leaving heavier Mg isotopes in the weathered residues (e.g., Teng et al., 2010b; Tipper et al., 2010; Huang et al., 2012; Liu et al., 2014). However, although chemical weathering produces a wide range of δ^{26}Mg in the secondary minerals, this process cannot solely account for the Mg isotopic heterogeneity (−1.64 to +0.92) seen in clastic sediments (Li et al., 2010, 2014; Huang et al., 2013; Wimpenny et al., 2014b). Huang et al. (2013) and Wimpenny et al. (2014b) found that the presence of carbonate minerals in loess deposits exerts a large impact on their bulk Mg isotopic compositions. Because carbonate minerals contain Mg that is characteristically light, the addition of carbonates potentially introduces light Mg isotopes to the bulk clastic sediments. However, the influence of carbonates on the Mg isotopic compositions of water-lain clastic sediments has yet to be characterized. Moreover, clastic sediments commonly experience low-temperature metamorphism (<300 ℃; Wintsch and Kvale, 1994; Sutton and Land,

1996; Milliken, 2003; Merriman et al., 2009). As Mg is water soluble, Mg isotope frationation due to metamorphic dehydration might occur at low temperatures. Although previous studies found high-temperature (>300 ℃) metamorphic dehydration causes limited Mg isotope fractionation (Li et al., 2010, 2014; Teng et al., 2013; Wang et al., 2014a, 2015), the behavior of Mg isotopes during low-temperature metamorphism (e.g., diagenesis and low-grade metamorphism) remains unknown.

To address these questions, we carried out leaching experiments and Mg isotopic analyses on three suites of low-grade metamorphosed mudrocks from lower Paleozoic basins within the British Caledonides that had previously been analyzed for major, trace element and Nd, Sr, and Li isotopic compositions (Qiu et al., 2009), as well as their metamorphic mineralogy (Merriman, 2006). Mudrocks make up ~50% of sedimentary rocks, and represent the most typical clastic sediments on the Earth's surface (e.g., Taylor and McLennan, 1985), thus, they are suitable for studying the behavior of Mg isotopes in clastic sediments. We found large (~1‰) Mg isotopic variations in the mudrocks regardless of the metamorphic grade, and up to 1.59‰ Mg isotopic differences between residues and leachates. These findings suggest that diagenesis and low-grade metamorphic dehydration do not cause significant Mg isotope fractionation in bulk mudrocks, but the addition of carbonate minerals may impart light Mg isotopic signatures to the clastic sediments. Thus, only Mg isotopic compositions of the silicate fraction reflect the signature of their provenance.

2 Geological Background and Samples

Mudrocks were collected from three Ordovician to Silurian sedimentary basins in the British Caledonides (Fig. 1): the northern Lake District, southern Lake District, and Southern Uplands (Merriman et al., 2009). The northern Lake District basin was formed in an extensional setting on the southern margin of the Iapetus Ocean during the early to mid-Ordovician (Stone and Merriman, 2004). The southern Lake District was formed following the flexural subsidence of the crust when the Iapetus Ocean was closed during the late Ordovician and early Silurian, while the Southern Uplands basin was developed as an accretionary thrust complex at the Laurentian continental margin (Leggett et al., 1979; Stone et al., 1987; Kneller, 1991). Deposits in these basins are commonly turbidite-dominated, mudrock sequences that were overprinted by diagenesis and low-grade metamorphism (Merriman et al., 2009). They consist mainly of clay assemblages dominated by illite, muscovite and chlorite, with non-clay fractions composed of quartz (<40%), albite (<15%), and minor amounts (<5%) of dolomite, calcite, K-feldspar, hematite, pyrite, and trace amounts (<1%) of rutile or anatase (Merriman et al., 2009). Samples were selected from previous collections that were used to study the metamorphic patterns of the three basins (Fortey, 1989; Merriman and Roberts, 2000; Johnson et al., 2001), and later used to investigate their major/trace elements and Sr, Nd, and Li isotopic geochemistry (Merriman et al., 2009; Qiu et al., 2009).

The northern Lake District mudrocks are from the Skiddaw Group (Fig. 1) that was metamorphosed under relatively high heat flow conditions (30–50 ℃ km^{-1}) in an extensional setting (Stone and Merriman, 2004). The lower-grade mudrocks from the deep diagenetic zone and low anchizone are carbonate-free and contain authigenic illite as the major mineral phase, accompanied by variable amounts of intermediate Na/K-mica, illite-smectite (I-S) and chlorite (Merriman, 2006). The higher-grade mudrocks from the high anchizone and epizone consist mainly of authigenic muscovite and chlorite, with paragonite and intermediate Na/K-mica, and minor pyrophyllite, albite, rutile, and quartz (Merriman, 2006). The provenance of the northern Lake District mudrocks is a highly weathered ancient upper continental crust (Qiu et al., 2009).

The southern Lake District mudrocks are from the Windermere Supergroup (Fig.1), which was metamorphosed under low heat flow conditions (<20 ℃ km^{-1}) (Soper and Woodcock, 2003). Clay minerals consist of K-mica and chlorite, with minor corrensite. Paragonite and pyrophyllite are not recorded in these mudrocks, but carbonate minerals are usually present (Merriman, 2006). Both weathered upper continental crust and juvenile arc volcanic materials are present in the provenance of the southern Lake District basin (Qiu et al., 2009).

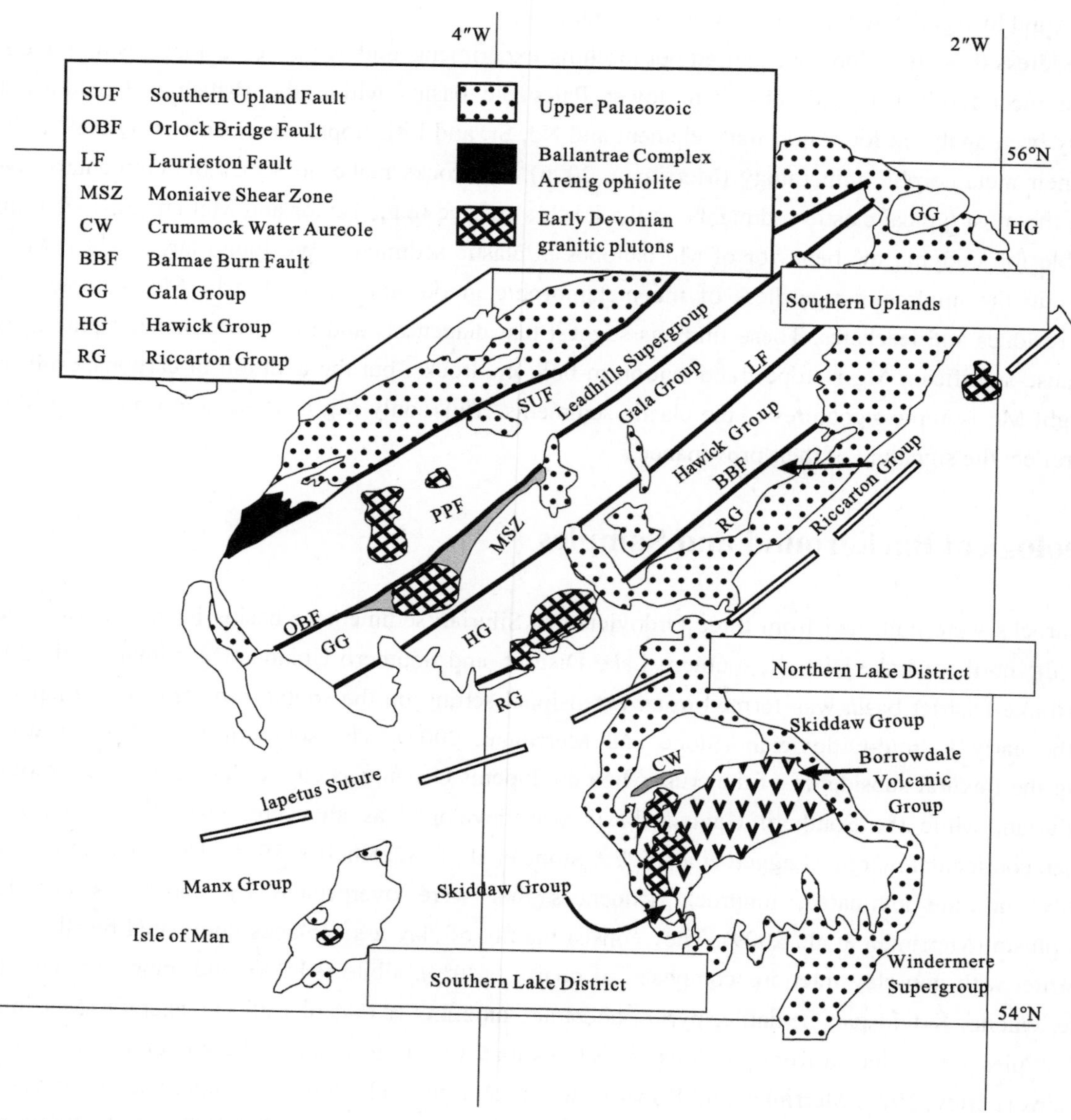

Fig. 1　Map showing the location of the three sedimentary basins from which the mudrocks derive: Southern Uplands, Northern Lake District, and Southern Lake District (after Merriman et al., 2009).

The Southern Uplands mudrocks were deposited in the Ordovician to Silurian (Fig.1), and were metamorphosed under a low geothermal gradient of <25 ℃ km^{-1} (Merriman and Roberts, 2000). Chlorite and K-mica are the dominant clay minerals. Minor amounts of albite, dolomite, corrensite, kaolinite, and intermediate Na/K-mica are also present (Merriman, 2006). The provenance of the Southern Uplands basin is, like that of the southern Lake District basin, composed of a mixture of arc lavas and weathered upper continental crust (Qiu et al., 2009).

3 Methods

All experiments were carried out at the Isotope Laboratory of the University of Washington, Seattle. For bulk rock powders, approximately two to six mg of samples was weighed, and treated sequentially with Optima-grade HF-HNO_3, HNO_3-HCl, and HNO_3. After complete dissolution, the samples were evaporated to dryness at ~160 ℃, and finally taken up in 1 N HNO_3 for chromatographic separation.

Leaching experiments were carried out on mudrocks from the southern Lake District and Southern Uplands to remove carbonate minerals. Dilute acetic and hydrochloric acids are commonly used in leaching experiments (e.g., Ostrom, 1961). Since diluted acetic acid may not completely dissolve dolomite minerals in sediments (Wimpenny et al., 2014b), we also used diluted hydrochloric acid. Previous studies suggest that 0.3 N HCl may have a negligible effect on either well or poorly crystallized illite and chlorite that are the two major Mg-bearing clays in the mudrocks (Ostrom, 1961). Therefore, we used 0.3 N HCl in our leaching experiments. For each sample, approximately 12 to 24 mg of rock powder was immersed in 10 ml of 0.3 N HCl at room temperature. The slurries were ultra-sonicated for ~45 minutes and then centrifuged to separate the supernatant and the residue. After separation, the supernatants were evaporated to dryness at ~160 ℃, and re-dissolved in 12 N Optima-grade HCl. The residues were cleaned using Milli-Q water (18.2 MΩ cm) three times before being dissolved in a mixture of Optima-grade HF-HNO_3-HCl acids.

The separation of Mg was achieved by cation exchange chromatography using Bio-Rad 200–400 mesh AG50W-X8 resin in 1 N HNO_3 (Teng et al., 2007b, 2010a; Yang et al., 2009; Li et al., 2010). An additional chromatographic step was processed for the leachates to separate Mg from Ca using Bio-Rad 200–400 mesh AG50W-X12 resin in 12 N HCl (Ling et al., 2013; Wang et al., 2014b). Three standards, Kilbourne Hole (KH) olivine, San Carlos (SC) olivine, and seawater, were processed together with samples for each batch of column chemistry. The same column procedure was performed twice to obtain pure Mg solutions for mass spectrometry. The total procedural blank is <10 ng, which represents <0.1% of the Mg loaded on the column (Teng et al., 2010a).

Magnesium isotopic ratios were determined using the standard-sample bracketing protocol on a *Nu Plasma* MC-ICPMS with a "wet" plasma introduction system (Teng and Yang, 2014). The ^{26}Mg, ^{25}Mg and ^{24}Mg were measured simultaneously in separate Faraday cups (H5, Ax, and L4). The background Mg signal for ^{24}Mg was $<10^{-4}$ V, which is negligible relative to the sample signals of 3-4 V. Magnesium isotopic results are reported in δ notation in per mil relative to DSM-3: $\delta^{x}Mg = [(^{x}Mg/^{24}Mg)_{sample}/(^{x}Mg/^{24}Mg)_{DSM\text{-}3}-1] \times 1000$, where x refers to mass 25 or 26. Three in-house standards were analyzed during the course of this study: KH olivine, SC olivine and seawater, and yielded average $\delta^{26}Mg$ values of -0.25 ± 0.03 (2SD; n=3), -0.25 ± 0.04 (2SD; n=3), and -0.84 ± 0.03 (2SD; n=6), respectively (Supplementary Table 1), which are in agreement with previously reported values (e.g., Yang et al., 2009; Foster et al., 2010; Li et al., 2010; Teng et al., 2010a, 2015; Ling et al., 2011).

The Ca/Al, Ca/K, and Mg/Al ratios of the leachates and residues were determined on the *Nu Plasma* MC-ICPMS. Fractions of the dissolved aliquots of the leachate and residue solutions were diluted in 3% HNO_3 prior to analysis. Four gravimetrically prepared SPEX Claritas™ ICPMS elemental standard solutions with a wide range of elemental ratios (covering the ratios of unknowing samples being analyzed) were analyzed to generate a calibration curve. This set of standard solutions was analyzed several times to verify that the calibration curve had remained unchanged during the course of the analyses of sample solutions. Five rock standards with known elemental ratios, including a basalt from Nancy, France (BR), three Chinese reference materials (GBW07112 gabbro, GBW07111 granodiorite, and GBW07103 granite), and a USGS shale standard from Wyoming, USA (SCo-1), were also analyzed together with the sample solutions, to monitor accuracy and precision. The uncertainty of Ca/Al, Ca/K and Mg/Al ratios are better than 10% (Supplementary Table 2).

4 Results

Major elemental ratios and Mg isotopic data of bulk rocks, leachates and residues are reported in Table 1. Magnesium isotopic compositions of all samples fall on a single mass-dependent fractionation line on the three-isotope diagram with a slope of 0.510 (not shown).

The $\delta^{26}Mg$ of the northern Lake District mudrocks vary from −0.17 to +0.25 (Fig. 2). Mudrocks from the southern Lake District and Southern Uplands have more variable $\delta^{26}Mg$, ranging from −0.74 to −0.09 and from −0.74 to −0.08 (Fig. 2), respectively.

The acid leachates contain considerable amounts of Al and K in addition to Mg and Ca (Table 1). Leachates of the southern Lake District mudrocks have Ca/Al of 0.36−8.40 and Ca/K of 0.79−59.3 (Fig. 3). The corresponding residues have significantly lower Ca/Al and Ca/K ratios of 0.01−0.14 and 0.03−0.47 (Fig. 3), respectively. Similarly, leachates of the Southern Uplands mudrocks have high Ca/Al of 0.10−4.77 and Ca/K of 0.13−11.6 (Fig. 3); whereas the residues have low Ca/Al of 0.004−0.10 and Ca/K of 0.01−0.27 (Fig. 3).

The Mg isotopic compositions of the leachates are always lighter than the residues (Fig. 4). Leachate $\delta^{26}Mg$ values range from −1.54 to −0.84 for the southern Lake District mudrocks and from −1.53 to −0.21 for the Southern Uplands mudrocks; residue $\delta^{26}Mg$ values vary from −0.39 to +0.05 for the southern Lake District mudrocks and from −0.32 to +0.09 for the Southern Uplands mudrocks (Fig. 4). Correspondingly, Mg isotopic differences between the residue and leachate (expressed as $\Delta^{26}Mg_{residue\text{-}leachate} = \delta^{26}Mg_{residue} - \delta^{26}Mg_{leachate}$) are in the range of 0.05 ~ 1.59‰.

Table 1　Magnesium isotopic compositions and elemental ratios for the bulk rock, leachate and residue of mudrocks from the northern Lake District, southern Lake District and Southern Uplands

Sample		$\delta^{26}Mg$	2SD	$\delta^{25}Mg$	2SD	KI	CIA	Mg/Al	Ca/Al	Ca/K
Northern Lake District										
LC348	Bulk-rock	−0.09	0.06	−0.03	0.03	0.2	79	0.08	0.02	0.06
LC199o	Bulk-rock	−0.13	0.06	−0.05	0.04	0.22	80	0.09	0.02	0.06
LC142	Bulk-rock	−0.04	0.06	−0.03	0.03	0.32	80	0.14	0.02	0.08
LC521r	Bulk-rock	+0.02	0.07	+0.02	0.05	0.39	81	0.09	0.02	0.08
LC482r	Bulk-rock	−0.13	0.07	−0.04	0.05	0.4	85	0.08	0.01	0.03
LC507	Bulk-rock	+0.25	0.07	+0.16	0.05	0.46	83	0.11	0.02	0.09
LC434	Bulk-rock	−0.17	0.06	−0.08	0.03	0.51	82	0.1	0.01	0.06
LC495r	Bulk-rock	−0.10	0.07	−0.06	0.05	0.63	81	0.07	0.01	0.06
Southern Lake District										
LC940	Bulk-rock	−0.36	0.04	−0.19	0.03	0.26	61	0.3	0.29	0.88
	Residue	−0.31	0.07	−0.19	0.07			0.27	0.05	0.16
	Leachate	−0.85	0.10	−0.43	0.08			0.34	3.4	12.24
	Leachate[R]	−0.83	0.07	−0.41	0.06					
	Average	−0.84	0.06	−0.42	0.05					
SH24o	Bulk-rock	−0.28	0.06	−0.15	0.03	0.27	73	0.31	0.02	0.06
	Residue	−0.26	0.07	−0.15	0.06			0.28	0.01	0.03
	Leachate	−1.54	0.10	−0.76	0.07			0.33	0.63	0.79

Continued

Sample		$\delta^{26}Mg$	2SD	$\delta^{25}Mg$	2SD	KI	CIA	Mg/Al	Ca/Al	Ca/K
SH19	Bulk-rock	−0.33	0.04	−0.15	0.03	0.29	71	0.3	0.06	0.17
	Residue	−0.25	0.07	−0.11	0.06			0.21	0.03	0.1
	Leachate	−0.76	0.10	−0.39	0.07			0.32	0.79	1.71
SH22	Bulk-rock	−0.46	0.06	−0.23	0.03	0.29	71	0.31	0.04	0.12
	Residue	−0.39	0.07	−0.2	0.07			0.31	0.02	0.06
	Leachate	−1.03	0.07	−0.53	0.06			0.32	0.71	1.79
SH59	Bulk-rock	−0.32	0.06	−0.12	0.03	0.33	66	0.3	0.21	0.61
	Residue	−0.09	0.07	−0.03	0.06			0.27	0.08	0.24
	Leachate	−1.40	0.10	−0.72	0.07			0.58	4.82	19.85
LC1606r,o	Bulk-rock	−0.31	0.07	−0.17	0.05	0.34	72	0.3	0.02	0.05
	Residue	−0.29	0.07	−0.14	0.06			0.28	0.01	0.04
	Leachate	−0.84	0.10	−0.42	0.07			0.32	0.36	0.54
LC1570r,o	Bulk-rock	−0.48	0.06	−0.24	0.03	0.4	74	0.15	2.54	10.18
	Residue	+0.05	0.10	+0.01	0.08			0.17	0.14	0.47
	Leachate	−1.54	0.07	−0.77	0.06			0.16	8.4	59.33
LC1617	Bulk-rock	−0.75	0.05	−0.39	0.02	0.45	68	0.37	0.53	1.48
	Bulk-rock [R]	−0.71	0.07	−0.36	0.04					
	Average	−0.74	0.04	−0.38	0.02					
	Residue	−0.25	0.07	−0.11	0.07			0.25	0.06	0.16
	Leachate	−1.31	0.10	−0.66	0.08			0.72	2.38	12.39
	Leachate [R]	−1.30	0.07	−0.62	0.06					
	Average	−1.30	0.06	−0.63	0.05					
LC1618r,o	Bulk-rock	−0.09	0.06	−0.03	0.04	0.66	74	0.13	0.32	1.24
	Residue	+0.03	0.07	+0.01	0.06			0.12	0.06	0.22
	Leachate	−1.00	0.10	−0.52	0.07			0.12	2.39	8.04
Southern Uplands										
BRS781r	Bulk-rock	−0.24	0.07	−0.12	0.05	0.2	72	0.34	0.02	0.05
	Residue	−0.21	0.07	−0.08	0.06			0.26	0.01	0.03
	Leachate	−0.25	0.10	−0.10	0.07			0.18	0.18	0.26
BRS790r	Bulk-rock	−0.26	0.07	−0.12	0.05	0.2	63	0.32	0.31	0.85
	Residue	−0.24	0.07	−0.12	0.06			0.3	0.04	0.11
	Leachate	−0.45	0.10	−0.21	0.07			0.31	4.77	9.25
	Leachate [R]	−0.47	0.07	−0.25	0.06					
	Average	−0.46	0.06	−0.23	0.05					
BRS807r	Bulk-rock	−0.12	0.06	−0.05	0.04	0.22	69	0.32	0.04	0.11
	Residue	−0.08	0.07	−0.07	0.06			0.22	0.03	0.09
	Leachate	−0.58	0.10	−0.29	0.07			0.2	1.58	2.24
	Leachate [R]	−0.56	0.07	−0.25	0.06					
	Average	−0.57	0.06	−0.27	0.05					
BRS824r	Bulk-rock	−0.20	0.06	−0.09	0.04	0.23	73	0.26	0.03	0.08
	Residue	−0.09	0.07	−0.03	0.06			0.25	0.01	0.04
	Residue [R]	−0.13	0.08	−0.05	0.05			0.22	0.4	0.68
	Average	−0.11	0.05	−0.04	0.04					
	Leachate	−0.52	0.07	−0.21	0.06					

Continued

Sample		$\delta^{26}Mg$	2SD	$\delta^{25}Mg$	2SD	KI	CIA	Mg/Al	Ca/Al	Ca/K
BRS882	Bulk-rock	−0.36	0.07	−0.18	0.05	0.28	69	0.3	0.03	0.07
	Residue	+0.03	0.07	+0.05	0.06			0.28	0.02	0.04
	Leachate	−1.17	0.10	−0.58	0.08			0.31	0.43	0.66
BRS879r	Bulk-rock	−0.16	0.06	−0.06	0.04	0.29	71	0.28	0.05	0.11
	Residue	−0.32	0.07	−0.15	0.06			0.28	0.05	0.12
	Leachate	−0.66	0.10	−0.34	0.07			0.16	0.22	0.71
BRS753r	Bulk-rock	−0.15	0.06	−0.06	0.04	0.32	73	0.25	0.02	0.06
	Residue	−0.24	0.07	−0.1	0.06			0.24	0.01	0.05
	Residue [R]	−0.30	0.08	−0.18	0.05					
	Average	−0.27	0.05	−0.15	0.04					
	Leachate	−0.44	0.10	−0.23	0.07			0.12	0.24	0.68
	Leachate [R]	−0.45	0.08	−0.22	0.05					
	Average	−0.45	0.06	−0.23	0.04					
BRS1028	Bulk-rock	−0.54	0.07	−0.29	0.05	0.44	74	0.22	0.24	0.61
	Residue	−0.20	0.07	−0.06	0.06			0.14	0.1	0.27
	Residue [R]	−0.14	0.08	−0.08	0.05					
	Average	−0.18	0.05	−0.07	0.04					
	Leachate	−1.02	0.10	−0.55	0.08			0.7	1.73	7.29
BRS742r,o	Bulk-rock	−0.11	0.06	−0.07	0.04	0.45	73	0.16	0.01	0.04
	Residue	+0.08	0.10	+0.01	0.07			0.16	0.01	0.02
	Residue [R]	+0.09	0.07	+0.02	0.06					
	Average	+0.09	0.06	+0.02	0.05					
	Leachate	−0.46	0.10	−0.30	0.07			0.2	0.12	0.25
BRS829r,o	Bulk-rock	−0.08	0.06	−0.02	0.04	0.48	73	0.18	0.01	0.02
	Residue	−0.07	0.10	−0.05	0.07			0.17	0	0.01
	Leachate	−0.21	0.07	−0.07	0.06			0.19	0.1	0.13
	Leachate [R]	−0.21	0.08	−0.05	0.05					
	Average	−0.21	0.05	−0.06	0.04					
BRS710r, o	Bulk-rock	−0.74	0.06	−0.38	0.04	0.5	56	0.39	0.4	1.45
	Residue	−0.12	0.10	−0.09	0.08			0.26	0.04	0.17
	Residue [R]	−0.07	0.07	−0.06	0.06					
	Average	−0.09	0.06	−0.07	0.05					
	Leachate	−1.49	0.10	−0.76	0.08			1.85	3.92	11.59
	Leachate [R]	−1.54	0.07	−0.76	0.06					
	Average	−1.53	0.06	−0.76	0.05					

KI values are from Merriman et al. (2009); Bulk-rock major elemental ratios are from Qiu et al. (2009), and elemental ratios of leachates and residues are from this study.

r: Re-sample from the original sample of Merriman et al. (2009)

o: Sample with organic carbon

Average = weighted average value;

2SD = 2 times the standard deviation of the population of n (n>20) repeated measurements of the standards during an analytical session.

[R] Repeat column chemistry and instrumental analysis.

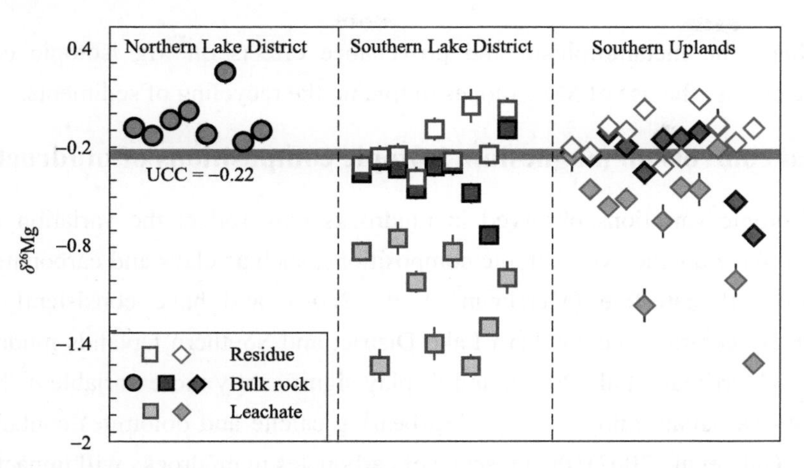

Fig. 2 δ^{26}Mg of bulk-rocks, leachates and residues of mudrocks from the northern Lake District, southern Lake District and Southern Uplands. The gray line represents the δ^{26}Mg of average upper continental crust (UCC = −0.22; Li et al., 2010). Data are reported in Table 1.

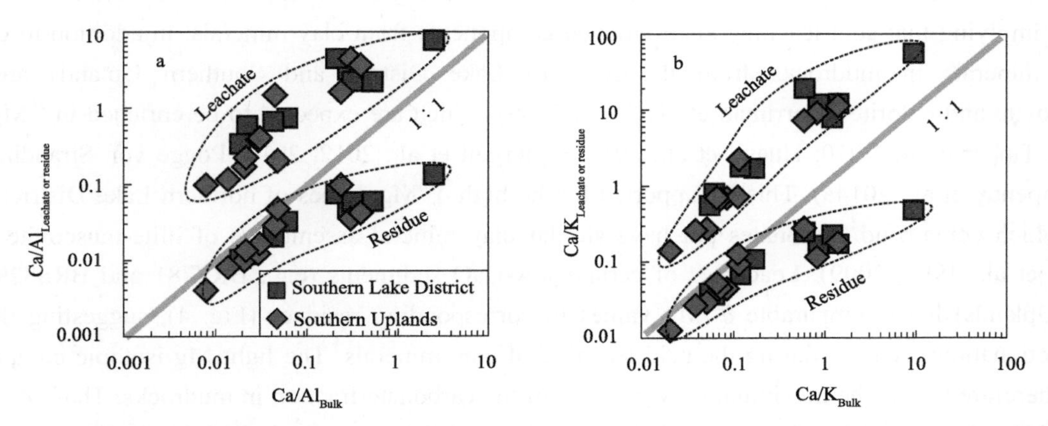

Fig. 3 Ca/Al (a) and Ca/K (b) ratios in leachates and residues versus that in the bulk rocks. Data are reported in Table 1.

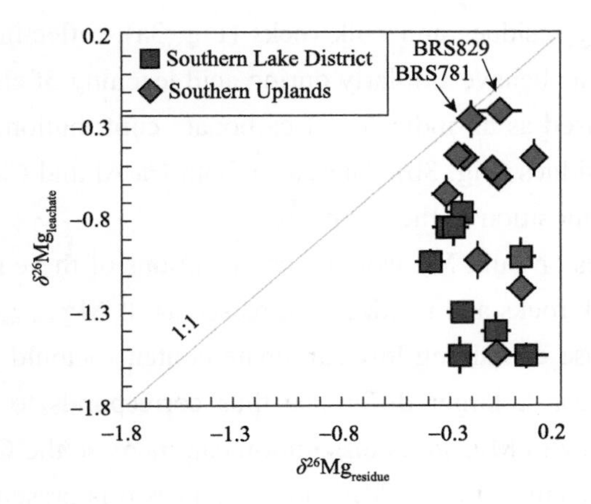

Fig. 4 Comparison of δ^{26}Mg values between leachates ($\delta^{26}Mg_{leachate}$) and residues ($\delta^{26}Mg_{residue}$). Data are reported in Table 1.

5 Discussion

In this section, we first examine the mineralogical controls on Mg isotopic compositions of the clastic sediments; then evaluate the metamorphism and provenance effects on Mg isotopic compositions of these mudrocks. Finally, we discuss the use of Mg isotopes in tracing the recycling of sediments.

5.1 Mineralogical controls on magnesium isotopic compositions of mudrocks

The large Mg isotopic variations observed in mudrocks may reflect the variation in the proportions of Mg-bearing phases that have distinct Mg isotopic compositions, such as clays and carbonates. The northern Lake District mudrocks are carbonate-free (Merriman et al., 2006), and have consistently heavy Mg isotopic compositions (Fig. 2). By contrast, the southern Lake District and Southern Uplands mudrocks contain variable amounts of carbonates (Merriman et al., 2006), and display significantly more variable δ^{26}Mg, extending to very low values (Fig. 2). As carbonate minerals (e.g., Mg-bearing calcite and dolomite) contain variable Mg that is isotopically light (e.g., Galy et al., 2002), the presence of carbonates in mudrocks will impact bulk δ^{26}Mg values.

Our acid leaching experiments show that Mg isotopic compositions of the leachates are always lighter than the residues (Fig. 4), consistent with isotopically light Mg leached from carbonates (Wimpenny et al., 2014b). Nonetheless, these leachates also contain considerable amounts of Al and K, which must derive from the clays (Table. 1), implying that acid leaching also removed components from clay minerals, in addition to carbonates. The clay minerals in mudrocks from the southern Lake District and Southern Uplands are Mg-rich illite-muscovite and chlorite (Merriman et al., 1995, 2009), which are expected to be enriched in ^{26}Mg (Teng et al., 2010b; Tipper et al., 2010; Huang et al., 2012; Opfergelt et al., 2012, 2014; Pogge von Strandmann et al., 2012; Wimpenny et al., 2014a). This is supported by the high δ^{26}Mg values of northern Lake District mudrocks (Fig. 2), which contain no carbonates but have similar clay mineral assemblage of illite-muscovite + chlorite (Merriman et al., 1995, 2009). Leachates of certain low-CaO sediments (e.g., BRS781 and BRS829 from the Southern Uplands) have comparable δ^{26}Mg values to corresponding residues (Fig. 4), suggesting that no Mg isotope fractionation occurred during the acid leaching of clay minerals. The light Mg isotopic compositions of leachates therefore reflect the maximum δ^{26}Mg values of the carbonate fraction in mudrocks. These δ^{26}Mg values also fall within the range of carbonate leachates from loess (Wimpenny et al., 2014b). Further support for the contribution of carbonate to the leachate comes from the elemental ratios. The leachates have significantly higher Ca/Al ratios than corresponding residues and bulk rocks (Fig. 3a), reflecting the preponderance of carbonates. Alkali and alkaline earth elements behave similarly during acid leaching of clay minerals (e.g., Grim, 1953); thus, the Ca/K ratio of leachates is used as an indicator of carbonate contribution. Likewise, leachate Ca/K ratios are significantly higher than the residues (Fig. 3b). Therefore, both Ca/Al and Ca/K ratios indicate the dominance of carbonate in controlling the composition of the leachates.

The influence of carbonates on bulk Mg isotopic compositions of these mudrocks is evaluated using the Mg isotopic difference between bulk rocks and residues (expressed as $\Delta^{26}\mathrm{Mg}_{\mathrm{residual\text{-}bulk}} = \delta^{26}\mathrm{Mg}_{\mathrm{residual}} - \delta^{26}\mathrm{Mg}_{\mathrm{bulk}}$). The carbonate-free sediments or those containing low carbonate contents should yield similar δ^{26}Mg values between bulk rocks and residues (Fig. 2). A larger difference thus corresponds to higher Mg fraction contributed by carbonate minerals. Calcite is low in Mg, so its dissolution can increase the $\mathrm{Ca/Al}_{\mathrm{leachate}}$ but may not significantly influence the bulk Mg isotopic composition or $\mathrm{Mg/Al}_{\mathrm{leachate}}$ unless it is present in large quantities (Fig. 5a, b). By contrast, dolomite, because of its high Mg and Ca concentrations, can have larger impacts on the bulk Mg isotopic composition, as well as $\mathrm{Mg/Al}_{\mathrm{leachate}}$ and $\mathrm{Ca/Al}_{\mathrm{leachate}}$ (Fig. 5a, b).

Residues from acid leaching represent the silicate fraction. As clay assemblages dominate the Mg budget of silicate fraction of these mudrocks, the Mg isotopic variations in residues reflect the relative proportions of illite/muscovite and chlorite that are produced by two parallel clay metamorphic mineral reaction series (Merriman, 2006): (1) smectite → mixed-layer illite/smectite → illite → muscovite; and (2) smectite →

mixed-layer chlorite/smectite → chlorite. Chlorite has high Mg/Al of 0.8–1.1, whereas illite and muscovite have low Mg/Al of 0.06–0.22 (Merriman et al., 1995). The $Mg/Al_{residue}$ is thus an indicator of the varying proportions of illite/muscovite vs. chlorite (Fig. 6). The carbonate-free mudrocks from the northern Lake District have the lowest Mg/Al ratios (Fig. 6), which is consistent with their high modal abundance of illite and muscovite. The heavy Mg isotopic compositions suggest enrichment of ^{26}Mg in illite and muscovite. By contrast, residues of mudrocks from southern Lake District and Southern Uplands have generally higher and more variable Mg/Al ratios, which show a negative correlation with δ^{26}Mg (Fig. 6). The correlation implies that illite and muscovite are isotopically heavier than chlorite in Mg isotopes.

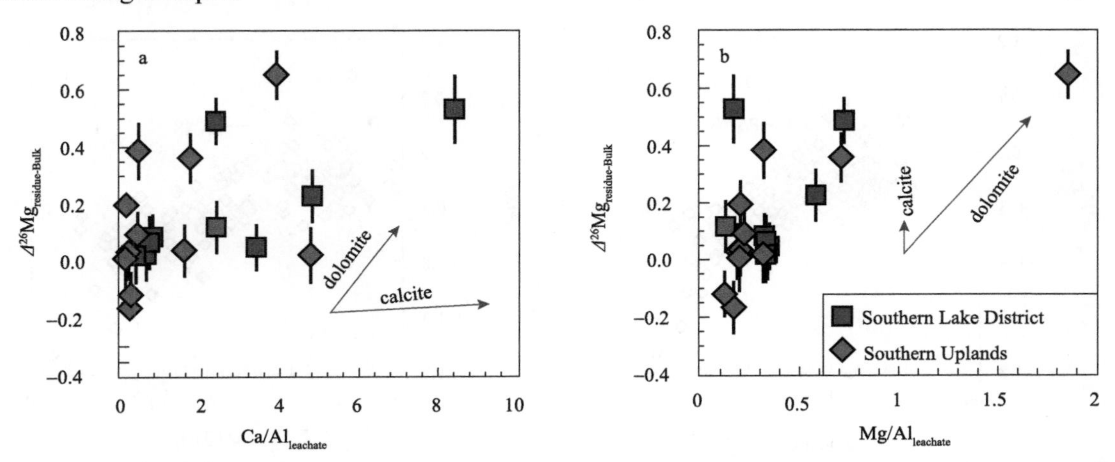

Fig. 5 Variation of Mg isotopic difference between bulk rock and residue ($\Delta^{26}Mg_{residue\text{-}bulk} = \delta^{26}Mg_{residue} - \delta^{26}Mg_{bulk}$) as a function of (a) leachate's Ca/Al ratios ($Ca/Al_{leachate}$), and (b) leachate's Mg/Al ratios ($Mg/Al_{leachate}$). The arrows indicate the trends created by calcite or dolomite dissolution. Data are reported in Table 1.

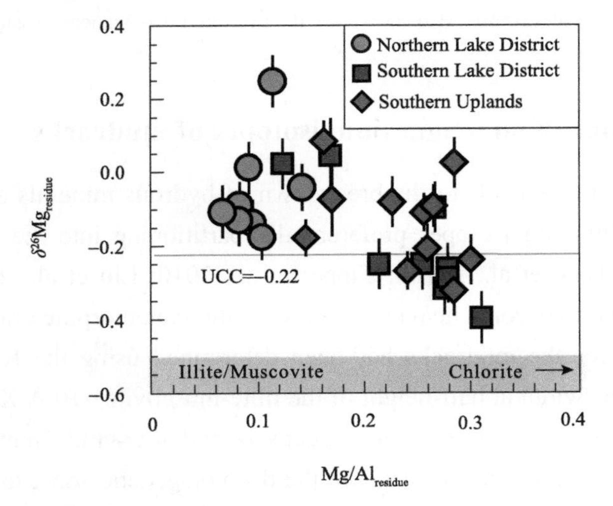

Fig. 6 Variation of $\delta^{26}Mg_{residue}$ as a function of residue's Mg/Al ratios ($Mg/Al_{residue}$). Average upper continental crust δ^{26}Mg shown by the gray line (Li et al., 2010). Data are reported in Table 1.

Overall, the relative abundances of illite/muscovite and chlorite control the Mg isotopic compositions of the silicate fraction of these mudrocks; and the presence of carbonate minerals introduces isotopically light Mg to the bulk sediments. This is further highlighted when all published δ^{26}Mg values for clastic sedimentary rocks are compiled (Fig. 7). The CaO/Al_2O_3 and CaO/TiO_2 ratios are indicators of carbonate proportions in sediments, as they are extremely high in carbonate minerals but low in clays. Globally, δ^{26}Mg values of clastic sedimentary rocks show rough, negative correlations with CaO/Al_2O_3 and CaO/TiO_2 (Fig. 7). In general, sediments with low

CaO/Al_2O_3 and CaO/TiO_2 ratios, for instance, lower than the average Post-Archean Australian Shales (PAAS), have variable Mg isotopic compositions, but are mostly heavier than the average value for bulk upper continental crust (−0.22; Fig. 7). The heterogeneity of $\delta^{26}Mg$ may be caused by the mineralogical variation of silicate phases. On the other hand, sediments with high CaO/Al_2O_3 and CaO/TiO_2 ratios have lighter Mg isotopic compositions, reflecting the incorporation of carbonate phases in these rocks (Fig. 7). While previous studies have shown clear evidence that Mg isotopic compositions of loess sediments are influenced by the presence of carbonate minerals (Huang et al., 2013; Wimpenny et al., 2014b), our results indicate that Mg isotopic compositions of water-lain sediments are also controlled by silicate-carbonate mixing.

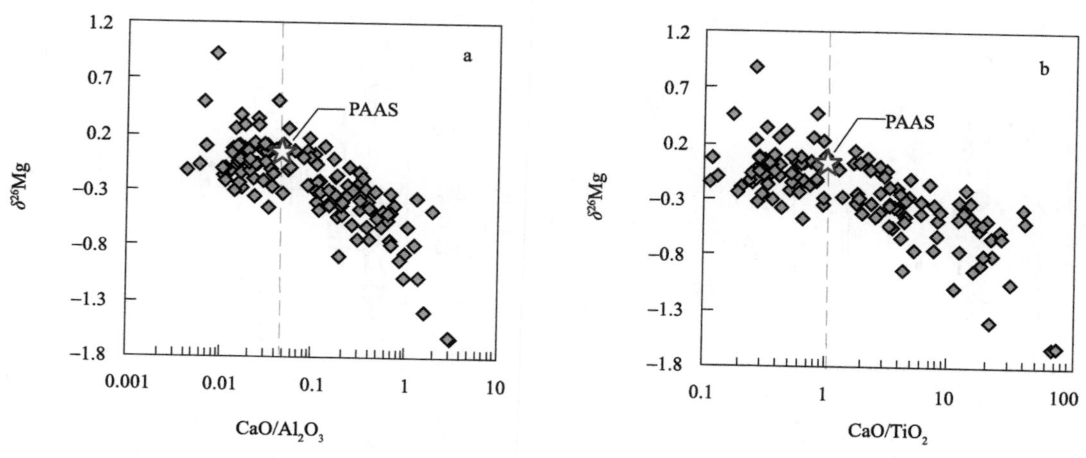

Fig. 7 Plot of $\delta^{26}Mg$ of clastic sedimentary rocks versus (a) CaO/Al_2O_3 and (b) CaO/TiO_2. The Mg isotopic data for clastic sedimentary rocks are from Wombacher et al. (2009), Li et al. (2010, 2014), Huang et al. (2013), Wimpenny et al. (2014b), and this study; The corresponding major elemental data are from Gallet et al. (1998), Gao et al. (1998), Jahn et al. (2001), Nance and Taylor (1976), Taylor et al. (1983), and Qiu et al. (2009). Also shown are the average Post-Archean Australian Shales (PAAS) from Li et al. (2010). See text for discussion.

5.2 Effects of metamorphism on magnesium isotopes of mudrocks

Metamorphism of crustal rocks leads to the breakdown of hydrous minerals and release of hydrous fluids. Magnesium is soluble with light Mg isotopes preferentially partitioning into the fluid during low-temperature water-rock interactions (e.g., Teng et al., 2010b; Tipper et al., 2010; Liu et al., 2014). Thus, $\delta^{26}Mg$ values of metamorphic rocks are expected to become heavier with increasing metamorphic grade.

The metamorphic grade for the mudrocks has been determined using the Kübler index (KI), which is a measure of small changes in the width at half-height of the illite-muscovite ~10 Å X-Ray Diffraction (XRD) peak (Peacor, 1992). This width varies in response to diagenesis and low-grade metamorphism. With increasing metamorphic grade, KI values decrease from >0.42 for the deep diagenetic zone, to 0.42–0.25 for the anchizone, and finally to <0.25 for the epizone/lower greenschist-facies (Merriman and Peacor, 1999). The KI values of the mudrocks from the British Caledonides range from 0.20 to 0.63 for the northern Lake District, from 0.26 to 0.66 for the southern Lake District, and from 0.20 to 0.50 for the Southern Uplands, consistent with metamorphism from the deep diagenetic zone to the high anchizone. Our results show that bulk Mg isotopic compositions of these mudrocks do not vary systematically with increasing metamorphic grade, as represented by the decrease of KI (Fig. 8a). Neither silicate fraction nor carbonate fraction shows a correlation between their $\delta^{26}Mg$ and KI (not shown). These observations suggest that the Mg isotopic compositions of the bulk mudrocks, as well as their components, may not be directly influenced by diagenesis or low-grade metamorphic dehydration, and that the original Mg is accommodated in newly formed mineral phases rather than being partitioned into the metamorphic fluids. A similar conclusion was also reached for Li (which is more soluble than Mg) and its isotopes (Teng et al., 2007a; Qiu et al., 2009, 2011a,b). This observation, together with the absence of Mg isotope fractionation during high-temperature metamorphism (>300 ℃; Li et al., 2011, 2014; Teng et al., 2013; Wang et al., 2014a, 2015), indicates that metamorphic dehydration has an insignificant influence on Mg isotopic compositions of metamorphic rocks.

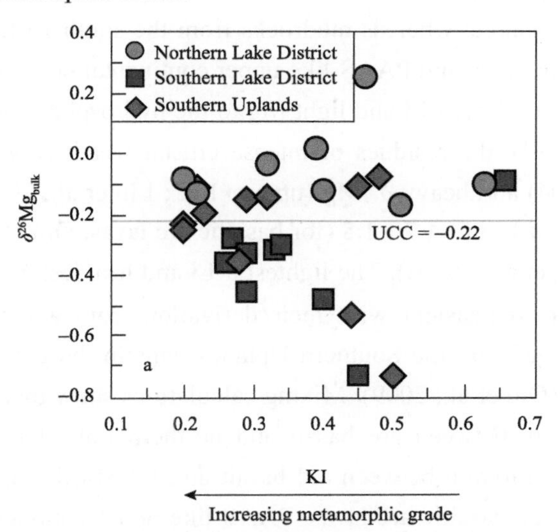

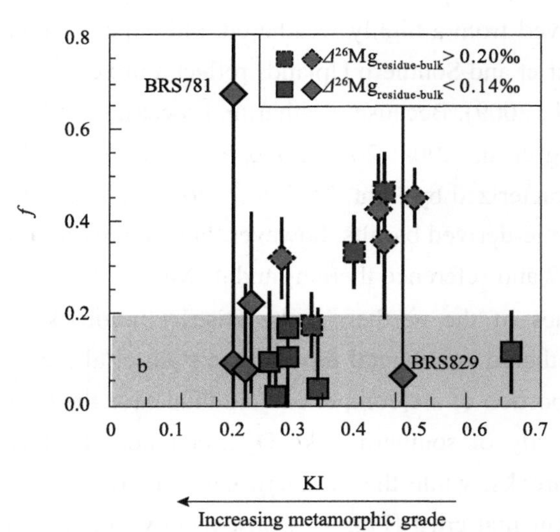

Fig. 8 (a) Variation of bulk δ^{26}Mg as a function of Kubler Index (KI); (b) Variation of the Mg fraction contributed by carbonate (f) as a function of KI. The KI value decreases as metamorphism progresses (arrows). KI values are from Merriman et al. (2009). The gray line in (a) represents the average upper continental crust δ^{26}Mg (Li et al., 2010). The gray area in (b) represents $f = 10 \pm 10\%$. δ^{26}Mg values are reported in Table 1. See text for discussion.

Mudrocks with relatively high δ^{26}Mg values (e.g., comparable to those of carbonate-free mudrocks from northern Lake District) are considered to contain no or very little carbonate. If such high-δ^{26}Mg samples from the southern Lake District and Southern Uplands are excluded, Mg isotopic compositions of the remainder tend to become more positive with increasing metamorphic grade (Fig. 8a). One possible explanation for this relationship is that the dissolution or decomposition of carbonate minerals during prograde metamorphism expelled light Mg isotopes from the mudrocks. As such, the Mg fraction contributed by the carbonate (f) is estimated using two end-memeber mixing: $f=(\delta^{26}Mg_{residue}-\delta^{26}Mg_{bulk})/(\delta^{26}Mg_{residue}-\delta^{26}Mg_{leachate})$, which ranges from $2\pm7\%$ to $47\pm9\%$ (Fig. 8b). Two samples from the Southern Uplands (BRS781 and BRS829) yield extremely large error bars for f because of the identical δ^{26}Mg value between residue and leachate. If these samples are excluded, the majority shows a rough, positive correlation between f and KI (Fig. 8b). Indeed, most mudrocks yield $\Delta^{26}Mg_{residue-bulk}<$ 0.14‰ (Figs. 2 and 5), which is insignificant given the external uncertainty of 0.10‰ for the ^{26}Mg/^{24}Mg, and thus the corresponding f cluster around $10\pm10\%$ (Fig. 8b). Consequently, sediments with relatively larger $\Delta^{26}Mg_{residue-bulk}$ (e.g., >0.20‰ in Fig. 2) give more reliable f, and again they show a positive correlation with KI for the mudrocks from each basin (Fig. 8b). This suggests that the influence of carbonates on the bulk Mg isotopic compositions is weakened with increasing metamorphic grade. However, this explanation is not unique and future Mg isotopic studies of carbonate-bearing clastic sediments are desirable.

5.3 Effects of provenance on magnesium and lithium isotopes of silicate fraction

The above discussion suggests that bulk Mg isotopic compositions of clastic sediments do not always reflect their source characteristics due to the influence of carbonates. By contrast, bulk Li isotopic compositions are controlled by the silicate δ^7Li values, as carbonates contain insignificant amount of Li compared to the clays. Because Mg and Li isotopes of silicate fractions are unaffected by metamorphic dehydration (Teng et al., 2007a, 2013; Qiu et al., 2009; 2011a,b; Li et al., 2011, 2014; Wang et al., 2014a, 2015), their variations may reflect the differences in the provenance.

A previous Sr-Nd-Li isotopic and trace elemental study found that northern Lake District mudrocks are derived from a highly weathered, old upper continental crust source; whereas mudrocks from the southern Lake District and Southern Uplands reflect a mixed provenance of arc lava and PAAS-like upper continental crust (Qiu et al., 2009). Because weathering processes preferentially release heavy Li and light Mg to the hydrosphere (e.g., Teng et al., 2004; 2010b; Tipper et al., 2010; Liu et al., 2014), the residues of intense crustal weathering are characterized by light δ^7Li (down to −20; Rudnick et al., 2004) and heavy δ^{26}Mg (up to +1.81; Liu et al., 2014). Mantle-derived basalts, however, have relatively homogenous δ^7Li of $+4.7\pm1.8$ (for basaltic arc lavas; Qiu et al., 2009 and reference therein) and δ^{26}Mg of -0.25 ± 0.07 (Teng et al., 2010a). The lightest δ^7Li and heaviest δ^{26}Mg values of the northern Lake District mudrocks (Fig. 9), are consistent with their derivation from a highly weathered continental provenance (Qiu et al., 2009). Mudrocks from the Southern Uplands contain the greatest proportion of arc component and thus have the heaviest δ^7Li (Qiu et al., 2009). Mixing calculations show that the majority of southern Lake District mudrocks fall on an array between arc basalt and northern Lake District mudrocks; while the Southern Uplands mudrocks represent a mixture between arc basalt and PAAS-like upper continental crust (Fig. 9). Consequently, mixing among average basaltic arc lavas, PAAS-like upper continental crust, and a highly weathered component, as represented by the northern Lake District mudrocks, can reproduce most of the δ^7Li and δ^{26}Mg variations observed in these mudrocks (Fig. 9). However, as pointed by Qiu et al., (2009), this is an oversimplified scenario, and different mixing end members most likely apply to different samples (Fig. 9).

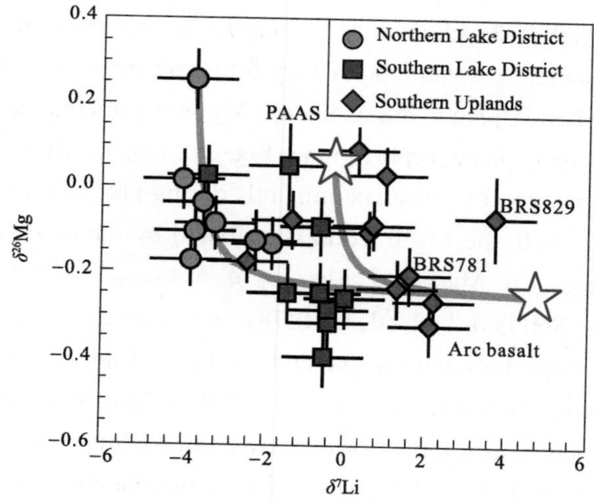

Fig. 9　Coupled δ^{26}Mg-δ^7Li variations for the silicate fraction of mudrocks from the three basins. The gray lines are mixing trends between basaltic arc lava and PAAS, and between basaltic arc lava and northern Lake District mudrocks, respectively. The [Li] and δ^7Li of arc basalts, PAAS and LC507 are from Qiu et al. (2009 and references therein). The MgO used for arc basalts and PAAS are ~6 wt.% and 2.3 wt.%, respectively; the δ^{26}Mg used for arc basalts and PAAS are −0.25 and +0.07, respectively. Sample LC507 from the northern Lake District mudrocks is taken to represent the end member of highly weathered upper continental crustal provenance. δ^{26}Mg values are reported in Table 1. See text for further details.

5.4　Implication for sediment recycling

Subducted sediments can influence on the compositions of oceanic island basalts as well as arc lavas (e.g.,

Plank and Langmuir, 1998; Plank, 2014); for example, sediment addition enriches highly fluid moible elements (such as Li, Be, Ba and Sr) in arc laves and may give rise to the EM I end member in the mantle (e.g., Hofmann, 1997; Ryan and Chauvel, 2014). Deeply subducted sedimentary rocks may largely retain their δ^{26}Mg signatures, as Mg isotopes are unaffected by either metamorphism or partial melting durnig crustal subduction (e.g., Li et al., 2011, 2014; Teng et al., 2010b, 2013; Wang et al., 2014a, 2015). The recycling of dolomite into the mantle might impact light Mg isotopic compositions to the mantle-derived rocks (Yang et al., 2012; Huang et al., 2015). However, addition of a low-MgO clastic sedimentary component to the mantle source may not significantly modify the mantle Mg isotopic compositions, owing to the large Mg budget of the mantle (~38 wt.%; McDonough and Sun, 1995). On the other hand, Mg isotopes may be potentially good tracers of sediment assimilation during magma ascent. This is due to the fact that basaltic magmas have much lower MgO content (~8 wt.%) than peridotite, and therefore their Mg isotopic compositions are relatively more sensitive to sediment or sediment-derived melt addition. Bulk mixing models suggest that 10~20% sediment addition to a basaltic magma would potentially produce δ^{26}Mg exceeding the range of oceanic basalts.

Burial of sedimentary rocks will produce an isotopically heterogenous middle-lower continental crust with repect to Mg isotopes (Teng et al., 2013; Wang et al., 2015; Yang et al., 2015). A wide range of δ^{26}Mg in a variety of granitoids (e.g., I, S and A types) as well as in granulite xenoliths bear witness to the sedimentary recycling process (Shen et al., 2009; Li et al., 2010; Telus et al., 2012; Teng et al., 2013; Wang et al., 2015; Yang et al., 2015). Although light-δ^{26}Mg signatures are generally seen in carbonate-bearing clastic sediments, it is also inferred that thermal evolution of a granulite-facies lower continental crust would induce carbonate-silicate Mg isotopic exchange or decarbonation of carbonate-bearing sediments leaving a silicate residue enriched in light Mg isotopes (Shen et al., 2013; Wang et al., 2014b; Yang et al., 2015).

6 Conclusions

Leaching experiments and Mg isotopic analyses carried out on three suites of low-grade metamorphosed mudrocks from lower Paleozoic basins within the British Caledonides (northern Lake District, southern Lake District, and Southern Uplands) demonstrate that:

(1) The δ^{26}Mg varies widely in these mudrocks, i.e., −0.17 to +0.25 in carbonate-free northern Lake District mudrocks, and −0.74 to −0.08 in carbonate-bearing mudrocks from southern Lake District and Southern Uplands.

(2) Large Mg isotope differences, up to 1.59‰, occur between leachates and residues. The δ^{26}Mg of the leachates are always lighter than that of the residues, due to the dissolution of leachable carbonate minerals. Mg isotopic compositions of global clastic sediments are controlled by silicate-carbonate mixing, while the silicate mineralogy determines the δ^{26}Mg of silicate fraction in sediments.

(3) The Mg isotopes are not directly affected by low-temperature (<300℃) metamorphic dehydration, based on the absence of correlations between bulk δ^{26}Mg and KI values for the mudrocks. However, δ^{26}Mg of carbonate-bearing mudrocks may become heavier as metamorphism progresses, due to carbonate dissolution or decomposition with increasing metamorphic grade.

(4) Magnesium isotopic compositions of the silicate fraction in clastic sediments more faithfully reflect provenance signatures, as bulk Mg isotopic compositions may be influenced by the presence of carbonate minerals.

(5) Burial of clastic sediments may produce Mg isotopic heterogeneity in the middle-lower continental crust, as well as their derivatives (e.g., S, I, and A type granites). Sediment assimilation in basaltic magmas may leave distinguishable Mg isotopic signatures in some extreme case.

Acknowledgements

We thank Richard Merriman for providing samples, Melissa Hornick for help in the clean lab, Jody Bourgeois, Charlotte Schreiber, and Aaron Brewer for their thoughtful discussions, Ed Tipper and two anonymous reviewers for insightful comments, and Shichun Huang for careful and efficient handling. This work was supported by the National Science Foundation (EAR-0838227, EAR-1056713 and EAR-1340160) and the National Nature Science Foundation of China (41230209 and 41090372).

References

Bizzarro M., Paton C., Larsen K., Schiller M., Trinquier A. and Ulfbeck D. (2011) High-precision Mg-isotope measurements of terrestrial and extraterrestrial material by HR-MC-ICPM Simplications for the relative and absolute Mg-isotope composition of the bulk silicate Earth. *J. Anal. Atomic Spec*. 26, 565-577.

Bourdon B., Tipper E. T., Fitoussi C. and Stracke A. (2010) Chondritic Mg isotope composition of the Earth. *Geochim. Cosmochim. Acta* 74, 5069-5083.

Brenot A., Cloquet C., Vigier N., Carignan J. and France-Lanord C. (2008) Magnesium isotope systematics of the lithologically varied Moselle river basin, France. *Geochim. Cosmochim. Acta* 72(20), 5070-5089.

Chakrabarti R. and Jacobsen S. B. (2010) The isotopic composition of magnesium in the inner Solar System. *Earth Planet. Sci. Lett.* 293, 349-358.

Dauphas N., Teng F.-Z. and Arndt N. T. (2010) Magnesium and iron isotopes in 2.7 Ga Alexo komatiites: Mantle signatures, no evidence for Soret diffusion, and identification of diffusive transport in zoned olivine. *Geochim. Cosmochim. Acta* 74, 3274-3291.

Fortey N. 1(989) Low grade metamorphism in the Lower Ordovician Skiddaw Group of the Lake District, England, Proceedings of the Yorkshire Geological and Polytechnic Society. Geological Society of London. pp. 325-337.

Foster G. L., Pogge von Strandmann P. A. E. and Rae J. W. B. (2010) Boron and magnesium isotopic composition of seawater. *Geochem. Geophys. Geosyst*. 11. http://dx.doi.org/10.1029/ 2010GC003201.

Gallet S., Jahn B.-M., Van Vliet LanoëB., Dia A. and Rossello E. (1998) Loess geochemistry and its implications for particle origin and composition of the upper continental crust. *Earth Planet. Sci. Lett*. 156, 157-172.

Galy A., Bar-Matthews M., Halicz L. and O'Nions R. K. (2002) Mg isotopic composition of carbonate: insight from speleothem formation. *Earth Planet. Sci. Lett*. 201, 105-115.

Gao S., Luo T.-C., Zhang B.-R., Zhang H.-F., Han Y.-W., Zhao Z.-D. and Hu Y.-K. (1998) Chemical composition of the continental crust as revealed by studies in East China. *Geochim. Cosmochim. Acta* 62, 1959-1975.

Grim R. E. (1953) *Clay Mineralogy*. McGraw-Hill Book Co., Inc,, New York, p. 384.

Handler M. R., Baker J. A., Schiller M., Bennett V. C. and Yaxley G. M. (2009) Magnesium stable isotope composition of Earth's upper mantle. *Earth Planet. Sci. Lett*. 282, 306-313.

Higgins J. A. and Schrag D. P. (2010) Constraining magnesium cycling in marine sediments using magnesium isotopes. *Geochim. Cosmochim. Acta* 74, 5039-5053.

Hofmann A. W. (1997) Mantle geochemistry: the message from oceanic volcanism. *Nature* 385, 219-229.

Huang F., Zhang Z. F., Lundstrom C. C. and Zhi X. C. (2011) Iron and magnesium isotopic compositions of peridotite xenoliths from Eastern China. *Geochim. Cosmochim. Acta* 75, 3318-3334.

Huang K.-J., Teng F.-Z., Wei G.-J., Ma J.-L. and Bao Z.-Y. (2012) Adsorption-and desorption-controlled magnesium isotope fractionation during extreme weathering of basalt in Hainan Island, China. *Earth Planet. Sci. Lett*. 359, 73-83.

Huang K.-J., Teng F.-Z., Elsenouy A., Li W.-Y. and Bao Z.-Y. (2013) Magnesium isotopic variations in loess: origins and implications. *Earth Planet. Sci. Lett*. 374, 60-70.

Huang J., Li S.-G., Xiao Y., Ke S., Li W.-Y. and Tian Y. (2015) Origin of low δ^{26}Mg Cenozoic basalts from South China Block and their geodynamic implications. *Geochim. Cosmochim. Acta* 164, 298-317.

Immenhauser A., Buhl D., Richter D., Niedermayr A., Riechelmann D., Dietzel M. and Schulte U. (2010) Magnesium-isotope fractionation during low-Mg calcite precipitation in a limestone cave–field study and experiments. *Geochim. Cosmochim. Acta*

74(15), 4346-4364.

Jahn B.-M., Gallet S. and Han J. (2001) Geochemistry of the Xining, Xifeng and Jixian sections, Loess Plateau of China: eolian dust provenance and paleosol evolution during the last 140 ka. *Chem. Geol.* 178, 71-94.

Johnson E. W., Soper N. J., Burgess I. C., Beddoe-Stephens B., Carruthers R. M., Fortey N. J., Hirrons S. R., Merritt J. W., Millward D., Molyneux S. G., Roberts B., Rushton A. W. A., Walker A. B. and Young B. (2001) Geology of the Country Around Ulverston: Memoir for 1: 50 000 Geological Sheet 48 (England and Wales). Stationery Office.

Kneller B. (1991) A foreland basin on the southern margin of Iapetus. *J. Geol. Soc.* 148, 207-210.

Lai Y.-J., Pogge von Strandmann P. A. E., Dohmen R., Takazawa E. and Elliott T. (2015) The influence of melt infiltration on the Li and Mg isotopic composition of the Horoman Peridotite Massif. *Geochim. Cosmochim. Acta* 164, 318-332.

Leggett J., McKerrow W. T. and Eales M. (1979) The Southern Uplands of Scotland: a lower Palaeozoic accretionary prism. *J.Geol. Soc.* 136, 755-770.

Li W.-Y., Teng F.-Z., Ke S., Rudnick R. L., Gao S., Wu F.-Y. and Chappell B. W. (2010) Heterogeneous magnesium isotopic composition of the upper continental crust. *Geochim. Cosmochim. Acta* 74, 6867-6884.

Li W.-Y., Teng F.-Z., Wing B. A. and Xiao Y. L. (2014) Limited magnesium isotopic variation in metapelites from the Onawa contact aureole, Maine: implications for Mg isotopes as new tracer for crustal recycling. *Geochem. Geophys. Geosyst.* 15, 408-415.

Ling M. X., Sedaghatpour F., Teng F.-Z., Hays P. D., Strauss J. and Sun W. (2011) Homogeneous magnesium isotopic composition of seawater: an excellent geostandard for Mg isotope analysis. *Rapid Commun. Mass Spectrom.* 25, 2828-2836.

Ling M.-X., Liu Y.-L., Williams I. S., Teng F.-Z., Yang X.-Y., Ding X., Wei G.-J., Xie L.-H., Deng W.-F. and Sun W.-D. (2013) Formation of the world's largest REE deposit through protracted fluxing of carbonatite by subduction-derived fluids. *Scientific reports* 3. http://dx.doi.org/10.1038/srep01776.

Liu X. M., Teng F.-Z., Rudnick R. L., McDonough W. F. and Cummings M. L. (2014) Massive magnesium depletion and isotope fractionation in weathered basalts. *Geochim. Cosmochim. Acta* 135(15), 336-349.

Ma L., Teng F.-Z., Jin L., Ke S., Yang W., Gu H. O. and Brantley S. L. (2015) Magnesium isotope fractionation during shale weathering in the Shale Hills Critical Zone Observatory: Accumulation of light Mg isotopes in soils by clay mineral transformation. *Chem. Geol.* 397, 37-50.

Merriman R. (2006) Clay mineral assemblages in British Lower Palaeozoic mudrocks. *Clay Miner.* 41, 473-512.

Merriman R. and Peacor D. (1999) Very low-grade metapelites: mineralogy, microfabrics and measuring reaction progress. In *Low-Grade Metamorphism* (eds. M. Frey and D. Robinson), Blackwell Publishing Ltd. pp. 10-60.

Merriman R. and Roberts B. (2000) Low-grade metamorphism in the Scottish Southern Uplands terrane: deciphering the patterns of accretionary burial, shearing and cryptic aureoles. *Trans. R.Soc. Edinb.: Earth Sci.* 91, 521-537.

Merriman R. J., Roberts B., Peacor D. R. and Hirons S. R. (1995)Strain-related differences in the crystal growth of white mica and chlorite: a TEM and XRD study of the development of metapelitic microfabrics in the Southern Uplands thrust terrane, Scotland. *J. Metamorp. Geol.* 13(5), 559-576.

Merriman R., Breward N., Stone P., Green K. and Kemp S. (2009) Element mobility and low-grade metamorphism of mudrocks in British Caledonian Basins. British Geological Survey Open Report, OR/09/017, http://nora.nerc.ac.uk/8146/1/OR09017.pdf.

Milliken K. (2003) Late diagenesis and mass transfer in sandstone shale sequences. *Treat. Geochem.* 7, 159-190.

Nance W. B. and Taylor S. (1976) Rare earth element patterns and crustal evolution–I. Australian post-Archean sedimentary rocks. *Geochim. Cosmochim. Acta* 40, 1539-1551.

Opfergelt S., Georg R. B., Delvaux B., Cabidoche Y. M., Burton K. W. and Halliday A. N. (2012) Mechanisms of magnesium isotope fractionation in volcanic soil weathering sequences, Guadeloupe. *Earth Planet. Sci. Lett.* 341-344, 176-185.

Opfergelt S., Burton K. W., Georg R. B., West A. J., Guicharnaud R. A., Sigfusson B., Siebert C., Gislason S. R. and Halliday A.N. (2014) Magnesium retention on the soil exchange complex controlling Mg isotope variations in soils, soil solutions and vegetation in volcanic soils, Iceland. *Geochim. Cosmochim. Acta* 125, 110-130.

Ostrom M. E. (1961) Separation of clay minerals from carbonate rocks by using acid. *J. Sediment. Res.* 31, 123-129.

Peacor D. R. (1992) Diagenesis and low-grade metamorphism of shales and slates. *Rev. Mineral. Geochem.* 27, 335-380.

Plank T. (2014) The chemical composition of subducting sediments. In *The Crust* (ed. R. L. Rudnick). In *Treatise on Geochemistry,* second ed., vol. 4 (eds. H. D. Holland and K. K. Turekian). Elsevier, Oxford, pp. 607-629.

Plank T. and Langmuir C. H. (1998) The chemical composition of subducting sediment and its consequences for the crust and mantle. *Chem. Geol.* 145, 325-394.

Pogge von Strandmann P. A. E. (2008a) Precise magnesium isotope measurements in core top planktic and benthic foraminifera.

Geochem. Geophys. Geosyst. 9, Q12015. http://dx.doi.org/10.1029/2008GC002209.

Pogge von Strandmann P. A., Burton K. W., James R. H., van Calsteren P., Gislason S. R. and Sigfússon B. (2008b) The influence of weathering processes on riverine magnesium isotopes in a basaltic terrain. *Earth Planet. Sci. Lett.* 276(1), 187-197.

Pogge von Strandmann P. A., Elliott T., Marschall H. R., Coath C., Lai Y. J., Jeffcoate A. B. and Ionov D. A. (2011) Variations of Li and Mg isotope ratios in bulk chondrites and mantle xenoliths. *Geochim. Cosmochim. Acta* 75(18), 5247-5268.

Pogge von Strandmann P. A. E., Opfergelt S., Lai Y.-J., Sigfússon B., Gislason S. R. and Burton K. W. (2012) Lithium, magnesium and silicon isotope behaviour accompanying weathering in a basaltic soil and pore water profile in Iceland. *Earth Planet. Sci. Lett.* 339-340, 11-23.

Qiu L., Rudnick R. L., McDonough W. F. and Merriman R. J. (2009) Li and δ^7Li in mudrocks from the British Caledonides: Metamorphism and source influences. *Geochim. Cosmochim. Acta* 73, 7325-7340.

Qiu L., Rudnick R. L., McDonough W. F. and Bea F. (2011a) The behavior of lithium in amphibolite- to granulite-facies rocks of the Ivrea-Verbano Zone, NW Italy. *Chem. Geol.* 289, 76-85.

Qiu L., Rudnick R. L., Ague J. J. and McDonough W. F. (2011b) A lithium isotopic study of sub-greenschist to greenschist facies metamorphism in an accretionary prism, New Zealand. *Earth Planet. Sci. Lett.* 301, 213-221.

Rudnick R. L., Tomascak P. B., Njo H. B. and Gardner L. R. (2004) Extreme lithium isotopic fractionation during continental weathering revealed in saprolites from South Carolina. *Chem. Geol.* 212, 45-57.

Ryan J. and Chauvel C. (2014) The subduction zone filter and the impact of recycled materials on the evolution of the mantle. In *The Mantle* (ed. R. W. Carlson). In *Treatise on Geochemistry*, second ed., vol. 3 (eds. H. D. Holland and K. K. Turekian). Elsevier, Oxford, pp. 479-508.

Shen B., Jacobsen B., Lee C. T., Yin Q. Z. and Morton D. M. (2009) The Mg isotopic systematics of granitoids in continental arcs and implications for the role of chemical weathering in crust formation. *Proc. Natl. Acad. Sci. U.S.A.* 106, 20652-20657.

Shen B., Wimpenny B., Lee C. T., Tollstrup D. and Yin Q. Z. (2013) Magnesium isotope systematics of endoskarns: Implications for wallrock reaction in magma chambers. *Chem. Geol.* 356, 209-214.

Soper N. and Woodcock N. (2003) The lost Lower Old Red Sandstone of England and Wales: a record of post-Iapetan flexure or Early Devonian transtension? *Geol. Mag.* 140, 627-647.

Stone P. and Merriman R. (2004) Basin thermal history favours an accretionary origin for the Southern Uplands terrane, Scottish Caledonides. *J. Geol. Soc.* 161, 829-836.

Stone P., Floyd J., Barnes R. and Lintern B. (1987) A sequential back-arc and foreland basin thrust duplex model for the Southern Uplands of Scotland. *J. Geol. Soc.* 144, 753-764.

Sutton S. and Land L. (1996) Postdepositional chemical alteration of Ouachita shales. Geol. Soc. *Am. Bull.* 108, 978-991.

Taylor S. R. and McLennan S. M. (1985) *The Continental Crust: its Composition and Evolution*. Blackwell, Oxford.

Taylor S., McLennan S. and McCulloch M. (1983) Geochemistry of loess, continental crustal composition and crustal model ages. *Geochim. Cosmochim. Acta* 47, 1897-1905.

Telus M., Dauphas N., Moynier F., Tissot F. L. H., Teng F. Z., Nabelek P. I., Craddock P. R. and Groat L. A. (2012) Iron, zinc, magnesium and uranium isotopic fractionation during continental crust differentiation: The tale from migmatites, granitoids, and pegmatites. *Geochim. Cosmochim. Acta* 97, 247-265.

Teng F.-Z. and Yang W. (2014) Comparison of factors affecting the accuracy of high-precision magnesium isotope analysis by multi-collector inductively coupled plasma mass spectrometry. *Rapid Commun. Mass Spec.* 28, 19-24.

Teng F.-Z., McDonough W. F., Rudnick R. L. and Wing B. A. (2007a) Limited lithium isotopic fractionation during progressive metamorphic dehydration in metapelites: a case study from the Onawa contact aureole, Maine. *Chem. Geol.* 239(1), 1-12.

Teng F.-Z., Wadhwa M. and Helz R. T. (2007b) Investigation of magnesium isotope fractionation during basalt differentiation: Implications for a chondritic composition of the terrestrial mantle. *Earth Planet. Sci. Lett.* 261, 84-92.

Teng F.-Z., Li W.-Y., Ke S., Marty B., Dauphas N., Huang S., Wu F.-Y. and Pourmand A. (2010a) Magnesium isotopic composition of the Earth and chondrites. *Geochim. Cosmochim. Acta* 74, 4150-4166.

Teng F.-Z., Li W.-Y., Rudnick R. L. and Gardner L. R. (2010b) Contrasting lithium and magnesium isotope fractionation during continental weathering. *Earth Planet. Sci. Lett.* 300, 63-71.

Teng F.-Z., Yang W., Rudnick R. L. and Hu Y. (2013) Heterogeneous magnesium isotopic composition of the lower continental crust: A xenolith perspective. *Geochem. Geophys. Geosyst.*. http://dx.doi.org/10.1002/ggge.20238.

Teng F.-Z., Li W. Y., Ke S., Yang W., Liu S. A., Sedaghatpour F., Wang S.-J., Huang K. J., Hu Y., Ling M. X., Xiao Y., Liu X. M., Li X. W., Gu H. O., Sio C., Wallace D., Su B. X., Zhao L., Harrington M. and Brewer A. (2015) Magnesium isotopic compositions of international geological reference materials. *Geostand. Geoanal. Res.* http://dx.doi.org/10.1111/j.1751-908X.2014.00326.x.

Tipper E., Galy A. and Bickle M. (2006a) Riverine evidence for a fractionated reservoir of Ca and Mg on the continents: Implications for the oceanic Ca cycle. *Earth Planet. Sci. Lett.* 247, 267-279.

Tipper E., Galy A., Gaillardet J., Bickle M., Elderfield H. and Carder E. (2006b) The magnesium isotope budget of the modern ocean: constraints from riverine magnesium isotope ratios. *Earth Planet. Sci. Lett.* 250, 241-253.

Tipper E. T., Galy A. and Bickle M. J. (2008) Calcium and magnesium isotope systematics in rivers draining the Himalaya-Tibetan-Plateau region: Lithological or fractionation control? *Geochim. Cosmochim. Acta* 72, 1057-1075.

Tipper E. T., Gaillardet J., Louvat P., Capmas F. and White A. F. (2010) Mg isotope constraints on soil pore-fluid chemistry: evidence from Santa Cruz, California. *Geochim. Cosmochim. Acta* 74, 3883-3896.

Wang S.-J., Teng F.-Z., Williams H. M. and Li S.-G. (2012) Magnesium isotopic variations in cratonic eclogites: origins and implications. *Earth Planet. Sci. Lett.* 359, 219-226.

Wang S.-J., Teng F.-Z., Li S.-G. and Hong J.-A. (2014a) Magnesium isotopic systematics of mafic rocks during continental subduction. *Geochim. Cosmochim. Acta* 143, 34-48.

Wang S.-J., Teng F.-Z. and Li S.-G. (2014b) Tracing carbonatesilicate interaction during subduction using magnesium and oxygen isotopes. *Nat. Commun.* http://dx.doi.org/10.1038/ ncomms6328.

Wang S.-J., Teng F.-Z. and Bea F. (2015) Magnesium isotopic systematics of metapelite in the deep crust and implications for granite petrogenesis. *Geochem. Perspect. Lett.* 1, 75-83.

Wimpenny J., Gíslason S. R., James R. H., Gannoun A., Pogge Von Strandmann P. A. and Burton K. W. (2010) The behaviour of Li and Mg isotopes during primary phase dissolution and secondary mineral formation in basalt. *Geochim. Cosmochim. Acta* 74(18), 5259-5279.

Wimpenny J., Burton K. W., James R. H., Gannoun A., Mokadem F. and Gíslason S. R. (2011) The behaviour of magnesium and its isotopes during glacial weathering in an ancient shield terrain in West Greenland. *Earth Planet. Sci. Lett.* 304(1), 260-269.

Wimpenny J., Colla C. A., Yin Q. Z., Rustad J. R. and Casey W. H. (2014a) Investigating the behaviors of Mg isotopes during the formation of clay minerals. *Geochim. Cosmochim. Acta* 128, 178-194.

Wimpenny J., Yin Q. Z., Tollstrup D., Xie L. W. and Sun J. (2014b) Using Mg isotope ratios to trace Cenozoic weathering changes: a case study from the Chinese loess plateau. *Chem. Geol.* 376, 31-43.

Wintsch R. P. and Kvale C. M. (1994) Differential mobility of elements in burial diagenesis of siliciclastic rocks. *J. Sediment. Res.* 64, 349-361.

Wombacher F., Eisenhauer A., Heuser A. and Weyer S. (2009) Separation of Mg, Ca and Fe from geological reference materials for stable isotope ratio analyses by MC-ICP-MS and double-spike TIMS. *J. Anal. At. Spectrom.* 24, 627-636.

Xiao Y., Teng F.-Z., Zhang H.-F. and Yang W. (2013) Large magnesium isotope fractionation in peridotite xenoliths from eastern North China craton: product of melt-rock interaction. *Geochim. Cosmochim. Acta* 115, 241-261.

Yang W., Teng F.-Z. and Zhang H.-F. (2009) Chondritic magnesium isotopic composition of the terrestrial mantle: a case study of peridotite xenoliths from the North China craton. *Earth Planet. Sci. Lett.* 288, 475-482.

Yang W., Teng F.-Z., Zhang H.-F. and Li S.-G. (2012) Magnesium isotopic systematics of continental basalts from the North China craton: implications for tracing subducted carbonate in the mantle. *Chem. Geol.* 328, 185-194.

Yang W., Teng F.-Z., Li W.-Y., Liu S.-A., Ke S., Liu Y.-S. Zhang H. F. and Gao S. (2015) Magnesium isotopic composition of the deep continental crust. American Mineralogist, accepted.

Magnesium isotopic heterogeneity across the cratonic lithosphere in eastern China and its origins*

Ze-Zhou Wang[1], Sheng-Ao Liu[1], Shan Ke [1], Yi-Can Liu[2] and Shu-Guang Li [1,2]

1. State Key Laboratory of Geological Processes and Mineral Resources, China University of Geosciences, Beijing 100083, China
2. CAS Key Laboratory of Crust-Mantle Materials and Environments, School of Earth and Space Sciences, University of Science and Technology of China, Hefei 230026, China

亮点介绍：本文研究了同一火成岩侵入体中来自岩石圈不同深度的镁铁质岩包体的镁同位素组成。所获数据显示下地壳 Mg 同位素组成是高度非均一的，平均比中地壳轻。在幔源包体中发现了极轻的镁同位素组成（−1.23‰～−0.73‰）。Mg 同位素在克拉通岩石圈中的不均一性是由碳酸盐交代作用引起的。

Abstract Available data in the literature have demonstrated a broad magnesium (Mg) isotope range for mantle and lower continental crustal rocks, implying an isotopically heterogeneous continental lithosphere, but its origin has not been thoroughly understood. Here, to investigate the primary cause of lithospheric Mg isotopic heterogeneity, we report major-trace elements, Sr and Mg isotope data for thirty deep-seated mafic xenoliths, which sampled different lithospheric depths in the southeastern North China Craton (NCC). The xenoliths are classified into three types based upon petrology and mineralogy, sampling from middle continental crust (Group I), lower continental crust (Group II) and lithospheric mantle (Group III), respectively. The Group I xenoliths have mantle-like to slightly high δ^{26}Mg values (−0.32‰ to +0.01‰), whereas some of the Group II xenoliths have very low δ^{26}Mg values (−0.93‰ to −0.07‰), reflecting substantial reaction with intracrustal carbonate-derived fluids. Combined with data in the literature, the results suggest that the Mg isotopic composition of the lower continental crust is much more heterogeneous and lighter on average relative to the middle continental crust. Except for one sample, the Group III xenoliths have extremely low δ^{26}Mg values (−1.23‰ to −0.73‰), the lightest among values already reported for mantle-derived rocks including peridotites and basalts. They also have highly variable ^{87}Sr/^{86}Sr ratios, of 0.70387 to 0.71675. The covariation of Mg and Sr isotopes in Group III xenoliths can be explained by Mg and Sr isotopic exchange reactions during mantle metasomatism, implying that the sub-continental mantle has been significantly modified by fluids derived from recycled carbonate-pelite bearing oceanic crust. Together with the metasomatism age of ~400 Ma obtained for one Group III xenolith, the results provide new evidence for the presence of extremely low-δ^{26}Mg rocks in the lithosphere and indicate ancient carbonate recycling during Paleo-Tethys slab subduction. Experimental and computational studies have demonstrated that the

* 本文发表在：Earth and Planetary Science Letters, 2016, 451: 77-88

solubility of carbonate in aqueous fluids increases with increasing pressures. Therefore, we suggest that the decreasing δ^{26}Mg values of mantle-derived rocks with increasing depths across cratonic lithosphere, as found in the NCC and other regions, have probably been caused by carbonate metasomatism to various extents at different lithospheric depths.

1 Introduction

Magnesium (Mg), having three stable isotopes of ^{24}Mg, ^{25}Mg and ^{26}Mg, is the second most abundant element in the mantle and eighth most abundant element in the crust. Investigations of oceanic basalts and mantle peridotites have shown that the terrestrial mantle has a homogeneous Mg isotopic composition (Handler et al., 2009; Yang et al., 2009; Bourdon et al., 2010; Teng et al., 2010; Liu et al., 2011; Pogge von Strandmann et al., 2011), with an average δ^{26}Mg of −0.25±0.07‰ (Teng et al., 2010). By contrast, a large range (>1‰) in δ^{26}Mg has been observed for the upper continental crust, although it has an average δ^{26}Mg value (−0.22‰) almost identical to that of the mantle (Li et al., 2010; Liu et al., 2010). Chemical weathering is proposed to be one of the main processes resulting in the Mg isotopic heterogeneity of the upper crust, which transports the light Mg isotopes into the hydrosphere, leaving the siliceous regoliths (e.g., soils, loess) enriched in heavy Mg isotopes (e.g., Li et al., 2010).

In contrast to silicates, sedimentary carbonates are characterized by highly variable and extremely light Mg isotopic compositions (−5.5‰ to −0.5‰, e.g., Young and Galy, 2004). The remarkable difference in Mg isotopic compositions between normal mantle-derived rocks and sedimentary carbonates makes Mg isotopes an efficient proxy for tracing recycled carbonates in mantle sources. In this context, recent studies have shown that late Cretaceous to Cenozoic (<110 Ma) continental basalts from eastern China have lighter Mg isotopic compositions (−0.6‰ to −0.3‰) relative to global oceanic basalts (Yang et al., 2012; Huang et al., 2015; Tian et al., 2016). The low-δ^{26}Mg signatures were interpreted as a result of interaction of the mantle with carbonate melts derived from subducted oceanic slab. A few wehrlites from the NCC also have slightly low δ^{26}Mg, of −0.4‰, which was also attributed to mantle metasomatism (Yang et al., 2009; Xiao et al., 2013).

To date, although abundant Mg isotope data have been reported for upper crustal materials, the Mg isotopic composition of the deep crust remains poorly constrained and controversial. Two suites of lower crustal xenoliths from Australia show a wide δ^{26}Mg range from −0.72‰ to +0.19‰ (Teng et al., 2013), which was attributed to protolith heterogeneity and preferential sampling of minerals such as garnet. By contrast, a ~0.5‰ isotopic variation (from −0.74‰ to −0.26‰) in granulite xenoliths from Hannuoba (northern NCC) was interpreted as a result of interaction with carbonate fluids (Yang et al., 2016). Composites from high-grade metamorphic terranes in Archean cratons of China have δ^{26}Mg values of −0.40‰ to +0.12‰ that were ascribed to protolith heterogeneity (Yang et al., 2016). To better understand the mechanisms controlling Mg isotopic heterogeneity in the deep crust and lithospheric mantle requires a detailed investigation of samples representing vertically different parts of the cratonic lithosphere in the same location.

Various types of deep-seated xenoliths captured by early Cretaceous dioritic intrusions occur widely in the Xu-Huai area on the southeastern NCC (Liu et al., 2009, 2013; Xu et al., 2009). Three types of xenoliths, originating from middle continental crust (MCC), lower continental crust (LCC) and sub-continental lithospheric mantle (SCLM), respectively, are recognized in this study and they provide an excellent opportunity to investigate the nature and origin of Mg isotopic variation of the deep lithosphere. Our results show that Mg isotopes, combined with Sr isotopes, can identify more widespread carbonate metasomatism in deep lithosphere than previously realized.

2 Geological setting and sample description

The NCC is one of the oldest and largest cratonic blocks in the world, with crustal remnants over 3.8 Ga exposed at surface or in the form of xenoliths (e.g., Zhai et al., 2000). It is bounded by the Central Asian Orogenic Belt to the north and the Qinling-Dabie-Sulu ultrahigh pressure metamorphic (UHPM) belt to the south and east. Separated by the Trans-North China Orogen (TNCO), the NCC can be divided into Western Block (WB) and Eastern Block (EB) (Fig. 1a). The Xuzhou-Huaibei (Xu-Huai) area is located on the southeastern margin of the Eastern Block (Fig. 1b). This area is covered by Neoproterozoic to Permian strata that were subjected to deformation during the Triassic and constitute the arcuate thrust nappe structure (Fig. 1c). Limestones and dolostones are the dominant lithologies in the Neoproterozoic and Paleozoic covers. Precambrian metamorphic basement, namely Wuhe complex, is exposed in Bengbu area, about 150 km south of the Xu-Huai area (Fig. 1b). The Wuhe complex is mainly composed of mafic lithologies (e.g., garnet amphibolites) and phlogopite-bearing marbles (Fig. S1), which have been suggested to have undergone high-pressure (HP) granulite-facies metamorphism at 1.8–1.9 Ga (Liu et al., 2009, 2013; Xu et al., 2009).

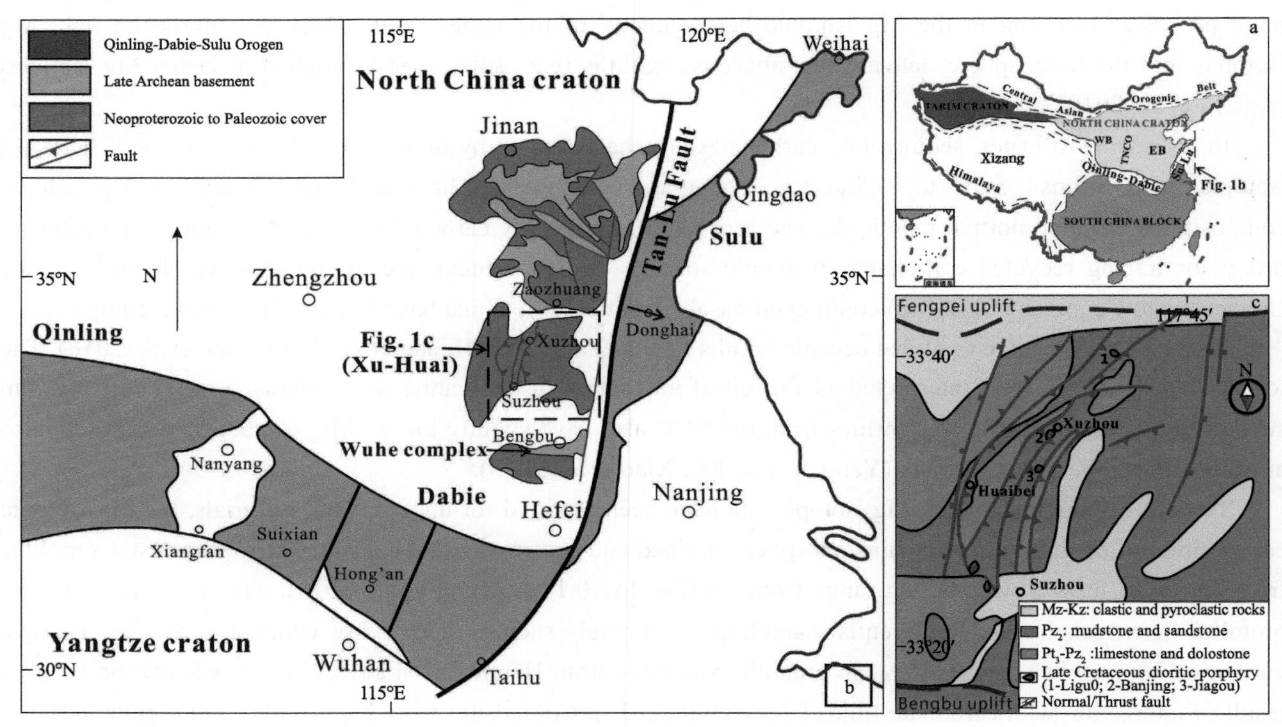

Fig. 1　Schematic geological map showing the major tectonic divisions of China (a), the location of Xu-Huai in the NCC (b), and the location of Cretaceous dioritic intrusions in the Xu-Huai area (c). Modified after Liu et al. (2013). Mz: Mesozoic, Kz: Cenozoic, Pz_2: Late Paleozoic, Pt_3: Neoproterozoic.

Late Mesozoic mafic to felsic magmatic rocks are widespread surrounding the NNE trending Tan-Lu fault in the southeastern NCC (e.g., Liu et al., 2012). In the Xu-Huai area, a suite of dioritic to monzodioritic porphyry bodies were intruded into the deformed strata at ~130 Ma. These rocks have previously been proposed to form via lower crust delamination (Xu et al., 2006) and the captured xenoliths originated from lithospheric mantle and deep crust (Liu et al., 2013). Various types of xenoliths have been recognized and well characterized by previous studies (Liu et al., 2009, 2013; Xu et al., 2009). A total of 30 xenoliths from the Jiagou intrusion in the Xu-Huai area were selected for major-trace elements, Sr and Mg isotopic analysis. These xenoliths range in diameters from

2 to 25 cm, and can be classified into three groups (I, II and III) based on their mineral assemblages (see below). In addition, four host rocks (dioritic porphyry) were also studied for comparison with the sample 11JG-5a(w) being collected next to the xenolith sample 11JG-5b.

Modal mineral compositions of these xenoliths are listed in Table1. The Group I xenoliths contain mainly amphibole (amp), plagioclase and quartz with a small amount of biotite, chlorite and ilmenite, and can be further divided into two subgroups: plagioclase amphibolite (Fig. 2c) and amphibolite (Fig. 2d) based on their modal proportions of amphibole. The Group I xenoliths are considered as amphibolite facies rocks derived from the MCC, due to the lack of garnet and pyroxene. The Group II xenoliths include garnet granulite and garnet amphibolite, and are composed of garnet, clinopyroxene, amphibole, plagioclase and quartz with minor rutile, ilmenite and sphene (Figs. 2e, g, f). The peak metamorphic assemblage of garnet, clinopyroxene, plagioclase, quartz and rutile suggests that the Group II xenoliths underwent HP granulite-facies metamorphism (Liu et al., 2009). The amphibole coronas on clinopyroxene and sphene coronas on rutile record amphibolite-facies retrogression during exhumation of these xenoliths (Figs. 2e, g). The Group III xenoliths include spinel-bearing garnet clinopyroxenite, phlogopite clinopyroxenite and spinel pyroxenite. Two Group III samples (11JG-5b and 14JG-36) were collected by this study and others are from Liu et al. (2013). Sample 11JG-5b contains abundant brown amphiboles (i.e. Ti-rich amphibole formed under high pressures; Liu et al., 2009) (Fig. 2h). Sample 14JG-36 is a spinel-bearing garnet clinopyroxenite with garnet coronas around spinel (Fig. 2i). The mineral assemblages and microtextures of the Group III xenoliths indicate that they were formed under conditions of spinel-garnet transitional facies in the SCLM (Liu et al., 2013).

Table 1 A summary of mineral assemblages, δ^{26}Mg (‰) and ^{87}Sr/^{86}Sr ratios of Xu-Huai deep-seated xenoliths

Sample	Lithology	Mineral assemblage	δ^{26}Mg	2SD	^{87}Sr/^{86}Sr	2SD
GroupI						
11JG-1	Amphibolite	0.98Amp + 0.02(Pl + Qz)	−0.32	0.03		
11JG-7	Amphibolite	0.95Amp +0.05(Pl +Qz)	0.01	0.08	0.70792	0.00002
11JG-11	Amphibolite	0.93Amp+0.07(Pl+Qz)	−0.25	0.03	0.70869	0.00001
11JG-13b	Amphibolite	0.97Amp+0.03(Pl+Qz)	−0.11	0.04	0.70702	0.00001
11JG-14	Amphibolite	0.99Amp+0.01(Pl+Qz)	0.00	0.02		
11JG-6a	Plag. amphibolite	0.26Amp +0.67(Pl +Qz) +0.07Bi	−0.31	0.03	0.70748	0.00006
11JG-6b	Plag. amphibolite	0.54Amp+0.45(Pl+Qz)+0.01Ilm	−0.21	0.06		
11JG-6c	Plag. amphibolite	0.25Amp+0.69(Pl+Qz)+0.06Bi	−0.28	0.01	0.70709	0.00003
11JG-8	Plag. amphibolite	0.60Amp+0.30(Pl+Qz)	−0.30	0.07		
11JG-12	Plag. amphibolite	0.87Amp+0.13(Pl+Qz)	−0.27	0.10	0.70345	0.00001
11JG-15	Plag. amphibolite	0.76Amp +0.24(Pl +Qz)	−0.23	0.06		
GroupII						
11JG-9a	Grt.amphibolite	0.03Grt+0.34Amp+0.63(Pl+Qz)	−0.24	0.08	0.70377	0.00001
11JG-9b	Grt.amphibolite	0.04Grt+0.31Amp+0.66(Pl+Qz)	−0.21	0.06	0.70317	0.00001
11JG-13c	Grt.amphibolite	0.14Grt+0.26Amp+0.07Cpx+0.53(Pl+Qz)	−0.20	0.07	0.70748	0.00001
14JG-23a	Grt.amphibolite	0.41Grt+0.49Amp+0.05Cpx+0.05(Pl+Qz)	−0.91	0.03	0.70589	0.00001

Continued

Sample	Lithology	Mineral assemblage	δ^{26}Mg	2SD	^{87}Sr/^{86}Sr	2SD
11JG-18	Grt.amphibolite	0.15Grt+0.81Amp+0.04(Pl+Qz)	−0.28	0.06		
11JG-10	Grt.granulite(r)	0.15Grt+0.43Amp+0.29Cpx+0.13(Pl+Qz)	−0.15	0.09	0.70692	0.00001
11JG-13a	Grt.granulite(r)	0.29Grt+0.26Amp+0.40Cpx+0.05(Pl+Qz)	−0.24	0.10	0.70703	0.00001
11JG-19a	Grt.granulite	0.30Grt+0.57Cpx+0.07(PL+Q)+0.06Ilm	−0.53	0.08	0.70764	0.00001
14JG-01	Grt.granulite(r)	0.22Grt+0.56Cpx+0.14Amp+0.08(Pl+Qz)	−0.07	0.08	0.70575	0.00001
14JG-02	Grt.granulite(r)	0.29Grt+0.16Amp+0.40Cpx+0.15(Pl+Qz)	−0.23	0.07	0.70577	0.00001
14JG-29	Grt.granulite(r)	0.15Grt+0.51Amp+0.22Cpx+0.12(Pl+Qz)	−0.16	0.06	0.70887	0.00001
14JG-35	Grt.granulite(r)	0.19Grt+0.67Amp+0.08Cpx+0.05(Pl+Qz)+0.01Ilm	−0.93	0.04	0.70825	0.00001
Group GroupIII						
07JG03[a]	Grt. clinopyroxenite	Grt+Cpx+Spl	−0.98	0.06	0.70409	0.00002
07JG17[a]	Spl.clinopyroxenite	Cpx+Spl	−1.07	0.08	0.70623	0.00001
07JG21[a]	Spl.clinopyroxenite	Cpx+Spl	−0.23	0.09	0.70387	0.00001
07JG30[a]	Grt. clinopyroxenite	Grt+Cpx+Amp+Spl	−0.73	0.07	0.71262	0.00001
08JG06[a]	Phl. clinopyroxenite	Cpx+Cpx+Phl	−0.76	0.07	0.71675	0.00001
11JG-5b	Grt. clinopyroxenite	0.11Grt+0.84Cpx+0.03Apm+0.02Spl	−0.75	0.03	0.70732	0.00001
14JG-36	Grt. clinopyroxenite	0.05Grt+0.94Cpx+0.01Spl	−1.23	0.00	0.70886	0.00001

Note: Modal mineral compositions were determined using point-counting method (1000 points) under optical microscope. Amp=amphibole, Pl+Qz=plagioclase+quartz, Bi=biotite, Ilm=ilmenite, Grt=garnet, Cpx=clinopyroxene, Spl=spinel, Phl=phlogopite. “r” refers to granulite samples which underwent amphibolite-facies retrogression.

[a] The samples are from Liu et al. (2013) and no modal data are obtained.

3　Analytical methods

3.1　Major and trace elements

Major elements were analyzed using wet-chemistry methods at the China University of Geosciences, Beijing (CUGB). Loss on ignition (LOI) was determined by gravimetric methods. Analytical uncertainties (2SD) for major elements were better than 2% and for the majority were better than 1% based on repeated measurement of rock standard GBW07105 (Table S1). Trace elements were measured using an ELAN DRCII inductively coupled plasma mass spectrometer (ICP-MS) at the University of Science and Technology of China (USTC) following the method established by Hou and Wang (2007). Whole-rock powder weighing ~50 mg was dissolved in a mixture of HF +HNO_3 at 190 ℃ using Parr bombs and complete sample dissolution was achieved after ~72 h. The dissolved samples were diluted to 80 ml using 1% HNO_3 before analysis. Reproducibility was better than 5% for elements with concentrations >10 μg/g and less than 10% for those <10 μg/g based on long-term analysis of standard materials (Table S1).

3.2　Strontium isotopes

Strontium (Sr) isotopic analysis was performed at the Isotope Geochemistry Laboratory of CUGB. About 50–70 mg whole-rock powders were digested in Savillex® screw top beakers with a mixture of HF, HNO_3 and

HCl. Dissolved solution was loaded on 3ml Bio-Rad 200–400 mesh AG50W-X12 resin in order to separate Sr from other elements. Sr isotopic analysis was conducted on a Thermo-Finnigan Neptune plus MC-ICP-MS. The mass fractionation correction for Sr isotopic ratios was based on $^{86}Sr/^{88}Sr$ =0.1194 using an exponential law. During the course of sample analysis, the NIST SRM 987 yielded an average $^{87}Sr/^{86}Sr$ ratio of 0.710274 ±25(2SD, n=4), in agreement with the long-term value of $^{87}Sr/^{86}Sr$ ratio of 0.710274±21(2SD, n =61) obtained in this lab.

3.3 Magnesium isotopes

Magnesium isotopic analysis was performed at the Isotope Geochemistry Laboratory of CUGB. All chemical procedures were carried out in laminar flow hoods (Class 100) in a clean room (Class 1000). Approximately 2–10 mg whole-rock powders, or 2–5 mg mineral separates containing over ten grains in order to avoid inter-grain heterogeneity, were dissolved in Savillex® screw-top beakers. About 10μg Mg in each sample can be obtained for isotopic analysis. Samples were first dissolved in a 3:1 (v/v) mixture of concentrated HF-HNO_3, followed by heating at a temperature of 160 ℃ on a hotplate in an exhaust hood until the solutions became transparent. Afterwards, the solutions were evaporated to dryness at 140 ℃. The dried residues were refluxed with a 3:1 (v/v) mixture of concentrated HCl-HNO_3 in order to remove residual fluorides. The solutions were then dried down at 80 ℃ overnight, and refluxed with concentrated HNO_3 so as to achieve 100% dissolution. The final samples were dissolved in 1 N HNO_3 in preparation for chromatographic separation.

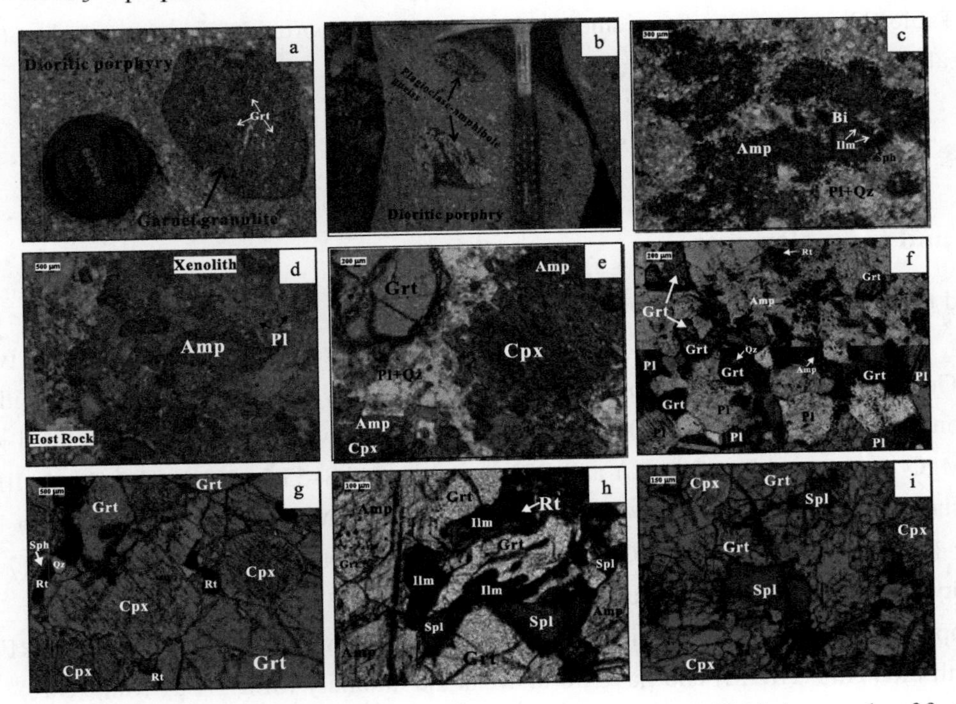

Fig. 2 Filed photographs (a, b) and photomicrographs (c–i) of Xu-Huai xenoliths. (a, b) Field photographs of fresh garnet granulite and plagioclase amphibolite; (c) Bi surrounding Rt and Ilm in plagioclase amphibolite (sample 11JG-6a; Group I); (d) amphibole cumulates in amphibolites (11JG-13b; Group I); (e) Amp corona surrounding Cpx in garnet granulite (11JG-10; Group II); (f) granular texture, Grt+Amp+Pl in garnet amphibolite (11JG-9a; Group II); (g) Grt+Cpx+Qz+Rt (rimmed by Sph) in garnet granulite (11JG-19a; GroupII); (h) Spl+Grt+Amp in spinel-bearing garnet clinopyroxenite (11JG-5b; Group III); (i) Grt coronae around Spl in spinel-bearing garnet clinopyroxenite (14JG-36; Group III). Amp=amphibole, Bi=biotite, Cpx=clinopyroxene, Grt=garnet, Ilm=ilmenite, Pl=plagioclase, Qz=Quartz, Rt=rutile, Sph=sphene, Spl=spinel.

Separation of Mg was achieved by cation exchange chromatography using Bio-Rad 200–400 mesh AG50W-X8 cation exchange resin. Prior to loading on the column, the resin was washed with 18.2 MΩ water, 6 N HCl and 1 N HNO_3 multiple times. Afterwards, the samples were loaded on 2 ml resin and eluted with 1 N HNO_3 following the procedure outlined in Teng et al. (2010) and Ke et al. (2016). The final solutions were heated to dryness in a vented laminar-flow hood and dissolved in 3% HNO_3 for mass spectrometry. The same procedure was repeated for some samples until all the interference elements (e.g., Ti) were removed completely. At least one USGS igneous rock standard was processed through column chemistry each time. The yield of Mg after column chemistry is better than 99% and total procedural blank is always less than 10 ng, leading to insignificant corrections to sample isotopic compositions.

Magnesium isotopic ratios were measured by the sample standard bracketing method using a Thermo-Finnigan Neptune plus MC-ICP-MS in low-resolution mode. Details are given by Ke et al. (2016). The data are reported in the δ-notation relative to DSM3(Galy et al., 2003): $\delta^{x}Mg=[(^{x}Mg/^{24}Mg)_{sample}/(^{x}Mg/^{24}Mg)_{DSM3}-1]\times10^{3}$, where x refers to 25 or 26. Long-term external reproducibility on the $^{26}Mg/^{24}Mg$ ratio is better than ±0.06(2SD) based on replicate analyses of natural and synthetic standards (Ke et al., 2016). The internal precision determined through more than three analyses of the same sample solutions during a single analytical session is better than 0.1‰. For each sample, when internal precision is larger than 0.06‰, the internal precision is always given. Two basalt standards (BHVO-2 and BCR-2) were analyzed during the course of sample analysis, yielding $\delta^{26}Mg$ values of −0.25±0.05‰ (n=5, 2SD) and −0.24±0.03‰ (n=3, 2SD), respectively (Table S3), in excellent agreement with the values reported by previous studies (e.g., An et al., 2014; Teng et al., 2015).

4 Results

4.1 Major and trace elements

Major and trace elemental compositions of the studied xenoliths are given in Table S1. The majority of the xenoliths have mafic compositions with SiO_2 ranging from 42.0 to 53.5 wt.%, except for two plagioclase amphibolites (Group I) with higher SiO_2 contents (55.4 wt.% and 60.9 wt.%). The Group I xenoliths have a large MgO range from 3.6 to 18.1wt.% and SiO_2 range from 48.7 to 60.9wt.%, whereas the Group II xenoliths show less variable MgO contents (4.9–9.5 wt.%) and SiO_2 contents (42.8–53.5 wt.%). The Group III xenoliths are generally enriched in MgO (8.0–14.6 wt.%) and Mg# (100 ×molar $Mg/(Mg+Fe^{2+})$; 54–76), but have low SiO_2 contents (42.0–48.1 wt.%) (Fig. 3a). The Group III xenoliths also have a large range of CaO (7.0–20.5 wt.%), with the majority more enriched in CaO relative to other groups (Fig. 3b).

The Group I xenoliths display significant enrichment of light rare earth elements (LREE) relative to heavy rare earth elements (HREE) and have no or weak Eu anomaly ($\delta Eu = Eu_N/(Sm_N \cdot Gd_N)^{0.5}$ = 0.48–1.03, where subscript N denotes normalization to CI chondritic value (Sun and McDonough, 1989))(Fig. 4a). They also have highly variable REE (28–294 μg/g), Cr (118–2857 μg/g) and Ni (52–445 μg/g) concentrations. The Group II xenoliths also show LREE enrichment relative to HREE (Fig. 4b), similar to the average LCC (Rudnick and Gao, 2003), but they display variable Eu anomalies (δEu=0.68–1.65). Except for sample 07JG21 with weak LREE enrichment, the Group III xenoliths have flat REE patterns and weak Eu anomalies (δEu=0.63–0.92), resembling E-MORB but differing from OIB and N-MORB (Fig.4c). Most of these xenoliths studied here have highly variable and extremely positive U and Pb anomalies, which are different from the global average LCC, N/E-MORB, OIB and mafic granulites or pyroxenites from other regions in the NCC (Figs. 4d–f).

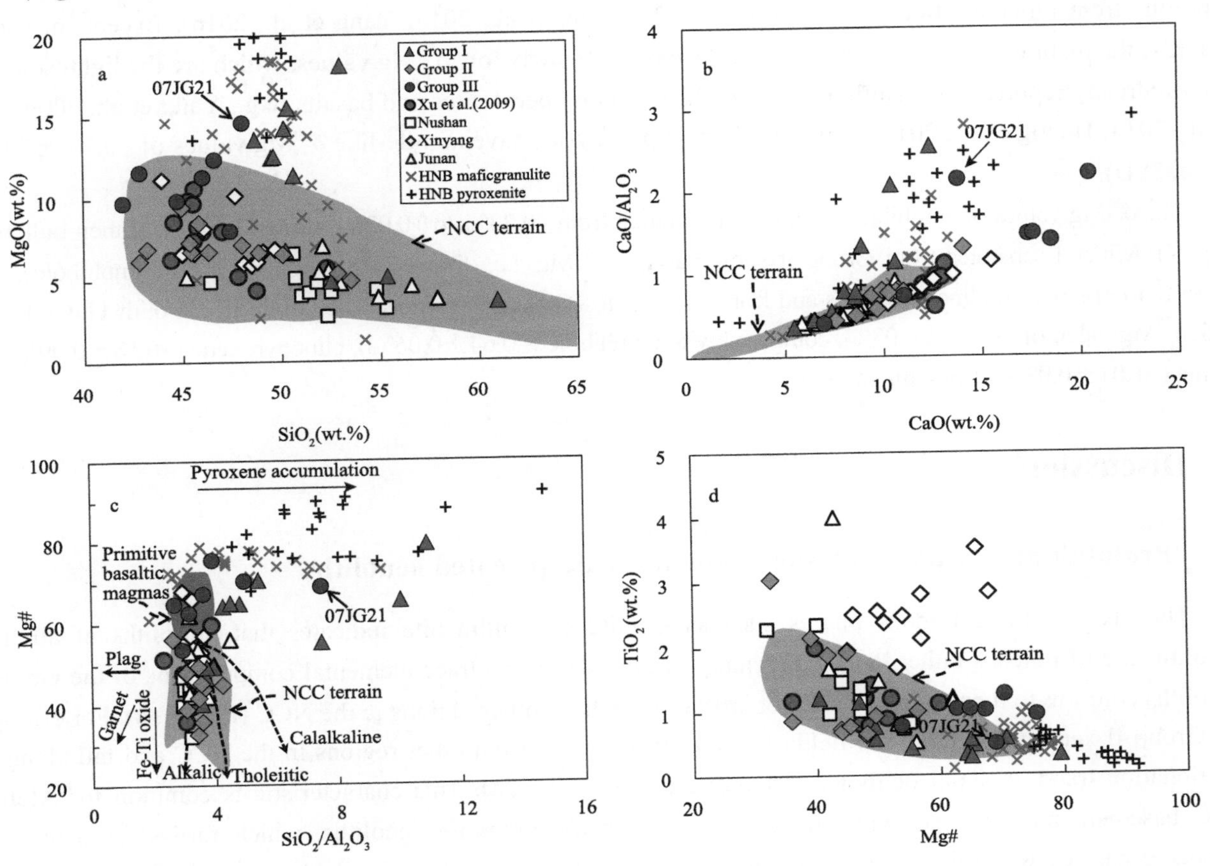

Fig. 3 Correlations of selected major elements for the Xu-Huai xenoliths. Mafic granulite and pyroxenite xenoliths from other regions and the fields for Archean granulite terrains in the NCC are also shown for comparison. Data sources: Nushan (Huang et al., 2004); Xinyang (Zheng et al., 2003); Junan (Ying et al., 2010); Hannuoba (Liu et al., 2001; Xu, 2002; Zhou et al., 2002); Table S1.

4.2 Strontium isotopes

Strontium isotopic compositions of xenoliths analyzed here are reported in Table S2. The Group I and II xenoliths have present-day $^{87}Sr/^{86}Sr$ ratios ranging from 0.70345 to 0.70869 and from 0.70317 to 0.70877, respectively. By contrast, the present-day $^{87}Sr/^{86}Sr$ ratios of the Group III xenoliths are highly variable and more radiogenic (0.70387 to 0.71675).

4.3 Magnesium isotopes

Magnesium isotopic compositions of whole rocks and mineral separates, together with reference materials (BHVO-2 and BCR-2) analyzed here are reported in Table S3. The inter-mineral Mg isotopic difference between two coexisting minerals is defined as $\Delta^{26}Mg_{X-Y}=\delta^{26}Mg_X-\delta^{26}Mg_Y$, where X and Y refer to different minerals. All samples including whole-rocks, minerals and reference materials analyzed in this study fall on a single mass fractionation line with a slope of $0.513\pm0.024(2\sigma)$. The overall range of $\delta^{26}Mg$ values of the studied xenoliths is up to 1.24‰ (−1.23 to +0.01‰; Figs. 5–6). In detail, the Group I xenoliths have a narrow $\delta^{26}Mg$ range and most of samples exhibit mantle-like values except for two amphibolites with heavy δ^{26}Mg around zero. The δ^{26}Mg values of Group II xenoliths are more variable and similar to or slightly lower than those of lower crustal xenoliths from other locations (−0.76‰ to +0.19‰; Teng et al., 2013; Yang et al., 2016). Except for sample 07JG21, the mantle-derived Group III xenoliths have extremely low δ^{26}Mg values, which are the lightest among values already reported for mantle-derived rocks including peridotites and basalts (e.g., Yang et al., 2009; Xiao et al., 2013; Huang et al., 2015). The host dioritic porphyries have mantle-like δ^{26}Mg values of −0.25 ±0.04‰ (n =4, 2SD).

The δ^{26}Mg values of amphiboles from Group I range from −0.33‰ to +0.02‰, similar to those of their bulk rocks (Fig. 7). Mineral separates from Group II show variable δ^{26}Mg, i.e., from −0.78‰ to +0.10‰ in amphiboles, from −0.85‰ to +0.16‰ in clinopyroxenes and from −1.34‰ to −0.41‰ in garnets. The Group III xenolith 11JG-5b has a bulk δ^{26}Mg value of −0.75 ±0.03‰, compared with amphibole (−0.64±0.06‰), clinopyroxene (−0.58±0.09‰) and garnet (−1.01±0.03‰) from this sample.

5 Discussion

5.1 Protolith nature and metasomatism of the deep-seated xenoliths

The absence of aluminous phases such as kyanite and sillimanite indicates that protoliths of all these xenoliths are of igneous rather than sedimentary origin. Major and trace elemental compositions of the Group II xenoliths conform to those of ancient lower crustal rocks from other regions in the NCC (Figs. 3, 4). For example, the Group II xenoliths overlap the field of lower crustal rocks from other regions in the NCC and fall along the fractionation trend of alkali or tholeiitic basaltic magma (Fig. 3c). This characteristic is common for Archean NCC basement but contrasts with Hannuoba (HNB) mafic granulite xenoliths, which formed as a result of Mesozoic basaltic underplating (e.g. Liu et al., 2001; Ying et al., 2010). Available zircon U–Pb ages of lower crustal xenoliths in this area suggest that their protoliths mainly formed in the late Archean (~2.5 Ga) and underwent HP granulite-facies metamorphism at around 1.8–1.9 Ga (Liu et al., 2009, 2013; Xu et al., 2009), coincident with the ages of two major thermal-tectonic-magmatic events occurring in the NCC (e.g., Zhai et al., 2000). By contrast, the Group I xenoliths with high Mg# are compositionally similar to pyroxenites and mafic granulites of cumulate origin from Hannuoba in the northern NCC (Fig. 3; Liu et al., 2001; Xu, 2002), suggesting that the Group I xenoliths formed as cumulates in the middle continental crust. The negative Nb–Ta–Zr–Hf anomalies and positive Ba–U–Pb anomalies suggest that precursors of the Group I xenoliths have arc magma signatures (Fig. 4).

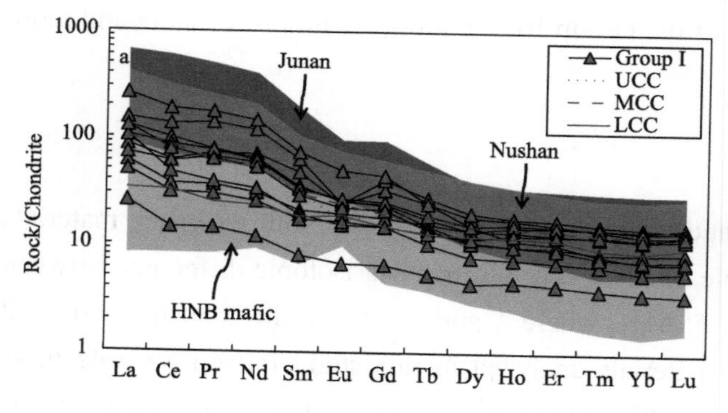

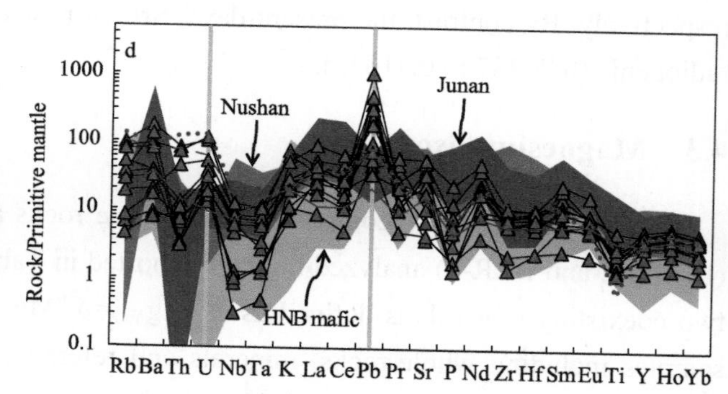

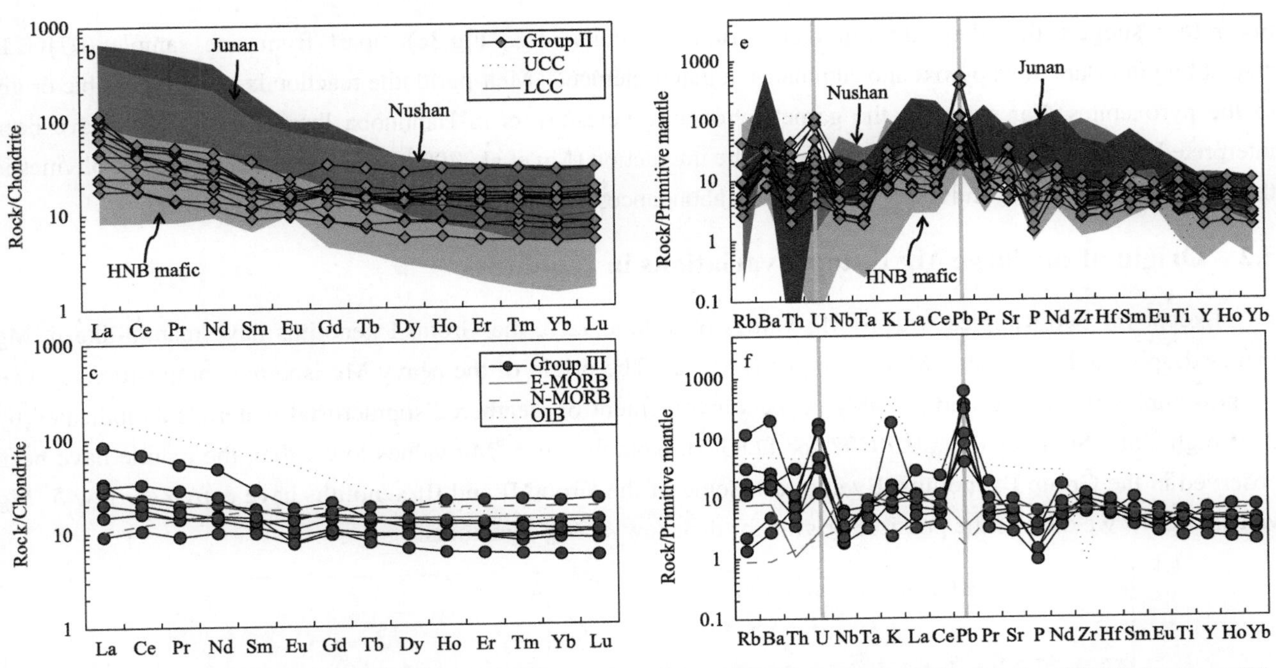

Fig. 4 Chondrite-normalized REE patterns and primitive mantle-normalized trace-element patterns for the Xu-Huai xenoliths. Mafic granulites from Junan, Nushan and Hannuoba as well as the global average UCC, MCC, LCC, OIB, E-MORB and N-MORB are shown for comparison. The normalized values and data for OIB, E-MORB and N-MORB are from Sun and McDonough(1989). Data of UCC, MCC and LCC are from Rudnick and Gao(2003). Data sources for mafic granulites are as the same as in Fig. 3.

Garnet commonly appears as coronae around spinel and exsolution lamellae in clinopyroxene in the Group III xenoliths. These textures have been interpreted as due to slow isobaric cooling at upper mantle depths at P–T conditions of 1.6–2.0 GPa and 900–1100 ℃ (e.g., O'Neill, 1981; Jerde et al., 1993). The Group III xenoliths also have relatively high Mg# (up to 76), flat REE patterns and positive zircon εHf(t) values (+0.5 to +6.7) (Liu et al., 2013). These characteristics indicate that their precursors are mantle-derived basaltic melts that stalled in the lithospheric mantle. Pyroxenites commonly originate from high-pressure cumulates (e.g., Bodinier et al., 1987), but no typical cumulate textures (like modal layer) have been observed in the Group III pyroxenites. The major and trace element features are also different from the HNB pyroxenites of cumulate origin (e.g., Figs.3, S2),

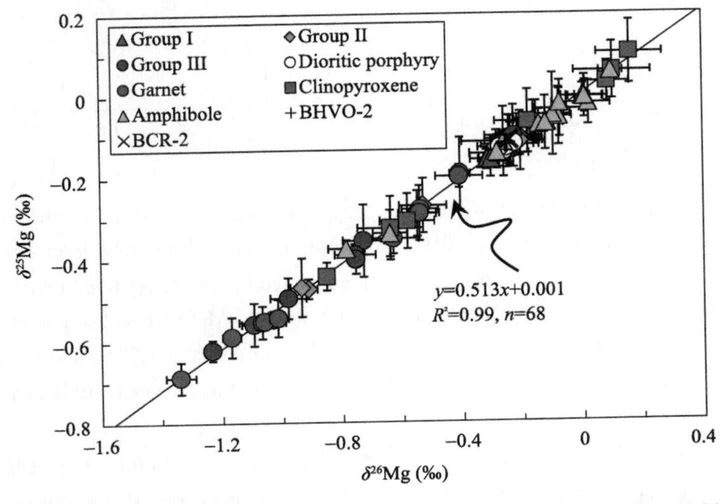

Fig. 5 Plot of δ^{26}Mg vs. δ^{25}Mg for all samples including standards reported in this study. Data are from Table 1 and Table S3.

but rather suggest that they are relatively primitive melts (e.g., Fig.3c), apart from one sample (07JG21) resembling the Hannuoba pyroxenite cumulates in major elements. Melt-peridotite reaction is another possible origin of the pyroxenites. For example, the garnet pyroxenite veins/layers in Hannuoba lherzolite xenoliths have been interpreted as the products of silicic melt-peridotite interaction (Liu et al., 2005). However, the absence of olivines in the Group III pyroxenites and their low Ni and Cr abundances do not support melt-peridotite interaction.

5.2 Origin of the large Mg isotopic variations in xenoliths

Except for two samples with δ^{26}Mg values of around zero, the Group I xenoliths have mantle-like δ^{26}Mg values despite highly variable MgO contents (Fig. 6a). The origin of the heavy Mg isotopic compositions in two of the samples is unclear, and possibly reflects involvement of weathered supracrustal materials as indicated by their high $^{87}Sr/^{86}Sr$ ratios (e.g., 0.70792 for 11JG-7). Notably, no δ^{26}Mg values lower than the mantle have been observed in the Group I xenoliths. By contrast, some of the Group II and III xenoliths have extremely low δ^{26}Mg values. Below we discuss the possible origins for their low δ^{26}Mg signatures.

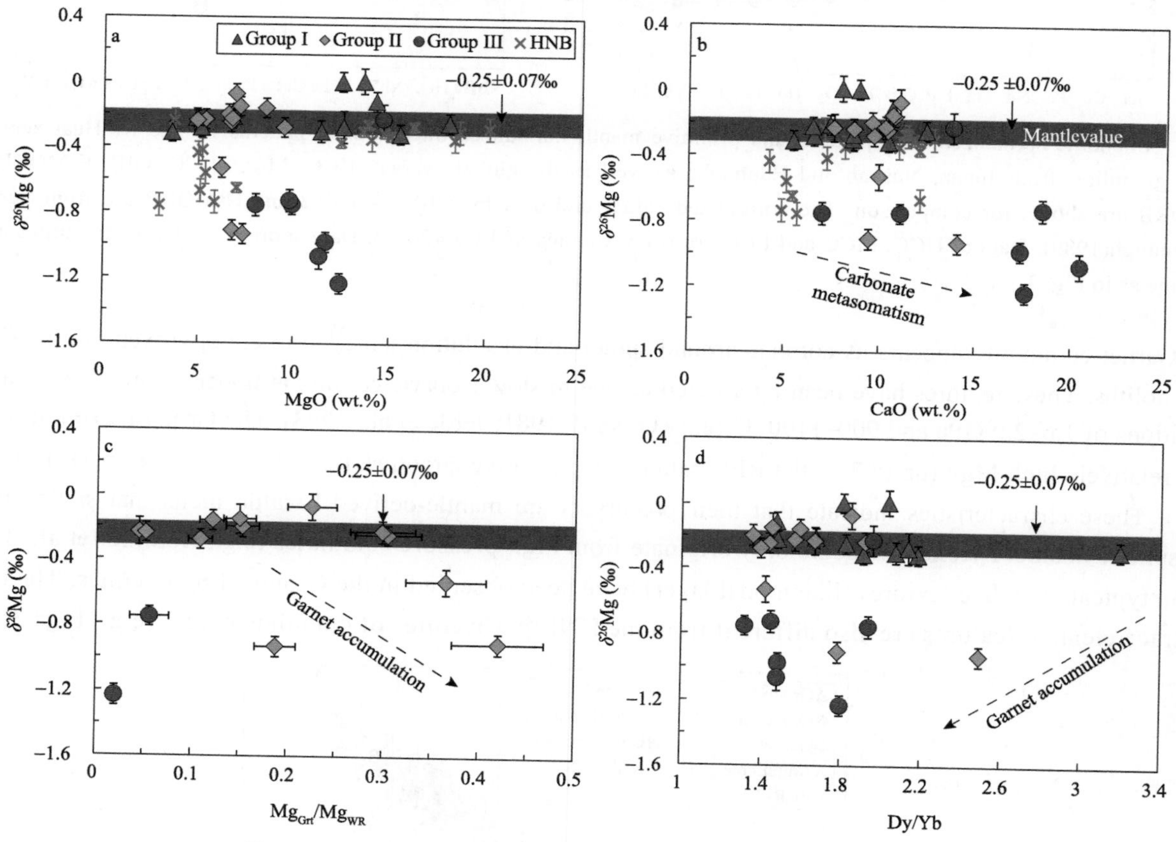

Fig. 6 Correlations between δ^{26}Mg and MgO (a), CaO (b), Mg_{Grt}/Mg_{WR}(c) and Dy/Yb (d). The grey band represents average δ^{26}Mg of global oceanic basalts (−0.25±0.07‰; Teng et al., 2010). Mg isotopic data of Hannuoba granulite xenoliths are also shown for comparison (Yang et al., 2016). Mg_{Grt}/Mg_{WR} refers to percent of Mg contained in garnets relative to bulk rocks, obtained on the basis of MgO contents and modal abundances of garnet, and MgO contents of bulk rock. MgO content of garnet is assumed to be 7±0.8 wt.% (2SD) for Group II and 5±1.8 wt.% (2SD) for Group III, based on the data of Xu et al. (2009) and Liu et al. (2009, 2013). Error bars represent 2SD uncertainties. Data are from Table 1 and Table S1. Note that the dashed arrows are only qualitatively indicative.

Since garnet has significantly lower δ^{26}Mg values than other silicate minerals (e.g., olivine, pyroxene, amphibole and mica) and oxides (e.g., spinel) (Li et al., 2011; Wang et al., 2012) and most of the studied xenoliths with low δ^{26}Mg values contain garnet, it is necessary to evaluate the influence of garnet accumulation on the bulk Mg isotopic compositions of the xenoliths. First, the absence of a correlation between bulk δ^{26}Mg values and relative percent of garnets in bulk rocks (Fig. 6c) or Dy/Yb ratios (Fig. 6d) suggests that garnet is not the major cause for bulk-rock Mg isotopic variations. Second, the major Mg-bearing minerals from the isotopically light samples (Groups II and III) also have low δ^{26}Mg values (Fig. 7). For example, clinopyroxene and amphibole from the isotopically light sample 11JG-5b (−0.75±0.03‰) also have low δ^{26}Mg values of −0.64±0.06‰ and −0.58±0.09‰, respectively. These observations strongly suggest that the low δ^{26}Mg values of bulk xenoliths are not a result of garnet enrichment.

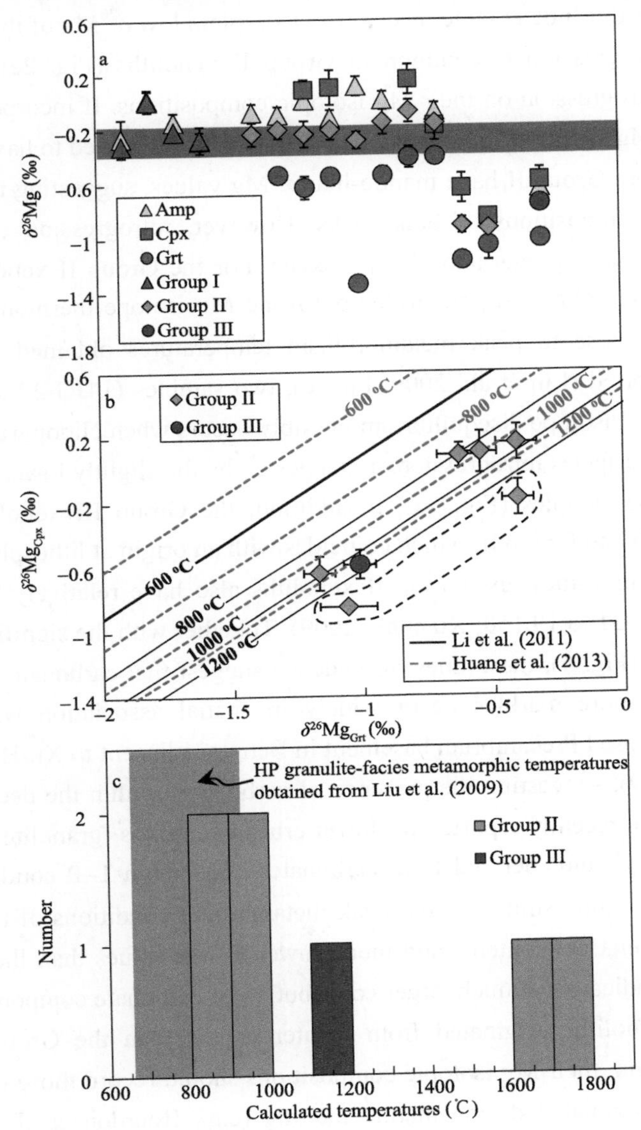

Fig. 7 (a) A comparison of δ^{26}Mg values between major Mg-bearing silicate minerals and whole rocks. (b) $\delta^{26}Mg_{Cpx}$ vs. $\delta^{26}Mg_{Grt}$ for Group II and III samples. (c) A histogram showing temperatures calculated from the Grt-Cpx Mg isotope thermometer (Li et al., 2011). The grey band represents global oceanic basalts (δ^{26}Mg=−0.25±0.07‰; Teng et al., 2010). Error bars represent 2SD uncertainties. The grey dashed line and solid black line represent the theoretically and empirically calculated equilibrium Mg isotope fractionation line between Cpx and Grt, respectively (Li et al., 2011; Huanget al., 2013). Data are from Table S3.

Diffusion-driven Mg isotope fractionation at high temperatures has been documented in many studies (e.g. Richter et al., 2008; Pogge von Strandmann et al., 2011). For example, δ^{26}Mg of peridotite xenoliths could drop to −0.36‰ due to diffusive replacement of H^+ by Mg^{2+} from basalts during xenolith outgassing (Pogge von Strandmann et al., 2011). The host rocks of the studied xenoliths have low MgO contents (<5 wt.%; Xu et al.,

2006) but normal δ^{26}Mg values (−0.25±0.04‰), making it difficult to generate such low δ^{26}Mg values (down to −1.23‰) in the xenoliths without significant isotope shift in the host magma, if large amounts of Mg diffused from host magma to xenoliths. In addition, many xenoliths with high MgO contents have normal δ^{26}Mg (Fig. 5), which is contradictory to that expected if light Mg preferentially incorporated into the xenoliths. The Soret effect (thermal diffusion process) can also be excluded as the light isotope prefers the hot end (host magma in this case) and the heavy isotope prefers the cold end (captured xenoliths in this case) (e.g., Richter et al., 2008; Huang et al., 2010). A diffusion process is thus not a viable mechanism to explain low δ^{26}Mg of the Group II and III xenoliths.

Amphibolite-facies retrogression is common in Group II xenoliths (Fig. 2e) and thus it is necessary to evaluate the influence of retrogression on their Mg isotopic compositions. If incorporation of exotic Mg-bearing fluids results in the low-δ^{26}Mg signatrue, all retrograded xenoliths are expected to have low δ^{26}Mg values. Most of the retrograded xenoliths from Group II have mantle-like δ^{26}Mg values, suggesting that retrogression had limited influence on Mg isotopic compositions of bulk rocks. However, retrogression could potentially disturb the equilibrium fractionation between garnet and clinopyroxene. For the Group II xenoliths, although temperatures obtained from the empirically calibrated garnet-clinopyroxene Mg isotope thermometer (825–998 ℃; Li et al., 2011) in four samples are close to peak metamorphism temperatures obtained using the Fe–Mg exchange thermometer previously reported (Liu et al., 2009; Fig.7c), two samples (14JG-29 and 14JG-35) show apparent discordance. The inter-mineral isotope disequilibrium possibly occurs when clinopyroxene or garnet transforms to amphibole during amphibolite-facies retrogression, as suggested by the slightly lower δ^{26}Mg values of amphiboles than clinopyroxenes in some samples (Fig. 7a). In addition, the Group III xenolith (11JG-5b) has a higher temperature of 1114 ℃ relative to Group II, which coincides with an origin at lithospheric mantle depth.

In addition to low δ^{26}Mg values, the Group II xenoliths also have relatively high radiogenic Pb isotopic composition (^{206}Pb/^{204}Pb=17.531 to 19.150; Xu et al., 2009). Together with the significantly positive anomalies in U and Pb of these xenoliths (Fig. 4), these lines of evidence suggest that carbonate is a potential source for the fluids in the deep crust. Impure marbles occur widely in spatial association with amphibolite and garnet amphibolite blocks in the exposed Precambrian basement in Bengbu adjacent to Xu-Huai in the southeastern NCC (Figs. 1b, S1; Liu et al., 2009), suggesting the presence of carbonates within the deep crust of the south-eastern NCC. Low δ^{26}Mg values were recently reported for lower crustal xenoliths (granulites) from Hannuoba and were attributed to metasomatism by fluids derived from carbonates under high T–P conditions (1 GPa, 800–950 ℃; Yang et al., 2016), conditions quite similar to the peak metamorphic conditions of the Xu-Huai granulites. The Group II xenoliths have higher CaO contents and much lower δ^{26}Mg values than the Hannuoba granulites with low δ^{26}Mg values (Fig. 6b), indicating a much larger contribution of carbonate components.

Since the Group III xenoliths originated from greater depths than the Group II xenoliths and have a mantle-derived magmatic origin, their Mg isotopic compositions should record those of the mantle sources, given that Mg isotopes are not fractionated during mantle melting (e.g., Bourdon et al., 2010; Teng et al., 2010). Sedimentary carbonates have extremely light Mg isotopic compositions (−5.5‰ to −0.5‰; e.g., Young and Galy, 2004) and act as the most probable metasomatic agent. This is supported by the broad negative correlation between δ^{26}Mg and bulk CaO contents (Fig. 6b). However, carbonatitic metasomatism fails to explain some trace element signatures of the Group III xenoliths. For example, carbonatitic metasomatism could generate negative Zr–Hf–Ti anomalies and positive Nb–Ta anomalies in mantle and mantle-derived rocks (e.g., Hoernle et al., 2002), e.g., low-δ^{26}Mg continental basalts from eastern China (Huang et al., 2015). However, these features are not observed in the Group III xenoliths (Fig. 4), suggesting that the metasomatic agent is not carbonatitic melt. Instead, the occurrence of hydrous minerals (phlogopite and non-retrogressive amphibole) in these xenoliths imply that their mantle source underwent fluid metasomatism (Fig. 2h and Liu et al., 2013). The negative Nb–Ta anomalies and positive Ba–K–U–Pb anomalies of the Group III xenoliths are also typical characteristics of subduction-zone fluid metasomatism (e.g., Kessel et al., 2005). Although decarbonation of carbonates in the subduction slab is difficult, carbonates are readily soluble in the released fluids during metamorphic dehydration (e.g., Frezzotti et al., 2011; Ague and Nicolescu, 2014). These carbonate-rich fluids may act as active agents in shifting the Mg isotopic composition of the mantle. In addition, the absence of significantly negative Zr–Hf–Ti anomalies requires addition of silicate sediments to the mantle source (e.g., Rudnick and Gao, 2003). The moderate to weak δEu anomalies of the Group III xenoliths, which are commonly observed in EM-type OIB, also support the involvement of sediments in their mantle sources (Sun and McDonough, 1989). Note that there are no clearly visible carbonate mineral phases in the Xu-Huai xenoliths (Fig. 2), indicating that the fluids are carbonate-undersaturated.

As Sr always replaces Ca in the lattice of carbonate that has high Sr contents (e.g., DePaolo, 1986), interaction of the pristine mantle with carbonates should generate elevated Sr concentrations in the mantle source and in melts from this source. However, the Group III xenoliths have no prominent positive Sr anomalies (57–197 μg/g), which seems to be inconsistent with that expected. This may indicate that Mg isotopic compositions and Sr concentrations of the Group III xenoliths are not a result of simple Mg and Sr mixing between carbonates and the mantle, but reflect Mg–Sr isotope exchange during fluid-rock interaction. Magnesium isotope exchange between silicate and carbonate has also been proposed to account for the extremely low δ^{26}Mg carbonated eclogites from Dabie-Sulu Orogen, eastern China (−0.65‰ to −1.93‰; Wang et al., 2014). Since Sr isotopes are not fractionated during melting processes, Sr isotopic ratios of these xenoliths could help constrain the nature of the metasomatic agent. Therefore, a Mg–Sr isotope exchange model, which was modified after a Mg–Sr isotope mixing model of Huang and Xiao (2016), was investigated to explain our data. The results show that most of the isotopically light samples can be explained by Mg–Sr isotope exchange between N-MORB and high-Mg calcite and magnesite (Fig. 8). Two samples having much more radiogenic Sr isotopic ratios were probably overprinted by strong clastic sediment metasomatism, which is also evidenced by their trace element signatures as discussed above. Overall, protoliths of the Group III xenoliths have been significantly metasomatized by carbonate-rich fluids derived from a subducted slab with contribution from clastic sediments.

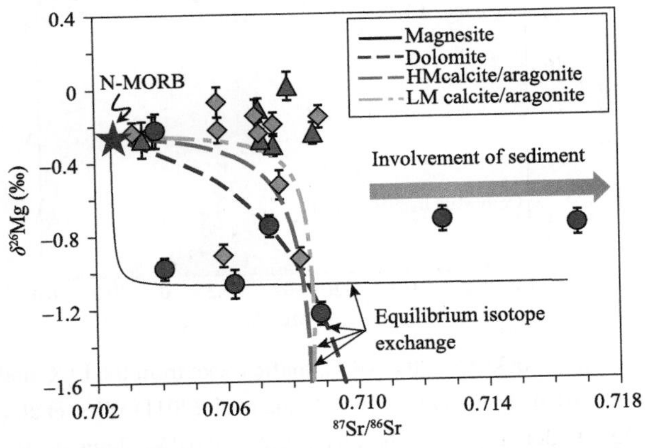

Fig. 8 Magnesium and strontium isotopic exchange between N-MORB and different carbonate species. The curves represent the Mg and Sr isotopic evolution of N-MORB due to an isotopic exchange reaction obtained following the equation: $(\varepsilon Sr_{N\text{-}MORB}^{f}-\varepsilon Sr_{N\text{-}MORB}^{i})/(\delta^{26}Mg_{N\text{-}MORB}^{f}-\delta^{26}Mg_{N\text{-}MORB}^{i})=(\varepsilon Sr_{carb.}^{i}-\varepsilon Sr_{N\text{-}MORB}^{f})/\delta^{26}Mg_{carb.}^{i}-\delta^{26}Mg_{N\text{-}MORB}^{f}+\Delta^{26}Mg_{N\text{-}MORB\text{-}carb.})\cdot([Sr]_{carb.}\cdot[Mg]_{N\text{-}MROB})/([Sr]_{N\text{-}MORB}\cdot[Mg]_{carb.})$, where f=final status after isotopic exchange, i=initial status before isotopic exchange, CARB. =carbonate (i.e., magnesite, dolomite, high-Mg (HM) calcite/aragonite and low-Mg (LM) calcite/aragonite), $\varepsilon Sr=[(^{87}Sr/^{86}Sr)/(^{87}Sr/^{86}Sr)_{UR}-1)]\times10^{4}$, [Sr]=Sr concentration and [Mg]=Mg concentration. $^{26}Mg_{N\text{-}MORB\text{-}carb.}$=0.03‰ at 1100 ℃ (Macris et al., 2013) was used in this model. Other parameters used in this model are from Huang and Xiao (2016) and given in Table S4. Mg isotopes, Sr concentration and Sr isotope data are listed in Table 1, Table S1 and Table S2, respectively.

5.3 The influence of carbonate solubility on deep crustal Mg isotopic composition

The current understanding of the nature and origin of Mg isotopic heterogeneity of the deep crust is still limited. Two suites of lower crustal xenoliths from Australia show a wide δ^{26}Mg range of −0.72‰ to +0.19‰ (Teng et al., 2013). A ~0.5‰ isotopic variation (from −0.74‰ to −0.26‰) in mafic granulite xenoliths from Hannuoba (NCC) has been interpreted as a result of interaction with carbonate fluids (Yang et al., 2016). By contrast, felsic to mafic rocks from the MCC generally have heavier δ^{26}Mg values of −0.40‰ to +0.12‰, and mafic composites have mantle-like δ^{26}Mg values from −0.26‰ to −0.16‰ (Yang et al., 2016). Overall, the data obtained in this study and from the literature suggest that the mafic LCC is more isotopically heterogeneous and lighter compared with mafic rocks from the MCC (Fig. 9).

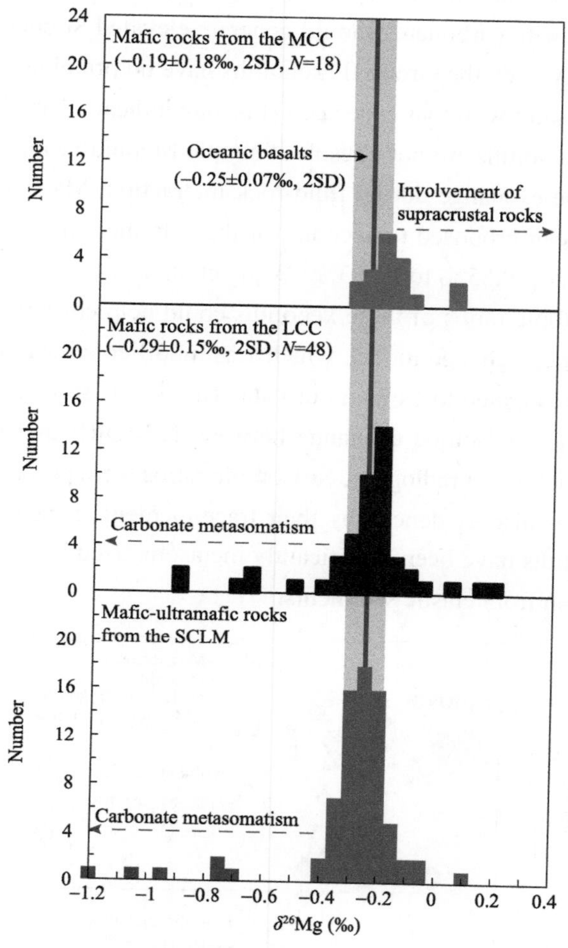

Fig. 9 Comparison of δ^{26}Mg between mafic rocks from the MCC, mafic rocks from the LCC and mafic to ultramafic rocks from the SCLM. Data are from Yang et al. (2009, 2016), Teng et al.(2010), Huang et al. (2011), Xiao et al. (2013) and this study (Table 1). The yellow bar represents average δ^{26}Mg of global oceanic basalts (−0.25±0.07‰; Teng et al., 2010). (For interpretation of the references to color in this figure legend, the reader is referred to the web version of this article.)

Numerous experimental studies have shown that the solubility of calcite in aqueous fluids decreases with increasing temperature at low pressure (<0.3 GPa) (Manning et al., 2013). On the contrary, at deep crustal and mantle pressure (>0.6 GPa), calcite solubility shows reverse trend, increasing with both increasing temperature and pressure. Similarly, Pan et al. (2013) predicted that the solubility products of carbonate species (including calcite, aragonite, dolomite and magnesite) in water increase substantially with increasing pressures using ab initio simulation method. These results indicate that carbonate-rich fluids are probably more pervasive in the deep crust than in the shallow crust. Therefore, metasomatic fluids derived from carbonate dissolution at different crustal depths may play an important role in controlling Mg isotopic compositions of the deep crust. Unlike the UCC, in which the Mg isotopic heterogeneity is mainly controlled by chemical weathering, carbonate metasomatism may be the most important process resulting in isotopic heterogeneity of the LCC.

5.4 Implications for tracing recycled ancient carbonates using Mg isotopes

As a major element in carbonates (e.g., dolomite and magnesite), the stable isotopes of Mg have received increasing attention in tracing recycled sedimentary carbonates in the deep Earth (Yang et al., 2012; Huang et al., 2015; Tian et al., 2016; Ke et al., 2016). For example, the low-δ^{26}Mg signatures of continental basalts from eastern China have been argued to indicate a widespread carbonated mantle source, which has been hypothesized to be related to the subduction of the west Pacific slab during the Mesozoic and Cenozoic (Yang et al., 2012; Huang et al., 2015; Tian et al., 2016). Considering the close spatial relationship with both the west Pacific Ocean to the east and the Paleo-Tethys Ocean to the south, the lithosphere beneath the Xu-Huai area was potentially affected by two episodes of subduction-related events. The formation age of the host dioritic porphyries roughly corresponds to the onset of west Pacific plate subduction beneath eastern China, which means that Mg isotopic variation in these captured xenoliths should record earlier metasomatism events. One of the Group III xenoliths (sample 07JG30 with δ^{26}Mg = −0.73‰) has been analyzed by zircon U–Pb dating and gives a magmatic age of 393±7 Ma (Liu et al., 2013). This age constraint indicates that the low-δ^{26}Mg Group III xenoliths probably record the fingerprint of recycled ancient carbonates related to the northward subduction of the Paleo-Tethys slab. A possible scenario is summarized below. Subduction of the Paleo-Tethys slab carried large amount of oceanic carbonates into the deep mantle, and fluids rich in dissolved carbonates migrated upwards from the slab and interacted with the asthenospheric mantle and/or lithospheric mantle. The flux of carbonate-rich fluids lowered the solidus of the mantle peridotite and generated the primitive melts of the Group III xenoliths, which ascended and then stalled at the lithospheric mantle.

6 Conclusion

Major-trace elements, Sr and Mg isotope data were reported for deep-seated xenoliths from the southeastern NCC, with the aim of characterizing Mg isotopic heterogeneity of the cratonic lithosphere in the same location and providing constraints on its origin. These xenoliths are from middle continental crust (Group I), ancient lower crust (Group II), and lithospheric mantle (Group III), respectively. The δ^{26}Mg values range from −0.32‰ to 0.01‰ for the Group I, from −0.93‰ to −0.07‰ for the Group II and from −1.23‰ to −0.23‰ for the Group III xenoliths. The low-δ^{26}Mg signatures of the Group II xenoliths are most likely due to metasomatism by carbonate-rich fluids in the lower crust, whereas the extremely low δ^{26}Mg values of the Group III xenoliths can be explained by fluid–mantle interaction. The metasomatic fluids derived from recycled carbonate-bearing oceanic crust. Together with data reported for middle and lower crustal rocks in other studies, it is concluded that the mafic LCC is more heterogeneous in Mg isotopic composition and lighter on average relative to the mafic rocks from the MCC, and rocks from the LCC commonly have light Mg isotopic compositions. The experimentally and computationally demonstrated, increasing solubility of carbonate in fluids with elevated pressure suggests that carbonate fluids are more pervasive in the lower crust relative to the middle crust. This may account for the presence of a low-δ^{26}Mg anomaly in the mafic rocks from the LCC but the general absence in mafic rocks from the MCC.

The extremely light Mg isotopic compositions of the Group III xenoliths provide important evidence for the presence of low-δ^{26}Mg mantle sources beneath the NCC. In combination with the previously-reported age of 393±7 Ma for one Group III xenolith sample, the results also provide evidence for ancient carbonate recycling related to Paleo-Tethys slab subduction during the Paleozoic. Therefore, Mg isotopes could be widely applied to trace recycled carbonates in the mantle.

Acknowledgements

We are grateful to Fangyue Wang, Jian Huang and Ji Shen for help during field work, Ting Gao, Zijian Li, Xunan Meng, and Lijuan Xu for the assistance for Mg and Sr isotope analysis in the lab and Zhenhui Hou for help during trace-element analysis. We thank the editor Prof. Derek Vance for comments and careful editorial handling, and Profs. Fang Huang, Joel Baker and an anonymous reviewer for their constructive comments, which significantly improved the manuscript. This work is supported by the National Natural Foundation of China (41203013 and 41473017) to S.A.L and (412302090) to S.G.L.

Appendix A. Supplementary material

Supplementary material related to this article can be found online at http://dx.doi.org/10.1016/j.epsl.2016.07.021.

References

Ague, J.J., Nicolescu, S., 2014. Carbon dioxide released from subduction zones by fluid-mediated reactions. Nat. Geosci. 7, 355-360.

An, Y., Wu, F., Xiang, Y., Nan, X., Yu, X., Yang, J., Yu, H., Xie, L., Huang, F., 2014. High precision Mg isotope analyses of low-Mg rocks by MC-ICP-MS. Chem. Geol. 390, 9-21.

Bodinier, J., Guiraud, M., Fabries, J., Dostal, J., Dupuy, C., 1987. Petrogenesis of layered pyroxenites from the Lherz, Freychinede and Prades ultramafic bodies (Ariege, French Pyrenees). Geochim. Cosmochim. Acta 51, 279-290.

Bourdon, B., Tipper, E.T., Fitoussi, C., Stracke, A., 2010. Chondritic Mg isotope com- position of the Earth. Geochim. Cosmochim. Acta 74, 5069-5083.

DePaolo, D.J., 1986. Detailed record of the Neogene Sr isotopic evolution of seawater from DSDP Site 590B. Geology 14, 103-106.

Frezzotti, M., Selverstone, J., Sharp, Z., Compagnoni, R., 2011. Carbonate dissolution during subduction revealed by diamond-bearing rocks from the Alps. Nat. Geosci. 4, 703-706.

Galy, A., Yoffe, O., Janney, P.E., Williams, R.W., Cloquet, C., Alard, O., Halicz, L., Wadhwa, M., Hutcheon, I.D., Ramon, E., 2003. Magnesium isotope heterogeneity of the isotopic standard SRM980 and new reference materials for magnesium–isotope–ratio measurements. J. Anal. At. Spectrom. 18, 1352-1356.

Handler, M.R., Baker, J.A., Schiller, M., Bennett, V.C., Yaxley, G.M., 2009. Magnesium stable isotope composition of Earth's upper mantle. Earth Planet. Sci. Lett. 282, 306-313.

Hoernle, K., Tilton, G., Le Bas, M.J., Duggen, S., Garbe-Schönberg, D., 2002. Geochemistry of oceanic carbonatites compared with continental carbonatites: mantle recycling of oceanic crustal carbonate. Contrib. Mineral. Petrol. 142, 520-542.

Hou, Z., Wang, C., 2007. Determination of 35 trace elements in geological samples by inductively coupled plasma mass spectrometry. J. Univ. Sci. Technol. China 37, 940-944 (in Chinese with English abstract).

Huang, F., Chakraborty, P., Lundstrom, C.C., Holmden, C., Glessner, J.J.G., Kieffer, S.W., Lesher, C.E., 2010. Isotope fractionation in silicate melts by thermal diffusion. Nature 464, 396-400.

Huang, F., Zhang, Z., Lundstrom, C.C., Zhi, X., 2011. Iron and magnesium isotopic compositions of peridotite xenoliths from

Eastern China. Geochim. Cosmochim. Acta 75, 3318-3334.

Huang, F., Chen, L., Wu, Z., Wang, W., 2013. First-principles calculations of equilibrium Mg isotope fractionations between garnet, clinopyroxene, orthopyroxene, and olivine: implications for Mg isotope thermometry. Earth Planet. Sci. Lett. 367, 61-70.

Huang, J., Li, S.-G., Xiao, Y., Ke, S., Li, W.-Y., Tian, Y., 2015. Origin of low δ^{26}Mg Cenozoic basalts from South China Block and their geodynamic implications. Geochim. Cosmochim. Acta 164, 298-317.

Huang, J., Xiao, Y., 2016. Mg–Sr isotopes of low-δ^{26}Mg basalts tracing recycled carbonate species: implication for the initial melting depth of the carbonated mantle in Eastern China. Int. Geol. Rev. 58, 1350-1362.

Huang, X.-L., Xu, Y.-G., Liu, D.-Y., 2004. Geochronology, petrology and geochemistry of the granulite xenoliths from Nushan, east China: implication for a heterogeneous lower crust beneath the Sino-Korean Craton. Geochim. Cosmochim. Acta 68, 127-149.

Jerde, E.A., Taylor, L.A., Crozaz, G., Sobolev, N.V., 1993. Exsolution of garnet within clinopyroxene of mantle eclogites: major- and trace-element chemistry. Contrib. Mineral. Petrol. 114, 148-159.

Ke, S., Teng, F.-Z., Li, S., Gao, T., Liu, S.-A., He, Y., Mo, X., 2016. Mg, Sr, and O isotope geochemistry of syenites from northwest Xinjiang, China: tracing carbonate recycling during Tethyan oceanic subduction. Chem. Geol. 437, 109-119.

Kessel, R., Schmidt, M.W., Ulmer, P., Pettke, T., 2005. Trace element signature of subduction-zone fluids, melts and supercritical liquids at 120–180 km depth. Nature 437, 724-727.

Li, W.-Y., Teng, F.-Z., Ke, S., Rudnick, R.L., Gao, S., Wu, F.-Y., Chappell, B., 2010. Heterogeneous magnesium isotopic composition of the upper continental crust. Geochim. Cosmochim. Acta 74, 6867-6884.

Li, W.-Y., Teng, F.-Z., Xiao, Y., Huang, J., 2011. High-temperature inter-mineral mag- nesium isotope fractionation in eclogite from the Dabie orogen, China. Earth Planet. Sci. Lett. 304, 224-230.

Liu, S.-A., Teng, F.-Z., He, Y.-S., Ke, S., Li, S.-G., 2010. Investigation of magnesium isotope fractionation during granite differentiation: implication for Mg isotopic composition of the continental crust. Earth Planet. Sci. Lett. 297, 646-654.

Liu, S.-A., Teng, F.-Z., Yang, W., Wu, F.-Y., 2011. High-temperature inter-mineral magnesium isotope fractionation in mantle xenoliths from the North China craton. Earth Planet. Sci. Lett. 308, 131-140.

Liu, S.-A., Li, S., Guo, S., Hou, Z., He, Y., 2012. The Cretaceous adakitic–basaltic–granitic magma sequence on south- eastern margin of the North China Craton: implications for lithospheric thinning mechanism. Lithos 134, 163-178.

Liu, Y., Gao, S., Jin, S.-Y., Hu, S.-H., Sun, M., Zhao, Z.-B., Feng, J.-L., 2001. Geochemistry of lower crustal xenoliths from Neogene Hannuoba Basalt, North China Craton: implications for petrogenesis and lower crustal composition. Geochim. Cosmochim. Acta 65, 2589-2604.

Liu, Y., Gao, S., Lee, C.-T.A., Hu, S., Liu, X., Yuan, H., 2005. Melt–peridotite interactions: links between garnet pyroxenite and high-Mg# signature of continental crust. Earth Planet. Sci. Lett. 234, 39-57.

Liu, Y.-C., Wang, A.D., Rolfo, F., Groppo, C., Gu, X.F., Song, B., 2009. Geochronological and petrological constraints on Palaeoproterozoic granulite facies metamorphism in southeastern margin of the North China Craton. J. Metamorph. Geol. 27, 125-138.

Liu, Y.-C., Wang, A.-D., Li, S.-G., Rolfo, F., Li, Y., Groppo, C., Gu, X.-F., Hou, Z.-H., 2013. Composition and geochronology of the deep-seated xenoliths from the southeastern margin of the North China Craton. Gondwana Res. 23, 1021-1039.

Macris, C.A., Young, E.D., Manning, C.E., 2013. Experimental determination of equilibrium magnesium isotope fractionation between spinel, forsterite, and magnesite from 600 to 800 ℃. Geochim. Cosmochim. Acta 118, 18-32.

Manning, C.E., Shock, E.L., Sverjensky, D., 2013. The chemistry of carbon in aqueous fluids at crustal and uppermantle conditions: experimental and theoretical constraints. Rev. Mineral. Geochem. 75, 109-148.

O'Neill, H.S.C., 1981. The transition between spinel lherzolite and garnet lherzolite, and its use as a geobarometer. Contrib. Mineral. Petrol. 77, 185-194.

Pan, D., Spanu, L., Harrison, B., Sverjensky, D.A., Galli, G., 2013. Dielectric properties of water under extreme conditions and transport of carbonates in the deep Earth. Proc. Natl. Acad. Sci. 110, 6646-6650.

Pogge von Strandmann, P.A., Elliott, T., Marschall, H.R., Coath, C., Lai, Y.-J., Jeffcoate, A.B., Ionov, D.A., 2011. Variations of Li and Mg isotope ratios in bulk chondrites and mantle xenoliths. Geochim. Cosmochim. Acta 75, 5247-5268.

Richter, F.M., Watson, E.B., Mendybaev, R.A., Teng, F.-Z., Janney, P.E., 2008. Magnesium isotope fractionation in silicate melts by chemical and thermal diffusion. Geochim. Cosmochim. Acta 72, 206-220.

Rudnick, R., Gao, S., 2003. Composition of the continental crust. Treatise Geochem. 3, 1-64.

Sun, S., McDonough, W., 1989. Chemical and isotopic systematics of oceanic basalts: implications for mantle composition and processes. J. Geol. Soc. 42, 313.

Teng, F.-Z., Li, W.-Y., Ke, S., Marty, B., Dauphas, N., Huang, S., Wu, F.-Y., Pourmand, A., 2010. Magnesium isotopic composition of the Earth and chondrites. Geochim. Cosmochim. Acta 74, 4150-4166.

Teng, F.-Z., Yang, W., Rudnick, R.L., Hu, Y., 2013. Heterogeneous magnesium isotopic composition of the lower continental crust: a xenolith perspective. Geochem. Geophys. Geosyst. 14, 3844-3856.

Teng, F.-Z., Li, W.-Y., Ke, S., Yang, W., Liu, S.-A., Sedaghatpour, F., Wang, S.J., Huang, K.J., Hu, Y., Ling, M.X., 2015. Magnesium isotopic compositions of international geological reference materials. Geostand. Geoanal. Res. 3, 329-339.

Tian, H., Yang, W., Li, S.-G., Ke, S., Chu, Z., 2016. Origin of low δ^{26}Mg basalts with EM-I component: evidence for interaction between enriched lithosphere and carbonated asthenosphere. Geochim. Cosmochim. Acta, in press.

Wang, S.-J., Teng, F.-Z., Williams, H.M., Li, S.-G., 2012. Magnesium isotopic variations in cratonic eclogites: origins and implications. Earth Planet. Sci. Lett. 359, 219-226.

Wang, S.-J., Teng, F.-Z., Li, S.-G., 2014. Tracing carbonate–silicate interaction during subduction using magnesium and oxygen isotopes. Nat. Commun. 5.

Xiao, Y., Teng, F.-Z., Zhang, H.-F., Yang, W., 2013. Large magnesium isotope fractionation in peridotite xenoliths from eastern North China craton: product of melt–rock interaction. Geochim. Cosmochim. Acta 115, 241-261.

Xu, W.-L., Gao, S., Wang, Q., Wang, D., Liu, Y., 2006. Mesozoic crustal thickening of the eastern North China craton: evidence from eclogite xenoliths and petrologic implications. Geology 34, 721-724.

Xu, W.-L., Gao, S., Yang, D.-B., Pei, F.-P., Wang, Q.-H., 2009. Geochemistry of eclogite xenoliths in Mesozoic adakitic rocks from Xuzhou-Suzhou area in central China and their tectonic implications. Lithos 107, 269-280.

Xu, Y., 2002. Evidence for crustal components in the mantle and constraintson crustal recycling mechanisms: pyroxenite xenoliths from Hannuoba, North China. Chem. Geol. 182, 301-322.

Yang, W., Teng, F.-Z., Zhang, H.-F., 2009. Chondritic magnesium isotopic composition of the terrestrial mantle: a case study of peridotite xenoliths from the North China craton. Earth Planet. Sci. Lett. 288, 475-482.

Yang, W., Teng, F.-Z., Zhang, H.-F., Li, S.-G., 2012. Magnesium isotopic systematics of continental basalts from the North China craton: implications for tracing subducted carbonate in the mantle. Chem. Geol. 328, 185-194.

Yang, W., Teng, F.-Z., Li, W.-Y., Liu, S.-A., Ke, S., Liu, Y.-S., Zhang, H.-F., Gao, S., 2016. Magnesium isotopic composition of the deep continental crust. Am. Mineral. 101, 243-252.

Ying, J.-F., Zhang, H.-F., Tang, Y.-J., 2010. Lower crustal xenoliths from Junan, Shan-dong province and their bearing on the nature of the lower crust beneath the North China Craton. Lithos 119, 363-376.

Young, E.D., Galy, A., 2004. The isotope geochemistry and cosmochemistry of magnesium. Rev. Mineral. Geochem. 55, 197-230.

Zhai, M., Bian, A., Zhao, T., 2000. The amalgamation of the supercontinent of North China Craton at the end of Neo-Archaean and its breakup during late Palaeo-proterozoic and Meso-Proterozoic. Sci. China, Ser. D-Earth Sci. 43, 219-232.

Zheng, J., Sun, M., Lu, F., Pearson, N., 2003. Mesozoic lower crustal xenoliths and their significance in lithospheric evolution beneath the Sino-Korean Craton. Tectonophysics 361, 37-60.

Zhou, X., Sun, M., Zhang, G., Chen, S., 2002. Continental crust and lithospheric mantle interaction beneath North China: isotopic evidence from granulite xenoliths in Hannuoba, Sino-Korean craton. Lithos 62, 111-124.

第三部分　地质过程中的金属稳定同位素分馏行为

Copper and iron isotope fractionation during weathering and pedogenesis: Insights from saprolite profiles*

Sheng-Ao Liu[1], Fang-Zhen Teng[2], Shuguang Li[1,3], Gang-Jian Wei[4], Jing-Long Ma[4] and Dandan Li[1]

1. State Key Laboratory of Geological Processes and Mineral Resources, China University of Geosciences, Beijing 100083, China
2. Isotope Laboratory, Department of Earth and Space Sciences, University of Washington, Seattle, WA 98195, USA
3. CAS Key Laboratory of Crust-Mantle Materials and Environments, School of Earth and Space Sciences, University of Science and Technology of China, Hefei 230026, China
4. Key Laboratory of Isotope Geochronology and Geochemistry, Guangzhou Institute of Geochemistry, Chinese Academy of Sciences, Guangzhou 510640, China

亮点介绍：对形成于热带和亚热带的两种玄武岩风化剖面进行了铜铁同位素研究，不同氧化还原条件下铁同位素的对比表明，氧化还原状态对铁的迁移率和同位素分馏起着关键的控制作用。突出了铁、铜同位素在化学风化和土壤易位过程中作为氧化还原条件、古气候和生物循环的重要示踪剂的潜在应用。

Abstract Iron and copper isotopes are useful tools to track redox transformation and biogeochemical cycling in natural environment. To study the relationships of stable Fe and Cu isotopic variations with redox regime and biological processes during weathering and pedogenesis, we carried out Fe and Cu isotope analyses for two sets of basalt weathering profiles (South Carolina, USA and Hainan Island, China), which formed under different climatic conditions (subtropical vs. tropical). Unaltered parent rocks from both profiles have uniform δ^{56}Fe and δ^{65}Cu values close to the average of global basalts. In the South Carolina profile, δ^{56}Fe values of saprolites vary from −0.01‰ to 0.92‰ in the lower (reduced) part and positively correlate with $Fe^{3+}/\Sigma Fe$ ($R^2 = 0.90$), whereas δ^{65}Cu values are almost constant. By contrast, δ^{56}Fe values are less variable and negatively correlate with $Fe^{3+}/\Sigma Fe$ ($R^2 = 0.88$) in the upper (oxidized) part, where large (4.85‰) δ^{65}Cu variation is observed with most samples enriched in heavy isotopes. In the Hainan profile formed by extreme weathering under oxidized condition, δ^{56}Fe values vary little (0.05–0.14‰), whereas δ^{65}Cu values successively decrease from 0.32‰ to −0.12‰ with depth below 3 m and increase from −0.17‰ to 0.02‰ with depth above 3 m. Throughout the whole profile, δ^{65}Cu positively correlate with Cu concentration and negatively correlate with the content of total organic carbon (TOC). Overall, the contrasting Fe isotopic patterns under different redox

* 本文发表在：Geochimica et Cosmochimica Acta, 2014, 146: 59-75

conditions suggest redox states play the key controls on Fe mobility and isotope fractionation. The negative correlation between δ^{56}Fe and $Fe^{3+}/\Sigma Fe$ in the oxidized part of the South Carolina profile may reflect addition of isotopically light Fe. This is demonstrated by leaching experiments, which show that Fe mineral pools extracted by 0.5 N HCl, representing poorly-crystalline Fe (hydr)-oxides, are enriched in light Fe isotopes. The systematic Cu isotopic variation in the Hainan profile reflects desorption and downward transport of isotopically heavy Cu, leaving the organically-bound Cu enriched in light isotope as supported by the negative correlation of δ^{65}Cu with TOC ($R^2 = 0.88$). The contrasting (mostly positive vs. negative) Cu isotopic signatures in the upper parts of these two profiles can be attributed to the different climatic conditions, e.g., high rainfall at a tropical climate in Hainan favors desorption and the development of organism, whereas relatively dry climate in South Carolina favors Cu re-precipitation from soil solutions and adsorption onto Fe (hydr)-oxides. Our results highlight the potential applications of Fe and Cu isotopes as great tracers of redox condition, ancient climate and biological cycling during chemical weathering and pedogenic translocation.

1 Introduction

Continental weathering governs the production of soils from rocks and is an important process controlling the distribution of trace metals in natural systems (Liaghati et al., 2004). This process can impact the ecosystems by releasing dissolved metals and controlling their distribution in porewaters and contaminated soils (Rubio et al., 2000). Stable isotopic systematics of Fe and Cu may be excellent tools that could be used to trace biological cycles in soils (Anbar, 2004; Johnson et al., 2004; Dauphas and Rouxel, 2006; Bigalke et al., 2011). Towards a comprehensive understanding of the mechanism of Fe and Cu isotope fractionation during continental weathering, large amounts of data have been reported for dissolved Fe and Cu in rivers and seawaters (Bermin et al., 2006; Borrok et al., 2008; Vance et al., 2008; Kimball et al., 2009; Radic et al., 2011), for soils and associated porewaters (Fantle and DePaolo, 2004; Emmanuel et al., 2005; Thompson et al., 2007; Wiederhold et al., 2007; Poitrasson et al., 2008; Bigalke et al., 2010a, 2011; Mathur et al., 2012; Yesavage et al., 2012), as well as for experimental leaching of primary sulfide and silicate minerals (Brantley et al., 2004; Chapman et al., 2009; Fernandez and Borrok, 2009) and adsorption of Cu to mineral or bacteria surface (Balistrieri et al., 2008; Pokrovsky et al., 2008; Navarrete et al., 2011). These pioneering studies have documented that significant fractionations of both Fe and Cu isotopes can occur during mineral dissolution and adsorption involved in soil formation.

Soils are key components in the biological cycling of elements and represent the interface between the solid Earth, hydrosphere and biosphere at the Earth's surface. Studies of soil profiles can provide direct constraints on the behavior of metal isotopes during weathering and pedogenesis, in which various factors (e.g. redox condition, acidity, climate and biological effect etc.) may play different roles. Previous studies have suggested that there are multiple mechanisms fractionating Fe and Cu isotopes during continental weathering, including redox transformation (Fantle and DePaolo, 2004; Chapman et al., 2009; Mathur et al., 2012), precipitation, ad/desorption (Balistrieri et al., 2008), and biological processes (Mathur et al., 2005, Thompson et al., 2007, Wiederhold et al., 2007, Bigalke et al., 2010b, Kiczka et al., 2011, Yesavage et al., 2012). In most cases, however, isotopic variations observed in natural soils resulted from more than one process. Therefore, it is difficult to unambiguously fingerprint the role of each process based on a single isotopic systematics.

Combined studies of Fe and Cu isotopic systematics have an important advantage in evaluating the roles of different mechanisms during continental weathering. This is because Fe and Cu behave differently in several ways although both are redox-sensitive. First, Fe (II) is generally more mobile and isotopically lighter than Fe (III) species whereas Cu (II) is more mobile and isotopically heavier than Cu(I) (Zhu et al., 2002; Fernandez and Borrok, 2009). The different mobility of Fe and Cu species can result in contrasting behaviors of Fe and Cu isotopes during mineral dissolution (Fernandez and Borrok, 2009). Thus, if only redox condition primarily controls element transformation and isotope fractionation, then opposite direction of Fe and Cu isotopic variations will be expected because higher valence state generally favors heavier isotopes. Second, Fe is a major constituent in most of silicate rocks primarily hosted by mafic minerals as well as in Fe-oxy-hydroxides, whereas Cu is a trace element and mainly hosted by easily-altered sulfide phases. This can result in different susceptibility of Fe and Cu transformation and isotope fractionation on the intensity of weathering (Mathur et al., 2012). Third, Cu can be strongly adsorbed onto clay minerals or Fe (hydr)-oxides (Dube et al., 2001), an additional process that may fractionate Cu isotopes relative to the dissolved or silicate-bound Cu (Balistrieri et al., 2008). Based on these potential differences, a combined study of Fe and Cu isotopes in the same soil samples can help demonstrate the role of different mechanisms in Fe or Cu isotope fractionation during continental weathering. To date, few studies have reported data for both Fe and Cu isotopes in the same soil samples.

In this study, we reported a combined study of Fe and Cu isotopes on two well-studied diabase and basalt weathering profiles from South Carolina, USA and Hainan Island, China, respectively (Gardner et al., 1981; Ma et al., 2007). Samples from these two profiles are suitable for this scientific subject because they are significantly different in the weathering intensity, oxidation conditions and climate conditions (subtropical vs. tropical). In addition, the Hainan profile displays significant variation in the amount of total organic carbon with depth, allowing the role of biological recycling on Cu or Fe isotope fractionation to be well evaluated. Finally, Mg isotopes have been analyzed for both profiles and Li isotopes have been studied for the South Carolina profile (Rudnick et al., 2004; Teng et al., 2010; Huang et al., 2012). Since both Mg and Li are redox-insensitive, comparisons of Fe and Cu isotopes with Mg and Li isotopes in the same profiles can further evaluate the role of redox conditions on Fe and Cu isotope fractionation, although these elements are probably controlled by different minerals. These two profiles thus provide natural examples documenting how Fe and Cu isotopes are fractionated during continental weathering associated with complicated variation of redox condition and formation of secondary minerals, as well as in the presence of organic matters.

2 Samples

2.1 Diabase weathering profile from South Carolina, USA

The studied saprolite samples were collected along a nearly vertical profile from Cayce, South Carolina. The saprolites developed on Mesozoic diabases that cut a granite quarry as a dike (N33°58.09′, W81°03.07′) (Gardner et al., 1981). These saprolites formed during the Tertiary in a humid, subtropical climate, overlain by a thin (~2 m) layer of Coastal Plain sediment. The diabase dike is ~7 m wide, with saprolites developed within the top 11 m (Gardner et al., 1981). The unaltered diabase crops out below 11 m and contains plagioclase (40%), clinopyroxene (29%) and opaque minerals (3%) (Gardner et al., 1981). In addition, it contains two unusual phases of talc (20%) and chlorite (8%). Green- and red-stained alteration haloes occur sequentially within the granite through that the dike cuts. These form a narrow (3–6 m thick) aureole in the upper part (0–6 m) of the profile, but increase in thickness up to ~30 m below 6 m depth.

Saprolite samples have been pulverized and analyzed for density, clay mineral proportions and bulk chemical compositions by Gardner et al. (1981). The weathering intensity gradationally increases with decreasing depth. The greatest amount of leaching occurred at the shallowest level due to rainwater infiltration as shown by the least density within the top 2 m. Kaolinite, smectite and Fe-oxides dominate secondary minerals in the saprolites.

The most striking feature of this weathering profile is a discontinuity that exists at ~2 m depth (Gardner et al., 1981). Both clay mineralogy and bulk chemistry show discontinuities and display different variation trends cross the discontinuity (Fig. 1). Bulk density generally increases with depth in the lower part (below 2 m) (Fig. 1b). Below 2 m, siderite veins were formed by weathering of original chlorite veins in the diabase. Towards the discontinuity, the ratio of kaolinite to smectite (K/S) increases reflecting transformation of smectite to kaolinite during progressing weathering, and bulk density decreases (Fig. 1a and b). Above 2 m, no siderite veins formed, the kaolinite/smectite ratio and bulk density are almost constant, and kaolinite and smectite contents decrease towards the surface (Gardner et al., 1981). Formation of the ~2 m discontinuity was interpreted by Gardner et al. (1981) as a result of an abrupt change in redox conditions. The upper part was oxidized, whereas the lower profile was reduced. There is also abrupt change of chemical compositions across the 2 m discontinuity (Fig. 1). Furthermore, below 2 m, chemical and lithium isotopic compositions show a discontinuity at 6 m depth, which has been interpreted to reflect the former presence of a water table at this depth (Gardner et al., 1981; Rudnick et al., 2004).

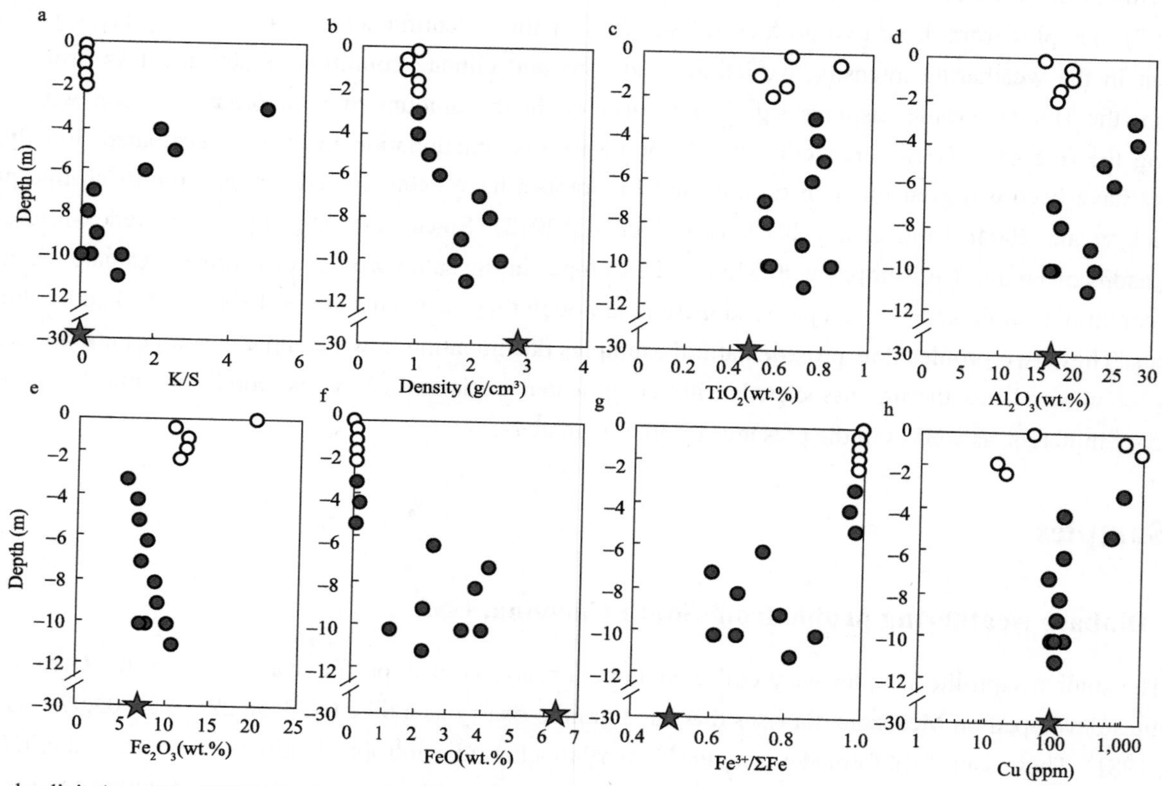

Fig. 1 kaolinite/smectite (K/S) ratio, bulk density and selected element concentrations of saprolites as a function of depth for the South Carolina profile. Star represents unaltered diabase. Open circles represent samples at or above 2 m depth and closed circles represent samples below 2 m. Data are reported in Table 2 and from Gardner et al. (1981). Cu concentrations were analyzed by solution ICP-MS in this study with typical uncertainty of ±10% (1σ).

Sixteen saprolite samples from the weathering profile and one unaltered diabase at 30 m were analyzed for Fe and Cu isotopic compositions. The mineralogy, major and trace element geochemistry as well as Li and Mg isotopes for these samples were previously reported (Gardner et al., 1981; Rudnick et al., 2004; Teng et al., 2010).

2.2 Basalt weathering profile from Hainan Island, China

The weathering profile is exposed at a small hill in the northeastern region of the Hainan province in southern China. This region has a tropical, moist, monsoonal style climate, with a mean annual air temperature of 25 ℃ and mean annual precipitation of 1500 mm (Ma et al., 2007). Samples were collected with an uninterrupted progression from unaltered basalts at the bottom to an extremely weathered laterite residue towards the surface. The top soil and the gravel layer at the upper 50 cm were not sampled in order to avoid the disruption of farming activities. A set of fine laterite with a homogeneous red color developed beneath the gravel layer. Seven samples (HK06-01 to -07) were collected at intervals of 30 cm. Three samples (HK06-08 to -10) were sampled in the section from 250 to 320 cm depth. Below 320 cm, the soil color becomes pistachio with unaltered core stones, and nine samples were collected at intervals of 10–15 cm. In addition, the unaltered tholeiitic basalt sample (HK06-R1) was collected from 5 m below. The tholeiitic basalts erupted during the Neogene, and contain 10% of pyroxene in the phenocryst, and 60% of plagioclase, 25% of clinopyroxene and few opaque minerals in the groundmass (Ma et al., 2007).

The saprolites developed in the profile are dominated by secondary minerals, such as kaolinite, halloysite, gibbsite and Fe-oxy-hydroxides. Primary minerals are absent in the saprolites. A transition in clay mineralogy exists at 3 m depth of the weathering profile (Fig. 2). Beneath 3 m, the clay mineralogy is dominated by halloysite with modal abundances ranging from 30.7% to 87.4%, and kaolinite is absent except in the sample HK06-20 (20%). Above 3 m, kaolinite dominants the secondary mineral assemblage with modal abundances ranging from 28.3% to 82.0%, and only one sample (HK06-7) contains halloysite (53.4%). Other clay minerals, such as illites, are absent except in the sample HK06-7 that contains 16% illite (Ma et al., 2007).

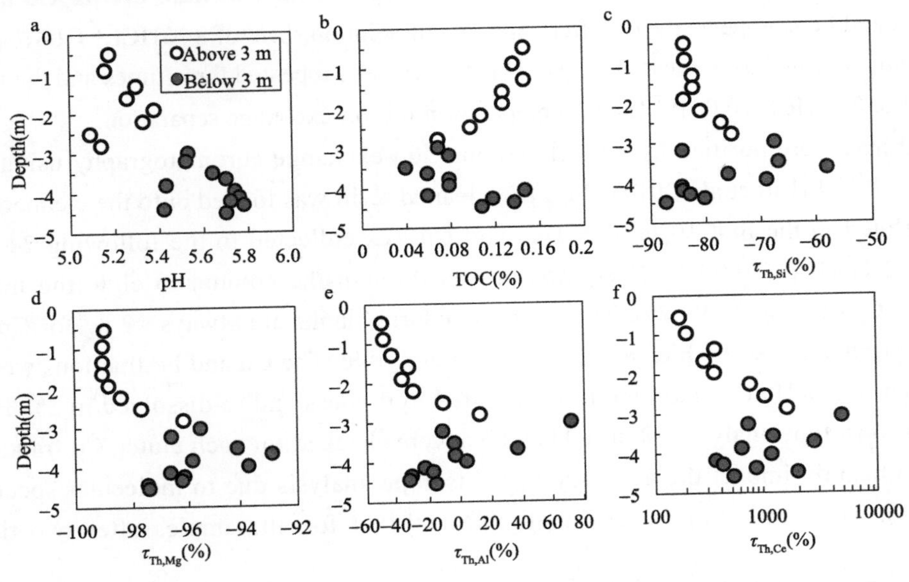

Fig. 2 Normalized element concentrations, pH values and total organic carbon (TOC) of saprolites as a function of depth for the Hainan profile. Data are from Table 4 and Ma et al. (2007).

Ma et al. (2007) found that the abundances of most elements, including major elements, rare earth elements (REEs) and immobile elements (Ti, Zr, Hf, Nb, and Ta) were significantly depleted above 3 m and gradually became enriched or less depleted in the section below 3 m. The depletion of REEs and immobile elements increases towards the surface, suggesting that weathering intensity increases towards the surface (Ma et al., 2007). CIA values (the chemical index of alteration) of saprolites in this profile are greater than 99%, and concentration of Al_2O_3 is up to 36 wt.%. These indicate that the chemical weathering intensity could be categorized as extreme (Nesbitt and Wilson, 1992). In addition, redox-sensitive elements such as Mn, Co, Ce, Cr and U are enriched in the middle profile, with maximum enrichment occurring at ~3 m. These enrichments are accompanied by significant depletion of total organic carbon and the absence of organic nitrogen, as well as higher water content in the middle profile. This suggests that organic colloids and redox conditions played an important role in transferring these elements during weathering (Ma et al., 2007). Studies of isotopic systematics with high masses such as Sr, Nd and Hf in the weathering saprolites found significant isotopic fractionations relative to the unaltered basalt, indicating extreme chemical weathering (Ma et al., 2010).

Twenty-one samples including one unaltered basalt and twenty saprolites from the Hainan profile were analyzed for Fe and Cu isotopes in this study. The mineralogy, major and trace element geochemistry, as well as Mg isotopes for these samples were previously reported (Ma et al., 2007, Huang et al., 2012).

3　Analytical methods

The detailed procedures for sample digestion, column chemistry and instrumental analysis in this study follow the methods of Liu et al. (2014). Only a brief description is given below.

3.1　Sample dissolution and chemical purification

The studied saprolites and unaltered basalt/diabase have Cu concentrations ranging from ~20 to >1000 ppm. Accordingly, 10–20 mg samples were weighted and digested to obtain at least 0.4 μg Cu and 20 μg Fe for high-precision Cu and Fe isotope analysis. After complete dissolution, 1 ml of 8 N HCl + 0.001% H_2O_2 was added to the beaker and then heated to dryness at 80 ℃. This process was repeated three times and the final material was dissolved in 1 ml of 8 N HCl + 0.001% H_2O_2 in preparation for ion-exchange separation.

Copper and iron were purified by a single column ion-exchange chromatography using Bio-Rad strong anion resin AG-MP-1M (Liu et al., 2014). 2 ml pre-cleaned resin was loaded onto the cleaned column. Matrix elements were eluted in the first 10 ml 8 N HCl and Cu was collected in the following 24 ml of 8 N HCl. Then, 18 ml of 2 N HCl + 0.001% H_2O_2 was passed through the column to elute the iron fraction. The recovery for both Cu and Fe is >99.7%. The total procedural blanks are always <2 ng for Cu and <10 ng for Fe based on long-term analyses, which are considered negligible. The Cu and Fe fractions were evaporated to dryness, dissolved in 3% HNO_3, and then re-evaporated to dryness and re-dissolved in 3% HNO_3 to remove all chlorine prior to isotope analysis. Matrix Ti and Na were checked for each eluted Cu fraction because they are found to significantly impact the accuracy of Cu isotope analysis due to molecular spectral interference (Liu et al., 2014). Their ratios to Cu are negligible (<0.1%) for all samples after two times of column chemistry.

In addition, two saprolite samples (M7 and M11) with extreme Fe isotopic compositions from the South Carolina profile were also purified using the AG-X8 resin following the general method of Dauphas et al. (2009). The results agree well (within ±0.02‰) with those obtained by the present single-column procedure and further confirm that our single column purification of Cu and Fe could produce precise and accurate Fe isotope data (Liu et al., 2014).

3.2 Instrumental analysis

Copper and iron isotopic ratios were measured using the Neptune plus MC-ICP-MS at the Isotope Geochemistry Laboratory of the China University of Geosciences, Beijing. Standard-sample bracketing (SSB) method was used in order to correct for instrumental mass fractionation (Zhu et al., 2002; Schoenberg and Von Blanckenburg, 2005; Borrok et al., 2007; Dauphas et al., 2009). External normalization using Zn or Ni as a spike is unnecessary here because it produced little improvement of precision and accuracy compared with the SSB method (Liu et al., 2014). A measurement consists of four blocks of 40 cycles of ~8 s each, and thus each value reported is average of 160 ratios. Cu isotopic data are reported in standard δ-notation in per mil relative to standard reference material (SRM) NIST 976. Iron isotopes were measured in high-resolution mode with mass resolution $M/\Delta M=\sim10{,}000$. ^{53}Cr was measured simultaneously to correct any ^{54}Cr interference on ^{54}Fe and the signal ratio of $^{53}Cr/^{54}Fe$ is at or below the level of 10^{-5}. Because Cr was undetected in the Fe faction after column chemistry, corrected results agree with the uncorrected values within ±0.02‰. Fe isotope data are reported in standard δ-notation in per mil relative to reference material IRMM-014.

The long-term external reproducibility for $\delta^{65}Cu$ and $\delta^{56}Fe$ measurements is better than ±0.05‰ (2SD) and ±0.049‰ (2SD), respectively, based on repeated analyses of natural samples and synthetic solutions (Liu et al., 2014). Several international rock standards (e.g., BHVO-2, BCR-2, BIR-1 and GSP-2) were analyzed for both Cu and Fe isotopes during the course of this study (Table 1). The results agree well with previous studies for Cu isotopes (Li et al., 2009; Bigalke et al., 2010a) and Fe isotopes (Carddock and Dauphas, 2010; Millet et al., 2012).

Table 1 Fe and Cu isotopic compositions of international rock standards analyzed in this study

Sample type	Name	$\delta^{65}Cu$	2SD	n^{a}	$\delta^{56}Fe$	2SD	$\delta^{57}Fe$	2SD	n
Basalt, Hawaiian, USA	BHVO-2	0.13	0.06	4	0.11	0.04	0.22	0.04	4
Basalt, Iceland	BIR-1	−0.01	0.05	4	0.08	0.03	0.13	0.07	4
Basalt, USGS	BCR-2	0.22	0.05	4	0.11	0.03	0.17	0.01	3
Granodoirite, USA	GSP-2	0.32	0.05	4	0.17	0.03	0.25	0.07	4
Andesite, USA	AGV-2	0.10	0.04	4	0.11	0.04	0.18	0.03	4

2SD = 2 times the standard deviation of the population of n repeat measurements of a sample solution.

[a] The times of repeat measurements of the same purification solution by MC-ICP-MS.

3.3 Sequential extraction of bulk soils from the South Carolina profile

In parallel, an aliquot of saprolite samples from the upper part of the South Carolina profile was subjected to a two-step extraction with 0.5 N HCl. The method was designed to operationally separate poorly-crystalline iron oxides from crystalline iron oxides and silicate-bound iron (e.g., Fantle and DePaolo, 2004; Wiederhold et al., 2007). Since saprolites from the entire Hainan profile have relatively homogeneous Fe isotopic compositions (see below), only samples from the South Carolina profile were subject to leaching. In detail, 5 ml of 0.5 N HCl was added to ~0.1 g of bulk soils in a PFA tube. The tubes were shaken vigorously for 24 h at room temperature (~20 ℃) for sufficient reaction and centrifuged for 10 min, and then the supernatants were filtered and separated from the residues. The residues were completely dissolved by following the digestion procedure of bulk saprolites and basalt protoliths. Both the extracted solutions and residues were processed through column and measured for Fe isotopic ratios.

4 Results

Fe and Cu isotopic compositions of the unaltered diabase and saprolites from the South Carolina profile are reported in Table 2. The Fe isotopic data for extracted solutions and residues are reported in Table 3. Fe and Cu isotopic compositions of the unaltered basalt and saprolites from the Hainan profile are reported in Table 4. All samples including geostandards analyzed in this study define the mass fractionation line in three-isotope space (δ^{57}Fe vs. δ^{56}Fe) with a slope of 1.484 ± 0.012 (1σ), indicating no analytical artifacts from unresolved isobaric interferences on Fe isotopes.

Table 2 Major element composition, bulk density and Fe and Cu isotopic compositions of saprolites and unaltered diabase from South Carolina

Sample	Depth (m)	Density g/cm^3	K/S	TiO_2 wt.%	Fe_2O_3B wt.%	Fe_2O_3E wt.%	FeO wt.%	$Fe^{3+}/\Sigma Fe$	$\tau_{Ti,Fe}$	Cu (ppm)	$\tau_{Ti,Cu}$	CIA	δ^{65}Cu	2SD	δ^{56}Fe	2SD	δ^{57}Fe	2SD
M1	0.1	1	0.04	0.66	16.19	4.04	0.00	1.00	4.0	48.2	−65	88	0.75	0.05	0.37	0.07	0.61	0.07
M3	0.5	0.8	0.03	0.88	10.75	0.13	0.06	0.99	−57	1103	504	88	−1.22	0.09	0.77	0.05	1.20	0.07
M4	1	0.8	0	0.52	12.29	0.10	0.09	0.99	−18	1936	1694	92	3.63	0.06	0.72	0.05	1.06	0.05
Repeat													3.65	0.08	0.67	0.04	1.00	0.07
M5	1.5	1	0.03	0.64	12.04	0.19	0.08	0.99	−35	14.2	−89	90	0.62	0.03	0.72	0.03	1.02	0.05
M6	2	1	0.07	0.58	11.27	0.24	0.08	0.99	−32	19.3	−84	87	1.48	0.05	0.61	0.06	0.93	0.05
Repeat													1.55	0.07	0.60	0.03	0.90	0.09
M7	3	1	5	0.77	5.27	0.19	0.14	0.98	−75	1065	567	95	−0.17	0.03	0.92	0.03	1.35	0.02
Repeat															0.90	0.05	1.35	0.06
M8	4	1	2.1	0.78	6.45	0.16	0.16	0.97	−70	140.9	−13	95	0.03	0.03	0.87	0.06	1.27	0.06
M9	5	1.2	2.5	0.81	6.57	0.27	0.10	0.98	−71	723.7	331	93	0.24	0.06	0.75	0.04	1.08	0.06
M10	6	1.4	1.7	0.76	7.40	0.50	2.46	0.74	−52	142.6	−10	91	0.15	0.05	0.52	0.06	0.75	0.06
M11	7	2.1	0.3	0.55	6.75	0.49	4.15	0.61	−27	86.8	−24	54	0.63	0.07	−0.01	0.06	−0.05	0.04
Repeat															0.01	0.06	0.01	0.06
M12	8	2.3	0.15	0.56	7.89	0.92	3.80	0.68	−21	126.2	9	55	−0.01	0.04	0.07	0.06	0.11	0.06
Repeat													0.02	0.06	0.03	0.03	0.07	0.06
M13	9	1.8	0.4	0.72	8.52	0.61	2.21	0.79	−45	116.4	−24	71	0.14	0.06	0.27	0.04	0.38	0.06
M14	10	1.7	1.1	0.85	9.56	0.64	1.24	0.88	−54	148.7	−16	90	0.05	0.03	0.57	0.06	0.84	0.06
L14-8	10	2.5	0.25	0.57	6.87	0.89	3.39	0.67	−31	95.8	−19	49	0.10	0.04	0.11	0.04	0.17	0.05
L14-9	10	2.5	0	0.58	5.95	1.14	3.98	0.62	−32	108.4	−10	46	0.02	0.06	0.08	0.03	0.13	0.06
M15	11	1.9	1	0.73	10.09	0.71	2.18	0.82	−38	110.8	−27	88	0.04	0.05	0.38	0.03	0.60	0.06
M20	30	3	0	0.48	3.13	0.37	6.26	0.50	0.00	99.6	0.00	45	0.01	0.04	0.04	0.07	0.09	0.06

Sample M20 at 30 m depth is the unaltered diabase protolith. Depth, bulk densities, K/S and major element data are from Gardner et al. (1981), where K/S = kaolinite/smectite intensity ratios were analyzed from XRD and FeO concentration was determined by titration. CIA values are from Rudnick et al. (2004), which is molar $Al_2O_3/(Al_2O_3 + CaO^* + Na_2O + K_2O)$, where CaO^* refers to Ca that is not contained in carbonate and phosphate. Fe_2O_3B = bound Fe_2O_3; Fe_2O_3E = extractable Fe_2O_3 determined by sodium dithionite–citrate–bicarbonate. Ti-normalized $\tau_{Ti,Fe(Cu)}$, $100\times[(X/Ti)_{saprolite}/(X/Ti)_{protolith} - 1]$, where X = Fe or Cu, is used to evaluate element mobility by assuming that Ti is most immobile during basalt weathering. Two samples (M-7 and M-11) with extreme Fe isotopic compositions are repeated using the general procedure following the method of Dauphas et al. (2009). Repeat indicates repeating sample dissolution, column chemistry and isotope ratio measurement.

Table 3 Fe isotopic compositions of leached mineral pools and residues by 0.5 N HCl for five bulk soils from the upper profile of the South Carolina profile

Sample no.	Leached pools					Residue					Bulk soils			
	Percent	δ^{56}Fe	2SD	δ^{57}Fe	2SD	Percent	δ^{56}Fe	2SD	δ^{57}Fe	2SD	δ^{56}Fe	2SD		
M-1	0.73	0.46	0.04	0.71	0.06	0.27	−0.15	0.04	−0.21	0.07	0.30	0.04		
M-3	0.29	0.61	0.05	0.94	0.07	0.72	0.88	0.05	1.37	0.06	0.80	0.05		
M-4	0.15	0.17	0.05	0.26	0.07	0.85	0.79	0.04	1.22	0.06	0.69	0.04		
M-5	0.13	0.10	0.05	0.15	0.05	0.87	0.78	0.05	1.18	0.05	0.69	0.05		
M-6	0.19	0.02	0.05	0.04	0.08	0.81	0.77	0.06	1.15	0.08	0.62	0.06		

The calculated Fe isotopic compositions of bulk soils were weighted mean of the leached mineral pools and residues. The calculated values are in agreement, within analytical uncertainty, with measured values of bulk soils reported in Table 2.

Table 4 Major- and trace elements, pH, TOC and Fe and Cu isotopic compositions of saprolites and unaltered basalt from the Hainan Island, China

Sample	Depth (m)	pH	TOC	Th (ppm)	Fe_2O_3T wt.%	$\tau_{Th,Fe}$	Cu (ppm)	$\tau_{Th,Cu}$	δ^{65}Cu	2SD	δ^{56}Fe	2SD	δ^{57}Fe	2SD
HK06-1	0.5	5.18	0.15	9.2	23.48	−39.2	118	−55.7	−0.17	0.05	0.12	0.06	0.18	0.09
HK06-2	0.9	5.16	0.14	9.19	23.45	−39.2	120	−54.9	−0.12	0.04	0.12	0.04	0.17	0.01
HK06-3	1.3	5.30	0.15	8.58	22.69	−37.0	119	−52.1	−0.14	0.05	0.12	0.03	0.19	0.03
HK06-4	1.6	5.26	0.13	7.9	21.62	−34.8	124	−45.8	−0.14	0.04	0.12	0.04	0.18	0.09
HK06-5	1.9	5.38	0.13	7.83	24.4	−25.8	146	−35.7	−0.12	0.04	0.13	0.04	0.18	0.07
HK06-6	2.2	5.33	0.11	7.1	23.41	−21.5	158	−23.2	−0.10	0.05	0.07	0.05	0.10	0.05
HK06-7	2.5	5.09	0.10	5.86	20.97	−14.8	160	−5.8	−0.05	0.05	0.06	0.05	0.11	0.06
HK06-8	2.8	5.14	0.07	5.02	20.43	−3.1	144	−1.0	−0.04	0.05	0.08	0.05	0.12	0.05
Repeat									−0.08	0.04				
HK06-9	3	5.53	0.07	3.72	15.74	0.8	173	60.5	0.02	0.05	0.10	0.03	0.16	0.06
HK06-10	3.2	5.52	0.08	5.99	25.52	1.5	237	36.5	0.22	0.04	0.09	0.06	0.16	0.09
HK06-12	3.5	5.64	0.04	4.9	19.32	−6.1	229	61.2	0.32	0.05	0.09	0.06	0.14	0.06
HK06-13	3.65	5.70	0.06	3.87	15.23	−6.3	163	45.3	0.23	0.05	0.09	0.06	0.15	0.04
HK06-14	3.8	5.43	0.08	5.21	23.4	7.0	237	56.9	0.08	0.05	0.05	0.05	0.09	0.02
HK06-15	3.95	5.74	0.08	4.74	19.44	−2.3	185	34.7	0.11	0.05	0.11	0.03	0.18	0.07
HK06-16	4.1	5.76	0.15	6.37	27.23	1.8	236	27.8	0.08	0.05	0.08	0.03	0.12	0.08
HK06-17	4.2	5.72	0.06	5.97	27.62	10.2	362	109.2	0.29	0.06	0.11	0.04	0.15	0.06
Repeat									0.28	0.03				
HK06-18	4.3	5.78	0.12	6.81	27.65	−3.3	234	18.6	−0.01	0.05	0.14	0.06	0.22	0.07
HK06-19	4.4	5.42	0.14	6.7	23.74	−15.6	239	23.1	−0.10	0.05	0.08	0.05	0.18	0.05
HK06-20	4.5	5.70	0.11	6.58	27.04	−2.1	233	22.2	−0.12	0.05	0.08	0.05	0.15	0.08
Repeat									−0.10	0.04				
HK06-R1	–	–	–	2.46	10.33	0	71.3	0	0.04	0.05	0.08	0.05	0.12	0.07

Sample HK-6-R1 is the unaltered diabase protolith. Depth, pH, total organic carbon (TOC) and major-trace elements are from Ma et al. (2007). Th-normalized $\tau_{Th,X}$, $100\times[(X/Th)_{saprolite}/(X/Th)_{protolith}-1]$, where X = Fe or Cu, is used to evaluate element mobility by assuming that Th is most immobile during extreme weathering.

4.1 Fe and Cu concentrations and isotopic compositions of the South Carolina profile

The unaltered diabase has δ^{56}Fe =+0.04‰, which falls within the range of global basaltic rocks (Beard et al., 2003a, Dauphas et al., 2010, Teng et al., 2013). The saprolites display a wide range of δ^{56}Fe from −0.01‰ to +0.92‰ (n=14). This 0.93‰ variation is one of the largest (0.3–0.9‰) among various parts of different bulk soils worldwide (Fantle and DePaolo, 2004, Emmanuel et al., 2005, Thompson et al., 2007, Wiederhold et al., 2007). In detail, Fe isotopic compositions of the saprolites are either similar to or skewed towards heavier values relative to the unaltered diabase and display a discontinuity at 2 m (Fig. 3). Fe isotopic variation in the lower part (below 2 m) is much larger (0.93‰) than that in the upper part (0.39‰) (above 2 m). The proportion of Fe in the extracted mineral pools is 13–73% of Fe in the bulk soils. The mineral pools extracted by 0.5 N HCl are isotopically lighter than the bulk soils by 0.2–0.5‰, except for the upmost sample (M-1). Fe mineral pool of sample M-1 has a slightly heavier isotopic composition by ~0.09‰ than the bulk soil. The leached residues display heavy δ^{56}Fe values relative to the extracted pools for four soil samples, whereas the extracted residue of sample M-1 is enriched in light Fe isotopes. For all five samples, the bulk δ^{56}Fe values calculated from the weighted average of the extracted mineral pools and residues agree with measured values of bulk digests (Table 3).

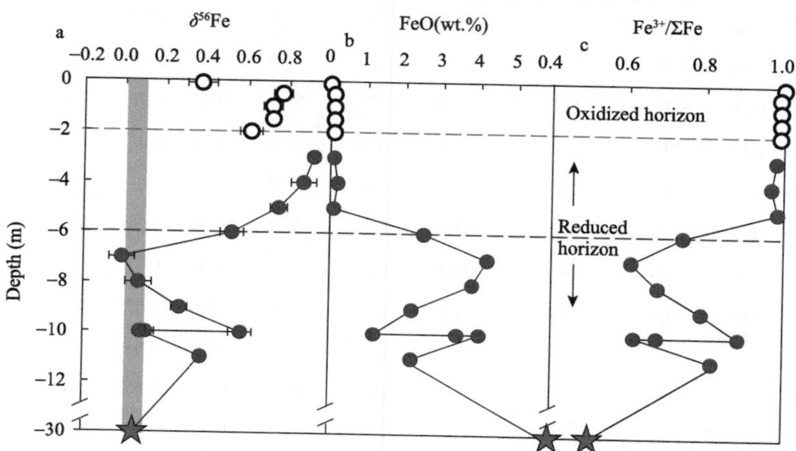

Fig. 3　Variation of δ^{56}Fe values (a), FeO contents (b) and $Fe^{3+}/\Sigma Fe$ (c) of saprolites from the South Carolina profile as a function of depth. Symbols are same as in Fig. 1. The two dashed lines denote two possible discontinuities at depths of 2 m and 6 m, respectively. The 2 m depth denotes a boundary of redox condition with the upper part oxidized and the lower part reduced (Gardner et al., 1981).

The Ti-normalized concentration ($\tau_{Ti,X}$) can quantitatively evaluate the relative element mobility during chemical weathering of basaltic rocks (Nesbitt and Young, 1982), described as $\tau_{Ti,X}=100\times[(X/Ti)_{saprolite}/(X/Ti)_{protolith}-1]$, where X = Fe or Cu. The saprolites in the lower part have $\tau_{Ti,Fe}$=−75 to −21(Table 2), suggesting that about 21–75% of Fe was leached out of the profile during weathering. In the upper part, $\tau_{Ti,Fe}$ values vary from −57 to 4.0, suggesting that some samples underwent significant Fe loss but there was extraneous Fe for the sample closest the surface.

The unaltered diabase of the South Carolina profile has δ^{65}Cu = +0.01‰. This value is consistent with values reported for igneous rocks, which are generally close to zero (Maréchal et al., 1999, Zhu et al., 2000, Li et al., 2009). δ^{65}Cu values of the saprolites range from −1.22‰ to +3.63‰ with an overall variation of up to 4.85‰, significantly larger than those previously reported in soil profiles (0.3–1.8‰) (Bigalke et al., 2010a, 2011, Mathur et al., 2012). Like Fe isotopes, Cu isotopes also show a clear discontinuity at 2 m, with saprolites below 2 m exhibiting less variation compared with those located above 2 m (Fig. 4a). Ti-normalized Cu concentration

($\tau_{Ti,Cu}$) of most samples located below 2 m is less variable and similar to that of the unaltered diabase. By contrast, over two orders of magnitude variation in $\tau_{Ti,Cu}$ values (−89 to 1694) is observed for saprolites above 2 m (Fig. 4b).

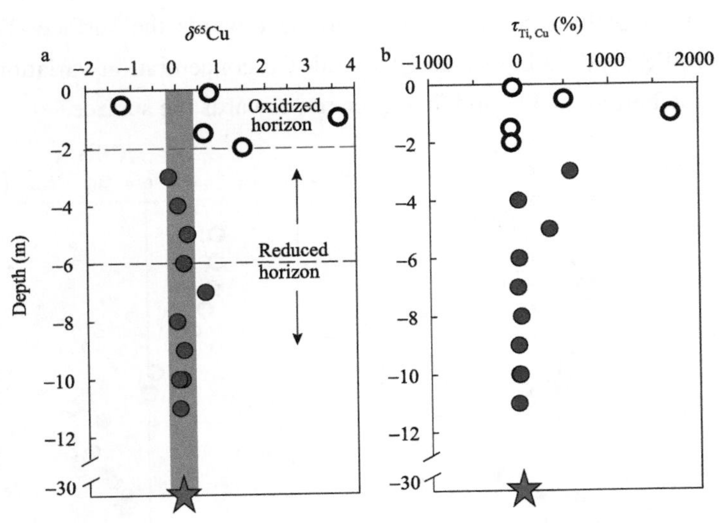

Fig. 4 Variation of Cu isotopic compositions and normalized Cu concentrations of saprolites from the South Carolina as a function of depth. The dashed line denotes the 2 m depth discontinuity. Symbols are same as in Fig. 1. The error bars of δ^{65}Cu values are smaller than the symbol size. Data are reported in Table 2.

4.2 Fe and Cu concentrations and isotopic compositions of the Hainan profile

The unaltered tholeiitic basalt from the Hainan profile has δ^{56}Fe = +0.08‰, falling within the range of global oceanic basalts (Teng et al., 2013). The saprolites display a narrow range of δ^{56}Fe values from +0.05‰ to +0.14‰ (Fig. 5a), which are indistinguishable from the parent basalt. Five samples closest to the surface seem to be isotopically heavier than samples below them, but the difference is barely beyond the analytical uncertainty. For the Hainan profile, relative mobility of elements is normalized to thorium (Th), since Th is the least mobile element during extreme weathering (Ma et al., 2007). The normalized Fe concentration ($\tau_{Th,Fe}$) of saprolites below 3 m is almost constant and similar to that of the unaltered basalt. Above 3 m, moderate variation of $\tau_{Th,Fe}$ occurs towards the surface.

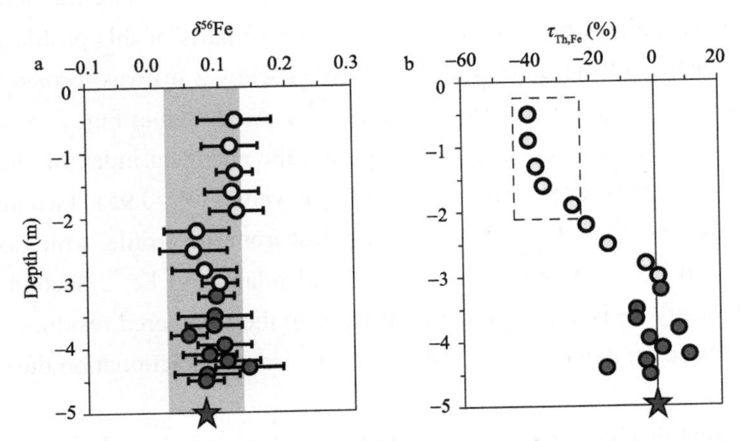

Fig. 5 The variation of Fe isotopic compositions and normalized Fe concentrations of saprolites from the Hainan profile as a function of depth. Fe isotopic data are reported in Table 4 and Fe concentration data are from Ma et al. (2007).

δ^{65}Cu value of the unaltered tholeiitic basalt is +0.04‰, falling within the range of global igneous rocks. δ^{65}Cu values of the saprolites vary significantly from −0.18‰ to +0.31‰ over the entire profile (Fig. 6a). The samples at the bottom (4.3–4.5 m) are depleted in ^{65}Cu relative to the unaltered basalt, but increase upwards from 4.3 to 3 m. Above 3 m, δ^{65}Cu values progressively decrease towards the surface (Fig. 6a). The Cu isotopic variation over the entire profile depth is highly coupled with Cu concentration variation (Fig. 6b). For example, $\tau_{Ti,Cu}$ values increase from the bottom to 3 m and then decrease towards the surface.

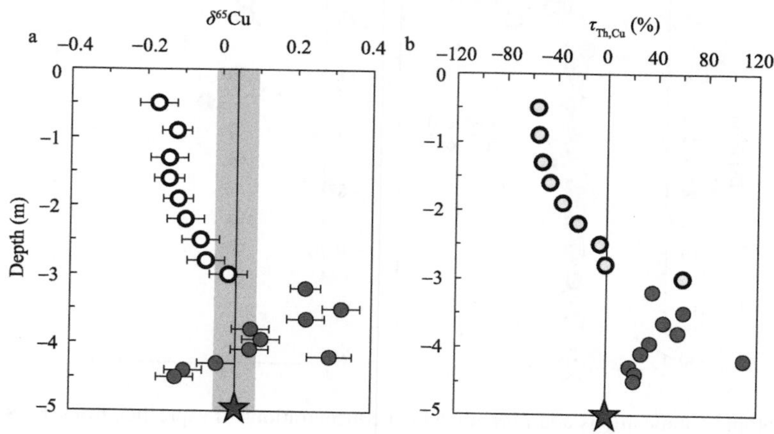

Fig. 6　The variation of Cu isotopic compositions and normalized Cu concentrations of saprolites from the Hainan profile as a function of depth. Cu isotopic data are reported in Table 4 and Cu concentration data are from Ma et al. (2007).

5　Discussion

In this section, we first discuss the possible mechanisms and processes causing Fe and Cu isotopic variations in these two weathering profiles. Then, we discuss the potential implications of Fe and Cu isotopes for tracing redox condition, paleoclimate and biological cycling during weathering and pedogenic translocation based on observations from the present study.

5.1　Iron isotope fractionation during weathering and pedogenesis

The South Carolina profile provides an opportunity for studying Fe isotope fractionation during continental weathering at different redox regimes because the upper and lower parts of this profile were formed at different redox conditions. The lower part of the South Carolina profile (below 2 m) was formed under reduced condition as evidenced by the occurrence of siderite (Gardner et al., 1981). As shown in Fig. 7, δ^{56}Fe values of saprolites in this part are positively correlated with $Fe^{3+}/\Sigma Fe$ (R^2=0.90) and the chemical index of alteration (CIA) (R^2=0.84), and negatively correlated with FeO contents (R^2=0.93) and $\tau_{Ti,Fe}$ values (R^2=0.95). Two important conclusions can be drawn from these correlations. First, Fe was significantly lost from the profile, which is closely associated with the weathering intensity, and Fe^{2+} was preferentially removed relative to Fe^{3+}. Second, light Fe isotopes were released to solutions with heavy Fe isotopes preferentially left in the weathered residues. The results thus provide the most direct evidence, based on natural observations, that Fe isotope fractionation during silicate weathering is redox-controlled.

Two completely different mechanisms have been proposed to explain the behavior of Fe isotope fractionating during weathering: (i) uptake of Fe from a single mineral with organic ligands and (ii) preferential decomposition of isotopically distinct phases. A combined effect of formation of the Fe-depleted layer of the silicate surface, uptake of isotopically heavy ferric Fe from solution by bacterial cells and preferential adsorption of isotopically light Fe^{2+} onto mineral was proposed to explain the light isotopic compositions of leaching Fe relative to starting hornblende (Brantley et al., 2004). Alternatively, leaching experiments of biotite granite and basalt using hydrochloric and oxalic acids indicate that Fe initially released into solution was isotopically light due to dissolution of chlorite and pigeonite, respectively (Chapman et al., 2009). A similar mechanism, relying on potential isotopic difference among solid phases of the parent rocks, has also been proposed for Mg isotope fractionation during granite dissolution (Ryu et al., 2011). The unaltered diabase from the South Carolina profile contains significant amounts of ferrous phases such as chlorite (8%), which was weathered prior to pyroxene (Gardner et al., 1981). Consequently, our results favor the explanation that initial dissolution of chlorite followed by pyroxene may rapidly release isotopically light Fe^{2+} into the solutions and leave isotopically heavy Fe in the residue (e.g., Kiczka et al., 2010).

Previous studies reported significant Li and Mg isotope fractionation in the South Carolina profile (Rudnick et al., 2004; Teng et al., 2010). As weathering intensity increases and density decreases towards the surface, Mg isotopes become progressively heavier whereas Li isotopes get lighter relative to the unaltered diabase. These results are generally consistent with the release of isotopically light Mg or heavy Li into porewater during weathering as observed from river waters (Huh et al., 1998; Tipper et al., 2006). However, unlike Fe, concentrations and isotopes of both Li and Mg show no discontinuity at the depth of 2 m that marks the redox boundary within the profile (Rudnick et al., 2004; Teng et al., 2010). This difference further manifests that redox conditions and formation of secondary Fe-bearing phases (see below) plays a key role in fractionating Fe isotopes during weathering.

The Fe isotopic variation in the Hainan profile is much smaller than the South Carolina profile (Fig. 5a). Most samples display limited variation in Fe concentrations ($\tau_{Th,Fe}=\pm 20\%$) compared to the reduced part of the South Carolina profile (−75% to −21%), although five samples above 2 m have moderately negative $\tau_{Th,Fe}$ (−39% to −26%) that are similar to the oxidized part of the South Carolina profile. Given the extreme weathering under oxidized condition, Fe would be transformed into immobile ferric Fe and can be re-precipitated as Fe (hydr)-oxides as observed in the Hainan profile where gibbsite and Fe (hydr)-oxides dominate the major Fe phases (Ma et al., 2007). The five samples with moderately negative $\tau_{Th,Fe}$ values undergone significant Fe lost and have slightly heavy $\delta^{56}Fe$ values similar to saprolites from the South Carolina profile in same $\tau_{Th,Fe}$ values, which also indicates redox-controlled Fe isotope fractionation during weathering. Nevertheless, the limited Fe isotopic variation over the whole profile suggests that extreme weathering induces limited Fe isotope fractionation.

After primary mineral dissolution, secondary processes may also further affect Fe isotopic compositions of the saprolites. Iron isotopic variation of saprolites at above 2 m (oxidized part) is smaller than that at below 2 m (reduced part) in the South Carolina profile (Fig. 3). In addition, $\delta^{56}Fe$ values of saprolites suddenly decrease towards the surface across the 2 m discontinuity but the values are still heavier than the unaltered diabase. All samples above 2 m except the one closest to the surface have $\tau_{Ti,Fe}$ values that are significantly negative (−57% to −18%). The heavy Fe isotopic compositions of saprolites above 2 m relative to the unaltered diabase and negative $\tau_{Ti,Fe}$ values indicate loss of isotopically light Fe(II). However, this process seems unlikely to explain the negative relationship of $\delta^{56}Fe$ with $Fe^{3+}/\Sigma Fe$ (R^2=0.88) (Fig. 7a) because it predicts that heavy $\delta^{56}Fe$ should positively correlate with elevated $Fe^{3+}/\Sigma Fe$ as observed in the lower part (below 2 m), if Fe isotopic variation is mainly a result of redox transformation. Additional processes must have affected samples in the upper part.

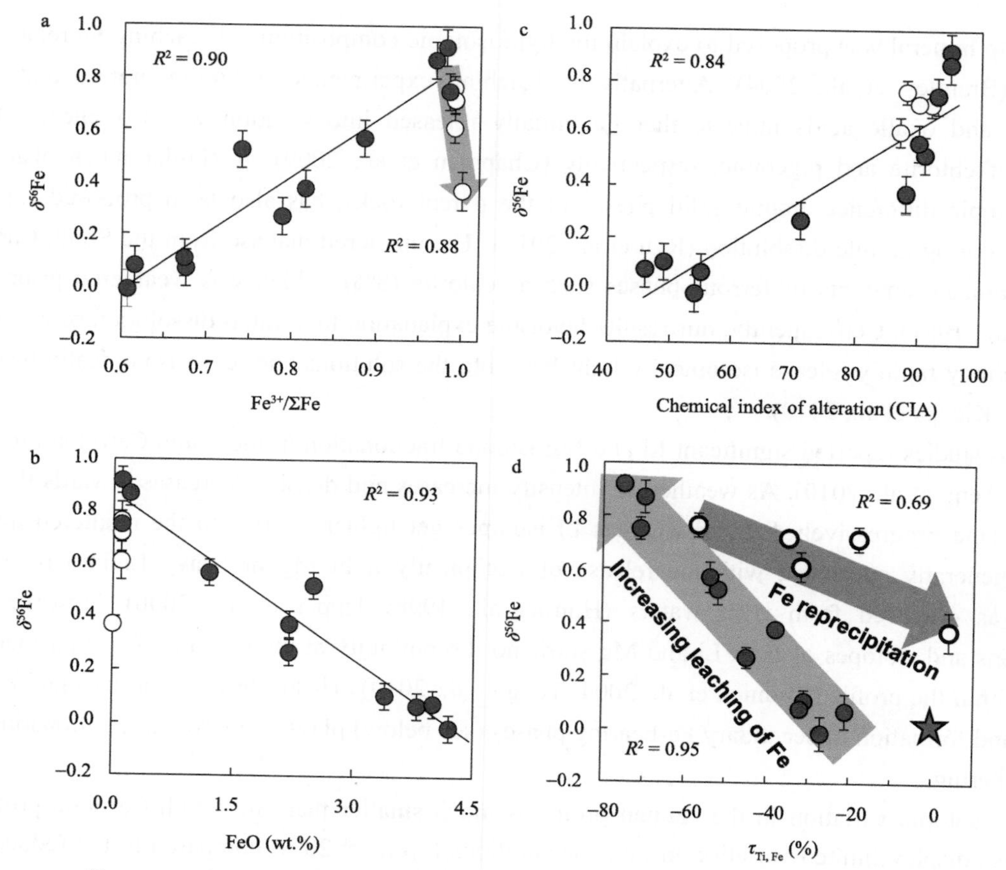

Fig. 7　Correlation of δ^{56}Fe with $Fe^{3+}/\Sigma Fe$, FeO concentrations, chemical index of alteration (CIA) and $\tau_{Ti,Fe}$ values for the saprolites from the South Carolina profile. The data are from Table 2 and Gardner et al. (1981).

The negative relationship of δ^{56}Fe with $Fe^{3+}/\Sigma Fe$ and $\tau_{Ti,Fe}$ values above 2 m indicates the addition of Fe with light isotopic compositions, e.g., re-precipitated Fe from waters or soil solutions. This is strongly supported by the fact that the sample closest to the surface has positive $\tau_{Ti,Fe}$ value (4.0) relative to the unaltered diabase. This hypothesis is also confirmed by sequential leaching of bulk soils which allows analysis of Fe isotopic compositions of different Fe fractions in soils without inducing Fe isotope fractionation (Fantle and DePaolo, 2004; Emmanuel et al., 2005; Wiederhold et al., 2007; Bigalke et al., 2011). The Fe mineral pools of saprolites from the upper part of the South Carolina profile extracted by 0.5 N HCl, representing the poorly-crystalline Fe oxides, are isotopically lighter than the bulk soils (Fig. 8) except for the sample closest to the surface (M-1). This demonstrates that saprolites in the upper profile contain isotopically light, poorly-crystalline Fe hydr-oxides. The formation of poorly-crystalline Fe hydr-oxides in the upper profile was suggested to have resulted from oxidation of the formerly existed siderite in the soils to Fe^{3+} ions and precipitation (Gardner et al., 1981). Siderite generally has light Fe isotopic composition because it is precipitated from groundwater or soil solutions and Fe isotope fractionation between $Fe(II)_{aq}$ and siderite is positive: $10^3\ln \alpha_{Fe(II)\text{-}Siderite} = +0.48‰$ at 20 ℃ (Wiesli et al., 2004). The Fe^{3+}-bearing phases are also enriched in light Fe isotopes similar to the original siderites, which were mixed with other parts of the saprolites to produce the light Fe isotopic signatures across the 2 m discontinuity. Collectively, these additional processes could have contributed to a greater extent of Fe isotope fractionation after primary mineral dissolution.

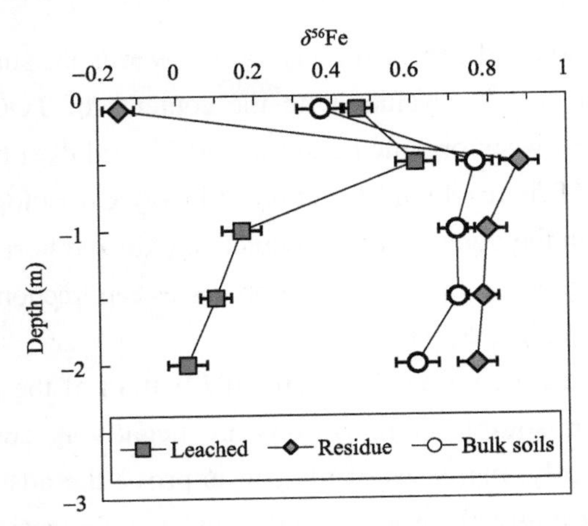

Fig. 8 The variation of δ^{56}Fe values of extracted Fe mineral pools and residues of the saprolites from the upper part of the South Carolina profile (above 2 m) as a function of depth. Data are reported in Table 3.

It is noteworthy that the Fe mineral pool extracted from sample M-1 has heavier δ^{56}Fe than the bulk soil (Fig. 8). 73% of total Fe has been leached from the sample compared to other samples (13–29%). There is thus a minor component in this sample that is isotopically light and is resistant to dissolution by 0.5 N HCl. This component may be organic if the organically-bound Fe is isotopically light (e.g., Brantley et al., 2001). Such a feature has been previously observed by Fantle and DePaolo (2004) but rarely observed in most of soil studies (e.g., Emmanuel et al., 2005; Wiederhold et al., 2007; Bigalke et al., 2011). Wiederhold et al. (2007) argued that the treatment of bulk soils at 600 ℃ to destroy organic matter as done in Fantle and DePaolo (2004) may also change the Fe mineralogy of the soil sample and thus influence the leaching data. Our samples of saprolites were not treated by similar procedure and the results might thus confirm that organic matters with isotopically light Fe relative to poorly-crystalline Fe oxides and silicate-bound iron can exist after extraction by 0.5 N HCl.

5.2 Copper isotope fractionation during weathering and pedogenesis

The variation of Cu concentration with depth is coupled with the variations of major elements such as Si, Mg and Al, and redox-sensitive trace elements such as Ce in the Hainan profile (Fig. 2). Given that saprolites in this profile formed under wet conditions, high rainfall occurred in association with formation of halloysite, Fe-oxides and kaolinite. Above 3 m, rainfall resulted in downward transformation of elements and formation of an enriched layer in the middle profile (~3–4 m) that was highly oxidized (Ma et al., 2007). The successive downward decrease of total organic carbon (TOC) above 3 m is also in association with the remarkable change of redox condition, as a result of decomposition of organic colloids in the oxidized layer. In addition, the presence of organic ligands during oxic conditions could enhance the mobilization of Cu because Cu forms strong complexes with chelating ligands (Neaman et al., 2005). The variations of both Cu (Si, Mg and Al, etc.) concentrations and the contents of TOC with depth can thus be due to redox-controlled transformation (Ma et al., 2007). Such a process can cause significant Cu isotope fractionation.

The positive correlation of δ^{65}Cu with $\tau_{Ti,Cu}$ values in the Hainan profile (Fig. 9a) indicates that Cu leached out of the profile is enriched in heavy Cu isotopes and the weathered residues are isotopically light. In detail, the increased δ^{65}Cu with depth above 3 m reflects the downward transformation of desorbed Cu, which is isotopically heavier than the organically-bound Cu. This transformation resulted in the enrichment of heavy Cu isotopes in the middle profile (3–4 m) and depletion in the upper profile (above 3 m). This can happen as the organic ligands are decomposed under oxidized condition and decreasing pH value towards the surface, which resulted in desorption (Fig. 2). The negative correlation of δ^{65}Cu values with the contents of TOC (Fig. 9b) in the Hainan profile indicates Cu isotope fractionation between organic ligand-bound Cu and dissolved (desorbed) Cu with the former isotopically lighter than the latter. This results in depletion of heavy Cu isotopes in the organic carbon-enriched layer (above 3 m) and enrichment in the organic carbon-depleted, oxidized horizon (~3 m). These observations are consistent with previous studies suggesting that biological processes can fractionate Cu isotopes in soils (Brantley et al., 2004; Mathur et al., 2005; Bigalke et al., 2011).

Below 3 m, δ^{65}Cu values decrease with depth towards the bottom of the profile, which is in contrast to the trend seen at above 3 m. Copper isotopic compositions are negatively correlated with pH values and the contents of TOC (Fig. 9a). Generally, increases of pH can improve the adsorption potential for cations onto the solid phases. Copper can be strongly absorbed onto the surface of negatively charged clay minerals (Dube et al., 2001). Experimental studies show that heavier Cu isotope preferentially adsorbs onto Fe hydr-oxide surface (Balistrieri et al., 2008; Pokrovsky et al., 2008). All samples below 3 m have positive $\tau_{Ti,Cu}$ values and most of them have heavy Cu isotopic compositions (Fig. 6), probably indicating preferential adsorption of heavy Cu isotopes onto Fe hydr-oxides (gibbsite, ilmenite and goethite). The decreased δ^{65}Cu and Cu concentration with depth at below 3 m may be explained as progressive decrease of dissolved, isotopically heavy Cu from upper profile with the distance to the surface, leaving the organic-rich soils enriched in light Cu isotope (^{63}Cu) with depth.

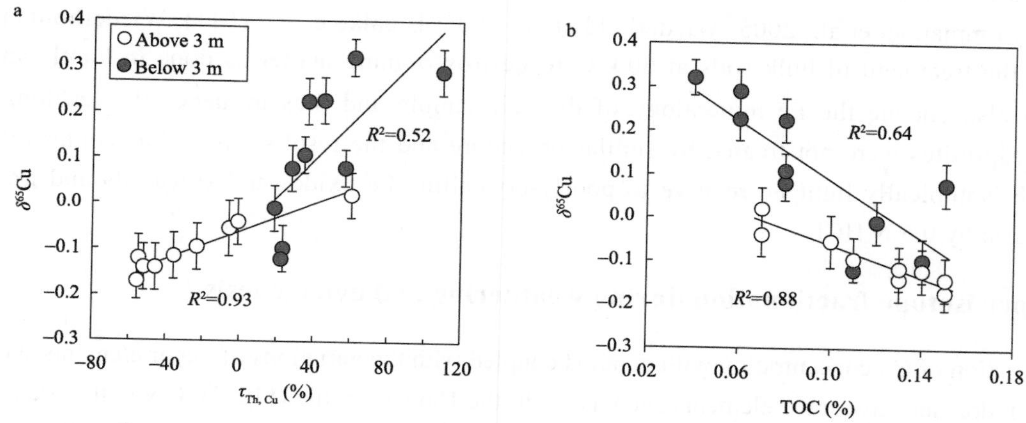

Fig. 9　Correlation of δ^{65}Cu with total organic carbon (TOC) and normalized Cu concentrations for the Hainan profile. Data are from Table 4 and Ma et al. (2007).

Huang et al. (2012) found that Mg isotopic compositions increase with depth above 3 m and decrease with depth below 3 m in the Hainan profile, which is very similar to Cu isotopic variation in the profile. Thorium-normalized Mg concentration also displays similar variation to δ^{26}Mg values. The variations of Mg concentration and isotopic compositions were interpreted as adsorption and desorption processes: adsorption of Mg to kaolin minerals, with preferential uptake of heavy Mg isotopes onto kaolin minerals; and desorption of Mg through cation exchange of Mg with the relatively lower hydration energy cations in the upper profile. Although δ^{65}Cu values show similar variation with the proportions of kaolin minerals in saprolites of the Hainan profile (Fig. 10), the correlation is much weaker than the correlation between δ^{65}Cu and TOC (Fig. 9b). Thus, although Cu adsorption onto clay minerals and desorption may fractionate Cu isotopes, biological cycling plays an important role in governing the Cu isotopic variation over the profile depth.

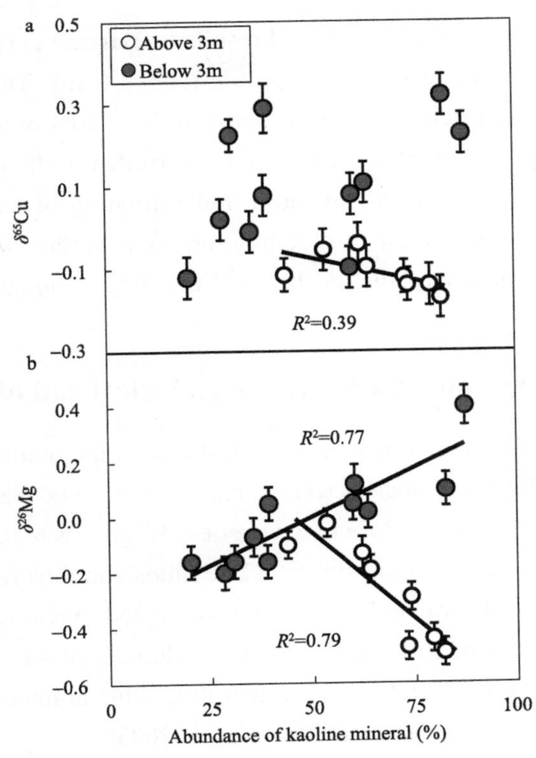

Fig. 10 Correlation of δ^{65}Cu and δ^{26}Mg with the abundance of kaolin minerals for the Hainan profile. The abundances of kaolin minerals are from Ma et al. (2007). δ^{26}Mg data are from Huang et al. (2012). Cu isotope data are reported in Table 4.

The overall variation of Cu isotopes (>4.8‰) in the South Carolina profile is the largest among soil profiles reported to date. The range of δ^{65}Cu (0.82‰) in the lower part (above 2 m) is much narrower than that in the upper part (4.85‰) (Fig. 4a). The Cu concentration variation is also different between the upper and lower profiles, with the upper profile two orders of magnitude ($\tau_{Ti,Cu}$ = −89 to 1649) larger than that in the lower profile ($\tau_{Ti,Cu}$ = −27 to 567) (Fig. 4b). Most of samples in the upper profile have high Cu concentrations (up to 1936 μg/g), which are similar to or even higher than those of contaminated soils (Bigalke et al., 2010a), indicating significant addition of Cu. Such an enrichment of Cu concentrations may be anthropogenic contamination such as contribution of smelters. Re-precipitation or adsorption of Cu onto Fe hydr-oxides could also increase Cu concentrations of the saprolites. Four out of five samples in the upper part of the South Carolina profile have positive δ^{65}Cu whereas one sample at 0.5 m has negative δ^{65}Cu (−1.22‰). This is similar to three zones with different Cu isotopic composition in porphyry deposits: the leach cap (oxidized zone) depleted in ^{65}Cu, enrichment blanket enriched in δ^{65}Cu and hypogene zone with ^{65}Cu close to zero or the average of igneous rocks (Mathur et al., 2009). In general, soils have depleted δ^{65}Cu values as a result of oxidative weathering (Mathur et al., 2005; 2012; Fernandez and Borrok, 2009) during which isotopically heavy Cu^{2+} was dissolved and released into solutions (Bermin et al., 2006; Vance et al., 2008; Mathur et al., 2012). Although most soils are depleted in δ^{65}Cu, soils enriched in ^{65}Cu were also reported (Bigalke et al., 2010a). Notably, the large Cu isotopic variation and heavy δ^{65}Cu in the upper profile is much different from the small variation and light δ^{65}Cu in the Hainan profile.

The different Cu isotopic patterns in these two studied profiles may be due to different climatic conditions under which the soils formed. The Hainan profile developed at a tropical climate with high rainfall and high contents of organic carbon, hence adsorption is relatively weak and downward transformation of desorbed Cu resulted in light Cu isotopic composition in the upper part (above 3 m). By contrast, at a subtropical climate in the South Carolina, adsorption is relatively strong and results in extreme enrichment of Cu concentration and heavy Cu isotopes of some saprolites in the upper profile (above 2 m). This is supported by experimental studies that heavy Cu isotopes are preferentially adsorbed onto Fe (hydr)-oxides with $\delta^{65}Cu_{solid}$-$\delta^{65}Cu_{solution}$ at +1‰ (Balistrieri et al., 2008; Pokrovsky et al., 2008). The saprolites in the upper part of the South Carolina (above 2 m) are enriched in Fe_2O_3 as a result of rapid transformation of kaolinite and siderite to Fe^{3+}-rich smectite and Fe hydr-oxides. Therefore, Cu re-precipitation from surface waters and/or soil solutions and adsorption onto Fe hydr-oxides may explain the heavy Cu isotopic signatures of saprolites in the upper profile.

5.3 Implications for using Fe and Cu isotopes as geological and biological tracers

Isotope geochemistry of Fe has gained particular interest due to its multiple redox states, which makes Fe isotopes widely used as redox indicator of ancient weathering environments. Nevertheless, most of post-Archean sedimentary rocks have $\delta^{56}Fe$ values that are close to the average of igneous rocks (Beard et al., 2003a; Beard and Johnson, 2004). Especially, despite large changes in $Fe^{3+}/\Sigma Fe$ ratios and thus redox condition, bulk sedimentary detritus remains unchanged in its $\delta^{56}Fe$ values in modern (oxygenated) environments (Yamaguchi et al., 2005). This relatively uniform $\delta^{56}Fe$ distribution reflects the very low solubility of Fe^{3+}-oxide minerals, which makes Fe act as a conservative element and difficult to be lost through fluid-mineral interactions, resulting in no change in Fe isotopic compositions in the weathering products (Beard et al., 2003b).

Compared with sedimentary rocks, redox conditions often change at short intervals and on a small scale in soils. Therefore, significant bulk-rock Fe isotopic variation was commonly observed in soil profiles (Fantle and DePaolo, 2004; Emmanuel et al., 2005; Wiederhold et al., 2007; Bigalke et al., 2011). However, Fe isotopic variation in soils may involve multiple processes including reductive dissolution, precipitation and biological cycling. It is therefore difficult to unambiguously fingerprint the role of each process. Our observations in the South Carolina profile strongly suggest that Fe isotope fractionation during basalt weathering is redox-controlled. Significant Fe isotopic variation can occur during mineral dissolution at reduced conditions and the magnitude of isotopic variations can be quantitatively related to the redox states in an isolated system. The linear variation of $\delta^{56}Fe$ values with the ratios of $Fe^{3+}/\Sigma Fe$ observed in the South Carolina profile (Fig. 7a) has crucial implications for Fe isotopes to be used as a predictor of long-term redox conditions, which are difficult to assess otherwise.

Like Fe, redox transformations between Cu(I) and Cu(II) species are also the principal process that fractionates Cu isotopes in natural systems (Zhu et al., 2002; Ehrlich et al., 2004). This nature of Cu evokes the great potential for using Cu isotopes as redox tracers in natural environments. However, Cu can be strongly adsorbed onto clay minerals or iron hydr-oxides relative to Fe (Dube et al., 2001), which strongly depends on solution pH values and much weakly depends on redox condition. In Fig. 11, we summarized published Cu isotope data for rivers and oceans (Vance et al., 2008), porewater (Mathur et al., 2012), soils (Bigalke et al., 2011; Mathur et al., 2012) and contaminated soils (Bigalke et al., 2010a). Rivers and oceans generally have positive $\delta^{65}Cu$ values relative to igneous rocks, probably due to preferential release of heavy Cu isotopes during continental weathering (Mathur et al., 2012). By contrast, $\delta^{65}Cu$ values of soils are not always light as expected and exhibit a large range with both positive and negative values. The heavy $\delta^{65}Cu$ values in soils may be attributed to adsorption as observed from contaminated soils (Bigalke et al., 2010a) as well as to re-precipitation from waters or soil solutions as observed from the South Carolina profile in this study.

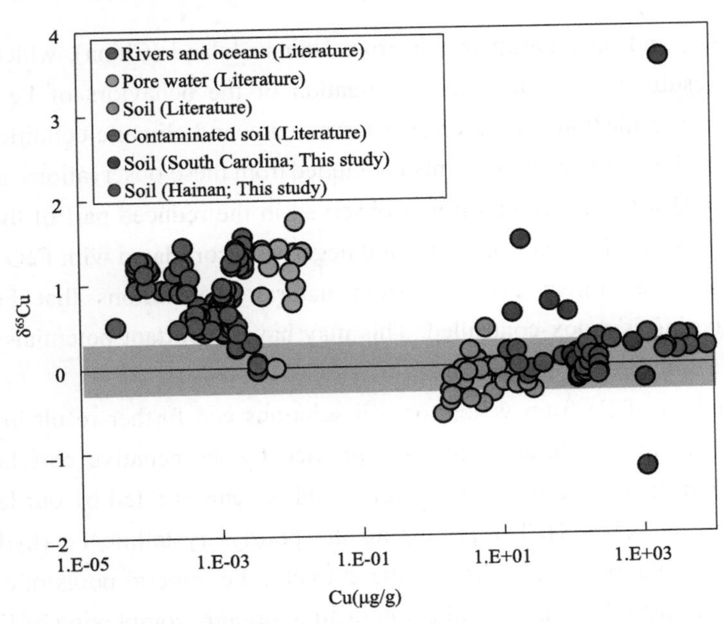

Fig. 11 A summary of Cu concentrations and isotopic compositions for seawater, river water, porewater and soils. The grey area represents the range of igneous rocks (Li et al., 2010). Data sources: rivers and oceans (Vance et al., 2008), pore water (Mathur et al., 2012), soils (Bigalke et al., 2011, Mathur et al., 2012), contaminated soils (Bigalke et al., 2010a). Data of soils from the South Carolina and Hainan profiles are from Tables 2 and 4.

In sharp contrast to the South Carolina profile, the Hainan profile exhibits much smaller Cu isotopic variation over depth and most samples in the upper part (above 3 m) have light Cu isotopic compositions. This difference indicates that Cu mobility and re-precipitation were probably favored by different climatic conditions. For example, at tropical climate re-precipitation is likely difficult to take place and desorption is strong due to high rainfall, resulting in isotopic depletion in the upper part of soil horizons. In summary, different climatic conditions may favor different capacity of adsorption as well as the development of organism, both inducing significant and distinct Cu isotope fractionation. The contrasting Cu isotopic signatures in the upper parts of these two studied profiles formed at different climatic conditions (subtropical vs. tropical) may have implications for using Cu isotopes to trace the ancient climate.

Previous studies have demonstrated that biological processes could significantly fractionate Cu isotopes. For example, Cu isotope fractionation has been observed during adsorption of Cu onto cell surfaces, co-precipitation and adsorption onto the mineral coatings (Mathur et al., 2005; Kimball et al., 2009; Navarrete et al., 2011). Zhu et al. (2002) observed Cu isotope variation during its incorporation into proteins synthesized by bacteria and yeast. Consistent with the results from previous Cu isotope studies on soils, our results indicate that complexation of Cu by organic matters is a key step for Cu isotope fractionation in the development of soils. Even in Cu-rich soils (>100 μg/g) formed by weathering of basalt in the Hainan profile, we observed that biological effects still played a key role in Cu transformation and isotope fractionation. This reveals the great potential of using Cu isotopes as a tracer of biological cycling, although further studies focusing on Cu isotope fractionation during adsorption on clay minerals and chemical reduction and reoxidation of soils are needed to improve our knowledge about the behaviors of Cu isotopes during long-term pedogenic processes.

6 Conclusion

We reported a comparative study of stable Fe and Cu isotopic ratios of saprolites developed on a diabase dike from South Carolina (USA) and on a basalt profile from Hainan Island (China), which formed under different climatic conditions. The results allow a detailed investigation of the behaviors of Fe and Cu isotopes during chemical weathering and pedogenic transformation at different redox and climatic conditions as well as at different amounts of organic matters. Several important points concluded from these observations are summarized below:

Firstly, large (up to 0.93‰) δ^{56}Fe variation was observed in the reduced part of the South Carolina profile, which is positively correlated with $Fe^{3+}/\Sigma Fe$ and CIA and negatively correlated with FeO concentrations and $\tau_{Ti,Fe}$. These results provide the most direct evidence from natural observations that Fe mobility and isotope fractionation during weathering is redox-controlled. This may have important potentials for using Fe isotopes as redox tracers for long-term soil formation processes.

Secondly, re-precipitation of Fe from waters or soil solutions can further result in Fe isotopic variation in bulk soils after primary mineral dissolution. This is supported by the negative correlation between δ^{56}Fe and $Fe^{3+}/\Sigma Fe$ in the oxidized part of the South Carolina profile and is demonstrated by our leaching experiments that Fe mineral pools extracted by 0.5 N HCl, representing the poorly-crystalline Fe (hydr)-oxides, are generally isotopically lighter than the bulk soils. Nevertheless, the extracted Fe mineral pools of one sample closest to the surface is isotopically heavier than the bulk soil, likely indicating organic complexing of Fe in this sample.

Thirdly, systematic Cu isotopic variation occurs in the Hainan profile formed under wet conditions with high rainfall, in contrast to the uniform Fe isotopic composition over the whole profile. The Cu isotopic variation with depth is coupled with Cu concentration and reflects downward transformation of desorbed/dissolved Cu that is enriched in heavy isotope. The organically-bound Cu is enriched in light Cu isotope relative to dissolved Cu as demonstrated by the positive correlation of δ^{65}Cu with the contents of TOC. This implies that ad/desorption and biological processes controlled Cu isotopic variation during weathering and soil formation, which has potential implications for using Cu isotopes as tracers of biogeochemical cycling.

Fourthly, saprolites in the upper part (above 2 m) of the South Carolina profile exhibit large Cu isotopic variation with most samples being isotopically heavy, which is in contrast to the small Cu isotopic variation and light isotopic signature in the upper part (above 3 m) of the Hainan profile. This difference indicates that Cu isotopic patterns in the upper parts of soil profiles were probably favored by different climatic conditions, under which rainfalls resulted in Cu desorption, transformation downward and re-precipitation. Therefore the different Cu isotopic patterns in saprolite profiles formed at different climatic conditions may have implications for using Cu isotopes to track the paleoclimate change.

Finally, our results show that although both Fe and Cu are redox-sensitive, they behave differently during redox-related weathering processes. These differences result in different dependence of Fe and Cu isotopes on weathering intensity, redox conditions and biological effects. Therefore, the combined utilization of Fe and Cu isotopes may be more efficient in fingerprinting the unique processes involved in weathering and pedogenesis compared with single isotopic systematics.

Acknowledgements

We are grateful to Drs. Roberta Rudnick and Bob Gardner for sharing saprolite samples from the South Carolina weathering profile. Shan Ke, Yongsheng He and Yinghuai Lu are thanked for help in the Lab. We thank Drs. J.-B, Chen, M.S. Fantle and an anonymous reviewer for constructive comments and the AE Jérôme Gaillardet for handling, which largely improved the manuscript. This work is supported by the National Natural Foundation of China (41203013, 41473017) and the fundamental research funds (2-9-2014-068) to S.A.L.

References

Anbar A. D. (2004) Iron stable isotopes: Beyond biosignatures. Earth Planet. Sci. Lett. 217, 223-236.

Balistrieri L. S., Borrok D. M., Wanty R. B. and Ridley W. I. (2008) Fractionation of Cu and Zn isotopes during adsorption onto amorphous Fe(III) oxyhydroxide: Experimental mixing of acid rock drainage and ambient river water. Geochim. Cosmo chim. Acta 72, 311-328.

Beard B. L. and Johnson C. M. (2004) Fe isotope variations in the modern and ancient earth and other planetary bodies. Rev. Mineral. Geochem. 55, 319-357.

Beard B. L., Johnson C. M., Skulan J. L., Nealson K. H., Cox L. and Sun H. (2003a) Application of Fe isotopes to tracing the geochemical and biological cycling of Fe. *Chem. Geol.* 195, 87-117.

Beard B. L., Johnson C. M., Von Damm K. L. and Poulson R. L. (2003b) Iron isotope constraints on Fe cycling and mass balance in oxygenated Earth oceans. Geology 31, 629-632.

Bermin J., Vance D., Archer C. and Statham P. J. (2006) The determination of the isotopic composition of Cu and Zn in seawater. Chem. Geol. 226, 280-297.

Bigalke M., Weyer S., Kobza J. and Wilcke W. (2010a) Stable Cu and Zn isotope ratios as tracers of sources and transport of Cu and Zn in contaminated soil. Geochim. Cosmochim. Acta 74, 6801-6813.

Bigalke M., Weyer S. and Wilcke W. (2010b) Stable Copper Isotopes: A Novel Tool to Trace Copper Behavior in Hydro- morphic Soils. Soil. Sci. Soc. Am. J. 74, 60-73.

Bigalke M., Weyer S. and Wilcke W. (2011) Stable Cu isotope fractionation in soils during oxic weathering and podzolization. Geochim. Cosmochim. Acta 75, 3119-3134.

Borrok D. M., Nimick D. A., Wanty R. B. and Ridley W. I. (2008) Isotopic variations of dissolved copper and zinc in stream waters affected by historical mining. Geochim. Cosmochim. Acta 72, 329-344.

Borrok D. M., Wanty R. B., Ridley W. I., Wolf R., Lamothe P. J. and Adams M. (2007) Separation of copper, iron, and zinc from complex aqueous solutions for isotopic measurement. Chem. Geol. 242, 400-414.

Brantley S., Liermann L. and Bullen T. (2001) Fractionation of Fe isotopes by soil microbes and organic acids. Geology 29, 535.

Brantley S. L., Liermann L. J., Guynn R. L., Anbar A., Icopini G. A. and Barling J. (2004) Fe isotopic fractionation during mineral dissolution with and without bacteria. Geochim. Cos- mochim. Acta 68, 3189-3204.

Carddock P. R. and Dauphas N. (2010) Iron isotopic composition of geological reference materials and chondrites. Geostand. Geoanal. Res. 35, 101-123.

Chapman J. B., Weiss D. J., Shan Y. and Lemburger M. (2009) Iron isotope fractionation during leaching of granite and basalt by hydrochloric and oxalic acids. Geochim. Cosmochim. Acta 73, 1312-1324.

Dauphas N. and Rouxel O. (2006) Mass spectrometry and natural variations of iron isotopes. Mass Spectrom. Rev., 25.

Dauphas N., Pourmand A. and Teng F.-Z. (2009) Routine isotopic analysis of iron by HR-MC-ICPMS: How precise and how accurate? Chem. Geol. 267, 175-184.

Dauphas N., Teng F.-Z. and Arndt N. T. (2010) Magnesium and iron isotopes in 2.7 Ga Alexo komatiites: Mantle signatures, no evidence for Soret diffusion, and identification of diffusive transport in zoned olivine. Geochim. Cosmochim. Acta 74, 3274-3291.

Dube A., Zbytniewski R., Kowalkowski T., Cukrowska E. and Buszewski B. (2001) Adsorption and migration of heavy metals in soil. Pol. J. Environ. Stud. 10, 1-10.

Ehrlich S., Butler I., Halicz L., Rickard D., Oldroyd A. and Matthews A. (2004) Experimental study of the copper isotope fractionation between aqueous Cu(II) and covellite, CuS. Chem. Geol. 209, 259-269.

Emmanuel S., Erel Y., Matthews A. and Teutsch N. (2005) A preliminary mixing model for Fe isotopes in soils. Chem. Geol. 222, 23-34.

Fantle M. S. and Depaolo D. J. (2004) Iron isotopic fractionation during continental weathering. Earth Planet. Sci. Lett. 228, 547-562.

Fernandez A. and Borrok D. M. (2009) Fractionation of Cu, Fe, and Zn isotopes during the oxidative weathering of sulfide-rich rocks. Chem. Geol. 264, 1-12.

Gardner L. R., Kheoruenromne I. and Chen H. S. (1981) Geochemistry and mineralogy of an unusual diabase saprolite near Columbia, South Carolina. Clays Clay Miner. 29, 184-190.

Huang K.-J., Teng F.-Z., Wei G.-J., Ma J.-L. and Bao Z.-Y. (2012) Adsorption-and desorption-controlled magnesium isotope fractionation during extreme weathering of basalt in Hainan Island, China. Earth Planet. Sci. Lett. 359-360, 73-83.
Huh Y., Chan L. H., Zhang L. and Edmond J. M. (1998) Lithium and its isotopes in major world rivers: Implications for weathering and the oceanic budget. Geochim. Cosmochim. Acta 62, 2039-2051.
Johnson C. M., Beard B. L., Roden E. E., Newman D. K. and Nealson K. H. (2004) Isotopic constraints on biogeochemical cycling of Fe. Geochem. Non-Traditional Stable Isotopes.
Kiczka M., Wiederhold J. G., Frommer J., Kraemer S. M., Bourdon B. and Kretzschmar R. (2010) Iron isotope fraction- ation during protonand ligand-promoted dissolution of primary phyllosilicates. Geochim. Cosmochim. Acta 74, 3112-3128.
Kiczka M., Wiederhold J. G., Frommer J., Voegelin A., Kraemer S. M., Bourdon B. and Kretzschmar R. (2011) Iron speciation and isotope fractionation during silicate weathering and soil formation in an alpine glacier forefield chronosequence. Geo-chim. Cosmochim. Acta 75, 5559-5573.
Kimball B. E., Mathur R., Dohnalkova A. C., Wall A. J., Runkel R. L. and Brantley S. L. (2009) Copper isotope fractionation in acid mine drainage. Geochim. Cosmochim. Acta 73, 1247-1263.
Li W., Jackson S. E., Pearson N. J., Alard O. and Chappell B. W. (2009) The Cu isotopic signature of granites from the Lachlan Fold Belt, SE Australia. Chem. Geol. 258, 38-49.
Li W., Jackson S. E., Pearson N. J. and Graham S. (2010) Copper isotopic zonation in the Northparkes porphyry Cu–Au deposit. SE Australia. Geochim. Cosmochim. Acta 74, 4078-4096.
Liaghati T., Preda M. and Cox M. (2004) Heavy metal distribution and controlling factors within coastal plain sediments, Bells Creek catchment, southeast Queensland, Australia. Environ. Int. 29, 935-948.
Liu S.-A., Li D.-D., Li S.-G., Teng F.-Z., Ke S., He Y.-S. and Lu Y.-H. (2014) High-precision copper and iron isotope analysis of igneous rock standards by MC-ICP-MS. J. Anal. At. Spectrom. 29, 122-133.
Ma J.-L., Wei G.-J., Xu Y.-G., Long W.-G. and Sun W.-D. (2007) Mobilization and re-distribution of major and trace elements during extreme weathering of basalt in Hainan Island, South China. Geochim. Cosmochim. Acta 71, 3223-3237.
Ma J., Wei G., Xu Y. and Long W. (2010) Variations of Sr–Nd–Hf isotopic systematics in basalt during intensive weathering. Chem. Geol. 269, 376-385.
Maréchal C.N. Télouk, P. and Albarède F. (1999) Precise analysis of copper and zinc isotopic compositions by plasma-source mass spectrometry Chem. Geol., 156, 251-273.
Mathur R., Ruiz J., Titley S., Liermann L., Buss H. and Brantley S. (2005) Cu isotopic fractionation in the supergene environment with and without bacteria. Geochim. Cosmochim. Acta 69, 5233-5246.
Mathur R., Titley S., Barra F., Brantley S., Wilson M., Phillips A., Munizaga F., Maksaev V., Vervoort J. and Hart G. (2009) Exploration potential of Cu isotope fractionation in porphyry copper deposits. J. Geochem. Explor. 102, 1-6.
Mathur R., Jin L., Prush V., Paul J., Ebersole C., Fornadel A., Williams J. Z. and Brantley S. (2012) Cu isotopes and concentrations during weathering of black shale of the Mar cellus Formation, Huntingdon County, Pennsylvania (USA). Chem. Geol. 304-305, 175-184.
Millet M.-A., Baker J. A. and Payne C. E. (2012) Ultra-precise stable Fe isotope measurements by high resolution multiple- collector inductively coupled plasma mass spectrometry with a 57Fe–58Fe double spike. Chem. Geol. 304-305, 18-25.
Navarrete J. U., Borrok D. M., Viveros M. and Ellzey J. T. (2011) Copper isotope fractionation during surface adsorption and intracellular incorporation by bacteria. Geochim. Cosmochim. Acta 75, 784-799.
Neaman A., Chorover J. and Brantley S. L. (2005) Element mobility patterns record organic. Geology 33, 117-120.
Nesbitt H. W. and Wilson R. E. (1992) Recent chemical weathering of basalts. Am. J. Sci. 292, 740-777.
Nesbitt H. W. and Young G. M. (1982) Early Proterozoic climates and plate Motions inferred from major element chemistry of lutites. Nature 299, 715-717.
Poitrasson F., Viers J., Martin F. and Braun J. J. (2008) Limited iron isotope variations in recent lateritic soils from Nsimi, Cameroon: Implications for the global Fe geochemical cycle. Chem. Geol. 253, 54-63.
Pokrovsky O. S., Viers J., Emnova E. E., Kompantseva E. I. and Freydier R. (2008) Copper isotope fractionation during its interaction with soil and aquatic microorganisms and metal oxy(hydr)oxides: Possible structural control. Geochim. Cosmo- chim. Acta 72, 1742-1757.
Radic A., Lacan F. and Murray J. W. (2011) Iron isotopes in the seawater of the equatorial Pacific Ocean: New constraints for the oceanic iron cycle. Earth Planet. Sci. Lett. 306, 1-10.
Rubio B., Nombela M. A. and Vilas F. (2000) Geochemistry of major and trace elements in sediments of the Ria de Vigo (NW

Spain): An assessment of metal pollution. Mar. Pollut. Bull. 40, 968-980.

Rudnick R. L., Tomascak P. B., Njo H. B. and Gardner L. R. (2004) Extreme lithium isotopic fractionation during continental weathering revealed in saprolites from South Carolina. Chem. Geol. 212, 45-57.

Ryu J.-S., Jacobson A. D., Holmden C., Lundstrom C. and Zhang Z. (2011) The major ion, $\delta^{44/40}$Ca, $\delta^{44/42}$Ca, and $\delta^{26/24}$Mg geochemistry of granite weathering at pH=1 and T=25。Power-law processes and the relative reactivity of minerals. Geochim. Cosmochim. Acta 75, 6004-6026.

Schoenberg R. and Von Blanckenburg F. (2005) An assessment of the accuracy of stable Fe isotope ratio measurements on samples with organic and inorganic matrices by high-resolution multicollector ICP-MS. Int. J. Mass. Spectrom. 242, 257–272.

Teng F.-Z., Li W.-Y., Rudnick R. L. and Gardner L. R. (2010) Contrasting lithium and magnesium isotope fractionation during continental weathering. Earth Planet. Sci. Lett. 300, 63-71.

Teng F.-Z., Dauphas N., Huang S. and Marty B. (2013) Iron isotopic systematics of oceanic basalts. Geochim. Cosmochim. Acta 107, 12-26.

Thompson A., Ruiz J., Chadwick O. A., Titus M. and Chorover J. (2007) Rayleigh fractionation of iron isotopes during pedogenesis along a climate sequence of Hawaiian basalt. Chem. Geol. 238, 72-83.

Tipper E. T., Galy A., Gaillardet J., Bickle M. J., Elderfield H. and Carder E. A. (2006) The magnesium isotope budget of the modern ocean: Constraints from riverine magnesium isotope ratios. Earth Planet. Sci. Lett. 250, 241-253.

Vance D., Archer C., Bermin J., Perkins J., Statham P. J., Lohan M. C., Ellwood M. J. and Mills R. A. (2008) The copper isotope geochemistry of rivers and the oceans. Earth Planet. Sci. Lett. 274, 204-213.

Wiederhold J. G., Teutsch N., Kraemer S. M., Halliday A. N. and Kretzschmar R. (2007) Iron isotope fractionation in oxic soils by mineral weathering and podzolization. Geochim. Cosmo chim. Acta 71, 5821-5833.

Wiesli R. A., Beard B. L. and Johnson C. M. (2004) Experimental determination of Fe isotope fractionation between aqueous Fe (II), siderite and "green rust" in abiotic systems. Chem. Geol. 211, 343-362.

Yamaguchi K. E., Johnson C. M., Beard B. L. and Ohmoto H. (2005) Biogeochemical cycling of iron in the Archean-Paleoproterozoic Earth: Constraints from iron isotope variations in sedimentary rocks from the Kaapvaal and Pilbara Cratons. Chem. Geol. 218, 135-169.

Yesavage T., Fantle M. S., Vervoort J., Mathur R., Jin L., Liermann L. J. and Brantley S. L. (2012) Fe cycling in the Shale Hills Critical Zone Observatory, Pennsylvania: An analysis of biogeochemical weathering and Fe isotope fractionation. Geo-chim. Cosmochim. Acta 99, 18-38.

Zhu X. K., O'nions R. K., Guo Y., Belshaw N. S. and Rickard D. (2000) Determination of natural Cu-isotope variation by plasma-source mass spectrometry: Implications for use as geochemical tracers. Chem. Geol. 163, 139-149.

Zhu X. K., Guo Y., Williams R. J. P., O'nions R. K., Matthews A., Belshaw N. S., Canters G. W., De Waal E. C., Weser U., Burgess B. K. and Salvato B. (2002) Mass fractionation processes of transition metal isotopes. Earth Planet. Sci. Lett. 200, 47-62.

Magnesium isotopic systematics of mafic rocks during continental subduction*

Shui-Jiong Wang [1,2,3], Fang-Zhen Teng [2,3], Shu-Guang Li [1,4], Ji-An Hong[4]

1. State Key Laboratory of Geological Processes and Mineral Resources, China University of Geosciences, Beijing 100083, China
2. Isotope Laboratory, Department of Geosciences and Arkansas Center for Space and Planetary Sciences, University of Arkansas, Fayetteville, AR 72701, USA
3. Isotope Laboratory, Department of Earth and Space Sciences, University of Washington, Seattle, WA 98195-1310, USA
4. CAS Key Laboratory of Crust-Mantle Materials and Environments, School of Earth and Space Sciences, University of Science and Technology of China, Hefei 230026, Anhui, China

亮点介绍：本文通过测定中国东部华南陆块北缘及俯冲陆壳的一套前进变质岩石（绿片岩、角闪岩和榴辉岩）的 Mg 同位素组成研究陆壳俯冲过程中镁同位素的地球化学行为。测定数据表明，绿片岩、角闪岩和榴辉岩 Mg 同位素组成类似，表明镁铁质地壳俯冲过程中的前进变质和脱水过程不造成显著的 Mg 同位素分馏。

Abstract Magnesium isotopic compositions of a set of prograde metamorphosed rocks (greenschists, amphibolites, and eclogites) from East China were measured in order to understand the behavior of Mg isotopes during continental subduction. The δ^{26}Mg values of the greenschists vary from –0.269±0.057‰ to –0.133±0.042‰ (2SD), and correlate negatively with the MgO contents and Nb/U ratios, possibly due to the crustal assimilation of isotopicallly heavy felsic schists (δ^{26}Mg up to –0.099‰) during the genesis of MORB-like protoliths. The two subgroups of amphibolites have slightly different Mg isotopic compositions. The group I amphibolites with cumulate origin have δ^{26}Mg values ranging from –0.243±0.061‰ to –0.192±0.050‰, which reflects clinopyroxene accumulation nature of the protolith. The group II amphibolites have relatively light δ^{26}Mg values varying from –0.358±0.061‰ to –0.224±0.056‰, which may result from either source contamination or crustal assimilation of isotopically light components (e.g., carbonates). Eclogitic garnet and clinopyroxene display large inter-mineral Mg isotope fractionations ($\Delta^{26}Mg_{Cpx\text{-}Grt}$=1.097±0.056‰ to 1.645±0.081‰) equilibrated at peak metamorphic temperatures. The δ^{26}Mg values of the eclogites vary from –0.348±0.041‰ to –0.137±0.063‰, overlapping the ranges of greenschists and amphibolites. Such Mg isotopic variations are not related to unrepresentative sampling nor host-eclogite chemical interactions, instead, they reflect the protolith heterogeneities. Collectively, the similarity of Mg isotopic compositions among the greenschists

* 本文发表在：Geochimica et Cosmochimica Acta, 2014, 143: 34-48

(–0.196±0.044‰), amphibolites (–0.271±0.042‰), and eclogites (–0.226±0.044‰), suggests that Mg isotope fractionation during continental subduction, if any, is limited.

1 Introduction

Magnesium (Mg) is a major element in earth's mantle, crust, and hydrosphere. Over geological time-scales, Mg is continuously extracted from the mantle to form crust by igneous processes, transferred from the continents to the hydrosphere via continental weathering, and returned to the mantle through subduction or delamination. These processes comprise a long-term Mg cycle, which contributes to the compositional evolution of continental crust (Rudnick, 1995). Magnesium has three stable isotopes (^{24}Mg, ^{25}Mg, and ^{26}Mg), with the relative mass difference >8% between ^{24}Mg and ^{26}Mg. Previous studies have found limited Mg isotope fractionations during high-temperature igneous process (Teng et al., 2007, 2010a; Handler et al., 2009; Yang et al., 2009; Bourdon et al., 2010; Huang et al., 2011; Pogge von Strandmann et al., 2011), but large during low-temperature processes, e.g., chemical weathering drives light Mg isotopes into the hydrosphere (e.g., Tipper et al., 2006a,b; Foster et al., 2010; Ling et al., 2011), leaving the weathered residues enriched in ^{26}Mg (Tipper et al., 2006a; Brenot et al., 2008; Pogge von Strandmann et al., 2008; Li et al., 2010; Teng et al., 2010b; Wimpenny et al., 2010; Huang et al., 2012, 2013b), and carbonate precipitation preferentially uptakes ^{24}Mg from the ambient aqueous solutions (e.g., Galy et al., 2002). Therefore, surface-processed materials may have Mg isotopic compositions significantly distinct from that of normal mantle (δ^{26}Mg = –0.25±0.07‰; Teng et al., 2010a), making Mg isotopes a potential tracer for crustal recycling (e.g., Ke et al., 2011). For example, previous studies have found δ^{26}Mg values of certain mantle derivates are deviated significantly from the "normal" mantle value of –0.25±0.07‰ (Teng et al., 2010a), and these Mg isotopic signatures were attributed to the incorporation of subducted materials in their mantle sources (Yang et al., 2012; Wang et al., 2012b; Xiao et al., 2013).

A primary assumption for the above geochemical hypotheses and interpretations is that Mg isotopic systematics of the subducting rocks suffer little change during metamorphic dehydration. However, to date, the extent to which Mg isotopes fractionate during this process remains largely unknown. Li et al. (2011) has found that bulk-rock δ^{26}Mg values of ten orogenic eclogites from Dabie orogen are relatively homogeneous (on average –0.32±0.08‰), although there is large inter-mineral Mg isotope fractionation between coexisting garnet and clinopyroxene. Based on the assumption that the gabbroic protoliths of the eclogites have δ^{26}Mg values around "normal" mantle value of –0.25±0.07‰, Li et al. (2011) concluded that eclogite-facies metamorphism produced limited Mg isotope fractionations. Despite all these, direct assessments on the behaviors of Mg isotopes during prograde metamorphism are still absent.

To address the above issue, a straightforward method is to compare the protoliths to their prograde metamorphic counterparts for Mg isotopic compositions. In this study, a set of genetically related meta-basaltic rocks including greenschists, amphibolites, and eclogites from South China Block and Dabie-Sulu orogen, East China, were analyzed for Mg isotopic ratios. The main purpose of this work is: (1) to assess the behaviors of Mg isotopes during continental subduction; and (2) to evaluate the origin and extent of Mg isotopic variations in metamorphic rocks.

2 Geological settings and samples

2.1 Geological background

The Dabie-Sulu orogen was formed by continent-continent collision between the South China Block (SCB) and the North China Block (NCB) in the Triassic (Li et al., 1989, 1993, 2000; Hacker et al., 1998; Zheng et al., 2003), and was later separated into two terranes by the left-lateral movement of the Tan-Lu fault (Fig. 1a). The Dabie orogen to the west of the Tan-Lu fault is divided into five litho-tectonic unites (Fig. 1a; Li et al., 2001; Zheng et al., 2005). They are, from south to north, the Susong-Hongan blueschist-facies metamorphic zone (SH zone), the south Dabie ultra-high pressure (UHP)/low-temperature eclogite-facies zone (SD zone), the central Dabie UHP/middle-temperature eclogite-facies zone (CD zone), the north Dabie UHP/high-temperature eclogite-facies zone (ND zone), and the Beihuaiyang greenschist- to amphibolite-facies zone (BHY zone). To the southwest of Dabie orogen are the SCB rocks that were not involved into the Triassic continental subduction, as represented by the Wudang terrane (Fig. 1a).

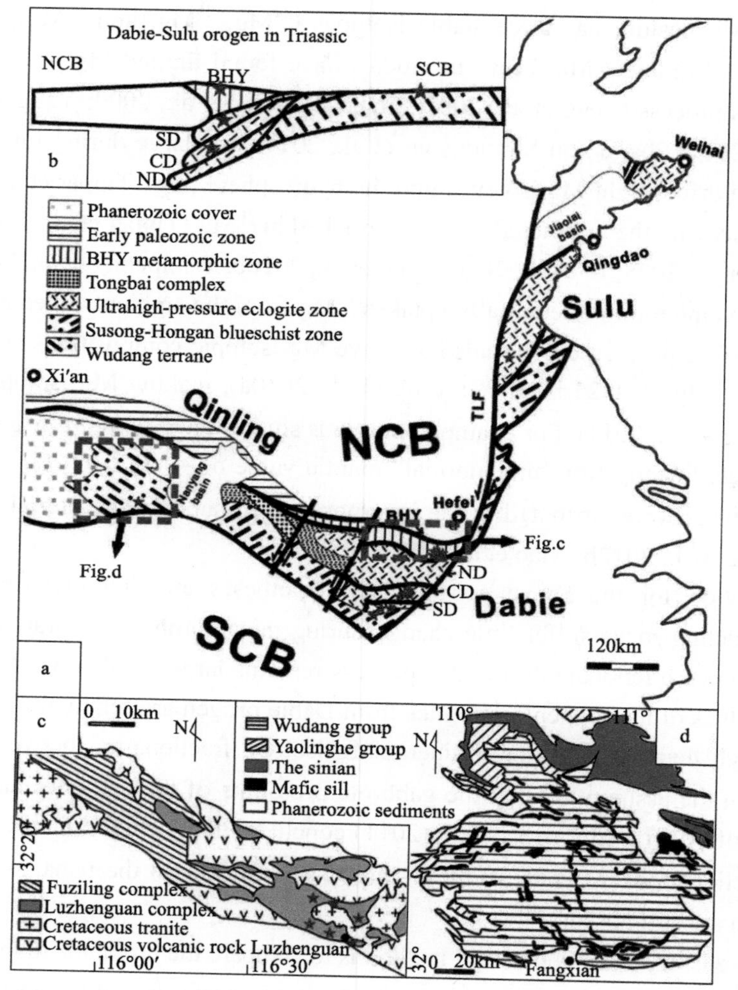

Fig. 1 (a) Sketch geological map of the Dabie-Sulu orogenic belt in central-east China (modified after Zheng et al., 2005); (b) Cartoon model illustrating the subduction of SCB beneath NCB created the Dabie-Sulu orogen, and the BHY zone as the accretion wedge during continental subduction (modified after Zheng et al., 2005); (c) Sketch geological map of the Beihuaiyang metamorphic zone in Dabie orogen; (d) Sketch geological map of the Wudangshan area in the South China Block. Also shown are the metamorphic zones in Dabie orogen. The green stars labeled the sample locations. NCB = North China Block; SCB = South China Block; BHY = Beihuaiyang metamorphic zone; SD = the south Dabie UHP/low-temperature eclogite-facies zone; CD = the central Dabie UHP/middle-temperature eclogite-facies zone; ND = the north Dabie UHP/high-temperature eclogite-facies zone; TLF = Tan-Lu fault. (For interpretation of the references to colour in this figure legend, the reader is referred to the web version of this article.)

The SH zone is composed of Proterozoic metamorphic volcano-sedimentary rocks, overprinted by Triassic blueschist-facies metamorphism (Li et al., 1993, 2001). The SD, CD and ND zones are mainly composed of UHP metamorphic rocks including eclogite, gneiss, marble and mafic-ultramafic rocks, and represent respectively independent crustal slices decoupled during continental subduction and exhumation (Li et al., 2003; Liu et al., 2007; Liu and Li, 2008; Wang et al., 2012a). Eclogites from SD and CD zones occur as fresh blocks or lenses within gneisses, marbles and ultramafic massifs. Coesite and/or diamond inclusions preserved in zircons and garnets from the eclogites support that these rocks were subducted to mantle depths of at least 90-120 km (Okay et al., 1989; Xu et al., 1992; Ye et al., 2000; Liu et al., 2001; Li et al., 2004). Previous studies show that the eclogites from the SD zone experienced peak metamorphism at temperature of ~650 ℃, whereas the eclogites from the CD zone recorded the peak metamorphic temperature of ~750 ℃ (e.g., Xiao et al., 2000; Li et al., 2004). Eclogites from the ND zone are overprinted by high-temperature granulite-facies retrograde metamorphism and later Cretaceous migmatization (Wang et al., 2012a, 2013), leading to the intensive retrogression of eclogitic minerals. Geochemical, stable and radiogenic isotopic evidence indicate that the protoliths of eclogites and surrounding gneisses are Neoproterozoic bimodal volcanic suites in SCB with formation ages of 750–800 Ma (e.g., Zheng et al., 2006; Liu and Liou., 2011).

The BHY zone is mainly composed of meta-sedimentary and meta-igneous rocks, and represents a passive-margin accretionary wedge formed during Triassic continental subduction (Fig. 1b; Li et al., 2001; Zheng et al., 2005). It is subdivided into two complexes: Fuziling and Luzhenguan complexes (Fig. 1c; Li et al., 2001; Zheng et al., 2005). The Fuziling complex is made of Paleozoic flysch sediments overprinted by greenschist-facies metamorphism, and the Luzhenguan complex is mainly composed of granitic gneisses, amphibolites, and graywackes that are overprinted by Triassic peak amphibolite-facies metamorphism (Li et al., 2001; Faure et al., 2003; Zheng et al., 2005). Previous geochronological studies have shown that amphibolites from Luzhenguan complex share the common feature of Neoproterozoic intrusive ages (~750 Ma; Wu et al., 2004), which are comparable to the protolith ages of UHP metamorphic rocks from the SD, CD and ND zones. Therefore, the protoliths of meta-igneous rocks exposed in Dabie orogen, including eclogites from the UHP metamorphic zones and amphibolites from the BHY zone, are associated with the Neoproterozoic bimodal volcanic rocks from SCB (Zheng et al., 2003, 2005).

Wudang terrane represents the un-subducted crust now presented in the northern margin of SCB, and is mainly composed of the Neoproterozoic Wudang and Yaolinghe groups (Fig. 1d), with zircon U-Pb ages of 755±3 Ma and 685±5 Ma, respectively (Ling et al., 2008). The Wudang group consists of fine-grain sedimentary beds intercalated with rift-related bimodal volcanic sequence (Fig. 1d), which were overprinted by Neoproterozoic up to greenschist-facies metamorphism (Huang, 1993; Ling et al., 2008, 2010). Trace elemental and zircon U-Pb-Hf isotopic studies suggest that Neoproterozoic basaltic rocks in SCB have experienced various degrees of crustal assimilation during their genesis in the continental rifting environment (e.g., Zheng et al., 2007; Xia et al., 2012).

2.2 Samples

Samples investigated here include Neoproterozoic bimodal volcanic rocks of Wudang group from the Wudang terrane, amphibolites from the Luzhenguan complex, and eclogites from the SD and CD zones (Fig. 1a-d).

2.2.1 Bimodal volcanic rocks from SCB

Neoproterozoic bimodal rocks in SCB comprise interlayered basaltic and felsic volcanic rocks. Due to the greenschist-facies metamorphic overprints, basaltic and felsic rocks are termed as greenschist and felsic schist, respectively. The greenschists are mainly composed of chlorite + epidote + plagioclase with minor quartz, amphibole, biotite and calcite. The abundance of chlorite and epidote can be >60 vol.%. On a total alkalis versus silica (TAS) plot (Fig. 2), greenschists straddle the boundary between the sub-alkalic and alkali fields. Felsic schists are composed of plagioclase + quartz + K-feldspar + chlorite + epidote with minor biotite and calcite, falling in the sub-alkalic side with dacite-rhyolitic compositions (Fig. 2).

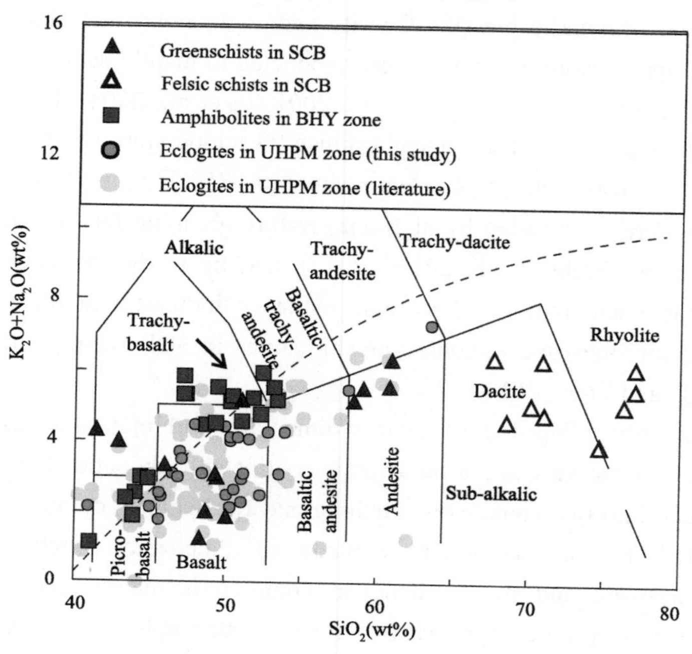

Fig. 2　Total alkalis versus SiO_2 diagram (Middlemost, 1994) for rocks in present study. Literature data of eclogites are from Jahn (1998), Zhang et al. (2006), and Zhao et al. (2007). Other data are given in Supplementary materials.

2.2.2 Amphibolites from BHY zone

Meta-basaltic rocks from Luzhenguan complex in the BHY zone are characterized by volcanic porphyritic structures, with mineral assemblage of amphibole + clinopyroxene + plagioclase + biotite + epidote + quartz + calcite for the matrix, and clinopyroxene/amphibole + plagioclase for the phenocrysts. Samples with low SiO_2 contents (<45 wt.%) are generally in the picro-basalt field, while those with relatively high SiO_2 contents (>45 wt.%) are within the basalt to trachy-basalt fields (Fig. 2).

2.2.3 Eclogites and country rocks from SD and CD zones

Dominated eclogitic minerals are garnet (10–80 vol%) and clinopyroxene (15–80 vol%) with minor phengite, amphibole, epidote, quartz, apatite, rutile and calcite. Eclogites display large major and trace elemental variations, indicating multiple origins of their protoliths (e.g., titanomagnetite/ilmenite-rich gabbroic cumulate, common gabbroic/dioritic cumulate, and basalts interlayered with sedimentary rocks; Jahn, 1998; Li et al., 2000; Zhang et al., 2006; Tang et al., 2007; Liu et al., 2008). They are mostly within the sub-alkalic field with basaltic compositions (Fig. 2). The typical host rocks of the eclogites are marble, peridotite, and gneiss. The impure marbles are composed of calcite and dolomite in various abundances with minor silicate minerals. The peridotite is a garnet-bearing lherzolite, and the paragneiss has a general mineral assemblage of plagioclase + quartz + K-feldspar + biotite + phengite + clinozoisite/epidote.

3 Analytical method

Magnesium isotopic analyses were performed at the Isotope Laboratory of the University of Arkansas, Fayetteville. Sample dissolution, column chemistry and instrumental analysis have been reported elsewhere (e.g., Yang et al., 2009; Li et al., 2010; Teng et al., 2010a, 2011; Teng and Yang, 2014) and are briefly described below.

Fresh garnet and clinopyroxene grains in eclogites were handpicked under a binocular microscope. Extra care was taken to avoid any alteration products during separation. The mineral separates and whole-rock powders were dissolved in Savillex screw-top beakers by using a combination of Optima-grade HF-HNO_3-HCl acids. Solutions were evaporated to dryness at 160 ℃ after complete dissolution, and the residues were dissolved in 1 N HNO_3 ready for ion exchange column chemistry. Chemical separation and purification of Mg were achieved by cation exchange chromatography with Bio-Rad 200–400 mesh AG50W–X8 resin in 1 N HNO_3 media. Each sample containing>50 μg Mg was loaded on the resin and eluted with 1 N HNO_3. At least 2 standards (KH-olivine and seawater) were processed together for each batch of column chemistry. The same column procedure was performed twice in order to obtain pure Mg solutions for mass spectrometry. The total procedural blank is < 10ng, which represents < 0.1% of Mg loaded on the column (Teng et al., 2010a).

Magnesium isotopic compositions were determined by the standard – sample bracketing method using a *Nu Plasma* MC-ICPMS at low resolution mode, with ^{26}Mg, ^{25}Mg and ^{24}Mg measured simultaneously in separated Faraday cups (H5, Ax, and L4) (Teng and Yang, 2014). The background Mg signals for the ^{24}Mg were $<10^{-4}$ V, which is negligible relative to sample signals of 3−4 V. During sample analysis, each sample solution together with standards treated as un-known was measured > 2 times per analytical session. The long-term external reproducibility is better than 0.07 on $^{26}Mg/^{24}Mg$ (Teng et al., 2010a). Magnesium isotopic results are reported in the conventional δ notation in per mil relative to DSM-3, $\delta^{x}Mg = [(^{x}Mg/^{24}Mg)_{sample}/(^{x}Mg/^{24}Mg)_{DSM\text{-}3} - 1] \times 1000$, where x refers to mass 25 or 26.

4 Results

Magnesium isotopic compositions of 74 whole-rock samples and 22 eclogitic minerals are reported in Tables 1 and 2, respectively. Major and trace elemental compositions for whole-rock samples are reported in the Supplementary materials.

Table 1 Magnesium isotopic compositions of samples and standards

Sample	MgO (wt.%)	$\delta^{26}Mg$	2SD	$\delta^{25}Mg$	2SD
Felsic schists from Wudang terrane in SCB					
HWD003	0.86	−0.135	0.051	−0.075	0.027
HWD006		−0.111	0.051	−0.043	0.027
Replicate		−0.123	0.048	−0.028	0.048
Average	1.01	−0.117	0.035	−0.040	0.023
HWD008		−0.187	0.051	−0.101	0.027
Replicate		−0.197	0.063	−0.110	0.060
Average	0.50	−0.191	0.040	−0.102	0.025
HWD009	0.88	−0.099	0.063	−0.073	0.060
HWD011		−0.167	0.051	−0.083	0.027
Replicate		−0.139	0.063	−0.061	0.060

Continued

Sample	MgO (wt.%)	δ^{26}Mg	2SD	δ^{25}Mg	2SD
Average	2.00	−0.156	0.040	−0.079	0.025
HWD016	1.32	−0.173	0.072	−0.065	0.039
HWD019	0.67	−0.159	0.057	−0.072	0.039
HWD022	1.19	−0.129	0.057	−0.045	0.039
HWD023	1.90	−0.290	0.049	−0.166	0.035
Greenschists from Wudang terrane in SCB					
HWD001	5.86	−0.217	0.051	−0.109	0.027
HWD005		−0.223	0.051	−0.113	0.027
Replicate		−0.235	0.048	−0.133	0.048
Average	8.60	−0.230	0.035	−0.118	0.023
HWD007		−0.203	0.051	−0.099	0.027
Replicate		−0.200	0.063	−0.105	0.060
Average	2.63	−0.202	0.040	−0.100	0.025
HWD010	5.75	−0.233	0.051	−0.111	0.027
HWD012		−0.173	0.057	−0.089	0.039
Replicate		−0.176	0.063	−0.080	0.060
Average	2.61	−0.174	0.042	−0.087	0.033
HWD013	8.39	−0.245	0.057	−0.148	0.039
HWD014		−0.140	0.057	−0.059	0.039
Replicate		−0.125	0.063	−0.059	0.060
Average	2.20	−0.133	0.042	−0.059	0.033
HWD015	4.57	−0.221	0.072	−0.116	0.039
HWD017	8.07	−0.198	0.057	−0.097	0.039
HWD018	3.60	−0.180	0.057	−0.102	0.039
HWD020	3.07	−0.150	0.057	−0.062	0.039
HWD021	6.09	−0.269	0.057	−0.147	0.039
HWD024		−0.163	0.037	−0.100	0.054
Replicate		−0.186	0.037	−0.079	0.054
Average	4.93	−0.175	0.026	−0.089	0.038
Amphibolites from Luzhenguan complex in BHY zone					
99CHC-1	5.42	−0.228	0.063	−0.131	0.054
99CHC-2		−0.223	0.063	−0.124	0.054
Replicate		−0.242	0.043	−0.103	0.039
Average	4.58	−0.236	0.035	−0.110	0.032
99CHC-3	5.15	−0.192	0.050	−0.096	0.049
99CHC-4	5.40	−0.204	0.061	−0.092	0.056
99CHC-5	4.93	−0.215	0.063	−0.105	0.054
99CHC-6	5.21	−0.243	0.061	−0.107	0.056
99SQ-1	3.67	−0.306	0.061	−0.099	0.056
99SQ-2	6.87	−0.286	0.061	−0.115	0.056
99SQ-3	4.54	−0.318	0.061	−0.146	0.056
99SQ-4	4.17	−0.358	0.061	−0.155	0.056

Continued

Sample	MgO (wt.%)	$\delta^{26}Mg$	2SD	$\delta^{25}Mg$	2SD
99SQ-5	4.43	−0.280	0.061	−0.137	0.056
99SQ-6	4.04	−0.303	0.063	−0.150	0.054
11LZ-3	8.54	−0.272	0.056	−0.060	0.047
11GF-3	4.16	−0.224	0.056	−0.126	0.047
11GF-5	3.98	−0.293	0.048	−0.171	0.048
11GF-11	8.15	−0.348	0.048	−0.196	0.048
11HJH-1	5.16	−0.260	0.048	−0.172	0.048
11HJH-2	4.62	−0.301	0.048	−0.135	0.048
11HJH-3	6.42	−0.276	0.048	−0.166	0.048
11HJH-4	5.87	−0.265	0.048	−0.148	0.048
Eclogites from UHPM zones in Dabie-Sulu orogen					
02QL-1#		−0.173	0.055	−0.078	0.050
Replicate		−0.219	0.061	−0.121	0.081
Average	7.60	−0.194	0.041	−0.090	0.043
02QL-4#		−0.340	0.075	−0.161	0.049
Replicate		−0.286	0.066	−0.101	0.070
Average	3.00	−0.310	0.050	−0.142	0.040
09QL-10#	5.39	−0.159	0.052	−0.094	0.040
09QL-13#	3.45	−0.257	0.075	−0.136	0.049
11HZ-03*	3.43	−0.240	0.061	−0.120	0.039
11HZ-04*	8.59	−0.252	0.061	−0.128	0.039
11HZ-05*		−0.170	0.054	−0.081	0.047
Replicate		−0.208	0.066	−0.140	0.070
Average	7.56	−0.185	0.042	−0.099	0.039
11HZ-06*	7.14	−0.265	0.054	−0.128	0.047
11SM-1*	4.07	−0.215	0.063	−0.129	0.035
11SM-2*	4.47	−0.230	0.056	−0.139	0.052
11SH-17	7.51	−0.214	0.063	−0.094	0.035
11SH-18	7.25	−0.209	0.052	−0.114	0.040
11SH-19		−0.156	0.055	−0.084	0.050
Replicate		−0.194	0.066	−0.046	0.070
Average	6.86	−0.172	0.042	−0.071	0.041
92HT-7	5.49	−0.277	0.063	−0.161	0.035
92HT-7-1		−0.139	0.075	−0.088	0.049
Replicate		−0.159	0.061	−0.072	0.081
Average	7.36	−0.151	0.047	−0.084	0.042
92HF-2	6.03	−0.137	0.063	−0.036	0.035
11BXL-001	8.72	−0.199	0.052	−0.086	0.040
11BXL-003	7.85	−0.297	0.054	−0.129	0.047
11BXL-004	6.98	−0.286	0.066	−0.160	0.070
11BXL-005	6.31	−0.235	0.054	−0.154	0.047
11BXL-006	6.57	−0.268	0.061	−0.128	0.039

Continued

Sample	MgO (wt.%)	δ^{26}Mg	2SD	δ^{25}Mg	2SD
11BXL-007	6.95	−0.257	0.055	−0.164	0.050
99MW-3	8.10	−0.191	0.054	−0.094	0.047
11SH-13	5.85	−0.218	0.055	−0.104	0.050
11SHE-1	5.13	−0.181	0.040	−0.089	0.041
11SHE-2	6.21	−0.246	0.075	−0.121	0.049
92HT-5		−0.338	0.052	−0.225	0.040
Replicate		−0.363	0.066	−0.134	0.070
Average	3.43	−0.348	0.041	−0.203	0.035
Paragneiss from UHPM zone in Dabie orogen					
92SH-5	0.41	+0.151	0.058	+0.065	0.042
Peridotite from UHPM zone in Dabie orogen					
11BXL-002	25.97	−0.312	0.052	−0.138	0.040
Marble from UHPM zone in Dabie orogen					
11SH-16	0.95	−0.864	0.058	−0.461	0.042
92HT-9	19.97	−2.443	0.058	−1.290	0.042
11SH-M	1.02	−0.633	0.058	−0.356	0.042
In-house standards					
KH-Ol (*n*=13)		−0.250	0.028	−0.128	0.024
Seawater(*n*=9)		−0.837	0.036	−0.430	0.026

2SD=2 times the standard deviation of the population of *n* (*n*>20) repeat measurement of the standards during an analytical session; Replicate represents the repeat sample dissolution, column chemistry and instrumental analysis. The eclogites labeled with "#" are from the Qinglongshan in Sulu orogen. Their peak metamorphic temperatures are similar to the eclogites from CD zone. The eclogites labeled with "*" are from SD zone, and others are from the CD zone. The Qinglongshan eclogites and the eclogites from the CD zone are termed as CD eclogites.

Table 2 Magnesium isotopic compositions of eclogitic minerals

Sample	Mineral	δ^{26}Mg	2SD	δ^{25}Mg	2SD
11BXL-006	Grt	−0.871	0.046	−0.463	0.039
	Cpx	+0.241	0.046	+0.130	0.039
	Replicate	+0.211	0.046	+0.134	0.066
	Average	+0.226	0.032	+0.131	0.033
11BXL-007	Grt	−0.852	0.046	−0.418	0.039
	Cpx	+0.303	0.046	+0.158	0.039
99QL-10	Grt	−0.801	0.046	−0.415	0.039
	Cpx	+0.325	0.046	+0.187	0.039
02QL-1	Grt	−0.824	0.046	−0.425	0.039
	Cpx	+0.325	0.046	+0.173	0.039
02QL-4	Grt	−0.770	0.046	−0.381	0.066
	Replicate	−0.831	0.046	−0.439	0.066
	average	−0.801	0.032	−0.410	0.046
	Cpx	+0.428	0.046	+0.193	0.066

Continued

Sample	Mineral	δ^{26}Mg	2SD	δ^{25}Mg	2SD
99QL-13	Grt	−0.967	0.046	−0.495	0.066
	Replicate	−0.951	0.046	−0.482	0.066
	Average	−0.959	0.032	−0.489	0.046
	Cpx	+0.267	0.046	+0.144	0.066
11HZ-06	Grt	−1.253	0.062	−0.643	0.040
	Cpx	+0.302	0.056	+0.151	0.047
11SH-13	Grt	−0.944	0.062	−0.482	0.040
	Cpx	+0.217	0.056	+0.109	0.047
11SH-18	Grt	−1.041	0.062	−0.499	0.040
	Cpx	+0.235	0.057	+0.107	0.045
92HT-7-1	Grt	−0.972	0.057	−0.510	0.045
	Replicate	−0.962	0.057	−0.494	0.045
	Average	−0.967	0.041	−0.502	0.032
	Cpx	+0.292	0.057	+0.208	0.045
11HZ-5	Grt	−1.398	0.057	−0.710	0.045
	Cpx	+0.247	0.057	+0.152	0.045

2SD=2 times the standard deviation of the population of n ($n > 20$) repeat measurement of the standards during an analytical session; Replicate represents the repeat sample dissolution, column chemistry and instrumental analysis.

4.1 Bimodal volcanic rocks from SCB

Thirteen Neoproterozoic greenschists yield δ^{26}Mg values ranging from −0.269±0.057‰ to −0.133±0.042‰ (Fig. 3a), with an average value of −0.196±0.044‰. The Mg isotopic compositions of the greenschists are negatively correlated with the MgO contents (Fig. 3a). The nine felsic schists have δ^{26}Mg values varying from −0.290±0.049‰ to −0.099±0.063 ‰ (Fig. 3a) with an average value of −0.161±0.084‰, falling within the range of global granites (Shen et al., 2009; Li et al., 2010; Liu et al., 2010; Telus et al., 2012; Ling et al., 2013a). With the exception of one sample (HWD023), Mg isotopic compositions of these felsic schists are systemically heavier than those of interlayered greenschists (Fig. 3a).

4.2 Amphibolites from BHY zone

δ^{26}Mg values of the 20 amphibolites range from −0.358±0.061‰ to −0.192±0.050‰, with an average value of −0.271±0.042‰ (Fig. 3b). No correlations exist between δ^{26}Mg and MgO (Fig. 3b).

4.3 Eclogites and country rocks from SD and CD zones

δ^{26}Mg values of the 27 eclogites vary from −0.348±0.041‰ to −0.137±0.063‰ (Fig. 3c), with an average value of −0.226±0.044‰, similar to those of greenschists and amphibolites (Fig. 3c). Host rocks of these eclogites (e.g., marble, gneiss and ultramafic rocks) have distinct Mg isotopic compositions (Table 1). The three marbles yield light δ^{26}Mg values ranging from −0.633±0.058‰ to −2.443±0.058‰. The peridotite has δ^{26}Mg value of −0.312±0.052‰, and the paragneiss has a heavy δ^{26}Mg value of +0.151±0.058‰, which overlaps those of pelites in eastern China (Li et al., 2010).

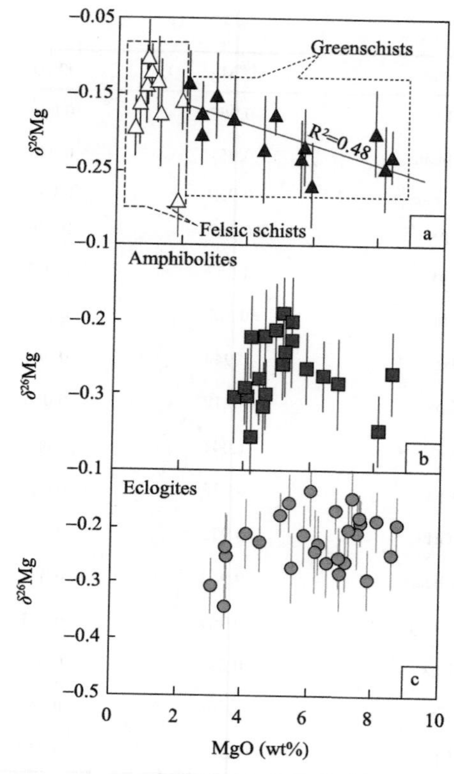

Fig. 3 Plots of δ^{26}Mg versus MgO for (a) greenschists and felsic schists in SCB, (b) amphibolites in BHY zone, and (c) eclogites in UHPM zones. Data are from Table 1.

4.4 Eclogitic minerals

δ^{26}Mg values of the eclogitic minerals vary from -0.801 ± 0.032 to -1.398 ± 0.057‰ in garnet, and from $+0.217\pm0.056$‰ to $+0.428\pm0.046$‰ in coexisting clinopyroxene (Table 2; Fig. 4). Garnet and clinopyroxene in eclogites from the SD zone display larger inter-mineral Mg isotope fractionations ($\Delta^{26}Mg_{Cpx\text{-}Grt}$=1.555–1.645 ‰) than those from the CD zone ($\Delta^{26}Mg_{Cpx\text{-}Grt}$ = 1.097–1.277‰) (Fig. 4).

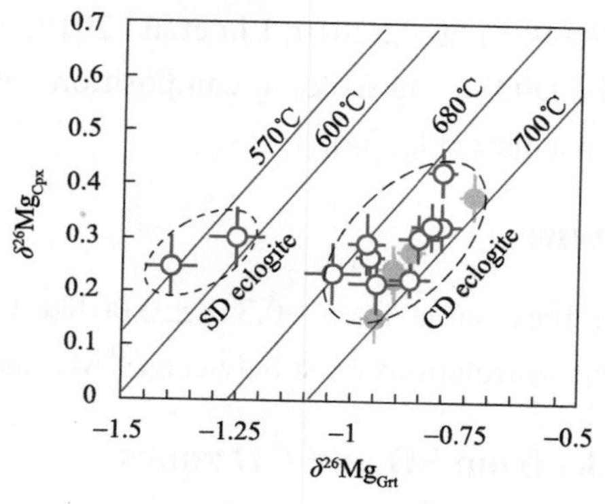

Fig. 4 Plot of $\delta^{26}Mg_{Grt}$ versus $\delta^{26}Mg_{Cpx}$. The equilibrium Mg isotope fractionation lines are from Huang et al. (2013a) by assuming that the peak metamorphic pressure of 3GPa. The grey circles are the data from Li et al. (2011). Other mineral data are from Table 2. (For interpretation of the references to colour in this figure legend, the reader is referred to the web version of this article.)

5 Discussion

Overall, the total Mg isotopic variation among greenschists, amphibolites, and eclogites is 0.259‰, which is significant compared to the isotope fractionation produced by close-system magmatic differentiation (≤ 0.07‰; Teng et al., 2007, 2010a). The inter-mineral Mg isotope fractionation in eclogites is larger (up to 1.645‰) than those previously reported (up to 1.14‰; Li et al., 2011; Wang et al., 2012b). In this section, we first assess the factors controlling the δ^{26}Mg of the greenschists and felsic schists, and then use the greenschists as the protolith of the high-grade metamorphic rocks to evaluate the behaviors of Mg isotopes during prograde metamorphism. Finally, we discuss the origin of Mg isotopic variations in amphibolites and eclogites.

5.1 Magnesium isotopic systematics of bimodal volcanic rocks from SCB

Studies on weathering profiles suggest that light Mg isotopes are preferentially released into the fluid, leaving the weathered residue enriched in heavy Mg isotopes (Teng et al., 2010b; Huang et al., 2012). During greenschist-facies metamorphism, Mg is also fluid-mobile (Coish, 1977). The subtle negative correlation between δ^{26}Mg and MgO (R^2=0.48) of these greenschists may thus result from Mg isotope fractionation induced by greenschist-facies metamorphic dehydration. However, δ^{26}Mg values of the greenschists and felsic schists do not systemically increase with the Rb/TiO_2 ratio (Fig. 5a), an index of the degree of metamorphic dehydration. Although some greenschists and felsic schists have extremely low Rb, K and Cs contents (Fig. 5a), indicating a high degree of metamorphic dehydration, their δ^{26}Mg values stay within the range of other samples (Fig. 5a). Therefore, the Mg isotopic variations observed in these Neoproterozoic greenschists and felsic schists cannot be caused by metamorphic dehydration but other processes.

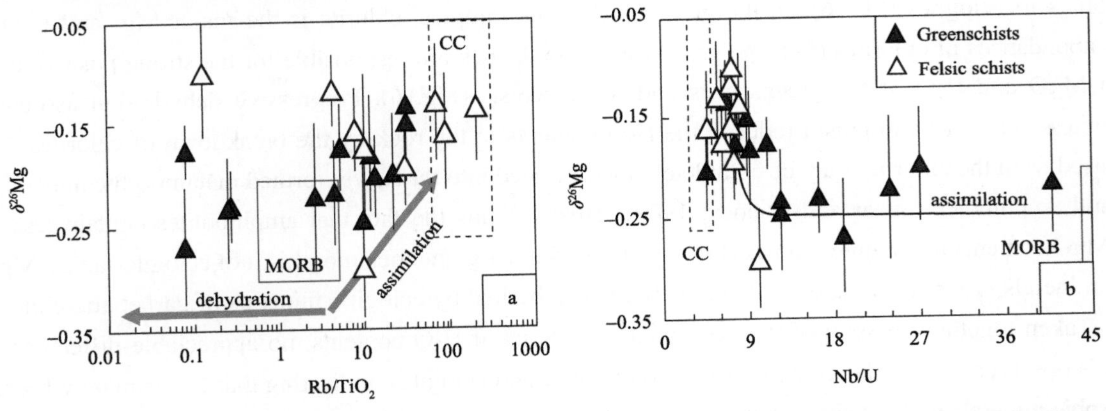

Fig. 5 Plots of δ^{26}Mg versus Rb/TiO_2 (a) and Nb/U (b) for greenschists and felsic schists in SCB. The bold line in (b) represents the bulk mixing curve between the average high-δ^{26}Mg felsic schists and MORB. Data for MORB are from Sun and McDonough (1989) and Teng et al. (2010a); Data for CC (continental crust) are from Rudnick and Gao (2003) and Li et al. (2010). The δ^{26}Mg values of greenschists and felsic schists are from Table 1, and the elemental compositions are given in Supplementary materials.

The δ^{26}Mg values of the felsic schists (−0.161±0.084‰ on average) are slightly heavier than those of I-type granites (Fig. 3a; −0.21±0.07‰; Liu et al., 2010). Heavy δ^{26}Mg values are generally seen in the sedimentary silicates such as shales and those heavily weathered regoliths (Li et al., 2010; Teng et al., 2010b). Recycling of these weathered components into the lower continental crust can cause the Mg isotopic heterogeneity of lower crust and produce high-δ^{26}Mg granites (Shen et al., 2009; Teng et al., 2013). Therefore, the high-δ^{26}Mg signatures preserved in these felsic schists may reflect various degrees of source contamination by recycled weathered materials. There is additional independent evidence for the source contamination of these felsic schists. Magmatic zircons in some Neoproterozoic granites preserve heavy δ^{18}O values (up to 10.4 ‰), which requires an ancient low-temperature hydrothermal alteration of their source region before anatexis (Zheng et al., 2007).

Continental crust typically has a Nb/U ratio of ~6.1 (Rudnick and Gao, 2003), significantly lower than that of MORBs (Nb/U=50; Sun and McDonough, 1989). The variable and low Nb/U ratios of 4.6–27 (except sample HWD001 with Nb/U=41) in greenschists (Fig. 5b) are taken as evidence for variable crustal assimilation during the genesis of these greenschists (e.g., Xia et al., 2012). Although the δ^{26}Mg values of the greenschists fall within the range of MORBs when uncertainties are considered, they do show subtle correlations with the Nb/U ratios and MgO contents (Fig. 3a and 5b). In particular, samples with the lowest Nb/U ratios have the heaviest δ^{26}Mg values (Fig. 5b), indicating that crustal assimilation process has driven the δ^{26}Mg of the greenschists towards heavier values. The coupled co-variation between δ^{26}Mg values and Nb/U ratios of the greenschists can be modeled by admixing of a MORB source with the high-δ^{26}Mg felsic schists (Fig. 5b). On the δ^{26}Mg – MgO plot, the δ^{26}Mg of the greenschists also continuously increase from the MORB-like value towards the end-member represented by the high-δ^{26}Mg felsic schists (Fig. 3a). Therefore, the Mg isotopic variations in these greenschists mainly reflect crustal assimilation of high-δ^{26}Mg felsic schists.

5.2 The behavior of Mg isotopes during prograde metamorphism

Comparison of Mg isotopic compositions of the greenschists to those of amphibolites and eclogites helps reveal the behaviors of Mg isotopes during prograde metamorphic dehydration. Change of water (H_2O^+) contents in rocks is a good indicator of the extent of metamorphic dehydration. With the increasing metamorphic grade, H_2O in the rock is progressively lost via dehydration. As seen in samples studied here, the H_2O contents decrease from greenschists, amphibolites to eclogites (Fig. 6). In the greenschist-facies samples, chlorite is the major Mg- and H_2O- carrier. Various abundances of chlorites (5–55%) present in the bulk-rock are responsible for the strong positive correlation between MgO and H_2O of the greenschists and felsic schists (Fig. 6). Progressive dehydration associated with amphibolites-facies metamorphism reduced the H_2O contents in the rock by the breakdown of chlorites. However, Mg released from the chlorites may be completely incorporated into the newly formed metamorphic minerals such as biotite and amphiboles (Spear and Cheney, 1989). This explains the fact that amphibolites contain less H_2O but similar MgO contents to the greenschists (Fig. 6). During higher grade metamorphism of eclogite-facies, Mg-bearing hydrous minerals broke down, and the Mg released was inherited by eclogitic minerals of garnet and clinopyroxene (Fig. 6). Taken together, in spite of the systematic variations of H_2O contents, no appreciable difference of MgO contents is observed among the greenschists, amphibolites, and eclogites, indicating that Mg is mainly hosted in the metamorphic minerals without significant loss during prograde metamorphism.

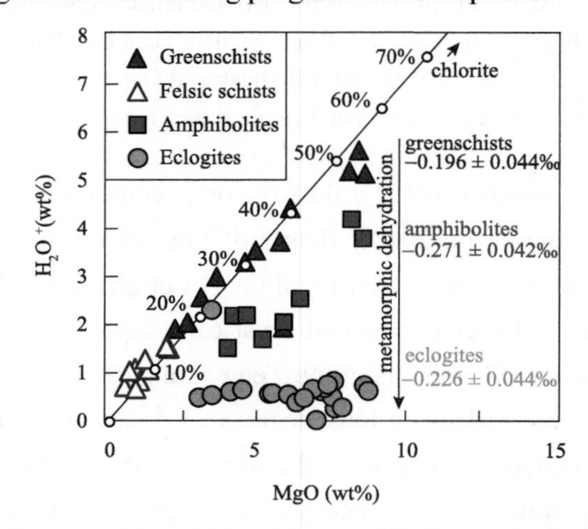

Fig. 6　Plot of H_2O^+ versus MgO for the greenschists, felsic schists, amphibolites and eclogites. The line represents water and MgO contents of rocks containing different abundances of chlorite. The mineral data of the chlorite is from the Deer et al. (1992). The data are given in the Supplementary materials.

The average δ^{26}Mg values of amphibolites and eclogites are −0.271±0.042‰ and −0.226±0.044‰, respectively, which are indistinguishable from that of the greenschists (−0.196±0.044‰), reflecting no systematic variations of δ^{26}Mg with the increasing metamorphic grade (Fig. 7). The lack of Mg isotopic differences among greenschists, amphibolites and eclogites, is consistent with the limited amounts of Mg loss during prograde metamorphic dehydration. Similarly, Teng et al. (2013) also shows that mafic granulites from Chudleigh retain Mg isotopic compositions of their protoliths of −0.243±0.058‰, even at higher metamorphic temperatures of 700–1000 ℃. In summary, Mg isotope fractionation during the subduction of basaltic rocks is negligible (Fig. 7), and Mg isotopic variations recorded in the meta-basaltic rocks reflect their protolith heterogeneities.

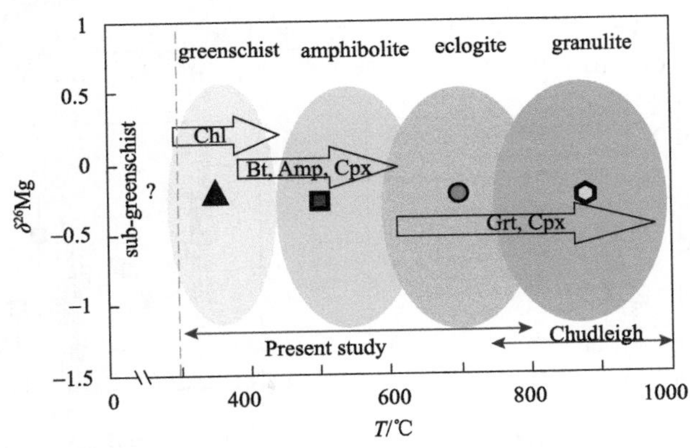

Fig. 7 Cartoon showing the evolution of Mg isotopes as a function of the metamorphic grade in mafic rocks. The ellipses indicate the sequence of metamorphic grade from greenschist to eclogite/granulite-facies. Different symbols represent the average δ^{26}Mg values of mafic rocks at different metamorphic grades. The Mg isotopic data of the mafic granulite in Chudleigh are from Teng et al. (2013). The other Mg isotopic data are from Li et al. (2011) and this study. Chl = chlorite, Bt = biotite, Amp = amphibole, Cpx = clinopyroxene, Grt = garnet. The behavior of Mg isotope during subgreenschist-facies metamorphism is currently unknown.

5.3 Magnesium isotopic systematics of amphibolites from BHY zone

Based on major and trace elemental compositions, amphibolites from BHY zone can be divided into two groups. The group I amphibolites have high CaO but low REE contents with positive Sr and Ti anomalies (Fig. 8a–c), whereas the group II amphibolites have low CaO but high REE contents without positive Sr and Ti anomalies (Fig. 8a–c). These two groups also display slightly different Mg isotopic compositions (Fig. 8b, c). The group I amphibolites have relatively heavy δ^{26}Mg values, with an average value of −0.215±0.048‰, whereas the group II amphibolites have relatively light δ^{26}Mg values, with an average value of −0.291±0.040‰. The potential processes responsible for the Mg isotopic compositions of these two groups of amphibolites are evaluated below in turn.

5.3.1 Group I amphibolites

The positive Sr and Ti anomalies, together with the high CaO contents (Fig. 8a, b), suggest that group I amphibolites contain high abundance of plagioclase and ilmenite. CIPW normalization shows that these rocks are mainly composed of An-rich plagioclase (44.74–55.05 wt%) and diopside (21.55–39.22 wt%) with minor ilmenite (3.15–3.91 wt%), supporting a cumulate origin (e.g., ilmenite-bearing gabbro). As the major Mg-carrier in gabbroic cumulates, the clinopyroxene plays an important role in controlling the bulk-rock δ^{26}Mg values. Theoretical calculation and studies of natural samples suggest that clinopyroxenes are slightly heavier than the coexisting melt in Mg isotopes (Handler et al., 2009; Yang et al., 2009; Young et al., 2009; Liu et al., 2011;

Schauble, 2011; Xiao et al., 2013). Accumulation of clinopyroxene and plagioclase can thus generate slightly heavier Mg isotopic compositions than the complementary basaltic melts. Therefore, the δ^{26}Mg value of −0.215 ± 0.048‰ for the group I amphibolites is of cumulate origin.

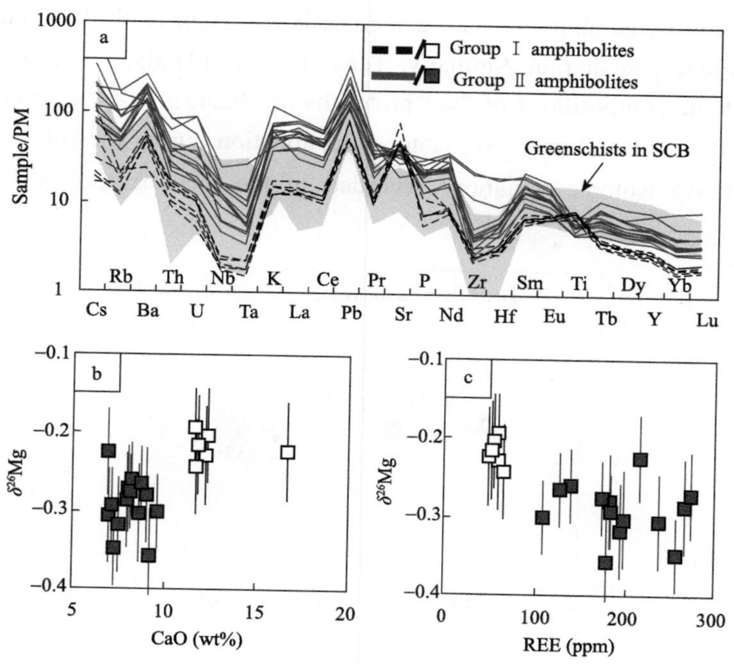

Fig. 8　Diagrams showing the two groups of amphibolites in BHY zone. Data for the primary mantle are from Sun and McDonough (1989). Major and trace elemental data are given in the Supplementary materials.

5.3.2　Group II amphibolites

The group II amphibolites have typical arc-like geochemical signatures, characterized by enrichment of LREE and LILE but depletion of HFSE (Fig. 8a). This feature can be explained if the protoliths were derived from (1) the asthenospheric mantle and later suffered crustal assimilation, or (2) the lithospheric mantle that was modified by subducted materials. The slightly lighter δ^{26}Mg value of the group II amphibolites can be caused by either of the above two processes.

The low Nb/U ratios (4.5–23) and enriched Nd isotopic compositions (εNd_{750Ma} =−11.2 for two samples; see Supplementary materials) of the group II amphibolites imply the incorporation of old crustal materials in their protoliths. Given that asthenospheric mantle-derived basalts have homogenous δ^{26}Mg value of −0.25±0.07‰ (e.g., Teng et al., 2010a), process of crustal assimilation is invoked to produce the light Mg isotopic characteristics of the group II amphibolites. Carbonate rocks have extremely light δ^{26}Mg values (e.g., Young and Galy, 2004; Ling et al. 2013b), and may serve as the most likely candidate. Alternatively, certain lithospheric mantle modified by subducted components may obtain δ^{26}Mg deviated from the "normal" mantle value (Yang et al., 2012; Wang et al., 2012b; Xiao et al., 2013). For example, cratonic eclogites from Kaapvaal Craton have δ^{26}Mg values as light as −0.798 ‰ (Wang et al., 2012b), while wehrlites from North China Craton have δ^{26}Mg values ranging from −0.39 to +0.09‰ (Xiao et al., 2013). Recent studies also found that Cenozoic continental basalts from North China Craton are characterized by light Mg isotopic compositions (−0.60 to −0.42‰; Yang et al., 2012). All these light δ^{26}Mg values in the mantle derivates are attributed to the involvement of recycled carbonates into their mantle source. Therefore, it is also possible that the group II amphibolites were produced by direct partial melting of the lithospheric mantle modified by recycled carbonates. Overall, the slightly light Mg isotopic compositions of the group II amphibolites could be produced by either crustal assimilation or source heterogeneity.

5.4 Magnesium isotopic systematics of eclogites from SD and CD zones

Eclogites in Dabie-Sulu orogen are divided into three types according to their host rock types (e.g., Jahn et al., 1998; Tang et al., 2007): type I eclogites occur as blocks or lenses within gneisses, type II eclogites occur as enclaves and interlayers within marbles, and type III eclogites occur as interlayers in mafic to ultramafic rocks. Host rocks of these three eclogite types have distinct δ^{26}Mg values (Fig. 9a), with paragneiss isotopically heavier and meta-carbonate lighter than the peridotite. Nonetheless, Mg isotopic variations among the three types of eclogites are indistinguishable and independent on the host rock types (Fig. 9a). It is thus suggested that eclogite-host interactions during continental subduction and exhumation have negligible effects on the whole-rock δ^{26}Mg values.

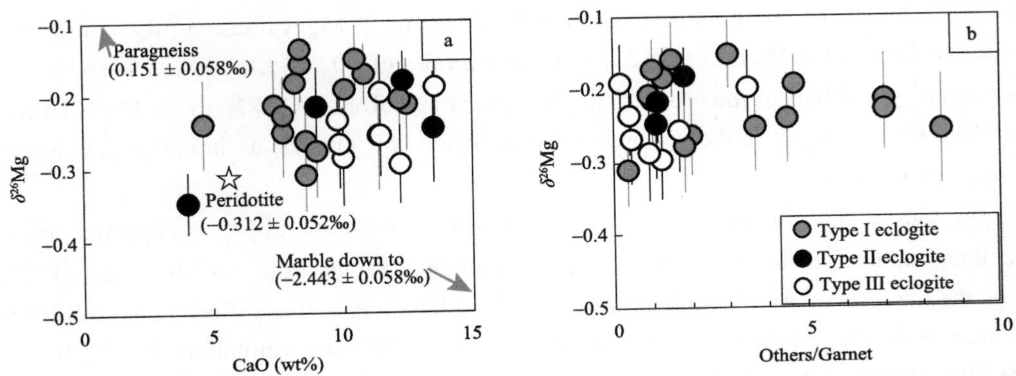

Fig. 9 Diagrams illustrating the δ^{26}Mg variations of the eclogites as a function of CaO content (a) and mineral abundance (b). δ^{26}Mg values of paragneiss, peridotite and marbles analyzed here from Dabie orogen are also shown in (a). The others/garnet ratio represents the abundance ratio of (clinopyroxene + phengite + amphibole)/garnet. The major elements and mineral abundance are reported in the supplementary materials.

Previous studies documented large equilibrium Mg isotope fractionation between coexisting garnet and clinopyroxene in both orogenic and cratonic eclogites (Li et al., 2011; Wang et al., 2012b). Cratonic eclogites formed at relatively higher temperatures display smaller inter-mineral Mg isotope fractionation than orogenic eclogites. Garnet-clinopyroxene pairs in eclogites from SD and CD zones record peak metamorphic temperatures of 570–600 ℃ and 680–750 ℃, respectively (Fig. 4), which are slightly lower than those obtained from Mg-Fe exchange thermometer (Xiao et al., 2000; Li et al., 2004). Large inter-mineral Mg isotope fractionation between garnet and clinopyroxene in theory, should not affect the bulk-rock Mg isotopic composition unless specimens were unrepresentatively sampled. Previous studies have found that unrepresentative sampling in small, banded, coarse-grained granulite xenoliths can cause large Mg isotopic variation in whole rocks (Teng et al., 2013). In consideration of the medium/large size of eclogitic garnet and clinopyroxene minerals, large rock specimens (normally 12 × 8 × 2 cm) were used to make whole-rock powder to avoid any potential unrepresentative sampling. Although SD eclogites exhibit relatively larger inter-mineral Mg isotope fractionations than CD eclogites (Fig. 4), they have similar whole-rock δ^{26}Mg values (−0.226±0.022‰ vs. −0.226±0.011‰; Table. 1). The absence of correlation between whole-rock δ^{26}Mg values and mineral proportions further excludes unrepresentative sampling as the cause for Mg isotopic variations in eclogites.

Overall, the Mg isotopic variations in eclogites were produced by neither host-eclogite interactions nor unrepresentative sampling, but likely reflect the protolith heterogeneity. Similar to the amphibolites and greenschists, eclogites from SD and CD zones were also enriched in Sr and Nd isotopic compositions (e.g.,

εNd_{230Ma} ranges from −20 to −0.1), indicating a strong crustal signature in the protoliths of eclogites (e.g., Jahn et al., 1998; Li et al., 2000; Li et al., 2004). Therefore, the Mg isotopic variations in eclogites from SD and CD zones also result from the incorporation of isotopically distinct crustal materials in their sources.

6 Conclusions

Based on systematic studies of Mg isotopes of the bimodal volcanic rocks from SCB, amphibolites from BHY zone, and eclogites from SD and CD zones, the following conclusions are drawn:

(1) δ^{26}Mg values range from −0.269‰ to −0.133‰ in 13 greenschists and from −0.290‰ to −0.099‰ in nine felsic schists. The subtle Mg isotopic variation results from the crustal assimilation of the isotopically heavy felsic schists.

(2) The two groups of amphibolites have slightly different δ^{26}Mg values: δ^{26}Mg values of the group I amphibolites range from −0.243‰ to −0.192‰ with an average of −0.215±0.048‰, caused by clinopyroxene accumulation; and δ^{26}Mg values of the group II amphibolites range from −0.358‰ to −0.224‰ with an average of −0.291±0.040‰, produced by either source contamination or crustal assimilation of isotopically light components (e.g., carbonates).

(3) Eclogitic garnet and clinopyroxene record equilibrium inter-mineral Mg isotope fractionations at peak metamorphic temperatures. Bulk δ^{26}Mg values of 28 eclogites range from −0.348‰ to −0.137‰, and are independent on the unrepresentative sampling or eclogite-host interaction, thus reflect the protolith heterogeneity.

(4) The average δ^{26}Mg values of the greenschists (−0.196±0.044‰), amphibolites (−0.271±0.042‰), and eclogites (−0.226±0.044‰) are similar within uncertainties, suggesting that Mg isotope fractionation during prograde metamorphic dehydration is limited.

Acknowledgements

We thank Hai-Ou Gu, Wang-Ye Li, Kang-Jun Huang, Yan Hu and Melissa Hornick for help in the clean lab, Ji Shen and Shi-Chao An for help in the field work. The constructive comments from the anonymous reviewers significantly improved the manuscript. The efficient editorial handling of Weidong Sun is greatly appreciated. This work was funded by National Science Foundation (EAR-0838227, EAR-1056713 and EAR-1340160) to Fang-Zhen Teng, and National Nature Science Foundation of China (41230209 and 41090372) to Shu-Guang Li.

Appendix A. supplementary data

Supplementary data associated with this article can be found, in the online version, at http://dx.doi.org/10.1016/j.gca.2014.03.029.

References

Bourdon B., Tipper E. T., Fitoussi C. and Stracke A. (2010) Chondritic Mg isotope composition of the Earth. Geochim. Cosmochim. Acta 74, 5069-5083.

Brenot A., Cloquet C., Vigier N., Carignan J. and Francelanord C. (2008) Magnesium isotope systematics of the lithologically varied Moselle river basin, France. Geochim. Cosmochim. Acta 72, 5070-5089.

Coish R. A. (1977) Ocean floor metamorphism in the Betts Cove Ophiolite, Newfoundland. Contrib. Miner. Petrol. 60, 255-270.

Deer W. A., Howie R. A. and Zussman J. (1992) An introduction to the rock-forming minerals. Longman Scientific & Technical, Hong Kong, 336.

Faure M., Lin W., Schärer U., Shu L., Sun Y. and Arnaud N. (2003) Continental subduction and exhumation of UHP rocks. Structural and geochronological insights from the Dabieshan (East China). Lithos 70, 213-241.

Foster G., von Strandmann P. P. and Rae J. (2010) Boron and magnesium isotope composition of seawater. Geochem. Geophys. Geosyst. 11, Q08015.

Galy A., Bar-Matthews M., Halicz L. and O'Nions R. K. (2002) Mg isotope composition of carbonate: insight from speleothem formation. Earth Planet. Sci. Lett. 201, 105-115.

Hacker B. R., Ratschbacher L., Webb L., Ireland T., Walker D. and Shuwen D. (1998) U/Pb zircon ages constrain the architecture of the ultrahigh-pressure Qinling-Dabie Orogen, China. Earth Planet. Sci. Lett. 161, 215-230.

Handler M. R., Baker J. A., Schiller M., Bennett V. C. and Yaxley G. M. (2009) Magnesium stable isotope composition of Earth's upper mantle. Earth Planet. Sci. Lett. 282(1–4), 306-313.

Huang W. F. (1993) Multiphase deformation and displacement within a basement complex on a contiental margin: the Wudang Complex in the Qinling Orogen, China. Tectonophysics 224, 305-326.

Huang F., Zhang Z. F., Lundstrom C. C. and Zhi X. C. (2011) Iron and magnesium isotope compositions of peridotite xenoliths from Eastern China. Geochim. Cosmochim. Acta 75, 3318-3334.

Huang K.-J., Teng F.-Z., Wei G.-J., Ma J.-L. and Bao Z.-Y. (2012) Adsorption- and desorption-controlled magnesium isotope fractionation during extreme weathering of basalt in Hainan Island, China. Earth Planet. Sci. Lett. 359, 73-83.

Huang F., Chen L., Wu Z. and Wang W. (2013a) First-principles calculations of equilibrium Mg isotope fractionations between garnet, clinopyroxene, orthopyroxene, and olivine: implications for Mg isotope thermometry. Earth Planet. Sci. Lett. 367, 61-70.

Huang K. J., Teng F. Z., Elsenouy A., Li W. Y. and Bao Z. Y. (2013b) Magnesium isotopic variations in loess: origins and implications. Earth Planet. Sci. Lett. 374, 60-70.

Jahn B. (1998) Geochemical and isotopic characteristics of UHP eclogites and ultramafic rocks of the Dabie orogen: implications for continental subduction and collisional tectonics. When Continents Collide: Geodynamics and Geochemistry of Ultrahigh-Pressure Rocks, 203-239.

Ke S., Liu S. A., Li W. Y., Yang W. and Teng F. Z. (2011) Advances and application in magnesium isotope geochemistry. Acta Petrol. Sin. 27, 383-397.

Li S. G., Hart S. R., Zheng S. G., Liu D. L., Zhang G. W. and Guo A. L. (1989) Timing of collision between the north and south China blocks—the Sm–Nd isotopic age evidence. Sci. China (Ser. B) 32, 1391-1400.

Li S. G., Xiao Y., Liou D., Chen Y., Ge N., Zhang Z., Sun S., Cong B., Zhang R. and Hart S. R. (1993) Collision of the North China and Yangtse Blocks and formation of coesite-bearing eclogites: timing and processes. Chem. Geol. 109, 89-111.

Li S. G., Jagoutz E., Chen Y. and Li Q. (2000) Sm–Nd and Rb–Sr isotopic chronology and cooling history of ultrahigh pressure metamorphic rocks and their country rocks at Shuanghe in the Dabie Mountains, Central China. Geochim. Cosmochim. Acta 64, 1077-1093.

Li S. G., Huang F., Nie Y. H., Han W. L., Long G., Li H. M., Zhang S. Q. and Zhang Z. H. (2001) Geochemical and geochronological constraints on the suture location between the North and South China blocks in the Dabie Orogen, Central China. Phys. Chem. Earth Part A. 26, 655-672.

Li S. G., Huang F., Zhou H. and Li H. (2003) U–Pb isotope compositions of the ultrahigh pressure metamorphic (UHPM) rocks from Shuanghe and gneisses from Northern Dabie zone in the Dabie Mountains, central China: constraint on the exhumation mechanism of UHPM rocks. Sci. China, Ser. D Earth Sci. 46, 200-209.

Li X. P., Zheng Y. F., Wu Y. B., Chen F., Gong B. and Li Y. L. (2004) Low-T eclogite in the Dabie terrane of China: petrological and isotopic constraints on fluid activity and radiometric dating. Contrib. Miner. Petrol. 148, 443-470.

Li W.-Y., Teng F.-Z., Ke S., Rudnick R. L., Gao S., Wu F.-Y. and Chappell B. W. (2010) Heterogeneous magnesium isotope composition of the upper continental crust. Geochim. Cosmochim. Acta 74, 6867-6884.

Li W.-Y., Teng F.-Z., Xiao Y. and Huang J. (2011) Hightemperature inter-mineral magnesium isotope fractionation in eclogite from the Dabie orogen, China. Earth Planet. Sci. Lett. 304, 224-230.

Ling W. L., Ren B. F., Duan R. C., Liu X. M., Mao X. W., Peng L. H., Liu Z. X., Cheng J. P. and Yang H. M. (2008) Timing of the

Wudangshan Yaolinghe volcanic sequences and mafic sill in South Qinling: U–Pb zircon geochronology and tectonic implication. China Sci. Bull. 53, 2192-2199.

Ling W. L., Duan R. C., Liu X. M., Cheng J. P., Mao X. W., Peng L. H., Liu Z. X., Yang H. M. and Ren B. F. (2010) U–Pb dating of detrital zircons from the Wudangshan Group in the South Qinling and its geological significance. Chin. Sci. Bull. 55, 2440-2448.

Ling M. X., Sedaghatpour F., Teng F. Z., Hays P. D., Strauss J. and Sun W. (2011) Homogeneous magnesium isotope composition of seawater: an excellent geostandard for Mg isotope analysis. Rapid Commun. Mass Spectrom. 25, 2828-2836.

Ling M. X., Li Y., Ding X., Teng F. Z., Yang X. Y., Fan W. M., Xu Y. G. and Sun W. (2013a) Destruction of the North China craton induced by ridge subductions. J. Geol. 121(2), 197-213.

Ling M. X., Liu Y. L., Williams I. S., Teng F. Z., Yang X. Y., Ding X., Wei G. J., Xie L. H., Deng W. F. and Sun W. D. (2013b) Formation of the world's largest REE deposit through protracted fluxing of carbonatite by subduction-derived fluids. Sci. Rep. http://dx.doi.org/10.1038/srep01776.

Liu Y. C. and Li S. G. (2008) Detachment within subducted continental crust and multi-slice successive exhumation of ultrahigh-pressure metamorphic rocks: evidence from the Dabie–Sulu orogenic belt. Chin. Sci. Bull. 53, 3105-3119.

Liu F. L. and Liou J. G. (2011) Zircon as the best mineral for P–T–time history of UHP metamorphism: a review on mineral inclusions an U–Pb SHRIMP ages of zircons from Dabie–Sulu UHP rocks. J. Asian Earth Sci. 40, 1-39.

Liu F., Xu Z., Katayama I., Yang J., Maruyama S. and Liou J. (2001) Mineral inclusions in zircons of para- and orthogneiss from pre-pilot drillhole CCSD-PP1, Chinese Continental Scientific Drilling Project. Lithos 59, 199-215.

Liu Y. C., Li S. G. and Xu S. T. (2007) Zircon SHRIMP U–Pb dating for gneisses in northern Dabie high T/P metamorphic zone, central China: implications for decoupling within subducted continental crust. Lithos 96, 170-185.

Liu Y., Zong K., Kelemen P. B. and Gao S. (2008) Geochemistry and magmatic history of eclogites and ultramafic rocks from the Chinese continental scientific drill hole: subduction and ultrahigh-pressure metamorphism of lower crustal cumulates. Chem. Geol. 247, 133-153.

Liu S.-A., Teng F.-Z., He Y., Ke S. and Li S. (2010) Investigation of magnesium isotope fractionation during granite differentiation: implication for Mg isotope composition of the continental crust. Earth Planet. Sci. Lett. 297, 646-654.

Liu S.-A., Teng F.-Z., Yang W. and Wu F.-Y. (2011) Hightemperature inter-mineral magnesium isotope fractionation in mantle xenoliths from the North China craton. Earth Planet. Sci. Lett. 308, 131-140.

Middlemost E. A. (1994) Naming materials in the magma/igneous rock system. Earth Sci. Rev. 37, 215-224.

Okay A. I., Xu S. and Sengor A. M. C. (1989) Coesite from the Dabie Shan eclogites, central China. Eur. J. Mineral. 1, 595-598.

Pogge von Strandmann P. A. E., Burton K. W., James R. H., van Calsteren P., Gislason S. R. and Sigfússon B. (2008) The influence of weathering processes on riverine magnesium isotopes in a basaltic terrain. Earth Planet. Sci. Lett. 276, 187-197.

Pogge von Strandmann P. A., Elliott T., Marschall H. R., Coath C., Lai Y.-J., Jeffcoate A. B. and Ionov D. A. (2011) Variations of Li and Mg isotope ratios in bulk chondrites and mantle xenoliths. Geochim. Cosmochim. Acta 75, 5247-5268.

Rudnick R. (1995) Making continental crust. Nature 378(6557), 571-577.

Rudnick R. and Gao S. (2003) Composition of the continental crust. Treatise Geochem. 3, 1-64.

Schauble E. A. (2011) First-principles estimates of equilibrium magnesium isotope fractionation in silicate, oxide, carbonate and hexaaquamagnesium(2+) crystals. Geochim. Cosmochim. Acta 75, 844-869.

Shen B., Jacobsen B., Lee C. T. A., Yin Q. Z. and Morton D. M. (2009) The Mg isotopic systematics of granitoids in continental arcs and implications for the role of chemical weathering in crust formation. Proc. Natl. Acad. Sci. USA 106, 20652-20657.

Spear F. S. and Cheney J. T. (1989) A petrogenetic grid for pelitic schists in the system SiO2–Al2O3–FeO–MgO–K2O–H2O. Contrib. Miner. Petrol. 101(2), 149-164.

Sun S.-S. and McDonough W. F. (1989) Chemical and isotopic systematics of oceanic basalts: implications for mantle composition and processes. Geol. Soc. Lond., Spec. Publ. 42, 313-345.

Tang H. F., Liu C. Q., Nakai S. and Orihashi Y. (2007) Geochemistry of eclogites from the Dabie–Sulu terrane, eastern China: new insights into protoliths and trace element behaviour during UHP metamorphism. Lithos 95, 441-457.

Telus M., Dauphas N., Moynier F., Tissot F. L., Teng F. Z., Nabelek P. I., Craddock P. R. and Groat L. A. (2012) Iron, zinc, magnesium and uranium isotopic fractionation during continental crust differentiation: the tale from migmatites, granitoids, and pegmatites. Geochim. Cosmochim. Acta 97, 247-265.

Teng F.-Z. and Yang W. (2014) Comparison of factors affecting the accuracy of high-precision magnesium isotope analysis by multi-collector inductively coupled plasma mass spectrometry. Rapid Commun. Mass Spectrom 28, 19-24.

Teng F.-Z., Wadhwa M. and Helz R. T. (2007) Investigation of magnesium isotope fractionation during basalt differentiation:

Implications for a chondritic composition of the terrestrial mantle. Earth Planet. Sci. Lett. 261, 84-92.

Teng F.-Z., Li W.-Y., Ke S., Marty B., Dauphas N., Huang S., Wu F.-Y. and Pourmand A. (2010a) Magnesium isotope composition of the Earth and chondrites. Geochim. Cosmochim. Acta 74, 4150-4166.

Teng F.-Z., Li W.-Y., Rudnick R. L. and Gardner L. R. (2010b) Contrasting lithium and magnesium isotope fractionation during continental weathering. Earth Planet. Sci. Lett. 300, 63-71.

Teng F.-Z., Dauphas N., Helz R. T., Gao S. and Huang S. (2011) Diffusion-driven magnesium and iron isotope fractionation in Hawaiian olivine. Earth Planet. Sci. Lett. 308, 317-324.

Teng F.-Z., Yang W., Rudnick R. L. and Hu Y. (2013) Heterogeneous magnesium isotopic composition of the lower continental crust: a xenolith perspective. Geochem. Geophys. Geosyst. 14, 3844-3856. http://dx.doi.org/10.1002/ggge.20238.

Tipper E., Galy A. and Bickle M. (2006a) Riverine evidence for a fractionated reservoir of Ca and Mg on the continents: Implications for the oceanic Ca cycle. Earth Planet. Sci. Lett. 247, 267-279.

Tipper E., Galy A., Gaillardet J., Bickle M., Elderfield H. and Carder E. (2006b) The magnesium isotope budget of the modern ocean: constraints from riverine magnesium isotope ratios. Earth Planet. Sci. Lett. 250, 241-253.

Wang S.-J., Li S.-G., An S.-C. and Hou Z.-H. (2012a) A granulite record of multistage metamorphism and REE behavior in the Dabie orogen: constraints from zircon and rock-forming minerals. Lithos 136-139, 109-125.

Wang S.-J., Teng F.-Z., Williams H. M. and Li S.-G. (2012b) Magnesium isotopic variations in cratonic eclogites: Origins and implications. Earth Planet. Sci. Lett. 359, 219-226.

Wang S.-J., Li S.-G., Chen L.-J., He Y.-S., An S.-C. and Shen J. (2013) Geochronology and geochemistry of leucosomes in the North Dabie Terrane, East China: implication for post-UHPM crustal melting during exhumation. Contrib. Miner. Petrol. 165, 1009-1029.

Wimpenny J., Gíslason S. R., James R. H., Gannoun A., Pogge Von Strandmann P. A. E. and Burton K. W. (2010) The behaviour of Li and Mg isotopes during primary phase dissolution and secondary mineral formation in basalt. Geochim. Cosmochim. Acta 74, 5259-5279.

Wu Y., Zheng Y., Gong B., Tang J., Zhao Z. and Zha X. (2004) Zircon U–Pb ages and oxygen isotope compositions of the Luzhenguan magmatic complex in the Beihuaiyang zone. Acta Petrol. Sin. 20, 1007.

Xia L. Q., Xia Z. C., Xu X. Y., Li X. M. and Ma Z. P. (2012) MidLate Neoproterozoic rift-related volcanic rocks in China: geological records of rifting and break-up of Rodinia. Geosci. Front. 3, 375-399.

Xiao Y., Hoefs J., van den Kerkhof A. M., Fiebig J. and Zheng Y. (2000) Fluid history of UHP metamorphism in Dabie Shan, China: a fluid inclusion and oxygen isotope study on the coesite-bearing eclogite from Bixiling. Contrib. Miner. Petrol. 139(1), 1-16.

Xiao Y., Teng F.-Z., Zhang H.-F. and Yang W. (2013) Large magnesium isotope fractionation in peridotite xenoliths from eastern North China craton: product of melt–rock interaction. Geochim. Cosmochim. Acta 115, 241-261.

Xu S. T., Okay A. I., Ji S. Y., Sengor A. M. C., Su W., Liu Y. C. and Jiang L. L. (1992) Diamond from the Dabie Shan metamorphic rocks and its implication for tectonic setting. Science 256, 80-82.

Yang W., Teng F.-Z. and Zhang H.-F. (2009) Chondritic magnesium isotope composition of the terrestrial mantle: a case study of peridotite xenoliths from the North China craton. Earth Planet. Sci. Lett. 288, 475-482.

Yang W., Teng F.-Z., Zhang H.-F. and Li S.-G. (2012) Magnesium isotopic systematics of continental basalts from the North China craton: implications for tracing subducted carbonate in the mantle. Chem. Geol. 328, 185-194.

Ye K., Cong B. and Ye D. (2000) The possible subduction of continental material to depths greater than 200 km. Nature 407, 734-736.

Young E. D. and Galy A. (2004) The isotope geochemistry and cosmochemistry of magnesium. Rev. Mineral. Geochem. 55, 197-230.

Young E. D., Tonui E., Manning C. E., Schauble E. and Macris C. A. (2009) Spinel-olivine magnesium isotope thermometry in the mantle and implications for the Mg isotope composition of Earth. Earth Planet. Sci. Lett. 288, 524-533.

Zhang Z., Xiao Y., Hoefs J., Liou J. and Simon K. (2006) Ultrahigh pressure metamorphic rocks from the Chinese Continental Scientific Drilling Project: I. Petrology and geochemistry of the main hole (0–2050 m). Contrib. Miner. Petrol. 152, 421-441.

Zhao Z.-F., Zheng Y.-F., Chen R.-X., Xia Q.-X. and Wu Y.-B. (2007) Element mobility in mafic and felsic ultrahigh-pressure metamorphic rocks during continental collision. Geochim. Cosmochim. Acta 71, 5244-5266.

Zheng Y. F., Fu B., Gong B. and Li L. (2003) Stable isotope geochemistry of ultrahigh pressure metamorphic rocks from the Dabie–Sulu orogen in China: implications for geodynamics and fluid regime. Earth Sci. Rev. 62, 105-161.

Zheng Y. F., Zhou J. B., Wu Y. B. and Xie Z. (2005) Low-grade metamorphic rocks in the Dabie-Sulu orogenic belt: a passivemargin

accretionary wedge deformed during continent subduction. Int. Geol. Rev. 47, 851-871.
Zheng Y. F., Zhao Z. F., Wu Y. B., Zhang S. B., Liu X. and Wu F. Y. (2006) Zircon U–Pb age, Hf and O isotope constraints on protolith origin of ultrahigh-pressure eclogite and gneiss in the Dabie orogen. Chem. Geol. 231, 135-158.
Zheng Y. F., Zhang S. B., Zhao Z. F., Wu Y. B., Li X. H., Li Z. X. and Wu F. Y. (2007) Constrasting zircon Hf and O isotopes in the two episodes of Neoproterozoic granitoids in South China: Implications for growth and reworking of contiental crust. Lithos 96, 127-150.

Tracing carbonate-silicate interaction during subduction using magnesium and oxygen isotopes*

Shui-Jiong Wang [1, 2], Fang-Zhen Teng [2] and Shu-Guang Li [1, 3]

1. State Key Laboratory of Geological Processes and Mineral Resources, China University of Geosciences, Beijing 100083, China
2. Isotope Laboratory, Department of Earth and Space Sciences, University of Washington, Seattle, Washington 98195, USA
3. CAS Key Laboratory of Crust-Mantle Materials and Environments, School of Earth and Space Sciences, University of Science and Technology of China, Hefei, Anhui 230026, China

亮点介绍：本文分析了大别-苏鲁超高压变质（UHPM）带的含柯石英大理岩及其包含的碳酸盐化榴辉岩的 Mg-O 同位素组成，以及该 UHPM 大理岩的原岩（华南陆块北缘新元古代沉积的碳酸盐岩）的 Mg 同位素组成。与原岩相比，UHPM 大理岩增加了介于方解石和白云石端元之间的含镁或富镁大理岩，它们具有相似和较重的 O 同位素组成，且较地幔值轻的 Mg 同位素组成；此外，碳酸盐化榴辉岩也较原岩极度富集重的 O 同位素和轻的 Mg 同位素。上述 Mg-O 同位素特征表明在俯冲作用过程中，碳酸盐和榴辉岩之间存在 Ca-Mg 元素和同位素交换。上述发现回答了深俯冲富镁碳酸盐的来源问题，也提出了如何区分同样具有轻 Mg 同位素组成的俯冲富镁碳酸盐和碳酸盐化榴辉岩问题。

Subduction of carbonates and carbonated eclogites into the mantle plays an important role in transporting carbon into deep Earth. However, to what degree isotopic exchanges occur between carbonate and silicate during subduction remains unclear. Here we report Mg and O isotopic compositions for ultrahigh pressure metamorphic marbles and enclosed carbonated eclogites from China. These marbles include both calcite- and dolomite-rich examples and display similar O but distinct Mg isotopic signatures to their protoliths. Their δ^{26}Mg values vary from −2.508 to −0.531‰, and negatively correlate with MgO/CaO ratios, unforeseen in sedimentary carbonates. Carbonated eclogites have extremely heavy δ^{18}O (up to +21.1‰) and light δ^{26}Mg values (down to −1.928‰ in garnet and −0.980‰ in pyroxene) compared with their protoliths. These unique Mg-O isotopic characteristics reflect differential isotopic exchange between eclogites and carbonates during subduction, making coupled Mg and O isotopic studies potential tools for tracing deep carbon recycling.

Deep carbon recycling plays a key role in the global carbon cycle by modulating the CO_2 budget of the Earth's atmosphere over geologic timescales[1,2]. Among different subducted lithologies, carbonates and carbonated eclogites are of particular importance in transporting carbon from Earth's surface into its interior[3–6]. As a subducting slab descends, some fractions of the carbon carried by carbonate minerals are liberated back at the

* 本文发表在：Nature Communications, 2014, 5: 5328

surface via volcanism[7]; however, most is delivered to great mantle depths together with silicates[8,9]. Carbonate-silicate interaction along the subduction pressure-temperature (*P-T*) paths is particularly important because it not only influences the long-term carbon flux in subduction zones[10] but also controls the isotopic systematics of coexisting carbonate and silicate that are ultimately recycled into the mantle. The latter in turn contributes fundamentally to the chemical and physical properties of the mantle[11–13], for example, initiation of melting by lowering the mantle solidus, generation of alkalic magma in oceanic islands and carbonate metasomatism in cratonic mantle.

Magnesium (Mg) and oxygen (O) are two essential elements closely associated with carbon in carbonate minerals. Their isotopic compositions differ greatly from normal mantle rocks, that is, unaltered oceanic basalts and lherzolites have δ^{26}Mg of -0.25 ± 0.07‰ (ref. 14) and δ^{18}O of $+5.5\pm0.2$‰(ref. 15), whereas sedimentary carbonates have extremely light δ^{26}Mg (down to −6‰)[16–18] and heavy δ^{18}O (up to +28 ‰)[19,20]. Thus, Mg-O isotopic systematics can provide insight into carbonate-silicate chemical interactions during the subduction of carbonates and carbonated eclogites.

Here, we present Mg, O, C and Nd isotopic data for a suite of ultrahigh pressure metamorphic (UHPM) marbles and enclosed carbonated eclogites from the Sulu orogen, east China. For comparison, we also report Mg isotopic data for Sinian sedimentary carbonates from the South China Block, and eclogites hosted in UHPM gneisses (termed as normal eclogites hereafter) from the Sulu orogen. Our results show that the carbonated eclogites have anomalously light δ^{26}Mg and heavy δ^{18}O as compared with the normal eclogites; moreover, UHPM marbles display negative correlation between δ^{26}Mg and MgO/CaO. These Mg-O isotopic characteristics reflect differential isotopic exchange between the marble and enclosed eclogite during subduction, and provide important constraints on deep carbon recycling.

1 Results

1.1 Geologic setting and samples

Triassic subduction of the South China Block beneath the North China Block resulted in the UHPM belts of Dabie and Sulu regions (Supplementary Fig. 1a)[21]. Eclogites are widespread as pods, discontinuous layers or blocks in gneisses, ultramafic massifs and marbles. Previous studies suggest that these eclogites were formed by UHP metamorphism of Neoproterozoic basaltic rocks during the Triassic subduction of the South China Block[21–24]. Diamond and coesite inclusions present in eclogites and their country rocks indicate that they were in-situ subducted and exhumed with peak metamorphic temperatures of 700–880 ℃ and pressures up to 5.5 GPa (refs 25-27). The marble masses in the Rongcheng area crop out as large-scale discontinuous lenses with lengths of more than 2,000m and are surrounded by orthogneisses containing layers or lenses of eclogites, amphibolites and serpentinized peridotites (Supplementary Fig. 1b). Samples studied here were collected from local marble quarries where many eclogites are present as centimetre- to metre-sized lenticular or spherical blocks in host marbles (Supplementary Fig. 1c). These marble-hosted eclogites are termed as carbonated eclogites hereafter. Field observation shows that the margins of carbonated eclogite blocks are invariably retrograded to amphibolites, while the cores are free of retrogression. The primary mineral assemblage of the carbonated eclogite is garnet, clinopyroxene, quartz, plagioclase and amphibole, with accessory minerals of calcite, dolomite, pyrrhotite, chalcopyrite, and rutile (Supplementary Table 1). Garnet and pyroxene minerals in carbonated eclogites are enriched in Ca relative to those in normal eclogites because of the UHPM metasomatism by host marbles[27]. The marbles range from calcite-rich to dolomite-rich in composition, and consist mainly of calcite and dolomite with silicate mineral assemblage of plagioclase, quartz, and other minor phases such as pyroxene, amphibole, olivine and talc. Calcite contains significant MgO up to 4.43 wt.%, and dolomite has MgO in the range of 16.57–21.84 wt.%(refs 28,29).

1.2 Magnesium isotopes

The δ^{26}Mg values of normal eclogites vary from −0.440 to +0.092 ‰, with an average value of −0.224± 0.048 ‰ (n=26; Fig. 1a and Supplementary Table 2), similar to that of their Neoproterozoic basaltic protoliths (δ^{26}Mg=−0.196±0.044‰; Fig. 1a)[30]. Garnets from these normal eclogites are systemically lighter than coexisting pyroxenes ($\Delta^{26}Mg_{Cpx-Grt}=\delta^{26}Mg_{Cpx}-\delta^{26}Mg_{Grt}$=0.887–1.099‰), with δ^{26}Mg ranging from −0.943 to −0.622 ‰ in garnets and from +0.151 to +0.438‰ in pyroxenes (Fig. 1b and Supplementary Table 2). By contrast, carbonated eclogites are highly depleted in ^{26}Mg as compared with the normal eclogites, with δ^{26}Mg varying from −1.928 to −0.648‰ (Fig. 1a and Supplementary Table 3). These Mg isotopic ratios are the lightest known values in silicate rocks. The amphibolitic margin (11RC-2E) and eclogitic core (11RC-1) of an individual retrograded eclogite block (with diameter of ~30 cm) have δ^{26}Mg of −0.648‰ and −0.684‰ (Supplementary Table 3), respectively. Garnet and pyroxene separates from carbonated eclogites also have the lightest δ^{26}Mg values (Fig. 1b and Supplementary Table 4), that is, from −1.937 to −1.191‰ in garnet and from −0.980 to −0.483‰ in pyroxene.

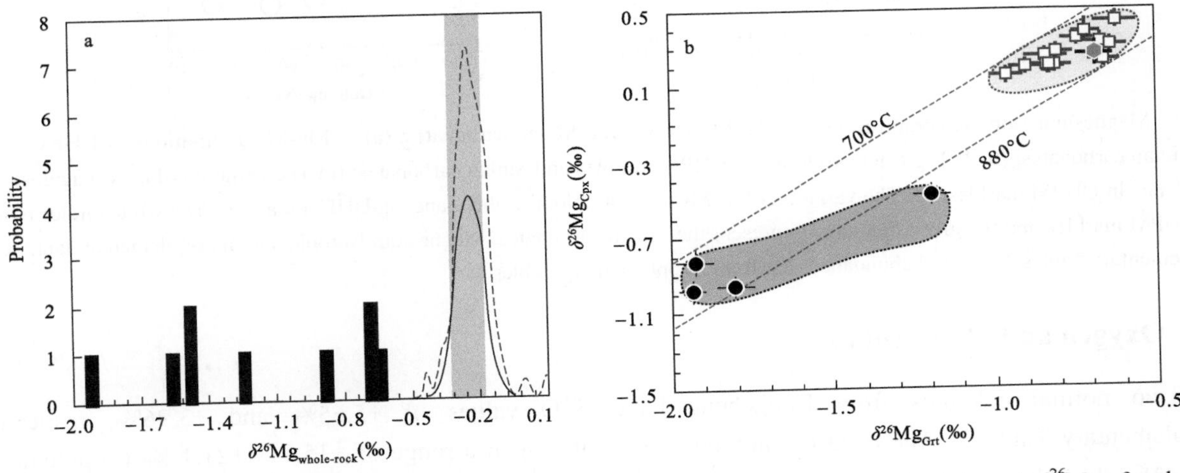

Fig. 1 Magnesium isotopic compositions of carbonated and normal eclogites. (a) Histogram of whole-rock δ^{26}Mg of carbonated eclogites. The grey band represents the 'normal' mantle value[14]; the red dashed and blue solid curves are kernel density estimates of δ^{26}Mg values of the normal eclogites from Sulu orogen and the Neoproterozoic basaltic protoliths from South China Block[30], respectively. δ^{26}Mg values of carbonated eclogites, represented by the histogram, are significantly lighter than both normal eclogites and their protoliths; (b) δ^{26}Mg values of garnet ($\delta^{26}Mg_{Grt}$) and pyroxene ($\delta^{26}Mg_{Cpx}$) separates from the normal and carbonated eclogites. The green hexagon represents the mineral separates from a normal eclogite close to the UHPM marble in the Rongcheng area. The red squares are mineral separates from normal eclogites in Chinese Continental Scientific Drill (CCSD). The black circles represent the mineral separates from carbonated eclogites. Green dotted lines are the equilibrium garnet-pyroxene Mg isotope fractionation lines[51]. Magnesium isotopic data are from Supplementary Tables 2-4.

Sinian carbonates comprise limestone (MgO/CaO=0.02) and dolostone (MgO/CaO=0.62–0.73), with δ^{26}Mg varying from −4.101 to −2.396‰ in limestone and from −2.210 to −1.009‰ in dolostone (Fig. 2a,b and Supplementary Table 3). The generally lighter δ^{26}Mg in limestone than in dolostone is consistent with previous studies of normal sedimentary carbonates[16–18]. By contrast, δ^{26}Mg values of the UHPM marbles vary from −2.508 to −0.531‰ and form a negative correlation with the dolomite abundance and a positive correlation with the

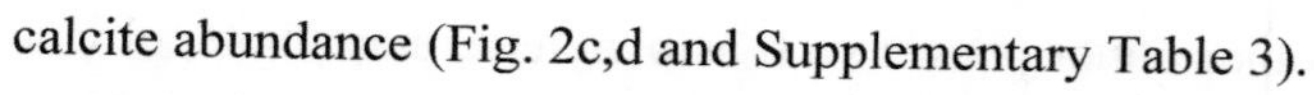

calcite abundance (Fig. 2c,d and Supplementary Table 3).

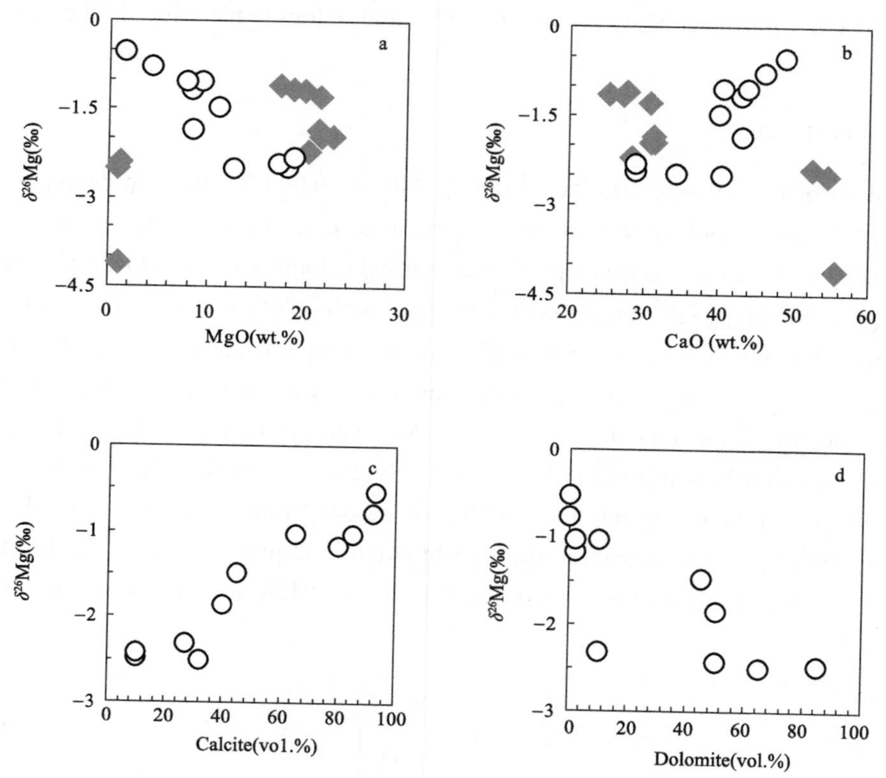

Fig. 2 Magnesium isotopic compositions of UHPM marbles and Sinian carbonates. (a) δ^{26}Mg-MgO variations of UHPM marbles and Sinian carbonates; (b) δ^{26}Mg-CaO variations of UHPM marbles and Sinian carbonates; (c) The variation of δ^{26}Mg against calcite abundance in UHPM marbles; (d) The variation of δ^{26}Mg against dolomite abundance in UHPM marbles. The white circles represent the UHPM marbles and the green diamonds represent the Sinian carbonates. Magnesium isotopic and major elemental data are from Supplementary Tables 3; mineral abundances are from Supplementary Tables 5.

1.3 Oxygen and other isotopes

Two normal eclogites from Rongcheng have δ^{18}O values of +5.65‰ and +3.96‰, respectively (Supplementary Table 3). The δ^{18}O of carbonated eclogites span a range of +15.9 – +21.1 ‰ (Supplementary Table 3), which is consistent with previous studies of carbonated eclogites[31]. Neodymium isotope analyses of these carbonated eclogites yield initial ε_{Nd} of −12.3 to −1.1 (Supplementary Table 6), falling in the range of normal eclogites (ε_{Nd}=−16.7 to −0.1)[32]. The δ^{18}O and δ^{13}C of the UHPM marbles are in the ranges +20.3 to +23.3 ‰ and −2.01 to +4.77‰ (Supplementary Table 3), respectively, similar to those of their protoliths[19,20].

2 Discussion

Garnet and pyroxene from both normal and carbonated eclogites reach Mg isotopic equilibrium at peak metamorphic temperatures of 700–880 ℃ (Fig. 1b). The bulk carbonated eclogites and mineral separates have anomalously light δ^{26}Mg and heavy δ^{18}O values when compared with the normal eclogites and Neoproterozoic basaltic protoliths (Fig. 1a,b). Sedimentary limestones generally have lighter δ^{26}Mg than dolostones[16–18], whereas the UHPM marbles display the opposite trend, that is, isotopically heavy marbles have high CaO and low MgO contents (Fig. 2a,b). As such, the Mg-O isotopic signatures preserved in the carbonated eclogites and UHPM marbles do not reflect their protolith heterogeneity, rather they reflect carbonate-eclogite interactions during subduction including carbonate-eclogite mixing, metamorphic dehydration or rehydration, decarbonation and isotopic exchange.

Carbonates are known to have extremely heavy $\delta^{18}O$ and light $\delta^{26}Mg$ values[16–20]; thus, carbonate-eclogite mixing might explain the bulk Mg-O isotopic characteristics of carbonated eclogites. If we assume that eclogite lenses in marbles have initial $\delta^{26}Mg$ and $\delta^{18}O$ values similar to the normal eclogites, then a mixing calculation with carbonates indicates that ~60–90% contribution by carbonate is required to fit the bulk Mg-O isotopic compositions. However, the low CO_2 contents of carbonated eclogites (0.11–6.43 wt.%; Supplementary Table 3) imply that the calculated carbonate proportion is unrealistically high. In addition, carbonate-eclogite mixing predicts lighter $\delta^{26}Mg$ values in those samples containing higher CO_2 contents, which is not seen in the carbonated eclogites (Supplementary Fig. 2). Most importantly, garnet and pyroxene minerals in carbonated eclogites are equally enriched in ^{24}Mg and ^{18}O. This requires a chemical exchange rather than simple physical mixing between eclogite lenses and marbles.

The Mg and O isotope fractionations between carbonated eclogites and their protoliths are far greater than predicted by closed-system metamorphic dehydration[30,33]. In addition, the $\delta^{26}Mg$ of low MgO/CaO marbles are significantly heavier than the values of their pre-metamorphosed counterpart ,Sinian limestones (Fig. 2a,b), which is opposite to the direction of isotope fractionation induced by metamorphic dehydration[34]. Thus, closed-system metamorphic dehydration is unlikely to account for the observed Mg-O isotopic variations in carbonated eclogites and UHPM marbles. Although the retrograded amphibolitic rim of an individual carbonated eclogite block has higher alkalies and CO_2 relative to the eclogitic core, they have similar Mg isotopic compositions (Supplementary Table 3), ruling out the possibility of retrogression-induced Mg isotope fractionation.

Metamorphic decarbonation involves the decomposition of a substantial amount of carbonate minerals by incorporating Mg and Ca oxides into the silicates and releasing CO_2(ref.35) . This may explain the light $\delta^{26}Mg$ and heavy $\delta^{18}O$ of silicate minerals in carbonated eclogites. However, metamorphic decarbonation cannot significantly fractionate Mg isotope, neither form the $\delta^{26}Mg$-CaO and $\delta^{26}Mg$-MgO correlations in UHMP marbles (Fig. 2a,b). Furthermore, previous studies have revealed that the subduction P-T trajectory of these carbonated eclogites and UHPM marbles does not intersect with experimentally determined carbonate-out boundaries (Supplementary Fig. 3)[8], suggesting limited decarbonation of these UHPM marbles. This is further supported by the similar O and C isotopic compositions between Sinian carbonates and UHPM marbles (Supplementary Fig. 4), as large O and C isotope fractionations are expected during decarbonation[36].

The most likely process causing the Mg and O isotopic variations in carbonated eclogites and UHPM marbles is thus through fluid-mediated isotopic exchange. Previous studies revealed that Mg isotope fractionation between silicate and carbonate decreases dramatically from low temperatures (⩾2‰ at T<300 ℃) to high temperatures (0.05–0.08‰ at T = 600–800 ℃)[37–40]. Similar is true for O isotopes[33]. Therefore, extensive isotopic exchanges between marbles and eclogites are expected during prograde metamorphism to reduce their Mg and O isotopic differences. Assuming that sedimentary limestone and dolostone have heavy $\delta^{18}O$ of +24‰(ref. 19) but light $\delta^{26}Mg$ around –4.00 ‰ and –2.50‰(refs 16–18),respectively, and eclogite lenses in carbonates have initial Mg and O isotopic compositions similar to the normal eclogites (with average $\delta^{18}O$=+5.5‰ and $\delta^{26}Mg$=–0.25‰), the eclogites would become depleted in ^{26}Mg and enriched in ^{18}O after UHP isotopic exchange with the marbles (Fig. 3). Magnesium isotopic exchange also elevated the $\delta^{26}Mg$ values of UHPM marbles in various degrees (Fig. 3). As limestone contains significantly lower Mg than dolostone, at a given weighted ratio between marble and eclogite the limestone is more susceptible than dolostone to Mg isotopic exchange with the eclogites (Fig. 3). Consequently, the calcite-rich carbonates exhibit a large shift in $\delta^{26}Mg$ toward heavier values, whereas the dolomite-rich carbonates tend to preserve their initial $\delta^{26}Mg$ values (Fig. 4). The negative co-variation of $\delta^{26}Mg$

versus MgO/CaO for UHPM marbles (Fig. 4) is therefore produced by the differential Mg isotopic exchange of compositionally varied carbonates with enclosed eclogite lenses.

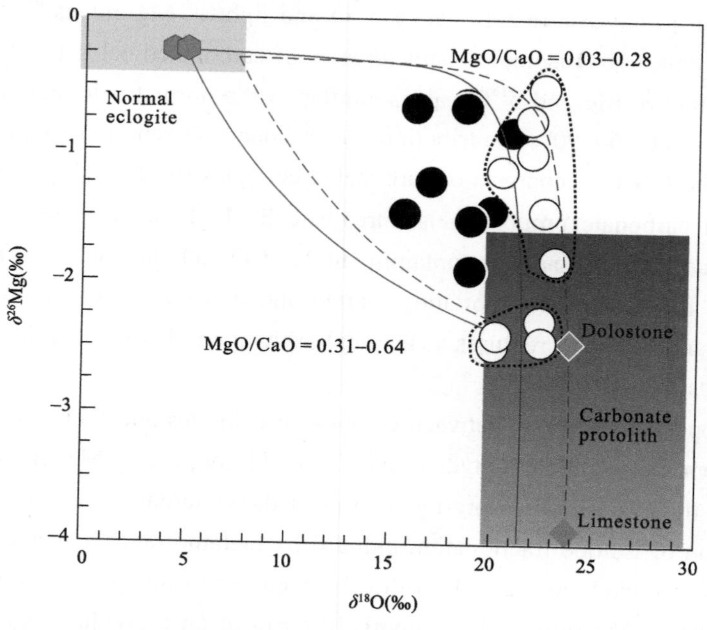

Fig. 3 Magnesium and O isotopic exchange between UHPM marbles and enclosed eclogites. The UHPM marbles and carbonated eclogites are represented by the white and black circles, respectively. Two end members of marble protoliths, limestone and dolostone are considered to have initial chemical compositions of $Mg_{0.001}Ca_{0.999}CO_3$ and $Mg_{0.5}Ca_{0.5}CO_3$, respectively. Their O and Mg isotopic compositions are: $\delta^{18}O$=+24 ‰ and $\delta^{26}Mg$=–4.00‰ for limestone, and $\delta^{18}O$=+24‰ and $\delta^{26}Mg$=–2.50‰ for dolostone. The eclogites are assumed to have initial $\delta^{18}O$ of +5.5‰, $\delta^{26}Mg$ of –0.25‰ and MgO=8 wt.%. The green hexagons represent two normal eclogites in gneisses close to the UHPM marble in the Rongcheng area. Equations that govern the isotopic exchange model are given in Supplementary Note 1. The red and blue curves represent the isotopic exchange of normal eclogites with limestone and with dolostone, respectively. The solid curves represent the Mg-O isotopic evolution of eclogites and the dashed curves represent the Mg-O isotopic evolution of marbles by isotopic exchange. The $\delta^{18}O$ ranges of Sinian carbonates[19] and normal eclogites[23] are shown for comparison. Magnesium and O isotopic data are from Supplementary Tables 2 and 3.

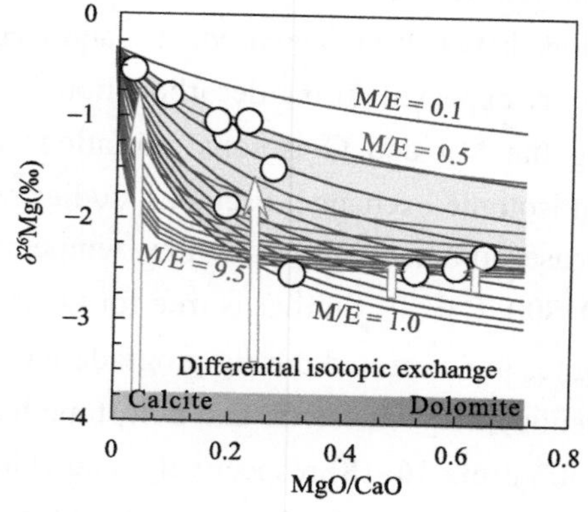

Fig. 4 Differential isotopic exchanges of eclogites with calcite-rich and dolomite-rich marbles. The eclogites are assumed to have initial $\delta^{26}Mg$ of –0.25‰ and MgO=8wt.%. Equations that govern differential isotopic exchange model are given in the Supplementary Note 2. At a given weighted ratio of carbonate to eclogite (M/E) during isotopic exchange, marbles with different compositions (limestone versus dolostone) shift their $\delta^{26}Mg$ in different degrees (shown by arrows). The red and blue curves assume that the initial $\delta^{26}Mg$ of carbonates are –4.00‰ and –2.50‰, respectively. Different curves represent the isotopic exchange under different M/E ratios. The increment of M/E ratio is 0.1 for the red and 1.0 for the blue. Data are from Supplementary Table 3.

Sedimentary carbonates are the main light-δ^{26}Mg sink of Earth's reservoirs[18]. Recycling of carbonates into the mantle has the potential to cause local mantle Mg isotope heterogeneity[41]. Knowledge on whether Mg isotopic compositions of sedimentary carbonates can be preserved during crustal subduction is fundamental to quantifying the amount of recycled carbonate in mantle sources using Mg isotopes. Our study suggests that calcite-rich carbonates are highly unlikely to retain their initial light δ^{26}Mg but become isotopically heavier towards the value of silicates, whereas dolomite-rich carbonates are more capable of preserving their initial values. On the other hand, the carbonated silicates (for example, eclogites) gain light δ^{26}Mg values that are significantly different from other normal eclogites and ambient mantle. These silicates may produce significant light-δ^{26}Mg components in the mantle, as sampled by cratonic eclogites[42] and highly metasomatized peridotite xenoliths[43].

3 Methods

3.1 Magnesium isotopic analyses

Magnesium isotopic analyses were carried out at the Isotope Laboratory of the University of Arkansas, Fayetteville, USA. Whole-rock powders of eclogites and carbonates, and the mineral separates were digested using Optima-grade mixed acid of HF-HNO_3-HCl. After complete dissolution, dried residues from carbonate solutions were added 12 N HCl, and those from silicate solutions were taken up in 1 N HNO_3 for ion column chemistry. Detail column chemistry procedures have been reported elsewhere[44,45].

Chemical separation and purification of Mg were achieved by cation exchange chromatography with Bio-Rad AG50W-X8 resin in 1 N HNO_3 media. Additional chromatographic step was processed for the carbonate to further separate Mg from Ca by using Bio-Rad 200−400 mesh AG50W-X12 resin in 12N HCl media. The same column procedure was performed twice in order to obtain the pure Mg recovery. The pure Mg solutions of silicate and carbonate were then dried down and re-dissolved in 3% HNO_3 ready for mass spectrometry.

The Mg isotopic compositions were analyzed by the sample-standard bracketing method using a *Nu Plasma* MC-ICPMS at low resolution mode[46]. Each batch of sample analysis contains at least one well-characterized standard. Sample solution was repeated on ratio measurements for more than four times within a session. The long-term precision is better than ±0.07‰ (2 s.d.) for the $^{26}Mg/^{24}Mg$ ratio. Magnesium isotopic results are reported in the conventional δ notation in per mil relative to DSM-3, $\delta^{26}Mg = [(^{26}Mg/^{24}Mg)_{sample}/(^{26}Mg/^{24}Mg)_{DSM\text{-}3}-1] \times 1{,}000$.

3.2 Oxygen-carbon and neodymium isotopic analyses

Oxygen and carbon isotopic analyses were conducted at the CAS Key Laboratory of Crust-mantle Materials and Environments in the University of Science and Technology of China, Hefei. Oxygen isotopes of eclogites were measured by the laser fluorination technique. The O_2 was extracted by a CO_2 laser and transferred to a Finigan Delta + mass spectrometer for the measurement of $^{18}O/^{16}O$ and $^{17}O/^{16}O$ ratios. The O and C isotopes of carbonates were analyzed by the GasBench II technique in the continuous flow mode and the extracted CO_2 gases were measured on a Finnigan MAT253 mass spectrometer. Detail procedures are described previously[47]. The O and C isotopic results are reported in the δ notation in per mil relative to Vienna Standard Mean Ocean Water and Vienna Peedee Belemnite, respectively.

Samarium-Nd isotopic analyses were performed using an IsoProbe-T thermal ionization mass spectrometer at the State Key Laboratory of Lithospheric Evolution, Institute of Geology and Geophysics, Chinese Academy of Sciences. Samarium isotopes were measured using single Ta filament. Neodymium isotopes were determined using single tungsten filament with TaF_5 as an ionization activator. Detail column chemistry and mass spectrometer operation condition have been described previously[48]. $^{143}Nd/^{144}Nd$ ratios were corrected for mass fractionation using $^{146}Nd/^{144}Nd$=0.7219. During the period of data collection, the measured value for the JNdi-Nd standard was 0.512117±10 (2 s.d.).

3.3 Whole rock chemistry and mineral abundances

Major elements (except FeO, H_2O^+ and CO_2) were analyzed at the Hebei Institute of Regional Geology and Mineral Resources, China, by wavelength dispersive X-Ray fluorescence spectrometry[49]. The analytical uncertainties are better than 1%. The FeO contents of the samples were determined by conventional wet chemical method[50]. After the quantitative oxidation of Fe^{2+} to Fe^{3+}, the ferrous iron is regenerated by back-titration with ammonium ferrous sulfate. Then, the original Fe^{2+} content can be calculated. H_2O^+ and CO_2 were determined by gravimetric methods and potentiometry, respectively. The relative abundances of carbonate minerals were determined by X-ray diffraction using a Rigaku Smart lab (9 kW) at the China University of Geosciences, Beijing. The X-ray source was a Cu anode operated at 45 kV and 200 mA using CuKα_1 radiation equipped with a diffracted beam graphite monochromator.

References

1. Dasgupta, R. & Hirschmann, M. M. The deep carbon cycle and melting in Earth's interior. Earth Planet Sci. Lett. 298, 1-13 (2010).
2. Huybers, P. & Langmuir, C. Feedback between deglaciation, volcanism, and atmospheric CO_2. Earth Planet Sci. Lett. 286, 479-491 (2009).
3. Yaxley, G. M. & Green, D. H. Experimental demonstration of refractory carbonate-bearing eclogite and siliceous melt in the subduction regime. Earth Planet Sci. Lett. 128, 313-325 (1994).
4. Hammouda, T. High-pressure melting of carbonated eclogite and experimental constraints on carbon recycling and storage in the mantle. Earth Planet Sci. Lett. 214, 357-368 (2003).
5. Dasgupta, R., Hirschmann, M. M. & Withers, A. C. Deep global cycling of carbon constrained by the solidus of anhydrous, carbonated eclogite under upper mantle conditions. Earth Planet Sci. Lett. 227, 73-85 (2004).
6. Connolly, J. Computation of phase equilibria by linear programming: a tool for geodynamic modeling and its application to subduction zone decarbonation. Earth Planet Sci. Lett. 236, 524-541 (2005).
7. Burton, M. R., Sawyer, G. M. & Granieri, D. Deep carbon emissions from volcanoes. Rev. Mineral Geochem. 75, 323-354 (2013).
8. Dasgupta, R. Ingassing, storage, and outgassing of terrestrial carbon through geologic time. Rev. Mineral Geochem. 75, 183-229 (2013).
9. Galvez, M. E. et al. Graphite formation by carbonate reduction during subduction. Nat. Geosci. 6, 473-477 (2013).
10. Ague, J. J. & Nicolescu, S. Carbon dioxide released from subduction zones by fluid-mediated reactions. Nat. Geosci. 7, 355-360 (2014).
11. Dasgupta, R. & Hirschmann, M. M. Melting in the Earth's deep upper mantle caused by carbon dioxide. Nature 440, 659-662 (2006).
12. Huang, S., Farkaš, J. & Jacobsen, S. B. Stable calcium isotopic compositions of Hawaiian shield lavas: evidence for recycling of ancient marine carbonates into the mantle. Geochim. Cosmochim. Acta 75, 4987-4997 (2011).
13. Kiseeva, E. S. et al. An experimental study of carbonated eclogite at 3-5–5-5GPa—implications for silicate and carbonate metasomatism in the cratonic mantle. J. Petrol. 53, 727-759 (2012).
14. Teng, F.-Z. et al. Magnesium isotopic composition of the Earth and chondrites. Geochim. Cosmochim. Acta 74, 4150-4166 (2010).
15. Eiler, J. M. Oxygen isotope variations of basaltic lavas and upper mantle rocks. Rev. Mineral. Geochem. 43, 319-364 (2001).
16. Galy, A., Bar-Matthews, M., Halicz, L. & O'Nions, R. K. Mg isotopic composition of carbonate: insight from speleothem formation. Earth Planet Sci. Lett. 201, 105-115 (2002).
17. Wombacher, F. et al. Magnesium stable isotope fractionation in marine biogenic calcite and aragonite. Geochim. Cosmochim. Acta 75, 5797-5818 (2011).

18. Young, E. D. & Galy, A. The isotope geochemistry and cosmochemistry of magnesium. Rev. Mineral. Geochem. 55, 197-230 (2004).
19. Lambert, I., Walter, M., Wenlong, Z., Songnian, L. & Guogan, M. Palaeoenvironment and carbon isotope stratigraphy of upper Proterozoic carbonates of the Yangtze platform. Nature 325, 140-142 (1987).
20. Schidlowski, M., Eichmann, R. & Junge, C. E. Precambrian sedimentary carbonates: carbon and oxygen isotope geochemistry and implications for the terrestrial oxygen budget. Precambrian Res. 2, 1-69 (1975).
21. Li, S. et al. Collision of the North China and Yangtse Blocks and formation of coesite-bearing eclogites: timing and processes. Chem. Geol. 109, 89-111 (1993).
22. Zheng, Y.-F. et al. Zircon U–Pb age, Hf and O isotope constraints on protolith origin of ultrahigh-pressure eclogite and gneiss in the Dabie orogen. Chem. Geol. 231, 135-158 (2006).
23. Zheng, Y.-F., Fu, B., Gong, B. & Li, L. Stable isotope geochemistry of ultrahigh pressure metamorphic rocks from the Dabie-Sulu orogen in China: implications for geodynamics and fluid regime. Earth Sci. Rev. 62, 105-161 (2003).
24. Liu, F. & Liou, J. Zircon as the best mineral for P–T–time history of UHP metamorphism: A review on mineral inclusions and U–Pb SHRIMP ages of zircons from the Dabie–Sulu UHP rocks. J. Asian Earth Sci. 40, 1-39 (2011).
25. Xu, S. et al. Diamond from the Dabie Shan metamorphic rocks and its implication for tectonic setting. Science 256, 80-82 (1992).
26. Ye, K., Cong, B. & Ye, D. The possible subduction of continental material to depths greater than 200 km. Nature 407, 734-736 (2000).
27. Kato, T., Enami, M. & Zhai, M. Ultra-high-pressure (UHP) marble and eclogite in the Su-Lu UHP terrane, eastern China. J. Metamorph. Geol. 15, 169-182 (1997).
28. Ye, K. & Hirajima, T. High-pressure marble at Yangguantun, Rongcheng county, Shandong province, eastern China. Mineral. Petrol. 57, 151-165 (1996).
29. Ogasawara, Y., Zhang, R. & Liou, J. Petrogenesis of dolomitic marbles from Rongcheng in the Su-Lu ultrahigh-pressure metamorphic terrane, eastern China. Island Arc. 7, 82-97 (1998).
30. Wang, S.-J., Teng, F.-Z., Li, S.-G. & Hong, J.-A. Magnesium isotopic systematics of mafic rocks during continental subduction. Geochim. Cosmochim. Acta. 10.1016/j.gca.2014.03.029 (2014).
31. Chu, X., Guo, J., Fan, H. & Jin, C. Oxygen isotope compositions of eclogites in Rongcheng, Eastern China. Chin. Sci. Bull. 48, 372-378 (2003).
32. Jahn, B. in When Continents Collide: Geodynamics and Geochemistry of Ultrahigh-Pressure Rocks Vol. 10 (eds Hacker, B. R. & Liou, J. G.) 203-239 (Kluwer Academic Publishers, 1998).
33. Matthews, A. Oxygen isotope geothermometers for metamorphic rocks. J. Metamorph. Geol. 12, 211-219 (1994).
34. Li, W., Chakraborty, S., Beard, B. L., Romanek, C. S. & Johnson, C. M. Magnesium isotope fractionation during precipitation of inorganic calcite under laboratory conditions. Earth Planet Sci. Lett. 333, 304-316 (2012).
35. Knoche, R., Sweeney, R. J. & Luth, R. W. Carbonation and decarbonation of eclogites: the role of garnet. Contrib. Mineral Petrol 135, 332-339 (1999).
36. Valley, J. W. Stable isotope geochemistry of metamorphic rocks. Rev. Mineral. Geochem. 16, 445-489 (1986).
37. Tipper, E. et al. The magnesium isotope budget of the modern ocean: Constraints from riverine magnesium isotope ratios. Earth Planet Sci. Lett. 250, 241-253 (2006).
38. Immenhauser, A. et al. Magnesium-isotope fractionation during low-Mg calcite precipitation in a limestone cave–Field study and experiments. Geochim. Cosmochim. Acta 74, 4346-4364 (2010).
39. Higgins, J. A. & Schrag, D. P. Constraining magnesium cycling in marine sediments using magnesium isotopes. Geochim. Cosmochim. Acta 74, 5039-5053 (2010).
40. Macris, C. A., Young, E. D. & Manning, C. E. Experimental determination of equilibrium magnesium isotope fractionation between spinel, forsterite, and magnesite from 600℃ to 800℃ Geochim. Cosmochim. Acta 118, 18-32 (2013).
41. Yang, W., Teng, F.-Z., Zhang, H.-F. & Li, S.-G. Magnesium isotopic systematics of continental basalts from the North China craton: Implications for tracing subducted carbonate in the mantle. Chem. Geol. 328, 185-194 (2012).
42. Wang, S.-J., Teng, F.-Z., Williams, H. M. & Li, S.-G. Magnesium isotopic variations in cratonic eclogites: origins and implications. Earth Planet Sci. Lett. 359, 219-226 (2012).
43. Xiao, Y., Teng, F.-Z., Zhang, H.-F. & Yang, W. Large magnesium isotope fractionation in peridotite xenoliths from eastern North China craton: product of melt-rock interaction. Geochim. Cosmochim. Acta 115, 241-261 (2013).
44. Yang, W., Teng, F.-Z. & Zhang, H.-F. Chondritic magnesium isotopic composition of the terrestrial mantle: a case study of

peridotite xenoliths from the North China craton. Earth Planet Sci. Lett. 288, 475-482 (2009).
45. Ling, M.-X. et al. Formation of the world's largest REE deposit through protracted fluxing of carbonatite by subduction-derived fluids. Sci. Rep. 3 (2013).
46. Teng, F. Z. & Yang, W. Comparison of factors affecting the accuracy of high-precision magnesium isotope analysis by multi-collector inductively coupled plasma mass spectrometry. Rapid Commun. Mass Spectrom. 28, 19-24 (2014).
47. Zha, X. P., Zhao, Y. Y. & Zheng, Y. F. An online method combining a Gasbench II with continuous flow isotope ratio mass spectrometry to determine the content and isotopic compositions of minor amounts of carbonate in silicate rocks. Rapid Commun. Mass Spectrom. 24, 2217-2226 (2010).
48. Chu, Z.-Y. et al. Evaluation of sample dissolution method for Sm-Nd isotopic analysis of scheelite. J. Anal. At. Spectrom. 27, 509-515 (2012).
49. Gao, S. et al. Silurian-Devonian provenance changes of South Qinling basins: implications for accretion of the Yangtze (South China) to the North China cratons. Tectonophysics 250, 183-197 (1995).
50. Wilson, A. The micro-determination of ferrous iron in silicate minerals by a volumetric and a colorimetric method. Analyst 85, 823-827 (1960).
51. Huang, F., Chen, L., Wu, Z. & Wang, W. First-principles calculations of equilibrium Mg isotope fractionations between garnet, clinopyroxene, orthopyroxene, and olivine: Implications for Mg isotope thermometry. Earth Planet Sci. Lett. 367, 61-70 (2013).

Acknowledgements

We would like to thank Y.-F. Zheng for support on oxygen isotopic analysis and for thoughtful discussion, Z.-Z. Han, S.-C. An, H.-M. Zhang and J.-A. Hong for field works, H.-O. Gu and W.-Y. Li for help in the clean lab. The work is supported by the National Science Foundation (EAR-0838227, EAR-1056713 and EAR1340160) to F.-Z.T. and the National Nature Scientific Foundation of China (No. 41230209, 41090372) to S.-G.L.

Author contributions

S.-J.W., F.-Z.T. and S.-G.L. designed the study, interpreted the data and wrote the paper. S.-J.W carried out all the laboratory work. All authors contributed equally to this work.

Additional information

Supplementary Information accompanies this paper at http://www.nature.com/ naturecommunications
Competing financial interests: The authors declare no competing financial interests. Reprints and permission information is available online at http://npg.nature.com/ reprintsandpermissions/
How to cite this article: Wang, S.-J. et al. Tracing carbonate–silicate interaction during subduction using magnesium and oxygen isotopes. Nat. Commun. 5:5328 doi: 10.1038/ncomms6328 (2014).

Copper isotope fractionation during adsorption onto kaolinite: Experimental approach and applications*

Dandan Li, Sheng-Ao Liu and Shuguang Li

State Key Laboratory of Geological Processes and Mineral Resources, China University of Geosciences, Beijing 100083, China

亮点介绍：首次对黏土矿物吸附过程 Cu 同位素分馏开展实验研究，发现：(i)吸附过程中 Cu 同位素分馏显著。(ii)离子强度对同位素分馏有显著影响，对解释河口地区或工业排水污染区的沉积物、土壤和水体同位素组成具有重要意义。(iii)Cu 同位素分馏值与初始铜浓度有关，pH 和温度影响很小，为解释全球不同温度和 pH 条件下样品的 Cu 同位素变化提供了重要依据。

Abstract The adsorption of copper and other heavy metals onto clay minerals is an important process that controls the distribution of trace metals in natural environments. Copper isotopes are a potentially useful tool to track the source of contaminated metals in soils formed in natural systems, but Cu isotope fractionation during adsorption onto clay minerals, the major component in soils, has not been thoroughly studied. In this study, we carried out for the first time a series of experiments to investigate the isotope fractionation of Cu during adsorption onto kaolinite for a wide range of conditions, including the contact time (t=10–360 min), temperature (T=1–50 ℃), initial Cu concentration of the starting solution (C_0 = 2–100 μg/g), pH value (4.0–6.0) and ionic strength ($NaNO_3$; I=0–0.1 mol/L). Our results indicate that Cu isotopes are significantly fractionated with preferential adsorption of the light isotope (^{63}Cu) onto the mineral surface. The fractionation factors ($\Delta^{65}Cu_{adsorbed\text{-}solution}=\delta^{65}Cu_{adsorbed}-\delta^{65}Cu_{solution}$) weakly depend on the pH and temperature with a constant value of approximately –0.27‰ at C_0=20 μg/g and in the absence of $NaNO_3$. Addition of $NaNO_3$ into the starting solution has a dramatic negative influence on the $\Delta^{65}Cu_{adsorbed\text{-}solution}$ values that range from –1.46 to –0.29‰. Such results are useful for interpreting Cu isotopic variations observed in sediments, soils and water from estuarine settings or industrial sewage pollution areas. The $\Delta^{65}Cu_{adsorbed\text{-}solution}$ values significantly increase with increasing initial Cu concentration of the starting solutions at C_0<30 μg/g, but approach a stable value of –0.17±0.10‰ (2SD) when the kaolinite has reached its maximum adsorption capacity at C_0>30 μg/g. The results imply that the isotopic compositions of the Cu adsorbed onto natural soils may vary greatly at relatively low Cu concentrations of the soil solutions. Furthermore, the pore waters after draining kaolinite-bearing rocks would become isotopically heavier due to the preferential adsorption of ^{63}Cu onto kaolinite. Given that no redox change occurred in all experiments, we propose that the most likely mechanism responsible for such Cu isotope fractionation is the different adsorption capacities of isotopically different species in aqueous solutions

* 本文发表在：Chemical Geology, 2015, 396: 74-82

and the formation of outer-sphere surface Cu(II) complexes. Our study represents one important step for future studies to use Cu isotopes to trace the source of metal contaminants in natural soils.
Keywords　Cu isotopes, isotope fractionation, adsorption, kaolinite, experiment

1 Introduction

Copper (Cu) is a common trace metal in sediments, soils and aquatic environments at relatively low concentration levels, and it acts as an important nutrient for vegetation and biota. However, at elevated concentrations Cu can become toxic. Due to the enhancement of its toxicity through accumulation in living organisms and consequent biomagnification in the food chain (An et al., 2001), excess Cu and other heavy metals can cause various diseases and disorders and have become important environmental and human health issues (Balistrieri and Mebane, 2014).

Weathering is the primary way for the formation of clays and clay minerals at the Earth's surface. It releases dissolved metals to natural systems and controls their distribution in pore waters and soils (Rubio et al., 2000; Liaghati et al., 2004). Soil, composed mainly of microorganisms, metal oxy(hydr)oxides and silicate clays, is a dominant carrier of Cu and other heavy metals. During weathering, the released Cu is incorporated into pedogenic minerals such as clay minerals and metal oxy(hydr)oxides (Contin et al., 2007) and partly bonded to organic matter during pedogenesis (Alcacio et al., 2001). Copper isotopes may act as a potentially useful tool to trace the source of Cu in soils. For example, significant Cu isotopic variations have been observed in soils (Bigalke et al., 2010a, 2011; Liu et al., 2014a). Several mechanisms have been proposed to explain Cu isotopic variations in soils, including redox weathering of Cu-bearing sulfides (Mathur et al., 2012), adsorption by soil microorganisms (Pokrovsky et al., 2008; Navarrete et al., 2011; Liu et al., 2014a) and adsorption onto metal oxy(hydr)oxides (Balistrieri et al., 2008; Pokrovsky et al., 2008).

Adsorption is an important physicochemical process retaining inorganic and organic substances in soils and concentrating trace metals in solutions (Essington, 2004). Natural clay minerals can be used as adsorbents for removal of heavy metals from contaminated water and soils (Ikhsan et al., 1999; Gupta and Bhattacharyya, 2005; Gu and Evans, 2008). Several studies have experimentally documented the Cu isotope fractionation during adsorption onto organic matter or metal oxy(hydr)oxides. Pokrovsky et al. (2008) observed a significant enrichment of the light Cu isotope on the cell surface of the soil bacterium *P. aureofaciens* at pH 1.8−3.5, and an enrichment of the heavy Cu isotope on the surface of metal oxy(hydr)oxides at pH 4-6. Balistrieri et al. (2008) also reported that the heavy Cu isotope is preferentially adsorbed onto the surface of amorphous ferric oxy(hydr)oxides. Navarrete et al. (2011) reported that lab strains and natural consortia preferentially incorporate the light Cu isotope with $\Delta^{65}Cu_{solution\text{-}solid}$ varying from 1.0 to 4.4‰.

In soil, Cu can also be adsorbed onto clay minerals (Yavuz et al., 2003; Veli and Alyüz, 2007; Šljivić et al., 2009; Jiang et al., 2010), in addition to metal oxy(hydr)oxides. Most clay minerals form where rocks are in contact with water, air, or steam (Foley, 1999). Kaolinite is one of the most abundant components in clay minerals. It occurs widely in tropical and sub-tropical soils and is the second most abundant clay mineral in ocean sediments (Grim, 1968). Kaolinite has high surface areas with cation-exchange capacities. It is a representative layered aluminosilicate mineral with a simple structure. It contains negatively charged functional groups, including the silanol (≡SiOH) and aluminol (≡AlOH) hydroxyl groups on the mineral edges, and permanently charged sites ($\equiv X^-$) on the basal surfaces (Yavuz et al., 2003; Gräfe et al., 2007; Gu and Evans, 2008). These functional groups readily form complexes with aqueous metal cations including Cu (Sen et al., 2002; Yavuz et al., 2003; Bhattacharyya and Gupta, 2008b). Previous studies suggested that the permanent structural charge of kaolinite is minor (Ferris and Jepson, 1975), and thus adsorption takes place mainly at the proton-bearing surface functional groups such as silanols and aluminols exposed at the edge of the sheets (Zachara et al., 1988) (Fig. 1). Despite recent advances on the understanding of such adsorption processes, little is known about the relationship between adsorption and isotope fractionation.

In this study, for the first time we carried out batch adsorption experiments under various experimental conditions to investigate the magnitude and direction of Cu isotope fractionation during adsorption onto clay mineral surfaces. Our results show that significant Cu isotope fractionation occurs in this process, and such fractionation should play a fundamental role in determining the isotopic composition of metals in natural soils and solutions. The results may provide insights into the mechanisms of metal adsorption that occurs in natural environments.

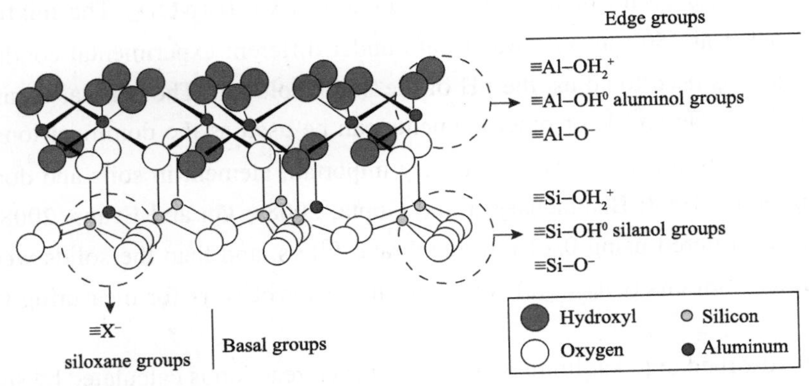

Fig. 1 Proposed charged sites on kaolinite used in the two site adsorption model (Gu and Evans, 2008).

2 Experimental methods

All experiments were performed in a clean laboratory fitted with HEPA-filtered Class 100 air supply and laminar-flow clean benches. All working acids were purified in-house by sub-boiling distillation in a Teflon® still, and > 18.2 MΩ H_2O was made from a Milli-Q (MQ) water system. All reaction labwares, including Teflon® and glassware, were cleaned with 1:1 (v) nitric acid, 1:1 (v) hydrochloric acid and MQ H_2O. Other materials, including columns, pipette tips, test tubes and caps, were washed in a heated bath of 3% HNO_3 and rinsed with MQ H_2O prior to use.

2.1 Starting materials

Kaolinite used in this study is the Chinese mineral standard of GBW 03121(200 mesh). It was dried at 105 ℃ for 2 hours using a partially enclosed hot plate (designed to limit airborne contamination) and finally stored in a Teflon bottle. The starting sample contains mainly kaolinite (~55 wt.%) and quartz (~35 wt.%) with other minerals in trace levels (<10 wt.%). The general formula of kaolinite is Al_2O_3-$2SiO_2$-$2H_2O$. the major elemental compositions of the used sample are: SiO_2 54.6 wt.%, Al_2O_3 31.4 wt.%, TiO_2 0.69 wt.%, TFe_2O_3 0.50 wt.%, K_2O 0.34 wt.%, MgO 0.12 wt.%, P_2O_5 0.10 wt.%, CaO 0.05 wt.%, Na_2O 0.02 wt.%. Note that natural kaolinite contains a certain amount of Cu, which must be taken into account in the adsorption experiments. The Cu concentration of the used kaolinite is 47.3 μg/g, and its δ^{65}Cu value determined in this study is 2.10±0.05‰ (*n*=6) relative to NIST 976. This amount of Cu will be deducted when calculating the actual amounts adsorbed onto kaolinite (see details below). The adsorbate is a stock solution of $Cu(NO_3)_2$ with a Cu concentration of 1000 μg/g prepared from an in-house mono-elemental Cu standard solution (GSB Cu; > 99.99%).

2.2 Experimental conditions and XRD analysis

All experiments were designed to provide a quantitative characterization of Cu isotope fractionation during adsorption onto the surface of kaolinite under different conditions. These conditions include: (a) contact time ranging from 10 to 360 min; (b) Cu concentration of the starting solutions ranging from 2 to 100 μg/g at constant pH; and (c) pH consisting of five parallel experiments at pH 4.0, 4.5, 5.0, 5.6 and 6.0. In addition, reaction temperature and ionic strength (background electrolyte) were also investigated. All experiments were conducted at an initial pH value equal to or less than 6.0, so that aqueous Cu(II) did not appreciably hydrolyze or reach saturation with formation of Cu oxide or Cu-oxy(hydr)oxide phases (Šljivić et al., 2009).

Batch adsorption experiments were carried out in 100 ml Erlenmeyer flasks by mixing a constant amount of kaolinite (500.0±0.1 mg) with 20 ml of the aqueous solution of $Cu(II)(NO_3)_2$. The mixture in the flasks was shaken using a rotary shaker at 150 rpm in a water bath under different experimental conditions. No buffer was used; instead, HNO_3 acid was used to adjust the pH of the initial solution. The pH was measured via a pH meter with an uncertainty of ±0.1. Na and K or other elements can be used as the dominant ions to change the ionic strength of the solution. We used Na because it is a very important element in soils and does not affect the clay mineral structure (Jung et al., 1998; Bhattacharyya and Gupta, 2008a; Gu and Evans, 2008; Jiang et al., 2010). The samples were vacuum filtered using 0.22 μm membrane filters, and then the solids were rinsed three times using MQ H_2O. The final solutions ($Cu_{measured}$) were kept in Teflon beakers for measuring Cu concentration and isotopic composition.

The amount of Cu adsorbed onto kaolinite ($Cu_{adsorbed}$) in each reaction is calculated by subtracting the amount of Cu in the final solution from the total amount of Cu in the initial solution, plus the amount of Cu released from the starting kaolinite. Blank experiments in Cu-free solutions were run as the terminal endpoint for each experimental condition. The amount of Cu released from kaolinite in the adsorption experiments is assumed to be equal to the amount of Cu released in the blank experiments under the same conditions. Cu concentrations were measured using inductively coupled plasma mass-spectrometry (ICP-MS), with an uncertainty of ±3% (2SD). The level of Cu contamination (<2 ng) from the filtration and transformation procedure is negligible.

X-ray diffraction analysis (XRD) was used to determine whether or not phase changes occurred during the adsorption processes. The starting kaolinite and residual solids after adsorption were powdered to ~200 mesh and analyzed by an X-ray power diffraction (SmartLab) with Cu Kα1 radiations at the China University of Geosciences, Beijing. The analyses were scanned from 3 to 70° 2θ using a step of 0.02° at a rate of 8°/min.

2.3 Cu isotope analysis

Copper isotopic ratios were measured using a sample-standard bracketing method on a *Neptune plus* MC-ICP-MS at the Isotope Geochemistry Laboratory of the China University of Geosciences, Beijing. Sample dissolution, column chemistry and instrumental analysis followed the method of Liu et al. (2014b). Only a brief description is given below.

The residual solids were dissolved in a 1:1 (v/v) mixture of double-distilled HF and HNO_3 in Savillex screw-top beakers and heated at 160 ℃ on a hotplate in an exhaust hood (Class 100). The solutions were evaporated to dryness at 150 ℃. The residue was then added with 1:3 (v/v) mixed HNO_3 and HCl and heated on a hot plate for a few hours before evaporation to dryness at 80 ℃. The samples were refluxed with concentrated HNO_3 and subsequently dried down at 80 ℃. 1 ml of 8 N HCl + 0.001% H_2O_2 was added to the beakers and the samples were heated to dryness. This process was repeated three times to ensure that all cations were converted to chloride species prior to ion-exchange separation.

All supernatant samples were dried down at 80 ℃, and then 1 ml 8 N HCl + 0.001% H_2O_2 was added to each beaker. Because matrix elements (e.g., Na) might be released from the starting kaolinite when it reacts with the $Cu(NO_3)_2$ solution, all supernatant solutions were subject to column chemistry. Cu was isolated from matrix elements using the strong anion exchange resin AG-MP-1M (Bio-Rad). 2 ml pre-cleaned resin was loaded into the column and then washed using 0.5 N HNO_3 and 8 N HCl + 0.001% H_2O_2 alternating with MQ water three times. Matrix elements were eluted in the first 10 ml of 8 N HCl + 0.001% H_2O_2, and Cu was collected in the following 24 ml of 8 N HCl + 0.001% H_2O_2. The total procedural blanks are < 2 ng. The Cu fractions were evaporated to dryness and dissolved in 3% HNO_3 (*m/m*). This process was repeated three times to ensure that all cations were converted to nitrate species prior to isotopic ratio analysis. Cu isotopic ratios of the supernatant solutions are reported relative to the starting Cu solution (GSB Cu). The filtration of the supernatant solutions in the adsorption experiments caused no Cu isotopic offset (± 0.05‰) by measuring the filtered known Cu solution. The Cu isotopic analysis in our working conditions is sensitive to the presence of matrix Na (Liu et al., 2014a). To overcome the matrix Na interference, chemical purification is necessary for Cu isotopic analysis. The ratios of Na to Cu are less than 0.01 in all supernatant solutions after column chemistry. The low signal of matrix Na yielded a negligible influence on Cu isotopic ratio analysis. The long-term external reproducibility of $\delta^{65}Cu$ is ± 0.05‰ (2SD) based on repeated analyses of natural rocks and in-house solutions (Liu et al., 2014b).

The final solutions in all adsorption experiments were analyzed for Cu isotopic ratios ($\delta^{65}Cu_{measured}$). The isotopic compositions of Cu adsorbed onto kaolinite ($\delta^{65}Cu_{adsorbed}$) were calculated based on mass balance using Eqs. (1) and (2), where $f_{adsorbed}$ and $f_{released}$ are the percentage of Cu absorbed onto kaolinite and the percentage of Cu released from kaolinite measured in the blank experiment, respectively. The $\delta^{65}Cu_{initial}$, $\delta^{65}Cu_{measured}$, $\delta^{65}Cu_{released}$ and $\delta^{65}Cu_{solution}$ are the Cu isotopic compositions of the starting solution, solution fraction including the released fraction, Cu released from kaolinite and solution fraction excluding the released fraction (residual solution), respectively.

$$\delta^{65}Cu = [\delta^{65}Cu_{measured} - \delta^{65}Cu_{released} \times f_{released}] / (1 - f_{released}) \quad (1)$$

$$\delta^{65}Cu_{adsorbed} = [\delta^{65}Cu_{initial} - \delta^{65}Cu_{solution} \times (1 - f_{adsorbed})] / f_{adsorbed} \quad (2)$$

Similarly, the isotopic compositions of Cu adsorbed onto kaolinite ($\delta^{65}Cu_{adsorbed}$) were also calculated based on the solid fraction, expressed as follows:

$$\delta^{65}Cu_{solid} = [\delta^{65}Cu_{kaolinite} - \delta^{65}Cu_{released} \times f_{released}] / (1 - f_{released}) \quad (3)$$

$$\delta^{65}Cu_{adsorbed} = [\delta^{65}Cu_{measured} - \delta^{65}Cu_{solid} \times (1 - f_{adsorbed})] / f_{adsorbed} \quad (4)$$

The $\delta^{65}Cu_{kaolinite}$ and $\delta^{65}Cu_{solid}$ represent Cu isotopic compositions of the starting kaolinite and the solid fraction including the adsorbed fraction, respectively.

The isotope fractionation factor ($\Delta^{65}Cu_{adsorbed\text{-}solution} = 1000 \cdot \ln\alpha$) between the adsorbed Cu (calculated based on the solution fraction) and residual Cu in the solutions is calculated as follows:

$$\Delta^{65}Cu_{adsorbed\text{-}solution} = \delta^{65}Cu_{adsorbed} - \delta^{65}Cu_{solution} \quad (5)$$

3 Results

3.1 Effects of experimental conditions on adsorption amount

The original data of Cu concentrations and isotopic ratios are reported in the Supplementary Tables s1–s7 and are plotted in Figs. 2–6. The percentage of adsorbed Cu is not constant, which quickly increases in the first 60 min and then slowly approaches equilibrium with a constant percentage of adsorbed Cu (~20%) (Fig. 2a). This is consistent with previous studies (Bhattacharyya and Gupta, 2006, 2008b; Jiang et al., 2010). Although apparent equilibrium was achieved at 60 min, we selected 240 min as the contact time for other experiments in order to ensure sufficient equilibrium.

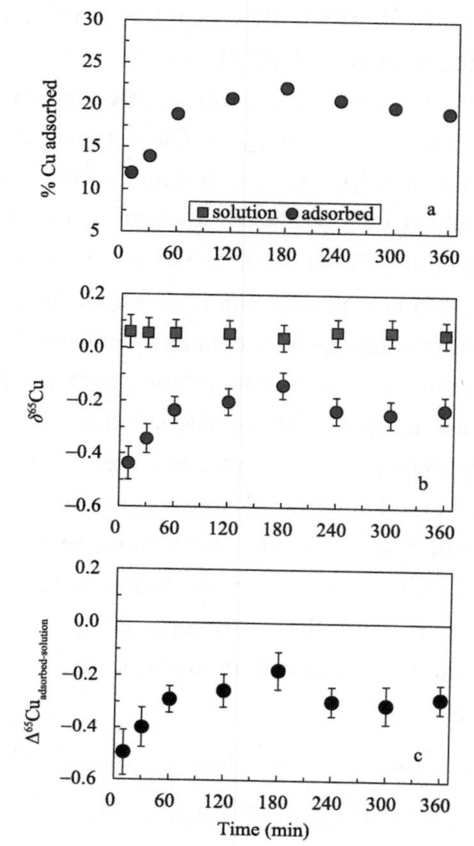

Fig. 2　The variation of fraction of Cu adsorbed onto kaolinite as a function of time (min) at C_0=20 μg/g, pH=5.6, T=25 ℃, without $NaNO_3$. a. The percentage of adsorbed Cu vs. time; b. δ^{65}Cu of measured residual Cu in the solutions (green squares) and of calculated adsorbed Cu (red circles); c. The fractionation values ($\Delta^{65}Cu_{\text{adsorbed-solution}}$) between adsorbed Cu and residual Cu in the solutions. All the following figures have the same plots of a, b and c except for changing the x-axis to the corresponding variable (Fig. 3, initial Cu concentration; Fig. 4, pH; Fig. 5, temperature; and Fig. 6, ionic strength).

The effect of the initial Cu concentration of the starting solution was examined and the results are shown in Fig. 3a. As the initial concentration increases from 2 to 100 μg/g, the percentage of adsorbed Cu rapidly decreases at concentrations up to 30 μg/g and then decreases slowly at higher concentrations. For example, at an initial Cu concentration of 2 μg/g, approximately 46.8% of Cu was adsorbed onto the kaolinite. At initial Cu concentrations larger than 60 μg/g, the percentage is less than 10%. An initial Cu concentration of 20 μg/g in the starting solutions was used in the following experiments.

Copper ions undergo hydrolysis, and the solubility of Cu is low at pH>6. For example, at pH>7, the consumption of OH^- ions results in the precipitation of $Cu(OH)_2$ (Bosso and Enzweiler, 2002; Šljivić et al., 2009). This process could cause large Cu isotope fractionation (Ehrlich et al., 2004). Therefore, all experiments were carried out at pH≤6 to prevent the production of insoluble Cu species. The percentage of adsorbed Cu increases as the initial pH increases from 4.0 to 5.0 and then becomes relatively constant (~20%) at pH = 5.0–6.0 (Fig. 4a). An initial pH of 5.6 was chosen for other experiments because apparent adsorption equilibrium is achieved at this pH value.

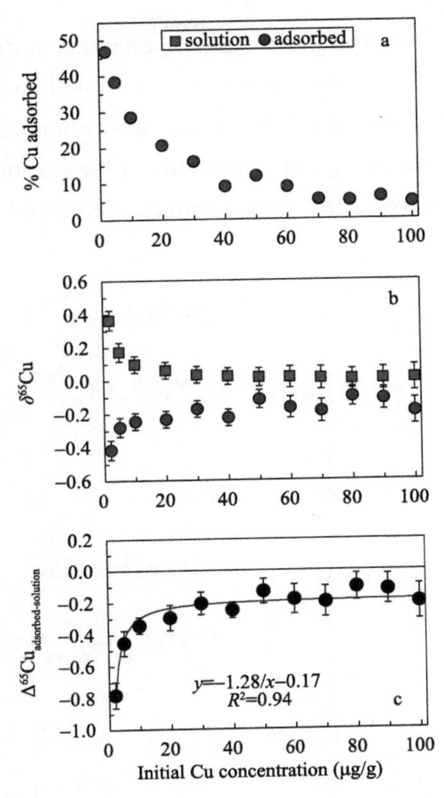

Fig. 3 The variation of fraction of Cu adsorbed onto kaolinite as a function of initial concentration (μg/g) at t=240 min, pH=5.6, T=25 ℃, without $NaNO_3$.

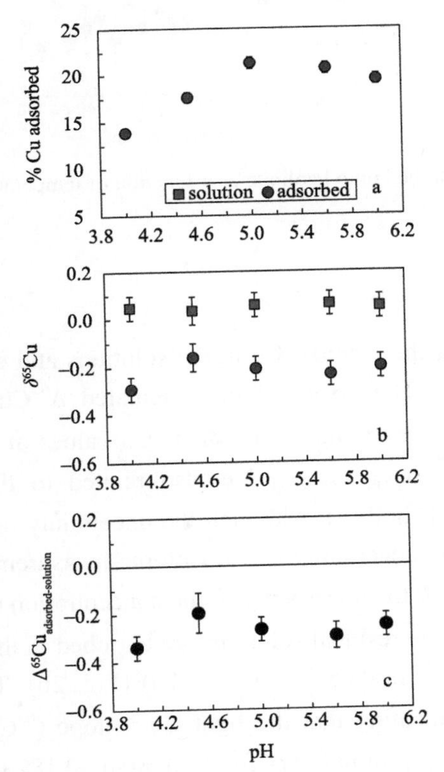

Fig. 4 The variation of fraction of Cu adsorbed onto kaolinite as a function of pH at t=240 min, C_0=20 μg/g, T=25 ℃, without $NaNO_3$.

To study thermodynamics of the adsorption process, experiments at different temperatures were carried out. The results are shown in Fig. 5a. As the temperature increases from 1 to 50 ℃, the percentage of adsorbed Cu slightly increases. The influences of ionic strength on Cu adsorption are shown in Fig. 6a. The adsorption capacities of kaolinite decrease with increasing concentrations of background electrolytes. For example, ~20% of Cu was adsorbed in the absence of $NaNO_3$, but the percentage decreased to ~6% when 0.1 mol/L $NaNO_3$ was added.

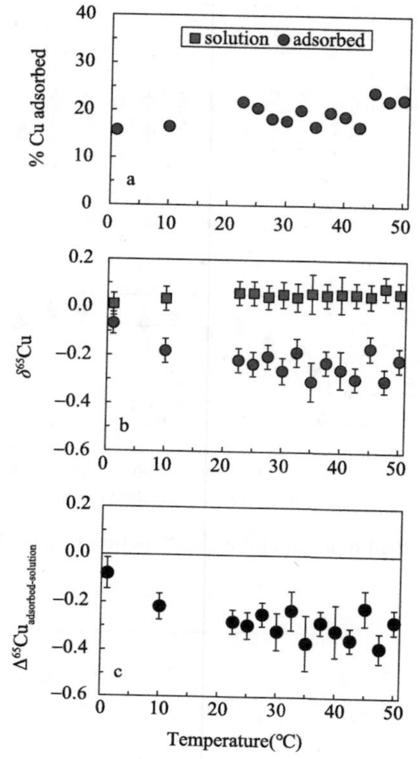

Fig. 5 The variation of fraction of Cu adsorbed onto kaolinite as a function of temperature (℃) at t=240 min, C_0=20 μg/g, pH=5.6, without $NaNO_3$.

3.2 Cu isotope results

Copper isotopic compositions of the residual Cu in the solutions and calculated values for adsorbed Cu are presented in Figs. 2–6, respectively. The errors of the calculated $\Delta^{65}Cu_{adsorbed-solution}$ values are based on the methods of Liu et al. (2011), by considering the errors of δ^{65}Cu values of both residual solutions and adsorbed Cu. Additionally, the errors of $\Delta^{65}Cu_{adsorbed-solution}$ are also related to the uncertainty of the measured Cu concentrations as illustrated in Eqs. (1) and (2). Although the uncertainty of the Cu concentration measurements (±3%) is significantly larger than the uncertainty of Cu isotopic measurements, the offset of $\Delta^{65}Cu_{adsorbed-solution}$ values is less than ±0.10‰ even though the uncertainty of the concentration measurements is considered.

In the contact time experiments, the residual solutions are enriched in the heavy Cu isotope (^{65}Cu) relative to the starting solution, with a mean δ^{65}Cu=0.05±0.01‰ (2SD) (Fig. 2b). The residual solution is significantly heavier than the starting solution, indicating that the light Cu isotope (^{63}Cu) was preferentially adsorbed onto kaolinite (Fig. 2b). The $\Delta^{65}Cu_{adsorbed-solution}$ values vary from −0.50 to −0.18‰ in the range of the contact time (10–360 min) and tend to be constant (−0.27±0.09‰; 2SD) after 60 min (Fig. 2c).

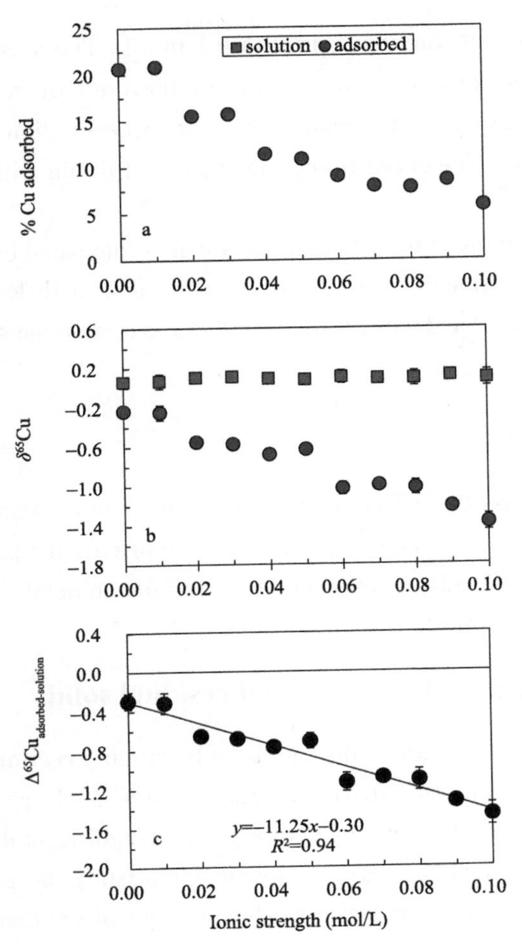

Fig. 6 The variation of fraction of Cu adsorbed onto kaolinite as a function of ionic strength (mol/L) at pH=5.6, C_0=20 μg/g, t= 240 min, 25 ℃.

In the initial concentration experiments, the residual solutions are also enriched in the heavy Cu isotope relative to the Cu adsorbed onto kaolinite. At a relatively low Cu concentration of 2 μg/g, the residual Cu in the solution has a δ^{65}Cu value of 0.37±0.06‰ (2SD). However, at higher Cu concentrations (>30 μg/g), δ^{65}Cu of the residual Cu in the solution is close to that of the starting solution, due to the small amounts of adsorbed Cu (< ~10%) (Fig. 3b). The $\Delta^{65}Cu_{adsorbed-solution}$ values vary from –0.78 to –0.21‰ as the initial Cu concentrations increase from 2 to 30 μg/g, and the values tend to be constant (–0.17‰) at higher initial Cu concentrations of 40 to 100 μg/g (Fig. 3b, c). The curve-fitting results yield a best fitting with $\Delta^{65}Cu_{adsorbed-solution} = -1.28/C_0 - 0.17$ and R^2=0.94 (Fig. 3c).

In the pH experiments, δ^{65}Cu values of the residual solutions remain constant at pH varying from 4.0 to 6.0. The average δ^{65}Cu value (0.05±0.02‰; 2SD) is significantly greater than zero (Fig. 4b), indicating that light Cu isotope is preferentially adsorbed onto kaolinite. The average of $\delta^{65}Cu_{adsorbed}$ is –0.22±0.10‰ (2SD) (Fig. 4b). The $\Delta^{65}Cu_{adsorbed-solution}$ values are also almost constant with an average of –0.27±0.11‰ (2SD) (Fig. 4c).

In the temperature experiments, the residual solutions are enriched in heavy Cu isotope and have an average $\delta^{65}Cu_{solution}$ = 0.05±0.03‰ (2SD). $\delta^{65}Cu_{adsorbed}$ is negatively correlated with the temperature at T< 10 ℃, but the correlation becomes weaker at higher temperatures (Fig. 5b). The $\Delta^{65}Cu_{adsorbed-solution}$ values vary from –0.38 to –0.08‰when temperatures increase from 1 to 50 ℃ (Fig. 5c).

In the ionic strength experiments, δ^{65}Cu values of the residual solutions show insignificant variations at background electrolyte concentrations varying from 0 to 0.1 mol/L. The average value (0.09±0.03‰; 2SD) is significantly larger than zero (Fig. 6b). The Cu isotopic compositions of the adsorbed Cu (−1.38 to −0.23‰) decrease significantly with increasing background electrolyte concentration (Fig. 6b). The $\Delta^{65}Cu_{adsorbed\text{-}solution}$ values vary from −1.46 to −0.29‰. The curve-fitting yielded a best fitting with $\Delta^{65}Cu_{adsorbed\text{-}solution}=-11.25\ I_{NaNO3} - 0.30$ and $R^2 = 0.94$ (Fig. 6c).

The copper isotopic compositions of the adsorbed Cu are also calculated based on the solid fraction. The data are listed in the Supplementary Table S7. The mean δ^{65}Cu value at different conditions is −0.27±0.03‰, consistent within error with the calculated value (−0.22±0.09‰) based on the solution fraction.

4 Discussion

In this section, we first discuss the effects of various experimental conditions on Cu isotope fractionation between adsorbed Cu and residual Cu in the solutions. Then, we discuss the possible mechanism responsible for the Cu isotope fractionation during adsorption onto the kaolinite mineral. Finally, we discuss the potential applications of our experimental findings to natural systems.

4.1 Characterization of the initial kaolinite and residual solids

Whether or not phase changes occurred during the adsorption experiments should be evaluated before discussing the influence of adsorption on Cu isotope fractionation. XRD patterns of the initial kaolinite and residual solids after adsorption experiments are given in Fig. 7. No significant differences of mineral assemblages between the initial kaolinite and residual solids were found. Importantly, no precipitation of copper phases and other minerals occurred during the adsorption process. Thus, we speculated that Cu adsorption onto kaolinite did not induce chemical reactions, and is only a physical absorption process with the formation of outer-sphere surface Cu(II) complexes.

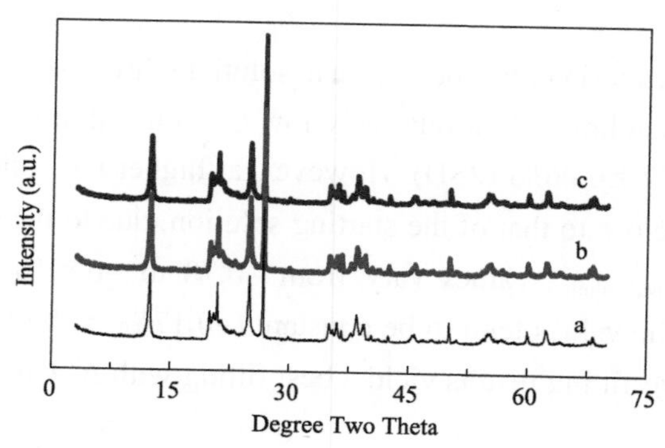

Fig. 7　X-ray Power Diffraction (XRD) patterns of the initial kaolinite and residual solids after the adsorption experiments: a. starting material, kaolinite; b. residual solid, without Cu(II) added into the solution; c. residual solid, with 20 μg/g Cu(II) in the solution. Other experimental conditions are pH=5.6, t=240 min, 25 ℃, without $NaNO_3$.

4.2 Cu isotope fractionation during adsorption onto kaolinite

4.2.1 The effect of contact time on Cu adsorption and isotope fractionation

The adsorption rate depends on the metal ions that are transported from the bulk solutions to the actual adsorption sites, and adsorption equilibrium is established rapidly (Dalang et al., 1984; Veglio and Beolchini, 1997; Yu et al., 2000; Yavuz et al., 2003; Jiang et al., 2010). The results of the contact time experiments show that maximum uptake was achieved within 60 min with constant percentage of adsorbed Cu (~20%) (Fig. 2a). This is due to the inhabitation of adsorption sites on the kaolinite surface through time until equilibrium is reached.

Isotopically, the $\Delta^{65}Cu_{adsorbed-solution}$ values initially increase with contact time and then become constant after 60 min (Fig. 2b, c). This is difficult to explain by thermodynamic equilibrium fractionation which is controlled by temperature and is not related to time. A possible explanation is that the transport of Cu species from the solutions to the actual adsorption sites needs some time and isotopic equilibrium had not been reached before apparent adsorption equilibrium. Other possible explanations may include (i) the desolvation kinetics of the Cu species or the surface adsorption sites and (ii) the diffusion kinetics through the interface between the solid surface and the bulk solution, etc.

4.2.2 The effect of initial concentration on Cu adsorption and isotope fractionation

At a certain weight of kaolinite, the percentage of metal adsorption onto kaolinite is determined by the adsorption capacity of the clay mineral (Fig. 3a). The percentage of adsorbed Cu decreases with increasing Cu concentration of the starting solutions, indicating that adsorption sites are widely available for the Cu ions at lower concentrations (C_0<30 μg/g). However, the competition between Cu and other elements becomes stronger at higher Cu concentrations (C_0>30 μg/g). Thus, a unit mass of kaolinite can adsorb a higher proportion of ions at a lower ion concentration. Given that the amounts of the adsorbate remain constant, the percentage of adsorption decreases with increasing initial Cu concentration of the solutions (Bhattacharyya and Gupta, 2008c).

The high percentage of adsorbed Cu at low initial Cu concentrations (<30 μg/g) dramatically changes the amount and isotopic composition of the Cu in the initial solution. Consequently, adsorption causes significant Cu isotope fractionation between adsorbed Cu and residual Cu in the solutions. However, when the initial Cu concentration increases, the maximum adsorption capacity of kaolinite will be reached and the percentage of adsorbed Cu is very low (<10%). Cu isotopic composition of the final solution mainly depends on $\delta^{65}Cu$ value of the initial solution. Thus, the $\delta^{65}Cu_{solution}$ values are close to the $\delta^{65}Cu_{initial}$ value (zero) and do not change substantially in the adsorption experiments; the $\Delta^{65}Cu_{adsorbed-solution}$ values also remain relatively constant.

4.2.3 The effect of pH value on Cu adsorption and isotope fractionation

The pH value of solutions is one key factor that affects the adsorption capacity of Cu and other metal ions (Bar-Yosef, 1979; Bolland et al., 1980; Zachara et al., 1988; Schulthess and Huang, 1990; Jung et al., 1998; Hizal and Apak, 2006; Gu and Evans, 2008; Jiang et al., 2010). The permanent structural charges of kaolinite are very minor, so metal adsorption takes place mainly at the proton-bearing surface functional groups (e.g., silanols and aluminols) exposed at the edge of sheets (Zachara et al., 1988). The pH-dependent surface charges of kaolinite can be explained by proton donor-acceptor reactions that occur simultaneously on these groups (Bolland et al., 1980; Zhou and Gunter, 1992). As illustrated in Fig. 4a, the percentage of adsorbed Cu is generally low (<15%) at low pH value (4.0), but increases at higher pH from 5.0 to 6.0. This occurs because the active sites on the kaolinite surface are weakly acidic and gradually deprotonated at higher pH, resulting in more negative surface charges and ultimately more bivalent metal cations (Cu(II)) adsorbed onto the surface.

Interestingly, the Cu isotope fractionation between adsorbed Cu and residual solutions ($\Delta^{65}Cu_{adsorbed-solution}$) is insensitive to the change of pH. This may be attributed to the fact that the concentration of H^+ in the solutions is significantly lower than the concentration of Cu(II) (20 μg/g) within the pH range (4.0 to 6.0), so the amounts of original Cu(II) species in the solutions do not change significantly. This could result in a stable Cu isotope fractionation between the adsorbed Cu and residual solutions (Fig. 4b, c).

4.2.4 The effect of temperature on Cu adsorption and isotope fractionation

The percentage of adsorbed Cu is the highest (22.6%) at high temperature (50 ℃) and the lowest (15.8%) at low temperature (1 ℃), indicating that adsorption is promoted by higher temperatures. The results indicate that the interactions between aqueous Cu and kaolinite are endothermic (Fig. 5a). There are three possible interpretations for this observation.

The first possible interpretation is that the Cu ions are well hydrated, and for Cu to be adsorbed onto kaolinite, the complex compounds have to lose the hydration shell, which consumes energy (Chen and Wang, 2006; Sheng et al., 2009). The second likely interpretation is that the adsorption process has to overcome a small activation barrier, and increasing the temperature facilitates Cu adsorption onto the kaolinite surface (Shukla et al., 2002; Ho, 2003). The third possibility is that the number of adsorption sites on the surface of clay minerals increase due to the dissociation of some surface components of the clay at higher temperatures which promotes Cu(II) uptake (Al-Asheh and Duvnjak, 1999).

We investigated Cu isotope fractionation in the range of temperatures (1–50 ℃), which can be expected for oxic depositional settings in the global ocean (except in the early Archean) (Knauth, 2005). The results show that $\delta^{65}Cu_{solution}$ is not significantly correlated with temperature; whereas $\delta^{65}Cu_{adsorbed}$ values have a weak negative correlation with temperature (Fig. 5b). This suggests that higher temperatures would promote the adsorption of light Cu isotope onto kaolinite. The $\Delta^{65}Cu_{adsorbed\text{-}solution}$ values also weakly depend on temperature, although there is a slightly negative correlation at T<10 ℃ (Fig. 5c).

4.2.5 The effect of ionic strength on Cu adsorption and isotope fractionation

As shown in Fig. 6a, increasing ionic strength has a negative influence on the percentage of adsorbed Cu. This may be due to competition between the background electrolyte ions and Cu(II), which decreases the number of active sites on the kaolinite as more electrolyte ions are adsorbed (Spark et al., 1995). The experimental results demonstrate that $\Delta^{65}Cu_{adsorbed\text{-}solution}$ is highly sensitive to ionic strength (Fig. 6c). The adsorbed Cu becomes isotopically much lighter with the increase of ionic strength of the solution. Similarly, the $\Delta^{65}Cu_{adsorbed\text{-}solution}$ values become lower as the ionic strength increases (Fig. 6b, c). The reason for this relationship is unclear. Generally, ionic strength significantly affects the amount of Cu adsorbed onto kaolinite at pH<6; at these pH values, outer-sphere complexation reactions occur (Hayes and Leckie, 1987; Gu and Evans, 2008). As shown in Fig. 6a, the adsorption process is greatly influenced by changes in the background electrolytes. Therefore, Cu(II) forms outer-sphere complexes with negatively charged sites on kaolinite at pH=5.6. The background electrolytes compete with Cu(II), which probably results in very low amounts and light Cu isotopic composition of the adsorbed Cu.

In summary, our results show that light Cu isotope (^{63}Cu) is always preferentially adsorbed onto the surface of kaolinite under various experimental conditions. Cu isotope fractionation between the adsorbed Cu and residual Cu in the solutions is weakly correlated with the pH values and temperature, moderately correlated with the reaction time, and strongly depends on the initial Cu concentration and ionic strength. The general Cu isotope fractionation factor ($\Delta^{65}Cu_{adsorbed\text{-}solution}$) under these conditions (t=240 min, C_0=20 μg/g, pH=4.0–6.0, T= 1–50 ℃ and without $NaNO_3$) is −0.29±0.06‰ (2SD).

4.3 Mechanism of Cu isotope fractionation during adsorption onto kaolinite

Redox transformation is a key mechanism for Cu isotope fractionation in nature. For example, Zhu et al. (2002) reported that the reduction of $Cu(II)_{aq}$ and $Cu(I)_{iodide}$ at 20 ℃ generates a large Cu isotope fractionation

($\Delta_{Cu(II)-Cu(I)}$=4.03±0.04‰). Ehrlich et al. (2004) also found that Cu(I)S precipitation from $Cu(II)_{aq}$ yields a fractionation of $\Delta_{Cu(II)-CuS}$=3.06±0.14‰ at 20 ℃. By contrast, the precipitation of insoluble Cu(II) species from Cu(II) solution generates a much smaller Cu isotope fractionation ($\Delta_{Cu(II)-Cu(OH)2}$=0.28±0.02‰; $\Delta_{Cu(II)-malachite}$ = 0.20±0.04‰) at 30 ℃ (Maréchal and Sheppard, 2002; Ehrlich et al., 2004). In addition, Cu adsorption onto amorphous ferric oxyhydroxide also yields a small Cu isotope fractionation with $\Delta_{solution-solid}$=−0.73±0.08‰ (Balistrieri et al., 2008). The Cu isotope fractionation during adsorption of Cu onto kaolinite obtained this study is also much smaller than that generated under redox reactions. This is consistent with the fact that our adsorption experiments were made in aerobic conditions; thus, no redox changes have occurred.

However, the mechanisms responsible for Cu isotope fractionation between solids and aqueous solutions remain unknown. The Cu isotope fractionation during adsorption onto kaolinite in our experiments is inconsistent with a kinetic isotope effect (Fig. 8), reflecting equilibrium fractionation between adsorbed Cu and residual Cu in the solutions. Previous studies on metal adsorption onto metal oxy(hydr)oxides suggest two potential mechanisms for equilibrium isotope fractionation during adsorption. The first one suggests that there is an equilibrium isotope fractionation between co-existing aqueous species of metal in the solution, and one of these species is preferentially adsorbed (Siebert et al., 2003; Barling and Anbar, 2004; Balistrieri et al., 2008; Pokrovsky et al., 2008). In this case, isotope fractionation is generated among different aqueous phases in the initial solution. The second mechanism suggests an equilibrium isotope fractionation between an aqueous species of metal and metal adsorbed onto the mineral surface (Siebert et al., 2003; Barling and Anbar, 2004; Balistrieri et al., 2008; Pokrovsky et al., 2008). In this case, isotope fractionation is generated during the adsorption process.

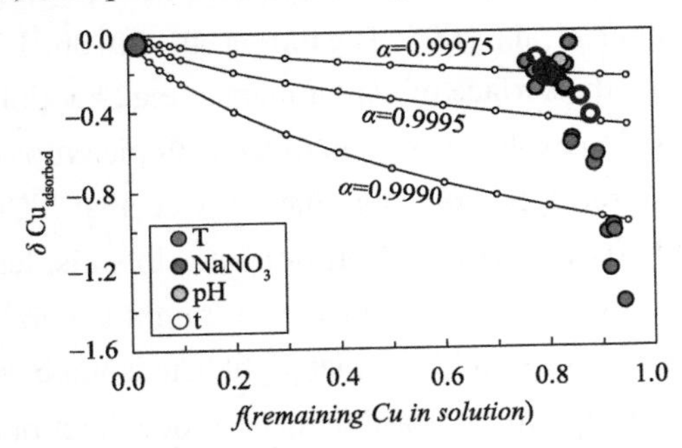

Fig. 8 Rayleigh modeling of the isotopic composition of adsorbed Cu as a function of remaining Cu (f) in the solution at various values of the fractionation factor (α). The α values represent the fractionation factor between adsorbed Cu and remaining Cu in the solution [$\alpha = ({}^{65}Cu/{}^{63}Cu)_{adsorbed} / ({}^{65}Cu/{}^{63}Cu)_{solution}$]. The Rayleigh equation: $\delta Cu_{adsorbed} = (\delta Cu_{initial} + 1000) \times (1 - f^{\alpha}) / (1 - f) - 1000$.

Pokrovsky et al. (2008) assumed that Al-OH sites in the kaolinite surface are an analogue of gibbsite surface centers. If this is correct, a similar direction of Cu isotope fractionation during adsorption onto gibbsite and kaolinite will be expected. Cu forms inner-sphere complexes with Cu-O during adsorption onto gibbsite or Fe(III) oxy(hydr)oxide, and the heavy Cu isotope is enriched in the solid phase (Balistrieri et al., 2008; Pokrovsky et al., 2008). This differs from our experimental studies, which show that light Cu isotope is preferentially absorbed onto kaolinite. Jung et al. (1998) found that Cu(II) forms inner-sphere complexes by strong bonds, but Cu adsorbed onto kaolinite may form inner-sphere complexes on Al sites and outer-sphere complex on Si sites. Previous work shows that Cd was not adsorbed by Si tetrahedral sheets of kaolinite, and Si tetrahedral sheets simultaneously impede Cd from binding to Al octahedral sheets (Gräfe et al., 2007). Thus, it is held as inner-sphere complexes on gibbsite and outer-sphere complexes on kaolinite. A study of Cu isotope fractionation during Cu interactions with microorganisms shows that outer-sphere complexes might change with pH, ionic strength and growth time

(Pokrovsky et al., 2008). In our adsorption experiments, Cu isotopes are significantly fractionated at different ionic strengths, suggesting that Cu(II) forms outer-sphere surface complexes.

The hydrated Cu(II) ions exist as $Cu(H_2O)_6^{2+}$ complex with six-fold coordination (Sherman, 2001) and $Cu(H_2O)_5^{2+}$ complex with five-fold coordination (Pasquarello et al., 2001). Calculations by Fujii et al. (2013) show that the Cu isotope fractionation between aqueous $Cu(H_2O)_6^{2+}$ and $Cu(H_2O)_5^{2+}$ species with optimized geometries is −0.26‰ at 25 ℃. This value is very similar to the general Cu isotope fractionation (−0.29‰ at 25 ℃) between adsorbed Cu and residual Cu measured in this study. Thus, the more likely mechanism for the Cu isotope fractionation is the preferential adsorption of the $Cu(H_2O)_6^{2+}$ species onto the surface of kaolinite. However, more studies are needed to constrain the processes by which outer-sphere complexes control isotopic shift and relevant structural information.

4.4 Implications for application to natural systems

The study of Cu isotopic compositions in soils has important implications in tracing the biological cycling of trace metals in nature. Kaolinite is one of the major phases in soils produced during continental weathering and pedogenesis. In highly weathered soils, kaolinite has high surface areas and cation-exchange capacities, which can greatly absorb Cu and other heavy metals. In most cases, variations of Cu isotopic ratios result from more than one process in natural soils. Previous studies have suggested that several processes could result in significant Cu isotope fractionation, including redox transformation (Zhu et al., 2002; Ehrlich et al., 2004), precipitation (Maréchal and Sheppard, 2002; Ehrlich et al., 2004), adsorption onto Fe(III) oxy(hydr)oxide (Balistrieri et al., 2008) and biological processes (Mathur et al., 2005; Bigalke et al., 2010b; Liu et al., 2014a). Our experimental results show that adsorption of Cu onto the surface of clay minerals (e.g., kaolinite) can also significantly fractionate Cu isotopes. The results may be applied to explain variation of Cu isotopic ratios observed in natural soils.

Bigalke et al. (2010a) studied three types of soils and showed that ^{65}Cu is preferentially adsorption onto oxy(hydr)oxides, organic matter and possibly other adsorption sites in soils, leaving the solution enriched in light Cu isotope. They also speculated that these adsorption processes might generally enrich the solid phases in heavy isotopes. However, our experimental results indicate that light Cu isotope is preferentially adsorbed onto the surface of kaolinite at various conditions, in contrast to Cu isotope fractionation during adsorption onto iron oxy(hydr)oxides. Liu et al. (2014a) recently studied a basalt weathering profile from Hainan Island, China. They found negative correlations of $\delta^{65}Cu$ values with both the contents of total organic carbon (TOC) and the proportions of kaolinite minerals in bulk soils. Although the negative correlation between $\delta^{65}Cu$ and TOC indicates biological uptake, adsorption of Cu onto kaolinite may also play an important role. Thus, a better understanding of the behavior of Cu isotopes during adsorption onto soil components is important to interpret the variations of Cu isotopic ratios in soils.

Pore waters, waters from acid mine drainage, river waters and seawaters have heavier Cu isotopic compositions than igneous rocks (Vance et al., 2008; Kimball et al., 2009; Mathur et al., 2012). This difference has been attributed to (1) oxidative leaching of sulfides with preferential release of isotopically heavy Cu under abiotic conditions or (2) isotope fractionation between a ligand-bound dissolved phase and a labile, adsorbed particulate phase. Kaolinite, which is one of the main clay minerals in soils and ocean sediments, may be an isotopically light source. For example, pore waters will become isotopically heavier after draining kaolinite.

Our results demonstrate that the $\Delta^{65}Cu_{adsorbed\text{-}solution}$ value is insensitive to pH and temperature. This is important for interpreting Cu isotopic compositions of the oceans in oxic depositional settings, regardless of variable temperatures in geological time. However, the $\Delta^{65}Cu_{adsorbed\text{-}solution}$ value is strongly affected by the ionic strength, which is critical for interpreting Cu isotopic variations in sediments from estuarine settings, where salinity varies substantially.

Our experimental data show that the $\Delta^{65}Cu_{adsorbed\text{-}solution}$ increases with the initial Cu concentration and then reaches a constant value. These results can be used to evaluate Cu isotopic compositions of soil solutions from which Cu was adsorbed and to trace the source of metals in natural soils, regardless of variations in pH and temperature. However, soils contain other phases, e.g., organisms and oxides. These phases preferentially absorb the heavy Cu isotope (^{65}Cu), which shifts the Cu isotopic compositions in the opposite direction. Further work is needed to evaluate the relative adsorption capacities of these phases in order to better apply Cu isotopes as a tracer of metal cycling in soils.

5 Conclusions

We report for the first time an experimental study of Cu isotope fractionation during adsorption onto kaolinite. The results show that light Cu isotopes are preferentially adsorbed onto the kaolinite surface, leaving the residual solutions enriched in heavy isotopes. Such isotope fractionation may have been caused by the preferential adsorption of isotopically different Cu species in aqueous solutions.

The results demonstrate that the effects of temperature and pH on Cu isotope fractionation during adsorption onto kaolinite are relatively weak. A consistent isotope fractionation of approximately –0.29‰ between adsorbed Cu and residual Cu in the solutions is obtained. The results can be applied to interpret Cu isotopic variations in the rock record in terms of global oxic conditions. By contrast, the isotope fractionation between adsorbed Cu and residual Cu in the solutions is largely related to ionic strength and Cu concentration of the starting solution. The significant effects of ionic strength (from 0 to 0.1 mol/L) on Cu isotope fractionation can be used to interpret the behavior of Cu isotopic signatures in sediments and soils from estuarine settings or industrial sewage pollution areas. The results have implications to trace the source of contaminated metals in natural soils using Cu isotopes. In addition, the isotope fractionation of Cu during adsorption onto kaolinite is important to explain Cu isotopic variations in natural soils and pore waters.

Acknowledgements

We are grateful to Dr. Jingao Liu for discussion on an early version of this manuscript. We thank R. Mathur and Rich Wanty for constructive comments and the Editor Jeremy Fein for efficient handling, which largely improved the manuscript. This work is supported by the National Natural Foundation of China (41473017) and the fundamental research funds (2-9-2014-068) to S.A.L.

Appendix A. Supplementary data

Supplementary data to this article can be found online at http://dx.doi.org/10.1016/j.chemgeo.2014.12.020.

References

Al-Asheh, S., Duvnjak, Z., 1999. Sorption of heavy metals by canola meal. Water Air Soil Pollut. 114 (3-4), 251-276.

Alcacio, T.E., Hesterberg, D., Chou, J.W., Martin, J.D., Beauchemin, S., Sayers, D.E., 2001. Molecular scale characteristics of Cu(II) bonding in goethite-humate complexes. Geochim. Cosmochim. Acta 65 (9), 1355-1366.

An, H.K., Park, B.Y., Kim, D.S., 2001. Crab shell for the removal of heavy metals from aqueous solution. Water Res. 35(15), 3551-3556.

Balistrieri, L.S., Mebane, C.A., 2014. Predicting the toxicity of metal mixtures. Sci. Total Environ. 466-467, 788-799.

Balistrieri, L.S., Borrok, D.M.,Wanty, R.B., Ridley,W.I., 2008. Fractionation of Cu and Zn isotopes during adsorption onto amorphous Fe(III) oxyhydroxide: experimental mixing of acid rock drainage and ambient river water. Geochim. Cosmochim. Acta 72 (2), 311-328.

Barling, J., Anbar, A., 2004. Molybdenum isotope fractionation during adsorption by manganese oxides. Earth Planet. Sci. Lett. 217 (3), 315-329.

Bar-Yosef, B., 1979. pH-dependent zinc adsorption by soils. Soil Sci. Soc. Am. J. 43 (6), 1095-1099.

Bhattacharyya, K.G., Gupta, S.S., 2006. Kaolinite, montmorillonite, and their modified derivatives as adsorbents for removal of Cu(II) from aqueous solution. Sep. Purif. Technol. 50 (3), 388-397.

Bhattacharyya, K.G., Gupta, S.S., 2008a. Adsorption of a few heavy metals on natural and modified kaolinite and montmorillonite: a review. Adv. Colloid Interf. Sci. 140 (2), 114-131.

Bhattacharyya, K.G., Gupta, S.S., 2008b. Influence of acid activation on adsorption of Ni(II) and Cu(II) on kaolinite and montmorillonite: kinetic and thermodynamic study. Chem. Eng. J. 136 (1), 1-13.

Bhattacharyya, K.G., Gupta, S.S., 2008c. Kaolinite and montmorillonite as adsorbents for Fe(III), Co(II) and Ni(II) in aqueous medium. Appl. Clay Sci. 41 (1), 1-9.

Bigalke, M., Weyer, S., Kobza, J., Wilcke, W., 2010a. Stable Cu and Zn isotope ratios as tracers of sources and transport of Cu and Zn in contaminated soil. Geochim. Cosmochim. Acta 74 (23), 6801-6813.

Bigalke, M.,Weyer, S.,Wilcke,W., 2010b. Stable copper isotopes: a novel tool to trace copper behavior in hydromorphic soils. Soil Sci. Soc. Am. J. 74 (1), 60-73.

Bigalke, M., Weyer, S., Wilcke, W., 2011. Stable Cu isotope fractionation in soils during oxic weathering and podzolization. Geochim. Cosmochim. Acta 75 (11), 3119-3134.

Bolland, M., Posner, A., Quirk, J., 1980. pH-independent and pH-dependent surface charges on kaolinite. Clays Clay Minerals 28(6), 412-418.

Bosso, S., Enzweiler, J., 2002. Evaluation of heavy metal removal from aqueous solution onto scolecite. Water Res. 36(19), 4795-4800.

Chen, C., Wang, X., 2006. Adsorption of Ni(II) from aqueous solution using oxidized multiwall carbon nanotubes. Ind. Eng. Chem. Res. 45 (26), 9144-9149.

Contin, M., Mondini, C., Leita, L., Nobili, M.D., 2007. Enhanced soil toxic metal fixation in iron (hydr)oxides by redox cycles. Geoderma 140 (1), 164-175.

Dalang, F., Buffle, J., Haerdi, W., 1984. Study of the influence of fulvic substances on the adsorption of copper (II) ions at the kaolinite surface. Environ. Sci. Technol. 18 (3), 135-141.

Ehrlich, S., Butler, I., Halicz, L., Rickard, D., Oldroyd, A., Matthews, A., 2004. Experimental study of the copper isotope fractionation between aqueous Cu(II) and covellite, CuS. Chem. Geol. 209 (3), 259-269.

Essington, M.E., 2004. Soil and Water Chemistry: An Integrative Approach. CRC Press. Ferris, A., Jepson, W., 1975. The exchange capacities of kaolinite and the preparation of homoionic clays. J. Colloid Interface Sci. 51 (2), 245-259.

Foley, N.K., 1999. Environmental Characteristics of Clays and Clay Mineral Deposits. US Department of the Interior, US Geological Survey.

Fujii, T., Moynier, F., Abe, M., Nemoto, K., Albarède, F., 2013. Copper isotope fractionation between aqueous compounds relevant to low temperature geochemistry and biology. Geochim. Cosmochim. Acta 110, 29-44.

Gräfe, M., Singh, B., Balasubramanian, M., 2007. Surface speciation of Cd(II) and Pb(II) on kaolinite by XAFS spectroscopy. J. Colloid Interface Sci. 315 (1), 21-32.

Grim, R.E., 1968. Clay Mineralogy. 2nd edn. McGraw-Hill, New York.

Gu, X., Evans, L.J., 2008. Surface complexation modelling of Cd(II), Cu(II), Ni(II), Pb(II) and Zn(II) adsorption onto kaolinite. Geochim. Cosmochim. Acta 72 (2), 267-276.

Gupta, S.S., Bhattacharyya, K.G., 2005. Interaction of metal ions with clays: I. A case study with Pb(II). Appl. Clay Sci. 30 (3-4), 199-208.

Hayes, K., Leckie, J., 1987. Modeling ionic strength effects on cation adsorption at hydrous oxide/solution interfaces. J. Colloid Interface Sci. 115 (2), 564-572.

Hizal, J., Apak, R., 2006. Modeling of copper(II) and lead(II) adsorption on kaolinite-based clay minerals individually and in the presence of humic acid. J. Colloid Interface Sci. 295 (1), 1-13.

Ho, Y.-S., 2003. Removal of copper ions from aqueous solution by tree fern.Water Res. 37(10), 2323-2330.

Ikhsan, J., Johnson, B.B.,Wells, J.D., 1999. A comparative study of the adsorption of transition metals on kaolinite. J. Colloid Interface Sci. 217 (2), 403-410.

Jiang, M.-Q., Jin, X.-Y., Lu, X.-Q., Chen, Z.-L., 2010. Adsorption of Pb(II), Cd(II), Ni(II) and Cu(II) onto natural kaolinite clay. Desalination 252 (1), 33-39.

Jung, J., Cho, Y.-H., Hahn, P., 1998. Comparative study of Cu^{2+} adsorption on goethite, hematite and kaolinite: mechanistic modeling approach. Bull. Korean Chem. Soc. 19(3), 324-327.

Kimball, B.E., Mathur, R., Dohnalkova, A.C., Wall, A.J., Runkel, R.L., Brantley, S.L., 2009. Copper isotope fractionation in acid mine drainage. Geochim. Cosmochim. Acta 73(5), 1247-1263.

Knauth, L.P., 2005. Temperature and salinity history of the Precambrian ocean: implications for the course of microbial evolution. Palaeogeogr. Palaeoclimatol. Palaeoecol. 219 (1), 53-69.

Liaghati, T., Preda, M., Cox, M., 2004. Heavy metal distribution and controlling factors within coastal plain sediments, Bells Creek catchment, southeast Queensland, Australia. Environ. Int. 29 (7), 935-948.

Liu, S.-A., Teng, F.-Z., Yang, W., Wu, F.-Y., 2011. High-temperature inter-mineral magnesium isotope fractionation in mantle xenoliths from the North China craton. Earth Planet. Sci. Lett. 308 (1), 131-140.

Liu, S.-A., Teng, F.-Z., Li, S., Wei, G.-J., Ma, J.-L., Li, D., 2014a. Copper and iron isotope fractionation during weathering and pedogenesis: insights from saprolite profiles. Geochim. Cosmochim. Acta 146, 59-75.

Liu, S.-A., Li, D., Li, S., Teng, F.-Z., Ke, S., He, Y., Lu, Y., 2014b. High-precision copper and iron isotope analysis of igneous rock standards by MC-ICP-MS. J. Anal. At. Spectrom. 29(1), 122-133.

Maréchal, C., Sheppard, S., 2002. Isotopic fractionation of Cu and Zn between chloride and nitrate solutions and malachite or smithsonite at 30 degrees and 50 degrees C. Geochimica et Cosmochimica Acta, Goldschmidt 2002 Conference, Davos, p. A484.

Mathur, R., Ruiz, J., Titley, S., Liermann, L., Buss, H., Brantley, S., 2005. Cu isotopic fractionation in the supergene environment with and without bacteria. Geochim. Cosmochim. Acta 69 (22), 5233-5246.

Mathur, R., Jin, L., Prush, V., Paul, J., Ebersole, C., Fornadel, A., Williams, J., Brantley, S., 2012. Cu isotopes and concentrations during weathering of black shale of the Marcellus Formation, Huntingdon County, Pennsylvania (USA). Chem. Geol. 304, 175-184.

Navarrete, J.U., Borrok, D.M., Viveros, M., Ellzey, J.T., 2011. Copper isotope fractionation during surface adsorption and intracellular incorporation by bacteria. Geochim. Cosmochim. Acta 75 (3), 784-799.

Pasquarello, A., Petri, I., Salmon, P.S., Parisel, O., Car, R., Tóth, É., Powell, D.H., Fischer, H.E., Helm, L.,Merbach, A.E., 2001. First solvation shell of the Cu(II) aqua ion: evidence for fivefold coordination. Science 291 (5505), 856-859.

Pokrovsky, O., Viers, J., Emnova, E., Kompantseva, E., Freydier, R., 2008. Copper isotope fractionation during its interaction with soil and aquatic microorganisms and metal oxy(hydr)oxides: possible structural control. Geochim. Cosmochim. Acta 72 (7), 1742-1757.

Rubio, B., Nombela, M., Vilas, F., 2000. Geochemistry of major and trace elements in sediments of the Ria de Vigo (NWSpain): an assessment of metal pollution. Mar. Pollut. Bull. 40 (11), 968-980.

Schulthess, C., Huang, C., 1990. Adsorption of heavymetals by silicon and aluminum oxide surfaces on clay minerals. Soil Sci. Soc. Am. J. 54 (3), 679-688.

Sen, T.K., Mahajan, S., Khilar, K.C., 2002. Adsorption of Cu^{2+} and Ni^{2+} on iron oxide and kaolin and its importance on Ni^{2+} transport in porous media. Colloids Surf. A Physicochem. Eng. Asp. 211 (1), 91-102.

Sheng, G., Wang, S., Hu, J., Lu, Y., Li, J., Dong, Y., Wang, X., 2009. Adsorption of Pb(II) on diatomite as affected via aqueous solution chemistry and temperature. Colloids Surf. A Physicochem. Eng. Asp. 339 (1), 159-166.

Sherman, D.M., 2001. Quantum chemistry and classical simulations of metal complexes in aqueous solutions. Rev. Mineral. Geochem. 42 (1), 273-317.

Shukla, A., Zhang, Y.-H., Dubey, P., Margrave, J., Shukla, S.S., 2002. The role of sawdust in the removal of unwanted materials from water. J. Hazard. Mater. 95 (1), 137-152.

Siebert, C., Nägler, T.F., von Blanckenburg, F., Kramers, J.D., 2003.Molybdenum isotope records as a potential new proxy for paleoceanography. Earth Planet. Sci. Lett. 211 (1), 159-171.

Šljivić, M., Smičiklas, I., Pejanović, S., Plećaš, I., 2009. Comparative study of Cu^{2+} adsorption on a zeolite, a clay and a diatomite from Serbia. Appl. Clay Sci. 43 (1), 33-40.

Spark, K., Wells, J., Johnson, B., 1995. Characterizing trace metal adsorption on kaolinite. Eur. J. Soil Sci. 46 (4), 633-640.

Vance, D., Archer, C., Bermin, J., Perkins, J., Statham, P.J., Lohan, M.C., Ellwood, M.J., Mills, R.A., 2008. The copper isotope geochemistry of rivers and the oceans. Earth Planet. Sci. Lett. 274 (1-2), 204-213.

Veglio, F., Beolchini, F., 1997. Removal of metals by biosorption: a review. Hydrometallurgy 44 (3), 301-316.

Veli, S., Alyüz, B., 2007. Adsorption of copper and zinc from aqueous solutions by using natural clay. J. Hazard. Mater. 149(1), 226-233.

Yavuz, Ö., Altunkaynak, Y., Güzel, F., 2003. Removal of copper, nickel, cobalt and manganese from aqueous solution by kaolinite. Water Res. 37 (4), 948-952.

Yu, B., Zhang, Y., Shukla, A., Shukla, S.S., Dorris, K.L., 2000. The removal of heavy metal from aqueous solutions by sawdust adsorption-removal of copper. J. Hazard. Mater. 80 (1), 33-42.

Zachara, J., Cowan, C., Schmidt, R., Ainsworth, C., 1988. Chromate adsorption by kaolinite. Clays Clay Minerals 36 (4), 317-326.

Zhou, Z., Gunter, W.D., 1992. The nature of the surface charge of kaolinite. Clay Clay Miner. 40 (3), 365-368.

Zhu, X., Guo, Y., Williams, R., O'nions, R., Matthews, A., Belshaw, N., Canters, G., De Waal, E., Weser, U., Burgess, B., 2002. Mass fractionation processes of transition metal isotopes. Earth Planet. Sci. Lett. 200 (1), 47-62.

Copper and zinc isotope fractionation during deposition and weathering of highly metalliferous black shales in central China*

Yiwen Lv[1], Sheng-Ao Liu[1], Jian-Ming Zhu[1,2] and Shuguang Li[1]

1. State Key Laboratory of Geological Processes and Mineral Resources, China University of Geosciences, Beijing 100083, China
2. The State Key Laboratory of Environmental Geochemistry, Institute of Geochemistry, Chinese Academy of Sciences, No. 46 Guanshui Road, Guiyang 550002, China

亮点介绍：恩施茅口组中等富金属黑色页岩及硅质岩原岩具有显著重的 Zn 同位素组成，认为其金属富集过程和硫化物沉淀和/或有机物吸附有关，提出 Zn 同位素可作为自然系统中金属富集过程的示踪剂。而风化页岩中巨大的 Cu 同位素变化，反映了多阶段氧化还原过程。风化过程中重的 Cu、Zn 同位素倾向于从硫化物溶解进入流体并从浅部向下淋滤迁移，后被 Fe 硫化物重新固定。该风化过程向海洋水体输入了部分 Cu、Zn，是 Cu、Zn 地球化学循环过程中不可忽视的一环。

Abstract Black shales represent one of the main reservoirs of metals released to hydrosphere via chemical weathering and play an important role in geochemical cycling of metals in the ocean. The stable isotope systematics of transitional metals (e.g., Cu and Zn) may be used as a proxy for evaluating their geochemical cycling. To investigate the behaviors of Cu and Zn isotopes during metal enrichment of black shales and the migration during weathering, in this study we reported Cu and Zn concentration and isotope data for unweathered and weathered metalliferous shales and siliceous interbeds from the Maokou Formation in central China. The unweathered shales and cherts have moderately enriched Cu and Zn concentrations with silicate-like δ^{65}Cu (+0.14 ± 0.09‰, 1σ) but heavy δ^{66}Zn (0.51 ± 0.11‰, 1σ). The elevated δ^{66}Zn values reflect an important contribution from seawater via sulfide precipitation and/or organic matter (OM) adsorption. The Zn isotopic compositions of these metalliferous shales are different from those of the 'normal' shales, highlighting the potential of Zn isotopes as a tracer for metal enrichment in natural systems.

The weathered shales and cherts have an extreme δ^{65}Cu range from −6.42‰ to +19.73‰ and a modest δ^{66}Zn range of +0.25‰ to +0.78‰. The strongly weathered samples have lower Cu and Zn concentrations and lighter isotopic compositions compared to the weakly weathered samples. The leaching of Cu- and Zn-rich sulfides in shallow depths and their downward transport and refixation by Fe-sulfide account for the Cu and Zn isotope fractionation, with the huge Cu isotope variation generated

* 本文发表在：Chemical Geology，2016, 445：24-35

by multistage redox leaching. In general, δ^{66}Zn values of the weathered shales shift towards light values compared to the unweathered protoliths, suggesting that shale weathering releases Zn which is isotopically heavier than igneous rocks and the global riverine average (+0.33‰). Our results therefore indicate that Cu isotopes can be extremely fractionated during weathering of Cu-rich shales and both heavy Cu and Zn isotopes are preferentially released into fluids during shale weathering. These results should be considered when evaluating geochemical cycling of Cu and Zn in the modern or past oceans.

1 Introduction

Shales account to about 25.4% proportion of exposed continents, linked to the cycling of carbon, nitrogen, phosphorous, sulfur, oxygen and heavy metals such as copper, zinc, iron, molybdenum, vanadium, and uranium (Huyck, 1990; Amiotte Suchet et al., 2003). Trace metals can be enriched in black shales, which become metalliferous so that they are often associated with economically important ore deposits. On the other hand, metalliferous black shales are sometimes the threat to natural environment (Woo et al., 2002). The metals in black shales are mainly in the forms of sulfide and complex incorporated with organic matter (OM) (Nijenhuis et al., 1999; Brumsack, 2006). The sulfide minerals and OM in metalliferous shales tend to decompose under oxidizing conditions and break down at low pH due to the leaching of sulfide without the buffing material. Subsequently, the release of heavy metals from these shales via weathering is often linked to water pollution, land contamination, instability of ecosystem and potential threat to human health (Woo et al., 2002).

Copper (Cu) and zinc (Zn) are two of the most important transitional metals and micronutrient elements. It has been well known that biological process significantly fractionates Cu and Zn isotopes (Maréchal et al., 2000). For example, Andersen et al. (2011) reported that the biogenic opal extracted from bulk marine sediment samples has heavy Zn isotopic composition (+0.7‰ to +1.5‰) relative to clastic sediments (~+0.25‰) in ocean (Maréchal et al., 2000). These findings make Cu and Zn isotopes a new tool for investigating the biological cycling of metals on the modern and ancient Earth (Maréchal et al., 2000; Little et al., 2014). A recent investigation sheds light on the global oceanic budget and modern oceanic mass balance of Cu and Zn using their isotopes (Little et al., 2014). The inputs from rivers and dust have light Cu isotope compositions (δ^{65}Cu relative to NIST 976; +0.63‰) compared to seawater (+0.9‰), and the oxic sediments are a complement reservoir with light Cu isotope composition (+0.31‰). For Zn, the riverine inputs have the same Zn isotopic composition (+0.33‰) as the lithogenic Zn. However, the Zn isotopic compositions of seawater and the outputs (Fe–Mn crusts and carbonates) are all heavier than that of the inputs. The present model of oceanic mass balance of Zn thus has problems with the missing of an isotopically light sink or an isotopically heavy input. Especially, Zn isotopic signatures of the sink in the euxinic sediments (e.g., shales) are still unclear but may be very important for understanding mass balance of Zn in the ocean.

Because of the biological toxicity with excess concentration (Påhlsson, 1989), the mobility and environmental behavior of Cu and Zn in the critical zone (soils and weathering horizons) have been recognized and the isotope fractionation mechanism has already drawn attention (Fernandez and Borrok, 2009; Liu et al., 2014a; Li et al., 2015). However, the Cu and Zn isotopic compositions of shales and the fractionation during weathering are not well studied or debated. Several experimental and field studies have investigated the isotopic behaviors of Cu and Zn during chemical weathering. For example, the leaching experiments of Cu and Zn sulfide-rich rocks show that the heavy Cu and Zn isotopes are preferentially enriched in fluids (Fernandez and Borrok, 2009; Wall et al., 2011). Experimental studies show that the dissolved Zn fractions are isotopically light during the early stage of biotite granite dissolution (Weiss et al., 2014). For Cu, Mathur et al. (2012) observed that Cu isotope fractionation during weathering of pyrite bearing soils developed on shales obeys the Rayleigh fractionation model. Liu et al. (2014a) reported that organic matters can play a key role in Cu isotope fractionation with organically-bound Cu enriched in light isotopes. Compared with Cu isotopes, the behavior of Zn isotopes during continental weathering is less constrained. Thus, the study on Cu and Zn isotopic compositions of fresh and weathering shales may hold new constraints on the inputs and outputs of oceanic Cu–Zn budget.

This present study aims to study Cu and Zn isotopic signatures of fresh shales and to investigate the behaviors of Cu and Zn isotopes during shale weathering. The fresh and weathered metal-rich shales and cherts in horizontal and vertical strata from Maokou Formation in central China are chosen for this study. The results shed lights on (1) oceanic mass balance of Cu and Zn and (2) the application of Cu and Zn isotopes as tracers of geochemical cycling and metal mobilization during continental weathering.

2 Samples

The regional geological setting for samples studied in this study was previously described in detail by Zhu et al. (2014) and is briefly summarized below. The fresh and weathered shales and cherts of the Lower Permian Maokou Formation (P_1m^3) were collected from the Yutangba (30°10′810″N, 109°46′728″E) and Shadi (30°20′304″N, 109°45′124″E) in the Enshi area, central China. Yutangba is located in the northwestern limb of the Shuanghe syncline in the northeastern part of the upper Yangtze Platform fold belt, sited about 81 km southwestern of the Enshi Prefecture in the Hubei Province (Fig. 1). The strata are horizontal at the Shadi and mainly consist of carbonaceous shales and cherts, whereas the strata at Yutangba strike ENE and dip 40°–70° SSE. The strata at Yutangba develop SSE-dipping normal faults. The target strata, carbonaceous shales of Maokou Formation (P_1m^3), are extremely polymetallic-enriched and several Se ore deposits occur in the studied area. At the same time, the incidence of Se poisoning is arising since 1960s in this area, leading to the displacement of population (Yang et al., 1983). Both the unaltered and weathered rocks from Yutangba and Shadi were selected and unaltered samples obtained from drill cores at Yutangba. Samples at Shadi are gathered vertically across the horizontal strata from a newly cut profile, sorted into the strongly weathered, the weakly weathered and the unweathered samples identified by observations of hand specimens, the distance from the top and the concentrations of total sulfur (TS), total organic carbon (TOC) and total iron (TFe). The weakly weathered and unweathered samples have high TOC, TFe and TS contents and significantly positive correlation between TFe and TS (R^2=0.85). For strongly weathered samples, TOC is heavily depleted; TS contents are variable due to some efflorescent salts; TFe is enriched and has negative correlation with TS (R^2=0.98) (Zhu et al., 2014). At Yutangba, two series of weathered samples were sampled horizontally across the vertical strata. Transect 1 (T1; cp0 prefix) was sampled at 2 m height above the floor at an interval of ca. 0.1 m. Roughly 7 m height above the floor, transect 2 (T2; YTB prefix) was sampled and samples from this transect are weathered more strongly than samples from transect 1 (Zhu et al., 2014). Samples from T2 have much lower TOC, TS and TFe contents than rocks from Shadi and T1. Fault crossed transects and weakened the strength of rocks, tending to accelerate the chemical weathering rate of rocks. Fault developed nearly around the samples YTB 25 and cp015. Totally, five unaltered samples and fifty-six weathered samples from Yutangba and Shadi were selected here for trace elements and Cu and Zn isotope analyses. These samples are well-characterized and have been studied for Se isotopes (Zhu et al., 2014).

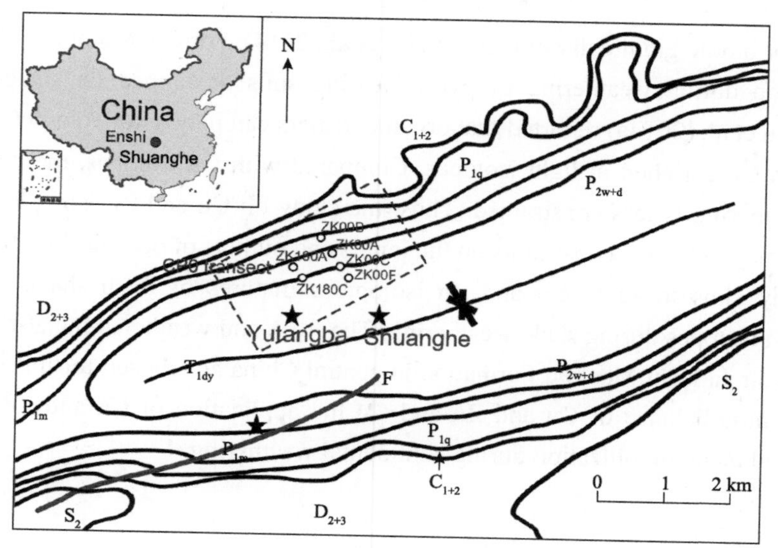

Fig. 1 Sample location map showing the location of the study area and sketch map of the geological features (after Zhu et al., 2014). Inset shows the location of study area in China.

3 Methods

3.1 Trace elements analysis

The concentrations of trace elements are analyzed on ELAN DRC-e inductively coupled plasma-mass spectrometry (ICP-MS) at the Institute of Geochemistry, Chinese Academy of Sciences (CAS). The method is outlined in Liang et al. (2000) and described briefly below in summary. Rock powders were completely dissolved in HF (38%) and HNO_3 (68%) at 200 ℃ using high-pressure PTFE bomb. The samples are redissolved in 1% HNO_3 for trace element analysis after evaporation to dryness. The measured trace elemental concentrations are corrected with the addition of Rh as an internal standard. The relative standard deviation (RSD) is less than ±10%.

3.2 Cu and Zn isotope analysis

All chemical procedures were performed in laminar flow hoods (Class 100) in the clean room (Class 1000) with filtered air. All beakers were PTFE (Savillex®). Double distilled reagents and≥18.2 MΩ water were used for sample dissolution and all other processes. Approximately 20 mg of powder was digested in Teflon vessels using high pressure Parr reactor bombs with a mixture of HF, HNO_3 and HCl at 190 ℃. Each portion, added with 2 to 3 ml of HNO_3, was then heated in oven over 36 h at 160 ℃ after decomposing carbonate minerals in concentrated HF (38 vol.%) and HNO_3. H_2O_2 was added to decompose the organic matter. If the samples were not dissolved completely, the residues were re-dissolved in 50% HNO_3. After complete digestion, 1 ml of 8 N HCl was added and the sample was heated to dryness. Dissolved samples were finally prepared in 1 ml of 8 N HCl + 0.001%H_2O_2 for chemical ion-exchange separation.

Copper and zinc were separated with a single-column procedure modified from Maréchal et al. (1999). The procedure for chemical purification and Cu isotopic analysis was reported in previous studies from our group (Liu et al., 2014a, 2014b, 2015; Li et al., 2015). The procedure for Zn isotope analysis is outlined in this study.

The samples were loaded on pre-cleaned column stuffed with 2 ml Bio-Rad strong anion resin AG-MP-1M. Copper was collected in the 24 ml of 8 N HCl after 10 ml of 8 N HCl is applied to elute the matrix. 18 ml of 2 N HCl + 0.001% H_2O_2 was then added to elute the Fe fraction. Zinc was collected in the following 10 ml of 0.5 N

HNO_3 after 2 ml of 0.5 N HNO_3 was discarded. After evaporating to dryness at 80 ℃ and dissolved in 3% (m/m) HNO_3 to drive the chloride ion away, samples in 3% HNO_3 were checked for the elimination of matrix elements before isotopic measurement. Mg and Al were detected in the Zn collected solution after once column chemistry. The presence of Mg and Al produces Argides ($Al^{27}Ar^+$, $Mg^{24}Ar^+$ and $Mg^{26}Ar^+$) which will significantly affect Zn isotope measurements (Mason et al., 2004). The in-house standard solutions with doping elements of Mg and Al are analyzed to check the possible influence of these matrix interferences on Zn isotope analysis (Table S1 in Supplementary data). The results show that Mg/Zn and Al/Zn of N1 would cause detectable Zn isotope shifts from the true values. The same procedure of column chemistry was thus repeated to reduce Mg/Zn and Al/Zn to less than 0.5. The removal of Cd from the Zn fraction could not be completely achieved by the second column step, whereas Cd has no influence on Zn isotopic analysis based on the spiking experiments (Chapman et al., 2006). In addition, Ti/Cu ratios were checked before Cu isotope analysis to exclude the possible Ti interference and the ratios for all samples are less than 0.008. The total procedural blanks are neglected and within the limit of ~2 ng and ~6 ng for Cu and Zn, respectively.

Isotopic analysis was carried out on a Neptune plus MC-ICP-MS in the Isotope Geochemistry Laboratory of the China University of Geosciences, Beijing. Sample-standard bracketing (SSB) method was used for mass bias correction. Cu and Zn isotopic determinations were performed in wet plasma mode. The samples and standards are diluted to ~200 ppb for Zn and 100 ppb for Cu in 3% HNO_3. The take-up time is 120 s. The high-sensitivity (X) cones made of Ni are used for the increase of transmission to ensure that the ^{64}Zn signals are usually N3 V/200 ppb. Cu and Zn isotopic analysis for each measurement is operated for three blocks of 40 cycles in low-resolution mode with ^{63}Cu in L3 Faraday cup, ^{64}Zn in the L2 Faraday cup, ^{65}Cu in L1 Faraday cup, ^{66}Zn in the central Faraday cup, ^{67}Zn in the H1 Faraday cup and ^{68}Zn in the H2 Faraday cup. Copper and zinc isotopic ratios are reported against SRM NIST 976 and JMC 3-0749L, respectively:

$$\delta^{65}Cu = ((^{65}Cu/^{63}Cu)_{sample}/(^{65}Cu/^{63}Cu)_{NIST\ 976} - 1) \times 1000$$

$$\delta^{66/68}Zn = ((^{66/68}Zn/^{64}Zn)_{sample} / (^{66/68}Zn/^{64}Zn)_{JMC\ 3\text{-}0749L} - 1) \times 1000.$$

The external reproducibility for $\delta^{65}Cu$ measurement is ±0.05‰ (2SD) based on long-term analyses of igneous rock standards (Liu et al., 2014b). $\delta^{66}Zn$ values of geological reference materials analyzed in this study are reported in Table 1. Long-term analysis of AAS (Alfa Aesar, Germany) Zn solution yields $\delta^{66}Zn$ value (+0.05±0.03‰, 2SD, n=40), which is well consistent with the value (+0.05±0.04‰; 2σ) calibrated by Chen et al. (2009) during two years (Fig. 2a). The USGS basalt standard BHVO-2 is analyzed repeatedly over a long period

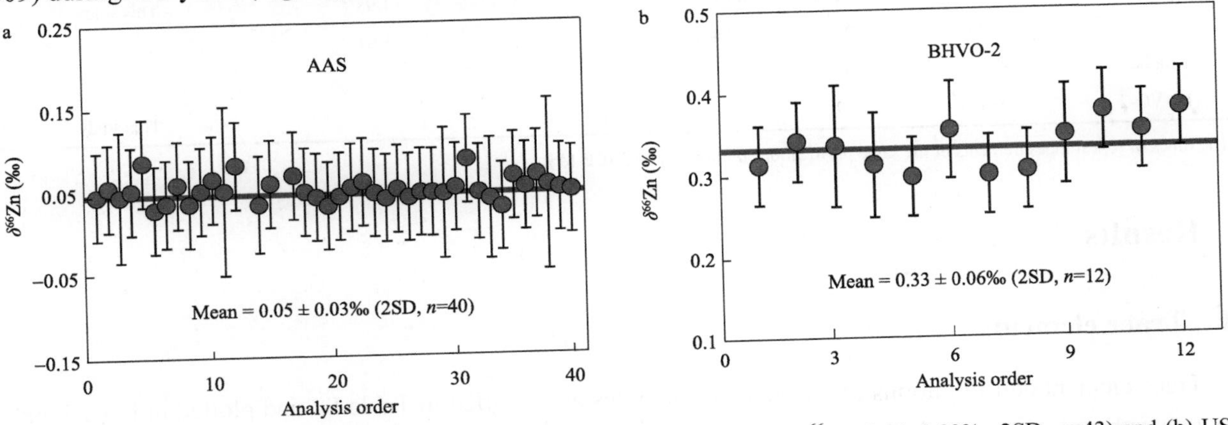

Fig. 2 Long-term analyses of Zn isotopic composition for (a) AAS Zn solutions ($\delta^{66}Zn$=0.05±0.03‰, 2SD, n=43) and (b) USGS standard BHVO-2 ($\delta^{66}Zn$=0.33±0.06‰, 2SD, n=12).

of one year longer, yielding an average δ^{66}Zn value of +0.33±0.06‰ (2SD, n=12; Fig. 2b). This value is in agreement within error with the value from previous studies (e.g., +0.33±0.04‰; Chen et al., 2013). Several other reference materials were also analyzed in this study and their δ^{66}Zn values are generally identical to those reported in the literature (Table 1). All samples analyzed in this study define the mass fractionation line in three-isotope space (δ^{66}Zn vs. δ^{68}Zn) with a slope of 2.019±0.037 (1σ), indicating no analytical artifacts from unresolved isobaric interferences on Zn isotopes.

Table 1 Cu and Zn isotopic compositions of geostandards reported in this study

Name	δ^{66}Zn	2sd	δ^{68}Zn	2sd	N	δ^{68}Zn/δ^{66}Zn	Reference
BCR-2	0.27	0.06	0.52	0.1	3	1.93	
	0.27	0.02	0.52	0.06	3	1.93	
	0.26	0.04	0.52	0.06	3	2	This study
	0.31	0.06	0.61	0.08	3	1.97	
Average	0.28	0.04	0.57	0.09		2.04	
BCR-2	0.25						Sossi et al. (2015)
AGV-2	0.3	0.01	0.56	0.1	3	1.87	This study
AGV-2	0.32	0.04	0.62	0.05	3	1.94	Chen et al. (2013)
BHVO-2	0.31	0.05	0.65	0.05	3	2.1	
	0.34	0.03	0.65	0	3	1.91	
	0.34	0.07	0.59	0.15	3	1.74	
	0.31	0.06	0.63	0.05	3	2.03	
	0.3	0.02	0.64	0.08	3	2.13	
	0.35	0.06	0.67	0.12	3	1.91	
	0.3	0.02	0.65	0.02	3	2.17	This study
	0.3	0.02	0.61	0.06	3	2.03	
	0.35	0.06	0.61	0.14	3	1.74	
	0.38	0.02	0.72	0.04	3	1.89	
	0.35	0.01	0.65	0.04	3	1.86	
	0.38	0.01	0.74	0.02	3	1.95	
Average	0.33	0.06	0.65	0.09		1.97	
BHVO-2	0.33	0.04	0.65	0.06		1.97	Chen et al. (2013)
GBW07105	0.44	0.05	0.87	0.1	3	1.98	This study
	0.44	0.06	0.84	0.1	3	1.91	
Name	δ^{65}Cu	2sd	n				
BHVO-2	0.12	0.09	3				This study

N: The times of repeat measurements of the same purification solutions by MC-ICP-MS.

4 Results

4.1 Trace elements

Trace element concentrations of unweathered samples are reported in Table S2 and plotted in Figs. 3 and 4a. The unweathered shales and cherts have relatively LREE-depleted patterns when normalized to the North American shale composite (NASC) (Gromet et al., 1984). All samples display positive anomalies of Y and Gd, strongly negative Ce anomalies and weak negative Eu anomalies, with depletion in total REE concentrations relative to NASC (Fig. 4a). Notably, the REE + Y pattern of the unweathered shales and cherts is similar to that of seawater except the less enriched HREEs.

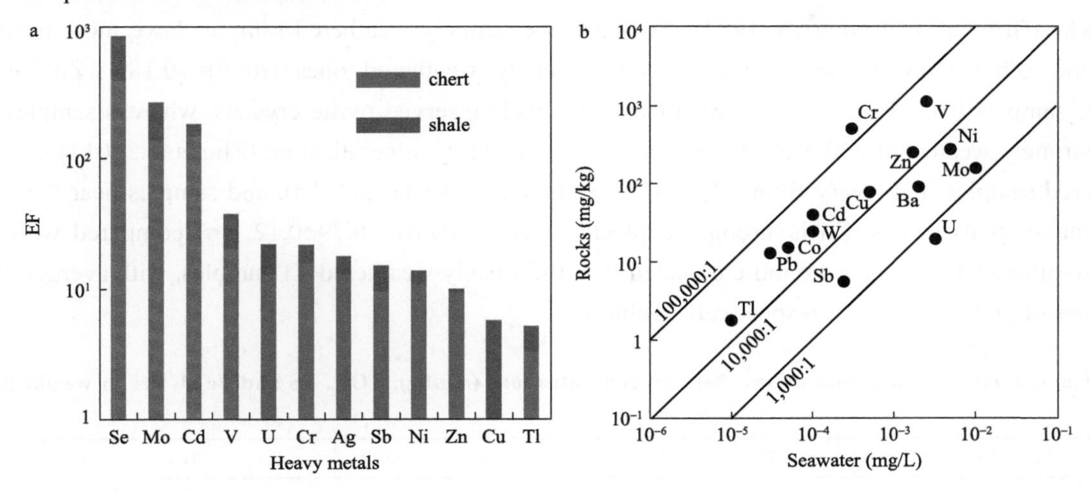

Fig. 3 (a) Enrichment factor (EF) of metals in unweathered samples from drill cores at Yutangba. EF = (X/Nb)sample/ (X/Nb)average shale (Rimmer et al., 2004). The average value of shales is based on that listed by Wedepohl (1971); (b) average concentrations of metals in unweathered samples compared to present-day seawater (Holland, 1979).

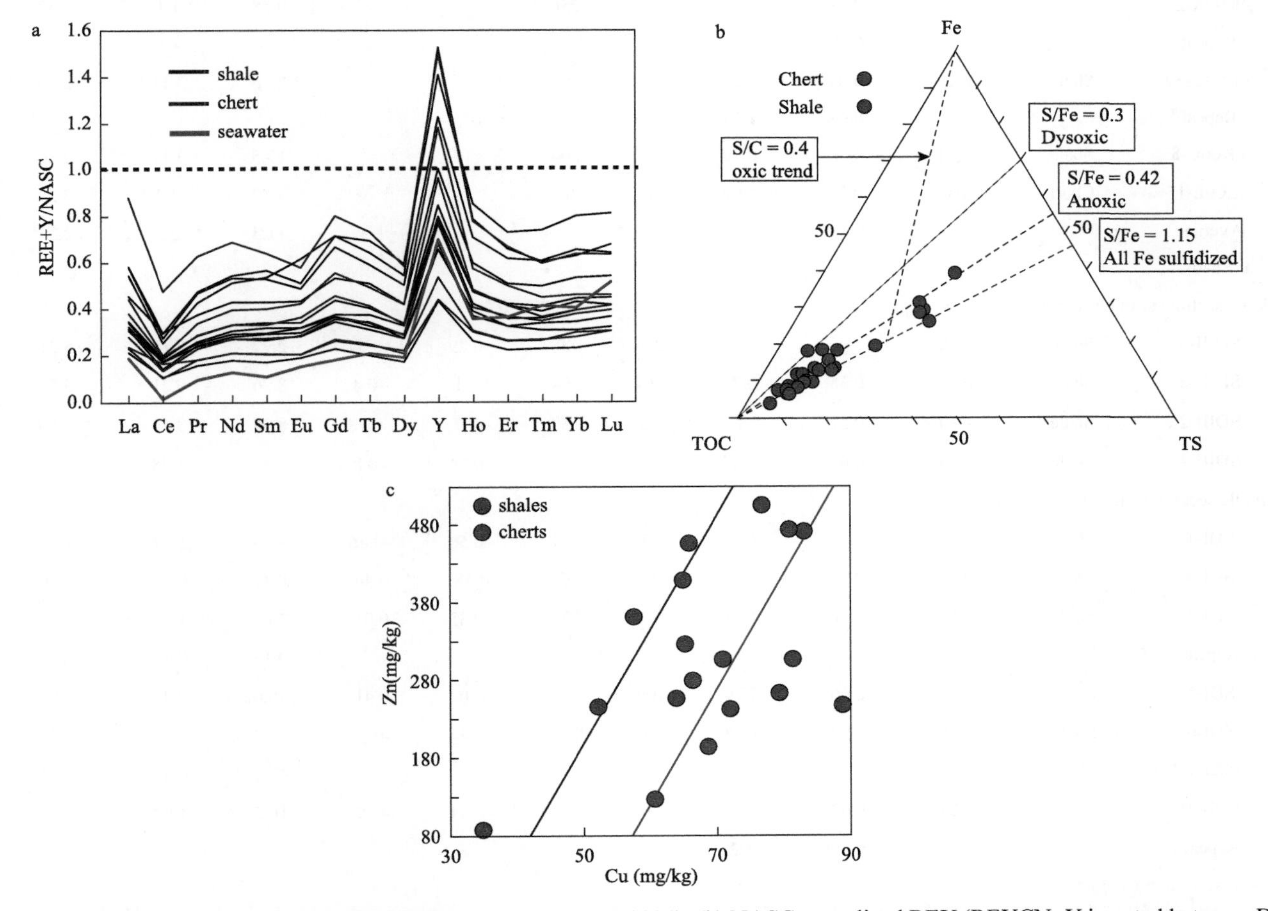

Fig. 4 (a) Fe–TOC–TS triangular diagram (after Rimmer et al., 2004); (b) NASCnormalized REY (REYCN; Y inserted between Dy and Ho) patterns for the unweathered samples. Data are reported by Zhu et al. (2014); (c) Cu concentrations vs. Zn concentrations of unweathered black shales and cherts.

To evaluate the degrees of depletion or enrichment of elements of interest during weathering, the mass transfer coefficient (τ) is applied: $\tau = ((C_{i,w}/C_{i,p}) / (C_{j,w}/C_{j,p}))-1$, where $C_{i,w}$ and $C_{j,w}$ refer to the concentrations of target and immobile element in weathered samples, respectively; $C_{i,p}$ and $C_{j,p}$ represent the concentrations in parent rocks (Brimhall and Dietrich, 1987). At Shadi, the strongly weathered samples have more negative τCu (−0.43) and τZn (−0.63) values compared to the weakly weathered ones (τ(Cu)=−0.18, τ(Zn)=−0.02). At Yutangba, samples from T1 are weakly weathered with wellpreserved pyrite crystals, whereas samples near the fault are strongly weathered with iron staining and other secondary mineralization (Zhu et al., 2014). τ(Cu) values of weathered samples in T1 vary from −0.78 to 11.56 (average=1.94±3.09, 1σ), and samples near the fault have more negative τ(Cu) values. Zn is strongly depleted (average τ(Zn)=−0.79±0.12, 1σ) compared with Cu. The strongly weathered T2 samples are more depleted than the weakly weathered T1 samples, with average τ(Cu) and τ(Zn) values of −0.09 and −0.95, respectively (Table 2).

Table 2 Cu and Zn isotopic compositions (‰) and concentrations (mg/kg). TOC, TS and Fe_2O_3 are in weight percentage (wt.%).

Sample	Type	δ^{65}Cu	δ^{66}Zn	δ^{68}Zn	Cu	Zn	τ(Cu)	τ(Zn)	TOC	TS	Fe_2O_3
The unaltered samples at Yutangba drill cores											
ZK00A-38	Chert	0.17	0.38	0.71	58	361			6.05	1.11	1.29
Repeat			0.36	0.71							
ZK00A-20	Chert	0.27	0.58	1.11	64	256			9.58	1.1	1.37
Repeat			0.59	1.11							
ZK00C-30	Shale	0.15	0.63	1.24	81	306			22.6	3.11	2.86
Repeat			0.63	1.24							
ZK00C-8	Shale	0.14	0.4	0.79	77	504			13.5	3.11	2.86
ZK00E-15	Chert	−0.01	0.55	1.07	66	82			4.87	3.2	3.29
Average									11.3	2.3	2.33
Shadi											
Upper section: strongly weathered											
SDIII-4	Shale	0.39	–	–	226	438	−0.29	−0.54	2.92	0.15	5.21
SDIII-3	Shale	0.83	0.38	0.75	269	434	0.11	−0.4	8.26	0.21	5.57
SDIII-2	Shale	−0.7	0.51	1	82	180	−0.7	−0.78	8.98	3.44	3.59
SDIII-1	Shale	0.24	0.65	1.27	40	157	−0.84	−0.8	7.55	3.48	3.35
Middle section: some weathering apparent											
SDI-1	Shale	0.2	0.64	1.27	53	124	−0.56	−0.66	13.3	1.77	1.39
SDI-3	Chert	0.09	0.56	1.14	34	92	0.58	0.44	8.41	0.56	0.36
SDI-4	Shale	0.12	0.6	1.27	73	324	−0.32	0.01	23.3	1.84	1.77
Repeat									0.17	0.63	1.2
SDI-5	Shale	0.27	0.66	1.29	109	481	−0.04	0.41	36.6	2.47	1.85
SDI-6	Shale	0.28	0.59	1.18	47	197	−0.44	−0.23	18.3	1.29	1.33
SDI-7	Chert	0.53	–	–	–	–	–	–	4.71	0.41	0.25
SDI-10	Shale	0.24	0.62	1.23	70	279	−0.34	−0.13	16.5	1.92	1.77
Repeat			0.59	1.2							
Lower section: no visible weathering											
SD3	Shale	0.26	0.46	0.89	126	1038	−0.01	1.7	35.5	2.12	1.57
Repeat			0.51	0.97							
Average									16.3	1.75	2.52
Yutangba											

Continued

Sample	Type	δ^{65}Cu	δ^{66}Zn	δ^{68}Zn	Cu	Zn	τ(Cu)	τ(Zn)	TOC	TS	Fe_2O_3
Transect 1											
cp01	Shale	6	0.56	1.25	222	82	1.41	−0.75	30.5	1.96	1.54
cp02	Shale	5.95	0.57	1.17	190	57	1.43	−0.79	24.2	1.6	1.29
Repeat			0.53	1.09							
cp03	Chert	8.2	0.61	1.22	226	46	5.97	−0.6	8.18	0.79	0.81
Repeat			0.58	1.24							
cp04	Mudstone	19.42	0.64	1.29	1962	136	11.56	−0.76	17.7	1.65	3.28
Repeat		20.04									
cp05	Chert	7.05	0.6	1.23	70	31	1.73	−0.66	7	0.59	0.66
Repeat			0.61	1.22							
cp06	Shale		0.64	1.46	240	31	2.39	−0.88	22.4	0.88	0.74
cp07	Chert	4.28	0.69	1.48	62	27	1.41	−0.7	12.2	0.67	0.6
cp09	Shale	1.71	0.78	1.52	166	79	0.46	−0.8	39.9	2.3	1.87
cp011	Chert	−0.06	0.6	1.28	82	15	5.71	−0.65	8.44	0.36	0.33
cp012	Chert	−0.51	0.46	1.94	31	10	0.67	−0.85	6.96	0.47	0.48
cp013	Shale		0.66	1.33	66	28	−0.34	−0.92	15.4	1.66	1.71
cp014	Shale	−0.41			63	31	−0.41	−0.92	26.3	0.52	0.99
cp015	Mudstone	0.2			104	56	−0.48	−0.92	12.1	0.31	0.88
cp016	Shale	−0.27	0.69	1.49	52	20	−0.51	−0.94	25.5	0.47	0.84
cp017	Chert	1.3	0.6	1.2	42	21	0.98	−0.72	7.81	0.37	0.6
cp018	Chert	−2.46			35	9	0.22	−0.92	4.67	0.38	0.5
cp019	Chert	−0.18			43	7	6.66	−0.62	6.91	0.18	0.18
cp020	Shale		0.57	1.21	181	19	1.32	−0.93	39.1	1.48	0.74
cp021	Shale	−3.35			26	9	−0.69	−0.97	29.6	0.59	0.27
cp021a	Mudstone	−2.18	0.65	1.25	54	243	−0.78	−0.68	17	0.44	5.26
Repeat			0.67	1.33							
Average									18.1	0.88	1.18
Transect 2											
YTB1	Shale	−2.7			87	149	−0.63	−0.79	1.3	0.11	2.63
YTB2	Shale	−0.44			38	9	−0.8	−0.98	4.36	0.14	0.23
YTB3	Shale	0.01			74	11	−0.67	−0.98	5.05	0.18	1.19
YTB4	Chert	−0.56			27	5	−0.63	−0.98	3.21	0.12	0.51
YTB5	Shale	−3.31	0.42	0.85	185	41	−0.07	−0.93	3.05	0.15	1.74
YTB7	Shale	−1.95			156	13	0	−0.97	1.95	0.18	1.86
YTB9	Shale	−0.9	0.35	0.82	114	15	−0.54	−0.98	1.62	0.27	0.65
YTB10	Chert	−2.25			38	5	2.4	−0.89	2.45	0.12	0.13
YTB11	Shale	−1.57			140	20	−0.44	−0.97	2.25	0.23	0.93
YTB13	Shale	−0.11			64	13	−0.72	−0.98	2.1	0.16	0.8
Repeat		−0.06									
YTB15	Chert	−0.63			42	12	1.35	−0.83	2.67	0.13	0.21
Repeat		−0.67									
YTB16	Shale	0.72			76	18	−0.7	−0.98	6.55	0.27	0.64
YTB18	Shale	1.14			176	18	0.65	−0.95	3.77	0.19	0.57

Continued

Sample	Type	$\delta^{65}Cu$	$\delta^{66}Zn$	$\delta^{68}Zn$	Cu	Zn	τ(Cu)	τ(Zn)	TOC	TS	Fe_2O_3
YTB20	Shale	−0.68			52	11	−0.78	−0.99	7.5	0.25	
YTB21	Shale	−0.12			60	12	−0.62	−0.97	5.74	0.18	
Repeat		−0.17									
YTB22	Shale	−6.42	0.25	0.57	53	4	1.26	−0.95	2.19	0.1	0.79
Repeat		−6.33									
YTB23	Shale	−3.25			267	12	0.21	−0.98	9.79	0.27	0.37
YTB24	Shale	−4.6			169	11	0.04	−0.98	3.85	0.16	0.21
YTB25	Shale	−2.49			72	15	−0.62	−0.97	3.67	0.1	0.26
YTB26	Chert	−0.89	0.34	0.62	67	11	−0.25	−0.97	4.47	0.19	0.27
YTB27	Chert	0.29	0.39	0.82	57	30	−0.53	−0.94	4.75	0.18	1.47
Repeat			0.36	0.86							
YTB28	Chert	−0.82			138	18	0.18	−0.96	4.45	0.21	0.94
Average									3.94	0.18	0.82

4.2 Cu and Zn isotopes

4.2.1 Unweathered rocks from drill cores

Copper and zinc isotopic compositions of unweathered and weathered samples are reported in Table 2. Unweathered shales and cherts from drill cores at Yutangba have similar Cu and Zn isotopic compositions. Their $\delta^{65}Cu$ values range from −0.01‰ to +0.17‰ with an average of +0.14±0.09‰ (1σ, n=5) consistent with the range of igneous rocks and marine sedimentary materials (Maréchal et al., 1999, 2000; Liu et al., 2015). $\delta^{66}Zn$ values of these unweathered samples vary from +0.40‰ to +0.67‰ with a mean of +0.51±0.10‰ (1σ, n=5), significantly heavier than igneous rocks (+0.28±0.05‰; Chen et al., 2013) and marine clastic sediments (+0.18±0.08‰) (Pons et al., 2011, 2013).

4.2.2 Weathered samples

Copper isotopic compositions of the strongly weathered samples from Shadi ($\delta^{65}Cu$=+0.19±0.56‰; 1σ) are more variable and slightly lighter compared to the weakly weathered samples ($\delta^{65}Cu$=+0.24±0.13‰). The latter have almost the same $\delta^{65}Cu$ values as those of the unweathered samples from drill cores (Table 2). The average $\delta^{66}Zn$ of the weathered samples at Shadi is +0.61 ± 0.08‰, 1σ), similar to the value of samples from drill cores (+0.40‰ to +0.67‰). The strongly weathered samples from T2 at Yutangba have extremely negative $\delta^{65}Cu$ values of −1.44±1.77‰ (1σ) compared to the unweathered samples (+0.14‰). Their $\delta^{66}Zn$ values display a small range with an average of +0.35±0.06‰ also lighter than the unweathered samples (+0.51‰) (Table 2). Copper isotopic compositions of the weathered samples from T1 vary from −3.35‰ to +19.42‰ with an average of +2.63±5.45‰ (Table 2). Especially, samples near the fault are strongly depleted in Cu and have lowest $\delta^{65}Cu$ values (Table 2). Zn isotopic compositions (+0.66±0.07‰) of weathered samples from transect 1 are slightly heavier than strongly weathered sample from transect 2 (+0.25‰ to +0.42‰) and unweathered samples from drill cores (+0.37‰ to +0.63‰).

5 Discussion

5.1 Mechanisms of metal enrichment in black shales: Cu and Zn isotopic constraints

The Maokou unweathered shales and imbedded cherts reveal the pronounced enrichment in organic matters (TOC>10 wt.%) and metals (i.e. V, Cr, Ni, Cu, Zn, Ag, U, Mo, Cd, Sb, Se and Tl), with the shales slightly more metal-enriched than the cherts (Table S2, Fig. 3a). For example, trace metals including Cu and Zn in the black shales and cherts are enriched by a factor of ~10^5 relative to present-day seawater (Bruland and Lohan, 2006; Fig. 3b). The enrichment extent of metals in shales is commonly assessed by the enrichment factor (EF) defined as EF = $(X/Nb)_{sample}$ / $(X/Nb)_{average\ shale}$ (Rimmer et al., 2004). The average EF values of Cu and Zn in the unweathered samples are 5.64 and 9.49, respectively, showing moderate enrichments (Fig. 3a). The mechanism of multi-metal enrichments in Maokou shales and cherts is debated. There are three possible mechanisms proposed, including (i) sulfide formation and organic complexation under anoxic/euxinic condition, (ii) biological assimilation, and/or (iii) extra metal supply by hydrothermal activity (Yao et al., 2002; Fan et al., 2008; Zhu et al., 2014).

The first possibility is a feasible explanation for the Cu–Zn enrichment. The redox-sensitive and sulfide-forming trace metals (e.g., Mo, V, Zn, Cu and U) and OM are commonly concentrated in marine euxinic environments during early digenesis of black shales (Nameroff et al., 2002; Kametaka et al., 2005). The euxinic condition for the studied samples is evidenced by the fact that almost all samples plot between S/Fe=0.42 line and S/Fe=1.15 line in the Fe–TOC–TS triangular diagram (Fig. 4b), which reflects that the deposition environment is anoxic or contains excess sulfur (Rimmer et al., 2004). In anoxic waters, Zn and Cu will mainly precipitate from seawater as independent sulfide phase or uptake by Fe-sulfides in solid solutions (Algeo and Maynard, 2004).

Zinc isotopic compositions (+0.51±0.10‰, 1σ) of the unweathered shales and cherts are heavier than those of both igneous rocks (+0.28±0.05‰; Chen et al., 2013) and marine clastic sediments including 'normal' shales reported in previous studies (+0.18±0.08‰) (Maréchal et al., 2000; Pons et al., 2011, 2013). The involvement of carbonates with heavy δ^{66}Zn (e.g., up to 1.34‰; Pichat et al., 2003) is unlikely because rare carbonate (<3%) exists in these samples based on X-ray diffraction analysis (XRD) and Zn isotopes are not correlated with the varying bulk CaO contents (1.3 to 4.1 wt.%, Fig. S1 and Tables S3, S4). Instead, it is more likely induced by the capture of seawater with Zn isotopic composition (+0.51‰) heavier than marine clastic sediments (Little et al., 2014). This hypothesis is further supported by the coupling of trace metal concentration of the unweathered shales and cherts with seawater (Fig. 3b). The shales and cherts also have similar REE + Y patterns with the seawater, in which the positive Y anomaly is the most striking signature of seawater (Fig. 4a). A similar contribution from seawater may also exist for Cu, as indicated by the positive correlation between Cu and Zn contents (Fig. 4c). Cu isotopic composition (+0.14±0.09‰) of the unweathered samples is similar to that of igneous sources (+0.06±0.20‰; Liu et al., 2015) and marine sediments (δ^{65}Cu = +0.18±0.14‰; Maréchal et al., 1999), but significantly lighter than seawater (0.9‰; Little et al., 2014). The lighter Cu isotopic composition of these black shales and cherts than seawater is likely induced by the Cu(II)/Cu(I) redox reaction of sulfide formation (Ehrlich et al., 2004). The lines of evidence thus suggest that metal enrichments in the Maokou shales and cherts were mainly derived from seawater.

Zinc isotope fractionation can also happen during sorption onto OM, Mn, Fe, Al oxides and hydroxides (Pokrovsky et al., 2005; Gélabert et al., 2006; Balistrieri et al., 2008; Jouvin et al., 2009). For example, experimental studies have shown that the sorption onto an analogue of OM induces a positive (~+0.24‰) Zn isotope fractionation (Jouvin et al., 2009). It seems that the sorption process, if present, produces limited Zn isotope variation in oceanic sediment traps and 'normal' shales that have silicate-like δ^{66}Zn of ca. +0.25‰

(Maréchal et al., 2000; Pons et al., 2011, 2013). However, the Maokou black shales and cherts have much higher TOC contents than "normal" shales, potentially resulting in sorption of isotopically heavy Zn from seawater. Furthermore, biological scavenging of isotopically heavy Zn onto sinking OM has been reported, which could elevate δ^{66}Zn values during the diagenesis of black shales and cherts (John et al., 2007; John and Conway, 2014).

The second possible mechanism is biological assimilation and its eventual burial, which has been used to explain Se enrichment and near-zero $\delta^{82/76}$Se values in the Maokou shales (Zhu et al., 2014). For Zn, phytoplankton uptakes isotopically light Zn (John et al., 2007), which is inconsistent with the relatively high δ^{66}Zn values of the Maokou shales. Thus, the second explanation is unlikely. The third possible mechanism is hydrothermal activity. The trace elements (Ba, Mn and Pb) that are commonly enriched in hydrothermal environments are weakly enriched in the Maokou shales (Table S2; Rona, 1978; Marchig et al., 1982; Fleet, 1983). In addition, the Maokou cherts have higher ratios of w(Al)/w(Al + Fe + Mn) (averaged 0.64) than hydrothermal cherts (<0.32; Table S3; Adachi et al., 1986). Hydrothermal input has a light Zn isotopic composition of +0.24‰ (Little et al., 2014), which is inconsistent with the data as observed. Therefore, hydrothermal input is probably not important for the enrichments of metals in the Maokou shales and cherts.

In summary, we conclude that the Cu and Zn isotope signatures of the Maokou shales and cherts indicate an important contribution of Cu and Zn from seawater. The mechanisms for Zn and Cu uptake from seawater may mainly include sulfide precipitation and/or adsorption by OM. These processes potentially result in enrichment of metals including Cu and Zn in these shales and cherts.

5.2 Copper isotopic variation during shale weathering: redox transport

At Yutangba, the weakly weathered samples (T1) are less depleted in Cu than the strongly weathered samples (T2) located above T1. τ(Cu) value of T1 samples is greater than zero except for the samples near the fault (Fig. 5). Samples from T1 also have heavier δ^{65}Cu (mean=+2.63±5.45‰) compared to samples from T2, and there is positive correlation between δ^{65}Cu and τ(Cu) in T1 samples (Fig. 6a). The plausible explanation of the systematic Cu isotopic variation is the oxidative leaching of Cu(I) sulfide and re-mineralization of Cu(II) minerals as previously observed in ore deposits and leaching experiments (Mathur et al., 2005, 2009; Fernandez and Borrok, 2009). The leachate has heavy δ^{65}Cu in the (Cu(II)/Cu(I) redox) reaction of sulfide during experimental leaching, resulting in ^{65}Cu-depletion in leached minerals and ^{65}Cu-enrichment during re-mineralization of Cu(II) minerals (Fernandez and Borrok, 2009). The dominate Cu-bearing fractions in shales and cherts are sulfide and OM, and sulfide apparently reacts faster with O_2 than OM does (Huerta-Diaz and Morse, 1992; Wildman et al., 2004; Tribovillard et al., 2006). Hence, the similar TOC contents but lower TS and TFe contents in the weakly weathered

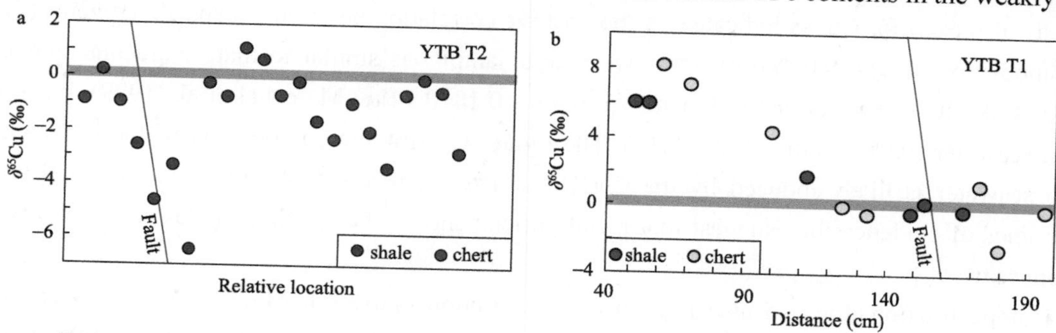

Fig. 5　Plots of δ^{65}Cu vs. distance for samples from T2 (a) and T1 (b) at Yutangba. No distance data are available for T2 and the data are plotted according to their relative location. The red lines denote faults. The average Cu isotopic composition of unweathered shales and cherts is shown in gray lines.

samples in comparison with the unweathered samples indicates that leaching of sulfide is a vital process of Cu mobilization during shale weathering. The significant Cu isotopic variation (–0.68‰ to +0.06‰) of soils developed on shales has been explained to be a result of pyrite weathering (Mathur et al., 2012).

The Cu isotopic variation (–6.38‰ to +19.73‰) of weathered samples at Yutangba is much larger than the average fractionation factor of $\Delta^{65}Cu_{(Cu(II)Aq-CuS)}$ (+3.06±0.14‰) at 20 ℃ (Ehrlich et al., 2004), indicating multistage redox leaching and redistribution. Repeated leaching and precipitation during weathering enlarges the offset of δ^{65}Cu compared to the starting value (Mathur et al., 2009; Palacios et al., 2011). Cu isotopic compositions of the weathered shales and cherts in T1 at Yutangba have moderately positive correlation with Se isotopic compositions (Fig. 6c). Particularly, the sample cp04 has both extremely heavy Cu and Se isotopic compositions (δ^{65}Cu = +19.74‰; $\delta^{82/76}$Se = +7.81‰). At Shadi, sample SDIII-2 has the most negative δ^{65}Cu value (–0.7‰) and also the lowest $\delta^{82/76}$Se value (–4.08‰) in the profile. The co-variation of Cu and Se isotopes further indicates that redox transport is the main mechanism of Cu and Se mobilization during shale weathering.

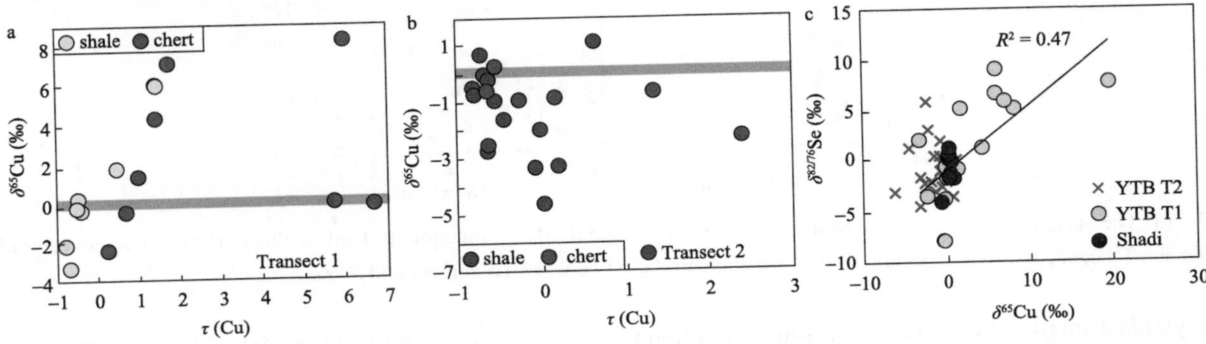

Fig. 6 Cu isotopic compositions vs. τ(Cu) in Yutangba T1 (a) and Yutangba T2 (b). (c) $\delta^{82/76}$Se vs. δ^{65}Cu of weathered samples. The average Cu isotopic composition of unweathered shales and cherts is shown in gray lines. Cu isotope data are reported in Table 2. The Se isotope data are from Zhu et al. (2014). $\tau = ((C_{i,w}/C_{i,p}) / (C_{j,w}/C_{j,p})) - 1$. Cu and Zn concentrations of parent rocks are taken as the average Cu concentration (77 mg/kg, n=15) and Zn concentration (301 mg/kg, n=11) of unweathered samples when calculating τ values. Ti is considered as the immobile element, and the average concentration of Ti (0.13±0.10 wt.%) of unweathered rocks is assumed as the initial Ti concentration of parent rocks.

Compared with the huge δ^{65}Cu variation in Yutangba, the samples at Shadi have more homogenous Cu isotopic composition (Fig. 7a). This difference reflects that samples at Shadi absent repeated leaching and redox reactions at Yutangba. The different Cu isotopic signature at different transects is similar to that of Se isotopes, induced by the dissimilar abilities of strata to conduct fluids (Zhu et al., 2014). The strata at Shadi and Yutangba develop horizontally and vertically, respectively, and the latter is much easier to conduct the fluid and to enhance seepage made by the faults at Yutangba.

5.3 Zinc redistribution and isotopic variation during black shale weathering

The weathered samples at Shadi have been sorted into the strongly weathered, the weakly weathered and the unweathered samples by Zhu et al. (2014). Based on τ(Zn) values, the three groups correspond to the three zones in weathering model (leached zone, enriched zone and proto-lith; Brimhall and Dietrich, 1987). Zn is lost in the leached zone and accumulates in the enriched zone (Fig. 7a). Samples in leached zone have relatively light Zn isotopic composition, indicating that heavy Zn isotopes tend to be released during leaching (Fig. 7a). Considering that Zn is mainly hosted by sulfide and the adsorptive phase as discussed in Section 5.1, the leaching of sulfide and desorption commonly control the loss of Zn during weathering (Nameroff et al., 2002; Berner et al., 2013), which probably induces Zn isotope fractionation. Experimental studies have shown that leaching of sphalerite-rich

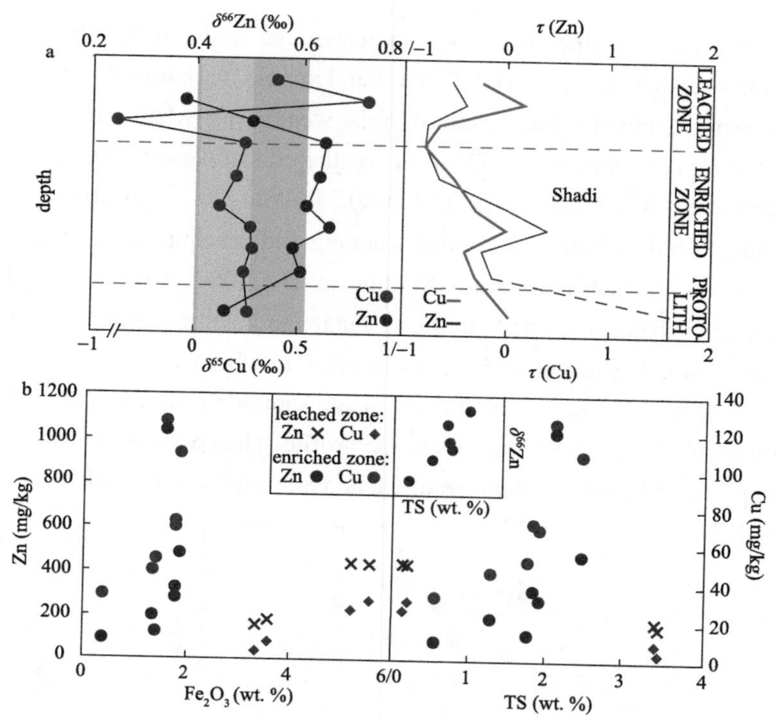

Fig. 7　(a) The variation of isotopic composition and τ values of Zn (Cu) as a function of depth at Shadi. (b) Zn (Cu) concentrations vs. TFe and TS concentrations at Shadi. Inset shows the correlation between TS content and δ^{66}Zn.

rocks yields a small but detectable Zn isotope fractionation, with the leaching fluids enriched in heavy Zn isotopes (by 0 to +0.2‰; Fernandez and Borrok, 2009). In addition, previous studies have reported that adsorption by OM, clays and oxides generates heavy Zn isotopic composition (Pokrovsky et al., 2005; Gélabert et al., 2006; Balistrieri et al., 2008; Jouvin et al., 2009). Thus, sulfide breakdown and desorption from OM are the likely processes to lower δ^{66}Zn values of the weathered samples.

The enriched zone has heavy Zn isotopic composition that is roughly coupled with the τ(Zn) variation (Fig. 7a). In other words, heavy Zn isotopes tend to be enriched in the enriched zone during Zn accumulation. The mechanism of Zn accumulation is perhaps similar to the enrichment mechanism of Cu which was observed in ore deposits and weathered shales and basalts (Mathur et al., 2009, 2012; Liu et al., 2014a). By contrast, no relationship between τ values and isotopic compositions exists for Zn as Cu does, probably due to the large variation of protolith Zn concentration (Table 2), which makes it difficult to evaluate the loss of Zn during weathering quantitatively. The re-precipitation and sorption onto OM and sulfide could be potentially involved in the enrichment process, whereas the refixation of Zn (Cu) may be mainly related to Fe sulfide based on the positive correlation between Zn (Cu) and TFe (TS) contents (Fig. 7b). Sulfides (e.g., pyrite and pyrrhotite) are excellent scavengers for Zn and Cu by adsorption or displacement in lattice (Jean and Bancroft, 1986; Simpson et al., 2000).

At Yutangba, samples from T2 have more negative τ(Zn) values (mean=−0.95) and lower δ^{66}Zn values (mean=+0.35‰) than those from T1 (τ=−0.8; δ^{66}Zn=+0.62‰). The upper transect (T2) and lower transect (T1) at Yutangba could correspond to the leached zone and the enriched zone at Shadi. In addition, the correlation between Zn and TFe, TS and Cu occurring in the enriched zone of Shadi is also observed in T1 at Yutangba (Fig. 8a, b and c). This indicates that except for leaching, accumulation also happened in T1 although the τ(Zn) value is low (<0). Overall, the results show that significant Zn isotope fractionation could be generated during shale weathering, with the released fluids enriched in heavy Zn isotopes. This is analogous to Cu isotopes, despite a much smaller magnitude of isotope fractionation due to the absence of multi valences of Zn.

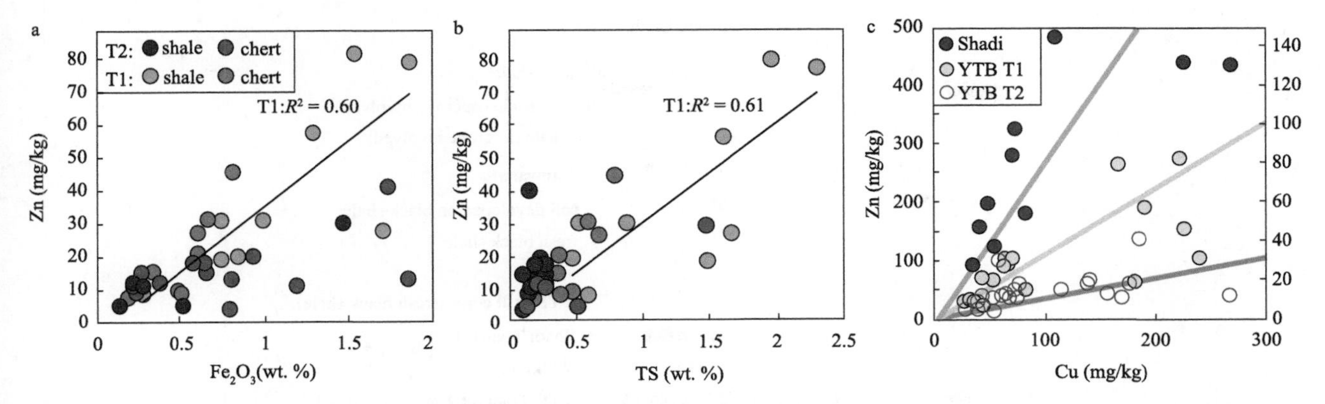

Fig. 8 The correlation of Zn concentrations with TFe (a), TS (b) and Cu (c) concentrations for the weathered samples. For Panel c, Cu concentrations at Shadi are read from the left axis and Cu concentrations of YTB T1 and T2 are read from the right axis. Data are reported in Table 2.

The weathered shales and cherts from both upper (T2) and lower (T1) transects at Yutangba have more negative τ(Zn) values (T1: –0.8; T2: –0.95) than τ(Cu) (T1: 1.93; T2: –0.12). Indeed, Zn is more mobile than Cu based on leaching tests and sequential chemical extraction of black shales (Lavergren et al., 2009). The leaching ability of sulfide minerals in black shales is in the order of the sphalerite (ZnFe)S > Cu-sulfide (Cu_2S, CuS, CuS_2, $CuFeS_2$) > pyrite (FeS_2) (Farbiszewska-Kiczma et al., 2004). The different leaching abilities can well explain the observations that the more strongly weathered transect (T2) has much lower Zn/Cu ratio compared to the transect (T1) due to strong leaching (Fig. 8c).

5.4 Implications for geochemical cycling of Cu and Zn

The development of Cu and Zn isotopes as a new tracer of biogeochemical information requires an understanding of the modern oceanic isotopic mass balance of these two metals. A recent investigation sheds light on the global oceanic budget and modern oceanic mass balance of Cu and Zn using their isotopes (Little et al., 2014). The initial data show that the inputs from rivers and dust have light Cu isotopic compositions (+0.63‰) compared to seawater (+0.9‰), and the oxic sediments are a complement reservoir with light Cu isotopic composition (+0.31‰). For Zn, the riverine inputs have the same Zn isotopic composition (+0.33‰) as the lithogenic Zn. The Zn isotopic compositions of seawater and the outputs (Fe–Mn crusts and carbonates), however, are all heavier than that of the inputs. The present model of oceanic mass balance of Zn thus has problems with the missing of an isotopically light sink or an isotopically heavy input (Little et al., 2014). Especially, Zn isotopic signatures of the sink in the euxinic sediments (e.g., shales) are still unclear but may be very important for understanding mass balance of Zn in the ocean.

The Zn and Cu isotopic compositions of unweathered black shales studied here may fill the knowledge about euxinic sediments. The Maokou shales were formed in euxinic conditions and they have lower δ^{65}Cu (+0.14±0.10‰) and higher δ^{66}Zn (+0.51±0.11‰) compared with the riverine inputs. Combined with the data in the literature (Maréchal et al., 2000; Asael et al., 2007), the δ^{65}Cu values of euxinic shales do not significantly drift off the igneous sources and oceanic clastic sediments (Fig. 9). The heterogeneous Zn isotopic composition of black shales is probably due to isotopic variation generated during diagenesis (Fig. 10). On the basis of the limited data available so far, the isotopic composition of euxinic sediments could not be defined. However, the Cu and Zn isotopic compositions of shales studied here and those reported in literature do not match with the isotopic compositions of projected missing sinks (δ^{65}Cu=+1.19‰; δ^{66}Zn=–0.3‰) in the oceans (Little et al., 2014), implying the presence of other isotopically light outputs or heavy inputs.

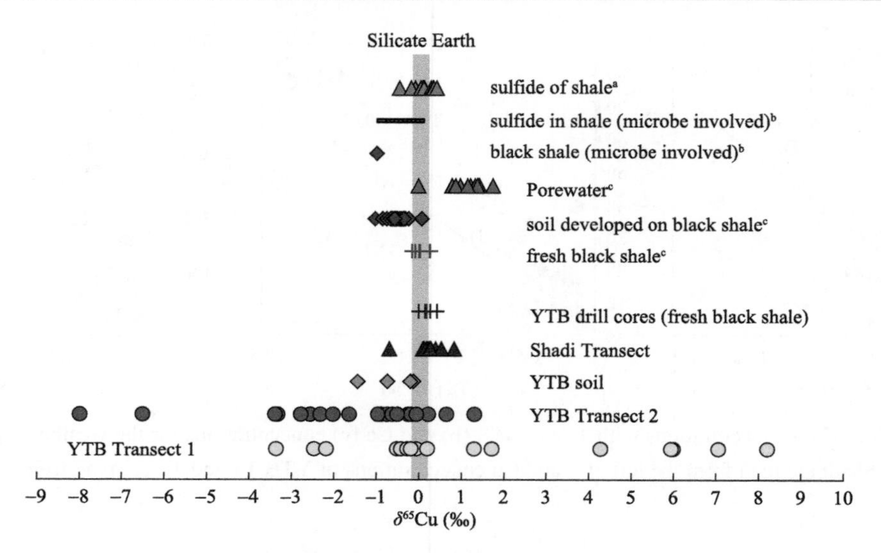

Fig. 9　Summary of Cu isotopic composition of shales and cherts reported in this study and sulfides in shales and pore waters reported in the reference. Data sources: (a) Asael et al. (2007), (b) Archer and Vance (2002), and (c) Mathur et al. (2012).

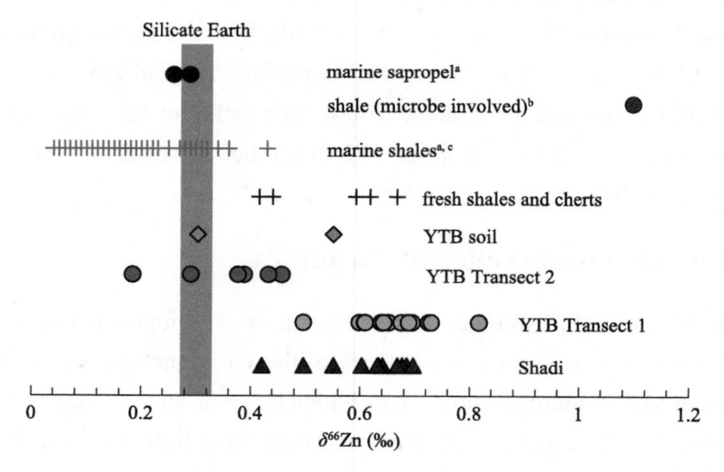

Fig. 10　Summary of Zn isotopic composition of shales and cherts reported in this study and those reported in reference, and oceanic sapropel reported in reference. Data sources: (a) Archer and Vance (2002), (b) Maréchal et al. (2000), and (c) Pons et al. (2011).

Given that black shales are highly enriched in heavy metals (Cu, Zn), weathering of the metalliferous black shales will play an important role in metal cycling in the ocean (Kametaka et al., 2005). The riverine inputs to the modern ocean are estimated to have Cu and Zn isotopic compositions of +0.68‰ and +0.33‰, respectively (Little et al., 2014). Although the proportion of dissolved Cu and Zn in river waters derived from shale weathering is unknown in the estimate of Little et al. (2014), our results indicate that extreme Cu isotope fractionation can be generated during weathering of shales and both heavy Cu and Zn isotopes are preferentially released from sulfide or OM into fluids during shale weathering. Similar fractionations can occur during sulfide weathering of copper deposits (e.g., Mathur et al., 2009). Black shale weathering, and other process including sulfides leaching (e.g., ore weathering) and OM decaying should thus be considered when evaluating the inputs of Cu and Zn in the modern or past oceans.

6　Conclusions

This paper reports Cu and Zn isotope data for unweathered and weathered metalliferous black shales and cherts from the Maokou Formation, Central China. The results are used to investigate the enrichment mechanism of metals in shales and their redistribution by chemical weathering of black shales.

Copper isotopic composition of the unweathered shales and cherts is similar to that of silicate rocks but lighter than seawater, whereas their Zn isotopic composition is heterogeneous and significantly heavier compared to the 'normal' shales and marine sediments reported in previous studies. The heavy δ^{66}Zn values are explained to reflect an important contribution of Zn in the shales from seawater via sulfide precipitation; the light δ^{65}Cu values compared with seawater is not conflict with this model and could be attributed to redox reaction during sulfide precipitation under anoxic environment. This is supported by the similar REE and Y patterns between the unweathered shales and cherts and seawater. Considering the prominent enrichment of organic matter (OM) in the Maokou black shales, the heavy δ^{66}Zn could also be induced by OM adsorption process. Overall, Zn isotopic compositions of these metalliferous shales are different from those of the 'normal' shales, highlighting the potential of Zn isotopes as a tracer for metal enrichment in metalliferous shales.

Both unweathered and weathered samples have positive correlations between Cu and Zn contents, indicating similar behaviors of Cu and Zn during the formation and weathering of black shales. Compared with the unweathered samples, the weathered shales and cherts have lower concentrations and heavier isotopic compositions of both Cu and Zn. The large variation of Cu isotopic composition reflects multistage redox cycling during the weathering processes. Our results indicate that extreme Cu isotope fractionation can be generated during weathering of Cu-rich shales and both heavy Cu and Zn isotopes are preferentially released into fluids during shale weathering. Typically, ~0.1‰ Zn isotope fractionation can be generated during the weathering of shales. These results should be considered when evaluating geochemical cycling of Cu and Zn in the modern or past oceans, especially the riverine flux and isotopic composition of Cu and Zn. The metalrich black shales, as an important sink from the oceans, also need to be considered when establishing the oceanic mass balance.

Acknowledgements

The authors would like to thank D. Li and Z.-Z. Wang for their assistance in the lab. We thank Profs. J.-B. Chen and F. Huang for providing the AAS and JMC Zn solutions and for their helpful discussion. We thank R. Mathur and two anonymous reviewers for their constructive comments and the editor for handling, which significantly improved the manuscript. This work is supported by the National Natural Foundation of China (41473017, 41203013) and the Fundamental Research Funds (2-9-2015-299) to SAL.

Appendix A. Supplementary data

Supplementary data to this article can be found online at http://dx. doi.org/10.1016/j.chemgeo.2015.12.017.

References

Adachi, M., Yamamoto, K., Sugisaki, R., 1986. Hydrothermal chert and associated siliceous rocks from the northern Pacific their geological significance as indication of ocean ridge activity. Sediment. Geol. 47 (1), 125-148.

Algeo, T.J., Maynard, J.B., 2004. Trace-element behavior and redox facies in core shales of Upper Pennsylvanian Kansas-type cyclothems. Chem. Geol. 206 (3), 289-318.

Amiotte Suchet, P., Probst, J.L., Ludwig, W., 2003. Worldwide distribution of continental rock lithology: implications for the atmospheric/soil CO2 uptake by continental weathering and alkalinity river transport to the oceans. Glob. Biogeochem. Cycles

17 (2).

Andersen, M.B., Vance, D., Archer, C., Anderson, R.F., Ellwood, M.J., Allen, C.S., 2011. The Zn abundance and isotopic composition of diatom frustules, a proxy for Zn availability in ocean surface seawater. Earth Planet. Sci. Lett. 301 (1-2), 137-145.

Archer, C., Vance, D., 2002. Large fractionations in Fe, Cu and Zn isotopes associated with Archean microbially-mediated sulphides. Geochim. Cosmochim. Acta 66, A26.

Asael, D., Matthews, A., Bar-Matthews, M., Halicz, L., 2007. Copper isotope fractionation in sedimentary copper mineralization (Timna Valley, Israel). Chem. Geol. 243 (3-4), 238-254.

Balistrieri, L.S., Borrok, D.M., Wanty, R.B., Ridley, W.I., 2008. Fractionation of Cu and Zn isotopes during adsorption onto amorphous Fe (III) oxyhydroxide: experimental mixing. of acid rock drainage and ambient river water. Geochim. Cosmochim. Acta 72 (2), 311-328.

Berner, Z.A., Puchelt, H., Noeltner, T., Kramar, U., 2013. Pyrite geochemistry in the Toarcian Posidonia Shale of south - west Germany: evidence for contrasting trace - element patterns of diagenetic and syngenetic pyrites. Sedimentology 60 (2), 548-573.

Brimhall, G.H., Dietrich, W.E., 1987. Constitutive mass balance relations between chemical composition, volume, density, porosity, and strain in metasomatic hydrochemical systems: results on weathering and pedogenesis. Geochim. Cosmochim. Acta 51 (3), 567-587.

Bruland, K.W., Lohan, M.C., 2006. Controls of trace metals in seawater. The Oceans and Marine Geochemistry. Elsevier, pp. 23-47.

Brumsack, H.-J., 2006. The trace metal content of recent organic carbon-rich sediments: implications for Cretaceous black shale formation. Palaeogeogr. Palaeoclimatol. Palaeoecol. 232 (2-4), 344-361.

Chapman, J.B., Mason, T.F., Weiss, D.J., Coles, B.J., Wilkinson, J.J., 2006. Chemical separation and isotopic variations of Cu and Zn from five geological reference materials. Geostand. Geoanal. Res. 30 (1), 5-16.

Chen, J.-B., Louvat, P., Gaillardet, J., Birck, J.-L., 2009. Direct separation of Zn from dilute aqueous solutions for isotope composition determination using multi-collector ICPMS. Chem. Geol. 259 (3-4), 120-130.

Chen, H., Savage, P.S., Teng, F.-Z., Helz, R.T., Moynier, F., 2013. Zinc isotope fractionation during magmatic differentiation and the isotopic composition of the bulk Earth. Earth Planet. Sci. Lett. 369-370 (0), 34-42.

Ehrlich, S., Butler, I., Halicz, L., Rickard, D., Oldroyd, A., Matthews, A., 2004. Experimental study of the copper isotope fractionation between aqueous Cu (II) and covellite, CuS. Chem. Geol. 209 (3), 259-269.

Fan, H., Wen, H., Hu, R., 2008. Enrichment of multiple elements and depositional environment of selenium-rich deposits in Yutangba, Western Hubei. Acta Sedimentol. Sin. 26 (2), 271.

Farbiszewska-Kiczma, J., Farbiszewska, T., Bąk, M., 2004. Bioleaching of metals from polish black shale in neutral medium. Physicochem. Probl. Miner. Process. 38, 273-280.

Fernandez, A., Borrok, D.M., 2009. Fractionation of Cu, Fe, and Zn isotopes during the oxidative weathering of sulfide-rich rocks. Chem. Geol. 264 (1), 1-12.

Fleet, A., 1983. Hydrothermal and hydrogenous ferro-manganese deposits: do they form a continuum? The rare earth element evidence. Hydrothermal Processes at Seafloor Spreading Centers 535-555. Springer, US.

Gélabert, A., Pokrovsky, O., Viers, J., Schott, J., Boudou, A., Feurtet-Mazel, A., 2006. Interaction between zinc and freshwater and marine diatom species: surface complexation and Zn isotope fractionation. Geochim. Cosmochim. Acta 70 (4), 839-857.

Gromet, L.P., Haskin, L.A., Korotev, R.L., Dymek, R.F., 1984. The "North American shale composite": its compilation, major and trace element characteristics. Geochim. Cosmochim. Acta 48 (12), 2469-2482.

Holland, H.D., 1979. Metals in black shales; a reassessment. Econ. Geol. 74 (7), 1676-1680.

Huerta-Diaz, M.A., Morse, J.W., 1992. Pyritization of trace metals in anoxic marine sediments. Geochim. Cosmochim. Acta 56 (7), 2681-2702.

Huyck, H.L., 1990. When is a metalliferous black shale not a black shale. Metalliferous black shales and related ore deposits. Proc. 1989 US Working Group Meet. 254. IGCP, pp. 42-56

Jean, G.E., Bancroft, G.M., 1986. Heavy metal adsorption by sulphide mineral surfaces. Geochim. Cosmochim. Acta 50 (7), 1455-1463.

John, S.G., Conway, T.M., 2014. A role for scavenging in the marine biogeochemical cycling of zinc and zinc isotopes. Earth Planet. Sci. Lett. 394, 159-167.

John, S.G., Geis, R.W., Saito, M.A., Boyle, E.A., 2007. Zinc isotope fractionation during highaffinity and low-affinity zinc transport by the marine diatom Thalassiosira oceanica. Limnol. Oceanogr. 52 (6), 2710-2714.

Jouvin, D., Louvat, P., Juillot, F., Maréchal, C.N., Benedetti, M.F., 2009. Zinc isotopic fractionation: why organic matters. Environ. Sci. Technol. 43 (15), 5747-5754.

Kametaka, M., Takebe, M., Nagai, H., Zhu, S., Takayanagi, Y., 2005. Sedimentary environments of the Middle Permian phosphorite-chert complex from the northeastern Yangtze platform, China; the Gufeng Formation: a continental shelf radiolarian chert. Sediment. Geol. 174 (3), 197-222.

Lavergren, U., Åström, M.E., Bergbäck, B., Holmström, H., 2009. Mobility of trace elements in black shale assessed by leaching tests and sequential chemical extraction. Geochem-Explor. Environ. Anal. 9 (1), 71-79.

Li, D., Liu, S.-A., Li, S., 2015. Copper isotope fractionation during adsorption onto kaolinite: experimental approach and applications. Chem. Geol. 396, 74-82.

Liang, Q., Jing, H., Gregoire, D.C., 2000. Determination of trace elements in granites by inductively coupled plasma mass spectrometry. Talanta 51 (3), 507-513.

Little, S.H., Vance, D., Walker-Brown, C., Landing, W.M., 2014. The oceanic mass balance of copper and zinc isotopes, investigated by analysis of their inputs, and outputs to ferromanganese oxide sediments. Geochim. Cosmochim. Acta 125 (0), 673-693.

Liu, S.-A., Teng, F.-Z., Li, S., Wei, G.-J., Ma, J.-L., Li, D., 2014a. Copper and iron isotope fractionation during weathering and pedogenesis: insights from saprolite profiles. Geochim. Cosmochim. Acta 146, 59-75.

Liu, S.-A., Li, D., Li, S., Teng, F.-Z., Ke, S., He, Y., Lu, Y., 2014b. High-precision copper and iron isotope analysis of igneous rock standards by MC-ICP-MS. J. Anal. At. Spectrom. 29 (1), 122-133.

Liu, S.-A., Huang, J., Liu, J.G., Wörner, G., Yang, W., Tang, Y.J., Chen, Y., Tang, L.-M., Zheng, J.P., Li, S.-G., 2015. Copper isotopic composition of the silicate Earth. Earth Planet. Sci. Lett. 427, 95-103.

Marchig, V., Gundlach, H., Möller, P., Schley, F., 1982. Some geochemical indicators for discrimination between diagenetic and hydrothermal metalliferous sediments. Mar. Geol. 50 (3), 241-256.

Maréchal, C.N., Télouk, P., Albarède, F., 1999. Precise analysis of copper and zinc isotopic compositions by plasma-source mass spectrometry. Chem. Geol. 156 (1), 251-273.

Maréchal, C.N., Nicolas, E., Douchet, C., Albarède, F., 2000. Abundance of zinc isotopes as a marine biogeochemical tracer. Geochem. Geophys. Geosyst. 1 (5).

Mason, T.F., Weiss, D.J., Horstwood, M., Parrish, R.R., Russell, S.S., Mullane, E., Coles, B.J., 2004. High-precision Cu and Zn isotope analysis by plasma source mass spectrometry Part 1. Spectral interferences and their correction. J. Anal. At. Spectrom. 19 (2), 209-217.

Mathur, R., Ruiz, J., Titley, S., Liermann, L., Buss, H., Brantley, S., 2005. Cu isotopic fractionation in the supergene environment with and without bacteria. Geochim. Cosmochim. Acta 69 (22), 5233-5246.

Mathur, R., Titley, S., Barra, F., Brantley, S., Wilson, M., Phillips, A., Munizaga, F., Maksaev, V., Vervoort, J., Hart, G., 2009. Exploration potential of Cu isotope fractionation in porphyry copper deposits. J. Geochem. Explor. 102 (1), 1-6.

Mathur, R., Jin, L., Prush, V., Paul, J., Ebersole, C., Fornadel, A., Williams, J., Brantley, S., 2012. Cu isotopes and concentrations during weathering of black shale of the Marcellus Formation, Huntingdon County, Pennsylvania (USA). Chem. Geol. 304, 175-184.

Nameroff, T.J., Balistrieri, L.S., Murray, J.W., 2002. Suboxic trace metal geochemistry in the Eastern Tropical North Pacific. Geochim. Cosmochim. Acta 66 (7), 1139-1158.

Nijenhuis, I.A., Bosch, H.J., Sinninghe Damsté, J.S., Brumsack, H.J., De Lange, G.J., 1999. Organic matter and trace element rich sapropels and black shales: a geochemical comparison. Earth Planet. Sci. Lett. 169 (3-4), 277-290.

Påhlsson, A.-M.B., 1989. Toxicity of heavy metals (Zn, Cu, Cd, Pb) to vascular plants. Water Air Soil Pollut. 47 (3-4), 287-319.

Palacios, C., Rouxel, O., Reich, M., Cameron, E.M., Leybourne, M.I., 2011. Pleistocene recycling of copper at a porphyry system, Atacama Desert, Chile: Cu isotope evidence. Mineral. Deposita 46 (1), 1-7.

Pichat, S., Douchet, C., Albarède, F., 2003. Zinc isotope variations in deep-sea carbonates from the eastern equatorial Pacific over the last 175 ka. Earth Planet. Sci. Lett. 210 (1), 167-178.

Pokrovsky, O., Viers, J., Freydier, R., 2005. Zinc stable isotope fractionation during its adsorption on oxides and hydroxides. J. Colloid Interface Sci. 291 (1), 192 200.

Pons, M.L., Quitté, G., Fujii, T., Rosing, M.T., Reynard, B., Moynier, F., Albarède, F., 2011. Early Archean serpentine mud volcanoes at Isua, Greenland, as a niche for early life. Proc. Natl. Acad. Sci. U. S. A. 108 (43), 17639-17643.

Pons, M.L., Fujii, T., Rosing, M., Quitté, G., Télouk, P., Albarède, F., 2013. A Zn isotope perspective on the rise of continents. Geobiology 11 (3), 201-214.

Rimmer, S.M., Thompson, J.A., Goodnight, S.A., Robl, T.L., 2004. Multiple controls on the preservation of organic matter in Devonian-Mississippian marine black shales: geochemical and petrographic evidence. Palaeogeogr. Palaeoclimatol. Palaeoecol. 215 (1), 125-154.

Rona, P.A., 1978. Criteria for recognition of hydrothermal mineral deposits in oceanic crust. Econ. Geol. 73 (2), 135-160.
Simpson, S.L., Rosner, J., Ellis, J., 2000. Competitive displacement reactions of cadmium, copper, and zinc added to a polluted, sulfidic estuarine sediment. Environ. Toxicol. Chem. 19 (8), 1992-1999.
Sossi, P.A., Halverson, G.P., Nebel, O., Eggins, S.M., 2015. Combined separation of Cu, Fe and Zn from rock matrices and improved analytical protocols for stable isotope determination. Geostand. Geoanal. Res. 39, 129-149.
Tribovillard, N., Algeo, T.J., Lyons, T., Riboulleau, A., 2006. Trace metals as paleoredox and paleoproductivity proxies: an update. Chem. Geol. 232 (1-2), 12-32.
Wall, A.J., Mathur, R., Post, J.E., Heaney, P.J., 2011. Cu isotope fractionation during bornite dissolution: an in situ X-ray diffraction analysis. Ore Geol. Rev. 42 (1), 62-70.
Wedepohl, K., 1971. Environmental influences on the chemical composition of shales and clays. Phys. Chem. Earth 8, 305-333.
Weiss, D.J., Boye, K., Caldelas, C., Fendorf, S., 2014. Zinc isotope fractionation during early dissolution of biotite granite. Soil Sci. Soc. Am. J. 78 (1), 171-179.
Wildman, R.A., Berner, R.A., Petsch, S.T., Bolton, E.W., Eckert, J.O., Mok, U., Evans, J.B., 2004. The weathering of sedimentary organic matter as a control on atmospheric O2: I. Analysis of a black shale. Am. J. Sci. 304 (3), 234-249.
Woo, N., Choi, M., Lee, K., 2002. Assessment of groundwater quality and contamination from uranium-bearing black shale in Goesan-Boeun areas, Korea. Environ. Geochem. Health 24 (3), 264-273.
Yang, G., Wang, S., Zhou, R., Sun, S., 1983. Endemic selenium intoxication of humans in China. Am. J. Clin. Nutr. 37 (5), 872-881.
Yao, L., Gao, Z., Yang, Z., Long, H., 2002. Origin of seleniferous cherts in Yutangba Se deposit, southwest Enshi, Hubei Province. Sci. China Ser. D Earth Sci. 45 (8), 741-754.
Zhu, J.-M., Johnson, T.M., Clark, S.K., Zhu, X.-K., Wang, X.-L., 2014. Selenium redox cycling during weathering of Se-rich shales: a selenium isotope study. Geochim. Cosmochim. Acta 126, 228-249.

Magnesium isotope fractionation during dolostone weathering*

Ting Gao[1], Shan Ke[2], Fang-Zhen Teng[2], Shouming Chen[1], Yongsheng He[1] and Shu-Guang Li[1]

1. State Key Laboratory of Geological Processes and Mineral Resources, School of Earth Science and Resources, China University of Geosciences, Beijing 100083, China
2. Isotope Laboratory, Department of Earth and Space Sciences, University of Washington, Seattle, WA 98195, USA

亮点介绍：本文发现了白云岩强风化阶段，以方解石和白云石溶解为主，风化残余具有非常重的 Mg 同位素组成；弱风化阶段，则伴随着方解石再沉淀和极少量白云石溶解，导致风化残余 Mg 同位素偏轻。该特点有望对风化环境和气候具有一定的指示意义，也为利用白云岩 Mg 同位素组成进行古环境演化提出了警示，必须识别和扣除风化的影响。

Abstract The element, Mg and Sr isotope ratios of a weathering profile from Hubei, China have been measured in order to document the behavior of Mg isotopes during dolostone weathering. The profile is developed in the joint system, and the weathering intensity increases from the least weathered dolostone towards the joint plane. According to the element and isotope ratios, the weathering profile can be divided into the weakly and intensely weathered zones. In the weakly weathered zone, Mg/Al (atomic) ratios decrease from 4.99 to 1.75 with increasing weathering intensity, while Ca/Al ratios slightly decrease from 13.70 to 9.02. These suggest massive loss of Mg and simultaneous conservation of Ca. δ^{26}Mg values in the weakly weathered zone slightly decrease from −1.90‰ to −2.22‰, lighter than the corresponding protolith (−1.90‰). These element and isotope variations are likely caused by the cooccurrence of dolomite dissolution and calcite re-precipitation, suggesting that the fluids co-existing with the dolostone are saturated with respect to calcite while not saturated with respect to dolomite. By contrast, both Mg/Al and Ca/Al ratios of intensely weathered samples display a decreasing trend towards the joint plane, ranging from 1.62 to 0.16 and 10.24 to 2.05, respectively. The significant loss of Mg and Ca indicates the considerable dissolution of both dolomite and calcite. δ^{26}Mg and ^{87}Sr/^{86}Sr values increase from −2.22‰ to −0.41‰ and from 0.71128 to 0.71368, respectively. These observations demonstrate that the isotope compositions of the intensely weathered residues are dominantly controlled by the increasing silicate fractions after carbonate dissolution, given that silicates have considerably high ^{87}Sr/^{86}Sr and δ^{26}Mg values relative to the carbonates. Overall, our study reveals large Mg isotope fractionation during dolostone weathering. The direction and extent of Mg isotope fractionation are mainly governed by the dissolution and re-precipitation of carbonate minerals, which are further controlled by the fluid's saturation with respect to carbonate minerals (calcite and dolomite). Therefore,

* 本文发表在：Chemical Geology，2016，445：14-23

cautions should be taken when using whole rock Mg isotope data of dolostones to trace paleoclimate change and dolostone genesis. Primary dolomite could be partially reserved in weakly weathered dolostones and step-leaching experiments may help to reveal its bearing Mg isotope signals.

Keywords　Mg isotopes, dolostone weathering, dolomite dissolution, calcite re-precipitation, saturation

1　Introduction

Magnesium isotope composition of dolostone has been increasingly employed to constrain the evolution of seawater chemistry, paleoclimate environments and the origin of dolostone (Pokrovsky et al., 2011; Kasemann et al., 2014; Liu et al., 2014a; Geske et al., 2015a, 2015b; Huang et al., 2015; Husson et al., 2015) due to the following geochemical characteristics. Firstly, a large fractionation between ^{26}Mg and ^{24}Mg is expected at low temperatures due to ~8% relative mass difference, which has been shown between natural/synthetic inorganic/biogenic carbonates and solutions ($\Delta^{26}Mg = \delta^{26}Mg_{carbonate} - \delta^{26}Mg_{solution} = -1$ to -3‰, e.g. Higgins and Schrag, 2010; Immenhauser et al., 2010; Wombacher et al., 2011; Li et al., 2012b, 2015; Mavromatis et al., 2013; Wang et al., 2013; Saenger and Wang, 2014). Secondly, Mg is a major element in dolomite, and the potential Mg sources for dolostone formation, including carbonate sediments, terrestrial silicates, and seawater, have very distinct Mg isotope compositions (Tipper et al., 2006b; Li et al., 2012b; Huang et al., 2013; Wang et al., 2015). Thirdly, Mg isotope composition of the modern seawater is homogeneous (Foster et al., 2010; Ling et al., 2011), consistent with the long residence time (~ 13 Ma) of Mg in the ocean (Li, 1982). Finally, the change of Mg isotope composition in marine carbonates on million-year scale is controlled by isotope composition of Mg in the oceans. The main source of Mg to the oceans is the weathering of carbonates and silicate rocks. These characteristics make Mg isotopes, combined with other stable isotopes (e.g. carbon, boron) and radiogenic strontium isotopes, potentially be an excellent tracer for chemical evolution of oceans and for paleoclimate environments, such as sea regression and transgression (Pokrovsky et al., 2011), ocean acidification event (Kasemann et al., 2014), seawater and climate evolution during post-'Snowball Earth' event (Liu et al., 2014a; Husson et al., 2015). In addition, Mg isotopes may provide a promising insight into the origin of dolostone (Geske et al., 2015a, 2015b; Huang et al., 2015).

Nonetheless, dolostone can be altered due to recrystallization and chemical weathering (Land, 1985; Machel, 1997; Al-Aasm and Packard, 2000; Warren, 2000), which are usually accompanied by the element migration (e.g. Mg, Ca and Sr) and possible isotope fractionation (e.g. Mg, O and C; Land, 1985; Montanez and Read, 1992; Malone et al., 1994; Yoo and Lee, 1998; Fantle and Higgins, 2014). Knowledge on Mg isotope fractionation during diagenesis and weathering processes thus is a prerequisite to use Mg isotope as a proxy to reconstruct paleoclimate environments and study the origin of dolostone. Thus far, only a few studies have been carried out to constrain Mg isotope fractionation during diagenesis of dolostone and the results are debated. Fantle and Higgins (2014) showed that limestone diagenesis and dolomitization significantly fractionate Mg isotopes, implying that the diagenesis of carbonates may greatly alter their $\delta^{26}Mg$ values. By contrast, Geske et al. (2012) and Azmy et al. (2013) found that diagenesis would alter the texture of the dolomite while the Mg isotope signals remain intact. With respect to Mg isotope fractionation during dolostone weathering, only one study has been carried out. Jacobson et al. (2010) investigated the fractionation mechanism during dedolomitization reactions such as dolomite dissolution, calcite precipitation and Mg-for-Na ion-exchange, mainly based on water samples. They concluded that dolomite recrystallization and calcite precipitation did not significantly fractionate Mg isotopes, while the preferential uptake of ^{24}Mg during Mg-for-Na ion-exchange was responsible for the observed isotope variation. In order to understand Mg isotope fractionation during dolostone weathering and its relevant controlling mechanisms, more case studies on natural weathering profiles are thus needed.

Here, we report Mg isotope data from a well-exposed weathering profile that developed on the Ca-dolostone from Cambrian strata in Changyang county of Hubei, China. The weathering intensity along the profile changes gradually. Thus the profile is well suitable for systematical investigation of the behaviors of Mg isotopes during dolostone weathering. We found large (up to 1.8‰) Mg isotope variation along the weathering profile and both the direction and extent of fractionation vary with the weathering intensity. The fluid's saturation with respect to carbonate minerals (calcite and dolomite) dominates the dissolution and re-precipitation of carbonates and thus controls the Mg isotope fractionation.

2 Samples

The profile studied here is from the lower Cambrian Shuijingtuo Formation in Xijiaao village, Changyang county, Hubei Province, South China (111°08′35.41″E, 30°31′49.82″N; Fig. 1A). This area is predominately covered by carbonates and shales. Carbonate rocks in this area are extensively developed with joints (Fig. 1B). The dolostone weathering profile is ~17 cm long (the least weathered zone is defined as 0 cm). The weathering degree increases gradually from the least weathered end towards joint plane with the color from black to yellow (Fig. 2). Petrographic study shows that the least weathered dolostone consists of dolomite, calcite and small amounts

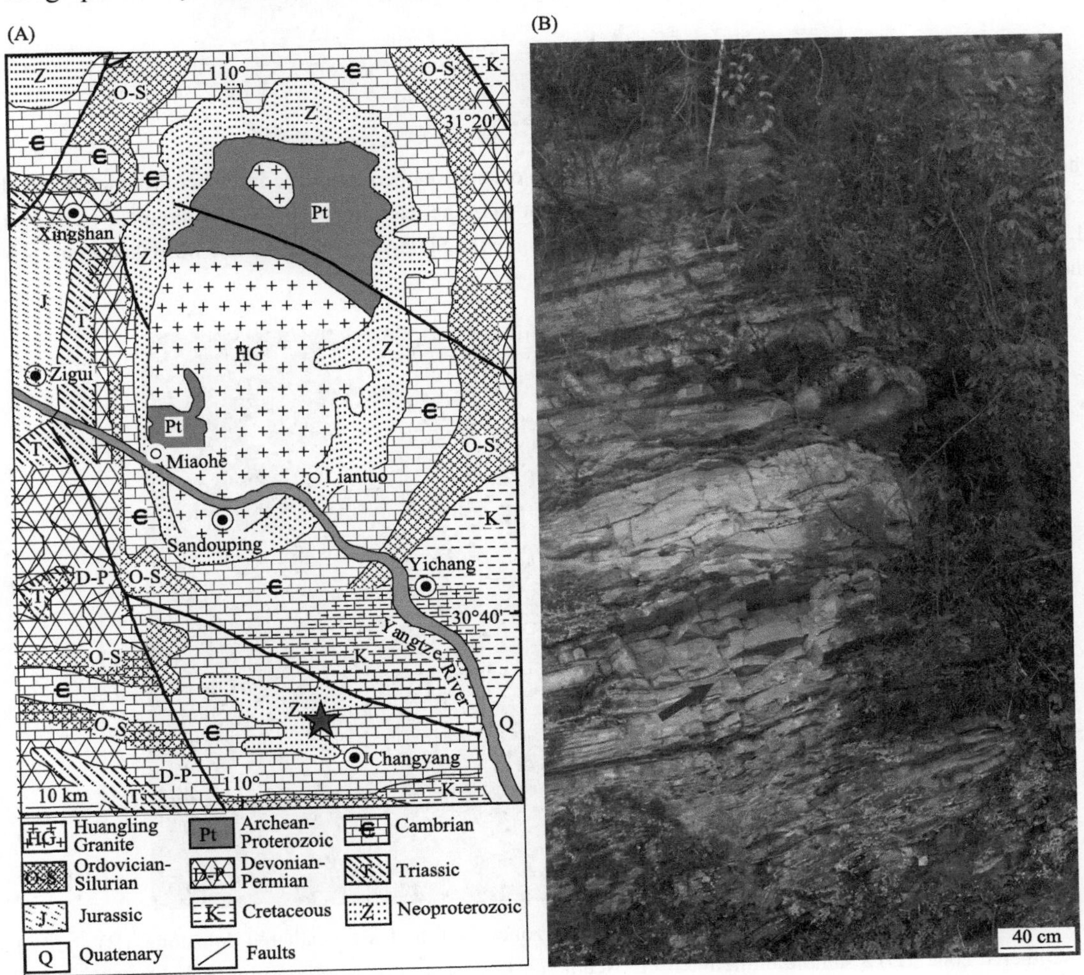

Fig. 1 (A) Geological map of the study area in Hubei Province, South China (modified after Jiang et al., 2012). The star represents the study area. (B) The photo of the weathering profile in the field. The arrow represents the sample location.

of silicates (dolomite + calcite = 90–95%, silicates = 5–10%, Fig. 3A). Alteration and carbonate recrystallization are not widely present in the least weathered dolostone. With the enhancement of weathering, the proportion of carbonate minerals decreases from 90–95% to 5–10% and that of silicate minerals increases from 5–10% to 90–95% (Fig. 3B, C, D). Besides, the grain diameter of carbonate minerals reduces and the clayzation becomes more intense on the mineral surface.

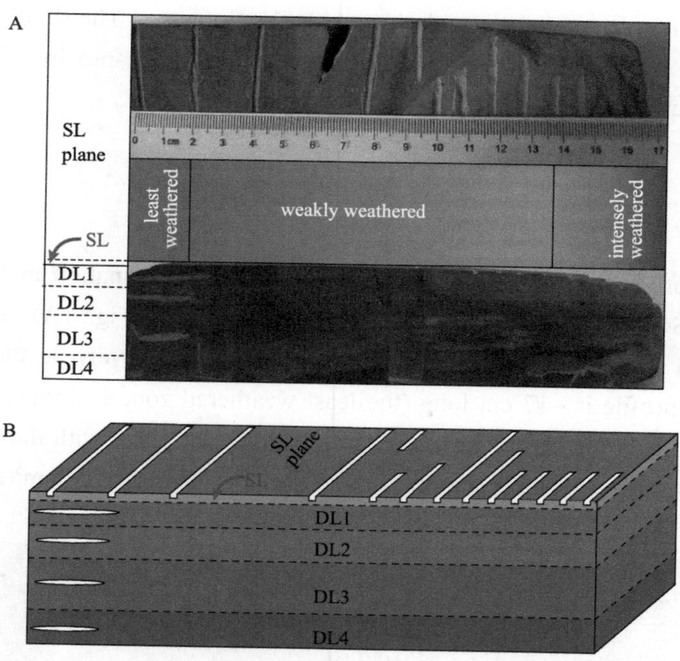

Fig. 2　Photograph of the specimen (A) and sketch (B) of the dolostone weathering profile in Hubei Province of South China. SL represents the sampling layer. 14 samples were successively taken from the least weathered end to the most weathered end in this layer. DL1, DL2, DL3 and DL4 represent different layers paralleling to the SL layer, and one sample was taken from each layer in the least weathered end.

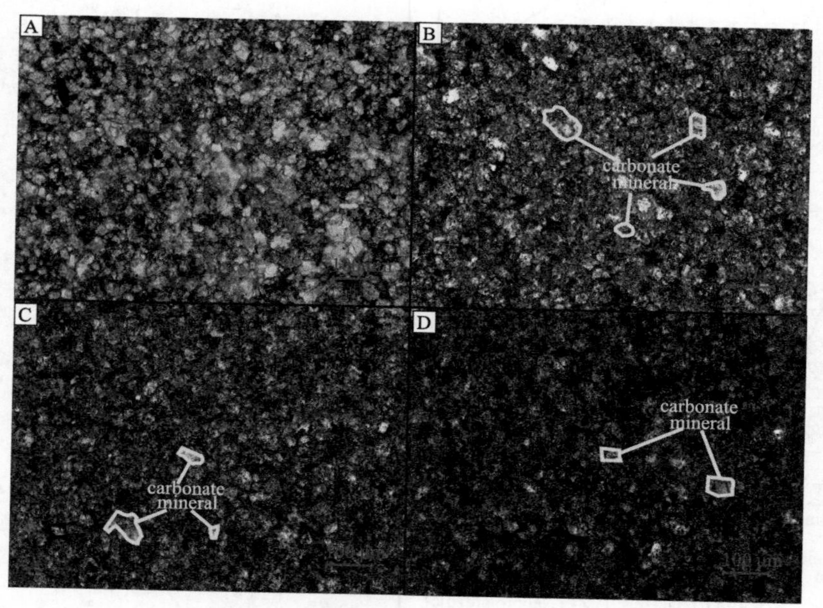

Fig. 3　Photomicrographs of the samples in different weathering zones. (A) the least weathered dolostone; (B) and (C) samples in the weakly weathered zone; (D) a sample in the intensely weathered zone. The relative proportions of each section are described as follows: (A) carbonates = 90–95%, silicates = 5–10%; (B) carbonates = 80–85%, silicates = 15–20%; (C) carbonates = 75–80%, silicates = 20–25%; (D) carbonates = 5–10%, silicates = 90–95%.

SL, DL1, DL2, DL3 and DL4 are five different deposition layers. To avoid the contamination from other deposition layers, fourteen samples (HY-PM3-3 to HY-PM3-16) were drilled from the same deposition layer (SL) (Fig. 2). To compare the geochemical composition of different deposition layers, another four samples (HY-1, HY-2, HY-3 and HY-4) were collected from four different deposition layers (DL1, DL2, DL3 and DL4) in the least weathered end (Fig. 2).

3 Analytical methods

The analyses of elemental ratios, Mg and Sr isotopes were performed at the Isotope Laboratory of the China University of Geosciences, Beijing. The procedures for sample digestion, column chemistry and instrumental analysis are described below.

3.1 Sample digestion

Based on the Mg content of samples, 1–10 mg of rock power was weighted in Savillex screw-top beakers and was digested by using a combination of single-distilled HCl, HF, HNO_3 and $HClO_4$. The digestion procedures are as follows: (1) sample powders were dissolved by concentrated HF-HNO_3 (3:1, v/v) as well as 3–4 drops of $HClO_4$ at 140 ℃ for 1–2 days, and then dried at 190 ℃; (2) the residues were refluxed by concentrated HCl-HNO_3 (3:1, v/v) and concentrated HNO_3 at 140 ℃ consequently; (3) three solution splits were prepared in 3% HNO_3, 1 N HNO_3, and 2.5 N HCl for elemental ratio analysis, Mg and Sr isotope column chemistry, respectively.

3.2 Elemental ratio (Mg/Al, Ca/Al and Mg/Ca) analysis

Magnesium, Ca and Al concentrations were determined and monitored by ^{24}Mg, ^{27}Al and ^{44}Ca intensities on a Neptune plus MC-ICP-MS. Each measurement comprises of 10 cycles. A 400 ppb Mg, Al and Ca mixed solution, prepared by pure single element standard solutions from "GSB standard", was used for calibration. The primary data are listed in Appendix A. Elemental ratios can be calculated by using the following formula:

$$M / N = (M_{smp} / N_{smp}) \times (N_{std} / M_{std})$$

M and N refer to intensities of two monitoring isotopes; smp and std refer to sample and standard, respectively.

3.3 Magnesium isotope analysis

Magnesium was purified for isotope analysis through cation exchange chromatograph with Bio-Rad 200–400 mesh AG50W-X8 resin (Ke et al., 2016), by following previously established methods (Teng et al., 2007, 2010a, 2015; Yang et al., 2009; Li et al., 2010; Liu et al., 2010). The resin was cleaned by 6 N HCl, 1 N HNO_3, 1 N HF and Milli-Q (18.2 MΩ) water alternatively before loading samples. Magnesium was eluted by 1 N HNO_3 and the Mg yield was between 99.5 and 99.9% based on the analyses of pure Mg solutions as well as reference materials (BHVO-2, BCR-2 and AGV-2). The whole procedure blank is <10 ng, having a negligible contribution to samples. To obtain a sufficiently pure Mg fraction, each sample was passed through the column twice.

Magnesium isotope compositions were measured by the standard-sample bracketing method using a Neptune Plus MC-ICP-MS. A "wet" sample introduction system, equipped with a quartz spray chamber and a 50 μl/min PFA nebulizer, and low-resolution mode were adopted for Mg isotope analysis. Samples were introduced as 400 ppb solutions in 3% HNO_3 and concentrations of samples and standards are matched within ±10%. A 400 ppb solution typically yields beam intensity of ~6 V for ^{24}Mg, with background Mg signals of $<10^{-4}$ V. The standard-sample sequence was repeated 4 times for each sample to get a better reproducibility. Magnesium isotope ratios of samples are expressed as deviations between sample and standard DSM3 (Galy et al., 2003):

$$\delta^{x}\text{Mg} = \left[\left({}^{x}\text{Mg}/{}^{24}\text{Mg}\right)_{\text{sample}} / \left({}^{x}\text{Mg}/{}^{24}\text{Mg}\right)_{\text{DSM3}} - 1\right] \times 1000$$

where x refers to mass 25 or 26. Long term external reproducibility is 0.06‰ for $\delta^{26}Mg$ (2SD; Ke et al., 2016). Duplicated measurement of the seawater standard (Hawaiian seawater) yields results (average $\delta^{26}Mg=-0.83\pm 0.03$‰, 2SD, n=2) consistent with the previously reported values (Foster et al., 2010; Ling et al., 2011; Teng et al., 2015).

3.4 Strontium isotope analysis

Strontium was purified by cation exchange chromatography with columns filled with 2 ml Bio-Rad AG50W-X12 resin (200–400 mesh). Before separation, the resin was cleaned by 20 ml 6 N HCl and 5 ml Milli-Q water. Samples were loaded on 0.5 ml 2.5 N HCl, followed by 1.5 ml 2.5 N HCl and 15 ml 4 N HCl. Thereafter, Sr was collected by 7 ml 4 N HCl and 4 ml 6 N HCl. Strontium isotope ratios were measured on the Neptune Plus MC-ICP-MS. Each measurement comprises 30 s baseline measurement at half mass positions (85.5 and 86.5) and 160 cycles of 4.198 s integration. ^{85}Rb and ^{83}Kr were monitored for the correction of the ^{87}Rb contribution on ^{87}Sr and ^{86}Kr on ^{86}Sr. Mass bias correction was corrected against $^{88}Sr/^{86}Sr$ = 8.375209, assuming an exponential law. The measured $^{87}Sr/^{86}Sr$ of NIST SRM987 during this measurement has an average value of 0.71028 ± 0.000031 (2SD, n=3), showing good agreement with published data (e.g. Li et al., 2012a and references therein).

4 Results

Elemental ratios (Mg/Al, Ca/Al and Mg/Ca), Mg and Sr isotope compositions are listed in Table 1 for all the samples.

Table 1 Elemental ratios and isotope compositions of standards and weathered dolostone samples from Hubei, China

Layer ID	Sample ID	Distance (cm)	Mg/Ca	Mg/Al	Ca/Al	$^{87}Sr/^{86}Sr$	$\delta^{25}Mg$(‰)	2SD	$\delta^{26}Mg$ (‰)	2SD
	HY-PM3-3*	0.0	0.46	5.64	12.22	0.71091	−1.02	0.04	−1.90	0.04
	HY-PM3-4	2.0	0.44	4.47	10.11	0.71118	−0.99	0.03	−1.90	0.03
	HY-PM3-5	4.0	0.40	4.99	12.54	0.71097	−0.98	0.08	−1.92	0.08
	HY-PM3-6	7.7	0.24	3.02	12.42	0.71161	−1.04	0.05	−2.02	0.05
	HY-PM3-7	9.4	0.32	3.41	10.77	0.71152	−1.02	0.05	−1.99	0.05
	HY-PM3-8	10.0	0.21	2.86	13.70	0.71130	−1.10	0.08	−2.11	0.08
SL	HY-PM3-9	10.7	0.21	2.78	12.98	0.71129	−1.07	0.04	−2.06	0.04
	HY-PM3-10	11.7	0.21	2.34	11.13	0.71125	−1.12	0.06	−2.17	0.06
	HY-PM3-11	12.7	0.19	1.75	9.02	0.71137	−1.15	0.06	−2.22	0.06
	HY-PM3-12	13.7	0.16	1.62	10.24	0.71128	−1.14	0.06	−2.22	0.06
	HY-PM3-13	14.7	0.13	1.04	8.06	0.71143	−1.12	0.01	−2.17	0.03
	HY-PM3-14	15.7	0.12	0.28	2.23	0.71353	−0.73	0.01	−1.41	0.01
	HY-PM3-15	16.2	0.07	0.16	2.22	0.71249	−0.21	0.03	−0.41	0.01
	HY-PM3-16	16.7	0.09	0.17	2.05	0.71368	−0.34	0.07	−0.64	0.03

Continued

Layer ID	Sample ID	Distance (cm)	Mg/Ca	Mg/Al	Ca/Al	$^{87}Sr/^{86}Sr$	$\delta^{25}Mg$(‰)	2SD	$\delta^{26}Mg$ (‰)	2SD
DL1	HY-1	/	0.41	6.00	14.58	0.71088	−1.01	0.04	−1.98	0.05
DL2	HY-2	/	0.11	2.88	25.83	0.70971	−1.06	0.09	−2.08	0.03
DL3	HY-3	/	0.34	8.15	23.88	0.71018	−0.83	0.03	−1.61	0.10
DL4	HY-4	/	0.30	3.77	12.38	0.71095	−1.00	0.09	−1.95	0.07
	Seawater	/	/	/	/	/	−0.43	0.07	−0.84	0.07
	Seawater	/	/	/	/	/	−0.40	0.07	−0.82	0.06
	Average	/	/	/	/	/	−0.42	0.07	−0.83	0.07
/	SRM987	/	/	/	/	0.71029	/	/	/	/
	SRM987	/	/	/	/	0.71028	/	/	/	/
	SRM987	/	/	/	/	0.71026	/	/	/	/
	Average	/	/	/	/	0.71028	/	/	/	/

$\delta^{x}Mg = [(^{x}Mg/^{24}Mg)_{sample} / (^{x}Mg/^{24}Mg)_{DSM3} - 1] \times 1000$, where x = 25 or 26 and DSM3 is Mg solution made from pure Mg metal (Galy et al., 2003). 2SD = 2 times the standard deviation of population analyzed. "*" represents the least weathered dolostone. "/" represents the inexistence or no measurements.

4.1 Compositions of the different deposition layers

The samples from the five different deposition layers (SL, DL1, DL2, DL3, DL4) represent five deposition characteristics and have different element and isotope compositions (Fig. 4). All the weathering profile samples were collected from the surface layer (SL). The least weathered dolostone from the SL has the Mg/Ca, Mg/Al and Ca/Al ratios of 0.46, 5.64 and 12.22, respectively. $\delta^{26}Mg$ ratio is −1.90‰ and falls in the range of $\delta^{26}Mg$ ratios of the typical dolostones (Pokrovsky et al., 2011; Azmy et al., 2013; Kasemann et al., 2014; Liu et al., 2014a; Geske et al., 2015a; Huang et al., 2015; Husson et al., 2015). $^{87}Sr/^{86}Sr$ ratio is 0.71091, higher than that of the seawater at 520−525 Ma (Kaufman et al., 1996; McArthur et al., 2012), reflecting the contribution of silicate fractions.

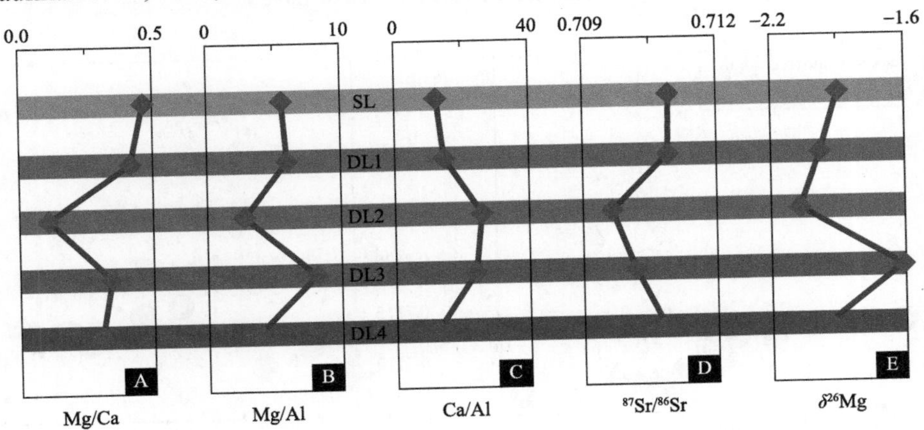

Fig. 4 The elemental ratios (Mg/Ca, Mg/Al, Ca/Al) and isotope compositions ($^{87}Sr/^{86}Sr$, $\delta^{26}Mg$) of different deposition layers (SL, DL1, DL2, DL3 and DL4). Data are reported in Table 1.

The layers DL1 and SL have similar elemental ratios (Mg/Ca=0.41−0.46, Mg/Al=5.64−6.00, Ca/Al=12.22−14.58) and isotope compositions ($\delta^{26}Mg$=−1.98‰ to −1.90‰, $^{87}Sr/^{86}Sr$=0.71088−0.71091) (Fig. 4). By contrast, DL2, DL3 and DL4 have rather different elemental ratios (Mg/Ca = 0.11−0.34, Mg/Al = 2.88−8.15, Ca/Al=12.38−25.83) and isotope compositions ($\delta^{26}Mg$=−1.61‰ to −2.08‰, $^{87}Sr/^{86}Sr$=0.70971−0.71095). These geochemical variations among different deposition layers suggest that it is important to collect samples within the same layer when studying Mg isotope fractionation mechanisms during dolostone weathering. All 14 samples of the weathering profile were strictly sampled along the layer SL to avoid any disturbance from the other deposition layers. Moreover, the layer SL has similar composition to the next layer DL1, which further minimizes the possible contamination effects.

4.2 Compositions of the weathering profile

Several immobile elements (e.g. Al, Ti, Zr and Nd) can be used to evaluate the relative depletion or enrichment of mobile elements during chemical weathering. Similar to a previous study (Rudnick et al., 2004), Al was chosen here because it is one of the least mobile elements and more prone to be enriched in weathered residues. The dolostone weathering profile is divided into two different weathering zones, namely weakly weathered and intensely weathered zones, based on elemental ratios (Mg/Al, Ca/Al and Mg/Ca, Fig. 5) and isotope compositions (δ^{26}Mg and ^{87}Sr/^{86}Sr, Fig. 6).

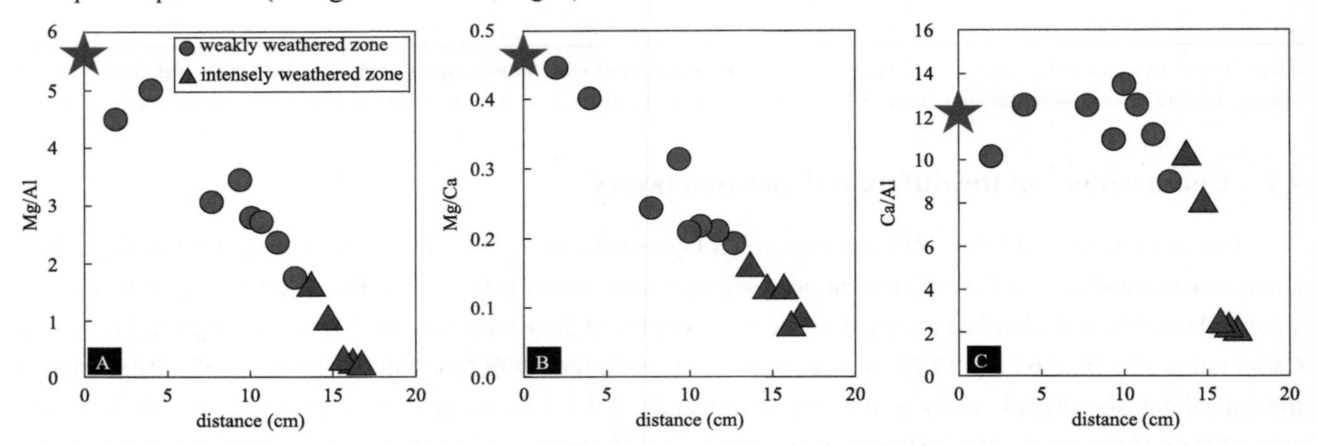

Fig. 5　Mg/Al (A), Mg/Ca (B), Ca/Al (C) as a function of distance in the dolostone weathering profile. The star represents the least weathered dolostone. Error bars represent 2SD uncertainties. Data are reported in Table 1.

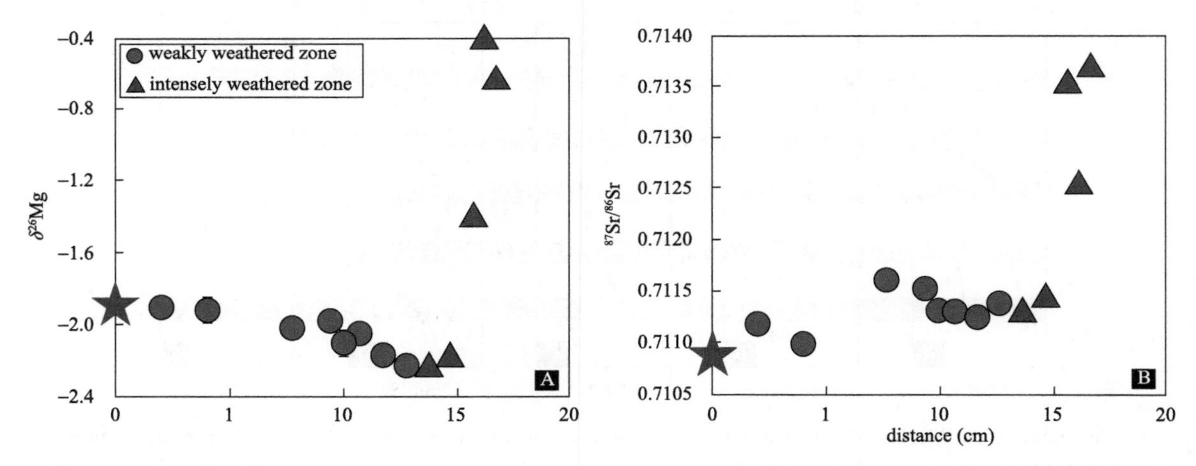

Fig. 6　δ^{26}Mg (A) and ^{87}Sr/^{86}Sr (B) as a function of distance in the dolostone weathering profile. The star represents the least weathered dolostone. Error bars represent 2SD uncertainties. Data are reported in Table 1.

In the weakly weathered zone, Mg/Al and Mg/Ca ratios are lower than the least weathered dolostone and show a decreasing trend with increasing weathering intensity, ranging from 4.99 to 1.75 and 0.44 to 0.19, respectively (Fig. 5A, B). By contrast, Ca/Al ratios slightly decrease from 13.70 to 9.02 (Fig. 5C). δ^{26}Mg values are lower than the least weathered dolostone and successively decrease from −1.90‰ to −2.22‰, while ^{87}Sr/^{86}Sr ratios slightly increase from 0.71097 to 0.71161 (Fig. 6).

In the intensely weathered zone, Mg/Al, Mg/Ca and Ca/Al ratios all decrease as weathering intensity increases, ranging from 1.62 to 0.16, 0.16 to 0.07 and 10.24 to 2.05, respectively (Fig. 5). δ^{26}Mg values and ^{87}Sr/^{86}Sr ratios increase significantly, varying from −2.22‰ to −0.41‰ and 0.71128 to 0.71368, respectively (Fig. 6).

5 Discussion

Chemical weathering is a process in which primary minerals are progressively dissolved and secondary minerals are formed. Mobile elements (e.g. Mg and Ca) during weathering are redistributed in primary minerals, secondary minerals and solutions (Nesbitt and Markovics, 1980). For example, the Mg/Al and Ca/Al ratios in the dolostone weathering profile here vary up to 97% and 83%, respectively, suggesting that nearly all Mg and most of Ca in the most intensely weathered samples have been removed to the solution and the rest of them may be absorbed or incorporated into secondary minerals. Large Mg isotope fractionation had been found during weathering of igneous rocks and shales (Tipper et al., 2006a, 2006b; Pogge von Strandmann et al., 2008; Teng et al., 2010b; Wimpenny et al., 2010; Ryu et al., 2011; Huang et al., 2012; Opfergelt et al., 2012, 2014; Pogge von Strandmann et al., 2012; Liu et al., 2014b; Ma et al., 2015). The dolostone weathering profile yields a δ^{26}Mg variation up to 1.81‰. Particularly, those two samples near the joint plane have rather heavy Mg isotope compositions (−0.41‰ to −0.64‰), which are close to those of igneous rocks (Teng et al., 2007, 2010a; Shen et al., 2009; Li et al., 2010; Yang et al., 2012). The large variations of elemental ratio and Mg isotope composition most likely reflect the dissolution of primary minerals and formation of secondary minerals. Below, we first discuss the behaviors of carbonate minerals and the nature of the fluids, and then reveal and model the Mg isotope fractionation mechanisms during dolostone weathering.

5.1 Behaviors of carbonate minerals during dolostone weathering

The weakly weathered zone is characterized by lower Mg/Al ratios (1.75 to 4.99) and slight variation of Ca/Al ratios (9.02 to 13.70) when compared to the least weathered dolostone (Fig. 5). The Mg/Al ratios decrease with increasing weathering intensity, suggesting that a large amount of initial Mg has been gradually lost as weathering progresses. For example, the sample HY-PM3-11 has the Mg/Al ratio of 1.75, indicating that approximately up to 70% of Mg was lost. Such huge loss of Mg suggests substantial dissolution of dolomite dissolution, the less variation of Ca/Al ratios and the significant decrease of Mg/Ca ratios indicate that calcite re-precipitation occurs simultaneously with dolomite dissolution to compensate Ca and keep Ca/Al ratios with less variation.

In contrast to the weakly weathered zone, both Mg/Al and Ca/Al ratios in the intensely weathered zone show a decreasing trend as weathering progresses, varying from 1.62 to 0.16 and from 10.24 to 2.05, respectively (Fig. 5). Again, dolomite dissolution is required to explain the decreasing Mg/Al ratios but it alone cannot explain the decrease in both Mg/Al and Ca/Al ratios. Even dolomite in the least weathered dolostone has been completely dissolved during weathering, only approximately 46% of the bulk Ca would be lost, which cannot explain the actual 80% loss of total Ca calculated by Ca/Al ratios (from 10.24 to 2.05) in the most intensely weathered samples. Therefore, calcite dissolution is also needed to further lower the Ca/Al ratios in the intensely weathered zone as calcite is the only dominant Ca hosting mineral other than dolomite. Dolomite and calcite dissolution are thus concurrent in the intensely weathered zone. These conclusions are consistent with the petrographic study that the content of carbonate minerals in the intensely weathered samples is much lower than the least weathered dolostone.

The fluid-dolostone reactions involve dissolution and re-precipitation of carbonate minerals depending on the saturation state of fluids with respect to carbonate minerals. The dolomite dissolution and calcite re-precipitation as suggested in the weakly weathered zone were also observed in Chinese Loess Plateau samples (Jeong et al., 2008; Wimpenny et al., 2014). The re-precipitation of secondary calcite is likely attributed to the saturation of fluids with respect to calcite given that large amounts of primary calcite and dolomite are dissolved causing high concentration of Ca^{2+} and CO_3^{2-} in fluids. In addition, this cooccurrence of dolomite dissolution and calcite re-precipitation occurs in carbonate regions where gypsum or anhydrite is present (Plummer and Back, 1980; Back et al., 1983; Jacobson and Wasserburg, 2005; Jacobson et al., 2010). The irreversible dissolution of gypsum or anhydrite would make the fluid reach saturation with respect to calcite and then drive the calcite re-precipitation. The calcite re-precipitation decreases the pH and removes CO_3^{2-} from fluids, thereby causing the dissolution of dolomite. As to the weathering profile in this study, no gypsum or anhydrite was present. Therefore, another process is needed to explain the behaviors of dolomite dissolution and calcite reprecipitation.

Jin et al. (2008) found that dolomite continuously dissolved after calcite equilibrium, because the solubility of dolomite is higher than calcite. Saturation index (SI) of the fluids with respect to calcite is defined as $SI = \log([Ca^{2+}] \times [CO_3^{2-}]/K_{calcite})$, where $[Ca^{2+}]$ and $[CO_3^{2-}]$ are activities of Ca^{2+} and CO_3^{2-}, and $K_{calcite}$ is the solubility constant of calcite. Saturation index of calcite is controlled by $K_{calcite}$ and concentrations of Ca^{2+} and CO_3^{2-}. The intensely weathered zone is near the joint plane. Fluids, like the rain waters, are poor in Ca^{2+}, Mg^{2+}, and CO_3^{2-} and thus could be undersaturated with respect to both dolomite and calcite, which could lead to substantial dissolution of dolomite and calcite in the intensely weathered zone close to the joint plane (Fig. 7). Most fluid after water-rock interaction may drain away along the joint and a small fraction of fluid infiltrates the dolostone because of the hydrostatic pressure, which would infiltrate the dolostone deeply and continue to dissolve carbonates. Massive dissolution of dolomite and calcite in the intensely weathered zone would concentrate Ca^{2+} and CO_3^{2-} in the fluids. This could drive the infiltration fluid saturated with respect to calcite in the boundary between the weakly weathered and intensely weathered zone, whereas keep undersaturated with respect to dolomite because of the higher equilibrium constant for dolomite than that for calcite at temperatures < 30 ℃ (Langmuir, 1997). Thus, dolomite could continue dissolving while calcite begins to re-precipitate in the weakly weathered zone. Once calcite re-precipitation begins, Ca^{2+} and CO_3^{2-} would be reduced, and thereby leading to a feedback of accelerating dolomite dissolution.

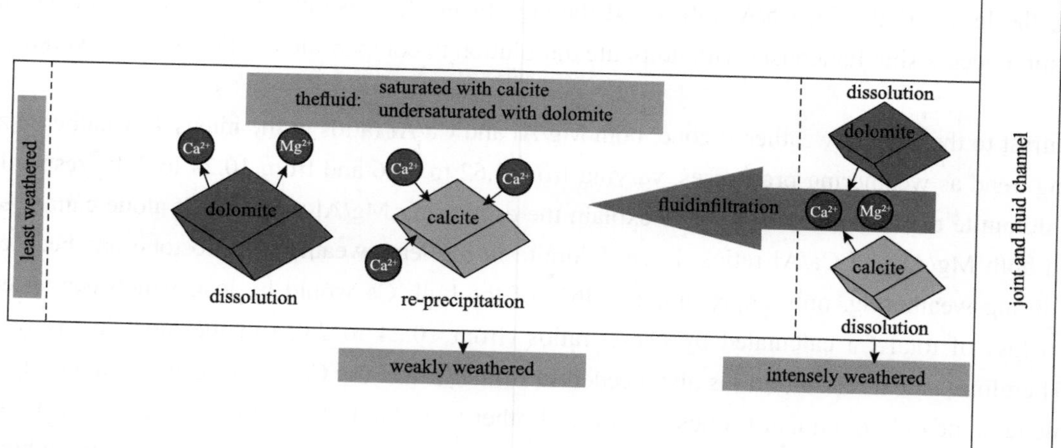

Fig. 7　Schematic diagram of carbonate mineral behaviors and fluid's properties along the dolostone weathering profile. In the intensely weathered zone, the fluids from the joint are undersaturated with respect to both dolomite and calcite. In the weakly weathered zone, the fluids become saturated with respect to calcite but undersaturated with respect to dolomite after interaction with dolomites and calcites in the intensely weathered zone.

5.2 Magnesium isotope fractionation during dolostone weathering

The chemical weathering of dolostone mainly involves the dissolution of primary carbonate minerals (e.g. dolomite and calcite) and formation of secondary minerals (e.g. calcite). These processes also control the Mg isotope fractionation in the dolostone weathering profile. The intensely and weakly weathered zones display different fractionation directions driven by different fractionation mechanisms. Below, we discuss the behaviors of Mg isotopes in these two zones, respectively.

5.2.1 The intensely weathered zone

The least weathered dolostone is mainly composed of carbonates (90–95%) with trace amounts of silicates (5–10%). The intensely weathered zone is characterized by great dissolution of both dolomite and calcite due to its direct contact with externally undersaturated fluids such as rain waters. As carbonates are more reactive and soluble than silicates (Horton et al., 1999; White et al., 2005), the carbonates to silicates ratio largely decreases with increasing weathering intensity. This is supported by the much higher $^{87}Sr/^{86}Sr$ ratios (0.71128–0.71368) of samples from the intensely weathered zone than the least weathered dolostone (0.71091). Since silicates have higher $\delta^{26}Mg$ values than carbonates (Huang et al., 2013; Wimpenny et al., 2014; Wang et al., 2015), $\delta^{26}Mg$ values of the weathered products therefore depend on the Mg proportion between carbonates and silicates as well as their $\delta^{26}Mg$ values. This interpretation agrees with the general increase of $\delta^{26}Mg$ (−2.22‰ to −0.41‰, Fig. 6) with the increasing degree of weathering. In particular, the two most intensely weathered samples (HY-PM3-15 and HY-PM3-16) have extremely heavy Mg isotope compositions up to −0.41‰ and −0.64‰ that are close to igneous rocks (Teng et al., 2007, 2010a; Shen et al., 2009; Li et al., 2010; Yang et al., 2012). This suggests the Mg budget of the whole rocks is dominated by silicates and a nearly complete loss of carbonates, which was also found in studies of loess and paleosol (Wimpenny et al., 2014).

5.2.2 The weakly weathered zone

In contrast to the intensely weathered zone, calcite re-precipitates simultaneously with dolomite dissolution in the weakly weathered zone. The positive correlation between $\delta^{26}Mg$ and Mg/Al (Fig. 8A) suggests that the heavy Mg isotopes are preferentially lost during chemical weathering. Four possible mechanisms could explain this fractionation of Mg isotopes: (1) preferential release of heavy Mg isotopes during dolomite dissolution; (2) preferential selection of ^{24}Mg over ^{26}Mg during Mg-for-Na ion-exchange; (3) decrease in ratio of dolomite to calcite; and (4) preferential incorporation of isotopically light Mg during secondary calcite precipitation.

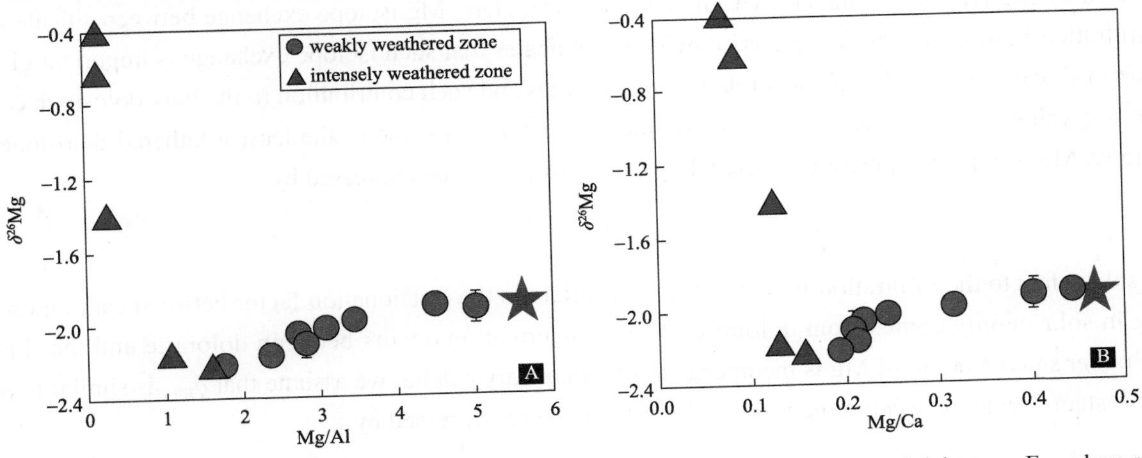

Fig. 8 Mg/Al (A) and Mg/Ca (B) as a function of $\delta^{26}Mg$. The star represents the least weathered dolostone. Error bars represent 2SD uncertainties. Data are reported in Table 1.

Previously published data indicated that dolomite tends to have lighter Mg isotope composition than its co-existing solution (Higgins and Schrag, 2010; Fantle and Higgins, 2014; Huang et al., 2015; Li et al., 2015). However, no isotope fractionation is expected during dissolution of dolomite as shown by previous studies. River waters draining dolostones were found to have Mg isotope composition similar to that of dolostones (Tipper et al., 2006b). Step leaching experiments revealed that the leacheates from dolostone have constant δ^{26}Mg values (Liu et al., 2014a). In this case, dolomite dissolution is not likely to be responsible for the isotope fractionation of Mg.

Mg for Na ion-exchange process has been tentatively called to explain Mg isotope variation in groundwater samples as fluids preferentially remove ^{24}Mg during Mg-for-Na ion-exchange (Jacobson et al., 2010). This process, however, cannot be the main process altering Mg isotope composition of dolostone given that (1) silicate content in dolostone is much lower than carbonate and (2) the abundance of Na in dolostone is extremely limited (Na/Al $\leq$0.06, see Appendix A) and (3) the abundance of Na in dolostone remain unchanged as weathering intensity changes.

Instead, δ^{26}Mg decline in the weakly weathered zone is likely caused by the decrease of dolomite to calcite ratio and precipitation of secondary calcite. Given that the calcite has lighter Mg isotope composition than dolomite (Galy et al., 2002; Tipper et al., 2006b), dolomite dissolution would decrease the dolomite to calcite ratio and thus lower δ^{26}Mg values of whole rocks. This agrees with the positive correlations between Mg/Al vs. δ^{26}Mg as well as Mg/Ca vs. δ^{26}Mg (Fig. 8). In addition, the secondary calcite that precipitates from the infiltration fluid preferentially incorporates ^{24}Mg (Galy et al., 2002; Immenhauser et al., 2010), thereby possibly further lowering δ^{26}Mg values of the dolostone.

Below, we quantitatively assess whether combined dolomite dissolution and calcite re-precipitation processes can produce the Mg isotope fractionation observed in the weakly weathered samples. In such a process, Mg isotope fractionation of weathered dolostone is determined by the input and output mass fluxes and their isotope compositions. As discussed above, the output flux of Mg from the weakly weathered dolostone is governed by dolomite dissolution while the input flux is due to calcite re-precipitation. Based on mass balance, the isotope deviation of a weathered dolostone from the least weathered dolostone ($\Delta_{\text{weathered-unweathered}}$) can be expressed by:

$$\Delta_{\text{weathered}-\text{unweathered}} = \frac{\delta_{\text{unweathered}} + F_{\text{cal.r}} \times \delta_{\text{cal.r}} - F_{\text{dol}} \times \delta_{\text{dol}}}{1 + F_{\text{cal.r}} - F_{\text{dol}}} - \delta_{\text{unweathered}} \tag{1}$$

where δ is the δ^{26}Mg value, F is the Mg fraction, cal.r is the secondary calcite, dol is the dissolved dolomite, weathered is the weathered dolostone, and protolith is the least weathered dolostone. The Mg fraction (F) is normalized by Mg/Al ratio of the least weathered dolostone. Here, Mg isotope exchange between silicate fraction and infiltration fluid is not considered, as no evidence indicates that such isotope exchange is important given that Mg sink in silicates is considerably low relative to dolomites and such contribution to the bulk dolostone could not alter δ^{26}Mg values of dolostone. $\delta_{\text{unweathered}}$ is defined as -1.90‰, a value of the least weathered dolostone in our field study. Mg isotope composition of secondary calcite ($\delta_{\text{cal.r}}$) can be expressed by:

$$\delta_{\text{cal.r}} = \delta_{\text{sol.r}} + \Delta_{\text{cal.r}-\text{sol.r}} \tag{2}$$

where sol.r refers to the infiltration fluid, and $\Delta_{\text{cal.r-sol.r}}$ refers to the fractionation factor between cal.r and sol.r. As (1) Mg in sol.r mainly comes from dolomite and no fractionation occurs between dolomite and the fluid, and (2) the rather small amount of Mg is incorporated into secondary calcite, we assume that $\delta_{\text{sol.r}}$ is similar to δ_{dol} and keeps constant during the weathering process. Eq. (2) can be re-expressed by:

$$\delta_{\text{cal.r}} = \delta_{\text{dol}} + \Delta_{\text{cal.r}-\text{sol.r}} \tag{3}$$

In Eq. (3), $\Delta_{cal.r-sol.r}$ is adopted as −2.6‰ based on the experimental calibration of temperature dependence by Li et al. (2012b) and local annual average temperature of 18 ℃ (Hu et al., 2008). Seasonal and/or annual temperature fluctuations have negligible influence on $\Delta_{cal.r-sol.r}$ due to its insensitive temperature dependence (0.011 ± 0.002‰/ ℃ in Li et al. 2012b). Given that Ca/Al ratios display limited variation in the weakly weathered zone, we assume that the Ca influx and outflux are balanced in this model, then $F_{cal.r}$ can be calculated by:

$$F_{cal.r} = \frac{F_{dol} \times X_{Mg/(Mg+Ca)}}{1 - X_{Mg/(Mg+Ca)}} \tag{4}$$

where $X_{Mg/(Mg+Ca)}$ is the proportion of $MgCO_3$ in calcite.

The modeling results indicate that the Mg isotope fractionation is mainly controlled by the ratio of influx to outflux Mg and the isotope composition of primary dolomite (Fig. 9): (1) the higher ratio of reprecipitated Mg to dissolved Mg, the larger magnitude of fractionation; (2) the heavier Mg isotope composition of dolomite, the easier to achieve the observed fractionation. Our data ideally fall in the modeled lines (Fig. 9), supporting the interpretation that the concurrence of dolomite dissolution and calcite re-precipitation could lead to the observed Mg isotope fractionation. In addition, these models appear to indicate that the secondary calcite is low Mg calcite ($X_{Mg/(Mg+Ca)} < 0.05$) and the δ_{dol} lies between −1.75‰ and −1.90‰.

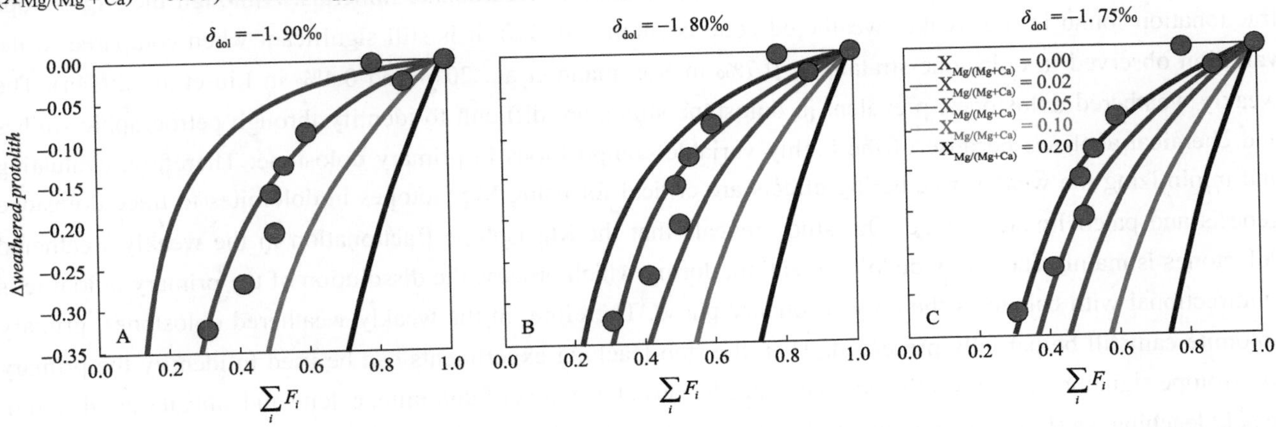

Fig. 9 $\sum F_i$ vs. $\Delta_{weathered-protolith}$ of the weakly weathered zone in dolostone weathering profile. $\sum F_i$ is the sum of Mg fraction: $\sum F_i = 1 + F_{cal.r} - F_{dol}$. $\Delta_{weathered-protolith} = \delta_{weathered} - \delta_{protolith}$, weathered and protolith represent the weathered and the least weathered dolostone, respectively. (A) δ_{dol} = −1.90‰; (B) δ_{dol} = −1.80‰; (C) δ_{dol} = −1.75‰.

The occurrence of both dolomite dissolution and calcite precipitation in the weakly weathered zone was termed as the "dedolomitization" which commonly occurs in the surface or near-surface settings (Raines and Dewers, 1997; Ronchi et al., 2004). This process is in opposite to dolomitization that is characterized by the formation of dolomite and decomposition of calcite. Fantle and Higgins (2014) found that dolomitization of limestone could lead to 1.3‰ Mg isotope fractionation, much larger than the 0.3‰ variation induced by dedolomitization observed in this study. The difference of Mg isotope fractionation between dolomitization and dedolomitization may depend on Mg influx and outflux, diagenetic fractionation factors, and fluid properties. For example, dolomitization will bring much external Mg into the Mg-depleted limestone and lead to a large diagenetic isotope fractionation up to −2‰, which can significantly change whole rock Mg isotope compositions (Fantle and Higgins, 2014). By contrast, dolomite dissolution does not appreciably fractionate Mg isotopes as discussed above. Additionally, samples in Fantle and Higgins (2014) were collected from an ocean drilling core and the dolomitization process occurred in an open system with a high fluid to solid mass ratio. By contrast, dedolomitization in this study took place under a surface environment with a low fluid to solid mass ratio. The fluids formed in these two different environments are probably distinct as well. The fluid during dolomitization of limestone in Fantle and Higgins (2014) is seawater-like, with Mg isotope composition much heavier than carbonates. However, dedolomitization fluid in this study has δ^{26}Mg value similar to dolostone as discussed above.

6 Implications

Our study of the dolostone weathering profile demonstrates that chemical weathering can significantly fractionate Mg isotopes even at the scale (~17 cm) of a hand specimen, with the direction and magnitude of isotope fractionation varying in different weathering stages. Therefore, the weathering influence has to be considered when using Mg isotopes of carbonates to trace paleoclimate change and the origin of dolostone.

The isotope fractionation produced in the intensely weathered zone is very large (~1.8‰), which can easily compromise the fractionation caused by other processes. Nonetheless, intensely weathered samples could be easily identified and avoided during sampling because of their distinct petrographic features such as color, hardness, and porosity. For example, the intensely weathered samples in this study have yellow color, are soft with very high porosity because of the substantial dissolution of carbonate minerals. Although the Mg isotope fractionation found in the weakly weathered zone is small (~0.3‰), it is still significant when compared to the variation observed in dolostone strata (e.g. 0.7‰ in Kasemann et al., 2014 and 0.4‰ in Liu et al., 2014a). The weakly weathered dolostones, prevalent in dolostone strata, are difficult to identify through petrographic studies and chemical analyses because of the highly variable compositions in primary dolostone. Therefore, evaluating and minimizing the weakly weathering effects are critical for using Mg isotopes in dolomites to trace dolostone genesis and paleoclimate change. Our study reveals that the Mg isotope fractionation in the weakly weathered dolostones is mainly caused by dedolomitization, during which process the dissolution of the primary dolomite is unidirectional with calcite as the only secondary phase. Therefore, in the weakly weathered dolostones, primary dolomite can still be partially preserved. Then the step-leaching experiments can be used to identify the primary Mg isotope signature of the weathered dolostones because behaviors of dolomite, calcite and silicates are different to acid leaching, as shown in previous studies (Liu et al., 2014a; Wen et al., 2015).

7 Conclusions

Extreme Mg depletion (~97%) and Mg isotope fractionation (δ^{26}Mg up to 1.8‰) are observed in a dolostone weathering profile from Hubei, China. The weathering profile is divided into the weakly and intensely weathered zones based on the element and isotope ratios. In the weakly weathered zone, whole rock δ^{26}Mg decreases from −1.90‰ to −2.22‰ with decreasing Mg/Al ratios and nearly constant Ca/Al ratios, reflecting the cooccurrence of dolomite dissolution and calcite re-precipitation, decrease in dolomite to calcite ratio, and preferential incorporation of ^{24}Mg into secondary calcite. Significant loss of both Mg and Ca is evident in the intensely weathered zone. Accordingly, δ^{26}Mg increases from −2.22‰ to −0.41‰, and ^{87}Sr/^{86}Sr from 0.71128 to 0.71368, which likely result from the increasing ratio of silicates to carbonates in the weathered residues. Our study suggests that both direction and magnitude of Mg isotope fractionation during dolostone weathering are controlled by dissolution and re-precipitation of carbonate minerals, as well as the saturation of carbonate minerals in the coexisting fluids. Possible weathering effect thus shall be carefully evaluated and minimized before using Mg isotopes of dolostone to trace dolostone genesis and paleoclimate change in the future work. Primary dolomite could be partially preserved in the weakly weathered dolostones and can be identified by step-leaching experiments.

Supplementary data to this article can be found online at http://dx. doi.org/10.1016/j.chemgeo.2016.07.012.

Acknowledgements

We thank Dr. Lijuan Xu, Zijian Li, Xu-Nan Meng and Ruiying Li for the help during sample analysis, and Zhengrong Wang, Lixin Jin, Jianming Zhu, Sheng-Ao Liu, Bing Shen, Kangjun Huang, Xianguo Lang, Yinghuai Lu, Pengfei Li, Chuanwei Zhu, and Ruoqi Wan for discussions. The constructive comments by Josh Wimpenny and two anonymous reviewers, as well as the careful and efficient editing of guest editor Lin Ma are greatly appreciated. This work was supported by the National Science Foundation of China (No. 41103010, No. 41202010 and No. 41328004), the Fundamental Research Funds for the Central Universities (No. 2652014035) and the National Science Foundation (EAR 1340160).

References

Al-Aasm, I.S., Packard, J.J., 2000. Stabilization of early-formed dolomite: a tale of divergence from two Mississippian dolomites. Sediment. Geol. 131, 97-108.

Azmy, K., Lavoie, D., Wang, Z., Brand, U., Al-Aasm, I., Jackson, S., Girard, I., 2013. Magnesium-isotope and REE compositions of Lower Ordovician carbonates from eastern Laurentia: implications for the origin of dolomites and limestones. Chem. Geol. 356, 64-75.

Back, W., Hanshaw, B.B., Plummer, L.N., Rahn, P.H., Rightmire, C.T., Rubin, M., 1983. Process and rate of dedolomitization: mass transfer and ^{14}C dating in a regional carbonate aquifer. Geol. Soc. Am. Bull. 94, 1415-1429.

Fantle, M.S., Higgins, J., 2014. The effects of diagenesis and dolomitization on Ca and Mg isotopes in marine platform carbonates: implications for the geochemical cycles of Ca and Mg. Geochim. Cosmochim. Acta 142, 458-481.

Foster, G.L., Pogge von Strandmann, P.A.E., Rae, J.W.B., 2010. Boron and magnesium isotopic composition of seawater. Geochem. Geophys. Geosyst. 11, Q08015. http://dx.doi. org/10.1029/2010GC003201.

Galy, A., Bar-Matthews, M., Halicz, L., O'Nions, R.K., 2002. Mg isotopic composition of carbonate: insight from speleothem formation. Earth Planet. Sci. Lett. 201, 105-115.

Galy, A., Yoffe, O., Janney, P.E., Williams, R.W., Cloquet, C., Alard, O., Carignan, J., 2003. Magnesium isotope heterogeneity of the isotopic standard SRM980 and new reference materials for magnesium-isotope-ratio measurements. J. Anal. At. Spectrom. 18, 1352-1356.

Geske, A., Zorlu, J., Richter, D.K., Buhl, D., Niedermayr, A., Immenhauser, A., 2012. Impact of diagenesis and low grade metamorphosis on isotope ($\delta^{26}Mg$, $\delta^{13}C$, $\delta^{18}O$ and $^{87}Sr/^{86}Sr$) and elemental (Ca, Mg, Mn, Fe and Sr) signatures of Triassic sabkha dolomites. Chem. Geol. 332, 45-64.

Geske, A., Goldstein, R.H., Mavromatis, V., Richter, D.K., Buhl, D., Kluge, T., Immenhauser, A., 2015a. The magnesium isotope ($\delta^{26}Mg$) signature of dolomites. Geochim. Cosmochim. Acta 149, 131-151.

Geske, A., Lokier, S., Dietzel, M., Richter, D.K., Buhl, D., Immenhauser, A., 2015b. Magnesium isotope composition of sabkha porewater and related (Sub-)Recent stoichiometric dolomites, Abu Dhabi (UAE). Chem. Geol. 393, 112-124.

Higgins, J., Schrag, D., 2010. Constraining magnesium cycling in marine sediments using magnesium isotopes. Geochim. Cosmochim. Acta 74, 5039-5053.

Horton, T.W., Chamberlain, C.P., Fantle, M., Blum, J.D., 1999. Chemical weathering and lithologic controls of water chemistry in a high-elevation river system: Clark's Fork of the Yellowstone River, Wyoming and Montana. Water Resour. Res. 35, 1643-1655.

Hu, C., Henderson, G., Huang, J., Chen, Z., Johnson, K., 2008. Report of a three-year monitoring programme at Heshang Cave, Central China. Int. J. Speleol. 37, 143-151.

Huang, K.-J., Teng, F.-Z., Wei, G.-J., Ma, J.-L., Bao, Z.-Y., 2012. Adsorption- and desorption-controlled magnesium isotope fractionation during extreme weathering of basalt in Hainan Island, China. Earth Planet. Sci. Lett. 359, 73-83.

Huang, K.-J., Teng, F.-Z., Elsenouy, A., Li, W.-Y., Bao, Z.-Y., 2013. Magnesium isotopic variations in loess: origins and implications. Earth Planet. Sci. Lett. 374, 60-70.

Huang, K.-J., Shen, B., Lang, X.G., Tang, W.B., Peng, Y., Ke, S., Li, F.B., 2015. Magnesium isotopic compositions of the Mesoproterozoic dolostones: implications for Mg isotopic systematics of marine carbonate. Geochim. Cosmochim. Acta 164, 333-351.

Husson, J.M., Higgins, J.A., Maloof, A.C., Schoene, B., 2015. Ca and Mg isotope constraints on the origin of Earth's deepest $\delta^{13}C$ excursion. Geochim. Cosmochim. Acta 160, 243-266.

Immenhauser, A., Buhl, D., Richter, D., Niedermayr, A., Riechelmann, D., Dietzel, M., Schulte, U., 2010. Magnesium-isotope fractionation during low-Mg calcite precipitation in a limestone cave-field study and experiments. Geochim. Cosmochim. Acta 74, 4346-4364.

Jacobson, A., Wasserburg, G., 2005. Anhydrite and the Sr isotope evolution of groundwater in a carbonate aquifer. Chem. Geol. 214, 331-350.

Jacobson, A.D., Zhang, Z., Lundstrom, C., Huang, F., 2010. Behavior of Mg isotopes during dedolomitization in the Madison Aquifer, South Dakota. Earth Planet. Sci. Lett. 297, 446-452.

Jeong, G.Y., Hillier, S., Kemp, R.A., 2008. Quantitative bulk and single-particle mineralogy of a thick Chinese loess-paleosol section: implications for loess provenance and weathering. Quat. Sci. Rev. 27, 1271-1287.

Jiang, G., Wang, X., Shi, X., Xiao, S., Zhang, S., Dong, J., 2012. The origin of decoupled carbonate and organic carbon isotope signatures in the early Cambrian (ca. 542-520 Ma) Yangtze platform. Earth Planet. Sci. Lett. 317, 96-110.

Jin, L., Williams, E.L., Szramek, K.J., Walter, L.M., Hamilton, S.K., 2008. Silicate and carbonate mineral weathering in soil profiles developed on Pleistocene glacial drift (Michigan, USA): mass balances based on soil water geochemistry. Geochim. Cosmochim. Acta 72, 1027-104.

Kasemann, S.A., Pogge von Strandmann, P.A.E., Prave, A.R., Fallick, A.E., Elliott, T., Hoffmann, K.H., 2014. Continental weathering following a Cryogenian glaciation: evidence from calcium and magnesium isotopes. Earth Planet. Sci. Lett. 396, 66-77.

Kaufman, A.J., Knoll, A.H., Semikhatov, M.A., Grotzinger, J.P., Jacobsen, S.B., Adams, W., 1996. Integrated chronostratigraphy of Proterozoic-Cambrian boundary beds in the western Anabar region, northern Siberia. Geol. Mag. 133, 509-533.

Ke, S., Teng, F.-Z., Li, S., Gao, T., Liu, S.-A., He, Y., Mo, X., 2016. Mg, Sr, and O isotope geochemistry of syenites from northwest Xinjiang, China: tracing carbonate recycling during Tethyan oceanic subduction. Chem. Geol. 37, 109-119.

Land, L.S., 1985. The origin of massive dolomite. J. Geol. Educ. 33, 112-125.

Langmuir, D., 1997. Aqueous Environmental Geochemistry. Prentice Hall, Upper Saddle River, NY (600 pp.).

Li, C.F., Li, X.H., Li, Q.L., Guo, J.H., Li, X.H., Yang, Y.H., 2012a. Rapid and precise determination of Sr and Nd isotopic ratios in geological samples from the same filament loading by thermal ionization mass spectrometry employing a single-step separation scheme. Anal. Chim. Acta 727, 54-60.

Li, W., Chakraborty, S., Beard, B.L., Romanek, C.S., Johnson, C.M., 2012b. Magnesium isotope fractionation during precipitation of inorganic calcite under laboratory conditions. Earth Planet. Sci. Lett. 333, 304-316.

Li, W., Beard, B.L., Li, C., Xu, H., Johnson, C.M., 2015. Experimental calibration of Mg isotope fractionation between dolomite and aqueous solution and its geological implications. Geochim. Cosmochim. Acta 157, 164-181.

Li, W.-Y., Teng, F.-Z., Ke, S., Rudnick, R.L., Gao, S., Wu, F.Y., Chappell, B.W., 2010. Heterogeneous magnesium isotopic composition of the upper continental crust. Geochim. Cosmochim. Acta 74, 6867-6884.

Li, Y.H., 1982. A brief discussion on the mean oceanic residence time of elements. Geochim. Cosmochim. Acta 46, 2671-2675.

Ling, M.X., Sedaghatpour, F., Teng, F.-Z., Hays, P.D., Strauss, J., Sun, W., 2011. Homogeneous magnesium isotopic composition of seawater: an excellent geostandard for Mg isotope analysis. Rapid Commun. Mass Spectrom. 25, 2828-2836.

Liu, C., Wang, Z., Raub, T.D., Macdonald, F.A., Evans, D.A., 2014a. Neoproterozoic capdolostone deposition in stratified glacial meltwater plume. Earth Planet. Sci. Lett. 404, 22-32.

Liu, S.-A., Teng, F.-Z., He, Y., Ke, S., Li, S.-G., 2010. Investigation of magnesium isotope fractionation during granite differentiation: implication for Mg isotopic composition of the continental crust. Earth Planet. Sci. Lett. 297, 646-654.

Liu, X.-M., Teng, F.-Z., Rudnick, R.L., McDonough, W.F., Cummings, M.L., 2014b. Massive magnesium depletion and isotope fractionation in weathered basalts. Geochim. Cosmochim. Acta 135, 336-349.

Ma, L., Teng, F.-Z., Jin, L., Ke, S., Yang, W., Gu, H.O., Brantley, S.L., 2015. Magnesium isotope fractionation during shale weathering in the Shale Hills Critical Zone Observatory: accumulation of light Mg isotopes in soils by clay mineral transformation. Chem. Geol. 397, 37-50.

Machel, H.G., 1997. Recrystallization versus neomorphism, and the concept of 'significant recrystallization' in dolomite research. Sediment. Geol. 113, 161-168.

Malone, M.J., Baker, P.A., Burns, S.J., 1994. Recrystallization of dolomite: evidence from the Monterey Formation (Miocene), California. Sedimentology 41, 1223-1239.

Mavromatis, V., Gautier, Q., Bosc, O., Schott, J., 2013. Kinetics of Mg partition and Mg stable isotope fractionation during its incorporation in calcite. Geochim. Cosmochim. Acta 114, 188-203.

McArthur, J.M., Howarth, R.J., Shields, G.A., 2012. Strontium isotope stratigraphy. In: Gradstein, F.M., Ogg, J.G., Schmitz, M.D., Ogg, G.M. (Eds.), The Geologic Time Scale. Elsevier, pp. 127-144.

Montanez, I.P., Read, J.F., 1992. Fluid-rock interaction history during stabilization of early dolomites, upper Knox Group (Lower Ordovician), US Appalachians. J. Sediment. Res. 62, 753-778.

Nesbitt, H.W., Markovics, G., 1980. Chemical processes affecting alkalis and alkaline earths during continental weathering. Geochim. Cosmochim. Acta 44, 1659-1666.

Opfergelt, S., Georg, R.B., Delvaux, B., Cabidoche, Y.M., Burton, K.W., Halliday, A.N., 2012. Mechanisms of magnesium isotope fractionation in volcanic soil weathering sequences, Guadeloupe. Earth Planet. Sci. Lett. 341, 176-185.

Opfergelt, S., Burton, K.W., Georg, R.B., West, A.J., Guicharnaud, R.A., Sigfusson, B., Halliday, A.N., 2014. Magnesium retention on the soil exchange complex controlling Mg isotope variations in soils, soil solutions and vegetation in volcanic soils, Iceland. Geochim. Cosmochim. Acta 125, 110-130.

Plummer, L.N., Back, W., 1980. The mass balance approach: application to interpreting the chemical evolution of hydrologic systems. Am. J. Sci. 280, 130-142.

Pogge von Strandmann, P.A.E., Burton, K.W., James, R.H., van Calsteren, P., Gislason, S.R., Sigfússon, B., 2008. The influence of weathering processes on riverine magnesium isotopes in a basaltic terrain. Earth Planet. Sci. Lett. 276, 187-197.

Pogge von Strandmann, P.A.E., Opfergelt, S., Lai, Y.J., Sigfússon, B., Gislason, S.R., Burton, K.W., 2012. Lithium, magnesium and silicon isotope behaviour accompanying weathering in a basaltic soil and pore water profile in Iceland. Earth Planet. Sci. Lett. 339, 11-23.

Pokrovsky, B.G., Mavromatis, V., Pokrovsky, O.S., 2011. Co-variation of Mg and C isotopes in late Precambrian carbonates of the Siberian Platform: a new tool for tracing the change in weathering regime? Chem. Geol. 290, 67-74.

Raines, M.A., Dewers, T.A., 1997. Dedolomitization as a driving mechanism for karst generation in Permian Blaine Formation, southwestern Oklahoma, USA. Carbonates Evaporites 12, 24-31.

Ronchi, P., Jadoul, F., Savino, R., 2004. Quaternary dedolomitization along fracture systems in a late triassic dolomitized platform (Western Southern Alps, Italy). Carbonates Evaporites 19, 51-66.

Rudnick, R.L., Tomascak, P.B., Njo, H.B., Gardner, L.R., 2004. Extreme lithium isotopic fractionation during continental weathering revealed in saprolites from South Carolina. Chem. Geol. 212, 45-57.

Ryu, J.-S., Jacobson, A.D., Holmden, C., Lundstrom, C., Zhang, Z., 2011. The major ion, $\delta^{44/40}$Ca, $\delta^{44/42}$Ca, and $\delta^{26/24}$Mg geochemistry of granite weathering at pH = 1 and T = 25 ℃: power-law processes and the relative reactivity of minerals. Geochim. Cosmochim. Acta 75, 6004-6026.

Saenger, C., Wang, Z., 2014. Magnesium isotope fractionation in biogenic and abiogenic carbonates: implications for paleoenvironmental proxies. Quat. Sci. Rev. 90, 1-21.

Schauble, E.A., 2011. First-principles estimates of equilibrium magnesium isotope fractionation in silicate, oxide, carbonate and hexaaquamagnesium (2+) crystals. Geochimica et Cosmochimica Acta, 75, 844-869.

Shen, B., Jacobsen, B., Lee, C.-T.A., Yin, Q.-Z., Morton, D.M., 2009. The Mg isotopic systematics of granitoids in continental arcs and implications for the role of chemical weathering in crust formation. Proc. Natl. Acad. Sci. 106, 20652-20657.

Teng, F.-Z., Wadhwa, M., Helz, R.T., 2007. Investigation of magnesium isotope fractionation during basalt differentiation: implications for a chondritic composition of the terrestrial mantle. Earth Planet. Sci. Lett. 261, 84-92.

Teng, F.-Z., Li, W.-Y., Ke, S., Marty, B., Dauphas, N., Huang, S., Pourmand, A., 2010a. Magnesium isotopic composition of the Earth and chondrites. Geochim. Cosmochim. Acta 74, 4150-4166.

Teng, F.-Z., Li, W.-Y., Rudnick, R.L., Gardner, L.R., 2010b. Contrasting lithium and magnesium isotope fractionation during continental weathering. Earth Planet. Sci. Lett. 300, 63-71.

Teng, F.-Z., Li, W.-Y., Ke, S., Yang, W., Liu, S.-A., Sedaghatpour, F., Xiao, Y., 2015. Magnesium isotopic compositions of international geological reference materials. Geostand. Geoanal. Res. 39, 329-339.

Tipper, E., Galy, A., Bickle, M., 2006a. Riverine evidence for a fractionated reservoir of Ca and Mg on the continents: implications for the oceanic Ca cycle. Earth Planet. Sci. Lett. 247, 267-279.

Tipper, E., Galy, A., Gaillardet, J., Bickle, M.J., Elderfield, H., Carder, E.A., 2006b. The magnesium isotope budget of the modern ocean: constraints from riverine magnesium isotope ratios. Earth Planet. Sci. Lett. 250, 241-253.

Wang, S.-J., Teng, F.-Z., Rudnick, R.L., Li, S.-G., 2015. The behavior of magnesium isotopes in low-grade metamorphosed mudrocks. Geochim. Cosmochim. Acta 165, 435-448.

Wang, Z., Hu, P., Gaetani, G., Liu, C., Saenger, C., Cohen, A., Hart, S., 2013. Experimental calibration of Mg isotope fractionation between aragonite and seawater. Geochim. Cosmochim. Acta 102, 113-123.

Warren, J., 2000. Dolomite: occurrence, evolution and economically important associations. Earth Sci. Rev. 52, 1-81.

Wen, B., Evans, D.A.D., Li, Y.X., Wang, Z., Liu, C., 2015. Newly discovered Neoproterozoic diamictite and cap carbonate (DCC) couplet in Tarim Craton, NW China: stratigraphy, geochemistry, and paleoenvironment. Precambrian Res. 271, 278-294.

White, A.F., Schulz, M.S., Lowenstern, J.B., Vivit, D.V., Bullen, T.D., 2005. The ubiquitous nature of accessory calcite in granitoid rocks: implications for weathering, solute evolution, and petrogenesis. Geochim. Cosmochim. Acta 69, 1455-1471.

Wimpenny, J., Gíslason, S.R., James, R.H., Gannoun, A., Pogge von Strandmann, P.A.E., Burton, K.W., 2010. The behaviour of Li and Mg isotopes during primary phase dissolution and secondary mineral formation in basalt. Geochim. Cosmochim. Acta 74, 5259-5279.

Wimpenny, J., Yin, Q.-Z., Tollstrup, D., Xie, L.-W., Sun, J., 2014. Using Mg isotope ratios to trace Cenozoic weathering changes: a case study from the Chinese Loess Plateau. Chem. Geol. 376, 31-43.

Wimpenny J, Colla C A, Yin Q Z, et al., 2014b. Investigating the behaviour of Mg isotopes during the formation of clay minerals. Geochimica et Cosmochimica Acta, 128, 178-194.

Wombacher, F., Eisenhauer, A., Böhm, F., Gussone, N., Regenberg, M., Dullo, W.C., Rüggeberg, A., 2011. Magnesium stable isotope fractionation in marine biogenic calcite and aragonite. Geochim. Cosmochim. Acta 75, 5797-5818.

Yang, W., Teng, F.-Z., Zhang, H.-F., 2009. Chondritic magnesium isotopic composition of the terrestrial mantle: a case study of peridotite xenoliths from the North China craton. Earth Planet. Sci. Lett. 288, 475-482.

Yang, W., Teng, F.-Z., Zhang, H.-F., Li, S.-G., 2012. Magnesium isotopic systematics of continental basalts from the North China craton: implications for tracing subducted carbonate in the mantle. Chem. Geol. 328, 185-194.

Yoo, C.M., Lee, Y.I., 1998. Origin and modification of early dolomites in cyclic shallow platform carbonates, Yeongheung Formation (Middle Ordovician), Korea. Sediment. Geol. 118, 141-157.

Mineral composition control on inter-mineral iron isotopic fractionation in granitoids*

Hongjie Wu, Yongsheng He, Leier Bao, Chuanwei Zhu and Shuguang Li

State Key Laboratory of Geological Processes and Mineral Resources, China University of Geosciences, Beijing 100083, China

亮点介绍：本研究对四个大别山的典型I型花岗岩中的矿物进行了成分和单矿物铁同位素分析。结果表明，铁同位素从重到轻的顺序为：长石>黄铁矿>磁铁矿>黑云母≈角闪石。本研究首次发现长石-云母-磁铁矿之间的铁同位素分馏尺度与矿物成分密切相关，且可以用矿物成分对Fe-O键强的影响来解释。此外，相对于共存的磁铁矿，长石具有非常重的铁同位素。本研究还测试了大别山混合岩淡色体的铁同位素，发现斜长石堆晶确实会造成全岩铁同位素偏重。文献中具有重铁同位素组成的高硅花岗岩类往往具有较低的Sr、Ba含量以及负Eu异常，所以不能用长石堆晶来解释。所以，岩浆演化过程中成分的变化导致的矿物间/矿物-熔体间分馏系数的变化伴随矿物分离结晶作用是导致高硅花岗岩类具有较重铁同位素的主要原因。

Abstract This study reports elemental and iron isotopic compositions of feldspar and its coexisting minerals from four Dabie I-type granitoids to evaluate the factors that control inter-mineral Fe isotopic fractionation in granitoids. The order of heavy iron isotope enrichment is feldspar > pyrite > magnetite > biotite ≈ hornblende. Feldspar has extraordinary heavier iron isotopic compositions than its co-existing biotite ($\Delta^{56}Fe_{plagioclase-magnetite}$ = +0.559‰ to +1.331‰, $\Delta^{56}Fe_{alkali\text{-}feldspar-biotite}$ = +0.698‰ to +1.092‰), which can be attributed to its high Fe^{3+}/Fe_{tot} ratio and low coordination number (tetrahedrally-coordinated) of Fe^{3+}. $\Delta^{56}Fe_{magnetite-biotite}$ of coexisting magnetite and biotite ranges from 0.090‰ to 0.246‰. Based on homogeneous major and iron isotopic compositions of mineral replicates, the inter-mineral fractionation in this study should reflect equilibrium fractionation. The large variations of inter-mineral fractionation among feldspar, magnetite and biotite are not induced by temperature variation, but strongly depend on mineral compositions. The $\Delta^{56}Fe_{plagioclase-biotite}$ and $\Delta^{56}Fe_{alkali\text{-}feldspar-biotite}$ are positively correlated with albite mode in plagioclase and orthoclase mode in alkali-feldspar, respectively. This could be explained by different Fe-O bond strength in feldspar due to different $Fe^{3+}/\sum Fe$ or different crystal parameters. The $\Delta^{56}Fe_{magnetite-biotite}$ increases with decreasing $Fe^{3+}/\sum Fe_{biotite}$ and increasing mole $(Na+K)/Mg_{biotite}$, indicating a decrease of β factor in low $Fe^{3+}/\sum Fe$ and high (Na+K)/Mg biotite.

High-silica leucosomes from Dabie migmatites with a feldspar accumulation petrogenesis have high $\delta^{56}Fe$ values ($\delta^{56}Fe$ = 0.42‰ to 0.567‰) than leucosome that represents pristine partial melt ($\delta^{56}Fe$ =0.117‰±0.016‰), indicating that accumulation of feldspar could account for high $\delta^{56}Fe$ values of these rocks. High $\delta^{56}Fe$ values are also predicted for other igneous rocks that are mainly composed of

* 本文发表在：Geochimica et Cosmochimica Acta, 2017, 198: 208-217

cumulate feldspar crystals, e.g., anorthosites. Feldspar accumulation, however, cannot explain high $\delta^{56}Fe$ values of most high-silica granitoids reported in the literature, based on their low Sr, Ba contents and negative Eu anomalies.

Keywords Fe isotopes, inter-mineral fractionation, compositional controls, feldspar accumulation, high-silica granitoids

1 Introduction

It is debated why high-silica granitic rocks (SiO_2 > 71 wt.%, $\delta^{56}Fe_{average}$=0.218±0.154‰, 2SD, N=84) have systematically heavier Fe isotopic compositions than dioritic-granodioritic rocks (53 wt.% < SiO_2 < 71 wt.%, $\delta^{56}Fe_{average}$=0.097±0.126‰, 2SD, N=103) (data from Dauphas et al., 2009; Foden et al., 2015; Heimann et al., 2008; Poitrasson and Freydier, 2005; Schuessler et al., 2009; Sossi et al., 2012; Telus et al., 2012). This phenomenon was firstly observed by Poitrasson and Freydier (2005) and interpreted as a result of exsolution of late magmatic aqueous fluids that preferentially dissolved ferrous iron rich in lighter isotopes. This idea was examined further by (Heimann et al., 2008), who proposed that the chloride-rich Fe^{2+}-bearing fluids exchanged with magnetite before exsolution. Li and Zn are fluid-mobile elements and their isotopic compositions can be fractionated during fluid-rock interaction (Schuessler et al., 2009; Telus et al., 2012 and references therein). However, correlated Li-Fe and Zn-Fe isotope fractionations were not observed (Schuessler et al., 2009; Telus et al., 2012). Soret diffusion was proposed to account for this phenomenon (Lundstrom, 2009; Zambardi et al., 2014), but the expected correlated heavy Mg and U isotopic compositions were not detected in other plutons (Telus et al., 2012). Therefore, multi-isotopic studies do not support fluid exsolution or diffusion explanations for granitic rocks.

Fractional crystallization is a possible explanation for the heavy Fe isotopic compositions of high-silica granitic rocks (Dauphas et al., 2014; Foden et al., 2015; Schuessler et al., 2009; Sossi et al., 2012; Telus et al., 2012). Some modelling performed by Schuessler et al. (2009) indicated that the rhyolites can be produced by fractional crystallization of a dacitic melt with a bulk mineral-melt fractionation factor $\Delta^{56}Fe_{melt\text{-}mineral}$ of about 0.1‰. However, some of the crystalline phases in granitic magma are Fe^{3+} rich oxides, such as magnetite. Crystallization of magnetite could reduce $\delta^{56}Fe$ values of melt because magnetite has heavier isotopic compositions relative to silicate phases (Heimann et al., 2008; Polyakov et al., 2007; Telus et al., 2012). Foden et al. (2015) found that granites with high $\delta^{56}Fe$ values are mainly A-type, corresponding to protracted fractional crystallization in which magnetite saturation is delayed. Dauphas et al. (2014) measured the mean force constants of iron bonds in silicate glasses and proposed that the β factor of iron is higher in rhyolitic glass than in basaltic-dacitic glass. In rhyolitic magma, the force constants of iron in melt may be even stronger than those of crystallizing magnetite. This will lead to a positive equilibrium isotopic fractionation between melt and crystals and drive the iron isotopic composition of the high-silica melts toward high $\delta^{56}Fe$ values. However, it is still unclear how the β factors of minerals change with the variation of their compositions, which are determined by the co-existing melts.

Feldspar is generally ignored when considering Fe isotope fractionation during magma processes due to its very low Fe content. However, mafic mineral is so rare while feldspar is abundant in high-silica granite that feldspar becomes an important iron carrier. For example, plagioclase from a Dabie high-silica granite T17 (SiO_2: 72.37 wt.%; FeO_t: 1.80 wt.%) contains 0.26 wt.% FeO (Ma et al., 1998). Considering the proportion of feldspar in the granite (more than 70%), over 10% iron is estimated to be carried in feldspar. In addition, an extremely heavy Fe isotope composition is expected for feldspar, as Fe^{3+} is dominant and tetrahedrally coordinated in feldspar

(Hofmeister and Rossman, 1984; Lundgaard and Tegner, 2004). An alkali-feldspar standard JF-2 from a pegmatite yields δ^{56}Fe as high as +0.816±0.025‰ (Millet et al., 2012). Therefore, feldspar may play a more important role in iron isotopic fractionation of high-silica magmas than previously thought. So far, iron isotopic fractionation factors between feldspar and its coexisting minerals are still lacking.

We carried out high-precision Fe isotopic analyses on four well-characterized I-type granitoids from the Dabie Orogen and minerals therein, including plagioclase, alkali-feldspar, biotite, hornblende, magnetite, pyrite and titanite. Our results show that feldspar has heavy (δ^{56}Fe: +0.557‰ to +1.326‰) iron isotopic compositions, with $\Delta^{56}Fe_{plagioclase-biotite}$ ranging from +0.559‰ to +1.331‰ and $\Delta^{56}Fe_{alkali\text{-}feldspar-biotite}$ = +0.698‰ to +1.092‰. Inter-mineral iron isotopic fractionations are most likely in equilibrium and controlled by chemical compositions of minerals. Based on feldspar data measured here, feldspar accumulation tend to produce high δ^{56}Fe rocks, which is evidenced by Dabie leucosomes. Leucosomes from Dabie migmatites with feldspar accumulation petrogenesis yield δ^{56}Fe up to 0.567‰, significantly higher than their pristine partial melt (~0.117‰).

2 Sample description

Four granites for this study (07ZB-6, 07XM-1, 07LD-2, and 06ZY-3) are Early Cretaceous post-collisional granitoids from the Dabie Orogen, East China. Their petrology and geochemistry have been previously studied and published (He et al., 2013; He et al., 2011). The four samples are typical I-type granites and have a wide range of SiO_2 (64.6–76.6wt.%, see appendix 2 Table S1). These samples contain various amount of quartz (20%–35%), plagioclase (35%–40%), alkali-feldspar (25%–40%), biotite (2%–8%) with accessory titanite, magnetite, apatite, pyrite and zircon (see Table 1 and appendix 1 Fig. S1). Hornblende was only found in 07LD-2 (6%) and 07ZB-6 (2%–3%).

Table 1 Iron isotopic compositions of granites, leucosomes, mineral seperates and geostandards in this study

Sample	Type	Modal (%)	$\delta^{56}Fe^{a}$	2se	$\delta^{57}Fe^{a}$	2se	n^{b}
Granites							
07LD-2	Plagioclase	35	0.770	0.052	1.132	0.066	4
	Replicate[c]		0.781	0.040	1.148	0.062	9
	K-feldspar	25	0.893	0.040	1.303	0.062	9
	Biotite	8	0.093	0.034	0.136	0.058	9
	Replicate[c]		0.091	0.060	0.122	0.077	3
	Replicate[c]		0.064	0.052	0.107	0.066	4
	Hornblende	6	0.106	0.034	0.121	0.058	9
	Replicate[c]		0.105	0.052	0.170	0.066	4
	Whole-rock		0.092	0.037	0.119	0.048	4
06ZY-3	Plagioclase	20	0.981	0.040	1.446	0.062	9
	K-feldspar	40	0.850	0.040	1.257	0.062	9
	Biotite	2	0.108	0.034	0.171	0.058	9
	Replicate[c]		0.142	0.060	0.234	0.077	3
	Replicate[c]		0.111	0.052	0.153	0.066	4
	Magnetite	3	0.198	0.034	0.305	0.058	9
	Replicate[c]		0.200	0.052	0.292	0.066	4
	Whole-rock		0.203	0.029	0.308	0.042	4

Continued

Sample	Type	Modal (%)	$\delta^{56}Fe^{a}$	2se	$\delta^{57}Fe^{a}$	2se	n^{b}
07ZB-6	Plagioclase	35	0.557	0.040	0.798	0.062	9
	K-feldspar	35	0.696	0.049	1.025	0.075	6
	Biotite	6	−0.002	0.034	−0.020	0.058	9
	Replicate[c]		0.025	0.060	0.108	0.077	3
	Replicate[c]		0.010	0.052	0.020	0.066	4
	Magnetite	2	0.181	0.034	0.285	0.058	9
	Titanite	1	0.217	0.034	0.318	0.058	9
	Whole-rock		0.135	0.037	0.215	0.048	4
07XM-1	Plagioclase	35	1.326	0.040	1.939	0.062	9
	K-feldspar	35	1.087	0.040	1.593	0.062	9
	Biotite	8	−0.005	0.040	0.010	0.062	9
	Replicate[c]		0.017	0.052	0.008	0.066	4
	Magnetite	2	0.241	0.034	0.361	0.058	9
	Pyrite	1	0.433	0.034	0.623	0.058	9
	Whole-rock		0.123	0.023	0.187	0.040	4
Leucosomes (Plagioclase accumulation petrogneiss)							
0907MSH-7-B	Whole-rock		0.424	0.016	0.650	0.036	9
0907MSH-6-B	Whole-rock		0.567	0.016	0.848	0.036	9
	Replicate[c]		0.568	0.016	0.842	0.036	9
0907QT-4-B	Whole-rock		0.482	0.016	0.695	0.036	9
	Replicate[c]		0.476	0.016	0.674	0.036	9
0907QT-1-B1	Whole-rock		0.468	0.016	0.685	0.036	9
Leucosome (None-plagioclase accumulation petrogneiss)							
0907MSH-8	Whole-rock		0.117	0.016	0.144	0.036	9
Geostandards							
BHVO-2	Whole-rock		0.131	0.052	0.198	0.066	4
GSR-1	Whole-rock		0.180	0.052	0.282	0.066	4
GSB	Solution		−0.709	0.052	−1.053	0.066	4

[a] Reported relative to IRMM-014.

[b] *n* denotes the total number of analyses per newly purified sample solution by MC-ICPMS.

[c] Replicate: repeat sample dissolution, column chemistry and instrumental analysis.

Five leucosomes from the Dabie Orogen were measured for whole rock iron isotopic composition to evaluate the effect of feldspar accumulation. Feldspar accumulation is evident for four of the five samples based on their extremely high Sr and Eu but low other REE concentrations (Wang et al., 2013). The other sample (0907MSH-8) may represent the pristine partial melt of dioritic to granodioritic gneisses (Wang et al., 2013). The five samples have SiO_2 contents ranging from 72.10 to 76.14 wt.%. Four leucosomes with feldspar accumulation have very low FeO_t concentrations (0.15 to 0.24 wt.%), while 0907MSH-8 has a FeO_t concentration of 1.90 wt.% (see appendix 2 Table S1).

3 Analytical methods

3.1 Electron analysis of minerals

Mineral compositions were measured at CAS Key Laboratory of Crust-Mantle Materials and Environments, University of Science and Technology of China. Major element concentrations of plagioclase, alkali-feldspar, biotite, titanite and hornblende were determined by a Shimadzu Electron Probe Micro analyzer (EMPA 1600) operating at 15 kV accelerating voltage with 20 nA beam current and <5 μm beam diameter. A series of natural minerals and synthetic oxides were used to calibrate the data based on a ZAF program. For major elements, the analytical precision is better than 2%. Iron content of feldspar was obtained with MC–ICP–MS by dissolution, since it's too low to get an accurate FeO_t content by EPMA.

3.2 Mineral separation

Pre-separation of minerals was conducted with the procedures of rough broken, fine crush, magnetic separation, manual extraction and density separation. After that, minerals were carefully handpicked under a binocular microscope twice to ensure purity and immunity of alteration. Special precautions were taken to guarantee feldspar separates are all colorless and without any magnetite or mafic mineral inclusions. In order to get enough iron for isotopic measurement, about 1000 small fragments (average total weight: 23 mg) of feldspar were prepared for each sample. The biotite grains are laminar in shape, dark green to black in color and almost transparent. Hornblende grains are generally brown or black, columnar and translucent to transparent. One titanite sample was picked based on its sphenoid shape and reddish brown color.

3.3 Fe isotope analysis

Iron isotopic analyses were performed at Isotope Geochemistry Lab in China University of Geosciences, Beijing. Procedures for sample dissolution, column chemistry and instrument analysis are after Dauphas et al. (2009b) and He et al. (2015). A brief description is given below.

Variable amount (0.35–39.55 mg) of samples were weighed into 6 ml Savillex PFA vials based on their Fe concentrations. Before dissolution, all minerals were leached by 1:1 mixture of BV-III grade ethanol and Milli-Q® water in an ultrasonic bath for 10 min and then carefully washed by Milli-Q® water for three times. After the samples were totally dissolved by step-wise acid digestion, iron was purified with a polypropylene column filled with 1 ml AG1-X8 resin in a HCl medium. Whole procedural blanks processed along with Dabie samples yield iron concentrations less than 2 ng, and thus is considered negligible. Iron isotopic compositions were determined by the standard–sample bracketing method using a Thermo–Finnigan Neptune Plus multiple collector–inductively coupled plasma–mass spectrometer (MC–ICP–MS) in HR mode. Typically, isotopic compositions reported here are the averages of nine times of repeat analyses. Iron isotopic variations are expressed in δ-notation, $\delta^{i}Fe = [({}^{i}Fe/{}^{54}Fe)_{sample} / ({}^{i}Fe/{}^{54}Fe)_{IRMM\text{-}014} - 1] \times 10^3$, where i refers to mass 56 or 57. All uncertainties in this study are reported as 2se (95% confidence intervals) and are calculated by $(2se)^2 = (2se_{MS})^2 + (2se_{Chem})^2$. Errors of bracketing calibrators measurement ($2se_{MS}$) are calculated as $2sd_{bracketing\ calibrator}/\sqrt{n}$. Accidental error from the chemistry ($2se_{Chem}$) adopted here is 0.005‰ and 0.012‰ for $\delta^{56}Fe$ and $\delta^{57}Fe$, respectively (He et al., 2015). Analyses over a 2-year period of reference materials demonstrate a long-term reproducibility and accuracy of < 0.03‰ for $^{56}Fe/^{54}Fe$ ratios. Samples analyzed here yield a linear trend with a slope of 1.466±0.028 (1se, R^2= 0.998) in the three isotope plot (Table 1 and Fig. 1), which agrees with theoretically expected slope of mass-dependent fractionation (Young et al., 2002).

Replicate analyses were applied for each biotite, one hornblende, one plagioclase and one magnetite sample from independent dissolution of separate splits of minerals to isotopic measurement. All the ten replicate analyses yield consistent values within current precision (Table 1). Two international geostandards (BHVO-2 and GSR-1) and one in-house standard (GSB Fe solution) were processed with Dabie samples to evaluate the accuracy and precision. GSR-1 (FeO_t : 1.93 wt.%) of 1.38 mg (contains c.a. 20 μg pure Fe) was processed to simulate feldspar samples (23 mg typically containing >12 μg Fe) with comparably low absolute Fe. The measured $\delta^{56}Fe$ values of the standards (BHVO-2: +0.131±0.052‰; GSR-1: +0.180±0.052‰; GSB Fe solution: –0.709±0.052‰).

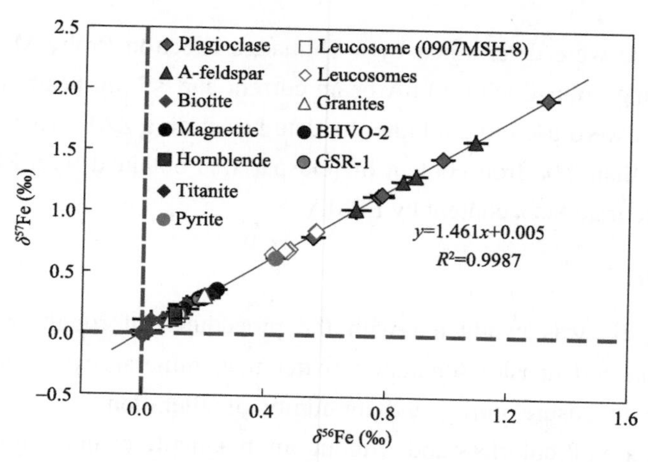

Fig. 1　Iron three-isotope plot of samples analyzed in this study. A-feldspar represents alkali-feldspar Standard samples BHVO-2 and GSR-1 are also plotted. The grey line represents a liner regression of δ^{56}Fe vs. δ^{57}Fe. The slope 1.461 is consistent with the values determined for mass-dependent fractionation (Young et al., 2002).

4　Results

Major elemental concentrations of minerals are reported in appendix 2 Table S2. At least three analyses have been done for each mineral phase in each rock sample. Compositions of each mineral are uniform for every rock but vary sample by sample. Biotite has MgO and FeO contents ranging from 11.51wt.% to 13.37wt.% and from 16.84 wt.% to 19.94 wt.%, respectively. $Fe^{3+}/\sum Fe$ ratio in biotite was calculated based on charge balance theory and ranges from 0.14 to 0.26. Based on albite – anorthite – orthoclase three end-member calculation, average albite mode (Ab%) of plagioclase and average orthoclase mode (Or%) of alkali-feldspar in four granites varies from 68.6% to 86.9% and from 75.4% to 88.0%, respectively. FeO contents of feldspar are 0.07%–0.15% for plagioclase and 0.03%–0.10% for alkali-feldspar. The hornblende from sample 07LD-2 has FeO content of 19.64 wt.% –19.87 wt.% and CaO contents of 10.85 wt.%–11.46 wt.%. The titanite from sample 07ZB-6 has FeO content of 1.67 wt.%–1.85 wt.%.

Iron isotopic data of granites and mineral separates, leucosomes, together with reference materials, are reported in Table 1. Four granites measured in this study have whole rock δ^{56}Fe ranging from +0.092±0.037‰ to +0.203±0.029‰, typical of intermediate to felsic rocks (Fig. 2a). Plagioclase and alkali-feldspar from the Dabie granites yield rather high δ^{56}Fe ranging from +0.557 ± 0.040‰ to +1.326 ± 0.040‰ and from +0.696±0.040‰ to +1.087±0.040‰, respectively, comparable to the K-feldspar reference sample from pegmatite (δ^{56}Fe: 0.816±0.025‰) (Millet et al., 2012). Biotite yields δ^{56}Fe ranging from −0.005±0.040‰ to +0.108±0.034‰, slightly below or comparable to their whole rocks ($\Delta^{56}Fe_{biotite-whole\ rock}$=−0.129‰ to −0.007‰) and consistent with the data previously reported (Fig. 2b) (Heimann et al., 2008; Telus et al., 2012). Magnetite was measured for only three granites, because no magnetite was separated from sample 07LD-2. They tend to have higher δ^{56}Fe values than their whole rocks (δ^{56}Fe=+0.181±0.034‰ to +0.241±0.034‰, $\Delta^{56}Fe_{magnetite-whole\ rock}$=−0.006‰ to +0.118‰) (Fig. 2b). Hornblende from 07LD-2 has δ^{56}Fe of +0.106 ± 0.034‰. Titanite from 07ZB-6 yields a δ^{56}Fe value of +0.217±0.033‰, slightly higher than its whole rock ($\Delta^{56}Fe_{titanite-whole\ rock}$=+0.082) (Fig. 2b). Pyrite from 07XM-1 has a high δ^{56}Fe value of +0.433±0.034‰. In summary, the observed order from highest to lowest $^{56}Fe/^{54}Fe$ is feldspar > pyrite > magnetite > titanite > biotite ≈ hornblende.

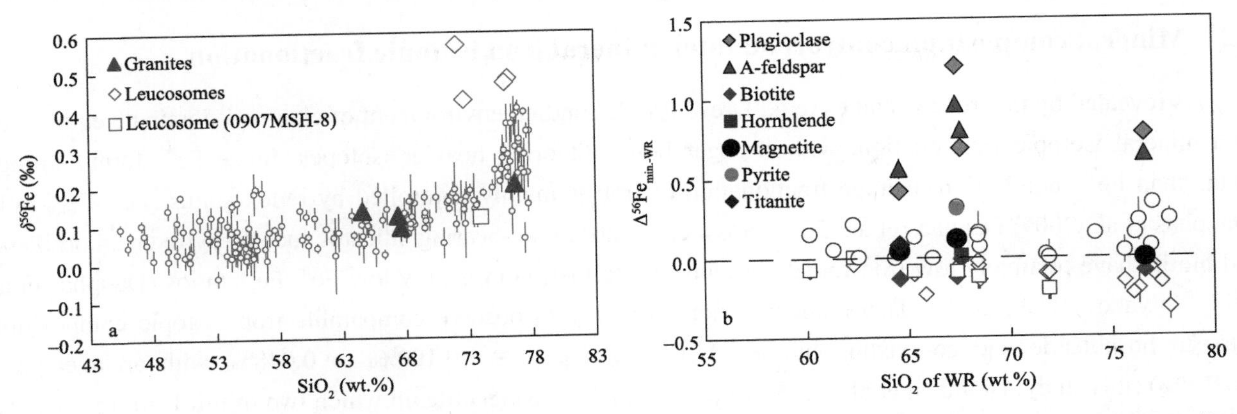

Fig. 2 (a) Wt.% SiO_2 vs. $\delta^{56}Fe$ of granites and leucosomes measured in this study compared to results of previous studies for igneous rocks. Data of intrusive rocks are from Poitrasson and Freydier. (2005), Heimann et al. (2008), Sossi et al. (2012) and Telus et al. (2012). Data of volcanic rocks are from Heimann et al. (2008), Dauphas et al. (2009), Schuessler et al. (2009), Sossi et al. (2012), Telus et al. (2012) and Teng et al. (2008). (b) $\Delta^{56}Fe_{mineral-whole\ rock}$ of mineral separates vs. SiO_2 wt.% of their whole rocks compared to literature data. WR=whole rock; $\Delta^{56}Fe_{mineral-WR} = \delta^{56}Fe_{mineral}-\delta^{56}Fe_{whole\ rock}$. Unfilled symbols represent literature data, including Heimann et al. (2008), Sossi et al. (2012) and Telus et al. (2012).

Four leucosomes from the Dabie Orogen with a plagioclase accumulation origin have rather heavy Fe isotopic compositions (+0.424±0.016‰ to +0.567±0.016‰, Table 1), their $\delta^{56}Fe$ values are even higher than published maximum $\delta^{56}Fe$ value of igneous rocks (Fig. 2a) (0.400 ± 0.020‰; Heimann et al., 2008). By contrast, leucosome 0907MSH-8, which may represent the pristine partial melt of dioritic to granodioritic gneisses (Wang et al., 2013), yields relatively low $\delta^{56}Fe$ value of +0.117‰±0.016‰ (Fig. 2a).

5 Discussion

5.1 Equilibrium or disequilibrium isotope fractionation?

The iron isotopic fractionation among mineral pairs from the Dabie granites shows significant variations (e.g., $\Delta^{56}Fe_{plagioclase-biotite}$ = +0.559‰ to +1.331‰, $\Delta^{56}Fe_{alkali-feldspar-biotite}$ = +0.698‰ to +1.092‰). It is essential to evaluate whether these variations reflect disequilibrium isotopic fractionation or represent equilibrium isotope fractionation controlled by varying parameters (e.g., mineral composition or temperature). Several lines of evidence indicate that chemical and isotopic equilibrium have been achieved within these granites.

Firstly, all the four granites studied here have coarse-grained granular texture (see appendix 1 Fig. S1), implying they had enough time to crystallize and equilibrate. Secondly, petrographic observations and compositional analyses do not show compositional zoning or heterogeneity in minerals of these granites (see electron probe results in appendix 2 Table S2). Thirdly, oxygen isotope studies of the Dabie granitoids have demonstrated that oxygen isotope compositions of minerals are in equilibrium at crystallization temperature (Chen et al., 2007; Xie et al., 2001; Xie et al., 2006; Zhao et al., 2004). In feldspar, iron diffuses faster than oxygen (Behrens et al., 1990; Farver, 2010). Therefore, the inter-mineral iron isotopic equilibrium between feldspar and other minerals should also have been achieved within these granites. Finally, mineral replicates, for which different separates of minerals were dissolved independently, yield consistent $\delta^{56}Fe$ values. This suggests homogeneous iron isotopic composition for each kind of mineral from the individual sample. Hence, data of minerals from the Dabie granites may record equilibrium iron isotopic fractionation.

5.2 Mineral composition controls on inter-mineral iron isotopic fractionation

As revealed by theoretical studies (e.g., Urey, 1947), bonding environment of elements dominate equilibrium inter-mineral isotopic fractionation, with stronger bonds favoring heavier isotopes. Since Fe^{3+} forms stronger bonds than Fe^{2+}, equilibrium isotope fractionation of iron is mainly controlled by redox status (Fe^{3+}/Fe_{tot} ratio) (Dauphas et al., 2009; Dauphas et al., 2014; Sossi et al., 2012). Fe-bearing silicates such as pyroxene, hornblende and biotite have relatively low iron force constants due to their comparably low Fe^{3+}/Fe_{tot} ratios (Dauphas et al., 2014; Polyakov et al., 2007). Therefore, it is not surprising to observe comparable iron isotopic compositions between hornblende and co-existing biotite ($\Delta^{56}Fe_{hornblende\text{-}biotite}$ = −0.100‰ to 0.048‰ with an average of −0.015‰) (this study, Heimann et al., 2008; Telus et al., 2012). Magnetite, in which two of thirds of iron occur as Fe^{3+} [1/3 of the total Fe as $^{VI}Fe^{2+}$, 1/3 as $^{IV}Fe^{3+}$, and 1/3 as $^{VI}Fe^{3+}$], has higher $\delta^{56}Fe$ values than their co-existing biotite ($\Delta^{56}Fe_{magnetite\text{-}biotite}$ = 0.083‰ to 0.560‰ with an average of 0.248‰) (this study, Heimann et al., 2008; Sossi et al., 2012; Telus et al., 2012). Although pyrite is a Fe^{2+}-bearing mineral, it is calculated to have heavier iron isotopes than coexisting magnetite and biotite due to its strong covalent Fe-S bond (Polyakov and Mineev, 2000; Blanchard et al., 2009). Pyrite from some hydrothermal deposits have very low $\delta^{56}Fe$ values, which reflects non-equilibrium fractionation (e.g., Rouxel et al., 2004). The pyrite from 07XM-1 has iron isotopic composition than its coexisting biotite and magnetite ($\Delta^{56}Fe_{pyrite\text{-}biotite}$ = 0.438‰; $\Delta^{56}Fe_{pyrite\text{-}magnetite}$ = 0.192‰). This supports the idea that inter-mineral fractionations here reflect the equilibrium conditions. Feldspar (especially orthoclase and albite-rich plagioclase) has high Fe^{3+}/Fe_{tot} ratios and all Fe^{3+} are tetrahedrally coordinated (Hofmeister and Rossman, 1984; Lundgaard and Tegner, 2004 and references therein). Therefore, strong Fe-O bond strength can explain why feldspar has higher $\delta^{56}Fe$ values than their coexisting magnetite and biotite ($\Delta^{56}Fe_{feldspar\text{-}magnetite}$ = +0.376‰ to +1.084‰, $\Delta^{56}Fe_{feldspar\text{-}biotite}$ = +0.559‰ to +1.331‰). In summary, the directions of inter-mineral fractionation among minerals (biotite, amphibole, and magnetite) are due to the different coordination environment of iron.

It is important to reveal the origin of the large variations of $\Delta^{56}Fe$ among minerals. Equilibrium Fe isotopic fractionation between mineral pairs was theoretically controlled by temperature and can be approximated as an equation in which $\Delta^{56}Fe$ between minerals is a linear function of $1/T^2$ (Bigeleisen and Mayer, 1947: Urey, 1947). To evaluate the equilibrium temperature of minerals, the thermodynamic models for ternary feldspar of Elkins-Tanton and Grove (1990) and Fuhrman and Lindsley (1988) were applied to samples in this study (Appendix 2). The two models give similar results (the difference <7 ℃) and obtained temperature values vary from 458 to 694 ℃. This temperature represents the equilibrium temperature for coexisting feldspars, which would be a good estimate of the equilibrium temperature of coexisting feldspar, biotite, and magnetite. $\Delta^{56}Fe$ values of every mineral pairs measured here do not follow the predicted, linear correlation with $10^6/T^2$ with a certain slope (Fig. 3). Therefore, the inter-mineral fractionation observed in our study cannot be simply explained by variation in the equilibrium temperature. Instead, inter-mineral Fe isotopic fractionations correlate with the mineral compositions, reflecting a dominant role of mineral composition on the varying fractionation scale.

5.2.1 Fractionation between magnetite and biotite

In this study, $\Delta^{56}Fe_{magnetite\text{-}biotite}$ increases with increasing of (Na + K)/Mg (mole ratio) and decreasing $Fe^{3+}/\sum Fe$ in biotite ($\Delta^{56}Fe_{magnetite\text{-}biotite}$=1.16×(Na+K)/Mg(mole ratio)−0.56, $\Delta^{56}Fe_{magnetite\text{-}biotite}$=−0.054×($Fe^{3+}/\sum Fe_{biotite}$)+1.29, Fig. 4). Since magnetite has a simple structure and the coordination state of Fe in magnetite is stable (Wilke et al., 2001), the $\Delta^{56}Fe_{magnetite\text{-}biotite}$ may primarily depend on the composition of biotite. Given that Fe^{3+} forms stronger bonds than Fe^{2+} with oxygen, the negative correlation between $\Delta^{56}Fe_{magnetite\text{-}biotite}$ and $Fe^{3+}/\sum Fe$

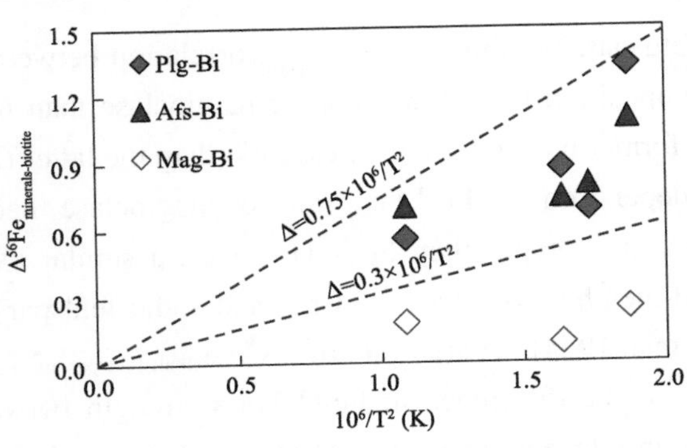

Fig. 3 Δ^{56}Fe versus $10^6/T^2$ of their corresponding whole rocks. Plg, Afs, Mag and Bi represent plagioclase, alkali-feldspar, magnetite and biotite, respectively. The temperature T was calculated from ternary feldspar geothermometer. On such diagram, the theoretical temperature dependence inter-minerals fractionations should be linear trends with intercepts of near-zero (e.g., the dotted lines). Obviously, the inter-minerals fractionations in this study are inconsistent with the temperature dependent prediction.

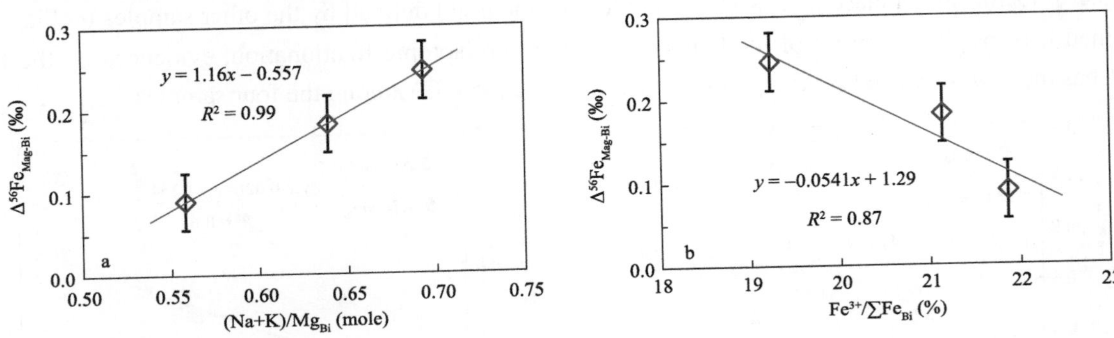

Fig. 4 (a) Plot of (Na+K)/Mg_{Bi} (mole) vs. $\Delta^{56}Fe_{Mag\text{-}Bi}$. Bi=biotite. (b) Plot of $\Delta^{56}Fe_{Mag\text{-}Bi}$ vs. $Fe^{3+}/\sum Fe$ ratios of biotite. The $Fe^{3+}/\sum Fe$ ratios were calculated from compositions of biotite based on charge balance theory.

ratio of biotite implies that less Fe^{3+} in biotite would induce decreasing of biotite's β factor. The following substitution reaction could explain why increase of (Na + K)/Mg (mole ratio) will lead to decrease of $Fe^{3+}/\sum Fe$ in biotite:

$$Fe^{3+} + Na^+/K^+ \leftrightarrow 2Mg^{2+}$$

This reaction is based on the knowledge that Fe^{3+} substitutes for Mg^{2+} in biotite ($K(Mg,Fe)_3(AlSi_3O_{10})(OH)_2$) and this substitution uses K^+ or Na^+ for charge balance. The equilibrium constant $\mathscr{K}$ of this reaction can be expressed by:

$$\mathscr{K} = ([Mg^{2+}]^2)/([Fe^{3+}] \times [Na^+ \text{ or } K^+])$$

This can be rewritten as $[Fe^{3+}] = ([Mg^{2+}]^2)/(\mathscr{K} \times [Na^+ \text{ or } K^+])$. At a given temperature and pressure, $\mathscr{K}$ is a constant. If (Na + K)/Mg ratio in biotite increases, the concentration of Fe^{3+} in biotite would decrease accordingly and resulting in lower β factor of Fe and larger fractionation between magnetite and biotite.

5.2.2 Fractionation between feldspar and magnetite

The fractionation factors between feldspar and magnetite show even larger variations ($\Delta^{56}Fe_{plagioclase\text{-}magnetite}$ = +0.376% to +1.084‰, $\Delta^{56}Fe_{alkali\text{-}feldspar\text{-}magnetite}$ = +0.516% to +0.846%). Mineral composition analyses show $\Delta^{56}Fe_{plagioclase\text{-}magnetite}$ and $\Delta^{56}Fe_{alkali\text{-}freldspar\text{-}magnetite}$ are positively correlated with albite mode of plagioclase and orthoclase mode of alkali-feldspar, respectively ($\Delta^{56}Fe_{plagioclase\ magnetite}$=0.022×Ab−1.15, $\Delta^{56}Fe_{alkali\text{-}reldspar\text{-}magnetite}$= 0.026×Or−1.46, Fig. 5a and b). This indicates that the iron isotopic fractionations between feldspar and magnetite depend on the composition of feldspar. Notably, the positive correlation between $\Delta^{56}Fe_{plagioclase\text{-}magnetite}$ and albite mode indicates iron isotopic composition is heavier in sodic plagioclase than in calcic plagioclase (Fig. 5a). A previous study revealed that the former has higher Fe^{3+}/Fe_{tot} ratio than the latter (Hofmeister and Rossman, 1984). The $\Delta^{56}Fe_{plagioclase\text{-}magnetite}$ could depend on the Fe^{3+}/Fe_{tot} ratio of plagioclase, because Fe^{3+} has low coordination number and favors more heavy iron isotopes than Fe^{2+}. However, a similar explanation cannot be applied on alkali-feldspar, since its two end-members (potassic feldspar and sodic feldspar) both have high Fe^{3+}/Fe_{tot} ratios (>95%, Hofmeister and Rossman.1984). Variation of $\Delta^{56}Fe_{alkali\text{-}feldspar\text{-}magnetite}$ with orthoclase mode of alkali-feldspar can be attributed to the difference in Fe-O bond strength between potassic feldspar and sodic feldspar. Fe^{3+} occupies the T-site in feldspar. The T-O interatomic distances in sanidine (T_{1O}-O: 1.651 Å, T_{1m}-O: 1.651 Å, T_{2O}-O: 1.638 Å, T_{2m}-O: 1.638 Å) are systematically shorter than those in albite (T_{1O}-O: 1.656 Å, T_{1m}-O: 1.657 Å, T_{2O}-O: 1.638 Å, T_{2m}-O: 1.643 Å) (e.g., Keefer and Brown, 1978). Accordingly, Fe-O bonds are likely shorter and thus stronger in potassic feldspar than in sodic feldspar. Alkali-feldspar with higher Or % thus tends to have higher β factor. Plagioclase in 07XM-1 is plotted off the trend defined by the other samples on Fig. 5a. This is explained here by the overprint of the temperature effect on isotopic fractionation, evidenced by the fact that 07XM-1 has the lowest two-feldspar equilibrium temperature (458 ℃) among the four samples.

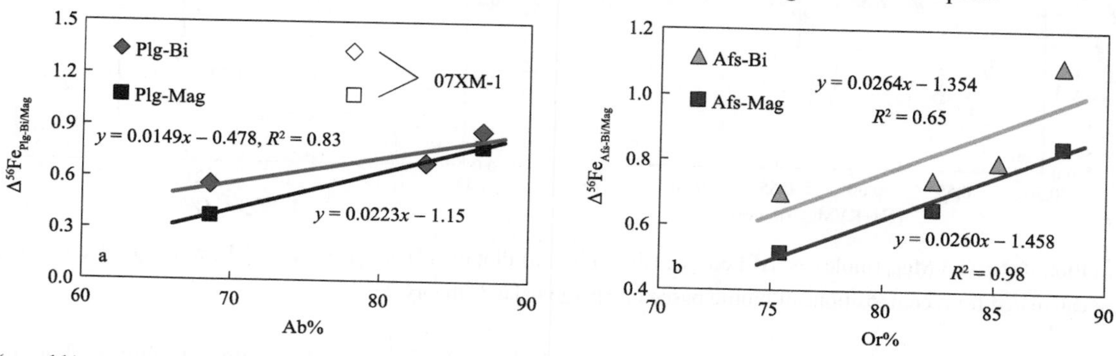

Fig. 5 (a and b) are composition of plagioclase and alkali-feldspar vs. their $\Delta^{56}Fe_{feldspar\text{-}biotite/magnetite}$ values, respectively. Ab = albite, Or = orthoclase.

5.2.3 Fractionation between feldspar and biotite

To facilitate future applications of inter-mineral Fe isotopic data here, the fractionation factors between feldspar and biotite are also plotted, and their correlations with the compositions of feldspar are given ($\Delta^{56}Fe_{plagioclase\text{-}biotite}$=0.015×Ab−0.48, $\Delta^{56}Fe_{alkali\text{-}feldspar\text{-}magnetite}$=0.026×Or−1.35, Fig. 5a and b). As noted in Section 5.2.1 and 5.2.2, variation of $\Delta^{56}Fe_{feldspar\text{-}biotite}$ reflects the combined compositional control from both biotite and feldspar.

5.3 Isotope effect of plagioclase accumulation

Rocks that experienced feldspar accumulation are common in nature, such as lunar and terrestrial anorthosite, some granites and some leucosomes from migmatites (e.g., Wiebe, 1980; Wu et al., 2005; Ohtake et al., 2009; Wang et al., 2013; Lee and Morton, 2015). Since feldspar has rather heavier iron isotopic composition than other rock-forming minerals, it is necessary to check whether feldspar accumulation can produce high $\delta^{56}Fe$ values of some igneous rocks. Leucosomes from Dabie migmatites with a feldspar accumulate origin yield high $\delta^{56}Fe$ values from 0.424±0.016‰ to 0.567±0.016‰. By contrast, the leucosome that represents pristine partial melt without substantial feldspar accumulation has a relatively normal $\delta^{56}Fe$ of 0.117±0.016‰, typical of granitic rocks. The four leucosomes with high $\delta^{56}Fe$ values have high SiO_2 contents (72.10–76.14 wt.%) and low FeO_t

concentrations (0.15–0.24 wt.%), indicating that a considerable amount of iron is present in feldspar. Combined with the iron isotopic composition results of feldspar in this study, feldspar accumulation can explain the high ^{56}Fe values of these leucosomes. The iron isotopic data of leucosome-melanosome pairs could be used to assess iron isotopic fractionation during partial melting of continental crust. However, the leucosomes may experience complex differentiation process such as fractional crystallization, thus they may not represent pristine melts (Telus et al., 2012). Our results suggest feldspar accumulation can produce high δ^{56}Fe leucosomes, and cautions shoud be taken when using migmatites to study iron isotope fractionation during crustal anatexis. Those intrusive rocks representing feldspar cumulates, e.g., anorthosites and some granites (e.g., Wiebe, 1980; Wu et al., 2005; Ohtake et al., 2009), are expected to have high δ^{56}Fe values. Anorthosites from Paleoproterozoic Bushveld Complex in South Africa indeed have δ^{56}Fe up to 0.27‰ (Stausberg et al., 2015), which can be explained by feldspar accumulation, considering the strong correlation between their bulk rock δ^{56}Fe and modal mineralogy. Lunar anorthosites have extremely high plagioclase proportions (>90%, Heiken et al., 1991), but they have an average δ^{56}Fe value of 0.14‰ (n=3, Poitrasson et al., 2004), which is only slightly higher than low-Ti basalts (average: 0.07‰, n=12; Weyer et al., 2005; Liu et al, 2010). Unlike terrestrial feldspar, lunar anorthosite only contains Fe^{2+}(Heiken et al., 1991), yielding $\Delta^{56}Fe_{plagioclase\text{-}pyroxene}$ smaller than 0.2‰ (Craddock et al., 2010). Therefore, plagioclase accumulation during lunar magmatism may only be able to produce minor iron isotopic fractionation.

The above observations demonstrate that heavy iron isotopic composition of feldspar could account for high δ^{56}Fe values of some cumulates. Lee and Morton (2015) suggested that crystal accumulation may be more common in the origin of granites than what has been previously estimated. Therefore, it is important to examine whether feldspar accumulation can account for high δ^{56}Fe values of high-silica granitic rocks. Theoretically, accumulation of plagioclase and alkali-feldspar would produce high δ^{56}Fe values, high Sr, Ba content and high Eu/Eu* simultaneously. However, most high δ^{56}Fe granitic rocks have low Sr, Ba contents and negative Eu anomalies, which doesn't fit the predicted accumulation trend (Fig. 6a-c). This implies accumulation of feldspar cannot account for high ^{56}Fe values of most high-silica granitic rocks.

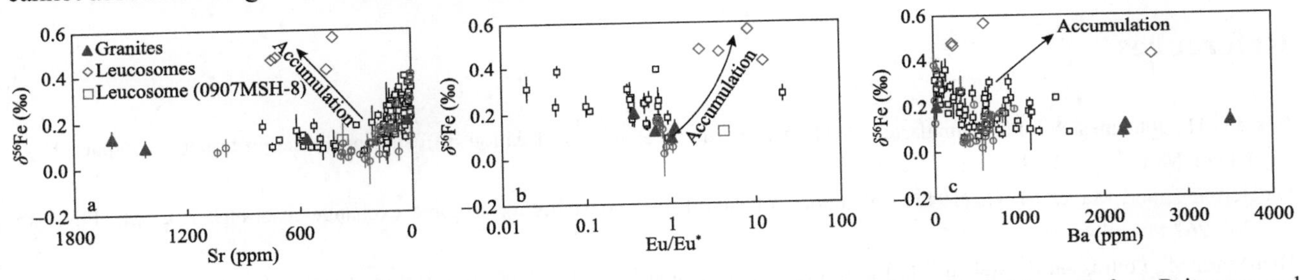

Fig. 6 Iron isotopic composition of bulk-rock δ^{56}Fe values vs. (a) Sr, (b) Eu/Eu*, (c) Ba. Data adopted here are from Poitrasson and Freydier (2005), Heimann et al. (2008), Dauphas et al. (2009a, b), Schuessler et al. (2009), Sossi et al. (2012), Telus et al. (2012), Foden et al. (2015) and references therein. Curve and arrow in each figure represent the direction of evolution according to accumulation of feldspar.

6 Conclusions

(1) Feldspar has very heavy iron isotopic compositions relative to the coexisting Fe-Mg minerals, with $\Delta^{56}Fe_{feldspar\text{-}biotite}$ = 0.56–1.33‰. The heavy iron isotopic signature of feldspar is due to their high Fe^{3+}/Fe_{tot} ratio and low coordination number (tetrahedrally-coordinated) of Fe^{3+}.

(2) Inter-mineral iron isotopic fractionations of magnetite-biotite, plagioclase-biotite and alkali-feldspar-biotite are correlated with the compositions of biotite, plagioclase and alkali-feldspar, respectively. This may reflect the

compositional effect of mineral compositions on the inter-mineral fractionation of Fe isotopes.

(3) Leucosomes from Dabie migmatites with a feldspar accumulate origin have high δ^{56}Fe values up to 0.567‰. High δ^{56}Fe values are also predicted for other igneous rocks that are mainly composed of cumulate feldspar crystals, e.g., anorthosites. Feldspar accumulation, however, cannot explain high δ^{56}Fe values of most high-silica granitoids reported in the literature, based on their low Sr, Ba contents and negative Eu anomalies.

Acknowledgements

Constructive comments from three anonymous reviewers, and efficient handling from Professor Nicolas Dauphas are greatly appreciated. We thank Fangzhen Teng for helpful comments. Dr.Shuijion Wang kindly provides leucosome samples from the DabieOrogen. We are grateful to Jiangling Xu and Min Feng for their help in maintaining the EMPA. This work was financially supported by Most of China, No.2016YFC0600408; theNational Natural Science Foundation of China (41473016): the Constructive comments from Fangzhen Teng are greatly appreciated. Shuijion Wang is thanked for the provision of leocosome samples from the Dabie Orogen. We are grateful to Jiangling Xu and Min Feng for their help in maintaining the EMPA. This work was financially supported by the National Natural Science Foundation of China (41473016); the Fundamental Research Funds for the Central Universities (2652014056) to YSH and State Key Lab. of Geological Processes and Mineral Resources. This is CUGB petrogeochemical contribution No.PGC-201515.

Appendix A. Supplementary data

Supplementary data associated with this article can be found, in the online version, at http://dx.doi.org/10.1016/j.gca.2016.11.008.

References

Behrens H., Johannes W. and Schmalzried H. (1990) On the mechanisms of cation diffusion processes in ternary feldspars. Phys. Chem. Miner. 17, 62-78.

Bigeleisen J. and Mayer M. G. (1947) Calculation of equilibrium constants for isotopic exchange reactions. J. Chem. Phys. 15, 261-267.

Blanchard M., Poitrasson F., Meheut M., Lazzeri M., Mauri F. and Balan E. (2009) Iron isotope fractionation between pyrite (FeS_2), hematite (Fe_2O_3) and siderite ($FeCO_3$): a first-principles density functional theory study. Geochim. Cosmochim. Acta 73, 6565-6578.

Chen J. F., Zheng Y.-F., Zhao Z.-F., Li B., Xie Z., Gong B. and Qian H. (2007) Relationships between O isotope equilibrium, mineral alteration and Rb-Sr chronometric validity in granitoids: implications for determination of cooling rate. Contrib. Mineral. Petrol. 153, 251-271.

Craddock P. R. and Dauphas N. (2011) Iron isotopic compositions of geological reference materials and chondrites. Geostand. Geoanal. Res. 35, 101-123.

Craddock P., Dauphas N., and Clayton R. (2010) Mineralogical control on iron isotopic fractionation during lunar differentiation and magmatism, Lunar and Planetary Science Conference, p. 1230.

Dauphas N., Craddock P. R., Asimow P. D., Bennett V. C., Nutman A. P. and Ohnenstetter D. (2009a) Iron isotopes may reveal the redox conditions of mantle melting from Archean to Present. Earth Planet. Sci. Lett. 288, 255-267.

Dauphas N., Pourmand A. and Teng F.-Z. (2009b) Routine isotopic analysis of iron by HR-MC-ICPMS: how precise and how accurate? Chem. Geol. 267, 175-184.

Dauphas N., Roskosz M., Alp E., Neuville D., Hu M., Sio C., Tissot F., Zhao J., Tissandier L. and Médard E. (2014) Magma redox and structural controls on iron isotope variations in Earth's mantle and crust. Earth Planet. Sci. Lett. 398, 127-140.

Elkins-Tanton L. and Grove T. L. (1990) Ternary feldspar experiments and thermodynamic models. Am. Miner. 75, 544-559.

Farver J. R. (2010) Oxygen and hydrogen diffusion in minerals. Rev. Mineral. Geochem. 72, 447-507.

Foden J., Sossi P. A. and Wawryk C. M. (2015) Fe isotopes and the contrasting petrogenesis of A-, I- and S-type granite. Lithos 212-215, 32-44.

Fuhrman M. L. and Lindsley D. H. (1988) Ternary-feldspar modeling and thermometry. Am. Miner. 73, 201-215.

He Y., Li S., Hoefs J., Huang F., Liu S.-A. and Hou Z. (2011) Postcollisional granitoids from the Dabie orogen: new evidence for partial melting of a thickened continental crust. Geochim. Cosmochim. Acta 75, 3815-3838.

He Y., Li S., Hoefs J. and Kleinhanns I. C. (2013) Sr Nd Pb isotopic compositions of early cretaceous granitoids from the Dabie orogen: constraints on the recycled lower continental crust. Lithos 156, 204-217.

He Y., Ke S., Teng F. Z., Wang T., Wu H., Lu Y. and Li S. (2015) High-precision iron isotope analysis of geological reference materials by high-resolution MC-ICP-MS. Geostand. Geoanal. Res..

Heiken G., Vaniman D., ans French B. M. (1991) Lunar Sourcebook: A User's Guide to the Moon. CUP Archive.

Heimann A., Beard B. L. and Johnson C. M. (2008) The role of volatile exsolution and sub-solidus fluid/rock interactions in producing high $^{56}Fe/^{54}Fe$ ratios in siliceous igneous rocks. Geochim. Cosmochim. Acta 72, 4379-4396.

Hofmeister A. M. and Rossman G. R. (1984) Determination of Fe^{3+} and Fe^{2+} concentrations in feldspar by optical absorption and EPR spectroscopy. Phys. Chem. Miner. 11, 213-224.

Keefer K. and Brown G. (1978) Crystal structures and compositions of sanidine and high albite in cryptoperthitic intergrowth. Am. Miner. 63, 1264-1273.

Lee C.-T. A. and Morton D. M. (2015) High silica granites: terminal porosity and crystal settling in shallow magma chambers. Earth Planet. Sci. Lett. 409, 23-31.

Liu Y., Spicuzza M. J., Craddock P. R., Day J. M., Valley J. W., Dauphas N. and Taylor L. A. (2010) Oxygen and iron isotope constraints on near-surface fractionation effects and the composition of lunar mare basalt source regions. Geochim. Cosmochim. Acta 74, 6249-6262.

Lundgaard K. L. and Tegner C. (2004) Partitioning of ferric and ferrous iron between plagioclase and silicate melt. Contrib. Mineral. Petrol. 147, 470-483.

Lundstrom C. (2009) Hypothesis for the origin of convergent margin granitoids and Earth's continental crust by thermal migration zone refining. Geochim. Cosmochim. Acta 73, 5709-5729.

Ma C., Li Z., Ehlers C., Yang K. and Wang R. (1998) A postcollisional magmatic plumbing system: Mesozoic granitoid plutons from the Dabieshan high-pressure and ultrahighpressure metamorphic zone, east-central China. Lithos 45, 431-456.

Millet M.-A., Baker J. A. and Payne C. E. (2012) Ultra-precise stable Fe isotope measurements by high resolution multiplecollector inductively coupled plasma mass spectrometry with a ^{57}Fe-^{58}Fe double spike. Chem. Geol. 304, 18-25.

Ohtake M., Matsunaga T., Haruyama J., Yokota Y., Morota T., Honda C., Ogawa Y., Torii M., Miyamoto H. and Arai T. (2009) The global distribution of pure anorthosite on the Moon. Nature 461, 236-240.

Poitrasson F. and Freydier R. (2005) Heavy iron isotope composition of granites determined by high resolution MC-ICP-MS. Chem. Geol. 222, 132-147.

Poitrasson F., Halliday A. N., Lee D.-C., Levasseur S. and Teutsch N. (2004) Iron isotope differences between Earth, Moon, Mars and Vesta as possible records of contrasted accretion mechanisms. Earth Planet. Sci. Lett. 223, 253-266.

Polyakov V. B. and Mineev S. D. (2000) The use of Mössbauer spectroscopy in stable isotope geochemistry. Geochim. Cosmochim. Acta 64, 849-865.

Polyakov V., Clayton R., Horita J. and Mineev S. (2007) Equilibrium iron isotope fractionation factors of minerals: reevaluation from the data of nuclear inelastic resonant X-ray scattering and Mössbauer spectroscopy. Geochim. Cosmochim. Acta 71, 3833-3846.

Rouxel O., Fouquet Y. and Ludden J. N. (2004) Subsurface processes at the Lucky Strike hydrothermal field, Mid-Atlantic Ridge: evidence from sulfur, selenium, and iron isotopes. Geochim. Cosmochim. Acta 68, 2295-2311.

Schuessler J. A., Schoenberg R. and Sigmarsson O. (2009) Iron and lithium isotope systematics of the Hekla volcano, Iceland—evidence for Fe isotope fractionation during magma differentiation. Chem. Geol. 258, 78-91.

Sossi P. A., Foden J. D. and Halverson G. P. (2012) Redoxcontrolled iron isotope fractionation during magmatic differentiation: an example from the Red Hill intrusion S. Tasmania. Contrib. Mineral. Petrol. 164, 757-772.

Stausberg N., Lesher C., Glessner J., Barfod G. and Tegne C. (2015) Iron Isotope Systematics of the Bushveld Complex. Goldschmidt Abstracts, South Africa, p. 2972.

Telus M., Dauphas N., Moynier F., Tissot F. L., Teng F.-Z., Nabelek P. I., Craddock P. R. and Groat L. A. (2012) Iron, zinc, magnesium and uranium isotopic fractionation during continental crust differentiation: the tale from migmatites, granitoids, and pegmatites. Geochim. Cosmochim. Acta 97, 247-265.

Teng F. Z., Dauphas N. and Helz R. T. (2008) Iron isotope fractionation during magmatic differentiation in Kilauea Iki lava lake. Science 320, 1620-1622.

Urey H. C. (1947) The thermodynamic properties of isotopic substances. J. Chem. Soc., 562-581.

Wang S.-J., Li S.-G., Chen L.-J., He Y.-S., An S.-C. and Shen J. (2013) Geochronology and geochemistry of leucosomes in the North Dabie Terrane, East China: implication for post-UHPM crustal melting during exhumation. Contrib. Mineral. Petrol. 165, 1009-1029.

Weyer S., Anbar A. D., Brey G. P., Münker C., Mezger K. and Woodland A. B. (2005) Iron isotope fractionation during planetary differentiation. Earth Planet. Sci. Lett. 240, 251-264.

Wiebe R. A. (1980) Anorthositic magmas and the origin of Proterozoic anorthosite massifs. Nature 286, 564-567.

Wilke M., Farges F., Petit P.-E., Brown G. E. and Martin F. (2001) Oxidation state and coordination of Fe in minerals: An Fe KXANES spectroscopic study. Am. Miner. 86, 714-730.

Wu F.-Y., Yang J.-H., Wilde S. A. and Zhang X.-O. (2005) Geochronology, petrogenesis and tectonic implications of Jurassic granites in the Liaodong Peninsula, NE China. Chem. Geol. 221, 127-156.

Xie Z., Wang Z.-R., Zheng Y.-F., Chen J.-F. and Zhang X. (2001) The mineral O isotopic equilibrium of Zhubuyuan granite and gneiss in the North Dabie Mountains and the Rb-Sr geochronologic affection. Geochimica 30, 95.

Xie Z., Zheng Y.-F., Zhao Z.-F., Wu Y.-B., Wang Z., Chen J., Liu X. and Wu F.-Y. (2006) Mineral isotope evidence for the contemporaneous process of Mesozoic granite emplacement and gneiss metamorphism in the Dabie orogen. Chem. Geol. 231, 214-235.

Young E. D., Galy A. and Nagahara H. (2002) Kinetic and equilibrium mass-dependent isotope fractionation laws in nature and their geochemical and cosmochemical significance. Geochim. Cosmochim. Acta 66, 1095-1104.

Zambardi T., Lundstrom C. C., Li X. and McCurry M. (2014) Fe and Si isotope variations at Cedar Butte volcano; insight into magmatic differentiation. Earth Planet. Sci. Lett. 405, 169-179.

Zhao Z.-F., Zheng Y.-F., Wei C.-S. and Gong B. (2004) Temporal relationship between granite cooling and hydrothermal uranium mineralization at Dalongshan in China: a combined radiometric and oxygen isotopic study. Ore Geol. Rev. 25, 221-236.

Diffusion-driven magnesium and iron isotope fractionation at a gabbro-granite boundary*

Hongjie Wu[1,2], Yongsheng He[1,2], Fang-Zhen Teng[2], Shan Ke[1], Zhenhui Hou[3] and Shuguang Li[1]

1. State Key Laboratory of Geological Processes and Mineral Resources, China University of Geosciences, Beijing 100083, China
2. Isotope Laboratory, Department of Earth and Space Sciences, University of Washington, Seattle, Washington 98195, USA
3. CAS Key Laboratory of Crust-Mantle Materials and Environments, School of Earth and Space Sciences, University of Science and Technology of China, Anhui 230026, China

亮点介绍：本研究对大别山的一处辉长岩-花岗岩接触边界进行了元素及 Mg、Fe 同位素的剖面研究。该接触边界 16cm 的范围内 δ^{26}Mg 从接触界面向花岗岩一侧逐渐降低(从–0.28±0.04‰到–0.63±0.08‰)，而 δ^{56}Fe 则逐渐升高（从–0.07±0.03‰到+0.25±0.03‰），二者呈现出明显的负相关性。该研究利用菲克第二定律扩散方程进行的模拟计算表明，Mg、Fe 互扩散可以解释实际观察的同位素剖面。该研究表明 Mg、Fe 互扩散至少可在分米尺度上产生显著的 Mg、Fe 同位素动力学分馏，具有作为地质速度计的潜力。该研究同时也提醒在研究镁铁质包体、基性岩脉、层状岩浆岩等的 Mg、Fe 同位素时应充分考虑扩散导致的影响。

Abstract Significant magnesium and iron isotope fractionations were observed in an adjacent gabbro and granite profile from the Dabie Orogen, China. Chilled margin and granitic veins at the gabbro side and gabbro xenoliths in the granite indicate the two intrusions were emplaced simultaneously. The δ^{26}Mg decreases from –0.28±0.04‰ to –0.63±0.08‰ and δ^{56}Fe increases from –0.07±0.03‰ to +0.25±0.03‰ along a ~16 cm traverse from the contact to the granite. Concentrations of major elements such as Al, Na, Ti and most trace elements also systematically change with distance to the contact. All the observations suggest that weathering, magma mixing, fluid exsolution, fractional crystallization and thermal diffusion are not the major processes responsible for the observed elemental and isotopic variations. Rather, the negatively correlated Mg and Fe isotopic compositions as well as co-variations of Mg and Fe isotopes with Mg# reflect Mg-Fe inter-diffusion driven isotope fractionation, with Mg diffusing from the chilled gabbro into the granitic melt and Fe oppositely. The diffusion modeling yields a characteristic diffusive transport distance of ~6 cm. Consequently, the diffusion duration, during which the granite may have maintained a molten state, can be constrained to ~2 My. The cooling rate of the granite is calculated to be 52–107 ℃/My. Our study suggests diffusion profiles can be a powerful geospeedometry. The observed isotope fractionations also indicate that Mg-Fe inter-diffusion can produce large stable isotope fractionations at least on a decimeter scale, with implications for Mg and Fe

* 本文发表在：Geochimica et Cosmochimica Acta，2018，222：671-684

isotope study of mantle xenoliths, mafic dikes, and inter-bedded lavas.

Keywords　Mg-Fe inter-diffusion, iron isotope, magnesium isotope, kinetic fractionation

1 Introduction

Magnesium and iron isotope fractionations at high temperature were considered limited but recent studies found large equilibrium (>0.5‰) and kinetic (up to several per mil) isotope fractionations (see recent reviews of Dauphas et al., 2017; Teng, 2017; Watkins et al., 2017). For example, large equilibrium inter-mineral Mg isotope fractionation among garnet, spinel, olivine and pyroxenes has been reported in igneous and metamorphic rocks (e.g., Liu et al., 2011; Wang et al., 2015; Li et al., 2016), making it a potential high-precision geothermometry. Large Mg and Fe isotope fractionation driven by chemical diffusion has been reported with applications toward deciphering the cooling history preserved in igneous samples (e.g., Dauphas et al., 2010; Teng et al., 2011; Chopra et al., 2012; Sio et al., 2013; Oeser et al., 2015; Pogge von Strandmann et al., 2015; Sio and Dauphas, 2017). Identifying the process that accounts for isotope fractionation is the prerequisite for using them as tracers, geospeedometry, or geothermometry. However, meaningful interpretation of fractionation signatures is often hampered by the fact that they could be ascribed to many processes such as fractional crystallization, partial melting and thermal diffusion (see recent reviews of Dauphas et al., 2017; Teng, 2017; Watkins et al., 2017).

The combination of multiple isotopic systems, where possible, can help distinguish processes responsible for isotopic variations. Fractional crystallization and partial melting can dramatically fractionate Fe isotopes (e.g., Weyer et al., 2005; Schoenberg and von Blanckenburg, 2006; Weyer and Ionov, 2007; Teng et al., 2008; 2013; Sossi et al., 2012; Telus et al., 2012), whereas the same processes lead to negligible fractionation of Mg isotopes (e.g., Teng et al., 2007; 2010a; Liu et al., 2010). Thermal diffusion can theoretically fractionate Mg and Fe isotopes in the same direction, yielding a positive correlation between $^{26}Mg/^{24}Mg$ and $^{56}Fe/^{54}Fe$ (e.g., Richter et al., 2008; 2009; Huang et al., 2010). In chemical diffusion experiments using juxtaposing basaltic and rhyolitic melts couples, both Fe and Mg diffuse from the basaltic to the rhyolitic melt, with the latter simultaneously concentrated light Fe and Mg isotopes and thus a positive correlations between $^{26}Mg/^{24}Mg$ and $^{56}Fe/^{54}Fe$ can be expected (Richter et al., 2008; 2009). Since Mg^{2+} and Fe^{2+} occupy the same lattice site in silicate minerals due to their identical charges and similar ionic radii, inter-diffusion of Mg and Fe ($Mg^{2+}_{phase1}+Fe^{2+}_{phase2}\leftrightarrow Mg^{2+}_{phase2}+Fe^{2+}_{phase1}$) have been observed among minerals and between co-existing minerals and magmas (e.g., olivine, spinel, perovskite; see the review of Zhang and Cherniak, 2010). The Fe and Mg would diffuse in opposite directions during inter-diffusion between solid and melt, which is different from the situation of diffusion within basaltic and rhyolitic melts that both Fe and Mg go from the basaltic melt to the rhyolitic melt. Recently, inter-diffusion of Mg and Fe was proposed to produce negatively correlated Mg-Fe isotope fractionation in olivine (Dauphas et al., 2010; Teng et al., 2011; Sio et al., 2013; Oeser et al., 2015). Therefore, the combined Mg-Fe isotopic effects are quite different during fractional crystallization, partial melting, thermal and chemical diffusion among or between melts and minerals.

Previous studies have found that Mg-Fe inter-diffusion profiles in olivine can be used as a geospeedometry (Dauphas et al., 2010; Teng et al., 2011; Sio et al., 2013; Oeser et al., 2015), nonetheless, whether such fractionation could happen between adjacent rocks is still unknown. Here, we report whole-rock Mg and Fe isotopic data for a gabbro-granite boundary profile from the Dabie Orogen, China. The results show a gradient of Mg and Fe isotopic compositions. Negatively correlated Mg and Fe isotopic compositions as well as co-variations of Mg and Fe isotopic compositions with Mg# likely reflect inter-diffusion of Mg and Fe between the chilled gabbro and its coexisting granitic melt. Results of diffusion modeling agree well with the measured profiles and yield a diffusion duration range of 1.5–2.2 My and a cooling rate range of 52–107 ℃/My. Our results demonstrate the potential application of the diffusion-controlled Mg-Fe isotope fractionation as a geospeedometry of geological events.

2 Sample description

The profile studied here was collected from the boundary between a granite intrusion and a gabbro intrusion, located in the Dabie Orogen, Central China. The gabbro intrusion was generally called Daoshichong pluton, which is one of the major post-collisional mafic-ultramafic plutons in Dabie Orogen (Dai et al., 2011; Xu et al., 2012). The granite intrusion belongs to a group of Early Cretaceous monzogranite plutons with ages younger than 130Ma (Xu et al., 2007). Both intrusions are wider than 20 m in the field. Mafic xenoliths in the granite near the boundary, small granitic veins in the gabbro and a chilled margin on the gabbro side indicate the two intrusions were emplaced almost simultaneously under liquid or semi-solid state (Fig. 1). One gabbro sample (13DSC-5) and one granite sample (13DSC-6) were collected 4 m and 0.5 m away from the boundary, respectively, to represent end-members unaffected by diffusion. The gabbro sample (13DSC-5) is medium-grained and contains pyroxene (25%), plagioclase (40%), hornblende (20%) and oxide minerals (10%–15%) (see photomicrographs in Appendix A). The granite sample (13DSC-6) is a medium to coarse-grained monzogranite, which consists of quartz (25%), plagioclase (45%), alkali-feldspar (25%), biotite (less than 2%) and accessory minerals such as oxides (see photomicrographs in Appendix A).

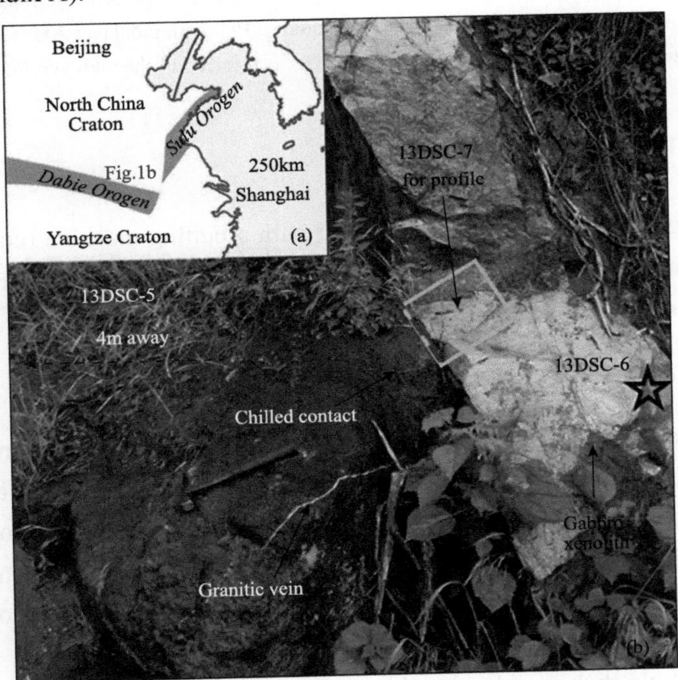

Fig. 1 (a) Simplified geological map showing eastern China and major tectonic units, with roughly marked sampling location. (b) Field photograph of our study profile, with detailed captions of sampling. Both intrusions are wider than 20 meters in the field.

One large hand specimen (13DSC-7) collected from the boundary was sliced into nine pieces (13DSC-7-1~13DSC-7-9) for a profile study (Figs.1 and 2). 13DSC-7-1 is the chilled margin of the gabbro, and the others are granite. The sliced granite samples have the same mineral assemblage with 13DSC-6, but their crystal size changes from fine-grained at the contact to medium- to coarse-grained in 13DSC-7-9 (Fig. 2). 13DSC-7-1 has a finely crystalline texture and contains hornblende, chlorite, oxide minerals with minor quartz and plagioclase (Fig. 2).

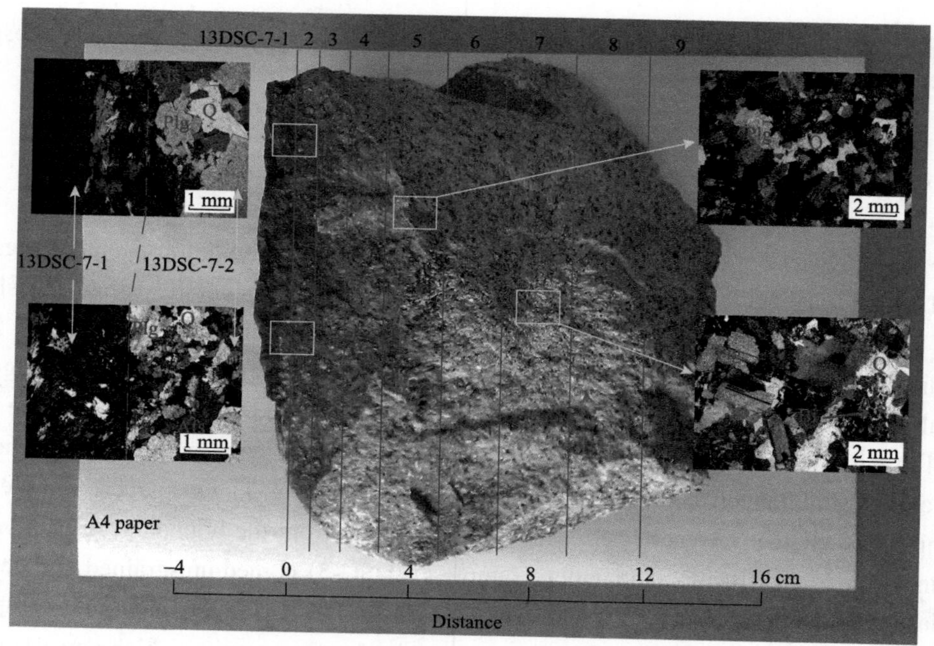

Fig. 2 Photograph of sample 13DSC-7 for profile study before cutting. It was cut into 9 pieces named 13DSC-7-1 to 13DSC-7-9 following the red lines in the figure. 13DSC-7-1 is the chilled margin of gabbro and the other eight are granite. The distance is defined as the distance to the contact of gabbro and granite. The yellow squares represent the locations of thin sections. Polarized micrographs of the thin sections are also shown in the figure. Q = quartz; Plg = plagioclase; Afs = alkali-feldspar; Bi = biotite. (For interpretaion of the refcrences to colour in this figure legend, the reader is referred to the web version of this article.)

3 Analytical methods

Surfaces of all samples were removed by ~0.5−1 cm with a cutting machine (corundum grinding wheel) to avoid the influence of weathering. After washing with Milli-Q® water, samples were roughly broken by a copper hammer and then crushed to a fine powder (<200 mesh in size) in an agate ball crusher.

3.1 Major and trace elements

Major elements were analyzed using inductively coupled plasma atomic emission spectroscopy at the China University of Geosciences, Beijing. Ferrous and ferric iron content was determined by wet-chemistry methods. Gravimetric methods were used to determine the loss on ignition (LOI) prior to major element analyses. Analytical uncertainties (2SD) were better than 1% for the majority of major elements and were better than 2% for MnO and P_2O_5 based on repeated measurement of USGS rock standards. Total carbon contents were measured using an Analytik-Jean multi-series analyzer with a furnace temperature of 1000 ℃.

For trace elements, whole-rock powder weighing ~50 mg was dissolved in a mixture of HF + HNO_3 at 190 ℃ using Parr bombs in the clean lab at the China University of Geosciences, Beijing for ~72 h to ensure complete sample dissolution. The dissolved samples were diluted to 80 ml using 1% HNO_3 before analysis. Trace element data were obtained by an ELAN DRCII inductively coupled plasma mass spectrometer (ICP-MS) at the University of Science and Technology of China (USTC) following the established method (Hou and Wang, 2007). Reproducibility was better than 5% for elements with concentrations >10 μg/g and better than 20% for most elements of <10 μg/g based on long-term analysis of standard materials (Hou and Wang, 2007).

3.2 Magnesium and iron isotope analyses

Magnesium and iron isotopic analyses were performed at the Isotope Geochemistry Lab, China University of Geosciences, Beijing. The procedures have been previously detailed in Ke et al. (2016) and He et al. (2015), largely similar to Teng et al. (2010a) and Dauphas et al. (2009), respectively. A brief description is given below.

All chemical procedures were carried out in class 100 laminar flow hoods in a class 1000 clean room. Approximately 3 mg gabbro powder, or 10 mg granite powders containing over 50 ug Fe and 40 ug Mg were first dissolved in a 3:1 (v/v) mixture of concentrated HF–HNO_3. After heating at a temperature of 160 ℃ on a hotplate until the solutions became transparent, the solutions were evaporated to dryness at 140 ℃. The dried residues were then refluxed twice with a 3:1 (v/v) mixture of concentrated HCl–HNO_3. After the samples were totally dissolved by the step-wise acid digestion, iron was purified with a polypropylene column filled with 1 ml AG1-X8 pre-cleaned resin in a HCl medium. The matrix elements during iron column chemistry were collected for Mg purification using AG50W-X8 pre-cleaned resin in 1 N HNO_3 media. The separated Mg and Fe were then repeated through the relevant column chemistry. The whole procedural blank is less than 5 ng for both Mg and Fe. Magnesium and iron isotopic compositions were determined by the standard–sample bracketing method using a Thermo–Finnigan Neptune Plus multiple-collector inductively coupled plasma mass spectrometer (MC–ICP–MS) in LR and HR mode, respectively. The Mg data are reported in the δ-notation relative to DSM3: δ^{i}Mg (‰)= $[(^{i}Mg/^{24}Mg_{sample}/(^{i}Mg/^{24}Mg)_{DSM3} -1] \times 10^3$, where i refers to mass 25 or 26. Magnesium isotopic compositions reported here are the averages of four times of repeat analyses and their uncertainties are reported as 2SD. Long-term external reproducibility on the $^{26}Mg/^{24}Mg$ ratio is better than ±0.06‰ based on replicate analyses of natural and synthetic standards (Ke et al., 2016). Iron isotopic variations are expressed in δ-notation relative to IRMM-014, δ^{i}Fe (‰)= $[(^{i}Fe/^{54}Fe)_{sample}/(^{i}Fe/^{54}Fe)_{IRMM-014} -1] \times 10^3$, where i refers to mass 56 or 57. Iron isotopic compositions reported here are the averages of several times (typically four times) of repeat analyses and their uncertainties in this study are reported as 2SE in 95% c.i. after Dauphas et al. (2009) and He et al. (2015) considering errors arising from both the instrumental analysis and the chemical procedures. Analyses over a 2-year period of reference materials demonstrate a long-term reproducibility and accuracy of ~0.04‰ for $^{56}Fe/^{54}Fe$ ratios. Samples analyzed here yield linear trends with slopes of 0.534±0.036 (1se, R^2 = 0.997) and 0.637±0.053 (1se, R^2=0.974) in the three-isotope plot (Fig. 3) of Mg and Fe, respectively. They agree with the theoretically expected slopes of mass-dependent fractionation ($\delta^{25}Mg/\delta^{26}Mg$=0.520, $\delta^{56}Fe/\delta^{57}Fe$=0.678, see Young et al. (2002)).

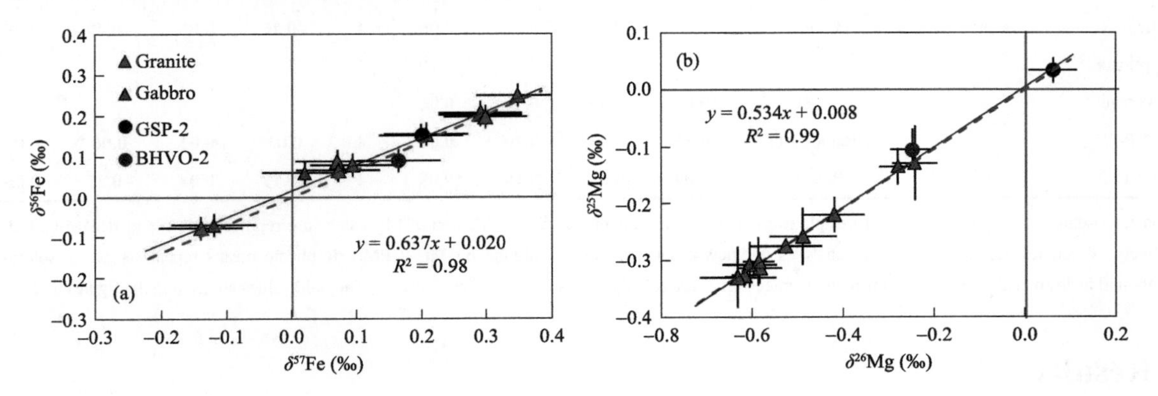

Fig. 3 (a) and (b) are iron and magnesium three-isotope plot of samples analyzed in this study, respectively. Standard samples BHVO-2 and GSP-2 are also plotted. The blue lines represent linear regressions of δ^{56}Fe vs. δ^{57}Fe or δ^{26}Mg vs. δ^{25}Mg. The slopes are consistent with the values determined for mass-dependent fractionation (Young et al., 2002). The theoretical lines from Young et al. (2002) are plotted for comparison (red dotted lines). (For interpretation of the references to colour in this figure legend, the reader is referred to the web version of this article.)

Two international geostandards (BHVO-2 and GSP-2) were processed with our samples to evaluate the accuracy. The measured δ^{26}Mg and δ^{56}Fe values of the standards (BHVO-2: δ^{26}Mg=–0.25±0.05‰, δ^{56}Fe=+0.09±0.03‰; GSP-2: δ^{26}Mg=+0.06±0.05‰, δ^{56}Fe=+0.15±0.03‰) agree well with their recommended values (BHVO-2: δ^{26}Mg=−0.20±0.07‰, δ^{56}Fe=0.109±0.002‰; GSP-2: δ^{26}Mg=+0.04±0.02‰, δ^{56}Fe=+0.157±0.005‰) within quoted errors (He et al., 2015; Teng et al., 2015). All replicate analyses yield consistent values within quoted errors (Table 1).

Table 1　Magnesium and iron isotopic compositions of samples in this study

Sample	Position (cm)	Rock Type	δ^{56}Fe	2se	δ^{57}Fe	2se	*N*	δ^{25}Mg	2sd	δ^{26}Mg	2sd	*N*
13DSC-5	~−400	Gabbro	**0.06**	**0.03**	**0.02**	**0.06**	4	**−0.13**	**0.07**	**−0.24**	**0.05**	4
13DSC-7-1	−1.0	Gabbro (chilled margin)	−0.07	0.03	−0.12	0.06	4	−0.14	0.02	−0.26	0.02	4
Replicate			−0.06	0.03	−0.09	0.06	4	−0.14	0.04	−0.30	0.07	4
Replicate								−0.12	0.03	−0.28	0.04	4
Average			**−0.06**	**0.02**	**−0.10**	**0.05**		**−0.13**	**0.02**	**−0.28**	**0.03**	
13DSC-7-2	0.5	Granite	−0.08	0.03	−0.14	0.06	4	−0.23	0.04	−0.41	0.07	4
Replicate			−0.03	0.08	−0.03	0.15	1	−0.21	0.03	−0.43	0.06	4
Average			**−0.07**	**0.03**	**−0.12**	**0.06**		**−0.22**	**0.02**	**−0.42**	**0.05**	
13DSC-7-3	2.5	Granite	0.08	0.03	0.09	0.06	4	−0.27	0.01	−0.53	0.08	4
Replicate			0.07	0.05	0.12	0.10	2					
Average			**0.08**	**0.03**	**0.10**	**0.06**						
13DSC-7-4	3.9	Granite	0.09	0.03	0.07	0.06	4	**−0.31**	**0.03**	**−0.58**	**0.05**	4
Replicate			0.08	0.05	0.13	0.10	2					
Average			**0.09**	**0.03**	**0.09**	**0.06**						
13DSC-7-5	5.5	Granite	**0.15**	**0.03**	**0.21**	**0.06**	4	**−0.31**	**0.04**	**−0.61**	**0.06**	4
13DSC-7-6	7.6	Granite	**0.20**	**0.03**	**0.29**	**0.06**	4	**−0.33**	**0.02**	**−0.62**	**0.03**	4
13DSC-7-7	9.9	Granite	**0.21**	**0.03**	**0.29**	**0.06**	4	**−0.33**	**0.05**	**−0.63**	**0.03**	4
13DSC-7-8	11.4	Granite	0.20	0.03	0.30	0.06	4	**−0.30**	**0.04**	**−0.59**	**0.03**	4
Replicate			0.21	0.05	0.30	0.10	2					
Average			**0.20**	**0.03**	**0.30**	**0.06**						
13DSC-7-9	13.4	Granite	**0.25**	**0.03**	**0.35**	**0.06**	4	**−0.33**	**0.04**	**−0.63**	**0.08**	4
13DSC-6	50	Granite	0.06	0.03	0.07	0.06	4	**−0.26**	**0.05**	**−0.49**	**0.07**	4
Replicate			0.05	0.05	0.15	0.10	2					
Average			**0.06**	**0.03**	**0.10**	**0.06**						
GSP-2		Granodiorite	**0.15**	**0.03**	**0.20**	**0.05**	4	**0.03**	**0.02**	**0.06**	**0.05**	4
BHVO-2		Basalt	**0.09**	**0.03**	**0.16**	**0.05**	4	**−0.11**	**0.04**	**−0.25**	**0.05**	4

Position represents the relative distance of sample to the gabbro-granite contact face. δ^{56}Fe and δ^{26}Mg values are reported relative to IRMM-014 and DSM-3, respectively. *N* denotes the total number of analyses per newly purified sample solution by MC-ICPMS. Replicate means repeat sample dissolution, column chemistry and instrumental analysis. The errors of average values are calculated based on weighted average. The bold values are used in the figures and main text.

4　Results

The Mg and Fe isotopic data as well as elemental data are reported in Tables 1 and 2, respectively. The variations of elemental and isotopic data along the profile are also plotted (Figs. 4 and 5). MgO and FeO contents of the gabbro end member (sample 13DSC-5, MgO: 15.45 wt.%, FeOt: 11.93 wt.%) are more than 10 times higher than those of the granite end member (sample 13DSC-6, MgO: 0.75 wt.%, FeO_t: 1.04 wt.%). The gabbro end member (13DSC-5) has δ^{26}Mg=–0.24±0.05‰ and δ^{56}Fe=+0.06±0.03‰, typical of mafic rocks (e.g., Dauphas et al., 2017; Teng, 2017). The granite end member (13DSC-6) yields δ^{26}Mg and δ^{56}Fe values of –0.49 ± 0.07‰ and +0.06±0.03‰, respectively.

Table 2 Major (wt. %) and trace element (ppm) compositions of samples in this study

Sample	13DSC-5	13DSC-7-1	13DSC-7-2	13DSC-7-3	13DSC-7-4	13DSC-7-5	13DSC-7-6	13DSC-7-7	13DSC-7-8	13DSC-7-9	13DSC-6	GSP-2	GSP-2 ref.	BHVO-2(1)	BHVO-2(2)	BHVO ref.	GSR-1	GSR-1 ref.
Rock Type	Gabbro	Gabbro (chilled margin)	Granite	Granite	Granite	Granite	Granite	Granite	Granite	Granite	Granite	USGS std.	USGS std.	USGS std.	USGS std.	USGS std.	GBW std.	GBW std.
Position (cm)	~–400	–1.0	0.5	2.5	3.9	5.5	7.6	9.9	11.4	13.4	50	na.	na.	na.	na.	na.	na.	na.
SiO_2	46.3	47.4	74.8	74.8	72.7	72.9	69.4	71.7	70.5	71.4	72.0	67.0	66.6	na.	na.		73.1	72.8
TiO_2	1.56	1.79	0.07	0.10	0.09	0.14	0.16	0.16	0.14	0.19	0.16	0.67	0.66	na.	na.		0.29	0.29
Al_2O_3	10.27	17.09	12.63	14.36	15.44	14.92	15.38	15.53	15.74	15.97	16.05	14.80	14.90	na.	na.		13.40	13.40
MnO	0.22	0.19	0.02	0.01	0.01	0.02	0.03	0.02	0.01	0.01	0.01	0.05	0.04	na.	na.		0.06	0.06
MgO	15.45	15.26	2.02	0.73	0.59	0.61	0.64	0.55	0.92	0.62	0.75	0.95	0.96	na.	na.		0.40	0.42
FeO	8.75	9.08	1.20	0.67	0.58	0.62	0.59	0.64	0.58	0.63	0.56	na.	na.	na.	na.		na.	na.
Fe_2O_3	3.54	6.82	0.48	0.28	0.22	0.41	0.65	0.47	0.46	0.70	0.53	4.89	4.90	na.	na.		2.13	2.14
CaO	10.76	0.46	1.39	0.19	0.42	0.89	3.42	0.94	1.29	0.27	0.43	2.10	2.10	na.	na.		1.57	1.55
Na_2O	1.41	1.71	4.35	5.03	5.05	4.46	5.18	5.40	5.46	5.56	6.39	2.80	2.78	na.	na.		3.11	3.13
K_2O	1.65	0.09	2.98	3.78	4.84	4.91	4.45	4.51	4.81	4.51	3.06	5.44	5.38	na.	na.		5.04	5.01
P_2O_5	0.10	0.11	0.05	0.03	0.03	0.12	0.09	0.03	0.03	0.08	0.08	0.29	0.29	na.	na.		0.09	0.07
LOI	1.80	9.34	1.51	0.82	0.69	0.94	2.89	1.23	1.10	1.31	0.50	0.69	na.	na.	na.		0.72	0.70
TC	0.52	0.22	0.17	0.23	0.19	0.38	0.67	0.27	0.17	0.27	nd.	na.	na.	na.	na.		na.	na.
CIA	30	82	49	53	52	51	44	50	49	52	52	na.	na.	na.	na.		na.	na.
Total FeO	11.93	15.22	1.64	0.93	0.78	0.99	1.18	1.06	1.00	1.25	1.04	na.	na.	na.	na.		na.	na.
Total Fe_2O_3	13.26	16.91	1.82	1.03	0.87	1.10	1.31	1.18	1.11	1.39	1.15	na.	na.	na.	na.		na.	na.
$Fe^{3+}/\Sigma Fe$	0.27	0.40	0.26	0.27	0.26	0.37	0.50	0.40	0.42	0.50	0.46	na.	na.	na.	na.		na.	na.
Mg#	70	64	69	59	57	53	49	48	62	47	57	28	28	na.	na.		27	28
Li	19	76	46	30	25	24	17	17	19	16	na.	34	36	4.54	4.02	4.74	na.	
Be	1.4	1.0	0.9	0.9	0.9	0.9	1.0	1.0	1.1	1.0	na.	1.5	1.5	1.06	0.95	na.	na.	
Sc	6.7	6.7	1.8	1.5	1.3	1.8	2.4	2.1	2.0	2.5	na.	6.0	6.3	32.1	30.3	32.8	na.	
V	325	321	42	27	24	27	26	22	30	33	na.	77	52	457	431	336	na.	
Cr	1028	770	5.4	2.0	1.4	0.9	1.4	1.8	1.4	2.5	na.	20	20	315	299	305	na.	
Co	67.2	23.7	2.7	1.3	1.1	1.2	1.2	1.3	1.4	1.5	na.	7.5	7.3	46.5	44.7	44.5	na.	
Ni	333	231	9.1	23.5	2.0	2.0	2.7	1.7	1.3	2.1	na.	16.5	17	127	121	121	na.	
Cu	87	18	20	730	2	2	19	3	8	13	na.	54	43	149	141	131	na.	
Zn	163	358	35	17	12	14	15	15	15	17	na.	134	120	133	131	112	na.	
Ga	21	55	16	14	15	15	15	16	16	16	na.	25	22	22.7	21.9	na.	na.	
Rb	42	2	96	109	139	147	128	134	143	129	na.	340	245	9.94	9.72	9.53	na.	
Sr	161	40	161	185	215	212	219	205	239	236	na.	195	240	425	417	404	na.	
Y	26	15	5	8	7	12	15	13	12	15	na.	26	28	25.0	24.4	28.3	na.	
Zr	169	83	54	119	88	125	139	151	161	181	na.	498	550	184	177	178	na.	

Continued

Sample	13DS C-5	13DSC -7-1	13DSC -7-2	13DSC -7-3	13DSC -7-4	13DSC -7-5	13DSC -7-6	13DSC -7-7	13DSC -7-8	13DSC -7-9	13DS C-6	GSP-2	GSP-2 ref.	BHVO -2(1)	BHVO -2(2)	BHVO ref.	GSR-1	GSR-1 ref.
Nb	11	68	4	8	10	15	17	17	16	18	na.	27	27	18.4	18.0	18.8	na.	
Cs	0.52	0.09	0.26	0.26	0.25	0.27	0.27	0.32	0.36	0.28	na.	1.24	1.2	0.1	0.1	0.1	na.	
Ba	921	61.2	871	985	1258	1259	1091	1093	1204	1114	na.	1424	1340	137	136	130	na.	
La	36	18	14	29	24	33	36	37	38	40	na.	174	180	15.7	15.6	15.1	na.	
Ce	96	42	25	51	40	57	63	63	66	70	na.	435	410	36.2	36.1	36.9	na.	
Pr	14	5	3	5	4	6	7	7	7	8	na.	56	51	5.46	5.35	5.34	na.	
Nd	57.6	19.8	9.0	18.5	15.0	21.3	23.2	22.7	24.1	26.3	na.	214	200	25.7	24.9	24.4	na.	
Sm	10.5	3.7	1.4	2.8	2.3	3.4	3.9	3.7	3.8	4.4	na.	26.4	27	6.16	5.95	6.05	na.	
Eu	3.02	0.76	0.39	0.61	0.59	0.79	0.87	0.84	0.84	0.86	na.	2.31	2.3	2.02	1.97	2.03	na.	
Gd	9.36	3.68	1.34	2.63	2.09	3.25	3.74	3.63	3.77	4.28	na.	19.9	12	6.28	6.38	6.04	na.	
Tb	1.14	0.52	0.16	0.29	0.24	0.40	0.47	0.44	0.43	0.52	na.	1.65	na.	0.92	0.90	0.95	na.	
Dy	5.65	2.82	0.85	1.42	1.25	2.04	2.48	2.25	2.19	2.73	na.	6.7	6.1	4.96	4.85	5.21	na.	
Ho	1.12	0.61	0.18	0.29	0.26	0.43	0.53	0.47	0.46	0.56	na.	1	1	1.02	0.98	0.99	na.	
Er	3.03	1.78	0.53	0.86	0.75	1.24	1.54	1.38	1.32	1.59	na.	3.0	2.2	2.60	2.50	2.55	na.	
Tm	0.41	0.27	0.08	0.13	0.11	0.19	0.23	0.21	0.20	0.24	na.	0.3	0.29	0.35	0.34	na.	na.	
Yb	2.53	1.85	0.53	0.85	0.76	1.21	1.58	1.36	1.29	1.57	na.	1.76	1.6	2.05	1.99	1.96	na.	
Lu	0.38	0.30	0.09	0.14	0.13	0.19	0.26	0.22	0.20	0.24	na.	0.25	0.23	0.30	0.29	0.28	na.	
Hf	5.16	3.36	2.19	3.95	3.52	4.05	4.36	4.67	4.90	5.52	na.	16.2	14.0	4.88	4.73	4.32	na.	
Ta	0.46	2.58	0.50	0.90	1.20	1.64	1.72	1.57	1.58	1.82	na.	0.86	na.	1.17	1.14	1.19	na.	
Pb	2.8	5.0	13.8	13.4	14.1	18.0	19.6	18.3	21.8	20.6	na.	49	42	1.49	1.47	1.54	na.	
Th	0.5	8.8	10.8	15.6	24.1	16.6	20.0	25.0	21.8	21.1	na.	114	105	1.28	1.26	1.24	na.	
U	0.16	3.68	1.04	1.22	2.12	1.43	1.56	1.86	1.74	1.65	na.	2.7	2.4	0.44	0.43	0.41	na.	

nd. – not detectable; na. – not available; LOI – Loss on ignition; TC – Total carbon; Ref. – reference values; std. – standard material; Total FeO = FeO + 0.9× Fe_2O_3; Total Fe_2O_3 = Total FeO/0.9; Mg# = $Mg^{2+}/(Mg^{2+}+Fe^{2+}+Fe^{3+})\times100$; $Fe^{3+}/\Sigma Fe = Fe^{3+}/(Fe^{2+}+Fe^{3+})$; CIA = $Al_2O_3/(Al_2O_3+CaO+Na_2O+K_2O)\times100$; All the major elements were normallized to a volatile-free basis using the LOI (except for rock standards). BHVO-2(1) and BHVO-2(2): repeat sample dissolution and instrumental analysis of BHVO-2.

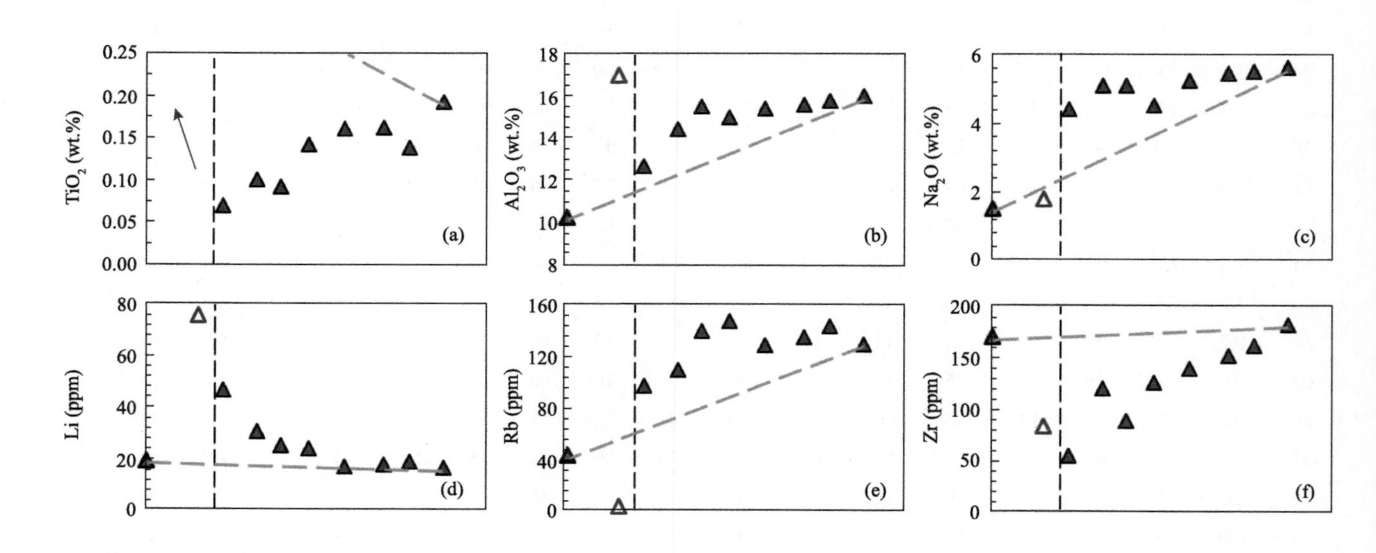

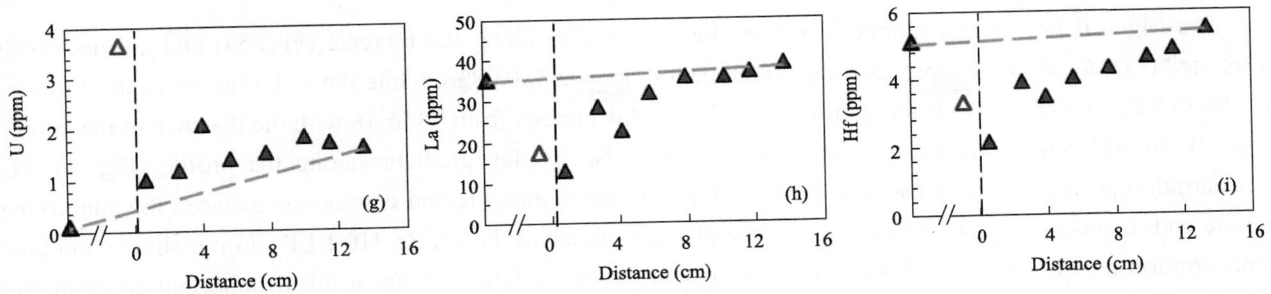

Fig. 4 Concentration profiles for representative major and trace elements. Distance x is given relative to the interface between the gabbro and granite. The red solid triangle represents granite profile samples. The green solid and hollow triangle symbols represent sample 13DSC-5 and 13DSC-7-1, respectively. The grey dotted lines represent the effect of physical mixing between the gabbro and granite magma. Sample 13DSC-5 is regarded as the gabbro end-member unaffected by diffusion. Sample 13DSC-7-1 is the chilled margin of the gabbro (adjacent to the granite). The abnormal element concentrations of 13DSC-7-1 could be attributed to the mineral effect of the chilled margin. As a chilled margin of gabbro, the mineral species and component could be altered and 13DSC-7-1 tends to concentrate or deplete certain elements. For example, the appearance of chlorite would elevate the Al_2O_3 content of the whole rock. The green arrow in (a) points toward the TiO_2 contents of the gabbro samples (13DSC-5 and 13DSC-7-1) because they are too high to be shown in the scale of granites' TiO_2 contents. (For interpretation of the relerences to colour in this figure legend, the reader is relerred to the web version of this article.)

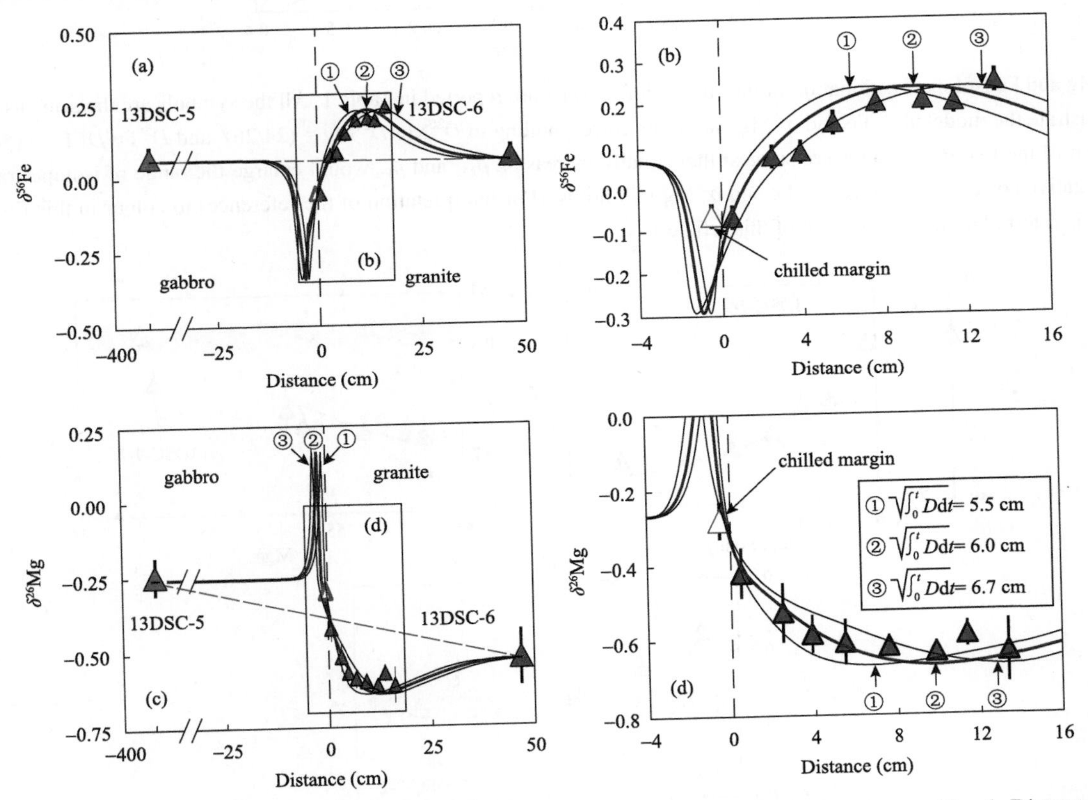

Fig. 5 Measured and modeled Mg and Fe isotopic diffusion profiles. All the symbols are the same as Fig. 4. Distance x is given relative to the interface between the gabbro and granite. The grey dotted lines represent the effect of physical mixing between the gabbro and granite magma. Blue and black curves are modeled lines based on Fick's second law. Model parameters: T = 750 °C; $C_{G,Mg}$ = 1.5 wt.%; $C_{R, Mg}$ = 0.5 wt.%; $C_{G, Fe}$ = 1.00 wt.%; $C_{R, Fe}$ = 0.50 wt.%; $D_{Mg\text{-}Fe, R} = 100\times D_{Mg\text{-}Fe, G}$; $\beta_{Mg}=\beta_{Fe}$ =0.05. The modeled curve yielding the best fit to the data is represented by the blue line. The $\sqrt{\int_0^t Ddt}$ in the figure represents $\sqrt{\int_0^t D_{Mg\text{-}Fe,R}dt}$. The two black curves represent the upper and lower limit $\sqrt{\int_0^t D_{Mg\text{-}Fe,R}dt}$ values yielding reasonable fits to the data. (For interpretation of the references to colour in this figure legend, the reader is referred to the web version of this article.)

The Mg and Fe isotopic compositions are highly variable along the traverse (Fig. 5a) and are negatively correlated (Fig. 6). δ^{26}Mg decreases from −0.28±0.04‰ to −0.63±0.08‰, while the δ^{56}Fe increases from −0.06±0.03‰ to 0.25±0.03‰ towards the granite end. The Mg# decreases from 70 to 48 with the distance to the contact (Fig. 7). In addition, concentrations of most elements also display gradients along the profile (Fig. 4). The concentration profiles of elements such as Al, Na, Rb, U are monotonic and continuous between the gabbro and granite end members (Fig. 4). By contrast, some elements (such as Ti, Li, Zr, Hf, REE) do not show monotonic concentration profiles (Fig. 4). It should be noted that gabbro 13DSC-7-1 has a lot of abnormal concentration values, the reason for which will be discussed in section 5.1.

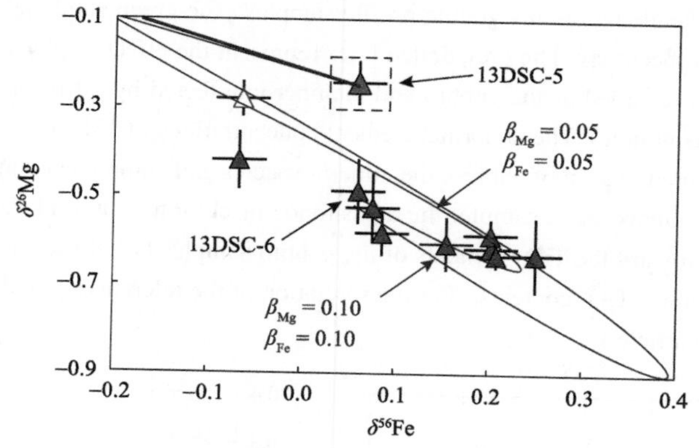

Fig. 6　Mg and Fe isotope fractionation of the study profile. Data are reported in Table 1. All the symbols are the same as Fig. 4. The curves illustrate the modeled δ^{56}Fe and δ^{26}Mg variations corresponding to D^{26}Mg/D^{24}Mg = $(24/26)^{\beta}$ and D^{56}Fe/D^{54}Fe = $(54/56)^{\beta}$. The comparison of the two modeled curves suggest that equally increasing β_{Mg} and β_{Fe} would enlarge the range of isotope fractionation, but the negative correlation between δ^{56}Fe and δ^{26}Mg still holds. (For interpretation of the references to colour in this figure legend, the reader is referred to the web version of this article.)

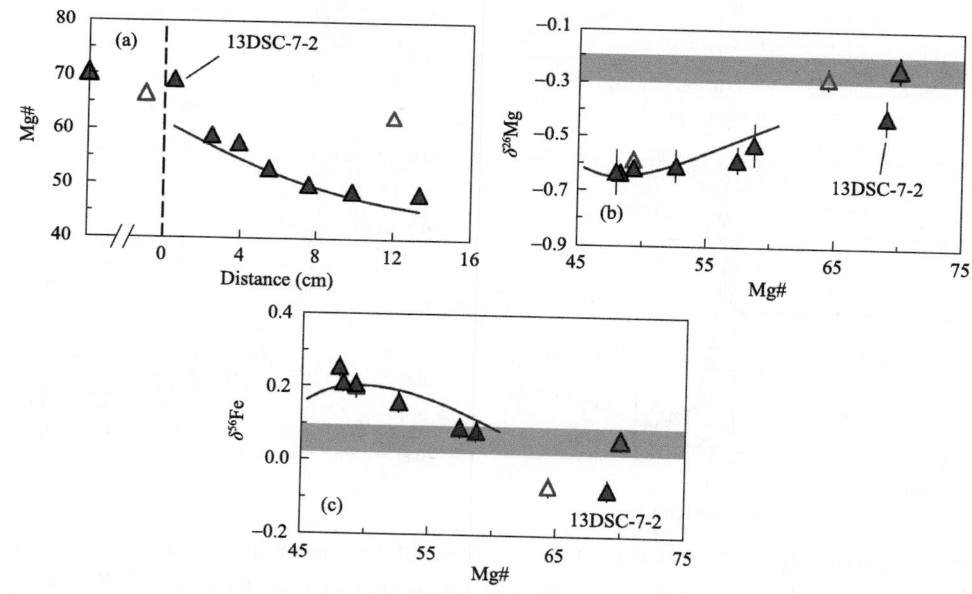

Fig. 7　(a) Measured and modeled Mg# profile. The Mg# of 13DSC-7-8 appears as an outlier, and the reason is not well understood. (b) and (c) are plots of δ^{56}Fe and δ^{26}Mg with Mg# in profile. Color shaded area in each panel represents the isotopic compositions of unaffected gabbro 13DSC-5. The blue lines are the modeled line of δ^{56}Fe and δ^{26}Mg varying with Mg#. The red hollow triangle represents sample 13DSC-7-8, which is excluded in the Mg# vs. Distance figure due to its abnormal Mg#. Obviously, Mg# of sample 13DSC-7-2 near the gabbro is higher than the modeling results. This could be due to a mixing between the rhyolitic melt and the interstitial basaltic melt near the interface or contamination of mafic minerals from the chilled margin. (For interpretation of the references to colour in this figure legend, the reader is referred to the web version of this article.)

5 Discussion

The highly variable elemental and Mg-Fe isotopic compositions along the profile could be induced by weathering, fluid exsolution, magma mixing, equilibrium fractionation or diffusion. In this section, we evaluate in detail the mechanisms that account for the isotope fractionations in turn.

5.1 Weathering, fluid exsolution, magma mixing or fractional crystallization

Both Mg and Fe isotope fractionation can occur during weathering process (e.g., Georg et al., 2007; Teng et al., 2010b; Huang et al., 2013; Liu et al., 2014), but several lines of evidence indicate the observed isotopic gradients in this study was not produced by weathering. Firstly, the samples are fresh (Fig. 2) and the surfaces of rock samples were removed before crushing and grinding. Secondly, LOI and CIA index ($100^*[Al_2O_3/(Al_2O_3 + CaO + Na_2O + K_2O)]$) for all samples (except 13DSC-7-1) are low and comparable to fresh igneous rocks (Table 2). Thirdly, the total carbon (TC) contents of the studied samples are low (<0.67 wt.% with an average of 0.31 wt.%), precluding the existence of secondary carbonates that may affect whole rock δ^{26}Mg values significantly (Teng, 2017). Therefore, the observed chemical and isotopic gradients in this study cannot be explained by weathering. It is noted that sample 13DSC-7-1 has a high LOI of 9.34 wt.% and CIA of 82. This is consistent with its high proportion (>30%) of chlorite, which is rich in water and Al_2O_3. Although chlorite could be a secondary mineral of weathering, it also commonly occurs in the chilled margin of mafic rocks as a result of high temperature hydrothermal alteration during intrusion and crystallization (e.g., Hayman et al., 2009). Chlorite in 13DSC-7-1 is well-crystallized. In addition, its coexisting plagioclase and adjacent granites (e.g., 13DSC-7-2) are unweathered. Therefore, abundant chlorite in 13DSC-7-1 is likely due to high-temperature hydrothermal alteration. Hydrothermal fluids may be from the adjacent granitic magma, which exsolves water during solidification. Mg and Fe isotopic compositions of 13DSC-7-1 seem not to be affected by the alteration, since they have not been decoupled (Figs. 5 and 6).

The Fe isotopes can also be strongly fractionated during fluid exsolution, with isotopically light Fe preferentially partitioning into fluids (e.g., Poitrasson and Freydier, 2005; Heimann et al., 2008). Although the influence of fluid exsolution on Mg isotopes is unclear, it is expected that light Mg isotopes also tend to be preferentially leached by the exsolving fluid (Brenot et al., 2008; Teng et al., 2010b). Negatively correlated δ^{26}Mg and δ^{56}Fe here thus rule out the possibility of fluid exsolution as the main cause of Mg and Fe isotopic variations in these granites.

Although partial melting and fractional crystallization may lead to significant iron isotope fractionation (Teng et al., 2008; Schuessler et al., 2009; Sossi et al., 2012; Foden et al., 2015; He et al., 2017; Wu et al., 2017; Xu et al., 2017). These processes, however, cannot account for the observed δ^{26}Mg variation and the negative correlation between δ^{26}Mg and δ^{56}Fe, because Mg isotope fractionation is sufficiently small during magmatic processes (e.g., Teng et al., 2007; Liu et al., 2010).

Mechanical mixing between gabbro and granite magmas is also unlikely to explain the observed chemical and isotopic variations because the variations of Ti, Li, Be, Zr, Hf, REE contents and Mg-Fe isotopic compositions of profile samples are far beyond the magma compositional differences represented by the two end-member samples 13DSC-5 and 13DSC-6 (Figs. 4 and 5). For instance, 13DSC-5 and 13DSC-6 have the same δ^{56}Fe within errors (0.06±0.03‰), mixing of which cannot explain the δ^{56}Fe variation of the profile samples from −0.07±0.03‰ to 0.25±0.03‰.

In summary, the Mg-Fe isotopic data of the profile samples cannot be explained by weathering, fluid exsolution, magma mixing, partial melting, and fractional crystallization.

5.2 Chemical diffusion-driven magnesium and iron isotope fractionation

The highly variable elemental and Mg-Fe isotopic compositions along the gabbro-granite profile may likely reflect diffusion-driven kinetic isotope fractionations. Both thermal and chemical diffusion could potentially produce Mg and Fe isotopic variations at high temperatures (Dauphas, 2007; Richter et al., 2008; 2009; Huang et al., 2010; Teng et al., 2011; Chopra et al., 2012; Sio et al., 2013; Oeser et al., 2015; Pogge von Strandmann et al., 2015). Thermal diffusion can be ruled out based on the negative δ^{26}Mg-δ^{56}Fe correlation (Fig. 6), because a positive correlation between δ^{26}Mg and δ^{56}Fe is expected (Richter et al., 2008; 2009). Chemical diffusion between the gabbro and granite as co-existing magmas/melts can also be ruled out, since both Mg and Fe are expected to diffuse from the mafic to the felsic magmas and thus a positive correlation between δ^{26}Mg and δ^{56}Fe would be also expected (Richter et al., 1999; 2008; 2009). The most likely mechanism is thus Mg-Fe inter-diffusion.

Mg^{2+} and Fe^{2+} usually occupy the same lattice site in silicate crystals due to their identical charge and similar ionic radii. Mg and Fe exchange ($Mg^{2+}_{phase1} + Fe^{2+}_{phase2} \leftrightarrow Mg^{2+}_{phase2} + Fe^{2+}_{phase1}$) via diffusion commonly occurs among minerals and between minerals and their co-existing magmas in disequilibrium (see the review of Zhang and Cherniak, 2010). This process is called 'Mg-Fe inter-diffusion'. The lattice limitation implies that at least one of the diffusing participants has been crystallized when Mg-Fe inter-diffusion occurs. Inter-diffusion of Mg-Fe can produce negatively correlated δ^{26}Mg and δ^{56}Fe, as Mg and Fe diffuse in opposite directions (Dauphas et al., 2010; Teng et al., 2011; Sio et al., 2013; Oeser et al., 2015). The negative correlation between δ^{26}Mg and δ^{56}Fe here most likely reflects Mg-Fe inter-diffusion between the gabbro and the granite. The gradually decrease of δ^{26}Mg and increase of δ^{56}Fe in the granite with the distance to the contact indicate Mg diffuses from the gabbro into the granite and Fe diffuses inversely. This is also evidenced by the significantly enhanced Mg# of the granite samples near the contact (Fig. 7a) as well as the strong correlations between Mg# and δ^{26}Mg or δ^{56}Fe values (Fig. 7b and c). The chilled margin at the gabbro side indicates the basaltic magma has experienced a rapid quenching at the early stage of emplacement, possibly when the basaltic magma reached a thermal equilibrium with the granitic intrusion. A temperature corresponding to >95% crystalline degree can be estimated to be about 750 ℃ for gabbro 13DSC-5 using MELTS, which is higher than the wet solidus line of granite under an intrusion depth (e.g., 10-15 km) (Chen and Grapes, 2007). Therefore, it is possible that the gabbro side was under solid state during element diffusion, with the granitic side simultaneously being still as a semi-solid or molten state. This could explain why Mg-Fe inter-diffusion occurred in the profile observed here. The inter-diffusion of Mg and Fe is driven by the Mg-Fe exchange reaction between the solid gabbro and the rhyolitic melt, which is analogous to the reaction between olivine and its surrounding melts. Diffusion after the solidification of both the gabbro and granite should be minimal, since Mg and Fe are dominantly hosted in few mafic mineral grains in the granite, isolated by abundant but almost Mg and Fe-free quartz and feldspar. It is noted that the boundary granite sample 13DSC-7-2 has higher Fe and Mg contents and Mg# than what are expected by inter-diffusion (Figs. 7 and 8). This could be due to mixing between the granitic and the interstitial basaltic melt near the interface. Mixing is also evidenced by the Co and V concentrations in the sample 13DSC-7-2 (Fig. 8). Contamination of mafic minerals from the chilled margin cannot be completely ruled out, since the contact interface is not an absolute plane and it is difficult to separate the chilled margin completely from the granite slices. The isotope profiles seem unaffected by this contamination since they show continues variations.

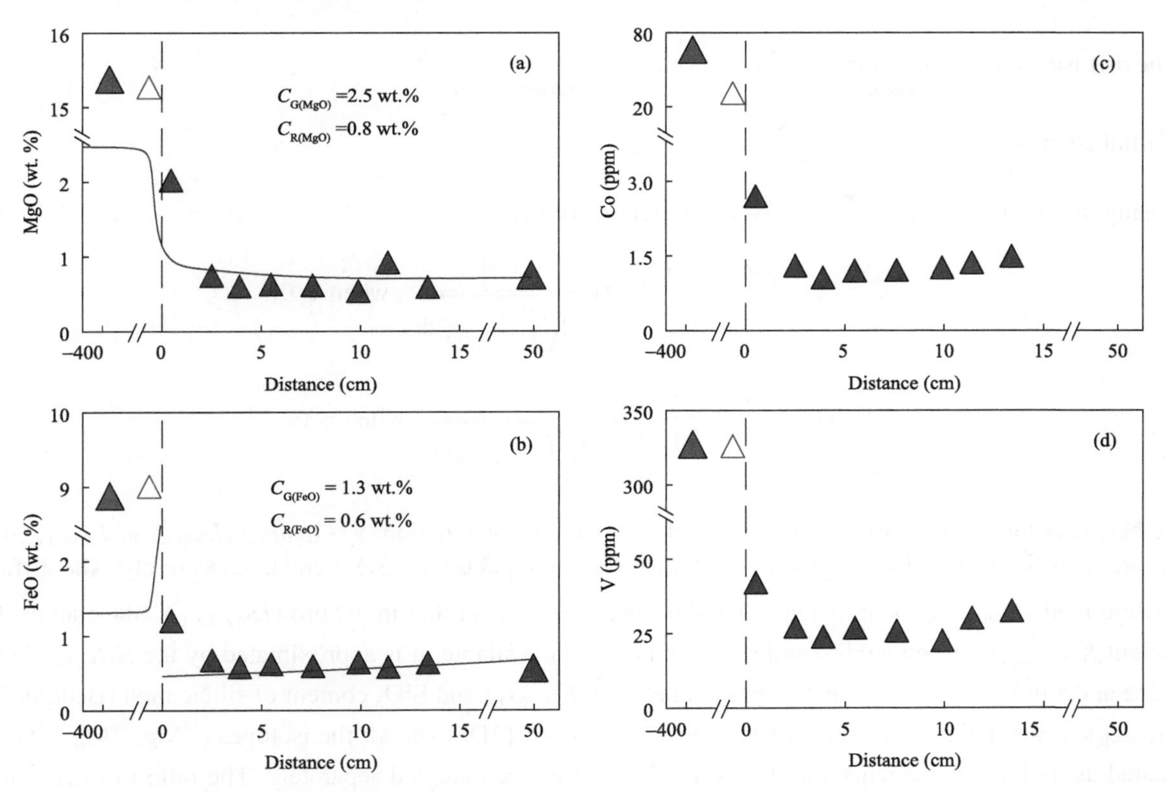

Fig. 8 Concentration profiles for MgO (a), FeO (b), Co (c) and V (d). All the symbols are the same as Fig. 4. The blue curves in (a) and (b) are the modeled lines of MgO and FeO activity variations based on best fit of the Mg#, δ^{56}Fe and δ^{26}Mg profiles. The activity parameters are displayed in the figures. (For interprelalion of the relerences to colour in this figure legend, the reader is relerred to the web version of this article.)

Diffusion between the gabbroic chilled margin and the granitic melt is also supported by elemental profiles. The diffusion of a component against its own concentration gradient has often been observed in systems involving solid phase, which is called uphill diffusion (see a review of Zhang et al., 1989; Liang, 2010). The compositions of Ti, Li, Zr, Hf and REE in this study do not show monotonic concentration profiles, which are very similar to those reported in Zhang et al. (1989) and Richter et al. (2003) and can also be attributed to uphill diffusion.

The elemental and isotopic data along the profile can be fitted by inter-diffusion of Fe and Mg between a solid gabbro and a molten rhyolite. Since the two sides belong to different phases, there would be both partitioning and diffusion of Mg and Fe. The diffusion in the two phases are modeled using a one-dimensional diffusion model separately, as each side requires different diffusivities.

The interface between the gabbro and granite is set as x=0. The initial effective element concentration (element activity) at x<0 (gabbro) is C_G, and that in the x>0 half (rhyolitic melt) is C_R. The partition coefficients of Mg and Fe between the two phases are K_{Mg} and K_{Fe}. The gabbro-rhyolite exchange coefficient $K_{D(Fe\text{-}Mg)}$ is K_{Fe}/K_{Mg}. The Mg and Fe have the same inter-diffusion coefficient ($D_{Mg\text{-}Fe}$). The Mg-Fe inter-diffusivity in gabbro half (x<0) is $D_{Mg\text{-}Fe,G}$, and in the rhyolitic melt half (x>0) is $D_{Mg\text{-}Fe,R}$. Time-dependent diffusion coefficients are adopted here to estimate the cooling rate of the rhyolitic melt by fitting the isotope profiles. The following equations are used:

Diffusion equation: $\dfrac{\partial C}{D\partial t}=\dfrac{\partial^2 C}{\partial x^2}$.

Define: $\alpha=\int_0^t D\mathrm{d}t$,

the diffusion equation becomes: $\dfrac{\partial C}{\partial \alpha}=\dfrac{\partial^2 C}{\partial x^2}$.

Initial condition: $C_{t=0}=\begin{cases}C_{\mathrm{G}} & x<0\\ C_{\mathrm{R}} & x>0\end{cases}$.

Using the solution from Zhang (2008) as reference, we get:

$$C=C_{\mathrm{G}}+\frac{\gamma(C_{\mathrm{R}}-KC_{\mathrm{G}})}{1+K\gamma}\operatorname{erfc}\frac{|x|}{2\sqrt{\int_0^t D_{\text{Mg-Fe,G}}\,\mathrm{d}t}}\text{, when } x<0;$$

$$C=C_{\mathrm{R}}+\frac{KC_{\mathrm{G}}-C_{\mathrm{R}}}{1+K\gamma}\operatorname{erfc}\frac{x}{2\sqrt{\int_0^t D_{\text{Mg-Fe,R}}\,\mathrm{d}t}}\text{, when } x>0.$$

Where here, t is the duration of the diffusion; erfc is Gauss error function; $\gamma=(\rho_{\mathrm{R}}/\rho_{\mathrm{G}})(D_{\text{Mg-Fe, R}}/D_{\text{Mg-Fe, G}})^{1/2}$, ρ_{R} and ρ_{G} are densities of the rhyolitic melt (~ 2.5g/cm^3) and the gabbro (~ 3.1 g/cm^3), respectively. The diffusivity in rhyolitic melt ($D_{\text{Mg-Fe, R}}$) is assumed to be 100 times higher than that in gabbro ($D_{\text{Mg-Fe, G}}$). Since an exchange coefficient $K_{D(\text{Fe-Mg})}$ between gabbro and rhyolitic melt is unavailable, it is approximated by the $K_{D(\text{Fe-Mg})}$ between pyroxene and silicic melt. Based on an equation between $K_{D(\text{Fe-Mg})}$ and SiO_2 content of silicic melt (Bédard, 2010), the $K_{D(\text{Fe-Mg})}$ is calculated to be 0.22 using the SiO_2 content of 13DSC-6. All the isotopes (^{24}Mg, ^{26}Mg, ^{54}Fe, ^{56}Fe) are treated as individual elements and their diffusion curves are modeled separately. The ratio of diffusivities of isotopes was expressed as $D_2/D_1=(\mathrm{m}_1/\mathrm{m}_2)^{\beta}$ (Richter et al., 1999). The δ^{26}Mg and δ^{56}Fe values of 13DSC-5 and 13DSC-6 are considered to represent the values of unaffected gabbro and granite, respectively. The values of β, C_{G}, C_{R} and $\sqrt{\int_0^t D_{\text{Mg-Fe,R}}\,\mathrm{d}t}$ are estimated to best fit the FeO, MgO, Mg# and isotopic profiles (Fig.5, Fig. 7). The best fit $C_{\text{G, Mg}}$, $C_{\text{R, Mg}}$, $C_{\text{G, Fe}}$ and $C_{\text{R, Fe}}$ values are 1.5 wt.%, 0.5 wt.%, 1.0 wt.% and 0.5 wt.%, respectively. Obviously, the fitted effective concentrations of Mg (1.5 wt.%) and Fe (1.0 wt.%) in the gabbro are lower than the actual Mg and Fe concentrations (see Table 2 and Appendix A), which may be due to the low element activity in the solid phase. The β is 0.05 for both Mg and Fe. The value for Mg is consistent with that from the experimental study (0.05±0.01, Richter et al., 2008), while the value for Fe is slightly higher than that from the experimental study (0.03±0.01, Richter et al., 2009). This inconsistence could be due to different compositions between our samples and the experiment materials and the uncertainties of our estimation of the element activities in the gabbro. The $\sqrt{\int_0^t D_{\text{Mg-Fe,R}}\,\mathrm{d}t}$ (characteristic diffusive transport distance) is 6 cm, with an accessible range of 5.5–6.7 cm (both lower or higher values did not yield a reasonable fit, Fig. 5). Although the inter-diffusion coefficient between gabbro and the granitic melt is unavailable, an estimation of $D_{\text{Mg-Fe,R,0}}$ ($D_{\text{Mg-Fe,R}}$ at 750 ℃) can be approached by the equation for binary ionic diffusion (Zhang, 2008):

$$D_{\text{Mg-Fe,R}}=\frac{D_{\mathrm{Mg}}\times D_{\mathrm{Fe}}\times(C'_{\mathrm{Mg}}+C'_{\mathrm{Fe}})}{C'_{\mathrm{Mg}}\times D_{\mathrm{Mg}}+C'_{\mathrm{Fe}}\times D_{\mathrm{Fe}}},$$

where C'_{Mg} and C'_{Fe} are molar concentrations of Mg and Fe in rhyolitic melt, D_{Mg} and D_{Fe} are self-diffusivities of Mg and Fe in rhyolitic melt. Based on a temperature and composition dependent diffusion coefficient function on Mg diffusivities in high silica melt (Zhang et al., 2010), the D_{Mg} is calculated to be 3.7×10^{-13} cm^2/s using a composition of 13DSC-6 under a temperature of 750 ℃. The diffusion coefficient of Fe is calculated from an experimental data-based Arrhenius equation, which yields a D_{Fe} of 1.1×10^{-12} cm^2/s (Todd and Ratliffe, 1990). C'_{Mg} and C'_{Fe} used here are molar concentrations of Mg and Fe in 13DSC-6. Hence, the inter-diffusivity of

$D_{\text{Mg-Fe,R,0}} \approx 6.5\times10^{-13}$ cm^2/s. If it is assumed that diffusion occurred at the initial condition without cooling, the diffusion timescale would be:

$$\tau = \int_0^t D_{\text{Mg-Fe,R}}\,\mathrm{d}t \,/\, D_{\text{Mg-Fe,R,0}}\ .$$

As a result, the duration of diffusion can be calculated to be 1.8 My with a range of 1.5 My and 2.2 My. Due to the assumption that diffusion occurred at the initial condition without cooling, the duration estimated above is the lower limit. In other word, the granitic magma was maintained under a semi-solid or molten state after intrusion for at least 1.5 My.

Given the diffusion property (the activation energy), the initial temperature T_0 and α obtained from fitting the profiles, a cooling rate may be obtained as follows:

Consider a thermal history of monotonic cooling represented by the asymptotic cooling model:

$$T = T_0 / (1 + \frac{t}{\tau_c}),$$

where τ_c is the cooling timescale for temperature to decrease from T_0 to $T_0/2$. The relation between τ and τ_c is (Zhang, 2008):

$$\tau = \tau_c (\frac{RT_0}{E}).$$

The cooling rate q is:

$$q = -\frac{\mathrm{d}T}{\mathrm{d}t\big|_{t=0}} = \frac{T_0}{\tau_c} = \frac{D_0 R T_0^2}{E\alpha} = \frac{RT_0^2}{E\tau},$$

where E is the activation energy, R is the universal gas constant. The T_0 used here is 750 ℃. Although the activation energy for Mg-Fe inter-diffusion in the granitic melt is unavailable, an approximate E is approached by assuming it is comparable to activation energy of Fe^{2+} or Mg^{2+} diffusion in rhyolitic melts. Based on the experimental data of ferrous iron diffusion in aluminosilicate melt (Dunn and Ratliffe, 1990), the activation energy of Fe^{2+} can be calculated to be 47.73 kJ/mol with uncertainly limites of 40.41 kJ/mol and 55.06 kJ/mol. The activation energy of Mg^{2+} diffusion in rhyolitic melts is unavailable. Considering the uncertainly range of τ, the cooling rate q can be calculated to be 76 ℃/My with an uncertainly range from 52 ℃/My to 107 ℃/My. Previous studies show the ages for the DSC gabbro intrusion is 129 ± 1 Ma (Dai et al., 2011). ^{40}Ar-^{39}Ar and fission track techniques revealed that North Dabie (where our samples collected) experienced a rapid lifting and cooled to 540 ℃ at ca. 125 Ma (Chen, 1995). The maximum duration of diffusion is 4 Ma based on the time interval between emplacement of intrusions and uplift-cooling of the Dabie Orogen, which is longer than the calculated duration (1.5–2.2 My). Therefore, the calculated duration of diffusion and cooling rate are consistent with the cooling history of North Dabie Orogen.

6 Summary and Implications

The negatively correlated $\delta^{26}Mg$ and $\delta^{56}Fe$ and their correlations with Mg# in the studied gabbro-granite boundary profile from the Dabie Orogen, China can be explained by isotope fractionations driven by Mg-Fe inter-diffusion, likely between the chilled gabbro margin and the granitic melt. Our studies demonstrate the potential application of the diffusion-controlled Mg-Fe isotope fractionation as geospeedometry of geological events by modeling the diffusion profiles. Since the inter-diffusion after the solidification of the granitic magma is negligible, the calculated diffusion duration 1.5–2.2 My indicates the granitic magma was maintained under a semi-solid or molten state magmas of quite different compositions or temperatures, e.g., inter-bedded mafic and felsic magmas, xenoliths/enclaves versus their hosting lavas, and cumulates in a cooling magma chamber, etc.

Acknowledgements

We deeply appreciate editorial handling by Weidong Sun and constructive comments from Stefan Weyer and two other anonymous reviewers who tremendously improved this manuscript. We are grateful to Qiang Zeng and Xiaolei Liu for their help in sample preparation. This work was financially supported by the DREAM project of Most China (No.2016YFC0600408); the National Natural Science Foundation of China (41473016); the Fundamental Research Funds for the Central Universities (2652014056) and State Key Lab. Of Geological Processes and Mineral Resources. This is CUGB petrogeochemical contribution No. PGC-201524.

References

Bédard J. H. (2010) Parameterization of the Fe=Mg exchange coefficient (Kd) between clinopyroxene and silicate melts. Chem. Geol. 274, 169-176.

Brenot A., Cloquet C., Vigier N., Carignan J. and France-Lanord C. (2008) Magnesium isotope systematics of the lithologically varied Moselle river basin, France. Geochim. Cosmochim. Acta 72, 5070-5089.

Chen J. (1995). Cooling age of Dabie Orogen, China, determined by 40Ar-39Ar and fission track techniques. Chinese Science Abstracts Series B. pp. 52.

Chen G.-N. and Grapes R. (2007) Granite genesis: in-situ melting and crustal evolution. Springer.

Chopra R., Richter F. M., Watson E. B. and Scullard C. R. (2012) Magnesium isotope fractionation by chemical diffusion in natural settings and in laboratory analogues. Geochim. Cosmochim. Acta 88, 1-18.

Dai L.-Q., Zhao Z.-F., Zheng Y.-F., Li Q., Yang Y. and Dai M. (2011) Zircon Hf-O isotope evidence for crust-mantle interaction during continental deep subduction. Earth Planet. Sci. Lett. 308, 229-244.

Dauphas N. (2007) Diffusion-driven kinetic isotope effect of Fe and Ni during formation of the Widmanstätten pattern. Meteorit. Planet. Sci. 42, 1597-1613.

Dauphas N., Pourmand A. and Teng F.-Z. (2009) Routine isotopic analysis of iron by HR-MC-ICPMS: How precise and how accurate? Chem. Geol. 267, 175-184.

Dauphas N., Teng F.-Z. and Arndt N. T. (2010) Magnesium and iron isotopes in 2.7 Ga Alexo komatiites: Mantle signatures, no evidence for Soret diffusion, and identification of diffusive transport in zoned olivine. Geochim. Cosmochim. Acta 74, 3274-3291.

Dauphas N., John S. G. and Rouxel O. (2017) Iron isotope systematics. Rev. Mineral. Geochem. 82, 415-510.

Dunn T. and Ratliffe W. A. (1990) Chemical diffusion of ferrous iron in a peraluminous sodium aluminosilicate melt: 0.1 MPa to 2.0 GPa. J. Geophys. Res.: Solid Earth 95, 15665-15673.

Foden J., Sossi P. A. and Wawryk C. M. (2015) Fe isotopes and the contrasting petrogenesis of A-, I- and S-type granite. Lithos 212-215, 32-44.

Georg R., Reynolds B., West A., Burton K. and Halliday A. (2007) Silicon isotope variations accompanying basalt weathering in Iceland. Earth Planet. Sci. Lett. 261, 476-490.

Hayman N., Anma R. and Veloso E. (2009) Data report: microstructure of chilled margins in the sheeted dike complex of IODP Hole 1256D. Proc. IODP 309, 312.

He Y., Ke S., Teng F. Z., Wang T., Wu H., Lu Y. and Li S. (2015) High precision iron isotope analysis of geological reference materials by high resolution MC-ICP-MS. Geostand. Geoanal. Res.

He Y., Wu H., Ke S., Liu S. A. and Wang Q. (2017) Iron isotopic compositions of adakitic and non-adakitic granitic magmas: Magma compositional control and subtle residual garnet effect. Geochim. Cosmochim. Acta 203, 89-102.

Heimann A., Beard B. L. and Johnson C. M. (2008) The role of volatile exsolution and sub-solidus fluid/rock interactions in producing high 56Fe/54Fe ratios in siliceous igneous rocks. Geochim. Cosmochim. Acta 72, 4379-4396.

Hou Z.-H. and Wang C.-X. (2007) Determination of 35 trace elements in geological samples by inductively coupled plasma mass spectrometry. J. Univ. Sci. Technol. China 37, 940-944.

Huang F., Chakraborty P., Lundstrom C., Holmden C., Glessner J., Kieffer S. and Lesher C. (2010) Isotope fractionation in silicate melts by thermal diffusion. Nature 464, 396-400.

Huang K.-J., Teng F.-Z., Elsenouy A., Li W.-Y. and Bao Z.-Y. (2013) Magnesium isotopic variations in loess: Origins and implications. Earth Planet. Sci. Lett. 374, 60-70.

Ke S., Teng F.-Z., Li S.-G., Gao T., Liu S.-A., He Y. and Mo X. (2016) Mg, Sr, and O isotope geochemistry of syenites from northwest Xinjiang, China: Tracing carbonate recycling during Tethyan oceanic subduction. Chem. Geol. 437, 109-119.

Li W. Y., Teng F. Z., Xiao Y., Gu H. O., Zha X. P. and Huang J. (2016) Empirical calibration of the clinopyroxene-garnet magnesium isotope geothermometer and implications. Contrib. Mineral. Petrol. 171, 1-14.

Liang Y. (2010) Multicomponent diffusion in molten silicates: theory, experiments, and geological applications. Rev. Mineral. Geochem. 72, 409-446.

Liu S.-A., Teng F.-Z., He Y., Ke S. and Li S. (2010) Investigation of magnesium isotope fractionation during granite differentiation: implication for Mg isotopic composition of the continental crust. Earth Planet. Sci. Lett. 297, 646-654.

Liu S.-A., Teng F.-Z., Yang W. and Wu F.-Y. (2011) Hightemperature inter-mineral magnesium isotope fractionation in mantle xenoliths from the North China craton. Earth Planet. Sci. Lett. 308, 131-140.

Liu S.-A., Teng F.-Z., Li S., Wei G.-J., Ma J.-L. and Li D. (2014) Copper and iron isotope fractionation during weathering and pedogenesis: insights from saprolite profiles. Geochim. Cosmochim. Acta 146, 59-75.

Oeser M., Dohmen R., Horn I., Schuth S. and Weyer S. (2015) Processes and time scales of magmatic evolution as revealed by Fe-Mg chemical and isotopic zoning in natural olivines. Geochim. Cosmochim. Acta 154, 130-150.

Pogge von Strandmann P. A. E., Dohmen R., Marschall H. R., Schumacher J. C. and Elliott T. (2015) Extreme magnesium isotope fractionation at outcrop scale records the mechanism and rate at which reaction fronts advance. J. Petrol. 56, 33-58.

Poitrasson F. and Freydier R. (2005) Heavy iron isotope composition of granites determined by high resolution MC-ICP-MS. Chem. Geol. 222, 132-147.

Richter F. M., Liang Y. and Davis A. M. (1999) Isotope fractionation by diffusion in molten oxides. Geochim. Cosmochim. Acta 63, 2853-2861.

Richter F. M., Davis A. M., DePaolo D. J. and Watson E. B. (2003) Isotope fractionation by chemical diffusion between molten basalt and rhyolite. Geochim. Cosmochim. Acta 67, 3905-3923.

Richter F. M., Watson E. B., Mendybaev R. A., Teng F.-Z. and Janney P. E. (2008) Magnesium isotope fractionation in silicate melts by chemical and thermal diffusion. Geochim. Cosmochim. Acta 72, 206-220.

Richter F. M., Watson E. B., Mendybaev R., Dauphas N., Georg B., Watkins J. and Valley J. (2009) Isotopic fractionation of the major elements of molten basalt by chemical and thermal diffusion. Geochim. Cosmochim. Acta 73, 4250-4263.

Schoenberg R. and von Blanckenburg F. (2006) Modes of planetary-scale Fe isotope fractionation. Earth Planet. Sci. Lett. 252, 342-359.

Schuessler J. A., Schoenberg R. and Sigmarsson O. (2009) Iron and lithium isotope systematics of the Hekla volcano, Iceland—evidence for Fe isotope fractionation during magma differentiation. Chem. Geol. 258, 78-91.

Sio C. K. I., Dauphas N., Teng F.-Z., Chaussidon M., Helz R. T. and Roskosz M. (2013) Discerning crystal growth from diffusion profiles in zoned olivine by in situ Mg-Fe isotopic analyses. Geochim. Cosmochim. Acta 123, 302-321.

Sio C. K. I. and Dauphas N. (2017) Thermal and crystallization histories of magmatic bodies by Monte Carlo inversion of Mg-Fe isotopic profiles in olivine. Geology 45, 67-70.

Sossi P. A., Foden J. D. and Halverson G. P. (2012) Redoxcontrolled iron isotope fractionation during magmatic differentiation: an example from the Red Hill intrusion. S. Tasmania. Contrib. Mineral. Petrol. 164, 757-772.

Telus M., Dauphas N., Moynier F., Tissot F. L., Teng F.-Z., Nabelek P. I., Craddock P. R. and Groat L. A. (2012) Iron, zinc, magnesium and uranium isotopic fractionation during continental crust differentiation: the tale from migmatites, granitoids, and pegmatites. Geochim. Cosmochim. Acta 97, 247-265.

Teng F.-Z. (2017) Magnesium isotope geochemistry. Rev. Mineral. Geochem. 82, 219-287.

Teng F.-Z., Wadhwa M. and Helz R. T. (2007) Investigation of magnesium isotope fractionation during basalt differentiation:

Implications for a chondritic composition of the terrestrial mantle. Earth Planet. Sci. Lett. 261, 84-92.

Teng F.-Z., Dauphas N. and Helz R. T. (2008) Iron isotope fractionation during magmatic differentiation in Kilauea Iki lava lake. Science 320, 1620-1622.

Teng F.-Z., Li W.-Y., Ke S., Marty B., Dauphas N., Huang S., Wu F.-Y. and Pourmand A. (2010a) Magnesium isotopic composition of the Earth and chondrites. Geochim. Cosmochim. Acta 74, 4150-4166.

Teng F.-Z., Li W.-Y., Rudnick R. L. and Gardner L. R. (2010b) Contrasting lithium and magnesium isotope fractionation during continental weathering. Earth Planet. Sci. Lett. 300, 63-71.

Teng F.-Z., Dauphas N., Helz R. T., Gao S. and Huang S. (2011) Diffusion-driven magnesium and iron isotope fractionation in Hawaiian olivine. Earth Planet. Sci. Lett. 308, 317-324.

Teng F.-Z., Dauphas N., Huang S. and Marty B. (2013) Iron isotopic systematics of oceanic basalts. Geochim. Cosmochim. Acta 107, 12-26.

Teng F. Z., Li W. Y., Ke S., Yang W., Liu S. A., Sedaghatpour F., Wang S. J., Huang K. J., Hu Y. and Ling M. X. (2015) Magnesium isotopic compositions of international geological reference materials. Geostand. Geoanal. Res. 39, 329-339.

Todd D. and Ratliffe W. A. (1990) Chemical diffusion of ferrous iron in a peraluminous sodium aluminosilicate melt: 0.1 MPa to 2.0 GPa. J. Geophys. Res. Atmos. 951, 15665-15673.

Wang S.-J., Teng F.-Z., Rudnick R. L. and Li S.-G. (2015) Magnesium isotope evidence for a recycled origin of cratonic eclogites. Geology 43, 1071-1074.

Watkins J. M., DePaolo D. J. and Watson E. B. (2017) Kinetic fractionation of non-traditional stable isotopes by diffusion and crystal growth reactions. Rev. Mineral. Geochem. 82, 85-125.

Weyer S., Anbar A. D., Brey G. P., Münker C., Mezger K. and Woodland A. B. (2005) Iron isotope fractionation during planetary differentiation. Earth Planet. Sci. Lett. 240, 251-264.

Weyer S. and Ionov D. A. (2007) Partial melting and melt percolation in the mantle: the message from Fe isotopes. Earth Planet. Sci. Lett. 259, 119-133.

Wu H., He Y., Bao L., Zhu C. and Li S. (2017) Mineral composition control on inter-mineral iron isotopic fractionation in granitoids. Geochim. Cosmochim. Acta 198, 208-217.

Xu H., Ma C., Song Y., Zhang J. and Ye K. (2012) Early Cretaceous intermediate-mafic dykes in the Dabie orogen, eastern China: Petrogenesis and implications for crust-mantle interaction. Lithos 154, 83-99.

Xu H., Ma C. and Ye K. (2007) Early cretaceous granitoids and their implications for the collapse of the Dabie orogen, eastern China: SHRIMP zircon U-Pb dating and geochemistry. Chem. Geol. 240, 238-259.

Xu L. J., He Y., Wang S. J., Wu H. and Li S. (2017) Iron isotope fractionation during crustal anatexis: constraints from migmatites from the dabie orogen, central china. Lithos 284, 171-179.

Young E. D., Galy A. and Nagahara H. (2002) Kinetic and equilibrium mass-dependent isotope fractionation laws in nature and their geochemical and cosmochemical significance. Geochim. Cosmochim. Acta 66, 1095-1104.

Zhang Y. (2008) Geochemical Kinetics. Princeton University Press. Zhang Y., Walker D. and Lesher C. E. (1989) Diffusive crystal dissolution. Contrib. Mineral. Petrol. 102, 492-513.

Zhang Y. and Cherniak D. J. (2010). Diffusion in minerals and melts.

Zhang Y., Ni H. and Chen Y. (2010) Diffusion data in silicate melts. Rev. Mineral. Geochem. 72, 311-408.

Diffusion-driven extreme Mg and Fe isotope fractionation in Panzhihua ilmenite: Implications for the origin of mafic intrusion*

Heng-Ci Tian[1,2], Chi Zhang[1,2], Fang-Zhen Teng[3], Yi-Jie Long[1,4], Shu-Guang Li[5], Yongsheng He[5], Shan Ke[5], Xin-Yang Chen[3] and Wei Yang[1,2]

1. Key Laboratory of Earth and Planetary Physics, Institute of Geology and Geophysics, Chinese Academy of Sciences, Beijing 100029, China
2. Innovation Academy for Earth Science, Chinese Academy of Sciences, Beijing 100029, China
3. Isotope Laboratory, Department of Earth and Space Sciences, University of Washington, Seattle, WA 98195, USA
4. University of Chinese Academy of Sciences, Beijing 100049, China
5. State Key Laboratory of Geological Processes and Mineral Resources, China University of Geosciences, Beijing 100083, China

亮点介绍：本文对攀枝花镁铁质层状岩体含矿层和非矿层中的橄榄石、辉石和钛铁矿单矿物开展了Mg-Fe同位素分析，发现钛铁矿具有明显的Mg-Fe同位素分馏，且含矿层和非矿层中钛铁矿的Mg-Fe同位素扩散方向是相反的。这很可能是粒间液态不混溶过程引起的同位素分馏。初始阶段的高钛玄武质熔体，经过硅酸盐矿物的结晶，形成了晶粥层的骨架，随后粒间熔体发生不混溶，分离出Fe-rich相和Si-rich相。前者具有高Fe含量，后者表现为低Mg特征。由于密度差异，富Si熔体逐渐上升，而富Fe熔体则逐渐下沉。不同化学成分的钛铁矿与这两相熔体之间存在较大的Mg活度梯度，从而引起极大的同位素分馏效应。

Abstract To investigate the petrogenesis of Fe-Ti oxide ore deposits, we report Mg and Fe isotopic compositions for coexisting olivine, clinopyroxene and ilmenite in Fe-Ti oxide ores, magnetite-bearing gabbros and gabbros from the Panzhihua Fe-Ti-oxide-bearing layered mafic intrusion, Southwest China. Olivine and clinopyroxene have δ^{26}Mg values ranging from −0.47 to −0.32‰ and −0.40 to −0.18‰, and δ^{57}Fe values from −0.21 to +0.23‰ and −0.16 to +0.20‰, respectively. Most of these mineral pairs display disequilibrium inter-mineral fractionation as they fall off the theoretically predicted equilibrium fractionation lines. Ilmenites from oxide ores and magnetite-bearing gabbros (Group 1) have mantle-like or lower δ^{26}Mg values of −0.80 to −0.13‰ and δ^{57}Fe values of −0.33 to −0.23‰, and those from gabbros (Group 2) display slightly to extremely higher δ^{26}Mg values of +0.48 to +23.10‰ and much lower δ^{57}Fe values of −0.59 to −0.33‰. The δ^{26}Mg negatively correlates with δ^{57}Fe in all ilmenites, which cannot be explained by simple extensive fractional crystallization or Soret diffusion. Instead, the

* 本文发表在：Geochimica et Cosmochimica Acta, 2020, 278: 361-375

negative correlations between δ^{26}Mg and δ^{57}Fe and between MgO and FeO in ilmenites result from Mg-Fe inter-diffusion among ilmenite, silicate minerals and high-Ti basaltic melts. The extremely large isotopic variations were produced by the large Mg activity gradient between different types of ilmenites and melts, which was enhanced by interstitial liquid immiscibility process. Our study therefore demonstrates that combined Mg-Fe isotopes can be used to trace the genesis of Fe-Ti-oxide-bearing ore.
Keywords Ilmenite, Mg-Fe isotopes, chemical diffusion, panzhihua layered intrusion, interstitial liquid immiscibility

1 Introduction

Understanding the origins of Fe-Ti oxide ore deposit is of primary importance for exploration, efficient mining operations and ore processing (Charlier et al., 2015). The economic concentrations of Fe and Ti are mainly carried by ilmenite and magnetite, which are commonly emplaced in the mafic layered intrusions that contain olivine, plagioclase, clinopyroxene and apatite (Klemm et al., 1985; Hunter and Sparks, 1987; Toplis and Carrol, 1996; Zhou et al., 2002, 2005; Cawthorn and Ashwal, 2009; Namur et al., 2010; Song et al., 2013). However, the petrogenesis of the giant Fe-Ti oxide ore is still controversial and several hypotheses have been proposed to explain its origin, including accumulation of Fe-Ti oxides in the late magmatic fractionation processes (Klemm et al., 1985; Song et al., 2013), an Fe-rich immiscible liquid segregated from mafic magma (Reynolds, 1985; Von Gruenewaldt, 1993; Zhou et al., 2005), or with magma addition and/or mixing (Robinson et al., 2003).

Magnesium and Fe isotopes can provide new constraints on the origins of Fe-Ti oxide ores because (1) fractional crystallization and partial melting do not significantly fractionate Mg isotopes but can cause large Fe isotopic variation (Teng et al., 2007, 2008, 2013); (2) combined Mg-Fe isotopic studies provide evidence to distinguish Soret diffusion from chemical diffusion in minerals (Richter et al., 2008, 2009a; Huang et al., 2010; Teng et al., 2011; Sio et al., 2013; Oeser et al., 2015; Sio and Dauphas, 2017); and (3) liquid immiscibility could lead to large Mg and Fe isotope fractionation with Si-rich phase enriched in heavy isotopes relative to the Fe-rich phase due to its high degree of polymerization (Zhu et al., 2015). In view of this, a few studies have combined Mg and Fe isotopes to explore the petrogenesis of Fe-Ti oxide ores although no consensus has been reached. For example, Liu et al. (2014b) suggested that the disequilibrium Fe isotopic fractionation among coexisting olivine, clinopyroxene and titanomagnetite from Fe-Ti oxide ores in Baima (Southwest China) was caused by changes in oxygen fugacity during phase segregation. By contrast, Chen et al. (2014, 2018) suggested that isotopic disequilibrium from the Baima intrusion resulted from chemical diffusion under sub-solidus conditions. These different interpretations lead to different implications on ore-forming processes. Hence, more systematic studies are required to better understand the mechanisms responsible for the large Mg and Fe isotopic variations in ilmenite and coexisting silicates in the mafic intrusion.

The Panzhihua layered mafic intrusion, located at the central part of the Emeishan large igneous province, Southwest China (Fig. 1), hosts a world-class Fe-Ti ore deposit, which is considered to be genetically related to the Emeishan mantle plume (e.g., Pang et al., 2008a). Here, we analyzed Mg and Fe isotopic compositions of olivine, clinopyroxene and ilmenite from a set of samples ranging from Fe-Ti-rich ore deposit layer to Fe-Ti-poor layers including oxide ores, magnetite-gabbros (Mt-gabbro), ilmenite-gabbros (ilm-gabbro) and gabbros from the Panzhihua layered mafic intrusion. Our results suggest that large Mg and Fe isotopic variations exist in ilmenites and coexisting silicates, which was mainly caused by chemical diffusion during the interstitial liquid immiscibility process.

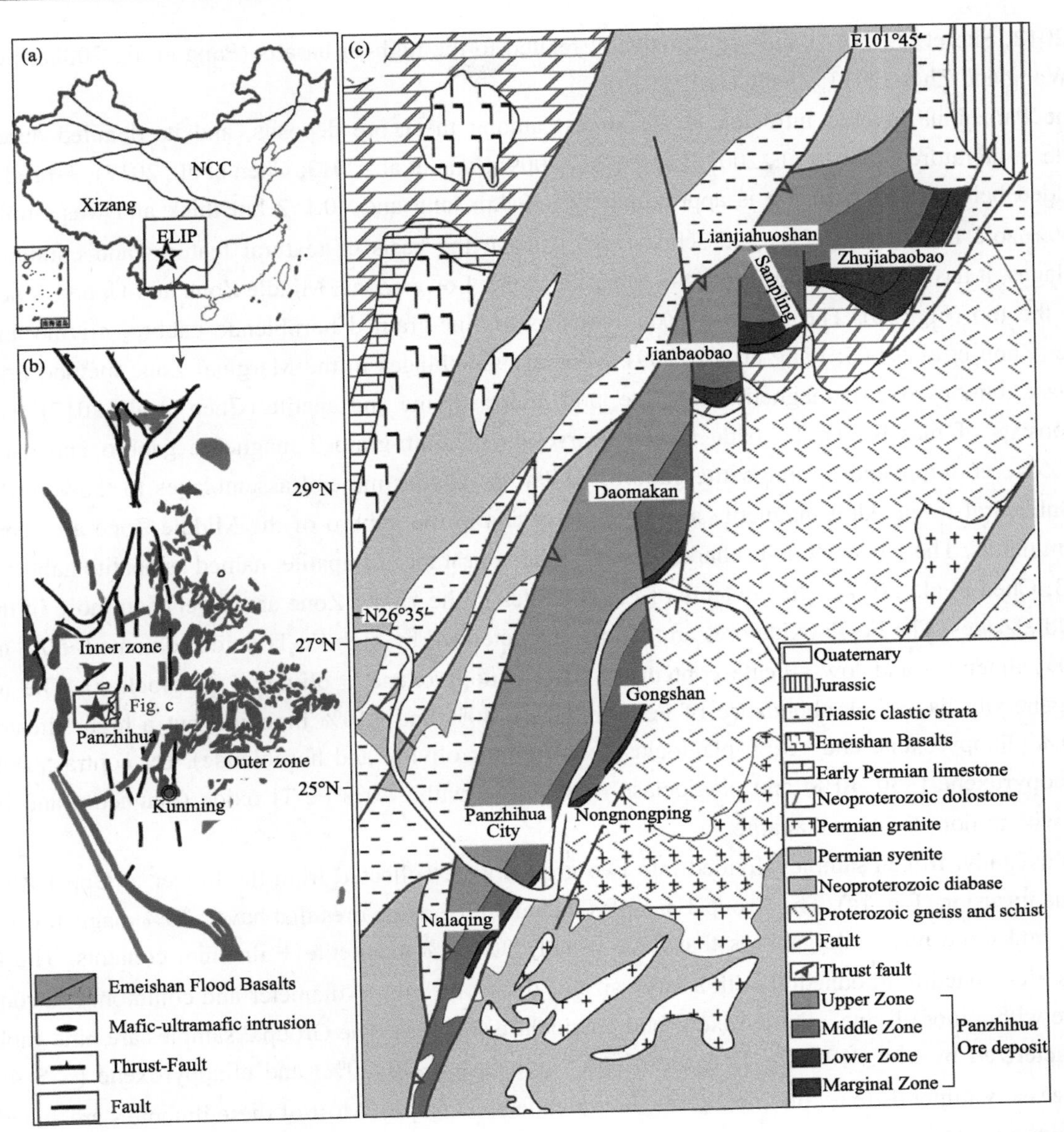

Fig. 1 Simplified geological maps of (a) major tectonic units of China, (b) distribution of the continental flood basalts and contemporaneous mafic intrusions within the Emeishan large igneous province and (c) Panzhihua layered mafic intrusion (modified from Zhou et al., 2002; Song et al., 2013). The samples investigated in this study were collected from the Lanjiahuoshan profile.

2 Geological background and samples

The Emeishan large igneous province (ELIP) is located in Southwest China, between the Tibetan Plateau and Yangtze block (Fig. 1a), and consists of flood basalts, minor picrite and a variety of mafic-ultramafic to felsic intrusive rocks with an area of more than 2.5 × 10^5 km^2 (Fig. 1b), which are interpreted to be the product of a mantle plume at the end of the Guadalupian (~260 Ma) (Chung and Jahn, 1995; Xu et al., 2001; Zhou et al., 2002). It is subdivided into a central part and an outer zone. Importantly, five world-class Fe-Ti-(V) oxide ore deposits also developed within the mafic-ultramafic intrusions along the N-S-trending Panzhihua- and Anninghe Fault in the central part of the ELIP (Fig. 1). They are the Panzhihua, Hongge, Xinjie, Baima, and Taihe intrusions, from south to north (Zhou et al., 2002, 2005; Zhong and Zhu, 2006; Song et al., 2013). These intrusions have SHRIMP zircon U-Pb ages of ~260 Ma (Zhou et al., 2002, 2005; Zhong and Zhu, 2006; Shellnutt et al., 2012; She et al., 2014) and are genetically related to the high-Ti basalts (Pang et al., 2008a; Song et al., 2013; Wang and Zhou, 2013; Zhang et al., 2013).

The Panzhihua layered intrusion is the largest among these ore deposits, and its detailed description is available in literature (Zhou et al., 2005; Pang et al., 2009; Song et al., 2013; Chen et al., 2017). A brief summary is provided below. This intrusion is approximately ~19 km long and ~0.1–2 km thick and was emplaced into Neoproterozoic limestone, gneiss and schist (Fig. 1c). On the basis of textural features and cumulus mineral assemblages, it has been subdivided into the Marginal Zone, Lower Zone, Middle Zone and Upper Zone from the base to the top (Fig. 1c). The Marginal Zone consists of fine-grained hornblende gabbro, olivine gabbro and wehrlite (Zhou et al., 2005; Pang et al., 2009). Mineral assemblages in the Marginal Zone include fine-grained plagioclase, clinopyroxene, magnetite, ilmenite, and minor olivine and apatite (Chen et al., 2017). The Lower Zone consists of massive Fe-Ti oxide layers, coarse- to medium-grained magnetite gabbro and gabbro. The Middle Zone is dominated by magnetite gabbro, which has similar mineral assemblages to those in the Lower Zone. Subhedral to euhedral magnetite and ilmenite grains in the gabbro of the Middle Zone are separated by silicate minerals. The Upper Zone is characterized by the appearance of apatite, named as apatite gabbro (Song et al., 2013; Chen et al., 2017). The contents of Fe-Ti oxides in the Upper Zone are lower than those in the Lower and Middle Zones. The Fe-Ti oxide ores are mainly characterized by high Fe-Ti oxide contents (>50% magnetite and ~10% ilmenite) and low silicate contents (<40%, clinopyroxene + olivine + plagioclase). The magnetite gabbro generally has a low abundance of Fe-Ti oxides (<40% and <10% ilmenite) but a high silicate content (~20–30% clinopyroxene and <30% plagioclase, with minor olivine and hornblende). By contrast, gabbro has high clinopyroxene (~30–40%) and plagioclase (~30–45%), with <20% Fe-Ti oxide (magnetite and ilmenite), coupled with minor olivine and apatite.

In this study, fifteen samples divided into two groups were collected from the Lower to Upper Zone in the Panzhihua intrusion (Fig. 1c). The Group 1 samples include three oxide ores that have >70% magnetite + ilmenite contents, and three magnetite-gabbros that have usually 30–60% magnetite + ilmenite contents. The Group 1 ilmenites are subhedral to euhedral with a large size of 0.7 to 1.5 mm in diameter and commonly surrounded by magnetite with minor olivine, clinopyroxene and plagioclase (Fig. 2a). The Group 2 samples are nine gabbros that are characterized by low <10% Fe-Ti oxides, high plagioclase (~30–70%) and clinopyroxene (~25–40%) and minor accessory minerals (e.g., olivine, biotite, hornblende and apatite). Most of these ilmenites are anhedral with a small size of <1 mm and they commonly occur as interstitial fillings between silicate minerals (Fig. 2b). In general, subhedral clinopyroxenes from these samples have abundant exsolution of Fe-Ti oxides that produce Schiller effects and dominantly comprise of titanomagnetite and ilmenite (Cao et al., 2019). The olivine grains are subhedral to euhedral, some of which contain a few Fe-Ti oxide inclusions. Most of the hornblende grains occurred as reaction rims between silicates and oxides, which are also observed in previous studies (Zhou et al., 2005; Pang et al., 2009). For chemical analyses, all these rocks were crushed to ~60 mesh to separate the Fe-Ti oxides and silicate minerals. Then, the separated minerals (olivine, clinopyroxene and ilmenite grains) were handpicked under a binocular microscope to a purity of approximately 100%. To avoid potential contamination from oxide inclusions in olivines, only the transparent green crystals were selected for Mg and Fe isotopes. For example, back scattered electron (BSE) images (Fig. S1) indicate that the separated ilmenite grains are very pure, close to 100%, with a few of them containing trace amounts of inclusions. Finally, all separated minerals were cleaned in an ultrasonic bath in Milli-Q water before isotopic analysis.

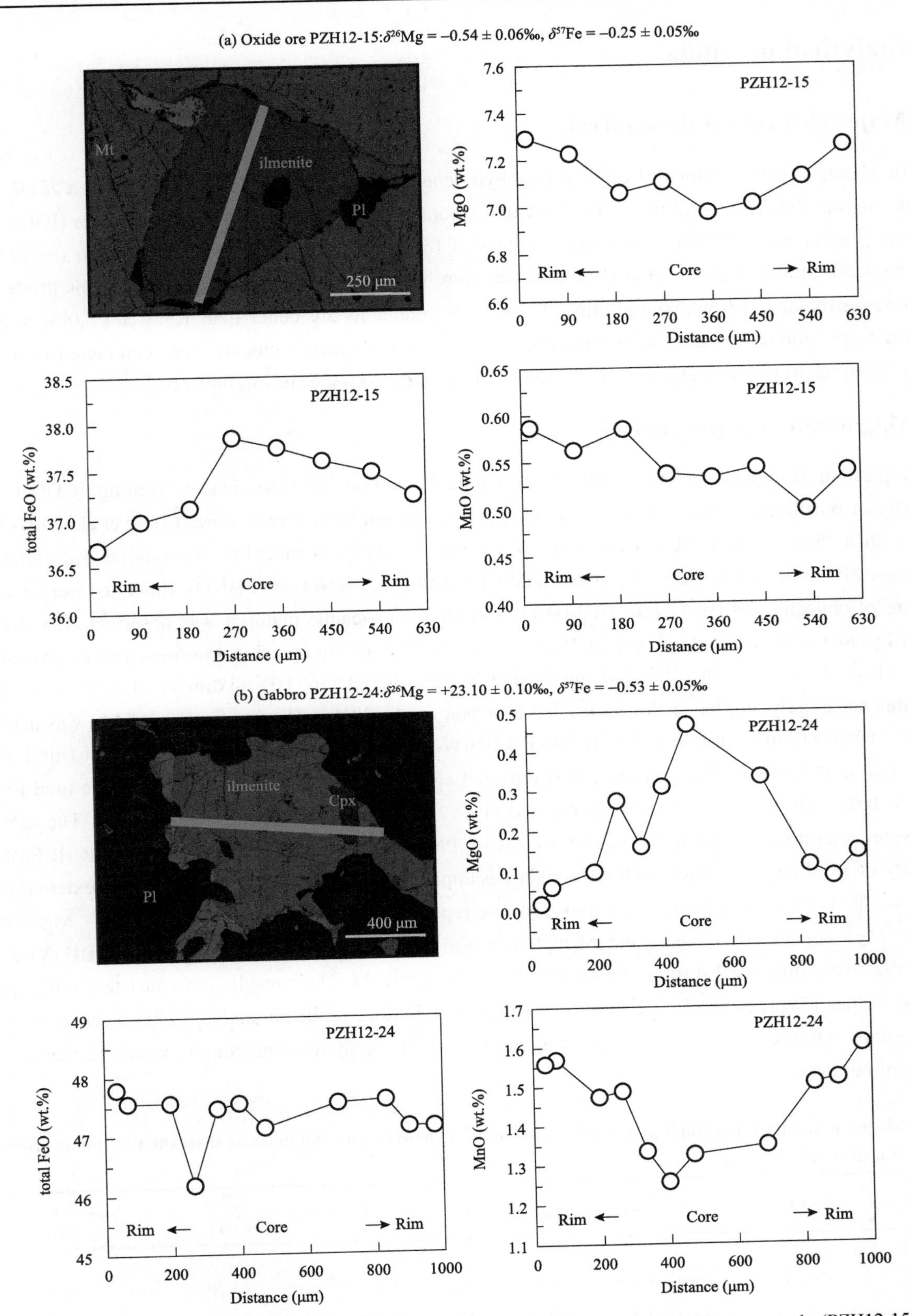

Fig. 2 (a) BSE image, MgO, FeO and MnO profile for a typical ilmenite grain from an oxide ore sample (PZH12-15). (b) BSE image, MgO, FeO and MnO profile for an ilmenite sample that is extremely enriched in heavy Mg isotopes in the Apatite-gabbro sample (PZH12-24). Note the sharp contract in MgO, FeO and MnO profiles between these two samples. Data are reported in Table S1.

3 Analytical methods

3.1 Major elements in the minerals

The chemical compositions of olivine, clinopyroxene and ilmenite were determined with a *JEOL JXA8100* electron microprobe at the Institute of Geology and Geophysics, Chinese Academy of Sciences (IGGCAS). The analytical conditions included an accelerating voltage of 15 kV, a beam current of 12 nA, a beam size of 5 μm and 10–30 s counting time. Natural minerals and oxides were used as standards to monitor the whole procedure. The precisions for major (>1.0 wt.%) and minor (<1.0 wt.%) elements are better than 1.5% and 5.0%, respectively. Ilmenites from fourteen samples were randomly selected and the measurements were conducted in their cores. Several ilmenites were also analyzed for concentration profiles from the rims to the cores.

3.2 Magnesium isotopic analysis

Magnesium isotopes were analyzed at the China University of Geosciences, Beijing (CUGB) and the University of Washington, Seattle (UW), following previously established procedures (Teng et al., 2007; Ke et al., 2016), with a slightly modified procedure for Mg isotope analysis in ilmenite. The olivine and clinopyroxene grains were dissolved in a mixture of concentrated HF-HNO_3, HCl-HNO_3, and HNO_3. Ilmenites were dissolved in a mixture of concentrated HF-HNO_3, HCl-HNO_3 and HCl. No opaque material was observed after dissolution. Before Mg separation, a Ti-column was applied to separate Ti from Mg by using anion-exchange chromatography with Bio-Rad AG1-X8 resin (200–400 mesh) and a 1 N HCl–0.5 N HF mixture acid as the eluent. These procedures are largely similar to those reported by Chen et al. (2018). The purification of Mg was achieved by cation-exchange chromatography with Bio-Rad AG50W-X8 pre-cleaned resin (200–400 mesh) in 1 N HNO_3 medium (Ke et al., 2016). This procedure was repeated again to obtain a pure Mg solution. The total procedural blank was below 10 ng for Mg, which represented <0.1% of the total amount of Mg processed. The Mg isotopic ratios were determined using a *Neptune Plus* MC-ICPMS at CUGB and *Nu Plasma II* MC-ICPMS at the University of Washington, Seattle with the standard–sample bracketing method. The long-term external precision is ±0.06‰ (2SD) for δ^{26}Mg. Isotopic compositions are reported as δ-notation: $\delta^{25,26}Mg(‰) = [(^{25,26}Mg/^{24}Mg)_{sample}/(^{25,26}Mg/^{24}Mg)_{DSM3}-1]\times 1000$. The standard materials San Carlos olivine, Hawaiian seawater, BHVO-2, BCR-2 and AGV-2, were processed during our sample analyses (Table 1). Their results are consistent with previously published values (Teng et al., 2015; Ke et al., 2016). In addition, replicate analyses have been carried out for ilmenites with extreme values and the results agree well within analytical uncertainty, which further assures the accuracy of our data.

Table 1　Magnesium and iron isotopic compositions of olivine, clinopyroxene and ilmenite from the Panzhihua layered mafic intrusion, Southwest China

Sample	Rock type	Minerals	δ^{26}Mg (‰)	2SD	δ^{25}Mg (‰)	2SD	δ^{57}Fe (‰)	2se	δ^{56}Fe (‰)	2se
Group 1										
PZH12-13	Oxide ore	Clinopyroxene	−0.40	0.08	−0.21	0.03	0.12	0.07	0.09	0.02
		Olivine	−0.41	0.07	−0.22	0.04	0.23	0.07	0.16	0.02
		Ilmenite								
		Batch 1	−0.73	0.05	−0.39	0.03	−0.24	0.05	−0.15	0.03
		Batch 2	−0.80	0.03	−0.41	0.02				
UW		Batch 3	−0.72	0.07	−0.36	0.05				

Continued

Sample	Rock type	Minerals	δ^{26}Mg (‰)	2SD	δ^{25}Mg (‰)	2SD	δ^{57}Fe (‰)	2se	δ^{56}Fe (‰)	2se
PZH12-15	Oxide ore	Clinopyroxene	−0.18	0.04	−0.10	0.03	0.20	0.07	0.11	0.03
		Ilmenite								
		Batch 1	−0.54	0.06	−0.27	0.01	−0.25	0.05	−0.20	0.03
		Batch 2	−0.49	0.14	−0.25	0.09				
UW		Batch 3	−0.53	0.07	−0.29	0.05				
PZH12-18	Oxide ore	Clinopyroxene	−0.30	0.07	−0.15	0.08	0.07	0.06	0.03	0.04
		Ilmenite								
		Batch 1	−0.25	0.04	−0.14	0.04	−0.33	0.05	−0.22	0.03
		Batch 2	−0.23	0.05	−0.11	0.02	−0.39	0.07	−0.26	0.05
		Batch 3	−0.32	0.01	−0.17	0.05				
PZH12-07	Mt-gabbro	Clinopyroxene	−0.37	0.04	−0.18	0.05	0.00	0.03	0.01	0.03
		Ilmenite								
		Batch 1	−0.38	0.06	−0.19	0.03	−0.30	0.07	−0.20	0.05
		Batch 2	−0.43	0.06	−0.22	0.04				
PZH12-09	Mt-gabbro	Olivine	−0.47	0.04	−0.24	0.03	0.07	0.03	0.06	0.03
		Ilmenite								
		Batch 1	−0.80	0.05	−0.41	0.06	−0.23	0.07	−0.17	0.05
		Batch 2	−0.79	0.03	−0.40	0.04				
PZH12-19	Mt-gabbro	Clinopyroxene	−0.34	0.02	−0.16	0.04	0.19	0.06	0.14	0.04
		Ilmenite	−0.13	0.04	−0.07	0.04	−0.28	0.05	−0.19	0.03
		Repeat	−0.13	0.01	−0.07	0.04				
Group 2										
PZH12-03	Hbl-gabbro	Ilmenite								
		Batch 1	8.19	0.03	4.16	0.05	−0.59	0.07	−0.36	0.05
		Batch 2	6.85	0.01	3.47	0.03				
UW		Batch 3	7.45	0.07	3.77	0.05				
UW		Batch 4	7.98	0.07	4.05	0.05				
PZH12-10	Ol-gabbro	Clinopyroxene	−0.22	0.06	−0.09	0.04	0.02	0.03	−0.01	0.03
		Olivine	−0.32	0.05	−0.16	0.07	−0.21	0.03	−0.15	0.03
		Ilmenite								
		Batch 1	3.77	0.03	1.95	0.00	−0.47	0.07	−0.31	0.05
		Batch 2	3.57	0.03	1.84	0.02				
PZH12-11	Ol-gabbro	Clinopyroxene	−0.33	0.09	−0.15	0.05	0.17	0.07	0.17	0.03
		Ilmenite								
		Batch 1	2.59	0.02	1.31	0.05	−0.33	0.07	−0.23	0.05
		Batch 2	2.31	0.02	1.17	0.04				
PZH12-12	Bt-gabbro	Clinopyroxene	−0.32	0.06	−0.17	0.06	0.05	0.07	0.05	0.03
		Ilmenite	2.94	0.06	1.49	0.05	−0.50	0.07	−0.31	0.05
PZH12-14	Bt-gabbro	Clinopyroxene	−0.23	0.04	−0.12	0.04	0.15	0.07	0.08	0.03
		Ilmenite	1.51	0.06	0.77	0.04	−0.41	0.05	−0.28	0.03
PZH12-16	Hbl-gabbro	Clinopyroxene	−0.20	0.03	−0.07	0.04	−0.16	0.07	−0.10	0.04
		Ilmenite								

Continued

Sample	Rock type	Minerals	δ^{26}Mg (‰)	2SD	δ^{25}Mg (‰)	2SD	δ^{57}Fe (‰)	2se	δ^{56}Fe (‰)	2se
		Batch 1	4.34	0.03	2.16	0.02				
		Batch 2	5.62	0.02	2.85	0.01	−0.44	0.07	−0.30	0.03
		Batch 3	6.70	0.03	3.39	0.01	−0.37	0.07	−0.29	0.03
		Batch 4	6.40	0.01	3.25	0.02	−0.47	0.07	−0.30	0.03
		Batch 5	5.04	0.02	2.54	0.03	−0.42	0.07	−0.30	0.03
		Batch 6	4.87	0.04	2.45	0.04	−0.44	0.07	−0.34	0.03
UW		Batch 7	6.23	0.07	3.16	0.05				
UW		Repeat	6.27	0.07	3.16	0.05				
UW		Batch 8	5.16	0.07	2.61	0.05				
PZH12-17	Ol-gabbro	Clinopyroxene	−0.29	0.03	−0.14	0.06	0.07	0.06	0.04	0.04
		Olivine	−0.33	0.00	−0.15	0.04	−0.07	0.06	−0.02	0.04
		Ilmenite								
		Batch 1	0.48	0.09	0.25	0.07	−0.36	0.05	−0.26	0.03
		Batch 2	0.25	0.06	0.12	0.05				
PZH12-22	Hbl-gabbro	Clinopyroxene	−0.29	0.05	−0.14	0.03	0.11	0.06	0.09	0.04
		Ilmenite								
		Batch 1	0.62	0.04	0.32	0.06	−0.36	0.05	−0.23	0.03
		Batch 2	0.46	0.05	0.24	0.03				
PZH12-24	Apat-gabbro	Clinopyroxene	−0.32	0.03	−0.15	0.02	0.03	0.05	0.04	0.02
		Ilmenite								
		Batch 1	20.88	0.04	10.43	0.04				
		Batch 2	23.10	0.10	11.60	0.07	−0.53	0.05	−0.35	0.03
		Batch 3	20.03	0.02	10.11	0.02	−0.56	0.07	−0.40	0.03
		Batch 4	21.88	0.02	11.03	0.01	−0.53	0.07	−0.40	0.03
		Batch 5	21.69	0.01	10.92	0.02	−0.55	0.07	−0.39	0.05
		Batch 6	22.67	0.02	11.42	0.03	−0.61	0.07	−0.40	0.05
		Batch 7	21.33	0.01	10.75	0.01	−0.52	0.07	−0.34	0.05
UW		Batch 8	21.04	0.07	10.61	0.05				
UW		Batch 9	22.23	0.07	11.18	0.05				
UW		Batch 10	22.68	0.07	11.45	0.05				
Standards										
		BHVO-2	−0.21	0.04	−0.09	0.03	0.14	0.07	0.12	0.03
		BCR-2	−0.16	0.04	−0.08	0.02	0.15	0.05	0.12	0.03
		AGV-2	−0.13	0.07	−0.07	0.06	0.17	0.07	0.13	0.05
		AGV-2[a]	−0.14	0.07	−0.07	0.07				
		BCR-2[a]	−0.18	0.03	−0.08	0.01				
UW		San Carlos[a]	−0.33	0.07	−0.16	0.05				
UW		H-Seawater[a]	−0.85	0.07	−0.43	0.05				

Note: The UW denotes that samples were determined at the University of Washington (Seattle) and the others were determined at CUGB. 2SD denotes two times standard deviation of 4 repeat measurements (each >20 ratios) of a sample solution. 2se (95% confidence intervals) represents the uncertainty of iron isotopic compositions, which is calculated based on the errors from the total procedure (Dauphas et al., 2009; He et al., 2015). Repeat means replicated measurement of Mg isotopic ratios on the same solutions. Superscript "a" denotes that these standards were also passed through the Ti-column before Mg-column. The term "batch" represents different batches of mineral grains. Abbreviations: Mt-gabbro = magnetite gabbro; Ol-gabbro = olivine gabbro; Hbl-gabbro = hornblende gabbro; Bt-gabbro = biotite gabbro; Apat-gabbro = apatite gabbro. H-Seawater = Hawaiian seawater.

3.3 Iron isotopic analysis

Iron isotopes were analyzed at the CUGB, following previously established procedures (He et al., 2015). An aliquot of the digested ilmenite solutions used for Mg isotopes was used for Fe isotopic analysis. The other minerals were weighed and digested in Teflon beakers using concentrated HF-HNO_3-HCl with several drops of $HClO_4$. The separation of Fe was achieved by anion-exchange chromatography with the resin of Bio-Rad AG1X-8 (200−400 mesh) in the HCl media (Dauphas et al., 2009; He et al., 2015). The same column procedure was performed twice in order to obtain a pure Fe solution for mass spectrometry. The Fe isotopic ratios were determined using a *Neptune Plus* MC-ICPMS with the standard–sample bracketing method. The long-term external precision is ±0.04‰ (2SE) for δ^{56}Fe (He et al., 2015). The Fe isotope data are reported in standard δ-notation: $\delta^{56,57}\text{Fe}\ (‰) = [(^{56,57}\text{Fe}/^{54}\text{Fe})_{\text{sample}}/(^{56,57}\text{Fe}/^{54}\text{Fe})_{\text{IRMM-014}}-1] \times 1000$. The standard reference materials yield the following results: δ^{57}Fe = 0.14±0.07‰ for BHVO-2, δ^{57}Fe = 0.15±0.05‰ for BCR-2, and δ^{57}Fe = 0.17± 0.07‰ for AGV-2 (Table 1), which are consistent with previously reported values (0.17±0.01‰ for BHVO-2, 0.12±0.01‰ for BCR-2 and 0.14 ± 0.01‰for AGV-2; Craddock and Dauphas, 2011; He et al., 2015).

4 Results

The chemical compositions of olivine, clinopyroxene and ilmenite are reported in Table S1. The Mg and Fe isotopic compositions of standard reference materials, four olivines, thirteen clinopyroxenes and fifteen ilmenites are presented in Table 1.

4.1 Major elemental compositions of ilmenites

Overall, ilmenites studied here display wider variations in MgO (0.05−8.01 wt.%), FeO (36.3−47.5 wt.%) (total Fe calculated as FeO) and MnO (0.46−1.63 wt.%) contents (Table S1) than those from the Baima layered intrusion (Chen et al., 2018). Group 1 ilmenites display higher MgO (4.25−8.01 wt.%) and lower FeO (36.31−41.73 wt.%) and MnO (0.46−0.59 wt.%) contents than those from Group 2 (3.53−0.05 wt.%, 39.3−47.5 wt.%, and 0.53−1.63 wt.%, respectively; Table S1). Additionally, most of ilmenites in Group 1 are mainly surrounded by magnetite, with minor olivine and/or clinopyroxene. For individual ilmenite grains, their MgO content increases, FeO content decreases, and MnO content remains invariant from core to rim (Fig. 2a). By contrast, most ilmenites from Group 2 are mainly in contact with plagioclase, and their MgO content decreases from the core to the rim, FeO content remains invariant, and MnO content increases (Fig. 2b). FeO and MgO are negatively correlated in both Groups 1 and 2 ilmenites (Fig. 3a) and MnO and MgO are negatively correlated in Group 2 ilmenites (Fig. 3b).

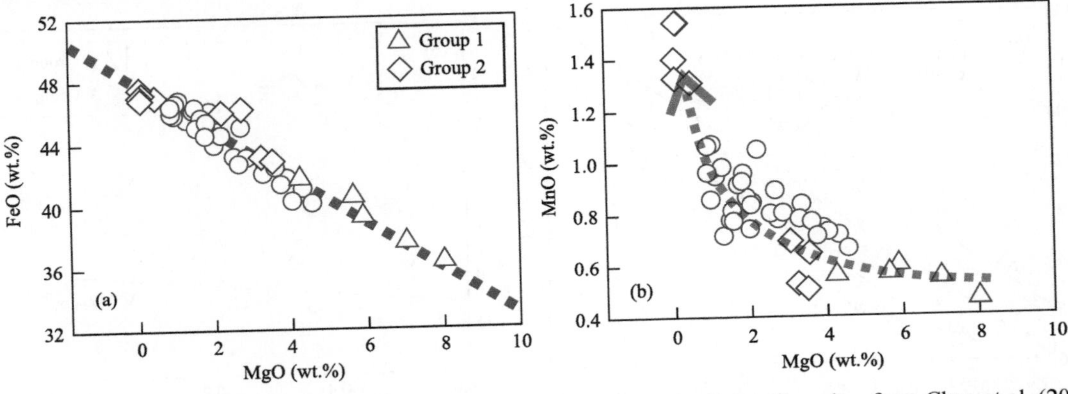

Fig. 3 (a) FeO vs. MgO and (b) MnO vs. MgO for the Panzhihua ilmenites. The Baima ilmenites from Chen et al. (2018) are also plotted as circles. Data are reported in Table S1.

4.2 Magnesium isotopic compositions

The olivine and clinopyroxene exhibit small Mg isotopic variations, with δ^{26}Mg ranging from −0.47 to −0.32‰ and from −0.40 to −0.18‰, respectively (Fig. 4a). Compared to the coexisting silicate minerals, ilmenites exhibit the most extreme variation in all terrestrial materials so far, with δ^{26}Mg ranging from −0.80 to −0.13‰ in Group 1 and from +0.48 to +23.10‰ in Group 2 (Fig. 4a). In addition, ilmenites from Group 1 display limited intra-mineral Mg isotopic variations, while those from Group 2 have highly variable intra-mineral Mg isotopic compositions (Table 1). For example, ilmenite grains separated from sample PZH12-24 can have up to +3‰ Mg isotopic fractionation among different batches.

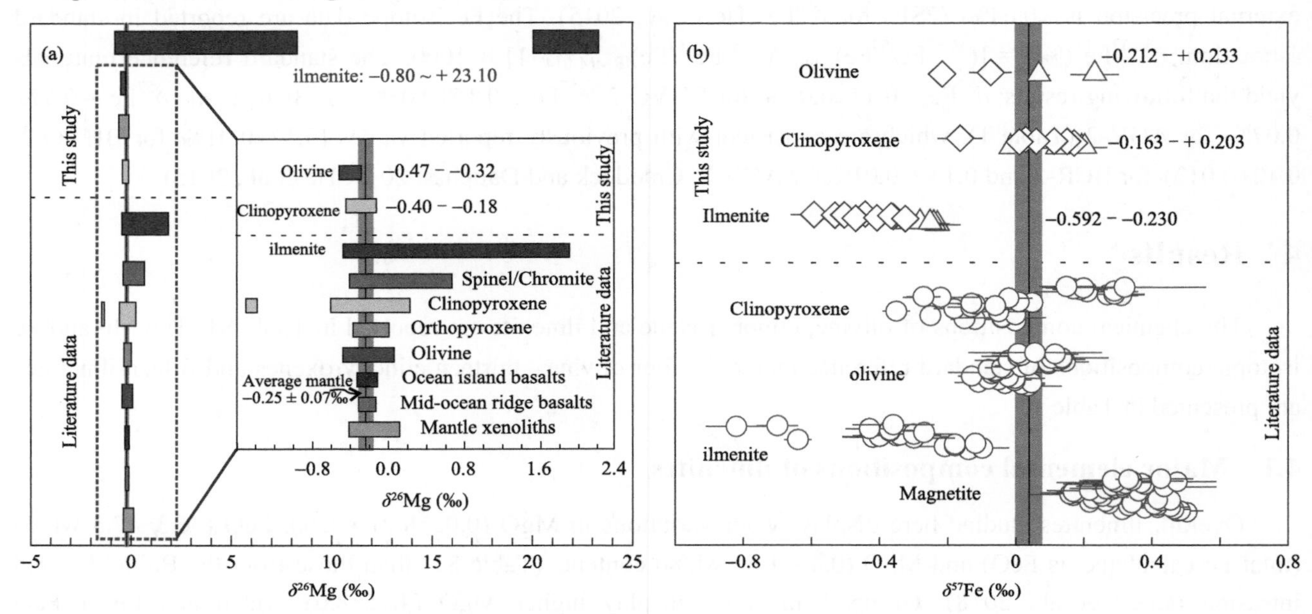

Fig. 4 (a) δ^{26}Mg of olivine, clinopyroxene and ilmenite, along with literature data (Teng, 2017 and references therein). Data are reported in Table 1. The vertical dashed line and gray band represent the average mantle Mg isotopic composition. (b) δ^{57}Fe of olivine, clinopyroxene and ilmenite, along with literature data for minerals from layered mafic intrusion (Chen et al., 2014; Liu et al., 2014b). The line and gray band represent the mantle Fe isotopic compositions (δ^{57}Fe=0.04±0.04‰, Weyer and Ionov, 2007; Craddock et al., 2013).

The inter-mineral Mg isotopic difference between two coexisting minerals is defined as $\Delta^{26}\text{Mg}_{X-Y} = \delta^{26}\text{Mg}_X - \delta^{26}\text{Mg}_Y$, where subscript X and Y refer to different minerals. The $\Delta^{26}\text{Mg}_{\text{Ol-Cpx}}$, $\Delta^{26}\text{Mg}_{\text{Ilm-Cpx}}$ and $\Delta^{26}\text{Mg}_{\text{Ilm-Ol}}$ vary from −0.11 to −0.01‰, −0.36 to +23.41‰, and −0.33 to +4.10‰, respectively (Fig. 5).

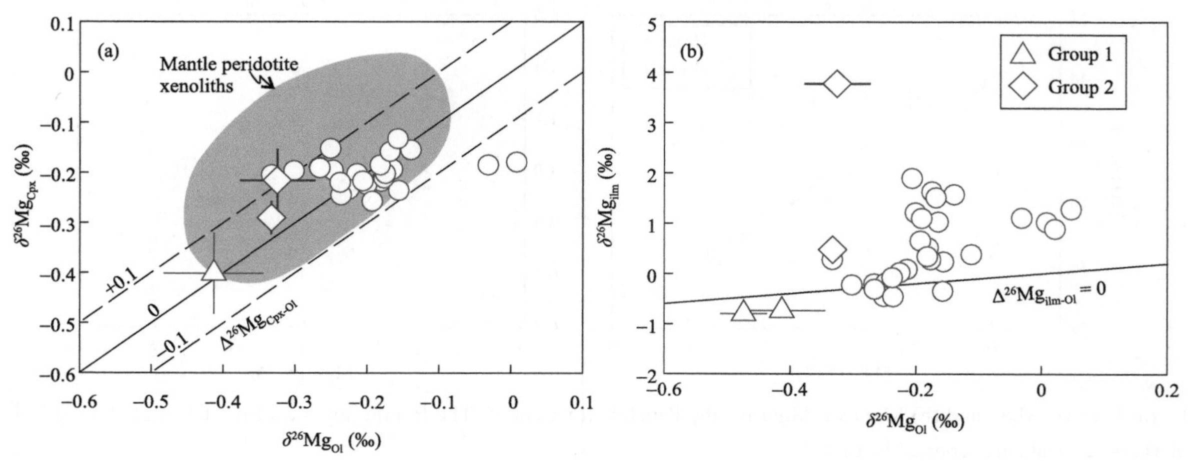

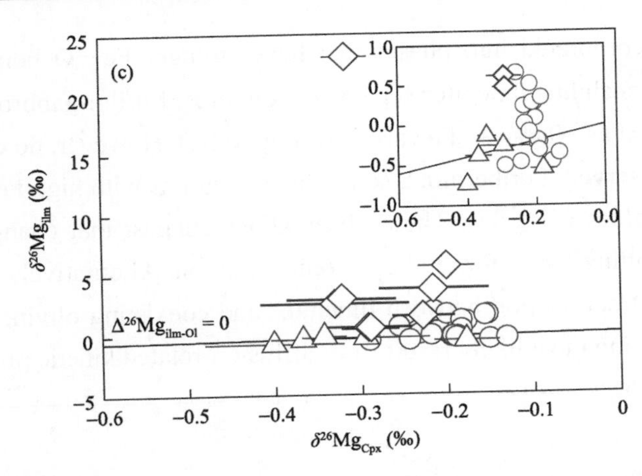

Fig. 5 Inter-mineral Mg isotope fractionation between olivine and clinopyroxene (a), between olivine and ilmenite (b), and between clinopyroxene and ilmenite (c) for the Panzhihua layered mafic intrusion. Data are reported in Table 1. The equilibrium isotope fractionation lines in the plots (a)–(c) are from Schauble (2011) and Chen et al. (2018). The light green area in panel (a) represents the δ^{26}Mg of olivine and clinopyroxene from mantle peridotite xenoliths (Teng, 2017 and reference therein). The white circles in panels (a)–(c) denote Baima ilmenites from Chen et al. (2018).

4.3 Iron isotopic compositions

Olivine, clinopyroxene and ilmenite have distinctive Fe isotopic compositions. The δ^{57}Fe values in olivine and clinopyroxene range from −0.21 to +0.23‰ and −0.16 to +0.20‰, respectively (Fig. 4b). The clinopyroxene range is larger than that reported in Cao et al. (2019), where they found clinopyroxene with abundant exsolution lamellae of Fe-Ti oxides have similar δ^{56}Fe range to those of the grains with less exsolution. The δ^{57}Fe values of ilmenites vary from −0.33 to −0.23‰ in Group 1 and from −0.59 to −0.33‰ in Group 2 (Fig. 4b).

The inter-mineral Fe isotopic difference ($\Delta^{57}Fe_{X\text{-}Y}=\delta^{57}Fe_X-\delta^{57}Fe_Y$, where subscript X and Y refer to different minerals) varies largely, with $\Delta^{57}Fe_{Ol\text{-}Cpx}$, $\Delta^{57}Fe_{Ilm\text{-}Cpx}$, and $\Delta^{57}Fe_{Ilm\text{-}Ol}$ ranging from −0.23 to +0.11‰, −0.56 to −0.28‰, and −0.47 to −0.26‰, respectively (Fig. 6).

5 Discussion

Our results indicate large disequilibrium isotope fractionation among these coexisting minerals because most of these mineral pairs fall off the equilibrium Mg and Fe isotope fractionation lines at magmatic temperature of ~950–1200 ℃ (Figs. 5 and 6). This disequilibrium isotopic fractionation is also evidenced from the large intra-mineral Mg isotope fractionation in Group 2 ilmenites. For example, δ^{26}Mg values in ilmenite grains vary from +4.34 to +6.70‰ in sample PZH12-16 and from +20.03 to +23.10‰ in sample PZH12-24 (Table 1). Previous studies suggest that fractional crystallization (Williams et al., 2005; Teng et al., 2008; Cao et al., 2019), change of oxygen fugacity (Liu et al., 2014b), or diffusion-related kinetic processes (Richter et al., 2008, 2009a; Huang et al., 2010; Chen et al., 2018) have the potential to produce Mg and/or Fe isotope fractionation. However, the former two processes can be ruled out based on the following observations. Though large inter-mineral isotopic fractionation can lead to the crystallization of isotopically light ilmenites and heavy magnetites (Cao et al., 2019), this process alone cannot cause significant Mg isotope fractionation because magnetite crystallization does not affect MgO contents in melts. On the other hand, oxygen fugacity can be an important factor controlling Fe isotope fractionation during magma differentiation (e.g., Williams et al., 2005; Teng et al., 2008), with heavy

Fe isotopes preferentially incorporated into phases that have stronger Fe—O bonds or higher $Fe^{3+}/\Sigma Fe$ ratios (Dauphas et al., 2017). The Panzhihua ilmenites separated from ores, Mt/ilm-gabbros and gabbros display a large range of $Fe^{3+}/\Sigma Fe$ ratios as well as ~0.6‰ $\delta^{57}Fe$ variation (Figure S2). However, no correlation between $\delta^{57}Fe$ and $Fe^{3+}/\Sigma Fe$ in ilmenites were observed. Furthermore, some ilmenite grains with high $Fe^{3+}/\Sigma Fe$ have lower $\delta^{57}Fe$ than those with lower $Fe^{3+}/\Sigma Fe$ values (Fig. S2). These observations suggest that changes in oxygen fugacity were unlikely to be the main controlling factor for the Fe isotopic variation. Alternatively, the extremely large Mg, and to a lesser extent, Fe isotope fractionation between ilmenites and coexisting olivine and clinopyroxene from the Panzhihua layered mafic intrusion most likely result from diffusion-related kinetic processes.

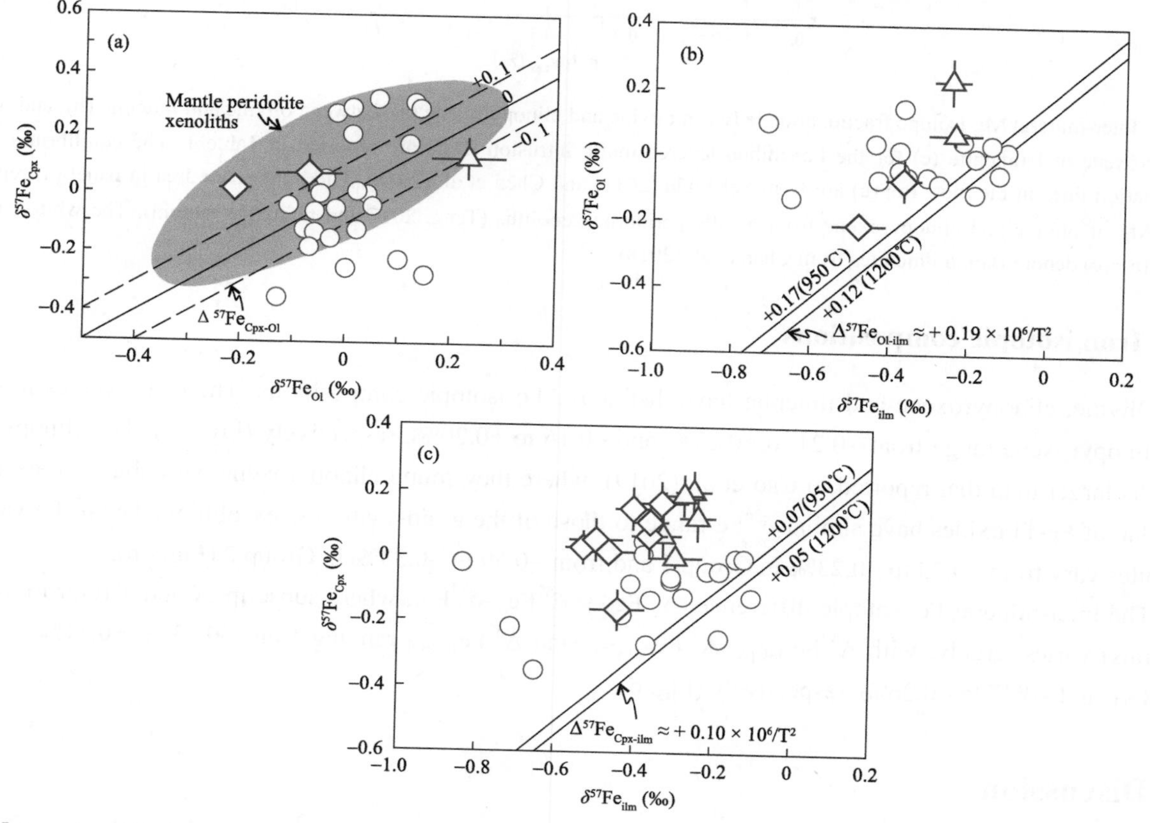

Fig. 6　Inter-mineral Fe isotope fractionation between olivine and clinopyroxene (a), between olivine and ilmenite (b), and between clinopyroxene and ilmenite (c) for the Panzhihua layered mafic intrusion. Data are reported in Table 1. The equilibrium fractionation lines in plots a-c are from Polyakov and Mineev (2000) and Chen et al. (2014). The calculated temperature range varies from 950 to 1200 ℃ (Zhang et al., 2012). The light green area in plot (a) represents the $\delta^{57}Fe$ of olivine and clinopyroxene from mantle peridotite xenoliths (Dauphas et al., 2017 and references therein). The symbols are the same as those in Fig. 5.

5.1 Diffusion-driven disequilibrium Mg-Fe isotope fractionations in cumulus minerals

Both chemical and Soret diffusion can induce large isotope fractionations (Richter et al., 2008, 2009a; Huang et al., 2010). Soret diffusion fractionates Mg and Fe isotopes in the same direction, hence yielding a positive correlation between $\delta^{57}Fe$ and $\delta^{26}Mg$. The negative correlation between $\delta^{57}Fe$ and $\delta^{26}Mg$ observed in ilmenites (Fig. 7) thus rules out thermal diffusion as the cause of the large Mg and Fe isotope fractionations. Instead, multiple lines of evidence point to chemical diffusion process, which will be discussed in detail below.

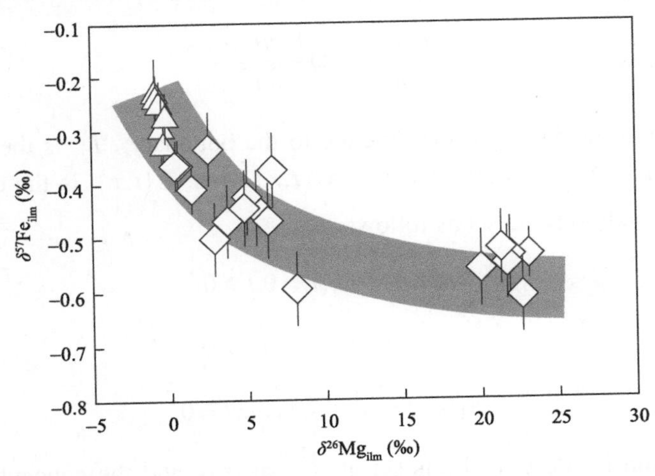

Fig. 7 The correlations between Mg and Fe isotopes in all ilmenites from Panzhihua layered mafic intrusion. The data are reported in Table 1.

5.1.1 Isotope fractionation during Mg-Fe inter-diffusion

The overall negative correlation between Mg and Fe isotopes in all ilmenites is consistent with diffusion-driven kinetic isotope fractionations during Mg-Fe inter-diffusion in zoned minerals (Dauphas et al., 2010; Teng et al., 2011; Sio et al., 2013; Oeser et al., 2015; Wu et al., 2018). The fluxes of Mg and Fe during inter-diffusion are identical in magnitude and opposite in direction. Since Fe content is (at least five times) higher than Mg content in ilmenite, the magnitude of Fe isotope fractionation is thus diluted by a larger background of Fe than Mg. Thus, the magnitude of Fe isotopic variation in ilmenite is smaller than that of Mg. Furthermore, a non-linear relationship between Mg and Fe isotopes occurs when almost all Mg in ilmenite is substituted by Fe, which is also consistent with the negative correlation between MgO and FeO (Fig. 3a).

Elemental profiles in ilmenite grains further reveal two contrasting inter-diffusion processes. MgO content increases from the core to the rim in ilmenite grain from sample PZH12-15 (Group 1), suggesting diffusion of Mg into the ilmenite. By contrast, MgO content decreases from the core to the rim in ilmenite grain PZH12-24 (Group 2), indicating diffusion of Mg out of the ilmenite. Since light isotopes always diffuse faster than heavy ones (Richter et al., 2009b), Group 1 ilmenites are expected to be enriched in light Mg isotopes and Group 2 ilmenites in heavy Mg ones (Fig. 5). Iron isotopes are expected to behave oppositely to Mg isotopes in both groups, though FeO content does not display significant zoning in Group 2 ilmenites. This mainly reflects the low MgO content (<0.5 wt.%) compared to FeO content (>46 wt.%) in the ilmenite grains from the sample PZH12-24. Nonetheless, MnO content displays opposite zonation to MgO content in this sample (Fig. 2a) and negatively correlates with MgO in Group 2 ilmenites (Fig. 3b). This might suggest that in addition to the inter-diffusion between Mg and Fe, substitution of Mg for Mn also occurred because of their similar ionic radii and lattice sizes (Shannon, 1976).

The disequilibrium Mg and Fe isotope fractionations in those olivines and clinopyroxenes were also likely caused by this chemical diffusion process. These silicate minerals have significantly higher MgO contents compared to ilmenites, thus their Mg isotopic variations were much less prominent (Table S1).

5.1.2 Isotope diffusion model

In this section, we present a detailed modeling calculation for the Mg-Fe diffusion in ilmenites. The chemical diffusion is modeled by assuming a spherical ilmenite in the melts. Considering that the diffusivity in melts is much higher than that in solids, the surface concentration is set as constant, and the non-steady diffusion process is characterized by Crank (1975):

$$\frac{\partial u(t,r)}{\partial t} = D\frac{\partial^2 u(t,r)}{\partial r^2}$$

where r is the distance from the center of the ilmenite to the boundary; D is the diffusion coefficient of the minerals; the function $u(t,r)$ is defined by $u(t,r)=rC(t,r)$, and $C(t,r)$ is the element concentration at r and time t. The boundary conditions are set as follows:

$$u(t,0)=C_1, r=0, t \geqslant 0$$

$$u(t,a)=C_0, r=a, t \geqslant 0$$

$$u(0,r)=rC, 0<r<a, t=0$$

where the initial concentration inside ilmenite is set as C_1 at r=0, and the concentration on the ilmenite-melt boundary r=a is set as C_0.

Isotopes are fractionated during chemical diffusion because of their slightly different diffusivities, which are directly related to atomic mass (Richter et al., 1999):

$$D_i / D_j = \left(m_j / m_i\right)^{\beta}$$

where m_i and m_j are the masses of isotope i and j. The value for β is considered as an empirical constant and generally depends on the diffusion medium and conditions. Moreover, for simplicity, Mg-Fe is assumed to be 1:1 substitution, as the degree of Mg-Mn substitution is negligible. The total diffusivity of Mg and Fe in ilmenite must satisfy:

$$\sum x_{^n\text{Mg}} D_{^n\text{Mg}} = \sum x_{^m\text{Fe}} D_{^m\text{Fe}}$$

where n and m are the index for different isotopes. The diffusion rates for Mg in synthetic ilmenite at ~1000 ℃ range from 10^{-16} to 10^{-13} m^2/s (Stenhouse et al., 2010). The average size of the ilmenite in this study is set to be 0.6 mm based on petrographic observations. Furthermore, the isotopic variations in Groups 1 and 2 ilmenites are modelled separately. The initial values and modelled parameters for diffusion are reported in Table 2. This quantitative calculation is computed using Mathematica program (the code is presented in the supporting information).

Table 2　Parameters for chemical diffusion in a sphere to model Mg and Fe isotopic variations in ilmenites

Mg mol.% in boundary surface (Co)[a]	Fe mol.% in boundary surface (Co)[a]	Initial Mg mol.% in crystal ilmenite (C_1)[a]	Initial Fe mol.% in crystal ilmenite (C_1)[a]	β_{Mg} in ilmenite[b]	β_{Fe} in ilmenite[b]	δ^{26}Mg (‰)[c]	δ^{57}Fe (‰)[c]
Model parameters for Group 1							
0.155	0.345	0.05	0.45	0.02, 0.04	0.035	−0.25	−0.33
Model parameters for Group 2							
(1) 0.0002	0.4998	0.08	0.42	0.10	0.30	−0.25	−0.33
(2) 0.0015	0.4985	0.07	0.43	0.06	0.25	−0.25	−0.33
(3) 0.006	0.494	0.06	0.44	0.05	0.16	−0.25	−0.33

Note: [a] We used two different boundary conditions and β-factors for inter-diffusion. According to the in situ major elemental data from EPMA, the total amount of Mg and Fe (mol.%) in ilmenite is nearly constant about 0.5. That is the molar ratio of (Mg + Fe)/(Mg + Fe + Ti). For Group 2 ilmenites, we use three sets of initial and boundary condition parameters. [b] These are the kinetic Mg and Fe isotopic fractionation factors used in ilmenites. [c] The initial ilmenite Mg and Fe isotopic compositions are represented by the sample PZH12-18 that has mantle-like Mg isotopic composition.

Our results indicate that the modelling curves fit the measured data best when β_{Mg} = ~0.02–0.04 and β_{Fe} = ~0.035 for Group 1 and β_{Mg} = ~0.05–0.10 and β_{Fe} = ~0.16–0.30 for Group 2 ilmenites (Figs. 8 and 9). Due to the absence of a well-constrained isotopic diffusion profile through the ilmenite, these values are not unique. Variable β_{Mg} and β_{Fe} values may come from different boundary environments for Mg and Fe in ilmenites because these samples were collected from a > 100-m length scale that could hold different initial elemental concentrations and temperatures. Our calculation also yields a kinetic diffusive timescale of 7 days to 19.1 years for Group 1 and of 55 days to 150.7 years for Group 2 ilmenites (Figs. 8 and 9), which may reflect the timescale for each intrusion cooling instead of the whole magma chamber cooling. In addition, based on our model calculations, the significantly high δ^{26}Mg values (+20.03 to +23.10‰) in Group 2 ilmenites would indicate an initial ilmenite MgO content of around 4.2 wt.%. This high MgO content suggests that those ilmenites crystallized from an early stage (e.g., Song et al., 2013; Zheng et al., 2014).

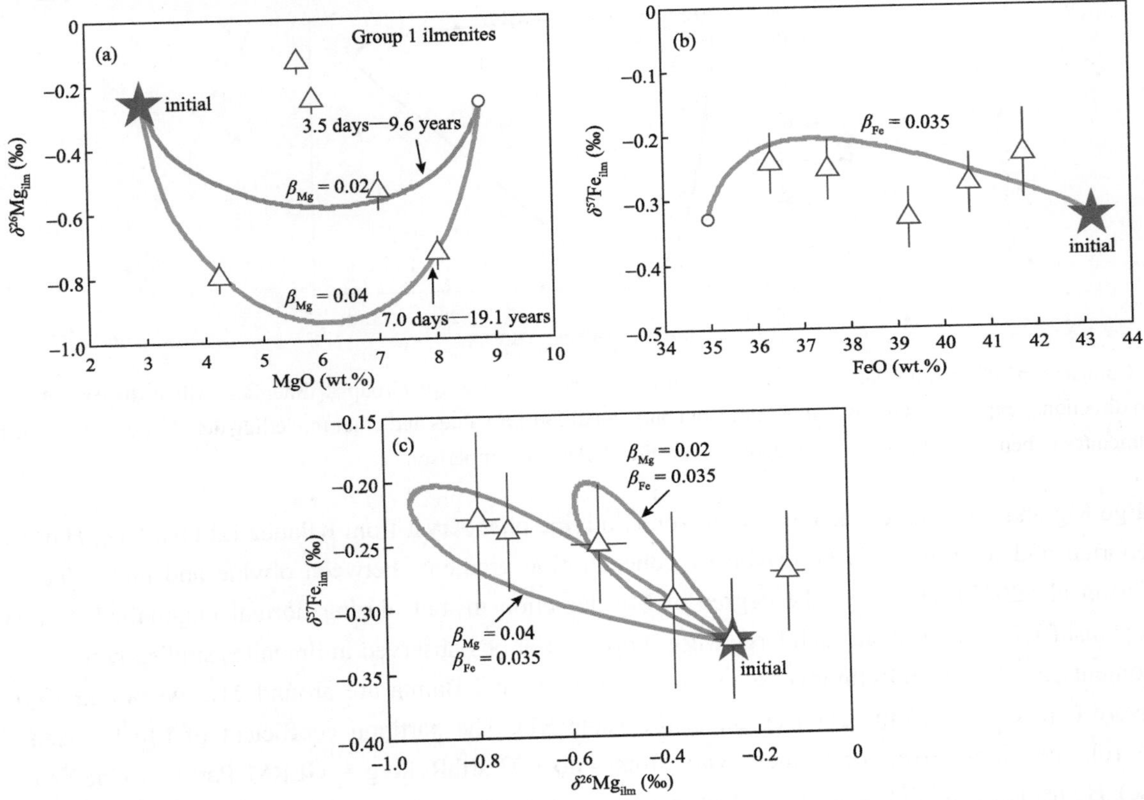

Fig. 8 Comparison of ilmenite data with modelled diffusion-driven curves for Group 1 ilmenites. Boundary conditions, β-values and other modelling details are reported in Table 2. The stars with jacinth color in panels represent the initial components.

5.2 Constraints on the petrogenesis of Panzhihua Fe-Ti layered mafic intrusion

The origin and petrogenesis of Panzhihua Fe-Ti layered mafic intrusion is still under debate. While some studies suggest that the mafic intrusion was mainly derived from an accumulation of Fe-Ti oxide crystals during magmatic differentiation (Pang et al., 2008a, 2008b, 2009; Shellnutt and Jahn, 2010; Zhang et al., 2012; Song et al., 2013), others imply that it could have formed from an Fe-Ti-rich immiscible liquid segregated from magmas (Zhou et al., 2005; Liu et al., 2014a, 2016; Cao et al., 2019). Our geochemical and modeling results suggest that interstitial liquid immiscibility probably plays the most important role in the genesis of Panzhihua layered mafic intrusion.

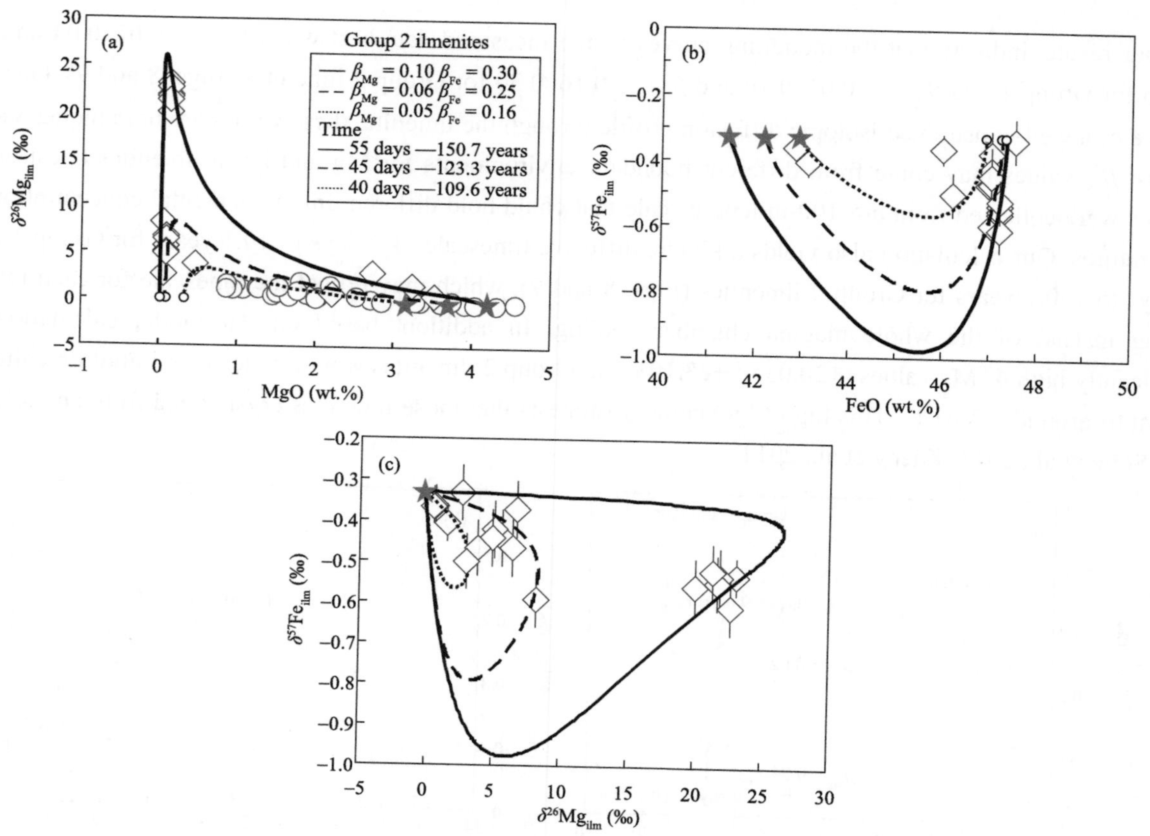

Fig. 9 Comparison of ilmenite data with modelled diffusion-driven curves for Group 2 ilmenites, which displays the opposite diffusion direction compared to Group 1 ilmenites. Boundary conditions, β-values and other modelling details are reported in Table 2. Baima ilmenites (Chen et al., 2018) are also reported as white circles for comparison.

Large Mg and Fe isotope fractionation in zoned olivine phenocryst from Kilauea Iki lava lake, Hawaii, have been reported and interpreted to be driven by concentration gradients between olivine and melts (Teng et al., 2011; Sio et al., 2013). However, the degree of Mg depletion in melt during normal magmatic differentiation process is insufficient to produce such large Mg isotopic variations observed in ilmenites studied here. The lowest MgO contents in olivine and clinopyroxene from Panzhihua and Baima are around 21.2 wt.% and 13.0 wt.%, respectively (Liu et al., 2014b; Chen et al., 2018; Table S1). The partition coefficient of Mg between silicate minerals (olivine/clinopyroxene) and melts vary from 4 to 6 (EarthRef.org – GERM Partition Coefficient (Kd) Database). By using the minimum MgO content in clinopyroxene, the melts should contain a minimum of ~2.2 wt.% MgO, which is inconsistent with the extremely low MgO contents of ilmenites from Group 2. This is because Mg is compatible in ilmenites (La Tourrette et al., 1991), which should have MgO content at least similar to the melt.

The most likely scenario to produce a large MgO activity gradient between ilmenites and melts involves interstitial immiscible liquid phase segregation (Holness et al., 2011; Wang et al., 2018; Cao et al., 2019). At an early stage, extensive crystallization of silicate minerals from the parental high-Ti basaltic magmas would develop a crystal mush in the intermittent layers of the Panzhihua chamber (Wang et al., 2018; Cao et al., 2019). The interstitial liquids between these minerals would progressively enrich in Fe due to the continued fractionation of silicate minerals and then intersected the immiscible field as the magma cools to approximately 1100 ℃. Finally, the interstitial melt would separate into an Fe-rich and a Si-rich phase, with the former having high-Mg contents and the latter having low-Mg contents (Zhou et al., 2005; Liu et al., 2014a). Due to the density differences in the interstitial liquid, the Si-rich components would separate from Fe-rich melt and migrate upward, while the Fe-rich melt remains in the lower part (Holness et al., 2011; Wang et al., 2018), as evidenced by the two conjugate Fe-rich and Si-rich immiscible phases in the apatite melt inclusions in the Panzhihua gabbroic-layered intrusion (Wang et al., 2018). Because of the convection induced by heat diffusion and crystallization (Huppert and Turner, 1981), some early crystallized ilmenites from parental basaltic magma or Fe-rich interstitial liquid would be trapped by the Si-rich melt during its migration. These two types of melts were in thermodynamic equilibrium initially but the interaction with the early crystallized ilmenites surrounded by these two types of melts would generate a large Mg activity gradient between ilmenite and melts.

Experimental study revealed that Si-rich phase could have a MgO content as low as 0.08 wt.%, while Fe-rich melt could have MgO content ten times higher (Charlier and Grove, 2012). In the Si-rich, Mg-poor melt phase (Group 2), Mg will diffuse out of early formed ilmenites into the melt and Fe will diffuse into the ilmenites due to the concentration gradients. As a result, ilmenites will have heavy Mg and light Fe isotopic compositions, as well as low MgO and high FeO contents. In addition, the decrease in D_{Mg} between ilmenite and melt during magma cooling (Toplis and Carroll, 1995) would also lead to the diffusion of Mg from ilmenites into melts, hence further increasing the magnitude of isotopic fractionation. By contrast, Mg would diffuse from the melt into ilmenites and Fe would diffuse out into the Fe-rich and Mg-rich melt phase (Group 1), producing ilmenites with low δ^{26}Mg and high δ^{57}Fe signatures.

The interstitial liquid phase segregation model not only explains the isotopic variation, but also provides insights for the lower FeO and higher MgO contents in ilmenites from the Fe-Ti-rich layer (Group 1) than those from gabbros (Group 2) (Fig. 10). In theory, ilmenites from Fe-Ti-oxide-bearing ore layer in the lower part crystallized from an Fe-rich magma and hence should have higher FeO and lower MgO contents than those from gabbros in the upper part. The opposite trends in Panzhihua and Baima ilmenites (Fig. 10) indicate the presence of

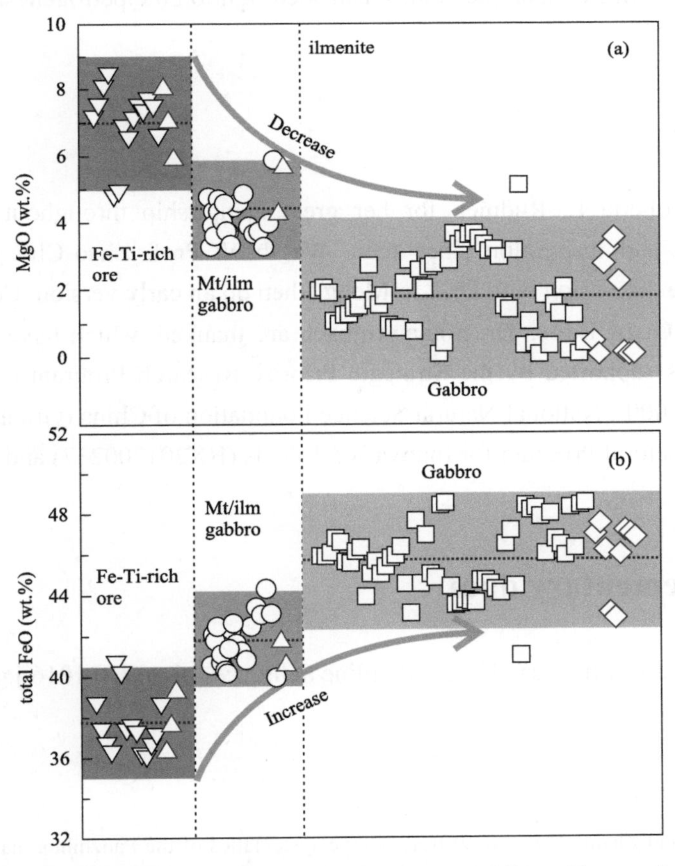

Fig. 10 (a) MgO and (b) total FeO contents of ilmenites from Fe-Ti-rich ores, Mt/ilm gabbros and gabbros in Panzhihua and Baima areas. The data are compiled in Table S2. The horizontal dashed lines represent the average value for each group, and the gray area represents the average value plus the error bar (2SD). Abbreviations: Mt = magnetite, and Ilm = ilmenite.

chemical diffusion that have changed their original characteristics. The large activity gradient between ilmenites and interstitial melts produced after the liquid phase segregation enhanced the Mg-Fe inter-diffusion. As such, with increasing degree of Mg-Fe inter-diffusion, Group 1 ilmenites continued to lose Fe, while Group 2 obtained more Fe, eventually yielding the contrasting FeO and MgO distribution between Group 1 and Group 2 ilmenites.

6 Conclusions

The main conclusions drawn from the high-precision Mg and Fe isotopic analyses of olivine, clinopyroxene and ilmenite from the Panzhihua layered mafic intrusion are:

(1) The δ^{26}Mg and δ^{57}Fe values range from −0.47 to −0.32‰ and −0.21 to +0.23‰ in olivine, from −0.40 to −0.18‰ and −0.16 to +0.20‰ in clinopyroxene, and from −0.80 to +23.10‰ and −0.59 to −0.23‰ in ilmenites. The δ^{26}Mg of +23.10‰ is the highest value reported to date.

(2) The negative correlation between δ^{26}Mg and δ^{57}Fe in ilmenites, as well as the disequilibrium Mg and Fe isotope fractionation in coexisting olivine and clinopyroxene points towards kinetic isotopic fractionation during inter-diffusion processes.

(3) The large Mg and Fe isotope fractionation in these minerals were derived from chemical diffusion between minerals and melts during phase segregation caused by interstitial liquid immiscibility.

(4) Our results suggest that the interstitial liquid immiscibility plays a crucial role in the origins of Fe-Ti-rich ore deposits, hence combined Mg-Fe isotopic studies can shed light on the petrogenesis of Fe-Ti ore deposits and layered mafic intrusions.

Acknowledgements

FZT is grateful to Roberta L. Rudnick for her great mentorship throughout his PhD studies and her continuous encouragement and inspiration since then. We thank Prof. Zhao-Chong Zhang for help in field sampling and appreciate the discussion with Dr. Lie-Meng Chen on an early version. Constructive comments from anonymous reviewers and Guest Editor Dr. Sonja Aulbach are thanked, which have significantly improved the manuscript. This work was supported by the Strategic Priority Research Program of the Chinese Academy of Sciences (Grant XDB18000000), National Natural Science Foundation of China (Grants 41730214, 41490633 and 41729001), National Postdoctoral Program for Innovative Talents (BX201700237) and China Scholarship Council program (201804910290).

Appendix A. Supplementary data

Supplementary data to this article can be found online at https://doi.org/10.1016/j.gca.2019.10.004.

References

Cao Y. H., Wang Y., Huang F. and Zhang Z. F. (2019) Iron isotope systematics of the Panzhihua mafic layered intrusion associated with giant Fe-Ti oxide deposit in the Emeishan large igneous province, SW China. J. Geophys. Res.

Cawthorn R. G. and Ashwal L. D. (2009) Origin of Anorthosite and Magnetitite Layers in the Bushveld Complex, Constrained by Major Element Compositions of Plagioclase. J. Petrol. 50, 1607-1637.

Charlier B. and Grove T. L. (2012) Experiments on liquid immiscibility along tholeiitic liquid lines of descent. Contrib. Mineral. Petrol. 164, 27-44.

Charlier B., Namur O., Bolle O., Latypov R. and Duchesne J. C. (2015) Fe-Ti-V-P ore deposits associated with Proterozoic massif-type anorthosites and related rocks. Earth Sci. Rev. 141, 56-81.

Chen L.-M., Teng F.-Z., Song X.-Y., Hu R.-Z., Yu S.-Y., Zhu D. and Kang J. (2018) Magnesium isotopic evidence for chemical disequilibrium among cumulus minerals in layered mafic intrusion. Earth Planet. Sci. Lett. 487, 74-83.

Chen L.-M., Song X.-Y., Hu R.-Z., Yu S.-Y., He H.-L., Dai Z.-H., She Y.-W. and Xie W. (2017) Controls on trace-element partitioning among co-crystallizing minerals: Evidence from the Panzhihua layered intrusion, SW China. Am. Mineral. 102, 1006-1020.

Chen L.-M., Song X.-Y., Zhu X.-K., Zhang X.-Q., Yu S.-Y. and Yi J.-N. (2014) Iron isotope fractionation during crystallization and sub-solidus re-equilibration: Constraints from the Baima mafic layered intrusion, SW China. Chem. Geol. 380, 97-109.

Chung S. L. and Jahn B. M. (1995) Plume-lithosphere interaction in generation of the Emeishan flood basalts at the Permian-Triassic boundary. Geology 23, 889-892.

Craddock P. R. and Dauphas N. (2011) Iron Isotopic Compositions of Geological Reference Materials and Chondrites. Geostand. Geoanal. Res. 35, 101-123.

Craddock P. R., Warren J. M. and Dauphas N. (2013) Abyssal peridotites reveal the near-chondritic Fe isotopic composition of the Earth. Earth Planet. Sci. Lett. 365, 63-76.

Crank J. T. (1975) The Mathematics of Diffusion, second ed. Clarendon Press, Oxford.

Dauphas N., Pourmand A. and Teng F. Z. (2009) Routine isotopic analysis of iron by HR-MC-ICPMS: How precise and how accurate? Chem. Geol. 267, 175-184.

Dauphas N., Teng F. Z. and Arndt N. T. (2010) Magnesium and iron isotopes in 2.7 Ga Alexo komatiites: Mantle signatures, no evidence for Soret diffusion, and identification of diffusive transport in zoned olivine. Geochim. Cosmochim. Acta 74, 3274-3291.

Dauphas, N., John, S. G. and Rouxel, O. (2017). Iron Isotope Systematics. Rev. Mineral. Geochem. 82, 415-510.

He Y.-S., Ke S., Teng F.-Z., Wang, T.-T., Wu H.-J., Lu Y.-H. and Li S.-G. (2015) High-Precision Iron Isotope Analysis of Geological Reference Materials by High-Resolution MC-ICP-MS. Geostand. Geoanal. Res. 39, 341-356.

Hess P. C. and Parmentier E. M. (1995) A Model for the Thermal and Chemical Evolution of the Moons Interior - Implications for the Onset of Mare Volcanism. Earth Planet. Sci. Lett. 134, 501-514.

Holness, M. B., Stripp, G., Humphreys, M. C.S., Veksler, I. V., Nielsen, T. F. D. and Tegner, C. (2011) Silicate Liquid Immiscibility within the Crystal Mush: Late-stage Magmatic Microstructures in the Skaergaard Intrusion, East Greenland. J. Petrol. 52, 175-222.

Huang F., Chakraborty P., Lundstrom C. C., Holmden C., Glessner J. J. G., Kieffer S. W. and Lesher C.E. (2010) Isotope fractionation in silicate melts by thermal diffusion. Nature 464, 396-400.

Huppert, H. and Turner, J. (1981).Double-diffusive convection. J. Fluid Mech. 106, 299-329.

Hunter R.H. and Sparks R. S. J. (1987) The Differentiation of the Skaergaard Intrusion. Contrib. Mineral. Petrol. 95, 451-461.

Ke S., Teng F. Z., Li S. G., Gao T., Liu S. A., He Y. S. and Mo X. X. (2016) Mg, Sr, and O isotope geochemistry of syenites from northwest Xinjiang, China: Tracing carbonate recycling during Tethyan oceanic subduction. Chem. Geol. 437, 109-119.

Klemm D. D., Henckel J., Dehm R. and Vongruenewaldt G. (1985) The Geochemistry of Titanomagnetite in Magnetite Layers and Their Host Rocks of the Eastern Bushveld Complex. Econ. Geol. 80, 1075-1088.

La Tourrette, T. Z., Burnett, D. S. and Bacon, C. R., (1991) Uranium and minor-element partitioning in Fe-Ti oxides and zircon from partially melted granodiorite, Crater Lake, Oregon. Geochim. Cosmochim. Acta 55, 457-469.

Liu P.-P., Zhou M.-F., Ren Z., Wang C. Y. and Wang K. (2016) Immiscible Fe- and Si-rich silicate melts in plagioclase from the Baima mafic intrusion (SW China): Implications for the origin of bi-modal igneous suites in large igneous provinces. J. Asian Earth Sci. 127, 211-230.

Liu P.-P., Zhou M.-F., Wang C. Y., Xing C.-M. and Gao J.-F. (2014a) Open magma chamber processes in the formation of the Permian Baima mafic-ultramafic layered intrusion, SW China. Lithos 184-187, 194-208.

Liu P.-P., Zhou M.-F., Luais B., Cividini D. and Rollion-Bard C. (2014b) Disequilibrium iron isotopic fractionation during the high-temperature magmatic differentiation of the Baima Fe-Ti oxide-bearing mafic intrusion, SW China. Earth Planet. Sci. Lett. 399, 21-29.

Namur O., Charlier B., Toplis M. J., Higgins M. D., Liegeois J. P. and Vander Auwera J. (2010) Crystallization Sequence and Magma

Chamber Processes in the Ferrobasaltic Sept Iles Layered Intrusion, Canada. J. Petrol. 51, 1203-1236.

Oeser M., Dohmen R., Horn I., Schuth S. and Weyer S. (2015) Processes and time scales of magmatic evolution as revealed by Fe-Mg chemical and isotopic zoning in natural olivines. Geochim. Cosmochim. Acta 154, 130-150.

Pang K. N., Li C. S., Zhou M. F. and Ripley E. M. (2009) Mineral compositional constraints on petrogenesis and oxide ore genesis of the late Permian Panzhihua layered gabbroic intrusion, SW China. Lithos 110, 199-214.

Pang K. N., Li C. S., Zhou M. F. and Ripley E. M. (2008a) Abundant Fe-Ti oxide inclusions in olivine from the Panzhihua and Hongge layered intrusions, SW China: evidence for early saturation of Fe-Ti oxides in ferrobasaltic magma. Contrib. Mineral. Petrol. 156, 307-321.

Pang, K. N., Zhou, M. F., Lindsley D., Zhao D. and Malpas J. (2008b) Origin of Fe-Ti oxide ores in mafic intrusions: Evidence from the Panzhihua intrusion, SW China. J. Petrol. 49, 295-313.

Polyakov V. B. and Mineev S. D. (2000) The use of Mossbauer spectroscopy in stable isotope geochemistry. Geochim. Cosmochim. Acta 64, 849-865.

Reynolds I. M. (1985) The Nature and Origin of Titaniferous Magnetite-Rich Layers in the Upper Zone of the Bushveld Complex - a Review and Synthesis. Econ. Geol. 80, 1089-1108.

Richter F. M., Liang Y. and Davis A. M. (1999) Isotope fractionation by diffusion in molten oxides. Geochim. Cosmochim. Acta 63, 2853-2861.

Richter F. M., Watson E. B., Mendybaev R., Dauphas N., Georg B., Watkins J. and Valley J. (2009a) Isotopic fractionation of the major elements of molten basalt by chemical and thermal diffusion. Geochim. Cosmochim. Acta 73, 4250-4263.

Richter F. M., Dauphas N. and Teng F. Z. (2009b). Non-traditional fractionation of non-traditional isotopes: Evaporation, chemical diffusion and Soret diffusion. Chem. Geol. 258, 92-103.

Richter F. M., Watson E.B., Mendybaev R.A., Teng F. Z. and Janney P. E. (2008) Magnesium isotope fractionation in silicate melts by chemical and thermal diffusion. Geochim. Cosmochim. Acta 72, 206-220.

Robinson P., Kullerud K., Tegner C., Robins B. and McEnroe S. A. (2003) Could the Tellnes ilmenite deposit have been produced by in-situ magma mixing? Norges Geol. Unders. Spec. Publ. 9, 107-108.

Schauble E. A. (2011) First-principles estimates of equilibrium magnesium isotope fractionation in silicate, oxide, carbonate and hexaaquamagnesium(2+) crystals. Geochim. Cosmochim. Acta 75, 844-869.

Shannon R. D. (1976) Revised effective ionic radii and systematic studies of interatomic distances in halides and chalcogenides. Acta Cryst. A32, 751-767.

She Y. W., Yu S. Y., Song X. Y., Chen L. M., Zheng W. Q. and Luan Y. (2014) The formation of P-rich Fe-Ti oxide ore layers in the Taihe layered intrusion, SW China: Implications for magma-plumbing system process. Ore Geol. Rev. 57, 539-559.

Shellnutt J. G., Denyszyn S. W. and Mundil R. (2012) Precise age determination of mafic and felsic intrusive rocks from the Permian Emeishan large igneous province (SW China). Gondwana Res. 22, 118-126.

Shellnutt J. G. and Jahn B. M. (2010) Formation of the Late Permian Panzhihua plutonic-hypabyssal-volcanic igneous complex: Implications for the genesis of Fe-Ti oxide deposits and A-type granites of SW China. Earth Planet. Sci. Lett. 289, 509-519.

Sio C. K. I. and Dauphas N. (2017) Thermal and crystallization histories of magmatic bodies by Monte Carlo inversion of Mg-Fe isotopic profiles in olivine. Geology 45, 67-70.

Sio C. K. I., Dauphas N., Teng F.Z., Chaussidon M., Helz R. T. and Roskosz M. (2013) Discerning crystal growth from diffusion profiles in zoned olivine by in situ Mg-Fe isotopic analyses. Geochim. Cosmochim. Acta 123, 302-321.

Song X.-Y., Qi H.-W., Hu R.-Z., Chen L.-M., Yu S.-Y. and Zhang J.-F. (2013) Formation of thick stratiform Fe-Ti oxide layers in layered intrusion and frequent replenishment of fractionated mafic magma: Evidence from the Panzhihua intrusion, SW China. Geochem. Geophys. Geosyst. 14, 712-732.

Stenhouse I., O'Neill H. and Lister G. (2010) Diffusion in natural ilmenite. In: EGU General Assembly, #343 (abstract).

Teng, F. Z. (2017). Magnesium isotope geochemistry. In: Teng, F.Z., Watkins, J., Dauphas, N. (Eds.), Non-Traditional Stable Isotopes. Rev. Mineral. Geochem. 82, 219-287.

Teng F.-Z., Wadhwa M. and Helz R. T. (2007) Investigation of magnesium isotope fractionation during basalt differentiation: Implications for a chondritic composition of the terrestrial mantle. Earth Planet. Sci. Lett. 261, 84-92.

Teng F.-Z., Dauphas N. and Helz R. T. (2008) Iron isotope fractionation during magmatic differentiation in Kilauea Iki Lava Lake. Science 320, 1620-1622.

Teng F.-Z., Dauphas N., Helz R. T., Gao S. and Huang S.-C. (2011) Diffusion-driven magnesium and iron isotope fractionation in Hawaiian olivine. Earth Planet. Sci. Lett. 308, 317-324.

Teng F.-Z., Dauphas N., Huang S.-C. and Marty B. (2013) Iron isotopic systematics of oceanic basalts. Geochim. Cosmochim. Acta 107, 12-26.

Teng F.-Z., Li W.-Y., Ke S., Yang W., Liu S.-A., Sedaghatpour F., Wang S.-J., Huang K.-J., Hu Y., Ling M.-X., Xiao Y., Liu X.-M., Li X.-W., Gu H.-O., Sio C. K., Wallace D. A., Su B.-X., Zhao L., Chamberlin J., Harrington M. and Brewer A. (2015) Magnesium Isotopic Compositions of International Geological Reference Materials. Geostand. Geoanal. Res. 39, 329-339.

Toplis M. J. and Carroll M. R. (1995) An Experimental-Study of the Influence of Oxygen Fugacity on Fe-Ti Oxide Stability, Phase-Relations, and Mineral-Melt Equilibria in Ferro-Basaltic Systems. J. Petrol. 36, 1137-1170.

Toplis M. J. and Carroll M. R. (1996) Differentiation of ferro-basaltic magmas under conditions open and closed to oxygen: Implications for the Skaergaard intrusion and other natural systems. J. Petrol. 37, 837-858.

Von Gruenewaldt G. (1993) Ilmenite-Apatite Enrichments in the Upper Zone of the Bushveld Complex: A Major Titanium-Rock Phosphate Resource. Int. Geol. Rev. 35, 987-1000.

Wang, C. Y. and Zhou M. F. (2013) New textural and mineralogical constraints on the origin of the Hongge Fe-Ti-V oxide deposit, SW China. Miner. Deposita 48, 787-798.

Wang, K., Wang, C. Y. and Ren, Z. Y. (2018) Apatite-hosted melt inclusions from the Panzhihua gabbroic-layered intrusion associated with a giant Fe-Ti oxide deposit in SW China: insights for magma unmixing within a crystal mush. Contrib. Mineral. Petrol. 173, 59.

Weyer S. and Ionov D. A. (2007) Partial melting and melt percolation in the mantle: The message from Fe isotopes. Earth Planet. Sci. Lett. 259, 119-133.

Williams H.M., Peslier A. H., McCammon C., Halliday A. N., Levasseur S., Teutsch N. and Burg J. P. (2005) Systematic iron isotope variations in mantle rocks and minerals: the effects of partial melting and oxygen fugacity. Earth Planet. Sci. Lett. 235, 435-452.

Wu H. J., He Y.-S., Teng F.-Z., Ke S., Hou Z.-H. and Li S.-G. (2018) Diffusion-driven magnesium and iron isotope fractionation at a gabbro-granite boundary. Geochim. Cosmochim. Acta 222, 671-684.

Xu Y.-G., Chung S.L., Jahn B. M. and Wu G.-Y. (2001) Petrologic and geochemical constraints on the petrogenesis of Permian-Triassic Emeishan flood basalts in southwestern China. Lithos 58, 145-168.

Zhang X. Q., Song X. Y., Chen L. M., Xie W., Yu S. Y., Zheng W. Q., Deng Y. F., Zhang J. F. and Gui S. G. (2012) Fractional crystallization and the formation of thick Fe-Ti-V oxide layers in the Baima layered intrusion, SW China. Ore Geol. Rev. 49, 96-108.

Zhang X. Q., Song X. Y., Chen L. M., Yu S. Y., Xie W., Deng Y., Zhang J. F. and Gui S. G. (2013) Chalcophile element geochemistry of the Baima layered intrusion, Emeishan Large Igneous Province, SW China: implications for sulfur saturation history and genetic relationship with high-Ti basalts. Contrib. Mineral. Petrol. 166, 193-209.

Zheng W. Q., Deng Y. F., Song X. Y., Chen L. M., Yu S. Y., Zhou G. F., Liu S. R. and Xiang J. X. (2014) Composition and genetic significance of the ilmenite of Panzhihua intrusion. Acta Petrol. Sin. 30, 1432-1442 (in Chinese with English abstract).

Zhong H. and Zhu W. G. (2006) Geochronology of layered mafic intrusions from the Pan-Xi area in the Emeishan large igneous province, SW China. Miner. Deposita 41, 599-606.

Zhou M.-F., Malpas J., Song X.-Y., Robinson P. T., Sun M., Kennedy A. K., Lesher C. M. and Keays R. R. (2002) A temporal link between the Emeishan large igneous province (SW China) and the end-Guadalupian mass extinction. Earth Planet. Sci. Lett. 196, 113-122.

Zhou M.-F., Robinson P. T., Lesher C. M., Keays R. R., Zhang C. J. and Malpas J. (2005) Geochemistry, petrogenesis and metallogenesis of the Panzhihua gabbroic layered intrusion and associated Fe-Ti-V oxide deposits, Sichuan Province, SW China. J. Petrol. 46, 2253-2280.

Zhu D., Bao H.-M. and Liu Y. (2015) Non-traditional stable isotope behaviors in immiscible silica-melts in a mafic magma chamber. Sci. Rep. 5, 17561.

Origins of two types of Archean potassic granite constrained by Mg isotopes and statistical geochemistry: Implications for continental crustal evolution*

Rui-Ying Li[1], Shan Ke[1], Shuguang Li[1,2], Shuguang Song[3], Chao Wang[1,3] and Chuntao Liu[1]

1. State Key Laboratory of Geological Processes and Mineral Resources, School of Earth Sciences and Resources, China University of Geosciences, Beijing 100083, China
2. CAS Key Laboratory of Crust-Mantle Materials and Environments, School of Earth and Space Sciences, University of Science and Technology of China, Hefei 230026, China
3. MOE Key Laboratory of Orogenic Belts and Crustal Evolution, School of Earth and Space Sciences, Peking University, Beijing 100871, China

亮点介绍：本文借助 Mg 同位素判别出华北克拉通冀东晚太古代存在 I 型和 S 型两类钾质花岗岩。通过统计 4066 个太古宙长英质火成岩的年龄和 K_2O/Na_2O，揭示出 S 型钾质花岗在太古宙中期初现、在太古宙晚期增多的发育规律；发现了陆壳的成熟演化需要壳源岩石熔融和大陆风化的共同作用，前者贡献比例大、后者贡献效率高。

Abstract The chemical composition of the upper continental crust (UCC) dramatically changed from predominantly sodic to more potassic during the late Archean (3000–2500 Ma), although how this change occurred remains poorly constrained. The origins of Neoarchean potassic granites may provide clues for this crucial change, as their formation dominated the compositional transition of the UCC. In this study, we conducted high-precision Mg isotopic analyses of Neoarchean potassic granites (2558–2520 Ma) and tonalite–trondhjemite–granodiorites (TTGs; 2595–2574 Ma) from the North China Craton and statistical geochemistry in K_2O/Na_2O ratios of global Archean granitoids. Results show that these TTGs, with K_2O/Na_2O ratios of ~0.61 and $\delta^{26}Mg$ values of –0.39‰ to –0.22‰, were generated from heterogeneous basaltic sources. In contrast, potassic granites yield two distinctive groups of $\delta^{26}Mg$ values and can be subdivided into I- and S-types. The I-type granite has mantle-like $\delta^{26}Mg$ values of – 0.28‰ to –0.22‰ and weak enrichment in K_2O (K_2O/Na_2O=1.0–1.5), and was likely derived by partial melting of igneous sources. The S-type is characterized by much higher $\delta^{26}Mg$ values of +0.60‰ to +0.91‰ and strong K enrichment (K_2O/Na_2O=1.57–23.26), with geochemical characteristics suggesting derivation by partial melting of sediments. Based on the geochemical characteristics of the S-type potassic granites in this study, we sort the S-type granites out of 4066 Archean felsic igneous rocks worldwide. Their temporal distribution suggests that S-type potassic granites have appeared in the middle Archean (3500–3000 Ma) and became widespread in the late Archean (3000–2500 Ma), but

* 本文发表在：Lithos, 2020, 368-369: 105570

were scarce in the early Archean (4000–3500 Ma). The onset of S-type granites indicates reworking of supracrustal rocks may have started since 3500 Ma. And widespread S-type potassic granites during 3000–2500 Ma play an important role for the high K_2O/Na_2O ratio of UCC at the end of Archean. A mass-balance calculation shows that 25.88 wt.% I-type, 4.12 wt.% S-type potassic granite and 70 wt.% TTGs could increase the K_2O/Na_2O ratio of the UCC to ~0.83 which is the global average value for felsic rocks at the end-Archean. To increase K_2O/Na_2O of the Archean UCC, contributions made by I- and S-type potassic granites could probably reach 3:2, respectively. Consequently, the evolved crustal composition from predominantly sodic to potassic results from the synergy of re-working of igneous (unaltered) and sedimentary (weathered) rocks, by which the former contributed more quantitatively, while the latter contributed more efficiently.

Keywords Late Archean, North China Craton, crustal evolution, potassic granites, TTGs, Mg isotopes

1 Introduction

The Archean was a period of major growth of continental crust, with 60%–70% of the present-day continental volume being formed before 2500 Ma (Dhuime et al., 2012; Goodwin, 1991). However, the crustal component was not always consistent with the primary crust, which evolved over time (Condie, 2014; Laurent et al., 2014). For example, the modern upper continental crust (UCC) is enriched in K and heavy rare earth elements (HREEs) (K_2O/Na_2O=0.86; $(La/Yb)_N$=11; Rudnick and Gao, 2003). In contrast, the primary felsic UCC in the early Archean was dominated by tonalite–trondhjemite–granodiorites (TTGs) (Clos et al., 2018; Johnson et al., 2019; Laurent et al., 2014), with high Na and low HREE contents (K_2O/Na_2O=0.35; $(La/Yb)_N$=32; Moyen and Martin, 2012). The rapid increase in K_2O/Na_2O of felsic igneous rocks occurred near the end of the Archean (Keller and Schoene, 2012) with decrease in TTGs and increase in potassic granites (Condie, 2014; Goodwin, 1991; Laurent et al., 2014; Moyen, 2019). Globally, potassic granitoids, the second most widespread Archean lithology, intruded after TTGs as the last Archean magmatic event in cratons, albeit asynchronously (Laurent et al., 2014; Nebel et al., 2018), representing the final stabilization of Archean cratons and marking progressive crustal maturation (Close et al., 2018; Laurent et al., 2014; Nebel et al., 2018; Whalen et al., 2004). Therefore, the potassic granites would provide important information to understand the crustal evolution in the Archean. Two types of Archean potassic granites have been proposed up till now. One is the I-type granites, which is produced by re-working of pre-existing TTGs (Champion and Sheraton, 1997; Nebel et al., 2018; Patiño Douce and Beard, 1995; Skjerlie and Johnston, 1996a; Watkins et al., 2007) or partial melting of K-rich basaltic rocks (Sisson et al., 2005; Xiao and Clemens, 2007). The other is S-type granites, pointing to the derivation of melting of metasediments (Day and Weiblen, 1986), however, minor in the Archean and less reported (Moyen, 2019). Although it is generally accepted that I-type potassic granites are more dominated in the Archean terranes than the S-type one, how the role and proportion of Archean I-type and S-type potassic granites to increasing K_2O/Na_2O of the UCC is still poorly constrained. To answer this issue is important for understanding the two mechanisms of making continental crust more potassic: (1) partial melting of igneous rocks, during which K is enriched in melts via dehydration melting of biotite and hornblende; and (2) surficial chemical weathering, during which K is mainly kept in residual clay minerals and Na is transported into oceans with aqueous solutions (e.g. Rudnick and Gao, 2003; and references therein). Besides, sediments, as the magmatic source for S-type granites, are products of continental weathering. Chemical weathering would consume CO_2 from the atmosphere and may related to a rise of atmospheric O_2 in the late Archean–early Proterozoic (e.g., Lowe and Tice, 2004). Thus, more researches on spatial and temporal distribution of Archean S-type potassic granites would offer fundamental information to constrain the timing and intensity of continental weathering in the Archean.

Because Mg isotopes fractionate limitedly during silicate magmatic differentiation but strongly in surficial weathering, they would perform distinctive signatures in different lithologies (Teng, 2017; and references therein). For example, non-contaminated mantle has uniform δ^{26}Mg values of –0.25±0.07‰ (Teng et al., 2010a), while carbonates have lighter δ^{26}Mg values as low as –5.57‰ (Wombacher et al., 2011) and weathered silicates have higher values of up to +1.8‰ (Liu et al., 2014). Therefore, Mg isotopes could be sensitive to sediments in magma sources and helpful to provide some constraints on distinguishing I- and S-type potassic granites, owing to the specific isotopic compositions of igneous and sedimentary rocks.

Here we investigate the Mg isotopic composition of Neoarchean potassic granites (2558–2520 Ma) and contemporaneous TTGs (2595–2574 Ma) from the North China Craton (NCC). Our data exhibit both I- and S-type Archean potassic granites occur in NCC, which have different geochemical characteristics, especially in their Mg isotopic compositions and K_2O/Na_2O ratios. With the end objective being to understand the mechanism inducing the continental crust more potassic, we compiled 4066 Archean felsic igneous rocks and compared with our samples, by which the development of Archean S-type potassic granites and the contributions of I- and S-type potassic granites to increasing K_2O/Na_2O of the Archean UCC are given in this study.

2 Regional geology and samples

2.1 Geological setting

The NCC is the largest and oldest craton in China and contains extensive exposures of Archean lithologies with different ages (4000–2500 Ma; Liu et al., 1992; Song et al., 1996; Wan et al., 2005), making it an ideal area for study of the evolution of the continental crust during the Archean.

The studied areas are located within the Suizhong granitic terrane along the eastern margin of the NCC (Fig. 1a). The Suizhong granites comprise mainly Neoarchean potassic granites and minor TTGs, and extend from Suizhong and Xingcheng through Huludao to Jinzhou (Fig. 1b, c) (Fu et al., 2017, 2018; Wang et al., 2016; Xie et al., 2019). The TTGs were intruded by potassic granites, and are typically exposed as xenoliths within the latter. Potassic granites in this terrane comprise predominantly monzodiorite and monzogranite to granite, with minor granodiorite and syenogranite (Fu et al., 2018; Xie et al., 2019). The Suizhong granitic terrane is partially covered by Proterozoic deposits and Mesozoic volcanic–sedimentary sequences, and intruded by Mesozoic igneous rocks (Wang et al., 2016).

2.2 Samples

Eighteen representative samples were analyzed, including four granodiorites from Juhuadao, six granites from Taili, and eight granites from Huludao (Fig. 1c). Here we refer to samples with high K_2O contents (≥3 wt.%) and K_2O/Na_2O ratios (≥1) as potassic granites, with granites from Taili and Huludao thus being so classified. Granodiorites from Juhuadao are of the TTG series because of their lower K_2O/Na_2O ratios (~0.61) and TTG characteristics in the An–Ab–Or diagram (Wang et al., 2016). Whole-rock major and trace element analyses, zircon U–Pb dating, and Nd–Hf isotopic analyses were undertaken by Wang et al. (2016). Here, we briefly review the characteristics of these samples.

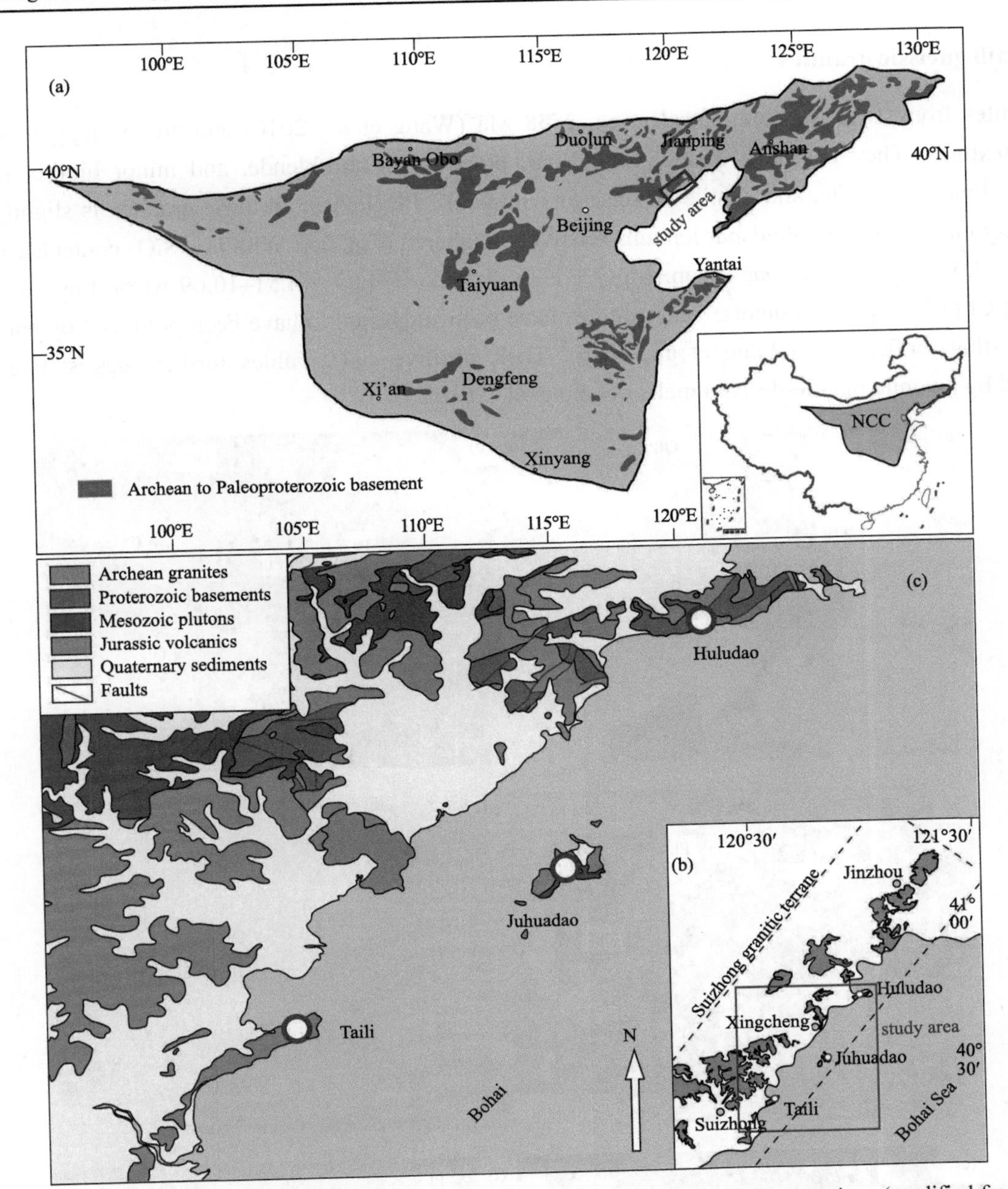

Fig. 1 (a) Geological map of the North China Craton, indicating major Precambrian basement regions (modified from Wan et al., 2015 and Wang et al., 2016). (b) Simplified geological map of Archean basement in the Suizhong granitic terrane (modified from Wang et al., 2016 and Fu et al., 2017). (c) Geological framework of the study area (modified from a 1:200000 geological map of Liaoning Province). Sample locations are represented by red circles.

2.2.1 Juhuadao granodiorites

TTGs from Juhuadao are represented by granodiorites emplaced at 2595–2574 Ma (Wang et al., 2016). They are medium- to coarse-grained rocks with massive structure and comprise quartz, plagioclase, K-feldspar, and biotite, with accessory zircon, titanite, magnetite, fluorite, and apatite. Plagioclase is weakly altered to clay, and occasionally biotite is weakly chloritized (Fig. 2a, b). Because of their enrichment in large-ion lithophile elements (LILEs) and K_2O/Na_2O ratios (average = 0.61; Fig. 3b) higher than those reported for typical TTGs (<0.5; Moyen and Martin, 2012), Wang et al. (2016) suggested that they were generated by partial melting of K-enriched mafic sources.

2.2.2　Taili gneissic granites

Granites from Taili were emplaced at ca. 2558 Ma (Wang et al., 2016) and are strongly deformed with gneissic textures. They comprise quartz, K-feldspar, plagioclase, hornblende, and minor biotite, with zircon, magnetite, titanite, fluorite, and apatite as accessory minerals. Plagioclase in these granites is slightly altered to clay and sericitized, while K-feldspar remains relatively unaltered (Fig. 2c). With low SiO_2 contents (60.16–69.43 wt.%; Fig. 3a) and high ferromagnesian ($MgO + FeO_t + TiO_2 + MnO$ = 4.51–10.69 wt.%; Fig. 3d), Ni (10–20 ppm), and Cr (17–31 ppm) contents, Taili granites have been suggested to have been produced by partial melting of pre-existing mafic crust (Wang et al., 2016). Their positive $\varepsilon_{Hf}(t)$ values further suggest that they were hybridized by juvenile mantle-derived melts (Wang et al., 2016).

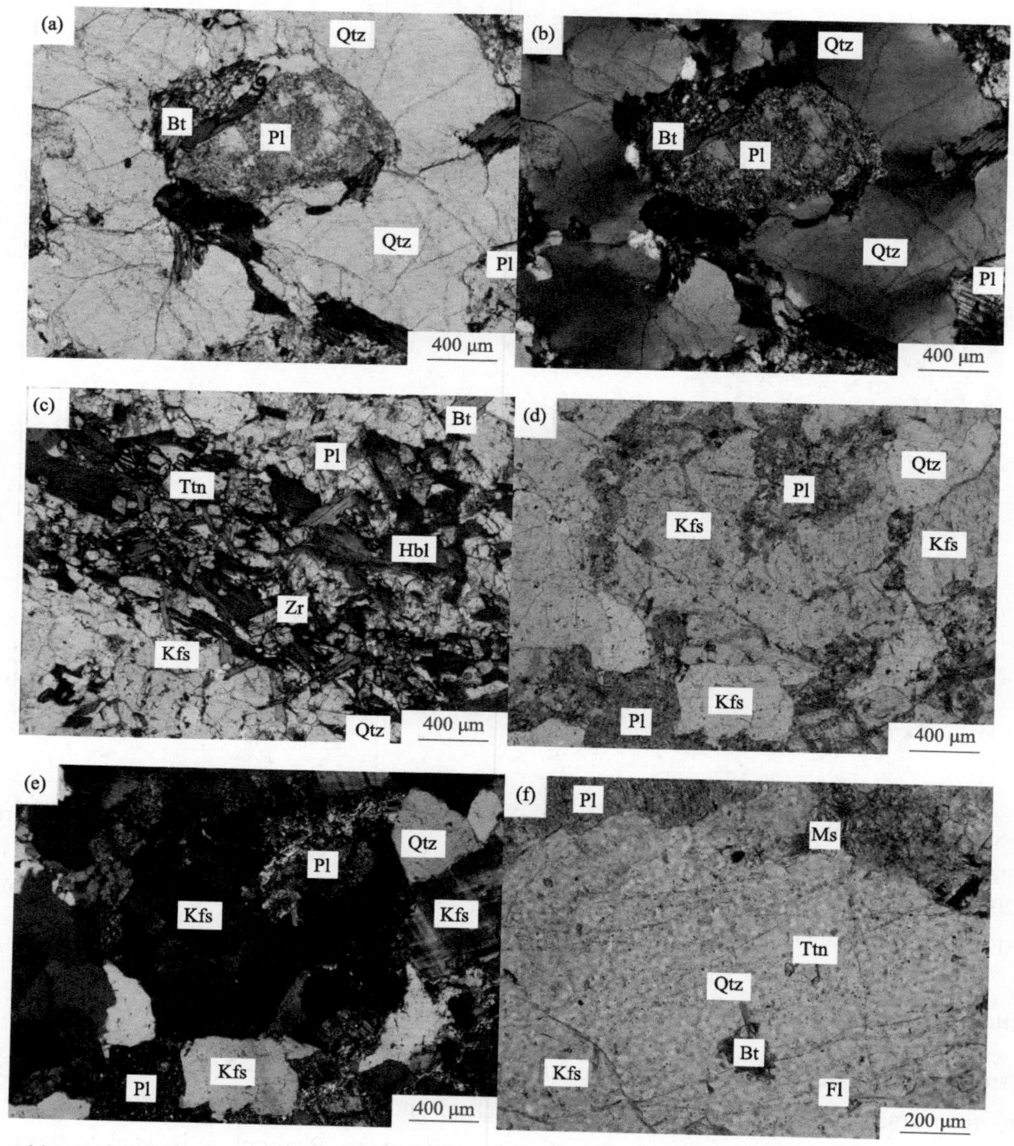

Fig. 2　Petrographic characteristics of Archean granitoids. (a, b) Granodiorites from Juhuadao showing the mineral assemblage of Qtz–Pl–Bt. (c) Gneissic granites from Taili showing the mineral assemblage of Qtz–Kfs–Hbl–Pl–Bt–Ttn–Zr. (d, e) Potassic granites from Huludao showing the mineral assemblage of Qtz–Kfs–Pl, with Pl presenting sericitization and clay alteration. (f) Photomicrograph showing Bt and Ms in Huludao potassic granites, with Qtz–Kfs–Pl as major minerals and Ttn–Fl as accessory minerals. Bt = biotite; Fl = fluorite; Hbl = hornblende; Kfs = K-feldspar; Ms = muscovite; Pl = plagioclase; Qtz = quartz; Ttn = titanite; Zr = zircon.

2.2.3 Huludao granites

Huludao granites were emplaced at 2545–2520 Ma (Wang et al., 2016) and are unconformably overlain by Paleoproterozoic metasandstone. These granites are fine- to medium-grained rocks with massive textures and a pinkish color. They contain quartz, K-feldspar, minor plagioclase, rare biotite, and muscovite (Fig. 2d–f). Accessory minerals include titanite, apatite, epidote, zircon, fluorite, and magnetite. Plagioclase shows sericitization and clay alteration, while K-feldspar remains unaltered as in the Taili gneissic granites (Fig. 2c–f). Wang et al. (2016) suggested that these granites were produced by re-melting of pre-existing TTGs, based on their high SiO_2 contents (69.69–72.32 wt.%; Fig. 3a), low ferromagnesian ($MgO+FeO_t+TiO_2+MnO$=1.66–3.05 wt.%; Fig. 3d), Ni (6–12 ppm), and Cr (4–22 ppm) contents and peraluminous nature (Fig. 3c).

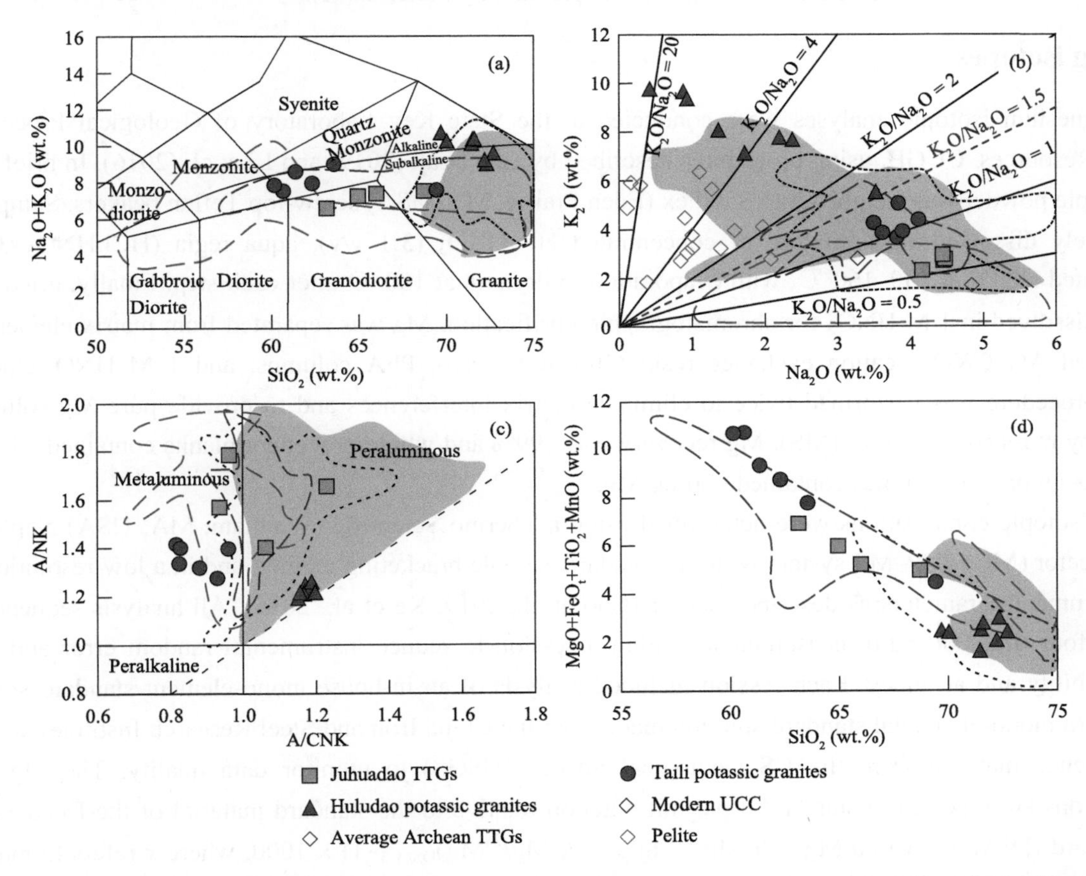

Fig. 3 Plots of (a) (K_2O+Na_2O)–SiO_2, (b) K_2O–Na_2O, (c) (A/NK)–(A/CNK), and (d) ($MgO+FeO_t+TiO_2+MnO$)–SiO_2 for Archean potassic granites and TTGs, compared with experimental melts of various protoliths, including low-K (K_2O<1 wt.%) mafic rocks (blue dashed line, Clemens, 2006; Patiño Douce and Beard, 1995; Rapp and Watson, 1995; Springer and Seck ,1997), high-K (K_2O>1 wt.%) mafic rocks (purple dash line, Sisson et al., 2005; Xiao and Clemens, 2007), TTGs (dotted line, Clemens, 2006; Conrad et al., 1988; Patiño Douce ,2005; Singh and Johannes, 1996a, b; Skjerlie and Johnston, 1992,1993; Watkins et al., 2007) and sediments (grey shaded area, Castro et al., 1999; García-Arias et al., 2012; Gardien et al., 1995; Grant, 2004; Nair and Chacko, 2002; Patiño Douce and Beard, 1995; Patiño Douce and Harris, 1998; Patiño Douce and Johnston, 1991; Pickering and Johnston, 1998; Spicer et al., 2004; Vielzeuf and Holloway, 1988; Vielzeuf and Montel, 1994). Data of modern UCC and average Archean TTGs are from Rudnick and Gao (2003) and Moyen and Martin (2012), respectively. Compositions of pelite are from Li et al. (2010).

3 Analytical methods

3.1 Major elements

Whole-rock major element compositions were determined at the China University of Geosciences, Beijing (CUGB), China. Samples were analyzed by Leeman Prodigy (Teledyne Leeman Labs, Hudson, NH, USA) inductively coupled plasma optical emission spectroscopy (ICP–OES). Loss-on-ignition (LOI) was determined gravimetrically after heating samples at 1000 ℃ for 30 min. Based on repeated measurements of rock reference standards GSR-1 (granite) and GSR-5 (shale) from National Research Center for Certified RMs of China, analytical uncertainties were better than 1% (1σ) (Supplementary materials A).

3.2 Mg isotopes

Magnesium isotopic analyses were conducted at the State Key Laboratory of Geological Processes and Mineral Resources, CUGB, using procedures described by Gao et al. (2019) and Li et al. (2016). In brief, 1–8 mg bulk sample powder was weighed into Savillex (Eden Prairie, MN, USA) screw-top Teflon beakers. Samples were successively dissolved in a mixture of concentrated HF–HNO_3 (3:1 v/v), aqua regia (HCl:HNO_3=3:1), and concentrated HNO_3 at 130–160 ℃, with evaporation to dryness at 100 ℃ after each step. Finally, dried residues were re-dissolved in 1 M HNO_3 for chromatographic purification. Mg was separated from matrix elements using pre-cleaned AG50W-X8 cation exchange resin (200–400 mesh), PFA columns, and 1 M HNO_3 eluent. The column procedure was performed twice to eliminate matrix interferences and to provide pure Mg solutions for analysis by mass spectrometry (MS). Mg recovery was >99% and whole-procedure blanks contained <10 ng Mg, which was ignored as samples contained ~20 μg Mg.

Mg isotopic compositions were determined using a Thermo Scientific (Waltham, MA, USA) Neptune Plus multicollector (MC)–ICP–MS system with the standard–sample bracketing method and in a low-resolution mode, with instrument parameters as described earlier (Gao et al., 2019; Ke et al., 2016). All analysis sequences were repeated four times on ratio measurements within a session to reduce instrumental random error and improve reproducibility and accuracy. Each session included analysis of an in-house mono-element standard solution of GSB Mg (a mono-elemental standard solution made from the China Iron and Steel Research Institute) and at least one reference material from the US Geological Survey (USGS) to monitor data quality. The Mg isotopic compositions are reported in standard δ per mil notation relative to the standard material of the Dead Sea metal Mg standard (DSM-3), with $\delta^{x}Mg = ((^{x}Mg/^{24}Mg)_{sample}/(^{x}Mg/^{24}Mg)_{DSM\text{-}3}-1) \times 1000$, where x refers to mass 25 or 26. Uncertainties are given as two standard deviations (2SD) computed from the four repeated analyses. Long-term external precision was ±0.06‰ (2SD) for $\delta^{26}Mg$ (Gao et al., 2019; Ke et al., 2016).

To ensure the accuracy of the entire chemical procedure, three USGS reference materials (BHVO-2, BCR-2, and GSP-2) were processed with granitic samples, yielding average $\delta^{26}Mg$ values of –0.23±0.02‰, –0.19±0.03‰, and 0.01±0.02‰, respectively, consistent with published values (Table 1; An et al., 2014; Baker et al., 2005; Huang et al., 2012; Huang et al., 2015; Lee et al., 2014; Pogge von Strandmann et al., 2011).

Table 1 Mg isotopic compositions of reference materials

Standards	$\delta^{26}Mg$ (‰)	2SD	$\delta^{25}Mg$ (‰)	2SD	References
BHVO-2	–0.22	0.03	–0.11	0.01	this study
repeat*	–0.24	0.02	–0.11	0.01	this study
average	–0.23	0.02	–0.11	0.01	this study

Continued

Standards	δ^{26}Mg (‰)	2SD	δ^{25}Mg (‰)	2SD	References
average	−0.24	0.08	−0.12	0.05	Huang et al. (2012)
	−0.26	0.04	−0.14	0.06	Huang et al. (2015)
	−0.20	0.07	−0.10	0.05	Lee et al. (2014)
	−0.26	0.06	−0.14	0.04	Pogge von Strandmann et al. (2011)
BCR-2	−0.18	0.06	−0.09	0.02	this study
repeat*	−0.20	0.04	−0.09	0.01	this study
average	−0.19	0.03	−0.09	0.01	this study
	−0.16	0.01	−0.08	0.02	An et al. (2014)
	−0.19	0.02	−0.09	0.07	Baker et al. (2005)
	−0.26	0.08	−0.13	0.05	Lee et al. (2014)
GSP-2	0.01	0.02	0.01	0.01	this study
	0.04	0.02	0.03	0.01	An et al. (2014)
	0.03	0.09	0.00	0.06	Huang et al. (2015)

2SD = 2 times the standard deviation of the population of n (= 4) repeat measurements of a single solution.
repeat* = repeat dissolution and column chemistry of individual samples.
average = average of the measured values in this study.

4 Results

4.1 Major element compositions

The major element compositions of all samples are given in Table 2, with three Huludao samples being analyzed in this study and data for the other samples being from Wang et al. (2016).

The three Huludao potassic granites measured in this study have high SiO_2 contents of 71.60–72.71 wt.% and are peraluminous with A/CNK (molar $Al_2O_3/(CaO+Na_2O+K_2O)$) ratios of 1.15–1.20 (Fig. 3a, c). Their ferromagnesian ($MgO+FeO_t+TiO_2+MnO$=2.10–2.83 wt.%) and CaO (0.11–0.34 wt.%) contents are low compared with those of Taili potassic granites (Fig. 3d; Table 2). They are relatively enriched in K_2O (5.54–9.37 wt.%) with high K_2O/Na_2O ratios of 1.57–10.08 (Fig. 3b). Their major element characteristics determined here are generally consistent with those of other Huludao samples reported by Wang et al. (2016).

Table 2 Major element and Mg isotopic compositions of Neoarchean potassic granites and TTGs measured in this study

Samples	SiO_2 (wt.%)	Al_2O_3 (wt.%)	FeO_t (wt.%)	MnO (wt.%)	MgO (wt.%)	CaO (wt.%)	Na_2O (wt.%)	K_2O (wt.%)	LOI (wt.%)	δ^{26}Mg (‰)	2SD	δ^{25}Mg (‰)	2SD
Potassic granites													
Taili gneissic granites													
13TL07 [a]	63.57	15.15	5.35	0.06	1.53	2.47	3.83	5.07	0.71	−0.26	0.04	−0.13	0.01
13TL08 [a]	61.37	15.30	6.16	0.08	1.97	3.21	4.11	4.40	0.70	−0.25	0.06	−0.12	0.03
15TL01 [a]	62.32	13.94	5.89	0.08	1.82	3.59	3.49	4.27	2.21	−0.28	0.03	−0.14	0.03
15TL02 [a]	60.66	14.58	7.37	0.10	2.12	4.18	3.81	3.67	0.73	−0.22	0.03	−0.13	0.00
15TL03 [a]	69.43	14.21	2.88	0.05	1.22	2.57	3.61	3.86	0.61	−0.27	0.02	−0.13	0.04
15TL05 [a]	60.16	14.99	7.21	0.09	2.25	4.10	3.89	3.92	0.67	−0.25	0.03	−0.13	0.01

Continued

Samples	SiO_2 (wt.%)	Al_2O_3 (wt.%)	FeO_t (wt.%)	MnO (wt.%)	MgO (wt.%)	CaO (wt.%)	Na_2O (wt.%)	K_2O (wt.%)	LOI (wt.%)	δ^{26}Mg (‰)	2SD	δ^{25}Mg (‰)	2SD
Huludao granites													
12XC29 [a]	69.69	14.98	1.51	0.01	0.63	0.45	0.88	9.67	1.03	0.90	0.03	0.47	0.02
repeat*										0.92	0.04	0.46	0.04
average										0.91	0.03	0.46	0.01
12XC30 [a]	70.03	15.05	1.67	0.01	0.42	0.32	2.38	7.68	1.26	0.75	0.03	0.40	0.02
repeat*										0.73	0.03	0.39	0.01
average										0.74	0.03	0.39	0.01
12XC31 [a]	71.45	14.99	0.97	0.01	0.47	0.30	2.20	7.77	0.88	0.63	0.03	0.34	0.02
repeat*										0.65	0.02	0.35	0.01
average										0.64	0.03	0.34	0.02
12XC32 [a]	72.32	13.37	2.16	0.03	0.43	0.36	1.72	7.13	1.08	0.88	0.03	0.46	0.01
repeat*										0.89	0.03	0.47	0.05
average										0.89	0.02	0.47	0.01
12XC33 [a]	71.50	13.75	1.82	0.01	0.48	0.25	0.42	9.77	0.98	0.77	0.04	0.40	0.03
repeat*										0.81	0.02	0.42	0.04
average										0.79	0.05	0.41	0.02
18HLD-01 [b]	72.71	13.46	1.98	0.01	0.34	0.11	1.36	8.08	1.05	0.78	0.09	0.41	0.06
18HLD-02 [b]	71.60	14.04	2.10	0.02	0.37	0.26	0.93	9.37	0.83	0.83	0.09	0.42	0.06
18HLD-03 [b]	72.21	14.41	1.56	0.01	0.33	0.34	3.52	5.54	0.92	0.60	0.09	0.32	0.06
TTGs													
Juhuadao granodiorites													
12XC19 [a]	64.95	16.17	4.00	0.07	1.51	3.85	4.45	2.71	0.77	–0.26	0.01	–0.13	0.01
12XC20 [a]	63.15	16.72	4.74	0.09	1.63	4.40	4.16	2.32	0.97	–0.22	0.03	–0.12	0.05
12XC24 [a]	66.00	17.21	3.49	0.06	1.09	1.99	4.47	2.78	0.32	–0.39	0.02	–0.19	0.02
repeat*										–0.39	0.06	–0.18	0.03
average										–0.39	0.01	–0.18	0.01
12XC27 [a]	68.72	14.81	3.53	0.06	0.94	1.89	4.44	2.95	1.17	–0.34	0.03	–0.16	0.03
repeat*										–0.38	0.05	–0.18	0.03
average										–0.36	0.05	–0.17	0.03

[a] Data of major elements were from Wang et al. (2016).

[b] Data of major elements are analyzed in this study at CUGB.

Total iron shown as FeOt.

2SD = 2 times the standard deviation of the population of n (= 4) repeat measurements of the sample during an analytical session.

repeat* = repeat dissolution and column chemistry of individual samples.

average = average of the measured values in this study.

4.2 Mg isotope compositions

Mg isotopic compositions of all granitoids are listed in Table 2 and plotted in Fig. 4. Four Juhuadao TTG samples have a narrow range of δ^{26}Mg values of –0.39‰ to –0.22‰ (Fig. 4), with two having values (–0.26‰ and –0.22‰) identical to mantle values (–0.25‰±0.07‰; Teng et al., 2010a), and two (–0.39‰ and –0.36‰) being slightly lighter. In contrast, the Taili and Huludao potassic granites have distinctive Mg isotopic compositions spanning a wide range of up to 1.19‰. Taili samples have mantle-like signatures with δ^{26}Mg values of –0.28‰ to –0.22‰ (Fig. 4), while the Huludao samples have high δ^{26}Mg values of +0.60‰ to +0.91‰, similar to those of pelite, saprolite, and paleo-weathered crust (Li et al., 2010; Liu et al., 2014;Tian et al., 2019) (Fig. 4).

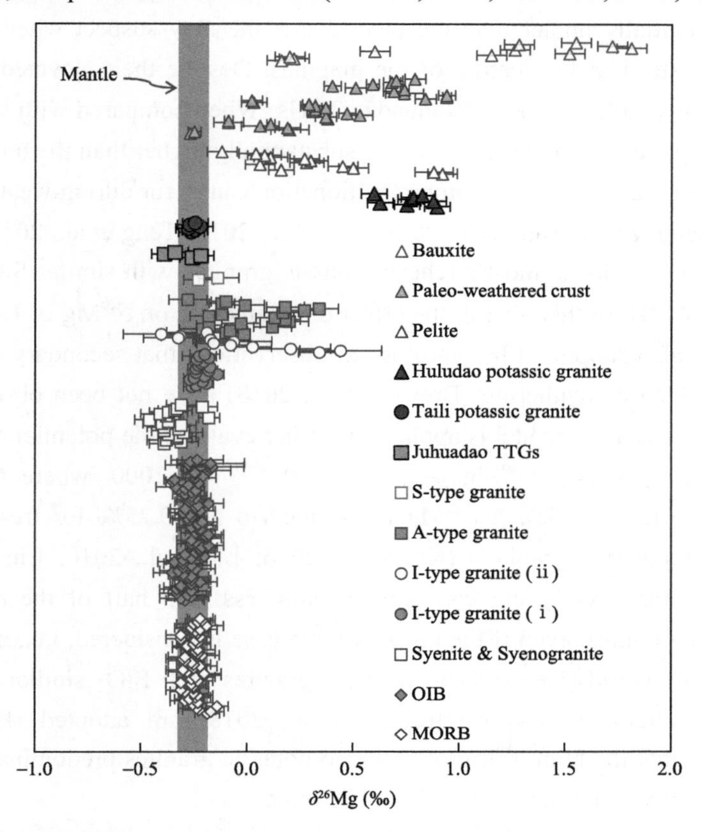

Fig. 4 δ^{26}Mg values of the Archean TTGs and potassic granites compared with those of granitoids and basalts in literatures. The mid-ocean ridge basalt (MORB) and ocean island basalt (OIB) data are from Bourdon et al. (2010) and Teng et al. (2010a). Data for syenite and syenogranite are from Ke et al. (2016); A- and S-type granite are from Li et al. (2010). I-type granite (i) represents typical I-type granites cited from Li et al. (2010) and Liu et al. (2010). I-type granite (ii) with a large continuous variation of δ^{26}Mg is from Shen et al. (2009), explained by assimilation with sedimentary wall rocks. The data for pelite, paleo-weathered crust, and bauxite are from Li et al. (2010), Tian et al. (2019) and Liu et al. (2014), respectively. The vertical shaded band represents a mantle δ^{26}Mg value of –0.25 ± 0.07‰ (Teng et al., 2010a).

5 Discussion

5.1 Influence of secondary alteration on Mg isotopes

Large Mg isotopic fractionation occurs during chemical weathering, with light isotope preferentially released into the hydrosphere and heavy one remaining in the regolith (Brewer et al., 2018; Liu et al., 2014; Teng et al., 2010b). In all samples, most plagioclase exhibits slight to moderate alteration (Fig. 2), pointing to various extent of weathering on these granites after emplacement. Therefore, to assess the influence of secondary alteration on Mg isotopes should be the priority.

The intensity of weathering on granites can be evaluated by Chemical Index of Alteration (CIA, molar ratio of 100 × $Al_2O_3/(Al_2O_3 + Na_2O + CaO + K_2O)$, Nesbitt and Young, 1982). As indicated by weathering profiles of granites and basalts, weathering tends to increase δ^{26}Mg as CIA increases (Fig. 5a). Most Juhuadao TTGs and Taili potassic granites have mantle-like δ^{26}Mg, typical of fresh igneous rocks, and show no correlation between δ^{26}Mg and CIA (Fig. 5a). Two Juhuadao samples (i.e. sample 12XC24 and 12XC27) have slightly lighter δ^{26}Mg

(–0.39‰ and –0.36‰), which also cannot be explained by weathering (e.g., Brewer et al., 2018). Weathering thus has not significantly affected δ^{26}Mg of Juhuadao TTGs and Taili potassic granites. By contrast, Huludao potassic granites have δ^{26}Mg substantially higher than the others, and one may suspect whether such high δ^{26}Mg were produced by weathering instead of the feature of the magmas. Despite their elevated δ^{26}Mg, Huludao potassic granites have low CIA comparable to those of Juhuadao TTGs. When compared with weathering profile samples in the literatures, Huludao potassic granites have δ^{26}Mg substantially higher than the most fractionated, weathered samples at a given CIA (Fig. 5a). Large Mg isotope fractionation can occur during weathering only after intensive removal of Mg from the samples (Brewer et al., 2018; Liu et al., 2014; Teng et al., 2010b). The Huludao potassic granites have MgO contents similar to most Archean potassic granites with similar SiO_2 (i.e. 70–74 wt.%) (Fig. 5b, Supplementary materials B). In this regard, the effect of weathering on δ^{26}Mg of Huludao potassic granites is likely minor. This view is also supported by petrological observations that secondary minerals such as illite and other clays generated by intense weathering (Brewer et al., 2018) have not been observed in Huludao granites (Fig. 2d–f). A Rayleigh fractionation model is applied to further evaluate the potential effect of weathering on the Huludao samples ($\delta^{26}Mg_{weathered\ granite}=(\delta^{26}Mg_{fresh\ granite}+1000)\ f^{\ (1/\alpha-1)}-1000$, where f refers to the fraction of bedrock MgO remaining in the sample). A δ^{26}Mg is assumed to be –0.25‰ for fresh granite, typical of most I-type granites (e.g., Taili potassic granites) (Ke et al., 2016; Li et al., 2010; Liu et al., 2010; this study). Modelling results indicate that weathering can only explain less than half of the δ^{26}Mg difference between Huludao and Taili potassic granites, even if the most extreme case is considered, i.e., an unusually high protolith MgO content (maximum of worldwide Archean potassic granites with SiO_2 similar to Huludao granites) and maximum of possible fractionation factor (Brewer et al., 2018) are adopted (Fig. 5b). Therefore, it is conservatively to conclude that the high δ^{26}Mg of Huludao potassic granites predominantly reflects the feature of their magmas, and the role of weathering, if any, should be minor.

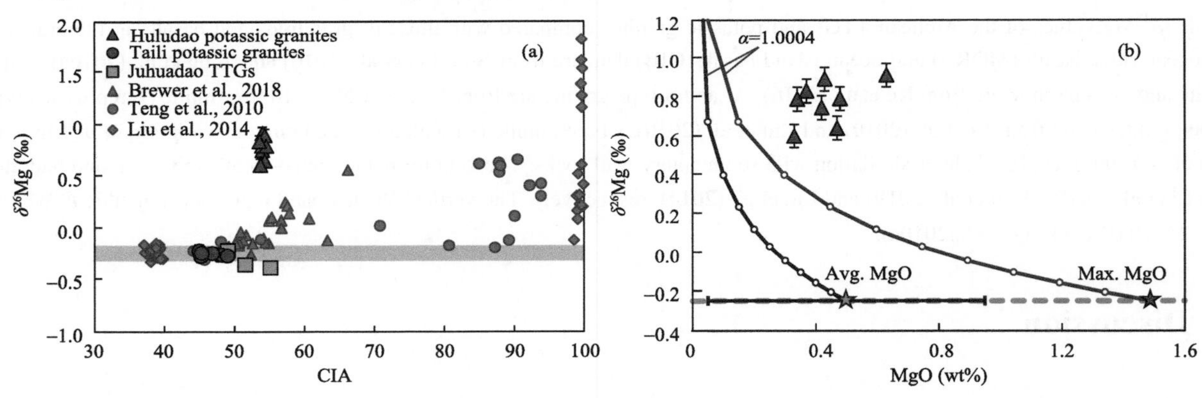

Fig. 5　(a) δ^{26}Mg correlation with Chemical Index of Alteration (CIA). The gray bands represent the average mantle δ^{26}Mg value of –0.25±0.07‰ (Teng et al., 2010a). The grey symbols represent literature data from weathering granites profiles (Brewer et al., 2018) and weathering basalt profile (Teng et al., 2010b; Liu et al., 2014). (b) Modeling of Mg isotopic fractionation during weathering in Huludao potassic granites. The horizontal dash line denotes the assumed δ^{26}Mg of fresh granite of –0.25‰. The red and blue stars represent the maximum MgO (1.49 wt.%) and average MgO (0.50±0.45 wt.%, 2sd) of an assumed fresh granite, obtained from the worldwide Archean potassic granites (Supplementary material B). With the maximum α of 1.0004 from Brewer et al. (2018), the red and blue curved line depicts Mg fractionation via Rayleigh distillation process with the maximum and average MgO assumed above, respectively.

5.2　δ^{26}Mg variations in TTGs

Previous studies have suggested that Mg isotopic fractionation during partial melting of the mantle and differentiation of basaltic (Teng et al., 2010a) or granitic magmas (Ke et al., 2016; Li et al., 2010; Liu et al., 2010) is insignificant. TTGs derived from partial melting of basaltic sources (Champion and Sheraton, 1997; Moyen and Martin, 2012; Rapp and Watson, 1995; Springer and Seck, 1997) would therefore inherit Mg isotopes from their basaltic precursors without significant fractionation. The small variations in δ^{26}Mg values of Juhuadao TTGs (– 0.39‰ to –0.22‰) may thus reflect the heterogeneity of their source, as they were produced by synchronous partial melting of garnet-free amphibolite and garnet amphibolite (Wang et al., 2016).

5.3 Potassic granites with mantle-like δ^{26}Mg signatures

The narrow range of δ^{26}Mg values of Taili potassic granites (–0.28‰ to –0.22‰) and their similarity to the mantle values (–0.25±0.07‰; Teng et al., 2010a) reflect their igneous origins. These potassic granites have lower SiO_2 (60.58–70.44 wt.%) and higher ferromagnesian ($MgO+FeO_t+TiO_2+MnO$=4.51–10.69 wt.%) contents (Fig. 3a, d), relative to the Huludao potassic granites. They are weakly enriched in K_2O (K_2O/Na_2O < 1.5; Fig. 3b) and plot in the metaluminous field in the (A/NK)–(A/CNK) diagram (Fig. 3c). Such features are consistent with those of experimental melts produced by partial melting of high-K mafic rocks (Fig.3, 6), such as hornblende gabbro (e.g., Sisson et al., 2005). On the basis of their positive zircon Hf isotopic compositions ($\varepsilon_{Hf}(t)$ > +2) and negative Eu–Sr anomalies, these potassic granites have been suggested to be produced by low-pressure melting of pre-existing enriched mafic crust mixed with juvenile mantle-derived melts (Wang et al., 2016).

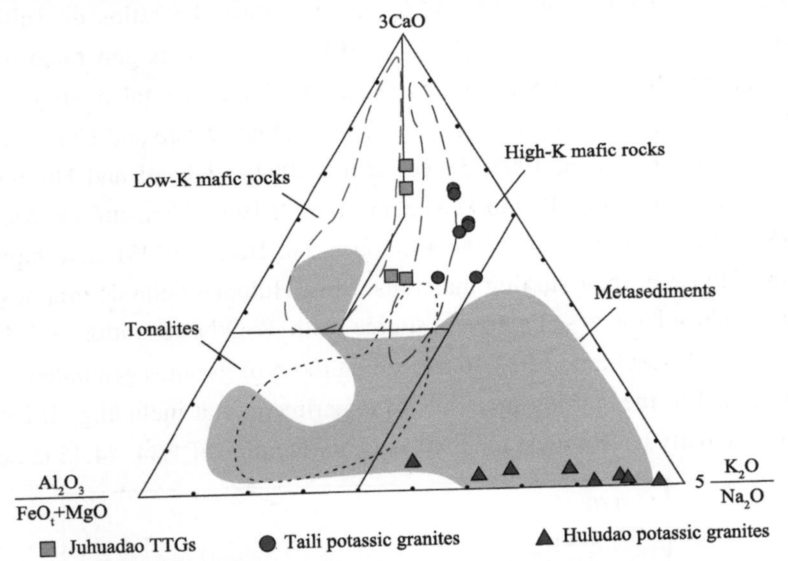

Fig. 6 $Al_2O_3/(FeO_t+MgO)$–3CaO–5(K_2O/Na_2O) ternary diagram (after Laurent et al., 2014). The fields represent average compositions of melts derived from various sources estimated by Laurent et al. (2014). The experimental melts of low-K mafic rocks (blue dash line), high-K mafic rocks (purple dash line), dotted line (TTGs) and sediments (grey shaded area) are cited from the same literatures referred in Fig.3.

The mantle-like Mg isotopes together with other geochemical features agree with the derivation from igneous protoliths of Taili potassic granites, which are essentially Archean I-type granites.

5.4 Potassic granites with high δ^{26}Mg signatures

Distinct from Taili potassic granites, Huludao potassic granites have higher SiO_2 (70.78–73.57 wt.%), K_2O/Na_2O (1.57–23.26), A/CNK (1.15–1.20) values and lower ferromagnesian ($MgO+FeO_t+TiO_2+MnO$=1.66–3.05 wt.%) contents, which match with the experimental melts generated by melting of both TTGs and sediments (Fig. 3). Combined with similar Hf-Nd isotopic features to those of the Juhuadao and Taili samples (Supplementary materials C), Huludao potassic granites was previously explained by re-melting of pre-existing

TTGs (Wang et al., 2016). However, based on comprehensive consideration of $Al_2O_3/(FeO_t+MgO)$, CaO and K_2O/Na_2O, these granites all fall within the source field of metasediments in the ternary discrimination diagram (Fig. 6), although the experimental granites derived by partial melting of sediments (grey area) partly overlap those source fields of TTGs and mafic rocks (Fig. 6). Such applications of major elements complicate the petrogenetic interpretation of Huludao potassic granites.

Strikingly, Huludao potassic granites have δ^{26}Mg up to +0.60‰ to +0.91‰, substantially higher than the Juhuadao TTGs (–0.39‰ to –0.22‰) (Fig. 4). With the effect of weathering being evaluated to be minor, the TTGs thus cannot be their sole source owing to limited Mg isotopic fractionation during magmatic processes (Teng, 2017; and references therein). High δ^{26}Mg values, similar to those of Huludao granites, are generally preserved in the weathered silicate relicts, such as pelite (Li et al., 2010), bauxite (Liu et al., 2014), and paleo-weathered crust (Tian et al., 2019) (Fig. 4), due to preferentially loss of light Mg isotopes into aqueous solutions during chemical weathering of silicates (Teng, 2017; and references therein). Partial melting of these sediments could produce the melt of S-type granites with high δ^{26}Mg values inherited from the sedimentary protoliths. The extremely high and limitedly varied δ^{26}Mg of Huludao potassic granites, convergent to those of the pelitic sediments (Fig. 4), does not support a mixing source of igneous rocks and sediments. Therefore, they are the S-type granites as partial melts from a sediments-dominated source. Furthermore, Huludao potassic granites have extremely high K_2O/Na_2O ratios of 1.57–23.26, far exceeding the ratios of Taili potassic granites and granitic melts derived from TTGs and mafic rocks (Fig. 3b). Instead, melts generated by melting of sediments could have large variable and high K_2O/Na_2O ratios (Fig. 3b). Experimental melts derived from sedimentary materials, such as pelite (Grant, 2004; Nair and Chacko, 2002; Patiño Douce and Harris, 1998; Patiño Douce and Johnston, 1991; Pickering and Johnston, 1998; Spicer et al., 2004; Vielzeuf and Holloway, 1988), graywacke (Castro et al., 1999; Conrad et al., 1988; Patiño Douce and Beard, 1995; Vielzeuf and Montel, 1994), and gneiss (Castro et al., 1999; García-Arias et al., 2012; Patiño Douce and Beard, 1995) have highly variable K_2O/Na_2O ratios of up to eight times those of their starting materials. Most Huludao potassic granites with K_2O/Na_2O ratios of 1.57–10.99 can be produced by a sedimentary source with K_2O/Na_2O ratios ≥7.4 (Fig. 7). But sample 12XC33 has much higher K_2O/Na_2O ratio of 23.26 exceeding those of granites generated by these sediments. This exception might be attributed to the starting materials of experiments not including all compositionally different sedimentary rocks which actually have a wide range of K_2O/Na_2O ratios of 1.34–44.45 (Li et al., 2010).

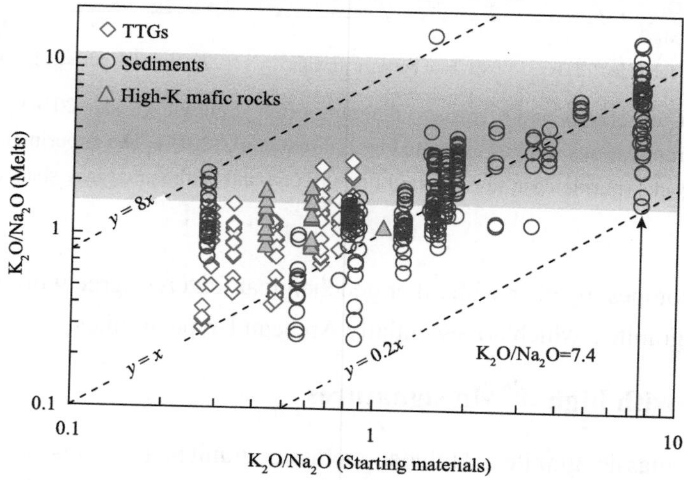

Fig. 7 Fractionation of K_2O/Na_2O ratios between experimental melts and relevant starting materials. The horizontal shaded bar represents a K_2O/Na_2O range of 1.5–10.9, covering most of Huludao potassic granites (except sample 12XC33 with a ratio of 23.26) when a sedimentary source has K_2O/Na_2O of 7.4. Melts derived from TTGs, sediments and high-K mafic rocks are from the same literatures referred in Fig.3. All experiments were conducted under the conditions of P = 1–15 kbar and T≤1250 ℃.

It is intriguing that Huludao potassic granites have similar Hf–Nd isotopes to those of Taili potassic granites and Juhuadao TTGs (Supplementary materials C; Wang et al., 2016), although they show greatly different Mg isotopes from each other (Fig. 4). Huludao potassic granites display positive zircon $\varepsilon_{HF}(t)$ of 0.0–6.7 and negative whole-rock $\varepsilon_{Nd}(t)$ values of −1.1 to −1.4 (Supplementary materials C) (Wang et al., 2016). Such decoupled Hf–Nd isotopes are also shown by the Taili (I-type) potassic granites and Juhuadao TTGs (Supplementary materials C). The decoupled variation of Hf–Nd isotopic signatures may reflect the shorter half-life of ^{176}Lu relative to that of ^{147}Sm, resulting the fact that the variation of $^{176}Hf/^{177}Hf$ is larger than that of $^{143}Nd/^{144}Nd$ in a given timespan (Wang et al., 2016; and reference therein). The indistinguishable Hf–Nd isotopes in TTGs, Taili and Huludao potassic granites suggest that their source rocks have short and similar residence time in continental crust before the re-working, supported by their similar Hf–Nd model ages (T_{DM2}) between 2600–2900 Ma (Supplementary materials C).

Taken all together, Huludao potassic granites are S-type granites produced by partial melting of sedimentary rocks. The above discussions highlight that Mg isotopes are helpful to identify S-type potassic granites, due to their sensitivity to chemical weathering.

At the end of the Archean, the NCC was undergoing the amalgamation of micro-continental blocks and cratonization (Wang et al., 2016), during which sediments may have been carried to deeper crustal levels and reworked to produce the S-type Huludao granites.

5.5 Temporal distribution of Archean S-type potassic granite

A compilation of data for 4066 global Archean granitoids was used for comparison with our samples to estimate when the S-type potassic granites appeared and extensively developed during the Archean. These data were taken from the compilation of Keller and Schoene (2012), filtered by SiO_2 contents of 56–75 wt.%, including 166 granitoids with ages of 4000–3500 Ma (Fig. 8a), 1169 granites with ages of 3500–3000 Ma (Fig. 8b), and 2731 granites with ages of 3000–2500 Ma (Fig. 8c). These comprehensive diagrams clearly show that most of Archean potassic granites with K_2O/Na_2O ratios of $\geqslant 1.5$ plot in the metasediment-origin area, similar to the S-type potassic granites from this study (Fig. 8d). Given that Archean granites with K_2O/Na_2O ratios of $\geqslant 1.5$ might be S-type granites, then, S-type potassic granites may have been scarce in the early Archean (4000–3500 Ma), present in the middle Archean (3500–3000 Ma), and widespread in the late Archean (3000–2500 Ma) globally (Fig. 8a, b, c).

5.6 Contributions of I- and S-type potassic granites to increasing K_2O/Na_2O ratio of the Archean UCC

Data compiled by Keller and Schoene (2012) indicate that the geochemical composition of Archean felsic rocks was transformed towards K-enrichment, with their average K_2O/Na_2O ratio increasing to ~0.83 by the end-Archean. Archean TTGs have an average K_2O/Na_2O ratio of only 0.47, according to the composition of 97 TTGs collected by Condie (2005). Thus, well-developed potassic granites since 3500 Ma, especially since 3000 Ma, play a key role in causing such an increase of K_2O/Na_2O ratio (Fig. 8). Goodwin (1991) estimated that 70% of Archean granitoids are TTGs, with potassic granites constituting ~30% in Archean cratons. The objective of this study is trying to investigate how much I- and S-type potassic granites are preserved in Archean terranes and how much contributions they made to making crust more potassic. Based on the statistical data mentioned above, S-type potassic granites with $K_2O/Na_2O \geqslant 1.5$ have an average K_2O/Na_2O ratio of 4.75, and I-type potassic granites ($1 \leqslant K_2O/Na_2O < 1.5$) exhibit an average K_2O/Na_2O of 1.18. Assuming that only I-type potassic granites and TTGs are preserved in Archean terranes, a composition of 50% I-type in the Archean basement would increase UCC K_2O/Na_2O ratio to 0.83. If only S-type potassic granites and TTGs are preserved, 8.5% of the former would be required to produce UCC with an average K_2O/Na_2O ratio of 0.83. Both assumptions contradict with the actual proportions of TTGs and potassic granites (70% and 30%, respectively, Goodwin, 1991) exposed in Archean cratons, so both I- and S-type granites are needed to account for the compositional transition of the UCC. Mass-balance calculations indicate that 25.88 wt.% I-type + 4.12 wt.% S-type potassic granites + 70 wt.% TTGs in Archean terranes are required to fit the UCC K_2O/Na_2O ratio of 0.83 at the end Archean (Fig. 9). Contribution of I- and S-type potassic granites to the increase of K_2O/Na_2O of Archean crust could reach 3 ∶ 2 by a semiquantitative estimation, indicating that a synergy of intracrustal melting and continental weathering induces the continental crust more potassic, with the former contributed more quantitatively and the latter contributed more efficiently.

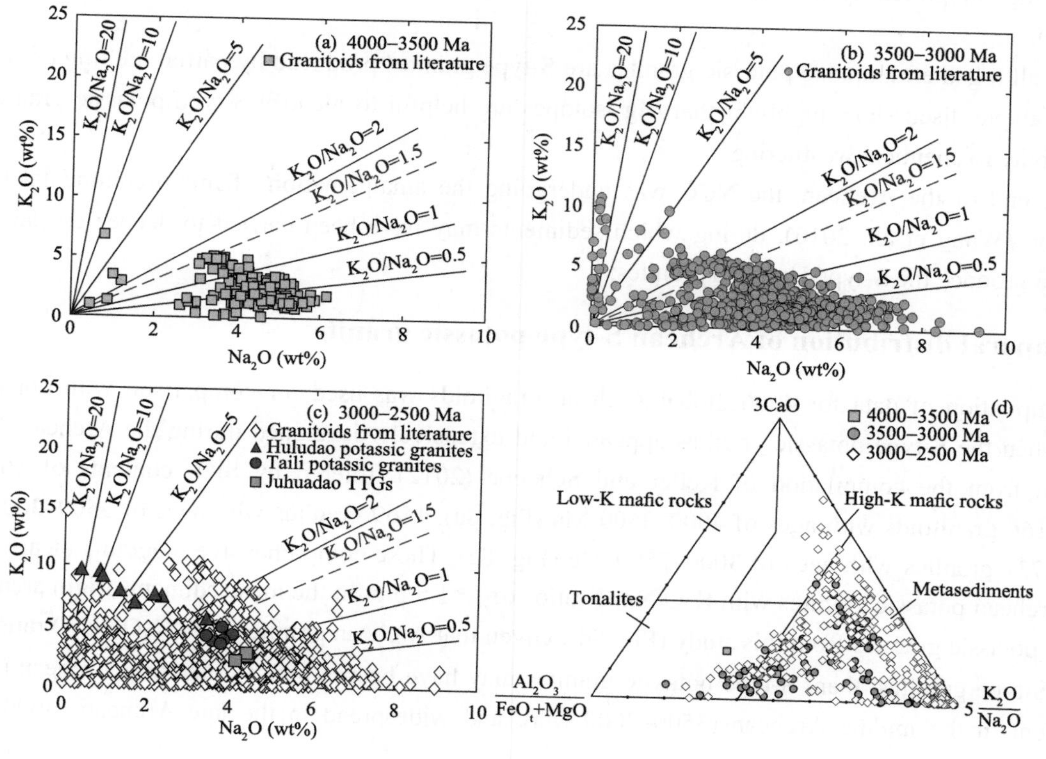

Fig. 8　(a, b, c) K_2O–Na_2O plots for global Archean felsic igneous rocks, displaying the development of potassic granitoids in different periods. Samples in this study show overlaps with late Archean granitoids on K_2O/Na_2O ratios. (d) Discrimination diagram for Archean granitoids with $K_2O/Na_2O \geqslant 1.5$, exhibiting more than 80% of them fall into the area of metasediments-origin.

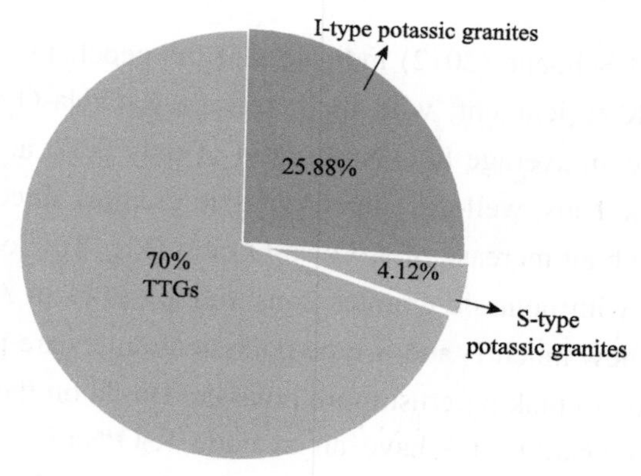

Fig. 9　The proportions of TTGs, I- and S-type potassic granites in the Archean.

6 Implications

Sediments as a source of S-type potassic granites are produced by continental weathering. Hence, temporal distribution of S-type potassic granites could mirror the timing and intensity of continental weathering in the Archean. As shown by statistical data and discussion above (Fig. 8), the occurrence of S-type potassic granites in the middle Archean reflect the onset of reworking of supracrustal sediments developed in the middle Archean, consistent with the previous study of Vezinet et al. (2019). While, the emergence of exposed land masses and their weathering may begin in the early Archean, if consider eroded crust is on average 300–400 Ma older than the deposition age of their sediments (Garçon et al., 2017). A great development of potassic granites in the late Archean were generated by reworking of continental crust (Laurent et al., 2014), possibly through amalgamation of micro-continental blocks to form the first large stable continent or supercontinent (Campbell and Allen, 2008; Laurent et al., 2014). Consistent with this terrane amalgamation model, the compiled geochemical data (Fig. 8) supports that plate subduction or continent–continent collision may not occurred in the early Archean, but may have begun in the middle Archean (Condie, 2008; Moyen and Van Hunen, 2012; Moyen and Laurent, 2018; Vezinet et al., 2019). Consequently, onset of plate tectonics, which triggered the reworking of both igneous and sedimentary Archean continental crust, was one of prime causes of the abrupt shift in UCC chemical composition from dominantly sodic to more potassic.

7 Conclusions

In this study, we report Mg isotope compositions of late Archean TTGs and potassic granites from North China Craton. Mg isotopes combined with major elements identified the occurrence of I- and S-type Archean potassic granites, which have made different contributions to increasing K_2O/Na_2O of the UCC. The statistic data of worldwide Archean felsic igneous rocks reveal an increased K_2O/Na_2O ratio of Archean granitoids and proportion of S-type potassic granites over the Archean era. Our conclusions are as follows.

(1) Neoarchean TTGs from Juhuadao have $\delta^{26}Mg$ values of –0.39‰ to –0.22‰, consistent with heterogeneous basaltic sources.

(2) Potassic granites produced by the reworking of igneous rocks (I-type) and sedimentary rocks (S-type) have distinctive Mg isotopic compositions. I-type potassic granites with weak K enrichment ($1.0 \leqslant K_2O/Na_2O < 1.5$) have mantle-like $\delta^{26}Mg$ values of –0.28‰ to –0.22‰, while S-type potassic granites with strong K enrichment ($K_2O/Na_2O \geqslant 1.5$) have substantially higher $\delta^{26}Mg$ values of +0.60‰ to +0.91‰.

(3) Compiled data of 4066 Archean felsic rocks indicate that S-type potassic granites have appeared in the middle Archean and became widespread by the late Archean, but were scarce in the early Archean.

(4) The evolved crustal composition from predominantly sodic to potassic needs a synergy of re-working of igneous and sedimentary rocks, corresponding to the geological processes of intracrustal melting and surficial weathering, respectively. The re-melting of igneous rocks contributed more quantitatively, while re-working of sediments contributed more efficiently to increasing the K_2O/Na_2O of the Archean UCC.

Declaration of Competing Interest

The authors declare that they have no known competing financial interests or personal relationships that could have appeared to influence the work reported in this paper.

Acknowledgements

Constructive comments from Yongsheng He, two anonymous reviewers and editor Xian-Hua Li are greatly appreciated, which highly improved the quality of this manuscript. We also thank Yang Wang and Wenning Lu for their comments and discussion. This work was supported by the National Natural Science Foundation of China (41873020 to SK and 41730214 to S-GL) and State Key Laboratory of Geological Processes and Mineral Resources.

Appendix A. Supplementary data

Supplementary data to this article can be found online at https://doi.org/10.1016/j.lithos.2020.105570.

References

An, Y., Wu, F., Xiang, Y., Nan, X., Yu, X., Yang, J., Yu, H., Xie, L., Huang, F., 2014. Highprecision Mg isotope analyses of low-Mg rocks by MC-ICP-MS. Chem. Geol. 390, 9-21.

Baker, J.A., Bizzarro, M.,Witting, M., Connelly, J., Haack, H., 2005. Early planetesimal melting from an age of 4.5662 Gyr for differentiated meteorites. Nature. 436, 1127-1131.

Bourdon, B., Tipper, E.T., Fitoussi, C., Stracke, A., 2010. Chondritic Mg isotope composition of the Earth. Geochim. Cosmochim. Acta 74 (17), 5069-5083.

Brewer, A., Teng, F.-Z., Dethier, D., 2018. Magnesium isotope fractionation during granite weathering. Chem. Geol. 501, 95-103.

Campbell, Ian H., Allen, Charlotte M., 2008. Formation of supercontinents linked to increases in atmpspheric oxygen. Nat. Geosci. 1, 554-558.

Castro, A., Patino Douce, A.E., Corretgé, L.G., Jesus, D.D.L.R., El-Biad, M., El-Hmidi, H., 1999. Origin of peraluminous granites and granodiorites, Iberian massif, Spain: an experimental test of granite petrogenesis. Contrib. Mineral. Petrol. 135, 255-276.

Champion, D.C., Sheraton, J.W., 1997. Geochemistry and Nd isotope systematics of Archaean granites of the Eastern Goldfields, Yilgarn Craton, Australia: implications for crustal growth processes. Precambrian Res. 83, 109-132.

Clemens, J.D., 2006.Melting of the continental crust: Fluid regimes, melting reactions, and source-rock fertility. In: Brown, M., Rushmer, T. (Eds.), Evolution and Differentiation of the Continental Crust. Cambridge University Press, Cambridge, pp. 297-331.

Clos, F., Weinberg, R., Zibra, I., 2018. Building the Archean Continental Crust: 300 Ma of Felsic Magmatism in the Yalgoo Dome (Yilgarn Craton). Geological Survey of Western Australia, Perth, Australia.

Condie, K.C., 2005. TTGs and adakites: are they both slab melts? Lithos. 80, 33-44.

Condie, K.C., 2008. Did the character of subduction change at the end of the Archean? Constraints from convergent-margin granitoids. Geology. 36, 611.

Condie, K.C., 2014. How to make a continent: thirty-five years of TTG research. Evolution of Archean Crust and Early Life. 2014. Springer, Dordrecht, pp. 179-193.

Conrad,W.K., Nicholls, I.A., Wall, V.J., 1988.Water-saturated and -undersaturated melting of metaluminous and peraluminous crustal compositions at 10 kb: evidence for the origin of silicic magmas in the Taupo volcanic zone, New Zealand, and other occurrences. J. Petrol. 29, 765-803.

Day, Warren C., Weiblen, P.W., 1986. Origin of late Archean granite: geochemical evidence from the Vermilion Granitic complex of northern Minnesota. Contib Mineral. Petrol. 93, 283-296.

Dhuime, B., Hawkesworth, C.J., Cawood, P.A., Storey, C.D., 2012. A change in the geodynamics of continental growth 3 billion years ago. Science. 335, 1334-1336.

Fu, J., Liu, S., Wang, M., Chen, X., Guo, B., Hu, F., 2017. Late Neoarchean monzogranitic-syenogranitic gneisses in the Eastern Hebei-Western Liaoning Province, North China Craton: Petrogenesis and implications for tectonic setting. Precambrian Res. 303, 392-413.

Fu, J., Liu, S., Cawood, P.A.,Wang, M., Hu, F., Sun, G., Gao, L., Hu, Y., 2018. Neoarchean magmatic arc in theWestern Liaoning Province, northern North China Craton: Geochemical and isotopic constraints from sanukitoids and associated granitoids. Lithos. 322, 296-311.

Gao, T., Ke, S., Li, R.-Y., Meng, X.N., He, Y., Liu, C., Wang, Y., Li, Z., Zhu, J.-M., 2019. High-Precision Magnesium Isotope Analysis of Geological and Environmental Reference Materials by Multiple-collector Inductively Coupled Plasma Mass Spectrometry. Rapid Commun. Mass Spectrom. 38, 767-777.

García-Arias, M., Corretgé, L.G., Castro, A., 2012. Trace element behavior during partial melting of Iberian orthogneisses: An experimental study. Chem. Geol. 292-293, 1-17.

Gardien, V., Thompson, A.B., Grujic, D., Ulmer, P., 1995. Experimental melting of biotite + plagioclase + quartz ± muscovite assemblages and implications for crustal melting. J. Geophys. Res. 100, 15581-15591.

Garçon,M., Carlson, R.W., Shirey, S.B., Arndt, N.T., Horan,M.F.,Mock, T.D., 2017. Erosion of Archean continents:the Sm-Nd and Lu-Hf isotopic record of Barberton sedimentary rocks. Geochim. Cosmochim. Acta 206, 216-235.

Goodwin, A.M., 1991. Precambrian Geology: The Dynamic Evolution of the Continental Crust. Academic Press, London.

Grant, J.A., 2004. Liquid compositions from low-pressure experimental melting of politic rock from Morton Pass, Wyoming, USA. J. Metamorphic Geol. 22, 65-78.

Huang, K.-J., Teng, F.-Z., Wei, G.-J., Ma, J.-L., Bao, Z.-Y., 2012. Adsorption- and desorptioncontrolled magnesium isotope fractionation during extreme weathering of basalt in Hainan Island, China. Earth Planet Sci. Lett. 359-360, 73-83.

Huang, J., Li, S.-G., Xiao, Y., Ke, S., Li,W.-Y., Tian, Y., 2015. Origin of low δ 26 Mg Cenozoic basalts from South China Block and their geodynamic implications. Geochim. Cosmochim. Acta 164, 298-317.

Johnson, T.E., Kirkland, C.L., Gardiner, N.J., Brown, M., Smithies, R.H., Santosh, M., 2019. Secular change in TTG compositions: Implications for the evolution of Archaean geodynamics. Earth Planet. Sci. Lett. 505, 65-75.

Ke, S., Teng, F.-Z., Li, S.-G., Gao, T., Liu, S.-A., He, Y.,Mo, X., 2016. Mg, Sr, and O isotope geochemistry of syenites from Northwest Xinjiang, China: Tracing carbonate recycling during Tethyan oceanic subduction. Chem. Geol. 437, 109-119.

Keller, C.B., Schoene, B., 2012. Statistical geochemistry reveals disruption in secular lithospheric evolution about 2.5Gyr ago. Nature. 485, 490-493.

Laurent, O., Martin, H., Moyen, J.F., Doucelance, R., 2014. The diversity and evolution of late-Archean granitoids: evidence for the onset of "modern-style" plate tectonics between 3.0 and 2.5Ga. Lithos. 205, 208-235.

Lee, S.-W., Ryu, J.-S., Lee, K.-S., 2014. Magnesium isotope geochemistry in the Han River, South Korea. Chem. Geol. 364, 9-19.

Li, W.-Y., Teng, F.-Z., Ke, S., Rudnick, R.L., Gao, S., Wu, F.-Y., Chappell, B.W., 2010. Heterogeneous magnesium isotopic composition of the upper continental crust. Geochim. Cosmochim. Acta 74, 6867-6884.

Li, R.-Y., Ke, S., He, Y., Gao, T., Meng, X.N., 2016. High precision Magnesium isotope measurement for high-Cr samples. Bull. Mineral. Petrol. Geochem. 35, 441-447.

Liu, D.Y., Nutman, A.P., Compston, W., Wu, J.S., Shen, Q.H., 1992. Remnants of ≥3800 Ma crust in the Chinese part of the Sino-Korean craton. Geology. 20, 339-342.

Liu, S.-A., Teng, F.-Z., He, Y., Ke, S., Li, S., 2010. Investigation ofmagnesium isotope fractionation during granite differentiation: Implication for Mg isotopic composition of the continental crust. Earth Planet. Sci. Lett. 297, 646-654.

Liu, X.-M., Teng, F.-Z., Rudnick, R.L., McDonough, W.F., Cummings, M.L., 2014. Massive magnesium depletion and isotope fractionation in weathered basalts. Geochim. Cosmochim. Acta 135, 336-349.

Lowe, Donald R., Tice, Michael M., 2004. Geologic evidence for Archean atmospheric and climatic evolution: Fluctuating levels of CO_2, CH_4, and O_2 with an overriding tectonic control. Geology. 32, 493-496.

Moyen, J.-F., 2019. Archean granitoids: classification, petrology, geochemistry and origin. Geological Society, London, Special Publications SP489-2018-34.

Moyen, J.-F., Laurent, O., 2018. Archaean tectonic systems: a view from igneous rocks. Lithos. 302-303, 99-125.

Moyen, J.-F., Martin, H., 2012. Forty years of TTG research. Lithos. 148, 312-336.

Moyen, J.-F., van Hunen, J., 2012. Short-term episodicity of Archaean plate tectonics. Geology. 40, 451-454.

Nair, R., Chacko, T., 2002. Fludi-absent melting of high-grade semipelites: P-T constraints on orthopyrozene formation and implications for granulite genesis. J. Petrol. 43, 2121-2142.

Nebel, O., Capitanio, F.A., Moyen, J.F.,Weinberg, R.F., Clos, F., Nebel-Jacobsen, Y.J., Cawood, P.A., 2018. When crust comes of age: on the chemical evolution of Archaean, felsic continental crust by crustal drip tectonics. Philos. Trans. Roy. Soc. A 376 20180103.

Nesbitt, H.W., Young, G.M., 1982. Early Proterozoic climates and plate motions inferred from major element chemistry of lutites. Nature. 299 (5885), 715.

Patiño Douce, A.E., 2005. Vapor-absent melting of tonalite at 15-32 kbar. J. Petrol. 46, 275-290.

Patiño Douce, A.E., Beard, J., 1995. Dehydration-melting of biotite gneiss and quartz amphibolite from 3 to 15 kbar. J. Petrol. 36, 707-738.

Patiño Douce, A.E., Harris, N., 1998. Experimental constraints on Himalayan anatexis. J. Petrol. 39, 689-710.

Patiño Douce, A.E., Johnston, A.D., 1991. Phase equilibria and melt productivity in the pelitic system: implications for the origin of peraluminous granitoids and aluminous granulites. Contrib. Mineral. Petrol. 107, 202-218.

Pickering, J.M., Johnston, A.D., 1998. Fluid-absent melting behavior of a two-mica metapelite: experimental constraints on the origin of Black Hills granite. J. Petrol. 39, 1787-1804.

Pogge von Strandmann, P.A.E., Elliott, T., Marschall, H.R., Coath, C., Lai, Y.-J., Jeffcoate, A.B., Ionov, D.A., 2011. Variations of Li and Mg isotope ratios in bulk chondrites and mantle xenoliths. Geochim. Cosmochim. Acta 75, 5247-5268.

Rapp, R.P., Watson, E.B., 1995. Dehydration Melting of Metabasalt at 8-32 kbar: Implications for Continental Growth and Crust-Mantle Recycling. J. Petrol. 36, 891-931.

Rudnick, R.L., Gao, S., 2003. Composition of the continental crust. Treatise on geochemistry. Elsevier. Amsterdam. 3, 1-64.

Shen, B., Jacobsen, B., Lee, C.T.A., Yin, Q.Z.,Morton, D.M., 2009. The Mg isotopic systematics of granitoids in continental arcs and implications for the role of chemical weathering in crust formation. Proc. Natl. Acad. Sci. 106, 20652-20657.

Singh, Jagmohan, Johannes, Wilhelm, 1996a. a. Dehydration melting of tonalites. Part I. beginning of melting. Contrib. Mineral. Petrol. 125, 16-25.

Singh, Jagmohan, Johannes, Wilhelm, 1996b. a. Dehydration melting of tonalites. Part II. Composition of melts and solids. Contrib. Mineral. Petrol. 125, 26-44.

Sisson, T.W., Ratajeski, K., Hankins,W.B., Glazner, A.F., 2005. Voluminous granitic magmas from common basaltic sources. Contrib. Mineral. Petrol. 148, 635-661.

Skjerlie, K.P., Johnston, A.D., 1992. Vapor-absent melting at 10 kbar of a biotitebearingand amphibole-bearing tonalitic gneiss: implications for the generation of A-typegranites. Geology. 20, 263-266.

Skjerlie, K.P., Johnston, A.D., 1993. Fluid-absent melting behavior of an F-rich tonaliticgneiss at mid-crustal pressures: implications for the generation of anorogenicgranites. J. Petrol. 34, 785-815.

Song, B., Nutman, A.P., Liu, D., Wu, J., 1996. 3800 to 2500 Ma crustal evolution in the Anshan area of Liaoning Province, northeastern China. Precambrian Res. 78, 79-94.

Spicer, E.M., Stevens, G., Buick, I.S., 2004. The low-pressure partial-melting behaviour of natural boron-bearing metapelites from the Mt. Stafford area, Central Australia. Contrib. Mineral. Petrol. 148, 160-179.

Springer, W., Seck, H.A., 1997. Partial fusion of basic granulites at 5 to 15 kbar: implications for the origin of TTG magmas. Contrib. Mineral. Petrol. 127, 30-45.

Teng, F.-Z., 2017. Magnesium Isotope Geochemistry. Rev.Mineral. Geochem. 82, 219-287.

Teng, F.-Z., Li, W.-Y., Ke, S., Marty, B., Dauphas, N., Huang, S., Wu, F.-Y., Pourmand, A., 2010a. Magnesium isotopic composition of the Earth and chondrites. Geochim. Cosmochim. Acta 74, 4150-4166.

Teng, F.-Z., Li, W.-Y., Rudnick, R.L., Gardner, L.R., 2010b. Contrasting lithium and magnesium isotope fractionation during continental weathering. Earth Planet. Sci. Lett. 300, 63-71.

Tian, H.-C., Yang,W., Li, S.-G.,Wei, H.-Q., Yao, Z.-S., Ke, S., 2019. Approach to trace hidden paleo-weathering of basaltic crust through decoupled Mg Sr and Nd isotopes recorded in volcanic rocks. Chem. Geol. 509, 234-248.

Vezinet, A., Thomassot, E., Pearson, D.G., Stern, R.A., Luo, Y., Sarkar, C., 2019. Extreme $\delta^{18}O$ signatures in zircon from the Saglek Block (North Atlantic Craton) document reworking of mature supracrustal rocks as early as 3.5 Ga. Geology. 47, 605-608.

Vielzeuf, D., Holloway, H.R., 1988. Experimental determination of the fluid-absent melting relations in the pelitic system: Consequences for crustal differentiation. Contrib. Mineral. Petrol. 98, 257-276.

Vielzeuf, D., Montel, J.M., 1994. Partialmelting of metagreywackes. Part I. Fluid-absent experiments and phase relationships. Contrib. Mineral. Petrol. 117, 375-393.

Wan, Y., Liu, D., Song, B., Wu, J., Yang, C., Zhang, Z., Geng, Y., 2005. Geochemical and Nd isotopic compositions of 3.8Ga meta-quartz dioritic and trondhjemitic rocks from the Anshan area and their geological significance. J. Asian Earth Sci. 24, 563-575.

Wan, Y.,Ma, M., Dong, C., Xie, H., Xie, S., Ren, P., Liu, D., 2015.Widespread late Neoarchean reworking of Meso- to Paleoarchean continental crust in the Anshan-Benxi area, North China Craton, as documented by U-Pb-Nd-Hf-O isotopes. Am. J. Sci. 315, 620-670.

Wang, C., Song, S., Niu, Y.,Wei, C., Su, L., 2016. TTG and potassic granitoids in the eastern North China Craton: making Neoarchean upper continental crust during microcontinental collision and post-collisional extension. J. Petrol. 57, 1775-1810.

Watkins, J.M., Clemens, J.D., Treloar, P.J., 2007. Archaean TTGs as sources of younger granitic magmas melting of sodic metatonalites at 0.61.2 GPa. Contrib. Mineral. Petrol. 154, 91-110.

Whalen, J.B., Percival, J.A., McNicoll, V.J., Longstaffe, F.J., 2004. Geochemical and isotopic (Nd-O) evidence bearing on the origin of late- to post-orogenic high-K granitoid rocks in the Western Superior Province: implications for late Archean tectonomagmatic processes. Precambrian Res. 132, 303-326.

Wombacher, F., Eisenhauer, A., Böhm, F., Gussone, N., Regenberg, M., Dullo, W.C., Rüggeberg, A., 2011. Magnesium stable isotope fractionation in marine biogenic calcite and aragonite. Geochim. Cosmochim. Acta 75, 5797-5818.

Xiao, L., Clemens, J.D., 2007. Origin of potassic (C-type) adakite magmas: Experimental and field constraints. Lithos. 95, 399-414.

Xie, H., Wan, Y., Dong, C., Kröner, A., Xie, S., Liu, S., Ma, M., Liu, D., 2019. Late Neoarchean synchronous TTG gneisses and potassic granitoids in southwestern Liaoning Province, North China Craton: Zircon U-Pb-Hf isotopes, geochemistry and tectonic implications. Gondw. Res. 70, 171-200.

Large Mg-Fe isotope fractionation in volcanic rocks from northeast China: The role of chemical weathering and magma compositional effect*

Haiquan Wei[1], Heng-Ci Tian[2], Shu-Guang Li[3], Wei Yang[2], Bing-Yu Gao[4,5], Shan Ke[3], Rui-Ying Li[2], Xiaowen Chen[6] and Hongmei Yu[1]

1. Jilin Changbaishan Volcano National Observation and Research Station, Institute of Geology, China Earthquake Administration (CEA), Beijing100029, China
2. Key Laboratory of Earth and Planetary Physics, Institute of Geology and Geophysics, Chinese Academy of Sciences, Beijing 100029, China
3. State Key Laboratory of Geological Processes and Mineral Resources, China University of Geosciences, Beijing 100083, China
4. Key laboratory of Mineral Resources, Institute of Geology and Geophysics, Chinese Academy of Sciences, Beijing 100029, China
5. University of Chinese Academy of Sciences, Beijing 100049, China
6. The Geological Museum of China, Beijing 100034, China

亮点介绍：为了进一步探讨长白山火山岩所记录的古气候事件是在有氧还是无氧环境下的化学风化，本研究对长白山玄武岩和粗面岩开展了 Fe 同位素分析研究。结果表明玄武岩与粗面岩 Fe 同位素组成比较相近（前者 δ^{56}Fe = 0.16–0.17‰；后者 δ^{56}Fe = 0.09–0.23‰），是由岩浆过程造成的。这一结果表明粗面岩混染的玄武岩风化产物并没有显著改造玄武岩的铁同位素组成，说明很可能是在有氧条件下的化学风化。

Abstract This study presents Mg and Fe isotopic data for a suite of volcanic rocks including basalt, trachyte and comendite from Changbaishan, northeast China. Our results show that the Millennium Eruption (ME) trachytes and comendites have δ^{26}Mg ranging from −0.37 to 0.14‰. Combined with literature data, the positive correlation between δ^{26}Mg and ^{87}Sr/^{86}Sr ratios and the almost constant value for Nd isotopes suggest that the heavy Mg isotopic composition most likely derived from involvement of weathered products of basalt instead of the old supracrustal/basement rocks. Iron isotopic composition is significantly varied throughout our suite of samples, with δ^{56}Fe values of 0.17±0.04‰ and 0.16±0.03‰ for basalts, 0.09 to 0.23‰ for cone-construction trachytes, and 0.23 to 0.37‰ for the ME samples. The slightly higher δ^{56}Fe than MORB observed in basalts reflect the role of subducted materials that could elevate source $Fe^{3+}/\Sigma Fe$ ratios via redox reaction. The trachytes from the cone construction, thought to be significantly affected by chemical weathered product, have a restricted Fe

* 本文发表在：Chemical Geology, 2021, 565: 120075

isotopic variation, while those ME samples with less weathered product input have overall higher δ^{56}Fe values. This contrasting Fe isotopic signature suggests that the paleo-weathering event likely occurred under oxidized environment, leaving the residue with little Fe isotope fractionation. By contrast, the good correlations between Fe isotopes and indicators of magmatic differentiation (e.g., tFe_2O_3, SiO_2, Mg#) in all rocks reflect that fractional crystallization under an open system most likely accounts for the Fe isotopic variation, which was enhanced by the compositional effect as suggested by the broadly positive relationship between δ^{56}Fe and (Na+K)/(Ca+Mg). This study therefore highlights the potential applications of Mg-Fe isotopes as great tracers of redox condition and ancient climate.

Keywords Mg-Fe isotopes, paleo-weathering event, Changbaishan Tianchi volcano, fractional crystallization, open magmatic system

1 Introduction

Magnesium and iron isotopes are proven to be largely fractionated at low temperature environments (e.g., Teng et al., 2010a; Liu et al., 2014). Studies have found over 1.5‰ Mg isotope variation in sedimentary rocks (e.g., shales, pelites, and loess) and residues of silicate rocks formed by chemical weathering (e.g., Li et al., 2010), making it as an effective probe for the recycled supracrustal materials (Li et al., 2017a; Teng, 2017). Iron is a redox-sensitive element and its isotopes are readily fractionated relevant to reducing weathering but limited fractionation under oxidizing conditions (Liu et al., 2014). At high temperature, both partial melting and subsequent fractional crystallization do not significantly fractionate Mg isotopes but can cause large Fe isotopic variation (Teng et al., 2008, 2010b, 2013). More importantly, the combination of Mg and Fe isotopes have been widely applied to illustrate/distinguish complicated geological processes such as tracing the deep oxygen cycling (He et al., 2019), tracing the crust evolution (Shen et al., 2009; Li et al., 2020), distinguishing the chemical heterogeneity derived from crystal growth or kinetic processes (e.g., Teng et al., 2011; Sio et al., 2013) or using them as geospeedometry (Teng et al., 2011; Wu et al., 2018). For example, Tian et al. (2020) reports a large-scale of diffusion-driven Mg-Fe isotope fractionation in ilmenites from Panzhihua layered mafic intrusion, suggesting that the interstitial liquid immiscibility plays a crucial role in the origins of Fe-Ti-rich ore deposits. Therefore, the Mg-Fe isotopes can help advance our understanding of what magmatic processes have contributed to the formation and evolution of the magmas.

Recently, the remarkably heavy Mg isotopic compositions (up to 0.94‰) have been reported in Changbaishan trachytes and attributed to the contribution of previously weathered residue of basaltic rocks based on their decoupled Mg-Sr and Nd isotopic ratios (Tian et al., 2019). If this explanation is correct, the Fe isotopic compositions of Changbaishan trachytes with heavy Mg isotopes should also record the geochemical fingerprints for such weathered components and can be used to further decipher the redox state of the chemical weathering during that period. This is important for the interpretation of the past climate in northeastern China during the period when the weathered products formed. To date, no available Fe isotopic data have been reported for Changbaishan volcanic rocks.

In this paper, Fe isotopic compositions for selected basalts, cone-construction trachytes and ME trachyte-comendite samples were investigated (Fig.1). These samples cover most of the geochemical characteristics and lithologies of the volcanic rocks distributed in Changbaishan volcanic region. We have also made Sr-Nd-Mg isotopic analyses for highly evolved ME products in order to examine whether a signal of the weathered residue occurs in these materials. Significant Fe (~0.3‰ for δ^{56}Fe) and Mg (~0.5‰ for δ^{26}Mg) isotopic variations are observed in these volcanic rocks. These isotopic data are then used to constrain whether the previously weathered basalt continuously participated in the Changbaishan volcanic rocks, the mechanism of Fe isotope fractionation, and to investigate any connections between Mg and Fe isotopes.

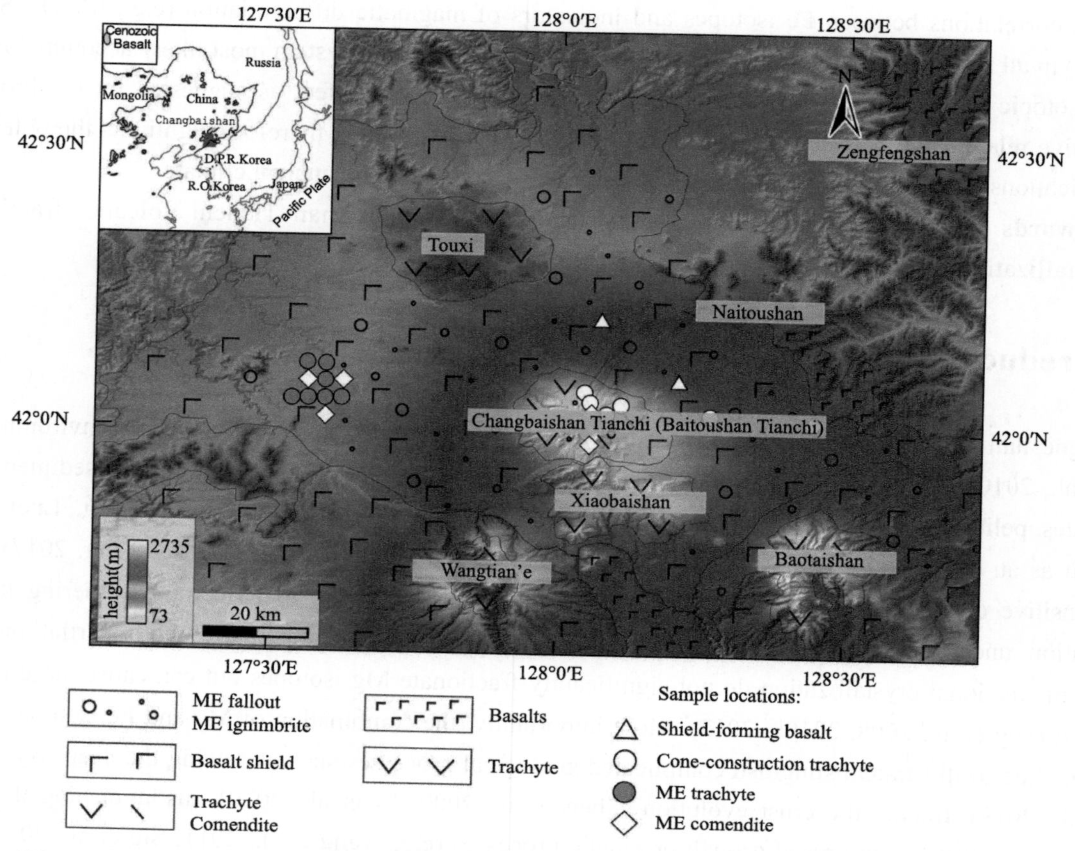

Fig. 1 Simplified map showing the distribution of the Late Cenozoic volcanic rocks in Changbaishan and surrounding regions. Sample localities are distinguished by different symbols with distinct colors, which are also used in the following figures. ME: Millennium Eruption.

2 Samples and methods

2.1 Samples

Changbaishan Tianchi volcano located on the China-North Korea boundary belongs to a shield volcano built on the Neogene basalt lavas (also named as the Gaima Plateau) (Fig. 1). This stratovolcano predominantly consists of pre-shield basalts, shield-forming basalts, cone-construction trachytes and Holocene comendite-trachyte series from the bottom to the top (see review of Zhang et al. 2018). Those basalts are generally younger than 10 Mys and widely spread around the volcano with an estimated area of ~7400 km^2 (Jin and Zhang, 1994). This shield includes Quanyang, Toudao, Baishan and Laofangzixiaoshan Formations. The Tianchi volcanic cone (i.e., trachyte lavas and pyroclastic equivalents) was formed between ca. 1.1 Ma and ca. 0.02 Ma and the youngest trachyte lavas (19–15 ka) at the Tianwenfeng are regarded as the final products of the cone-construction stage. The ME (946 CE) was one of the largest eruptions during the past 2000 years (Xu et al., 2013; Oppenheimer et al., 2017), creating a ~5 km diameter caldera and generating a large amount of comenditic ignimbrite around the volcano (Wei et al., 2007, 2013; Pan et al., 2017). Ash from this eruption is even detected in

Japan Sea, Russian Far East and the northern Greenland (e.g., Andreeva et al., 2011; Sun et al., 2014; McLean et al., 2016).

In this study, seven ME trachytes and six ME comendites are analyzed for trace elemental and Sr-Nd-Mg isotopic compositions (see sample locations in Fig. 1). These samples were collected from the ME Plinian fallout and ignimbrite deposits (~50 km west and east of the caldera lake) and from the caldera rim. Some pumices compose of alternate dark and light streaks, providing direct evidence for mixing between the ME felsic and more mafic melts (Wei et al., 2013). The dark and light streaks in pumices were separated and then pulverized for chemical analyses. Sample names with "S" and "Q" refer to the components of dark and light streaks, respectively. Their major elemental compositions are reported in a previous work (Chen et al., 2017) and compiled in Table S1. Phenocrysts in ME comendites are mostly alkali feldspar, alkali pyroxene and fayalitic olivine. Altered minerals such as kaolinization of feldspars are rare in these rocks. For Fe isotopes, two shield-forming basalts, five cone-construction trachytes from the Xiaobaishan and Baitoushan Formations, three ME trachytes and three ME comendites were selected for analyses (Fig. 1) because they cover most of the chemical and isotopic characteristics of the Changbaishan Tianchi volcano.

2.2 Analytical methods

Analytical procedures of Sr and Nd isotopes are similar to previous works (Li et al., 2015, 2016), which are measured on a *Thermo-fisher* Triton Plus TIMS at the Institute of Geology and Geophysics, Chinese Academy of Sciences. About ~100 mg whole rock powders were dissolved by concentrated HF-HNO_3-$HClO_4$ acids mixture. A chromatographic separation of Sr and Nd was achieved using a two-step ion exchange technique. During the measurement, $^{86}Sr/^{88}Sr=0.1194$ and $^{146}Nd/^{144}Nd=0.7219$ values were employed for correcting instrumental mass fractionation. Measurements of the NBS-987 and JNdi-1 standards yield $^{87}Sr/^{86}Sr=0.710257\pm0.000013$ and $^{143}Nd/^{144}Nd=0.512098\pm0.000008$, respectively. Simultaneously, the international rock standard BCR-2 was also analyzed to monitor the whole procedures, which present $^{87}Sr/^{86}Sr=0.704989\pm0.000013$ and $^{143}Nd/^{144}Nd=0.512625\pm0.000009$, consistent with the data in other labs (e.g., Raczek et al., 2003; Jweda et al., 2016).

Magnesium isotopic analysis was documented using a *Nu Plasma* MC-ICP-MS at the University of Washington (Seattle) and a *Neptune plus* MC-ICP-MS at the China University of Geosciences (Beijing) following methods previously established (Li et al., 2010; Teng et al., 2010b, 2015; Teng and Yang, 2014; Ke et al., 2016). For chemical procedure, powdered rocks were dissolved in Teflon beakers using three steps. The first step is in HF-HNO_3, and then in HCl-HNO_3 and finally in HNO_3. No opaque residue was observed after these steps. The dried residues were re-dissolved in 1 mol/L HNO_3 in preparation for Mg separation. Then, a cation exchange column containing Bio-Rad 200–400 mesh AG50W-X8 resin was adopted and the eluent was 1 mol/L HNO_3. Due to the high ratios of Na and Al to Mg in rocks, all samples including the standards were then passed through the columns three times to obtain a purer Mg solution. Instrumental mass bias was corrected using standard-sample-standard bracketing method. The data are expressed in standard δ-notation relative to standard DSM3: $\delta^{25,26}Mg = [(^{25,26}Mg/^{24}Mg)_{sample}/(^{25,26}Mg/^{24}Mg)_{DSM3} - 1] \times 1000$. Replicate analyses have been done for some samples and the results agree well within current uncertainty. In addition, $\delta^{26}Mg$ values of AGV-2, GSP-2, Hawaii seawater and San Carlos olivine reference materials were also analyzed during the course, which are indistinguishable from the values published in other studies (e.g., Ke et al., 2016; Tian et al., 2020).

Iron isotopic analysis was done at the University of Washington (Seattle), USA, following the steps described in Dauphas et al. (2009) and Teng et al. (2013). The solutions used for Mg isotopic analysis were not used for Fe isotopes here. Powdered samples were weighted and then digested by Optima-grade HF, HNO_3 and

HCl acids. To separate Fe from the solution, an anion exchange chromatography with the resin of Bio-Rad AG1X-8 (200–400 mesh) in HCl medium was chosen. The first step was 6 M HCl acid to remove the matrix elements and the second step was 0.4 M HCl that was used as the eluent to release Fe. This process was repeated again to achieve a much pure Fe solution. Iron isotopic ratios were determined on a *Nu Plasma II* MC-ICP-MS with the aforementioned method (for Mg). The Fe isotopic data are reported in per mil deviation relative to the international standard IRMM-014: $\delta^{56,57}$Fe = [(^{56,57}Fe/^{54}Fe)$_{sample}$/(^{56,57}Fe/^{54}Fe)$_{IRMM-014}$ −1] × 1000. Iron isotopic data will be discussed in δ^{56}Fe below. Two standards (JB-1 and BHVO-1; listed in Table 1) were also analyzed, which yielded Fe isotopic data similar to the recommended values (He et al., 2015).

Table 1　Mg, Fe, Sr and Nd isotopic compositions of the volcanic rocks studied in this work

Samples		Lab	Location	^{87}Sr/^{86}Sr	2SE	^{143}Nd/^{144}Nd	2SE	δ^{26}Mg	2SD	δ^{56}Fe	2SE	*N*
Shield-forming basalts												
I-98-1	Basalt		42°10.2′N, 128°4.8′E							0.16	0.03	6
T9-2	Basalt		42°6.2′N, 128°12.7′E							0.17	0.04	7
Cone-construction trachytes												
8-24-15	Trachyte		42°1.9′N, 128°16.1′E							0.09	0.03	9
I-25-1	Trachyte		42°4.9′N, 128°3.1′E							0.23	0.05	4
2-15-2	Trachyte		42°3.3′N, 128°6.8′E							0.18	0.04	7
2(1)S	Trachyte		42°3.5′N, 128°3.5′E							0.22	0.04	7
I-7-2	Trachyte		42°1.9′N, 128°6.1′E							0.20	0.04	8
ME trachytes-comendites												
30-P-S	Trachyte	CUGB	42°3.3′N, 127°40.4′E	0.705079	0.000014	0.512583	0.000008	−0.30	0.02			
28-P-S	Trachyte	CUGB	42°3.3′N, 127°40.4′E	0.705038	0.000012	0.512582	0.000008	−0.37	0.04	0.24	0.04	8
Replicate		CUGB						−0.36	0.05			
31-P-S	Trachyte	CUGB	42°3.3′N, 127°40.4′E	0.705053	0.000014	0.512585	0.000009	−0.25	0.02			
Replicate		CUGB						−0.28	0.07			
29-P-S	Trachyte	CUGB	42°3.3′N, 127°40.4′E	0.705130	0.000014	0.512581	0.000009	−0.30	0.04			
Replicate		CUGB						−0.31	0.08			
33-P	Trachyte	CUGB	42°3.3′N, 127°40.4′E	0.705333	0.000014	0.512571	0.000009	0.14	0.05	0.23	0.04	8
		CUGB						0.08	0.03			
32-P-S	Trachyte	CUGB	42°3.3′N, 127°40.4′E	0.705323	0.000012	0.512599	0.000008	0.05	0.06			
24-P-Q	Trachyte	UW	42°3.3′N, 127°40.4′E	0.705292	0.000013	0.512587	0.000009	−0.03	0.08	0.26	0.04	8
T24-5	Comendite	CUGB	42°3.3′N, 127°40.4′E					−0.14	0.06	0.30	0.05	8
28-P-Q	Comendite	CUGB	42°3.3′N, 127°40.4′E	0.705242	0.000013	0.512579	0.000008	−0.32	0.04			
31-P-Q	Comendite	UW	42°3.3′N, 127°40.4′E	0.705174	0.000014	0.512574	0.000008	−0.16	0.07			

Continued

Samples		Lab	Location	$^{87}Sr/^{86}Sr$	2SE	$^{143}Nd/^{144}Nd$	2SE	$\delta^{26}Mg$	2SD	$\delta^{56}Fe$	2SE	*N*
Duplicate		UW						−0.14	0.08			
30-P-Q	Comendite	UW	42°3.3′N, 127°40.4′E	0.705319	0.000014	0.512583	0.000012	−0.30	0.07			
Duplicate		UW						−0.28	0.08			
YME	Comendite	UW	42°1.5′N, 128°26.8′E	0.705133	0.000012	0.512581	0.000008	−0.24	0.08	0.37	0.04	6
NME	Comendite	UW	41°59.3′N, 128°3.5′E	0.705512	0.000012	0.512582	0.000009	0.12	0.08	0.34	0.05	8
Duplicate		UW						0.07	0.06			
Standards												
BCR-2				0.704989	0.000013	0.512625	0.000009					
JB-1										0.13	0.03	
BHVO-1										0.14	0.04	
GSP-2		CUGB						0.07	0.02			
Duplicate		CUGB						0.04	0.13			
AGV-2		CUGB						−0.14	0.07			
Hawaiian Seawater		UW						−0.85	0.07			
Duplicate		UW						−0.87	0.06			
San Carlos olivine		UW						−0.33	0.07			

Notes: The Sr and Nd isotope were measured in IGGCAS. For Mg isotopic analysis, the UW denotes that samples were determined at the University of Washington (Seattle) and the others were determined at CUGB. All Fe isotopic data were done at UW. Replicate means repeated column chemistry and instrumental analysis. Duplicate means repeated measurement of isotopic ratios on the same solutions. *N* denotes the times of Fe isotopic measurement.

3 Results

The trace elemental concentrations for ME samples along with major oxides data from literature are summarized in Table S1 and the isotopic compositions analyzed here are listed in Table 1. The ME trachyte-comendite series exhibit a negative relationship between total alkali and silica contents (Figure S1). Together with literature data, most studied trachytes and comendites lay along the fractional crystallization trends (Figure S1). They display weakly to extremely negative Sr, P, Eu and Ti anomalies (Figure S2). Their $^{87}Sr/^{86}Sr$ ratios range from 0.705038 to 0.705512, with most of them higher than the underlying shield-forming basalts (e.g., Liu et al., 2015; Zhang et al., 2018). By contrast, Nd isotopic ratios (0.512571−0.512599) of these samples remain invariant from trachyte to highly differentiated comendite and are similar to those of both the shield-forming basalts and cone-construction trachytes. The ME trachytes have $\delta^{26}Mg$ values (−0.37 to 0.14‰) comparable to the comendites (−0.32 to 0.12‰), which fall within the field between basalt and cone-construction trachyte (Fig. 2a and Table 1). For Fe isotopes, the two basalts have $\delta^{56}Fe$ values of 0.16±0.03‰ and 0.17±0.04‰, respectively, slightly higher than the normal MORB (~0.105; Teng et al., 2013). The $\delta^{56}Fe$ values of cone-construction trachytes range between 0.09 and 0.23‰, while ME trachytes and comendites have the highest Fe isotopic compositions varying from 0.23 to 0.37‰ (Table 1). All studied samples broadly follow the $\delta^{56}Fe$ versus SiO_2 trend defined by previously reported data for igneous rocks, i.e., an increase in $\delta^{56}Fe$ with increasing SiO_2 (Fig. 2b). Additionally, $\delta^{56}Fe$ values inversely vary with tFe_2O_3 (Fig. 2c) and do not correlate with $\delta^{26}Mg$ (Fig. 2d).

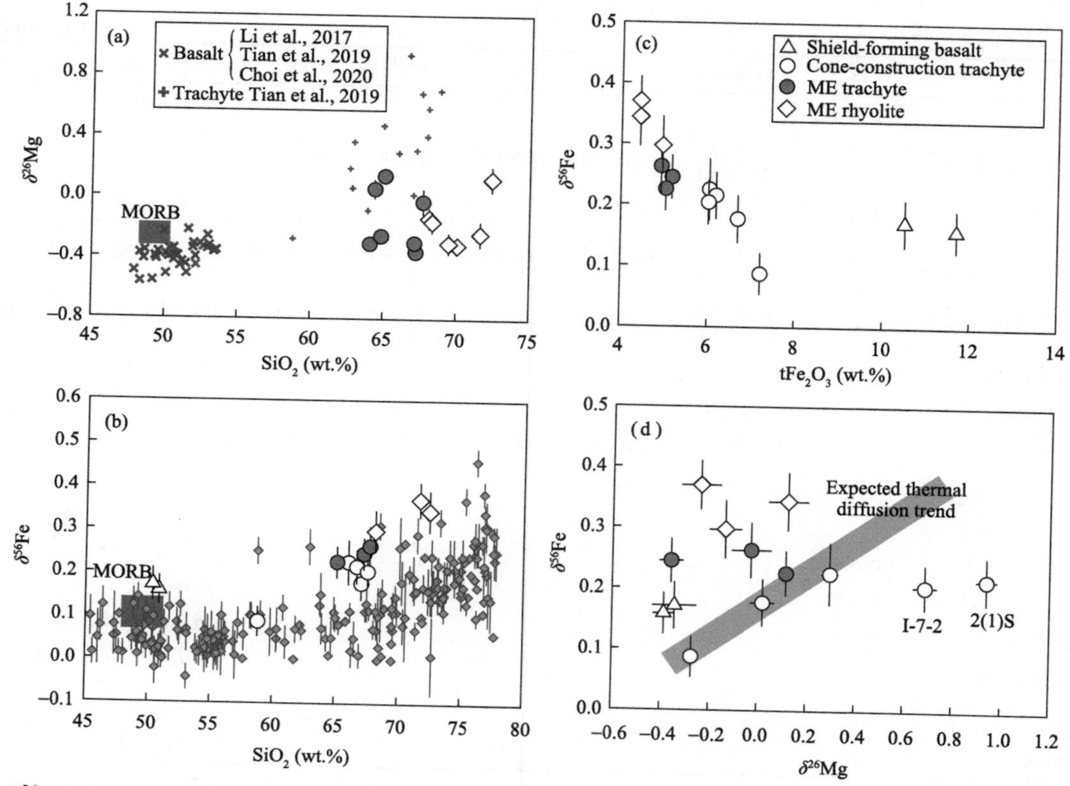

Fig. 2 (a) δ^{26}Mg versus SiO$_2$. The average MORB of −0.25±0.06 is from Teng et al. (2010b). The triangles denote data from literature data (Li et al., 2017a; Tian et al., 2019; Choi et al., 2020). (b) δ^{56}Fe versus SiO$_2$. Gray diamonds denote literature data (Dauphas et al., 2009; Poitrasson and Freydier, 2005; Heimann et al., 2008; Schuessler et al., 2009; Telus et al., 2012; Zambardi et al., 2014; Foden et al., 2015). Uncertainties are 2 standard deviations of the mean (2SD). Iron isotopic composition of the MORB is from Teng et al. (2013). (c) δ^{56}Fe versus tFe$_2$O$_3$; (d) δ^{56}Fe versus δ^{26}Mg. The Mg and Fe isotopic data are from this study.

4 Discussion

The newly analyzed ME trachyte-comendite series show a substantial variation in δ^{26}Mg and are generally enriched in heavy Mg isotopes (Fig. 2a). This isotopic variation is unlikely induced by surface chemical weathering since the degree of isotopic fractionation is independent on the LOI (loss on ignition) (Fig. 3a). In a plot of δ^{56}Fe versus LOI (Fig. 3b), the ME trachyte-comendite series have higher LOI than the basalt and cone-construction trachyte. The former seems to have heavy Fe isotopic compositions with increasing LOI (Fig. 3b), suggesting a possible influence from low-temperature alteration. In general, both Nb and Ti show element immobility during low-temperature alteration processes (e.g., Ma et al., 2007). The good correlations between δ^{56}Fe with Nb and Ti for the ME trachyte-comendite series as well as cone-construction trachytes (Fig. 3c and d), however, argue against the presence of low-temperature alteration process. Such large Mg-Fe isotopic variations observed in these samples therefore primarily reflect their source regions and will be discussed in detail.

4.1 Magnesium isotopic variation in ME trachyte-comendite series

Many studies have suggested that the Changbaishan Cenozoic shield-forming basalts are generated by partial melting of the asthenospheric mantle (e.g., Zou et al., 2008; Kuritani et al., 2011; Li et al., 2017a; Choi et al., 2020), which is imaged by the high-precision tomography (Fan and Chen, 2019). The intermediate-silicic magmatic rocks were interpreted to source from a basaltic composition by fractional crystallization although there is

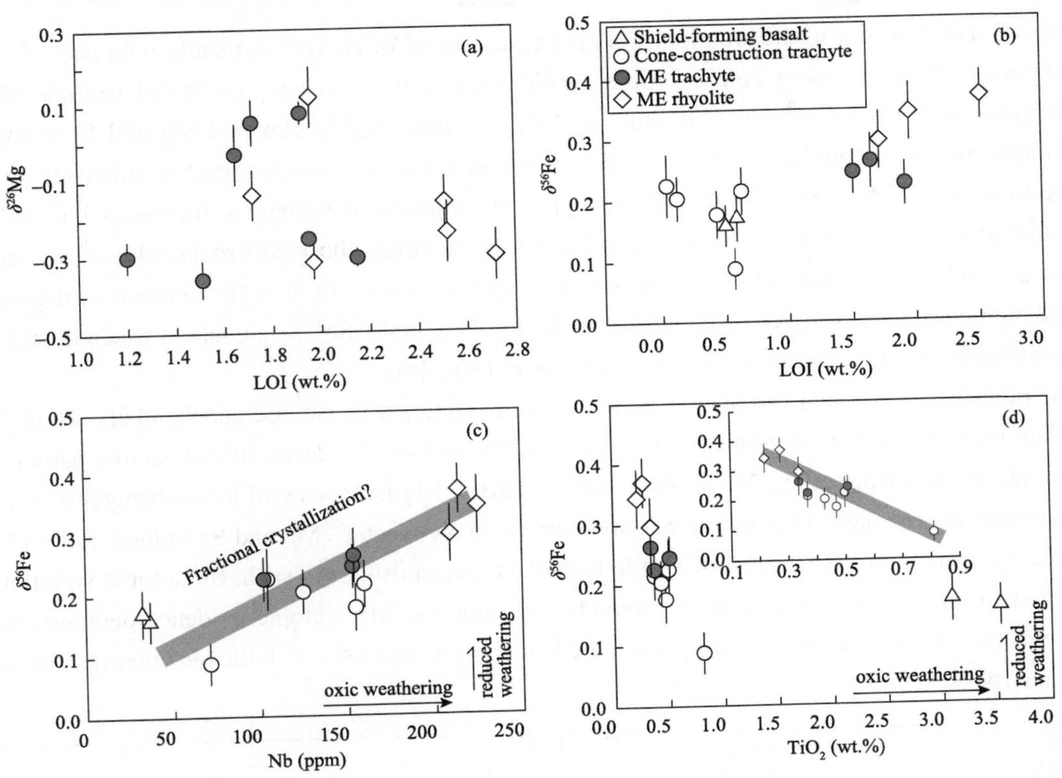

Fig. 3 δ^{26}Mg versus LOI (a); δ^{56}Fe versus LOI (b); δ^{56}Fe versus Nb (c) and δ^{56}Fe versus TiO_2 (d). The trends of Fe isotope fractionation during chemical weathering are deduced from Liu et al. (2014).

still debate on the origin of their variable ratios of Sr, Nd, Pb and Mg isotopes: (1) whole rock compositional variations are consistent with the expected results from crystallization (e.g., Liu et al., 2015; Li et al., 2004; Andreeva et al., 2018; Zhang et al., 2018; Tian et al., 2019). For example, Changbaishan basalts, trachytes and rhyolites have nearly identical Nd isotopic compositions (Fig. 4a) and Sr abundances decrease from ~800 ppm in basalts down to several ppm in high-SiO_2 comendites (Table S1); (2) the relatively higher $^{230}Th/^{232}Th$ ratios (>0.65) in ME trachytes and comendites than the mafic rocks (~0.63) (Kuritani et al., 2020); and (3) lack of any old zircon crystals (e.g., >1 Ma) or crustal xenoliths in trachytic and comenditic rocks (e.g., Zou et al., 2010; Cheong et al., 2019). Quantitatively, the trachytic and comenditic magmas can be produced via ~70% and ~90% crystallization, respectively (Kuritani et al., 2020).

In current work, the newly measured δ^{26}Mg for young trachytes and comendites overlap with the upper part of the basalts and the lower part of the cone-construction trachytes (Fig.4b). Shield-forming basalts were previously reported to have heterogeneous Mg isotopic variation ranging from −0.56 to −0.22‰ (Li et al., 2017a; Tian et al., 2019; Choi et al., 2020), which was attributed to the source metasomatism rather than from exogenous effects of shallow processes. In this case, the variation of the Mg isotopic ratios of some felsic ME products would be explained essentially by the Mg isotope heterogeneity of their parental primary magmas. Here, the fractional crystallization model was employed to calculate the Mg isotopic variation based on the rhyolite-MELTS results and melt-mineral fractionation factors (Tian et al., 2019). As plotted in Fig. 4b, the black dashed curve and gray area represent the modelled results as the δ^{26}Mg of parental melts are set to be −0.37±0.05‰. Notably, some trachytes and ME comendites fall on the gray area, meaning that fractional crystallization can explain part of the Mg isotopic variation.

The decoupled Mg-Sr and Nd isotopic signature of ME samples is unlikely derived from the contamination by supracrustal materials during its eruption. This is because input of ancient (typically with ε_{Nd}<−6, McCulloch and Wasserburg, 1978) or young country rocks would have resulted in coupled Sr-Nd isotopic ratios. Since intense chemical weathered products of basaltic rocks could have highly elevated Mg and Sr isotopes, it was invoked to interpret the decoupled Mg-Sr and Nd isotopes observed in cone-construction trachytes by Tian et al. (2019). We found that δ^{26}Mg values of the ME trachyte and comendite lavas tend to increase with ^{87}Sr/^{86}Sr ratios, similar to that of the cone-construction trachyte (Fig. 4c). By analogy, those ME rocks with δ^{26}Mg and ^{87}Sr/^{86}Sr higher than the underlying basalts (Fig. 4c) more likely have the same origin to the cone-construction trachytes. These weathered components therefore could impart high δ^{26}Mg and ^{87}Sr/^{86}Sr signals in trachyte and comendite through assimilation and fractional crystallization processes (Fig. 4c).

On the other hand, our data can also provide new constraints on Sr isotope geochemistry. High ^{87}Sr/^{86}Sr in Changbaishan trachytes and comendites are previously explained by the decay of radioactive parent ^{87}Rb (e.g., Andreeva et al., 2018; Zhang et al., 2018). As a stable element, Mg isotopes will be unchanged with time unless external processes are invoked. The finding of good correlation between δ^{26}Mg and Sr isotopic ratios (Fig. 4c) are likely caused by a same process instead of two independent mechanisms. As such, Sr isotopic variation was also closely related to the addition of weathered products. In summary, Mg isotopic evidence demonstrates that the previously weathered basalt continuously participated in the petrogenesis of both old intermediate and young felsic volcanic rocks.

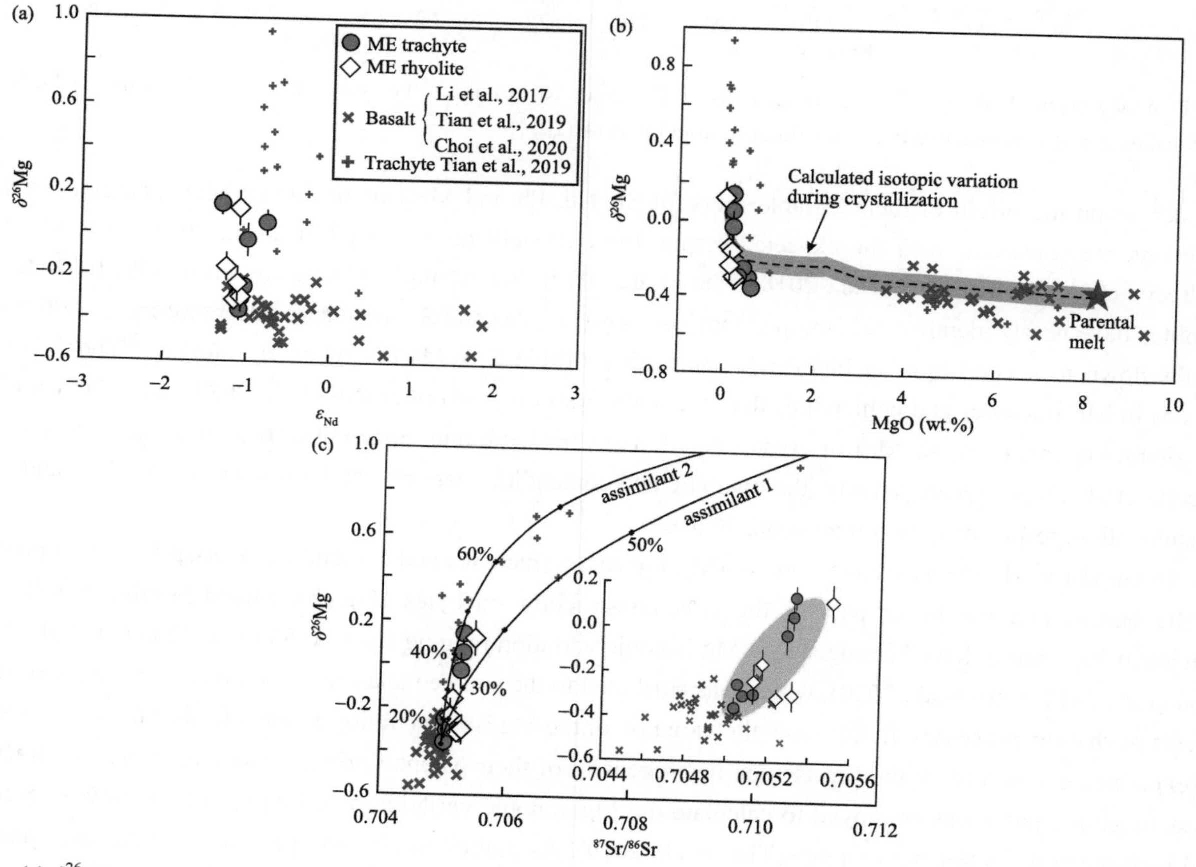

Fig. 4 (a) δ^{26}Mg versus ε_{Nd}. (b) Plot of δ^{26}Mg against MgO. The parental melt is assumed to be δ^{26}Mg=−0.37±0.05‰. (c) δ^{26}Mg versus ^{87}Sr/^{86}Sr. Concentrations and isotopic compositions of elements in end members are as follow: MgO (wt.%): 1.1 (parental melt), 0.15 (assimilant 1), 0.6 (assimilant 2); δ^{26}Mg (‰): −0.37 (parental melt), 1.8 (assimilant 1), 1.5 (assimilant 2); Sr (ppm): 430 (parental melt), 7 (assimilant 1), 10 (assimilant 2); ^{87}Sr/^{86}Sr: 0.705 (parental melt), 0.718 (assimilant 1), 0.715 (assimilant 2). The ticks on curves indicate 10% increment.

4.2 Iron isotopic variation in Changbaishan volcanic rocks

All measured $\delta^{56}Fe$ values for Changbaishan volcanic rocks gradually increase with increasing SiO_2 content, in agreement with the literature data where those two datasets overlap (Fig. 2b). The two basalts studied here (0.16±0.03 for I-98-1 and 0.17±0.04 for T9-2; Table 1) are isotopically heavier than both the Earth's upper mantle (~0.02; Weyer and Ionov, 2007) and MORB (~0.105; Teng et al., 2013). They hold Fe/Mn ratios of 67 to 74 (Table S1), higher than that of the MORB (~56; Gale et al., 2013), suggesting a significant contribution from pyroxenite to the genesis of primary basaltic magma (e.g., Sobolev et al., 2007). This is consistent with a recent study that identified a hybrid source comprising of eclogite/pyroxenite and garnet peridotite using trace elemental indices (e.g., Gb/Yb vs Yb; Choi et al., 2020). Mantle pyroxenites were shown to have heterogeneous and heavier Fe isotopic compositions relative to the upper mantle. Garnet-pyroxenite xenoliths from Oahu, Hawaii display $\delta^{56}Fe$ values varying from 0.05 to 0.16 (Williams and Bizimis, 2014). In this case, their subsequent melting could acquire a higher $\delta^{56}Fe$ (Konter et al., 2016), which can readily cover the observed Fe isotopic data in these two basalts. Moreover, if garnet is a residual phase during melting, it would also impart a heavy Fe isotopic signature to the remaining melts as garnet commonly has lighter Fe isotopic composition than coexisting olivine and pyroxene (He et al., 2017). In view of this, exotic processes such as thermal diffusion or recycled sedimentary carbonates are not necessary. In fact, recycled crustal components such as sedimentary carbonates, terrigenous silicate sediments and carbonated eclogites have been invoked to interpret the EM1-like isotopic features and large Mg isotopic variations (Li et al., 2017a; Choi et al., 2020). The subducted materials, such as sedimentary carbonates can oxide Fe^0 and/or Fe^{2+} to Fe^{3+} via redox reactions in the mantle. Partial melting of Fe^{3+}-rich metasomatized mantle will generate basaltic melt with heavy Fe isotopic composition. For instance, an unusual high $\delta^{56}Fe$ signature (up to 0.29) seen in nephelinitic component from South China Block was partially attributed to the hybridization by recycled carbonates (He et al., 2019). Carbonatitic metasomatism can reduce the Fe/Mn ratios due to its high MnO content based on experimental results (Thomson et al., 2016). The high Fe/Mn (67–74) and low $\delta^{56}Fe$ shown by our two basalts relative to the nephelinites (~58 for Fe/Mn; He et al., 2019) can be explained by less input of recycled carbonates than that in nephelinitic source. Indeed, the only two Fe isotopic data analyzed here are insufficient to determine/discriminate the above mentioned scenarios.

For the trachytes and comendites, the more evolved igneous rocks generally show heavier Fe isotopes (Fig. 2b). Late-stage magmatic fluid exsolution was introduced to interpret the isotopic variations in silicic igneous rocks (e.g., Poitrasson and Freydier, 2005; Heimann et al., 2008). Isotopically light Fe was considered to preferentially partition in late magmatic aqueous fluids exsolved from highly differentiated magmas (Poitrasson and Freydier, 2005). The Th/U and Rb/La ratios, which have been used as sensitive proxies for the presence of fluids (e.g., He et al., 2017; Xia et al., 2017), were chosen here to test whether the observed Fe isotopic variation was affected by deuteric fluids. As illustrated in Fig. 5a, two trachytes with high Th/U ratios do not have the expected heavy Fe isotopic compositions, while almost the highest $\delta^{56}Fe$ values were observed at a given low Th/U value. Furthermore, $\delta^{56}Fe$ of all trachytes and comendites are positively correlated with Rb/La, opposite to the predicted trend (Fig. 5b), which is taken as a further evidence that the heavy Fe isotopes are not related to the fluid exsolution. Open-system mixing process between exogenous crustal component with high $\delta^{56}Fe$ and mafic magma component with low $\delta^{56}Fe$ appears to be unlikely either. For the Changbaishan volcanic rocks investigated here, there is no clear relationship between Fe isotopes and ε_{Nd} (Table 1). Although physical mixing of trachytic and comenditic magmas was observed for the pumices (Pan et al., 2017; this study), this process was not responsible for Fe isotopic variation as they are the products of different stages of the differentiation process.

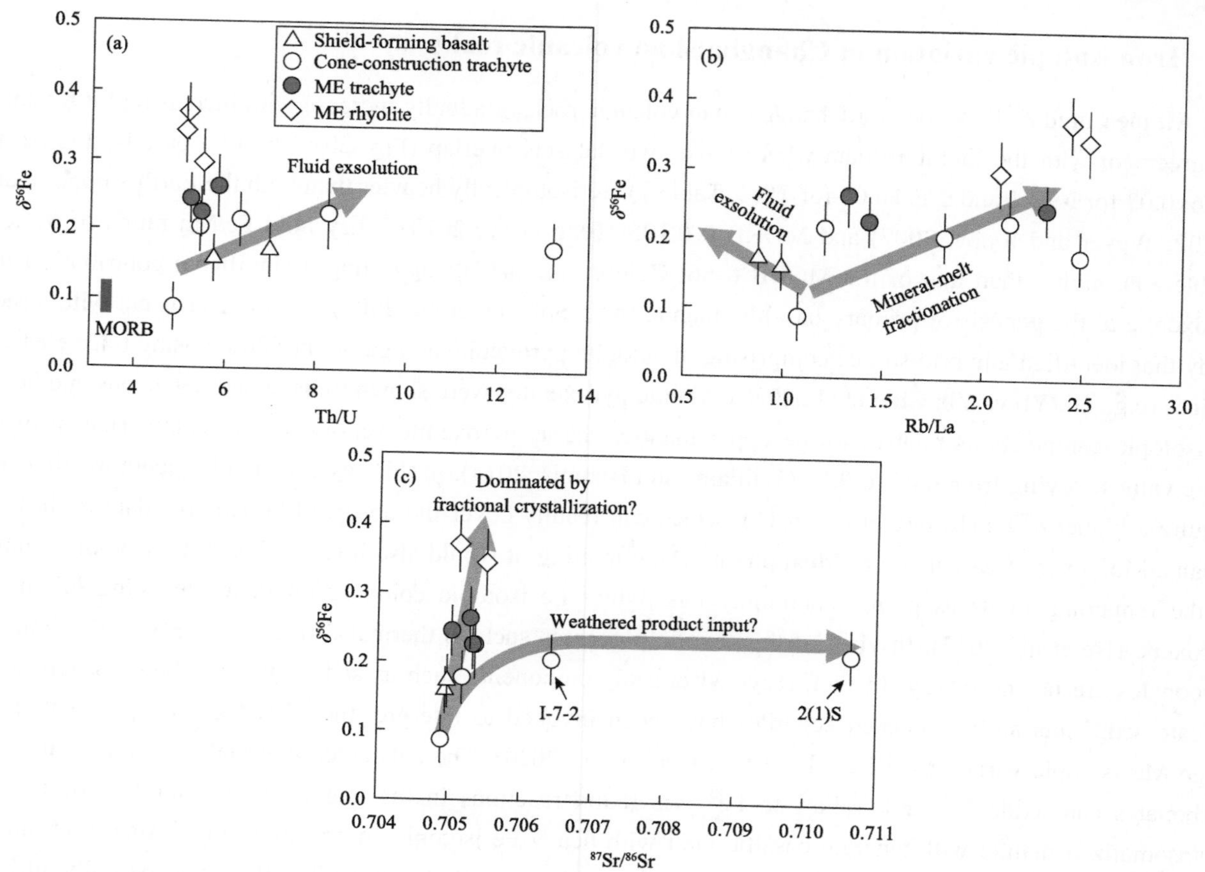

Fig. 5　Plot of δ^{56}Fe versus Th/U (a), Rb/La (b) and ^{87}Sr/^{86}Sr (c) of the studied samples from Changbaishan volcano. The Th/U ratio of MORB is from Gale et al. (2013). Iron isotopic composition of the MORB comes from Teng et al. (2013).

The most likely driver points to the aforementioned weathered products and fractional crystallization mechanism as was first suggested for the lavas from Kilauea Iki lava lake (Teng et al., 2008). Overall, all Changbaishan trachytes and comendites show two trends in the plot of δ^{56}Fe versus ^{87}Sr/^{86}Sr (Fig. 5c), which are distinct from their Mg isotopic signatures. The five cone-construction trachytes with higher proportions of weathered products involvement display a non-linear relationship between ^{87}Sr/^{86}Sr and Fe isotopes, whereas ME samples with less addition of weathered products have the heaviest Fe isotopic compositions (Fig. 5c). This observation demonstrates that recycled weathered products mainly controlled the Fe isotopes of the cone-construction samples but had a limited shift (<0.1‰). In other words, it means that the speculated paleo-weathering event did not result in apparent Fe isotope fractionation in the weathered residue when compared with source-like Fe isotopes (~0.15‰). Studies on saprolite profiles indicate redox states were thought to be the first order in controlling Fe isotope fractionation (Liu et al., 2014; Li et al., 2017b). Under oxidizing conditions, Fe is transformed into immobile ferric Fe and rapidly precipitated in the form of Fe oxides, which was the dominant Fe species in the profiles. By contrast, Fe isotope fractionation is considerable under reducing conditions. As such, one possibility is that the paleo-weathering of basalts was subject to oxidized conditions, thus had negligible influence on Fe isotopes of Changbaishan volcanic rocks. This also implies that Fe isotope fractionation may be only relevant to oxidation-reduction reactions regardless of the weathering rate, humidity, temperature and/or latitude. For ME samples, their δ^{56}Fe values vary with SiO_2 and tFe_2O_3 contents (Fig. 2). Furthermore, they fall on an open-system evolutionary trend defined by the Hekla lavas where ME samples move towards higher δ^{56}Fe as Mg# decreases down to approximately 2 (Fig. 6a). This reflects a more extensive scale of differentiation process and the main fractional crystallization phases are Fe^{2+}-rich silicate and oxide minerals. Our results coincide with the view from inclusion data where Andreeva et al. (2019) found an overall decreasing trend in oxygen fugacity from Changbaishan basaltic to felsic magmas. That trend is mainly caused by magma degassing in the form of SO_2 at decompression and spinel precipitation rather than the much segregation of Fe^{3+} (i.e., magnetite) from the melt. Under this condition open to oxygen, crystallization of these mineral assemblages therefore would increase the δ^{56}Fe of remaining melt.

Iron isotope fractionation can be made using a Rayleigh fractionation model as described below:

$$\delta^{56}\mathrm{Fe}_{\mathrm{melt}} = (\delta^{56}\mathrm{Fe}_{\mathrm{source}} + 1000) \times f^{(\alpha-1)} - 1000$$

where f denotes the fraction of Fe remaining in the evolving melt and α refers to the equilibrium isotope fractionation factor. Due to the changes of $Fe^{3+}/\Sigma Fe$ during magmatic processes, it is difficult to obtain precise Fe isotope fractionation factors between the melt and minerals. A bulk isotope fractionation factor with $\alpha \sim$ 0.99981–0.99995 (Fig. 6b) can match the Fe isotopic variation observed in our samples. The maximum value of $\Delta^{56}\mathrm{Fe}_{\mathrm{melt\text{-}mineral}} \sim 0.19‰$ used here is a little larger than that in previous studies (e.g., 0.13‰ in He et al., 2017). This may be derived from the compositional effect or a little low initial Fe content used in this model. Nonetheless, future work on experimental determination of Fe isotopic fractionation factor in highly differentiated magmas remains important to test this value.

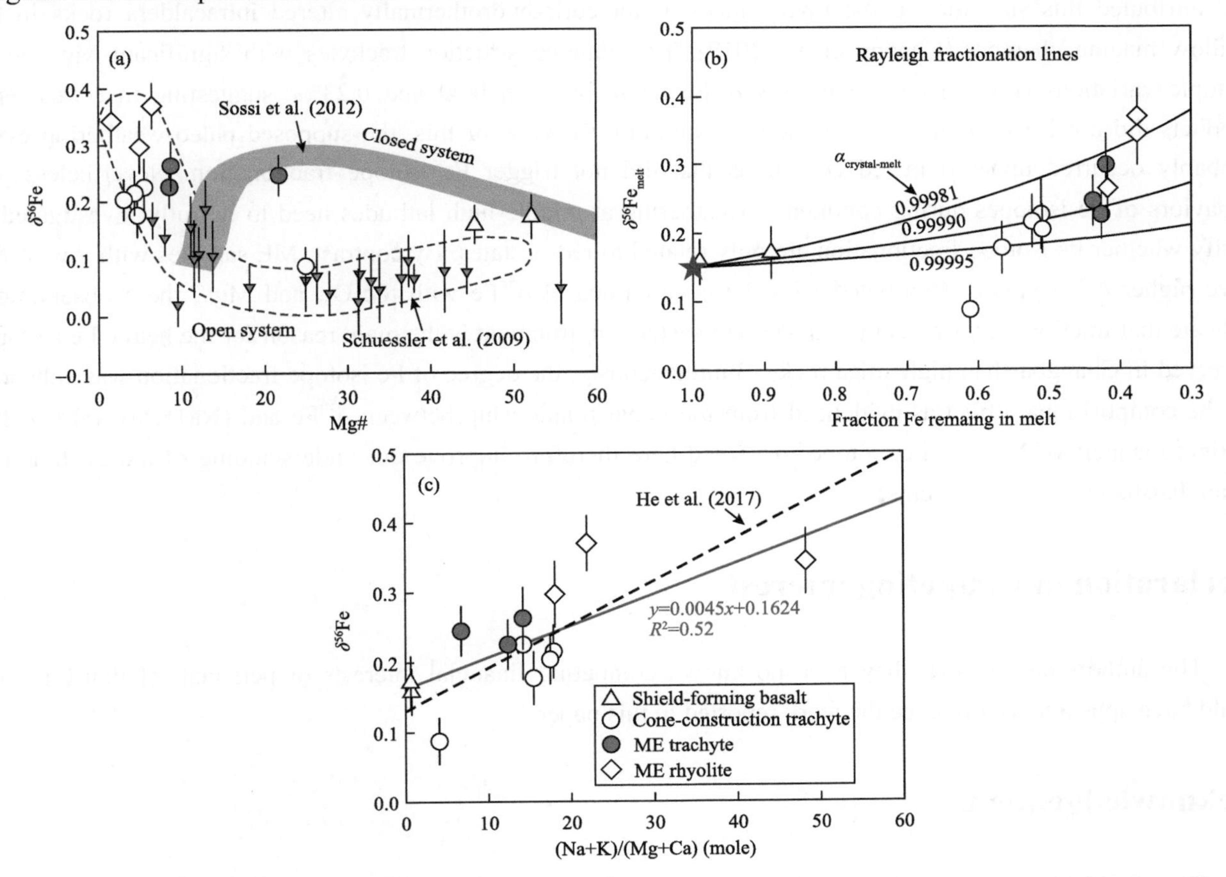

Fig. 6 (a) δ^{56}Fe versus Mg#. The gray area denotes the Fe isotopic variation from Sossi et al. (2012) for comparison. (b) Iron isotopic variation against Fe fraction remaining in the melt for the samples from Changbaishan volcano with a Rayleigh fractionation model ($\delta^{56}\mathrm{Fe}_{\mathrm{melt}} = (\delta^{56}\mathrm{Fe}_{\mathrm{source}} + 1000) \times f^{(\alpha-1)} - 1000$), using different bulk isotope fractionation factors ($\alpha_{\mathrm{crystal\text{-}melt}}$) between crystallizing minerals and melt. The f is the ratio of Fe remaining in the melt and red star represents the initial value (assumed δ^{56}Fe = 0.15 ‰ as the initial value). (c) δ^{56}Fe versus (Na+K)/(Mg+Ca) for the Changbaishan volcanic rocks. The orange line is the regression line. (For interpretation of the references to colour in this figure legend, the reader is referred to the web version of this article.)

On the other hand, the changes of melt compositions during magmatic evolution could lead to the modification of the structure or speciation of Fe. Enrichment of alkalis has the potential to enlarge Fe isotope fractionation, while alkaline earth cations impart opposite effect as manifested by Foden et al. (2015) and He et al. (2017). The analyzed trachytes and comendites here have a broadly positive relationship between δ^{56}Fe and (Na + K)/(Ca + Mg), with a slope close to the maximum estimate of the magma compositional effect (Fig. 6c). This phenomenon demonstrates that the increased concentrations of Na and K and lower Ca and Mg in the melt enhance the Fe isotope fractionation. Therefore, it can be concluded that the Fe isotopes of Changbaishan magmas were controlled by chemical weathered products, fractional crystallization and compositional effect.

5 Summary and implications

In present work, the ME trachyte-comendite series display a large Mg isotopic fractionation with δ^{26}Mg values ranging from −0.37 to +0.14‰, similar to that of the cone-construction trachytes. Such an isotopic variation observed here and the good correlations between Mg isotopes and ^{87}Sr/^{86}Sr ratios indicate the continuous participation of formerly weathered residue of basalts in the magmatic evolution. This interpretation is somewhat similar to the oxygen isotopic study that reported low δ^{18}O in zircons from the trachydacitic pumices (ca. 12−9 ka) and attributed this signature to the involvement of meteoric-hydrothermally altered intracaldera rocks in the shallow magma chamber (Cheong et al., 2017). The cone-construction trachytes with significant Mg and Sr isotopic variations (Fig. 4) have a narrow δ^{56}Fe range between 0.09 and 0.23‰, suggesting that weathered products did not cause remarkable Fe isotopic variation. In view of this, the supposed paleo-weathering event probably occurred under oxidized conditions that did not trigger Fe isotope fractionation. Nevertheless, the behaviors of Fe isotopes during continental weathering at middle-high latitudes need to be fully investigated to verify whether its isotope fractionation are only related to redox states. By contrast, ME samples with low δ^{26}Mg have higher δ^{56}Fe values. Combined with the covariations of δ^{56}Fe with tFe_2O_3, and Mg#, these observations indicate that fractional crystallization in an open-system environment is the main reason for the heavy Fe isotopes observed in Changbaishan high silica rocks. Simultaneously, the degree of Fe isotope fractionation was enhanced by the compositional effect as evidenced from the linear relationship between δ^{56}Fe and (Na+K)/(Ca+Mg). The distinct manners of Mg and Fe isotopes observed here therefore improve our understanding of the evolution of Changbaishan magmatic systems.

Declaration of competing interest

The authors declare that they have no known competing financial interests or personal relationships that could have appeared to influence the work reported in this paper.

Acknowledgements

We would like to thank the associate editor Christian France-lanord, Dr. Jian Huang and an anonymous reviewer for their insightful comments, which greatly improved the manuscript. We also thank Prof. Fang-Zhen Teng for the discussion and Dr. Xin-Yang Chen for help during Mg and Fe isotopic analyses. This work is financially funded by the National Natural Science Foundation of China (41730214, 41861144025), National Postdoctoral Program (2018M631566), Fundamental Scientific Research Projects of Institute of Geology, China

Earthquake Administration (IGCEA1603).

Appendix A. Supplementary data

Supplementary data to this article can be found online at https://doi.org/10.1016/j.chemgeo.2021.120075.

References

Andreeva, O.A., Naumov, V.B., Andreeva, I.A., Kovalenko, V.I., 2011. Basaltic melts in olivine from alkaline pumice of primor'e: evidence from the study of melt inclusions. Dokl. Earth Sci. 438, 656-660.

Andreeva, O.A., Yarmolyuk, V.V., Andreeva, I.A., Borisovskiy, S.E., 2018. Magmatic evolution of Changbaishan Tianchi Volcano, China-North Korea: evidence from mineral-hosted melt and fluid inclusions. Petrol. 26, 515-545.

Andreeva, O.A., Andreeva, I.A., Yarmolyuk, V.V., 2019. Effect of redox conditions on the evolution of magmas of Changbaishan Tianchi volcano, China–North Korea. Chem. Geol. 508, 225-233.

Chen, X.W., Wei, H.Q., Yang, L.F., Chen, Z.Q., 2017. Petrological and Mineralogical Characteristics of Tianchi Volcano, Changbai Mountain: Implications for Crystallization Differentiation and Magma Mixing. Acta Geol. Sin. 38,177-192.

Cheong, A.C.S., Jeong, Y.J., Jo, H.J., Sohn, Y.K., 2019. Recurrent Quaternary magma generation at Baekdusan (Changbaishan) volcano: New zircon U-Th ages and Hf isotopic constraints from the Millennium Eruption. Gondwana Res. 68, 13-21.

Cheong, A.C.S, Sohn, Y.K., Jeong, Y.J., Jo, H.J., Park, K-H., Lee, Y.S., Li, X-H., 2017. Latest Pleistocene crustal cannibalization at Baekdusan (Changbaishan) as traced by oxygen isotopes of zircon from the Millennium Eruption. Lithos. 284, 132-137.

Choi, H.O., Choi, S.H., Lee, Y.S., Ryu, J-S., Lee, D.C., Lee, S.G., Sohn, Y.K., Liu, J.Q., 2020. Petrogenesis and mantle source characteristics of the late Cenozoic Baekdusan (Changbaishan) basalts, North China Craton. Gondwana Res. 78, 156-171.

Dauphas, N., Pourmand, A., Teng, F.Z., 2009. Routine isotopic analysis of iron by HR-MC-ICPMS: how precise and how accurate? Chem. Geol. 267, 175-184.

Fan, X., Chen, Q.F., 2019. Seismic constraints on the magmatic system beneath the Changbaishan Volcano: insight into its origin and regional tectonics. J. Geophys. Res. 124, 2003-2024.

Foden, J., Sossi, P.A., Wawryk, C.M., 2015. Fe isotopes and the contrasting petrogenesis of A-, I- and S-type granite. Lithos 212-215, 32-44.

Gale, A., Dalton, C.A., Langmuir, C.H., Su, Y.J., Schilling, J.G., 2013. The mean composition of ocean ridge basalts. Geochem. Geophys. Geosyst. 14, 489-518.

He, Y.S., Ke, S., Teng, F.Z., Wang, T.T., Wu, H.J., Lu, Y.H., Li, S.G., 2015. High-Precision Iron Isotope Analysis of Geological Reference Materials by High-Resolution MC-ICP-MS. Geostand. Geoanal. Res. 39, 341-356.

He, Y.S., Wu, H.J., Ke, S., Liu, S.A., Wang, Q., 2017. Iron isotopic compositions of adakitic and non-adakitic granitic magmas: Magma compositional control and subtle residual garnet effect. Geochim. Cosmochim. Acta 203, 89-102.

He, Y., Meng, X., Ke, S., Wu, H., Zhu, C., Teng, F.-Z., Hoefs, J., Huang, J., Yang, W., Xu, L., Hou, Z., Ren, Z.-Y., Li, S., 2019. A nephelinitic component with unusual δ^{56}Fe in Cenozoic basalts from eastern China and its implications for deep oxygen cycle. Earth Planet. Sci. Lett. 512, 175-183.

Heimann, A., Beard, B.L., Johnson, C.M., 2008. The role of volatile exsolution and sub-solidus fluid/rock interactions in producing high ^{56}Fe/^{54}Fe ratios in siliceous igneous rocks. Geochim. Cosmochim. Acta 72, 4379-4396.

Jin, B., Zhang, X., 1994. Researching Volcanic Geology in Changbai Mountain. Northeast Korea Nation Education Press, Changchun, 1-233 (in Chinese).

Jweda, J., Bolge, L., Class, C., Goldstein, S.L., 2016. High Precision Sr-Nd-Hf-Pb Isotopic Compositions of USGS Reference Material BCR-2. Geostand. Geoanal. Res. 40, 101-115.

Ke, S., Teng, F.Z., Li, S.G., Gao, T., Liu, S.A., He, Y.S., Mo, X.X., 2016. Mg, Sr, and O isotope geochemistry of syenites from northwest Xinjiang, China: Tracing carbonate recycling during Tethyan oceanic subduction. Chem. Geol. 437, 109-119.

Konter, J.G., Pietruszka, A.J., Hanan, B.B., Finlayson, V.A., Craddock, P.R., Jackson, M.G., Dauphas, N., 2016. Unusual δ^{56}Fe values in Samoan rejuvenated lavas generated in the mantle. Earth Planet. Sci. Lett. 450, 221-232.

Kuritani, T., Ohtani, E., Kimura, J.I., 2011. Intensive hydration of the mantle transition zone beneath China caused by ancient slab stagnation. Nat. Geosci. 4, 713-716.

Kuritani, T., Nakagawa, M., Nishimoto, J., Yokoyama, T., Miyamoto, T., 2020. Magma plumbing system for the Millennium Eruption at Changbaishan volcano, China: constraints from whole-rock U–Th disequilibrium. Lithos, 366-367, 105564.

Li, N., Fan, Q.C., Sun, Q., Zhang, W.L., 2004. Magma evolution of Changbaishan Tianchi volcano: evidences from the main phenocrystal minerals. Acta Petrol. Sin. 20, 575-582 (in Chinese with English abstract).

Li, W.Y., Teng, F.Z., Ke, S., Rudnick, R.L., Gao, S., Wu, F.Y., Chappell, B.W., 2010. Heterogeneous magnesium isotopic composition of the upper continental crust. Geochim. Cosmochim. Acta 74, 6867-6884.

Li, C.F., Chu, Z.Y., Guo, J.H., Li, Y.L., Yang, Y.H., Li, X.H., 2015. A rapid single column separation scheme for high-precision Sr-Nd-Pb isotopic analysis in geological samples using thermal ionization mass spectrometry. Anal. Methods 7, 4793-4802.

Li, C.F., Wang, X.C., Guo, J.H., Chu, Z.Y., Feng, L.J., 2016. Rapid separation scheme of Sr, Nd, Pb, and Hf from a single rock digest using a tandem chromatography column prior to isotope ratio measurements by mass spectrometry. J. Anal. At. Spectrom. 31, 1150-1159.

Li, S.G., Yang, W., Ke, S., Meng, X., Tian, H., Xu, L., He, Y., Huang, J., Wang, X.-C., Xia, Q., Sun, W., Yang, X., Ren, Z.-Y., Wei, H., Liu, Y., Meng, F., Yan, J., 2017a. Deep carbon cycles constrained by a large-scale mantle Mg isotope anomaly in eastern China. Natl. Sci. Rev. 4, 111-120.

Li, M., He, Y.-S., Kang, J.-T., Yang, X.-Y., He, Z.-W., Yu, H.-M., Huang, F., 2017b. Why was iron lost without significant isotope fractionation during the lateritic process in tropical environments? Geoderma 290, 1-9.

Li, R.Y., Ke, S., Li, S., Song, S., Wang, C., Liu, C., 2020. Origins of two types of Archean potassic granite constrained by Mg isotopes and statistical geochemistry: Implications for continental crustal evolution. Lithos 368-369, 105570.

Liu, J.Q, Chen, S.S., Guo, Z.F., Guo, W.F., He, H.Y., You, H.T., Kim, H.M., Sung, G.H., Kim, H., 2015. Geological background and geodynamic mechanism of Mt. Changbai volcanoes on the China-Korea border. Lithos 236, 46-73.

Liu, S.A., Teng, F.Z., Li, S.G., Wei, G.J., Ma, J.L., Li, D.D., 2014. Copper and iron isotope fractionation during weathering and pedogenesis: Insights from saprolite profiles. Geochim. Cosmochim. Acta 146, 59-75.

Ma, J.L., Wei, G.J., Xu, Y.G., Long, W.G., Sun, W.D., 2007. Mobilization and re-distribution of major and trace elements during extreme weathering of basalt in Hainan Island, South China. Geochim. Cosmochim. Acta 71, 3223-3237.

McLean, D., Albert, P.G., Nakagawa, T., Staff, R., Suzuki, T., Suigetsu 2006 Project Members., Smith, V., 2016. Identification of the Changbaishan 'Millennium' (B-Tm) eruption deposit in the Lake Suigetsu (SG06) sedimentary archive, Japan: Synchronisation of hemispheric-wide palaeoclimate archives. Quat. Sci. Rev. 150, 301-307.

Oppenheimer, C., Wacker, L., Xu, J.D., Galvan, J.D., Stoffel, M., Guillet, S., Corona, C., Sigl, M., Cosmo, N.D., Hajdas, L., Pan, B., Breuker, R., Schineider, L., Esper, J., Fei, J., Hammond, J.O.S., Buntgen, U., 2017. Multi-proxy dating the 'Millennium Eruption' of Changbaishan to late 946 CE. Quat. Sci. Rev. 158, 164-171.

Pan, B., de Silva, S.L., Xu, J.D., Chen, Z.Q., Miggins, D.P., Wei, H.Q., 2017. The VEI-7 Millennium eruption, Changbaishan-Tianchi volcano, China/DPRK: New field, petrological, and chemical constraints on stratigraphy, volcanology, and magma dynamics. J. Volcanol. Geotherm. Res. 343, 45-59.

Poitrasson, F., Freydier, R., 2005. Heavy iron isotope composition of granites determined by high resolution MC-ICP-MS. Chem. Geol. 222, 132-147.

Raczek, I., Jochum, K.P., Hofmann, A.W., 2003. Neodymium and strontium isotope data for USGS reference materials BCR-1, BCR-2, BHVO-1, BHVO-2, AGV-1, AGV-2F GSP-1, GSP-2 and eight MPI-DING reference glasses. Geostand. Newsl. 27, 173-179.

Schuessler, J.A., Schoenberg, R., Sigmarsson, O., 2009. Iron and lithium isotope systematics of the Hekla volcano, Iceland: evidence for Fe isotope fractionation during magma differentiation. Chem. Geol. 258, 78-91.

Shen, B., Jacobsen, B., Lee, C.T.A., Yin, Q.Z., Morton, D.M., 2009. The Mg isotopic systematics of granitoids in continental arcs and implications for the role of chemical weathering in crust formation. P. Natl. Acad. Sci. USA 106, 20652-20657.

Sio, C.K.I., Dauphas, N., Teng, F.Z., Chaussidon, M., Helz, R.T., Roskosz, M., 2013. Discerning crystal growth from diffusion profiles in zoned olivine by in situ Mg-Fe isotopic analyses. Geochim. Cosmochim. Acta 123, 302-321.

Sobolev, A.V., Hofmann, A.W., Kuzmin, D.V., Yaxley, G.M., Arndt, N.T., Chung, S.L., Danyushevsky, L.V., Elliott, T., Frey, F.A., Garcia, M.O., Gurenko, A.A., Kamenetsky, V.S., Kerr, A.C., Krivolutskaya, N.A., Matvienkov, V.V., Nikogosian, I.K., Rocholl, A., Sigurdsson, I.A., Sushchevskaya, N.M., Teklay, M., 2007. The amount of recycled crust in sources of mantle-derived melts. Science 316, 412-417.

Sossi, P.A., Foden, J.D., Halverson, G.P., 2012. Redox-controlled iron isotope fractionation during magmatic differentiation: an example from the Red Hill intrusion, S. Tasmania. Contrib. Mineral. Petrol. 164, 757-772.

Sun, C.Q., Plunkett, G., Liu, J.Q., Zhao, H., Sigl, M., McConnell, J.R., Pilcher, J.R., Vinther, B., Steffensen, J.P., Hall, V., 2014. Ash from Changbaishan Millennium eruption recorded in Greenland ice: implications for determining the eruption's timing and impact. Geophys. Res. Lett. 41, 694-701.

Telus, M., Dauphas, N., Moynier, F., Tissot, F.L.H., Teng, F.Z., Nabelek, P.I., Craddock, P.R., Groat, L.A., 2012. Iron, zinc, magnesium and uranium isotopic fractionation during continental crust differentiation: the tale from migmatites, granitoids, and pegmatites. Geochim. Cosmochim. Acta 97, 247-265.

Teng, F.Z., 2017. Magnesium isotope geochemistry. Rev. Mineral. Geochem. 82, 219-287.

Teng, F.Z., Yang, W., 2014. Comparison of factors affecting the accuracy of high-precision magnesium isotope analysis by multi-collector inductively coupled plasma mass spectrometry. Rapid Commun. Mass. Spectrom. 28, 19-24.

Teng, F.Z., Dauphas, N., Helz, R.T., 2008. Iron isotope fractionation during magmatic differentiation in Kilauea Iki Lava Lake. Science 320, 1620-1622.

Teng, F.Z., Li, W.Y., Rudnick, R.L., Gardner, L.R., 2010a. Contrasting lithium and magnesium isotope fractionation during continental weathering. Earth Planet. Sci. Lett. 300, 63-71.

Teng, F.Z., Li, W.Y., Ke, S., Marty, B., Dauphas, N., Huang, S.C., Wu, F.Y., Pourmand, A., 2010b. Magnesium isotopic composition of the Earth and chondrites. Geochim. Cosmochim. Acta 74, 4150-4166.

Teng, F.Z., Dauphas, N, Helz, R.T., Gao, S, Huang, S.C., 2011. Diffusion-driven magnesium and iron isotope fractionation in Hawaiian olivine. Earth Planet. Sci. Lett. 308, 317-324.

Teng, F.Z., Dauphas, N., Huang, S.C., Marty, B., 2013. Iron isotopic systematics of oceanic basalts. Geochim. Cosmochim. Acta 107, 12-26.

Teng, F.Z., Li, W.Y., Ke, S., Yang, W., Liu, S.A., Sedaghatpour, F., Wang, S.J., Huang, K.J., Hu, Y., Ling, M.X., Xiao, Y., Liu, X.M., Li, X.W., Gu, H.O., Sio, C.K., Wallace, D.A., Su, B.X., Zhao, L., Chamberlin, J., Harrington, M., Brewer, A., 2015. Magnesium isotopic compositions of international geological reference materials. geostand. Geoanal. Res. 39, 329-339.

Thomson, A.R., Walter, M.J., Kohn, S.C., 2016. Slab melting as a barrier to deep carbon subduction. Nature 529, 76-79.

Tian, H.C., Yang, W., Li, S.G., Wei, H.Q., Yao, Z.S., Ke, S., 2019. Approach to trace hidden paleo-weathering of basaltic crust through decoupled Mg-Sr and Nd isotopes recorded in volcanic rocks. Chem. Geol. 509, 234-248.

Tian, H.C., Zhang, C., Teng, F.Z., Long, Y.J., Li, S.G., He, Y., Ke, S., Chen, X.Y., Yang, W., 2020. Diffusion-driven extreme Mg and Fe isotope fractionation in Panzhihua ilmenite: Implications for the origin of mafic intrusion. Geochim. Cosmochim. Acta 278, 361-375.

McCulloch ,M.T., Wasserburg, G.J., 1978. Sm–Nd and Rb–Sr chronology of continental crust formation. Science 200, 1003-1011.

Wei, H.Q., Wang, Y., Jin, J.Y., Gao, L., Yun, S.Y., Jin, B.L., 2007. Timescale and evolution of the intracontinental Tianchi volcanic shield and ignimbrite-forming eruption, Changbaishan, Northeast China. Lithos 96, 315-324.

Wei, H.Q., Liu, G.M., Gill, J., 2013. Review of eruptive activity at Tianchi volcano, Changbaishan, northeast China: implications for possible future eruptions. B. Volcanol. 75, 706.

Weyer, S., Ionov, D.A., 2007. Partial melting and melt percolation in the mantle: The message from Fe isotopes. Earth Planet. Sci. Lett. 259, 119-133.

Williams, H.M., Bizimis, M., 2014. Iron isotope tracing of mantle heterogeneity within the source regions of oceanic basalts. Earth Planet. Sci. Lett. 404, 396-407.

Wu, H., He, Y., Teng, F.-Z., Ke, S., Hou, Z., Li, S., 2018. Diffusion-driven magnesium and iron isotope fractionation at a gabbro-granite boundary. Geochim. Cosmochim. Acta 222, 671-684.

Xia, Y., Li, S., Huang, F., 2017. Iron and Zinc isotope fractionation during magmatism in the continental crust: evidence from bimodal volcanic rocks from Hailar basin, NE China. Geochim. Cosmochim. Acta 213, 35-46.

Xu, J.D., Pan, B., Liu, T.Z., Hajdas, I., Zhao, B., Yu, H., Liu, R., Zhao, P., 2013. Climatic impact of the Millennium eruption of Changbaishan volcano in China: new insights from high-precision radiocarbon wiggle-match dating. Geophys. Res. Lett. 40, 54-59.

Zambardi, T., Lundstrom, C.C., Li, X.X., McCurry, M., 2014. Fe and Si isotope variations at Cedar Butte volcano; insight into magmatic differentiation. Earth Planet. Sci. Lett. 405, 169-179.

Zhang, M.L., Guo, Z.F., Liu, J.Q., Liu, G.M., Zhang, L.H., Lei, M., Zhao, W.B., Ma, L., Sepe, V., Ventura, G., 2018. The intraplate Changbaishan volcanic field (China/North Korea): a review on eruptive history, magma genesis, geodynamic significance, recent dynamics and potential hazards. Earth Sci. Rev. 187, 19-52.

Zou, H.B., Fan, Q.C., Yao, Y.P., 2008. U-Th systematics of dispersed young volcanoes in NE China: Asthenosphere upwelling caused by piling up and upward thickening of stagnant Pacific slab. Chem. Geol. 255, 134-142.
Zou, H.B., Fan, Q.C., Zhang, H.F., 2010. Rapid development of the great Millennium eruption of Changbaishan (Tianchi) Volcano, China/North Korea: evidence from U-Th zircon dating. Lithos 119, 289-296.

Zinc isotopic behavior of mafic rocks during continental deep subduction*

Li-Juan Xu, Sheng-Ao Liu and Shuguang Li

State Key Laboratory of Geological Processes and Mineral Resources, School of Scientific Research, China University of Geosciences, Beijing 100083, China

亮点介绍： 对华南陆块北缘和大别-苏鲁造山带陆壳俯冲形成的不同变质程度的基性变质岩，如武当群的绿片岩，北淮阳的斜长角闪岩和中大别-北大别的榴辉岩的 Zn 同位素研究发现绿片岩、角闪岩和榴辉岩具有相似的 Zn 同位素组成，且与全球玄武岩在误差范围内一致，表明俯冲玄武质火成岩的进变质脱水过程不会导致明显的 Zn 同位素分馏。

Abstract Zinc isotopes may act as a new tool of tracking recycling of crustal materials that causes compositional heterogeneity of the mantle. This application relies on an investigation of Zn isotopic behaviors during slab subduction. In this study, we report Zn isotopic compositions for a suite of metabasalts (greenschists, amphibolites, and coesite-bearing eclogites) from the Dabie Orogen (China), which were formed via the subduction of mafic rocks into different depths and up to > 200 km. Three out of eight greenschists are characterized by lighter $\delta^{66}Zn_{JMC\text{-}Lyon}$ (0.10‰–0.16‰) than those of global basalts (0.28‰±0.05‰), which may be caused by crustal assimilation of the protoliths by sedimentary rocks due to their extremely high $^{87}Sr/^{86}Sr$ (up to 0.7130) and low ε_{Nd} values (down to −12.3). The remaining greenschists have relatively low $^{87}Sr/^{86}Sr$ and their $\delta^{66}Zn$ values (0.21‰–0.38‰) overlap the ranges of amphibolites (0.18‰–0.32‰) and coesite-bearing eclogites (0.18‰–0.36‰). There is no correlation between $\delta^{66}Zn$ and sensitive indicators of dehydration (Rb/TiO_2, Ba/Yb, and H_2O^+), suggesting that no detectable Zn isotope fractionation has occurred during the deep subduction of mafic rocks even into > 200 km, which is attributed to the limited loss of Zn during prograde metamorphism and dehydration. Thus, Zn isotopic compositions of the deeply subducted mafic rocks are inherited from their protoliths. Considering that these metamorphosed rocks have higher $\delta^{66}Zn$ than that of the mantle value by up to 0.2‰, the recycled/subducted mafic crust can incorporate isotopically heavy Zn into the mantle. The subducted slabs may partially melt and generate a metasomatized mantle, resulting in changes of Zn isotopic composition of the hybridized mantle as have been observed in some mantle xenoliths and basaltic lavas.

Keywords Zn isotopes, metamorphosed rocks, dehydration, crustal deep subduction

1 Introduction

Subduction zones are key regions for recycling of surface materials from the crust into the mantle, and for a

* 本文发表在：Geoscience Frontiers, 2021, 5: 182-191

series of geological processes that lead to new crustal growth (Turcotte and Schubert, 2014). In recent years, zinc stable isotopes have been developed as a new tool of identifying recycled crustal materials in the mantle sources of basaltic lavas. That is, the mantle has a relatively homogeneous Zn isotopic composition based on analysis of mantle peridotites and ultramafic igneous rocks ($\delta^{66}Zn_{JMC\text{-}Lyon}$=0.18‰±0.05‰; Wang et al., 2017; McCoy-West et al., 2018; Sossi et al., 2018; Liu et al., 2019), and Zn isotope fractionation during magmatic differentiation is limited (<0.1‰) (Chen et al., 2013; Wang et al., 2017). By contrast, Zn isotopic compositions of crustal components (e.g., sediment, carbonate, altered oceanic crust and abyssal peridotite) are highly heterogeneous, with $\delta^{66}Zn$ values varying widely from −0.15‰ to +1.43‰ (Pichat et al., 2003; Bentahila et al., 2008; Pons et al., 2011; Little et al., 2014, 2016; Huang et al., 2016; Liu et al., 2017, 2019). These differences suggest that Zn isotopes could be a useful tool of tracing the potential contribution of recycled crustal components to the mantle via subduction (Liu et al., 2016; Pons et al., 2016; Liu and Li, 2019; Beunon et al., 2020). A recent study also proposed that Zn isotopes may identify recycled oceanic crust in the sources of large igneous provinces (Yang and Liu, 2019), given the detectable Zn isotopic difference between altered/un-altered oceanic crust (0.19‰–0.55‰) (Huang et al., 2016; Wang et al., 2017) and the terrestrial mantle.

A primary assumption for the application of Zn isotopes to trace recycled crustal materials is that Zn isotope fractionation during meta-morphic dehydration processes in subduction zones is negligible. To our knowledge, three studies have investigated the behavior of Zn isotopes in subduction zones, two on ultramafic rocks and one on metabasalts. Pons et al. (2016) documented a gradual decrease in $\delta^{66}Zn$ (0.32‰±0.08‰ to 0.16‰±0.06‰) with increasing metamor-phic grade in the subducted Alpine serpentinites, and suggested that the heavy, serpentinite-derived Zn isotopic signatures may be recorded in the mantle wedge. Debret et al. (2018) suggested that Zn isotope fractionation occurs at several stages of subduction via carbonate precipitation and dissolution at metasomatic interfaces between metasedimentary and ultramafic rocks in the slab. However, the MORB-like Zn isotopic compositions of arc basalts (Huang et al., 2018a) suggest that the amount of isotopically heavy Zn released at early stages of subduction—if any—is probably small. Inglis et al. (2017) found limited Zn isotopic variation in metagabbros and metabasalts (greenschist- to eclogite-facies) representing mafic oceanic crust that had experienced increasing prograde metamorphism in a subduction zone. They suggested that relatively a small amount of Zn have been mobilized from these rocks in S- and C-bearing fluids. Notably, these metagabbros and metabasalts recorded a peak metamorphic pressure of less than 90 km (Bucher et al., 2005; Bucher and Grapes, 2009). Given that many alkaline basaltic lavas (e.g., intraplate basalts, OIBs) originated from the asthenospheric mantle or even the lower mantle (Keith, 1999; Deschamps et al., 2011), whether or not the deeply (e.g., >90 km) subducted crust can preserve its primary Zn isotopic signature is a key for improving the application of Zn isotopes. In other words, it remains unknown regarding whether or not Zn isotopes are significantly fractionated during deep subduction of mafic rocks into the mantle, even up to >200 km.

One approach to characterize the behavior of Zn isotopes during crustal deep subduction is to analyze rocks with different degrees of metamorphism. In this study, high-precision Zn isotope data for a suite of well-characterized metamorphosed rocks from the Dabie Orogen, central China are reported. These rocks have undergone various degrees of metamorphism from greenschist-facies through amphibolite-facies to eclogite-facies. In particular, the Dabie eclogites are coesite-bearing and represent the products of deeply subducted continental crust to an ultra-depth of > 200 km (Ames et al., 1996; Cong, 1996; Ye et al., 2000). The new Zn isotope data, together with the Sr-Nd isotopes and other existing elemental data, allow us to evaluate the extent of Zn isotope fractionation in subduction zones, and then to assess the influence of recycled crust on the Zn isotopic heterogeneity of the mantle and derivative igneous rocks.

2 Geological setting and samples

The Dabie-Sulu Orogen in east-central China was formed by the Triassic continental collision related to subduction of the South China Craton (SCC) beneath the North China Craton (NCC) (e.g., Li et al., 1993, 2000; Ames et al., 1996; Cong, 1996) (Fig. 1a). The Dabie Orogen is divided into five litho-tectonic units based on petrology, mineralogy, geochemistry, and *P-T-t* evolution (e.g., Xu et al., 1992; Li et al., 2001). From its north to south (Fig. 1), it includes the Beihuaiyang greenschist-amphibolite-facies terrane (BHY), the northern high-T/ultrahigh-pressure (UHP) Dabie terrane (NDT), the central mid-T/UHP Dabie terrane (CDT), the southern low-T/UHP Dabie terrane (SDT), and the Susong-Hongan metamorphic terrane (SHT).

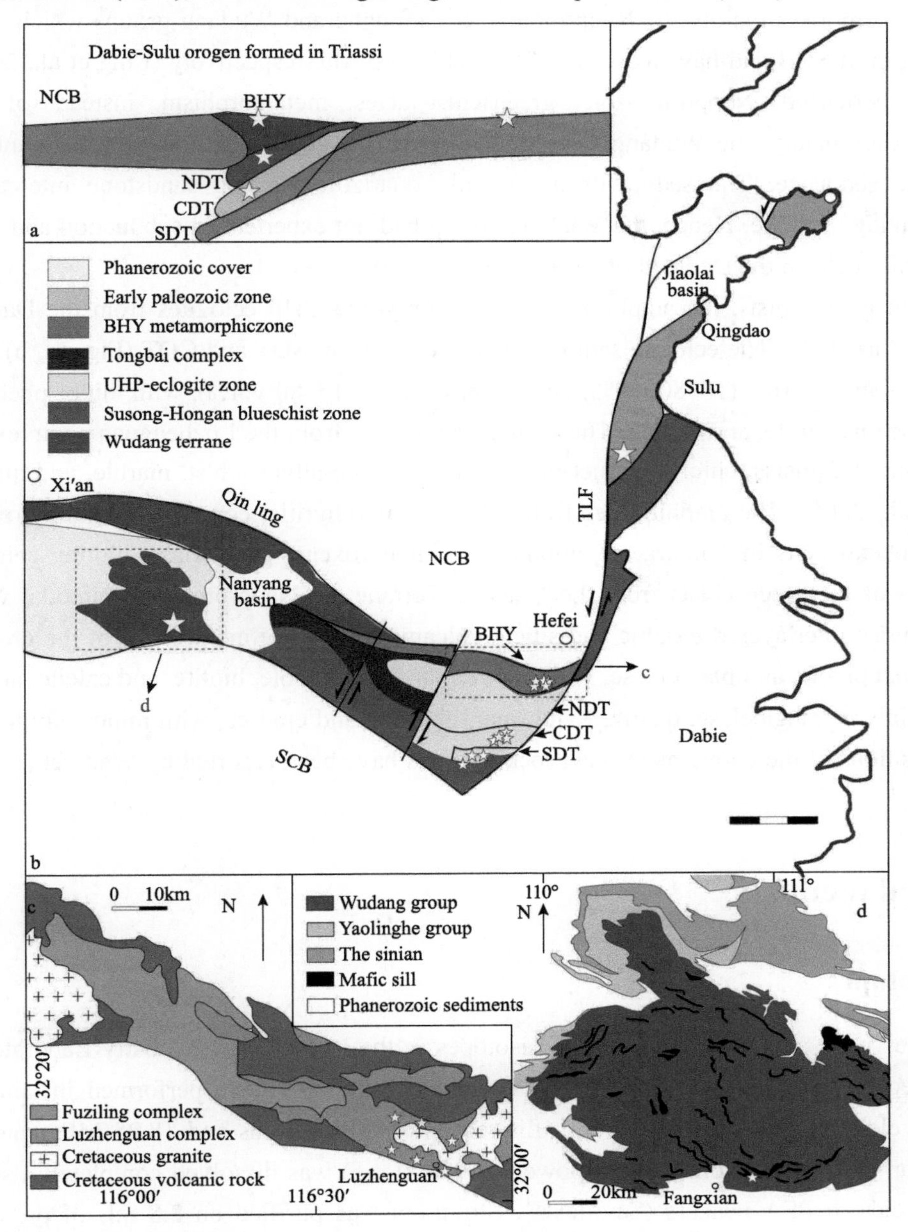

Fig. 1 (a, b) Simplified geological map of the Dabie-Sulu Orogen in China modified after Wang et al. (2014) and Zheng et al. (2005). (c) Beihuaiyang metamorphic zone ineastern China. (d) Sketch geological map of the Wudangshan area. NCB=North China Block; SCB=South China Block; BHY = Beihuaiyang greenschist-amphibolite-facies terrane; SDT=southern low-T/UHP Dabie Terrane; CDT = central mid-T/UHP Dabie Terrane; NDT=northern high-T/UHP Dabie Terrane.

Previous studies documented that eclogites from the Dabie Orogen have been subducted into >200 km (Ye et al., 2000). Coesite and diamond inclusions have also been found in these eclogites (Ames et al., 1996; Cong, 1996; Ye et al., 2000). The peak eclogite-facies metamorphism temperature is ~650 ℃ and ~750 ℃ for SDT and CDT, respectively (e.g., Okay et al., 1989; Xiao et al., 2000; Li et al., 2004). The rocks in the BHY have experienced low-grade greenschist- to amphibolite-facies metamorphism during the Triassic (Li et al., 2001; Zheng et al., 2005), represented by the Fuziling and Luzhenguan complexes. The Fuziling complex is a metaclastic flysch formation that was overprinted by greenschist-facies metamorphism (Li et al., 2001; Zheng et al., 2005). The amphibolites and eclogites share common Neoproterozoic protolith ages (mainly 830–740 Ma) associated with Neoproterozoic rift magmatism in SCB (Zheng et al., 2003, 2005; Wu et al., 2007). The Wudang Terrane (Fig. 1d) comprises mainly the Neoproterozoic Yaolinghe and Wudang groups which are located along the northern margin of SCB and have ages of 685±5 and 755±5 Ma, respectively (Ling et al., 2008, 2010). Both groups have experienced Neoproterozoic greenschist-facies metamorphism instead of Triassic UHP metamorphism. Additionally, the Wudang Group consists of fine-grained sedimentary beds intercalated with a bimodal volcanic sequence. The sediments are mainly quartzofeldspathic sandstone intercalated with silty mudstone and muddy siltstone. Hence, the Wudang Group had not experienced subduction and it is the protolith of the metamorphic rocks in the Dabie Orogen (Ling et al., 2010).

Totally, eight greenschists, ten amphibolites, and twenty-one UHP eclogites from the Dabie Orogen have been analyzed in this study. The eclogite samples analyzed are from SDT and CDT (Fig. 1a, b). The minerals in the eclogites are mainly garnet (10–80 vol%) and clinopyroxene (15–80 vol%), with minor phengite, amphibole, epidote, quartz, apatite, rutile, and calcite. These amphibolites are from the Luzhenguan complex (Fig. 1c) which comprises amphibolite, gneiss, chlorite-garnet-mica schist, chlorite-albite schist, marble, and quartzite (Li et al., 2001; Zheng et al., 2005). The amphibolites from BHY are porphyritic, comprising clinopyroxene, amphibole, and plagioclase phenocrysts in a matrix of amphibole, clinopyroxene, plagioclase, biotite, epidote, quartz, and calcite. Protoliths of the greenschists from the Wudang Terrane are Neoproterozoic bimodal volcanic rocks in SCB which comprise interlayered basaltic and silicic volcanic rocks. The metabasites of the greenschists consist mainly of chlorite, epidote, and plagioclase, with minor quartz, amphibole, biotite, and calcite, and the metasilicic rocks consist mainly of plagioclase, quartz, K-feldspar, chlorite, and epidote, with minor biotite and calcite. The chemical compositions of the above-mentioned rock samples have been reported by Wang et al. (2014) and Shen et al. (2015).

3 Analytical methods

3.1 Sr-Nd isotopes

Rock samples were analyzed for Sr and Nd isotopes in the Isotope Geochemistry Laboratory of the China University of Geosciences, Beijing (CUGB). All chemical procedures were performed in laminar flow hoods (Class 100) in a clean room (Class 1000) with filtered air. Distilled acids and 18.25 MΩ water were used for sample dissolution and purification. Sample powder (50–100 mg) was dissolved completely using a mixture of HF-HNO_3-HCl acids in PFA beakers (Savillex®). Strontium was purified on 2.8 mL of pre-cleaned Bio-Rad cation resin AG50W-X12 (200–400 mesh chloride form; USA) by eluting 11 mL of 4 N HCl, and Nd was purified on 2 mL of LN resin by eluting 4 mL of 0.18 N HCl after being recovered from the AG50W-X12 resin. The purified solutions were evaporated to dryness and then dissolved in 3% HNO_3 for isotopic analysis on the

Thermo-Finnigan Neptune Plus multi-collector-inductively coupled plasma-mass spectrometer (MC-ICP-MS). The Sr and Nd isotopes were corrected to $^{86}Sr/^{88}Sr$ = 0.1194 and $^{146}Nd/^{144}Nd$ = 0.7219 for instrumental mass bias correction, respectively. Alfa Aesar Nd was analyzed as an in-house reference standard. In the past several years, the long-term measured value for this standard is $^{143}Nd/^{144}Nd$=0.512425±30 on average. The basalt standard BHVO-2 yielded $^{87}Sr/^{86}Sr$ = 0.703518±37 (2SD; N = 4) over a period of two years. These values are identical within uncertainties with those of Beire et al. (2015) (0.703490±52, 2SD) and DuFrane et al. (2009) (0.703496± 36, 2SD). The long-term measured value for BHVO-2 is $^{143}Nd/^{144}Nd$=0.512974±28 (2SD; N=25), which is identical with that of Li et al. (2007) (0.512984±11; 2SD) and Wiesmaier et al. (2011) (0.512980±50, 2SD).

3.2 Zn isotopes

Zinc isotopic compositions were analyzed at the Isotope Geochemistry Laboratory of CUGB, following the procedures reported by Liu et al. (2016) and Lv et al. (2016). Rock sample powder containing >0.4 μg Zn was completely dissolved using double-distilled HF-HNO_3-HCl acids in Savillex beakers. After evaporation, the residue was dissolved in 1 mL of 8 N HCl. Chromatographic separation of Zn was achieved by using pre-cleaned AG-MP-1M resin (Bio-Rad strong anion resin). The matrix was removed by eluting 31 mL of 8 N HCl and 10 mL of 2 N HCl. Then, separation of Zn was achieved in the final 10 mL of 0.5 N HNO_3. In this process, the yield of Zn is near to 100% and the total procedural blank for Zn is <6 ng which is negligible compared with the total amount of sample Zn (>400 ng). Finally, the purified Zn was analyzed on the MC-ICP-MS at a low-resolution mode. The sample–standard bracketing method was used for instrumental mass bias correction. The Zn standard JMC Lyon 3-0749L and in-house mono-element standard solution GSB Zn were analyzed. Zinc isotopic data are reported in the standard δ-notation in per mil relative to the Zn standard JMC Lyon 3-0749L:

$$\delta^{66}\text{Zn or } \delta^{68}\text{Zn} = ((^{66,68}\text{Zn}/^{64}\text{Zn})_{\text{sample}} / (^{66,68}\text{Zn}/^{64}\text{Zn})_{\text{JMC 3-0749L}} - 1) \times 1000$$

Two international basalt standards (BHVO-2 and BCR-2), which are compositionally close to the studied metamorphic rocks (metabasalts), were processed through the column chemistry along with the samples to assess the accuracy of data (Table 1). The obtained values ($\delta^{66}Zn_{BHVO-2}$=0.31±0.05‰, 2SD, n=3; $\delta^{66}Zn_{BCR-2}$ =0.27± 0.05‰, 2SD, n=3) are identical within uncertainty with the recommended values ($\delta^{66}Zn_{BHVO-2}$ = 0.27±0.05‰ to 0.33±0.01‰; $\delta^{66}Zn_{BCR-2}$ =0.25±0.01‰ to 0.27±0.03‰; Chen et al., 2013; Sossi et al., 2015; Liu et al., 2016; Wang et al., 2017; Huang et al., 2018a, b).

Table 1 δ^{66}Zn (‰), $^{87}Sr/^{86}Sr_{(i)}$ (t = 220 Ma) and $\varepsilon_{Nd}(t)$ (t = 220 Ma) data for metamorphosed rocks of different metamorphic grades from the Dabie Orogen

Samples	δ^{66}Zn	2SD	δ^{68}Zn	2SD	$^{87}Sr/^{86}Sr_{(i)}$	$^{143}Nd/^{144}Nd_{(i)}$	$\varepsilon_{Nd}(t)$
Greenschists from Wudang Terrane							
HWD-11	0.38	0.05	0.75	0.10	0.70597		
HWD-23	0.29	0.05	0.57	0.10			
HWD-9	0.30	0.05	0.59	0.10	0.70520	0.51218	−3.35
HWD-18	0.16	0.05	0.31	0.10	0.70744	0.51175	−11.8
HWD-21	0.26	0.05	0.51	0.10	0.70631		
HWD-13	0.21	0.05	0.41	0.10		0.51174	−12.1
HWD-26	0.10	0.05	0.19	0.10			
HWD-5	0.12	0.05	0.24	0.10	0.71299	0.51173	−12.3

Continued

Samples	δ^{66}Zn	2SD	δ^{68}Zn	2SD	^{87}Sr/^{86}Sr$_{(i)}$	^{143}Nd/^{144}Nd$_{(i)}$	ε_{Nd} (t)
Amphibolites from BHY							
99SQ-5	0.18	0.05	0.36	0.10	0.70813	0.51150	−16.6
99SQ-6	0.23	0.05	0.46	0.10	0.70811	0.51151	−16.5
99SQ-1	0.24	0.05	0.48	0.10	0.70890	0.51147	−17.2
99SQ-3	0.26	0.05	0.51	0.10	0.70879	0.51148	−17.1
99SQ-2	0.30	0.05	0.60	0.10	0.70648	0.51148	−17.0
99SQ-4	0.32	0.05	0.62	0.10		0.51168	−13.1
99CHC-4	0.24	0.05	0.48	0.10	0.70704	0.51180	−10.9
99CHC-6	0.21	0.05	0.41	0.10		0.51166	−13.5
99CHC-5	0.22	0.05	0.43	0.10		0.51173	−12.2
99CHC-2	0.25	0.05	0.49	0.10	0.70718		
Eclogites from CDT and NDT							
11HZ-4	0.25	0.05	0.50	0.10	0.70631		
92HT-5	0.23	0.05	0.44	0.10	0.71016	0.51244	1.80
92HT-2	0.22	0.05	0.43	0.10	0.71020		
11HZ-5	0.22	0.05	0.43	0.10	0.70584		
11HZ-2	0.21	0.05	0.41	0.10			
11HZ-6	0.18	0.05	0.36	0.10			
11HZ-6	0.24	0.05	0.48	0.10		0.51150	−16.7
99WM-3	0.23	0.06	0.45	0.12	0.70494	0.51215	−3.9
11SH-17	0.28	0.05	0.55	0.10	0.70487	0.51201	−6.71
92HT-7-1	0.20	0.05	0.40	0.10	0.71014	0.51132	−20.1
11BXL-4	0.31	0.05	0.61	0.10	0.70435	0.51249	2.64
11SH-E1	0.27	0.05	0.54	0.10	0.70524	0.51145	−17.7
02QL-4	0.36	0.05	0.70	0.10			
11SM-1	0.26	0.05	0.51	0.10	0.70542	0.51182	−10.5
99QL-10	0.21	0.05	0.41	0.10			
09-SH-7-3	0.22	0.05	0.44	0.10		0.51133	−20.0
09-SH-6-4	0.33	0.05	0.66	0.10		0.51149	−16.9
09-SH-6-3a	0.32	0.07	0.65	0.14		0.51167	−13.3
09-SH-7-1	0.23	0.05	0.46	0.10		0.51151	−16.5
09-SH-7-2	0.31	0.05	0.63	0.10		0.51182	−10.5
09-SH-6-3b	0.29	0.05	0.58	0.10		0.51123	−21.9
International Standards							
BHVO-2	0.31	0.05	0.61	0.10	0.70353	0.51297	
BCR-2	0.27	0.05	0.52	0.10	0.70503	0.51261	

4　Results

The zinc isotopic compositions of thirty-nine whole-rock samples and the Sr-Nd isotopic ratios of twenty-nine samples in this study are reported in Table 1. The bulk major and trace element data for these greenschists, amphibolites, and coesite-bearing eclogites (Fig. 2) are available in Wang et al. (2014) and Shen et al. (2015).

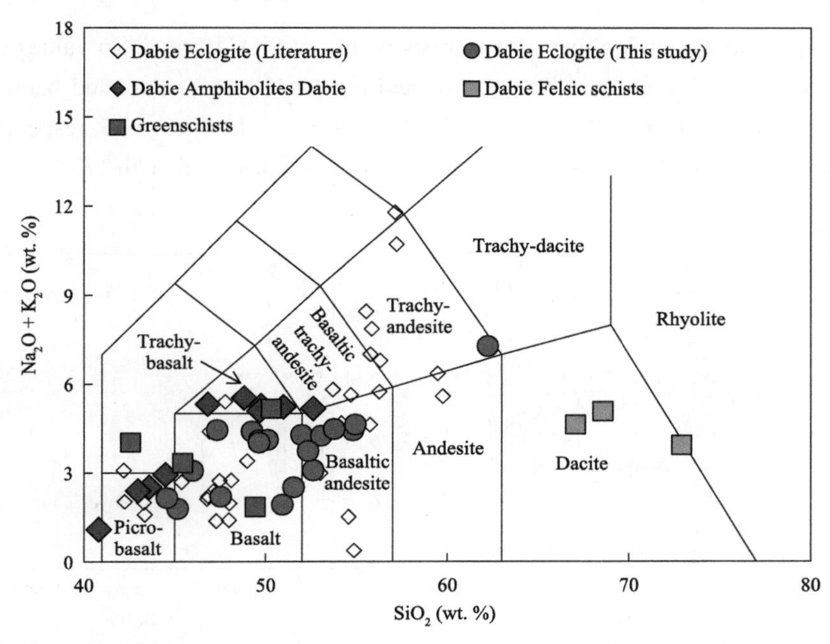

Fig. 2 The diagram of total alkalis (Na_2O + K_2O; wt.%) versus SiO_2 (wt.%) (Le Bas et al., 1986) for metamorphic rocks in this study. The literature data of eclogites are from Zhang et al. (2006) and Zhao et al. (2007).

4.1 Bimodal volcanic rocks from Wudang complex

Eight Neoproterozoic greenschists have δ^{66}Zn values varying widely from 0.10‰ to 0.38‰ (average = 0.23 ±0.19‰, 2SD, n=8), which are negatively correlated with Zn and MgO contents (Fig. 3a, b). The protoliths of these greenschists are bimodal volcanic rocks. The felsic schists possess relatively high δ^{66}Zn values ranging from

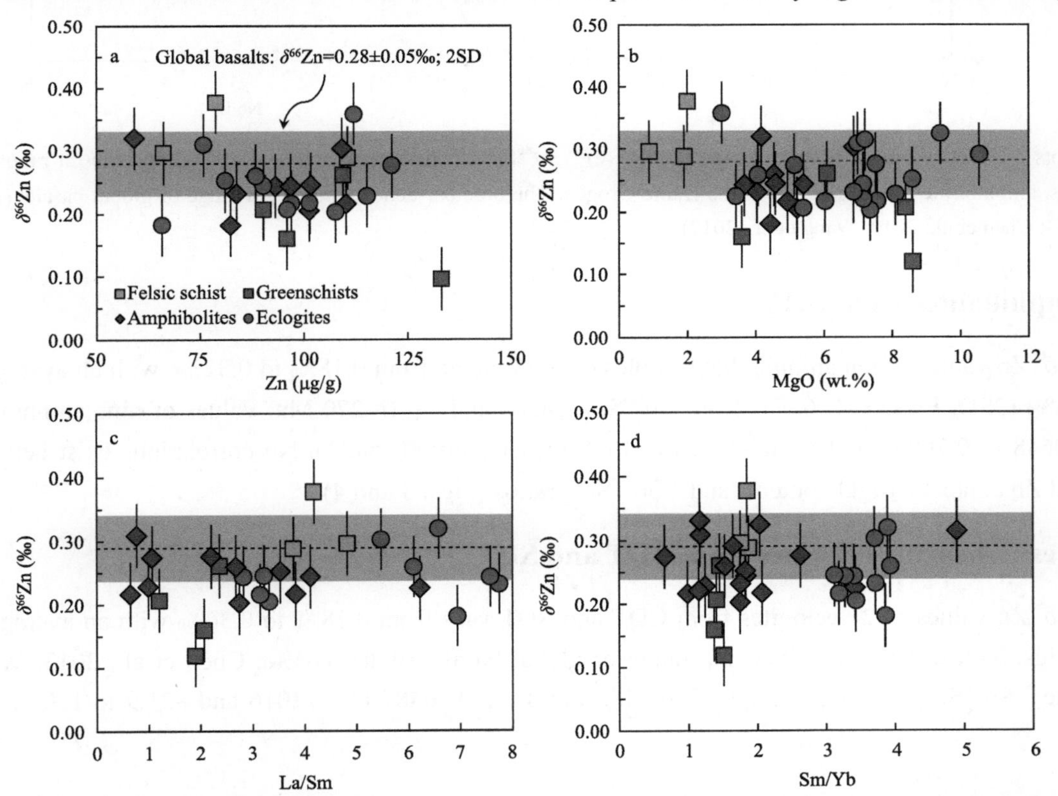

Fig. 3 Plots of δ^{66}Zn versus (a) Zn, (b) MgO, (c) La/Sm and (d) Sm/Yb for the greenschists, amphibolites, and eclogites of the Dabie Orogen. The blue bar denotes the δ^{66}Zn range of global basalts (δ^{66}Zn=0.28±0.05‰, 2SD; Chen et al., 2013; Wang et al., 2017).

0.29‰ to 0.38‰ (Figs. 2 and 3), while other greenschists with low-SiO_2 have $\delta^{66}Zn$ values ranging from 0.10‰ to 0.26‰ (Figs. 2 and 3). The initial $^{87}Sr/^{86}Sr_{(i)}$ ratios and ε_{Nd} (t=220 Ma) calculated back to 220 Ma of these Neoproterozoic greenschists range from 0.70520 to 0.71299 and from –12.3 to –3.35, respectively (Fig. 4c; Table 1). Additionally, $\delta^{66}Zn$ values of these greenschists are negatively correlated with $^{87}Sr/^{86}Sr_{(i)}$ (t=220 Ma) ratios (Fig. 4c).

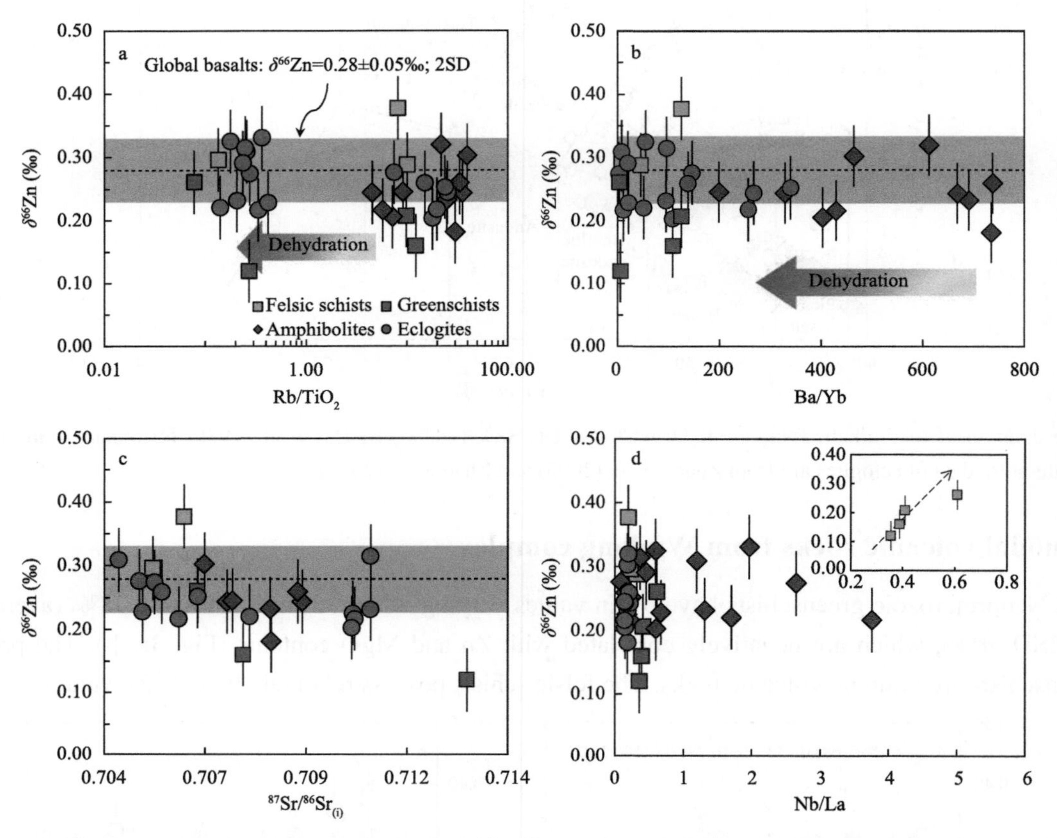

Fig. 4　Plots of $\delta^{66}Zn$ versus (a) Rb/TiO_2, (b) Ba/Yb, (c) $^{87}Sr/^{86}Sr_{(i)}$ (calculated back to 220 Ma) and Nb/La ratios (d) for the greenschists, amphibolites, and eclogites of the Dabie Orogen. The blue bar denotes the $\delta^{66}Zn$ range of global basalts ($\delta^{66}Zn$=0.28± 0.05‰; 2SD; Chen et al., 2013; Wang et al., 2017).

4.2　Amphibolites from BHY

The $\delta^{66}Zn$ values of ten amphibolite samples of BHY range from 0.18‰ to 0.32‰, with an average value of 0.25±0.08‰ (2SD; Figs. 3–4, 6, 7). The $^{87}Sr/^{86}Sr_{(i)}$ ratios and ε_{Nd} (t=220 Ma) values of nine amphibolites vary from 0.70648 to 0.70890 and from –17.2 to –13.2, respectively (Table 1). No correlations exist between $\delta^{66}Zn$ values and Zn contents, MgO contents and $^{87}Sr/^{86}Sr_{(i)}$ ratios (Figs. 3 and 4).

4.3　Coesite-bearing eclogites from CDT and SDT

The $\delta^{66}Zn$ values of 21 eclogites from CDT and SDT vary from 0.18‰ to 0.36‰ with an average of 0.26± 0.10‰ (Figs. 3–4, 6, 7), falling into the range of global basalts (0.28±0.05‰; Chen et al., 2013; Wang et al., 2017). The $^{87}Sr/^{86}Sr_{(i)}$ ratios and ε_{Nd} (t=220 Ma) values are 0.70487 to 0.71016 and –21.9 to 1.75, respectively (Table 1).

5 Discussion

Here we first access the factors causing Zn isotopic variations in the greenschists. Then, we apply new Zn isotope data of the greenschists, amphibolites, and coesite-bearing eclogites to discuss the behaviors of Zn isotopes during metamorphism and dehydration. Finally, this study would discuss the implications of these subducted metamorphic rocks for Zn isotopic heterogeneity of the mantle.

5.1 Zinc isotopic systematics of bimodal volcanic rocks from BHY

The BHY greenschists display a larger range of δ^{66}Zn (0.10‰–0.38‰) compared with that of the amphibolites (0.18‰–0.32‰) and coesite-bearing eclogites (0.18‰–0.36‰) (Fig. 3). Two possible causes for the large δ^{66}Zn range are (i) Zn isotope fractionation during greenschist-facies metamorphism and/or (ii) protolith heterogeneity, as indicated by the negative correlation between δ^{66}Zn and Zn contents in the greenschists (Fig. 3a). Given the low solubility of Ti and Yb relative to Rb and Ba in aqueous fluids (e.g., Kessel et al., 2005; Zack and John, 2007), the ratios of Rb/TiO_2 and Ba/Yb can be used as indicators of fluid-rock interaction and metamorphic dehydration. As shown in Fig. 4a and 4b, the δ^{66}Zn values of greenschists do not systemically vary with the geochemical indices of metamorphic dehydration (Rb/TiO_2; Ba/Yb). Although some of the greenschists have very low contents of Cs, K and Rb, their δ^{66}Zn values are similar to those of other greenschist samples (Fig. 4a, b). Hence, the Zn isotopic variations of these greenschists may not result from greenschist-facies metamorphic dehydration and should reflect protolith heterogeneity.

The protoliths of the greenschists are bimodal volcanic rocks with a large range of SiO_2 contents (42.6–72.9 wt.%) (Fig. 2). Three felsic schists display higher δ^{66}Zn values (0.29‰–0.38‰) relative to other mafic greenschists (0.10‰–0.26‰; Figs. 2 and 3) and the δ^{66}Zn values are negatively correlated with Zn and MgO contents (Fig. 3a, b), suggesting that magmatic differentiation of their protoliths led to Zn isotopic variations in the greenschists. In addition, these felsic schists have δ^{66}Zn values similar to those of granites of the similar SiO_2 (wt.%) contents (Xu et al., 2019). It has been widely reported that felsic igneous rocks have higher δ^{66}Zn compared with mafic igneous rocks, related to crustal melting, magmatic differentiation, and/or fluid exsolution (e.g., Telus et al., 2012; Doucet et al., 2018; Xu et al., 2019; Wang et al., 2020). Thus, the slightly high δ^{66}Zn of three BHY felsic schists may be inherited from their felsic protoliths.

Zinc isotope fractionation produced by closed-system magmatic differentiation is generally less than 0.1‰ (Chen et al., 2013; Moynier et al., 2017; Wang et al., 2017; Sossi et al., 2018; Huang et al., 2018a). However, three mafic greenschists have δ^{66}Zn values of 0.10‰–0.16‰ which are lower than those of global basalts (0.28± 0.05‰), and even slightly lower than that of the mantle (0.18±0.05‰; Wang et al., 2017; Sossi et al., 2018; Liu et al., 2019). Previous studies showed that sedimentary carbonates usually have high δ^{66}Zn of up to 1.34‰ (e.g., Pichat et al., 2003; Liu et al., 2016, 2017). The low δ^{66}Zn of these metabasites thus exclude significant influence of these samples by recycled sedimentary carbonates during the formation of their igneous protoliths. Notably, these mafic greenschists have extremely high $^{87}Sr/^{86}Sr_{(i)}$ (up to 0.7130) and low ε_{Nd}(t) (down to −12.3) (Fig. 4c, Table 1), suggesting that crustal contamination has occurred during the formation of their protoliths (e.g., Jahn, 1998; Li et al., 2000, 2004). The continental crust has lower Nb/La (~0.39) than that of the MORBs (Nb/La=1.04; Sun and McDonough, 1989; Rudnick and Gao, 2003). Thus, the low Nb/La ratios (Fig. 4d) of most of these metabasites indicate a strong crustal signature in the sources of their igneous protoliths (Wang et al., 2014b; Xia et al., 2012; Lu et al., 2019). These mafic greenschists (except the felsic greenschists) display positive correlations of δ^{66}Zn with Nb/La (Fig. 4d), further implying that the δ^{66}Zn variations may be related to crustal contamination of their protoliths. Studies on sedimentary lithologies in the Western Alps have shown that the metasedimentary rocks of different metamorphic grades typically have low δ^{66}Zn (0.00‰–0.13‰; Inglis et al., 2017). Some continental margin sediments even have δ^{66}Zn of down to –0.05‰ (Little et al., 2016). Abundant metasedimentary rocks are also present in the Dabie-Sulu Orogen (Li et al., 2001; Zheng et al., 2005). Thus, three mafic greenschists with low δ^{66}Zn may have been involved in crustal assimilation. After excluding the samples significantly affected by sedimentary components, the average δ^{66}Zn of the greenschists becomes 0.29±0.12‰ (2SD, *n*=5).

5.2 Possible causes for Zn isotopic heterogeneity of amphibolites and eclogites

The Wudang Group was overprinted by greenschist-facies meta-morphism and not experienced subduction. They could be considered as the protolith of the amphibolites and eclogites in the Dabie Orogen (Ling et al., 2010). The Zn isotopic compositions of amphibolites (0.18‰–0.32‰) and eclogites (0.18‰–0.36‰) are slightly heteroge-neous and may also mirror their protolith heterogeneity. Previous stud-ies proposed three possible mechanisms to explain the Zn isotopic heterogeneity of basalts: (1) partial melting; (2) fractional crystalliza-tion; (3) crustal contamination/source heterogeneity.

Partial melting would induce non-negligible Zn isotopic variations not only in the crust (Doucet et al., 2018; Xu et al., 2019) but also in the mantle (Doucet et al., 2016; Wang et al., 2017; Sossi et al., 2018; Huang et al., 2018a; Liu et al., 2019). Doucet et al. (2016) suggested that there was an up to 0.16‰ Zn isotope fractionation at high degrees of melt extraction in garnet-facies mantle partial melting. However, Wang et al., (2017) suggested that detectable Zn isotope fractionation (~0.1‰) occurred during spine-facies mantle partial melting. Recently, Sossi et al. (2018) proposed that the Zn isotope fractionation (~0.08‰) between mantle and melt is driven by the fractionation of $^{VI}Zn^{2+}$ and $^{IV}Zn^{2+}$ between them. Thus, partial melting would induce heavy Zn isotopic composition in the melt and light Zn isotopic composition in the residual. there is no correlation between δ^{66}Zn and MgO (wt.%), Zn (μg/g), La/Sm, Sm/Yb for the amphibolites and coesite-bearing eclogites in this study (Fig. 3), hinting that the δ^{66}Zn variations of these samples may not be caused by different degrees of partial melting of their igneous protoliths.

Crystallization of olivine or Fe-Ti oxides would increase the δ^{66}Zn of basalts, as Fe-Ti oxides and olivine phenocrysts are usually characterized by light δ^{66}Zn (Chen et al., 2013; Yang and Liu, 2019). Considering that zinc is an incompatible element and has a partition coefficient $DZn_{ol/melt}$ of 0.96 in olivine (Davis et al., 2013; Kohn and Schofield, 1994), fractional crystallization of olivine will lead to a negative relationship of Zn and MgO in the liquid. By contrast, accumulation of olivine will drive Zn and MgO in the liquid to an opposite direction. There are no correlations between Zn and MgO, Zn and δ^{66}Zn for the amphibolites and eclogites, arguing against significant olivine accumulation and/or fractional crystallization. In addition, there is a lack of correlation between CaO/Al_2O_3 or FeO_T and MgO or δ^{66}Zn (Fig. 5), indicating that crystallization of clinopyxene or Fe-Ti oxides has a negligible influence on the protoliths of these metamorphic rocks.

For the amphibolites and eclogites, $^{87}Sr/^{86}Sr_{(i)}$ gradually increase with a decrease of $\varepsilon_{Nd}(t)$ (Supplementary Data, Fig. S1) and are accompanied by elevating SiO_2 contents. This suggests that crustal contamination may have occurred for their igneous protoliths during magma ascending or recycled crustal materials have been involved into the mantle sources. However, the absence of relationship between δ^{66}Zn and $^{87}Sr/^{86}Sr_{(i)}$ (Fig. 4c) implies that crustal contamination only has a negligible influence on Zn isotopic systematics of amphibolites and eclogites. In addition, there is no correlation between δ^{66}Zn and Nb/La for eclogites, and these metamorphic rocks have δ^{66}Zn values similar to those of terrestrial basalts (Chen et al., 2013; Wang et al., 2017; Huang et al., 2018a; Liu et al., 2019). Thus, the δ^{66}Zn values of most metamorphic rocks (except three greenschists, HWD-5, 18, 26) are inherited from their igneous protoliths.

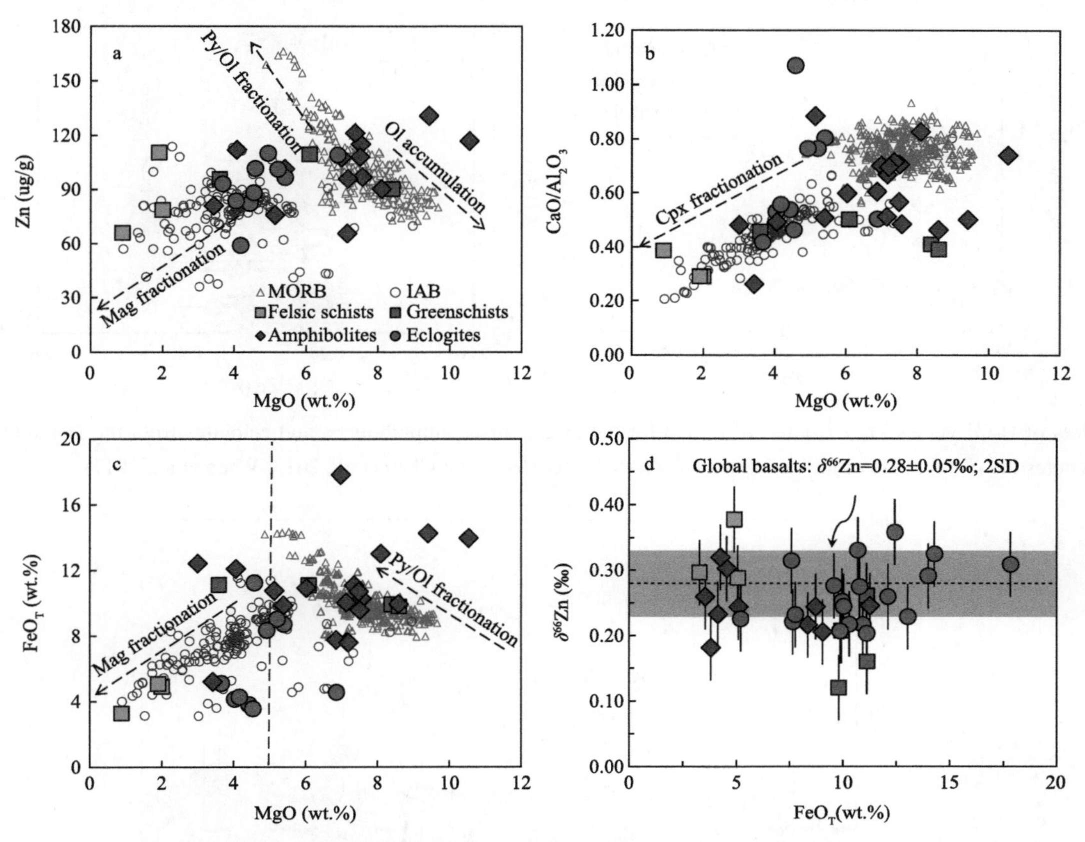

Fig. 5 Plots of (a) Zn, (b) CaO/Al_2O_3 and (c) FeO_T versus MgO for the MORBs, Arc-basalts, Dabie greenschists, Dabie amphibolites, and Dabie eclogites. (d) Plots of FeO_T versus δ^{66}Zn for the greenschists, amphibolites, and eclogites. The data of the MORBs, Arc-basalts are from GEOROC (http://georoc.mpch-mainz.gwdg.de/georoc/).

5.3 Behaviors of Zn isotopes during prograde and retrograde metamorphism

Here zinc isotope fractionation during prograde metamorphism and dehydration of subducted mafic crust is evaluated by comparing the Zn isotopic compositions of greenschists, amphibolites, and eclogites. The metamorphic dehydration processes typically release H_2O and volatiles (e.g., CO_2, S, F and Cl), as well as water-soluble elements, into the mantle wedge and overlying slabs (e.g., Kessel et al., 2005; Kelley and Cottrell, 2009; Debret et al., 2014). H_2O in these rocks would be progressively lost via dehydration with an increase of the metamorphic grade. Water contents (H_2O^+) in the Dabie rocks gradually decrease from greenschists through amphibolites to eclogites (Fig. 6a, b). However, the systematic decrease in H_2O contents is not associated with significant changes in Zn/Y ratios and Zn isotopic compositions (Fig. 6a, b). This suggests that there is limited Zn loss and Zn isotope fractionation associated with metamorphic dehydration of subducted mafic crust. Chlorite-epidote and biotite-amphibole groups are the major Zn- and H_2O-carriers in greenschist-and amphibolite-facies rocks, respectively (Spear and Cheney, 1989). It appears that Zn released from chlorite-epidote is completely retained in the newly formed minerals with increasing metamorphic grades, such as biotite and amphibole. During eclogite-facies metamorphism, Zn-bearing hydrous minerals are dehydrated, and the released Zn is further inherited by eclogitic minerals, such as garnet and clinopyroxene (Fig. 6a). This explains the similarity of Zn contents among greenschists, amphibolites, and coesite-bearing eclogites (Table 1; Fig. 3a). In addition, the δ^{66}Zn values of rocks of different metamorphic facies do not change with Rb/TiO_2 and Ba/Yb ratios (Fig. 4a, b). Thus, it is concluded that Zn isotope fractionation during continental subduction, even to the UHP eclogite-facies (>200 km; Ye et al., 2002), is negligible (Fig. 7).

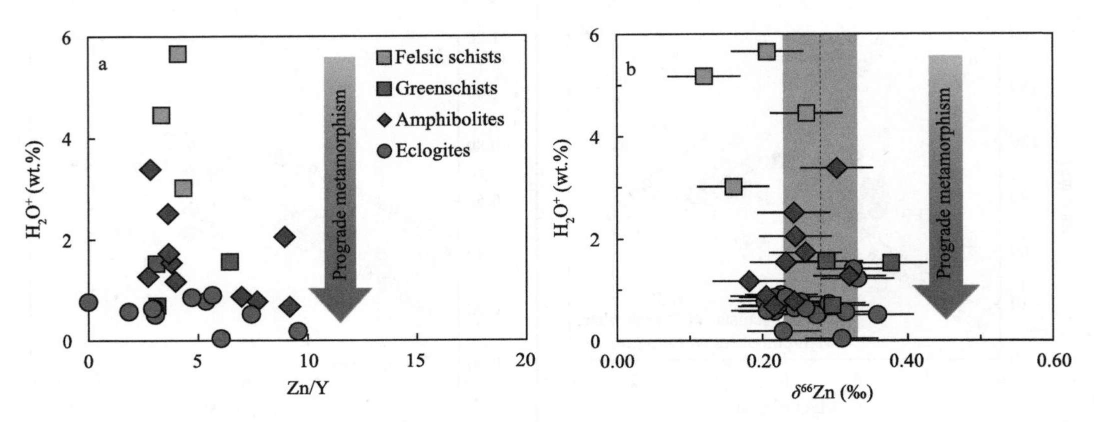

Fig. 6　Plots of H_2O^+ versus Zn/Y (a) and δ^{66}Zn (b) for the greenschists, amphibolites, and eclogites from the Dabie Orogen. The blue bar denotes the δ^{66}Zn range of global basalts (δ^{66}Zn=0.28±0.05‰, 2SD; Chen et al., 2013; Wang et al., 2017).

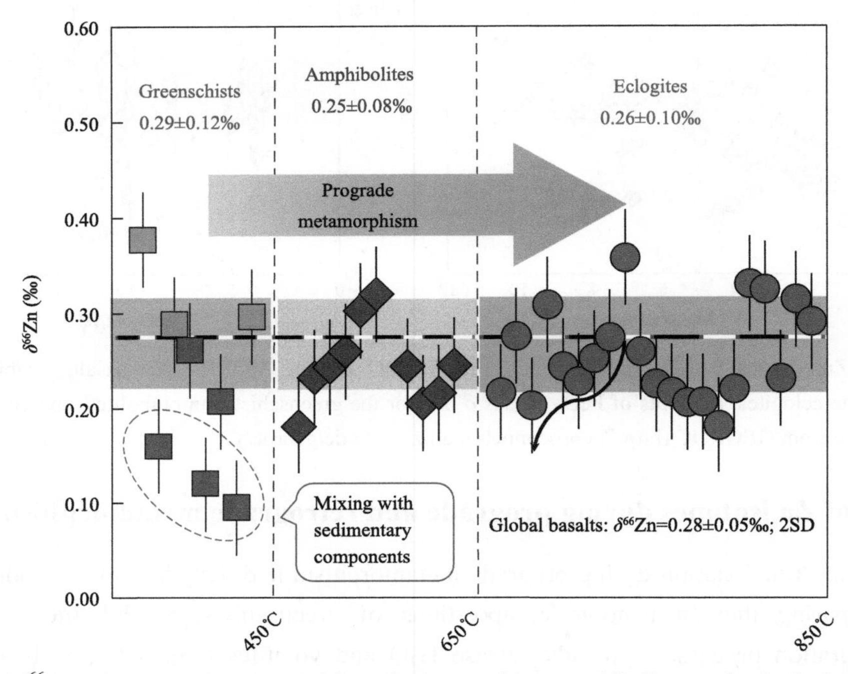

Fig. 7　comparison of δ^{66}Zn values among metamorphic rocks (greenschists, amphibolites, and eclogites) from the Dabie Orogen formed at different temperatures and pressures.

Three eclogite samples from Shuanghe in the Dabie Orogen that experienced retrograde metamorphism were also analyzed in this study, including peak eclogite-facies (09-SH-6-3b), transition-facies (09-SH-6-3a), and amphibolite facies (09-SH-6-4) (Shen et al., 2015). The LOI contents of the three samples are apt to decrease, hinting the missing of significant amounts of water. However, the δ^{66}Zn values and Zn contents of these samples are similar within errors regardless of the notable variations in LOI (Supplementary Data, Fig. S2). This finding further suggests the lack of Zn isotope fractionation associated with the degree of metamorphic dehydration.

Inglis et al. (2017) previously reported the absence of systematic Zn isotopic variations among metamorphosed oceanic crust samples formed by various grades of metamorphism, including twelve metabasaltic and metagabbroic rocks formed in eclogite-facies conditions. The metamorphism climax of these eclogites is 24–26 kbar and 550–600 ℃ (Bucher et al., 2005; Bucher and Grapes, 2009). Thus, the subduction depth of these metamorphic rocks in the study of Inglis et al. (2017) is smaller than ~90 km. Typically, the eclogite samples in this study represent much deeper subduction of mafic rocks to a depth of >200 km, and the results suggest that Zn isotope fractionation induced by metamorphic dehydration during ultra-deep subduction of mafic crust is still insignificant.

In summary, the average δ^{66}Zn value of amphibolites (0.25±0.08‰, n=10) and coesite-bearing eclogites (0.26±0.10‰, n=21) is indistinguishable from that of the greenschists (0.29±0.12‰, n=5) from the Dabie Orogen (Fig. 7), reflecting limited variations of Zn isotopic compositions induced by prograde metamorphism and dehydration of the mafic crust at different temperatures and pressures (Fig. 7). The lack of appreciable Zn isotopic differences among greenschists, amphibolites and eclogites is in accordance with the limited amount of Zn loss during prograde metamorphic dehydration.

5.4 Implications for Zn isotopic heterogeneity of the mantle

Subduction delivers slab-derived components (melts and fluids) to the mantle and potentially modifies chemical composition of the mantle wedge, which leads to mantle heterogeneity (e.g., McCulloch and Gamble, 1991; Elliott et al., 1997; Metrich and Wallace, 2008). Recent studies have shown that the potential subducting components in the crust are highly heterogeneous in term of Zn isotopic composition. For example, serpentinites and altered oceanic peridotites have δ^{66}Zn varying from –0.05‰ to 0.62‰ (Pons et al., 2011, 2016; Debret et al., 2018; Liu et al., 2019) and sedimentary materials have δ^{66}Zn from –0.15‰ to 1.43‰ (Maréchal et al., 2000; Pichat et al., 2003; Bentahila et al., 2008; Pons et al., 2011; Little et al., 2014, 2016; Liu et al., 2016, 2017; Inglis et al., 2017; Lv et al., 2016, 2018). In contrast, the mantle has a homogeneous Zn isotopic composition of 0.18± 0.05‰ (2SD). Therefore, Zn isotopes may be a powerful tool to investigate the recycling of crustal components into the Earth's mantle (Liu et al., 2016, 2020; Wang et al., 2018; Liu and Li, 2019; Yang and Liu, 2019; Beunon et al., 2020).

The results presented in this study indicate that limited Zn isotope fractionation occurs during ultra-deep (>200 km) subduction of mafic crust. The δ^{66}Zn of amphiboles and coesite-bearing eclogite are all higher than those of the mantle. When subducted crustal rocks partially melt and generate a metasomatized mantle, a heavy Zn isotopic signature of the hybridized mantle and their extrusive basalts can be observed. Yang and Liu (2019) reported that picrites from the Emeishan large igneous province have higher δ^{66}Zn (0.33‰ for the parental magma, Fig. 8) than that of the mantle, which is unlikely to be the consequence of large-degree partial melting of a normal mantle source by using non-modal melting modelling. Subsequently, a hybridized mantle source containing ~15% of recycled basaltic crust is required, suggesting that the source of Emeishan picritic melts contains recycled oceanic crust (Yang and Liu, 2019). The heavier Zn isotopic compositions of some ocean island basalts (OIBs) relative to those of the mantle and MORBs have also been taken as evidence for the presence of recycled carbonate-bearing eclogites in OIB's sources (Beunon et al., 2020).

To better understand the influence of recycled mafic crust on Zn isotopic composition of the mantle, possible ways of mantle metasomatism are evaluated by a binary mixing model. As shown in Fig. 8, addition of ~20% recycled eclogite would change the δ^{66}Zn values and Zn contents of spinel-facies and garnet-facies peridotites to 0.19‰, 63 μg/g and 0.22‰, 67 μg/g, respectively. In detail, the δ^{66}Zn values and Zn contents of starting peridotites and recycled mafic crust are set to be 0.16‰ and 55 μg/g for spinel-facies peridotite, 0.20‰ and 60 μg/g for garnet-facies peridotites and 0.26‰ and 96 μg/g for coesite-bearing eclogites, respectively. The incremental non-modal melting mode was used to evaluate the influence of recycled mafic crust on Zn isotope fractionation during mantle melting, which has been previously used to simulate Mg, Fe and Zn isotope fractionation during mantle melting (Williams and Bizimis, 2014; Zhong et al., 2017; Yang and Liu, 2019). The Zn isotope fractionation factor between melt and residue ($\alpha_{\text{melt-residue}}$) can be calculated as follows:

$$\alpha_{\text{melt-residue}} = \alpha_{\text{melt-spl}} \times \left[\sum_{i=1}^{\text{mineral}} \left(X_i \cdot C_{\text{mineral}} \right) / \sum_{i=1}^{\text{mineral}} \left(X_i \cdot C_{\text{mineral}} \cdot \alpha_{\text{mineral-spl}} \right) \right]$$

where C_{mineral} is the concentration of Zn in the mineral, and can be calculated by using D_{Zn} (Le Roux et al., 2011; Davis et al., 2013; Yang and Liu, 2019; Sossi et al., 2018). The $\alpha_{\text{mineral-spl}}$ is the isotope fractionation factor between spinel and its co-existing mineral assemblage. The X is the normalized mineral modal abundance, and the starting fraction for the minerals of spinel-facies and garnet-facies peridotite are from Yang and Liu (2019) and Sossi et al. (2018). The calculated δ^{66}Zn value for the primary basalt from the normal mantle is 0.24‰. As shown in Fig. 8, if 10%–20% of eclogite is added to the mantle sources, the Zn isotopic ratios of the melt will significantly increase after mantle melting, supposing that the source lithology will not be significantly changed (e.g., Lambart, 2017). This simulation strengthens the application of Zn isotopes as a tool of tracing recycled mafic crust in the mantle sources of basaltic lavas.

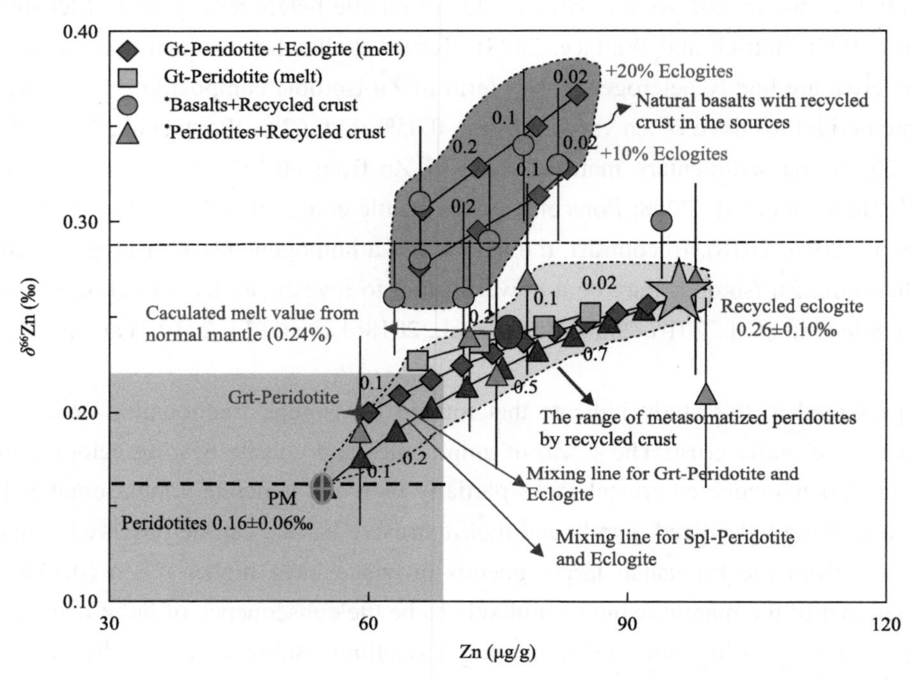

Fig. 8 Plots of δ^{66}Zn versus Zn concentrations for the metasomatized peridotites (gray circle, Wang et al., 2017; Huang et al., 2018a) and picrites (gray triangle) from a hybridized mantle in previous studies (Yang and Liu, 2019). The black lines with blue diamonds and blue triangles are part of the mixing lines for garnet-facies peridotites (red star: 0.20±0.04‰, McCoy-West et al., 2018) and spinel-facies peridotites (yellow circle: 0.16±0.04‰, Sossi et al., 2018) with eclogites (yellow star: 0.26±0.10‰), respectively. Both two lines show the melt fractions with blue words. The black line with yellow squares is the melt generated by partial melting of garnet-facies peridotites. The two black lines with orange diamonds are the melt generated by partial melting of hybridized garnet-facies peridotites with 10% or 20% recycled eclogites. All of these lines show the melt fractions of 0.02, 0.1, 0.2 and 0.3 (yellow words).

6 Conclusions

Based on a combined study of Zn, Sr and Nd isotopes for greenschists, amphibolites, and eclogites from the Dabie Orogen, which were formed by prograde metamorphism of mafic rocks at various degrees, the conclusions are as follows:

(1) The low δ^{66}Zn (0.10‰–0.16‰) of three mafic greenschists may be produced by crustal assimilation of

their igneous protoliths, as indicated by the low Nb/La, Nb/U ratios, low ε_{Nd} and high $^{87}Sr/^{86}Sr$. After excluding the samples significantly affected by sedimentary components, the average $\delta^{66}Zn$ of greenschists is 0.29±0.12‰ (2SD).

(2) The modest $\delta^{66}Zn$ variations in greenschists (0.21‰–0.38‰), amphibolites (0.18‰–0.32‰) and coesite-bearing eclogites (0.18±0.05‰ to 0.36±0.04‰) may mirror their isotopically inhomogeneous igneous protoliths.

(3) The similar Zn isotopic compositions of greenschists (0.29±0.12‰), amphibolites (0.25±0.08‰), and eclogites (0.26 ± 0.10‰) suggest that no detectable Zn isotope fractionation has occurred during continental deep subduction, which is attributed to limited loss of Zn during prograde metamorphism.

(4) Modelling suggests that the subducted mafic crust would result in significant changes of the Zn isotopic composition of the hybridized mantle and its extrusive basalts.

Declaration of Competing Interest

The authors declare that they have no known competing financial interests or personal relationships that could have appeared to influence the work reported in this paper.

Acknowledgements

We greatly appreciated the Dr. Jian Huang and one anonymous reviewer for their constructive comments and efficient editorial handling of Chief Editor M. Santosh. We are grateful to Dr. S.J. Wang and Dr. J. Shen for providing some of the analyzed samples and to D. Liu, Z.-Z. Wang, C.Y. Liu, and C. Yang for help in the lab. This work is supported by the National Key R and D Program of China (2019YFA0708400), Fundamental Research Funds for the Central Universities (Grant No. 292018049), the National Natural Science Foundation of China (Grant No. 41730214), and the Strategic Priority Research Program (B) of the Chinese Academy of Sciences (Grant XDB18000000).

Appendix A. Supplementary data

Supplementary data to this articlecan be found onlineathttps://doi.org/10.1016/j.gsf.2021.101182.

References

Ames, L., Zhou, G., Xiong, B., 1996. Geochronology and isotopic character of ultrahigh pressure metamorphism with implications for collision of the Sino-Korean and Yangtze cratons, Central China. Tectonics IS 472-489.

Bentahila, Y., Ben Othman, D., Luck, J.-M., 2008. Strontium, lead and zinc isotopes in marine cores as tracers of sedimentary provenance: a case study around Taiwan orogen. Chem. Geol. 248, 62-82.

Beunon, H., Mattielli, N., Doucet, L.S., Moine, B., Debret, B., 2020. Mantle heterogeneity through Zn systematics in oceanic basalts: Evidence for a deep carbon cycling. Earth-Sci. Rev. 205, 103174.

Bucher, K., Grapes, R., 2009. The eclogite-facies Allalin Gabbro of the Zermatt-Saas ophiolite, Western Alps: a record of subduction zone hydration. J. Petrol. 50 (8), 1405-1442.

Bucher, K., Fazis, Y., Capitani, C., Grapes, R., 2005. Blueschists, eclogites, and decompression assemblages of the Zermatt-Saas ophiolite: High-pressure metamorphism of subducted Tethys lithosphere. Am. Mineral. 90 (5-6), 821-835.

Chen, H., Savage, P., Teng, F.-Z., Helz, R., Moynier, F., 2013. Zinc isotope fractionation during magmatic differentiation and the isotopic composition of the bulk Earth. Earth Planet. Sci. Lett. 369-370, 34-42.

Cong, B., 1996. Ultrahigh-Pressure Metamorphic Rocks in the Dabieshan-Sulu Region of China. Science Press, Beijing, pp. 1-224.

Davis, F., Humayun, M., Hirschmann, M., Cooper, R.S., 2013. Experimentally determined mineral/melt partitioning of first-row transition elements (FRTE) during partial melting of peridotite at 3 GPa. Geochim. Cosmochim. Acta 104, 232-260.

Debret, B., Koga, K., Nicollet, C., Andreani, M., Schwartz, S., 2014. F, Cl and S input via serpentinite in subduction zones: implications for the nature of the fluid released at depth. Terra Nova 26, 96-101.

Debret, B., Beunon, H., Mattielli, M., RibeirodaCosta, I., Escartin, J., 2018. Ore component mobility, transport and mineralization at mid-oceanic ridges: a stable isotopes (Zn, Cu and Fe) study of the Rainbow massif (Mid-Atlantic Ridge 36°14′N). Earth Planet. Sci. Lett. 503, 170-180.

Deschamps, F., Kaminski, E., Tackley, P., 2011. A deep mantle origin for the primitive signature of ocean island basalt. Nat. Geosci. 4 (12), 879-882.

Doucet, L., Laurent, O., Mattielli, N., Debouge, W., 2018. Zn isotope heterogeneity in the continental lithosphere: new evidence from Archean granitoids of the northern Kaapvaal craton, South Africa. Chem. Geol. 476, 260-271.

Doucet, L.S., Mattielli, N., Ionov, D.A., Debouge,W., Golovin, A.V., 2016. Zn isotopic heterogeneity in the mantle: A melting control? Earth. Planet. Sci. Lett. 451, 232-240.

DuFrane, S., Turner, S., Dosseto, A., van Soest, M., 2009. Reappraisal of fluid and sediment contributions to Lesser Antilles magmas. Chem. Geol. 265, 272-278.

Elliott, T., Plank, T., Zindler, A.,White,W., Bourdon, B., 1997. Element transport from slab to volcanic front at the Mariana arc. J. Geophys. Res. 102, 14991-15019.

Huang, J., Liu, S.A., Gao, Y.J., Xiao, Y.L., Chen, S., 2016. Copper and zinc isotope systematics of altered oceanic crust at IODP Site 1256 in the eastern equatorial Pacific. J. Geophys. Res-Sol. Ea. 121, 7086-7100.

Huang, J., Zhang, X.-C., Chen, S., Tang, L.M.,Worner, G., Yu, H.M., Huang, F., 2018a. Zinc isotopic systematics of Kamchatka-Aleutian arc magmas controlled by mantle melting. Geochim. Cosmochim. Acta 238, 85-101.

Huang, J., Chen, S., Zhang, X.C., Huang, F., 2018b. Effects of melt percolation on Zn isotope heterogeneity in the mantle: constraints from peridotite massifs in Ivrea-Verbano Zone, Italian Alps. J. Geophys. Res-Solid Earth 123 (4), 2706-2722.

Inglis, E., Debret, B., Burton,W.K., Millet,M.-A., Pons, M.-L., Dale, C.W., Bouilhol, P., Cooper, M., Nowell, G.M., McCoy-West, A.J., Williams, H.M., 2017. The behavior of iron and zinc stable isotopes accompanying the subduction of mafic oceanic crust: a case study from Western Alpine ophiolites. Geochem. Geophys. Geosyst. 18 (7), 2562-2579.

Jahn, B., 1998. Geochemical and isotopic characteristics of UHP eclogites and ultramafic rocks of the Dabie orogen: implications for continental subduction and collisional tectonics. In: Hacker, B.R., Liou, J.G. (Eds.), When Continents Collide: geodynamics and Geochemistry of Ultrahigh-Pressure Rocks. Petrology and Structural Geology. vol 10. Springer, Dordrecht, pp. 203-239. https://doi.org/10.1007/978-94-015-9050-1_8.

Keith, P., 1999. Melting depths and mantle heterogeneity beneath Hawaii and the East Pacific Rise: constraints from Na/Ti and rare earth element ratios. J. Geophys. Res-Solid Earth 104 (B2), 2817-2829.

Kelley, K., Cottrell, E., 2009. Water and the oxidation state of subduction zone magmas. Science 325, 605-607.

Kessel, R., Schmidt, M.W., Ulmer, P., Pettke, T., 2005. Trace element signature of subduction-zone fluids, melts and supercritical liquids at 120-180 km depth. Nature 437, 724-727.

Kohn, S.C., Schofield, P.F., 1994. The importance of melt composition in controlling traceelement behaviour: an experimental study of Mn and Zn partitioning between forsterite and silicate melts. Chem. Geol. 117 (1-4), 73-87.

Lambart, S., 2017. No direct contribution of recycled crust in Icelandic basalts. Geochem. Perspect. Let. 7-12.

Le Bas, M.J., Le,M.R., Streckeisen, A., Zanettin, B., 1986. A chemical classification of volcanic rocks based on the total alkali-silica diagram. Jour. Petrol. 27, 745-750.

Le Roux, V., Dasgupta, R., Lee, C.-T., 2011. Mineralogical heterogeneities in the Earth's mantle: constraints from Mn, Co, Ni and Zn partitioning during partial melting. Earth Planet. Sci. Lett. 307 (3-4), 395-408.

Li, C.-F., Chen, F.-K., Li, X.-H., 2007. Precise isotopic measurements of sub-nanogram Nd of standard reference material by thermal ionization mass spectrometry using the NdO + technique. Int. J. Mass Spectrom. 266, 34-41.

Li, S., Jagoutz, E., Chen, Y., Li, Q., 2000. Sm-Nd and Rb-Sr isotopic chronology and cooling history of ultrahigh pressure metamorphic rocks and their country rocks at Shuanghe in the Dabie Mountains, Central China. Geochim. Cosmochim. Acta 64, 1077-1093.

Li, S.G., Xiao, Y.L., Liou, D., Chen, Y., Ge, N., Zhang, Z., Sun, S.S., Cong, B., Zhang, R., Hart, S.R., 1993. Collision of the North China and Yangtse Blocks and formation of coesitebearing eclogites: timing and processes. Chem. Geol. 109, 89-111.

Li, S.G., Huang, F., Nie, Y.H., Han,W.L., Zhang, Z.H., 2001. Geochemical and geochronological constraints on the suture location between the North and South China blocks in the Dabie Orogen, Central China. Phys. Chem. Earth Pt. A. 26, 655-672.

Li, X.P., Zheng, Y.F., Wu, Y.B., Chen, F., Gong, B., Li, Y.L., 2004. Low-T eclogite in the Dabie terrane of China: petrological and isotopic constraints on fluid activity and radiometric dating. Contrib. Mineral. Petrol. 148, 443-470.

Ling,W.L., Ren, B.F., Duan, R.C., Liu, X.M., Mao, X.W., Peng, L.H., Liu, Z.X., Cheng, J.P., Yang, H.M., 2008. Timing of theWudangshan Yaolinghe volcanic sequences and mafic sill in South Qinling: U-Pb zircon geochronology and tectonic implication. Chin. Sci. Bull. 53, 2192-2199.

Ling,W.L., Duan, R.C., Liu, X.M., Cheng, J.P.,Mao, X.W., Peng, L.H., Liu, Z.X., Yang, H.M., Ren, B.F., 2010. U-Pb dating of detrital zircons from the Wudangshan Group in the South Qinling and its geological significance. Chin. Sci. Bull. 55, 2440-2448.

Little, S., Vance, D.,Walker-Brown, C., Landing,W., 2014. The oceanicmass balance of copper and zinc isotopes, investigated by analysis of their inputs, and outputs to ferromanganese oxide sediments. Geochim. Cosmochim. Acta 125, 673-693.

Little, S., Vance, D., McManus, J., Severmann, S., 2016. Key role of continental margin sediments in the oceanic mass balance of Zn and Zn isotopes. Geology 44, 207-210.

Liu, S.-A., Li, S.-G., 2019. Tracing deep carbon cycle using metal stable isotopes: opportunities and challenges. Engineering 5, 448-457.

Liu, S.-A.,Wang, Z.-Z., Li, S.G., Huang, J., Yang,W., 2016. Zinc isotope evidence for a largescale carbonated mantle beneath eastern China. Earth Planet. Sci. Lett. 444, 169-178.

Liu, S.-A., Wu, H., Shen, S.Z., Jiang, G., Zhang, S., Lv, Y., Zhang, H., Li, S., 2017. Zinc isotope evidence for intensive magmatismim mediately before the end-Permian mass extinction. Geology 45 (4), 343-346.

Liu, S.-A., Liu, P.-P., Lv, Y.,Wang, Z.-Z., Dai, J.-G., 2019. Cu and Zn isotope fractionation during oceanic alteration: implications for oceanic Cu and Zn cycles. Geochim. Cosmochim. Acta 257, 191-205.

Liu, S.-A., Wang, Z.-Z., Yang, C., Li, S.-G., Ke, S., 2020. Mg and Zn isotope evidence for two types of mantle metasomatism and deep recycling of magnesium carbonates. J. Geophys. Res-Solid Earth 125, e2020JB020684.

Lv, Y., Liu, S.-A., Zhu, J., Li, S., 2016. Copper and zinc isotope fractionation during deposition and weathering of highly metalliferous black shales in Central China. Chem. Geol. 445, 24-35.

Lv, Y., Liu, S.-A.,Wu, H.C., Simon, V.H., Chen, S.M., Li, S., 2018. Zn-Sr isotope records of the Ediacaran Doushantuo Formation in South China: diagenesis assessment and implications. Geochim. Cosmochim. Acta 239, 330-345.

Maréchal, C.N., Nicolas, E., Douchat, C., Albaréde, F., 2000. Abundance of zinc isotopes as a marine biogeochemical tracer. Geochem. Geophys. Geosyst. 1, 1015.

McCoy-West, A.J., Fitton, G.J., Pons, M.-L., Inglis, E.C.,Williams, H.M., 2018. The Fe and Zn isotope composition of deep mantle source regions: Insights from Baffin Island picrites. Geochim. Cosmochim. Acta 238, 542-562.

McCulloch, M.T., Gamble, J.A., 1991. Geochemical and geodynamical constraints on subduction zone magmatism. Earth Planet. Sc. Lett. 102, 358-374.

Metrich, N., Wallace, P.J., 2008. Volatile abundances in basaltic magmas and their degassing paths tracked by melt inclusions. Rev. Mineral. Geochem. 69, 363-402.

Moynier, F., Vance, D., Fujii, T., Savage, P., 2017. The isotope geochemistry of zinc and copper. Rev. Mineral. Geochem. 82, 543-600.

Okay, A.I., Xu, S., Sengor, A.M.C., 1989. Coesite from the Dabie Shan eclogites, Central China. Eur. J. Mineral. 1, 595-598.

Pichat, S., Douchet, C., Albaréde, F., 2003. Zinc isotope variations in deep-sea carbonates from the eastern equatorial Pacific over the last 175 ka. Earth Planet. Sci. Lett. 210, 167-178.

Pons, M.-L., Quitte, G., Fujii, T., Rosing, M.T., Reynard, B., Moynier, F., Douchet, C., Albarée, F., 2011. Early Archean serpentine mud volcanoes at Isua, Greenland, as a niche for early life. Proc. Natl. Acad. Sci. U. S. A. 108, 17639-17643.

Pons, M.-L., Debret, B., Bouilhol, P., Delacour, A.,Williams, H., 2016. Zinc isotope evidence for sulfate-rich fluid transfer across subduction zones. Nat. Commun. 7, 13794.

Rudnick, R.L., Gao, S., 2003. 3.01 - Composition of the Continental Crust. In: Holland, H.D., Turekian, K.K. (Eds.), Treatise on Geochemistry. Vol. 3. Elsevier Science, pp. 1-64.

Shen, J., Liu, J., Qin, L., Wang, S.J., Li, S., Xia, J., Yu, H.M., Yang, J., 2015. Chromium isotope signature during continental crust subduction recorded in metamorphic rocks. Geochem. Geophys. Geosyst. 16 (11), 3840-3854.

Sossi, P., Halverson, G., Nebel, O., Eggins, S., 2015. Combined separation of Cu, Fe and Zn from rock matrices and improved analytical protocols for stable isotope determination. Geostand. Geoanal. Res. 39 (2), 129-149.

Sossi, P., Nebel, O., O'Neill, H.,Moynier, F., 2018. Zinc isotope composition of the Earth and its behaviour during planetary accretion. Chem. Geol. 477, 73-84.

Spear, F., Cheney, J.T., 1989. A petrogenetic grid for pelitic schists in the system SiO_2-Al_2O_3-FeO-MgO-K_2O-H_2O. Contrib. Mineral. Petrol. 101 (2), 149-164.

Sun, S.S., McDonough, W.-S., 1989. Chemical and isotopic systematics of oceanic basalts: implications for mantle composition and processes. Geol. Soc. Lond., Spec. Publ. 42, 313-345.

Telus, M., Dauphas, N., Frédéric, M., Tissot, F.L.H., Teng, F.Z., Nabelek, P.I., Craddock, P.R., Groat, L.A., 2012. Iron, zinc, magnesium and uranium isotope fractionation during continental crust differentiation: the tale from migmatites, granitoids, and pegmatites. Geochim. Cosmochim. Acta 97, 247-265.

Turcotte, D., Schubert, G., 2014. Geodynamics. 3rd. Cambridge University Press, Cambridge, p. 626.

Wang, Z.-Z., Liu, S.-A., Liu, Z.-C., Zheng, Y.-C.,Wu, F.-Y., 2020. Extreme Mg and Zn isotope fractionation recorded in the Himalayan leucogranites. Geochim. Cosmochim. Acta 278, 305-321.

Wang, Z.-Z., Liu, S.-A., Liu, J., Huang, J., Xiao, Y., Chu, Z.-Y., Zhao, X.-M., Tang, L., 2017. Zinc isotope fractionation during mantle melting and constraints on the Zn isotope composition of Earth's upper mantle. Geochim. Cosmochim. Acta 198, 151-167.

Wang, Z.-Z., Liu, S.-A., Chen, L., Li, S., Zeng, G., 2018. Compositional transition in natural alkaline lavas through silica-undersaturated melt-lithosphere interaction. Geology 46(9), 771-774.

Wang, S., Teng, F., Li, S., Hong, J., 2014. Magnesium isotopic systematics of mafic rocks during continental subduction. Geochim. Cosmochim. Acta 143, 34-48.

Wiesmaier, S., Deegan, F.M., Troll, V.R., Carracedo, J.C., Chadwick, J.P., Chew, D.M., 2011. Magma mixing in the 1100 AD Montaña Reventada composite lava flow, Tenerife, Canary Islands: interaction between rift zone and central volcano plumbing systems. Contrib. Mineral. Petrol. 162, 651-669.

Williams, H.M., Bizimis, M., 2014. Iron isotope tracing of mantle heterogeneity within the source regions of oceanic basalts. Earth Planet. Sci. Lett. 404, 396-407.

Wu, Y.-B., Zheng, Y.-F., Tang, J., Gong, B., Zhao, Z.-F., Liu, X., 2007. Zircon U-Pb dating of water-rock interaction during Neoproterozoic rift magmatism in South China. Chem. Geol. 246, 65-86.

Xia, L., Xia, Z., Xu, X., Li, X., Ma, Z., 2012. Mid-late Neoproterozoic rift-related volcanic rocks in China: geological records of rifting and break-up of Rodinia. Geosci. Front. 3, 375-399.

Xiao, Y., Hoefs, J., van den Kerkhof, A.M., Fiebig, J., Zheng, Y., 2000. Fluid history of UHP metamorphism in Dabie Shan, China: a fluid inclusion and oxygen isotope study on the coesite-bearing eclogite from Bixiling. Contrib. Mineral. Petrol. 139 (1), 1-16.

Xu, L.-J., Liu, S.-A.,Wang, Z.-Z., Liu, C., Li, S., 2019. Zinc isotopic compositions of migmatites and granitoids from the Dabie Orogen, Central China: implications for zinc isotopic fractionation during differentiation of the continental crust. Lithos 324, 454-465.

Xu, S.T., Okay, A.I., Ji, S.Y., Sengor, A.M.C., Su, W., Liu, Y.C., Jiang, L.L., 1992. Diamond from the Dabie Shan metamorphic rocks and its implication for tectonic setting. Science 256, 80-82.

Yang, C., Liu, S.-A., 2019. Zinc isotope constraints on recycled oceanic crust in the mantle sources of the Emeishan large igneous province. J. Geophys. Res-Solid Earth 124, 12537-12555.

Ye, K., Cong, B., Ye, D., 2000. The possible subduction of continental material to depths greater than 200 km. Nature 407, 734-736.

Zack, T., John, T., 2007. An evaluation of reactive fluid flow and trace element mobility in subducting slabs. Chem. Geol. 239, 199-216.

Zhang, Z., Xiao, Y., Hoefs, J., Liou, J., Simon, K., 2006. Ultrahigh pressure metamorphic rocks from the Chinese Continental Scientific Drilling Project: I. Petrology and geochemistry of the main hole (0-2050 m). Contrib. Miner. Petrol. 152, 421-441.

Zhang, G., Liu, Y., Moynier, Frèdèric, Zhu, Y., Wang, Z., Hu, Z., Zhang, L., Li, M., Chen, H., 2020. Zinc isotopic composition of the lower continental crust estimated from lower crustal xenoliths and granulite terrains. Geochim. Cosmochim. Acta 276, 92-108.

Zhao, Z.-F., Zheng, Y.-F., Chen, R.X., Xia, Q.X., Wu, Y.-B., 2007. Element mobility in mafic and felsic ultrahigh-pressure metamorphic rocks during continental collision. Geochim. Cosmochim. Acta 71, 5244-5266.

Zheng, Y., Fu, B., Gong, B., Li, L., 2003. Stable isotope geochemistry of ultrahigh pressure metamorphic rocks from the Dabie-Sulu orogen in China: implications for geodynamics and fluid regime. Earth-Sci. Rev. 62, 105-161.

Zheng, Y., Zhou, J.,Wu, Y.B., Xie, Z., 2005. Low-grade metamorphic rocks in the Dabie-Sulu orogenic belt: a passive margin accretionary wedge deformed during continent subduction. Int. Geol. Rev. 47, 851-871.

Zhong, Y., Chen, L.-H., Wang, X.-J., Zhang, G.-L., Xie, L.-W., Zeng, G., 2017. Magnesium isotopic variation of oceanic island basalts generated by partial melting and crustal recycling. Earth Planet. Sci. Lett. 463, 127-135.

Zinc isotope fractionation between Cr-spinel and olivine and its implications for chromite crystallization during magma differentiation*

Chun Yang[1,2], Sheng-Ao Liu[1], Long Zhang[3,4], Ze-Zhou Wang[1,5], Ping-Ping Liu[6] and Shu-Guang Li[1]

1. State Key Laboratory of Geological Processes and Mineral Resources, China University of Geosciences, Beijing, 100083, China
2. Institut für Planetologie, University of Münster, Wilhelm-Klemm-Straße 10, Münster, 48149, Germany
3. State Key Laboratory of Isotope Geochemistry, Guangzhou Institute of Geochemistry, Chinese Academy of Sciences, Guangzhou 510640, China
4. CAS Center for Excellence in Deep Earth Science, Guangzhou, 510640, China
5. Isotope Laboratory, Department of Earth and Space Sciences, University of Washington, Seattle, WA, USA
6. Key Laboratory of Orogenic Belts and Crustal Evolution, School of Earth and Space Sciences, Peking University, Beijing, 100871, China

亮点介绍： 本文选取了北祁连造山带的玉石沟蛇绿岩及北冰洋 Gakkel 海沟的深海橄榄岩中尖晶石-橄榄石矿物作为研究对象，厘定了铬尖晶石与橄榄石之间的锌同位素分馏，并估计铬铁矿分异对于玄武质熔体锌同位素分馏的影响。首次验证了尖晶石中的锌同位素分馏主要受控于“成分效应”。研究表明铬铁矿的结晶分异可使熔体锌同位素比值升高，锌含量显著降低，二者呈负相关关系；而全球洋岛玄武岩及大陆板内碱性玄武岩则具有更重的锌同位素组成和更高的锌含量，二者为正相关关系。这一结果不支持前人提出的铬铁矿分离结晶导致熔体镁同位素变轻的模型。

Abstract We report the first zinc isotope data (expressed as δ^{66}Zn relative to the JMC-Lyon standard) for chromian spinels (Cr-spinels) and coexisting olivines in oceanic peridotites. All spinel-olivine pairs fall on the 1:1 fractionation line in the diagram of $\delta^{66}Zn_{spinel}$ versus $\delta^{66}Zn_{olivine}$, suggesting equilibrium isotope fractionation. Cr-spinels are always isotopically lighter than coexisting olivines ($\Delta^{66}Zn_{spinel\text{-}olivine} = \delta^{66}Zn_{spinel} - \delta^{66}Zn_{olivine}$=–0.50‰ to –0.33‰; n=13), which is opposite to the positive Zn isotope fractionation between Al-spinel and olivine observed in cratonic peridotites. The “inverse” Zn isotope fractionation between Cr-spinel and olivine is unlikely to have been caused by low-temperature alteration of the oceanic peridotites, given the lack of correlation between $\Delta^{66}Zn_{spinel\text{-}olivine}$ values and chemical indices of serpentinization and weathering (e.g., LOIs, MgO/SiO_2). Ionic model calculation

* 本文发表在：Geochimica et Cosmochimica Acta, 2021, 313: 277-294

indicates that even considering that Zn occurs at both tetrahedral and octahedral sites in the crystal lattice of spinel, the magnitude of equilibrium fractionation is still inconsistent with the observed Zn isotope fractionation between Cr-spinel and olivine. We suggest a "chemical effect" in which the Zn-O bond length (tetrahedral site) in spinel increases when Cr substitutes Al in octahedral site, which is corroborated by the striking negative correlation of δ^{66}Zn with Cr# [molar $Cr^{3+}/(Cr^{3+}+Al)$] in natural spinels. Therefore, Zn isotopic compositions of natural spinels can be highly variable depending on their chemical compositions.

During magma differentiation, zinc is moderately incompatible in silicate minerals (olivine and pyroxene) but highly compatible in Cr-spinels/chromites that have Zn contents tens of times higher than those of basaltic melts. Given its light Zn isotopic composition, chromite crystallization–if any–can evidently elevate δ^{66}Zn and lower Zn contents of the residual melts. Lunar mare basalts are typically characterized by higher δ^{66}Zn and lower Zn contents relative to terrestrial basalts, but modelling suggests that chromite crystallization during magmatic differentiation is unlikely to account for Zn isotopic and elemental data of lunar mare basalts. Global ocean island basalts (OIBs) and some intraplate alkali basalts have systematically higher δ^{66}Zn and higher Zn contents than those of the normal mantle-derived melts (e.g., mid-ocean ridge basalts; MORBs), which contradicts with a chromite crystallization model. Instead, such signatures can reflect recycling of surface carbonates into the Earth's deep mantle, reinforcing the application of zinc isotopes as a tracer of deep carbonate recycling.

Keywords Zinc isotope, chromite, olivine, chemical effect, fractional crystallization

1 Introduction

With the significant advances of analytical techniques in recent years, a detectable Zn isotopic variation has been widely observed in high-temperature geological samples. For instance, terrestrial basalts exhibit a wide range of Zn isotopic composition (expressed as δ^{66}Zn relative to the JMC-Lyon standard) from 0.21±0.07‰ to 0.77± 0.04‰ (Chen et al., 2013; Liu et al., 2016, 2020; Moynier et al., 2017; McCoy-West et al., 2018; Huang et al., 2018; Wang et al., 2017, 2018; Yang and Liu, 2019; Beunon et al., 2020; Li et al., 2021), in comparison with the estimate of the Bulk Silicate Earth (BSE) (δ^{66}Zn=0.18±0.05‰; Wang et al., 2017; Sossi et al., 2018; Liu et al., 2019). The significantly heavy Zn isotopic composition recorded in some ocean island basalts (OIBs) (up to 0.40 ±0.04‰) and intra-plate alkali basalts (up to 0.63 ± 0.05‰) have been taken as evidence for the presence of recycled carbonate-bearing components (with δ^{66}Zn up to 1.34‰; Pichat et al., 2003) in the Earth's deep mantle (Liu et al., 2016, 2020; Liu and Li, 2019; Beunon et al., 2020). These findings mark Zn isotopes a novel geochemical tracer of unraveling the final fate of subducted surface carbon, which is mainly in the form of carbonate sediments (Plank and Manning, 2019). In addition to terrestrial basalts, extremely heavy Zn isotopic compositions (δ^{66}Zn=~1.4±0.50‰) have also been reported for lunar mare basalts. In a completely distinct way, the high δ^{66}Zn values of lunar mare basalts have been interpreted to reflect volatilization-induced loss of isotopically light Zn as a result of impact during formation and differentiation of the Moon (Paniello et al., 2012; Kato et al., 2015; Day et al., 2020). This finding suggests that Zn isotopes can be a powerful tool for deciphering volatilization-related events during planet formation and evolution.

Nevertheless, interpretations for any isotopic variations in mantle-derived melts must take crystal-melt differentiation during magma evolution into account. In particular, evaluating the influence of fractional crystallization of various types of minerals on Zn isotopic systematics of magmas is fundamental to the applications of Zn isotopes mentioned above. Previous study on a suite of cogenetic samples from the Kilauea Iki lava lake (Hawaii) found that the more differentiated lithologies tend to be slightly enriched in heavier Zn isotopes as a result of olivine and/or Fe-Ti oxide crystallization (Chen et al., 2013). Analysis of mineral phenocrysts in terrestrial basalts supports this interpretation, with olivine and clinopyroxene phenocrysts being isotopically lighter than the host melt by ~0.15‰ (McCoy-West et al., 2018; Yang and Liu, 2019) and ~0.10‰ (Yang and Liu, 2019), respectively. Besides, light Zn isotopic composition of Ti-oxide has been deduced by fractional crystallization trend in mafic granulites (Zhang et al., 2020). This is attributed to the different bonding environments between basaltic melt and silicate minerals, with Zn in the former preferring the four-coordinated site and Zn in the latter favoring the six-coordinated site (Ducher et al., 2016; Wang et al., 2017; Sossi et al., 2018). Accordingly, Zn isotope fractionation induced by fractional crystallization of silicate minerals (olivine and pyroxene) in the basaltic melts is detectable but generally does not exceed 0.1‰ (Chen et al., 2013; Moynier et al., 2017). This magnitude of isotope fractionation is considered to be insignificant in comparison with the large Zn isotopic variation observed in terrestrial and lunar mare basalts (e.g., Herzog et al., 2009; Paniello et al., 2012; Liu et al., 2016; Beunon et al., 2020; Li et al., 2021).

Notably, oxides like Cr-spinels or chromites commonly occur as ubiquitous accessory phases in igneous rocks and are extremely Zn-rich due to the high partition coefficient of Zn^{2+} in spinel ($D_{Zn}^{Spl\text{-}Melt}$=5.3; Davis et al., 2013; Horn et al., 1994). For example, Cr-spinels in ultramafic and mafic rocks have Zn contents of up to ~3000 μg/g (e.g., Griffin et al., 1993; Yu et al., 2019), in sharp contrast to silicate minerals and melts that commonly contain only tens of μg/g Zn (Le Roux et al., 2010; Wang et al., 2017). In particular, basaltic magmas may experience early fractional crystallization of spinels, in which Cr-spinel and olivine are often the first phases to crystallize from primary magmas containing >6 wt.% MgO (Thy, 1983; Kamenetsky et al., 2001; Barnes and Roeder, 2001). Given a high compatibility of Zn in spinel, fractional crystallization of Cr-spinels will substantially change Zn contents of the residual melts. Nevertheless, Zn isotopic compositions of different types of spinels and the influence of fractional crystallization of these Zn-rich oxides on Zn isotopic systematics of basaltic melts remain poorly constrained. Luck et al. (2005) presented Zn isotopic composition of separated spinel phase in meteorite by chemical leaching method, which has a lower δ^{66}Zn value than that of olivine. Bridgestock et al. (2014) reported Zn isotope fractionation between a chromite-rich inclusion and coexisting metal and silicate phases in iron meteorites, obtaining a negative $\Delta^{66}Zn_{chromite\text{-}metal}$ value of ~ −1.5‰. By contrast, Wang et al. (2017) reported that the Zn isotopic composition of Al-spinels is 0.13‰ heavier than that of olivines in cratonic peridotites on average. As yet, no Zn isotopic data is available for Cr-spinels or chromites in terrestrial rocks.

In this study, we present Zn elementary and isotopic data for Cr-spinels in oceanic peridotites. These new data are compared with existing Zn isotope data for Al-spinels and will help to assess the factors controlling Zn isotopic variation in natural spinels as well as the influence of spinel crystallization on Zn systematics in basaltic melts. Our results reveal an "inverse" Zn isotope fractionation between Cr-spinels and silicate minerals, which has crucial significance for understanding inter-mineral Zn isotope fractionation at high temperatures and the origin of Zn isotopic variation in terrestrial and extraterrestrial basalts.

2 Sample description

2.1 Petrology and mineralogy of peridotites

Chromian spinels and coexisting olivines in this study are from ultramafic rocks including both abyssal and ophiolitic peridotites. The abyssal peridotites are from two areas in the Sparsely Magmatic Zone of the ultraslow-spreading Gakkel Ridge (85.44°N, 14.52°E; 84.64°N, 4.22°E). Sample locations are shown in Fig. 1a.

The majority of peridotites from this area have been altered to clay and serpentinite, containing secondary minerals including serpentine, hornblende, andradite, chlorite, and oxides (Liu et al., 2017, 2019). We selected two least altered samples: one lherzolite (sample PS59 235-4) which has only a few percent serpentine and slightly weathered olivine and one harzburgite (sample HLY 40-56) with mostly tiny and fresh spinels. Detailed petrology and chemistry of the abyssal peridotites are available in Liu et al. (2017), and whole-rock Zn isotope data have been reported in Liu et al. (2019). The contents of loss-on-ignitions (LOIs) of the two selected abyssal peridotites are lower than 1 wt.%, indicating that olivine and spinel separates in these samples should preserve their relatively original chemical characteristics.

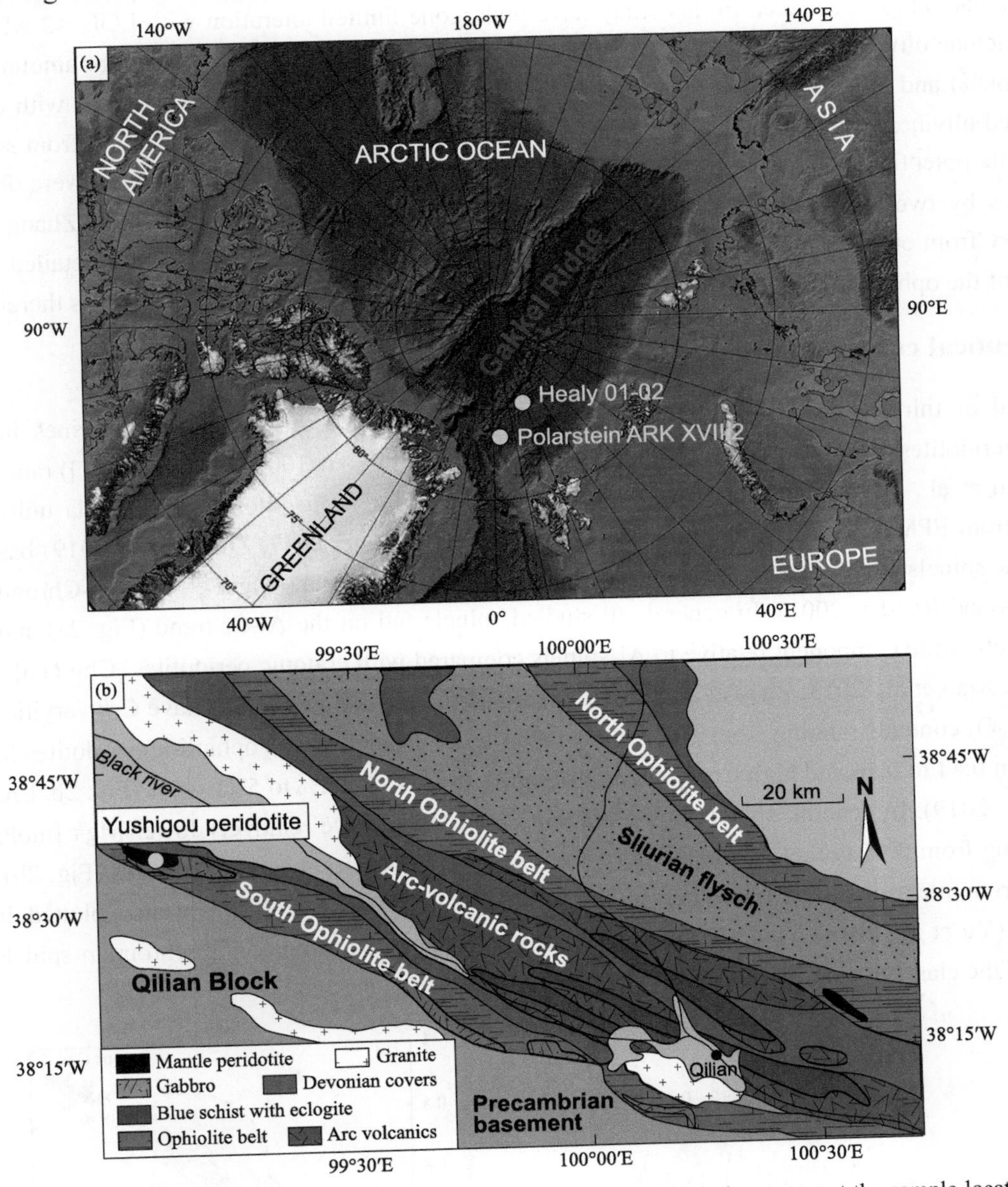

Fig. 1 (a) Physiographic map of the Gakkel Ridge in Arctic Ocean, where yellow circles represent the sample locations (modified from Liu et al., 2017). (b) Schematic geological map of the North Qilian Orogen, showing the ophiolitic peridotites within the Qilian suture zone. Yellow circle represents the sample location, Yushigou (Song et al., 2009; Zhang et al., 2019). (For interpretation of the references to colour in this figure legend, the reader is referred to the web version of this article.)

The ophiolitic peridotites are from the Yushigou ophiolite complex in the North Qilian orogen on the northeastern margin of the Qinghai-Tibetan plateau (Fig. 1b). The North Qilian orogen represents an Early Paleozoic oceanic suture zone containing ophiolites, arc magmatic rocks, and blueschist–eclogite facies high-pressure metamorphic rocks (Song et al., 2009, 2013). The Yushigou ophiolite complex in the south of the North Qilian orogen exposes well-developed rock assemblages of peridotite, gabbros, pillow lavas, and sediments (Song et al., 2009, 2013). The majority of ophiolitic peridotites are harzburgites and have undergone various extents of serpentinization, but some fresh harzburgites with limited alteration are preserved (Song et al., 2009; Zhang et al., 2019). Geochemical and petrological data suggest that the harzburgites were a highly depleted residue after a great extent of melt extraction probably in the forearc mantle (Song et al., 2009; Zhang et al., 2019). The harzburgites selected in this study have undergone limited alteration with LOIs <2 wt.%. The major minerals include olivine (ol, ~70–85 vol.%), orthopyroxene (opx, ~10–25 vol.%), and minor amounts of Cr-spinel (spl, <1 vol.%) and clinopyroxene (cpx, <2–3 vol.%). They display a protogranular texture with coarse-grained kink-banded olivines, which are the fingerprint of diffusion-controlled dislocation resulting from solid-state flow at the mantle potential temperature (~1000 ℃). Equilibrium temperatures of ~900–1000 ℃ were obtained for the harzburgites by two-pyroxene thermometer at a pressure of 1.5 GPa (Song et al., 2009; Zhang et al., 2019). Spinels vary from euhedral, subeuhedral to anhedral in the matrix or occur as inclusions. Detailed petrology and chemistry of the ophiolitic peridotites are available in Zhang et al. (2019, 2021) and references therein.

2.2 Chemical compositions of spinels and olivines

A total of thirteen spinel-olivine mineral pairs are selected for this study. The olivines in abyssal and ophiolitic peridotites display a limited compositional variation with Fo (=100×Mg/[Mg + Fe^{2+}]) ranging from 89.6 to 93.1 (Liu et al., 2017; Zhang et al. 2019). For spinels, the Fe^{3+} apfu (atoms per formula unit) contents are calculated from EPMA (electron probe microanalyzer) data (Liu et al., 2017; Zhang et al. 2019) based on charge balance. The spinels are plotted on the compositional triangle of Al-Spinel ((Mg, Fe^{2+})Al_2O_4)-Chromite-Magnetite (e.g. Barnes and Roeder, 2001). In general, all studied spinels fall on the Al-Cr trend (Fig. 2a) and have higher contents of chromite component relative to Al-spinels compared with cratonic peridotites (Chu et al., 2009; Wang et al., 2017; Xiao et al., 2013; Zhao et al., 2015). Spinels in the abyssal peridotites have Cr# varying from 0.30 to 0.54 and Cr_2O_3 contents ranging from 26.2 wt.% to 45.3 wt.%. Spinels in the ophiolitic peridotites have high Cr# varying from 0.64 to 0.76 and high Cr_2O_3 contents ranging from 45.6 wt.% to 55.1 wt.% (Fig. 2b, Liu et al., 2017; Zhang et al., 2019). In general, spinels in the ophiolitic peridotites have relatively lower Mg# [molar Mg/(Mg + Fe^{2+})] varying from 0.45 to 0.60 but higher Cr# than those of spinels in abyssal peridotites (Fig. 2b). The porous chromites from metasomatized ophiolites in the North Qinling Orogenic Belt are also plotted in Fig. 2 for comparison (Yu et al., 2019). Overall, the investigated spinels can be grouped into chromian spinels (Cr-spinel) according to the classification by Cassard et al. (1981).

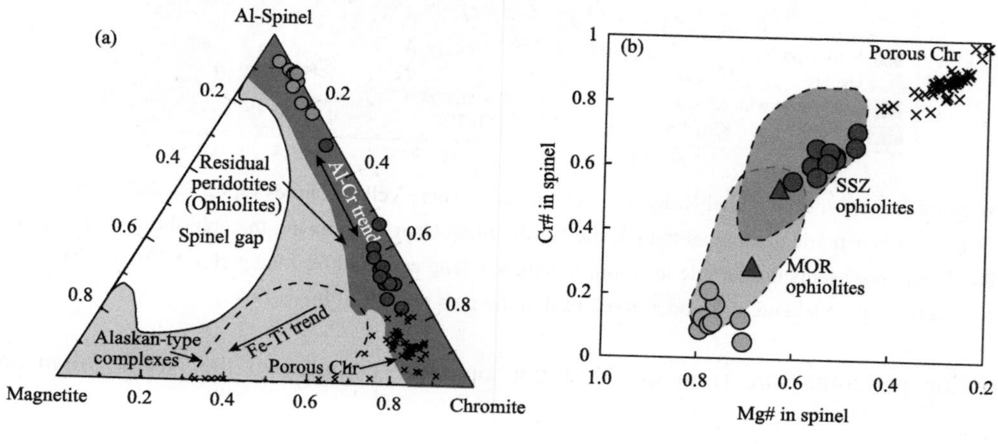

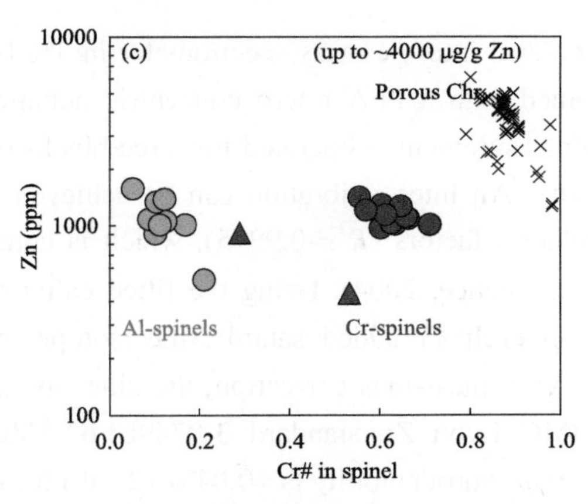

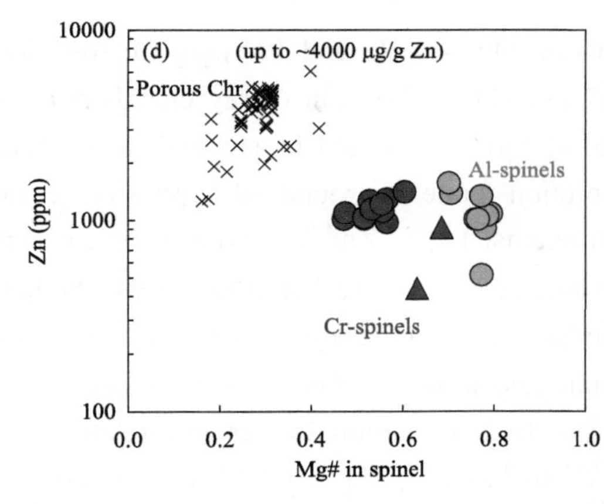

Fig. 2 Variation in chemical compositions of spinels in this study. (a) Triangular plot of Spinel-Chromite-Magnetite for the classification of spine group minerals (Barnes and Roeder, 2001). The Fe^{3+} contents of spinels are calculated from the Electron Probe Microanalysis (EPMA) data based on charge balance (Table S1). The components of spinel, chromite and magnetite are calculated on the basis of Cr, Al, and calculated Fe^{3+} contents. (b) The plot of Cr# (molar Cr/(Cr+Al)) versus Mg# (molar Mg/(Mg+Fe^{2+})) in studied spinels. Oceanic spinels from different tectonic setting are shown in this plot (MOR ophiolites, SSZ ophiolites and Boninite; Warren, 2016). (c-d) Diagrams of Zn contents against Cr# and Mg# in the studied spinels. The Y-axis is in logarithmic scale. Red triangles are analysed spinels from abyssal peridotites and red dots denote ophiolitic peridotites. The yellow dots are Al-spinels from cratonic peridotites (Wang et al., 2017), and crosses represent porous chromites from highly altered ophiolites (Yu et al., 2019). (For interpretation of the references to colour in this figure legend, the reader is referred to the web version of this article.)

3 Analytical methods

The selected olivine and spinel grains are relatively unaltered, although the majority of bulk peridotites have been slightly altered. Spinel and olivine grains were first handpicked under binocular microscope, and olivines with fluid inclusions or oxides on the edge were not selected. All of the mineral grains were cleaned in an ultrasonic bath with Milli-Q water, and then crushed into powders using an agate mortar and pestle prior to digestion. To ensure complete dissolution, powdered spinels were digested with a 1:1 (v/v) mixture of double-distilled concentrated HCl + HF in high-pressure bombs at 190 ℃ for 48 h, following the procedure of Liu et al. (2011). The samples were then digested in a mixture of double-distilled concentrated HNO_3 + HCl (Aqua Regia) in PTFE beakers (Savillex®) at 150 ℃, and finally dissolved into 1 ml 8 N HCl in preparation for chromatographic separation. Olivine separates can be fully dissolved in 1:3 (v/v) mixture of double-distilled high concentrated HNO_3 + HF in PTFE beakers at 140 ℃ for 48 h, and then dried down at hot plate around 150 ℃. After the digestion by Aqua Regia and concentrated HCl acids, olivines were finally digested into 1 ml 8 M HCl prior to column chemistry. Zinc was purified following the protocols outlined in Maréchal et al. (1999) and Liu et al. (2016). Dissolved samples in 1 ml solution were loaded on 2 ml pre-cleaned AG-MP-1M resin and Zn was collected in 10 ml 0.5 N HNO_3 after the elution of matrices. Due to the high Cr contents of spinels, an additional column procedure was performed in order to eliminate matrix interference and improve analytical accuracy and precision. The final elution was dried down and dissolved in 1 ml 3% (m/m) HNO_3. Firstly, Zn concentration of the purified solution was obtained using ICP-MS by measuring ^{66}Zn signal against standards with known Zn concentrations. The recovery of Zn during column chemistry is larger than 99% and the whole procedure blank is less than 4 ng. The uncertainties were better than ±5% based on repeated analyses of rock standards with recommended values.

Zinc isotopic ratios were measured using a Thermo Scientific *Neptune Plus* MC-ICP-MS at the Isotope Geochemistry Lab of the China University of Geosciences, Beijing. Instrumental mass bias correction was achieved through the combination of sample-standard bracketing and Cu-doping methods. Solutions with concentrations of 200 ng/g Zn and 100 ng/g Cu were introduced into the mass spectrometer in 3% HNO_3 using "wet plasma" introduction system (spray chamber) combined with a PFA micro concentric nebulizer with an uptake rate of ~50 μL/min. Zinc isotopic analysis for each measurement is operated for three blocks of 40 cycles in a low-resolution mode at around 10 V/ppm of sensitivity. An inter-calibration can be achieved by plotting $\ln(^{65}Cu/^{63}Cu)$ against $\ln(^{66,68}Zn/^{64}Zn)$ with excellent correlation factors ($R^2 \geqslant 0.9975$), which is consistent with those in previous studies (Maréchal et al., 1999; Archer and Vance, 2004). Using the fitted calibration line of in-house standards and the analyzed $\ln(^{65}Cu/^{63}Cu)$ ratio in each Cu-added sample, the isotope fractionation between sample and in-house standard can be calculated. After mass-bias correction, the zinc isotopic data are reported in standard δ-notation in per mil relative to JMC Lyon Zn standard 3-0749L: $\delta^{66,68}Zn=[(^{66,68}Zn/^{64}Zn)_{sample}/(^{66,68}Zn/^{64}Zn)_{Lyon}-1]\times 1000$. The long-term external reproducibility is ±0.04‰ (2 s.d.) for $\delta^{66}Zn$ ratios based on repeated analysis of igneous rock standards and synthetic solutions (Wang et al., 2017). Several reference materials (BHVO-2, BCR-2) were analyzed to assess accuracy, and the results (Table 1) agree well with the recommended values (Sossi et al., 2015; Moynier et al., 2017; Wang et al., 2017; Debret et al., 2018).

Table 1　Zinc isotopic data for Cr-spinels (Spl) and olivines (Ol) in oceanic peridotites in this study

Sample	Mineral	$\delta^{66}Zn$	2sd	$\delta^{68}Zn$	2sd	$\Delta^{66}Zn_{Spl-Ol}$	2sd	Mg#	Cr#	Al_2O_3	Cr_2O_3	Zn (μg/g)	T (℃)
Abyssal peridotites													
PS59 235-4	Ol	0.33	0.04	0.65	0.06	−0.33	0.06	0.897				41.3	1032
	Spl	−0.05	0.04	−0.10	0.05			0.655	0.30	41.4	26.2	915	
HLY 40-56	Ol	0.19	0.03	0.38	0.05	−0.35	0.05	0.918				37.5	1041
	Spl	−0.16	0.04	−0.33	0.06			0.628	0.54	26.3	45.3	439	
Ophiolitic peridotites peridotites													
17QH419	Ol	0.19	0.04	0.38	0.06	−0.36	0.06	0.909				32.1	826
	Spl	−0.17	0.04	−0.34	0.05			0.453	0.68	15.5	50.2	1015	
17QH422	Ol	0.35	0.04	0.70	0.06	−0.48	0.06	0.913					-
	Spl	−0.13	0.04	−0.26	0.06			0.406	0.73	12.9	51.4	1088	
17QH423	Ol	0.31	0.04	0.62	0.05	−0.50	0.06	0.912				33.2	779
	Spl	−0.19	0.04	−0.38	0.05			0.420	0.76	11.6	55.1	1009	
17QH425	Ol	0.29	0.04	0.59	0.04	−0.45	0.06	0.920				30.1	829
	Spl	−0.16	0.04	−0.33	0.06			0.503	0.64	17.9	48.5	971	
17QH415	Ol	0.27	0.05	0.54	0.04	−0.32	0.06	0.913				32.3	889
	Spl	−0.05	0.03	−0.05	0.08			0.455	0.67	16.3	49.8	1030	
17QH416	Ol	0.32	0.04	0.64	0.08	−0.41	0.04	0.914				30.0	925
	Spl	−0.09	0.02	−0.17	0.06			0.510	0.64	18.4	48.8	1289	
17QH418	Ol	0.21	0.02	0.42	0.01	−0.39	0.04	0.923				29.2	912
	Spl	−0.18	0.03	−0.28	0.09			0.507	0.69	15.4	52.0	1109	
17QH420	Ol	0.27	0.02	0.53	0.03	−0.39	0.04	0.919				30.9	922
	Spl	−0.12	0.04	−0.25	0.02			0.550	0.59	21.6	45.5	1408	
17QH427	Ol	0.29	0.04	0.58	0.02	−0.42	0.05	0.913				31.6	916
	Spl	−0.12	0.03	−0.25	0.10			0.464	0.66	15.4	51.0	1264	
17QH428	Ol	0.27	0.04	0.53	0.00	−0.43	0.05	0.911				33.0	916
	Spl	−0.16	0.03	−0.31	0.08			0.492	0.61	16.6	48.9	1138	
17QH429	Ol	0.28	0.07	0.55	0.08	−0.47	0.07	0.912				31.9	899
	Spl	−0.19	0.02	−0.45	0.11			0.455	0.67	19.8	45.9	1205	

Continued

Sample	Mineral	δ^{66}Zn	2sd	δ^{68}Zn	2sd	$\Delta^{66}Zn_{Spl-Ol}$	2sd	Mg#	Cr#	Al_2O_3	Cr_2O_3	Zn (μg/g)	T(℃)
Standards													
BCR-2		0.27	0.04	0.54	0.05								
BHVO-2		0.33	0.04	0.65	0.04								
IRMM 3702		0.28	0.03	0.56	0.05								

Mg# = molar Mg/(Mg+Fe^{2+}), Cr# = molar Cr/(Cr+Al). The units for δ^{66}Zn, δ^{68}Zn and $\Delta^{66}Zn_{Spl-Ol}$ are per mil (‰). Data of major elements (wt.%) and temperatures (calculated using pyroxene thermometer) are from Liu et al. (2017) and Zhang et al. (2019).

4 Results

Zinc concentrations and isotopic compositions of Cr-spinel and olivine separates are listed in Table 1. Olivine separates have Zn contents of 29.2 to 41.3 μg/g, lower than those of olivines in fertile mantle peridotites (e.g., Le Roux et al., 2010) and that of the primitive mantle (~55 μg/g; McDonough and Sun, 1995). δ^{66}Zn values of olivines vary from 0.19‰ to 0.35‰ (n=13), which falls within the δ^{66}Zn range already reported for the unaltered ophiolitic peridotites (0.17‰ to 0.33‰, Liu et al., 2019). Compared with olivines, Cr-spinels have extremely high Zn contents and the $[Zn]_{Cr\text{-}Spl}/[Zn]_{Ol}$ ratios are up to 12–45 (Fig. 2). δ^{66}Zn values of Cr-spinels vary from −0.19‰ to −0.05‰ (n=13). All analyzed olivines and spinels display mass-dependent Zn isotope fractionation with positive-correlated δ^{68}Zn and δ^{66}Zn values with a slope of 1.993 (R^2=0.998, Fig. S1). The apparent Zn isotope fractionation ($\Delta^{66}Zn_{Spl-Ol}=\delta^{66}Zn_{Spl}-\delta^{66}Zn_{Ol}$) ranges from −0.50‰ to −0.33‰ (n=13). This is opposite to the positive Zn isotope fractionation ($\Delta^{66}Zn_{Spl-Ol}$=0.13±0.09‰) between Al-spinels and olivines observed in cratonic peridotite xenoliths (Wang et al., 2017). That is, an unexpected, "inverse" direction of spinel-olivine Zn isotope fractionation is observed in oceanic peridotites (Fig. 3).

5 Discussion

In this section, we firstly discuss possible mechanisms for the "inverse" spinel-olivine Zn isotope fractionation in oceanic peridotites. Then, we discuss the influence of spinel crystallization on Zn isotopic systematics of basaltic magmas based on the apparent isotope fractionation factors between Cr-spinels and silicates. Finally, we test the influence of Cr-spinel crystallization on Zn isotopic systematics of basaltic magmas.

5.1 Effects of low-temperature alteration

Compared with cratonic peridotites, oceanic (both abyssal and ophiolitic) peridotites are readily subjected to fluid-rock interaction and significant Zn isotopic variations have been observed during this process (Pons et al., 2011, 2016; Liu et al., 2019; Debret et al., 2020). Given that the studied abyssal and ophiolitic peridotites have undergone various extents of serpentinization (Liu et al., 2017; Zhang et al., 2019), secondary low-temperature alteration must be considered for any Zn isotopic variations in minerals within them.

5.1.1 Serpentinization and weathering of olivines

Olivine is sensitive to serpentinization on the seafloor and in the forearc mantle, which typically results in the formation of serpentine minerals like lizardite and chrysotile, together with other minerals (brucite ± talc ± magnetite; Evans et al., 2013). Olivines in the studied oceanic peridotites underwent only slight serpentinization as revealed by petrographic observations (Fig. S2; Zhang et al., 2019), and are isotopically heavier than olivines in unaltered cratonic peridotites by ~0.10‰ (Fig. 3). Petrographic observations revealed the presence of a small amount of carbonate-bearing fluid inclusion in olivine from the Yushigou harzburgites (Zhang et al., 2021), and such fluids may have heavy Zn isotopic composition (Fujii et al., 2014; Debret et al., 2020). However, there is a lack of correlation between δ^{66}Zn values and CaO contents in olivines (Fig. 4a, 4b), and no carbonate-bearing fluid inclusion has been identified in olivines from the studied abyssal peridotites. In general, olivine serpentinization and weathering will lead to an increase in whole-rock LOIs, and a decrease trend of MgO/SiO_2 but nearly constant Al_2O_3/SiO_2 (Liu et al., 2017; Hart and Zinder, 1986). This process can be assessed by the proxy of MgO/SiO_2^* that is a measurement of departure of the analyzed MgO/SiO_2 from the unaltered geochemical fractionation array (Hart and Zindler, 1986). An appreciable drop in MgO/SiO_2 would imply extensive Mg loss due to alteration or weathering (Snow and Dick, 1995), although fore-arc peridotites may have elevated SiO_2 content due to melt extraction (Ionov, 2010; Zhang et al., 2019). No correlation between $\Delta^{66}Zn_{Spl\text{-}Ol}$ and MgO/SiO_2^* has been observed (Fig. 4c, 4d), indicating that the inverse Zn isotope fractionation between Cr-spinel and olivine cannot result from olivine serpentinization and weathering. In particular, the observed δ^{66}Zn difference between Cr-spinel and olivine (up to −0.50‰) is much larger than the isotopic difference between serpentinite and unaltered mantle peridotites (Debret et al., 2018; Liu et al., 2019). In one abyssal peridotite sample (PS59 235-4) that is least altered and contains only a few percent serpentine, Cr-spinel is also isotopically lighter than coexisting olivine (Fig. 3). Thus, olivine serpentinization and interaction with external CO_2-bearing fluids are difficult to account for the inverse isotope fractionation between spinel and olivine in oceanic peridotites, although the weak serpentinization may slightly elevate the primary δ^{66}Zn value of olivines by ~0.10‰

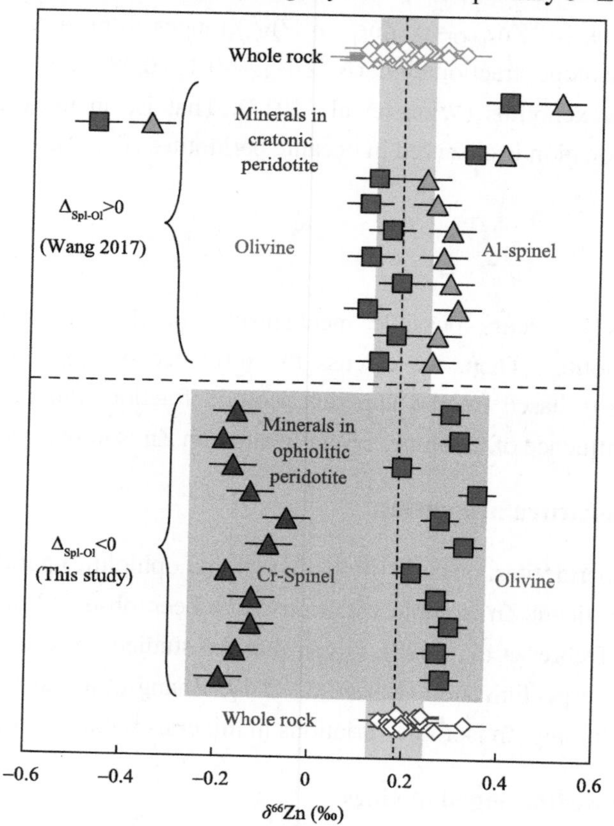

Fig. 3 Zinc isotope fractioantion between spinel and olivine ($\Delta_{Spl\text{-}Ol}$) in peridotites from continental and oceanic settings. The data for continental (cratonic) peridotites are from Wang et al. (2017), and data for oceanic (both abyssal and ophiolitic) peridotites are from this study (Table 1). For the whole rock data, cratonic peridotites are from Doucet et al. (2016), Wang et al. (2017), Sossi et al. (2018) and Huang et al. (2019), and oceanic peridotites are from Liu et al. (2019). The dashed area represents the estimated δ^{66}Zn value for Bulk Silicate Earth (0.18±0.06‰, Wang et al., 2017; Sossi et al., 2018; Liu et al., 2019).

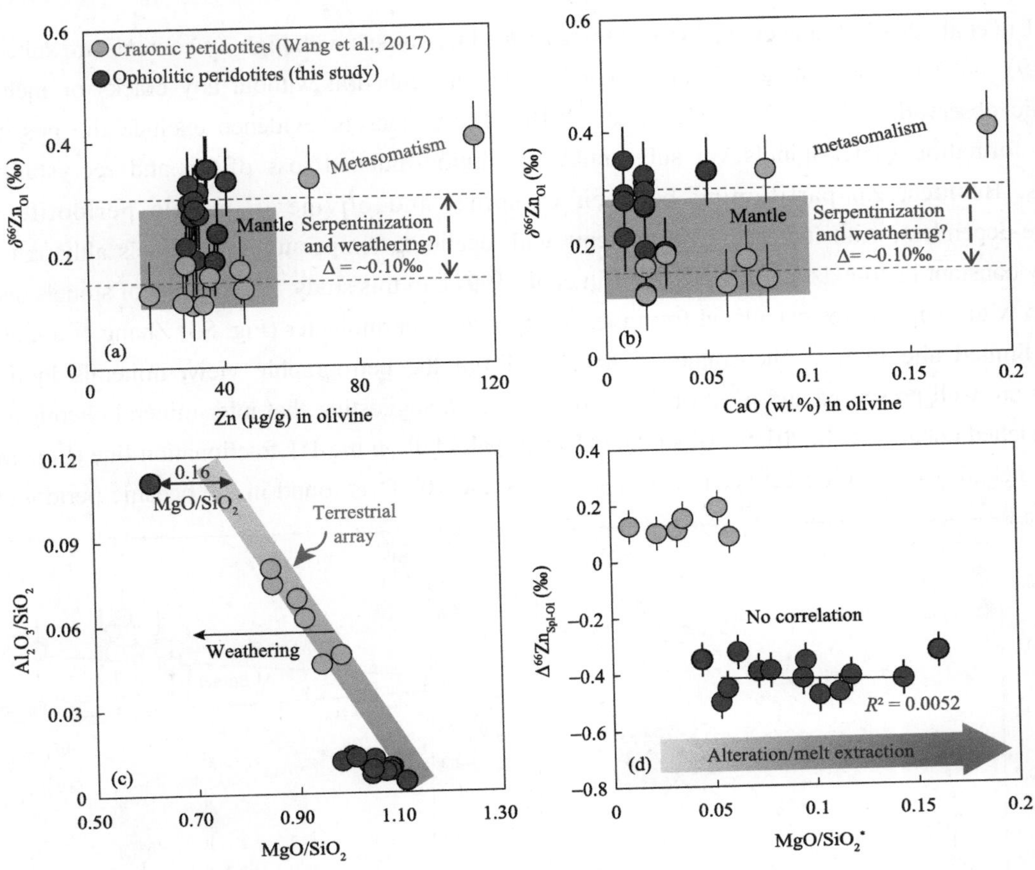

Fig. 4 Plots of $\delta^{66}Zn$ against indices of low temperature alteration. (a–b)$\delta^{66}Zn_{ol}$ against Zn and CaO contents of olivines in ophiolitic peridotites. Only the metasomatised olivines in wehrites have distinct chemical and isotopic compositions. The black and red dashed lines denote the average $\delta^{66}Zn$ values of olivines in cratonic and ophiolitic peridotites, respectively, which shows a small $\delta^{66}Zn$ difference of 0.10‰ possibly induced by serptinization and/or weathering of ophiolitic olivines. (c) Plot of Al_2O_3/SiO_2 against MgO/SiO_2^* values for the studied peridotites; the terrestrial array is from Hart and Zinder (1986). (d) Plot of $\Delta^{66}Zn_{Spl-Ol}$ against MgO/SiO_2^*, indicating a limited effect of low-temperature alteration on the Zn isotopic systematics. (For interpretation of the references to colour in this figure legend, the reader is referred to the web version of this article.)

relative to the unaltered olivines possibly due to the preferential adsorption of heavy Zn onto secondary mineral phases (Liu et al., 2019).

5.1.2 Secondary Cr-spinels formed by seafloor alteration?

Spinel is more resistant to seafloor alteration relative to silicate minerals such as olivine (Barnes and Roeder, 2001), but serpentinization may involve the dissolution and recrystallization of mantle spinels into magnetite, Cr-spinel, and ferrichromite (e.g., Marques et al., 2007). Zoned porous chromites with high Cr# (>0.80) formed via metasomatism and re-mineralization typically have high Zn contents of up to 4000 μg/g (Fig. 2c, 2d), due to the preferential incorporation of divalent elements like Zn, Mn and Co into spinels during low-grade greenschist-amphibolite facies metamorphism and serpentinization (Singh and Singh, 2013; Yu et al., 2019). However, Zn concentrations of spinels in this study are much lower than those of metasomatized Cr-spinels (Fig. 2c, 2d), and the analyzed Cr-spinels are all tiny and not zoned (Fig. S2; Liu et al., 2017; Zhang et al., 2019). Considering the possible existence of metasomatism-induced zoning, only Cr-spinel grains with multiple spots of analysis can be used as the homogeneous petrogenetic indicator (El Dien et al. 2019). There are more than one analysis for spinels in studied peridotites, and no obvious variation in Cr# among different spinel grains in individual samples is observed (Liu et al., 2017; Zhang et al., 2019). Most of zoned Cr-spinels with high Zn contents are anhedral (El Dien et al., 2019), but Cr-spinels in the Yushigou harzburgites are euhedral without any cracks or inclusions under petrographic observation (Fig. S2; Zhang et al., 2019). These lines of evidence exclude the possibility of the secondary formation of Cr-spinels via substantial dissolution-induced loss of Zn and recrystallization from Al-spinels. Besides, Zn partitioning between Cr-spinel and olivine in mantle peridotite is strongly temperature-dependent, with increasing Zn contents with declining temperatures in spinels although Zn contents keep nearly constant in co-existing olivines (Griffin et al., 1994). In this study, Zn contents of spinels display a good relationship with temperatures calculated from the two-pyroxene thermometer (Fig. S3; Zhang et al., 2019), further indicating limited alteration of the studied Cr-spinels. From the petrographic view, minerals in the Yushigou harzbugites are well-preserved and most of them are not zoned, suggesting that inter-mineral chemical equilibrium has been reached (Zhang et al., 2019). All studied mineral pairs fall on the 1:1 fractionation line (Fig. 5a), however, no obvious negative correlation between $\Delta^{66}Zn_{Spl-Ol}$ values and $10^6/T^2$ is found in the oceanic peridotites (Fig. 5b).

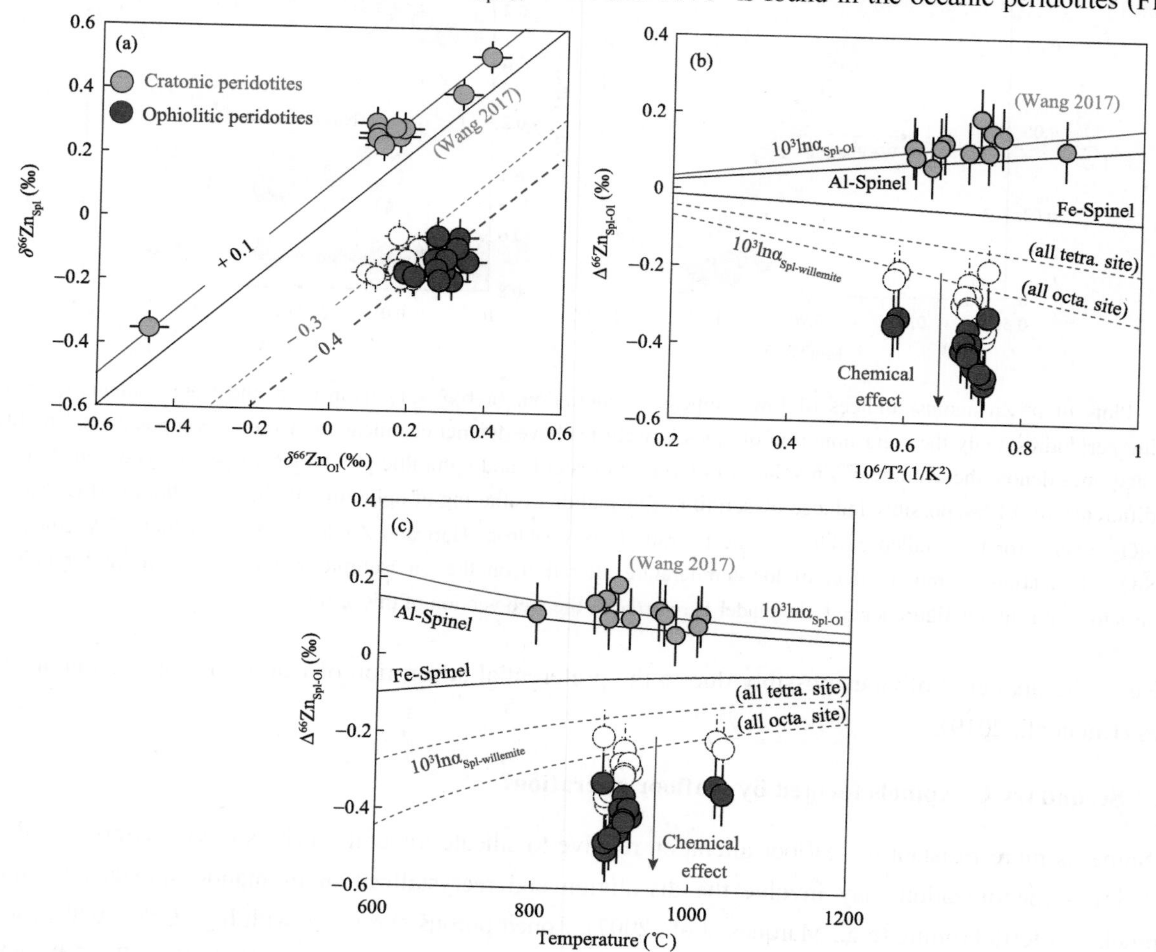

Fig. 5 (a) Plot of $\delta^{66}Zn_{Spl}$ against $\delta^{66}Zn_{Ol}$. Data ploted on the 1:1 line is indicative of inter-mineral equilibrium isotope fractionation. (b-c) Plots of the measured spinel-olivine Zn isotope fractionation against equilibrium temperature (℃) in peridotite. The fractionation of ~0.10‰ induced by olivine serpentinization and weathering is subtracted from the measured values. Theoretical calculations for the spinel-olivine and spinel-willemite (Zn_2SiO_4) fractionation by the ionic model are shown as red solid and dashed lines, respectively. Theoretically calculated values between spinel with different travalent ion compositions (gahnite, franklinite) and silicate minerals (e.g., hemimorphite) are also shown for comparson (Ducher et al., 2016). The red arrow reprensents potential "chemical effect" on Zn isotopic composition of spinels at the same temperatures. See text for more details. Equilibrium temperatures for oceanic peridotites are listed in Table 1, and those for cratonic peridotites are from Wang et al. (2017) and reference therein. (For interpretation of the references to colour in this figure legend, the reader is referred to the web version of this article.)

These observations in turn indicate that an equilibrium Zn isotope fractionation between Cr-spinel and olivine may have been achieved at high temperatures, but the fractionation is not exclusively controlled by temperature. Similar phenomenon is also observed for Mg isotope fractionation between spinel and forsterite at the same temperature that is highly correlated with Cr# of spinel (Tang et al., 2021). Notably, Cr-spinels have a larger chemical variation compared with olivines (Fig. S3), indicating a possible mechanism induced by "chemical effect" in spinel for the "inverse" Zn isotope fractionation in oceanic peridotites.

5.2 A chemical control on spinel-silicate Zn isotope fractionation

The general chemical formula of spinel-group minerals is $A^{2+}B_2^{3+}O_4^{2-}$, with variable cation occupancies in the octahedral and tetrahedral sites. Normally, divalent metals (such as Mg^{2+}, Fe^{2+}, Ni^{2+}, Zn^{2+}, etc.) occupy the A site, whereas trivalent cations (e.g. Al^{3+}, Cr^{3+}, Fe^{3+}, etc.) occupy the B site. However, spinel can vary from the normal state with only A^{2+} in the tetrahedral site (T-site, four-fold coordination) to the inverse state with only B^{3+} in the tetrahedral site (M-site, six-fold coordination) and all the A^{2+} and the rest of B^{3+} in the octahedral site. The spinel formula can be re-written as $A_{1-i}B_i[A_iB_{2-i}]O_4$, where the disorder is defined by the inversion parameter, i (O'Neill, 1992). Zn^{2+} (0.74 Å) has a similar ionic radii to that of Mg^{2+} (0.72 Å) and thus mainly substitutes for Mg^{2+} in the tetrahedral site of normal spinel. The inversion parameter (i) of zinc ferrite ($ZnFe_2O_4$) is up to 0.19 at 900 ℃ (O'Neill, 1992), indicating potential octahedral site occupation of Zn in inverse spinel. The structures of normal and inverse spinels depend on the crystal field stabilization energy (CFSE) of ions in tetrahedral and octahedral sites. In zinc ferrite, the small Zn^{2+} ion with d^{10} of metal d-orbital energy fits more easily into the tetrahedral site comparing with $Fe^{3+}(d^5)$, and both Zn^{2+} and Fe^{3+} have zero CFSE. In addition, Cr-spinels are always normal spinels because Cr^{3+} strongly prefers the octahedral site. Overall, from $ZnAl_2O_4$ through $ZnCr_2O_4$ to $ZnFe_2O_4$, the normal spinel structure will remain unchanged based on the crystal field theory.

5.2.1 Ionic model prediction for Zn isotope fractionation between spinel and silicate

Theoretical studies suggest that equilibrium inter-mineral isotope fractionation is mainly controlled by the bonding environment of the element of interest, with stronger bonding favoring incorporation of heavy isotopes (Bigeleisen and Mayer, 1947; Urey, 1947). In general, stronger bonding is associated with lower coordination number (CN). Accordingly, spinel-group minerals should be enriched in heavier Zn isotopes relative to silicate minerals (e.g., olivines) due to the lower CN of Zn^{2+} in spinel when equilibrium isotope fractionation is reached, which is indeed observed in cratonic peridotites (Wang et al., 2017) (Fig. 3). Unexpectedly, an inverse direction of Zn isotope fractionation between Cr-spinel and olivine is obtained in oceanic peridotites in this study. A similar phenomenon was originally mentioned in Ben Othman et al. (2006), but unexplained and simply attributed to the different geodynamic environments (continental vs. oceanic). Here we calculate the equilibrium Zn isotope fractionation factor between spinel and olivine according to the ionic model (Macris et al., 2015; Shen et al., 2018; Young et al., 2009, 2015). The equation for Zn isotope fractionation factors between two minerals A and B ($\alpha_{A\text{-}B}$), can be simply expressed as:

$$\delta^{66}\mathrm{Zn}_A - \delta^{66}\mathrm{Zn}_B \approx 10^3 \alpha_{A-B}^{\frac{66}{64}} = \frac{10^3}{24}\left(\frac{h}{k_b T}\right)^2 \left(\frac{1}{m_{64}} - \frac{1}{m_{66}}\right)\left[\frac{K_{f,A}}{4\pi^2} - \frac{K_{f,B}}{4\pi^2}\right]$$

Where m_{64} and m_{66} represent the atomic masses of ^{64}Zn and ^{66}Zn, respectively. In addition, k_b is Boltzmann's constant, h is the Planck's constant, and $K_{f,A}$ and $K_{f,B}$ are the average force constants for phases A and B, respectively. For instance, A is spinel and B represents olivine in this model. We adopt a well-developed approach, which is similar to that used for Fe isotopes by Macris et al. (2015) and Mg isotopes by Young et al. (2015), to qualify the isotopic effects driven by changes in coordination. The equation for average force constants $K_{f,ij}$ is:

$$K_{f,ij} = \frac{z_i z_j e^2 (1-n)}{4\pi\varepsilon_0 r_{ij}^3}$$

Where z_i and z_j represent the valences of cation and anion, respectively, e is he charge of an electron, ε_0 is the vacuum permittivity, and n is Born-Mayer constant for the repulsion term which empirically equals to 12 (Young et al., 2015). Based on the effective cation and anion radii of Shannon (1976), the equilibrium inter-ion distance r_{ij} is calculated for a given coordination environment and cation valence. In general, heavy isotopes preferentially concentrate in sites with lower coordination number and higher valence (Young et al., 2015). In the case of Zn isotopes, the different coordination environment is the main factor in the ionic model due to the absence of valence changes of Zn in silicate minerals. For olivine with a general formula (X_{M1}, Y_{M2})SiO_4, the relatively small-sized Zn^{2+} ion is generally incorporated into the M1 octahedral site (e.g. Redfern, 1998). Zinc occupies tetrahedral site in pure zinc silicates such as willemite (Zn_2SiO_4) at ambient pressure (e.g., Klaska et al., 1978). In addition, Zn prefers tetrahedral site in normal spinel such as Al-spinel and Cr-spinel, and Zn in octahedral site, if any, is limited in both normal and inverse spinels due to crystal field energy theory.

The estimated spinel-olivine Zn isotope fractionation factors between 600 ℃ and 1200 ℃ are shown in Fig. 5, together with the calculated values between spinels with different octahedral ion compositions (gahnite, franklinite) and silicate minerals (e.g., hemimorphite) based on ab initio calculation (Ducher et al., 2016). The $\Delta^{66}Zn_{Spl\text{-}Ol}$ variation induced by olivine serpentinization and weathering is subtracted by ~0.10‰ from the measured value as discussed above (Fig. 4, 5). Unexpectedly, even if one consider that Zn occurs in both tetrahedral and octahedral sites in spinel, the magnitude of equilibrium fractionation is still smaller than the observed Zn isotope fractionation between Cr-spinel and olivine at the same temperature (Fig. 5). Only the $\Delta^{66}Zn_{Spl\text{-}willemite}$ (willemite: Zn_2SiO_4) value is close to the fractionation values between Cr-spinel and olivine in oceanic peridotites. It should be noted that Cr-spinels have much larger chemical variations compared with olivines (Fig. S3), and the effect of compositional change of spinel on Zn isotope fractionation between minerals is not able to be obtained in the current ionic model. Therefore, additional effect is required to account for the inverse Zn isotope fractionation between Cr-spinel and olivine.

5.2.2 The "chemical effect" on spinel-silicate Zn isotope fractionation

Natural spinel-group minerals are structurally similar but chemically diverse. Theoretical calculations and experiment results show that partitioning of Mg isotopes in spinel minerals at similar temperatures strongly depends on the octahedral ion composition (Schauble, 2011; Tang et al., 2021). Namely, the reduced partition function ratios (β-factors) for $^{26}Mg/^{24}Mg$ decrease significantly from aluminum spinel ($MgAl_2O_4$) through magnesiochromite ($MgCr_2O_4$) to magnesioferrite ($MgFe_2O_4$). Similar chemical effect also exists for Zn isotopes. The β-factors for $^{66}Zn/^{64}Zn$, expressed as $1000\cdot\ln(\beta_{66\text{-}64})$, are found to remarkably decrease from gahnite ($ZnAl_2O_4$) to franklinite ($ZnFe_2O_4$), although Zn atoms are always tetrahedrally coordinated to oxygen atoms and both Al and Fe atoms occupy the octahedral sites (Ducher et al., 2016). It is noteworthy that the $1000\cdot\ln(\beta_{66\text{-}64})$ value of franklinite can become lower than that of zinc silicate (e.g., hemimorphite). This indicates that Fe^{3+}-rich spinel ($A^{2+}Fe_2^{3+}O_4$), in contrast to Al-spinel ($A^{2+}Al_2O_4$), can be enriched in lighter Zn isotopes relative to silicate minerals in equilibrium. However, no β-factor data are available yet for $ZnCr_2O_4$ in the literature, but a decreasing β-factor could be expected when Cr for Al substitutions are made by referring to the models of Schauble (2011) and Ducher et al. (2016). Thereupon, further evidence from the crystal structure perspective is needed.

Notably, the changing of β-factors reflects the bonding environment variation in crystal cell. Zn isotopic variation in natural spinels is mainly controlled by the Zn—O bond length variation in the tetrahedral site in equilibrium (1.950 to 1.999 Å, O'Neill, 1994; Andreozzi et al., 2001). Bosi et al. (2011) suggested that the variation of T−O bond length in Zn-rich spinel ($ZnAl_xCr_{2-x}O_4$) is mainly controlled by cation occupancy in both octahedral and tetrahedral sites. Here we collect previous bond length data for T—O, M—O sites in Zn-rich spinels and the chemical compositions of spinels as well. The bonding environments of Zn-rich and Mg-rich spinels are shown in Fig. 6. There are excellent positive correlations in the plots of M—O bond length against Cr# and T—O bond length in spinels (Fig. 6a, 6b). The ionic radii of Cr^{3+} (0.615) is larger than that of Al^{3+} (0.535) in octahedral site, resulting in longer Cr—O bond in M—O site as Cr# increases in spinels. These correlations indicate that Zn—O bond length extends with increasing Cr# in spinel as a result of Al^{3+} substitution by Cr^{3+} in the M—O site. During the substitution of Cr^{3+} into Al^{3+}, the octahedral bond angle decreases (Fig. 6c), resulting in longer bond length of $(O—O)_{shared}$ in MO_6 compared with ideal octahedron. This increase provides a reduced oxygen shielding effect that enhances the interaction of the octahedral cations and leads to an energetic stability of the spinel structure, which in turn increases Zn—O bond length in tetrahedral site (Bosi et al., 2011). Typically, Zn—O bond length increases from ~1.950 Å in gahnite ($ZnAl_2O_4$) to ~1.975 Å in Zn-chromite ($ZnCr_2O_4$), which is consistent with a decrease in $\delta^{66}Zn$ from Al-spinels to Cr-spinels. Besides, since Zn and Mg can share an oxygen atom in tetrahedral site, Zn substation for Mg in Al-spinel can also extend the T—O bond length based on experimental data (Andreozzi et al., 2001). By contrast, Zn—O bond lengths in both M1 and M2 sites in Zn-rich silicate minerals such as willemite (Zn_2SiO_4) are around ~1.96 Å (Klaska et al., 1978), which is at the intermediate value of the variation ranges for Zn-rich spinels (Fig. 6b). This is also consistent with an inverse $\Delta^{66}Zn_{Spl\text{-}Ol}$ observed in this study. When the Zn—O bond length of spinel slightly exceeds the bond length of Zn—O in

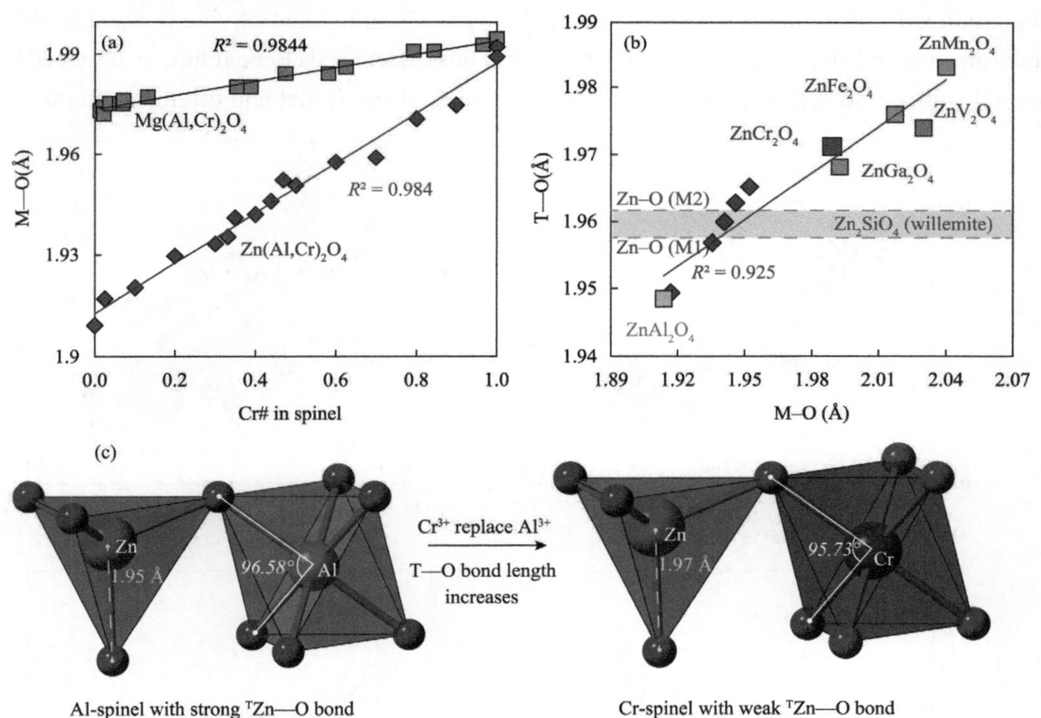

Fig. 6 Diagrams for bonding environment variations in spinels. (a) Plot of bond length in octahedral site (M—O, in unit Å) against Cr# in spinel. Data for Zn-rich spinel and Mg-rich spinel are shown (Bosi et al., 2011; Hålenius et al. 2010; Verger et al., 2016). (b) Diagram of bond length changes in tetradral site (T—O) and octrahedral site (M—O), including experimental synthetic spinels and natural spinels (Bosi et al., 2011; O'Neill and Dollase., 1994; O'Neill et al., 1992; Ebbinghaus et al. 2004; Åsbrink et al. 1999). A positive trend is seen; the bond length data for willemite (Zn_2SiO_4) is shown for comparsion (Klaska et al., 1978). (c) A cartoon illustration of crystall struatures from Al-spinel to Cr-spinel. The smaller bond angle and increasing Zn—O bond lengths are shown (Bosi et al., 2011).

olivine as spinel Cr# increases, the $\Delta^{66}Zn_{Spl-Ol}$ then becomes lower than zero and the inverse fractionation direction begins to appear.

Natural observations support the theoretical expectation. The $\delta^{66}Zn$ of natural spinels gradually decrease with increasing ($Cr+Fe^{3+}$) and ($Mn+Zn+Fe^{2+}$) contents (R^2=0.923, 0.737; Fig. 7a, 7b), which are the indices of increasing substitution of Al^{3+} by Cr^{3+} and Fe^{3+} in the octahedral site, together with substitution of Mg by divalent element (such as Zn, Fe^{2+}, etc.) in tetrahedral site. In addition, well-fitted positive and negative correlations between temperature-corrected $\Delta^{66}Zn_{Spl-Ol}$ value and Mg# or Cr# are shown in Fig.7c, 7d, respectively. Using the calculated apfu value of spinel composition, we adopt chromite and Al-spinel proportions as indicators of compositional change in spinel (chromite%=100×($Fe^{2+}+Cr^{3+}$)/total cation; Al-spinel%=100×($Mg^{2+}+Al^{3+}$)/total cation). The $\Delta^{66}Zn_{Spl-Ol}$ values linearly decrease with increasing chromite proportion and decreasing proportion of Al-spinel (Fig.7e, 7f). Due to the small variation of Fe^{3+} contents in studied spinels, the compositional effect induced by Fe^{3+} substitution should be much smaller than that of Cr^{3+}. Only one sample (PS59 235-4) from Gakkel Ridge does not fall within the trend, and it may result from slightly low-temperature weathering and serpentinization of olivine as indicated by high MgO/SiO_2^* value of this sample (Fig. 4). These observations support the prediction that the various Zn isotopic compositions of natural spinels are a result of different octahedral and tetrahedral ion compositions, rather than the difference in geodynamic environments (oceanic vs. continental) (Ben Othman et al., 2006). Similar inverse olivine-spinel fractionation is also observed in Fe isotope systematics of the studied ophiolitic peridotites (Zhang et al., 2019), and this fractionation may be induced by increasing chromite component in spinel that leads to increasing Fe—O bond length (Lenaz and Skogby, 2013). Notably, the predicted Zn isotope fractionation factor between Cr-spinel and olivine is considerably larger than that of in inter-mineral Fe isotope fractionation (Fig. S4), and this offset needs to be future investigated by theoretical or experimental studies. To sum up, Zn isotope fractionation between spinel and olivine is mainly controlled by

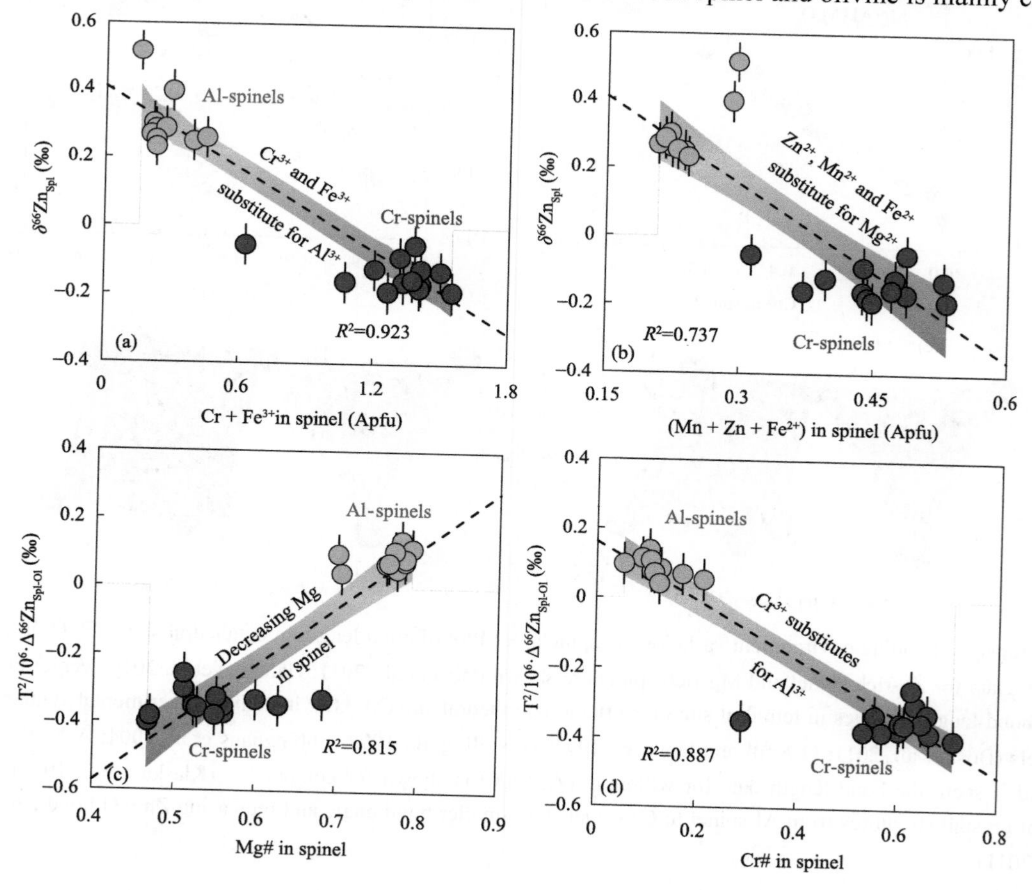

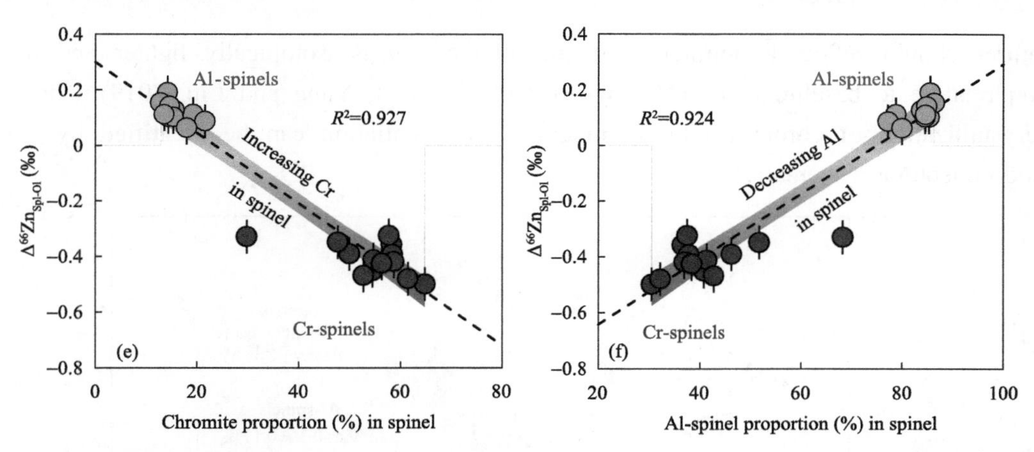

Fig. 7 (a–b) Plots of $\delta^{66}Zn$ against ($Cr+Fe^{3+}$) and ($Mn + Zn + Fe^{2+}$) content in octahedral site and tetrahedral site of spinel (in apfu, atom per formula unit), respectively, showing decreasing $\delta^{66}Zn$ as the transition of Al-spinel ($MgAl_2O_4$) to Cr-spinel ($FeCr_2O_4$). (c–d) Plots of temperature-corrected Zn isotope fractionation (expressed as $T^2/10^6 \cdot \Delta^{66}Zn_{Spl-Ol}$) against Mg# and Cr# of spinel, respectively. (e–f) Plots of $\Delta^{66}Zn_{Spl-Ol}$ against the proportion of chromite component (%) and the proportion of $MgAl_2O_4$ component (%) in spinel, respectively. The content of chromite and $MgAl_2O_4$ in spinel is equivalent to 100 × ($Cr^{3+} + Fe^{2+}$)/total cation and ($Al^{3+} + Mg^{2+}$)/total cation, respectively (Table S1). Trend in these diagrams represents the substituting of Al into Cr in the octahedral site of spinel. Major element data of spinels are from Liu et al. (2017), Zhang et al. (2019), Wang et al. (2017) and reference therein. Uncertainties (2s.d.) of $\Delta^{66}Zn$ are calculated as the square root of the sum of the square of individual errors for $\delta^{66}Zn$ values, and the red shadow area represents 95% confidence intevial of the dataset.

changing bonding environments in spinels. Therefore, Zn isotopes may be utilized as a novel indicator for compositional variation in natural spinels and associated ultramafic rocks.

5.3 Zinc and its isotopic behaviors during chromite crystallization

The direct pathway to evaluate Zn isotope fractionation during chromite crystallization is to analyze crystallized chromite grains in basaltic melts. However, chromite in basaltic melts commonly occurs as tiny inclusions within or at the edge of phenocrysts like olivine or clinopyroxene, which are crystallized at different evolving stages (e.g. Scowen et al., 1991). Chromite that occurs in peridotites commonly has a distinct boundary with coexisting silicate minerals (e.g. Song et al., 2009), which can be easily separated. Thus, the $\Delta^{66}Zn_{Spl-Ol}$ values obtained in this study allows us to assess the role of chromite crystallization in Zn isotopic systematics of mantle-derived melts. Given the high partition coefficient of Zn in spinels/chromites relative to silicate melts ($D_{Zn}^{Spl/melt}$ = 5.3; Davis et al., 2013), significant Zn isotope fractionation is expected to occur during this process.

Using MELTS and following a Rayleigh distillation process, we modelled Zn isotopic variation of the residual melt as a function of chromite crystallization. The isotope fractionation factor (α) between Cr-spinel and silicate melt is assumed to be slightly larger than the apparent fractionation between Cr-spinel and olivine (−0.50‰ to −0.33‰), considering that basaltic melt is isotopically heavier by ~0.10‰ than olivine in equlibrium (Wang et al., 2017; Sossi et al., 2018). Variable fractionation factors ($\alpha_{Spl\text{-}melt}$) of 0.9996 to 0.99995 are set for chromite crystallization. The Zn content and $\delta^{66}Zn$ value of the initial melt in this model are assumed to be close to those of the average MORB ([Zn] = 80 μg/g, $\delta^{66}Zn$ = 0.28‰; Wang et al., 2017; Sossi et al., 2018). The results show that chromite crystallization will lead to a moderate increase of $\delta^{66}Zn$ values but a drastic decline of Zn concentrations in the residual melts (Fig. 8). Chen et al. (2013) attributed the slightly elevated $\delta^{66}Zn$ with increasing Zn concentrations at late-stage of magma evolution of the Kilauea Iki lava lake (Hawaii) to fractional crystallization of olivine and/or Fe-Ti oxides. Our results show that fractional crystallization of Zn-bearing oxides (e.g., chromite, magnetite) is unlikely to have played an important role in Zn systematics of these samples (Fig. 8).

Such signatures should reflect a dominant role of olivine that is isotopically lighter and has lower Zn concentration relative to basaltic melts (McCoy-West et al., 2018; Yang and Liu, 2019). Thus, the role of fractional crystallization of chromites during magmatic differentiation can be identified by combining Zn elemental and Zn isotopic systematics.

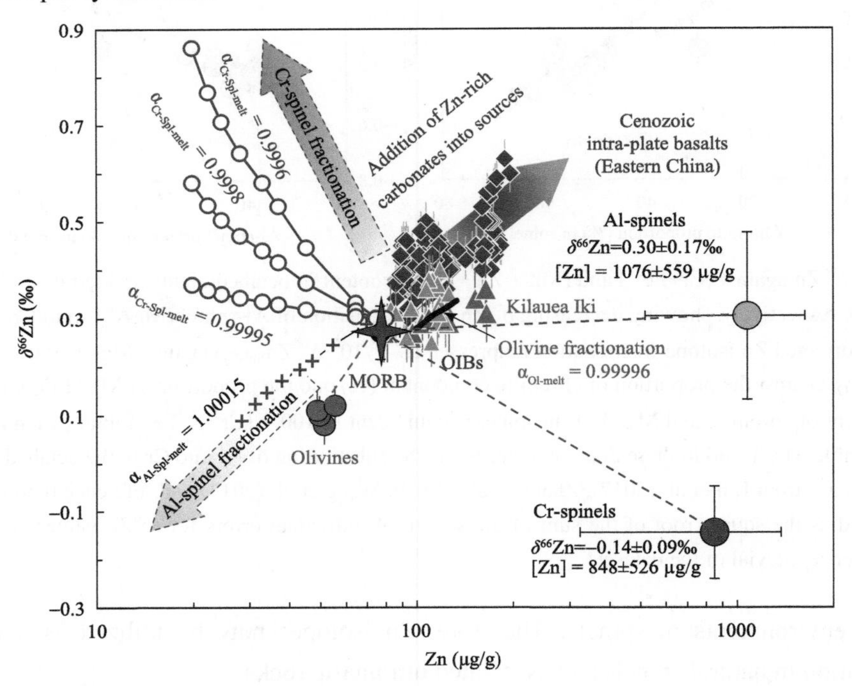

Fig. 8 Modelling results for the variations of δ^{66}Zn and Zn contents of melts as a function of fractional crystallization of chromite, Al-rich spinel, and olivine in magmas, calculated by assuming a Rayleigh distillation process. The equation is: $\delta^{66}Zn_{melt} = (\delta^{66}Zn_{initial} + 1000)f^{(\alpha-1)} - 1000$, where f is the fraction of melts remaining calculated using MELTS. The average MORB (δ^{66}Zn = 0.28 ± 0.03‰, Zn = 80 μg/g; Wang et al., 2017; Huang et al., 2018) is set as the initial melt. The mineral-melt isotope fractionation factors (α) are assumed to be the apparent fractionation factors between spinel/chromite and silicate minerals. Data for olivine phenocrysts are from Yang and Liu (2019). Kilauea Iki lavas (Chen et al., 2013), global OIBs (Wang et al., 2017; Beunon et al., 2020), and intraplate alkali basalts in eastern China (Liu et al., 2016; Wang et al., 2018) are also shown for comparision.

The finding of high δ^{66}Zn values (up to 0.77‰) in some intraplate alkali basalts has led to the primary proposal of Zn isotopes as a novel tool of tracking recycling of marine carbonates into the mantle (Liu et al., 2016, 2020; Wang et al., 2018; Wang and Liu, 2021; Jin et al., 2020; Li et al., 2021). Recently, Beunon et al. (2020) collected a complete Zn isotopic dataset for global OIBs and also attributed their higher δ^{66}Zn values relative to those of MORBs to the widespread presence of recycled carbonate-bearing oceanic crustal component (carbonate + eclogite) in OIB's sources. Since chromite is a Mg-rich mineral (MgO>10 wt.%) and has heavier Mg isotopic composition than that of silicate minerals (Xiao et al., 2016; Su et al., 2017), a recent study evoked chromite crystallization as a primary cause for the light Mg isotopic compositions of mantle-derived melts (Su et al., 2019) As shown in Fig. 8, intraplate alkali basalts and global OIBs display an excellent positive correlation between δ^{66}Zn values and Zn contents, both of which are much higher than those of MORBs (Liu et al., 2016; Wang et al., 2018; Wang and Liu, 2021; Jin et al., 2020; Beunon et al., 2020). The coupling of high δ^{66}Zn and high Zn contents is opposite to the trend as expected for chromite crystallization. Instead, recycling of magnesium carbonates (e.g. magnesite) with heavy Zn isotopic composition and high Zn concentrations into the mantle sources accounts for the coupling of high δ^{66}Zn and Zn contents as well as high Zn/Fe ratios (Liu et al., 2016; Liu and Li, 2019; Beunon et al., 2020).

In comparison with terrestrial basalts (e.g., MORBs) with δ^{66}Zn value of ~0.3‰ and Zn concentrations of

~80 μg/g, lunar mare basalts have extremely high δ^{66}Zn (+1.4 ± 0.5‰) but very low Zn abundances (~1.5 μg/g), which is proposed to have mainly been caused by evaporation processes (Herzog et al., 2009; Paniello et al., 2012; Kato et al., 2015; Day et al., 2020). Experimental studies showed that evaporation with a large α value of 0.9959–0.9997 can explain the measured Zn elemental and isotopic systematics of mare basalts, indicating approximately 92–99% Zn loss of its initial value (Day et al., 2017a; Wimpenny et al., 2019; Sossi et al., 2020). However, whether or not late-stage magmatic differentiation has contributed to the loss of Zn and other volatile elements in lunar mare basalts is still debated, particularly considering that Cr-spinel fractionation can also result in high δ^{66}Zn value and Zn concentrations in the melts. The chromium and magnesium isotopic compositions of mare basalts are correlated with differentiation indices, which has been proposed to reflect fractional crystallization or cumulation of spinel and pyroxene during magma evolving (Bonnand et al., 2016; Sedaghatpour and Jocobsen, 2018).

Here we carried out modelling to assess to what extent the Zn systematics of mare basalts may have been influenced by chromite crystallization (Fig. 9a). For other extraterrestrial samples, the chromite crystallization trend obtained by chromite-rich inclusions in iron meteorites (Bridgestock et al., 2014) is also plotted in Fig. 9 for comparison. The modelling results show that >7% chromite crystallization is needed to account for the extremely low Zn content and heavy δ^{66}Zn of mare basalts, which will strongly lower Cr contents of mare basalts. However, the Cr contents of mare basalts are much higher than those of terrestrial basalts (Bonnand et al., 2016), which argues against significant fractional crystallization of chromite (Fig. 9b). It is proposed that the control of evaporation/condensation of Zn isotopes is also recorded by samples with clear surficial condensation such as orange glasses (Herzog et al., 2009) or the lunar breccias 66095 (Day et al., 2017b), and the fractionated Mg-suite lithologies have extremely high δ^{66}Zn exceeding +9‰ due to evaporation (Day et al., 2020; Van Kooten et al., 2020). It should be pointed out that the control of volatilization on the isotopic composition of mare basalts is also clearly reflected in other volatile elements, such as K (Wang and Jacobson, 2016; Tian et al., 2021), Rb (Pringle and Moynier, 2017), Ga (Kato and Moynier, 2017), Cl (Gargano et al., 2020) or Sn (Wang et al., 2019). Thus, the Zn elemental and isotopic systematics of lunar basalts are dominantly controlled by volatilization-related Zn loss during the giant impact (e.g., Paniello et al., 2012; Kato et al., 2015; Day et al., 2020), and the effect of chromite fractionation, if any, is limited or not sampled in the current lunar lithologies.

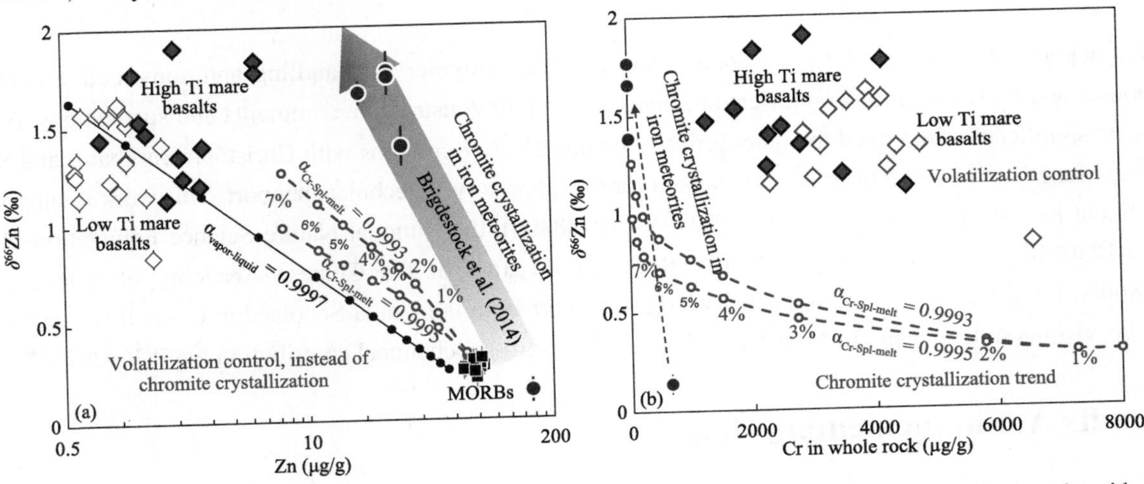

Fig. 9 (a) The plot of δ^{66}Zn against Zn content for lunar mare basalts, iron meteorites, and MORBs. Modelling results with different chromite-melt Zn isotope fractionation factors (α) are shown, together with the Rayleigh fractionation line for evaporation with $\alpha_{vapor-liquid}$ = 0.9997 (Wimpenny et al., 2019; Sossi et al., 2020). (b) the diagram of δ^{66}Zn against Cr content for lunar mare basalts and iron meteorites. Modelling results for chromite crystallization are shown with an assumed initial composition (Cr = 8000 μg/g and δ^{66}Zn = 0.28‰). Data are from Paniello et al. (2012), Kato et al. (2015), and Day et al. (2020) for mare basalts, from Bridgestock et al. (2014) for iron meteorites, and from Wang et al. (2017) and Huang et al. (2018) for MORBs.

6 Conclusion

We present the first Zn isotopic data for Cr-spinels and obtained Zn isotope fractionation between Cr-spinels and silicate minerals in oceanic peridotites. Along with previously reported data for Al-spinels, we show that Al-spinel is isotopically heavier whereas Cr-spinel is isotopically lighter than coexisting silicate minerals. Ionic model calculations show that the equilibrium fractionation is inconsistent with the observed zinc isotopic fractionation between Cr-spinel and olivine, even if we consider the presence of Zn at both tetrahedral and octahedral sites in spinel. The "inverse" $\Delta^{66}Zn_{spl\text{-}ol}$ value is attributed to the changing bonding environment of tetrahedral site in spinel which is mainly controlled by chemical composition variation in octahedral site in spinel. The negative correlation of $\delta^{66}Zn$ with Cr# in spinels from oceanic peridotites in this study strongly supports a chemical control on Zn isotopic systematics of spinels.

Cr-spinel has extremely high Zn contents relative to silicate melts, consistent with high partition coefficient of Zn in spinels. In conjunction with the lighter Zn isotopic composition of Cr-spinels relative to silicates, a negative correlation between $\delta^{66}Zn$ and Zn concentrations in basaltic melts is expected for chromite crystallization. This is in contrast to the positive correlation of $\delta^{66}Zn$ and Zn contents observed in some intraplate alkali basalts and global OIBs, contradicting the chromite crystallization model. By contrast, lunar mare basalts have typically high $\delta^{66}Zn$ and low Zn contents, but such signatures are also unlikely to have been induced by Cr-spinel fractionation. Overall, zinc elemental and isotopic systematics have the ability to identify the crystallization of chromite during magmatic differentiation.

Declaration of Competing Interest

The authors declare that they have no known competing financial interests or personal relationships that could have appeared to influence the work reported in this paper.

Acknowledgements

We acknowledge editors Jeffrey Catalano and Frédéric Moynier for handling and constructive comments. Luc Doucet and two anonymous reviewers are thanked for their constructive comments and suggestions. All these comments significantly improved the manuscript. We appreciate discussions with Christoph Burkhardt and Stephan Klemme. Meng-Lun Li and Yiwen Lv are thanked for the laboratory or technical support. This work is supported by the National Key R&D Program of China (2019YFA0708400), the National Nature Science Foundation of China (grant 41730214), and the Strategic Priority Research Program of the Chinese Academy of Sciences (grant XDB18000000) to S.A.L. Chun Yang acknowledges support from the China Scholarship Council (202006400075) during his visiting in University of Münster. This is CUGB petro-geochemical contribution No.PGC-201577.

Appendix A. Supplementary data

Supplementary data to this article can be found online at https://doi.org/10.1016/j.gca.2021.08.005.

References

Åsbrink S., Waśkowska A., Gerward L., Olsen J. S. and Talik E. (1999) High-pressure phase transition and properties of spinel

$ZnMn_2O_4$. Physical Review B 60, 12651.

Andreozzi G. B., Lucchesi S., Skogby H. and della Giusta A. (2001) Compositional dependence of cation distribution in some synthetic (Mg, Zn)(Al, Fe^{3+})$_2O_4$ spinels. European Journal of Mineralogy 13, 391-402.

Archer, C. and Vance, D. (2004). Mass discrimination correction in multiple-collector plasma source mass spectrometry: an example using Cu and Zn isotopes. Journal of Analytical Atomic Spectrometry, 19(5), 656-665.

Barnes S. J. and Roeder P. L. (2001) The range of spinel compositions in terrestrial mafic and ultramafic rocks. Journal of petrology 42, 2279-2302.

Beunon H., Mattielli N., Doucet L. S., Moine B. and Debret B. (2020) Mantle heterogeneity through Zn systematics in oceanic basalts: Evidence for a deep carbon cycling. Earth-Science Reviews 205, 103174.

Bigeleisen J. and Mayer M. G. (1947) Calculation of equilibrium constants for isotopic exchange reactions. The Journal of Chemical Physics 15, 261-267.

Bonnand P., Parkinson I. J. and Anand M. (2016) Mass dependent fractionation of stable chromium isotopes in mare basalts: Implications for the formation and the differentiation of the Moon. Geochimica et Cosmochimica Acta 175, 208-221.

Bosi F., Andreozzi G. B., Hålenius U. and Skogby H. (2011) Zn-O tetrahedral bond length variations in normal spinel oxides. American Mineralogist 96, 594-598.

Boyce J. W., Treiman A. H., Guan Y., Ma C., Eiler J. M., Gross J., Greenwood J. P. and Stolper E. M. (2015) The chlorine isotope fingerprint of the lunar magma ocean. Science Advances 1, e1500380.

Bridgestock L. J., Williams H., Rehkämper M., Larner F., Giscard M. D., Hammond S., Coles B., Andreasen R., Wood B. J., Theis K. J., Smith C. L., Benedix G. K. and Schönbächler M. (2014) Unlocking the zinc isotope systematics of iron meteorites. Earth and Planetary Science Letters 400, 153-164.

Cassard, D., Nicolas, A., Rabinovitch, M., Moutte, J., Leblanc, M., & Prinzhofer, A. (1981). Structural classification of chromite pods in southern New Caledonia. Economic Geology, 76(4), 805-831.

Chen H., Savage P. S., Teng F.-Z., Helz R. T. and Moynier F. (2013) Zinc isotope fractionation during magmatic differentiation and the isotopic composition of the bulk Earth. Earth and Planetary Science Letters 369-370, 34-42.

Chu Z.-Y., Wu F.-Y., Walker R. J., Rudnick R. L., Pitcher L., Puchtel I. S., Yang Y.-H. and Wilde S. A. (2009) Temporal Evolution of the Lithospheric Mantle beneath the Eastern North China Craton. J Petrology 50, 1857-1898.

Davis F. A., Humayun M., Hirschmann M. M. and Cooper R. S. (2013) Experimentally determined mineral/melt partitioning of first-row transition elements (FRTE) during partial melting of peridotite at 3GPa. Geochimica et Cosmochimica Acta 104, 232-260.

Day J. M. D., Moynier F., Meshik A. P., Pradivtseva O. V. and Petit D. R. (2017a) Evaporative fractionation of zinc during the first nuclear detonation. Science Advances 3, e1602668.

Day J. M. D., Moynier F. and Shearer C. K. (2017b) Late-stage magmatic outgassing from a volatile-depleted Moon. PNAS 114, 9547-9551.

Day J. M. D., van Kooten E. M. M. E., Hofmann B. A. and Moynier F. (2020) Mare basalt meteorites, magnesian-suite rocks and KREEP reveal loss of zinc during and after lunar formation. Earth and Planetary Science Letters 531, 115998.

Debret B., Bouilhol P., Pons M. L. and Williams H. (2018) Carbonate Transfer during the Onset of Slab Devolatilization: New Insights from Fe and Zn Stable Isotopes. Journal of Petrology 59, 1145-1166.

Debret B., Garrido C. J., Pons M.-L., Bouilhol P., Inglis E., López Sánchez-Vizcaíno V. and Williams H. (2021) Iron and zinc stable isotope evidence for open-system high-pressure dehydration of antigorite serpentinite in subduction zones. Geochimica et Cosmochimica Acta. 296, 210-225.

Ducher M., Blanchard M. and Balan E. (2016) Equilibrium zinc isotope fractionation in Zn-bearing minerals from first-principles calculations. Chemical Geology 443, 87-96.

Ebbinghaus S. G., Hanss J., Klemm M. and Horn S. (2004) Crystal structure and magnetic properties of ZnV_2O_4. Journal of alloys and compounds 370, 75-79.

Fujii T., Moynier F., Blichert-Toft J. and Albarède F. (2014) Density functional theory estimation of isotope fractionation of Fe, Ni, Cu, and Zn among species relevant to geochemical and biological environments. Geochimica et Cosmochimica Acta 140, 553-576.

Gamal El Dien H., Arai S., Doucet L.-S., Li Z.-X., Kil Y., Fougerouse D., Reddy S. M., Saxey D. W. and Hamdy M. (2019) Cr-spinel records metasomatism not petrogenesis of mantle rocks. Nature Communications 10, 5103.

Gargano A., Sharp Z., Shearer C., Simon J. I., Halliday A. and Buckley W. (2020) The Cl isotope composition and halogen contents of Apollo-return samples. PNAS 117, 23418-23425.

Griffin W. L., Sobolev N. V., Ryan C. G., Pokhilenko N. P., Win T. T. and Yefimova E. S. (1993) Trace elements in garnets and chromites: Diamond formation in the Siberian lithosphere. Lithos 29, 235-256.

Hålenius U., Andreozzi G. B. and Skogby H. (2010) Structural relaxation around Cr^{3+} and the red-green color change in the spinel (sensu stricto)-magnesiochromite ($MgAl_eO_4$-$MgCr_2O_4$) and gahnite-zincochromite ($ZnAl_2O4$-$ZnCr_2O_4$) solid-solution series. American Mineralogist 95, 456-462.

Hart S. R. and Zindler A. (1986) In search of a bulk-Earth composition. Chemical Geology 57, 247-267.

Haskin L. and Warren P. (1991) Lunar chemistry. Lunar sourcebook 4, 357-474.

Herzog G. F., Moynier F., Albarède F. and Berezhnoy A. A. (2009) Isotopic and elemental abundances of copper and zinc in lunar samples, Zagami, Pele's hairs, and a terrestrial basalt. Geochimica et Cosmochimica Acta 73, 5884-5904.

Horn I., Foley S. F., Jackson S. E. and Jenner G. A. (1994) Experimentally determined partitioning of high field strength- and selected transition elements between spinel and basaltic melt. Chemical Geology 117, 193-218.

Huang J., Zhang X.-C., Chen S., Tang L., Wörner G., Yu H. and Huang F. (2018) Zinc isotopic systematics of Kamchatka-Aleutian arc magmas controlled by mantle melting. Geochimica et Cosmochimica Acta 238, 85-101.

Ionov D. A. (2010) Petrology of mantle wedge lithosphere: new data on supra-subduction zone peridotite xenoliths from the andesitic Avacha volcano, Kamchatka. Journal of Petrology 51, 327-361.

Jin Q.-Z., Huang J., Liu S.-C. and Huang F. (2020) Magnesium and zinc isotope evidence for recycled sediments and oceanic crust in the mantle sources of continental basalts from eastern China. Lithos 370-371, 105627.

Kamenetsky V. S., Crawford A. J. and Meffre S. (2001) Factors Controlling Chemistry of Magmatic Spinel: an Empirical Study of Associated Olivine, Cr-spinel and Melt Inclusions from Primitive Rocks. J Petrology 42, 655-671.

Kato C., Moynier F., Valdes M. C., Dhaliwal J. K. and Day J. M. D. (2015) Extensive volatile loss during formation and differentiation of the Moon. Nature Communications 6, 7617.

Kato C. and Moynier F. (2017) Gallium isotopic evidence for extensive volatile loss from the Moon during its formation. Science Advances 3(7), e1700571.

Klaska K.-H., Eck J. C. and Pohl D. (1978) New investigation of willemite. Acta Crystallographica Section B: Structural Crystallography and Crystal Chemistry 34, 3324-3325.

Kooten E. M. M. E. van, Moynier F. and Day J. M. D. (2020) Evidence for Transient Atmospheres during Eruptive Outgassing on the Moon. The Planetary Science Journal. 1(3), 67.

Le Roux V., Lee C.-T. A. and Turner S. J. (2010) Zn/Fe systematics in mafic and ultramafic systems: Implications for detecting major element heterogeneities in the Earth's mantle. Geochimica et Cosmochimica Acta 74, 2779-2796.

Lenaz D. and Skogby H. (2013) Structural changes in the $FeAl_2O_4$-$FeCr_2O_4$ solid solution series and their consequences on natural Cr-bearing spinels. Physics and Chemistry of Minerals 40, 587-595.

Li M.-L., Liu S.-A., Lee H.-Y., Yang C. and Wang Z.-Z. (2021) Magnesium and zinc isotopic anomaly of Cenozoic lavas in central Myanmar: Origins and implications for deep carbon recycling. Lithos 386-387, 106011.

Liu P.-P., Teng F.-Z., Dick H. J. B., Zhou M.-F. and Chung S.-L. (2017) Magnesium isotopic composition of the oceanic mantle and oceanic Mg cycling. Geochimica et Cosmochimica Acta 206, 151-165.

Liu S.-A. and Li S.-G. (2019) Tracing the Deep Carbon Cycle Using Metal Stable Isotopes: Opportunities and Challenges. Engineering 5, 448-457.

Liu S.-A., Liu P.-P., Lv Y., Wang Z.-Z. and Dai J.-G. (2019) Cu and Zn Isotope Fractionation during Oceanic Alteration: Implications for Oceanic Cu and Zn Cycles. Geochimica et Cosmochimica Acta. 257, 191-205.

Liu S.-A., Teng F.-Z., Yang W. and Wu F.-Y. (2011) High-temperature inter-mineral magnesium isotope fractionation in mantle xenoliths from the North China craton. Earth and Planetary Science Letters 308, 131-140.

Liu S.-A., Wang Z.-Z., Li S.-G., Huang J. and Yang W. (2016) Zinc isotope evidence for a large-scale carbonated mantle beneath eastern China. Earth and Planetary Science Letters 444, 169-178.

Liu S.-A., Wang Z.-Z., Yang C., Li S.-G. and Ke S. (2020) Mg and Zn Isotope Evidence for Two Types of Mantle Metasomatism and Deep Recycling of Magnesium Carbonates. Journal of Geophysical Research: Solid Earth 125, e2020JB020684.

Luck J.-M., Othman D. B. and Albarède F. (2005) Zn and Cu isotopic variations in chondrites and iron meteorites: Early solar nebula reservoirs and parent-body processes. Geochimica et Cosmochimica Acta 69, 5351-5363.

Macris C. A., Manning C. E. and Young E. D. (2015) Crystal chemical constraints on inter-mineral Fe isotope fractionation and implications for Fe isotope disequilibrium in San Carlos mantle xenoliths. Geochimica et Cosmochimica Acta 154, 168-185.

Marques A. F. A., Barriga F. J. and Scott S. D. (2007) Sulfide mineralization in an ultramafic-rock hosted seafloor hydrothermal system: From serpentinization to the formation of Cu-Zn-(Co)-rich massive sulfides. Marine Geology 245, 20-39.

Maréchal C. N., Télouk P. and Albarède F. (1999) Precise analysis of copper and zinc isotopic compositions by plasma-source mass spectrometry. Chemical Geology 156, 251-273.

McCoy-West A. J., Godfrey Fitton J., Pons M.-L., Inglis E. C. and Williams H. M. (2018) The Fe and Zn isotope composition of deep mantle source regions: Insights from Baffin Island picrites. Geochimica et Cosmochimica Acta. 238, 542-562

McDonough W. F. and Sun S. -s. (1995) The composition of the Earth. Chemical Geology 120, 223-253.

Moynier F., Vance D., Fujii T. and Savage P. (2017) The Isotope Geochemistry of Zinc and Copper. Reviews in Mineralogy and Geochemistry 82, 543-600.

O'neill H. S. C., Annersten H. and Virgo D. (1992) The temperature dependence of the cation distribution in magnesioferrite ($MgFe_2O_4$) from powder XRD structural refinements and Mössbauer spectroscopy. American Mineralogist 77, 725-740.

O'neill H. S. C. (1992) Temperature dependence of the cation distribution in zinc ferrite ($ZnFe_2O_4$) from powder XRD structural refinements. European Journal of Mineralogy, 571-580.

O'neill H. S. C. and Dollase W. A. (1994) Crystal structures and cation distributions in simple spinels from powder XRD structural refinements: $MgCr_2O_4$, $ZnCr_2O_4$, Fe_3O_4 and the temperature dependence of the cation distribution in $ZnAl_2O_4$. Physics and Chemistry of Minerals 20, 541-555.

Othman D. B., Luck J. M., Bodinier J. L., Arndt N. T. and Albarede F. (2006) Cu-Zn isotopic variations in the Earth's mantle. Geochimica et Cosmochimica Acta 18, A46.

Paniello R. C., Day J. M. D. and Moynier F. (2012) Zinc isotopic evidence for the origin of the Moon. Nature 490, 376-379.

Pichat S., Douchet C. and Albarède F. (2003) Zinc isotope variations in deep-sea carbonates from the eastern equatorial Pacific over the last 175 ka. Earth and Planetary Science Letters 210, 167-178.

Plank T. and Manning C. E. (2019) Subducting carbon. Nature 574, 343-352.

Pons M.-L., Debret B., Bouilhol P., Delacour A. and Williams H. (2016) Zinc isotope evidence for sulfate-rich fluid transfer across subduction zones. Nature Communications 7, 1-8.

Pons M.-L., Quitté G., Fujii T., Rosing M. T., Reynard B., Moynier F., Douchet C. and Albarède F. (2011) Early Archean serpentine mud volcanoes at Isua, Greenland, as a niche for early life. Proceedings of the National Academy of Sciences 108, 17639-17643.

Pringle E. A. and Moynier F. (2017) Rubidium isotopic composition of the Earth, meteorites, and the Moon: Evidence for the origin of volatile loss during planetary accretion. Earth and Planetary Science Letters 473, 62-70.

Redfern S. a. T., Knight K. S., Henderson C. M. B. and Wood B. J. (1998) Fe-Mn cation ordering in fayalite-tephroite (Fe_xMn_{1-x})2SiO4 olivines: a neutron diffraction study. Mineralogical Magazine 62, 607-615.

Schauble E. A. (2011) First-principles estimates of equilibrium magnesium isotope fractionation in silicate, oxide, carbonate and hexaaquamagnesium(2+) crystals. Geochimica et Cosmochimica Acta 75, 844-869.

Scowen P. A. H., Roeder P. L. and Helz R. T. (1991) Reequilibration of chromite within Kilauea Iki lava lake, Hawaii. Contributions to Mineralogy and Petrology 107, 8-20.

Sedaghatpour F. and Jacobsen S. B. (2019) Magnesium stable isotopes support the lunar magma ocean cumulate remelting model for mare basalts. Proc Natl Acad Sci USA 116, 73-78.

Shannon R. D. (1976) Revised effective ionic radii and systematic studies of interatomic distances in halides and chalcogenides. Acta crystallographica section A: crystal physics, diffraction, theoretical and general crystallography 32, 751-767.

Sharp Z. D., Shearer C. K., McKeegan K. D., Barnes J. D. and Wang Y. Q. (2010) The chlorine isotope composition of the Moon and implications for an anhydrous mantle. Science 329, 1050-1053.

Shen J., Qin L., Fang Z., Zhang Y., Liu J., Liu W., Wang F., Xiao Y., Yu H. and Wei S. (2018) High-temperature inter-mineral Cr isotope fractionation: A comparison of ionic model predictions and experimental investigations of mantle xenoliths from the North China Craton. Earth and Planetary Science Letters 499, 278-290.

Singh A. K. and Singh R. B. (2013) Genetic implications of Zn- and Mn-rich Cr-spinels in serpentinites of the Tidding Suture Zone, eastern Himalaya, NE India. Geological Journal 48, 22-38.

Snow J. E. and Dick H. J. (1995) Pervasive magnesium loss by marine weathering of peridotite. Geochimica et Cosmochimica Acta 59, 4219-4236.

Song S., Niu Y., Su L. and Xia X. (2013) Tectonics of the north Qilian orogen, NW China. Gondwana Research 23, 1378-1401.

Sossi P. A., Nebel O., O'Neill H. St. C. and Moynier F. (2018) Zinc isotope composition of the Earth and its behaviour during planetary accretion. Chemical Geology 477, 73-84.

Sossi P. A. and O'Neill H. St. C. (2017) The effect of bonding environment on iron isotope fractionation between minerals at high temperature. Geochimica et Cosmochimica Acta 196, 121-143.

Sossi P. A., Moynier F., Treilles R., Mokhtari M., Wang X. and Siebert J. (2020) An experimentally-determined general formalism for evaporation and isotope fractionation of Cu and Zn from silicate melts between 1300 and 1500 ℃ and 1 bar. Geochimica et Cosmochimica Acta 288, 316-340.

Su B.-X., Hu Y., Teng F.-Z., Xiao Y., Zhang H.-F., Sun Y., Bai Y., Zhu B., Zhou X.-H. and Ying J.-F. (2019) Light Mg isotopes in mantle-derived lavas caused by chromite crystallization, instead of carbonatite metasomatism. Earth and Planetary Science Letters 522, 79-86.

Su B.-X., Hu Y., Teng F.-Z., Xiao Y., Zhou X.-H., Sun Y., Zhou M.-F. and Chang S.-C. (2017) Magnesium isotope constraints on subduction contribution to Mesozoic and Cenozoic East Asian continental basalts. Chemical Geology 466, 116-122.

Tang H., Szumila I., Trail D. and Young E. D. (2021) Experimental determination of the effect of Cr on Mg isotope fractionation between spinel and forsterite. Geochimica et Cosmochimica Acta 296. 152-169.

Tian Z., Jolliff B. L., Korotev R. L., Fegley B., Lodders K., Day J. M. D., Chen H. and Wang K. (2020) Potassium isotopic composition of the Moon. Geochimica et Cosmochimica Acta 280, 263-280.

Thy, P., 1983. Spinel minerals in transitional and alkali basaltic glasses from Iceland. Contri. Mineral. Petrol. 83, 141-149.

Urey H. C. (1947) The thermodynamic properties of isotopic substances. Journal of the Chemical Society (Resumed), 562-581.

Verger L., Dargaud O., Rousse G., Rozsályi E., Juhin A., Cabaret D., Cotte M., Glatzel P. and Cormier L. (2016) Spectroscopic properties of Cr^{3+} in the spinel solid solution $ZnAl_2$-$xCrxO_4$. Phys Chem Minerals 43, 33-42.

Wang K. and Jacobsen S. B. (2016) Potassium isotopic evidence for a high-energy giant impact origin of the Moon. Nature 538, 487-490.

Wang X., Fitoussi C., Bourdon B., Fegley B. and Charnoz S. (2019) Tin isotopes indicative of liquid-vapour equilibration and separation in the Moon-forming disk. Nat. Geosci. 12, 707-711.

Wang Z.-Z. and Liu S.-A. (2021) Evolution of Intraplate Alkaline to Tholeiitic Basalts via Interaction Between Carbonated Melt and Lithospheric Mantle. Journal of Petrology 4, 4.

Wang Z.-Z., Liu S.-A., Chen L.-H., Li S.-G. and Zeng G. (2018) Compositional transition in natural alkaline lavas through silica-undersaturated melt-lithosphere interaction. Geology 46, 771-774.

Wang Z.-Z., Liu S.-A., Liu J., Huang J., Xiao Y., Chu Z.-Y., Zhao X.-M. and Tang L. (2017) Zinc isotope fractionation during mantle melting and constraints on the Zn isotope composition of Earth's upper mantle. Geochimica et Cosmochimica Acta 198, 151-167.

Wimpenny J., Marks N., Knight K., Rolison J. M., Borg L., Eppich G., Badro J., Ryerson F. J., Sanborn M., Huyskens M. H. and Yin Q. (2019) Experimental determination of Zn isotope fractionation during evaporative loss at extreme temperatures. Geochimica et Cosmochimica Acta 259, 391-411.

Xiao Y., Teng F.-Z., Zhang H.-F. and Yang W. (2013) Large magnesium isotope fractionation in peridotite xenoliths from eastern North China craton: Product of melt-rock interaction. Geochimica et Cosmochimica Acta 115, 241-261.

Yang C. and Liu S.-A. (2019) Zinc Isotope Constraints on Recycled Oceanic Crust in the Mantle Sources of the Emeishan Large Igneous Province. Journal of Geophysical Research: Solid Earth 124, 12537-12555.

Young E. D., Tonui E., Manning C. E., Schauble E. and Macris C. A. (2009) Spinel-olivine magnesium isotope thermometry in the mantle and implications for the Mg isotopic composition of Earth. Earth and Planetary Science Letters 288, 524-533.

Young E. D., Manning C. E., Schauble E. A., Shahar A., Macris C. A., Lazar C. and Jordan M. (2015) High-temperature equilibrium isotope fractionation of non-traditional stable isotopes: Experiments, theory, and applications. Chemical Geology 395, 176-195.

Yu H., Zhang H.-F., Zou H.-B. and Yang Y.-H. (2019) Minor and trace element variations in chromite from the Songshugou dunites, North Qinling Orogen: Evidence for amphibolite-facies metamorphism. Lithos 328-329, 146-158.

Zhang G., Liu Y., Moynier F., Zhu Y., Wang Z., Hu Z., Zhang L., Li M. and Chen H. (2020) Zinc isotopic composition of the lower continental crust estimated from lower crustal xenoliths and granulite terrains. Geochimica et Cosmochimica Acta 276, 92-108.

Zhang L., Sun W., Zhang Z., An Y. and Liu F. (2019) Iron isotopic composition of supra-subduction zone ophiolitic peridotite from northern Tibet. Geochimica et Cosmochimica Acta 258, 274-289.

Zhang L., Wang Q., Ding X. and Li W.-C. (2021) Diverse serpentinization and associated abiotic methanogenesis within multiple types of olivine-hosted fluid inclusions in orogenic peridotite from northern Tibet. Geochimica et Cosmochimica Acta 296, 1-17.

Zhao X.-M., Zhang H.-F., Zhu X.-K., Zhu B. and Cao H. (2015) Effects of melt percolation on iron isotopic variation in peridotites from Yangyuan, North China Craton. Chemical Geology 401, 96-110.

Magnesium isotope geochemistry of the carbonate-silicate system in subduction zones*

Shui-Jiong Wang[1] and Shu-Guang Li[1, 2]

1. State Key Laboratory of Geological Processes and Mineral Resources, China University of Geosciences (Beijing), Beijing 100083, China
2. CAS Key Laboratory of Crust-Mantle Materials and Environments, School of Earth and Space Sciences, University of Science and Technology of China, Hefei 230026, China

亮点介绍：本文系统综述了板块俯冲过程中硅酸盐和碳酸盐体系经历的俯冲板片变质脱水、变质脱碳、碳酸盐溶解、部分熔融和碳酸盐-硅酸盐相互作用过程中的 Mg 同位素地球化学行为，为利用 Mg 同位素示踪深部碳循环奠定了重要理论基础。

Abstract The lighter magnesium (Mg) isotopic signatures observed in intraplate basalts are commonly thought to result from deep carbonate recycling, provided that the sharp difference in Mg isotopic composition between surface carbonates and the normal mantle is preserved during plate subduction. However, deep subduction of carbonates and silicates could potentially fractionate Mg isotopes and change their chemical compositions. Subducting silicate rocks that experienced metamorphic dehydration lose a small amount of Mg, and preserve the original Mg isotopic signature of their protoliths. When the dehydrated fluids dissolve carbonate minerals, they may evolve to lighter Mg isotopic compositions. The solubility of carbonate minerals in fluids decreases in the order of calcite, aragonite, dolomite, magnesite and siderite, leading to selective and partial dissolution of carbonate minerals along subduction path. At the island arc depth (70–120 km), the metamorphic fluid dissolves mainly Mg-poor calcites, and thus the fluid has difficulty modifying the Mg isotopic system of the mantle wedge and associated arc basalts. At greater depth of the back arc system or continental margin (> 150 km), the supercritical fluid can dissolve Mg-rich carbonate minerals, and its interaction with the mantle wedge could significantly imprint the light Mg isotopic signature onto the mantle rocks and derivatives. Meanwhile, the carbonate and silicate remaining within the subducting slab could experience elemental and isotopic exchange, during which the silicate can obtain a light Mg isotopic signature and high CaO/Al_2O_3, whereas the carbonates, particularly the Ca-rich limestone, shift Mg isotopes and MgO contents towards higher values. If this isotopic and elemental exchange event occurs widely during crustal subduction, subducted Ca-rich carbonates can partially transform into being Mg-rich, and a portion of recycled silicates (e.g. carbonated eclogites) can have light Mg isotopic composition alongside carbonates. Both serve as the low-δ^{26}Mg endmember recycled back into the deep mantle, but the latter is not related to deep carbonate recycling. Therefore, it is important to determine

* 本文发表在：National Science Review, 2022, 9(6): nwac036

whether the light Mg isotopic signatures observed in intraplate basalts are linked to deep carbonate recycling, or alternatively, recycling of carbonated eclogites.

Keywords Magnesium isotope, fractionation, deep carbon cycle, subduction, carbonate

1 Introduction

A magnesium (Mg) isotopic system has been applied to trace the deep recycling of carbonates [1] for three broad reasons. First, surface carbonates, regardless of inorganic or organic origin, have remarkably lighter Mg isotopic compositions than terrestrial silicates [2, 3]. This suggests that an injection of carbonates into the mantle has the ability to cause mantle Mg isotopic heterogeneity. Second, igneous processes such as mantle melting, degassing, and crystallization produce negligible Mg isotope fractionations [4-6], such that the Mg isotopic signature of mantle sources can be directly seen from their derivative basalts. Finally, crustal subduction seems not to erase the contrasting Mg isotopic signature between sedimentary carbonates and silicates [7]. While the last statement is empirically accepted [1], the behavior of Mg isotopes in a subduction zone is complicated and relatively less well constrained. Attempts have been made over the last decade to decipher the magnitude and mechanism of Mg isotope fractionation by subduction-related processes [7, 8]. It helps to answer some fundamental questions, for example, (i) Is carbonate the only low-δ^{26}Mg carrier among those recycled into the mantle? (ii) Can the composition and solubility of carbonate be changed during subduction? (iii) Why can the low-δ^{26}Mg signature be observed in intraplate basalts but not in island arc basalts?

This contribution, built upon materials presented in previous reviews and incorporating the findings of most recent studies [9-11], aims to provide an overview of the behavior of carbonate-silicate systems and their Mg isotopes in subduction zones. We examine the physical and chemical properties of the subducting silicate-carbonate package during crustal subduction in the first section of this article. In the second section, we evaluate how subduction-related processes could affect the Mg isotopic system and the chemical composition of subducting carbonate and silicate. In the third section, we put these fractionation events in the context of a plate tectonic framework to explore the robustness of linking Mg isotopic anomalies in mantle-derived rocks to carbonate recycling.

2 Carbonate-silicate package in subduction zone

The carbonates that enter into the trench are mainly from the platform carbonates on the overriding plate and marine carbonates precipitated on the oceanic floor [12]. They are carried by the subducting plate, together with the silicates, to the deep mantle. The subducting carbonate-silicate package experiences significant changes in physical and chemical properties, leading to carbon mobility and potential isotope fractionations. Processes of particular interest are summarized below (Fig. 1).

2.1 Metamorphic dehydration

At elevated pressures and temperatures in subduction zones, the fluid in the pore spaces of rocks, boundaries between crystals, or in hydrous minerals, will be liberated due to compression and metamorphism. With increasing metamorphic grade, for example, from sub-greenschist- to greenschist-, amphibolite- and eclogite-facies, the amount and chemical composition of the dehydrated fluid vary as a function of the lithologies and dehydration reactions [13]. The dehydrated fluid then migrates upwards due to its low density and viscosity compared to the surrounding rocks. The importance of metamorphic dehydration is 2-fold. First, metamorphic dehydration may cause loss of Mg along with phase changes. If subducting rocks after metamorphic dehydration, display different Mg isotopic compositions from their protolith, the extent and magnitude of such isotope fractionation must be calibrated. Second, the fluids may change from aqueous fluid to supercritical fluid as pressure and temperature increase during crustal subduction, and they play a key role in mass transfer and elemental/isotopic exchanges in subduction zones [14, 15]. Extensive fluid-rock interactions in subduction zones could facilitate carbon mobility through a series of reactions such as decarbonation, carbonate dissolution and carbonate-silicate reaction [16-18]. We are interested in whether the fluid has a similar Mg isotopic composition as its protolith, whether Mg isotopes fractionate during carbon-mobility events and whether chemical composition and solubility of the subducting carbonate change during plate subduction.

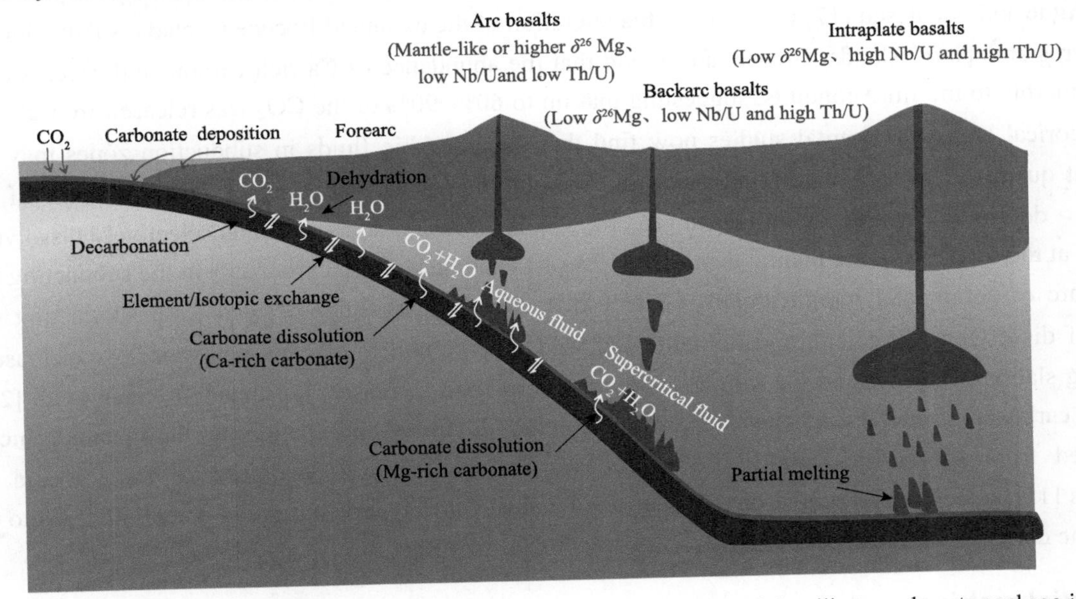

Fig. 1 Cartoons showing some of the important processes related to the subduction of the silicate-carbonate package into the deep mantle (not to scale). Please see the main text for details.

2.2 Decarbonation

Metamorphism of carbonate rocks may cause decarbonation via reactions of $CaCO_3 + SiO_2 = CaSiO_3 + CO_2$, or $CaMg(CO_3)_2 + 2SiO_2 = CaMg(SiO_3)_2 + 2CO_2$. Given the strong dependence of decarbonation on pressure, temperature and composition, the degree of decarbonation changes among subducted lithologies, varies from one subduction to another, and may differ significantly between Precambrian and modern subductions. Based on phase equilibria computation of the metamorphic decarbonation of subducting slabs, Kerrick and Connolly [19] proposed that along typical subduction geotherms metamorphic decarbonation is unlikely to happen and its effects on transferring CO_2 from subducting slabs to arc magmas is negligible. In general, decarbonation is more efficient in carbonated sediments relative to carbonated basaltic rocks and siliceous limestones [19, 20]. In rare cases where subduction geotherms are high, for example the subduction of young oceanic crust at slow convergent rates, decarbonation is feasible in the forearc regions (because of low pressure and relatively high temperatures) [20]. This raises the possibility that decarbonation may have been high or complete at Precambrian subduction zones at which the temperature was as much as ~100 ℃ higher than the hottest subduction zone [21]. The infiltration of H_2O-rich fluid could also promote decarbonation of subducted marine sediments [19, 22]. Stewart and Ague [18] predicts that 40%–65% of the CO_2 in subducting crust can be released via metamorphic decarbonation at forearc depths, which is remarkably higher than previously thought. In addition, natural observations and experimental studies found that carbonate minerals can be reduced to form hydrocarbons in subduction-zone settings under low oxygen fugacity, with the residual mineral assemblage consisting of iron-bearing dolomite, magnetite, and graphite [23]. According to the decarbonation reactions, Mg isotopic signatures of carbonates would be inherited by newly formed silicates during decarbonation.

2.3 Carbonate dissolution

Metamorphic fluid is probably the most important agent mobilizing carbon from subducting slab to the arc mantle [17]. Fluid inclusions in subduction-related ultrahigh metamorphic rocks contain a range of carbonate minerals, suggesting that a substantial amount of carbonate minerals can be dissolved in metamorphic fluids [17, 24, 25]. Ague and Nicolescu [17] investigated the alteration of the exhumed Eocene Cycladic subduction complex on the Syros and Tinos islands, Greece, and found that the abundance of Ca-rich carbonate decreases drastically from the marble to the fluid conduits, suggesting that up to 60%–90% of the CO_2 was released from the rocks by fluid. Theorical and experimental studies now find that metamorphic fluids in subduction zones may transport significant quantities of carbonate minerals, with the solubility of carbonate minerals in the order of calcite > aragonite > dolomite > magnesite > siderite [16, 26, 27]. As a result, aqueous fluids selectively dissolve Ca-rich carbonate at the forearc and arc mantle depth (70–120 km) leaving Mg-rich carbonates in the subducting slabs [1]. At back arc or continental margin depths (⩾150 km), supercritical fluids derived from subducting slabs are capable of dissolving Mg-calcite and dolomite [24, 25]. The solubility of carbonate minerals increases as the subducting slab goes deeper and the salinity of fluid composition increases. For example, Shen et al. [25] found abundant carbonate mineral inclusions including calcite, dolomite and magnesite, in metamorphic zircons precipitated from supercritical fluids. Given the distinct Mg isotopic signature of Ca-rich and Mg-rich carbonates [11], selective and partial dissolution of subducted Ca-rich carbonates can potentially lead to different Mg isotopic compositions of evolving fluids.

2.4 Partial melting

Another important process that could mobilize carbon in subduction zones is the melting of carbonate-bearing rocks. The fate of subducted carbonates that survived decarbonation and dissolution at forearc and arc depth hinges on the location of the solidus of carbonated rocks relative to the thermal structure of the subduction zone. Experimental studies have determined the solidi of three dominant carbonated lithologies: carbonated ocean floor sediments, carbonated altered basalt and carbonated peridotite [28-35]. Carbonated sediments have the lowest solidi and thus are more prone to losing carbon during subduction. Grassi and Schmidt [28] suggested that carbonated sediments may melt at two depths of the subducting slabs: 6–9 GPa and 20–22 GPa. At any given pressure, the solidi of carbonated oceanic floor basalts and carbonated peridotites are on average higher than that of carbonated sediments, and remain hotter than the slab-top condition of most modern subduction zones [36]. Pure carbonate rocks have an even higher melting temperature than carbonated silicates [37]. Therefore, carbonated oceanic floor basalts and carbonates are the two major carbon carriers in the subducting slabs. Isotopic studies suggest that the carbon in the subducting slabs could be introduced to the mantle transition zone (410–660 km) [1]. Recently, Thomson et al. [33] determined the melting phase relations of a synthetic mid-ocean ridge basalt (MORB) composition containing 2.5 wt% CO_2 between 3 and 21 GPa, and found that the melting curve of carbonated oceanic crust will intersect the majority of slab geotherms at depths of 300-700 km, leading to the idea that melting at this depth would create a barrier to direct carbonate recycling into the lower mantle. Melting of recycled carbonated rocks could contribute to the formation of intraplate basalts, which in turn could impart their distinct Mg isotopic signature to the mantle melts.

2.5 Carbonate-silicate reaction

As previously discussed, thermodynamic modeling of the devolatilization of carbonate-bearing subducting slab and melting experiments point towards the preservation of solid carbonates along geotherms of modern subduction zones [38]. The carbonate minerals interact chemically with silicates during subduction and undergo changes in both physical and chemical properties. Kushiro [39] studied carbonate-silicate interaction at pressures between 2.3 and 7.7 GPa and temperatures between 800 and 1400 ℃. They found that calcite is unstable in the presence of enstatite, and reacts with enstatite to form dolomite and diopside. Studies predict that the carbonate mineral stable at shallow depth is calcite-rich, at intermediate depths is dolomite-rich and at greater depth is magnesite-rich [36], suggesting that carbonate carried by subducted plate mainly resides in $MgCO_3$ throughout much of the mantle via forward reaction $CaCO_3+MgSiO_3=MgCO_3+CaSiO_3$. Therefore, the above silicate-carbonate interaction could transform calcite to magnesite so that it fixes carbon in the subducting slabs in form of more stable Mg-rich carbonate minerals. The carbonate-silicate interaction during crustal subduction may induce massive elemental and isotopic exchange.

3 Magnesium isotope fractionation during subduction

The preceding discussion introduces the subduction-related processes that could potentially mobilize carbon and fractionate Mg isotopes. In this section, we review recent advances with regard to the behavior of Mg isotopes in these processes.

3.1 Mg isotopic compositions of metamorphic rocks

Mg preferentially partitions into a solid during metamorphic dehydration, leading to lower Mg concentrations of the fluid relative to the source. Typical subduction-zone fluids have Mg concentrations (0 to 125 mmol/kg) lower than the seawater (averagely 50 mmol/kg) [13]. This is consistent with the results of many experimental studies on elemental partitioning between fluid and minerals during dehydration of sedimentary and basaltic rocks [40-42], which yield distribution coefficients of Mg ($D_{solid/fluid}$) in a range of 0.7 to 70 [40-42]. Assuming that a rock contains 5% of fluid that is sequentially lost during dehydration, the dehydrated fluid takes away only <7% of the bulk-rock Mg. The isotope fractionation through metamorphic dehydration ($\varepsilon_{fluid\text{-}solid}$) has not been experimentally determined yet. However, mass balance calculation suggests that a variation of $\varepsilon_{fluid\text{-}solid}$ from −1.0‰ to +1.0‰ would result in a dehydration-induced shift of δ^{26}Mg in solid smaller than 0.04 ‰, within current analytical uncertainties for δ^{26}Mg.

The inferred lack of Mg isotope fractionation in rocks during metamorphic dehydration is supported by Mg isotopic analyses on metamorphic rocks. Wang et al. [43] measured Mg isotopic compositions of a suite of metabasalts from the Dabie-Sulu orogen, Eastern China, all of which share the same protolith. These samples include greenschist, amphibolite and eclogites, representing products of prograde metamorphism in the subduction zone. Despite the decreasing loss of ignition (LOI) with increasing metamorphic grade, the metabasalts have similar Mg isotopic compositions. Therefore, metamorphic dehydration has a limited effect on Mg isotopic systematics in metabasalts. A similar conclusion has been reached by Li et al. [44] and Teng et al. [45] who found that high-grade metamorphic granulite and eclogites have homogeneous mantle-like δ^{26}Mg values as their protoliths. Metamorphic dehydration of sedimentary rocks causes limited Mg isotopic changes in bulk rock as well. Metapelites exposed in the Irvea zone, Italy, represent a typical prograde metamorphic sequence from middle amphibolite- to granulite-facies. The Mg isotopic compositions of these metapelites do not vary with metamorphic grade but are inherited from the source heterogeneity [46]. Li et al. [47] studied metapelites from the Onawa contact aureole, Maine. They documented that metapelites across the aureole, with increasing metamorphic grade, from the outmost chlorite zone to the andalusite-cordierite zone, potassium feldspar zone sillimanite zone and leucocratic-vein zone, have similar Mg isotopic compositions. Both studies conclude that prograde metamorphic dehydration causes limited Mg isotopic changes in metapelites. Additionally, Wang et al. [48] studied the behavior of Mg isotopes at even lower grade metamorphic conditions where devolatilization might be larger. The mudrocks studied by Wang et al. [48] experienced diagenesis to sub-greenschist metamorphism. The generally heavy Mg isotopic compositions of the mudrocks are not related to metamorphic dehydration but are inherited from their sources [48].

Despite a limited shift in Mg isotopic composition at the bulk-rock scale during prograde metamorphic dehydration, massive redistribution of Mg occurs among metamorphic minerals accompanied by large Mg isotope fractionations. One typical example comes from the metapelites in the Irvea zone [46]. Biotite and garnet are the two major Mg hosts in these metapelites. During prograde metamorphic reaction of biotite + sillimanite + plagioclase + quartz → garnet + K-feldspar + rutile + melt, the mineralogy of metapelites changes from biotite dominated at amphibolite-facies to garnet-dominated at granulite-facies (Fig. 2). Due to large inter-mineral Mg isotope fractionation between biotite and garnet ($\Delta^{26}Mg_{bt\text{-}grt} = 0.96\times10^6/T^2$), the mineral δ^{26}Mg values, as expected from mass balance, increase with increasing metamorphic prograde (Fig. 2).

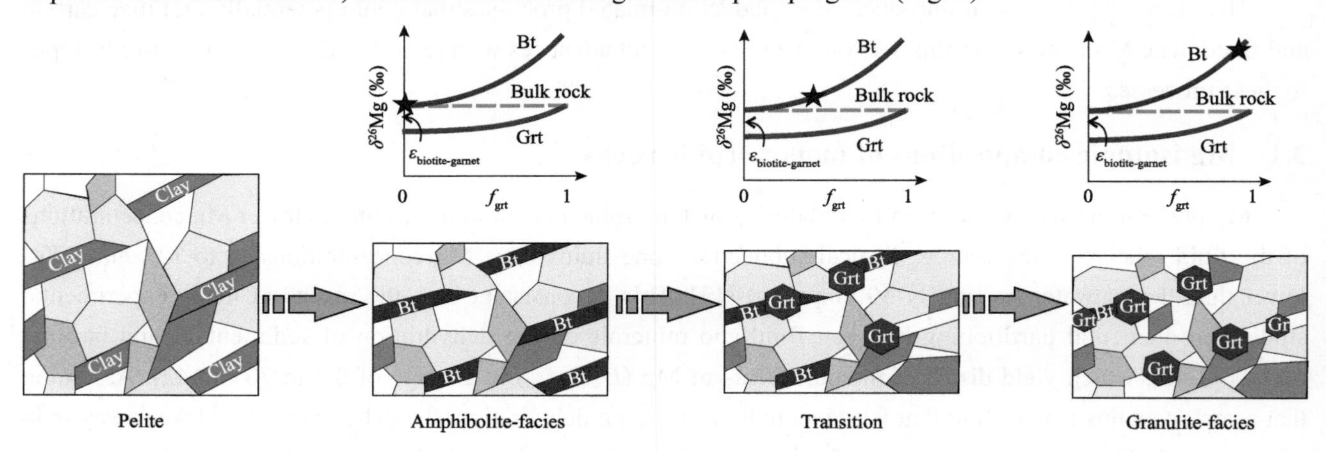

Fig. 2　Cartoon showing the redistribution of Mg isotopes during prograde metamorphism of metapelites from the Irvea zone. $\varepsilon_{biotite\text{-}garnet}$ represents the Mg isotope fractionation between biotite and garnet. At amphibolite-facies, the biotite (Bt) is the dominant Mg host of the metapelites, and its δ^{26}Mg value is close to the bulk-rock value. At granulite-facies, the garnet (Grt) becomes the dominant Mg host of the metapelites, and its δ^{26}Mg value approaches to the bulk-rock value with biotite having δ^{26}Mg offset from the garnet by $\varepsilon_{biotite\text{-}garnet}$.

Overall, the Mg isotopic compositions of subducted silicate rocks preserve their protolith's signature. Subducted carbonates after metamorphism may experience decarbonation, dissolution and isotopic exchange, whose effects on the Mg isotopic system will be discussed below.

3.2　Mg isotopic compositions of dehydrated fluids

Dehydrated fluids from silicate rocks have highly variable and generally heavier Mg isotopic compositions relative to their source rocks under high pressure-temperature (*P-T*) metamorphic conditions. The fluid-precipitated high pressure (HP) quartz veins in the Dabie orogen represent the fluids derived from dehydration of the metabasalt. They have higher $\delta^{26}Mg$ values (from +0.08 to +0.15‰) relative to their basaltic protolith (−0.25±0.04‰) [49]. The coesite-bearing white schist at Dora-Maira in the Western Alps is characterized by strong Mg enrichment, which could be caused by infiltration of Mg-rich fluid derived from dehydration of serpentinites. Chen et al. [50] found that these white schists (T = 730 ℃ and 4.0 GPa) have extremely heavy Mg isotopic compositions ($\delta^{26}Mg$ up to +0.72‰), and suspected that the fluid could be derived from the breakdown of Mg-rich hydrous minerals such as talc and antigorite in serpentinite at the slab-mantle interface. The heavy Mg isotopic compositions (with $\delta^{26}Mg$ up to +0.61‰) of the coesite-bearing jadeite quartzites from the Dabie orogen are also interpretated as being a result of the infiltration of fluid dehydrated from the breakdown of biotite in subducted metasedimentary rocks [51]. The retrograde eclogites and blueschists from southwestern Tianshan have been interacted with metamorphic fluids mainly derived from subducting sediments in the subduction channel [52]. Geochemical proxies of the eclogites and blueschists allow us to distinguish two components of the fluid. One is high-large-ion lithosphile (LILEs) fluid derived from dehydration of mica-group minerals. The other has higher Pb concentration and $^{87}Sr/^{86}Sr$ relative to typical oceanic basalts, suggesting that the fluid is likely released from dehydration of epidote-group minerals. The fluid derived from mica dehydration contains a considerable amount of Mg that is isotopically heavy [46], and thus shifted the $\delta^{26}Mg$ of retrograde eclogites towards higher values. The fluid derived from epidote dehydration has little Mg so as not to influence the Mg isotopic system of the retrograde eclogites.

Given the similar octahedral coordination environment of Mg-O in between common metamorphic hydrous minerals (e.g. biotite and hornblende) and fluid $[Mg(H_2O)_6)]^{2+}$, the isotope fractionation by dehydration of hydrous minerals might be small. It is suggested that the Mg isotopic systematics of these fluids are mainly determined by the hydrous minerals from which they derive. If the hydrous minerals control the Mg isotopic composition of the dehydrated fluids, a further question arises: why do the hydrous minerals in metamorphic rocks have heterogeneous and generally heavier Mg isotopic compositions? First, most hydrous minerals in metamorphic rocks were transferred from clay minerals that are products of surface water-rock interaction. Surface chemical weathering produces large Mg isotope fractionations, leading to the incorporation of heavy Mg isotopes into clays in the weathering residue [53]. Second, as inferred by the Irvea zone metapelites, it is highly likely that the hydrous mineral's $\delta^{26}Mg$ value increases with increasing metamorphic grade (Fig. 2), given the massive redistribution of Mg among metamorphic minerals and potentially large inter-mineral isotope fractionation between hydrous minerals and newly formed metamorphic minerals such like garnet. Take the biotite in metapelites from the Ivrea zone as an example: the $\delta^{26}Mg$ of biotite increases from −0.08 ‰ at amphibolite-facies to +1.10‰ at granulite-facies. As a result, metamorphic fluids derived from biotite dehydration could have highly variable Mg isotopic compositions that are closely correlated to the metamorphic grade (Fig. 2).

As a note, this section only mentions the primary fluid dehydrated from silicates. When the fluid travels and interacts with carbonates, the Mg isotopic composition will change as discussed below.

3.3 Carbonate dissolution on the Mg isotopic systematics of metamorphic fluids

When the metamorphic fluid dissolves carbonate minerals, the Mg isotopic composition of the fluid may become lighter. Sedimentary carbonate rocks range from Ca-rich limestone to Mg-rich dolomite. Ca-rich carbonate minerals have generally lighter Mg isotopic compositions than Mg-rich carbonate minerals (Fig. 3). At forearc and island arc depths, metamorphic aqueous fluids dissolve mainly calcite and, to a less extent, dolomite [1, 16, 17, 26]. As temperature and pressure increase at the back arc or continental margin depth (>150 km), the supercritical fluids with high solubilities of trace elements and carbonate minerals are able to dissolve dolomite and magnesite [25]. Given selective dissolution of carbonate minerals during subduction, the fluid may have different Mg isotopic compositions depending on the solute.

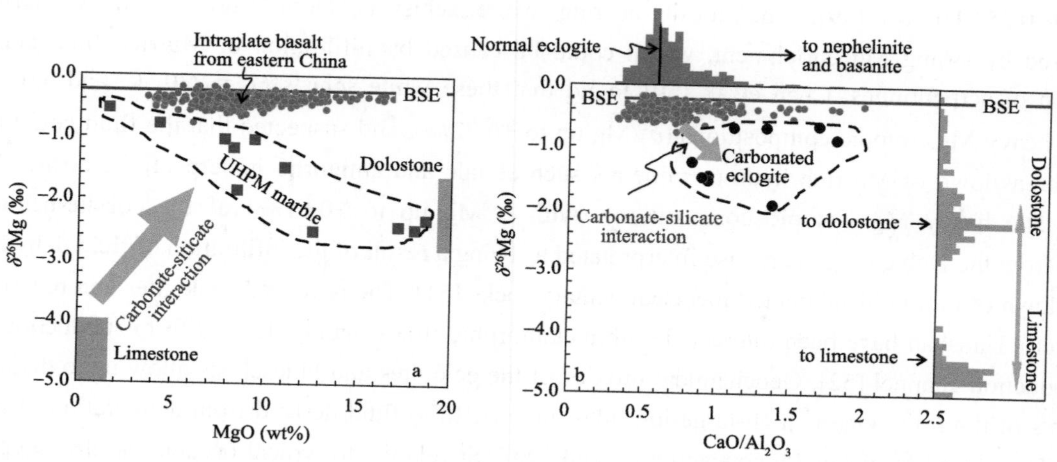

Fig. 3 A compilation of Mg isotopic composition of sedimentary carbonates, intraplate basalts from eastern China, normal and carbonated eclogites, and ultra-high pressure metamorphic (UHPM) marbles. The UHPM marbles display a negative correlation between δ^{26}Mg and MgO after interacting with the concomitant silicate during crustal subduction. The carbonated eclogites after interacting with the carbonate during crustal subduction have low δ^{26}Mg and high CaO/Al_2O_3, which is distinct from that of the normal eclogites. The histogram of sedimentary carbonate is modified from; Mg isotopic data of normal/carbonated eclogites and UHPM marbles are from Ref. [54]; Data for intraplate basalts are from Refs. [1, 55-59]. The Bulk Silicate Earth value is from Ref. [4].

Chen et al. [60] measured Mg isotopic compositions of the jadeitites from Myanmar. These white jadeitites were precipitated from Na-Al-Si fluids at the forearc slab-mantle interface (1−1.5 GPa, and 300−500 ℃). They are characterized by extremely light Mg isotopic compositions (δ^{26}Mg=−0.55 to −0.92‰) that are negatively correlated with CaO/TiO_2 and CaO/Al_2O_3 ratios. Chen et al. [60] proposed that the high-salinity reduced fluid dehydrated from subducting slabs enhanced the dissolution of Ca-rich carbonates that eventually lowered the δ^{26}Mg values of fluids. Chen et al. [61] studied high-pressure metamorphic leucophyllites from Eastern Alps. They were formed under similar pressure but higher temperature than the jadeitites in Myanmar (500−600 ℃), and thus are capable of dissolving Mg-rich calcite at the forearc depth. Two types of fluids are recognized in terms of Mg isotopes. One has high δ^{26}Mg values (>0.3‰) which is likely from dehydration of talc-rich serpentinite, and the other has extremely low δ^{26}Mg values (<−1.3‰), likely produced by dissolution of mainly Mg-calcite at forearc conditions. Shen et al. [25] studied the Maowu ultramafic massif, which represents an exhumed fragment of mantle wedge from the Dabie orogen, with peak metamorphism of 5.3−6.3 GPa and 800 ℃. The garnet pyroxenite within the Maowu ultramafic massif was formed by mantle metasomatism of supercritical fluids derived from subducting slabs, and is characterized by high Th/U ratios (up to 23) and light Mg isotopic compositions (δ^{26}Mg down −0.99‰). Abundant carbonate mineral inclusions, including calcite, dolomite and magnesite, have been found in metamorphic zircons formed from supercritical fluids. The supercritical fluids have high Mg content and light Mg isotopic compositions as they dissolve a considerable amount of Mg-rich carbonate minerals [25]. When traveling and interacting with the mantle wedge, the supercritical fluids impart light Mg isotopic signatures and high Th/U to the Maowu garnet pyroxenite [25].

3.4 Mg isotope fractionation during decarbonation

Decarbonation during modern subduction may be negligible, but could be facilitated in the presence of fluid [18]. Decarbonation releases CO_2 while leaving Mg and Ca to the silicate. This reaction could lead to the newly formed silicate enriched in light Mg isotopes and high CaO/Al_2O_3. Shen et al. [62] analyzed the Mg isotopic composition of endoskarn xenoliths from Sierra Nevada batholith in California, and found that the pyroxenite rim, which is the product of the decarbonation reaction, is characterized by light Mg isotopic composition and high CaO/Al_2O_3. The Mg isotopic anomalies can be explained by mixing of Mg between granodioritic magma and dolomitic wallrock. Decarbonation of the dolomitic wallrock transfers the Mg isotopic signature from carbonate to silicate. In the Precambrian subduction where the geothermal gradient was higher, decarbonation may have been significant. It is possible that subducted carbonates would have been completely decarbonated leaving light Mg isotopes to the subducting silicates.

3.5 Mg isotopic exchange between carbonate and silicate

The large Mg isotopic difference between surface carbonate and silicate will be reduced at elevated temperatures during crustal subduction, if the equilibrium isotope fractionation rule ($\Delta^{26}Mg=A\times10^6/T^2$) applies. The experimental study using a three-isotope method found that equilibrium Mg isotope fractionation between magnesite and forsterite follows the equation of $\Delta^{26}Mg_{forsterite\text{-}magnesite} = 0.06\ (\pm0.04)\times10^6/T^2$ at high temperatures [63], that is, 0.44±0.10‰ at 600 ℃. These experimental determined high-temperature equilibrium fractionation values are significantly smaller than the apparent isotopic difference observed at the surface environment (Fig. 3). Whether or not complete isotopic equilibrium between coexisting carbonate and silicate can be achieved during crustal subduction is uncertain, but massive diffusion-induced isotopic exchange between the two lithologies is expected.

Eclogite boudins enclosed in the ultrahigh metamorphic marbles in the Rongcheng area, Sulu orogenic belt, have chemically interacted with the host marble during high-pressure metamorphism. Wang et al. [54] found that the eclogite boudins have extremely low $\delta^{26}Mg$ and high $\delta^{18}O$ values, which is in sharp contrast to the normal eclogites in the Sulu orogen. The ultrahigh-pressure metamorphic marbles show negative correlation between $\delta^{26}Mg$ and MgO/CaO, which is opposite to their protoliths, in which dolostones have heavier Mg isotopic composition than the limestones (Fig. 3). These Mg and O isotopic anomalies, observed in both eclogite boudins and marbles, are interpreted as a result of elemental and isotopic exchange during crustal subduction. The big difference in Mg content between limestone and dolostone results in differential Mg isotopic exchange against eclogites boudins. The Mg-poor limestone suffered extensive elemental and isotopic exchange has its $\delta^{26}Mg$ and MgO contents elevated significantly, whereas the Mg-rich dolostone retains its original $\delta^{26}Mg$ values because of the high-Mg nature (Fig. 3a). The eclogitic minerals, after elemental and isotopic exchange, obtain light Mg isotopic and high CaO/Al_2O_3 signatures (Fig. 3b). The carbonate-silicate interaction during crustal subduction is of particular consequence. First, the carbonated eclogites, after isotopic exchange, can have low $\delta^{26}Mg$ values down to −1.93‰ and high CaO/Al_2O_3 up to 1.81 (Fig. 3b). Recycling of these components can produce Mg isotopic heterogeneity of the mantle domains but it is not directly related to carbonate recycling. Second, the carbonates, after isotopic exchange, rearrange the $\delta^{26}Mg$ vs. MgO array (Fig. 3a), and thus the endmember of carbonates recycled into the deep mantle is mainly Mg-rich dolostone and magnesite.

4 Linking the mg isotopic system to recycled carbonate

The above-mentioned Mg isotopic geochemistry in subduction zone proves that multiple subduction-related processes can change the Mg isotopic system of subducting silicate and carbonate. Understanding the behavior of Mg isotopes at different stages of crustal subduction can place constraints on the robustness of linking Mg isotopic anomalies in mantle-derived rocks to carbonate recycling.

From trench to the island arc depth (70–120 km), the fluids dehydrated from meta-sediments, metabasalts or metaperidotite are mainly aqueous fluids containing only a small amount of Mg compared to their sources. Thus, the loss of Mg by metamorphic dehydration does not cause any Mg isotopic changes in the metamorphic products. The aqueous fluids selectively dissolve calcite while leaving Mg-rich carbonate minerals in the subducting slab. Fluid infiltration also facilitates decarbonation and isotopic exchange between subducting silicate and carbonate at forearc and island arc depths. At this stage, most calcites in subducting slabs are either decarbonated or dissolved in aqueous fluids, releasing CO_2 into arc volcanism, and some are transferred to Mg-rich carbonate minerals due to Ca-Mg exchange between silicate and carbonate. The latter can be delivered to the deep mantle by subducting slabs. The subducting silicates, when interacting with the carbonates (for example, carbonated eclogites), can obtain light Mg isotopic signatures. Although the fluid may evolve to be highly enriched in light Mg isotopes because of carbonate dissolution, its impact on the Mg isotopic system of the mantle wedge source of arc basalts is limited due to the remarkably lower Mg concentration compared to the peridotitic mantle. This can explain why most arc basalts with a source that has been modified by infiltration of such CO_2-rich fluids do not usually display light Mg isotopic signatures (Fig. 4). The involvement of subducting sediments or sediment-derived melts in the mantle source gives the arc basalts have mantle-like or slightly heavier Mg isotopic compositions [1, 64].

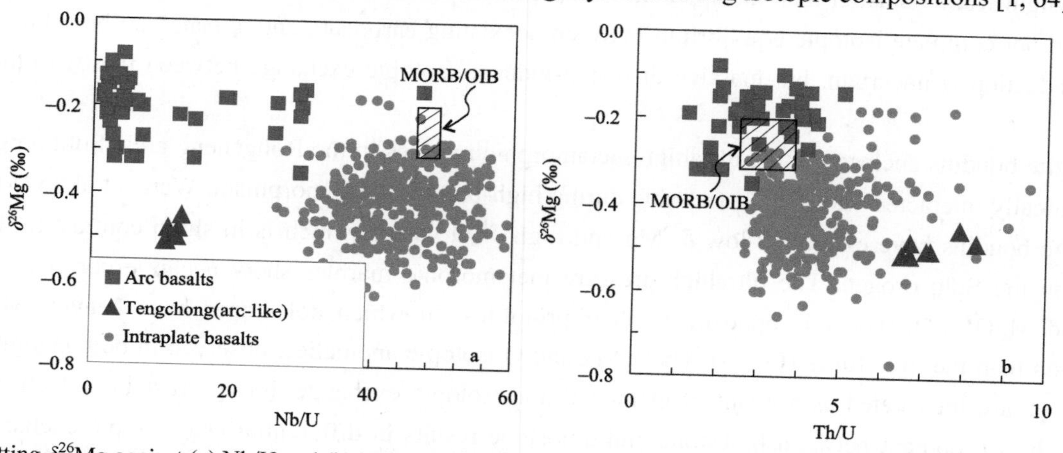

Fig. 4 Plotting δ^{26}Mg against (a) Nb/U and (b) Th/U for arc basalts, arc-like Tengchong basalts, and intraplate basalts from eastern China. The mantle source of arc basalts has been metasomatized by the aqueous fluids dehydrated from subducting slabs. The aqueous fluids have low Th/U, and low Nb/U because of rutile residual in the source during dehydration. The arc-like Tengchong basalts are derived from a back arc or continental margin mantle source modified by supercritical fluid with high Th/U but low Nb/U. The supercritical fluid can dissolve Mg-rich carbonate minerals and thus have high Mg and low δ^{26}Mg features. The intraplate basalts from eastern China have higher Nb/U and relatively high Th/U. They are from a mantle source that has been modified by melts of carbonates or carbonated eclogites from the subducting slab. Data for arc basalts are from Refs. [1, 64]; data for Tengchong basalts are from Ref. [65]; data for intraplate basalts from eastern China are from Refs. [1, 55-59].

Within increasing temperature and pressure to the back arc or active continental margin system (>150 km to <410 km), supercritical fluid appears. In contrast to the aqueous fluid, the supercritical fluid has higher solubility of trace elements and carbonate minerals. In particular, the $D_{Th}^{liquid/solid}$ is > $D_U^{liquid/solid}$ in supercritical fluids but is < $D_U^{liquid/solid}$ in aqueous fluids [42]. Therefore, the supercritical fluid has higher Th/U than the aqueous fluid. In addition, a number of carbonate mineral inclusions, including calcite, dolomite and magnesite observed in metamorphic zircons precipitated from supercritical fluids in the Maowu massif of Dabie orogen, suggest that supercritical fluid at mantle depth >150 km can dissolve more Mg-rich carbonate minerals but not rutile in eclogites [25]. The mantle wedge metasomatized by the supercritical fluids have δ^{26}Mg down to −0.99‰ and Th/U ratio up to 23 [25]. The arc-like basaltic rocks generated from such mantle source (e.g., Tengchong basalt)

[65] can be distinguished from typical arc basalts in terms of lower δ^{26}Mg and higher Th/U (Fig. 4b). Both have low Nb/U ratios because of rutile residual in the subducted eclogites (Fig. 4b).

When the silicate-carbonate package is subducted to the mantle transition zone (410–660 km), both Mg-rich carbonates and carbonated eclogites can melt. Recycling and involvement of these components in the mantle source can account for the light Mg isotopic signatures observed in ocean island basalt (OIB)-like intraplate basalts. They are distinguishable from arc or arc-like basalts by their high Nb/U ratio and variably low δ^{26}Mg values (Fig. 4a). However, it is still a puzzle whether the light Mg isotopic signatures result from recycling of carbonates or carbonated eclogites. Previous studies revealed that the low-δ^{26}Mg basalts from New Zealand, Eastern China, Hainan Island, Vietnam and Pitcairn Island are related to the recycling of carbonated eclogites [1, 55, 58, 66-68]. Wang et al. [66] first proposed that carbonated eclogite-derived melts are involved in the genesis of low-δ^{26}Mg Antipodes Volcano basalts from New Zealand, based on the negative correlation between δ^{26}Mg and Gd/Yb ratios. Li et al. [1] concluded that there are two low-δ^{26}Mg components in the mantle of Eastern China and Hainan Island. The mantle of Eastern China is characterized by low Fe/Mn and high CaO/Al_2O_3 ratios that are consistent with carbonated peridotite-derived melts; the mantle of Hainan Island features high Fe/Mn and low CaO/Al_2O_3 ratios, which is indicative of recycled eclogite-derived melts. Future work coupling Mg isotopes to other major/trace element proxies is needed to further constrain how the low-δ^{26}Mg signature is related to deep carbon cycle. As a consequence, using the Mg isotopic system to quantify the proportion of recycled carbonate component in the mantle source is still in early stages.

Other stable isotopic systems, such as Ca and Zn isotopes, have been increasingly applied to trace deep carbonate recycling [11], and shed more light to the nature and fate of deep recycled carbon. However, the geochemical behavior of the silicate-carbonate system during crustal subduction remains poorly known. Before the stable isotopic systems of divalent metals in carbonates are put together to provide better constraints on deep carbonate cycling, the behavior of metal stable isotopes during subduction needs to be evaluated.

Acknowledgements

We thank Xu-Han Dong for compiling the data for intraplate basalts from Eastern China.

Funing

This work was reported by the National Nature Science Foundation of China (41730214) and the National Key R&D Program of China (2019YFA0708404).

References

1. Li S-G, Yang W and Ke S *et al.* Deep carbon cycles constrained by a large-scale mantle Mg isotope anomaly in eastern China. Natl Sci Rev 2017; 4: 111-20.
2. Tipper E, Galy A and Gaillardet J *et al.* The magnesium isotope budget of the modern ocean: constraints from riverine magnesium isotope ratios. Earth Planet Sci Lett 2006; 250: 241-53.
3. Galy A, Bar-Matthews M and Halicz L *et al.* Mg isotopic composition of carbonate: insight from speleothem formation. Earth Planet Sci Lett 2002; 201: 105-15.
4. Teng F-Z, Li W-Y and Ke S *et al.* Magnesium isotopic composition of the Earth and chondrites. Geochim Cosmochim Acta 2010; 74: 4150-66.

5. Chen C, Dai W and Wang Z *et al.* Calcium isotope fractionation during magmatic processes in the upper mantle. Geochim Cosmochim Acta 2019; 249: 121-37.
6. Kang J-T, Ionov DA and Liu F *et al.* Calcium isotopic fractionation in mantle peridotites by melting and metasomatism and ca isotope composition of the Bulk Silicate Earth. Earth Planet Sci Lett 2017; 474: 128-37.
7. Wang S-J, Teng F-Z and Li S-G *et al.* Magnesium isotopic systematics of mafic rocks during continental subduction. Geochim Cosmochim Acta 2014; 143: 34-48.
8. Lu W-N, He Y and Wang Y *et al.* Behavior of calcium isotopes during continental subduction recorded in meta-basaltic rocks. Geochim Cosmochim Acta 2020; 278: 392-404.
9. Teng F-Z. Magnesium isotope geochemistry. Rev Mineral Geochem 2017; 82: 219-87.
10. Antonelli MA and Simon JI. Calcium isotopes in hightemperature terrestrial processes. Chem Geol 2020; 548: 119651.
11. Liu S-A and Li S-G. Tracing the deep carbon cycle using metal stable isotopes: opportunities and challenges. Engineering 2019; 5: 448-57.
12. Plank T and Manning CE. Subducting carbon. Nature 2019; 574: 343-52.
13. Manning CE. The chemistry of subduction-zone fluids. Earth Planet Sci Lett 2004; 223: 1-16.
14. Bebout GE and Penniston-Dorland SC. Fluid and mass transfer at subduction interfaces—the field metamorphic record. Lithos 2016; 240: 228-58.
15. Shen J, Wang S-J and Qin L *et al.* Tracing serpentinite dehydration in a subduction channel: chromium element and isotope evidence from subducted oceanic crust. Geochim Cosmochim Acta 2021; 313: 1-20.
16. Kelemen PB and Manning CE. Reevaluating carbon fluxes in subduction zones, what goes down, mostly comes up. Proc Natl Acad Sci USA 2015; 112: E3997-4006.
17. Ague JJ and Nicolescu S. Carbon dioxide released from subduction zones by fluid-mediated reactions. Nat Geosci 2014; 7: 355-60.
18. Stewart E and Ague JJ. Pervasive subduction zone devolatilization recycles CO_2 into the forearc. Nat Commun 2020; 11: 6220.
19. Kerrick D and Connolly J. Metamorphic devolatilization of subducted marine sediments and the transport of volatiles into the Earth's mantle. Nature 2001; 411: 293-6.
20. Molina JF and Poli S. Carbonate stability and fluid composition in subducted oceanic crust: an experimental study on H_2O-CO_2-bearing basalts. Earth Planet Sci Lett 2000; 176: 295-310.
21. Sizova E, Gerya T and Brown M *et al.* Subduction styles in the Precambrian: insight from numerical experiments. Lithos 2010; 116: 209-29.
22. GormanPJ, Kerrick D and Connolly J. Modeling open system metamorphic decarbonation of subducting slabs. Geochem Geophys Geosyst 2006; 7: Q04007.
23. Tao R, Zhang L and Tian M *et al.* Formation of abiotic hydrocarbon from reduction of carbonate in subduction zones: constraints from petrological observation and experimental simulation. Geochim Cosmochim Acta 2018; 239: 390-408.
24. Frezzotti M, Selverstone J and Sharp Z *et al.* Carbonate dissolution during subduction revealed by diamond-bearing rocks from the Alps. Nat Geosci 2011; 4: 703-6.
25. Shen J, Li S-G andWang S-J *et al.* Subducted Mg-rich carbonates into the deep mantle wedge. Earth Planet Sci Lett 2018; 503: 118-30.
26. Pan D, Spanu L and Harrison B *et al.* Dielectric properties of water under extreme conditions and transport of carbonates in the deep Earth. Proc Natl Acad Sci USA 2013; 110: 6646-50.
27. Manning CE, Shock EL and Sverjensky D. The chemistry of carbon in aqueous fluids at crustal and uppermantle conditions: experimental and theoretical constraints. Rev Mineral Geochem 2013; 75: 109-48.
28. Grassi D and Schmidt MW. The melting of carbonated pelites from 70 to 700 km depth. J Petrol 2011; 52: 765-89.
29. Tsuno K and Dasgupta R. Melting phase relation of nominally anhydrous, carbonated pelitic-eclogite at 2.5-3.0 GPa and deep cycling of sedimentary carbon. Contrib Mineral Petrol 2011; 161: 743-63.
30. Thomsen TB and Schmidt MW. Melting of carbonated pelites at 2.5-5.0 GPa, silicate-carbonatite liquid immiscibility, and potassium-carbon metasomatism of the mantle. Earth Planet Sci Lett 2008; 267: 17-31.
31. Dasgupta R, Hirschmann MM and Dellas N. The effect of bulk composition on the solidus of carbonated eclogite from partial melting experiments at 3 GPa. Contrib Mineral Petrol 2005; 149: 288-305.
32. Dasgupta R, Hirschmann MM and Withers AC. Deep global cycling of carbon constrained by the solidus of anhydrous, carbonated eclogite under upper mantle conditions. Earth Planet Sci Lett 2004; 227: 73-85.

33. Thomson AR, Walter MJ and Kohn SC *et al.* Slab melting as a barrier to deep carbon subduction. Nature 2016; 529: 76-9.
34. Hammouda T. High-pressure melting of carbonated eclogite and experimental constraints on carbon recycling and storage in the mantle. Earth Planet Sci Lett 2003; 214: 357-68.
35. Hammouda T and Keshav S. Melting in the mantle in the presence of carbon: review of experiments and discussion on the origin of carbonatites. Chem Geol 2015; 418: 171-88.
36. Dasgupta R and Hirschmann MM. The deep carbon cycle and melting in Earth's interior. Earth Planet Sci Lett 2010; 298: 1-13.
37. Suito K, Namba J and Horikawa T *et al.* Phase relations of CaCO3 at high pressure and high temperature. Am Mineral 2001; 86: 997-1002.
38. Dasgupta R. Ingassing, storage, and outgassing of terrestrial carbon through geologic time. Rev Mineral Geochem 2013; 75: 183-229.
39. Kushiro I. Carbonate-silicate reactions at high pressures and possible presence of dolomite and magnesite in the upper mantle. Earth Planet Sci Lett 1975; 28: 116-20.
40. Johnson MC and Plank T. Dehydration and melting experiments constrain the fate of subducted sediments. Geochem Geophys Geosyst 1999; 1: 1007.
41. Martin LA, Wood BJ and Turner S *et al.* Experimental measurements of trace element partitioning between lawsonite, zoisite and fluid and their implication for the composition of arc magmas. J Petrol 2011; 52: 1049-75.
42. Kessel R, Schmidt MW and Ulmer P *et al.* Trace element signature of subduction-zone fluids, melts and supercritical liquids at 120-180 km depth. Nature 2005; 437: 724-7.
43. Wang S-J, Teng F-Z and Li S-G *et al.* Magnesium isotopic systematics of mafic rocks during continental subduction. Geochim Cosmochim Acta 2014; 143: 34-48.
44. Li W-Y, Teng F-Z and Xiao Y *et al.* High-temperature inter-mineral magnesium isotope fractionation in eclogite from the Dabie orogen, China. Earth Planet Sci Lett 2011; 304: 224-30.
45. Teng FZ, Yang W and Rudnick RL *et al.* Heterogeneous magnesium isotopic composition of the lower continental crust: a xenolith perspective. Geochem Geophys Geosyst 2013; 14: 3844-56.
46. Wang S, Teng F and Bea F. Magnesium isotopic systematics of metapelite in the deep crust and implications for granite petrogenesis. Geochem Persp Let 2015; 1: 75-83.
47. Li WY, Teng FZ and Wing BA *et al.* Limited magnesium isotope fractionation during metamorphic dehydration in metapelites from the Onawa contact aureole, Maine. Geochem Geophys Geosyst 2014; 15: 408-15.
48. Wang S-J, Teng F-Z and Rudnick RL *et al.* The behavior of magnesium isotopes in low-grade metamorphosed mudrocks. Geochim Cosmochim Acta 2015; 165: 435-48.
49. Huang J, Guo S and Jin Q-Z *et al.* Iron and magnesium isotopic compositions of subduction-zone fluids and implications for arc volcanism. Geochim Cosmochim Acta 2020; 278: 376-91.
50. Chen Y-X, Schertl H-P and Zheng Y-F *et al.* Mg-O isotopes trace the origin of Mg-rich fluids in the deeply subducted continental crust ofWestern Alps. Earth Planet Sci Lett 2016; 456: 157-67.
51. Gao X-Y, Wang L and Chen Y-X *et al.* Geochemical evidence from coesitebearing jadeite quartzites for large-scale flow of metamorphic fluids in a continental subduction channel. Geochim Cosmochim Acta 2019; 265: 354-70.
52. Wang S-J, Teng F-Z and Li S-G *et al.* Tracing subduction zone fluid-rock interactions using trace element and mg-sr-nd isotopes. Lithos 2017; 290: 94-103.
53. Teng F-Z, Li W-Y and Rudnick RL *et al.* Contrasting lithium and magnesium isotope fractionation during continental weathering. Earth Planet Sci Lett 2010; 300: 63-71.
54. Wang S-J, Teng F-Z and Li S-G. Tracing carbonate-silicate interaction during subduction using magnesium and oxygen isotopes. Nat Commun 2014; 5: 5328.
55. Zeng G, Chen L-H and Hofmann AW *et al.* Nephelinites in eastern China originating from the mantle transition zone. Chem Geol 2021; 576: 120276.
56. Huang J, Li S-G and Xiao Y *et al.* Origin of low δ^{26}Mg Cenozoic basalts from South China Block and their geodynamic implications. Geochim Cosmochim Acta 2015; 164: 298-317.
57. Su B-X, Hu Y and Teng F-Z *et al.* Magnesium isotope constraints on subduction contribution to Mesozoic and Cenozoic East Asian continental basalts. Chem Geol 2017; 466: 116-22.
58. Sun Y, Teng FZ and Ying JF *et al.* Magnesium isotopic evidence for ancient subducted oceanic crust in LOMU-like potassium-rich volcanic rocks. J Geophys Res Solid Earth 2017; 122: 7562-72.

59. Wang X-J, Chen L-H and Hofmann AW *et al.* Mantle transition zone-derived EM1 component beneath NE China: geochemical evidence from Cenozoic potassic basalts. Earth Planet Sci Lett 2017; 465: 16-28.
60. Chen Y, Huang F and Shi GH *et al.* Magnesium isotope composition of subduction zone fluids as constrained by jadeitites from Myanmar. J Geophys Res Solid Earth 2018; 123: 7566-85.
61. Chen Y-X, Démny A and Schertl H-P *et al.* Tracing subduction zone fluids with distinct Mg isotope compositions: insights from high-pressure metasomatic rocks (leucophyllites) from the Eastern Alps. Geochim Cosmochim Acta 2020; 271: 154-78.
62. Shen B, Wimpenny J and Lee C-TA *et al.* Magnesium isotope systematics of endoskarns: implications for wallrock reaction in magma chambers. Chem Geol 2013; 356: 209-14.
63. Macris CA, Young ED and Manning CE. Experimental determination of equilibrium magnesium isotope fractionation between spinel, forsterite, and magnesite from 600 ℃ to 800 ℃. Geochim Cosmochim Acta 2013; 118: 18-32.
64. Teng F-Z, Hu Y and Chauvel C. Magnesium isotope geochemistry in arc volcanism. Proc Natl Acad Sci USA 2016; 113: 7082-7.
65. Tian H-C, Yang W and Li S-G *et al.* Low δ^{26}Mg volcanic rocks of Tengchong in Southwestern China: a deep carbon cycle induced by supercritical liquids. Geochim Cosmochim Acta 2018; 240: 191-219.
66. Wang S-J, Teng F-Z and Scott JM. Tracing the origin of continental HIMUlike intraplate volcanism using magnesium isotope systematics. Geochim Cosmochim Acta 2016; 185: 78-87.
67. Hoang THA, Choi SH and Yu Y *et al.* Geochemical constraints on the spatial distribution of recycled oceanic crust in the mantle source of late Cenozoic basalts, Vietnam. Lithos 2018; 296: 382-95.
68. Wang X-J, Chen L-H and Hofmann AW *et al.* Recycled ancient ghost carbonate in the Pitcairn mantle plume. Proc Natl Acad Sci USA 2018; 115: 682-7.

Chromium isotope fractionation during magmatic processes: Evidence from mid-ocean ridge basalts*

Haibo Ma[1], Li-Juan Xu[1], Ji Shen[2, 3], Sheng-Ao Liu[1] and Shuguang Li[1, 2]

1. State Key Laboratory of Geological Processes and Mineral Resources, China University of Geosciences, Beijing 100083, China
2. CAS Key Laboratory of Crust-Mantle Materials and Environments, School of Earth and Space Sciences, University of Science and Technology of China, Hefei 230026, P.R. China
3. CAS Center for Excellence in Comparative Planetology, Hefei 230026, P.R. China

亮点介绍：本文首次对洋中脊的玄武岩进行了高精度的Cr同位素分析，结果显示洋中脊玄武岩的Cr同位素变化主要受控于含铬矿物的分离结晶作用。相比于洋岛玄武岩，本文发现洋中脊玄武岩岩浆分异过程中Cr同位素分馏系数要更大，可能原因是后者具有更低的氧逸度导致的，其更低的V/Sc也支持这个解释。

Abstract Chromium (Cr) isotopes represent a powerful tool for tracing the redox conditions during planetary magmatic evolution. However, so far, the systematic investigation of Cr isotope variation has been only performed on ocean island basalts (OIBs) in terrestrial magmatic rocks. Therefore, the Cr stable isotope compositions (expressed as δ^{53}Cr relative to NIST SRM 979) of other magmatic rocks, formed under different oxygen fugacity, have remained unconstrained. In this paper, we present the first Cr stable isotopic data of mid-ocean ridge basalts (MORBs) from the Eastern Pacific Ocean ridge, the Indian ridge, and the Atlantic Ocean ridge. The oxygen fugacity of such basalts is different from that of the above-mentioned OIBs. The Rhyolite-MELTS model shows that the chemical composition variations in the studied basalts are induced by varying extents of fractional crystallization of olivine, clinopyroxene, plagioclase, and spinel. The δ^{53}Cr values of the MORBs range from −0.27±0.03‰ to −0.07±0.02‰ (n=28), and two distinct groups of basalts are identified based on the correlation between δ^{53}Cr and MgO. On the one hand, the δ^{53}Cr of group I basalts (−0.27±0.03‰ to −0.14±0.03‰; n=24) are systematically lower than the established average value of the Bulk Silicate Earth (BSE) (δ^{53}Cr= −0.12±0.04‰; 2SD), which are positively correlated with their MgO and Cr concentrations, indicating that Cr isotopes are fractionated during magmatic differentiation. Moreover, Rayleigh fractionation modelling suggests that the crystallization of olivine, clinopyroxene, and spinel gives rise to the Cr depletion and thereby decreases δ^{53}Cr values. On the other hand, group II basalts (−0.10±0.03‰ to −0.07±0.02‰; n=4) exhibit higher δ^{53}Cr values than group I with identical MgO and Cr concentrations. This is possibly associated with the crystallization of clinopyroxene under low pressure.

* 本文发表在：Geochimica et Cosmochimica Acta, 2022, 327: 79-95

The average Cr isotope composition (δ^{53}Cr=−0.16±0.02‰, n=3) of the primitive basalts (MgO> 9%) represents that of the primary MORB melt, which is lighter than the average value of BSE. Using the non-modal melting equations, the δ^{53}Cr of the MORB mantle source is estimated to be −0.12±0.02‰ (2σ), which is consistent with that of BSE. Compared with the Cr isotopic data of OIBs from Fangataufa island, we find that the equilibrium fractionation factors ($\Delta^{53}Cr_{crystal-melt}$ = +0.04‰ to +0.13‰) of MORBs during fractional crystallization are larger than that of Fangataufa island lavas ($\Delta^{53}Cr_{crystal-melt}$ = 0.010±0.005‰ for low-K suite, and 0.020±0.010‰ for high-K suite; Bonnand et al., 2020a), indicating that the basalts from Fangataufa island have higher oxygen fugacity than those of MORBs analyzed in this study, which is strongly supported by their higher V/Sc ratios.

Keywords Cr isotopes, mid-ocean ridge basalts, redox conditions, magmatic differentiation

1 Introduction

Cr stable isotopes have been widely used to trace planetary evolution and differentiation processes (Moynier et al., 2011; Schoenberg et al., 2016; Bonnand et al., 2016a, 2016b, 2018; Sossi et al., 2018; Zhu et al., 2019, 2021a, 2021b, 2021c, 2022; Shen et al., 2020). In particular, the Cr stable isotope differences between chondrites and differentiated planets, e.g., the Earth, the Moon and the Vesta, and their precursors have remained largely debated. Two major mechanisms have been proposed to explain such Cr isotope differences: (1) loss of heavy isotopes by oxidizing gas during the magma ocean stage of the Moon, the Vesta, and the Earth (Sossi et al., 2018; Zhu et al., 2019, 2021a), and (2) the dominant influence of the differences in redox condition in generating Cr isotope fractionation during partial melting and fractional crystallization of different silicate planets (Shen et al., 2020; Bonnand et al., 2020a). Although both processes could account for the observed Cr isotope differences among planets, the latter process has been hypothesized only based on the analyses OIBs. Hence, more work on Cr isotope fractionation during magmatic evolutions under different redox conditions is required.

Previous studies have shown that the oxygen fugacity of MORBs is significantly different from that of OIBs (Rhodes and Vollinger, 2005; Mallmann and O'Neill, 2007; Cottrell and Kelley, 2011; Moussallam et al., 2014, 2016; Shorttle et al., 2015; Brounce et al., 2017). For instance, Mallmann and O'Neill (2007) found that OIBs ($\Delta\log fO_2$ (QFM) = +0.54; oxygen fugacity (fO_2) reported in log-bar units relative to the quartz-fayalite-magnetite (QFM) buffer) are more oxidized than MORBs (−0.67). Their difference in oxygen fugacity has been attributed to incorporating ancient and more oxidized crustal materials into the mantle source of OIBs (Hofmann and White, 1982), the temperature and pressure of melting (Shorttle et al., 2015), or degassing of certain volatile elements (Moussallam et al., 2014, 2016). To better estimate the effect of the redox conditions on the observed difference in Cr isotope compositions, MORBs are the best available samples. In addition, MORBs produced by decompressed melting of asthenosphere mantle under mid-ocean ridge are the most abundant igneous rocks on the Earth because they cover most of the global ocean floor, forming ~60,000 km long globe-encircling ocean ridges (Niu, 2016). MORBs are also the main component of oceanic crust. Therefore, the composition of the MORBs provides insights into the Earth's most extensive crustal reservoir and the best available constraints on the composition of oceanic crust (Gale et al., 2013).

Contrary to radiogenic isotopes, mass-dependent stable Cr isotopes can be fractionated by high-temperature processes (Xia et al., 2017; Shen et al., 2018, 2020; Bonnand et al., 2020a; Jerram et al., 2021; Zhu et al. 2022). Therefore, in order to better use Cr isotopes as efficient tracers, we need to constrain the fractionation behavior of Cr isotopes during high-temperature processes. Most recent Cr isotope analyses of peridotites, mantle-derived chromites, and basaltic rocks have been used to demonstrate Cr isotope fractionation during partial melting of mantle (Xia et al., 2017; Shen et al., 2018, 2020; Bonnand et al., 2020a; Jerram et al., 2021; Zhu et al., 2022). For example, Xia et al. (2017) suggested that refractory peridotites possess higher δ^{53}Cr values relative to fertile peridotites, implying that Cr isotopes are fractionated during partial melting. Similarly, the data of natural samples and theoretical calculations have provided a strong indication for equilibrium inter-mineral Cr isotope fractionation in mantle peridotites, which highlights the feasibility of Cr isotope fractionation during partial melting of the mantle (Shen et al., 2018; Jerram et al., 2021). The above studies suggested that the δ^{53}Cr values of the basaltic melt produced by partial melting of the mantle should be isotopically lower than those of the mantle source. This result has been reported in the later work on OIBs (Shen et al., 2020; Bonnand et al., 2020a). However, these basalts were considered derived from the heterogeneous mantle influenced by the recycled component (e.g., altered oceanic crust and sediment carbonates) (Huang and Frey, 2005; Huang et al., 2009, 2011; Shen et al., 2020; Bonnand et al., 2020a). In fact, a recent study has shown the effect of mantle heterogeneity on Cr isotope compositions of basalts (Shen et al., 2020). Compared with these basalts from Kilauea Iki and Mauna Kea, the basalts from Koolau display lighter Cr isotopes (the Kilauea Iki, Mauna Kea, and Koolau are located on the island of Hawaii). One possible interpretation is that the incorporation of altered oceanic crust causes the mantle source to possess slightly isotopically lighter than those of Kilauea Iki and Mauna Kea (Shen et al., 2020). Therefore, the individual contributions of mantle source heterogeneity and mantle melting leading to Cr isotope fractionation between basalts and peridotites remain elusive. More high-precision Cr stable isotope analyses of basalts are necessary to better constrain the behavior of Cr isotope fractionation during magmatic processes.

Using double spike techniques, we report high-precision Cr stable isotopic data for twenty-eight fresh MORBs from the Eastern Pacific Ocean ridge, the Indian ridge, and the Atlantic Ocean ridge. Since these MORBs derive from the uniform asthenosphere mantle, they effectively avoid the influence of mantle source heterogeneity. In addition, two enriched MORB (E-MORB) derived from the interaction of enriched plumes from deep mantle with normal midocean ridge basalts (N-MORB) or low degree melting of enriched mantle (Schilling, 1973) are also included in this study. They have the potential to further understand the influence of mantle heterogeneity on the Cr isotope composition of basalts. The objectives of this study are to (i) estimate the magnitude of Cr isotope fractionation during partial melting of mantle on Earth, (ii) evaluate the Cr isotope composition of oceanic crust (e.g., with this result, we can establish a benchmark for using Cr isotopes as a tracer of magmatic processes), and (iii) evaluate whether Cr isotopes can trace the difference in oxygen fugacity between MORBs and OIBs by combining the published OIB data (Shen et al., 2020; Bonnand et al., 2020a) with our MORB data.

2 Geological settings and samples

Twenty-eight MORB samples were analyzed from a wide range of geographic locations, including fifteen lavas from the Eastern Pacific Ocean ridge (EPR) (3° 6.59′ S, 102° 32.68′ W), seven from the Carlsberg ridge (CR) (63° 49.09′ E, 3° 42.06′ N), four from the South Mid-Atlantic Ocean ridge (SMAR) (63° 49.09′ E, 3° 42.06′ N), two from the North Atlantic Ocean ridge (AR) (5° 12.40′ S, 33° 20.14′ W) and one from the Southwest Indian ridge (SWIR) (9–25 °E) (Fig. 1).

The EPR is a fast-spreading ridge with a full spreading rate in excess of 80 mm/yr (White and Klein, 2013), while the SMAR is slow-spreading (~35 mm/yr) (Chen et al., 2020). Some MORBs from the EPR and all MORBs from SMAR have been studied (Chen et al., 2020). The MORBs from EPR display significant variations in MgO (8.2–6.6 wt.%) and CaO content (12.1–10.5 wt.%), which have been explained by the fractional crystallization of olivine, clinopyroxene, and plagioclase (Chen et al., 2020). In comparison, the MORBs from SMAR show homogeneity in MgO (8.1–7.7 wt.%) and CaO (11.8–11.3 wt.%). All MORBs mentioned above are fresh and typically the normal mid-ocean ridge basalts (N-MORB), except for SMAR-04 (Chen et al., 2020).

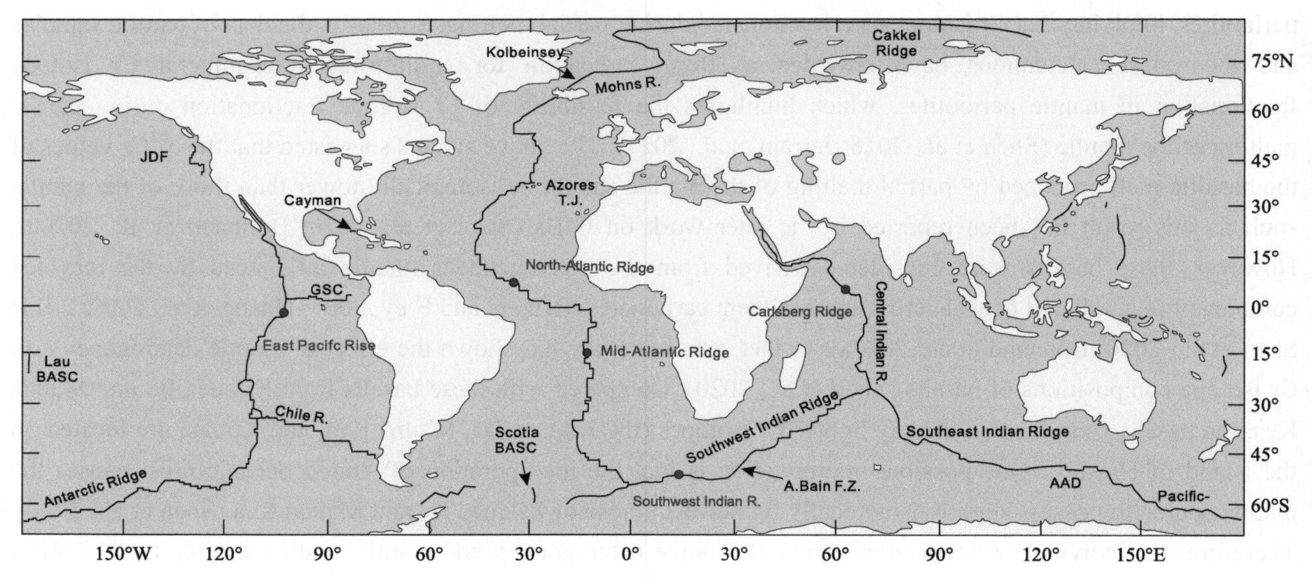

Fig. 1 Map of sampling locations (red cycle) in the global mid-ocean ridge system, modified from White and Klein, 2013.

The CR, which is a typical slow-spreading ridge (22–36 mm/yr; full spreading rate) in the northwest Indian Ocean, defines the plate boundary between the Indian and Somalian plates (Ray et al., 2012). AR is also a slow-spreading ridge (10–20 mm/yr; full spreading rate) likewise (Xiao et al., 2002). The major element composition of MORBs from CR and AR have been reported by Liu et al. (2015). These MORBs from CR display notable variations in their chemical compositions with MgO and CaO ranging from 9.6 to 7.9 wt.% and 12.7 to 10.9 wt.%, respectively. The variations of MgO (8.5–7.0 wt.%) and CaO (11.3–10.9 wt.%) in MORBs from AR, however, are limited (Liu et al., 2015). The SWIR, which is situated between the African and Antarctic plates, measures 7700 km in length and is one of the slowest spreading mid-ocean ridges with an average full spreading rate of 14–15 mm/yr (Standish et al., 2008). The major elements data of MORBs from SWIR have been reported by Standish et al. (2008).

3 Analytical methods

3.1 Major and trace element analysis

The major elements were analyzed using the Inductively Coupled Plasma-Atomic Emission Spectrometry (ICPAES) at the State Key Laboratory of Geological Processes and Mineral Resources, the China University of Geosciences, Beijing (CUGB). Approximately 1000 mg of sample powder was weighed into a Pt-(Au) crucible. The loss on ignition (LOI) values of the samples were obtained by heating the crucible in a pre-heated furnace at ~1000 ℃ for more than one hour. Sample powders (~400 mg) were mixed with Li-tetraborate ($Li_2B_4O_7$) and fused in a Pt- (Au) crucible at ~1000 ℃. The uncertainties of major element analysis are better than ±2%.

Whole-rock trace element composition analyses were conducted using the Thermo-Fisher Element XR at the State Key Laboratory of Geological Processes and Mineral Resources (RIG lab), CUGB. The analytical procedures followed the previously established methods (Chen et al., 2017). Powder aliquots (~50 mg) of both samples and standards (BHVO-2) were dissolved in a distilled HF + HNO_3 + HCl concentrated acids in bombs at

190 ℃ in an oven for 18 h. Several procedural blanks were processed simultaneously with the samples. After complete dissolution, the HF + HNO_3 + HCl mixture was evaporated to dryness and repeatedly treated with HNO_3 to break down the fluoride compounds. The samples were subsequently dissolved in concentrated HNO_3, followed by the addition of Milli-Q in bombs at 190 ℃ for 2 h. After drying, the samples were dissolved in 2% HNO_3. The samples were further diluted by 2% HNO_3 and doped with indium as an internal standard to monitor instrumental drift during analysis. The precision of all trace elements was estimated to be ±10%, and the accuracy was better than ±5% for most elements based on the analysis of the BHVO-2 rock standard.

3.2 Strontium and Neodymium isotope analysis

Strontium (Sr) and Neodymium (Nd) isotope analyses of basalts were performed in the Isotope Geochemistry Laboratory of CUGB. Sample powders (~30 mg) were dissolved completely in a mixture of HF and HNO_3 in Teflon Beakers. After the solutions were completely evaporated, the dried samples were re-dissolved using aqua regia until complete dissolution. Sr and Nd were purified following the protocols outlined in Xu et al. (2021). Sr was purified using 2.8 ml of pre-cleaned AG50W-X12 (400 mesh) by eluting 11 ml of 4 M HCl, and Nd was purified using 2 ml of LN resin by eluting 4 mL of 0.18 M HCl after being recovered from the AG50W-X12 resin. The Sr and Nd purified solutions were evaporated to dryness for isotopic analysis. Sr and Nd isotopic ratios were measured using a Thermo Fisher Scientific TRITON Plus thermal ionization mass spectrometer (TIMS) at the Isotope Geochemistry Laboratory of CUGB.. The Sr and Nd isotopes were corrected to $^{86}Sr/^{88}Sr$ = 0.1194 and $^{146}Nd/^{144}Nd$=0.7219, respectively, for instrumental mass bias correction. Furthermore, Alfa Aesar Nd was analyzed as an in-house reference standard. Based on isotope measurements carried out in past several years, the long-term measured value for this standard is $^{143}Nd/^{144}Nd$=0.512425±30 (2σ, standard deviation) on average. The long-term analysis of reference material BHVO-2 in this lab yielded $^{87}Sr/^{86}Sr$=0.703475±25 (2σ) and $^{143}Nd/^{144}Nd$ =0.512979±19 (2σ). The standard basalt BHVO-2 yielded $^{87}Sr/^{86}Sr$=0.703477±15 (2σ) and $^{143}Nd/^{144}Nd$=0.512974 ±13 (2σ) in this study. These values are identical to the previously published values within the range of uncertainty (Weis et al., 2005, 2006; Li et al., 2007; DuFrane et al., 2009; Wiesmaier et al., 2011; Xu et al., 2021).

3.3 Cr isotope analysis

Sample powders containing at least ~1 μg of Cr were weighed and digested in a mixture of double-distilled HF and HNO_3 in Teflon Beakers. After the powders were completely digested, the sample aliquots were evaporated to dryness. Aqua regia (HCl: HNO_3=3:1) was then added to the sample and heated until the aliquots were completely dissolved. The aliquots were heated again, evaporated to dryness, and refluxed with double-distilled HCl. Subsequently, the sample aliquots were mixed with ^{50}Cr-^{54}Cr double-spike according to the optimized ratio of Cr_{spike}/Cr_{total} = 0.25 (Cr_{total} = Cr_{spike} + Cr_{sample}). Following previous studies (Wu et al., 2020; Zhu et al., 2021b), the sample-spike mixtures were placed on a hot plate overnight at 120 ℃ to homogenize the double-spike and aliquots. After evaporation to dryness, the final residues were re-dissolved in 6 M HCl for chromatographic separation. Cr was purified according to the protocol established by Liu et al. (2019). Aliquots dissolved in 6 M HCl were heated for three or four hours and diluted with high purity water to reach a concentration of 1 M HCl.

The aliquot was then loaded in 1 ml pre-cleaned AG50W-X12 (400 mesh) on cation exchange resins and Cr was directly eluted using 1 M HCl. The aliquots were loaded on 0.33 ml pre-cleaned AG50W-X12 resins, and pure Cr was collected in 2 M HCl. The total yield of Cr through two-step procedure was great than 80% for silicate samples (Liu et al., 2019). The total procedural blank was 3–4 ng, which is negligible relative to the total amount of sample considered, i.e., ~1000 ng. The purified Cr was evaporated in concentrated HNO_3 three to four times to convert the HCl medium to HNO_3. To remove organics (from the resin), 500 μl of concentrated HNO_3 along with 30 μl $HClO_4$ was added and heated at 120 ℃ for two weeks. The Cr isotope compositions were determined using TIMS at the Isotope Geochemistry Laboratory of CUGB. 2 μl of 3% HNO_3 containing ~1 μg of Cr, was loaded on outgassed Re single filaments with Al-doped silica gel and saturated boric acid (Zhu et al., 2021a). Each analysis consists of fourteen blocks of twelve cycles with an integration time of 8 s, obtained using an amplifier rotation. Measurements were made with an ion current of 4–8×10^{-11} A for ^{52}Cr. Each sample was measured two or four times. Cr isotope composition of the samples analyzed in this study has been reported relative to the National Institute of Standards and Technology (NIST) Standard Reference Material 979 (SRM 979), following the delta notation:

$$\delta^{53}\mathrm{Cr}(‰)=\left(\frac{(^{53}\mathrm{Cr}/^{52}\mathrm{Cr})_{\mathrm{sample}}}{(^{53}\mathrm{Cr}/^{52}\mathrm{Cr})\mathrm{NIST\ SRM\ 979}}-1\right)\times 1000 \qquad (1)$$

The reproducibility of NIST SRM 979 and NIST SRM 3112a was found to be 0.00±0.02‰ (2SD; n=42) and −0.07±0.02‰ (2SD; n=32), respectively. The long-term external precision of unprocessed NIST SRM 979 and NIST SRM 3112a was better than 0.03‰. The δ^{53}Cr of USGS standards such as JP-1, BHVO-1, and BHVO-2, are in agreement with the published data for these standards, Table 1 (Schoenberg et al., 2008, 2016; Zhu et al., 2018; Sossi et al., 2018; Liu et al., 2019; Zhu et al., 2021a, 2021b, 2021c), confirming the accuracy of our new data.

Table 1 The Cr, Sr, Nd isotopes and selected element data of the MORBs analyzed in this study

Sample	Location	Rock type	MgO (wt.%)	Cr (μg/g)	δ^{53}Cr (‰)	2SD	N	^{87}Sr/^{86}Sr	^{143}Nd/^{144}Nd
MORB glasses from the East Pacific Ocean Ridge									
TVG01	EPR	Basalt	6.39	87	−0.08	0.02	4	0.70251±2	0.51319±2
Replicate					−0.10	0.02	3		
Replicate					−0.08	0.02	2		
Average					−0.09				
TVG03-9	EPR	Basalt	7.31	174	−0.22	0.02	2	0.70269±1	0.51318±2
TVG04	EPR	Basalt	7.87	342	−0.18	0.02	3	0.70249±2	0.51317±1
Replicate					−0.19	0.02	2		
Average					−0.19				
TVG07	EPR	Basalt	7.3	169	−0.24	0.02		0.70256±1	0.51318±1
TVG09	EPR	Basalt	6.79	168	−0.22	0.02	3	0.70251±2	0.51317±1
Replicate					−0.22	0.02	2		
Average					−0.22				
TVG14	EPR	Basalt	6.34	96	−0.24	0.02	3	0.70240±1	0.51319±2
TVG18	EPR	Basalt	6.43	89	−0.08	0.02	3	0.70249±2	0.51318±1
Replicate					−0.08	0.03	3		
Average					−0.08				
EPR-01	EPR	Basalt	7.03[a]	153[a]	−0.21	0.02	2		
EPR-03	EPR	Basalt	8.18[a]	325[a]	−0.21	0.02	2		
EPR-04	EPR	Basalt	6.57[a]	77[a]	−0.07	0.02	3		
Replicate					−0.07	0.02	2		
Average					−0.07				

Continued

Sample	Location	Rock type	MgO (wt.%)	Cr (μg/g)	δ^{53}Cr (‰)	2SD	N	^{87}Sr/^{86}Sr	^{143}Nd/^{144}Nd
EPR-05	EPR	Basalt	6.65[a]	88[a]	−0.24	0.02	2		
EPR-07	EPR	Basalt	7.44[a]	166[a]	−0.22	0.03	2		
Replicate					−0.24	0.02	3		
Average					−0.23				
EPR-08	EPR	Basalt	7.42[a]	164[a]	−0.20	0.03	2		
EPR-09	EPR	Basalt	7.37[a]	165[a]	−0.24	0.02	2		
EPR-11	EPR	Basalt	6.55[a]	78[a]	−0.09	0.02	2		
Replicate					−0.10	0.03	3		
Average					−0.10				
MORB glasses from the Atlantic Ocean Ridge									
SMAR-01	SMAR	Basalt	8.09[a]	286[a]	−0.21	0.03	4		
SMAR-04	SMAR	Basalt	7.99[a]	247[a]	−0.18	0.02	2		
SMAR-05	SMAR	Basalt	7.69[a]	253[a]	−0.27	0.03	3		
SMAR-09	SMAR	Basalt	7.61[a]	260[a]	−0.25	0.02	2		
AR04-1	AR	Basalt	8.52[b]	284	−0.22	0.02	2	0.70289±1	0.51306±2
AR05-1	AR	Basalt	7.03[b]	195	−0.20	0.02	3	0.70321±2	0.51296±1
MORB glasses from the Carlsberg Ocean Ridge									
CR01-1	CR	Basalt	8.76[b]	315	−0.20	0.02	3	0.70303±1	0.51305±2
CR02-1	CR	Basalt	7.86[b]	247	−0.25	0.02	4	0.70286±1	0.51306±2
CR02-3	CR	Basalt	7.86[b]	247	−0.22	0.02	4	0.70300±2	0.5130±1
CR03-1	CR	Basalt	9.57[b]	391	−0.14	0.02	2	0.70289±1	0.51302±2
CR03-3	CR	Basalt	9.56[b]	391	−0.16	0.03	4	0.70310±1	0.51296±2
CR04-5	CR	Basalt	8.90[b]	328	−0.21	0.03	2	0.70305±2	0.51308±1
MORB glasses from the Southwest Indian ridge									
KN162-9 48-4	SWIR	Basalt	9.07[c]	384	−0.17	0.02	3		
Replicate					−0.17	0.02	3		
Average					−0.17				
JP-1		Peridotite			−0.08	0.02	5		
BHVO-1		Basalt			−0.11	0.02	8		
BHVO-2		Basalt			−0.12	0.02	15	0.70348±2	0.51300±2

Note: Replicate-replicate digestion of the same powder.

[a] The major and trace element compositions of basalts from Chen et al. (2020).

[b] The major element compositions of basalts from Liu et al. (2015).

[c] The major element compositions of basalts from Standish et al. (2008).

4 Results

The major and trace elements data of all samples are reported in Table S1, combined with the major and trace elements of MORBs from EPR and SMAR (Chen et al., 2020) as well as the major elements from CR, AR, and SWIR (Liu et al., 2015; Standish et al., 2008). All MORBs selected for study are fresh, as indicated by the low LOI (<2%) (Fig. 2a and c). The negative LOI values could be attributed to oxidation of ferric iron in the MORBs during experimental analysis (Chen et al., 2020). Most of our samples belong to N-MORB (Fig. 3d) and have a wide range of MgO contents from 6.34 to 9.57 wt.%, while SiO_2 shows a narrow range from 49.1 to 51.3 wt.%. In terms of compatible element abundances (e.g., Cr), such basalts display moderate variations from 77 to 391 μg/g (Table 1).

The Sr and Nd isotopic data of MORBs are listed in Table 1 and plotted in Fig. 3a. These basalts have $^{87}Sr/^{86}Sr$ and $^{143}Nd/^{144}Nd$ ratios ranging from 0.702102±10 (2σ) to 0.703206±16 (2σ) and 0.512957±21 to 0.513188 ± 16 (2σ), respectively. The results overlap with the published data for MORBs (Fig. 3a).

The $\delta^{53}Cr$ of twenty-eight MORBs samples are provided in Table 1, which ranged from −0.07 ± 0.02‰ to −0.27±0.03‰, with an average value of −0.20±0.11‰ (2SD, n=28). The $\delta^{53}Cr$ values of the studied samples are divided into two distinct groups based on the correlation between $\delta^{53}Cr$ and MgO content (Fig. 6a). The MORBs of group I (−0.14±0.03‰ to −0.27±0.03‰, n=24) are characterized by lower $\delta^{53}Cr$ than the BSE ($\delta^{53}Cr$=−0.12± 0.04‰, 2SD, Jerram et al., 2020; −0.10±0.12‰, 2SD, Schoenberg et al., 2008; −0.11±0.06‰, 2SD, Sossi et al., 2018). These MORBs also display co-variations between $\delta^{53}Cr$ and both MgO and Cr concentrations. In comparison, the group II (−0.07±0.02‰ to −0.10±0.03‰, n=4) have heavier Cr isotope compositions than those of group I with identical MgO and Cr concentrations. There is no systematic difference in the Cr isotope compositions between N-MORB and E-MORB (Fig. 3d).

5 Discussion

Our isotope measurements reveal resolvable differences in the Cr isotope compositions of MORBs and BSE. Several possible geological processes can account for the variation in the $\delta^{53}Cr$ values in the investigated samples. The effects of alteration and source heterogeneity are assessed first. Subsequently, magmatic processes, especially fractional crystallization processes, are evaluated. The magnitude of fractionation caused by partial melting and Cr isotope composition of the MORB mantle source is estimated. Finally, we discuss the implications of Cr isotope fractionation related to the evolution of terrestrial planets associated with redox conditions.

5.1 Secondary alteration

After the formation of MORBs, they are transported off-axis to each side of the spreading center, accumulating sediments, and gradually altering with age (White and Klein, 2013). The extent of secondary alteration was assessed based on LOI contents. In addition, the ratio of K_2O/P_2O_5 is also regarded as an alteration index because seawater alteration may cause leaching of K (relative to P) from basalts into fluid (Huang and Frey, 2003; Huang et al., 2016). As shown in Fig. 2a and c, the LOI values of all samples are < 2%, indicating that these MORBs experience limited or no seafloor alteration. Besides, the Cr concentration and isotope composition of MORBs do not correlate with LOI or K_2O/P_2O_5 (Fig. 2), suggesting that secondary alteration does not affect the Cr concentration or isotope composition of the studied MORBs.

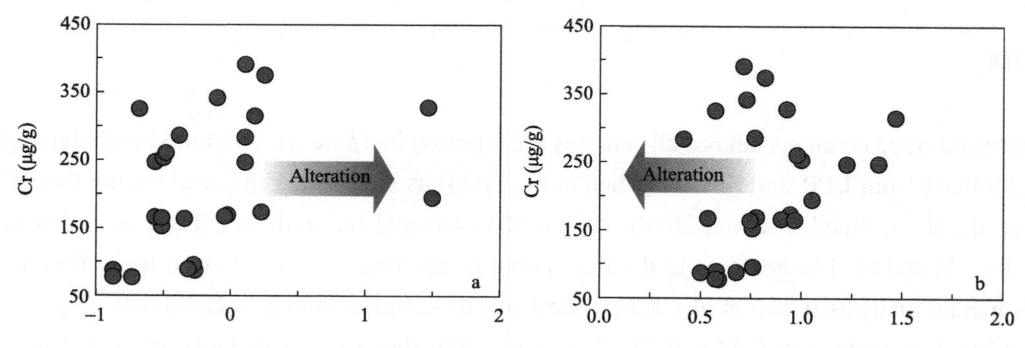

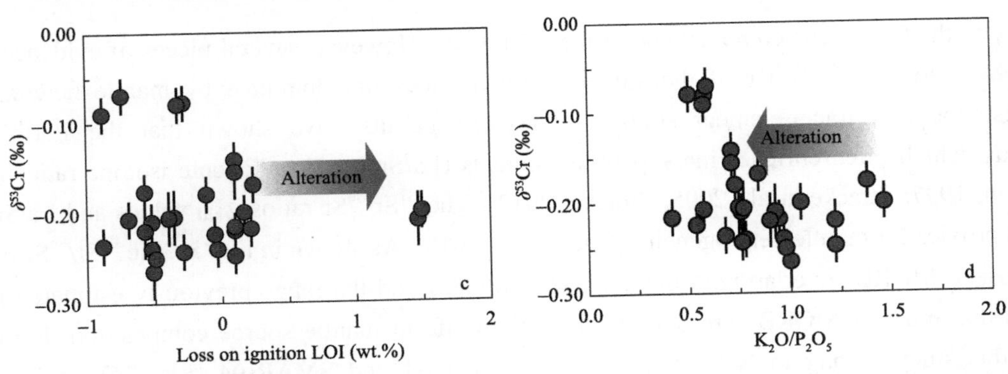

Fig. 2 Plots (a) and (b) show the relationship between Cr concentrations and the post-magmatic alteration indices (LOI and K_2O/P_2O_5). Plots (c) and (d) show the Cr isotope compositions versus LOI and K_2O/P_2O_5, respectively.

Compared with the Cr concentration of studied MORBs (77–391 μg/g), seawater contains much less Cr, making it hardly changes the Cr isotope budget of MORBs (0.05 to 1 ng/g; Bonnand et al., 2013; Pereira et al., 2016; Reinhard et al., 2013). In addition, the reported natural seawater samples usually have δ^{53}Cr values ranging from 0.40 to 1.60 ‰ (Bonnand et al., 2013; Pereira et al., 2016; Scheiderich et al., 2015) which are higher than those of the MORBs. It is impossible that seawater alteration leads to lower δ^{53}Cr of the MORBs. Previous work has shown that Cr^{2+} is not retained on quenching to the glass because of electron exchange reaction ($Cr^{2+} + Fe^{3+} \rightarrow Cr^{3+} + Fe^{2+}$) on cooling (Berry et al., 2006), suggesting that the Cr oxidation state is Cr^{3+} in these MORBs. The Cr^{3+} is insoluble and immobile in natural pH range (Bonnand et al., 2013). These lines of evidence demonstrate that the δ^{53}Cr variation of the studied MORBs does not dominate by secondary alteration. This agrees with earlier conclusions that mild alteration did not modify the Cr isotope composition of the oceanic crust (Wang et al., 2016).

5.2 Mantle source heterogeneity

Recent studies have found that the Cr isotope composition of the mantle is modified by mantle metasomatism (Xia et al., 2017; Shen et al., 2018; Jerram et al., 2021). For instance, Xia et al. (2017) and Jerram et al. (2021) both found that metasomatized peridotites possess a wider range of δ^{53}Cr variations than non-metasomatized peridotites, which was attributed to kinetic isotope fractionation of Cr associated with interactions between silicate melts and peridotites. The inter-mineral Cr isotope fractionations (e.g., $\Delta^{53}Cr_{Spinel\text{-}Olivine}$, $\Delta^{53}Cr_{Spinel\text{-}Orthopyroxene}$, and $\Delta^{53}Cr_{Spinel\text{-}Clinopyroxene}$) of clinopyroxene (Cpx) -rich lherzolites and wehrlites metasomatized by the evolution of silicate melt and recycled sedimentary carbonates, respectively, are larger than lherzolites, where all mineral pairs have reached Cr isotope equilibrium. These studies suggested that Cr isotope disequilibrium of Cpx-rich lherzolites likely implies Cr isotope heterogeneity inherited from the mantle source. At the same time, the intermineral Cr isotope disequilibrium of wehrlites has been attributed to kinetic diffusion of isotopically light Mg and Cr from olivine to carbonatitic melt (Shen et al., 2018). In summary, mantle metasomatism may lead to heterogeneity in Cr isotope composition of mantle peridotites. It is expected that if the mantle source has been metasomatized by recycled materials or silicate melts, basalts derived from such a source might have more various Cr isotope compositions relative to those from the normal mantle. Such Cr isotope heterogeneity has been observed in basalts from mantle plumes, i.e., samples from Koolau, Hawaii (Shen et al., 2020). The lower δ^{53}Cr of basalts from Koolau compared to rest of the Hawaii samples with similar MgO content, has been attributed to the incorporation of the recycled altered oceanic crust with light Cr isotopes into the mantle source of Koolau (Shen et al., 2020). Considering that there are two E-MORB in this study, we have to assess the contribution of source heterogeneity to the Cr isotope compositions of these MORBs. However, several pieces of evidence suggest that the measurable variation in δ^{53}Cr of the studied MORBs does not dominate by mantle heterogeneity here. Numerous isotope and trace element studies of oceanic basalts have shown that the Earth's mantle is heterogeneous, which often enriches incompatible elements (La/Sm) and radiogenic isotope ratios (Sr, Nd, and Pb) (Hofmann, 1997; Stracke et al., 2005; Stracke, 2012). The ^{87}Sr/^{86}Sr ratios, ε_{Nd} values and La/Sm ratios can also serve as proxies for mantle heterogeneity (Chen et al., 2019). As shown in Fig. 3a, the ^{87}Sr/^{86}Sr ratios and ε_{Nd} values of most of MORBs overlap with the average N-MORB and the range previously estimated for MORBs. This implies that most of MORBs are derived from the uniform mantle source composition. Furthermore, all MORBs in this study belong to N-MORB, except for AR05-1 and SMAR-04 (Fig. 3d), indicating that the chemical compositions of these N-MORB are not affected by mantle source heterogeneity (Zindler and Hart, 1986; Sun and McDonough, 1989; Hofmann and Hémend, 2006). The ^{87}Sr/^{86}Sr ratios, ε_{Nd} values, and (La/Sm)$_N$ ratios of E-MORB are different from most of the N-MORB, suggesting that they are influenced by mantle source heterogeneity (Fig. 3b–d). However, E-MORB and N-MORB have identical Cr isotope compositions within the uncertainty range (Fig. 3b–d), which implies that the individual contribution of source heterogeneity to the change in δ^{53}Cr of the E-MORBs is insignificant. Alternatively, the signal induced by source heterogeneity may be overwhelmed by subsequent geological processes. This indicates that their Cr isotope compositions are not significantly affected by mantle source heterogeneity. The two E-MORB samples are included in the subsequent discussion. Hence, we propose that the dominant cause for the Cr isotope variation of the studied MORBs is likely Cr isotope fractionation during magmatic differentiation or partial melting of the mantle.

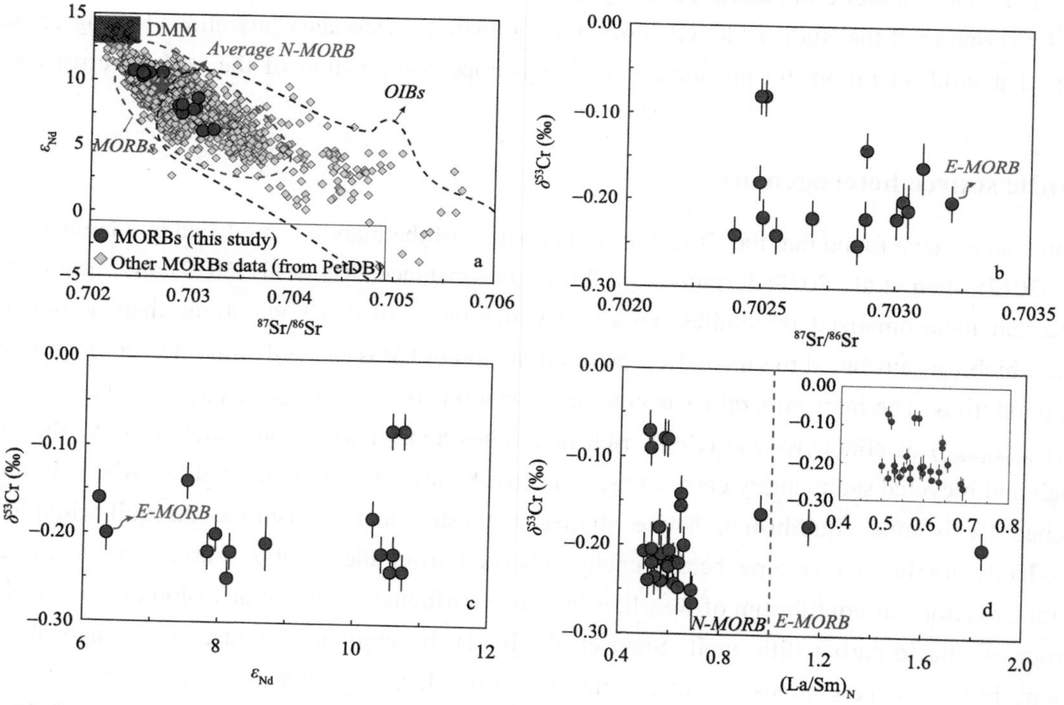

Fig. 3　Plot (a) displays the Sr and Nd isotope composition of MORBs analyzed in this study and others from PetDB database (https://search.earthchem.org/). The shaded area and blue star are depleted MORB mantle and average Sr and Nd isotope composition of the N-MORB, respectively (Zindler and Hart, 1986; Workman and Hart, 2005). MORBs and OIBs fields are derived from Zindler and Hart (1986). Plots (b) to (d) show the Cr isotope compositions versus ^{87}Sr/^{86}Sr (b), ε_{Nd} (c) and (La/Sm)$_N$ (d). Plot (d) the MORBs with (La/Sm)$_N$ <1 belong to the normal MORB (N-MORB), and the reverse is the enriched MORB (E-MORB).

5.3 Cr isotope fractionation during fractional crystallization and the average Cr isotope composition of oceanic crust

5.3.1 Rhyolite-MELTS Models and Cr isotope fractionation during fractional crystallization

It has been recently suggested that Cr isotopes show significant fractionation during basaltic magma differentiation, which is associated with fractional crystallization of minerals (olivine, pyroxene, and spinel) with different Cr isotope compositions from magmatic melts (Bonnand et al., 2016a, 2020a; Shen et al., 2020; Wagner et al., 2021; Zhu et al., 2019). Shen et al. (2020) found that the δ^{53}Cr of Kilauea Iki positively correlates with MgO and Cr concentrations, suggesting that basaltic magma differentiation generates isotopically lighter residual melts. In subsequent studies, significant positive correlations between δ^{53}Cr and magma differentiation indices (MgO and Cr) were also observed in basalts from Fangataufa island (Bonnand et al., 2020a) and komatiite-tholeiite suites (Wagner et al., 2021), which are explained by the removal of the Cr^{3+}-rich phase (e.g., spinel and clinopyroxene) (Bonnand et al., 2020a; Wagner et al., 2021). In conclusion, it can be suggested that fractional crystallization of minerals affects Cr isotope fractionation of basaltic melts.

Before discussing the processes that might have generated Cr isotope variations in the studied MORBs, it is necessary to identify the dominant processes that are responsible for the variations in the major and trace element compositions, including the variation of MgO contents from 6.34 to 9.57 wt.%, Cr concentrations from 77 to 391 μg/g, and Ni concentrations from 45 to 153 μg/g. Partial melting of the mantle and fractional crystallization could generate such variations in MgO contents. However, the variations of trace elements, such as Cr, Ni, and La, reflect that these MORBs are more likely influenced by fractional crystallization. As shown in Fig. 4, the observed trends in Cr, La and $(La/Sm)_N$ ratios in this study are significantly different from the theoretically calculated ones induced by variation of partial melting degree. In addition, the ratio of $(La/Sm)_N$ (i.e., primitive mantle-normalized La/Sm) is also regarded as an index of the degree of partial melting owing to more incompatible nature of La compared to that of Sm. However, most of the studied MORBs are characterized by a narrow range of $(La/Sm)_N$ ratios (0.50–0.69; Fig. 4b), except for three samples. This implies that the effect of the variation of partial melting degree on the composition of the studied MORBs is insignificant. In summary, we suggest that the chemical compositions of the studied MORBs are mainly controlled by fractional crystallization process.

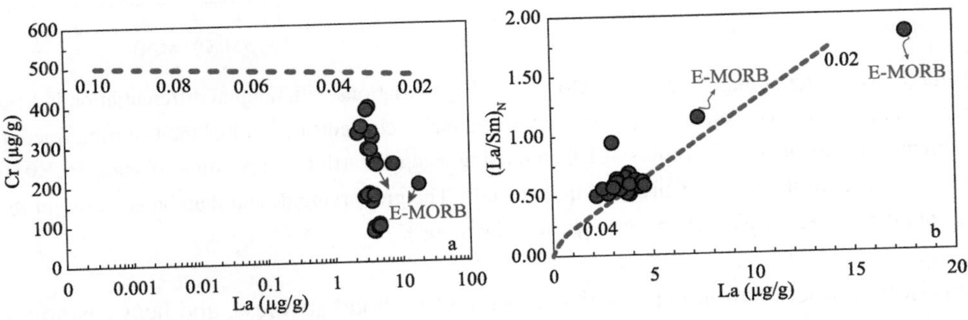

Fig. 4 La versus (a) Cr and (b) $(La/Sm)_N$ for the samples analyzed in this study. The dotted line is the partial melting model discussed in the text in the section 5.3. The numbers are the degrees of partial melting. Derailed calculation methods and parameters are given in Appendix.

To better constrain crystallization sequence and crystallizing assemblages, a matrix Rhyolite-MELTS version 1.0.2 (for anhydrous condition) (Gualda et al., 2012; Ghiorso and Gualda, 2015) model was performed. The run was performed using an QFM redox buffer under isobaric pressure (2 kbars). A less evolved MORBs (CR03-1) was regarded as the starting composition in the Rhyolite-MELTS model. The initial parameters for Rhyolite-MELTS model using the studied MORBs starting composition were: fO_2=QFM and anhydrous condition with isobaric pressure. The model was run from 1250 ℃ down to 1150 ℃ in 10 ℃ increments. For each 10 ℃ increment, Rhyolite-MELTS provides the proportion of each crystallizing phase (Fig. 5c).

Rhyolite-MELTS isobaric model was calculated at fO_2=QFM using the MORBs (CR03-1) as a parental composition predicts a crystallization sequence from first to last crystallizing phase of spinel, clinopyroxene, olivine and plagioclase (Fig. 5c). At≥1240 ℃, spinel is the only solidus phase, and clinopyroxene, olivine and plagioclase are in the liquidus phase. Clinopyroxene is crystallized when the melt reaches 1230 ℃, and subsequently, plagioclase produces under 1220 ℃ before olivine. In addition, the variations of major elements (Al_2O_3 and TiO_2) calculated by the model are consistent with the observed ones (Fig. 5a and b). In conclusion, the chemical compositions of MORBs are attributed to the crystallization of olivine, clinopyroxene, plagioclase, and spinel. Thus, evaluating the effect of crystallization of these minerals on Cr isotope compositions of the whole rock is necessary for the studied samples (we exclude plagioclase because D_{Cr} for plagioclase/melt is less than 0.1; Bédard, 2006).

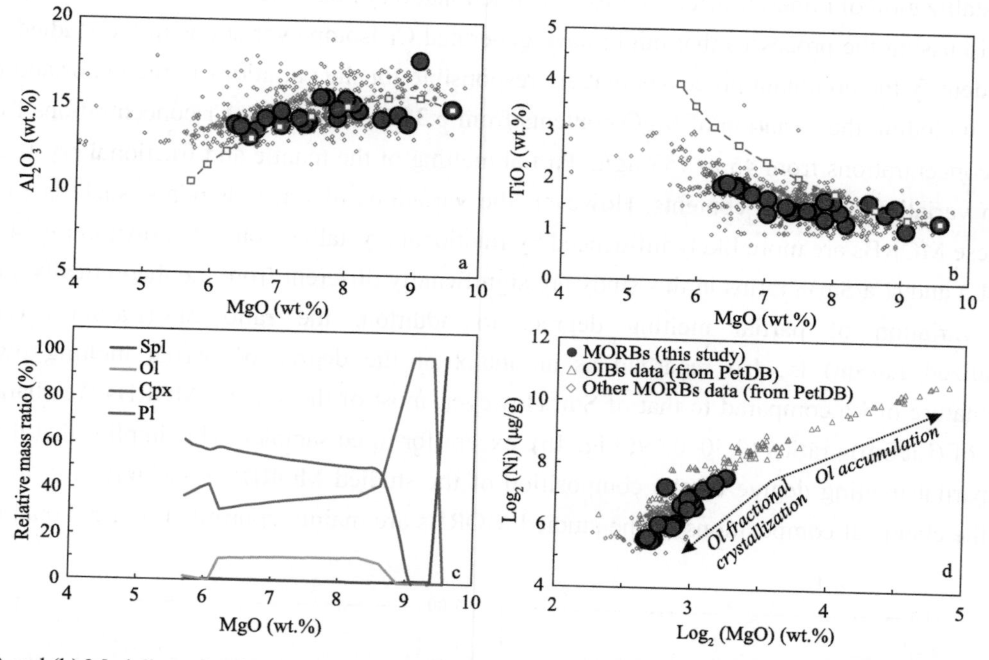

Fig. 5　plots (a) and (b) Modeling of major elements (Al_2O_3 and TiO_2) variations with magma differentiation indicator of MgO using Rhyolite-MELTS. The dotted lines with squares represent calculated major elements of residual melts during fractional crystallization with a 10 ℃ increment. (c) Outcomes of Rhyolite-MELTS modeling using a starting composition of sample CR03-1. (d) The plot of Ni (μg/g) versus MgO (wt.%) of the MORBs analyzed in this study. The grey triangles and diamonds represent the published OIBs and MORBs data from PetDB, respectively (https://search.earthchem.org/).

Equilibrium isotope fractionation is primarily dependent on bond stiffness, and heavy isotopes typically tend to concentrate in substances where that element forms the stiffest bonds (Urey, 1974; Schauble et al., 2004; Young et al., 2015). Previous studies have documented that Cr—O bond stiffness is often related to bond length, coordination number (CN), and oxidation state of Cr in minerals (Shen et al., 2018). In basaltic melts and minerals, Cr^{2+} and Cr^{3+} are the predominant species, and $Cr^{3+}/\Sigma Cr$ strongly depends on the redox environment of the magmatic system (Berry and O'Neill, 2004; Berry et al., 2006; Bell et al., 2014, 2017; Papike et al., 2016). The Cr isotope fractionation factors between major minerals and melts have not been well constrained for basaltic systems. Fortunately, recent work allows us to assess the behavior of Cr isotope fractionation between these minerals and whole-rock (Moynier et al., 2011; Shen et al., 2018, 2020; Bonnand et al., 2020b; Berry et al., 2021a). As discussed above, the three minerals that host Cr in the MORBs analyzed here are olivine, clinopyroxene, and spinel. In previous work, the ab initio calculation of equilibrium Cr isotope fractionation has predicted that spinel tends to concentrate in heavy Cr isotopes relative to olivine (Moynier et al., 2011). Moreover, theoretical ionic modeling has been proposed to predict Cr isotope fractionation of inter-mineral and mineral-melt (Shen et al., 2018, 2020; Berry et al., 2021a). Based on the model, Cr isotope fractionation of inter-mineral or mineral-melt essentially depends on the difference in valence states, Cr—O bond length, and coordination environment. Thus, we evaluate the behavior of Cr isotopes during MORB melt evolution using the ionic model of Shen et al. (2020) and Berry et al. (2021a). The $Cr^{3+}/\Sigma Cr$ ratios of MORBs in this study have been quantitatively calculated using the results of Berry et al. (2006). Based on the accurate determinations of Cr^{2+}—O and Cr^{3+}–O bond lengths by Berry et al. (2021a), the Cr—O bond force constants of minerals and melts can be calculated (Shen et al., 2020; Berry et al., 2021a):

$$K_{f,ij} = \frac{z_i z_j e^2 (1-n)}{4\pi\varepsilon_0 r_{ij}^3} \tag{2}$$

where z_i and z_j are the valences of cations and anions, e is the charge of electron, ε_0 is the electric constant, r_{ij} is inter-ionic distance between cations i and anion j, and n is the Born-Mater constant for the repulsion term (Young et al., 2015). The oxidation fugacity of MORBs has been well defined (QFM−0.67), corresponding to $Cr^{3+}/\Sigma Cr$ of ~0.5 (Mallmann and O'Neill, 2007; Berry et al., 2006). According to Eq. (2), the force constants (K) for minerals and melts are calculated. The K value of MORB melts is 1585 Nm^{-1}, while that of the olivine, clinopyroxene and spinel crystallized from the melts are 1395 Nm^{-1}, 1854 Nm^{-1}, and 1918 Nm^{-1}, respectively. Compared to equilibrium melt, spinel and clinopyroxene tend to concentrate in heavy Cr isotopes, but olivine is lighter isotopes in this study.

As shown in Fig. 6a, MORBs can be divided into two groups based on correlation between $\delta^{53}Cr$ and MgO. The $\delta^{53}Cr$ values of group I show a systematic decrease with reducing MgO and Cr concentrations, similar to the data of Fangataufa island (Bonnand et al., 2020a), suggesting that the variation of Cr isotope composition in our studied samples is dominated by fractional crystallization. To better explain the relationships between $\delta^{53}Cr$ values and the major and trace elements data, we present a Rayleigh fractionation model to reproduce the significant $\delta^{53}Cr$ variation, using the following equation:

$$\delta^{53}Cr_{melt} = \delta^{53}Cr_{melt0} + \left[\Delta^{53}Cr_{crystal\text{-}melt} \times \ln(f_{Cr})\right] \tag{3}$$

where $\delta^{53}Cr_{melt}$ and $\delta^{53}Cr_{melt0}$ refer to $\delta^{53}Cr$ of residual and initial melt (parental magma), respectively; $\Delta^{53}Cr_{crystal\text{-}melt}$ ($\Delta^{53}Cr_{crystal\text{-}melt} = \delta^{53}Cr_{crystal} - \delta^{53}Cr_{melt}$) represents Cr isotope fractionation between the crystallized mineral assemblage and residual melt. f_{Cr} is the fraction of Cr remaining in the melt, which can be calculated using $f_{Cr} = F \times C_{melt,\,Cr}/C_{melt0,\,Cr}$ ($C_{melt,\,Cr}$ and $C_{melt0,\,Cr}$ refer to Cr in residual and initial melt, respectively). F is the fraction of remaining melt. The $\delta^{53}Cr$ value of the initial MORB melt in this study must be constrained because the choice of this value may affect the magnitude of the crystal-melt fractionation factor ($\Delta^{53}Cr_{crystal\text{-}melt}$). The variation in the chemical compositions of MORB glasses can also provide information on the characteristics of the parental magma of MORBs. The plots of logarithmic values of the incompatible element abundance versus log of whole-rock MgO content shows exponential relationships during fractional crystallization. In contrast, accumulation follows a simple linear relation (Nebel et al., 2014). Therefore, the plot can be utilized to identify the parental magma composition. The slope breakpoint corresponds to the parental magma composition on the log–log plot of incompatible versus MgO abundance. As shown in Fig. 5d, there is no breakpoint in the array, suggesting that no olivine accumulation occurred. This implies that these MORBs contained the highest MgO abundance (most primitive samples) are the representative of the initial melts. Accordingly, the Cr isotope composition of the initial melt for the MORBs studied is −0.16±0.02‰ (2SD; n=3). As shown in Fig. 7a–c, the trends of δ^{53}Cr versus Cr, MgO, and $^{T}Fe_2O_3$ contents of MORBs can be well modeled using the result of Rhyolite-MELTS with a fractionation factor of $\Delta^{53}Cr_{crystal-melt}$ of +0.04‰ to +0.13‰.

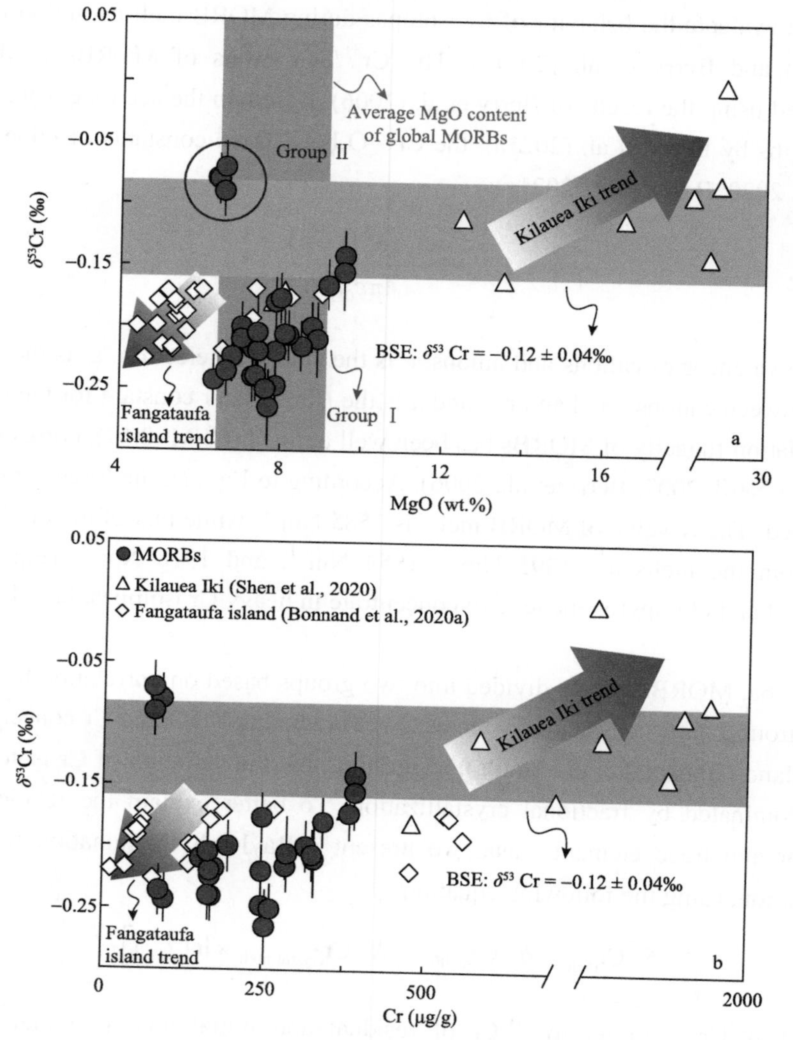

Fig. 6　The plots of δ^{53}Cr versus (a) MgO and (b) Cr of the MORBs analyzed in this study. For comparison, data from Kilauea Iki (Shen et al., 2020) and Fangataufa island (Bonnand et al., 2020a) are also plotted. The colored arrows indicate the trends of δ^{53}Cr variation with magma differentiation (notable, the purple allow is the trend of fractional crystallization, while the green is the accumulation of olivine and spinel). The blue (δ^{53}Cr=−0.12±0.04‰; 2SD) and orange (MgO=7.71±1.31 wt.%; 2σ) areas represent the Bulk Silicate Earth estimated by Jerram et al. (2020) and the average MgO content of global MORBs, respectively.

The fractionation factors used in this model for MORBs are larger than the Fangataufa island lavas ($\Delta^{53}Cr_{crystal-melt}$=0.010±0.005‰ for low-K suite, and 0.020±0.010‰ for high-K suite; Bonnand et al., 2020a) due to the different redox conditions of the basaltic melts (see Section 5.5 for a detailed discussion).

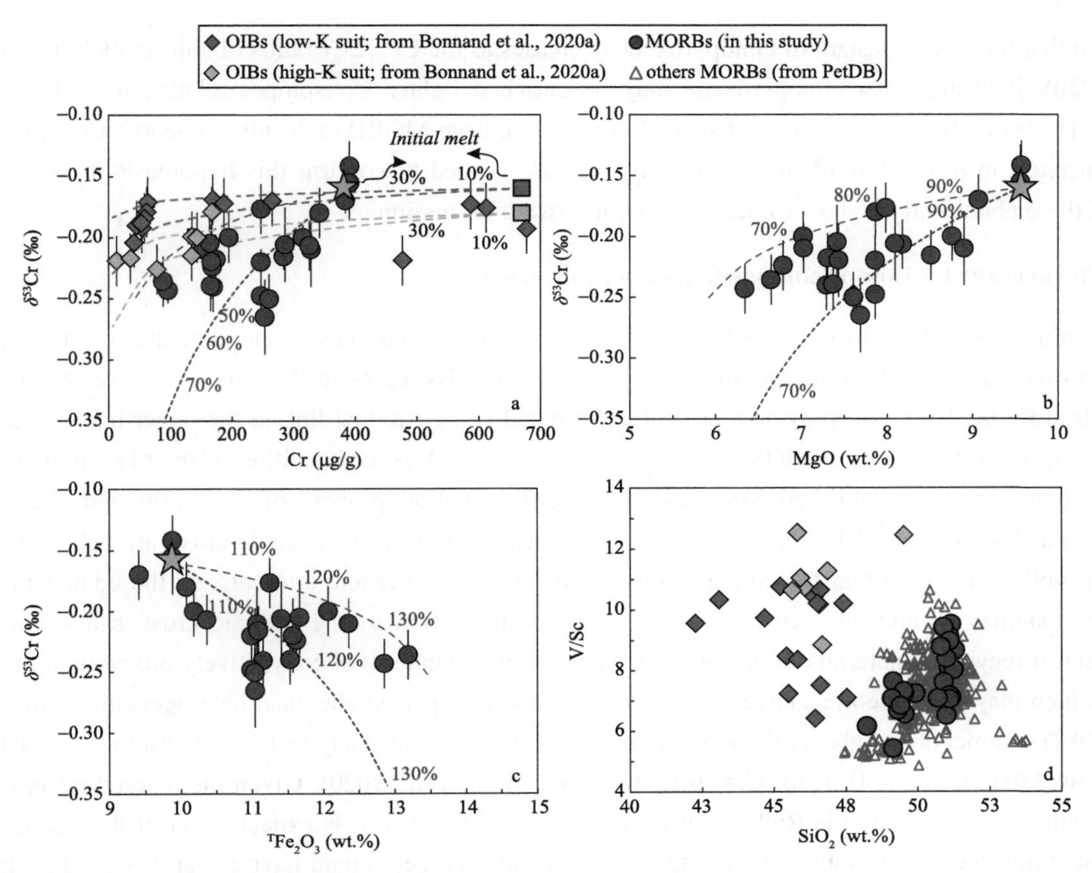

Fig. 7 Plots (a) to (c) show model calculations on Cr isotope fractionation during fractional crystallization. Plot (a) shows that δ^{53}Cr versus the amount of Cr in the residual melt during fractional crystallization of the MORBs analyzed in this study and basalts from Fangataufa island (Bonnand et al., 2020a), while plots (b) and (c) show that δ^{53}Cr vs. MgO and $^{T}Fe_2O_3$ in the residual melt during fractional crystallization. The red circles and colored (the blue and yellow are low-K and high-K suit basalts, respectively) diamonds represent the MORBs in this study and basalts from Fangataufa island, respectively. The blue and yellow quadrilaterals and light blue stars denote the estimated compositions of initial melt of the basalts from Fangataufa island and the MORBs analyzed in this study, respectively. The red, blue and yellow dotted lines are the Rayleigh fractionation model for the studied MORBs and Fangataufa basalts (blue and yellow), respectively. The numbers on the curves represent the Cr/MgO/$^{T}Fe_2O_3$ left in the residual melt (the $^{T}Fe_2O_3$ content in the residual melt increases with magma evolution because the Fe is incompatible during early magma evolution). The isotopic fractionation ($\Delta^{53}Cr_{crystal\text{-}melt}$) used for Fangataufa basalts are 0.010±0.005‰ (low-K) and 0.020 ± 0.010‰ (high-K). For the studied MORBs, the fractionation is 0.04 to 0.13‰. (d) The plot of V/Sc ratios versus SiO_2 contents. The red circles and colored (the blue and yellow are low-K and high-K suit basalts, respectively) diamonds and green hollow triangles represented the studied MORBs and basalts from Fangataufa island and published MORBs from PetDB database. Derailed calculation methods and parameters are given in Appendix.

However, the δ^{53}Cr values of group II are deviated from the trend of Fangataufa island, and have heavier Cr isotope compositions relative to others with similar MgO and Cr concentrations (Fig. 6). Although fractional crystallization of olivine leads to the preferential incorporation of heavy Cr isotopes in the residual melts, the sole removal of olivine cannot account for the decrease in Cr concentrations due to its slight incompatibility in olivine ($D_{Ol/melt}$=0.85; Mallmann and O'Neill, 2009). In addition, similar δ^{53}Cr values have also been previously reported in basalts from Kilauea Iki, which links to the re-dissolution of late-stage spinel (Shen et al., 2020). The higher δ^{53}Cr values in this study may be related to another process because the formation of late-stage spinel requires persistent high temperature (Shen et al., 2020). A recent study on experimental petrology has demonstrated that a decrease in pressure increases the $Cr^{2+}/\Sigma Cr$ ratio of silicate melt (Berry et al., 2021b). Theoretical ionic modeling has shown that the force constant of clinopyroxene decreases as the $Cr^{2+}/\Sigma Cr$ ratios of silicate melt increase (Shen et al., 2020), indicating that clinopyroxene may be enriched lighter Cr isotopes relative to melts under low pressure. Therefore, the crystallization of such clinopyroxene from MORB melts gives rise to a Cr depletion trend and an increase in δ^{53}Cr. Significantly, more samples are needed to confirm this hypothesis in the future. We excluded these abnormally high δ^{53}Cr values in the following discussion.

5.3.2 The average Cr isotope composition of oceanic crust

The mid-ocean ridge constitutes ~75% of the total global erupted magma and generates most oceanic crust (White and Klein, 2013). As a product of mantle melting, MORBs represent the most extensive crustal reservoir on terrestrial Earth. They also provide a critical genetic and compositional link to the upper mantle (Gale et al., 2013). Thus, the δ^{53}Cr values of MORBs can constrain the Cr isotope composition of the oceanic mantle, and a reference point can be established for tracking the differentiation process of the Earth and rocky planets. Subduction and subsequent dehydration, and in some cases, melting of the oceanic crust, are believed to trigger island arc volcanism, such that the composition of the subducting oceanic crust affects the composition of the island arc magmas (White and Klein, 2013). The deep subduction of the oceanic crust is also the primary mechanism of recycling materials from the surface back to the mantle. It is convectively mixed with the ambient mantle, which may form the the source of some hot spots and dispersed chemical heterogeneities (Hofmann and White, 1982). As mentioned above, the δ^{53}Cr values of MORBs in this study (−0.27±0.03‰ to −0.14±0.03‰; n=24) are lower than those of BSE (δ^{53}Cr=−0.12±0.04‰; Jerram et al., 2020). Given the marked difference in Cr isotope composition between the Bulk Earth mantle and oceanic crust, it is expected that if the oceanic crust is incorporated into the mantle source, basalt derived from such sources should have lower δ^{53}Cr values than from normal mantle, such as the basalts from Koolau, Hawaii (Shen et al., 2020).

As discussed above, owing to the Cr isotope fractionation of MORBs during fractional crystallization, the δ^{53}Cr of oceanic crust is difficult to constrain. However, the simple linear correlation between MgO and δ^{53}Cr allows us to indirectly estimate the δ^{53}Cr of the oceanic crust. Previous work has proposed the average major element chemical composition of global MORBs using 2010 complete whole-rock analyses in PetDB (White and Klein, 2013). The statistical results of the study suggested that the average MgO content of global MORBs is 7.71 ±1.31wt.% (2σ). The 20 studied samples fall in the average MgO content of the global MORB range (6.40 to 9.02 wt.%) (Fig. 6a). All remaining samples with mean δ^{53}Cr=−0.22±0.05‰ (2SD, n=20) are not found for statistical outliers at the 95% confidence level using Grubbs' test. In conclusion, the average δ^{53}Cr value of oceanic crust is −0.22±0.05‰ (2SD, n=20).

The two-tailed Student's t-test yielded a t statistic of 19.10, comparing to the 95% confidence level of 2.09. This p-value (9.09×10^{-9}) is less than our chosen alpha (0.05), indicating that the difference in δ^{53}Cr between oceanic crust and the BSE is statistically significant. This value is remarkably lower than that of BSE (δ^{53}Cr=−0.12±0.04‰; Jerram et al., 2020), which may be caused by partial melting of mantle (see section 5.4 for the detailed discussion). The Cr isotopes display great potential for tracing the mantle-crust interaction process. However, it is worth noting that the number of MORBs analyzed in this work is limited relative to the entire oceanic crust. Therefore, more samples are required for accurate estimation of the Cr isotope composition of oceanic crust.

5.4 Estimation of the Cr isotope composition of the depleted mantle source

The Cr isotope composition of the depleted mantle can be estimated using the primary MORB melts, which experience partial melting only during their formation. Therefore, the Cr isotope composition of the primary melts reflects that of their mantle source to the greatest extent. The most primitive MORBs (MgO contents of 9.57, 9.56, and 9.07 wt.%) analyzed in this study are comparable with that of primary MORB melt (9.74 wt.%) estimated by Workman and Hart (2005). In addition, the $Mg^{\#}$ ($Mg^{\#} = Mg / (Mg + Fe^{2+}) \times 100$) of such MORBs (69–71) are very similar to those of primary MORB melts ($Mg^{\#}$=72) in equilibrium with mantle olivine of $Fo_{89.6}$ (Niu and O'Hara, 2008; Niu, 2016). This implies that the MORBs with the highest MgO content can represent primary MORB melt. Accordingly, the Cr isotope composition of the primary melt of MORBs analyzed in this study is −0.16±0.02‰ (2SD; n=3). Notably, the δ^{53}Cr value of the primary MORB melt has been estimated based on only three MORB samples; therefore, additional MORB data are required for confirming the reported values.

Analyses of mantle peridotites and minerals demonstrated the occurrence of significant Cr isotope fractionation during partial melting of the mantle (Xia et al., 2017; Shen et al., 2018; Jerram et al., 2021). Since Cr isotope fractionation has been well constrained based on combined the theoretical model and experimental data of natural samples (Shen et al., 2018; Jerram et al., 2021), Cr isotope fractionation during partial melting can be modeled well. The redox conditions have been previously shown to significantly dominate Cr isotope fractionation during magmatic processes, so the redox state of the mantle source must be considered (Shen et al., 2018, 2020; Bonnand et al., 2020b). Previous work has been well defined the redox state of the MORB mantle source (QFM+0.07±0.14; Cottrell and Kelley, 2011). Based on the experimental results, the $Cr^{3+}/\Sigma Cr$ ratio of the MORB mantle source is ~0.55 (Berry et al., 2006). We use the non-modal melting equations outlined in previous studies (Shen et al., 2018; Sossi and O'Neill, 2017) to model Cr isotope fractionation during partial melting of the mantle. In detail, we consider the spinel-facies peridotite as the mantle source of the MORBs. The starting mineral fraction and melting mode for spinel peridotite are considered from Workman and Hart (2005) and Wasylenki et al. (2003), and the partition coefficient values of Cr between mantle-derived minerals and basaltic melt are obtained from Mallmann and O'Neill (2009) and Roeder et al. (2006).

The non-modal melting model used in this study is similar to that used in previous studies for modelling Fe and Cr isotope fractionation of mantle melting (Shen et al., 2018; Sossi and O'Neill, 2017). In the melting model of spinel peridotite, the fractionation factor between melt and residue can be calculated using the equation:

$$\alpha \mathrm{Cr}_{1-2}^{53/52} = \frac{D_{1-2}^{53}}{D_{1-2}^{52}} = \frac{({}^{53}C_{\mathrm{Cr}})_1 / ({}^{53}C_{\mathrm{Cr}})_2}{({}^{53}C_{\mathrm{Cr}})_1 / ({}^{53}C_{\mathrm{Cr}})_2} \tag{4}$$

where $\alpha \mathrm{Cr}_{1-2}^{53/52}$ is the fractionation factor between residue (1) and the melt (2) D_{1-2}^{53} and D_{1-2}^{52} are the partition coefficients between phases 1 and 2, respectively.

At a partial melting degree equivalent to that for the MORBs analysed in this study (~10%; it is commonly agreed that MORBs represent near-fractional melting of ~10% mantle based on the relative enrichments of highly incompatible elements in MORBs (Hofmann, 1988; Langmuir et al., 1992; Lee et al., 2005)), modelling results predict that the melts from the spinel-facies peridotite source are ~0.040±0.002‰ lighter than that of the mantle source under QFM+0.07±0.14 ($Cr^{3+}/\Sigma Cr$=0.55) and 1300 ℃ (see Shen et al., 2020). Based on the δ^{53}Cr value of the primary melt estimated above, the δ^{53}Cr value of the MORB mantle source is −0.12±0.02‰ (2σ). This value agrees with the previous estimates using peridotites (δ^{53}Cr=−0.12±0.10‰, Schoenberg et al., 2008; δ^{53}Cr=−0.14± 0.12‰, Xia et al., 2017; δ^{53}Cr=−0.11±0.06‰, Sossi et al., 2018; δ^{53}Cr=−0.12±0.04‰, Jerram et al., 2020).

5.5 Implication for the oxygen fugacity difference of MORBs and OIBs

Till now, no consensus has been reached regarding the potential processes responsible for generating the difference in the Cr isotope composition of silicate rocks of different planets (the Earth, Moon, and Vesta) in the solar system. Among diverse models, the most widely used are linked to the loss of heavy Cr isotopes via oxidizing gas (Sossi et al., 2018; Zhu et al., 2019, 2021a) and differences in redox condition dominating Cr isotope fractionation during magmatic differentiation (Bonnand et al., 2020a; Shen et al., 2020). Both lunar basalts and HED meteorites from Vesta are isotopically lighter relative to the Earth, which has been linked to the loss of heavy Cr isotopes by an oxidizing gas in the magma ocean (Bonnand et al., 2016a; Sossi et al., 2018; Zhu et al., 2019). The Cr isotope difference between the Bulk silicate Earth and enstatite chondrites also has been explained by volatile loss of heavy Cr isotopes from magma oceans on Earth (Zhu et al., 2021a). However, these hypotheses remain debatable (Bonnand et al., 2020a; Shen et al., 2020). In contrast, the fractional crystallization of Cr-bearing minerals (spinel, pyroxene, and olivine) under different oxygen fugacity has been proposed to explain the difference of Cr isotopes in subsequent studies of basalts from Fangataufa and Hawaii (Bonnand et al., 2020a; Shen et al., 2020) and theoretical calculations (Shen et al., 2020). The $Cr^{3+}/\Sigma Cr$ values of minerals and melts are mainly controlled by the redox state of the environment, resulting in differences in their corresponding Cr–O bonding force constants. Such differences further generate different fractionation factors between the crystal and melt phases. In brief, the magnitude of the fractionation factor during magmatic differentiation is inversely related to the oxygen fugacity of the environment and leads to different Cr isotope compositions.

To test this hypothesis, we have compared the Cr isotope composition of MORBs (this work) with OIBs published in the literature (Bonnand et al., 2020a). In the first place, the oxygen fugacity of OIBs is more variable compared to that of MORBs (Brounce et al., 2017). For example, the least degassed magmas from Mauna Kea are more oxidized than mid-ocean ridge basalt magmas, indicating that the upper mantle sources of Hawaiian magmas have higher oxygen fugacity than MORB sources (Brounce et al., 2017). This difference might be attributed to the recycling material from the oxidized surface to the deep mantle, which subsequently returned to the surface as a component of buoyant plumes (Brounce et al., 2017).

Compared with the data of OIBs from Fangataufa island (here, we have not considered basalts from Hawaii because most of them are controlled by spinel and olivine accumulation), the variation in $\delta^{53}Cr$ of the studied MORBs are significant relative to that of basalts from Fangataufa island within given range of MgO (Fig. 6a). This difference attributed to the equilibrium fractionation factor ($\Delta^{53}Cr_{crystal\text{-}melt}$ = +0.04‰ to +0.13‰) of MORBs during fractional crystallization is larger than that of Fangataufa island lavas ($\Delta^{53}Cr_{crystal\text{-}melt}$ = +0.005‰ to +0.030‰). Before discussing what causes such Cr isotopes differences, the contribution of mantle heterogeneity on Cr isotope compositions of Fangataufa island basalts must be evaluated. To avoid the effects of fractional crystallization (see above discussion), we use the initial melts to reflect them of mantle source. The $\delta^{53}Cr$ value of initial melt from MORBs (−0.16±0.02‰ from this work) is identical to that of OIBs (−0.18±0.02‰; Bonnand et al., 2020a) within the range of uncertainty, suggesting that the effect of mantle heterogeneous on the Cr isotope compositions of basalts from Fangataufa island is negligible. One possible explanation is that the proportion of recycling materials is too low (2%; Bonnand et al., 2020a) to modify the Cr isotope composition of mantle. The difference in variation of $\delta^{53}Cr$ between MORBs and OIBs (Fangataufa island) is most likely due to the large equilibrium fractionation factor of the former. The variation in $\delta^{53}Cr$ of MORBs can be modelled with a Rayleigh fractionation model with a bulk isotopic fractionation of ($\Delta^{53}Cr_{crystal-melt}$) +0.04‰ to +0.13‰ (Fig. 7a–c). Furthermore, the fractionation factors used in our models are larger than those of Fangataufa basalts ($\Delta^{53}Cr_{crystal-melt}$ = +0.005‰ to +0.030‰). This implies that the oxygen fugacity of MORBs in this study is lower than that of OIBs from Fangataufa island. To test this hypothesis, we use the V/Sc ratios in MORBs and OIBs as indicators of oxygen fugacity (Lee et al., 2005). We chose V and Sc for the following reasons. First, the behaviors of V and Sc during magmatic processes are mostly similar to each other than to any other element (Lee et al., 2005). Second, the partitioning of V is redox-sensitive (V^{3+}, V^{4+}, and V^{5+}), whereas that of Sc is not. As the oxygen fugacity of magmatic systems increases, V becomes more incompatible owing to the significantly smaller ionic radii of V^{4+} and V^{5+} relative to V^{3+}. Finally, the application of V/Sc ratio rather than V alone helps in reducing the effects of magmatic differentiation processes that might have diluted V and Sc concentrations. As shown in Fig. 7d, the V/Sc ratios of Fangataufa basalts are remarkably higher than those of MORBs in our study, indicating that the Fangataufa island lavas are more oxidized than MORBs. Our conclusion is that redox conditions can markedly affect equilibrium Cr isotope fractionation during magmatic processes, and can provide a potential tool for understanding the redox conditions of magmatic differentiation.

6 Conclusion

This study presents high-precision Cr stable isotope data for MORBs. Based on the analyzed data, the following conclusions can be drawn:

(1) The δ^{53}Cr of MORBs are positively correlated with the whole-rock Cr and MgO concentrations, indicating that Cr isotope fractionation is controlled by fractional crystallization of olivine, spinel and clinopyroxene. The effects of secondary alteration and source heterogeneity on δ^{53}Cr values of MORBs are limited. Moreover, the estimated δ^{53}Cr values of primitive MORB melt (δ^{53}Cr=−0.16±0.02‰; n=3) are lower than the average value of BSE (−0.12±0.04‰; 2SD). Theoretical calculations indicate that the Cr isotope composition of oceanic mantle is −0.12±0.02‰. In addition, the average Cr isotope composition of oceanic crust is −0.22±0.05‰.

(2) Compared with the data of Fangataufa island lavas, the equilibrium fractionation factors ($\Delta^{53}Cr_{crystal-melt}$ = 0.03 to 0.14‰) of the MORBs analyzed in this study are larger during fractional crystallization. This implies that the oxygen fugacity of the MORBs is lower than that of basalts from Fangataufa island, which is strongly supported by the lower V/Sc ratios than those of basalts of Fangataufa island. Thus, Cr isotopes are expected to be utilized as an oxybarometer to understand the redox conditions during the magmatic evolution.

Declaration of Competing Interest

The authors declare that they have no known competing financial interests or personal relationships that could have appeared to influence the work reported in this paper.

Acknowledgements

We are grateful to Jeffrey Catalano (Editor-in-Chief) and Shichun Huang (AE) for handing and constructive comments, and to four anonymous reviewers for their careful comments and suggestions that helped to improve the manuscript. We also thank Y.-W Su and W.-R Liu. for their assistance in the lab. We also thank Prof. Zaicong Wang (Department of Earth Sciences, China University of Geosciences, Wuhan), Jixi Li (Institute of Oceanology, Chinese Academy of Sciences, Hangzhou), and Chunhui Tao (Institute of Oceanology, Chinese Academy of Sciences, Qingdao) for providing us samples. This work is supported by the National Natural Science Foundation of China (Grants 41730214), National Key R&D Program of China (2019YFA0708400) and the National Natural Science Foundation of China (Grants 41403010 and 41622303). This work is also supported by the State Key Laboratory of Geological Processes and Mineral Resources, China University of Geosciences (MSFGPMR201811) to XLJ, the Strategic Priority Research Program (B) of the Chinese Academy of Sciences (grant XDB18000000 to SAL), and the MOST Special Fund from the State Key Laboratory of Geological Processes and Mineral Resources, China University of Geosciences (Grant No. 2652018049 to LJX).

Appendix A. Supplementary material

Supplementary data to this article can be found online at https://doi.org/10.1016/j.gca.2022.04.018.

References

Bédard J. H. (2006) Trace element partitioning in plagioclase feldspar. Geochim. Cosmochim. Acta 70, 3717-3742.

Bell A. S., Burger P. V., Le L., Shearer C. K., Papike J. J., Sutton S. R., Newville M. and Jones J. (2014) XANES measurements of Cr valence in olivine and their applications to planetary basalts. Am. Mineral. 99, 1404-1412.

Bell A. S., Shearer C., Burger P., Ren M., Newville M. and Lanzirotti A. (2017) Quantifying and correcting the effects of anisotropy in XANES measurements of chromium valence in olivine: Implications for a new olivine oxybarometer. Am. Mineral. 102, 1165-1172.

Berry A. J., Miller L. A., O'Neill H. S. C. and Foran G. J. (2021a) The coordination of Cr^{2+} in silicate glasses and implications for mineral-melt fractionation of Cr isotopes. Chem. Geol. 586, 120483.

Berry A. J., O'Neill H. S. C. and Foran G. J. (2021b) The effects of temperature and pressure on the oxidation state of chromium in silicate melts. Contrib. to Mineral. Petrol. 176, 1-14.

Berry A. J. and O'Neill H. S. C. (2004) A XANES determination of the oxidation state of chromium in silicate glasses. Am. Mineral. 89, 790-798.

Berry A. J., O'Neill H. S. C., Scott D. R., Foran G. J. and Shelley J. M. G. (2006) The effect of composition on Cr^{2+}/Cr^{3+} in silicate melts. Am. Mineral. 91, 1901-1908.

Bonnand P., Doucelance R., Boyet M., Bachèlery P., Bosq C., Auclair D. and Schiano P. (2020a) The influence of igneous processes on the chromium isotopic compositions of Ocean Island basalts. Earth Planet. Sci. Lett. 532.

Bonnand P., Bruand E., Matzen A. K., Jerram M., Schiavi F., Wood B. J., Boyet M. and Halliday A. N. (2020b) Redox control on chromium isotope behaviour in silicate melts in contact with magnesiochromite. Geochim. Cosmochim. Acta 288, 282-300.

Bonnand P. and Halliday A. N. (2018) Oxidized conditions in iron meteorite parent bodies. Nat. Geosci. 11, 401-404.

Bonnand P., James R. H., Parkinson I. J., Connelly D. P. and Fairchild I. J. (2013) The chromium isotopic composition of seawater and marine carbonates. Earth Planet. Sci. Lett. 382, 10-20.

Bonnand P., Parkinson I. J. and Anand M. (2016a) Mass dependent fractionation of stable chromium isotopes in mare basalts: Implications for the formation and the differentiation of the Moon. Geochim. Cosmochim. Acta 175, 208-221.

Bonnand P., Williams H. M., Parkinson I. J., Wood B. J. and Halliday A. N. (2016b) Stable chromium isotopic composition of meteorites and metal-silicate experiments: Implications for fractionation during core formation. Earth Planet. Sci. Lett. 435, 14-21.

Brounce M., Stolper E. and Eiler J. (2017) Redox variations in Mauna Kea lavas, the oxygen fugacity of the Hawaiian plume, and the role of volcanic gases in Earth's oxygenation. Proc. Natl. Acad. Sci. U. S. A. 114, 8997-9002.

Chen C., Ciazela J., Li W., Dai W., Wang Z., Foley S. F., Li M., Hu Z. and Liu Y. (2020) Calcium isotopic compositions of oceanic crust at various spreading rates. Geochim. Cosmochim. Acta 278, 272-288.

Chen S., Niu Y., Guo P., Gong H., Sun P., Xue Q., Duan M. and Wang X. (2019) Iron isotope fractionation during mid-ocean ridge basalt (MORB) evolution: Evidence from lavas on the East Pacific Rise at 10°30′N and its implications. Geochim. Cosmochim. Acta 267, 227-239.

Chen S., Wang X., Niu Y., Sun P., Duan M., Xiao Y., Guo P., Gong H., Wang G. and Xue Q. (2017) Simple and cost-effective methods for precise analysis of trace element abundances in geological materials with ICP-MS. Sci. Bull. 62, 277-289.

Cottrell E. and Kelley K. A. (2011) The oxidation state of Fe in MORB glasses and the oxygen fugacity of the upper mantle. Earth Planet. Sci. Lett. 305, 270-282.

DuFrane S. A., Turner S., Dosseto A. and van Soest M. (2009) Reappraisal of fluid and sediment contributions to Lesser Antilles magmas. Chem. Geol. 265, 272-278.

Gale A., Dalton C. A., Langmuir C. H., Su Y. and Schilling J. G. (2013) The mean composition of ocean ridge basalts. Geochemistry,

Geophys. Geosystems 14, 489-518.

Ghiorso M. S. and Gualda G. A. R. (2015) An H_2O-CO_2 mixed fluid saturation model compatible with rhyolite-MELTS. Contrib. to Mineral. Petrol. 169, 1-30.

Gualda G. A. R., Ghiorso M. S., Lemons R. V. and Carley T. L. (2012) Rhyolite-MELTS: A modified calibration of MELTS optimized for silica-rich, fluid-bearing magmatic systems. J. Petrol. 53, 875-890.

Hofmann A. W. (1988) Chemical differentiation of the Earth: the relationship between mantle, continental crust, and oceanic crust. Earth Planet. Sci. Lett. 90, 297-314.

Hofmann A. W. (1997) Mantle geochemistry: the message from oceanic volcanism. Nature 385, 219-229.

Hofmann A. W. and Hémend C. (2006) The origin of E-MORB. Geochim.Cosmochim.Acta 70, A257.

Hofmann A. W. and White W. M. (1982) Mantle plumes from ancient oceanic crust. Earth Planet. Sci. Lett. 57, 421-436.

Huang S., Abouchami W., Blichert-Toft J., Clague D. A., Cousens B. L., Frey F. A. and Humayun M. (2009) Ancient carbonate sedimentary signature in the Hawaiian plume: evidence from Mahukona volcano,Hawaii. Geochem. Geophys. Geosyst. 10(8).

Huang S., Farkaš J. and Jacobsen S. B. (2011) Stable calcium isotopic compositions of Hawaiian shield lavas: evidence for recycling of ancient marine carbonates into the mantle. Geochim. Cosmochim. Acta 75, 4987-4997.

Huang S. and Frey F. (2003) Trace element abundances of Mauna Kea basalt from phase 2 of the Hawaii Scientific Drilling Project: Petrogenetic implications of correlations with major element content and isotopic ratios. Geochem. Geophys. Geosyst., 4.

Huang S. and Frey F. A. (2005) Recycled oceanic crust in the Hawaiian Plume: evidence from temporal geochemical variations within the Koolau Shield. Contrib. Miner. Petrol. 149, 556-575.

Huang S., Frey F. A., Blichert-Toft J., Fodor R., Bauer G. R. and Xu G. (2005) Enriched components in the Hawaiian plume: evidence from Kahoolawe Volcano, Hawaii. Geochem. Geophys. Geosyst. 6(11).

Huang S., Vollinger M. J., Frey F. A., Rhodes J. M. and Zhang Q. (2016) Compositional variation within thick (>10 m) flow units of Mauna Kea Volcano cored by the Hawaii Scientific Drilling Project. Geochim. Cosmochim. Acta 185, 182-197.

Jerram M., Bonnand P., Harvey J., Ionov D. and Halliday A. N. (2021) Stable chromium isotopic variations in peridotite mantle xenoliths: metasomatism versus partial melting. Geochim. Cosmochim. Acta 317, 138-154.

Jerram M., Bonnand P., Kerr A. C., Nisbet E. G., Puchtel I. S. and Halliday A. N. (2020) The δ^{53}Cr isotope composition of komatiite flows and implications for the composition of the bulk silicate Earth. Chem. Geol. 551, 119761.

Langmuir C. H., Klein E. M., and Plank T. (1992) Petrological systematics of mid-ocean ridge basalts: Constraints on melt generation beneath ocean ridges. Geophysical Monograph, American Geophysical Union 71, 183-280.

Lee C. T. A., Leeman W. P., Canil D. and Li Z. X. A. (2005) Similar V/Sc systematics in MORB and arc basalts: Implications for the oxygen fugacities of their mantle source regions. J. Petrol. 46, 2313-2336.

Li C. F., Chen F. and Li X. H. (2007) Precise isotopic measurements of sub-nanogram Nd of standard reference material by thermal ionization mass spectrometry using the NdO+ technique. Int. J. Mass Spectrom. 266, 34-41.

Liu C. Y., Xu L. J., Liu C. T., Liu J., Qin L. P., Zhang Z. Da, Liu S. A. and Li S. G. (2019) High-Precision Measurement of Stable Cr Isotopes in Geological Reference Materials by a Double-Spike TIMS Method. Geostand. Geoanalytical Res. 43, 647-661.

Liu S. A., Huang J., Liu J., Wörner G., Yang W., Tang Y. J., Chen Y., Tang L., Zheng J. and Li S. (2015) Copper isotopic composition of the silicate Earth. Earth Planet. Sci. Lett. 427, 95-103.

Mallmann G. and O'Neill H. S. C. (2007) The effect of oxygen fugacity on the partitioning of Re between crystals and silicate melt during mantle melting. Geochimi. Cosmochimi. Acta 71, 2837-2857.

Mallmann G. and O'Neill H. S. C. (2009) The crystal/melt partitioning of V during mantle melting as a function of oxygen fugacity compared with some other elements (Al, P, Ca, Sc, Ti, Cr, Fe, Ga, Y, Zr and Nb). J. Petrol. 50, 1765-1794.

Moussallam Y., Edmonds M., Scaillet B., Peters N., Gennaro E., Sides I. and Oppenheimer C. (2016) The impact of degassing on the oxidation state of basaltic magmas: A case study of Kīlauea volcano. Earth Planet. Sci. Lett. 450, 317-325.

Moussallam Y., Oppenheimer C., Scaillet B., Gaillard F., Kyle P., Peters N., Hartley M., Berlo K. and Donovan A. (2014) Tracking the changing oxidation state of Erebus magmas, from mantle to surface, driven by magma ascent and degassing. Earth Planet. Sci. Lett. 393, 200-209.

Moynier F., Yin Q. Z. and Schauble E. (2011) Isotopic evidence of Cr partitioning into Earth's core. Science (80-.). 331, 1417-1420.

Nebel O., Campbell I. H., Sossi P. A. and Van Kranendonk M. J. (2014) Hafnium and iron isotopes in early Archean komatiites record a plume-driven convection cycle in the Hadean Earth. Earth Planet. Sci. Lett. 397, 111-120.

Niu Y. (2016) The meaning of global ocean ridge basalt major element compositions. J. Petrol. 57, 2081-2103.

Niu Y. and O'Hara M. J. (2008) Global correlations of ocean ridge basalt chemistry with axial depth: A new perspective. J. Petrol. 49,

633-664.

Papike J. J., Simon S. B., Burger P. V., Bell A. S., Shearer C. K. and Karner J. M. (2016) Chromium, vanadium, and titanium valence systematics in Solar System pyroxene as a recorder of oxygen fugacity, planetary provenance, and processes. Am. Mineral. 101, 907-918.

Pereira N. S., Voegelin A. R., Paulukat C., Sial A. N., Ferreira V. P. and Frei R. (2016) Chromium-isotope signatures in scleractinian corals from the Rocas Atoll, Tropical South Atlantic. Geobiology 14, 54-67.

Ray D., Kamesh Raju K. A., Baker E. T., Srinivas Rao A., Mudholkar A. V., Lupton J. E., Surya Prakash L., Gawas R. B. and Vijaya Kumar T. (2012) Hydrothermal plumes over the Carlsberg Ridge, Indian Ocean. Geochemistry, Geophys. Geosystems 13.

Reinhard C. T., Planavsky N. J., Robbins L. J., Partin C. A., Gill B. C., Lalonde S. V., Bekker A., Konhauser K. O. and Lyons T. W. (2013) Proterozoic ocean redox and biogeochemical stasis. Proc. Natl. Acad. Sci. U. S. A. 110, 5357-5362.

Rhodes J. M. and Vollinger M. J. (2005) Ferric/ferrous ratios in 1984 Mauna Loa lavas: A contribution to understanding the oxidation state of Hawaiian magmas. Contrib. to Mineral. Petrol. 149, 666-674.

Roeder P., Gofton E. and Thornber C. (2006) Cotectic proportions of olivine and spinel in olivine-tholeiitic basalt and evaluation of pre-eruptive processes. J. Petrol. 47, 883-900.

Schauble E., Rossman G. R. and Taylor H. P. (2004) Theoretical estimates of equilibrium chromium-isotope fractionations. Chem. Geol. 205, 99-114.

Scheiderich K., Amini M., Holmden C. and Francois R. (2015) Global variability of chromium isotopes in seawater demonstrated by Pacific, Atlantic, and Arctic Ocean samples. Earth Planet. Sci. Lett. 423, 87-97.

Schilling J G. (1973). Iceland mantle plume: Geochemical study of Reykjanes Ridge. Nature, 242 (5400): 565-571.

Schoenberg R., Merdian A., Holmden C., Kleinhanns I. C., Haßler K., Wille M. and Reitter E. (2016) The stable Cr isotopic compositions of chondrites and silicate planetary reservoirs. Geochim. Cosmochim. Acta 183, 14-30.

Schoenberg R., Zink S., Staubwasser M. and von Blanckenburg F. (2008) The stable Cr isotope inventory of solid Earth reservoirs determined by double spike MC-ICP-MS. Chem. Geol. 249, 294-306.

Shen J., Qin L., Fang Z., Zhang Y., Liu J., Liu W., Wang F., Xiao Y., Yu H. and Wei S. (2018) High-temperature inter-mineral Cr isotope fractionation: A comparison of ionic model predictions and experimental investigations of mantle xenoliths from the North China Craton. Earth Planet. Sci. Lett. 499, 278-290.

Shen J., Xia J., Qin L., Carlson R. W., Huang S., Helz R. T. and Mock T. D. (2020) Stable chromium isotope fractionation during magmatic differentiation: Insights from Hawaiian basalts and implications for planetary redox conditions. Geochim. Cosmochim. Acta 278, 289-304.

Shorttle O., Moussallam Y., Hartley M. E., Maclennan J., Edmonds M. and Murton B. J. (2015) Fe-XANES analyses of Reykjanes Ridge basalts: Implications for oceanic crust's role in the solid Earth oxygen cycle. Earth Planet. Sci. Lett. 427, 272-285.

Sossi P. A., Moynier F. and Van Zuilen K. (2018) Volatile loss following cooling and accretion of the Moon revealed by chromium isotopes. Proc. Natl. Acad. Sci. U. S. A. 115, 10920-10925.

Sossi P. A. and O'Neill H. S. C. (2017) The effect of bonding environment on iron isotope fractionation between minerals at high temperature. Geochim. Cosmochim. Acta 196, 121-143.

Standish J. J., Dick H. J. B., Michael P. J., Melson W. G. and O'Hearn T. (2008) MORB generation beneath the ultraslow spreading Southwest Indian Ridge (9-25°E): Major element chemistry and the importance of process versus source. Geochemistry, Geophys. Geosystems 9.

Stracke A. (2012) Earth's heterogeneous mantle: A product of convection-driven interaction between crust and mantle. Chem. Geol. 330-331, 274-299.

Stracke A., Hofmann A. W. and Hart S. R. (2005) FOZO, HIMU, and the rest of the mantle zoo. Geochemistry, Geophys. Geosystems 6.

Sun S.-S. and McDonough W. F. (1989) Chemical and isotopic systematics of oceanic basalts: Implications for mantle composition and processes. Geol.Soc.Lond.Spec.Public. 42, 313-345.

Urey, H.C. (1947). The thermodynamic properties of isotopic substances. J. Chem. Soc. 562-581.

Wagner L. J., Kleinhanns I. C., Weber N., Babechuk M. G., Hofmann A. and Schoenberg R. (2021) Coupled stable chromium and iron isotopic fractionation tracing magmatic mineral crystallization in Archean komatiite-tholeiite suites. Chem. Geol. 576, 120121.

Wang X., Planavsky N. J., Reinhard C. T., Zou H., Ague J. J., Wu Y., Gill B. C., Schwarzenbach E. M. and Peucker-Ehrenbrink B. (2016) Chromium isotope fractionation during subduction-related metamorphism, black shale weathering, and hydrothermal alteration. Chem. Geol. 423, 19-33.

Wasylenki L. E., Baker M. B., Kent A. J. R. and Stolper E. M. (2003) Near-solidus melting of the Shallow Upper Mantle: Partial melting experiments on depleted peridotite. J. Petrol. 44, 1163-1191.

Weis D., Kieffer B., Maerschalk C., Barling J., De Jong J., Williams G. A., Hanano D., Pretorius W., Mattielli N., Scoates J. S., Goolaerts A., Friedman R. M. and Mahoney J. B. (2006) High-precision isotopic characterization of USGS reference materials by TIMS and MC-ICP-MS. Geochemistry, Geophys. Geosystems 7.

Weis D., Kieffer B., Maerschalk C., Pretorius W. and Barling J. (2005) High-precision Pb-Sr-Nd-Hf isotopic characterization of USGS BHVO-1 and BHVO-2 reference materials. Geochemistry, Geophys. Geosystems 6.

White W. M. and Klein E. M. (2013) Composition of the Oceanic Crust., Elsevier Ltd.

Wiesmaier S., Deegan F. M., Troll V. R., Carracedo J. C., Chadwick J. P. and Chew D. M. (2011) Magma mixing in the 1100 AD Montaña Reventada composite lava flow, Tenerife, Canary Islands: Interaction between rift zone and central volcano plumbing systems. Contrib. to Mineral. Petrol. 162, 651-669.

Workman R. K. and Hart S. R. (2005) Major and trace element composition of the depleted MORB mantle (DMM). Earth Planet. Sci. Lett. 231, 53-72.

Wu G., Zhu J. M., Wang X., Johnson T. M. and Han G. (2020) High-Sensitivity Measurement of Cr Isotopes by Double Spike MC-ICP-MS at the 10 ng Level. Anal. Chem. 92, 1463-1469.

Xia J., Qin L., Shen J., Carlson R. W., Ionov D. A. and Mock T. D. (2017) Chromium isotope heterogeneity in the mantle. Earth Planet. Sci. Lett. 464, 103-115.

Xiao W., Windley B. F., Hao J. and Li J. (2002) Arc-ophiolite obduction in the western Kunlun Range (China): Implications for the Palaeozoic evolution of central Asia. J. Geol. Soc. London. 159, 517-528.

Xu L. J., Liu S. A. and Li S. (2021) Zinc isotopic behavior of mafic rocks during continental deep subduction. Geosci. Front. 12, 101182.

Young E. D., Manning C. E., Schauble E. A., Shahar A., Macris C. A., Lazar C. and Jordan M. (2015) High-temperature equilibrium isotope fractionation of non-traditional stable isotopes: Experiments, theory, and applications. Chem. Geol. 395, 176-195.

Zhu J. M., Wu G., Wang X., Han G. and Zhang L. (2018) An improved method of Cr purification for high precision measurement of Cr isotopes by double spike MC-ICP-MS. J. Anal. At. Spectrom. 33, 809-821.

Zhu K., Barrat J., Yamaguchi A., Rouxel O., Germain Y., Langlade J. and Moynier F. (2022) Nickel and chromium stable isotopic composition of ureilites: implications for the Earth's core formation and differentiation of the ureilite parent body. Geophys. Res. Lett., 1-11.

Zhu K., Moynier F., Alexander C. M. O., Davidson J., Schrader D. L., Zhu J.-M., Wu G.-L., Schiller M., Bizzarro M. and Becker H. (2021a) Chromium Stable Isotope Panorama of Chondrites and Implications for Earth Early Accretion. Astrophys. J. 923, 94.

Zhu K., Moynier F., Schiller M., Alexander C. M. O. D., Barrat J. A., Bischoff A. and Bizzarro M. (2021b) Mass-independent and mass-dependent Cr isotopic composition of the Rumuruti (R) chondrites: Implications for their origin and planet formation. Geochim. Cosmochim. Acta 293, 598-609.

Zhu K., Moynier F., Schiller M., Becker H., Barrat J. A., Bizzarro M., Zhu K., Moynier F., Schiller M., Alexander C. M. O. D., Davidson J., Schrader D. L., van Kooten E. and Bizzarro M. (2021c) Tracing the origin and core formation of the enstatite achondrite parent bodies using Cr isotopes. Geochim. Cosmochim. Acta 308, 158-186.

Zhu K., Sossi P. A., Siebert J. and Moynier F. (2019) Tracking the volatile and magmatic history of Vesta from chromium stable isotope variations in eucrite and diogenite meteorites. Geochim. Cosmochim. Acta 266, 598-610.

Zindler A. and Hart S. (1986) CHEMICAL GEODYNAMICS. Ann. Rev. Earth Planet. Sci. 14, 493-571.

第四部分　金属稳定同位素示踪深部碳循环

Magnesium isotopic systematics of continental basalts from the North China craton: Implications for tracing subducted carbonate in the mantle*

Wei Yang[1,2], Fang-Zhen Teng[2], Hong-Fu Zhang[1] and Shu-Guang Li[3,4]

1. State Key Laboratory of Lithospheric Evolution, Institute of Geology and Geophysics, Chinese Academy of Sciences, P.O. Box 9825, Beijing 10029, China
2. Isotope Laboratory, Department of Geosciences and Arkansas Center for Space and Planetary Sciences, University of Arkansas, Fayetteville, AR 72701, U.S.A.
3. CAS Key Laboratory of Crust-Mantle Materials and Environments, School of Earth and Space Sciences, University of Science and Technology of China, Hefei 230026, China
4. State Key Laboratory of Geological Processes and Mineral Resources, China University of Geosciences, Beijing 100083, China

亮点工作：本研究是世界上首次报道的应用 Mg 同位素示踪深部碳循环的实例，发现华北> 120 Ma 义县组玄武岩具有类似于地幔的 Mg 同位素组成（δ^{26}Mg = −0.31 ~ −0.25），而<110 Ma 阜新和太行山玄武岩均具有轻的 Mg 同位素组成（δ^{26}Mg = −0.60 ~ −0.42）。本文论证了后者的轻 Mg 同位素组成只能来自西太平洋俯冲板片携带的再循环的碳酸盐岩混入的岩浆源区。

Abstract To explore the possibility of tracing recycled carbonate by using Mg isotopes and to evaluate the effects of the western Pacific oceanic subduction on the upper mantle evolution of the North China craton, Mg isotopic compositions of the Mesozoic–Cenozoic basalts and basaltic andesites from the craton have been investigated. The samples studied here come from a broad area in the craton with variable ages of 125–6 Ma, and can be divided into two groups based on geochemical features: the >120 Ma Yixian basalts and basaltic andesites, and the <110 Ma Fuxin and Taihang basalts. Our results indicate that these two groups have distinct Mg isotopic compositions. The >120 Ma Yixian basalts and basaltic andesites, with low Ce/Pb, Nb/U ratios and lower-crust like Sr-Nd-Pb isotopic compositions, have a mantle-like Mg isotopic composition, with δ^{26}Mg values ranging from −0.31‰ to −0.25‰ and an average of −0.27±0.05‰ (2SD, *n*=5). This suggests that continental crust contamination may strongly modify their Ce/Pb, Nb/U ratios and Sr-Nd-Pb isotopic compositions but do not influence their Mg isotopic compositions. By contrast, the <110 Ma basalts from Fuxin and Taihang exhibit lower δ^{26}Mg values of −0.60‰ to −0.42‰, with an average of −0.46±0.10‰ (2SD). Since these basalts still preserve mantle-like Sr-Nd-Pb isotopic compositions and Ce/Pb, Nb/U ratios and have high U/Pb and Th/Pb ratios similar to those of HIMU basalts, the light Mg isotopic composition is most likely derived

* 本文发表在：Chemical Geology, 2012, 328: 185–194

from interaction of their mantle source with isotopically light recycled carbonate melt. Since the Tethys and Mongolia oceanic subductions from south and north toward the North China craton were terminated in the Triassic and light Mg isotopic signature in basalts did not appear before 120 Ma, the subducted Pacific oceanic crust could be the major source of the recycled carbonate. Therefore, this study not only presents an example to trace recycled carbonate using Mg isotopes but also confirms the important role of the western Pacific oceanic subduction in generating the <110 Ma basalts in the North China craton.

1 Introduction

The "carbon cycle" constitutes one of the most important areas of the Earth science research in this century. The deep carbon cycle refers to carbon ingassing to the mantle through subduction and outgassing through magmatic and volcanic processes. Altered oceanic crust contains a significant fraction of carbonate (Alt and Teagle, 1999), and subduction zone dehydration do not significantly remove this carbonate (Yaxley and Green, 1994; Molina and Poli, 2000; Kerrick and Connolly, 2001). This carbonate delivered to the mantle then significantly affects chemical and physical properties of the mantle.

Carbon isotopes are generally used to trace the subducted carbonate, which are efficient to identify recycled organic carbon but not very sensitive to inorganic carbon. However, the subducted carbon is dominated by inorganic carbonate (Plank and Langmuir, 1998; Alt and Teagle, 1999). Therefore, additional tracers are required to better identify such recycled inorganic carbonate in the mantle.

Magnesium (Mg) isotopes could be one of such tracers. Magnesium isotope fractionation is limited during high temperature processes (Teng et al., 2007; 2010a; Handler et al., 2009; Yang et al., 2009; Bourdon et al., 2010; Liu et al., 2010), but is significant during low temperature processes (Galy et al., 2002; Young and Galy, 2004; Tipper et al., 2006a; 2006b; 2008; 2010; Pogge von Strandmann et al., 2008; Higgins and Schrag, 2010; Li et al., 2010; Teng et al., 2010b). The mantle is homogeneous in its Mg isotopic composition, with an average δ^{26}Mg value of −0.25±0.07‰ (Teng et al., 2007; 2010a; Handler et al., 2009; Yang et al., 2009; Bourdon et al., 2010; Liu et al., 2011). The upper continental crust is heterogeneous and on average similar to the mantle (Shen et al., 2009; Li et al., 2010; Liu et al., 2010). Notably, carbonates have significant light Mg isotopic compositions with δ^{26}Mg value ranging from −5.31 to −1.09 (Young and Galy, 2004; Tipper et al., 2006a; Brenot et al., 2008; Pogge von Strandmann, 2008; Hippler et al., 2009; Higgins and Schrag, 2010), which makes Mg isotopes a potential tracer of recycled carbonate in the mantle.

The Mesozoic and Cenozoic basalts from the North China craton (NCC) are ideal samples to explore Mg isotopes as a potential tracer of recycled carbonate in the mantle. There is an abrupt change in the chemical compositions between basalts formed before 120 Ma and those formed after 110 Ma. The >120 Ma basalts with strong continental geochemical signatures could be derived from the mantle that has been contaminated by delaminated lower continental crust (Huang et al., 2007; Liu et al., 2008), whereas the <110 Ma basalts with mantle-like Sr-Nd-Pb isotope and HIMU-like trace element signatures could be originated from the mantle affected by subducted oceanic crust (Yang and Li, 2008; Zhang et al., 2009; Xu et al., 2010; Wang et al., 2011). Recent studies suggest that the source of the <110 Ma basalts could be metasomatized by carbonate (Chen et al., 2009; Zeng et al., 2010). Therefore, Mg isotopic composition of these basalts could help us understand how recycled carbonate affects the mantle.

Here, we investigated Mg isotopic compositions of the >120 Ma basalts and basaltic andesites, and <110 Ma basalts from the North China craton. Our results indicate that Mg isotopic compositions of basaltic lavas change abruptly from >120 Ma to <110 Ma. The >120 Ma basalts and basaltic andesites have mantle-like Mg isotopic composition whereas the <110 Ma basalts are isotopically light (δ^{26}Mg = −0.60 to −0.42‰). This light Mg isotopic composition most likely results from metasomatism by recycled carbonate melt, which could be derived from subducted Pacific oceanic crust. This study not only presents an example to trace recycled carbonate by using Mg isotopes but also confirms the important role of the western Pacific oceanic subduction in generating the <110 Ma basalts in the North China craton.

2 Geological settings and sample description

2.1 Geological settings

The North China craton is one of the world's oldest Archean cratons, which preserves >3.8 Ga crustal remnants (Liu et al., 1992). It can be divided into three regions: the Eastern Block, the Western Block and the Trans-North China Orogen. The Trans-North China Orogen was formed by the collision between the Eastern Block and Western Block at ~1.85 Ga, marking the final amalgamation of the NCC (Zhao et al., 2008). Two linear geological and geophysical zones, the NEE strike Tan-Lu fault zone and Daxing'anling–Taihangshan gravity lineament crosscut the craton (Fig. 1). The NCC experienced multiple circumcraton subductions and collisions, which are manifested by the Paleozoic to Triassic Qinling-Dabie-Sulu ultrahigh-pressure belt in south, the Central Asian Orogenic Belt in north and the Mesozoic–Cenozoic subduction of Pacific plate in the east (Fig. 1).

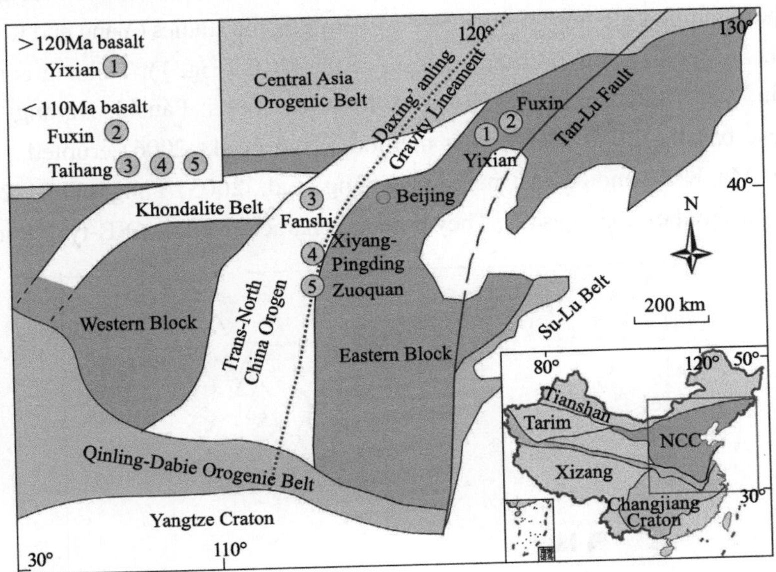

Fig. 1 Simplified geological map showing sample localities and subdivisions of the North China craton (modified from Tang et al., 2010). The North China craton is cut by two major linear zones, Tan-Lu fault zone to the east and Daxing'anling gravity lineament to the west. The basalt samples from Fanshi, Xiyang-Pingding and Zuoquan are all called Taihang basalts in this paper. Inset shows location of the NCC relative to other blocks and fold belts.

The Mesozoic and Cenozoic basalts occurred widely in the NCC. The basalts erupted before 120 Ma (e.g., Yixian basalts) have distinctive chemical compositions from those erupted after 110 Ma (e.g. Fuxin and Taihang basalts). The >120 Ma basalts are characterized by negative ε_{Nd} (down to −20), variable $^{87}Sr/^{86}Sr$, low radiogenic Pb isotope ratios, and typical "continental" geochemical signatures, such as enrichment of large ion lithophile elements (LILE, e.g., Rb and Ba) and depletion of high field strength element (HFSE, e.g., Nb and Ta) (Guo et al., 2001; 2003; Qiu et al., 2002; Zhang et al., 2002; 2003; Zhang, 2007; Liu et al., 2008; Yang and Li, 2008). By contrast, the <110 Ma basalts have relatively depleted Nd and Sr isotopic compositions and HIMU-like trace element patterns, e.g. depletion of Rb and Pb and enrichment of Nb (Peng et al., 1986; Song, 1990; Zhi et al., 1990; Zou et al., 2000; Zhang et al., 2003; 2009; Xu et al., 2005; Tang et al., 2006; Yang and Li, 2008; Chen et al., 2009; Zeng et al., 2010; 2011; Wang et al., 2011).

2.2 Sample description

The samples studied here are collected from Yixian and Fuxin in the Eastern Block, and Taihang in Trans-North China Orogen, respectively (Fig. 1). Their petrology, chemical and Sr-Nd-Pb isotopic compositions have been presented in previous studies (Zhang et al., 2003; Tang et al., 2006; Yang and Li, 2008). Only a brief description is given below.

Yixian is located to the west of Tan-Lu fault zone at the northern margin of the NCC (Fig. 1). The Yixian lavas mainly consist of basalts, andesites and rhyolites dated to 125–120 Ma by zircon U-Pb method (Yang and Li, 2008). Like many >120 Ma basalts in the craton, the Yixian basalts and basaltic andesites studied here have typical "lower crustal" signatures, such as depletion of HFSEs, low radiogenic Pb ($^{206}Pb/^{207}Pb$=16.21–16.66, $^{207}Pb/^{204}Pb$=15.22–15.29) and low $\varepsilon_{Nd}(T)$ (−13.4 to −9.7) (Yang and Li, 2008). Their low Ce/Pb ratios (4.6 to 8.2) are similar to the average Ce/Pb ratio (~5.0) of the lower continental crust (Rudnick and Gao, 2003) (Fig. 2a), suggesting the involvement of lower crustal materials in producing these basalts. The samples investigated here include three basalts and two basaltic andesites, and have relatively high SiO_2 contents (52.8–56.7%) and low CaO/Al_2O_3 ratios (0.36–0.46; Fig. 2c). These samples are fresh as indicated by petrographic studies (Yang and Li, 2008).

Fuxin is located near Yixian at the northern margin of the NCC (Fig. 1). Taihang is more than 700 km west from Yixian and Fuxin, and Taihang samples are from three locations: Fanshi, Xiyang–Pingding and Zuoquan (Fig. 1). Although these basalts (26–6 Ma, Ar–Ar method; Tang et al., 2006) erupted >70 Myrs later than the Fuxin basalts (106–100 Ma, K-Ar and Ar–Ar methods; Zhang et al., 2003; Yang and Li, 2008), they have similar geochemical characteristics to the Fuxin basalts. They both are characterized by OIB-type trace element distribution

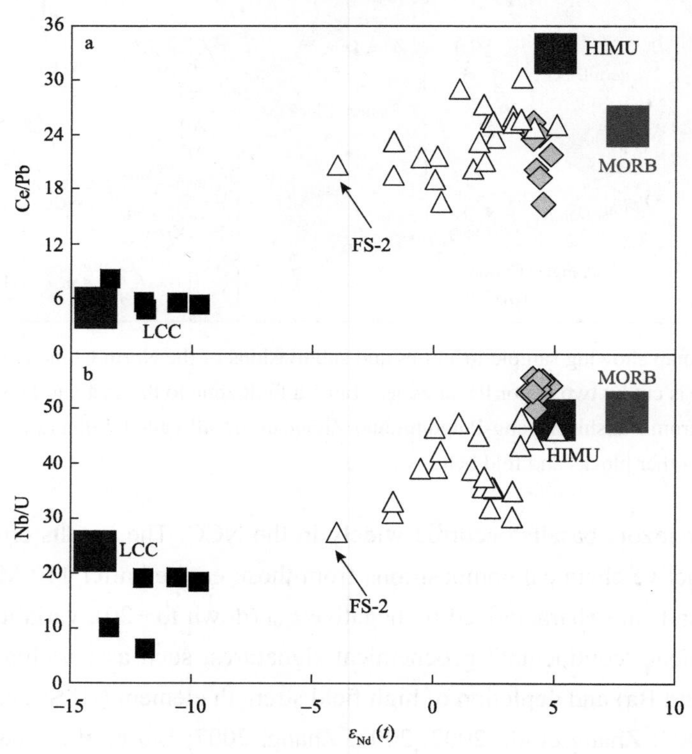

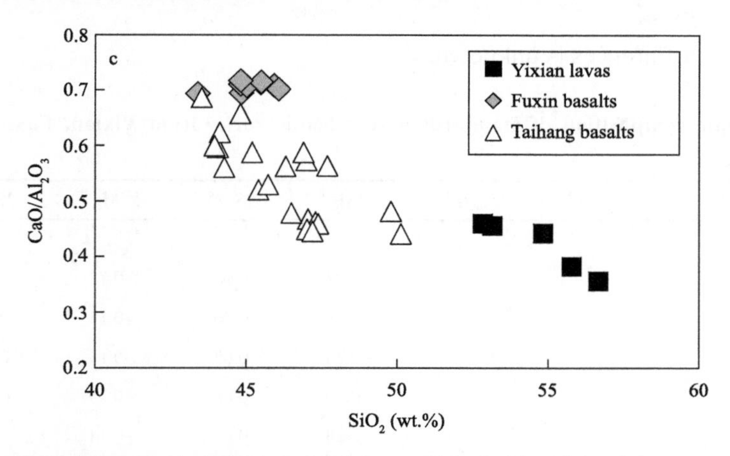

Fig. 2 (a) Ce/Pb vs. $\varepsilon_{Nd}(t)$, (b) Nb/U vs. $\varepsilon_{Nd}(t)$, and (c) CaO/Al_2O_3 vs. SiO_2 for the Yixian lavas, and the Fuxin and Taihang basalts studied here. Database: MORB and HIMU from Sun and McDonough (1989), lower continental crust (LCC) from Rudnick and Gao (2003), the Yixian lavas, and Fuxin and Taihang basalts from Zhang et al. (2003), Tang et al. (2006) and Yang and Li (2008).

patterns, with enriched light rare earth elements (LREEs) and Nb and depleted Rb and Pb. They have lower $^{87}Sr/^{86}Sr$ (0.7036–0.7054), higher $\varepsilon_{Nd}(T)$ (−4.0–5.1) and Pb isotopic ratios ($^{206}Pb/^{207}Pb$=16.76–18.32, $^{207}Pb/^{204}Pb$= 15.28–15.46) than those of Yixian basalts (Zhang et al., 2003; Tang et al., 2006). Their Ce/Pb and Nb/U ratios vary from 16.3 to 30.3 and 27.5 to 55.6, respectively (Fig. 2a, b), close to those of MORB and HIMU basalts (Sun and McDonough, 1989). They have low SiO_2 contents (43.4%–50.1%) and high CaO/Al_2O_3 ratios (0.44–0.72; Fig. 2c). Nine Fuxin basalts and 21 Taihang basalts were analyzed for Mg isotopes here.

3 Analytical methods

Magnesium isotopic analyses were performed at the Isotope Laboratory of the University of Arkansas, Fayetteville, following the established procedures (Teng et al., 2007; 2010a; Yang et al., 2009; Li et al., 2010). Only a brief description is given below.

All chemical procedures were carried out in a clean laboratory environment. Depending on Mg concentration, 1 to 25 mg of sample powder was weighted in Savillex screw-top beakers in order to have >50 μg Mg in the solution and was then dissolved in a mixture of concentrated HF-HNO_3-HCl solution. Separation of Mg was achieved by cation exchange chromatography with Bio-Rad 200–400 mesh AG50W-X8 resin in 1 N HNO_3 media following the established procedures (Teng et al., 2007; 2010a; Yang et al., 2009; Li et al., 2010). Magnesium isotopic compositions were analyzed by the standard bracketing method using a *Nu Plasma* MC-ICP-MS. Magnesium isotope data are reported in standard δ-notation relative to DSM3: $\delta^{X}Mg$= $[(\delta^{X}Mg/^{24}Mg)sample/(\delta^{X}Mg/^{24}Mg)_{DSM3}-1]\times1000$, where X=25 or 26. $\Delta^{25}Mg'= \delta^{25}Mg'-0.512\times\delta^{26}Mg'$, where $\delta^{X}Mg'=1000\times\ln[(\delta^{X}Mg+ 1000)/1000]$ (Young and Galy, 2004).

The internal precision of the measured $^{26}Mg/^{24}Mg$ ratio based on ≥4 repeat runs of the same sample solution during a single analytical session is <±0.1‰ (Teng et al., 2010a). Multiple analyses of olivine KH-1 yielded $\delta^{26}Mg$ values of −0.33 to −0.27 (Table 1), which is in agreement with that reported by Teng et al. (2010a) ($\delta^{26}Mg$= −0.27±0.07‰; 2SD, n=16). The synthetic granite standard (IL-granite) yielded a $\delta^{26}Mg$ value of −0.06‰ (Table 1), within uncertainties of the reported value of −0.01±0.06‰ (2SD, n=13; Teng et al., 2010a). San Carlos olivine standard studied by Chakrabarti and Jacobsen (2010) was also analyzed here for inter-laboratory comparison, which yielded $\delta^{26}Mg$ values of −0.33 to −0.30‰ (Table 1), similar to those reported by Young et al. (2009) and Liu et al (2010), but heavier than the value ($\delta^{26}Mg$=−0.55‰) reported by Chakrabarti and Jacobsen

(2010). The cause for these differences is unknown.

Table 1 Magnesium isotopic composition of standards and volcanic rocks from Yixian, Fuxin and Taihang Mountains, North China craton

Sample	SiO_2 [a]	MgO [a]	$\varepsilon_{Nd}(t)$ [a]	$\delta^{26}Mg$ [b]	2SD [c]	$\delta^{25}Mg$ [b]	2SD [c]	$\Delta^{25}Mg'$ [d]
Standards								
KH-1 Olivine [e]				−0.27	0.06	−0.17	0.10	−0.03
Repeat [f]				−0.28	0.06	−0.15	0.05	0.00
Repeat				−0.28	0.07	−0.13	0.05	+0.02
Repeat				−0.30	0.06	−0.15	0.04	+0.01
Repeat				−0.28	0.05	−0.14	0.07	+0.01
Repeat				−0.33	0.09	−0.13	0.05	+0.04
Repeat				−0.31	0.06	−0.15	0.05	+0.01
Repeat				−0.32	0.06	−0.17	0.05	0.00
Average				−0.30	0.04	−0.15	0.04	
San Calos Ol				−0.33	0.07	−0.17	0.05	0.00
Repeat				−0.30	0.07	−0.16	0.05	0.00
Average				−0.32	0.04	−0.17	0.01	
Il-granite [g]				−0.06	0.06	−0.02	0.05	+0.01
Repeat				−0.06	0.05	−0.05	0.07	−0.02
Average				−0.06	0.01	−0.04	0.04	
Yixian basalts and basaltic andesites (*125–120 Ma*)								
HBJ4-1	54.83	6.45	−10.6	−0.25	0.06	−0.14	0.05	−0.01
HBJ4-2	53.16	6.23	−12.0	−0.28	0.06	−0.14	0.06	+0.01
HBJ4-3	52.84	7.03	−9.7	−0.27	0.06	−0.15	0.06	−0.01
SHT-14	55.78	5.21	−13.4	−0.26	0.06	−0.15	0.06	−0.01
SHT-3	56.67	5.92	−11.9	−0.31	0.07	−0.19	0.05	−0.03
Average				−0.27	0.05	−0.15	0.04	
Fuxin basalts(*106–100 Ma*)								
JG-01	44.84	8.31	4.8	−0.46	0.06	−0.27	0.05	−0.03
JG-02	45.48	8.25	3.9	−0.54	0.06	−0.26	0.05	+0.02
Repeat				−0.52	0.05	−0.29	0.07	−0.02
JG-03	44.82	8.38	4.1	−0.48	0.06	−0.25	0.05	0.00
JG-04	45.92	8.09	4.4	−0.53	0.06	−0.28	0.05	0.00
JG-05	45.01	8.30	4.5	−0.46	0.06	−0.24	0.05	0.00
JG-06	46.07	8.06	4.2	−0.47	0.06	−0.25	0.05	−0.01
JG-07	43.40	8.39	4.4	−0.60	0.06	−0.32	0.05	−0.01
Repeat				−0.55	0.06	−0.27	0.04	+0.02
JG-08	45.50	8.16	4.4	−0.49	0.06	−0.25	0.05	+0.01
JG-09	44.82	8.30	4.1	−0.49	0.07	−0.24	0.05	+0.02
Average				−0.51	0.09	−0.27	0.05	
Taihang basalts(*26–6 Ma*)								
FS-1	44.07	8.81	3.3	−0.44	0.04	−0.22	0.06	+0.01
FS-8	44.28	8.30	2.3	−0.46	0.04	−0.24	0.06	0.00
FS-10	44.13	9.09	2.5	−0.46	0.04	−0.22	0.06	+0.02

Continued

Sample	SiO_2 [a]	MgO [a]	$\varepsilon_{Nd}(t)$ [a]	$\delta^{26}Mg$ [b]	2SD [c]	$\delta^{25}Mg$ [b]	2SD [c]	$\Delta^{25}Mg'$ [d]
Taihang basalts(*26–6 Ma*)								
FS-30	43.53	9.93	3.3	−0.44	0.04	−0.25	0.06	−0.02
FS-32	45.20	8.74	2.1	−0.49	0.04	−0.27	0.06	−0.01
FS-33	43.96	8.81	2.4	−0.48	0.04	−0.22	0.06	+0.03
HHL-1	45.41	8.10	1.6	−0.43	0.04	−0.22	0.06	0.00
HHL-2	45.73	8.26	2.1	−0.52	0.06	−0.26	0.05	+0.01
FS-2	47.01	8.36	−4.0	−0.30	0.06	−0.16	0.05	0.00
FS-3	47.06	7.43	−1.7	−0.53	0.06	−0.26	0.05	+0.02
FS-9	44.81	9.85	−1.7	−0.46	0.06	−0.24	0.05	0.00
FS-36	46.31	8.47	0.2	−0.51	0.06	−0.25	0.05	+0.02
FS-38	46.91	8.91	−0.5	−0.45	0.06	−0.23	0.05	0.00
FHS-1	47.70	7.83	1.1	−0.44	0.07	−0.25	0.04	−0.02
GB-2	47.31	6.72	3.6	−0.50	0.07	−0.27	0.04	−0.01
JD-1	47.03	6.48	4.1	−0.46	0.07	−0.26	0.04	−0.02
JX-1	47.40	6.72	1.9	−0.49	0.07	−0.25	0.04	+0.01
JX-3	46.51	6.94	5.1	−0.46	0.07	−0.25	0.04	−0.01
MAS-1	47.21	6.47	3.6	−0.54	0.07	−0.29	0.04	−0.01
ZQ-1	49.80	7.17	0.3	−0.42	0.07	−0.22	0.04	0.00
ZQ-4	50.13	6.67	0.1	−0.42	0.07	−0.22	0.04	0.00
Average				−0.46	0.10	−0.24	0.05	

[a] Data from Zhang et al. (2003), Tang et al. (2006), Yang and Li (2008).

[b] $\delta^{X}Mg = [(^{X}Mg/^{24}Mg)_{sample}/(^{X}Mg/^{24}Mg)_{DSM3}-1] \times 1000$, where $X = 25$ or 26 and DSM3 is Mg solution made from pure Mg metal (Galy et al., 2003).

[c] 2SD = 2 times the standard deviation of the population of >20 repeat measurements of the standard during a session.

[d] $\Delta^{25}Mg' = \delta^{25}Mg' - 0.512 \times \delta^{26}Mg'$, where $\delta^{X}Mg' = 1000 \times \ln[(\delta^{X}Mg + 1000)/1000]$ (Young and Galy, 2004).

[e] KH-1 olivine is an in-house standard that has been analyzed through whole-procedural column chemistry and instrumental analysis with $\delta^{26}Mg = -0.27 \pm 0.07$ and $\delta^{25}Mg = -0.14 \pm 0.04$ (n=16; 2SD) relative to DSM-3 (Teng et al., 2010a).

[f] Repeat of column chemistry and measurement of different aliquots of a stock solution.

[g] IL-granite is a synthetic solution with Mg: Al: Fe: Ca: K: Na: Ti: Ni = 1:30:5:5:20:10:0.1:0.1.

4 Results

Magnesium isotopic compositions of reference materials and the Mesozoic and Cenozoic basalts from the NCC are reported in Table 1 along with their chemical compositions (Zhang et al., 2003; Tang et al., 2006; Yang and Li, 2008). In a plot of $\delta^{25}Mg$ vs. $\delta^{26}Mg$ (Fig. 3), all samples fall along the terrestrial equilibrium mass fractionation curve with a slope of 0.521. The $\Delta^{25}Mg'$ values of all samples are within ± 0.04‰ (Table 1). Overall, Mg isotopic compositions of these basalts from NCC vary significantly, with $\delta^{26}Mg$ values ranging from –0.60‰ to –0.25‰ and $\delta^{25}Mg$ values from –0.32‰ to –0.14‰ (Table 1).

The Yixian basalts and basaltic andesites have homogenous Mg isotopic compositions (Fig. 4 and Table 1), with $\delta^{26}Mg$ ranging from –0.31 to –0.25‰ and an average of –0.27±0.05‰ (2SD, n=5), which is within analytical uncertainties of the average $\delta^{26}Mg$ of the mantle (–0.25±0.07‰; Teng et al., 2010a).

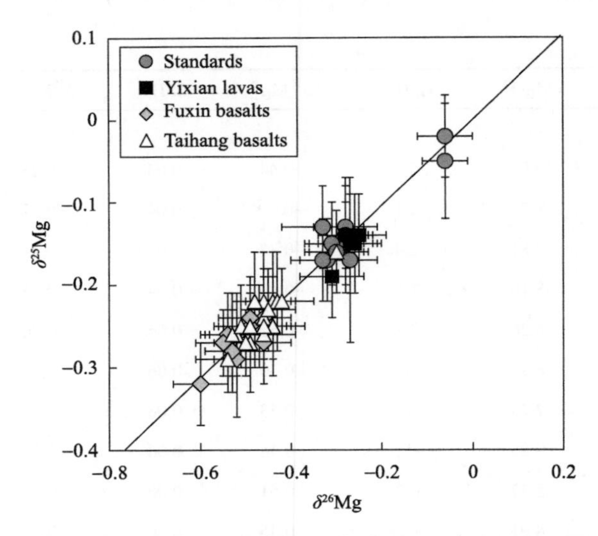

Fig. 3　Magnesium isotope plot of the samples in this study. The solid line represents the terrestrial fractionation line with a slope equal to 0.521. Error bars represent 2SD uncertainties. Data are from Table 1.

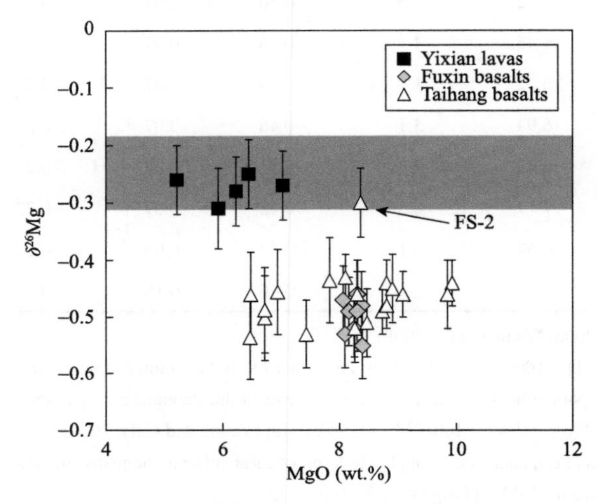

Fig. 4　δ^{26}Mg vs. MgO (wt.%) for the Yixian lavas, and the Fuxin and Taihang basalts. The grey bar represents global oceanic basalts (δ^{26}Mg=–0.25±0.07‰; Teng et al., 2010a). Error bars represent 2SD uncertainties. Data are from Table 1.

29 out of the 30 Fuxin and Taihang basalts exhibit a variation of 0.18‰ in δ^{26}Mg values from –0.60 to −0.42‰, which are obviously lighter than those of the average mantle (–0.25±0.07‰; Teng et al., 2010a) but close to those of some wehrlites (–0.48 to –0.43‰) in the North China craton (Yang et al., 2009). Sample FS-2 deviates from the population of the others with a δ^{26}Mg value of –0.30±0.06‰ (Fig. 4). Notably, this sample also has the lowest $\varepsilon_{Nd}(t)$ value of –4.0 and lowest Nb/U ratio of 27.5 among all Fuxin and Taihang basalts (Fig. 2b).

5　Discussion

Magnesium isotopic composition of basalts is generally expected to be mantle-like because basaltic differentiation does not fractionate Mg isotopes (Teng et al., 2007; 2010a) and Mg isotopic composition of the mantle is difficult to be modified by recycled materials, which generally have low-Mg content compared to the mantle (MgO=~37.8%; McDonough and Sun, 1995). For example, the >120 Ma Yixian basalts are considered to be derived from the mantle source hybridized by continental crustal materials based on their low Ce/Pb and Nb/U ratios and enriched Sr-Nd isotopic compositions (Yang and Li, 2008). However, they still preserve mantle-like Mg isotope composition, with δ^{26}Mg ranging from –0.31 to –0.25‰ and an average of –0.27 ± 0.05‰ (2SD, n=5).

By contrast, most of the <110 Ma basalts have lighter Mg isotopic composition than the average mantle (Teng et al., 2010a), with δ^{26}Mg values from –0.60 to –0.42‰. Similar light Mg isotopic composition has been previously observed in the wehrlites from the Beiyan area, NCC (Yang et al., 2009). In general, Mg isotopic compositions of basalts and peridotites from the NCC can be divided into two groups (Fig. 5). One group, including the basalts and two wehrlites, has relatively light Mg isotopic composition (Yang et al., 2009 and this study). The other group has mantle-like Mg isotopic composition, represented by the majority of peridotites (Yang et al., 2009; Teng et al., 2010a). This may suggest that the isotopically light wehrlites could be either the mantle source of or cumulates from the <110 Ma basalts. Three potential processes may account for such a light Mg isotopic composition in the <110 Ma basalts and the two wehrlites: (1) Kinetic Mg isotope fractionation in the mantle source; (2) Interaction with silicate melts (either crustal contamination or mantle metasomatism by recycled crustal material); (3) Interaction with carbonate rocks. Below, we will first evaluate which process could produce the observed light Mg isotopic composition in these basalts, and then discuss their origin together with their other geochemical characteristics.

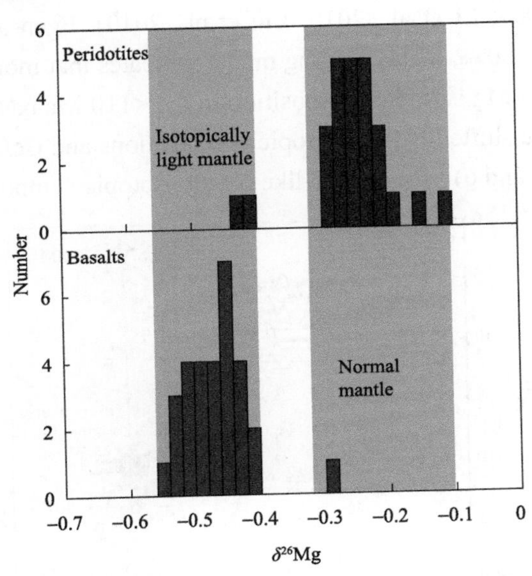

Fig. 5 Histogram for δ^{26}Mg values of the <110 Ma basalts and peridotites from the North China craton. The grey bars represent two kind of mantle sources with distinct Mg isotopic composition. Data are from Yang et al. (2009), Teng et al. (2010a) and Table 1.

5.1 Kinetic Mg isotope fractionation

Kinetic Mg isotope fractionation driven by chemical and thermal diffusion can occur at high temperatures as shown in experimental studies (Richter et al., 2008; Huang et al., 2009; 2010) but unlikely be the mechanisms that generate such light Mg isotopic compositions in the mantle source of the <110 Ma basalts. One recent study shows co-variation of Mg and Li isotopes in mantle xenoliths (Pogge von Strandmann et al., 2011), which is explained by chemical diffusion driven kinetic isotope fractionations during the transport of mantle xenoliths. However, such kinetic isotope fractionation by chemical diffusion could be ruled out to produce the isotopically light mantle source of the <110 Ma basalts from the NCC. First, lighter isotopes diffuse faster than heavier ones during chemical diffusion. When Mg diffuses from peridotites to host basalts, Mg isotopic composition of peridotites should be shifted towards heavy values, as shown in Pogge von Strandmann et al. (2011). This is the opposite to the wehrlites from the NCC. In addition, peridotites affected by diffusion also display large inter-mineral Mg isotope fractionation, with $\Delta^{26}Mg_{cpx\text{-}ol}$ up to 0.31‰ (Pogge von Strandmann et al., 2011). By contrast, the $\Delta^{26}Mg_{cpx-ol}$ values of all peridotite xenoliths from the NCC are <0.10‰, indicating that Mg isotopes reach equilibrium in these peridotite xenoliths (Yang et al., 2009). This further rules out chemical diffusion driven kinetic fractionation as causes of those isotopically light wehrlites.

Kinetic isotope fractionation by thermal diffusion could also be ruled out since the <110 Ma basalts studied here come from a very broad area and have variable ages (106–6 Ma). It is unlikely that thermal diffusion could operate on such a large scale and continue over such a long period of time (>70 Myrs from the Fuxian basalts to the Taihang basalts). Therefore, both chemical and thermal diffusion appear unable to create the light Mg isotopic composition in the <110 Ma basalts from the NCC.

5.2 Interactions with silicate melts

Crustal contamination by silicate wall rocks and mantle metasomatism by recycled silicate crustal material are both unlikely to produce such a light Mg isotopic composition in the <110 Ma basalts because crustal materials generally have low-Mg content compared to the mantle (MgO=~37.8%; McDonough and Sun, 1995). The silicate rocks in continental crust have a relatively heavier Mg isotopic composition with $\delta^{26}Mg$ ranging from –0.44 to 0.90‰ (Shen et al., 2009; Li et al., 2010; Liu et al., 2010). Even assuming that they could have an average $\delta^{26}Mg$ value as low as –0.60‰, binary mixing model indicates that more than 60% crustal contamination is required to produce such a light Mg isotopic composition in the <110 Ma basalts (Fig. 6). This large amount of crustal contamination should have shifted Nd-Pb isotopic compositions and Ce/Pb, Nb/U ratios of these basalts to crustal-like compositions (Figs. 2 and 6). The mantle-like Nd-Pb isotopic compositions and Ce/Pb, Nb/U ratios in

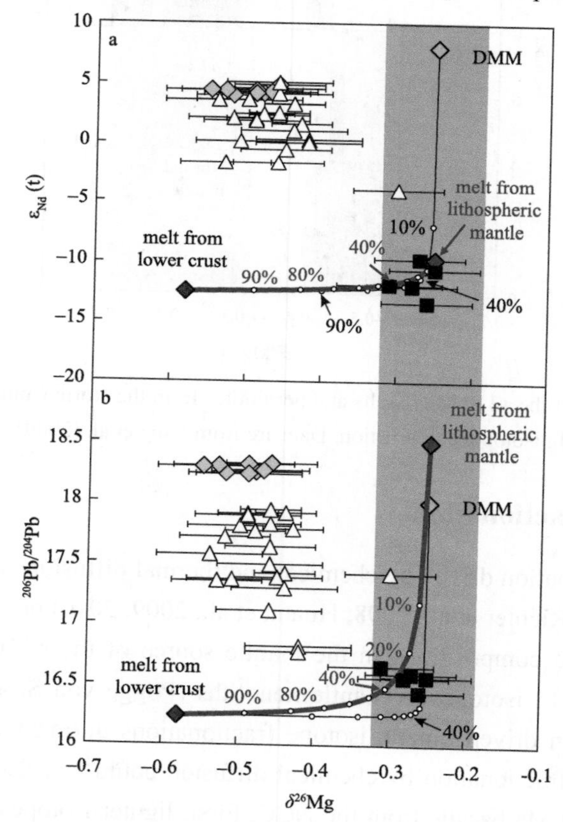

Fig. 6　Plots of $\varepsilon_{Nd}(t)$ and $^{206}Pb/^{204}Pb$ vs. $\delta^{26}Mg$ for the Yixian lavas, and the Fuxin and Taihang basalts. The symbols are the same as Fig. 4. The grey bar represents global oceanic basalts ($\delta^{26}Mg$=–0.25±0.07‰; Teng et al., 2010a). Error bars represent 2SD uncertainties. The curves represent mixing of two end members. Numbers at small circles represent fraction of the melt derived from the lower continental crust. Data are from Table 1. See Table 2 for modeling parameters.

Table 2 Parameters for model calculations

	MgO (wt.%)	δ^{26}Mg	FeO (wt.%)	Nd (ppm)	ε_{Nd}	Pb (ppm)	$^{206}Pb/^{204}Pb$
DMM [a]	37.8	−0.25		1.1	8	0.2	18
Primitive magma from SCLM [b]	12.0	−0.25		7.8	−9.7	1.2	18.5
Melt from LCC [c]	3.5	−0.60		28.9	−12.5	15.3	16.24
Melt from OC [d]	4.6	−0.83	7.3				
Mantle harzburgites [e]	45.2	−0.24	8.5				
Kd $^{Fe\text{-}Mg}$ between peridotite and melt: 0.3 [f]							

[a] DMM = Depleted MORB mantle. Database: MgO (McDonough and Sun, 1995); δ^{26}Mg (Teng et al., 2010a); ε_{Nd} (Jahn et al., 1999); Nd, Pb and $^{206}Pb/^{204}Pb$ (Sun and McDonough, 1989).

[b] SCLM = sub-continental lithospheric mantle. Database: δ^{26}Mg (Teng et al., 2010a); ε_{Nd} and $^{206}Pb/^{204}Pb$ (Zheng and Lu, 1999); MgO, Nd and Pb based on data of oceanic basalt (Sun and McDonough, 1989).

[c] LCC = lower continental crust of the northern margin of the NCC. δ^{26}Mg is assumed as light as -0.6 to demonstrate the light isotopic composition of the basalts could not derive from crustal contamination. Other data are from (Yang and Li, 2008).

[d] OC = oceanic crust. Database: MgO and FeO from partial melt of an oceanic gabbro G108 (Sobolev et al., 2005); δ^{26}Mg is assumed as light as -0.83 to demonstrate the light isotopic composition of the basalts could not derive from silicate melt metasomatism.

[e] Data are from two harzburgites of the NCC (Wu et al., 2006; Teng et al., 2010a).

[f] Data are estimated based on Kd $^{Fe\text{-}Mg}$ of olivine and pyroxene with melt (Roeder and Emslie, 1970; Brey and Kohler, 1990).

the <110 Ma basalts hence suggest that the contamination from silicate rocks in continental crust is limited and can be ruled out as causes of such light Mg isotopic compositions of these basalts.

For mantle metasomatism, the potential metasomatic agent could be originated from the delaminated continental crust (Liu et al., 2008; Zeng et al., 2011), the asthenospheric mantle (Xiao et al., 2010) or subducted oceanic crust (Yang and Li, 2008; Zhang et al., 2009; Xu et al., 2010; Wang et al., 2011). As mentioned above, the continental crust has a relatively heavier Mg isotopic composition with δ^{26}Mg ranging from –0.44 to 0.90 (Shen et al., 2009; Li et al., 2010; Liu et al., 2010), and the asthenosphere has a homogenous Mg isotopic composition as sampled by global MORBs (δ^{26}Mg=–0.25±0.06, 2SD; Teng et al., 2010a). Their Mg isotopic compositions are both heavier than those of the <110 Ma basalts (–0.60 to –0.42).

Alternatively, the oceanic crust may have a relatively lighter Mg isotopic composition. First, seawater has a homogeneous and light Mg isotopic composition with δ^{26}Mg=–0.832±0.068 (Ling et al., 2011 and references therein). Water–rock interactions should hence lead to an isotopically light oceanic crust. In addition, the fluxweighted average δ^{26}Mg value of global runoff is –1.09 (Tipper et al., 2006b), which is even lighter than that of the seawater. Assuming a steady-state Mg isotopic composition for the oceans, the light Mg in the global runoff must balance the flux from the oceans to the oceanic crust, which suggests the oceanic crust has to be isotopically light (Teng et al., 2010b). Moreover, δ^{26}Mg value of a marine shale standard (SCo-1) is –0.94±0.08 (Li et al., 2010).

However, even assuming that the silicate melt from the subducted oceanic crust has δ^{26}Mg value as low as seawater (–0.8‰), model calculation suggests that a total melt/rock ratio of 3.2–4.4 is required to form the isotopically light mantle source in the NCC (Fig. 7). Such a high melt/rock ratio may be possible when large volume of melt migrates through the mantle peridotite in a local area. However, the <110 Ma basalts studied here come from four localities in the NCC. In addition, migration of large volume of melt derived from subducted oceanic crust through the mantle should result in adakitic magma eruption (Rapp et al., 1999), which is inconsistent with current observation that nearly no <110 Ma adakitic rock has been found in the NCC.

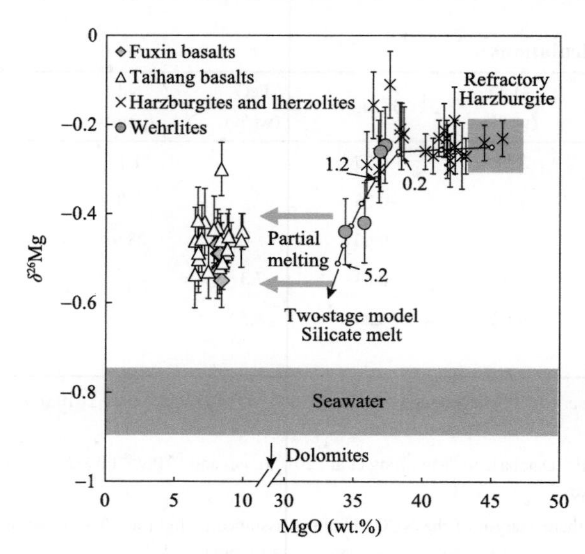

Fig. 7 δ^{26}Mg vs. MgO (wt.%) for the Fuxin and the Fanshi basalts with the peridotites in the North China craton. The grey bar represents seawater (δ^{26}Mg=−0.83±0.07‰, 2SD) (Ling et al., 2011 and references therein). The grey box represents refractory harzburgite from the NCC (Teng et al., 2010a). Average δ^{26}Mg of dolomite is −2.2‰ (Higgins and Schrag, 2010). Numbers at small circles represent total melt/rock ratios. Data are from Yang et al. (2009), Teng et al. (2010a) and Tables 1 and 2. Error bars represent 2SD uncertainties. For modeling reaction of peridotite with silicate melt, two ideal processes are assumed. At the beginning the peridotite simply mixes with the melt, and then just exchange Fe-Mg with the melt. The latter process assumes that peridotite exchanges Fe-Mg with melt to achieve elemental and isotopic equilibrium but never mix with melts.

Although mantle metasomatism by silicate melt cannot generate the light Mg isotopic composition in the <110 Ma basalts, this does not mean the mantle source of the <110 Ma basalts would have never metasomatized by silicate melt. The mantle could preserve its original Mg isotopic composition even though it had been significantly metasomatized by silicate melts.

5.3 Interactions with high-Mg carbonate melts

High-Mg carbonate (e.g., dolomite) melt metasomatism however could easily modify Mg isotopic composition of the mantle or the basalts. Dolomite can have high MgO contents and light Mg isotopic composition, with δ^{26}Mg values ranging from −5.31 to −1.09‰ (Young and Galy, 2004; Tipper et al., 2006a; Brenot et al., 2008; Pogge von Strandmann, 2008; Hippler et al., 2009; Higgins and Schrag, 2010). For example, by assuming that dolomite has an average MgO of 20% and δ^{26}Mg of −2.2‰ (Higgins and Schrag, 2010). Both mixing with ~20% dolomitic melt in the mantle and contamination by ~5% dolomite wallrock can shift the basalts to such a light Mg isotopic composition.

In addition, the isotopically light wehrlites in the NCC have higher CaO than those of other peridotites (see Fig. 5b in Yang et al., 2009). This CaO/Al_2O_3 variation in the wehrlites does not result from the mode abundances of clinopyroxene, because the isotopically light wehrlites have similar mode abundance of clinopyroxene (25%) to the "normal" wehrlites (27%–28%) (Yang et al., 2009). Thus, the isotopically light wehrlites may be affected by carbonate. And if these wehrlites are cumulates from or source of the <110 Ma basalts, the basalts may also be modified by carbonates. Coincidentally, the isotopically light <110 Ma basalts have relatively higher CaO/Al_2O_3 than those of >110 Ma basalts (Fig. 8c), although this high CaO/Al_2O_3 may partially result from crystal fractionation of plagioclase.

5.3.1 Contamination by carbonate wallrock

Dolomites in the upper continental crust generally have light Mg isotopic composition (Higgins and Schrag, 2010). However, it is unlikely that the light Mg isotopic composition in the <110 Ma basalts could result from interactions with dolomite wallrock. First, the <110 Ma basalts studied here come from a very broad area and have variable ages (106–6 Ma). However, they have nearly identical Mg isotopic composition with δ^{26}Mg values ranging from −0.60 to −0.42‰. It is difficult to explain why they were all contaminated by ~5% dolomite wallrock. Second, dolomite only occupies a very small part of the upper crust. If all lavas have been contaminated by dolomite, they could also be largely contaminated by other crustal materials, which would significantly shift Nd isotopic compositions, Ce/Pb and Nb/U ratios. However, this is not observed in the <110 Ma basalts. Furthermore, the Yixian (~125 Ma) and the Fuxin (~106 Ma) basalts both come from western Liaoning area, thus have similar wallrocks. Although the former have enriched Nd isotopic composition (Yang and Li, 2008), their mantle-like Mg isotopic composition indicates that they maybe escaped from dolomite contamination. On the contrary, the latter with light Mg isotopic composition appears to be significantly contaminated by dolomite, still preserves depleted Nd isotopic composition (Yang and Li, 2008).

In addition, the two isotopically light wehrlites from Beiyan area could not be cumulates from the basalts based on their petrological and geochemical features (Xiao et al., 2010). Their olivines show undulatory extinction and kink bands which cannot be found in cumulates (Xiao et al., 2010). Furthermore, these wehrlites have lower Cr# in spinel (<40) than those of cumulates (>50) (Xiao et al., 2010). Finally, their cpxs have sieve textures (Xiao et al., 2010), which could be a feature for melt-rock reactions as indicated by the experimental study (Shaw and Dingwell, 2008). This also indicates that there could be an isotopically light mantle source beneath the NCC after 110 Ma, and that a contamination model by dolomite wallrock is not necessary.

5.3.2 Mantle metasomatism by carbonate melt

To date, Mg isotopic data for mantle carbonatites are still not available. Thus, it is difficult to distinguish where the metasomatic agent (carbonatitic melt) was derived from. However, the observations below suggest that the metasomatic agent (carbonatitic melt) is more likely derived from oceanic crust than from the asthenosphere. First, if the carbonatitic melt from asthenosphere could modify the Mg isotopic composition of the mantle to light values, some of MORBs and OIBs should have light Mg isotopic composition. However, none of MORBs and OIBs has such a light Mg isotopic composition. Second, the geochemical data of the <110 Ma basalts suggest involvement of subducted oceanic crust materials in their mantle source. They are characterized by HIMU features, e.g., relatively higher U/Pb, Th/Pb, Ce/Pb and Nb/U ratios than those of >120 Ma basalts (Figs. 2 and 8). The dehydration of subducted oceanic crust may result in the increases of U/Pb, Th/Pb, Ce/Pb and Nb/U ratios (Hofmann, 1997; Kogiso et al., 1997). Some <110 Ma basalts have U/Pb ratios up to 0.5 and Th/Pb ratios up to 1.8, which are typical characteristics of the HIMU mantle end member (Fig. 8), and may reflect the signature of recycled subducted oceanic crust (Hofmann, 1997). Thus, the source of the <110 Ma basalts may also involve the carbonate in the recycled subducted oceanic crust.

The proportion of dolomite in subducted oceanic crust is not well constrained. Since modern deep water carbonates are dominated by calcite (Plank and Langmuir, 1998; Alt and Teagle, 1999) and dolomite is generally formed at continental shelves, it seems that no dolomite could be brought into the mantle by oceanic crust subduction. However, the subducted oceanic crust may indeed contain a lot of high-Mg carbonates. First, substantial quantities of the terrigenous sediments are known to enter the mantle at subduction zones as suggested by some oceanic island basalts (e.g., Samoa) with low Nd, very high Sr isotopic compositions and other trace element

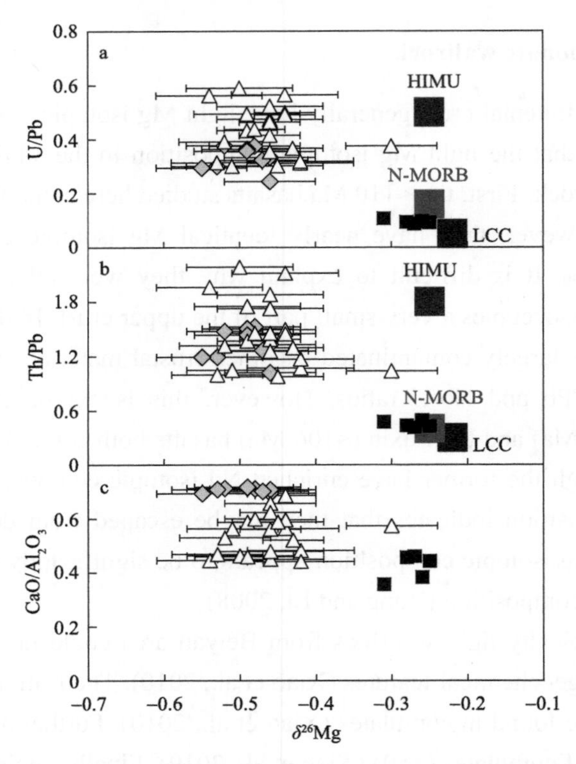

Fig. 8 (a) U/Pb, (b) Th/Pb and (c) CaO/Al_2O_3 vs. $\delta^{26}Mg$ for the Yixian lavas, and the Fuxin and Taihang basalts. The symbols are the same as Fig. 4. Error bars represent 2SD uncertainties. Database: U/Pb and Th/Pb ratios of HIMU and MORB from Sun and McDonough (1989), U/Pb and Th/Pb ratios of LCC from Rudnick and Gao (2003), Mg isotopic composition of HIMU, MORB and LCC from Teng et al. (2010a) and Li et al. (2010), other data from Table 1, Zhang et al. (2003), Tang et al. (2006) and Yang and Li (2008).

features (Hofmann, 2007; Jackson et al., 2007). Thus, it is possible these recycle terrigenous sediments contain continental shelf dolomite. In addition, as the oceanic crust subducting, low Mg carbonates can be transformed to high-Mg carbonates. The Ca-rich carbonate (e.g., limestone) exchange Ca-Mg with silicates to form dolomite or magnesite (Dasgupta and Hirschmann, 2010 and references therein). During this process, Mg diffuses from silicate minerals into carbonate which may also generate a light Mg isotopic composition in the carbonate. Future experimental studies are required for revealing the Mg isotope fractionation between carbonate and silicate during subduction.

5.4 Implications

5.4.1 Tracing recycled carbonate in the mantle by using magnesium isotopes

Cycling of carbon into and out of the mantle plays a key role in the global carbon cycle and influences the CO_2 budget of the Earth's atmosphere. Subducted oceanic crust contains a significant fraction of carbonate (Alt and Teagle, 1999), and subduction zone dehydration do not significantly remove this carbonate (Yaxley and Green, 1994; Molina and Poli, 2000; Kerrick and Connolly, 2001). These carbonated eclogites advected into the upper mantle would first produce carbonate melts at depths of 280–400 km (Dasgupta et al., 2004), which may then significantly modify chemical and physical properties of the mantle.

Carbon isotopes are efficient to identify recycled organic carbon but not very sensitive to inorganic carbonate. However, the subducted carbon is dominated by inorganic carbonate (Plank and Langmuir, 1998; Alt and Teagle, 1999). Huang et al. (2011) observed light Ca isotopic compositions of Hawaiian basalts with $\delta^{44}Ca_{SRM915a}$ ranging from 0.75‰ to 1.02‰, indicating recycled ancient marine carbonate involved in their mantle source. Thus, Ca isotopes could be a new tracer for Ca and C cycling (Huang et al., 2011).

Similarly, we observed light Mg isotopic compositions of the <110 Ma basalts from the NCC, indicating recycled carbonate in their mantle source. Therefore, Mg isotopes could be another useful tracer for interactions between mantle rocks and recycled carbonate. Since Mg isotopes cannot distinguish organic carbonate (δ^{26}Mg= −5.31 to −1.88‰; Pogge von Strandmann, 2008; Hippler et al., 2009) from inorganic carbonate (δ^{26}Mg= −5.09 to −1.09‰; Young and Galy, 2004; Tipper et al., 2006a; Brenot et al., 2008; Higgins and Schrag, 2010), combined Mg-Ca-C isotopic studies could shed light on tracing recycled carbonate in the mantle.

5.4.2 Constraints on the role of the Pacific subduction in generating continental basalts in the NCC

The influence of western Pacific subduction on the geological evolution of the south China block has been recognized (Li and Li, 2007; Sun et al., 2007; Ling et al., 2009). However, it remains uncertain whether Pacific subduction-related component was contributed to the late Cretaceous and Cenozoic basalts in the NCC or not. The NCC has experienced three circum-craton subductions, which are manifested by the Paleozoic to Triassic Qinling-Dabie-Sulu ultrahigh-pressure belt in south, the Central Asian Orogenic Belt in the north and the Mesozoic–Cenozoic subduction of Pacific plate in the east (Fig. 1). Although several authors propose that the upper mantle beneath the NCC might have been affected by subduction (Yang and Li, 2008; Zhang et al., 2009; 2010; Xu et al., 2010), it is difficult to tell whether these signatures come from Pacific subduction or not. Magnesium isotopic data of the Mesozoic and Cenozoic basalts could provide spatial and temporal constraints on how the Pacific subducted crust affected the upper mantle beneath the NCC.

Magnesium isotopic data of the Mesozoic and Cenozoic basalts display an abrupt Mg isotopic variation between the >120 Ma basalt and <110 Ma basalt. This is consistent with the abrupt Nd isotopic variation in these basalts (Fig. 9), indicating that the upper mantle beneath the NCC has been significantly modified by melt from the recycled carbonate. Because carbonation lowers the solidus of peridotite (Dasgupta et al., 2007), if carbonated peridotites with light Mg isotopic composition existed in the upper mantle, they would melt first and generate isotopically light basalts. Light Mg isotopic signature in basalts did not appear before 120 Ma, suggesting that the isotopically light mantle source didn't form till 120 Ma. Only the Pacific plate was subducting beneath the NCC during the Mesozoic–Cenozoic, thus, is expected to play an important role in generating the <110 Ma basalts of the NCC.

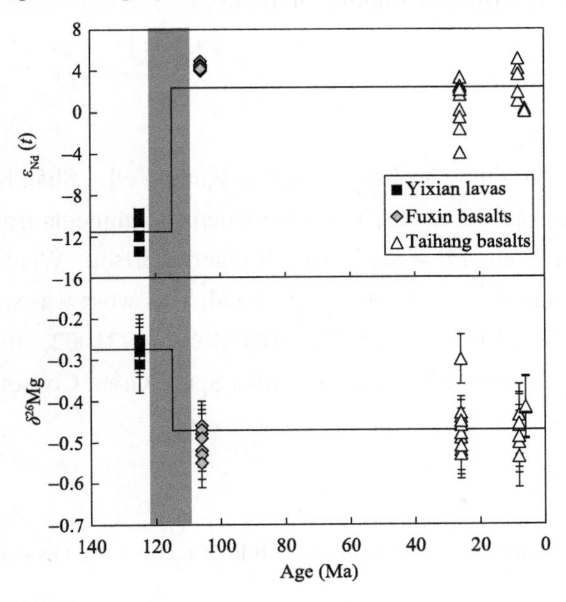

Fig. 9 Plots of $\varepsilon_{Nd}(t)$ and δ^{26}Mg vs. age for the Yixian lavas, the Fuxin and Taihang basalts. Error bars represent 2SD uncertainties. The grey box indicates the time period when the abrupt change occurred. Data are from Table 1.

The <110 Ma basalts investigated here are from both the Eastern Block and Trans-North China Orogen of the NCC (Fig. 1), indicating that the subduction of Pacific plate has affected the Taihang Mountain to the west, which is about 2000 km far from the Japan trench. Based on the seismic study, the Pacific slab is subducting beneath the Japan Islands and Japan Sea and becomes stagnant in the mantle transition zone under East China (Zhao and Ohtani, 2009). The upper mantle above the stagnant Pacific slab under East Asia may have formed a big mantle wedge which exhibits significantly low seismic velocity (Zhao and Ohtani, 2009). Slab melting may take place in the big mantle wedge. As a result, the upper mantle beneath the NCC has been metasomatized by the slab melt and shifts to HIMU geochemical features and light Mg isotopic composition.

6 Conclusions

Based on Mg isotopic compositions of the Mesozoic and Cenozoic basalts and basaltic andesites from the North China craton, we conclude:

(1) The >120 Ma Yixian basalts and basaltic andesites have mantle-like Mg isotopic composition, with δ^{26}Mg values varying from −0.31 to −0.25‰ and an average of −0.27±0.05‰ (2SD). This indicates that the lithospheric mantle beneath the NCC have a "normal" mantle-like Mg isotopic composition before 120 Ma.

(2) 29 out of 30 <110 Ma basalts from the NCC have light Mg isotopic composition, with δ^{26}Mg ranging from −0.60 to −0.42‰. The light Mg isotopic composition of the <110 Ma basalts reflects that of their mantle source rather than contamination of crustal materials because they still preserve mantle like Sr-Nd-Pb isotopic compositions and Ce/Pb, Nb/U ratios.

(3) The light Mg isotopic composition in the upper mantle beneath the North China craton is most likely derived from metasomatism of isotopically light recycled carbonate melt, which is also supported by other geochemical features of the <110 Ma basalts, e.g. high U/Pb and Th/Pb ratios similar to those of HIMU basalts. Our study presents an example to trace recycled carbonate by using Mg isotopes.

(4) The temporal change of the Mg isotopic compositions between >120 Ma and <110 Ma basalts indicates that the isotopically light mantle component could not form till 120 Ma, thus the subducted Pacific Plate might play an important role during the lithospheric thinning in the NCC.

Acknowledgement

We thank Yan-Jie Tang for providing Taihang samples, Wang-Ye Li, Shan Ke, Sheng-Ao Liu for assistances in the lab and Weifu Guo, Jingao Liu for discussions. Constructive comments from Shichun Huang, Fang Huang, Bin Shen, Jun-Jun Zhang, Yan Xiao, Edward Tipper, Richard Carlson, Weidong Sun, Cin-Ty Lee and four anonymous reviewers on the manuscript are greatly appreciated. This work was supported by the National Science Foundation of China (Grants 41173012, 90714008, 91014007, 40921002, 40803011), the National Science Foundation (EAR-0838227 and EAR-1056713) and Arkansas Space Grant Consortium (SW19002).

References

Alt, J.C., Teagle, D.A.H., 1999. The uptake of carbon during alteration of ocean crust. Geochimica et Cosmochimica Acta 63 (10), 1527-1535.

Bourdon, B., Tipper, E.T., Fitoussi, C., Stracke, A., 2010. Chondritic Mg isotope composition of the Earth. Geochimica et

Cosmochimica Acta 74 (17), 5069-5083.

Brenot, A., Cloquet, C., Vigier, N., Carignan, J., France-Lanord, C., 2008. Magnesium isotope systematics of the lithologically varied Moselle river basin, France. Geochimica et Cosmochimica Acta 72 (20), 5070-5089.

Brey, G.P., Kohler, T., 1990. Geothermobarometry in four-phase lherzolites II: New thermobarometers, and practical assessment of existing thermobarometers. Journal of Petrology 31 (6), 1353-1378.

Chakrabarti, R., Jacobsen, S.B., 2010. The isotopic composition of magnesium in the inner Solar System. Earth and Planetary Science Letters 293 (3-4), 349-358.

Chen, L.-H., Zeng, G., Jiang, S.-Y., Hofmann, A.W., Xu, X.-S., Pan, M.-B., 2009. Sources of Anfengshan basalts: subducted lower crust in the Sulu UHP belt, China. Earth and Planetary Science Letters 286 (3-4), 426-435.

Dasgupta, R., Hirschmann, M.M., 2010. The deep carbon cycle and melting in Earth's interior. Earth and Planetary Science Letters 298 (1-2), 1-13.

Dasgupta, R., Hirschmann, M.M., Withers, A.C., 2004. Deep global cycling of carbon constrained by the solidus of anhydrous, carbonated eclogite under upper mantle conditions. Earth and Planetary Science Letters 227, 73-85.

Dasgupta, R., Hirschmann, M.M., Smith, N.D., 2007. Partial melting experiments of peridotite+CO_2 at 3 GPa and genesis of alkalic ocean island basalts. Journal of Petrology 48 (11), 2093-2124.

Galy, A., Bar-Matthews, M., Halicz, L., O'Nions, R.K., 2002. Mg isotopic composition of carbonate: insight from speleothem formation. Earth and Planetary Science Letters 201 (1), 105-115.

Galy, A., Yoffe, O., Janney, P.E., Williams, R.W., Cloquet, C., Alard, O., Halicz, L., Wadhwa, M., Hutcheon, I.D., Ramon, E., Carignan, J., 2003. Magnesium isotope heterogeneity of the isotopic standard SRM980 and new reference materials for magnesiumisotope-ratio measurements. Journal of Analytical Atomic Spectrometry 18, 1352-1356.

Guo, F., Fan, W.M., Wang, Y.J., Lin, G., 2001. Late Mesozoic mafic intrusive complexes in North China Block: constraints on the nature of subcontinental lithospheric mantle. Physics and Chemistry of the Earth (A) 26, 759-771.

Guo, F., Fan, W.M., Wang, Y.J., Lin, G., 2003. Geochemistry of late mesozoic mafic magmatism in west Shandong Province, eastern China: characterizing the lost lithospheric mantle beneath the North China Block. Geochemical Journal 37 (1), 63-77.

Handler, M.R., Baker, J.A., Schiller, M., Bennett, V.C., Yaxley, G.M., 2009. Magnesium stable isotope composition of Earth's upper mantle. Earth and Planetary Science Letters 282 (1-4), 306-313.

Higgins, J.A., Schrag, D.P., 2010. Constraining magnesium cycling in marine sediments using magnesium isotopes. Geochimica et Cosmochimica Acta 74 (17), 5039-5053.

Hippler, D., Buhl, D., Witbaard, R., Richter, D., Immenhauser, A., 2009. Towards a better understanding of magnesium-isotope ratios from marine skeletal carbonates. Geochimica et Cosmochimica Acta 73 (20), 6134-6146.

Hofmann, A., 1997. Mantle geochemistry: the message from oceanic volcanism. Nature 385 (6613), 219-229.

Hofmann, A.W., 2007. Geochemistry: the lost continents. Nature 448 (7154), 655-656.

Huang, F., Li, S., Dong, F., Li, Q., Chen, F., Wang, Y., Yang, W., 2007. Recycling of deeply subducted continental crust in the Dabie Mountains, central China. Lithos 96 (1-2), 151-169.

Huang, F., Lundstrom, C.C., Glessner, J., Ianno, A., Boudreau, A., Li, J., Ferre, E.C., Marshak, S., DeFrates, J., 2009. Chemical and isotopic fractionation of wet andesite in a temperature gradient: experiments and models suggesting a new mechanism of magma differentiation. Geochimica et Cosmochimica Acta 73 (3), 729-749.

Huang, F., Chakraborty, P., Lundstrom, C.C., Holmden, C., Glessner, J.J.G., Kieffer, S.W., Lesher, C.E., 2010. Isotope fractionation in silicate melts by thermal diffusion. Nature 464 (7287), 396-400.

Huang, S., Farkaš, J., Jacobsen, S.B., 2011. Stable calcium isotopic compositions of Hawaiian shield lavas: evidence for recycling of ancient marine carbonates into the mantle. Geochimica et Cosmochimica Acta 75 (17), 4987-4997.

Jackson, M.G., Hart, S.R., Koppers, A.A.P., Staudigel, H., Konter, J., Blusztajn, J., Kurz, M., Russell, J.A., 2007. The return of subducted continental crust in Samoan lavas. Nature 448 (7154), 684-687.

Jahn, B.M., Wu, F.Y., Lo, C.H., Tsai, C.H., 1999. Crust-mantle interaction induced by deep subduction of the continental crust: geochemical and Sr-Nd isotopic evidence from post-collisional mafic-ultramafic intrusions of the northern Dabie complex, central China. Chemical Geology 157 (1-2), 119-146.

Kerrick, D.M., Connolly, J.A.D., 2001. Metamorphic devolatilization of subducted marine sediments and the transport of volatiles into the Earth's mantle. Nature 411 (6835), 293-296.

Kogiso, T., Tatsumi, Y., Nakano, S., 1997. Trace element transport during dehydration processes in the subducted oceanic crust: 1. Experiments and implications for the origin of ocean island basalts. Earth and Planetary Science Letters 148 (1-2), 193-205.

Li, Z.X., Li, X.H., 2007. Formation of the 1300-km-wide intracontinental orogen and postorogenic magmatic province in Mesozoic South China: a flat-slab subduction model. Geology 35 (2), 179-182.

Li, W.Y., Teng, F.Z., Ke, S., Rudnick, R.L., Gao, S., Wu, F.Y., Chappell, B.W., 2010. Heterogeneous magnesium isotopic composition of the upper continental crust. Geochimica et Cosmochimica Acta 74 (23), 6867-6884.

Ling, M.X., Wang, F.Y., Ding, X., Hu, Y.H., Zhou, J.B., Zartman, R.E., Yang, X.Y., Sun, W., 2009. Cretaceous ridge subduction along the lower Yangtze River belt, eastern China. Economic Geology 104 (2), 303.

Ling, M.-X., Sedaghatpour, F., Teng, F.-Z., Hays, P.D., Strauss, J., Sun, W., 2011. Homogeneous magnesium isotopic composition of seawater: an excellent geostandard for Mg isotope analysis. Rapid Communications in Mass Spectrometry 25 (19), 2828-2836.

Liu, D.Y., Nutman, A.P., Compston, W., Wu, J.S., Shen, Q.H., 1992. Remnants of >=3800 Ma crust in the Chinese part of the Sino-Korean craton. Geology 20 (4), 339.

Liu, Y., Gao, S., Kelemen, P.B., Xu, W., 2008. Recycled crust controls contrasting source compositions of Mesozoic and Cenozoic basalts in the North China Craton. Geochimica et Cosmochimica Acta 72 (9), 2349-2376.

Liu, S.-A., Teng, F.-Z., He, Y., Ke, S., Li, S., 2010. Investigation of magnesium isotope fractionation during granite differentiation: implication for Mg isotopic composition of the continental crust. Earth and Planetary Science Letters 297 (3-4), 646-654.

Liu, S.-A., Teng, F.-Z., Yang, W., Wu, F.-Y., 2011. High-temperature inter-mineral magnesium isotope fractionation in mantle xenoliths from the North China craton. Earth and Planetary Science Letters 308 (1-2), 131-140.

McDonough, W.F., Sun, S.S., 1995. The composition of the earth. Chemical Geology 120, 223-253.

Molina, J.F., Poli, S., 2000. Carbonate stability and fluid composition in subducted oceanic crust: an experimental study on H2O-CO2-bearing basalts. Earth and Planetary Science Letters 176 (3-4), 295-310.

Peng, Z.C., Zartman, R.E., Futa, K., Chen, D.G., 1986. Pb-, Sr- and Nd- isotopic systematics and chemical characteristic of Cenozoic basalts, eastern China. Chemical Geology 59, 3-33.

Plank, T., Langmuir, C.H., 1998. The chemical composition of subducting sediment and its consequences for the crust and mantle. Chemical Geology 145 (3-4), 325-394.

Pogge von Strandmann, P.A.E., 2008. Precise magnesium isotope measurements in core top planktic and benthic foraminifera. Geochemistry, Geophysics, Geosystems 9, Q12015. http://dx.doi.org/10.1029/2008GC002209.

Pogge von Strandmann, P.A.E., James, R.H., van Calsteren, P., Gislason, S.R., Burton, K.W., 2008. Lithium, magnesium and uranium isotope behaviour in the estuarine environment of basaltic islands. Earth and Planetary Science Letters 274 (3-4), 462-471.

Pogge von Strandmann, P.A.E., Elliott, T., Marschall, H.R., Coath, C., Lai, Y.-J., Jeffcoate, A.B., Ionov, D.A., 2011. Variations of Li and Mg isotope ratios in bulk chondrites and mantle xenoliths. Geochimica et Cosmochimica Acta 75 (18), 5247-5268.

Qiu, J.S., Xu, X.S., Lo, C.H., 2002. Potash-rich volcanic rocks and lamprophyres in western Shandong province: 40Ar-39Ar dating and source tracing from Luxi area. Chinese Science Bulletin 47, 91-99.

Rapp, R.P., Shimizu, N., Norman, M.D., Applegate, G.S., 1999. Reaction between slabderived melts and peridotite in the mantle wedge: experimental constraints at 3.8 GPa. Chemical Geology 160 (4), 335-356.

Richter, F.M., Watson, E.B., Mendybaev, R.A., Teng, F.-Z., Janney, P.E., 2008. Magnesium isotope fractionation in silicate melts by chemical and thermal diffusion. Geochimica et Cosmochimica Acta 72, 206-220.

Roeder, P.L., Emslie, R.F., 1970. Olivine-liquid equilibria. Contributions to Mineralogy and Petrology 29, 275-289.

Rudnick, R.L., Gao, S., 2003. Composition of the continental crust. In: Rudnick, R.L. (Ed.), The Crust. Treatise on Geochemistry. Elsevier-Pergamon, Oxford, pp. 1-64.

Shaw, C.S.J., Dingwell, D.B., 2008. Experimental peridotite-melt reaction at one atmosphere: a textural and chemical study. Contributions to Mineralogy and Petrology 155 (2), 199-214.

Shen, B., Jacobsen, B., Lee, C.T.A., Yin, Q.Z., Morton, D.M., 2009. The Mg isotopic systematics of granitoids in continental arcs and implications for the role of chemical weathering in crust formation. Proceedings of the National Academy of Sciences 106 (49), 20652-20657.

Sobolev, A.V., Hofmann, A.W., Sobolev, S.V., Nikogosian, I.K., 2005. An olivine-free mantle source of Hawaiian shield basalts. Nature 434 (7033), 590-597.

Song, Y., 1990. Isotopic characteristics of Hannuoba basalts, eastern China: implications for their petrogenesis and the composition of subcontinental mantle. Chemical Geology 88, 35-52.

Sun, S., McDonough, W., 1989. Chemical and isotopic systematics of oceanic basalts: implications for mantle composition and processes. Geological Society of London. Special Publication 42 (1), 313.

Sun, W., Ding, X., Hu, Y.-H., Li, X.-H., 2007. The golden transformation of the Cretaceous plate subduction in the west Pacific.

Earth and Planetary Science Letters 262 (3-4), 533-542.

Tang, Y.-J., Zhang, H.-F., Ying, J.-F., 2006. Asthenosphere-lithospheric mantle interaction in an extensional regime: Implication from the geochemistry of Cenozoic basalts from Taihang Mountains, North China Craton. Chemical Geology 233 (3-4), 309-327.

Tang, Y.-J., Zhang, H.-F., Nakamura, E., Ying, J.-F., 2010. Multistage melt/fluid-peridotite interactions in the refertilized lithospheric mantle beneath the North China Craton: constraints from the Li-Sr-Nd isotopic disequilibrium between minerals of peridotite xenoliths. Contributions to Mineralogy and Petrology 1-17.

Teng, F.-Z., Wadhwa, M., Helz, R.T., 2007. Investigation of magnesium isotope fractionation during basalt differentiation: implications for a chondritic composition of the terrestrial mantle. Earth and Planetary Science Letters 261 (1-2), 84-92.

Teng, F.-Z., Li, W.-Y., Ke, S., Marty, B., Dauphas, N., Huang, S., Wu, F.-Y., Pourmand, A., 2010a. Magnesium isotopic composition of the Earth and chondrites. Geochimica et Cosmochimica Acta 74 (14), 4150-4166.

Teng, F.-Z., Li, W.Y., Rudnick, R.L., Gardner, L.R., 2010b. Contrasting lithium and magnesium isotope fractionation during continental weathering. Earth and Planetary Science Letters 300 (1-2), 63-71.

Tipper, E.T., Galy, A., Bickle, M.J., 2006a. Riverine evidence for a fractionated reservoir of Ca and Mg on the continents: implications for the oceanic Ca cycle. Earth and Planetary Science Letters 247 (3-4), 267-279.

Tipper, E.T., Galy, A., Gaillardet, J., Bickle, M.J., Elderfield, H., Carder, E.A., 2006b. The magnesium isotope budget of the modern ocean: constraints from riverine magnesium isotope ratios. Earth and Planetary Science Letters 250 (1-2), 241-253.

Tipper, E., Galy, A., Bickle, M., 2008. Calcium and magnesium isotope systematics in rivers draining the Himalaya-Tibetan-Plateau region: lithological or fractionation control? Geochimica et Cosmochimica Acta 72 (4), 1057-1075.

Tipper, E.T., Gaillardet, J., Louvat, P., Capmas, F., White, A.F., 2010. Mg isotope constraints on soil pore-fluid chemistry: evidence from Santa Cruz, California. Geochimica et Cosmochimica Acta 74 (14), 3883-3896.

Wang, Y., Zhao, Z.-F., Zheng, Y.-F., Zhang, J.-J., 2011. Geochemical constraints on the nature of mantle source for Cenozoic continental basalts in east-central China. Lithos 125 (3-4), 940-955.

Wu, F.-Y., Walker, R.J., Yang, Y.-H., Yuan, H.-L., Yang, J.-H., 2006. The chemical-temporal evolution of lithospheric mantle underlying the North China Craton. Geochimica et Cosmochimica Acta 70 (19), 5013-5034.

Xiao, Y., Zhang, H.-F., Fan, W.-M., Ying, J.-F., Zhang, J., Zhao, X.-M., Su, B.-X., 2010. Evolution of lithospheric mantle beneath the Tan-Lu fault zone, eastern North China Craton: evidence from petrology and geochemistry of peridotite xenoliths. Lithos 117 (1-4), 229-246.

Xu, Y.-G., Ma, J.-L., Frey, F.A., Feigenson, M.D., Liu, J.-F., 2005. Role of lithosphereasthenosphere interaction in the genesis of Quaternary alkali and tholeiitic basalts from Datong, western North China Craton. Chemical Geology 224 (4), 247-271.

Xu, Y.G., Yu, S.Y., Zheng, Y.F., 2010. Evidence from pyroxenite xenoliths for subducted lower oceanic crust in subcontinental lithospheric mantle. Geochimica et Cosmochimica Acta 74 (12, supplement 1), A1164.

Yang, W., Li, S.G., 2008. Geochronology and geochemistry of the Mesozoic volcanic rocks in Western Liaoning: implications for lithospheric thinning of the North China Craton. Lithos 102 (1-2), 88-117.

Yang, W., Teng, F.-Z., Zhang, H.-F., 2009. Chondritic magnesium isotopic composition of the terrestrial mantle: a case study of peridotite xenoliths from the North China craton. Earth and Planetary Science Letters 288 (3-4), 475-482.

Yaxley, G.M., Green, D.H., 1994. Experimental demonstration of refractory carbonatebearing eclogite and siliceous melt in the subduction regime. Earth and Planetary Science Letters 128 (3-4), 313-325.

Young, E.D., Galy, A., 2004. The isotope geochemistry and cosmochemistry of magnesium. In: Hohnson, C.M., Beard, B.L., Albarede, F. (Eds.), Geochemistry of Nontraditional Stable Isotopes, Reviews in Mineralogy & Geochemistry. Mineralogical Society of America, Washington, pp. 197-230.

Young, E.D., Tonui, E., Manning, C.E., Schauble, E., Macris, C.A., 2009. Spinel-olivine magnesium isotope thermometry in the mantle and implications for the Mg isotopic composition of Earth. Earth and Planetary Science Letters 288 (3-4), 524-533.

Zeng, G., Chen, L.-H., Xu, X.-S., Jiang, S.-Y., Hofmann, A.W., 2010. Carbonated mantle sources for Cenozoic intra-plate alkaline basalts in Shandong, North China. Chemical Geology 273 (1-2), 35-45.

Zeng, G., Chen, L.-H., Hofmann, A.W., Jiang, S.-Y., Xu, X.-S., 2011. Crust recycling in the sources of two parallel volcanic chains in Shandong, North China. Earth and Planetary Science Letters 302 (3-4), 359-368.

Zhang, H., Deloule, E., Tang, Y., Ying, J., 2010. Melt/rock interaction in remains of refertilized Archean lithospheric mantle in Jiaodong Peninsula, North China Craton: Li isotopic evidence. Contributions to Mineralogy and Petrology 1-17.

Zhang, H.-F., 2007. Temporal and spatial distribution of Mesozoic mafic magmatism in the North China Craton and implications for secular lithospheric evolution. Geological Society of London. Special Publication 280, 35-54.

Zhang, H.F., Sun, M., Zhou, X.H., Fan, W.M., Zhai, M.G., Ying, J.F., 2002. Mesozoic lithosphere destruction beneath the North China Craton: evidence from major-, trace-element and Sr-Nd-Pb isotope studies of Fangcheng basalts. Contributions to Mineralogy and Petrology 144, 241-253.

Zhang, H.F., Sun, M., Zhou, X.H., Zhou, M.F., Fan, W.M., Zheng, J.P., 2003. Secular evolution of the lithosphere beneath the eastern North China Craton: evidence from Mesozoic basalts and high-Mg andesites. Geochimica et Cosmochimica Acta 67, 4373-4387.

Zhang, J.-J., Zheng, Y.-F., Zhao, Z.-F., 2009. Geochemical evidence for interaction between oceanic crust and lithospheric mantle in the origin of Cenozoic continental basalts in east-central China. Lithos 110 (1-4), 305-326.

Zhao, D.P., Ohtani, E., 2009. Deep slab subduction and dehydration and their geodynamic consequences: evidence from seismology and mineral physics. Gondwana Research 16 (3-4), 401-413.

Zhao, G., Wilde, S.A., Sun, M., Li, S., Li, X., Zhang, J., 2008. SHRIMP U-Pb zircon ages of granitoid rocks in the Lüliang Complex: implications for the accretion and evolution of the Trans-North China Orogen. Precambrian Research 160 (3-4), 213-226.

Zheng, J.P., Lu, F.X., 1999. Petrologic characteristics of kimberlite-borne mantle xenoliths from the Shandong and Liaoning Peninsula: paleozoic lithosphere mantle and its heterogeneity. Acta Petrologica Sinica 15 (1), 65-74.

Zhi, X., Song, Y., Frey, F.A., Feng, J., Zhai, M., 1990. Geochemistry of Hannuoba basalts, eastern China: Constraints on the origin of continental alkalic and tholeiitic basalt. Chemical Geology 88 (1-2), 1-33.

Zou, H.B., Zindler, A., Xu, X.S., Qi, Q., 2000. Major, trace element, and Nd, Sr and Pb isotope studies of Cenozoic basalts in SE China: mantle sources, regional variations, and tectonic significance. Chemical Geology 171, 33-47.

Magnesium isotopic variations in cratonic eclogites: Origins and implications*

Shui-Jiong Wang[1,2], Fang-Zhen Teng[2], Helen M. Williams[3] and Shu-Guang Li[1,4]

1. State Key Laboratory of Geological Processes and Mineral Resources, China University of Geosciences, Beijing 100083, China
2. Isotope Laboratory, Department of Geosciences and Arkansas Center for Space and Planetary Sciences, University of Arkansas, Fayetteville, AR 72701, USA
3. Department of Earth Sciences, Durham University, Science Labs, Durham, DH1 3LE, UK
4. CAS Key Laboratory of Crust-Mantle Materials and Environments, School of Earth and Space Sciences, University of Science and Technology of China, Hefei 230026, Anhui, China

亮点介绍：本文报道了来自南非卡瓦勒和贝尔斯班克金伯利岩管道的克拉通榴辉岩的全岩及单矿物 Mg 同位素组成(δ^{26}Mg)。分析结果表明克拉通榴辉岩的 Mg 同位素组成变化较大，且显著低于全球地幔橄榄岩 Mg 同位素组成，为克拉通榴辉岩源自于俯冲的蚀变洋壳的理论提供了进一步的证据。

Abstract Cratonic eclogites play an important role in the formation and dynamic evolution of the subcontinental lithospheric mantle. However, their origin, whether as fragments of subducted oceanic crust or high-pressure mantle cumulates, remains controversial. Here, we report Mg isotopic compositions (δ^{26}Mg) for cratonic eclogites from Kaalvallei and Bellsbank kimberlite pipes, South Africa. We find that clinopyroxene is 0.375 ± 0.069 to 0.676 ± 0.075 ‰ heavier than coexisting garnet, which reflects equilibrium isotope fractionation between these phases, primarily driven by the difference in Mg coordination between clinopyroxene and garnet. Bulk eclogites have strikingly variable Mg isotopic compositions, which range from -0.797 ± 0.075 to -0.139 ± 0.061‰, values that are significantly lighter than the range displayed by global mantle peridotites to date (-0.25 ± 0.07‰, 2SD). As significant Mg isotope fractionation is only known to occur during low-temperature water-rock interaction, our results provide further evidence for the derivation of cratonic eclogites from subducted altered oceanic crust. In addition, the lack of correlation between Δ^{26}Mg and Δ^{57}Fe provides evidence for redox control on equilibrium inter-mineral Fe isotope fractionation.

Keywords Magnesium isotopes, iron isotopes, isotope fractionation, cratonic eclogite, mantle

* 本文发表在：Earth and Planetary Science Letters, 2012, 359-360: 219-226

1 Introduction

Cratonic eclogites hosted by kimberlite are samples of relict subcontinental lithospheric mantle preserved beneath ancient continental cratons. As such, they have the potential to provide fundamental information on the genesis of continental cratons and the recycling and preservation of subducted oceanic crust into the deep mantle. Their origin, however, is still a subject of controversy (Smyth et al., 1989; Caporuscio and Smyth, 1990; Ireland et al., 1994; Jacob et al., 1994, Jacob, 2004; Schulze et al., 1997; Barth et al., 2001, 2002; Griffin and O'Reilly, 2007; Gréau et al., 2011). The "mantle hypothesis" interprets cratonic eclogites as products of high-pressure cumulates from mantle melts based on cumulate textures, mineralogical layering, and exsolution of garnet from pyroxene in cratonic eclogites (Smyth et al., 1989; Caporuscio and Smyth, 1990). By contrast, the "crustal hypothesis" states that cratonic eclogites represent products of ancient subducted oceanic crust (either basaltic melts or cumulates) and their protoliths have experienced surface processes such as water-rock interaction and/or hydrothermal alteration prior to subduction and stacking beneath pre-existing continental lithosphere (Ireland et al., 1994; Jacob et al., 1994; Barth et al., 2001, 2002). The most compelling evidence in support of the "crustal hypothesis" comes from studies of oxygen (O) and carbon (C) isotopes. The large variations in O isotopic composition in many cratonic eclogites have been interpreted to inherit from altered oceanic crust, where O isotopes deviated from those of the Earth's mantle are produced through seawater-rock interactions (Jacob et al., 1994) and similar arguments apply to C isotopes (Schulze et al., 1997). Recent studies show, however, that partial melting and fluid-solid reaction processes could also produce the large O isotopic variations in cratonic eclogites (Griffin and O'Reilly, 2007; Williams et al., 2009), and C isotopes can be as well fractionated by different degrees of oxidation at mantle environment (Cartigny et al., 1998). Accordingly, new tracers are needed in order to resolve the debate on the origin of cratonic eclogites.

Magnesium (Mg) stable isotopes have strong potential as such a tracer. Magnesium has only one oxidation state hence Mg isotopes do not fractionate during redox-related processes. High-temperature fractionation of Mg isotopes is generally limited, as peridotites and unaltered oceanic basalts have quite homogenous Mg isotopic compositions (average δ^{26}Mg=−0.25±0.07 ‰, 2SD; Teng et al., 2007, 2010a; Handler et al., 2009; Yang et al., 2009; Bourdon et al., 2010; Pogge von Strandmann et al., 2011). On the contrary, Mg isotopes are highly fractionated during low-temperature surface processes (Fig.1). For example, both continental and oceanic crustal rocks have highly heterogeneous Mg isotope compositions (Fig.1; Shen et al., 2009; Higgins and Schrag, 2010; Li et al., 2010; Liu et al., 2010; Wombacher et al., 2011), due to various degrees of water-rock interaction (Tipper et al., 2006a; Pogge von Strandmann et al., 2008; Teng et al., 2010b; Wimpenny et al., 2010). On these grounds, cratonic eclogites are expected to have mantle-like Mg isotopic compositions if the "mantle hypothesis" holds true, whereas the opposite is true should the "crustal hypothesis" account for the origin of cratonic eclogites.

Specific conditions (e.g., diffusion) can, however, generate large Mg isotopic variations in igneous rocks (Richter et al., 2008; Huang et al., 2009, 2010; Dauphas et al., 2010; Pogge von Strandmann et al., 2011; Teng et al., 2011). For example, diffusion-related process has produced isotope fractionations up to 0.4 ‰ in Mg and 1.6‰ in Fe from Hawaii olivine (Teng et al., 2011). Significant high-temperature equilibrium inter-mineral Mg isotope fractionations can also take place between coexisting minerals (e.g., between garnet and omphacite in eclogite; spinel and olivine in peridotite; Young et al., 2009; Li et al., 2011; Liu et al., 2011). These effects have been predominantly attributed to the distinct bonding environment of Mg in these minerals (coordination number of Mg is 8 and 4 in garnet and spinel, respectively compared to 6 in pyroxene and olivine), as lower Mg coordination with stronger Mg-O bonds prefers the heavier Mg isotopes (Bigeleisen and Mayer, 1947; Urey, 1947).

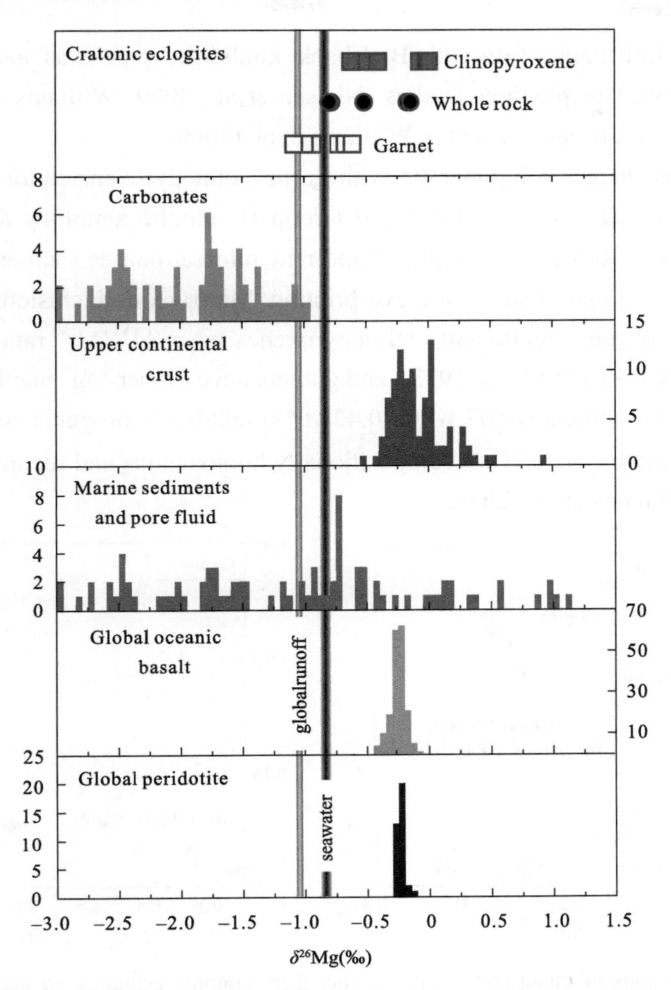

Fig. 1 Magnesium isotopic compositions of major reservoirs: carbonates (Young and Galy., 2004; Tipper et al., 2006a; Brenot et al., 2008; Higgins and Schrag., 2010; Jacobson et al., 2010), upper continental crust (Li et al., 2010), marine sediment & pore fluid (Higgins and Schrag., 2010; Wombacher et al., 2011), global oceanic basalt (Teng et al., 2007, 2010a; Bourdon et al., 2010), global peridotite (Bourdon et al., 2010; Teng et al., 2010a;), seawater and global runoff (Tipper et al., 2006b). δ^{26}Mg values for carbonates and marine sediments can be as low as −5.29‰ and −5.57‰, respectively.

Here, we present Mg isotopic data for a set of well-characterized cratonic eclogites from Kaalvallei and Bellsbank kimberlite pipes in Kaapvaal craton, South Africa (Nielsen et al., 2009; Williams et al., 2009). Our results yield up to 0.676±0.075 ‰ Mg isotope fractionation (calculated as $\delta^{26}Mg_{clinopyroxene}-\delta^{26}Mg_{garnet}$; errors are propagated using the standard sum-of-squares approach) between coexisting clinopyroxene and garnet. The bulk eclogites have Mg isotopic compositions (recalculated from mineral separate data) ranging from −0.797 ± 0.075 to −0.139 ± 0.061 ‰, values that are significantly lighter than the range displayed by global mantle peridotites, hence supporting the "crustal hypothesis" for the origin of cratonic eclogites. Decoupled Mg and Fe isotope variations suggest that the observed Fe isotope fractionation is highly redox-controlled.

2 Geological setting and samples

The Cretaceous Kaalvallei and Bellsbank kimberlite pipes are located on the Kaapvaal craton, South Africa. Five cratonic eclogite samples (382, 402, 423, Kaalvallei-A, and 375) were collected from concentrates at the Kaalvallei kimberlite pipe. They are from the same suite as samples studied by Viljoen (2005). Additionally one cratonic eclogite sample (Bellsbank) from the Bellsbank kimberlite pipe was analysed. These six cratonic eclogites have been the subject of previous studies (Nielsen et al., 2009; Williams et al., 2009) and a detailed description of these samples has been provided in Williams et al. (2009).

All the samples studied here are bimineralic with garnet: clinopyroxene ratios ranging from 0.20:0.80 to 0.65:0.35 and are texturally identical to Group I and Group II eclogite xenoliths described in Roberts Victor (Macgregor and Carter, 1970). Replacement textures and new mineral phases such as amphibole and phlogopite are absent, indicating limited modal metasomatic overprinting. Minor oxide inclusions are occasionally observed in both garnet and clinopyroxene. Significantly, clinopyroxenes have Al^{VI}/Al^{IV} ratios >2 (Fig.2), which imply high-pressure equilibration (Aoki and Shiba, 1973), and garnets have higher $Mg^{\#}$ numbers [0.62–0.83, $Mg^{\#}$=molar MgO/(MgO+FeO)] and Cr_2O_3 contents (0.07 wt %–0.42wt %) relative to orogenic eclogitic garnets (Fig.2; Li et al., 2011). Both garnet and clinopyroxene are compositionally homogenous and record equilibration temperatures from 935 °C to 1401 °C (Williams et al., 2009).

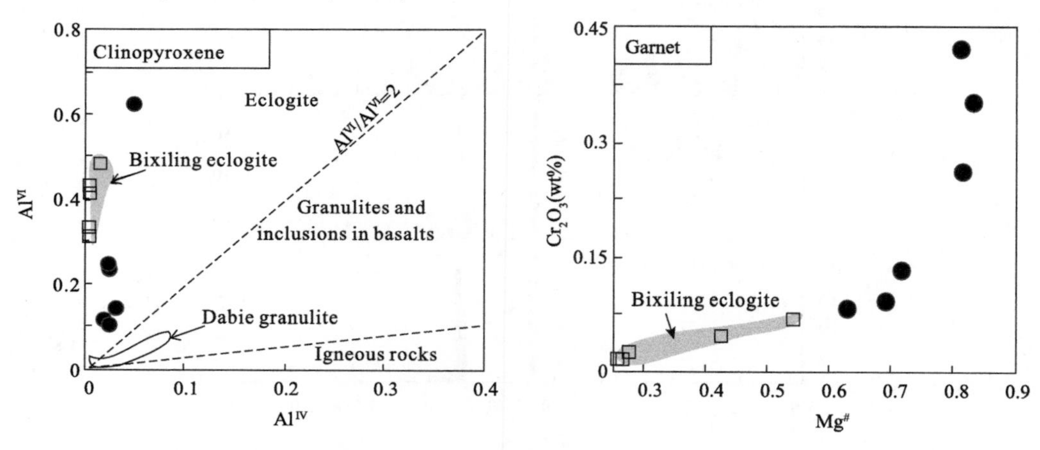

Fig. 2　Major element compositions of clinopyroxene and garnet from cratonic eclogites. In the Al^{VI} vs. Al^{IV} diagram, all the clinopyroxene from cratonic eclogites plot in the eclogite field defined by Al^{VI}/Al^{IV} >2. Fields and ratio are from Aoki and Shiba (1973). Data of Dabie granulites are from Wang et al. (2012); In the Cr_2O_3-$Mg^{\#}$ diagram, all garnets from cratonic eclogites are distinct from orogenic eclogite in higher Cr_2O_3 content and $Mg^{\#}$. Data of Bixiling orogenic eclogite are from Li et al. (2011)

Because these eclogite xenoliths are too small to provide representative samples for whole-rock chemical analysis, as well as the potential existence of contamination by kimberlite infiltration, bulk powder samples of the eclogites were not prepared and whole-rock compositions of these eclogites were instead calculated by combining the chemistry of both garnet and clinopyroxene according to the mineral abundance estimation (Williams et al., 2009). The re-constructed whole-rock compositions are characterized by higher MgO (8.46 wt %–17.9 wt %) and lower SiO_2 (45.9 wt %–52.1 wt %) contents compared to those of eclogites from orogenic massifs (Li et al., 2011) and broadly similar to those of gabbroic cumulates (Barth et al., 2002).

3　Analytical methods

Eclogite samples were provided as coarse gravels and were subsequently crushed in an agate mortar with an agate pestle to ~50 mesh size. Fresh garnet and clinopyroxene grains were handpicked under a binocular microscope with extra care taken to avoid any alteration products during separation. Before dissolution, 0.34–2.1 mg separated minerals were ultrasonicated for 3 times with each 10 minutes in Milli-Q water (18.2 MΩ cm) at room temperature. All chemical procedures including mineral dissolution and column chemistry were carried out in the clean laboratory at the Isotope Laboratory of the University of Arkansas, Fayetteville, following established procedures (Yang et al., 2009; Li et al., 2010; Teng et al., 2010a).

Minerals were digested in Savillex screw-top beakers in a 6:1 (v/v) mixture of Optima-grade HF and HNO_3 acids on a hotplate in an exhaust hood. After about one week, the sample solutions were dried at 120 °C and the dried residues were refluxed with a 1:3 (v/v) mixture of Optima-grade HNO_3 and HCl acid, followed by heating at 160 °C to dryness. Concentrated HNO_3 was then added at 160 °C to ensure complete dissolution. Finally, the solution was evaporated to dryness at 160 °C, and the dried residue was dissolved in 1 N HNO_3 for ion exchange column chemistry.

Chemical separation and purification of Mg were achieved by cation exchange chromatography with Bio-Rad 200–400 mesh AG50W-X8 resin in 1 N HNO_3 media (Teng et al., 2007). The same column procedure was performed twice in order to obtain pure Mg solutions for mass spectrometry. The total procedural blank is <10 ng, which represented <0.1 % of Mg loaded on the column (Teng et al., 2010a).

Magnesium isotopic compositions were analyzed by the sample-standard bracketing method using a *Nu Plasma* MC-ICPMS at low resolution mode, with ^{26}Mg, ^{25}Mg and ^{24}Mg measured simultaneously in separated Faraday cups (H5, Ax, and L4). No molecular interferences or double charge interferences were observed during analyses. The background Mg signals for the ^{24}Mg were $<10^{-4}$ V, which are negligible relative to sample signals of 3–4 V. Each batch of sample analysis contains at least 1 well-characterized standard. Sample solutions were re-measured >4 times per analytical session. The long-term precision was better than 0.07‰ (Teng et al., 2010a), based on replicate analyses of synthetic solution, mineral and rock standards. Magnesium isotope results are reported in the conventional δ notation in per mil relative to DSM-3 (Galy et al., 2003), $\delta^{x}Mg=[(^{x}Mg/^{24}Mg)_{sample}/(^{x}Mg/^{24}Mg)_{DSM\text{-}3}-1]\times1000$, where x refers to 25 or 26.

4 Results

Magnesium isotopic compositions of the mineral separates and whole rocks, together with the well-characterized international (seawater) and in-house (KH-olivine) standards are reported in Table 1. Different batches of garnets and clinopyroxenes from the same eclogite sample have identical Mg isotopic compositions within analytical uncertainty (Table 1).

Table 1 Magnesium isotopic composition of mineral separates and constructed whole-rock for cratonic eclogites and reference materials (KH-olivine and seawater)

Sample	Mineral/rock[a]	Mode[h]	Wt.(mg)	$\delta^{26}Mg$	2SD[g]	$\delta^{25}Mg$	2SD
375	Cpx	0.65	1.26	–0.622	0.061	–0.336	0.051
	Grt[1 b]	0.35	1.07	–1.014	0.051	–0.541	0.052
	Grt[2]		1.44	–0.973	0.082	–0.504	0.054
	Duplicate[c]			–0.990	0.051	–0.533	0.052
	Ave. Grt[d]			–0.997	0.033	–0.526	0.030
	Whole-rock[e]			–0.779	0.069	–0.416	0.059
382	Cpx[1]	0.8	0.93	0.001	0.051	–0.002	0.052
	Cpx[2]		2.1	–0.025	0.051	–0.025	0.052
	Ave. Cpx			–0.012	0.036	–0.014	0.037
	Grt[1]	0.2	1.57	–0.681	0.061	–0.328	0.039
	Grt[2]		0.34	–0.652	0.061	–0.340	0.039
	Ave. Grt			–0.666	0.043	–0.334	0.028
	Whole-rock			–0.164	0.056	–0.088	0.046

Continued

Sample	Mineral/rock[a]	Mode[h]	Wt.(mg)	δ^{26}Mg	2SD[g]	δ^{25}Mg	2SD
402	Cpx[1]	0.8	1.32	−0.015	0.061	−0.002	0.051
	Cpx[2]		0.72	0.040	0.061	0.010	0.051
	Ave. Cpx			0.013	0.043	0.004	0.036
	Grt[1]	0.2	0.9	−0.647	0.061	−0.338	0.051
	Grt[2]		0.63	−0.568	0.061	−0.306	0.051
	Ave. Grt			−0.608	0.043	−0.322	0.036
	Whole-rock			−0.139	0.061	−0.076	0.051
423	Cpx[1]	0.35	1.84	−0.101	0.061	−0.030	0.051
	Cpx[2]		0.62	−0.044	0.051	−0.055	0.052
	Ave. Cpx			−0.068	0.039	−0.042	0.036
	Grt	0.65	1.64	−0.723	0.061	−0.377	0.051
	Whole-rock			−0.520	0.072	−0.273	0.062
Bellsbank	Cpx	0.55	1.4	−0.407	0.061	−0.206	0.039
	Grt[1]	0.45	0.78	−1.091	0.061	−0.555	0.051
	Grt[2]		0.86	−1.075	0.061	−0.565	0.051
	Ave. Grt			−1.083	0.043	−0.560	0.036
	Whole-rock			−0.798	0.075	−0.411	0.053
Kaalvallei-A	Cpx[1]	0.4	0.5	−0.565	0.082	−0.259	0.054
	Duplicate			−0.523	0.051	−0.280	0.052
	Cpx[2]		0.34	−0.579	0.082	−0.285	0.054
	Duplicate			−0.511	0.051	−0.279	0.052
	Ave. Cpx			−0.533	0.031	−0.276	0.026
	Grt[1]	0.6	1.27	−0.920	0.082	−0.450	0.054
	Duplicate			−0.939	0.051	−0.484	0.052
	Grt[2]		1.79	−0.962	0.082	−0.461	0.054
	Duplicate			−0.911	0.051	−0.488	0.052
	Ave. Grt			−0.929	0.031	−0.471	0.026
	Whole-rock			−0.794	0.044	−0.404	0.037
KH-Olivine	Ol			−0.276	0.061	−0.151	0.039
	Replicate[f]			−0.274	0.043	−0.124	0.039
	Replicate			−0.236	0.043	−0.150	0.039
	Replicate			−0.251	0.075	−0.135	0.049
	Replicate			−0.235	0.055	−0.116	0.050
	Replicate			−0.237	0.052	−0.128	0.040
	Replicate			−0.250	0.054	−0.133	0.047
	Replicate			−0.227	0.050	−0.111	0.049
	Replicate			−0.248	0.063	−0.142	0.054
	Replicate			−0.270	0.043	−0.106	0.039
	Replicate			−0.289	0.061	−0.155	0.051
	Ave. Ol			−0.253	0.016	−0.132	0.013
Seawater	Seawater			−0.879	0.082	−0.431	0.054
	Replicate			−0.827	0.051	−0.413	0.052

Continued

Sample	Mineral/rock[a]	Mode[h]	Wt.(mg)	δ^{26}Mg	2SD[g]	δ^{25}Mg	2SD
	Replicate			−0.825	0.063	−0.412	0.035
	Replicate			−0.849	0.056	−0.415	0.052
Seawater	Replicate			−0.861	0.061	−0.418	0.056
	Replicate			−0.842	0.043	−0.410	0.039
	Ave. Seawater			−0.844	0.023	−0.415	0.019

[a] Grt=garnet; Cpx=clinopyroxene; Ol=olivine.

[b] The number on the upper-right corner denotes different batches of mineral grains from the same eclogite sample.

[c] Duplicate: repeated measurement of Mg isotopic ratios on the same solution.

[d] Ave.=weighted average value.

[e] Whole-rock Mg isotopic compositions that are calculated based on the mineral data. Variations in the modal abundance of garnet and clinopyroxene by ±10 % can shift the whole-rock δ^{26}Mg values within ±0.07 ‰, which is still within the analytical uncertainty.

[f] Replicate: repeat sample dissolution, column chemistry and instrumental analysis.

[g] 2SD=2 times the standard deviation of the population of (n >20) repeat measurements of the standards during an analytical session.

[h] Mineral modal abundances are taken from Williams et al. (2009).

The δ^{26}Mg values span a considerable range from −1.083 to −0.608‰ in garnet and from −0.622 to +0.013 ‰ in clinopyroxene (Fig.3). Clinopyroxene is systematically heavier than coexisting garnet, with $\Delta^{26}Mg_{clinopyroxene-garnet}$ ($=\delta^{26}Mg_{clinopyroxene} - \delta^{26}Mg_{garnet}$) ranging from 0.375 ± 0.069 to 0.676 ± 0.075‰. The δ^{26}Mg values of whole-rock eclogites, calculated based on δ^{26}Mg values and modal abundance of clinopyroxene and garnet, vary from −0.798 ± 0.075 to −0.139 ± 0.061‰ (Fig.3, Table.1), of which 4 exhibit significantly lighter Mg isotopic compositions than the mantle value (Fig.3; Teng et al., 2010a). The constructed whole-rock δ^{26}Mg values are not susceptible to variation of the mineral modes, as changes in the modal abundance of garnet and clinopyroxene by ± 10% can only shift the whole-rock δ^{26}Mg values within ± 0.07‰, which is still within analytical uncertainty.

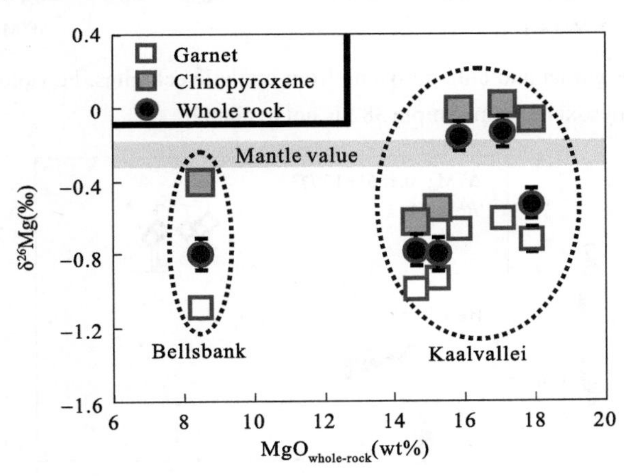

Fig. 3 Variations of δ^{26}Mg values as a function of whole-rock MgO contents. The horizontal yellow band represents the Mg isotopic composition of the mantle (−0.25 ± 0.07‰, 2SD, Teng et al., 2010a).

5 Discussion

5.1 Equilibrium inter-mineral Mg isotope fractionation

Both equilibrium (Young et al., 2009; Li et al., 2011; Liu et al., 2011) and chemical diffusion-driven kinetic (Richter et al., 2008; Huang et al., 2009; Dauphas et al., 2010; Teng et al., 2011) Mg isotope fractionation could occur at high temperature and potentially produce large Mg isotope fractionations between clinopyroxene and garnet. Kinetic isotope fractionation during inter-diffusion of Mg and Fe should generate a negative linear correlation between $\delta^{26}Mg$ and $\delta^{56}Fe$ (Dauphas et al., 2010; Teng et al., 2011). No such correlations were observed (Fig.4). Furthermore, no elemental variations exist within single mineral grain or among different grains from an individual eclogite sample (Williams et al., 2009). Analysis of different mineral fragments from the same sample yield identical Mg isotope compositions (Table 1), suggesting that there is minimal intra-mineral Mg isotopic variation at the mineral scale (<0.2 mm) and implying equilibrium Mg isotope fractionation. The final support for equilibrium inter-mineral Mg isotope fractionation comes from the linear correlation between $\Delta^{26}Mg_{clinopyroxene\text{-}garnet}$ and $1/T^2$ (Fig. 5). An equilibrium fractionation line is defined on $\Delta^{26}Mg$-$1/T^2$ diagram for all cratonic eclogites studied here and orogenic eclogites that formed at lower temperatures from Dabie Mountain, China (Li et al., 2011), with the following equilibrium fractionation equation $\Delta^{26}Mg_{clinopyroxene\text{-}garnet}=0.86\times10^6/T^2$ (R^2=0.84, Fig.5). This equilibrium inter-mineral Mg isotope fractionation may reflect the distinct bonding environment of Mg in garnet [coordination number (CN) of Mg is 8] compared to clinopyroxene (CN of Mg is 6), where stronger Mg-O bonds in lower Mg coordination sites prefer heavier Mg isotopes to lighter ones (Li et al., 2011).

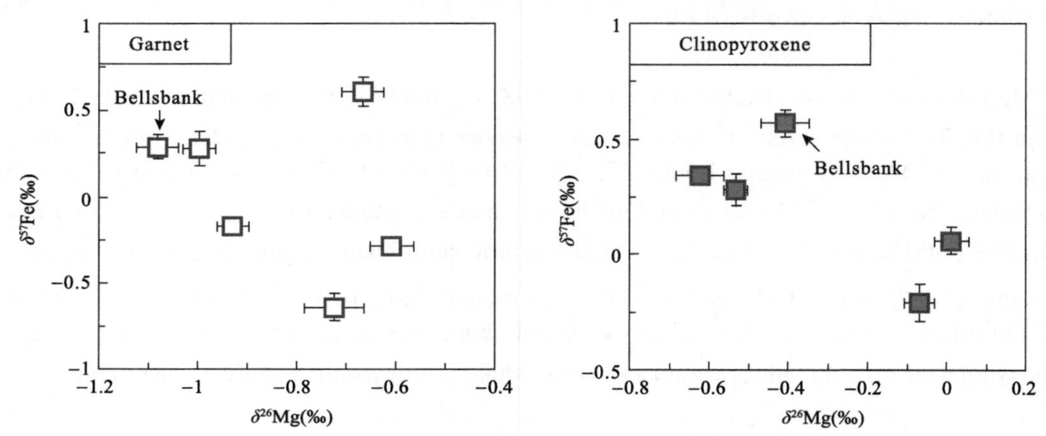

Fig. 4　$\delta^{26}Mg$ vs. $\delta^{57}Fe$ diagrams for garnet and clinopyroxene from cratonic eclogites. Fe isotopic compositions are from Williams et al. (2009). $\delta^{57}Fe$ value of the clinopyroxene from sample 382 is not available.

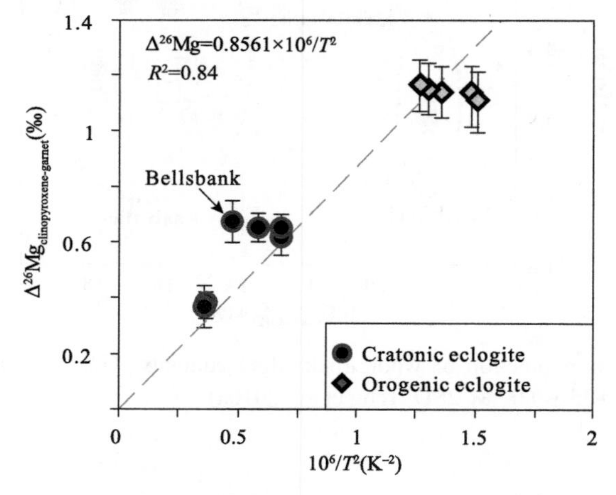

Fig. 5　Equilibrium Mg isotope fractionation between clinopyroxene and garnet ($\Delta^{26}Mg_{clinopyroxene\text{-}garnet}=\delta^{26}Mg_{clinopyroxene}-\delta^{26}Mg_{garnet}$) as a function of $1/T^2$. The orogenic eclogites data are from Li et al (2011).

5.2　Crustal origin of cratonic eclogites

When compared to mantle peridotites and unaltered oceanic basalts (Teng et al., 2007, 2010a; Handler et al., 2009; Yang et al., 2009; Bourdon et al., 2010; Pogge von Strandmann et al., 2011), cratonic eclogites have distinct, extremely light Mg isotopic compositions (Fig.3). Provided that prograde metamorphism does not significantly fractionate Mg isotopes (Li et al., 2011), and the rapid eruption of kimberlitic magma does not allow for extensive interaction between eclogitic minerals and host magma (Kelley and Wartho, 2000), the large Mg isotopic variations in these eclogites must therefore result from other, open-system processes. These include melt extraction and diffusion processes associated with eclogite-peridotite interaction and melt/fluid metasomatism, or, alternatively protolith heterogeneity.

Diffusion processes can generate large Mg isotope fractionations due to the faster diffusion of lighter Mg isotopes relative to heavier isotopes (e.g., Richter et al., 2008). It is, however, unlikely to be the main process responsible for producing the light δ^{26}Mg values observed in the bulk eclogites as the eclogite sample with the highest Mg$^{\#}$ also has the heaviest Mg isotopic composition, opposite to what would be expected from diffusion. In addition, diffusion processes alone (i.e. without any mineral recrystallization) should generate disequilibrium inter-mineral Mg isotope fractionation. This is inconsistent with the equilibrium Mg isotope fractionation between clinopyroxene and garnet observed here.

Metasomatism can also alter the chemical and isotopic compositions of eclogites at different scales and obscure their primary features. Although metasomatic overprinting has been demonstrated to be of limited importance in producing variations in stable Fe, O and Tl isotope signatures observed in these samples (Viljoen, 2005; Nielsen et al., 2009; Williams et al., 2009), it is necessary to evaluate its impact on Mg isotope systematics. Mg isotopic compositions of global oceanic basalts are homogenous (Teng et al., 2010a), thus silicate metasomatism, in the absence of any additional fractionation processes, cannot produce the light Mg isotopic compositions of these cratonic eclogites. Mantle carbonatite melt is another important metasomatic agent with the potential to change the Mg isotopic compositions of target rocks towards light values, as both theoretical studies and analysis of crustal carbonates suggest that they have isotopically light Mg isotope signatures (Galy et al., 2002; Young and Galy, 2004; Brenot et al., 2008; Higgins and Schrag, 2010; Jacobson et al., 2010; Schauble, 2011). Carbonatite metasomatism should, however, be readily identifiable through an increase in the values of metasomatic indices such as Sr, Sr/Y, and LREE. There are however no correlations between δ^{26}Mg and such indices of metasomatism (Fig. 6). For example, the sample with the highest ratio of Sr/Y has a relative unfractionated

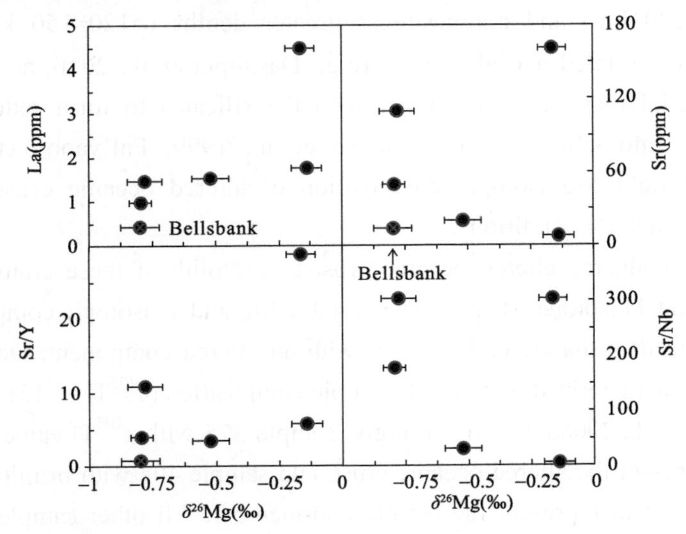

Fig. 6 Variations of bulk eclogite δ^{26}Mg values as a function of elemental indices of metasomatism. Trace element data are from Williams et al. (2009). Sample labeled with "×" is the Bellsbank eclogite. The high Sr/Nb ratio of Bellsbank resulted from the low Nb content, which is under the detect limitation.

δ^{26}Mg value of −0.164‰. Furthermore, the eclogites studied here have δ^{18}O values ranging from 4.49‰ to 5.58‰ (Williams et al., 2009), slightly below the mantle value (5.5‰ to 5.9‰), but much lower than values exhibited by mantle carbonatites (Deines, 1989), which provides further evidence against carbonatite metasomatism as the primary cause of Mg isotope heterogeneity.

The most likely explanation for these isotopically light cratonic eclogites is therefore protolith heterogeneity. As Mg isotopes can only be significantly fractionated during surface processes such as carbonate precipitation (Galy et al., 2002; Higgins and Schrag, 2010) and silicate weathering (Tipper et al., 2006a; Pogge von Strandmann et al., 2008; Teng et al., 2010b; Wimpenny et al., 2010), these protoliths must have had a surface origin. For example, oceanic dolomite is deeply depleted in heavy Mg isotopes (<−2‰) with δ^{26}Mg values 2.0‰ to 2.7‰ lighter than the precipitating pore fluid (Higgins and Schrag, 2010). Marine sediments have highly variable but isotopically extremely light Mg isotopic compositions (Higgins and Schrag, 2010). While fresh oceanic basalts such as MORBs and OIBs have homogenous Mg isotopic compositions around the mantle value (defined by unaltered peridotites), the bulk altered oceanic crust is inferred to have a lighter Mg isotopic composition. This is based on the systematics of Mg isotopes during weathering: the δ^{26}Mg value of seawater (δ^{26}Mg=−0.83±0.09‰, Foster et al., 2010; Ling et al., 2011 and references therein) is higher than that of the global runoff (δ^{26}Mg=−1.09±0.05‰, Tipper et al., 2006b), which requires the uptake of light Mg isotopes by alteration of oceanic crust as suggested by Tipper et al (2006b). During oceanic alteration, carbonates form by the uptake of Mg and Ca in seawater that were originally released from oceanic basalts (Wilkinson and Algeo, 1989). A significant portion of carbonate resides in the top 300 m of altered oceanic crust with an average CO_2 content of ~3 wt% (Alt and Teagle, 1999). Local enrichment of carbonate veins (as high as 75% modal abundance) can account for >10 wt% of the CO_2 content of altered basalt (Staudigel et al., 1996). Considering the extremely light Mg isotopic compositions of carbonates, at least parts of the altered oceanic crust should have a distinct, light Mg isotopic composition relative to the primitive mantle. Subduction should subsequently transform this isotopically light altered oceanic crustal material into carbonated eclogites during prograde metamorphism (Green and Ringwood, 1967). Carbonates in eclogites, due to their higher solidus than the geothermal gradient of even hot subduction, can be preserved in the form of dolomite at intermediate depths (60–120 km) and magnesite at greater depths (>120–150 km), potentially surviving processes such as subduction-related partial melting (e.g., Dasgupta et al., 2010, and references therein). This subducted carbonate material may ultimately react with the silicates to form reduced carbon (diamond or graphite) and release Mg into silicates (e.g., Knoche et al., 1999; Pal'vanov et al., 2002), providing a mechanism by which the light Mg isotopic composition of altered oceanic crust can be inherited by its metamorphic equivalents in mantle conditions.

Further evidence for subducted altered oceanic crust as protolith of these cratonic eclogites comes from coupled variations of Mg and Tl isotopes (Fig.7). The variable Mg and Tl isotopic compositions of these eclogites can be explained by mixing of a mantle end-member with an altered component. Oceanic crust altered at low temperature by seawater is characterized by light Tl isotopic composition (ε^{205}Tl= −15), in great contrast to mantle rocks (ε^{205}Tl= −2; Nielsen et al., 2006a, b). Accordingly, sample 375 with ε^{205}Tl value of −5 and δ^{26}Mg value of −0.779 ‰ is chosen to represent the altered oceanic crust, and sample 402 with primitive mantle-like Tl and Mg isotopic signatures is selected to represent the mantle end-member. All other samples fall on this mixing line between two end-members, further suggesting the involvement of altered oceanic crust into the formation of cratonic eclogites (Fig.7).

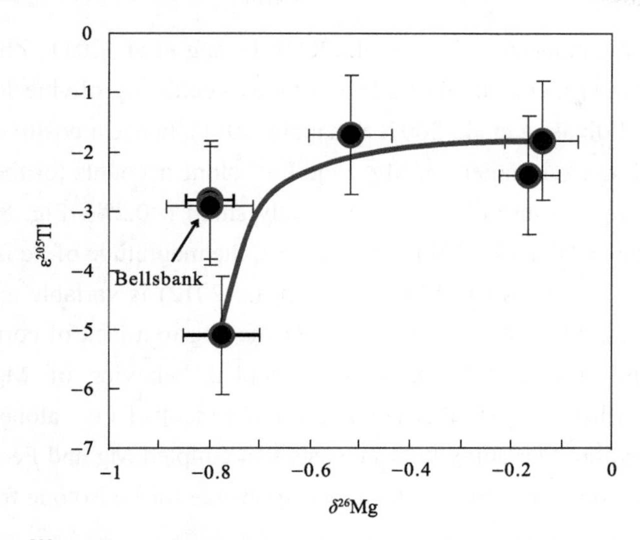

Fig. 7 Correlation of δ^{26}Mg and ε^{205}Tl values for eclogite samples. Sample 402 and 375 are taken to represent the end-member of mantle and altered oceanic crust, respectively. Tl contents and Tl isotopic compositions of the end-member components are taken from Nielsen et al. (2009) and as follows. Sample 402: Tl=80 ppb, ε^{205}Tl=−1.8; Sample 375: Tl=464 ppb, ε^{205}Tl=−5.1. Mg contents are taken from Williams et al. (2009) and Mg isotopic compositions are from Table 1. Mixing line represents simple mixing between the mantle and altered oceanic that have lost 90 % original Tl during subduction-related dehydration.

5.3 Redox-controlled inter-mineral Fe isotope fractionation

Large inter-mineral Fe isotope fractionations between garnet and clinopyroxene have been previously observed in mantle rocks such as garnet peridotites, pyroxenites and eclogites (Beard and Johnson, 2004; Williams et al., 2005, 2009; Weyer et al., 2007). Although the observed fractionations can be ascribed to equilibrium processes (Beard and Johnson, 2004; Williams et al., 2009), the underlying mechanisms are not yet fully resolved. As in the case of Mg isotopes, contrasts in the bonding environment of Fe, here involving both Fe redox state and coordination, in minerals and melts are predicted to be responsible for the inter-mineral Fe isotope fractionation (Polyakov and Mineev, 2000; Polyakov et al., 2007; Schuessler et al., 2007; Hill and Schauble, 2008; Shahar et al., 2008). Resolving the effects of Fe redox state (Williams et al., 2004, 2005, 2012; Weyer et al., 2007; Weyer and Ionov, 2007; Dauphas et al., 2009; Teng et al., 2008; Sossi et al., 2012) from coordination has so far proved difficult.

Combined Mg and Fe isotopic analyses of the same samples may help to evaluate the relative roles of oxidation state and coordination on Fe isotope fractionation, as Mg isotopes do not fractionate during redox reactions and Mg^{2+} and Fe^{2+} have identical charge and similar ionic radii, and occupy the same site in common rock-forming minerals. For example, in garnet with general chemical formula of $X_3Y_2Z_3O_{12}$, Mg^{2+} and Fe^{2+} have ionic radii of 0.890 Å and 0.920 Å (Shannon and Prewitt, 1969; Shannon, 1976), respectively, and occupy the same X site with 8-fold coordination, whereas the Fe^{3+} has ionic radii of only 0.645 Å and goes into the Y site with six-fold coordination. It then follows that if inter-mineral Fe isotope fractionation is controlled purely by the coordination and bonding environment of Fe^{2+}, then positive correlations between δ^{26}Mg and δ^{57}Fe for an individual mineral and between Δ^{26}Mg and Δ^{57}Fe are expected. No such correlations however are observed in samples investigated here (Fig.8). This implies that Fe^{3+} also plays a critical role in inter-mineral Fe isotope fractionation, while the proportions of Fe^{3+} and Fe^{2+} in minerals are largely related to the oxidation state. Considering that equilibrium Fe isotope fractionation has been achieved for the studied cratonic eclogites (Williams et al., 2009), the change of redox state of Fe thus accounts for the fractionation of Fe isotopes. This conclusion is further supported by literature studies where coupled Mg and Fe isotopic analyses of clinopyroxene and olivine in peridotites were conducted (Yang et al., 2009; Huang et al., 2011; Zhao et al., 2012). Theoretical calculation suggests that, clinopyroxene should be heavier than coexisting olivine for both Mg and Fe isotopes (Polyakov and Mineev, 2000; Polyakov et al., 2007; Schauble, 2011), hence a positive correlation between Δ^{26}Mg and Δ^{57}Fe is expected if bonding environment of Mg^{2+} and Fe^{2+} alone accounts for the effect. The fractionation of Mg isotopes between clinopyroxene and olivine is, relatively small (<0.2‰, Fig. 8) and within the theoretical prediction at mantle temperature (Schauble, 2011). By contrast, the magnitude of Fe isotope fractionation between clinopyroxene and olivine (up to 0.54‰ for Δ^{57}Fe, Zhao et al., 2012) is variable and beyond that predicted by theory (Polyakov and Mineev, 2000; Polyakov et al., 2007), leading to a lack of correlation between Δ^{26}Mg and Δ^{57}Fe for clinopyroxene and olivine (Fig.8). This decoupled behavior of Mg and Fe isotopes during high-temperature processes further suggests that bonding environment of Fe^{2+} alone is unlikely to be the major control on Fe isotope fractionation. Our study thus suggests that coupled Mg and Fe isotopic analyses on mineral pairs may provide a novel way to identify the mechanism responsible for Fe isotope fractionation.

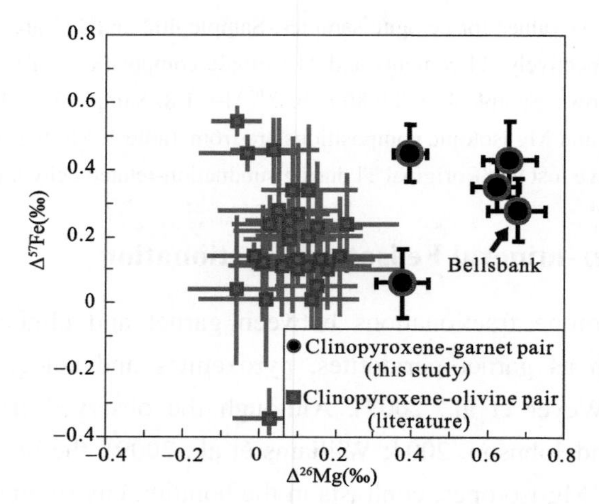

Fig. 8　Δ^{26}Mg-Δ^{57}Fe diagram for minerals from the cratonic eclogites and peridotites. Mg and Fe isotopic compositions for the clinopyroxene and garnet in cratonic eclogites are from this study and Williams et al. (2009), respectively. Mg and Fe isotopic compositions for the clinopyroxene and olivine pairs in peridotites are from Huang et al. (2011), Zhao et al. (2012) and Yang et al. (2009).

Acknowledgements

We thank Kangjun Huang for help in the clean lab, Bill Griffin and Fanus Viljoen for the access to samples. We also thank Fang Huang and Wang-Ye Li for helpful discussion. Constructive comments from Ed Tipper and an anonymous reviewer, and careful and efficient handling for Bernard Marty are greatly appreciated. This work was funded by the National Science Foundation (EAR-0838227 and EAR-1056713) to Fang-Zhen Teng, NERC Advanced Fellowship (NE/F014295/2) to Helen Williams, and the National Nature Science Foundation of China (No. 41230209 and 40973016) to Shu-Guang Li.

References

Alt, J. C., Teagle, D. A. H., 1999. The uptake of carbon during alteration of ocean crust. Geochim. Cosmochim. Acta 63, 1527-1535.

Aoki, K. I., Shiba, I., 1973. Pyroxenes from lherzolite inclusions of Itinome-gata, Japan. Lithos 6, 41-51.

Barth, M. G., Rudnick, R. L., Horn, I., McDonough, W. F., Spicuzza, M. J., Valley, J. W., Haggerty, S. E., 2001. Geochemistry of xenolithic eclogites from West Africa, Part I: a link between low MgO eclogites and Archean crust formation. Geochim. Cosmochim. Acta 65, 1499-1527.

Barth, M. G., Rudnick, R. L., Horn, I., McDonough, W. F., Spicuzza, M. J., Valley, J. W., Haggerty, S. E., 2002. Geochemistry of xenolithic eclogites from West Africa, Part 2: origins of the high MgO eclogites. Geochim. Cosmochim. Acta 66, 4325-4345.

Beard, B., Johnson, C., 2004. Inter-mineral Fe isotope variations in mantle-derived rocks and implications for the Fe geochemical cycle. Geochim. Cosmochim. Acta 68, 4727-4743.

Bigeleisen, J., Mayer, M. G., 1947. Calculation of equilibrium constants for isotopic exchange reactions. J. Chem. Phys. 15, 261.

Bourdon, B., Tipper, E. T., Fitoussi, C., Stracke, A., 2010. Chondritic Mg isotope composition of the Earth. Geochim. Cosmochim. Acta 74, 5069-5083.

Brenot, A., Cloquet, C., Vigier, N., Carignan, J., Francelanord, C., 2008. Magnesium isotope systematics of the lithologically varied Moselle river basin, France. Geochim. Cosmochim. Acta 72, 5070-5089.

Caporuscio, F. A., Smyth, J. R., 1990. Trace element crystal chemistry of mantle eclogites. Contrib. Mineral. Petrol. 105, 550-561.

Cartigny, P., Harris, J. W., Javoy, M., 1998. Eclogitic diamond formation at Jwaneng: no room for a recycled component. Science 280, 1421-1424.

Dasgupta, R., Hirschmann, M. M., 2010. The deep carbon cycle and melting in the Earth's interior. Earth Planet. Sci. Lett. 298, 1-13.

Dauphas, N., Craddock, P. R., Asimow, P. D., Bennett, V. C., Nutman, A. P., Ohnenstetter, D., 2009. Iron isotopes may reveal the redox conditions of mantle melting from Archean to Present. Earth Planet. Sci. Lett. 288, 255-267.

Dauphas, N., Teng, F.-Z., Arndt, N. T., 2010. Magnesium and iron isotopes in 2.7 Ga Alexo komatiites: mantle signatures, no evidence for Soret diffusion, and identification of diffusive transport in zoned olivine. Geochim. Cosmochim. Acta 74, 3274-3291.

Deines, P., 1989. Stable Isotope Variations in Carbonatites. Carbonatites: Genesis and Evolution. Unwin Hyman, London (pp. 301-359).

Foster, G. L., Pogge von Strandmann, P. A. E., Rae, J. W. B., 2010. Boron and magnesium isotopic composition of seawater. Geochem. Geophys. Geosyst., 11, http://dx.doi.org/10.1029/2010GC003201.

Galy, A., Bar-Matthews, M., Halicz, L., O'Nions, R. K., 2002. Mg isotopic composition of carbonate: insight from speleothem formation. Earth Planet. Sci. Lett. 201, 105-115.

Galy, A., Yoffe, O., Janney, P. E., Williams, R. W., Cloquet, C., Alard, O., Halicz, L., Wadhwa, M., Hutcheon, I. D., Ramon, E., Carignan, J., 2003. Magnesium isotope heterogeneity of the isotopic standard SRM980 and new reference materials for magnesium-isotope-ratio measurements. J. Anal. Atom. Spectrom. 18, 1352-1356.

Gré au, Y., Huang, J.-X., Griffin, W. L., Renac, C., Alard, O., O'Reilly, S. Y., 2011. Type I eclogites from Roberts Victor kimberlites: products of extensive mantle metasomatism. Geochim. Cosmochim. Acta 75, 6927-6954.

Green, D., Ringwood, A., 1967. An experimental investigation of the gabbro to eclogite transformation and its petrological applications. Geochim. Cosmochim. Acta 31, 767-833.

Griffin, W. L., O'Reilly, S. Y., 2007. Cratonic lithospheric mantle: Is anything subducted? Episodes-Newsmag. Int. Union Geol. Sci. 30, 43-53.

Handler, M. R., Baker, J. A., Schiller, M., Bennett, V. C., Yaxley, G. M., 2009. Magnesium stable isotope composition of Earth's upper mantle. Earth Planet. Sci. Lett. 282, 306-313.

Higgins, J. A., Schrag, D. P., 2010. Constraining magnesium cycling in marine sediments using magnesium isotopes. Geochim. Cosmochim. Acta 74, 5039-5053.

Hill, P., Schauble, E., 2008. Modeling the effects of bond environment on equilibrium iron isotope fractionation in ferric aquo-chloro complexes. Geochim. Cosmochim. Acta 72, 1939-1958.

Huang, F., Chakraborty, P., Lundstrom, C. C., Holmden, C., Glessner, J. J. G., Kieffer, S. W., Lesher, C. E., 2010. Isotope fractionation in silicate melts by thermal diffusion. Nature 464, 396-400.

Huang, F., Lundstrom, C. C., Glessner, J., Ianno, A., Boudreau, A., Li, J., Ferre, E. C., Marshak, S., DeFrates, J., 2009. Chemical and isotopic fractionation of wet andesite in a temperature gradient: experiments and models suggesting a new mechanism of magma differentiation. Geochim. Cosmochim. Acta 73, 729-749.

Huang, F., Zhang, Z., Lundstrom, C. C., Zhi, X., 2011. Iron and magnesium isotopic compositions of peridotite xenoliths from Eastern China. Geochim. Cosmochim. Acta 75, 3318-3334.

Ireland, T. R., Rudnick, R. L., Spetsius, Z., 1994. Trace elements in diamond inclusions from eclogites reveal link to Archean granites. Earth Planet. Sci. Lett. 128, 199-213.

Jacob, D., Jagoutz, E., Lowry, D., Mattey, D., Kudrjavtseva, G., 1994. Diamondiferous eclogites from Siberia: remnants of Archean oceanic crust. Geochim. Cosmochim. Acta 58, 5191-5207.

Jacob, D. E., 2004. Nature and origin of eclogite xenoliths from kimberlites. Lithos 77, 295-316.

Jacobson, A. D., Zhang, Z., Lundstrom, C., Huang, F., 2010. Behavior of Mg isotopes during dedolomitization in the Madison Aquifer, South Dakota. Earth Planet. Sci. Lett. 297, 446-452.

Kelley, S., Wartho, J., 2000. Rapid kimberlite ascent and the significance of Ar-Ar ages in xenolith phlogopites. Science 289, 609-611.

Knoche, R., Sweeney, R. J., Luth, R. W., 1999. Carbonation and decarbonation of eclogites: the role of garnet. Contrib. Mineral. Petrol. 135, 332-339.

Li, W.-Y., Teng, F.-Z., Ke, S., Rudnick, R. L., Gao, S., Wu, F.-Y., Chappell, B. W., 2010. Heterogeneous magnesium isotopic composition of the upper continental crust. Geochim. Cosmochim. Acta 74, 6867-6884.

Li, W.-Y., Teng, F.-Z., Xiao, Y., Huang, J., 2011. High-temperature inter-mineral magnesium isotope fractionation in eclogite from the Dabie orogen, China. Earth Planet. Sci. Lett. 304, 224-230.

Ling, M.-X., Sedaghatpour, F., Teng, F.-Z., Hays, P. D., Strauss, J., Sun, W., 2011. Homogeneous magnesium isotopic composition of seawater: an excellent geostandard for Mg isotope analysis. Rapid Commun. Mass Spectrom. 25, 2828-2836.

Liu, S.-A., Teng, F.-Z., He, Y., Ke, S., Li, S., 2010. Investigation of magnesium isotope fractionation during granite differentiation: implication for Mg isotopic composition of the continental crust. Earth Planet. Sci. Lett. 297, 646-654.

Liu, S.-A., Teng, F.-Z., Yang, W., Wu, F.-Y., 2011. High-temperature inter-mineral magnesium isotope fractionation in mantle xenoliths from the North China craton. Earth Planet. Sci. Lett. 308, 131-140.

Macgregor, I. D., Carter, J., 1970. The chemistry of clinopyroxenes and garnets of eclogite and peridotite xenoliths from the Roberts Victor Mine, South Africa. Phys. Earth Planet. Int. 3, 391-397.

Nielsen, S. G., Rehkämper, M., Norman, M. D., Halliday, A. N., Harrison, D., 2006a. Thallium isotopic evidence for ferromanganese sediments in the mantle source of Hawaiian basalts. Nature 439, 314-317.

Nielsen, S. G., Rehkämper, M., Teagle, D. A. H., Butterfield, D. A., Alt, J. C., Halliday, A. N., 2006b. Hydrothermal fluid fluxes calculated from the isotopic mass balance of thallium in the ocean crust. Earth Planet. Sci. Lett. 251, 120-133.

Nielsen, S. G., Williams, H. M., Griffin, W. L., O'Reilly, S. Y., Pearson, N., Viljoen, F., 2009. Thallium isotopes as a potential tracer for the origin of cratonic eclogites. Geochim. Cosmochim. Acta 73, 7387-7398.

Pal'yanov, Y. N., Sokol, A. G., Borzdov, Y. M., Khokhryakov, A. F., Sobolev, N. V., 2002. Diamond formation through carbonate-silicate interaction. Am. Mineral. 87, 1009-1013.

Pogge von Strandmann, P. A. E., Burton, K. W., James, R. H., van Calsteren, P., Gislason, S. R., Sigfússon, B., 2008. The influence of weathering processes on riverine magnesium isotopes in a basaltic terrain. Earth Planet. Sci. Lett. 276, 187-197.

Pogge von Strandmann, P. A. E., Elliott, T., Marschall, H. R., Coath, C., Lai, Y.-J., Jeffcoate, A. B., Ionov, D. A., 2011. Variations of Li and Mg isotope ratios in bulk chondrites and mantle xenoliths. Geochim. Cosmochim. Acta 75, 5247-5268.

Polyakov, V., Clayton, R., Horita, J., Mineev, S., 2007. Equilibrium iron isotope fractionation factors of minerals: reevaluation from the data of nuclear inelastic resonant X-ray scattering and Mössbauer spectroscopy. Geochim. Cosmochim. Acta 71, 3833-3846.

Polyakov, V. B., Mineev, S. D., 2000. The use of Möossbauer spectroscopy in stable isotope geochemistry. Geochim. Cosmochim. Acta 64, 849-865.

Richter, F. M., Watson, E., Mendybaev, R., Teng, F.-Z., Janney, P., 2008. Magnesium isotope fractionation in silicate melts by chemical and thermal diffusion. Geochim. Cosmochim. Acta 72, 206-220.

Schauble, E. A., 2011. First-principles estimates of equilibrium magnesium isotope fractionation in silicate, oxide, carbonate and hexaaquamagnesium(2+) crys-tals. Geochim. Cosmochim. Acta 75, 844-869.

Schuessler, J., Schoenberg, R., Behrens, H., Blanckenburg, F., 2007. The experi-mental calibration of the iron isotope fractionation factor between pyrrhotite and peralkaline rhyolitic melt. Geochim. Cosmochim. Acta 71, 417-433.

Schulze, D., Valley, J., Viljoen, K., Stiefenhofer, J., Spicuzza, M., 1997. Carbon isotope composition of graphite in mantle eclogites. J. Geol. 105, 379-386.

Shahar, A., Young, E. D., Manning, C. E., 2008. Equilibrium high-temperature Fe isotope fractionation between fayalite and magnetite: an experimental cali-bration. Earth Planet. Sci. Lett. 268, 330-338.

Shannon, R., 1976. Revised effective ionic radii and systematic studies of interatomic distances in halides and chalcogenides. Acta Crystallogr. A 32, 751-767.

Shannon, R. D., Prewitt, C. T., 1969. Effective ionic radii in oxides and fluorides. Acta Crystallogr. B 25, 925-946.

Shen, B., Jacobsen, B., Lee, C. T. A., Yin, Q. Z., Morton, D. M., 2009. The Mg isotopic systematics of granitoids in continental arcs and implications for the role of chemical weathering in crust formation. Proc. Natl. Acad. Sci. 106, 20652-20657.

Smyth, J. R., Caporuscio, F. A., McCormick, T. C., 1989. Mantle eclogites: evidence of igneous fractionation in the mantle. Earth Planet. Sci. Lett. 93, 133-141.

Sossi, P. A., Foden, J. D., Halverson, G. P., 2012. Redox-controlled iron isotope fractionation during magmatic differentiation: an example from the Red Hill intrusion, S. Tasmania. Contrib. Mineral. Petrol. 10. 1007/s00410-012-0769-x.

Staudigel, H., Plan, T., White, B., Schmincke, H. U., 1996. Geochemical fluxes during seafloor alteration of the basaltic upper oceanic crust: DSDP sites 417 and 418. Geophys. Monogr. 96, 19-38.

Teng, F.-Z., Dauphas, N., Helz, R. T., Gao, S., Huang, S., 2011. Diffusion-driven magnesium and iron isotope fractionation in Hawaiian olivine. Earth Planet. Sci. Lett. 308, 317-324.

Teng, F.-Z., Li, W.-Y., Ke, S., Marty, B., Dauphas, N., Huang, S., Wu, F.-Y., Pourmand, A., 2010a. Magnesium isotopic composition of the Earth and chondrites. Geochim. Cosmochim. Acta 74, 4150-4166.

Teng, F.-Z., Li, W.-Y., Rudnick, R. L., Gardner, L. R., 2010b. Contrasting lithium and magnesium isotope fractionation during continental weathering. Earth Planet. Sci. Lett. 300, 63-71.

Teng, F.-Z., Wadhwa, M., Helz, R. T., 2007. Investigation of magnesium isotope fractionation during basalt differentiation: implications for a chondritic composition of the terrestrial mantle. Earth Planet. Sci. Lett. 261, 84-92.

Teng, F. Z., Dauphas, N., Helz, R. T., 2008. Iron isotope fractionation during magmatic differentiation in Kilauea Iki lava lake. Science 320, 1620-1622.

Tipper, E., Galy, A., Bickle, M., 2006a. Riverine evidence for a fractionated reservoir of Ca and Mg on the continents: implications for the oceanic Ca cycle. Earth Planet. Sci. Lett. 247, 267-279.

Tipper, E., Galy, A., Gaillardet, J., Bickle, M., Elderfield, H., Carder, E., 2006b. The magnesium isotope budget of the modern ocean: constraints from riverine magnesium isotope ratios. Earth Planet. Sci. Lett. 250, 241-253.

Urey, H. C., 1947. The thermodynamic properties of isotopic substances. J. Chem. Soc., 562-581.

Viljoen, K. S., 2005. Contrasting Group I and Group II eclogite xenolith petrogen-esis: petrological, trace element and isotopic evidence from eclogite, garnet-websterite and alkremite xenoliths in the Kaalvallei Kimberlite, South Africa. J. Petrol. 46, 2059-2090.

Wang, S.-J., Li, S.-G., An, S.-C., Hou, Z.-H., 2012. A granulite record of multistage metamorphism and REE behavior in the Dabie orogen: constraints from zircon and rock-forming minerals. Lithos 136-139, 109-125.

Weyer, S., Anbar, A. D., Brey, G. P., Münker, C., Mezger, K., Woodland, A. B., 2007. Fe-isotope fractionation during partial melting on Earth and the current view on the Fe-isotope budgets of the planets (reply to the comment of F. Poitrasson and to the comment of B. L. Beard and C. M. Johnson on "Iron isotope fractionation during planetary differentiation" by S. Weyer, A. D. Anbar, G. P. Brey, C. Münker, K. Mezger and A. B. Woodland). Earth Planet. Sci. Lett. 256, 638-646.

Weyer, S., Ionov, D. A., 2007. Partial melting and melt percolation in the mantle: the message from Fe isotopes. Earth Planet. Sci. Lett. 259, 119-133.

Wilkinson, B. H., Algeo, T. J., 1989. Sedimentary carbonate record of calcium-magnesium cycling. Am. J. Sci. 289, 1158-1194.

Williams, H., Peslier, A., McCammon, C., Halliday, A., Levasseur, S., Teutsch, N., Burg, J., 2005. Systematic iron isotope variations in mantle rocks and minerals: the effects of partial melting and oxygen fugacity. Earth Planet. Sci. Lett. 235, 435-452.

Williams, H. M., McCammon, C. A., Peslier, A. H., Halliday, A. N., Teutsch, N., Levasseur, S., Burg, J. P., 2004. Iron isotope fractionation and the oxygen fugacity of the mantle. Science 304, 1656-1659.

Williams, H. M., Nielsen, S. G., Renac, C., Griffin, W. L., O'Reilly, S. Y., McCammon, C. A., Pearson, N., Viljoen, F., Alt, J. C., Halliday, A. N., 2009. Fractionation of oxygen and iron isotopes by partial melting processes: implications for the interpreta-tion of stable isotope signatures in mafic rocks. Earth Planet. Sci. Lett. 283, 156-166.

Williams, H. M., Wood, B. J., Wade, J., Frost, D. J., Tuff, J., 2012. Isotopic evidence for internal oxidation of the Earth's mantle during accretion. Earth Planet. Sci. Lett. 321, 54-63.

Wimpenny, J., Gíslason, S. R., James, R. H., Gannoun, A., Pogge Von Strandmann, P. A. E., Burton, K. W., 2010. The behaviour of Li and Mg isotopes during primary phase dissolution and secondary mineral formation in basalt. Geochim. Cosmochim. Acta 74, 5259-5279.

Wombacher, F., Eisenhauer, A., Böhm, F., Gussone, N., Regenberg, M., Dullo, W. C., Rüggeberg, A., 2011. Magnesium stable isotope fractionation in marine biogenic calcite and aragonite. Geochim. Cosmochim. Acta 75, 5797-5818.

Yang, W., Teng, F.-Z., Zhang, H.-F., 2009. Chondritic magnesium isotopic composi-tion of the terrestrial mantle: a case study of peridotite xenoliths from the North China craton. Earth Planet. Sci. Lett. 288, 475-482.

Young, E. D., Galy, A., 2004. The isotope geochemistry and cosmochemistry of magnesium. Rev. Mineral. Geochem. 55, 197-230.

Young, E. D., Tonui, E., Manning, C. E., Schauble, E., Macris, C. A., 2009. Spinel-olivine magnesium isotope thermometry in the mantle and implications for the Mg isotopic composition of Earth. Earth Planet. Sci. Lett. 288, 524-533.

Zhao, X., Zhang, H., Zhu, X., Tang, S., Yan, B., 2012. Iron isotope evidence for multistage melt-peridotite interactions in the lithospheric mantle of eastern China. Chem. Geol. 292-293, 127-139.

深部碳循环的 Mg 同位素示踪*

李曙光 [1,2]

1. 中国地质大学(北京)，地质过程与矿产资源国家重点实验室，北京，100083
2. 中国科学技术大学，地球与空间科学学院，中国科学院壳幔物质与环境重点实验室，安徽，合肥，230026

亮点介绍：本文系统介绍了 Mg 同位素示踪深部碳循环的原理，同时对高温高压条件下碳酸盐稳定性和相转变、板片俯冲过程中镁同位素行为以及再循环碳酸盐对地幔镁同位素组成的作用进行了详细论述。本文指出在多数大洋板块俯冲条件下，俯冲碳酸盐通过变质分解而脱碳不可能发生，但富钙碳酸盐可被俯冲板片析出流体溶解，其溶解度随温度升高而增大，而菱镁矿溶解度下降，且成为高压下能稳定存在碳酸盐矿物，白云石在高压下分解为菱镁矿+文石。因此，能俯冲进入深部地幔的碳酸盐主要是菱镁矿，它可以局部改变深部碳酸盐化地幔的镁同位素组成，为应用 Mg 同位素示踪深部碳循环提供了理论基础。

摘要 大洋板块俯冲导致的深部碳循环可影响地球历史的大气 CO_2 的收支情况及气候变化。沉积碳酸盐岩是地球中轻镁同位素的主要储库，它通过板块俯冲再循环进入地幔有可能引起地幔局部的 Mg 同位素组成不均一性。因此，在这样一个基本假设基础上，即俯冲岩石的镁同位素在变质脱水和岩浆过程中不发生显著变化，镁同位素有可能成为深部碳循环的示踪剂。前人研究已经证明岩浆过程不会发生显著镁同位素分馏。然而，至今对俯冲、变质过程镁同位素的分馏程度以及低 δ^{26}Mg 玄武岩成因还属未知。为此，本研究聚焦在高温高压条件下碳酸盐的稳定性和相转换、板块俯冲过程中的镁同位素行为、循环碳酸盐对地幔镁同位素组成可能产生的影响。

关键词 深部碳循环，碳酸盐，镁同位素，板块俯冲，地幔地球化学

0 引言

大气 CO_2 是最重要的温室气体之一。因此大气圈的碳收支情况(CO_2 budget)是影响地球历史上气候变化的关键因素。为此人们需要研究地球各大碳储库之间的碳循环情况。近几十年来，人们研究最多的是地球表层碳循环(气圈-水圈-生物圈-土壤圈-表层岩石圈之间的碳循环)。其中涉及表层岩石圈的主要是：大规模火山作用向大气喷发的大量 CO_2，它是大气 CO_2 的一个重要“源”；生物光合作用和地表风化作用吸收大气 CO_2，海水还吸收溶解大气 CO_2，它们最终构成生物圈，或以碳酸盐形式沉积，是大气 CO_2 的一个重要“汇”。问题是这些“汇”能够平衡掉地球内部通过火山喷气排向大气的 CO_2 吗？Javoy 等[1]通过岩浆岩的碳同位素研究指出仅靠地球表层碳循环无法消耗掉地球历史向地球外部圈层排放的碳量，并提出这多余的碳的出路只能通过板块俯冲将它们再循环进入地幔来解释。

既然地幔成为全球碳循环的一个重要“汇”，对平衡大气 CO_2 含量起重要作用，每年进出地幔的碳通量对估算大气 CO_2 的收支是重要参数。我们面临的科学问题是：(1)如何验证地幔含有再循环的沉积碳酸

* 本文发表在：地学前缘，2015，22(5)：143-159

盐，以及它们的通量有多大？（2）再循环碳酸盐进入地幔后的地球化学行为如何？应用同位素示踪这一深部碳循环是一个可行的验证途径[2]。

根据碳酸盐组成，C-O-Ca-Mg 同位素是潜在的示踪剂。金刚石碳同位素是最早被应用示踪深部碳循环的。碳同位素对于区分有机碳和无机碳的效果很好，有机碳 $\delta^{13}C$ 小于–15‰，而无机碳大于–10‰。因此金刚石的碳同位素研究为深部碳循环提供了可靠证据。如巴西地区 Juina-5 金伯利岩管中有一部分金刚石 $\delta^{13}C$ 小于–15‰，最低可达–24.1‰，其内部矿物包裹物的岩石学分析表明这些金刚石来源于下地幔，为洋壳俯冲到下地幔提供了直接的地球化学证据[3]。由于地幔原始火成碳酸岩的 O 同位素（$\delta^{18}O$ = +6‰~+10‰）和沉积碳酸盐岩的 O 同位素（$\delta^{18}O$ = +15‰~+30‰）有显著差别，C-O 同位素结合也常被用来示踪它们是源自地幔原始碳，还是再循环的地壳碳酸盐岩。如川西新生代火成碳酸岩具有地幔原始碳酸岩的 C-O 同位素特征[4]；而东喜马拉雅火成碳酸岩脉具有很高的，类似沉积碳酸盐岩的 C-O 同位素特征[5]。但是 C 同位素示踪深部碳循环有两大缺点：（1）由于板块俯冲携带的碳酸盐相当部分是无机碳酸盐，它与地幔碳差异不大，C 同位素对鉴别它们无能为力；（2）低 $\delta^{13}C$ 值存在多解，它有可能是 CO_2 去气分馏的结果[1]。因此，我们有必要开展 Ca、Mg 同位素示踪深部碳循环研究。

夏威夷玄武岩 Ca 同位素研究是目前发表的 Ca 同位素示踪深部碳循环实例[6]。该文观察到夏威夷玄武岩的 $\delta^{44/40}Ca$–Sr/Nb 与 $\delta^{44/40}Ca$–$^{87}Sr/^{86}Sr$ 呈明显负相关关系，表明含碳酸盐的海洋沉积物再循环进入其源区，且混入的越多，$\delta^{44/40}Ca$ 值越低。通过拟合 $\delta^{44/40}Ca$ 与 Sr/Nb 和 $^{87}Sr/^{86}Sr$ 相关曲线，进行碳酸盐混入量的模拟计算，结果表明形成夏威夷玄武岩的地幔柱混入 4%的古老海洋沉积碳酸盐。然而由于 Ca 同位素测量的高成本，目前已发表用于示踪深部碳循环的研究仅此一例。相比之下，Mg 同位素示踪深部碳循环研究更活跃得多。

1　Mg 同位素示踪深部碳循环的原理

Mg 是地球的主要组成元素，广泛分布于地球各大储库中。Mg 有 ^{24}Mg（78.99%），^{25}Mg（10.00%），^{26}Mg（11.01%）3 个稳定同位素。^{26}Mg 它可以由短寿命核素 ^{26}Al 衰变而来。放射成因 ^{26}Mg 仅存在于太阳系早期物质。通常我们用 $\delta^{26}Mg$ 表达样品的 Mg 同位素组成。

$$\delta^{26}Mg=[(^{26}Mg/^{24}Mg)_{样品}/(^{26}Mg/^{24}Mg)_{标样}-1]\times 1000$$

现在可利用 MC-ICP-MS 精确测定 Mg 同位素组成。低 $\delta^{26}Mg$ 值代表具有较轻的 Mg 同位素组成，反之具有较重 Mg 同位素组成。

Mg 同位素可以用来示踪板块俯冲过程和深部碳循环基于以下假设：

（1）Mg 同位素在表生过程可产生较大分馏，故沉积碳酸盐岩与地幔的 Mg 同位素组成有较大差异。

（2）高温岩浆过程 Mg 同位素不产生显著分馏，故幔源岩浆岩的 Mg 同位素组成可代表其地幔源区的组成。

（3）沉积碳酸盐随板块俯冲而经历的变质和脱水过程所导致的 Mg 同位素分馏不能抹杀它与地幔 Mg 同位素的差异。故再循环碳酸盐可改造地幔的 Mg 同位素组成。

已有的研究已表明前 2 条基本原理是成立的，而板块俯冲过程中的 Mg 同位素地球化学行为则需要做深入研究（详见第 3 部分）。

1.1　地幔及碳酸盐岩的 Mg 同位素组成

对幔源岩石，如洋中脊玄武岩（MORB）和橄榄岩（peridotite）包体，研究表明地幔的 Mg 同位素平均组成为 $\delta^{26}Mg$=–0.25±0.07‰[7-8]。

碎屑沉积岩和土壤作为风化作用的固相产物相对地幔富集重 Mg 同位素。流经硅酸岩的河水含有风化作用可溶性产物相对富集轻 Mg 同位素（图 1）。这说明硅酸岩矿物风化过程中 Mg 同位素发生显著分馏，轻 Mg 同位素优先进入流体。河水每年输入的 1.6×10^{12} kg $CaCO_3$ 进入海洋，它们 $\delta^{26}Mg < 0.5‰$，其平均值为–1.09‰。海水的 Mg 同位素较均一，其 $\delta^{26}Mg$=–0.8‰。海洋沉积钙质碳酸盐的 $\delta^{26}Mg$=–1‰~–5.5‰，远低于海水和河水；白云岩的 $\delta^{26}Mg$=–1‰~–2.5‰（图 1）。深海沉积碳酸盐的非常低 $\delta^{26}Mg$ 值源自两个原因：

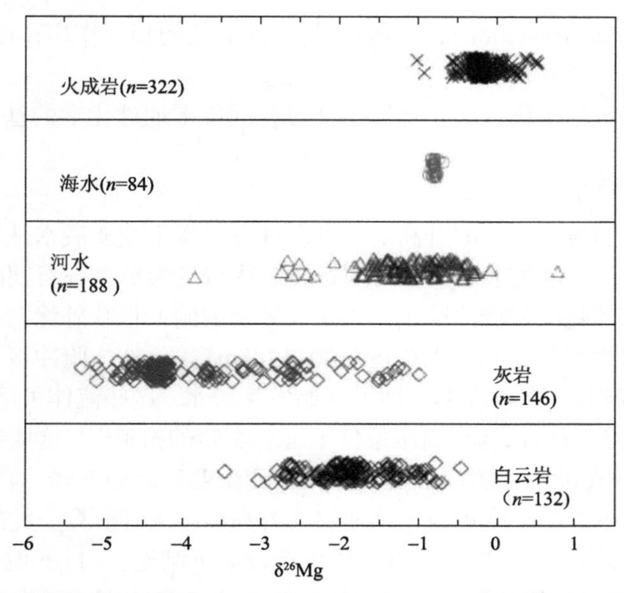

图 1　海水碳酸盐岩(灰岩、白云岩)的 Mg 同位素组成及与河水、海水和火成岩的比较(李伟强私人通信，据文献[9]修改)

(1)碳酸盐与水之间的平衡同位素分馏。其中常温下含镁方解石与水之间的平衡同位素分馏为 Δ^{26}Mg=–2.6‰(22°C)[9]或–3.5‰(25°C)[10]；白云石与水之间的平衡同位素分馏略低，根据最近的热水实验结果推算其常温下 Δ^{26}Mg =–1.8‰[11]。该机制可以解释大部分碳酸盐的低 δ^{26}Mg 值，但难于解释 δ^{26}Mg < –4.3‰的碳酸盐。

(2)生物分馏。远洋深水沉积的 $CaCO_3$ 绝大部分是有孔虫、颗石藻，其 δ^{26}Mg 值可低至–6.19‰[12-13].

1.2　高温岩浆过程 Mg 同位素无显著分馏

高温岩浆过程 Mg 同位素无显著分馏已被众多研究所证实。如夏威夷 Kilauea 火山玄武岩的研究表明基性岩浆的分离结晶作用不导致 Mg 同位素分馏[7]。大别山，澳大利亚和中东部花岗岩研究表明中酸性岩浆分异过程也不导致 Mg 同位素显著分馏[14-16](图 2)。

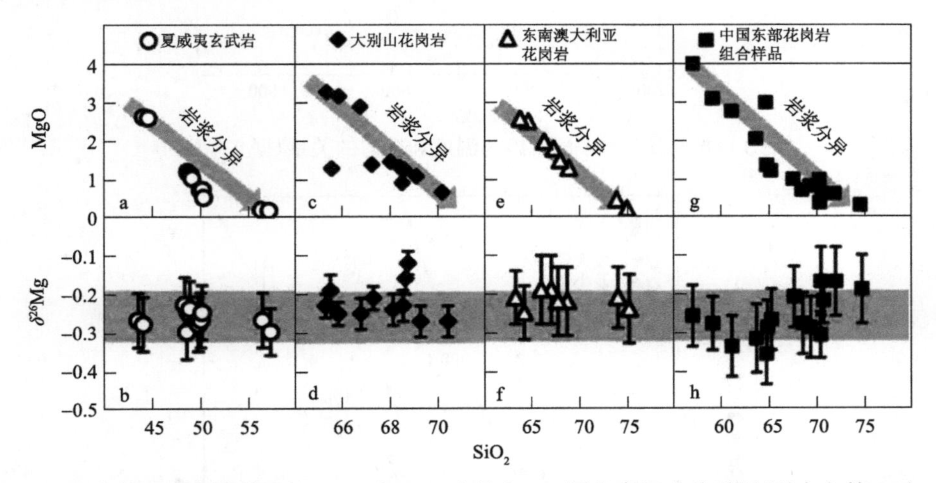

图 2　岩浆岩分离结晶过程 SiO_2 对 MgO 含量和 Mg 同位素组成关系图(引自文献[16])

2　板块俯冲条件下碳酸盐矿物的稳定性及转化

为了了解板块俯冲过程中再循环碳酸盐岩的 Mg 同位素变化，我们需要先了解碳酸盐岩在俯冲过程中可能发生的变化。碳酸盐随俯冲板块进入地幔可能发生五类反应：(1)碳酸盐的溶解；(2)碳酸盐矿物的相转换；(3)碳酸盐高温分解(decarbonation)；(4)碳酸盐与硅酸盐的相互作用；(5)碳酸盐化岩石的脱碳熔融(decarbonation melting)。

研究这些反应是实验岩石学工作，它是理解俯冲过程同位素地球化学的基础。

2.1　俯冲带碳酸盐的溶解

板片在俯冲过程中由于温度、压力的升高。可以多种方式发生变质脱水从而产生流体。流体可来自蚀变洋壳及沉积物的变质脱水，大量脱水发生在角闪石类矿物转变为单斜辉石期间，约在 50~70km 深度。流体还可来自岩石圈地幔上部的蛇纹岩变质脱水。该蛇纹岩是因俯冲板片外缘弯曲断裂引入流体产生。由于叶蛇纹石的高稳定性，它的脱水发生在较大深度(120~130km)。此外，俯冲板片顶部与地幔楔接触界面的沉积物，蚀变洋壳和地幔楔碎块，在较浅部位即可开始发生脱水。这些流体可溶解俯冲板片携带的碳酸盐。

目前仅 Ca 质碳酸盐(方解石)有高温-高压条件下溶解实验的报道[17]。这些实验表明，在上地壳低压(<0.3GPa)条件下 Ca 质碳酸盐在纯水中的溶解度很低，且随温度升高而下降。在中、下地壳高压(>0.5GPa)条件下 Ca 质碳酸盐在纯水中的溶解度较高，且随温度升高而增大(图 3)。此外，流体含盐(NaCl)度增大可提高 Ca 质碳酸盐的溶解度(图 4)。根据已有实验可获得如下结论：(1)下地壳变质流体比中上地壳流体可导致碳酸盐更高的活动性；(2)俯冲板块析出流体有较高的 T, P 和较高含盐度，因而可导致俯冲板块的 Ca 质碳酸盐较多溶解。

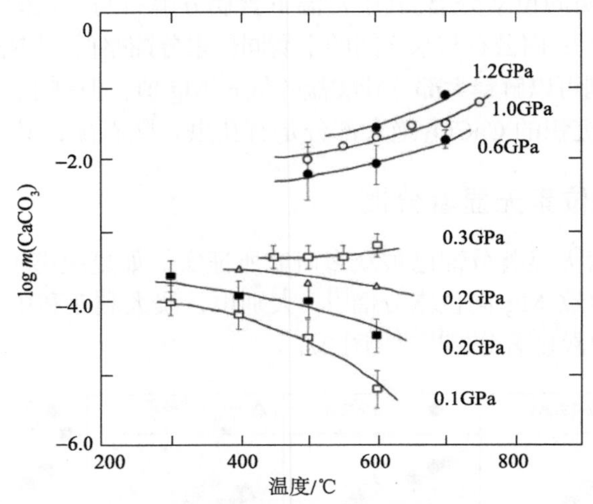

图 3　方解石在水中溶解度与温度和压力的关系(据文献[17])

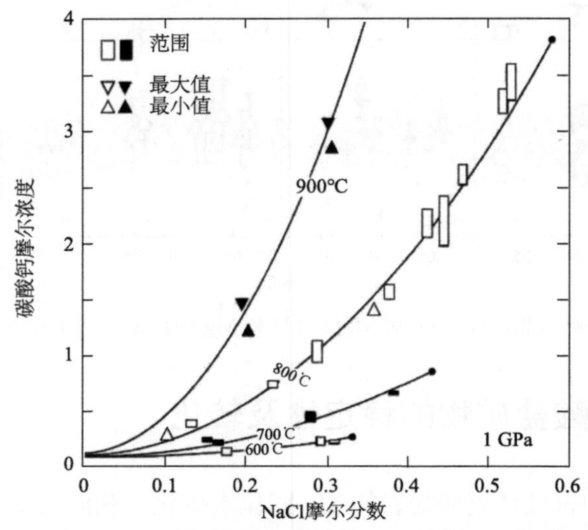

图 4　方解石在 1GPa 压力和不同温度条件下水中溶解度与盐度的关系(据文献[17])

目前镁质碳酸盐仅有低压（< 0.1GPa）条件下的溶解实验报道[18]。该文揭示了菱镁矿在水流体中的溶解速率随温度升高而下降，并随流体的碳酸根离子活度和PH值增高而下降。对镁质碳酸盐今后特别需要做高温高压条件下的溶解实验，检查高压下其溶解度与温度关系是否也像钙质碳酸盐一样出现翻转，以及相同条件下菱镁矿和方解石的溶解度对比实验。

除了实验外，人们也揭示了许多俯冲过程中碳酸盐溶解的地质证据。如在西天山洋壳俯冲成因榴辉岩中广泛发育的含碳酸盐的高压脉体证明这一碳酸盐溶解反应过程的存在[19]。此外，在西阿尔卑斯超高压变质岩金刚石的流体包体中有大量碳酸根离子并发现碳酸盐晶体，证明了俯冲带流体对碳酸盐的溶解作用[20]。

这样，俯冲板片携带进入地幔的碳通量中有一部分在较浅部位就溶解于俯冲板片析出流体并进入岛弧火山系统中，并最终伴随岛弧火山岩释放出来。

为描述这一岛弧浅部碳循环所占比例，人们提出一个“脱碳效率”的概念，即洋壳俯冲产生的岛弧火山排碳量与通过俯冲洋壳进入地幔碳通量的比值[21]。全球岛弧火山脱碳效率统计显示年轻的热洋壳俯冲脱碳效率很低（约20%）。这可能是因为它的脱水作用发生在弧前浅部，它使地幔楔蛇纹石化而不能熔融产生岩浆；对于俯冲洋壳年龄>50Ma的岛弧火山，脱碳效率可高达50%。溶解的碳酸盐可进入火山系统。另有约50%的未溶解碳酸盐可进入地幔深部。

2.2 高 *T-P* 条件下碳酸盐的稳定域和相转换

在板块俯冲过程中，随着温度和压力的升高，一些碳酸盐可能变得不稳定而发生相转变。Sato 和 Katsura[22]的实验表明，白云石可以在 700°C，P < 6GPa（深度 < 180km）的深度保持稳定；在 *P*>6GPa（深度>180km）时分解为菱镁矿和文石（图5）。这个压力高于石墨-金刚石相变压力。

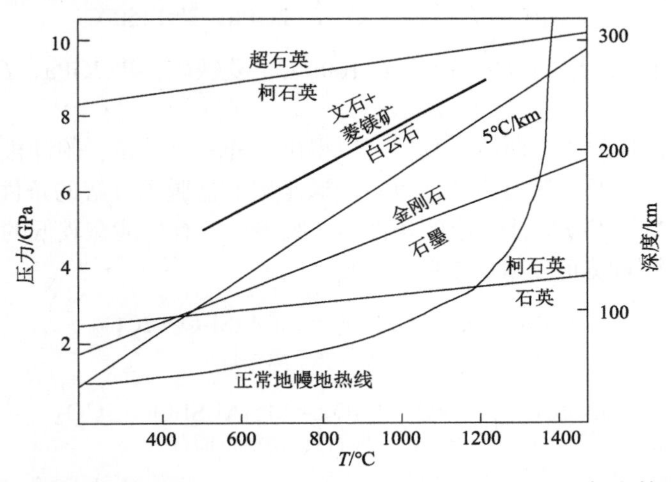

图5 白云石分解为菱镁矿+文石及若干关键相变边界 *P-T* 相图（据文献[22]修改）

然而Tao等[23]的实验给出了与 *S-K* 平行但压力相对较低的的白云石分解相变线（DD反应），白云石可以在700°C，*P* < 5.3GPa（深度 < 160km）的深度保持稳定。在深度>160km时分解为菱镁矿和文石（图6(a)）。造成这一差异的原因可能是不同实验压力标定有误差[23]。此外，Tao等[23]还发现当白云石为铁质白云石时，它因含Fe增高而使其稳定域相变压力剧烈下降（AD反应）（图6(a)）。因此，对典型冷俯冲板片，其 *P-T* 往往处于DD反应线和AD反应线之间，为白云石和菱镁矿共存区。热力学计算也表明且在同一压力条件下，随白云石铁含量不同会导致产生白云石（Dol）、白云石+菱镁矿+文石（Dol+Mgs+Argue）和菱镁矿+文石（Mgs+Argue）3个稳定区（图6(b)）[23]。

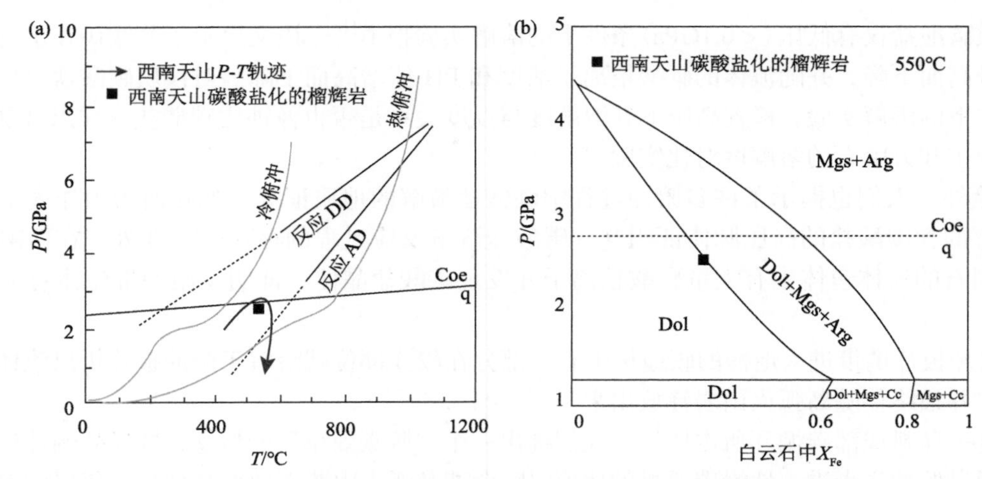

图 6 白云石分解为菱镁矿+文石的 *P-T* 相变线（DD 反应）和铁白云石分解为文石+菱铁矿的 *P-T* 相变线（AD 反应）并与典型俯冲板片的 *P-T* 路径比较（a）和含铁白云石的热力学计算 $P\text{-}X_{Fe}$ 相图（b）（据文献[23]）

2.3 碳酸盐的脱碳分解反应

我们关注的一个科学问题是碳酸盐在板块俯冲时可否直接发生变质脱碳分解反应（decarbonation）？实验表明菱铁矿最易发生脱碳分解反应[24]。

$$\text{菱铁矿} \rightarrow \text{磁铁矿} + CO_2 + \text{石墨}（P=2\text{GPa}，T=800°\text{C}）$$

在相同压力下，菱镁矿的分解温度远高于菱铁矿。

$$\text{菱镁矿} \rightarrow \text{方镁石} + CO_2（P=2\text{GPa}，T=1450°\text{C}）$$

在更高的压力和温度条件下（菱镁矿：*P*>3GPa，*T*>1600°C；菱铁矿：*P*>7GPa，*T*>1500°C）碳酸盐可直接发生熔融。

根据这些实验和俯冲板片的 *P-T* 曲线，我们可以得出这样一个结论：俯冲板片携带的碳酸盐岩不可能直接发生脱碳分解反应。因为即使热板块俯冲也不能满足碳酸盐脱碳分解的条件[24]。

然而，在有二氧化硅参与和碳酸盐反应的条件下，脱碳反应有可能在较低的温压条件下发生。Knoche 等[25]指出碳酸盐与石英或柯石英可发生如下脱碳反应。

$$\underset{\text{白云石}}{CaMg(CO_3)_2} + \underset{\text{柯石英}}{2SiO_2} = \underset{\text{透辉石}}{CaMgSi_2O_6} + 2CO_2$$

$$\underset{\text{菱镁矿}}{3MgCO_3} + \underset{\text{蓝晶石}}{Al_2SiO_5} + \underset{\text{柯石英}}{2SiO_2} = \underset{\text{镁铝石榴石}}{Mg_3Al_2Si_3O_{12}} + 3CO_2$$

$$\underset{\text{方解石}}{3CaCO_3} + \underset{\text{蓝晶石}}{Al_2SiO_5} + \underset{\text{柯石英}}{2SiO_2} = \underset{\text{钙铝石榴石}}{Ca_3Al_2Si_3O_{12}} + 3CO_2$$

在相同压力下，其反应温度低于直接的脱碳分解反应。因此在热板块俯冲条件下，沉积物的碳酸盐岩和硅质岩夹层可发生此类脱碳反应[25]。这可能对太古代板块俯冲有重要意义。

2.4 高温高压条件下碳酸盐和硅酸盐矿物的反应

一个值得关注的科学问题是，俯冲板块携带的碳酸盐是与俯冲洋壳和泥质沉积物一起俯冲到深部的，在此升温升压过程中，碳酸盐和钙镁硅酸盐之间是否发生 Ca-Mg 元素交换反应？它对碳酸盐稳定性有何影响？

Kushiro 等[26]实验表明在高压下（*P*>2300MPa），有斜方辉石存在时，方解石不稳定，它与斜方辉石发生 Ca-Mg 交换反应，生成白云石+透辉石。在 *P*>4500MPa，*T*>1000°C时（或外推 *P*>3200MPa，*T*>700°C），白云石不稳定，而形成菱镁矿+透辉石。这表明随压力增高，Ca 更倾向进入硅酸盐。然而菱镁矿在很高压力下都是稳定的，Isshiki 等[27]实验表明菱镁矿的稳定域可到 P=110GPa，即在大部分下地幔区域菱镁矿都是稳定的。

2.5 碳酸盐化俯冲洋壳和橄榄岩的脱碳熔融

我们关注的科学问题是：(1)碳酸盐化俯冲洋壳能否发生脱碳熔融？(2)碳酸盐再循环进入地幔深部形成碳酸盐化橄榄岩，它们被熔融的条件是什么？

碳酸盐化俯冲洋壳在发生熔融反应前已经变质为榴辉岩，因此 Dasgupta 和 Hirschmann[28]总结了碳酸盐化榴辉岩的发生部分熔融的固相线实验资料并和冷、热俯冲板片顶部与地幔接触界面的 P-T 轨迹进行了比较。图 7 显示一般情况下现代俯冲洋壳的碳酸盐化榴辉岩不可能发生脱碳熔融，因为平均俯冲板块地热增温线无法与碳酸盐化榴辉岩的固相线相交。仅少数热板块俯冲在深度>150km 时碳酸盐岩化榴辉岩会部分熔融。

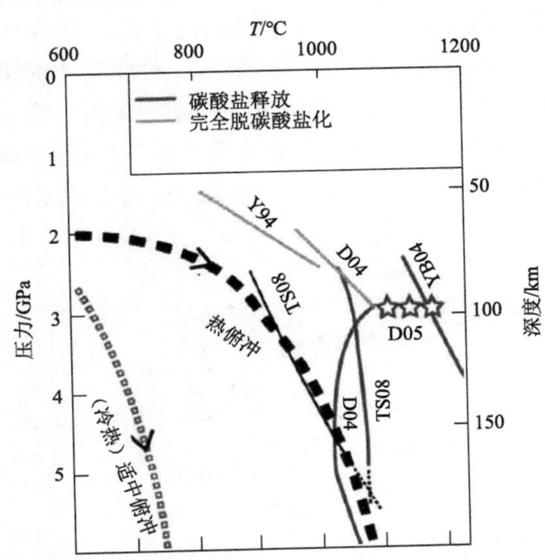

图 7 俯冲碳酸盐化榴辉岩的初始熔融固相线及与俯冲板片顶部 P-T 轨迹比较(据文献[28])

没有被俯冲板块脱水流体溶解的碳酸盐可再循环进入地幔深部与地幔橄榄岩混杂形成碳酸盐化橄榄岩，Dasgupta[29]总结了有关实验资料绘制了碳酸盐化橄榄岩的固相线，并与非碳酸盐化的干橄榄岩固相线比较(图 8)。图 8 表明碳酸盐化橄榄岩可大大降低其固相线温度，在 70km 以下，它比无挥发分橄榄岩

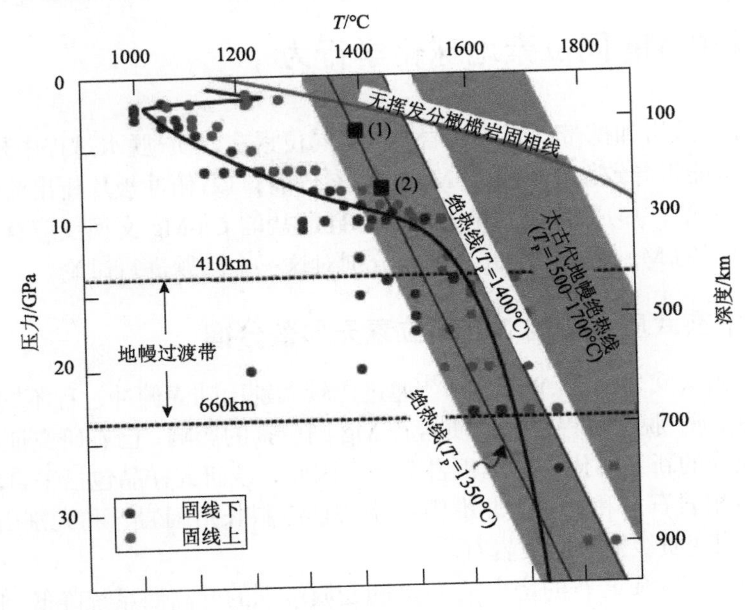

图 8 实验测定的正常干地幔橄榄岩和碳酸盐化橄榄岩初始熔融的固相线图(据文献[29])

固相线(红色)低 300~500℃。根据上地幔绝热线与碳酸盐化橄榄岩固相线的交点，现代地幔上涌至 300km 深度，其碳酸盐化橄榄岩就开始熔融。这个深度释放的 C 以氧化态形式(CO_2或碳酸盐)存在。而对无挥发分橄榄岩则要上升至 70km 才开始减压熔融。因此，碳酸盐化地幔的脱碳熔融效率高于普通地幔。据此我们可知再循环到地幔深部的碳酸盐，其脱碳熔融导致的 CO_2 再逸出主要发生在洋中脊玄武岩(MORB，大陆玄武岩)和与地幔热柱有关的海岛玄武岩(OIB)喷发过程。

2.6　小结

综合所有实验数据，俯冲板片进入上地幔后，碳酸盐的稳定区如下：在P>2300MPa(>69 km)在有Opx存在情况下，方解石不稳定，反应形成白云石+透辉石。在一定温度下，压力升高可导致白云石不稳定，并分解为菱镁矿+文石；其分解压力随白云石Fe含量增高而下降。故在P-T图上，DD(白云石分解线)和AD(菱铁矿分解线)线之间是白云石+菱镁矿的共存区(图6a)；对冷俯冲板块，T=400~850℃，P=2.5~6.0GPa，是白云石+菱镁矿的共存区(图9)。当P>6GPa(>180km)，T<850℃时，白云石不稳定，含镁碳酸盐仅有菱镁矿存在(图 9)。需要指出的是，图 9 显示的是在冷俯冲板片的P-T条件下碳酸盐相变模型，对于 T>1000℃的地幔楔情况则有不同，对应 T=1000℃，P>7GPa(>210km)白云石才变得不稳定(图6a)。

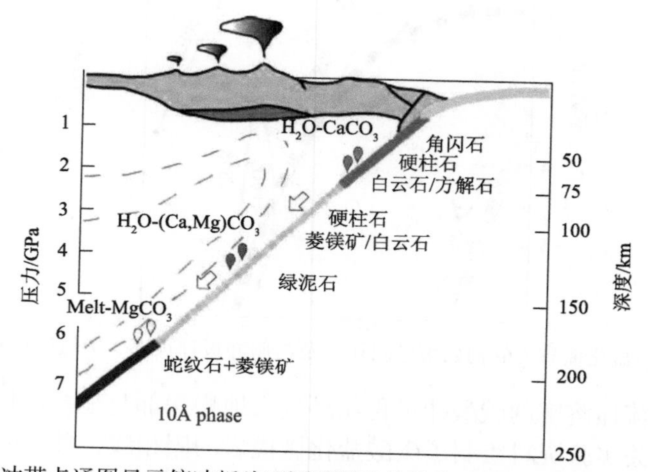

图 9　俯冲带卡通图显示俯冲板片不同碳酸盐的稳定 P-T 区间(据文献[30]修改)

3　板块俯冲过程中 Mg 同位素地球化学行为

我们关注俯冲洋壳玄武岩和携带的碳酸盐岩的 Mg 同位素在变质–脱水过程中是否发生变化。这包括 3 方面问题：(1)变质脱水能否导致俯冲洋壳的 Mg 同位素分馏；(2)俯冲板片析出流体对碳酸盐的溶解反应可能导致的 Mg 同位素分馏；(3)俯冲板片的碳酸盐和硅酸盐的 Ca-Mg 交换反应有可能伴随有 Mg 同位素交换作用，从而改变它们的 Mg 同位素组成。下面分别对这三个问题进行讨论。

3.1　板块俯冲过程中玄武质洋壳的 Mg 同位素无显著分馏

俯冲大洋地壳主要由玄武岩组成。Wang 等[31]通过比较大别山陆壳俯冲过程不同程度变质基性岩的 Mg 同位素组成，探讨俯冲变质–脱水过程对玄武质岩石 Mg 同位素的影响。已有研究证明，大别山超高压变质岩的源岩是华南陆块北缘的新元古代火山沉积岩系[32]。因此，该研究样品包括采自华南陆块(SCB)北缘武当群新元古代的绿片岩相岩石、和大别山北淮阳地体(BHY)斜长角闪岩、和大别山超高压变质带(CD 和 SD)的榴辉岩等 3 种不同变质级别玄武质岩石。

通过比较发现三类变玄武质岩石的烧失量(LOI)随变质级别的升高而系统降低，但是MgO 含量和δ^{26}Mg 值并未表现出系统变化(图 10)。绿片岩、角闪岩和榴辉岩的 Mg 同位素均位于正常地幔值((–0.25±0.07)‰)

范围之内，三者不存在系统分异。这说明在前进变质脱水过程中，Mg 同位素没有发生显著分馏。变基性岩的 Mg 同位素的一定变化范围主要反映了源岩不均一性[31]。

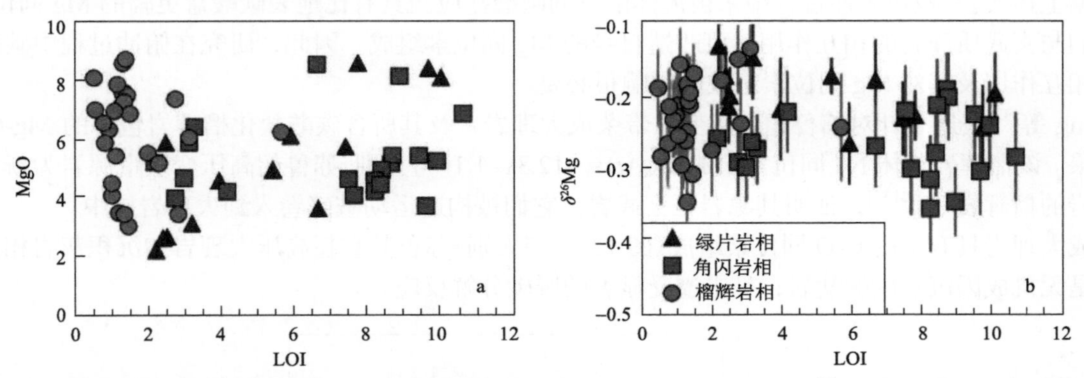

图 10 大别造山带绿片岩、斜长角闪岩、和榴辉岩的烧失量(LOI)对 MgO 含量和 δ^{26}Mg 值图(数据引自文献[31])

图 11 显示低级变玄武质岩石的 H_2O 含量与 MgO 含量呈正相关，表明 H_2O 主要与含 Mg 矿物结合(如绿泥石、角闪石)。随变质级别的升高 H_2O 含量系统降低，但是 MgO 含量并未表现出系统变化，说明在前进变质脱水过程中，Mg 元素被新生变质矿物(辉石和石榴子石)继承而未发生显著丢失。因此，变质脱水过程不会发生 Mg 同位素分馏。

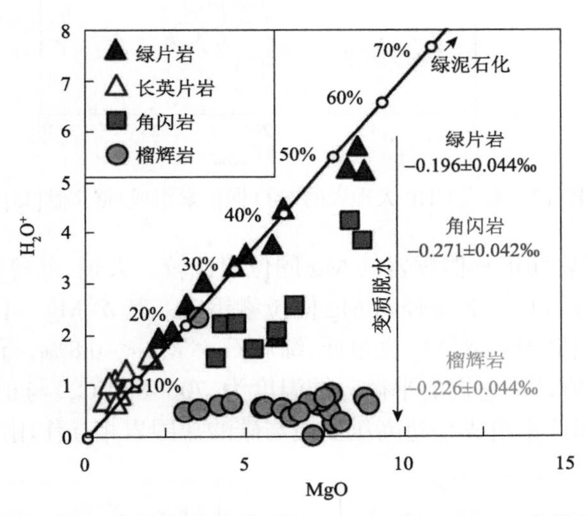

图 11 大别造山带绿片岩、斜长角闪岩和榴辉岩的 H_2O 含量对 MgO 含量关系图(据文献[31])

图中−0.196‰，−0.271‰和−0.226‰分别对应绿片岩，斜长角闪岩和榴辉岩的平均 δ^{26}Mg 值

3.2 俯冲板片碳酸盐部分溶解可能导致的 Mg 同位素分馏

目前只有常温常压下的碳酸盐溶解实验。该实验表明，溶于水的 Mg 倾向重同位素，其分馏系数 Δ^{26}Mg= −2.80+0.0107×T(℃)[9]。尽管该实验是在常温条件(T≤40℃)下做的，不好外推到俯冲板块变质温度(T=450~500℃)，如果溶于水的 Mg 倾向重同位素这一趋势不变，可以推测交代地幔楔的俯冲板片脱水流体应该具有较碳酸盐重的 Mg 同位素组成。因此，岛弧玄武岩的 Mg 同位素不会显著偏离地幔值，Mg 同位素示踪岛弧岩浆系统的碳循环可能不如示踪进入地幔深部的再循环碳酸盐敏感。显然，这一根据常温试验所做的推论有很大不确定性，我们特别需要进行高温、高压条件下的碳酸盐溶解实验以便观察其 Mg 同位素分馏情况。

3.3 俯冲过程中碳酸盐与硅酸盐相互作用对 Mg-O 同位素的影响

根据上述实验，板块俯冲过程中未被流体溶解的碳酸盐应当具有比地表碳酸盐更轻的 Mg 同位素组成，然而它们与玄武质洋壳的相互作用又可改造自身的 Mg 同位素组成。因此，研究在俯冲过程中碳酸盐与玄武岩的相互作用及其对 Mg 同位素组成的影响很必要。

Wang 等[33]报道了针对苏鲁超高压变质带荣成大理岩，及其所含碳酸盐化榴辉岩包体的 Mg-O 同位素研究结果。该榴辉岩包体 Nd 同位素组成（$\varepsilon_{Nd}(t)=-12.3\sim-1.1$）与大别-苏鲁超高压变质带源岩为新元古代基性火山岩的榴辉岩一致[33]，证明其源岩是玄武岩，它们被构造运动破碎卷入到大理岩层中。

荣成大理岩具有重的 C-O 同位素组成(图 12)，与大别-苏鲁其它超高压大理岩和沉积灰岩相同，这说明它们是无机成因沉积碳酸盐岩，且未遭受显著的脱碳分解反应。

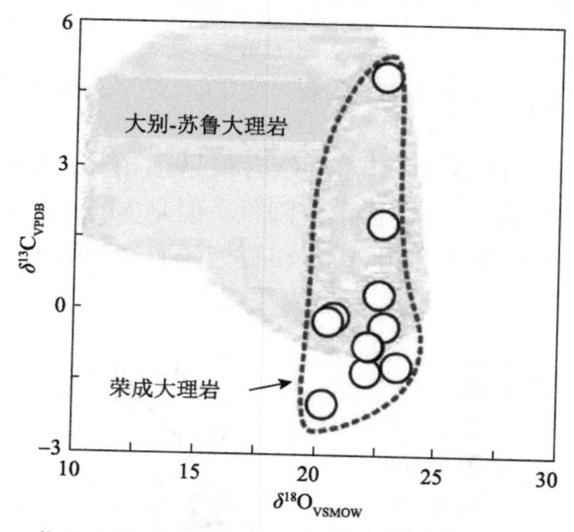

图 12　荣成 UHP 大理岩的 C-O 同位素组成(据文献[33])

图 13 显示碳酸盐化榴辉岩和正常榴辉岩的 Mg 同位素比较。大别-苏鲁带普通榴辉岩的 Mg 同位素接近地幔值。荣成碳酸盐化榴辉岩具有非常轻的 Mg 同位素组成。其 δ^{26}Mg=−1.9‰~−0.6‰(图 13a)。荣成碳酸盐化榴辉岩的石榴石和单斜辉石的 δ^{26}Mg 也很低，绿辉石 δ^{26}Mg <-0.8‰，石榴石 δ^{26}Mg <-1.8‰(图 13b)。根据它们之间 Mg 同位素差值计算它们的平衡分馏温度为 700~880°C，与正常榴辉岩的变质温度相同(图 13b)，说明其非常低的 Mg 同位素组成是超高压变质与碳酸盐围岩相互作用的结果。

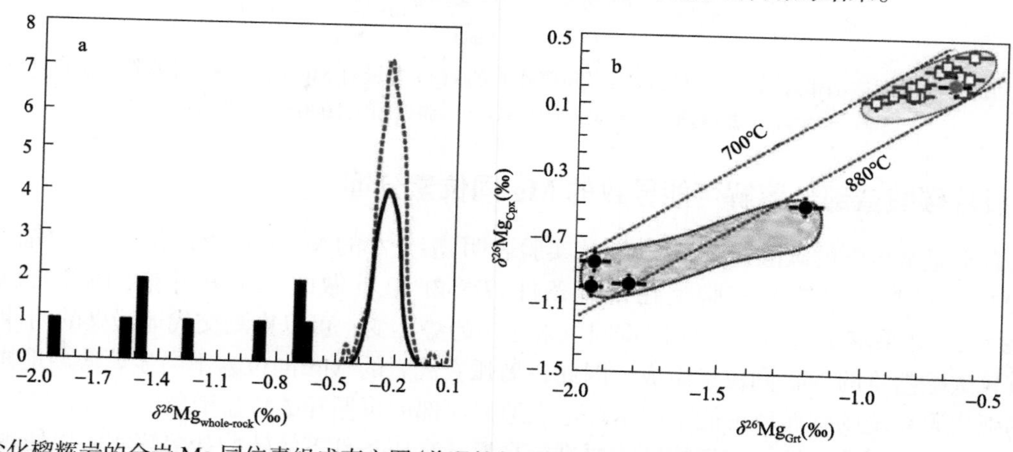

图 13　碳酸盐化榴辉岩的全岩 Mg 同位素组成直方图(普通榴辉岩和其源岩(华南新元古代玄武岩)进行比较)(a)和碳酸盐化榴辉岩、普通榴辉岩的主要矿物 Mg 同位素组成(石榴石(Grt)和绿辉石(Cpx)的 Mg 同位素组成均落在 700~800°C平衡分馏线上)(b)(据文献[33])

荣成 UHPM 大理岩和扬子陆块北缘新元古代浅变质大理岩 Mg 同位素比较显示新元古代浅变质大理岩由钙质灰岩和白云岩两个端元组成，缺少富镁方解石中间态(图 14a，14b)。它们的 Mg 同位素与地表沉积碳酸盐岩相同：其 Mg 同位素值与 MgO 含量呈正相关。Ca 质灰岩 δ^{26}Mg 低至–4.2‰；白云岩最低仅–2.5‰(图 14a)。与此相反，荣成 UHPM 大理岩由方解石大理岩–富镁方解石大理岩–白云质大理岩组成，其 MgO 含量呈连续变化。它们的 Mg 同位素值与 MgO 含量呈负相关，与地表沉积碳酸盐岩相反。Ca 质大理岩 δ^{26}Mg 升高至–0.5‰；而白云质大理岩最低仍维持在–2.5‰(图 14)。

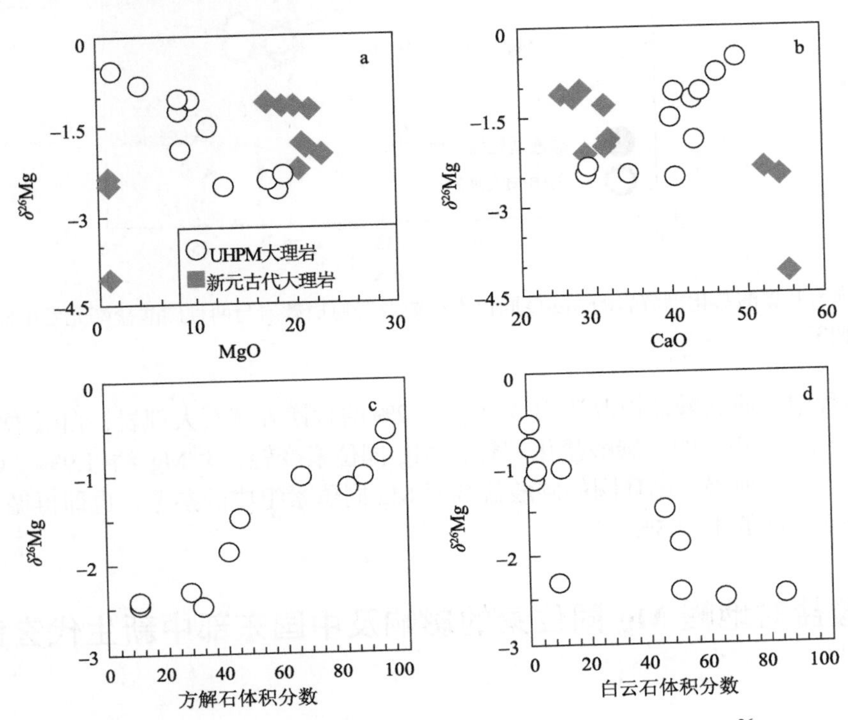

图 14 荣成 UHPM 大理岩和扬子陆块北缘新元古代沉积碳酸盐岩的 δ^{26}Mg–MgO 图(a)；δ^{26}Mg–CaO 图(b)；δ^{26}Mg–UHPM 大理岩的方解石含量图(c)和 δ^{26}Mg–UHPM 大理岩的白云石含量图(d)(据文献[33])

此外图 15 还显示，荣成碳酸盐化榴辉岩还具有重的氧同位素组成，其 δ^{18}O=16‰~22‰，远高于地幔值，而接近 UHP 大理岩的氧同位素组成。荣成碳酸盐化榴辉岩这种低 δ^{26}Mg，高 δ^{18}O 特征不能用碳酸盐与榴辉岩简单机械混合来解释，因为它们的绿辉石和石榴石同样具有低的 δ^{26}Mg 值，且模拟计算表明简单混合需要太多的(60~90%)碳酸盐加入，因而不可能。Wang 等[33]论证了上述碳酸盐化榴辉岩和 UHPM 大理岩 Mg 同位素变化只能用碳酸盐岩与硅酸岩在高温–高压条件下相互发生同位素交换反应的结果。由于白云岩和 Ca 质灰岩的 Mg 含量有很大差异，但氧含量相同，它们与榴辉岩在 Mg-O 同位素图上的同位素交换(混合)曲线具有完全不同的曲率(图 15)，这可被用来判断反应机制。

图 15 显示，白云岩(MgO/CaO=0.31~0.64)位于白云岩端元附近，这说明白云岩的 Mg-O 同位素基本没有发生变化，它基本没与榴辉岩发生同位素交换反应；而钙质碳酸盐(MgO/CaO=0.03~0.28)沿着 Ca 质灰岩与普通榴辉岩的同位素交换混合线分布(图 15)，这说明是钙质碳酸盐吸收了硅酸盐的 Mg 而变成富镁方解石，且使 Mg 同位素变重。相应的，由于同位素交换，碳酸盐化榴辉岩 Mg 同位素变轻，O 同位素变重。这一同位素示踪结果与前述高温高压条件下碳酸盐与硅酸盐反应实验结果一致[26]，即在高压条件下，方解石的 Ca 更容易进入斜方辉石，从而变为白云石+透辉石组合。

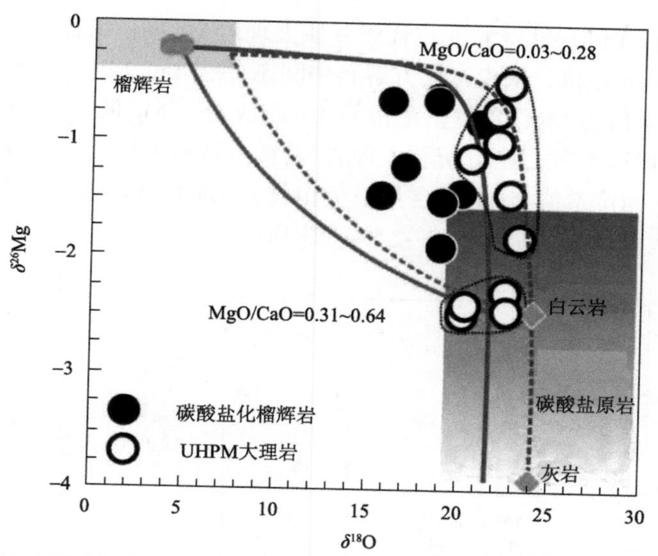

图 15　荣成 UHP 大理岩和碳酸盐化榴辉岩的 Mg-O 同位素组成及普通榴辉岩与两种碳酸盐端元发生同位素交换反应的混合模拟计算曲线(据文献[33])

结论：俯冲过程中钙质碳酸盐被改造为 δ^{26}Mg < −2‰的含镁方解石大理岩；白云岩镁同位素组成保持不变，δ^{26}Mg 约−2.5‰ 。相应的，碳酸盐化榴辉岩 Mg 同位素变轻，δ^{26}Mg 约−1.9‰。O 同位素变重。这些值可作为判断进入深部地幔再循环圈的碳酸盐端元 Mg 同位素组成的参考，也即再循环进入地幔的碳酸盐其端元 δ^{26}Mg 值不可能低于−2.5‰。

4　再循环碳酸盐对地幔 Mg 同位素的影响及中国东部中新生代玄武岩镁同位素研究

根据前述碳酸盐在板块俯冲过程中可能发生的反应深部碳循环基本可划分为两个循环圈：

(1) 岛弧岩浆系统循环圈(island arc recycling)。俯冲板块脱水部分溶解碳酸盐(partial dissolution of carbonate)，使其进入岛弧火山系统。前面讨论已表明碳酸盐部分溶解过程可导致俯冲板片析出流体所溶解的镁的 Mg 同位素变重，因此 Mg 同位素示踪岛弧岩浆系统的碳循环可能不太敏感。

(2) 深部地幔循环圈(deep mantle recycling)。俯冲板片中未被析出流体溶解的碳酸盐可再循环进入地幔深部，最终通过碳酸盐化橄榄岩部分熔融进入 MORB 或板内火山系统。前面已介绍，这部分再循环碳酸盐和碳酸盐化榴辉岩具有较低的 δ^{26}Mg 值，它可以造成局部地幔具有轻的 Mg 同位素异常值。因此，我们关注俯冲板块中未溶解碳酸盐再循环对深部地幔碳循环的示踪效果。

中东部晚白垩世和新生代玄武岩源自受西太平洋板块俯冲影响的上地幔。一个令人感兴趣的问题是西太平洋板块俯冲是否导致巨量碳酸盐再循环进入该区地幔，并影响其 Mg 同位素组成？已有两项工作[34-35]分别测定了华北和华南部分中新生代玄武岩 Mg 同位素。它们均具有较轻的 Mg 同位素组成(图 16)。

4.1　华北中新生代玄武岩 Mg 同位素研究

Yang 等[34]研究了华北克拉通中、新生代玄武岩的 Mg 同位素地球化学。根据岩石的年龄和地球化学特征可将其分成两组：

(1) >120 Ma 的辽西义县组玄武岩，具有 LREE 和大离子亲石元素富集和低 ε_{Nd} 同位素特征。它们的 ε_{Nd}=−13.4~−9.7，显示其地幔源区长期富集 LREE 或受古老下地壳混染。但是它们的 Mg 同位素组成平均值 δ^{26}Mg=(−0.27±0.05)‰，具有类似于地幔的 Mg 同位素组成特征(图 17)。

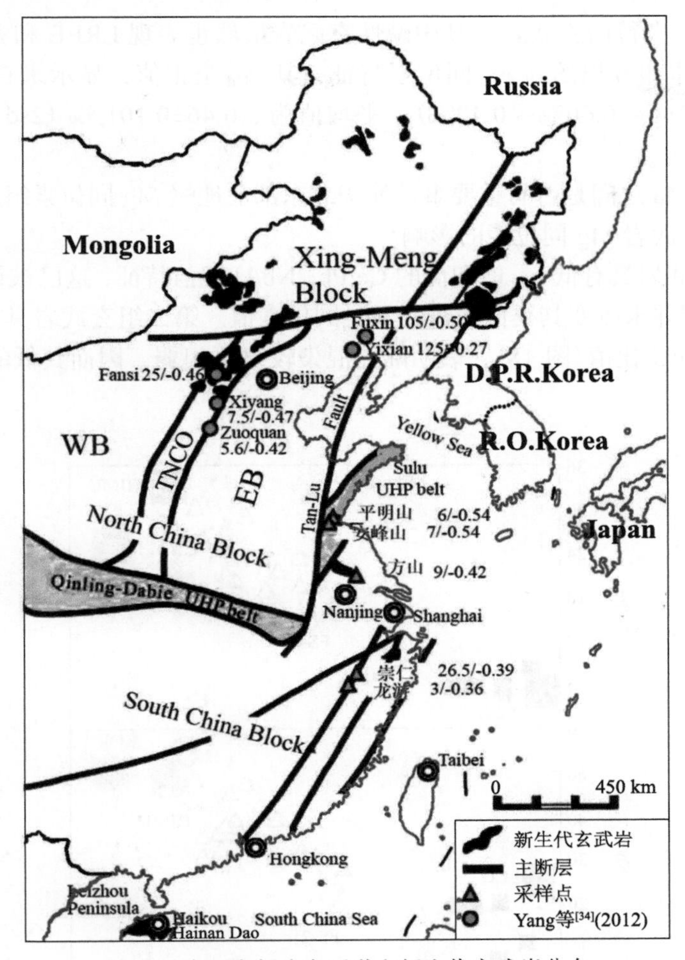

图 16 中国东部晚白垩世和新生代玄武岩分布

（据文献[34-35]修改）黄圈和黄三角分别指示 Yang 等[34]（2012）和 Huang 等[35]（2015）分析了 Mg 同位素组成的采样地区。数值含义为该地玄武岩平均“年龄 Ma/δ^{26}Mg 值”

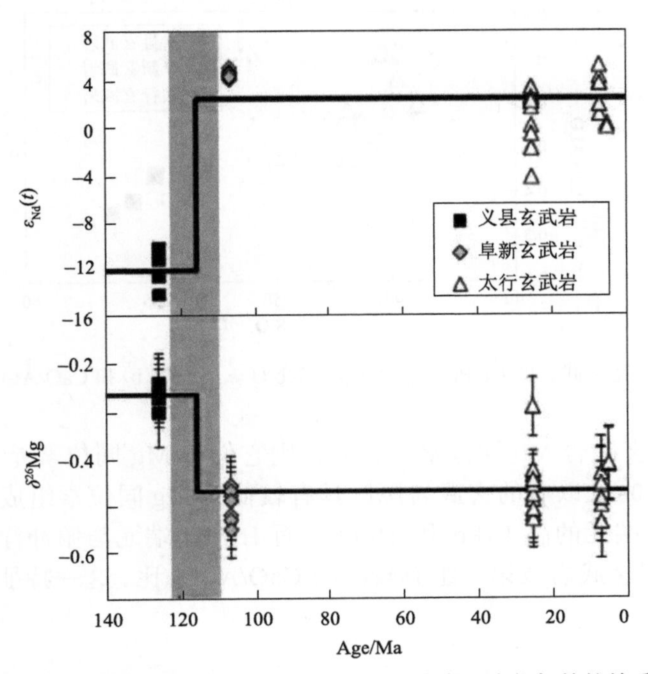

图 17 华北义县，阜新和太行玄武岩的 Nd，Mg 同位素组成与年龄的关系（据文献[34]）

(2) < 110 Ma 阜新和太行山玄武岩，其中碱性玄武岩虽然也表现 LREE 和大离子亲石元素富集，但它们有含地幔橄榄岩包体和亏损的高 ε_{Nd} 同位素特征，其 ε_{Nd} 是正值，显示来自亏损地幔。但是它们的 Mg 同位素组成偏低(δ^{26}Mg=–0.60‰~–0.42‰)，平均值为(–0.46±0.10) ‰ (2sd)，显著低于地幔平均值(图 17)。

为了判明在 120-110Ma 之间是否有重要事件使华北东部上地幔 Mg 同位素组成发生了显著变化，我们还需要排除地壳混染对玄武岩 Mg 同位素的影响。

图 18 显示第一组玄武岩具有低 ε_{Nd} 值和低的 Ce/Pb，Nb/U 比值特征，这已被认为是拆沉大陆下地壳混杂上地幔的结果[36-37]，但并未改变其类似地幔的 Mg 同位素值。第二组玄武岩具有亏损地幔型的 Nd 同位素特征和高的 Ce/Pb，Nb/U 比值(图 18)，表明陆壳混染较第 1 组弱，因而其低的 δ^{26}Mg 值不是陆壳混染结果。

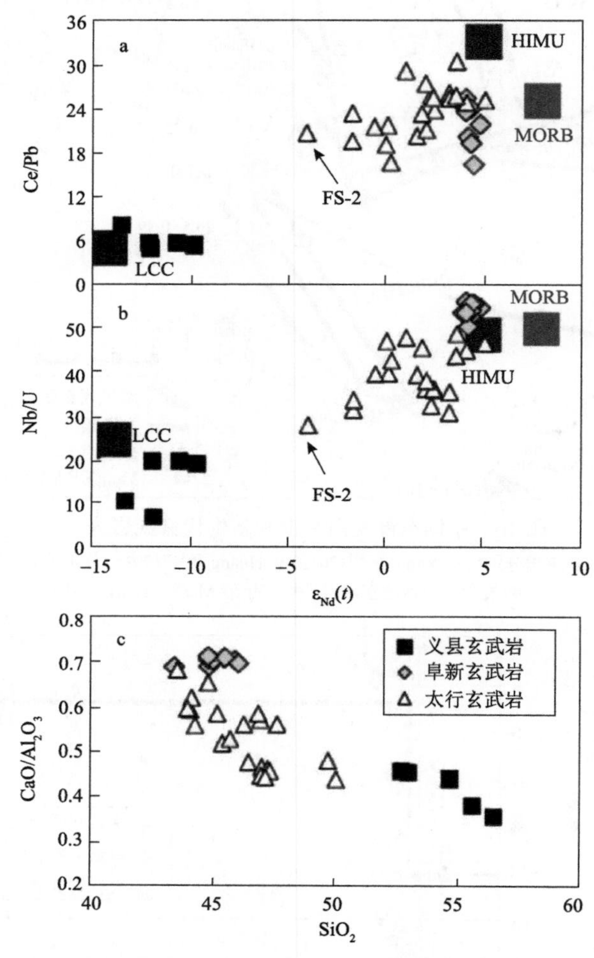

图 18　华北中新生代两类玄武岩的 Ce/Pb–$\varepsilon_{Nd}(t)$ (a) 和 Nb/U–$\varepsilon_{Nd}(t)$ 图(b) 和 CaO/Al_2O_3–SiO_2 图(c) (据文献[34])

排除了陆壳混染对玄武岩 Mg 同位素的影响，则它们的 Mg 同位素差异主要是地幔源区的差异。那么是什么事件导致 110Ma 以后的玄武岩源区具有较低的 Mg 同位素组成呢？图 19 显示第二组玄武岩具有类似地幔 HIMU 端元的高 U/Pb 和 Th/Pb，而 HIMU 端元与俯冲洋壳再循环有关。此外图 18 和图 19 均还显示第二组玄武岩较第一组有较高的 CaO/Al_2O_3 比，这一特征与富 CaO 碳酸盐再循环推论相符。

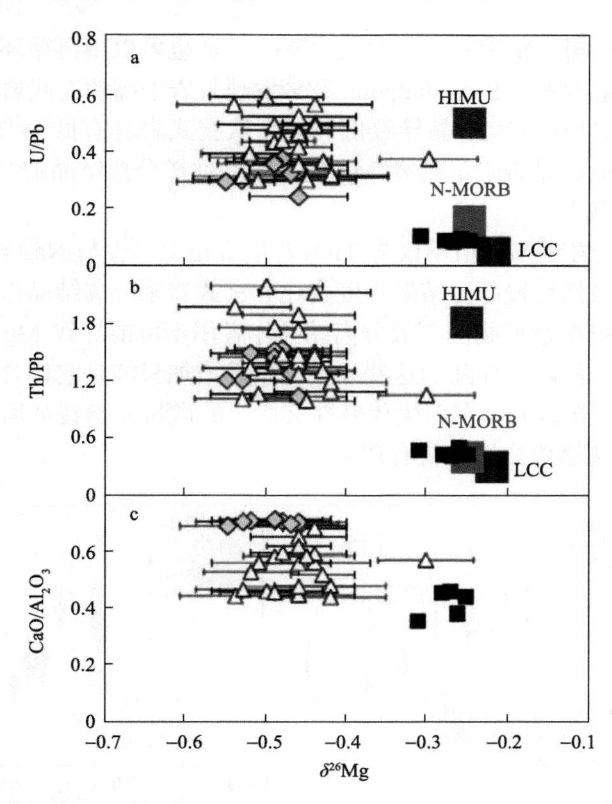

图 19 华北中新生代两类玄武岩的 U/Pb (a)、Th/Pb (b) 和 CaO/Al_2O_3 (c) 对 $\delta^{26}Mg$ 图 (据文献[34])

因此，Yang 等[34]的结论是第二组玄武岩的轻 Mg 同位素组成最可能是俯冲洋壳携带再循环沉积物中碳酸盐对上地幔的混染造成的。

4.2 华南部分新生代玄武岩 Mg 同位素研究和低 $\delta^{26}Mg$ 玄武岩成因

Huang 等[35]分析了山东、苏北、浙江部分新生代玄武岩的 Mg，Nd 同位素和主、微量元素，发现它们与华北新生代玄武岩一样具有正的 ε_{Nd} 值和低的 $\delta^{26}Mg$ 值(–0.6‰~–0.35‰)(图 20)，以及高的 Ce/Pb，Nb/U，Th/Pb，U/Pb 和 CaO/Al_2O_3 比值。

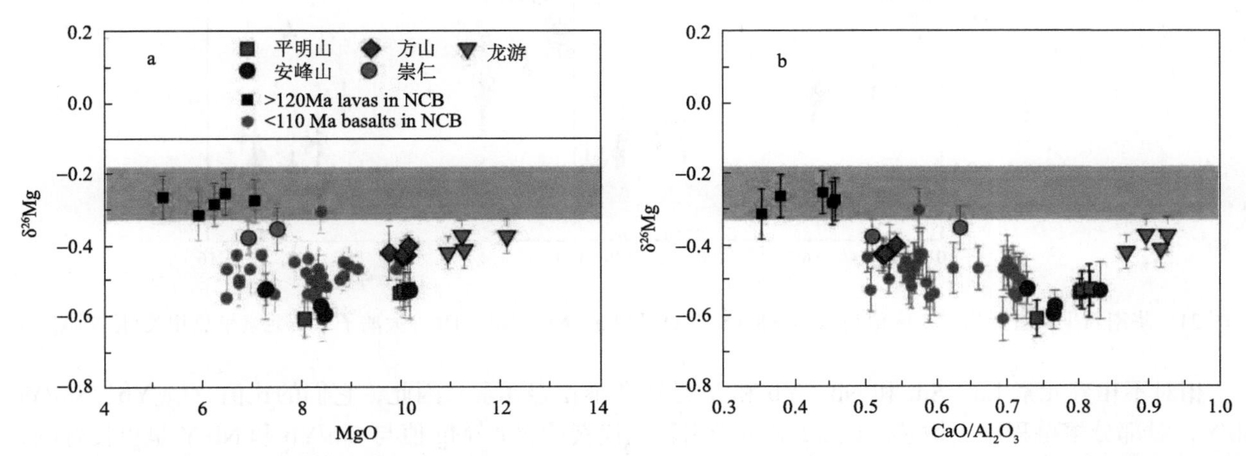

图 20 华南新生代玄武岩(样品采自平明山、方山、安峰山、龙游、崇仁)的 $\delta^{26}Mg$ 值对 MgO 含量和 CaO/Al_2O_3 比值图(数据引自文献[35])

灰色圆点代表华北中新生代玄武岩[34]；灰色带是正常地幔值(–0.25±0.07)[8]

按照 Yang 等[34]的论证逻辑，华南玄武岩的低 $\delta^{26}Mg$ 特征也可以用再循环洋壳及碳酸盐混染地幔来解释。但是，这一解释遇到了新挑战。Sedaghatpour 等[38]发现月岩中高钛玄武岩的 $\delta^{26}Mg$ 值可低至–0.61‰。该文推测是月球岩浆海的钛铁矿分离结晶导致低 Mg#高钛玄武岩具有低 $\delta^{26}Mg$ 值。华南大陆玄武岩的 $\delta^{26}Mg$-TiO_2 呈负相关(图 21b)，是否暗示其 $\delta^{26}Mg$ 值也是钛铁矿分离结晶的结果？这对根据玄武岩 $\delta^{26}Mg$ 值示踪深部碳循环是一个挑战。

图 21 显示华南大陆玄武岩 $\delta^{26}Mg$ 值不仅与 TiO_2 含量负相关，也与 Na_2O+K_2O，La，Nb，Nd，Th 等大离子亲石元素呈负相关。虽然地幔部分熔融程度变化和玄武岩浆分离结晶作用都可导致大离子亲石元素的变化，但是已有研究证明硅酸盐矿物的高温分离结晶作者用不可能导致 Mg 同位素的显著分馏[7, 14-16]。因此，华南大陆玄武岩 $\delta^{26}Mg$ 的降低伴随的这些元素含量的增加只能是它们含碳酸盐的地幔源区部分熔融程度降低的结果。Zeng 等[39]在研究山东新生代碱性玄武岩的微量元素特征后指出这些元素含量随部分熔融程度降低而增加是碳酸盐化橄榄岩熔融特有现象。

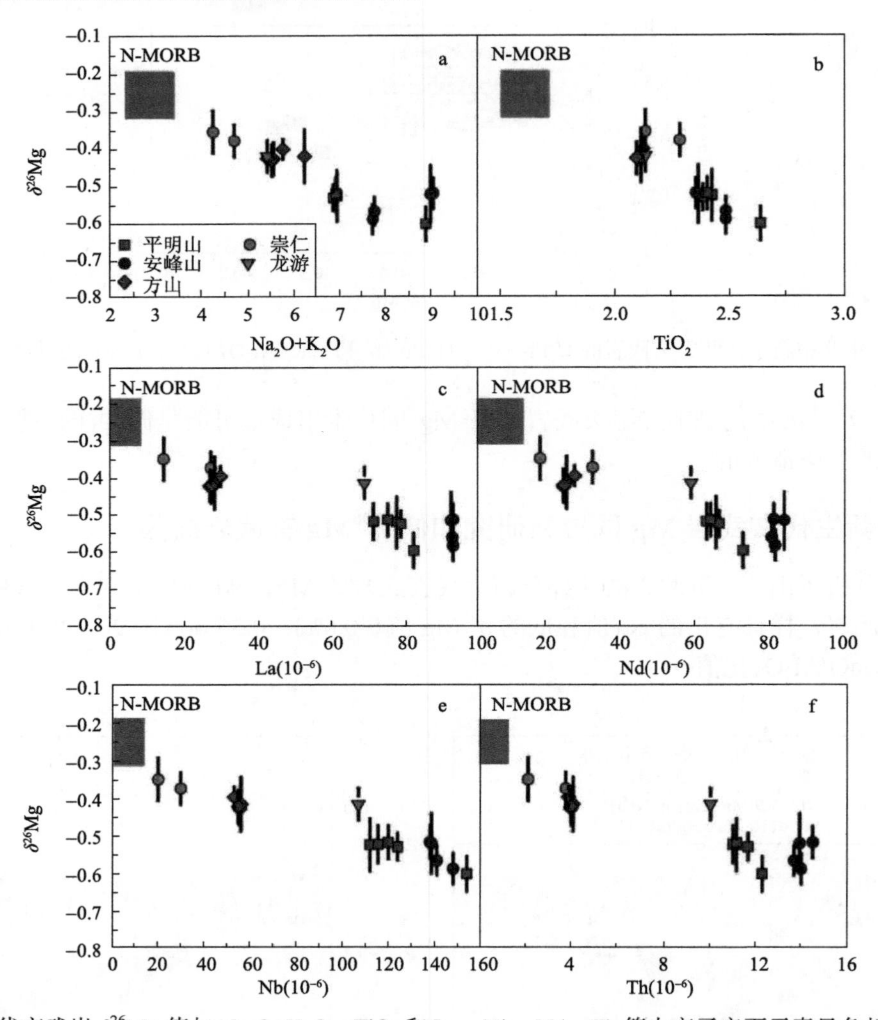

图 21　华南新生代玄武岩 $\delta^{26}Mg$ 值与 Na_2O+K_2O、TiO_2 和 La、Nb、Nd、Th 等大离子亲石元素呈负相关(据文献[35])

相对不相容元素 La、Sm 和 Nb，Yb 和 Y 是轻微不相容元素。因此，它们的比值，La/Yb、Sm/Yb 和 Nb/Y，对部分熔融程度更敏感。图 22 显示华南新生代玄武岩 $\delta^{26}Mg$ 值与 Sm/Yb 和 Nb/Y 呈更良好的线性关系，平衡部分熔融模拟计算表明，它们的比值随部分熔融程度减低而增加(图 22)。如果该部分熔融程度变化不仅影响了熔体的 Sm/Yb 和 Nb/Y，而且如图 22 所示，还同步影响了华南新生代玄武岩的 $\delta^{26}Mg$ 值，则该地幔源区必须是碳酸盐化的橄榄岩，因为如果源区是不含碳酸盐的二辉橄榄岩，已有的研究表明橄榄石、斜方辉石和单斜辉石与玄武岩熔体之间不会产生显著 Mg 同位素分馏[7, 14-16]。图 23 的石榴石橄榄岩平衡部分熔融模拟计算可很好地拟合 La/Yb-Sm/Yb 演化趋势，并显示随着熔融程度降低，La/Yb 和 Sm/Yb 比值升高。

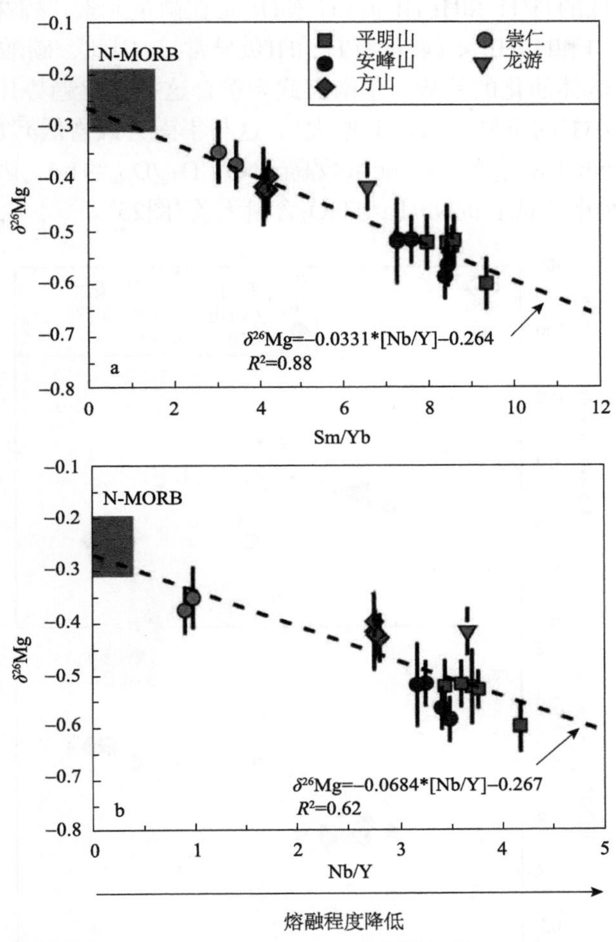

图 22 华南新生代玄武岩 δ^{26}Mg 值与 Sm/Yb 和 Nb/Y 呈良好线性关系（据文献[35]）

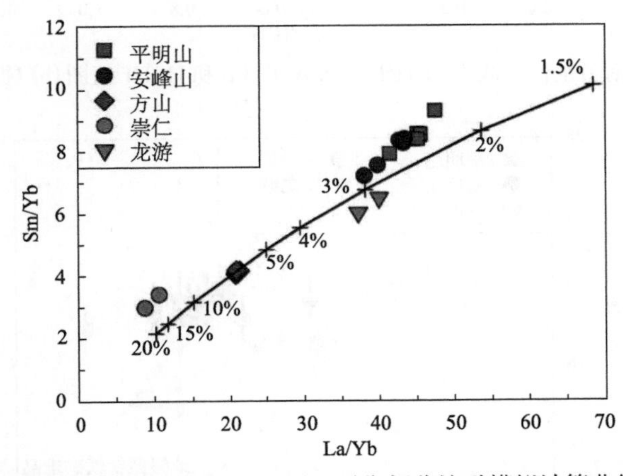

图 23 华南新生代玄武岩 Sm/Yb-Nb/Y 图及平衡部分熔融模拟计算曲线（据文献[35]）

碳酸盐化橄榄岩部分熔融实验研究表明，其低比例熔融的熔体中含碳酸盐比例和 TiO_2、Na_2O 含量均高于高比例熔融熔体[40]，这会导致低比例熔融熔体 δ^{26}Mg 值降低，而高比例熔融熔体 δ^{26}Mg 值升高并接近地幔平均值。这与 Huang 等[35]的结论一致。

Ti-Hf负异常，高Ca/Al和Nb/Ta与δ^{26}Mg的相关性也可用来证明华南玄武岩的低δ^{26}Mg值与钛铁矿堆晶无关，而与地幔源区含沉积碳酸盐有关。碳酸盐熔体具有富Ca、REE，贫Ti、Hf特征，因而具有高Ca/Al，非常大的Ti、Hf负异常或极低的Ti/Ti*和Hf/Hf*值（Ti*和Hf*是在微量元素“蜘蛛”图上不显示Ti和Hf异常应对应的Ti和Hf含量，Ti/Ti$^* < 1$和Hf/Hf$^* < 1$表示有Ti、Hf负异常）。因此，碳酸盐化橄榄岩低比例熔融熔体应当具有从MORB向碳酸盐熔体演化的趋势。华南玄武岩符合这一演化趋势（图24）。反之，钛铁矿的堆晶提高岩石TiO_2含量，从而减小Ti的负异常（Ti/Ti*增大）。这与华南玄武岩低δ^{26}Mg值样品具有低Ti/Ti*矛盾。此外，钛铁矿能够强烈分异Nb/Ta，它的Ta分配系数高于Nb，D_{Ta}/D_{Nb}约1.3。因此，钛铁矿堆晶可降低熔体的Nb/Ta，呈负相关系，但华南玄武岩的Nb/Ta与TiO_2含量无关（图25）。

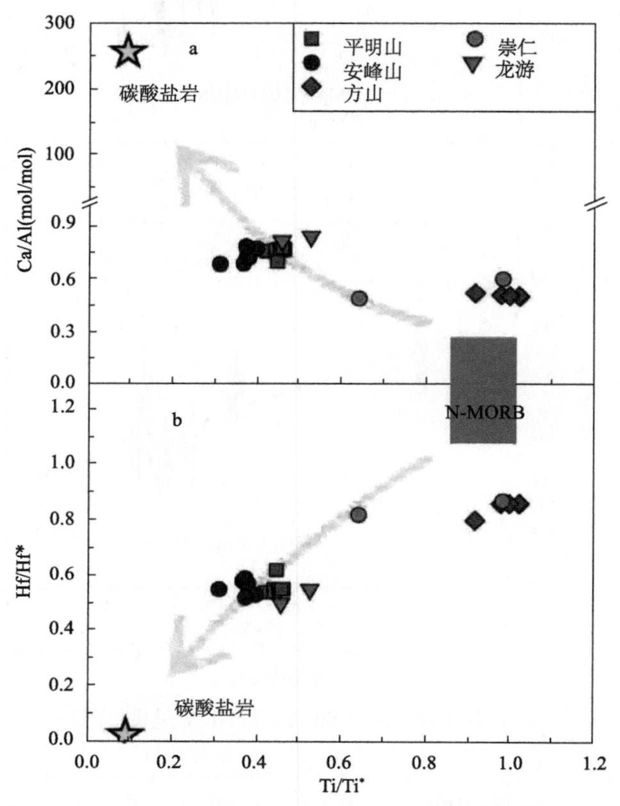

图 24　华南新生代玄武岩的 Ti/Ti*-Ca/Al 图(a)和 Hf/Hf*值图(b)（据文献[35]）

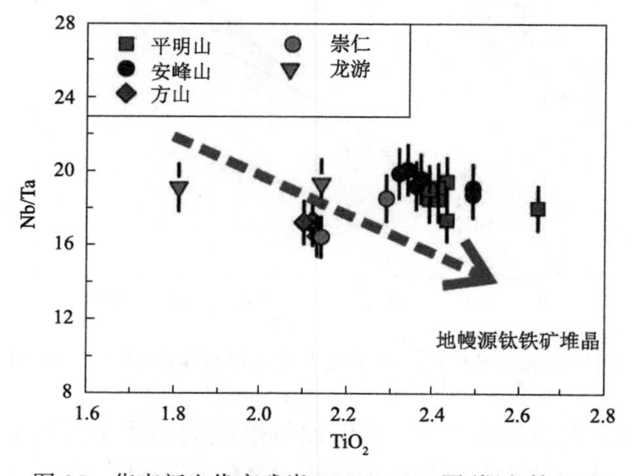

图 25　华南新生代玄武岩 Nb/Ta-TiO_2图（据文献[35]）

Zeng 等[39]在研究山东新生代碱性玄武岩时指出，碳酸盐化橄榄岩熔体在 $Na_2O+K_2O-TiO_2$ 图上有不同于碳酸盐化辉石岩或碳酸盐化榴辉岩以及角闪岩的趋势，后者都有较高的 TiO_2 含量。图 26 显示华南玄武岩的是源于碳酸盐化橄榄岩而非源于碳酸盐化俯冲洋壳(榴辉岩)的部分熔融。它是被俯冲洋壳带到地幔深部的再循环碳酸盐交代地幔橄榄岩而后发生部分熔融的结果。Xiao 等[41]报道的山东昌乐北岩的单辉橄榄岩(wehrlite)包体是这样被再循环碳酸盐交代的实例。该类包体是熔体交代二辉橄榄岩成因，含方解石、磷灰石、角闪石、金云母等交代矿物。Xiao 等[41]发现该地二辉橄榄岩包体具有正常地幔 Mg 同位素组成，但交代成因的单辉橄榄岩具有低的 δ^{26}Mg 值(低至–0.39‰)。这被解释为是岩石圈地幔被具有低 δ^{26}Mg 值的熔体交代成因。

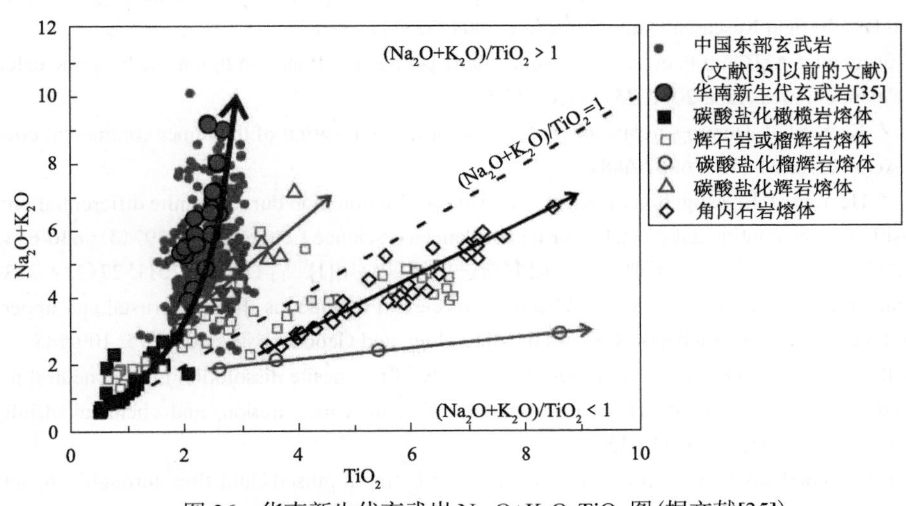

图 26 华南新生代玄武岩 $Na_2O+K_2O-TiO_2$ 图(据文献[35])

4.3 中国东部晚白垩世-新生代玄武岩低 δ^{26}Mg 值的地球动力学意义

华北和华南晚白垩世-新生代玄武岩低 δ^{26}Mg 值特征的发现及成因研究证明整个中国东部上地幔存在被再循环碳酸盐碳交代的碳酸盐化橄榄岩，它是中国东部晚白垩世-新生代玄武岩的源区，其分布范围南北跨度大于 1800km。这一空间分布特点说明影响该地幔的再循环碳酸盐只能是与太平洋板块俯冲有关。因此，中国东部晚白垩世-新生代玄武岩低 δ^{26}Mg 值特征证明了太平洋板块俯冲可造成巨量沉积碳酸盐的再循环，同时它也是太平洋板块俯冲影响中国东部陆下上地幔的有力证据。

感谢李伟强对本文的建议和提供修改图1图件。

参考文献

[1] Javoy M, Pineau F, Allegre C J. Carbone geodynamic cycle[J]. Nature, 1982, 300(11): 171-173.

[2] 张宏铭，李曙光. 深部碳循环及同位素示踪：回顾与展望[J]. 中国科学(D 辑)：地球科学, 2012, 42(10):1459-1472.

[3] Walter M J, Kohn S.C., Araujo D, et al. Deep Mantle Cycling of Oceanic Crust: Evidence from Diamonds and their mineral inclusions[J]. Science, 2011, 334: 54-57.

[4] Hou Z, Tian S, Yuan Z, et al. The Himalayan collision zone carbonatites in western Sichuan, SW China: Petrogenesis, mantle source and tectonic implication[J]. Earth and Planetary Science Letters, 2006, 244 (1): 234-250.

[5] Liu Y, Berner Z, Massonne H J, et al. Carbonatite-like dykes from the eastern Himalayan syntaxis: geochemical, isotopic, and petrogenetic evidence for melting of metasedimentary carbonate rocks within the orogenic crust[J]. Journal of Asian Earth Sciences, 2006, 26(1): 105-120.

[6] Huang S, Farkaš J, Jacobsen S B. Stable calcium isotopic compositions of Hawaiian shield lavas: Evidence for recycling of ancient marine carbonates into the mantle[J]. Geochimica et Cosmochimica Acta, 2011, 75(17): 4987-4997.

[7] Teng F Z, Wadhwa M, Helz R T. Investigation of magnesium isotope fractionation during basalt differentiation: implications for a chondritic composition of the terrestrial mantle[J]. Earth and Planetary Science Letters, 2007, 261 (1): 84-92.

[8] Teng F Z, Li W Y, Ke S, et al. Magnesium isotopic composition of the Earth and chondrites[J]. Geochimica et Cosmochimica Acta, 2010, 74 (14): 4150-4166.

[9] Li W, Chakraborty S, Beard B L, et al. Magnesium isotope fractionation during precipitation of inorganic calcite under laboratory conditions[J]. Earth and Planetary Science Letters, 2012, 333: 304-316.

[10] Mavromatis V, Gautier Q, Bosc O, et al. Kinetics of Mg partition and Mg stable isotope fractionation during its incorporation in calcite[J]. Geochimica et Cosmochimica Acta, 2013, 114: 188-203.

[11] Li W, Beard B L, Li C, et al. Experimental calibration of Mg isotope fractionation between dolomite and aqueous solution and its geological implications[J]. Geochimica et Cosmochimica Acta, 2015, 157: 164-181.

[12] Chang V T C, Williams R J P, Makishima A, et al. Mg and Ca isotope fractionation during $CaCO_3$ biomineralisation[J]. Biochemical and Biophysical Research Communication, 2004, 323 (1): 79-85.

[13] Wombacher F, Eisenhauer A, Böhm F, et al. Magnesium stable isotope fractionation in marine biogenic calcite and aragonite[J]. Geochimica et Cosmochimica Acta, 2011, 75 (19): 5797-5818.

[14] Li W Y, Teng F Z, Ke S, et al. Heterogeneous magnesium isotopic composition of the upper continental crust[J]. Geochimica et Cosmochimica Acta, 2010, 74 (23): 6867-6884.

[15] Liu S A, Teng F Z, He Y, et al. Investigation of magnesium isotope fractionation during granite differentiation: implication for Mg isotopic composition of the continental crust[J]. Earth and Planetary Science Letters, 2010, 297 (3): 646-654.

[16] 柯珊，刘胜遨，李王晔，等. 镁同位素地球化学研究新进展及其应用[J]. 岩石学报, 2011, 27 (2)：383-397.

[17] Manning C E, Shock E L, Sverjensky D A. The chemistry of carbon in aqueous fluids at crustal and upper-mantle conditions: experimental and theoretical constraints[J]. Reviews of Mineralogy and Geochemistry, 2013, 75: 109-148.

[18] Saldi G D, Schott J, Pokrovsky O S, et al. An experimental study of magnesite dissolution rates at neutral to alkaline conditions and 150 and 200 °C as a function of pH, total dissolved carbonate concentration, and chemical affinity[J]. Geochimica et Cosmochimica Acta, 2010, 74 (22): 6344-6356.

[19] John T, Gussone N, Podladchikov Y Y, et al. Volcanic arcs fed by rapid pulsed fluid flow through subducting slabs[J]. Nature Geoscience, 2012, 5 (7): 489-492.

[20] Frezzotti M L, Selverstone J, Sharp Z D, et al. Carbonate dissolution during subduction revealed by diamond-bearing rocks from the Alps[J]. Nature Geoscience, 2011, 4 (10): 703-706.

[21] Johnston F K B, Turchyn A V, Edmonds M. Decarbonation efficiency in subduction zones: Implications for warm Cretaceous climates[J]. Earth and Planetary Science Letters, 2011, 303 (1): 143-152.

[22] Sato K, Katsura T. Experimental investigation on dolomite dissociation into aragonite+magnesite up to 8.5 GPa[J]. Earth and Planetary Science Letters, 2001, 184 (2): 529-534.

[23] Tao R, Zhang L, Fei Y, et al. The effect of Fe on the stability of dolomite at high pressure: Experimental study and petrological observation in eclogite from southwestern Tianshan, China[J]. Geochimica et Cosmochimica Acta, 2014, 143: 253-267.

[24] Tao R, Fei Y, Zhan L. Experimental determination of siderite stability at high pressure[J]. American Mineralogist, 2013, 98(8/9): 1565-1572.

[25] Knoche R, Sweeney R J, Luth R W. Carbonation and decarbonation of eclogites: the role of garnet[J]. Contribution Mineralogy and Petrology, 1999, 135 (4): 332-339.

[26] Kushiro I. Carbonate-silicate reactions at high pressures and possible presence of dolomite and magnasite in the upper mantle[J]. Earth and Planetar Science Letters, 1975, 28 (2): 116-120.

[27] Isshiki M, Irifune Tetsuo, Hirose K, et al. Stability of magnesite and its high-pressure form in the lowermost mantle[J]. Nature, 2004, 427: 60-63.

[28] Dasgupta R, Hirschmann M M. The deep carbon cycle and melting in Earth's interior[J]. Earth and Planetary Science Letters, 2010, 298 (1):1-13.

[29] Dasgupta R. Ingassing, Storage, and Outgassing of Terrestrial Carbon through Geologic Time[J]. Reviews of Mineralogy and Geochemistry, 2013, 75(1): 183-229.

[30] Poli S, Franzolin E, Fumagalli P, et al. The transport of carbon and hydrogen in subducted oceanic crust: An experimental study to 5 GPa[J]. Earth and Planetary Science Letters, 2009, 278 (3): 350-360.

[31] Wang S J, Teng F Z, Li S G, et al. Magnesium isotopic systematics of mafic rocks during continental subduction[J]. Geochimica et Cosmochimica Acta, 2014, 143: 34-48.

[32] Zheng Y F, Fu B, Gong B, et al. Stable isotope geochemistry of ultrahigh pressure metamorphic rocks from the Dabie-Sulu orogen in China: Implications for geodynamics and fluid regime[J]. Earth-Science Reviews, 2003, 62 (1): 105-161.

[33] Wang S J, Teng F Z, Li S G. Tracing carbonate-silicate interaction during subduction using magnesium and oxygen isotopes[J]. Nature communications, 2014.

[34] Yang W, Teng F Z, Zhang H F, et al. Magnesium isotopic systematics of continental basalts from the North China craton: Implications for tracing subducted carbonate in the mantle[J]. Chemical Geology, 2012, 328: 185-194.

[35] Huang J, Li S G, Xiao Y L, et al. Origin of low δ^{26}Mg Cenozoic basalts from South China Block and their geodynamic implications[J]. Geochimica et Cosmochimica Acta, 2015 (in press).

[36] Yang W, Li S G. Geochronology and geochemistry of the Mesozoic volcanic rocks in Western Liaoning: implications for lithospheric thinning of the North China Craton[J]. Lithos, 2008, 102 (1): 88-117.

[37] Huang F, Li S G, Yang W. Contributions of the lower crust to Mesozoic mantle-derived mafic rocks from the North China Craton: implications for lithospheric thinning[J]. Geological Society, London, Special Publication, 2007, 280 (1): 55-75.

[38] Sedaghatpour F, Teng F Z, Liu Y, et al. Magnesium isotopic composition of the Moon[J]. Geochimica et Cosmochimica Acta, 2013, 120, 1-16.

[39] Zeng G, Chen L H, Xu X S, et al. Carbonated mantle sources for Cenozoic intra-plate alkaline basalts in Shandong, North China[J]. Chemical Geology, 2010, 273 (1): 35-45.

[40] Dasgupta R, Hirschmann M M, Smith N D. Partial melting experiments of peridotite+CO_2 at 3 GPa and genesis of alkalic ocean island basalts[J]. Journal of Petrology, 2007: 48 (11): 2093-2124.

[41] Xiao Y, Teng F Z, Zhang H F, et al. Large magnesium isotope fractionation in peridotite xenoliths from eastern North China craton: Product of melt-rock interaction[J]. Geochimica et Cosmochimica Acta, 2013, 115: 241-261.

Origin of low δ^{26}Mg Cenozoic basalts from South China Block and their geodynamic implications*

Jian Huang[1], Shu-Guang Li[2, 1], Yilin Xiao[1], Shan Ke[2], Wang-Ye Li[1] and Ye Tian[1]

1. CAS Key Laboratory of Crust-Mantle Materials and Environments, School of Earth and Space Sciences, University of Science and Technology of China, Hefei 230026, China
2. State Key Laboratory of Geological Processes and Mineral Resources, China University of Geosciences, Beijing 100083, China

亮点介绍： 本文首次发现华南新生代玄武岩相比地幔富集轻镁同位素，玄武岩的 δ^{26}Mg 值与全碱含量、流体不活动微量元素组成和比值以及 Ca/Al 比值呈现负相关性，而与 Hf 和 Ti 异常值呈现正相关性。上述特征与富集轻镁同位素的碳酸盐橄榄岩的不一致熔融趋势一致，说明中国东部新生代玄武岩的低 δ^{26}Mg 值是古太平洋板块俯冲导致的碳酸盐再循环的结果。

Abstract Origin of low δ^{26}Mg basalts is a controversial subject and has been attributed to interaction of isotopically light carbonatitic melts derived from a subducted oceanic slab with the mantle (Yang et al., 2012), or alternatively, to accumulation of isotopically light ilmenite ($FeTiO_3$) in their mantle source (Sedaghatpour et al. 2013). To study the origin of low δ^{26}Mg basalts and evaluate whether Mg isotope ratios of basalts can be used to trace deeply recycled carbon, high-precision major and trace element and Mg isotopic analyses on the Cenozoic alkaline and tholeiitic basalts from the South China Block (SCB), eastern China have been carried out in this study. The basalts show light Mg isotopic compositions, with δ^{26}Mg ranging from –0.60 to –0.35‰. The relatively low TiO_2 contents (<2.7 wt.%) of our basalts, roughly positive correlations between δ^{26}Mg and Ti/Ti* and their constant Nb/Ta ratios (16.4–20) irrespective of variable TiO_2 contents indicate no significant amounts of isotopically light ilmenite accumulation in their mantle source. Notably, the basalts display negative correlations between δ^{26}Mg and the amounts of total alkalis (i.e., Na_2O+K_2O) and incompatible trace elements (e.g., Ti, La, Nd, Nb, Th) and trace element abundance ratios (e.g., Sm/Yb, Nb/Y). Generally, with decrease of δ^{26}Mg values, their Hf/Hf* and Ti/Ti* ratios decrease, whereas Ca/Al and Zr/Hf ratios increase. These features are consistent with incongruent partial melting of an isotopically light carbonated mantle, suggesting that large variations in Mg isotope ratios occurred during partial melting of such carbonated mantle under high temperatures. The isotopically light carbonated mantle were probably formed by interaction of the mantle with low δ^{26}Mg carbonatitic melts derived from the deeply subducted low δ^{26}Mg carbonated eclogite transformed from carbonate-bearing oceanic crust during plate subduction. As only the Pacific slab has an influence on both the North China Block (NCB) and SCB, our results together with the study of Yang et al. (2012) demonstrate that the recycled carbonatitic melts might have originated from

* 本文发表在：Geochimica et Cosmochimica Acta, 2015, 164: 298-317

the stagnant Pacific slab beneath East Asia in the Cretaceous and Cenozoic and that a widespread carbonated upper mantle exists beneath eastern China, which may serve as the main source for the <110Ma basalts in this area. Thus, our study demonstrates that Mg isotope ratios of basalts are a powerful tool to trace deeply recycled carbon.

Keywords Magnesium isotopes, deep carbon cycle, basalts, pacific slab subduction, carbonated mantle

1 Introduction

Magnesium (Mg) is a major constituent of the mantle (MgO=37.8 wt.%, McDonough and Sun, 1995) and the continental crust (MgO=4.66 wt.%, Rudnick and Gao, 2003) and the second most important cation in the seawater (MgO=~2.2 wt.%, Millero, 1974). It has three stable isotopes, ^{24}Mg, ^{25}Mg and ^{26}Mg, with natural abundances of 78.99%, 10.00% and 11.01%, respectively (Rosman and Taylor, 1998). The relatively mass difference between ^{24}Mg and ^{26}Mg is ~8%, large enough to produce significant Mg isotope fractionation in cosmochemical, geochemical and biological processes. Mg isotope fractionation has been used to play constraints on a variety of scientific issues such as the evolution of the early solar system (e.g., Lee et al., 1976; Bizzarro et al., 2004), equilibrium temperatures of metamorphic and mantle rocks (e.g., Li et al., 2011; Huang et al., 2013), continental weathering (e.g., Tipper et al., 2006; Teng et al., 2010a; Huang et al., 2012a; Liu et al., 2014; Wimpenny et al., 2014a), plant growth (e.g., Black et al., 2006; Bolou-Bi et al., 2010), enzyme synthesis (Buchachenko et al., 2008) and paleoclimate changes (Saenger and Wang, 2014).

With respect to the cosmochemistry and geochemistry of Mg isotopes, one of the most important findings is that some high-Ti lunar basalts and some terrestrial basalts have very light Mg isotopic compositions with δ^{26}Mg as low as –0.60 (Yang et al., 2012; Sedaghatpour et al., 2013). The interpretations for the origins of the low δ^{26}Mg basalts are inconsistent. Because carbonate rocks (e.g., dolostone and limestone) and minerals (e.g., calcite, aragonite, dolomite and magnesite) have extremely light Mg isotopic compositions, with δ^{26}Mg varying widely from –5.54 to –0.47 (e.g., Galy et al., 2002; Young and Galy, 2004; Tipper et al., 2006; Pogge von Strandmann et al., 2008; Higgins and Schrag, 2010; Jacobson et al., 2010; Ke et al., 2011; Pokrovsky et al., 2011; Wombacher et al., 2011), the low δ^{26}Mg terrestrial basalts from the North China Block (NCB), eastern China have been suggested to result from interaction of their mantle source with isotopically light carbonatitic melts derived from the subducted oceanic slab (Yang et al., 2012). However, based on the large Mg isotope fractionation recorded in high-Ti and low-Ti lunar basalts, with the former generally having much lower δ^{26}Mg values than the latter (–0.59 to –0.37 vs. –0.33 to –0.02, Sedaghatpour et al., 2013), the low δ^{26}Mg high-Ti lunar basalts have been suggested to originate from an isotopically light mantle source produced by crystallization of ilmenite ($FeTiO_3$) with low δ^{26}Mg values at the late stage in the lunar magma ocean (Sedaghatpour et al., 2013). To evaluate wheather Mg isotope ratios of basalts can be used to trace deeply recycled carbon, it is necessary to re-evaluate whether the low δ^{26}Mg basalts from eastern China were caused by mixture of isotopically light carbonatitic melts into their mantle source as suggested by Yang et al. (2012), or alternatively, by accumulation of isotopically light ilmenite in their mantle source (Sedaghatpour et al., 2013), although it is not sure whether ilmenite has a light Mg isotopic composition, because so far no Mg isotopic data for it has been reported, .

Here, we present high-precision Mg isotopic analyses on a suite of well-characterized Cenozoic alkaline and tholeiitic basalts from the South China Block (SCB), eastern China. The basalts show a large variation in major and trace element geochemistry as well as similar depleted Sr-Nd isotopic compositions (e.g., Zou et al., 2000; Chen et al., 2009; Wang et al., 2011). Previous studies have demonstrated that a carbonated mantle may be the main source for the alkaline basalts (Chen et al., 2009; Zeng et al., 2010; Yang et al., 2012; Sakuyama et al.,

2013). Our results show that these basalts have low δ^{26}Mg values. Interestingly, a negative correlation between δ^{26}Mg and TiO_2 exists in the basalts studied here. Thus, the Mg isotopic compositions of these basalts firstly allow us to re-evaluate whether the low δ^{26}Mg basalts were caused by recycled carbonates through oceanic plate subduction or accumulation of isotopically light ilmenite in their mantle source. Secondly, we refer the geodynamic implications deduced from the low δ^{26}Mg basalts from eastern China.

2 Geological settings and sample descriptions

In eastern China, the Cenozoic volcanic rocks are widely distributed along the coastal provinces and adjacent offshore shelf extending over 4000 km from Heilongjiang Province in the north to Hainan island in the south in the eastern edge of the Eurasian continent (Fig. 1). They constitute an important part of the volcanic belt of the

Fig. 1　Simplified geological map of eastern China that mainly consists of the SCB, the NCB, and NE China (i.e., the Xing-Meng Block). The NCB consists of the Western Block (WB), the Trans-North China Orogen (TNCO) and the Eastern Block (EG). The regions where basalts were sampled for Mg isotope investigation in this study are marked as yellow triangles.

western circum-Pacific rim and are one of the world's presently active tectono-magmatic regions (e.g., Zhou and Armstrong, 1982). The Cenozoic volcanic rocks are mainly alkaline basalts that are thought to represent melts derived from the upper mantle, given their depleted Sr-Nd isotopic compositions and OIB-like trace-element signatures in spidergram (e.g., enrichment in Nb, Ta and LREEs, and negative K and Pb anomalies; e.g., Zhou and Armstrong, 1982; Peng et al., 1986; Liu et al., 1994; Zou et al., 2000; Xu et al., 2005; Tang et al., 2006; Chen et al., 2009; Zeng et al., 2010, 2011; Wang et al., 2011).

The samples investigated in this study were collected from Pingmingshan, Anfengshan, Fangshan, Chongren, and Longyou from the SCB (Fig. 1). K-Ar dating results show that basalts from Pingmingshan and Anfengshan have ages of 7.3–12.3 and 4.0–6.4 Ma, respectively (Chen and Peng, 1988; Jin et al., 2003), whereas those from Fangshan, Chongren and Longyou have ages of 2.9–3.5, ~26.4, and 9.0–9.4 Ma, respectively (Chen and Peng, 1988; Ho et al., 2003). Mantle peridotite xenoliths are common in all localities (e.g., Qi et al., 1995; Jin et al., 2003; Reisberg et al., 2005).

Twenty-three basaltic samples were selected for investigations, and all of them are of the porphyritic texture. Most of the studied samples are fresh and unaltered, exceptions are samples 13AFS9-10 from Anfengshan, 10LYSK11 from Longyou, which are altered with iddingsitization of olivine phenocrysts. The phenocrysts consist predominantly of olivine in the Anfengshan basalts, of olivine and clinopyroxene in the Fangshan and Longyou basalts, and of olivine, clinopyroxene and plagioclase in the Pingmingshan and Chongren basalts. The groundmass in these basalts is variable and mainly consists of plagioclase, olivine, augite, nepheline, magnetite and glass.

3 Analytical methods

3.1 Major and trace element analysis

The samples were sawed into slices and only central fresh parts were used for bulk-rock analyses. The pieces were crushed in a corundum jaw crusher to 60 mesh, and then ~60 g of each crushed sample was powdered in an agate ring mill to <200 mesh in size. Bulk rock abundances of major elements were determined using an X-ray fluorescence spectrometer (XRF) on glass disks at the laboratory of ALS minerals at Guangzhou. Pre-ignition was used to determine the loss on ignition (LOI) prior to major elements analyses. Accuracy and precision for major oxides are generally better than 1% based on replicate analyses of certified USGS rock standards. Bulk rock trace element data were obtained by an ELAN DRCII inductively coupled plasma-mass spectrometry (ICP-MS) at the University of Science and Technology of China (USTC) after ultrapure acid digestion (HNO_3+HF+$HClO_4$) of sample powders (~50 mg) in Teflon bombs. Analytical procedures were described in detail by Huang et al. (2012b, 2014). The measured values of international USGS standards (BHVO-2 and BIR-1) are in satisfactory agreement with the recommended values within error, and the precision and accuracy for majority of trace elements analyzed are better than 6% (Table S1 in Supplementary Materials).

3.2 Mg isotope analysis

Magnesium isotopic analyses were performed at the State Key Laboratory of Geological Processes and Mineral Resources, China University of Geosciences (Beijing) (CUGB), following the procedures very similar to those established by Yang et al. (2009), Teng et al. (2010b) and Teng and Yang (2014). A brief description is given below.

All chemical procedures were conducted in a clean laboratory environment at CUGB. Sample powders were dissolved in Savillex screw-up beakers in a mixture of concentrated HF–HCl–HNO_3. Chemical separation of Mg was achieved by cation exchange chromatography with Bio-Rad 200–400 mesh AG50W-X8 pre-cleaned resin in 1 N HNO_3 media. The same column procedure was processed twice for all samples in order to obtain a pure Mg solution for mass spectrometry and to check the efficiency of our column to separate Mg from interference cations. The eluted solutions were firstly evaporated to dryness and then re-dissolved in 3% HNO_3, ready for final dilution immediately prior to analysis. Three USGS reference materials (e.g., BHVO-2, AGV-2 and GSP-2) were processed through column chemistry with each batch of the investigated samples. The total procedural blanks during the course of this study were ~8 ng, comparable to that of Teng et al. (2010b).

Magnesium isotopic compositions were measured by the sample-standard bracketing method on a *Neptune Plasma* MC-ICPMS in a low-resolution mode. The in-run precision on the $^{26}Mg/^{24}Mg$ ratio for a single block of 40 ratios is <±0.02‰ (2SD). The internal precision on the measured $^{26}Mg/^{24}Mg$ ratio based on 4 repeated analyses of the same sample solution during analytical sessions of this study, is ⩽±0.08‰ (2SD, Table. 2). The results are reported in the conventional δ notation that is defined as $\delta^{X}Mg=[(^{X}Mg/^{24}Mg)_{sample}/(^{X}Mg/^{24}Mg)_{DSM3}-1]\times1000$, where X=25 or 26, and DSM3 is an international reference of solution made from pure Mg metal (Galy et al., 2003). Analyses of the well-characterized USGS Mg rock standards in this study yielded $\delta^{26}Mg$=–0.25±0.05‰ (2SD, n=2) for BHVO-2, –0.15±0.03‰ (2SD, n=2) for AGV-2 and –0.03±0.09‰ for GSP-2 (Table 2), which are in excellent agreement with previously published values within error (e.g., Teng et al., 2007, 2010a, b, 2015; Huang et al., 2009; An et al., 2014).

4 Results

Results for bulk-rock major and trace element concentrations and Mg isotopic compositions of the studied basalts are listed in Tables 1 and 2, respectively.

Table 1　Major and trace element concentrations of the Cenozoic basalts from the South China Block

Sample [a]	13PMS1	13PMS1R	13PMS2	13PMS3	13PMS4	13PMS5	13PMS6	13PMS7
Location	Pingmingshan							
Major element (wt.%)								
SiO_2	41.02		40.86	41.08	41.05	41.14	41.07	40.91
TiO_2	2.43		2.39	2.43	2.41	2.41	2.41	2.39
Al_2O_3	11.71		11.70	11.74	11.86	11.73	11.66	11.58
Fe_2O_3	13.34		13.18	13.23	13.00	13.20	13.35	13.24
MnO	0.20		0.20	0.20	0.20	0.20	0.20	0.20
MgO	10.03		9.93	10.00	9.71	10.11	10.10	10.28
CaO	9.52		9.36	9.43	9.58	9.47	9.50	9.53
Na_2O	4.73		4.50	4.63	4.29	4.43	4.60	4.22
K_2O	2.23		2.38	2.32	2.46	2.27	2.33	2.18
P_2O_5	1.25		1.25	1.29	1.23	1.26	1.28	1.26
LOI	3.05		3.22	3.24	3.36	3.35	3.00	3.61
Total	99.51		98.97	99.59	99.15	99.57	99.50	99.40
Na_2O+K_2O	6.96		6.88	6.95	6.75	6.70	6.93	6.40
Na_2O/K_2O	2.12		1.89	2.00	1.74	1.95	1.97	1.94
Mg# [b]	0.64		0.64	0.64	0.64	0.64	0.64	0.65

Continued

Sample [a]	13PMS1	13PMS1R	13PMS2	13PMS3	13PMS4	13PMS5	13PMS6	13PMS7
Location	Pingmingshan							
Trace element (ppm)								
Li	10.4	11.2	11.9	9.76	11.7	11.7	12.2	12.2
Sc	12.9	12.8	12.6	14.4	12.5	12.7	13.0	13.2
V	119	122	130	136	125	130	133	139
Cr	197	203	217	225	198	216	223	232
Ni	193	197	202	217	197	206	211	220
Cu	56.1	57.9	51.8	45.1	54.5	51.6	53.5	53.7
Rb	34.7	35.9	38.5	39.3	45.9	37.9	38.9	41.7
Sr	1211	1275	1490	1557	1583	1348	1417	1420
Y	30.2	31.8	32.9	33.6	33.6	32.7	33.3	33.3
Zr	307	316	335	312	336	328	333	334
Nb	112	113	124	116	126	119	120	119
Cs	1.49	1.41	1.49	1.32	1.49	1.47	1.50	1.36
Ba	642	613	672	592	697	648	634	634
La	77.6	73.0	78.7	71.4	79.1	75.9	75.2	75.5
Ce	139	131	141	129	127	137	127	135
Pr	15.6	14.8	15.9	14.1	16.0	15.6	15.4	15.3
Nd	64.3	60.5	65.1	61.8	65.5	63.9	62.9	62.6
Sm	14.7	13.7	14.9	13.8	15.0	14.6	14.3	14.2
Eu	4.58	4.28	4.68	4.43	4.67	4.57	4.46	4.49
Gd	13.1	12.0	13.0	14.6	13.1	12.8	12.5	12.6
Tb	1.80	1.65	1.79	1.62	1.80	1.75	1.70	1.73
Dy	8.67	8.00	8.65	7.51	8.66	8.38	8.24	8.28
Ho	1.27	1.17	1.28	1.16	1.26	1.22	1.20	1.20
Er	2.85	2.64	2.87	2.60	2.91	2.78	2.73	2.75
Tm	0.32	0.29	0.32	0.28	0.32	0.31	0.30	0.31
Yb	1.75	1.60	1.75	1.74	1.79	1.70	1.67	1.69
Lu	0.22	0.20	0.21	0.26	0.22	0.21	0.21	0.21
Hf	6.59	6.14	6.71	6.14	6.65	6.53	6.45	6.43
Ta	6.46	5.99	6.71	5.96	6.68	6.44	6.28	6.27
Pb	5.86	5.57	5.47	5.71	5.58	5.23	5.34	5.31
Th	11.1	10.8	11.6	11.0	11.7	11.3	11.1	11.2
U	2.81	2.73	3.13	3.18	3.09	3.14	3.09	3.06
Ca/Al [c]	0.74		0.73	0.73	0.74	0.74	0.74	0.75
Nb/U	39.9	41.5	39.7	36.4	40.8	37.9	38.9	38.9
Ce/Pb	23.7	23.5	25.8	22.6	22.8	26.2	23.8	25.4
La/Yb	44.3	45.6	45.0	41.0	44.2	44.6	45.0	44.7
Sm/Yb	8.4	8.6	8.5	7.9	8.4	8.6	8.6	8.4
Nb/Ta	17.3	18.9	18.5	19.4	18.9	18.5	19.1	19.0
Zr/Hf	46.6	51.5	49.9	50.9	50.5	50.2	51.6	51.9
Ti/Ti* [d]	0.46		0.45	0.43	0.45	0.46	0.47	0.47
Hf/Hf* [e]	0.54		0.54	0.53	0.53	0.54	0.54	0.54

Continued

Sample	13PMS8	13AFS1	13AFS2	13AFS3	13AFS4	13AFS9	13AFS10	10FS6
Location	Pingmingshan	Anfengshan						Fangshan
Major element (wt.%)								
SiO_2	41.17	44.46	44.41	42.41	42.36	40.86	40.65	46.56
TiO_2	2.64	2.36	2.37	2.49	2.49	2.34	2.32	2.12
Al_2O_3	12.48	12.05	12.04	11.89	11.86	11.73	11.51	13.89
Fe_2O_3	14.58	12.48	12.50	13.06	13.04	12.57	12.55	11.35
MnO	0.22	0.20	0.20	0.21	0.21	0.20	0.20	0.15
MgO	8.03	7.26	7.24	8.44	8.36	10.54	10.13	9.72
CaO	9.24	8.77	8.74	9.08	9.08	9.61	9.54	7.43
Na_2O	6.44	6.19	6.22	5.59	5.61	3.73	4.20	3.66
K_2O	2.46	2.87	2.78	2.15	2.17	1.88	2.01	2.60
P_2O_5	1.13	1.43	1.42	1.27	1.27	1.35	1.44	0.54
LOI	1.39	1.63	1.60	2.90	2.88	**4.66**	**4.11**	1.60
Total	99.78	99.70	99.52	99.49	99.33	99.47	98.66	99.62
Na_2O+K_2O	8.90	9.06	9.00	7.74	7.78	5.61	6.21	6.26
Na_2O/K_2O	2.62	2.16	2.24	2.60	2.59	1.98	2.09	1.41
Mg#	0.56	0.58	0.58	0.60	0.60	0.66	0.66	0.67
Trace element (ppm)								
Li	13.5	13.8	17.8	14.3	14.4	14.7	16.6	9.28
Sc	11.4	10.4	12.8	11.0	10.9	14.1	14.3	15.9
V	119	112	120	164	209	131	132	145
Cr	104	107	114	111	113	186	172	323
Ni	113	97.9	108	106	103	168	160	296
Cu	55.2	43.0	36.2	47.5	47.5	60.9	51.5	54.8
Rb	29.7	42.0	39.3	37.4	40.0	32.0	50.5	45.4
Sr	1468	1429	1632	1596	1566	1518	1823	585
Y	36.8	42.4	43.7	42.4	41.3	39.5	40.5	20.4
Zr	363	372	423	385	371	372	343	197
Nb	154	138	139	148	141	130	133	56.2
Cs	1.60	1.91	1.86	1.66	1.81	1.28	1.66	0.82
Ba	585	728	701	699	675	633	701	437
La	82.2	92.6	92.1	92.6	92.4	84.3	87.4	28.8
Ce	125	140	169	141	140	125	132	54.6
Pr	17.1	19.4	18.5	19.5	19.1	17.3	18.0	6.25
Nd	71.4	79.3	82.1	79.7	78.8	71.0	73.3	25.1
Sm	16.3	17.8	17.7	18.3	18.1	16.3	17.0	5.76
Eu	5.10	5.08	5.09	5.39	5.35	5.01	5.23	1.92
Gd	14.2	15.3	19.4	15.9	15.6	14.0	14.8	5.39
Tb	1.95	2.14	2.10	2.18	2.14	1.94	2.02	0.82
Dy	9.36	10.30	9.57	10.40	10.30	9.30	9.66	4.34
Ho	1.35	1.54	1.49	1.51	1.49	1.36	1.41	0.71
Er	3.01	3.51	3.35	3.45	3.39	3.08	3.20	1.78
Tm	0.33	0.42	0.39	0.39	0.38	0.35	0.36	0.23
Yb	1.75	2.35	2.45	2.19	2.15	1.97	2.04	1.39

Continued

Sample	13PMS8	13AFS1	13AFS2	13AFS3	13AFS4	13AFS9	13AFS10	10FS6
Location	Pingmingshan	Anfengshan						Fangshan
Lu	0.21	0.31	0.35	0.28	0.27	0.25	0.26	0.19
Hf	8.30	8.50	8.11	8.47	8.36	6.99	7.20	4.04
Ta	8.56	7.19	7.08	7.76	7.53	6.48	6.70	3.29
Pb	5.66	7.72	7.55	7.28	7.38	6.16	6.27	3.09
Th	12.2	14.4	13.8	13.9	13.6	13.1	13.3	4.0
U	4.12	3.10	3.27	2.87	2.89	3.45	3.34	1.18
Ca/Al	0.67	0.66	0.66	0.70	0.70	0.75	0.75	0.49
Nb/U	37.4	44.6	42.4	51.6	48.9	37.7	39.9	47.8
Ce/Pb	22.1	18.1	22.4	19.4	19.0	20.3	21.1	17.7
La/Yb	47.0	39.4	37.6	42.3	43.0	42.8	42.8	20.7
Sm/Yb	9.3	7.6	7.2	8.4	8.4	8.3	8.3	4.1
Nb/Ta	18.0	19.2	19.6	19.1	18.7	20.1	19.9	17.1
Zr/Hf	43.7	43.8	52.1	45.5	44.4	53.2	47.6	48.8
Ti/Ti*	0.46	0.38	0.32	0.38	0.39	0.41	0.38	0.98
Hf/Hf*	0.61	0.57	0.54	0.56	0.56	0.52	0.51	0.85

Sample	10FS8	10FS9	10FS10	10FS11	10CR1	10CR2	10LYSK11	10LYSK13
Location	Fangshan	Fangshan			Chongren		Longyou	
Major element (wt.%)								
SiO_2	46.66	46.73	46.48	46.44	51.10	51.13	43.4	42.8
TiO_2	2.10	2.12	2.10	2.13	2.29	2.14	1.81	2.14
Al_2O_3	13.72	13.84	13.72	13.88	11.87	13.47	10.1	11.4
Fe_2O_3	11.12	11.22	11.13	11.33	13.08	12.23	10.78	12.2
MnO	0.15	0.15	0.15	0.16	0.18	0.17	0.16	0.18
MgO	10.08	9.99	9.99	10.11	6.91	7.49	13.05	12
CaO	7.20	7.29	7.27	7.56	6.07	8.54	9.02	9.87
Na_2O	3.90	3.94	3.71	3.67	3.27	3.22	2.99	4.01
K_2O	1.68	1.61	1.74	2.12	1.46	1.04	2.07	1.44
P_2O_5	0.53	0.53	0.53	0.53	0.58	0.34	0.91	1.11
LOI	2.23	2.31	2.48	1.89	3.06	–0.08	**4.94**	2.15
Total	99.37	99.73	99.30	99.82	99.87	99.69	99.23	99.30
Na_2O+K_2O	5.58	5.55	5.45	5.79	4.73	4.26	5.06	5.45
Na_2O/K_2O	2.32	2.45	2.13	1.73	2.24	3.10	1.44	2.78
Mg#	0.68	0.68	0.68	0.68	0.55	0.59	0.74	0.70
Trace element (ppm)								
Li	9.13	9.38	9.49	8.55	7.10	6.47	16.9	9.47
Sc	15.6	15.8	15.7	15.7	13.4	19.0	19.3	19.4
V	130	145	144	140	135	161	143	152
Cr	347	355	351	397	108	168	504	375
Ni	307	304	303	303	170	157	451	349
Cu	60.9	62.4	63.4	58.3	112	79.7	55.7	56.5
Rb	32.6	32.9	32.5	41.6	33.9	20.6	64.4	38.3
Sr	608	609	630	557	298	344	974	1194
Y	19.9	19.8	19.9	19.3	33.4	20.7	25.1	29.2

Continued

Sample	10FS8	10FS9	10FS10	10FS11	10CR1	10CR2	10LYSK11	10LYSK13
Location	Fangshan	Fangshan			Chongren		Longyou	
Zr	191	192	190	184	221	125	219	244
Nb	55.0	55.6	55.1	53.2	30.0	20.2	88.8	107
Cs	2.10	1.73	1.51	1.10	1.03	0.41	1.1	1.06
Ba	421	425	427	455	544	320	648	915
La	27.6	27.9	28.1	30.4	27.8	15.2	57.1	68.9
Ce	52.9	53.3	53.6	58.3	55.7	31.0	106	126
Pr	6.03	6.07	6.09	6.66	7.01	3.89	11.7	14.2
Nd	24.3	24.4	24.5	27.2	31.9	18.2	48.0	57.8
Sm	5.53	5.56	5.67	6.23	9.15	5.35	9.43	11.4
Eu	1.83	1.84	1.84	2.08	2.76	1.89	2.82	3.44
Gd	5.08	5.15	5.20	5.74	8.78	5.48	8.13	9.55
Tb	0.77	0.78	0.78	0.87	1.43	0.91	1.08	1.27
Dy	4.11	4.15	4.13	4.69	7.72	4.99	5.41	6.28
Ho	0.68	0.68	0.68	0.75	1.30	0.86	0.87	1.02
Er	1.68	1.69	1.69	1.88	3.24	2.11	2.15	2.46
Tm	0.22	0.22	0.22	0.25	0.43	0.28	0.27	0.3
Yb	1.30	1.33	1.35	1.46	2.61	1.73	1.55	1.74
Lu	0.18	0.18	0.18	0.20	0.35	0.24	0.21	0.23
Hf	3.92	3.93	3.99	4.06	5.47	3.36	4.56	4.99
Ta	3.20	3.21	3.20	3.23	1.62	1.23	4.65	5.54
Pb	2.69	2.69	2.84	3.09	4.10	2.46	4.34	3.75
Th	3.9	4.0	4.0	3.8	3.7	2.1	9.0	10.0
U	1.15	1.15	1.16	1.09	0.78	0.49	1.98	2.18
Ca/Al	0.48	0.48	0.48	0.50	0.47	0.58	0.81	0.79
Nb/U	48.0	48.5	47.7	49.0	38.7	41.6	44.8	49.1
Ce/Pb	19.7	19.8	18.9	18.9	13.6	12.6	24.4	33.6
La/Yb	21.2	21.0	20.8	20.8	10.7	8.8	36.8	39.6
Sm/Yb	4.3	4.2	4.2	4.3	3.5	3.1	6.1	6.6
Nb/Ta	17.2	17.3	17.2	16.5	18.5	16.4	19.1	19.3
Zr/Hf	48.7	48.9	47.6	45.3	40.4	37.2	48.0	48.9
Ti/Ti*	1.03	1.02	1.00	0.92	0.65	0.99	0.53	0.47
Hf/Hf*	0.85	0.85	0.85	0.79	0.81	0.86	0.54	0.49

[a] 13PMS1R is the replicated analysis of sample 13PMS1 for trace element concentrations.

[b] $Mg\#=Mg/(Mg+0.85Fe^{tot})$.

[c] $Ca/Al=([CaO]_{wt.\%}/56)/(2*[Al_2O_3]_{wt.\%}/102)$.

[d] $Ti/Ti^*=Ti_N/(Nd_N^{-0.055}\times Sm_N^{0.333}\times Gd_N^{0.722})$.

[e] $Hf/Hf^*=Hf_N/(Sm_N\times Nd_N)^{0.5}$.

Table 2　Magnesium isotopic compositions of the Cenozoic basalts from the South China Block

Sample	$\delta^{26}Mg$[a]	2SD[b]	$\delta^{25}Mg$	2SD[b]	N	$\Delta^{25}Mg'$[c]
13PMS1	−0.52	0.07	−0.25	0.09	4	0.03
13PMS2	−0.53	0.04	−0.26	0.03	4	0.02
13PMS3	−0.52	0.05	−0.24	0.03	4	0.03
Replicate[d]	−0.52	0.05	−0.24	0.03	4	0.03

Continued

Sample	$\delta^{26}Mg^a$	2SD[b]	$\delta^{25}Mg$	2SD[b]	*N*	$\Delta^{25}Mg'^c$
13PMS6	–0.52	0.05	–0.26	0.03	4	0.01
13PMS8	–0.60	0.05	–0.31	0.05	4	0.01
13AFS1	–0.52	0.04	–0.26	0.02	4	0.01
Replicate	–0.51	0.08	–0.26	0.05	4	0.00
13AFS2	–0.52	0.08	–0.25	0.07	4	0.02
13AFS3	–0.59	0.04	–0.29	0.04	4	0.02
Replicate	–0.58	0.05	–0.29	0.04	4	0.02
13AFS4	–0.56	0.04	–0.28	0.02	4	0.02
10FS6	–0.41	0.07	–0.22	0.03	4	0.00
Replicate	–0.42	0.06	–0.21	0.02	4	0.01
10FS8	–0.42	0.05	–0.20	0.03	4	0.01
10FS9	–0.43	0.05	–0.23	0.05	4	–0.01
10FS10	–0.42	0.04	–0.22	0.04	4	–0.01
10FS11	–0.40	0.03	–0.20	0.05	4	0.00
10CR1	–0.37	0.05	–0.19	0.04	4	0.00
10CR2	–0.35	0.06	–0.18	0.05	4	0.00
10LYSK13	–0.41	0.05	–0.21	0.05	4	0.01
BHVO-2	–0.26	0.04	–0.14	0.06	4	0.00
Replicate	–0.24	0.06	–0.12	0.02	4	0.01
AGV-2	–0.16	0.03	–0.10	0.06	4	–0.02
Replicate	–0.15	0.02	–0.06	0.04	4	0.01
GSP-2	0.03	0.09	0.00	0.06	4	–0.01

[a] $\delta^{X}Mg=\{(^{X}Mg/^{24}Mg)_{sample}/(^{X}Mg/^{24}Mg)_{DSM3}-1)\}\times 1000$, where X=25 or 26 and DSM3 is solution made from pure Mg metal (Galy et al., 2003).

[b] 2SD indicates twice the standard deviation of the population of 4 repeat measurements of a sample solution.

[c] $\Delta^{25}Mg'=\delta^{25}Mg'-0.521\delta^{26}Mg'$, where $\delta^{X}Mg'=1000\times\ln[(\delta^{X}Mg+1000)/1000]$ (Young and Galy, 2004). It is reported largely as a quality control on the data, with values that should be close to zero.

[d] Replicate denotes repeating sample dissolution, column chemistry and instrumental analysis.

4.1 Major and trace elements

The investigated samples have a large compositional range of SiO_2 (40.6–51.1 wt.%) and high contents of MgO (6.91–13.1 wt.%) and TiO_2 (1.81–2.64 wt.%). They also show high total alkalis (Na_2O+K_2O=3.71–9.06 wt.%), with Na_2O/K_2O ratios ranging from 1.4 to 3.1, indicating their alkali-rich and high-sodium nature. Following the nomenclature of La Bas et al. (1986), samples from Anfengshan, Pingmingshan and Longyou are basanites, whereas those from Fangshan are trachybasalt, and those from Chongren are normal basalts (Fig. 2). All basalts are alkaline except for samples from Chongren, which are tholeiites (Fig. 2). No clear correlation was observed between MgO and SiO_2, TiO_2, Fe_2O_3, CaO and K_2O, while Al_2O_3 displays a slightly negative correlation with MgO (Fig. 3).

In the chondrite-normalized REE diagram (Fig. 4a), all basalts show enrichment of LREEs over HREEs ($[La/Yb]_N$=6.0–32), with no significant Eu or Ce anomaly. In primitive mantle-normalized spidergram (Fig. 4b), all but rocks from Chongren resemble many ocean island basalts in terms of enrichment in Nb and Ta, and depletion in K and Pb relative to LREEs. Additionally, the basalts generally show negative Zr, Hf and Ti (Hf/Hf*=0.49–0.86, Ti/Ti*=0.32–1.03, indexes for Hf and Ti anomalies, respectively) and high Ca/Al ratios (0.47–0.81) (Fig. 5). Nb/U and Ce/Pb ratios of the Cenozoic basalts range from 36 to 52 and 13 to 34, respectively

(Table 1), similar to those of MORBs and OIBs (Nb/U=47±10, Ce/Pb=25±5, Hofmann et al, 1986), but much higher than those of continental crust (Nb/U=6.2, Ce/Pb=4, Rudnick and Gao, 2003).

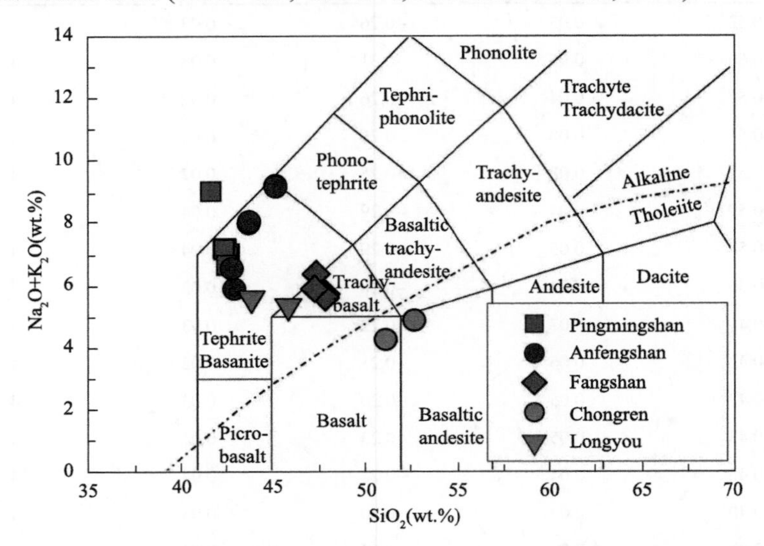

Fig. 2 Na_2O+K_2O vs. SiO_2 diagram (Le Bas, 1986) for the SCB Cenozoic basalts.

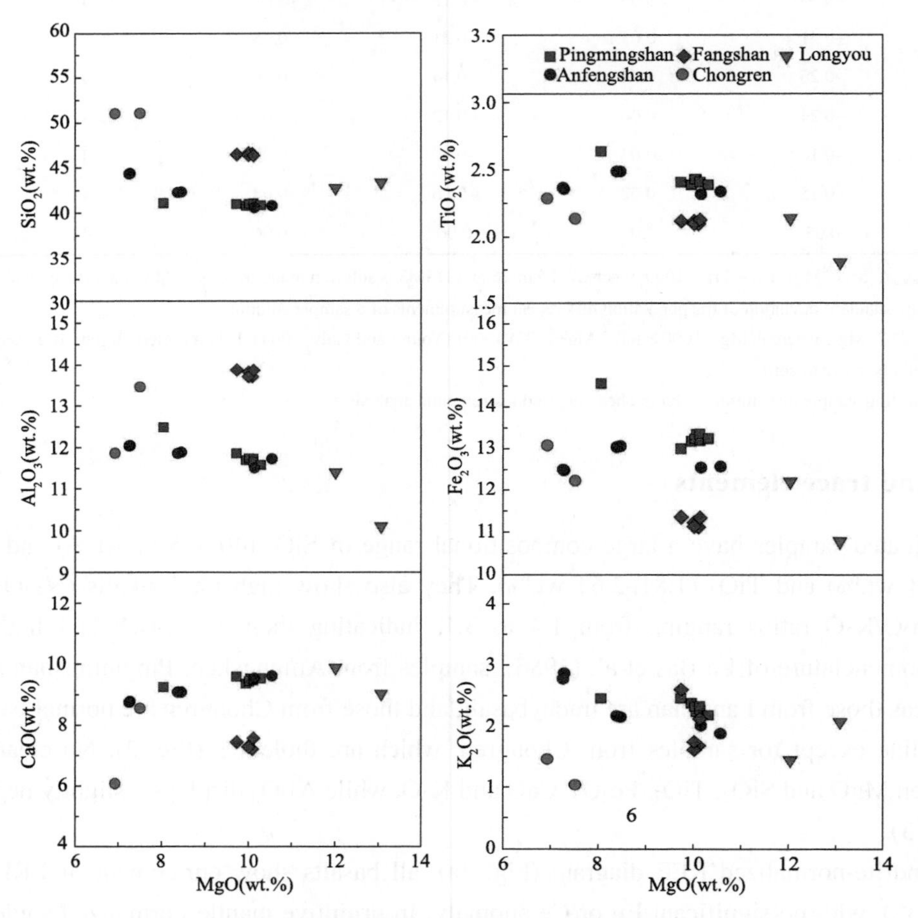

Fig. 3 MgO vs. other oxides diagrams for the SCB Cenozoic basalts.

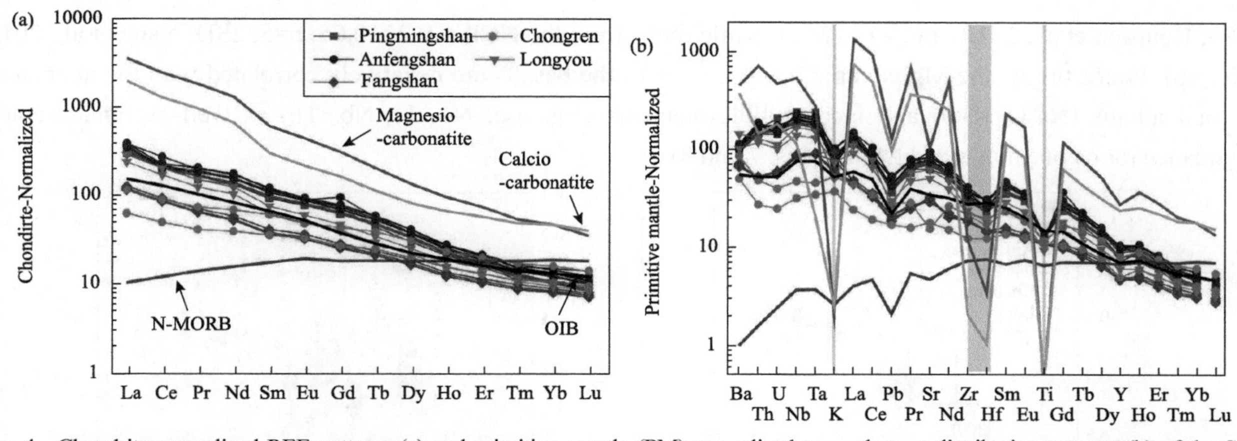

Fig. 4 Chondrite-normalized REE patterns (a) and primitive mantle (PM)-normalized trace element distribution patterns (b) of the SCB Cenozoic basalts. Normalized values are from McDonough and Sun (1995), and data for N-MORB and OIB are from Sun and McDonough (1989). The average values for magnesio- and calico-carbonatites are taken from references (Hoernle et al., 2002; Bizimis et al., 2003).

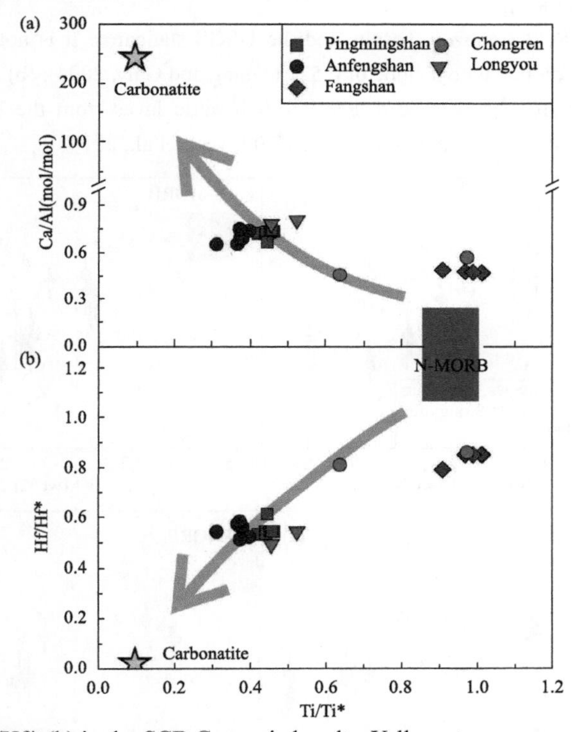

Fig. 5 Ti/Ti* vs. Ca/Al (a) and Hf/Hf* (b) in the SCB Cenozoic basalts. Yellow stars represent the average ratios of Ca/Al (248.4), Ti/Ti* (0.106) and Hf/Hf* (0.016) for magnesio- and calico-carbonatites (Hoernle et al., 2002; Bizimis et al., 2003). Those ratios for N-MORB are calculated based on major and trace element data presented by Hofmann (1988).

4.2 Magnesium isotopes

In the plot of δ^{25}Mg vs. δ^{26}Mg (Fig. 6a), the basalts and the USGS standards fall along the terrestrial equilibrium mass fractionation line with a slope of 0.521, similar to previous studies of natural and synthetic samples (e.g., Young and Galy, 2004; Yang et al., 2009, 2012; Li et al., 2010; Teng et al., 2010a, b). The Δ^{25}Mg′ ($\Delta^{25}\text{Mg}'=\delta^{25}\text{Mg}' - 0.521\delta^{26}\text{Mg}'$, where $\delta^{X}\text{Mg}'=1000\times\ln[(\delta^{X}\text{Mg}+1000)/1000]$, X=25 or 26, Young and Galy, 2004) values for all basalts range from –0.01 to 0.04, very close to zero (Table 2). In general, our Cenozoic basalts have highly variable but overall light Mg isotopic compositions, with δ^{26}Mg values ranging from –0.60 to –0.35 (Fig. 6). These δ^{26}Mg values are similar to those of the <110 Ma basalts from the NCB (–0.60 to –0.42, Yang et al., 2012) but obviously lighter than that of the normal mantle (~–0.25, Teng et al., 2007, 2010b; Handler et al.,

2009; Bourdon et al., 2010) and >120 Ma basaltic rocks from the NCB (−0.27±0.05, n=5, 2SD, Yang et al., 2012) (Fig. 6b). Interestingly, the Mg isotopic compositions of the basalts are negatively correlated with the abundance of total alkalis (Na_2O+K_2O) and incompatible elements (e.g., La, Nd, Ti, Nb, Th) as well as trace element abundance ratios of Sm/Yb and Nb/Y (Figs. 7 and 8).

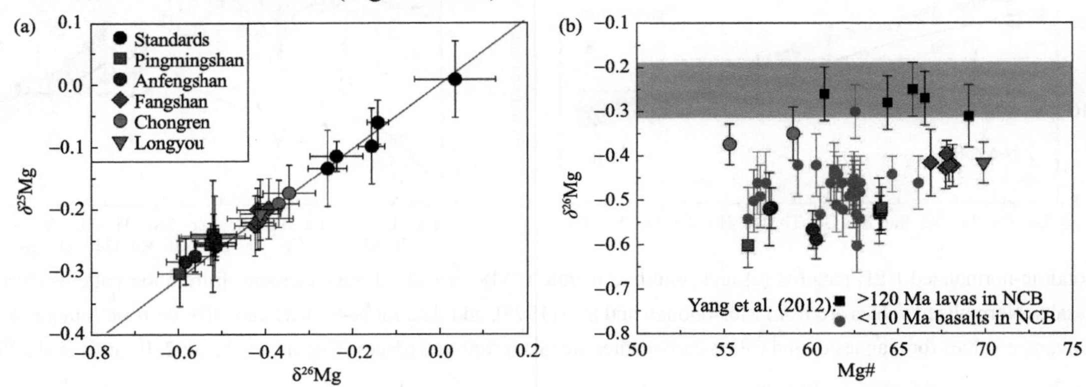

Fig. 6 (a) δ^{26}Mg vs. δ^{25}Mg in the SCB Cenozoic basalts and the USGS standards. It is noted that all data distribute along the terrestrial equilibrium mass fractionation line with a slope of 0.521 (Young and Galy, 2004); (b) δ^{26}Mg vs. Mg# in the SCB Cenozoic basalts. Also shown for comparison are Mg isotopic data of fresh basaltic lavas from the NCB (Yang et al., 2012). Gray bar represents the widely accepted δ^{26}Mg of the normal mantle (−0.25±0.07, Teng et al., 2010b).

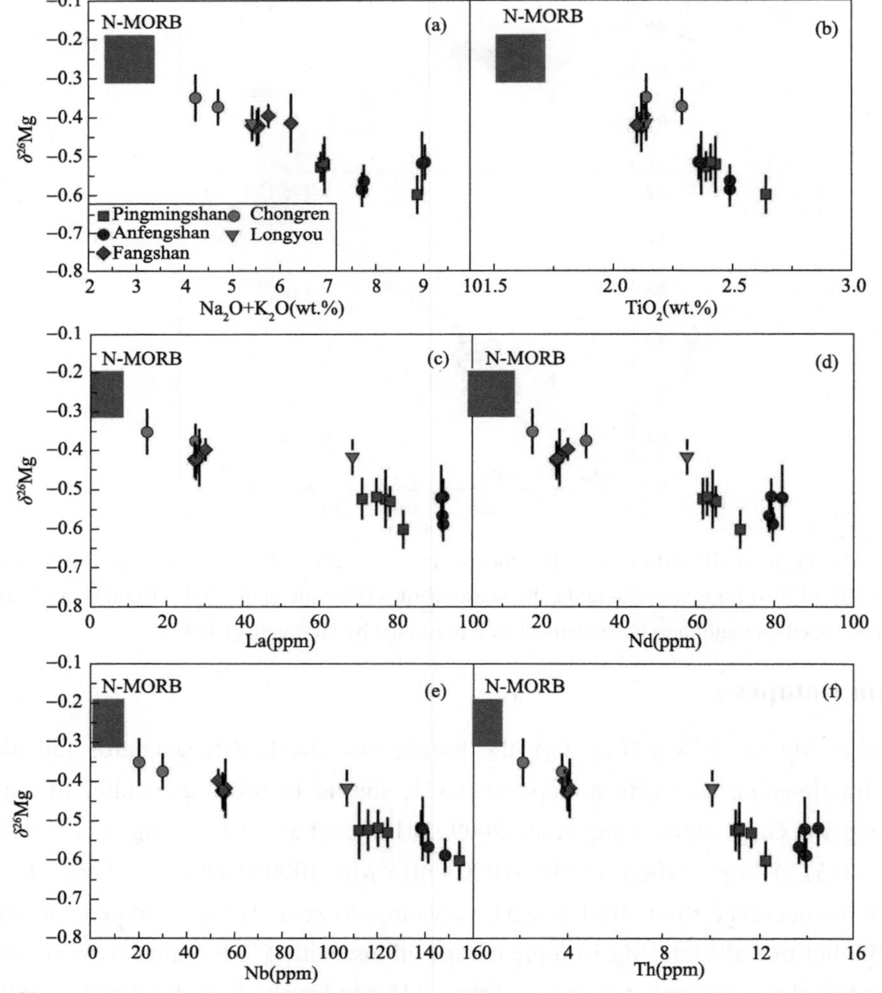

Fig. 7 Variations of δ^{26}Mg with abundances of incompatible elements in the SCB Cenozoic basalts. The δ^{26}Mg of N-MORB is cited as −0.25±0.07 (Teng et al., 2010b) and the values for incompatible elements are taken from Hofmann (1988).

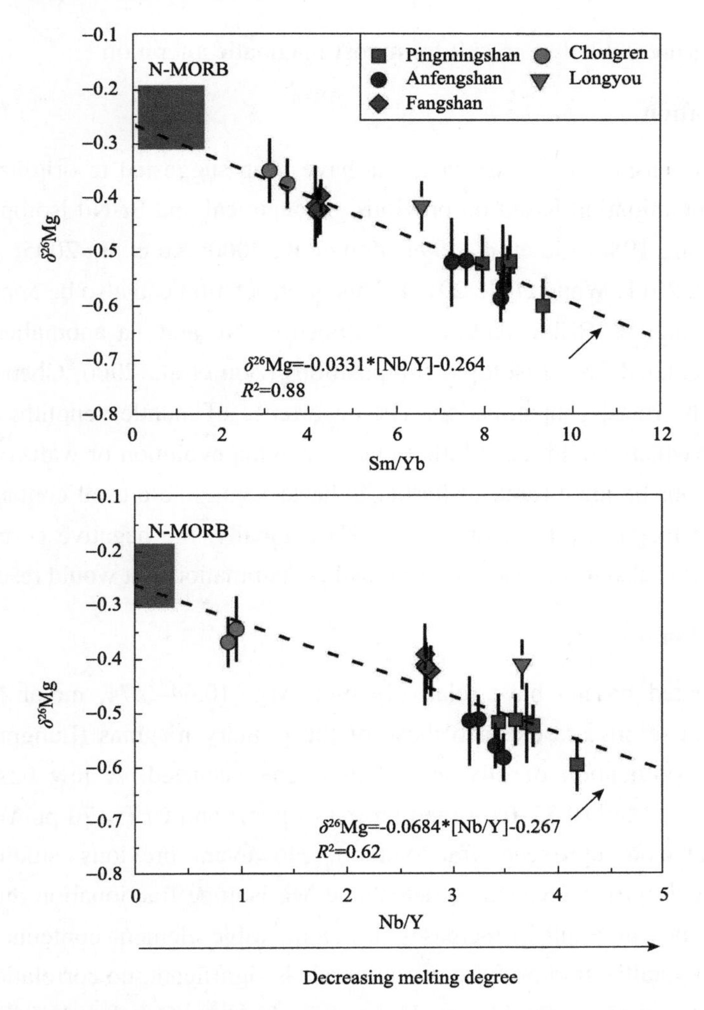

Fig. 8 δ^{26}Mg vs. Sm/Yb (a) and Nb/Y (b) in the SCB Cenozoic basalts. The negative correlations between δ^{26}Mg and Sm/Yb and Nb/Y, trace element abundance ratios that are sensitive to partial melting, suggest that the observed δ^{26}Mg variation in the SCB Cenozoic basalts is caused by partial melting of a carbonated mantle (See text for details).

5 Discussion

5.1 Effects of shallow-level processes

5.1.1 Post-magmatic alteration

Loss on ignition (LOI) of our basalts varies from –0.08 to 4.94 wt.%, suggesting that some samples have experienced alterations, manifested by the transformation of olivine phenocrysts to low-T iddingsite in samples 13AFS9-10 and 10LYSK11. However, good correlations between abundances of fluid-mobile elements (e.g., Ba, Sr, Pb, Th, U, La, Nd) and the fluid-immobile element Nb (Fig. 9) suggest that the effect of alteration, which would markedly disturb such correlations, is not significant. In addition, residual products of basalts (e.g., saprolite and soil) usually have heavier Mg isotopic compositions compared to their protoliths due to preferential incorporation of ^{26}Mg into the secondary Mg-bearing clay minerals (e.g., Pogge von Strandmann et al., 2008; Teng et al., 2010a; Huang et al., 2012a; Liu et al., 2014; Wimpenny et al., 2014b). However, the studied basalts have lighter Mg isotopic compositions relative to normal mantle-derived magmatic rocks (e.g., MORBs and OIBs, –0.25±0.07, Teng et al., 2010b) (Figs. 6-8), opposite to the expected results of alteration. Therefore, the low

δ^{26}Mg values of our basalts are unlikely to result from post-magmatic alteration.

5.1.2 Crustal contamination

The Cenozoic alkaline basalts from eastern China have been suggested to originate from the upper mantle with negligible crustal contamination based on previous geochemical and Sr-Nd isotopic studies (e.g., Zhou and Armstrong, 1982; Peng et al., 1986; Liu et al., 1994; Zou et al., 2000; Xu et al., 2005; Tang et al., 2006; Chen et al., 2009; Zeng et al., 2010, 2011; Wang et al., 2011). This interpretation can also be applied to the alkaline basalts investigated here because their OIB-like features (e.g., positive Nb and Ta anomalies and negative K and Pb anomalies, Fig. 4b) and depleted Sr-Nd isotopic compositions (Zou et al., 2000; Chen et al., 2009; Wang et al., 2011) are inconsistent with crustal contamination. The occurrence of mantle xenoliths indicates that the basaltic magmas ascended rapidly, which would leave little time for magma evolution or wall-rock assimilations. OIB-like Ce/Pb and Nb/U ratios of our basalts (Table 1) further indicate negligible crustal contamination that would lower these ratios of the basaltic magmas (Hofmann et al., 1986). Finally, the negative correlation between SiO_2 and Na_2O+K_2O contents (Fig. 2) is also inconsistent with crustal contamination that would result in an opposite trend.

5.1.3 Crystal fractionation

Most of the investigated basalts have relatively high Mg# (0.64–0.74, molar Mg/[Mg+Fe^{+2}], Table 1), suggesting that their compositions are close to those of the primary magmas (Langmuir et al., 1977) and that insignificant fractional crystallization of olivine and pyroxene occurred. A few basalts (i.e., 13AFS1-4 and 10CR1-2) with relatively low Mg# (0.55–0.60) and Ni (<170 ppm) and Cr (<170 ppm) contents (Table 1) might have experienced olivine and pyroxene fractionation. However, previous studies show that fractional crystallization of olivine and pyroxene cause no detectable Mg isotope fractionation during basalt differentiation (e.g., Teng et al., 2007), but can result in increases in incompatible element contents of basalts (Allègre et al., 1977). Thus, if fractional crystallization of these two minerals is significant, no correlations between δ^{26}Mg values and incompatible element contents should be observed in basalts. Our basalts display obviously negative correlations between δ^{26}Mg values and contents of incompatible elements, such as Ti, La, Nd, Nb and Th (Fig. 7), suggesting negligible fractional crystallization of olivine and pyroxene. Furthermore, since fractional crystallization of minerals such as olivine and pyroxene has a very limited effect on two incompatible element abundance ratios (e.g., Sun and Hanson, 1975; Minster and Allègre, 1978), the negative correlations between δ^{26}Mg values and Sm/Yb and Nb/Y ratios in our basalts (Fig. 8) must record the original features of the primary magmas. Finally, no negative Eu anomalies (Fig. 4a) are indicative of negligible removal of plagioclase. Thus, the observed variations in geochemistry and δ^{26}Mg values of our basalts are largely related to mantle processes.

5.2 Geochemical variations with degree of partial melting in the upper mantle

Since shallow-level processes (including post-magmatic alteration, crustal contamination and crystal fractionation) have insignificant effects, the compositions of the studied basalts are close to the primary magmas. Thus, the different features in geochemistry are likely to result from different mantle sources or from different degrees of partial melting of similar mantle sources. Given that the Sr-Nd isotopic compositions of these basalts are depleted and vary in a narrow range (Zou et al., 2000; Chen et al., 2009; Wang et al., 2011), their LILE and LREE enrichments suggest that they were probably derived from similar sources, which were recently enriched by mantle metasomatism after a long-term depletion. Since La and Sm are more incompatible than Yb during partial melting of the mantle rocks, La/Yb and Sm/Yb ratios are sensitive to partial melting degree. As illustrated in Fig. 10, variable degrees (2–20‰) of batch melting of a hypothetical light REE-enriched mantle source ($[La/Yb]_N$ >1) in the garnet stability field can produce the La/Yb vs. Sm/Yb systematics of the alkaline and tholeiitic basalts investigated here. Specifically, alkaline basalts with high La/Yb and Sm/Yb ratios from Pingmingshan have the lowest degree of partial melting (~2%), while tholeiitic basalts with lower La/Yb and Sm/Yb ratios from Chongren have the highest degree of melting (~20%). These estimates are consistent with the different contents of incompatible elements such as Ba, Th, U, La, Sr, Nd and Nb (Fig. 9), because the amount of incompatible elements in basaltic melts increases with decreasing degree of partial melting. In addition, the Cenozoic basalts investigated here have highly variable Nb/Y ratios that are also strongly dependent on the degree of partial melting, because Nb is much more incompatible than Y during partial melting of the mantle rocks (Fig. 8b).

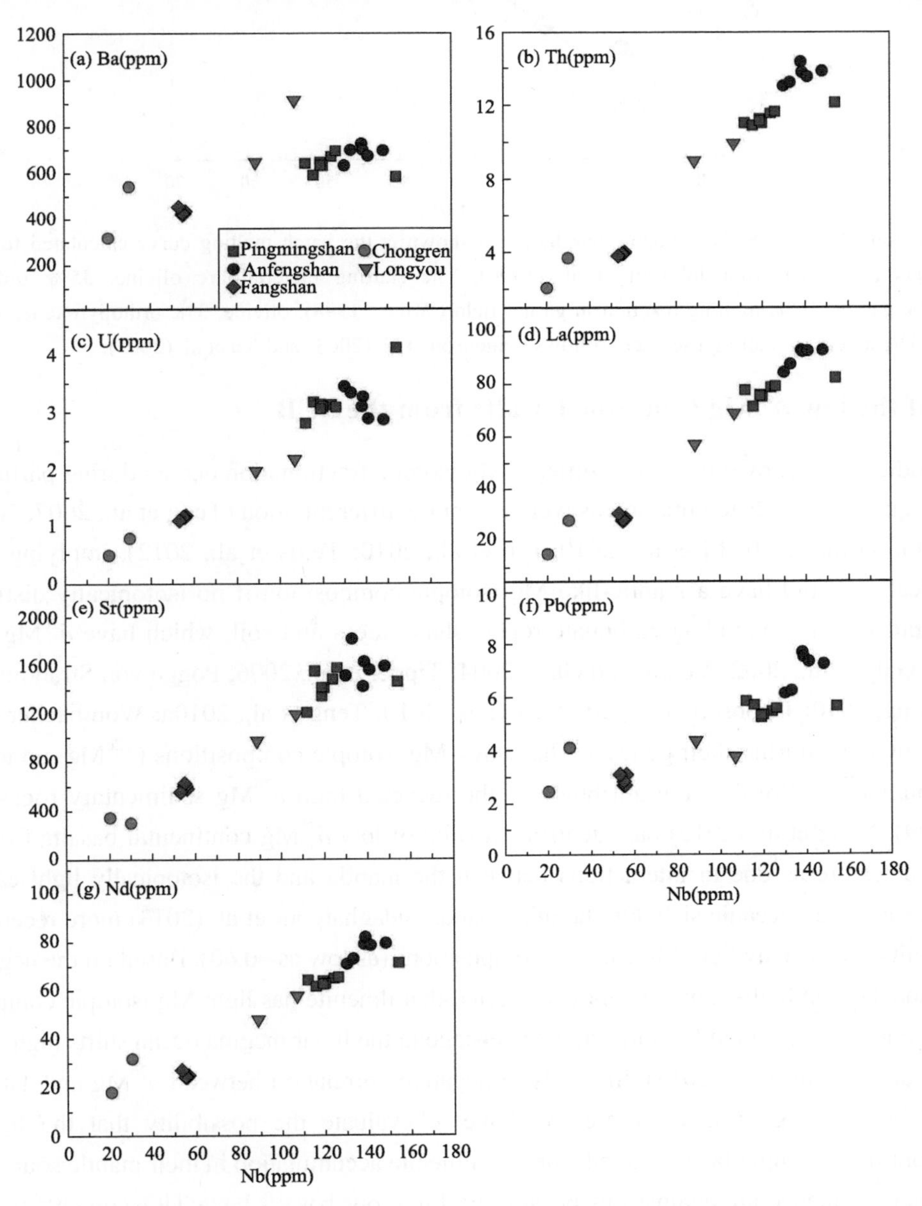

Fig. 9 Variations of selected elements versus Nb in the SCB Cenozoic basalts.

Therefore, the data and discussions above lead us to suggest that the basalts studied here originated from the upper mantle and their geochemical differences closely reflect variable degrees of partial melting. With decreasing degree of partial melting, the basalts display higher total alkalis and TiO_2 contents, higher concentrations of incompatible elements (e.g., Ba, Sr, La, Nd, Nb, U, Th), increased La/Yb, Sm/Yb and Nb/Y ratios, more pronounced negative K, Zr, Hf and Ti anomalies in primitive mantle (PM)-normalized trace element spidergram, and much lighter Mg isotopic compositions, as shown in Figs. 4 and 7-10.

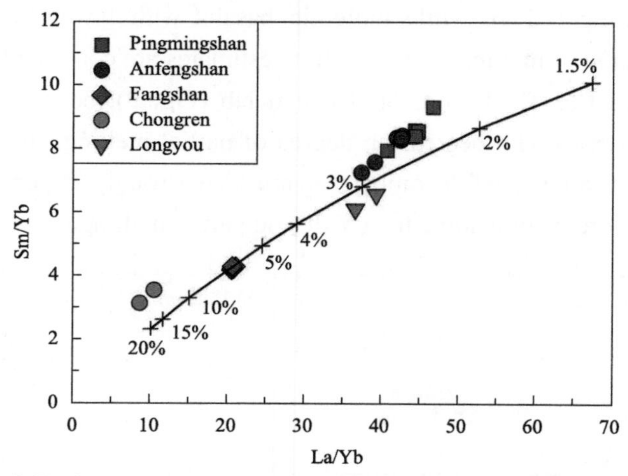

Fig. 10　La/Yb vs. Sm/Yb in the SCB Cenozoic basalts. Also shown is the batch melting curve calculated for garnet peridotite. Partition coefficients are taken from Johnson et al. (1990). The starting material are olivine, 55%; orthopyroxene, 20%; clinopyroxene, 15%; garnet, 10%; melting reaction in garnet field (Walter, 1998): olivine, 3%; orthopyroxene, 3%; clinopyroxene, 70%; garnet, 24%. The inverse modeling used here follows Feigenson et al. (2003) and Xu et al. (2005).

5.3　Origin of the low δ^{26}Mg Cenozoic basalts from the SCB

Previous studies have shown that no significant Mg isotope fractionation occurs during partial melting of the mantle and subsequent basalt differentiation as well as granite differentiation (Teng et al., 2007, 2010b; Handler et al., 2009; Bourdon et al., 2010; Li et al., 2010; Liu et al., 2010; Telus et al., 2012), implying that basalts and granites theoretically should have a mantle-like Mg isotopic composition if no isotopically distinct components (such as sedimentary rocks, including carbonate rocks, shale, loess and soil, which have δ^{26}Mg ranging from –5.57 to 1.8, e.g., Galy et al., 2002; Young and Galy, 2004; Tipper et al., 2006; Pogge von Strandmann et al., 2008; Higgins and Schrag, 2010; Jacobson et al., 2010; Li et al., 2010; Teng et al., 2010a; Wombacher et al., 2011; Liu et al., 2014) are involved during their genesis. The heavy Mg isotopic compositions (δ^{26}Mg=up to 0.44) of I-type granites from southern California were attributed to the recycled high δ^{26}Mg sedimentary rocks in their source (Shen et al., 2009). Yang et al. (2012) has identified a suite of low δ^{26}Mg continental basalts from the NCB and interpreted such a feature as due to interactions between the mantle and the isotopically light carbonatitic melts derived from the subducted oceanic slab. On the other hand, Sedaghatpour et al. (2013) more recently reported that high-Ti lunar basalts also display light Mg isotopic compositions (as low as –0.60). Based on the negative correlation between δ^{26}Mg and TiO_2 in high-Ti basalts, they suggested that ilmenite has light Mg isotopic compositions and the accumulation of ilmenite in the mantle source at the late stage in the lunar magma ocean shifts high-Ti basalts to low δ^{26}Mg values (Sedaghatpour et al., 2013). Similarly, a negative correlation between δ^{26}Mg and TiO_2 has also been observed in our basalts (Fig. 7b). Therefore, we have to evaluate the possibility that the light Mg isotopic compositions of our basalts might have resulted from the ilmenite accumulation in their mantle source.

Several lines of evidence are against this possibility. First, our basalts have TiO_2 contents (<2.7 wt.%) much lower than those of high-Ti lunar basalts (>6 wt.%, Sedaghatpour et al., 2013), implying that no significant accumulation of ilmenite in the mantle source for our basalts. Second, as ilmenite generally displays an enrichment of Nb-Ta and preferentially incorporates Ta relative to Nb with D_{Ta}/D_{Nb}=~1.3 between ilmenite and mafic melts (Dygert et al., 2013), accumulation of ilmenite in the mantle source would cause a negative correlation between Nb/Ta and TiO_2 rather other constant Nb/Ta ratios in our basalts irrespective of variable TiO_2

(Fig. 11). Third, the roughly positive correlation between δ^{26}Mg and Ti/Ti* (Fig. 12d) in our basalts also doesn't stand for this interpretation, because if large amounts of isotopically light and Ti-rich ilmenites are present in the mantle source, an opposite trend should be observed. Thus, the light Mg isotopic compositions of the investigated basalts cannot be resulted from ilmenite accumulation in their mantle source.

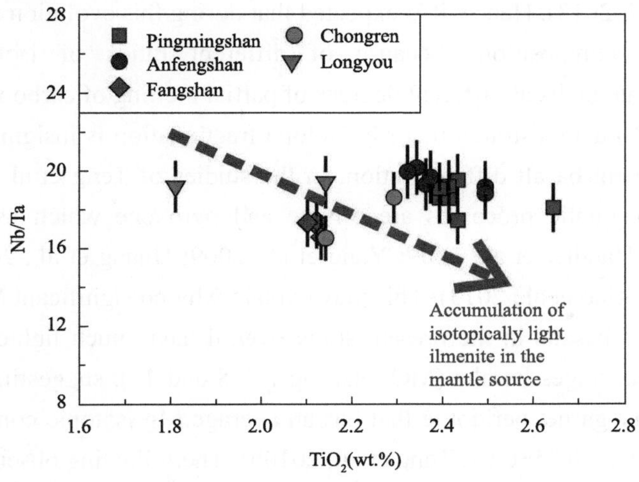

Fig. 11 TiO_2 vs. Nb/Ta in the SCB Cenozoic basalts.

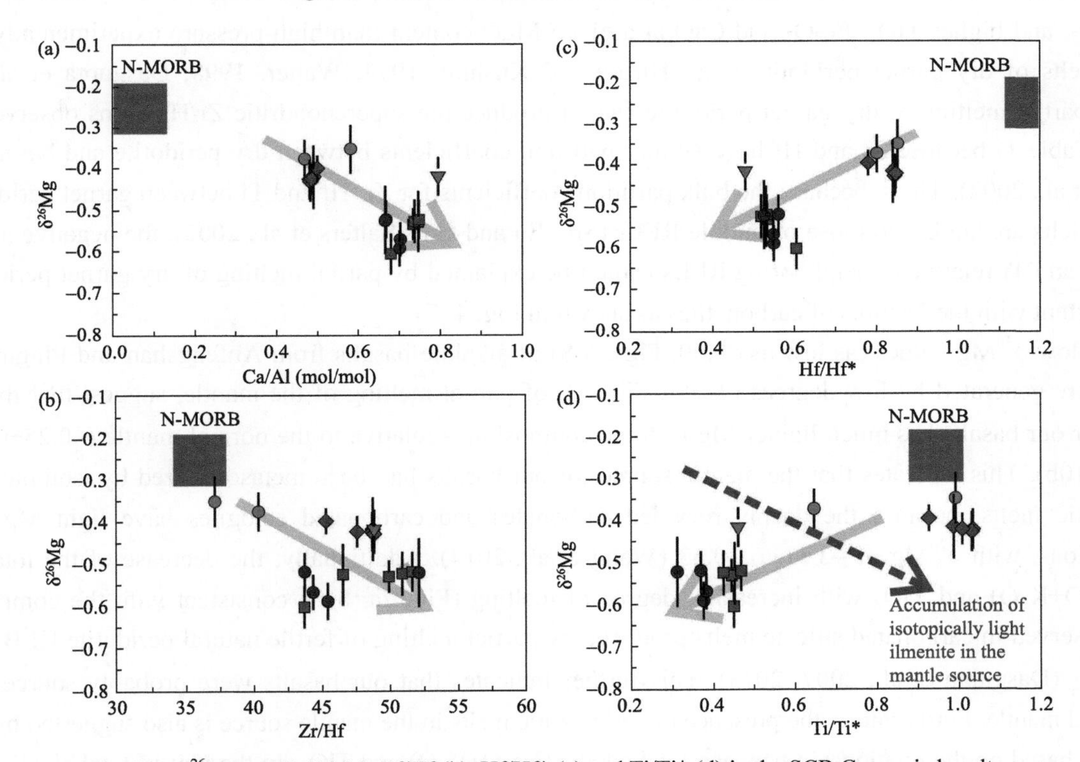

Fig. 12 δ^{26}Mg vs. Ca/Al (a), Zf/Hf (b), Hf/Hf* (c) and Ti/Ti* (d) in the SCB Cenozoic basalts.

In addition to the negative correlation between δ^{26}Mg and TiO_2, the δ^{26}Mg values of our basalts also decrease with the amounts of other incompatible elements (e.g., La, Nd, Nb, Th) and trace element abundance ratios (e.g., Sm/Yb, Nb/Y) (Figs. 7, 8) that are sensitive to partial melting. This suggests that large variations in Mg isotope ratios have occurred during partial melting of the mantle under high temperatures and pressures, with melts produced by low degrees of melting having lighter Mg isotopic compositions relative to melts produced by high degrees of melting. A dissolution experiment on Boulder Creek Granodiorite has shown that concomitant variations in δ^{26}Mg values of reactive fluids reflect conservative mixing of Mg released from isotopically distinct minerals (e.g., chlorite, biotite and hornblende) rather than Mg isotope fractionation. It is experimentally determined that during partial melting of carbonated peridotite, the first incipient melts near solidus are carbonatitic melts, which evolve to carbonated silicate melts (≤25 wt. % CO_2 in melts) with increasing degree of melting (Dasgupta et al., 2007, 2013). Hence, it is expected that during this evolution path, there should exist large variations in Mg isotopic compositions because of different ratios of isotopically light carbonatitic melts/isotopically heavy silicate melts at different degrees of partial melting of carbonated peridotite.

Teng et al. (2007, 2010b) demonstrated that Mg isotope fractionation is insignificant during partial melting of dry peridotite and subsequent basalt differentiation. In the studies of Teng et al. (2007, 2010b), the Mg-rich minerals involved during magmatic processes are olivine and pyroxene which usually have mantle-like Mg isotopic compositions (e, g., Handler et al., 2009; Yang et al., 2009; Huang et al., 2011; Liu et al., 2011; Pogge von Strandmann et al., 2011; Xiao et al., 2013). This may explain why no significant Mg isotope fractionation was observed in their studies. The basalts in the present study overall have much lighter Mg isotopic compositions compared to global mid-ocean ridges basalts (MORBs, Figs. 7, 8 and 13), suggesting that the mantle source for the studied basalts is not a dry garnet peridotite that has an average Mg isotopic composition identical to that of MORBs (δ^{26}Mg=–0.26±0.07 vs. –0.25±0.04, Teng et al., 2010b). The following observations also suggest that dry garnet peridotite is not a suitable source for our alkaline basalts. First, alkaline basalts usually have lower SiO_2 and Al_2O_3, and higher TiO_2, Fe_2O_3 and CaO at a given MgO content than high-pressure experimentally-derived partial melts of dry garnet peridotite (e.g., Hirose and Kushiro, 1993; Walter, 1998; Dasgupta et al., 2007). Second, partial melting of dry garnet peridotite cannot produce the superchondritic Zr/Hf ratios observed in our basalts (Table 1) because Zr and Hf have similar partition coefficients between dry peridotite and basaltic melts (Salters et al., 2002). Third, because the bulk partition coefficients for Zr, Hf and Ti between garnet peridotite and basaltic melts are similar to those of middle REEs (Sm, Eu and Gd) (Salters et al., 2002), the negative anomalies of Zr, Hf and Ti relative to neighboring REEs cannot be explained by partial melting of dry garnet peridotite but are consistent with the features of carbonatites as shown in Fig. 4.

The low δ^{26}Mg values (as low as –0.60, Figs. 6-8) of alkaline basalts from Anfengshan and Pingmingshan, which were generated by low degrees (2–3%, Fig. 10) of partial melting of the mantle, suggest that the mantle source for our basalts has much lighter Mg isotopic compositions relative to the normal mantle (–0.25±0.7, Teng et al., 2010b). This indicates that the mantle source for our basalts had been metasomatized by isotopically light carbonatitic melts, because the deeply recycled carbonates and carbonated eclogites have light Mg isotopic compositions, with δ^{26}Mg of –2.51 to –0.53 (Wang et al., 2014). Additionally, the decrease of the total alkalis (i.e., Na_2O+K_2O) and TiO_2 with increasing degree of melting (Fig. 7a, b) is consistent with the compositional trends observed in carbonated silicate melts produced by partial melting of fertile natural peridotite KLB-1+1–2.5 wt.% CO_2 (Dasgupta et al., 2007, 2013). This further indicates that our basalts were probably sourced from a carbonated mantle. Furthermore, the presence of carbonatitic melts in the mantle source is also suggested by Zeng et al. (2010) based on the relationship between total alkalis (Na_2O+K_2O) and TiO_2. In the plot of total alkalis vs. TiO_2 (Fig. 13), the Cenozoic basalts from eastern China fall along the trend defined by experimentally-derived melts of carbonated peridotite, implying that carbonated peridotite is probably the main source for them.

As carbonation significantly lowers the solidus of mantle peridotite, carbonated mantle may melt before anything else and contribute more when the degrees of melting are low. It has been experimentally determined that the incipient melts from carbonated mantle near solidus are carbonatitic melts (Dasgupta et al., 2007; 2013). Carbonatitic melts will evolve to carbonated silicate melts with increasing temperature and degree of melting as the dissolution of clinopyroxene and/or olivine into carbonatitic melts becomes significant, melt fraction increases and the concentration of CO_2 in melts gets diluted (Dasgupta et al., 2007; 2013). Therefore, under high degrees of melting, the so-called "carbonatitic fingerprints" (e.g., high Ca/Al and Zr/Hf ratios, extremely low Ti/Ti* and Hf/Hf* ratios, and strongly negative anomalies of Zr, Hf and Ti in spidergram, Hoernle et al., 2002; Bizimis et al., 2003) will be diluted as shown in the studied basalts from Fangshan and Chongren (Figs. 3 and 4). The features that the δ^{26}Mg values in our basalts increase with increasing degree of melting could be attributed to the incongruent melting of the carbonated mantle. The basalts produced by low degrees of melting were contributed more from istopically light carbonatitic melts and thus show much lighter Mg isotopic compositions, as observed in basalts from Anfengshan and Pingmingshan; while the basalts produced by higher degrees of melting were contributed largely from isotopically heavy silicate melts and thus display heavy Mg isotopic compositions, as observed in basalts from Chongren. As illustrated in Fig. 12, the basalts with lower δ^{26}Mg values generally have higher Ca/Al and Zr/Hf ratios, and lower Hf/Hf* and Ti/Ti* ratios. Because carbonatitic melts have high ratios of Ca/Al and Zr/Hf and low ratios of Hf/Hf* and Ti/Ti* (Hoernle et al., 2002; Bizimis et al., 2003), these features also imply that our basalts probably record the compositional evolution trend from carbonatitic melts with light Mg isotopic compositions to silicate melts with heavy isotopic compositions (represented by N-MORB in Fig. 12) as observed in partial melting experiments on carbonated peridotite (e.g., Dasgupta et al., 2007, 2013). Thus, the mantle source for our basalts from the SCB is a carbonated mantle with light Mg isotopic compositions that formed by incorporation of isotopically light carbonatitic melts into the upper mantle. As the studied basalts are enriched in LILE and LREE and have depleted Sr-Nd isotopic compositions (e.g., Zou et al., 2000; Chen et al., 2009; Wang et al., 2011), the incorporation of carbonatitic components into the upper mantle must take place recently without a long time-integrated ingrowth of Sr-Nd isotopic systems.

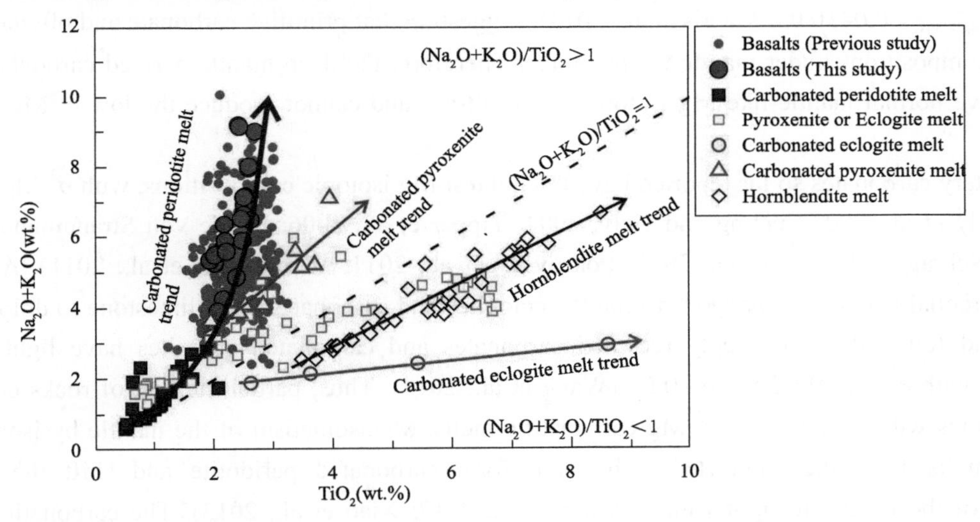

Fig. 13 Na_2O+K_2O vs. TiO_2 for the Cenozoic basalts in eastern China. Data sources for the Cenozoic basalts in eastern China are from the references (Zhi et al., 1990; Zou et al., 2000; Xu et al., 2005; Tang et al., 2006; Liu et al., 2008; Chen et al., 2009; Zhang et al., 2009; Zeng et al., 2010, 2011; Wang et al., 2011; Xu et al., 2012). Also shown for comparison are experimentally-derived melts from carbonated peridotite (Hirose, 1997; Dasgupta et al., 2007), carbonated pyroxenite (Gerbode and Dasgupta, 2010), pyroxenite or eclogite (Hirschmann et al., 2003; Kogiso et al., 2003; Pertermann and Hirschmann, 2003; Kogiso and Hirschmann, 2006), hornblendite (Pilet et al., 2008), and carbonated eclogite (Dasgupta et al., 2006).

5.4 Did the carbonatitic components come from the subducted oceanic slab?

The carbonatitic melts that metasomatized the upper mantle to form carbonated peridotite for generating the alkaline basalts might come from the deep mantle at great depths or the subducted slabs (Dasgupta et al., 2007;

2013). It is experimentally shown that the carbonatitic melts from the deep mantle can be generated by partial melting or redox melting of the primitive carbon-bearing peridotite at depths of greater than 200 km (Dasgupta and Hirschman, 2006; Dasgupta et al., 2013; Stagno et al., 2013), where carbon is stored chiefly in carbonates (e.g., dolomite and magnesite), graphite/diamond and carbides (e.g., Luth, 1999; Dasgupta and Hirschmann, 2010). The graphite or diamond in the deep mantle would react with silicates to form dolomite or magnesite during their ascent through the following redox reactions proposed by Eggler (1982) and Luth (1993):

$$\underset{\text{graphite/diamond}}{2C} + 2O_2 + \underset{\text{olivine}}{2Mg_2SiO_4} = \underset{\text{enstatite}}{Mg_2Si_2O_6} + \underset{\text{magnesite}}{2MgCO_3} \tag{1}$$

and

$$\underset{\text{dioposide}}{CaMgSi_2O_6} + \underset{\text{olivine}}{2Mg_2SiO_4} + \underset{\text{graphite/diamond}}{2C} + 2O_2 = \underset{\text{enstatite}}{2Mg_2Si_2O_6} + \underset{\text{dolomite}}{CaMg(CO_3)_2} \tag{2}$$

Both reactions are involved with free O_2 as oxidants, which could originate from the mantle-derived fluids through reduction of oxidized species, such as CO_2, H_2O and sulfates (e.g., Luth, 1993). These oxidized species are usually abundant in mantle-derived fluids, manifested by the study of fluid inclusions in diamond and minerals of spinel and garnet-peridotite xenoliths (e.g., Navon et al., 1988; Frezzotti et al., 2012). According to these reactions, the element Mg in these carbonates is mainly derived from olivine and pyroxene that have mantle-like Mg isotopic compositions in the normal mantle (e, g., Hardler et al., 2009; Yang et al., 2009; Huang et al., 2011; Liu et al., 2011; Pogge von Strandmann et al., 2011; Xiao et al., 2013). A recent experimental study shows that Mg isotope fractionation between olivine and carbonate (e.g., magnesite) is limited at temperatures of ⩾800 ℃ ($\Delta^{26}Mg_{olivine-magnesite} \leqslant 0.04 \pm 0.04$, Macris et al., 2013), suggesting that primitive carbonate and olivine have similar Mg isotopic compositions under mantle temperatures. Therefore, the deep mantle-derived carbonatitic melts are inferred to have normal mantle-like Mg isotopic compositions and cannot produce the low $\delta^{26}Mg$ values of our basalts.

Sedimentary carbonates so far reported have the lightest Mg isotopic compositions, with $\delta^{26}Mg$ of –5.54 to –0.47 (e.g., Galy et al., 2002; Young and Galy, 2004; Tipper et al., 2006; Pogge von Strandmann et al., 2008; Higgins and Schrag, 2010; Ke et al., 2011; Pokrovsky et al., 2011; Wombacher et al., 2011). A recent study revealed differential isotopic exchange between the eclogites and carbonates (e.g., limestone to dolostone) during subduction and found that the deeply recycled carbonates and carbonated eclogites have light Mg isotopic compositions, with $\delta^{26}Mg$ of –2.51 to –0.53 (Wang et al., 2014). Thus, partial melting of rocks containing low $\delta^{26}Mg$ carbonates will generate low $\delta^{26}Mg$ carbonatitic melts. Metasomatism of the mantle by isotopically light carbonatitic melts from the subducted slab could form carbonated peridotite and shift the Mg isotopic compositions of the mantle to light values (Yang et al., 2012; Xiao et al., 2013). The carbonatitic melts were probably derived from partial melting of the recycled carbonated eclogite transformed from the subducted carbonate-bearing oceanic crust during plate subduction. This can be inferred from the following observations. First, the OIB-like trace element distribution patterns (Fig. 4) and key element ratios (e.g., U/Pb, Th/Pb, Nb/U, Ce/Pb, Table 1) for most of the basalts imply the involvement of a few percent of recycled oceanic crust in their mantle source (e.g., Wang et al., 2011; Xu et al., 2012; Sakuyama et al., 2013). Second, carbonated eclogite has been reported to have light Mg isotopic compositions, with $\delta^{26}Mg$ as low as –1.93 (Wang et al., 2012, 2014).

5.5 Geodynamic implications

The light Mg isotopic compositions of the <110 Ma basalts from the NCB have been attributed to the interaction of mantle peridotite with isotopically light carbonatitic melts derived from the subducted oceanic slab

(Yang et al., 2012). However, the NCB suffered from three circum-craton oceanic subductions since the Paleozoic era, including the Paleo-Tethys oceanic subduction from south, the Mongolia oceanic subduction from north and the Pacific oceanic subduction from East (Windley et al., 2010). It is thus difficult to judge which oceanic subduction supplied recycled carbonated eclogite in the upper mantle of the NCB. Previous studies show that on the northern side of the NCB, southwards subduction of the Mongolia oceanic plate started in the Ordovician and ceased in the Permo-Triassic (Xiao et al., 2003), while northwards subduction of the Paleo-Tethys oceanic plate below the southern margin of the NCB began in the Paleozoic and ceased in the Triassic (Li et al., 1993, 2000, 2001). On the basis of plate reconstruction, westwards subduction of the Izanaghi-Pacific plate beneath the eastern Asian continent was suggested to start as early as early Cretaceous (Müller et al., 2008). Meanwhile, the mantle-like Mg isotopic compositions of the >120 Ma basalts in the NCB (Fig. 6) suggest that the isotopically light mantle source didn't form till 120 Ma (Yang et al., 2012). Considering only the Pacific plate was subducting beneath the NCB in the Mesozoic-Cenozoic, Yang et al. (2012) further pointed out that subduction of Pacific plate plays an important role in generating the light Mg isotopic compositions of the <110 Ma basalts of the NCB. In this study, the light Mg isotopic compositions of the Cenozoic basalts from the SCB provide a convincing evidence to support that the recycled carbonated eclogite in the upper mantle of the NCB were derived from the Pacific slab, because only the Pacific slab has an influence on both blocks of North and South China. In addition, our results combined with Yang's study also suggest that the upper mantle of eastern China may be significantly metasomatized by carbonatitic melts formed by partial melting of carbonated eclogite transformed from the subducted Pacific slab. Such carbonatitic melts might be responsible for the abrupt changes of Mg and Nd isotopic compositions between the >120 and <110 Ma continental basalts from eastern China (Yang et al., 2012).

High-resolution seismic tomography revealed that the Pacific slab is subducting beneath the Japan Islands and becomes stagnant in the mantle transition zone (410–660 km) beneath eastern China, with its western edge ~2000 km away from the Japan Trench (e.g., Zhao et al., 2011). The stagnant Pacific slab might bring large amounts of carbonated eclogites into the mantle transition zone as they can survive from subduction-zone dehydration and melting at modern subduction zones (Dasgupta, 2013 and references therein). Thus, we propose that the carbonatitic melts for creating the light Mg isotopic compositions of the upper mantle of eastern China probably originate from the stagnant Pacific slab. Partial molten carbonatitic melts would metasomatize the upper mantle and lead to the formation of a carbonated mantle above the stagnant Pacific slab beneath East Asia, which have formed a big mantle wedge (e.g., Zhao et al., 2011). Carbonated peridotite with light Mg isotopic compositions existed in the upper mantle would melt first and generate the isotopically light basalts, because carbonation lowers the solidus of peridotite (e.g., Dasgupta et al., 2007, 2013).

6 Conclusions

High-precision major and trace element data and Mg isotopic analyses on the Cenozoic alkaline and tholeiitic basalts from the SCB, eastern China lead us to make conclusions as follows:

(1) The Mg isotopic compositions of the studied basalts are much lighter relative to the normal mantle, with δ^{26}Mg values ranging from –0.60 to –0.35. The possibility of isotopically light ilmenite accumulation in their mantle source for causing the light Mg isotopic compositions of our basalts can be ruled out, because (i) their relatively lower TiO_2 contents (<2.5 wt.%) compared to high-Ti lunar basalts (>6.5 wt.%) investigated by Sedaghatpour et al. (2013) suggest no significant abundance of ilmenite in the mantle source of our basalts; and (ii) the roughly positive correlation between their δ^{26}Mg values and Ti/Ti* as well as their constant Nb/Ta ratios irrespective of variable TiO_2 contents suggests no isotopically light ilmenite accumulation in their mantle source,

which would result in negative correlations between Nb/Ta and TiO_2, δ^{26}Mg and Ti/Ti*. Thus, the low δ^{26}Mg basalts from the SCB were probably sourced from a carbonated mantle that formed by interaction of the mantle with isotopically light carbonatitic melts, and our results confirm that Mg isotope ratios can be used as a powerful tool to trace recycled carbonates.

(2) The δ^{26}Mg values of our basalts decrease with the amounts of incompatible elements (e.g., Ti, La, Nd, Nb, Th) and trace element abundance ratios (e.g., Sm/Yb, Nb/Y) that are sensitive to partial melting, suggesting that large variations in Mg isotope ratios occurred during partial melting of the mantle under high temperatures and pressures. Additionally, their Hf/Hf* and Ti/Ti* ratios increase, and Ca/Al and Zr/Hf ratios decrease with increasing degrees of partial melting. These features can be ascribed to the incongruent partial melting of the carbonated mantle. At low degrees of melting, the partial melts are contributed more from isotopically light carbonatitic melts that have high Ca/Al and Zr/Hf ratios, low Hf/Hf* and Ti/Ti* ratios, while at higher degrees of partial melting, the partial melts are contributed more from isotopically heavy silicate melts that have low Ca/Al and Zr/Hf ratios, high Hf/Hf* and Ti/Ti* ratios. Thus, the large Mg isotopic variations in our basalts represent conservative mixing of isotopically distinct materials rather than isotope fractionation at mantle pressures and temperatures.

(3) The carbonatitic melts probably originate from the stagnant Pacific slab beneath East Asia, which is consistent with the results of seismic tomography (e.g., Zhao et al., 2011). Thus, our results combined with the study of Yang et al. (2012) demonstrate that the subducted Pacific slab provides the recycled carbonates and that there exists a widespread carbonated upper mantle beneath eastern China, which serves as the main source for the <110 Ma alkaline basalts.

Acknowledgement

This work is financially supported by grants from the National Science Foundation of China (NOs. 41090372, 41230209, 41328004 and 41273037) and the Fundamental Research Funds for the Central Universities (WK2080000068). We are grateful to Zhen-Hui Hou and Hai-Yang Liu for assistance during analyses of trace elements and to Ting Gao, Zi-Jian Li, Ze-Zhou Wang for Mg isotopic analyses. Grateful thanks are due to four anonymous reviewers whose constructive and critical reviews greatly improve this contribution. We acknowledge the executive editor Dr. Marc Norman and the associated editor Dr. Weidong Sun for their punctuality and dedication during the editing work.

Appendix A. Supplementary data

Supplementary data associated with this article can be found, in the online version, at http://dx.doi.org/10.1016/j.gca.2015.04.054.

References

Allègre C. J., Treuil M., Minster J.-F., Minster B. and Albarède F. (1977) Systematic use of trace element in igneous process. Contrib. Mineral. Petrol. 60, 57-75.

An Y. J., Wu F., Xiang Y., Nan X. Y., Yu X., Yang J., Yu H. M., Xie L. W. and Huang F. (2014) High-precision Mg isotope analyses of low-Mg rocks by MC-ICP-MS. Chem. Geol. 390, 9-21.

Bizimis M., Salters V. M. and Dawson J. B. (2003) The brevity of carbonatite sources in the mantle: evidence from Hf isotopes.

Contrib. Mineral. Petrol. 145, 281-300.

Bizzarro M., Baker J. A. and H aack H. (2004) Mg isotope evidence for contemporaneous formation of chondrules and refractory inclusions. Nature 431, 275-278.

Black J. R., Yin Q.-Z. and Casey W. H. (2006) An experimental study of magnesium-isotope fractionation in chlorophyll-a photosynthesis. Geochim. Cosmochim. Acta 70, 4072-4079.

Bolou-Bi E. B., Poszwa A., Leyval C. and Vigier N. (2010) Experimental determination of magnesium isotope fractionation during higher plant growth. Geochim. Cosmochim. Acta 74, 2523-2537.

Bourdon B., Tipper E. T., Fitoussi C. and Stracke A. (2010) Chondritic Mg isotope composition of the Earth. Geochim. Cosmochim. Acta 74, 5069-5083.

Buchachenko A. L., Kouznetsov D. A., Breslavskaya N. N. and Orlova M. A. (2008) Magnesium isotope effects in enzymatic phosphorylation. J. Phys. Chem. B 112, 2548-2556.

Chen D. G. and Peng Z. C. (1988) K-Ar ages and Pb, Sr isotopic characteristics of some Cenozoic volcanic rocks from Anhui and Jiangsu provinces, China. Act. Petrol. Sin. 5, 3-12 (In Chinese with English abstract).

Chen L.-H., Zeng G., Jiang S.-Y., Hofmann A. W., Xu X.-S. and Pan M.-B. (2009) Sources of Anfengshan basalts: Subducted lower crust in the Sulu UHP belt, China. Earth Planet. Sci. Lett. 286, 426-435.

Dasgupta R. and Hirschmann M. M. (2006) Melting in the Earth's deep upper mantle caused by carbon dioxide. Nature 440, 659-662.

Dasgupta R., Hirschmann M. M. and Stalker K. (2006) Immiscible transition from carbonate-rich to silicate-rich melts in the 3 GPa melting interval of eclogite+CO_2 and genesis of silicaundersaturated ocean island lavas. J. Petrol. 47, 647-671.

Dasgupta R., Hirschmann M. M. and Smith N. D. (2007) Partial melting experiments of peridotite+CO_2 at 3 GPa and genesis of alkalic ocean island basalts. J. Petrol. 48, 2093-2124.

Dasgupta R. and Hirschmann M. M. (2010) The deep carbon cycle and melting in Earth's interior. Earth Planet. Sci. Lett. 298, 1-13.

Dasgupta R. (2013) Ingassing, storage, and outgassing of terrestrial carbon through geologic time. Rev. Mineral. Geochem. 75, 183-229.

Dasgupta R., Mallik A., Tsuno K., Withers A. C., Hirth G. and Hirschmann M. M. (2013) Carbon-dioxide-rich silicate melt in the Earth/'s upper mantle. Nature 493, 211-215.

Dygert N., Liang Y. and Hess P. (2013) The importance of melt TiO_2 in affecting major and trace element partitioning between Fe-Ti oxides and lunar picritic glass melts. Geochim. Cosmochim. Acta 106, 134-151.

Eggler D. H. and Baker, D. R (1982). Reduced volatiles in the system C-O-H: Implications to mantle melting, fluid, formation, and diamond genesis. In Akimoto S. and others (eds.), High-pressure in geophysics. Adv. Earth Planet. Sci. 12, 237-250.

Feigenson M. D., Bolge L. L., Carr M. J. and Herzberg C. T. (2003) REE inverse modeling of HSDP2 basalts: Evidence for multiple sources in the Hawaiian plume. Geochem. Geophys. Geosyst. 4, 8706.

Frezzotti M. L., Ferrando S., Tecce F. and Castelli D. (2012) Water content and nature of solutes in shallow-mantle fluids from fluid inclusions. Earth Planet. Sci. Lett. 351-352, 70-83.

Galy A., Bar-Matthews M., Halicz L. and O'Nions R. K. (2002) Mg isotopic composition of carbonate: insight from speleothem formation. Earth Planet. Sci. Lett. 201, 105-115.

Galy A., Yoffe O., Janney P. E., Williams R. W., Cloquet C., Alard O., Halicz L., Wadhwa M., Hutcheon I. D., Ramon E. and Carignan J. (2003) Magnesium isotope heterogeneity of the isotopic standard SRM980 and new reference materials for magnesium-isotope-ratio measurements. J. Anal. At. Spectrom. 18, 1352-1356.

Gerbode C. and Dasgupta R. (2010) Carbonate-fluxed melting of MORB-like pyroxenite at 2.9 GPa and genesis of HIMU ocean island basalts. J. Petrol. 51, 2067-2088.

Handler M. R., Baker J. A., Schiller M., Bennett V. C. and Yaxley G. M. (2009) Magnesium stable isotope composition of Earth's upper mantle. Earth Planet. Sci. Lett. 282, 306-313.

Higgins J. A. and Schrag D. P. (2010) Constraining magnesium cycling in marine sediments using magnesium isotopes. Geochim. Cosmochim. Acta 74, 5039-5053.

Hirose K. and Kushiro I. (1993) Partial melting of dry peridotites at high pressures: Determination of compositions of melts segregated from peridotite using aggregates of diamond. Earth Planet. Sci. Lett. 114, 477-489.

Hirose K. (1997) Partial melt compositions of carbonated peridotite at 3 GPa and role of CO_2 in alkali-basalt magma generation. Geophys. Res. Lett. 24, 2837-2840.

Hirschmann M. M., Kogiso T., Baker M. B. and Stolper E. M. (2003) Alkalic magmas generated by partial melting of garnet pyroxenite. Geology 31, 481-484.

Ho K.-S., Chen J.-C., Lo C.-H. and Zhao H.-L. (2003) ^{40}Ar-^{39}Ar dating and geochemical characteristics of late Cenozoic basaltic rocks from the Zhejiang-Fujian region, SE China: eruption ages, magma evolution and petrogenesis. Chem. Geol. 197, 287-318.

Hoernle K., Tilton G., Le Bas M., Duggen S. and Garbe-Schönberg D. (2002) Geochemistry of oceanic carbonatites compared with continental carbonatites: mantle recycling of oceanic crustal carbonate. Contrib. Mineral. Petrol. 142, 520-542.

Hofmann A. W., Jochum K. P., Seufert M. and White W. M. (1986) Nb and Pb in oceanic basalts: new constraints on mantle evolution. Earth Planet. Sci. Lett. 79, 33-45.

Hofmann A. W. (1988) Chemical differentiation of the Earth: the relationship between mantle, continental crust, and oceanic crust. Earth Planet. Sci. Lett. 90, 297-314.

Huang F., Glessner J., Ianno A., Lundstrom C. and Zhang Z. (2009) Magnesium isotopic composition of igneous rock standards measured by MC-ICP-MS. Chem. Geol. 268, 15-23.

Huang F., Zhang Z., Lundstrom C. C. and Zhi X. (2011) Iron and magnesium isotopic compositions of peridotite xenoliths from Eastern China. Geochim. Cosmochim. Acta 75, 3318-3334.

Huang K.-J., Teng F.-Z., Wei G.-J., Ma J.-L. and Bao Z.-Y. (2012a) Adsorption- and desorption-controlled magnesium isotope fractionation during extreme weathering of basalt in Hainan Island, China. Earth Planet. Sci. Lett. 359-360, 73-83.

Huang J., Xiao Y. L., Gao Y. J., Hou Z. H. and Wu W. P. (2012b) Nb-Ta fractionation induced by fluid-rock interaction in subduction-zones: constraints from UHP eclogite- and vein- hosted rutile from the Dabie orogen, Central-Eastern China. J. Metamorph. Geol. 30, 821-842.

Huang F., Chen L., Wu Z. and Wang W. (2013) First-principles calculations of equilibrium Mg isotope fractionations between garnet, clinopyroxene, orthopyroxene, and olivine: Implications for Mg isotope thermometry. Earth Planet. Sci. Lett. 367, 61-70.

Jacobson A. D., Zhang Z., Lundstrom C. and Huang F. (2010) Behavior of Mg isotopes during dedolomitization in the Madison Aquifer, South Dakota. Earth Planet. Sci. Lett. 297, 446-452.

Jin Z. M., Yu R., Yang W. and Ou X. (2003) Mantle-derived Xenoliths of Peridotite from Pingmingshan, Donghai County, Jiangsu Province and Their Implications for Deep Structures. Act. Geol. Sin. 77, 451-462 (In Chinese with English abstract).

Johnson K. T. M., Dick H. J. B. and Shimizu N. (1990) Melting in the oceanic upper mantle: An ion microprobe study of diopsides in abyssal peridotites. J. Geophys. Res.: Solid Earth 95, 2661-2678.

Ke S., Liu S.-A., Li W.-Y., Yang W. and Teng F.-Z. (2011) Advances and application in magnesium isotope geochemistry. Act. Petrol. Sin. 27, 383-397 (In Chinese with English abstract).

Kogiso T., Hirschmann M. M. and Frost D. J. (2003) High- pressure partial melting of garnet pyroxenite: possible mafic lithologies in the source of ocean island basalts. Earth Planet. Sci. Lett. 216, 603-617.

Kogiso T. and Hirschmann M. M. (2006) Partial melting experiments of bimineralic eclogite and the role of recycled mafic oceanic crust in the genesis of ocean island basalts. Earth Planet. Sci. Lett. 249, 188-199.

Langmuir C. H., Bender J. F., Bence A. E., Hanson G. N. and Taylor S. R. (1977) Petrogenesis of basalts from the FAMOUS area: Mid-Atlantic Ridge. Earth Planet. Sci. Lett. 36, 133-156.

Le Bas M. J. (1986) A chemical classification of volcanic rocks based on the total alkali-silica diagram. J. Petrol. 27, 745-750.

Lee T., Papanastassiou D. A. and Wasserburg G. J. (1976) Demonstration of ^{26}Mg excess in Allende and evidence for ^{26}Al. Geophys. Res. Lett. 3, 41-44.

Li S. G., Xiao Y. L., Liou D., Chen Y., Ge N., Zhang Z., Sun S.-S., Cong B., Zhang R. Y., Hart S. R. and Wang S. (1993) Collision of the North China and Yangtse Blocks and formation of coesite-bearing eclogites: Timing and processes. Chem. Geol. 109, 89-111.

Li S. G., Jagoutz E., Chen Y. and Li Q. L. (2000) Sm-Nd and Rb- Sr isotopic chronology and cooling history of ultrahigh pressure metamorphic rocks and their country rocks at Shuanghe in the Dabie Mountains, Central China. Geochim. Cosmochim. Acta 64, 1077-1093.

Li S. G., Huang F., Nie Y. H., Han W. L., Long G., Li H. M., Zhang S. Q. and Zhang Z. H. (2001) Geochemical and geochronological constraints on the suture location between the North and South China blocks in the Dabie Orogen, Central China. Phys. Chem. Earth (A): Solid Earth Geodesy. 26, 655-672.

Li W.-Y., Teng F.-Z., Ke S., Rudnick R. L., Gao S., Wu F.-Y. and Chappell B. W. (2010) Heterogeneous magnesium isotopic composition of the upper continental crust. Geochim. Cosmochim. Acta 74, 6867-6884.

Li W.-Y., Teng F.-Z., Xiao Y. and Huang J. (2011) High- temperature inter-mineral magnesium isotope fractionation in eclogite from the Dabie orogen, China. Earth Planet. Sci. Lett. 304, 224-230.

Liu C.-Q., Masuda A. and Xie G.-H. (1994) Major- and traceelement compositions of Cenozoic basalts in eastern China: Petrogenesis and mantle source. Chem. Geol. 114, 19-42.

Liu Y., Gao S., Kelemen P. B. and Xu W. (2008) Recycled crust controls contrasting source compositions of Mesozoic and Cenozoic basalts in the North China Craton. Geochim Cosmochim Acta 72, 2349-2376.

Liu S.-A., Teng F.-Z., He Y., Ke S. and Li S. (2010) Investigation of magnesium isotope fractionation during granite differentiation: Implication for Mg isotopic composition of the continental crust. Earth Planet. Sci. Lett. 297, 646-654.

Liu S.-A., Teng F.-Z., Yang W. and Wu F.-Y. (2011) High-temperature inter-mineral magnesium isotope fractionation in mantle xenoliths from the North China craton. Earth Planet. Sci. Lett. 308, 131-140.

Liu X.-M., Teng F.-Z., Rudnick R. L., McDonough W. F. and Cummings M. L. (2014) Massive magnesium depletion and isotope fractionation in weathered basalts. Geochim. Cosmochim. Acta 135, 336-349.

Luth R. W. (1993) Diamonds, eclogites, and the oxidation state of the Earth's mantle. Science 261, 66-68.

Luth R. W. (1999) Carbon and carbonates in the mantle. In: Fei Y., Bertka C. M., Mysen B. O. (eds.), Mantle petrology: field observations and high pressure experimentation: a tribute to francis R. (Joe) Boyd, The Geochem. Society 6. 297-316.

Müller R. D., Sdrolias M., Gaina C., Steinberger B. and Heine C. (2008) Long-term sea-level fluctuations driven by ocean basin dynamics. Science 319, 1357-1362.

Macris C. A., Young E. D. and Manning C. E. (2013) Experimental determination of equilibrium magnesium isotope fractionation between spinel, forsterite, and magnesite from 600 to 800 °C. Geochim. Cosmochim. Acta 118, 18-32.

McDonough W. F. and Sun S. S. (1995) The composition of the Earth. Chem. Geol. 120, 223-253.

Millero F. J. (1974) The Physical Chemistry of Seawater. Annu Rev. Earth Planet. Sci. 2, 101-150.

Minster J. F. and Allègre C. J. (1978) Systematic use of trace elements in igneous processes. Contrib. Mineral. Petrol. 68, 37-52.

Navon O., Hutcheon I. D., Rossman G. R. and Wasserburg G. J. (1988) Mantle-derived fluids in diamond micro-inclusions. Nature 335, 784-789.

Peng Z. C., Zartman R. E., Futa K. and Chen D. G. (1986) Pb-, Sr- and Nd-isotopic systematics and chemical characteristics of Cenozoic basalts, eastern China. Chem. Geol. 59, 3-33.

Pertermann M. and Hirschmann M. M. (2003) Partial melting experiments on a MORB-like pyroxenite between 2 and 3 GPa: Constraints on the presence of pyroxenite in basalt source regions from solidus location and melting rate. J. Geophys. Res.: Solid Earth 108, 2125.

Pilet S., Baker M. B. and Stolper E. M. (2008) Metasomatized lithosphere and the origin of alkaline lavas. Science 320, 916-919.

Pogge von Strandmann P. A. E. (2008) Precise magnesium isotope measurements in core top planktic and benthic foraminifera. Geochem. Geophys. Geosyst. 9, Q12015.

Pogge von Strandmann P. A. E., Burton K. W., James R. H., van Calsteren P., Gislason S. R. and Sigfú sson B. (2008) The influence of weathering processes on riverine magnesium isotopes in a basaltic terrain. Earth Planet. Sci. Lett. 276, 187-197.

Pogge von Strandmann P. A. E., Elliott T., Marschall H. R., Coath C., Lai Y.-J., Jeffcoate A. B. and Ionov D. A. (2011) Variations of Li and Mg isotope ratios in bulk chondrites and mantle xenoliths. Geochim. Cosmochim. Acta 75, 5247-5268.

Pokrovsky B. G., Mavromatis V. and Pokrovsky O. S. (2011) Covariation of Mg and C isotopes in late Precambrian carbonates of the Siberian Platform: A new tool for tracing the change in weathering regime? Chem. Geol. 290, 67-74.

Qi Q. U., Taylor L. A. and Zhou X. (1995) Petrology and Geochemistry of Mantle Peridotite Xenoliths from SE China. J. Petrol. 36, 55-79.

Reisberg L., Zhi X., Lorand J.-P., Wagner C., Peng Z. and Zimmermann C. (2005) Re-Os and S systematics of spinel peridotite xenoliths from east central China: Evidence for contrasting effects of melt percolation. Earth Planet. Sci. Lett. 239, 286-308.

Rosman K. J. R. and Taylor P. D. P. (1998) Isotopic compositions of the elements 1997. Pure Appl. Chem. 70, 217-235.

Rudnick R. L. and Gao S. (2003) Composition of the continental crust. Treatise Geochem. 3, 1-64.

Saenger C. and Wang Z. (2014) Magnesium isotope fractionation in biogenic and abiogenic carbonates: implications for paleoenvironmental proxies. Quaternary Sci. Rev. 90, 1-21.

Sakuyama T., Tian W., Kimura J.-I., Fukao Y., Hirahara Y., Takahashi T., Senda R., Chang Q., Miyazaki T., Obayashi M., Kawabata H. and Tatsumi Y. (2013) Melting of dehydrated oceanic crust from the stagnant slab and of the hydrated mantle transition zone: Constraints from Cenozoic alkaline basalts in eastern China. Chem. Geol. 359, 32-48.

Salters V. J. M., Longhi J. E. and Bizimis M. (2002) Near mantle solidus trace element partitioning at pressures up to 3.4 GPa. Geochem. Geophys. Geosyst. 3, 1038.

Sedaghatpour F., Teng F.-Z., Liu Y., Sears D. W. G. and Taylor L. A. (2013) Magnesium isotopic composition of the Moon. Geochim. Cosmochim. Acta 120, 1-16.

Shen B., Jacobsen B., Lee C.-T. A., Yin Q.-Z. and Morton D. M. (2009) The Mg isotopic systematics of granitoids in continental arcs and implications for the role of chemical weathering in crust formation. Proc. Natl. Acad. Sci. U.S.A. 106, 20652-20657.

Stagno V., Ojwang D. O., McCammon C. A. and Frost D. J. (2013) The oxidation state of the mantle and the extraction of carbon from Earth's interior. Nature 493, 84-88.

Sun S. s. and Hanson G. (1975) Origin of Ross Island basanitoids and limitations upon the heterogeneity of mantle sources for alkali basalts and nephelinites. Contrib. Mineral. Petrol. 52, 77-106.

Tang Y.-J., Zhang H.-F. and Ying J.-F. (2006) Asthenosphere- lithospheric mantle interaction in an extensional regime: Implication from the geochemistry of Cenozoic basalts from Taihang Mountains, North China Craton. Chem. Geol. 233, 309-327.

Telus M., Dauphas N., Moynier F., Tissot F. L. H., Teng F.-Z., Nabelek P. I., Craddock P. R. and Groat L. A. (2012) Iron, zinc, magnesium and uranium isotopic fractionation during continental crust differentiation: The tale from migmatites, granitoids, and pegmatites. Geochim. Cosmochim. Acta 97, 247-265.

Teng F.-Z., Wadhwa M. and Helz R. T. (2007) Investigation of magnesium isotope fractionation during basalt differentiation: Implications for a chondritic composition of the terrestrial mantle. Earth Planet. Sci. Lett. 261, 84-92.

Teng F.-Z., Li W.-Y., Rudnick R. L. and Gardner L. R. (2010a) Contrasting lithium and magnesium isotope fractionation during continental weathering. Earth Planet. Sci. Lett. 300, 63-71.

Teng F.-Z., Li W.-Y., Ke S., Marty B., Dauphas N., Huang S., Wu F.-Y. and Pourmand A. (2010b) Magnesium isotopic composition of the Earth and chondrites. Geochim. Cosmochim. Acta 74, 4150-4166.

Teng F.-Z. and Yang W. (2014) Comparison of factors affecting the accuracy of high-precision magnesium isotope analysis by multi-collector inductively coupled plasma mass spectrometry. Rapid Commun. Mass Spectrom. 28, 19-24.

Teng F.-Z., Li W.-Y., Ke S., Yang W., Liu S.-A., Sedaghatpour F., Wang S.-J., Huang K.-J., Hu Y., Ling M.-X., Xiao Y., Liu X.- M., Li X.-W., Gu H.-O., Sio C. K., Wallace D. A., Su B.-X., Zhao, Chamberlin J., Harrington M. and Brewer A. Magnesium isotopic compositions of international geological reference materials. Geostand. Geoanalyt. Res., 10.1111/j.1751-908X.2014.00326.x.

Tipper E. T., Bickle M. J., Galy A., West A. J., Pomiès C. and Chapman H. J. (2006) The short term climatic sensitivity of carbonate and silicate weathering fluxes: Insight from seasonal variations in river chemistry. Geochim. Cosmochim. Acta 70, 2737-2754.

Walter M. J. (1998) Melting of garnet peridotite and the origin of komatiite and depleted lithosphere. J. Petrol. 39, 29-60.

Wang S.-J., Teng F.-Z., Williams H. M. and Li S.-G. (2012) Magnesium isotopic variations in cratonic eclogites: Origins and implications. Earth Planet. Sci. Lett. 359-360, 219-226.

Wang S.-J., Teng F.-Z. and Li S.-G. (2014) Tracing carbonatesilicate interaction during subduction using magnesium and oxygen isotopes. Nat. Commun. 5. http://dx.doi.org/10.1038/ ncomms6328.

Wang Y., Zhao Z.-F., Zheng Y.-F. and Zhang J.-J. (2011) Geochemical constraints on the nature of mantle source for Cenozoic continental basalts in east-central China. Lithos 125, 940-955.

Wimpenny J., Yin Q.-Z., Tollstrup D., Xie L.-W. and Sun J. (2014a) Using Mg isotope ratios to trace Cenozoic weathering changes: A case study from the Chinese loess plateau. Chem. Geol. 376, 31-43.

Wimpenny J., Colla C. A., Yin Q., Rustad J. R. and Casey W. H. (2014b) Investigating the behaviour of mg isotopes during the formation of clay minerals. Geochim. Cosmochim. Acta 128, 178-194.

Windley B. F., Maruyama S. and Xiao W. J. (2010) Delamination/ thinning of sub-continental lithospheric mantle under Eastern China: The role of water and multiple subduction. Am. J. Sci. 310, 1250-1293.

Wombacher F., Eisenhauer A., Böhm F., Gussone N., Regenberg M., Dullo W. C. and Rüggeberg A. (2011) Magnesium stable isotope fractionation in marine biogenic calcite and aragonite. Geochim. Cosmochim. Acta 75, 5797-5818.

Xiao W. J., Windley B. F., Hao J. and Zhai M. (2003) Accretion leading to collision and the Permian Solonker suture, Inner Mongolia, China: Termination of the central Asian orogenic belt. Tectonics 22, 1069-1088.

Xiao Y., Teng F.-Z., Zhang H.-F. and Yang W. (2013) Large magnesium isotope fractionation in peridotite xenoliths from eastern North China craton: Product of melt-rock interaction. Geochim. Cosmochim. Acta 115, 241-261.

Xu Y.-G., Ma J.-L., Frey F. A., Feigenson M. D. and Liu J.-F. (2005) Role of lithosphere-asthenosphere interaction in the genesis of Quaternary alkali and tholeiitic basalts from Datong, western North China Craton. Chem. Geol. 224, 247-271.

Xu Y.-G., Zhang H.-H., Qiu H.-N., Ge W.-C. and Wu F.-Y. (2012) Oceanic crust components in continental basalts from Shuangliao, Northeast China: Derived from the mantle transition zone? Chem. Geol. 328, 168-184.

Yang W., Teng F.-Z. and Zhang H.-F. (2009) Chondritic magnesium isotopic composition of the terrestrial mantle: A case study of peridotite xenoliths from the North China craton. Earth Planet. Sci. Lett. 288, 475-482.

Yang W., Teng F.-Z., Zhang H.-F. and Li S.-G. (2012) Magnesium isotopic systematics of continental basalts from the North China craton: Implications for tracing subducted carbonate in the mantle. Chem. Geol. 328, 185-194.

Young E. D. and Galy A. (2004) The isotope geochemistry and cosmochemistry of magnesium. Rev. Mineral. Geochem. 55, 197-230.

Zeng G., Chen L.-H., Xu X.-S., Jiang S.-Y. and Hofmann A. W. (2010) Carbonated mantle sources for Cenozoic intra-plate alkaline basalts in Shandong, North China. Chem. Geol. 273, 35-45.

Zeng G., Chen L.-H., Hofmann A. W., Jiang S.-Y. and Xu X.-S. (2011) Crust recycling in the sources of two parallel volcanic chains in Shandong, North China. Earth Planet. Sci. Lett. 302, 359-368.

Zhao D. P., Yu S. and Ohtani E. (2011) East Asia: Seismotectonics, magmatism and mantle dynamics. J. Asian Earth Sci. 40, 689-709.

Zhi X., Song Y., Frey F. A., Feng J. and Zhai M. (1990) Geochemistry of Hannuoba basalts, eastern China: Constraints on the origin of continental alkalic and tholeiitic basalt. Chem. Geol. 88, 1-33.

Zhou X. M. and Armstrong R. L. (1982) Cenozoic volcanic rocks of eastern China—secular and geographic trends in chemistry and strontium isotopic composition. Earth Planet. Sci. Lett. 58, 301-329.

Zou H. B., Zindler A., Xu X. and Qi Q. (2000) Major, trace element, and Nd, Sr and Pb isotope studies of Cenozoic basalts in SE China: mantle sources, regional variations, and tectonic significance. Chem. Geol. 171, 33-47.

Magnesium isotope evidence for a recycled origin of cratonic eclogites*

Shui-Jiong Wang[1, 2], Fang-Zhen Teng[1], Roberta L. Rudnick[3] and Shu-Guang Li[2, 4]

1. Isotope Laboratory, Department of Earth and Space Sciences, University of Washington, Seattle, Washington 98195-1310, USA
2. State Key Laboratory of Geological Processes and Mineral Resources, China University of Geosciences, Beijing 100083, China
3. Geochemical Laboratory, Department of Geology, University of Maryland, College Park, Maryland 20742, USA
4. CAS Key Laboratory of Crust-Mantle Materials and Environments, School of Earth and Space Sciences, University of Science and Technology of China, Hefei 230026, Anhui, China

亮点介绍： 西非塞拉利昂的太古宙地盾内的科伊杜金伯利岩杂岩中石榴石和单斜辉石矿物的Mg同位素组成表明，低Mg或高Mg榴辉岩中的石榴石和单斜辉石存在Mg同位素平衡分馏；围岩金伯利岩对捕虏体的Mg同位素组成影响较小。低Mg榴辉岩的Mg同位素特征表明其源自于再循环洋壳。高Mg榴辉岩的Mg同位素和Mg、Fe含量特征表明其成因是部分熔融的低Mg榴辉岩与橄榄岩围岩进行Mg-Fe交换的结果。综上，克拉通榴辉岩的Mg同位素组成可以用来示踪再循环洋壳，并且表明了蚀变洋壳的再循环可能导致地幔的Mg同位素组成不均一性

Abstract The Mg isotopic compositions of garnet and clinopyroxene mineral separates and whole rocks from 21 xenolithic eclogites (11 low-MgO eclogites and 10 high-MgO eclogites) from the Koidu kimberlite complex, erupted within the Archean Man Shield, Sierra Leone, West Africa, provide new evidence bearing on the origin of cratonic eclogites. Garnet and clinopyroxene in both low-MgO and high-MgO eclogites generally record equilibrium inter-mineral Mg isotope partitioning, with δ^{26}Mg varying from –2.15‰ to –0.46‰ in garnets and from –0.49‰ to +0.35‰ in clinopyroxenes. Bulk δ^{26}Mg values (–1.38‰ to +0.05‰), constructed from garnet and clinopyroxene data, are similar to results from rock powders (–1.60‰ to +0.17‰), suggesting that kimberlite infiltration has had negligible influence on the Mg isotopic compositions of the xenoliths. The δ^{26}Mg values of low-MgO eclogites (–0.80‰ to +0.05‰) exceed the range of mantle peridotite xenoliths (–0.25‰±0.04‰), consistent with the eclogite's derivation from recycled altered oceanic crust. Similarly variable δ^{26}Mg values in high-MgO eclogites (–0.95‰ to –0.13‰), together with their high MgO and low FeO contents, suggest that high-MgO eclogites were produced by Mg-Fe exchange between partially molten low-MgO eclogites and surrounding peridotites. Our study shows that cratonic xenolithic eclogites preserve a

* 本文发表在：Geology, 2015, 43(12): 1071-1074

record of Mg isotopic compositions produced by low-pressure, surficial isotope fractionations. The recycling of oceanic crust will therefore increase the Mg isotope heterogeneity of the mantle.

1 Introduction

Mantle eclogite xenoliths sample the cratonic lithospheric mantle and have garnered significant interest due to their possible origin via ancient subduction cycles and the fact that a significant proportion of kimberlite-borne diamonds have grown in an eclogitic medium (e.g., Jacob, 2004; Stachel and Harris, 2008). However, there is a broad spectrum of eclogite compositions, and debate continues as to whether all xenolithic eclogites, especially those containing very high MgO contents, reflect ancient recycling of basaltic or picritic crust (e.g., Griffin and O'Reilly, 2007).

Cratonic eclogites from Sierra Leone, West Africa, have been divided into two groups based on their MgO contents (e.g., Hills and Haggerty, 1989): low-MgO eclogites (with measured bulk-rock MgO <15 wt%) and high-MgO eclogites (with measured bulk-rock MgO >15 wt%). Garnets from the low-MgO eclogites display variable $\delta^{18}O$ values (+4.7‰ to +6.8‰, Barth et al., 2001), corresponding to whole rock compositions that extend beyond the range of peridotites (+5.5‰±0.4 ‰; Mattey et al., 1994), suggesting their protoliths may be altered oceanic crust (e.g., Barth et al., 2001). By contrast, the high-MgO eclogites have garnets with mantle-like $\delta^{18}O$ values (+5.1‰ to +5.7‰; Barth et al., 2002). The origin of high MgO eclogites the world over is enigmatic, with hypotheses ranging from in situ high-pressure mantle cumulates (e.g., Hills and Haggerty, 1989), low-pressure oceanic gabbroic cumulates or melting residues (e.g., Ireland et al., 1994; Barth et al., 2002; Tappe et al., 2011), to subducted oceanic crust that has been modified by interaction with peridotite (e.g., Smart et al., 2009).

Magnesium isotopes can provide insights on the origin of cratonic eclogites because they are readily fractionated, by up to 7‰, during low-temperature processes (Tipper et al., 2006; Liu et al., 2014), but display only limited variations (⩽0.07‰) during high-temperature igneous differentiation (Teng et al., 2007, 2010; Handler et al., 2009; Yang et al., 2009; Pogge von Strandmann et al., 2011; Xiao et al., 2013). For example, altered oceanic crust has a heterogeneous Mg isotopic composition due to carbonate precipitation and clay formation during seafloor alteration (Higgins and Schrag, 2010), whereas pristine oceanic basalts and mantle peridotite xenoliths display a narrow range of $\delta^{26}Mg$ (–0.25‰±0.07‰; Teng et al., 2010). To date, Mg isotopic data have been reported for only six cratonic eclogites from South Africa. The calculated bulk $\delta^{26}Mg$ values vary widely from –0.80‰ to –0.14‰, which were attributed to the incorporation of subducted altered oceanic crust in their protoliths (Wang et al., 2012).

Here we present Mg isotopic data for 21 well-characterized xenolithic eclogites and 40 corresponding mineral separates from Koidu, Sierra Leone, West Africa. Both low-MgO and high-MgO eclogites display large Mg isotopic variations (as much as 1.43‰), suggesting that the low-MgO eclogites represent fragments of subducted altered oceanic crust and that the high-MgO eclogites are products of Mg-Fe exchange between low-MgO eclogites and surrounding peridotites.

2 Koidu eclogites

The xenolithic eclogites derive from the Mesozoic Koidu kimberlite complex located on the Man Shield of the West Africa Craton (Fig. 1; Hills and Haggerty, 1989; Fung and Haggerty, 1995; Rollinson, 1997). The majority of Koidu eclogites are bimineralic with the modal abundance of garnet and clinopyroxene each ranging from 30% to 60%. Fine-grained alteration assemblages constitute between 2 modal% and 17 modal% of the rocks (Hills and Haggerty, 1989). Two groups of eclogites are distinguishable in terms of their measured and calculated bulk MgO contents: low-MgO eclogites have MgO of 6–13 wt%, and high-MgO eclogites have MgO>16 wt% (Hills and Haggerty, 1989). Garnets in low-MgO eclogites are compositionally similar to the residual garnets produced in high-pressure experiments on mid-oceanic ridge basalts (MORBs) or gabbros with MgO of 7.3–11.7 wt% (Fig. DR1 in the GSA Data Repository①). Coexisting clinopyroxenes vary from jadeitic diopside to omphacite (Na_2O=3.65–7.22 wt%; Hills and Haggerty, 1989). By contrast, garnets in high-MgO eclogites have significantly higher MgO of 16.8–20.3 wt%, extending beyond the range of those formed in high-pressure experiments on MORBs or gabbros, but overlapping the garnets formed during eclogite-peridotite interaction experiments (Fig. DR1). The coexisting clinopyroxenes are mainly diopsidic in composition (Na_2O=1.71–3.72 wt%; Hills and Haggerty, 1989).

Some Koidu eclogites preserve partial melting textures (e.g., presence of spongy, symplectic textures in clinopyroxenes) and contain zoned garnets (Hills and Haggerty, 1989; Fung and Haggerty, 1995). These garnets have Mg-poor cores (9.04–10.04 wt%) that are compositionally akin to low-MgO garnets, and Mg-rich rims (13.22–17.38 wt%) that are similar to high-MgO garnets or fall within the Mg gap between low- and high-MgO garnets (Hills and Haggerty, 1989; Fung and Haggerty, 1995). Garnet $\delta^{18}O$ values are generally much more variable in low-MgO eclogites (+4.7‰ to +6.8‰) than in high-MgO eclogites (+5.1‰ to +5.7‰; Barth et al., 2001, 2002). Equilibrium temperatures estimated using garnet-clinopyroxene Mg-Fe exchange thermometry range from 850 to 1100 °C for the low-MgO eclogites and from 950 to 1150 °C for the high-MgO eclogites (Hills and Haggerty, 1989; Fung and Haggerty, 1995).

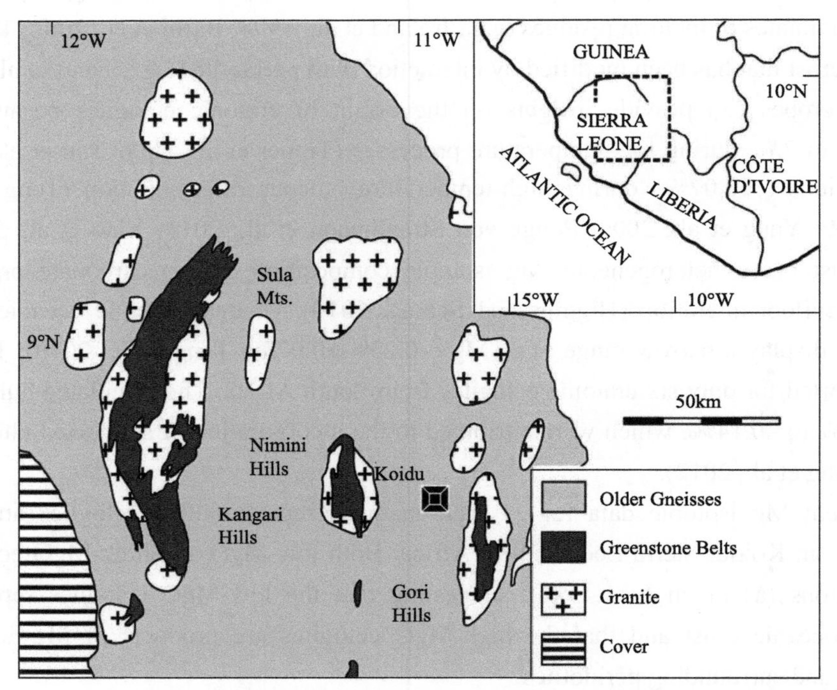

Fig. 1 Generalized geological map of northeastern Sierra Leone (from Rollinson, 1999), showing location of Koidu kimberlites (8°37′52″N to 8°37′58″N latitude and 10°57′55″W to 10°58′19″W longitude).

3 Results

The Mg isotopic data of in-house reference materials and eclogite samples are reported in Tables DR1 and

① GSA Data Repository item 2015356, analytical methods, and Mg isotopic data of bulk rock powders and mineral separates, is available online at www.geosociety.org/pubs/ft2015.htm, or on request from editing@geosociety.org or Documents Secretary, GSA, P.O. Box 9140, Boulder, CO 80301, USA.

DR2 in the Data Repository, respectively. Garnet δ^{26}Mg values range from –2.15‰ to –0.48‰ in low-MgO eclogites, and from –1.22‰ to –0.46‰ in high-MgO eclogites (Fig. 2). Clinopyroxene δ^{26}Mg values range from –0.40‰ to +0.28‰ in low-MgO eclogites, and from –0.49‰ to +0.35‰ in high-MgO eclogites (Fig. 2). Altered garnets and clinopyroxenes yield Mg isotopic compositions that are within uncertainty of values of their fresh counterparts (Fig. DR2). The δ^{26}Mg values of garnet and clinopyroxene are positively correlated (R=0.92, excluding one low-MgO eclogite sample, KEC81-7), and fall on equilibrium fractionation lines corresponding to temperatures of 900–1350 °C (Fig. 2). Temperatures estimated using inter-mineral Mg isotope fractionation are in rough agreement with these results (Fig. DR3). The sample KEC81-7 defines an anomalously low temperature of ~550 °C based on Mg isotopes (Fig. 2).

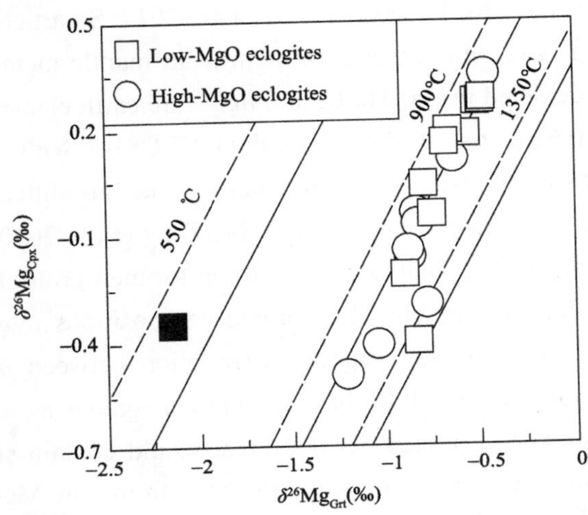

Fig. 2 $\delta^{26}Mg_{Grt}$ (garnet) versus $\delta^{26}Mg_{Cpx}$ clinopyroxene for low- and high-MgO eclogites. Equilibrium Mg isotope fractionations for different temperatures are calculated from Huang et al. (2013), assuming pressures of 1 GPa (dashed line) and 6 GPa (solid line). Sample KEC81-7 is shown as filled square.

The bulk δ^{26}Mg values of 15 whole rock powders ($\delta^{26}Mg_{measured}$) vary from –1.60‰ to –0.26‰ in low-MgO eclogites, and from –1.07‰ to +0.17 ‰ in high-MgO eclogites (Table DR2). The bulk δ^{26}Mg values calculated from garnet and clinopyroxene data ($\delta^{26}Mg_{calculated}$) range from –1.38‰ to +0.05‰ in low-MgO eclogites and from –0.95‰ to –0.13‰ in high-MgO eclogites. The $\delta^{26}Mg_{measured}$ and $\delta^{26}Mg_{calculated}$ values generally agree with each other and fall on or near a 1:1 line (Fig. DR4). Nonetheless, $\delta^{26}Mg_{calculated}$ values are used hereafter.

4 Discussion

The range of Mg isotopic compositions in cratonic eclogites cannot be caused by unrepresentative sampling of mineral phases, as the Mg isotopic variations in garnet and clinopyroxene surpass that which can be generated by varying mineral modal abundance by even±20%. They are also unlikely to be due to exchange with the host kimberlite. The diffusion coefficient of Mg in garnet is ~10^{-13} mm^2/s at a temperature of ~1000 °C (Perchuk et al., 2009), thus a grain with a diameter of 4 mm needs at least 1 m.y. to be homogenized. The rapid eruption of kimberlitic magma (within hours and days; Kelley and Wartho, 2000) rules out resetting of mineral δ^{26}Mg values by kimberlite-eclogite isotopic exchange. Even in the sample that shows evidence of garnet-clinopyroxene isotopic disequilibrium (KEC81-7), different garnet or clinopyroxene grains yield identical Mg isotopic compositions (Table DR2), suggesting that the observed disequilibrium was not caused by kimberlite-induced diffusion, but rather by earlier processes that took place at mantle depths. Therefore, it appears that exchange with or infiltration of kimberlite has had negligible influence on the Mg isotopic compositions of eclogite xenoliths. The large Mg isotopic variations thus reflect isotopic heterogeneities at mantle depths that existed prior to the eclogite's entrainment in the kimberlite magma.

4.1 Origin of the Low-MgO Eclogites

Low-MgO eclogites in general are too high in Al_2O_3 and too low in MgO content to match the compositions of high-pressure melts derived from peridotites. They contain no olivine, but many have positive or negative Eu anomalies; $\delta^{18}O$ values also extend beyond the range for mantle peridotites. All of these observations are indicative of a low-pressure origin for the low-MgO eclogites (e.g., Jacob et al., 1994; Jacob, 2004; Barth et al., 2001; Nielsen et al., 2009; Tappe et al., 2011; Pernet-Fisher et al., 2014; Smart et al., 2014).

The low-MgO Koidu eclogites did not experience significant mantle metasomatism, as evidenced by their trace elemental patterns that are depleted in Ba, Th, U, and light rare earth elements (LREEs) (Barth et al., 2001). Their variable $\delta^{26}Mg$ values (–0.62‰ to +0.05‰, excluding KEC81-7 with disequilibrium inter-mineral Mg isotope fractionation) must reflect the Mg isotopic heterogeneity of the protoliths. Given that pristine mantle rocks display a narrow range of bulk $\delta^{26}Mg$ values (–0.25‰±0.07‰; Teng et al., 2010), the Mg isotopic heterogeneity in the low-MgO Koidu eclogites supports a shallow crustal origin for their protoliths.

Altered oceanic crust with highly variable Mg isotopic compositions may be the best candidate for the protoliths of low-MgO eclogites. The roughly negative correlation between $\delta^{26}Mg$ and Na_2O/CaO (Fig. 3a) suggests that the $\delta^{26}Mg$ values of the protoliths become more negative as seafloor spilitization progresses. However, the low-$\delta^{26}Mg$ end member of altered oceanic crust would contain significant amounts of carbonate (Higgins and Schrag, 2010), which is compositionally different from low-MgO eclogites. The absence of a correlation between $\delta^{26}Mg$ and $\delta^{18}O$ values for the low-MgO eclogites (Fig. 3b) further implies that processes other than seafloor alteration may have had taken place to produce the Mg isotopic variation in low-MgO eclogites. Differential Mg isotopic exchange between carbonates and silicate rocks during crustal subduction may lower the $\delta^{26}Mg$ value of the upper part of oceanic crust (Wang et al., 2014). Exhumed orogenic eclogites that have experienced isotopic exchange with surrounding carbonate are characterized by low $\delta^{26}Mg$ values down to –1.93‰ (Wang et al., 2014). Therefore, it is likely that the large Mg isotopic variations in low-MgO eclogites are not only caused by ancient seawater alteration, but also enhanced by carbonate-silicate isotopic exchange during subduction.

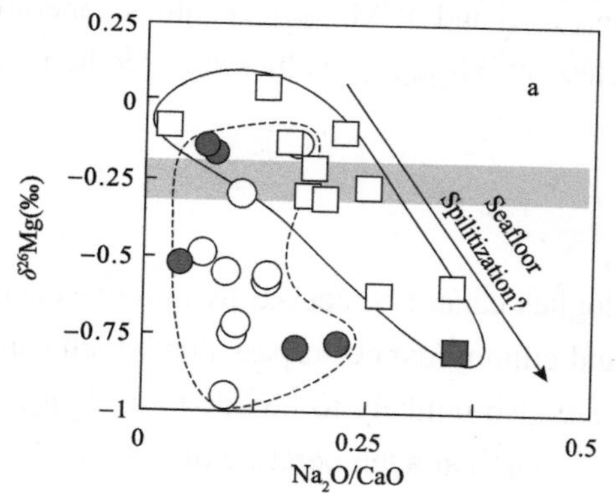

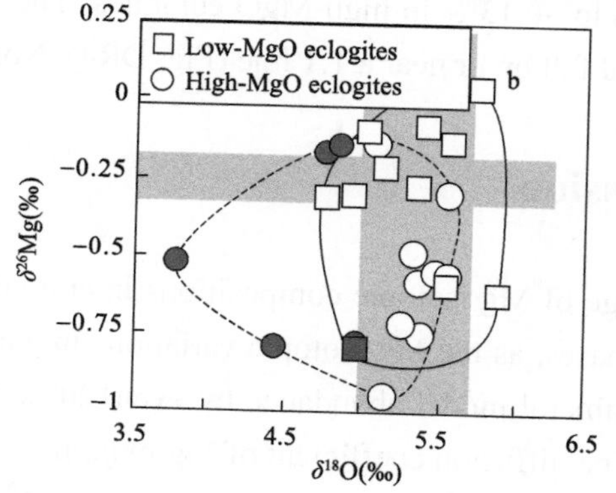

Fig. 3 Calculated bulk rock $\delta^{26}Mg$ values as function of Na_2O/CaO ratio (a) and garnet $\delta^{18}O$ value (b) for low- and high-MgO eclogites, Koidu kimberlite complex, Sierra Leone. The $\delta^{18}O$ is from Barth et al. (2001, 2002); bulk Na_2O/CaO is calculated based on mineral abundance and composition reported in Hills and Haggerty (1989). Gray bars represent the normal mantle $\delta^{26}Mg$ and $\delta^{18}O$ values (Mattey et al., 1994; Teng et al., 2010). Filled symbols are cratonic eclogites from South Africa (Williams et al., 2009; Wang et al., 2012).

4.2 Origin of the High-MgO Eclogites

While many studies have suggested an origin for low-MgO eclogites via subduction of altered oceanic crust, the petrogenesis of high-MgO eclogites is still unresolved. Hypotheses proposed for the formation of high-MgO eclogites include their origin as high-pressure mantle cumulates (Hills and Haggerty, 1989), metamorphic counterparts of shallow cumulates or melt residues formed during Archean subduction (Ireland et al., 1994; Rollinson, 1997; Barth et al., 2002; Tappe et al., 2011), and melt-facilitated interaction of low-MgO eclogites with surrounding peridotite (Smart et al., 2009; Smit et al., 2014).

High-pressure mantle cumulates should have δ^{26}Mg around the normal mantle value of –0.25‰ (e.g., Teng et al., 2010). Low-pressure high-MgO counterparts located in the deeper portions of oceanic lithosphere or at the base of continental crust are unlikely to have interacted with seawater or carbonate, and therefore should also display mantle-like δ^{26}Mg values. These hypotheses, therefore, cannot explain the large Mg isotopic variation in high-MgO Koidu eclogites (–0.95‰ to –0.13‰). An experimental study of high-temperature (600–800 °C) equilibrium Mg isotope fractionation between olivine and carbonate by Macris et al. (2013) has shown a small inter-mineral Mg isotope fractionation of 0.04‰–0.44‰, which implies that primary mantle carbonatite melts should have Mg isotopic compositions that do not deviate far from the normal mantle value. This is consistent with observations that δ^{26}Mg values reported for the most metasomatised mantle rocks (e.g., wehrlites) fall in a narrow range of –0.44‰ to +0.06‰ (Yang et al., 2009; Pogge von Strandmann et al., 2011; Xiao et al., 2013). In addition, carbonatitic metasomatism does not sufficiently enrich SiO_2 and MgO to match the composition of the high-MgO eclogites (e.g., Smart et al., 2009). Therefore, carbonatitic metasomatism does not appear to be a viable explanation for the Mg isotopic systematics in high-MgO eclogites.

The most likely mechanism for the origin of high-MgO elcogites is thus melt-facilitated Mg-Fe exchange between low-MgO eclogite and surrounding peridotite. This petrogenesis has been proposed for high-MgO eclogites from Jericho in the Slave craton (Smart et al., 2009), where diamonds hosted in the high-MgO eclogites contain mineral inclusions that are characteristic of low-MgO eclogites (Smart et al., 2009, 2012). High-pressure experiments on a mixed eclogite-peridotite lithology have shown that melt-facilitated diffusional Mg-Fe exchange between eclogite and refractory peridotite produces a residual eclogite that is enriched in MgO and depleted in FeO (Yaxley and Green, 1998). Eclogites in the lithospheric mantle have a lower solidus than surrounding peridotite (Hisrchmann and Stolper, 1996), and thus will melt before peridotites, generating melts that facilitate rapid diffusional Fe-Mg exchange between restitic eclogite and peridotite (Yaxley and Green, 1998; Smart et al., 2009). The high-MgO Koidu eclogites display a negative correlation between FeO and MgO (Barth et al., 2002). Garnets in these high-MgO eclogites also have similar compositions to those formed in the eclogite-peridotite equilibration experiments (Fig. DR1). Some Koidu eclogites preserve partial melting textures, with zoned garnets showing evidence for transformation from Mg-poor cores to Mg-rich rims (Hills and Haggerty, 1989; Fung and Haggerty, 1995). All of these observations are consistent with Mg-Fe exchange between low-MgO eclogite and peridotite. The Mg in the high-MgO eclogites is therefore a mixture of Mg from peridotite and low-MgO eclogite. However, high-MgO eclogites display similar Mg isotopic variations as the low-MgO eclogites (Fig. 4a), which, in turn, suggests that the range in δ^{26}Mg in low-MgO eclogites may be underestimated due to the relatively small sample size. Low-MgO eclogites are likely to have even lighter Mg isotopic compositions than that reported in the present study, given the fact that their proposed protoliths display much larger Mg isotopic variations (as much as 3‰; Higgins and Schrag, 2010; Wang et al., 2014). A bulk mixing calculation between exhumed orogenic carbonated eclogites and peridotite suggests a peridotite contribution of 20%–30% by mass for the high-MgO Koidu eclogites (Fig. 4a). The lack of deviation of δ^{18}O from the normal mantle range for the high-MgO Koidu eclogites is also consistent with this mixing scenario. Adding 20% −30% of a peridotitic component to the low-MgO eclogites (δ^{18}O=+4.7‰ to +6.8‰) will dampen the original δ^{18}O variability, producing high-MgO eclogites with δ^{18}O in the range of +5.1‰ to +5.9‰ (Fig. 4b). Therefore, the O isotopic compositions of high-MgO Koidu eclogites, though mantle-like, probably do not represent a pristine mantle signature, but are likely the products of hybridization between recycled crustal component and mantle.

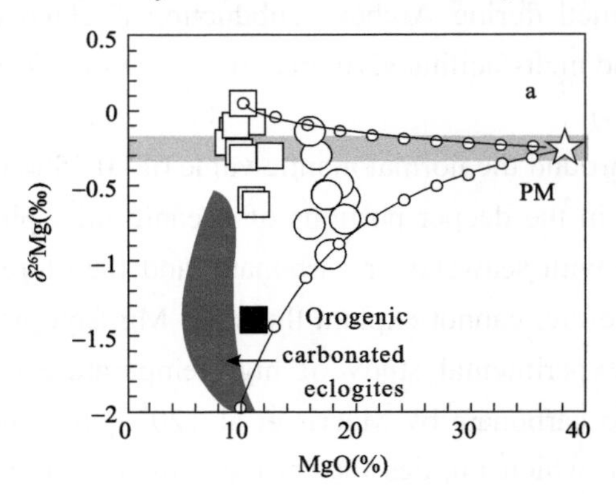

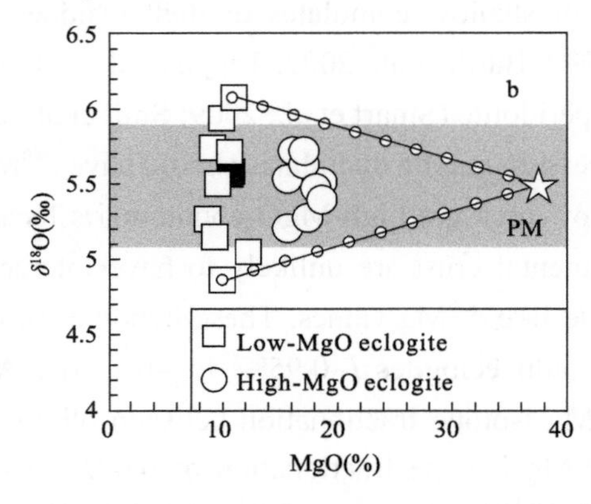

Fig. 4 Calculated bulk rock δ^{26}Mg values (a), and garnet δ^{18}O values (b), as function of MgO content, Koidu kimberlite complex, Sierra Leone. δ^{18}O is from Barth et al. (2001, 2002); MgO is calculated based on mineral abundance and composition reported in Hills and Haggerty (1989). Gray bars represent the normal mantle δ^{26}Mg and δ^{18}O values (Mattey et al., 1994; Teng et al., 2010). Dark-gray field represents orogenic carbonated eclogites from Wang et al. (2014). Mixing between low-MgO eclogite (including orogenic carbonated eclogite) and primitive mantle (PM) are represented by red curves, with 10% increments. Sample KEC81-7 is shown as a black square.

Acknowledgements

We are very grateful to Steve Haggerty for providing the samples. Constructive comments from Katie Smart, Sune Nielsen, and an anonymous reviewer, and careful and efficient handling from Ellen Thomas, are greatly appreciated. This work was financially supported by the National Science Foundation grants EAR-0838227, EAR-1056713, and EAR-1340160 to Teng and EAR-980467 to Rudnick.

References

Barth, M.G., Rudnick, R.L., Horn, I., McDonough, W.F., Spicuzza, M.J., Valley, J.W., and Haggerty, S.E., 2001, Geochemistry of xenolithic eclogites from West Africa, Part I: A link between low MgO eclogites and Archean crust formation: Geochimica et Cosmochimica Acta, v. 65, p. 1499-1527, doi:10.1016/S0016-7037(00)00626-8.

Barth, M.G., Rudnick, R.L., Horn, I., McDonough, W.F., Spicuzza, M.J., Valley, J.W., and Haggerty, S.E., 2002, Geochemistry of xenolithic eclogites from West Africa, Part 2: Origins of the high MgO eclogites: Geochimica et Cosmochimica Acta, v. 66, p. 4325-4345, doi:10.1016/S0016-7037(02)01004-9.

Fung, A.T., and Haggerty, S.E., 1995, Petrography and mineral compositions of eclogites from the Koidu Kimberlite Complex, Sierra Leone: Journal of Geophysical Research, v. 100, p. 20451-20473, doi:10.1029/95JB01573.

Griffin, W.L., and O'Reilly, S.Y., 2007, Cratonic lithospheric mantle: is anything subducted?: Episodes, v. 30, p. 43-53.

Handler, M.R., Baker, J.A., Schiller, M., Bennett, V.C., and Yaxley, G.M., 2009, Magnesium stable isotope composition of Earth's upper mantle: Earth and Planetary Science Letters, v. 282, p. 306-313, doi:10.1016/j.epsl.2009.03.031.

Higgins, J.A., and Schrag, D.P., 2010, Constraining magnesium cycling in marine sediments using magnesium isotopes: Geochimica et Cosmochimica Acta, v. 74, p. 5039-5053, doi:10.1016/j.gca.2010.05.019.

Hills, D.V., and Haggerty, S.E., 1989, Petrochemistry of eclogites from the Koidu kimberlite complex, Sierra Leone: Contributions to

Mineralogy and Petrology, v. 103, p. 397-422, doi:10.1007/BF01041749.

Hirschmann, M.M., and Stolper, E.M., 1996, A possible role for garnet pyroxenite in the origin of the "garnet signature" in MORB: Contributions to Mineralogy and Petrology, v. 124, p. 185-208, doi:10.1007/s004100050184.

Huang, F., Chen, L., Wu, Z., and Wang, W., 2013, First-principles calculations of equilibrium Mg isotope fractionations between garnet, clinopyroxene, orthopyroxene, and olivine: Implications for Mg isotope thermometry: Earth and Planetary Science Letters, v. 367, p. 61-70, doi:10.1016/j.epsl.2013.02.025.

Ireland, T.R., Rudnick, R.L., and Spetsius, Z., 1994, Trace elements in diamond inclusions from eclogites reveal link to Archean granites: Earth and Planetary Science Letters, v. 128, p. 199-213, doi:10.1016/0012-821X(94)90145-7.

Jacob, D., 2004, Nature and origin of eclogite xenoliths from kimberlites: Lithos, v. 77, p. 295-316, doi:10.1016/j.lithos.2004.03.038.

Jacob, D., Jagoutz, E., Lowry, D., Mattey, D., and Kudrjavtseva, G., 1994, Diamondiferous eclogites from Siberia: remnants of Archean oceanic crust: Geochimica et Cosmochimica Acta, v. 58, p. 5191-5207, doi:10.1016/0016-7037(94)90304-2.

Kelley, S., and Wartho, J., 2000, Rapid kimberlite ascent and the significance of Ar-Ar ages in xenolith phlogopites: Science, v. 289, p. 609-611, doi:10.1126/science.289.5479.609.

Liu, X.-M., Teng, F.-Z., Rudnick, R.L., McDonough, W.F., and Cummings, M.L., 2014, Massive magnesium depletion and isotope fractionation in weathered basalts: Geochimica et Cosmochimica Acta, v. 135, p. 336-349, doi:10.1016/j.gca.2014.03.028.

Macris, C.A., Young, E.D., and Manning, C.E., 2013, Experimental determination of equilibrium magnesium isotope fractionation between spinel, forsterite, and magnesite from 600 to 800 °C: Geochimica et Cosmochimica Acta, v. 118, p. 18-32, doi:10.1016/j.gca.2013.05.008.

Mattey, D., Lowry, D., and Macpherson, C., 1994, Oxygen isotope composition of mantle peridotite: Earth and Planetary Science Letters, v. 128, p. 231-241, doi:10.1016/0012-821X(94)90147-3.

Nielsen, S.G., Williams, H.M., Griffin, W.L., O'Reilly, S.Y., Pearson, N., and Viljoen, F., 2009, Thallium isotopes as a potential tracer for the origin of cratonic eclogites: Geochimica et Cosmochimica Acta, v. 73, p. 7387-7398, doi:10.1016/j.gca.2009.09.001.

Perchuk, A., Burchard, M., Schertl, H.-P., Maresch, W., Gerya, T., Bernhardt, H.-J., and Vidal, O., 2009, Diffusion of divalent cations in garnet: multi-couple experiments: Contributions to Mineralogy and Petrology, v. 157, p. 573-592, doi:10.1007/s00410-008-0353-6.

Pernet-Fisher, J.F., Howarth, G.H., Liu, Y., Barry, P.H., Carmody, L., Valley, J.W., Bodnar, R.J., Spetsius, Z.V., and Taylor, L.A., 2014, Komsomolskaya diamondiferous eclogites: evidence for oceanic crustal protoliths: Contributions to Mineralogy and Petrology, v. 167, p. 1-17, doi:10.1007/s00410-014-0981-y.

Pogge von Strandmann, P.A.E., Elliott, T., Marschall, H.R., Coath, C., Lai, Y.-J., Jeffcoate, A.B., and Ionov, D.A., 2011, Variations of Li and Mg isotope ratios in bulk chondrites and mantle xenoliths: Geochimica et Cosmochimica Acta, v. 75, p. 5247-5268, doi:10.1016/j.gca.2011.06.026.

Rollinson, H.R., 1997, Eclogite xenoliths in west African kimberlites as residues from Archaean granitoid crust formation: Nature, v. 389, p. 173-176, doi:10.1038/38266.

Rollinson, H.R., 1999, Petrology and geochemistry of metamorphosed komatiites and basalts from the Sula Mountains greenstone belt, Sierra Leone: Contributions to Mineralogy and Petrology, v. 134, p. 86-101, doi:10.1007/s004100050470.

Smart, K.A., Heaman, L.M., Chacko, T., Simonetti, A., Kopylova, M., Mah, D., and Daniels, D., 2009, The origin of high-MgO diamond eclogites from the Jericho Kimberlite, Canada: Earth and Planetary Science Letters, v.284, p.527-537, doi:10.1016/j.epsl.2009.05.020.

Smart, K.A., Chacko, T., Stachel, T., Tappe, S., Stern, R.A., Ickert, R.B., and EIMF (Edinburgh Ion Microprobe Facility), 2012, Eclogite formation beneath the northern Slave craton constrained by diamond incluisons: Oceanic lithosphere origin without a crustal signature: Earth and Planetary Science Letters, v. 319-320, p. 165-177, doi:10.1016/j.epsl.2011.12.032.

Smart, K.A., Chacko, T., Simonetti, A., Sharp, Z., and Meaman, L.M., 2014, A record of Paleoproterozoic subduction preserved in the northern Slave cratonic mantle: Sr-Pb-O isotope and trace-element investigations of eclogite xenoliths from the Jericho and Muskox kimberlites: Journal of Petrology, v. 55, p. 549-583, doi:10.1093/petrology/egt077.

Smit, K.V., Stachel, T., Creaser, R.A., Ickert, R.B., DuFrane, S.A., Stern, R.A., and Seller, M., 2014, Origin of eclogite and pyroxenite xenoliths from the Victor kimberlite, Canada, and implications for Superior craton formation: Geochimica et Cosmochimica Acta, v. 125, p. 308-337, doi:10.1016/j.gca.2013.10.019.

Stachel, T., and Harris, J.W., 2008, The origin of cratonic diamonds—Constraints from mineral inclusions: Ore Geology Reviews, v. 34, p. 5-32, doi:10.1016/j.oregeorev.2007.05.002.

Tappe, S., Smart, K.A., Pearson, D.G., Steenfelt, A., and Simonetti, A., 2011, Craton formation in Late Archean subduction zones revealed by first Greenland eclogites: Geology, v. 39, p. 1103-1106, doi:10.1130/G32348.1.

Teng, F.-Z., Li, Y.-Y., Ke, S., Marty, B., Dauphas, N., Huang, S., Wu, F.Y., and Pourmand, A., 2010, Magnesium isotopic composition of the Earth and chondrites: Geochimica et Cosmochimica Acta, v. 74, p. 4150-4166, doi:10.1016/j.gca.2010.04.019.

Teng, F.-Z., Wadhwa, M., and Helz, R.T., 2007, Investigation of magnesium isotope fractionation during basalt differentiation: implications for a chondritic composition of the terrestrial mantle: Earth and Planetary Science Letters, v. 261, p. 84-92, doi:10.1016/j.epsl.2007.06.004.

Tipper, E., Galy, A., Gaillarde, J., Bickle, M.J., Elderfield, H., and Carder, E.A., 2006, The magnesium isotope budget of the modern ocean: Constraints from riverine magnesium isotope ratios: Earth and Planetary Science Letters, v. 250, p. 241-253, doi:10.1016/j.epsl.2006.07.037.

Wang, S.-J., Teng, F.-Z., Williams, H.M., and Li, S.-G., 2012, Magnesium isotopic variations in cratonic eclogites: Origins and implications: Earth and Planetary Science Letters, v. 359, p. 219-226, doi:10.1016/j.epsl.2012.10.016.

Wang, S.-J., Teng, F.-Z., and Li, S.-G., 2014, Tracing carbonate-silicate interaction during subduction using magnesium and oxygen isotopes: Nature Communications, v. 5, doi:10.1038/ncomms6328.

Williams, H.M., Nielsen, S.G., Renac, C., Griffin, W.L., O'Reilly, S.Y., McCammon, C.A., Pearson, N., Viljoen, F., Alt, J.C., and Halliday, A.N., 2009, Fractionation of oxygen and iron isotopes by partial melting processes: Implications for the interpretation of stable isotope signatures in mafic rocks: Earth and Planetary Science Letters, v. 283, p. 156-166, doi:10.1016/j.epsl.2009.04.011.

Xiao, Y., Teng, F.-Z., Zhang, H.-F., and Yang, W., 2013, Large magnesium isotope fractionation in peridotite xenoliths from eastern North China craton: Product of melt-rock interaction: Geochimica et Cosmochimica Acta, v. 115, p. 241-261, doi:10.1016/j.gca.2013.04.011.

Yang, W., Teng, F.-Z., and Zhang, H.-F., 2009, Chondritic magnesium isotopic composition of the terrestrial mantle: a case study of peridotite xenoliths from the North China craton: Earth and Planetary Science Letters, v. 288, p. 475-482, doi:10.1016/j.epsl.2009.10.009.

Yaxley, G., and Green, D., 1998, Reactions between eclogite and peridotite: mantle refertilisation by subduction of oceanic crust: Schweizerische Mineralogische und Petrographische Mitteilungen, v. 78, p. 243-255.

Mg, Sr, and O isotope geochemistry of syenites from northwest Xinjiang, China: Tracing carbonate recycling during Tethyan oceanic subduction*

Shan Ke[1], Fang-Zhen Teng[2], Shuguang Li[1], Ting Gao[1], Sheng-Ao Liu[1], Yongsheng He[1] and Xuanxue Mo[1]

1. State Key Laboratory of Geological Processes and Mineral Resources, China University of Geosciences, Beijing 100083, China
2. Isotope Laboratory, Department of Earth and Space Sciences, University of Washington, Seattle, WA 98195, USA

亮点介绍：应用 Mg 同位素对青藏高原西北部帕米尔构造结新生代苦子干碱性正长岩和正长花岗岩进行研究，发现其具有显著轻的 Mg 同位素组成（−0.46‰~−0.17‰），并显示出 Mg-O 同位素的负相关性，证明其源区混有再循环沉积碳酸盐。这一工作为确定特提斯带地幔存在碳酸盐岩再循环提供了有力的证据。

Abstract Magnesium isotopic compositions of igneous rocks could be potentially used to trace recycling of supracrustal materials. High-δ^{26}Mg granitoids have been previously reported and explained to reflect the involvement of surface weathered materials in their sources. Low-δ^{26}Mg granitoids, however, have not been reported. In this study, we report high-precision Mg isotopic analyses of Cenozoic alkaline syenites and syenogranites from the Kuzigan and Zankan plutons, northwest Xinjiang, China. The Kuzigan syenites were originated from the mantle metasomatized by recycled supracrustal materials, and the syenogranites are differentiated products of the syenites. Both syenites and syenogranites have δ^{26}Mg values (−0.46 to −0.26‰ and −0.41 to −0.17‰, respectively) significantly lighter than the mantle (−0.25±0.07‰, 2SD). No correlation of δ^{26}Mg with either SiO_2 or MgO is observed, indicating limited Mg isotope fractionation during alkaline magmatic differentiation. The low δ^{26}Mg of the syenites and syenogranites thus reflect a light Mg isotopic source. This, combined with high $^{87}Sr/^{86}Sr$ ratios (0.70814 to 0.71105) and negative correlation between δ^{26}Mg and δ^{18}O, suggests that the magma source contains recycled marine carbonates. Modeling of the Mg-O-Sr isotopic data indicates that the recycled carbonate is mainly limestone with minor dolostone, suggesting that the metasomatism occurred at depths shallower than 60 to 120 km. Given that the plutons are located at the India-Eurasia collision zone, the carbonate recycling was most likely derived from the subducted Tethyan oceanic crust during the Mesozoic-Cenozoic. Our study suggests that the combined Mg, O, and Sr isotopic studies are powerful for tracing recycled carbonates and identifying their species in mantle sources.

* 本文发表在：Chemical Geology, 2016, 437: 109-119

Keywords Magnesium isotopes, syenites, magmatic differentiation, carbonate recycling, Tibetan and Pamir plateau

1 Introduction

Warm global climate at Cretaceous-early Cenozoic era was attributed to a high pCO_2 level possibly maintained by subduction of Tethyan oceanic crust with pelagic carbonate and subsequent release of CO_2 from arc volcanism (Kent and Muttoni, 2008). However, according to the carbon budget in subduction zones, the expelled carbon from arc volcanism is much less than that subducted, implying that a mass of undissolved carbonate is subducted to the deep mantle (Kerrick and Connolly, 2001). Tracing the recycled carbonates and evaluating their proportions in the mantle becomes an important issue (e.g., Li, 2015).

Recent studies show that Mg isotopic compositions of igneous rocks could be used to trace the recycling of supracrustal materials in the magma source (Shen et al., 2009; Li et al., 2010; Yang et al., 2012; Huang et al., 2015; Wang et al., 2016). Compared with the upper mantle (δ^{26}Mg=−0.25±0.07‰, 2SD; Teng et al., 2010), clastic sedimentary rocks (δ^{26}Mg of −1.64 to +0.92‰; Li et al., 2010; Huang et al., 2013) and carbonates (δ^{26}Mg of −5.60 to −0.66‰; Beinlich et al., 2014; Saenger and Wang, 2014), on average, have heavy and light Mg isotopic compositions, respectively. Therefore, Mg isotopes are an effective tool in distinguishing recycled clastic sediments and carbonates, given that Mg isotopes are not significantly fractionated during high temperature magmatic processes. Especially, stability of carbonate species (calcite, dolomite and magnesite) in the mantle via subduction depends on the P-T conditions (Dasgupta and Hirschmann, 2010). Thus, constraining the species of recycled carbonates has a crucial implication for understanding the depth of the recycled carbonates. However, the species of recycled carbonate is difficult to identify by using Mg isotopes alone because all carbonate minerals (e.g., calcite, dolomite, and magnesite) have light Mg isotopic composition (Saenger and Wang, 2014). Nevertheless, carbonate minerals have significantly different Mg and Sr contents but similar O contents, then mixing of calcite, dolomite and magnesite with the mantle will generate different curves in the plots of Mg versus O and Sr isotopes. Thus, a combination of Mg, Sr, and O isotopes may be an effective approach to identify the species of carbonates and to explore their potential application in tracing deep carbon recycling.

The Pamir syntaxis belongs to the westernmost part of the Himalayan-Tibetan orogen, formed by the Tethyan oceanic subduction and the India-Eurasia collision during Paleozoic to Cenozoic. Cenozoic post-collisional potassic and ultrapotassic igneous rocks are widely distributed in the middle Pamir (Hacker et al., 2005; Ke et al., 2008; Jiang et al., 2012). Previous studies proposed that these potassic and ultrapotassic rocks in Pamir originated from the enriched mantle source related to the subducted India continent (Pan, 2000; Jiang et al., 2012). Contemporaneous ultrapotassic magmatic rocks in Xizang were also generally proposed to be the products of partial melting of the enriched mantle (Turner et al., 1996; Miller et al., 1999; Nomade et al., 2004; Williams et al., 2004; Chung et al., 2005; Zhao et al., 2009; Jiang et al., 2012; Wang et al., 2014a). Therefore, if huge amounts of carbonates were carried into the deep mantle by subduction of the Tethyan oceanic slabs, the chemical properties of mantle beneath the Tibetan plateau would have been modified by recycled carbonates and then may imprint the signature on mantle-derived igneous rocks, a hypothesis which can be examined using Mg-O-Sr isotopes as discussed above.

In this study, we report Mg, O, and Sr isotopic data for ultrapotassic alkaline syenites and evolved syenogranites from two Cenozoic plutons (Kuzigan and Zankan) in the Pamir syntaxis, northwestern Xinjiang, China. The aim of this study is to (i) investigate the behavior of Mg isotopes during alkaline magmatic differentiation which is not addressed yet, and the origin of the low δ^{26}Mg syenites and syenogranites; (ii) evaluate the imprint of carbonate recycling in the mantle beneath the Pamir plateau during the subduction of the Tethyan oceanic slabs; and (iii) explore a way using Mg-O-Sr isotopes to identify recycled carbonate species.

2 Geological setting and samples

The Pamir syntaxis, located in the western Himalayan-Tibetan orogen, is composed of several distinctive terrenes and suture zones that accreted to the southern margin of Asia during the late Paleozoic-Mesozoic closure of the Paleo-Tethys Ocean and the Mesozoic-Cenozoic closure of the Neo-Tethys Ocean (Burtman and Molnar, 1993; Schwab et al., 2004). The Pamir area is divided into the northern, central and southern zone based on the Paleozoic and Mesozoic sutures (Burtman and Molnar, 1993). To the south, the Indus suture and Shyok suture were the marks of the final India-Eurasia collision in Cenozoic (Bouilhol et al., 2013).

Cenozoic igneous rocks were widely distributed in the central and southern Pamir (Ducea et al., 2003; Hacker et al., 2005). The Kuzigan alkaline pluton, paralleling the Karakorum fault, is one of this Cenozoic magmatic belt (Fig. 1a). It is bounded by a Miocene granite pluton (Karibasheng) (Fig. 1a) to the northeast and gneissic granites of unknown age to the southwest (Fig. 1b). The Kuzigan pluton is made up of alkaline syenites, syenitoids and syenogranites, with medium- to coarse-grained aegirine-augite syenites and syenogranites as their representative rocks. The different lithologies have gradual contacts without clear boundaries, indicating a synchronous magmatic activity. SHIRMP U-Pb zircon ages of syenites and syenogranites confirm that they belong to the same magmatic episode at ca. 11 Ma (Ke et al., 2008; Jiang et al., 2012). The Zankan syenitic pluton is another intrusive body of the

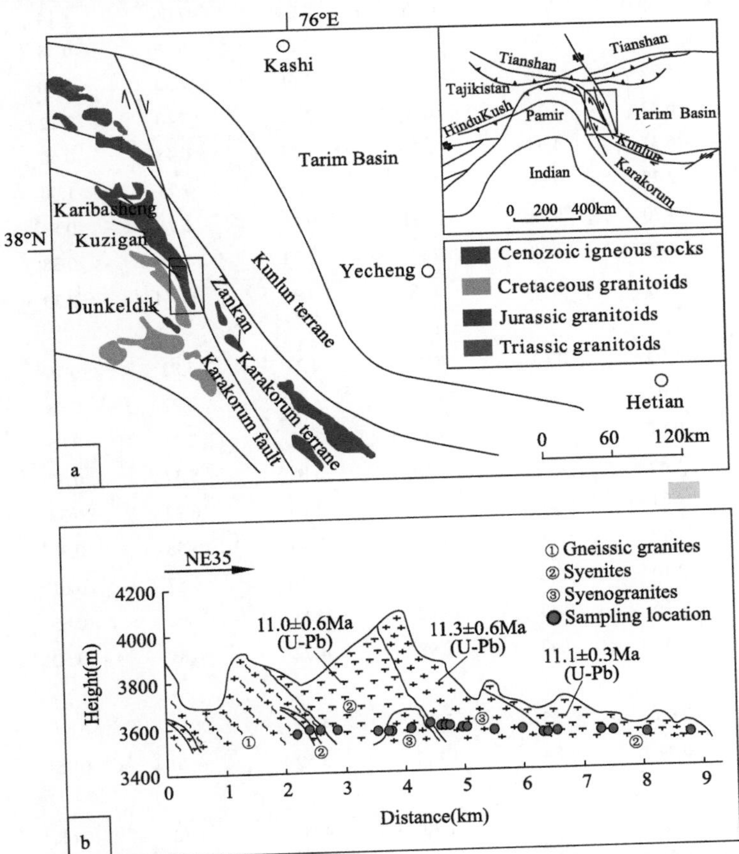

Fig.1 Simplified geological map of the eastern Pamir and sample section. (a) Geological map modified from Jiang et al. (2012), emphasizing Cenozoic igneous rocks. Inset map shows the location of Kuzigan and Zankan plutons in the Pamir syntaxis. (b) Section map of the Kuzigan alkaline pluton, emphasizing the rock types, contact relationships, and sampling locations.

Cenozoic magmatism belt to the southeast along the Karakorum fault (Fig. 1a). Kuzigan and Zankan plutons were considered to be the products of the synchronous magmatic activity and tectonic process in the Pamir syntaxis based on the same age (ca. 11 Ma; Pan, 2000), rock type and geochemical feature (Pan, 2000).

Syenites and syenogranites from the Kuzigan pluton share the same mineral assemblage of potassium feldspar, oligoclase, quartz, aegirine-augite, minor biotite and rare hornblende. Mineral modes, however, are different, e.g. aegirine-augite are enriched in syenites (6–20% in volume) but depleted in syenogranites (less than 2% in volume). In addition, syenites and syenogranites have similar Sr, Nd and Pb isotopic compositions (see Table 1 for Sr isotopes, and Ke et al., 2006 and Jiang et al., 2012 for Nd-Pb isotopes). Hence, the same spatial and temporal distribution, the similar mineral assemblage and Sr-Nd-Pb isotopic compositions suggest that syenites and syenogranites from the Kuzigan pluton were from the same source (Pan, 2000; Ke et al., 2006; Jiang et al., 2012). Syenites from the Zankan pluton are medium-grained aegirine-augite syenites with similar texture and mineral assemblage as syenites from the Kuzigan pluton. The gneissic biotite granites from the wall rock are poorly studied, which are composed of potassium feldspar, oligoclase, quartz and minor biotite.

Table 1 Bulk-rock compositions O and Sr isotopic compositions of syenites and syenogranites from the Kuzigan and Zankan plutons

Sample	SiO_2	FeO	MgO	CaO	Na_2O	K_2O	LOI	$\delta^{18}O$	$^{87}Sr/^{86}Sr$
Kuzigan syenites									
017E [a, b]	55.41	6.03	2.49	9.29	2.30	7.97	0.43		0.708143
017Q3	58.10	5.80	3.41	9.24	2.80	6.69	0.33	9.3	0.709822
017U1	61.32	3.82	2.36	6.07	3.24	6.76	0.49		0.709923
018B1 [b]	62.81	3.49	2.28	6.55	3.28	7.06	1.14	9.6	0.709895
018L [b]	63.32	3.57	0.40	4.50	5.55	6.39	0.58	8.6	0.709752
018P	56.30	5.72	2.36	6.71	2.61	7.33	0.80		0.709164
018R2	55.20	6.43	3.71	10.06	2.56	5.43	2.12		0.710032
018R4	51.60	6.31	4.64	12.13	1.25	6.45	0.60		0.710098
019A2 [b]	65.81	2.78	0.60	3.22	4.30	7.17	1.00	10.4	
019H [b]	54.18	4.70	5.54	14.31	0.58	6.71	0.53		0.709960
019J [a, b]	58.38	4.48	4.18	7.91	2.33	7.98	0.38	9.4	0.711045
019L1	61.49	2.01	1.95	5.18	0.96	11.74	0.33		0.710426
Kuzigan syenogranites									
018A	74.60	1.03	0.10	0.96	4.27	5.82	0.76	10.0	0.710284
018E [b]	74.16	1.02	0.24	1.34	4.02	4.92	0.58	11.6	0.709402
018M [b]	74.81	0.81	0.17	1.42	5.11	3.74	0.55	11.3	0.709582
018O [a, b]	73.75	1.86	0.23	1.19	3.88	5.57	0.57	11.2	0.709368
018S	72.49	1.51	0.25	1.61	3.36	6.33	0.38	10.5	0.710027
018U [b]	71.72	1.77	0.39	1.91	3.76	5.58	0.37	10.2	0.709914
019A3	66.62	2.10	0.43	2.41	4.27	6.57	0.61		0.709916
019E	67.43	2.46	0.53	3.02	4.13	6.60	0.65	10.9	0.709883
019I [b]	66.35	2.46	0.47	2.98	4.72	6.87	0.52	11.2	0.709914
Zankan syenites									
024A	64.68	2.60	0.35	3.05	4.25	8.24	0.67	10.1	0.709910
024B	57.05	6.44	1.84	10.14	3.59	6.70	0.75	11.1	0.710254
Wall rocks									
017E2 [b]	72.57	1.38	0.47	1.70	4.00	4.72	0.68		
017I	71.60	1.41	0.48	1.73	4.01	5.02	0.79	8.50	

[a] Sr isotopic compositions are from Ke et al. (2006).

[b] SiO_2, FeO (total FeO), MgO, CaO, Na_2O, K_2O, and LOI (loss in ignition) contents are from Ke et al. (2006).

Previous studies suggested that the Kuzigan pluton were derived from the hybridized lithospheric mantle and could not be produced by partial melting of the thickened lower crust or subducted slab based on the following arguments (Jiang et al., 2012): (1) Melts derived by partial melting of thickened lower crust or of subducted slab would show the adakitic affinity, which, however, are not observed in the Kuzigan pluton; (2) the Kuzigan pluton has much higher Ba (2200–9100 ppm) and Sr (840–3100 ppm, most>1100 ppm) contents than those of the common granites, suggesting that the source of Kuzigan pluton is more enriched in Ba and Sr than the source of common granites; and (3) high Ba/Sr (10–34) and low Rb/Sr (0.08–0.26) ratios suggest a metasomatized mantle source with the presence of both phlogopite and amphibole.

Twelve syenites and nine syenogranites were collected along a geological section from the Kuzigan pluton. Two syenites samples from the Zankan pluton for comparison and two gneissic granites from the wall rock are selected. All samples collected in this study are fresh without secondary alteration.

3 Analytical methods

3.1 Major elements

Major elements were determined by a Prodigy inductively coupled plasma-optical emission spectroscopy (ICP-OES) using standard methods at the China University of Geosciences, Beijing (CUGB). Loss on ignition (LOI) was determined gravimetrically after heating the samples at 1000°C for half hour. The analytical uncertainties were better than 1% after repeated measurements of rock reference standards AGV-2 and GBW07103.

3.2 Oxygen and strontium isotopes

Oxygen isotopes were analyzed using a *Finnigan MAT-251EM* mass spectrometer at the Key Laboratory for Isotope Geology, Ministry of Land and Resources. Oxygen isotopes were processed by the BrF_5 method described by Clayton and Mayeda (1963) and then converted into CO_2 by reacting with graphite rod at 700 °C. Analytical reproducibility in this study is ±0.2‰ (2SD). The oxygen isotopic data are reported by δ values relative to Vienna Standard Mean Ocean Water (SMOW).

Strontium isotopes were analyzed by using a *Neptune Plus* MC-ICP-MS at the Isotope Geochemistry Laboratory of the China University of Geosciences, Beijing. Strontium isotope fractionation is corrected to $^{86}Sr/^{88}Sr=0.1194$. Data are corrected for ^{87}Rb interference by the measured ^{85}Rb abundance. A 200 ppb Sr solution typically produced 5.0 V for ^{88}Sr and 0.004 V for ^{85}Rb signal (with a 10^{11}-Ω resistor for the Faraday cups) with $^{85}Rb/^{88}Sr$ less than 8×10^{-4}. Correction of the interference of ^{87}Rb breaks down only when $^{85}Rb/^{88}Sr$ is less than 1×10^{-3}. Therefore, separation of Sr from Rb was efficient and $^{87}Sr/^{86}Sr$ of samples are reliable in this study. NBS987 measured during the course of this study yields $^{87}Sr/^{86}Sr=0.710262\pm16$ (2SD, n=7). The BHVO-2 gives a value of $^{87}Sr/^{86}Sr=0.703458\pm10$ (2SD). The total blank for Sr isotope analysis is less than 1.0 ng.

3.3 Magnesium isotopes

Magnesium isotopes were analyzed using a *Nu Plasma* MC-ICP-MS at the Isotope Laboratory of the University of Arkansas (UA), USA, and a *Neptune Plus* MC-ICP-MS at the Isotope Geochemistry Laboratory of the China University of Geosciences, Beijing. The experiment parameters applied in the two laboratories at UA and CUGB are shown in Tables 2. All geostandards and three samples that were measured in both of the labs yield consistent results within quoted errors (Tables 3 and 4).

3.3.1 Analytical method at UA

The analytical method in UA has been previously reported (e.g. Yang et al., 2009; Li et al., 2010; Teng and Yang, 2014; Teng et al., 2015a). Briefly, the whole rock powder (200 mesh) was dissolved in a 3:1 mixture of concentrated HF:HNO_3, then a concentrated HCl:HNO_3 mixture (3:1) and re-dissolved by concentrated HNO_3. The chemical separation of Mg was achieved using ion exchange resin AG50W-X8 (Bio-Rad 200-400 mesh) following the procedure described by Teng et al. (2007). The same column procedure was repeated two times with the whole procedure blank less than 10 ng. The long-term external precision was ±0.06‰ for δ^{25}Mg and ±0.07‰ for δ^{26}Mg (2SD) (Teng et al., 2010, 2015a, b).

3.3.2 Analytical method at the CUGB

The chemical procedure used in CUGB is the same as that in UA. The recovery of Mg is up to 99.5%–99.9% and whole procedure blank is less than 10 ng as well. Mg isotope analysis, however, was conducted using a *Neptune* MC-ICP-MS (Table 2). The detailed instrument parameters at the CUGB have been previously reported by Liu et al. (2014) and He et al. (2015). Hence, we report the parts relevant to Mg isotope analysis in details below.

All samples and standards were introduced to the Ar plasma in 3% HNO_3 using a Scott double pass quartz glass spray chamber and a low-flow 50 μL/min PFA self-aspiration micro nebulizer, with help of a Cetac ASX-110 autosampler. The sample introduction system was washed with 3% HNO_3 for two minutes after each measurement to avoid cross-contamination. Mg isotopes were determined in a low-mass resolution mode under wet plasma conditions. The isobaric interferences, such as $^{12}C^{12}C^+$, $^{12}C^{14}N^+$, $^{23}NaH^+$ and $^{25}MgH^+$ have been checked at the low resolution mode. They are unresolvable by signal peaks of 3% HNO_3 (the blank solution) and thus could be negligible. ^{24}Mg, ^{25}Mg and ^{26}Mg ion beams are simultaneously collected in L3, C, and H3 Faraday cups. Each measurement consists of a 3 s idle time and 30 cycles of 4.19 or 8.39 s integration time. About 400 ppb Mg concentration typically produced 4.5~8 V ^{24}Mg signal (with a 10^{11}-Ω resistor for the Faraday cups) (Table 2). The Mg isotope ratios are reported related to the Dead Sea metal Mg standard (DSM3) using the following expression:

$$\delta^x\text{Mg}\ (‰)=[(^x\text{Mg}/^{24}\text{Mg})_{sample}/(^x\text{Mg}/^{24}\text{Mg})_{DSM3}-1]\times 1000,\ \text{where}\ x=25\ \text{or}\ 26.$$

Table 2　Instrumental operating parameters at UA and CUGB

Lab	UA	CUGB
Instrument	*Nu Plasma* MC-ICP-MS	*Neptune Plus* MC-ICP-MS
Spray chamber	Quartz cyclonic spray chamber	Scott double pass quartz glass spray chamber
Nebulizer	100 μl/minute glass concentric self-aspiration nebulizer	50 μl/minute PFA self-aspiration micro nebulizer
Autosampler	SC-2 from ESI	Cetac ASX-110
Interface cones	Nickel	Nickel
Faraday cups	L4, Ax, and H5	L3, C, H3
Instrument resolution	Low resolution mode	Low resolution mode
Typical ^{24}Mg sensitivity	2.5 to 4.5 V/100 ppb	4.5 to 8 V/400ppb
Analytical section	4 blocks and 40 cycles for each block	4 or 6 blocks and 30 cycles for each block
Background ^{24}Mg signal	$<10^{-4}$ V	$<2\times10^{-4}$ V
Blank[a]	<10 ng	<10 ng
In-run precision[b]	± 0.02‰ (2SD)	± 0.02‰ (2SD)
Internal precision[c]	<0.1‰ (2SD)	<0.1‰ (2SD)
External precision	0.07‰ (2SD) for δ^{26}Mg	0.06‰ (2SD) for δ^{26}Mg

The parameters of UA are from Yang et al. (2009), Teng et al. (2010), Teng and Yang (2014).

[a] Blank is the whole-procedure blank through dissolution, column processing and measurement.

[b] In-run precision is the precision on ^{26}Mg/^{24}Mg ratio for a single measurement of one block.

[c] Internal precision is the precision on ^{26}Mg/^{24}Mg ratio based on≥4 repeat runs of the same sample solution.

The sample-standard sequence for each sample was repeated 4 or 6 times to get a better reproducibility and accuracy. The internal precision, calculated by 4 or 6 times duplicate measurement within an analytical session, is generally better than ±0.10‰ (2SD). During a typical analytical session, the in-house standard GSB Mg with known values (an ultrapure single elemental standard solution from China Iron and Steel Research Institute) and at least one rock reference standard (e.g. BHVO-2, BCR-2, AGV-2) were analyzed with unknown samples to check long-term reproducibility of the method.

The long-term external precision (2SD) is ±0.06‰ for δ^{26}Mg and ±0.05‰ for δ^{25}Mg, based on the following analyses since June 2012 (Fig. 2): (1) Mg isotope analyses of pure Mg solutions without column chemistry (GSB Mg: δ^{26}Mg=–2.04±0.04‰, δ^{25}Mg=–1.05±0.03‰, n=58) and (2) analyses of reference materials from USGS processed through column chemistry twice, yielding δ^{26}Mg=–0.25±0.06‰ and δ^{25}Mg=–0.13±0.04‰ for BHVO-2 (n=48), δ^{26}Mg=–0.19±0.05‰ and δ^{25}Mg=–0.10±0.03‰ for BCR-2 (n=19), δ^{26}Mg=–0.17±0.05‰ and δ^{25}Mg=–0.08±0.03‰ for AGV-2 (n=13) (Table 3, Fig. 2). δ^{26}Mg values of BCR-2 are identical to the literature data within the analytical uncertainty (e.g., Huang et al., 2011). δ^{26}Mg values of AGV-2 obtained in this study are identical to the data reported by An et al. (2014), but slightly heavier than that reported by Lee et al. (2014). δ^{26}Mg values of BHVO-2, PCC-1, DST-1 and seawater analyzed in CUGB are well consistent with the recommended values within uncertainty (Teng et al., 2015a) (Table 3).

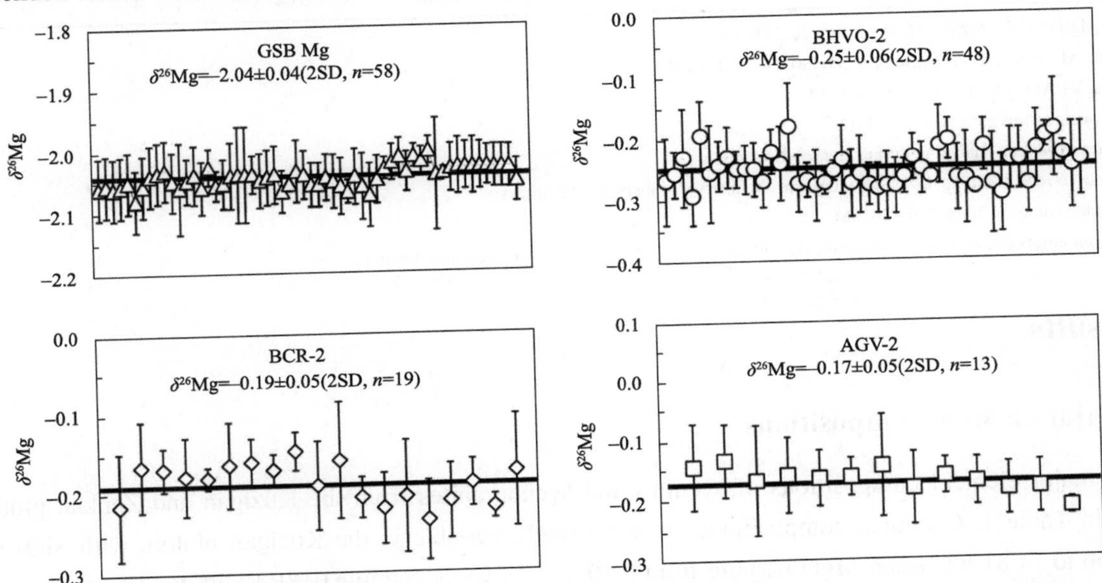

Fig. 2 Magnesium isotopic compositions of BHVO-2, BCR-2, AGV-2, and GSB Mg (an ultrapure single elemental standard solution from China Iron and Steel Research Institute) since 06/2012. The black horizontal line represents the average δ^{26}Mg. Error bars represent 2SD uncertainties. Data are reported in Table 3.

Table 3 Mg isotopic compositions of synthetic solutions and reference materials measured at UA and CUGB

Lab	Lab ID	δ^{26}Mg	2SD	δ^{25}Mg	2SD	n^a	N^b
	KH Olivine	–0.27	0.08	–0.15	0.05	4	
	KH Olivine	–0.31	0.06	–0.15	0.05	4	
	KH Olivine	–0.27	0.06	–0.17	0.06	4	
UA	IL-Granite	0.02	0.08	0.01	0.05	4	
	IL-Granite	–0.04	0.07	–0.02	0.05	4	
	IL-Granite	–0.03	0.07	–0.01	0.06	4	

Continued

Lab	Lab ID	$\delta^{26}Mg$	2SD	$\delta^{25}Mg$	2SD	n^a	N^b
UA	IL-Granite	−0.07	0.05	−0.03	0.01	4	
	IL-Chondrite	−0.07	0.07	−0.03	0.04	4	
	IL-Mg-1	−0.08	0.07	−0.01	0.04	4	
	IL-Clinopyroxene	−0.02	0.04	−0.02	0.05	4	
	BHVO-2 [c]	−0.20	0.07	−0.10	0.04		7
	Seawater [c]	−0.83	0.09	−0.43	0.06		
	PCC-1 [c]	−0.23	0.06	−0.10	0.01		3
	DST-1 [c]	−0.30	0.01	−0.13	0.01	4	5
CUGB	GSB Mg [d]	−1.05	0.03	−2.04	0.04	4	58
	BHVO-2	−0.25	0.06	−0.13	0.04	4	48
	BCR-2	−0.19	0.05	−0.10	0.03	4	19
	AGV-2	−0.17	0.05	−0.08	0.03	4	13
	Seawater	−0.83	0.02	−0.42	0.03	4	2
	PCC-1	−0.25	0.04	−0.13	0.02	4	2
	DST-1	−0.32	0.04	−0.16	0.04	4	2

IL-Granite: Mg:Fe:Al:Ca:Na:K:Ti:Ni=1:5:30:5:10:20:0.1:0.1.

IL-Chondrite: Mg:Fe:Ca:Ni:Al:Na:K:Ti=1:3:0.2:0.2:0.2:0.1:0.1:0.1.

IL-Mg-1: Mg:Fe:Al:Ca:Na:K:Ti=1:1:1:1:1:1:0.1.

IL-Clinopyroxene: Mg:Fe:Ca:Al=1:0.5:2:0.1.

a *n* is the times of sample analyzing during an analytical session.

b *N* is the times to repeat sample dissolution, column chemistry and instrumental analysis.

c Recommended values in Teng et al. (2015a).

d GSB Mg is an ultrapure single elemental standard solution from China Iron and Steel Research Institute.

4 Results

4.1 Major element compositions

The major element compositions of syenites and syenogranites from the Kuzigan and Zankan plutons are reported in Table 1. Chemical compositions are extensively variable in the Kuzigan pluton, with SiO_2 ranging from 51.60 to 74.81 wt.% and MgO varying from 0.10 to 5.54 wt.%. Sample 018R4 with the lowest SiO_2 content (51.60 wt.%) is classified as syenite based on the mineral assemblage (60% potassium feldspar, 3% oligoclase, 22% aegirine-augite, 10% biotite, and 5% melanite). The Kuzigan and Zankan plutons are characterized by high K_2O contents (3.74 to 11.74 wt.%). In particular, the syenites are highly enriched in K_2O with K_2O/Na_2O mostly >2, showing affinities with shoshonitic and ultrapotassic igneous rocks in Xizang (Jiang et al., 2012; Zhao et al., 2009; Liu et al., 2014a).

4.2 Oxygen and strontium isotopic compositions

Oxygen and strontium isotopic data are reported in Table 1. Samples from the Kuzigan pluton have high $\delta^{18}O$ values from 8.6 to 11.6‰ with an average of 10.2‰. $\delta^{18}O$ values of the syenite (8.6 to 10.4‰) are slightly lighter than those of the syenogranites (10.0 to 11.6‰). Syenites from the Zankan pluton also display heavy O isotopic compositions (10.1 and 11.1‰), slightly heavier than those of the Kuzigan pluton. The wall rock, gneissic granites, is the lightest among samples analyzed in this study ($\delta^{18}O$=8.5‰).

The syenites and syenogranites have relatively constant Sr isotopic ratios over the wide SiO_2 range. In detail, the syenites show a larger range of $^{87}Sr/^{86}Sr$ ratios (0.708143 to 0.711045) than that of syenogranites (0.709368 to 0.710284). Two syenites from the Zankan pluton have similar Sr isotopic ratios as those from the Kuzigan pluton (Table 1).

4.3 Magnesium isotopic compositions

Magnesium isotopic data of syenites and syenogranites are presented in Table 4. In a $\delta^{25}Mg$ versus $\delta^{26}Mg$ plot, samples and reference materials together yield a slope of 0.513. Several samples analyzed at UA were also analyzed at CUGB for comparison. The results from the two labs are well consistent within analytical uncertainty (Table 4). Therefore, the Mg isotope data obtained from the two labs are comparable.

Table 4 Mg isotopic compositions of syenites and syenogranites from the Kuzigan and Zankan plutons

Lab	Samples	$\delta^{26}Mg$	2SD	$\delta^{25}Mg$	2SD
Kuzigan syenites					
CUGB	017E	−0.39	0.07	−0.17	0.05
UA	017Q3	−0.31	0.08	−0.20	0.05
CUGB	017U1	−0.38	0.09	−0.18	0.04
UA	018B1	−0.33	0.07	−0.15	0.05
UA	018L	−0.33	0.09	−0.15	0.05
	Replicate [a]	−0.26	0.05	−0.14	0.04
	Duplicate [b]	−0.26	0.07	−0.13	0.06
CUGB	018P	−0.37	0.02	−0.20	0.07
CUGB	018R2	−0.30	0.09	−0.17	0.05
CUGB	018R4	−0.39	0.06	−0.20	0.02
UA	019A2	−0.32	0.05	−0.17	0.05
	Replicate	−0.37	0.07	−0.19	0.05
UA	019H	−0.39	0.08	−0.20	0.05
UA	019J	−0.31	0.07	−0.14	0.05
UA	019L1	−0.40	0.09	−0.21	0.05
	Replicate	−0.38	0.05	−0.20	0.04
Kuzigan syenogranites					
UA	018A	−0.18	0.05	−0.10	0.05
	Replicate	−0.17	0.05	−0.08	0.04
CUGB	Replicate	−0.20	0.08	−0.09	0.08
UA	018E	−0.38	0.07	−0.20	0.06
UA	018M	−0.41	0.09	−0.21	0.05
CUGB	Replicate	−0.39	0.05	−0.20	0.04
UA	018O	−0.37	0.09	−0.15	0.05
	Replicate	−0.28	0.05	−0.14	0.04
	Duplicate	−0.30	0.07	−0.16	0.06
UA	018S	−0.34	0.08	−0.17	0.05
UA	018U	−0.33	0.05	−0.18	0.04
CUGB	Replicate	−0.32	0.05	−0.17	0.04
CUGB	019A3	−0.39	0.02	−0.19	0.03

Continued

Lab	Samples	δ^{26}Mg	2SD	δ^{25}Mg	2SD
UA	019E	−0.21	0.07	−0.11	0.06
UA	019I	−0.31	0.07	−0.15	0.05
Zankan syenites					
CUGB	024A	−0.37	0.07	−0.18	0.05
CUGB	024B	−0.46	0.07	−0.20	0.05
Wall rocks, gneissic granites					
UA	017E2	−0.26	0.07	−0.13	0.05
UA	017I	−0.27	0.05	−0.15	0.04

[a] Replicate: repeated sample dissolution, column chemistry and instrumental analysis.

[b] Duplicate: repeated measurement of Mg isotopic ratios on the same solutions.

Magnesium isotopic compositions of syenites and syenogranites from the Kuzigan pluton vary from −0.40 to −0.26‰ with an average of −0.35±0.08‰ and from −0.40 to −0.17‰ with an average of −0.32±0.14‰, respectively. This is in contrast to the heavy δ^{26}Mg in I- and A-type granitoids (Shen et al. 2009; Li et al., 2010), and the mantle-like Mg isotopic compositions in I-type granites (Liu et al., 2010) (Fig. 3). Despite overlapping, δ^{26}Mg values of the syenogranites are on average slightly heavier than those of the syenites (Fig. 3). Two syenites from the Zankan pluton have slightly lighter Mg isotopic compositions (−0.37 and −0.46‰) than those from the Kuzigan region (Table 4). The wall rocks, gneissic granites, have mantle-like δ^{26}Mg values of −0.26‰ and −0.27‰ (Table 4).

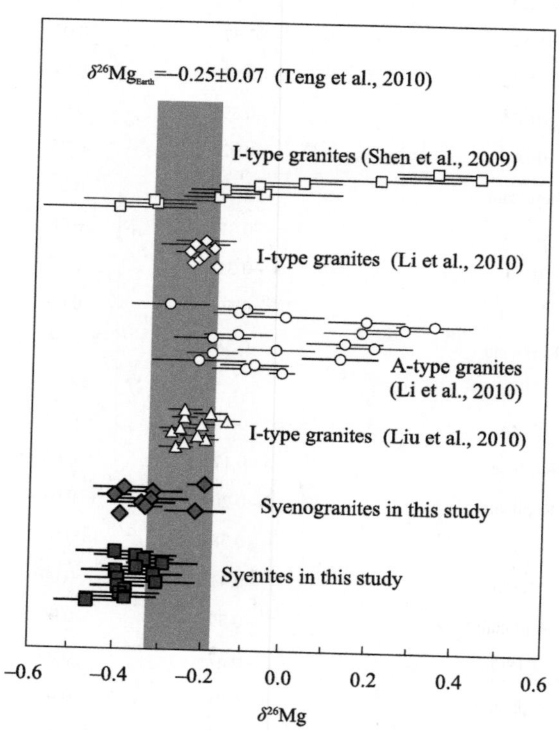

Fig. 3　Magnesium isotopic compositions of granitoids rocks in literatures and the Kuzigan and Zankan plutons in this study. Mg isotopic data for A-type and I-type granitoids are from published literatures (Shen et al., 2009; Li et al., 2010; Liu et al., 2010). The gray bar is δ^{26}Mg value of the Earth (−0.25±0.07‰; Teng et al., 2010).

5 Discussion

The light Mg isotopic compositions in the syenites and syenogranites from the Kuzigan and Zankan plutons may result from magmatic differentiation, assimilation of wall rocks during ascent, kinetic isotopic fractionation (thermal and chemical diffusion), or source heterogeneity. In this section, we first explore the possible mechanisms responsible for the light Mg isotopic signatures. Then we discuss the carbonate recycling during subduction of Tethyan oceanic crust based on the observations from Mg-O-Sr isotope geochemistry of these alkaline rocks.

5.1 The lack of Mg isotope fractionation during alkaline magmatic differentiation

Equilibrium Mg isotope fractionation is limited during crystal-melt differentiation of komatiitic, basaltic and granitic magmas, consistent with the undetectable Mg isotopic differences among common silicate minerals (olivine, pyroxene, hornblende and biotite) (Handler et al., 2009; Yang et al., 2009; Young et al., 2009; Dauphas et al., 2010; Liu et al., 2010, 2011; Pogge von Strandmann et al., 2011; Xiao et al., 2013; Hu et al., 2015). To date, the behavior of Mg isotope fractionation during magmatic differentiation in the alkaline syenitic system is unclear.

The primary magma of the Kuzigan alkaline pluton had undergone fractional crystallization evolving from syenites to syenogranites as illustrated in the Hacker diagrams (Fig.4). Fractional crystallization of aegirine-augite would enrich the melt in Al_2O_3, K_2O, and Na_2O and deplete the melt in MgO, FeO (total FeO) and CaO from syenites to syenogranites based on the chemical composition of aegirine-augites. The SiO_2 is negatively correlated with MgO, CaO, and total FeO and is positively correlated with Na_2O+K_2O in the syenite series (Fig. 4), indicating the fractional crystallization of aegirine-augites from the original syenitic magma. This is supported by the gradually decreased contents of aegirine-augites in the Kuzigan pluton from syenites to syenogranites. However, there is no significant correlation between Mg isotopic compositions and the indicators of fractional crystallization of aegirine-augites, e.g., MgO, CaO, and total FeO (Fig.5). Although fractional crystallization of aegirine-augites seems to shift the evolved syenogranites towards slightly heavier δ^{26}Mg values with an average of −0.32‰, the Mg isotopic difference between syenites and syenogranites falls within the analytical uncertainty

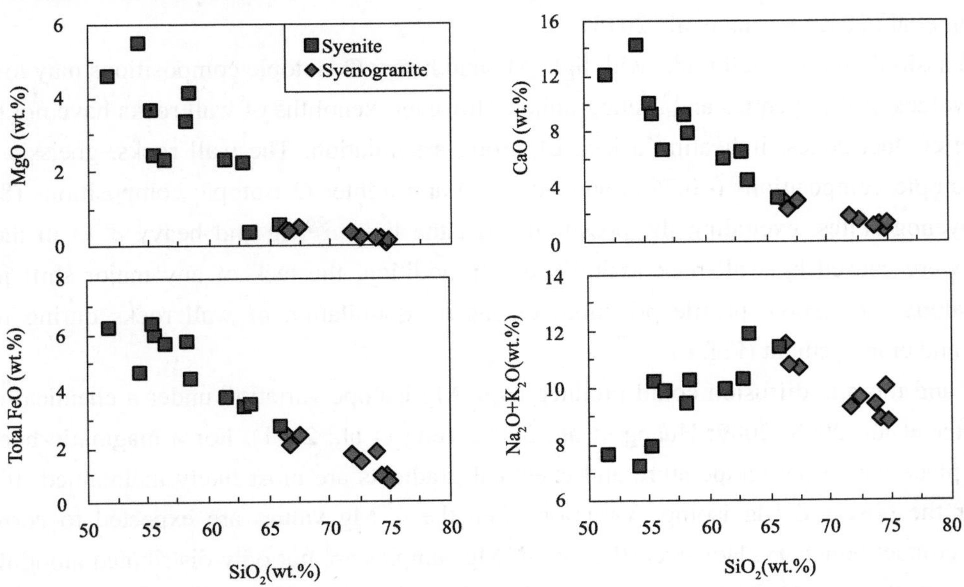

Fig. 4 Plots of SiO_2 contents versus MgO, total FeO, CaO, and Na_2O+K_2O contents. The data are reported in Table 1.

(Fig. 5). Therefore, the effect of crystal-melt fractionation on Mg isotopic compositions in alkaline syenitic system from the Kuzigan pluton is insignificant, consistent with the previous studies on I- and A-type granite systems (Shen et al., 2009; Li et al., 2010; Liu et al., 2010).

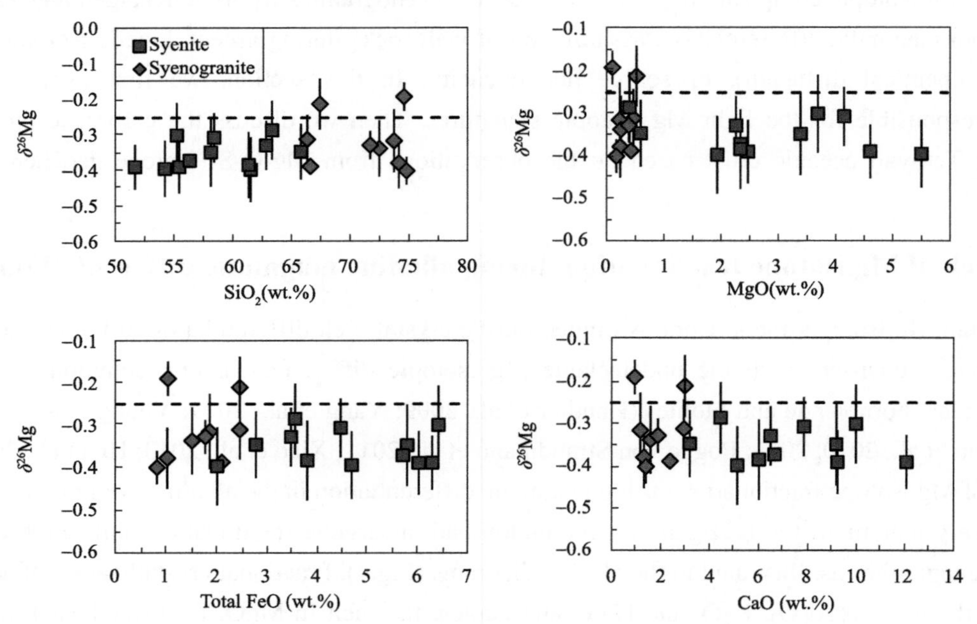

Fig. 5　Plots of δ^{26}Mg versus SiO_2, MgO, Total FeO, and CaO contents. The dashed line represents δ^{26}Mg value of the Earth (−0.25‰; Teng et al., 2010). Data are reported in Tables 1 and 4.

5.2　Origin of low δ^{26}Mg and high δ^{18}O in syenites and syenogranites.

Given the limited Mg isotope fractionation during alkaline magmatic differentiation, other processes that could potentially produce low-δ^{26}Mg signatures include: (1) assimilation by wall rocks, (2) thermal and chemical diffusion within the magma chamber and/or around the contact boundaries (Richter et al., 2008, 2009; Teng et al., 2008, 2011; Huang et al., 2009; Lundstrom, 2009), and/ or (3) heterogeneous source involved by low-δ^{26}Mg materials (Yang et al., 2012; Huang et al., 2015).

Extensive assimilation by wall rocks with light Mg and heavy O isotopic compositions may lower δ^{26}Mg and increase δ^{18}O values of the syenites and syenogranites. However, xenoliths of wall rocks have not been found yet even along the contact zones, indicating a lack of strong assimilation. The wall rocks, gneissic granites, have heavier Mg isotopic compositions (−0.26‰ and −0.27‰) and lighter O isotopic compositions (8.5‰) than the syenites and syenogranites, excluding the possibility that the light δ^{26}Mg and heavy δ^{18}O in the syenites and syenogranites were caused by wall rock assimilation. In addition, the lack of any major shift in Mg isotopic compositions along the sample profile precludes extensive assimilation of wall rocks during magma ascent, differentiation and emplacement (Fig. 6).

Chemical and thermal diffusion could produce large Mg isotope variation under a chemical or temperature gradient (Richter et al., 2008, 2009; Huang et al., 2009; Teng et al., 2011). For a magmatic body, the contact boundary is a place where the temperature and chemical gradients are most likely maintained. If diffusion was responsible for the observed Mg isotope variation, then the δ^{26}Mg values are expected to correlate with the distance to the contact boundary. However, the low-δ^{26}Mg samples are not only distributed along the boundaries, but also occur over the whole ~9 km long geological section cross the Kuzigan pluton (Fig. 6). Furthermore, the negative correlation between Mg and O isotopes is difficult to explain by thermal diffusion (Fig. 7). During thermal diffusion, heavy isotopes always tend to be enriched at the cold end relative to light isotopes, resulting in two different stable isotopic systems that correlate positively with each other (Richter et al., 2008, 2009; Huang et al., 2009; Teng et al., 2011). Therefore, the negative correlation between δ^{26}Mg and δ^{18}O rules out chemical or thermal diffusion as the cause of the low δ^{26}Mg in the Kuzigan syenites and syenogranites.

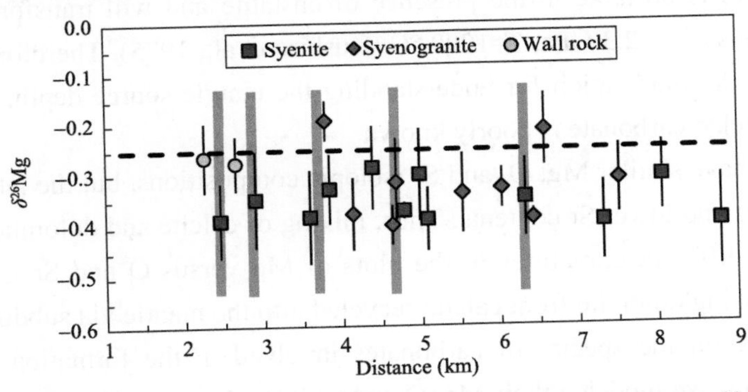

Fig. 6 Relationship of δ^{26}Mg values and sample spatial distribution. The dashed line represents δ^{26}Mg value of the Earth (−0.25‰; Teng et al., 2010). The gray strips are contact boundaries based on the sample section (Fig. 1b).

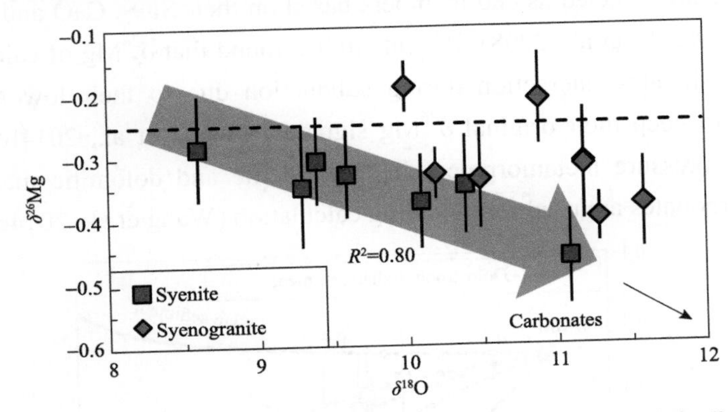

Fig. 7 Plot of δ^{26}Mg versus δ^{18}O of syenites and syenogranites from the Kuzigan and Zankan alkaline intrusions. Data are from Tables 1 and 4.

The most likely mechanism for the low-δ^{26}Mg in the Kuzigan syenites and syenogranites is source heterogeneity, as shown by the O isotopic data. Syenites and syenogranites from the Kuzigan and Zankan plutons have heavy O isotopic compositions compared with the normal mantle (δ^{18}O=5.5‰; Hoefs, 2009). Magmatic differentiation could increase δ^{18}O of syenogranites because quartz is the most ^{18}O-rich mineral compared to feldspar and clinopyroxene (Hoefs, 2009). Nonetheless, the heavy O isotopic compositions of syenites should not be affected by magmatic differentiation and thus reflect either crustal contamination or source heterogeneity. Crustal contamination would increase δ^{18}O and ^{87}Sr/^{86}Sr as well as SiO_2 and Sr contents, and lead to positive correlations between δ^{18}O and SiO_2 and between ^{87}Sr/^{86}Sr and Sr, which are not observed here. Therefore, the high δ^{18}O values of these syenites must inherit from their mantle source that had been affected by sedimentary or metasedimentary protoliths. The negative correlation between δ^{26}Mg and δ^{18}O strongly suggests that carbonates are the most likely component in their mantle source based on their extremely light Mg and heavy O isotopic compositions (Fig. 7). The sedimentary signature is also reflected in the high ^{87}Sr/^{86}Sr (Table 1) and extreme LILE enrichment (e.g., Ba and Sr) (Ke et al., 2006; Jiang et al., 2012).

5.3 Species of recycled carbonate and mantle source depth

Stability of subducted carbonate species in the mantle strongly depends on the *P-T* conditions. Dolomite could dissociate into aragonite + magnesite under high *P-T* conditions (e.g., at T=700 °C, P > 5.3−6.0 GPa) (Sato and Katsura, 2001). Calcite is unstable in the presence of enstatite and will transform to dolomite + diopside assemblage under conditions of P=2.3 GPa, T=1000 °C (Kushiro et al., 1975). Therefore, constraining the species of recycled carbonates has an implication for understanding the mantle source depth. However, to date, how to identify the species of recycled carbonate is poorly known.

Calcite and dolomite have similar Mg, O, and Sr isotopic compositions, but the latter has heavy Mg isotopic composition and higher Mg and lower Sr contents. Thus, mixing of calcite and dolomite with the depleted mantle will generate curves with different curvatures in the plots of Mg versus O and Sr isotopes, which could be a potential tool for distinguishing dolomite from calcite recycled into the mantle via subduction.

To identify and constrain the species of carbonates involved in the formation of the syenites from the Kuzigan and Zankan plutons, we modeled their Mg, O and Sr isotopic compositions (Table 5, Fig. 8, 9). In order to eliminate the effect of siliceous-clastic component, the relatively pure dolostones and limestones from Cretaceous strata in Xizang are selected as end members based on their SiO_2, CaO and MgO contents (Chen and Dong, 1982; Zong et al., 1982; Li et al., 2008). Recent studies found that δ^{26}Mg of calcite-rich carbonates varies greatly through carbonate-silicates interaction during subduction due to their low MgO contents, while the dolomite carbonates tend to keep their original δ^{26}Mg signature (Wang et al., 2014b). Therefore, Mg isotopic compositions of ultrahigh pressure metamorphic (UHPM) calcitic and dolomitic marbles in subduction zone rather than sedimentary carbonates are used for modeling calculation (Wang et al., 2014b).

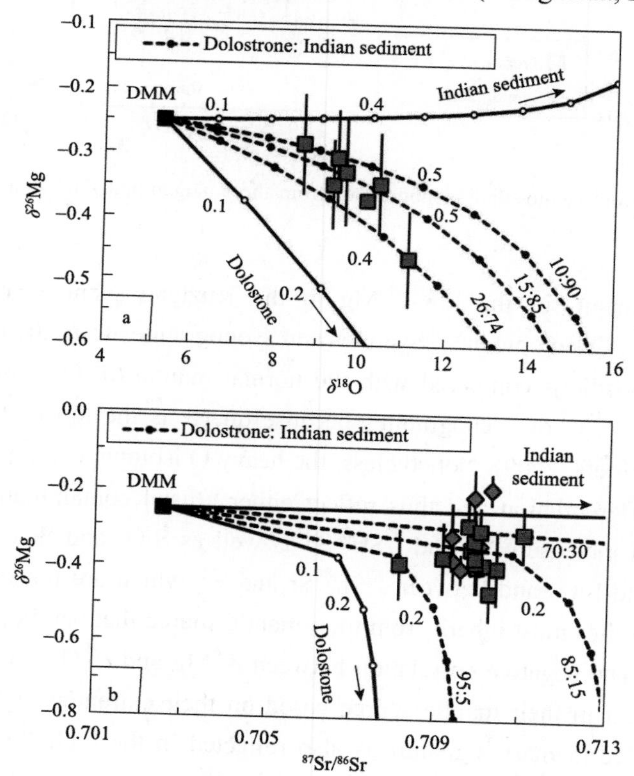

Fig. 8 (a) Mg-O isotopes modeling for mixing of the depleted mantle (DMM), dolostone, and the Indian sediment. (b) Mg-Sr isotopes modeling for mixing of the DMM, dolostone, and the Indian sediment. The label on each curve is proportions of dolostone: Indian sediment. Parameters and data source are reported in Tables 1, 4, and 5. Points on mixing curves are at 10% intervals.

Table 5 Magnesium, oxygen and strontium isotopic compositions and MgO, Sr contents of end members for mixing modeling

End member	δ^{26}Mg	δ^{18}O	^{87}Sr/^{86}Sr	MgO (wt.%)	Sr (ppm)
DMM	–0.25	5.5	0.7025	38	7.7
Dolostone	–2.17	22.0	0.7080	23	222
Limestone	–0.66	22.0	0.7080	0.5	340
Indian sediments	–0.05	17.0	0.8000	2.3	88

DMM is the depleted mantle. The δ^{18}O and δ^{26}Mg (–0.25±0.07‰) values of DMM are from Hoefs (2009) and Teng et al. (2010), respectively. Other data of the DMM are taken from Workman and Hart (2005).

Dolostone is the Cretaceous dolostones of Cuoqin basin, Qinhai-Xizang Plateau (Li et al., 2008). In order to avoid the post-diagenetic effect on Sr isotopic composition of the dolostone, ^{87}Sr/^{86}Sr ratio of the Cretaceous sea water in the south Xizang (0.708; Huang et al., 2004) is used to represent that of the dolostones. δ^{26}Mg value of dolostone is an average Mg isotopic composition of ultrahigh pressure metamorphic (UHPM) dolomitic marbles from Wang et al. (2014b). Other data of the dolostone are taken from Li et al. (2008).

Limestone is the Cretaceous pure biogenic limestones of Nyanang-Gamba, Xizang. The δ^{18}O value of Gamba limestones is from Li et al. (2005). ^{87}Sr/^{86}Sr ratio of the Cretaceous sea water in the south Xizang (0.708; Huang et al., 2004) is used to represent that of the limestones. MgO content is from Chen and Dong (1982). The Sr concentration is from Zong et al. (1982). The δ^{26}Mg value is the average of Mg isotopic compositions of UHPM calcitic marbles from Wang et al. (2014b).

Indian sediment is clastic sediments from the Lesser, High and Tethyan Himalayan sediments from the Sutlej Valley (Richards et al., 2005). MgO content, Sr concentration and ^{87}Sr/^{86}Sr ratio are from Richards et al. (2005) and Bouilhol et al. (2013). δ^{18}O values of clastic sediments span a large range and we select δ^{18}O=17‰ as its average for modeling. The δ^{26}Mg value is the Mg isotopic composition of sediments from Li et al. (2010).

The Mg-O isotope modeling shows that mixing of a small amount of dolostone with the depleted mantle (DMM) could significantly lower the δ^{26}Mg values and increase the δ^{18}O values. However, dolostone alone makes the slope of the mixing line too sharp to fit the Mg and O isotopic compositions of syenites. Involvement of the Indian sediments (clastic sediments) or limestone could overlap the observed δ^{26}Mg variation (Fig. 8 and 9). However, unrealistically large amount of the Indian sediments (from 74% to 90% relative to dolostone) is required in order to fit the data by mixing dolostone, Indian sediments and DMM (Fig. 8a). This amount of Indian sediments is contradictory to the modeling result based on Mg and Sr isotopic compositions in Fig. 8b, in which a maximum 30% of Indian sediments is needed.

If the recycled carbonates contain both dolostone and limestone, then the limestone must be a major component (65% to 90%) to fit the Mg and O isotopic data (Fig. 9a). However, involvement of carbonates (limestone + dolostone) alone could not account for Sr isotopic compositions of the samples (Fig. 9b). Hence, the clastic sediments (e.g., Indian sediments) with heavy Mg isotopic composition and high ^{87}Sr/^{86}Sr are required. These clastic sediments were probably carried into the mantle source by subducted India continent during the India-Eurasia collision. The Mg-Sr isotope modeling shows that all samples lie along mixing lines between the DMM and a mixture of dolostone (18%–34%), limestone (60%–67%) and Indian sediment (6%–15%) (Fig. 9b). A mixure with similar proportions of dolostone, limestone and clastic sediments also explains the co-variation of Mg and O isotopes (Fig. 9c). Therefore, the variations of Mg, O and Sr isotopic compositions can be best explained by mixing dolostone, limestone and clastic sediments in the magma source region of the Kuzigan and Zankan plutons.

The contamination by both limestone and dolostone indicates that the hybridized depth in the mantle is shallow, because stability of deeply recycled carbonates strongly depends on their depth in the mantle (Kushiro et al., 1975; Sato and Katsura, 2001; Dasgupta and Hirschmann, 2010). Therefore, the mantle source of the syenites and syenogranites in Kuzigan and Zankan areas had been hybridized at a depth shallower than 60 to 120 km, consistent with a previous study that suggested a <70 to 100 km metasomatized mantle source inferred from the metasomatic volatile phase relation (phlogopite and amphibole) (Jiang et al., 2012).

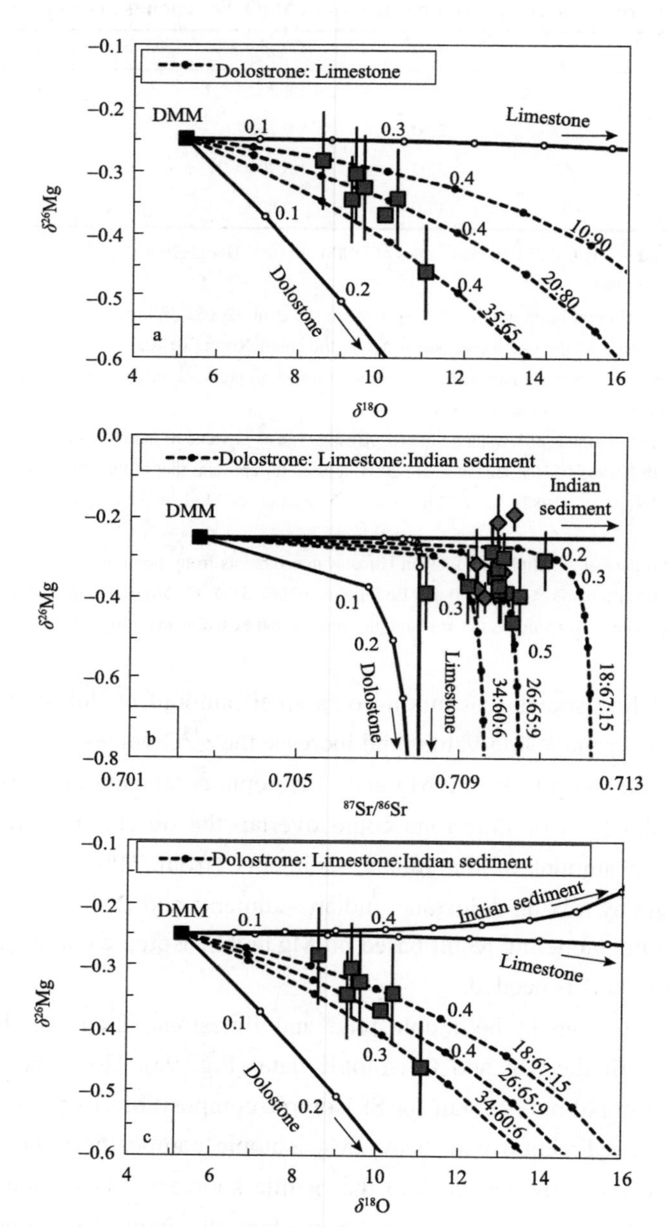

Fig. 9　(a) Mg-O isotopes modeling for mixing of the DMM, dolostone, and limestone. (b).Mg-Sr isotopes modeling for mixing of the DMM, dolostone, limestone and the Indian sediment. (c) Mg-O isotopes modeling for mixing of the DMM, dolostone, limestone and Indian sediment. Parameters and data source are listed in Tables 1, 4, and 5. Points on mixing curves are at 10% intervals. The label on each curve is proportions of dolostone:limestone and dolostone:limestone:Indian sediment, respectively. Mixing calculations illustrate different sedimentary rocks recycled and then hybridized the upper mantle in the Pamir region.

5.4 Implication for carbonate recycling during subduction of the Tethyan oceanic slabs

Pamir syntaxis is a complex subduction accretion system formed by northward subduction of the Tethyan ocean to Asia during Paleozoic to early Cenozoic. Particularly, huge amounts of carbonates were transported into the deep mantle along with the subduction and closure of the Neo-Tethys ocean. According to the carbon budget, a part of carbon was released to the surface by volcanic emission of CO_2, which was suggested to be responsible for the global warming during Cretaceous to early Cenozoic (Kerrick and Connolly, 2001; Kent and Muttoni, 2008). Another part of carbon was stored in carbonate minerals and recycled into the deep mantle. In this case, the recycling carbonates would be recorded in mantle-derived post-collisional magmatic rocks distributed in Xizang-Pamir region.

Ultrapotassic and potassic samples from the Kuzigan alkaline complex reveal that carbonates were involved into the upper mantle source based on their low δ^{26}Mg isotopic compositions. Such a signature has also been observed in the Zankan syenites to the south in the Pamir region. Coincidentally, the Oligocene-Miocene (24 to 10Ma) ultrapotassic volcanic rocks from the central and southern Lhasa subterranes also present exhibit the similar low δ^{26}Mg values, implying the carbonate contamination in the mantle underneath the Tibetan plateau (Liu et al., 2015). It is reasonable to speculate that recycled carbonate might be broadly existent in the mantle sources of the Cenozoic potassic and ultrapotassic igneous rocks widely distributed in the Tibetan-Pamir Plateau. More studies are needed to constrain the spatial-temporal scale of carbonate metasomatism in the upper mantle underneath the Tibetan Plateau.

6 Conclusion

We report the Mg, O and Sr isotopic compositions for the syenites and evolved syenogranites from the Kuzigan and Zankan plutons of the Pamir syntaxis in NW Xinjiang, China. The main conclusions are:

(1) The alkaline syenites and syenogranites from the Kuzigan and Zankan regions have light Mg isotopic compositions with δ^{26}Mg varying from –0.46 to –0.17‰. Syenites and syenogranites have similar Mg isotopic compositions, indicating limited Mg isotope fractionation during alkaline magma fractional crystallization.

(2) The syenites display a negative correlation between Mg and O isotopic compositions, which suggest the involvement of recycled sedimentary carbonates in their mantle source.

(3) The combined Mg-O-Sr isotopic data could be used to identify species of recycled carbonates and thus constrain the metasomatized depth. The carbonated peridotite source for the Kuzigan and Zankan alkaline plutons could be mainly in mantle lithosphere (60 to 120 km).

(4) In view of the tectonic setting of the Pamir syntaxis, the recycling of carbonate was likely related to subduction of the Tethyan oceanic slab during Mesozoic-Cenozoic.

Acknowledgement

We thank Dr. Shihong Tian, Zijian Li, Xunan Meng, Ruiying Li, and Lijuan Xu for help during sample analysis. We are grateful to Drs. Shuijiong Wang, Wangye Li, Yang Sun for comments and discussion, and the anonymous reviewers for formal reviews. This work is supported by the National Nature Scientific Foundation of China (No. 41103010, No. 2011CB403102, No.41230209 and No. 41328004) and National Science Foundation (EAR-0838227 and EAR-1340160).

References

An, Y.J.,Wu, F., Xiang, Y.X., Nan, X.Y., Yu, X., Yang, J.H., Yu, H.M., Xie, L.W., Huang, F., 2014. High-precision Mg isotope analyses of low-Mg rocks by MC-ICP-MS. Chem. Geol. 390, 9-21.

Beinlich, A., Mavromatis, V., Austrheim, H., Oelkers, E.H., 2014. Inter-mineral Mg isotope fractionation during hydrothermal ultramafic rock alteration-implications for the global Mg-cycle. Earth Planet. Sci. Lett. 392, 166-176.

Bouilhol, P., Jagoutz, O., Hanchar, J.M., Dudas, F.O., 2013. Dating the India-Eurasia collision through arc magmatic records. Earth Planet. Sci. Lett. 366, 163-175.

Burtman, V.S., Molnar, P., 1993. Geological and geophysical evidence for deep subduction of continental crust beneath the Pamir. Geol. Soc. Am. Spec. Pap. 281, 1-76.

Chen, N.S., Dong, Z.S., 1982. Sedimentological geochemistry of the strata from Ordovician to Tertiary in Nyanang-Gamba area, southern Xizang. Geochimica 1, 103-111.

Chung, S.-L., Chu, M.-F., Zhang, Y., Xie, Y., Lo, C.-H., Lee, T.-Y., Lan, C.-Y., Li, X., Zhang, Q., Wang, Y., 2005. Tibetan tectonic evolution inferred from spatial and temporal variations in post-collisional magmatism. Earth Sci. Rev. 68, 173-196.

Clayton, R.N., Mayeda, T.K., 1963. The use of bromine pentafluoride in the extraction of oxygen from oxides and silicates for isotopic analysis. Geochim. Cosmochim. Acta 27, 43-52.

Dasgupta, R., Hirschmann, M.M., 2010. The deep carbon cycle and melting in Earth's interior. Earth Planet. Sci. Lett. 298, 1-13.

Dauphas, N., Teng, F.-Z., Arndt, N.T., 2010. Magnesium and iron isotopes in 2.7 Ga Alexo komatiites: mantle signatures, no evidence for Soret diffusion, and identification of diffusive transport in zoned olivine. Geochim. Cosmochim. Acta 74, 3274-3291.

Ducea, M.N., Lutkov, V., Minaev, V.T., Hacker, B., Ratschbacher, L., Luffi, P., Schwab, M., Gehrels, G.E., McWilliams, M., Vervoort, J., 2003. Building the Pamirs: the view from the underside. Geology 31, 849-852.

Hacker, B., Luffi, P., Lutkov, V., Minaev, V., Ratschbacher, L., Plank, T., Ducea, M., Patiño-Douce, A., McWilliams, M., Metcalf, J., 2005. Near-ultrahigh pressure processing of continental crust: Miocene crustal xenoliths from the Pamir. J. Petrol. 46, 1661-1687.

Handler, M.R., Baker, J.A., Schiller, M., Bennett, V.C., Yaxley, G.M., 2009. Magnesium stable isotope composition of Earth's upper mantle. Earth Planet. Sci. Lett. 282, 306-313.

He, Y.S., Ke, S., Teng, F.-Z., Wang, T.T.,Wu, H.J., Lu, Y.H., Li, S.G., 2015. High-precision iron isotope analysis of geological reference materials by high-resolution MC-ICP-MS. Geostand. Geoanal. Res. 39, 341-356.

Hoefs, J., 2009. Stable Isotope Geochemistry. Springer.

Hu, Y., Teng, F.-Z., Zhang, H.-F., Xiao, Y., Su, B.-X., 2015. Metasomatism-induced mantle magnesium isotopic heterogeneity: evidence from pyroxenites. Geochim. Cosmochim. Acta. http://dx.doi.org/10.1016/j.gca.2015.11.001

Huang, S.J., Shi, H., Shen, L.C., Zhang, M.,Wu, W.H., 2004. Global correlation for strontium isotope curve in the late Cretaceous of Tibet and dating marine sediment. Sci. China Ser. D Earth Sci. 34, 335-344.

Huang, F., Lundstrom, C.C., Glessner, J., Ianno, A., Boudreau, A., Li, J., Ferré, E.C., Marshak, S., DeFrates, J., 2009. Chemical and isotopic fractionation of wet andesite in a temperature gradient: experiments and models suggesting a new mechanism of magma differentiation. Geochim. Cosmochim. Acta 73, 729-749.

Huang, F., Zhang, Z., Lundstrom, C.C., Zhi, X., 2011. Iron and magnesium isotopic compositions of peridotite xenoliths from Eastern China. Geochim. Cosmochim. Acta 75, 3318-3334.

Huang, K.-J., Teng, F.-Z., Elsenouy, A., Li, W.-Y., Bao, Z.-Y., 2013. Magnesium isotopic variations in loess: origins and implications. Earth Planet. Sci. Lett. 374, 60-70.

Huang, J., Li, S.-G., Xiao, Y., Ke, S., Li, W.-Y., Tian, Y., 2015. Origin of low δ_{26}Mg Cenozoic basalts from South China Block and their geodynamic implications. Geochim. Cosmochim. Acta 164, 298-317.

Jiang, Y.-H., Liu, Z., Jia, R.-Y., Liao, S.-Y., Zhou, Q., Zhao, P., 2012. Miocene potassic granite-syenite association in western Tibetan Plateau: implications for shoshonitic and high Ba-Sr granite genesis. Lithos 134-135, 146-162.

Ke, S., Mo, X.X., Luo, Z.H., Liang, T., Zhan, H.M., Li, L., Li,W.T., 2006. Petrogenesis and geochemistry of Cenozoic Taxkorgan alkalic complex and its geological significance. Acta Petrol. Sin. 221, 905-915.

Ke, S., Luo, Z.H., Mo, X.X., Zhang, W.H., Liang, T., Zhan, H.M., 2008. The geochronology of Taxkorgan alkalic complex, Northeast Pamir. Acta Petrol. Sin. 24, 315-324.

Kent, D.V., Muttoni, G., 2008. Equatorial convergence of India and early Cenozoic climate trends. PNAS 105, 16065-16070.

Kerrick, D., Connolly, J., 2001. Metamorphic devolatilization of subducted marine sediments and the transport of volatiles into the Earth's mantle. Nature 411, 293-296.

Kushiro, I., 1975. Carbonate-silicate reactions at high presures and possible presence of dolomite and magnesite in the upper mantle. Earth Planet. Sci. Lett. 28, 116-120.

Lee, S.-W., Ryu, J.-S., Lee, K.-S., 2014. Magnesium isotope geochemistry in the Han River, South Korea. Earth Planet. Sci. Lett. 364, 9-19.

Li, S.G., 2015. Tracing deep carbon recycling by Mg isotopes. Earth Sci. Front. 22, 143-159.

Li, X.H.,Wang, C.S., Cui, J., 2005. Deviation analysis of the comparable carbon and oxygen isotopic values of the Upper

Cretqaceous limestones form Gamba, Tibet, China. Bull. Mineral. Petrol. Geochem. 24, 190-194.
Li, Y.L., Huang, Y.J.,Wang, C.S., Zhu, L.D., Wang, L.C., Wei, Y.S., 2008. Geochemical characteristics and genetic analysis of the Cretaceous dolomite in the Cuoqin Basin, Qinghai- Tibet Plateau. Acta Petrol. Sin. 24, 609-615.
Li,W.-Y., Teng, F.-Z., Ke, S., Rudnick, R.L., Gao, S.,Wu, F.Y., Chappell, B.W., 2010. Heterogeneous magnesium isotopic composition of the upper continental crust. Geochim. Cosmochim. Acta 74, 6867-6884.
Liu, S.-A., Teng, F.-Z., He, Y.-S., Ke, S., Li, S.-G., 2010. Investigation of magnesium isotope fractionation during granite differentiation: implication for Mg isotopic composition of the continental crust. Earth Planet. Sci. Lett. 297, 646-654.
Liu, S.-A., Teng, F.-Z., Yang, W., Wu, F.-Y., 2011. High-temperature inter-mineral magnesium isotope fractionation in mantle xenoliths from the North China craton. Earth Planet. Sci. Lett. 308, 131-140.
Liu, D., Zhao, Z.D., Zhu, D.-C., Niu, Y.L., DePaolo, D.J., Harrison, T.M., Mo, X.X., Dong, G.C., Zhou, S., Sun, C.G., Zhang, Z.C., Liu, J.L., 2014a. Postcollisional potassic and ultrapotassic rocks in southern Tibet: mantle and crustal origins in response to India-Asia collision and convergence. Geochim. Cosmochim. Acta 143, 207-231.
Liu, S.-A., Li, D.-D., Li, S.-G., Teng, F.-Z., Ke, S., He, Y.-S., Lu, Y.-H., 2014b. High-precision copper and iron isotope analysis of igneous rock standards by MC-ICP-MS. J. Anal. At. Spectrom. 29, 122-133.
Liu, D., Zhao, Z.D., Zhu, D.-C., Niu, Y.L., Widom, E., Teng, F.-Z., DePaolo, D.J., Ke, S., Xu, J.-F., Wang, Q., Mo, X.X., 2015. Identifying mantle carbonatite metasomatism through Os-Sr-Mg isotopes in Tibetan ultrapotassic rocks. Earth Planet. Sci. Lett. 430, 458-469.
Lundstrom, C., 2009. Hypothesis for the origin of convergent margin granitoids and Earth's continental crust by thermal migration zone refining. Geochim. Cosmochim. Acta 73, 5709-5729.
Miller, C., Schuster, R., Klötzli, U., Frank,W., Purtscheller, F., 1999. Post-collisional potassic and ultrapotassic magmatism in SW Tibet: geochemical and Sr-Nd-Pb-O isotopic constraints for mantle source characteristics and petrogenesis. J. Petrol. 40, 1399-1424.
Nomade, S., Renne, P.R., Mo, X.X., Zhao, Z.D., Zhou, S., 2004. Miocene volcanism in the Lhasa block, Tibet: spatial trends and geodynamic implications. Earth Planet. Sci. Lett. 221, 227-243.
Pan, Y.S., 2000. Geological evolution of he Karakorum-Kunlun Mountains. Sciences Press, Beijing.
Pogge von Strandmann, P.A.E., Elliott, T., Marschall, H.R., Coath, C., Lai, Y.-J., Jeffcoate, A.B., Ionov, D.A., 2011. Variations of Li and Mg isotope ratios in bulk chondrites and mantle xenoliths. Geochim. Cosmochim. Acta 75, 5247-5268.
Richards, A., Argles, T., Harris, N., Parrish, R., Ahmad, T., Darbyshire, F., Draganits, E., 2005. Himalayan architecture constrained by isotopic tracers from clastic sediments. Earth Planet. Sci. Lett. 236, 773-796.
Richter, F.M., Watson, E.B., Mendybaev, R.A., Teng, F.-Z., Janney, P.E., 2008. Magnesium isotope fractionation in silicate melts by chemical and thermal diffusion. Geochim. Cosmochim. Acta 72, 206-220.
Richter, F.M., Dauphas, N., Teng, F.-Z., 2009. Non-traditional fractionation of non-traditional isotopes: evaporation, chemical diffusion and Soret diffusion. Earth Planet. Sci. Lett. 258, 92-103.
Saenger, C.,Wang, Z.R., 2014. Magnesium isotope fractionation in biogenic and abiogenic carbonates: implications for paleoenvironmental proxies. Quat. Sci. Rev. 90, 1-21.
Sato, K., Katsura, T., 2001. Experimental investigation on dolomite dissociation into aragonite+magnesite up to 8.5 GPa. Earth Planet. Sci. Lett. 184, 529-534.
Schwab, M., Ratschbacher, L., Siebel, W., McWilliams, M., Minaev, V., Lutkov, V., Chen, F., Stanek, K., Nelson, B., Frisch, W., Wooden, J.L., 2004. Assembly of the Pamirs: age and origin of magmatic belts from the southern Tien Shan to the southern Pamirs and their relation to Tibet. Tectonics 23, TC4002.
Shen, B., Jacobsen, B., Lee, C.A., Yin, Q.Z., Morton, D.M., 2009. The Mg isotopic systematics of granitoids in continental arcs and implications for the role of chemical weathering in crust formation. PNAS 106, 20652-20657.
Teng, F.-Z., Yang,W., 2014. Comparison of factors affecting the accuracy of high-precision magnesium isotope analysis by multi-collector inductively coupled plasma mass spectrometry. Rapid Commun. Mass Spectrom. 28, 19-24.
Teng, F.-Z., Wadhwa, M., Helz, R.T., 2007. Investigation of magnesium isotope fractionation during basalt differentiation: implications for a chondritic composition of the terrestrial mantle. Earth Planet. Sci. Lett. 261, 84-92.
Teng, F.-Z., Rudnick, R.L., McDonough, W.F., Gao, S., Tomascak, P.B., Liu, Y., 2008. Lithium isotopic composition and concentration of the deep continental crust. Chem. Geol. 255, 47-59.
Teng, F.-Z., Li, W.-Y., Ke, S., Marty, B., Dauphas, N., Huang, S., Wu, F.-Y., 2010. Magnesium isotopic composition of the Earth and chondrites. Geochim. Cosmochim. Acta 74, 4150-4166.

Teng, F.-Z., Dauphas, N., Helz, R.T., Gao, S., Huang, S., 2011. Diffusion-driven magnesium and iron isotope fractionation in Hawaiian olivine. Earth Planet. Sci. Lett. 308, 317-324.

Teng, F.-Z., Li, W.-Y., Ke, S., Yang, W., Liu, S.-A., Sedaghatpour, F., Wang, S.-J., Huang, K.-J., Hu, Y., Ling, M.-X., Xiao, Y., Liu, X.-M., Li, X.-W., Gu, H.-O., Sio, C.K., Wallace, D.A., Su, B.-X., Zhao, L., Chamberlin, J., Harrington, M., Brewer, A., 2015a. Magnesium isotopic compositions of international geological reference materials. Geostand. Geoanal. Res. 39, 329-339.

Teng, F.-Z., Yin, Q.-Z., Ullmann, C.V., Chakrabarti, R., Pogge von Strandmann, P.A.E., Yang, W., Li, W.-Y., Ke, S., Sedaghatpour, F., Wimpenny, J., Meixner, A., Romer, R.L., Wiechert, U., Jacobsen, S.B., 2015b. Interlaboratory comparison of magnesium isotopic compositions of 12 felsic to ultramafic igneous rock standards analyzed by MCICPMS. Geochem. Geophys. Geosyst. 16, 3197-3209.

Turner, S., Arnaud, N., Liu, J., Rogers, N., Hawkesworth, C., Harris, N., Kelley, S., Van Calsteren, P., Deng,W., 1996. Post-collision, shoshonitic volcanismon the Tibetan Plateau: implications for convective thinning of the lithosphere and the source of ocean island basalts. J. Petrol. 37, 45-71.

Wang, B.D., Chen, J.L., Xu, J.F.,Wang, L.Q., 2014a. Geochemical and Sr-Nd-Pb-Os isotopic compositions of Miocene ultrapotassic rocks in southern Tibet: petrogenesis and implications for the regional tectonic history. Lithos 208-209, 237-250.

Wang, S.-J., Teng, F.-Z., Li, S.-G., 2014b. Tracing carbonate-silicate interaction during subduction using magnesium and oxygen isotopes. Nat. Commun. 5, 5328.

Wang, S.-J., Teng, F.-Z., Scott, J.M., 2016. Tracing the origin of continental HIMU-like intraplate volcanism using magnesium isotope systematics. Geochim. Cosmochim. Acta http://dx.doi.org/10.1016/j.gca.2016.01.007 (in press).

Williams, H., Turner, S., Pearce, J., Kelley, S., Harris, N., 2004. Nature of the source regions for post-collisional, potassic magmatism in southern and northern Tibet from geochemical variations and inverse trace element modelling. J. Petrol. 45, 555-607.

Workman, R.K., Hart, S.R., 2005. Major and trace element composition of the depleted MORB mantle (DMM). Earth Planet. Sci. Lett. 231, 53-72.

Xiao, Y., Teng, F.-Z., Zhang, H.-F., Yang,W., 2013. Large magnesium isotope fractionation in peridotite xenoliths from eastern North China craton: product of melt-rock interaction. Geochim. Cosmochim. Acta 115, 241-261.

Yang, W., Teng, F.-Z., Zhang, H.-F., 2009. Chondritic magnesium isotopic composition of the terrestrial mantle: a case study of peridotite xenoliths from the North China craton. Earth Planet. Sci. Lett. 288, 475-482.

Yang,W., Teng, F.-Z., Zhang, H.-F., Li, S.-G., 2012. Magnesium isotopic systematics of continental basalts from the North China craton: implications for tracing subducted carbonate in the mantle. Chem. Geol. 328, 185-194.

Young, E.D., Tonui, E., Manning, C.E., Schauble, E.A., Macris, C., 2009. Spinel-olivine magnesium isotope thermometry in the mantle and implications for the Mg isotopic composition of Earth. Earth Planet. Sci. Lett. 288, 524-533.

Zhao, Z.D., Mo, X.X., Dilek, Y., Niu, Y.L., DePaolo, D.J., Robinson, P., Zhu, D.C., Sun, C.G., Dong, G.C., Zhou, S., Luo, Z.H., Hou, Z.Q., 2009. Geochemical and Sr-Nd-Pb-O isotopic compositions of the post-collisional ultrapotassic magmatism in SW Tibet: petrogenesis and implications for India intra-continental subduction beneath southern Tibet. Lithos 113, 190-212.

Zong, P.H., Zhou, X.X., Zhao, Z.H., Yu, S.H., Wang, C.S., Yang, Y., Yi,W.X., Zhang, Z.Q., 1982. REE of Nyalm strata, South Xizang analyzed by Neutron activation. Chin. Sci. Bull. 17, 1062-1065.

Zinc isotope evidence for a large-scale carbonated mantle beneath eastern China*

Sheng-Ao Liu[1], Ze-Zhou Wang[1], Shu-Guang Li[1, 2], Jian Huang[2] and Wei Yang[3]

1. State Key Laboratory of Geological Processes and Mineral Resources, China University of Geosciences, Beijing 100083, China
2. CAS Key Laboratory of Crust-Mantle Materials and Environments, University of Science and Technology of China, Hefei, Anhui 230026, China
3. State Key Laboratory of Lithospheric Evolution, Institute of Geology and Geophysics, Chinese Academy of Sciences, P.O. Box 9825, Beijing 100029, China

亮点介绍： 对中国东部大陆玄武岩锌同位素的系统研究，与正常地幔相比，发现 <110 Ma 具有轻镁同位素组成的玄武岩有较重的锌同位素组成，且 δ^{66}Zn 值与 Sm/Yb, Nb/Y, Zn, Nb, CaO, CaO/Al_2O_3 正相关。故首次提出锌同位素是地幔深部碳酸盐岩循环的一种新的示踪剂。

Abstract A large set of zinc (Zn) stable isotope data for continental basalts from eastern China were reported to investigate the application of Zn isotopes as a new tracer of deep carbonate cycling. All of the basalts with ages of <110 Ma have systematically heavy δ^{66}Zn (relative to JMC 3-0749L) ranging from 0.30‰ to 0.63‰ (n=44) compared to the mantle (0.28±0.05‰; 2sd) and >120 Ma basalts from eastern China (0.27±0.06‰; 2sd). Given that Zn isotope fractionation during magmatic differentiation is limited (⩽0.1‰), the elevated δ^{66}Zn values reflect the involvement of isotopically heavy crustal materials (e.g., carbonates with an average δ^{66}Zn of ~0.91‰) in the mantle sources. SiO_2 contents of the <110 Ma basalts negatively correlate with parameters that are sensitive to the degree of partial melting (e.g., Sm/Yb, Nb/Y, [Nb]) and with the concentration of Zn, which also behaves incompatibly during mantle melting. This is inconsistent with a volatile-poor peridotite source and instead suggests partial melting of carbonated peridotites which, at lower degree of melting, generates more Si-depleted (and more Ca-rich) melts. Zinc isotopic compositions are positively correlated with Sm/Yb, Nb/Y, [Nb] and [Zn], indicating that melts produced by lower degrees of melting have heavier Zn isotopic compositions. Carbonated peridotites have a lower solidus than volatile-poor peridotites and therefore at lower melting extents, contribute more to the melts, which will have heavier Zn isotopic compositions. Together with the positive relationships of δ^{66}Zn with CaO and CaO/Al_2O_3, we propose that the heavy Zn isotopic compositions of the <110 Ma basalts were generated by incongruent partial melting of carbonated peridotites. Combined with previously reported Mg and Sr isotope data, we suggest that the large-scale Zn isotope anomaly indicates the widespread presence of recycled Mg (Zn)-rich carbonates in the mantle beneath eastern China since the Late Mesozoic. Since Zn is a trace element in the mantle

* 本文发表在：Earth and Planetary Science Letters, 2016, 444: 169-178

and Zn isotopic compositions of marine carbonates and the mantle differ markedly, we highlight Zn isotopes as a new and useful tool of tracing deep carbonate cycling in the Earth's mantle.

1 Introduction

Cycling of carbon into and out of the mantle plays an important role in the global carbon cycle, which is important for the CO_2 budget of the Earth's atmosphere and thus influences Earth's climate. For example, warm global climate in the Cretaceous-early Cenozoic era has been attributed to a high ρCO_2 level, possibly maintained by subduction of the Tethyan oceanic crust with pelagic carbonate and subsequent release of CO_2 from arc volcanism (Kerrick and Connolly, 2001; Kent and Muttoni, 2008). Nevertheless, an estimate of the carbon budget in subduction zones shows that the expelled carbon from arc volcanism is much less than that subducted (Kerrick and Connolly, 2001; Johnston et al., 2011). Therefore, a significant mass of non-dissolved carbonates can survive subduction-related dehydration and melting and must be subducted/recycled to the mantle via slab subduction at convergent margins (e.g., Dasgupta and Hirschmann, 2006, 2010). The recycled carbonates were located in the mantle and then transferred to the atmosphere in the form of CO_2 via decompression melting of carbonated mantle under oxidizing conditions, under which carbon behaves as an incompatible element in magmatic systems (see review by Dasgupta and Hirschmann, 2010). This has evoked an important scientific issue called deep carbon cycling.

At present, research on deep carbon recycling is still at its early stage. Study of deep carbon cycling involves a series of scientific issues, including the total storage of deep carbon, fluxes during deep carbon cycling, phases containing deep carbon, partial melting of the carbon-bearing mantle, the release of CO_2 to the atmosphere by volcanism, and the proportion of subduction-related carbon and primary mantle-derived carbon in total CO_2 released by volcanism. Among those, tracing recycled carbonates and evaluating their amounts in the mantle are essential to a complete understanding of deep carbon cycling. Carbon isotopes can easily distinguish organic carbon from inorganic carbon, but about 95% of subduction-related and primary mantle-derived carbon released by volcanism is inorganic carbon; hence, carbon isotopes cannot be used to distinguish subduction-related carbon from primary mantle-derived carbon (Deines, 2002).Recently, stable isotopes of metals like Ca and Mg, the major ions in marine carbonates, have shown a great potential for tracing recycled crustal materials (in particular, carbonates) in the mantle (Huang et al., 2011; Yang et al., 2012; Huang et al., 2015). However, both Mg and Ca are major elements in the mantle so that it is doubtful whether recycled carbonates, commonly in the form of carbonated oceanic crust, could substantially modify Mg and Ca isotopic composition of the mantle. New efficient geochemical proxies need to be developed.

Zinc is a lithophile and trace element in the mantle (~55 μg/g; McDonough and Sun, 1995). It has five stable isotopes: ^{64}Zn (49.17%), ^{66}Zn (27.73%), ^{67}Zn (4.04%), ^{68}Zn (18.45%), and ^{70}Zn (0.61%). The terrestrial mantle is estimated to have an average $\delta^{66}Zn$ of 0.28±0.05‰ (2sd) (Chen et al., 2013), although some peridotites and basalts have slightly more variable Zn isotopic compositions (Fig. 1) (Herzog et al., 2009; Bigalke et al., 2010; Moeller et al., 2012; Telus et al., 2012; Chen et al., 2013; Makishima and Nakamura, 2013; Sossi et al., 2015; Chen et al., in press). Trace Zn can be incorporated into the crystal lattice of carbonate minerals (Reeder et al., 1999), which have heavier Zn isotopic composition (avg. $\delta^{66}Zn$=0.91±0.47‰, 2sd; Pichat et al., 2003) compared to the mantle (Fig. 1). During partial melting of the mantle, Zn behaves moderately incompatibly and the mineral/melt partition coefficients are 0.96 (ol/melt), 0.451 (opx/melt) and 0.333 (cpx/melt), respectively (Davis et al., 2013). Compared to the upper mantle, primitive basaltic melts (MgO >8.5 wt.%) from mid-ocean ridges, ocean islands and arc settings generally contain around 100 μg/g Zn (Le Roux et al., 2010). Given the marked difference of Zn isotopic composition between marine carbonates and the mantle, it is expected that if the mantle source contains recycled carbonates, basalts from such a source would have heavier Zn isotopic compositions than those from normal mantle. Thus, the fingerprint of recycled carbonates in the mantle can be investigated by measuring a series of basalts with a wide range of chemical compositions.

To test this application, in this study we have measured Zn isotopic compositions for Late Mesozoic to Cenozoic continental basalts from a wide range of locations in eastern China. Together with previously reported magnesium and strontium isotope data on the same samples, our study provides a strong support for the presence of recycled carbonates in the deep mantle beneath east-ern China and demonstrates the potential of zinc isotopes as a new tracer of deep carbonate cycling.

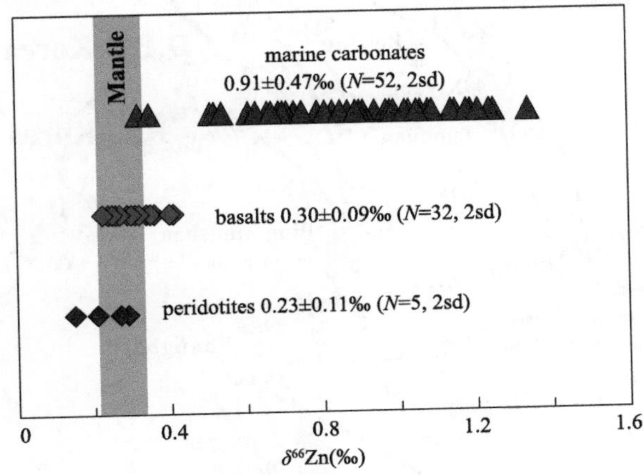

Fig.1 A summary of δ^{66}Zn values of peridotites, basalts and marine carbonates reported in the literature. Data sources: the mantle (0.28±0.05‰, 2sd; Chen et al., 2013), basalts (Herzog et al., 2009; Bigalke et al., 2010; Moeller et al., 2012; Telus et al., 2012; Chen et al., 2013; Sossi et al., 2015), peridotites (Makishima and Nakamura, 2013; Chen et al., 2013; Sossi et al., 2015; Chen et al., in press), marine carbonates (Pichat et al., 2003).

2 Geological settings and samples

Eastern China includes two major blocks: the North China Block (NCB) and the South China Block (SCB) (Fig. 2). They are separated by the Dabie-Sulu orogen, formed during Triassic continent-continent collision between these two blocks (Li et al., 1993). The basement of the NCB consists primarily of Early to Late Archean high- and low-grade TTG gneisses and syntectonic granitoids, and supracrustal rocks as well as some Early Proterozoic magma-tectonic belts (Zhao et al., 2001). The SCB is a major continental block in East Asia with a complex tectonic history. It was formed by the amalgamation of the Yangtze block (in the northwest) with the Cathaysia block (in the southeast) in the early Neoproterozoic (e.g., Charvet et al., 1996).

Late Mesozoic to Cenozoic continental basalts are widely distributed in eastern China along the coastal provinces and adjacent offshore shelf, extending over 4000 km from Heilongjiang Province in the north to Hainan island in the south on the eastern edge of the Eurasian continent (Fig. 2). They constitute an important part of the volcanic belt of the western circum-Pacific rim and are one of the world's presently active tectono-magmatic regions. The basalts erupted before 120 Ma are characterized by very negative $\varepsilon_{Nd}(t)$ values and are proposed to have originated from the enriched lithospheric mantle (e.g., Zhang et al., 2002; Yang and Li, 2008). The Late Mesozoic (<110 Ma) to Cenozoic volcanic rocks are mainly alkaline basalts and have ocean island basalt (OIB)-like trace element distribution patterns, which are thought to represent melts derived from the asthenospheric mantle (e.g., Zhou and Armstrong, 1982; Tang et al., 2006; Yang and Li, 2008; Chen et al., 2009).

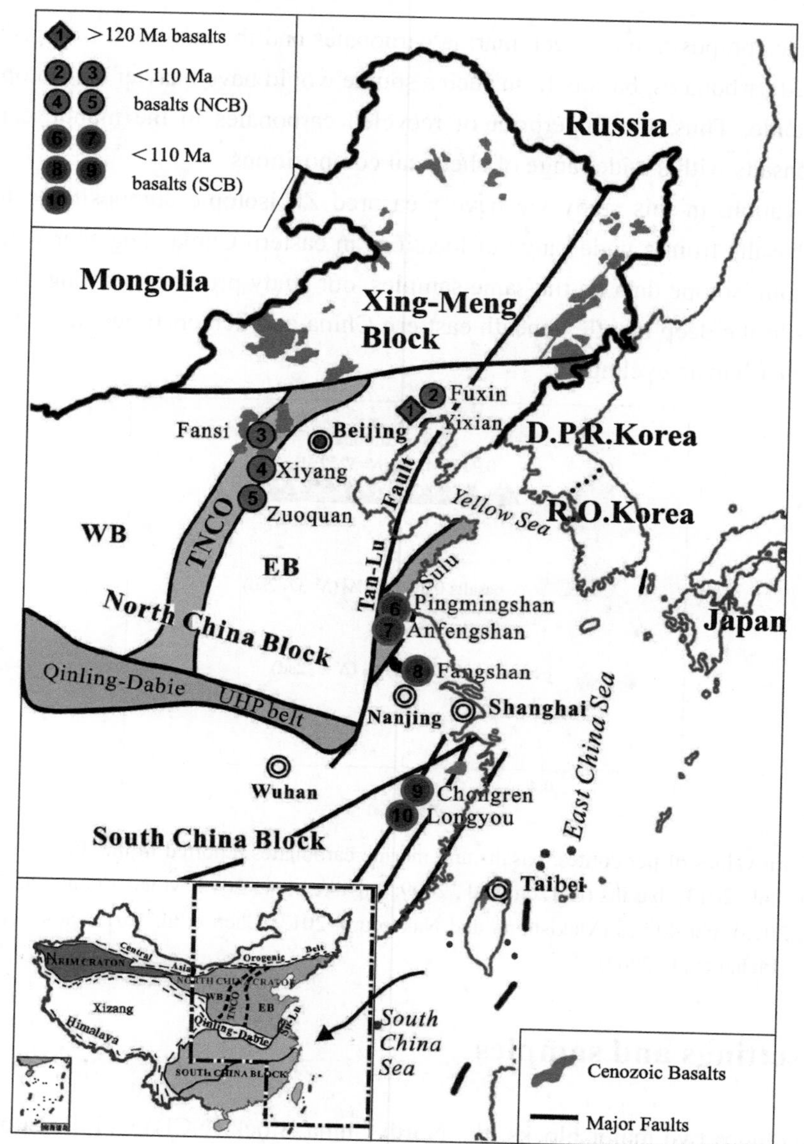

Fig. 2 Simplified geological map of eastern China, showing the South China Block (SCB), the North China Block (NCB), and the Xing-Meng Block. The NCB and SCB are separated by the Triassic Dabie-Sulu orogen formed during continent-continent collision between these two blocks (Li et al., 1993). The NCB consists of the Western Block (WB), the Trans-North China Orogen (TNCO) and the Eastern Block (EB). Sample localities are indicated by different symbols of different colours, and inset shows location of the study area in China. (For interpretation of the references to colour in this figure legend, the reader is referred to the web version of this article.)

The samples investigated in this study are from several locations in eastern China (Fig. 2). The Yixian basalts were formed at 132 Ma, and others (Fuxian, Fansi, Xiyang, Zuoquan, Pingmingshan, Anfengshan, Fangshan, Chongren, Longyou) were formed during 110 Ma to 4 Ma, classified as <110 Ma basalts here. These samples have been well characterized in previous studies for petrology, major and trace elements, Sr–Nd isotopes as well as for Mg and Cu isotopes (Tang et al., 2006; Yang and Li, 2008; Yang et al., 2012; Liu et al., 2015; Huang et al., 2015). In detail, the phenocrysts in these samples consist of olivine, clinopyroxene and plagioclase, and the groundmass is variable and mainly consists of plagioclase, olivine, augite, nepheline, magnetite and glass. They have SiO_2 contents ranging from 40.9 to 49.8 wt.% and Zn concentrations varying widely from 86 to 196 μg/g. All of the <110 Ma basalts are characterized by OIB-type trace element distribution patterns, with enriched light rare earth elements (LREEs) and Nb and depleted Rb and Pb. Their Ce/Pb and Nb/U ratios vary from 16.3 to 33.6 and 27.5 to 55.6, respectively, similar to those of MORBs and OIBs, but much higher than those of continental crust (Yang and Li, 2008; Huang et al., 2015). The initial $^{87}Sr/^{86}Sr$ ratios vary from 0.70328 to 0.70537 and $\varepsilon_{Nd}(t)$ values vary from −4.0 to +6.8 (Yang and Li, 2008; Huang and Xiao, in press).

3 Analytical methods

All chemical procedures were performed in laminar flow hoods (Class 100) in a clean room (Class 1000) with filtered air. All beakers were PTFE (Savillex®). Double distilled reagents and ≥18.2 MΩ water were used for sample dissolution and all other processes. To obtain ~1 μg Zn for high-precision isotope analysis, approximately 10 mg of sample was digested with a mixture of HF, HNO_3 and HCl. After complete digestion, 1 ml of 8 N HCl was added and the sample was evaporated to dryness. Dissolved samples were finally prepared in 1 ml of 8 N HCl + 0.001% H_2O_2 for ion-exchange separation.

Zinc was separated with a single-column procedure modified from Maréchal et al. (1999). The procedure for chemical purification has been reported in previous studies from our group (Liu et al., 2014, 2015; Lv et al., 2016). Briefly, the samples were loaded on a column containing 2 ml pre-cleaned Bio-Rad strong anion resin AG-MP-1M. Zinc was collected in 10 ml of 0.5 N HNO_3 after matrix, copper and iron were eluted. After evaporation to dryness at 80 °C and dissolution in 3% HNO_3 to drive the chloride ion away, the residues were dissolved in 3% HNO_3. The samples were analyzed using ICP-MS to check for the elimination of matrix elements before isotopic measurement. Mg and Al were checked for each collected Zn solution after column chemistry given that the presence of Mg and Al produces argides ($Al^{27}Ar^+$, $Mg^{24}Ar^+$ and $Mg^{26}Ar^+$) that significantly influence Zn isotope measurements (Mason et al., 2004). The recovery of Zn is close to 100% and the total procedural blanks are negligible and ≤ ~6 ng.

Isotopic analysis was carried out on the Thermo Scientific Neptune plus MC-ICP-MS instruments at the Isotope Geochemistry Laboratory of the China University of Geosciences, Beijing. Samples and standards are diluted to obtain 200 ppb Zn solutions in 3% HNO_3 and sample-standard bracketing (SSB) method was used for mass bias correction. High-sensitivity (X) cones made of Ni are used to ensure that the ^{64}Zn signals are usually >3 V/200 ppb. Zinc isotopic analysis for each measurement is operated for three blocks of 40 cycles in the low-resolution mode. The data are reported in δ-notation in per mil against JMC3-0749L: $\delta^{66}Zn$ or $\delta^{68}Zn = ((^{66,68}Zn/^{64}Zn)_{sample}/(^{66,68}Zn/^{64}Zn)_{JMC\ 3\text{-}0749L} - 1) \times 1000$. The external reproducibility for $\delta^{66}Zn$ measurement is better than ±0.05‰ (2sd) based on long-term analyses of international basalt standard BHVO-2 (0.31±0.05‰; n=22; 2sd) and standard solution IRMM 3702 (0.27±0.03‰; n=47; 2sd) over a period of two years (Table S1, Fig. S1).

4 Results

Zinc isotope data for continental basalts from eastern China, together with previously reported major-trace elements and Mg and Sr isotope data on the same samples, are listed in Tables S2, S3. All samples fall on a mass fractionation line in three-isotope space ($\delta^{66}Zn$ vs. $\delta^{68}Zn$; Fig. 3) with a slope of 1.998±0.008 (1σ; R^2=0.997; n=54), indicating that there are no analytical artifacts from unresolved isobaric interferences on measured Zn isotopic ratios. Analyses of the USGS basalt standards (BHVO-2, BCR-2 and BIR-1a) yield results in good agreement with published values (e.g., Herzog et al., 2009; Bigalke et al., 2010; Moeller et al., 2012; Telus et al., 2012; Chen et al., 2013; Sossi et al., 2015; Chen et al., in press).

Zinc isotopic compositions of the >120 Ma basalts are significantly different from those of <110 Ma basalts (Fig.4). The >120 Ma basalts have δ^{66}Zn values ranging from 0.23‰ to 0.32‰, with a mean of 0.27±0.06‰ (n=5; 2sd) identical to the values of global OIB and MORB. The <110 Ma basalts have significantly heavy δ^{66}Zn values, ranging from 0.30‰ to 0.63‰ (n=48). Their δ^{66}Zn values positively correlate with Zn concentrations (Fig.4), Sm/Yb, Nb/Y and Nb concentrations (Fig.5) and correlate negatively with SiO_2 contents (Fig.6). δ^{66}Zn values are also positively correlated with Zn/Fe ratios (Fig.7). The Zn isotopic compositions are negatively correlated with the Mg isotopic compositions of the same samples reported in previous studies (Yang et al., 2012; Huang et al., 2015) (Fig.8a).

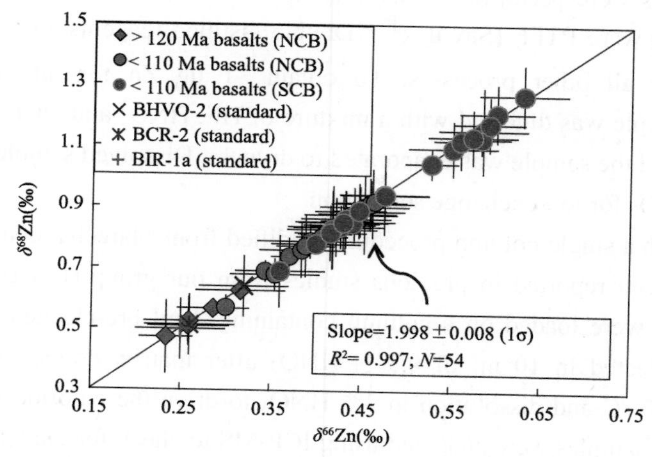

Fig. 3　Relationship between measured δ^{66}Zn and δ^{68}Zn values for basalts from eastern China and standard materials. All samples analyzed define the mass fractionation line with a slope of 1.998±0.008, indicating that there are no analytical artifacts from unresolved isobaric interferences on measured Zn isotopic ratios. The data are reported in Table S3. The symbol type and colour refer to those for the sample locations in Fig. 2. (For interpretation of the references to colour in this figure legend, the reader is referred to the web version of this article.)

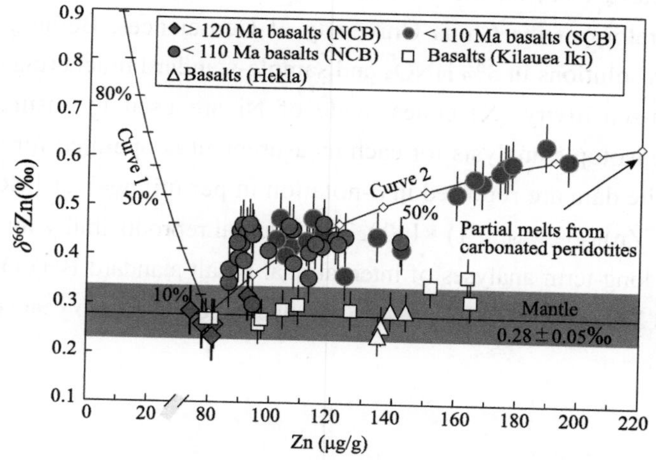

Fig.4　Variation of Zn isotopic compositions with Zn concentrations for continental basalts from eastern China. Data for Kilauea Iki and Hekla basalts are from Chen et al. (2013). Curve 1 represents the trend expected for contamination of basaltic magmas by sedimentary carbonates within the crust. The basalt end member used for modeling is taken to have a Zn concentration of 80 μg/g and a δ^{66}Zn value of 0.28‰ (Chen et al., 2013). The carbonate end member used for modeling is assumed to have a Zn concentration of ~10 μg/g and a δ^{66}Zn value of 0.91‰ (Pichat et al., 2003). Curve 2 represents mixing of melts from normal mantle melting (Zn=80 μg/g; δ^{66}Zn=0.28‰) with melts from low-degree melting of carbonated peridotites (assuming Zn=220 μg/g; δ^{66}Zn=0.65‰). The symbol type and colour refer to those for the sample locations in Fig.2. (For interpretation of the references to colour in this figure legend, the reader is referred to the web version of this article.)

5 Discussion

5.1 The origin of high δ^{66}Zn basalts in eastern China

Zinc isotopic compositions of basalts including MORB, OIB and continental flood basalts published previously range from 0.24‰ to 0.44‰ (Herzog et al., 2009; Chen et al., 2013; Zhou et al., 2014; Sossi et al., 2015). Most of these basalts have identical δ^{66}Zn values of around 0.28‰ (Fig.1) but some of them have slightly heavier values whose origin has not been thoroughly understood. There are four possible mechanisms that could give rise to heavy Zn isotopic compositions in basalts, including (1) crystal-melt differentiation, (2) magmatic degassing, (3) crustal contamination, and/or (4) isotopic heterogeneity caused by recycled materials in the mantle sources.

Research on the Kilauea Iki lava lake, Hawaii, shows that the more differentiated lithologies have heavier Zn isotopic compositions, which was explained as the result of crystallization of olivine and/or Fe–Ti oxides at the very end of the differentiation sequence (Chen et al., 2013). The relatively light δ^{66}Zn in two San Carlos olivines (0.07‰ and 0.12‰; Sossi et al., 2015) supports the speculation that crystallization of olivine may enrich the residual melts in heavy Zn isotopes. There is no extensive fractional crystallization of olivine in our samples, given their high MgO contents (average=8.9 wt.%; Table S2) close to primary mantle-derived magmas (~8.5 wt.%). In particular, the magnitude of Zn isotope fractionation in the strongly fractionated Kilauea Iki lavas is typically smaller than 0.1‰ (varying from 0.26‰ to 0.36‰), and thus any Zn isotope variation in basalts by more than 0.1‰ should reflect other processes. Our samples have an overall δ^{66}Zn range of 0.23‰ to 0.63‰ and define a much steeper trend than the Kilauea Iki suite in the diagram of δ^{66}Zn vs. [Zn] (Fig.4), which is unlikely to be explained by crystal-melt fractionation at magmatic temperatures.

Zinc is a moderately volatile element and volatile loss of Zn could result in Zn isotope fractionation, with light Zn preferentially partitioning into the volatile phase (e.g., John et al., 2008). One typical example is lunar igneous rocks, which have heavier Zn isotopic compositions and lower Zn concentrations than terrestrial igneous rocks, interpreted to reflect volatile depletion of the Moon through evaporation (Paniello et al., 2012; Kato et al., 2015). The <110 Ma basalts from eastern China display a positive correlation between δ^{66}Zn and Zn abundance (Fig. 4), which is thus inconsistent with that expected for volatilization. In addition, volatile loss of Zn via magmatic degassing in the studied samples is inconsistent with the co-variation of δ^{66}Zn with abundance of the barely volatile elements (e.g., Nb; Fig. 5f). In fact, a study on fumarolic gases, rocks and condensate samples from Merapi volcano (Indonesia) shows significant Zn isotope fractionation during magmatic degassing, with gas condensates enriched in the heavy isotopes of Zn, but rock samples from associated lava flows having “normal” Zn isotopic composition (Toutain et al., 2008). A study of Hekla lavas that underwent strong degassing, as indicated by up to 78–95% loss of initial sulphur, also shows limited Zn isotope variation (Chen et al., 2013). Therefore, volatile loss of Zn via magmatic degassing is not a viable mechanism to explain the enrichment of the <110 Ma basalts in heavy Zn isotopes.

Two lines of evidence suggest that crustal contamination is also unlikely to be the process generating the heavy Zn isotopic compositions. First, most of the <110 Ma basalts have depleted Sr–Nd isotopic compositions ($^{87}Sr/^{86}Sr_i$ <0.7045; $\varepsilon_{Nd}(t) > 0$) and positive Nb and Ta anomalies in their primitive mantle normalized trace element distribution patterns (Tang et al., 2006; Yang and Li, 2008; Huang et al., 2015), indicating that crustal contamination, if any, is negligible. Second, marine carbonates have much lower Nb contents compared to basaltic rocks (Turekian and Wedepohl, 1961), and if a basaltic magma underwent contamination by sedimentary carbonates while it ascended through the crust, a negative correlation between δ^{66}Zn and Nb concentrations should occur. This is in conflict with the positive relationship illustrated in Fig. 5f.

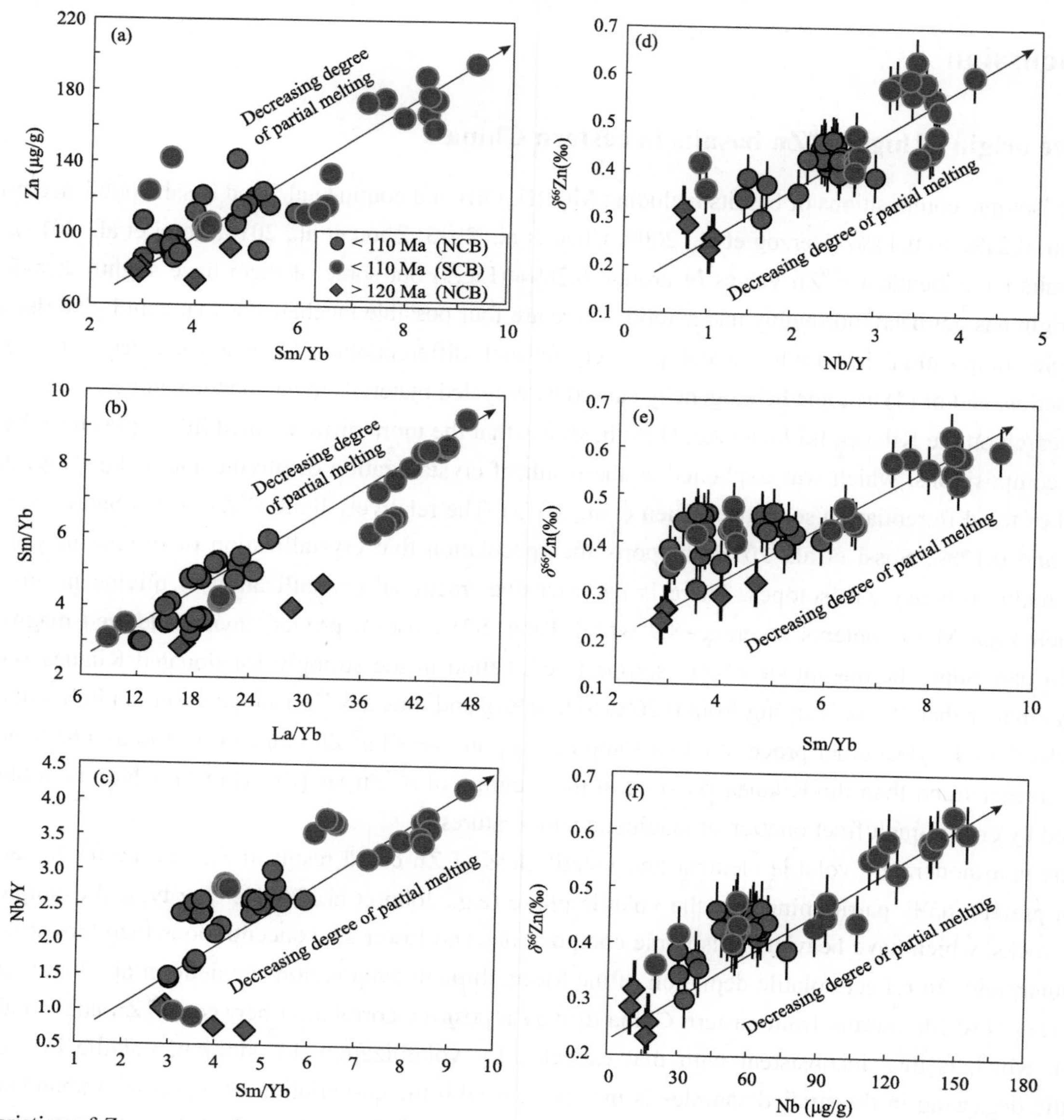

Fig. 5　Variation of Zn concentrations with Sm/Yb (a), Sm/Yb with La/Yb (b), Nb/Y vs. Sm/Yb (c) and variation of δ^{66}Zn with Nb/Y (d), Sm/Yb (e), and Nb concentrations (f). Data are from Table S3. It is shown that these samples were formed by various degrees of partial melting and that Zn isotopic compositions increase with decreasing degrees of melting. The degree (%) of partial melting for each sample can be found in Huang et al. (2015). Note that the arrows in this figure and in Fig. 6 represent the trends of decreasing degree of partial melting and are not true linear regressions of the data.

The above three processes (crystal-melt differentiation, magmatic degassing, and crustal contamination) are therefore unlikely. Because Nb is commonly immobile in fluids during slab dehydration (Kogiso et al., 1997), the positive relationship between δ^{66}Zn and Nb contents also excludes the possibility that the heavy δ^{66}Zn values are related to slab-derived fluids. Thus the elevated Zn isotopic compositions of the <110 Ma basalts should reflect the isotopically heavy mantle sources that could have been caused by recycled crustal materials via plate subduction. Data for altered basaltic oceanic crust are not yet available but they should fall between deep seawater (~0.51‰; Zhao et al., 2014) and MORB (~0.28‰). Thus, the recycling of altered oceanic crust is unlikely to account for the heavy δ^{66}Zn of up to 0.63‰ in the <110 Ma basalts. Consequently, marine carbonates with an average δ^{66}Zn of 0.91‰ (Pichat et al., 2003) carried by the subducted plate are the more realistic candidates. Below we apply several key parameters to constrain the origin.

Given that Zn behaves moderately incompatibly during mantle melting (Le Roux et al., 2010; Davis et al., 2013), lower degrees of partial melting generate basaltic melts with higher Zn concentrations. Since La and Sm are more incompatible than Yb and Nb is more incompatible than Y during partial melting of mantle rocks, La/Yb, Sm/Yb and Nb/Y ratios of basalts are also strongly dependent on the degree of partial melting. In contrast to partial melting, fractional crystallization of olivine and/or pyroxene should not significantly change the ratios of two incompatible element abundances (e.g., La/Yb, Sm/Yb and Nb/Y). Thus, the co-variations of Zn concentrations, La/Yb, Sm/Yb and Nb/Y ratios (Fig. 5a, b, c) indicate that the <110 Ma basalts have been formed by a wide range of degrees of mantle partial melting (ca. 2% to 15%; Huang et al., 2015).

The positive correlations of $\delta^{66}Zn$ with [Zn], [Nb], Sm/Yb and Nb/Y (Figs. 4, 5d, e, f) suggest that large Zn isotope variations have also occurred during the melting processes that generated these basalts, with melts produced by low degrees of melting having heavier Zn isotopic compositions relative to melts produced by high degrees of melting. Because Zn has only one valence state (Zn^{2+}), such a large $\delta^{66}Zn$ variation of 0.4‰ is unlikely to be explained by equilibrium fractionation between peridotite and melt at mantle temperatures. For example, alkalic OIBs are commonly generated at lower degrees of melting (<5%) than MORBs (10%–15%) (e.g., Klein and Karsten, 1995), but most of OIB and other alkaline basalts have indistinguishable Zn isotopic composition from MORB (Ben Othman et al., 2006; Herzog et al., 2009; Chen et al., 2013; Zhou et al., 2014; Fig. 1). Thus, partial melting of a common mantle source parental to MORB and most OIB is unlikely to account for the large variation in $\delta^{66}Zn$ and its relationships with [Zn], [Nb], Sm/Yb, Nb/Y and La/Yb in the <110 Ma basalts (Figs. 4, 5). Instead, they are best accounted for by incongruent partial melting of a carbonated mantle source. Carbonation significantly lowers the solidus of peridotites, allowing carbonated peridotites to melt before volatile-poor peridotites (Dasgupta et al., 2007, 2013). When the melting degree is low, carbonated peridotites contribute more to the basaltic melts that will have heavier Zn isotopic compositions.

Commonly, lower degrees of partial melting of volatile-poor peridotites generate higher SiO_2, lower CaO and lower CaO/Al_2O_3 ratios in melts (e.g., Laporte et al., 2014). By contrast, experimental studies show that SiO_2 contents of carbonated silicate melts from carbonated peridotites diminish significantly with increasing dissolved CO_2 in the melt, whereas the CaO contents and CaO/Al_2O_3 ratios increase markedly (Dasgupta et al., 2007). That is, at low melt fractions, partial melts of carbonated peridotites would have enhanced CaO contents and CaO/Al_2O_3 ratios and diminished (unsaturated) SiO_2 contents (Dasgupta et al., 2007). The negative correlation of SiO_2 with Sm/Yb (Fig. 6a) coincides with the melting trend of carbonated peridotites, but contrasts with peridotite melting in the absence of CO_2. The positive correlation between CaO/Al_2O_3 and CaO contents (Fig. 6c) also matches with the partial melting of carbonated peridotites as a result of preferential transfer of Ca into the melts relative to Al (Dasgupta et al., 2007). At lower extents of melting, carbonated peridotites contribute more to the melts, which would have lower SiO_2 contents, higher CaO/Al_2O_3 and heavier $\delta^{66}Zn$ values. This agrees well with the observations depicted in Fig.6; there is a negative correlation between $\delta^{66}Zn$ and SiO_2 and a positive correlation between $\delta^{66}Zn$ and CaO/Al_2O_3, which strongly supports a carbonated peridotite source. Several samples seem to deviate from the trend (Fig.6c, d) due to their low Al_2O_3 contents. Interestingly, experimental melts of carbonated peridotites also display an inverse variation trend of Al_2O_3 at very low degree of partial melting (Fig.3 in Dasgupta et al., 2007). This inverse trend possibly corresponds to the starting of garnet consumption during mantle partial melting (Dasgupta et al., 2007).

Zn/Fe (Zn/Mn, Fe/Mn) ratios of basalts are not significantly affected by modal variations in peridotites but will fractionate if garnet and/or clinopyroxene (cpx) are the main phases in the residue (Le Roux et al., 2010). This is based on the lower mineral/melt partition coefficient of Zn/Fe in garnet and cpx than that in olivine and orthopyroxene (Le Roux et al., 2010). The <110 Ma basalts have high ratios of Zn/Fe and Zn/Mn, which are partly due to their high Zn concentrations (see Section 5.2 for a possible explanation). Together with the positive correlation between Zn/Mn and Zn/Fe (Fig. 7a), these results imply an important role of garnet or cpx in the residue during partial melting. Basalts with higher Zn/Fe ratios also have heavier δ^{66}Zn values (Fig. 7b), suggesting that the melts with heavy δ^{66}Zn values produced by lower degree melting contain more garnet or cpx in the residue.

In summary, Zn isotopic compositions of the <110 Ma basalts vary with parameters that are sensitive to melting degree (Sm/Yb, Nb/Y, [Zn], [Nb]), carbonation of peridotites (SiO_2, CaO, CaO/Al_2O_3) and garnet or cpx in the residue (Zn/Fe, Zn/Mn). These relationships are unlikely explained by the varying degrees of partial melting of peridotites that generate most MORB and OIB, and instead point out carbonated peridotite as a more likely proximal source for the isotopically heavy basalts. A carbonated mantle source for the Cenozoic basalts from Shandong province (NCB) and the SCB is also suggested on the basis of the relationship between total alkalis ($Na_2O + K_2O$) and TiO_2 as well as Hf–Nd isotopic decoupling and Mg–Sr isotopes (Chen et al., 2009; Zeng et al., 2010; Yang et al., 2012; Huang et al., 2015).

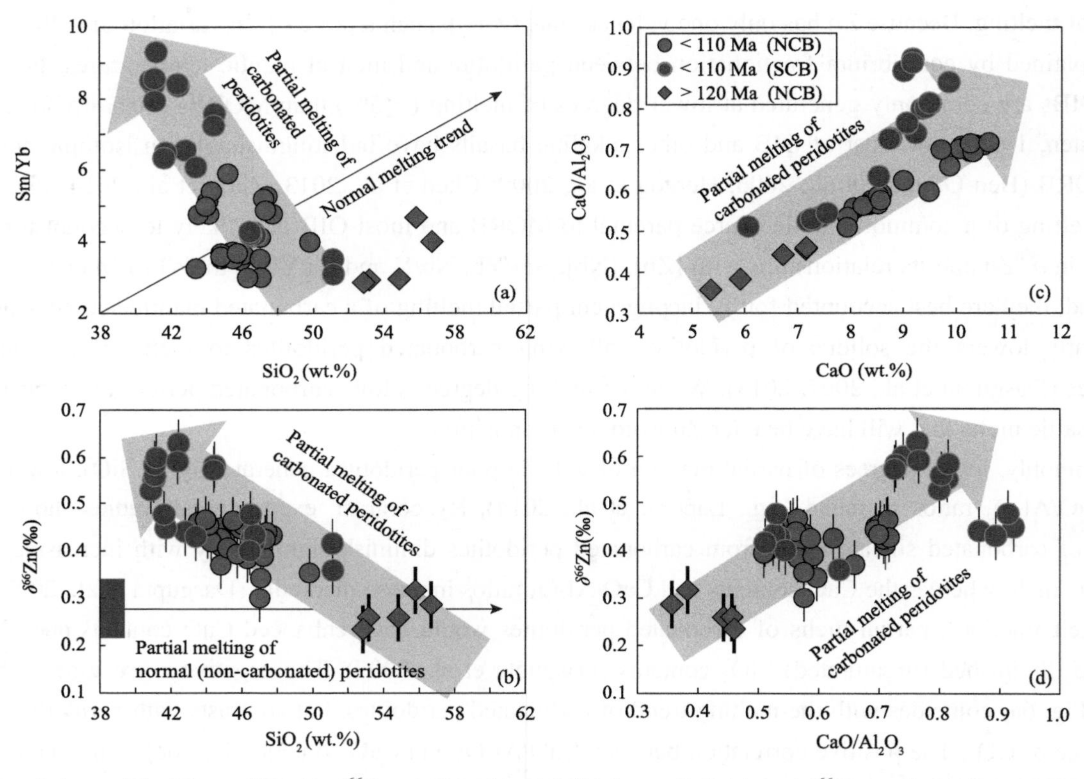

Fig. 6　Correlations of Sm/Yb vs. SiO_2 (a), δ^{66}Zn vs. SiO_2 (b), CaO/Al_2O_3 vs. CaO (c), and δ^{66}Zn vs. CaO/Al_2O_3 (d) for the basalts from eastern China. Data are from Table S3. The trend for partial melting of carbonated peridotites is based on the experimental studies of Dasgupta et al. (2007).

5.2 Possible constraints on nature of the recycled carbonates

An important characteristic of the <110 Ma basalts in eastern China is that they have relatively low initial $^{87}Sr/^{86}Sr$ ratios of 0.70328 to 0.70537 (Table S3). As shown in Fig. 8b, there is a rapid increase of δ^{66}Zn values but little change of $^{87}Sr/^{86}Sr$ ratios in most of the <110 Ma basalts. Modelling suggests that, to produce this trend, the recycled carbonates are required to have high Zn but low Sr concentrations, since marine carbonates commonly have high $^{87}Sr/^{86}Sr$ ratios. Different species of carbonate minerals have dramatically different Sr contents, decreasing from calcite/aragonite (avg. 1311 μg/g), through dolomite (avg. 1151 μg/g), to magnesite (avg. 1.84 μg/g; Huang and Xiao, in press and reference therein). By contrast, dolomite and magnesite in ultrahigh pressure metamorphic rocks, representing the deeply subducted slab, could have very high Zn concentrations of up to 147 μg/g and 449 μg/g, respectively (Li et al., 2014), in comparison with calcite (taken to be similar to that of limestone, avg. ~20 μg/g; Turekian and Wedepohl, 1961). That is, magnesite is an important mineral carried by the subducted plate to the deep mantle, and it potentially has high Zn but low Sr contents. The Zn-Sr isotope data therefore suggest that the recycled carbonates in the mantle source of the <110 Ma from eastern China are most likely composed of magnesite+dolomite. This hypothesis is consistent with the results from seismic tomography that showed the location of the Pacific slab in the mantle transition zone (410–660 km) beneath eastern China (e.g., Zhao et al., 2011); under such *P-T* conditions magnesite would be a stable phase of the recycled carbonates (e.g., Sato and Katsura, 2001). The presence of magnesite in the mantle source may also partly contribute to the high Zn concentrations and high Zn/Fe ratios of the <110 Ma basalts as discussed above.

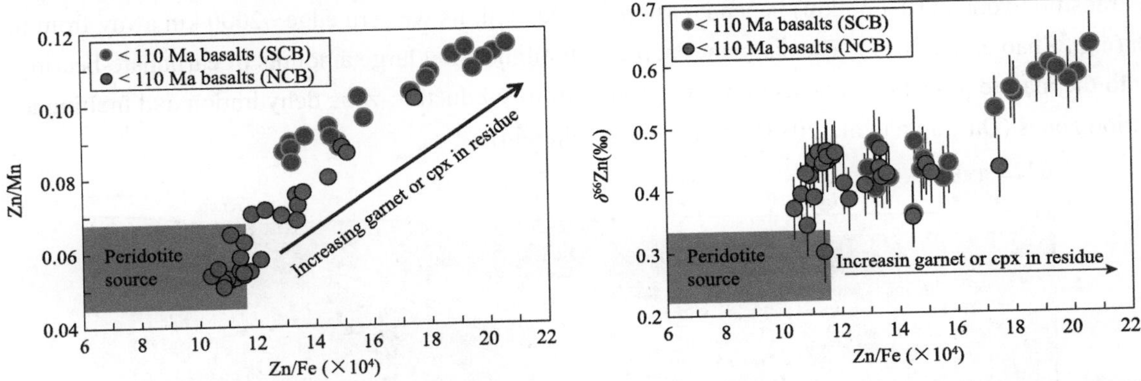

Fig.7 Correlations of Zn/Mn vs. Zn/Fe (upper) and δ^{66}Zn vs. Zn/Fe (lower) for the <110 Ma basalts from eastern China. Zn concentration and isotope data are from this study (Tables S2, S3). The contents of Fe (wt.%) and Mn (μg/g), listed in Table S2, are from Tang et al. (2006), Yang et al. (2012) and Huang et al. (2015).

5.3 Evidence for a large-scale carbonated mantle beneath eastern China

Our Zn isotope results provide new and solid evidence for a carbonated mantle located beneath eastern China (including NCB and SCB) during the late Mesozoic to Cenozoic. Zinc isotopic compositions of the <110 Ma basalts are negatively correlated with Mg isotopic compositions of the same samples (Fig. 8a). No known high-temperature processes could explain the relationship and, to our knowledge, it can only be explained by the presence of recycled marine carbonates (having heavy δ^{66}Zn and light δ^{26}Mg relative to the mantle) in the mantle sources. A cartoon model (Fig. 9) gives a simple description on the recycling of the subducting carbonate-bearing plate and the generation of high δ^{66}Zn and low δ^{26}Mg basalts.

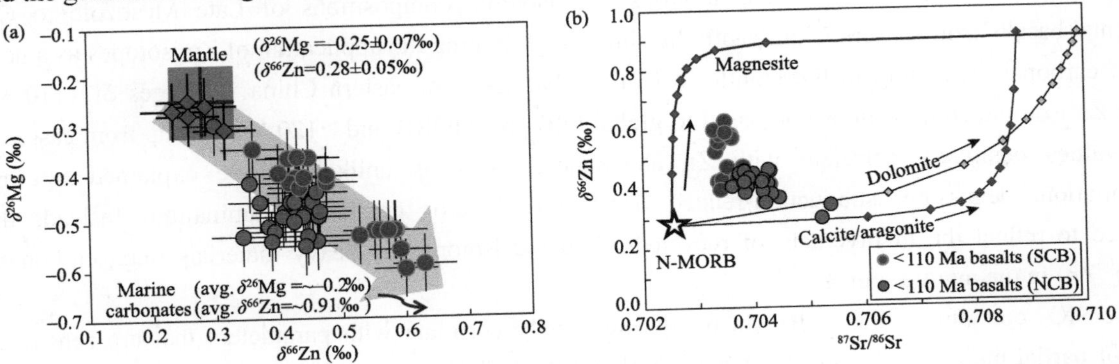

Fig. 8 Correlations between δ^{26}Mg and δ^{66}Zn (a) and between δ^{66}Zn and ^{87}Sr/^{86}Sr (b) for the <110 Ma basalts from eastern China. MgO and Sr isotope data for the basalts, listed in Tables S2, S3, are from Yang and Li (2008), Yang et al. (2012), Huang et al. (2015) and Huang and Xiao (in press). The average Mg isotopic composition of marine carbonates is from the compiled data by Huang and Xiao (in press). The parameters used in this modelling are summarized in Table S4. Mixing hyperbolas are marked in 10% increments.

The mantle-like Zn and Mg isotopic compositions of the >120 Ma basalts in the NCB (Fig. 8) suggest that the isotopically light mantle source did not form before 120 Ma (Yang et al., 2012; Huang et al., 2015; this study). After 110 Ma, the large-scale Zn and Mg isotope anomaly appeared abruptly in the <110 Ma basalts in eastern China. The abrupt shift in Zn and Mg isotopic compositions suggests that the recycled carbonates in the mantle beneath eastern China were plausibly derived from the subducting Pacific slab, given that only the Pacific slab has an influence on both NCB and SCB during the Late Mesozoic. The recycled carbonate-bearing oceanic crust might be responsible for the abrupt changes of Mg, Zn, as well as Nd isotopic compositions between >120 and <110 Ma basalts from eastern China (Yang et al., 2012; Huang et al., 2015; this study). High-resolution seismic tomography has revealed that the Pacific slab is subducting beneath the Japan Islands and becomes stagnant in the mantle transition zone (410–660 km) beneath eastern China, with its western edge ~2000 km away from the Japan Trench (e.g., Zhao et al., 2011). The stagnant Pacific slab might bring large amounts of carbonate-bearing oceanic crust into the mantle transition zone as they can survive from subduction-zone dehydration and melting at modern subduction zones (Dasgupta et al., 2013) and references therein).

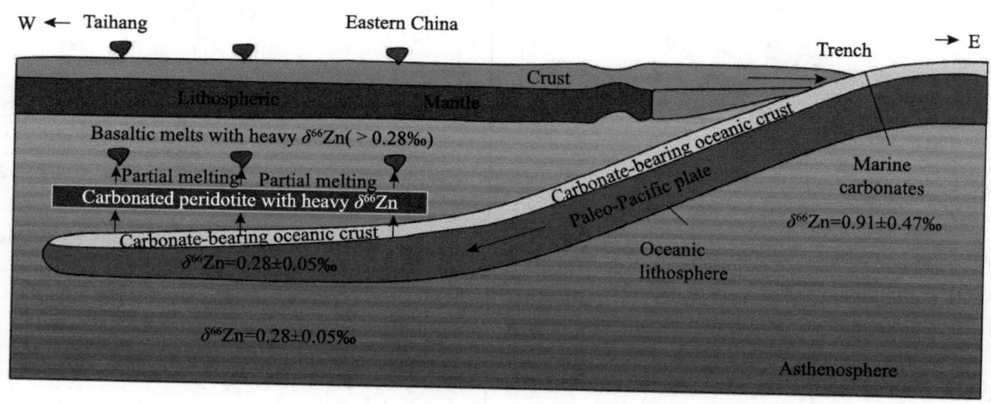

Fig.9　Cartoon model showing the subduction of the west Pacific plate beneath eastern China during the Late Mesozoic. The subducted plate brought marine carbonates with heavy δ^{66}Zn (mean=~0.91‰) into the deep mantle and formed carbonated peridotites. The carbonated peridotites have a lower solidus than volatile-poor peridotites. Thus they melted earlier and contributed more to the basaltic melts, generating positive correlations of basalt δ^{66}Zn with [Zn], [Nb], Sm/Yb and Nb/Y as observed (Figs. 4, 5).

6　Conclusion

We have measured the Zn concentrations and isotopic compositions of Late Mesozoic to Cenozoic continental basalts from eastern China, with the aim of highlighting the application of Zn isotopes as a new tracer of deep carbonate recycling in the mantle. All of the basalts from eastern China with ages of <110 Ma have heavier Zn isotopic compositions compared to global OIB and MORB and >120 Ma basalts from eastern China. δ^{66}Zn values positively correlate with Zn abundance, which is unlikely to be explained by magmatic differentiation, degassing, subduction-related dehydration and/or crustal contamination. Instead, they are suggested to reflect the involvement of recycled/subducted isotopically heavy materials (e.g., carbonates with heavy δ^{66}Zn) in the mantle sources.

The SiO_2 contents of the <110 Ma basalts negatively correlate with parameters that are sensitive to the degree of partial melting, e.g., Sm/Yb, Nb/Y, [Nb] and [Zn]. This contrasts with partial melting of volatile-poor peridotites and instead suggests partial melting of carbonated peridotites which, at lower degree of melting, generates more Si-depleted (and more Ca-rich) melts. δ^{66}Zn values are positively correlated with these parameters, indicating that the melts produced by lower degrees of melting have heavier Zn isotopic compositions. Carbonated peridotites have a lower solidus than volatile-poor peridotites, which at lower melting extents, contribute more to the basaltic melts. As a result, these melts will have heavier Zn isotopic compositions compared with those produced by partial melting of normal mantle. This explanation is also strongly supported by the positive relationship between $\delta^{66}Zn$ and CaO/Al_2O_3 in the <110 Ma basalts. Thus, Zn isotopes could be a powerful tool for tracing recycled carbonate in the Earth's deep mantle.

Combined with the light Mg isotopic signatures in the same samples (Yang et al., 2012; Huang et al., 2015), we suggest that the large-scale Zn isotope anomaly indicates the widespread presence of recycled carbonates in the mantle beneath eastern China, which is possibly linked to subduction of the west Pacific plate during the Late Mesozoic to Cenozoic. Modelling on the basis of Zn and Sr isotope data further suggests that the recycled carbonates in the mantle beneath eastern China could be mainly composed of Mg and Zn enriched species (e.g., magnesite±dolomite).

Acknowledgement

We are grateful to Yanjie Tang for providing some of the basalt samples used in this study. We thank the Editor Prof. Derek Vance for comments, efficient handling and language editing and three anonymous reviewers for constructive comments, all of which significantly improved the manuscript. This work is supported by the National Key Project for Basic Research (project 2015CB452606), the National Natural Science Foundation of China (No. 41473017) and the Fundamental Research Funds (2-9-2014-068) to SAL and the National Natural Science Foundation of China (No. 41230209) to SGL and (No. 41573018) to JH. This is CUGB petro-geochemical contribution No. PGC-201504.

Appendix A. Supplementary material

Supplementary material related to this article can be found online at http://dx.doi.org/10.1016/j.epsl.2016.03.051.

References

Ben Othman, D., Luck, J.M., Bodinier, J.L., Arndt, N.T., Albarede, F., 2006. Cu-Zn isotope variation in the Earth's mantle. Geochim. Cosmochim. Acta 70 (18), A46.

Bigalke, M., Weyer, S., Kobza, J., Wilcke, W., 2010. Stable Cu and Zn isotope ratios as tracers of sources and transport of Cu and Zn in contaminated soil. Geochim. Cosmochim. Acta 74, 6801-6813.

Charvet, J., Shu, L.S., Shi, Y.S., Guo, L.Z., Faure, M., 1996. The building of south China: collision of Yangzi and Cathaysia blocks, problems and tentative answers. J. Southeast Asian Earth Sci. 13, 223-235.

Chen, H., Savage, P.S., Teng, F.-Z., Helz, R.T., Moynier, F., 2013. Zinc isotope fractionation during magmatic differentiation and the isotopic composition of the bulk Earth. Earth Planet. Sci. Lett. 369-370, 34-42.

Chen, L.-H., Zeng, G., Jiang, S.-Y., Hofmann, A.W., Xu, X.-S., Pan, M.-B., 2009. Sources of Anfengshan basalts: subducted lower crust in the Sulu UHP belt, China. Earth Planet. Sci. Lett. 286, 426-435.

Chen, S., Liu, Y., Hu, J., Zhang, Z., Hou, Z., Huang, F., Yu, H., in press. Zn isotopic compositions of NIST SRM683 and whole-rock reference materials. Geostand. Geoanal. Res. http://dx.doi.org/10.1111/j.1751-908X.2015.00377.x.

Dasgupta, R., Hirschmann, M.M., 2006. Melting in the Earth's deep upper mantle caused by carbon dioxide. Nature 440, 659-662.

Dasgupta, R., Hirschmann, M.M., Smith, N.D., 2007. Partial melting experiments of peridotite+CO_2 at 3 GPa and genesis of alkalic ocean island basalts. J. Petrol. 48, 2093-2124.

Dasgupta, R., Hirschmann, M.M., 2010. The deep carbon cycle and melting in Earth's interior. Earth Planet. Sci. Lett. 298, 1-13.

Dasgupta, R., Mallik, A., Tsuno, K., Withers, A.C., Hirth, G., Hirschmann, M.M., 2013. Carbon-dioxide-rich silicate melt in the Earth's upper mantle. Nature 493, 211-215.

Davis, F.A., Humayun, M., Hirschmann, M.M., Cooper, R.S., 2013. Experimentally determined mineral/melt partitioning of first-row transition elements (FRTE) during partial melting of peridotite at 3 Gpa. Geochim. Cosmochim. Acta 104, 232-260.

Deines, P., 2002. The carbon isotope geochemistry of mantle xenoliths. Earth-Sci. Rev. 58, 247-278.

Herzog, G.F., Moynier, F., Albarede, F., Berezhnoy, A.A., 2009. Isotopic and elemental abundances of copper and zinc in lunar samples, Zagami, Pele's hairs, and a terrestrial basalt. Geochim. Cosmochim. Acta 73, 5884-5904.

Huang, J., Li, S.-G., Xiao, Y., Ke, S., Li, W.-Y., Tian, Y., 2015. Origin of low δ^{26}Mg Cenozoic basalts from South China Block and their geodynamic implications. Geochim. Cosmochim. Acta 164, 298-317.

Huang, J., Xiao, Y., in press. Mg-Sr isotopes of low δ^{26}Mg basalts tracing the recycled carbonate species: implication for the initial melting depth of the carbonated mantle in eastern China. Int. Geol. Rev. http://dx.doi.org/10.1080/00206814. 2016.1157709.

Huang, S.C., Farkaš, J., Jacobsen, S.B., 2011. Stable calcium isotopic compositions of Hawaiian shield lavas: evidence for recycling of ancient marine carbonates into the mantle. Geochim. Cosmochim. Acta 75, 4987-4997.

John, S.G., Rouxel, O.J., Craddock, P.R., Engwall, A.M., Boyle, E.A., 2008. Zinc stable isotopes in seafloor hydrothermal vent fluids and chimneys. Earth Planet. Sci. Lett. 269, 17-28.

Johnston, F.K.B., Turchyn, A.V., Edmonds, M., 2011. Decarbonation efficiency in subduction zones: implications for warm Cretaceous climates. Earth Planet. Sci. Lett. 303 (1-2), 143-152.

Kato, C., Moynier, F., Valdes, M.C., Dhaliwal, J.K., Day, J.M.D., 2015. Extensive volatile loss during formation and differentiation of the Moon. Nat. Commun. 6, 7617. http://dx.doi.org/10.1038/ncomms8617.

Kent, D.V., Muttoni, G., 2008. Equatorial convergence of India and early Cenozoic climate trends. Proc. Natl. Acad. Sci. USA 105 (42), 16065-16070.

Kerrick, D., Connolly, J., 2001. Metamorphic devolatilization of subducted marine sediments and the transport of volatiles into the Earth's mantle. Nature 411 (6835), 293-296.

Klein, E.M., Karsten, J.L., 1995. Ocean-ridge basalts with convergent-margin geochemical affinities from the Chile Ridge. Nature 374, 52-57.

Kogiso, T., Tatsumi, Y., Nakano, S., 1997. Trace element transport during dehydration processes in the subducted oceanic crust: 1. Experiments and implications for the origin of ocean island basalts. Earth Planet. Sci. Lett. 148, 193-205.

Laporte, D., Lambart, S., Schiano, P., Ottolini, L., 2014. Experimental derivation of nepheline syenite and phonolite liquids by partial melting of upper mantle peridotites. Earth Planet. Sci. Lett. 404, 319-331.

Le Roux, V., Lee, C.T.A., Turner, S.J., 2010. Zn/Fe systematics in mafic and ultramafic systems: implications for detecting major element heterogeneities in the Earth's mantle. Geochim. Cosmochim. Acta 74, 2779-2796.

Li, J.L., Klemd, Reiner, Gao, Jun, Meyer, Melanie, 2014. Compositional zoning in dolomite from lawsonite-bearing eclogite (SW Tianshan, China): evidence for prograde metamorphism during subduction of oceanic crust. Am. Mineral. 99, 206-217.

Li, S.G., Xiao, Y.L., Liou, D.L., Chen, Y.Z., Ge, N.J., Zhang, Z.Q., Sun, S.S., Cong, B.L., Zhang, R.Y., Hart, S.R., Wang, S.S., 1993. Collision of the North China and Yangtse blocks and formation of coesite-bearing eclogitestiming and processes. Chem. Geol. 109, 89-111.

Liu, S.-A., Li, D., Li, S., Teng, F.-Z., Ke, S., He, Y., Lu, Y., 2014. High-precision copper and iron isotope analysis of igneous rock standards by MC-ICP-MS. J. Anal. At. Spectrom. 29 (1), 122-133.

Liu, S.-A., Huang, J., Liu, J.G., Wörner, G., Yang, W., Tang, Y.J., Chen, Y., Tang, L.-M., Zheng, J.P., Li, S.-G., 2015. Copper isotopic composition of the silicate Earth. Earth Planet. Sci. Lett. 427, 95-103.

Lv, Y.-W., Liu, S.-A., Zhu, J.-M., Li, S.-G., 2016. Copper and zinc isotope fractionation during deposition and weathering of highly metalliferous black shales in central China. Chem. Geol. 422, 82-93.

Makishima, A., Nakamura, E., 2013. Low-blank chemistry for Zn stable isotope ratio determination using extraction chromatographic resin and double spike- multiple collector-ICP-MS. J. Anal. At. Spectrom. 28, 127-133.

Maréchal, C.N., Télouk, P., Albarède, F., 1999. Precise analysis of copper and zinc isotopic compositions by plasma-source mass spectrometry. Chem. Geol. 156 (1), 251-273.

Mason, T.F., Weiss, D.J., Horstwood, M., Parrish, R.R., Russell, S.S., Mullane, E., Coles, B.J., 2004. High-precision Cu and Zn isotope analysis by plasma source mass spectrometry Part 1. Spectral interferences and their correction. J. Anal. At. Spectrom. 19 (2), 209-217.

McDonough, W.F., Sun, S.S., 1995. The composition of the earth. Chem. Geol. 120, 223-253.

Moeller, K., Schoenberg, R., Pedersen, R.-B., Weiss, D., Dong, S., 2012. Calibration of the new certified reference materials ERM-AE633 and ERM-AE647 for copper and IRMM-3702 for zinc isotope amount ratio determinations. Geostand. Geoanal. Res. 36, 177-199.

Paniello, R.C., Day, J.M.D., Moynier, F., 2012. Zinc isotopic evidence for the origin of the Moon. Nature 490, 376-379.

Pichat, S., Douchet, C., Albarède, F., 2003. Zinc isotope variations in deep-sea carbonates from the eastern equatorial Pacific over the last 175 ka. Earth Planet. Sci. Lett. 210, 167-178.

Reeder, R.J., Lamble, G.M., Northrup, P.A., 1999. XAFS study of the coordination and local relaxation around Co^{2+}, Zn^{2+}, Pb^{2+}, and Ba^{2+} trace elements in calcite. Am. Mineral. 84, 1049-1060.

Sato, K., Katsura, T., 2001. Experimental investigation on dolomite dissociation into aragonite+magnesite up to 8.5 GPa. Earth Planet. Sci. Lett. 184, 529-534.

Sossi, P.A., Halverson, G.P., Nebel, O., Eggins, S.M., 2015. Combined separation of Cu, Fe and Zn from rock matrices and improved analytical protocols for stable isotope determination. Geostand. Geoanal. Res. 39, 129-149.

Tang, Y.-J., Zhang, H.-F., Ying, J.-F., 2006. Asthenosphere-lithospheric mantle interaction in an extensional regime: implication from the geochemistry of Cenozoic basalts from Taihang Mountains, North China Craton. Chem. Geol. 233, 309-327.

Telus, M., Dauphas, N., Moynier, F., Tissot, F.L.H., Teng, F.-Z., Nabelek, P.I., Craddock, P.R., Groat, L.A., 2012. Iron, zinc, magnesium and uranium isotopic fractionation during continental crust differentiation: the tale from migmatites, granitoids, and pegmatites. Geochim. Cosmochim. Acta 97, 247-265.

Toutain, J.P., Sonke, J., Munoz, M., Nonell, A., Polve, M., Viers, J., Freydier, R., Sortino, F., Joron, J.L., Sumarti, S., 2008. Evidence for Zn isotopic fractionation at Merapi volcano. Chem. Geol. 253, 74-82.

Turekian, K.K., Wedepohl, K.H., 1961. Distribution of the elements in some major units of the earth's crust. Geol. Soc. Am. Bull. 72 (2), 175-192.

Yang, W., Li, S.G., 2008. Geochronology and geochemistry of the Mesozoic volcanic rocks in Western Liaoning: implications for lithospheric thinning of the North China Craton. Lithos 102, 88-117.

Yang, W., Teng, F.-Z., Zhang, H.-F., Li, S.-G., 2012. Magnesium isotopic systematics of continental basalts from the North China craton: implications for tracing subducted carbonate in the mantle. Chem. Geol. 328, 185-194.

Zeng, G., Chen, L.-H., Xu, X.-S., Jiang, S.-Y., Hofmann, A.W., 2010. Carbonated mantle sources for Cenozoic intra-plate alkaline basalts in Shandong, North China. Chem. Geol. 273, 35-45.

Zhang, H.F., Sun, M., Zhou, X.H., Fan, W.M., Zhai, M.G., Ying, J.F., 2002. Mesozoic lithosphere destruction beneath the North China Craton: evidence from major-, trace-element and Sr-Nd-Pb isotope studies of Fangcheng basalts. Contrib. Mineral. Petrol. 144, 241-253.

Zhao, D.P., Yu, S., Ohtanic, E., 2011. East Asia: seismotectonics, magmatism and mantle dynamics. J. Asian Earth Sci. 40, 689-709.

Zhao, G.C., Wilde, S.A., Cawood, P.A., Sun, M., 2001. Archean blocks and their boundaries in the North China Craton: lithological, geochemical, structural and P-T path constraints and tectonic evolution. Precambrian Res. 107, 45-73.

Zhao, Y., Vance, D., Abouchami, W., De Baar, H.J.W., 2014. Biogeochemical cycling of zinc and its isotopes in the Southern Ocean. Geochim. Cosmochim. Acta 125, 653-672.

Zhou, J.X., Huang, Z.L., Zhou, M.F., Zhu, X.K., Muchez, P., 2014. Zinc, sulfur and lead isotope variations in carbonate-hosted Pb-Zn sulfide deposits, southwest China. Ore Geol. Rev. 58, 41-54.

Zhou, X.M., Armstrong, R.L., 1982. Cenozoic volcanic rocks of eastern China—secular and geographic trends in chemistry and strontium isotopic composition. Earth Planet. Sci. Lett. 58, 301-329.

Origin of low δ^{26}Mg basalts with EM-I component: Evidence for interaction between enriched lithosphere and carbonated asthenosphere*

Heng-Ci Tian[1, 2], Wei Yang[1], Shu-Guang Li[3], Shan Ke[3], Zhu-Yin Chu[4]

1. Key Laboratory of Earth and Planetary Physics, Institute of Geology and Geophysics, Chinese Academy of Sciences, Beijing 100029, China
2. University of Chinese Academy of Sciences, Beijing 100049, China
3. State Key Laboratory of Geological Processes and Mineral Resources, China University of Geosciences, Beijing 100083, China
4. State Key Laboratory of Lithospheric Evolution, Institute of Geology and Geophysics, Chinese Academy of Sciences, Beijing 100029, China

亮点介绍：本文对具有富集 EM-I 特征的中国东北五大连池和二克山玄武岩进行 Mg 同位素分析，发现这些玄武岩与中国东部其他地区新生代具有亏损地幔特征的玄武岩一样，也具有显著轻的 Mg 同位素组成。因而它的源区很可能是西太平洋板片携带再循环碳酸盐进入地幔过渡带熔融产生的富含碳酸盐熔体交代上覆对流软流圈地幔的结果；然而该玄武岩的 EM-I 特征又要求其源区曾经历过古老的富轻稀土流体交代作用。依据该玄武岩的高 Ba/Th 和低 Ce/Pb 特征，本文论证了这一古老的 EM-I 富集地幔应位于岩石圈。因此，该玄武岩是碳酸盐化软流圈地幔来源的具有低 δ^{26}Mg 的低 SiO_2 熔体与具有 EM-I 特征的富集岩石圈地幔相互作用的产物。

Abstract This study presents stable Mg isotopic data for Cenozoic potassic basalts from Wudalianchi and Erkeshan in northeastern China to determine the interactions between upwelling carbonated asthenosphere and enriched lithospheric mantle. Although the Wudalianchi and Erkeshan basalts have variable MgO contents of 4.45 to 9.47 wt.%, they exhibit a homogeneous Mg isotopic composition with δ^{26}Mg values ranging from −0.57 to −0.46‰ and averaging −0.51±0.06‰ (2SD, n=18). This Mg isotopic composition is lighter than that of the average mantle (δ^{26}Mg=−0.25±0.07‰) but similar to late Cretaceous (<110 Ma) and Cenozoic basalts from the North China Craton and the South China Block (δ^{26}Mg=−0.60 to −0.35‰). The high CaO/Al_2O_3 and Ba/Rb, and low Hf/Hf* ratios of the Wudalianchi and Erkeshan basalts are typical characteristics of carbonatitic metasomatism, suggesting that the light Mg isotopic composition could derive from involvement of recycled sedimentary carbonates in the mantle source. The high Dy/Er ratios (2.55 to 2.75) and excess of ^{230}Th (^{230}Th/^{238}U=1.24 to 1.33) suggest presence of garnet in a relatively deep mantle source. Additionally, the seismic tomographic observations show the existence of the stagnant Pacific slab in the mantle transition zone (410−660 km)

* 本文发表在：Geochimica et Cosmochimica Acta, 2016, 188: 93-105

under eastern China. This carbonated mantle source should be located in the asthenosphere. However, compared to other low δ^{26}Mg basalts from eastern China with MORB-like Sr-Nd-Pb isotopic compositions and OIB-like trace element features, the Wudalianchi and Erkeshan basalts exhibit EMI-like Sr-Nd-Pb isotopic compositions combined with high SiO_2, Ba, K, Pb and LREE contents, high Ba/Th and low Ce/Pb ratios. These geochemical features require a contribution from another mantle source, most likely an EM-I lithospheric mantle. Therefore, an asthenosphere-lithosphere interaction model is proposed for determine the origin of the Wudalianchi and Erkeshan basalts. The original melt was derived from partial melting of the carbonated asthenospheric mantle that was metasomatized by carbonate melt from the stagnant Pacific slab in the mantle transition zone. This ascending melt with a low δ^{26}Mg signature subsequently interacted with the EM-I lithospheric mantle. The interaction dissolved pyroxene and crystallized olivine, releasing LILE, REE, Sr and Pb from dissolved minerals into the melt. Simultaneously, Mg was inherited by newly formed olivine. Therefore, this interaction modified the trace elements and the Sr-Nd-Pb isotopic compositions of the melt toward EM-I geochemical features but preserved the light Mg isotope signatures.

1 Introduction

It is widely accepted that the heterogeneity of the mantle observed in the composition of mantle-derived magmas is largely attributable to subduction and recycling of continental or oceanic crust into the deep mantle (Cohen and O'Nions, 1982; White and Hofmann, 1982; Zindler and Hart, 1986). Many studies have focused on the interactions between recycled materials and depleted mantle to explain the origins of EM and HIMU components (e.g., Cohen and O'Nions, 1982; White and Hofmann, 1982; Menzies et al., 1983; Zindler and Hart, 1986; Kogiso et al., 1997; Jackson et al., 2007). However, little is known about the interactions between recycled materials and enriched mantle and the associated consequences, e.g., late recycled crustal material overprints mantle previously enriched by earlier recycling events of the crust.

Recently, the light Mg isotopic compositions (δ^{26}Mg=−0.60 to −0.35‰) of the <110 Ma basalts from the North China Craton (NCC) and the South China Block (SCB) have been attributed to the interaction between mantle peridotites and isotopically light carbonate melts derived from the subducted Pacific slab (Yang et al., 2012; Huang et al., 2015). This relationship is supported by seismic tomographic observations that suggest the stagnant Pacific slab existed in the mantle transition zone under eastern China (Huang and Zhao, 2006; Zhao et al., 2011). If this explanation is correct, the Cenozoic basalts from Northeast (NE) China should also contain light Mg isotopic compositions. The basalts from NE China have EM-I Sr-Nd-Pb isotopic features that are distinct from the NCC and SCB basalts, indicating the contribution of an enriched mantle source (Choi et al., 2006; Chu et al., 2013). Therefore, Mg isotopic analyses of the basalts from NE China will help to determine the interaction between recycled sedimentary carbonates and an enriched mantle.

Here, we investigate the Mg isotopic compositions of Cenozoic potassic basalts with EM-I geochemical features from Wudalianchi and Erkeshan in NE China. Our results show that these basalts have homogeneous and light Mg isotopic compositions, with δ^{26}Mg values varying from −0.57 to −0.46‰. We argue that such a light Mg isotopic composition can only be explained by the involvement of recycled sedimentary carbonates in the asthenospheric mantle source. The melt originated from the carbonated mantle and subsequently interacted with an EM-I lithospheric mantle. Thus, this study provides a specific example of carbonated asthenosphere interaction with enriched lithospheric mantle.

2 Geological background and sample description

Cenozoic continental basalts are widespread in eastern China (Fig. 1). A considerable number of studies on these basalts have been conducted since the 1980s (e.g., Zhou and Armstrong, 1982; Peng et al., 1986; Zhou et al., 1988; Song et al., 1990; Zhi et al., 1990; Liu et al., 1994; Zou et al., 2000; Zhang et al., 2003, 2009; Xu et al., 2005; Tang et al., 2006; Yang and Li, 2008; Chen et al., 2009; Zeng et al., 2010, 2011; Wang et al., 2011; Yang et al.. 2012; Huang et al., 2015). The Cenozoic basalts in NE China, including Wudalianchi, Erkeshan, Keluo and Xiaogulihe volcanic fields, are located on the northern margin of the Songliao graben, encompassing an area of > 1400 km^2 (Fig. 1). These basalts are highly potassic with EM-I geochemical features (Basu et al., 1991; Chen et al., 2007; Zhang et al., 1991, 1995; Zou et al., 2003; Kuritani et al., 2013; Sun et al., 2014, 2015), distinguishing them from other Cenozoic continental basalts in eastern China (Zhou and Armstrong, 1982; Song et al., 1990; Zhi et al., 1990; Basu et al., 1991; Fan and Hooper, 1991; Liu et al., 1994; Chen et al., 2007). They represent an enriched mantle endmember in eastern China (Basu et al., 1991; Zhang et al., 1991, 1995; Zou et al., 2003; Chen et al., 2007; Kuritani et al., 2013; Sun et al., 2014, 2015).

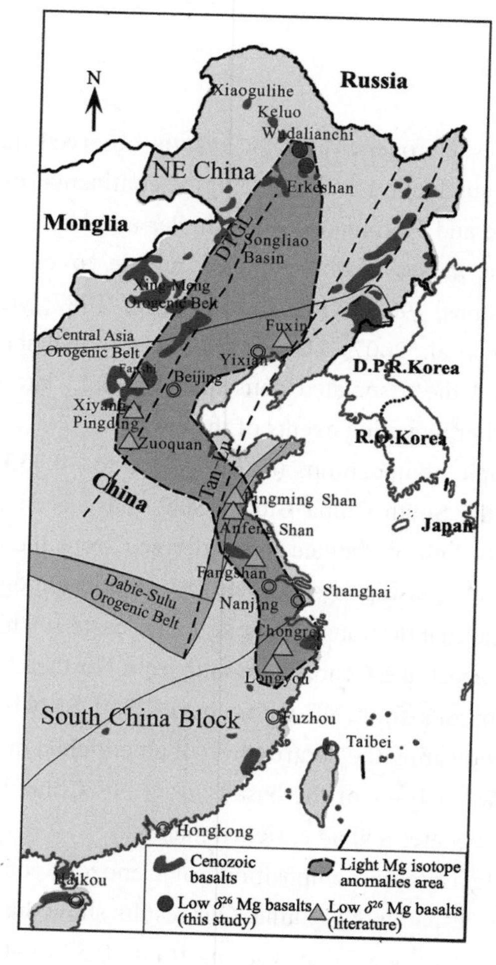

Fig. 1 Simplified geological map of eastern China showing samples locations (modified from Yang et al., 2009). Wudalianchi and Erkeshan are denoted by red circles. The yellow triangles represent the locations of the previously reported low δ^{26}Mg basalts from the North China Craton and the South China Block (Yang et al., 2012; Huang et al., 2015). The dark grey area is the distribution of the low δ^{26}Mg basalts, indicating a large-scale light Mg isotope anomaly in the mantle beneath eastern China. (For interpretation of the references to colour in this figure legend, the reader is referred to the web version of this article.)

The samples studied here were collected from lava flows related to the eruptions (0.25 Ma to 1721 AD, Zhang et al., 1995) from the Wudalianchi volcanic area, and lava flows related to the Erkeshan eruption (0.56 Ma, Liu et al., 2001) from the Kedong volcanic area (Fig. 1). Their petrology, major and trace element compositions, Re-Os and Sr-Nd-Hf isotopic compositions have been reported in a previous study (Chu et al., 2013). Only a brief description is given below.

The Wudalianchi and Erkeshan basalts are characterized by high K_2O contents of 4.50% to 6.09%, with K_2O/Na_2O ratios of 1.09 to 1.67 (Chu et al., 2013). They have variable MgO (4.45 to 9.47 wt.%), SiO_2 (49.02 to 53.54 wt.%), Al_2O_3 (12.15 to 14.39 wt.%), CaO (5.09 to 7.44 wt.%), total FeO (8.19 to 9.64 wt.%) and TiO_2 (2.26 to 2.77 wt.%) contents. They have enriched LREEs and exhibit strong positive Rb, Ba, K and Pb anomalies based on a spider diagram (Fig. 2). Accordingly, they display lower Ce/Pb (9.4 to 13.5) and higher Ba/Th (198.2 to 316.1) ratios than those of basalts from the NCC and SCB (Fig. 3a, b), reflecting input of sediment-derived fluid (Sakuyama et al., 2013). The high Nb/U ratios (38.9 to 54.2) of the Wudalianchi and Erkeshan basalts resemble those of basalts from the NCC and SCB (Fig. 3a), suggesting that their low Ce/Pb ratios are not caused by crustal contamination. The geochemistry of the Wudalianchi and Erkeshan basalts suggests contradictory sources. The high Dy/Er ratios (2.55 to 2.75) and significant excess of ^{230}Th ($^{230}Th/^{238}U$=1.24 to 1.33) suggest the presence of garnet in a relatively deep mantle source (Zou et al., 2003; Chen et al., 2007). Based on thermodynamic calculations, the temperatures of the Wudalianchi magmas were estimated to have been ~1250 °C, higher than the projected maximum temperature of the lithospheric mantle beneath the Wudalianchi volcanic field (Kuritani et al., 2013). However, high SiO_2 contents (49.02 to 53.54 wt.%) of the Wudalianchi basalts require a shallow melting depth (Chen et al., 2007).

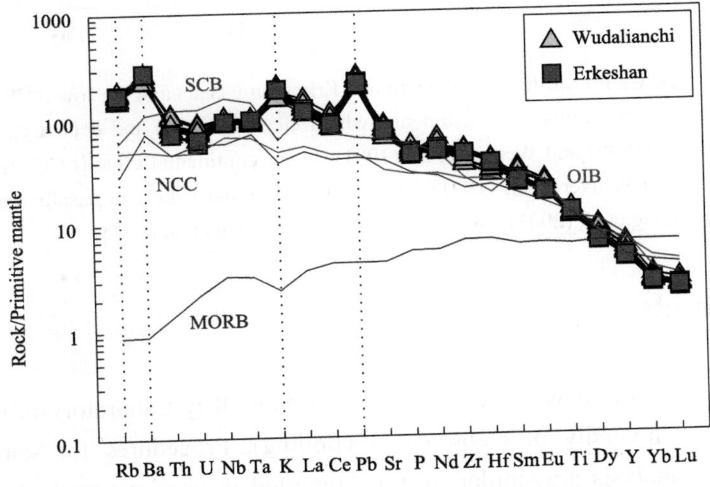

Fig. 2 Primitive mantle-normalized trace element patterns for the Wudalianchi and Erkeshan basalts and the low $\delta^{26}Mg$ basalts from NCC and SCB. The Wudalianchi and Erkeshan basalts show strong positive Rb, Ba, K and Pb anomalies compared to the trace element patterns of the NCC (North China Craton) and SCB (South China Block) basalts. Primitive mantle normalized values, MORB (mid-ocean ridge basalts) and OIB (ocean island basalts) data are from Sun and McDonough (1989). The NCC and SCB basalt data are from Tang et al. (2006) and Huang et al. (2015), respectively.

The EM-I Sr-Nd-Pb isotopic compositions of the Wudalianchi and Erkeshan basalts with high $^{87}Sr/^{86}Sr$ of 0.7051 to 0.7057, low $^{143}Nd/^{144}Nd$ of 0.5123 to 0.5125 and unradiogenic Pb ($^{206}Pb/^{204}Pb$=16.7 to 17.1) isotope compositions have been well documented (Basu et al., 1991; Zhang et al., 1991; Zou et al., 2003; Chu et al. 2013). However, their mantle source is still a subject of debate (Zhang et al., 1995, 1998, 2000; Zou et al., 2003; Choi et al., 2006; Chen et al., 2007; Chu et al., 2013; Kuritani et al., 2013). Several models may explain the mantle source: (1) subcontinental lithospheric mantle (SCLM) enriched by past metasomatism during the Proterozoic

(Zhang et al., 1995, 1998, 2000), likely phlogopite-bearing garnet peridotites (Zou et al., 2003; Sun et al., 2015); (2) shallow asthenosphere containing fragments of delaminated SCLM due to the Sr–Nd–Pb–Hf isotopic characteristics resembling a DMM-EM1 array (Choi et al., 2006); (3) mixed FOZO (Focal Zone; Hart et al., 1992) and LoMu (low U/Pb; Douglass et al., 1999) mantle sources (Chen et al., 2007); (4) mantle transition zone metasomatized by K-rich sediment-derived fluids approximately 1.5 Ga ago, due to the stagnation of an ancient slab (Kuritani et al., 2013) and (5) SCLM metasomatized by delaminated lower continental crust based on low Ce/Pb ratios and unradiogenic Pb isotopic compositions (Chu et al., 2013).

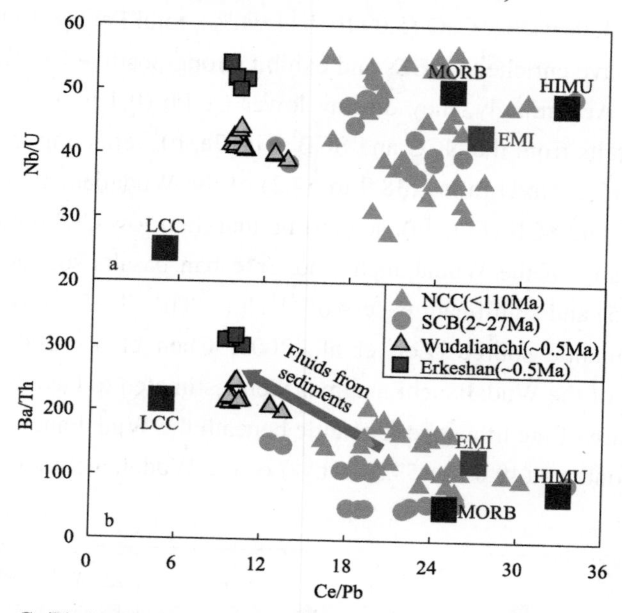

Fig. 3　(a) Nb/U and (b) Ba/Th vs. Ce/Pb for the Wudalianchi and Erkeshan basalts and the low δ^{26}Mg basalts from NCC and SCB. The arrow represents metasomatism by fluids from sediments, which elevates Ba/Th and decreases Ce/Pb. Database: MORB and HIMU from Sun and McDonough (1989) and Weaver et al. (1991), lower continental crust (LCC) from Rudnick and Gao (2003), EM-I from Leroex et al. (1990) and Weaver et al. (1991). The Wudalianchi and Erkeshan basalts are from Chu et al. (2013). The NCC and SCB basalts are from Zhang et al. (2003), Tang et al. (2006) and Huang et al. (2015).

3　Analytical methods

Stable Mg isotopic compositions were determined at the State Key Laboratory of Geological Processes and Mineral Resources, China University of Geosciences (Beijing). Procedures for sample dissolution, column chemistry and instrumental analysis are similar to those reported in previous studies (Teng et al., 2007, 2010; Huang et al., 2015). A brief description is given below.

Sample dissolution and column chemistry were performed in an ultra-clean laboratory. Approximately 1–5 mg sample powders were dissolved in Savillex screw-top beakers with HF-HNO_3 (~3:1) mixed acids followed by HCl-HNO_3 (~3:1) mixed acids. Then, the samples were refluxed with concentrated HNO_3. Finally, all dried residues were dissolved in 1M HNO_3 prior to ion exchange column chemistry. Magnesium was separated by cation exchange chromatography with Bio-Rad 200–400 mesh AG50W-X8 resin following the established procedures (Teng et al., 2007, 2010; Yang et al., 2009). The prepared samples were eluted using 1M HNO_3 from the column. The same column procedure was performed twice to obtain pure Mg solutions for mass spectrometry. The eluted solutions were evaporated to dryness and dissolved in 3% HNO_3 prior to instrumental analysis. The concentrations of cations (e.g., Na, Al, K, Ti and Fe) in the solutions were measured by ICP-MS, with cation/Mg (mass/mass) ratios of <0.02, which is negligible for the Mg isotope analyses. The BHVO-2, BCR-2 and GSR-3 standard materials and a blank were prepared and analysed following the same procedure to evaluate the precision, accuracy and blank.

Magnesium isotopic compositions were measured by the sample-standard bracketing method to correct for instrumental mass bias and drift using a Thermo-Finnigan Neptune plus MC-ICP-MS. Solutions containing 400 ppb Mg were introduced into the plasma (~100 μL/min) via a standard H-skimmer cone and an ESI PFA MicroFlow nebulizer with a quartz Scott-type spray chamber. Stable Mg isotopes were analysed in low-resolution mode, with ^{26}Mg, ^{25}Mg and ^{24}Mg measured simultaneously in separate Faraday cups (H3, C and L3). The duration of each sample run was approximately 6–7 min. Repeated analyses of synthetic solution, mineral and rock standards are indicative of a long-term reproducibility of ≤±0.10‰ for δ^{26}Mg values (2SD, Table 1). The Mg content of the blank was <10 ng, which was negligible for the Mg isotope analyses. The results are reported in standard δ-notation relative to DSM3, where $\delta^{25,26}\text{Mg}=[(^{25,26}\text{Mg}/^{24}\text{Mg})_{\text{sample}}/(^{25,26}\text{Mg}/^{24}\text{Mg})_{\text{DSM3}}-1] \times 1000$ (Galy et al., 2003). The standard materials were also analysed, with δ^{26}Mg=–0.20±0.09‰ for BCR-2, δ^{26}Mg=–0.27±0.01‰ for BHVO-2 and δ^{26}Mg=–0.48±0.03‰ for GSR-3. These values are consistent with previously published data (–0.36 to –0.16‰ for BCR-2 and –0.32 to –0.16‰ for BHVO-2; Teng et al., 2007, 2015; Huang et al., 2009, 2011, 2015; An et al., 2014).

Table 1 Mg isotopic compositions of volcanic rocks from Wudalianchi and Erkeshan, Northeast China

Sample[a]	Location[a]	MgO[a]	TiO_2[a]	CIA[a]	δ^{26}Mg[b]	2SD[c]	δ^{25}Mg[b]	2SD[c]	Δ^{25}Mg′[d]
YQ1	Wudalianchi	6.17	2.45	44.4	–0.52	0.08	–0.27	0.04	0.00
YQ3	Wudalianchi	6.18	2.45	44.5	–0.49	0.05	–0.24	0.02	0.02
YQ4	Wudalianchi	9.47	2.39	41.0	–0.48	0.04	–0.24	0.06	0.01
YQ6	Wudalianchi	4.45	2.72	43.8	–0.51	0.03	–0.27	0.04	0.00
YQ7	Wudalianchi	7.40	2.35	41.9	–0.54	0.08	–0.28	0.07	0.01
YQ8	Wudalianchi	7.69	2.34	41.4	–0.48	0.10	–0.25	0.05	0.00
LHS1	Wudalianchi	7.45	2.30	43.7	–0.49	0.06	–0.26	0.04	–0.01
LHS3	Wudalianchi	7.46	2.29	43.8	–0.52	0.05	–0.25	0.03	0.02
LHS4	Wudalianchi	6.21	2.77	44.1	–0.52	0.05	–0.26	0.03	0.01
LHS6	Wudalianchi	5.96	2.34	44.9	–0.47	0.10	–0.24	0.06	0.00
HSS2	Wudalianchi	5.93	2.42	44.8	–0.49	0.03	–0.26	0.05	0.00
HSS5	Wudalianchi	5.99	2.42	45.0	–0.46	0.07	–0.26	0.08	–0.02
HSS6	Wudalianchi	8.10	2.26	41.0	–0.48	0.07	–0.25	0.04	0.00
HSS8	Wudalianchi	6.01	2.32	44.9	–0.48	0.07	–0.23	0.05	0.02
KD1	Erkeshan	6.87	2.57	44.5	–0.55	0.08	–0.28	0.05	0.01
KD3	Erkeshan	6.88	2.60	44.4	–0.57	0.10	–0.28	0.05	0.02
KD4	Erkeshan	6.83	2.61	44.5	–0.55	0.10	–0.29	0.05	0.00
KD5	Erkeshan	6.90	2.59	44.6	–0.54	0.10	–0.27	0.04	0.01
Standards									
BCR-2					–0.20	0.09	–0.12	0.06	–0.02
BHVO-2					–0.27	0.01	–0.12	0.03	0.03
GSR-3					–0.48	0.03	–0.24	0.02	0.01

[a] The samples are from Wudalianchi and Erkeshan (Chu et al., 2013). CIA=Chemical Indexes of Alteration, defined as=molecular ratios of $Al_2O_3/(Al_2O_3 + CaO + Na_2O + K_2O) \times 100$ Nesbitt and Young (1982).

[b] $\delta^{25,26}\text{Mg}=\{(^{25,26}\text{Mg}/^{24}\text{Mg})_{\text{sample}}/(^{25,26}\text{Mg}/^{24}\text{Mg})_{\text{DSM3}} - 1) \times 1000$, DSM3 is a Mg solution made from pure Mg metal (Galy et al., 2003).

[c] 2SD=two times the standard deviation of the population of 4 repeat measurements of a sample solution.

[d] $\Delta^{25}\text{Mg}'=\delta^{25}\text{Mg}' - 0.521 \times \delta^{26}\text{Mg}'$, where $\delta^{25,26}\text{Mg}'=1000 \times \ln[(\delta^{25,26}\text{Mg} + 1000)/1000]$ Young and Galy (2004).

4 Results

Stable Mg isotopic compositions of the Wudalianchi and Erkeshan basalts and standard reference materials are listed in Table 1. The $\Delta^{25}Mg'$ ($\Delta^{25}Mg'=\delta^{25}Mg'-0.521\times\delta^{26}Mg'$, where $\delta^{25,26}Mg'=1000\times\ln[(\delta^{25,26}Mg+1000)/1000]$; Young and Galy, 2004) values for the Wudalianchi and Erkeshan basalts are within 0±0.02‰ (Table 1).

Overall, the Wudalianchi and Erkeshan basalts have a homogeneous Mg isotopic composition (Fig. 4a and Table 1), with $\delta^{25}Mg$ values ranging from −0.29 to −0.23‰ and $\delta^{26}Mg$ values ranging from −0.57 to −0.46‰. These Mg isotopic compositions are similar to those of <110 Ma basalts from NCC ($\delta^{26}Mg$=−0.60 to −0.42‰; Yang et al., 2012) and Cenozoic basalts from SCB ($\delta^{26}Mg$=−0.60 to −0.35‰; Huang et al., 2015), but lighter than that of average mantle ($\delta^{26}Mg$=−0.25±0.07‰; Teng et al., 2010) and > 120 Ma basaltic rocks from NCC ($\delta^{26}Mg$=−0.27±0.05‰, n=5, 2SD; Yang et al., 2012) (Fig. 4a).

The $\delta^{26}Mg$ values of the Wudalianchi and Erkeshan basalts are invariant with MgO and TiO_2 contents, in accordance with those of the low $\delta^{26}Mg$ basalts from the NCC and SCB (Fig. 4). They have low Hf/Hf* (0.62 to 1.07; $Hf/Hf^*=Hf_N/(Sm_N\times Nd_N)^{0.5}$), high Ba/Rb (14.4 to 19.9) and CaO/Al_2O_3 (0.36 to 0.57) ratios compared to MORB (Hofmann, 1988), but similar to the low $\delta^{26}Mg$ basalts from the NCC and SCB (Fig. 5).

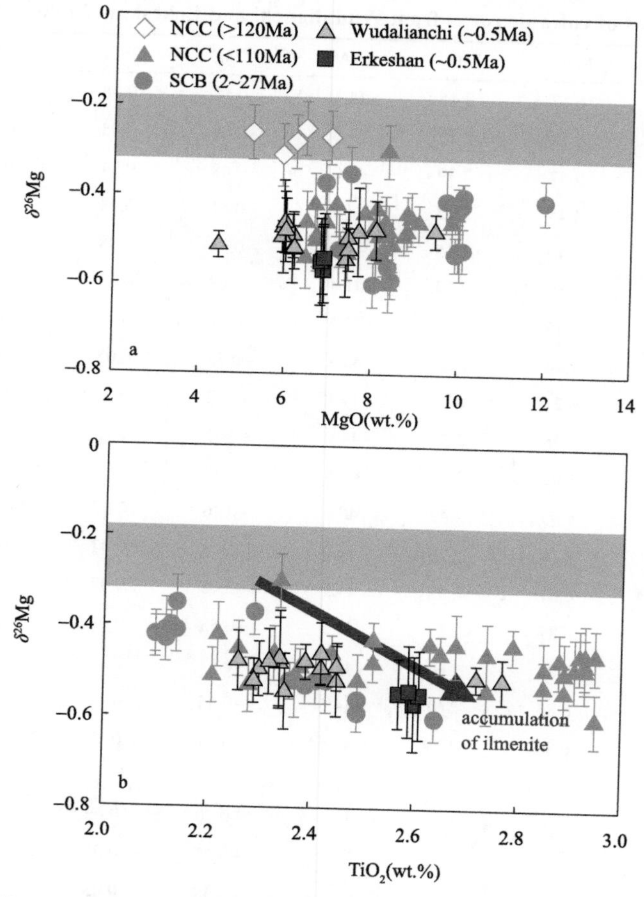

Fig. 4 $\delta^{26}Mg$ vs. (a) MgO (wt.%) and (b) TiO_2 (wt.%) for the Wudalianchi and Erkeshan basalts and the low $\delta^{26}Mg$ basalts from NCC and SCB. The arrow represents the trend for accumulation of ilmenite. Database: the NCC basalts from Yang et al. (2012), the SCB basalts from Huang et al. (2015), the Wudalianchi and Erkeshan basalts from Table 1 and Chu et al. (2013). The grey bar represents the average mantle ($\delta^{26}Mg$=−0.25±0.07‰; Teng et al., 2010). Error bars represent 2SD uncertainties.

5 Discussion

5.1 Characteristics of the primary melt

The chemical index of alteration (CIA) values of the Wudalianchi and Erkeshan basalts vary from 41.0 to 45.0 (Table 1), which fall within the fresh basalt range of 30 to 45 (Nesbitt and Young, 1982). These basalts exhibit a relatively large Mg# variation of 55.3 to 70.6, which may reflect compositional evolution from the primary melts (Langmuir et al., 1977). To discuss their source characteristics, the geochemical features of the primary melt should be first evaluated.

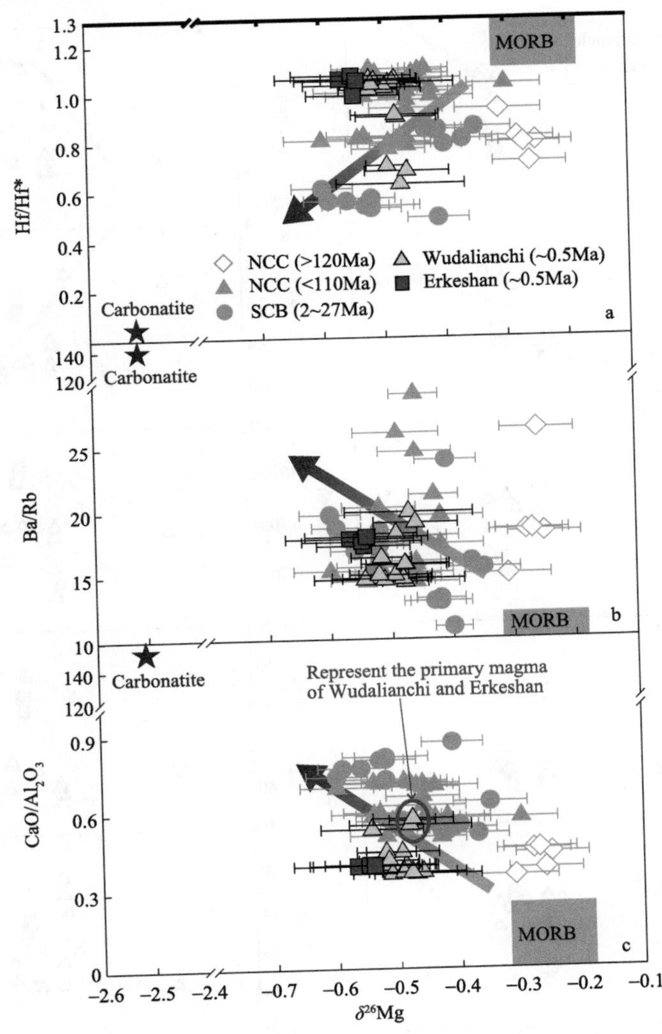

Fig. 5 (a) Hf/Hf*, (b) Ba/Rb and (c) CaO/Al_2O_3 vs. δ^{26}Mg for the Wudalianchi and Erkeshan basalts with the low δ^{26}Mg basalts from NCC and SCB. The pink star represents the average ratios of CaO/Al_2O_3 (152.4), Hf/Hf* (0.01) and Ba/Rb (140.5) for magnesio- and calico-carbonatites (Hoernle et al., 2002; Bizimis et al., 2003). The arrows represent the involvement of recycled sedimentary carbonate in the mantle source. The NCC and SCB basalt data are from Yang et al. (2012) and Huang et al. (2015), respectively. MORB data are from Hofmann (1988). The δ^{26}Mg values of MORB and carbonatite are from Teng et al. (2010) and Wang et al. (2014), respectively. Error bar represents 2SD uncertainties. $Hf/Hf^*=Hf_N/(Sm_N \times Nd_N)^{0.5}$, where subscript N means normalized to chondrites (Sun and McDonough, 1989). (For interpretation of the references to colour in this figure legend, the reader is referred to the web version of this article.)

The Wudalianchi and Erkeshan basalts show positive correlations between Mg# and both TFeO (Fig. 6a) and

CaO (Fig. 6b), indicating olivine and pyroxene fractional crystallization. Accordingly, CaO/Al_2O_3 ratios of these basalts display a large variation and show a positive correlation with Mg# (Fig. 6c). This correlation could be attributed to pyroxene fractional crystallization rather than plagioclase crystallization, because no obvious Eu anomaly is observed in these basalts (Fig. 2). In addition, the negative correlation between TiO_2 and Mg# (Fig. 6d) with no negative Ti anomaly (Fig. 2) indicates limited ilmenite fractional crystallization. The variations of major oxides (e.g., MgO, K_2O and CaO) can be modelled by fractional crystallization processes using MELTS software (Fig. 6a-d; Gualda and Ghiorso, 2015). The results indicate fractional crystallization of olivine, orthopyroxene and clinopyroxene (Fig. 6a-d). The major elements of the primary melt associated with the Wudalianchi and Erkeshan basalts are estimated from the samples with the highest MgO/Al_2O_3 ratios (YQ4 and HSS6).

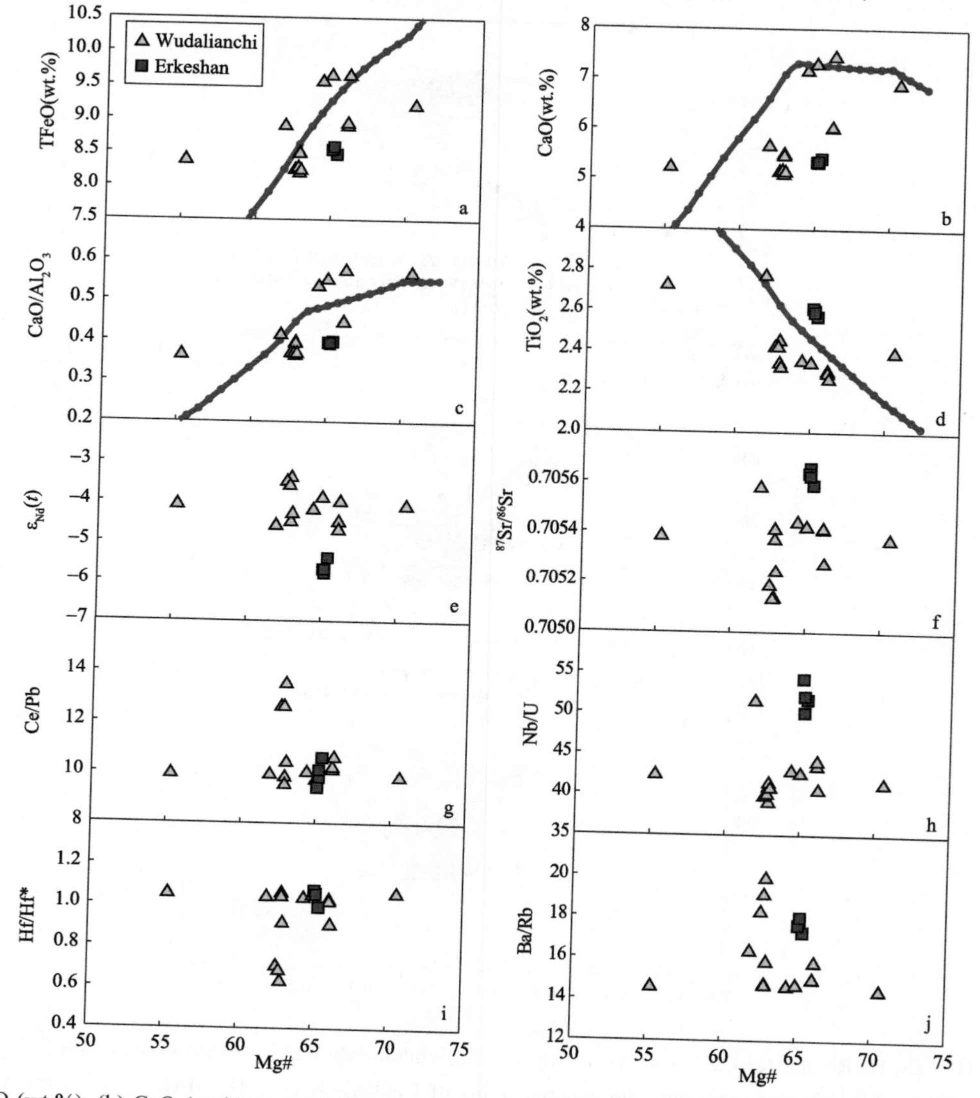

Fig. 6 (a) TFeO (wt.%), (b) CaO (wt.%), (c) CaO/Al_2O_3, (d) TiO_2 (wt.%), (e) $\varepsilon_{Nd}(t)$, (f) $^{87}Sr/^{86}Sr$, (g) Ce/Pb, (h) Nb/U, (i) Hf/Hf* and (j) Ba/Rb vs. Mg# for the Wudalianchi and Erkeshan basalts. The red curves represent the fractional crystallization (olivine and/or pyroxene) processes modelled using MELTS software. The parental magma compositions of SiO_2 (45.90 wt.%), TiO_2 (1.65 wt.%), Al_2O_3 (10.10 wt.%), Fe_2O_3 (3.30 wt.%), FeO (7.05 wt.%), MgO (18.50 wt.%), CaO (5.60 wt.%), Na_2O (3.00 wt.%), K_2O (3.70 wt.%) and P_2O_5 (1.08 wt.%) is based on a large number of simulations. The given conditions are ~15 kbar and $\log f_{O_2}$=NNO – 2.0 (NNO corresponding to Ni-NiO buffer). Red circles on the fractional crystallization curves indicate 10 °C temperature decrements. Major oxides, trace elements and Sr-Nd isotopic data of the Wudalianchi and Erkeshan basalts are from Chu et al. (2013). (For interpretation of the references to color in this figure legend, the reader is referred to the web version of this article.)

Although the Wudalianchi and Erkeshan basalts have undergone significant olivine and pyroxene fractional crystallization, their Ce/Pb, Nb/U, Hf/Hf* and Ba/Rb ratios and Sr-Nd isotopes do not vary with Mg# values (Fig. 6e-j), indicating limited influence on these values during fractional crystallization. The trace element ratios (e.g., Ce/Pb, Nb/U, Hf/Hf* and Ba/Rb) and Sr-Nd-Pb isotopic compositions of the primary melt are denoted with the variability of the complete dataset (see Table 2 for summarized data). For comparison, data for the low δ^{26}Mg basalts from the NCC and SCB are also listed in Table 2.

Table 2 Composition of primary melt of the Wudalianchi and Erkeshan basalts, and the NCC and SCB basalts

	Wudalianchi & Erkeshan[a]		NCC[a]		SCB[a]	
	Average[b]	1SD[c]	Average[b]	1SD[c]	Average[b]	1SD[c]
SiO_2	49.4	0.5	44.2	0.9	41.9	1.2
MgO	8.8	1.0	9.9	0.1	11.1	1.3
CaO	7.2	0.4	9.0	1.3	9.7	0.3
Al_2O_3	12.6	0.6	14.7	0.4	11.5	0.2
TiO_2	2.3	0.1	2.5	0.2	2.3	0.2
TFeO	9.4	0.3	12.3	0.6	12.8	0.8
Na_2O	4.0	0.3	3.1	0.1	4.3	0.4
K_2O	4.5	0.1	1.2	0.5	1.9	0.6
CaO/Al_2O_3	0.57		0.61		0.84	
Ba/Rb	14.4 – 19.9		9.5 – 67.8		9.6 – 23.9	
Hf/Hf*[d]	0.63 –1.07		0.77 – 1.10		0.49 – 0.86	
Ba/Th	198.2 – 316.1		75.2 – 203.5		48.0 – 152.4	
Ce/Pb	9.4 – 13.5		16.3 – 30.3		12.6 – 33.6	
Nb/U	38.9 – 54.2		30.3 – 55.6		36.5– 51.6	
Dy/Er	2.55 – 2.75		1.91 – 2.55		2.36 – 3.11	
$^{87}Sr/^{86}Sr$	0.7051 – 0.7057		0.7033 – 0.7054			
ε_{Nd}	–5.8 – –3.2		–1.7 – 5.1			
$^{206}Pb/^{204}Pb$	16.7 – 17.1		16.8 – 18.1			

[a] Wudalianchi and Erkeshan data from Basu et al., (1991), Zhang et al., (1991), Zou et al., (2003), Chu et al. (2013); NCC data from Zhang et al. (2003), Tang et al. (2006); SCB data from Huang et al. (2015).

[b] The major element composition of primary melt is estimated from the samples with the highest MgO/Al_2O_3 ratios; the trace element ratios denote the variability of the complete dataset.

[c] 1SD=one standard deviation.

[d] $Hf/Hf^*=Hf_N/(Sm_N \times Nd_N)^{0.5}$.

5.2 Origin of the light Mg isotopic composition

The δ^{26}Mg values of the Wudalianchi and Erkeshan basalts do not vary with MgO content (Fig. 4a), indicating limited Mg isotope fractionation during magma differentiation. This observation is consistent with previous studies of Mg isotope fractionation during magma differentiation (Teng et al., 2007, 2010; Liu et al., 2010) and of the low δ^{26}Mg basalts from the NCC and SCB (Fig. 4a; Yang et al., 2012; Huang et al., 2015). In addition, the high Nb/U (38.9 to 54.2) ratios of the Wudalianchi and Erkeshan basalts suggest limited input of crustal materials. Therefore, the light Mg isotopic composition of the Wudalianchi and Erkeshan basalts is inherited from the mantle source, which may result from two processes: accumulation of ilmenite and/or involvement of carbonates.

A recent study on lunar high-Ti basalts indicates that ilmenite could have low δ^{26}Mg values (<–0.61±0.03‰), and the accumulation of ilmenite may hence result in light Mg isotopic compositions (Sedaghatpour et al., 2013). However, this origin of low δ^{26}Mg basalts can be ruled out for the Wudalianchi and Erkeshan lavas. First, the Wudalianchi and Erkeshan basalts have lower TiO_2 (2.3 to 2.8 wt.%) contents than the lunar high-Ti (TiO_2 > 6.0 wt.%) basalts. Second, the trace elements Nb and Ta are compatible in ilmenite and the partition coefficients D_{Ta}/D_{Nb} are approximately ~1.3 between ilmenite and mafic melts (Dygert et al., 2013). Accumulation of ilmenite in the mantle source should generate a negative correlation between Nb/Ta ratios and TiO_2 (wt.%). However, the Wudalianchi and Erkeshan basalts display a positive correlation between Nb/Ta and TiO_2 (wt.%) (Fig. 7). Third, unlike the lunar high-Ti basalts, no correlation between δ^{26}Mg and TiO_2 is observed for the Wudalianchi and Erkeshan basalts (Fig. 4b). Thus, the accumulation of isotopically light ilmenite is unlikely to have affected the mantle source of the Wudalianchi and Erkeshan basalts.

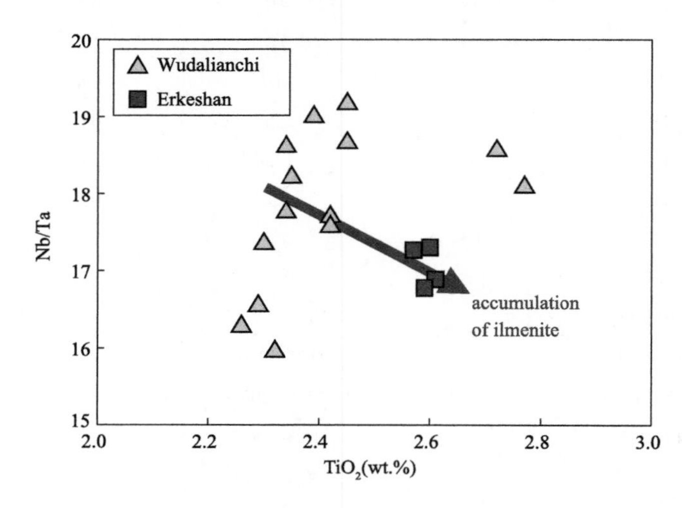

Fig. 7 Nb/Ta vs. TiO_2 (wt.%) for the Wudalianchi and Erkeshan basalts. The accumulation of isotopically light ilmenite should generate a negative correlation between Nb/Ta and TiO_2 (wt.%), which is not observed for the Wudalianchi and Erkeshan basalts.

The light Mg isotopic compositions (δ^{26}Mg=0.60 to –0.35‰) of the <110 Ma basalts from the North China Craton (NCC) and South China Block (SCB) have been attributed to the interaction between mantle peridotites and isotopically light carbonate melts derived from the subducted Pacific slab (Yang et al., 2012; Huang et al., 2015).The high CaO/Al_2O_3 (0.57) and Ba/Rb (14.4 to 19.9), and low Hf/Hf* ratios (0.63 to 1.07) of the Wudalianchi and Erkeshan basalts (Fig. 5) are similar to those of the low δ^{26}Mg basalts from the NCC and SCB (Table 2). These are typical characteristics of carbonatitic metasomatism because carbonatite is enriched in Ba, Th and La and depleted in Zr, Hf and Ti (Hoernle et al., 2002; Bizimis et al., 2003). A binary mixing calculation based on N-MORB and carbonate melts indicates that the involvement of ~10% carbonate melts can form the light Mg isotopic composition in the basalts from eastern China (Huang and Xiao, 2016). Therefore, the light Mg isotopic composition of the Wudalianchi and Erkeshan basalts could also derive from a mantle source involving sedimentary carbonates (Yang et al., 2012; Huang et al., 2015).

The distribution of low δ^{26}Mg basalts defines a large area with a light Mg isotope anomaly in the mantle of eastern China (Fig. 1), which coincides with the subducted Pacific slab in the mantle transition zone, as revealed by seismic tomography (Fukao et al., 1992; Huang and Zhao, 2006; Zhao et al., 2011). Based on experimental studies, the subducted carbonated oceanic crust underwent slab melting to produce carbonatitic melts in the mantle transition zone (410–660 km) because the majority of slabs will intersect the melting curve of carbonated oceanic crust at depths of approximately 300–700 km (Dasgupta and Hirshmann, 2010; Thomson et al., 2016). Therefore, the low δ^{26}Mg mantle source was formed by interaction between the mantle and the carbonate melt from the subducted oceanic crust above the mantle transition zone (Fig. 8). The carbonated asthenospheric mantle could partially melt at a depth of ~300 km (Dasgupta and Hirschmann, 2010) to generate the Wudalianchi and Erkeshan basalts, which is supported by their high Dy/Er ratios (2.55 to 2.75) and significant excess of ^{230}Th (^{230}Th/^{238}U=1.24 to 1.33) (Zou et al., 2003; Chen et al., 2007; Kuritani et al., 2013).

5.3 The contribution of an EM-I mantle source

Although the origin of the Wudalianchi and Erkeshan basalts is still a subject of debate, many studies suggested contributions of at least two components with different geochemical features (Zhang et al., 1995, 1998, 2000; Zou et al., 2003; Choi et al., 2006; Chen et al., 2007; Kuritani et al., 2013; Chu et al., 2013). The debate focuses on the characteristics of these two components and their origins.

As discussed above, one mantle source was the carbonated asthenospheric mantle, which is similar to that inferred for the NCC and SCB basalts. However, the Wudalianchi and Erkeshan basalts exhibit many distinct geochemical features from those of the NCC and SCB basalts (Table 2): e.g., high SiO_2 (49.4 wt.%) and K_2O (4.5 wt.%), low MgO (8.8 wt.%) and TFeO (9.4 wt.%) contents, high Ba/Th (198.2 to 316.1) and low Ce/Pb (9.4 to 13.5) ratios, and EM-I Sr-Nd-Pb isotopic compositions. Thus, the second mantle source should feature an EM-I signature.

Several sources for the EM-I signatures were previously proposed: the mantle transition zone metasomatized by K-rich sediment-derived fluids ~1.5 Ga ago (Kuritani et al., 2013), or subcontinental lithospheric mantle (SCLM) metasomatized by delaminated lower continental crust (Chu et al., 2013) or by sediment-derived fluid during the Proterozoic (Zhang et al., 1995, 1998, 2000; Zou et al., 2003).

Based on the high temperature of the Wudalianchi Magmas (~1250 °C), Kuritani et al. (2013) proposed that the EM-I signature of the Wudalianchi basalts may be derived from the mantle transition zone. However, the following evidence argues against this view. First, the continental basalts from NCC and SCB, which were also derived from the mantle transition zone (Yang et al., 2012; Huang et al., 2015), do not show EM-I signatures. This indicates that the mantle transition zone may not exhibit EM-I signatures. In addition, K-hollandite is required in the model of Kuritani et al. (2013) to explain the high K, Rb, Sr and Ba contents in the Wudalianchi basalts. However, this mineral is highly depleted in Nb, which will result in extremely low Nb/U ratios. This is not consistent with the high Nb/U (38.9 to 54.2) ratios observed in the Wudalianchi and Erkeshan basalts.

Chu et al. (2013) suggested the EM-I signatures originated from a subcontinental lithospheric mantle contaminated by delaminated lower continental crust because the low Ce/Pb ratios and unradiogenic Pb isotopic compositions are typical features of the lower continental crust. However, this process decreases not only Ce/Pb, but also Nb/U ratios, which is not supported by the high Nb/U (38.9 to 54.2) ratios of the Wudalianchi and Erkeshan basalts.

Alternatively, the EM-I signatures are most likely derived from a SCLM EM-I signatures are during the Proterozoic (Zhang et al., 1995, 1998, 2000; Zou et al., 2003). The enrichments of Rb, Ba and Pb and high Ba/Th ratios are typical characteristics of sediment-derived fluid inputs (Sakuyama et al., 2013), which must have occurred at least ~1.5 Ga ago to produce such low ^{206}Pb/^{204}Pb ratios (16.7 to 17.1) (Basu et al., 1991; Zhang et al., 1991; Zou et al., 2003; Kuritani et al., 2013).

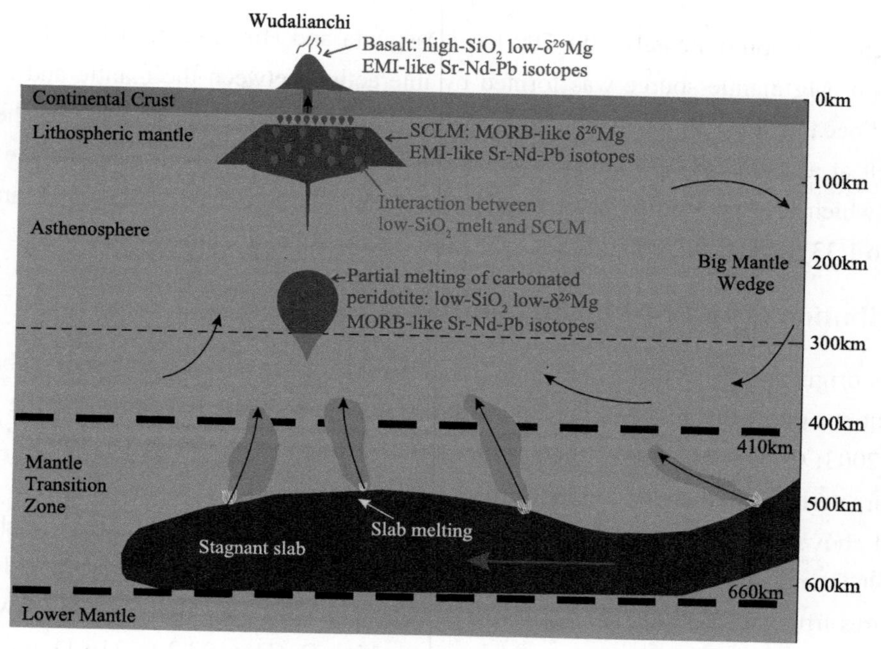

Fig. 8 Cartoon showing the origin of the Wudalianchi and Erkeshan basalts. The thickness of the stagnant slab is based on Zou et al. (2008), Kameyama and Nishioka, (2012) and Li et al. (2013). The original melt derived from partial melting of the carbonated asthenospheric mantle, interacted with the EM-I lithospheric mantle during melt ascent. This interaction dissolved pyroxenes and crystallized olivine, significantly modifying the trace element and Sr-Nd-Pb isotopic composition of the melt towards EM-I geochemical signatures. However, this interaction had little effect on Mg isotopes.

5.4 Interaction between carbonated asthenosphere and enriched lithosphere

Both carbonated asthenosphere and enriched lithosphere contributed to the Wudalianchi and Erkeshan basalts. Two stages are, therefore, required: partial melting of the carbonated peridotite to produce low δ^{26}Mg basaltic melt (Stage 1), and interaction of the basaltic melt with the enriched lithospheric mantle to obtain EM-I Sr-Nd-Pb isotopic compositions (Stage 2) (Fig. 9).

However, why did interaction with the EM-I SCLM significantly modify the trace element and Sr-Nd-Pb isotopic composition of the melt, but have little effect on the Mg isotopes? Two possible mechanism for interaction between melt derived from the asthenosphere and the EM-I SCLM should be considered.

(1) Mixing between melts derived from the asthenosphere and the EM-I SCLM. Given the limited fractionation of Mg isotopes during partial melting of the mantle (Teng et al., 2007, 2010), the basaltic melt from the EM-I SCLM should exhibit a mantle-like Mg isotopic composition. Thus, when the low δ^{26}Mg melt derived from the carbonated asthenosphere mixed with the melt from the EM-I SCLM, the Sr-Nd-Pb and Mg isotopic compositions were synchronously modified based on the relevant mixing proportions. However, although the Wudalianchi and Erkeshan basalts have typical EM-I Sr-Nd-Pb isotopic features, they still preserve light Mg isotopic compositions, similar to those of basalts from the NCC and SCB (Yang et al., 2012; Huang et al., 2015). This suggests that the interaction was more complex than simple melt mixing.

(2) A reaction between the low SiO_2 basaltic melt derived from the asthenosphere and the EM-I SCLM. Partial melting of carbonated peridotite produces low SiO_2 alkaline melts (e.g., nephelite, basanite, alkaline basalt) (Dasgupta et al., 2007). The primary melt of the Wudalianchi and Erkeshan basalts exhibits higher SiO_2 (49.4 wt.%) and K_2O (4.5 wt.%), and lower MgO (8.8 wt.%) and TFeO (9.4 wt.%) contents than those of low δ^{26}Mg basalts from the NCC and SCB (Table 2). This indicates that the subsequent interaction elevated SiO_2 and K_2O, and decreased MgO and TFeO concentrations of the melt. This inference is consistent with the interaction

between low SiO_2 melt and peridotite, which converts pyroxene to olivine, as reported in previous studies (Kelemen et al., 1992; Edwards and Malpas, 1996; Zhou et al., 1996, 2014; Suhr et al., 1998):

$$\text{Opx, Cpx + low SiO}_2\text{ melt} \rightarrow \text{Ol (solid) + high SiO}_2\text{ melt}$$

Such an interaction can explain all petrological and geochemical features of the Wudalianchi and Erkeshan basalts. First, the melt was derived from a deep source (carbonated asthenosphere) with garnet present, which is consistent with the high Dy/Er ratios (2.55 to 2.75), significant excess ^{230}Th ($^{230}Th/^{238}U$=1.24 to 1.33) (Zou et al., 2003; Chen et al., 2007), and the high temperature of the melt (~1250 °C; Kuritani et al., 2013). In addition, the high SiO_2 contents, previously considered to require a shallow melting depth (Chen et al., 2007), resulted from interaction with the SCLM. Moreover, the trace elements of the EM-I SCLM largely entered the melt during the interaction because the solid product was olivine. Consequently, the Ce/Pb and Ba/Th ratios were significantly modified, and the Sr-Nd-Pb-Hf isotopic characteristics resemble a DMM-EMI array (Choi et al., 2006). Finally, all Mg was transferred from Opx and Cpx into the newly formed olivine during the interaction. Thus, the Mg isotopic composition of the melt was not significantly affected.

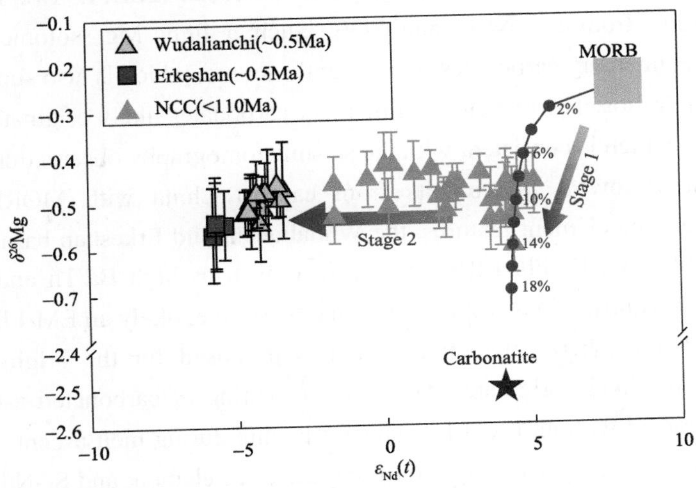

Fig. 9 $\delta^{26}Mg$ vs. $\varepsilon_{Nd}(t)$ for the Wudalianchi and Erkeshan basalts and the low $\delta^{26}Mg$ basalts from NCC. Stage 1: involvement of recycled sedimentary carbonate in the mantle source to produce low $\delta^{26}Mg$ basaltic melts. Stage 2: interaction of the melt with enriched lithospheric mantle to obtain melts with EM-I Sr-Nd-Pb isotopic compositions. The red curve represents binary mixing between depleted MORB (mid-ocean ridge basalts) and the carbonates, and circles on the binary modelling curves indicate 2% increments. Mg and Nd isotopic data for the NCC basalts are from Zhang et al. (2003), Tang et al. (2006) and Yang et al. (2012). Parameters for MORB ($\delta^{26}Mg$ of −0.25‰, MgO of 11.2 wt.%, $\varepsilon_{Nd}(t)$ of +8 and Nd of 7.3 ppm) are from Langmuir et al. (1977), Zindler and Hart (1986), Sun and McDonough (1989) and Teng et al. (2010), respectively; Carbonatite values ($\delta^{26}Mg$ of −2.508‰, MgO of 12.6 wt.%, $\varepsilon_{Nd}(t)$ of +4 and Nd of 500 ppm) are from Wang et al. (2014), Hoernle et al. (2002) and Bizimis et al. (2003), respectively. Error bar represents 2SD uncertainties. (For interpretation of the references to color in this figure legend, the reader is referred to the web version of this article.)

Therefore, the following model is proposed (Fig. 8): The subduction of the western Pacific plate subduction was at started in the Jurassic (Maruyama et al., 1997; Wu et al., 2007), whereby the slab may have reached the bottom of the transition zone and become stagnant slab in the Cretaceous (Niu, 2005). The convective upper mantle was hybridized by a partial melt from the recycled carbonates of the stagnant oceanic slab, forming carbonated peridotites above the mantle transition zone. Due to convection in the big mantle wedge that were proposed for the active tectonic and volcanic processes in northeast Asia based on multiscale tomographic images (Zhao et al., 2009), diapirs of carbonated peridotite ascended. When they reached the solidus of carbonated peridotite at a depth of approximately ~300 km (Dasgupta and Hirschmann, 2010), partial melting generated a low δ^{26}Mg melt with high Ce/Pb (~25) and Nb/U (~50), low Ba/Th (~53) and MORB-like Sr-Nd-Pb isotopic compositions. Because there is no EM-I lithospheric mantle beneath NCC and SCB, the geochemical features of this melt did not change in these regions during interaction between the asthenosphere-derived melt and lithosphere, producing the low δ^{26}Mg basalts with MORB-like Sr-Nd-Pb isotopic compositions of the NCC and SCB. In NE China, however, the melt interacted with the EM-I SCLM, forming the Wudalianchi and Erkeshan basalts with low δ^{26}Mg and EM-I Sr-Nd-Pb isotopic compositions (Fig. 8).

6 Conclusions

The stable Mg isotope compositions of the Wudalianchi and Erkeshan basalts from Northeast China were determined at high precision. A number of conclusions for the origin of the basalts follow from the new data.

(1) The Wudalianchi and Erkeshan basalts have homogeneous and light Mg isotopic compositions, with δ^{26}Mg ranging from −0.57‰ to −0.46‰ and averaging −0.51±0.06‰ (2SD, n=18). These values are similar to those of low δ^{26}Mg basalts from the NCC and SCB. Such a light Mg isotopic composition reflects the involvement of recycled sedimentary carbonates in the mantle source, which is also supported by high CaO/Al_2O_3 and Ba/Rb, and low Hf/Hf* ratios. The recycled sedimentary carbonates likely originate from the stagnant Pacific slab beneath eastern China, which is consistent with the seismic tomography observations.

(2) Compared to other low δ^{26}Mg basalts from eastern China with MORB-like Sr-Nd-Pb isotopic compositions and OIB-like trace element features, the Wudalianchi and Erkeshan basalts exhibit EM-I Sr-Nd-Pb isotope signatures, high SiO_2, Ba, K, Pb, LREE contents, as well as high Ba/Th and low Ce/Pb ratios. These geochemical features require contributions from another mantle source, likely an EM-I lithospheric mantle.

(3) An asthenosphere-lithosphere interaction model is proposed for the origin of the Wudalianchi and Erkeshan basalts, which most likely originated from partial melting of carbonated asthenospheric mantle. This material subsequently interacted with an EM-I lithospheric mantle during melt ascent. This interaction dissolved pyroxenes and crystallized olivine, significantly modifying the trace element and Sr-Nd-Pb isotopic compositions of the melt, but with little effect on Mg isotopes.

Acknowledgements

We thank Mark Rehkämper, Hai-Bo Zou and an anonymous reviewer, whose constructive and critical reviews greatly improved this manuscript. We are grateful to Xu-Nan Meng, Rui-Ring Li, Yan Xiao, Sheng-Ao Liu, Ting Gao and Zi-Jian Li for assistances in the lab at the State Key Laboratory of Geological Processes and Mineral Resources, China University of Geosciences (Beijing), and Zhuo-Sen Yao for guidance using the MELTS software and Fang-Zhen Teng for constructive suggestions. This work is financially supported by the National Science Foundation of China (Grants 41230209, 41322022, 41328004, 41221002 and 41173012) and Youth Innovation Promotion Association, Chinese Academy of Sciences.

References

An Y.-J., Wu F., Xiang Y.-X., Nan X.-Y., Yu X., Yang J.-H., Yu H.-M., Xie L.-W. and Huang F. (2014) High-precision Mg isotope analyses of low-Mg rocks by MC-ICP-MS. Chem. Geol. 390, 9-21.

Basu A. R., Wang J.-W., Huang W.-K., Xie G.-H. and Tatsumoto M. (1991) Major element, REE, and Pb, Nd and Sr isotopic geochemistry of Cenozoic volcanic rocks of eastern China: implications for their origin from suboceanic-type mantle reservoirs. Earth Planet. Sci. Lett. 105, 149-169.

Bizimis M., Salters V. J. M. and Dawson J. B. (2003) The brevity of carbonatite sources in the mantle: evidence from Hf isotopes. Contrib. Mineral. Petrol. 145, 281-300.

Chen Y., Zhang Y.-X., Graham D., Su S.-G. and Deng J.-F. (2007) Geochemistry of Cenozoic basalts and mantle xenoliths in Northeast China. Lithos 96, 108-126.

Chen L.-H., Zeng G., Jiang S.-Y., Hofmann A. W. and Xu X.-S. (2009) Sources of Anfengshan basalts: subducted lower crust in the Sulu UHP belt, China. Earth Planet. Sci. Lett. 286, 426-435.

Choi S. H., Mukasa S. B., Kwon S. T. and Andronikov A. V. (2006) Sr, Nd, Pb and Hf isotopic compositions of late Cenozoic alkali basalts in South Korea: Evidence for mixing the two dominant asthenospheric mantle domains beneath East Asia. Chem. Geol. 232, 134-151.

Chu Z.-Y., Harvey J., Liu C.-Z., Guo J.-H., Wu F.-Y., Tian W., Zhang Y.-L. and Yang J.-H. (2013) Source of highly potassic basalts in northeast China: Evidence from Re-Os, Sr-Nd-Hf isotopes and PGE geochemistry. Chem. Geol. 357, 52-66.

Cohen R. S. and O'Nions R. K. (1982) Identification of Recycled Continental Materials in the Mantle from Sr, Nd and Pb Isotope Investigations. Earth Planet. Sci. Lett. 61, 73-84.

Dasgupta R., Hirschmann M. M. and Smith N. D. (2007) Partial melting experiments of peridotite + CO_2 at 3 GPa and genesis of alkali ocean island basalts. J. Petrol. 48, 2093-2124.

Dasgupta R. and Hirschmann M. M. (2010) The deep carbon cycle and melting in Earth's interior. Earth Planet. Sci. Lett. 298, 1-13.

Douglass J., Schilling J.-G. and Fontignie D. (1999) Plume-ridge interactions of the Discovery and Shona mantle plumes with the southern Mid-Atlantic Ridge (40°-55°S). J. Geophys. Res. 104, 2941-2962.

Dygert N., Liang Y. and Hess P. (2013) The importance of melt TiO_2 in affecting major and trace element partitioning between Fe-Ti oxides and lunar picritic glass melts. Geochim. Cosmochim. Acta 106, 134-151.

Edwards S. J. and Malpas J. (1996) Melt-perdotite interactions in shallow mantle at the East Pacific rise: Evidence from ODP site 895 (Hess deep). Mineral. Maga. 60, 191-206.

Fan Q.-C. and Hooper P. R. (1991) The Cenozoic Basaltic Rocks of Eastern China: Petrology and Chemical Composition. J. Petrol. 32, 765-810.

Fukao Y., Obayashi M., Inoue H. and Nenbai M. (1992) Subducting slab stagnant in the mantle transition zone. J. Geophys. Res. 97, 4809-4822.

Galy A., Yoffe O., Janney P. E., Williams R. W., Cloquet C., Alard O., Halicz L., Wadhwa M., Hutcheon I. D., Ramon E. and Carignan J. (2003) Magnesium isotope heterogeneity of the isotopic standard SRM980 and new reference materials for magnesium-isotope-ratio measurements. J. Anal. At. Spectrom. 18, 1352-1356.

Gualda G. A. R. and Ghiorso M. S. (2015) MELTS_Excel: A Microsoft Excel-based MELTS interface for research and teaching of magma properties and evolution. Geochem. Geophys. Geosyst. 16, 315-324.

Hart S. R., Hauri E. H., Oschmann L. A. and Whitehead J. A. (1992) Mantle Plumes and Entrainment: Isotopoc Evidence. Science 256, 517-520.

Hoernle K., Tilton G., Le Bas M., Duggen S. and Garbe-Schonberg D. (2002) Geochemistry of oceanic carbonatites compared with continental carbonatites: mantle recycling of oceanic crustal carbonate. Contrib. Mineral. Petrol. 142, 520-542.

Hofmann A. W. (1988) Chemical differentiation of the Earth: the relationship between mantle, continental crust, and oceanic crust. Earth Planet. Sci. Lett. 90, 297-314.

Huang F., Glessner J., Ianno A., Lundstrom C. and Zhang Z.-F. (2009) Magnesium isotopic composition of igneous rock standards measured by MC-ICP-MS. Chem. Geol. 268, 15-23.

Huang F., Zhang Z.-F., Lundstrom C. C. and Zhi X.-C. (2011) Iron and magnesium isotopic compositions of peridotite xenoliths from Eastern China. Geochim. Cosmochim. Acta 75, 3318-3334.

Huang J., Li S.-G., Xiao Y.-L., Ke S., Li W.-Y. and Tian Y. (2015) Origin of low δ^{26}Mg Cenozoic basalts from South China Block and their geodynamic implications. Geochim. Cosmochim. Acta 164, 298-317.

Huang J.-L. and Zhao D.-P. (2006) High-resolution mantle tomography of China and surrounding regions. J. Geophys. Res. 111, B09305.doi:10.1029/2005JB004066.

Jackson M. G., Kurz M. D. and Workman R. K. (2007) New Samoan lavas from Ofu Island reveal a hemispherically heterogeneous high $^3He/^4He$ mantle. Earth Planet. Sci. Lett. 264, 360-374.

Kameyama M. and Nishioka R. (2012) Generation of ascending flows in the Big Mantle Wedge (BMW) beneath northeast Asia induced by retreat and stagnation of subducted slab. Geophys. Res. Lett. 39, L10309.
Kelemen P. B., Dick H. J. B. and Quick J. E. (1992) Formation of harzburgite by pervasive melt/rock reaction in the upper mantle. Nature 358, 635-641.
Kogiso T., Tatsumi Y., Shimoda G. and Barsczus H. G. (1997) High μ (HIMU) ocean island basalts in southern Polynesia: New evidence for whole mantle scale recycling of subducted oceanic crust. J. Geophys. Res. 102, 8085-8103.
Kuritani T., Kimura J. I., Ohtani E., Miyamoto H. and Furuyama K. (2013) Transition zone origin of potassic basalts from Wudalianchi volcano, northeast China. Lithos 156, 1-12.
Langmuir C. H., Bender J. F., Bence A. E., Hanson G. N. and Taylor S. R. (1977) Petrogenesis of Basalts from Famous Area: Mid-Atlantic Ridge. Earth Planet. Sci. Lett. 36, 133-156.
Leroex A. P., Cliff R. A. and Adair B. J. I. (1990) Tristan da Cunha, South Atlantic: Geochemistry and Petrogenesis of a Basanite-Phonolite Lava Series. J. Petrol. 314, 779-812.
Li J., Wang X., Wang X. -J. and Yuen D. A. (2013) P and SH velocity structure in the upper mantle beneath Northeast China: Evidence for a stagnant slab in hydrous mantle transition zone. Earth Planet. Sci. Lett. 367, 71-81.
Liu C.-Q., Masuda A. and Xie G.-H. (1994) Major- and trace-element compositions of Cenozoic basalts in eastern China: Petrogenesis and mantle source. Chem. Geol. 114, 19-42.
Liu J.-Q., Han J.-T. and Fyfe W. S. (2001) Cenozoic episodic volcanism and continental rifting in northeast China and possible link to Japan Sea development as revealed from K-Ar geochronology. Tectonophysics 339, 385-401.
Liu S.-A., Teng F.-Z., He Y.-S., Ke S. and Li S.-G. (2010) Investigation of magnesium isotope fractionation during granite differentiation: Implication for Mg isotopic composition of the continental crust. Earth Planet. Sci. Lett. 297, 646-654.
Maruyama S., Isozaki Y., Kimura G. and Terabayashi M. (1997) Paleogeographic maps of the Japanese islands: Plate tectonic synthesis from 750Ma to the present. Isl Arc 6, 121-142.
Menzies M. A., Leeman W. P. and Hawkesworth C. J. (1983) Isotope geochemistry of Cenozoic volcanic rocks reveals mantle heterogeneity below western USA. Nature 303, 205-209.
Nesbitt H. W. and Young G. M. (1982) Early Proterozoic climates and plate motions inferred from major element chemistry of lutites. Nature 299, 715-717.
Niu Y. L. (2005) Generation and evolution of basaltic magmas: some basic concepts and a new view on the origin of mesozoic-cenozoic basaltic volcanism in Eastern China. Geol. J. China Univ. 11, 9-46.
Peng Z.-C., Zartman R. E., Futa K. and Chen D.-G. (1986) Pb-, Sr- and Nd-isotopic systematics and chemical characteristics of Cenozoic basalts, Eastern China. Chem. Geol. 59, 3-33.
Rudnick R. L. and Gao S. (2003) Composition of the continental crust. Treatise Geochem. 3, 1-64.
Sakuyama T., Tian W., Kimura J. I., Fukao Y., Hirahara Y., Takahashi T., Senda R., Chang Q., Miyazaki T., Obayashi M., Kawabata H. and Tatsumi Y. (2013) Melting of dehydrated oceanic crust from the stagnant slab and of the hydrated mantle transition zone: Constraints from Cenozoic alkaline basalts in eastern China. Chem. Geol. 359, 32-48.
Sedaghatpour F., Teng F.-Z., Liu Y., Sears D. W. G. and Taylor L. A. (2013) Magnesium isotopic composition of the Moon. Geochim. Cosmochim. Acta 120, 1-16.
Song Y., Frey F. A. and Zhi X.-C. (1990) Isotopic characteristics of Hannuoba basalts, eastern China: implications for their petrogenesis and the composition of subcontinental mantle. Chem. Geol. 88, 35-52.
Suhr G., Seck H. A., Shimizu N., Gunther D. and Jenner G. (1998) Infiltration of refractory melts into the lowermost oceanic crust: evidence from dunite- and gabbro-hosted clinopyroxenes in the Bay of Islands ophiolite. Contrib. Mineral. Petrol. 131, 136-154.
Sun S.-S. and McDonough W. F. (1989) Chemical and isotopic systematics of oceanic basalts: implications for mantle composition and processes. Geol. Soc. London, Special Publications 42, 313-345.
Sun Y., Ying J.-F., Su B.-X., Zhou X.-H. and Shao J.-A. (2015) Contribution of crustal materials to the mantle sources of Xiaogulihe ultrapotassic volcanic rocks, Northeast China: New constraints from mineral chemistry and oxygen isotopes of olivine. Chem. Geol. 405, 10-18.
Sun Y., Ying J.-F., Zhou X.-H., Shao J.-A., Chu Z.-Y. and Su B.-X. (2014) Geochemistry of ultrapotassic volcanic rocks in Xiaogulihe NE China: Implications for the role of ancient subducted sediments. Lithos 208, 53-66.
Tang Y.-J., Zhang H.-F. and Ying J.-F. (2006) Asthenosphere-lithospheric mantle interaction in an extensional regime: Implication from the geochemistry of Cenozoic basalts from Taihang Mountains, North China Craton. Chem. Geol. 233, 309-327.
Teng F.-Z., Li W.-Y., Ke S., Yang W., Liu S.-A., Sedaghatpour F., Wang S.-J., Huang K.-J., Hu Y., Ling M.-X., Xiao Y., Liu X.-M., Li

X.-W., Gu H.-O., Sio C. K., Wallace D. A., Su B.-X., Zhao L., Chamberlin J., Harrington M. and Brewer A. (2015) Magnesium Isotopic Compositions of International Geological Reference Materials. Geostand. Geoanal. Res. 39, 329-339.

Teng F.-Z., Li W.-Y., Ke S., Marty B., Dauphas N., Huang S.-C., Wu F.-Y. and Pourmand A. (2010) Magnesium isotopic composition of the Earth and chondrites. Geochim. Cosmochim. Acta 74, 4150-4166.

Teng F.-Z., Wadhwa M. and Helz R.T. (2007) Investigation of magnesium isotope fractionation during basalt differentiation: Implications for a chondritic composition of the terrestrial mantle. Earth Planet. Sci. Lett. 261, 84-92.

Thomson A. R., Walter M. J., Kohn S.C. and Brooker R. A. (2016) Slab melting as a barrier to deep carbon subduction. Nature 529, 76-79.

Wang S.-J., Teng F.-Z. and Li S.-G. (2014) Tracing carbonate-silicate interaction during subduction using magnesium and oxygen isotopes. Nat. Commun. 5. http://dx.doi.org/10.1038/ncomms6328.

Wang Y., Zhao Z.-F., Zheng Y.-F. and Zhang, J.-J. (2011) Geochemical constraints on the nature of mantle source for Cenozoic continental basalts in east-central China. Lithos 125, 940-955.

Weaver B. L. (1991) The origin of ocean island basalt end-menber compositions: trace element and isotopic constraints. Earth Planet. Sci. Lett. 104, 381-397.

White W. M. and Hofmann A. W. (1982) Sr and Nd isotope geochemistry of oceanic basalts and mantle evolution. Nature 296, 821-825.

Wu F.-Y., Yang J.-H., Lo C.-H., Wilde S.-A., Sun D.-Y. and Jahn B. M. (2007) The Heilongjiang Group: A Jurassic accretionary complex in the Jiamusi Massif at the western Pacific margin of northeastern China. Isl Arc 16, 156-172.

Xu Y. G., Ma J.-L., Frey F. A., Feigenson M. D. and Liu J.-F. (2005) Role of lithosphere-asthenosphere interaction in the genesis of Quaternary alkali and tholeiitic basalts from Datong, western North China Craton. Chem. Geol. 224, 247-271.

Yang W. and Li S.-G. (2008) Geochronology and geochemistry of the Mesozoic volcanic rocks in Western Liaoning: Implications for lithospheric thinning of the North China Craton. Lithos 102, 88-117.

Yang W., Teng F.-Z. and Zhang H.-F. (2009) Chondritic magnesium isotopic composition of the terrestrial mantle: A case study of peridotite xenoliths from the North China craton. Earth Planet. Sci. Lett. 288, 475-482.

Yang W., Teng F.-Z., Zhang H.-F. and Li S.-G. (2012) Magnesium isotopic systematics of continental basalts from the North China craton: Implications for tracing subducted carbonate in the mantle. Chem. Geol. 328, 185-194.

Young E. D. and Galy A. (2004) The isotope geochemistry and cosmochemistry of magnesium. Rev. Mineral. Geochem. 55, 197-230.

Zeng G., Chen L.-H., Xu X.-S., Jiang S.-Y. and Hofmann A. W. (2010) Carbonated mantle sources for Cenozoic intra-plate alkaline basalts in Shandong, North China. Chem. Geol. 273, 35-45.

Zeng, G., Chen, L.H., Hofmann, A.W., Jiang, S.Y., Xu, X.S., 2011. Crust recycling in the sources of two parallel volcanic chains in Shandong, North China. Earth Planet. Sci. Lett. 302, 359-368.

Zhang H.-F., Sun M., Zhou X.-H., Zhou M.-F., Fan W.-M. and Zheng J.-P. (2003) Secular evolution of the lithosphere beneath the eastern North China Craton: Evidence from Mesozoic basalts and high-Mg andesites. Geochim. Cosmochim. Acta 67, 4373-4387.

Zhang J.-J., Zheng Y.-F. and Zhao Z.-F. (2009) Geochemical evidence for interaction between oceanic crust and lithospheric mantle in the origin of Cenozoic continental basalts in east-central China. Lithos 110, 305-326.

Zhang M., Menzies M. A., Suddaby P. and Thirlwall M. F. (1991) EM1 signature from within the post-Archean subcontinental lithospheric mantle: Isotopic evidence from the potassic volcanic rocks in NE China. Geochem. J. 25, 387-398.

Zhang M., Suddaby P., Thompson R. N., Thirlwall M. F. and Menzies M. A. (1995) Potassic Volcanic Rocks in NE China: Geochemical Constraints on Mantle Source and Magma Genesis. J. Petrol. 36, 1275-1303.

Zhang M., Suddaby P., O'Reilly S. Y., Norman M. and Qiu J.-X. (2000) Nature of the lithospheric mantle beneath the eastern part of the Central Asian fold belt: mantle xenolith evidence. Tectonophysics 328, 131-156.

Zhang M., Zhou X.-H. and Zhang J.-B. (1998) Nature of the lithospheric mantle beneath NE China: evidence from potassic volcanic rocks and mantle xenolithos. In: Flower, M., Chuang S.L., Lo C. H., Lee T. Y. (Eds.), Mantle Dynamics and Plate Interactions in East Asia. Am. Geophys. Union, Geohys. Monogr. 27, 197-219.

Zhao D.-P., Tian Y., Lei J.-S., Liu L.-C. and Zheng S.-H. (2009) Seismic image and origin of the Changbai intraplate volcano in East Asia: Role of big mantle wedge above the stagnant Pacific slab. Phys. Earth. Planet. In. 173, 197-206

Zhao D.-P., Yu S. and Ohtani E. (2011) East Asia: Seismotectonics, magmatism and mantle dynamics. J. Asian Earth Sci. 40, 689-709.

Zhi X.-C., Song Y., Frey F. A., Feng J.-L. and Zhai M.-Z. (1990) Geochemistry of Hannuoba Basalts, Eastern China: Constraints on the Origin of Continental Alkalic and Tholeiitic Basalt. Chem. Geol. 88, 1-33.

Zhou X.-H. and Armstrong R. L. (1982) Cenozoic volcanic rocks of eastern China: Secular and geographic trends in chemistry and strontium isotopic composition. Earth Planet. Sci. Lett. 58, 301-329.

Zhou X.-H., Zhu B.-Q., Liu R.-X. and Chen W.-J. 1988. Cenozoic basaltic rocks in eastern China. In: Macdougall, J.D. (Ed.), Continental Flood Basalts. Springer. Netherlands. pp. 311-330.

Zhou M.-F., Robinson P. T., Malpas J. and Li, Z.-J. (1996) Podiform chromitites in the Luobusha ophiolite (southern Tibet): Implications for melt-rock interaction and chromite segregation in the upper mantle. J. Petrol. 37, 3-21.

Zhou M.-F., Robinson P. T., Su B.-X., Gao J.-F., Li J.-W., Yang J.-S. and Malpas J. (2014) Compositions of chromite, associated minerals, and parental magmas of podiform chromite deposits: the role of slab contamination of asthenospheric melts in suprasubduction zone environments. Gondwana Res. 26, 262-283.

Zindler A. and Hart S. (1986) Chemical Geodynamics. Annu Rev. Earth Planet. Sci. 14, 493-571.

Zou H.-B., Fan Q.-C. and Yao Y.-P. (2008) U-Th systematics of dispersed young volcanoes in NE China: Asthenosphere upwelling caused by piling up and upward thickening of stagnant Pacific slab. Chem. Geol. 255, 134-142.

Zou H.-B., Reid M. R., Liu Y.-S., Yao Y.-P., Xu X.-S. and Fan Q.-C. (2003) Constraints on the origin of historic potassic basalts from northeast China by U-Th disequilibrium data. Chem. Geol. 200, 189-201.

Zou H.-B., Zindler A., Xu X.-S. and Qi Q. (2000) Major, trace element, and Nd, Sr and Pb isotope studies of Cenozoic basalts in SE China: mantle sources, regional variations, and tectonic significance. Chem. Geol. 171, 33-47.

Deep carbon cycles constrained by a large-scale mantle Mg isotope anomaly in eastern China*

Shu-Guang Li[1,2], Wei Yang[3], Shan Ke[1], Xunan Meng[1], Hengci Tian[3], Lijuan Xu[1], Yongsheng He[1], Jian Huang[2], Xuan-Ce Wang[4], Qunke Xia[5], Weidong Sun[6], Xiaoyong Yang[2], Zhong-Yuan Ren[6], Haiquan Wei[7], Yongsheng Liu[8], Fancong Meng[9] and Jun Yan[10]

1. State Key Laboratory of Geological Processes and Mineral Resources, China University of Geosciences, Beijing 100083, China
2. CAS Key Laboratory of Crust-Mantle Materials and Environments, School of Earth and Space Sciences, University of Science and Technology of China, Hefei 230026, China
3. Key Laboratory of Earth and Planetary Physics, Institute of Geology and Geophysics, Chinese Academy of Sciences, Beijing 100029, China
4. The Institute for Geoscience Research, Department of Applied Geology, Curtin University, GPO Box U1987, Perth, WA 6845, Australia
5. School of Earth Sciences, Zhejiang University, Hangzhou 310027, China
6. Key Laboratory of Mineralogy and Metallogeny, Guangzhou Institute of Geochemistry, Chinese Academy of Sciences, Guangzhou 510640, China
7. Institute of Geology, China Earthquake Administration, Beijing 100029, China
8. State Key Laboratory of Geological Processes and Mineral Resources, Faculty of Earth Sciences, China University of Geosciences, Wuhan 430074, China
9. Institute of Geology, Chinese Academy of Geological Sciences, Beijing 100037, China
10. School of Resources and Environmental Engineering, Hefei University of Technology, Hefei 230009, China

亮点介绍：本文首次厘定出中国东部对流上地幔存在一年龄＜110Ma 大规模轻 Mg 同位素异常区，其空间分布区与东亚地幔过渡带的滞留西太平洋俯冲板片分布区高度重合，证明该大尺度轻 Mg 同位素异常与西太平洋板块俯冲导致的深部碳循环有关。与此相反，本文还发现与环太平洋板块俯冲有关的岛弧玄武岩具有类似地幔或略重的 Mg 同位素组成，并提出了俯冲板片在较浅深度（70~120km）析出水流体仅能选择溶解富钙碳酸盐，故其不能改变被交代的岛弧地幔楔的 Mg 同位素组成的理论。据此，西太平洋板块俯冲形成的大地幔楔有两个深部碳循环圈，分别为在较浅部位俯冲板片析出的水流体选择溶解钙质碳酸盐交代岛弧地幔楔的碳循环，以及位于地幔过渡带深俯冲板片熔体含镁质碳酸盐交代上覆对流上地幔的深部碳循环。

Abstract Although deep carbon recycling plays an important role in the atmospheric CO_2 budget and climate changes through geological time, the precise mechanisms remain poorly understood. Since recycled sedimentary carbonate through plate subduction is the main light-δ^{26}Mg reservoir within

* 本文发表在：National Science Review, 2017, 4:111-120

deep-Earth, Mg isotope variation in mantle-derived melts provides a novel perspective when investigating deep carbon cycling. Here, we show that the Late Cretaceous and Cenozoic continental basalts from 13 regions covering the whole of eastern China have low δ^{26}Mg isotopic compositions, while the Early Cretaceous basalts from the same area and the island arc basalts from circum-Pacific subduction zones have mantle-like or heavy Mg isotopic characteristics. Thus, a large-scale mantle low δ^{26}Mg anomaly in eastern China has been delineated, suggesting the contribution of sedimentary carbonates recycled into the upper mantle, but limited into the lower mantle. This large-scale spatial and temporal variation of Mg isotopes in the mantle places severe constraints on deep carbon recycling via oceanic subduction.

Keywords　Mg isotopes, deep carbon cycling, continental basalts, eastern China, circum-Pacific subduction zones, mantle geochemistry

1　Introduction

Deep carbon cycling is one of the controlling factors for the distribution of carbon in all terrestrial reservoirs [1–3]. Carbon is transported into the mantle by subducted slabs and returned to the surface by degassing of volcanoes [2, 4]. However, the details of this deep carbon cycling and the residence time of carbon in the mantle are poorly understood.

Although global carbon recycling was proposed based on C isotopes [4], it is difficult to elucidate the detailed deep carbon recycling processes because of the significant fractionation of C isotopes during magma degassing [4]. Alternatively, Mg isotopes constitute a powerful tracer for recycled carbonates in the mantle [5–7] due to the distinctive Mg isotopic compositions between sedimentary carbonates and mantle peridotites (δ^{26}Mg=–5.31 to –1.09 vs –0.25±0.07) [8–10].

Low δ^{26}Mg signatures documented in some Late Cretaceous and Cenozoic continental basalts from eastern China were attributed to recycled sedimentary carbonates via the subducted Pacific oceanic slab into their mantle source [5–7]. However, how large the low δ^{26}Mg mantle domain beneath eastern China is and whether it coincides with the western Pacific subducted slab stagnated in the mantle transition zone beneath eastern China, revealed by seismic tomography [11], have not yet been determined. In addition, several important questions regarding the deep carbon recycling processes remain unanswered or are under debate: (i) How can the effects on mantle Mg isotope compositions by recycled sedimentary carbonates or carbonated eclogites be distinguished? (ii) What is the Mg isotopic behavior of recycled carbonates during slab subduction? (iii) What is the proportions of carbon retained in the subducted slab and carried into the deep mantle [12–14]? (iv) Where is the recycled carbon stored in the mantle and what is its residence time [2, 15, 16]?

Given that both island arc volcanisms in the circum-Pacific and continental volcanisms in eastern China are related to the subduction of Pacific slabs, in order to better understand the above issues, here we systematically investigated the spatial and temporal variations of Mg isotopes in the basaltic rocks from 21 localities (Fig. 1).

2　A large-scale low δ^{26}Mg anomaly

The basaltic samples investigated here and those from the extant literature can be divided into four groups based on their tectonic settings, ages and geochemical features. (i) Island Arc Basalt (IAB): the island arc or back arc basin basalts from the circum- Pacific, including Kamchaka, the Philippines, Costa Rica and the Lau Basin (Fig. 1A), are characterized by low ^{87}Sr/^{86}Sr (0.703344–0.703800), high ε_{Nd} (6.8–9.2) and low Nb/U ratios (2.65–29.65) (Fig. 2) [9, 17–19]. (ii) Continental Basalt Old (CBO): these rocks are collected from three localities in eastern China, including Yixian, Feixian-Fangcheng and Dabie Orogen (Fig. 1B). They have ages older than 115

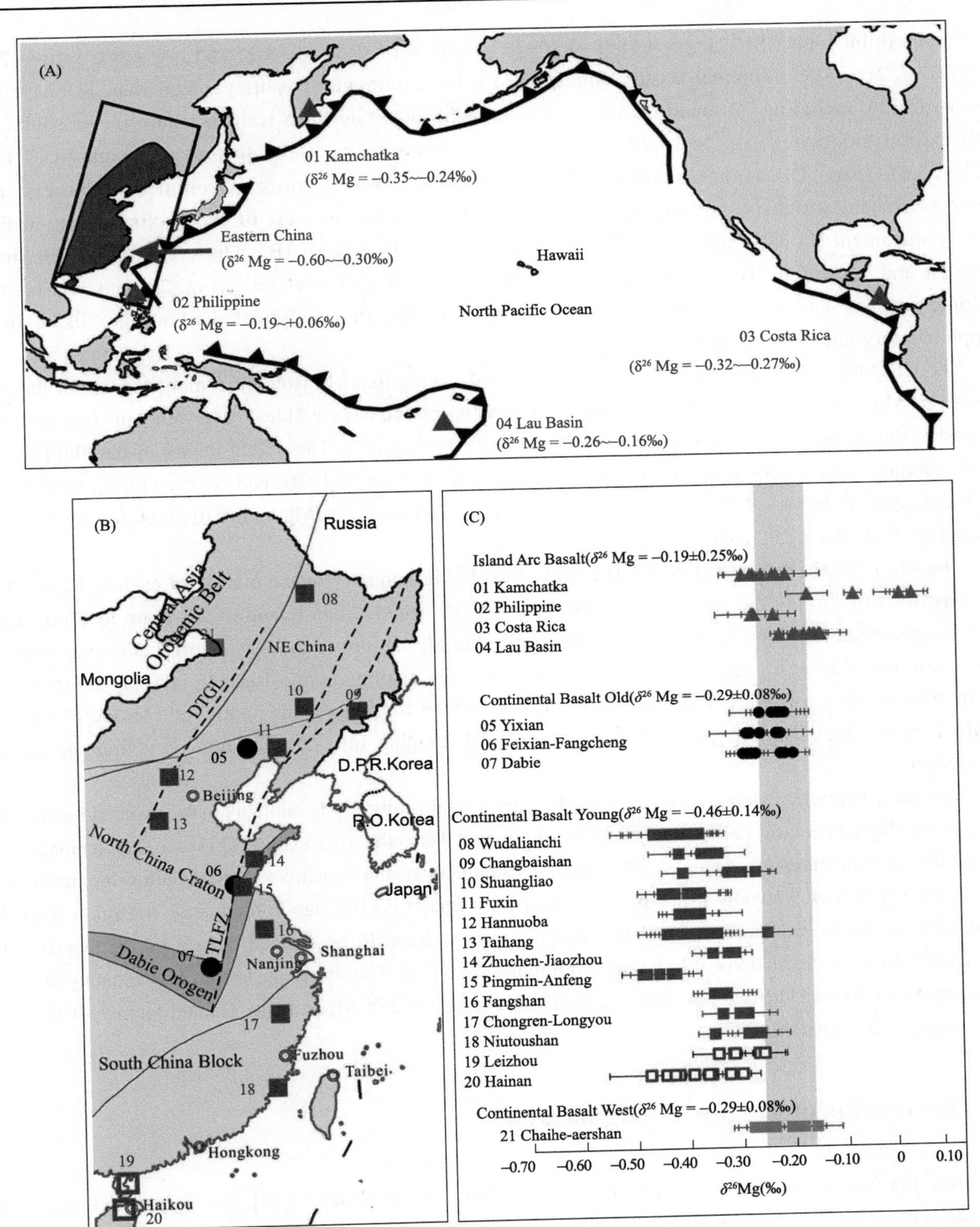

Fig. 1 (A) Schematic map of East Asia and North Pacific Ocean. The red area represents the low δ^{26}Mg anomaly in the mantle beneath eastern China. The blue triangles indicate the locations of the island arc samples in Circum- Pacific. (B) Locations of continental basalt samples on the map of eastern China. DTGL and TLFZ represent Daxing'anling-Taihang Gravity Lineament and Tan-Lu Fault Zone, respectively. (C) δ^{26}Mg variations of basaltic samples from 21 locations. The gray bar represents the average mantle (δ^{26}Mg=–0.25±0.07)[9]. Data source: Wudalianchi from Tian et al. [7]; Yixian, Fuxin and Taihang from Yang et al. [5]; Pingming-Anfeng, Fangshan and Chongren-Longyou from Huang et al. [6]; other data from this study.

Myr and exhibit high $^{87}Sr/^{86}Sr$ (0.705868–0.709767), low ε_{Nd} (−19.8 to −9.7) and low Nb/U ratios (2.67–19.5) (Fig. 2). (iii) Continental Basalt Young (CBY): the continental basalts younger than 110 Myr from eastern China, including 13 localities (from Wudalianchi in the north to Hainan Island in the south, Fig. 1B), exhibit moderate $^{87}Sr/^{86}Sr$ (0.703140–0.705637), moderate ε_{Nd} (−5.7 to 6.8) and high Nb/U ratios (27.5–55.6) (Fig. 2). (iv) Continental Basalt West (CBW): Cenozoic continental basalts from Chaihe-Aershan, which is the only sample location that lies to the west of the Daxing'anling-Taihang gravity lineament (DTGL) and in the Central Asia Orogenic Belt (Fig. 1B). These basalts exhibit similar $^{87}Sr/^{86}Sr$ and ε_{Nd} to the CBY group, but have higher Nb/U ratios of 46.9–65.5 (Fig. 2). The details of the tectonic settings and geochemical features of these samples and analytical methods are listed in the Supplementary data (available at *NSR* online).

The CBO and CBW groups have homogeneous and mantle-like Mg isotopic compositions with the same average $\delta^{26}Mg$ value of −0.29±0.08 (2SD) (Fig. 1C). The IAB group has variable $\delta^{26}Mg$ values of −0.35 to 0.06, in contrast to the narrow range of the mantle (−0.25±0.04, 2SD) (Fig. 1C). The $\delta^{26}Mg$ values of the Philippine arc basalts ranging from −0.19 to 0.06 are the highest values reported so far for IABs, and are even higher than those of the Martinique arc lavas (−0.25 to −0.1) [20]. The CBY group has low $\delta^{26}Mg$ values of −0.60 to −0.30, with an average of −0.46±0.14 (2SD) (Fig. 1C).

The Mg isotopic results define a large-scale low $\delta^{26}Mg$ anomaly in the mantle of eastern China, from the coastline of China in the east to the DTGL in the west, from Hainan Island in the south to Wudalianchi and Changbaishan in the north (Fig. 1B). Temporally, the Mg isotopic anomaly occurred abruptly after 110 Myr. Prior to 115 Myr, all basaltic rocks from the same area display mantle-like Mg isotopic compositions. To the best of our knowledge, this study is the first to reveal the large-scale spatial and temporal variation in Mg isotopes in the mantle, and then provide novel insights into deep carbon recycling via oceanic subduction.

With the exception of Hainan and Leizhou, the portion of this low $\delta^{26}Mg$ anomaly in the mainland completely overlaps the stagnant Pacific slab in the mantle transition zone beneath eastern China [11] (see Supplementary Fig. 2, available as Supplementary data at *NSR* online). This coincidence documents a causal relationship between the large-scale Mg isotopic anomaly in the mainland and the stagnant Pacific slab in the mantle transition zone. The continental basalts from Hainan and Leizhou have been considered to be derived from a young mantle plume associated with deep subductions of India and Philippines slabs [21]. Therefore, the low $\delta^{26}Mg$ anomaly in eastern China can be subdivided into two domains: the mainland anomaly (CBY-ML) and the Hainan anomaly (CBY-HN), which may have different origins.

3 Two origins of the low $\delta^{26}Mg$ basalts

Given that Mg isotopes are little fractionated during basalt differentiation [22], the higher Nb/U ratios and higher ε_{Nd} but lower $\delta^{26}Mg$ values of the CBY group than those of the CBO group (Fig. 2B and C) suggest that the low $\delta^{26}Mg$ feature in the CBY group must be derived from their mantle sources instead of the magmatic processes, e.g. fractional crystallization and crustal contamination. Previous studies suggested that such light Mg isotopic compositions ($\delta^{26}Mg$=−0.60 to −0.35) can only be explained by the involvement of recycled sedimentary carbonates in their mantle source [5–7]. However, the previous studies ignored another possible origin for low $\delta^{26}Mg$ basalts, which could be produced by recycled eclogites with light Mg isotopic compositions ($\delta^{26}Mg$ down to −1.928) due to carbonate–silicate isotopic exchange during subduction [23–25]. Since recycled eclogites may contain only a small amount of carbon [24], it is necessary to re-evaluate the origin of the low δ^{26}Mg feature of the CBY group.

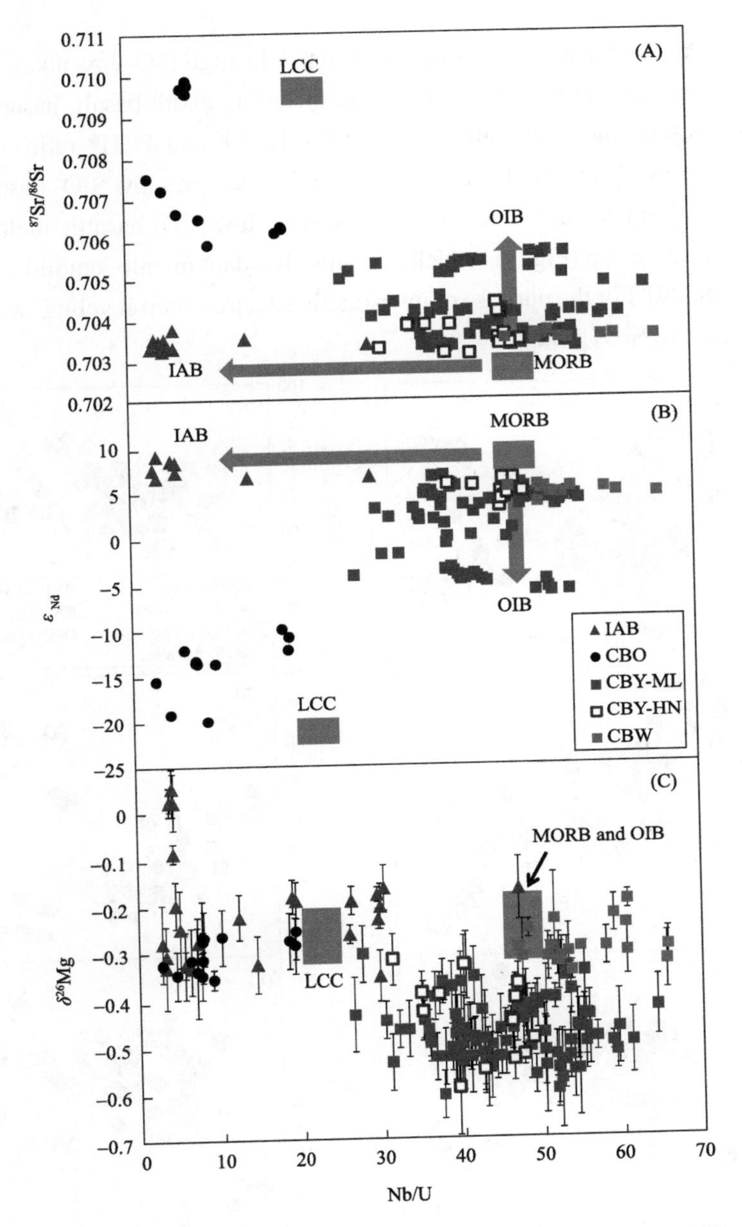

Fig. 2 ^{87}Sr/^{86}Sr (A), ε_{Nd} (B) and δ^{26}Mg (C) vs Nb/U plots for basaltic rocks. Data source: Mid-ocean ridge basalts (MORB), Oceanic island basalts (OIB), and Island arc basalts (IAB) from Hofmann et al. [17] and Teng et al. [9]; Lower continental crust (LCC) from Rudnick and Gao [18] and Yang et al. [19]; basaltic samples from Supplementary Table S1 (available as Supplementary data at *NSR* online). The four groups have distinct geochemical features and can be distinguished each other in these diagrams. The CBY group exhibits low δ^{26}Mg features, which is distinct from the CBW, IAB and CBO groups. The CBY-ML (red solid squares) shows the samples from the Mainland of eastern China and the CBY-HN (red open squares) shows the samples from the Hainan island and the Leizhou Peninsula.

Partial melting of typical recycled carbonated oceanic crust (in the form of carbonated eclogite) at a pressure of 3 GPa can produce silicate melt and carbonate-rich melt coexisting over a wide temperature interval [26]. Their proportion is dependent on carbonate contents in the recycled oceanic crust. Reactions of these two immiscible melts with peridotite can produce olivine-free pyroxenite [27] and carbonated (or diamond-bearing) peridotite, respectively. These two mantle sources may produce melts with low δ^{26}Mg values but different major and trace element features.

The CBY-ML group can be divided into three sub-groups: the high SiO_2-low alkali group (tholeiite), the high SiO_2-high alkali group (potassic) and the low SiO_2 group, including alkali basalt, basanite and nephelinite based on their SiO_2 and Na_2O+ K_2O contents as well as CaO/Al_2O_3, La/Yb and Ti/Ti* ratios (Fig. 3). Both the potassic basalts and the tholeiites are considered to derive from interaction between low SiO_2 basaltic melt with lithosphere mantle [7, 28]. Their low δ^{26}Mg feature could be inherited from low SiO_2 basaltic melt [7]. The majority of the CBY-ML group is a low SiO_2 group (Fig. 3), which contains abundant mantle xenoliths and was directly derived from the asthenosphere [28, 29]. For the purpose of investigating deep carbon recycling, we will focus on the origin of the low δ^{26}Mg feature of the low SiO_2 group.

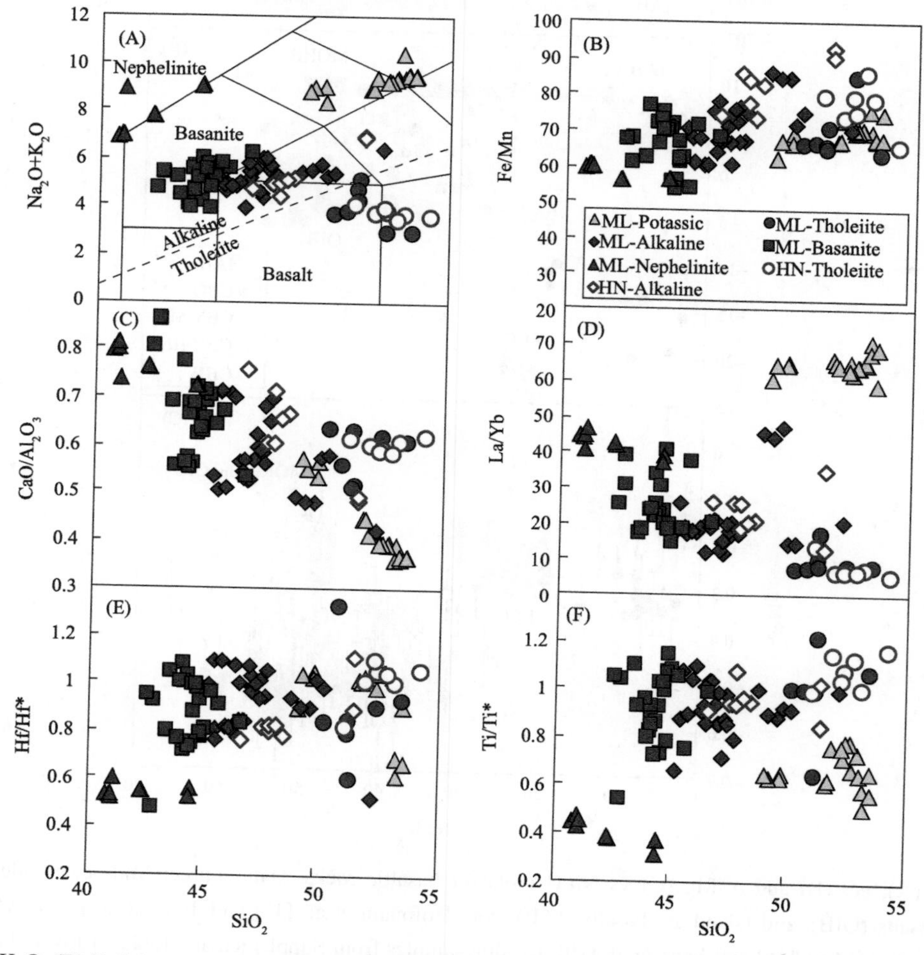

Fig. 3　(A) Na_2O+K_2O, (B) Fe/Mn, (C) CaO/Al_2O_3, (D) La/Yb, (E) Hf/Hf* and (F) Ti/Ti* vs SiO_2 for the CBY group. The CBY-ML is subdivided into ML-Potassic (yellow solid triangles), ML-tholeiite (blue solid circles), ML-Alkaline (red solid diamonds), ML-Basanite (red solid squares), ML-Nephelinite (red solid triangles). The CBY-HN is subdivided into HN-Tholeiite (blue open circles) and HN-Alkaline (red open diamonds). Data are from Supplementary Table S1 (available as Supplementary data at *NSR* online).

The nephelinite with the lowest SiO_2 contents in the low SiO_2 group exhibits significant K, Zr, Hf and Ti negative anomalies (and thus, having the lowest Hf/Hf* and Ti/Ti* ratios) in primitive mantle-normalized trace element spider diagrams (Supplementary Fig. 1 and Fig. 3E and F, available as Supplementary data at NSR online), as well as high CaO/Al_2O_3, high La/Yb and low Fe/Mn ratios (Fig. 3C and D), which constitute typical characteristics of carbonate melt [30–33]. The correlations between SiO_2 with Fe/Mn, CaO/Al_2O_3, La/Yb, Hf/Hf* and Ti/Ti* ratios (Fig. 3B–F) and the correlations between $\delta^{26}Mg$ with CaO/Al_2O_3, Fe/Mn, Hf/Hf* and Ti/Ti* ratios (Fig. 4A–D) for the CBY-ML low SiO_2 group could be a result of partial melting instead of fractional crystallization. This is because fractional crystallizations can influence both CaO/Al_2O_3 and Fe/Mn ratios of magma but cannot significantly change their La/Yb ratios and $\delta^{26}Mg$ values [22, 34]. A low degree of partial melting of carbonated peridotite can produce low SiO_2 melt with high CaO/Al_2O_3, La/Yb and low Fe/Mn ratios [6, 29–31]. Therefore, we concluded that the nephelinite, basanite and alkaline basalts could be derived from carbonated peridotite with various degrees of partial melting. The nephelinites with the lowest SiO_2 content and $\delta^{26}Mg$ (Fig. 4) can represent the low $\delta^{26}Mg$ component (LMC) in the mantle source of the CBY-ML group. This LMC is characterized by low Fe/Mn, Hf/Hf*, Ti/Ti* and high CaO/Al_2O_3 ratios (Fig. 4A–D), which are typical features of carbonated peridotite-derived melt. Consequently, carbonated peridotite is the main source of the low $\delta^{26}Mg$ basalts in the mainland of eastern China.

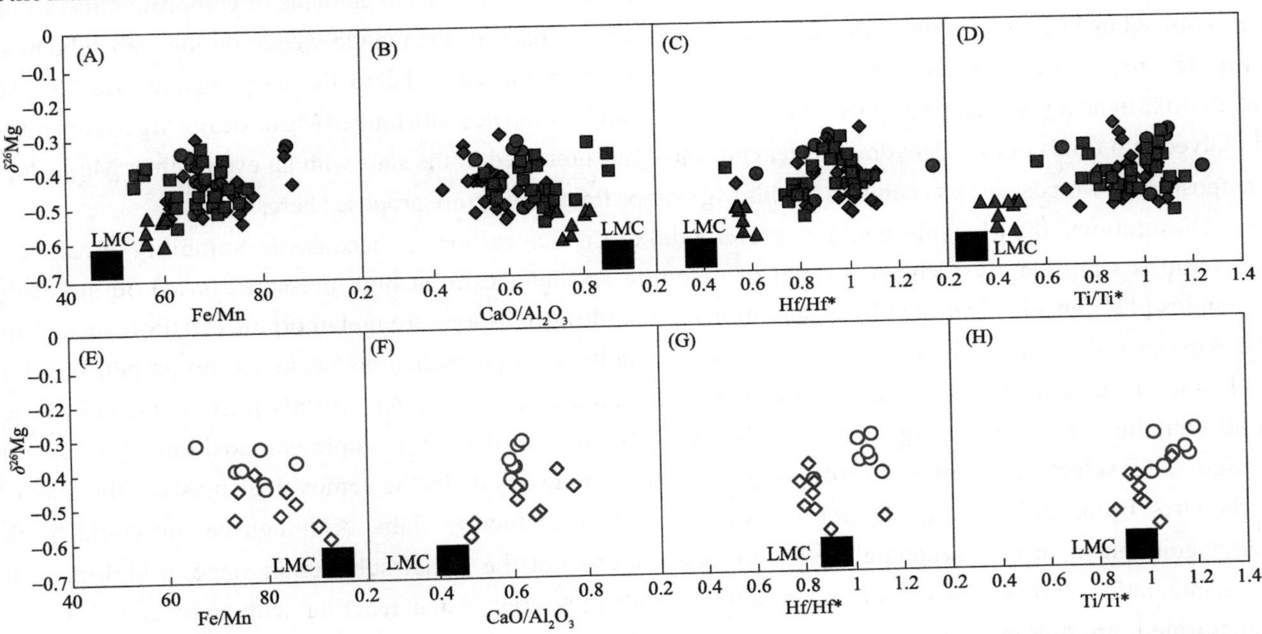

Fig. 4 (A–D) Fe/Mn, CaO/Al_2O_3, Hf/Hf*, and Ti/Ti* vs $\delta^{26}Mg$ plots of the CBY-ML and (E-H) Fe/Mn, CaO/Al_2O_3, Hf/Hf*, and Ti/Ti* vs $\delta^{26}Mg$ plots of the CBY-HN, showing difference of the low $\delta^{26}Mg$ components (LMC) between the CBY-ML and CBY-HN. The symbols are same as in Fig. 3. The contrast correlations between CaO/Al_2O_3 and Fe/Mn with $\delta^{26}Mg$ for the CBY-ML and CBY-HN suggest that the light Mg isotopic composition in the CBY-ML and CBY-HN groups probably resulted from the involvement of recycled sedimentary carbonates or low $\delta^{26}Mg$ eclogite, respectively, in their mantle sources. The potassic basalts from Wudalianchi are excluded in these plots because they have been significantly affected by interactions with an ancient enriched lithosphere as indicated by their EM-I Sr-Nd isotopic features [7].

Distinct from the LMC of the CBY-ML group, the LMC of the CBY-HN group as indicated by the low SiO_2 alkaline basalts exhibits high Fe/Mn, Hf/Hf*, Ti/Ti* and low CaO/Al_2O_3 ratios (Fig. 4E–H), which suggest that carbonate was little involved in the mantle source of the Hainan low SiO_2 alkaline basalts. Although the high Fe/Mn ratios could be derived from either core–mantle interaction [35] or the presence of pyroxenite or eclogite in the mantle source [26, 30, 31, 33], core–mantle interaction cannot result in low $\delta^{26}Mg$ values. Therefore, the low $\delta^{26}Mg$ features in the Hainan basalts could be related to recycled carbonated eclogites instead of sedimentary carbonates.

The low δ^{26}Mg anomaly of CBY-ML coincides with the stagnant Pacific slab in the mantle transition zone under eastern China, as revealed by seismic tomography [11]. In addition, the low δ^{26}Mg anomaly of CBY-HN coincides with the Hainan plume [21] (see Supplementary Fig. 2, available as Supplementary data at NSR online). The different origins between the CBY-ML and CBY-HN suggest that the major recycled carbon reservoir is located in the upper mantle, but carbon recycling into the lower mantle is limited in eastern China. This is consistent with a recent experimental study, which concluded that direct recycling of carbon into the lower mantle may have been highly restricted throughout most of Earth's history [16].

4 Behavior of subducted magnesium

Although the mantle sources of the IAB samples have been metasomatized by fluid derived from subducting slabs [36–39], they, in contrast to the CBY group, have mantle-like or heavy Mg isotopes (δ^{26}Mg=–0.35 to 0.06), indicating Mg isotope fractionation during subduction dehydration. Significant amounts of carbonate minerals can be dissolved by infiltrating fluids, aiding the transfer of carbon back to the mantle wedge during slab subduction [40]. The rest of the carbonates are then retained in the slab and carried into the deep mantle [12, 13]. The mantle-like or heavy Mg isotopes of the IAB indicate that only carbonates with limited Mg or heavy Mg isotopes were dissolved into fluids during dehydration, leaving carbonates retained in the slab with an even lighter Mg isotopic composition. Two possible mechanisms for this Mg isotope fractionation are proposed here.

Dissolutions of carbonates could be preferential to Ca-rich carbonate, because the solubility of calcite or aragonite is significantly higher than that of dolomite and magnesite at high pressures based on molecular dynamics [41]. This is supported by observations in the ultra-high-pressure metamorphic (UHPM) rocks from the Alps that show that, although magnesite and dolomite as solid inclusions occur within garnet, the fluid inclusions in garnet and diamond contain only Mg-calcite/calcite daughter crystals [42]. Thus, dehydrated fluid contains Ca, but little Mg, that has a limited influence on the Mg isotopic composition of the mantle wedge. This selective dissolution process of carbonates may result in the removal of most of the Ca-rich carbonates from, and leaving Mg-rich carbonates in, the subducting slabs. Although calcite could be the major component in the carbonate carried by oceanic crust at the initial subduction stage, a high-pressure experimentation shows that calcite is unstable at high pressures; when reacting with pyroxene, it can be transformed into dolomite at pressures between 23 and 45 kbars or magnesite at higher pressures [43]. Thus, the proportion of Mg-rich carbonate retained in the subducted slabs and carried into the deep mantle could be significant. This explanation is supported by Mg-Sr isotopes of the CBY-ML samples. The basalts of the CBY-ML group display light Mg isotopic compositions (δ^{26}Mg=–0.60 to –0.30) and relatively low ^{87}Sr/^{86}Sr ratios (0.703140–0.705637) (Fig. 5), which indicate that their mantle source had been hybridized by recycled carbonates with low Sr content. This is probably a mixture of magnesite with minor dolomite, as indicated by modeling calculations (Fig. 5).

However, this process alone cannot induce the heavy Mg isotopes observed in some IABs, e.g. the Philippines, with δ^{26}Mg up to 0.06±0.04, because Mg-calcite also has a light Mg isotopic composition. These heavy Mg isotopes cannot be explained by recycled argillaceous sediments because of the low ^{87}Sr/^{86}Sr ratios (0.7033–0.7035) of Philippine arc basalts. An alternative mechanism is the preferential addition of heavy Mg isotopes from fluids released from the altered MORB and altered abyssal peridotites [20]. Abyssal peridotites can have high Mg concentrations and heavy Mg isotopic compositions (δ^{26}Mg=–0.25 to 0.10) [20].

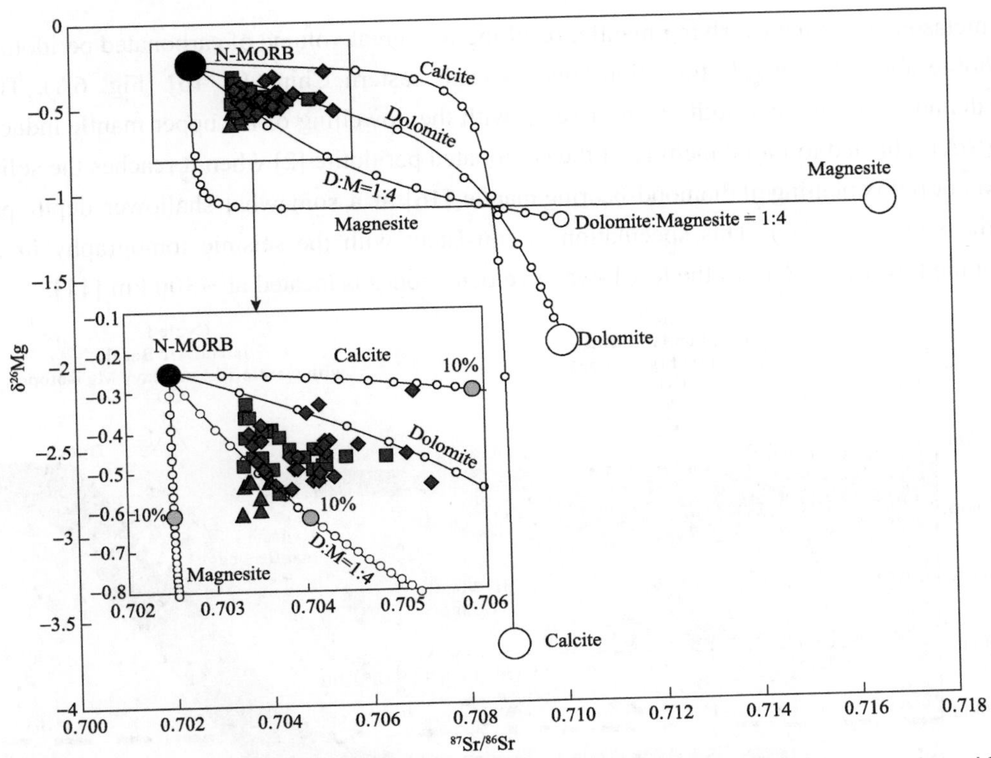

Fig. 5 Diagram of Mg and Sr isotope ratios of the basalts of the low SiO_2 group of the CBY-ML, together with mixing curves between peridotite-derived melt (N-MORB) and calcite-, magnesite- or dolomite-derived melts, given that the Mg-Sr contents and isotopic compositions of carbonate-derived melts are same as carbonates. This diagram shows that the species of recycled carbonates in the upper mantle of Eastern China are mainly Mg-rich carbonites and the initial melt of the CBY-ML group contained 1–10% carbonate-derived melt. The symbols are same as in Fig. 3. The step-sizes represented by small circles are 10% or 1% in the insert, which are the mass ratios of carbonates/(carbonates + N-MORB). See Supplementary data (available as Supplementary data at NSR online) for details of the compositions of the end members and mixing calculation.

Nevertheless, the heavy Mg isotopes in the IAB and the light Mg isotopes in the CBY-ML indicate that a large amount of carbonates are retained in the slab and carried into the mantle transition zone (e.g. 50%) [12, 13]. Otherwise, the IAB should exhibit a light Mg isotopic composition, and the CBY-ML should show mantle-like Mg isotopic composition.

5 Two deep carbon recycles and their residence times

Combining all observations on the IAB, CBY-ML and CBY-HN, deep carbon recycling can be subdivided into two cycles (Fig. 6A) [44]: cycle 1, the partially dissolved Ca-rich carbonates during the slab dehydration process were injected into the mantle wedge, and then released by arc volcanism; and cycle 2, the undissolved Mg-rich carbonates during slab dehydration were carried into the mantle transition zone by subducted slabs, melted and metasomatized the upper mantle, and were released by intraplate volcanism. Carbonate subduction through the transition zone and into the lower mantle seems to be limited, as suggested by experimental studies [16]. Cycle 1 has attracted considerable attention [40, 42] and its residence time is approximately 5–10 Myr [14]. Cycle 2, however, has not yet been well studied.

A melting experimental study on carbonated oceanic crust reveals that, when oceanic slabs were subducted into the mantle transition zone, the majority of slabs intersected the solidus for carbonated recycled MORB, producing carbonatite melt [16]. As the carbonatitic melts exhibit low viscosity and high mobility, they migrated upward and metasomatized the overlying mantle, resulting in a great volume of carbonated peridotite or diamond-bearing peridotite above the mantle transition zone beneath eastern China [2, 16] (Fig. 6A). The carbonated peridotite or diamond-bearing peridotite could rise up with the upwelling of the upper mantle induced by the slab rollback (Fig. 6A). This led to initial melting of the carbonated peridotite [2] when it reaches the solidus at a depth of 300–330 km or redox melting of diamond-bearing mantle [16] at a somewhat shallower depth, produc ing low δ^{26}Mg basaltic melts (Fig. 6A). This speculation is consistent with the seismic tomography in eastern China indicating that the lower boundary of the low P-wave velocity zones is located at ~300 km [11].

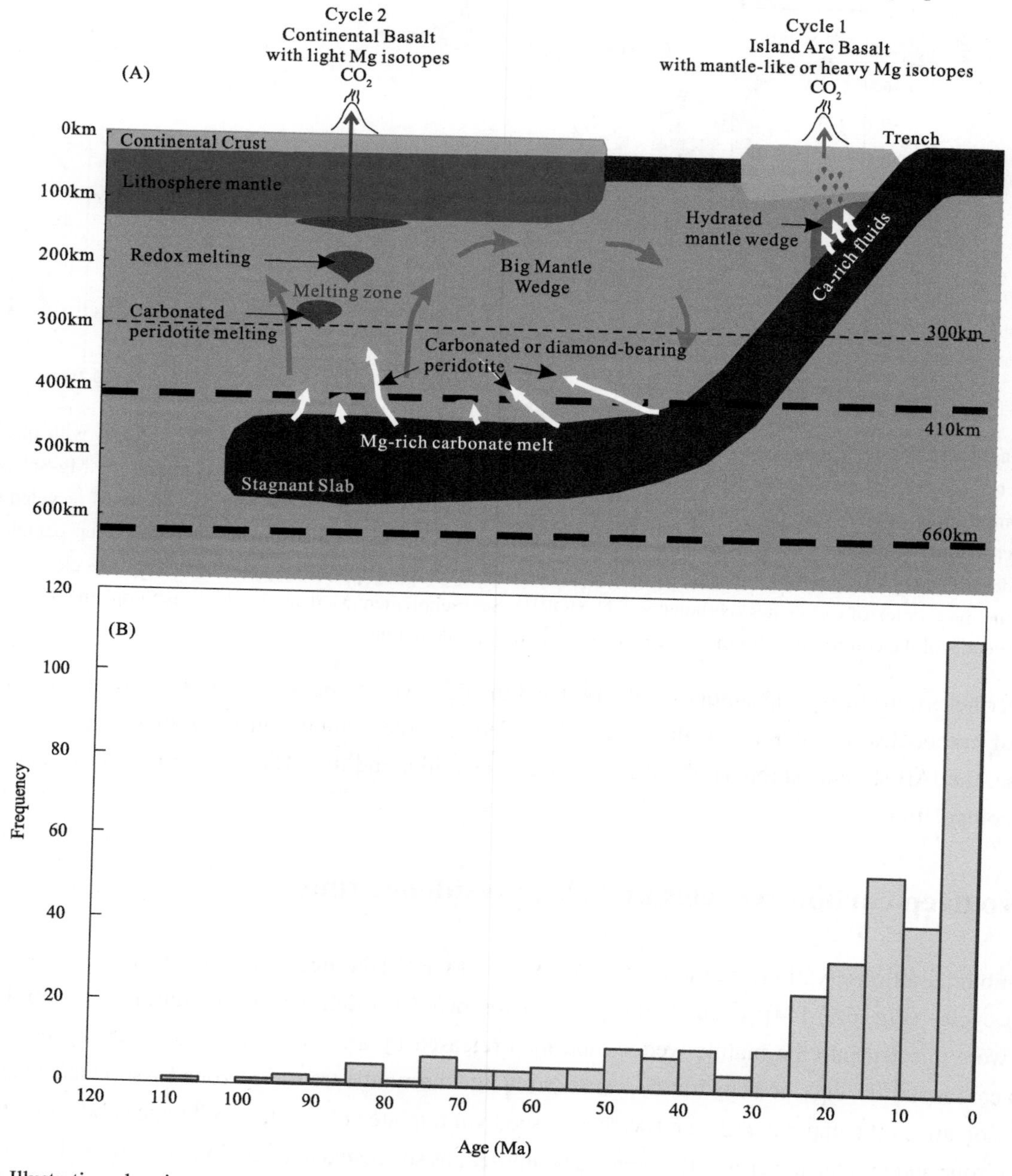

Fig. 6 (A) Illustration showing two cycles of the deep carbon cycling. In cycle 1, the partially dissolved Ca-rich carbonates in slab-derived fluid were injected into the mantle wedge, and then CO_2 was released by island arc volcanism. The island arc basalts have mantle-like or even heavier Mg isotopes. In cycle 2, the undissolved Mg-rich carbonates during slab dehydration, were carried into mantle transition zone by the subducted slab, partially melted and metasomatized the upper mantle, and then CO_2 was released by intraplate volcanism. These basalts have a light Mg isotopic composition. (B) Histogram of the ages of the basalts younger than 110Ma in the mainland of eastern China. Data are from Liu [44] and references therein.

The resident time of carbon for cycle 2 can be estimated based on the following geophysical and geo-chronological data. Seismic tomography shows that the subducted flat part of the slab in the mantle transition zone in eastern China is approximately 1500 km long, whereas the dipping part is approximately 1200 km long [11]. 165 Myr is a possible initial time for the west Pacific subduction, which is supported by much substantial geological evidence showing the initiation of western Pacific plate subduction in the late Jurassic period (180–150 Myr) [45–47]. Thus, when the slab reached the bottom of the transition zone, slab rollback and continental rifting in eastern China could occur [48]. The age statistics of Early Cretaceous igneous rocks in eastern China show that the peak age of 125 Myr should be a rifting time in eastern China [49]. Thus, approximately 40 Myr is required for the dipping slab subduction, which means that the average subduction speed is 3 cm/y. Since a low δ^{26}Mg signal has not been observed in basaltic rocks until 106 Myr, another 20 Myr is needed for subsequent processes: partial melting of subducted slabs, carbonate metasomatism above the mantle transition zone, mantle upwelling and its partial melting, and basalt eruption. Thus, the minimum resident time for cycle 2 is approximately 60 Myr (Fig. 6A), which is considerably shorter than previous estimations [2, 15].

6 Implications for paleoclimate

If the above estimation for the resident time of recycled carbon by Pacific subduction in eastern China is correct, given that approximately 50% of the subducted carbon is removed from the down-going plate and returned to the Earth's surface by arc volcanisms in east Asia [2, 12], another 50% of recycled carbon was stored in the mantle transition zone and upper mantle for a time period of 60−100 Myr. Thus, the large mantle wedge under eastern China is a huge carbon reservoir, which may have influenced the paleoclimate when strong degassing occurred. Based on mixing modeling calculations in the Mg-Sr isotopic diagram (Fig. 5 and see Supplementary data for details, available as Supplementary data at NSR online), the initial melt of the CBY-ML contains carbon ranging from 3, 400 to 24, 000 ppm, which is twice as much as that of the MORB [4]. The age statistics of the CBY-ML in eastern China show that the recycled carbon stored in the upper mantle began to return to the surface through continental volcanisms beginning in 106 Ma. Its activity increased quickly after 23 Ma and became much stronger during the Quaternary (Fig. 6B). Based on digital-image processing of a geological map, up to 90% of Cenozoic volcanic rocks in eastern China erupted during the Quaternary [50]. Therefore, the Quaternary volcanisms in eastern China could constitute one of the important sources of liberating carbon from the Earth's interior to the exosphere, which might have played a critical role in global climate and environmental changes. Therefore, it is important to monitor the CO_2 emissions in dormant volcanoes, hot springs and the Tan-Lu fault zone in eastern China.

Supplementary data

Supplementary data are available at *NSRSCP* online.

Acknowledgement

We sincerely thank Prof. Gerhard Wörner for providing IAB samples from Kamchatka and Costa Rica. We are also grateful to anonymous reviewers for their constructive comments. SGL initiated and organized this research. S.G.L. and W.Y. wrote the manuscript. S.K., X.M., H.T. and Y.H. analysed the samples for Mg isotopes. L.X. and X.M. analysed the samples for Sr and Nd isotopes. J.H. analysed the samples from Leizhou, and Fangcheng for Sr-Nd isotopes and major-trace elements. The samples are provided by G.W. for Kamchatka and Costa Rica; Q.X. for Shuangliao, Niutoushan, Leizhou and Chaihe-Aershan; X.C.W. for Hainan; X.Y. for the Philippines; W.S. for the Lau Basin; Z.Y.R. for Hannuoba; H.W. for Changbaishan; Y.L. for Feixian, and F.M. and J.Y. for Zhucheng. All authors discussed the results and implications.

Funding

This project is supported by the National Natural Science Foundation of China (41230209 to S.G.L., 41322022 and 41430105 to W.Y.).

References

1. Hazen RM and Schiffries CM. Why deep carbon? Rev Mineral Geochem 2013; 75: 1-6.
2. Dasgupta R and Hirschmann MM. The deep carbon cycle and melting in Earth's interior. Earth Planet Sci Lett 2010; 298: 1-13.
3. Dasgupta R. Ingassing, storage, and outgassing of terrestrial carbon through geologic time. Rev Mineral Geochem 2013; 75: 183-229.
4. Javoy M, Pineau F and Allegre CJ. Carbon geodynamic cycle. Nature 1982; 300: 171-3.
5. Yang W, Teng F-Z and Zhang H-F et al. Magnesium isotopic systematics of continental basalts from the North China craton: implications for tracing subducted carbonate in the mantle. Chem Geol 2012; 328: 185-94.
6. Huang J, Li S-G and Xiao Y et al. Origin of low δ^{26}Mg Cenozoic basalts from South China Block and their geodynamic implications. Geochim Cosmochim Acta 2015; 164: 298-317.
7. Tian H, Yang W and Li S-G et al. Origin of low δ^{26}Mg basalts with EM-I component: evidence for interaction between enriched lithosphere and carbonated asthenosphere. Geochim Cosmochim Acta 2016; 188: 93-105.
8. Higgins JA and Schrag DP. Constraining magnesium cycling in marine sediments using magnesium isotopes. Geochim Cosmochim Acta 2010; 74: 5039-53.
9. Teng F-Z, Li W-Y and Ke S et al. Magnesium isotopic composition of the Earth and chondrites. Geochim Cosmochim Acta 2010; 74: 4150-66.
10. Wombacher F, Eisenhauer A and Böhm F et al. Magnesium stable isotope fractionation in marine biogenic calcite and aragonite. Geochim Cosmochim Acta 2011; 75: 5797-818.
11. Huang J and Zhao D. High-resolution mantle tomography of China and surrounding regions. J Geophys Res: Solid Earth 2006; 111: B09305.
12. Johnston FKB, Turchyn AV and Edmonds M. Decarbonation efficiency in subduction zones: implications for warm Cretaceous climates. Earth Planet Sci Lett 2011; 303: 143-52.
13. Wallace PJ. Volatiles in subduction zone magmas: concentrations and fluxes based on melt inclusion and volcanic gas data. J Volcanol Geotherm Res 2005; 140: 217-40.
14. Kelemen PB and Manning CE. Reevaluating carbon fluxes in subduction zones, what goes down, mostly comes up. Proc Natl Acad Sci USA 2015; 112: E3997- 4006.
15. Sleep NH and Zahnle K. Carbon dioxide cycling and implications for climate on ancient Earth. J Geophys Res: Planets 2001; 106: 1373-99.
16. Thomson AR, Walter MJ and Kohn SC et al. Slab melting as a barrier to deep carbon subduction. Nature 2016; 529: 76-9.
17. Hofmann AW. Mantle geochemistry: the message from oceanic volcanism. Nature 1997; 385: 219-29.
18. Rudnick RL and Gao S. Composition of the continental crust. In: Rudnick RL (ed), The Crust. Oxford: Elsevier-Pergamon, 2003, 1-64.
19. Yang W, Teng F-Z and Li W-Y et al. Magnesium isotopic composition of the deep continental crust. Am Mineral 2016; 101: 243-52.
20. Teng FZ, Hu Y and Chauvel C. Magnesium isotope geochemistry in arc volcanism. Proc Natl Acad Sci USA 2016; 113: 7082-7.
21. Wang X-C, Li Z-X and Li X-H et al. Identification of an ancient mantle reservoir and young recycled materials in the source

region of a young mantle plume: implications for potential linkages between plume and plate tectonics. Earth Planet Sci Lett 2013; 377-378: 248-59.

22. Teng FZ, Wadhwa M and Helz RT. Investigation of magnesium isotope fractionation during basalt differentiation: implications for a chondritic composition of the terrestrial mantle. Earth Planet Sci Lett 2007; 261: 84-92.
23. Wang S-J, Teng F-Z and Williams HM et al. Magnesium isotopic variations in cratonic eclogites: origins and implications. Earth Planet Sci Lett 2012; 359- 360: 219-26.
24. Wang S-J, Teng F-Z and Li S-G. Tracing carbonate-silicate interaction during subduction using magnesium and oxygen isotopes. Nature Commun 2014; 5: 5328.
25. Wang S-J, Teng F-Z and Rudnick RL et al. Magnesium isotope evidence for a recycled origin of cratonic eclogites. Geology 2015; 43: 1071-4.
26. Dasgupta R, Hirschmann MM and Stalker K. Immiscible transition from carbonate-rich to silicate-rich melts in the 3 GPa melting interval of eclogite + CO_2 and genesis of silica-undersaturated ocean island lavas. J Petrol 2006; 47: 647-71.
27. Sobolev AV, Hofmann AW and Sobolev SV et al. An olivine-free mantle source of Hawaiian shield basalts. Nature 2005; 434: 590-7.
28. Xu Y-G, Ma J-L and Frey FA et al. Role of lithosphere-asthenosphere interaction in the genesis of Quaternary alkali and tholeiitic basalts from Datong, western North China Craton. Chem Geol 2005; 224: 247-71.
29. Zeng G, Chen L-H and Xu X-S et al. Carbonated mantle sources for Cenozoic intra-plate alkaline basalts in Shandong, North China. Chem Geol 2010; 273: 35-45.
30. Dasgupta R, Hirschmann MM and Smith ND. Partial melting experiments of peridotite + CO_2 at 3 GPa and genesis of alkalic ocean island basalts. J Petrol 2007; 48: 2093-124.
31. Sobolev AV, Hofmann AW and Kuzmin DV et al. The amount of recycled crust in sources of mantle-derived melts. Science 2007; 316: 412-17.
32. Hauri EH, Shimizu N and Dieu JJ et al. Evidence for hotspot-related carbonatite metasomatism in the oceanic upper mantle. Nature 1993; 365: 221-7.
33. Walter MJ, Bulanova GP and Armstrong LS et al. Primary carbonatite melt from deeply subducted oceanic crust. Nature 2008; 454: 622-5.
34. Alle`gre CJ and Minster JF. Quantitative models of trace element behavior in magmatic processes. Earth Planet Sci Lett 1978; 38: 1-25.
35. Humayun M, Qin L and Norman MD. Geochemical evidence for excess iron in the mantle beneath Hawaii. Science 2004; 306: 91-4.
36. Deng J, Yang X and Zhang Z-F et al. Early Cretaceous arc volcanic suite in Cebu Island, Central Philippines and its implications on paleo-Pacific plate subduction: constraints from geochemistry, zircon U-Pb geochronology and Lu-Hf iso- topes. Lithos 2015; 230: 166-79.
37. Churikova T, Dorendorf F and Wörner G. Sources and fluids in the mantle wedge below Kamchatka: evidence from across-arc geochemical variation. J Petrol 2001; 42: 1567-93.
38. Sun W, Bennett VC and Kamenetsky VS. The mechanism of Re enrichment in arc magmas: evidence from Lau Basin basaltic glasses and primitive melt in- clusions. Earth Planet Sci Lett 2004; 222: 101-14.
39. Abratis M and Worner G. Ridge collision, slab-window formation, and the flux of Pacific asthenosphere into the Caribbean realm. Geology 2001; 29: 127-30.
40. Ague JJ and Nicolescu S. Carbon dioxide released from subduction zones by fluid-mediated reactions. Nature Geos 2014; 7: 355-60.
41. Pan D, Spanu L and Harrison B et al. Dielectric properties of water under extreme conditions and transport of carbonates in the deep Earth. Proc Natl Acad Sci USA 2013; 110: 6646-50.
42. Frezzotti ML, Selverstone J and Sharp ZD et al. Carbonate dissolution during subduction revealed by diamond-bearing rocks from the Alps. Nature Geos 2011; 4: 703-6.
43. Kushiro I, Satake H and Akimoto S. Carbonate-silicate reactions at high- pressures and possible presence of dolomite and magnesite in upper mantle. Earth Planet Sci Lett 1975; 28: 116-20.
44. Liu J. Volcanoes in China. Beijing: Science Press, 1999.
45. Engebretson DC, Cox A and Gordon RG. Relative motions between oceanic and continental plates in the Pacific Basin. Geological Society of America Special Papers 1985; 206: 1-60.

46. Maruyama S, Isozaki Y and Kimura G et al. Paleogeographic maps of the Japanese islands: plate tectonic synthesis from 750 Ma to the present. Island Arc 1997; 6: 121-42.
47. Wu F-Y, Yang J-H and Lo C-H et al. The Heilongjiang Group: a Jurassic accre- tionary complex in the Jiamusi Massif at the western Pacific margin of northeastern China. Island Arc 2007; 16: 156-72.
48. Kusky TM, Windley BF and Wang L et al. Flat slab subduction, trench suction, and craton destruction: comparison of the North China, Wyoming, and Brazilian cratons. Tectonophysics 2014; 630: 208-21.
49. Wu F-Y, Lin J-Q and Wilde SA et al. Nature and significance of the Early Cretaceous giant igneous event in eastern China. Earth Planet Sci Lett 2005; 233: 103-19.
50. Chen X, Chen L and Chen Y et al. Distribution summary of Cenozoic basalts in central and eastern China. Geo J China Univ 2014; 20: 507-19.

Could sedimentary carbonates be recycled into the lower mantle? Constraints from Mg isotopic composition of Emeishan basalts*

Heng-Ci Tian[1,2], Wei Yang[1], Shu-GuangLi[3], and Shan Ke[3]

1. Key Laboratory of Earth and Planetary Physics, Institute of Geology and Geophysics, Chinese Academy of Sciences, Beijing 100029, China
2. University of Chinese Academy of Sciences, Beijing 100049, China
3. State Key Laboratory of Geological Processes and Mineral Resources, China University of Geosciences, Beijing 100083, China

亮点介绍：在碳循环过程中，碳酸盐岩能否被俯冲进入下地幔，是一个有争议的问题。本文通过对地幔柱成因的峨眉山大火成岩省两类玄武岩系统的Mg同位素研究，发现其低钛和高钛两类玄武岩都具有与地幔相似的Mg同位素组成。模拟计算表明，其源区混入的碳酸盐岩的量≤2%。因此，再循环沉积碳酸盐岩很可能难以进入到下地幔，这与高温高压实验岩石学的结果一致。

Abstract Whether or not sedimentary carbonates can be recycled into the lower mantle through subduction remains unclear. To further elucidate this issue, we investigate the Mg isotopic composition of Permian basalts from the Emeishan large igneous province (ELIP). Emeishan basalts can be divided into two major groups: the low-Ti and high-Ti basalts, which exhibit distinct major, trace element, and Sr-Nd-Pb isotopic compositions. However, they both possess mantle-like Mg isotopic compositions with δ^{26}Mg values of −0.33 to −0.19‰ and −0.35 to −0.19‰, respectively. Both the low-Ti and high-Ti basalts have experienced compositional evolution, e.g., fractional crystallization or crustal contamination, because their Mg$^{\#}$ (39.8−61.1) values are significantly lower than that of primary melt (~72.0). However, their δ^{26}Mg values do not vary with Mg$^{\#}$, indicating that the Mg isotopic compositions of the basalts were inherited from their mantle source. The low-Ti basalts are enriched in light rare earth elements with extremely low $\varepsilon_{Nd}(t)$ values (mostly ranging from −10.5 to −8.8) and low Ce/Pb, high Ba/Th ratios, suggesting that they were derived from the lithospheric mantle that was metasomatized by slab-derived fluids. Their mantle-like Mg isotopic composition suggests that the metasomatism of slab-derived fluids did not affect their Mg isotopic composition. This is consistent with most of the reported Arc basalts with mantle-like Mg isotopic compositions. On the other hand, the high-Ti basalts display an OIB-like trace element pattern and positive $\varepsilon_{Nd}(t)$ values (+0.2 −+2.1), which were possibly derived from a mantle plume that originated from the lower mantle. Their mantle-like Mg isotopic composition indicates that sedimentary carbonates recycled into the lower mantle beneath the ELIP were limited (<2%).

* 本文发表在：Lithos, 2017, 292–293：250-261

1 Introduction

Although carbon cycling theory requires that substantial quantities of sedimentary carbonates are recycled into the mantle by slab subduction (Dasgupta, 2013; Javoy et al., 1982), their precise fate in the mantle remains unknown. For instance, it is possible that they are stored in the mantle wedge (Kelemen and Manning, 2015), continental lithospheric mantle (Hu et al., 2016; Huang et al., 2011a; Ke et al., 2016; Wang et al., 2016a; Xiao et al., 2013; Yang et al., 2009), convective upper mantle (Huang et al., 2015; Li et al., 2017; Tian et al., 2016; Yang et al., 2012), lower mantle (Huang et al., 2011b) or even returned to the Earth's atmosphere by volcanism at island arcs, continents, or oceanic islands. The residence time of recycled carbonates in the mantle could be 5 to 10 Ma in the mantle wedge beneath island arcs (Kelemen and Manning, 2015), 60 to several hundred Ma in the upper mantle (Li et al., 2017) or 1–2 Ga in the lower mantle (Dasgupta and Hirschmann, 2010; Sleep and Zahnle, 2001) depending on where the carbonates are stored. Therefore, it is critical to determine how deep the sedimentary carbonates can be recycled into mantle by plate subduction.

Magnesium isotopes are demonstrated as effective tracers for the recycling of sedimentary carbonates in the mantle source (e.g., Huang et al., 2015; Li et al., 2017; Tian et al., 2016; Yang et al., 2012). This is due to the distinctive Mg isotopic compositions between sedimentary carbonates (δ^{26}Mg=–5.60 to –0.66‰; Beinlich et al., 2014; Saenger and Wang, 2014) and mantle peridotites (δ^{26}Mg=–0.25±0.07‰; Teng et al., 2010). Other processes (e.g. crustal contamination, magma evolution, lithosphere-asthenosphere interaction) can hardly shift Mg isotopic composition of basalts. For example, the crust has low MgO content with an average of ~2.48 wt.% (Rudnick and Gao, 2003) and mantle-like average Mg isotopic composition (δ^{26}Mg=0.24±0.07‰; Yang et al., 2016). Thus, crustal contamination cannot modify the Mg isotopic compositions of basalts. In addition, Mg isotope fractionation during magmatic evolution (Teng et al., 2007, 2010) and prograde metamorphism dehydration in the subducted slab (Wang et al., 2014) are limited. Moreover, the lithosphere-asthenosphere interaction cannot modify the Mg isotopic composition of upwelling basaltic melts (Tian et al., 2016).

Recent Mg isotopic studies of continental basalts in eastern China revealed that sedimentary carbonates can be recycled into the mantle transition zone (MTZ, 410–660 km) (Huang et al., 2015; Li et al., 2017; Tian et al., 2016;Yang et al., 2012). However, whether or not sedimentary carbonates could be carried into the lower mantle via subduction is a topic of debate. Combined Ca-Sr isotopes of Hawaiian basalts suggest that approximately 4% ancient marine carbonates were added into the source of the Hawaiian plume (Huang et al., 2011b). However, experimental studies have pointed out that lower mantle carbonates recycling may be limited (Hammouda and Keshav, 2015; Kelemen and Manning, 2015; Thomson et al., 2016). Even though low δ^{26}Mg features were observed in the plume-related Hainan basalts and HIMU-like New Zealand basalts, they are considered to result from recycled carbonated eclogites rather than sedimentary carbonates based on correlations between Mg isotopes and major/trace element ratios (Li et al., 2017; Wang et al., 2016b).

In order to determine whether subduction into the lower mantle can carry sedimentary carbonates, more plume-related basalts derived from the lower mantle should be investigated by Mg isotopes. The Permian basalts from the Emeishan large igneous province (ELIP), southwestern China, are widely accepted to be mantle plume in origin (e.g., Chuang and Jahn, 1995; Xiao et al., 2004; Xu et al., 2001, 2004). The Emeishan basalts can be divided into the low-Ti and high-Ti basalts based on their distinctive geochemical characteristics (e.g., Xu et al., 2001; Xiao et al., 2004). Although the petrogenesis of the low-Ti basalts remains controversial (e.g., Song et al., 2008; Wang et al., 2007; Xiao et al., 2004; Xu et al., 2001, 2004, 2007), the high-Ti basalts are considered to originate from a mantle plume (e.g., Chung et al., 1998; Huang et al., 2014; Song et al., 2001, 2008; Xiao et al., 2003, 2004; Xu et al., 2001, 2004; Zhang and Wang, 2002). Due to the high potential mantle temperature

(>1550 °C) as inferred by REE inversion in the basalts (Xu et al., 2001) and picrites (Zhang et al., 2006), the mantle plume could be derived from the lower mantle (Song et al., 2001). The data presented in this study show that both types of basalts from the ELIP have mantle-like Mg isotopic compositions, with δ^{26}Mg values of −0.33 to −0.19‰ and −0.35 to −0.19‰, respectively. This suggests that recycled carbonate is limited to less than 2% in the lower mantle source of the Emeishan plume.

2 Geological background and sample descriptions

The Emeishan large igneous province (ELIP) is located in the western margin of the Yangtze craton, southwest China (Chung and Jahn, 1995; Xu et al., 2001) (Fig. 1a), covering an area of ~250, 000 km^2. The thickness of the volcanic lavas varies from ~5 km in the west to ~100 m in the east (Chung and Jahn, 1995; Xu et al., 2001). The ELIP can be divided into a central part and an outer zone, based on the thickness of the volcanic lavas, their chemical compositions, occurring rock types, etc. (Song et al., 2005). The exposed volcanic rocks include picrites, basalts, basaltic andesites, rhyolite-trachyte, intrusions, and basaltic pyroclastics (Xu et al., 2001).

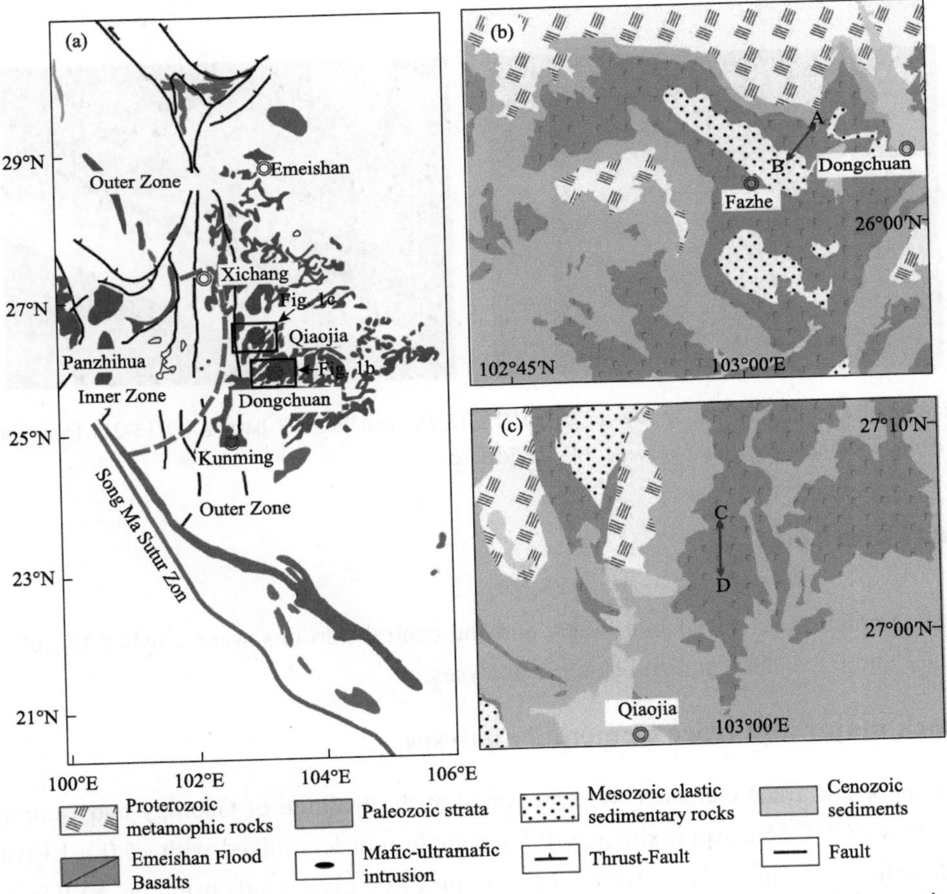

Fig. 1 (a) Geological map of the Emeishan large igneous province, showing the distribution of the continental flood basalts (modified from Song et al., 2008). (b) Simplified map of the Dongchuan area. (c) Simplified map of the Qiaojia area. Lines A–B and C–D represent the profiles where samples were collected from Dongchuan and Qiaojia, respectively.

The Emeishan basalts unconformably overlie late Permian limestone formations and are overlain by Triassic sedimentary rocks (Chung and Jahn, 1995). They erupted at approximately 260 Ma ago based on zircon U-Pb and whole rock Ar-Ar dating (Ali et al., 2004; Guo et al., 2004; Tao et al., 2009; Wang and Zhou, 2006; Xu et al., 2008; Zhong and Zhu, 2006; Zhou et al., 2002, 2008), coinciding with the end-Guadalupian mass extinction (Zhou et al., 2002). The Emeishan basalts include two major groups: low-Ti basalts (TiO_2 <2.5 wt.%; Ti/Y <500) and high-Ti basalts (TiO_2>2.5 wt.%; Ti/Y>500) (e.g., Xiao et al., 2004; Xu et al., 2001). The central part of ELIP consisted of the low-Ti and the high-Ti basalts, whereas the outer zone is dominated by high-Ti basalts (e.g., Song et al., 2006). Although the mantle source of the low-Ti basalts remains controversial (e.g., Chung and Jahn, 1995; Song et al., 2008; Xu et al., 2001, 2007), most studies suggested that they were derived from partial melting of subcontinental lithospheric mantle (e.g., Song et al., 2008; Xiao et al., 2004). The high-Ti basalts with OIB-like trace elements and highly positive $\varepsilon_{Nd}(t)$ and low $^{87}Sr/^{86}Sr(t)$ isotopic features may have originated from a mantle plume (e.g., Song et al., 2008; Xiao et al., 2003, 2004; Xu et al., 2001).

In this study, 21 samples were collected from two well-preserved lava successions from Dongchuan and Qiaojia in the central part of the ELIP (Fig. 1a). The Dongchuan samples (Fig. 1b) include both high-Ti and low-Ti basalts (Song et al., 2008), whereas the Qiaojia samples (Fig. 1c) are all high-Ti basalts. Based on the petrographic observations, the low-Ti basalts mainly contain clinopyroxene with occasionally minor plagioclase phenocrysts in a fine-grained groundmass, whereas the high-Ti basalts are dominated by plagioclase with minor clinopyroxene phenocrysts (Fig. 2).

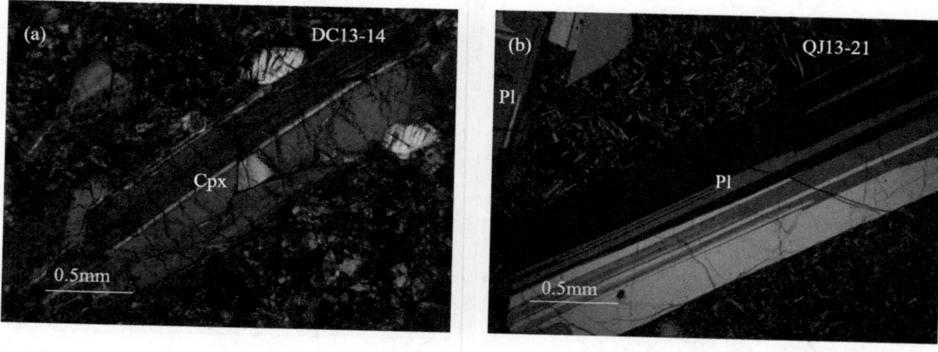

Fig. 2 Photo micrographs for representative low-Ti basalt (DC13-14) and high-Ti basalts (QJ13-21) from the Emeishan large igneous province. Abbreviations: Cpx=clinopyroxene; Pl=plagioclase.

3 Analytical methods

The selected samples were sawed into chips, and the central portions were crushed to 200 mesh using an agate mill for major and trace elemental and isotopic analyses.

3.1 Whole rock major and trace elemental analyses

Whole rock major and trace elements were measured at the Institute of Geology and Geophysics, Chinese Academy of Sciences (IGGCAS). Approximately 0.5 g sample powders mixed with ~5.0 g $Li_2B_4O_7$ were fused into a glass bead. Major element analyses were measured on fused glass beads using an AXIOS Minerals X-ray fluorescence (XRF) spectrometer. The uncertainties for major elements are <1.0 wt.%. Loss on ignition (LOI) was determined after 2 h of baking under a constant temperature of 1000°C and 30 min cooling to ambient temperature.

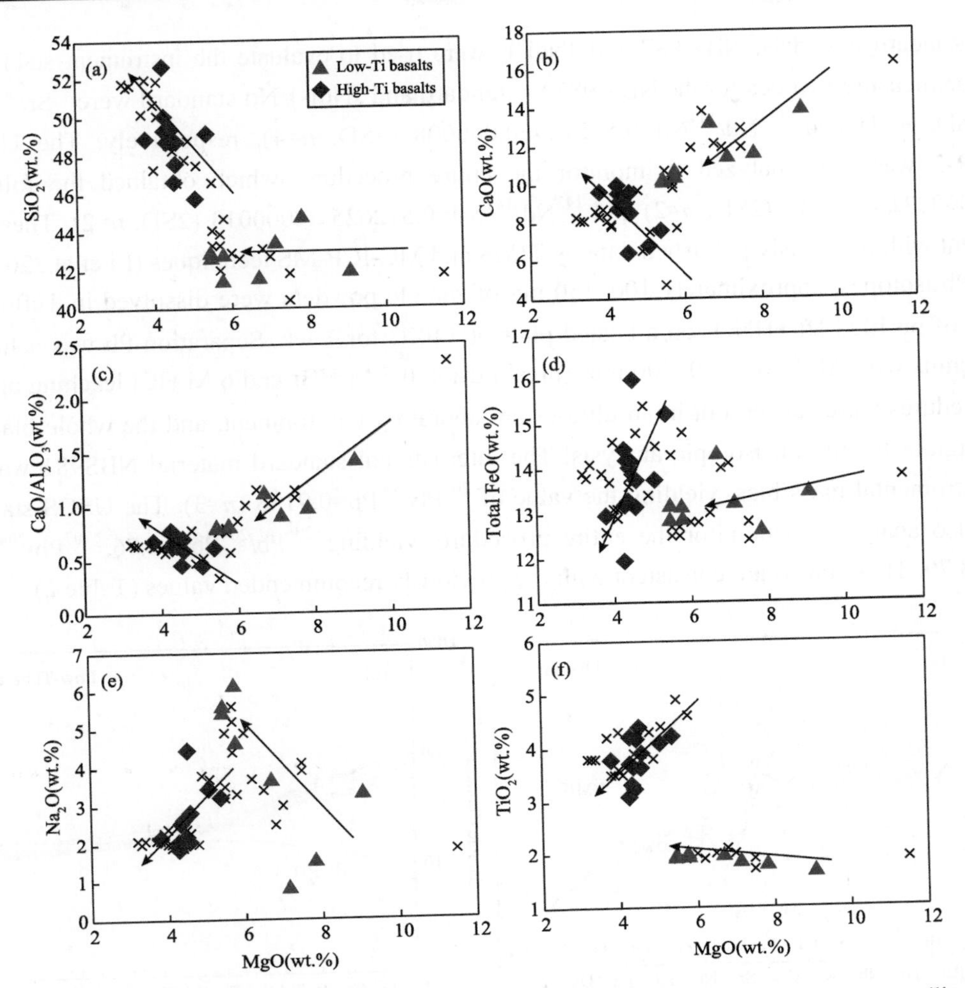

Fig. 3 Major oxide variations for the Emeishan basalts. The arrows represent the trends of fractional crystallization. The symbol crosses represent literature data from Song et al. (2008).

For trace element analyses, approximately 40 mg sample powders were dissolved in a mixture of $HF+HNO_3$ acids in Teflon bombs (see Chu et al., 2009 for details). The trace elements were measured using an ELEMENT instrument at IGGCAS. Two reference materials, GSR1 and GSR3, were measured to evaluate the chemical preparation procedure and the condition of the instrument. GSR1 and GSR3 data are listed in Supplementary Table 1, which are consistent with the reference values. Both precision and accuracy are <5% for most of the trace elements.

3.2 Sr-Nd-Pb isotopic analyses

The whole-rock Sr-Nd-Pb isotopic compositions were analyzed at IGGCAS using a Finnigan MAT-262 thermal ionization mass spectrometer (TIMS) instrument. The Rb-Sr and Sm-Nd isotopic measurements were made following the procedures described by Li et al. (2012a) and Yang et al. (2010). Sample powders for Sr and Nd isotopic analyses were dissolved in a Savillex Teflon screw-top capsule after being spiked with mixed ^{87}Rb-^{84}Sr and ^{149}Sm-^{150}Nd tracers prior to $HF+HNO_3+HClO_4$ dissolution. Rb, Sr, Sm and Nd were separated using the classical two-step ion exchange chromatographic method. The whole procedural blank was <300 pg for Rb-Sr and <100 pg for Sm-Nd systems, which was negligible for Rb-Sr and Sm-Nd analyses. The isotopic ratios corrected for mass fractionation were normalized to $^{86}Sr/^{88}Sr=0.1194$ and $^{146}Nd/^{144}Nd=0.7219$, respectively. The international standard materials, NBS-987 and JNdi-1, were used to evaluate the instrument stability during data collection. The measured values for the NBS-987 Sr standard and JNdi-1 Nd standard were $^{87}Sr/^{86}Sr=0.710250\pm0.000026$ (2SD, $n=4$) and $^{143}Nd/^{144}Nd=0.512120\pm0.000008$ (2SD, $n=4$), respectively. The USGS reference material BCR-2 was also analyzed to monitor the entire procedure, which obtained the following results: $^{87}Sr/^{86}Sr=0.705023\pm0.000011$ (2SD, $n=2$) and $^{143}Nd/^{144}Nd=0.512625\pm0.000013$ (2SD, $n=2$). These results are in good agreement with previously published data by TIMS and MC-ICP-MS techniques (Li et al., 2012a, b).

For the Pb isotopes, approximately 100–150 mg of sample powders were dissolved in Teflon beakers with mixture acids of purified $HF+HNO_3$ on a heated plate at 140°C for 1 wk. Separation Pb was achieved by anion exchange columns with AG1-X8 (100–200 mesh) resin using 0.6 M HBr and 6 M HCl leaching agent. All of the chemical procedures were carried out in an ultra-clean laboratory environment, and the whole blank is <200 pg, which is negligible for the Pb isotopic analysis. The international standard material NBS981 was measured to correct for instrumental mass bias, yielding the value of $^{207}Pb/^{206}Pb=0.9138$ ($n=3$). The USGS standard material BCR-2 was also analyzed to monitor the entire procedure, yielding $^{206}Pb/^{204}Pb=18.76$, $^{207}Pb/^{204}Pb=15.62$ and $^{208}Pb/^{204}Pb=38.70$. These ratios are consistent with the previously recommended values (Table 2).

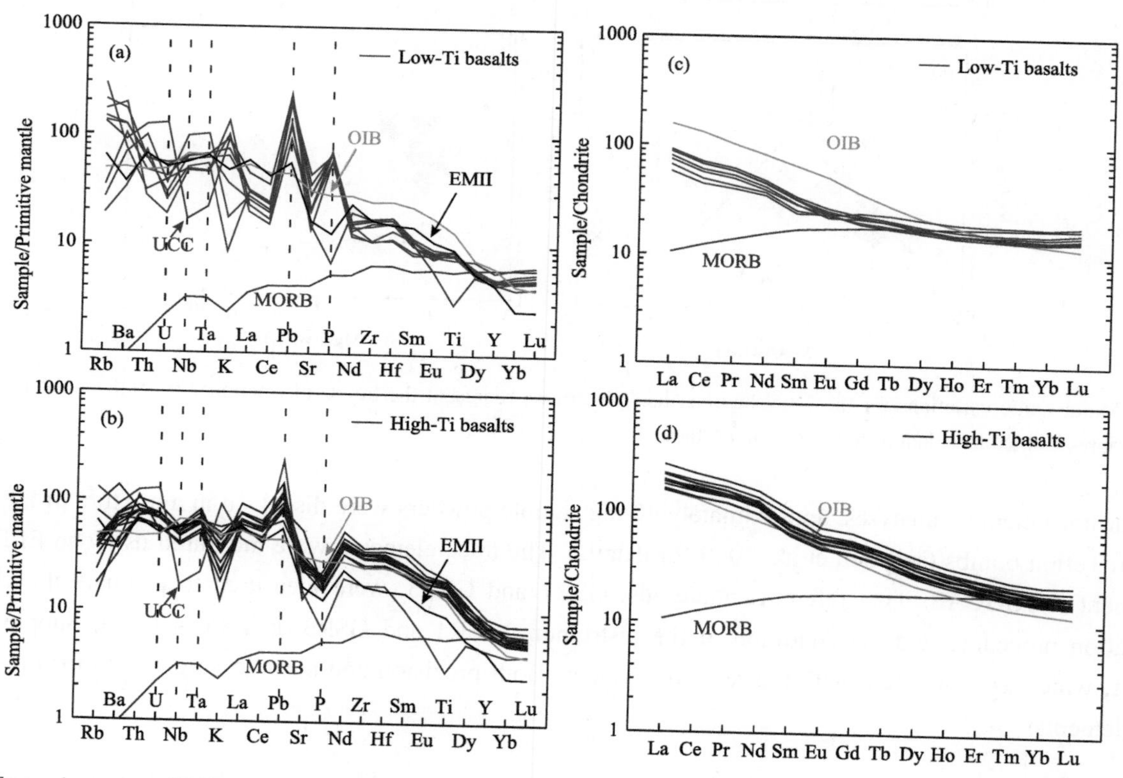

Fig. 4 Trace elements and REE abundances of the Emeishan basalts. (a) and (b) are trace element diagrams normalized by primitive mantle values. (c) and (d) are REE patterns normalized by chondrite values (Sun and McDonough, 1989). Data source: UCC from Rudnick and Gao (2003), OIB and MORB from Sun and McDonough (1989), and Samoan lavas (an EMII-like endmember) from Workman et al. (2004).

3.3 Magnesium isotopic analysis

Stable Mg isotopic compositions were determined at the State Key Laboratory of Geological Processes and Mineral Resources, China University of Geosciences (Beijing) (CUGB), following the methods described in previous studies (Huang et al., 2015; Ke et al., 2016; Tian et al., 2016; Yang et al., 2012). The chemistry procedure for the Mg separation was performed in an ultra-clean laboratory. Based on the Mg content, one to ten mg of sample powders were dissolved in beakers with the best mixture acids of HF and HNO_3 (~3:1). These capped beakers were heated at 160°C, and then the solutions were evaporated to dryness. Following this step, the dried samples were refluxed in HCl-HNO_3 (~3:1) and then again evaporated to dryness at 80°C. The dried samples were then re-dissolved in HNO_3 and heated to dryness at 70°C. Finally, the residues were dissolved in 1 M HNO_3 in preparation for chromatographic separation. Separation of Mg was obtained using cation exchange columns loaded with Bio-Rad AG50W-X8 (200–400 mesh) pre-cleaned resin using 1 M HNO_3 as the leaching agent (Teng et al., 2010). The separation columns were used twice to achieve the required level of purification. Prior to Mg isotopic measurements, the ratios of the abundances of matrix elements (e.g., Na, Al, Ca, Ti, and Fe) to Mg were measured in the solutions using ICP-MS, with cation/Mg (mass/mass) ratios of <0.04, which is negligible for the Mg isotope analyses. The total procedure blank was ~8.5 ng, and is negligible compared to the Mg loaded on the column.

Magnesium isotopic compositions were analyzed by the standard bracketing method using a Neptune Plasma MC-ICPMS at CUGB. The instrument parameters for Mg analyses are similar to those described in Ke et al. (2016). The long-term precision based on the repeated analyses of synthetic pure Mg solution and natural materials is ≤0.10‰ for δ^{26}Mg (Huang et al., 2015; Ke et al., 2016; Tian et al., 2016). The results are reported in the conventional δ notation, which is defined as: $\delta^{25, 26}Mg=[(^{25, 26}Mg/^{24}Mg)_{sample}/(^{25, 26}Mg/^{24}Mg)_{DSM3}-1] \times 1000$, and DSM3 is an international reference of solution made from pure Mg metal (Galy et al., 2003). The analyses of standard materials BCR-2, BHVO-2, AGV-2, and GSP-2 yielded δ^{26}Mg values (Table 3) in good agreement with previously published values (−0.36 to −0.16‰ for BCR-2, −0.32 to −0.16‰ for BHVO-2, −0.22 to −0.10‰ for AGV-2, and +0.02 to +0.06‰ for GSP-2; An et al., 2014; Huang et al., 2015; Ke et al., 2016; Teng et al., 2015; Tian et al., 2016).

4 Results

4.1 Major and trace elements

The low-Ti basalts exhibit limited variation of SiO_2 (41.38–44.72 wt.%) but a large variation of MgO (5.45–9.07 wt.%). On the other hand, the high-Ti basalts display large SiO_2 variation (45.78–52.67 wt.%) and small MgO variation (3.80–5.38 wt.%) (Fig. 3a). MgO contents of the low-Ti basalts show positive correlation with CaO, CaO/Al_2O_3, and total FeO contents, and negative correlation with Na_2O and TiO_2 contents (Fig. 3). In contrast, MgO contents of the high-Ti basalts display positive correlations with total FeO, Na_2O and TiO_2, and negative correlations with CaO and CaO/Al_2O_3 (Fig. 3).

In the primitive-mantle normalized diagrams, the low-Ti basalts exhibit weak positive anomalies of Nb and Ta, as well as strong positive anomalies of K, Pb and P, which is distinctive to both the Oceanic Island Basalts (OIB) and Upper Continental Crust (UCC), but similar to the Samoan lavas-like signatures (an EM-II mantle endmember), with exception to the strong positive P anomaly (Fig. 4a). These characteristics imply that volatile-bearing minerals were in the mantle source (Song et al., 2008). In contrast, high-Ti basalts show an OIB-like trace element composition (Fig. 4b). They are enriched in light rare earth elements relative to the heavy rare earth elements (Fig. 4c, d), with $(La/Yb)_N$ values of 9.5 to 12.7 (Table 1 and Supplementary Table 1). In addition, the low-Ti basalts exhibit higher Nb/U (35.9–92.5) and Ba/Th (37.3–580.4), but lower Ce/Pb (2.5–4.8) ratios than the high-Ti basalts (Fig. 5).

Table 1　Representative major and trace-element data for the Dongchuan and Qiaojia lavas, Emeishan large igneous province

Samples	DC13-07	DC13-08	DC13-12	DC14-04	DC13-16	DC13-17	QJ13-17	QJ13-21
Types	Low-Ti			High-Ti				
Locality	Dongchuan			Dongchuan			Qiaojia	
Major elements (wt.%)								
SiO_2	41.54	41.86	41.38	45.78	49.42	46.66	48.76	49.10
TiO_2	1.95	1.64	1.98	4.09	4.17	4.39	3.32	3.24
Al_2O_3	11.84	9.85	13.17	15.00	13.40	13.89	13.40	14.38
Fe_2O_3[a]	14.29	13.44	13.12	13.75	14.17	15.99	14.19	14.03
MnO	0.27	0.30	0.25	0.19	0.18	0.17	0.17	0.18
MgO	6.70	9.07	5.76	5.07	4.46	4.53	4.34	4.41
CaO	13.15	13.84	10.62	6.77	9.00	6.41	9.13	9.26
Na_2O	3.64	3.28	6.12	3.45	2.19	4.46	2.55	2.64
K_2O	0.53	0.26	1.09	1.73	1.06	1.05	0.81	0.69
P_2O_5	1.44	1.51	1.57	0.43	0.46	0.51	0.34	0.35
LOI	3.46	4.00	4.14	3.40	1.54	2.42	3.26	1.44
Total	98.81	99.05	99.20	99.7	100.05	100.48	100.27	99.72
Trace elements								
Sc	26.7	31.3	20.4	23.5	26.2	30.7	25.1	26.7
V	427.5	375.1	298.1	371.5	398.8	450.6	383.8	326.6
Co	49.2	53.0	44.6	35.3	45.2	46.5	45.7	42.8
Zn	154.7	187.1	137.6	133.6	151.6	179.8	135.2	133.8
La	19.0	20.9	17.6	45.0	53.5	64.7	37.1	37.2
Ce	36.9	41.2	34.7	103.6	117.3	137.2	82.2	82.3
Pr	5.1	5.6	4.7	13.7	16.0	18.2	11.3	11.5
Nd	19.5	21.0	18.3	58.8	63.4	72.8	44.7	44.8
Sm	4.5	4.9	4.4	11.5	14.3	16.3	9.8	9.7
Eu	1.3	1.4	1.4	3.3	3.7	4.2	2.8	2.8
Gd	4.5	5.0	4.1	10.4	11.5	13.5	8.8	8.8
Tb	0.8	0.9	0.7	1.5	1.7	1.9	1.3	1.3
Dy	4.5	5.4	4.1	8.0	9.0	10.3	6.9	7.0
Ho	0.9	1.1	0.9	1.5	1.7	1.9	1.3	1.3
Er	2.5	3.2	2.3	3.6	4.2	4.7	3.2	3.3
Tm	0.4	0.5	0.4	0.5	0.6	0.6	0.4	0.5
Yb	2.4	3.1	2.4	2.9	3.4	3.8	2.6	2.8
Lu	0.4	0.5	0.4	0.4	0.5	0.6	0.4	0.4
Rb	16.8	12.0	21.6	57.1	23.3	26.4	49.4	14.3
Sr	930.1	424.8	444.0	679.7	583.7	270.6	605.7	1144.6
Ba	475.5	227.5	711.6	947.1	513.3	278.8	391.6	471.4
Y	21.4	26.4	20.1	35.8	35.2	41.1	29.1	29.8
Nb	70.0	36.2	45.6	37.7	40.2	41.4	29.1	29.3
Ta	4.2	2.2	2.8	2.7	3.1	3.4	2.2	2.2
Zr	167.2	123.3	177.0	374.9	398.6	425.7	295.4	281.3
Hf	4.9	4.2	4.9	9.6	11.1	12.1	8.3	8.0
Th	8.4	6.1	4.0	6.9	8.7	9.3	5.3	5.6
U	0.8	0.5	0.9	1.6	1.9	1.1	1.2	1.2
Pb	14.0	8.5	13.9	7.7	7.6	10.4	6.1	6.0

[a] Total FeO as Fe_2O_3.

4.2 Sr-Nd-Pb isotopes

The initial Sr-Nd-Pb isotopic ratios were all corrected to 260 Ma (Zhou et al., 2008; Zhou et al., 2002) and data is presented in Table 2. The low-Ti basalts exhibit high initial $^{87}Sr/^{86}Sr(t)$ values varying from 0.7078 to 0.7088, extremely low $\varepsilon_{Nd}(t)$ values ranging from −10.5 to −8.8 and high $^{206}Pb/^{204}Pb(t)$ from 18.48 to 18.55 and $^{207}Pb/^{204}Pb(t)$ from 15.60 to 15.62. These low-Ti basalts define a trend that lie between enriched mantle 1 (EMI) and enriched mantle 2 (EMII) (Fig. 6). In contrast, the high-Ti basalts yield low $^{87}Sr/^{86}Sr(t)$ (0.7051−0.7065) and Pb isotopic ratios, but high initial $\varepsilon_{Nd}(t)$ (+0.2−+2.1) (Table 2 and Fig. 6a), similar to those of the ECFB elsewhere in the ELIP (Fig. 6; Song et al., 2008; Wang et al., 2007; Xiao et al., 2004; Xu et al., 2001). In addition, in the Pb isotopic ratio diagram (Fig. 6b), both high- and low-Ti basalts plot upon the Northern Hemisphere Reference Line (NHRL; Hart, 1984).

4.3 Magnesium isotopes

Stable Mg isotopic compositions of the low- and high-Ti samples are listed in Table 3. The $\Delta^{25}Mg'$ values ($\Delta^{25}Mg'=\delta^{25}Mg' - 0.521\times\delta^{26}Mg'$, where $\delta^{25,26}Mg'=1000 \times \ln(\delta^{25,26}Mg/1000+1)$; Young and Galy, 2004) range from −0.02 to 0.03 (Table 3). In the plot of $\delta^{25}Mg'$ vs. $\delta^{26}Mg'$, all samples and standards fall along the terrestrial equilibrium mass fractionation curve with a slope of 0.521 (Fig. 7a), and thus we only use $\delta^{26}Mg$ in the subsequent discussion.

The $\delta^{26}Mg$ values of the low-Ti basalts vary from −0.33 to −0.19‰, with an average of −0.25±0.08‰ (2SD, n=8), and the high-Ti basalts display similar variations with $\delta^{26}Mg$ ranging from −0.35 to −0.19‰ and an average of −0.25±0.09‰ (2SD, n=13) (Table 3). Overall, both the low- and high-Ti basalts have identical Mg isotopic compositions, similar to the average mantle ($\delta^{26}Mg$=−0.25±0.07‰, 2SD, n=139, Teng et al., 2010). In addition, $\delta^{26}Mg$ values of both the low- and high-Ti basalts do not vary with increasing of MgO content (Fig. 7b), Ce/Pb and Ba/Th ratios (Fig. 8).

Table 2 Rb-Sr, Sm-Nd, and Pb isotopic ratios and abundances for the Dongchuan and Qiaojia lavas

Sample	Rb (ppm)	Sr (ppm)	$^{87}Sr/^{86}Sr$ (measured)	$2\sigma_m$	$^{87}Sr/^{86}Sr(t)^a$	Sm (ppm)	Nd (ppm)	$^{143}Nd/^{144}Nd$ (measured)	$2\sigma_m$	$^{143}Nd/^{144}Nd (t)^a$	$\varepsilon_{Nd}(t)^b$	$^{206}Pb/^{204}Pb (t)^a$	$^{207}Pb/^{204} (t)^a$	$^{208}Pb/^{204}Pb (t)^a$
Low-Ti type														
DC13-07	16.0	914.5	0.707994	0.000009	0.707807	4.1	17.5	0.512002	0.000013	0.511764	−10.5	18.50	15.60	38.65
DC13-08	11.7	413.1	0.708298	0.000012	0.707997	4.4	19.3	0.512090	0.000013	0.511856	−8.8	18.52	15.60	38.54
DC13-12	20.3	423.4	0.709324	0.000012	0.708813	3.4	15.2	0.512025	0.000012	0.511798	−9.9	18.48	15.62	38.67
DC13-13	119.8	709.7	0.710590	0.000010	0.708789	4.1	18.9	0.512018	0.000012	0.511796	−9.9	18.55	15.61	38.68
DC13-14	86.3	586.5	0.710281	0.000013	0.708712	4.3	19.9	0.512023	0.000011	0.511802	−9.8	18.55	15.61	38.70
DC13-15	82.1	595.9	0.709962	0.000011	0.708492	4.2	19.6	0.512012	0.000012	0.511790	−10.0	18.53	15.62	38.73
High-Ti type														
DC13-16	22.4	571.6	0.705985	0.000010	0.705568	11.8	58.7	0.512586	0.000011	0.512380	1.5	18.19	15.59	38.36
DC13-17	24.0	248.4	0.707255	0.000012	0.706226	12.9	64.8	0.512580	0.000011	0.512376	1.4	18.36	15.58	38.36
DC13-21	9.6	455.2	0.706746	0.000012	0.706522	11.3	51.9	0.512627	0.000013	0.512402	1.9	18.46	15.61	38.73
DC13-22	22.6	496.1	0.706270	0.000011	0.705784	11.5	53.1	0.512635	0.000012	0.512413	2.1	18.50	15.61	38.73
QJ13-03	18.1	529.9	0.706373	0.000011	0.706008	10.4	50.7	0.512528	0.000011	0.512317	0.2	18.18	15.58	38.63
QJ13-04	21.2	535.2	0.706394	0.000011	0.705972	10.5	51.6	0.512524	0.000012	0.512315	0.2	18.15	15.57	38.58
QJ13-07	20.6	552.4	0.706271	0.000009	0.705874	10.6	51.9	0.512528	0.000010	0.512318	0.3	17.96	15.56	38.32
QJ13-12	23.9	608.4	0.706407	0.000014	0.705988	11.5	56.7	0.512524	0.000010	0.512316	0.2	18.15	15.56	38.55
QJ13-13	7.1	481.0	0.706020	0.000009	0.705863	9.0	42.9	0.512551	0.000010	0.512336	0.6	18.09	15.58	38.52
QJ13-14	25.8	565.5	0.706332	0.000009	0.705845	11.0	54.4	0.512535	0.000011	0.512328	0.5	18.09	15.56	38.45
QJ13-17	47.7	587.5	0.705979	0.000012	0.705114	9.1	43.4	0.512613	0.000011	0.512398	1.8	18.09	15.53	38.34

Continued

Sample	Rb (ppm)	Sr (ppm)	$^{87}Sr/^{86}Sr$ (measured)	$2\sigma_m$	$^{87}Sr/^{86}Sr(t)$[a]	Sm (ppm)	Nd (ppm)	$^{143}Nd/^{144}Nd$ (measured)	$2\sigma_m$	$^{143}Nd/^{144}Nd$ (t)[a]	$\varepsilon_{Nd}(t)$[b]	$^{206}Pb/^{204}Pb$ (t)[a]	$^{207}Pb/^{204}$ (t)[a]	$^{208}Pb/^{204}Pb$ (t)[a]
QJ13-21	13.5	1137.9	0.706002	0.000012	0.705876	9.0	42.9	0.512592	0.000012	0.512376	1.4	18.12	15.54	38.36
BCR-2 (measured)	46.8	336.1	0.705038	0.000011		6.55	28.7	0.512625	0.000013			18.76	15.62	38.70
Replicate[c]	47.0	336.0	0.705008	0.000011		6.58	28.79	0.512626	0.000011					
BCR-2 (reported)[d]	46.9	340	0.705000			6.58	28.7	0.512636				18.75	15.62	38.72

a The $^{87}Sr/^{86}Sr(t)$, $^{143}Nd/^{144}Nd(t)$, $^{206}Pb/^{204}Pb(t)$, $^{207}Pb/^{204}Pb(t)$, and $^{208}Pb/^{204}Pb(t)$ were calculated for 260 Ma (Zhou et al., 2002, 2008).

b $\varepsilon_{Nd}(t)$ values were calculated using $(^{143}Nd/^{144}Nd)_{CHUR(0)}$=0.512638.

c Repeated sample dissolution, column chemistry, and instrumental analysis.

d Reported values for the reference materials are from GeoREM(http://georem.mpch-mainz.gwdg.de/).

Table 3　Magnesium isotopic composition of the Dongchuan and Qiaojia lavas, as well as reference standard materials

Type	Sample	$\delta^{26}Mg$[a]	2SD[b]	$\delta^{25}Mg$[a]	2SD[b]	$\Delta^{25}Mg'$[c]
Low-Ti	DC13-07	−0.19	0.04	−0.08	0.03	0.02
Low-Ti	DC13-08	−0.33	0.06	−0.15	0.04	0.02
Low-Ti	DC13-12	−0.25	0.02	−0.12	0.04	0.01
Low-Ti	DC13-13	−0.25	0.05	−0.12	0.01	0.01
Low-Ti	DC13-14	−0.26	0.07	−0.14	0.03	−0.01
Low-Ti	DC13-15	−0.26	0.03	−0.13	0.05	0.01
	Replicate[d]	−0.21	0.07	−0.13	0.05	−0.02
Low-Ti	DC14-05	−0.24	0.07	−0.13	0.06	0.00
Low-Ti	DC14-06	−0.24	0.08	−0.10	0.08	0.02
High-Ti	DC14-04	−0.35	0.04	−0.18	0.05	0.00
High-Ti	DC13-16	−0.20	0.02	−0.10	0.04	0.00
High-Ti	DC13-17	−0.23	0.03	−0.13	0.01	−0.01
High-Ti	DC13-21	−0.25	0.05	−0.12	0.04	0.00
High-Ti	DC13-22	−0.21	0.06	−0.12	0.04	−0.01
High-Ti	QJ13-03	−0.19	0.04	−0.10	0.07	0.00
High-Ti	QJ13-04	−0.23	0.05	−0.14	0.01	−0.02
High-Ti	QJ13-07	−0.21	0.08	−0.11	0.07	0.00
High-Ti	QJ13-12	−0.29	0.01	−0.12	0.03	0.03
High-Ti	QJ13-13	−0.30	0.09	−0.13	0.00	0.03
High-Ti	QJ13-14	−0.26	0.02	−0.14	0.04	0.00
High-Ti	QJ13-17	−0.30	0.02	−0.15	0.03	0.00
High-Ti	QJ13-21	−0.22	0.06	−0.12	0.03	−0.01
Standards	BHVO-2	−0.30	0.06	−0.15	0.04	0.01
	Replicate[d]	−0.27	0.01	−0.12	0.03	0.03
	average	−0.29		−0.13		
	BCR-2	−0.19	0.01	−0.11	0.02	−0.01
	Replicate[d]	−0.20	0.09	−0.12	0.06	−0.02
	average	−0.19		−0.12		
	AGV-2	−0.21	0.01	−0.09	0.04	0.02
	GSP-2	0.00	0.04	−0.02	0.05	−0.02

a $\delta^{25,26}Mg=\{(^{25,26}Mg/^{24}Mg)_{sample}/(^{25,26}Mg/^{24}Mg)_{DSM3}-1\}\times 1000$, DSM3 is Mg solution made from pure Mg metal (Galy et al., 2003).

b 2SD=two times the standard deviation of the population of four repeat measurements of a sample solution.

c $\Delta^{25}Mg'=\delta^{25}Mg'-0.521\delta^{26}Mg'$, where $\delta^{25,26}Mg'=1000\times\ln[(\delta^{25,26}Mg+1000)/1000]$ (Young and Galy, 2004).

d Replicate denotes repeating sample dissolution, column chemistry, and instrumental analysis.

5 Discussion

5.1 Fractional crystallization and crustal contamination

Both the low-Ti and high-Ti basalts may have experienced compositional evolution, e.g., fractional crystallization or crustal contamination, because their Mg# (39.8–61.1) (Table 1 and Supplementary Table 1) values are lower than that of primary melt (~72.0) (Langmuir et al., 1977). The Ce/Pb and Nb/U ratios, and Sr-Nd isotopic compositions of both the low-Ti and high-Ti basalts do not vary with increasing SiO_2 (Fig. 9), indicating that crustal contamination of both the low-Ti and high-Ti basalts was limited.

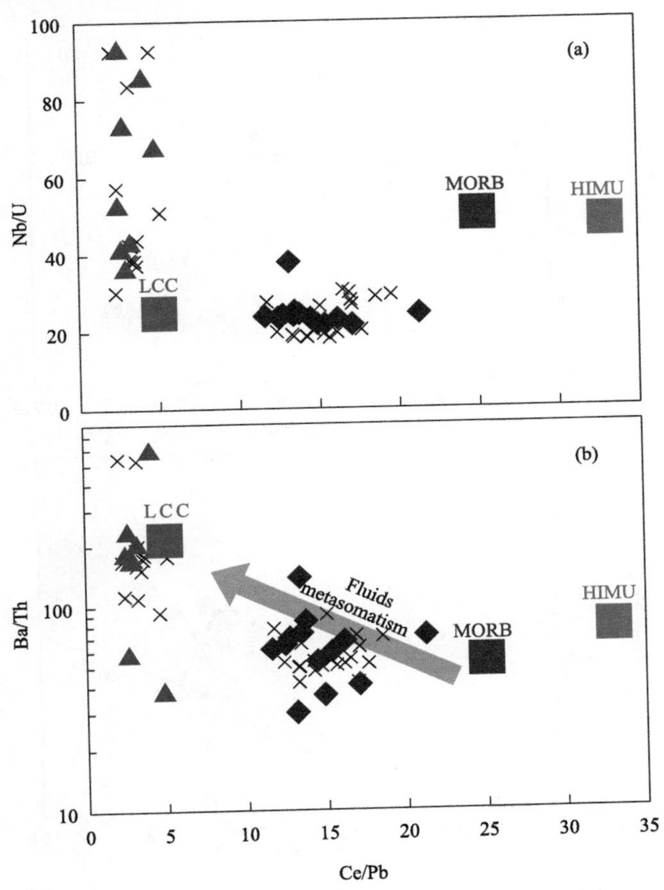

Fig. 5 (a) Nb/U and (b) Ba/Th vs. Ce/Pb for the Emeishan basalts. The arrow represents metasomatism by fluids from subducting slabs, which elevates Ba/Th and decreases Ce/Pb ratios. The symbol crosses represent literature data from Song et al. (2008). Database: MORB and HIMU from Sun and McDonough (1989) and Weaver (1991), and lower continental crust (LCC) from Rudnick and Gao (2003).

MgO contents of the low-Ti and high-Ti basalts follow different trends with CaO, CaO/Al_2O_3, total FeO, Na_2O and TiO_2 (Fig. 3), suggesting different mineral assemblages of fractional crystallization. For the low-Ti basalts, the positive correlation between MgO and CaO suggests the crystal fractionation of either pyroxene or plagioclase (Fig. 3b). However, CaO/Al_2O_3 ratios of these basalts exhibit large variation and show a positive correlation with MgO (Fig. 3c). This correlation could be attributed to pyroxene fractional crystallization rather than plagioclase crystallization, which is also supported by the lack of any negative Eu anomaly (Fig. 4c) as well as petrographic observations (Fig. 2a). Although no correlation between SiO_2 and total FeO was observed for the low-Ti basalts (Fig. 3a, d), the low Ni concentrations (as low as 18.8 ppm) imply crystal fractionation of olivine and pyroxene (Song et al., 2008).

The high-Ti basalts may have mainly experienced plagioclase fractional crystallization based on the negative correlation between MgO and Na_2O (Fig. 3). A small amount of olivine and pyroxene fractional crystallization could have also occurred according to the positive correlations between Ni and MgO (Song et al., 2008). The positive correlations between MgO and total FeO, and TiO_2 contents (Fig. 3d, f) suggest that Fe-Ti oxides (e.g., ilmenite) was significantly fractionated. This observation is also supported by the negative correlation between Nb/Ta and TiO_2 content (Table 1 and Supplementary Table 1), because elemental Ta is more compatible than Nb in ilmenite (Dygert et al., 2013).

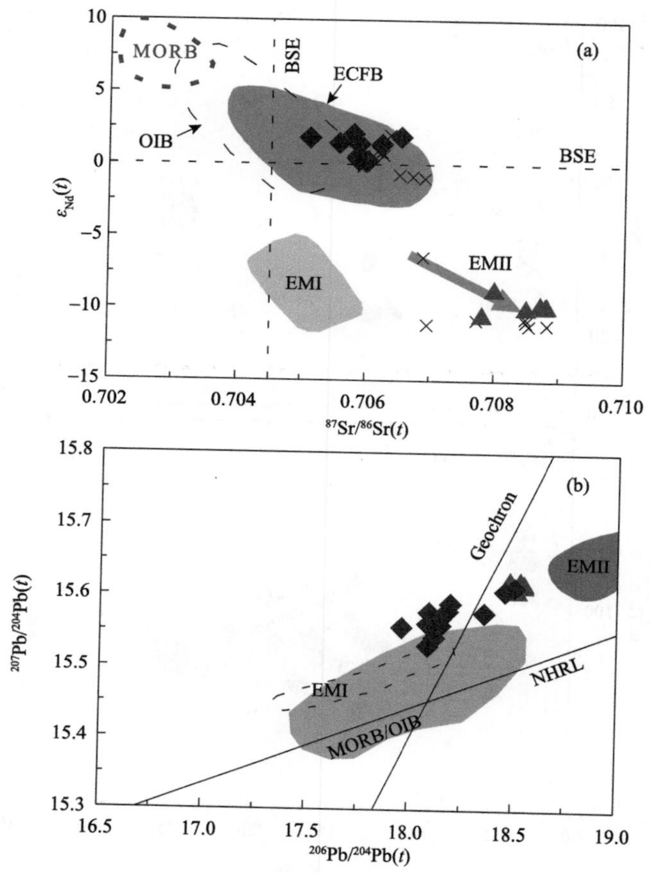

Fig. 6 (a) Initial $^{87}Sr/^{86}Sr(t)$ vs. $\varepsilon_{Nd}(t)$ for the low- and high-Ti Emeishan basalts. The symbols are the same as in Fig. 5. ECFB represents other Emeishan high-Ti continental flood basalts from Xu et al. (2001), Xiao et al. (2004), and Wang et al. (2007). (b) Initial Pb isotopic correlation diagram for the Emeishan basalts. NHRL stands for the Northern Hemisphere Reference Line (Hart, 1984). $(^{207}Pb/^{206}Pb)_{NHRL}=0.1084\times(^{206}Pb/^{204}Pb)+13.491$. The symbol crosses represent literature data from Song et al. (2008). Data source: OIB, MORB, EMI, EMII, and LCC (lower continental crust) are from Zindler and Hart (1986) and Hofmann (1997).

5.2 Magnesium isotopic compositions of the ELIP

Although both the low-Ti and high-Ti samples have undergone significant fractional crystallization, their $\delta^{26}Mg$ values do not vary with MgO contents (Fig. 7b), indicating limited Mg isotope fractionation during magma differentiation. This observation agrees well with previous studies (e.g., Huang et al., 2015; Teng et al., 2007, 2010; Tian et al., 2016). Therefore, Mg isotopic compositions of both the low- and high-Ti basalts are inherited from their mantle sources.

The low-Ti and high-Ti basalts exhibit distinct geochemical characteristics (Figs. 3, 5 and 6) and the petrogenesis of the low-Ti basalts are still controversial (e.g., Huang et al., 2014; Song et al., 2001, 2004, 2008; Xiao et al., 2004; Xu et al., 2001; Zhang and Wang, 2002). Some studies suggested the Emeishan CFBs were products of a mantle plume (Chung and Jahn, 1995; Xu et al., 2001), whereas others have more recently suggest that only the high-Ti basalts derived from the mantle plume (e.g., Song et al., 2008; Xiao et al., 2003, 2004). Compared with other low-Ti basalts exposed in the ELIP elsewhere (e.g., Xiao et al., 2004; Xu et al., 2001), the low-Ti samples studied here are atypical types because they show Samoan lavas-like trace elemental abundances and low $\varepsilon_{Nd}(t)$ values (mostly ranging from −10.5 to −8.8), suggesting possible derivation from an enriched lithospheric mantle (e.g., Song et al., 2008; Xiao et al., 2004). The mantle source of the low-Ti basalts investigated here was possibly metasomatized by fluids derived from subducting slabs during the Archaean (Song et al., 2008). This is supported by their enrichment of the fluid-mobile elements (e.g., Rb, Ba, Th, and Pb) (Fig. 4a) and high Ba/Th and low Ce/Pb ratios (Fig. 8). Meanwhile, an in-situ oxygen isotopic study of the primitive olivine grains in the Emeishan picrites also suggested that the lithospheric mantle was possibly modified by a subduction-related event in the Newproterozoic (Yu et al., 2017). This event most likely elevated Rb/Sr and U/Pb but decreased Sm/Nd ratios. The long-term residence in the mantle would have significantly modified the depleted MORB-like Sr-Nd-Pb isotopic compositions into more enriched Sr-Nd-Pb as observed in these low-Ti samples.

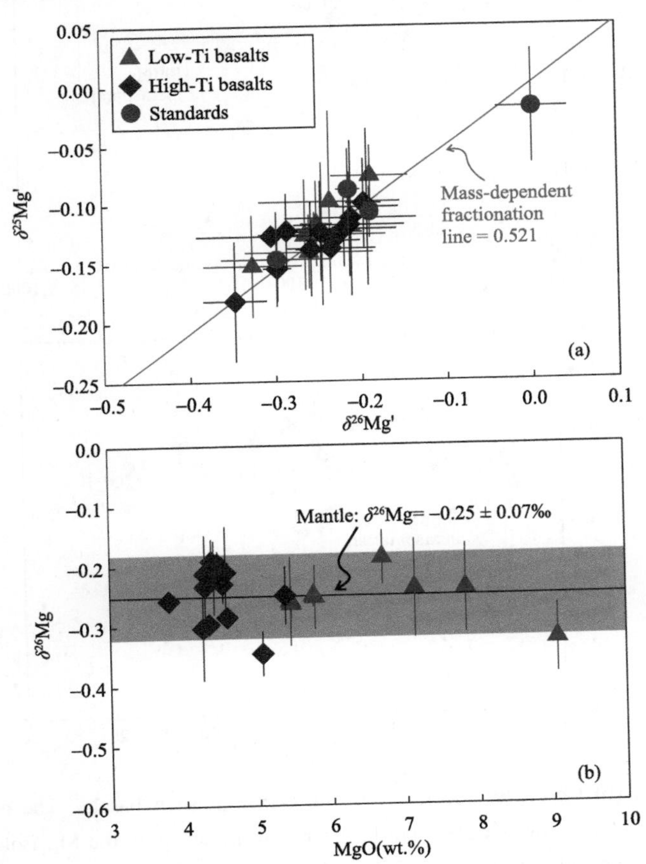

Fig. 7 (a) Magnesium isotope plot of the Emeishan basalts. The solid line represents the terrestrial fractionation with a slope of 0.521. (b) δ^{26}Mg vs. MgO for the low- and high-Ti basalts. Error bars represent 2SD uncertainties. The green yellow band represents the average mantle (δ^{26}Mg=−0.25±0.07‰; Teng et al., 2010).

The high-Ti basalts exhibit typical OIB-like trace element patterns, highly positive $\varepsilon_{Nd}(t)$ and low $^{87}Sr/^{86}Sr(t)$

isotopic compositions, which are consistent with the high-Ti basalts elsewhere in the ELIP, indicating origin from partial melting of a mantle plume (e.g., Chung et al., 1998; Fan et al., 2008; Huang et al., 2014; Song et al., 2008; Wang et al., 2007; Xiao et al., 2004; Xu et al., 2001; Zhou et al., 2006).

However, both types of basalts exhibit Mg isotopic compositions similar to that of the average mantle (−0.25±0.07‰, Teng et al., 2010). As discussed above, the mantle source of the low-Ti basalts was possibly influenced by slab-derived fluids. However, the ancient fluidic metasomatism in the lithospheric mantle modified the trace elements and Sr-Nd-Pb isotopic compositions, but had little effect on the Mg isotopic composition in the low-Ti basalts. These results are also consistent with the Mg isotopic compositions of the Arc basalts from circum-Pacific, which have mantle-like Mg isotopic compositions (Li et al., 2017). This can be explained by the following reasons. Firstly, the fluids may contain low carbonate abundances because of their high Ti/Ti* and Hf/Hf* ratios (indexes for Ti and Hf anomalies; Huang et al., 2015), and the low-Ti basalts do not follow the trend of carbonate metasomatism (Fig. 10d). Second, the slab-dehydrated fluids may only contain low Mg content compared to the mantle (Frezzotti et al., 2011; Pan et al., 2013).

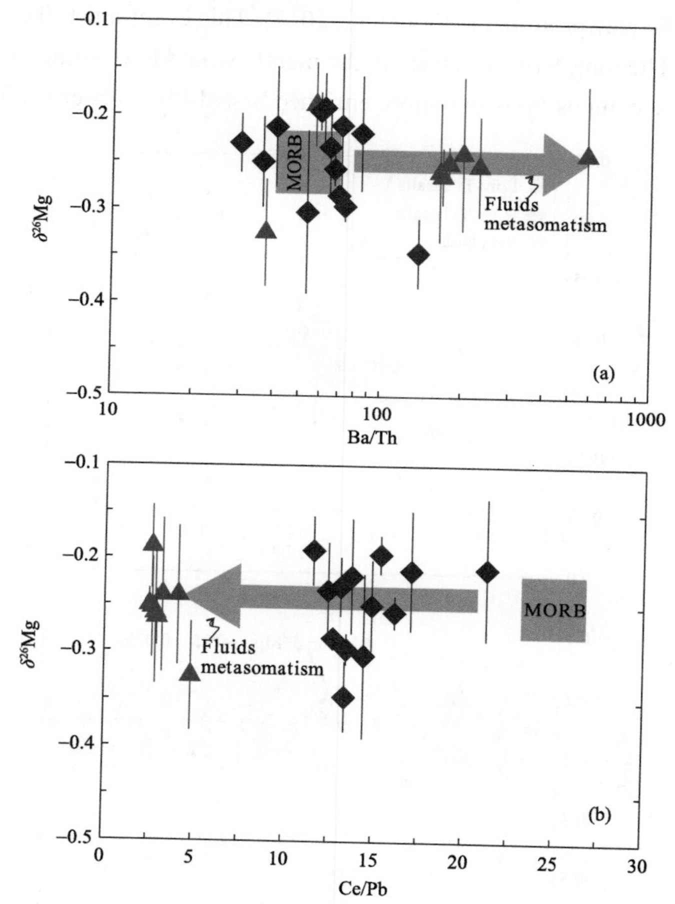

Fig. 8 δ^{26}Mg vs. (a) Ba/Th and (b) Ce/Pb for the low- and high-Ti Emeishan basalts. The arrows represent trends of fluid metasomatism, which increase Ba/Th and lower Ce/Pb, but have limited influence on the Mg isotopic composition of the mantle. Parameters of the MORB are from Sun and McDonough (1989) and Teng et al. (2010).

The high-Ti basalts also exhibit mantle-like Mg isotopic compositions with δ^{26}Mg values of −0.35 to −0.19‰ (average of −0.25±0.09‰, 2SD, n=13). These data suggest that recycled carbonates in the lower mantle source of the Emeishan mantle plume were limited, probably less than 2%, because>2% involvement of recycled sedimentary carbonates into the mantle source should be identified by Mg isotopes under the current analytical precision. This is also supported by their trace element abundances. Typical carbonates are enriched in Ba and Th, and depleted in Zr, Hf, and Ti (Bizimis et al., 2003; Hoernle et al., 2002). As shown in Fig. 10, the high-Ti basalts do not follow the trend of carbonate metasomatism. In addition, such a limited amount of recycled sedimentary carbonates cannot shift Sr, Nd and Pb isotopic isotopic compositions of the high-Ti basalts (Fig. 6).

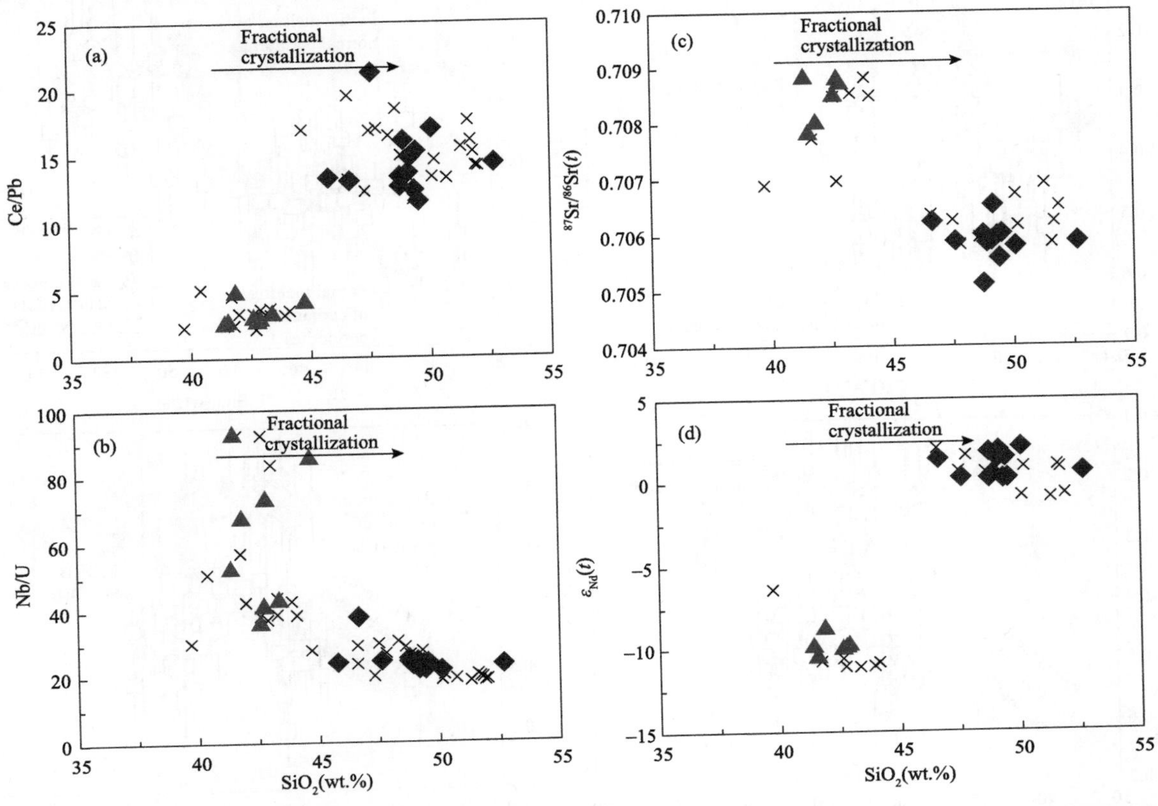

Fig. 9 SiO_2 vs. Ce/Pb (a), Nb/U (b), $^{87}Sr/^{86}Sr(t)$ (c), and $\varepsilon_{Nd}(t)$ (d) for the Emeishan basalts. The symbols are the same as in Fig. 5. The arrows indicate the trend of crustal contamination and fractional crystallization, respectively.

5.3 Geochemical implications for deep carbon cycling

It is important to identify how deep the sedimentary carbonates can be recycled into the mantle by plate subduction. This determines the residence time of recycled carbon. The high-Ti basalts from the Emeishan mantle plume exhibit a mantle-like Mg isotopic composition, indicating that recycled carbonates in their lower mantle source are limited (<2%). This result is consistent with the studies of oceanic basalts from Hawaii and other plumes, which also yield mantle-like Mg isotopic compositions with δ^{26}Mg values of –0.35 to –0.18‰ (Teng et al., 2010). Although the plume-related Hainan basalts and HIMU-like New Zealand basalts exhibit low δ^{26}Mg features (Li et al., 2017; Wang et al., 2016b), their light Mg isotopic compositions are considered to have originated from recycled carbonated eclogites, which do not contain much recycled carbon (Li et al., 2017; Wang et al., 2016b). Thus, this study, as well as previous Mg isotopic studies of basalts, indicate that sedimentary carbonates were unlikely to have been directly recycled into the low mantle. The majority of slab geotherms could intersect a deep depression along the melting curve of carbonated oceanic crust at depths of approximately 300 to 700 km, creating a barrier for direct carbonate recycling into the deep mantle (Thomson et al., 2016).

If sedimentary carbonates can barely be recycled into the lower mantle, their residence time in the mantle could be either 5 to 10 Ma in the mantle wedge beneath island arcs (Kelemen and Manning, 2015) or 60 to several hundred Ma in the upper mantle (Li et al., 2017), which are considerably shorter than the estimation of 1–2 Ga when they recycled into the lower mantle in higher abundances (Sleep and Zahnle, 2001; Dasgupta and Hirschmann, 2010).

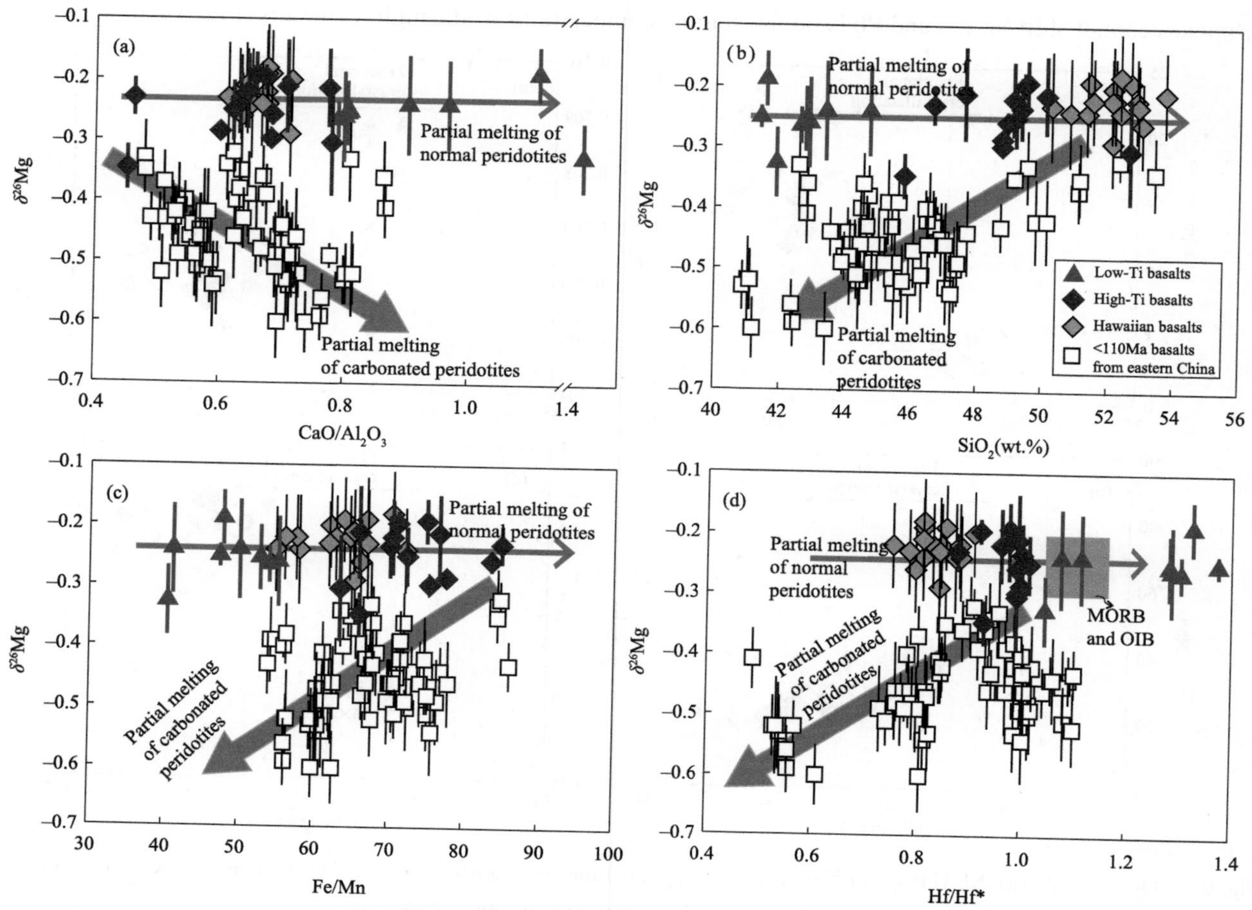

Fig. 10 Correlations of δ^{26}Mg vs. CaO/Al$_2$O$_3$ (a), SiO$_2$ (b), Fe/Mn (c), and Hf/Hf* (d) for the Emeishan basalts. The arrows represent partial melting of carbonated peridotites (Dasgupta et al., 2007) and normal mantle peridotites (Laporte et al., 2014). Hf/Hf*=Hf$_N$/(Sm$_N$×Nd$_N$)$^{0.5}$, where the subscript N means normalized to primitive mantle (Sun and McDonough, 1989). Database: MORB and OIB from Hofmann (1988) and Sun and McDonough (1989); Hawaiian basalts from Frey et al. (1994) and Teng et al. (2010); and eastern China basalts from Yang et al. (2012), Huang et al. (2015), Tian et al. (2016), and Li et al. (2017).

6 Conclusions

This study presents high-precision Mg isotopic measurements of low- and high-Ti basalts from the Emeishan large igneous province (ELIP), southwest China. The main conclusions are:

(1) Although the low-Ti and high-Ti basalts exhibit distinct major, trace elemental and Sr-Nd-Pb isotopic compositions, they both have mantle-like Mg isotopic compositions with δ^{26}Mg value of –0.33 to –0.19‰ and –0.35 to –0.19‰, respectively.

(2) Both the low-Ti and high-Ti basalts have experienced compositional evolution, e.g., fractional crystallization or crustal contamination, because their Mg$^{\#}$ (39.8–61.1) are significantly lower than that of the primary melt (~72.0). However, their δ^{26}Mg values do not vary with Mg$^{\#}$, indicating that the Mg isotopic compositions of the basalts were inherited from their mantle source.

(3) The low $\varepsilon_{Nd}(t)$ values, combined with enriched LILEs (e.g., Rb, Ba, Th, and Pb) and low Ce/Pb and high

Ba/Th ratios in these low-Ti basalts imply that the lithospheric mantle was metasomatized by fluids during the Archaean. Their mantle-like Mg isotopic compositions indicate that the slab-derived fluids contained little Mg-rich carbonates, which could not have modified its original Mg isotopic signature.

(4) The high-Ti basalts display typical OIB-like trace elemental patterns and positive $\varepsilon_{Nd}(t)$ and low $^{87}Sr/^{86}Sr(t)$ isotopic compositions, which most likely derived from a mantle plume originating from the lower mantle. Their mantle-like Mg isotopic compositions suggest that sedimentary carbonate abundances in the lower mantle source of the Emeishan plume were limited (<2%).

Acknowledgements

We would like to thank Hong-Yue Wang, Yan-Hong Liu, Wen-Jun Li, Bing-Yu Gao, Hong-Yan Li, and Wei-Yi Li for their generous help with major, trace elements, and Sr-Nd-Pb isotopic analyses at IGGCAS. Ting Gao, Zi-Jian Li, Xu-Nan Meng and Rui-Ying Li are sincerely thanked for their help in Mg isotope analyses at CUGB and Hitesh Changela for carefully polishing the English writing. The manuscript benefited greatly from constructive reviews by the chief editor, Dr. Andrew Kerr and two anonymous reviewers. This work was financially supported by the National Science Foundation of China (Grants 41322022, 41230209, and 41221002).

References

Ali, J.R., Lo, C.H., Thompson, G.M., Song, X.Y., 2004. Emeishan Basalt Ar-Ar overprint ages define several tectonic events that affected the western Yangtze Platform in the Mesozoic and Cenozoic. Journal of Asian Earth Sciences 23, 163-178.

An, Y.J., Wu, F., Xiang, Y.X., Nan, X.Y., Yu, X., Yang, J.H., Yu, H.M., Xie, L.W., Huang, F., 2014. High-precision Mg isotope analyses of low-Mg rocks by MC-ICP-MS. Chemical Geology 390, 9-21.

Beinlich, A., Mavromatis, V., Austrheim, H., Oelkers, E.H., 2014. Inter-mineral Mg isotope fractionation during hydrothermal ultramafic rock alteration - Implications for the global Mg-cycle. Earth and Planetary Science Letters 392: 166-176.

Bizimis, M., Salters, V.J.M., Dawson, J.B., 2003. The brevity of carbonatite sources in the mantle: evidence from Hf isotopes. Contributions to Mineralogy and Petrology145, 281-300.

Chu, Z.Y., Wu, F.Y., Walker, R.J., Rudnick, R.L., Pitcher, L., Puchtel, I.S., Yang, Y.H., Wilde, S.A., 2009. Temporal Evolution of the Lithospheric Mantle beneath the Eastern North China Craton. Journal of Petrology 50, 1857-1898.

Chung, S.L., Jahn, B.M., 1995. Plume-Lithosphere Interaction in Generation of the Emeishan Flood Basalts at the Permian-Triassic Boundary. Geology 23, 889-892.

Chung, S.L., Jahn, B.M., Genyao, W., Lo, C.H., Bolin, C., 1998. The Emeishan flood Basalt in SW China: A mantle plume initiation model and its connection with continental breakup and mass extinction at the Permian-Triassic boundary. Mantle Dynamics and Plate Interactions in East Asia 27, 47-58.

Dasgupta, R., 2013. Ingassing, Storage, and Outgassing of Terrestrial Carbon through Geologic Time. Reviews in Mineralogy and Geochemistry 75, 183-229.

Dasgupta, R., Hirschmann, M.M., 2010. The deep carbon cycle and melting in Earth's interior. Earth and Planetary Science Letters 298, 1-13.

Dasgupta, R., Hirschmann, M.M., Smith, N.D., 2007. Partial melting experiments of peridotite CO_2 at 3 GPa and genesis of alkalic ocean island basalts. Journal of Petrology 48, 2093-2124.

Dygert, N., Liang, Y., Hess, P., 2013. The importance of melt TiO_2 in affecting major and trace element partitioning between Fe-Ti oxides and lunar picritic glass melts. Geochimica et Cosmochimica Acta 106, 134-151.

Frey, F.A., Garcia, M.O., Roden, M.F., 1994. Geochemical characteristics of Koolau Volcano: Implications of intershield geochemical differences among Hawaiian volcanos. Geochimica et Cosmochimica Acta 58, 1441-1462.

Frezzotti, M.L., Selverstone, J., Sharp, Z.D., Compagnoni, R., 2011. Carbonate dissolution during subduction revealed by diamond-bearing rocks from the Alps. Nature Geoscience 4, 703-706.

Galy, A., Yoffe, O., Janney, P.E., Williams, R.W., Cloquet, C., Alard, O., Halicz, L., Wadhwa, M., Hutcheon, I.D., Ramon, E., Carignan, J., 2003. Magnesium isotope heterogeneity of the isotopic standard SRM980 and new reference materials for magnesium-isotope-ratio measurements. Journal of Analytical Atomic Spectrometry 18, 1352-1356.

Guo, F., Fan, W.M., Wang, Y.J., Li, C.W., 2004. When did the emeishan mantle plume activity start? Geochronological and geochemical evidence from ultramafic-mafic dikes in southwestern China. International Geology Review 46, 226-234.

Hammouda, T., Keshav, S., 2015. Melting in the mantle in the presence of carbon: Review of experiments and discussion on the origin of carbonatites. Chemical Geology 418, 171-188.

Hart, S.R., 1984. A large-scale isotope anomaly in the Southern Hemisphere Mantle. Nature 309, 753-757.

Hoernle, K., Tilton, G., Le Bas, M.J., Duggen, S., Garbe-Schonberg, D., 2002. Geochemistry of oceanic carbonatites compared with continental carbonatites: mantle recycling of oceanic crustal carbonate. Contributions to Mineralogy and Petrology 142, 520-542.

Hofmann, A.W., 1988. Chemical differentiation of the Earth: the relationship between mantle, continental crust, and oceanic crust. Earth and Planetary Science Letters 90, 297-314.

Hofmann, A.W., 1997. Mantle geochemistry: the message from oceanic volcanism. Nature 385, 219-229.

Hu, Y., Teng, F.Z., Zhang, H.F., Xiao, Y., Su, B.X., 2016. Metasomatism-induced mantle magnesium isotopic heterogeneity: Evidence from pyroxenites. Geochimica et Cosmochimica Acta 185, 88-111.

Huang, F., Zhang, Z.F., Lundstrom, C.C., Zhi, X.C., 2011a. Iron and magnesium isotopic compositions of peridotite xenoliths from Eastern China. Geochimica et Cosmochimica Acta 75, 3318-3334.

Huang, S.C., Farkas, J., Jacobsen, S.B., 2011b. Stable calcium isotopic compositions of Hawaiian shield lavas: Evidence for recycling of ancient marine carbonates into the mantle. Geochimica et Cosmochimica Acta, 75, 4987-4997.

Huang, H., Du, Y.S., Yang, J.H., Zhou, L., Hu, L.S., Huang, H.W., Huang, Z.Q., 2014. Origin of Permian basalts and clastic rocks in Napo, Southwest China: Implications for the erosion and eruption of the Emeishan large igneous province. Lithos 208: 324-338.

Huang, J., Li, S.G., Xiao, Y.L., Ke, S., Li, W.Y., Tian, Y., 2015. Origin of low δ^{26}Mg Cenozoic basalts from South China Block and their geodynamic implications. Geochimica et Cosmochimica Acta 164, 298-317.

Javoy, M., Pineau, F., Allegre, C.J., 1982. Carbon Geodynamic Cycle. Nature 300, 171-173.

Ke, S., Teng, F.Z., Li, S.G., Gao, T., Liu, S.A., He, Y.S., Mo, X.X., 2016. Mg, Sr, and O isotope geochemistry of syenites from northwest Xinjiang, China: Tracing carbonate recycling during Tethyan oceanic subduction. Chemical Geology 437, 109-119.

Kelemen, P.B., Manning, C.E., 2015. Reevaluating carbon fluxes in subduction zones, what goes down, mostly comes up. Proceedings of the National Academy of Sciences of the United States of America 112, E3997-E4006.

Langmuir, C.H., Bender, J.F., Bence, A.E., Hanson, G.N., Taylor, S.R., 1977. Petrogenesis of Basalts from the Famous Area: Mid-Atlantic Ridge. Earth and Planetary Science Letters 36, 133-156.

Laporte, D., Lambart, S., Schiano, P., Ottolini, L., 2014. Experimental derivation of nepheline syenite and phonolite liquids by partial melting of upper mantle peridotites. Earth and Planetary Science Letters 404, 319-331.

Li, C.F., Li, X.H., Li, Q.L., Guo, J.H., Li, X.H., Yang, Y.H., 2012a. Rapid and precise determination of Sr and Nd isotopic ratios in geological samples from the same filament loading by thermal ionization mass spectrometry employing a single-step separation scheme. Analytica Chimica Acta, 727: 54-60.

Li, C.F., Li, X.H., Li, Q.L., Guo, J.H., Li, X.H., Feng, L.J., Chu, Z.Y., 2012b. Simultaneous Determination of ^{143}Nd/^{144}Nd and ^{147}Sm/^{144}Nd Ratios and Sm-Nd Contents from the Same Filament Loaded with Purified Sm-Nd Aliquot from Geological Samples by Isotope Dilution Thermal Ionization Mass Spectrometry. Analytical Chemistry 84, 6040-6047.

Li, S.G., Yang, W., Ke, S., Meng, X.N., Tian, H.C., Xu, L.J., He, Y.H., Huang, J., Wang, X.C., Xia, Q.K., Sun, W.D., Yang, X.Y., Ren, Z.Y., Wei, H.Q., Liu, Y.S., Meng, F.C., Yan, J., 2017. Deep carbon cycles constrained by a large-scale mantle Mg isotope anomaly in eastern China. National Science Review 4, 111-120.

Pan, D., Spanu, L., Harrison, B., Sverjensky, D.A., Galli, G., 2013. Dielectric properties of water under extreme conditions and transport of carbonates in the deep Earth. Proceedings of the National Academy of Sciences of the United States of America 110, 6646-6650.

Rudnick, R.L., Gao, S., 2003. Composition of the continental crust. Treatise on Geochemistry 3, 1-64.

Saenger, C., Wang, Z.R., 2014. Magnesium isotope fractionation in biogenic and abiogenic carbonates: implications for paleoenvironmental proxies. Quaternary Science Reviews 90: 1-21.

Sleep, N.H., Zahnle, K., 2001. Carbon dioxide cycling and implications for climate on ancient Earth. Journal of Geophysical Research 106, 1373-1399.

Song, X.Y., Zhou, M.F., Hou, Z.Q., Cao, Z.M., Wang, Y.L., Li, Y.G., 2001. Geochemical Constraints on the Mantle Source of the

Upper Permian Emeishan Continental Flood Basalts, Southwestern China. International Geology Review, 43, 213-225.
Song, X.Y., Zhou, M.F., Cao, Z.M., Robinson, P.T., 2004. Late Permian rifting of the South China Craton caused by the Emeishan mantle plume? Journal of the Geological Society 161, 773-781.
Song, X.Y., Zhang, C.J., Hu, R.Z., Zhong, H., Zhou, M.F., Ma, R.Z., Li, Y.G., 2005. Genetic Links of Magmatic Deposits In The Emeishan Large Igneous Province With Dynamics of Mantle Plume. Journal of Mineral Petrol 25, 35-44 (in Chinese with English abstract).
Song, X.Y., Zhou, M.F., Keays, R.R., Cao, Z.M., Sun, M., Qi, L., 2006. Geochemistry of the Emeishan flood basalts at Yangliuping, Sichuan, SW China: implications for sulfide segregation. Contributions to Mineralogy and Petrology 152, 53-74.
Song, X.Y., Qi, H.W., Robinson, P.T., Zhou, M.F., Cao, Z.M., Chen, L.M., 2008. Melting of the subcontinental lithospheric mantle by the Emeishan mantle plume; evidence from the basal alkaline basalts in Dongchuan, Yunnan, Southwestern China. Lithos 100, 93-111.
Sun, S.S., McDonough, W.F., 1989. Chemical and isotopic systematics of oceanic basalts: implications for mantle composition and processes. Geological Society, London, Special Publications 42, 313-345.
Tao, Y., Ma, Y.S., Miao, L.C., Zhu, F.L., 2009. SHRIMP U-Pb zircon age of the Jinbaoshan ultramafic intrusion, Yunnan Province, SW China. Chinese Science Bulletin 54, 168-172.
Teng, F.Z., Wadhwa, M., Helz, R.T., 2007. Investigation of magnesium isotope fractionation during basalt differentiation: Implications for a chondritic composition of the terrestrial mantle. Earth and Planetary Science Letters 261, 84-92.
Teng, F.Z., Li, W.Y., Ke, S., Marty, B., Dauphas, N., Huang, S.C., Wu, F.Y., Pourmand, A., 2010. Magnesium isotopic composition of the Earth and chondrites. Geochimica et Cosmochimica Acta 74, 4150-4166.
Teng, F.Z., Li, W.Y., Ke, S., Yang, W., Liu, S.A., Sedaghatpour, F., Wang, S.J., Huang, K.J., Hu, Y., Ling, M.X., Xiao, Y., Liu, X.M., Li, X.W., Gu, H.O., Sio, C.K., Wallace, D.A., Su, B.X., Zhao, L., Chamberlin, J., Harrington, M., Brewer, A., 2015. Magnesium Isotopic Compositions of International Geological Reference Materials. Geostandards and Geoanalytical Research 39, 329-339.
Thomson, A.R., Walter, M.J., Kohn, S.C., Brooker, R.A., 2016. Slab melting as a barrier to deep carbon subduction. Nature 529, 76-79.
Tian, H.C., Yang, W., Li, S.G., Ke, S., Chu, Z.Y., 2016. Origin of lowδ^{26}Mg Mg basalts with EM-I component: Evidence for interaction between enriched lithosphere and carbonated asthenosphere. Geochimica et Cosmochimica Acta, 188: 93-105.
Wang, C.Y., Zhou, M.F., 2006. Genesis of the Permian Baimazhai magmatic Ni-Cu-(PGE) sulfide deposit, Yunnan, SW China. Mineralium Deposita 41, 771-783.
Wang, C.Y., Zhou, M.-F., Qi, L., 2007. Permian flood basalts and mafic intrusions in the Jinping (SW China)-Song Da (northern Vietnam) district: Mantle sources, crustal contamination and sulfide segregation. Chemical Geology, 243(3-4): 317-343.
Wang, S.J., Teng, F.Z., Li, S.G., Hong, J.A., 2014. Magnesium isotopic systematics of mafic rocks during continental subduction. Geochimica et Cosmochimica Acta 143: 34-48.
Wang, S.J., Teng, F.Z., Scott, J.M., 2016b. Tracing the origin of continental HIMU-like intraplate volcanism using magnesium isotope systematics. Geochimica et Cosmochimica Acta 185, 78-87.
Wang, Z.Z., Liu, S.A., Ke, S., Liu, Y.C., Li, S.G., 2016a. Magnesium isotopic heterogeneity across the cratonic lithosphere in eastern China and its origins. Earth and Planetary Science Letters 451, 77-88.
Weaver, B.L., 1991. The Origin of Ocean Island Basalt End-Member Compositions - Trace-Element and Isotopic Constraints. Earth and Planetary Science Letters 104, 381-397.
Workman, R.K., Hart, S.R., Jackson, M., Regelous, M., Farley, K.A., Blusztajn, J., Kurz, M., Staudigel, H., 2004. Recycled metasomatized lithosphere as the origin of the enriched mantle II (EM2) end-member: Evidence from the Samoan volcanic chain. Geochemistry, Geophysics, Geosystems 5, Q04008, doi:10.1029/2003GC000623.
Xiao, L., Xu, Y.G., Chung, S.L., He, B., Mei, H., 2003. Chemostratigraphic Correlation of Upper Permian Lavas from Yunnan Province, China: Extent of the Emeishan Large Igneous Province. International Geology Review 45, 753-766.
Xiao, L., Xu, Y.G., Mei, H.J., Zheng, Y.F., He, B., Pirajno, F., 2004. Distinct mantle sources of low-Ti and high-Ti basalts from the western Emeishan large igneous province, SW China: implications for plume-lithosphere interaction. Earth and Planetary Science Letters 228, 525-546.
Xiao, Y., Teng, F.Z., Zhang, H.F., Yang, W., 2013. Large magnesium isotope fractionation in peridotite xenoliths from eastern North China craton: Product of melt-rock interaction. Geochimica et Cosmochimica Acta 115, 241-261.
Xu, J.F., Suzuki, K., Xu, Y.G., Mei, H.J., Li, J., 2007. Os, Pb, and Nd isotope geochemistry of the Permian Emeishan continental flood basalts: Insights into the source of a large igneous province. Geochimica et Cosmochimica Acta 71: 2104-2119.

Xu, Y.G., Luo, Z.Y., Huang, X.L., He, B., Xiao, L., Xie, L.W., Shi, Y.R., 2008. Zircon U-Pb and Hf isotope constraints on crustal melting associated with the Emeishan mantle plume. Geochimica et Cosmochimica Acta 72, 3084-3104.

Xu, Y.G., Chung, S.L., Jahn, B.M., Wu, G.Y., 2001. Petrologic and geochemical constraints on the petrogenesis of Permian-Triassic Emeishan flood basalts in southwestern China. Lithos 58, 145-168.

Xu, Y.G., He, B., Chung, S.L., Menzies, M.A., Frey, F.A., 2004. Geologic, geochemical, and geophysical consequences of plume involvement in the Emeishan flood-basalt province. Geology 32, 917-920.

Yang, W., Teng, F.Z., Zhang, H.F., 2009. Chondritic magnesium isotopic composition of the terrestrial mantle: A case study of peridotite xenoliths from the North China craton. Earth and Planetary Science Letters 288, 475-482.

Yang, W., Teng, F.Z., Zhang, H.F., Li, S.G., 2012. Magnesium isotopic systematics of continental basalts from the North China craton: Implications for tracing subducted carbonate in the mantle. Chemical Geology 328, 185-194.

Yang, W., Teng, F.Z., Li, W.Y., Liu, S.A., Ke, S., Liu, Y.S., Zhang, H.F., Gao, S., 2016. Magnesium isotopic composition of the deep continental crust. American Mineralogist 101: 243-252.

Yang, Y.H., Zhang, H.F., Chu, Z.Y., Xie, L.W., Wu, F.Y., 2010. Combined chemical separation of Lu, Hf, Rb, Sr, Sm and Nd from a single rock digest and precise and accurate isotope determinations of Lu-Hf, Rb-Sr and Sm-Nd isotope systems using Multi-Collector ICP-MS and TIMS. International Journal of Mass Spectrometry 290, 120-126.

Young, E.D., Galy, A., 2004. The isotope geochemistry and cosmochemistry of magnesium. Reviews in Mineralogy and Geochemistry 55, 197-230.

Yu, S.Y., Shen, N.P., Song, X.Y., Ripley, E.M., Li, C., Chen, L.M., 2017. An integrated chemical and oxygen isotopic study of primitive olivine grains in picrites from the Emeishan Large Igneous Province, SW China: Evidence for oxygen isotope heterogeneity in mantle sources. Geochimica et Cosmochimica Acta 215, 263-276.

Zhang, Z.C., Wang, F.S., 2002. Geochemistry of two types of basalts in the Emeishan basaltic province: Evidence for mantle plume-lithosphere interaction. Acta Geologica Sinica 76, 229-237.

Zhang, Z.C., J.Mahoney, J., Mao, J.W., Wang, F.S., 2006. Geochemistry of Picritic and Associated Basalt Flows of the Western Emeishan Flood Basalt Province, China. Journal of Petrology 47, 1997-2019.

Zhong, H., Zhu, W.G., 2006. Geochronology of layered mafic intrusions from the Pan-Xi area in the Emeishan large igneous province, SW China. Mineralium Deposita 41 599-606.

Zhou, M.F., Malpas, J., Song, X.Y., Robinson, P.T., Sun, M., Kennedy, A.K., Lesher, C.M., Keays, R.R., 2002. A temporal link between the Emeishan large igneous province (SW China) and the end-Guadalupian mass extinction. Earth and Planetary Science Letters 196, 113-122.

Zhou, M.-F., Arndt, N.T., Malpas, J., Wang, C.Y., Kennedy, A.K., 2008. Two magma series and associated ore deposit types in the Permian Emeishan large igneous province, SW China. Lithos 103, 352-368.

Zindler, A., Hart, S., 1986. Chemical geodynamics. Annual Review of Earth and Planetary Sciences 14, 493-571.

Tracing subduction zone fluid-rock interactions using trace element and Mg-Sr-Nd isotopes*

Shui-Jiong Wang[1, 2], Fang-Zhen Teng[2], Shu-Guang Li[1], Li-Fei Zhang[3], Jin-Xue Du[1, 3], Yong-Sheng He[1] and Yaoling Niu[4, 5]

1. State Key Laboratory of Geological Processes and Mineral Resources, China University of Geosciences, Beijing 100083, China
2. Isotope Laboratory, Department of Earth and Space Sciences, University of Washington, Seattle, WA 98195-1310, USA
3. School of Earth Space Sciences, Peking University, Beijing, China
4. Institute of Oceanology, Chinese Academy of Sciences, Qingdao 266071, China
5. Department of Earth Sciences, Durham University, Durham DH1 3LE, UK

亮点介绍：西南天山榴辉岩的 Mg 同位素，主微量元素及 Sr-Nd 同位素数据表明，俯冲通道中的变质流体主要来源于俯冲板块的沉积物，少量来源于蚀变洋壳或蛇纹石化橄榄岩。流体成分主要有两种：富含 LILEs 的流体可能来自于云母类矿物的变质脱水，它能够使榴辉岩 Mg 同位素系统变重；而富含 Sr 和 Pb 的流体可能来自于帘石类矿物的变质脱水，由于帘石类矿物贫 Mg，其变质脱水形成的流体对榴辉岩 Mg 同位素基本不产生影响。以上发现揭示了俯冲区域的 Mg 同位素地球化学行为，并解释了岛弧玄武岩相较地幔具有重 Mg 同位素组成的原因。

Abstract Slab-derived fluids play a key role in mass transfer and elemental/isotopic exchanges in subduction zones. The exhumation of deeply subducted crust is achieved via a subduction channel where fluids from various sources are abundant, and thus the chemical/isotopic compositions of these rocks could have been modified by subduction-zone fluid-rock interactions. Here, we investigate the Mg isotopic systematics of eclogites from southwestern Tianshan, in conjunction with major/trace element and Sr-Nd isotopes, to characterize the source and nature of fluids and to decipher how fluid-rock interactions in subduction channel might influence the Mg isotopic systematics of exhumed eclogites. The eclogites have high LILEs (especially Ba) and Pb, high initial $^{87}Sr/^{86}Sr$ (up to 0.7117; higher than that of coeval seawater), and varying Ni and Co (mostly lower than those of oceanic basalts), suggesting that these eclogites have interacted with metamorphic fluids mainly released from subducted sediments, with minor contributions from altered oceanic crust or altered abyssal peridotites. The positive correlation between $^{87}Sr/^{86}Sr$ and Pb* (an index of Pb enrichment; $Pb^*=2*Pb_N/[Ce_N+Pr_N]$), and the decoupling relationships and bidirectional patterns in $^{87}Sr/^{86}Sr$-Rb/Sr, Pb*-Rb/Sr and Pb*-Ba/Pb spaces imply the presence of two compositionally different components for the fluids: one enriched in

* 本文发表在：Lithos, 2017, 290-291: 94-103

LILEs, and the other enriched in Pb and $^{87}Sr/^{86}Sr$. The systematically heavier Mg isotopic compositions ($\delta^{26}Mg$= −0.37 to +0.26) relative to oceanic basalts (−0.25±0.07) and the roughly negative correlation of $\delta^{26}Mg$ with MgO for the southwestern Tianshan eclogites, cannot be explained by inheritance of Mg isotopic signatures from ancient seafloor alteration or prograde metamorphism. Instead, the signatures are most likely produced by fluid-rock interactions during the exhumation of eclogites. The high Rb/Sr and Ba/Pb but low Pb* eclogites generally have high bulk-rock $\delta^{26}Mg$ values, whereas high Pb* and $^{87}Sr/^{86}Sr$ eclogites have mantle-like $\delta^{26}Mg$ values, suggesting that the two fluid components have diverse influences on the Mg isotopic systematics of these eclogites. The LILE-rich fluid component, possibly derived from mica-group minerals, contains a considerable amount of isotopically heavy Mg that has shifted the $\delta^{26}Mg$ of the eclogites towards higher values. By contrast, the $^{87}Sr/^{86}Sr$- and Pb-rich fluid component, most likely released from epidote-group minerals in metasediments, has little Mg so as not to modify the Mg isotopic composition of the eclogites. In addition, the influence of talc-derived fluid might be evident in a very few eclogites that have low Rb/Sr and Ba/Pb but slightly heavier Mg isotopic compositions. These findings represent an important step toward a broad understanding of the Mg isotope geochemistry in subduction zones, and contributing to understanding why island arc basalts have averagely heavier Mg isotopic compositions than the normal mantle.

Keywords Mg isotopes, subduction channel, fluid-rock interaction, eclogite, Tianshan

1 Introduction

Subduction channel is a highly reactive interface between subducting oceanic lithosphere and mantle wedge, in which mass transfer as well as elemental and isotopic exchanges actively occur (e.g., Bebout and Penniston-Dorland, 2016). In this region, fluids released from various subducting slab lithologies (e.g., sediments, altered oceanic crust, and altered abyssal peridotites) can be mixed and penetrate into exhuming rocks, inducing extensive fluid-rock interactions (John et al., 2008; van der Straaten et al., 2008, 2012; Zack and John, 2007). The fluids, when emanating from the interface into the mantle wedge, can further impart their chemical/isotopic signatures to the juxtaposed mantle rocks and associated arc volcanism.

Trace elements in conjunction with Sr-Nd-O isotopic systematics have been widely used to identify and understand fluid-rock interactions in subduction channels (Glodny et al., 2003; Halama et al., 2011; John et al., 2004, 2012; King et al., 2006). The magnesium (Mg) isotopic systematics might be a useful tracer of subduction-zone fluid-rock interactions, potentially providing insights into the source and nature of fluids. Magnesium is fluid-mobile at low temperatures, which leads to large Mg isotope fractionations as much as 7 ‰ during Earth's surface processes (Teng, 2017 and references therein). Recent studies also documented high mobility of Mg during subduction-zone metamorphism (Chen et al., 2016; Horodyskyj et al., 2009; Pogge von Strandmann et al., 2015; van der Straaten et al., 2008). Chen et al. (2016) found high $\delta^{26}Mg$ values (up to +0.72) for white schists from Western Alps, and linked them to infiltration of Mg-rich fluids derived from dehydration of serpentinites. Recent studies also documented generally heavier Mg isotopic compositions in arc volcanic rocks relative to normal peridotitic sources ($\delta^{26}Mg$=−0.25±0.07), which were explained as the addition of heavy Mg isotopes from subducting slabs to the mantle wedge (Li et al., 2017;Teng et al., 2016). A general conclusion derived from these studies is that the subduction-zone fluids might be isotopically heavy in terms of Mg isotopes. Nevertheless, the interpretation of any Mg isotopic variations in subduction-related rocks requires the knowledge of how Mg isotopes behave in subduction channels, and how fluid-rock interactions could affect the Mg isotopic systematics of a rock.

Orogenic eclogites of seafloor protolith may be the best choice to study subduction channel processes. Oceanic crust undergoes seawater alteration prior to subduction and is, therefore, more hydrated relative to the continental crust (Miller et al., 1988). It experiences extensive dehydration together with the sediment veneer during subduction (Gerya et al., 2002). In addition, the exhumation of oceanic crust via the subduction channel proceeds at relatively slower rate (mm/yr; Agard et al., 2009). All of these allow eclogites of seafloor protolith to preserve a record of extensive fluid-rock interactions during exhumation. An increasing number of studies have shown that fluid-rock interactions can readily modify the chemical and isotopic compositions of exhumed eclogites (e.g., Bebout, 2007; Klemd, 2013; Xiao et al., 2012), although how the chemical/isotopic composition shift depends on the nature and abundance of fluids with which the eclogites have interacted.

In this study, we investigate a suite of well-characterized eclogites/blueschists and mica schists from southwestern Tianshan, China. We present the first Mg isotopic data for the orogenic eclogites of seafloor protolith, and in combination with Sr-Nd isotopic and trace elemental data, we explore the influence of subduction-zone fluid-rock interactions on the Mg isotopic systematics of eclogites. Our results show that these eclogites are variably enriched in heavy Mg isotopes, which may result from interactions of the eclogites with both high-MgO and low-MgO fluids released from different hydrous minerals in the subduction channel.

2 Geological settings and samples

The high-pressure to ultrahigh-pressure (HP-UHP) metamorphic belt of Chinese southwestern Tianshan, located along the suture between the Yili and the Tarim blocks, was formed during the northward subduction of the Palaeo-South Tianshan oceanic crust beneath the Yili block (Gao et al., 1999; Windley et al., 1990; Zhang et al., 2002, 2008). The eclogites and retrograded blueschists in southwestern Tianshan occur as interlayers or lenticular bodies in mica schists, representing the relic oceanic crust that experienced subduction and exhumation in response to later continental collision. The protoliths of eclogites and associated blueschists range from MORBs to OIBs as indicated by the geochemical data and their preserved pillow structures in the field (Ai et al., 2006; Gao and Klemd, 2003; Zhang et al., 2002, 2008). The eclogites and their host rocks have experienced peak coesite-bearing eclogite-facies metamorphism at 324–312 Ma (Zhang et al., 2005; Su et al., 2010; Klemd et al., 2011; Q.-L. Li et al., 2011a; Yang et al., 2013), followed by a slow exhumation rate to amphibolite-facies between 320 Ma and 240 Ma (e.g., Zhang et al., 2013). The peak and retrograde metamorphic temperatures estimated for the southwestern Tianshan eclogites vary from 450 to 630 °C (e.g., Du et al., 2014a, 2014b). The retrograde metamorphic temperatures are slightly higher than the peak-eclogite facies temperatures as a result of thermal relaxation during the exhumation (e.g., Zhang et al., 2013). The presence of abundant millimeter to decimeter-wide and centimeter to meter-long veins in southwestern Tianshan blueschists and eclogites indicates extensive fluid-rock interactions and fluid-mediated mass transport during crustal subduction and exhumation (Beinlich et al., 2010; Gao et al., 2007; Gao and Klemd, 2001; John et al., 2008, 2012; Lü et al., 2012).

The petrology and metamorphic evolution of the studied eclogites and mica schists have been well characterized (Zhang et al., 2003; Ai et al., 2006; Lü et al., 2009; Du et al., 2011, 2014b; Xiao et al., 2012). The eclogites consist mainly of garnet, omphacite, glaucophane, paragonite, epidote, calcite, dolomite and quartz/coesite; the mica schists are mainly composed of garnet, glaucophane, phengite, epidote, paragonite, plagioclase and quartz/coesite. A detail description of the studied eclogites and mica schists including the sample localities has been given in Supplementary Table S1.

3 Analytical methods

3.1 Major and trace elements

Major elements were analyzed at the Hebei Institute of Regional Geology and Mineral Resources, China, by wavelength dispersive X-Ray fluorescence spectrometry (Gao et al., 1995). Analytical uncertainties are generally better than 1%. The H_2O^+ and CO_2 were determined by gravimetric methods and potentiometry, respectively. Trace elements were analyzed using an Elan 6100 DRC ICP-MS at the CAS key laboratory of crust-material and environments, University of Science and Technology of China, Hefei. Samples were analyzed with aliquots of USGS standards BHVO-2, BIR-1, AGV-2 and GSP-2. Results for the USGS standards together with the reference values are reported in Supplementary Table S2. Analytical uncertainties are better than 5% for most of the elements.

3.2 Strontium and Nd isotopic analysis

The Sr and Nd were separated from the matrix with cation exchange chromatography with Bio-Rad AG50W-X12 resin using the method described by Chu et al. (2009). The Sr and Nd isotopes were performed using an Isoprobe-T thermal ionization mass spectrometer (TIMS) at the State Key Laboratory of Lithospheric Evolution, Institute of Geology and Geophysics, Chinese Academy of Sciences. Measured $^{87}Sr/^{86}Sr$ and $^{143}Nd/^{144}Nd$ ratios were corrected for mass-fractionation using $^{86}Sr/^{88}Sr=0.1194$ and $^{146}Nd/^{144}Nd=0.7219$, respectively. During the course of this study, standards of NBS987-Sr and jNdi-Nd yielded a value of $^{87}Sr/^{86}Sr=0.710245\pm20$ and $^{143}Nd/^{144}Nd=0.512117\pm10$, respectively.

3.3 Magnesium isotopic analysis

Magnesium isotopic ratios were analyzed for bulk rock powders and mineral separates at the University of Washington, Seattle. The separation of Mg was achieved by cation exchange chromatography using Bio-Rad AG50W-X8 resin in 1 N HNO_3 media (Li et al., 2010; Teng et al., 2007, 2010, 2015; Yang et al., 2009). Two standards, Kilbourne Hole (KH) olivine and seawater, were processed together with samples for each batch of column chemistry. The Mg isotopic ratios were determined using the standard-sample bracketing protocol on a *Nu* plasma MC-ICPMS (Teng and Yang, 2014). The blank Mg signal for ^{24}Mg was $<10^{-4}$ V, which is negligible relative to the sample signals of 3-5 V. The KH olivine and seawater yielded average $\delta^{26}Mg$ of -0.25 ± 0.05 and -0.82 ± 0.06, respectively, consistent with previous reported values (Foster et al., 2010; Li et al., 2010; Teng et al., 2010; Ling et al., 2011; Wang et al., 2016).

4 Results

Major and trace elemental compositions of the eclogites and mica schists are summarized in Supplementary Table S3. The eclogites have SiO_2 ranging from 39.82 to 52.47 wt.% and MgO ranging from 3.19 to 9.68 wt.% (Supplementary Table S3), and plot in subalkalic basalt field in Zr/Ti versus Nb/Y diagram

(Supplementary Fig. S1; Pearce, 1996). The high contents of H_2O^+ (0.58 to 3.38 wt.%) and CO_2 (0.08 to 8.96 wt.%) are consistent with the presence of water- and/or carbon oxide-bearing minerals such as zoisite/clinozoisite and calcite/dolomite. The eclogites have variably high LILEs (e.g., Ba, Rb, Cs, and K) and Pb, but low Ni and Co concentrations (Supplementary Table S3). The mica schists are felsic with SiO_2 ranging from 59.53 to 76.66 wt.% and MgO ranging from 1.81 to 3.60 wt.% (Supplementary Table S3). They are characterized by variable contents of LILEs, Sr and Pb, which may be controlled by different proportions of mica-group minerals (host of LILEs) and epidote-group minerals (major host of Sr and Pb) in southwestern Tianshan metasediments (e.g., Xiao et al., 2012).

The Sr and Nd isotopic compositions of the eclogites are reported in Table 1. The eclogites have positive age-corrected ε_{Nd}(320 Ma) value ranging from +2.8 to +10.1 (with one exception of -2.4; Fig. 1). They have extremely high and variable initial Sr isotopic compositions [$^{87}Sr/^{86}Sr$(320Ma)] varying from 0.7058 to 0.7117 (Fig. 1), a range that is even higher than that of Ordovican to Carboniferous seawater ($^{87}Sr/^{86}Sr$=0.7075–0.7090; Veizer, 1989). As a result, the eclogites plot rightward far from the field defined by depleted MORB and OIB in $\varepsilon_{Nd}(t)$ − $^{87}Sr/^{86}Sr(t)$ diagram (Fig. 1).

The $\delta^{26}Mg$ values of southwestern Tianshan eclogites vary widely from −0.37±0.05 to +0.26±0.04 (Table 2), equal to or higher than unaltered oceanic basalts and eclogites of continental basalt protolith, both of which have homogeneous Mg isotopic compositions around the normal mantle value (−0.25± 0.07; Fig. 2). Garnets in southwestern Tianshan eclogites yield $\delta^{26}Mg$ values varying from −1.75±0.07 to −1.10±0.07, and omphacites have $\delta^{26}Mg$ values ranging from −0.04± 0.05 to +0.46± 0.07 (Table 2), with corresponding inter-mineral Mg isotope fractionation ($\Delta^{26}Mg_{Cpx\text{-}Grt} = \delta^{26}Mg_{Cpx} - \delta^{26}Mg_{Grt}$) in the range of 1.23−1.98. Temperatures estimated using garnet-clinopyroxene Mg isotope geothermometer range from 485 °C to 675 °C(Huang et al., 2013; W. Y. Li et al., 2016b), which are in rough agreement with the peak and retrograde metamorphic temperatures for the Tianshan eclogites (e.g., Du et al., 2014a, 2014b). Six mica schists from southwestern Tianshan have bulk $\delta^{26}Mg$ values ranging from −0.11±0.05 to +0.23±0.02 (Table 2).

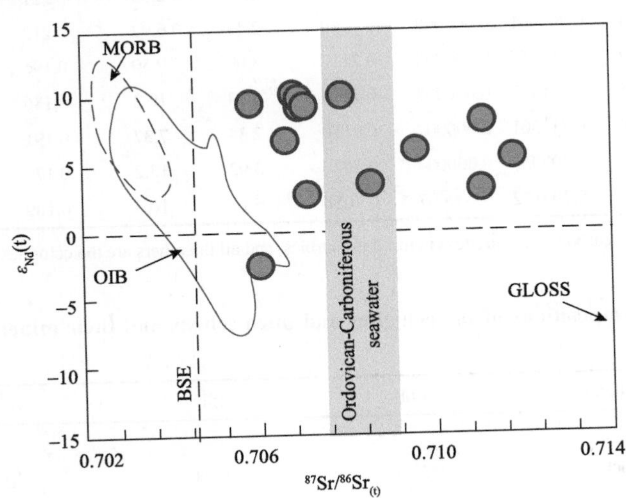

Fig. 1 The Sr and Nd isotopic compositions of the eclogites from southwestern Tianshan. MORB and OIB fields are from Zindler and Hart (1986); $^{87}Sr/^{86}Sr$ ratio of the Ordovican to Carboniferous (O−C) seawater is from Veizer (1989), and $^{87}Sr/^{86}Sr$ ratio of the global subducting sediments (GLOSS) can be high as much as 0.73 (Plank and Langmuir, 1998).

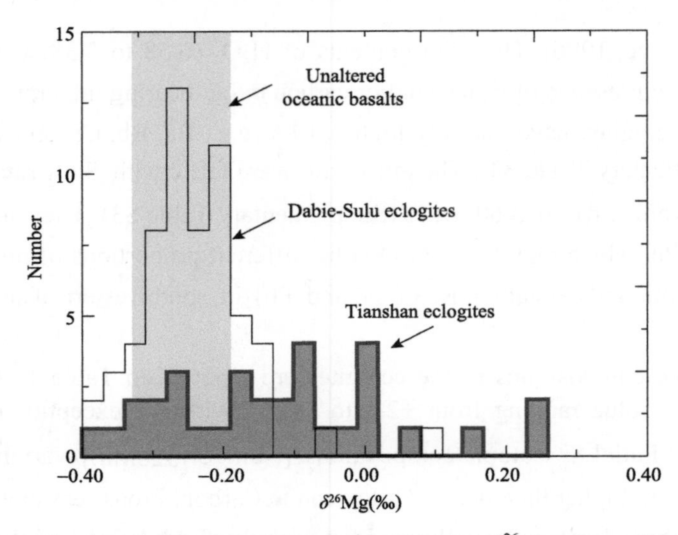

Fig. 2　Histogram of δ^{26}Mg values for the eclogites from southwestern Tianshan. δ^{26}Mg values of the eclogites with continental origin are from Li et al. (2010) and Wang et al. (2014a, 2014 b). δ^{26}Mg values of the unaltered oceanic crust are from Teng et al. (2010).

Table 1　The Sr and Nd isotopic compositions of the eclogites from southwestern Tianshan

Sample	Rb(ppm)	Sr(ppm)	^{87}Rb/^{86}Sr	^{87}Sr/^{86}Sr	2sigma	^{87}Sr/^{86}Sr(320 Ma)	Sm(ppm)	Nd(ppm)	^{147}Sm/^{144}Nd	^{143}Nd/^{144}Nd	2sigma	ε_{Nd}(320 Ma)
H902-7	31.1	332	0.271	0.707198	0.000006	0.7060	9.35	37.6	0.150	0.512418	0.000007	−2.4
300-1	89.2	203	1.275	0.712858	0.000045	0.7071	5.74	23.1	0.150	0.512682	0.000011	2.8
H902-4	3.9	115	0.099	0.709964	0.000004	0.7095	3.35	11.5	0.176	0.512903	0.000016	6.0
H902-5	8.5	47.0	0.523	0.710863	0.000004	0.7085	0.67	1.94	0.207	0.512841	0.000026	3.5
305-1a	0.8	92.3	0.024	0.706988	0.000006	0.7069	1.51	4.47	0.204	0.513122	0.000012	9.2
305-2 a	1.4	181	0.022	0.706855	0.000006	0.7068	2.01	5.46	0.222	0.513209	0.000016	10.1
305-3 a	9.1	273	0.097	0.707012	0.000007	0.7066	1.32	4.23	0.188	0.512966	0.000024	6.8
305-4 a	5.0	42.8	0.338	0.709384	0.000008	0.7078	2.4	6.56	0.221	0.513201	0.000016	10.0
X3-1 a	1.1	175	0.018	0.706961	0.000008	0.7069	2.43	6.92	0.212	0.513172	0.000007	9.8
8-12	4.0	297	0.039	0.711889	0.000008	0.7117	3.09	9.50	0.196	0.512917	0.000009	5.4
H608-6	5.4	149	0.105	0.711467	0.000003	0.7110	3.14	10.2	0.186	0.512775	0.000019	3.1
8-4	2.6	128	0.058	0.711301	0.000007	0.7110	2.34	7.37	0.191	0.513043	0.000034	8.1
8-9	0.4	112	0.011	0.707053	0.000004	0.7070	3.92	13.2	0.179	0.513076	0.000013	9.2
8-20	4.4	170	0.075	0.706122	0.000007	0.7058	5.11	16.3	0.189	0.513108	0.000017	9.5

Samples marked with a superscript "a" are carbonated eclogites enclosed in marbles, and all the others are the eclogites enclosed in mica schists.

Table 2　Magnesium isotopic compositions of the eclogites and mica schists and their mineral separates from southwestern Tianshan

Sample	Rock/Mineral	δ^{26}Mg	2SD	δ^{25}Mg	2SD
Eclogites/blueschists					
H902-7	Bulk rock	0.27	0.08	0.16	0.05
	Replicate	0.24	0.06	0.14	0.08
	average	0.25	0.05	0.15	0.04
	Grt	−1.37	0.07	−0.69	0.06
	Cpx	0.46	0.07	0.27	0.06
300-1	Bulk rock	0.25	0.05	0.16	0.04
	Replicate	0.28	0.07	0.15	0.05
	average	0.26	0.04	0.16	0.04

Continued

Sample	Rock/Mineral	δ^{26}Mg	2SD	δ^{25}Mg	2SD
H902-4	Bulk rock	−0.10	0.05	−0.03	0.04
	Grt	−1.61	0.07	−0.88	0.06
	Duplicate	−1.70	0.07	−0.87	0.05
	average	−1.66	0.05	−0.87	0.04
	Cpx	0.09	0.05	0.02	0.07
	Replicate	0.05	0.05	0.04	0.07
	average	0.07	0.04	0.03	0.05
H902-5	Bulk rock	−0.10	0.06	−0.04	0.05
	Grt	−1.58	0.07	−0.78	0.07
	Duplicate	−1.47	0.09	−0.77	0.06
	average	−1.54	0.06	−0.78	0.04
	Cpx	0.06	0.05	0.01	0.07
H907-21	Bulk rock	−0.19	0.06	−0.11	0.05
	Grt	−1.45	0.09	−0.76	0.06
	Cpx	−0.04	0.05	−0.05	0.07
305-1a	Bulk rock	−0.09	0.05	−0.07	0.05
	Grt	−1.10	0.07	−0.58	0.05
	Cpx	0.14	0.07	0.09	0.05
305-2 a	Bulk rock	0.02	0.08	0.02	0.05
	Grt	−1.17	0.06	−0.60	0.05
	Duplicate	−1.16	0.06	−0.64	0.05
	Replicate	−1.16	0.09	−0.59	0.06
	average	−1.16	0.04	−0.62	0.03
	Cpx	0.11	0.07	0.04	0.05
305-3 a	Bulk rock	−0.19	0.06	−0.10	0.03
305-4 a	Bulk rock	−0.28	0.05	−0.16	0.04
X3-1 a	Bulk rock	0.01	0.06	−0.01	0.05
	Grt	−1.16	0.06	−0.63	0.04
	Duplicate	−1.15	0.09	−0.60	0.06
	average	−1.16	0.05	−0.62	0.03
8-12	Bulk rock	−0.26	0.05	−0.13	0.04
8-19	Bulk rock	−0.33	0.05	−0.19	0.05
	Grt	−1.54	0.07	−0.79	0.06
	Replicate	−1.52	0.06	−0.75	0.06
	Duplicate	−1.51	0.07	−0.80	0.06
	average	−1.52	0.04	−0.78	0.03
	Cpx	0.27	0.07	0.13	0.05
8-26	Bulk rock	−0.31	0.05	−0.16	0.05
H608-6	Bulk rock	−0.25	0.05	−0.11	0.04
	Grt	−1.56	0.07	−0.81	0.05
8-3	Bulk rock	−0.24	0.05	−0.13	0.04
	Grt	−1.51	0.09	−0.83	0.06
8-4	Bulk rock	−0.17	0.05	−0.10	0.05

Continued

Sample	Rock/Mineral	δ^{26}Mg	2SD	δ^{25}Mg	2SD
	Replicate	−0.19	0.06	−0.09	0.05
	average	−0.18	0.04	−0.10	0.04
	Grt	−1.67	0.07	−0.87	0.07
	Duplicate	−1.65	0.09	−0.87	0.06
	average	−1.66	0.06	−0.87	0.04
8-5	Bulk rock	−0.26	0.05	−0.13	0.04
	Grt	−1.34	0.07	−0.70	0.06
8-7	Bulk rock	−0.34	0.06	−0.15	0.05
8-9	Bulk rock	−0.12	0.06	−0.05	0.05
8-20	Bulk rock	−0.37	0.05	−0.16	0.05
	Grt	−1.75	0.07	−0.89	0.05
	Cpx	−0.02	0.07	−0.03	0.05
H710-3	Bulk rock	−0.16	0.05	−0.03	0.05
A314-3[a]	Bulk rock	−0.15	0.08	−0.08	0.05
	Grt	−1.53	0.07	−0.80	0.05
	Cpx	0.45	0.05	0.26	0.07
105-1	Bulk rock	−0.03	0.07	−0.03	0.06
105-12	Bulk rock	0.00	0.07	0.02	0.06
106-14[a]	Bulk rock	−0.19	0.07	−0.12	0.06
110-3	Bulk rock	0.08	0.07	0.01	0.06
Q316-10	Bulk rock	−0.09	0.04	−0.04	0.02
A300-3	Bulk rock	0.00	0.03	0.00	0.03
a300-16	Bulk rock	−0.11	0.04	−0.05	0.02
H902-10	Bulk rock	−0.33	0.02	−0.16	0.02
k984 - 1	Bulk rock	−0.05	0.06	−0.02	0.03
H902-2 - 1	Bulk rock	0.15	0.01	0.07	0.01
Mica schist					
106-3B	Bulk rock	−0.18	0.07	−0.08	0.06
	Duplicate	−0.13	0.08	−0.05	0.05
	average	−0.16	0.05	−0.06	0.04
986-1	Bulk rock	−0.11	0.08	−0.02	0.05
305-5	Bulk rock	−0.16	0.05	−0.08	0.04
Q314-1	Bulk rock	0.23	0.02	0.13	0.03
Q316-4	Bulk rock	−0.25	0.02	−0.13	0.01
H865 - 1	Bulk rock	−0.13	0.03	−0.06	0.01

Samples marked with a superscript "a" are carbonated eclogites enclosed in marbles, and all the others are the eclogites enclosed in mica schists; Grt=garnet; Cpx=clinopyroxene; 2SD=two times the standard deviation of the population of n (n>20) repeat measurments of the standard during an analytical session;Replicate: repeat sample dissolution, column chemistry and instrument analysis of Mg isotopic ratios; Duplicate: repeat measurement of Mg isotopic ratios on the same solution.

5 Discussion

The overprint of fluid-rock interactions on the southwestern Tianshan eclogites/blueschists has been confirmed by many petrological and geochemical studies (John et al., 2008; van der Straaten et al., 2008, 2012; Beinlich et al., 2010; Lü et al., 2012; J.-L. Li et al., 2016a; Zhang et al., 2016). Depending on the nature and abundance of fluids in a subduction channel, the initial composition of an eclogite can be altered to various degrees after fluid-rock interactions. In this section, we first focus on the trace element and Sr-Nd isotopes to characterize the source and nature of the fluids, and then decipher how fluid-rock interactions may have influenced the Mg isotopic systematics of the eclogites. Finally, we discuss the Mg isotope geochemistry of slab-derived fluids in the subduction channel and their influences on the sub-arc peridotites.

5.1 Geochemical evidence for fluid-rock interactions

Trace element and Sr-Nd isotope geochemistry suggest interactions of eclogites with metamorphic fluids. The fluids are mainly derived from subducted sediments, with limited contributions from serpentinites or altered oceanic crusts. Most eclogites are variably enriched in LILEs (e.g., Ba, Cs, Rb, and K) and Pb (Fig. 3), which can be produced during either ancient seafloor alteration or subduction-zone fluid-rock interactions. Bebout (2007) documented that significant enrichments of Ba and Pb in metabasaltic rocks can be most directly associated with metasomatism because these two elements are only slightly enriched in altered oceanic basalts during seafloor alteration relative to other LILEs. The consistently high Ba/Rb, high Ba/K and low Ce/Pb of our eclogites are thus indicative of HP/UHP fluid-rock interactions rather than ancient seawater alteration (Fig. 3a, b, c). Furthermore, these eclogites have extremely high initial $^{87}Sr/^{86}Sr$ ratio up to 0.7117 (Fig. 1), a signature that cannot be attributed to pre-subduction seawater alteration because the Ordovican-Carboniferous seawater has much lower $^{87}Sr/^{86}Sr$ ratios of 0.7075–0.7090 (Veizer, 1989). The high $^{87}Sr/^{86}Sr$(320Ma) ratios thus must have resulted from interactions of the eclogites with fluids during metasomatism, and the fluids might be derived from subducted sediments whose $^{87}Sr/^{86}Sr$ ratios can be as high as 0.73 (Plank and Langmuir, 1998). In contrast to Sr isotopes, Nd isotopes appear to behave conservatively during the metasomatism (King et al., 2006). Due to the low mobility of REE during metamorphic dehydration under relatively low *P-T* conditions (Kessel et al., 2005), slab-derived fluids would contain too little Nd to affect the Nd isotopic systematics of eclogites (van der Straaten et al., 2012), such that the eclogites retain their depleted Nd isotopic signatures (Fig. 1). In accordance with the high $^{87}Sr/^{86}Sr$ ratios, most eclogites contain very low concentrations of Co and Ni evolving from oceanic basalts towards the GLOSS (global subducting sediments; Fig. 3d), pointing towards again interactions of the eclogites with sediment-derived fluids. Some eclogites however have Ni and Co contents overlapping or slightly higher than oceanic basalts (Fig. 3d). This indicates the possible contributions of altered oceanic crust-derived or serpentinite-derived fluids (e.g., van der Straaten et al., 2012), although subducted sediments must be the dominant source for fluids that have interacted with the eclogites.

The geochemical signatures of sediment-derived fluids might vary significantly in response to the mineralogical heterogeneity of subducting sediments. The eclogites display a series of geochemical features indicative of two compositionally different fluid components (Fig. 4). As shown in Rb/Sr vs. Pb* (an index of enrichment of Pb; $Pb^*=2*Pb_N/[Ce_N+Pr_N]$) and Ba/Pb vs. Pb* diagrams, the enrichment of Pb in eclogites is not always associated with the enrichment of LILEs (Fig. 4a and b). The observed decoupling patterns may indicate two major fluid components: one enriched in LILEs relative to Pb (e.g., high Rb/Sr and Ba/Pb but low Pb*), and the other enriched in Pb relative to LILEs (e.g., high Pb* but low Rb/Sr or Ba/Pb). The roughly positive correlation between Pb* and $^{87}Sr/^{86}Sr$(320Ma) (Fig. 4c), suggests that the high-Pb component also contains a significant amount of radiogenic Sr that has elevated the $^{87}Sr/^{86}Sr$ value of eclogites. Some carbonated eclogites are extremely enriched in elemental Sr but have relatively low $^{87}Sr/^{86}Sr$ values of 0.7066–0.7078 (Supplementary Fig. S2), suggesting that the surrounding marbles are not the source of high-$^{87}Sr/^{86}Sr$ fluid. Instead, the high-$^{87}Sr/^{86}Sr$

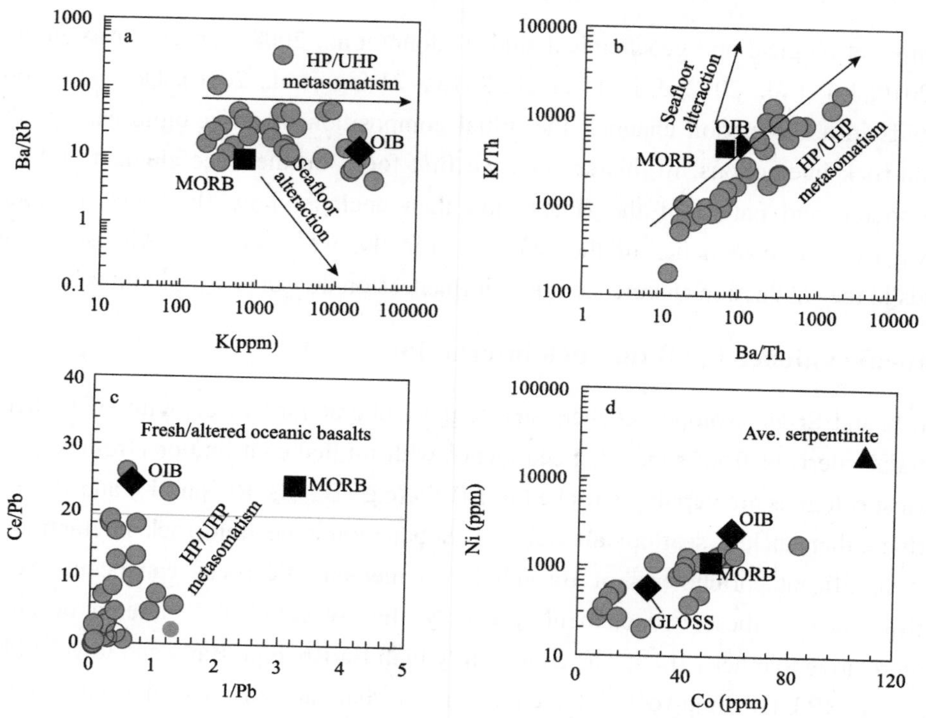

Fig. 3 Ba/Rb vs. K (a), K/Th vs. Ba/Th (b), and Ce/Pb vs. 1/Pb (c) diagrams to differentiate between ancient seawater alteration and metamorphic metasomatism after Bebout (2007). The Ni vs. Co diagram (d) indicates that most eclogites have lower Ni and Co concentration than oceanic basalts. Data of MORB and OIB are from Sun and McDonough (1989); the Ni and Co of average serpentinite are from data compiled by van der Straaten et al. (2008).

fluid component must be sourced from other metasediments, such as mica schists. The high-LILEs component, on the other hand, might contain too little Sr to modify the Sr isotopic composition of eclogites, as reflected by the decoupling relationship between $^{87}Sr/^{86}Sr$(320Ma) and Rb/Sr (Fig. 4d): the high-Rb/Sr eclogites display low $^{87}Sr/^{86}Sr$(320Ma) values, whereas the low-Rb/Sr samples are characterized by highly radiogenic Sr isotopic compositions (Fig. 4d). All these observations support that the eclogites were infiltrated by two fluid components. The distinct geochemical signatures of the two fluid components are consistent with the fact that LILEs and Sr-Pb are hosted in different hydrous minerals in subducted sediments: mica-group minerals are the dominant host for LILEs, whereas epidote-group minerals (and to a less extent carbonate minerals and paragonite) are the major host of Pb and Sr (e.g., Bebout et al., 2007, 2013; Busigny et al., 2003; Xiao et al., 2012). As a result, fluid dehydrated from mica-group minerals would have high Rb/Sr and Ba/Pb ratios, whereas fluid released from epidote-group minerals in metasediments could be enriched in Pb and Sr (as well as $^{87}Sr/^{86}Sr$). It is possible that varying modal mineralogy in the subducted sediments (e.g., mica-group minerals are abundant in metapelites and epidote-group minerals are abundant in greywackes) can result in decomposition of mica- and epidote-group minerals in different proportions along the subduction *P-T* path and generate the two fluid components in the subduction channel. During crustal subduction, biotite is thought to be completely decomposed at P= 1.3−1.5 GPa, at which the epidote-group minerals such as epidote and zoisite are still stable (Poli and Schmidt, 2002). Therefore, decomposition of biotite at the early stage during crustal subduction could release a significant amount of fluid that is enriched in LILEs. At a higher pressure above 2.5 GPa, epidote and zoisite might become unstable (Carswell, 1990; Poli and Schmidt, 2002). Metamorphic dehydration at this stage could thus release abundant Sr and Pb to the fluids. Such fluids, when released from subducting oceanic crust, would migrate upward along the subduction channel, infiltrate the exhuming eclogites and impart their distinct geochemical signatures to the eclogites via fluid-rock interactions.

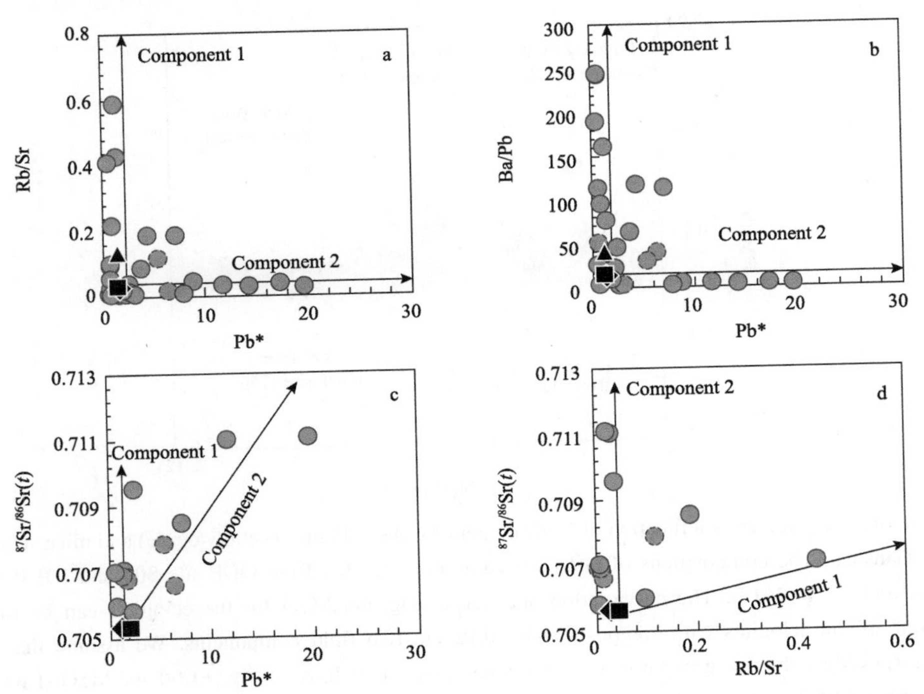

Fig. 4 Rb/Sr vs. Pb* (a), Ba/Pb vs. Pb* (b), $^{87}Sr/^{86}Sr$ (t) vs. Pb* (c) and $^{87}Sr/^{86}Sr$ (t) vs. Rb/Sr (d)diagrams to indicate the two fluid components. The Pb* represents an indices of enrichment of Pb in the eclogites: $Pb^*=2*Pb_N/(Ce_N+Pr_N)$. The higher the Pb*, the more enrichment of Pb for the eclogites. The carbonated eclogites are marked as dashed outline. The black triangle in panels a and b represents the average altered oceanic crust (super composite of Ocean Drilling Program Site 801) in Kelley et al. (2003). Black square and diamond represent the composition of MORB and OIB, respectively. The component 1 is enriched in LILEs, which might be derived from dehydration of mica-group minerals. The component 2 is enriched in Pb and $^{87}Sr/^{86}Sr$, likely released from epidote-group minerals.

5.2 Constraining the mechanisms of Mg isotopic variations in the eclogites

The eclogites have varying Mg contents (MgO=3.2 to 9.7 wt.%) at a given SiO_2 content, and more variable and systemically heavier Mg isotopic composition than fresh oceanic basalts (Fig. 5).The simplest explanation for the low MgO and high $\delta^{26}Mg$ of eclogites is physical/mechanical mixing with a high-$\delta^{26}Mg$ sedimentary component at some point before or during the exhumation of the eclogites. However, this is very unlikely because binary mixing calculation, using the highest $\delta^{26}Mg$ value of the six mica schists as an endmember (Q-314), suggests that at least >60% of sedimentary component is required to produce the Mg isotopic compositions of most eclogites (Fig. 5), such that the eclogites would have anomalously high SiO_2 contents (>55 wt.%). In addition, the SiO_2 of eclogites does not correlate with neither $^{87}Sr/^{86}Sr$(320Ma) nor $\varepsilon_{Nd}(t)$ (Supplementary Fig. S3), further supporting that binary mixing between basalt (or eclogite) and sediment (or metasediment) might not be the case. Magnesium is fluid-mobile, thus, processes like ancient seawater alteration, prograde metamorphism (e.g., release of Mg into metamorphic fluids and eclogite-host isotopic exchanges), and retrograde fluid-rock interactions (e.g., interaction with metamorphic fluid during exhumation), could potentially account for the observed Mg isotopic variations. Next, we endeavor to explore how Mg isotopes behave during these processes, based on which, we highlight the importance of subduction channel fluids in generating Mg isotopic variations in exhumed eclogites.

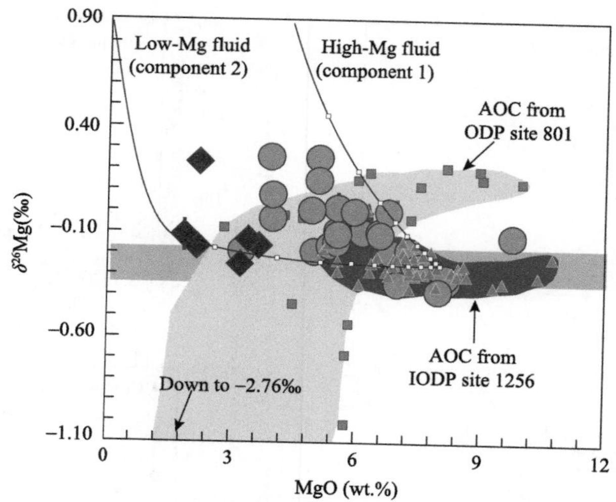

Fig. 5 The variation of δ^{26}Mg values as a function of MgO content for the eclogites (yellow circle) and mica schists (blue diamond) from southwestern Tianshan. The compositions of altered oceanic crust (AOC) from ODP site 801 and IODP site 1256 are from Huang et al. (2015) and Teng (2017). The co-variation between δ^{26}Mg and MgO for the eclogites can be roughly modeled as fluid-rock interactions of the eclogites with compositionally different two fluid components. We assume that the component 1, because of its origin from Mg-rich mica-group minerals or to a less extent talc, have δ^{26}Mg=+1.00 and MgO=1 wt.%; the component 2, released from Mg-poor epidote-group minerals, contain very little Mg (assuming MgO=0.05 wt.%). Although we assign a value of +1.00 for the δ^{26}Mg of the low-MgO component 2, the change of this value will not affect the modelling significantly, as the component 2 contains too little Mg so as not to influence the Mg isotopic composition of the eclogites. Thus, the two purple curves with increment of 10% represent the fluid-rock interaction of an eclogite (δ^{26}Mg=−0.25; MgO=8 wt.%) with high-MgO and low-MgO fluid components, with the partition coefficient of MgO between fluid and eclogite, $D_{\mathrm{eclogite/fluid}}$=4. The black dotted curve represents binary mixing between sediments and basalts, which suggests that >60% of sedimentary component is required to produce the Mg isotopic composition of the eclogites. The green bar represents the normal mantle δ^{26}Mg value (Teng et al., 2010). (For interpretation of the references to color in this figure legend, the reader is referred to the web version of this article.)

5.2.1 Seafloor alteration cannot explain the Mg isotopic signatures

Seafloor alteration produces even larger Mg isotopic variations, with Mg isotopes likely fractionated in a different manner from that observed in the eclogites, as shown in Fig. 5. Altered oceanic crusts (AOC) from two different sites have been reported for Mg isotopic compositions (Huang et al., 2015; Teng, 2017). Carbonate-barren AOC samples recovered from IODP site 1256 in the eastern equatorial Pacific retain a mantle-like δ^{26}Mg value as for fresh oceanic basalts (Fig. 5; Huang et al., 2015), based on which Huang et al. (2015) concluded that seafloor alteration causes limited Mg isotope fractionation, regardless of alteration temperature and water/rock ratio. At the other site (ODP site 801) in western Pacific, extensively altered AOC samples have highly variable δ^{26}Mg values ranging from −2.76 to +0.21 (Fig. 5), with low δ^{26}Mg values being associated with carbonate enriched samples and high δ^{26}Mg values associated with clay-rich samples (Teng, 2017). Due to carbonate dilution effect (Tipper et al., 2006), the AOC samples from ODP site 801 are distributed in a trend in which δ^{26}Mg values decrease as MgO decreases (Fig. 5). Different from AOC, none of the studied eclogites (32 in total) show enrichment of light Mg isotopes, although they contain variable abundances of carbonate minerals. Furthermore, neither heavily nor less altered AOC could account for the roughly negative correlation between δ^{26}Mg and MgO for the eclogites (Fig. 5). Thus, ancient seawater alteration is unlikely to be the cause of the variable and systemically heavier Mg isotopic compositions of the eclogites.

5.2.2 The role of prograde metamorphic dehydration and eclogite-host isotopic exchange

Magnesium isotope fractionation during prograde metamorphic dehydration or eclogite-host isotopic exchange cannot account for the Mg isotopic variations in our eclogites. It is possible that dehydrated fluids have distinct Mg isotopic compositions from the rock where the fluids are from. However, since the fraction of Mg partitioning into the fluid phases is so small compared to that inherited by metamorphic minerals during prograde metamorphism, metamorphic dehydration causes insignificant Mg isotope fractionation (<±0.07) on a bulk-rock scale (Li et al., 2014, 2014; Teng et al., 2013; W.-Y. Li et al., 2011b; Wang et al., 2014b, 2015a, 2015b). Local isotopic exchange between eclogite and its host rock can potentially change the original mantle-like Mg isotopic compositions of the eclogites (Wang et al., 2014a). To which direction the Mg isotopes of the eclogites fractionate depends on the types of host rock. For example, eclogite-host isotopic exchange would make eclogite boudins in carbonates/marbles isotopically lighter, whereas those enclosed in mica schists heavier (Wang et al., 2014a). However, no systemic relationship between δ^{26}Mg and host rock type was observed for the southwestern Tianshan eclogites. On the opposite, the carbonated eclogites (those enclosed in marbles) in our study are enriched in heavy Mg isotopes (δ^{26}Mg= –0.28 to +0.02; Table 2), which we interpret below as a result of infiltration of external fluids derived from metasediments.

5.2.3 Response of Mg isotopic systematics in the eclogites to fluid-rock interactions.

Thus, our favored interpretation of the Mg isotopic variation is fluid-rock interaction in a subduction channel. The fluids must be enriched in heavy Mg isotopes, and pervasively reactive in interacting with the eclogites because the eclogites have systemically heavier Mg isotopic compositions (Fig. 2), regardless of their diverse host rock types. Below, we discuss how the two fluid components may have affected the Mg isotopic compositions of the eclogites.

The two fluid components, due to their derivation from different hydrous minerals, have different impacts on the Mg isotopic systematics of eclogites. In the plots of δ^{26}Mg vs. Pb* and δ^{26}Mg vs. ^{87}Sr/^{86}Sr(320Ma) (Fig. 6a and b), the high-Pb* and ^{87}Sr/^{86}Sr(320Ma) samples retain a mantle-like δ^{26}Mg value, suggesting that the infiltration of high-Pb and ^{87}Sr/^{86}Sr fluid component had limited influences on the Mg isotopic composition of eclogites. Being the dominant source of high-Pb and ^{87}Sr/^{86}Sr component, the epidote-group minerals contain little Mg (e.g., Guo et al., 2012), and thus the fluid dehydrated from them is unable to modify the Mg isotopic composition of the eclogites (although the exact δ^{26}Mg value of any epidote-group mineral has not been reported so far). The low-Pb* and ^{87}Sr/^{86}Sr(320Ma) samples, on the other hand, have variably high δ^{26}Mg values (Fig. 6a and b). As expected, eclogites with high-Rb/Sr and Ba/Pb ratios have high δ^{26}Mg values (Fig. 6c and d). Because of the complexity of the fluid system and the uncertainty of its Mg concentration and Mg isotopic composition, we are not expecting to see good correlations between δ^{26}Mg and indices of enrichment of LILEs (such as Rb/Sr and Ba/Pb). However, the general patterns shown in Fig. 6c and d suggest that the high-LILEs component carries a significant amount of isotopically heavy Mg that has elevated the δ^{26}Mg values of the eclogites. Mica-group minerals, as the major source of high-LILE component, are enriched in MgO, and in addition their δ^{26}Mg values are characteristically high. For instance, biotites in metapelites from the Ivrea Zone in NW Italy have δ^{26}Mg values ranging from –0.08 to +1.10 (Wang et al., 2015b), and phengites in eclogites from the Dabie orogen have δ^{26}Mg values of +0.30 to +0.59 (W.-Y. Li et al., 2011). Therefore, eclogites metasomatized by the mica-derived fluid could gain high-δ^{26}Mg signatures. It is also important to note that in Fig. 6c and d, a part of low-Rb/Sr and Ba/Pb samples has slightly high δ^{26}Mg value, which we interpret as the possible influence of the talc-derived fluid, as the talc in serpentinite is depleted in LILEs but enriched in heavy Mg (δ^{26}Mg=+0.06 to

+0.30; Beinlich et al., 2014). Without good constraints on the Mg isotopic composition of fluid and the partition coefficient of MgO between fluid and eclogite, it is not yet possible to give a perfect fluid-rock interaction model for the whole dataset of the eclogites. However, the rough negative correlation between δ^{26}Mg and MgO for the eclogites can be generally modeled as interactions of eclogites with high-MgO (e.g., dehydrated from mica-group minerals or talc) and low-MgO fluid components (e.g., dehydrated from epidote-group minerals) under a variety of water/rock (fluid/eclogite) ratios (Fig. 5).

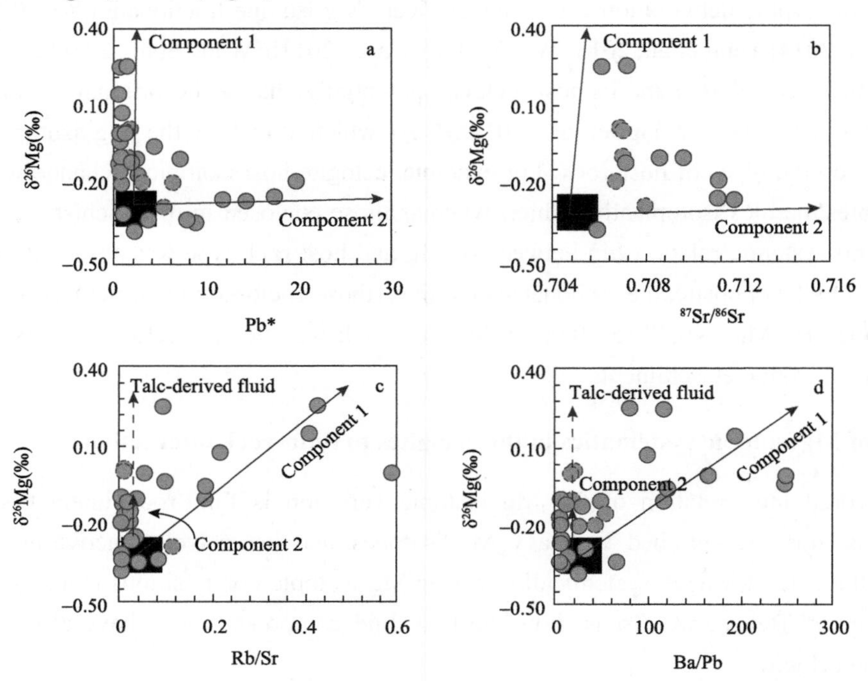

Fig. 6　δ^{26}Mg vs. Pb* (a), δ^{26}Mg vs. ^{87}Sr/^{86}Sr (t) (b), δ^{26}Mg vs. Rb/Sr (c), and δ^{26}Mg vs. Ba/Pb (d) diagrams showing the influence of the two fluid components on the Mg isotopic systematics of eclogites (shown as solid arrows). The carbonated eclogites are marked as dashed outline. The high-LILE fluid component contains a considerable amount of isotopically heavy Mg to shift the δ^{26}Mg of eclogites towards a higher value, whereas the high-^{87}Sr/^{86}Sr and -Pb fluid component contains little heavy Mg to influence the Mg isotopic systematics of eclogites. Some low-Rb/Sr and -Ba/Pb samples also have slightly heavy Mg isotopic compositions, which might point towards the contributions of fluids dehydrated from talc in serpentinite (shown as dashed arrows; Beinlich et al., 2014).

5.3　The origins of isotopically heavy fluids in subduction channel

Fluids in subduction channels are likely to have heavy Mg isotopic compositions, although different subducted lithologies themselves show highly variable δ^{26}Mg values. The subducted abyssal peridotites have slightly high δ^{26}Mg values of −0.25 – +0.10 (Liu et al., 2017). The subducted sediments and altered oceanic crusts have large variations in δ^{26}Mg values (−2.76‰ to +0.92‰), with low δ^{26}Mg associated with carbonated rocks and with high δ^{26}Mg associated with carbonate-free rocks (Hu et al., 2017; Huang et al., 2015; Li et al., 2010; Teng et al., 2016; Teng, 2017; Wang et al., 2015a;). One might expect that subsolidus decarbonation or carbonate dissolution during metamorphism could release light Mg isotopes, making the sediment-derived fluids isotopically light. However, decarbonation is an inefficient process for carbonated sediments/basalts along the P-T paths of oceanic subduction (Dasgupta and Hirschmann, 2010; Gorman et al., 2006). Carbonate species dissolved in metamorphic fluid is thought to be mainly $CaCO_3$ (Ague and Nicolescu, 2014; Kelemen and Manning, 2015; Li et al., 2017). Therefore, the presence of carbonate minerals in subducted rocks has negligible influence on the Mg isotopic composition of dehydrated fluids (Li et al., 2017). By contrast, breakdown of hydrous minerals might control the Mg concentration and Mg isotopic composition of dehydrated fluids. Reported δ^{26}Mg values for Mg-rich hydrous minerals, such like mica-group minerals and talc, are higher than the normal mantle value (Beinlich et al., 2014; W.-Y. Li et al., 2011; Wang et al., 2015b), and thus it is very likely that the dehydrated fluids are enriched in heavy Mg isotopes. For example, a recent study suggested that the fluid derived from talc and antigorite in serpentinite is likely characterized by high-Mg and high-δ^{26}Mg, and could be responsible for the high δ^{26}Mg values of white schists in Western Alps (Chen et al., 2016).

5.4 Implications on Mg isotopic systematics in sub-arc peridotites

Fluids in subduction channels can infiltrate the mantle wedge, inducing fluid-peridotite interactions and potentially modifying the Mg isotopic composition of associated peridotites. Only a few Mg isotopic data have been reported so far for mantle wedge peridotites, and they are indeed enriched in heavy Mg isotopes: six arc peridotites from Avacha Volcano in Kamchatka analyzed by Pogge von Strandmann et al. (2011) have δ^{26}Mg values ranging from –0.25 to –0.06, higher than the normal mantle value (–0.25±0.07; Teng et al., 2010). Although the actual mechanism responsible for the Mg isotopic variations in these peridotites is still uncertain, their high δ^{26}Mg values are consistent with petrological and geochemical evidence suggesting that these peridotites have been affected by upward fluid migration from the subducting slab (Ionov and Seitz, 2008). Most recently, Li et al. (2017) found that island arc or back arc basin basalts from circum-Pacific arcs, including Kamchatka, Philippines, Costa Rica and Lau Basin have generally high δ^{26}Mg values ranging from –0.35 to +0.06. Teng et al. (2016) reported the Martinique arc lava δ^{26}Mg values of –0.25 to –0.10. Those values overlap the Avacha peridotites and are systemically higher than normal oceanic basalts and peridotites, consistent with the interpretation that isotopically heavy fluids released from the subducted slab incorporate into the mantle wedge (Li et al., 2017; Teng et al., 2016). All the three cases suggest that massive flux of dehydrated fluid into the sub-arc mantle could facilitate extensive fluid-peridotite interaction and shift the δ^{26}Mg of sub-arc peridotite towards higher values.

6 Conclusions

To reveal the nature of fluid-rock interactions in subduction channels and the influence of subduction-zone fluids on the Mg isotopic systematics in exhumed rocks, we present major and trace elements, and Sr-Nd-Mg isotopic data for the eclogites and mica schists from southwestern Tianshan, China. The following conclusions can be drawn:

(1) The eclogites have high Ba/Rb and Ba/K but low Ce/Pb ratios, suggesting the overprint of subduction-zone metamorphic metasomatism. The highly radiogenic Sr isotopic composition [^{87}Sr/^{86}Sr(320Ma)=0.7058–0.7117; higher than that of coeval seawater], together with the varying and mostly low Ni and Co concentrations, further indicate that the eclogites have interacted with fluids mainly released from subducted sediments, with limited contributions from altered oceanic crust-or serpentinite-derived fluids.

(2) The positive correlation between ^{87}Sr/^{86}Sr and Pb*, and the bidirectional patterns in ^{87}Sr/^{86}Sr-Rb/Sr, Pb*-Rb/Sr, and Pb*-Ba/Pb spaces, suggest interaction of the eclogites with compositionally different two fluid components: the high-LILEs component which could be derived from dehydration of mica-group minerals, and the high-Pb and ^{87}Sr/^{86}Sr component likely released from epidote-group minerals in subducted sediments.

(3) The highly variable and systemically heavy Mg isotopic compositions of eclogites (δ^{26}Mg=–0.37 to +0.26) resulted from fluid-rock interactions in the subduction channel. The high-LILE component, dehydrated

from Mg-rich mica-group minerals or to a less extent from talc, contains a considerable amount of Mg that has shifted the δ^{26}Mg of the eclogites towards higher values. The high-Pb and ^{87}Sr/^{86}Sr component, dehydrated from Mg-poor epidote-group minerals, has little Mg so as not to influence the Mg isotopic composition of the eclogites.

Acknowledgement

The authors would like to thank two anonymous reviewers for insightful comments and Editor Sun-Lin Chung for careful and efficient handling. This study was financially supported by National Natural Science Foundation of China (41230209) to SGL, National Science Foundation (EAR-1340160) to FZT, and National Science Foundation of China (Grants 41330210, 41520104004) and Major State Basic Research Development Program (Grant 2015CB856105) to LFZ.

Appendix A. Supplementary data

Supplementary data to this article can be found online at http://dx. doi.org/10.1016/j.lithos.2017.08.004.

References

Agard, P., Yamato, P., Jolivet, L., Burov, E., 2009. Exhumation of oceanic blueschists and eclogites in subduction zones: timing and mechanisms. Earth-Science Reviews 92, 53-79.

Ague, J.J., Nicolescu, S., 2014. Carbon dioxide released from subduction zones by fluidmediated reactions. Nature Geoscience 7, 355-360.

Ai, Y.L., Zhang, L.F., Li, X.P., Qu, J.F., 2006. Geochemical characteristics and tectonic implications of HP-UHP eclogites and blueschists in southwestern Tianshan, China. Progress in Natural Science 16, 624-632.

Bebout, G.E., 2007. Metamorphic chemical geodynamics of subduction zones. Earth and Planetary Science Letters 260, 373-393.

Bebout, G.E., Penniston-Dorland, S.C., 2016. Fluid and mass transfer at subduction interfaces—the field metamorphic record. Lithos 240-243, 228-258.

Bebout, G.E., Bebout, A.E., Graham, C.M., 2007. Cycling of B, Li, and LILE (K, Cs, Rb, Ba, Sr) into subduction zones: SIMS evidence from micas in high-P/T metasedimentary rocks. Chemical Geology 239, 284-304.

Bebout, G.E., Agard, P., Kobayashi, K., Moriguti, T., Nakamura, E., 2013. Devolatilization history and trace element mobility in deeply subducted sedimentary rocks: evidence from western alps HP/UHP suites. Chemical Geology 342, 1-20.

Beinlich, A., Klemd, R., John, T., Gao, J., 2010. Trace-element mobilization during Cametasomatism along a major fluid conduit: eclogitization of blueschist as a consequence of fluid-rock interaction. Geochimica et Cosmochimica Acta 74, 1892-1922.

Beinlich, A., Mavromatis, V., Austrheim, H., Oelkers, E.H., 2014. Intermineral Mg isotope fractionation during hydrothermal ultramafic rock alteration-implications for the global Mg-cycle. Earth and Planetary Science Letters 392, 166-176.

Busigny, V., Cartigny, P., Philippot, P., Ader, M., Javoy, M., 2003. Massive recycling of nitrogen and other fluid-mobile elements (K, Rb, Cs, H) in a cold slab environment: evidence from HP to UHP oceanic metasediments of the Schistes Lustrés nappe (western Alps, Europe). Earth and Planetary Science Letters 215, 27-42.

Carswell, D.A., 1990. Eclogite Facies Rocks. Blackie and Son Ltd., pp. 14-49.

Chen, Y.-X., Schertl, H.P., Zheng, Y.-F., Huang, F., Zhou, K., Gong, Y.-Z., 2016. Mg-O isotopes trace the origin of Mg-rich fluids in the deeply subdcuted continental crust of Western Alps. Earth and Planetary Science Letters 456, 157-167.

Chu, Z., Chen, F., Yang, Y., Guo, J., 2009. Precise determination of Sm, Nd concentrations and Nd isotopic compositions at the nanogram level in geological samples by thermal ionization mass spectrometry. Journal of Analytical Atomic Spectrometry 24, 1534-1544.

Dasgupta, R., Hirschmann, M.M., 2010. The deep carbon cycle and melting in Earth's interior. Earth and Planetary Science Letters 298, 1-13.

Du, J., Zhang, L., Lü, Z., Chu, X., 2011. Lawsonite-bearing chloritoid-glaucophane schist from SW Tianshan, China: phase equilibria and P-T path. Journal of Asian Earth Sciences 42, 684-693.

Du, J.-X., Zhang, L.-F., Shen, X.-J., Bader, T., 2014a. A new PTt path of eclogites from Chinese southwestern Tianshan: constraints from PT pseudosections and Sm-Nd isochron dating. Lithos 200, 258-272.

Du, J., Zhang, L., Bader, T., Chen, Z., Lü, Z., 2014b. Metamorphic evolution of relict lawsonite-bearing eclogites from the (U) HP metamorphic belt in the Chinese southwestern Tianshan. Journal of Metamorphic Geology 32, 575-598.

Foster, G.L., Pogge von Strandmann, P.A.E., Rae, J.W.B., 2010. Boron and magnesium isotopic composition of seawater. Geochemistry, Geophysics, Geosystems 11 (8). http:// dx.doi.org/10.1029/2010GC003201.

Gao, J., Klemd, R., 2001. Primary fluids entrapped at blueschist to eclogite transition: evidence from the Tianshan meta-subduction complex in northwestern China. Contributions to Mineralogy and Petrology 142, 1-14.

Gao, J., Klemd, R., 2003. Formation of HP-LT rocks and their tectonic implications in the western Tianshan Orogen, NW China: geochemical and age constraints. Lithos 66, 1-22.

Gao, J., Klemd, R., Zhang, L., Wang, Z., Xiao, X., 1999. PT path of high-pressure/low-temperature rocks and tectonic implications in the western Tianshan Mountains, NW China. Journal of Metamorphic Geology 17, 621-636.

Gao, J., John, T., Klemd, R., Xiong, X., 2007. Mobilization of Ti-Nb-Ta during subduction: evidence from rutile-bearing dehydration segregations and veins hosted in eclogite, Tianshan, NW China. Geochimica et Cosmochimica Acta 71, 4974-4996.

Gao, S., Zhang, B.-R., Gu, X.-M., Xie, Q.-L., Gao, C.-L., Guo, X.-M., 1995. Silurian-Devonian provenance changes of South Qinling basins: implications for accretion of the Yangtze (South China) to the North China cratons. Tectonophysics 250, 183-197.

Gerya, T.V., Stöckhert, B., Perchuk, A.L., 2002. Exhumation of high-pressure metamorphic rocks in a subduction channel: a numerical simulation. Tectonics 21 (6-1-6-19).

Glodny, J., Austrheim, H., Molina, J.F., Rusin, A.I., Seward, D., 2003. Rb/Sr record of fluidrock interaction in eclogites: the Marun-Keu complex, Polar Urals, Russia. Geochimica et Cosmochimica Acta 67, 4353-4371.

Gorman, P.J., Kerrick, D., Connolly, J., 2006. Modeling open system metamorphic decarbonation of subducting slabs. Geochemistry, Geophysics, Geosystems 7.

Guo, S., Ye, K., Chen, Y., Liu, J.-B., Mao, Q., Ma, Y.-G., 2012. Fluid-rock interaction and element mobilization in UHP metabasalt: constraints from an omphacite-epidote vein and host eclogites in the Dabie orogen. Lithos 136-139, 145-167.

Halama, R., John, T., Herms, P., Hauff, F., Schenk, V., 2011. A stable (Li, O) and radiogenic (Sr, Nd) isotope perspective on metasomatic processes in a subducting slab. Chemical Geology 281, 151-166.

Horodyskyj, U., Lee, C.-T.A., Luffi, P., 2009. Geochemical evidence for exhumation of eclogite via serpentinite channels in ocean-continent subduction zones. Geosphere 5, 426-438.

Hu, Y., Teng, F.-Z., Plank, T., Huang, H.-J., 2017. Magnesium isotopic composition of subducting marine sediments. Chemical Geology http://dx.doi.org/10.1016/j.chemgeo.2017.06.010.

Huang, F., Chen, L.,Wu, Z.,Wang,W., 2013. First-principles calculations of equilibrium Mg isotope fractionations between garnet, clinopyroxene, orthopyroxene, and olivine: implications for Mg isotope thermometry. Earth and Planetary Science Letters 367, 61-70.

Huang, J., Ke, S., Gao, Y., Xiao, Y., Li, S., 2015. Magnesium isotopic compositions of altered oceanic basalts and gabbros from IODP site 1256 at the East Pacific rise. Lithos 231, 53-61.

Ionov, D.A., Seitz, H.-M., 2008. Lithium abundances and isotopic compositions in mantle xenoliths from subduction and intra-plate settings: mantle sources vs. eruption histories. Earth and Planetary Science Letters 266, 316-331.

John, T., Scherer, E.E., Haase, K., Schenk, V., 2004. Trace element fractionation during fluidinduced eclogitization in a subducting slab: trace element and Lu-Hf-Sm-Nd isotope systematics. Earth and Planetary Science Letters 227, 441-456.

John, T., Klemd, R., Gao, J., Garbe-Schönberg, C.-D., 2008. Trace-element mobilization in slabs due to non steady-state fluid-rock interaction: constraints from an eclogitefacies transport vein in blueschist (Tianshan, China). Lithos 103, 1-24.

John, T., Gussone, N., Podladchikov, Y.Y., Bebout, G.E., Dohmen, R., Halama, R., Klemd, R., Magna, T., Seitz, H.-M., 2012. Volcanic arcs fed by rapid pulsed fluid flow through subducting slabs. Nature Geoscience 5, 489-492.

Kelemen, P.B., Manning, C.E., 2015. Reevaluating carbon fluxes in subduction zones, what goes down, mostly comes up. Proceedings of the National Academy of Sciences 112, E3997-E4006.

Kelley, K.A., Plank, T., Ludden, J., Staudigel, H., 2003. Composition of altered oceanic crust at ODP sites 801 and 1149. Geochemistry, Geophysics, Geosystems 4:890. http://dx.doi.org/10.1029/2002GC000435.

Kessel, R., Schmidt, M.W., Ulmer, P., Pettke, T., 2005. Trace element signature of subduction-zone fluids, melts and supercritical liquids at 120-180 km depth. Nature 437, 724-727.

King, R.L., Bebout, G.E., Moriguti, T., Nakamura, E., 2006. Elemental mixing systematics and Sr-Nd isotope geochemistry of mélange formation: obstacles to identification of fluid sources to arc volcanics. Earth and Planetary Science Letters 246, 288-304.

Klemd, R., 2013. Metasomatism During High-Pressure Metamorphism: Eclogites and Blueschist-Facies Rocks, Metasomatism and the Chemical Transformation of Rock. Springer, pp. 351-413.

Klemd, R., John, T., Scherer, E., Rondenay, S., Gao, J., 2011. Changes in dip of subducted slabs at depth: petrological and geochronological evidence from HP-UHP rocks (Tianshan, NW-China). Earth and Planetary Science Letters 310, 9-20.

Li, W.-Y., Teng, F.-Z., Ke, S., Rudnick, R.L., Gao, S., Wu, F.-Y., Chappell, B., 2010. Heterogeneous magnesium isotopic composition of the upper continental crust. Geochimica et Cosmochimica Acta 74, 6867-6884.

Li, Q.-L., Lin,W., Su,W., Li, X.-h., Shi, Y.-H., Liu, Y., Tang, G.-Q., 2011a. SIMS U-Pb rutile age of low-temperature eclogites from southwestern Chinese Tianshan, NW China. Lithos 122, 76-86.

Li, W.-Y., Teng, F.-Z., Xiao, Y., Huang, J., 2011b. High-temperature inter-mineral magnesium isotope fractionation in eclogite from the Dabie orogen, China. Earth and Planetary Science Letters 304, 224-230.

Li, W.Y., Teng, F.Z., Wing, B.A., Xiao, Y., 2014. Limited magnesium isotope fractionation during metamorphic dehydration in metapelites from the Onawa contact aureole, Maine. Geochemistry, Geophysics, Geosystems 15, 408-415.

Li, J.-L., Klemd, R., Gao, J., John, T., 2016a. Poly-cyclic metamorphic evolution of eclogite: evidence for multistage burial-exhumation cycling in a subduction channel. Journal of Petrology 57, 119-146.

Li, W.Y., Teng, F.Z., Xiao, Y., Gu, H.O., Zha, X.P., Huang, J., 2016b. Empirical calibration of the clinopyroxene-garnet magnesium isotope geothermometer and implications. Contributions to Mineralogy and Petrology 171 (7), 1-14.

Li, S.-G., Yang,W., Ke, S., Meng, X.-N., Tian, H.-C., Xu, L.-J., He, Y.-S., Huang, J.,Wang, X.-C., Xia, Q.-K., Sun, W.-D., Yang, X.-Y., Ren, Z.-Y., Wei, H.-Q., Liu, Y.-S., Meng, F.-C., Yan, J., 2017. Deep carbon cycles constrained by a large-scalemantle Mg isotope anomaly in eastern China. National Science Review 4, 111-120.

Ling, M.X., Sedaghatpour, F., Teng, F.Z., Hays, P.D., Strauss, J., Sun, W.D., 2011. Homogeneous magnesium isotopic composition of seawater: an excellent geostandard for Mg isotope analysis. Rapid Communications in Mass Spectrometry 25 (19), 2828-2836.

Liu, P.-P., Teng, F.-Z., Dick, H.J.B., Zhou, M.-F., Chung, S.-L., 2017.Magnesium isotopic composition of the oceanic mantle and oceanic Mg cycling. Geochimica et Cosmochimica Acta 206, 151-165.

Lü, Z., Zhang, L., Du, J., Bucher, K., 2009. Petrology of coesite-bearing eclogite from Habutengsu Valley, western Tianshan, NW China and its tectonometamorphic implication. Journal of Metamorphic Geology 27, 773-787.

Lü, Z., Zhang, L., Du, J., Yang, X., Tian, Z., Xia, B., 2012. Petrology of HP metamorphic veins in coesite-bearing eclogite from western Tianshan, China: fluid processes and elemental mobility during exhumation in a cold subduction zone. Lithos 136, 168-186.

Miller, C., Stosch, H.-G., Hoernes, S., 1988. Geochemistry and origin of eclogites from the type locality Koralpe and Saualpe, Eastern Alps, Austria. Chemical Geology 67, 103-118.

Pearce, J.A., 1996. A user's guide to basalt discrimination diagrams. Trace element geochemistry of volcanic rocks: applications for massive sulphide exploration. Geological Association of Canada, Short Course Notes 12, 113.

Plank, T., Langmuir, C.H., 1998. The chemical composition of subducting sediment and its consequences for the crust and mantle. Chemical Geology 145, 325-394.

Pogge von Strandmann, P.A.E., Elliott, T., Marschall, H.R., Coath, C., Lai, Y.-J., Jeffcoate, A.B., Ionov, D.A., 2011. Variations of Li and Mg isotope ratios in bulk chondrites and mantle xenoliths. Geochimica et Cosmochimica Acta 75, 5247-5268.

Pogge von Strandmann, P.A.P., Dohmen, R., Marschall, H.R., Schumacher, J.C., Elliott, T., 2015. Extreme magnesium isotope fractionation at outcrop scale records the mechanism and rate at which reaction fronts advance. Journal of Petrology 56, 33-58.

Poli, S., Schmidt, M.W., 2002. Petrology of subducted slabs. Annual Review of Earth and Planetary Sciences 30, 207-235.

Su,W., Gao, J., Klemd, R., Li, J.-L., Zhang, X., Li, X.-H., Chen, N.-S., Zhang, L., 2010. U-Pb zircon geochronology of Tianshan eclogites in NW China: implication for the collision between the Yili and Tarim blocks of the southwestern Altaids. European Journal of Mineralogy 22, 473-478.

Sun, S.-S., McDonough, W., 1989. Chemical and isotopic systematics of oceanic basalts: implications for mantle composition and processes. Geological Society, London, Special Publications 42, 313-345.

Teng, F.-Z., 2017. Magnesium isotope geochemistry. Reviews in Mineralogy and Geochemistry 82, 219-287.

Teng, F.-Z., Yang, W., 2014. Comparison of factors affecting the accuracy of high-precision magnesium isotope analysis by multi-collector inductively coupled plasma mass spectrometry. Rapid Communications in Mass Spectrometry 28, 19-24.

Teng, F.-Z., Wadhwa, M., Helz, R.T., 2007. Investigation of magnesium isotope fractionation during basalt differentiation: implications for a chondritic composition of the terrestrial mantle. Earth and Planetary Science Letters 261, 84-92.

Teng, F.-Z., Li,W.-Y., Ke, S., Marty, B., Dauphas, N., Huang, S.,Wu, F.-Y., Pourmand, A., 2010. Magnesium isotopic composition of the earth and chondrites. Geochimica et Cosmochimica Acta 74, 4150-4166.

Teng, F.Z., Yang, W., Rudnick, R.L., Hu, Y., 2013. Heterogeneous magnesium isotopic composition of the lower continental crust: a xenolith perspective. Geochemistry, Geophysics, Geosystems 14, 3844-3856.

Teng, F.-Z., Li,W.Y., Ke, S., Yang,W., Liu, S.A., Sedaghatpour, F.,Wang, S.J., Huang, K.J., Hu, Y., Ling,M.X., Xiao, Y., Liu, X.M., Li, X.W., Gu, H.O., Sio, C.,Wallace, D., Su, B.X., Zhao, L., Harrington, M., Brewer, A., 2015. Magnesium isotopic compositions of international geological reference materials. Geostandards and Geoanalytical Research 39, 329-339.

Teng, F.-Z., Hu, Y., Chauvel, C., 2016. Magnesium isotope geochemistry in arc volcanism. Proceedings of the National Academy of Sciences (201518456).

Tipper, E., Galy, A., Gaillardet, J., Bickle, M., Elderfield, H., Carder, E., 2006. The magnesium isotope budget of the modern ocean: constraints from riverine magnesium isotope ratios. Earth and Planetary Science Letters 250, 241-253.

van der Straaten, F., Schenk, V., John, T., Gao, J., 2008. Blueschist-facies rehydration of eclogites (Tian Shan, NW-China): implications for fluid-rock interaction in the subduction channel. Chemical Geology 255, 195-219.

van der Straaten, F., Halama, R., John, T., Schenk, V., Hauff, F., Andersen, N., 2012. Tracing the effects of high-pressure metasomatic fluids and seawater alteration in blueschistfacies overprinted eclogites: implications for subduction channel processes. Chemical Geology 292, 69-87.

Veizer, J., 1989. Strontium isotopes in seawater through time. Annual Review of Earth and Planetary Sciences 17, 141-167.

Wang, S.-J., Teng, F.-Z., Li, S.-G., 2014a. Tracing carbonate-silicate interaction during subduction using magnesium and oxygen isotopes. Nature Communications 5.

Wang, S.-J., Teng, F.-Z., Li, S.-G., Hong, J.-A., 2014b. Magnesium isotopic systematics of mafic rocks during continental subduction. Geochimica et Cosmochimica Acta 143, 34-48.

Wang, S.-J., Teng, F.-Z., Rudnick, R.L., Li, S.-G., 2015a. The behavior of magnesium isotopes in low-grade metamorphosed mudrocks. Geochimica et Cosmochimica Acta 165, 435-448.

Wang, S.-J., Teng, F.-Z., Bea, F., 2015b. Magnesium isotopic systematics of metapelite in the deep crust and implications for granite petrogenesis. Geochemical Perspectives Letters 1, 75-83.

Wang, S.-J., Teng, F.-Z., Scott, J., 2016. Tracing the origin of continental HIMU-like intraplate volcanism using magnesium isotope systematics. Geochimica et Cosmochimica Acta 185, 78-87.

Windley, B., Allen,M., Zhang, C., Zhao, Z.,Wang, G., 1990. Paleozoic accretion and cenozoic redeformation of the Chinese Tien Shan range, central Asia. Geology 18, 128-131.

Xiao, Y., Lavis, S., Niu, Y., Pearce, J.A., Li, H., Wang, H., Davidson, J., 2012. Trace-element transport during subduction-zone ultrahigh-pressure metamorphism: evidence from western Tianshan, China. Geological Society of America Bulletin 124, 1113-1129.

Yang, W., Teng, F.-Z., Zhang, H.-F., 2009. Chondritic magnesium isotopic composition of the terrestrial mantle: a case study of peridotite xenoliths from the North China craton. Earth and Planetary Science Letters 288, 475-482.

Yang, X., Zhang, L.F., Tian, Z.L., Bader, T., 2013. Petrology and U-Pb zircon dating of coesitebearing metapelites from the Kebuerte Valley, western Tianshan, China. Journal of Asian Earth Sciences 70-71, 295-307.

Zack, T., John, T., 2007. An evaluation of reactive fluid flow and trace element mobility in subducting slabs. Chemical Geology 239, 199-216.

Zhang, L., Ellis, D.J., Jiang, W., 2002. Ultrahigh-pressure metamorphism in western Tianshan, China: part I. Evidence from inclusions of coesite pseudomorphs in garnet and from quartz exsolution lamellae in omphacite in eclogites. American Mineralogist 87, 853-860.

Zhang, L.-F., Ellis, D.,Williams, S., Jiang,W.-B., 2003. Ultrahigh-pressure metamorphism in eclogites from the western Tianshan, China—reply. American Mineralogist 88, 1157-1160.

Zhang, L., Song, S., Liou, J.G., Ai, Y., Li, X., 2005. Relict coesite exsolution in omphacite from Western Tianshan eclogites, China. American Mineralogist 90, 181-186.

Zhang, L., Lü, Z., Zhang, G., Song, S., 2008. The geological characteristics of oceanic-type UHP metamorphic belts and their tectonic implications: case studies from Southwest Tianshan and North Qaidam in NW China. Chinese Science Bulletin 53, 3120-3130.

Zhang, L., Du, J., Lü, Z., Yang, X., Gou, L., Xia, B., Chen, Z., Wei, C., Song, S., 2013. A huge oceanic-type UHP metamorphic belt in southwestern Tianshan, China: peak metamorphic age and PT path. Chinese Science Bulletin 58, 4378-4383.

Zhang, L., Zhang, L., Lü, Z., Bader, T., Chen, Z., 2016. Nb-Ta mobility and fractionation during exhumation of UHP eclogite from southwestern Tianshan, China. Journal of Asian Earth Sciences 122, 136-157.

Zindler, A., Hart, S., 1986. Chemical geodynamics. Annual Review of Earth and Planetary Sciences 14, 493-571.

中国东部大地幔楔形成时代和华北克拉通岩石圈减薄新机制——深部再循环碳的地球动力学效应*

李曙光[1,2]，汪　洋[1]

1. 地质过程与矿产资源国家重点实验室，中国地质大学(北京)，北京，10008
2. 中国科学院壳幔物质与环境重点实验室，中国科学技术大学地球和空间科学学院，合肥，230026

亮点介绍： 根据中国东部玄武岩的年代学和 Mg-Sr 同位素，限定了太平洋板片俯冲深度和大地幔楔形成时代。同时揭示了中国东部低 δ^{26}Mg 碱性玄武岩可能是碳酸盐化橄榄岩熔融产生的含碳酸盐-硅酸盐熔体交代上覆的岩石圈地幔的结果。基于这个发现，提出了富碳酸盐熔体与岩石圈相互作用可能是华北克拉通岩石圈减薄过程的一种新机制。

摘要　地震 P 波速度成像显示，西太平洋俯冲板片滞留在地幔过渡带，在中国东部形成“大地幔楔”结构。中国东部大陆玄武岩的 Mg 同位素调查揭示了该俯冲滞留板片携带大量碳酸盐交代地幔过渡带上覆的对流地幔，形成了碳酸盐化橄榄岩，它是中国东部晚白垩世和新生代大陆强碱性玄武岩的源岩。该玄武岩的 Mg-Sr 同位素组成与年龄关系显示自 106Ma 以来，该地幔源区的碳酸盐种属为菱镁矿+少量白云石，发生初熔的深度为 300~360km。因此，该地幔源区的碳酸盐化交代作用应发生在大于 360km 的深度，即在 106Ma 以前俯冲板片已经开始滞留地幔过渡带，形成了大地幔楔结构。这一时代支持大地幔楔的俯冲板片后撤形成机制。根据高温高压实验结果，碳酸盐化橄榄岩熔融产生的含碳酸盐硅酸盐熔体在到达至华北克拉通初步减薄的岩石圈底部(180~120km)时仍可以有高达 25~18wt%的 CO_2 含量，其 SiO_2 和 Al_2O_3 含量及 CaO/Al_2O_3 类似于强碱性的霞石岩和碧玄岩，并具有较高的 ε_{Nd} 值(2~6)。该熔体向上渗透并交代底部岩石圈地幔可形成碳酸盐化橄榄岩。由于克拉通地热增温线与碳酸盐化橄榄岩固相线相交于 130km 深度，则华北克拉通岩石圈底部的碳酸盐橄榄岩将发生部分熔融导致其物理性质类似软流圈且较容易被对流上地幔置换。其新生成的碳酸盐-硅酸盐熔体又可以向上渗透交代上覆的岩石圈地幔。如此重复这一碳酸盐化交代-熔融过程，可以使岩石圈减薄，而碳酸盐-硅酸盐熔体也转变呈较富硅和具有较低 ε_{Nd} 值(低至−2)的碱性玄武岩。随着华北克拉通岩石圈的减薄，其地热增温线逐步向大洋岩石圈靠近，岩石圈碳酸盐化橄榄岩的初熔深度可小于 130 km 和逐步接近 70 km。因此，富碳酸盐熔体与岩石圈相互作用是华北克拉通岩石圈减薄过程在晚白垩世和新生代的一种可能机制。据中国东部低 δ^{26}Mg 玄武岩的年龄统计，106~25Ma 区间岩石圈的碳酸盐化交代-熔融引发的减薄过程仅零星存在，而在 25Ma 以后才大规模发生。因此，华北克拉通岩石圈的减薄可能存在两个峰期：与克拉通破坏峰期(135~115Ma)同时发生的岩石圈减薄，和 25Ma 以后由富碳酸盐硅酸盐熔体与岩石圈相互作用引发的岩石圈进一步减薄，它使得华北克拉通东部岩石圈减薄至现今的约 70km 左右。

关键词　大地幔楔，华北克拉通，岩石圈减薄，深部碳循环，碱性玄武岩

* 本文发表在：中国科学：地球科学，2018，48(7)：809-824

1　引言

地球物理的地震P波速度成像显示，西太平洋俯冲板片在到达约600km深度时可以滞留在地幔过渡带。随着俯冲板片的后撤，这一俯冲板片滞留带可扩展到宽达1500km的高速带。这一壳-幔结构被称为“大地幔楔”（Huang和Zhao，2006）。中国东部“大地幔楔”壳幔结构的形成机制和时代对理解东亚西太平洋俯冲板片与陆缘的相互作用是十分重要的科学问题。关于其形成机制，目前存在平俯冲板片下沉至过渡带（Li和Li，2007）和陡俯冲板片后撤（Kusky等，2014）两种模型的争议，而关于俯冲板片在地幔过渡带滞留的起始时代还没有很好的约束。Liu等（2017）根据西太平洋古海底年龄和高精度P波速度层析成像揭示了现代滞留地幔过渡带的太平洋俯冲板片不老于30Ma，然而他们也推测最初的大地幔楔结构可能形成于140~110 Ma，但没有给出证据。

西太平洋板块俯冲还导致了华北克拉通的破坏，岩石圈在中-新生代减薄（厚度减少超过120 km）是其主要表现之一（朱日祥等（2012）及其中参考文献）。它的减薄机制和时代长期存在争议（朱日祥和郑天愉快, 2009；朱日祥等, 2011）。已提出的主要减薄机制有，迅速的加厚岩石圈拆沉说（Gao等，2004，2008），克拉通周边造山带山根拆沉说（Li等，2013；Liu等，2012a），长时间持续的流软流圈热侵蚀说（Xu，2001；Xu等，2004），太平洋俯冲板块后撤引张说（Kusky等，2014），底部岩石圈地幔被俯冲板片析出流体水化并软化说（Niu，2005；Windley等，2010）和橄榄岩-熔体相互作用改造岩石圈地幔说（Zhang，2005；Zheng等，2007）。关于减薄时代，岩石圈究竟是在早白垩世通过快速拆沉减薄（Gao等，2008；Meng等，2015），还是从白垩纪到新生代通过软流圈热侵蚀持续地减薄（Xu等，2004，2005；Zhang，2005；Zheng等，2007）是另外一个争议点。Liu等（2017）进一步指出Izanagi板块俯冲导致早白垩世华北克拉通的破坏和岩石圈减薄，而太平洋板块俯冲及伴随的广泛板内火山作用和弧后引张导致东亚大陆岩石圈的新生代进一步破坏与减薄。所有这些模型都忽略了再循环碳酸盐对地幔固相线的影响，也没有解释为什么华北克拉通岩石圈的最薄厚度是约70km?

近年来我们对中国东部白垩纪-新生代大陆玄武岩和部分环太平洋岛弧玄武岩进行了系统的Mg同位素地球化学研究，发现早白垩世玄武岩和岛弧玄武岩Mg同位素都类似地幔值（δ^{26}Mg=–0.25±0.07‰）或重一些；而晚白垩世和新生代玄武岩均具有轻的Mg同位素组成（δ^{26}Mg=–0.60~–0.30‰；Yang等，2012；Huang等，2015；Tian等，2016；Li等，2017）。其中碱性玄武岩具有的高CaO/Al_2O_3，低FeO/MnO比值，和在微量元素蜘蛛图上显著的K、Pb、Hf、Ti负异常是典型的碳酸盐熔体特征，证明它们的轻Mg同位素组成是再循环碳酸盐熔体交代地幔的结果（Li等，2017）。这些Mg同位素结果厘定了北从黑龙江省中俄边境、南到海南岛，东从中国海岸线西到大兴安岭-太行山重力梯度带，这样一个大尺度低δ^{26}Mg异常区。该异常的大陆部分较好地与地球物理揭示的地幔过渡带滞留俯冲板片的分布范围一致（Li等，2017）。这证明中国东部上地幔的轻镁同位素异常与地幔过渡带滞留的西太平洋俯冲板片携带的碳酸盐交代对流上地幔有关。它说明洋壳俯冲可以将大量碳酸盐带入对流上地幔，从而使得中国东部上地幔成为一个大的再循环碳库。这一研究成果不仅对深部碳循环的若干重要科学问题给出了重要制约（Li等，2017），由于碳酸盐化橄榄岩的固相线温度大幅度下降（Dusgupta等，2007，2013；Dusgupta和Hirschmann，2010；Dusgupta，2013），它将会显著影响我们对地幔的动力学过程的理解。本文将依据碳酸盐和碳酸盐化橄榄岩的高温高压实验结果和中国东部中-新生代低δ^{26}Mg玄武岩的Mg-Sr-Nd同位素数据对中国东部大地幔楔结构的形成时代和机制以及华北克拉通岩石圈减薄的过程与机制做深入讨论。

2　中国东部大地幔楔结构的形成时代与机制

2.1　中国东部低δ^{26}Mg碱性玄武岩源区的初熔深度

前人的研究一致认为中国东部碱性玄武岩来源于软流圈，其熔体迅速穿过岩石圈很少陆壳混染（Zhi

等，1990；Song 等，1990；Basu 等，1991；Liu 等，1994；Zou 等，2000；Xu 等，2005；Zeng 等，2010）。它们具有较低的 SiO_2 和 Al_2O_3 含量，和较高的 Na_2O+K_2O、CaO 和 TiO_2 含量，以及 K，Zr，Hf 和 Ti 负异常（Zeng 等，2010）。它们的高 La/Yb 说明源区熔融时有残留石榴子石（Xu 等，2005；Zeng 等，2010）。然而它们的初始熔融深度在 2016 年以前没有给出约束。

玄武岩熔融深度或压力估计通常使用熔体 SiO_2 压力计。它依据的是橄榄岩熔融时的缓冲反应：Mg_2SiO_4(ol) + SiO_2(liq) =$Mg_2Si_2O_6$(opx)。该反应前后摩尔体积变化很大，故对压力敏感（Lee 等，2009）。该压力计计算公式基本依据非碳酸盐化橄榄岩的熔融实验资料拟合给出（Lee 等，2009）。但是，碳酸盐化橄榄岩的固相线与无水橄榄岩固相线相差甚远（Dasgupta，2013），这使得基于熔体 SiO_2 的压力计不适用于碳酸盐化橄榄岩熔融产生的低 δ^{26}Mg 玄武岩。

Huang 和 Xiao（2016）和 Li 等（2017）基于中国东部晚白垩世以来碱性玄武岩的 Mg-Sr 同位素组成，以及该碱性玄武岩是由碳酸盐化橄榄岩部分熔融产生的，它可以视为该地幔源部分熔融产生的碳酸盐熔体和硅酸盐熔体（以洋中脊玄武岩 MORB 代表）混合组成，通过 MORB 与再循环碳酸盐的混合计算模拟，获得进入源区的碳酸盐主要是白云石+菱镁矿的结论（Huang 和 Xiao，2016；Li 等，2017）。由于该模型忽略了俯冲蚀变洋壳和泥质沉积物的影响，且由于 MORB 代表的硅酸盐熔体与地幔橄榄岩的 Mg/Sr 存在巨大差异，它可导致的混合曲线曲率不同有可能影响拟合线。本文建立了各种俯冲再循环物质，包括再循环俯冲碳酸盐类（方解石、白云石和菱镁矿）、蚀变洋壳和泥质沉积物与亏损上地幔的混合模型。该模型涵盖了如下两个过程：

（1）俯冲再循环物质与地幔的交代或混杂过程。俯冲板片部分熔融产生的含碳酸盐富硅熔体上升，伸入并与上覆地幔发生反应，形成碳酸盐化橄榄岩。该碳酸盐化橄榄岩含有富硅熔体交代形成的辉石岩脉，因而不相容元素较普通橄榄岩富集（Prytulak 和 Elliott，2007）。由于硅酸盐矿物对碳的相容性很低，碳或碳酸盐可在碳酸盐化橄榄岩中呈游离态存在，其中部分碳酸盐可氧化深部地幔的金属铁或 Fe^{2+}形成 Fe^{3+}而将自身还原成金刚石。其中碳酸盐的存在形态（即碳酸盐种属：方解石、文石、白云石、菱镁矿）取决于该碳酸盐化橄榄岩所处地幔的温压条件。这一碳酸盐化橄榄岩的 Mg 和 Sr 同位素组成应当是亏损上地幔和俯冲板片部分熔融碳酸盐化硅酸盐熔体的混合结果。

（2）中国东部碱性玄武岩形成过程。当碳酸盐化橄榄岩随俯冲板片后撤导致的上涌上地幔上升减压至与碳酸盐化橄榄岩固相线相交，导致该碳酸盐化橄榄岩发生部分熔融。该初始熔融产生富含碳酸盐和贫 SiO_2 的碳酸岩熔体，这意味着其主要贡献者是源区的碳酸盐。随着进一步地幔上涌和减压，熔体中石榴石、单斜辉石和橄榄石的贡献比例增大，导致该碳酸岩熔体的硅酸盐组分比例增大，而熔体中 CO_2 组分被稀释，从而过渡到碳酸盐化的硅酸盐熔体（Dasgupta 等，2013）。因此，最终产生类似霞石岩或碱玄岩、碧玄岩的碱性玄武质熔体是初始碳酸岩熔体和后来橄榄岩或辉石岩熔融产生的硅酸盐熔体的混合物。如果忽略该熔体穿越岩石圈时与岩石圈地幔相互作用对熔体 Mg-Sr 同位素的影响，则中国东部低 δ^{26}Mg 玄武岩的 Mg-Sr 同位素组成就可代表碳酸盐化橄榄岩部分熔融熔体的组成。

由于在上述的地幔交代和碳酸盐化橄榄岩熔融过程中，碳酸盐和硅酸盐组分之间可能发生复杂的元素和同位素交换，因此我们很难准确确定这些中间产物（也即上述初始碳酸岩熔体和橄榄岩或辉石岩熔融产生的硅酸盐熔体）的 Mg-Sr 含量和同位素组成。如果碳酸盐化橄榄岩部分熔融过程不导致 Mg-Sr 同位素的分馏，则该部分熔融熔体（也即中国东部低 δ^{26}Mg 碱性玄武岩）必然继承了其地幔源区（碳酸盐化橄榄岩）的 Mg-Sr 同位素组成。因此，我们可以用过程（1）的俯冲再循环物质，包括俯冲碳酸盐（方解石、白云石和菱镁矿）、泥质沉积物和蚀变洋壳，利用亏损上地幔的混合模型模拟中国东部低 δ^{26}Mg 碱性玄武岩的 Mg-Sr 同位素数据，以判断再循环进入该玄武岩源区的碳酸盐种属（图 1）。

图 1 显示中国东部大陆的晚白垩世和新生代碱性玄武岩普遍具有异常低的 δ^{26}Mg 值，和不太高的 ^{87}Sr/^{86}Sr 比值（<0.706）。由于蚀变洋壳或泥质沉积物具有较重的 Mg 同位素组成（Wang 等，2015，2017），亏损地幔仅与蚀变洋壳或泥质沉积物混合不可能解释中国东部大陆玄武岩的低 δ^{26}Mg 特征（图 1）。由于亏损地幔 DMM 的高 Mg 低 Sr 含量，因此所有相对具有低 Mg 高 Sr 含量的再循环碳酸盐（方解石，白云石），

以及它们与蚀变洋壳和泥质沉积物的组合物(如蚀变洋壳：沉积物：白云石=86：1：13)和亏损地幔的混合线都是向上凸的混合线(图 1)，在 $^{87}Sr/^{86}Sr<0.706$ 的范围内具有较高的 $\delta^{26}Mg$ 值，也不能拟合中国东部碱性玄武岩的 Mg-Sr 同位素数据。与方解石和白云石相反，菱镁矿具有比亏损地幔更高的 MgO 含量和更低的 Sr 含量，它与亏损地幔的混合线是向下凹的，且当菱镁矿的 $\delta^{26}Mg$ 值介于−1.11‰(菱镁矿[1])～−1.9‰(菱镁矿[2])时才能拟合霞石岩的 Mg-Sr 同位素组成(图 1)。图 1 还显示当俯冲再循环物质中以菱镁矿为主体，与较少量的白云石+蚀变洋壳+泥质沉积物组合也可以与亏损地幔混合解释部分中国东部碱性玄武岩的 Mg-Sr 同位素组成。因此，中国东部碱性玄武岩低 $\delta^{26}Mg$ 特征是由其地幔源区混入的再循环菱镁矿+少量白云石引起的。考虑到只有霞石岩较少受岩石圈的影响(详见 3.3.2 节)，最能代表其软流圈地幔源区的同位素组成，则其碳酸盐化橄榄岩地幔源区的碳酸盐基本是菱镁矿。

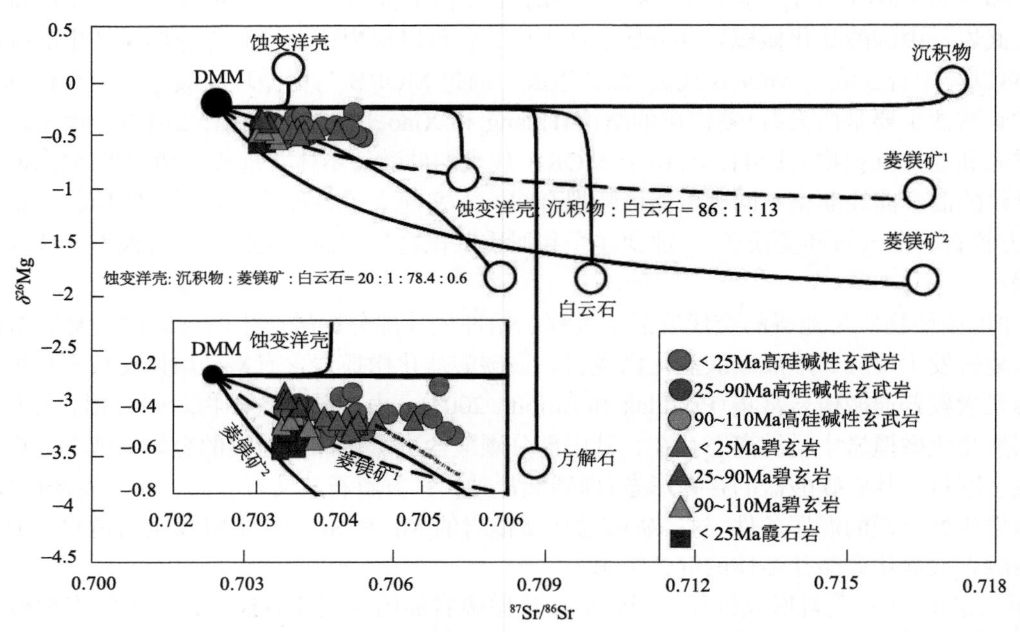

图 1　中国东部霞石岩，碧玄岩和碱性玄武岩的 Mg-Sr 同位素图，以及亏损地幔(DMM)与俯冲再循环地壳物质的 Mg-Sr 同位素混合计算模拟曲线

再循环物质包括：蚀变洋壳、泥质沉积物、方解石、白云石、菱镁矿[1]和菱镁矿[2]；同位素端元组成见表 1；玄武岩数据来自 Li 等(2017)

表 1　图 1 混合计算端元组成 a)

	MgO	$\delta^{26}Mg$	Sr	$^{87}Sr/^{86}Sr$
亏损地幔	38	−0.25	7.7	0.7025
白云石	22	−1.89	1311	0.7099
菱镁矿-1	47.6	−1.11	1.84	0.7164
菱镁矿-2	47.6	−1.9	1.84	0.7164
方解石/文石	1	−3.65	1151	0.7088
蚀变洋壳	4	0.05	180	0.7039
沉积物	2.6	−0.02	327	0.717

a) 数据来源：DMM 亏损地幔端元组成据 Ke 等(2016)，其中 $\delta^{26}Mg$ 来自 Hoefs(2009)和 Teng 等(2010)，其它数据据 Workman 和 Hart(2005)。白云石、菱镁矿-1 和方解石/文石来自 Li 等(2017)，和 Huang 和 Xiao(2016)。菱镁矿-2 的 $\delta^{26}Mg$ 值是拟合霞石岩最佳曲线所需数值。蚀变洋壳的 Mg 和 $\delta^{26}Mg$ 来自 Huang 等(2018)，Sr 和 $^{87}Sr/^{86}Sr$ 据 Hauff 等(2003)。沉积物中的 Mg 来自 Wang 等(2017)的云母片岩，$\delta^{26}Mg$ 据 Wang 等(2015，2017)的泥岩和云母片岩的 Mg 同位素的加权平均值，Sr 以及 $^{87}Sr/^{86}Sr$ 据全球俯冲沉积物(GLOSS)(Plank 和 Langmuir，1998)

实验研究表明不同的碳酸盐矿物种属（方解石、白云石、菱镁矿）有不同的稳定 *P-T* 条件（例如 Poli 等，2009）。根据碳酸盐化地幔发生部分熔融时的碳酸盐矿物种属就可以查明其初始熔融发生的 *P-T* 条件。已有若干实验测定了白云石=菱镁矿+文石的固相转化反应 *P-T* 界限（Sato 和 Katsura，2001；Morlidge 等，2006；Tao 等，2014）（图 2）。该固相转化反应 *P-T* 条件的变化（图 2 绿、红、蓝线）取决于白云石的铁含量（Tao 等，2014）。图 2 显示，碳酸盐化橄榄岩的固相线与这些白云石与菱镁矿相转化 *P-T* 界限的上下边界线分别相交于 *A* 点和 *B* 点，该交汇位置恰好落在现代地幔的绝热线区（图 2 灰色带），*A* 点对应的地幔绝热线起始 1350℃，而 *B* 点对应的地幔绝热线起始 1400℃（图 2）。因此，在 *A*、*B* 点均可发生碳酸盐化橄榄岩的初始熔融，它们对应的深度分别是 300~360km，并意味着在此深度发生初熔时，而地幔源区所含碳酸盐是菱镁矿+白云石；如果在更大深度发生初熔，则地幔源区所含碳酸盐是菱镁矿。这一理论推测结果与上述 Mg-Sr 同位素约束的中国东部碱性玄武岩源区的碳酸盐种属（菱镁矿为主+少量白云石）完全吻合。因此，300~360km 应该是中国东部碱性玄武岩源区的最小初熔深度。

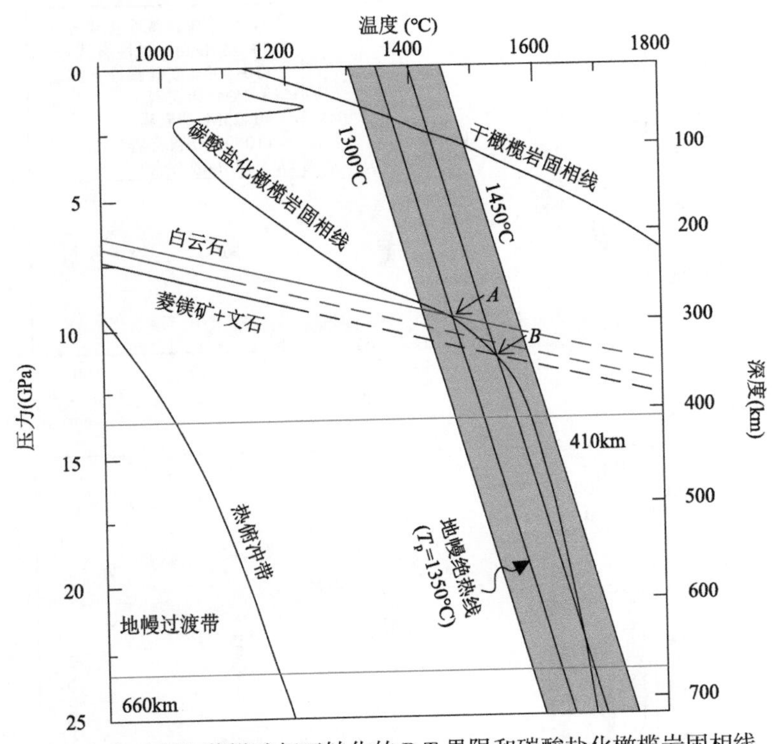

图 2　白云石与菱镁矿相互转化的 P-T 界限和碳酸盐化橄榄岩固相线

P-T 界限：蓝线，Sato 和 Katsura (2001)；红线，Morlidge 等(2006)；绿线，Tao 等(2014)。碳酸盐化橄榄岩固相线据 Dasgupta (2013)。灰色带指示及现代海洋地幔绝热线变化范围。该固相线与 T_P=1350℃和 1450℃地幔绝热线分别相交于 A 和 B 点，它们恰好落在白云石与菱镁矿相互转化的 *P-T* 界限（蓝线和绿线），据 Huang 和 Xiao (2016)修改

2.2　俯冲板片滞留地幔过渡带的起始时代

如果中国东部碱性玄武岩的初熔深度在 300~360km，则碳酸盐化地幔的交代作用应发生在 360km 以下深度。如果该地幔碳酸盐化交代深度在 360~410 km，由于地幔过渡带的上界深度为 410km，则碳酸盐化地幔交代熔体应来自地幔过渡带（410~660 km）的滞留俯冲板片。据此可以推测，中国东部最早的低 δ^{26}Mg 碱性玄武岩的出现就意味着西太平洋俯冲板片已在那时开始滞留在地幔过渡带，如果其源区的碳酸盐种属主要是菱镁矿。因此，根据中国东部碱性玄武岩的 Mg-Sr 同位素组成随时间演化的特征，我们可以估计不同时代中国东部低 δ^{26}Mg 值的碱性玄武岩源区的碳酸盐种属，从而约束俯冲板片开始滞留地幔过渡带的最早时代。

图 3a 显示中国东部碱性玄武岩的 Sr 同位素组成在 106~25Ma 期间均具有较低的 ^{87}Sr/^{86}Sr 比值和较小

变化。25Ma 以后，火山活动加剧，并出现具有较高 $^{87}Sr/^{86}Sr$ 比值的样品。25Ma 前后的这一变化很可能反映了早期 Izanagi 俯冲板片和晚期太平洋俯冲板片对中国东部上地幔的不同影响（Liu 等，2017）。图 3b 显示中国东部不同年龄的碱性玄武岩具有相似的 Mg 同位素组成，其 $\delta^{26}Mg$ 值不随其年龄变化而发生变化。如前所述，年龄为 106Ma 的最古老碱性玄武岩具有低 $^{87}Sr/^{86}Sr$ 比值和低 $\delta^{26}Mg$ 值，因而其源区碳酸盐种属基本为菱镁矿（图 1 绿色符号）。年龄在 80~40Ma 之间的碱性玄武岩源区也有很高比例的菱镁矿（菱镁矿：白云石=78.4：0.6）（图 1 红色符号）。因此，自 106Ma 以来，中国东部碱性玄武岩的源区均是含菱镁矿+少量白云石的碳酸盐化橄榄岩，表明它们初熔深度在 300~360km，以及自 106Ma 以来，俯冲板片就已滞留地幔过渡带，其部分熔融熔体向上迁移并交代了上覆地幔。这一结论与 Thomson 等（2016）的碳酸盐化俯冲洋壳熔融实验相吻合，该实验固相线在 13~16GPa 发生温度下降了约 200℃，并与大多数俯冲板块的地热增温线在 400~700km 深度相交。这意味着大多数俯冲板块在地幔过渡带均会发生部分熔融。

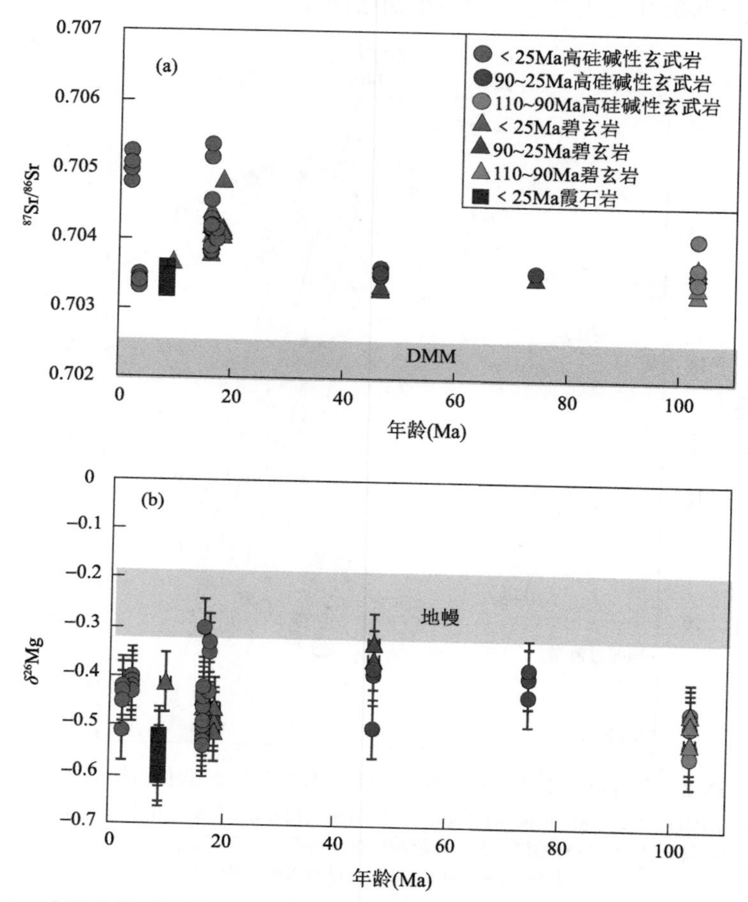

图 3　中国东部碱性玄武岩的 Sr（a）和 Mg（b）同位素组成与年龄的关系
数据据 Li 等，2017

如果上述结论是正确的，则大地幔楔结构至少在 106Ma 以前已经形成。由于 106Ma 是中国东部最古老的低 $\delta^{26}Mg$ 碱性玄武岩的喷发年龄（Yang 和 Li，2008），不是其地幔源区的交代年龄，考虑到地幔交代作用，地幔上涌和部分熔融，直到穿越岩石圈发生火山喷发等过程至少需要 20Ma 时间（以从 400km 深度上升至地表按平均速率 2cm/a 计），则实际俯冲板片滞留地幔过渡带的大地幔楔结构可能在早白垩世（约 125Ma）就已经形成。如果在早白垩世向中国东部尤其华北克拉通陆下俯冲的是 Izanagi 板块（Sun 等，2007；Liu 等，2017），则 Izanagi 俯冲板块可能同样滞留在地幔过渡带，并在东亚形成了大地幔楔结构。

由此得出，中国东部大地幔楔结构至少在 106Ma 以前，很可能在 125Ma 左右已经形成。早白垩世大地幔楔的形成可能与 Izanagi 板块俯冲有关，现在的大地幔楔结构由太平洋板块俯冲形成。

2.3 对大地幔楔形成机制的制约

前已述，关于大地幔楔的形成机制存在平俯冲板片下沉至过渡带（Li 和 Li，2007）和陡俯冲板片后撤（Kusky 等，2014）两种模型的争议。如果本文对俯冲板片滞留过渡带的时代大于 106Ma，很可能是 125Ma 的估计是正确的话，依据平俯冲模型，则 Izanagi 板片在晚侏罗-早白垩世时期应该是低角度平俯冲于华北克拉通岩石圈之下，其俯冲深度应该略大于 200km。在此深度下，该平俯冲板片可析出超临界流体（Kessel 等，2005），它可以溶解镁质碳酸盐并交代岩石圈地幔（Tian 等，2018）。平俯冲板片在早白垩世的下沉将导致的大规模地幔上涌从而引起该碳酸盐化交代岩石圈地幔的部分熔融和形成年龄约为 130~120Ma 的低 δ^{26}Mg 玄武岩浆。然而这一时期华北发育的 130~120Ma 的大陆玄武岩浆均具有类似地幔的 Mg 同位素组成，没有任何低 δ^{26}Mg 印记（Yang 等，2012；Li 等，2017）。因此，平俯冲模型不能解释中国东部低 δ^{26}Mg 玄武岩年龄均小于 110Ma 的事实。

相反，陡俯冲板片后撤模型可以合理解释这上述低 δ^{26}Mg 玄武岩年龄。该模型要求 Izanagi 板块在 125 Ma 时俯冲进入地幔过渡带并开始发生俯冲板片后撤，从而导致华北岩石圈伸展和大地幔楔的形成（Kusky 等，2014），并伴随大规模花岗岩事件的发生（Wu 等，2005）。Izanagi 板片俯冲如果起始于 165Ma，以 30° 倾角俯冲，当俯冲到 600km 深度时滞留在地幔过渡带，这样其倾斜的俯冲板片长度约 1200km（Kusky 等，2014）。只要平均俯冲速度为 3cm/a，俯冲的 Izanagi 板片就可在 125Ma 实现在地幔过渡带滞留。如上所述，随后的一系列过程，包括滞留俯冲板片的部分熔融及其向上渗透熔体交代上覆地幔，被交代碳酸盐化地幔的上涌和部分熔融，以碱性玄武岩熔体穿越岩石圈到达地表喷出，需要约 20Ma，从而使最早的低 δ^{26}Mg 玄武岩在~105Ma 喷出。这一依据陡俯冲板片后撤模型估算的结果与中东部低 δ^{26}Mg 玄武岩的最老年龄（106Ma）一致。

3 华北克拉通东部岩石圈减薄机制与时代

如前言所述，华北克拉通东部岩石圈减薄的机制与发生时代一直是地质界讨论的热点和有争议问题。针对这一问题已有众多的模型提出（Gao 等，2004，2008；Kusky 等，2014；Li 等，2013；Liu S A 等，2012；Liu 等，2015；Meng 等，2015；Niu，2005；Windley 等，2010；Xu ，2001；Xu 等，2004；Zhang，2005；Zheng 等，2007；朱日祥和郑天愉，2009；朱日祥等，2012）。然而已提出的所有模型都忽视了如下两个基本事实：(1) 西太平洋板块俯冲不仅携带了水而且携带了大量碳酸盐再循环进入对流地幔，引发地幔的碳酸盐化交代作用（Yang 等，2012；Huang 等，2015；Tian 等，2016；Li 等，2017）。碳酸盐化橄榄岩显著低于干橄榄岩的固相线温度会对中国东部地幔动力学和软流圈与岩石圈相互作用产生重要影响；(2) 华北克拉通东部最大减薄的岩石圈厚度约为 70km，什么因素限制了这一减薄厚度尚属未知。基于已获得的西太平洋板块俯冲导致大量碳酸盐再循环进入并交代中国东部上地幔，及该交代地幔的碳酸盐化橄榄岩部分熔融产生具低 δ^{26}Mg 特征的碱性玄武岩的认识（Li 等，2017），本文下面一节将聚焦富含碳酸盐熔体与岩石圈的相互作用并讨论其对华北克拉通岩石圈地幔减薄的影响和高 SiO_2 碱性玄武岩的成因。

3.1 碳酸盐化橄榄岩的部分熔融产生的富碳酸盐熔体

碳酸盐化橄榄岩的固相线已被高压实验很好测定了（Dasgupta 等，2007，2013）。该实验的起始物料是均匀混合的，CO_2 含量为 1 和 2.5wt%的碳酸盐化橄榄岩（Dasgupta 等，2007，2013），因此该固相线适用于被均匀渗滤熔体交代产生的碳酸盐化橄榄岩。

碳酸盐化橄榄岩在固相线附近的低比例熔融（F=2%~6%）产生的熔体 SiO_2 含量很低（<10wt%）和有很高的 CO_2 含量（38~45wt%），反映了它主要是碳酸盐熔体的贡献（Dasgupta 等，2007）。随着温度升高，熔融

比例增高，熔体中 CO_2 含量被稀释，熔体转化为碳酸盐化硅酸盐熔体，SiO_2 含量>25wt%，熔体中的 CO_2 含量 <25wt%(图 4)(Dasgupta 等，2013)，反映硅酸盐矿物，如单斜辉石或石榴石，熔融对熔体的贡献比例增大。图 4 显示，碳酸盐化橄榄岩固相线在 10~12GPa 区间与地幔绝热线相交发生初熔，这与前述讨论的中国东部大地幔楔的碳酸盐化橄榄岩在 300~360km 深度发生初熔是一致的。在该深度的初始熔体为碳酸盐熔体(SiO_2 含量<10wt%，含 38~45wt%CO_2)。当上涌地幔上升至 6GPa(180km)或 4GPa(120km)深度时，随着熔融比例的增大，熔体演化为含有 25wt%或 18wt%的 CO_2 含量(图 4)。高压实验研究表明，在压力>2.5GPa 时，游离 CO_2 不能与橄榄岩同时平衡存在，它们之间要发生如下反应生成白云石(Dalton 和 Wood，1993)：

$$CaMgSi_2O_6 + 2Mg_2SiO_4 + 2CO_2 = CaMg(CO_3)_2 + 4MgSiO_3 \tag{1}$$

单斜辉石　橄榄石　流体　白云石　斜方辉石

在高压(>4.5GPa)下，进一步反应形成菱镁矿(Dalton 和 Wood，1993)：

$$CaMg(CO_3)_2 + 2MgSiO_3 = CaMgSi_2O_6 + 2MgCO_3 \tag{2}$$

白云石　斜方辉石　单斜辉石　菱镁矿

因此，在软流圈的温压条件下，碳酸盐化橄榄岩部分熔融产生的碳酸盐化硅酸盐熔体是白云石或菱镁矿的碳酸盐熔体与硅酸盐熔体的混合物。

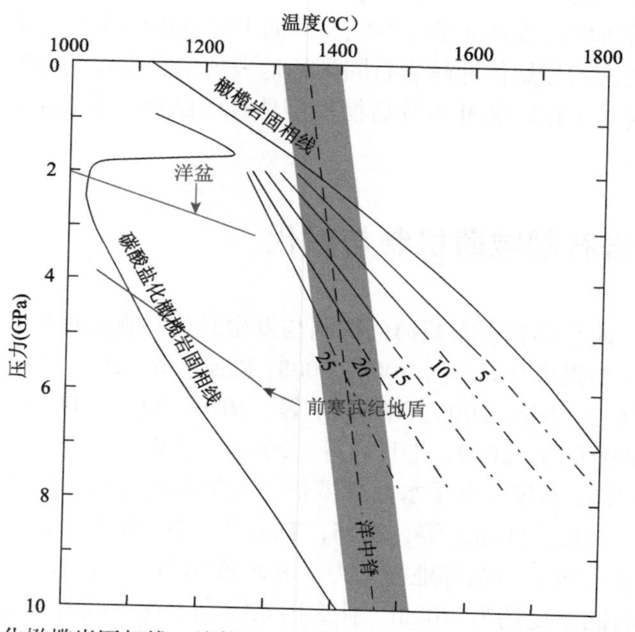

图 4　干橄榄岩及碳酸盐化橄榄岩固相线、熔体 CO_2 含量(重量百分比)等值线和大陆地盾及洋壳地热增温线

据 Dasgupta 等(2013)及 Mckenzie 和 Bickle(1988)

3.2　碳酸盐化的硅酸盐熔体与岩石圈地幔的相互作用和岩石圈减薄

高温高压实验和地幔包体研究都证明碳酸盐化硅酸盐熔体作用于岩石圈地幔可发生岩石圈地幔的碳酸盐化交代作用(Kushiro 等，1975；Hauri 等，1993；Ionov，1998；Coltorti 等，1999)。最近，Wang 等(2016)报道了碳酸盐熔体在岩石圈斜方辉橄岩中的渗滤实验，它可以在岩石圈下部产生富菱镁矿交代地幔层，并指出由于碳酸盐化岩石圈地幔强度弱，很易被对流软流圈剥蚀掉而导致岩石圈减薄(Wang 等，2016)。然而需要强调指出的是碳酸盐化岩石圈地幔交代作用除了导致岩石圈强度降低外，还可使交代作用产生的碳酸盐化橄榄岩熔点大幅下降，导致克拉通岩石圈地热增温线与碳酸盐化橄榄岩的固相线在 P=4.3GPa 时相交，从而引发岩石圈下部碳酸盐化交代地幔的部分熔融和岩石圈减薄(图 4)。前已述，当上涌的碳酸盐化

橄榄岩在到达 180km 或 120km 深度而部分熔融产生的熔体含 CO_2 可分别达 25wt%和 18wt%。当这样的熔体渗滤进入上覆岩石圈地幔，必然引发岩石圈地幔的碳酸盐化交代作用。如果渗滤熔体重量比列只要占被交代橄榄岩的 1/10，则被交代的碳酸盐化橄榄岩的 CO_2 含量即可达 2.5 或 1.8 wt%，这已经与 Dasgupta 等(2007，2013)实验测定碳酸盐化橄榄岩固相线的实验物料 CO_2 含量(2.5~1.0%)相当。由于前寒武纪克拉通地热增温线与碳酸盐化橄榄岩固相线相交于 4.3GPa(130km)(图 4，前寒武纪地盾)，故深度大于 130km 的碳酸盐化交代岩石圈地幔都会发生部分熔融(图 5a，b)。实验表明只要熔融比例 F 达到 0.2%，该部分熔融的岩石圈物性接近软流圈地幔(Chantel 等，2016)。一般情况下，如果熔融比例 $F>0.5\%$，则会有熔体从熔融源区析出，而如果 $F\leqslant 0.5\%$则熔体仍残留源区不会析出(Pilet 等，2011)。因此，只要碳酸盐化岩石圈地幔的熔融比列 $F>0.5\%$，就会有碳酸盐化硅酸盐熔体向上渗透引发上部岩石圈地幔的碳酸盐化交代作用，而熔体析出后残留的岩石圈地幔仍可圈闭<0. 5%的熔体从而具有类似软流圈的物理性质(图 5c)。由于该部分熔融岩石圈地幔的强度下降，它很易被对流上地幔置换。这可以用来解释华北克拉通东部已减薄岩石圈存在新形成的饱满的岩石圈地幔(Zheng 等，2007)。随着岩石圈减薄和与软流圈界面的抬升，大陆岩石圈的地热增温线也向上抬升而接近大洋岩石圈增温线，它与碳酸盐化橄榄岩固相线相交于 2.2GPa(相当 66km 深度)(图 4，洋盆)。因此，碳酸盐化交代岩石圈地幔部分熔融所产生富 CO_2 熔体可持续向上运移，引发上部岩石圈地幔的碳酸盐化交代和部分熔融，至接近 70km 的深度(图 5d)。图 4 显示，当 P=2.5GPa(深度 75km)时，碳酸盐化硅酸盐熔体的 CO_2 含量降至 6~7%左右，对应 SiO_2 升至 43%左右(Dasgupta 等，2007)，熔体以高 SiO_2 碱性玄武岩浆形式喷出地表(图 5d)。因此，70~66km(2.3~2.2GPa)是碳酸盐化岩石圈地幔部分熔融导致其减薄的上限，它取决于减薄的岩石圈地热增温线与碳酸盐化橄榄岩固相线相交的深度。这一理论推测与华北克拉通现代最大岩石圈减薄处的岩石圈厚度一致(朱日祥等，2009)。这种一致性表明来自上涌软流圈的碳酸盐化熔体交代岩石圈地幔是导致其晚白垩世以来，特别是新生代减薄的主导因素之一。

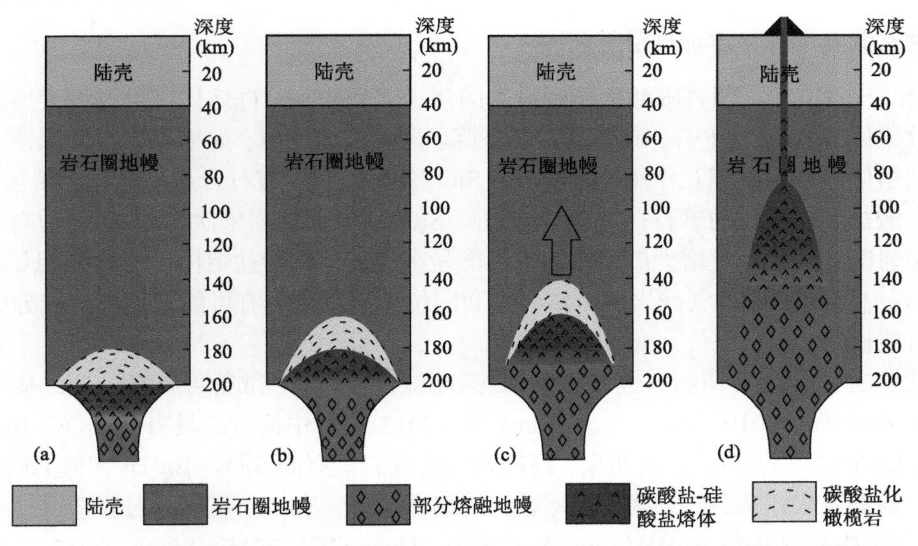

图 5　上涌碳酸盐-硅酸盐熔体与岩石圈地幔相互作用和熔融过程的卡通图

3.3　富含碳酸盐熔体与岩石圈相互作用的岩石化学证据

3.3.1　岩石圈地幔包体证据

来自软流圈的富碳酸盐熔体与岩石圈地幔的相互作用已被众多地幔橄榄岩包体研究所证实(例如 Hauri 等，1993；Ionov，1998；Coltorti 等，1999)。中国东部大陆玄武岩携带的岩石圈地幔包体的岩石学研究证明中国东部岩石圈地幔同样也遭受了来自软流圈的富碳酸盐熔体的交代作用(Zhang，2005；Xiao 等，2013；Hu 等，2016)。当富含碳酸盐的硅酸盐熔体与岩石圈橄榄岩在压力>2.5GPa 相互作用时可发生两类反应：一类如上面化学反应式(1)和(2)所示的碳酸盐组分与橄榄岩反应，生成白云石+斜方辉石或单斜辉石+菱镁矿；另一类是贫 SiO_2 的硅酸盐熔体与橄榄岩相互作用可发生如下反应(Zhou 等，2014)：

$$\text{Opx, Cpx} + \text{melt} \rightarrow \text{Ol (solid)} + SiO_2 \text{ (liquid)} \quad (3)$$

该反应生成富硅玄武质熔体和与之平衡的橄榄石。在压力≤2.5GPa 的地幔条件下，反应(3)仍可进行(Zhou 等，2014)，但反应(1)和(2)将不能发生，导致在尖晶石相碳酸盐化橄榄岩中只有方解石或富含 CO_2 的熔流体包裹体产出(Dalton 和 Wood，1993)。因此，富含碳酸盐的硅酸盐熔体与岩石圈橄榄岩相互作用可生成单斜辉石岩和富单斜辉石二辉橄榄岩或异剥橄榄岩，其橄榄石中含有新生的具低 Mg#值的橄榄石。由于反应(2)和(3)的产物单斜辉石较斜方辉石富 Sr，新生橄榄石较斜方辉石富 Mg，故该反应可改变被交代橄榄岩的 Mg-Sr 同位素组成，但不会显著改变熔体的 Mg-Sr 同位素组成。这类熔体-岩石圈相互作用证据已在华北多处观察到。如山东方城玄武岩携带的橄榄岩包体中观察到橄榄石化学组成呈环带结构，大颗粒橄榄石核部 $Mg^{\#}$=88~92，而边部橄榄石 $Mg^{\#}$=76~83，与玄武岩的橄榄石斑晶类似(Zhang，2005)。这一观察证明来自软流圈的玄武质熔体与岩石圈橄榄岩发生了化学反应，生成 $Mg^{\#}$较低的橄榄石。新生代玄武岩橄榄岩包体镁同位素研究表明，与岩石圈地幔相互作用的熔体是具有低 δ^{26}Mg 特征的含碳酸盐的硅酸盐熔体(Xiao 等，2013；Hu 等，2016)。如山东北岩新生代碱性玄武岩携带的异剥橄榄岩大多数具轻镁同位素组成(δ^{26}Mg 低至−0.39‰)，表明它是二辉橄榄岩与低 δ^{26}Mg 熔体相互作用的产物(Xiao 等，2013)；来自河北汉诺坝碱性玄武岩的含石榴石单斜辉石岩和不含石榴石单斜辉石岩分别具有异常低的 δ^{26}Mg 值(分别低至−0.48‰和−1.51‰)，证明是与含碳酸盐熔体相互作用的产物(Hu 等，2016)。需要指出的是上述低 δ^{26}Mg 包体是来自较浅部位尖晶石二辉橄榄岩相岩石圈的地幔包体，因而北岩异剥橄榄岩仅发现有方解石但未发现白云石或菱镁矿(Xiao 等，2013)，它们不能代表寄主玄武岩源区的特征。寄主玄武岩地球化学有可能提供其源区的碳酸盐化地幔交代作用证据。

3.3.2　寄主碱性玄武岩地球化学证据及其成因

根据图 4 和反应式(3)，源自碳酸盐化橄榄岩的富含碳酸盐和贫硅熔体可通过增大熔融比例或与岩石圈相互作用导致熔体 SiO_2 含量上升，碳酸盐含量下降。因此可以推测，如果中国东部碱性玄武岩是经过软流圈富含碳酸盐熔体与岩石圈相互作用的产物，则 SiO_2 含量最低的霞石岩是与岩石圈相互作用最少，最接近软流圈富含碳酸盐熔体组成的岩石；而碧玄岩和高 SiO_2 碱性玄武岩依次 SiO_2 含量增高，有可能是软流圈碳酸盐化橄榄岩部分熔融程度增大的结果，也可能是富碳酸盐硅酸盐熔体与岩石圈相互作用的结果。研究霞石岩-碧玄岩-高 SiO_2 碱性玄武岩的微量元素和同位素地球化学有可能帮助我们区分碧玄岩和碱性玄武岩的这两种可能机制。

山东新生代的玄武岩由早期弱碱性岩石系列(高硅碱性玄武岩)和晚期强碱性岩石系列(霞石岩-碧玄岩)组成(图 6a)(Zeng 等，2010，2011；Sakuyama 等，2013)。其中霞石岩具有低 SiO_2、Ba/Th 和高 FeO*、Na_2O、Ce/Pb 和 La/Yb 特点；与霞石岩相反，高硅碱性玄武岩具有高 SiO_2、Ba/Th 和低 FeO*、Na_2O、Ce/Pb 和 La/Yb 特点；而碧玄岩组成介于二者之间(图 6)。如上所述，有 2 个可能的过程造成这些主微量元素变化。(1)碳酸盐化橄榄岩部分熔融过程(Zeng 等，2010；Huang 等，2015；Li 等，2017)。高压实验表明碳酸盐化橄榄岩固相线附近的低比例熔融产生富 CO_2 贫 SiO_2 熔体具有高 Na_2O、CaO 和 FeO 含量特征，随着熔融比例升高，CO_2、Na_2O、CaO 和 FeO*含量下降，SiO_2 含量升高(Dasgupta 等，2007)。这可以解释霞石岩，碧玄岩和高硅碱性玄武岩的 Na_2O+K_2O 和 FeO*与 SiO_2 的负相关关系(图 6a，b)。但是单一的部分熔融过程不能解释这三类岩石的微量元素变化。(2)这些玄武岩的 2 个不同的地幔源区。一个是被来自在地幔过渡带滞留脱水俯冲板片熔融的高 Ce/Pb，低 Ba/Th 熔体交代的上覆碳酸盐化橄榄岩地幔；另一个是俯冲沉积物脱水流体交代的高 Ba/Th，低 Ce/Pb 富集地幔。如此，这些玄武岩的 Ba/Th 和 Ce/Pb 等对流体活动性敏感的微量元素比值变化就很容易被解释(图 6c)。

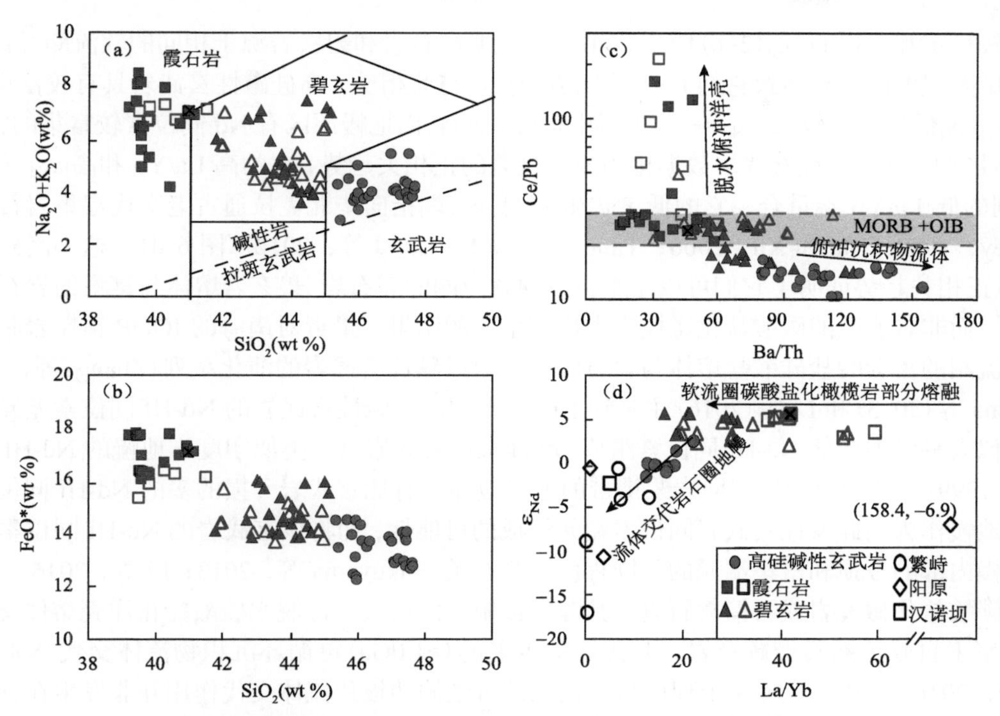

图 6　山东新生代霞石岩、碧玄岩、高硅碱性玄武的主-微量元素和 Nd 同位素图

(a) SiO_2-Na_2O+K_2O 图；(b) SiO_2-FeO*图；(c) Ba/Th-Ce/Pb 图；(d) ε_{Nd}-La/Yb 图。山东玄武岩数据来自 Zeng 等(2010，2011)和 Sakuyama 等(2013)；地幔包体数据据 Song 和 Frey(1989)、Ma 和 Xu(2006)、Tang 等(2008)和 LiuJ 等(2012)；MORB+OIB 的 Ce/Pb 据 Hofmann 等(1986)

Ce/Pb-Ba/Th 图显示，霞石岩有低的 Ba/Th 和很高而且变化范围很大的 Ce/Pb(20~210)(图 6c)，其最低值与全球的 MORB 和 OIB 的稳定 Ce/Pb 比值(25±5)相当(Hofmann 等，1986)，说明其地幔源区遭受了富 Ce 贫 Pb 介质的交代改造作用。Sakuyama 等(2013)指出地幔过渡带滞留的俯冲洋壳在俯冲过程脱水时因遭受强烈的 Ba、Pb 丢失而具有低 Ba/Th 和高 Ce/Pb 特征，该俯冲洋壳部分熔融产生的熔体向上运移并交代上覆地幔可形成具有低的 Ba/Th 和很高而且变化很大 Ce/Pb 的碳酸盐化橄榄岩地幔。当该地幔源区发生部分熔融时，由于 Ba 和 Th，以及 Ce 和 Pb，有类似的地幔硅酸盐矿物/熔体分配系数，且都属于高不相容元素，因此熔体的 Ba/Th 和 Ce/Pb 较地幔橄榄岩不发生显著分异。因此，该碳酸盐化橄榄岩熔融产生的富碳酸盐硅酸盐熔体(相当于霞石岩)可继承源区的低 Ba/Th 和高而且变化的 Ce/Pb 特征。与霞石岩相反，高硅碱性玄武岩具有异常低的 Ce/Pb(10~15)，显著低于全球 MORB 和 OIB 的 Ce/Pb(25±5)(Hofmann 等，1986)，和非常高的 Ba/Th(80~160)(图 6c)，反映了其地幔源区遭受了富 Ba 和 Pb，相对贫 Th 和 Ce 的介质的交代改造。因为 Ba 在水流体中较 Th 有更高的溶解度和活动性，源区高 Ba/Th 特征可能反映了流体交代而非熔体交代的结果。碧玄岩在 Ce/Pb-Ba/Th 图中处于霞石岩和高硅碱性玄武岩之间，因此它们的 Ce/Pb-Ba/Th 变化较难用来有效区分上述两种过程。前人认为上述产生低 Ce/Pb 和高 Ba/Th 特征的流体地幔交代作用发生在软流圈(Sakuyama 等，2013；Li 等，2016)；然而结合应用 Nd 同位素和微量元素数据可以证明俯冲沉积物脱水流体交代地幔应该发生在岩石圈，并非发生在软流圈。

具有判别意义的同位素应当是软流圈碳酸盐化橄榄岩与古老流体交代的岩石圈地幔具有不同的同位素组成。Mg-Sr 同位素可能不适合这一研究目的，因为如果交代岩石圈地幔的流体或超临界流体溶解有富镁碳酸盐(如白云石)，则被流体交代的岩石圈地幔可以和软流圈碳酸盐化橄榄岩地幔有类似的 Mg-Sr 同位素组成。Nd 同位素适合这一研究目的，因为中国东部的碳酸盐化橄榄岩上地幔是被较年轻的晚侏罗世西太平洋俯冲板片交代的(Li 等，2017)，而且俯冲海相沉积碳酸盐含 REE 很低(Turekian 和 Wedepohl，1961；Paula-Santos 等，2018)，其交代作用不会显著影响软流圈地幔的 Nd 同位素组成；而古老流体交代的岩石圈地幔因 LREE 富集，经长时间演化具有低的 ε_{Nd} 值(Song 和 Frey，1989；Ma 和 Xu，2006；Tang 等，2008；Liu J 等，2012)。ε_{Nd}-La/Yb 图显示尽管山东新生代霞石岩和碧玄岩的 La/Yb 和 Nd/Zr 比值有较大变化范围，但它们具有较稳定和亏损的 ε_{Nd}(2~6)值(图 6d)。这说明霞石岩和碧玄岩源于相同的软流圈亏损地幔源区，但由于部分熔融程度的差异导致它们 La/Yb 比值变化。与此相反，高硅碱性玄武岩具有较富集的 Nd 同位素组成，其 ε_{Nd} 偏低且变化较大(ε_{Nd}=+3~−2)(图 6d)，说明它们地幔源区有 Nd 同位素较富集的物质的贡献。在 ε_{Nd}-La/Yb 图中，高硅碱性玄武岩数据排列成一显著的正相关趋势，它的高 La/Yb 和高 ε_{Nd} 端元与霞石岩-碧玄岩系列的低 La/Yb 端重合，它的低 La/Yb 和低 ε_{Nd} 端指向华北克拉通古老交代富集岩石圈地幔包体(Song 和 Frey，1989；Ma 和 Xu，2006；Tang 等，2008；Liu J 等，2012)(图 6 d)。这一高硅碱性玄武岩的 ε_{Nd}-La/Yb 正相关趋势证明了它们可以是来自亏损软流圈的霞石岩-碧玄岩熔体与富集的岩石圈地幔相互作用的产物，而非软流圈的碳酸盐化橄榄岩大比例熔融的结果。最近对南海的 IODP 钻探岩芯的研究已观察到来自软流圈的富碳酸盐硅酸盐熔体与岩石圈反应形成碱性玄武岩的演化系列(Zhang 等，2017)。

Sakuyama 等(2013)和 Li 等(2016)注意到了霞石岩和高硅碱性玄武岩的 Nd-Hf 同位素差异，并指出霞石岩具有类似太平洋地幔的 Nd-Hf 同位素组成，高硅碱性玄武岩具有类似印度洋地幔的 Nd-Hf 同位素组成(Pearce 等，1999)。然而他们根据同岩浆携带的地幔包体具有比玄武岩亏损的多的 Nd-Hf 同位素组成，排除了岩石圈地幔作为高硅碱性玄武岩同位素端元组成的可能性，而将该玄武岩的 Nd-Hf 同位素变化都归结与软流圈地幔内部的亏损和富集端元的不同程度贡献有关(Sakuyama 等，2013；Li 等，2016). 即霞石岩浆来自软流圈碳酸盐化橄榄岩或石榴辉石岩，其源区被俯冲的西太平洋脱水碳酸盐化洋壳熔体交代；而高硅碱性玄武岩浆来自软流圈石榴辉石岩，其源区被古老的(>1.0Ga)再循环沉积物流体交代(Sakuyama 等，2013；Li 等，2016)。但是下列 3 个理由可以论证该古老俯冲板片流体交代作用并非发生在软流圈地幔，而应发生在岩石圈地幔。(1)寄主岩浆携带的地幔橄榄岩包体不是来自该岩浆的源区，而是在岩浆上升过程中从上覆的岩石圈俘虏来的，不能用寄主岩浆携带的地幔橄榄岩包体 Nd 同位素亏损来排除源区有交代富集的岩石圈地幔端元。事实上华北汉诺坝、繁峙、阳原等地的地幔包体低 ε_{Nd} 特征证明华北克拉通岩石圈存在古老交代富集的地幔(图 6d)(Song 和 Frey，1989；Ma 和 Xu，2006；Tang 等，2008；Liu J 等，2012)(2)高硅碱性玄武岩具有低 Ce/Pb 和高 Ba/Th 特征要求它的源区经历的是在较浅深度(80-120 公里)俯冲板片析出流体交代而非熔体交代(Ringwood，1990；Kessel 等，2005)。然而华北克拉通奥陶纪代金伯利岩的地幔岩包体证明该岩石圈在古生代及以前厚度可达 200km(Fan 和 Menzies，1992；Zheng 和 Lu，1999)，因此，新元古代-早古生代华北克拉通下的软流圈深度应大于 200km。高压实验研究表明，俯冲含碳酸盐泥质沉积物在深度 240km(P=8.0GPa)，固相线温度在 1050-950℃之间，低于地幔绝热线 150~300℃(Grassi 和 Schmidt，2011)，使得俯冲含碳酸盐泥质沉积物在这一软流圈深度必然发生熔融而不是脱水。(3)如上所述，根据 Nd-Hf 同位素组成(Sakuyama 等，2013；Li 等，2016)，高硅碱性玄武岩具有类似印度洋地幔的 Nd-Hf 同位素组成。如果其地幔源区位于软流圈，我们很难解释新元古代交代形成的印度洋软流圈地幔如何在新生代进入到北半球的太平洋地幔对流体系。然而，如果该古老流体交代的地幔位于岩石圈内部，它随华北陆块的向北漂移，我们很容易理解为什么在中国东部，即西太平洋陆缘的地幔源区会具有印度洋型地幔的地球化学特征。

3.4　华北克拉通岩石圈减薄历史

前人研究已经指出早白垩世(130~120 Ma)是华北克拉通破坏峰期，导致其破坏的过程可能包括岩石圈引张和大规模花岗岩岩浆作用(Wu 等，2005；Kusky 等，2014)，对流地幔的热和机械侵蚀(Xu，2001；Xu 等，2004)，软流圈熔体与岩石圈的相互作用(Zhang，2005；Zheng 等，2007)，岩石圈底部的水化作用(Niu，2005；Windley 等，2010)，以及加厚岩石圈的拆沉(Gao 等，2004, 2008)或克拉通边缘造山带的山根拆沉作用(Li 等，2013)。它们造成了华北克拉通岩石圈的强烈减薄，但并没有达到现今减薄的程度(图 7a)。中国东部晚白垩世到新生代期间板内玄武岩的发育说明华北克拉通东部的岩石圈减薄过程在早白垩世以后一直在持续(Xu 等，2004，2005)。根据对中国东部大陆低 δ^{26}Mg 玄武岩的年龄统计(Li 等，2017)，华北克拉通岩石圈在 106Ma 之后开始遭受局部碳酸盐化交代和熔融作用。但从 106~25Ma 之间，中国东部大陆玄武岩火山活动仅零星分布，表明它仅使已经初步减薄的岩石圈局部发生隧道状破坏(图 7b)。在 25Ma 以后，中国东部大陆玄武岩喷发活动增长快速并在第四纪达到高峰，反映这一时期碳酸盐化橄榄岩地幔大规模上涌，从而导致大规模富碳酸盐熔体与岩石圈相互作用和减薄发生(图 7c)。值得注意的是中东部霞石岩仅发育在 10Ma 以后的新生代晚期(图 3)。如前述，霞石岩是很少与岩石圈发生作用的直接来自软流圈地幔的岩浆，它的出现说明华北克拉通岩石圈在新生代晚期已进一步减薄。

综上，华北克拉通岩石圈的减薄可能存在两个峰期：(1)在白垩纪早期(135~115Ma)伴随克拉通引张和岩浆作用峰期发生的岩石圈减薄；(2)25Ma 以后，由富碳酸盐硅酸盐熔体与岩石圈相互作用引发的岩石圈减薄使得华北克拉通东部岩石圈厚度减薄至约 70km 左右。

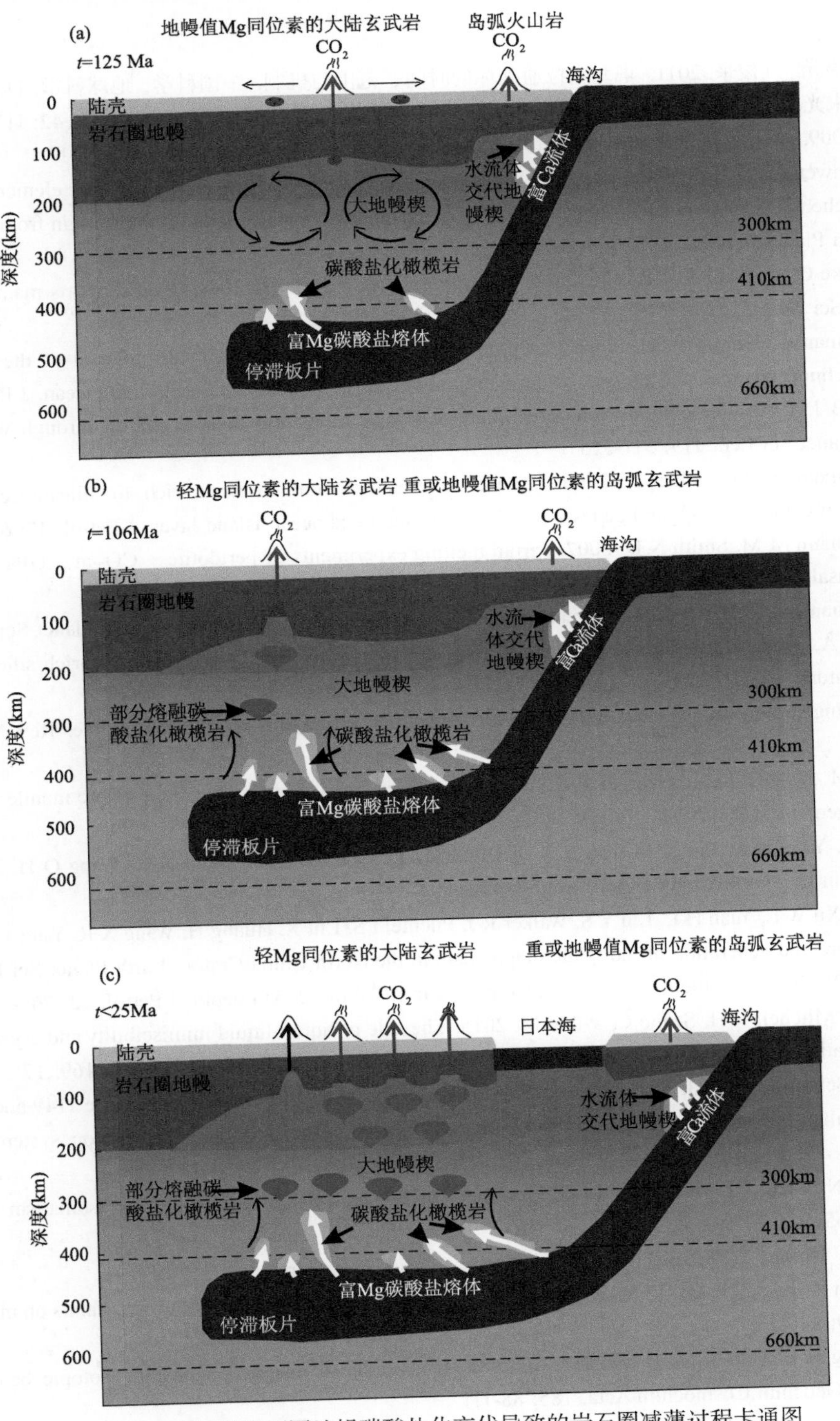

图 7 华北克拉通岩石圈地幔碳酸盐化交代导致的岩石圈减薄过程卡通图

致谢

感谢朱日祥院士邀请撰写此文。感谢刘金高教授、徐丽娟博士及匿名审稿人提出的修改意见，感谢黄金水教授提供的地球物理文献。

参考文献

朱日祥, 陈凌, 吴福元, 刘俊来. 2011. 华北克拉通破坏的时间、范围与机制. 中国科学: 地球科学, 41: 583-592
朱日祥, 徐义刚, 朱光, 张宏福, 夏群科, 郑天愉. 2012. 华北克拉通破坏. 中国科学: 地球科学, 42: 1135-1159
朱日祥, 郑天愉. 2009. 华北克拉通破坏机制与古元古代板块构造体系. 科学通报, 54: 1950-1961
Basu A R, Wang Junwen A R, Huang Wankang A R, Xie Guanghong A R, Tatsumoto M. 1991. Major element, REE, and Pb, Nd and Sr isotopic geochemistry of Cenozoic volcanic rocks of eastern China: Implications for their origin from suboceanic-type mantle reservoirs. Earth Planet Sci Lett, 105: 149-169
Chantel J, Manthilake G, Andrault D, Novella D, Yu T, Wang Y. 2016. Experimental evidence supports mantle partial melting in the asthenosphere. Sci Adv, 2: e1600246
Coltorti M, Bonadiman C, Hinton R W, Siena F, Upton B G J. 1999. Carbonatite metasomatism of the oceanic upper mantle: Evidence from clinopyroxenes and glasses in ultramafic xenoliths of grande comore, Indian Ocean. J Petrol, 40: 133-165
Dalton J A, Wood B J. 1993. The compositions of primary carbonate melts and their evolution through wallrock reaction in the mantle. Earth Planet Sci Lett, 119: 511-525
Dasgupta R, Hirschmann M M, Stalker K. 2006. Immiscible transition from carbonate-rich to silicate-rich melts in the 3 GPa melting interval of eclogite + CO_2 and genesis of silica-undersaturated ocean island lavas. J Petrol, 47: 647-671
Dasgupta R, Hirschmann M M, Smith N D. 2007. Partial melting experiments of peridotite + CO_2 at 3 GPa and genesis of Alkalic Ocean Island Basalts. J Petrol, 48: 2093-2124
Dasgupta R, Hirschmann M M. 2010. The deep carbon cycle and melting in Earth's interior. Earth Planet Sci Lett, 298: 1-13
Dasgupta R, Mallik A, Tsuno K, Withers A C, Hirth G, Hirschmann M M. 2013. Carbon-dioxide-rich silicate melt in the Earth’s upper mantle. Nature, 493: 211-215
Dasgupta R. 2013. Ingassing, storage, and outgassing of terrestrial carbon through geologic time. Rev Mineral Geochem, 75: 183-229
Fan W M, Menzies M A. 1992. Destruction of aged lower lithosphere and accretion of asthenosphere mantle beneath eastern China. Geotect Metallogen, 16: 171-179
Gao S, Rudnick R L, Yuan H L, Liu X M, Liu Y S, Xu W L, Ling W L, Ayers J, Wang X C, Wang Q H. 2004. Recycling lower continental crust in the North China craton. Nature, 432: 892-897
Gao S, Rudnick R L, Xu W L, Yuan H L, Liu Y S, Walker R J, Puchtel I S, Liu X, Huang H, Wang X R, Yang J. 2008. Recycling deep cratonic lithosphere and generation of intraplate magmatism in the North China Craton. Earth Planet Sci Lett, 270: 41-53
Grassi D, Schmidt M W. 2011. The melting of carbonated pelites from 70 to 700 km depth. J Petrol, 52: 765-789
Guzmics T, Zajacz Z, Mitchell R H, Szabó C, Wälle M. 2015. The role of liquid-liquid immiscibility and crystal fractionation in the genesis of carbonatite magmas: Insights from Kerimasi melt inclusions. Contrib Mineral Petrol, 169: 17
Hauff F, Hoernle K, Schmidt A. 2003. Sr-Nd-Pb composition of Mesozoic Pacific oceanic crust (Site 1149 and 801, ODP Leg 185): Implications for alteration of ocean crust and the input into the Izu- Bonin-Mariana subduction system. Geochem Geophys Geosyst, 4: 8913
Hauri E H, Shimizu N, Dieu J J, Hart S R. 1993. Evidence for hotspot-related carbonatite metasomatism in the oceanic upper mantle. Nature, 365: 221-227
Hoefs J. 2009. Stable Isotope Geochemistry. Springer-Verlag Berlin Heidelberg
Hofmann A W, Jochum K P, Seufert M, White W M. 1986. Nb and Pb in oceanic basalts: New constraints on mantle evolution. Earth Planet Sci Lett, 79: 33-45
Hu Y, Teng F Z, Zhang H F, Xiao Y, Su B X. 2016. Metasomatism- induced mantle magnesium isotopic heterogeneity: Evidence from pyroxenites. Geochim Cosmochim Acta, 185: 88-111
Huang J, Zhao D. 2006. High-resolution mantle tomography of China and surrounding regions. J Geophys Res, 111: B09305

Huang J, Li S G, Xiao Y, Ke S, Li W Y, Tian Y. 2015. Origin of low δ^{26}Mg Cenozoic basalts from South China Block and their geodynamic implications. Geochim Cosmochim Acta, 164: 298-317
Huang J, Xiao Y. 2016. Mg-Sr isotopes of low-δ^{26}Mg basalts tracing recycled carbonate species: Implication for the initial melting depth of the carbonated mantle in Eastern China. Int Geol Rev, 58: 1350-1362
Huang K J, Teng F Z, Plank T, Staudigel H, Hu Y, Bao Z Y. 2018. Magnesium isotopic composition of the altered oceanic crust: Implications for the magnesium geochemical cycle. Geochim Cosmochim Acta, Submitted
Ionov D. 1998. Trace element composition of mantle-derived carbonates and coexisting phasesin peridotite xenoliths from Alkali Basalts. J Petrol, 39: 1931-1941
Ke S, Teng F Z, Li S G, Gao T, Liu S A, He Y, Mo X. 2016. Mg, Sr, and O isotope geochemistry of syenites from northwest Xinjiang, China: Tracing carbonate recycling during Tethyan oceanic subduction. Chem Geol, 437: 109-119
Kessel R, Schmidt M W, Ulmer P, Pettke T. 2005. Trace element signature of subduction-zone fluids, melts and supercritical liquids at 120-180 km depth. Nature, 437: 724-727
Kogiso T, Tatsumi Y, Nakano S. 1997. Trace element transport during dehydration processes in the subducted oceanic crust: 1. Experiments and implications for the origin of ocean island basalts. Earth Planet Sci Lett, 148: 193-205
Kushiro I, Satake H, Akimoto S. 1975. Carbonate-silicate reactions at high presures and possible presence of dolomite and magnesite in the upper mantle. Earth Planet Sci Lett, 28: 116-120
Kusky T M, Windley B F, Wang L, Wang Z, Li X, Zhu P. 2014. Flat slab subduction, trench suction, and craton destruction: Comparison of the North China, Wyoming, and Brazilian cratons. Tectonophysics, 630: 208-221
Lee C T A, Luffi P, Plank T, Dalton H, Leeman W P. 2009. Constraints on the depths and temperatures of basaltic magma generation on Earth and other terrestrial planets using new thermobarometers for mafic magmas. Earth Planet Sci Lett, 279: 20-33
Li H Y, Xu Y G, Ryan J G, Huang X L, Ren Z Y, Guo H, Ning Z G. 2016. Olivine and melt inclusion chemical constraints on the source of intracontinental basalts from the eastern North China Craton: Discrimination of contributions from the subducted Pacific slab. Geochim Cosmochim Acta, 178: 1-19
Li S G, He Y S, Wang S J. 2013. Process and mechanism of mountain-root removal of the Dabie Orogen—Constraints from geochronology and geochemistry of post-collisional igneous rocks. Chin Sci Bull, 58: 4411-4417
Li S G, Yang W, Ke S, Meng X, Tian H, Xu L, He Y, Huang J, Wang X C, Xia Q, Sun W, Yang X, Ren Z Y, Wei H, Liu Y, Meng F, Yan J. 2017. Deep carbon cycles constrained by a large-scale mantle Mg isotope anomaly in eastern China. Nat Sci Rev, 4: 111-120
Li Z X, Li X H. 2007. Formation of the 1300-km-wide intracontinental orogen and postorogenic magmatic province in Mesozoic South China: A flat-slab subduction model. Geology, 35: 179-182
Liu C Q, Masuda A, Xie G H. 1994. Major- and trace-element compositions of Cenozoic basalts in eastern China: Petrogenesis and mantle source. Chem Geol, 114: 19-42
Liu J, Carlson R W, Rudnick R L, Walker R J, Gao S, Wu F. 2012. Comparative Sr-Nd-Hf-Os-Pb isotope systematics of xenolithic peridotites from Yangyuan, North China Craton: Additional evidence for a Paleoproterozoic age. Chem Geol, 332-333: 1-14
Liu J, Rudnick R L, Walker R J, Xu W, Gao S, Wu F. 2015. Big insights from tiny peridotites: Evidence for persistence of Precambrian lithosphere beneath the eastern North China Craton. Tectonophysics, 650: 104-112
Liu S A, Li S, Guo S, Hou Z, He Y. 2012. The Cretaceous adakitic- basaltic-granitic magma sequence on south-eastern margin of the North China Craton: Implications for lithospheric thinning mechanism. Lithos, 134-135: 163-178
Liu X, Zhao D, Li S, Wei W. 2017. Age of the subducting Pacific slab beneath East Asia and its geodynamic implications. Earth Planet Sci Lett, 464: 166-174
Ma J, Xu Y. 2006. Old EMI-type enriched mantle under the middle North China Craton as indicated by Sr and Nd isotopes of mantle xenoliths from Yangyuan, Hebei Province. Chin Sci Bull, 51: 1343-1349
McKenzie D, Bickle M J. 1988. The volume and composition of melt generated by extension of the lithosphere. J Petrol, 29: 625-679
Meng F, Gao S, Niu Y, Liu Y, Wang X. 2015. Mesozoic-Cenozoic mantle evolution beneath the North China Craton: A new perspective from Hf-Nd isotopes of basalts. Gondwana Res, 27: 1574-1585
Morlidge M, Pawley A, Droop G. 2006. Double carbonate breakdown reactions at high pressures: An experimental study in the system CaO-MgO-FeO-MnO-CO_2. Contrib Mineral Petrol, 152: 365-373
Niu Y L. 2005. Generation and evolution of basaltic magmas: Some basic concepts and a new view on the origin of Mesozoic-Cenozoic basaltic volcanism in Eastern China. Geol J China Univ, 11: 9−46

De Paula-Santos G M, Caetano-Filho S, Babinski M, Enzweiler J. 2018. Rare earth elements of carbonate rocks from the Bambuí Group, southern São Francisco Basin, Brazil, and their significance as paleoenvironmental proxies. Precambrian Res, 305: 327-340

Pearce J A, Kempton P D, Nowell G M, Noble S R. 1999. Hf-Nd element and isotope perspective on the nature and provenance of mantle and subduction components in Western Pacific Arc-Basin Systems. J Petrol, 40: 1579-1611

Pilet S, Baker M B, Muntener O, Stolper E M. 2011. Monte carlo simulations of metasomatic enrichment in the lithosphere and implications for the source of Alkaline Basalts. J Petrol, 52: 1415-1442

Plank T, Langmuir C H. 1998. The chemical composition of subducting sediment and its consequences for the crust and mantle. Chem Geol, 145: 325-394

Poli S, Franzolin E, Fumagalli P, Crottini A. 2009. The transport of carbon and hydrogen in subducted oceanic crust: An experimental study to 5 GPa. Earth Planet Sci Lett, 278: 350-360

Prytulak J, Elliott T. 2007. TiO_2 enrichment in ocean island basalts. Earth Planet Sci Lett, 263: 388-403

Ringwood A E. 1990. Slab-mantle interactions. Chem Geol, 82: 187-207

Sakuyama T, Tian W, Kimura J I, Fukao Y, Hirahara Y, Takahashi T, Senda R, Chang Q, Miyazaki T, Obayashi M, Kawabata H, Tatsumi Y. 2013. Melting of dehydrated oceanic crust from the stagnant slab and of the hydrated mantle transition zone: Constraints from Cenozoic alkaline basalts in eastern China. Chem Geol, 359: 32-48

Sato K, Katsura T. 2001. Experimental investigation on dolomite dissociation into aragonite+magnesite up to 8.5 GPa. Earth Planet Sci Lett, 184: 529-534

Song Y, Frey F A. 1989. Geochemistry of peridotite xenoliths in basalt from Hannuoba, Eastern China: Implications for subcontinental mantle heterogeneity. Geochim Cosmochim Acta, 53: 97-113

Song Y, Frey F A, Zhi X. 1990. Isotopic characteristics of Hannuoba basalts, eastern China: Implications for their petrogenesis and the composition of subcontinental mantle. Chem Geol, 88: 35-52

Sun W, Ding X, Hu Y H, Li X H. 2007. The golden transformation of the Cretaceous plate subduction in the west Pacific. Earth Planet Sci Lett, 262: 533-542

Tang Y J, Zhang H F, Ying J F, Zhang J, Liu X M. 2008. Refertilization of ancient lithospheric mantle beneath the central North China Craton: Evidence from petrology and geochemistry of peridotite xenoliths. Lithos, 101: 435-452

Tao R, Zhang L, Fei Y, Liu Q. 2014. The effect of Fe on the stability of dolomite at high pressure: Experimental study and petrological observation in eclogite from southwestern Tianshan, China. Geochim Cosmochim Acta, 143: 253-267

Teng F Z, Li W Y, Ke S, Marty B, Dauphas N, Huang S, Wu F Y, Pourmand A. 2010. Magnesium isotopic composition of the Earth and chondrites. Geochim Cosmochim Acta, 74: 4150-4166

Thomson A R, Walter M J, Kohn S C, Brooker R A. 2016. Slab melting as a barrier to deep carbon subduction. Nature, 529: 76-79

Tian H, Yang W, Li S G, Ke S, Duan X Z. 2018. Low δ^{26}Mg volcanic rocks of Tengchong in Southwestern China: A dep carbon cycle induced by surprcritical liquids. Geochim Cosmochim Acta, Submitted

Tian H C, Yang W, Li S G, Ke S, Chu Z Y. 2016. Origin of low δ^{26}Mg basalts with EM-I component: Evidence for interaction between enriched lithosphere and carbonated asthenosphere. Geochim Cosmochim Acta, 188: 93-105

Turekian K K, Wedepohl K H. 1961. Distribution of the elements in some major units of the Earth's crust. Geol Soc Am Bull, 72: 175-192

Wang C, Liu Y, Zhang J, Jin Z. 2016. Carbonate melt form subduction zone: The key for Craton destruction. Goldschmidt Conference Abstracts, 3307

Wang S J, Teng F Z, Rudnick R L, Li S G. 2015. The behavior of magnesium isotopes in low-grade metamorphosed mudrocks. Geochim Cosmochim Acta, 165: 435-448

Wang S J, Teng F Z, Li S G, Zhang L F, Du J X, He Y S, Niu Y. 2017. Tracing subduction zone fluid-rock interactions using trace element and Mg-Sr-Nd isotopes. Lithos, 290-291: 94-103

Windley B F, Maruyama S, Xiao W J. 2010. Delamination/thinning of sub-continental lithospheric mantle under Eastern China: The role of water and multiple subduction. Am J Sci, 310: 1250-1293

Workman R K, Hart S R. 2005. Major and trace element composition of the depleted MORB mantle (DMM). Earth Planet Sci Lett, 231: 53-72

Wu F Y, Lin J Q, Wilde S A, Zhang X, Yang J H. 2005. Nature and significance of the Early Cretaceous giant igneous event in eastern China. Earth Planet Sci Lett, 233: 103-119

Xiao Y, Teng F Z, Zhang H F, Yang W. 2013. Large magnesium isotope fractionation in peridotite xenoliths from eastern North China craton: Product of melt-rock interaction. Geochim Cosmochim Acta, 115: 241-261

Xu Y G. 2001. Thermo-tectonic destruction of the archaean lithospheric keel beneath the sino-korean craton in china: Evidence, timing and mechanism. Phys Chem Earth Part A-Solid Earth Geodesy, 26: 747-757

Xu Y G, Huang X L, Ma J L, Wang Y B, Iizuka Y, Xu J F, Wang Q, Wu X Y. 2004. Crust-mantle interaction during the tectono-thermal reactivation of the North China Craton: Constraints from SHRIMP zircon U-Pb chronology and geochemistry of Mesozoic plutons from western Shandong. Contrib Mineral Petrol, 147: 750-767

Xu Y G, Ma J L, Frey F A, Feigenson M D, Liu J F. 2005. Role of lithosphere-asthenosphere interaction in the genesis of Quaternary alkali and tholeiitic basalts from Datong, western North China Craton. Chem Geol, 224: 247-271

Yang W, Li S. 2008. Geochronology and geochemistry of the Mesozoic volcanic rocks in Western Liaoning: Implications for lithospheric thinning of the North China Craton. Lithos, 102: 88-117

Yang W, Teng F Z, Zhang H F, Li S G. 2012. Magnesium isotopic systematics of continental basalts from the North China craton: Implications for tracing subducted carbonate in the mantle. Chem Geol, 328: 185-194

Zeng G, Chen L H, Xu X S, Jiang S Y, Hofmann A W. 2010. Carbonated mantle sources for Cenozoic intra-plate alkaline basalts in Shandong, North China. Chem Geol, 273: 35-45

Zeng G, Chen L H, Hofmann A W, Jiang S Y, Xu X S. 2011. Crust recycling in the sources of two parallel volcanic chains in Shandong, North China. Earth Planet Sci Lett, 302: 359-368

Zhang G L, Chen L H, Jackson M G, Hofmann A W. 2017. Evolution of carbonated melt to alkali basalt in the South China Sea. Nat Geosci, 10: 229-235

Zhang H F. 2005. Transformation of lithospheric mantle through peridotite-melt reaction: A case of Sino-Korean craton. Earth Planet Sci Lett, 237: 768-780

Zheng J P, Griffin W L, O'Reilly S Y, Yu C M, Zhang H F, Pearson N, Zhang M. 2007. Mechanism and timing of lithospheric modification and replacement beneath the eastern North China Craton: Peridotitic xenoliths from the 100 Ma Fuxin basalts and a regional synthesis. Geochim Cosmochim Acta, 71: 5203-5225

Zheng J P, Lu F X. 1999. Mantle xenoliths from kimberlites, Shandong and Liaoning: Paleozoic lithospheric mantle character and its heterogeneity (in Chinese). Acta Petrol Sin, 15: 65-74

Zhi X, Song Y, Frey F A, Feng J, Zhai M. 1990. Geochemistry of Hannuoba basalts, eastern China: Constraints on the origin of continental alkalic and tholeiitic basalt. Chem Geol, 88: 1-33

Zou H, Zindler A, Xu X, Qi Q. 2000. Major, trace element, and Nd, Sr and Pb isotope studies of Cenozoic basalts in SE China: Mantle sources, regional variations, and tectonic significance. Chem Geol, 171: 33-47

Zhou M F, Robinson P T, Su B X, Gao J F, Li J W, Yang J S, Malpas J. 2014. Compositions of chromite, associated minerals, and parental magmas of podiform chromite deposits: The role of slab contamination of asthenospheric melts in suprasubduction zone environments. Gondwana Res, 26: 262-283

Subducted Mg-rich carbonates into the deep mantle wedge*

Ji Shen, [1] Shu-Guang Li, [2,1] Shui-Jiong Wang, [2] Fang-Zhen Teng, [3] Qiu-Li Li[4] and Yong-Sheng Liu[5]

1. CAS Key Laboratory of Crust-Mantle Materials and Environments, School of Earth and Space Sciences, University of Science and Technology of China, Hefei 230026, China
2. State Key Laboratory of Geological Processes and Mineral Resources, China, University of Geosciences, Beijing 100083, China
3. Isotope Laboratory, Department of Earth and Space Sciences, University of Washington, Seattle, WA 98195, USA
4. State Key Laboratory of Lithospheric Evolution, Institute of Geology and Geophysics, Chinese Academy of Sciences, Beijing 100029, China
5. State Key Laboratory of Geological Processes and Mineral Resources, Faculty of Earth Sciences, China University of Geosciences, Wuhan 430074, China

亮点介绍：本文结合矿物学、元素和Mg-O同位素地球化学、以及同位素年代学对大别造山带中部毛屋超镁铁质岩体进行系统研究，揭示该岩体为交代地幔楔碎片，并记录了早期洋壳俯冲阶段的富Mg碳酸盐交代作用。本工作发现了板片俯冲深度超过150km，俯冲板片析出的超临界流体中可以溶解富Mg碳酸盐。

Abstract Recent studies have concluded that subducted calcium (Ca-) rich carbonates could be dissolved in slab-derived aqueous fluids and transported upwards into the shallow mantle wedge (75−120 km), while magnesium (Mg-) rich carbonates could be delivered to a greater depth (i.e., the mantle transition zone, ~410 km), melted, and recycled into the convective upper mantle. However, it remains unknown whether or not Mg-rich carbonates can be transferred to the deep mantle wedge (>~120 km) by subduction-zone fluids, which, if true, is important for tracing deep carbon. In this paper, we report a comprehensive mineralogical, geochemical, stable (Mg and O) and radiogenic isotopic (zircon U-Pb) study of garnet clinopyroxenites from the Maowu ultra-mafic massif (a slice of the mantle wedge) in the Dabie orogeny, Central China. Whole-rock and mineral trace elemental features and zircon U-Pb ages reveal evidence of mantle wedge metasomatism by a slab-derived melt or supercritical fluid from the subducted rutile-bearing eclogitic Paleo-Tethys oceanic crust, in addition to subsequent metamorphism occurring during the Triassic collision between the South and North China blocks. Combined with the results of previous works, the high Th/U ratios of both whole rocks and metasomatized zircons with no oscillatory zoning lead us to infer that a supercritical liquid rather than a melt was the metasomatic agent during oceanic subduction at peak conditions (5.3−6.3 GPa and ~800°C,

* 本文发表在：Earth and Planetary Science Letters, 2018, 503: 118-130

160−190 km). Abundant carbonate mineral inclusions (including calcite, dolomite and magnesite) and the high $\delta^{18}O_{VSMOW}$ values of the metasomatized zircons (up to 12.2‰) indicate that sedimentary carbonates were leached by the supercritical fluid. Furthermore, whole-rock $\delta^{26}Mg$ values (−0.99‰ to −0.65‰) that are lower than normal mantle values (−0.25±0.07‰) imply that the incorporated carbonates contain not only calcites but also a certain amount of dolomites (approximately 1−10 wt.% of the metasomatic supercritical liquid). The dissolved Mg-rich carbonates in the slab-derived supercritical liquid could effectively modify the Mg isotope composition of the deep mantle wedge. Our study represents a critical step towards achieving a broad understanding of the behaviours of recycled carbonate during slab subduction.

Keywords Subducted carbonates, mantle wedge metasomatism, Mg isotope, O isotope, Supercritical liquid

1 Introduction

The recycling of sedimentary carbonates into the mantle via slab subduction is believed to be the predominant form of deep carbon cycling (Dasgupta and Hirschmann, 2010). However, the fate of subducted carbonates is a highly controversial issue. Some studies have concluded that some subducting carbon is returned to the convecting mantle (e.g., 20%−80%, Dasgupta and Hirschmann, 2010; 18%−70%, Johnston et al., 2011), whereas Kelemen and Manning (2015) proposed that almost all carbon in the slab was transferred to the mantle wedge in a subduction zone, limiting the efficiency of carbon recycling to the deep mantle. These discrepancies emphasize the difficulty in assessing the nature and composition of C-O-H fluids in deep subduction zones (Ferrando et al., 2017). For example, it has been widely accepted that a large amount of sedimentary calcium (Ca-) rich carbonates (i.e., calcites and aragonite) could be dissolved in slab-derived fluids and transported into the shallow mantle wedge during slab subduction (Ague and Nicolescu, 2014; Facq et al., 2014; Pan et al., 2013), in addition to decarbonation (Gorman et al., 2006). This promotes the higher estimated flux of carbon released from subducted slabs presented by Kelemen and Manning (2015). However, compared with Ca-rich carbonates, magnesium (Mg-) rich carbonates (i.e., dolomite and magnesite), which are another important carbon reservoir in surface sediments, behave conservatively during subduction, i.e., they are more stable and insoluble in subduction fluids (Luth 2001; Pan et al., 2013), so that they might be delivered to the deep mantle by subduction (Li et al., 2017). The question of whether subducted Mg-rich carbonates can be transferred back into the mantle wedge remains poorly constrained. Recently, Foley and Fischer (2017) proposed that the continental lithosphere was a vast carbon reservoir and that carbon remobilization can lead to variations in CO_2 outgassing through magmas from the continental lithosphere over geological timescales. The behaviours of dolomites (and magnesites) in subduction fluids play an important role in understanding the carbon stored in the continental lithosphere because some of the continental lithosphere was formed by the accumulation of sub-arc lithosphere during the island arc accretions.

Although a wide range of stable metal isotope systems (Mg, Fe, Zn, Ca, etc.) have been used to discuss the nature of fluid transfer in subduction zones (Debret et al., 2016; Huang et al., 2011; Pons et al., 2016; Wang et al., 2014a, 2017), with respect to recycling carbonates, Mg isotopes have displayed tracing potentials because sedimentary carbonates have lighter Mg isotope compositions relative to other terrestrial reservoirs (Li et al., 2017; Teng, 2017). Furthermore, because the Mg contents in dolomites (or magnesites) are tens to hundreds of times higher than those in calcites, the incorporation of Mg-rich carbonates may have remarkable effects on the Mg isotope compositions of subduction fluids and the mantle relative to Ca-rich carbonates. For example, intra-plate basalts and peridotite xenoliths with light Mg isotopes have been determined to be the result of recycling sedimentary Mg-rich carbonates or carbonated eclogites into the deep mantle (i.e., the mantle transition zone or convective upper mantle) via slab subduction (Hu et al., 2016; Huang et al., 2015; Li et al., 2017; Wang et al., 2016a, b; Yang et al., 2012), which melted and metasomatized the overlying convecting upper mantle or craton lithosphere. However, the relatively shallow mantle wedge (<120 km) and associated island arc volcanic rocks, including the Avacha metasomatized mantle wedge peridotites and island arc basalts from the circum-Pacific (e.g., Kamchatka, the Philippines, Costa Rica, and the Lau Basin), as well as the Lesser Antilles regions, display mantle-like and even heavier Mg isotope features (Li et al., 2017; Pogge von Strandmann et al., 2011; Teng et al., 2016). These results were interpreted to reflect the preferential dissolution of Ca-rich carbonates in slab-derived aqueous fluids (Li et al., 2017), while such fluids predominantly inherit heavy Mg isotope features from altered abyssal peridotites or sediments (Teng et al., 2016; Wang et al., 2017). These observations raise a question regarding whether sedimentary Mg-rich carbonates could be incorporated into the overlying mantle wedge at depths of greater than 120 km and whether they can influence the Mg isotope compositions of the mantle.

Experimental and theoretical studies have mainly focused on the dissolutions of calcites and CO_3^{2-} in subduction zone fluids (Ague and Nicolescu, 2014; Facq et al., 2014; Frezzotti et al., 2011), whereas those performed on dolomite (or magnesite) have been limited. Two factors, i.e., the increases of pressure and temperature via prograde subduction and high solutions (salinity and silicate components) in the fluids, could effectively enhance the solubilities of calcites and CO_3^{2-} (Kelemen and Manning, 2015 and references therein; Tumiati et al., 2017). The same reason was employed to interpret the first observation of dolomite dissolution in the Western Alps calcite-dolomite marble (Ferrando et al., 2017). A supercritical liquid from a subduction oceanic crust, which is saline and has higher silicate components and a deeper subduction slab source (>160 km) relative to shallow aqueous fluids (Kessel et al., 2005), was proposed to be capable of dissolving a great deal of CO_2 (~5 wt.%, Kelemen and Manning, 2015). Currently, whether Mg-rich carbonates can be dissolved by supercritical liquids and delivered into the mantle wedge remains undetermined. Investigating the magnesium isotopes of a carbonated deep mantle wedge could provide critical information for these issues.

The Maowu ultra-mafic massif is considered to be a fragment of the mantle wedge in the Dabie orogen, which has experienced mantle metasomatism and ultra-high-pressure metamorphism based on detailed studies of garnet orthopyroxenites (GO) (Chen et al., 2013, 2017; Malaspina et al., 2006, 2009). In contrast, few garnet clinopyroxenites (GC) within the massif have been reported, which display higher enrichments of Th, U and light rare earth elements (LREE) (Jahn et al., 2003) and thus likely record information about metasomatism and metamorphism. In this paper, we report an integrated study of the major and trace elements and Mg isotope compositions of the Maowu garnet clinopyroxenites, as well as the secondary ion mass spectrometry (SIMS) in situ zircon dating, oxygen (O) isotopes, and Raman spectroscopy analyses of mineral inclusions, to facilitate a better understanding of the nature of the slab-derived metasomatic fluid and carbonatitic metasomatism of the mantle wedge. For comparison, two quartz eclogites from the country rock gneisses of the Maowu massif were also investigated here, which have been identified as ultra-high-pressure metamorphic (UHPM) basaltic rocks belonging to the subducted continental crust of the South China Block (SCB), thus providing further constraint on the effects of Triassic UHP metamorphism.

2 Geology and samples

The Dabie-Sulu orogen is a Triassic continental collision zone formed by the subduction of the SCB underneath the North China Block (NCB) (Li et al., 1993). The observation of mineral assemblage exsolutions (i.e., clinopyroxene, rutile, and apatite) in garnets within coesite- or diamond-bearing eclogites indicates subduction depths of greater than 200 km (Ye et al., 2000). Before the Triassic collision, the Paleo-Tethys ocean, which was located between the NCB and the SCB, was subducted northward underneath the NCB (Li et al., 2001). The Maowu UHPM mafic-ultramafic massif occurred as a lenticular body (approximately 250 × 50 m in size) within the UHPM paragneisses located in the central Dabie zone (Supplementary Fig. S1a). The massif was dominated by GO, garnet websterites and GC, with minor garnet harzburgites and garnet dunites (Supplementary Fig. S1b, Chen et al., 2013, 2017; Malaspina et al., 2006). The GC investigated here occur as layers associated with GO and garnet websterites (Supplementary Fig. S1b, c). The boundaries between various types of rocks are sharp (Chen et al., 2017). The detailed field survey of the Maowu massif displays a geologically symmetrical structure between variable lithologies suggesting a channel metasomatism petrogenesis for the garnet orthopyroxenites, websterites and clinopyroxenites (Chen, 2011), rather than a metrical structure, which represents an important piece of evidence for an accumulative crustal origin (Fan et al., 1996). Although the Maowu massif was previously proposed to be a crustal cumulate in the SCB (Fan et al., 1996; Jahn et al., 2003), recent works have presented a series of geochemical evidence and outcrops demonstrating that this massif represents a rock slice from the precursor NCB mantle wedge located near the interface between the subducted slab and the mantle wedge, which suffered slab-derived fluid/melt metasomatism and UHP metamorphism (Chen et al., 2013, 2017; Malaspina et al., 2006, 2009). The peak P-T conditions determined using both THERMOCALC calculations and a garnet-orthopyroxene thermometer and barometer were estimated to be 5.3–6.3 GPa and ~800 °C (Chen et al., 2013), which indicates that the Maowu massif was delivered to a depth of at least ~160 km via mantle convections during oceanic crust subduction; it was then trapped in the continental crust of the SCB during continental subduction. The eight GC samples investigated here were sampled from different locations of two layers (00MW-2, 00MW-2(re) from one layer and 11MW-1, 3, 4, 5, 6, 8 from the other one), both of which were interlayered within the surrounding GO. These two layers were located only ~10 m from each other.

The country rocks of the Maowu massif are coesite-bearing paragneisses. Two quartz eclogite lenses (99MW-2 and 99MW-3) within these paragneisses, located ~1 km away from the massif, were selected for comparison (Supplementary Fig. S1b).

3 Analytical methods

Fresh mineral grains including garnets and clinopyroxenes were handpicked under a binocular microscope before dissolution. The chemical separation and purification of Mg was performed following the procedures reported in previous studies (Teng et al., 2010). A brief description is provided here: both mineral grains and whole-rock powders were dissolved in Savillex beakers in a combination of HF-HCl-HNO_3. After complete digestion, the separation of Mg was achieved by cation exchange chromatography using Bio-Rad 200–400 mesh AG50W-X8 resin in 1 N HNO_3. Two standards (KH-olivine and Hawaiian seawater) were processed along with each batch of samples. The procedural blanks were generally <10 ng, which was negligible relative to the amount of Mg (~50 μg) loaded on the column for samples and standards.

The magnesium isotope compositions of whole rocks and minerals were determined by the sample-standard bracketing method using a *Nu Plasma* multi-collector (MC-) ICP-MS at the Isotope Laboratory of the University of Arkansas, Fayetteville, in low-resolution mode with ^{26}Mg, ^{25}Mg and ^{24}Mg measured simultaneously in separate Faraday cups. The long-term precision was better than 0.07‰ (2SD) based on replicate analyses of synthetic solutions and mineral and rock standards (Teng et al., 2010). The internal precision of the measured ^{26}Mg/^{24}Mg ratio, based on four repeat runs of the same sample solution performed during analytical sessions, was

0.05‰–0.08‰ (2SD). The in-house standards (KH-olivine and Hawaiian seawater) yielded δ^{26}Mg values of −0.25±0.03‰ and −0.84±0.04‰ (Table 1), respectively, which were consistent with previous works (e.g., Wang et al., 2014a, b). Magnesium isotope results are reported in conventional δ notation in per mil relative to the DSM-3 standard, expressed as $\delta^{x}Mg=[(^{x}Mg/^{24}Mg)_{sample}/(^{x}Mg/^{24}Mg)_{DSM\text{-}3}-1]\times1000$, where x refers to 25 or 26. Magnesium isotope fractionation between two mineral phases (X and Y) is denoted by $\Delta^{26}Mg_{X\text{-}Y}=\delta^{26}Mg_{X}-\delta^{26}Mg_{Y}$.

Table 1　Magnesium isotope compositions of whole rocks and mineral separates (garnet and clinopyroxene) for Maowu garnet clinopyroxenites and quartz eclogites

Sample ID	Whole rock or minerals[a]	Column times[b]	δ^{26}Mg (‰)	2SD[c]	δ^{25}Mg (‰)	2SD
99MW-2	WR	2	−0.47	0.07	−0.25	0.05
	Grt	2	−0.79	0.08	−0.42	0.06
	Cpx	2	0.29	0.08	0.17	0.06
99MW-3	WR	2	−0.22	0.07	−0.11	0.04
	Grt	2	−0.77	0.08	−0.43	0.06
	Cpx	2	0.31	0.08	0.16	0.06
11MW-1	WR	2	−0.70	0.07	−0.36	0.04
	Grt	2	−1.31	0.07	−0.69	0.04
	Cpx	2	−0.31	0.07	−0.16	0.04
11MW-3	WR	2	−0.71	0.07	−0.38	0.06
	Grt	2	−1.58	0.07	−0.77	0.06
	Cpx	2	−0.56	0.07	−0.30	0.06
11MW-4	WR	2	−0.65	0.07	−0.34	0.05
	Grt	2	−1.51	0.08	−0.77	0.05
	Cpx	2	−0.39	0.08	−0.19	0.05
11MW-5	WR	2	−0.99	0.07	−0.49	0.06
	Grt	2	−1.78	0.08	−0.91	0.05
	Cpx	2	−0.68	0.08	−0.34	0.05
11MW-8	WR	2	−0.88	0.07	−0.44	0.05
	Grt	2	−1.53	0.08	−0.79	0.05
	Cpx	2	−0.56	0.08	−0.25	0.05
00MW-2	WR	2	−0.77	0.08	−0.40	0.05
	Grt	2	−1.40	0.08	−0.73	0.05
	Cpx	2	−0.40	0.08	−0.16	0.06
00MW-2(re)	WR	2	−0.95	0.08	−0.48	0.04
Reference materials						
BHVO-2		4	−0.25	0.07	−0.12	0.04
San Carlos olivine		4	−0.27	0.07	−0.13	0.04
In-house standards						
KH-Ol		13	−0.25	0.03	−0.13	0.02
Hawaiian Seawater		9	−0.84	0.04	−0.43	0.03

[a] WR-whole rock; Cpx-clinopyroxene; Grt-garnet.

[b] Two-processed through the column two times.

[c] 2SD-the 2SD uncertainties quoted for the samples were 2SE of single sample measurement, or 2SD reproducibility of repeated samples, or 2SD reproducibility of long-term standards, whichever is largest.

The analytical methods used for the major and trace element contents of whole rocks and minerals, zircon mineral inclusions, SIMS U-Pb dating and O isotope compositions are presented in detail in the Supplementary Materials section.

4 Results

4.1 Major and trace element contents of whole rocks

The major and trace element compositions of eight GC and two quartz eclogites are presented in Supplementary Table S1. Their magnesium numbers [Mg#=Mg/(Mg+Fe)] vary from 56 to 74 for the GC and are 42 and 52 for the two quartz eclogites. In terms of the major and trace element abundances of the GC, as MgO decreases, general increases in Fe and Ca and decreases in compatible elements (Cr, Ni) are observed (Supplementary Fig. S5). Thorium, U and LREE increase with enhanced MgO contents (Fig. 1a-c), which is inconsistent with magmatic differentiation or accumulation.

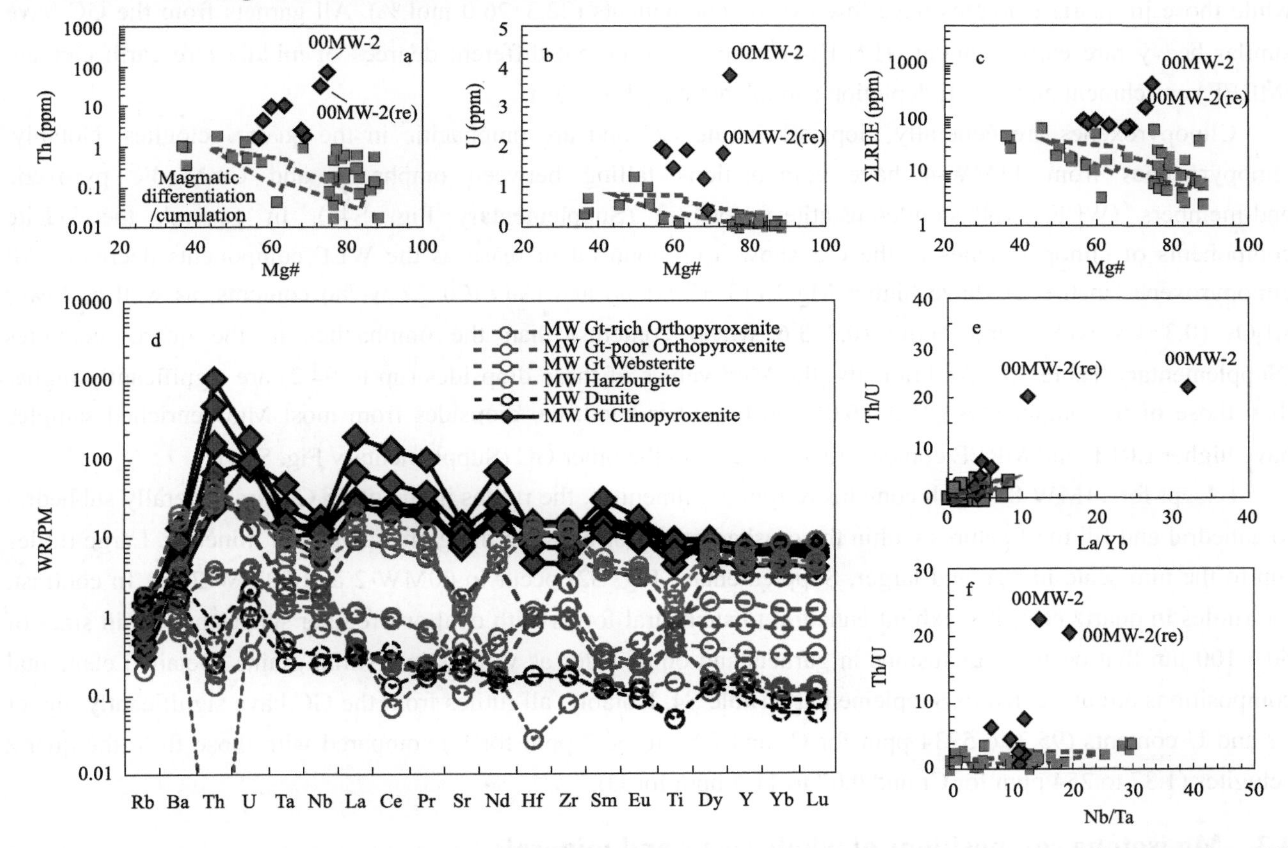

Fig. 1 Relationships between Th, U, ΣLREE contents and Mg# values of whole rocks for the Maowu garnet clinopyroxenites (GC) (a, b, c). Two GC with the highest Mg# [00MW-2 and 00MW-2(re)] have the most enriched Th, U, and ΣLREE. The primitive mantle-normalized spider diagram in panel (d) shows the trace element differences among GC, GO, garnet websterites, and harzburgites from the Maowu UHPM ultra-mafic massif. The trace element data for the MW GO, garnet websterites, and harzburgites are from Chen et al. (2017) and Jahn et al. (2003). Panels (e) and (f) show Th/U ratios vs. La/Yb and Nb/Ta ratios for the investigated GC. The grey squares were mantle pyroxenites with crystallization/accumulation petrogenesis in plots (a)-(c), (e) and (f) from Lee et al. (2006), Wang et al. (2001) and Xu (2002). Primitive mantle (PM) values are from McDonough and Sun (1995). (For interpretation of the colours in the figure(s), the reader is referred to the web version of this article.)

4.2 Petrography and major and trace elements of minerals

The Maowu GC consist of variable amounts of diopside (40–60 vol.%), garnet (30–55 vol.%), minor rutile (1–5 vol.%), orthopyroxene (<3 vol.%), amphibole (<3 vol.%), monazite (<1 vol.%), and apatite (<1 vol.%) (Supplementary Fig. S2). These GC have porphyroblastic textures, and the coarse-grained garnet porphyroblasts contains abundant polyphase solid inclusions (e.g., orthopyroxene, phlogopite, rutile, talc, apatite, kyanite, amphibolite, Supplementary Fig. S2a-c), which were also observed in coarse garnets from the Maowu GO and garnet websterites (Chen et al., 2013; Malaspina et al., 2006, 2009). Two quartz eclogite samples (99MW-2 and 99MW-3) contain garnet (40–50 vol.%), omphacite (40–50 vol.%), quartz (~5 vol.%), rutile (~2 vol.%), and muscovite (~1 vol.%), as well as some retrograde metamorphic symplektites of amphibole and plagioclase (Supplementary Fig. S2d).

The results of the representative analyses of minerals (i.e., garnet, clinopyroxene and rutile) are listed in Supplementary Tables S2, S3 and S4, respectively. The garnets in the GC have a large variation in pyrope components ranging from a low pyrope component (20.5 mol.%) to an enriched pyrope component (61.5 mol.%), while those in quartz eclogites have low pyrope components (22.3–26.0 mol.%). All garnets from the GC have similar heavy rare earth element (HREE)-enriched patterns but different degrees of middle rare earth element (MREE) enrichment and LREE depletion (Supplementary Fig. S3c).

Clinopyroxenes are generally diopsidic in the GC and are omphacitic in the quartz eclogites. Notably, clinopyroxenes from 11MW-4 have compositions falling between omphacite and Ca-Mg-Fe pyroxene end-members (WEF, wollastonite-enstatite-ferrosilite) (Supplementary Fig. S4a). In general, the jadeite components of clinopyroxenes in the GC show a pronounced increase as the WEF components decrease. All clinopyroxenes in the GC have higher MgO (13–17 wt.%) and CaO (20–27 wt.%) contents, as well as lower Al_2O_3 (0.3–4.5 wt.%) and Na_2O (0.3–3.6 wt.%) contents, than the omphacites in the quartz eclogites (Supplementary Table S3). Additionally, the Mg# values of these diopsides (up to 94.2) are significantly higher than those of the omphacites (81.2–86.0). On the other hand, the diopsides from most MgO-enriched samples have higher LREE and MREE contents than those from the other GC (Supplementary Fig. S4b).

Except for 11MW-8, which contains retrograde ilmenites, the rutiles in the other GC are generally subhedral to anhedral and fill the fractures within the crushed garnet or between the garnet and other minerals. Large rutiles (up to the mm-scale in size and larger, Supplementary Fig. S2) occur in 00MW-2 and 00MW-2(re). In contrast, the rutiles in quartz eclogites exhibit euhedral to subhedral forms with oval or irregular shapes and grain sizes of 50 - 100 μm that occur as inclusions in garnets and omphacites as well as interstitial grains. Detailed elemental compositions are presented in Supplementary Table S4. Notably, all rutiles from the GC have significantly higher Cr and U contents (95.3 to 5734 ppm for Cr and 24.5 to 66.2 ppm for U) compared with those from the quartz eclogites (1.37 to 254 ppm for Cr and 0.07 to 11.4 ppm for U).

4.3 Mg isotope compositions of whole rocks and minerals

The magnesium isotope compositions of the GC and the quartz eclogites, as well as reference samples, are reported in Table 1. The Maowu GC have significantly lighter Mg isotope compositions for both the whole rocks (−0.99 to −0.65‰) and major minerals (−1.78 to −1.31‰ for garnets and −0.68 to −0.31‰ for diopsides) relative to the widespread Dabie-Sulu eclogites as well as global basalts and mantle peridotites (Fig. 2a, Li et al., 2011; Teng et al., 2010; Wang et al., 2014b). In contrast, the whole-rock δ^{26}Mg values of the two quartz eclogites are −0.47‰ and −0.22‰, which are within the reported Mg isotope range of the Dabie-Sulu eclogites within error (−0.44‰ to −0.14‰, Li et al., 2011; Wang et al., 2014b).

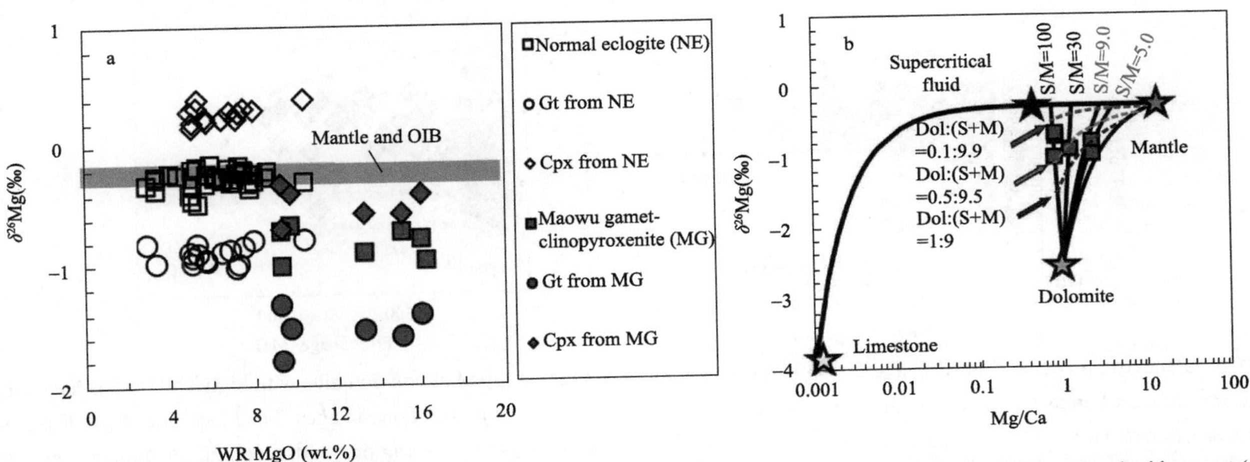

Fig. 2 (a) δ^{26}Mg vs. whole-rock (WR) MgO (wt.%) for widespread Dabie-Sulu eclogites (DE) and Maowu GC associated with garnet (Gt) and clinopyroxene (Cpx) separates from this and previous works (Li et al., 2011; Wang et al., 2014b). The grey bar represents the average Mg isotope composition of the mantle and ocean island basalt (OIB) (δ^{26}Mg=−0.25±0.07‰, 2SD, Teng et al. 2010). (b) Mixing modelling of the different sources in terms of Mg/Ca (molar ratio) and δ^{26}Mg. The mantle is assumed to have an initial δ^{26}Mg of −0.25‰, MgO=38 wt.%, and CaO=3.8 wt.%. The limestone and dolostone are assumed to have initial chemical compositions of $Mg_{0.001}Ca_{0.999}CO_3$ and $Mg_{0.5}Ca_{0.5}CO_3$, respectively, and to have initial δ^{26}Mg=−4.00‰ and δ^{26}Mg=−2.50‰ (Teng, 2017 and references therein), respectively. The composition of the slab-derived supercritical fluid is estimated to be similar to that of the experimental supercritical fluid (MgO=0.68 wt.% and CaO=1.93 wt.% with H_2O contents of 72.6 wt.%) at 6 Gpa and 800 °C (Kessel et al., 2005). During this calculation, the water content was deducted. The δ^{26}Mg value of such agent is assumed to be equal to metamorphic dehydration fluid and granitic melt, which have mantle-like δ^{26}Mg values of −0.25‰ (Liu et al., 2010; Wang et al., 2014b). Different weighted ratios (S/M) of mixing between the supercritical fluid (S) and the mantle peridotite (M) lead to different MgO and CaO contents, as well as Mg/Ca ratios, but constant mantle-like Mg isotope compositions of the metasomatized mantle. The incorporation of dolomite and limestone shifts the Mg isotope compositions of the metasomatized mantle along different curves; the former appears to be more significant, accounting for the light Mg isotope compositions of the metasomatized mantle. The dashed curves represent the mixing lines generated using different mass ratios of dolomite/(supercritical fluid + peridotite).

4.4 Mineral inclusions, zircon U-Pb dating and O isotope compositions

The U-Pb ages and O isotope compositions of the zircons from two eclogites and four Maowu GC are presented in Supplementary Table S5. The zircons from the GC are subhedral and anhedral and free of obvious magmatic cores with rounded to irregular shapes of various sizes (30–200 μm). Cathodoluminescence (CL) imaging with a scanning electron microscope (SEM) shows that the zircons from the GC have complex textures with interval multi-stage dark and bright CL domains and no oscillatory zoning. In contrast, zircons from the quartz eclogites have inherited magmatic cores with clear oscillatory zoning.

The investigated zircons from four GC have large variations in U (3 to 3486 ppm) and Th (1 to 2867 ppm), as well as Th/U ratios (0.007 to 1.185). Based on their Th/U ratios and mineral inclusions, these zircons could be classified into two types: type I zircon domains are characterized by high Th/U ratios (0.1 to 1.2, Fig. 3a) and different types of mineral inclusions, such as garnet, clinopyroxene, clinohumite, rutile, and multiphase solid inclusions, including carbonates (calcite + magnesite + dolomite, Fig. 4a, c-h) and silicates (clinopyroxene + amphibole + quartz), phosphates (apatite + monazite) and rutiles (Supplementary Fig. S7a-h'). Type I zircons define a discordant line with lower- and upper- intercept ages of 232.8±7.9 Ma and 457±55 Ma, respectively (n=106, mean square weighted deviation (MSWD)=1.2, Fig. 5c). Additionally, type I zircon domains display highly variable and heavier O isotope compositions ($\delta^{18}O_{VSMOW}$=4.3‰ to 12.2‰). In contrast, type II zircon domains are mainly characterized by low Th/U ratios (<0.1) (Fig. 3a) and carbonate-free high-pressure mineral inclusions (garnet, clinopyroxene, amphibole and rutile, Fig. 4b, i-k). They yield a relatively narrow range of concordant ages ranging from 215.1−265.1 Ma with a weighted mean age of 230.7±2.2 Ma (n=67, MSWD=6.5, Fig. 5c). Type II zircon domains also have variable δ^{18}O values (3.9‰−9.2‰).

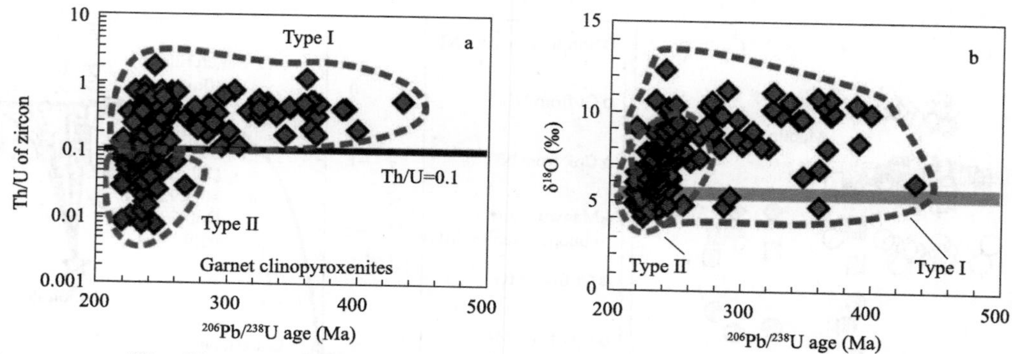

Fig. 3 Panel (a), a zircon ^{206}Pb/^{238}U age vs. Th/U ratio diagram showing that type I zircon domains with high Th/U ratios have large age variations, whereas type II zircon domains with low Th/U ratios yield concentrated Triassic ages. The black line in (a) displays a critical value of Th/U=0.1. Panel (b), a zircon ^{206}Pb/^{238}U age vs. O isotope diagram showing that both type I zircon domains and type II zircon domains have large δ^{18}O variations of 4.3‰–12.2‰ and 3.9‰–9.2‰, respectively. The grey area in (b) represents the δ^{18}O value of average mantle zircons (5.3±0.3‰, Valley, 2003).

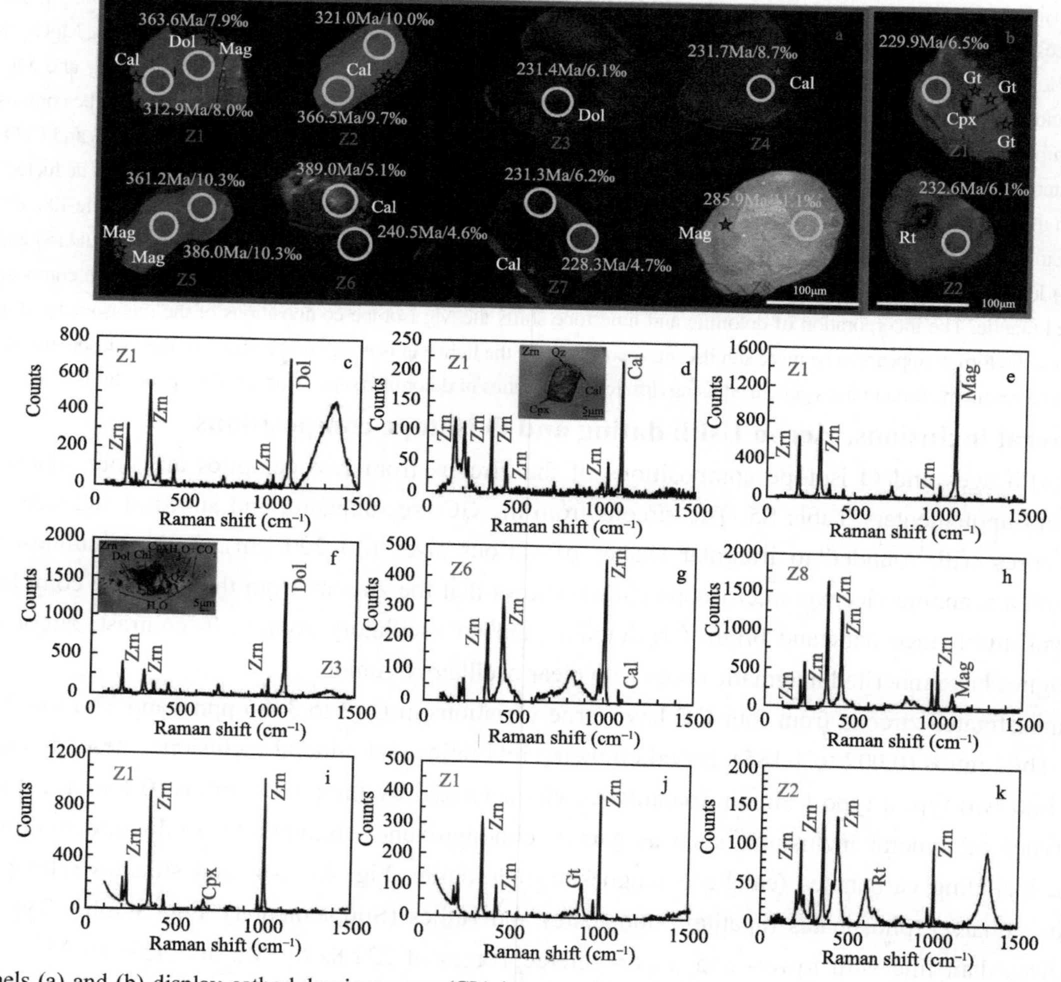

Fig. 4 Panels (a) and (b) display cathodoluminescence (CL) images of type I and type II zircons from Maowu GC, respectively, combined with in situ ^{206}Pb/^{238}U ages, O isotope compositions, and mineral inclusions. Panels (c) – (h) present Raman spectra of carbonate mineral inclusions in type I zircon domains, and Raman shifts of dolomite (1098 cm^{-1}), magnesite (1095 cm^{-1}), and calcite (1085 cm^{-1}) can be observed. In addition, the silicate mineral inclusions in type I zircon domains are presented in Supplementary Fig. S7. Photomicrographs of preserved multiphase solid inclusions in type I zircons are presented in Panels (d) and (f). The Raman spectra of the silicate mineral inclusions (garnet, clinopyroxene, and rutile) in type II zircon domains are displayed in panels (i)–(k). Cal-calcite, Dol-dolomite, Mag-magnesite, Gt-garnet, Cpx-clinopyroxene, Rt-rutile, Zrn-zircon, Qz-quartz, Chl-chlorite.

The in situ U-Pb dating results of inclusion-free zircon cores from the two quartz eclogites define a discordant line with upper- and lower-intercept ages of 737±27 Ma and 228±34 Ma [n=6, MSWD=0.81, Fig. 5a], respectively. These zircon cores have high Th/U ratios of 0.35–1.24 and slightly lighter O isotope compositions ($\delta^{18}O$ values=3.2‰ to 4.3‰) than the mantle zircons (5.3±0.3‰, Valley, 2003) (Supplementary Table S5). The zircon rims are characterized by low Th/U <0.1 and concordant ages ranging from 218–237 Ma (Fig. 5b), as well as UHPM mineral inclusions (garnet + clinopyroxene + rutile). The oxygen isotope compositions of these zircon domains exhibit a large range of −1.2 to 5.7‰ (Supplementary Table S5).

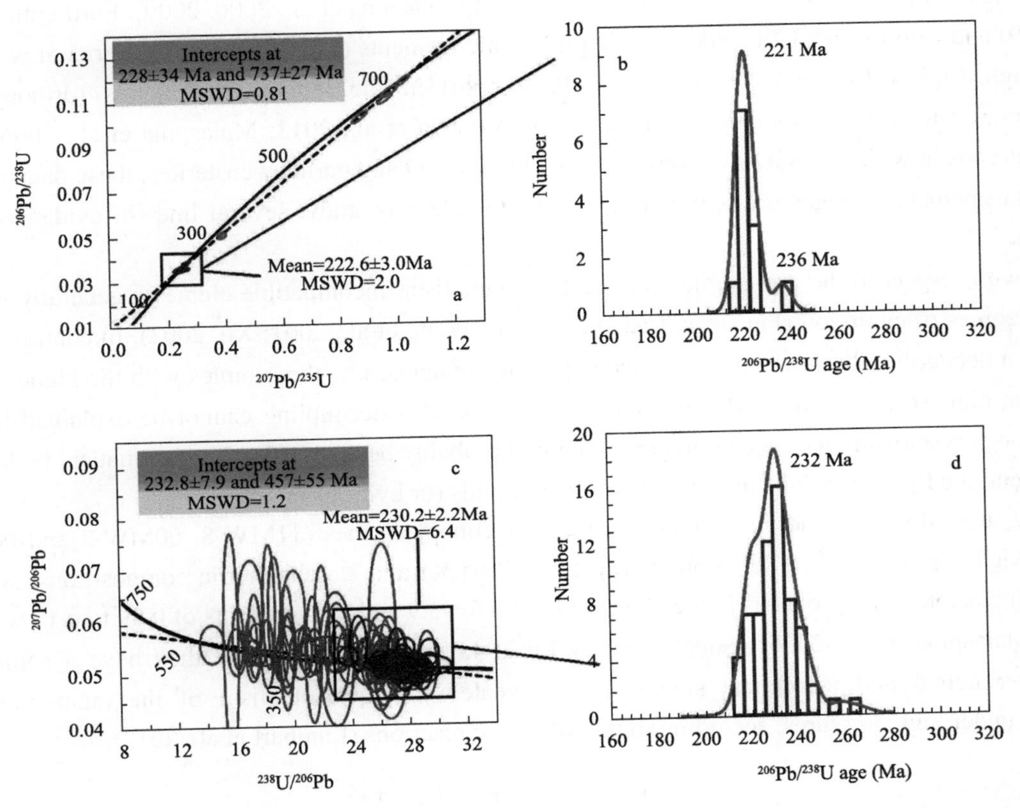

Fig. 5 (a) Concordia plots of two quartz eclogites (99MW-2, 3) within country paragneisses. The $^{206}Pb/^{238}U$-$^{207}Pb/^{235}U$ diagram gives discordia upper- and lower-intercept ages of 737±27 Ma and 228±34 Ma, respectively. Two age spikes for the zircon rims of 221 Ma and 236 Ma in the program (b) might reflect metamorphism in different stages during the Triassic subduction of continental crust. (c) Concordia plots of zircon U-Pb dating results for type I metasomatized zircon domains from four Maowu GC (11MW-1, 11MW-3, 11MW-8, and 00MW-2). The $^{207}Pb/^{206}Pb$-$^{238}U/^{206}Pb$ diagram shows discordia upper- and lower-intercept ages of 457±55 Ma and 232.8±7.9 Ma, respectively. Type II zircon domains give a range of concordant ages from 215.1–265.1 Ma, with a weighted mean age of 230.7±2.2 Ma (n=67, MSWD=6.5), which is similar to the average age of 232 Ma shown in histogram (d) and likely records the effects of the metamorphic fluid continuously released from the continental crust subducted during the Triassic.

5 Discussion

5.1 The nature of the protolith of the Maowu GC

Several processes could account for the formation of garnet pyroxenites within peridotites, including metasomatism and magmatism, i.e., mantle peridotites metasomatized by slab-derived silica-fluids or melts (Porreca et al., 2006) and cumulates/crystallizations derived from alkali basaltic magmas or mantle-derived melt mixed with subducted slab-derived materials (Lee et al., 2006; Wang et al., 2001; Xu, 2002).

The Maowu dunites were documented to be a relict mantle wedge of the NCB (Chen et al., 2017), while the GO within the massif were considered the products of metasomatism interactions between precursor mantle wedge dunites/harzburgites and subducted slab-derived hydrous melts or silica-fluids, which is supported by several petrological and geochemical observations (Chen et al., 2013; Malaspina et al., 2006, 2009). For example, the Maowu garnet orthopyroxenites have porphyroblastic textures, including coarse-grained garnet porphyroblasts and a medium-grained garnet, orthopyroxene, and clinopyroxene matrix, while relict olivines were preserved in both matrix and porphyroblast minerals (Chen et al., 2013; Malaspina et al., 2006, 2009). Furthermore, high Mg# values (>89) and significant LREE- and large ion lithophile elements (LILE)-enriched patterns of whole rocks, as well as high Mg# (>94) and Ni contents (>1000 ppm) in the individual minerals (orthopyroxene and clinopyroxene) reflect a metasomatized mantle protolith (Chen et al., 2013; Malaspina et al., 2006). Combined with the outcrops in which the GC are interlayered within the GO and garnet websterites, these data indicate that a mantle metasomatism petrogenesis is preferred for the GC. In this study, several lines of evidence support this speculation.

First, with respect to the pyroxenites formed by magmatism, incompatible elements generally increase with enhanced degrees of evolution (Fig. 1a-c, Lee et al., 2006; Wang et al., 2001; Xu, 2002). In contrast, the Maowu GC exhibit a decoupling between MgO and incompatible elements, i.e., the samples with the highest Mg# (up to 74) have the highest Th, U and LREE contents (Fig. 1a-c). This decoupling cannot be explained by magmatic processes (i.e., crystallization or accumulation) but could probably be the result of metasomatism by Mg-, Th-, U-, and LREE-enriched slab-derived fluids or supercritical liquids (or hydrous melts).

Second, the clinopyroxenes from high-Mg# garnet clinopyroxenes (11MW-8, 00MW-2 and 00MW-2(re)) have high Mg# (up to ~94.2), NiO contents (up to 0.12 wt.%) and stoichiometric compositions, similar to the matrix and inclusion clinopyroxenes in the GO (Mg# of 93.6−93.9 and NiO contents of 0.09 to 0.18 wt.%, Chen et al., 2013; Malaspina et al., 2006) (Supplementary Fig. S4a), which implies that they have a common mantle origin. Experimental and theoretical studies have revealed the metasomatism of the mantle peridotite by silica-fluids under high pressures, according to the following reactions (Lambart et al., 2012):

$$\text{Liquid} + \text{Ol} \rightarrow \text{Opx} + \text{Cpx} + \text{Gt}, \tag{1}$$

and

$$\text{Liquid} + \text{Opx} \rightarrow \text{Cpx} + \text{Gt}. \tag{2}$$

Therefore, the continuous incorporation of fluid along the channel in the mantle wedge could lead to more production of clinopyroxene and garnet at a location closer to the channel. All clinopyroxenes and garnets of the GC have composition variations ranging from those of ultra-mafic to mafic rocks (Supplementary Fig. S3a and S4a), suggesting that the GC seem to be produced by the metasomatism interactions of GO with slab-derived alkaline (Na- and Ca- rich) fluids or melts.

Third, the rutiles from high-Mg# GC have higher Cr contents (up to ~5800 ppm) compared to those from metamorphic cumulates, mafic rocks and felsic/pelitic rocks (see the review by Meinhold, 2010, Supplementary Fig. S5a), which indicates that their protoliths are enriched in compatible elements, likely from the mantle, which is also consistent with the mantle-like Mg-Al features in these rutiles (Supplementary Fig. S5b).

Fourth, similar polyphase inclusions, which represent trapped metasomatic fluids in the garnets from the GO (Chen et al., 2013; Malaspina et al. 2006, 2015), were also observed in the garnets and zircons of the GC (Fig. 4

and Supplementary Fig. S2b). Finally, the GC display similar trace element patterns but more enriched LREE, Th and U features relative to the metasomatized GO (Fig. 1d, Chen et al., 2013, 2017).

Therefore, we propose that the investigated GC share a common protolith with the GO, i.e., the harzburgites or the dunites in the Maowu massif, which represent a Paleoproterozoic fragment of the subcontinental lithospheric mantle beneath the southeastern margin of the NCB (Chen et al., 2017). These GC should have also undergone multi-stage metasomatic and metamorphic evolutions similar to the GO, including at least pre-Triassic metasomatism and Triassic metamorphism.

5.2 The origin and nature of the metasomatic melt/fluid

5.2.1 Constraints from whole-rock and mineral element compositions

If the entire Maowu massif was collectively affected by Triassic metamorphic fluid and only the GC and GO specifically suffered pre-Triassic metasomatism, the trace elemental differences between GC or GO and other ultra-mafic rocks (harzburgites and dunites) might provide a critical constraint on the geochemical features of the metasomatic agent during the pre-Triassic subduction of oceanic crust.

With respect to the Maowu garnet-poor orthopyroxenites, harzburgites and dunites, the GC display remarkably positive Th and U anomalies, negative high field strength element (HFSE, such as Nb, Ta, Zr and Hf) anomalies, and moderate to high enrichments of LREEs relative to HREEs (such as Dy, Y, Yb, and Lu) (Fig. 1d). These features require the metasomatic agent to be derived from subducted rutile-bearing eclogite-facies crust that suffered monazite/allanite decomposition because monazite/allanite decomposition can release abundant Th and LREE (Hermann et al., 2009) and residual rutiles and garnets can retain HFSEs and HREEs (Spandler et al., 2003). This conclusion is also supported by high Nb/Ta (up to 20) and high La/Yb (up to 30) ratios (Fig. 1e and f) of the GC due to the higher partitioning of Ta than Nb in rutile (Klemme et al., 2005) and the higher partitioning of Yb than La in garnet (Spandler et al., 2003). Several works have emphasized that Th behaves like a HFSE in subduction fluids, and it is more than ten orders of magnitude less soluble than U (Bailey and Ragnarsdottir, 1994; Brenan et al., 1995). However, when accompanied by enhanced temperatures and pressures as well as higher solute concentrations, the solubility of Th increases much more than that of U in fluids (Brenan et al., 1995; Hermann et al., 2006, 2009). For example, slab-derived melts/supercritical fluids could have significantly higher Th solubilities, such that high Th concentrations and Th/U ratios are often viewed as constituting a unique characteristic of them (Hermann et al., 2006, 2009; Kessel et al., 2005). The GC with higher La/Yb or Nb/Ta ratios have higher Th/U ratios, suggesting that the metasomatic agent was likely a supercritical silica-rich fluid or a hydrous melt (Fig. 1e and f). This U-enriched feature cannot be the result of Triassic UHP metamorphism, as documented by the obviously higher U contents in all rutile grains from the GC (24.5 to 66.2 ppm, Supplementary Table S4) relative to those of the rutiles from surrounding quartz eclogites and general meta-basic eclogites (<10 ppm, Meinhold, 2010). A zirconium-in-rutile thermometer yields equilibrium temperatures of 750 °C to 800 °C for the GC (6 GPa, Tomkins et al., 2007), which are similar to the proposed peak condition of the Maowu ultra-mafic massif (5.3–6.3 GPa and ~800 °C, Chen et al., 2013), thus indicating that this metasomatism took place under peak conditions. Furthermore, this *P-T* condition, which falls beyond the second critical end point of the basalt-water system (Kessel et al., 2005), is capable of generating the slab-derived supercritical fluid that was recorded in the trapped multiphase solid mineral inclusions within the garnet grains from the Maowu GO (Malaspina et al., 2006, 2015). Therefore, we propose that the pre-Triassic metasomatic agent for the Maowu massif was most likely a supercritical fluid with Th-, U-, LREE-enriched characteristics and high Th/U ratios derived from a rutile-bearing eclogitic-facies crust that suffered monazite/allanite-free decomposition. It should be noted that although different lithologies of the Maowu ultra-mafic massif displayed remarkable trace elemental differences, the similar fluid-mobile element contents (Rb and Ba) of all lithologies (Fig. 1d) might reflect overprints by Triassic metamorphic fluids.

5.2.2 Constraints from zircon textures and U-Th-Pb data

Because Zr is generally unsaturated in the geochemical reservoirs of primitive and depleted mantle, in which primary zircon rarely forms, the formation of zircon in ultra-mafic rocks is often associated with the addition of crustal material (Zheng et al., 2006). Due to the higher U-Th-Pb closure temperatures of zircons (>900 °C) relative to the peak temperature, the zircons from these Maowu GC could be a powerful tool for characterizing the timing of melt/fluid activities and the compositions of metasomatic and metamorphic agents (Hoskin and Schaltegger, 2003 and references therein).

The internal textures and Th/U ratio of zircons are generally used to distinguish their igneous and metamorphic petrogenesis. Igneous zircons exhibit obvious oscillatory zoning structures and high Th/U ratios, while metamorphic zircons usually lack oscillatory zoning and have remarkably low Th/U ratios (Hoskin and Schaltegger, 2003). The differences between their Th/U ratios are probably attributed to the different Th/U features of their original melts and fluids (Ayers and Peters, 2018; Hoskin and Schaltegger, 2003). For example, the inherited magmatic zircon cores with clear oscillatory zoning from two quartz eclogites have high Th/U ratios (0.35−1.24) and yield upper- and lower-intercept ages of 737±27 Ma and 228±34 Ma (Fig. 5a), which represent the Neoproterozoic protolith age and Triassic UHP metamorphic age, respectively, corresponding to the widespread formation of eclogites in the Dabie-Sulu UHPM zone (Zheng et al., 2004). The zircon rims, which are characterized by low Th/U ratios (<0.1), concordant ages ranging from 218−237 Ma (Fig. 5b), and representative eclogitic-facies metamorphic mineral inclusions (garnet + clinopyroxene + rutile, not shown here), reflect Triassic overgrowths associated with metamorphic fluids with low Th/U ratios.

With respect to the GC, type I zircons with an upper-intercept age of 457±55 Ma also have high Th/U ratios (0.1−1.2), indicating that their crystallization was related to the pre-Triassic metasomatism by a supercritical fluid or melt. Because the increased water contents in melts probably have negative effects on the formation of oscillatory zoning in magmatic zircons, zircons that crystallized from melts often have clear oscillatory zoning patterns in CL images, while those that crystallized from fluids lack oscillatory zoning (Hoskin and Schaltegger, 2003 and references therein). Based on the CL investigations of up to ~1000 zircon grains from the GC, the absence of oscillatory zoning in type I zircons further documents that the metasomatic agent was a silica-supercritical fluid, rather than a melt. The upper intercept age of 457±55 Ma defined by these metasomatized zircons is different from the Neoproterozoic protolith age of quartz eclogites and widespread eclogites in the Dabie-Sulu orogen, but it is similar to the upper-intercept age of 447+82/−79 Ma for zircons from one Maowu GC dated by the ID-TIMS method (Rowley et al., 1997). This age was interpreted to record the Paleo-Tethys oceanic crustal subduction during the Paleozoic prior to the Triassic continental collision, which is supported by the contemporary island arc basalts developed from the south margin of the NCB (Li et al., 2001). The lower intercept age of 232.8±7.9 Ma for type I zircons and weighted mean age of 230.7±2.2 Ma for the type II zircons characterized by low Th/U ratios (<0.1) document that these GC indeed underwent Triassic metamorphism during the subduction of the SCB continental crust. In agreement with previous works (Chen et al., 2013, 2017), our result suggests that the Maowu massif, as a relict mantle wedge of the NCB, underwent both Paleozoic metasomatism during the subduction of the Paleo-Tethys oceanic crust and UHP metamorphism during the continental collision between the NCB and the SCB.

5.3 Sedimentary carbonate transfers in the Paleozoic supercritical fluid

5.3.1 Mineral inclusion and oxygen isotope evidence

The Paleozoic metasomatized zircon domains (Type I) have unique mineral inclusions (multiphase solid inclusions) that are different from those within the Triassic metamorphic zircons (Type II) of the GC and the metamorphic zircons of the quartz eclogites. These multiphase solid inclusions show obviously small sizes (i.e., one to several μm in diameter) and polygonal or negative crystal shapes (Fig. 4). Similar multiphase solid inclusions were also observed in the garnets from the investigated GC (Supplementary Fig. S2b) and from the Maowu GO, which are considered to be the daughter minerals crystallized from trapped supercritical fluids (Malaspina et al., 2006, 2015). Here, we propose that a similar process generated these inclusions in the metasomatized zircons. Additionally, abundant carbonate minerals (including dolomite, magnesite and calcite) within the multiphase inclusions occur in the Paleozoic metasomatized zircons from all four GC, probably implying that the metasomatic supercritical fluid was enriched in carbonate components (Frezzotti et al., 2011). In contrast, the carbonate-free inclusions in the Triassic metamorphic zircons from both the Maowu GC and country quartz eclogites indicates that no obvious carbonate transfers occurred in Triassic metamorphic fluid.

In situ O isotope analyses provide further evidence for the above conclusion. Slightly lighter O isotope values ($\delta^{18}O_{VSMOW}$ values of 3.2‰ to 4.3‰, Fig. 6) of the zircon cores from the quartz eclogites relative to the mantle zircon value (5.3±0.3‰, Valley, 2003) are within the O isotope range for zircons from the Dabie-Sulu UHPM eclogites and orthogneisses (Fig. 6), which might represent the primary magmatic value affected by meteoric water (Zheng et al., 2004). The metamorphic zircon rims from quartz eclogites are also characterized by

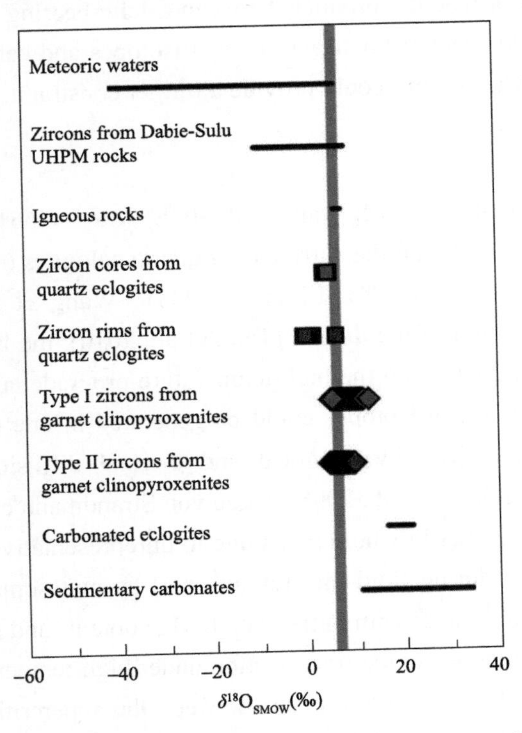

Fig. 6 Oxygen isotope compositions of type I and type II zircons from Maowu GC, as well as zircon cores and rims from country quartz eclogites, respectively (data are reported in Supplementary Table S5). For comparison, the $\delta^{18}O_{VSMOW}$ values of meteoric waters (Craig, 1961), zircons from the Dabie-Sulu UHPM rocks (Chen et al., 2011; Zheng et al., 2004), igneous rocks (Taylor, 1968), carbonated eclogites (Wang et al., 2014a) and sedimentary carbonates (Shields and Veizer, 2002) are shown. The grey bar represents the $\delta^{18}O$ values of mantle zircons.

lighter O isotopes ($\delta^{18}O$=–1.2‰ to 5.7‰, Fig. 6), indicating that zircon overgrowths were associated with the isotopically light-O metamorphic fluids during the Triassic (Chen et al., 2011; Zheng et al., 2004). In contrast, the Paleozoic metasomatized zircon domains of the Maowu GC have higher $\delta^{18}O$ values with large variations of 4.3‰ to 12.2‰ (Fig. 3b and Fig. 6). Considering their long-term retention in a mantle wedge lasting from the Paleozoic to Triassic, O isotope equilibrium between the zircons and other co-existing minerals should have been achieved under mantle *P-T* conditions (Valley, 2003). According to empirical calibrations, limited equilibrium fractionation factors of $\Delta^{18}O_{zircon-diopside}$ (~0.1‰ at 800 °C) and $\Delta^{18}O_{zircon-garnet}$ (0.06–0.34‰ at 800 °C) imply that the equilibrium O isotope compositions of zircons were slightly heavier than those of their whole rocks and could only vary within a narrow range, which is inconsistent with the large isotope variations in the metasomatized zircons observed from all individual samples (>8‰). Therefore, these metasomatized zircons were out of O isotope equilibrium, which might be attributed to the incorporation of Triassic metamorphic fluids during the rapid exhumation of the SCB. Given that the Triassic metamorphic fluids have light O isotope compositions, the heavy O isotope features of the Paleozoic metasomatized zircons should represent the initial isotope compositions of the metasomatic supercritical fluids, thereby supporting the idea of carbonate incorporations based on carbonate inclusions (Fig. 6). Metamorphic zircon domains also show high and variable $\delta^{18}O$ values (3.9‰ to 9.2‰, Fig. 3b), indicating that their formations were related to the partial crystallization and overgrowth of metasomatized zircons from the Triassic metamorphic fluids.

The heavy oxygen isotopes and carbonate inclusions of the metasomatized zircons confirm that the supercritical fluids contain sedimentary carbonate components. However, it remains difficult to assess the Mg-rich carbonate species origins (i.e., dolomite and magnesite) because they could be derived from dissolved sedimentary Mg-rich carbonates or reaction products between calcite-bearing fluid and mafic minerals (Frezzotti et al., 2011), both of which could account for heavy oxygen isotopes and carbonate inclusions. Thus, to further diagnose the carbonate species, Mg isotopes could provide a further constraint.

5.3.2　Mg isotope evidence

The GC exhibit lower whole-rock $\delta^{26}Mg$ values of –0.99‰ to –0.65‰ relative to those of the mantle (–0.25±0.07‰, 2SD, Teng et al., 2010) and the surrounding quartz eclogites (–0.22‰ to –0.47‰), as well as the Dabie UHPM eclogites (–0.44‰ to –0.14‰, Li et al., 2011; Wang et al., 2014b) (Fig. 2a). Given the homogeneous Mg isotope composition of the mantle (Teng et al., 2010), the limited isotope fractionation during magma differentiation (Liu et al., 2010), and the high-temperature prograde and retrograde metamorphism (Li et al., 2011; Wang et al., 2014b), light Mg isotopes could be generated in one of the following ways: 1) through kinetic Mg isotope fractionation driven by chemical and thermal diffusion during metasomatism at high temperatures (Huang et al., 2010; Richter et al., 2008; Pogge von Strandmann et al., 2015; Wu et al., 2018); 2) via enrichment of isotopically lighter garnet in whole rocks due to unrepresentative sampling (Li et al., 2011); 3) by leaching of isotopically heavy Mg during fluid infiltration; or 4) by metasomatism with the supercritical liquid containing sedimentary Mg-rich carbonates with initially high Mg contents and low $\delta^{26}Mg$ values.

Kinetic diffusion could cause large isotope fractionation under high-temperature conditions based on thermal and chemical gradients. Thermal diffusion might occur between the supercritical fluid and residual ultra-mafic rocks during Paleozoic metasomatism or subsequently between the GC and the garnet websterites/GO. In both cases, thermal diffusion should lead to the same fractionation trends for different isotope systems (Huang et al., 2010; Richter et al., 2008). Chemical diffusion between mantle rocks and the infiltrating supercritical fluid is dominant during metasomatism, and ^{24}Mg diffuses faster than ^{26}Mg during chemical diffusion. Simple solid-fluid diffusion would lead to light Mg isotopes diffusing into the supercritical fluid, thereby leaving isotopically heavier metasomatized mantle rocks (i.e., garnet pyroxenites), which contradicts our observations. However, interactions between channelled fluids after diffusion and lateral peridotites seem to be plausible because this mechanism has been proposed to explain the large isotope fractionations of Mg across a serpentinite-metapelite interface (Pogge von Strandmann et al., 2015) and Mg and Fe across a gabbro-granite boundary (Wu et al., 2018) at the centimetre to metre scales. The investigated GC occur as decimetre-scale veins and layers within the GO, which falls within the scale range of diffusion. However, chemical diffusion could also induce similar Mg and O isotope fractionation trends and lead to the infiltration of supercritical fluid characterized by light Mg and O isotopes (Pogge von Strandmann et al., 2015), which is inconsistent with our observations, i.e., low δ^{26}Mg and high δ^{18}O values of the GC and the negative correlation between them (Fig. 7). Consequently, kinetic diffusion cannot account for the Mg-O isotopes of the Maowu GC.

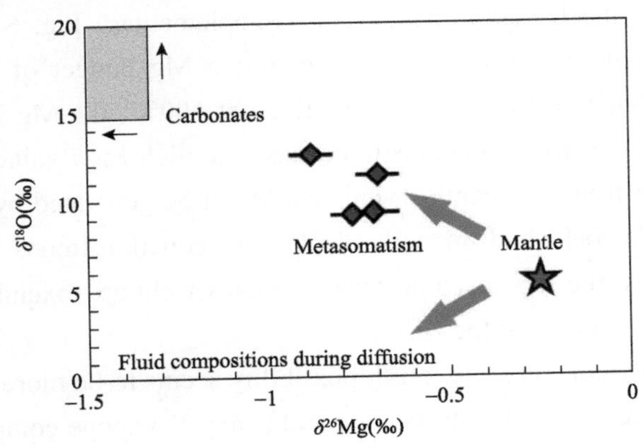

Fig. 7 Whole-rock $\delta^{18}O_i$ and δ^{26}Mg diagram for the Maowu GC, reflecting that these isotope features are most likely generated by metasomatism, rather than kinetic diffusion, in which the $\delta^{18}O_i$ values represent the whole-rock isotope compositions prior to Triassic metamorphism. Considering offsets by Triassic metamorphic fluids with low δ^{18}O, the $\delta^{18}O_i$ values are calculated according to the relation: $\Delta^{18}O_{(Zrc\text{-}WR)}=\delta^{18}O_{(Zrc)}-\delta^{18}O_{(WR)}\approx-0.0612(\text{wt.\% } SiO_2)+2.5$, as proposed by Valley et al. (2005). There is a basic assumption that the metasomatized zircons were in O isotope fractionation equilibrium with other co-existing minerals due to their long-term retention in the sub-arc mantle wedge from the Paleozoic to the Triassic, and the heaviest O isotopes of the zircons in the individual samples were used to estimate the $\delta^{18}O_i$. The Mg and O isotope compositions of the mantle and sedimentary carbonates are from Shields and Veizer (2002), Valley (2003), Teng et al. (2010) and Teng (2017). The variation trend of the infiltration fluid is according to Huang et al. (2010) and Pogge von Strandmann et al. (2015). Error bars are the 2SD uncertainties for δ^{26}Mg, and 1σ uncertainties for $\delta^{18}O_i$ are propagated from Cameca O isotope analyses.

In previous work, eclogites with mantle-like Mg isotope compositions displayed large equilibrium inter-mineral Mg fractionations between garnet and clinopyroxene, with garnet δ^{26}Mg values that were ~1.1‰ lower than those of their co-existing clinopyroxenes at ~600°C (Li et al., 2011). Therefore, unrepresentative sampling with the enrichment of isotopically light garnet can lead to a negative shift of whole-rock Mg isotope compositions. In our samples, we also observed considerable and equilibrium inter-mineral Mg isotope fractionations between garnet and clinopyroxene at eclogitic-facies metamorphic temperatures (600−680°C); however, both garnets and clinopyroxenes have consistently lower δ^{26}Mg values relative to the minerals in the eclogites, thus ruling out the possibility of unrepresentative sampling as the cause of the light Mg isotope compositions of the GC (Fig. 2a and Table 1).

Some isotopes could be largely fractionated during leaching by fluids, such as Mo, Fe, Zn (Debret et al., 2016; Greber et al., 2014; Pons et al., 2016). The dominant mechanism operating during isotope fractionation depends on the speciation of isotopes in the fluids and residuals, such as redox states and crystallographic sites. Magnesium has only a +2 valance in both fluid and rock systems; thus, a change in redox state cannot induce significant isotope fractionation. Although the exact nature of the Mg complexation in fluids at subduction zones is poorly constrained, Schauble (2011) calculated the equilibrium Mg isotope fractionation factors for a series of hexahydrate Mg^{2+} salts [i.e., $Mg(H_2O)_6SO_3$, $Mg(H_2O)_6HPO_3$, $Mg(H_2O)_6Zn_2Br_6$, $Mg(H_2O)_6SO_4 \cdot 5H_2O$ and $Mg(H_2O)_6SiF_6$] in fluids, all of which varied within a narrow range under high-temperature conditions (i.e., <~0.2‰ at 800°C). Furthermore, theoretical predications showed that the fractionation factor of clinopyroxene was similar to those of the Mg^{2+} salts (Schauble, 2011), while garnet was remarkably isotopically lighter than both of them (Huang et al., 2013). Thus, isotopically heavy Mg leaching by fluids seems to be capable of explaining the low-δ^{26}Mg features of the GC. Here, we present a simple Rayleigh distillation model to assess the evolution of the δ^{26}Mg values of the residual GC during Mg leaching (Supplementary Fig. S8a, b; see Supplementary Materials for detailed parameters and calculations). Among different models, $Mg(H_2O)_6SiF_6$ in fluids could always generate residuals with the lowest δ^{26}Mg values (Supplementary Fig. S8a, b). However, to match the observed Mg isotope features of the garnet clinopyroxenites, a Mg budget of >80 wt.% Mg in protoliths is required for a loss at 800°C (Supplementary Fig. S8a). Even at 600°C, the Mg loss remained high (>70 wt.%, Supplementary Fig. S8b). This scenario is unrealistic because the high Mg# values of the investigated diopsides and garnets are similar to those from the mantle, which could not be generated by obvious Mg loss. In addition, the δ^{26}Mg values of both whole rocks and minerals display no reduction trends with decreasing MgO contents (Supplementary Fig. S8c). Thus, the Mg leaching from the garnet clinopyroxenites to infiltration fluids cannot explain the observed low δ^{26}Mg features of the GC.

In view of the above three arguments, the fourth possibility seems to be more plausible because sedimentary carbonation reservoirs have extremely light Mg isotope and heavy O isotope compositions. The incorporation of calcite with a low MgO content cannot significantly influence the Mg isotope compositions of the mantle wedge and thus cannot explain our Mg isotope data (Fig. 2b). Instead, the Mg-rich carbonates should be taken into consideration. Due to the extremely low solubility of magnesite relative to dolomite (Pan et al., 2013), the most likely potential carbonate species is dolomite. Based on the observed Mg/Ca ratios and δ^{26}Mg values, a modelling calculation shows that approximately 1–10 wt.% of a dolomite component in the supercritical fluid is required to generate low δ^{26}Mg values in the range of –0.65‰ to –0.99‰ (Fig. 2b). Notably, although the melting of subducted carbonated eclogites could also account for the observed Mg-O isotopes and inclusions, the estimated peak P-T conditions (~800 °C and 5.3–6.3 GPa) for metasomatism are incapable of melting the carbonated eclogites (Dasgupta and Hirschmann, 2010). Therefore, we propose that the metasomatic supercritical fluid should contain dissolved Mg-rich sedimentary carbonates. The infiltration of such agents into the overlying mantle wedge will efficiently shift the Mg isotope compositions of metasomatized peridotites and give rise to local mantle heterogeneity in terms of Mg isotopes.

5.4 Implications for Mg isotope systems in the subduction zone

Our work presents critical insight into understanding Mg isotope systematics in a subduction zone (Fig. 8). Island arc volcanism is often induced by dehydration fluids from a subduction slab into the mantle wedge at relatively shallow depths (75–120 km). At this stage, the solubility of Ca-rich carbonate (i.e., calcite or aragonite) is significantly higher than those of dolomite and magnesite based on molecular dynamics (Pan et al., 2013); thus, dissolved carbonate components in dehydrated fluid have abundant Ca but little Mg, which has a limited influence on the Mg isotope composition of the mantle wedge (Li et al., 2017). Instead, at this stage, dehydration fluid from altered MORB and altered abyssal peridotites may preferentially contain heavy Mg isotopes (Wang et al., 2017; Li et al., 2017; Teng et al., 2016). Based on our study, however, it can be expected that, via enhanced slab subduction depths (up to ~160 km), Mg-rich carbonates (dolomite or magnesite) in the subducted slab become active and could be dissolved in a supercritical fluid. Thus, this agent could effectively influence the Mg isotope compositions of the mantle wedge or lithospheric mantle in the continental margin at a depth of ⩾ 160 km, which suggests that Mg isotopes can be used to trace carbonated metasomatism in the continental lithosphere, which is a potential vast store for carbon (Foley and Fischer, 2017; Kelemen and Manning, 2015). For example, the mantle xenoliths with extremely low δ^{26}Mg values (−1.23‰ to −0.73‰) and an age of ~400 Ma from the south margin of the North China craton showed that the sub-continental lithospheric mantle, which could be thicker than 180 km, has been significantly modified by supercritical fluids derived from a subducted carbonate-pelite-bearing oceanic crust (Wang et al., 2016b). When carbonated oceanic crust is subducted to deeper zones (probably the mantle transition zone) (Li et al., 2017; Thomson et al., 2016), the melting of carbonates and carbonated eclogites could produce carbonatitic melts, which could metasomatize the overlying mantle and imprint light Mg isotope features on intraplate basalts and peridotite xenoliths, as has been reported in previous works (Hu et al., 2016; Huang et al., 2015; Li et al., 2017; Teng et al., 2016; Wang et al., 2016a; Yang et al., 2012) (Fig. 8).

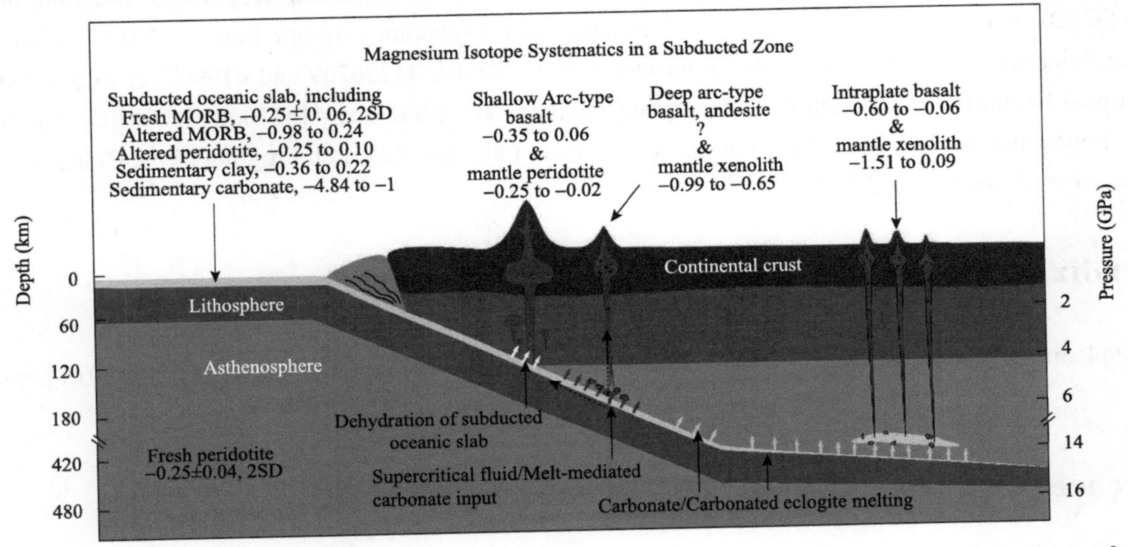

Fig. 8 Illustration showing magnesium isotope systematics in a subduction zone, in which there are three-stages of carbonate released from the subducted slab and input into the overlying mantle at different depths: (1) At shallow arc depth (75 km to 120 km), the dehydration of the subducted oceanic slab transfers a predominately Ca-rich carbonatitic component (calcite) into the mantle wedge; (2) At deep arc depth (approximately ~160 km), the dehydration or partial melting of the subducted oceanic slab could promote the incorporation of Mg-rich carbonates (dolomite, magnesite) into the mantle wedge; (3) At the mantle transition zone, the direct partial melting of the subducted oceanic slab with remaining Mg-rich carbonates produces carbonatitic melt that enters the overlying mantle. Three stages of carbonate liberation constitute the essential carbon recycling structure in a subduction zone. Combined with the variable Mg isotope compositions of a subducted oceanic slab, the Mg isotope features in the metasomatized mantle wedge and associated arc lava and intraplate basalts are distinct for each stage. Data sources: fresh MORB and fresh mantle peridotite (Teng et al., 2010), altered MORB, altered abyssal peridotite, shallow arc-type basalt, shallow sub-arc peridotite, sedimentary clay and carbonate (Teng et al., 2016 and references therein; Teng et al., 2017; Li et al., 2017) , deep sub-arc mantle xenolith data (this work), intraplate basalt (Huang et al., 2015; Li et al., 2017; Wang et al., 2016a; Yang et al., 2012), intraplate mantle xenolith (Hu et al., 2016; Wang et al., 2016b).

6 Conclusion

Here, we present a comprehensive study of the GC from the Maowu ultra-mafic massif in the Dabie orogen, Central China. Combined with the results of previous works, the major and trace elemental features and zircon U-Pb ages of these GC indicate that they were formed by the interaction between a mantle wedge and a supercritical fluid derived from the subducted Paleo-Tethys oceanic crust. Light Mg and heavy O isotopes, as well as carbonate inclusions in the metasomatized zircons, constitute strong evidence of the transfer of sedimentary Mg-rich carbonate in supercritical fluids at a deeper sub-arc depth (~160 km) during subduction. Our work clarifies the behaviours of subducted sedimentary Mg-rich carbonates and presents the first observation of an isotopically light-Mg mantle wedge. Based on these data, combined with those of previous works, we can construct the essential carbon recycling structure and integral Mg isotope systematics in this subduction zone.

Acknowledgements

We are grateful for discussions with Yi-Lin Xiao, Wei-Dong Sun, Sheng-Ao Liu and Yi-Xiang Chen. Constructive comments given by Editor Frederic Moynier, and two anonymous reviewers have improved the quality of the manuscript greatly. This work is supported by the National Key R&D Program of China (2016YFC0600404), the Chinese Ministry of Science and Technology (grant number 2015CB856102), the National Scientific Foundation of China (grant numbers 41730214, 41230209 and 41090372) to Shu-Guang Li, the National Scientific Foundation of China (grant numbers 41403001 and 41673006) to Ji Shen, the National Science Foundation (EAR-1340160) to Fang-Zhen Teng, and the Key Research Program of the Chinese Academy of Science (grant number XDPB11).

Appendix A. Supplementary material

Supplementary material related to this article can be found on-line at https://doi.org/10.1016/j.epsl.2018.09.011.

References

Ague, J.J., Nicolescu, S., 2014. Carbon dioxide released from subduction zones by fluid-mediated reactions. Nat. Geosci. 7, 355-360.

Ayers, J.C., Peters, T.J., 2018. Zircon/fluid trace element partition coefficients mea-sured by recrystallization of Mud Tank zircon at 1.5GPa and 800-1000°C. Geochim. Cosmochim. Acta 223, 60-74.

Bailey, E.H., Ragnarsdottir, K.V., 1994. Uranium and thorium solubilities in subduc-tion zone fluids. Earth Planet. Sci. Lett. 124, 119-129.

Brenan, J.M., Shaw, H.F., Ryerson, F.J., Phinney, D.L., 1995. Mineral-aqueous fluid partitioning of trace elements at 900°C and 2.0GPa: constraints on the trace element chemistry of mantle and deep crustal fluids. Geochim. Cosmochim. Acta 59, 3331-3350.

Chen, Y., 2011. Metamorphic Evolutions of Maowu Garnet Orthopyroxenites from the Dabie Orogen: Implication for the Matle Wedge Convection and Metasoma-tism. Postdoctoral Report. Institute of Geology and Geophysics of the Chinese Academy of Sciences.

Chen, Y., Su, B., Chu, Z., 2017. Modification of an ancient subcontinental lithospheric mantle by continental subduction: insight from the Maowu garnet peridotites in the Dabie UHP belt, eastern China. Lithos 278, 54-71.

Chen, Y., Ye, K., Guo, S., Wu, T.-F., Liu, J.-B., 2013. Multistage metamorphism of garnet orthopyroxenites from the Maowu mafic-ultramafic complex, Dabieshan UHP terrane, eastern China. Int. Geol. Rev. 55, 1239-1260.

Chen, Y.-X., Zheng, Y.-F., Chen, R.-X., Zhang, S.-B., Li, Q., Dai, M., Chen, L., 2011. Meta-morphic growth and recrystallization of zircons in extremely ^{18}O-depleted rocks during eclogite-facies metamorphism: evidence from U-Pb ages, trace elements, and O-Hf isotopes. Geochim. Cosmochim. Acta 75, 4877-4898.

Craig, H., 1961. Isotopic variations in meteoric waters. Science 133, 1702-1703.

Dasgupta, R., Hirschmann, M.M., 2010. The deep carbon cycle and melting in Earth's interior. Earth Planet. Sci. Lett. 298, 1-13.

Debret, B., Millet, M.-A., Pons, M.-L., Bouilhol, P., Inglis, E., Williams, H., 2016. Iso-topic evidence for iron mobility during subduction. Geology 44, 215-218.

Facq, S., Daniel, I., Montagnac, G., Cardon, H., Sverjensky, D.A., 2014. In situ Raman study and thermodynamic model of aqueous carbonate speciation in equilib-rium with aragonite under subduction zone conditions. Geochim. Cosmochim. Acta 132, 375-390.

Fan, Q.C., Liu, R., Ma, B., Zhao, D., Zhang, Q., 1996. The protolith and ultrahigh-pressure metamorphism of Maowu mafic-ultramafic rock block in Dabieshan Mountains. Acta Petrol. Sin. 12, 29-47 (in Chinese with English abstract).

Ferrando, S., Groppo, C., Frezzotti, M.L., Castelli, D., Proyer, A., 2017. Dissolving dolomite in a stable UHP mineral assemblage: evidence from Cal-Dol marbles of the Dora-Maira Massif (Italian Western Alps). Am. Mineral. 102, 42-60.

Foley, S.F., Fischer, T.P., 2017. An essential role for continental rifts and lithosphere in the deep carbon cycle. Nat. Geosci. 10, 897-902.

Frezzotti, M., Selverstone, J., Sharp, Z., Compagnoni, R., 2011. Carbonate dissolu-tion during subduction revealed by diamond-bearing rocks from the Alps. Nat. Geosci. 4, 703-706.

Gorman, P.J., Kerrick, D., Connolly, J., 2006. Modeling open system metamorphic de-carbonation of subducting slabs. Geochem. Geophys. Geosyst. 7.

Greber, N.D., Pettke, T., Nägler, T.F., 2014. Magmatic-hydrothermal molybdenum iso-tope fractionation and its relevance to the igneous crustal signature. Lithos 190, 104-110.

Hermann, J., Rubatto, D., 2009. Accessory phase control on the trace element signa-ture of sediment melts in subduction zones. Chem. Geol. 265, 512-526.

Hermann, J., Spandler, C., Hack, A., Korsakov, A.V., 2006. Aqueous fluids and hydrous melts in high-pressure and ultra-high pressure rocks: implications for element transfer in subduction zones. Lithos 92, 399-417.

Hoskin, P.W., Schaltegger, U., 2003. The composition of zircon and igneous and meta-morphic petrogenesis. Rev. Mineral. Geochem. 53, 27-62.

Hu, Y., Teng, F.-Z., Zhang, H.-F., Xiao, Y., Su, B.-X., 2016. Metasomatism-induced mantle magnesium isotopic heterogeneity: evidence from pyroxenites. Geochim. Cosmochim. Acta 185, 88-111.

Huang, F., Chakraborty, P., Lundstrom, C.C., Holmden, C., Glessner, J.J.G., Kieffer, S.W., Lesher, C.E., 2010. Isotope fractionation in silicate melts by thermal diffusion. Nature 464, 396-400.

Huang, F., Chen, L., Wu, Z., Wang, W., 2013. First-principles calculations of equi-librium Mg isotope fractionations between garnet, clinopyroxene, orthopyroxene, and olivine: implications for Mg isotope thermometry. Earth Planet. Sci. Lett. 367, 61-70.

Huang, J., Li, S.-G., Xiao, Y., Ke, S., Li, W.-Y., Tian, Y., 2015. Origin of low δ^{26}Mg Cenozoic basalts from South China Block and their geodynamic implications. Geochim. Cosmochim. Acta 164, 298-317.

Huang, S., Farkaš, J., Jacobsen, S.B., 2011. Stable calcium isotopic compositions of Hawaiian shield lavas: evidence for recycling of ancient marine carbonates into the mantle. Geochim. Cosmochim. Acta 75, 4987-4997.

Jahn, B., Fan, Q., Yang, J.J., Henin, O., 2003. Petrogenesis of the Maowu pyroxenite-eclogite body from the UHP metamorphic terrane of Dabieshan: chemical and isotopic constraints. Lithos 70, 243-267.

Johnston, F.K., Turchyn, A.V., Edmonds, M., 2011. Decarbonation efficiency in sub-duction zones: implications for warm Cretaceous climates. Earth Planet. Sci. Lett. 303, 143-152.

Kelemen, P.B., Manning, C.E., 2015. Reevaluating carbon fluxes in subduction zones, what goes down, mostly comes up. Proc. Natl. Acad. Sci. USA112, E3997-E4006.

Kessel, R., Schmidt, M.W., Ulmer, P., Pettke, T., 2005. Trace element signature of subduction-zone fluids, melts and supercritical liquids at 120-180km depth. Nature 437, 724-727.

Klemme, S., Prowatke, S., Hametner, K., G ü nther, D., 2005. Partitioning of trace elements between rutile and silicate melts: implications for subduction zones. Geochim. Cosmochim. Acta 69, 2361-2371.

Lambart, S., Laporte, D., Provost, A., Schiano, P., 2012. Fate of pyroxenite-derived melts in the peridotitic mantle: thermodynamic and experimental constraints. J. Petrol. 53, 451-476.

Lee, C.-T.A., Cheng, X., Horodyskyj, U., 2006. The development and refinement of continental arcs by primary basaltic magmatism, garnet pyroxenite accumula-tion, basaltic recharge and delamination: insights from the Sierra Nevada, Cali-fornia. Contrib. Mineral. Petrol. 151, 222-242.

Li, S.G., Huang, F., Nie, Y.H., Han, W.L., Long, G., Li, H.M., Zhang, S.Q., Zhang, Z.H., 2001. Geochemical and geochronological constraints on the suture location be-tween the North and South China blocks in the Dabie Orogen, Central China. Phys. Chem. Earth, Part A, Solid Earth Geod. 26, 655-672.

Li, S.G., Xiao, Y.L., Liou, D.L., Chen, Y.Z., Ge, N.J., Zhang, Z.Q., Sun, S.S., Cong, B.L., Zhang, R.Y., Hart, S.R., Wang, S.S., 1993.

Collision of the North China and Yangtse Blocks and formation of coesite-bearing eclogites: timing and processes. Chem. Geol. 109, 89-111.

Li, S.-G., Yang, W., Ke, S., Meng, X., Tian, H., Xu, L., He, Y., Huang, J., Wang, X.-C., Xia, Q., 2017. Deep carbon cycles constrained by a large-scale mantle Mg isotope anomaly in eastern China. Nat. Sci. Rev. 4, 111-120.

Li, W.-Y., Teng, F.-Z., Xiao, Y., Huang, J., 2011. High-temperature inter-mineral mag-nesium isotope fractionation in eclogite from the Dabie orogen, China. Earth Planet. Sci. Lett. 304, 224-230.

Liu, S.-A., Teng, F.-Z., He, Y., Ke, S., Li, S., 2010. Investigation of magnesium isotope fractionation during granite differentiation: implication for Mg isotopic compo-sition of the continental crust. Earth Planet. Sci. Lett. 297, 646-654.

Luth, R.W., 2001. Experimental determination of the reaction aragonite + magnesite =dolomite at 5 to 9GPa. Contrib. Mineral. Petrol. 141, 222-232.

Malaspina, N., Alvaro, M., Campione, M., Wilhelm, H., Nestola, F., 2015. Dynamics of mineral crystallization from precipitated slab-derived fluid phase: first in situ synchrotron X-ray measurements. Contrib. Mineral. Petrol. 169, 1-12.

Malaspina, N., Hermann, J., Scambelluri, M., 2009. Fluid/mineral interaction in UHP garnet peridotite. Lithos 107, 38-52.

Malaspina, N., Hermann, J., Scambelluri, M., Compagnoni, R., 2006. Polyphase in-clusions in garnet-orthopyroxenite (Dabie Shan, China) as monitors for metaso-matism and fluid-related trace element transfer in subduction zone peridotite. Earth Planet. Sci. Lett. 249, 173-187.

McDonough, W.F., Sun, S.S., 1995. The composition of the Earth. Chem. Geol. 120, 223-253.

Meinhold, G., 2010. Rutile and its applications in earth sciences. Earth-Sci. Rev. 102, 1-28.

Pan, D., Spanu, L., Harrison, B., Sverjensky, D.A., Galli, G., 2013. Dielectric proper-ties of water under extreme conditions and transport of carbonates in the deep Earth. Proc. Natl. Acad. Sci. USA110, 6646-6650.

Pogge von Strandmann, P.A.E., Dohmen, R., Marschall, H.R., Schumacher, J.C., Elliott, T., 2015. Extreme magnesium isotope fractionation at outcrop scale records the mechanism and rate at which reaction fronts advance. J. Petrol. 56, 33-58.

Pogge von Strandmann, P.A.E., Elliott, T., Marschall, H.R., Coath, C., Lai, Y.-J., Jeffcoate, A.B., Ionov, D.A., 2011. Variations of Li and Mg isotope ratios in bulk chondrites and mantle xenoliths. Geochim. Cosmochim. Acta 75, 5247-5268.

Pons, M.-L., Debret, B., Bouilhol, P., Delacour, A., Williams, H., 2016. Zinc isotope evidence for sulfate-rich fluid transfer across subduction zones. Nat. Commun. 7, 13794.

Porreca, C., Selverstone, J., Samuels, K., 2006. Pyroxenite xenoliths from the Rio Puerco volcanic field, New Mexico: melt metasomatism at the margin of the Rio Grande rift. Geosphere 2, 333-351.

Richter, F.M., Watson, E.B., Mendybaev, R.A., Teng, F.-Z., Janney, P.E., 2008. Magne-sium isotope fractionation in silicate melts by chemical and thermal diffusion. Geochim. Cosmochim. Acta 72, 206-220.

Rowley, D.B., Xue, F., Tucker, R.D., Peng, Z.X., Baker, J., Davis, A., 1997. Ages of ul-trahigh pressure metamorphism and protolith orthogneisses from the eastern Dabie Shan: U/Pb zircon geochronology. Earth Planet. Sci. Lett. 151, 191-203.

Schauble, E.A., 2011. First-principles estimates of equilibrium magnesium isotope fractionation in silicate, oxide, carbonate and hexaaquamagnesium(2+) crystals. Geochim. Cosmochim. Acta 75, 844-869.

Shields, G., Veizer, J., 2002. Precambrian marine carbonate isotope database: version 1.1. Geochem. Geophys. Geosyst.3.

Spandler, C., Hermann, J., Arculus, R., Mavrogenes, J., 2003. Redistribution of trace elements during prograde metamorphism from lawsonite blueschist to eclog-ite facies: implications for deep subduction-zone processes. Contrib. Mineral. Petrol. 146, 205-222.

Taylor Jr, H.P., 1968. The oxygen isotope geochemistry of igneous rocks. Contrib. Mineral. Petrol. 19, 1-71.

Teng, F.-Z., 2017. Magnesium isotope geochemistry. Rev. Mineral. Geochem. 82, 219-287.

Teng, F.-Z., Hu, Y., Chauvel, C., 2016. Magnesium isotope geochemistry in arc volcan-ism. Proc. Natl. Acad. Sci. USA 113, 7082-7087.

Teng, F.-Z., Li, W.-Y., Ke, S., Marty, B., Dauphas, N., Huang, S., Wu, F.-Y., Pourmand, A., 2010. Magnesium isotopic composition of the Earth and chondrites. Geochim. Cosmochim. Acta 74, 4150-4166.

Thomson, A.R., Walter, M.J., Kohn, S.C., Brooker, R.A., 2016. Slab melting as a barrier to deep carbon subduction. Nature 529, 76.

Tomkins, H.S., Powell, R., Ellis, D.J., 2007. The pressure dependence of the zirconium-in-rutile thermometer. J. Metamorph. Geol. 25, 703-713.

Tumiati, S., Tiraboschi, C., Sverjensky, D., Pettke, T., Recchia, S., Ulmer, P., Miozzi, F., Poli, S., 2017. Silicate dissolution boosts the CO_2concentrations in subduction fluids. Nat. Commun. 8, 616.

Valley, J.W., 2003. Oxygen isotopes in zircon. Rev. Mineral. Geochem. 53, 343-385.

Valley, J., Lackey, J., Cavosie, A., Clechenko, C., Spicuzza, M., Basei, M., Bindeman, I., Ferreira, V., Sial, A., King, E., 2005. 4.4

billion years of crustal maturation: oxygen isotope ratios of magmatic zircon. Contrib. Mineral. Petrol. 150, 561-580.

Wang, S.-J., Teng, F.-Z., Li, S.-G., 2014a. Tracing carbonate-silicate interaction during subduction using magnesium and oxygen isotopes. Nat. Commun. 5, 5328.

Wang, S.-J., Teng, F.-Z., Li, S.-G., Hong, J.-A., 2014b. Magnesium isotopic systematics of mafic rocks during continental subduction. Geochim. Cosmochim. Acta 143, 34-48.

Wang, S.-J., Teng, F.-Z., Li, S.-G., Zhang, L.-F., Du, J.-X., He, Y.-S., Niu, Y., 2017. Trac-ing subduction zone fluid-rock interactions using trace element and Mg-Sr-Nd isotopes. Lithos 290, 94-103.

Wang, S.-J., Teng, F.-Z., Scott, J.M., 2016a. Tracing the origin of continental HIMU-like intraplate volcanism using magnesium isotope systematics. Geochim. Cos-mochim. Acta 185, 78-87.

Wang, Z.z., Liu, S.-A., Ke, S., Liu, Y.-C., Li, S.-G., 2016b. Magnesium isotopic hetero-geneity across the cratonic lithosphere in eastern China and its origins. Earth Planet. Sci. Lett. 451, 77-88.

Wang, Z., Sun, S., Hou, Q., Li, J., 2001. Effect of melt-rock interaction on geochem-istry in the Kudi ophiolite (western Kunlun Mountains, northwestern China): implication for ophiolite origin. Earth Planet. Sci. Lett. 191, 33-48.

Wu, H., He, Y., Teng, F.-Z., Ke, S., Hou, Z., Li, S., 2018. Diffusion driven magne-sium and iron isotope fractionation at a gabbro-granite boundary. Geochim. Cosmochim. Acta 222, 671-684.

Xu, Y., 2002. Evidence for crustal components in the mantle and constraints on crustal recycling mechanisms: pyroxenite xenoliths from Hannuoba, North China. Chem. Geol. 182, 301-322.

Yang, W., Teng, F.-Z., Zhang, H.-F., Li, S.-G., 2012. Magnesium isotopic systematics of continental basalts from the North China craton: implications for tracing sub-ducted carbonate in the mantle. Chem. Geol. 328, 185-194.

Ye, K., Cong, B., Ye, D., 2000. The possible subduction of continental material to depths greater than 200km. Nature 407, 734-736.

Zheng, Y.F., Wu, Y.-B., Chen, F.-K., Gong, B., Li, L., Zhao, Z.-F., 2004. Zircon U-Pb and oxygen isotope evidence for a large-scale ^{18}O depletion event in igneous rocks during the Neoproterozoic. Geochim. Cosmochim. Acta 68, 4145-4165.

Zheng, Y.F., Zhao, Z.F., Wu, Y.B., Zhang, S.B., Liu, X., Wu, F.Y., 2006. Zircon U-Pb age, Hf and O isotope constraints on protolith origin of ultrahigh-pressure eclogite and gneiss in the Dabie orogen. Chem. Geol. 231, 135-158.

Low δ^{26}Mg volcanic rocks of Tengchong in Southwestern China: A deep carbon cycle induced by supercritical liquids*

Heng-Ci Tian[1,2], Wei Yang[1,2], Shu-Guang Li[3], Shan Ke[3] and Xian-Zhe Duan[4]

1. Key Laboratory of Earth and Planetary Physics, Institute of Geology and Geophysics, Chinese Academy of Sciences, Beijing 100029, China
2. Institutions of Earth Science, Chinese Academy of Sciences, Beijing 100029, China
3. State Key Laboratory of Geological Processes and Mineral Resources, China University of Geosciences, Beijing 100083, China
4. The School of Nuclear Resource Engineering, University of South China, Hengyang 421001, China

亮点介绍：云南腾冲玄武岩的 Sr-Nd-Pb-Mg 同位素研究表明这些玄武岩具有轻 Mg 同位素组成，并具有高的 Th/U 比值，显著高于正常地幔橄榄岩和碳酸盐化橄榄岩部分熔融的产物(通常<4.5)。该研究揭示了当俯冲板片俯冲至 120~400km 深度处时，其释放的超临界流体能够溶解以白云石为主的碳酸盐，并交代地幔，进而导致地幔熔融。这一碳循环周期大约为 48~53Ma，长于岛弧地幔楔深度碳循环的周期（5~10Ma），但短于中国东部地幔过渡带碳循环的周期（>60Ma）。

Abstract Oceanic subduction zones are important channels for carbon exchange between the Earth's crust and mantle. However, the nature of carbon cycles at depths from 120 to 410 km in the subduction zone remains unknown. To decipher this issue, high-precision stable Mg isotopes of arc-like volcanic rocks from Tengchong, Southwestern China, have been investigated. The Tengchong volcanic rocks comprise basalts and andesites, with MgO content varying from 2.41 to 8.48 wt.%. Both the basalts and andesites exhibit homogeneous and light Mg isotopic compositions with δ^{26}Mg ranging from −0.51 to −0.45‰ and −0.49 to −0.33‰, respectively. Their δ^{26}Mg values are lower than the average mantle (δ^{26}Mg=−0.25 ± 0.07‰) and island arc lavas (δ^{26}Mg = −0.35 to +0.06‰), but similar to the <110 Ma intra-continental basalts from eastern China (δ^{26}Mg = −0.60 to 0.30‰). This light Mg isotopic composition could not originate from the accumulation of ilmenite in their mantle source because both the Nb/Ta and δ^{26}Mg values of the basalts are invariant with TiO_2 content. The recycling of carbonated eclogites is also unlikely because of the lack of any correlation between δ^{26}Mg and either $(Gd/Yb)_N$ or Fe/Mn ratios in the Tengchong basalts. Alternatively, the most probable explanation for the light Mg isotopic composition is the recycling of sedimentary carbonates in the mantle source. This is supported by the high Na_2O+K_2O/TiO_2 ratios (3.3−4.1), low Ti/Ti* and Hf/Hf* values in the basalts, which are consistent with the partial melting trend of carbonated peridotite. Additionally, the high Ba/Th, low Rb/Cs and enriched Sr-Nd isotopes (EMII-like endmember) of the Tengchong basalts indicate the presence of Indian sediments in their mantle source. Furthermore, the extremely high Th/U (6.5−8.3)

* 本文发表在：Geochimica et Cosmochimica Acta, 2018, 240: 191-219

ratios in basalts suggest the higher mobility of Th than U, which is a unique characteristic of slab-derived supercritical liquids in subduction zones with pressures greater than 6 GPa. Based on the quantitative modeling of Mg-Sr-Nd-Pb isotopes and trace elements, the mantle source of the Tengchong basalts lies along mixing lines of the DMM with 1% supercritical liquids and 15–19% recycled mixture containing 66–72% dolomite, 22–26% calcite and 2–12% Indian sediments. Given that the recycled mixture is mainly dolomite (66–72%), we speculated that the slab-derived supercritical liquids can dissolve dolomite and then metasomatized the overlying mantle to form carbonated peridotite. Partial melting of this mantle source should be located at a depth of ~120 to 300 km under which supercritical liquids can occur, which is supported by seismic tomographic observations.

The Tengchong andesites display many geochemical features similar to the basalts, such as enrichments in LILEs (Large Ion Lithophile Elements), LREEs (Light Rare Earth Elements) and Sr-Nd-Pb isotopes, as well as low δ^{26}Mg values, suggesting a petrogenetic link between these two rock types. Our detailed study suggests that the andesites evolved from the Tengchong basalts via assimilation and fractional crystallization (AFC) processes. This interpretation is also supported by the geophysical tomography, which reveals a low-velocity anomalous zone in the continental crust.

This study reveals a new carbon cycle in which Mg-rich carbonate–dolomite–can be dissolved by supercritical liquids and subducted into a deep mantle wedge to depths of 120–300 km in the oceanic subduction zone. This deep metasomatic mantle wedge mixed with the upwelling mantle beneath the Tengchong volcano and partially melted to form the low δ^{26}Mg volcanic rocks.

1 Introduction

The deep carbon cycle is an important yet poorly understood process in Earth science. It describes the ingassing of carbon to the mantle through subduction, followed by outgassing to the atmosphere via magmatic and volcanic processes (Dasgupta, 2013; Hazen and Schiffries, 2013). Some studies estimate that significant quantities (about half) of the subducted carbon can be recycled into the convective mantle (Wallace, 2005; Dasgupta and Hirschmann, 2010; Johnston et al., 2011; Hofmann, 2017; Li et al., 2017). However, other studies suggest that the high dissolution proportion of calcite in slab-derived fluids can only lead to minor amounts of carbon recycled into the convective mantle, the rest being stored in the lithospheric mantle and crust (Kelemen and Manning, 2015). At such depths and elevated pressures, these fluids will become 'supercritical liquids' where the fluid and melt converge along a miscibility gap. This gap will eventually disappear with increasing temperature and pressure and then intersect the solidus to define an end point (Shen and Keppler, 1997; Stalder et al., 2001; Schmidt et al., 2004). Any fluid or melt beyond this point can be termed 'supercritical liquid' and it can be generated over 6 GPa based on experimental studies (e.g., Kessel et al., 2005). Understanding carbon transport in the mantle as well as carbonate dissolution in slab-derived or supercritical fluids is therefore paramount for understanding deep carbon cycles.

Due to the large Mg isotopic differences between sedimentary carbonates (δ^{26}Mg = –5.60 to –0.66‰; Beinlich et al., 2014; Saenger and Wang, 2014) and the mantle (δ^{26}Mg =–0.25±0.07 ‰; Teng et al., 2010a), Mg isotopes have proven to be a powerful tracer of sedimentary carbonate recycling in the mantle (e.g., Yang et al., 2012; Huang et al., 2015a; Tian et al., 2016, 2017; Hu et al., 2017; Li et al., 2017; Liu et al., 2017; Wang et al., 2017). Two deep carbon cycles were recently discovered by Li et al. (2017): cycle 1 (<120 km), during slab dehydration a fluid of partially dissolved Ca-rich carbonates was injected into a mantle wedge, and then released by arc volcanism without low δ^{26}Mg signature; and cycle 2 (>410 km), un-dissolved Mg-rich carbonates during slab dehydration were carried into the mantle transition zone by subducted slabs, melted and metasomatized the upper mantle, and were released by intraplate volcanism with low δ^{26}Mg feature (Li et al., 2017). However, whether a carbon cycle occurs at depths between 120 and 410 km is still unknown. An experimental study suggested that supercritical liquids would occur at pressures around 6 GPa (at depths of 180 km), probably with higher solubility than fluids at shallower depths (e.g., 4 GPa) (Kessel et al., 2005). Thus, supercritical liquids at such depths could potentially dissolve recycled Mg-rich carbonates, inducing volcanism with low δ^{26}Mg feature.

The Cenozoic basalts from the Tengchong arc-like volcanic field (TVF) in Southwestern China are ideal samples to investigate a deep carbon cycle. It is located near the border between China and Burma, and directly above the subducted Indian Oceanic plate at subduction depths ~150–400 km based on the seismic tomography (Figs. 1a and b). Many geochemical studies suggest that the Tengchong volcanic rocks with EM-II Sr-Nd isotopic features formed by the partial melting of an enriched mantle source metasomatized by recycled continental crustal materials (e.g., Zhu et al., 1983; Turner et al., 1996; Chen et al., 2002; Chung et al., 2005; Wang et al., 2006; Mo et al., 2007; Zhao and Fan, 2010; Li and Liu, 2012; Zhang et al., 2012; Zhou et al., 2012; Zou et al., 2014). The Mg isotopic compositions of the Tengchong volcanic rocks could therefore provide constraints on carbon cycles at depths from 150 and 400 km if supercritical liquids can dissolve Mg-rich carbonates.

This study presents major and trace elements as well as the Sr-Nd-Pb-Mg isotopic compositions of Pliocene basalts and Holocene andesites from the Tengchong volcanic field. The δ^{26}Mg values of the Tengchong volcanic rocks vary from −0.51 to −0.33‰, similar to those of the low δ^{26}Mg intra-continental basalts from eastern China (δ^{26}Mg = −0.60 to −0.30‰; Yang et al., 2012; Huang et al., 2015a; Tian et al., 2016; Li et al., 2017; Su et al., 2017), but lower than that of the island arc basalts (δ^{26}Mg = −0.35 to +0.06‰; Teng et al., 2016; Li et al., 2017). Combined with their major and trace elemental abundances, this light Mg isotopic composition probably resulted from the recycling of sedimentary carbonate (calcite and dolomite) that could have been dissolved by supercritical liquids. Thus, this study provides a specific example of carbon cycles into the deep mantle wedge relating to supercritical liquids during subduction.

2 Geological background and sample descriptions

The Tengchong volcanic field is located in southwestern China near the border between China and Burma, and the southeastern margin of the Tibetan Plateau (Fig. 1a). Following the closure of the Neo-Tethyan Ocean and the collision between the India and Asia continents, the India ocean plate was subducted eastward beneath Burma (e.g., Yin and Harrison, 2000). High-resolution tomographic images (Fig. 1b) show a low velocity zone with a width of approximately 100 km dipping eastward, extending to depths of ~400 km beneath the Tengchong volcanic field (Lei et al., 2009; Zhao and Liu, 2010; Zhao et al., 2011; Zhang et al., 2017). This suggests a geological link between the Tengchong volcanic field and the Burma plate (or India plate) subduction.

The basement beneath TVF was mainly composed of metamorphic rocks, including gneisses, migmatites and migmatitic granites (Chen et al., 2002), which was intruded by Mesozoic-Cenozoic granites and overlain by pre-Quaternary sedimentary rocks (YBGMR, 1979). The crustal thickness of the TVF is 33.5 to 38.0 km (Zhang et al., 2015) and magma chambers may exist in the crust (Qin et al., 2000; Jiang et al., 2004; Zhao et al., 2006).

Cenozoic volcanic rocks are widely distributed in TVF, with an area of about 90 km long from south to north and about 50 km wide from west to east, consisting of more than 60 volcanoes (e.g., Jiang, 1998a) (Fig. 1c). The Tengchong volcanic rocks erupted long after the collision with ages ranging from ~17.8 Ma to ~3000 years based on the U-Th, K/Ar and thermoluminescence dating (Mu et al., 1987; Li et al., 2000; Wang et al., 2006; Yin and

Li, 2000; Zou et al., 2010). These rocks were previously divided into 4 different stages: (1) Middle-Late Pliocene olivine basalt, (2) Early Pleistocene trachyandesite, (3) Middle-Late Pleistocene olivine basalt, basaltic andesite, and (4) Holocene basaltic trachyandesite, trachyandesite and trachydacite (Jiang, 1998a, b).

Thirty-nine volcanic rocks from south to north across the Tengchong area were collected, including basalts and andesites from the Pliocene (N_2) and Holocene, respectively (Fig. 1c). The basalts comprise of 5 samples from Qingliangshan and 2 samples from Wuhe. The major phenocryst in basalts is olivine. The phenocrysts are set in a fine-grained, intergranular groundmass, consisting of olivine, pyroxene, plagioclase, opaque oxides, etc. The andesites include 4 samples from Tuanshan, 5 samples from Yujiadashan, 2 samples from Laoguipo, 9 samples from Heikongshan and 12 samples from Dayingshan. Major phenocrysts in the andesites are clinopyroxene and plagioclase combined with minor olivine and opaque Fe-Ti oxides. Based on the eruption period and rock types (Mu et al., 1987; Jiang et al., 1998a), these andesites can be classified into two groups. Group 1 belongs to early Holocene components of the Tuanshan, Yujiadashan and Laoguipo samples. Group 2 belongs to the late Holocene recorded in the Heikongshan and Dayingshan samples.

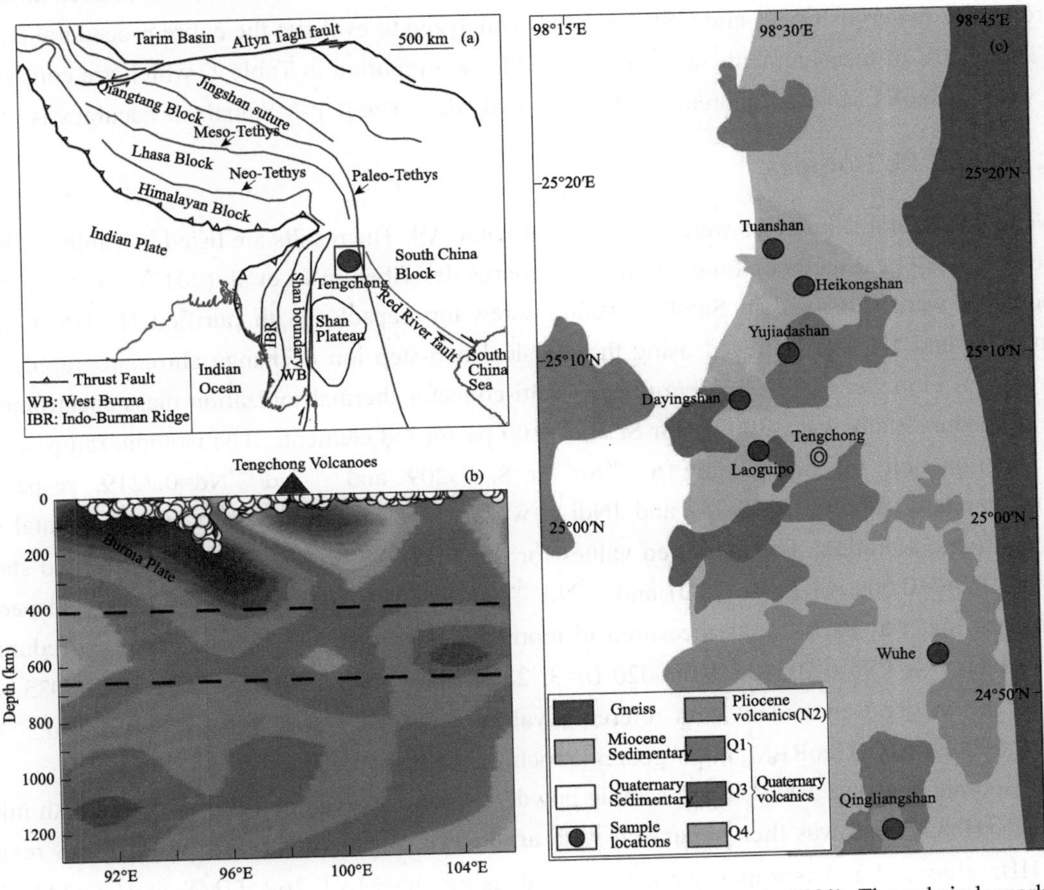

Fig. 1 (a) Map of the Tengchong area and its major tectonic setting (after Tapponnier et al., 1990). The red circle marks the sample locations from this study. (b) A high-resolution P wave tomography shows the vertical profile beneath the Tengchong area (modified from Huang and Zhao, 2006). (c) A schematic map showing the distribution of Tengchong Cenozoic volcanic rocks (after Wang et al., 2006).

3 Analytical methods

By petrological examination, fresh and representative volcanic rocks were chosen for chemical analyses.

After the surface contamination was completely removed, the inner chips were selected and crushed to powders (75 μm) for major, trace elemental and Sr-Nd-Pb-Mg isotopic analyses.

3.1 Major and trace elements

Whole rock major and trace elemental analyses were carried out at the Institute of Geology and Geophysics, Chinese Academy of Sciences (IGGCAS). For the major element analysis, approximately 0.5 g sample powders were mixed with ~5.0 g $Li_2B_4O_7$ powders, and then heated to fuse a glass bead. Major elements were measured on the fused glass with an Axios-Minerals sequential X-ray fluorescence (XRF) spectrometer. The analytical uncertainty is <1% for major elements >1 wt.% and ~10% uncertainty for elements <1 wt.%. In addition, loss on ignition (LOI) for the investigated samples was determined after two hours of baking under a constant temperature of 1000 °C and thirty minutes cooling to ambient temperature.

Trace elements, including the rare earth elements (REE), were analyzed using an ELEMENT inductively coupled plasma-mass spectrometer (ICP-MS), following the procedure described in Chu et al. (2009). Two standard reference materials GSR1 and GSR3 were also analyzed to evaluate the entire procedural and instrument stability. The results of measurements on GSR1 and GSR3 are reported in Table 1, which are consistent with the reference values (GeoREM, http://georem.mpchmainz.gwdg.de/). The ICP-MS analysis accuracy is <5%.

3.2 Sr, Nd and Pb isotopes

Sr, Nd and Pb isotopic analyses were carried out at IGGCAS. The results are listed in Table 2. The Rb-Sr and Sm-Nd isotopic analytical methods followed the procedures described in Li et al. (2012) and Yang et al. (2010). Sample powders were dissolved in Savillex Teflon screw-top capsules with purified HF-HNO_3-$HClO_4$ acids. Separation of Sr and Nd was achieved using the classical two-step ion exchange chromatographic method and measured using a Finnigan MAT-262 *Triton Plus* multi-collector thermal ionization mass spectrometer (TIMS). The entire procedure blank was <300 pg for Sr and <100 pg for Nd elements. The isotopic ratios were corrected for mass fractionation by normalizing to $^{88}Sr/^{86}Sr$=8.375209 and $^{146}Nd/^{144}Nd$=0.7219, respectively. The international standard samples, NBS-987 and JNdi-1, were also measured to evaluate instrumental stability for mass fractionation calibration. The measured values for the NBS-987 Sr standard and JNdi-1 Nd standard were $^{87}Sr/^{86}Sr$ =0.710240±0.000043 (n=5, 2SD) and $^{143}Nd/^{144}Nd$=0.512116±0.000014 (n=5, 2SD), respectively. The USGS reference material BCR-2 was measured to monitor the accuracy of the analytical procedures, with the following results: $^{87}Sr/^{86}Sr$=0.705005±0.000020 (n=3, 2SD) and $^{143}Nd/^{144}Nd$=0.512634 ±0.000035 (n=3, 2SD). These values are in good agreement with reference values ($^{87}Sr/^{86}Sr$ = 0.70492 ± 0.00055 and $^{143}Nd/^{144}Nd$ = 0.512635 ± 0.000029 from GeoREM, http://georem.mpch-mainz.gwdg.de/).

For Pb isotopic analysis, 100−150 mg sample powders were dissolved in Teflon beakers with mixture acids of purified HF+HNO_3. Lead was then separated with anion exchange columns using AG1-X8 resin (75−150 μm) with HBr eluent. Lead isotopes were measured using the MAT-262 TIMS at IGGCAS. The entire procedural blank for Pb concentration was <150 pg. The NBS981standard with $^{207}Pb/^{206}Pb$ =0.9132 (n=6) was used to calibrate isotopic mass fractionation. Repeated analyses of the BCR-2 standard yielded $^{206}Pb/^{204}Pb$=18.759 ± 0.003 (2SD, n=2), $^{207}Pb/^{204}Pb$=15.622 ± 0.001 (2SD, n=2) and $^{208}Pb/^{204}Pb$ = 38.707 ± 0.012 (2SD, n=2).

3.3 Mg isotopes

Magnesium isotopes were analyzed at the State Key Laboratory of Geological Processes and Mineral

Resources, China University of Geosciences (Beijing) (CUGB), following the methods previously reported by Huang et al. (2015a); Ke et al. (2016) and Tian et al. (2016, 2017). Chemical procedures were conducted in a clean lab. About 1-5 mg of each sample was dissolved in Savillex screw-top beakers in a HF-HNO_3 (3:1 v/v) acids mixture. The capped beakers were heated at 130–160 °C and subsequently evaporated to dryness at ~160 °C. The sample residues were then refluxed with HCl-HNO_3 (3:1 v/v) acids and evaporated to dryness at ~ 80 °C. The residues were refluxed again in concentrated HNO_3 media and heated to dryness at ~70 °C. Finally, all sample residues were dissolved in 1M HNO_3 for Mg separation. Samples containing about 10 μg Mg were subsequently separated on an analytical column by cation exchange chromatography with Bio-Rad AG50W-X8 pre-cleaned resin (38–75 μm), and eluted using 1M HNO_3 at a flow rate of 1 mL/30 min. This was carried out twice to obtain a pure Mg solution. The solutions were evaporated to dryness and then re-dissolved in 3% HNO_3. Prior to Mg isotopic analyses, the abundances of Ti, Al, Fe, Ca, Na and K relative to Mg in each sample solution were measured. The total recovery of Mg is more than 99.5% (Ke et al., 2016) and the cation/Mg (mass/mass) ratios measured by ICP-MS were less than 0.05. Ultra-pure acids and 18.2 MΩ·cm deionized water were used in all steps to decrease the background. The procedural blank is about 6.8 ng.

Magnesium isotopes were measured using a Thermo-Finnigan *Neptune plus* MC-ICP-MS at CUGB. The sample-standard bracketing method was used for instrumental mass bias correction (Ke et al., 2016). Sample solutions containing 400 ppb Mg were introduced into the plasma (~ 50 μL/min) via a standard H-skimmer cone, and an ESI PFA MicroFlow nebulizer with a quartz Scott-type spray chamber under Low-resolution mode. Sample solution ratio measurements were made for ⩾4 repeat runs during a session. The long-term reproducibility of measurements of both natural and synthetic pure Mg solution is better than 0.10‰ for δ^{26}Mg (Liu and Zhou, 2005; Huang et al., 2015a; Tian et al., 2016). The results are reported in the conventional δ-notation in per mil relative to DSM3: $\delta^{n}Mg = [(^{n}Mg/^{24}Mg)_{sample}/(^{n}Mg/^{24}Mg)_{DSM3}-1] \times 1000(‰)$, where n is either 25 or 26 (Galy et al., 2003). Two well-characterized USGS and 2 Chinese rock standards were analyzed, with average values of δ^{26}Mg=–0.20±0.09‰ (2SD) for BCR-2, δ^{26}Mg =–0.27±0.02‰ (2SD, n=2) for BHVO-2, δ^{26}Mg=–0.55±0.07‰ (2SD) for GSR-2 and δ^{26}Mg=–0.45±0.03‰ (2SD) for GSR-3. All of these values are indistinguishable from previously published data (–0.36 to –0.16‰ for BCR-2, –0.32 to –0.16‰ for BHVO-2) (Teng et al., 2007, 2015; Huang et al., 2009, 2011; An et al., 2014, 2015a; Tian et al., 2016; Li et al., 2017).

4 Results

4.1 Major and trace elements

The major and trace elemental compositions of the Tengchong volcanic rocks as well as the 2 standard materials are listed in Table 1. These volcanic rocks fall within the field of trachy-basalt, basaltic trachy-andesite and trachy-andesite in the TAS diagram (Fig. 2a). They exhibit a large variation in SiO_2 (48.36 to 62.53 wt.%) and MgO contents (2.41 to 8.48 wt.%) (Table 1). The Tengchong basalts have limited variation in SiO_2 content and show no obvious correlation with other major oxides (Fig. 2). In contrast, CaO, total FeO (TFeO), CaO/Al_2O_3 and TiO_2 decrease with increasing SiO_2 (Figs. 2b-e), whereas K_2O of the andesites correlates positively with SiO_2 (Fig. 2f). This is consistent with fractionation of the observed phenocryst assemblages (olivine ± pyroxene ± plagioclase ± Fe-Ti oxides). Notably, most of the Group 1 and 2 andesites fall along the magma evolution trends of the Tengchong basalts (Fig. 2).

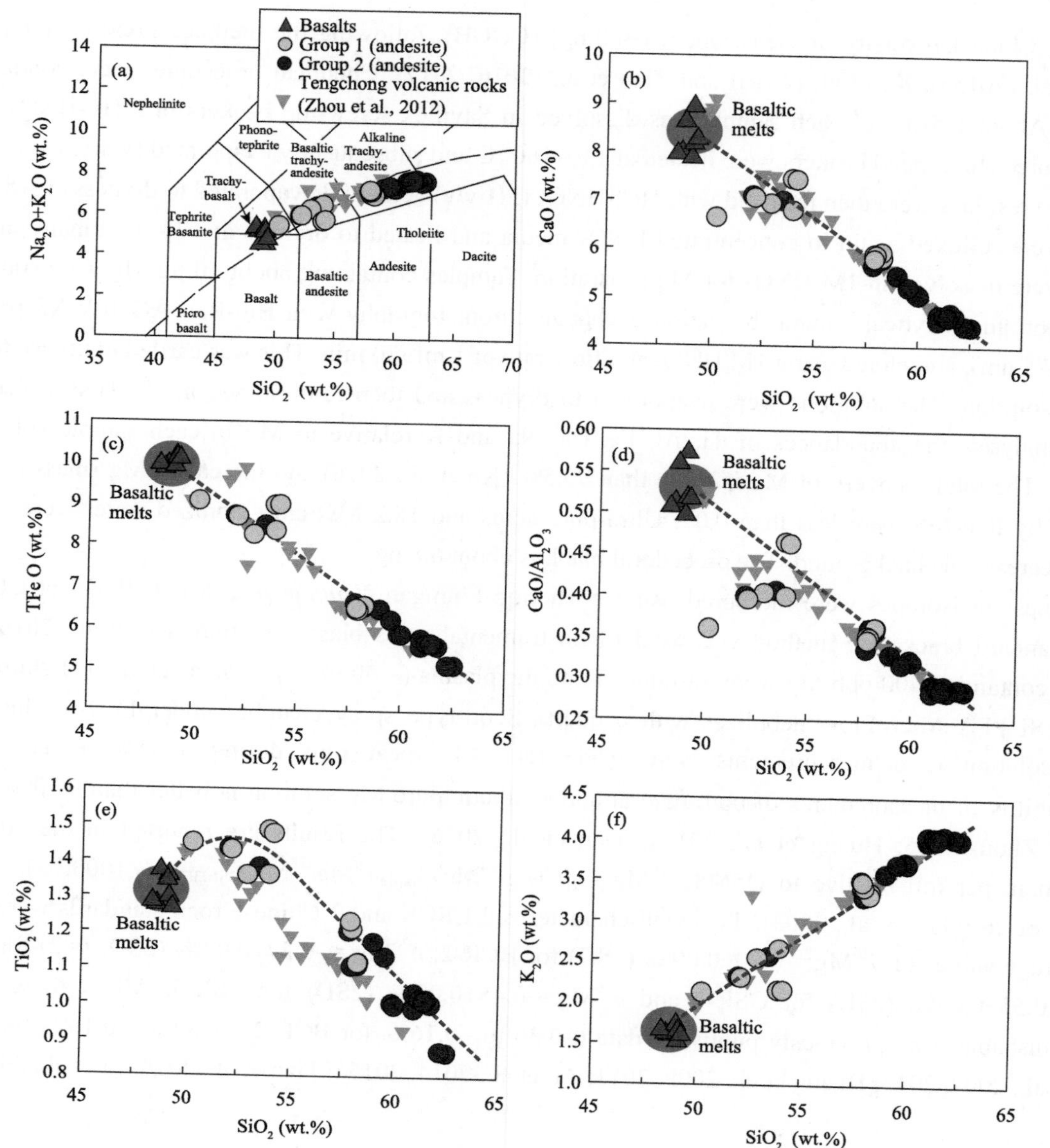

Fig. 2 (a) Total alkalis Na_2O+K_2O (wt.%) vs. SiO_2 (wt.%) (TAS) diagrams for the Tengchong volcanic rocks. The reference fields are from Le Bas (1986). Variations of (b) CaO (wt.%), (c) total FeO (TFeO) (wt.%), (d) CaO/Al_2O_3 (wt.%), (e) TiO_2 (wt.%) and (f) K_2O (wt.%) vs. SiO_2 (wt.%) for the Tengchong basalts and andesites. The orange red circle and the dashed red line represent the basaltic melt and magma evolution trends, respectively. Previously reported data are displayed in the gray triangle (Zhou et al., 2012). Note that the gray triangles in this figure and the Fig. 5 are the same. (For interpretation of the references to colour in this figure legend, the reader is referred to the web version of this article.)

Both the Tengchong basalts and andesites are enriched in LILEs (e.g., K, Rb and Ba) and depleted in HFSEs (e.g., Nb, Ta, Zr, Hf and P), similar to the trace element patterns of the Indian sediments and continental and island arc basalts, but are distinct from the OIB and Samoan lavas that display an EM-II endmember Sr-Nd isotopic feature (Figs. 3a and b). In the chondrite-normalized rare earth element diagrams (Fig. 3c and d), the basalts and andesites are enriched in light rare earth elements (LREE), with $(La/Yb)_N$ values ranging from 10.1 to 26.2 (Table 1). The Tengchong basalts exhibit low Nb/U (10.5–12.8) and Ce/Pb (7.7–9.6) ratios, similar to the continental and island arc basalts (Fig. 4a). Notably, these basalts yield extremely high Th/U (6.5–8.3) (Fig. 4b), which is different from continental and island arc basalts and the Cenozoic continental basalt in eastern China.

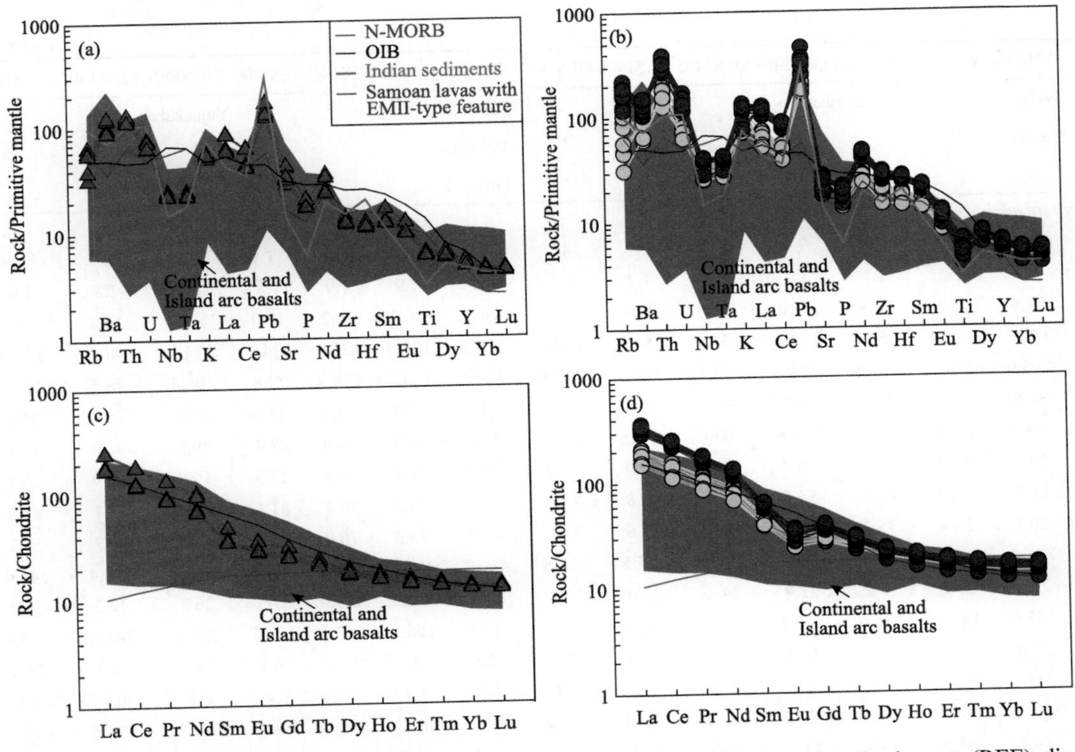

Fig. 3 Primitive mantle-normalized trace element patterns and chondrite-normalized rare earth element (REE) diagram of the Tengchong basalts (a and c), as well as the Tengchong andesites (b and d). The database are from Sun and McDonough (1989) for the N-MORB and OIB, Plank and Langmuir (1998) for the Indian sediments, and Workman et al. (2004) for the Samoan lavas with the EMII-type feature. Oceanic arc basalts (Turner et al., 2001; Elburg and Foden, 2005; Teng et al., 2016; Li et al., 2017) and continental arc basalts are plotted for comparison. Note: Continental arc basalts include Cascades (Borg et al., 2000) and Andean (data compiled by Geochemistry of Rocks of the Oceans and Continents database). Note that the data of the <110 Ma basalts from eastern China and oceanic and continental arc basalts in the following figures are the same as in this figure.

Table 1 Major oxides (wt.%) and trace elements (ppm) of the Tengchong Cenozoic basalts and andesites

Sample	WH1401	WH1402	QLS1403	QLS1404	QLS1405	QLS1406	QLS1407	TS1403	TS1404	TS1405	TS1406	YJDS1401	YJDS1402	YJDS1403	YJDS1404
Location	Wuhe		Qingliangshan					Tuanshan				Yujiadashan			
	Basalts							Andesites							
Groups								Group 1							
Major element (wt.%)															
SiO_2	49.14	48.74	48.58	48.36	49.07	49.15	49.31	58.27	58.23	57.87	57.96	53.94	50.26	52.9	53.91
TiO_2	1.36	1.37	1.28	1.29	1.34	1.30	1.27	1.12	1.12	1.22	1.21	1.36	1.45	1.36	1.49
Al_2O_3	15.61	15.42	15.59	15.69	15.84	15.77	16.00	16.14	16.43	16.78	16.79	17.03	18.45	17.20	16.03
$TFe_2O_3^b$	10.14	9.92	9.84	9.93	9.88	10.09	10.07	6.55	6.55	6.46	6.44	8.35	9.06	8.25	8.97
MnO	0.15	0.14	0.15	0.15	0.15	0.15	0.15	0.11	0.10	0.10	0.10	0.13	0.14	0.13	0.13
MgO	8.10	8.05	8.04	7.99	7.78	8.33	8.48	4.01	4.06	3.42	3.40	4.76	5.33	4.70	5.15
CaO	8.93	8.62	7.99	7.95	7.87	8.16	8.26	5.84	5.95	5.91	5.89	6.80	6.65	6.94	7.45
Na_2O	3.01	2.93	3.53	3.66	3.50	3.47	3.16	3.58	3.58	3.67	3.68	3.78	3.31	3.69	3.51
K_2O	1.52	1.60	1.62	1.66	1.73	1.65	1.60	3.36	3.32	3.42	3.42	2.66	2.12	2.54	2.15
P_2O_5	0.44	0.45	0.38	0.38	0.39	0.38	0.37	0.42	0.43	0.46	0.46	0.42	0.45	0.43	0.30
LOI	0.90	1.60	2.66	2.60	1.64	1.12	1.08	−0.20	0.34	−0.20	1.38	0.18	2.38	0.72	0.10
Total	99.29	98.84	99.65	99.64	99.19	99.56	99.75	99.19	100.11	99.12	100.73	99.41	99.60	98.86	99.20
$Mg\#^c$	65.1	65.4	65.6	65.2	64.7	65.8	66.3	58.8	59.1	55.2	55.2	57.1	57.8	57.0	57.2
Na_2O+K_2O/TiO_2	3.34	3.30	4.03	4.14	3.89	3.95	3.75	6.21	6.18	5.80	5.86	4.72	3.74	4.58	3.81
CIA^d	40.6	40.9	41.3	41.3	41.9	41.4	42.1	44.4	44.7	45.0	45.0	44.2	48.2	44.5	42.5

Continued

Sample	WH1401	WH1402	QLS1403	QLS1404	QLS1405	QLS1406	QLS1407	TS1403	TS1404	TS1405	TS1406	YJDS1401	YJDS1402	YJDS1403	YJDS1404
Location	Wuhe		Qingliangshan					Tuanshan				Yujiadashan			
	Basalts							Andesites							
Groups								Group 1							
Trace element (ppm)															
Li	5.8	6.4	7.6	7.9	8.7	7.7	8.3	14.3	14.6	13.6	11.1	12.4	10.3	11.9	11.0
Be	1.3	1.2	1.1	1.1	1.3	1.2	1.1	1.9	1.9	1.9	1.8	1.8	1.8	1.9	1.6
Sc	21.8	22.6	21.8	21.9	22.9	22.0	21.9	16.5	16.2	16.2	17.0	17.8	16.8	17.2	17.2
V	145.2	145.9	150.7	152.3	157.9	151.5	148.8	124.8	128.3	125.4	134.3	134.1	140.7	128.9	129.7
Cr	295.9	250.1	262.8	272.5	273.5	284.5	273.8	119.6	134.7	135.9	79.8	105.4	78.5	92.3	113.0
Co	38.9	36.5	37.9	38.2	36.9	39.4	39.6	24.5	25.3	24.6	25.9	25.6	26.9	25.3	24.8
Ni	108.6	107.9	117.7	114.8	103.6	134.5	124.9	41.6	43.5	48.9	42.4	49.8	45.0	45.1	40.5
Cu	29.7	32.9	30.7	30.9	31.9	34.9	31.0	27.0	27.4	24.8	17.3	19.8	20.9	25.1	21.9
Zn	88.2	111.6	84.1	85.5	86.2	86.8	86.6	69.4	73.5	70.1	81.1	81.2	81.1	81.3	77.6
Rb	20.3	24.4	35.3	37.3	40.9	37.6	35.1	81.3	78.3	76.9	30.3	57.9	29.3	59.7	52.1
Sr	876.9	908.0	709.6	709.4	601.0	695.5	663.8	410.2	420.7	412.8	554.8	520.2	498.4	541.4	527.8
Y	23.2	23.6	21.2	21.4	24.2	23.0	22.8	24.9	25.3	25.2	27.5	26.2	27.2	26.9	25.7
Zr	153.6	143.1	136.6	140.7	140.8	133.0	131.5	185.8	206.1	187.9	241.7	225.3	244.8	233.9	221.7
Nb	17.0	16.7	15.9	16.4	17.6	16.5	16.2	20.9	22.1	21.2	26.1	24.3	26.3	25.4	24.0
Cs	0.6	0.6	1.4	1.0	1.4	1.1	0.7	1.5	1.1	1.3	0.4	0.8	0.5	1.1	0.8
Ba	821.9	862.7	624.5	622.2	692.1	673.4	662.0	445.2	462.3	445.1	700.7	651.8	671.5	655.1	634.8
La	59.7	57.9	40.5	41.1	43.7	41.7	41.2	41.3	43.5	41.6	50.4	49.5	50.8	50.3	48.0
Ce	113.7	111.2	73.5	74.7	78.4	74.9	75.0	80.1	83.6	80.9	93.1	94.9	96.8	97.1	90.2
Pr	12.9	12.7	8.4	8.6	8.9	8.5	8.4	9.3	9.5	9.3	11.5	10.8	11.4	10.9	10.3
Nd	48.0	44.7	31.8	32.0	33.5	31.9	31.5	33.7	35.1	34.3	43.4	40.3	42.9	41.1	38.3
Sm	7.3	7.3	5.4	5.5	5.7	5.4	5.3	5.9	6.1	6.0	7.4	6.8	7.3	7.1	6.4
Eu	2.1	2.1	1.6	1.6	1.7	1.7	1.6	1.4	1.5	1.4	1.9	1.8	1.9	1.8	1.8
Gd	6.3	6.4	5.1	5.1	5.3	5.0	5.0	5.5	5.7	5.6	6.7	6.3	6.6	6.4	6.1
Tb	0.9	0.9	0.8	0.8	0.8	0.8	0.7	0.8	0.9	0.9	1.0	0.9	1.0	0.9	0.9
Dy	4.5	4.8	4.3	4.2	4.4	4.2	4.3	4.8	5.0	4.9	5.5	5.1	5.4	5.3	4.8
Ho	0.9	1.0	0.9	0.9	0.9	0.9	0.9	1.0	1.0	1.0	1.1	1.0	1.1	1.1	1.0
Er	2.4	2.6	2.2	2.3	2.4	2.3	2.3	2.6	2.6	2.7	2.9	2.7	2.9	2.8	2.5
Tm	0.3	0.4	0.3	0.3	0.3	0.3	0.3	0.4	0.4	0.4	0.4	0.4	0.4	0.4	0.4
Yb	2.2	2.2	2.1	2.2	2.1	2.1	2.1	2.4	2.5	2.4	2.6	2.5	2.6	2.6	2.3
Lu	0.3	0.3	0.3	0.3	0.3	0.3	0.3	0.4	0.4	0.4	0.4	0.4	0.4	0.4	0.3
Hf	3.8	3.8	3.5	3.5	3.5	3.4	3.4	5.0	5.1	5.1	5.9	5.7	5.9	5.8	5.4
Ta	0.9	0.9	0.9	0.9	1.0	1.0	0.9	1.3	1.4	1.3	1.5	1.4	1.5	1.5	1.4
Pb	11.9	12.1	9.3	9.6	10.0	9.0	8.8	13.8	13.9	17.1	14.9	13.3	14.2	13.8	11.7
Th	10.4	11.1	9.5	9.7	10.8	10.2	10.1	20.6	19.0	20.2	15.1	15.2	15.9	15.8	12.7
U	1.3	1.3	1.3	1.3	1.6	1.6	1.5	2.0	1.9	2.0	1.6	1.7	1.6	1.7	1.5
Ratios															
$(La/Yb)_N{}^e$	19.7	18.7	13.9	13.7	14.6	14.4	14.3	12.6	12.7	12.4	13.9	14.2	13.8	14.0	14.9
Ce/Pb	9.6	9.2	7.9	7.7	7.9	8.3	8.5	5.8	6.0	4.7	6.2	7.1	6.8	7.0	7.7
Nb/U	12.8	12.4	11.8	12.2	10.9	10.5	10.7	10.5	11.4	10.8	16.6	14.1	16.0	14.8	16.4
$(Gd/Yb)_N{}^e$	2.40	2.36	2.00	1.96	2.06	2.00	2.01	1.91	1.92	1.92	2.15	2.07	2.07	2.05	2.17
Nb/Ta	18.8	18.4	17.2	17.6	17.3	17.0	17.1	16.0	16.4	16.1	17.1	16.9	17.1	17.2	17.3
Hf/Hf^{*f}	0.51	0.52	0.67	0.67	0.64	0.66	0.65	0.89	0.87	0.89	0.82	0.86	0.84	0.85	0.86
Ti/Ti^{*f}	0.55	0.55	0.65	0.65	0.65	0.66	0.65	0.52	0.50	0.56	0.46	0.56	0.56	0.54	0.63
Th/U	7.9	8.3	7.1	7.2	6.7	6.5	6.7	10.3	9.8	10.3	9.6	8.9	9.6	9.2	8.7
Ba/Th	78.7	77.7	65.5	64.2	63.9	66.1	65.6	21.6	24.3	22.0	46.3	42.8	42.3	41.5	49.9
Rb/Cs	35.6	41.1	25.9	38.5	29.7	32.9	47.3	55.5	71.2	59.4	83.0	75.4	64.2	52.5	65.6
U/Pb	0.11	0.11	0.14	0.14	0.16	0.17	0.17	0.14	0.14	0.11	0.11	0.13	0.12	0.12	0.13
Th/Pb	0.88	0.92	1.03	1.00	1.09	1.13	1.15	1.49	1.37	1.18	1.01	1.14	1.12	1.14	1.09

Continued

Sample	YJDS1405	LGP1401	LGP1402	HKS1402	HKS1403	HKS1404	HKS1406	HKS1408	HKS1409	HKS1414	HKS1415	HKS1416	DYS1401	DYS1402
Location		Laoguipo		Heikongshan									Dayingshan	
	Andesite													
Groups	Group 1			Group 2										
Major elements (wt.%)														
SiO_2	52.13	54.13	57.88	62.53	60.01	58.23	57.95	57.86	53.46	52.02	62.22	59.90	61.54	60.98
TiO_2	1.43	1.48	1.23	0.87	1.00	1.11	1.11	1.21	1.38	1.44	0.87	1.01	1.02	1.00
Al_2O_3	17.85	16.05	16.72	15.43	15.75	16.21	16.31	16.68	17.23	17.77	15.39	15.73	16.14	15.88
TFe_2O_3	8.69	8.98	6.49	5.12	5.86	6.52	6.53	6.45	8.48	8.71	5.12	5.87	5.63	5.64
MnO	0.13	0.13	0.10	0.09	0.10	0.10	0.11	0.10	0.13	0.13	0.09	0.10	0.09	0.09
MgO	5.06	5.16	3.35	2.72	3.38	4.00	4.05	3.40	4.89	5.05	2.73	3.37	2.50	2.45
CaO	7.06	7.43	5.81	4.42	5.06	5.80	5.89	5.84	7.00	7.10	4.44	5.10	4.68	4.66
Na_2O	3.63	3.56	3.68	3.60	3.55	3.65	3.67	3.71	3.81	3.61	3.53	3.52	3.62	3.60
K_2O	2.30	2.15	3.48	4.00	3.70	3.34	3.30	3.40	2.56	2.31	4.02	3.73	4.03	3.98
P_2O_5	0.45	0.30	0.47	0.32	0.38	0.43	0.43	0.46	0.43	0.44	0.32	0.37	0.39	0.38
LOI	0.70	0.10	0.85	0.22	0.64	0.06	0.04	−0.10	0.58	0.04	0.48	0.28	−0.02	0.20
Total	99.43	99.47	100.06	99.32	99.43	99.45	99.38	99.02	99.96	98.62	99.21	98.97	99.60	98.86
Mg#	57.6	57.3	54.6	55.3	57.3	58.8	59.1	55.1	57.3	57.5	55.4	57.2	50.8	50.3
Na_2O+K_2O/TiO_2	4.14	3.86	5.80	8.75	7.25	6.31	6.30	5.88	4.60	4.12	8.66	7.21	7.52	7.59
CIA	45.6	42.5	45.0	45.7	45.2	44.5	44.5	44.9	44.2	45.4	45.7	45.1	46.1	45.9
Trace elements (ppm)														
Li	11.1	12.3	9.3	25.5	20.2	16.1	16.4	14.9	15.5	15.9	16.0	20.9	19.2	19.8
Be	1.8	1.5	1.5	2.5	2.2	1.9	2.0	1.9	1.9	2.0	2.0	2.1	2.4	2.3
Sc	11.2	20.6	22.5	10.9	13.4	14.5	15.4	14.2	14.4	14.7	14.7	12.9	12.0	11.0
V	134.2	137.0	149.6	84.1	94.3	106.0	107.5	112.0	114.5	116.1	115.6	98.8	90.2	89.5
Cr	81.1	175.5	188.4	135.2	129.4	204.8	195.2	105.4	99.7	76.2	93.1	155.3	64.2	143.1
Co	24.7	28.7	32.3	13.8	16.6	19.0	19.7	17.6	17.4	17.6	17.7	16.2	13.1	13.2
Ni	46.7	44.4	49.7	45.7	48.2	79.5	85.6	67.1	40.2	40.9	35.8	40.6	21.7	25.8
Cu	24.8	29.4	34.4	22.7	20.2	40.2	35.4	30.4	23.9	21.4	21.1	21.2	14.6	19.9
Zn	78.9	86.8	95.6	68.4	72.6	76.2	77.9	75.1	78.5	78.7	81.9	73.2	69.1	68.9
Rb	36.3	65.5	20.0	119.3	124.2	99.8	102.0	99.9	107.9	112.2	104.5	131.0	138.1	114.8
Sr	538.2	404.1	440.8	378.7	453.4	515.8	534.4	569.0	578.7	587.5	577.5	460.0	451.2	408.2
Y	24.8	25.4	26.3	26.2	25.9	25.3	26.3	27.0	27.4	27.6	27.9	22.7	27.2	26.8
Zr	231.6	164.6	164.0	279.7	289.5	298.3	299.9	310.0	330.3	333.9	337.9	272.5	307.9	304.7
Nb	25.0	18.1	18.4	27.3	27.1	26.9	27.9	27.8	28.1	28.8	28.9	20.3	27.0	26.9
Cs	0.8	1.3	0.3	2.5	1.9	1.3	1.3	1.2	1.3	1.4	1.3	1.9	1.9	2.1
Ba	634.2	427.2	453.6	747.8	878.7	932.5	946.0	1032.9	1013.6	1051.1	1032.5	730.4	889.5	805.2
La	44.3	37.4	35.7	70.0	75.4	72.9	74.0	80.0	81.7	82.4	84.0	73.6	82.0	78.4
Ce	89.6	71.9	68.8	136.1	143.8	139.5	140.8	150.8	152.9	155.9	157.2	145.9	154.0	149.2
Pr	10.0	8.5	8.2	14.8	16.0	15.5	15.5	16.5	17.0	17.2	17.4	16.3	16.7	16.4
Nd	37.0	32.3	31.8	53.7	58.0	57.0	57.5	60.2	61.9	63.5	64.1	59.5	60.7	58.8
Sm	6.4	6.0	6.1	8.6	9.1	8.9	9.0	9.4	9.5	9.9	9.9	9.0	9.3	9.2
Eu	1.8	1.6	1.7	1.6	1.8	1.9	1.9	2.1	2.1	2.1	2.1	1.6	1.8	1.8
Gd	6.1	5.8	5.9	7.0	7.4	7.4	7.4	8.0	8.0	8.1	8.2	7.1	7.6	7.6
Tb	0.9	0.9	0.9	1.0	1.0	1.0	1.0	1.1	1.1	1.1	1.1	0.9	1.0	1.0
Dy	4.9	5.2	5.5	5.1	5.2	5.1	5.3	5.4	5.5	5.6	5.7	4.5	5.4	5.3
Ho	1.0	1.0	1.1	1.0	1.0	1.0	1.0	1.1	1.1	1.1	1.1	0.9	1.1	1.0
Er	2.6	2.6	2.8	2.7	2.7	2.6	2.7	2.7	2.8	2.8	2.8	2.3	2.8	2.7
Tm	0.4	0.4	0.4	0.4	0.4	0.4	0.4	0.4	0.4	0.4	0.4	0.3	0.4	0.4
Yb	2.3	2.4	2.5	2.5	2.5	2.4	2.5	2.4	2.4	2.5	2.5	2.0	2.5	2.5
Lu	0.3	0.4	0.4	0.4	0.4	0.4	0.4	0.4	0.4	0.4	0.4	0.3	0.4	0.4
Hf	5.5	4.5	4.4	7.4	7.6	7.4	7.5	8.0	8.1	8.4	8.4	7.4	7.9	7.9
Ta	1.4	1.1	1.1	1.8	1.7	1.6	1.6	1.6	1.6	1.7	1.7	1.3	1.7	1.7
Pb	15.0	12.1	12.7	32.2	24.4	25.7	21.9	18.2	23.8	21.5	21.7	22.8	22.8	22.7
Th	10.7	15.7	13.2	32.2	28.8	23.4	23.8	23.2	22.8	23.1	23.3	31.7	29.7	28.9
U	1.5	1.5	1.3	3.6	3.1	2.4	2.5	2.5	2.5	2.5	2.5	2.8	3.0	2.9

Continued

Sample	YJDS1405	LGP1401	LGP1402	HKS1402	HKS1403	HKS1404	HKS1406	HKS1408	HKS1409	HKS1414	HKS1415	HKS1416	DYS1401	DYS1402
Location		Laoguipo		Heikongshan									Dayingshan	
	Andesite													
Groups	Group 1			Group 2										
Ratios														
$(La/Yb)_N$	14.0	11.0	10.1	19.8	21.7	21.7	21.6	24.3	24.4	23.9	24.0	26.2	23.8	22.8
Ce/Pb	6.0	6.0	5.4	4.2	5.9	5.4	6.4	8.3	6.4	7.3	7.2	6.4	6.8	6.6
Nb/U	16.4	11.8	14.0	7.5	8.8	11.1	11.3	11.2	11.3	11.6	11.7	7.3	9.0	9.1
$(Gd/Yb)_N$	2.23	1.96	1.93	2.30	2.46	2.53	2.49	2.79	2.77	2.71	2.71	2.92	2.55	2.55
Nb/Ta	17.5	16.1	16.5	15.4	15.9	16.8	17.0	17.1	17.3	17.1	17.2	15.8	16.0	16.0
Hf/Hf*	0.90	0.80	0.80	0.86	0.83	0.83	0.83	0.84	0.84	0.84	0.84	0.80	0.83	0.86
Ti/Ti*	0.61	0.66	0.54	0.31	0.34	0.38	0.37	0.38	0.43	0.44	0.27	0.35	0.33	0.33
Th/U	7.0	10.2	10.1	8.9	9.3	9.7	9.7	9.4	9.2	9.3	9.4	11.4	9.9	9.8
Ba/Th	59.2	27.3	34.2	23.2	30.5	39.8	39.8	44.5	44.5	45.4	44.3	23.0	30.0	27.9
Rb/Cs	44.8	48.6	63.9	48.1	64.5	76.4	77.8	83.9	82.9	81.9	78.8	68.7	72.7	54.0
U/Pb	0.10	0.13	0.10	0.11	0.13	0.09	0.11	0.14	0.10	0.12	0.11	0.12	0.13	0.13
Th/Pb	0.71	1.30	1.04	1.00	1.18	0.91	1.09	1.28	0.95	1.08	1.07	1.39	1.30	1.27

Sample	DYS1403	DYS1406	DYS1407	DYS1408	DYS1410	DYS1412	DYS1413	DYS1415	DYS1416	DYS1419	GSR1 (measured)	GSR1[a]	GSR3 (measured)	GSR3[a]
Location	Dayingshan													
	Andesites													
Groups	Group 2													
Major element (wt.%)														
SiO_2	60.96	61.43	61.74	60.96	61.25	61.54	61.43	57.81	58.98	59.52				
TiO_2	0.99	1.01	1.00	1.03	1.02	1.02	1.01	1.20	1.18	1.13				
Al_2O_3	15.89	16.08	16.24	15.98	16.00	16.09	16.08	16.86	16.57	16.44				
TFe_2O_3	5.57	5.66	5.58	5.77	5.65	5.63	5.64	6.61	6.47	6.21				
MnO	0.09	0.09	0.09	0.09	0.09	0.09	0.09	0.11	0.10	0.10				
MgO	2.41	2.50	2.46	2.49	2.47	2.42	2.49	3.05	2.99	2.83				
CaO	4.63	4.66	4.63	4.54	4.65	4.71	4.61	5.67	5.47	5.22				
Na_2O	3.58	3.62	3.64	3.55	3.60	3.64	3.61	3.63	3.67	3.65				
K_2O	4.00	4.03	4.05	4.04	4.03	4.03	4.00	3.50	3.59	3.70				
P_2O_5	0.37	0.38	0.38	0.38	0.39	0.38	0.38	0.48	0.47	0.44				
LOI	0.08	0.20	0.14	0.42	0.08	0.10	0.18	0.44	0.13	0.08				
Total	98.57	99.67	99.96	99.26	99.22	99.65	99.52	99.36	99.59	99.32				
Mg#	50.2	50.7	50.7	50.1	50.5	50.0	50.7	51.8	51.8	51.5				
Na_2O+K_2O/TiO_2	7.65	7.55	7.66	7.35	7.50	7.56	7.53	5.92	6.17	6.48				
CIA	46.0	46.1	46.3	46.4	46.0	45.9	46.3	45.6	45.5	45.7				
Trace elements (ppm)														
Li	19.1	19.8	18.8	20.3	18.7	19.5	19.6	15.9	17.5	16.7	127.9	131	9.1	9.5
Be	2.5	2.5	2.5	2.4	2.5	2.5	2.5	2.1	2.2	2.2	11.9	12.4	2.5	2.5
Sc	11.2	11.5	10.7	11.7	11.4	12.2	11.6	13.1	13.7	13.7	6.2	6.1	14.9	15.2
V	93.3	95.3	93.8	93.7	91.8	90.8	91.6	106.1	108.7	103.5	26.1	24	165.6	167
Cr	95.5	93.5	71.8	103.3	56.9	79.4	92.2	87.3	105.7	76.8	5.0	5	132.5	134
Co	13.3	13.7	13.2	14.1	13.7	13.5	13.8	16.0	16.5	15.5	2.9	3.4	43.6	46.5
Ni	86.3	24.3	24.1	25.2	23.6	24.5	25.4	29.6	35.8	89.9	1.8	2.3	133.9	140
Cu	43.1	16.4	15.7	14.1	18.7	16.3	14.4	20.1	18.9	59.0	2.0	3.2	48.1	48.6
Zn	71.3	71.7	69.0	71.4	73.0	72.3	69.5	79.8	78.5	76.8	25.2	28	150.4	150
Rb	118.6	121.7	102.7	126.9	129.3	141.3	120.6	74.2	102.9	119.3	470.6	466	38.9	37

Continued

Sample	DYS1403	DYS1406	DYS1407	DYS1408	DYS1410	DYS1412	DYS1413	DYS1415	DYS1416	DYS1419	GSR1 (measured)	GSR1[a]	GSR3 (measured)	GSR3[a]
Location	Dayingshan													
	Andesites													
Groups	Group 2													
Trace elements (ppm)														
Sr	425.2	417.3	409.4	416.3	444.2	459.0	423.6	548.4	545.8	532.6	104.8	106	1108.8	1100
Y	28.0	28.4	28.0	27.6	28.6	28.5	28.0	29.8	28.9	28.9	63.3	62	22.1	22
Zr	331.2	319.6	319.2	304.3	302.6	301.7	302.2	312.5	323.6	322.4	169.6	167	279.6	277
Nb	28.0	27.9	28.2	27.9	28.1	28.0	28.2	28.9	29.7	29.1	39.5	40	67.6	68
Cs	1.7	2.2	2.0	2.0	2.1	2.2	1.8	1.3	1.3	1.4	37.6	38.4	0.49	0.49
Ba	876.4	887.9	837.6	906.0	967.3	953.2	909.4	1047.6	1063.3	1058.2	335.2	343	526.8	526
La	78.8	79.3	81.1	77.9	81.3	87.2	81.1	81.4	80.5	83.1	52.5	54	55.6	56
Ce	150.2	154.7	157.7	148.7	153.9	159.6	155.8	152.1	153.7	156.6	103.5	108	108.7	105
Pr	16.4	16.8	16.9	16.5	16.9	17.5	17.3	17.3	17.0	17.4	12.5	12.7	13.2	13.2
Nd	59.6	60.9	61.3	60.0	61.4	63.7	62.3	63.6	62.7	63.7	46.1	47	54.3	54
Sm	9.4	9.4	9.5	9.3	9.5	9.8	9.6	10.0	9.8	9.9	9.4	9.7	10.4	10.2
Eu	1.8	1.9	1.8	1.9	1.9	1.9	1.9	2.1	2.1	2.1	0.83	0.85	3.2	3.2
Gd	7.7	7.9	7.8	7.8	7.9	8.0	7.9	8.3	8.2	8.2	9.1	9.3	9.0	8.5
Tb	1.1	1.1	1.1	1.1	1.1	1.1	1.1	1.2	1.1	1.1	1.62	1.65	1.2	1.2
Dy	5.5	5.6	5.5	5.5	5.6	5.6	5.6	6.0	5.8	5.8	10.1	10.2	5.6	5.6
Ho	1.1	1.1	1.1	1.1	1.1	1.1	1.1	1.2	1.2	1.2	2.02	2.05	0.89	0.88
Er	2.8	2.9	2.8	2.9	2.9	2.9	2.9	3.1	3.0	3.0	6.5	6.5	2.0	2
Tm	0.4	0.4	0.4	0.4	0.4	0.4	0.4	0.4	0.4	0.4	1.06	1.06	0.24	0.28
Yb	2.6	2.6	2.6	2.6	2.7	2.7	2.7	2.8	2.7	2.7	7.3	7.4	1.3	1.5
Lu	0.4	0.4	0.4	0.4	0.4	0.4	0.4	0.4	0.4	0.4	1.14	1.15	0.17	0.19
Hf	8.4	8.2	8.1	8.0	7.9	7.9	8.0	7.8	8.0	8.1	6.4	6.3	6.5	6.5
Ta	1.7	1.8	1.7	1.8	1.8	1.8	1.8	1.7	1.7	1.8	7.1	7.2	4.3	4.3
Pb	21.8	22.9	21.3	25.2	25.4	24.5	25.2	20.9	21.4	23.8	30.2	31	5.2	4.7
Th	30.2	31.2	31.1	33.1	32.5	33.1	32.9	25.2	24.6	27.6	52.5	54	6.1	6
U	3.1	3.1	3.1	3.3	3.2	3.2	3.2	2.4	2.5	2.8	19.2	18.8	1.4	1.4
Ratios														
$(La/Yb)_N$	21.9	21.8	22.4	21.1	21.8	23.5	21.9	21.1	21.5	22.1				
Ce/Pb	6.9	6.8	7.4	5.9	6.1	6.5	6.2	7.3	7.2	6.6				
Nb/U	9.0	8.9	9.2	8.6	8.7	8.8	8.8	12.0	11.8	10.5				
$(Gd/Yb)_N$	2.47	2.50	2.48	2.44	2.45	2.49	2.46	2.49	2.52	2.52				
Nb/Ta	16.1	15.8	16.2	15.9	15.9	15.9	16.0	16.8	17.1	16.54				
Hf/Hf*	0.89	0.86	0.84	0.85	0.82	0.79	0.82	0.78	0.81	0.81				
Ti/Ti*	0.32	0.32	0.32	0.33	0.32	0.32	0.32	0.36	0.36	0.35				
Th/U	9.8	10.0	10.1	10.2	10.1	10.3	10.3	10.4	9.8	9.9				
Ba/Th	29.0	28.5	26.9	27.4	29.7	28.8	27.6	41.6	43.3	38.4				
Rb/Cs	70.0	55.4	51.3	64.0	61.9	64.5	67.5	58.9	78.6	87.74				
U/Pb	0.14	0.14	0.14	0.13	0.13	0.13	0.13	0.12	0.12	0.12				
Th/Pb	1.39	1.36	1.46	1.31	1.28	1.35	1.31	1.20	1.15	1.16				

[a] Reported values for the reference materials are from GeoREM (http://georem.mpch-mainz.gwdg.de/).

[b] Total iron as Fe_2O_3

[c] Mg# = 100 × Mg^{2+} / (Mg^{2+} + Fe^{2+}), assuming Fe^{3+}/ (Fe^{2+}+Fe^{3+}) = 0.15.

[d] CIA = Chemical Indexes of Alteration, defined as molecular ratios of Al_2O_3/(Al_2O_3 + CaO + Na_2O +K_2O) × 100 (Nesbitt and Young, 1982).

[e] $(La/Yb)_N$, and $(Gd/Yb)_N$ where subscript N means normalized to chondrite (Sun and McDonough, 1989).

[f] $Ti/Ti^* = Ti_N/(Sm_N^{-0.055} \times Nd_N^{0.333} \times Gd_N^{0.722})$ and $Hf/Hf^* = Hf_N/(Sm_N \times Nd_N)^{0.5}$, where subscript N means normalized to primitive mantle (Sun and McDonough, 1989).

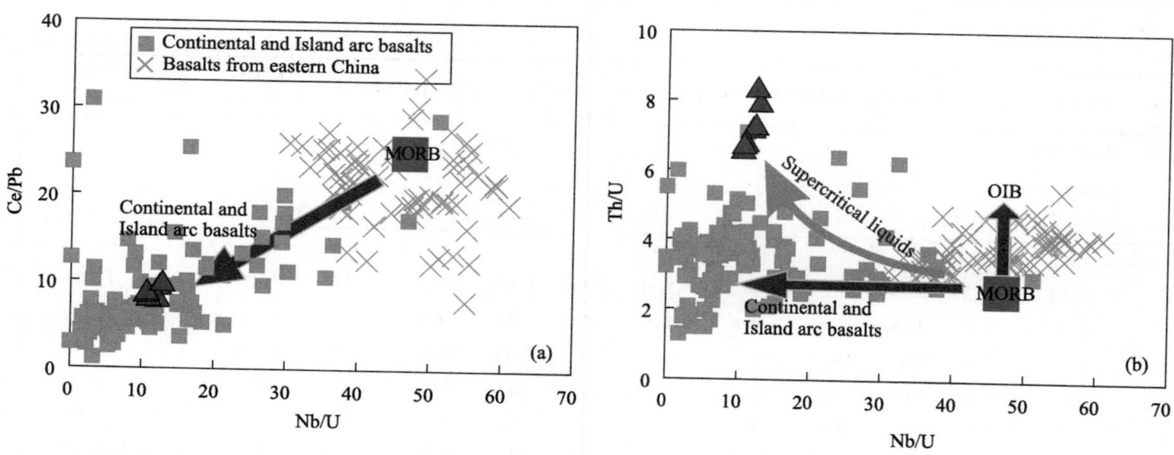

Fig. 4 Ce/Pb (a) and Th/U (b) versus Nb/U for the Tengchong basalts. Data for the MORB are from Sun and McDonough (1989). The gray line with arrow indicates the addition of supercritical liquids. The continental and island arc basalts and intra-plate basalts from eastern China are plotted for comparison.

4.2 Sr-Nd-Pb isotopic compositions

The Sr, Nd and Pb isotopic compositions of the Tengchong basalts are presented in Table 2. These basalts exhibit higher $^{87}Sr/^{86}Sr$ ratios of 0.7069 to 0.7083 and lower ε_{Nd} values (from −4.3 to −3.2) than those of the continental and arc basalts, but fall within the OIB field, pointing to an EM-II endmember signature (Fig. 5a and Table 2). They also display high radiogenic Pb isotopic compositions, with $^{206}Pb/^{204}Pb$ of 18.05 to 18.29 and $^{208}Pb/^{204}Pb$ of 38.90 to 39.13 (Table 2). The enriched Sr-Nd and relatively high Pb isotope ratios possibly suggested the addition of Indian sediments (Fig. 5). In addition, these basalts plot on the left of the 'Geochron' line and the upper side of the NHRL defined by Hart (1984) (Figs. 5c and d).

Table 2 Sr, Nd and Pb isotopic compositions of the Tengchong Cenozoic basalts and andesites

Sample	$^{206}Pb/^{204}Pb$	$2\sigma_m$	$^{207}Pb/^{204}Pb$	$2\sigma_m$	$^{208}Pb/^{204}Pb$	$2\sigma_m$	$^{87}Sr/^{86}Sr$	$2\sigma_m$	$^{143}Nd/^{144}Nd$	$2\sigma_m$	ε_{Nd}[a]
Basalts											
WH1401	18.046	0.001	15.661	0.001	38.923	0.003	0.708013	0.000011	0.512417	0.000007	−4.3
WH1402	18.046	0.001	15.655	0.001	38.904	0.003	0.708259	0.000012	0.512416	0.000009	−4.3
QLS1403	18.289	0.002	15.686	0.002	39.130	0.004	0.706943	0.000011	0.512468	0.000008	−3.3
QLS1404	18.286	0.001	15.679	0.001	39.107	0.003	0.706883	0.000011	0.512469	0.000009	−3.3
QLS1405	18.281	0.002	15.676	0.002	39.094	0.005	0.706906	0.000010	0.512474	0.000009	−3.2
QLS1406	18.281	0.002	15.672	0.002	39.084	0.004	0.706906	0.000012	0.512473	0.000008	−3.2
QLS1407	18.284	0.001	15.673	0.001	39.086	0.004	0.706871	0.000010	0.512470	0.000009	−3.3
Andesites											
TS1403	18.160	0.002	15.673	0.002	39.134	0.005	0.705763	0.000012	0.512348	0.000008	−5.7
TS1404	18.177	0.001	15.690	0.001	39.191	0.003	0.705734	0.000012	0.512352	0.000008	−5.6
TS1405	18.175	0.002	15.688	0.002	39.187	0.004	0.705775	0.000011	0.512349	0.000009	−5.6
TS1406	18.167	0.002	15.666	0.002	39.075	0.004	0.706735	0.000015	0.512368	0.000009	−5.3
YJDS1401	18.168	0.000	15.670	0.002	39.100	0.004	0.706756	0.000010	0.512351	0.000009	−5.6
YJDS1402	18.177	0.002	15.681	0.001	39.111	0.004	0.706728	0.000011	0.512370	0.000008	−5.2
YJDS1403	18.185	0.002	15.693	0.001	39.128	0.004	0.706764	0.000013	0.512351	0.000008	−5.6
YJDS1404	18.181	0.002	15.683	0.002	39.134	0.003	0.706767	0.000011	0.512371	0.000008	−5.2
YJDS1405	18.174	0.001	15.668	0.002	39.068	0.004	0.706766	0.000010	0.512378	0.000008	−5.1
LGP1401	18.185	0.002	15.687	0.002	39.203	0.005	0.706499	0.000011	0.512359	0.000009	−5.4
LGP1402	18.191	0.001	15.675	0.001	39.101	0.003	0.706393	0.000012	0.512401	0.000009	−4.6
HKS1402	18.122	0.002	15.693	0.001	39.205	0.003	0.708587	0.000011	0.512207	0.000008	−8.4
HKS1403	18.100	0.001	15.690	0.001	39.178	0.004	0.708584	0.000015	0.512201	0.000009	−8.5

Continued

Sample	^{206}Pb/^{204}Pb	$2\sigma_m$	^{207}Pb/^{204}Pb	$2\sigma_m$	^{208}Pb/^{204}Pb	$2\sigma_m$	^{87}Sr/^{86}Sr	$2\sigma_m$	^{143}Nd/^{144}Nd	$2\sigma_m$	ε_{Nd} [a]
HKS1404	18.076	0.001	15.676	0.001	39.017	0.003	0.708507	0.000011	0.512207	0.000008	−8.4
HKS1406	18.067	0.002	15.676	0.002	39.078	0.005	0.708445	0.000009	0.512224	0.000009	−8.1
HKS1408	18.005	0.001	15.642	0.001	39.026	0.004	0.708529	0.000012	0.512203	0.000008	−8.5
HKS1409	18.017	0.002	15.628	0.002	38.868	0.005	0.708563	0.000016	0.512204	0.000008	−8.5
HKS1414	18.026	0.002	15.657	0.002	39.068	0.004	0.708556	0.000010	0.512204	0.000008	−8.5
HKS1415	18.054	0.002	15.698	0.002	39.153	0.004	0.708538	0.000008	0.512208	0.000009	−8.4
HKS1416	17.912	0.002	15.658	0.002	39.116	0.004	0.708706	0.000011	0.512057	0.000009	−11.3
DYS1401	18.088	0.002	15.679	0.001	39.165	0.004	0.708812	0.000010	0.512183	0.000008	−8.9
DYS1402	18.099	0.001	15.674	0.001	39.084	0.004	0.708793	0.000011	0.512177	0.000009	−9.0
DYS1403	18.089	0.002	15.683	0.001	39.179	0.004	0.708871	0.000011	0.512176	0.000008	−9.0
DYS1406	18.103	0.002	15.696	0.002	39.218	0.005	0.708818	0.000011	0.512183	0.000008	−8.9
DYS1407	18.103	0.002	15.697	0.002	39.225	0.004	0.708803	0.000012	0.512169	0.000007	−9.1
DYS1408	18.114	0.002	15.706	0.002	39.244	0.005	0.708791	0.000012	0.512193	0.000008	−8.7
DYS1410	18.092	0.002	15.685	0.002	39.186	0.004	0.708795	0.000011	0.512192	0.000009	−8.7
DYS1412	18.087	0.002	15.678	0.001	39.155	0.004	0.708855	0.000011	0.512172	0.000009	−9.1
DYS1413	18.101	0.002	15.698	0.002	39.232	0.004	0.708855	0.000012	0.512186	0.000009	−8.8
DYS1415	18.077	0.002	15.680	0.002	39.127	0.005	0.708826	0.000010	0.512209	0.000008	−8.4
DYS1416	18.048	0.002	15.656	0.001	39.045	0.003	0.708877	0.000012	0.512202	0.000009	−8.5
DYS1419	18.056	0.001	15.654	0.001	39.048	0.004	0.708909	0.000010	0.512216	0.000009	−8.2

[a] ε_{Nd} values were calculated using $(^{143}Nd/^{144}Nd)_{CHUR} = 0.512638$.

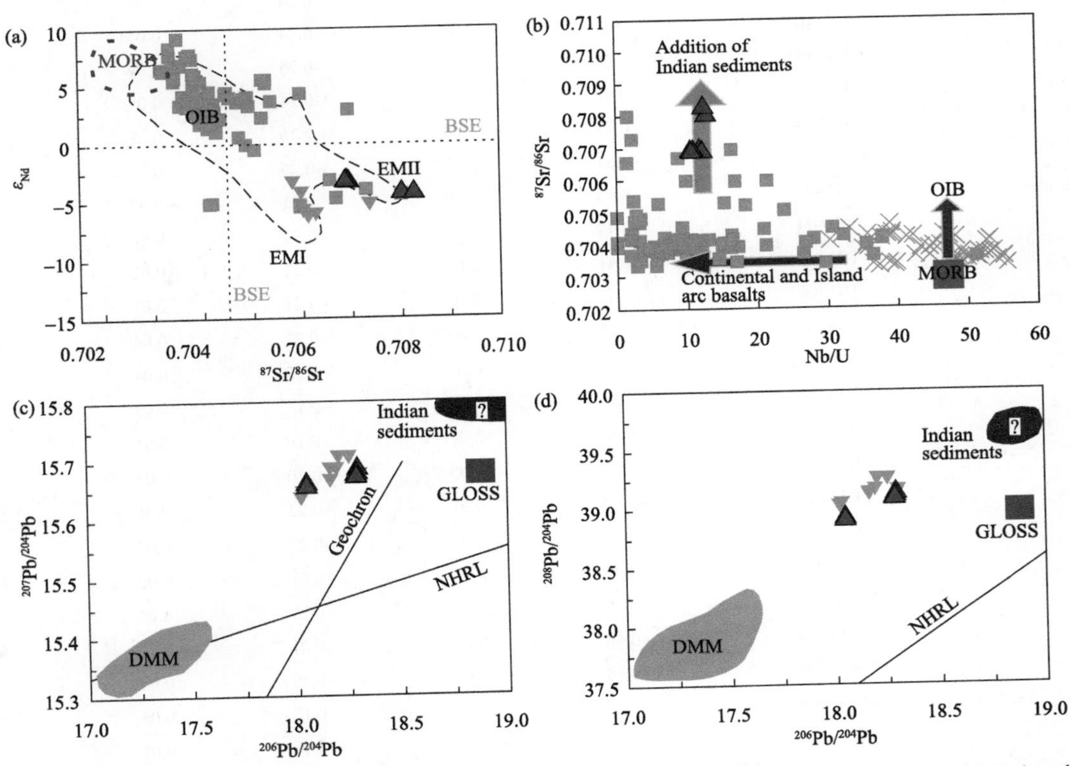

Fig. 5 ^{87}Sr/^{86}Sr vs. ε_{Nd} (a) and Nb/U (b), ^{206}Pb/^{204}Pb vs. ^{207}Pb/^{204}Pb (c) and ^{208}Pb/^{204}Pb (d) for the Tengchong basalts. The gray arrow in panel b possibly reflected the addition of Indian sediments. Data sources: trace elements and isotopic endmember of the MORB and OIB are from Zindler and Hart (1986), Sun and McDonough (1989) and Hofmann (1997). GLOSS and Indian sediments are from Plank and Langmuir (1998). NHRL stands for Northern Hemisphere Reference Line (Hart, 1984), with the equation of $(^{207}Pb/^{204}Pb)_{NHRL} = 0.1084\times(^{206}Pb/^{204}Pb)+13.491$ and $(^{208}Pb/^{204}Pb)_{NHRL} = 1.209\times(^{206}Pb/^{204}Pb)+15.627$. The line 'Geochron' is from Patterson (1956).

4.3 Mg isotopic composition

The stable Mg isotopic compositions of the Tengchong basalts and andesites and standard reference materials are listed in Table 3. The $\Delta^{25}Mg'$ values range from −0.01 to 0.02 (Table 3) ($\Delta^{25}Mg'=\delta^{25}Mg'-0.521\times\delta^{26}Mg'$, where $\delta^{25,26}Mg'=1000\times\ln(\delta^{25,26}Mg/1000+1)$; Young and Galy, 2004). The Mg isotopic composition of $\delta^{26}Mg$ values are described below.

Overall, $\delta^{26}Mg$ values of the Tengchong volcanic rocks vary from −0.51 to −0.33‰ (Table 3). As shown in Fig. 6a, their $\delta^{26}Mg$ values are lower than that of the average mantle (−0.25±0.07‰; Teng et al., 2010a) and the island arc basalts (−0.35 to +0.06‰; Teng et al., 2016; Li et al., 2017), but similar to those of the low $\delta^{26}Mg$ Cenozoic basalts from eastern China (−0.60 to −0.30‰;Yang et al., 2012; Huang et al., 2015a; Tian et al., 2016; Li et al., 2017; Su et al., 2017) (Fig. 6a).

The Tengchong basalts exhibit $\delta^{26}Mg$ values ranging from −0.51 to −0.45‰ with an average of −0.49± 0.04‰ (2SD, n=7), whereas the andesites (Group 1 and 2) yield slightly heavier $\delta^{26}Mg$ values of −0.49 to −0.33‰, with an average of −0.40±0.08‰ (2SD, n=32) (Fig. 6a). The $\delta^{26}Mg$ values of the basalts do not vary with either increasing MgO, $(Gd/Yb)_N$ or Fe/Mn ratios (Figs. 6b-d).

Table 3 Magnesium isotopic compositions of the Tengchong volcanic rocks as well as standard reference materials

Types	Sample	$\delta^{26}Mg^a$	2SD[b]	$\delta^{25}Mg^a$	2SD[b]	$\Delta^{25}Mg'^c$
Basalts	WH1401	−0.45	0.03	−0.22	0.02	0.01
	WH1402	−0.48	0.03	−0.24	0.03	0.01
	QLS1403	−0.50	0.08	−0.24	0.05	0.02
	QLS1404	−0.50	0.05	−0.25	0.04	0.01
	QLS1405	−0.48	0.09	−0.24	0.05	0.01
	QLS1406	−0.51	0.08	−0.26	0.08	0.01
	QLS1407	−0.51	0.02	−0.26	0.01	0.01
Andesites	TS1403	−0.44	0.10	−0.21	0.05	0.02
	TS1404	−0.49	0.03	−0.27	0.02	−0.01
	TS1405	−0.41	0.05	−0.21	0.02	0.00
	TS1406	−0.41	0.01	−0.21	0.01	0.00
	YJDS1401	−0.45	0.03	−0.22	0.04	0.01
	YJDS1402	−0.43	0.05	−0.22	0.04	0.00
	YJDS1403	−0.36	0.02	−0.18	0.03	0.01
	YJDS1404	−0.38	0.07	−0.19	0.01	0.01
	YJDS1405	−0.41	0.05	−0.20	0.03	0.01
	LGP1401	−0.44	0.01	−0.23	0.01	0.00
	LGP1402	−0.45	0.02	−0.22	0.02	0.01
	HKS1402	−0.43	0.06	−0.21	0.02	0.01
	HKS1403	−0.39	0.01	−0.21	0.03	0.00
	HKS1404	−0.43	0.01	−0.21	0.01	0.01
	HKS1406	−0.38	0.01	−0.19	0.04	0.00
	HKS1408	−0.34	0.06	−0.17	0.05	0.01
	HKS1409	−0.35	0.02	−0.18	0.01	0.01
	HKS1414	−0.45	0.06	−0.23	0.05	0.00
	HKS1415	−0.36	0.03	−0.19	0.02	−0.01
	HKS1416	−0.39	0.08	−0.20	0.10	0.01
	DYS1401	−0.39	0.03	−0.20	0.02	0.00
	DYS1402	−0.33	0.07	−0.17	0.04	0.00
	DYS1403	−0.38	0.07	−0.19	0.08	0.01

Continued

Types	Sample	δ^{26}Mg[a]	2SD[b]	δ^{25}Mg[a]	2SD[b]	Δ^{25}Mg'[c]
Andesites	DYS1406	−0.41	0.10	−0.20	0.11	0.02
	DYS1407	−0.40	0.10	−0.20	0.05	0.01
	DYS1408	−0.40	0.08	−0.21	0.04	0.00
	DYS1410	−0.36	0.00	−0.19	0.02	0.00
	DYS1412	−0.34	0.01	−0.17	0.02	0.01
	DYS1413	−0.38	0.05	−0.17	0.06	0.03
	DYS1415	−0.40	0.06	−0.21	0.04	0.00
	DYS1416	−0.37	0.00	−0.20	0.04	0.00
	DYS1419	−0.37	0.07	−0.18	0.03	0.01
Standards	GSR-2	−0.55	0.07	−0.28	0.04	0.01
	GSR-3	−0.45	0.03	−0.22	0.05	0.01
	BCR-2	−0.20	0.09	−0.10	0.03	0.00
	BHVO-2	−0.27	0.06	−0.12	0.10	0.02
	Replicate[d]	−0.26	0.08	−0.16	0.05	−0.02

[a] $\delta^{25,26}Mg = \{(^{25,26}Mg/^{24}Mg)_{sample}/(^{25,26}Mg/^{24}Mg)_{DSM3} - 1\} \times 1000$, DSM3 is a Mg solution made from pure Mg metal (Galy et al., 2003).

[b] 2SD = two times the standard deviation of the population of 4 repeat measurements of a sample solution.

[c] $\Delta^{25}Mg' = \delta^{25}Mg' - 0.521 \times \delta^{26}Mg'$, where $\delta^{25,26}Mg' = 1000 \times \ln[(\delta^{25,26}Mg + 1000)/1000]$ (Young and Galy, 2004).

[d] Replicate denotes repeating sample dissolution, column chemistry and instrumental analysis

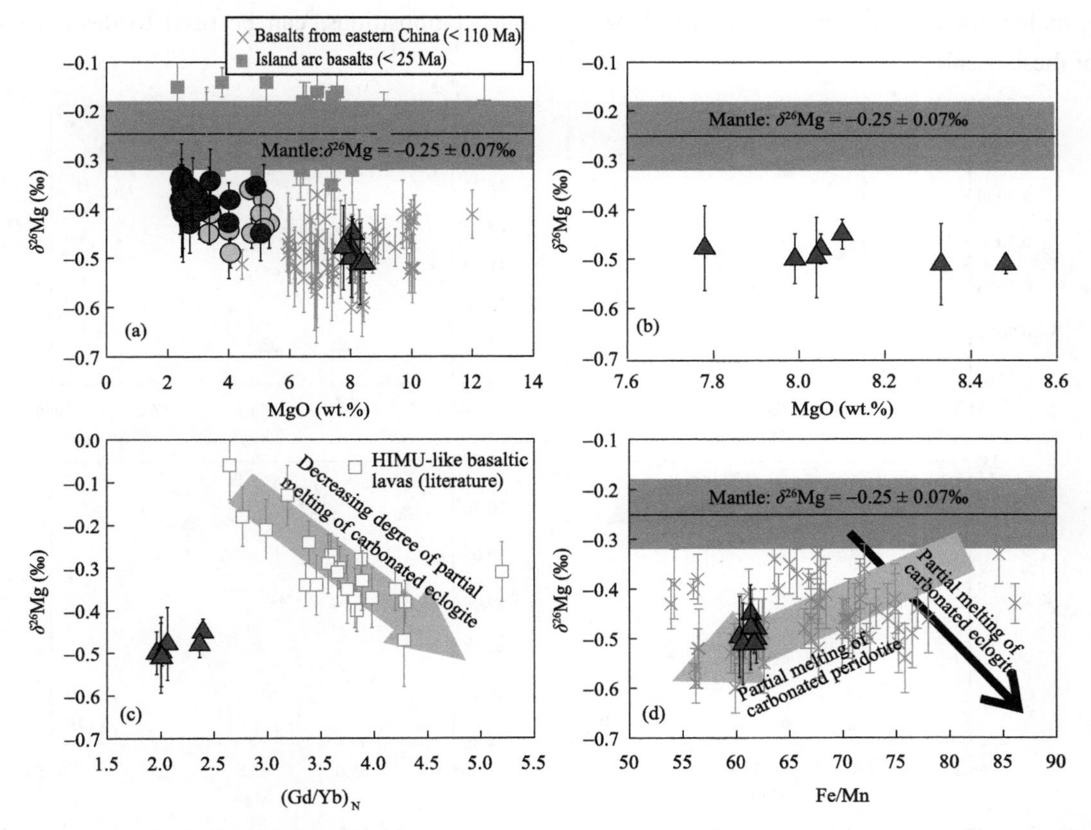

Fig. 6 δ^{26}Mg vs. (a) MgO (wt.%). The Tengchong basalt samples of panel (a) is expanded in panel (b). (c) δ^{26}Mg vs. $(Gd/Yb)_N$, and (d) δ^{26}Mg vs. Fe/Mn for the Tengchong volcanic rocks. Error bars represent 2SD uncertainties. Data source: Island arc basalts from Teng et al. (2016) and Li et al. (2017). HIMU-like basaltic lavas with Mg isotopic variation are from Wang et al. (2016). Note: subscript N means normalized to chondrite values (Sun and McDonough, 1989). The green bars in (a) and (c) represent the average mantle (δ^{26}Mg=−0.25±0.07‰; Teng et al., 2010a). (For interpretation of the references to colour in this figure legend, the reader is referred to the web version of this article.)

5 Discussion

The Tengchong volcanic rocks studied here exhibit homogeneous and light Mg isotopic compositions (δ^{26}Mg = −0.51 to −0.33‰), similar to the <110 Ma intra-continental basalts from eastern China (−0.60 to −0.30‰; Yang et al., 2012; Huang et al., 2015a; Tian et al., 2016; Li et al., 2017; Su et al., 2017). We will first focus on the origin of the Tengchong basalts with light Mg isotopic composition and EM-II signature and then the genetic links between basalts and andesites. Finally, we evaluate the geochemical implications for the deep carbon cycling at depths from 150 to 400 km.

5.1 Origin of the Tengchong basalts

The Chemical Index of Alteration (CIA) values of the Tengchong basalts range from 40.6 to 42.1 (Table 1), within the range of un-weathered basalts (30–45) as noted by Nesbitt and Young (1982). The Tengchong basalts display high and restricted Mg# values, ranging from 64.7 to 66.3 (Table 1 and Fig. 2), which is slightly lower than that of the primary magma of the normal MORB (Langmuir et al., 1977), possibly reflecting a low-degree of partial melting of mantle peridotite (~5%; Zhou et al., 2012). Furthermore, there is little or no crustal contamination in the Tengchong basalts, as suggested by the lack of any correlations between either $^{87}Sr/^{86}Sr$, ε_{Nd}, $^{206}Pb/^{204}Pb$ or $^{208}Pb/^{204}Pb$ with Mg# (Fig. 7). Thus, geochemical characteristics of the Tengchong basalts, including major, trace elemental and Sr-Nd-Pb-Mg isotopic compositions, can be used to discuss the mantle sources of these basalts.

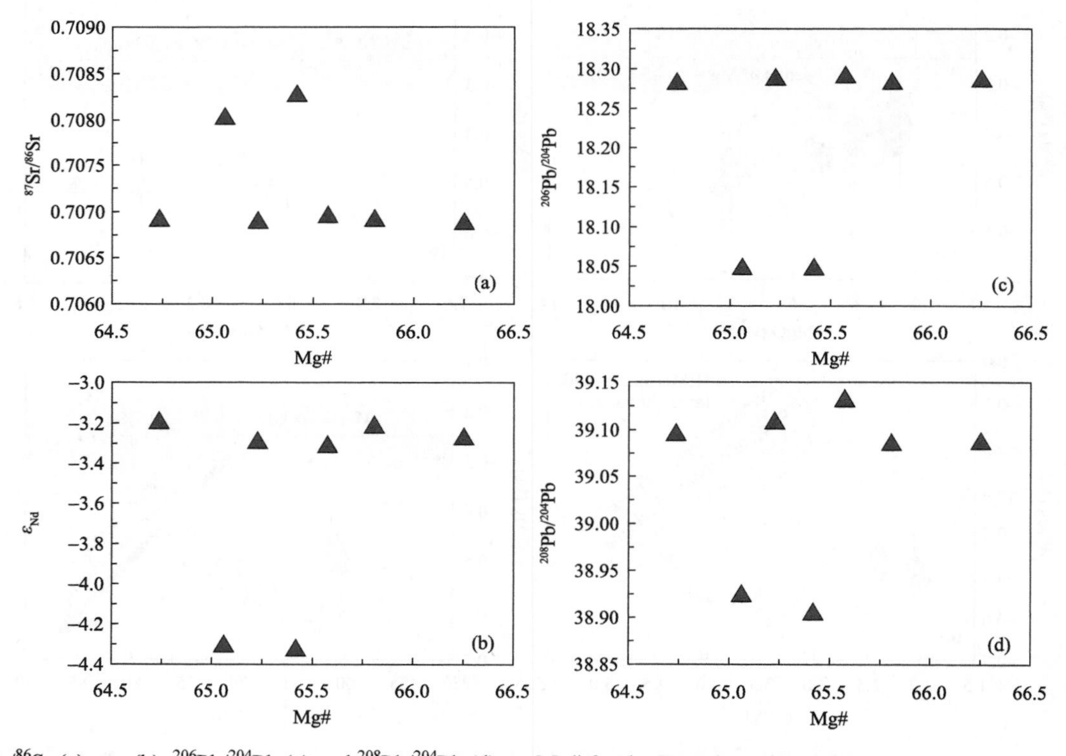

Fig. 7　$^{87}Sr/^{86}Sr$ (a), ε_{Nd} (b), $^{206}Pb/^{204}Pb$ (c) and $^{208}Pb/^{204}Pb$ (d) vs. Mg# for the Tengchong basalts in this study.

5.1.1 Contribution of recycled sedimentary carbonates

The δ^{26}Mg values of the Tengchong basalts do not vary with MgO content (Fig. 6a), indicating limited Mg isotope fractionation during magma differentiation. This agrees with previous work showing that magmatic processes had little effect on Mg isotopes (e.g., Teng et al., 2007, 2010a). Furthermore, little or the lack of crustal contamination should not have led to any Mg isotopic variation. Four potential candidates of the light Mg isotopic composition in the mantle are: (1) kinetic Mg isotope fractionation (e.g., Richter et al., 2008; Huang et al., 2010), (2) the accumulation of ilmenite with isotopically light Mg isotopes (Sedaghatpour et al., 2013), (3) the recycling of carbonated eclogites (Wang et al., 2016; Sun et al., 2017), or (4) the recycling of sedimentary carbonates (e.g., Yang et al., 2012; Huang et al., 2015a).

Kinetic processes can clearly result in Mg isotope fractionation by both the thermal and chemical diffusion at high temperatures, as demonstrated by melting experiments (e.g., Richter et al., 2008; Huang et al., 2010). As with all stable isotopes, the lighter isotopes, such as ^{6}Li, ^{24}Mg, and ^{54}Fe, always diffuse faster than the heavier ones during both chemical and thermal diffusion (Dauphas, 2007; Richter et al., 2008, 2009; Huang et al., 2010). For the thermal diffusion, a positive correlation between δ^{26}Mg and $\delta^{44/40}$Ca would be expected. However, the Tengchong volcanic rocks do not show positive correlation between $\delta^{44/40}$Ca and δ^{26}Mg (see Fig. 5 in Liu et al. 2017). Chemical diffusion can also be ruled out by the following evidences. First, during melt-peridotite reaction, Mg tends to diffuse from peridotite to melt while Ca diffuses from melt to peridotite, resulting in heavy Mg and light Ca isotopes in the peridotite (Zhao et al., 2017). Thus, the melt should have heavy Ca and light Mg isotopic compositions. However, this is inconsistent with the light Ca isotopic composition ($\delta^{44/40}$Ca = 0.67–0.80‰) in the Tengchong volcanic rocks (Liu et al., 2017). In addition, no correlations of δ^{26}Mg and $\delta^{44/40}$Ca with either MgO (Fig. 6b) or CaO content varying from 4.65 to 7.99 wt.% (Liu et al., 2017) were observed in Tengchong volcanic rocks. Moreover, the samples studied here come from a very broad area (>20 km, Fig. 1c). However, it is impossible that chemical diffusion could operate on such a large scale to form a homogeneous Mg isotopic composition (δ^{26}Mg = –0.51 to –0.45‰) in the Tengchong basalts. Therefore, the low δ^{26}Mg values in Tengchong basalts are unlikely controlled by kinetic fractionation during chemical or thermal diffusion.

The accumulation of isotopically light ilmenite in the mantle source was suggested to produce low δ^{26}Mg values in some lunar high-Ti basalts (Sedaghatpour et al., 2013). If this process occurred in the source of Tengchong basalts, the TiO_2 content of the basalts should negatively correlate with Nb/Ta ratios and δ^{26}Mg values (Huang et al., 2015a; Tian et al., 2016). However, slightly positive correlations occur (Fig. 8) thus ruling this process out.

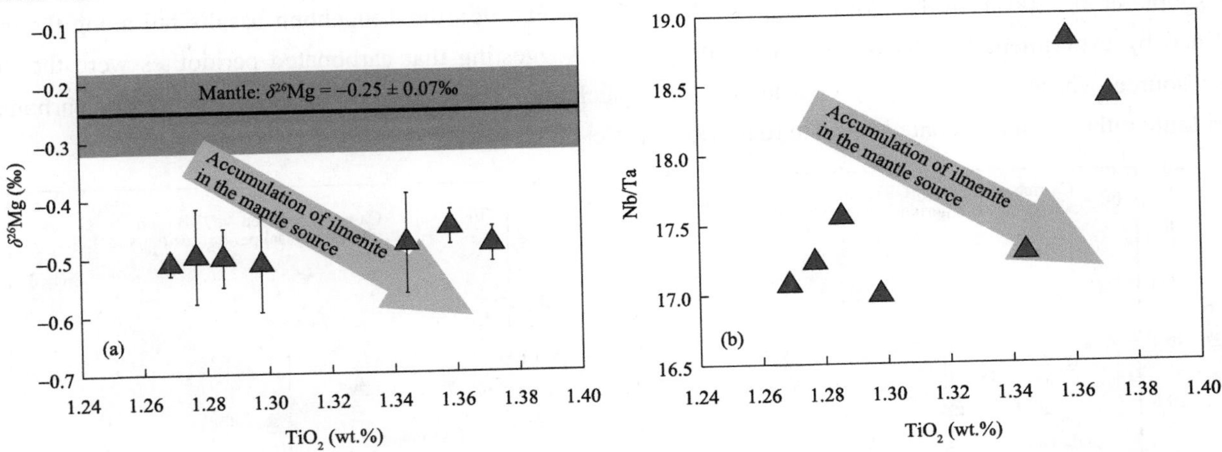

Fig. 8 δ^{26}Mg (a) and Nb/Ta (b) vs. TiO_2 (wt.%) for the Tengchong basalts. The arrow represents the accumulation of isotopically light ilmenite in the mantle source. Error bars represent 2SD uncertainties. The green bar represent the average mantle (δ^{26}Mg = –0.25 ± 0.07‰; Teng et al., 2010a). (For interpretation of the references to colour in this figure legend, the reader is referred to the web version of this article.)

Both recycled carbonated eclogites and recycled sedimentary carbonates have light Mg isotopic compositions, and have been used to explain the origins of some low δ^{26}Mg basalts and potassic-rich basalts (Yang et al., 2012; Huang et al., 2015a; Wang et al., 2016, 2017; Li et al., 2017; Sun et al., 2017). Eclogite-derived melts are silicate-rich, producing olivine-free pyroxenite via the interaction with peridotite (Sobolev et al., 2007). In contrast, the partial melting of recycled sedimentary carbonates generates carbonate-rich melt, resulting in carbonated peridotite. These two mantle sources produce low δ^{26}Mg melts but with different major and trace elemental compositions (Li et al., 2017).

Compared with carbonated peridotite-derived melts, pyroxenite-derived or eclogite-derived melts have high Fe/Mn, Hf/Hf*, Ti/Ti* and low CaO/Al_2O_3 ratios (e.g., Humayun et al., 2004; Dasgupta et al., 2006, 2007; Sobolev et al., 2007; Walter et al., 2008). In addition, the pyroxenites/eclogites are more enriched in Yb relative to Gd (EarthRef.org–GERM Partition Coefficient (Kd) Database). Thus, a low-degree of partial melting of pyroxenites or eclogites could have led to higher Gd/Yb ratios than in the peridotite melts. Recently, Wang et al. (2016) and Sun et al. (2017) used the negative relationship between $(Gd/Yb)_N$ and δ^{26}Mg to identify the occurrence of carbonated eclogites in their mantle source. Additionally, experiments demonstrate that clinopyroxene, orthopyroxene and garnet generally have $D_{Fe/Mn}$<1 (where $D_{Fe/Mn}=\dfrac{Fe^{min}/Fe^{melt}}{Mn^{min}/Mn^{melt}}$) and olivine has $D_{Fe/Mn}$ >1(Humayun et al., 2004 and compiled data in Liu et al. 2008). According to these partition coefficients, pyroxenites and eclogites should have a little lower $D_{Fe/Mn}$ value than that of peridotites. Thus, Fe/Mn ratios of pyroxenites and eclogites melts are commonly higher than in peridotite melts. However, the δ^{26}Mg values of the Tengchong basalts do not vary with either $(Gd/Yb)_N$ or Fe/Mn ratios (Figs. 6c and d). This means that the partial melting of recycled carbonated eclogites probably did not occur. Instead, the basalts have low Ti/Ti* and Hf/Hf* ratios relative to the MORB and fall along the trends of carbonate metasomatism (Figs. 9a and b). This feature is similar to the <110 Ma basalts from eastern China (e.g., Yang et al., 2012; Huang et al., 2015a), but distinct from the island arc basalts with MORB-like or heavier δ^{26}Mg values (Teng et al., 2016; Li et al., 2017), which were affected by Ca-rich or altered MORB- and altered abyssal peridotites-derived fluids (Figs. 9a and b). Furthermore, pyroxenite-derived or eclogite-derived melts have high TiO_2 content (up to 19.4 wt.%) (Dasgupta et al., 2006). In contrast, melts from carbonated peridotites are enriched in Na_2O and K_2O, but also depleted in TiO_2 (Dasgupta et al., 2007). In the plot of total alkalis vs. TiO_2 (Fig. 9c), the Tengchong basalts fall along the trend defined by experimental melts of carbonated peridotites, suggesting that carbonated peridotites were the main melt source. These geochemical properties of the Tengchong basalts strongly support a source of carbonated peridotite rather than carbonated eclogite (e.g., Dasgupta et al., 2007).

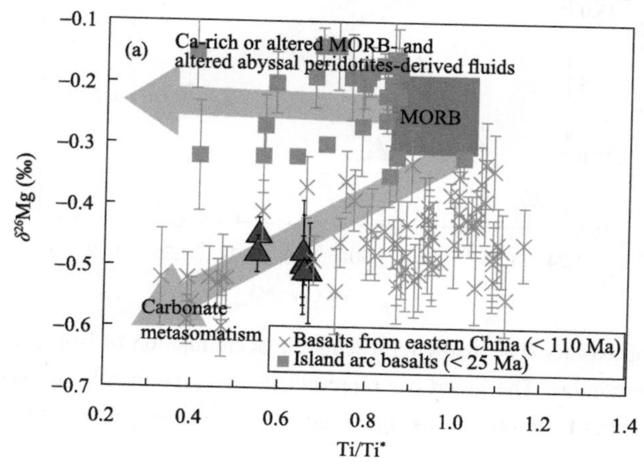

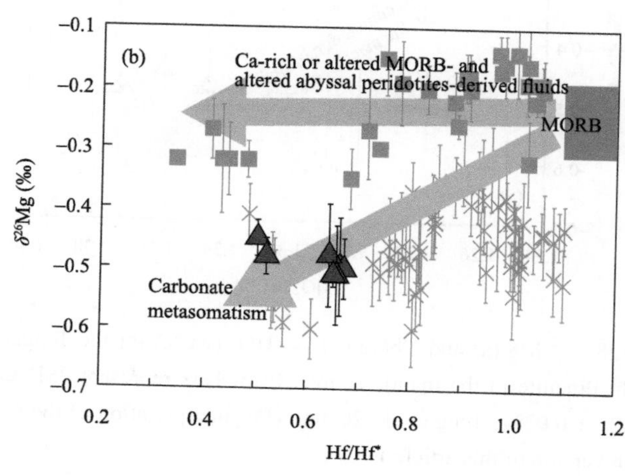

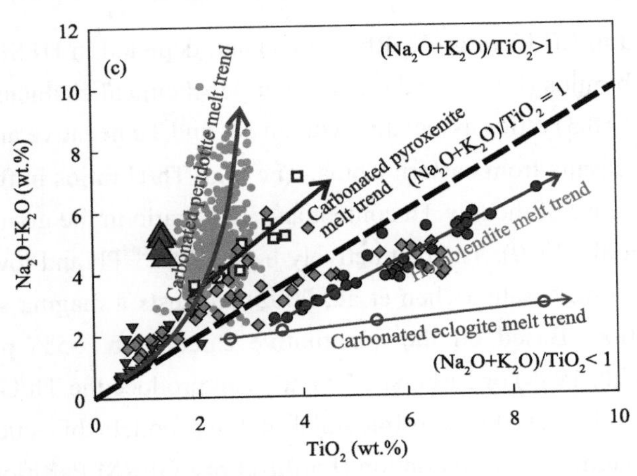

Fig. 9 (a) Ti/Ti* and (b) Hf/Hf* vs. δ^{26}Mg (wt.%) for the Tengchong basalts. Error bars represent 2SD uncertainties. Mg isotopic composition (δ^{26}Mg=−0.25±0.07‰) and other parameters of MORB are from Hofmann (1988) and Teng et al. (2010a). Elemental anomalies are calculated as follows: $Ti/Ti^* = Ti_N/(Sm_N^{-0.055} \times Nd_N^{0.333} \times Gd_N^{0.722})$, $Hf/Hf^* = Hf_N/(Sm_N \times Nd_N)^{0.5}$, where the subscript N means normalized to primitive mantle (Sun and McDonough, 1989). The arrows in the panel (a) and (b) indicate Ca-rich or altered MORB, as well as altered abyssal peridotites-derived fluids for the island arc basalts, carbonate metasomatism for Tengchong basalts and <110 Ma intra-continental basalts from eastern China. (c) Na_2O+K_2O (wt.%) vs. TiO_2 for the Tengchong basalts. Data sources for the Cenozoic basalts in eastern China are from the references (Zhi et al., 1990; Zou et al., 2000; Xu et al., 2005, 2012; Tang et al., 2006; Liu et al., 2008; Chen et al., 2009; Zeng et al., 2010, 2011; Wang et al., 2011; Huang et al., 2015a). Also displayed for comparison are experimentally-derived melts from carbonated peridotite (Hirose, 1997; Dasgupta et al., 2007), carbonated pyroxenite (Gerbode and Gasgupta, 2010), pyroxenite or eclogite (Hirschmann et al., 2003; Kogiso et al., 2003; Pertermann and Hirschmann, 2003; Kogiso and Hirschmann, 2006), hornblendite (Pilet et al., 2008), and carbonated eclogite (Dasgupta et al., 2006).

5.1.2 Contribution of the Indian sediments

When compared with the island arc basalts, the Tengchong basalts show enriched Sr and Nd isotopes toward the Indian sediments (Figs. 5a and b). Furthermore, they also exhibit both high Pb (8.8–12.1 ppm) and Th (9.5–11.1 ppm) content, high Ba/Th (63.9–78.7) and low Rb/Cs (25.9–47.3), Ce/Pb (7.7–9.6) ratios (Table 1), which are typical features of the Indian sediments (e.g., White et al., 1985; Ben Othman et al., 1989; Hart and Reid, 1991; Miller et al., 1994; Plank and Langmuir, 1998). It is difficult to distinguish the Indian sediments from the upper continental crustal materials because their trace element patterns are similar (Rudnick and Gao, 2003). However, the high $^{207}Pb/^{204}Pb$ and $^{208}Pb/^{204}Pb$ ratios for a given $^{206}Pb/^{204}Pb$ value of the Tengchong basalts (Figs. 5c and d) do not follow the mixing trends between DMM and the oceanic sediments (GLOSS). Such high radiogenic Pb isotopes could only derive from an old upper crust with high U/Pb (0.11–0.17) and Th/Pb (0.88–1.15) ratios, suggesting the recycling of Indian sediments into their source mantle (Figs. 5c and d). Several previous studies have observed a similar signature on mid-ocean ridge basalts from the Indian Ocean (e.g., Dupré and Allègre, 1983; Hamelin et al., 1986; Ito et al., 1987). Their investigations confirmed that the Indian Ocean MORBs are characterized by on average higher $^{87}Sr/^{86}Sr$ and lower $^{143}Nd/^{144}Nd$ and $^{206}Pb/^{204}Pb$ isotopes ratios, but contain higher $^{207}Pb/^{204}Pb$ and $^{208}Pb/^{204}Pb$ for a given $^{206}Pb/^{204}Pb$ value, than those from other oceans (e.g., Atlantic, Pacific). Therefore, the recycled materials in the mantle source of the Tengchong basalts should be Indian sediments containing ancient upper crustal materials.

5.1.3 Characteristics of the metasomatic agent

Although the Tengchong basalts have low δ^{26}Mg values similar to the intraplate basalts from eastern China (e.g., Yang et al., 2012; Li et al., 2017), they exhibit continental and island arc lavas-like geochemical features.

For instance, they are enriched in LILEs (e.g., K, Rb and Ba) and depleted in HFSEs (e.g., Nb, Ta, Zr, Hf and P) as well as low Nb/U and Ce/Pb ratios (Figs. 3 and 4). Another geochemical characteristic of the Tengchong basalt is the extremely high Th/U (6.5–8.3) ratios associated with an Nb and Ta negative anomaly (Figs. 3a and 4b). This is different from the OIB-like basalts from eastern China. The high Th/U ratios in the basalts are unlikely derived from crustal contamination because of the low Th content and Th/U ratio in the granulitic xenoliths exposed in the Tengchong volcanic area (Ji et al., 2000). The distinctively high $^{208}Pb/^{204}Pb$ and low $^{206}Pb/^{204}Pb$ ratios relative to the crustal rocks in the Tengchong basalts (Chen et al., 2002) suggests a magma source with a high Th/U ratio instead of crustal contamination. Based on the quantitative calculation, ~5% partial melting of the mantle peridotites (Sun and McDonough, 1989) or carbonated mantle can produce the Th/U ratios and Th abundance less than 4.5 and 8.5 ppm, respectively, based on the fractional melting model (this study). Because U behaves more incompatible than Th during magmatic differentiation (EarthRef.org-GERM Partition Coefficient (Kd) Database), the residual melt is depleted Th relative to U and possibly shows a small decreased Th/U ratios. This scene is consistent with the global MORB with Th/U ranging from 2.2 to 3.5 (e.g., Langmuir et al., 1977; Hofmann, 1988; Gale et al., 2013). In addition, although the recycling of continentally derived clay-rich mature sediments or mudstones can explain the high Th/U ratios (Zou et al., 2014), two lines of evidence are inconsistent with this explanation. One is that continental and island arc basalts from other localities worldwide do not have such high Th/U ratios relative to the Tengchong basalts (Fig. 4c). Another reason is that the clay or mudstones are possibly enriched heavy $\delta^{26}Mg$ (e.g., Teng et al., 2010b; Huang et al., 2012; Opfergelt et al., 2012; Liu et al., 2014; Wang et al., 2015) and thus the addition of clay-rich sediments or mudstones should have shifted the Mg isotopic compositions of the Tengchong basalts to slightly higher values than that of the mantle. Thus, the high Th (9.5–11.1 ppm) and Th/U (6.5–8.3) ratios in the Tengchong basalts must record their mantle source. Alternatively, these geochemical features suggest the higher mobility of Th over U in the mantle metasomatic agent, which is thought to be a unique characteristic of slab-derived supercritical liquids in subduction zones generated over 6 GPa pressures (Kessel et al., 2005).

5.1.4 Recycled carbonate species and mantle source depth

One of the key issues of deep carbon cycling is constraining the recycling depth of sedimentary carbonates during subduction, which can provide constraints on the residence time of the recycled carbon in the mantle (Li et al., 2017). The stability of carbonate species (calcite, dolomite, and magnesite) in the mantle strongly depends on the *P-T* conditions (Kushiro et al., 1975; Sato and Katsura, 2001). Different carbonate species have different Sr and Mg content. The Mg-Sr isotopic compositions of the low $\delta^{26}Mg$ basalts can distinguish different carbonate species in the mantle source to determine the recycled depth of carbonates (Huang and Xiao, 2016; Ke et al., 2016).

Since the mantle source of the Tengchong basalts have been affected by both recycled carbonates and other crustal materials, the Mg-Sr isotope system alone cannot determine the species of carbonates. Here, Mg-Sr-Nd-Pb isotopes and trace elements have been modeled (Figs. 10 and 11). As shown in Fig. 10a-c, pure dolomite mixed with the DMM + 1% supercritical liquids cannot match the Sr, Nd and Pb isotopic variations of the Tengchong basalts. Pure recycled calcite mixed with the DMM + 1% supercritical liquids can match Mg, Sr and Nd isotopes (Fig. 10a and b), but do not satisfy the Pb isotopic variation (Fig. 10c). In addition, because of the low MgO content of calcite (~2 wt.%; Ke et al., 2016), an unrealistic large proportion of calcite (60%) is required (Fig. 10a and b). Hence, the crustal materials (e.g., Indian sediments) with high $^{87}Sr/^{86}Sr$ and $^{206}Pb/^{204}Pb$ ratios are required. Our results show that Tengchong basalts lay along mixing lines of the DMM with 1% supercritical liquids and 15–19% recycled mixture containing 66–72% dolomite, 22–26% calcite and 2–12% Indian sediments (Fig. 10d-f). Approximately 3.0-10.0% degree of partial melting of this hybrid mantle well matches the trace elemental variation of the Tengchong basalts (Fig. 11).

A noteworthy question is that the Tengchong basalts exhibit low CaO/Al_2O_3 ratios and high SiO_2 contents compared with the experimental results of carbonated peridotite (Dasgupta et al., 2007) and MORB (Gale et al., 2013). Two possible mechanisms could be responsible for these features. One is the addition of Indian sediments, which have high SiO_2 (66.6 wt.%), Al_2O_3 (15.4 wt.%) and low CaO (3.59 wt.%) content similar to UCC (Rudnick and Gao, 2003). When the melting degree is low (<6%), the Indian sediments should be melt in the early stage and form the melt with low CaO, high Al_2O_3 and high SiO_2 characteristics. The other is the interaction between carbonated silicate melt and lithospheric mantle proposed by Li and Wang (2018). This interaction could convert lithospheric orthopyroxene to clinopyroxene, olivine and CO_2 (fluid), resulting in low low CaO/Al_2O_3 ratios and high SiO_2 contents in the melt.

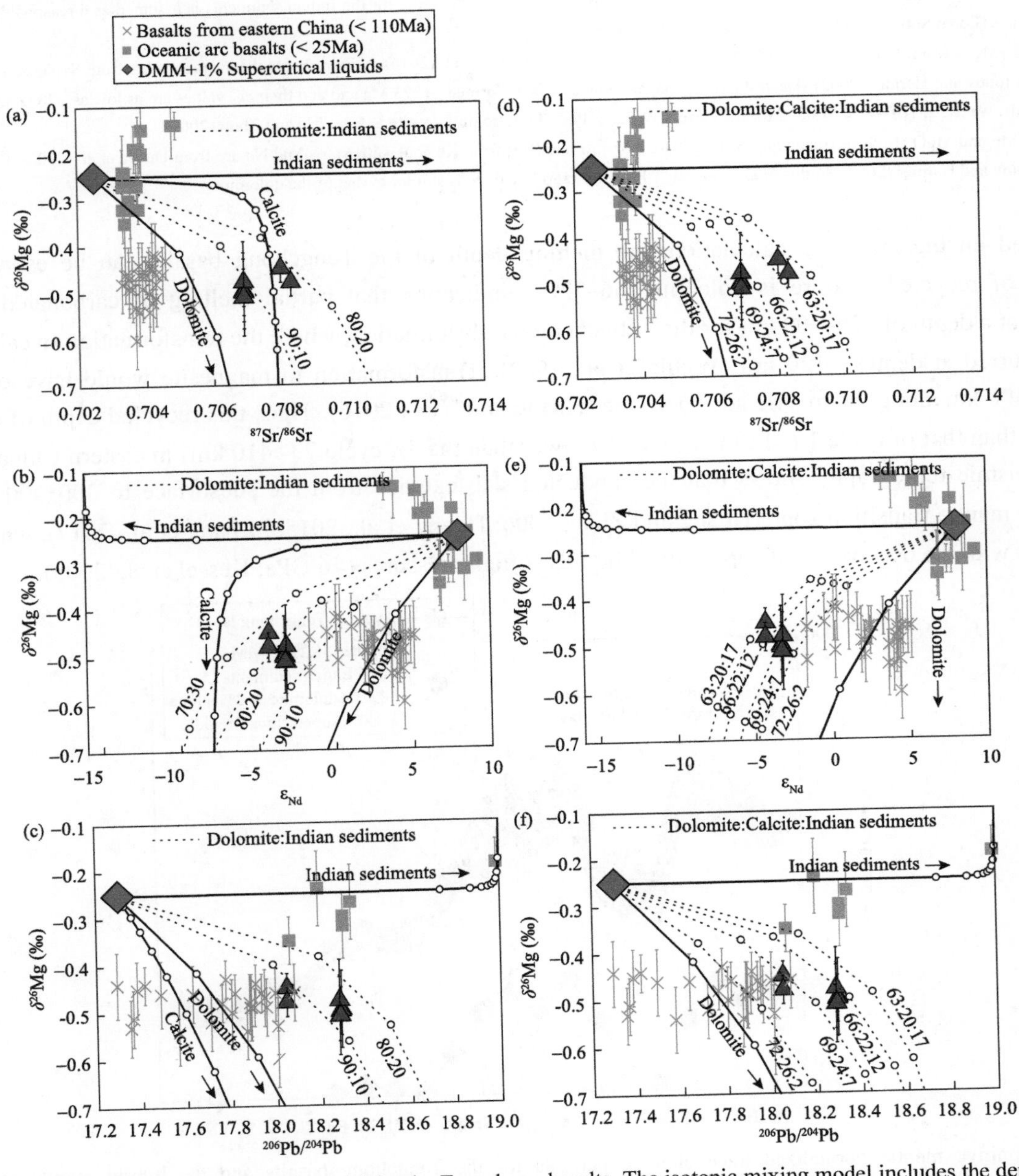

Fig. 10 Mg-Sr-Nd-Pb mixing calculation model for the Tengchong basalts. The isotopic mixing model includes the depleted mantle (DMM), supercritical liquid, dolomite, calcite and Indian sediments. The <110 Ma basalts from eastern China and the island arc basalts are also plotted. Mixing hyperbolas are marked in 10% increments. The parameters used in this modelling are summarized in Table 4.

Table 4 Parameters for model calculation

Endmembers	δ^{26}Mg	MgO	^{87}Sr/^{86}Sr	Sr(ppm)	ε_{Nd}	Nd (ppm)	^{206}Pb/^{204}Pb	Pb(ppm)
DMM[a]	−0.25	37.8	0.7025	21.1	8	1.35	17.3	0.5
Indian sediments[b]	−0.05	2.07	0.731	251	−16.0	28.4	19.0	24.5
Dolomite[c]	−2.86	23	0.708	220	−11.0	3.6	18.9	1.5
Calcite[d]	−3.65	2	0.708	340	−8.3	21.3	18.8	0.1

[a] DMM denotes the depleted mantle. Database: δ^{26}Mg value are from Teng et al. (2010) and MgO from McDonough and Sun (1995). ^{87}Sr/^{86}Sr, ε_{Nd} and ^{206}Pb/^{204}Pb isotopic compositions are from Zindler and Hart (1986) and Hofmann (1997). The concentration of Sr and Nd are from Sun and McDonough (1989). Based on the uncertainties for the source of Tengchong basalts, Pb concentration is set as 0.5, which is slightly high than the primitive mantle.

[b] Indian sediments: δ^{26}Mg is from Ke et al. (2016). Here, the Indian sediments are represented by Sumatra and Andaman sediments (Indian sediments) from Plank and Langmuir (1998). MgO is the average of Sumatra and Andaman sediments (Plank and Langmuir, 1998). Based on the variation ranges, ^{87}Sr/^{86}Sr, ε_{Nd}, Sr, Nd and Pb contents are from Plank and Langmuir (1998). Here, ^{206}Pb/^{204}Pb value for the Indian sediments is assumed as a reasonable value of 19.0, similar to the Global Subducting Sediments (Plank and Langmuir, 1998).

[c] Dolomite: δ^{26}Mg is from Blättler et al. (2015). MgO, ^{87}Sr/^{86}Sr and Sr is from Ke et al. (2016). Nd represents the average value from Sapienza et al. (2009) and Pb is from Ionov and Harmer (2002). The dolomite in Bayan Obo have been formed ~1273 Ma ago and their ε_{Nd} values are as low as −20 (Zhang et al., 2001). In this study, we use ε_{Nd} of −11.0 to stand for the current value. ^{206}Pb/^{204}Pb ratio of dolomite is from Liu and Zhou (2005).

[d] Calcite: δ^{26}Mg and MgO are from Huang and Xiao (2016), ^{87}Sr/^{86}Sr and Sr is from Ke et al. (2016). ε_{Nd} and Nd are from Ducea et al. (2005). Pb concentration is from Ionov and Harmer (2002). In this model, we set ^{206}Pb/^{204}Pb ratio of calcite is similar to that of the dolomite.

Based on the modeling calculation, the melting depth of the Tengchong basalts can be estimated. The majority of recycled mixture is dolomite (66–72%), indicating that partial melting of carbonated peridotite occurred at a depth of 120–300 km for the Tengchong basalt formation, where the transformation to calcite would have occurred at depths <120 km (Wyllie et al., 1983). Transformation to magnesite would have occurred at depths >300 km along the mantle adiabatic line (Huang and Xiao, 2016). Thus, this recycled depth of carbonates is deeper than that of cycle 1 (<120 km), but shallower than that of cycle 2 (>410 km) in eastern China (Li et al., 2017). Seismic tomography shows that the subducted slab extends from the subsurface to 300−400 km depth above the mantle transition zone (Huang and Zhao, 2006; Huang et al., 2015b; Zhang et al., 2017), which is also consistent with the occurrence of supercritical liquids at high pressures (>6 GPa; Kessel et al., 2005).

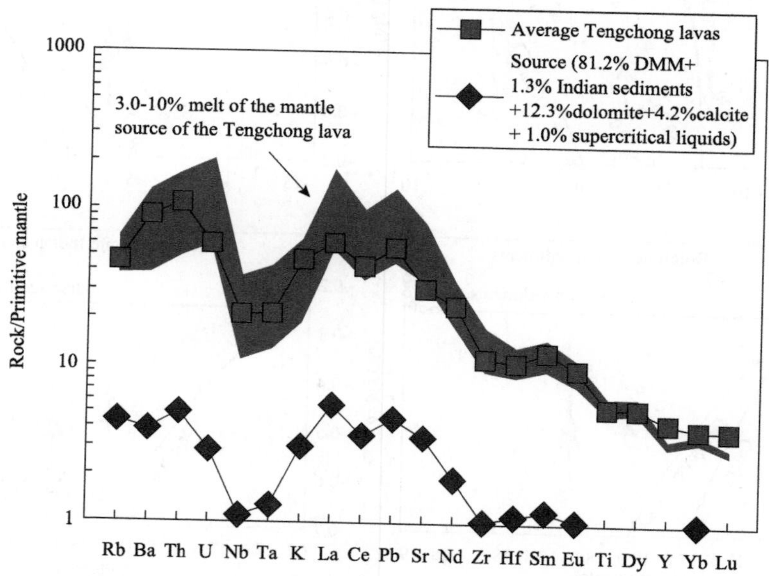

Fig. 11 Primitive mantle normalized trace element patterns for the Tengchong basalts and the hybrid mantle source. The compositions are calculated in a carbonate and Indian sediments recycling model for the Tengchong basaltic mantle source. Average Tengchong lavas are the average value of the 7 Tengchong basalts, which have been corrected for olivine fractionation in equilibrium at Fo90, and then trace element corrected. The parameters and calculations used in the model are summarized in Table 5.

Table 5 Calculation of the source component of the Tengchong volcanic rocks and modelling the composition evolution

Elements	Primitive mantle[a]	Dolomite[a]	Calcite[a]	Indian Sediments[a]	Supercritical Liquids[a]	Source[b] (81.2%DMM +1.3% Indian sediments +12.3%dolomite +4.2%calcite+1.0% Supercritical liquids)	Bulk partition coefs (garnet field)[c]	Avg. Tengchong lava (olivine corrected to Fo90)+1SD[d]	3.0% (model) melt of the source[e]	10.0% (model) melt of the source[e]
Rb	0.6	0.045	17.4	63.9	59.6	2.65	0.051615	27.5	39.4	23.1
Ba	6.6	22	180	517	331.8	25.68	0.000138	591.7	856.1	256.8
Th	0.0795	0.0012	1.14	8.9	16.7	0.40	0.002001	8.6	13.2	4.0
U	0.0203	0.0075	0.023	2.3	7.3	0.12	0.000701	1.2	4.0	1.2
Nb	0.658	0.0060	0.13	10.7	4.3	0.72	0.002846	13.9	24.0	7.2
Ta	0.037	0.0052	0.13	0.8	0.1	0.05	0.002846	0.8	1.6	0.5
K	240	0	0	18628.7	41.5	435.0	0.001442	11267.6	14500	4350
La	0.648	1.05	55	29.6	23.8	3.61	0.010884	38.9	113.2	36.1
Ce	1.675	0.8017	76	63.6	46.8	5.98	0.017746	71.8	163.5	59.6
Pb	0.15	1.5	0.10	21.1	10.5	0.69	0.014836	8.4	20.0	6.9
Sr	19.9	222	340	294.5	770.2	69.37	0.026226	616.2	1588.5	681.2
Nd	1.25	0.086	21.3	26.3	10.1	2.37	0.039905	30.2	42.2	22.0
Zr	10.5	0.20	1.51	166.0	3.1	10.8	0.042840	116.8	182.8	98.6
Hf	0.283	0.008	0.04	6.25	0.1	0.31	0.069360	3.0	3.7	2.5
Sm	0.406	0.033	1.81	4.8	1.2	0.48	0.065788	5.0	6.0	3.9
Eu	0.154	0.012	0.42	1.0	0.2	0.16	0.079808	1.5	1.7	1.2
Ti	1205	0.0032	0.02	4140	986.6	1041.37	0.122600	6588.5	7636.2	6004.3
Dy	0.674	0.014	1.37	3.63	0.2	0.66	0.139360	3.7	4.3	3.5
Y	4.3	0.087	7.06	20.5	1.1	4.08	0.262120	19.0	14.9	13.5
Yb	0.441	0.059	1.22	2.0	0.1	0.44	0.254720	1.8	1.7	1.5
Lu	0.0675	0.01	0.12	0.3	0.1	0.07	0.302600	0.3	0.21	0.19

[a] DMM is from McDonough and Sun (1995). The trace elements of dolomite and calcite are from carbonates in mantle peridotites (Ionov and Harmer, 2002; Ducea et al., 2005; Sapienza et al., 2009; Ke et al., 2016). Trace elements (except for Sr and Pb) of dolomite are the average values of the dolomite minerals. The Indian sediments are from Plank and Langmuir (1998). In addition, we assume the following additions and modifications: K concentrations of the dolomite and calcite are assumed as zero. The trace elemental composition of supercritical liquids were calculated from trace element of eclogite (11SHE-1 from eastern China, Wang et al., 2014) divided by the partition coefficient between rock and supercritical liquid (at 800 °C, Kessel et al., 2005).

[b] This list represents the source composition of Tengchong basalts: calculated by adding 1.3% Indian sediments + 12.3% dolomite + 4.2% calcite + 1.0% supercritical liquids to 81.2% DMM. These weight percent are based on the mixture of dolomite: calcite: Indian sediments = 69:24:7, and the optimal mixing amount is set as 17.8%.

[c] Bulk partition coefficients are calculated by assuming the following modal abundances: garnet stability field (2.0% garnet, 20.0% clinopyroxene, 14.0% orthopyroxene, 64.0% olivine). The Rb partition coefficient is from EarthRef.org–GERM Partition Coefficient (Kd) Database. Mineral-melt partition coefficients are from Kelemen et al. (2003).

[d] All the Tengchong basalts have MgO > 7.0 wt.% and can be used to correct for olivine fractionation to be in equilibrium with a mantle olivine composition of Fo90, trace elements corrected, and then averaged. We assumed that all the Fe^{3+} in Tengchong basalts are from Fe^{2+}.

[e] The composition listed is calculated by melting the Tengchong basalts source by 3.0–10.0%. The modal aggregated fractional melting model is assumed. The equation is $\frac{C_l^i}{C_o^i}=\frac{1}{F}*\left[1-(1-F)^{\left(\frac{1}{D_i}\right)}\right]$. The symbol C_l^i and C_o^i represent the element concentration of i in the melt and solid, respectively. F is the degree of partial melting and D is the bulk partition coefficient of element i. Bulk partition coefficients in this table are used for the model calculation.

5.2 Origin of the Tengchong andesites

Many previous studies have been dedicated to interpret the genesis of the andesitic rocks and the link between the basalts and andesites (e.g., Wang et al., 2006; Zhao and Fan, 2010; Li and Zhang, 2011; Li et al., 2012a, b; Yu et al., 2012; Zhang et al., 2012; Zhou et al., 2012; Huang et al., 2013; Zou et al., 2014, 2017). Most of these works favored that the andesitic rocks were derived from combined assimilation and fractional crystallization (AFC) processes. Here, the Tengchong andesites exhibit many geochemical features similar to the basalts, e.g., enrichment of LILEs and LREEs and Sr-Nd-Pb isotopic compositions (an EMII-like feature), as well as low δ^{26}Mg values (Figs. 3, 6, and 12), suggesting a petrogenetic link between the basalts and andesites. According to the Section 5.1, the Tengchong basalts can be regarded as the primary melts. As shown in Fig. 2, both the Group 1 and 2 andesites fall along the magma evolution trends initiated from the Tengchong basalts. The CaO, total FeO (TFeO), CaO/Al_2O_3 and TiO_2 contents decrease with increasing SiO_2 (Figs. 2b-e), whereas K_2O in the andesites positively correlates with SiO_2 content (Fig. 2f), suggesting olivine ± pyroxene ± plagioclase ± Fe-Ti oxides fractional crystallization. This is consistent with the observed phenocryst assemblages in the photomicrographs. In addition, both the Tengchong basalts and andesites show similar trace and rare earth elements patterns, such as enriched in LILEs (e.g., K, Rb and Ba), LREEs and depletion in HFSEs (e.g., Nb, Ta, Zr, Hf and P) (Fig. 3), mainly caused by fractional crystallization process.

However, in the plots of $^{87}Sr/^{86}Sr$ versus SiO_2 and Mg# (Figs. 12a and b), and $^{143}Nd/^{144}Nd$ and $^{206}Pb/^{204}Pb$ versus SiO_2 (Figs. 12c and d), the Tengchong volcanic rocks show two independent assimilation and fractional crystallization (AFC) trends. The basalts from Qingliangshan and the Group 1 andesites belong to the low-$^{87}Sr/^{86}Sr$, high-$^{143}Nd/^{144}Nd$ and high-$^{206}Pb/^{204}Pb$ group, while basalts from Wuhe and the Group 2 andesites belong to the high-$^{87}Sr/^{86}Sr$, low-$^{143}Nd/^{144}Nd$ and low-$^{206}Pb/^{204}Pb$ group (Fig. 12). Among the high-$^{87}Sr/^{86}Sr$ group, the more evolved samples have higher $^{87}Sr/^{86}Sr$ and $^{206}Pb/^{204}Pb$ ratios than the less-evolved samples. In contrast, among the low-$^{87}Sr/^{86}Sr$ group, the more evolved samples show lower $^{87}Sr/^{86}Sr$ and $^{206}Pb/^{204}Pb$ ratios than the less-evolved samples. These correlations imply that the two independent groups were sourced from distinctive wallrocks before their eruption.

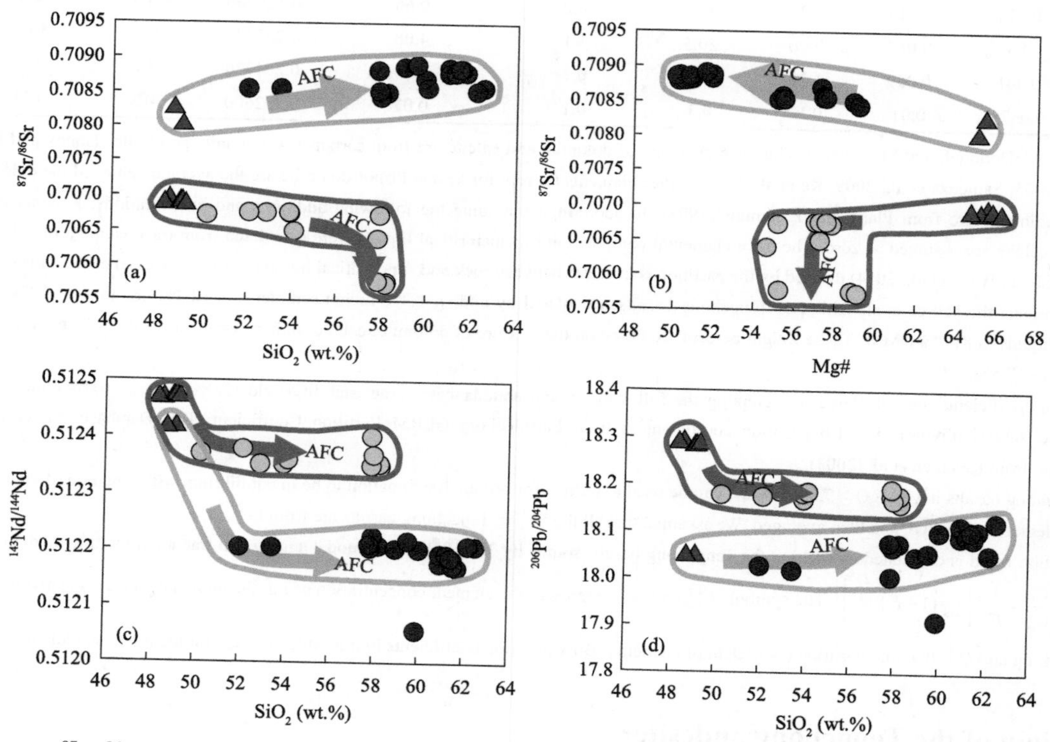

Fig. 12　Plots of $^{87}Sr/^{86}Sr$ versus SiO_2 (a) and Mg# (b), and $^{143}Nd/^{144}Nd$ (c) and $^{206}Pb/^{204}Pb$ (d) versus SiO_2 for the Tengchong basalts. The basalts from Qingliangshan and Group 1 andesites defined an AFC process (blue arrow) and the basalts from Wuhe and Group 2 andesites defined another AFC process (light yellow arrow). (For interpretation of the references to colour in this figure legend, the reader is referred to the web version of this article.)

In addition, the AFC process is also supported by the Mg isotopic compositions. The Group 1 and 2 andesites have slightly heavier Mg isotopic composition (average: −0.40 ± 0.08‰) than the Tengchong basalts (average: −0.49 ± 0.04‰), because of the continental crust having a mantle-like Mg isotopic composition (Yang et al., 2016). Thus, we propose that these andesites were derived from AFC processes before their eruption, which is consistent with previous studies (e.g., Zhao and Fan, 2010; Li and Liu, 2012; Li et al., 2012; Zhou et al., 2012; Zou et al., 2017).

This origin of the Tengchong andesite is also supported by geophysical studies, which revealed a low-velocity anomaly zone reflecting a magma chamber in the continental crust (e.g., Wang and Huangfu, 2004; Lei et al., 2009) as well the geochronological studies indicating that the Tengchong andesites erupted several Ma after the Tengchong basalts (e.g., Li et al., 2000; Wang et al., 2006; Zou et al., 2010).

5.3 Geodynamic implication for the deep carbon recycling

The Tengchong volcanic rocks provide important insights into the deep carbon cycle at depths from 120 to 410 km (Figs. 1a and b). The light Mg isotopic composition (δ^{26}Mg = −0.51 to −0.45‰) of the Tengchong basalts resulted from recycled carbonate metasomatism according to geochemical features (Fig. 9). Based on model calculations with supercritical liquids and seismic tomography, the recycled depth of carbonates is estimated from 180 to 300 km.

Here, we propose a model to illustrate carbon recycling in the deep oceanic subduction zone (Fig. 13). (1) ~55–50 Ma: After closure of the Neo-Tethyan Ocean, the Indian continental crust subducted beneath Eurasia (Donaldson et al., 2013; Ding et al., 2016; Yin and Harrison 2000; Aitchison et al. 2007) and was then replaced by the subduction of the Indian Oceanic plate beneath Burma and SW China (Zhou et al., 2012). The Indian Oceanic plate carried sediments including carbonates into the mantle. It had experienced slab dehydration twice: the first slab dehydration occurred under the sub-arc at low pressure hybridizing the shallow mantle wedge (70–120 km).

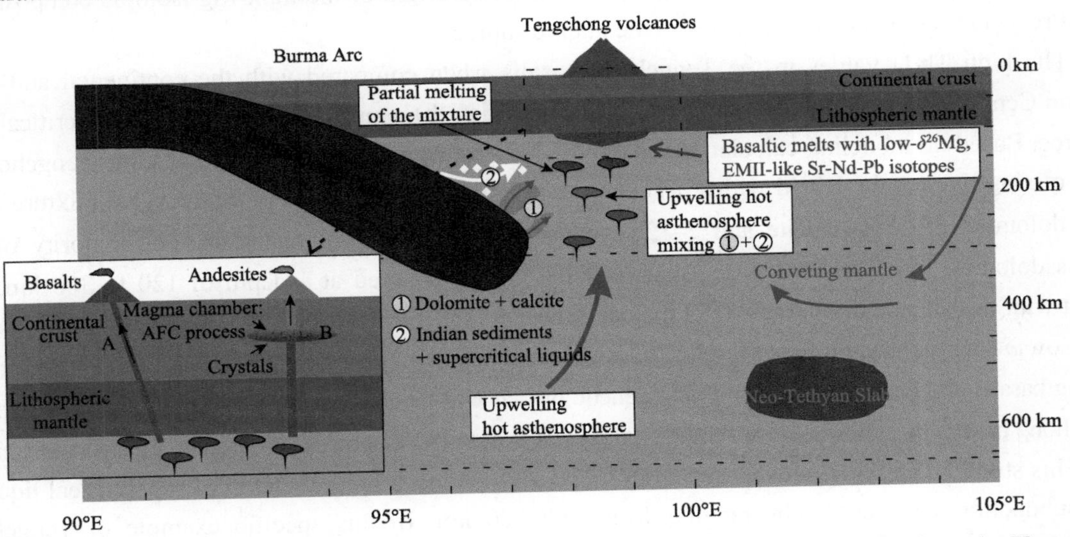

Fig. 13 A schematic model illustrating the origin of the Tengchong volcanic rocks based on the seismic study (Huang and Zhao, 2006; Zhang et al., 2017) and discussion in the text. The mantle source of the Tengchong volcanic rocks are metasomatized by a recycled mixture including dolomite, calcite, Indian sediments and the supercritical liquids. (A) Observations by seismic tomography, model calculations and the presence of supercritical liquids indicate that the partial melting source should be located from 120 to 300 km. Because of the break-off of the Neo-Tethyan slab, the upwelling hot asthenosphere provides the heat for partial melting of the metasomatized mantle to produce basaltic melts with low δ^{26}Mg, high Th/U and enriched Sr-Nd-Pb isotopic characteristics. (B) The basaltic melt entered into the continental crust, and then experienced assimilation and fractional crystallization (AFC) process. Finally, the evolved magmatic melts erupted to the Earth's surface to form the Tengchong andesites.

The second slab dehydration event occurred at high pressure generating the supercritical liquids that metasomatized the deep mantle wedge (120–300 km). (2) ~20 Ma: Based on seismic observations (Huang and Zhao, 2006), break-off of the Neo-Tethyan oceanic slab at ~400 km depths from the more buoyant Indian continental lithosphere caused the mantle upwelling through the slab window (Fig. 1b). The metasomatized deep mantle wedge was transported into the upwelling mantle and partially melted to produce basaltic melts with low δ^{26}Mg, high Th/U values and enriched Sr-Nd-Pb isotopic characteristics (Mahéo et al., 2002; Negredo et al., 2007; Lee et al., 2009; Ji et al., 2016). (3) ~4–2 Ma (Mu et al., 1987; Li et al., 2000): Part of this basaltic melt directly erupted to the Earth's surface to form Tengchong basalts (A) and some of this melt entered into the continental crust and evolved. (4) < 2 Ma (Mu et al., 1987; Jiang, 1998a): The evolved magma in the continental crust erupted to the surface to form the Tengchong andesites (B). The residence time of this cycle was ~48–53 Ma, which is longer than that of the sub-arc (5–10 Mys, Kelemen and Manning, 2015), but shorter than that of the mantle transition zone (>60Ma; Li et al., 2017).

Our study together with previous carbon cycle studies (Yang et al., 2012; Huang et al., 2015a; Tian et al., 2016; Li et al., 2017) provides three carbon cycles in the entire upper mantle: cycle 1: >120 km, 5–10 Ma (Kelemen and Manning, 2015); cycle 2: 120–410 km, 48–53 Ma and cycle 3: 410–660 km, >60 Ma (Li et al., 2017). These results are critical to our understanding of the deep carbon cycle.

6 Conclusions

The following conclusions are drawn:

(1) The Tengchong basalts exhibit low δ^{26}Mg values from –0.51 to –0.45‰, lower than that of the average mantle (–0.25±0.07‰; Teng et al., 2010a) and the island arc lavas (–0.35 to +0.06‰; Teng et al., 2016; Li et al., 2017). Based on the geochemical correlations, the most likely origin of the light Mg isotopic composition is the mixing of recycled sedimentary carbonates in the mantle source.

(2) The high Th/U values in the Tengchong basalts when compared with the continental and island arc basalts and Cenozoic continental basalts in eastern China (Fig. 4b), emphasizes the role of supercritical liquids in their source. Based on modeling calculations of Mg-Sr-Nd-Pb isotopes and trace elements, the Tengchong basalts lie along mixing lines of the DMM with 1% supercritical liquids with 15–19% of the recycled mixture containing 66–72% dolomite, 22–26% calcite and 2–12% upper continental crustal materials. The majority of recycled mixture is dolomite (66–72%) indicating that recycled mantle sourced at a depth of 120 to 300 km, which is consistent with the seismic tomography and the occurrence of supercritical liquids.

(3) Low δ^{26}Mg values, enriched Sr-Nd-Pb isotopic compositions, enrichment of LILEs and LREEs for both the Tengchong basalts and andesites suggest a petrogenetic link. It was highly possible these andesites were evolved from the Tengchong basalts by assimilation and fractional crystallization (AFC) processes based on their relationships.

(4) This study reveals a new deep carbon cycle possibly induced by the release of supercritical liquids in the oceanic subduction zone at depths ranging from 180–300 km. In this specific example of Tengchong, the residence time of carbon at 48–53 Mys is shorter than that in the mantle transition zone (>60 Mys) but longer than that of the sub-arc zone (5–10 Mys).

Acknowledgements

The authors would like to thank editorial handling by Dr. Shi-Chun Huang and insightful comments from three anonymous reviewers who greatly improved the original manuscript. We also thank Yan-Hong Liu, Bing-Yu Gao, Wei-Yi Li, Hong-Yan Li and Wen-Jun Li for their assistance with the major and trace elements and Sr-Nd-Pb isotopic analyses at IGGCAS, Xu-Nan Meng, Rui-Ying Li, Yan Xiao and Ting Gao for assistances in Mg isotopic analyses at CUGB. We also thank Hitesh Changela for carefully polishing the English writing on the manuscript. This work was supported by the National Natural Science Foundation of China (41730214, 41322022), "Strategic Priority Research Program" of the Chinese Academy of Sciences (Grant No. XDB18030603), Youth Innovation Promotion Association, Chinese Academy of Sciences (No. 2014056), National Postdoctoral Program for Innovative Talents (BX201700237) to Heng-Ci Tian and the Ministry of Science and Technology of People's Republic of China (2016YFE0203000).

References

Aitchison J. C., Ali J. R. and Davis A. M. (2007) When and where did India and Asia collide? J. Geophys. Res. 112, B05423, doi:10.1029/2006JB004706.

An Y.-J., Wu F., Xiang Y.-X., Nan X.-Y., Yu X., Yang J.-H., Yu H.-M., Xie L.-W. and Huang F. (2014) High-precision Mg isotope analyses of low-Mg rocks by MC-ICP-MS. Chem. Geol. 390, 9-21.

Beinlich A., Mavromatis V., Austrheim H. and Oelkers E. H. (2014) Inter-mineral Mg isotope fractionation during hydrothermal ultramafic rock alteration: implications for the global Mg-cycle. Earth Planet. Sci. Lett. 392, 166-176.

Ben Othman D., White W. M. and Patchett J. (1989). The Geochemistry of marine-sediments, island-arc magma genesis, and crust mantle recycling. Earth Planet. Sci. Lett. 94, 1-21.

Blättler C. L., Miller N. R. and Higgins J. A. (2015) Mg and Ca isotope signatures of authigenic dolomite in siliceous deep-sea sediments. Earth Planet. Sci. Lett. 419, 32-42.

Borg L. E., Brandon A. D., Clynne M. A. and Walker R. J. (2000) Re-Os isotopic systematics of primitive lavas from the Lassen region of the Cascade arc, California. Earth Planet. Sci. Lett. 177, 301-317.

Chen F., Satir M., Ji J. and Zhong D. (2002) Nd-Sr-Pb isotopes of Tengchong Cenozoic volcanic rocks from western Yunnan, China: evidence for an enriched-mantle source. J. Asian Earth Sci. 21, 39-45.

Chen L.-H., Zeng G., Jiang S.-Y., Hofmann A. W., Xu X.-S. and Pan M.-B. (2009) Sources of Anfengshan basalts: Subducted lower crust in the Sulu UHP belt, China. Earth Planet. Sci. Lett. 286, 426-435.

Chu Z.-Y., Wu F.-Y., Walker R. J., Rudnick R. L., Pitcher L., Puchtel I. S., Yang Y.-H. and Wilde S. A. (2009) Temporal evolution of the lithospheric mantle beneath the Eastern North China Craton. J. Petrol. 50, 1857-1898.

Chung S.-L. Chu M.-F., Zhang Y.-Q., Xie Y.-W., Lo C.-H., Lee T.-Y., Lan C.-Y., Li X.-H., Zhang Q. and Wang Y.-Z. (2005) Tibetan tectonic evolution inferred from spatial and temporal variations in post-collisional magmatism. Earth Sci. Rev. 68, 173-196.

Dasgupta R. (2013) Ingassing, storage, and outgassing of terrestrial carbon through geologic time. Rev. Mineral. Geochem. 75, 183-229.

Dasgupta R., Hirschmann M. M. and Smith N. D. (2007) Partial melting experiments of peridotite + CO_2 at 3 GPa and genesis of alkalic ocean island basalts. J. Petrol. 48, 2093-2124.

Dasgupta R., Hirschmann M. M. and Stalker K. (2006) Immiscible transition from carbonate-rich to silicate-rich melts in the 3 GPa melting interval of eclogite + CO_2 and genesis of silica-undersaturated ocean island lavas. J. Petrol. 47, 647-671.

Dasgupta R. and Hirschmann M. M. (2010) The deep carbon cycle and melting in Earth's interior. Earth Planet. Sci. Lett. 298, 1-13.

Dauphas N. (2007). Diffusion-driven kinetic isotope effect of Fe and Ni during formation of the Widmanstatten pattern. Meteorit. Planet. Sci 42, 1597-1613.

Ding H.-X., Zhang Z.-M., Dong X., Tian Z.-L., Xiang H., Mu H.-C., Gou Z.-B., Shui X.-F., Li W.-C. and Mao L.-J. (2016) Early Eocene (c. 50 Ma) collision of the Indian and Asian continents: constraints from the North Himalayan metamorphic rocks, southeastern Tibet. Earth Planet. Sci. Lett. 435, 64-73.

Donaldson D. G., Webb A. A. G., Menold C. A., Kylander-Clark A. R. C. and Hacker B. R. (2013) Petrochronology of Himalayan ultrahigh-pressure eclogite. Geology 41, 835-838.

Ducea M. N., Saleeby J., Morrison J. and Valencia V. A. (2005) Subducted carbonates, metasomatism of mantle wedges, and possible connections to diamond formation: An example from California. Am. Mineral. 90, 864-870.

Dupré B. and Allègre C. J. (1983) Pb-Sr isotope variation in Indian-Ocean basalts and mixing phenomena. Nature 303, 142-146.
Gale A., Dalton C. A., Langmuir C. H., Su Y. J. and Schilling J. G. (2013) The mean composition of ocean ridge basalts. Geochem. Geophys. Geosyst, 14, 489-518.
Galy A., Yoffe O., Janney P. E., Williams R. W., Cloquet C., Alard O., Halicz L., Wadhwa M., Hutcheon I. D., Ramon E. and Carignan J. (2003) Magnesium isotope heterogeneity of the isotopic standard SRM980 and new reference materials for magnesium-isotope-ratio measurements. J. Anal. At. Spectrom. 18, 1352-1356.
Gerbode C. and Dasgupta R. (2010) Carbonate-fluxed melting of MORB-like pyroxenite at 2.9 GPa and genesis of HIMU ocean island basalts. J. Petrol. 51, 2067-2088.
Hamelin B., Dupré B. and Allègre C.J. (1986) Pb-Sr-Nd Isotopic Data of Indian-Ocean Ridges: New Evidence of Large-Scale Mapping of Mantle Heterogeneities. Earth Planet. Sci. Lett. 76, 288-298.
Hart S. R. (1984) A large-scale isotope anomaly in the SouthernHemisphere mantle. Nature 309, 753-757.
Hart S. R. and Reid M. R. (1991) Rb/Cs fractionation: A link between granulite metamorphism and the S-process. Geochim. Cosmochim. Acta 55, 2379-2383.
Hazen R. M. and Schiffries C. M. (2013) Why Deep Carbon? Rev. Mineral. Geochem. 75, 1-6.
Hirose K. (1997) Partial melt compositions of carbonated peridotite at 3 GPa and role of CO_2 in alkali-basalt magma generation. Geophys. Res. Lett. 24, 2837-2840.
Hirschmann M. M., Kogiso T., Baker M. B. and Stolper E. M. (2003) Alkalic magmas generated by partial melting of garnet pyroxenite. Geology 31, 481-484.
Hofmann A. W. (1988) Chemical differentiation of the Earth: the relationship between mantle, continental crust, and oceanic crust. Earth Planet. Sci. Lett. 90, 297-314.
Hofmann A. W. (1997) Mantle geochemistry: the message from oceanic volcanism. Nature 385, 219-229.
Hofmann A. W. (2017) A store of subducted carbon beneath Eastern China. Nat. Sci. Rev. 4, 2-2.
Hofmann A. W., Jochum K. P., Seufert, M. and White W. M. 1986. Nb and Pb in oceanic basalts: new constraints on mantle evolution. Earth Planet. Sci. Lett. 79, 33-45.
Hu Y., Teng F.-Z., Plank T. and Huang K.-J. (2017). Magnesium isotopic composition of subducting marine sediments. Chem. Geol. 466, 15-31.
Huang F., Chakraborty P., Lundstrom C. C., Holmden C., Glessner J. J. G., Kieffer S. W. and Lesher C. E. (2010). Isotope fractionation in silicate melts by thermal diffusion. Nature 464, 396-400.
Huang F., Glessner J., Ianno A., Lundstrom C. and Zhang Z.-F. (2009) Magnesium isotopic composition of igneous rock standards measured by MC-ICP-MS. Chem. Geol. 268, 15-23.
Huang F., Zhang Z.-F., Lundstrom C. C. and Zhi X.-C. (2011) Iron and magnesium isotopic compositions of peridotite xenoliths from Eastern China. Geochim. Cosmochim. Acta 75, 3318-3334.
Huang J., Li S.-G., Xiao Y.-L., Ke S., Li W.-Y. and Tian Y. (2015a) Origin of low δ^{26}Mg Cenozoic basalts from South China Block and their geodynamic implications. Geochim. Cosmochim. Acta 164, 298-317.
Huang J. and Xiao Y.-L. (2016) Mg-Sr isotopes of low-δ26Mg basalts tracing recycled carbonate species: Implication for the initial melting depth of the carbonated mantle in Eastern China. Int. Geol. Rev. 58, 1350-1362.
Huang J.-L. and Zhao D.-P. (2006) High-resolution mantle tomography of China and surrounding regions. J. Geophys. Res. 111, http://dx.doi.org/10.1029/2005JB004066.
Huang K.-J., Teng F.-Z., Wei G.-J., Ma J.-L. and Bao Z.-Y. (2012). Adsorption- and desorption-controlled magnesium isotope fractionation during extreme weathering of basalt in Hainan Island, China. Earth Planet. Sci. Lett. 359, 73-83.
Huang Z.-C., Wang P., Xu M.-J., Wang L.-S., Ding Z.-F., Wu Y., Xu M.-J., Mi N., Hu D.-Y. and Li H. (2015b). Mantle structure and dynamics beneath SE Tibet revealed by new seismic images. Earth Planet. Sci. Lett. 411, 100-111.
Huang X.-W., Zhou M.-F., Wang C.-Y., Robinson P. T., Zhao J.-H. and Qi L. (2013). Chalcophile element constraints on magma differentiation of Quaternary volcanoes in Tengchong, SW China. J. Asian Earth Sci. 76, 1-11.
Humayun M., Qin L.-P. and Norman M. D. (2004). Geochemical evidence for excess iron in the mantle beneath Hawaii. Science 306, 91-94.
Ionov D. and Harmer R. E. (2002) Trace element distribution in calcite-dolomite carbonatites from Spitskop: inferences for differentiation of carbonatite magmas and the origin of carbonates in mantle xenoliths. Earth Planet. Sci. Lett. 198, 495-510.
Ito E., White W. M. and Gopel C. (1987) The O, Sr, Nd and Pb isotope geochemistry of MORB. Chem. Geol. 62, 157-176.
Ji J.-Q., Zhong D.-L. and Chen C.-Y. (2000). Geochemistry and genesis of Nabang metamorphic basalts, southwest Yunnan, China:

implications for the subducted slab break-off. Acta Petrol. Sin. 16, 433-442 (in Chinese with English abstract).

Jiang C.-S. (1998a) Distribution characteristics of Tengchong volcano in the Cenozoic Era. J. Seismol. Res. 21, 309-319 (in Chinese with English abstract).

Jiang C.-S. (1998b) Period division of volcano activities in the Cenozoic Era of Tengchong. J. Seismol. Res. 21, 320-329 (in Chinese with English abstract).

Jiang C.-S., Zhou Z.-H. and Zhao C.-P. (2004) The structure characteristics of the crust and upper mantle in the area of Tengchong volcano. J. Seismol. Res. 27, 1-6 (in Chinese with English abstract).

Johnston F. K. B., Turchyn A. V. and Edmonds M. (2011) Decarbonation efficiency in subduction zones: Implications for warm Cretaceous climates. Earth Planet. Sci. Lett. 303, 143-152.

Ke S., Teng F.-Z., Li S.-G., Gao T., Liu S.-A., He Y.-S. and Mo X.-X. (2016) Mg, Sr, and O isotope geochemistry of syenites from northwest Xinjiang, China: tracing carbonate recycling during Tethyan oceanic subduction. Chem. Geol. 437, 109-119.

Kelemen P. B. and Manning C. E. 2015 Reevaluating carbon fluxes in subduction zones, what goes down, mostly comes up. Proc. Natl. Acad. Sci. U. S. A. 112, E3997-E4006.

Kelemen P. B., Yogodzinski G. M. and Scholl D.W. (2003) Along-strike variation in the Aleutian Island Arc: Genesis of high Mg# andesite and implications for continental crust, Inside the Subduction Factory. Geophysical Monograph Series, pp. 223-276.

Kessel R., Schmidt M. W., Ulmer P. and Pettke T. (2005) Trace element signature of subduction-zone fluids, melts and supercritical liquids at 120-180 km depth. Nature 437, 724-727.

Kogiso T., Hirschmann M. M. and Frost D. J. (2003) Highpressure partial melting of garnet pyroxenite: possible mafic lithologies in the source of ocean island basalts. Earth Planet. Sci. Lett. 216, 603-617.

Kogiso T. and Hirschmann M. M. (2006) Partial melting experiments of bimineralic eclogite and the role of recycled mafic oceanic crust in the genesis of ocean island basalts. Earth Planet. Sci. Lett. 249, 188-199.

Kushiro I., Satake H. and Akimoto S. (1975). Carbonate-silicate reactions at high pressures and possible presence of dolomite and magnesite in the upper mantle. Earth Planet. Sci. Lett. 28, 116-120.

Langmuir C. H., Bender J. F., Bence A. E., Hanson G. N. and Taylor S. R. (1977) Petrogenesis of basalts from famous area: Mid-Atlantic Ridge. Earth Planet. Sci. Lett. 36, 133-156.

Le Bas M. J. (1986) A chemical classification of volcanic rocks based on the total alkali-silica diagram. J. Petrol. 27, 745-750.

Lee H.-Y., Chung S.-L., Lo C.-H., Ji J.-Q., Lee T.-Y., Qian Q. and Zhang Q. (2009) Eocene Neotethyan slab breakoff in southern Tibet inferred from the Linzizong volcanic record. Tectonophysics 477, 20-35.

Lei J.-S., Zhao D.-P. and Su Y.-J. (2009) Insight into the origin of the Tengchong intraplate volcano and seismotectonics in southwest China from local and teleseismic data. J. Geophys. Res. 114, B05302.

Li C.-F., Li X.-H., Li Q.-L., Guo J.-H., Li X.-H. and Yang Y.-H. (2012) Rapid and precise determination of Sr and Nd isotopic ratios in geological samples from the same filament loading by thermal ionization mass spectrometry employing a single-step separation scheme. Anal. Chimi. Acta 727, 54-60.

Li D.-M., Li Q. and Chen W.-J. (2000) Volcanic activities in the Tengchong volcano area since Pliocene. Acta Petrol. Sin. 16, 362-370.

Li D.-P., Luo Z.-H., Liu J.-Q., Chen Y.-L. and Jin Y. (2012). Magma Origin and Evolution of Tengchong Cenozoic Volcanic Rocks from West Yunnan, China: Evidence from Whole Rock Geochemistry and Nd-Sr-Pb Isotopes. Acta Geol. Sin. 86, 867-878.

Li N. and Zhang L.-Y. (2011). A study on volcanic minerals and hosted melt inclusions in newly-erupted Tengchong volcanic rocks, Yunnan Province. Acta Petrol. Sin. 27, 2842-2854 (in Chinese with English abstract).

Li X. and Liu J.Q. (2012) A study on the geochemical characteristics and petrogenesis of Holocene volcanic rocks in the Tengchong volcanic eruption field, Yunnan Province, SW China. Acta Petrol. Sin. 28, 1507-1516 (in Chinese with English abstract).

Li S. G. and Wang Y. (2018) Formation time of the big mantle wedge beneath eastern China and a new lithospheric thinning mechanism of the North China craton: Geodynamic effects of deep recycled carbon. Science China Earth Sciences, 61, 853-868.

Li S.-G., Yang W., Ke S., Meng X.-N., Tian H.-C., Xu L.-J., He Y.-S., Huang J., Wang X.-C., Xia Q.-K., Sun W.-D., Yang X.-Y., Ren Z.-Y., Wei H.-Q., Liu Y.-S., Meng F.-C. and Yan J. (2017) Deep carbon cycles constrained by a large-scale mantle Mg isotope anomaly in eastern China. Nat. Sci. Rev. 4, 111-120.

Liu F. and Zhou H.-W. (2005) Information of interaction of crust and mantle during continent subduction and exhumation in Dabie orogenic belt: evidence from Pb isotopes of marbles. Geological Science and Technology Information 24, 14-18 (in Chinese with English abstract).

Liu F., Li X., Wang G.-Q., Liu Y.-F., Zhu H.-L., Kang J.-T., Huang F., Sun W.-D., Xia X.-P. and Zhang Z.-F. (2017). Marine

Carbonate Component in the Mantle Beneath the Southeastern Tibetan Plateau: Evidence From Magnesium and Calcium Isotopes. J. Geophys. Res. 122, 9729-9744.

Liu X.-M., Teng F.-Z., Rudnick R. L., McDonough W. F. and Cummings M. L. (2014). Massive magnesium depletion and isotope fractionation in weathered basalts. Geochim. Cosmochim. Acta 135, 336-349.

Liu Y.-S., Gao S., Kelemen P. B. and Xu W.-L. (2008). Recycled crust controls contrasting source compositions of Mesozoic and Cenozoic basalts in the North China Craton. Geochim. Cosmochim. Acta 72, 2349-2376.

Mahéo G., Guillot S., Blichert-Toft J., Rolland Y. and Pecher A. (2002) A slab breakoff model for the Neogene thermal evolution of South Karakorum and South Tibet. Earth Planet. Sci. Lett. 195, 45-58.

McDonough, W.F., Sun, S.S., 1995. The Composition of the Earth. Chem. Geol. 120, 223-253.

Miller D. M., Goldstein S. L. and Langmuir C. H. (1994) Cerium/lead and lead isotope ratios in arc magmas and the enrichment of lead in the continents. Nature 368, 514-520.

Mo X.-X., Hou Z.-Q., Niu Y.-L., Dong G.-C., Qu X.-M., Zhao Z.-D. and Yang Z.-M. (2007) Mantle contributions to crustal thickening during continental collision: Evidence from Cenozoic igneous rocks in southern Tibet. Lithos 96, 225-242.

Mu Z.-G., Curtis G. H., Liao Z.-J. and Tong W. (1987) K-Ar Age and Strontium Isotopic Composition of the Tengchong Volcanic-Rocks, West Yunnan Province, China. Geothermics 16, 283-297.

Negredo A. M., Replumaz A., Villasenor A. and Guillot S. (2007) Modeling the evolution of continental subduction processes in the Pamir-Hindu Kush region. Earth Planet. Sci. Lett. 259, 212-225.

Nesbitt H. W. and Young G. M. (1982) Early Proterozoic climates and plate Motions Inferred from Major Element Chemistry of Lutites. Nature 299, 715-717.

Opfergelt S., Georg R. B., Delvaux B., Cabidoche Y. M., Burton K. W. and Halliday A. N. (2012). Mechanisms of magnesium isotope fractionation in volcanic soil weathering sequences, Guadeloupe. Earth Planet. Sci. Lett. 341, 176-185.

Patterson C. (1956) Age of meteorites and the Earth. Geochim. Cosmochim. Acta 10, 230-237.

Pertermann M. and Hirschmann M. M. (2003) Partial melting experiments on a MORB-like pyroxenite between 2 and 3 GPa: Constraints on the presence of pyroxenite in basalt source regions from solidus location and melting rate. J. Geophys. Res.: Solid Earth 108, 2125.

Pilet S., Baker M. B. and Stolper E. M. (2008) Metasomatized lithosphere and the origin of alkaline lavas. Science 320, 916-919.

Plank T. and Langmuir C. H. (1998) The chemical composition of subducting sediment and its consequences for the crust and mantle. Chem. Geol. 145, 325-394.

Pogge von Strandmann P. A. E., Elliott T., Marschall H. R., Coath C., Lai Y.-J., Jeffcoate A. B. and Ionov D. A. (2011). Variations of Li and Mg isotope ratios in bulk chondrites and mantle xenoliths. Geochim. Cosmochim. Acta 75, 5247-5268.

Qin J.-Z., Huang F.-G., Li Q., Qian X.-D. Su Y.-J. and Cai M.-J. (2000) 3-D chromatography of velocity structure in Tengchong volcano areas and nearby. J. Seismol. Res. 23, 157-164 (in Chinese with English abstract).

Richter F. M., Watson E. B., Mendybaev R., Dauphas N., Georg B., Watkins J. and Valley J. (2009). Isotopic fractionation of the major elements of molten basalt by chemical and thermal diffusion. Geochim. Cosmochim. Acta 73, 4250-4263.

Richter F. M., Watson E. B., Mendybaev R. A., Teng F.-Z. and Janney P. E. (2008). Magnesium isotope fractionation in silicate melts by chemical and thermal diffusion. Geochim. Cosmochim. Acta 72, 206-220.

Rudnick R. L. and Gao S. (2003) Composition of the continental crust. Treatise Geochem. 3, 1-64.

Saenger C. and Wang Z.-R. (2014) Magnesium isotope fractionation in biogenic and abiogenic carbonates: implications for paleoenvironmental proxies. Quat. Sci. Rev. 90, 1-21.

Sapienza G. T., Scambelluri M. and Braga R. (2009) Dolomite-bearing orogenic garnet peridotites witness fluid-mediated carbon recycling in a mantle wedge (Ulten Zone, Eastern Alps, Italy). Contrib. Mineral. Petrol. 158, 401-420.

Sato K. and Katsura T. (2001). Experimental investigation on dolomite dissociation into aragonite plus magnesite up to 8.5 GPa. Earth Planet. Sci. Lett. 184, 529-534.

Schmidt M. W., Vielzeuf D. and Auzanneau E. (2004) Melting and dissolution of subducting crust at high pressures: the key role of white mica. Earth Planet. Sci. Lett. 228, 65-84.

Sedaghatpour F., Teng F.-Z., Liu Y., Sears D. W. G. and Taylor L. A. (2013) Magnesium isotopic composition of the Moon. Geochim. Cosmochim. Acta 120, 1-16.

Shen A. H. and Keppler H. (1997) Direct observation of complete miscibility in the albite-H_2O system. Nature 385, 710-712.

Sobolev A.V., Hofmann A. W., Kuzmin D. V., Yaxley G. M., Arndt N. T., Chung S. L., Danyushevsky L. V., Elliott T., Frey F. A., Garcia M. O., Gurenko A. A., Kamenetsky V. S., Kerr A. C., Krivolutskaya N. A., Matvienkov V. V., Nikogosian I. K., Rocholl

A., Sigurdsson I. A., Sushchevskaya N. M. and Teklay M. (2007) The amount of recycled crust in sources of mantle-derived melts. Science 316, 412-417.

Stalder R., Ulmer P., Thompson A. B. and Gunther D. (2001) High pressure fluids in the system $MgO-SiO_2-H_2O$ under upper mantle conditions. Contrib. Mineral. Petrol. 140, 607-618.

Su B.-X., Hu Y., Teng F.-Z., Xiao Y., Zhou X.-H., Sun, Y., Zhou M.-F. and Chang S. C. (2017). Magnesium isotope constraints on subduction contribution to Mesozoic and Cenozoic East Asian continental basalts. Chem. Geol. 466, 116-122.

Sun S.-S. and McDonough W. F. (1989) Chemical and isotopic systematics of oceanic basalts: implications for mantle composition and processes. Geol. Soc. (Lond.) Spec. Publ. 42, 313-345.

Sun Y., Teng F.-Z., Ying J.-F., Su B.-X., Hu Y., Fan Q.-C. and Zhou X.-H. (2017). Magnesium Isotopic Evidence for Ancient Subducted Oceanic Crust in LOMU-Like Potassium-Rich Volcanic Rocks. J. Geophys. Res. 122, 7562-7572.

Tang Y.-J., Zhang H.-F. and Ying J.-F. (2006) Asthenosphere–lithospheric mantle interaction in an extensional regime: Implication from the geochemistry of Cenozoic basalts from Taihang Mountains, North China Craton. Chem. Geol. 233, 309-327.

Tapponnier P., Lacassin R., Leloup P. H., Scharer U., Zhong D.-L., Wu H.-W., Liu X.-H., Ji S.-C., Zhang L.-S. and Zhong J.-Y. (1990) The Ailao Shan/Red River metamorphic belt: Tertiary left-lateral shear between Indochina and South China. Nature 343, 431-437.

Teng F.-Z., Hu Y. and Chauvel C. (2016) Magnesium isotope geochemistry in arc volcanism. Proc. Natl. Acad. Sci. U. S. A. 113, 7082-7087.

Teng F.-Z., Li W.-Y., Ke S., Yang W., Liu S.-A., Sedaghatpour F., Wang S.-J., Huang K.-J., Hu Y., Ling M.-X., Xiao Y., Liu X.-M., Li X.-W., Gu H.-O., Sio C. K., Wallace D.-A., Su B.-X., Zhao L., Chamberlin J., Harrington M. and Brewer A. (2015) Magnesium isotopic compositions of international geological reference materials. Geostand. Geoanalyt. Res. 39, 329-339.

Teng F.-Z., Li W.-Y., Ke S., Marty B., Dauphas N., Huang S.-C., Wu F.-Y. and Pourmand A. (2010a) Magnesium isotopic composition of the Earth and chondrites. Geochim. Cosmochim. Acta 74, 4150-4166.

Teng F.-Z., Li W.-Y., Rudnick R. L. and Gardner L. R. (2010b). Contrasting lithium and magnesium isotope fractionation during continental weathering. Earth Planet. Sci. Lett. 300, 63-71.

Teng F.-Z., Wadhwa M. and Helz R. T. (2007) Investigation of magnesium isotope fractionation during basalt differentiation: implications for a chondritic composition of the terrestrial mantle. Earth Planet. Sci. Lett. 261, 84-92.

Tian H.-C., Yang W., Li S.-G., Ke S. and Chu Z.Y. (2016) Origin of low δ^{26}Mg basalts with EM-I component: Evidence for interaction between enriched lithosphere and carbonated asthenosphere. Geochim. Cosmochim. Acta 188, 93-105.

Tian H.-C., Yang W., Li S-G. and Ke S. (2017) Could sedimentary carbonates be recycled into the lower mantle? Constraints from Mg isotopic composition of Emeishan basalts. Lithos 292-293 250-261.

Turner S., Arnaud N., Liu J., Rogers N., Hawkesworth C., Harris N., Kelley S., VanCalsteren P. and Deng W. (1996) Post-collision, shoshonitic volcanism on the Tibetan plateau: implications for convective thinning of the lithosphere and the source of ocean island basalts. J. Petrol. 37, 45-71.

Turner S. and Foden J. (2001) U, Th and Ra disequilibria, Sr, Nd and Pb isotope and trace element variations in Sunda arc lavas: predominance of a subducted sediment component. Contrib. Mineral. Petrol. 142, 43-57.

Wallace P. J. (2005) Volatiles in subduction zone magmas: concentrations and fluxes based on melt inclusion and volcanic gas data. J. Volcanol. and Geotherm. Res. 140, 217-240.

Walter M. J., Bulanova G. P., Armstrong L. S., Keshav S., Blundy J. D., Gudfinnsson G., Lord O. T., Lennie A. R., Clark S. M., Smith C. B. and Gobbo L. (2008). Primary carbonatite melt from deeply subducted oceanic crust. Nature 454, 622-U630.

Wang C.-Y. and Huangfu G. (2004) Crustal structure in Tengchong Volcano-Geothermal Area, western Yunan, China. Tectonophysics 380, 69-87.

Wang F., Peng Z.-C., Zhu R.-X., He H.-Y. and Yang L.-K. (2006) Petrogenesis and magma residence time of lavas from Tengchong volcanic field (China): evidence from U series disequilibria and $^{40}Ar/^{39}Ar$ dating. Geochem. Geophy. Geosyst. 7, Q01002, doi:10.1029/2005GC001023.

Wang S.-J., Teng F.-Z. and Scott J. M. (2016) Tracing the origin of continental HIMU-like intraplate volcanism using magnesium isotope systematics. Geochim. Cosmochim. Acta 185, 78-87.

Wang S.-J., Teng F.-Z., Rudnick R. L. and Li S.-G. (2015). The behavior of magnesium isotopes in low-grade metamorphosed mudrocks. Geochim. Cosmochim. Acta 165, 435-448.

Wang S.-J., Teng F.-Z., Li S.-G. and Hong J.-A. (2014) Magnesium isotopic systematics of mafic rocks during continental subduction. Geochim. Cosmochim. Acta 143, 34-48.

Wang X.-J., Chen L.-H., Hofmann A. W., Mao F.-G., Liu J.-Q., Zhong Y., Xie L.-W. and Yang Y.-H. (2017). Mantle transition zone-derived EM1 component beneath NE China: Geochemical evidence from Cenozoic potassic basalts. Earth Planet. Sci. Lett. 465, 16-28.

Wang Y., Zhao Z.-F., Zheng Y.-F. and Zhang J.-J. (2011) Geochemical constraints on the nature of mantle source for Cenozoic continental basalts in east-central China. Lithos125, 940-955.

White W. M., Dupré B. and Vidal P. (1985) Isotope and trace element geochemistry of sediments from the Barbados Ridge-Demerara Plain Region, Atlantic Ocean. Geochim. Cosmochim. Acta 49, 1875-1886.

Workman R. K., Hart S. R., Jackson M., Regelous M., Farley K. A., Blusztajn J., Kurz M. and Staudigel H. (2004) Recycled metasomatized lithosphere as the origin of the enriched mantle II (EM2) end-member: evidence from the Samoan volcanic chain. Geochem. Geophys. Geosyst. 5. Q04008, doi:10.1029/2003GC000623.

Xu Y.-G., Ma J.-L., Frey F. A., Feigenson M. D. and Liu J.-F. (2005) Role of lithosphere–asthenosphere interaction in the genesis of Quaternary alkali and tholeiitic basalts from Datong, western North China Craton. Chem. Geol. 224, 247-271.

Xu Y.-G., Zhang H.-H., Qiu H.-N., Ge W.-C. and Wu F.-Y. (2012) Oceanic crust components in continental basalts from Shuangliao, Northeast China: Derived from the mantle transition zone? Chem. Geol. 328, 168-184.

Yang W., Teng F.-Z., Li W.-Y., Liu S.-A., Ke S., Liu Y.-S., Zhang H.-F. and Gao S. (2016) Magnesium isotopic composition of the deep continental crust. Am. Mineral. 101, 243-252.

Yang W., Teng F.-Z., Zhang H.-F. and Li S.-G. (2012) Magnesium isotopic systematics of continental basalts from the North China craton: Implications for tracing subducted carbonate in the mantle. Chem. Geol. 328, 185-194.

Yang Y.-H., Zhang H.-F., Chu Z.-Y., Xie L.-W. and Wu F.-Y. (2010) Combined chemical separation of Lu, Hf, Rb, Sr, Sm and Nd from a single rock digest and precise and accurate isotope determinations of Lu-Hf, Rb-Sr and Sm-Nd isotope systems using Multi-Collector ICP-MS and TIMS. Int. J. Mass Spectrom. 290, 120-126.

Yaxley G. M. and Brey G. P. (2004) Phase relations of carbonate-bearing eclogite assemblages from 2.5 to 5.5 GPa: implications for petrogenesis of carbonatites. Contrib. Mineral. Petrol. 146, 606-619.

YBGMR (Yunnan Bureau of Geology and Mineral Resources) (1979) Tengchong geologic map (1:200000) (in Chinese).

Yin A. and Harrison T. M. (2000) Geologic evolution of the Himalayan-Tibetan orogen. Annu. Rev. Earth Planet. Sci. 28, 211-280.

Yin G.-M. and Li S.-H. (2000) Thermoluminescence age of last volcanic eruption in TengchongMaanshan of Yunnan Province. J. Seismol. Res. 23, 388-391.

Young E. D. and Galy A. (2004) The isotope geochemistry and cosmochemistry of magnesium. Rev. Mineral. Geochem. 55, 197-230.

Yu H.-M., Xu J.-D., Lin C.-Y., Shi L.-B. and Chen X. D. (2012). Magmatic processes inferred from chemical composition, texture and crystal size distribution of the Heikongshan lavas in the Tengchong volcanic field, SW China. J. Asian Earth Sci. 58, 1-15.

Zeng G., Chen L.-H., Hofmann A. W., Jiang S.-Y. and Xu X.-S. (2011). Crust recycling in the sources of two parallel volcanic chains in Shandong, North China. Earth Planet. Sci. Lett. 302, 359-368.

Zeng G., Chen L.-H., Xu X.-S., Jiang S.-Y. and Hofmann A. W. (2010) Carbonated mantle sources for Cenozoic intra-plate alkaline basalts in Shandong, North China. Chem. Geol. 273, 35-45.

Zhang L., Hu Y.-L., Qin M., Duan Y., Duan Y.-Z., Peng H.-C. and Zhao H. (2015) Study on crustal and lithosphere thicknesses of Tengchong volcanic area in Yunnan. Chinese J. Geophys. 58, 1622-1633 (in Chinese with English abstract).

Zhang R.-Q., Wu Y., Gao Z.-Y., Fu Y. Y. V., Sun L., Wu Q. J. and Ding Z.-F. (2017). Upper mantle discontinuity structure beneath eastern and southeastern Tibet: New constraints on the Tengchong intraplate volcano and signatures of detached lithosphere under the western Yangtze Craton. J. Geophys. Res. 122, 1367-1380.

Zhang Y.-T., Liu J.-Q. and Meng F.-C. (2012) Geochemistry of Cenozoic volcanic rocks in Tengchong, SW China: relationship with the uplift of the Tibetan Plateau. Isl. Arc 21, 255-269.

Zhang Z.-Q., Tang S.-H., Yuan Z.-X., Bai G. and Wang J.-H. (2001) The Sm-Nd and Rb-Sr isotopic systems of the dolomites in the Bayan Obo ore deposit, Inner Mongolia, China. Acta Petrol. Sin. 17, 637-642 (in Chinese with English abstract).

Zhao C.-P., Ran H. and Chen K.-H. (2006) Present-day magma chambers in Tengchong volcano area inferred from relative geothermal gradient. Acta Petrol. Sin. 22, 1517-1528 (in Chinese with English abstract).

Zhao D.-P. and Liu L. (2010) Deep structure and origin of active volcanoes in China. Geosci. Frontiers 1, 31-44.

Zhao D.-P., Yu S. and Ohtani E. (2011) East Asia: seismotectonics, magmatism and mantle dynamics. J. Asian Earth Sci. 40, 689-709.

Zhao X.-M., Zhang Z.-F., Huang S.-C., Liu Y.-F., Li X. and Zhang H.-F. (2017) Coupled extremely light Ca and Fe isotopes in peridotites. Geochim. Cosmochim. Acta 208, 368-380.

Zhao Y.-W. and Fan Q.-C. (2010) Magma origin and evolution of Maanshan volcano, Dayingshan volcano and Heikongshan volcano in Tengchong area. Acta Petrol. Sin. 26, 1133-1140 (in Chinese with English abstract).

Zhi X., Song Y., Frey F. A., Feng J. and Zhai M. (1990) Geochemistry of Hannuoba basalts, eastern China: Constraints on the origin of continental alkalic and tholeiitic basalt. Chem. Geol. 88, 1-33.

Zhou M.-F., Robinson P. T., Wang C.-Y., Zhao J.-H., Yan D.-P., Gao J.-F. and Malpas J. (2012) Heterogeneous mantle source and magma differentiation of quaternary arc-like volcanic rocks from Tengchong, SE margin of the Tibetan Plateau. Contrib. Mineral. Petrol. 163, 841-860.

Zhu B.-Q., Mao C.-X., Lugmair G. W. and Macdougall J. D. (1983) Isotopic and geochemical evidence for the origin of Plio-Pleistocene volcanic rocks near the Indo-Eurasian collisional margin at Tengchong, China. Earth Planet. Sci. Lett. 65, 263-275.

Zindler A. and Hart S. (1986) Chemical Geodynamics. Annu. Rev. Earth Planet. Sci. 14, 493-571.

Zou H. B., Zindler A., Xu X. and Qi Q. (2000) Major, trace element, and Nd, Sr and Pb isotope studies of Cenozoic basalts in SE China: mantle sources, regional variations, and tectonic significance. Chem. Geol. 171, 33-47.

Zou H.-B., Fan Q.-C., Schmitt A. K. and Sui J.-L. (2010) U-Th dating of zircons from Holocene potassic andesites (Maanshan volcano, Tengchong, SE Tibetan Plateau) by depth profiling: time scales and nature of magma storage. Lithos 118, 202-210.

Zou H.-B., Ma M.-J., Fan Q.-C., Xu B., Li S.-Q., Zhao Y.-W. and King D. T. (2017). Genesis and open-system evolution of Quaternary magmas beneath southeastern margin of Tibet: Constraints from Sr-Nd-Pb-Hf isotope systematics. Lithos 272-273, 278-290.

Zou H.-B., Shen C.-C., Fan Q.-C. and Lin K. (2014) U-series disequilibrium in young Tengchong volcanics: recycling of mature clay sediments or mudstones into the SE Tibetan mantle. Lithos 192, 132-141

Compositional transition in natural alkaline lavas through silica-undersaturated melt—lithosphere interaction*

Ze-Zhou Wang[1], Sheng-Ao Liu[1], Li-Hui Chen[2], Shu-Guang Li[1,3] and Gang Zeng[2]

1. State Key Laboratory of Geological Processes and Mineral Resources, China University of Geosciences, Beijing 100083, China
2. State Key Laboratory for Mineral Deposits Research, School of Earth Sciences and Engineering, Nanjing University, Nanjing 210093, China
3. Chinese Academy of Sciences Key Laboratory of Crust-Mantle Materials and Environments, University of Science and Technology of China, Hefei, Anhui 230026, China

亮点介绍：本文应用 Zn 同位素研究自然碱性岩浆的霞石岩-碧玄岩-碱性橄榄石玄武岩的系列化学组成演化过程，其中霞石岩、高碱碧玄岩的重 Zn 同位素组成表明岩浆形成于碳酸盐化地幔的部分熔融，而低碱碧玄岩、碱性橄榄石玄武岩中等的 δ^{66}Zn 及 δ^{66}Zn 随 SiO_2 升高而下降，随不相容元素(Zn、Nb、Th)含量变低而降低的线性趋势均揭露了硅不饱和碧玄质熔体与富硅岩石圈地幔熔体的相互作用，提出硅不饱和熔体与岩石圈反应是造成碱性岩浆序列演化的重要因素。

Abstract Natural alkaline lavas have diverse compositions—varying widely from nephelinite through basanite to alkali olivine basalt—the origin of which is controversial. In particular, identifying the roles of recycling carbonates in the source and evolution of natural alkaline lavas is commonly difficult. Zinc isotope ratios (δ^{66}Zn) have great potential due to the strong δ^{66}Zn contrast between marine carbonates and the mantle. Here we present a systematic variation of Zn isotopes with Sr-Nd isotopes and incompatible elements (e.g., Nb, Th, and Zn) in nephelinites, basanites, and alkali olivine basalts from eastern China. The elevated δ^{66}Zn of nephelinites and high-alkali basanites relative to the mantle demonstrates that the silica-undersaturated melts were derived from a carbonated mantle. Alkali olivine basalts and low-alkali basanites show a gradual decline of δ^{66}Zn with SiO_2 and have Zn-Sr-Nd isotopic and chemical compositions shifted toward that of an enriched lithospheric mantle. Infiltration of silica-undersaturated basanitic melts and reaction with the lithospheric mantle account for the transition of strongly alkaline melts into weakly alkaline melts via consumption of orthopyroxene and mixing with silica-rich melt derived from lithospheric mantle. High-δ^{66}Zn wehrlite xenoliths found in these alkaline lavas record metasomatism of the lithospheric mantle by basanitic melts. Thus, silica-undersaturated melt–lithosphere interaction could be one of the most common causes of compositional diversity in natural alkaline lavas.

* 本文发表在：Geology, 2018, 46(9): 771-774

1 Introduction

Intraplate alkaline lavas are typically silica-undersaturated and alkali-rich relative to tholeiitic or calc-alkaline lavas erupted at plate boundaries. Their unique chemical and isotope signatures manifest the compositional heterogeneities of Earth's mantle, associated with recycling of crust materials (e.g., Hofmann, 1997). Natural alkaline lavas commonly exhibit a wide range of chemical compositions, from nephelinite through basanite to alkali olivine basalt, with a respective increase in silica and a decrease in alkalinity. The origin of this compositional transition, observed in intraplate volcanoes, is fundamental in petrology and has long been debated.

Trace element modeling indicates that increasing degree of melting of garnet peridotite or fractional crystallization facilitates the transition from nephelinite to alkali olivine basalt (e.g., O'Hara and Yoder, 1967; Frey et al., 1978; Caroff et al., 1997). However, experimental results show that natural alkaline lavas can hardly be produced by melting of volatile-free peridotite (Kogiso et al., 2003). As a possible resolution, melting of CO_2- or carbonate-bearing peridotites has reproduced a series of liquids that overlap part of the natural trend (Hirose, 1997; Dasgupta et al., 2007). Nevertheless, the melting degrees (5%–33%) obtained in experiments (e.g., Dasgupta et al., 2007) are too high to match those (1%–8%) calculated on the basis of trace elements (e.g., Caroff et al., 1997). An alternative mechanism for the nephelinite–alkali basalt transition is reaction of silica-undersaturated melt with subsolidus peridotite, during which melts are diluted by the progressive dissolution of orthopyroxene leading to an enrichment of silica and a decrease of alkalis and incompatible trace elements (Lundstrom, 2000; Pilet et al., 2008; Zhang et al., 2017; Li and Wang, 2018). Discriminating the above hypotheses requires an effective tool to distinguish a carbonated source from other sources (e.g., hybridization by recycled siliciclastic sediments), which is commonly difficult using traditional isotope approaches (e.g., Sr, Pb, and O isotopes). Zinc stable isotopes have such a potential owing to the remarkable δ^{66}Zn difference, but the same orders of magnitude of Zn concentrations between silicate reservoirs (mantle, oceanic mafic crust, and siliciclastic sediments; δ^{66}Zn = ~0.2‰–0.3‰) and marine carbonates (average δ^{66}Zn = ~0.91‰) (Fig. DR1 in the GSA Data Repository[1]; Liu et al., 2016).

Cenozoic alkaline basaltic lavas in the Shandong Peninsula of eastern China encompass a broad range of compositions from nephelinite through basanite to alkali olivine basalt, and are regarded as one of classic examples for studying the origin of intraplate alkaline lavas (Sakuyama et al., 2013; Zeng et al., 2010; H.-Y. Li et al., 2017; S.-G. Li et al., 2017). Here we report new Zn isotope data and compile available Sr-Nd isotope and major-trace element data on these alkaline lavas. These results shed light on the important role of interaction of carbonated mantle-derived silica-undersaturated melt with the overlying lithosphere in generating the compositional diversity of intraplate alkaline lavas.

2 Geological background and samples

Alkaline basaltic lavas have been emplaced over a distance of >4000 km from north to south in eastern China since ca. 106 Ma, constituting an important part of the western circum-Pacific volcanic belt (Fig. DR2). Seismic tomography has revealed that subducting Pacific slabs are stagnant in the mantle transition zone (MTZ) beneath eastern China (Fukao et al., 1992). These basalts have chemical and radiogenic isotopic compositions comparable to those of ocean island basalts (OIBs) and represent typical intraplate alkaline lavas derived from an

[1] GSA Data Repository item 2018280, analytical methods, modeling details, Figures DR1–DR5, and Table DR1, is available online at http://www.geosociety.org/datarepository/2018/ or on request from editing@geosociety.org.

asthenospheric source (e.g., Zhou and Armstrong, 1982; S.-G. Li et al., 2017), which is probably located above the MTZ (S.-G. Li et al., 2017). A total of 57 representative samples from Shandong Peninsula were selected for this study. Their eruption lasted from 24.0 to 0.3 Ma, with alkali olivine basalt at the early stage and nephelinite at the late stage (Zeng et al., 2010). The presence of abundant mantle xenoliths reflects rapid emplacement of the host lavas. All samples have minor olivine as phenocrysts set in a groundmass of olivine, Ti-magnetite, nepheline, and glass, but are devoid of plagioclase and pyroxene phenocrysts. Petrology and geochemistry of these samples have been well characterized (Zeng et al., 2010, 2011).

3 Results

Zinc isotope data are available in Table DR1 in the Data Repository, together with data for selected elements and Sr-Nd isotopes. The overall range of $\delta^{66}Zn$ is from 0.31‰ to 0.48‰, greatly exceeding the range of the upper mantle (0.20‰±0.05‰; 2σ) and mid-oceanic ridge basalts (MORBs) (0.28‰±0.03‰; 2σ) (Wang et al., 2017). Nephelinites (0.45‰±0.04‰; 2σ) have higher $\delta^{66}Zn$ than alkali olivine basalts (0.34‰±0.06‰; 2σ), the latter of which has values that lie between those of nephelinites and MORBs. Alkali olivine basalts show a striking $\delta^{66}Zn$ decline with increasing SiO_2 and decreasing alkalis and incompatible elements such as Nb, Th, and Zn, whereas nephelinites do not (Figs. 1 and 2). The $\delta^{66}Zn$ values of high-alkali basanites overlap those of nephelinites, but some low-alkali basanites have $\delta^{66}Zn$ values within the range of alkali olivine basalts.

4 Origin of strongly alkaline basalts

The lack of correlation of $\delta^{66}Zn$ with loss on ignition, MgO, Cr, and Ni (Fig. DR3) indicates that secondary alteration, crustal contamination, and fractional crystallization negligibly affect Zn isotopes. Primary melts from a volatile-free peridotite source have $\delta^{66}Zn<0.3‰$ (Wang et al., 2017), and thus the elevated $\delta^{66}Zn$ must reflect mantle hybridization by ^{66}Zn-enriched agents. Trace element characteristics (e.g., high Ce/Pb and U/Pb, low Ba/Th and Rb/Nb; Figs. 2E and 2F; Fig. DR4) suggest that the nephelinite and high-alkali basanites have a strong high-μ (HIMU) affinity (high $^{238}U/^{204}Pb$), indicating a source dominated by dehydrated oceanic crust (e.g., Hofmann, 1997; Kogiso et al., 1997; Willbold and Stracke, 2006), probably together with subducted carbonates (Castillo, 2015). Because dehydrated oceanic crust has slightly lower $\delta^{66}Zn$ relative to MORBs due to preferential release of ^{66}Zn-enriched fluids during slab dehydration (Pons et al., 2016), marine carbonates are the unique candidate for the ^{66}Zn enriched agents. Even if carbonate is a minor phase of oceanic crust, deep subducted (ultrahigh-pressure) carbonate species (e.g., magnesite) could have very high Zn concentration (up to ~450 μg/g; Li et al., 2014) relative to the upper mantle (~55 μg/g; Le Roux et al., 2010). Mixing of ~4.4% carbonates with the mantle can result in an elevated $\delta^{66}Zn$ of up to 0.45‰. Thereupon, melting of carbonate-bearing dehydrated oceanic crust accounts for the high-$\delta^{66}Zn$ mantle source for strongly alkaline lavas.

5 Transition from basanite to weakly alkaline basalts

The covariations of $\delta^{66}Zn$ with Nd isotopes and key elements can be separated into two distinct trends (Figs. 1 and 2). One trend is defined by nephelinites and high-alkali basanites that have variable concentrations of incompatible elements but uniform $\delta^{66}Zn$, reflecting the modest Zn isotope fractionation during mantle melting (Wang et al., 2017). Trace element modeling suggests 1.5%–4% melting in the garnet stability field (Zeng et al.,

2010). Another trend is defined by alkali olivine basalts and low-alkali basanites that show covariations of δ^{66}Zn with $\varepsilon_{Nd}(t)$ and incompatible elements. Because isotopic ratios are almost unfractionated during equilibrium melting, this trend demands mixing/interaction of two chemically and isotopically distinct sources. Extrapolation of these trendlines (Fig. 2) separates the two sources into basanite and an enriched mantle component, respectively.

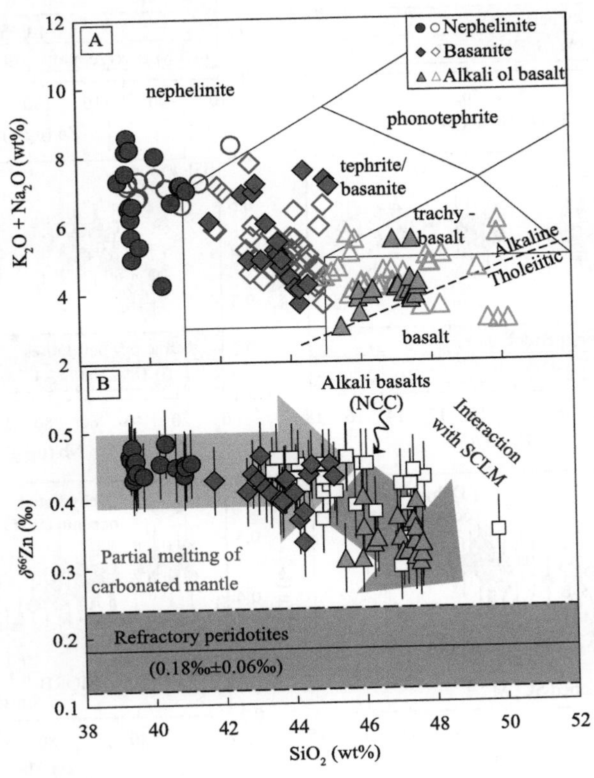

Fig. 1 A: Total alkali versus SiO_2 diagram for classification of alkaline lavas from Shandong Peninsula, eastern China. Solid symbols denote samples in this study; open symbols represent literature data (see data sources in Fig. DR5 [see footnote 1]). B: Plot of δ^{66}Zn versus SiO_2. Error bars represent 2σ. Data for alkali basalts from the North China craton (NCC) and refractory peridotites are from Liu et al. (2016) and Wang et al. (2017), respectively. ol—olivine; SCLM—sub-continental lithospheric mantle.

Recycled ancient eclogite and garnet pyroxenite were previously presumed to be the enriched component in the mantle (Zeng et al., 2011; H.-Y. Li et al., 2017). However, the alkali olivine basalts and low-alkali basanites have low Zn/Fe ratios typical of a peridotitic rather than a pyroxenitic source lithology (Fig. 2G). The enriched component is most likely the subcontinental lithospheric mantle (SCLM) modified by slab-derived fluids and/or melts as indicated by the high Ba/Th and low Ce/Pb of alkali olivine basalts and low-alkali basanites (Figs. 2E and 2F). Although most peridotite xenoliths entrained in these alkaline lavas have low $^{87}Sr/^{86}Sr$ and high $\varepsilon_{Nd}(t)$, they sampled the shallow SCLM that is located above and thus may differ in composition from the source region of alkali olivine basalts capturing the xenoliths. In fact, peridotite xenoliths with extremely enriched Sr-Nd isotope compositions are widely found in mafic intrusions from Shandong Peninsula (Fig. 3; Fig. DR5). These "enriched" peridotites also contain high abundances of fluid-mobile elements like Ba and Pb (Xu et al., 2008) and have low La/Yb ratios (Fig. 3). Highly enriched mantle signals [e.g., low $\varepsilon_{Nd}(t)$] are also recorded in the SCLM-derived lavas (>120 Ma basalts; Fig. 2H). These lines of evidence suggest that substantial reaction of basanitic melts with an enriched SCLM is highly probable. The existence of high-δ^{66}Zn wehrlite xenoliths is the direct record of metasomatism of the SCLM by ^{66}Zn-rich basanitic melts (Fig. 3A).

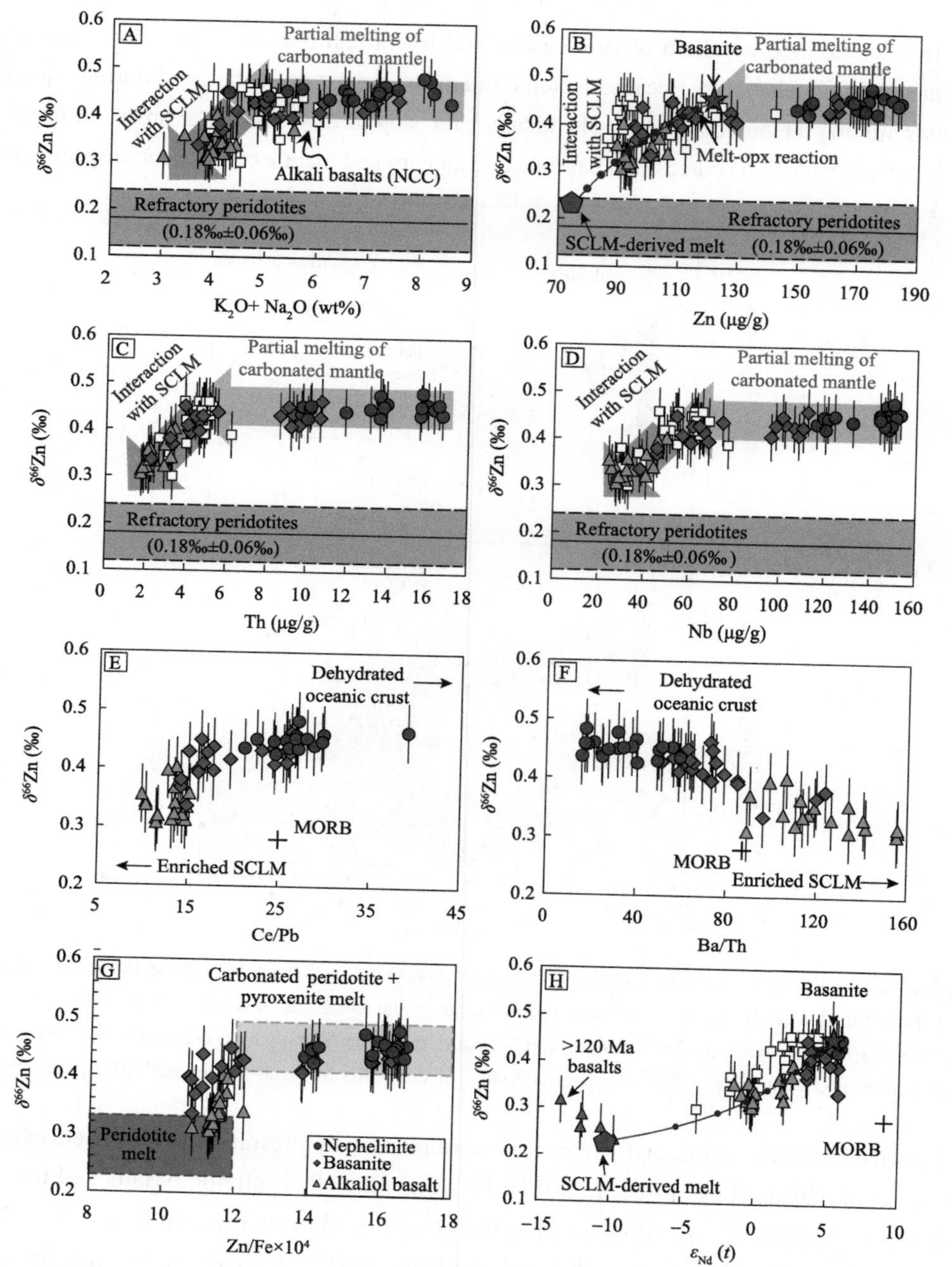

Fig. 2　Variations of δ^{66}Zn with $K_2O + Na_2O$ (A), Zn (B), Th (C), Nb (D), Ce/Pb (E), Ba/Th (F), Zn/Fe (G), and $\varepsilon_{Nd}(t)$ (H). Error bars represent 2σ. Data for alkali basalts and >120 Ma basalts from North China craton (NCC) are from Liu et al. (2016). δ^{66}Zn of refractory peridotite and midoceanic ridge basalt (MORB) are from Wang et al. (2017). The Zn/Fe ranges of peridotite and pyroxenite melts are from Le Roux et al. (2010). Solid curves in B and H represent evolution of basanite during reaction with orthopyroxene (opx) (blue) or mixing with sub-continental lithospheric mantle (SCLM)–derived melt (purple). Increment of mixing curves (small dot) is 0.1. Details of calculations are shown in the Data Repository (see footnote 1).

Considerable CO_2 can be dissolved in primary melts of basanite and nephelinite (Dasgupta et al., 2007), but reaction of CO_2-rich basanitic melts with peridotite would dissolve olivine and produce strongly silica-undersaturated melilitite instead of less silica-undersaturated melts (Mallik and Dasgupta, 2013). In fact, the primary basanitic melts derived from melting of carbonated mantle could have undergone extensive degassing in the upper mantle (1–4 GPa) (Boudoire et al., 2018). Degassing may also be enhanced along deep faults such as the Tan-Lu fault where weakly alkaline basalts erupted (Fig. DR2) during continental rifting (Lee et al., 2016). Subsequently, reactive infiltration of the CO_2-poor silica-undersaturated melt into the SCLM would gradually transform silica-undersaturated melts into less silica-undersaturated ones via olivine crystallization at the expense of orthopyroxene (Shawet al., 1998; Lundstrom, 2000; Pilet et al., 2008), with a concomitant decrease in incompatible elements and δ^{66}Zn by dilution (Figs. 1 and 2). However, reaction of melts with orthopyroxene alone cannot generate the δ^{66}Zn-[Zn] trend in alkali olivine basalts (Fig. 2B), as orthopyroxene is too poor in Zn. Lundstrom (2000) found that infiltration of sodium into peridotite not only causes dissolution of orthopyroxene but can also change the manner by which the peridotite melts by a process that is chemically, not thermally, activated. Hence, silica-rich melts with lower δ^{66}Zn and La/Yb ratios than the basanite can be produced by melting of SLCM at shallow depth during basanite-peridotite interaction. A binary mixing model suggests that mixing of the percolated basanite with ~30% to 70% of SCLM melt is able to reproduce the δ^{66}Zn-[Zn] and δ^{66}Zn-$\varepsilon_{Nd}(t)$ trends observed in the alkali olivine basalts (Figs. 2B and 2H). Such amounts of SCLM melts can be generated near the interface between basanitic melt and impregnated peridotite (Lundstrom, 2000).

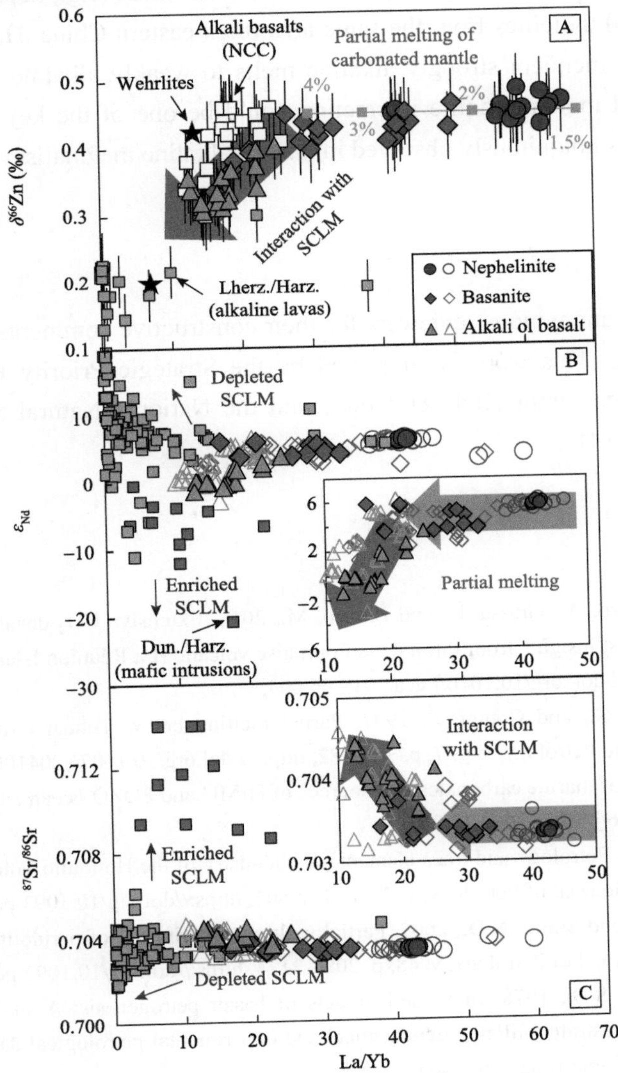

Fig. 3 Variations of La/Yb ratios with isotopic ratios of Zn (A), Nd (B), and Sr (C). δ^{66}Zn of wehrlite, lherzolite (Lherz.), and harzburgite (Harz.) are from Wang et al. (2017), and $^{87}Sr/^{86}Sr$, ε_{Nd}, and La/Yb are compiled from literature (see the Data Repository [see footnote 1]). Error bars represent 2σ. Orange line refers to degree of partial melting as function of La/Yb (after Zeng et al., 2010). NCC—North China craton; SCLM—sub-continental lithospheric mantle; Dun.—dunite.

6 Compositional diversity in intraplate alkaline lavas

The compositional continuum from nephelinites to alkali olivine basalts observed in many other intraplate alkaline lava suites (e.g., Hawaii, southeastern Australia, and Tubuai [French Polynesia]) has been commonly attributed to increasing degrees of mantle melting (e.g., Frey et al., 1978; Clague and Frey, 1982; Caroff et al., 1997). However, systematic isotopic variations found in some alkaline lavas from a geographically and temporally limited volcanic province indicate multiple sources (e.g., Zhang et al., 2017). Our study supports the experimental results showing that melting of a carbonated mantle can only generate the nephelinite-basanite series (Hirose, 1997; Dasgupta et al., 2007), whereas reaction of silica undersaturated basanitic melts with lithosphere is required for a complete transition from basanite into alkali olivine basalts and even tholeiites (Pilet et al., 2008). The composition of alkaline lavas could be efficiently modified when they ascend through the overlying lithosphere. Such processes are perhaps frequently identified where the lithospheric mantle is isotopically distinct from the percolated melts. It is interesting that strongly silica-undersaturated lavas (e.g., nephelinite) commonly occur later than alkali olivine basalts and tholeiites from the same area (e.g., eastern China, Hawaii). The temporal sequence may reflect transition of the incipient strongly alkaline melts to weakly alkaline melts via interaction with the SCLM. Silica-undersaturated melt–peridotite interaction could be one of the key processes responsible for the compositional diversity that is ubiquitously observed in natural alkaline magmatism (Pilet et al., 2008).

Acknowledgements

We are grateful to three anonymous reviewers for their constructive comments and to editor Chris Clark for his careful editorial handling. This work is supported by the Strategic Priority Research Program (B) of the Chinese Academy of Sciences (grant XDB18000000) and the National Natural Science Foundation of China (grants 41622303 and 41730214).

References

Boudoire, G., Rizzo, A.L., Di Muro, A., Grassa, F., and Liuzzo, M., 2018, Extensive CO_2 degassing in the upper mantle beneath oceanic basaltic volcanoes: First insights from Piton de la Fournaise volcano (La Réunion Island): Geochimica et Cosmochimica Acta, v. 235, p. 376-401, https://doi .org/10.1016/j.gca.2018.06.004.

Caroff, M., Maury, R.C., Guille, G., and Cotten, J., 1997, Partial melting below Tubuai (Austral Islands, French Polynesia): Contributions to Mineralogy and Petrology, v. 127, p. 369-382, https:// doi.org/10.1007/s004100050286.

Castillo, P.R., 2015, The recycling of marine carbonates and sources of HIMU and FOZO ocean island basalts: Lithos, v. 216-217, p. 254-263, https://doi.org/10.1016/j.lithos.2014.12.005.

Clague, D.A., and Frey, F.A., 1982, Petrology and trace element geochemistry of the Honolulu Volcanics, Oahu: Implications for the oceanic mantle below Hawaii: Journal of Petrology, v. 23, p. 447-504, https://doi.org/10.1093/petrology/23 .3.447.

Dasgupta, R., Hirschmann, M.M., and Smith, N.D., 2007, Partial melting experiments of peridotite + CO_2 at 3 GPa and genesis of alkalic ocean island basalts: Journal of Petrology, v. 48, p. 2093-2124, https://doi.org/10.1093/petrology/egm053.

Frey, F.A., Green, D.H., and Roy, S.D., 1978, Integrated models of basalt petrogenesis: A study of quartz tholeiites to olivine melilitites from south eastern Australia utilizing geochemical and experimental petrological data: Journal of Petrology, v. 19, p. 463-513, https://doi.org/10.1093 /petrology/19.3.463.

Fukao, Y., Obayashi, M., Inoue, H., and Nenbai, M., 1992, Subducting slabs stagnant in the mantle transition zone: Journal of Geophysical Research, v. 97, p. 4809-4822, https://doi.org/10 .1029/91JB02749.

Hirose, K., 1997, Partial melt compositions of carbonated peridotite at 3 GPa and role of CO_2 in alkali- basalt magma generation:

Geophysical Research Letters, v. 24, p. 2837-2840, https://doi.org/10 .1029/97GL02956.
Hofmann, A.W., 1997, Mantle geochemistry: The message from oceanic volcanism: Nature, v. 385, p. 219-229, https://doi.org/10.1038/385219a0.
Kogiso, T., Tatsumi, Y., and Nakano, S., 1997, Trace element transport during dehydration processes in the subducted oceanic crust: 1. Experiments and implications for the origin of ocean island basalts: Earth and Planetary Science Letters, v. 148, p. 193-205, https://doi.org/10.1016/S0012-821X (97)00018-6.
Kogiso, T., Hirschmann, M.M., and Frost, D.J., 2003, High-pressure partial melting of garnet pyroxenite: Possible mafic lithologies in the source of ocean island basalts: Earth and Planetary Science Letters, v. 216, p. 603-617, https://doi.org /10.1016/S0012-821X(03)00538-7.
Le Roux, V., Lee, C.-T.A., and Turner, S.J., 2010, Zn/ Fe systematics in mafic and ultramafic systems: Implications for detecting major element heterogeneities in the Earth's mantle: Geochimica et Cosmochimica Acta, v. 74, p. 2779-2796, https:// doi.org/10.1016/j.gca.2010.02.004.
Lee, H., Muirhead, J.D., Fischer, T.P., Ebinger, C.J., Kattenhorn, S.A., Sharp, Z.D., and Kianji, G., 2016, Massive and prolonged deep carbon emissions associated with continental rifting: Nature Geoscience, v. 9, p. 145-149, https://doi.org/10 .1038/ ngeo2622.
Li, H.-Y., Xu, Y.-G., Ryan, J.G., and Whattam, S.A., 2017, Evolution of the mantle beneath the eastern North China Craton during the Cenozoic: Linking geochemical and geophysical observations: Journal of Geophysical Research: Solid Earth, v. 122, p. 224-246, https://doi.org/10.1002 /2016JB013486.
Li, J.-L., Klemd, R., Gao, J., and Meyer, M., 2014, Compositional zoning in dolomite from lawsonite-bearing eclogite (SW Tianshan, China): Evidence for prograde metamorphism during subduction of oceanic crust: The American Mineralogist, v. 99, p. 206-217, https://doi.org/10 .2138/am.2014.4507.
Li, S.-G., and Wang, Y., 2018, Formation time of the big mantle wedge beneath eastern China and a new lithospheric thinning mechanism of the North China craton—Geodynamic effects of deep recycled carbon: Science China: Earth Sciences, https:// doi.org/10.1007/s11430-017-9217-7 (in press).
Li, S.-G., et al., 2017, Deep carbon cycles constrained by a large-scale mantle Mg isotope anomaly in eastern China: National Science Review, v. 4, p. 111-120, https://doi.org/10.1093/nsr/nww070.
Liu, S.-A., Wang, Z.-Z., Li, S.-G., Huang, J., and Yang, W., 2016, Zinc isotope evidence for a large-scale carbonated mantle beneath eastern China: Earth and Planetary Science Letters, v. 444, p. 169-178, https://doi.org/10.1016/j.epsl.2016.03.051.
Lundstrom, C.C., 2000, Rapid diffusive infiltration of sodium into partially molten peridotite: Nature, v. 403, p. 527-530, https:// doi.org/10.1038 /35000546.
Mallik, A., and Dasgupta, R., 2013, Reactive infiltration of MORB-eclogite-derived carbonated silicate melt into fertile peridotite at 3 GPa and genesis of alkalic magmas: Journal of Petrology, v. 54, p. 2267-2300, https://doi.org/10.1093 /petrology/egt047.
O'Hara, M.J., and Yoder, H.S., 1967, Formation and fractionation of basic magmas at high pressures: Scottish Journal of Geology, v. 3, p. 67-117, https://doi.org/10.1144/sjg03010067.
Pilet, S., Baker, M.B., and Stolper, E.M., 2008, Metasomatized lithosphere and the origin of alkaline lavas: Science, v. 320, p. 916-919, https://doi.org /10.1126/science.1156563.
Pons, M.-L., Debret, B., Bouilhol, P., Delacour, A., and Williams, H., 2016, Zinc isotope evidence for sulfate-rich fluid transfer across subduction zones: Nature Communications, v. 7, 13794, https://doi.org/10.1038/ncomms13794.
Sakuyama, T., et al., 2013, Melting of dehydrated oceanic crust from the stagnant slab and of the hydrated mantle transition zone: Constraints from Cenozoic alkaline basalts in eastern China: Chemical Geology, v. 359 p.32-48, https://doi .org/10.1016/ j.chemgeo.2013.09.012.
Shaw, C.S., Thibault, Y., Edgar, A.D., and Lloyd, F.E., 1998, Mechanisms of orthopyroxene dissolution in silica- undersaturated melts at 1 atmosphere and implications for the origin of silica-rich glass in mantle xenoliths: Contributions to Mineralogy and Petrology, v. 132, p. 354-370, https://doi.org /10.1007/s004100050429.
Wang, Z.-Z., Liu, S.-A., Liu, J., Huang, J., Xiao, Y., Chu, Z.-Y., Zhao, X.-M., and Tang, L., 2017, Zinc isotope fractionation during mantle melting and constraints on the Zn isotope composition of Earth's upper mantle: Geochimica et Cosmochimica Acta, v. 198, p. 151-167, https://doi.org /10.1016/j.gca.2016.11.014.
Willbold, M., and Stracke, A., 2006, Trace element composition of mantle end-members: Implications for recycling of oceanic and upper and lower continental crust: Geochemistry Geophysics Geosystems, v. 7, Q04004, https://doi.org/10.1029/2005GC001005.
Xu, W., Hergt, J.M., Gao, S., Pei, F., Wang, W., and Yang, D., 2008, Interaction of adakitic melt-peridotite: Implications for the

high-Mg# signature of Mesozoic adakitic rocks in the eastern North China Craton: Earth and Planetary Science Letters, v. 265, p. 123-137, https://doi.org/10.1016 /j.epsl.2007.09.041.

Zeng, G., Chen, L.-H., Xu, X.-S., Jiang, S.-Y., and Hofmann, A.W., 2010, Carbonated mantle sources for Cenozoic intra-plate alkaline basalts in Shandong, North China: Chemical Geology, v. 273, p. 35-45, https://doi.org/10.1016/j.chemgeo.2010.02.009.

Zeng, G., Chen, L.-H., Hofmann, A.W., Jiang, S.- Y., and Xu, X.-S., 2011, Crust recycling in the sources of two parallel volcanic chains in Shandong, North China: Earth and Planetary Science Letters, v. 302, p. 359-368, https://doi.org/10.1016/j.epsl.2010.12.026.

Zhang, G.-L., Chen, L.-H., Jackson, M.G., and Hofmann, A.W., 2017, Evolution of carbonated melt to alkali basalt in the South China Sea: Nature Geoscience, v. 10, p. 229-235, https://doi.org/10 .1038/ngeo2877.

Zhou, X., and Armstrong, R.L., 1982, Cenozoic volcanic rocks of eastern China—Secular and geographic trends in chemistry and strontium isotopic composition: Earth and Planetary Science Letters, v. 58, p. 301-329, https://doi.org/10.1016/0012-821X(82)90083-8.

A nephelinitic component with unusual δ^{56}Fe in Cenozoic basalts from eastern China and its implications for deep oxygen cycle*

Yongsheng He[1], Xunan Meng[1], Shan Ke[1], Hongjie Wu[1], Chuanwei Zhu[1], Fang-Zhen Teng[2], Jochen Hoefs[3], Jian Huang[4], Wei Yang[5], Lijuan Xu[1], Zhenhui Hou[4], Zhong-Yuan Ren[6] and Shuguang Li[1,4]

1. State Key Laboratory of Geological Processes and Mineral Resources, China University of Geosciences, Beijing 100083, China
2. Isotope Laboratory, Department of Earth and Space Sciences, University of Washington, Seattle, WA 98195, USA
3. Department of Geosciences, University of Göttingen, Goldschmidtstr. 1, Göttingen 37077, Germany
4. CAS Key Laboratory of Crust-Mantle Materials and Environments, School of Earth and Space Sciences, University of Science and Technology of China, Hefei 230026, China
5. State Key Laboratory of Lithospheric Evolution, Institute of Geology and Geophysics, Chinese Academy of Sciences, P.O. Box 9825, Beijing 100029, China
6. Key Laboratory of Mineralogy and Metallogeny, Guangzhou Institute of Geochemistry, Chinese Academy of Sciences, Guangzhou 510640, China

亮点介绍：本研究发现中国东部新生代玄武岩中普遍存在一个具有高碱度、高 $Fe^{3+}/\Sigma Fe$、重Fe同位素组成的岩浆端元组分。该重Fe同位素端元应来自氧化性橄榄岩源区，与再循环碳酸盐在深部（> 300 km）地幔发生的歧化反应有关：即俯冲碳酸盐的部分碳酸根还原成单质碳（金刚石），同时氧化源区硅酸盐的 Fe^{2+} 和金属Fe成 Fe^{3+}，在地幔熔融时形成高 $Fe^{3+}/\Sigma Fe$ 岩浆组分。本文证实了现今地球存在深部碳循环驱动的地球深部向浅表的净氧迁移机制。

Abstract Cycling of elements with multiple valences (e.g., Fe, C, and S) through subduction and magmatism may dictate the redox evolution of the deep mantle and atmosphere. To investigate the potential of Fe isotopes as a tracer of such cycles, here we report Fe isotopic compositions of thirty-seven Cenozoic basalts from eastern China. A nephelinitic melt component with δ^{56}Fe up to 0.29 has been identified, which cannot be explained by weathering, alteration, magma differentiation, or chemical diffusion. Its low Fe/Mn ~58, relatively low TiO_2 and high Na_2O+K_2O argue against a significant contribution of pyroxenite melting. Instead, the heavy Fe component requires enhanced isotope fractionation during partial melting of a peridotitic source with $Fe^{3+}/\Sigma Fe \geqslant 0.15$. Low Ba/Th ~ 50 and depleted $^{87}Sr/^{86}Sr(i)$ and $\varepsilon_{Nd}(t)$ suggest that the source was insignificantly affected by hydrous fluids and recycled terrigenous sediments. The heavy Fe component is known to be unique in its low

* 本文发表在：Earth and Planetary Science Letters, 2019, 512: 175-183

δ^{26}Mg and high δ^{66}Zn and indicates hybridization by recycled carbonates. The source $Fe^{3+}/\Sigma Fe$ was most likely enhanced at cost of reduction of recycled carbonates to diamonds in a mantle depth≥300 km. The origin of the heavy Fe component illustrates a pathway with net transportation of oxidizer back to Earth's surface: CO_2 (in carbonates) → C (as diamond frozen in the deep mantle) + O_2 (ferric Fe being scavenged by melt extraction). Secular cooling of global subduction zones may have stepwisely increased the efficiency of this carbon driven deep oxygen cycle in the past, providing an alternative explanation for the rise of atmospheric O_2.

Keywords iron isotopes, oxidized peridotitic source, deep carbon freezing, deep oxygen cycle, atmospheric oxygenation

1 Introduction

Cycling of elements with multiple valences, dominated by Fe, C and S, via subduction and magmatism can exchange the oxidizing potential between the upper mantle and Earth's surface. For example, modern subduction zones can bring a net redox budget (e.g., the ability to transfer O^{2-} to O_2) ~1.2×10^{19} mol O_2 / Ma into the upper mantle (Evans, 2012) that is significant compared to the present O_2 volume of the atmosphere (~4.0×10^{19} mol) (Trenberth and Smith, 2005). Influx of oxidizing agents may explain mantle domains with high oxygen fugacities (e.g., Kelley and Cottrell, 2009; Konter et al., 2016). Therefore, tracing deep oxygen cycles driven by subduction and magmatism is crucial to understand the redox evolution of both, the deep mantle and atmosphere, which, however, remains challenging.

Iron is the most abundant multiple valence element (naturally Fe^0, Fe^{2+} and Fe^{3+}) in the solid Earth. It has four stable isotopes: ^{54}Fe (5.85%), ^{56}Fe (91.75%), ^{57}Fe (2.12%), and ^{58}Fe (0.28%). The terrestrial mantle has an average δ^{56}Fe of ~0.02, while mid-ocean ridge (MORB) and oceanic island basalts (OIB) have systematically higher δ^{56}Fe with the majority clustering ~0.10, reflecting significant isotope fractionation during partial melting (e.g., Teng et al., 2013; Weyer and Ionov, 2007). Ferric Fe usually forms stronger bonds and thus is enriched in heavy isotopes compared to ferrous Fe, both in minerals and melts (e.g., Dauphas et al., 2014; Sossi and O'Neil, 2017). Furthermore, ferric Fe behaves incompatible and ferrous Fe distributes approximately equal between the melt and residue during peridotite partial melting (Canil et al., 1994). Therefore, δ^{56}Fe of basalts can reflect their source $Fe^{3+}/\Sigma Fe$ and the extent of preferential extraction of Fe^{3+} by the melt (e.g., Dauphas et al., 2014; Weyer and Ionov, 2007). Contribution of pyroxenites in the source and magma differentiation may also affect δ^{56}Fe of basalts (e.g., Konter et al., 2016; Teng et al., 2008; Williams and Bizimis, 2014). Partial melts from pyroxenitic sources can be well identified by indices, e.g., Fe/Mn (Sobolev et al., 2007). A significant increase in melt δ^{56}Fe only happens after substantial olivine and pyroxene crystallization (e.g., Teng et al., 2008), which can be corrected by back-calculations of crystallization (e.g., Teng et al., 2013). With contribution of pyroxenitic sources and magma differentiation being identified or corrected, δ^{56}Fe of basalts have the potential to trace their source $Fe^{3+}/\Sigma Fe$ and outflux of ferric Fe from the mantle. Together with other indices for recycled water, terrigenous sediments, and carbonates (e.g., Chen et al., 2015; Liu et al., 2016; Li et al., 2017; Plank and Langmuir, 1998), Fe isotopes have the potential to trace deep oxygen cycles induced by cycling of supercrustal components.

Cenozoic basalts distributed throughout the whole eastern China (Fig. S1) have been documented to be derived from mantle sources hybridized by variable subducted materials, ranging from water, terrigenous sediments, to carbonates, likely released from the subducted Pacific plate beneath the Eurasian continent (e.g., Chen et al., 2015; Li et al., 2016; Li et al., 2017). Cenozoic basalts from eastern China have δ^{26}Mg (−0.46±0.14, 2SD) systematically lower and δ^{66}Zn (0.45±0.15, 2SD) higher than MORB, indicating that their sources could have been hybridized by recycled carbonates (e.g., Liu et al., 2016; Li et al., 2017). Particularly, basalts with the lowest δ^{26}Mg and highest δ^{66}Zn are characterized by a low $^{87}Sr/^{86}Sr(i)$ ~0.703. The hybridizing carbonate agent should be Mg- and Zn-rich and Sr-depleted, likely dominated by magnesite. Accordingly, carbonates may have been recycled into depths probably > 300 km in the upper mantle beneath eastern China (e.g., Li et al., 2017), across the iron-carbon redox boundary where reduction of recycled carbonates by Fe^0 resulted from Fe^{2+} disproportion can occur (e.g., Rohrbach and Schmidt, 2011). These observations highlight Cenozoic basalts from eastern China as ideal samples to study deep oxygen cycles induced by cycling of supercrustal components.

To investigate the potential of Fe isotopes to trace deep oxygen cycles and evaluate the net redox budget for cycling of variable crustal components, here we report Fe isotopic compositions of thirty-seven Cenozoic basalts from eastern China. A nephelinitic melt component with δ^{56}Fe up to 0.29 has been identified, which represents the heaviest basalts so far found from peridotite melting. Together with previously reported and newly obtained elemental and Sr-Nd-Mg-Zn isotopic data on the same samples, this heavy Fe component was most likely derived from a peridotitic source with high $Fe^{3+}/\Sigma Fe$ that had been hybridized by recycled carbonates. Our study illustrates a massive deep oxygen cycle with a net redox budget, through releasing ferric Fe to the Earth's surface via recycling of carbonates into the deep mantle that are presumably reduced to diamonds. Considering the evolution of global plate tectonics, this carbon driven deep oxygen cycle may help to explain the oxygenation of the atmosphere.

2 Samples and methods

The samples measured in this study are from 12 localities widespread in eastern China, including Shuangliao, Hannuoba, Fansi, Xiyang-Pingding, Zuoquan, Qixia-Penglai, Pingmingshan, Anfengshan, Fangshan, Chongren, and Longyou (Fig. S1). Petrological, elemental and Sr-Nd-Mg-Zn isotopic compositions of the same set of samples, except those from Qixia-Penglai, have been previously documented (Supplementary Materials and Table S5). Nine basalts from Qixia-Penglai were newly collected in this study. Thirty-seven samples measured here range from 41.4–53.5 wt.% SiO_2 (discussed on a volatile-free bias) and 0.08–0.22 $(Na_2O+K_2O)/SiO_2$ (weight ratio), spanning the entire compositional range of Cenozoic basalts from eastern China. According to the enrichment of alkalis, indexed by $(Na_2O+K_2O)/SiO_2$, these samples can be divided into three subgroups: the nephelinitic (>0.16), tholeiitic (≤0.10–0.11), and transitional group (Fig. 1A). Olivine and pyroxenes commonly occur as the only Fe-rich phenocrysts, while magnetite is absent or exists in the matrix. All samples are relatively fresh, except that olivine in the basalts from Zuoquan and Qixia-Penglai has experienced slight iddingsitization.

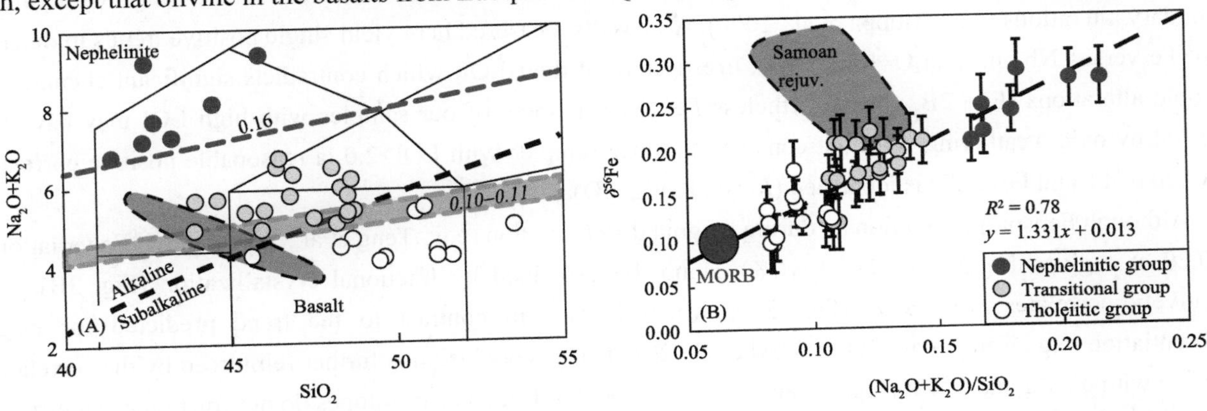

Fig. 1 Total alkaline vs. SiO_2 (A) and δ^{56}Fe vs. $(Na_2O+K_2O)/SiO_2$ (B) diagrams. Cenozoic basalts from Table S5, MORB from Teng et al. (2013) and rejuvenated Samoan lavas from Konter et al. (2016). Major elements are plotted on a volatile free bias.

Iron isotopic analysis was conducted at the Isotope Geochemistry Lab at China University of Geosciences, Beijing (CUGB). Samples were dissolved in a concentrated HF-HNO_3-HCl, and Fe was purified using AG1X-8 (200–400 mesh in chloride form). Iron isotopic ratios were obtained on a Thermo Finnigan Neptune Plus MC-ICPMS with a standard-sample-bracketing technique. The long term reproducibility and accuracy are better than 0.03 for δ^{56}Fe. To support the interpretation of Fe isotopic data, elemental and Sr-Nd-Mg isotopic data have also been obtained for the Qixia-Penglai samples and other samples that lack Sr-Nd-Mg isotopic data (Tables S1-S3). The ferrous Fe contents were determined by redox titration using $K_2Cr_2O_7$ solution; $Fe^{3+}/\Sigma Fe$ was then calculated with an accuracy <10%. Analytical details are given in Supplementary Materials.

3 Result

Nine Qixia-Penglai basalts have SiO_2 from 41.4 wt.% to 48.3 wt.% and $(Na_2O+K_2O)/SiO_2$ from 0.11 to 0.18 (Table S1). Six of these nine samples were measured for trace element concentrations (Table S2). Although a few samples (e.g., 14TST-2) have high $(Na_2O+K_2O)/SiO_2$, all the Qixia-Penglai basalts display trace element patterns similar to the other transitional samples, characterized by moderate enrichment of Nb and Ta without strong depletion in K and Ti (Fig. S2). The Qixia-Penglai basalts have $^{87}Sr/^{86}Sr(i)$ from 0.7037 to 0.7049 and $\varepsilon_{Nd}(t)$ from 1.3 to 4.3. Mg isotopic compositions of twenty three Cenozoic basalts were measured, yielding δ^{26}Mg from −0.66 to −0.29 with a mean value of −0.48±0.17 (2SD, N=23) comparable to those previously reported (−0.46±0.14; Li et al., 2017; and references therein).

Thirty-seven Cenozoic basalts from eastern China were measured for Fe isotopic compositions (Table S4) and yield highly variable but overall high δ^{56}Fe (0.10–0.29) increasing with $Fe^{3+}/\Sigma Fe$ (0.13–0.73) and the alkaline enrichment indexed by $(Na_2O+K_2O)/SiO_2$ (0.08–0.22) (Fig. 1 and Table S5). The heaviest samples have δ^{56}Fe substantially higher than MORB (~0.10) (Teng et al., 2013) approaching the highest δ^{56}Fe values identified in terrestrial basalts (Konter et al., 2016).

4 Discussion

4.1 Limited effect of weathering, differentiation, and diffusion on δ^{56}Fe of Cenozoic basalts

Samples with high δ^{56}Fe also have overall high LOI (Fig. 2A), which may suggest a possible role of weathering and secondary alterations. Niobium is immobile, and alkali elements are mobile during weathering and secondary alterations (e.g., Aiuppa et al., 2000). All basalts measured here yield single positive trends in diagrams of δ^{56}Fe versus Nb and $(Na_2O+K_2O)/SiO_2$, irrespective of their LOI, which contradicts significant chemical and isotopic alterations (Fig. 2B, C). Nevertheless $Fe^{3+}/\Sigma Fe$ in some of our samples with high LOI may have been affected by oxic weathering. Without considering those samples with LOI>2.0, a reasonable positive correlation between δ^{56}Fe and $Fe^{3+}/\Sigma Fe$ is identified (R^2=0.58, Fig. 2D).

Although Fe isotopes fractionate during magma differentiation (e.g., Teng et al., 2008), the δ^{56}Fe variation up to 0.20 at a given MgO (e.g., 8–10 wt.%) cannot be explained by fractional crystallization (Fig. 3B). δ^{56}Fe negatively correlates with SiO_2 (Fig. 3A), which is also in contrast to the trend predicted for magma differentiation (e.g., Teng et al., 2008; Sossi et al., 2012). This conclusion is further reinforced by the correlations of δ^{56}Fe with δ^{26}Mg and δ^{66}Zn (Fig. 3C and Fig. 4), given that Mg and Zn isotopes do not fractionate significantly during magma differentiation (e.g., Teng et al., 2010; Liu et al., 2016). Olivine grains from the Kilauea Iki lava lake samples record large, negatively correlated Fe and Mg isotope fractionation caused by Fe-Mg inter-diffusion

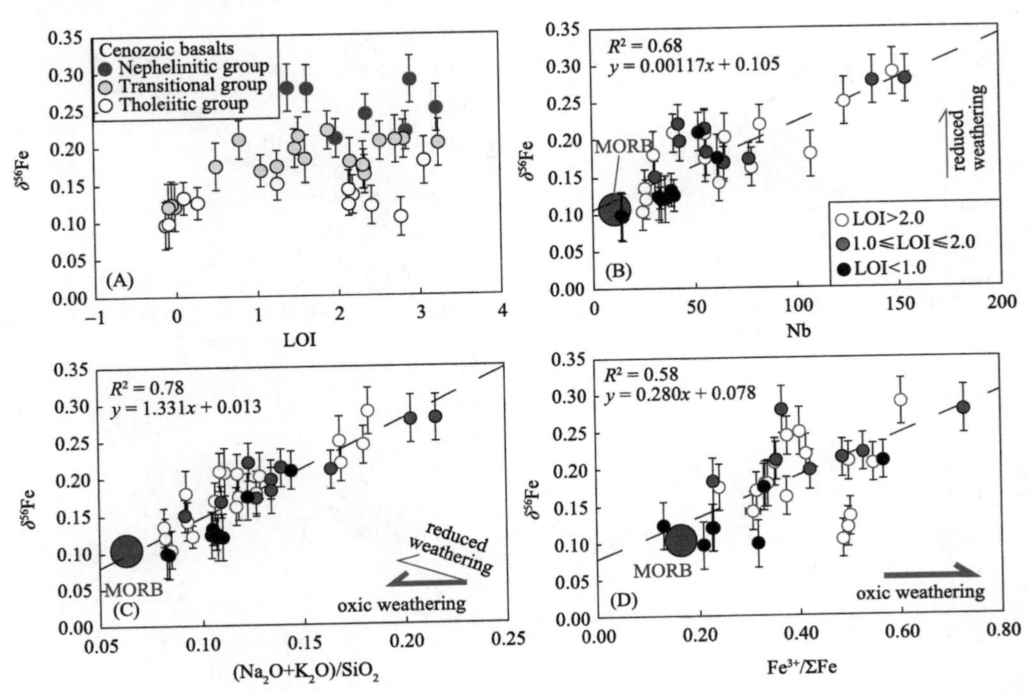

Fig. 2 Diagrams of δ^{56}Fe vs. LOI (A), δ^{56}Fe vs. Nb (B), δ^{56}Fe vs. $(Na_2O+K_2O)/SiO_2$ (C), and δ^{56}Fe vs. $Fe^{3+}/\Sigma Fe$ (D) for basalts from eastern China. MORB from Cottrell and Kelley (2011) and Teng et al. (2013). The regression in Fig. 2D only considers samples with LOI ≤ 2.0.

between crystallizing olivine and evolved basalts (e.g., Teng et al., 2011). Neither fractional crystallization nor accumulation of olivine with highly variable δ^{56}Fe (−1.10 to 0.49) and δ^{26}Mg (−0.40 to 0.03) seems to be able to produce a negatively correlated δ^{56}Fe and δ^{26}Mg in the differentiated magmas as observed in Cenozoic basalts from eastern China (Fig. 3C). Back-calculations of olivine crystallization suggest the effect of magma differentiation on δ^{56}Fe values is likely within or comparable to the error of our measurements (Fig. S3; Supplementary Materials).

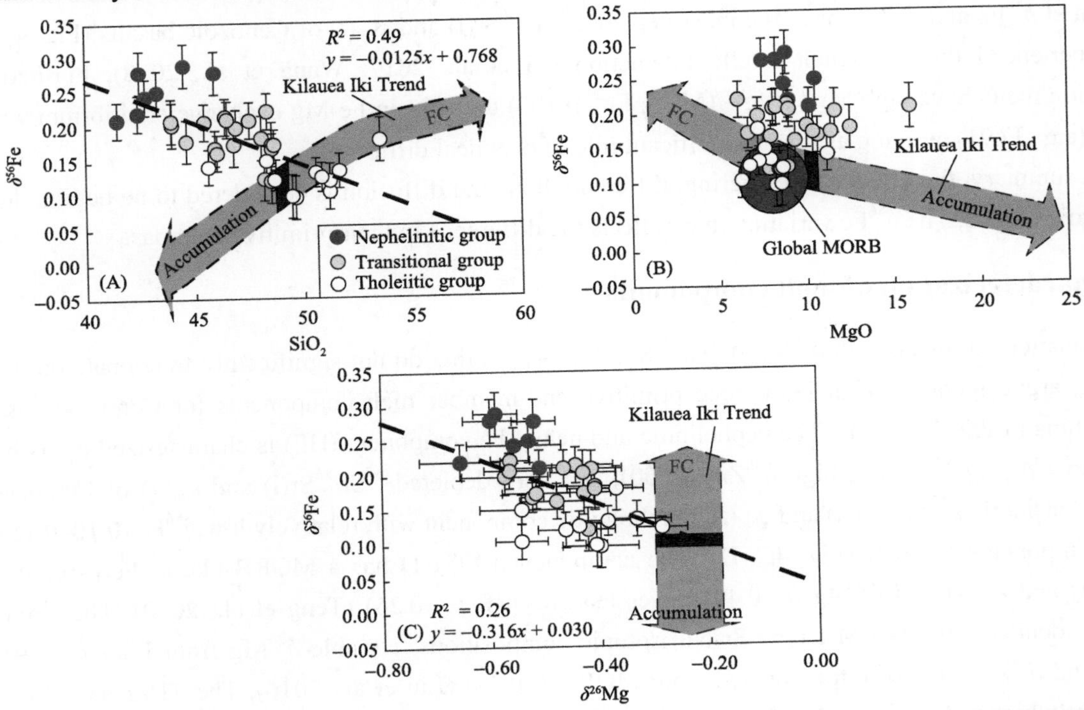

Fig. 3 δ^{56}Fe vs. SiO_2 (A), MgO (B) and δ^{26}Mg (C) diagrams. The differentiation trend of the Kilauea Iki lava lake, Hawaii (Teng et al., 2008, 2010) is plotted for comparison.

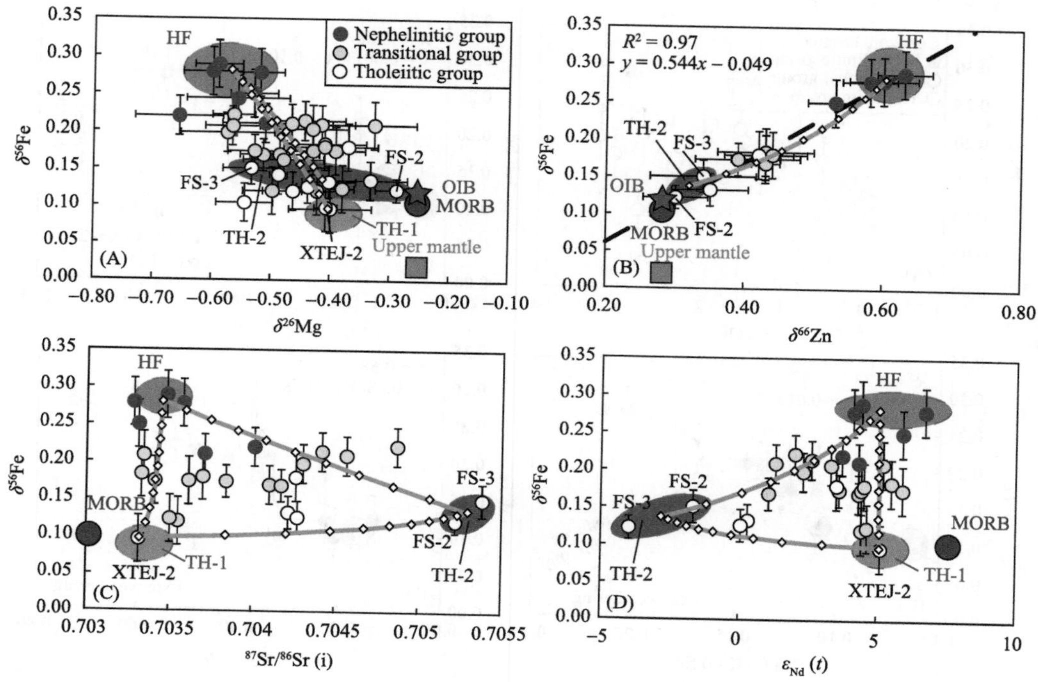

Fig. 4 δ^{56}Fe vs. δ^{26}Mg (A), δ^{66}Zn (B), ^{87}Sr/^{86}Sr(i) (C) and $\varepsilon_{Nd}(t)$ (D) diagrams. Isotopic variations in Cenozoic basalts can be explained by mixing among three magma components: the nephelinitic and heavy Fe component (HF), tholeiitic component I (TH-1) and II (TH-2) (Table 1). Diamonds indicate an increment of 10%. Mean upper mantle, MORB, and OIB are after Liu et al. (2016) and Teng et al. (2010, 2013).

Some Cenozoic basalts from eastern China may have experienced melt-lithospheric mantle interaction to variable extents (Xu et al., 2005; Wang et al., 2018), thereby initiating diffusion which may produce negatively fractionated δ^{56}Fe and δ^{26}Mg (Teng et al., 2011) (Fig. 4A). However, the nephelinitic basalts with the most fractionated δ^{56}Fe and δ^{26}Mg have the most depleted ^{87}Sr/^{86}Sr(i) and $\varepsilon_{Nd}(t)$ of Cenozoic basalts (Fig. 4), and thus have experienced the least melt-mantle interaction (Li et al., 2017; Wang et al., 2018). Furthermore, the nephelinitic melts have high Mg# ($Mg^{2+}/(Mg^{2+}+Fe^{2+})\times100$) up to 81 in Fe-Mg exchange equilibrium with mantle olivines (e.g., Fo91), arguing against a significant role of chemical diffusion.

In a summary, the effect of weathering, differentiation, and diffusion is considered to be negligible on δ^{56}Fe of Cenozoic basalts. The δ^{56}Fe variation thus reflects the difference in their primitive magmas.

4.2 Characterization of melt components

Correlations of δ^{56}Fe with δ^{26}Mg-δ^{66}Zn-^{87}Sr/^{86}Sr(i)-$\varepsilon_{Nd}(t)$ that do not significantly fractionate during magma processes, argue in favor of at least three primitive end-member melt components for Cenozoic basalts from eastern China (Table 1, Fig. 4). The nephelinitic and heavy Fe component (HF) is characterized by its high δ^{56}Fe (~0.28), low δ^{26}Mg (~−0.57), high δ^{66}Zn (~0.60), and very depleted ^{87}Sr/^{86}Sr(i) and $\varepsilon_{Nd}(t)$ of Cenozoic basalts (Fig. 4). Considering ^{87}Sr/^{86}Sr(i) and $\varepsilon_{Nd}(t)$, the tholeiitic component with relatively low δ^{56}Fe (0.10–0.15) consists of two components I and II (Fig. 4C, D). The component I (TH-1) has a MORB-like δ^{56}Fe (~0.10), depleted ^{87}Sr/^{86}Sr(i) and $\varepsilon_{Nd}(t)$ and δ^{26}Mg (~−0.40) lower than MORB (~−0.26) (Teng et al., 2010). The component II (TH-2) is identified by its EM-I type Sr-Nd isotopic compositions, variable δ^{26}Mg from FS-2 (−0.29) to FS-3 (−0.53), and δ^{66}Zn (~0.30) unfractionated from MORB (~0.28) (Liu et al., 2016). The TH-2 has δ^{56}Fe (0.12 to 0.15) slightly higher than that of MORB.

Table 1 Typical geochemical parameters of melt components identified among Cenozoic basalts from eastern China. The three melt components, identified among Cenozoic basalts from eastern China, are given as mean compositions of representative samples: the component HF (13PMS8, 13AFS1, and 13AFS3), TH-1 (XTEJ-2 and XETJ-4) and TH-2 (FS-2 and FS-3). Ti/Ti* = $Ti_N/(Nd_N^{-0.055} \times Sm_N^{0.333} \times Gd_N^{0.722})$. Data sources refer to Table S5

Component	HF	TH-1	TH-2
SiO_2	44.30	49.55	48.48
$Fe^{3+}/\Sigma Fe$	0.57	0.26	
CaO/Al_2O_3	0.74	0.65	0.59
$(Na_2O+K_2O)/SiO_2$	0.20	0.08	0.09
Ti/Ti*	0.43	1.54	0.89
Fe/Mn	58	72	67
Ba/Th	50	105	195
δ^{26}Mg	−0.57	−0.40	−0.41
δ^{66}Zn	0.60		0.32
δ^{56}Fe	0.28	0.10	0.14
$^{87}Sr/^{86}Sr(i)$	0.7034	0.7033	0.7053
$\varepsilon_{Nd}(t)$	5.1	5.1	−2.8
$(Dy/Yb)_N$	3.2	1.6	1.8

The three melt components also represent end-members in the TAS (Fig. 1A) and primitive mantle normalized trace element pattern (Fig. S2). The heavy Fe component possesses the most enriched alkalis and incompatible trace element abundances in Cenozoic basalts, and is characterized by strong depletion of Rb-Ba, K, and Zr-Hf-Ti. The TH-1 is most depleted and featured by its positive anomalies in Pb, Sr and Zr-Hf-Ti. The TH-2 is intermediate between the other two components. Both of the TH-1 and TH-2 are subalkaline in composition (Table 1).

The chemical and isotopic variation of Cenozoic basalts from eastern China can be explained by mixing of the three melt components identified above (Fig. 4). The correlations among indices sensitive and insensitive to magma processes (e.g., δ^{56}Fe and Nb versus Sr-Nd-Mg-Zn isotopic ratios) prefers magma mixing rather than source mixing (Fig. 2 and Fig. 4).

4.3 Origins of the tholeiitic melt components

Both the tholeiitic melt components (TH-1 and TH-2) possess Fe/Mn≥67, significantly higher than MORB and even OIB (Fig.5A). Given that melts from pyroxenitic sources tend to have higher Fe/Mn than those from peridotitic sources (e.g., Sobolev et al., 2007), the high Fe/Mn of TH-1 and TH-2 suggests a significant contribution from pyroxenite melting. Their petrogenesis nevertheless remains debated, involving either derivation from pyroxenitic sources in the asthenospheric mantle (e.g., Li et al., 2016) or interaction of the heavy Fe component with the lithospheric mantle (e.g., Xu et al., 2005; Wang et al., 2018). In the first scenario, the diversity in elemental and Sr-Nd-Mg isotopic compositions, e.g., $^{87}Sr/^{86}Sr(i)$ from 0.7033 to 0.7054, $\varepsilon_{Nd}(t)$ from −4.0 to 5.1 and δ^{26}Mg from −0.53 to −0.29 (Fig.4 and Fig.S2), indicates diverse kinds of pyroxenitic sources in the asthenospheric mantle (e.g., Li et al., 2016). Mantle pyroxenites usually have δ^{56}Fe variably higher than the upper mantle. For example, pyroxenite xenoliths from the rejuvenated Hawaii basalts yield δ^{56}Fe from 0.07 to 0.18 (Williams and Bizimis, 2014). Melt components from pyroxenitic sources thus tend to have δ^{56}Fe higher than those from peridotitic sources (Williams and Bizimis, 2014; Konter et al., 2016). TH-1 and TH-2, however, have δ^{56}Fe comparable to (~0.10) and only slightly higher (0.12–0.15) than MORB (~0.105; Teng et al., 2013) respectively. This requires i) pyroxenitic sources with low δ^{56}Fe, ii) limited Fe isotope fractionation during high degree pyroxenite melting (e.g., He et al., 2017), and/or iii) that these two melt components may not be pristine melts from pyroxenites and be diluted by partial melts from peridotites (e.g., Sobolev et al., 2007). TH-1 and TH-2 have Fe/Mn ranging from 67 to 72, significantly lower than experimental and predicted low degree partial melts of pyroxenites (e.g., Fe/Mn>76 for melting degree<42% in experiments on a secondary pyroxenite from Sobolev et al. (2007); Fig.S5). In the second scenario, mixing trends in Fig.4 could indicate variable extent of dilution by silica-rich melts derived from and/or re-equilibrium with the lithospheric mantle. Iron isotopic variation during interaction between adakitic melts and the upper mantle has been previously investigated (He et al., 2017). The homogeneous and MORB-like δ^{56}Fe of high-Mg# adakites from the Dabie orogen and central America indicates that the hybridizing melts may finally have been in re-equilibrium with the ambient mantle with their δ^{56}Fe determined by $\delta^{56}Fe_{mantle\ source}+\Delta^{56}Fe_{melt\text{-}solid}$ (He et al., 2017). δ^{56}Fe of the TH-1 and TH-2 approaching that of MORB likely also records re-equilibrium with the lithospheric mantle. In this scenario, the compositional variation of the TH-1 and TH-2 must reflect diversity in the lithospheric mantle beneath eastern China (e.g., Xu et al., 2005; Wang et al., 2018). The Fe isotopic data reported here, however, can not discriminate the above two scenarios.

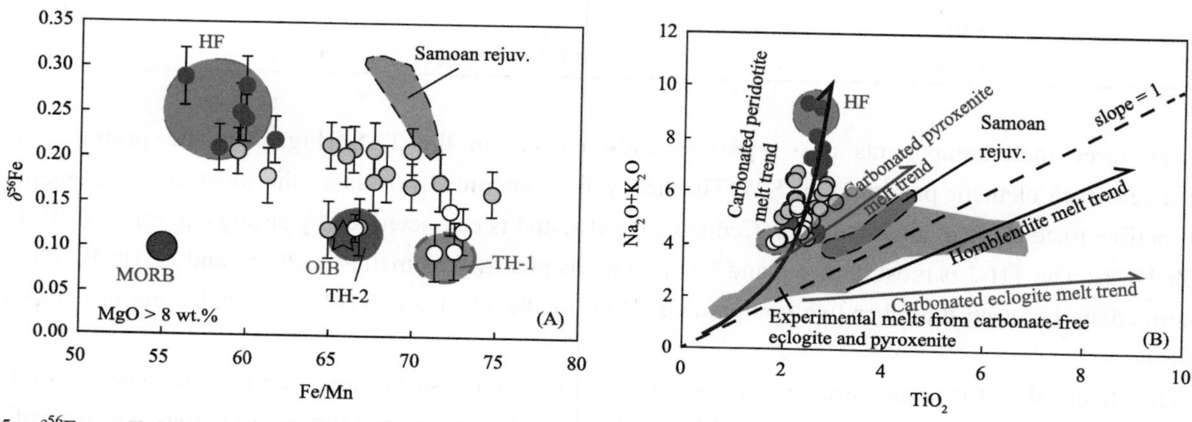

Fig. 5　δ^{56}Fe vs. Fe/Mn (A) and Na_2O+K_2O vs. TiO_2 (B) diagrams. Only samples with MgO>8 wt.% were plotted in (A) to avoid the effect of differentiation. MORB and OIB are from Qin and Humayun (2008) and Teng et al. (2013). Rejuvenated Samoan lavas (Konter et al., 2016) are also plotted for comparison. The field of experimental melts from carbonate-free eclogite and pyroxenite and melting trends of hornblendite and carbonated eclogite, pyroxenite and peridotite are taken from Huang et al. (2015), and the original data source is referred therein.

4.4　Origin of the heavy Fe component

The δ^{56}Fe values of primitive basalts are determined by source compositions and isotope fractionations during partial melting (Dauphas et al., 2014; Konter et al., 2016). Basalts may inherit high δ^{56}Fe from their mantle sources that have been previously hybridized by silicate melts commonly with high δ^{56}Fe, and such mantle sources mainly include fertile peridotites and pyroxenites (e.g., Weyer and Ionov, 2007; Konter et al., 2016). These fertile peridotites with high δ^{56}Fe tend to also have low Mg#. For example, lherzolite and wehrlite xenoliths from Tok, Siberia have δ^{56}Fe up to 0.17 and Mg# down to 84 (e.g., Weyer and Ionov, 2007). The high Mg# up to 81 in Cenozoic basalts representative of the heavy Fe component (e.g., 13AFS3; Table S5) suggests in Fe-Mg exchange equilibrium with mantle olivines (e.g., Fo91) and thus argues against derivation from a fertile peridotite source with high δ^{56}Fe but low Mg#. The highest δ^{56}Fe (~0.33) in terrestrial basalts has been observed in rejuvenated lavas from Samoan, and partially attributed to pyroxenite sources with high δ^{56}Fe values (Konter et al., 2016). The high δ^{56}Fe end-member in Samoan samples is characterized by their subalkaline compositions with

Fe/Mn≥67 (Fig. 1 and Fig. 5A). By contrast, the heavy Fe component is nephelinitic and has a low Fe/Mn ~58. Given that high degree partial melting can reduce Fe/Mn of the melts from pyroxenite melting (e.g., Sobolev et al., 2007; Davis et al., 2013), one may suspect that the heavy Fe component may be derived from high degree partial melting of pyroxenites with high $\delta^{56}Fe$. Since recycled oceanic eclogites and mantle secondary pyroxenites with high $\delta^{56}Fe$ commonly also have high Fe/Mn, partial melting of known pyroxenite sources can not simultaneously produce the high $\delta^{56}Fe$ and low Fe/Mn of the heavy Fe component (Fig. S5). Depletion of Zr, Hf and Ti in the heavy Fe component (Fig. S6) suggests that carbonatitic metasomatism in the source is necessary (e.g., Huang et al., 2015). Carbonatitic melts are also enriched in MnO with low Fe/Mn. For example, Mg-rich carbonatites from the Cape Verde Island (Hoernle et al., 2002) and experimental carbonatitic melts from carbonated MORB-like eclogites (Thomson et al., 2016) have MnO up to 1.68 wt.% and Fe/Mn down to 3.0. In this regard, a special pyroxenite source overprinted by carbonatitic metasomatism may not be completely ruled out for the heavy Fe component. However, the heavy Fe component has relatively low TiO_2 and high Na_2O+K_2O at a given TiO_2 (Fig. 5B; Table S5), comparable to partial melts from carbonated peridotites rather than carbonated pyroxenites (e.g., Huang et al., 2015; and reference therein). Therefore, we prefer the heavy Fe component was likely derived from a carbonated peridotite source (e.g., Dasgupta et al., 2007; Huang et al., 2015; Li et al., 2017). Source hybridization by carbonatitic melts preferentially enriched in lighter Fe isotopes (Johnson et al., 2009) also cannot explain the heavy Fe component. The unusually high $\delta^{56}Fe$ of the heavy Fe component most likely reflect enhanced Fe isotope fractionation during peridotite partial melting.

Iron isotope fractionation during peridotite partial melting is possibly enhanced by: i) presence of residual garnet having an Fe isotopic composition lighter than olivine, pyroxenes and spinel (e.g., Williams and Bizimis, 2014; Sossi and O'Neill, 2017); ii) high alkalis in the melt that stabilize melt Fe^{2+} and Fe^{3+} in low coordinations (e.g., Sossi et al., 2012); iii) high source $Fe^{3+}/\Sigma Fe$ (e.g., Dauphas et al., 2014). The heavy Fe component has a high $(Dy/Yb)_N$ up to 3.6, indicating abundant (up to 10%) residual garnet in its source (Huang et al., 2015). Calculations using a batch partial melting model and current available fractionation factors, however, suggest the presence of 10% residual garnet has negligible effect on the melt $\delta^{56}Fe$ (<0.01) during peridotite melting (Sossi and O'Neill, 2017). This is supported by the limited $\delta^{56}Fe$ variation (~0.07) observed in low-Mg adakites that represent crustal melts leaving up to 30% garnet at a lower temperature ~1100 °C (He et al., 2017). The heavy Fe component also appears as the highest Na_2O+K_2O end of Cenozoic basalts from eastern China (Fig. 1A). Previous studies on felsic igneous rocks indicate a preferential enrichment of heavy Fe isotopes in alkalis-enriched granitic magmas (Foden et al., 2015; He et al., 2017; Sossi et al., 2012). A global compilation of high silica granitic magmas reveals a maximum estimate of the magma compositional effect indexed by (Na+K)/(Ca+Mg): $\delta^{56}Fe = 0.0062 \times (Na+K)/(Ca+Mg) + 0.130$, reflecting that alkaline earth cations counteract the role of alkaline cations (He et al., 2017). All samples measured here are also enriched in CaO and MgO, with (Na+K)/(Ca+Mg) varying from 0.25 to 0.77, corresponding to a potential magma compositional effect on $\Delta^{56}Fe_{melt\text{-}residue} < 0.003$. This estimate is reinforced by observations that basaltic to dacitic glasses with (Na+K)/(Ca+Mg) ranging from 0.05 to 0.91 have identical mean force constants of Fe-O bonds (Dauphas et al., 2014). Therefore, the enhancement in Fe isotope fractionation, required to explain high $\delta^{56}Fe$ of the heavy Fe component, cannot be attributed to either its high Na_2O+K_2O or residual garnet in its sources. Its high Na_2O+K_2O and $(Dy/Yb)_N$ simply reflect the genetic link of the heavy Fe component to partial melting of a carbonated, garnet peridotitic source (e.g., Dasgupta et al., 2007). The heavy Fe component was most likely derived from a source with high $Fe^{3+}/\Sigma Fe$, which is also supported by its high V/Sc up to 14.9 (Fig. S7; e.g., Li and Lee, 2004).

Considering the Fe^{3+} and Fe^{2+} distribution, a batch partial melting model of peridotites (Dauphas et al., 2014), is adopted to assess the source $Fe^{3+}/\Sigma Fe$ (Supplementary Materials). A difficulty of such calculation comes from Fe isotope fractionation between the melt and minerals (e.g., $\Delta^{56}Fe_{melt\text{-}residue}$) changing with $Fe^{3+}/\Sigma Fe$ of both (e.g., Dauphas et al., 2014). Prediction on $\Delta^{56}Fe_{melt\text{-}residue}$ based on the mean force constants of Fe^{3+}-O and Fe^{2+}-O bonds has been established, but can only account for 1/3 of the fractionation between MORB and the upper mantle (e.g., Dauphas et al., 2014). Therefore, here we empirically enlarge $\Delta^{56}Fe_{melt\text{-}residue}$ calculated after Dauphas et al. (2014) by a factor of 3 to reconcile the $\delta^{56}Fe$ difference between MORB and the upper mantle (Teng et al., 2013; Supplementary Materials). Assuming a mean upper mantle source (Weyer and Ionov, 2007), a $Fe^{3+}/\Sigma Fe$ mantle source of 0.15, significantly higher than for MORB (~0.036; Canil et al., 1994), is necessary to explain the heavy Fe component (Fig. 6). The corresponding $Fe^{3+}/\Sigma Fe$ in the primitive basalt is about 0.58. A source value of 0.15 appears to be a minimum estimate, because i) the source may have a $\delta^{56}Fe$ lower than the mean upper mantle due to carbonatitic hybridization (Johnson et al., 2009; Li et al., 2017), and ii) $\delta^{56}Fe$ of basalts might be less sensitive to the source $Fe^{3+}/\Sigma Fe$ given incompatibility of ferric Fe decreasing with pressure (Fig. S8).

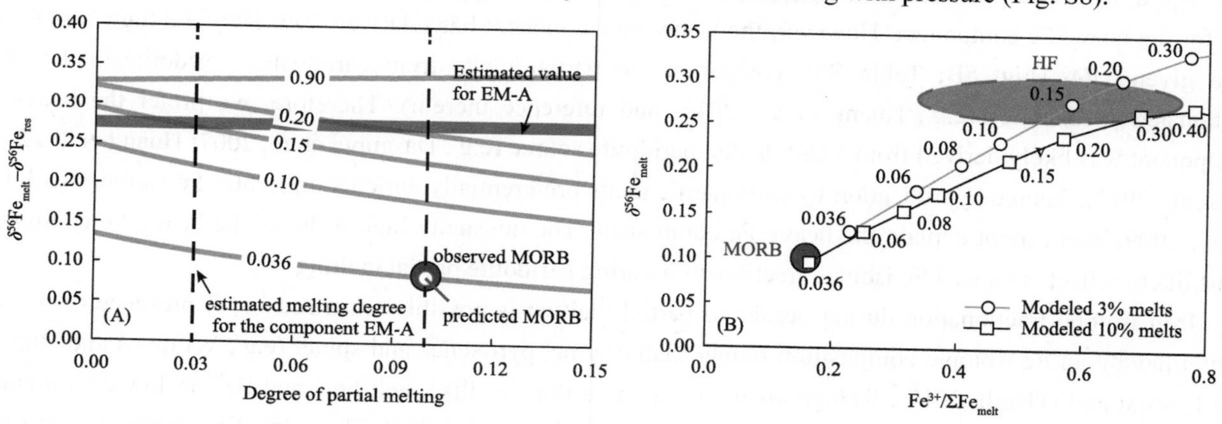

Fig. 6　Predicted fractionation factors between melts and residua (A) during and the modeled 3% and 10% partial melts (B) of peridotite partial melting with diverse source $Fe^{3+}/\Sigma Fe$. Numbers marked on curves in (A) and besides symbols in (B) denote $Fe^{3+}/\Sigma Fe$ of the source. The modelling details refer to Supplementary Materials.

A high source $Fe^{3+}/\Sigma Fe$ will increase the incompatibility of FeOt and thus lead to elevated melt Fe/Mn relative to the source during peridotite melting (e.g., Davis et al., 2013). Therefore, Fe/Mn of the source must have been decreased prior to partial melting to reconcile high $\delta^{56}Fe$ and low Fe/Mn of the heavy Fe component. Carbonatitic metasomatism in the source is supported by the low $\delta^{26}Mg$ and high $\delta^{66}Zn$ of the heavy Fe component that indicate involvement of recycled carbonates likely as carbonatitic melts released from the stagnated Pacific plate in the mantle transition zone beneath eastern China (Li et al., 2017; Liu et al., 2016). Hybridization of the source by ca. 11% MnO-rich carbonatitic melts (e.g., MnO = 1.57 wt.% and Fe/Mn = 4.4; Thomson et al., 2016) can account for the low Fe/Mn of the heavy Fe component, while all other parameters keep constant and phase relationship from Dasgupta et al. (2007) and partition coefficients for Fe^{2+} and Mn from Davis et al. (2013) are adopted.

Oxygen fugacity of mantle sources can be elevated by recycled terrigenous sediments (Evans, 2012) and influx of hydrous fluids (Kelley and Cottrell, 2009). The depleted $^{87}Sr/^{86}Sr(i)$ ~0.7034 of the heavy Fe component (Table 1 and Fig. 4) indicates contribution of recycled terrigenous sediments in its mantle source must be very low (e.g., much less than 1%; Fig. S9). Given the comparable incompatibility of H_2O and Ce during mantle melting, H_2O/Ce can be used to constrain H_2O enrichment in the mantle source (e.g., Chen et al., 2015). Water contents in basalts from Shuangliao have been estimated by measuring H_2O in clinopyroxene phenocryst (Chen et al., 2015). A negative correlation between $\delta^{56}Fe$ and H_2O/Ce in the Shuangliao samples contradicts with influx of hydrous fluids as the origin of the high $Fe^{3+}/\Sigma Fe$ source for the heavy Fe component(Fig.S9). In addition, a low Ba/Th ~50 of the heavy Fe component indicates its mantle source has been metasomatized by dehydrated crustal materials rather than hydrous fluids (Fig. S9 and reference therein). Therefore, the elevated source $Fe^{3+}/\Sigma Fe$ of the heavy Fe component was not caused by recycled terrigenous sediments and hydrous fluid influx, but instead by hybridization of recycled carbonates indicated by its low $\delta^{26}Mg$ and high $\delta^{66}Zn$ (Li et al., 2017; Liu et al., 2016). Deep derivation of the heavy Fe component is supported by its low $^{87}Sr/^{86}Sr$ (i) ~ 0.703, which favors hybridization dominantly by Sr-depleted magnesite at depths >300 km (Li et al., 2017). In such a deep mantle, recycled carbonates can be reduced to diamonds ("redox freezing"), while Fe^0, resulting from Fe^{2+} disproportionation ($3Fe^{2+} \rightarrow Fe^0 + 2Fe^{3+}$), is consumed but the Fe^{3+} preserved (e.g., Rohrbach and Schmidt, 2011; Xu et al., 2017). Mass balance calculation suggests reduction of only 1400 ppm CO_2 in magnesite to 390 ppm C will enhance the source $Fe^{3+}/\Sigma Fe$ from an ambient mantle value ~0.036 (Canil et al., 1994) to 0.15 that is required to explain the heavy Fe component. This estimate is ten times lower than the recycled CO_2 present in the source indicated by $\delta^{26}Mg$ and $\delta^{66}Zn$ (Li et al., 2017; Liu et al., 2016). The capability of mantle peridotites to "freeze" C depends on pressure and is limited by the Fe^0 saturation (Rohrbach and Schmidt, 2011). Free carbonates in excess to deep carbon freezing thus may have existed in the source of the heavy Fe component, and facilitated partial melting at depth >300 km below the Fe-C redox boundary at 250 km, where diamonds were still stable and $Fe^{3+}/\Sigma Fe$ of the source remained high at the solidus (Rohrbach and Schmidt, 2011).

5 Insight to oxygenation of paleo-atmosphere

The Earth has an atmosphere with 21 vol.% of free oxygen (O_2) that was stepwisely oxygenated through two major events: the Great Oxidation Event (GOE) at 2.5−2.2 Ga and the Neoproterozoic Oxidation Event (NOE) at 0.8−0.55 Ga (Kasting, 2013; Lyons et al., 2014; and references therein). The mechanism of oxygenation remains a matter of debate. Free oxygen production by photosynthesis may have started to operate well before GOE, and other processes that buffer the O_2 concentration thus may have dictated the evolution of the atmosphere (Kasting, 2013; Lyons et al., 2014; and references therein). Sedimentary burial of organic carbon is widely considered as a mechanism for the rise of O_2 (Kasting, 2013; Lyons et al., 2014; and references therein). This mechanism alone may not be sufficient to sustain the elevated O_2 level after oxygenation, since buried organic carbon could be efficiently remobilized and released back into the atmosphere due to oxidative weathering (e.g., Kump et al., 2011). The role of the solid Earth has been increasingly considered important in the atmosphere O_2 evolution (e.g., Lee et al., 2016; Mao et al., 2017).

The origin of the heavy Fe component identified here illustrates a deep oxygen cycle driven by deep carbon cycling, which could provide insights into the oxygenation of paleo-atmosphere. The oxygen fugacity of the heavy Fe component is calculated after Kress and Carmichael (1991) and using a melt composition of 13AFS3 and $Fe^{3+}/\Sigma Fe$ of 0.58 estimated by Fe isotopes. Referenced to 1200 °C and 1 atm, the heavy Fe component from this study has an fO_2 of ~4.1 log units above the quartz-fayalite-magnetite (QFM) buffer, substantially higher than that of MORB near the QFM buffer (e.g., Cottrell and Kelley, 2011). Here we emphasize eruption of such oxidizing basalts can ultimately promote net atmosphere O_2 production via decreasing volcanic O_2 sink after Gaillard et al. (2011) and Keller and Schoene (2012), providing that the real interaction between volcanism and the atmosphere could be complex (e.g., Gaillard et al., 2011; Kasting, 2013; Lee et al., 2016; reference therein). With Fe^{3+} being preferentially scavenged during melt extraction (Canil et al., 1994), the residual mantle domains became reduced, and recycled carbonates tend to be, at least partially, "frozen" as diamonds. Collectively, burying sedimentary carbonates into the deep mantle >250 km through subduction as immobile diamonds may ultimately promote O_2 production in the atmosphere: CO_2 (in carbonates) → C (as diamond frozen in the deep mantle) + O_2

(ferric Fe being preferentially scavenged by melt extraction), which is referred as the carbon driven deep oxygen cycle hereafter. A compilation of global basalts with ages from Archean to the present reveals a positive correlation between V/Sc and alkalinity, similar to what is observed in Cenozoic basalts from eastern China (Fig. 7). Given the genetic link between high alkalinity basalts and partial melting of carbonated sources (e.g., Dasgupta et al., 2007), the carbon driven deep oxygen cycle may have been timely and globally important.

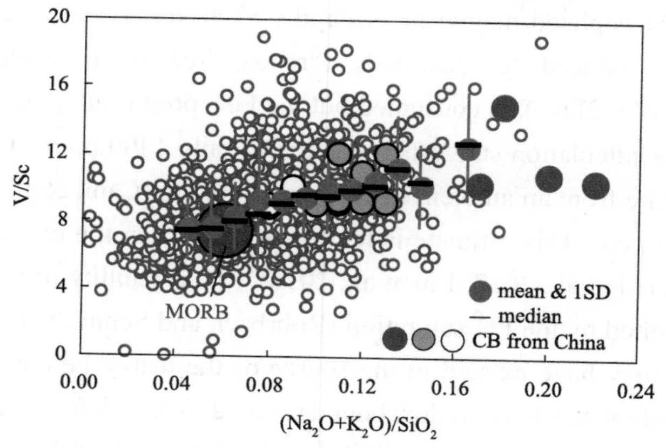

Fig. 7　V/Sc vs. $(Na_2O + K_2O)/SiO_2$ diagram of global basalts (n=1411). Global basalts with different ages are from GEOROC (http://georoc.mpch-mainz.gwdg.de/georoc/; accessed by Jan 22, 2018). All data with MgO<8.0 wt.%, Mg#>75, and CIA>55 have been excluded to reduce potential influences on magma compositions by fractional crystallization, accumulation, chemical weathering, and alteration. Arithmetical mean, median values, and one standard deviations are calculated for samples with $(Na_2O + K_2O)/SiO_2$ intervals between 0.04 and 0.15 by steps of 0.01 as well as for those with $(Na_2O + K_2O)/SiO_2 \geqslant 0.15$. The compilation indicates V/Sc of basalts, and thus oxidizing capability, increases with $(Na_2O+K_2O)/SiO_2$. Cenozoic basalts (CB) from eastern China are also plotted for a comparison.

The efficiency (or the net redox budget flux) of the carbon driven deep oxygen cycle heavily depends on the amount of carbonates subducted into depth greater than 250 km. On the one hand, the amount of carbonates that escaped from decarbonation beneath the arc system and were subducted into deeper mantle could have been stepwisely increased at the Archean-Proterozoic and Neoproterozoic-Cambrian boundary, due to development/secular cooling of global subduction zones indicated by vanishing of widespread slab melting and the occurrence of lawsonite-bearing blueschists and eclogites respectively (Fig. 8) (e.g., Brown, 2014; Moyen and Martin, 2012; and reference therein). This is supported by recycled crustal components occasionally revealed in Paleoproterozoic carbonatites (e.g., those from Borden, Canada, ~1.9 Ga) and universally in carbonatites younger than Neoproterozoic, e.g., by higher $\delta^{11}B$ compared to MORB (Hulett et al., 2016). On the other hand, statistics on global mafic rocks reveal that their mean alkalinity stepwisely increased across the Archean-Proterozoic and Neoproterozoic-Cambrian boundary (Keller and Schoene, 2012), simultaneous to the transitions of global plate tectonics (Fig. 8). Such changes in mean alkalinity of global mafic rocks could be attributed to eruption of more magmas with high alkalinity and likely oxygen fugacity, supported by a burst of detrital zircons from carbonatite-alkaline parent magmas at the Neoproterozoic-Cambrian boundary (Paulsen et al., 2017). The efficiency of the carbon driven deep oxygen cycle thus could have increased stepwisely with decreasing geothermal gradient of subduction zones in the past, which may provide an explanation for the striking temporal coincidence among the development/transition of the plate tectonic regime, geochemical evolution of global mafic rocks, and two major oxidation events of the paleo-atmosphere (Fig. 8).

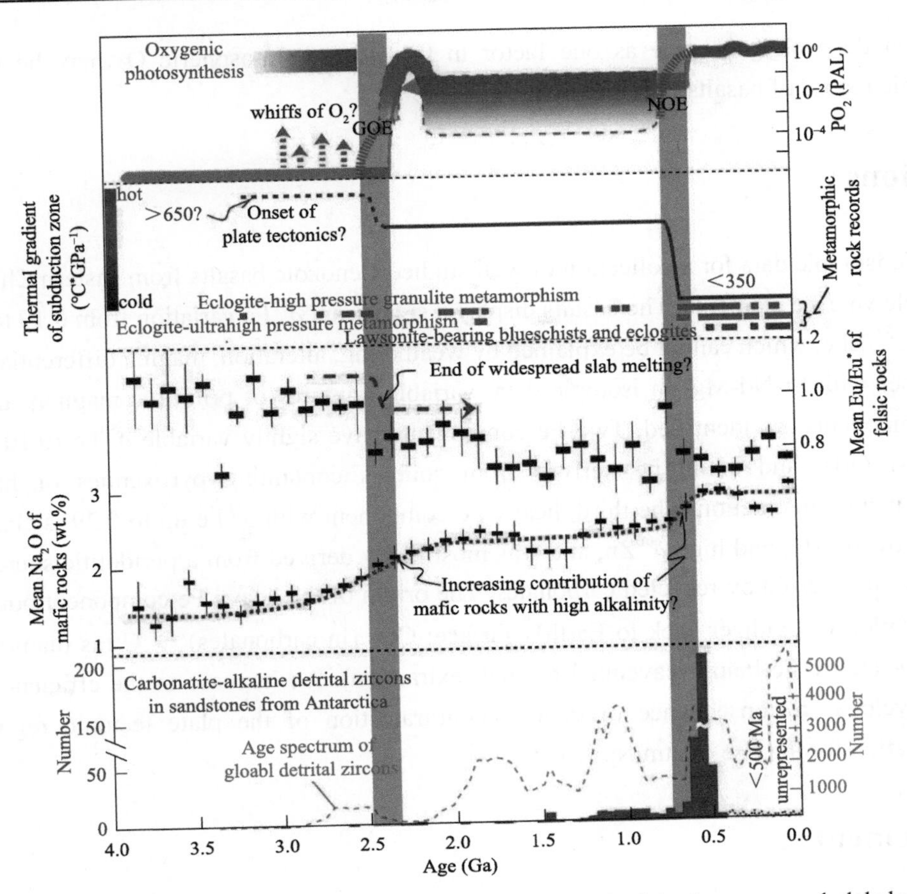

Fig. 8 Atmospheric O_2 evolution curves are compared with thermal gradient of subduction zones and global magmatism evolution in the past. O_2 curves are from Lyons et al. (2014). Thermal gradient of subduction zones is estimated according to metamorphic rock records (Brown, 2014; Weller and St-Onge, 2017) and the condition of slab melting (e.g., Moyen and Martin, 2012; and reference therein). Secular evolution of alkalinity of continental mafic (43–51 wt.% SiO_2) lithologies and Eu/Eu* of felsic (62–73 wt.% SiO_2) lithologies are also shown (Keller and Schoene, 2012). Mean alkalinity of continental mafic lithologies stepwisely increased across the Archean-Proterozoic and Neoproterozoic-Cambrian boundary, which is reinforced by a burst of detrital zircons from carbonatite-alkaline parent magmas (Paulsen et al., 2017) compared to the spectrum of all detrital zircons (Lee et al., 2016).

A quantitative evaluation on the impact of the carbon driven deep oxygen cycle on atmospheric oxygenation remains challenging. Difficulties come from evaluation on i) the amount of subducted carbonates escaped from decarbonation beneath the arc system, ii) to what extent the deep subducted carbonates were converted to diamonds, and iii) the time scale of the carbon driven deep oxygen cycle from carbonate subduction to eruption of oxidizing basalts. Nevertheless the net redox budget of the present carbon driven deep oxygen cycle can be roughly estimated by an adoption of the deep subducted carbonate flux of Evans (2012) (~4.64×10^{18} mol/Ma). Given the difference between influx of deep subducted carbonates and outflux of carbon by MORB and OIB (Evans, 2012), 30% of deep subducted carbonates are assumed to be converted to diamonds. The current carbon driven deep oxygen cycle thus can approximately promote a net atmospheric O_2 production rate of 1.4×10^{18} mol/Ma. Compared to the present O_2 volume of the atmosphere (~4.0×10^{19} mol), transition of the carbon driven deep oxygen cycle from a much less efficient state to the present one across the Neoproterozoic-Cambrian boundary may have potential to drive atmospheric oxygenation to the present atmospheric level in tens of Ma. Assuming that deep carbon cycling dictates the long-term fluctuation of the atmospheric CO_2 (e.g., Li et al., 2017; Paulsen et al., 2017), the impact of the carbon driven deep oxygen cycle might be recorded by the atmospheric O_2 level negatively correlated with CO_2 in the Phanerozoic (Fig. S10; Berner, 2006). A change in the efficiency of the carbon driven deep oxygen cycle as one factor in the rise of atmospheric O_2 may be tested by secular Fe-Mg-Zn isotopic record of basalts in future investigations.

6 Conclusions

We report Fe isotopic data for a collection of well studied Cenozoic basalts from eastern China to investigate cycling of multiple valence elements. The basalts display a significant δ^{56}Fe variation from 0.10 to 0.29, positively correlated with $Fe^{3+}/\Sigma Fe$, which cannot be explained by weathering, alteration, magma differentiation, or chemical diffusion. Together with Sr-Nd-Mg-Zn isotopic data, variable mixtures of primitive magmas are indicated, and three magma components are identified. Two Fe components have sightly variable δ^{56}Fe (0.10–0.15) typical of MORB and most OIB, and could be derived from sources containing pyroxenites or have experienced melt/lithospheric mantle interaction. The third, heavy Fe component with δ^{56}Fe up to 0.29 is characterized by its low Fe/Mn ~58, low δ^{26}Mg, and high δ^{66}Zn, and was most likely derived from a peridotitic source with $Fe^{3+}/\Sigma Fe \geqslant 0.15$ that had been oxidized by recycled carbonates. The origin of the heavy Fe component points to a pathway with net transportation of oxidizer back to Earth's surface: CO_2 (in carbonates) → C (as diamond frozen in the deep mantle) + O_2 (ferric Fe being scavenged by melt extraction). An increase in the efficiency of this carbon driven oxygen cycle as a consequence of development/transition of the plate tectonic regime provides an alternative explanation for the rise of atmospheric O_2.

Acknowledgement

Constructive reviews from Stefan Weyer and one anonymous reviewer and efficient editorial efforts by Dr. Frederic Moynier are greatly appreciated. Prof. Qun-Ke Xia kindly provided the basalts from Shuangliao. This work was financially supported by the National Natural Science Foundation of China (41730214, 41473016, 41230209, 41688103), the National(Key R&D Program) of China (No.2016YFC0600408), and State Key Laboratory of Geological Processes and Mineral Resources.

Appendix A. Supplementary material

Supplementary material related to this article can be found online at https://doi.org/10.1016/j.epsl.2019.02.009.

References

Aiuppa, A., Allard, P., D'Alessandro, W., Michel, A., Parello, F., Treuil, M., Valenza, M., 2000. Mobility and fluxes of major, minor and trace metals during basalt weathering and groundwater transport at Mt Etna volcano (Sicily). Geochim. Cosmochim. Acta 64, 1827-1841.

Berner, R.A., 2006. GEOCARBSULF: a combined model for Phanerozoic atmospheric O_2 and CO_2. Geochim. Cosmochim. Acta 70, 5653-5664.

Brown, M., 2014. The contribution of metamorphic petrology to understanding lithosphere evolution and geodynamics. Geosci. Front. 5, 553-569.

Canil, D., O'Neill, C., Pearson, D.G., Rudnick, R.L., McDonough, W.F., Carswell, D.A., 1994. Ferric iron in peridotites and mantle oxidation states. Earth Planet. Sci. Lett. 123, 205-220.

Chen, H., Xia, Q.-K., Ingrin, J., Jia, Z.-B., Feng, M., 2015. Changing recycled oceanic components in the mantle source of the Shuangliao Cenozoic basalts, NE China: new constraints from water content. Tectonophysics 650, 113-123.

Cottrell, E., Kelley, K.A., 2011. The oxidation state of Fe in MORB glasses and the oxygen fugacity of the upper mantle. Earth Planet. Sci. Lett. 305, 270-282.

Dasgupta, R., Hirschmann, M.M., Smith, N.D., 2007. Partial melting experiments of peridotite +CO_2 at 3 GPa and genesis of alkalic ocean island basalts. J. Petrol. 48, 2093-2124.

Dauphas, N., Roskosz, M., Alp, E.E., Neuville, D., Hu, M.Y., Sio, C.K., Tissot, F.L.H., Zhao, J., Tissandier, L., Medard, E., Cordier, C., 2014. Magma redox and structural controls on iron isotope variations in Earth's mantle and crust. Earth Planet. Sci. Lett. 398, 127-140.

Davis, F.A., Humayun, M., Hirschmann, M.M., Cooper, R.S., 2013. Experimentally determined mineral/melt partitioning of first-row transition elements (FRTE) during partial melting of peridotite at 3 GPa. Geochim. Cosmochim. Acta 104, 232-260.

Evans, K.A., 2012. The redox budget of subduction zones. Earth-Sci. Rev. 113, 11-32.

Foden, J., Sossi, P.A., Wawryk, C.M., 2015. Fe isotopes and the contrasting petrogen-esis of A-, I-and S-type granite. Lithos 212-215, 32-44.

Gaillard, F., Scailler, B., Arndt, N.T., 2011. Atmospheric oxygenation caused by a change in volcanic degassing pressure. Nature 478, 229-232.

He, Y., Wu, H., Ke, S., Liu, S.-A., Wang, Q., 2017. Iron isotopic compositions of adakitic and non-adakitic granitic magmas: magma compositional control and subtle residual garnet effect. Geochim. Cosmochim. Acta 203, 89-102.

Hoernle, K., Tilton, G., Le Bas, M.J., Duggen, S., Schönberg, D.G., 2002. Geochemistry of oceanic carbonatites compared with continental carbonatites: mantle recycling of oceanic crustal carbonate. Contrib. Mineral. Petrol. 142, 520-542.

Huang, J., Li, S.-G., Xiao, Y., Ke, S., Li, W.-Y., Tian, Y., 2015. Origin of low δ^{26}Mg Cenozoic basalts from South China Block and their geodynamic implications. Geochim. Cosmochim. Acta 164, 298-317.

Hulett, S.R.W., Simonetti, A., Rasbury, E.T., Hemming, N.G., 2016. Recycling of subducted crustal components into carbonatite melts revealed by boron isotopes. Nat. Geosci. 9, 904-908.

Johnson, C.M., Bell, K., Beard, B.L., Shultis, A.I., 2009. Iron isotope compositions of carbonatites record melt generation, crystallization, and late-stage volatile-transport processes. Mineral. Petrol. 98, 91-110.

Kasting, J.F., 2013. What caused the rise of atmospheric O_2? Chem. Geol. 362, 13-25.

Keller, C.B., Schoene, B., 2012. Statistical geochemistry reveals disruption in secular lithospheric evolution about 2.5 Gyr ago. Nature 485, 490-493.

Kelley, K.A., Cottrell, E., 2009. Water and the oxidation state of subduction zone magmas. Science 325, 605-607.

Konter, J.G., Pietruszka, A.J., Hanan, B.B., Finlayson, V.A., Craddock, P.R., Jackson, M.G., Dauphas, N., 2016. Unusual δ^{56} Fe values in Samoan rejuvenated lavas generated in the mantle. Earth Planet. Sci. Lett. 450, 221-232.

Kress, V.C., Carmichael, I.S.E., 1991. The compressibility of silicate liquids containing Fe_2O_3 and the effect of composition, temperature, oxygen fugacity and pressure on their redox states. Contrib. Mineral. Petrol. 108, 82-92.

Kump, L.R., Junium, C., Arthur, M.A., Brasier, A., Fallick, A., Melezhik, V., Lepland, A., Crne, A.E., Luo, G.M., 2011. Isotopic evidence for massive oxidation of organic matter following the Great Oxidation Event. Science 334, 1694-1696.

Lee, C.T.A., Yeung, L.Y., McKenzie, N.R., Yokoyama, Y., Qzaki, K., Lenardic, A., 2016. Two-step rise of atmospheric oxygen linked to the growth of continents. Nat. Geosci. 9, 417-424.

Li, H.-Y., Xu, Y.-G., Ryan, J.G., Huang, X.-L., Ren, Z.-Y., Guo, H., Ning, Z.-G., 2016. Olivine and melt inclusion chemical constraints on the source of intracontinental basalts from the eastern North China Craton: discrimination of contributions from the subducted Pacific slab. Geochim. Cosmochim. Acta 178, 1-19.

Li, S.G., Yang, W., Ke, S., Meng, X.N., Tian, H.C., Xu, L.J., He, Y.S., Huang, J., Wang, X.C., Xia, Q.K., Sun, W.D., Yang, X.Y., Ren, Z.Y., Wei, H.Q., Liu, Y.S., Meng, F.C., Yan, J., 2017. Deep carbon cycles constrained by a large-scale mantle Mg isotope anomaly in eastern China. Nat. Sci. Rev. 4, 111-120.

Li, Z.X., Lee, C.T., 2004. The constancy of upper mantle fO_2 through time inferred from V/Sc ratios in basalts. Earth Planet. Sci. Lett. 228, 483-493.

Liu, S.-A., Wang, Z.-Z., Li, S.-G., Huang, J., Yang, W., 2016. Zinc isotope evidence for a large-scale carbonated mantle beneath eastern China. Earth Planet. Sci. Lett. 444, 169-178.

Lyons, T.W., Reinhard, C.T., Planavsky, N.J., 2014. The rise of oxygen in Earth's early ocean and atmosphere. Nature 506, 307-315.

Mao, H.K., Hu, Q.Y., Yang, L.X., Liu, J., Kim, D.Y., Meng, Y., Zhang, L., Prakapenka, V.B., Yang, W., Mao, W.L., 2017. When water

meets iron at Earth's core-mantle boundary. Nat. Sci. Rev. 4, 870-878.
Moyen, J.-F., Martin, H., 2012. Forty years of TTG research. Lithos 148, 312-336.
Paulsen, T., Deering, C., Sliwinski, J., Bachmann, O., Guillong, M., 2017. Evidence for a spike in mantle carbon outgassing during the Ediacaran period. Nat. Geosci. 10, 930-934.
Plank, T., Langmuir, C.H., 1998. The chemical composition of subducting sediment and its consequences for the crust and mantle. Chem. Geol. 145, 325-394.
Qin, L., Humayun, M., 2008. The Fe/Mn ratio in MORB and OIB determined by ICP-MS. Geochim. Cosmochim. Acta 72, 1660-1677.
Rohrbach, A., Schmidt, M.W., 2011. Redox freezing and melting in the Earth's deep mantle resulting from carbon-iron redox coupling. Nature 472, 209-212.
Sobolev, A.V., Hofmann, A.W., Kuzmin, D.V., Yaxley, G.M., Arndt, N.T., Chung, S.L., Danyushevsky, L.V., Elliott, T., Frey, F.A., Garcia, M.O., Gurenko, A.A., Kamenetsky, V.S., Kerr, A.C., Krivolutskaya, N.A., Matvienkov, V.V., Nikogosian, L.K., Rocholl, A., Sigurdsson, I.A., Sushchevskaya, N.M., Teklay, M., 2007. The amount of recycled crust in sources of mantle-derived melts. Science 316, 412-417.
Sossi, P.A., Foden, J.D., Halverson, G.P., 2012. Redox-controlled iron isotope fractionation during magmatic differentiation: an example from the Red Hill intrusion, S. Tasmania. Contrib. Mineral. Petrol. 164, 757-772.
Sossi, P.A., O'Neill, C., 2017. The effect of bonding environment on iron isotope fractionation between minerals at high temperature. Geochim. Cosmochim. Acta 196, 121-143.
Teng, F.-Z., Dauphas, N., Helz, R.T., Gao, S., Huang, S., 2011. Diffusion-driven magnesium and iron isotope fractionation in Hawaiian olivine. Earth Planet. Sci. Lett. 308, 317-324.
Teng, F.-Z., Dauphas, N., Huang, S., Marty, B., 2013. Iron isotopic systematics of oceanic basalts. Geochim. Cosmochim. Acta 107, 12-26.
Teng, F.-Z., Li, W.-Y., Ke, S., Marty, B., Dauphas, N., Huang, S., Wu, F.-Y., Pourmand, A., 2010. Magnesium isotopic composition of the Earth and chondrites. Geochim. Cosmochim. Acta 74, 4150-4166.
Teng, F.Z., Dauphas, N., Helz, R.T., 2008. Iron isotope fractionation during magmatic differentiation in Kilauea Iki lava lake. Science 320, 1620-1622.
Thomson, A.R., Walter, M.J., Kohn, S.C., Brooker, R.A., 2016. Slab melting as a barrier to deep carbon subduction. Nature 529, 76-79.
Trenberth, K.E., Smith, L., 2005. The mass of the atmosphere: a constraint on global analyses. J. Climate 18, 864-875.
Wang, Z.Z., Liu, S.A., Chen, L.H., Li, S.G., Zeng, G., 2018. Compositional transition in natural alkaline lavas through silica-undersaturated melt-lithosphere interaction. Geology 46, 771-774.
Weller, O.M., St-Onge, M.R., 2017. Record of modern-style plate tectonics in the Palaeoproterozoic Trans-Hudson orogen. Nat. Geosci. 10, 305-311.
Weyer, S., Ionov, D.A., 2007. Partial melting and melt percolation in the mantle: the message from Fe isotopes. Earth Planet. Sci. Lett. 259, 119-133.
Williams, H.M., Bizimis, M., 2014. Iron isotope tracing of mantle heterogeneity within the source regions of oceanic basalts. Earth Planet. Sci. Lett. 404, 396-407.
Xu, C., Kynicky, J., Tao, R., Liu, X., Zhang, L., Pohanka, M., Song, W., Fei, Y., 2017. Recovery of an oxidized majorite inclusion from Earth's deep asthenosphere. Sci. Adv. 3, e1601589.
Xu, Y.G., Ma, J.L., Frey, F.A., Feigenson, M.D., Liu, J.F., 2005. Role of lithosphere-asthenosphere interaction in the genesis of Quaternary alkali and tholeiitic basalts from Datong, western North China Craton. Chem. Geol. 224, 247-271.

Tracing the deep carbon cycle using metal stable isotopes: Opportunities and challenges*

Sheng-Ao Liu and Shu-Guang Li

State Key Laboratory of Geological Processes and Mineral Resources, China University of Geosciences, Beijing 100083, China

亮点介绍：由于大洋碳酸盐与地幔间显著的 Ca、Mg、Zn 同位素组成差异，这三种二价金属稳定同位素已普遍用于示踪碳酸盐岩俯冲进入地幔的深部碳循环过程。本文综述了近年的相关研究，细致排除了部分熔融、碳酸盐化榴辉岩再循环、金属与碳解耦、扩散等干扰过程，论述中国东部<110Ma 玄武岩大尺度 Mg-Zn 同位素异常起因于富 Mg、Zn 碳酸盐岩再循环进入其地幔源区。

Abstract The subduction of marine carbonates and carbonated oceanic crust to the Earth's interior and the return of recycled carbon to the surface via volcanism may play a pivotal role in governing Earth's atmosphere, climate, and biosphere over geologic time. Identifying recycled marine carbonates and evaluating their fluxes in Earth's mantle are essential in order to obtain a complete understanding of the global deep carbon cycle (DCC). Here, we review recent advances in tracing the DCC using stable isotopes of divalent metals such as calcium (Ca), magnesium (Mg), and zinc (Zn). The three isotope systematics show great capability as tracers due to appreciable isotope differences between marine carbonate and the terrestrial mantle. Recent studies have observed anomalies of Ca, Mg, and Zn isotopes in basalts worldwide, which have been interpreted as evidence for the recycling of carbonates into the mantle, even into the mantle transition zone (410–660 km). Nevertheless, considerable challenges in determining the DCC remain because other processes can potentially fractionate isotopes in the same direction as expected for carbonate recycling; these processes include partial melting, recycling of carbonated eclogite, separation of metals and carbon, and diffusion. Discriminating between these effects has become a key issue in the study of the DCC and must be considered when interpreting any isotope anomaly of mantle-derived rocks. An ongoing evaluation on the plausibility of potential mechanisms and possible solutions for these challenges is discussed in detail in this work. Based on a comprehensive evaluation, we conclude that the large-scale Mg and Zn isotope anomalies of the Eastern China basalts were produced by recycling of Mg- and Zn-rich carbonates into their mantle source.

1 Introduction

The fate of marine carbonates during subduction has profound significance for the global carbon cycle, as

* 本文发表在: Engineering, 2019, 5: 448-457

they represent the major carbon-bearing phases that are recycled into our planet's interior. In the early Earth's mantle, the presence of both high geothermal gradients and mantle convection rates resulted in variable subduction rates that reduced the likelihood of adding crustal carbon into the deep mantle [1]. As Earth's mantle cooled, the mantle environment of subduction zones could permit the influx of crustal carbonates beyond the depth of arc magma generation, resulting in the more efficient addition of crustal carbon to the deep mantle. The transport of Earth's surface carbon into the mantle by subducted slabs and its return to the surface through the degassing of volcanoes constitute the global deep carbon cycle (DCC)[2], which is one of the major factors controlling the distribution of carbon in all terrestrial reservoirs [3,4]. The study of the DCC has invoked a number of scientific issues, including: identification of recycled crustal carbon; the storage, speciation, and fluxes of deep carbon; partial melting of carbonated mantle; and the proportion of subduction-related carbon and primary mantle-derived carbon in the total carbon dioxide (CO_2) released by volcanism. Among these issues, a definite identification of recycled marine carbonates in the Earth's mantle is a first and key step in DCC research.

Carbon isotopes have been widely applied to trace the DCC [2], but strong fractionation induced by magma degassing makes it difficult to elucidate recycled carbonate in volcanic rock sources based on carbon isotopes alone. Although the efficient subduction of organic carbon may play an important role in the rise of Earth's atmospheric oxygen [5], about 95% of subduction-related and primary mantle-derived carbon is inorganic carbon [6], and their carbon isotopic compositions are not significantly different. Thus, it is also difficult to identify recycled surface carbon in Earth's mantle using carbon isotopes alone. The stable isotopes of divalent metals such as calcium (Ca), magnesium (Mg), and zinc (Zn), which occur as major or trace constituents in carbonate, can be used to identify recycled carbonate in the mantle, primarily due to the limited influence of degassing on isotopic composition of the magmas and the large contrast in isotopic composition between marine carbonates and the terrestrial mantle. In past years, detectable anomalies of Ca, Mg, and Zn isotope compositions (expressed as $\delta^{44/40}Ca = [(^{44}Ca/^{40}Ca)_{sample}/(^{44}Ca/^{40}Ca)_{SRM\text{-}915a}-1]\times1000$, $\delta^{26}Mg = [(^{26}Mg/^{24}Mg)_{sample}/(^{26}Mg/^{24}Mg)_{DSM\text{-}3}-1]\times1000$, and $\delta^{66}Zn = [(^{66}Zn/^{64}Zn)_{sample}/(^{66}Zn/^{64}Zn)_{JMC3\text{-}0749L}-1]\times1000$, respectively) have been widely reported in mantle-derived rocks worldwide, including Hawaii [7], Eastern China including Hainan Island [8–15], the Tibetan Plateau [16–18], the Emeishan large igneous province (ELIP), and Tarim large igneous province (TLIP) [19–21], New Zealand [22], and Vietnam [23]. With the exception of New Zealand, Vietnam, and Hainan Island of China, the studies on most of these locations have speculated that the storage of a large amount of recycling marine carbonates in the mantle is related to subduction of carbonate-bearing oceanic crust.

The application of metal stable isotopes in tracing DCC is innovative and has been rapidly developed in recent years; however, it is critically challenged by the following important facts or potential difficulties: ① Ca, Mg, and Zn isotopes may be fractionated during partial melting of the mantle and the subducted oceanic crust; ② carbonated eclogites in subduction zones also possess extremely light Mg isotope compositions but commonly contain little carbon; ③ isotope anomalies of mantle-derived rocks may be dependent on the species of carbonates being subducted/recycled; ④ carbon and metal cations (Mg, Ca, and Zn) in carbonate may be decoupled during both slab subduction and return of CO_2-bearing magmas into the surface; and ⑤ diffusion-driven kinetic isotope effects may occur in magma sources or when magmas ascend through the overlying lithosphere. In view of potentially causing isotope fractionation in the same direction as expected for carbonate recycling, these processes must be taken into account while applying metal isotopes to DCC research. In this paper, we review the major advances in the application of metal stable isotopes to DCC; more importantly, we provide an ongoing evaluation of the plausibility of potential mechanisms and possible solutions for these difficult challenges.

2 Principle of the use of metal stable isotopes for tracing DCC

To apply the stable isotopes of metals of interest to the study of DCC, three key prerequisites are needed: ① a distinctive isotopic offset must exist between marine carbonates and the terrestrial mantle; ② the isotopic compositions cannot be significantly changed or fractionated during the subduction processes, partial melting, and magmatic differentiation; and ③ the elemental concentration ratios of carbonate to the mantle must be large enough to modify the isotopic composition of the hybridized mantle.

The efficiency of carbon subduction is usually largely controlled by the carbonate contents of the sediment column, and is partly linked to the latitude of the trench. The distribution of carbonate in the surface sediments of oceans is mainly controlled by surface productivity, dissolution, and dilution by noncarbonate biogenic and nonbiogenic sediment [24]. At the Earth's surface, divalent metal carbonates (i.e., Ca, Mg, Sr, Ba, Mn, Fe, Co, Ni, Zn, Cu, Cd, and Pb) constitute the main species of carbonate [25], such as calcite ($CaCO_3$), magnesite ($MgCO_3$), dolomite ($CaMg(CO_3)_2$), rhodochrosite ($MnCO_3$), smithsonite ($ZnCO_3$), and siderite ($FeCO_3$). Calcium and magnesium are the two most abundant cations in carbonates and constitute the endmember components of Ca-rich and Mg-rich carbonates, respectively. The Mg concentration (C) ratio of magnesite to the mantle ($C(\mathrm{Mg})_{\mathrm{magnesite}}/C(\mathrm{Mg})_{\mathrm{mantle}}$) is close to 1, and the Ca concentration of calcite is one order of magnitude higher than that of the fertile mantle (~3.5 wt%, [26]). Thus, the Mg and Ca concentration ratios of carbonate to the mantle could be sufficiently large to significantly alter the Mg–Ca isotope composition of the mantle. Trace Zn^{2+} can replace Mg^{2+} and incorporate into the crystal lattice of carbonate minerals [27]. Although Zn is trace in calcic carbonate (highly variable, but with an average of ~20 $\mu g \cdot g^{-1}$ [28]), the low Zn concentration of the mantle (~55 $\mu g \cdot g^{-1}$; [29]) results in a high $C(\mathrm{Zn})_{\mathrm{carbonate}}/C(\mathrm{Zn})_{\mathrm{mantle}}$ ratio (~0.4), which is sufficient to modify the Zn isotope composition of the mantle. In particular, dolomite and magnesite in ultrahighpressure metamorphic (UHPM) rocks, such as the deeply subducted slab, can have very high Zn concentrations of up to 147 and 449 $\mu g.g^{-1}$, respectively [30].

There are significant isotopic offsets between carbonate and the terrestrial mantle for Ca, Mg, and Zn isotopes (Fig. 1) [31–58]. The large isotope differences between marine carbonate and silicate reservoirs primarily reflect isotope fractionation generated during the release of metals into seawater by continental weathering and during carbonate precipitation from seawater. Boron isotopes ($\delta^{11}B$) have also shown potential in tracing recycled crustal materials in sources of carbonatitic melts [59]; however, in this paper, we focus on the stable isotopes of divalent metals (Ca, Mg, and Zn). For Ca and Mg, carbonates are enriched in the lighter isotopes in comparison with the terrestrial mantle (i.e., $\delta^{44/40}\mathrm{Ca}_{\mathrm{carbonate}} < \delta^{44/40}\mathrm{Ca}_{\mathrm{mantle}}$ [60]; $\delta^{26}\mathrm{Mg}_{\mathrm{carbonate}} < \delta^{26}\mathrm{Mg}_{\mathrm{mantle}}$ [31,61]), although it has been demonstrated that the $\delta^{44/40}\mathrm{Ca}$ values of sedimentary carbonates varied with time and some carbonates have higher $\delta^{44/40}\mathrm{Ca}$ ratios relative to the mantle [32]. By contrast, carbonates are more enriched in the heavier Zn isotopes in comparison with the mantle ($\delta^{66}\mathrm{Zn}_{\mathrm{carbonate}} > \delta^{66}\mathrm{Zn}_{\mathrm{mantle}}$) [10,33], and no evidence has been found to indicate that the $\delta^{66}\mathrm{Zn}$ values of sedimentary carbonates varied over geologic time [33,34]. Therefore, negative anomalies of Ca and Mg isotopes (i.e., lower $\delta^{44/40}\mathrm{Ca}$ and $\delta^{26}\mathrm{Mg}$ relative to those of the mantle) and positive anomalies of Zn isotopes (i.e., higher $\delta^{66}\mathrm{Zn}$ relative to that of the mantle) in igneous rocks—if any—may imply the incorporation of recycled sedimentary carbonates in the rock sources.

Due to the distinctive calcium and magnesium contents in Ca-rich and Mg-rich carbonates, the recycling of different species of marine carbonate would impose different Mg–Ca isotopic effects on mantle-derived rocks, such that a combination of Mg and Ca isotopes can quantitatively constrain the speciation of subducted carbonates (see Section 4.3 for details). Since Zn^{2+} (0.74 Å) has a similar ionic radii to Mg^{2+} (0.72 Å) and mainly substitutes for Mg^{2+} in silicates [62], high-temperature processes (e.g., magmatic differentiation and thermal diffusion) are expected to fractionate Mg and Zn isotopes in an identical direction. Therefore, whether light Mg or heavy Zn isotope signatures of igneous rocks reflect recycled carbonates in the sources or isotope fractionation induced by high-temperature processes may be effectively distinguished by combining Mg isotopes with Zn isotopes.

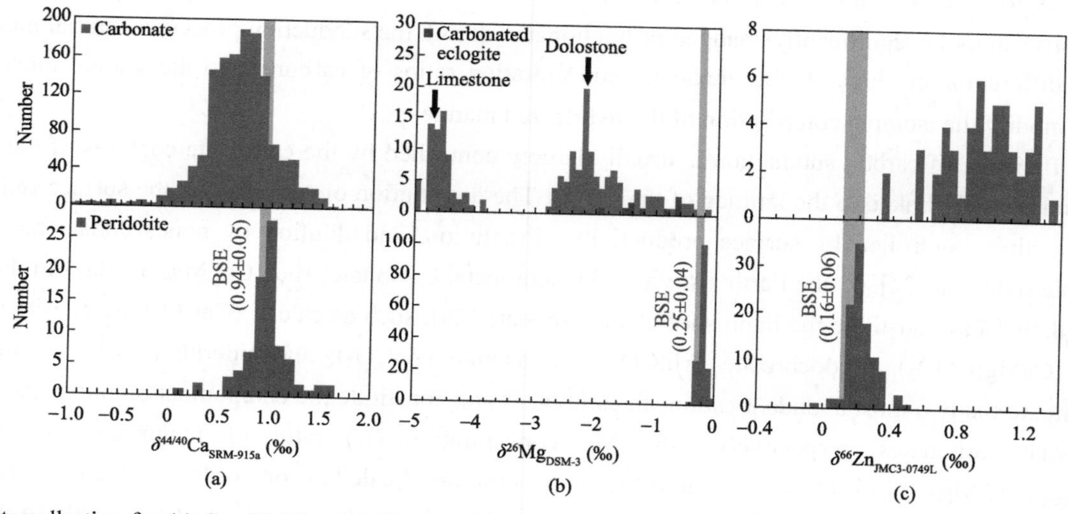

Fig. 1　Data collection for (a) Ca, (b) Mg, and (c) Zn isotopic compositions of sedimentary carbonates, carbonated eclogite, and mantle peridotites. Bulk silicate Earth (BSE) values (represented by the grey band) are from Ref. [38] for $\delta^{44/40}$Ca (relative to SRM-915a), Ref. [31] for δ^{26}Mg (relative to DSM-3), and Ref. [43] for δ^{66}Zn (relative to JMC3-0749L). Data are compiled from the following sources: sedimentary carbonates: $\delta^{44/40}$Ca [32] and references therein, δ^{26}Mg [45–50], and δ^{66}Zn [33,34]; carbonated eclogite: δ^{26}Mg [44]; peridotites: $\delta^{44/40}$Ca [36–41], δ^{26}Mg [31,51–57], and δ^{66}Zn [35,42,43,58].

During plate subduction, Mg isotopes are found to be limitedly fractionated [63]. The subducted mafic crust can preserve its primary isotope signature even after deep subduction to the ultrahigh-pressure eclogite facies. Magnesium isotopes are also not significantly fractionated during magmatic differentiation [31,64]. Siliciclastic sediments have heavy Mg isotope composition due to preferential release of light Mg during continental weathering [65], which is in contrast to carbonate sediments; thus, Mg isotopes can be used to discriminate recycled carbonates from siliciclastic sediments. Pons et al. [66] reported progressive decline of Zn isotope ratios during subduction of ultramafic crust (serpentinites), with heavy Zn isotopes preferentially being released into the fluids by incorporating with $ZnCO_3$ and/or $ZnSO_4$. Dehydrated serpentinites have a slightly lighter Zn isotope composition (δ^{66}Zn=0.16‰±0.06‰) than pre-dehydrated ones (0.32‰±0.08‰). By contrast, during subduction of the mafic oceanic crust, δ^{66}Zn value do not vary across metamorphic facies and the eclogites retain a mid-ocean ridge basalt (MORB)-like isotope composition [67]. Siliciclastic sediments have δ^{66}Zn values similar to or slightly lower than those of the mantle [68], which is also in contrast to the heavy δ^{66}Zn of carbonate; thus, Zn isotopes can be used to discriminate between recycled carbonates and siliciclastic sediments. Magmatic differentiation is found to fractionate Zn isotopes, but the magnitude of fractionation is considerably limited (< 0.1‰) [35,69]. No data is available for Ca isotopes in subduction-related samples, so this clearly needs future investigation. In particular, it is unclear whether selective dissolution of different species of carbonates within the subducted slab causes calcium isotope fractionation, so this topic deserves investigation. Overall, isotope fractionation of Mg and Zn (and possibly of Ca) during subduction-related dehydration is much smaller than the large isotope offset between carbonate and the mantle, making these isotopes powerful tools for tracing the circulation of sedimentary carbonates from the Earth's surface to the mantle.

3 Advances on application of Ca, Mg, and Zn isotopes to DCC

All data available in the literature for basalts with Ca, Mg, and Zn isotope anomalies were compiled, are plotted in Fig. 2 [7–12,14–16,18,19,21–23,31,38,43,70–74], and are briefly summarized below. This dataset is large ($n > 600$) and represents an important step of applying metal stable isotopes to the study of DCC. During the review process of this paper, at least nine papers reporting new data and detectable variations of calcium, magnesium, and zinc isotopes in mantle-derived rocks (peridotites and basaltic rocks) were published [36,37,70–76], highlighting the rapid development of these three isotope systematics in tracing DCC.

3.1 Calcium isotopes

Two studies have reported detectable variations of Ca isotopes in basaltic rocks and linked these variations to recycling of carbonates. Huang et al. [7] found light Ca isotope signatures in Hawaiian basalts ($\delta^{44/40}$Ca=0.75‰–1.02‰) relative to the fertile mantle (0.94‰±0.05‰ [38]). They observed decreasing $\delta^{44/40}$Ca with increasing ^{87}Sr/^{86}Sr and Sr/Nb ratios, and attributed the light Ca isotope compositions to recycling of ancient carbonate component (~4‰) in the Hawaii plume. Recently, Liu et al. [18] reported lower $\delta^{44/40}$Ca values of volcanic rocks from Tengchong in Southwest China (0.67‰–0.80‰) compared with the fertile mantle. A binary mixing model using Mg–Ca isotopes shows that 5%–8% carbonates dominated by dolostone have been recycled into the mantle sources. The recycled carbonates were proposed to have originated from the Indian oceanic crust.

Significant calcium isotope variations have also been found in mantle peridotite xenoliths [36–38,77], but the results and explanations differ considerably. Zhao et al. [77] reported extreme $\delta^{44/40}$Ca variation (−0.08‰ to 0.92‰) in peridotitic xenoliths from the North China Craton, and attributed the coupled light Ca–Fe isotopic signatures in these Fe-rich peridotites to kinetic isotope fractionation during melt-peridotite reaction. No clear role of recycled carbonate was illustrated in that study. Chen et al. [36] reported relatively high $\delta^{44/40}$Ca ratios for clinopyroxene (+0.84‰ to +1.17‰) and orthopyroxene (+0.82‰ to +1.22‰) in peridotite xenoliths from Fanshi area in the North China Craton. The heavy Ca isotope signature in the metasomatic clinopyroxenes, which were formed via metasomatism by both carbonatite melt and carbonate-rich silicate melt, was proposed to reflect a characteristic of evolved sedimentderived hydrous carbonate melts [36]. These authors further predicted that significant Ca isotope fractionation may have occurred during the subduction of carbonate sediments. Recently, Ionov et al. [37] reported calcium isotope data for mantle xenoliths containing metasomatic calcite from the Siberian Craton. These authors found that the $\delta^{44/40}$Ca ratios of the metasomatized peridotites (+0.81‰ to +0.83‰) do not significantly deviate from that of the fertile mantle (<0.15‰), and the acid-leachates (carbonates) and leaching residues have similar $\delta^{44/40}$Ca values. The authors therefore asserted that Ca isotopes have little use as a tracer of carbonate recycling [37]. Overall, it seems that the calcium isotope data of mantle peridotites is immensely complex, making it difficult to simply relate it to carbonate recycling.

3.2 Magnesium isotopes

An important superiority of Mg isotopes relative to other isotope systems (e.g., Ca and Zn) is the huge difference of Mg isotope composition between marine carbonates and the terrestrial mantle (up to 5‰; Fig. 1). In addition, compared with the large contrast in isotope abundances between ^{40}Ca and ^{42}Ca or ^{44}Ca, and the strong interference of ^{40}Ar on ^{40}Ca that represents a significant obstacle to the precise measurement of Ca isotope ratios [78], the analysis of magnesium isotopes is more convenient. These superiorities make Mg isotopes the most commonly applied tracer for the DCC to date. Below, we review the studies that reported Mg isotope anomalies in basalts separately on the basis of regional occurrence of these rocks.

3.2.1 East Asia

Yang et al. [8] reported the first Mg isotope anomaly in alkaline basalts from the north and central North China Craton. They found an abrupt decline of δ^{26}Mg ratios in<110 Ma basalts relative to>125 Ma basalts, and attributed this transition to subduction of the West Pacific slab beneath Eastern China in the late Mesozoic. Huang et al. [9] expanded the low-δ^{26}Mg anomaly to Cenozoic alkaline basalts from South China. Tian et al. [11] and Wang et al. [15] further found low-δ^{26}Mg values in Cenozoic alkaline basalts from Northeast China. Collectively, these studies have expanded the large-scale Mg isotope anomalies of Eastern China basalts to an area of more than 1 000 000 km^2. Li et al. [12] presented a more comprehensive Mg isotope dataset including arc basalts from the circle-Pacific region, basalts from the western part of the North China Craton, and basalts from the Hainan hot spot. These authors found that low-δ^{26}Mg anomalies only occur in Eastern China and are not present in the western North China Craton; based on Mg–Sr isotope modeling, they proposed that the recycled carbonate is mainly composed of dolomite ± magnesite [12]. Highresolution seismic tomography showed that the Pacific slab is subducting beneath the Japan Islands and becomes stagnant in the mantle transition zone (410–660 km) beneath Eastern China [79]. The low-δ^{26}Mg anomaly of basalts in the mainland of Eastern China coincides with the stagnant Pacific slab in the mantle transition zone beneath Eastern China, suggesting that the large-scale low-δ^{26}Mg anomalies indicate the location of recycled Mg-rich carbonates in the mantle transition zone beneath Eastern China [12]. Su et al. [14] also reported a large Mg isotope dataset on a suite of continental basalts from East Asia, and found a decreasing δ^{26}Mg trend with increasing distance from the present subduction trench. They attributed these variations to westward subduction of the Pacific oceanic plate beneath East Asia. More recently, Kim et al. [71] extended the low-δ^{26}Mg anomaly in alkaline volcanic rocks to Jeju Island, Korea, and similarly linked it to the stagnant, subducted, carbonate-bearing Pacific slab in the mantle transition zone beneath East Asia.

3.2.2 Tethyan subduction zone

The Tethyan subduction zone is another important area where low-δ^{26}Mg signatures have been widely observed in mantlederived rocks. Liu et al. [16] found light Mg isotope compositions of Cenozoic ultrapotassic rocks from south Xizang, and attributed the observations to the recycling of carbonated pelites during subduction of the Neo-Tethyan oceanic crust. Ke et al. [17] reported low δ^{26}Mg signatures and negative correlation between δ^{26}Mg and δ^{18}O values for Cenozoic syenites from the Pamir syntaxis, located in the western Himalayan–Tibetan orogen. Modeling on the basis of Mg–O–Sr isotope data indicated that the recycled carbonate is mainly limestone with minor dolostone, suggesting that the metasomatism occurred at depths shallower than 60–120 km. Liu et al. [18] also found low-δ^{26}Mg signatures together with low $\delta^{44/40}$Ca values for the Tengchong volcanic rocks in the southeastern Tibetan Plateau. These studies indicate that recycling of marine carbonate into the mantle may have been widespread during subduction of the Tethyan oceanic crust before and during the Cenozoic. Recently, Tian et al. [72] reported low Mg isotope ratios (−0.51‰ to −0.33‰) in basalts and andesites from Tengchong, and attributed the low-δ^{26}Mg anomalies as well as the extremely high Th/U (6.5–8.3) to dolomite dissolution by slab-derived supercritical liquids in subduction zones.

3.2.3 Large igneous provinces

Cheng et al. [21] reported low-δ^{26}Mg anomalies (−0.45‰ to −0.28‰) in basalts, mafic–ultramafic layered intrusions, diabase dykes, carbonatites, and mantle xenoliths in kimberlitic rocks from the TLIP. These authors proposed two distinct mantle domains as the sources of the TLIP basalts: ① a lithospheric mantle source for basalts and mafic–ultramafic layered intrusions, which was modified by calcite/dolomite and eclogite-derived high-silicon (Si) melts, as evidenced by enriched Sr–Nd, heavy oxygen, and light Mg isotope compositions; and ② a plume source for carbonatite, nephelinite, and kimberlitic rocks, with the involvement of magnesite or periclase/perovskite. Tian et al. [19] reported the mantlelike Mg isotope composition of Permian basalts from the ELIP. They proposed that sedimentary carbonates recycled into the lower mantle beneath the ELIP were very limited (<2%). Thus, whether or not marine carbonates can be carried into the lower mantle remains open to debate and requires further study.

3.2.4 Low-δ^{26}Mg anomalies proposed to be unrelated to the recycling of carbonate

Five studies have proposed different explanations for low-δ^{26}Mg anomalies observed in basalts from other regions, rather than recycling of marine carbonates. Wang et al. [22] reported Mg isotope data for high-time-integrated μ=^{238}U/^{204}Pb mantle (HIMU)-like intraplate basalts from New Zealand (–0.47‰ to –0.06‰). Based on the negative relationship between δ^{26}Mg and $(Gd/Yb)_{normalized}$ these authors attributed the observed low-δ^{26}Mg anomalies to recycling of carbonated oceanic crust (eclogites) in the mantle source. On the basis of high Fe/Mn and Ca/Al, and the absence of negative Hf and Ti anomalies, Li et al. [12] argued that the low-δ^{26}Mg anomalies of basalts from Hainan Island were caused by recycled oceanic eclogites. Sun et al. [70] presented negative correlations between δ^{26}Mg values and TiO_2 contents for potassium-rich basalts from Northeast China, and attributed the low-δ^{26}Mg anomalies to recycling of carbonated eclogites. More recently, Hoang et al. [23] reported Mg isotope data for late Cenozoic intraplate basaltic rocks from Central and Southern Vietnam (–0.62‰ to –0.28‰), and proposed a hybridized source containing peridotite and recycled oceanic crust (carbonated eclogites). Wang et al. [74] found low-δ^{26}Mg anomalies in Pitcairn Island basalts with a typical EM1 affinity. Considering the low Ca/Al ratios, these authors proposed that carbonate–silicate interaction in the subducted sediments exhausted the carbonates but incorporated the isotopically light magnesium of the carbonate into silicates. The latter were subducted into the deep mantle and resulted in the observed low-δ^{26}Mg anomalies.

3.3 Zinc isotopes

Zinc isotopes are a novel tracer for the DCC recently developed by Liu et al. [10]. These authors analyzed Zn isotope data for Mesozoic to Cenozoic basalts from Eastern China that had been previously analyzed for Mg isotopes. The <110 Ma basalts were shown to have elevated δ^{66}Zn values and display a negative correlation between δ^{66}Zn and δ^{26}Mg, which were interpreted as a result of the recycling of marine carbonates in the mantle source. Given the high Zn concentration of these basalts and the positive relationship between δ^{66}Zn and Zn concentration, Liu et al. [10] argued that the recycled carbonate is Zn-rich and is most likely dominated by magnesite. This argument provides independent evidence for determining the species of recycled carbonate during subduction of the West Pacific slab beneath Eastern China.

Recently, Wang et al. [73] presented a systematic variation of Zn isotopes with Sr–Nd isotopes and incompatible elements (e.g., Nb, Th, and Zn) in nephelinites, basanites, and alkali olivine basalts from the Shandong Peninsula in Eastern China (Fig. 2). Compared with nephelinites and high-alkali basanites that have elevated δ^{66}Zn ratios reflecting carbonate metasomatism of the sources, alkali olivine basalts and low-alkali basanites show a gradual decline of δ^{66}Zn with SiO_2, and have Zn–Sr–Nd isotopic and chemical compositions that are shifted toward those of an enriched lithospheric mantle. The authors proposed that infiltration and reaction of silica-undersaturated melts with the lithospheric mantle result in the transition of strongly alkaline melts into weakly alkaline melts, and could be one of the most common causes of compositional diversity in natural alkaline lavas.

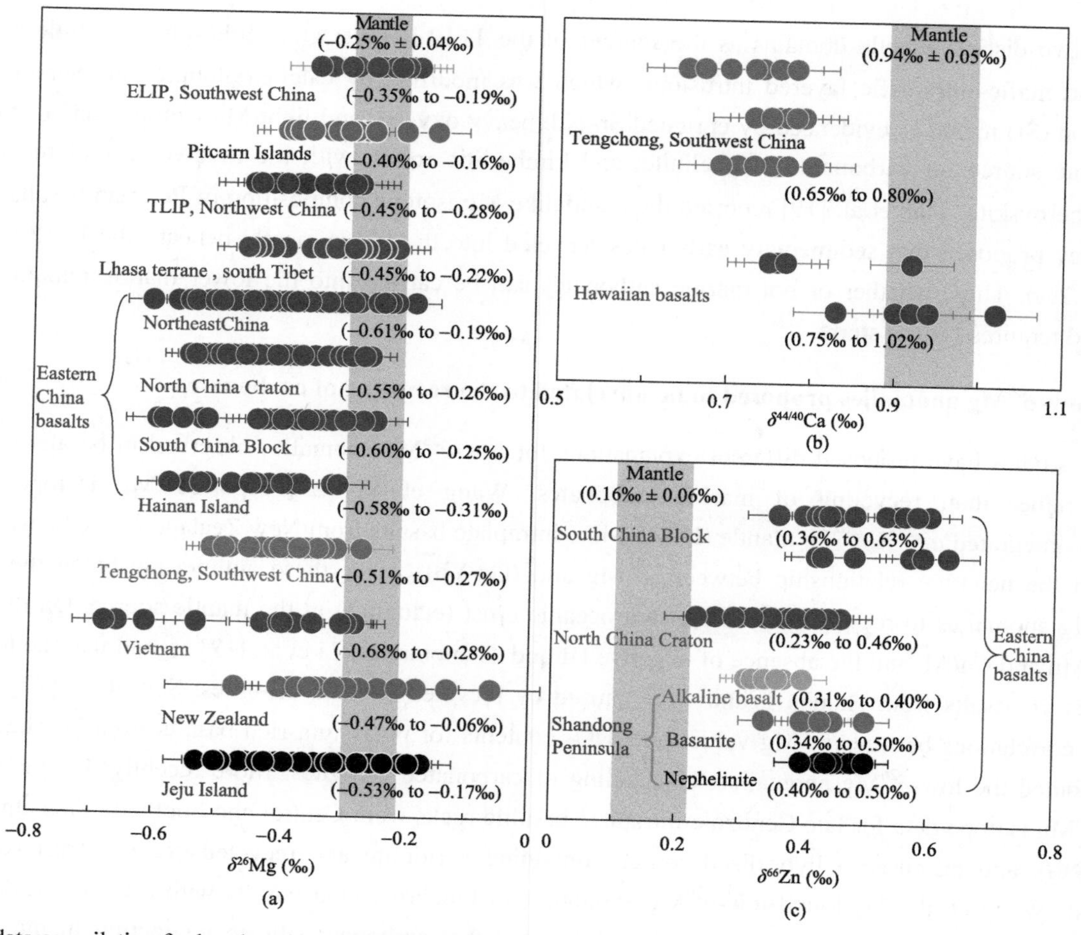

Fig. 2 A data compilation for basalts with isotope anomalies of (a) Mg, (b) Ca, and (c) Zn reported in the literature. Mantle values (represented by the grey band) are from Ref. [38] for $\delta^{44/40}$Ca, Ref. [31] for δ^{26}Mg, and Ref. [43] for δ^{66}Zn. Data are compiled from the following sources: δ^{26}Mg, Northeast China [11,15,70], North China Craton [8,12,14], South China Block [9,12,14], Hainan Island [12], ELIP [19], Pitcairn Islands [74], TLIP [21], Tengchong [18,72], Lhasa terrane [16], Vietnam [23], New Zealand [22], and Jeju Island [71]; $\delta^{44/40}$Ca, Tengchong [18] and Hawaii [7]; δ^{66}Zn, South China Block [10], North China Craton [10], and Shandong Peninsula [73].

4 Challenges and questions

As discussed above, many studies have already reported Ca, Mg, and Zn isotope data for basalts, and have linked the observed isotope anomalies to possible recycling of marine carbonate. However, considerable challenges exist and call into question the unique linkage between the isotope anomalies observed in basalts and carbonate recycling in mantle sources. In particular, several processes can potentially fractionate isotopes in the same direction as expected by carbonate recycling. In this section, particular attention is paid to evaluating the plausibility of these potential mechanisms and providing possible solutions for these challenges–aspects that have been commonly overlooked in earlier work. Our aim is to call attention to the interpretation of isotope data instead of simply linking any isotope anomaly to the carbonate effect without evaluating other possibilities in detail in future studies.

4.1 Partial melting effect

The improved precision of isotope analysis has led to the finding that calcium, magnesium, and zinc isotopes may be fractionated during mantle melting. Data on mantle minerals show that in equilibrium, olivine and orthopyroxene tend to become enriched in heavy Ca isotopes in comparison to clinopyroxene [39,40]. During partial melting of the mantle, light Ca isotopes preferentially enter into melt because clinopyroxene is consumed first [80,81]. Indeed, the $\delta^{44/40}$Ca of ultramafic rocks is negatively correlated with CaO and positively correlated with MgO [38,40,41], probably reflecting isotope fractionation during the partial melting of mantle peridotites. Kang et al. [38] found that the $\delta^{44/40}$Ca in moderately depleted peridotites (1.07‰±0.04‰) is slightly higher than that in fertile spinel and garnet peridotites (0.94‰±0.05‰), indicating that light Ca isotopes are preferentially partitioned into melt. Ionov et al. [37] also found heavy Ca isotope signatures in refractory, non-metasomatized off-craton peridotites (1.10‰±0.03‰) in comparison to fertile lherzolites due to isotope fractionation during melt extraction. Therefore, partial melting of mantle peridotites may result in light Ca isotope composition in basaltic melts. Since marine carbonates commonly have a lighter $\delta^{44/40}$Ca relative to the fertile mantle, application of Ca isotopes to trace recycled marine carbonates in the source of mantle derived melts must take this fractionation effect into account.

Magnesium isotopes have generally been suggested to be unfractionated during mantle partial melting [31,82–84], since global peridotite xenoliths and mantle-derived melts (e.g., basalt, komatiite) commonly have identical Mg isotope composition within analytical errors. Recently, Zhong et al. [85] showed that the δ^{26}Mg values of most ocean island basalts (OIBs) are negatively correlated with melting-sensitive trace-element ratios (e.g., Nb/Zr) but uncorrelated with source-sensitive elemental ratios. Modeling calculations suggest that preferential melting of garnet in garnetbearing sources (e.g., recycled oceanic crust and garnet peridotite) can generate melts with slightly low δ^{26}Mg values (about −0.35‰) at low-degree partial melting [85], since garnet is more enriched in light Mg isotopes in comparison with other coexisting silicate minerals (e.g., clinopyroxene). Zhang et al. [86] evoked a similar mechanism to interpret the low-δ^{26}Mg anomaly observed in the basalts of Eastern China. However, the maximum difference between the δ^{26}Mg values of melts and residue is not greater than 0.1‰ [85], which is unlikely to reconcile with the up to 0.4‰ δ^{26}Mg offset between the Eastern China basalts and the terrestrial mantle [8,9,12].

Recent studies show that zinc isotopes can also be fractionated during mantle melting. A study on peridotites from Siberia and Mongolia found that refractory peridotites have Zn isotope compositions (δ^{66}Zn=+0.14‰± 0.05‰) lighter than fertile peridotites (+0.30‰±0.07‰). The authors attributed this difference to pressure-dependent isotope fractionation induced by mantle melting [42]. Based on new data for mantle minerals and oceanic basalts, Wang et al. [35] argued that spinel consumption during mantle melting plays a key role in generating high Zn concentrations and heavy Zn isotope compositions of global basalts. More recently, Sossi et al. [43] suggested that the Zn–O bonding difference between melts and residue may account for the isotope fractionation of Zn during mantle melting. Huang et al. [87] reported that some arc volcanic rocks display a δ^{66}Zn higher than that of the depleted MORB mantle, also suggesting a small Zn isotope fractionation during mantle melting. Although a detailed evaluation of these different hypotheses is beyond the scope of this paper, it is clear that Zn isotope fractionation generated during partial melting of the mantle has a maximum of <0.1‰, which is unlikely to have resulted in the up to 0.4‰ δ^{66}Zn difference between the Eastern China basalts and the terrestrial mantle [10].

It is very surprising that the isotope effect induced by mantle melting is always in the same direction as the recycling of marine carbonates for all three isotope systematics of Ca, Mg, and Zn. While the magnitude of melting-induced isotope fractionation may be smaller than the isotope offset between carbonate and the mantle, this effect would become outstanding when quantifying the fluxes of recycling carbonate into the mantle. Especially for strongly alkaline basalts, which are commonly generated by extremely low degrees of mantle melting (1%–8%) [88], isotope fractionation produced by partial melting may exert significant influence on the interpretation and quantification of light $\delta^{44/40}$Ca and δ^{26}Mg and heavy δ^{66}Zn values in the mantle sources of basalts. Further research on the application of these isotope systematics to the DCC is required to consider this effect.

4.2 The carbonated eclogite "problem"

As discussed above, five case studies have attributed low-δ^{26}Mg anomalies in basalts from New Zealand, Northeast China, Hainan Island, Vietnam, and Pitcairn Island to recycling of carbonated silicates (eclogites). Distinguishing between carbonated eclogite and marine carbonate in sources of basalts has become a urgent problem, since most carbonated eclogites with extremely low δ^{26}Mg values (e.g., −1.92‰; Fig. 1) contain very little carbon (CO_2 < 0.22 wt%) [44]. The low-δ^{26}Mg signature of carbonated eclogites could have been caused by substantial carbonate-silicate isotopic exchange within the subducted slab [44]. Therefore, if the low-δ^{26}Mg anomalies observed in basalts were produced by the recycling of CO_2-poor carbonated eclogites, their presence would fail to indicate carbon recycling and would only imply the recycling of isotopically light crustal Mg.

Experimental studies have shown that partial melting of typical recycled carbonated oceanic crust (in the form of carbonated eclogite) at a pressure of 3 GPa can produce coexisting silica-rich melt and carbonate-rich melt over a wide temperature range [89]. Metasomatism of the mantle by carbonated eclogite-derived silicate melts will generate pyroxenite, and carbonatitic melts from the melting of recycled carbonate will form carbonated peridotites [90]. Although both metasomatized sources have low δ^{26}Mg values, their diverse roles in the sources of basalts may be distinguished by plotting δ^{26}Mg against major and trace elements that are sensitive to either a peridotitic or pyroxenitic source lithology (e.g., Fe/Mn, Hf/Hf*, Ti/Ti*, CaO/Al_2O_3, etc.) [12]. For example, high Fe/Mn ratios could be derived from a pyroxenite or eclogite source [91]. Melts from a mantle source hybridized by sedimentary carbonate (i.e., carbonated peridotite) are expected to typically have low Fe/Mn, Hf/Hf*, and Ti/Ti* and high CaO/Al_2O_3. Li et al. [12] concluded that the nephelinites in Eastern China with the lowest SiO_2 and δ^{26}Mg can represent the low δ^{26}Mg component (LMC) in the mantle source, which is characterized by low Fe/Mn, Hf/Hf*, and Ti/Ti* and high CaO/Al_2O_3 ratios, consistent with the typical features of carbonated peridotite-derived melt. By contrast, a negative relationship of δ^{26}Mg with Fe/Mn ratios and a positive relationship of δ^{26}Mg with CaO/Al_2O_3 have been observed in Hainan basalts from South China [12]. Compared with partial melts of mantle peridotite, the Hainan basalts have lower Na_2O/TiO_2, CaO/Al_2O_3, and Co/Fe, and higher TiO_2, Fe/Mn, Zn/Fe, and Zn/Mn [92]. These characteristics were argued to have been caused by a mixing source of recycled oceanic crust (carbonated eclogite) and peridotite [12,92].

There is no available data on Ca and Zn isotopes for carbonated eclogites. Nevertheless, isotopic exchange between sedimentary carbonate and basaltic crust within the subducted slab, which has significantly modified Mg isotopes, should have influenced Ca and Zn isotopes of the recycled oceanic crust as well. The exchange of light Ca and heavy Zn isotopes from carbonate into the basaltic crust is expected to form low-$\delta^{44/40}$Ca and high δ^{66}Zn carbonated eclogites. Since these components may be carbon-poor but would generate the same isotope effect as the recycling of carbonates, their isotopic compositions need to be determined in future studies and must be considered when applying Ca and Zn isotopes to trace recycled carbonates in sources of basalts.

4.3 Dependence of isotope anomalies on speciation of recycled carbonates

The great variation of Ca, Mg, and Zn contents in different species of carbonates means that their isotopes are sensitive to the speciation of recycled carbonates in the mantle. For example, Ca-rich carbonate (i.e., calcite) has low Mg and Zn concentrations, such that the recycling of such carbonates will cause a modest change in the

Mg and Zn isotope compositions of the carbonated mantle. By contrast, the recycling of Mg-rich carbonate (e.g., magnesite) could result in a remarkable anomaly of Mg and Zn isotopes but limited change in Ca isotope composition due to low Ca concentration. In other words, normal (mantle-like) δ^{26}Mg or δ^{66}Zn values of basaltic rocks do not necessarily reflect the absence of recycled carbonates in the rocks' mantle sources, because the recycled carbonate species may be poor in Mg and Zn (e.g., calcite or aragonite). Similarly, normal $\delta^{44/40}$Ca values of basalts cannot exclude the possibility of a substantial amount of carbonate recycling if the recycled species of carbonate is Ca-poor (e.g., magnesite and siderite). An evaluation of the dependence of Ca, Mg, and Zn isotope anomalies of basalts on the species of recycled carbonate is important, because normal (mantle-like) isotope values may result in an incorrect judgment that there is no recycled crustal carbon in the sources of the studied basalts. The most direct solution to this conundrum may be a combined utilization of all three isotopic systematics.

Theoretical studies based on molecular dynamics [93] have predicted that the solubility of calcite or aragonite is significantly higher than that of dolomite and magnesite at high pressures. This is supported by observations on UHPM rocks from the Alps that showed that while magnesite and dolomite occur as solid inclusions within garnet, the fluid inclusions in garnet and diamond contain only Mg-calcite/calcite daughter crystals [94]. Therefore, the diverse solubility of different carbonate species in dehydrated fluids may play an important role in controlling the Ca, Mg, and Zn isotope compositions of the metasomatized mantle. For example, dehydrated fluids dominated by calcite or aragonite dissolution contain a large amount of Ca but little Mg, which has a limited influence on the Mg isotope composition of the mantle but may cause significant Ca isotope variation. Island arc basalts have a mantle-like or heavier Mg isotope composition in comparison with the mantle, which is best explained by selective dissolution and removal of Ca-rich carbonates from—and leaving Mg-rich carbonates in—the subducting slabs at relatively shallow depths (75–120 km) [12]. By contrast, at depths of > 120 km, Mg-rich carbonates (e.g., dolomite) could be dissolved by the slab-derived supercritical liquid, resulting in significantly low Mg isotope ratios in the mantle and in mantle-derived rocks [72,76].

Although calcite is the major species of carbonate carried by oceanic crust at the initial subduction stage, high-pressure experimentation shows that calcite is unstable at high pressures. When reacting with pyroxene, calcite can be transformed into dolomite at pressures between 23 and 45 kbars (1 bar=0.1 MPa), or into magnesite under higher pressure [95]. More recent experimental and theoretical studies show that the high-pressure metastable phase of calcite (e.g., $MgCO_3$, $ZnCO_3$, $FeCO_3$) can be stable at pressures greater than 62 GPa, and may play an important role in carbon storage and transport in the deep Earth [96]. As the cationic radius decreases (Ca>Zn>Mg), divalent carbonates can be stable under higher pressure. Thus, the proportion of Mg-rich carbonate retained in the subducted slabs and carried into the deep mantle could be significant. The Mg and Zn isotope systematics have been demonstrated to be effective in tracing the recycling of Mg-rich carbonate [10,12], such as in alkaline basalts from Eastern China, where the recycled carbonates were most likely located in the mantle transition zone (410–660 km). Recently, Cheng et al. [20] reported a low-δ^{26}Mg anomaly in deeply derived rocks from the TLIP, and attributed it to the recycling of magnesite in the plume source in the lower mantle. By contrast, it could be speculated that calcium isotopes may be more effective in tracing carbon recycling in island arc settings than Mg and Zn isotopes, considering the higher solubility of Ca-rich carbonate (calcite or aragonite) relative to Mg-rich carbonate (e.g., magnesite).

4.4 Decoupling of metals and carbon during the DCC

The most severe challenge for tracing the DCC using metal isotopes is that metals (e.g., Ca, Mg, and Zn) and carbon may be thoroughly separated during the circulation process. This problem has cast doubt on whether or not metal stable isotopes can effectively trace the DCC. ''Separation'' may happen during slab subduction, mantle melting, magma ascent, and final eruption. During early subduction, the geothermal gradient is high and the decarbonization reaction will release CO_2 but may leave divalent metals in the ongoing subducting plate. This process results in the inefficient deep subduction of carbon, with most carbon being stripped off from the downgoing slab and released back to the exosphere by arc magmatism [1]. Separation of metals and carbon is also reflected in CO_2-poor carbonated eclogites formed via carbonate–silicate interaction within the subducting plate [44]. However, in modern subduction zones, sedimentary carbonates may survive decarbonization, and the downgoing lithologies may experience temperatures cooler than the key decarbonation reactions. The crustal carbon is then expected to avoid the magmatic release in the volcanic arcs and participate in the deeper cycling and mantle processes, as has been observed in East Asia and the TLIP. As discussed in Section 4.3, the recycling of carbonated eclogites and carbonates into mantle sources of basalts can be effectively discriminated by major and trace element indexes.

During melting of the carbonated mantle and the ascent and eruption of magmas, CO_2 and CO_2-rich fluids may escape from magmas even at mantle depths, partition into vapor, and thus separate from metals [97,98]. Therefore, using the metal stable isotopes of the eruptive volcanic rocks to quantify carbon fluxes of the magma sources is questionable and almost impossible. In this case, carbon isotopes serve as the unique tool of the quantification of carbon fluxes in the original magmas and sources [99,100]. As mentioned above, however, strong fractionation induced by magma degassing makes it difficult to use carbon isotopes to identify recycled carbonates in the pre-eruptive magma source of volcanic rocks. Thus, we recommend a combined utilization of carbon and metal isotopes for studying carbon circulation in the surface-mantle-surface system; the metal stable isotopes can be used to trace (and possibly quantify) recycled carbonates into the mantle, and the carbon isotopes can help to quantify the fluxes of the return of recycled carbon to the surface.

4.5 Diffusion effect

A noticeable phenomenon is that almost all of the studied basaltic rocks with Ca, Mg, and Zn isotope anomalies are alkalirich, such as those in East China [8–15], Southwest China [18], the Tibetan Plateau [16,17], New Zealand [22], Vietnam [23], the TLIP [21], and Pitcairn Island [74]. Most of these alkaline basalts originated from the asthenospheric mantle and had ascended through the sub-continental lithospheric mantle (SCLM) or the oceanic lithospheric mantle before they erupted on the surface. During this process, both thermal and chemical diffusion processes may happen and potentially cause isotope fractionation. Since the heavier isotopes always concentrate in cold regions and light isotopes concentrate in hot regions during thermal diffusion [101], the ascending of hot (e.g., plume-related) magmas through the SCLM may result in light isotope enrichment of the resultant melts. The diffusion-driven isotope effect thus challenges the interpretation of low-δ^{26}Mg anomalies in basalts as a consequence of carbonate recycling in sources (Fig. 3(a)). Since Zn^{2+} mainly replaces Mg^{2+} in silicate, thermal diffusion between peridotite and silicate melts is expected to fractionate Zn and Mg isotopes in the same direction [102,103]. This expectation is in contrast to the strongly negative relationship between δ^{26}Mg and δ^{66}Zn ratios observed in Cenozoic basalts from Eastern China (Fig. 3(a)). Therefore, we conclude that the large-scale δ^{26}Mg and δ^{66}Zn anomalies of Eastern China basalts are unlikely to have been caused by thermal diffusion.

A potential magnesium (calcium, zinc) concentration difference exists between ascending basaltic melts and surrounding peridotites. Since light isotopes diffuse faster than heavy isotopes [101], a chemical gradient will drive the diffusion of light Mg isotopes from peridotite to melts [104,105], forming low-δ^{26}Mg melts. The diffusion-driven isotope effect, again, challenges the interpretation of low-δ^{26}Mg anomalies in basalts being due to carbonate recycling (Fig. 3(a)). Unlike Mg, Zn is moderately incompatible during mantle melting and basalts have a higher Zn concentration (commonly>80 $\mu g \cdot g^{-1}$) than mantle peridotites (~55 $\mu g \cdot g^{-1}$) [62]. Chemical diffusion may thus result in elevated $\delta^{66}Zn$ and a negative relationship between $\delta^{26}Mg$ and $\delta^{66}Zn$ ratios in basaltic melts. However, the diffusion of zinc from basaltic melts to ambient peridotites is strongly unsupported by the positive relationship between Zn concentration and $\delta^{66}Zn$ ratios in Eastern China basalts (Fig. 3(b)) [10]. In other words, the samples with higher $\delta^{66}Zn$ ratios are unlikely to have lost more Zn via diffusion because these samples have relatively high Zn concentrations (Fig. 3(b)). Liu et al. [10] has argued that the elevated $\delta^{66}Zn$ ratios coupled with the high Zn concentrations of Eastern China basalts were caused by the addition of Zn-rich carbonate (e.g., magnesite) to the mantle sources.

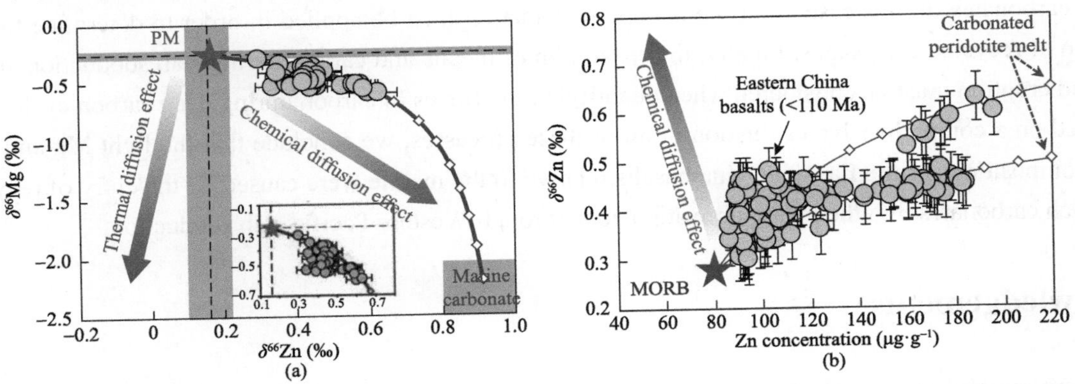

Fig. 3 Plots of (a) $\delta^{26}Mg$ versus $\delta^{66}Zn$ and (b) $\delta^{66}Zn$ versus Zn concentration for < 110 Ma basalts from Eastern China. Recycling of marine carbonate in the mantle source of basalts will produce a negative relationship between $\delta^{26}Mg$ and $\delta^{66}Zn$ and a positive relationship between $\delta^{66}Zn$ and Zn concentration [10]. The modeling curve in (a) refers to the mixing of primitive mantle ($C(MgO)$ = 38 wt%, $C(Zn)$ = 55 $\mu g \cdot g^{-1}$, $\delta^{26}Mg$ = −0.25‰, $\delta^{66}Zn$ = 0.16‰) with recycled Mg- and Zn-rich carbonate ($C(MgO)$ = 25 wt%, $C(Zn)$ = 280 $\mu g \cdot g^{-1}$, $\delta^{26}Mg$ = −0.22‰, $\delta^{66}Zn$ = 0.91‰), and the modeling curve in (b) refers to the mixing of MORB ($C(Zn)$ = 80 $\mu g \cdot g^{-1}$, $\delta^{66}Zn$ = 0.28‰) with assumed melts derived from carbonated peridotites ($C(Zn)$ = 220 $\mu g \cdot g^{-1}$, $\delta^{66}Zn$ = 0.50‰–0.65‰). The increment (diamond symbol) is 10%. The thermal diffusion-driven isotope effect is expected to produce a positive relationship between $\delta^{26}Mg$ and $\delta^{66}Zn$. The chemical diffusion-driven isotope effect is expected to produce a negative relationship between $\delta^{26}Mg$ and $\delta^{66}Zn$ and between $\delta^{66}Zn$ and Zn concentration, the latter of which is inconsistent with the observations (see text for details). Data of < 110 Ma basalts from Eastern China are given in Refs. [8–10]. PM: primitive mantle.

Diffusion may also happen in the source regions of basalts. The recycled carbonate-bearing oceanic crust may partially melt and produce SiO_2-rich melts, which could react with mantle peridotites to form pyroxenite veins. A magnesium concentration gradient will occur between surrounding peridotites and the newly formed pyroxenite veins. The diffusion of light Mg from peridotite to pyroxenite may result in the latter becoming enriched in light Mg isotopes (Fig. 3(a)). Subsequent partial melting of the low-$\delta^{26}Mg$ sources may produce low-$\delta^{26}Mg$ melts. Again, this model is in conflict with the high Zn concentrations and $\delta^{66}Zn$ ratios in Eastern China basalts (Fig. 3), since melts from the pyroxenites with higher $\delta^{66}Zn$ ratios, if formed via diffusion, should have relatively lower Zn concentrations in the absence of a significant difference in melting degrees. On the other hand, since diffusion will be very fast at high temperatures [101,106], it is unclear whether the possible diffusion-driven isotope effects of such sources could be preserved at mantle temperatures. Future simulation studies are required in order to quantitatively evaluate whether the predicted, isotopically abnormal sources could be preserved over the interval between pyroxenite formation and subsequent melting to produce basaltic melts.

5 Conclusions

The stable isotopes of divalent metals such as calcium, magnesium, and zinc have been widely applied to research on the DCC in recent years. Most of these studies have linked the observed isotope anomalies to the fingerprint of recycled marine carbonates in the mantle source of basalts, which strengthens the application of Ca, Mg, and Zn isotopes to trace Earth's DCC. Nevertheless, several processes can also fractionate the stable isotopes of Ca, Mg, and Zn to varying extents during the generation of mantle-derived rocks, including isotope fractionation induced by partial melting, the recycling of carbonated eclogite, and thermal/chemical diffusion. These effects create a barrier to the interpretation of low $\delta^{44/40}$Ca and δ^{26}Mg and high δ^{66}Zn values in basalts as the unique consequence of carbonate recycling, and demand consideration when interpreting any isotope anomalies of mantle-derived melts. In addition, isotope anomalies of mantle-derived rocks are dependent on the speciation of recycled carbonates. In this case, multiple isotope systematics must be applied in order to determine the possible speciation of recycled carbonates. Finally, the separation of metals and carbon during slab subduction and magma ascent and eruption must be considered when quantifying the fluxes of carbon during deep carbon cycling.

Based on a comprehensive evaluation of all of these processes, we conclude that the light Mg and heavy Zn isotope compositions of the Eastern China basalts relative to the mantle were caused by the entry of recycled Mg- and Zn-rich carbonates into the mantle transition zone through Western Pacific slab subduction.

Acknowledgements

We are grateful to Prof. Ho-Kwang Mao for inviting us to write this paper. We thank Prof. Terry Plank for discussion before we began drafting this manuscript. We also thank three reviewers for their comments and suggestions that helped to improve this paper. Finally, we thank Ze-Zhou Wang and Chun Yang for data collection. This paper is supported by the National Nature Science Foundation of China (41730214 and 41622303) and the Strategic Priority Research Program of the Chinese Academy of Sciences (XDB18030603).

Compliance with ethics guidelines

Sheng-Ao Liu and Shu-Guang Li declare that they have no conflict of interest or financial conflicts to disclose.

References

[1] Dasgupta R. Ingassing, storage, and outgassing of terrestrial carbon through geologic time. Rev Mineral Geochem 2013;75:183-229.

[2] Javoy M, Pineau F, Allègre CJ. Carbon geodynamic cycle. Nature 1982;300 (5888):171.

[3] Dasgupta R, Hirschmann MM. The deep carbon cycle and melting in Earth's interior. Earth Planet Sci Lett 2010;298:1-13.

[4] Hazen RM, Schiffries CM. Why deep carbon? Rev Mineral Geochem 2013;75:1-6.

[5] Duncan MS, Dasgupta R. Rise of Earth's atmospheric oxygen controlled by efficient subduction of organic carbon. Nat Geosci 2017;10:387-92.

[6] Deines P. The carbon isotope geochemistry of mantle xenoliths. Earth Sci Rev 2002;58(3-4):247-78.

[7] Huang S, Farkaš J, Jacobsen SB. Stable calcium isotope compositions of Hawaiian shield lavas: evidence for recycling of ancient marine carbonates into the mantle. Geochim Cosmochim Acta 2011;75(17):4987-97.

[8] Yang W, Teng FZ, Zhang HF, Li SG. Magnesium isotopic systematics of continental basalts from the North China craton: implications for tracing subducted carbonate in the mantle. Chem Geol 2012;328:185-94.

[9] Huang J, Li SG, Xiao Y, Ke S, Li WY, Tian Y. Origin of low δ^{26}Mg Cenozoic basalts from South China Block and their geodynamic implications. Geochim Cosmochim Acta 2015;164:298-317.

[10] Liu SA, Wang ZZ, Li SG, Huang J, Yang W. Zinc isotope evidence for a largescale carbonated mantle beneath Eastern China. Earth Planet Sci Lett 2016;444:169-78.

[11] Tian HC, Yang W, Li SG, Ke S, Chu ZY. Origin of low δ^{26}Mg basalts with EM-I component: evidence for interaction between enriched lithosphere and carbonated asthenosphere. Geochim Cosmochim Acta 2016;188:93-105.

[12] Li SG, Yang W, Ke S, Meng X, Tian H, Xu L, et al. Deep carbon cycles constrained by a large-scale mantle Mg isotope anomaly in Eastern China. Natl Sci Rev 2017;4(1):111-20.

[13] Hofmann AW. A store of subducted carbon beneath Eastern China. Natl Sci Rev 2017;4(1):2.

[14] Su BX, Hu Y, Teng FZ, Xiao Y, Zhou XH, Sun Y, et al. Magnesium isotope constraints on subduction contribution to Mesozoic and Cenozoic East Asian continental basalts. Chem Geol 2017;466:116-22.

[15] Wang XJ, Chen LH, Hofmann AW, Mao FG, Liu JQ, Zhong Y, et al. Mantle transition zone-derived EM1 component beneath NE China: geochemical evidence from Cenozoic potassic basalts. Earth Planet Sci Lett 2017;465:16-28.

[16] Liu D, Zhao Z, Zhu DC, Niu Y, Widom E, Teng FZ, et al. Identifying mantle carbonatite metasomatism through Os-Sr-Mg isotopes in Tibetan ultrapotassic rocks. Earth Planet Sci Lett 2015;430:458-69.

[17] Ke S, Teng FZ, Li SG, Gao T, Liu SA, He Y, et al. Mg, Sr, and O isotope geochemistry of syenites from northwest Xinjiang, China: tracing carbonate recycling during Tethyan oceanic subduction. Chem Geol 2016;437:109-19.

[18] Liu F, Li X, Wang G, Liu Y, Zhu H, Kang J, et al. Marine carbonate component in the mantle beneath the southeastern Tibetan Plateau: evidence from magnesium and calcium isotopes. J Geophys Res Solid Earth 2017;122:9729-44.

[19] Tian HC, Yang W, Li SG, Ke S. Could sedimentary carbonates be recycled into the lower mantle? Constraints from Mg isotopic composition of Emeishan basalts. Lithos 2017;292-293:250-61.

[20] Cheng Z, Zhang Z, Hou T, Santosh M, Chen L, Ke S, et al. Decoupling of Mg-C and Sr-Nd-O isotopes traces the role of recycled carbon in magnesio carbonatites from the Tarim Large Igneous Province. Geochim Cosmochim Acta 2017;202:159-78.

[21] Cheng Z, Zhang Z, Xie Q, Hou T, Ke S. Subducted slab-plume interaction traced by magnesium isotopes in the northern margin of the Tarim Large Igneous Province. Earth Planet Sci Lett 2018;489:100-10.

[22] Wang SJ, Teng FZ, Scott JM. Tracing the origin of continental HIMU-like intraplate volcanism using magnesium isotope systematics. Geochim Cosmochim Acta 2016;185:78-87.

[23] Hoang THA, Choi SH, Yu Y, Pham TH, Nguyen KH, Ryu JS. Geochemical constraints on the spatial distribution of recycled oceanic crust in the mantle source of late Cenozoic basalts, Vietnam. Lithos 2018;296:382-95.

[24] Biscaye PE, Kolla V, Turekian KK. Distribution of calcium carbonate in surface sediments of the Atlantic Ocean. J Geophys Res 1976;81:2595-603.

[25] Brečević L, Nöthig-Laslo V, Kralj D, Popović S. Effect of divalent cations on the formation and structure of calcium carbonate polymorphs. J Chem Soc, Faraday Trans 1996;92:1017-22.

[26] McDonough WF, Sun SS. The composition of the Earth. Chem Geol 1995;120 (3-4):223-53.

[27] Reeder RJ, Lamble GM, Northrup PA. XAFS study of the coordination and local relaxation around Co^{2+}, Zn^{2+}, Pb^{2+}, and Ba^{2+} trace elements in calcite. Am Mineral 1999;84(7-8):1049-60.

[28] Turekian KK, Wedepohl KH. Distribution of the elements in some major units of the earth's crust. Geol Soc Am Bull 1961;72(2):175-92.

[29] Salters VJM, Stracke A. Composition of the depleted mantle. Geochem Geophys Geosyst 2004;5:5.

[30] Li JL, Klemd R, Gao J, Meyer M. Compositional zoning in dolomite from lawsonite-bearing eclogite (SW Tianshan, China): evidence for prograde metamorphism during subduction of oceanic crust. Am Mineral 2014;99 (1):206-17.

[31] Teng FZ, Li WY, Ke S, Marty B, Dauphas N, Huang S, et al. Magnesium isotopic composition of the Earth and chondrites. Geochim Cosmochim Acta 2010;74 (14):4150-66.

[32] Fantle MS, Tipper ET. Calcium isotopes in the global biogeochemical Ca cycle: implications for development of a Ca isotope proxy. Earth Sci Rev 2014;129:148-77.

[33] Pichat S, Douchet C, Albarède F. Zinc isotope variations in deep-sea carbonates from the eastern equatorial Pacific over the last 175 ka. Earth Planet Sci Lett 2003;210:167-78.

[34] Liu SA, Wu H, Shen SZ, Jiang G, Zhang S, Lv Y, et al. Zinc isotope evidence for intensive magmatism immediately before the end-Permian mass extinction. Geology 2017;45:343-6.

[35] Wang ZZ, Liu SA, Liu J, Huang J, Xiao Y, Chu ZY, et al. Zinc isotope fractionation during mantle melting and constraints on the Zn isotope composition of Earth's upper mantle. Geochim Cosmochim Acta 2017;198:151-67.

[36] Chen C, Liu Y, Feng L, Foley SF, Zhou L, Ducea MN, et al. Calcium isotope evidence for subduction-enriched lithospheric

mantle under the northern North China Craton. Geochim Cosmochim Acta 2018;238:55-67.

[37] Ionov DA, Qi YH, Kang JT, Golovin AV, Oleinikov OB, Zheng W, et al. Calcium isotopic signatures of carbonatite and silicate metasomatism, melt percolation and crustal recycling in the lithospheric mantle. Geochim Cosmochim Acta 2019;248(1):1-13.

[38] Kang JT, Ionov DA, Liu F, Zhang CL, Golovin AV, Qin LP, et al. Calcium isotopic fractionation in mantle peridotites by melting and metasomatism and Ca isotope composition of the Bulk Silicate Earth. Earth Planet Sci Lett 2017;474:128-37.

[39] Huang S, Farkaš J, Jacobsen SB. Calcium isotopic fractionation between clinopyroxene and orthopyroxene from mantle peridotites. Earth Planet Sci Lett 2010;292(3-4):337-44.

[40] Kang JT, Zhu HL, Liu YF, Liu F, Wu F, Hao YT, et al. Calcium isotopic composition of mantle xenoliths and minerals from Eastern China. Geochim Cosmochim Acta 2016;174:335-44.

[41] Amini M, Eisenhauer A, Böhm F, Holmden C, Kreissig K, Hauff F, et al. Calcium isotopes ($\delta^{44/40}$Ca) in MPI-DING reference glasses, USGS rock powders and various rocks: evidence for Ca isotope fractionation in terrestrial silicates. Geostand Geoanal Res 2009;33(2):231-47.

[42] Doucet LS, Mattielli N, Ionov DA, Debouge W, Golovin AV. Zn isotopic heterogeneity in the mantle: a melting control? Earth Planet Sci Lett 2016;451:232-40.

[43] Sossi PA, Nebel O, O'Neill HSC, Moynier F. Zinc isotope composition of the Earth and its behaviour during planetary accretion. Chem Geol 2018;477:73-84.

[44] Wang SJ, Teng FZ, Li SG. Tracing carbonate-silicate interaction during subduction using magnesium and oxygen isotopes. Nat Commun 2014;5:5328.

[45] Pokrovsky BG, Mavromatis V, Pokrovsky OS. Co-variation of Mg and C isotopes in late Precambrian carbonates of the Siberian Platform: a new tool for tracing the change in weathering regime? Chem Geol 2011;290 (1):67-74.

[46] Geske A, Zorlu J, Richter D, Buhl D, Niedermayr A, Immenhauser A. Impact of diagenesis and low grade metamorphosis on isotope (δ^{26}Mg, δ^{13}C, δ^{18}O and ^{87}Sr/^{86}Sr) and elemental (Ca, Mg, Mn, Fe and Sr) signatures of Triassic sabkha dolomites. Chem Geol 2012;332:45-64.

[47] Riechelmann S, Buhl D, Schröder-Ritzrau A, Riechelmann D, Richter D, Vonhof H, et al. The magnesium isotope record of cave carbonate archives. Clim Past 2012;8(6):1849-67.

[48] Higgins JA, Schrag DP. Records of Neogene seawater chemistry and diagenesis in deep-sea carbonate sediments and pore fluids. Earth Planet Sci Lett 2012;357:386-96.

[49] Higgins JA, Schrag DP. The Mg isotopic composition of Cenozoic seawater- evidence for a link between Mg-clays, seawater Mg/Ca, and climate. Earth Planet Sci Lett 2015;416:73-81.

[50] Blättler CL, Miller NR, Higgins JA. Mg and Ca isotope signatures of authigenic dolomite in siliceous deep-sea sediments. Earth Planet Sci Lett 2015;419:32-42.

[51] Yang W, Teng FZ, Zhang HF. Chondritic magnesium isotopic composition of the terrestrial mantle: a case study of peridotite xenoliths from the North China Craton. Earth Planet Sci Lett 2009;288(3):475-82.

[52] Bourdon B, Tipper ET, Fitoussi C, Stracke A. Chondritic Mg isotope composition of the Earth. Geochim Cosmochim Acta 2010;74(17):5069-83.

[53] von Strandmann PAP, Elliott T, Marschall HR, Coath C, Lai YJ, Jeffcoate AB, et al. Variations of Li and Mg isotope ratios in bulk chondrites and mantle xenoliths. Geochim Cosmochim Acta 2011;75:5247-68.

[54] Huang F, Zhang Z, Lundstrom CC, Zhi X. Iron and magnesium isotopic compositions of peridotite xenoliths from Eastern China. Geochim Cosmochim Acta 2011;75:3318-34.

[55] Liu SA, Teng FZ, Yang W, Wu F. High-temperature inter-mineral magnesium isotope fractionation in mantle xenoliths from the North China Craton. Earth Planet Sci Lett 2011;308(1):131-40.

[56] Xiao Y, Teng FZ, Zhang HF, Yang W. Large magnesium isotope fractionation in peridotite xenoliths from eastern North China Craton: product of melt-rock interaction. Geochim Cosmochim Acta 2013;115:241-61.

[57] An Y, Huang JX, Griffin W, Liu C, Huang F. Isotopic composition of Mg and Fe in garnet peridotites from the Kaapvaal and Siberian Cratons. Geochim Cosmochim Acta 2017;200:167-85.

[58] Huang J, Chen S, Zhang XC, Huang F. Effects of melt percolation on Zn isotope heterogeneity in the mantle: constraints from peridotite massifs in Ivrea-Verbano Zone, Italian Alps. J Geophys Res Solid Earth 2018;123 (4):2706-22.

[59] Hulett SRW, Simonetti A, Rasbury ET, Hemming NG. Recycling of subducted crustal components into carbonatite melts revealed by boron isotopes. Nat Geosci 2016;9:904-8.

[60] Fantle MS, DePaolo DJ. Variations in the marine Ca cycle over the past 20 million years. Earth Planet Sci Lett

2005;237(1):102-17.
[61] Higgins JA, Schrag DP. Constraining magnesium cycling in marine sediments using magnesium isotopes. Geochim Cosmochim Acta 2010;74 (17):5039-53.
[62] Le Roux V, Lee CA, Turner SJ. Zn/Fe systematics in mafic and ultramafic systems: implications for detecting major element heterogeneities in the Earth's mantle. Geochim Cosmochim Acta 2010;74(9):2779-96.
[63] Wang SJ, Teng FZ, Li SG, Hong JA. Magnesium isotopic systematics of mafic rocks during continental subduction. Geochim Cosmochim Acta 2014;143:34-48.
[64] Liu SA, Teng FZ, He YS, Ke S, Li SG. Investigation of magnesium isotope fractionation during granite differentiation: implication for Mg isotopic composition of the continental crust. Earth Planet Sci Lett 2010;297:646-54.
[65] Teng FZ, Li WY, Rudnick RL, Gardner LR. Contrasting lithium and magnesium isotope fractionation during continental weathering. Earth Planet Sci Lett 2010;300(1):63-71.
[66] Pons ML, Debret B, Bouilhol P, Delacour A, Williams H. Zinc isotope evidence for sulfate-rich fluid transfer across subduction zones. Nat Commun 2016;7:13794.
[67] Inglis EC, Debret B, Burton KW, Millet MA, Pons ML, Dale CW, et al. The behaviour of iron and zinc stable isotopes accompanying the subduction of mafic oceanic crust: a case study from Western Alpine Ophiolites. Geochem Geophys Geosyst 2017;18(7):2562-79.
[68] Little SH, Vance D, McManus J, Severmann S. Key role of continental margin sediments in the oceanic mass balance of Zn and Zn isotopes. Geology 2016;44:207-10.
[69] Chen H, Savage PS, Teng FZ, Helz RT, Moynier F. Zinc isotope fractionation during magmatic differentiation and the isotopic composition of the bulk Earth. Earth Planet Sci Lett 2013;369:34-42.
[70] Sun Y, Teng F, Ying JF, Su BX, Hu Y, Fan QC, et al. Magnesium isotopic evidence for ancient subducted oceanic crust in LOMU-like potassium-rich volcanic rocks. J Geophys Res Solid Earth 2017;122(10):7562-72.
[71] Kim JI, Choi SH, Koh GW, Park JB, Ryu JS. Petrogenesis and mantle source characteristics of volcanic rocks on Jeju Island, South Korea. Lithos 2019;326- 327:476-90.
[72] Tian HC, Yang W, Li SG, Ke S, Duan XZ. Low δ^{26}Mg volcanic rocks of Tengchong in Southwestern China: a deep carbon cycle induced by supercritical liquids. Geochim Cosmochim Acta 2018;240:191-219.
[73] Wang ZZ, Liu SA, Chen LH, Li SG, Zeng G. Compositional transition in natural alkaline lavas through silica-undersaturated melt-lithosphere interaction. Geology 2018;46(9):771-4.
[74] Wang XJ, Chen LH, Hofmann AW, Hanyu T, Kawabata H, Zhong Y, et al. Recycled ancient ghost carbonate in the Pitcairn mantle plume. Proc Natl Acad Sci USA 2018;115(35):8682-7.
[75] Debret B, Bouilhol P, Pons ML, Williams H. Carbonate transfer during the onset of slab devolatilization: new insights from Fe and Zn stable isotopes. J Petrol 2018;59(6):1145-66.
[76] Shen J, Li SG, Wang SJ, Teng FZ, Li QL, Liu YS. Subducted Mg-rich carbonates into the deep mantle wedge. Earth Planet Sci Lett 2018;503:118-30.
[77] Zhao X, Zhang Z, Huang S, Liu Y, Li X, Zhang H. Coupled extremely light Ca and Fe isotopes in peridotites. Geochim Cosmochim Acta 2017;208:368-80.
[78] Schiller M, Paton C, Bizzarro M. Calcium isotope measurement by combined HR-MC-ICPMS and TIMS. J Anal At Spectrom 2012;27:38-49.
[79] Zhao D, Yu S, Ohtani E. East Asia: seismotectonics, magmatism and mantle dynamics. J Asian Earth Sci 2011;40(3):689-709.
[80] Green DH. Experimental melting studies on a model upper mantle composition at high pressure under water-saturated and waterundersaturated conditions. Earth Planet Sci Lett 1973;19(1):37-53.
[81] Jaques AL, Green DH. Anhydrous melting of peridotite at 0-15 kb pressure and the genesis of tholeiitic basalts. Contrib Mineral Petrol 1980;73 (3):287-310.
[82] Teng FZ. Magnesium isotope geochemistry. Rev Mineral Geochem 2017;82 (1):219-87.
[83] Handler MR, Baker JA, Schiller M, Bennett VC, Yaxley GM. Magnesium stable isotope composition of Earth's upper mantle. Earth Planet Sci Lett 2009;282 (1-4):306-13.
[84] Dauphas N, Teng FZ, Arndt NT. Magnesium and iron isotopes in 2.7 Ga Alexo komatiites: mantle signatures, no evidence for Soret diffusion, and identification of diffusive transport in zoned olivine. Geochim Cosmochim Acta 2010;74(11):3274-91.
[85] Zhong Y, Chen LH, Wang XJ, Zhang GL, Xie LW, Zeng G. Magnesium isotopic variation of oceanic island basalts generated by partial melting and crustal recycling. Earth Planet Sci Lett 2017;463:127-35.

[86] Zhang J, Liu Y, Ling W, Gao S. Pressure-dependent compatibility of iron in garnet: insights into the origin of ferropicritic melt. Geochim Cosmochim Acta 2017;197:356-77.

[87] Huang J, Zhang XC, Chen S, Tang L, Wörner G, Yu H, et al. Zinc isotopic systematics of Kamchatka-Aleutian arc magmas controlled by mantle melting. Geochim Cosmochim Acta 2018;238:85-101.

[88] Caroff M, Maury RC, Guille G, Cotten J. Partial melting below Tubuai (Austral Islands, French Polynesia). Contrib Mineral Petrol 1997;127:369-82.

[89] Dasgupta R, Hirschmann MM. Melting in the Earth's deep upper mantle caused by carbon dioxide. Nature 2006;440(7084):659-62.

[90] Li SG, Wang Y. Formation time of the big mantle wedge beneath eastern China and a new lithospheric thinning mechanism of the North China Craton-geodynamic effects of deep recycled Carbon. Sci China Earth Sci 2018;61:853-68.

[91] Sobolev AV, Hofmann AW, Kuzmin DV, Yaxley GM, Arndt NT, Chung SL, et al. The amount of recycled crust in sources of mantle-derived melts. Science 2007;316:412-7.

[92] Liu JQ, Ren ZY, Nichols ARL, Song MS, Qian SP, Zhang Y, et al. Petrogenesis of Late Cenozoic basalts from north Hainan Island: constraints from melt inclusions and their host olivines. Geochim Cosmochim Acta 2015;152:89-121.

[93] Pan D, Spanu L, Harrison B, Sverjensky DA, Galli G. Dielectric properties of water under extreme conditions and transport of carbonates in the deep Earth. Proc Natl Acad Sci USA 2013;110(17):6646-50.

[94] Frezzotti ML, Selverstone J, Sharp ZD, Compagnoni R. Carbonate dissolution during subduction revealed by diamond-bearing rocks from the Alps. Nat Geosci 2011;4:703-6.

[95] Kushiro I, Satake H, Akimoto S. Carbonate-silicate reactions at high pressures and possible presence of dolomite and magnesite in upper mantle. Earth Planet Sci Lett 1975;28:116-20.

[96] Boulard E, Pan D, Galli G, Liu Z, Mao WL. Tetrahedrally coordinated carbonates in Earth's lower mantle. Nat Commun 2015;6:6311.

[97] Longpré MA, Stix J, Klügel A, Shimizu N. Mantle to surface degassing of carbon-and sulphur-rich alkaline magma at El Hierro. Canary Islands. Earth Planet Sci Lett 2017;460:268-80.

[98] Boudoire G, Rizzo AL, Di Muro A, Grassa F, Liuzzo M. Extensive CO_2 degassing in the upper mantle beneath oceanic basaltic volcanoes: first insights from Piton de la Fournaise volcano (La Réunion Island). Geochim Cosmochim Acta 2018;235:376-401.

[99] Aubaud C, Pineau F, Hékinian R, Javoy M. Degassing of CO_2 and H_2O in submarine lavas from the Society hotspot. Earth Planet Sci Lett 2005;235:511-27.

[100] Barry PH, Hilton DR, Füri E, Halldórsson SA, Grönvold K. Carbon isotope and abundance systematics of Icelandic geothermal gases, fluids and subglacial basalts with implications for mantle plume-related CO_2 fluxes. Geochim Cosmochim Acta 2014;134:74-99.

[101] Richter FM, Watson EB, Mendybaev RA, Teng FZ, Janney PE. Magnesium isotope fractionation in silicate melts by chemical and thermal diffusion. Geochim Cosmochim Acta 2008;72(1):206-20.

[102] Dominguez G, Wilkins G, Thiemens MH. The Soret effect and isotopic fractionation in high-temperature silicate melts. Nature 2011;473:70.

[103] Huang F, Chakraborty P, Lundstrom CC, Holmden C, Glessner JJG, Kieffer SW, et al. Isotope fractionation in silicate melts by thermal diffusion. Nature 2010;464:396-400.

[104] Teng FZ, Dauphas N, Helz RT, Gao S, Huang S. Diffusion-driven magnesium and iron isotope fractionation in Hawaiian olivine. Earth Planet Sci Lett 2011;308(3-4):317-24.

[105] Sio CKI, Dauphas N, Teng FZ, Chaussidon M, Helz R, Roskosz M. Discerning crystal growth from diffusion profiles in zoned olivine by in situ Mg-Fe isotopic analyses. Geochim Cosmochim Acta 2013;123:302-21.

[106] Lai YJ, von Strandmann PAP, Dohmen R, Takazawa E, Elliott T. The influence of melt infiltration on the Li and Mg isotopic composition of the Horoman Peridotite Massif. Geochim Cosmochim Acta 2015;164:318-32.

Approach to trace hidden paleo-weathering of basaltic crust through decoupled Mg-Sr and Nd isotopes recorded in volcanic rocks*

Heng-Ci Tian[1,2], Wei Yang[1,2], Shu-Guang Li[3], Hai-Quan Wei[4], Zhuo-Sen Yao[5], Shan Ke[3]

1. Key Laboratory of Earth and Planetary Physics, Institute of Geology and Geophysics, Chinese Academy of Sciences, Beijing 100029, China
2. Institutions of Earth Science, Chinese Academy of Sciences, Beijing 100029, China
3. State Key Laboratory of Geological Processes and Mineral Resources, China University of Geosciences, Beijing 100083, China
4. Institute of Geology, China Earthquake Administration, Beijing 100029, China
5. Key Laboratory of Mineral Resources, Institute of Geology and Geophysics, Chinese Academy of Sciences, Beijing 100029, China

亮点介绍： 研究玄武岩的古风化壳是我们重建古气候以及预测未来气候变化的关键所在，但这方面的研究鲜有报道。主要是因为它常被新形成的火山岩和沉积物覆盖以及受后期地质作用地改造影响。本研究创新性地对东北长白山火山锥下部层位的玄武岩和上部层位的中酸性岩石开展研究，这是因为如果早期硅酸盐经历过化学风化，那么晚期中酸性火山岩在其岩浆上升和侵位过程中极有可能通过混染记录该事件。结果表明粗面岩与玄武岩 Sr-Mg 与 Nd 同位素解耦，前者具有明显重的 Mg 同位素组成。进一步研究表明粗面岩在形成过程中混染了早期经过强烈风化的玄武岩风化产物。这是因为玄武岩风化过程中，残余物的 Sr 和 Mg 同位素会逐渐变重，而 Nd 同位素却几乎保持不变。本文提出了一种示踪"隐藏"古风化壳的新方法。

Abstract On Earth, chemical weathering of basaltic rocks not only modulates the climate change, but also plays a crucial role on the ocean oxygenation and ecosystem evolution. Records of paleo-weathered crusts are a challenge to identify because they are obscured by subsequent alteration processes and buried by the newly formed volcanic lavas. However, late formed volcanic rocks theoretically could record previously weathered crust via their assimilation and fractional crystallization (AFC) processes. Here, we present a systematic study on major, trace elements and Sr-Nd-Mg isotopes in a suite of Changbaishan early formed basalts and late erupted trachytes from Northeast China. Detailed evidence of field observation, geochemical signature and rhyolite-MELTS simulations suggest that trachytes mainly originated from deep fractional crystallization of basaltic melts. The low abundant trachytes from the stratigraphic layer above the basalts display decoupled Mg-Sr and Nd isotopes. δ^{26}Mg and ^{87}Sr/^{86}Sr in trachytes range from −0.27 up to +0.94‰ (the heaviest isotopic composition ever reported

* 本文发表在：Chemical Geology, 2019; 509: 234-248

for Mg in the igneous rocks) and 0.70490 to 0.71065, respectively, which are significantly elevated relative to that of their mantle source. However, they preserved mantle source-like $^{143}Nd/^{144}Nd$ (0.512584 to 0.512656) ratios. These unusual characteristics cannot be explained by thermal diffusion, fractional crystallization or simple crustal contamination processes. Specifically, $\delta^{26}Mg$ correlates positively with SiO_2 and $^{87}Sr/^{86}Sr$, and negatively with MgO, CaO, Al_2O_3, Sr and Nb/U indicators. The highly elevated Mg and Sr isotopes but limited Nd isotopic changes in trachytes require a component to be the relatively younger weathering residue of basaltic rocks. Taking this component as a wall rock endmember, the well-developed above correlations can be best modelled by AFC processes. This reflects that some trachytes in the aftermath of AFC processes almost fully lost their primary source characteristics, but inherited the intense chemical weathering signature of basalts. This event probably occurred in the middle Miocene as studies of paleosols showed humid and warm climate during that period after the initial eruption of Changbaishan basalts. Therefore, our study as a specific case for the first time provides an efficient way to explore paleo-weathering of basaltic crusts buried in the Earth's surface, through comprehensive investigating Mg, Sr and Nd isotopes of volcanic rocks developed in the late stage of a volcano cone.

Keywords　Hidden paleo-weathering of crust, Sr-Nd-Mg isotopes, AFC processes, chemical weathering, Changbaishan volcanic rocks

1　Introduction

Chemical weathering of silicates, especially the large continental flood basalt provinces around the world, serves as a long-term sink of atmospheric CO_2 (Louvat and Allègre, 1997, 1998; Dessert et al., 2001, 2003). It regulates climate through the greenhouse effect of CO_2 over geological time, providing a negative climate feedback (e.g. Berner et al., 1983; Dessert et al., 2003; Ibarra et al., 2016; Li et al., 2016a). For example, previous studies have demonstrated that basaltic rocks with faster weathering rate than other silicates account for as much as 30 to 35% of modern CO_2 consumption by global silicate weathering, while they are estimated to occupy less than ~5% of the global land surface (e.g. Gaillardet et al., 1999; Dessert et al., 2003; Hartmann et al., 2009; Hartmann and Moosdorf, 2012; Moon et al., 2014; Balagizi et al., 2015). Additionally，their weathering rates are high in tropical and subtropical environment, and gradually decrease towards middle- and high-latitude (e.g. Kent and Muttoni, 2013; Balagizi et al., 2015; Dessert et al., 2015). This implies that basalt weathering plays a significant role for the long-term global carbon cycle. However, most studies on continental basalts weathering focus on the modern environment (Dessert et al., 2003; Li et al., 2016a). The paleo-weathered basaltic events received much less attention, which have been linked directly or indirectly to the paleoclimate change, ocean oxygenation, ecosystem changes or mass extinction (Dessert et al., 2003; Hartmann and Moosdorf, 2012; Sun et al., 2018; Yang et al., 2018). Knowledge of the past variability of climate in continental crust could be obtained through directly studying the paleo-weathered crust, which can help to inform our understanding of past climate change and provide the baseline for ongoing and future climate change. However, one of the main difficulties of studying basaltic rocks paleo-weathering is that weathered crusts were always covered by the newly formed volcanic lavas or sediments and were also modified by later geological processes. Many volcanoes worldwide not only have accumulated a lot of basaltic lavas in the deeper region, but also commonly accompanied with small amounts of intermediate-felsic rocks at the shallow part (e.g. Chung and Jahn, 1995; Bryan and Ernst, 2008; Dufek and Bachmann, 2010). Theoretically, the late formed intermediate-felsic rocks can record wallrocks information via crustal contamination processes during magma ascent and emplacement. More importantly, selecting sensitive proxies are the keys to ascertain the nature and intensity of past silicates weathering events.

Magnesium is a water-soluble major element that occurs in continental crust and mantle (e.g. McDonough and Sun, 1995; Rudnick and Gao, 2003). Mg can easily fractionate at low-temperature, while Mg will minimally fractionate in high-temperature magmas (e.g. Teng, 2017 and references therein). For example, during low-temperature continental weathering, the residual of sedimentary rocks (e.g. shales, pelites, and loess) and silicate rocks generally enrich heavy Mg isotopes relative to the un-weathered protolith, making the light Mg isotopes enter into the hydrosphere (Li et al., 2010; Opfergelt et al., 2012; Liu et al., 2014; Wimpenny et al., 2014). Furthermore, during the chemical weathering, Sr behaves much more active than Nd and is easily removed from the primary rocks (e.g. Nesbitt et al., 1980; McCulloch et al., 1981; Ma et al., 2007). During chemical weathering, significant changes of $^{87}Sr/^{86}Sr$ have been reported by most studies (e.g. McCulloch et al., 1981; Ma et al., 2010). This can be attributed to differential decomposition of minerals containing heterogeneous $^{87}Sr/^{86}Sr$ ratios (Blum and Erel, 1995; Negrel, 2006) or input of extraneous Sr with high $^{87}Sr/^{86}Sr$ ratios from aeolian deposits or groundwater (Price et al., 1991; Stewart et al., 2001; Kurtz et al., 2001; Dia et al., 2006; Ma et al., 2010). In contrast, due to the much high atomic numbers and weak immobility, Nd isotopes are unlikely to be fractionated by incongruent weathering unless some aeolian dusts involved into the weathering saprolite (Ma et al., 2010; Liu et al., 2013). Additionally, the best efficient approach to explore paleo-weathered crusts may be via investigating the late-erupted intermediate-felsic lavas laying on top of a volcano cone. That is because these volcanic lavas will inevitably assimilate the early erupted basalts during their ascent (Depaolo, 1981), and generally have low volume relative to early erupted volcanic lavas. Specifically, if the early basalts experienced intense weathering, it will be inherited by the late formed volcanic lavas.

The volcanic lavas exposed in Changbaishan and surrounding areas from Northeast (NE) China (Fig. 1a), are an excellent natural laboratory to resolve this issue. That is because (1) Changbaishan stratovolcano is composed of multi-stage eruptions from ~20 Ma until one thousand years ago (Zhang et al., 2018 and references therein) and (2) previous studies demonstrated that East Asia experienced several strong summer monsoon and tropical-like climate during the Miocene (e.g. Zou et al., 2004; Sun et al., 2015; Meng et al., 2018). In this paper, we investigate the Sr-Nd-Mg isotopes and trace elements of twelve early erupted alkali basalts and fourteen late-erupted trachytes (Fig. 1b), to understand how the paleo-weathered basaltic crust affects isotopic behaviors, to reveal the paleoclimate change in NE China and more importantly to obtain a general pathway to study the paleo-weathered crust.

2 Geological setting and samples

The Changbaishan volcanic area is located in NE China, between the Japan Sea back-arc basin and the Songliao Basin (Fig. 1a) and belongs to active Cenozoic volcanoes in China (e.g. Liu, 1999, 2000; Wei et al., 2003). Many faults have occurred in this area, such as Mishan-Dunhua Fault, Tianchi-Tumenjiang Fault and Tianchi-Baishan Fault (Zhang et al., 2002). Three large volcanic systems compose the Changbaishan volcanic area: Tianchi (China-North Korea), Wangtian'e (China) and Baotaishan (North Korea). The Changbaishan volcanic area is composed of an early basaltic shield (ca. ~2.8 to 0.31 Ma), middle trachyte composite cone (ca. 1.1 to 0.1 Ma) and the final eruption of the Holocene ignimbrite (<0.1 Ma) rocks (or alkali rhyolites) based on the Ar-Ar, K-Ar and U-Th dating (e.g. Liu and Wang, 1982; Wei et al., 2003, 2007; Fan et al., 2006; Zou et al., 2010, 2014; Liu et al., 2015; Ramos et al., 2016), on top of Archean metamorphic rocks and Mesozoic granites. This basaltic shield is composed of almost entirely of lavas flows. Based on the statistics, the lava shield covers an area of ~ 20,000 km^2, extending to 50 and 100 km in different directions (Wei et al., 2003).

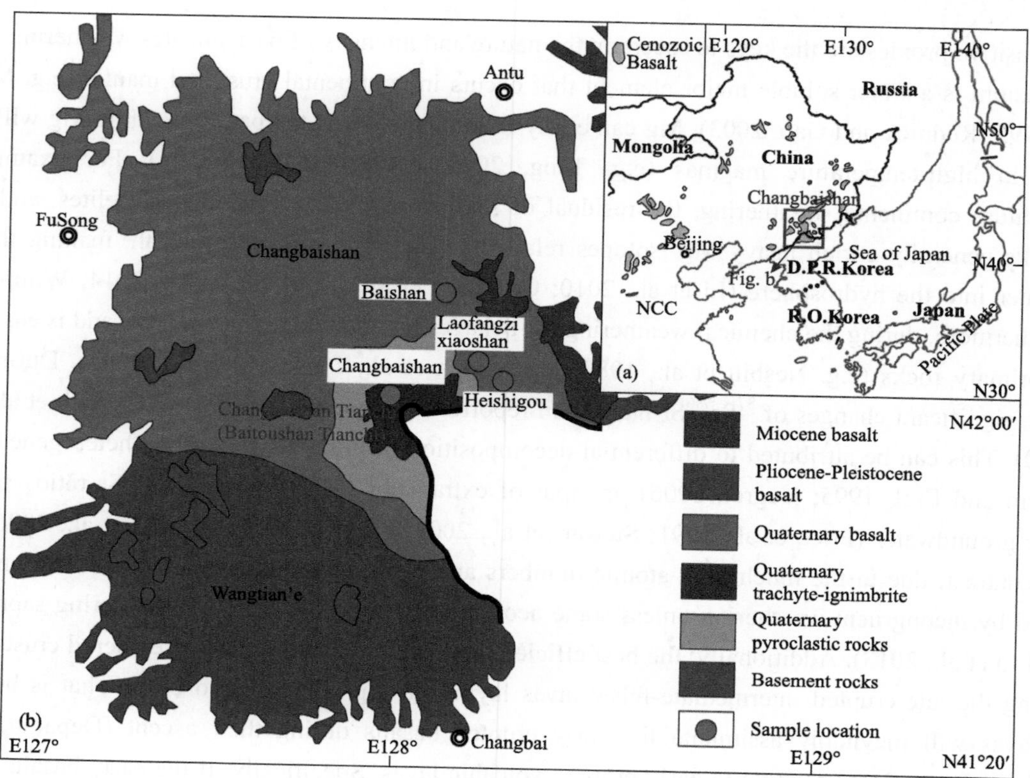

Fig. 1 (a) Map showing the location of Changbaishan and the distribution of Cenozoic volcanic rocks in Northeast China (modified from Wei et al., 2003). (b) The distribution of Cenozoic basalts and trachyte-ignimbrite in Changbaishan area (modified from Fan et al., 2006 and Guo et al., 2014). Four orange circles represent the locations of basalts and trachytes collected in this study.

There are still controversies over the origin of the volcanic rocks. Cenozoic magmatic activity in NE China has been interpreted as the result of decompression melting of the asthenosphere (e.g. Zou et al., 2003, 2008; Zhao et al., 2009; Kuritani et al., 2011; Tang et al., 2014; Choi et al., 2017). This explanation is supported by the high-resolution seismic tomography of a plume-like shape beneath the Changbaishan area (Zhao et al., 2009; Tang et al., 2014). Many models plausibly explain the slightly enriched Sr-Nd-Pb isotopes in Changbaishan basalts, including metasomatism from ancient stagnant slab (Kuritani et al., 2011), subduction of the Pacific slab in the mantle transition zone (410–660 km; Zhao et al., 2009), assimilated crustal materials (Zou et al., 2008) or the enriched lithospheric mantle (Chen et al., 2007). In contrast, most previous studies have suggested that trachytes formed from the prolonged fractional crystallization of minerals (olivine ± pyroxene ± plagioclase) from the alkali basalts, together with assimilation of some crustal materials (e.g. Xie et al., 1988; Liu et al., 1998, 2015; Li et al., 2004; Guo et al., 2015, 2016; Park et al., 2016).

In this study, twelve alkali basalts (ca. 1.91 to 0.18 Ma) and fourteen trachytes (ca. 1.14 to 0.12 Ma) (Table 1) were collected from Changbaishan volcanic area (Fig. 1b). Their lithology and major elements are reported by previous studies (Wei et al., 2007; Sun et al., 2008; Chen, 2013). Except for one sample with slightly high LOI (2.57 wt.%), all the other basalts and trachytes display low LOI ranging from –0.39 to 0.95 wt.% (Table 1), reflecting they are not affected by alteration. These alkali basalts are mainly porphyritic or holocrystalline, containing olivine (<5%), augite (3–5%) and plagioclase (<15%) phenocrysts (Chen, 2013). They have low MgO (3.60–5.53 wt.%), high SiO_2 (48.67–52.14 wt.%), Na_2O (3.64–4.05 wt.%) and K_2O (1.88–2.34 wt.%) contents. In contrast, the trachytes have characteristically trachytic texture, with the phenocrysts of plagioclase (10–30%) and minor pyroxene (<3%). They are characterized by low MgO (0.05–1.02 wt.%) and CaO (0.55–2.40 wt.%), variable Al_2O_3 (12.25–17.91 wt.%) and total FeO (4.81–7.21 wt.%) (Chen, 2013).

3 Analytical Methods

3.1 Trace elemental analyses

The chemical procedure was conducted in an ultra-clean laboratory at the State Key Laboratory of Geological Process and Mineral Resources, China University of Geosciences (Beijing) (CUGB). Trace element concentrations were determined with inductively coupled plasma mass spectrometry (ICP-MS) using an *ELEMENT* instrument at the Institute of Geology and Geophysics, Chinese Academy of Sciences (IGGCAS). Two reference materials GSR1 and GSR3 were also analyzed to monitor the entire chemical procedure and instrument stability. The measured values of GSR1 and GSR3 are consistent with the recommended values (Table 2). Precision and accuracy is better than 5% for most of the trace elements.

3.2 Sr-Nd-Mg isotopic analyses

Whole-rock Sr-Nd isotopic compositions were analyzed at IGGCAS using a *Triton Plus* thermal ionization mass spectrometer (TIMS) instrument, following the previously established methods (Li et al., 2015, 2016b). Sr and Nd were separated using the classical two-step ion exchange chromatographic method. Due to the high Rb/Sr (>6) in some samples (008-7, I-15-1, I-25-1, 2-15-2, 2(1)S, I-7-2, I-8-1, I-8-2 and I-101-1), the same chemical separation column was achieved twice. Isotopic ratios corrected for mass fractionation were normalized to $^{86}Sr/^{88}Sr = 0.1194$ and $^{146}Nd/^{144}Nd = 0.7219$, respectively. The standard materials NBS-987 and JNdi-1, were also used to evaluate the instrument stability during data collection. The results were $^{87}Sr/^{86}Sr=0.710256\pm0.000020$ (2SD, n=2) for NBS-987 and $^{143}Nd/^{144}Nd=0.512112\pm0.000007$ (2SD, n=2) for JNdi-1. Basalt standard (BCR-2) was also analyzed and yielded $^{87}Sr/^{86}Sr=0.705000\pm0.000014$ (2σ) and $^{143}Nd/^{144}Nd=0.512652\pm0.000008$ (2σ). These results are in good agreement with recommended values on GeoREM (http://georem.mpch-mainz.gwdg.de/).

Magnesium isotope analyses were carried out using a *Neptune* multi collector inductively coupled plasma mass spectrometer (MC-ICP-MS) at CUGB, based on the method described in previous studies (e.g. Teng et al., 2010; Ke et al., 2016). For chemical procedure, powdered samples in Savillex beakers were totally dissolved in concentrated HF-HNO_3-HCl. All dried sample residues were dissolved in 1 N HNO_3 for Mg separation. Samples containing ~10 μg Mg were separated by cation exchange chromatography using Bio-Rad 200–400 mesh AG50W-X8 pre-clean resin and eluted using 1 N HNO_3 at a rate of 1 mL/30 min. Considering higher Na_2O and Al_2O_3 content than MgO content in some of the trachytes (samples I-15-1, 2(1)S, I-7-2, I-8-2 and I-101-1), the same chemical separation column was achieved three times. This was also applied to the two standards of AGV-2 and GSP-2. After three passes, the abundances of Ti, Al, Fe, Ca, Na and K relative to Mg in each sample solution was measured and the cation/Mg (mass/mass) ratios were less than 0.05. The Mg yield after three passes was > 99.0%.

Purified Mg solutions were analyzed with the sample-standard bracketing method to correct for instrumental mass bias. The long-term reproducibility based on analyses of natural materials and synthetic pure Mg solution is better than 0.06‰ for $\delta^{26}Mg$ and 0.05‰ for $\delta^{25}Mg$ (Ke et al., 2016). The Mg isotopes are expressed as $\delta^{25}Mg$ and $\delta^{26}Mg$, where $\delta^{25,26}Mg_{sample} = [(^{25,26}Mg/^{24}Mg)_{sample}/(^{25,26}Mg/^{24}Mg)_{DSM3}-1] \times 1000$ (‰), where DSM3 is a solution of pure Mg metal (Galy et al., 2003). The data is presented in Table 3. The $\Delta^{25}Mg'$ ($\Delta^{25}Mg'=\delta^{25}Mg'-0.521\times\delta^{26}Mg'$, where $\delta^{25,26}Mg'=1000\times\ln[(\delta^{25,26}Mg+1000)/1000]$; Young and Galy, 2004) values are within 0±0.05‰ (Table 3). Thus, only the $\delta^{26}Mg$ values will be used in the discussion. Four USGS rock standards analyzed during the course of this study yielded $\delta^{26}Mg$ values of −0.27±0.06‰ for BHVO-2, −0.20±0.09‰ for BCR-2, −0.15±0.04‰ for AGV-2 (2SD, n=2) and +0.05±0.06‰ for GSP-2 (2SD, n=2), which are consistent with those reported in previous studies (e.g. Teng et al., 2007, 2015; Opfergelt et al., 2012; An et al., 2014; Ke et al., 2016).

4 Results

4.1 Trace elements

Trace elements of the Changbaishan basalts and trachytes are presented in Table 2. The basalts are enriched in light rare earth elements (LREEs) relative to heavy rare earth elements and display OIB-like REE patterns (Fig. 2a). In the primitive mantle-normalized element spider diagram, these basalts show positive Ba, K and Pb, as well as negative U anomalies (Fig. 2b). In contrast, the trachytes also display enrichment in LREEs relative to heavy rare earth elements (Fig. 2c). And most of them have higher total REE abundances (158.3–225.2 ppm) than that of the OIB, as well as no or strong Eu anomaly (δEu = $Eu_N/(Sm_N \times Gd_N)^{0.5}$ = 0.04–1.58, where subscript N denotes normalization to chondritic value (Sun and McDonough, 1989)) (Fig. 2c). Furthermore, the trachytes display weakly to extremely negative Sr, P and Ti anomalies, and weakly positive to negative Ba and Eu anomalies (Fig. 2d), probably reflecting significant minerals fractional crystallization.

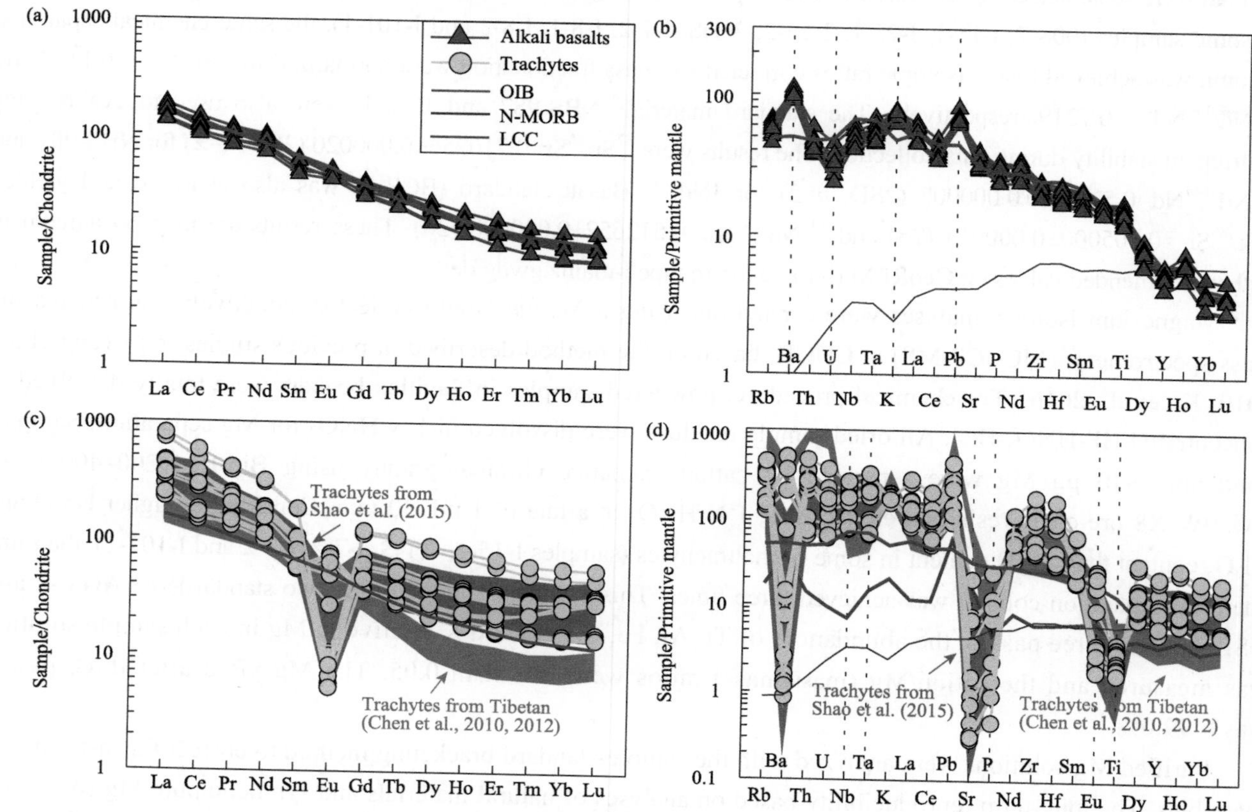

Fig. 2　Chondrite-normalized rare earth element (REE) diagram (a and c) and primitive mantle-normalized trace patterns (b and d) for the Changbaishan basalts and trachytes in this study. The primitive mantle and chondrite values are from Sun and McDonough (1989). Data source: OIB (ocean island basalt) and N-MORB (mid-ocean-ridge basalt) from Sun and McDonough (1989) and LCC (lower continental crust) is from Rudnick and Gao (2003). The colorful area represents trachytes from Tibetan (Chen et al., 2010, 2012) and southeast Queensland, Australia (Shao et al., 2015) for comparison.

4.2 Isotopic geochemistry

4.2.1 Sr and Nd isotopic compositions

Strontium and Nd isotopic data of the measured samples, as well as reference standard materials are presented in Table 3. The initial $^{87}Sr/^{86}Sr$ and $^{143}Nd/^{144}Nd$ ratios in the Changbaishan basalts and trachytes are corrected based on their ages (Wei et al., 2007; Sun et al., 2008) and Rb/Sr ratios, although they are very young. As a whole, the basalts have slightly enriched and limited variations in $^{87}Sr/^{86}Sr$ (0.70493–0.70506) and $^{143}Nd/^{144}Nd$ (0.512584–0.512619) ratios (Fig. 3), possibly reflecting the mixing trend between depleted mantle and EM-I component (Basu et al., 1991). In contrast, the trachytes have decoupled Sr-Nd isotopic compositions. They show a large $^{87}Sr/^{86}Sr$ range varying from 0.70490 to 0.71065, but display Changbaishan basalts-like $^{143}Nd/^{144}Nd$ ratios varying from 0.512584 to 0.512656 (Fig. 3).

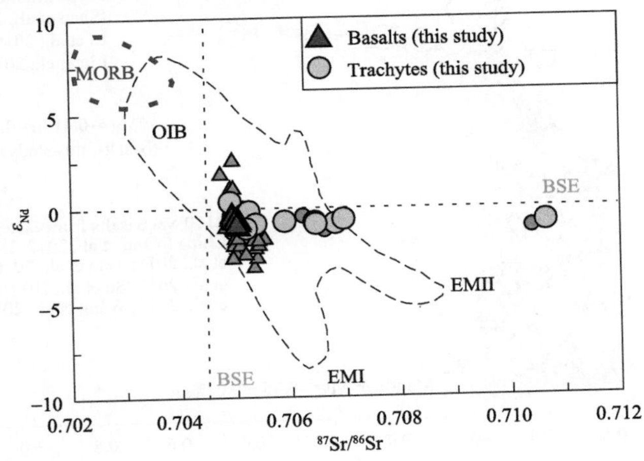

Fig. 3 $^{87}Sr/^{86}Sr$ vs. ε_{Nd} for the Changbaishan volcanic rocks. The grey triangle and circle represent the Changbaishan basalts and trachytes from literature, respectively. Data sources: isotopic endmember of the MORB and OIB are based on Zindler and Hart (1986). The compiled basalts and trachytes are from Xie et al. (1988), Fan et al. (2007), Zou et al. (2008), Kuritani et al. (2009), Guo et al. (2014, 2016) and Ma et al. (2015).

4.2.2 Mg isotopic composition

Compared with the terrestrial mantle ($\delta^{26}Mg$=–0.25±0.07‰; Teng et al., 2010), Changbaishan basalts studied here have lower $\delta^{26}Mg$ and limited variation from –0.41 to –0.34‰ (Table 3). As shown in Fig. 4, these newly analyzed values are similar to the previously published data on Changbaishan basalts (–0.51 to –0.42 ‰) and most of <110 Ma basalts from other parts of eastern China ($\delta^{26}Mg$ = –0.60 to 0.23‰) reported by Yang et al. (2012), Huang et al. (2015), Tian et al. (2016), Li et al. (2017), Su et al. (2017), Sun et al. (2017) and Wang et al. (2017).

However, $\delta^{26}Mg$ in the trachytes varies between –0.27 and +0.94‰ (Table 3). This exceeds the range measured for terrestrial igneous rocks or minerals reported thus far (Fig. 4). For example, their $\delta^{26}Mg$ values are heavier than I-, S- and A-type granitoids (–0.40 to +0.44‰, Shen et al., 2009; Li et al., 2010; Liu et al., 2010), syenites and syenogranites (–0.46 to –0.17‰, Ke et al., 2016). The most positive value is up to 1.1‰ heavier than those of the terrestrial mantle values ($\delta^{26}Mg$=–0.25±0.07‰, Teng et al., 2010) and 1.2‰ heavier than those of the underlying Changbaishan basalts (Fig. 4). Using previously published major and newly analyzed trace element and Sr-Nd data, we find that $\delta^{26}Mg$ inversely correlates with MgO, Al_2O_3, CaO, TiO_2, Sr (ppm), Nb/U and Zr/Hf values, but positively correlates with whole-rock SiO_2 and $^{87}Sr/^{86}Sr$ (Fig. 5).

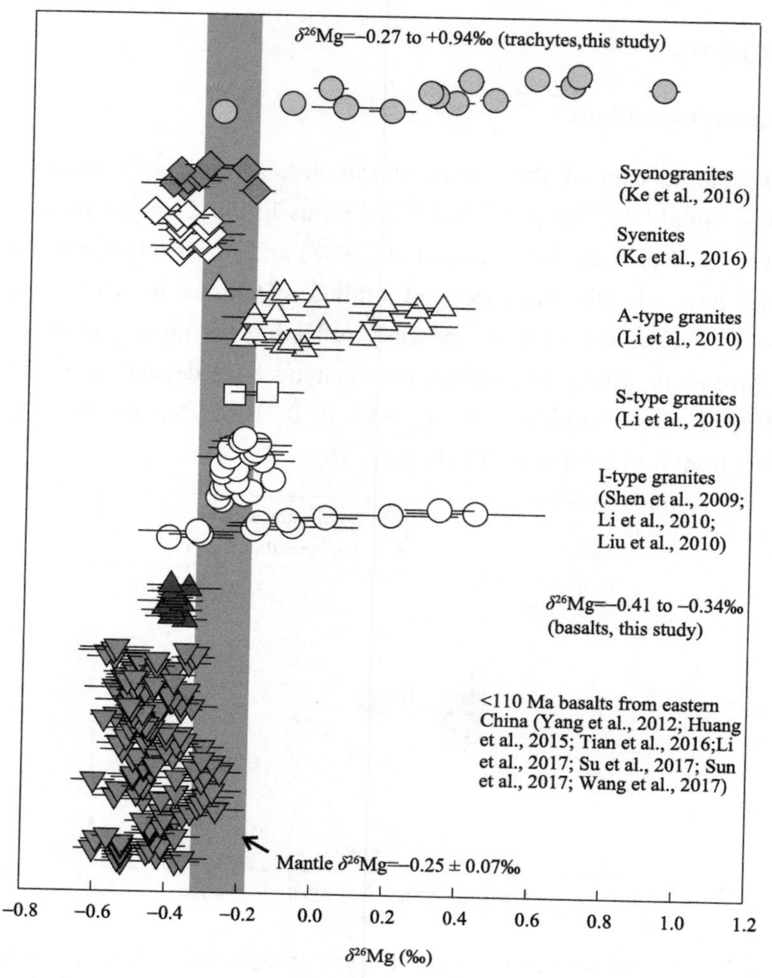

Fig. 4 Magnesium isotopic compositions of volcanic rocks. Data source: <110 Ma basalts from eastern China (Yang et al., 2012; Huang et al., 2015; Tian et al., 2016; Li et al., 2017; Su et al., 2017; Sun et al., 2017; Wang et al., 2017), granitoids rocks (Shen et al., 2009; Li et al., 2010; Liu et al., 2010), and syenites (Ke et al., 2016). The grey band is δ^{26}Mg range of the average Earth (−0.25± 0.07‰; Teng et al., 2010). Error bars on δ^{26}Mg represent 2 SD of repeated measurements and some error bars are smaller than symbols.

5 Discussion

The chemical index of alteration (CIA) values for the Changbaishan basalts and trachytes vary from 40.8 to 44.0 and 42.7 to 49.1 (Table 1), within the range of un-weathered basalts (30–45) and granites and granodiorites (45–55) respectively, as noted by Nesbitt and Young (1982). This demonstrates the samples are not affected by surface alterations. We found the heaviest Mg isotopic composition ever reported in fresh igneous rocks also containing highly enriched ^{87}Sr/^{86}Sr but mantle source-like Nd isotopes, showing decoupled Mg-Sr and Nd isotopes signature. Below, we first discuss the source and petrogenesis of the trachytes and then assess the role of chemical weathering in the formation of trachytes on the basis of their decoupling of Mg-Sr and Nd isotopes. Finally, we evaluate their implications for the paleo-weathering of basaltic crust in NE China.

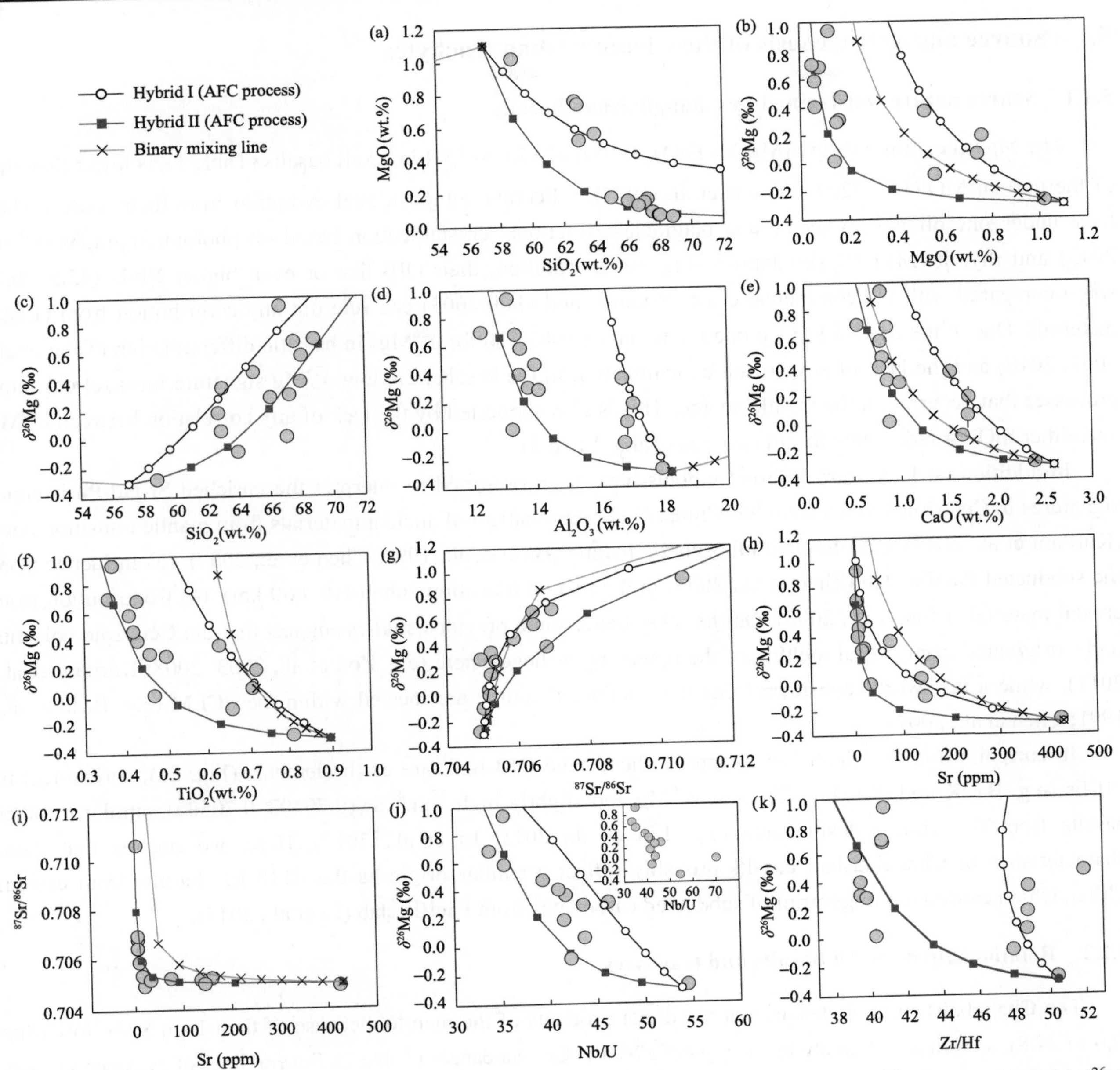

Fig. 5 Geochemical data on Changbaishan basalts and trachytes samples. (a) SiO_2 versus MgO. (b) $\delta^{26}Mg$ versus MgO. (c) $\delta^{26}Mg$ versus SiO_2. (d) $\delta^{26}Mg$ versus Al_2O_3. (e) $\delta^{26}Mg$ versus CaO. (f) $\delta^{26}Mg$ versus TiO_2. (g) $\delta^{26}Mg$ versus $^{87}Sr/^{86}Sr$. (h) $\delta^{26}Mg$ versus Sr (ppm). (i) $^{87}Sr/^{86}Sr$ versus Sr (ppm). (j) $\delta^{26}Mg$ versus Nb/U and (k) $\delta^{26}Mg$ versus Zr/Hf. Most of the error bars associated with an individual data is within the size of the symbol. Grey line with black cross represents binary mixing line between the initial magma endmember and continental weathered residue, characterized by heavy $\delta^{26}Mg$, low SiO_2 (wt.%), MgO (wt.%), CaO (wt.%) and TiO_2 (wt.%) and high Al_2O_3 (wt.%). The two colored curves represent two different "hybrid" models describing evolution of a magma chamber formed by mixing of initial magma and weathered residue and simultaneously undergoing fractional crystallization (AFC; Depaolo, 1981). For these calculations, the initial magma chamber was assumed to be similar to the geochemical characteristics of sample 8-24-15. Two different endmembers of weathered residue were employed based on the Mg isotopic, major and trace element variations (Ma et al., 2007, 2010; Ma, 2011; Huang et al., 2012; Liu et al., 2013, 2014). Circles on the curves represent 10% AFC increments. The modelled parameters are listed in Table 4.

5.1 Source and petrogenesis of the Changbaishan trachytes

5.1.1 Source nature constrained by Changbaishan basalts

The Mg# (i.e. molar [MgO/(MgO + FeO)] × 100) of 37.6 to 53.4 in alkali basalts (Table 1), is lower than that of the normal MORB (~72; Langmuir et al., 1977), reflecting compositional evolution from their source. They have undergone olivine, pyroxene and plagioclase fractional crystallization based on photomicrographs (Chen, 2013) and rhyolite-MELTS simulations (Fig. 6). In addition, their OIB-like or even higher Nb/U (42.5–70.7) when compared with the continental crust (Rudnick and Gao, 2003) can rule out any contribution from crustal materials. Due to the limited Mg isotope fractionation (<0.07‰ for δ^{26}Mg) in basaltic differentiation (Teng et al., 2007, 2010) and the lack of any crustal contamination in the basalts, the low δ^{26}Mg signature must reflect some processes that occurred in the mantle source. This is also supported by the lack of any correlation between δ^{26}Mg and either SiO_2 or radiogenic Sr-Nd isotopes (Table 1 and 3).

In addition, at least four different models have been proposed to interpret the enriched Sr-Nd-Pb isotopic signatures in Cenozoic basalts from NE China. (1) Metasomatism of ancient materials from mantle transition zone (Kuritani et al., 2011). (2) Enriched lithospheric mantle (Basu et al., 1991; Chen et al., 2007). (3) Influence from the subducted Pacific slab (Zhao et al., 2009) in the mantle transition zone (410–660 km). (4) Contribution from crustal materials (Zou et al., 2008). On the other hand, some previous studies suggest that the Cenozoic volcanic rocks originated from partial melting of the upwelling asthenosphere (e.g. Zou et al., 2003, 2008; Kuritani et al., 2011), while a few works considered that the dominated source are located within the SCLM (e.g. Basu et al., 1991; Chen et al., 2007).

In current work, the alkali basalts studied here have OIB-like rare earth elements (Fig. 2a), enrichment in LILEs (e.g. Ba, K and Pb) (Fig. 2b), low δ^{26}Mg and slightly high $^{87}Sr/^{86}Sr$ (0.70493–0.70506), similar to other basalts from the same volcanic area (e.g. Liu et al., 2015; Li et al., 2017). Here, we suggest that these characteristics of Changbaishan basalts probably reflect a similar origin as the <110 Ma basalts from eastern China, which carried the fingerprint of subducted carbonates from Pacific slab (Li et al., 2017).

5.1.2 Relationship between basalts and trachytes

The Changbaishan trachytes are not the direct products of the mantle, because of their high SiO_2, low Mg# (2.0 to 24.8), as well as extremely low compatible Ni and Cr abundances (Table 2). Several lines of evidence strongly support that trachytes mainly formed by the protracted fractional crystallization of alkali basaltic magmas accompanied with some contamination. Firstly, field observations show that the trachytes are from a stratigraphic layer above the basalts and are less abundant than the Changbaishan basalts. Secondly, their trace element patterns (Fig. 2c and d) are clearly distinguishable from the Tibetan trachyte lavas that were interpreted to be originated from crust-derived melts (Chen et al., 2010, 2012). They do however display a high degree of similarity to the trachytes from southeast Queensland, Australia that formed by extensive fractional crystallization of basaltic melts (Shao et al., 2015). Thirdly, phenocrysts of olivine, pyroxene and feldspar in basalts, trachytes and Holocene volcanic eruptions display continuous compositional evolution (Li et al., 2004). Finally, we use the rhyolite-MELTS software (Gualda et al., 2012; Ghiorso and Gualda, 2015) to model the fractional crystallization trend of the basaltic melts. As illustrated in Fig. 6, most of the major elements fall along the modelling curves, although the SiO_2 data is slightly high. This agrees well with previous studies (e.g. Xie et al., 1988; Liu et al., 1998, 2015; Li et al., 2004; Guo et al., 2015, 2016; Park et al., 2016).

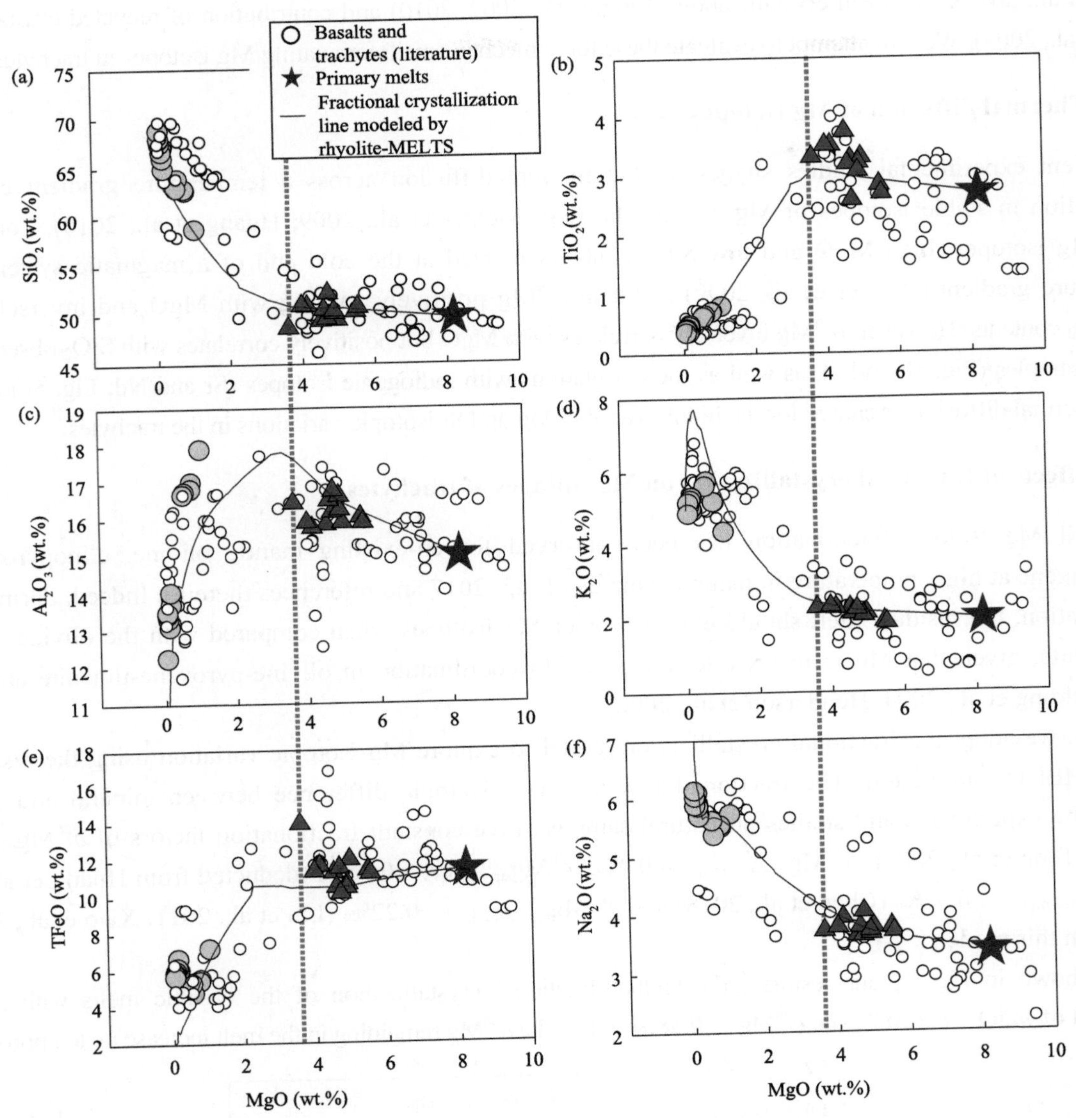

Fig. 6 Major oxide variations of the Changbaishan volcanic rocks. The white circles represent Changbaishan basalts and trachytes from literatures (Xie et al., 1988; Hsu et al., 2000; Fan et al., 2006; Chen et al., 2007; Zou et al., 2008; Chen, 2013; Guo et al., 2014; Ma et al., 2015). The blue curves represent the fractional crystallization (e.g., olivine, pyroxene, ilmenite) processes modelled using rhyolite-MELTS software (Gualda et al., 2012; Ghiorso and Gualda, 2015). The primary melts of SiO_2 (49.7 wt.%), TiO_2 (2.7 wt.%), Al_2O_3 (15.0 wt.%), Fe_2O_3 (1.1 wt.%), FeO (9.3 wt.%), MnO (0.15 wt.%), MgO (8.2 wt.%), CaO (7.4 wt.%), Na_2O (3.4 wt.%), K_2O (2.0 wt.%), P_2O_5 (0.4 wt.%) and H_2O (0.6 wt.%) is based on a large number of simulations. The given conditions are ~4.2 kbar and oxygen fugacity (ΔQFM-1). The evolution line represents the temperature decrements. The right part separated by the dash orange line is basaltic composition and the left part is intermediate-felsic composition. (For in-terpretation of the references to color in this figure legend, the reader is referred to the web version of this article.)

However, given the highly variable $^{87}Sr/^{86}Sr$ (0.70490 to 0.71065) ratios in trachytes relative to that of the basalts, this characteristic may reflect the presence of crustal assimilation in the late magma evolution during the formation of trachytes. The following part will discuss the nature and evolution of the crustal material in detail.

5.2 Chemical weathering inducing decoupled Mg-Sr and Nd isotopes of trachytes

These trachytes display the highest $\delta^{26}Mg$ (up to 0.94±0.04‰) among the previously reported data for fresh igneous rocks (Fig. 4). Interpretations of the largely variable $\delta^{26}Mg$ include thermal diffusion (Richter et al., 2009;

Huang et al., 2010), fractional crystallization (Teng et al., 2007, 2010) and contribution of recycled crustal materials (Shen et al., 2009). We will attempt to evaluate these three mechanisms fractionating Mg isotopes in trachytes.

5.2.1 Thermal diffusion of Mg isotopes

Recent experimental studies suggested that thermal diffusion across a temperature gradient can induce fractionation in stable isotopes of Mg, Ca and Fe (e.g. Richter et al., 2009; Huang et al., 2010). For instance, heavy Mg isotopes, high MgO and low SiO_2 contents located at the cold end of a magmatic system along a temperature gradient (Richter et al., 2009) result in δ^{26}Mg positively varying with MgO and inversely varying with SiO_2 contents. However, δ^{26}Mg inversely correlates with MgO but positively correlates with SiO_2 observed in our trachyte samples (Fig. 5b and c), as well as the correlations with radiogenic isotopes (Sr and Nd; Fig. 5g) thus argue against thermal diffusion as causes for the highly variable Mg and Sr isotopic variations in the trachytes.

5.2.2 Effects of fractional crystallization on Mg isotopes of trachytes

Small Mg isotope fractionation has been observed for co-existing mantle olivine, clinopyroxene and orthopyroxene at high temperature in natural samples (Teng, 2017 and references therein). Indeed, during magma differentiation, the residual melt should enrich heavier Mg isotopes when compared with the olivine, pyroxene and ilmenite, given that Mg is in six-fold and five-fold coordination in olivine-pyroxene-ilmenite and silicate melts (Wilding et al., 2004; Henderson et al., 2006).

Here, we adopt the fractional crystallization model to explore Mg isotopic variation using the results from rhyolite-MELTS simulation. The fractionation factors (i.e. isotopic difference between mineral and melt) are obtained by experiments and studies on natural samples. Five constant fractionation factors of $\delta^{26}Mg_{olivine\text{-}melt}$ = −0.07‰ (Teng et al., 2007), $\delta^{26}Mg_{cpx\text{-}melt}$ = −0.04‰, $\delta^{26}Mg_{opx\text{-}melt}$ = −0.05‰ (deducted from Huang et al., 2013), $\delta^{26}Mg_{ilmenite\text{-}melt}$ = −0.32‰ (Chen et al., 2018) and $\delta^{26}Mg_{spinel\text{-}melt}$ = +0.22‰ (Liu et al., 2011; Xiao et al., 2013) are assumed in this model.

As shown in Fig. 7, our results indicate that fractional crystallization of the basaltic melts with a starting component of MgO = 8.2 wt.% and δ^{26}Mg= −0.38‰ will make δ^{26}Mg remaining in the melt increase up to approximately

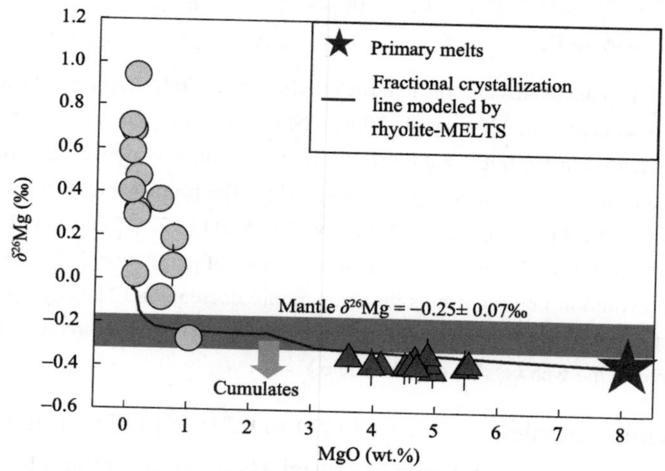

Fig. 7 Modelling the evolution of δ^{26}Mg during crystallization from the parental magma by removal of mafic minerals (olivine, clinopyroxene, spinel and ilmenite) based on the rhyolite-MELTS output result. Here, we adopted the fractional crystallization equation to model the δ^{26}Mg variation. The amount of fractional crystallization of each mineral is in accordance with the predicted results from rhyolite-MELTS. The blue line is the evolutionary trend of Mg isotopes. Some error bars on δ^{26}Mg are smaller than symbols. (For interpretation of the references to color in this figure legend, the reader is referred to the web version of this article.)

+0.05‰, when MgO decreases as low as 0.05 wt.% (Fig. 7). Notably, this value still cannot account for most of the δ^{26}Mg in trachyte rocks. Hence, the isotope offset between alkali basalts and trachytes (up to 1.2‰) is too large to be explained by a single fractional crystallization process. Therefore, although fractional crystallization process has slightly elevated the δ^{26}Mg in trachytes, it is not the main mechanism that highly fractionated the Mg isotopes.

5.2.3 The assimilation and fractional crystallization (AFC) model

From the above discussion, crustal materials are required in this volcanic system. However, the decoupling of Mg-Sr and Nd isotopes in trachytes rule out the assimilation of common silicate materials. Because input of ancient or young crustal rocks (i.e. the underlying Archean-Proterozoic basement rocks) would have resulted in coupled Sr-Nd isotopic variation. Considering the correlations between δ^{26}Mg and both the major and trace element variations and $^{87}Sr/^{86}Sr$ ratios (Fig. 5), at least one component is needed to produce the high δ^{26}Mg values in Changbaishan trachytes, which is characterized by heavy δ^{26}Mg (>+0.94‰), high $^{87}Sr/^{86}Sr$ (>0.7110), basalts-like $^{143}Nd/^{144}Nd$ (~0.512600), low MgO, Nb/U and Zr/Hf ratios. It has been reported that the continental crustal rocks (including TTG gneisses, granulites, I-, S- and A-type granites, syenites and syenogranites) yield a large δ^{26}Mg variation of −0.76 to +0.44‰ (Shen et al., 2009; Li et al., 2010; Liu et al., 2010; Teng et al., 2013; Ke et al., 2016; Yang et al., 2016). However, they are still lower than the heaviest value (+0.94‰) recorded in trachytic rocks. In contrast, although the supracrustal materials show larger Mg isotopic variations, e.g. pelites (−0.52 to +0.92‰; Li et al., 2010), shales (−0.27 to +0.49‰; Li et al., 2010), and mudrocks (−0.17 to +0.25‰; Wang et al., 2015), they are still not the appropriate candidates. To date, continental weathering of the diabase and basalts produce extremely variable δ^{26}Mg varying from −0.49‰ to +1.81‰ (Teng, 2017 and references therein), which can cover the largest value of +0.94‰ in trachytes. Furthermore, previous studies suggested that the Cenozoic basalts during intense weathering could induce apparent $^{87}Sr/^{86}Sr$ fractionation, but limited influence on Nd isotopes (Ma et al., 2010; Ma, 2011). This can be attributed to differential decomposition of minerals containing heterogeneous $^{87}Sr/^{86}Sr$ ratios (Blum and Erel, 1995; Negrel, 2006) or input of extraneous Sr with high $^{87}Sr/^{86}Sr$ ratios from aeolian deposits or groundwater (Price et al., 1991; Stewart et al., 2001; Kurtz et al., 2001; Dia et al., 2006; Ma et al., 2010). Additionally, the unique structure of the Changbaishan shield basalts provide favorable conditions for the formation of weathering basalts. Thus, the early-forming basalts that experienced intense weathering would be covered by the late fresh basalts that eventually formed a volcano with a layered structure. This explanation agrees well with a recent study of high δ^{18}O (3.68–5.03‰) in zircons from the trachytes, indicating the occurrence of altered rocks in the shallow magma chamber beneath the Changbaishan volcanic area (Cheong et al., 2017). Thus, the weathered basalts recycling into the trachytic magma chamber provide a reasonable and feasible explanation of the data.

Here, we adopt the AFC equation (Depaolo, 1981) to model the δ^{26}Mg versus major and trace elements and $^{87}Sr/^{86}Sr$ variations using reasonable parameters (Table 4) on the basis of residue of weathered basalts. As shown in Fig. 5, the simple binary mixing between the initial component and preexisting residue of weathered basalts is unlikely to match most of the major elemental variations. For example, the major elements Fe_2O_3 and Al_2O_3 will be increasing, whereas SiO_2, MgO, Na_2O, K_2O and CaO dramatically decrease in the weathered residual when compared with their parent rocks (e.g. Ma et al., 2007; Liu et al., 2013). Binary mixing between the initial component and weathered products will elevate Fe_2O_3 and Al_2O_3 content, but lower the SiO_2, MgO and CaO concentrations. However, correlations between δ^{26}Mg and both the SiO_2 and Al_2O_3 contents in the trachytes show opposing binary mixing trend lines (Fig. 5c and d). Furthermore, although some elements can be modelled by binary mixing, an unrealistically high proportion of weathered materials (>70%) is required to fully fit δ^{26}Mg variations in the trachyte lavas (Fig. 5b and e-h).

The AFC model in contrast treats Mg as a compatible element with a partition coefficient D_{Mg}>1. Silicon is incompatible with D_{Si} <1 in the fractional assemblages. Thus, Mg is rapidly depleted from the magma whilst Si gradually increases. In this case, two weathered endmember products are required to fit the trachyte variations falling on the AFC calculation curves (Fig. 5). This result shows that up to 50% AFC can reach the Mg-Sr isotopes and other elemental variations. That may be ascribed to the much lower volume of trachytes relative to the shield basalts, so that minor volume of intensely weathered products can easily reach this proportion. Therefore, the AFC process associated with the weathered basalts most likely resulted in the decoupled Mg-Sr and Nd isotopes, accompanied by major- and trace elemental variations in trachytes.

5.3 Approach to trace the paleo-weathering of basaltic crust in Earth's surface

Our study of the trachytes clearly show remarkable decoupling of Mg-Sr and Nd isotopes, as well as enriched Mg-Sr isotopic compositions relative to their primary sources (Figs. 4 and 5). This finding probably has the implication to reconstruct the Earth's climate system by using natural records of past climate change coupled with theoretical models. In combination with the major and trace element variations, these characteristics strongly point to the existence of intensely-weathered residue of basalts in the volcano shield. This extremely chemical weathering requires strong precipitation, wind or high temperature conditions. Additionally, based on previous studies of weathered basalts from South China (Ma et al., 2007; Huang et al., 2012) and Columbia River (Liu et al., 2014), we speculate that ~4–5 Mys continued weathering may be required to generate such a high degree of weathering (δ^{26}Mg up to +1.8‰). Based on K-Ar dating and field observations (Zhang et al., 2018 and references therein), basalts distributed in Changbaishan area mainly included the early pre-shield basalts with the age of ~22.6 to 10.4 Ma, and the late shield basalts with the age of ~5.02 to 1.05 Ma. Hence, it is likely that the pre-shield basalts would be intensely weathered during the long time span of the pre-shield stage (~12 Mys), the interval of pre-shield stage and the shield-forming stage (~5 Mys), and during the shield-forming stage (~4 Mys). An early work by Zou et al. (2004) proposed hot and humid conditions (with annual average temperature of >19 ℃ and the annual rainfall of >165 cm) existed in SE China during 17–15 Ma, based on Sr-Nd isotope systematics of xenoliths in SE China basalts at Fujian. Their interpretation was supported by subsequent studies (Kurschner et al., 2008; Jiang and Ding, 2008). In addition, because of the expansion of the eastern Antarctica ice sheet during the middle Miocene and the formation of the Greenland ice sheet in the late Miocene (Larsen et al., 1994; Zachos et al., 2001), global temperature gradually cooled since ~15 Ma. Taking the above two factors into account, we suggested that there probably existed a warm and humid condition in NE China during 17–15 Ma, which induced intense chemical weathering for pre-shield basalts. Subsequently, they were also affected by weakened weathering until the end of Miocene. This ancient weathered basaltic crust would be covered by the late fresh volcanic flow during Pliocene and became a part of the volcanic cone. As the basalt continued to erupt, the volcanic cone increased and grew, and the magma chamber also rose. When this melt rose to the surface, it would assimilate the ancient weathered crust in the volcanic cone. Finally, the latest formed trachytes from the volcano cone can obtain this weathering signature through AFC processes when they passed through the magma vent (Fig. 8). That is ascribed to two aspects (1) the very low abundant (<5%) of trachytes accounts for the total basalts (Jin and Zhang, 1994) and (2) the relatively low upwelling rate and long store time for the intermediate-felsic magma in the crustal chamber. The trachytes can obtain higher Mg and Sr isotopes through the AFC processes. Hence, our results suggest that the Northeast Asia once had a warm and humid climate during the middle Miocene.

This study provides a specific case that systematic studies of Sr-Nd-Mg isotopes, and conventional elements on late stage volcanic rocks can provide a potentially efficient way to explore the paleo-weathered basaltic crust.

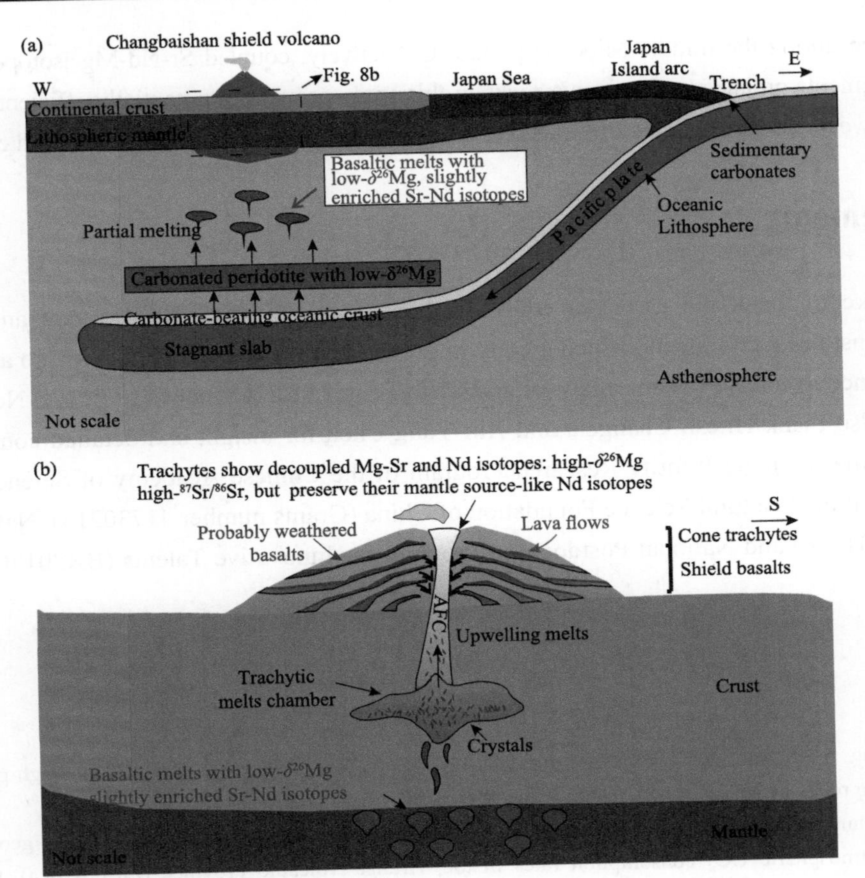

Fig. 8 A cartoon model illustrating the origin of the Changbaishan basalts and trachytes. (a) A subducted Pacific stagnant slab existed in the mantle transition zone underneath eastern China (modified from Huang and Zhao, 2006). This plate brought sedimentary carbonates with low δ^{26}Mg into the asthenosphere and formed carbonated peridotites that can generate basaltic melts with low δ^{26}Mg (e.g. Li et al., 2017). (b) A part of the basaltic melts with low-δ^{26}Mg values upwelled and entered into the magma chamber in the crust. Then, this melt continued to rise through the magmatic channel and involved of some weathered basalts residue (characterized by extreme high δ^{26}Mg and enriched ^{87}Sr/^{86}Sr ratios) via AFC processes. At last, the late erupted trachytes obtain heavy Mg and enriched Sr isotopes but preserved their original Nd isotopes compared to their source melts (Figs. 4, 5).

6 Conclusions

We present high-precision Sr-Nd-Mg isotopes, as well as trace elemental abundance on a suite of early erupted basalts and late trachytes from Changbaishan stratovolcano, NE China. Combined with previous studies, alkali basalts have lower δ^{26}Mg (−0.51 to −0.34‰) than that of the mantle, but similar to most of the late Cretaceous and Cenozoic basalts from eastern China (δ^{26}Mg = −0.60 to −0.23‰). These basalts have the same origins with recycled carbonates metasomatism as the <110 Ma basalts from other eastern China. In contrast, the trachytes show decoupled Mg-Sr and Nd isotopes, i.e. largely variable δ^{26}Mg values from −0.27 up to +0.94‰, high ^{87}Sr/^{86}Sr (0.70490 to 0.71065), but mantle source-like ^{143}Nd/^{144}Nd (0.512584 to 0.512656). Their decoupled Sr-Nd-Mg isotopes cannot be explained by thermal diffusion, fractional crystallization or simple crustal contamination processes. Instead, based on the correlations between δ^{26}Mg and ^{87}Sr/^{86}Sr, trace elemental ratios (e.g. Nb/U and Zr/Hf) and major oxides (e.g. MgO, SiO_2, CaO), they are suggested to reflect the role of chemical weathering residue. These observations together strongly support the existence of paleo-weathered crusts in Changbaishan volcanic cone. Combined with the paleo-environmental evidence, intense chemical weathering probably took place during the middle Miocene period. Collectively, coupled Sr-Nd-Mg isotopes and conventional major and trace elements on late stage volcanic rocks in this study provides a potentially efficient way to explore the existence of paleo-weathered crusts. This study shed new light on the research on paleo-weathered crust.

Acknowledgements

We would like to thank the associate editor Anthony Dosseto, Prof. Hai-Bo Zou and one anonymous reviewer for their insightful comments, which greatly improved the manuscript. Bing-Yu Gao and Wen-Jun Li are thanked for assistance in trace elements analyses and Chao-Feng Li for assistance in Sr and Nd isotopic analyses at IGGCAS. We also thank Hitesh Changela and Xin-Yang Chen for insight and detailed comments. This work was supported by the Strategic Priority Research Program of the Chinese Academy of Sciences (Grant number XDB18000000), National Natural Science Foundation of China (Grants number 41730214), National Postdoctoral Program (2018M631566) and National Postdoctoral Program for Innovative Talents (BX201700237) to Heng-Ci Tian.

References

An, Y.J., Wu, F., Xiang, Y.X., Nan, X.Y., Yu, X., Yang, J.H., Yu, H.M., Xie, L.W., Huang, F., 2014. High-precision Mg isotope analyses of low-Mg rocks by MC-ICP-MS. Chem. Geol. 390, 9-21.

Balagizi, C.M., Darchambeau, F., Bouillon, S., Yalire, M.M., Lambert, T., Borges, A.V., 2015. River geochemistry, chemical weathering, and atmospheric CO_2 consumption rates in the Virunga Volcanic Province (East Africa). Geochem. Geophys. Geosyst. 16, 2637-2660.

Basu, A.R., Wang, J.W., Huang, W.K., Xie, G.H., Tatsumoto, M., 1991. Major Element, Ree, and Pb, Nd and Sr Isotopic Geochemistry of Cenozoic Volcanic Rocks of Eastern China: implications for Their Origin from Suboceanic-Type Mantle Reservoirs. Earth Planet. Sci. Lett. 105, 149-169.

Berner, R.A., Lasaga, A.C., Garrels, R.M., 1983. The carbonate-silicate geochemical cycle and its effect on atmospheric carbon dioxide over the past 100 millions years. Am. J. Sci. 284, 641-683.

Blum, J.D., Erel, Y., 1995. A silicate weathering mechanism linking increases in marine $^{87}Sr/^{86}Sr$ with global glaciation. Nature 373, 415-418.

Bryan, S.E., Ernst, R.E., 2008. Revised definition of large igneous provinces (LIPs). Earth Sci. Rev. 86, 175-202.

Chen, J.L., Zha, W.X., Xu, J.F., Wang, B.D., Kang, Z.Q., 2012. Geochemistry of Miocene trachytes in Bugasi, Lhasa block, Tibetan Plateau: Mixing products between mantle- and crust-derived melts? Gondwana Res. 21, 112-122.

Chen, J.L., Xu, J.F., Wang, B.D., Kang, Z.Q., Jie, L., 2010. Origin of Cenozoic alkaline potassic volcanic rocks at KonglongXiang, Lhasa terrane, Tibetan Plateau: Products of partial melting of a mafic lower-crustal source? Chem. Geol. 273, 286-299.

Chen, L.M., Teng, F.-Z., Song, X.Y., Hu, R.Z., Yu, S.Y., Zhu, D., Kang, J., 2018. Magnesium isotopic evidence for chemical disequilibrium among cumulus minerals in layered mafic intrusion. Earth Planet. Sci. Lett. 487, 74-83.

Chen, X.W., 2013. Crystallization differentiation and magma mixing process of the Tianchi volcano, Changbaishan. Master Thesis Institute of Geology, China Earthquake Administration, Beijing 1-121 (in Chinese).

Chen, Y., Zhang, Y.X., Graham, D., Su, S.G., Deng, J.F., 2007. Geochemistry of Cenozoic basalts and mantle xenoliths in Northeast China. Lithos 96, 108-126.

Cheong, A.C.S., Sohn, Y.K., Jeong, Y. J., Jo, H. J., Park, K.H., Lee, Y.S., Li, X.H., 2017. Latest Pleistocene crustal cannibalization at Baekdusan (Changbaishan) as traced by oxygen isotopes of zircon from the Millennium Eruption. Lithos 284, 132-137.

Choi, H.O., Choi, S.H., Schiano, P., Cho, M., Cluzel, N., Devidal, J.-L., Ha, K., 2017. Geochemistry of olivine-hosted melt inclusions in the Baekdusan (Changbaishan) basalts: Implications for recycling of oceanic crustal materials into the mantle source. Lithos 284-285, 194-206.

Chung, S.L., Jahn, B.M., 1995. Plume-Lithosphere Interaction in Generation of the Emeishan Flood Basalts at the Permian-Triassic Boundary. Geology 23, 889-892.

Depaolo, D.J., 1981. Trace-Element and Isotopic Effects of Combined Wallrock Assimilation and Fractional Crystallization. Earth Planet. Sci. Lett. 53, 189-202.

Dessert, C., Dupré, B., Francois, L.M., Schott, J., Gaillardet, J., Chakrapani, G.J., Bajpai, S., 2001. Erosion of Deccan Traps determined by river geochemistry: impact on the global climate and the $^{87}Sr/^{86}Sr$ ratio of seawater. Earth Planet. Sci. Lett. 188, 459-474.

Dessert, C., Dupré, B., Gaillardet, J., Francois, L.M., Allegre, C.J., 2003. Basalt weathering laws and the impact of basalt weathering on the global carbon cycle. Chem. Geol. 202, 257-273.

Dessert, C., Lajeunesse, E., Lloret, E., Clergue, C., Crispi, O., Gorge, C., Quidelleur, X., 2015. Controls on chemical weathering on a mountainous volcanic tropical island: Guadeloupe (French West Indies). Geochim. Cosmochim. Acta 171, 216-237.

Dia, A., Chauvel, C., Bulourde, M., Gerard, M., 2006. Eolian contribution to soils on Mount Cameroon: Isotopic and trace element records. Chem. Geol. 226, 232-252.

Dufek, J., Bachmann, O., 2010. Quantum magmatism: Magmatic compositional gaps generated by melt-crystal dynamics. Geology 38, 687-690.

Fan, Q.C., Sui, J.L., Wang, T.H., Li, N., Sun, Q., 2006. Eruption history and magma evolution of the trachybasalt in the Tianchi volcano, Changbaishan. Acta Petrol. Sin. 22, 1449-1457 (in Chinese).

Fan, Q.C., Sui, J.L., Wang, T.H., Li N., Sun Q., 2007. History of volcanic activity, magma evolution and eruptive mechanisms of the Changbai volcanic province. Geological journal of China Universities 13, 175-190 (in Chinese).

Gaillardet, J., Dupré, B., Louvat, P., Allegre, C.J., 1999. Global silicate weathering and CO_2 consumption rates deduced from the chemistry of large rivers. Chem. Geol. 159, 3-30.

Galy, A., Yoffe, O., Janney, P.E., Williams, R.W., Cloquet, C., Alard, O., Halicz, L., Wadhwa, M., Hutcheon, I.D., Ramon, E., Carignan J., 2003. Magnesium isotope heterogeneity of the isotopic standard SRM980 and new reference materials for magnesium-isotope-ratio measurements. J. Anal. At. Spectrom. 18, 1352-1356.

Ghiorso, M.S., Gualda, G.A.R., 2015. An H_2O-CO_2 mixed fluid saturation model compatible with rhyolite-MELTS. Contrib. Mineral. Petrol. 169.

Gualda, G.A.R., Ghiorso, M.S., Lemons, R.V., Carley, T.L., 2012. Rhyolite-MELTS: a Modified Calibration of MELTS Optimized for Silica-rich, Fluid-bearing Magmatic Systems. J. Petrol. 53, 875-890.

Guo, W.F., Liu, J.Q., Guo, Z.F., 2014. Temporal variations and petrogenetic implications in Changbai basaltic rocks since the Pliocene. Acta Petrol. Sin. 30, 3595-3611 (in Chinese).

Guo, W.F., Liu, J.Q., Wu, C.L., Lei, M., Qin, H.P., Wang, N., Zhang, X., Chen, H.J., Wang, Z., 2016. Petrogenesis of trachyte and the felsic magma system at Tianchi volcano: trace elements and isotopic constraints. Geol. Rev. 62, 617-630 (in Chinese).

Guo, W.F., Liu, J.Q., Xu, W.G., Li, W., Lei, M., 2015. Reassessment of the magma system beneath Tianchi volcano, Changbaishan: Phase equilibria constraints. Chin. Sci. Bull. 60, 3489-3500 (in Chinese).

Hartmann, J., Jansen, N., Durr, H.H., Kempe, S., Kohler, P., 2009. Global CO_2-consumption by chemical weathering: What is the contribution of highly active weathering regions? Glob. Planet. Change 69, 185-194.

Hartmann, J., Moosdorf, N., 2012. The new global lithological map database GLiM: a representation of rock properties at the Earth surface. Geochem. Geophys. Geosyst. 13.

Henderson, G.S., Calas, G., Stebbins, J.F., 2006. The structure of silicate glasses and melts. Elements 2, 269-273.

Hsu, C.N., Chen, J.C., Ho, K.S., 2000. Geochemistry of Cenozoic volcanic rocks from Kirin Province, northeast China. Geochem. J. 34, 33-58.

Huang, F., Chakraborty, P., Lundstrom, C.C., Holmden, C., Glessner, J.J.G., Kieffer, S.W., Lesher, C.E., 2010. Isotope fractionation in silicate melts by thermal diffusion. Nature 464, 396-400.

Huang, F., Chen, L.J., Wu, Z.Q., Wang, W., 2013. First-principles calculations of equilibrium Mg isotope fractionations between garnet, clinopyroxene, orthopyroxene, and olivine: Implications for Mg isotope thermometry. Earth Planet. Sci. Lett. 367, 61-70.

Huang J., Ke, S., Gao, Y.J., Xiao, Y.L., Li, S.G., 2015. Magnesium isotopic compositions of altered oceanic basalts and gabbros from IODP site 1256 at the East Pacific Rise. Lithos 231, 53-61.

Huang, J., Zhao, D., 2006. High-resolution mantle tomography of China and surrounding regions. J. Geophys. Res. 111.

Huang, K.J., Teng, F.-Z., Wei, G.J., Ma, J.L., Bao, Z.Y., 2012. Adsorption- and desorption-controlled magnesium isotope fractionation during extreme weathering of basalt in Hainan Island, China. Earth Planet. Sci. Lett. 359, 73-83.

Ibarra, D.E., Caves, J.K., Moon, S., Thomas, D.L., Hartmann, J., Chamberlain, C.P., Maher, K., 2016. Differential weathering of basaltic and granitic catchments from concentration–discharge relationships. Geochim. Cosmochim. Acta 190, 265-293.

Jiang, H.C., Ding, Z.L., 2008. A 20 Ma pollen record of East-Asian summer monsoon evolution from Guyuan, Ningxia, China. Palaeogeogr. Palaeoclimatol. Palaeoecol. 265, 30-38.

Jin, B.L., Zhang, X.Y., 1994. Researching Volcanic Geology in Changbai Mt Northeast Korea (in Chinese). Yanbian: Nation Education Press, 1-223 (in Chinese).

Ke, S., Teng, F.-Z., Li, S.G., Gao, T., Liu, S.A., He, Y.S., Mo, X.X., 2016. Mg, Sr, and O isotope geochemistry of syenites from northwest Xinjiang, China: Tracing carbonate recycling during Tethyan oceanic subduction. Chem. Geol. 437, 109-119.

Kent, D.V., Muttoni, G., 2013. Modulation of Late Cretaceous and Cenozoic climate by variable drawdown of atmospheric pCO_2 from weathering of basaltic provinces on continents drifting through the equatorial humid belt. Clim. Past 9, 525-546.

Kuritani, T., Kimura, J.I., Miyamoto, T., Wei, H.Q., Shimano, T., Maeno, F., Jin, X., Taniguchi, H., 2009. Intraplate magmatism related to deceleration of upwrelling asthenospheric mantle: Implications from the Changbaishan shield basalts, northeast China. Lithos 112, 247-258.

Kuritani, T., Ohtani, E., Kimura, J.I., 2011. Intensive hydration of the mantle transition zone beneath China caused by ancient slab stagnation. Nat. Geosci. 4, 713-716.

Kurschner, W.M., Kvacek, Z., Dilcher, D.L., 2008. The impact of Miocene atmospheric carbon dioxide fluctuations on climate and the evolution of terrestrial ecosystems. Proc. Natl. Acad. Sci. 105, 449-453.

Kurtz, A.C., Derry, L.A., Chadwick, O.A., 2001. Accretion of Asian dust to Hawaiian soils: Isotopic, elemental, and mineral mass balances. Geochim. Cosmochim. Acta 65, 1971-1983.

Langmuir, C.H., Bender, J.F., Bence, A.E., Hanson, G.N., Taylor, S.R., 1977. Petrogenesis of Basalts from the Famous Area: Mid-Atlantic Ridge. Earth Planet. Sci. Lett. 36, 133-156.

Larsen, H.C., Saunders, A.D., Clift, P.D., Beget, J., Wei, W., Spezzaferri, S., 1994. Seven million years of glaciation in Greenland. Science 264, 952-955.

Li, C.F., Chu, Z.Y., Guo, J.H., Li, Y.L., Yang, Y.H., Li, X.H., 2015. A rapid single column separation scheme for high precision Sr–Nd–Pb isotopic analysis in geological samples using thermal ionization mass spectrometry. Anal. Methods 7, 4793-4802.

Li, C.F., Wang, X.C., Guo, J.H., Chu, Z.Y., Feng, L.J., 2016b. Rapid separation scheme of Sr, Nd, Pb and Hf from a single rock digest using a tandem chromatography column prior to isotope ratio measurements by mass spectrometry. J. Anal. At. Spectrom. 31, 1150-1159.

Li, G.J., Hartmann, J., Derry, L.A., West, A.J., You, C.F., Long, X., Zhan, T., Li, L., Li, G., Qiu, W., Li, T., Liu, L., Chen, Y., Ji, J., Zhao, L., Chen, J., 2016a. Temperature dependence of basalt weathering. Earth Planet. Sci. Lett. 443, 59-69.

Li, N., Fan, Q.C., Sun, Q., Zhang, W.L., 2004. Magma evolution of Changbaishan Tianchi volcano: Evidences from the main phenocrystal minerals. Acta Petrol. Sin. 20, 575-582 (in Chinese).

Li, S.G., Yang, W., Ke, S., Meng, X.N., Tian, H.C., Xu, L.J., He, Y.S., Huang, J., Wang, X. C., Xia, Q.K., Sun, W.D., Yang, X.Y., Ren, Z.Y., Wei, H.Q., Liu, Y.S., Meng, F.C., Yan, J., 2017. Deep carbon cycles constrained by a large-scale mantle Mg isotope anomaly in eastern China. Natl. Sci. Rev. 4, 111-120.

Li, W.Y., Teng, F.-Z., Ke, S., Rudnick, R.L., Gao, S., Wu, F.Y., Chappell, B.W., 2010. Heterogeneous magnesium isotopic composition of the upper continental crust. Geochim. Cosmochim. Acta 74, 6867-6884.

Liu, J.Q., Chen, S.S., Guo, Z.F., Guo, W.F., He, H.Y., You, H.T., Kim, H.M., Sung, G.H., Kim, H., 2015. Geological background and geodynamic mechanism of Mt. Changbai volcanoes on the China–Korea border. Lithos 236-237, 46-73.

Liu, J.Q., 1999. Volcanoes in China, Beijing: Science Press 1-219 (in Chinese).

Liu, J.Q., Wang, S.S., 1982. Age of Changbaishan Volcano and Tianchi Lake. Chin. Sci. Bull. 1312-1315 (in Chinese).

Liu, R.X., 2000. Active Volcanoes in China (in Chinese), Beijing: Seismological Press. 1-114 (in Chinese).

Liu, R.X., Qiu, S.H., Cai, L.Z., Wei, H.Q., Yang, Q.F., Xian, Z.Q., Bo, G.C., Zhong, J., 1998. The date of last large eruption of Changbaishan-Tianchi volcano and its significance. Sci. China 41, 69-74 (in Chinese).

Liu, S.A., Teng, F.-Z., He, Y.S., Ke, S., Li, S.G., 2010. Investigation of magnesium isotope fractionation during granite differentiation: Implication for Mg isotopic composition of the continental crust. Earth Planet. Sci. Lett. 297, 646-654.

Liu, S.A., Teng, F.-Z., Yang, W., Wu, F.Y., 2011. High-temperature inter-mineral magnesium isotope fractionation in mantle xenoliths from the North China craton. Earth Planet. Sci. Lett. 308, 131-140.

Liu, X.M., Rudnick, R.L., McDonough, W.F., Cummings, M.L., 2013. Influence of chemical weathering on the composition of the continental crust: Insights from Li and Nd isotopes in bauxite profiles developed on Columbia River Basalts. Geochim. Cosmochim. Acta 115, 73-91.

Liu, X.M., Teng, F.-Z., Rudnick, R.L., McDonough, W.F., Cummings, M.L., 2014. Massive magnesium depletion and isotope

fractionation in weathered basalts. Geochim. Cosmochim. Acta 135, 336-349.
Louvat, P., Allègre, C.J., 1997. Present denudation rates at Réunion island determined by river geochemistry: basalt weathering and mass budget between chemical and mechanical erosions. Geochim. Cosmochim. Acta 61, 3645-3669.
Louvat, P., Allègre, C.J., 1998. Riverine erosion rates on Sao Miguel volcanic island, Azores archipelago. Chem. Geol. 148, 177-200.
Ma, H.R., Yang, Q.F., Pan, X.D., Wu, C.Z., Chen, C., 2015. Origin of early Pleistocene basaltic lavas in the Erdaobaihe river basin, Changbaishan region. Acta Petrol. Sin. 31, 3484-3494 (in Chinese).
Ma, J.L., 2011. Elemental and isotopic systematics during intensive chemical weathering of basalts: variabilities and their mechanisms. PHD thesis. Guangzhou Institute of Geochemistry, Chinese Academy of Sciences, Guangzhou 1-108 (in Chinese).
Ma, J.L., Wei, G.J., Xu, Y.G., Long, W.G., 2010. Variations of Sr–Nd–Hf isotopic systematics in basalt during intensive weathering. Chem. Geol. 269, 376-385.
Ma, J.L., Wei, G.J., Xu, Y.G., Long, W.G., Sun, W.D., 2007. Mobilization and re-distribution of major and trace elements during extreme weathering of basalt in Hainan Island, South China. Geochim. Cosmochim. Acta 71, 3223-3237.
McCulloch, M.T., Gregory, R.T., Wasserburg, G.J., Taylor, H.P., 1981. Sm-Nd, Rb-Sr, and $^{18}O/^{16}O$ isotopic systematics in an oceanic crustal section: Evidence from the Samail Ophiolite. J. Geophys. Res. 86, 2721-2735.
McDonough, W.F., Sun, S.S., 1995. The Composition of the Earth. Chem. Geol. 120, 223-253.
Meng, X.Q., Liu, L.W., Wang, X.T., Balsam W., Chen J., Ji J. F., 2018. Mineralogical evidence of reduced East Asian summer monsoon rainfall on the Chinese loess plateau during the early Pleistocene interglacials. Earth Planet. Sci. Lett. 486, 61-69.
Moon, S., Chamberlain, C.P., Hilley, G.E., 2014. New estimates of silicate weathering rates and their uncertainties in global rivers. Geochim. Cosmochim. Acta 134, 257-274.
Negrel, P., 2006. Water-granite interaction: clues from strontium, neodymium and rare earth elements in soil and waters. Appl. Geochem. 21, 1432-1454.
Nesbitt, H.W., Markovics, G., Price, R.C., 1980. Chemical processes affecting alkalis and alkaline earths during continental weathering. Geochim. Cosmochim. Acta 44, 1659-1666.
Nesbitt, H.W., Young, G.M., 1982. Early Proterozoic Climates and Plate Motions Inferred from Major Element Chemistry of Lutites. Nature 299, 715-717.
Opfergelt, S., Georg, R.B., Delvaux, B., Cabidoche, Y.M., Burton, K.W., Halliday, A.N., 2012. Mechanisms of magnesium isotope fractionation in volcanic soil weathering sequences, Guadeloupe. Earth Planet. Sci. Lett. 341, 176-185.
Park, S.C., Kim, T.U., Cho, I.W., Kim, S.C., Kim, I.C., 2016. Scoriae magma evolution at Paekdu volcano, Democratic People's Republic of Korea. Acta Petrol. Sin. 32, 3214-3224. (in Chinese).
Price, R.C., Gray, C.M., Wilson, R.E., Frey, F.A., Taylor, S.R., 1991. The effects of weathering on rare-earth element, Y and Ba abundances in tertiary basalts from southeastern Australia. Chem. Geol. 93, 245-265.
Ramos, F.C., Heizler, M.T., Buettner, J.E., Gill, J.B., Wei, H.Q., Dimond, C.A., Scott, S.R., 2016. U-series and $^{40}Ar/^{39}Ar$ ages of Holocene volcanic rocks at Changbaishan volcano, China. Geology 44, 511-514.
Richter, F.M., Watson, E.B., Mendybaev, R., Dauphas, N., Georg, B., Watkins, J., Valley, J., 2009. Isotopic fractionation of the major elements of molten basalt by chemical and thermal diffusion. Geochim. Cosmochim. Acta 73, 4250-4263.
Rudnick, R.L., Gao, S., 2003. Composition of the continental crust. Treatise Geochem. 3, 1-64.
Shao, F.L., Niu, Y.L., Regelous, M., Zhu, D.-C., 2015. Petrogenesis of peralkaline rhyolites in an intra-plate setting: Glass House Mountains, southeast Queensland, Australia. Lithos 216-217, 196-210.
Shen, B., Jacobsen, B., Lee, C.T.A., Yin, Q.Z., Morton, D.M., 2009. The Mg isotopic systematics of granitoids in continental arcs and implications for the role of chemical weathering in crust formation. P. Natl. Acad. Sci. 106, 20652-20657.
Stewart, B.W., Capo, R.C., Chadwick, O.A., 2001. Effects of rainfall on weathering rate, base cation provenance, and Sr isotope composition of Hawaiian soils. Geochim. Cosmochim. Acta 65, 1087-1099.
Su, B.X., Hu, Y., Teng, F.Z., Xiao, Y., Zhou, X.H., Sun, Y., Zhou, M.F., Chang, S.C., 2017. Magnesium isotope constraints on subduction contribution to Mesozoic and Cenozoic East Asian continental basalts. Chem. Geol. 466, 116-122.
Sun, C.Q., Wei, H.Q., Liu, Q., Pan, X.D., 2008. K-Ar dating on the trachytes composing the Tianchi composite cone in Changbaishan. Seismol. Geol. 30, 484-496 (in Chinese).
Sun, H., Xiao, Y., Gao, Y., Zhang, G., Casey, J.F., Shen, Y., 2018. Rapid enhancement of chemical weathering recorded by extremely light seawater lithium isotopes at the Permian–Triassic boundary. P. Natl. Acad. Sci. 115, 3782-3787.
Sun, S.S., McDonough, W.F., 1989. Chemical and isotopic systematics of oceanic basalts: implications for mantle composition and processes. Geol. Soc. London, Special Publications 42, 313-345.

Sun, Y., Teng, F.-Z., Ying, J.F., Su, B.X., Hu, Y., Fan, Q.C., Zhou, X.H., 2017. Magnesium isotopic evidence for ancient subducted oceanic crust in LOMU-Like potassium-rich volcanic rocks. J. Geophys. Res. 122, 7562-7572.
Sun, Y.B., Ma, L., Bloemendal, J., Clemens, S., Qiang, X.K., An, Z.S., 2015. Miocene climate change on the Chinese Loess Plateau: Possible links to the growth of the northern Tibetan Plateau and global cooling. Geochem. Geophys. Geosyst. 16, 2097-2108.
Tang, Y.C., Obayashi, M., Niu, F.L., Grand, S.P., Chen, Y.J., Kawakatsu, H., Tanaka, S., Ning, J.Y., Ni, J.F., 2014. Changbaishan volcanism in northeast China linked to subduction-induced mantle upwelling. Nat. Geosci. 7, 470-475.
Teng, F.-Z., 2017. Magnesium Isotope Geochemistry. Rev. Mineral. Geochem. 82, 219-287.
Teng, F.-Z., Li, W.Y., Ke, S., Marty, B., Dauphas, N., Huang, S.C., Wu, F.Y., Pourmand, A., 2010. Magnesium isotopic composition of the Earth and chondrites. Geochim. Cosmochim. Acta 74, 4150-4166.
Teng, F.-Z., Li, W.Y., Ke, S., Yang, W., Liu, S.A., Sedaghatpour, F., Wang, S.J., Huang, K.J., Hu, Y., Ling, M.X., Xiao, Y., Liu, X.M., Li, X.W., Gu, H.O., Sio, C.K., Wallace, D.A., Su, B.X., Zhao, L., Chamberlin, J., Harrington, M., Brewer, A., 2015. Magnesium Isotopic Compositions of International Geological Reference Materials. Geostand. Geoanal. Res. 39, 329-339.
Teng, F.-Z., Wadhwa, M., Helz, R.T., 2007. Investigation of magnesium isotope fractionation during basalt differentiation: Implications for a chondritic composition of the terrestrial mantle. Earth Planet. Sci. Lett. 261, 84-92.
Teng, F.-Z., Yang, W., Rudnick, R.L., Hu, Y., 2013. Heterogeneous magnesium isotopic composition of the lower continental crust: A xenolith perspective. Geochem. Geophys. Geosyst. 14, 3844-3856.
Tian, H.-C., Yang, W., Li, S.G., Ke, S., Chu, Z.Y., 2016. Origin of low δ^{26}Mg basalts with EM-I component: Evidence for interaction between enriched lithosphere and carbonated asthenosphere. Geochim. Cosmochim. Acta 188, 93-105.
Wang, S.J., Teng, F.-Z., Rudnick, R.L., Li, S.G., 2015. The behavior of magnesium isotopes in low-grade metamorphosed mudrocks. Geochim. Cosmochim. Acta 165, 435-448.
Wang, X.J., Chen, L.H., Hofmann, A.W., Mao, F.G., Liu, J.Q., Zhong, Y., Xie, L.W., Yang, Y.H., 2017. Mantle transition zone-derived EM1 component beneath NE China: Geochemical evidence from Cenozoic potassic basalts. Earth Planet. Sci. Lett. 465, 16-28.
Wei, H., Sparks, R.S.J., Liu, R., Fan, Q., Wang, Y., Hong, H., Zhang, H., Chen, H., Jiang, C., Dong, J., Zheng, Y., Pan, Y., 2003. Three active volcanoes in China and their hazards. J. Asian Earth Sci. 21, 515-526.
Wei, H.Q., Wang, Y., Jin, J.Y., Gao, L., Yun, S.H., Jin, B.L., 2007. Timescale and evolution of the intracontinental Tianchi volcanic shield and ignimbrite-forming eruption, Changbaishan, Northeast China. Lithos 96, 315-324.
Wilding, M.C., Benmore, C.J., Tangeman, J.A., Sampath, S., 2004. Evidence of different structures in magnesium silicate liquids: coordination changes in forsterite- to enstatite-composition glasses. Chem. Geol. 213, 281-291.
Wimpenny, J., Yin, Q.Z., Tollstrup, D., Xie, L.W., Sun, J., 2014. Using Mg isotope ratios to trace Cenozoic weathering changes: A case study from the Chinese Loess Plateau. Chem. Geol. 376, 31-43.
Xiao, Y., Teng, F.-Z., Zhang, H.F., Yang, W., 2013. Large magnesium isotope fractionation in peridotite xenoliths from eastern North China craton: Product of melt-rock interaction. Geochim. Cosmochim. Acta 115, 241-261.
Xie, G.H., Wang, J.W., Basu, A.R., Tatsumoto, M., 1988. Petrochemistry and Sr, Nd, Pb isotopic geochemistry of Cenozoic volcanic rocks, Changbaishan area, northeast China. Acta Petrol. Sin. 1-13 (in Chinese).
Yang, W., Teng, F.-Z., Li, W.Y., Liu, S.A., Ke, S., Liu, Y.S., Zhang, H.F., Gao, S., 2016. Magnesium isotopic composition of the deep continental crust. Am. Mineral. 101, 243-252.
Yang, W., Teng, F.-Z., Zhang, H.F., Li, S.G., 2012. Magnesium isotopic systematics of continental basalts from the North China craton: Implications for tracing subducted carbonate in the mantle. Chem. Geol. 328, 185-194.
Yang, J.H., Cawood P. A., Du Y. S., Condon D. J., Yan J. X., Liu J. Z., Huang Y., Yuan D. X., 2018. Early Wuchiapingian cooling linked to Emeishan basaltic weathering? Earth Planet. Sci. Lett. 492, 102-111.
Young, E.D., Galy, A., 2004. The isotope geochemistry and cosmochemistry of magnesium. Rev. Mineral. Geochem. 55, 197-230.
Zachos, J., Pagani, M., Sloan, L., Thomas, E., Billups, K., 2001. Trends, rhythms and aberrations in global climate 65 Ma to Present. Science 292, 686-693.
Zhang, M.L., Guo, Z.F., Liu, J.Q., Liu, G.M., Zhang, L.H., Lei, M., Zhao, W.B., Ma, L., Sepe, V., Ventura, G., 2018. The intraplate Changbaishan volcanic field (China/North Korea): A review on eruptive history, magma genesis, geodynamic significance, recent dynamics and potential hazards. Earth Sci. Rev. 187, 19-52.
Zhao, D.P., Tian, Y., Lei, J. ., Liu, L.C., Zheng, S.H., 2009. Seismic image and origin of the Changbai intraplate volcano in East Asia: Role of big mantle wedge above the stagnant Pacific slab. Phys. Earth Planet. In. 173, 197-206.
Zindler, A., Hart, S., 1986. Chemical Geodynamics. Annu. Rev. Earth Pl. Sc.14, 493-571.

Zou, H.B., Fan, Q.C., Yao, Y.P., 2008. U-Th systematics of dispersed young volcanoes in NE China: Asthenosphere upwelling caused by piling up and upward thickening of stagnant Pacific slab. Chem. Geol. 255, 134-142.

Zou, H.B., Fan, Q.C., Zhang, H.F., 2010. Rapid development of the great Millennium eruption of Changbaishan (Tianchi) Volcano, China/North Korea: Evidence from U-Th zircon dating. Lithos 119, 289-296.

Zou, H.B., Fan, Q.C., Zhang, H.F., Schmitt A. K., 2014. U-series zircon age constraints on the plumbing system and magma residence times of the Changbai volcano, China/North Korea border. Lithos 200, 169-180.

Zou, H.B., McKeegan, K.D., Xu, X.S., Zindler, A., 2004. Fe-Al-rich tridymite–hercynite xenoliths with positive cerium anomalies: preserved lateritic paleosols and implications for Miocene climate. Chem. Geol. 207, 101-116.

Zou, H.B., Reid, M.R., Liu, Y.S., Yao, Y.P., Xu, X.S., Fan, Q.C., 2003. Constraints on the origin of historic potassic basalts from northeast China by U-Th disequilibrium data. Chem. Geol. 200, 189-201.

Mg and Zn isotope evidence for two types of mantle metasomatism and deep recycling of magnesium carbonates*

Sheng-Ao Liu[1], Ze-Zhou Wang[1, 2], Chun Yang[1], Shu-Guang Li[1] and Shan Ke[1]

1. State Key Laboratory of Geological Processes and Mineral Resources, China University of Geosciences, Beijing 100083, China
2. Isotope Laboratory, Department of Earth and Space Sciences, University of Washington, Seattle, WA 98195, USA

亮点介绍：西藏东南部的新生代超钾岩和富钠的碱性玄武岩的比较研究表明超钾岩具有类似地幔的δ^{26}Mg，且δ^{66}Zn较全球大洋玄武岩略轻，而富钠碱性玄武岩拥有明显较轻的δ^{26}Mg和较重的δ^{66}Zn值。这些特征表明前者的源区为再循环泥硅质沉积物交代的地幔；而后者源区存在富Zn的再循环成因菱镁矿和辉石岩熔体。因此，沉积泥质硅酸盐和碳酸盐交代深部地幔分别促成了西藏东南部超钾岩和富钠碱性玄武岩的形成，后者发育区与观测到的地幔过渡带（410~660 km）的深俯冲成因的滞留板块在空间上保持一致，为俯冲碳酸盐循环进入深地幔提供了证据。

Abstract To test the ability of Mg and Zn isotopes in discriminating between different types of mantle metasomatism and identifying deep carbon cycling, here we present a comparative study on two types of Cenozoic lavas in SE Xizang, that is, K-rich (potassic-ultrapotassic) lavas and Na-rich alkali basalts. The contrasting bulk rock chemical compositions, Sr-Nd isotopic ratios, and olivine chemistry between them suggest distinct sources in the lithospheric mantle and asthenosphere, respectively. The K-rich lavas have mantle-like δ^{26}Mg, slightly lighter δ^{66}Zn relative to global oceanic basalts, high ^{87}Sr/^{86}Sr, and low ^{143}Nd/^{144}Nd, indicating source metasomatism by recycled siliciclastic sediments. By contrast, the alkali basalts possess remarkably lighter δ^{26}Mg and heavier δ^{66}Zn values relative to the mantle that are typically characterized by carbonates. The coupling of high δ^{66}Zn with high-Zn contents and Zn/Fe ratios further suggests a pyroxenite source containing recycled Zn-rich magnesium carbonates. This is strongly corroborated by the similarity in major elements between the alkali basalts and experimental partial melts of pyroxenite + CO_2. Thus, mantle silicate and carbonate metasomatism contributed to the origin of K-rich and Na-rich lavas in SE Xizang, respectively. Notably, the occurrence of the alkali basalts is spatially consistent with a stagnant slab in the mantle transition zone (410–660 km), the latter of which is interpreted to represent the deeply subducted oceanic slab. These observations provide evidence for recycling of carbonates into the deep mantle, which represents a long-term circulation of subducted carbon compared with that of arc-trench systems and has crucial significance for global deep carbon cycling.

Keywords Magnesium isotopes, zinc isotopes, basaltic lavas, mantle metasomatism, deep carbon cycle

* 本文发表在：Journal of Geophysical Research-solid Earth, 2020, 125

1 Introduction

Mantle metasomatism is a vital process changing the mineralogical, chemical, and physical properties of the mantle through infiltration and percolation of melts and/or aqueous fluids within the upper mantle (Bailey, 1982; Roden & Murthy, 1985; Green & Wallace, 1988). Silicate and carbonatite melts constitute two important agents commonly involved in mantle metasomatism. Identifying mantle metasomatism by recycled carbonates and discriminating it from silicate metasomatism has profound significance for understanding global deep carbon cycles, as carbonates represent the major carbon-bearing phase that is delivered to our planet's interior via subduction (e.g., Plank and Manning, 2019). In subduction zones, a considerable mass of nondissolved carbonates can survive shallow dehydration and melting beyond the subarc regime (Yaxley & Green, 1994; Kerrick & Connolly, 2001) and may have been recycled into the deeper upper mantle and the mantle transition zone (MTZ) (Thomson et al., 2016; Li et al., 2017; Mazza et al., 2019) or are stored in the subcontinental lithospheric mantle (Foley & Fischer, 2017). Owing to the lower solidus temperature than nominally volatile-free peridotite (Dasgupta et al., 2007; Yaxley et al., 1991), the carbonate-bearing peridotite would be apt to partially melt prior to volatile-poor peridotite and then result in mantle carbonatite metasomatism. The imprint of recycled carbonate and its final storage depth in the mantle, however, still require to be deciphered from natural observations, in particular based on new geochemical tracers.

Carbon isotopes can easily distinguish organic carbon from inorganic carbon, but approximately 95% of subduction-related carbon is inorganic carbon that is not readily distinguished from primary mantle-derived carbon based on C isotopes solely (Deines, 2002). In addition, degassing-induced strong isotope fractionation of carbon during magma eruption (e.g., Aubaud et al., 2005) makes it difficult to track recycled carbonates in the sources of mantle-derived lavas. The isotopes of divalent metals (e.g., Mg and Zn), which occur as major or trace constituents in carbonate and silicate, may be promising tools of discriminating between different types of mantle metasomatism. For Mg, marine carbonates are isotopically lightest (δ^{26}Mg low to −5.6‰; Wombacher et al., 2011) and weathered silicates/sediments are isotopically heaviest (δ^{26}Mg up to +1.8‰; Liu et al., 2014) in terrestrial reservoirs. The Mg isotopic compositions of different types of carbonates (e.g., limestone and dolostone) are significantly distinct but all of them are much lighter than the mantle value (Fig. 1). For Zn, marine carbonates (e.g., Mesozoic-Cenozoic, δ^{66}Zn=0.99±0.25‰; Pichat et al., 2003; Sweere et al., 2018) are isotopically

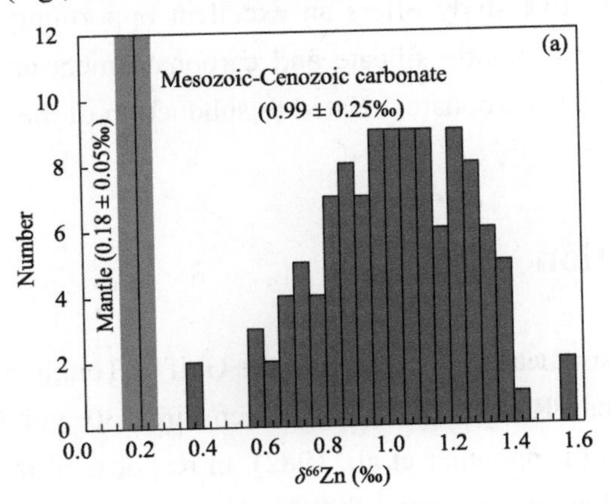

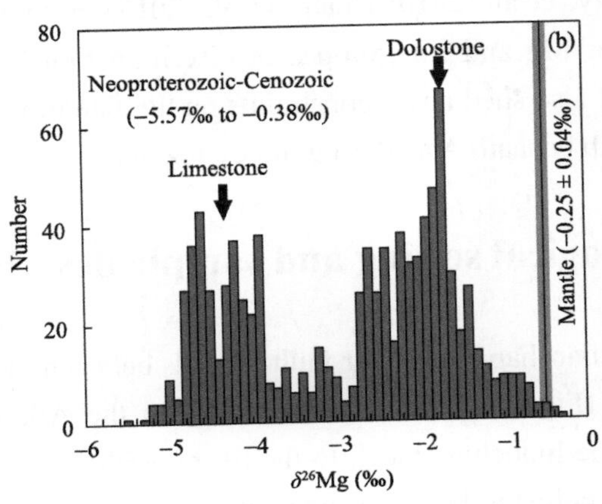

Fig. 1 Plots showing strong Mg and Zn isotopic contrast between marine carbonates and the mantle. Zn isotope data of Mesozoic-Cenozoic carbonates are from Pichat et al. (2003) and Sweere et al. (2018). Mg isotope data of carbonates and the mantle are from Teng (2017) and reference therein. The mantle δ^{66}Zn is from Wang et al. (2017), Sossi et al. (2018), McCoy-West et al. (2018), and Liu et al. (2019).

heavier by up to ~0.8‰ than the mantle (δ^{66}Zn=0.18±0.05‰; Liu et al., 2019; McCoy-West et al., 2018; Sossi et al., 2018; Wang et al., 2017). The heterogeneity of Zn isotopic compositions of marine carbonates is related to variation in seawater δ^{66}Zn value (Kunzmann et al., 2013; Liu et al., 2017; Pichat et al., 2003; Sweere et al., 2018), temperature, and/or pH value (Mavromatis et al., 2019) at which the carbonates precipitated. Siliciclastic sediments are isotopically similar to or lighter than igneous rock average (δ^{66}Zn=~0.28‰), especially organic-rich margin sediments that possess δ^{66}Zn values low to −0.05‰ (Little et al., 2016).

During slab subduction and dehydration, Mg isotope fractionation is limited through different degrees of metamorphism and the deeply subducted carbonates have extremely light δ^{26}Mg (Wang et al., 2014). Zinc isotope fractionation during the subduction of mafic oceanic crust is also limited (Inglis et al., 2017). It is proposed that isotopically heavy Zn may be preferentially released during dehydration of altered ultramafic rocks (Pons et al., 2016), but the small variation of Zn contents in subducted rocks (Debret et al., 2018; Inglis et al., 2017) indicates that this process probably results in limited Zn loss. The MORB-like Zn isotopic compositions of arc basalts (Huang et al., 2018) provide further evidence that the amount of isotopically heavy Zn released at early stages of subduction—if any—is small. Thereupon, lighter Mg and heavier Zn isotopic compositions of basaltic melts relative to the mantle values may indicate the incorporation of carbonate into the mantle via subduction (Li et al., 2017; Liu et al., 2016). For example, significant higher-than-mantle Zn isotopic compositions of basaltic rocks, such as late Mesozoic to Cenozoic intraplate basalts in Eastern China (Liu et al., 2016; Wang et al., 2018) and global ocean island basalts (Beunon et al., 2020), have been explained as a result of recycled carbonate-bearing components in their mantle sources. By contrast, melts derived from a mantle source hybridized by recycled siliciclastic sediments should have heavier Mg and possibly lighter Zn isotopic compositions.

In order to illustrate the possible differences in Mg and Zn isotopic composition induced by different types of mantle metasomatism, here we present a comparative study on two groups of Cenozoic mafic volcanism (K-rich versus Na-rich) widely distributed in the southeastern Tibetan Plateau. Bulk rock major-trace elements and Sr-Nd isotopes are presented together with in-situ chemical compositions of olivine phenocrysts in order to further characterize the nature of their mantle sources. Given that the two groups of volcanism have been inferred to originate from different mantle source domains (e.g., Flower et al., 2013; Guo et al., 2005; F. Huang, Chakraborty, et al., 2010; Huang et al., 2013), a comparative study offers an excellent opportunity to test the potential of Mg and Zn isotopes in discriminating between mantle silicate and carbonate metasomatism. The results will also shed novel constraints on the fate of recycled carbonates related to subduction of the Neo-Tethys oceanic slab beneath Asia during the Cenozoic.

2 Geological setting and sample description

The Ailaoshan-Red River fault extends between the southeastern Xizang and the Gulf of Tonkin in the South China Sea (Fig. 2). Left-lateral shearing on the Ailaoshan-Red River fault between circa 30 and 17 Ma has displaced the Indochina block to the SE by ~500–700 km (Tapponnier et al., 1982), in response to the Cenozoic continental collision between India and Asia (Fig. 2a). This fault is over 1,000 km long and represents the most pronounced morphologic and geologic discontinuity in Southeast Asia (Fig. 2a). The Ailaoshan-Red River Shear Zone is the plate boundary separating the Yangtze Craton to the east from Indochina Block to the west (Wang et al., 2000; Fig. 2a, b). Two types of mafic magmas are exposed along the Ailaoshan-Red River Shear Zone prior to or following the left-lateral slip on this major fault, respectively. The first type is exposed in the Dali-Lijiang and adjacent regions of western Yunnan and Sichuan and comprises a potassic to ultrapotassic episode between Late Eocene and Early Oligocene (Flower et al., 2013; Guo et al., 2005; Huang, Chakraborty, et al., 2010; Wang et al.,

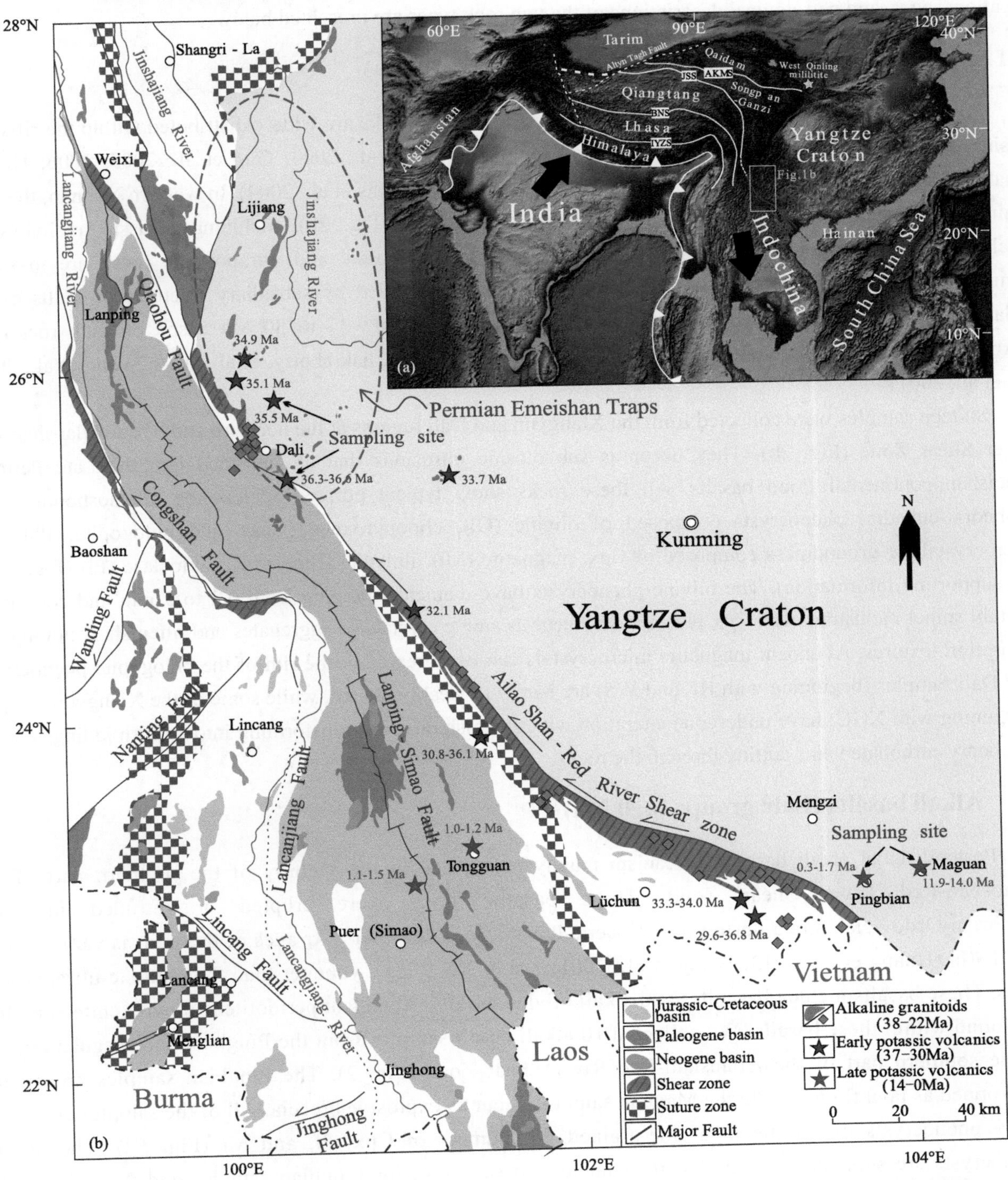

Fig. 2 (a) Tectonic framework of the Xizang an Plateau and its surrounding regions. IYZS: Indus-Yarlung-Zangbo Sutrue, BNS: Bangong-Nujiang Suture, JSS: Jinshajiang suture, AKMS: Anyimaqin-Kunlun-Muztagh Suture, ARSZ: Ailaoshan-Red River Shear Zone. (b) Simplified map showing the tectonic structure of Ailaoshan-Red River Shear Zone and its surrounding regions and the distributions of Cenozoic magmatic rocks (after Deng et al., 2014).

2001; Xu et al., 2001). The second type comprises Na-rich basalt and basanite phase between the Pliocene (or earlier) and Pleistocene (Flower et al., 2013; Huang et al., 2013; Wang et al., 2001; Xia & Xu, 2005). The locations, ages, and petrological descriptions of the two rock types are presented below.

2.1 Potassic-ultrapotassic rocks (Early group; 37−30 Ma)

Postcollisional Cenozoic potassic and ultrapotassic volcanic rocks are widely distributed within the rifts and basins of the Tibetan Plateau and its peripheral regions (Ding et al., 2003; Guo et al., 2005, 2006; Huang, Chakraborty, et al., 2010; Liu et al., 2015; Turner et al., 1996; Williams et al., 2004). In western Yunnan, they are mainly distributed along the Ailaoshan-Red River Shear Zone (Fig. 2b), including potassic trachybasalts, trachyandesites, shoshonites, and lamprophyres. Most of the potassic and ultrapotassic rocks are mafic in composition and occur as volcanic pipes or dykes closely controlled by subsidiary fractures or faults of the Ailaoshan-Red River Shear Zone. Their formation ages vary from ~37 to 30 Ma based on bulk rock/mica $^{40}Ar/^{39}Ar$ or zircon U-Pb dating methods (Guo et al., 2005; Huang, Chakraborty, et al., 2010; Wang et al., 2001; Xu et al., 2001).

Sixteen samples were collected from the Xiangyun and Dali regions at the northern end of the Ailaoshan-Red River Shear Zone (Fig. 2b). They occur as subvolcanic intrusions that are intruded into the Late Permian Emeishan continental flood basalts. All these rocks show typical porphyritic textures of shoshonites with subhedral-euhedral phenocrysts composed of olivine (Ol), clinopyroxene (Cpx), and phlogopite (Phl) and microcrystalline groundmass composed of Cpx, magnetite (Mt), ilmenite (Ilm), and plagioclase (Pl) (Fig. S1 in the supporting information). The olivine phenocrysts have diameters ranging from 0.2 to 5 mm and commonly contain spinel inclusions. The Cpx phenocrysts occur as single crystals or aggregates and often show zoning and resorption textures. Abundant magnetite microcrystals can be observed at the rim of the phlogopite phenocrysts. The Dali samples (beginning with HZ and WS) are barely to weakly altered, while some of the Xiangyun samples (beginning with XGC) have undergone alteration with greenish olivines transforming into reddish iddingsites and secondary carbonate veins cutting through the rocks.

2.2 Alkali basalts (Late group; 14−0 Ma)

Postcollisional alkali basaltic volcanism mainly took place in the vicinity of the southern part of the Ailaoshan-Red River Shear Zone. These basaltic lavas were erupted or intruded into the Cambrian-Ordovician strata during the Miocene-Pleistocene. The $^{40}Ar/^{39}Ar$ ages of these lavas vary from 14 to 0.1 Ma (Huang et al., 2013; Wang et al., 2001; Xia & Xu, 2005), younger than the potassic-ultrapotassic lavas. These basalts include basanites and alkali olivine basalts. Abundant peridotite and pyroxenite xenoliths are found within these basalts. We collected 10 alkali basalt samples from the Pingbian and Maguan regions in the southern part of the Ailaoshan-Red River Shear Zone (Fig. 2). The Pingbian samples are mainly outcropped as lava flows, while the Maguan samples occur as explosive breccia. All of the samples consist of 2–5% phenocrysts of olivine in a fine-grained groundmass of Cpx, Pl, and Mt (Fig. S2). The olivine phenocrysts are subhedral-euhedral with diameters of <0.2 mm for Pingbian samples and 0.2–0.5 mm for Maguan samples. In addition, abundant xenocrysts (1–3 mm) of Ol and Cpx have also been observed in the Maguan samples but are absent in the Pingbian samples. All of the alkali basalt samples collected in this study are unaltered.

3 Analytical methods

3.1 Major and trace elements

Rock chips were crushed to powders of 200 mesh (74 μm) in an agate mortar. The sample powders were weighed and dissolved separately for elemental, Sr-Nd and Mg-Zn isotopic analysis. The major element concentrations were determined by wet chemistry methods at the State Key Laboratory of Geological Processes and Mineral Resources, China University of Geosciences, Beijing (CUGB). The loss on ignition (LOI) was obtained by heating the samples at 980 °C over 30 min. The uncertainties are better than ±1% (1σ). For trace element analysis, samples were completely digested in a mixture of HF + HNO_3 + HCl acids using high-pressure bombs at 190 °C. Elemental analysis was accomplished using an ELAN DRCII inductively coupled plasma mass spectrometer (ICP-MS). The USGS basalt standard BHVO-2 was adopted to monitor the whole analytical procedure, with reproducibility better than 5% (1σ) for elements with concentrations of > 10 μg/g and ~10% (1σ) for those with concentrations of <10 μg/g. The major and trace elemental concentrations of BHVO-2 analyzed in this study are reported in Table S1. Typically, Zn concentration of BHVO-2 obtained (105.3 ppm; Table S1) is consistent with the recommended value (103 ± 6 ppm; USGS, http://www.usgs.gov/) within uncertainty.

3.2 Sr and Nd isotopes

The analysis of Sr and Nd isotopic ratios was performed on a Thermo-Fisher *Neptune Plus* multicollector ICP-MS (MC-ICP-MS) at the Isotope Geochemistry Lab of CUGB. Rock powders were sufficiently digested in the Teflon beakers with a mixture of HF-HNO_3-HCl acids. Samples were passed through a column loaded with precleaned Bio-Rad® cation resin AG50W-X12 (200–400 mesh) for collecting Sr first. Nd was then purified using LN resin. The instrumental mass fractionation was corrected by $^{86}Sr/^{88}Sr$=0.1194 and $^{146}Nd/^{144}Nd$=0.7219 based on the exponential law. Data for BHVO-2 obtained in this study (0.703510±10 for $^{87}Sr/^{86}Sr$ and 0.512976±10 for $^{143}Nd/^{144}Nd$) are consistent with the published values (e.g., Weis et al., 2005).

3.3 Mg and Zn isotopes

All chemical procedures were carried out in laminar flow hoods (Class 100) in a clean room (Class 1,000) with filtered air. Double distilled reagents and ⩾18.2 MΩ water were always used for sample dissolution and all other processes. Details of sample digestion, column chemistry, and isotope ratio analysis follow the established procedures for Mg (Ke et al., 2016) and Zn isotopes (Liu et al., 2016). A brief description is presented below. After complete digestion of ~20 mg sample powder, the solution of each sample was separated into two fractions that were evaporated to dryness and then transferred into 1 ml of 1 M nitric acid or 8 M hydrochloric acid for Mg and Zn purification, respectively. Magnesium was purified from matrix elements using Bio-Rad® AG50W-X8 resin. Zinc was separated with a single-column procedure using 2 ml precleaned Bio-Rad® strong anion resin AG-MP-1 M (100–200 mesh; chloride form) and collected in 10 ml of 0.5 N HNO_3 after matrices, copper, and iron were eluted in turn. After evaporation to dryness at 80 °C, the residues were dissolved in 3% HNO_3 by mass. In order to ensure complete separation from matrix, the eluted Mg and Zn solutions were subject again to column chemistry. The recovery of both Mg and Zn after two times of chemical purification is >99.5%. The whole procedure blank was less than 10 ng for Mg and less than 6 ng for Zn, which accounts for negligible portions of the loaded Mg (>20 μg) and Zn (1−3 μg) in the studied samples.

Magnesium and zinc isotopic ratios were measured separately using a Thermo Scientific *Neptune Plus* MC-ICP-MS at the Isotope Geochemistry Lab of the CUGB. Samples and in-house standards are diluted to obtain

100 ppb Mg or 200 ppb Zn solutions in 3% HNO_3. The samples and standards were introduced to the Ar plasma using a Scott double pass quartz glass spray chamber and a low-flow 50 μl/min PFA self-aspiration micronebulizer, with help of a Cetac ASX-110 autosampler. The sample introduction system was washed with 3% HNO_3 for 2 min after each measurement to avoid cross contamination. Mass bias correction was achieved using sample standard bracketing (SSB) method, and each measurement was operated for three blocks of 40 cycles. The isobaric interferences have been checked in the low-resolution mode. High-sensitivity (X) cones made of Ni were used to ensure that the ^{24}Mg and ^{64}Zn signals are typically >3 V/200 ppb with a 10^{11}-Ω resistor for the Faraday cups. The results are reported in delta notation against Mg isotope standard DSM-3 (Galy et al., 2003) and Zn isotope standard Lyon JMC (Maréchal et al., 1999), respectively, expressed as follows:

$$\delta^{26,25}\mathrm{Mg}\ (‰) = [(^{26,25}\mathrm{Mg}/^{24}\mathrm{Mg})_{\mathrm{sample}}/(^{26,25}\mathrm{Mg}/^{24}\mathrm{Mg})_{\mathrm{DSM3}} - 1] \times 1000$$

$$\delta^{68,66}\mathrm{Zn}\ (‰) = [(^{68,66}\mathrm{Zn}/^{64}\mathrm{Zn})_{\mathrm{sample}}/(^{68,66}\mathrm{Zn}/^{64}\mathrm{Zn})_{\mathrm{JMC\ 3\text{-}0749L}} - 1] \times 1000$$

The long-term external reproducibility (2SD) is better than ±0.06‰ for δ^{26}Mg (Ke et al., 2016) and ±0.04‰ for δ^{66}Zn (Liu et al., 2016; Wang et al., 2017) based on repeated analyses of seawater for Mg isotopes, IRMM 3702 for Zn isotopes and international rock standards (BHVO-2, BCR-2, and AGV-2) for both Mg and Zn isotopes. Two USGS reference materials BHVO-2 (δ^{26}Mg = −0.21±0.05‰, n=4; δ^{66}Zn = 0.31±0.04‰, n =4) and BCR-2 (δ^{26}Mg=−0.25±0.04‰, n=4; δ^{66}Zn=0.28±0.03‰, n=4) running through the whole procedure yielded results resembling the literature values (e.g., Sossi et al., 2015; Teng, 2017; Wang et al., 2017).

4 Results

4.1 Major-trace elements and Sr-Nd isotopes

Major and trace elements are reported in Table S1. The two lava groups have high-MgO and total alkali (K_2O + Na_2O) contents that mostly fall within the alkali series (Fig. 3a). The K-rich rocks belong to basaltic trachyandesite and trachyandesite (Le Bas et al., 1986; Fig. 3a) and have high K_2O/Na_2O of 0.62–3.51 (wt.%), with some being classified as ultrapotassic rocks following the definition of Foley et al. (1987), which is MgO>3 wt.%, K_2O>3 wt.%, and K_2O/Na_2O>2 (Fig. 3b). The alkali basalts are mainly sodic with K_2O/Na_2O of mostly <1. The K-rich rocks have Mg numbers ($Mg^{\#} = 100\times Mg/(Mg + Fe^{2+})$) between 63 and 77, which are slightly higher than those of the Na-rich basalts (59–67). The K-rich rocks are selectively enriched in large ion lithophile elements (LILEs), particularly Rb, Ba, and K, and exhibit strong negative anomalies of high field strength elements (HFSEs; such as Ta, Nb, and Ti) and positive anomalies of Pb in primitive mantle-normalized trace element diagram (Fig. 3c). In contrast, the alkali basalts are enriched in both LILE and HFSE and display positive Ta-Nb anomalies and negative Pb-Ti anomalies (Fig. 3d). Both the two groups of rocks are more enriched in light rare earth elements (LREEs) relative to heavy REEs (HREEs) (Fig. 3e, f).

The potassic-ultrapotassic rocks have high initial ^{87}Sr/^{86}Sr ratios of 0.70595 to 0.70773 and negative $\varepsilon_{\mathrm{Nd}}(t$ = 37–30 Ma) values of −4.81 to −1.11. The Na-rich alkali basalts have relatively low ^{87}Sr/^{86}Sr$_{(i)}$ varying from 0.70375 to 0.70483 and high-$\varepsilon_{\mathrm{Nd}}(t)$ values varying from + 2.70 to + 6.34 (Fig. 4). The ^{87}Sr/^{86}Sr and $\varepsilon_{\mathrm{Nd}}(t)$ values of sodic basalts are akin to global OIBs and alkali basalts from adjacent regions (e.g., Hainan Island, South China Sea; Wang et al., 2011).

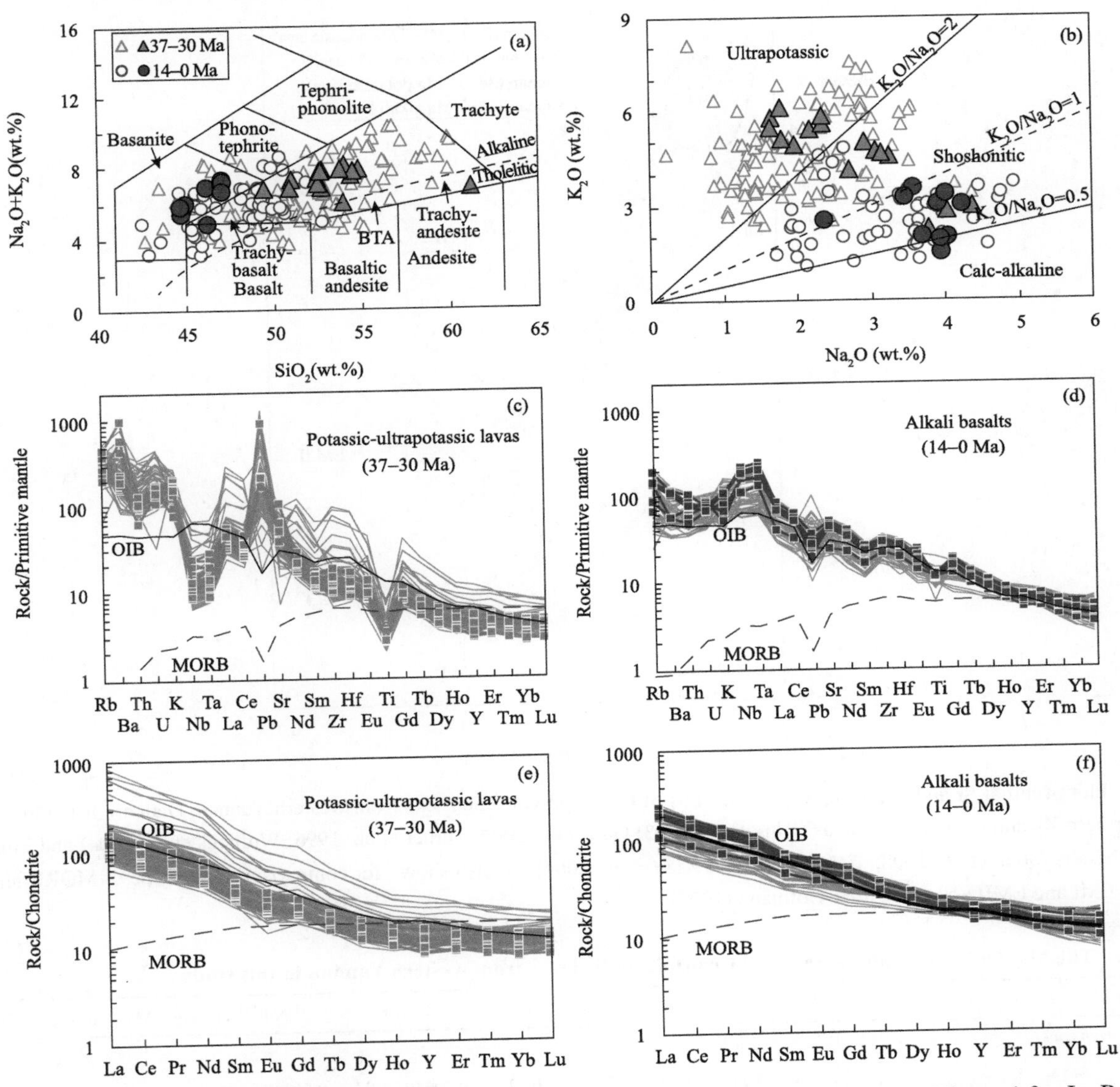

Fig. 3 (a and b) Geochemical classification of Cenozoic volcanic rocks from western Yunnan in SE Xizang (after Le Bas et al., 1986, and Foley et al., 1987). Solid symbols represent samples in this study, and open symbols represent those reported in the literature (Chen et al., 2014; Guo et al., 2005; Huang et al., 2013; Huang, Chakraborty, et al., 2010; Li et al., 2002; Lu et al., 2015; Wang et al., 2001; Xia & Xu, 2005; Xu et al., 2001). (c and d) Primitive mantle-normalized trace element patterns of the two groups of lavas. (e and f) Chondrite-normalized rare earth element patterns. Compositions of the primitive mantle and chondrites are from McDonough and Sun (1995), and average MORB (solid black line) and OIB (dashed gray line) are from Sun and McDonough (1989).

4.2 Mg and Zn isotopes

The Mg and Zn isotopic data of all samples are reported in Table 1. The potassic-ultrapotassic rocks display a wide range of δ^{26}Mg from −1.01‰ to −0.21‰, and below the narrow δ^{26}Mg range of the mantle (−0.25±0.04‰; Teng, 2017). The K-rich Xiangyun subgroup have extremely variable and low δ^{26}Mg values of −1.01‰ to −0.38‰. Notably, the Xiangyun samples have LOIs of 2.5–11.1 wt.% (Table S1), reflecting variable degrees of alteration after magma eruption. Mg and Zn isotope data of the Xiangyun samples may have been modified and are excluded from the following discussions. The remaining samples have low LOIs (mostly <1.0 wt.%), are almost unaltered, and display a narrow δ^{26}Mg range (−0.31‰ to −0.21‰; n=9). The alkali basalts have low LOIs (mostly <1.5 wt.%), and their δ^{26}Mg values (−0.44‰ to −0.33‰; n=10) are also systematically below the mantle value of −0.25±0.04‰.

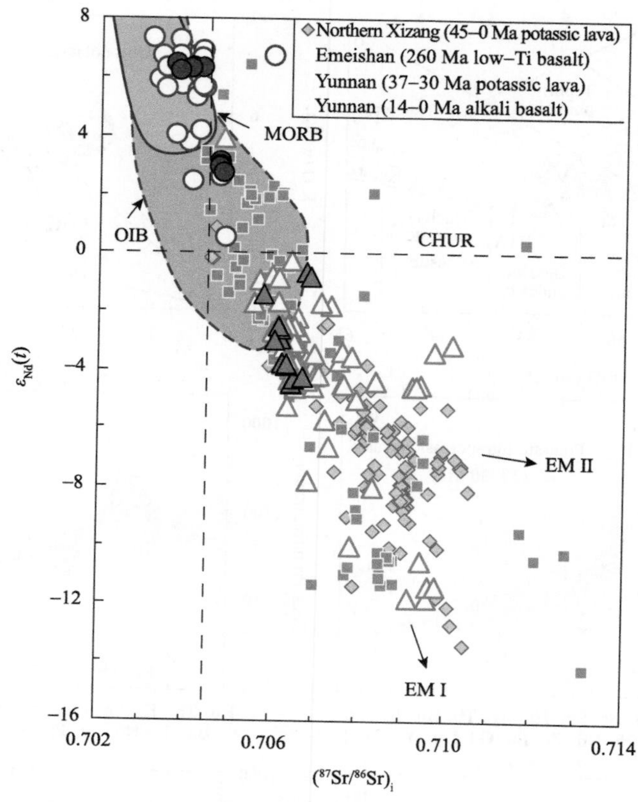

Fig. 4　Plot of initial Sr and Nd isotopic compositions of Cenozoic volcanic rocks from western Yunnan. Data are listed in Table 1. The northern Xizang potassic-rich lavas (Ding et al., 2003; Guo et al., 2006; Turner et al., 1996; Williams et al., 2004) and Emeishan low-Ti basalts (Song et al., 2008; Tian et al., 2017; Xiao et al., 2004) are also shown for comparison. The scope of MORB and OIB and the EMI and EMII arrays are after Hofmann (1997).

Table 1　The Mg-Zn-Sr-Nd isotopic ratios of Cenozoic mafic lavas from western Yunnan in this study

Sample	δ^{25}Mg	2SD	δ^{26}Mg	2SD	δ^{66}Zn	2SD	δ^{68}Zn	2SD	^{87}Sr/^{86}Sr	^{143}Nd/^{144}Nd	Age (Ma)	^{87}Sr/^{86}Sr$_i$	$\varepsilon_{Nd}(t)$
Potassic-rich rocks													
HZ04	—	—	—	—	0.24	0.01	0.47	0.02	0.706493 ± 9	0.512478 ± 10	35.5	0.706162	−2.84
HZ06	−0.09	0.01	−0.22	0.02	0.22	0.05	0.44	0.11	0.706769 ± 9	0.512392 ± 9	35.5	0.706441	−4.51
HZ09	−0.10	0.01	−0.22	0.02	0.18	0.02	0.36	0.03	0.706547 ± 9	0.512471 ± 9	35.5	0.706196	−2.98
HZ10	−0.12	0.03	−0.21	0.01	0.21	0.06	0.42	0.13	0.706542 ± 8	0.512487 ± 9	35.5	0.706189	−2.64
HZ13	−0.12	0.02	−0.24	0.01	0.18	0.05	0.35	0.11	0.706396 ± 10	0.512493 ± 8	35.5	0.706146	−2.55
HZ14	−0.12	0.01	−0.23	0.02	0.20	0.06	0.39	0.10	0.706778 ± 10	0.512393 ± 9	35.5	0.706443	−4.47
HZ15	−0.14	0.02	−0.27	0.02	0.23	0.01	0.45	0.00	0.706846 ± 9	0.512400 ± 9	35.5	0.706499	−4.34
WS05	−0.16	0.02	−0.31	0.03	0.30	0.02	0.58	0.08	0.706870 ± 10	0.512581 ± 8	35.5	0.706780	−0.70
WS06	−0.15	0.02	−0.27	0.01	0.26	0.05	0.52	0.11	0.706976 ± 9	0.512573 ± 9	35.5	0.706874	−0.85
WS09	−0.12	0.02	−0.24	0.01	0.21	0.06	0.42	0.12	0.705952 ± 9	0.512547 ± 8	35.5	0.705796	−1.44
WS10	—	—	—	—	0.24	0.02	0.47	0.05	0.706321 ± 9	0.512469 ± 10	35.5	0.706058	−3.02
XGC02	−0.51	0.02	−1.01	0.01	0.32	0.03	0.63	0.10	0.706470 ± 8	0.512427 ± 10	36.5	0.706226	−3.80
XGC03	−0.19	0.02	−0.38	0.01	0.24	0.03	0.51	0.04	0.707232 ± 10	0.512403 ± 9	36.5	0.706704	−4.28
XGC05	−0.25	0.01	−0.47	0.04	0.30	0.04	0.60	0.02	0.706509 ± 8		36.5	0.706306	
XGC06	−0.27	0.02	−0.53	0.02	0.31	0.01	0.62	0.02	0.706504 ± 9	0.512423 ± 9	36.5	0.706322	−3.89
XGC08	−0.29	0.02	−0.52	0.02	0.24	0.01	0.48	0.04	0.706511 ± 8	0.512426 ± 10	36.5	0.706330	−3.82

Continued

Sample	$\delta^{25}Mg$	2SD	$\delta^{26}Mg$	2SD	$\delta^{66}Zn$	2SD	$\delta^{68}Zn$	2SD	$^{87}Sr/^{86}Sr$	$^{143}Nd/^{144}Nd$	Age (Ma)	$^{87}Sr/^{86}Sr_i$	$\varepsilon_{Nd}(t)$
Na-rich alkali basalt													
LC01	−0.17	0.01	−0.33	0.02	0.48	0.03	0.94	0.01	0.704143 ± 10	0.512957 ± 10	13.0	0.704108	6.31
LC03	−0.18	0.01	−0.39	0.01	0.46	0.03	0.91	0.05	0.703747 ± 10	0.512963 ± 9	13.0	0.703722	6.43
LC04	−0.20	0.03	−0.39	0.01	0.49	0.02	0.97	0.01	0.704359 ± 8	0.512955 ± 10	13.0	0.704328	6.29
LC05	−0.20	0.02	−0.37	0.04	0.51	0.03	1.00	0.08	0.704093 ± 10	0.512954 ± 10	13.0	0.704066	6.27
ZJS01	−0.21	0.01	−0.44	0.03	0.39	0.03	0.79	0.03	0.703902 ± 9	0.512951 ± 10	13.0	0.703808	6.19
PZ01	−0.20	0.02	−0.40	0.01	0.45	0.02	0.89	0.02	0.704756 ± 10	0.512796 ± 10	1.0	0.704752	3.09
AJW01	−0.21	0.00	−0.40	0.01	0.42	0.03	0.84	0.07	0.704773 ± 9	0.512788 ± 10	1.0	0.704769	2.93
AJW02	−0.19	0.01	−0.40	0.04	0.43	0.02	0.85	0.03	0.704753 ± 9	0.512790 ± 9	1.0	0.704749	2.96
AJW03	−0.18	0.01	−0.39	0.02	0.45	0.03	0.88	0.01	0.704832 ± 9	0.512776 ± 9	1.0	0.704828	2.70
AJW06	−0.17	0.01	−0.35	0.04	0.44	0.06	0.87	0.10	0.704743 ± 9	0.512786 ± 10	1.0	0.704738	2.90

Note: $^{87}Sr/^{86}Sr_i = {}^{87}Sr/^{86}Sr - {}^{87}Rb/^{86}Sr \times (e^{\lambda t} - 1)$. $^{143}Nd/^{144}Nd\ (i) = {}^{143}Nd/^{144}Nd - {}^{147}Sm/^{144}Nd \times (e^{\lambda t} - 1)$. $\varepsilon_{Nd}(t) = [(^{143}Nd/^{144}Nd\ (i))_{sample}/(^{143}Nd/^{144}Nd\ (t))_{CHUR}$. $^{87}Rb/^{86}Sr$ and $^{43}Nd/^{144}Nd$ ratios (±2SE) are calculated based on Rb, Sr, Sm, and Nd contents. Data of the eruption age (t) is from Wang et al. (2001) and Huang, Chakraborty, et al. (2010). λ refers to the decay constant of ^{87}Rb (1.42×10^{-11}) or ^{147}Sm (0.654×10^{-11}); $(^{143}Nd/^{144}Nd)_{CHUR} = 0.512638$; $(^{147}Sm/^{144}Nd)_{CHUR} = 0.1967$.

The K-rich rocks have $\delta^{66}Zn$ ranging from 0.18‰ to 0.30‰ (n=11). Most of these values overlap the reported $\delta^{66}Zn$ values for mid-ocean ridge basalts (MORBs, 0.28±0.03‰; Huang et al., 2018; Wang et al., 2017), while some samples are isotopically lighter than MORBs. In contrast, the Na-rich alkali basalts have higher $\delta^{66}Zn$ values than both the potassic-ultrapotassic rocks and MORBs, ranging from 0.39±0.03‰ to 0.51±0.03‰ (n= 10). The high-$\delta^{66}Zn$ values of the alkali basalts are similar to those of the <110 Ma intraplate alkali basalts from Eastern China ($\delta^{66}Zn$ = 0.30‰ to 0.63‰; Liu et al., 2016; Wang et al., 2018).

5 Discussion

In this section, we first discuss the mantle source domains for the two groups of mafic lavas based on bulk major-trace elements, Sr-Nd isotope data, and olivine chemistry. Then, we attempt to identify different types of metasomatic agents in their mantle sources using Mg and Zn isotope data. Finally, we discuss the broad significance of these observations for global deep carbon cycling.

5.1 Contrasting mantle domains for two types of volcanism in SE Xizang

The two groups of Cenozoic basaltic volcanism in SE Xizang have been extensively investigated in terms of petrology, mineralogy, chemical, and Sr-Nd isotopic compositions in previous studies (Chen et al., 2014; Flower et al., 2013; Guo et al., 2005; Huang, Chakraborty, et al., 2010; Huang et al., 2013; Li et al., 2002; Lu et al., 2015; Wang et al., 2001; Xia & Xu, 2005; Xu et al., 2001). In addition to the large difference in K_2O and K_2O/Na_2O, the two groups of lavas also display strong contrasts in the trace element patterns (Fig. 3), Sr-Nd isotopic ratios (Fig. 4), and olivine phenocryst composition (Fig. 5). These differences imply different mantle source domains for the two groups of volcanism.

Several lines of evidence suggest that the potassic-ultrapotassic rocks originated from a metasomatized, refractory lithospheric mantle. First, olivine phenocrysts are more forsteritic (Fo up to ~92; Table S2) than those of the alkali basalts (Fig. 5). To produce olivine with Fo=92, a melt with $Mg^{\#}$=77.5 is required using the exchange partition coefficient ($[Fe^{2+}/Mg]^{ol}/[Fe^{2+}/Mg]^{melt}$) = 0.30±0.03 (Roeder & Emslie, 1970). The high-$Mg^{\#}$ values of melts in equilibrium with olivine phenocrysts suggest a refractory (previously molten) mantle source (e.g., a harzburgite). A harzburgite source has also been recognized by the presence of high-Cr$^{\#}$ [Cr$^{\#}$=Cr^{3+}/(Cr^{3+}+Al), mostly >0.75] spinels in the K-rich rocks (Huang, Chakraborty, et al., 2010). The finding of phlogopite-bearing harzburgite xenoliths in these rocks supports the presence of a metasomatized, refractory lithospheric mantle beneath the study area (Lu et al., 2015). Second, the K-rich rocks display pronounced primitive mantle-normalized HFSE depletions and LILE enrichments and have Nb/U (1.8–5.2) and Ce/Pb (1.4–2.9) ratios much lower than those of MORBs and OIBs, which have uniform Nb/U ratios of ~47 and Ce/Pb ratios of ~25 (Hofmann, 2003). These features suggest an enriched mantle source hybridized by subduction-related fluids/melts. Hydrous metasomatism by subducting materials typically results in the formation of a potassium-rich phase, for example, phlogopite and amphibole (Förster et al., 2019; Prelević et al., 2012; Roden & Murthy, 1985). A potassium-rich phase is also required to plausibly account for high-K_2O contents and K_2O/Na_2O ratios (0.6–3.5) of the potassic-ultrapotassic lavas. Third, Sr and Nd isotopic compositions of these K-rich rocks resemble those of Cenozoic potassic-ultrapotassic lavas from northern Xizang that were widely proposed to have originated from the enriched lithospheric mantle (Ding et al., 2003; Guo et al., 2006; Turner et al., 1996; Williams et al., 2004) (Fig. 4), suggesting a similar mantle source. The K-rich lavas also have Sr and Nd isotopic ratios similar to those of the Permian Emeishan low-Ti basalts in SE Xizang (Song et al., 2008; Tian et al., 2017; Xiao et al., 2004) (Fig. 4). To summarize, the mantle source of the K-rich lavas has been subjected to strong depletion via previous melt extraction and became later overprinted by metasomatism (e.g., Guo et al., 2005; Huang, Chakraborty, et al., 2010; Lu et al., 2015; Xu et al., 2001).

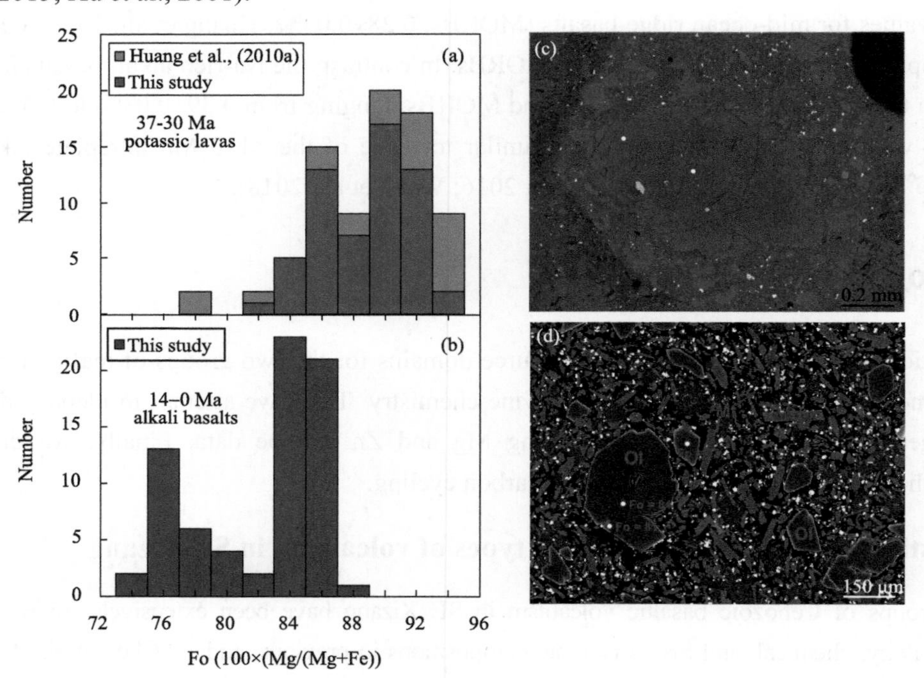

Fig. 5 (a and b) Comparison of Fo values in olivine phenocrysts from the two groups of lavas in this study. Data are listed in Table S2, and those reported in Huang, Chakraborty, et al. (2010) are also presented. Representative backscattered electron images of the potassium-rich lava (c) and the Na-rich alkali basalt (d). Ol = olivine; Cpx = clinopyroxene; Spl = spinel; Pl = plagioclase; and Mt = magnetite.

Apart from their relative enrichment in sodium, the alkali basalts show typical HFSE-rich trace element patterns of intraplate basalts (e.g., OIB-like). Their Nb/U ~53–88 and Ce/Pb ~15–29 are much higher than those of the potassic-ultrapotassic lavas and resemble those of oceanic basalts (Hofmann, 2003). The olivine phenocrysts are less forsteritic (73 to 87) than those of potassic-ultrapotassic rocks (82 to 92) (Fig. 5), suggesting a relatively fertile mantle source (e.g., pyroxenite). The depleted mantle-like Sr and Nd isotopic ratios also strongly contrast with the K-rich lavas (Fig. 4) and point to a magma source in the underlying asthenospheric mantle. Considering the lithosphere thickness of 80–100 km in this region (Liu et al., 1990; Yuan, 1989), the source depth of the alkali basalts must then be at least ~100 km. Overall, the contrasting chemical and Sr-Nd isotopic characteristics suggest that the two groups of volcanism in SE Xizang have distinct mantle source domains, in the enriched lithospheric mantle and the asthenospheric mantle, respectively. Both K-rich and Na-rich basaltic lavas also widely occur in the Circum-Mediterranean region. The Circum-Mediterranean K-rich lavas were interpreted to originate from the shallow mantle wedges metasomatized by subducting slabs (Lustrino et al., 2011), which are genetically similar to the K-rich lavas in SE Xizang. For the Circum-Mediterranean Na-rich igneous rocks, Lustrino and Wilson (2007) proposed a common sublithospheric mantle reservoir, with additional contributions from the HIMU component, the European Asthenospheric Reservoir and the Low Velocity Component.

5.2 Mg and Zn isotopic systematics of K-rich lavas

Magnesium isotopic composition of the K-rich lavas is indistinguishable from that of the mantle within the analytical uncertainty (Fig. 6). It is noteworthy that a large contrast in MgO content exists between peridotites

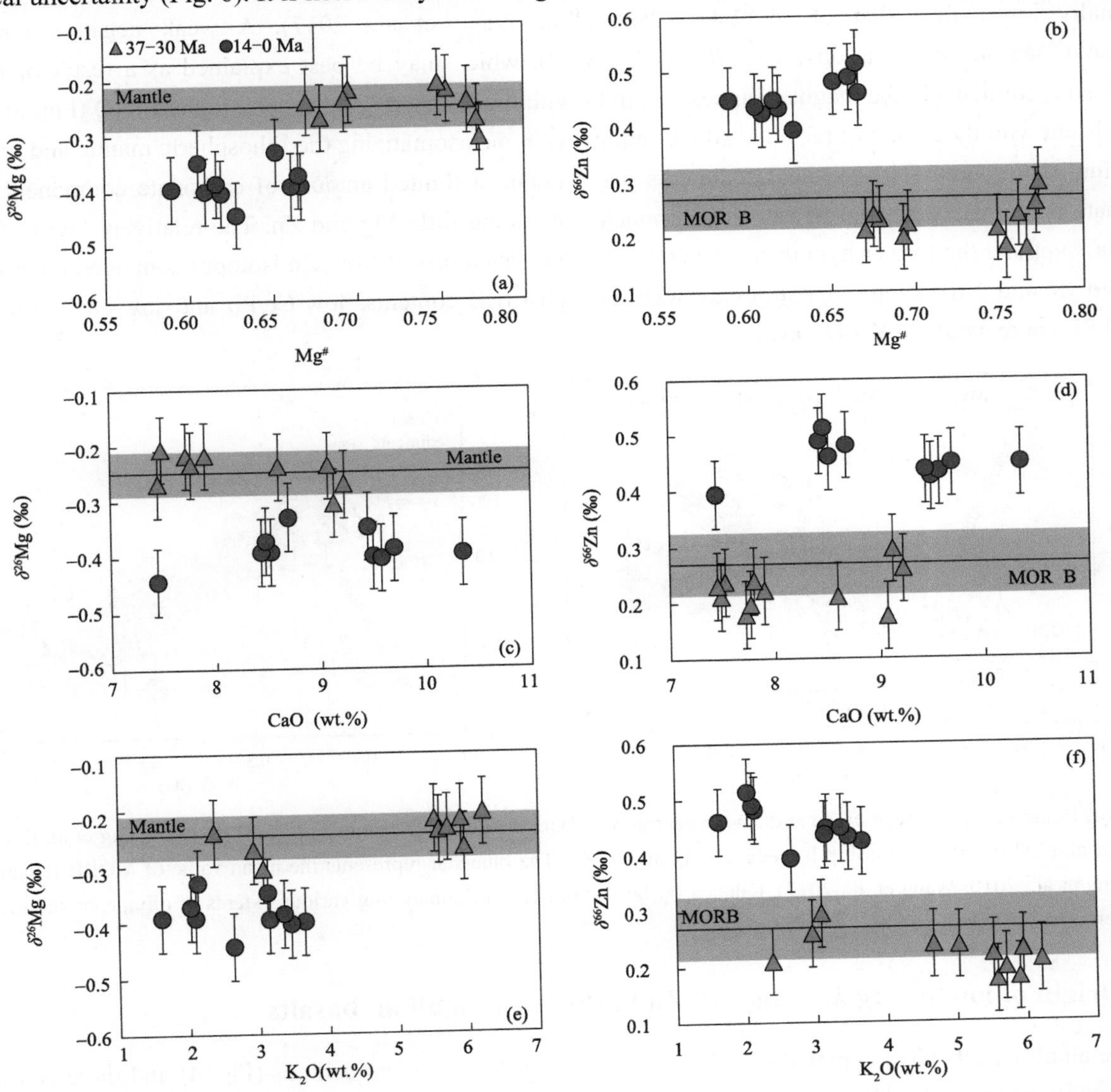

Fig. 6 Correlations of δ^{26}Mg and δ^{66}Zn values with Mg#, CaO and K_2O contents. The mantle δ^{26}Mg value (−0.25±0.04‰) is from Teng (2017) and the δ^{66}Zn range of MORBs is from Wang et al. (2017) and Huang et al. (2018).

(~38 wt.%) and siliciclastic sediments (e.g., Global Subducting Sediments, GLOSS-II, MgO = 2.75 wt.%; Plank, 2014), although the estimated average δ^{26}Mg for GLOSS-II (−0.34‰; Hu et al., 2017) is lower than the mantle value of −0.25±0.04‰ (Teng, 2017). Thus, binary mixing between mantle and siliciclastic sediment is unlikely to significantly modify Mg isotopic composition of the hybridized mantle even if substantial sediment volume has been involved (e.g., up to ~70%). Thus, despite extensive evidence for K-rich (phlogopite) metasomatism, this process has minimal effect on the Mg isotopic composition of the resulting ultrapotassic lavas. In contrast to Mg, Zn is a trace element in the mantle (~55 μg/g). The GLOSS-II has an average Zn concentration of 93±5.2 μg/g (Plank, 2014), resulting in a high $[Zn]_{sediment}/[Zn]_{mantle}$ ratio of ~1.7. To date, no Zn isotopic data is available for GLOSS, but existing data show that continental margin sediments can have light δ^{66}Zn values of low to −0.05‰ (−0.05‰ to 0.36‰; n= 36; Little et al., 2016). The large Zn concentration contrast between sediments and mantle, in conjunction with the light Zn isotopic compositions of some margin sediments, implies that Zn isotopic composition of the mantle can be modified if a large (>20%) amount of sediments is added.

Some of the K-rich lavas have δ^{66}Zn shifted toward slightly lighter values relative to MORB, although most samples are indistinguishable within analytical uncertainty (Fig. 6). The average δ^{66}Zn (0.22±0.03‰; n=11) is marginally lower than that of MORB (0.28±0.03‰; Wang et al., 2017). A weak negative correlation between δ^{26}Mg and δ^{66}Zn is observed (R^2=0.65, Fig. 7), which may be best explained as a result of recycled siliciclastic sediments in the magma sources. Samples with the lowest δ^{66}Zn have the highest K_2O (Fig. 6f), which is consistent with the model of recycled silicic sediments in metasomatizing the lithospheric mantle and creating a potassium-rich source. The recycled sediments may contain a limited amount of carbonate components or the carbonate is mainly composed of calcium carbonate containing little Mg and Zn. The relatively low CaO/Al_2O_3 (Fig. 8a) supports the former hypothesis. Overall, the conclusion driven from Zn isotopes is in agreement with the enriched Sr and Nd isotope signatures as well as high-LILE contents, low-Ce/Pb and low-Nb/U ratios, and low-HFSE contents of the K-rich lavas.

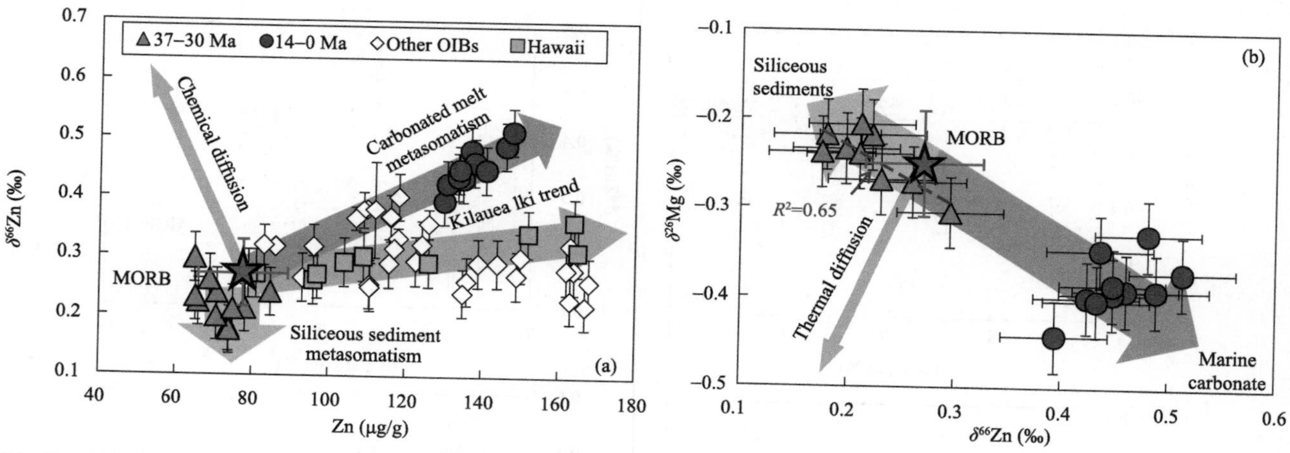

Fig. 7　(a) Relationship between δ^{66}Zn and Zn concentrations. Data for OIBs are from Chen et al. (2013), Wang et al. (2017), and Beunon et al. (2020). (b) Relationship between δ^{26}Mg and δ^{66}Zn. The blue star represents the mean value of MORB (Huang et al., 2018; Teng et al., 2010; Wang et al., 2017). Kilauea Iki lavas from Hawaii undergoing various extents of olivine crystallization are from Chen et al. (2013).

5.3　Origin of low δ^{26}Mg and high δ^{66}Zn in the Na-rich alkali basalts

The alkali basalts have depleted mantle-like Sr and Nd isotopic compositions (Fig. 4) and there is a lack of correlation between $^{87}Sr/^{86}Sr$ or $\varepsilon_{Nd}(t)$ and Mg number (not shown), indicating negligible, if any, crustal contamination. The positive Nb and Ta anomalies in the primitive mantle-normalized trace element patterns (Fig. 3) also exclude significant contamination by crustal materials. Contamination by crustal carbonate during magma ascending is also unlikely because sedimentary carbonates have high $^{87}Sr/^{86}Sr$ and much lower Nb contents compared with basaltic rocks (Turekian & Wedepohl, 1961). Thus, the significantly lighter-than-mantle Mg isotopic composition and heavier-than-mantle Zn isotopic composition of the alkali basalts (Fig. 6) should reflect isotopic variations of the primary, mantle-derived magmas. Several processes can potentially generate low $\delta^{26}Mg$ and high $\delta^{66}Zn$ in mantle-derived magmas, including partial melting, magmatic differentiation, diffusion effect, and/or source heterogeneity caused by recycled carbonates (Liu & Li, 2019). These processes must be taken into account because they potentially induce isotope fractionation in the same direction. Below we evaluate these possibilities.

5.3.1 Partial melting of a normal peridotite source and magmatic differentiation

Magnesium stable isotope fractionation during partial melting and fractional crystallization is limited (Liu et al., 2010; Teng et al., 2010). This is supported by the identical Mg isotopic composition among mantle peridotites, komatiites, and oceanic basalts (OIB, MORB) which are within the analytical uncertainty (Dauphas et al., 2010; Handler et al., 2009; Teng et al., 2010). The $\delta^{26}Mg$ values of the studied alkali basalts (−0.44±0.03‰ to −0.33±0.02‰) are detectably lower than those of the mantle (−0.25±0.04‰) and, thus, unable to have been caused by partial melting or fractional crystallization. Zinc isotopes may be fractionated during crystal-melt differentiation involving olivines that are isotopically lighter than the melts (McCoy-West et al., 2018; Sossi et al., 2015; Yang & Liu, 2019), but detectable fractionation (~0.1‰) only occurs at the very end of the differentiation sequence when MgO falls below 3 wt.% as observed in the Kilauea Iki lavas (Chen et al., 2013). The studied alkali basalts have near-primary compositions (MgO=8.7−12.2 wt.%) and have $\delta^{66}Zn$ values above the trend of Kilauea Iki lavas (Chen et al., 2013) as well as of global OIBs (Beunon et al., 2020; Wang et al., 2017) (Fig. 7a). There is also no visible correlation of $\delta^{66}Zn$ or $\delta^{26}Mg$ with indices of magmatic differentiation (e.g., $Mg^{\#}$ and CaO, Fig. 6; SiO_2, not shown), which is consistent with a negligible influence of magmatic differentiation on Mg and Zn isotopic systematics of the alkali basalts.

Partial melting induces non-negligible Zn isotope fractionation with the melts being slightly enriched in heavier Zn isotopes relative to the residue by 0.04–0.10‰ (Doucet et al., 2016; McCoy-West et al., 2018; Sossi et al., 2018; Wang et al., 2017). The fractionation is best attributed to a result of isotope difference between $^{VI}Zn^{2+}$ (sixfold coordination) in mantle minerals and $^{IV}Zn^{2+}$ (fourfold coordination) in silicate melts (Sossi et al., 2018), which accounts for the subtle difference of ~0.1‰ in $\delta^{66}Zn$ between the mantle (~0.18‰) and global basalt average (~0.28‰). The $\delta^{66}Zn$ of the alkali basalts (0.39‰ to 0.51‰) is up to 0.33‰ higher than the mantle value, and the difference is fairly beyond that induced by partial melting of a normal mantle source. Thereupon, the high $\delta^{66}Zn$ that is coupled with low $\delta^{26}Mg$ in the same samples (Fig. 7) is unlikely to have been produced by partial melting of a normal mantle source and/or igneous differentiation.

5.3.2 Diffusion effects

Large isotope fractionation of metals (e.g., Ca, Mg, and Fe) in silicate melts induced by thermal or chemical diffusion has been frequently observed under experimental conditions (e.g., Dominguez et al., 2011; X.-L. Huang, Niu, et al., 2010; Richter et al., 2008). During thermal diffusion (also known as Soret effect), the heavier isotopes always concentrate in cold regions and the lighter isotopes concentrate in hot regions. Thus, the ascending of hot basaltic magmas through the overlying mantle may drive diffusion of Mg and Zn and result in enrichments of both light Mg and Zn isotopes in the resultant melts (i.e., fractionated in the same direction; Dominguez et al., 2011; Huang, Niu, et al., 2010). However, the Na-rich alkali basalts from SE Xizang are characterized by lighter Mg but heavier Zn isotopic ratios relative to the mantle (Fig. 7). Thus, the Mg and Zn isotopic anomalies observed in these rocks are unlikely to have been caused by thermal diffusion between an isotopically normal basaltic melt with surrounding peridotites during ascent.

A chemical gradient exists between melt and surrounding peridotite when basaltic magmas ascend through the mantle. Since light isotopes diffuse faster than heavy isotopes (Richter et al., 2008), a gradient in chemical potential may drive the preferential diffusion of lighter Mg isotopes from surrounding peridotite (higher MgO) to melt (lower MgO), forming low-δ^{26}Mg melts. Unlike Mg, although Zn is moderately incompatible during partial melting of the mantle (Davis et al., 2013), the Zn concentration of normal basaltic melts is only marginally higher than that of mantle peridotites (~80 vs. 55 μg/g; Le Roux et al., 2010). Even if chemical diffusion happens that drives the diffusion of light Zn isotopes from melt to peridotite and result in elevated δ^{66}Zn of the melts, this process should generate a negative correlation between δ^{26}Mg and δ^{66}Zn in the alkali basalts, which is not observed (Fig. 7b). In addition, provided that the high δ^{66}Zn of a basaltic melt was produced via diffusion of Zn from an isotopically normal melt (e.g., MORB, δ^{66}Zn=0.28‰, [Zn]=~80 μg/g) to ambient peridotite, the resultant melt would have a Zn concentration lower than 80 μg/g due to diffusion-induced loss of Zn. This is contradicted by the high-Zn concentrations (130−148 μg/g) of the alkali basalts and their positive correlation with δ^{66}Zn (Fig. 7a). It is also unlikely that the high-δ^{66}Zn ratios were produced via chemical diffusion from an initial melt with Zn concentration higher than that of the basalt samples (>148 μg/g) to peridotites, because this process should similarly generate a negative correlation between δ^{66}Zn and Zn concentration. Thus, the coherent high-Zn concentrations and Zn isotopic ratios of the alkali basalts are unlikely to have been caused by chemical diffusion. Also, if diffusion-driven isotopic effect was a primary cause for isotopic anomalies of the Na-rich basaltic lavas, a similar effect (light δ^{26}Mg and heavy δ^{66}Zn) should also occur in the K-rich lavas, which is not observed (Fig. 6). Here we call for a careful treatment on chemical and thermal diffusion effects when treating Mg isotope data of mantle-derived rocks alone, and such effects may be qualitatively resolved by combining with Zn concentration and Zn isotope data in same samples (Liu & Li, 2019). Future quantitative simulations of the diffusion timescales may help to better identify the influence of diffusion effects on isotopic systematics of Mg and Zn in mantle-derived melts.

5.3.3 Source heterogeneity caused by recycled carbonates

Given that magmatic processes and diffusion effects are unlikely to account for the coupling of low-δ^{26}Mg, high-δ^{66}Zn, and high-Zn contents observed in the alkali basalts, the significant deviation of δ^{26}Mg and δ^{66}Zn of these rocks from the normal mantle value should reflect a hybridized mantle source. Marine carbonates are typically characterized by extremely light Mg and heavy Zn isotopic compositions (Kunzmann et al., 2013; Liu et al., 2017; Pichat et al., 2003; Wombacher et al., 2011) and serve as the robust candidate for a traceable metasomatic agent in the mantle. Experimental studies showed that silicate-undersaturated alkalic basaltic melts can be produced by partial melting of peridotite, eclogite, or pyroxenite with CO_2 being added (e.g., Dasgupta et al., 2007; Gerbode & Dasgupta, 2010; Kiseeva et al., 2013). The most striking differences of chemical compositions among these partial melts include CaO, SiO_2, Na_2O + K_2O, and TiO_2. As illustrated in Fig. 8a, the Na-rich alkali basalts bear compositional similarities to experimental partial melts of CO_2 + pyroxenite in term of CaO/Al_2O_3 and SiO_2. In the diagram of Na_2O+K_2O versus TiO_2 (Fig. 8b), the samples have higher TiO_2 compared with partial melts of CO_2 + peridotite and higher Na_2O + K_2O compared with partial melts of CO_2 + eclogite. Instead, they almost follow the trend of partial melts of CO_2 + pyroxenite, although the samples have relatively lower TiO_2 at a given K_2O + Na_2O compared with carbonated pyroxenite-derived melts. We hypothesize that such discrepancies might be resolved if a hybrid peridotite-carbonated pyroxenite source is considered or the melt underwent reaction with the lithospheric mantle during ascent. For simplification, here we consider only a role of pyroxenite and carbonate in the magma sources.

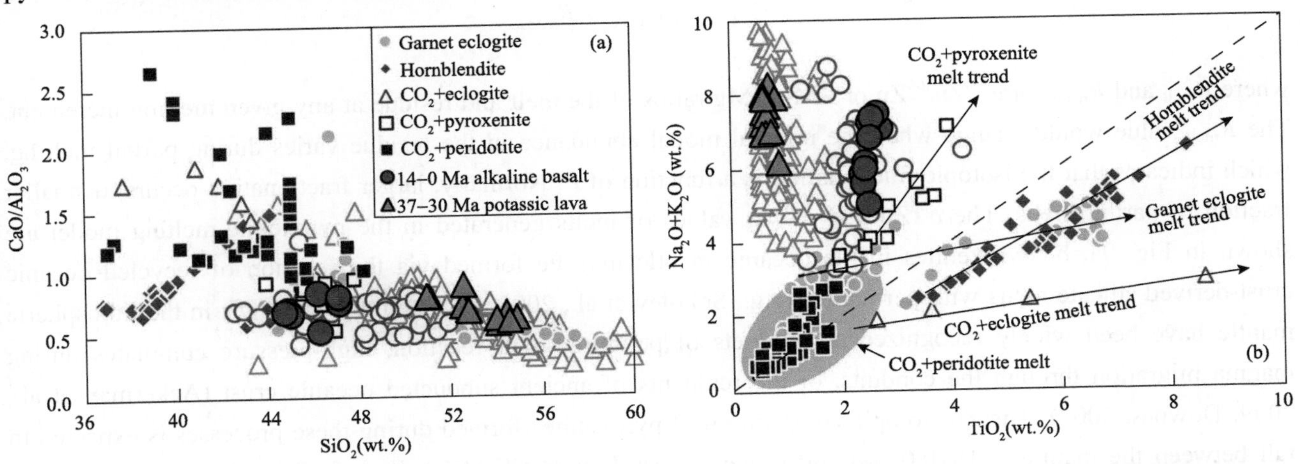

Fig. 8 Plots of CaO/Al_2O_3 versus SiO_2 (a) and $Na_2O + K_2O$ versus TiO_2 (b). Experimental melts from partial melting of garnet pyroxenite (Hirschmann et al., 2003; Keshav et al., 2004; Kogiso et al., 2003; Kogiso & Hirschmann, 2006; Pertermann & Hirschmann, 2003), hornblendite (Pilet et al., 2008), CO_2 + eclogite (Dasgupta et al., 2006), CO_2 + pyroxenite (Gerbode & Dasgupta, 2010), and CO_2 + peridotite (Dasgupta et al., 2007, 2013) are also shown for comparison. The open circles denote literature data as same as in Fig. 3.

Here we choose an incremental melting model in order to simulate the chemical and isotopic variation during pyroxenite melting (Williams & Bizimis, 2014; Yang & Liu, 2019; Zhong et al., 2017). The modeling parameters are listed in Table 2. In detail, the modal mineralogy of the residue in this incremental melting model is variable during the melting process, and the mass of melt at each step is constantly set as 2% of the bulk mass. Each step n ($n = 1, 2, 3, \ldots$) of melting degree (F) is calculated according to

$$F_n = \frac{2\%}{1-(n-1)\times 2\%}$$

The Zn, Mg, and Fe contents of the pyroxenite-derived melt are calculated using the non-modal melting equation (Shaw, 1970), which involved the bulk partition coefficients for each element (Le Roux et al., 2015), the melting mode and the initial mineral modal abundance (Pertermann et al., 2004). Then, the isotope fractionation factor between melt and residue is calculated using the equation:

$$\alpha_{\text{melt-residue}} = \alpha_{\text{melt-cpx}} \times \left(\sum_{i=1}^{\text{mineral}} \left[M_i \cdot \text{Zn(Mg)}_{\text{mineral}} \right] \Big/ \sum_{i=1}^{\text{mineral}} \left[M_i \cdot \text{Zn(Mg)}_{\text{mineral}} \cdot \alpha_{\text{mineral-cpx}} \right] \right)$$

The "M_i" represents the normalized mineral modal abundance in each incremental melt and the $\alpha_{\text{mineral-cpx}}$ equals the isotope fractionation factor between coexisting garnet and clinopyroxene in source pyroxenite. The Zn isotope fractionation factor of $\alpha_{\text{melt-cpx}}$ and $\alpha_{\text{grt-cpx}}$ can be calculated using the equations of $\Delta^{66}Zn_{\text{cpx-melt}} = -0.17\times 10^6/T^2$ and $\Delta^{66}Zn_{\text{grt-cpx}} = -0.03\times10^6/T^2$ (McCoy-West et al., 2018; Sossi et al., 2018) where the temperature is set as 1573 K. Then, the $\alpha_{\text{melt-residue}}$ value is ~1.000075 during the pyroxenite melting at 1573 K. For Mg, clinopyroxene is isotopically heavier than melt and garnet (Wang et al., 2012), meaning that the $\alpha_{\text{melt-cpx}}$ and $\alpha_{\text{grt-cpx}}$ values are smaller than 1. Here we choose a constant $\alpha_{\text{melt-residue}}$ of 1.00006 at 1573 K. Given that, Mg and Zn isotopic compositions of the melt and residue can be calculated as follows:

$$R_{\text{melt}} = \alpha_{\text{melt-residue}} \times R_{\text{residue}}$$

where R_{melt} and R_{residue} are $^{66}Zn/^{64}Zn$ or $^{26}Mg/^{24}Mg$ ratios of the melt and residue at any given melting increment. The R_{melt} value would change when the mineral modal abundance in the residue varies during partial melting, which indicates that the isotopic fractionation is a function of F. Normally, larger fractionation occurs in smaller fraction of partial melts. The $\delta^{66}Zn$ and $\delta^{26}Mg$ values of melts generated in the pyroxenite-melting model are shown in Fig. 9a, b. Pyroxenites in the oceanic mantle may be formed via the reaction of recycled oceanic crust-derived silicate melts with peridotites (e.g., Sobolev et al., 2005, 2007), and pyroxenites in the lithospheric mantle have been widely recognized as products of peridotite-melt reaction, high-pressure cumulates during magma migration through the conduits, or the remnants of ancient subducted oceanic crust (Ackerman et al., 2009; Downes, 2007). The Zn isotopic composition of pyroxenites formed during these processes is expected to fall between the mantle (0.18±0.05‰) and oceanic crust-derived silicate melts (~0.30‰). Here we assume an intermediate $\delta^{66}Zn$ value of 0.23‰, which is consistent with the value reported for natural pyroxenites in massif ultramafic rocks (Sossi et al., 2018), as the initial source pyroxenite isotopic composition in this melting model. The Zn and FeO concentrations of the pyroxenite end-member are calculated as weighted mean of minerals (clinopyroxene and garnet) on the basis of their modal abundance. For Mg isotopes and given the similar $\delta^{26}Mg$ between MORB and the mantle (Teng et al., 2010), a mantle-like $\delta^{26}Mg$ value of −0.25‰ and MgO content of 17.8 wt.% for the source pyroxenite are used in this melting model (Wang et al., 2012). The results show that partial melting of pyroxenite alone is unlikely to produce melts with $\delta^{66}Zn$ higher than 0.30‰ (Fig. 9a) and high-Zn/Fe ratios of the alkali basalts. Partial melting of pyroxenite alone would yield melts with higher $\delta^{26}Mg$ relative to the source (Fig. 9b), which is also in conflict with low $\delta^{26}Mg$ of the alkali basalts. This inconsistency is further reinforced by the differences in the major element oxides as shown in Fig. 8b.

Table 2　The parameters utilized in partial melting modeling of pyroxenite and mixing calculations

Mineral	Modal	Mode	Zn	FeO	D_{Zn}	D_{Fe}	D_{Mg}	$\Delta^{66}Zn_{\text{mineral-melt}}$	$\Delta^{26}Mg$
Cpx	0.75	0.10	55	5.3	0.19	0.94	2.79	$-0.17\times10^6/T^2$	
Grt	0.25	0.90	45	12	0.01	2.47	4.49	$-0.20\times10^6/T^2$	
Sum	1.00	1.00	52.5	6.97	0.73	1.32	3.21	$-0.18\times10^6/T^2$	$-0.12\times10^6/T^2$

End-member	Zn	$\delta^{66}Zn$ (‰)	Sr	$^{87}Sr/^{86}Sr$	MgO	$\delta^{26}Mg$ (‰)
Pyroxenite	52.5	0.23	113	0.7025	19.98	−0.25
Dolomite	132	0.9	1151	0.7099	22.00	−2.00
Magnesite	449	0.9	1.84	0.7164	47.60	−1.00
Calcite/Aragonite	20	0.9	1311	0.7088	0.80	−3.00
Sediments	93	0.11	302	0.7124	2.75	−0.34

Note: The melting mode and modal mineral abundances for pyroxenite melting are from Pertermann et al. (2004). Partition coefficients of Zn, Fe, and Mg are from Le Roux et al. (2015) and Pertermann et al. (2004). The $\delta^{66}Zn$ value of pyroxenite is from Sossi et al. (2018), and the contents of Zn (μg/g) and FeO (wt.%) are calculated from the weighted mean of minerals in pyroxenite. The $\Delta^{66}Zn_{\text{mineral-melt}}$ (‰) values are from Sossi et al. (2018) and McCoy-West et al. (2018), and $\Delta^{26}Mg_{\text{residue-melt}}$ (‰) is from Wang et al. (2020). The Sr content (μg/g) and $^{87}Sr/^{86}Sr$ and $\delta^{26}Mg$ values of pyroxenite are assumed to be similar to those of N-MORB (Hofmann, 1988; Teng et al., 2010). The Zn concentrations (μg/g) of dolomite and magnesite are from Li et al. (2014), and the Zn concentration of calcite/aragonite is from Turekian and Wedepohl (1961). The $\delta^{66}Zn$ of carbonate phases is the average of Mesozoic-Cenozoic sedimentary carbonates (Pichat et al., 2003; Sweere et al., 2018), and Sr contents and $^{87}Sr/^{86}Sr$ ratios are from Huang and Xiao (2016). The MgO contents (wt.%) and $\delta^{26}Mg$ of carbonates are reasonably assumed from data of sedimentary carbonates (Teng, 2017, and reference therein). The MgO, Zn, and Sr contents and $^{87}Sr/^{86}Sr$ of sediments are from Plank (2014). The $\delta^{26}Mg$ and $\delta^{66}Zn$ values of sediments are from Little et al. (2016) and Hu et al. (2017).

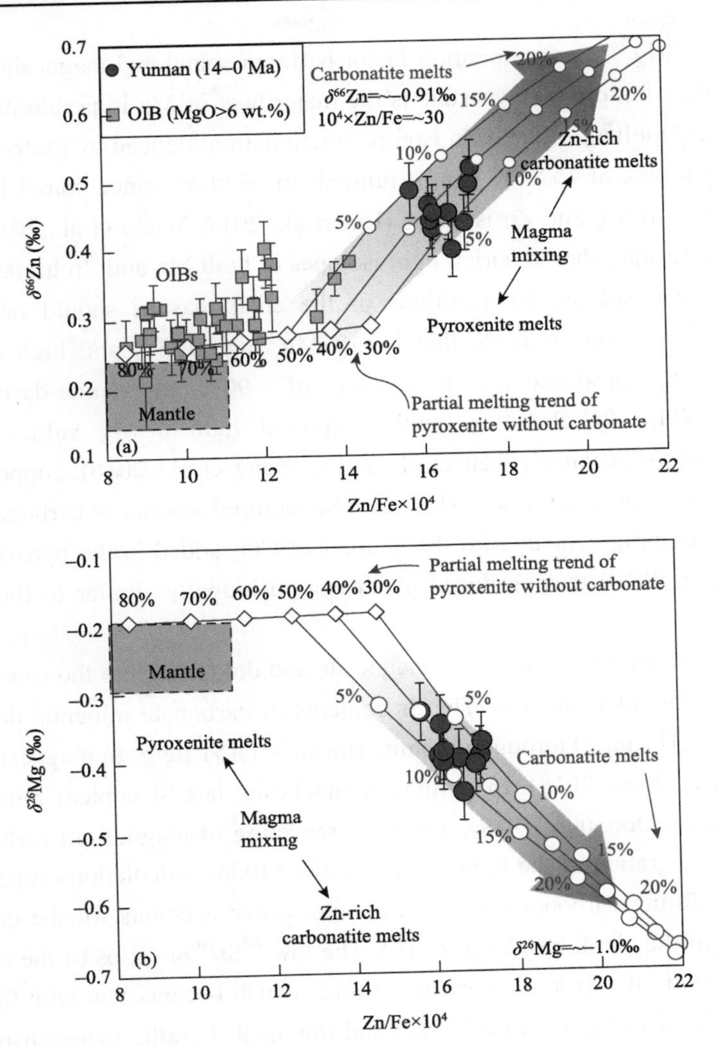

Fig. 9 Plots of δ^{66}Zn and δ^{26}Mg against Zn/Fe ($\times10^4$). Modeling results for pyroxenite melting are shown as blue curves, and diamonds denote 10% increment in degree of partial melting from 30% to 80%. Modeling parameters are given in Table 2. The OIBs with high MgO (>6 wt.%) are from Chen et al. (2013), Wang et al. (2017), and Beunon et al. (2020). Red curves represent mixing of Zn-rich carbonatite melts with pyroxenite-derived melts formed at various melting degrees (30%, 40%, and 50%). The results show that the Na-rich alkali basalts are the products of mixing of pyroxenite melts and carbonatite melts. See text for more details.

All of the above discrepancies can be consistently resolved by adding carbonate components into the sources. Carbonatite melts can gain high Zn/Fe during partial melting ($D^{Zn}_{Carb.\text{-}Melt}$ = 0.44, $D^{Fe}_{Carb.\text{-}Melt}$ = 0.63; Martin et al., 2012) and high MgO when magnesite or dolomite dominates the carbonate species (Merlini et al., 2012). Calcic carbonate (e.g., calcite/aragonite) is commonly poor in Zn (~20 μg/g; Turekian & Wedepohl, 1961), but dolomite and magnesite in ultrahigh-pressure (deeply subducted) metamorphic rocks have up to 147 and 449 μg/g Zn, respectively (Li et al., 2014). To quantify the amount of recycled carbonate in the magma source, here the δ^{66}Zn and $10^4\times$Zn/Fe of the magnesium carbonate are set as ~0.9‰ and ~30 (Schultz et al., 2004; Walter et al., 2018), and the δ^{26}Mg of dolomite and magnesite is set as –2.0‰ and –1.0‰, respectively (Table 2). The degree of melting for pyroxenite varies from 30% to 50%, and the carbonate is assumed to melt completely. Theoretical calculations show that Zn carbonate ($ZnCO_3$) is isotopically lighter by ~1‰ than silicate at room temperatures (Ducher et al., 2016), but when calculated to 1573 K, the fractionation is reduced to near zero (<–0.04‰). Similarly, experimental studies reported the Mg isotope fractionation between magnesite ($MgCO_3$) and forsterite (Mg_2SiO_4) to be close to zero (–0.04 ± 0.04‰) at 800 °C (Macris et al., 2013), which will become even less at higher temperatures. The $^{26}Mg/^{24}Mg$ fractionation factor between spinel and magnesite is large (+0.90±0.28‰ at 800 °C; Macris et al., 2013), but olivine/pyroxene is the major host of Mg in peridotite/pyroxenite. Thus, even if carbonate is not completely molten, Mg and Zn isotope fractionation induced by preferential melting of carbonate in the carbonate-bearing source at 1573 K is still limited. In addition, since spinel is the isotopically heaviest mineral in peridotites for both Mg and Zn isotopes (Liu et al., 2011; Wang et al., 2017), preferential melting of carbonate with spinel as a residue should enrich light isotopes of both Mg and Zn in melts, which is not observed. Thereupon, the Mg and Zn isotopic compositions of the alkali basalts should nearly represent the source compositions. The modeling results indicate that low-$\delta^{26}Mg$, high-$\delta^{66}Zn$, and high Zn/Fe ratios of the alkali basalts can be consistently produced by a mixture of ~90% pyroxenite-derived melts and ~8–10% carbonate-derived melts (Fig. 9). Previous studies reported light $\delta^{26}Mg$ values of down to −1.2‰ in mantle-derived, carbonated pyroxenites (Shen et al., 2018; Wang et al., 2016), supporting a carbonate-bearing pyroxenite source for the Na-rich alkali basalts (Fig. 9). The required amount of carbonate in this model (8–10%), equivalent to 4–5% CO_2, is in agreement with the amount of CO_2 added in the pyroxenite-melting experiments (5%, Gerbode & Dasgupta, 2010) that produced melts compositionally similar to the studied alkali basalts as discussed above (Fig. 8).

The addition of magnesium carbonates (e.g., magnesite and dolomite) into the sources is also consistent with the low-$^{87}Sr/^{86}Sr$ ratios of the alkali basalts. The Sr contents of carbonate minerals dramatically decrease from calcite/aragonite (mean = 1,311 μg/g) through dolomite (mean = 1,151 μg/g) to magnesite (mean = 1.84 μg/g; see data compiled by Huang & Xiao, 2016). In particular, magnesite has Sr content much lower than that of the mantle (21.1 μg/g; Sun & McDonough, 1989), and thus, recycling of magnesium carbonates is not expected to significantly modify $^{87}Sr/^{86}Sr$ ratios of the hybridized mantle. Mixing calculations supports this assumption that addition of ~8–10% magnesium carbonates to a pyroxenite source accounts for the covariation of Mg and Zn isotopes with Sr isotopes of the alkali basalts (Fig. 10). The low-$^{87}Sr/^{86}Sr$ ratios of the alkali basalts also imply a limited time-integrated growth of ^{87}Sr in the enriched source, which is consistent with the short duration between Cenozoic subduction of the Neo-Tethys oceanic slab and the alkali basaltic magmatism starting at 14 Ma. The recycled carbonate-bearing oceanic crust is expected to partially melt and produce carbonated silicate melts, which could react with peridotites to form silica-deficient pyroxenites (Fig. 11). Subsequent partial melting of the carbonate-bearing pyroxenites would produce the alkali basaltic melts. Collectively, the major elemental compositions and Mg-Zn-Sr isotopic ratios of the alkali basalts are consistent with a pyroxenite source that was hybridized by recycled magnesium carbonates from the subducted oceanic crust.

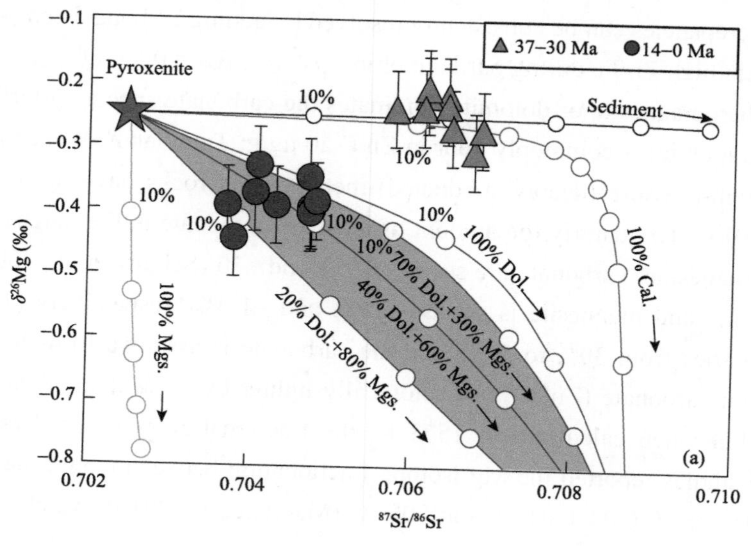

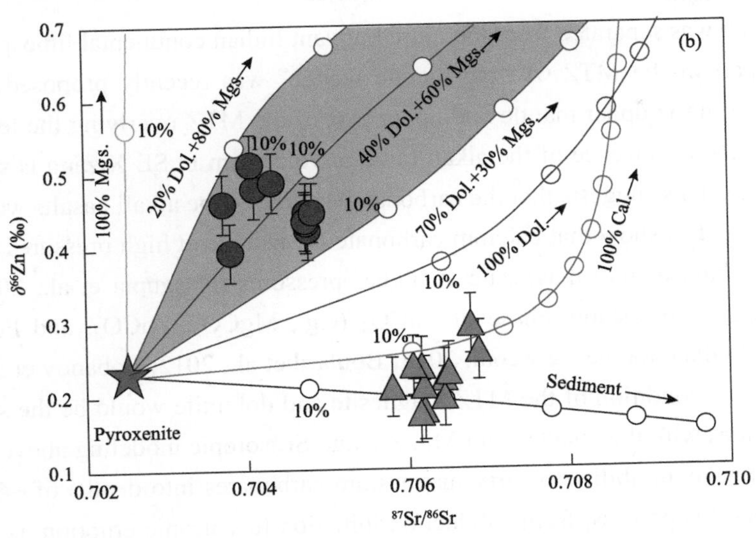

Fig. 10 Plots of δ^{26}Mg (a) or δ^{66}Zn (b) versus ^{87}Sr/^{86}Sr. The lines denote binary mixing between pyroxenite (blue star) and different species of carbonates as well as siliciclastic sediments (black curve). Modeling parameters are given in Table 2. The amount of carbonates (~8–10%) obtained in this modeling is in agreement with that obtained in Fig. 9. See text for more details.

5.4 Implications for deep carbon recycling

The carbon that enters subduction zones includes carbonate and reduced organic carbon that resides within the sedimentary and oceanic crust and the mantle layer of the incoming plate, but carbonate sediments dominate the subducting flux of carbon (Plank and Manning, 2019). Thus, carbon recycled into the mantle is mainly in the form of carbonate. Estimates of total annual carbonate subduction vary from ~66 to 103 megatons of carbon per year (Mt C yr^{-1}), with about 60% present in sediments and 40% in the altered oceanic lithosphere (Galvez and Pubellier, 2019). Dehydration and melting in subduction zones, however, do not completely remove the subducted carbonate. For example, an estimate of the C budget in subduction zones showed that the expelled carbon from arc lavas is much less than that subducted, leaving ~20 Mt C yr^{-1} unaccounted for (Kelemen and Manning, 2015). More recent estimates by Plank and Manning (2019) suggest that arc outputs currently represent a lower proportion of the inputs (~27%), roughly half of that derived from previous estimates (Kelemen and Manning, 2015). This is mainly due to a significant increase of new C flux estimate in marine carbonates since the mid-Cretaceous (Dutkiewicz et al., 2018). Therefore, a considerable mass of carbonates can survive beyond the subarc regime and has been recycled into the deeper mantle, but the final fate of the subducted carbonates is still open. There are two geodynamically distinct reservoirs of subducted carbon in the mantle: the deep lithospheric keels beneath continents and the more voluminous, convecting upper mantle, MTZ, and upper part of the lower mantle (Plank and Manning, 2019). A considerable portion of subducted carbonates may have been released into the mantle wedge or finally delivered into the continental lithosphere through subduction or continent-arc collision (Ague and Nicolescu, 2014; Foley and Fischer, 2017; Kelemen and Manning, 2015). Recent studies demonstrated that the subducted carbonated oceanic crust will undergo melting at depths of approximately 300 to 700 km, implying that carbonates can be delivered to the convective upper mantle (Thomson et al., 2016). However, evidence from natural observations on mantle-derived rocks for the deep subduction of carbonates beyond the subarc regime into the upper mantle or the MTZ is still rare (e.g., Li et al., 2017; Mazza et al., 2019).

At ~600 km depth, a large-scale high-velocity anomaly exists in SE Xizang that is composed of rigid rocks and possibly associated with the subducted Neo-Tethys oceanic slab (Huang & Zhao, 2006) (Fig. 11). Based on seismic observations (Huang & Zhao, 2006), the break off of the subducted Neo-Tethys slab happened at ~400 km depth and the oceanic slab was separated from the more buoyant Indian continental lithosphere and finally became stagnant in the lower part of the MTZ. A "big mantle wedge" was recently proposed to exist in SE Xizang, defined as a broad region in the upper mantle and upper part of the MTZ overlying the long stagnant slab (Lei et al., 2019). Interestingly, the occurrence of the alkali basaltic volcanism in SE Xizang is spatially coincident with the stagnant slab (Fig. 11). This suggests that the carbonated source of the alkali basalts was most likely located in the MTZ. Experimental studies show that calcium carbonate is unstable at high pressures and can be transformed into dolomite at 23–45 kbars or into magnesite at higher pressures (Dasgupta et al., 2004; Kushiro, 1975). In particular, the high-pressure metastable phase of $CaCO_3$ (e.g., $MgCO_3$, $ZnCO_3$, and $FeCO_3$) can be stable at pressures greater than 62 GPa (Bouibes & Zaoui, 2014; Boulard et al., 2012; Lobanov et al., 2015; Merlini et al., 2012). Thus, under the P–T condition of the MTZ magnesite and dolomite would be the stable phase of recycled carbonates, which coincides with the results from Mg, Zn, and Sr isotopic modeling above. This indicates that the subducted oceanic slabs have an ability to carry magnesium carbonates into depths of ~410–660 km. The cycle duration of carbon during this process, from subduction initiation to volcanic eruption, is estimated to be tens of millions of years or longer. This duration of this cycle is much longer than that of arc-trench systems (5–10 Ma; Kelemen & Manning, 2015) due to the longer pathway of carbon in the surface-MTZ-surface cycle.

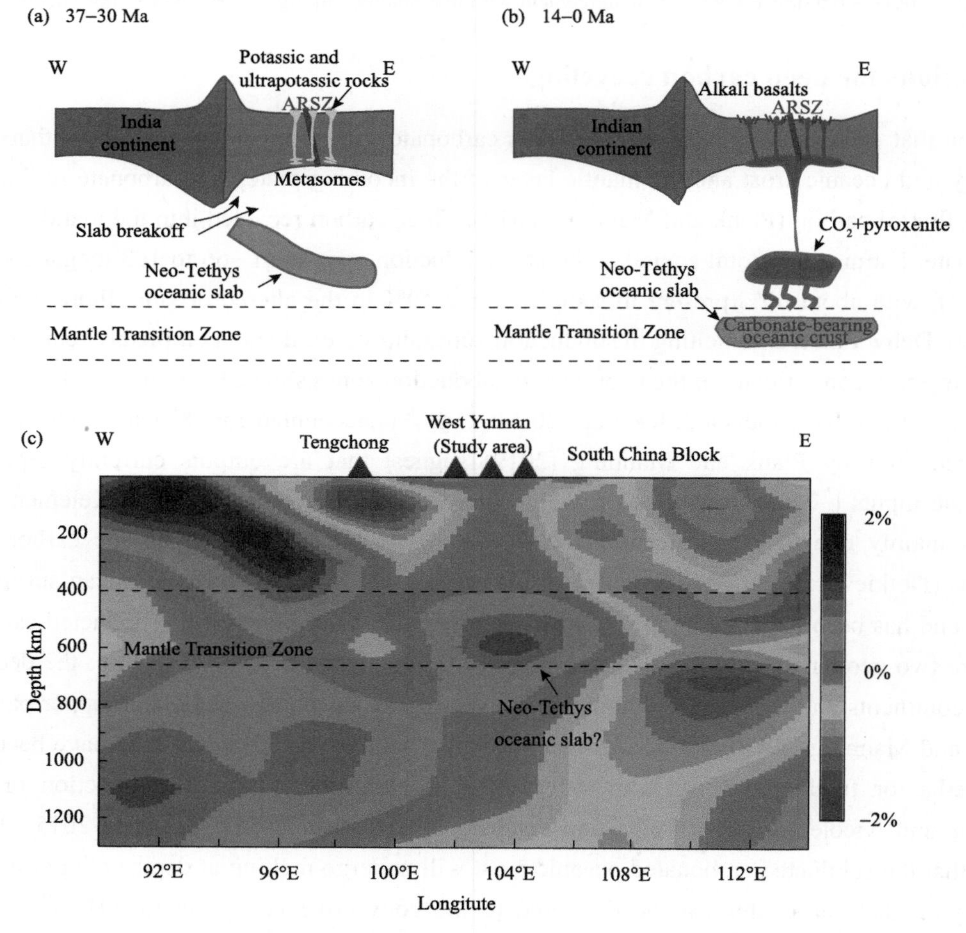

Fig. 11 (a and b) Schematic cartoon models illustrating the generation of Cenozoic mafic lavas in SE Xizang. (c) A vertical cross section of P wave velocity perturbations along the Tengchong volcanic field, western Yunnan and South China Block (after Huang & Zhao, 2006). The occurrence of the Na-rich alkali basalts is spatially coincident with a stagnant slab in the MTZ, suggesting a carbonate-bearing source in the MTZ.

Observations on ultrahigh-pressure metamorphic rocks from the Alps show that magnesite and dolomite occur as major solid inclusions within garnet (Frezzotti et al., 2011), which supports the deep subduction of Mg-rich carbonates. Several studies have also reported the presence of a deep storage of the subducted carbonate at the lower part of the transition zone (>580 km) or even the lower mantle (>670 km) based on investigations on mineral inclusions within super-deep diamonds or garnets (Anzolini et al., 2019; Brenker et al., 2007; Nestola et al., 2018). These studies, together with the evidence for recycled carbonates in the MTZ carried by the Neo-Tethys oceanic slab in this study, imply a significant flux of recycled carbonate in the Earth's mantle. Globally, carbon input flux at subduction zone may vary significantly for each downgoing plate according to their formation, evolution, and sedimentation history (Galvez & Pubellier, 2019; Plank & Manning, 2019). Future investigations of magmas from deep mantle sources in other subduction zones will test the hypothesis of a significant flux of recycled carbonates in the deep mantle and improve our understanding of the fate of subducted carbon and the Earth's deep carbon cycle over geologic time.

6 Conclusions

Aiming to discriminate between carbonate and silicate metasomatism and to effectively identify recycled carbonates in the mantle, we studied two groups of Cenozoic mafic volcanism (K-rich vs. Na-rich) widely distributed in the SE Tibetan Plateau. The K-rich rocks have an enriched lithospheric mantle origin, whereas the Na-rich alkali basalts represent partial melting products of the asthenospheric mantle. The slightly light Zn isotopic compositions and enriched Sr-Nd isotopic ratios of the K-rich lavas indicate that their mantle source was modified by siliciclastic sediments. By contrast, the alkali basalts have low-δ^{26}Mg, high-δ^{66}Zn, and high Zn/Fe ratios that cannot be attributed to magmatic processes or diffusion effects and that are interpreted to reflect source heterogeneity caused by recycled carbonates. Along with the compositional similarity with experimental partial melts of CO_2 + pyroxenite, the alkali basalts are proposed to have been produced by melting of a hybridized source of ~90% pyroxenite and ~8–10% magnesium carbonate (magnesite + dolomite). Provided that magnesite is stable, the recycled carbonates can be subducted into depths of over ~300 km. Seismic tomography revealed a stagnant plate in the MTZ in the study area, which could represent the deeply recycled carbonate-bearing oceanic crust. Collectively, our studies provide evidence from natural observations for the potential of oceanic subduction to transport surface carbon into the MTZ (410–660 km) at timescales that significantly exceed those of arc-trench cycle.

Acknowledgements

We thank Paolo A. Sossi and an anonymous reviewer for their careful comments and suggestions that helped to improve this manuscript. The editorial handing and comments by the editors are greatly acknowledged. We thank Xun Wang, Meng-Lun Li, and Ning Ma for their assistance with sampling collection, Li-Juan Xu for her assistance with Sr–Nd isotope analysis, and Jinhua Hao and Dexing Kong for their help with olivine phenocryst analysis. This work is supported by the "Strategic Priority Research Program" of the Chinese Academy of Sciences (Grant XDB18000000), the National Natural Science Foundation of China (Grant 41730214), and the National Key Research and Development Project of China (Grants 2016YFC0600310) to S. A. L. Analytical results for whole rocks and olivine chemistry are listed in the supporting information and can be found on figshare (doi:10.6084/m9.figshare.12745106.v1).

References

Ackerman, L., Jelínek, E., Medaris, G. Jr., Ježek, J., Siebel, W., & Strnad, L. (2009). Geochemistry of Fe-rich peridotites and associated pyroxenites from Horní Bory, Bohemian Massif: Insights into subduction-related melt-rock reactions. Chemical Geology, 259, 15-167. https://doi.org/10.1016/j.chemgeo.2008.10.042.

Ague, J. J. & Nicolescu, S. (2014) Carbon dioxide released from subduction zones by fluid-mediated reactions. Nature Geoscience 7, 355. https://doi.org/10.1038/ngeo2143.

Anzolini, C., Nestola, F., Mazzucchelli, M. L., Alvaro, M., Nimis, P., Gianese, A., et al. (2019) Depth of diamond formation obtained from single periclase inclusions. Geology, 47, 219-222. https://doi.org/10.1130/g45605.1.

Aubaud, C., Pineau, F., Hékinian, R., Javoy, M. (2005) Degassing of CO_2 and H_2O in submarine lavas from the Society hotspot. Earth and Planetary Science Letters 235, 511-527. https://doi.org/10.1016/j.epsl.2005.04.047.

Aubaud, C., Pineau, F., Hékinian, R., & Javoy, M. (2005). Degassing of CO_2 and H_2O in submarine lavas from the Society hotspot. Earth and Planetary Science Letters, 235, 511-527. https://doi.org/10.1016/j.epsl.2005.04.047.

Bailey, D. K. (1982) Mantle metasomatism—continuing chemical change within the Earth. Nature 296, 525. https://doi.org/10.1038/296525a0.

Beunon, H., Mattielli, N., Doucet, L.S., Moine, B., Debret, B. (2020) Mantle heterogeneity through Zn systematics in oceanic basalts: Evidence for a deep carbon cycling. Earth Science Reviews, https://doi.org/10.1016/j.earscirev.2020.103174.

Bouibes, A. & Zaoui, A. (2014) High-pressure polymorphs of $ZnCO_3$: Evolutionary crystal structure prediction. Scientific Report 4, 5172. https://doi.org/10.1038/srep05172.

Boulard, E., Menguy, N., Auzende, A.-L., Benzerara, K., Bureau, H., Antonangeli, D., et al. (2012) Experimental investigation of the stability of Fe-rich carbonates in the lower mantle. Journal of Geophysical Research: Solid Earth 117, 148-227 https://doi.org/10.1029/2011jb008733.

Brenker, F. E., Vollmer, C., Vincze, L., Vekemans, B., Szymanski, A., Janssens, K., et al. (2007) Carbonates from the lower part of transition zone or even the lower mantle. Earth and Planetary Science Letters, 260, 1-9. https://doi.org/10.1016/j.epsl.2007.02.038.

Chen, H., Savage, P. S., Teng, F.-Z., Helz, R. T. & Moynier, F. (2013) Zinc isotope fractionation during magmatic differentiation and the isotopic composition of the bulk Earth. Earth and Planetary Science Letters 369, 34-42. https://doi.org/10.1016/j.epsl.2013.02.037.

Chen, Y., Yao, S. & Pan, Y. (2014) Geochemistry of lamprophyres at the Daping gold deposit, Yunnan Province, China: constraints on the timing of gold mineralization and evidence for mantle convection in the eastern Tibetan Plateau. Journal of Asian Earth Sciences 93, 129-145. https://doi.org/10.1016/j.jseaes.2014.07.033.

Dasgupta, R., Hirschmann, M. M, Withers, A. C. (2004) Deep global cycling of carbon constrained by the solidus of anhydrous, carbonated eclogite under upper mantle conditions. Earth and Planetary Science Letters 227, 73-85. https://doi.org/10.1016/j.epsl.2004.08.004.

Dasgupta R., Hirschmann M.M., Stalker K. (2006) Immiscible transition from carbonate-rich to silicate-rich melts in the 3 GPa melting interval of eclogite+CO_2 and genesis of silica-undersaturated ocean island lavas. Journal of Petrology 47 (4), 647-671. https://doi.org/10.1093/petrology/egi088.

Dasgupta, R., Hirschmann, M. M. & Smith, N. D. (2007) Partial melting experiments of peridotite+ CO_2 at 3 GPa and genesis of alkalic ocean island basalts. Journal of Petrology 48, 2093-2124. https://doi.org/10.1093/petrology/egm053.

Dasgupta R., Mallik A., Tsuno K., Withers A.C., Hirth G., Hirschmann M.M. (2013) Carbon-dioxide-rich silicate melt in the Earth's upper mantle. Nature 493 (7431), 211. https://doi.org/10.1038/nature11731.

Dauphas N., Teng F. Z. and Arndt N. T. (2010) Magnesium and iron isotopes in 2.7 Ga Alexo komatiites: Mantle signatures, no evidence for Soret diffusion, and identification of diffusive transport in zoned olivine. Geochimica et Cosmochimica Acta 74, 3274-3291. https://doi.org/10.1016/j.gca.2010.02.031.

Davis, F. A., Humayun, M., Hirschmann, M. M. & Cooper, R. S. (2013) Experimentally determined mineral/melt partitioning of first-row transition elements (FRTE) during partial melting of peridotite at 3 GPa. Geochimica et Cosmochimica Acta 104, 232-260. https://doi.org/10.1016/j.gca.2012.11.009.

Debret, B., Bouilhol, P., Pons, M. L., & Williams, H. (2018). Carbonate Transfer during the Onset of Slab Devolatilization: New Insights from Fe and Zn Stable Isotopes. Journal of Petrology. 59(6): 1145-1166. https://doi.org/10.1093/petrology/egy057.

Deines, P. (2002) The carbon isotope geochemistry of mantle xenoliths. Earth Science Reviews 58, 247-278. https://doi.org/10.1016/s0012-8252(02)00064-8.

Deng, J., Wang, Q., Li, G. & Santosh, M. (2014) Cenozoic tectono-magmatic and metallogenic processes in the Sanjiang region, southwestern China. Earth Science Reviews 138, 268-299. https://doi.org/10.1016/j.earscirev.2014.05.015.

Ding, L., Kapp, P., Zhong, D. & Deng, W. (2003) Cenozoic volcanism in Tibet: evidence for a transition from oceanic to continental subduction. Journal of Petrology 44, 1833-1865. https://doi.org/10.1093/petrology/egg061.

Dominguez, G., Wilkins, G., Thiemens, M.H. (2011) The Soret effect and isotopic fractionation in high-temperature silicate melts. Nature, 473, 70-73. https://doi.org/10.1038/nature09911.

Doucet, L. S., Mattielli, N., Ionov, D. A., Debouge, W., & Golovin, A. V. (2016). Zn isotopic heterogeneity in the mantle: A melting control? Earth and Planetary Science Letters 451, 232-240. https://doi.org/10.1016/j.epsl.2016.06.040.

Downes, H. (2007). Origin and significance of spinel and garnet pyroxenites in the shallow lithospheric mantle: Ultramafic massifs in orogenic belts in western Europe and NW Africa. Lithos, 99(1-2), 1-24. https://doi.org/10.1016/j.lithos.2007.05.006 https://doi.org/10.1016/j.lithos.2007.05.006.

Ducher, M., Blanchard, M., Balan, E. (2016) Equilibrium zinc isotope fractionation in Zn-bearing minerals from first-principles calculations. Chemical Geology 443, 87-96. https://doi.org/10.1016/j.chemgeo.2016.09.016.

Dutkiewicz, A., Müller, R. D., Cannon, J., Vaughan, S. & Zahirovic, S (2018) Sequestration and subduction of deep-sea carbonate in the global ocean since the Early Cretaceous. Geology 47, 91-94. https://doi.org/10.1130/G45424.1 .

Flower, M. F., Hoàng, N., Lo, C., Quốc Cu'ò'ng, N., Liu, F., Deng, J. & Mo, X. (2013) Potassic magma genesis and the Ailao Shan-Red River fault. Journal of Geodynamics. 69, 84-105. https://doi.org/10.1016/j.jog.2012.06.008.

Foley, S., Venturelli, G., Green, D. H. & Toscani, L. (1987) The ultrapotassic rocks: characteristics, classification, and constraints for petrogenetic models. Earth Science Reviews 24, 81-134. https://doi.org/10.1016/0012-8252(87)90001-8.

Foley, S. F. & Fischer, T. P. (2017) An essential role for continental rifts and lithosphere in the deep carbon cycle. Nature Geoscience 10, 897-902. https://doi.org/10.1038/s41561-017-0002-7.

Förster, M. W., Buhre, S., Xu, B., Prelević, D., Mertz-Kraus, R. & Foley, S. F. (2019). Two-Stage Origin of K-Enrichment in Ultrapotassic Magmatism Simulated by Melting of Experimentally Metasomatized Mantle. Minerals 10, 41. https://doi.org/10.3390/min10010041.

Frezzotti, M. L., Selverstone, J., Sharp, Z. D. & Compagnoni, R. (2011) Carbonate dissolution during subduction revealed by diamond-bearing rocks from the Alps. Nature Geoscience 4, 703. https://doi.org/10.1038/ngeo1246.

Galvez, M. E. & Pubellier, M. (2019) How Do Subduction Zones Regulate the Carbon Cycle? Cambridge Univ. Press, 276-309. https://doi.org/10.1017/9781108677950.010.

Galy, A., Yoffe, O., Janney, P.E., Williams, R.W., Cloquet, C., Alard, O., Halicz, L., Wadhwa, M., Hutcheon, I.D., Ramon, E., Carignan, J. (2003) Magnesium isotope heterogeneity of the isotopic standard SRM980 and new reference materials for magnesium isotope-ratio measurements. Journal of Analytical Atomic Spectrometry 18, 1352-1356. https://doi.org/10.1039/b309273a.

Gerbode C., Dasgupta R. (2010) Carbonate-fluxed melting of MORB-like pyroxenite at 2.9 GPa and genesis of HIMU ocean island basalts. Journal of Petrology 51, 2067-2088. https://doi.org/10.1093/petrology/egq049.

Green, D. H. & Wallace, M. E. (1988) Mantle metasomatism by ephemeral carbonatite melts. Nature 336, 459. https://doi.org/10.1038/336459a0.

Guo, Z., Hertogen, J. A. N., Liu, J., Pasteels, P., Boven, A., Punzalan, L. E. A., He, H., Luo, X. & Zhang, W. (2005) Potassic magmatism in western Sichuan and Yunnan provinces, SE Tibet, China: petrological and geochemical constraints on petrogenesis. Journal of Petrology 46, 33-78. https://doi.org/10.1093/petrology/egh061.

Guo, Z., Wilson, M., Liu, J. & Mao, Q. (2006) Post-collisional, potassic and ultrapotassic magmatism of the northern Tibetan Plateau: Constraints on characteristics of the mantle source, geodynamic setting and uplift mechanisms. Journal of Petrology 47, 1177-1220. https://doi.org/10.1093/petrology/egl007.

Handler, M.R., Baker, J.A., Schiller, M., Bennett, V.C., Yaxley, G.M., 2009. Magnesium stable isotope composition of the Earth's upper mantle. Earth and Planetary Science Letters 282, 306-313. https://doi.org/10.1016/j.epsl.2009.03.031.

Hirschmann, M. M., Kogiso, T., Baker, M. B., & Stolper, E. M. (2003). Alkalic magmas generated by partial melting of garnet pyroxenite. Geology, 31 (6), 481-484. https://doi.org/10.1130/0091-7613(2003)031<0481:AMGBPM>2.0.CO;2.

Hofmann A.W. (1988) Chemical differentiation of the Earth: the relationship between mantle, continental crust, and oceanic crust. Earth and Planetary Science Letters 90, 297-314. https://doi.org/10.1016/0012-821x(88)90132-x.

Hofmann, A. W. (1997) Mantle geochemistry: the message from oceanic volcanism. Nature 385, 219. https://doi.org/10.1038/385219a0.

Hofmann, A. W. (2003) Sampling mantle heterogeneity through oceanic basalts: isotopes and trace elements. Treatise on geochemistry 2, 568. https://doi.org/10.1016/b0-08-043751-6/02123-x.

Hu, Y., Teng, F.-Z., Plank, T. & Huang, K.-J. (2017) Magnesium isotopic composition of subducting marine sediments. Chemical Geology 466, 15-31. https://doi.org/10.1016/j.chemgeo.2017.06.010.

Huang, F., Chakraborty, P., Lundstrom, C. C., Holmden, C., Glessner, J. J. G., Kieffer, S. W. & Lesher, C. E. (2010b) Isotope fractionation in silicate melts by thermal diffusion. Nature 464, 396. https://doi.org/10.1038/nature08840.

Huang, J. & Xiao, Y. (2016) Mg-Sr isotopes of low-δ^{26}Mg basalts tracing recycled carbonate species: Implication for the initial melting depth of the carbonated mantle in Eastern China. International Geology Reviews 58, 1350-1362. https://doi.org/10.1080/00206814.2016.1157709.

Huang, J., Zhang, X.-C., Chen, S., Tang, L., Wörner, G., Yu, H. & Huang, F. (2018) Zinc isotopic systematics of Kamchatka-Aleutian arc magmas controlled by mantle melting. Geochimica et Cosmochimica Acta 238, 85-101. https://doi.org/10.1016/j.gca.2018.07.012.

Huang, J. & Zhao, D. (2006) High-resolution mantle tomography of China and surrounding regions. Journal of Geophysical Research: Solid Earth 111(B9). https://doi.org/10.1029/2005jb004066.

Huang, X. K., Mo, X. X., Yu, X. H., Li, Y. & He, W. Y. (2013) Geochemical characteristics and geodynamic significance of Cenozoic basalts from Maguan and Pingbian, southeastern Yunnan Province. Acta Petrologica Sinica. 29, 1325-1337.

Huang, X.-L., Niu, Y., Xu, Y.-G., Chen, L.-L. & Yang, Q.-J. (2010a) Mineralogical and geochemical constraints on the petrogenesis of post-collisional potassic and ultrapotassic rocks from western Yunnan, SW China. Journal of Petrology 51, 1617-1654. https://doi.org/10.1093/petrology/egq032.

Inglis, E. C., Debret, B., Burton, K. W., Millet, M.-A., Pons, M.-L., Dale, C. W., et al. (2017) The behavior of iron and zinc stable isotopes accompanying the subduction of mafic oceanic crust: A case study from Western Alpine ophiolites. Geochemistry, Geophysics, Geosystems, 18(7), 2562-2579. https://doi.org/10.1002/2016GC006735 .

Ke, S., Teng, F.-Z., Li, S.-G., Gao, T., Liu, S.-A., He, Y. & Mo, X. (2016) Mg, Sr, and O isotope geochemistry of syenites from northwest Xinjiang, China: Tracing carbonate recycling during Tethyan oceanic subduction. Chemical Geology 437, 109-119. https://doi.org/10.1016/j.chemgeo.2016.05.002.

Kelemen, P. B. & Manning, C. E. (2015) Reevaluating carbon fluxes in subduction zones, what goes down, mostly comes up. Proceedings of the National Academy of Sciences 112, E3997-E4006. https://doi.org/10.1073/pnas.1507889112.

Kerrick, D. M. & Connolly, J. A. D. (2001) Metamorphic devolatilization of subducted oceanic metabasalts: implications for seismicity, arc magmatism and volatile recycling. Earth and Planetary Science Letters 189, 19-29. https://doi.org/10.1016/s0012-821x(01)00347-8.

Keshav S., Gudfinnsson G.H., Sen G., Fei Y. (2004) High-pressure melting experiments on garnet clinopyroxenite and the alkalic to tholeiitic transition in ocean-island basalts. Earth and Planetary Science Letters 223 (3-4), 365-379. https://doi.org/10.1016/j.epsl.2004.04.029.

Kiseeva, E. S., Litasov, K. D., Yaxley, G. M., Ohtani, E. & Kamenetsky, V. S. (2013) Melting and phase relations of carbonated eclogite at 9-21 GPa and the petrogenesis of alkali-rich melts in the deep mantle. Journal of Petrology 54, 1555-1583. https://doi.org/10.1093/petrology/egt023.

Kogiso T., Hirschmann M.M., Frost D.J. (2003) High-pressure partial melting of garnet pyroxenite: possible mafic lithologies in the source of ocean island basalts. Earth and Planetary Science Letters 216, 603-617. https://doi.org/10.1016/s0012-821x(03)00538-7.

Kogiso T., Hirschmann M.M. (2006) Partial melting experiments of bimineralic eclogite and the role of recycled mafic oceanic crust in the genesis of ocean island basalts. Earth and Planetary Science Letters 249, 188-199. https://doi.org/10.1016/j.epsl.2006.07.016.

Kunzmann M., Halverson G. P., Sossi P. A., Raub T. D., Payne J. L. and Kirby J. (2013) Zn isotope evidence for immediate resumption of primary productivity after snowball Earth. Geology 41, 27-30. https://doi.org/10.1130/g33422.1.

Kushiro, I. (1975) Carbonate-silicate reactions at high pressures and possible presence of dolomite and magnesite in the upper mantle. Earth and Planetary Science Letters 28, 116-120. https://doi.org/10.1016/0012-821x(75)90218-6.

Le Bas, M.J., LeMaitre, R.W., Strecheisen, A.L. and Zanettin, B. (1986) A chemical classification of volcanic rocks based on the total alkali-silica diagram. Journal of Petrology 27, 745-750. https://doi.org/10.1093/petrology/27.3.745.

Le Roux, V., Lee, C.-T. & Turner, S. J. (2010) Zn/Fe systematics in mafic and ultramafic systems: Implications for detecting major element heterogeneities in the Earth's mantle. Geochimica et Cosmochimica Acta 74, 2779-2796. https://doi.org/10.1016/j.gca.2010.02.004.

Le Roux, V., Dasgupta, R., & Lee, C.-T. A. (2015). Recommended mineral-melt partition coefficients for FRTEs (Cu), Ga, and Ge during mantle melting. American Mineralogist 100, 2533-2544. https://doi.org/10.2138/am-2015-5215.

Lei, J., Zhao, D., Xu, X., Xu, Y.-G., Du, M., 2019. Is there a big mantle wedge under eastern Tibet? Physics of the Earth and Planetary Interiors. 292, 100-113. https://doi.org/10.1016/j.pepi.2019.04.005.

Li, J.-L., Klemd, R., Gao, J. & Meyer, M. (2014) Compositional zoning in dolomite from lawsonite-bearing eclogite (SW Tianshan, China): Evidence for prograde metamorphism during subduction of oceanic crust. American Mineralogist 99, 206-217. https://doi.org/10.2138/am.2014.4507.

Li, S.-G., Yang, W., Ke, S., Meng, X., Tian, H., Xu, L., et al. (2017) Deep carbon cycles constrained by a large-scale mantle Mg isotope anomaly in eastern China. National Science Reviews 4, 111-120. https://doi.org/10.1093/nsr/nww070.

Li, X.-H., Zhou, H., Chung, S.-L., Lo, C.-H., Wei, G., Liu, Y. & Lee, C. (2002) Geochemical and Sr-Nd isotopic characteristics of late Paleogene ultrapotassic magmatism in southeastern Tibet. International Geology Review 44, 559-574. https://doi.org/10.2747/0020-6814.44.6.559.

Little S. H., Vance D., McManus J. and Severmann S. (2016) Key role of continental margin sediments in the oceanic mass balance of Zn and Zn isotopes. Geology 44, 207-210. https://doi.org/10.1130/g37493.1.

Liu, D., Zhao, Z., Zhu, D.-C., Niu, Y., Widom, E., Teng, F.-Z., et al. (2015) Identifying mantle carbonatite metasomatism through Os-Sr-Mg isotopes in Tibetan ultrapotassic rocks. Earth and Planetary Science Letters, 430, 458-469. https://doi.org/10.1016/j.epsl.2015.09.005.

Liu, F., Wu, H., Liu, J., Hu, G., Li, Q., and Qu, K. (1990) 3-D velocity images beneath the Chinese continent and adjacent regions. Geophysical Journal International 101, 379-394. https://doi.org/10.1111/j.1365-246x.1990.tb06576.x.

Liu, S.-A., Teng, F.-Z., He, Y., Ke, S. & Li, S. (2010) Investigation of magnesium isotope fractionation during granite differentiation: implication for Mg isotopic composition of the continental crust. Earth and Planetary Science Letters 297, 646-654. https://doi.org/10.1016/j.epsl.2010.07.019.

Liu S.-A., Teng F.-Z., Yang W. and Wu F. (2011) High-temperature inter-mineral magnesium isotope fractionation in mantle xenoliths from the North China craton. Earth and Planetary Science Letters 308, 131-140. https://doi.org/10.1016/j.epsl.2011.05.047 .

Liu, S.-A., Wang, Z.-Z., Li, S.-G., Huang, J. & Yang, W. (2016) Zinc isotope evidence for a large-scale carbonated mantle beneath eastern China. Earth and Planetary Science Letters 444, 169-178. https://doi.org/10.1016/j.epsl.2016.03.051.

Liu, S.-A., Wu, H., Shen, S., Jiang, G., Zhang, S., Lv, Y., Zhang, H. & Li, S. (2017) Zinc isotope evidence for intensive magmatism immediately before the end-Permian mass extinction. Geology 45, 343-346. https://doi.org/10.1130/g38644.1.

Liu, S.-A. & Li, S.-G. (2019) Tracing the Deep Carbon Cycle Using Metal Stable Isotopes: Opportunities and Challenges. Engineering, 5, 448-457. https://doi.org/10.1016/j.eng.2019.03.007.

Liu, S.-A., Liu, P.-P., Lv, Y., Wang, Z.-Z., Dai, J.-G. (2019) Cu and Zn Isotope Fractionation during Oceanic Alteration: Implications for Oceanic Cu and Zn Cycles. Geochimica et Cosmochimica Acta 257, 191-205. https://doi.org/10.1016/j.gca.2019.04.026.

Liu, X.-M., Teng, F.-Z., Rudnick, R.L., McDonough, W.F., Cummings, M. (2014) Massive magnesium depletion and isotopic fractionation in weathered basalts. Geochimica et Cosmochimica Acta 135, 336-349. https://doi.org/10.1016/j.gca.2014.03.028.

Lobanov, S. S., Goncharov, A. F. & Litasov, K. D. (2015) Optical properties of siderite ($FeCO_3$) across the spin transition: Crossover to iron-rich carbonates in the lower mantle. American Mineralogist 100, 1059-1064. https://doi.org/10.2138/am-2015-5053.

Lu, Y.-J., McCuaig, T. C., Li, Z.-X., Jourdan, F., Hart, C. J., Hou, Z.-Q. & Tang, S.-H. (2015) Paleogene post-collisional lamprophyres in western Yunnan, western Yangtze Craton: Mantle source and tectonic implications. Lithos 233, 139-161. https://doi.org/10.1016/j.lithos.2015.02.003.

Lustrino, M., Duggen, S., & Rosenberg, C. L. (2011). The Central-Western Mediterranean: Anomalous igneous activity in an anomalous collisional tectonic setting. Earth-Science Reviews, 104(1-3), 1-40.https://doi.org/10.1016/j.earscirev.2010.08.002.

Lustrino, M., and Wilson, M. (2007). The circum-Mediterranean anorogenic Cenozoic igneous province. Earth-Sci. Rev. 81, 1-65. https://doi.org/10.1016/j.earscirev.2006.09.002 .

Macris, C. A., Young, E. D., & Manning, C. E. (2013) Experimental determination of equilibrium magnesium isotope fractionation between spinel, forsterite, and magnesite from 600 to 800 C. Geochimica et Cosmochimica Acta 118, 18-32. https://doi.org/10.1016/j.gca.2013.05.008.

Maréchal, C. N., Télouk, P., Albarède, F. (1999) Precise analysis of copper and zinc isotopic compositions by plasma-source mass spectrometry. Chemical Geology. 156, 251-273. https://doi.org/10.1016/s0009-2541(98)00191-0.

Martin, L. H. J., Schmidt, M. W., Mattsson, H. B., Ulmer, P., Hametner, K., & Günther, D. (2012). Element partitioning between immiscible carbonatite-kamafugite melts with application to the Italian ultrapotassic suite. Chemical Geology, 320-321, 96-112. https://doi.org/10.1016/j.chemgeo.2012.05.019.

Mavromatis, V., González, A.G., Dietzel, M., Schott, J., 2019. Zinc isotope fractionation during the inorganic precipitation of calcite - Towards a new pH proxy. Geochimica et Cosmochimica Acta, 244: 99-112. https://doi.org/10.1016/j.gca.2018.09.005.

Mazza, S.E., Gazel, E., Bizimis, M., Moucha, R., Béguelin, P., Johnson, E.A., McAleer, R.J., Sobolev, A.V. (2019) Sampling the volatile-rich transition zone beneath Bermuda. Nature 569 (7756), 398. https://doi.org/10.1038/s41586-019-1183-6.

McCoy-West A. J., Fitton J. G., Pons M.-L., Inglis E. C. and Williams H. M. (2018) The Fe and Zn isotope composition of deep mantle source regions: insights from Baffin Island picrites. Geochimica et Cosmochimica Acta 238, 542-562. https://doi.org/10.1016/j.gca.2018.07.021.

McDonough, W. F. & Sun, S.-S. (1995) The composition of the Earth. Chemical Geology 120, 223-253. https://doi.org/10.5772/intechopen.88100.

Merlini, M., Crichton, W. A., Hanfland, M., Gemmi, M., Müller, H., Kupenko, I. & Dubrovinsky, L. (2012) Structures of dolomite at ultrahigh pressure and their influence on the deep carbon cycle. Proc. Natl. Acad. Sci. U.S.A. 109, 13509-13514. https://doi.org/10.1073/pnas.1201336109.

Nestola, F., Korolev, N., Kopylova, M., Rotiroti, N., Pearson, D. G., Pamato, M. G., et al. (2018) $CaSiO_3$ perovskite in diamond indicates the recycling of oceanic crust into the lower mantle. Nature, 555, 237. https://doi.org/10.1038/nature25972.

Pertermann, M., & Hirschmann, M. M. (2003). Anhydrous partial melting experiments on MORB-like eclogite: Phase relations, phase compositions and mineral-melt partitioning of major elements at 2-3 GPa. Journal of Petrology 44, 2173-2201. https://doi.org/10.1093/petrology/egg074.

Pertermann, M., Hirschmann, M. M., Hametner, K., Günther, D., & Schmidt, M. W. (2004). Experimental determination of trace element partitioning between garnet and silica-rich liquid during anhydrous partial melting of MORB-like eclogite. Geochemistry, Geophysics, Geosystems 5(5). https://doi.org/10.1029/2003gc000638.

Pichat, S., Douchet, C. & Albarède, F. (2003) Zinc isotope variations in deep-sea carbonates from the eastern equatorial Pacific over the last 175 ka. Earth and Planetary Science Letters 210, 167-178. https://doi.org/10.1016/s0012-821x(03)00106-7.

Pilet S., Baker M.B., Stolper E.M. (2008) Metasomatized lithosphere and the origin of alkali lavas. Science 320 (5878), 916-919. https://doi.org/10.1126/science.1156563.

Plank, T. (2014) The chemical composition of subducting sediments. Elsevier. https://doi.org/10.1016/b978-0-08-095975-7.00319-3.

Plank, T & Manning, C.E. (2019) Subducting carbon. Nature 574, 343-352. https://doi.org/10.1038/s41586-019-1643-z .

Pons, M.-L., Debret, B., Bouilhol, P., Delacour, A., & Williams, H. (2016). Zinc isotope evidence for sulfate-rich fluid transfer across subduction zones. Nature Communications, 7(1). https://doi.org/10.1038/ncomms13794.

Prelević, D., Akal, C., Foley, S. F., Romer, R. L., Stracke, A. & Van Den Bogaard, P. (2012) Ultrapotassic mafic rocks as geochemical proxies for post-collisional dynamics of orogenic lithospheric mantle: the case of southwestern Anatolia, Turkey. Journal of Petrology 53, 1019-1055. https://doi.org/10.1093/petrology/egs008.

Richter, F. M., Watson, E. B., Mendybaev, R. A., Teng, F.-Z. & Janney, P. E. (2008) Magnesium isotope fractionation in silicate melts by chemical and thermal diffusion. Geochimica et Cosmochimica Acta 72, 206-220. https://doi.org/10.1038/nature09954.

Roden, M.F., Murthy, V.R. (1985) Mantle metasomatism. Annu Rev. Earth. Pl. Sc. 13, 169-296. https://doi.org/10.1146/annurev.ea.13.050185.001413.

Roeder, P. L. & Emslie, Rf. (1970) Olivine-liquid equilibrium. Contributions to Mineralogy and Petrology 29, 275-289. https://doi.org/10.1007/bf00371276.

Schultz, F., Lehmann, B., Tawackoli, S., Rössling, R., Belyatsky, B., & Dulski, P. (2004) Carbonatite diversity in the Central Andes: The Ayopaya alkali province, Bolivia. Contributions to Mineralogy and Petrology 148, 391-408. https://doi.org/10.1007/s00410-004-0612-0.

Shaw, D. M. (1970). Trace element fractionation during anatexis. Geochimica et Cosmochimica Acta, 34, 237-243. https://doi.org/10.1016/0016-7037(70)90009-8.

Shen, J., Li, S.-G., Wang, S.-J., Teng, F.-Z., Li, Q.-L., Liu, Y.-S. (2018) Subducted Mg-rich carbonates into the deep mantle wedge. Earth and Planetary Science Letters 503, 118-130. Shen, J., Li, S.-G., Wang, S.-J., Teng, F.-Z., Li, Q.-L., Liu, Y.-S. (2018) Subducted Mg-rich carbonates into the deep mantle wedge. Earth and Planetary Science Letters 503, 118-130.

Sobolev, A. V., Hofmann, A. W., Sobolev, S. V. & Nikogosian, I. K. (2005) An olivine-free mantle source of Hawaiian shield basalts.

Nature 434, 590. https://doi.org/10.1038/nature03411.
Sobolev, A. V., Hofmann, A. W., Kuzmin, D. V., Yaxley, G. M., Arndt, N. T., Chung, S.-L., et al. (2007) The amount of recycled crust in sources of mantle-derived melts. Science, 316, 412-417.
Song, X.-Y., Qi, H.-W., Robinson, P. T., Zhou, M.-F., Cao, Z.-M. & Chen, L.-M. (2008) Melting of the subcontinental lithospheric mantle by the Emeishan mantle plume; evidence from the basal alkali basalts in Dongchuan, Yunnan, Southwestern China. Lithos 100, 93-111. https://doi.org/10.1016/j.lithos.2007.06.023.
Sossi, P. A., Halverson, G. P., Nebel, O., & Eggins, S. M. (2015) Combined Separation of Cu, Fe and Zn from Rock Matrices and Improved Analytical Protocols for Stable Isotope Determination. Geostandards and Geoanalytical Research 39, 129-149. https://doi.org/10.1111/j.1751-908x.2014.00298.x.
Sossi, P. A., Nebel, O., O'Neill, H. S. C., & Moynier, F. (2018). Zinc isotope composition of the Earth and its behaviour during planetary accretion. Chemical Geology, 477, 73-84. https://doi.org/10.1016/j.chemgeo.2017.12.006.
Sun, S.-S. & McDonough, W. F. (1989) Chemical and isotopic systematics of oceanic basalts: implications for mantle composition and processes. Geological Society, London, Special Publications 42, 313-345. https://doi.org/10.1144/gsl.sp.1989.042.01.19.
Sweere T.C., Dickson A.J., Jenkyns H.C., Porcelli D., Elrick M., van den Boorn S.H.J.M., Henderson G.M. (2018) Isotopic evidence for changes in the zinc cycle during Oceanic Anoxic Event 2 (Late Cretaceous). Geology 46 (5), 463-466. https://doi.org/10.1130/g40226.1.
Tapponnier, P., Peltzer, G., Le Dain, A. Y., Armijo, R. & Cobbold, P. (1982) Propagating extrusion tectonics in Asia: New insights from simple experiments with plasticine. Geology 10, 611-616. https://doi.org/10.1130/0091-7613(1982)10%3C611:petian%3E2.0.co;2.
Teng, F.-Z., Li, W.-Y., Ke, S., Marty, B., Dauphas, N., Huang, S., Wu, F.-Y. & Pourmand, A. (2010) Magnesium isotopic composition of the Earth and chondrites. Geochimica et Cosmochimica Acta 74, 4150-4166. https://doi.org/10.1016/0012-821x(67)90044-1.
Teng, F.-Z. (2017) Magnesium isotope geochemistry. Reviews in Mineralogy and Geochemistry 82, 219-287. https://doi.org/10.2138/rmg.2017.82.7.
Thomson, A. R., Walter, M. J., Kohn, S. C. & Brooker, R. A. (2016) Slab melting as a barrier to deep carbon subduction. Nature 529, 76. https://doi.org/10.1038/nature16174.
Tian, H.-C., Yang, W., Li, S.-G. & Ke, S. (2017) Could sedimentary carbonates be recycled into the lower mantle? Constraints from Mg isotopic composition of Emeishan basalts. Lithos 292, 250-261. https://doi.org/10.1016/j.lithos.2017.09.007.
Turekian, K. K. & Wedepohl, K. H. (1961) Distribution of the elements in some major units of the earth's crust. Geological Society of America Bulletin 72, 175-192. https://doi.org/10.1130/0016-7606(1961)72[175:doteis]2.0.co;2.
Turner, S., Arnaud, N., LIU, J., Rogers, N., Hawkesworth, C., Harris, N., Kelley, S. v, Van Calsteren, P. & Deng, W. (1996) Post-collision, shoshonitic volcanism on the Tibetan Plateau: implications for convective thinning of the lithosphere and the source of ocean island basalts. Journal of Petrology 37, 45-71. https://doi.org/10.1093/petrology/37.1.45.
Walter, B. F., Parsapoor, A., Braunger, S., Marks, M. A. W., Wenzel, T., Martin, M., & Markl, G. (2018) Pyrochlore as a monitor for magmatic and hydrothermal processes in carbonatites from the Kaiserstuhl volcanic complex (SW Germany). Chemical Geology 498, 1-16. https://doi.org/10.1016/j.chemgeo.2018.08.008.
Wang, J.-H., Yin, A., Harrison, T. M., Grove, M., Zhang, Y.-Q. & Xie, G.-H. (2001) A tectonic model for Cenozoic igneous activities in the eastern Indo-Asian collision zone. Earth and Planetary Science Letters 188, 123-133. https://doi.org/10.1016/s0012-821x(01)00315-6.
Wang, S.-J., Teng, F.-Z., Williams, H. M., & Li, S.-G. (2012). Magnesium isotopic variations in cratonic eclogites: Origins and implications. Earth and Planetary Science Letters, 359-360, 219-226. https://doi.org/10.1016/j.epsl.2012.10.016.
Wang, S.-J, Teng, F.-Z, Li, S.-G (2014). Tracing carbonate-silicate interaction during subduction using magnesium and oxygen isotopes. Nature Communication, 5:5328. https://doi.org/10.1038/ncomms6328.
Wang, X., Metcalfe, I., Jian, P., He, L. & Wang, C. (2000) The Jinshajiang-Ailaoshan suture zone, China: tectonostratigraphy, age and evolution. Journal of Asian Earth Sciences 18, 675-690. https://doi.org/10.1016/s1367-9120(00)00039-0.
Wang, X.-C., Li, Z.-X., Li, X.-H., Li, J., Liu, Y., Long, W.-G., Zhou, J.-B. & Wang, F. (2011) Temperature, pressure, and composition of the mantle source region of Late Cenozoic basalts in Hainan Island, SE Asia: a consequence of a young thermal mantle plume close to subduction zones? Journal of Petrology 53, 177-233. https://doi.org/10.1093/petrology/egr061.
Wang, Y., He, Y., & Ke, S. (2020). Mg isotope fractionation during partial melting of garnet-bearing sources: An adakite perspective. Chemical Geology, 537, 119478. https://doi.org/10.1016/j.chemgeo.2020.119478.
Wang Z.-Z., Liu S.-A., Ke S., Liu Y.-C. and Li S.-G. (2016) Magnesium isotopic heterogeneity across the cratonic litho- sphere in

eastern China and its origins. Earth and Planetary Science Letters 451, 77-88. https://doi.org/10.1016/j.epsl.2016.07.021.

Wang, Z.-Z., Liu, S.-A., Liu, J., Huang, J., Xiao, Y., Chu, Z.-Y., Zhao, X.-M. & Tang, L. (2017) Zinc isotope fractionation during mantle melting and constraints on the Zn isotope composition of Earth's upper mantle. Geochimica et Cosmochimica Acta 198, 151-167. https://doi.org/10.1016/j.gca.2016.11.014.

Wang, Z.-Z., Liu, S.-A., Chen, L.-H., Li, S.-G. & Zeng, G. (2018) Compositional transition in natural alkaline lavas through silica-undersaturated melt-lithosphere interaction. Geology 46, 771-774. https://doi.org/10.1130/g45145.1.

Weis, D., Kieffer, B., Maerschalk, C., Pretorius, W. & Barling, J. (2005) High-precision Pb-Sr-Nd-Hf isotopic characterization of USGS BHVO-1 and BHVO-2 reference materials. Geochemistry, Geophysics, Geosystems 6(2). https://doi.org/10.1029/2004gc000852.

Williams, H. M., Turner, S. P., Pearce, J. A., Kelley, S. P. & Harris, N. B. W. (2004) Nature of the source regions for post-collisional, potassic magmatism in southern and northern Tibet from geochemical variations and inverse trace element modelling. Journal of Petrology 45, 555-607. https://doi.org/10.1093/petrology/egg094.

Williams, H.M., Bizimis, M. (2014) Iron isotope tracing of mantle heterogeneity within the source regions of oceanic basalts. Earth and Planetary Science Letters 404, 396-407. https://doi.org/10.1016/j.epsl.2014.07.033.

Wang, Y., He, Y., & Ke, S. (2020). Mg isotope fractionation during partial melting of garnet-bearing sources: An adakite perspective. Chemical Geology, 537, 119478. https://doi.org/10.1016/j.chemgeo.2020.119478.

Xia, P. & Xu, Y. (2005) Domains and enrichment mechanism of the lithospheric mantle in western Yunnan: A comparative study on two types of Cenozoic ultrapotassic rocks. Science China Series D: Earth Science 48, 326-337. https://doi.org/10.1360/03yd0488.

Xiao, L., Xu, Y. G., Mei, H. J., Zheng, Y. F., He, B. & Pirajno, F. (2004) Distinct mantle sources of low-Ti and high-Ti basalts from the western Emeishan large igneous province, SW China: implications for plume-lithosphere interaction. Earth and Planetary Science Letters 228, 525-546. https://doi.org/10.1016/j.epsl.2004.10.002.

Xu, Y.-G., Menzies, M. A., Thirlwall, M. F. & Xie, G.-H. (2001) Exotic lithosphere mantle beneath the western Yangtze craton: petrogenetic links to Tibet using highly magnesian ultrapotassic rocks. Geology 29, 863-866. https://doi.org/10.1130/0091-7613(2001)029<0863:elmbtw>2.0.co;2.

Yang, C., Liu, S.-A. (2019) Zinc isotope constraints on recycled oceanic crust in the mantle sources of the Emeishan large igneous province. Journal of Geophysical Research: Solid Earth 124, 12,537-12,555. https://doi.org/10.1029/2019jb017405.

Yaxley, G. M., Crawford, A. J. & Green, D. H. (1991) Evidence for carbonatite metasomatism in spinel peridotite xenoliths from western Victoria, Australia. Earth and Planetary Science Letters 107, 305-317. https://doi.org/10.1016/0012-821x(91)90078-v.

Yaxley, G. M., Green, D. H. (1994) Experimental demonstration of refractory carbonate-bearing eclogite and siliceous melt in the subduction regime. Earth and Planetary Science Letters 128, 313-325. https://doi.org/10.1016/0012-821x(94)90153-8.

Yuan, X. C. (1989) On the deep structure of the Kang-dian rift. Acta Geologica Sinica 63, 1-13. Yuan, X. C. (1989) On the deep structure of the Kang-dian rift. Acta Geologica Sinica 63, 1-13.

Zhong, Y., Chen, L.-H., Wang, X.-J., Zhang, G.-L., Xie, L.-W., & Zeng, G. (2017). Magnesium isotopic variation of oceanic island basalts generated by partial melting and crustal recycling. Earth and Planetary Science Letters 463, 127-135. https://doi.org/10.1016/j.epsl.2017.01.040.

Oxidation of the deep big mantle wedge by recycled carbonates: Constraints from highly siderophile elements and osmium isotopes *

Ronghua Cai [1], Jingao Liu [1*], D. Graham Pearson [2], Dongxu Li [1], Yong Xu [1], Sheng-Ao Liu [1], Zhuyin Chu [3], Li-Hui Chen [4] and Shuguang Li [1,5]

1. State Key Laboratory of Geological Processes and Mineral Resources, China University of Geosciences, Beijing 100083, China
2. Department of Earth and Atmospheric Sciences, University of Alberta, Edmonton, Alberta T6G 2E3, Canada
3. State Key Laboratory of Lithospheric Evolution, Institute of Geology and Geophysics, Chinese Academy of Sciences, Beijing 100029, PR China
4. State Key Laboratory of Continental Dynamics, Department of Geology, Northwest University, Xi'an 710069, China
5. CAS Key Laboratory of Crust-Mantle Materials and Environments, School of Earth and Space Sciences, University of Science and Technology of China, Hefei 230026, China

亮点介绍：本研究首次利用强亲铁性元素及Os同位素证实滞留板片所释放的碳酸盐熔体能够氧化少量的地幔硫化物并且释放强亲铁性元素到碳酸盐熔体当中，对于理解大地幔楔的氧化还原状态以及再循环碳酸盐-软流圈相互作用有重大意义。

Abstract Widespread Cenozoic intraplate basalts from eastern China offer the opportunity to investigate the consequences of interaction between the stagnant Pacific slab and overlying asthenosphere and chemical heterogeneity within this "big mantle wedge". We present and compile a comprehensive study of highly siderophile elements and Mg-Zn isotopes of this magmatic suite (60 samples including nephelinites, basanites, alkali basalts and tholeiites). The large-scale Mg-Zn isotopic anomalies documented in these basalts have been ascribed to mantle hybridization by recycled Mg-carbonates from the stagnant western Pacific plate. Our results reveal that the nephelinites and basanites are characterized by unfractionated platinum-group element (PGE) patterns normalized to primitive upper mantle (PUM) (e.g., Pd_N/Ir_N normalized to PUM=1.1±0.8, 1σ), relatively high total PGE contents (e.g., Ir=0.25±0.14 ppb) and modern mantle-like $^{187}Os/^{188}Os$ (0.142±0.020). These characteristics are coupled with lighter Mg isotope ($\delta^{26}Mg$=−0.48±0.07‰) and heavier Zn isotope ($\delta^{66}Zn$=+0.46±0.06‰) compositions compared to the mantle values ($\delta^{26}Mg$: −0.25±0.07‰; $\delta^{66}Zn$: +0.18±0.05‰). Together, these data are interpreted to reflect the oxidative breakdown of low

* 本文发表在：Geochimica et Cosmochimica Acta, 2021, 295: 207-223

proportions of mantle sulfides in the sources of these small-degree melts, likely caused by recycled carbonates, which then release chalcophile-siderophile elements into carbonatitic melts. By contrast, the contemporaneous alkali basalts and tholeiites are characterized by highly fractionated PGE patterns (e.g., Pd_N/Ir_N=4.4±3.3; Ir = 0.037±0.027 ppb) and radiogenic $^{187}Os/^{188}Os$ (0.279±0.115) coupled with less fractionated Mg-Zn isotope compositions (δ^{26}Mg: −0.39±0.05‰; δ^{66}Zn: +0.35±0.03‰). In combination with other isotopic (e.g., Sr-Nd) and chemical (SiO_2, Ce/Pb, Ba/Th, Fe/Mn) constraints, the alkali basalts and tholeiites were derived from higher degree melting of ancient pyroxenite-bearing mantle in addition to mixing with the aforementioned nephelinitic and basanitic melts. Collectively, we suggest that deep recycled carbonates promoted melting within the "big mantle wedge" leading to the generation of Cenozoic intraplate basalts across eastern China and the "redox freezing of carbonates" may cause the oxidation of Fe^0 and S^{2-}. This process may provide an important mechanism to oxidize mantle sulfides and transfer precious metals from deep mantle to crust.

Keywords　Highly siderophile elements, mantle sulfides, carbonate recycling, intraplate basalts, big mantle wedge, Mg-Zn isotopes

1　Introduction

The origin of continental intraplate basalts remains a controversial subject in solid Earth science (Day, 2013; Farmer, 2014; Zheng et al., 2019). Compared to oceanic basalts, continental basaltic magmas are more diverse, ranging from small-volume potassic and sodic basaltic melts to large-volume continental flood basalts (Farmer, 2014). Studying the origin of continental basalts can help understanding the interaction of mantle with recycled crustal materials (Zheng et al., 2019). Here, we focus on the widespread and voluminous Cenozoic intraplate basalts across eastern China (Fig. 1). These basalts have chemical and isotopic compositions comparable to those of oceanic island basalts (OIBs) and represent typical continental intraplate basaltic rocks primarily derived from asthenosphere (Zou et al., 2000; Xu et al., 2018). Recent geophysical and tectonic studies have shown that the deeply subducted Pacific slab in this region stagnates in the mantle transition zone beneath eastern China, forming a "sandwich-like" mantle structure in the upper mantle—the so called "big mantle wedge" (Fig. 1; Zhao et al., 2004; Huang and Zhao, 2006). The distribution of Cenozoic basalts across eastern China coincides with the region underlain by the stagnant slab (Fig. 1), thus these basalts are regarded as melting products of the big mantle wedge (Li et al., 2017; Xu et al., 2018).

It is generally accepted that the dehydration or melting of the subducted slab at varying depths within the arc mantle wedge plays a major role in the formation of arc basalts (Xu et al., 2020a). By contrast, no definite theory is available for the origin of continental intraplate basalts within the deep big mantle wedge and associated crust-mantle interaction (Xu et al., 2018). Previous studies have emphasized that the big mantle wedge beneath eastern China is vertically heterogeneous, with the upper part containing ancient enriched mantle components and the lower part containing recycled carbonate components derived from the subducting Pacific plate (e.g., Li et al., 2016; Xu et al., 2018; Li and Wang, 2018). Arc mantle wedge metasomatized by fluids/melts derived from the subducted slab is more oxidized than the depleted mantle, with arc basalts characterized by higher oxygen fugacity than MORB (mid-oceanic ridge basalts) (Kelley and Cottrell, 2009). Yet, it is unclear whether fluids/melts (e.g., carbonatite melts) derived from the stagnant slab can cause oxidation of the big mantle wedge and mobilize precious metals (He et al., 2019, 2020; Hong et al., 2020). Tracing redox reactions as well as recycled materials in the mantle sources of the major Cenozoic basalt province in eastern China is vital to understanding the interaction between the subducted stagnant slab and the big mantle wedge.

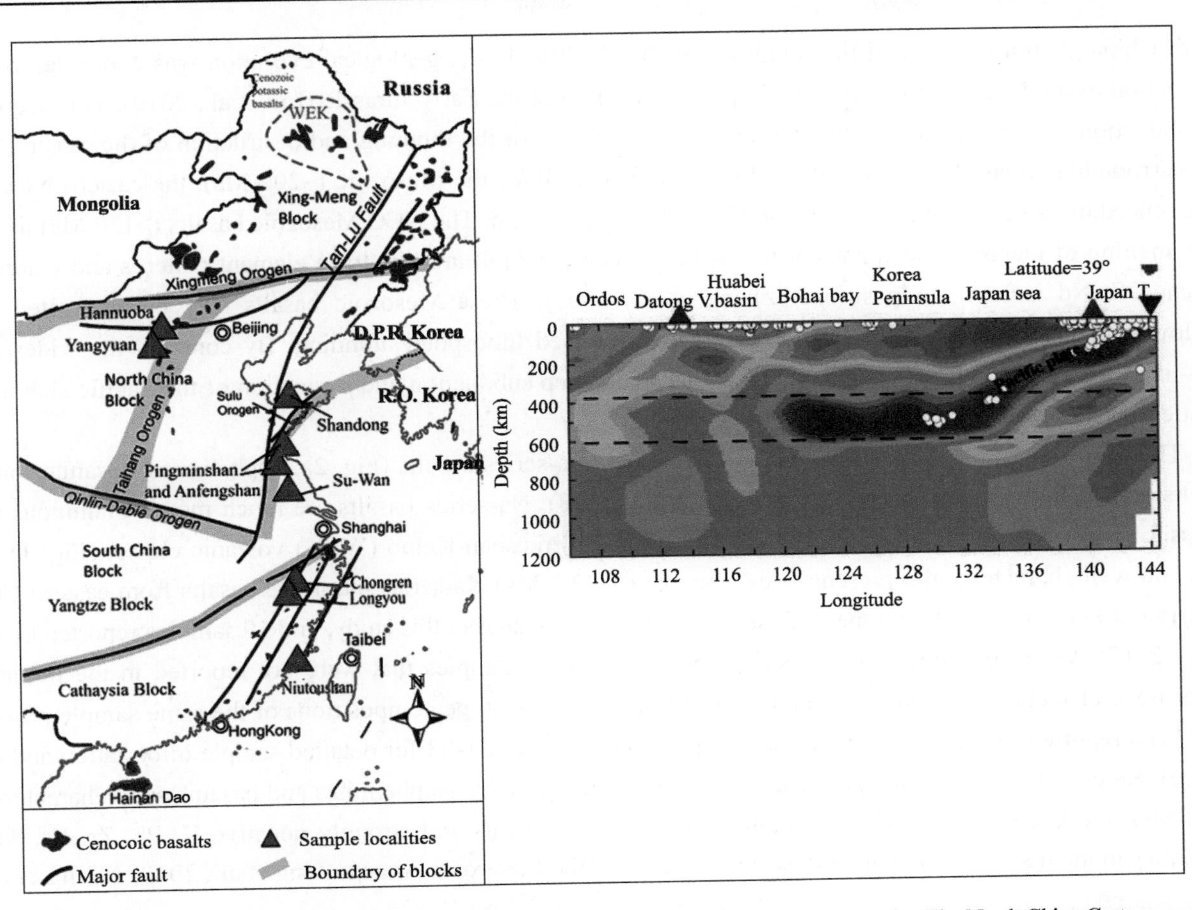

Fig. 1 Simplified geological map of eastern China and stagnant slab revealed by seismic tomography. The North China Craton consists of the Eastern and Western Blocks with the intervening central orogenic belt. The Xing-Meng Block and the North China Craton are separated by the Late Paleozoic Xing-Meng orogenic belt. The North China Craton and the South China Block are separated by the Triassic Qinling-Dabie orogen formed during continent-continent collision between these two blocks. WEK indicates Cenozoic Wudalianchi-Erkeshan-Keluo volcanic cluster. Seismic tomography and topographic map are modified from Huang and Zhao (2006).

Highly siderophile elements (HSE here including Os, Ir, Ru, Rh, Pt, Pd, Re), copper and the Re-Os isotope system have the potential to trace the behavior of mantle sulfides and recycled materials in the mantle source. The HSE of melts may record the redox evolution of mantle source due to the sensitivity of mantle sulfides to oxidation (He et al., 2020). During normal relatively low-degree mantle melting, the platinum-like platinum group elements (PPGE:Pt, Pd), Re and Cu behave moderately incompatible, whereas the iridium-like PGE (IPGE: Os, Ir, Ru) act as compatible elements (Pearson et al., 2004; Day, 2013; Gannoun et al., 2016). The HSE and Cu elemental and isotopic systematics can record some deep processes that are not easily revealed by lithophile elements and isotopes. We analyzed and compiled HSE and Cu concentrations, Os and Mg-Zn-Sr-Nd isotopic data for an extensive suite (60 samples) of eastern China Cenozoic basalts. The systematic variations in HSE patterns and Os isotope ratios of these rocks are coupled with their silicate-hosted isotope compositions and major-trace element contents. We propose that the variations of HSE, Cu and Os isotope systematics are associated with mantle oxidation induced by deep carbonate cycling and heterogeneity of the "big mantle wedge".

2 Geological setting and sample description

Eastern China consists of several major blocks including, from north to south, the Xing-Meng Block, the

North China Craton (NCC) and the South China Block (Fig. 1). Its geological evolution was controlled by the subduction of the Paleo-Pacific and Pacific plate since at least the Early Jurassic (Liu et al., 2019). This sequence of subduction is also regarded as the dominant driving force for the thinning and destruction of the eastern NCC and surrounding areas (Liu et al., 2019; Wu et al., 2019). Since the Mesozoic (~200 Ma), the eastern NCC has experienced massive regional extension and intraplate magmatism. The NCC Mesozoic basalts (>120 Ma) derived from melting of metasomatized ancient lithospheric mantle exhibit arc-like trace element patterns and extremely enriched Sr-Nd isotope compositions (Zhang et al., 2002). These Mesozoic basalts may be associated with shallow subduction of the Paleo-Pacific slab which induced lithosphere thinning. By contrast, the widespread OIB-like Cenozoic intraplate basalts are associated with deep subduction and stagnation of the Pacific slab in the mantle transition zone (Liu et al., 2019; Zheng et al., 2019).

The Cenozoic intraplate basalts can be divided into Na-series basalts (Fig. 2a; nephelinites, basanites, alkali basalts and tholeiites) and potassic basalts (Xu et al., 2018). Na-series basalts are much more voluminous than potassic basalts which are only found in the Wudalianchi-Erkeshan-Keluo (WEK) volcanic cluster (Fig. 1). We chose 60 well-characterized representative Cenozoic (0–32.2 Ma) Na-series intraplate basalts from eastern China to carry out PGE and Re-Os isotope analyses (41 samples measured in this study, and 19 samples reported by Chu et al., 2017). We also conducted Mg isotope analyses for 27 samples that were not reported in the literature. Major-trace element concentrations and most of Sr-Nd-Mg-Zn isotope compositions of the same sample powders have been reported in previous studies (See supplementary tables S1-S3 for detailed sample information and data source). Here we briefly summarize the key features of these basalts. Nephelinites and basanites are characterized by HIMU-like trace element patterns, high trace element contents and strongly negative K, Pb, Zr, Hf, Ti, Y anomalies in the trace element spider diagrams (Fig. S1 in the Electronic Annex; Zeng et al., 2010; Huang et al.,

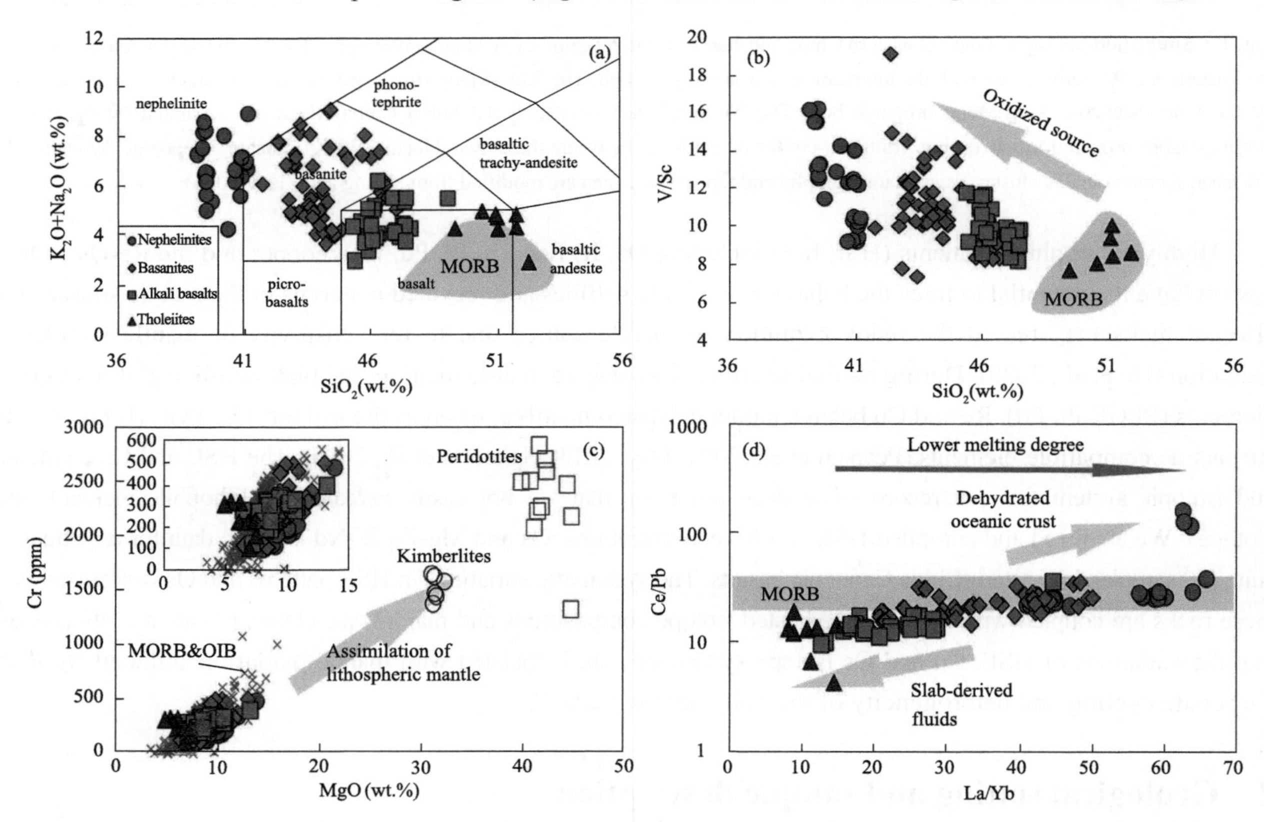

Fig. 2 Total alkali (K_2O+Na_2O) vs. SiO_2 (a), V/Sc vs. SiO_2 (b), Cr vs. MgO contents (c), Ce/Pb vs. La/Yb (d) diagrams of Cenozoic basalts from eastern China. See Appendix 3 in the Electronic Annex for data sources.

2015; Chu et al., 2017). They show high CaO/Al_2O_3 (0.74±0.16; 1σ, here and elsewhere in text unless specified), La_N/Yb_N (28±10) ratios and higher total FeO (13.3±1.5 wt. %) contents (Fig. 2, S2, S3; Huang et al., 2015; Chu et al., 2017). These basalts also have light Mg isotope ($\delta^{26}Mg = [(^{26}Mg/^{24}Mg)_{sample} / (^{26}Mg/^{24}Mg)_{DSM3} - 1] \times 1000$: −0.48±0.07‰) and heavy Zn isotope compositions ($\delta^{66}Zn = [(^{66}Zn/^{64}Zn)_{sample} /(^{66}Zn/^{64}Zn)_{JMC\ 3\text{-}0749L} - 1)] \times 1000$: +0.46±0.06‰) relative to the accessible upper mantle ($\delta^{26}Mg$: −0.25±0.07‰; $\delta^{66}Zn$: +0.18±0.05‰; Fig. 3c, 4), coupled with depleted Sr-Nd isotope compositions ($^{87}Sr/^{86}Sr$: 0.7035±0.0003, ε_{Nd}: 5.4±1.5) compositions and low Ba/Th (69±31), high Ce/Pb (25±5) ratios (Fig. 3a; Huang et al., 2015; Chu et al., 2017). By contrast, alkali basalts and tholeiites are characterized by slightly negative Pb anomalies, positive Ba anomalies (Fig. S1), higher Ba/Th ratios (117±27), lower Ce/Pb (14±5), La_N/Yb_N (14±5) and CaO/Al_2O_3 (0.62±0.05) ratios (Fig. 2, S3) and near EM-I type Sr-Nd isotope compositions (Fig. 3a; $^{87}Sr/^{86}Sr$: 0.7041±0.0004; ε_{Nd}:1.1±2.7; Huang et al., 2015; Chu et al., 2017). Their Mg-Zn isotope compositions (Fig. 4; $\delta^{26}Mg$: −0.39±0.05‰; $\delta^{66}Zn$: +0.35±0.03‰) lie between those of the nephelinites and mantle values (Huang et al., 2015; Wang et al., 2018a).

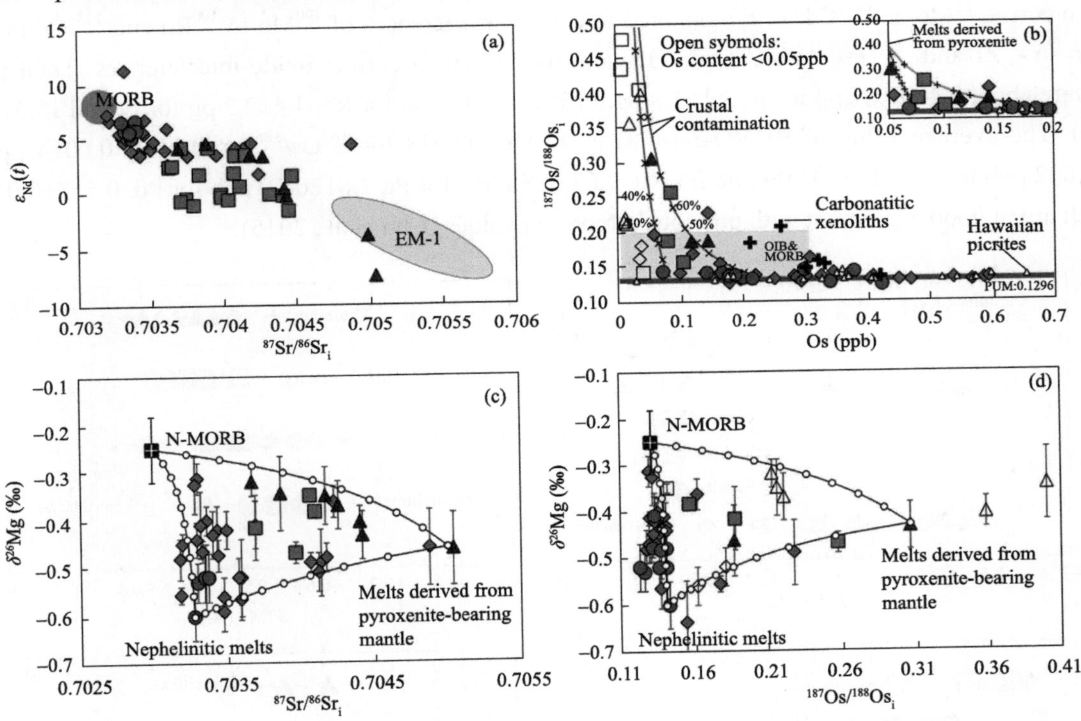

Fig. 3 (a) $^{87}Sr/^{86}Sr$ vs. $\varepsilon_{Nd}(t)$ diagram. (b) $^{187}Os/^{188}Os$ vs. Os contents diagram. The curve is modeled in line of the mixture of melts and crust. The Os content and $^{187}Os/^{188}Os$ for crust are 0.01ppb and 1.5, respectively. The Os content and $^{187}Os/^{188}Os$ for parental melts are 0.2 ppb or 0.075 ppb and 0.14, respectively. Data sources of the carbonatite xenoliths: He et al., 2020. (c, d) $\delta^{26}Mg$ vs. $^{87}Sr/^{86}Sr$ and $^{187}Os/^{188}Os$ diagrams.

3 Analytical methods

The analyses were conducted at State Key Laboratory of Geological Processes and Mineral Resources, China University of Geosciences, Beijing. We followed the protocols developed by Pearson and Woodland (2000) and modified by Xu et al. (2020b) for Re-Os isotope compositions and HSE abundances. About 1 g of sample powders along with a mixed HSE spike (^{99}Ru, ^{106}Pd, ^{185}Re, ^{190}Os, ^{191}Ir and ^{194}Pt) were digested by *aqua regia* (3 ml purified concentrated 10 N HCl and 6 ml purified concentrated 14 N HNO_3) within a Carius tube at 240 °C for 72h. Osmium was extracted from the *aqua regia* solution into CCl_4, then followed by back-extraction into HBr and further purified via microdistillation. The purified Os fraction was then loaded onto a platinum filament using HBr with a Ba $(OH)_2$ activator and measured as OsO_3^- on a Thermo Triton Plus N-TIMS (negative thermal ionization mass spectrometer) using the SEM in a peak-jumping mode. Osmium isotope compositions were corrected for mass-dependent isotope fractionation and oxygen isotope interferences. During the measurement campaign, 250 pg loads of the Johnson-Matthey UMD and DROsS standards were measured repeatedly. Our analyses of $^{187}Os/^{188}Os$ over a year period yielded mean values of 0.11375±0.00027 (2σ; n=10) for UMD and 0.16076±0.00035 (2σ; n=10) for DROsS which are both at the range of accepted values (Luguet et al., 2008). The total procedural Os blank was 0.89±0.25 pg (n=4) with $^{187}Os/^{188}Os$ of ~0.1627. The blank contributions on total Os were <1% for most samples, thus the blank corrections were negligible. The other PGEs and Re were purified and concentrated by cation exchange resin (AG50W-X8). The detailed separation chemistry procedures are described by Li et al. (2013). To further separate Zr-Hf-Mo-W from PGE-Re, solvent extraction using BPHA in chloroform was applied (Shinotsuka and Suzuki, 2007). All Re and PGE were measured via SEM on a Thermo Element 2 XR-ICP-MS. During the measurement, ^{95}Mo and ^{111}Cd were monitored to correct interferences of ^{100}Mo to ^{100}Ru and ^{108}Cd to ^{108}Pd, and ^{85}Rb-^{88}Sr-^{89}Y-^{90}Zr and ^{169}Tm-^{172}Yb-^{175}Lu-^{178}Hf were monitored to correct oxide interferences. Total procedural blanks were about 0.40±0.27 pg for Ir, 3.1±1.8 pg for Ru, 1.4±0.8 pg for Re, 4.4±2.3 pg for Pt and 15.3±8.2 pg for Pd (n=7). The average value of BHVO-2 (n=3) is 0.1476±0.0038 for $^{187}Os/^{188}Os$, 0.1156±0.0014 ppb for Os, 0.070±0.002 ppb for Ir, 0.156±0.009 ppb for Ru, 9.63±0.86 ppb for Pt, 2.81±0.10 ppb for Pd, 0.555±0.016 ppb for Re, which are in good agreement with previously reported values (Chu et al., 2015).

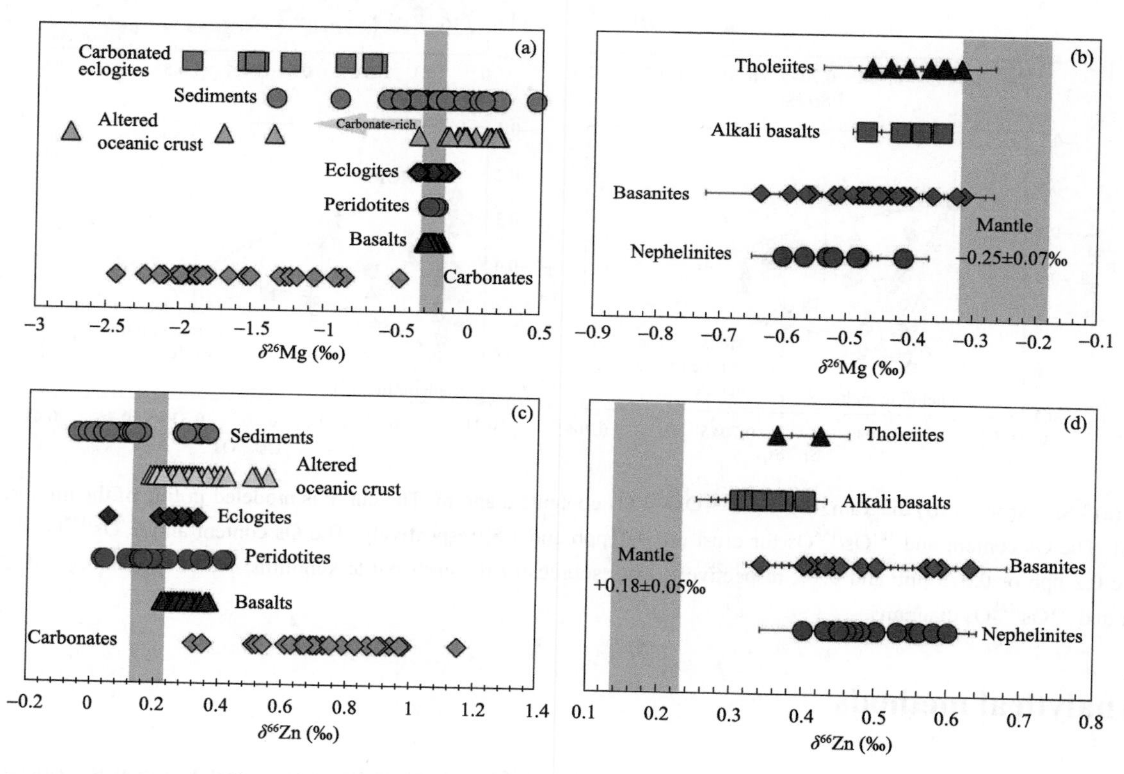

Fig. 4　(a, c) a summary of $\delta^{26}Mg$ and $\delta^{66}Zn$ values of peridotites, basalts, altered oceanic crust, eclogite, marine carbonates, carbonated eclogite. (b, d) a summary of $\delta^{26}Mg$ and $\delta^{66}Zn$ values of Cenozoic basalts from eastern China. See Appendix 3 for data sources.

For Mg isotope analyses, 1 to 5 mg sample powders were dissolved using concentrated acids of HF-HNO_3 (3:1), followed by HCl-HNO_3 (3:1), and converted to 1 N HNO_3 for column chemistry. Magnesium was purified from matrices using AG50W-X8 (200–400 mesh chloride form, Bio-Rad, USA) resin. The whole procedure Mg blank was <10 ng which is negligible compared to >10 μg Mg in samples. Magnesium isotope measurements were conducted on a Thermo-Finnigan Neptune Plus MC-ICP-MS using the sample-standard bracketing method. Each sample was measured four times and finally, the average of those values is reported. More detailed analytical procedures are described by Gao et al. (2019). Two separate analyses of BHVO-2 gave δ^{26}Mg values (−0.26 ± 0.04‰ and −0.20 ± 0.02‰, respectively; 2SD) that are in good agreement with the recommended value (−0.24 ± 0.08‰; Teng et al., 2017).

4 Results

Nephelinites and basanites have high IPGE abundances (Ir: 0.20±0.09 ppb and 0.25±0.16 ppb, respectively; Table S3), low Cu contents (52±5 ppm and 50±11 ppm, respectively) and exhibit flat PGE patterns normalized to primitive upper mantle (PUM) (Fig. 5; Pd_N/Ir_N: 0.9±0.4 and 1.1±0.9, respectively). These samples essentially show modern mantle-like $^{187}Os/^{188}Os_i$ (0.142±0.020) and high Os contents (Fig. 3b, 6; 0.27±0.15 ppb). By contrast, alkali basalts and tholeiites are characterized by highly fractionated PGE patterns (Fig. 5; Pd_N/Ir_N: 5.7±4.2

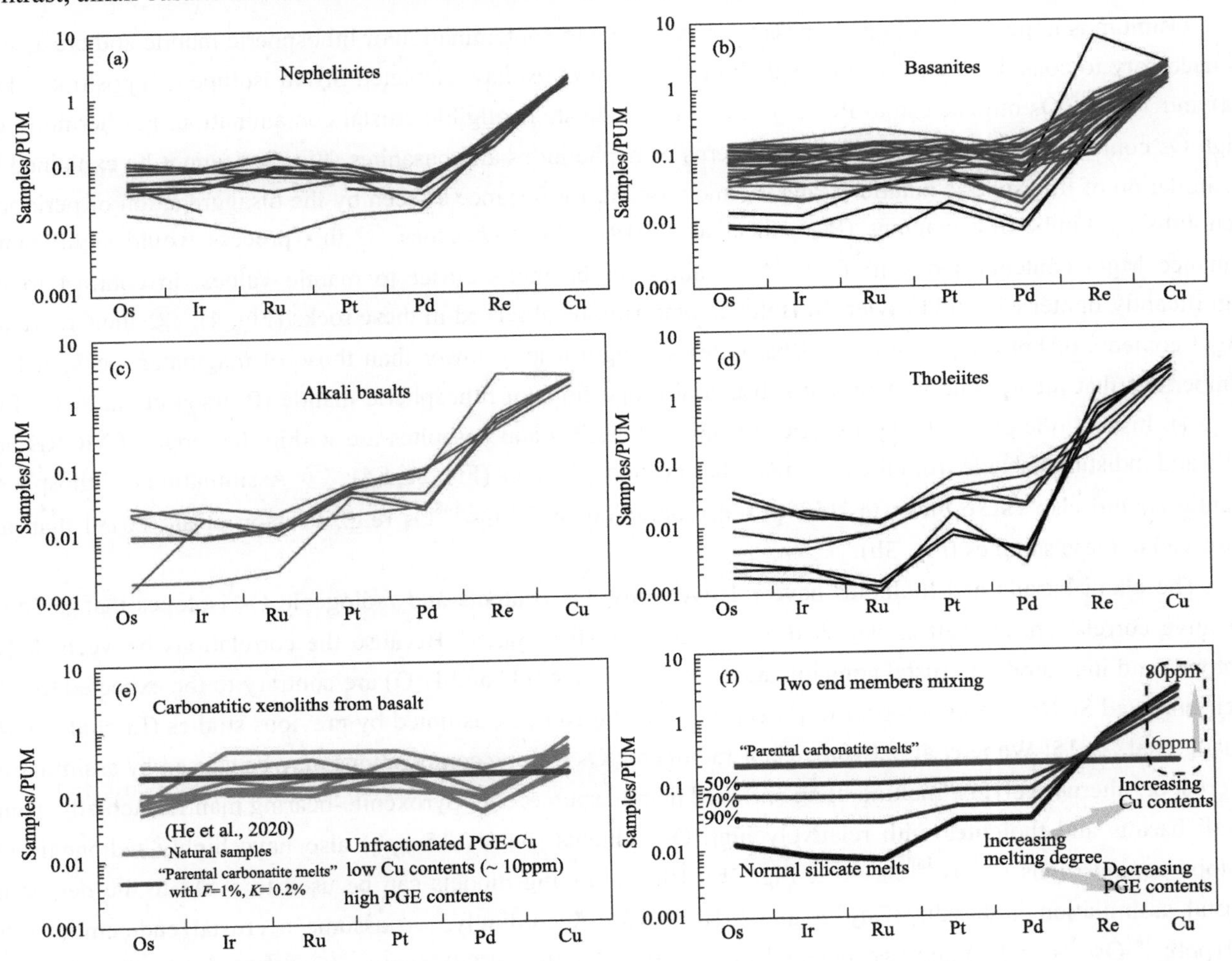

Fig. 5 PGE, Re and Cu contents of Cenozoic Na-series basalts from eastern China normalized to PUM. Data of the carbonatite xenoliths is from He et al. (2020), whereas rhenium contents were not reported. The grey lines in "Two end members mixing" diagram are mixtures of parental carbonatite melts and normal silicate melts in varying proportions. The element contents of normal silicate melts are average values of tholeiites. Element concentrations of parental carbonatite melts are calculated based on $C^l_{melt} = \frac{C_0^i \times K}{F}$ assuming K=0.2% and F=1%. The data of PUM is from Becker et al. (2006).

and 3.2±1.8, respectively), lower IPGE abundances (Ir: 0.04±0.03 ppb and 0.03±0.02 ppb, respectively; Table S3) and higher Cu contents (64±4 ppm and 80±19 ppm, respectively). These samples have more radiogenic $^{187}Os/^{188}Os_i$ (0.279±0.115) and lower Os contents (Fig. 3b, 6; 0.049±0.044 ppb).

Our new Mg isotope results (Su-Wan, Yangyuan and Hannuoba basalts; 27 samples) are consistent with those of previous studies (Huang et al., 2015; Li et al., 2017) having Mg isotope compositions that are clearly distinct from the mantle value (Table S2). Generally, the Cenozoic basalts (nephelinites and basanites) with limited inter-element PGE fractionation are characterized by lighter Mg and heavier Zn isotope compositions and lower Cu contents than the alkali basalts and tholeiites that have highly fractionated PGE patterns (Fig. 4, 5, 7).

5 Discussion

5.1 Evaluation of secondary processes and identification of source signatures

5.1.1 Assimilation processes

Osmium isotope compositions of basalts are sensitive to contamination of lithospheric mantle and crust, so it is necessary to consider their effects. Nephelinites and basanites have depleted Sr-Nd isotope compositions (Fig. 3a) and $^{187}Os/^{188}Os$ ratios close to PUM (Fig. 3b) that indicate negligible crustal contamination. Furthermore, the high Os contents and unfractionated PGE patterns of nephelinites and basanites (Fig. 2c) cannot be explained by assimilation of lithospheric mantle during the melt ascent, for instance as seen by the disaggregation of peridotite xenoliths by kimberlitic magmas (Pearson et al., 2019). This is because ① this process would significantly enhance MgO contents and shift the $\delta^{26}Mg$ values of the melts closer to mantle values, in contrast to the significantly lighter Mg and heavier Zn isotope compositions observed in these rocks (Fig. 4); ② the Cr, Ni and MgO contents of both nephelinites and basanites are significantly lower than those of magmatic rocks such as kimberlites that incorporate varying but substantial proportions of lithospheric mantle (Pearson et al., 2019; Fig. 2c, S4). Instead, the elemental systematics of the nephelinites and basanites are within the range of MORB and OIB and indistinguishable from those of alkali basalts and tholeiites (Fig. 2c, S4); ③ Assimilation of lithospheric mantle should also cause melts to shift to more unradiogenic $^{187}Os/^{188}Os$ (e.g., Pearson et al., 2019) than are observed in these samples (Fig. 3b).

The alkali basalts and tholeiites have relatively low Os contents and radiogenic Os isotopes that define a negative correlation, as well as enriched Sr-Nd isotopes (Fig. 3a, b). Because the correlations between Sr-Nd isotopes and indicators of crustal contamination (e.g., Th/Nb, Nb/U and K_2O) are contrary to the expected trends, their enriched Sr-Nd isotopes appear to reflect their mantle sources, as noted by previous studies (Li et al., 2016; Huang et al., 2015; Wang et al., 2018a). Their radiogenic Os isotope compositions may be caused by assimilation of crust or, alternatively, by melting of an enriched mantle source, e.g., pyroxenite-bearing mantle. Actually, some alkali basalts and tholeiites with relatively high Os contents (0.05–0.15 ppb) also have highly radiogenic Os isotope compositions ($^{187}Os/^{188}Os$ ~0.2; Fig. 3b). Binary mixing models can be used to evaluate the degree of crustal assimilation in basalts (Day et al., 2010; 2013). Accordingly, we assume a crustal endmember (Os: 0.01ppb; $^{187}Os/^{188}Os$: 1.5) and two parental melts (Os: 0.2 ppb and 0.075ppb; $^{187}Os/^{188}Os$: 0.14). Based on our modeling, a high proportion of crustal material (at least 20–30%) is needed to explain those samples with radiogenic Os isotope compositions and high Os contents-proportions which seem unrealistic because they would significantly change the major element chemistry and critical trace element ratios (e.g., Nb/U) of these magmas, rendering them rather unlike the primitive melts that they resemble (Fig. 3b; Chu et al., 2017). Therefore, we propose that the alkali basalts and tholeiites were primarily derived from an enriched mantle source. However, it is difficult to evaluate the degree of crustal contamination for those samples with Os contents <0.05 ppb (Day, 2013; Gannoun et al., 2016) which may experience significant AFC (assimilation and fractional crystallization) modification. Hence, we have excluded those samples with Os <0.05 ppb from further discussion (Fig. 3, 6).

5.1.2 Sulfide segregation and source signatures

Sulfides control the siderophile-chalcophile element content and fractionation in basaltic melts (Jenner, 2017). Correlation between Cu and MgO contents is usually used to identify the segregation of sulfides during magma differentiation (Lee et al., 2012; Reekie et al., 2019). In contrast to MORB magmas that are normally sulfide-saturated, OIB-type magmas usually have higher Cu contents due to sulfide-undersaturation or less sulfide segregation (Reekie et al., 2019). The Cenozoic basalts from eastern China generally have lower Cu contents than MORB and OIB (Fig. 7) and lack robust correlations between Cu and MgO, which indicates that they did not experience considerable sulfide segregation (Fig. 7). In detail, the nephelinites and basanites have lower Cu contents than the alkali basalts and tholeiites, with the latter group having Cu contents close to MORB (Fig. 7), which implies that the degree of melting dominates the Cu concentration variation of these magmas.

During the evolution of basaltic magmas, sulfide or PGE alloy segregation will modify the PGE-Re-Cu patterns of parental magmas. However, this process cannot account for the PGE systematics of the nephelinites and basanites, because the fO_2 of these melts is high (1.6 log units above the FMQ buffer; Hong et al., 2020), and as fO_2 increases, the S content at sulfide saturation (SCSS) will increase exponentially hampering the saturation of sulfide phases (Jugo et al., 2010). Fractionation of PGE alloy from such oxidized melts is also unlikely. Furthermore, any sulfide or PGE alloy segregation would dramatically reduce the PGE contents of their host melts and drive PGE-Re-Cu patterns towards greater fractionation, clearly inconsistent with their high IPGE contents and unfractionated PGE patterns (Fig. 5). Hence, we propose that the unique PGE-Re-Cu patterns of nephelinites and basanites mainly reflect the characteristics of their mantle sources, while a few samples with lower IPGE contents and slightly fractionated PGE patterns (black lines in Fig. 5) may reflect heterogeneity among the parental melts or unidentified magma evolution processes. The PGE-Re-Cu patterns of alkali basalts and tholeiites are generally comparable to those normal OIB and MORB. Most of the alkali basalts and tholeiites have relatively homogeneous PGE contents. Of note, the alkali basalts with significantly lower IPGE contents may experience IPGE-bearing alloy segregation, and the three tholeiites with lower total PGE contents and depleted Pd may reflect sulfide segregation (black lines in Fig. 5).

5.2 Recycled carbonates in the big mantle wedge and the generation of intraplate basalts

Global electrical conductivity-depth profiles indicate that the asthenosphere beneath eastern China at 250–600 km depths has anomalously high conductivity (σ: ~1 S/m; Karato, 2011). Carbonatitic melts have significantly higher electrical conductivity than hydrous olivine and molten silicates (Gaillard et al. 2008; Yoshino et al., 2010), which implies that the asthenosphere above the stagnant slab may have been carbonated. Based on the melting curve of carbonated oceanic crust (Thomson et al., 2016) and oxygen isotope evidence from inclusions within super-deep (~410 km) diamonds (Regier et al., 2020), the subducted oceanic slab will melt and release carbonatitic melts when reaching mantle transition zone depths.

The large-scale Mg-Zn isotopic anomalies documented in Cenozoic basalts from eastern China have been previously attributed to recycled Mg-Zn-rich carbonates (magnesite or dolomite) from the stagnant western Pacific plate that infiltrated their mantle sources (Fig. 4; Liu et al., 2016; Li et al., 2017). Furthermore, the nephelinites and basanites with the lightest Mg, heaviest Zn isotope compositions and more obvious “carbonatitic signals" (Fig. S1, S3; high CaO/Al_2O_3, La/Yb, negative Zr-Hf-Ti anomalies, suprachondritic Zr/Hf) may represent, or approach, pristine melts derived from deep carbonated mantle (Liu et al., 2016; Li et al., 2017). Collectively, the big mantle wedge beneath eastern China may contain large amounts of carbonates which promote mantle melting and the generation of intraplate basalts (Liu et al., 2016; Li et al., 2017; Xu et al., 2018). Since the rationale of tracing recycled carbonates using Mg-Zn isotopes has been thoroughly discussed in previous studies (e.g., Yang et al., 2012; Huang et al., 2015; Liu et al., 2016; Li et al., 2017; Xu et al., 2018; Liu and Li, 2019), we summarized the details only in the Electronic Annex (Appendix 1).

5.3 Ancient pyroxenite-bearing mantle source of alkali basalts and tholeiites

The alkali basalts and tholeiites have bulk compositions that are typical of melts derived from the upper 100 km of Earth's mantle (Li et al., 2016). Unlike the scenario of the nephelinites and basanites which will be further discussed below, partial melting of single carbonated peridotite at shallow mantle cannot account for the enriched Sr-Nd and radiogenic Os isotope compositions observed in these alkali basalts and tholeiites (Fig. 3a, b). This is because recycled sedimentary carbonates have extremely low Nd-Os contents (e.g., Nd: 0.1–2 ppm; He et al., 2020) and magnesites also have low Sr contents. Recycled Mg-carbonates cannot significantly modify the Nd-Os-Sr isotope compositions of the mantle sources (Turekian et al., 1961; Li and Wang, 2018; Jin et al., 2020); this is also the reason why both nephelinites and basanites have relatively depleted Sr-Nd isotope compositions (Fig. 3a). Hence, the alkali basalts and tholeiites require a different source, with enriched Sr-Nd and radiogenic Os isotope characteristics (Fig. 3a, b). Three groups of models have been invoked to explain the source characteristics: ①recycled ancient sediments (Sakuyama et al., 2013; Liu et al., 2015a; Jin et al., 2020), ②ancient garnet pyroxenite in shallow asthenosphere with EM-I like signature (Li et al., 2016; Hong et al., 2020), and ③ ancient enriched subcontinental lithospheric mantle (SCLM; Wang et al., 2018a; Li and Wang, 2018).

Since terrigenous sediments mostly have very low Os contents, the presence of terrigenous sediments in the source region of the Na-series basalts cannot explain such radiogenic Os isotope compositions of these rocks, unless unrealistic amounts of such sediments are incorporated (Eisele et al., 2002; Class et al., 2009; Jackson and Shirey, 2011). Of note, this suite of Cenozoic intraplate basalts from eastern China possess significantly higher Fe/Mn (72±6) than MORB (55±5; Chu et al., 2017; He et al., 2019), and the olivine phenocrysts of the alkali basalts also have high Ni and low Mn, Ca contents (Li et al., 2016). These two lines of evidence suggest their derivation from olivine-poor pyroxenitic sources (Sobolev et al., 2007). Thus, Li et al. (2016) proposed that the mantle source of alkali basalts entrained ancient EM-I garnet pyroxenites, of which the origin is controversial. Due to extensive and long-lived melt/fluid metasomatism, ancient SCLM is characterized by overall enriched Sr-Nd isotope compositions, high Ba/Th and low Ce/Pb ratios (Fig. 6; Pearson and Wittig, 2014). The contribution of melts derived from enriched SCLM (for example from beneath the NCC) is therefore proposed to be responsible for the elemental and isotopic compositions of alkali basalts and tholeiites (Fig. 6; Wang et al., 2018a; Li and Wang, 2018). The intrinsic melting mechanism of enriched SCLM is indeed dominated by melting of non-peridotite mantle lithology (i.e., pyroxenite dykes; Zhang et al., 2002; Kogiso et al., 2003; Gao et al., 2008; Zhao et al., 2013). Metasomatic melts/fluids derived from slabs are usually enriched in Re relative to Os, with elevated Re/Os ratios in the pyroxenite veins within the SCLM that would develop radiogenic $^{187}Os/^{188}Os$ over time (Reisberg et al., 1991; Pearson and Nowell, 2004; Becker, 2000; Dale et al., 2007, 2009). Such pyroxenites, particularly clinopyroxenites, usually have extremely radiogenic $^{187}Os/^{188}Os$ coupled with enriched Sr-Nd isotope compositions (Reisberg et al., 1991; Pearson and Nowell, 2004; Zhang et al., 2008; Acken et al., 2010; Ackerman et al., 2013; Wang and Becker, 2015b). Collectively, the radiogenic Os isotope compositions and other chemical characteristics of the alkali basalts and tholeiites can be attributed to melting of ancient pyroxenite-bearing mantle sources (Fig. 3a, b), and these ancient pyroxenite components may come from enriched SCLM (Wang et al., 2018a; Li and Wang, 2018) or asthenosphere (Li et al., 2016).

Based on Sr-Nd-Mg-Os isotopic mixing calculations, Cenozoic intraplate basalts from eastern China can be well accounted for by three melt endmembers: ① low-silica nephelinitic melts (melting of carbonated peridotite), ②high-silica tholeiites (melting of ancient pyroxenite-bearing mantle), and ③ N-MORB melts (melting of the depleted mantle). The compositional transition from nephelinites, basanites, through alkali basalts to tholeiites depends on the varying proportions of mixing of these three endmembers (Fig. 3c, d), which is further discussed below.

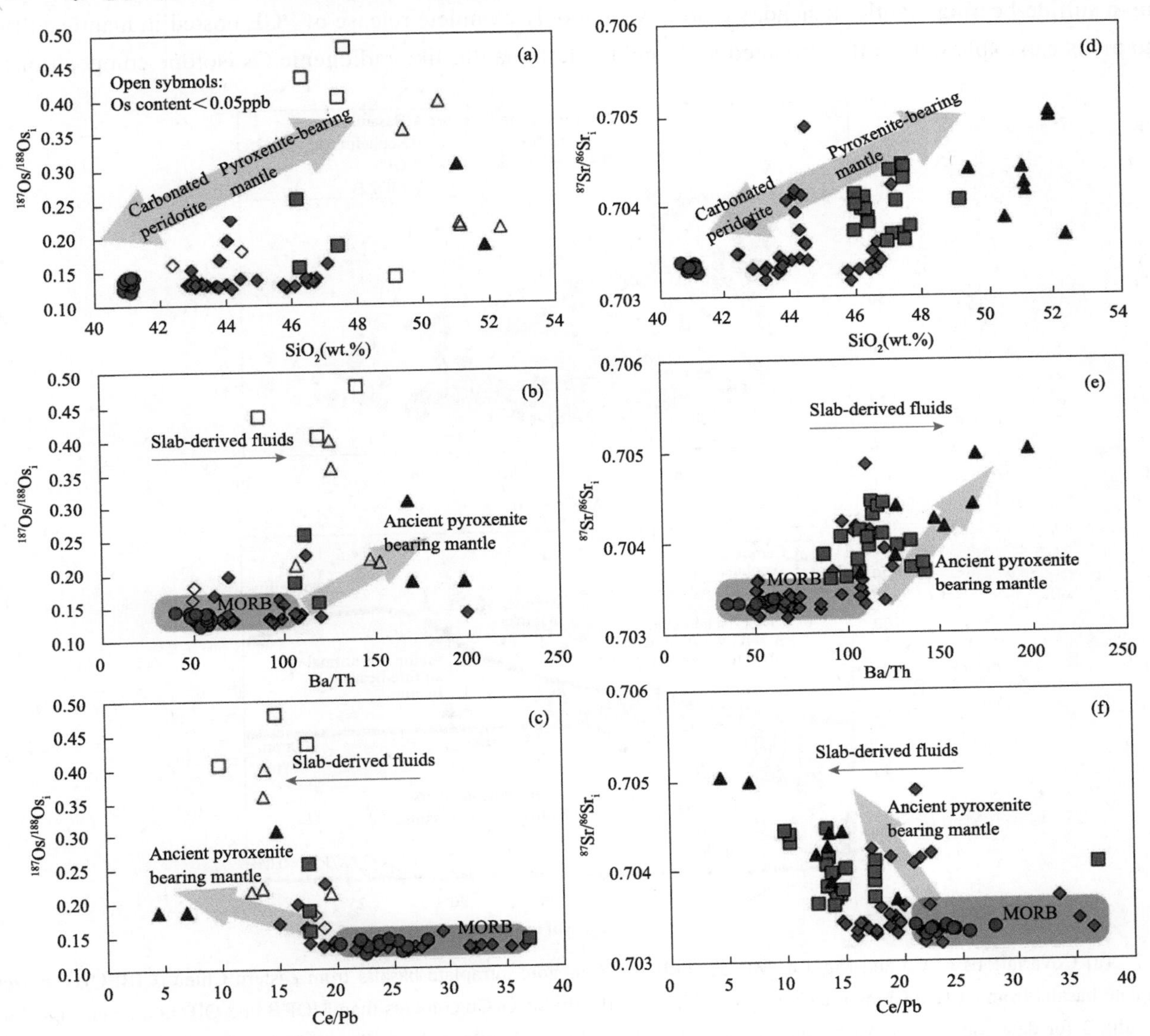

Fig. 6 $^{187}Os/^{188}Os$ and $^{87}Sr/^{86}Sr$ vs. SiO_2 (a, d), Ba/Th (b, e) and Ce/Pb (c, f) diagrams of the Cenozoic basalts from eastern China.

5.4 Generation of unfractionated PGE patterns of nephelinites and basanites and its implications

The current recognition about the partitioning of siderophile-chalcophile elements between residual mantle and basaltic melts is in debate (we summarize the details of this debate in the Electronic Annex), including: ①MSS-sulfide melt partitioning (incongruent partial melting of mantle sulfides; Alard et al., 2000; Bockrath et al., 2004; Ballhaus et al., 2006; Liu and Brenan, 2015) and ②sulfide melt-silicate melt partitioning (Wang et al.,

2013, 2018b; Wang and Becker, 2015a, b; Mungall and Brenan, 2014) (see details in Appendix 2 in the Electronic Annex). The highly fractionated PGE-Re-Cu patterns and Cu contents of the eastern China alkali basalts and tholeiites are comparable with normal mafic melts which can be essentially explained by partial melting of common sulfide-bearing mantle sources (purple curves in Fig. 7b). However, the strikingly unfractionated PGE patterns of the nephelinites and basanites, with 5–10 times enrichment of IPGE and lower Cu contents (Fig. 5, 7b) cannot be accounted for by the same PGE-partitioning environment that produces the drastically different, much more fractionated PGE patterns observed in MORB or OIB, i.e., partitioning between silicate melts in equilibrium with a sulfide-bearing mantle. Instead, we propose that only complete release of PGE hosted in mantle sulfides into melts can explain the unfractionated PGE and modern mantle-like radiogenic Os isotope compositions of

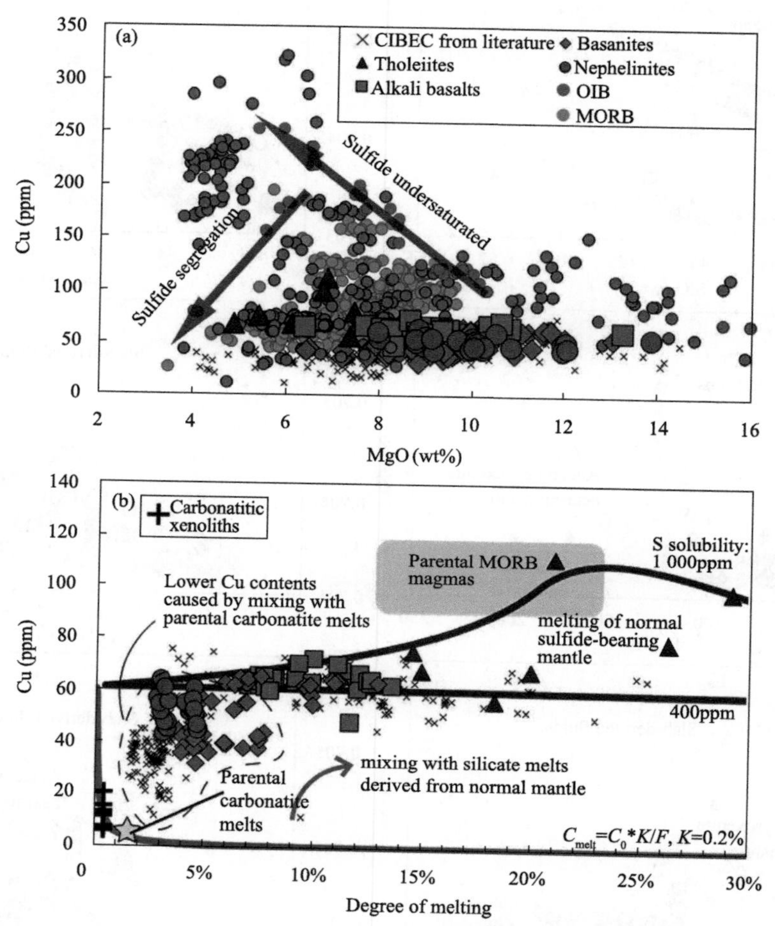

Fig. 7　(a) Covariations of Cu and MgO in MORB, OIB and Cenozoic intraplate basalts from eastern China (CIBEC). Cenozoic intraplate basalts from eastern China are characterized by remarkably lower Cu contents than MORB and OIB at a given MgO. See Appendix 3 for data sources. (b) Modelling the behavior of Cu during mantle melting. Purple curves denote the evolution of the melts during near-fractional melting of normal sulfide-bearing mantle (Lee et al., 2012). The parameters are the same as Lee et al. (2012), but the solubility of S in melts is set as 1000 ppm and 400 ppm, respectively. Red curve models the evolution of carbonate-rich melts calculated using $C^i_{melt}=\frac{C^i_0 * K}{F}$ (the concentration of metal elements i in the melts is C^i_{melt}, C^i_0 is the concentration of element i in ambient mantle, F and K are the degree of partial melting and the proportion of oxidized sulfides in the mantle source, respectively), assuming K=0.2%, $C^{Cu}_0=30$ ppm. Melting degrees of samples are calculated based on La/Yb ratios, with detailed calculation provided in the Fig. S3c. The data of the carbonatite xenoliths (He et al., 2020) are plotted in (b) for comparison, assuming their melting degree is 0.5%. (For interpretation of the references to colour in this figure legend, the reader is referred to the web version of this article.)

these rocks. This process may proceed by four mechanisms: ① the source sulfides are exhausted completely and melts are sulfide-undersaturated within mantle source, ② sulfide-undersaturated melts dissolve mantle sulfides directly *en route* to the surface, ③ sulfides are molten within peridotite throughout the asthenosphere and sulfide melts behave physically incompatibly during silicate melt extraction, and ④ extracted melts are in equilibrium with oxidized sulfides (S_2 or sulfate) which no longer host PGE-Cu, so that these metals are released into silicate melt phase (Mungall et al., 2006; He et al., 2020).

The unfractionated PGE-Re-Cu patterns and extremely high PGE contents documented in komatiites and picrites (Fig. 8) can be well explained by mechanism ① (Connolly et al., 2011; Day, 2013; Gannoun et al., 2016). However, this process does not fit the case of the nephelinites and basanites because they are products of low degree mantle melting (<5–10%).

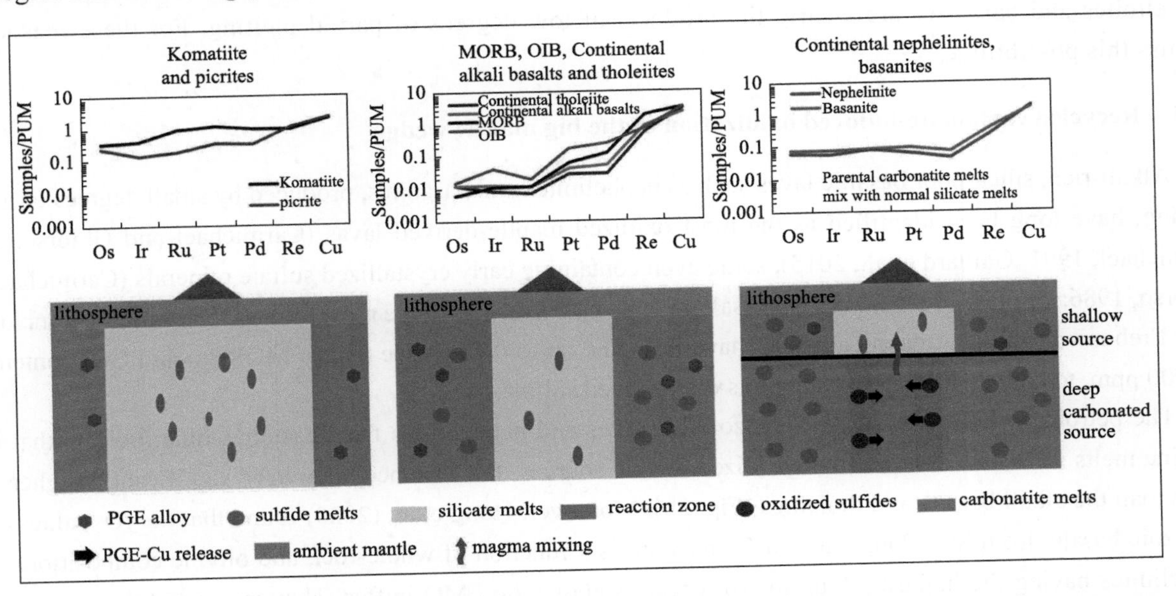

Fig. 8 PGE-Re-Cu patterns of different mantle melts. Komatiites and picrites are products of extremely high degree melting (30–50%) of the mantle, where source sulfides are exhausted completely and only PGE alloys are detained in residues. MORB, OIB and some continental basalts (alkali basalts and tholeiites) are products of lower degree (5–30%) melting of the mantle, where source sulfides cannot be exhausted completely, both PGE alloys and sulfides are detained in residues. Continental alkaline magmas (nephelinites, basanites) are products of extremely low degree (<5–10%) melting of the mantle. We propose a two end-members (carbonatite melts-normal silicate melts) mixing model to account for the unfractionated PGE and relatively enriched Cu-Re patterns. In this model, low proportions (<1%) of source sulfides may be oxidized and then the oxidized sulfides release metal to carbonatite melts. See Appendix 3 for data sources.

Based on studies of peridotite xenoliths, some peridotites experience sulfide breakdown or dissolution as a result of percolation of S-undersaturated melts (Liu et al., 2010, 2020; Tassara et al., 2018; Alard et al., 2011). Through such a process, i.e., mechanism ②, the melts might inherit the PGE-Re-Cu characteristics of the lithospheric mantle or even asthenospheric mantle. Interaction of ascending nephelinitic melts with lithospheric mantle is an unlikely explanation of the observed PGE-Re-Cu data, because interstitial sulfides (with high Pd/Ir and radiogenic $^{187}Os/^{188}Os$) relative to sulfides armoured within silicates (low Pd/Ir and unradiogenic Os isotope) in lithospheric mantle should dissolve preferentially into ascending melts (Harvey et al., 2011). It remains unknown whether S-undersaturated melts can dissolve sulfides directly within the asthenosphere to produce melts with unfractionated PGE patterns. If this process could operate in the asthenosphere, unfractionated PGE-Cu patterns should be observed in sulfide undersaturated OIBs, which is inconsistent with their observed highly fractionated PGE patterns in OIBs (Day, 2013; Fig. 8). In addition, this process cannot explain why unfractionated PGE patterns are only observed in nephelinites and basanites not in the alkali basalts which are also S-undersaturated (Fig. 5, 7a).

If sulfides in the asthenosphere are completely molten, their PGE-Re-Cu will be quantitatively retained in sulfide melts (Zhang and Hirschmann, 2016). In this case, if silicate melts derived from mantle sources entrain molten sulfides in suspension (Ballhaus et al., 2006), the melts could have unfractionated PGE-Re-Cu patterns. However, the feasibility of this physical process is questioned by experiments (Mungall and Su, 2005), because it is difficult to force sulfide droplets to mobilize through pore throats in partially molten peridotite (Mungall and Su, 2005; Mungall et al., 2006; Wang et al., 2020). Also, if physically permissible, this process ought to be most effective during high degrees of mantle melting (Balhaus et al., 2006) whereas the eastern China nephelinites and basanites are clearly the products of low degrees of partial melting. For these reasons we dismiss this possibility.

5.4.1 Recycled carbonate-induced oxidization of the big mantle wedge

Alkali-rich, silica-poor basaltic lavas such as nephelinites and basanites, produced by small degrees of mantle melting, have long been identified as the most oxidized mantle-derived lavas (Carmichael and Ghiorso, 1986; Carmichael, 1991; Gaillard et al., 2015), some even containing early crystallized sulfate minerals (Carmichael and Ghiorso, 1986; Gaillard et al., 2015). Moussallam et al. (2014) reported the most primary basanitic melt inclusions from Erebus volcano in Antarctica, which have $Fe^{3+}/\Sigma Fe$ up to 0.36 (twice that of MORBs) and CO_2 contents up to 7000 ppm, to contain measurable amounts of dissolved sulfate.

The petrological features of the Cenozoic basanites and nephelinites from eastern China indicate that these alkaline melts were also derived from oxidized mantle sources. For instance, they have significantly higher V/Sc ratios than the alkali basalts and tholeiites (Fig. 2b). Moreover, Hong et al. (2020) found that the fO_2 values of the Cenozoic basalts from Shandong, eastern China vary as a function of whole-rock and olivine compositions, with nephelinites having the highest fO_2 up to 1.6 log units above the FMQ buffer. They suggested that the high fO_2 values were inherited from asthenospheric mantle oxidized by recycled carbonates. Similarly, He et al. (2019) showed that the Mg-Zn isotope data of the Cenozoic basalts from eastern China are coupled with Fe isotope data. Also, the nephelinites have high $\delta^{56}Fe$ (+0.29) coupled with $Fe^{3+}/\Sigma Fe$ (~0.73) which indicates that the $Fe^{3+}/\Sigma Fe$ of the mantle source was enhanced by the partial reduction of carbonates (redox freezing of carbonates; Rohrbach and Schmidt, 2011), while the native iron may be exhausted (He et al., 2019).

The high fO_2, V/Sc and $\delta^{56}Fe$ of the nephelinites indicate that their source is oxidized and free of native iron (He et al., 2019), that could host PGEs in the deep asthenosphere (Frost et al., 2004; Rohrbach et al., 2007). Majoritic garnet inclusions within diamonds, from mantle transition zone depths, have been shown to have high $Fe^{3+}/\Sigma Fe$(~0.3), showing that parts of the mantle transition zone have fO_2 well beyond the stability of native iron (Kiseeva et al., 2018). This, together with the extreme oxygen isotope compositions of these majorites (Regier et al., 2020) implies that carbonated melts/fluids are the oxidizing agents responsible for the high $Fe^{3+}/\Sigma Fe$ of majoritic garnet. As such, metallic Fe, generated by disproportionation of Fe^{2+}, is likely absent within the carbonated asthenosphere modified by the subducted slab (Kiseeva et al., 2018) and should not affect PGE partitioning.

5.4.2 Platinum group elements mobilization within carbonated mantle

Based on above discussion, the eastern China nephelinites and basanites most likely represent melts derived from the mantle hybridized and oxidized by recycled carbonates above the stagnant slab. Furthermore, Mungall et al., (2006) reported some meimechites with extremely high PGE contents and relatively unfractionated PGE patterns, which they suggested that the unfractionated PGE patterns in melts resulting from low degree mantle melting can be attained when source sulfides are oxidized to sulfates. This evidence, combined with our new observations from eastern China indicates that the effect of recycled carbonates on stability of mantle sulfides and the consequences for mobilization of PGE-Re-Cu needs to be evaluated.

The scarce published PGE-Re-Cu data of carbonatites indicate that they have relatively low PGE-Re-Cu contents and fractionated PGE patterns which may be caused by segregation of immiscible sulfides during evolution of carbonatitic melts or low degree partial melting (Xu et al., 2008b; Ackerman et al., 2019). Based on sulfur solubility in silicate-carbonate melts, sulfur can be effectively transported by carbonate-silicate melts as sulfate, under oxidizing conditions (Woodland et al., 2019). Frezzotti et al., (2011) found $MgSO_4$ as fluid inclusions within diamonds from ultra-high pressure metamorphic rocks in the Alps. They proposed that carbonate reduction occurred by sulfide oxidation to form diamonds. High P-T experiments in the $MgCO_3$-FeS-oxide system also show that diamonds can form when mantle sulfides are oxidized during carbonate-oxide-sulfide interaction, reflecting the role of sulfides as a reducing agent for carbonatitic melts (Gun and Luth, 2006; Palyanov et al., 2007). Although the S species produced by oxidation in the experiments was suggested to be S_2 via the reaction:

$3SiO_2+Al_2O_3+3FeS+1.5CO_2=Fe_3Al_2Si_3O_{12}+1.5C+1.5S_2$ (Marx, 1972; Palyanov et al., 2007), the enhanced ferric iron proportion of the total iron within the carbonated mantle source produced by "redox freezing of carbonates" (He et al., 2019) may also play an important role in promoting the oxidation of sulfides. This is because ferric iron is regarded as an effective agent in oxidizing sulfides, for instance, by the reaction: $FeS+4Fe_2O_3+9FeSiO_3(opx)=SO_3+9Fe_2SiO_4(ol)$ suggested by Mungall (2002). Based on the above discussion, it is likely that "redox freezing of carbonates" cause not only oxidation of Fe^0 but also S^{2-}.

Recently, some carbonatite xenoliths (carbonatite melt pockets in mantle peridotite xenoliths) in Neogene basalts from Dalihu, eastern China were found to contain diamond, graphite and barite. They have been suggested to represent sedimentary carbonates recycled into the mantle given their limestone-like trace element and $^{87}Sr/^{86}Sr$ signatures. Also, they provide evidence for sulfide oxidation and release of PGE-Cu by recycled carbonates in mantle (Liu et al., 2015b; He et al., 2020), including: ① these carbonatite xenoliths have extremely high PGE contents and unfractionated PGE-Cu patterns (slightly higher PGE contents and lower Cu contents than the nephelinites and basanites from eastern China; Fig. 5), and their Os isotope compositions are close to PUM (Fig. 3b); and ② $\delta^{34}S_{sulfate}$ values are lower than those of sulfates in protolith sedimentary carbonates (i.e., limestone) , and ③ carbonate groundmass shows elevated Cu and Ni contents.

5.4.3 Two-endmember mixing model

Here, we invoke a two-endmember mixing model, between normal silicate melts (those with highly fractionated PGE patterns typical of silicate moderate to small-degree mantle melts) and parental carbonatite melts, to account for the PGE-Re-Cu systematics of the nephelinites and basanites. We established a simplified model of buoyant carbonatitic melts derived from the stagnant slab that react with ambient mantle to form carbonated mantle (Fig. 8). Within the reaction zone, Fe^0 is exhausted by carbonates to enhance the source $Fe^{3+}/\Sigma Fe$, and mantle sulfides within the reaction zone which no longer host PGE-Cu-Re due to their oxidation. Consequently, all PGE-Re-Cu are released into the carbonatitic melts, producing unfractionated PGE-Re-Cu patterns (Fig. 5e, 7b). We define the almost pure carbonatitic melts produced by this process as "parental carbonatite melts" (Fig. 5e, 7b). Based on mass balance, the concentration of chalcophile/siderophile element i in the melts is

$$C^i_{\text{melt}} = \frac{C^i_{\text{sulfide}} * m_{\text{sulfide-zone}}}{m_{\text{melt}}} \tag{1}$$

Where C^i_{sulfide} is the concentration of element *i* in sulfides, $m_{\text{sulfide-zone}}$ is the mass of sulfides in the reaction zone.

Given: $$C^i_0 * m_{\text{zone}} = C^i_{\text{sulfide}} * m_{\text{sulfide-zone}} \tag{2}$$

where m_{zone} is the mass of the reaction zone, C^i_0 is the concentration of element *i* in the ambient mantle. We can obtain

$$\begin{aligned} C^i_{\text{melt}} &= \frac{C^i_0 * m_{\text{zone}}}{m_{\text{melt}}} \\ m_{\text{melt}} &= F * m_0 \\ m_{\text{zone}} &= K * m_0 \\ C^i_{\text{melt}} &= \frac{C^i_0 * K}{F} \end{aligned} \tag{3}$$

where F and K are the mass proportions of melts and reaction zone in the mantle source, respectively (K is equivalent to the proportion of oxidized sulfides in the mantle source and F is equivalent to the melting degree of carbonated peridotite). Hence, the PGE-Re-Cu contents of the carbonatite xenoliths (assuming that they represent "parental carbonatite melts") mentioned above can be well explained by relatively low levels of oxidation of sulfides (K=0.1% to 0.5% at F=1%) (Fig. 5e, 7b). The release of metals is not controlled by the partition coefficient because oxidized sulfides within the reaction zone cannot host these elements. Although previous studies suggested that the oxidation of mantle sulfides and the release of metals should cause Cu-enrichment of basaltic melts (Mungall et al., 2002), we argue that the Cu content of the resulting melt is strongly controlled by the proportion of oxidized sulfides and the degree of melting (Fig. 5e, 7b). A low proportion of sulfide oxidation does not necessarily cause Cu-enrichment of melts (Fig. 5e, 7b) but can lead to significant PGE-enrichment (Fig. 5e).

With the upwelling of the heterogeneous mantle source of the Cenozoic eastern China intraplate basalts from the mantle transition zone to shallow mantle, the source materials (carbonatite veins, ancient pyroxenites and depleted peridotites) would cross their solidi and melt (Li et al., 2016; Xu et al., 2018). When reaching ~300 km, the carbonated peridotite would experience initial partial melting to produce almost pure carbonatitic melts (<10 wt% SiO_2 and ~40 wt% CO_2; Dasgupta and Hirschmann, 2006). Subsequently these carbonatitic melts will evolve to carbonated silicate melts (>25 wt% SiO_2 and <25 wt% CO_2) as silicate melt fractions increase at shallower mantle conditions (Dasgupta et al., 2007; 2013). At ~100 km, the ancient pyroxenite would melt to produce silica-rich melts (Kogiso et al., 2003; Li et al., 2016). Obviously, the more fusible carbonatitic veins will be preferentially sampled by nephelinites and basanites that comprise the lower-degree melts, whereas more refractory pyroxenite will be preferentially tapped by alkali basalts and tholeiites during melting at much shallower levels. Thus, the carbonatitic component will be diluted by silica-rich melts with increasing degrees of partial melting (Fig. 4, S3). In terms of PGE-Re-Cu systematics, we propose that the "parental carbonatite melts" produced by initial partial melting of this carbonated peridotite have unfractionated PGE-Re-Cu patterns (Fig. 5e), high PGE contents and unradiogenic Os isotope compositions while the silica-rich melts produced by melting of ancient pyroxenite-bearing mantle exhibit highly fractionated PGE-Re-Cu patterns (Fig. 5f) and radiogenic Os isotope compositions (Fig. 3c, d). Mixing between the "parental carbonatitic melts" and normal silicate melts (average tholeiites) in varying ratios from ~9:1 to 1:1 can essentially maintain the unfractionated PGE patterns with high IPGE contents and relatively enriched Cu-Re contents (slightly lower Cu contents than normal silicate melts, 30–60 ppm) as observed in the nephelinites and basanites (Fig. 5f, 7b, 8).

Here we briefly summarize the evolution of PGE-Cu-Re of the nephelinites and basanites: ① Deep carbonated mantle and formation of parental carbonatitic melts: Within the mantle reaction zone affected by infiltration of carbonates released from the stagnant slab in the mantle transition zone, ambient native Fe is exhausted to enhance $Fe^{3+}/\Sigma Fe$ and mantle sulfides are oxidized by carbonates to release their bearing PGE-Cu-Re into carbonatitic melts. The proportion of oxidized sulfides in the mantle source is ~0.1%–0.5%. The parental carbonatitic melts produced by this process should have unfractionated PGE-Re-Cu patterns (Fig. 5e). ②Parental carbonatitic melts evolve to carbonate-silicate melts: With higher degrees of melting of a now heterogeneous source at shallow mantle conditions, the silica-rich melts derived from sulfide-bearing mantle (pyroxenite veins or depleted peridotites) will dilute the PGE-Cu-Re contents and other carbonatitic signals of the parental carbonatite melts (e.g., Mg-Zn isotopes, CaO/Al_2O_3, Ti/Ti*). This process will reduce the PGE contents and enhance Cu contents of the resulting melts to produce similar PGE-Re-Cu concentrations and patterns as nephelinites and basanites (Fig. 5f).

5.4.4 Broader implications

Our data indicate that mantle sulfides may be oxidized by recycled carbonates to release metals. Some studies have revealed that the solubility of metals in carbonate melts is much higher than silicate melts (Blaine, 2010; Wolf, 2015; He et al., 2020). Here we emphasize that the deep recycled carbonates that enter the deep big mantle wedge beneath eastern China may provide an important mechanism to oxidize mantle sulfides and transfer precious metals from the deep mantle to Earth's continental crust, ultimately facilitating the formation of precious metal ore deposits in intraplate systems. The Phalaborwa carbonatite is an exceptional metal deposit that may be associated with recycled sedimentary carbonates (Rudashevsky et al., 2004; He et al., 2020). The eastern China intraplate alkaline lavas offer an enrichment mechanism for precious metals, pointing to the formation of precious metal ore deposits associated with small-volume carbonate-rich alkaline lavas (Graham et al., 2017; Fiorentini et al., 2018). Furthermore, the role of carbonates in affecting the stability of sulfides in the mantle wedge needs further investigation, for example, high P-T experiments on the carbonate-sulfide-oxide-silicate system would help us understand the details of oxidation of sulfides driven by carbonates, and their broader geochemical consequences.

6 Conclusions

(1) Cenozoic intraplate basalts across eastern China can be well explained by varying mixtures of three melt endmembers: ①low-silica nephelinitic melts, ②high-silica tholeiites, and ③ N-MORB melts. The alkali basalts and tholeiites with highly fractionated PGE patterns and radiogenic Os isotope compositions are mainly derived from melting of ancient pyroxenite-bearing mantle.

(2) The unfractionated PGE patterns documented in the nephelinites and basanites with modern mantle-like Os isotope, lighter Mg and heavier Zn isotope compositions were generated by oxidative breakdown of low levels

of source sulfides caused by deep recycled carbonates, releasing chalcophile-siderophile elements into the resulting carbonatite melts.

(3) The deep big mantle wedge above the mantle transition zone beneath eastern China is oxidized by recycled carbonates. "Redox freezing of carbonates" may cause the oxidation of Fe^0 and S^{2-}. Deep recycled carbonates may provide an important mechanism to oxidize mantle sulfides and transfer precious metals from deep mantle to Earth's continental crust.

Declaration of Competing Interest

The authors declare that they have no known competing financial interests or personal relationships that could have appeared to influence the work reported in this paper.

Acknowledgements

This research was financially supported by the National Key R&D Program of China (2019YFA0708400, 2020YFA0714800, and 2019YFC0605403), the National Natural Science Foundation of China (No. 41730214, 41822301 and 41790451), China "1000 Youth Talents Program", the Second Tibetan Plateau Scientific Expedition and Research Program (STEP) (2019QZKK0801) and the 111 project (B18048), as well as by pre-research Project on Civil Aerospace Technologies No. D 020202 funded by Chinese National Space Administration. We thank Xue Xiao, Xue Gao and Wenran Liu for help in the Mg isotope analyses, Jian Huang for providing some basalt samples. James Scott is thanked for editing an earlier version of this manuscript. We also thank Associate Editor Rich Walker for efficient handling, Lukáš Ackerman and two anonymous reviewers for constructive comments. This is CUGB petro-geochemical contribution no. PGC20150059 (RIG-no. 7).

Appendix A. Supplementary data

Supplementary data to this article can be found online at https://doi.org/10.1016/j.gca.2020.12.019.

References

Ackerman L., Pitcher L., Strnad L., Puchtel I. S., Jelinek E., Walker R. J. and Rohovec J. (2013) Highly siderophile element geochemistry of peridotites and pyroxenites from Horni Bory, Bohemian Massif: implications for HSE behaviour in subduction-related upper mantle. Geochim. Cosmochim. Acta 100, 158-175.

Ackerman L., Polák L., Magna T., Rapprich V., Durišová J. and Upadhyay D. (2019) Highly siderophile element geochemistry and Re-Os isotopic systematics of carbonatites: insights from Tamil Nadu, India. Earth Planet. Sci. Lett. 520, 175-187.

Alard O., Griffin W. L., Lorand J. P., Jackson S. E. and O'Reilly S. Y. (2000) Non-chondritic distribution of the highly siderophile elements in mantle sulphides. Nature 407, 891-894.

Alard O., Lorand J. P., Reisberg L., Bodinier J. L., Dautria J. M. and O'Reilly S. Y. (2011) Volatile-rich Metasomatism in Montferrier Xenoliths (Southern France): implications for the abundances of chalcophile and highly siderophile elements in the subcontinental mantle. J. Petrol. 52, 2009-2045.

Ballhaus C., Bockrath C., Wohlgemuth-Ueberwasser C., Laurenz V. and Berndt J. (2006) Fractionation of the noble metals by physical processes. Contrib. Mineral. Petrol. 152, 667-684.

Becker H. (2000) Re-Os fractionation in eclogites and blueschists and the implications for recycling of oceanic crust into the mantle. Earth Planet. Sci. Lett. 177, 287-300.

Becker H., Horan M. F., Walker R. J., Gao S., Lorand J.-P. and Rudnick R. L. (2006) Highly siderophile element composition of the Earth's primitive upper mantle: constraints from new data on peridotite massifs and xenoliths. Geochim. Cosmochim. Acta 70, 4528-4550.

Blaine F. A. (2010) The Effect of Volatiles (H_2O, Cl and CO_2) on the Solubility and Partitioning of Platinum and Iridium in Fluid-Melt Systems Ph.D. thesis. University of Waterloo.

Bockrath C., Ballhaus C. and Holzheid A. (2004) Fractionation of the platinum-group elements during mantle melting. Science 305, 1951-1953.

Carmichael I. S. E. (1991) The redox states of basic and silicic magmas: a reflection of their source regions?. Contrib. Mineral. Petrol. 106, 129-141.

Carmichael I. S. E. and Ghiorso M. S. (1986) Oxidation-reduction relations in basic magma: a case for homogeneous equilibria. Earth Planet. Sci. Lett. 78, 200-210.

Chu Z., Yan Y., Zeng G., Tian W., Li C., Yang Y. and Guo J. (2017) Petrogenesis of Cenozoic basalts in central-eastern China: Constraints from Re-Os and PGE geochemistry. Lithos 278-281, 72-83.

Chu Z. Y., Yan Y., Chen Z., Guo J. H., Yang Y. H., Li C. F. and Zhang Y. B. (2015) A comprehensive method for precise determination of Re, Os, Ir, Ru, Pt, Pd concentrations and Os isotopic compositions in geological samples. Geostand. Geoanal. Res. 39, 151-169.

Class C., Goldstein S. L. and Shirey S. B. (2009) Osmium isotopes in Grand Comore lavas: a new extreme among a spectrum of EM-type mantle endmembers. Earth Planet. Sci. Lett. 284, 219-227.

Connolly B. D., Puchtel I. S., Walker R. J., Arevalo R., Piccoli P. M., Byerly G., Robin Popieul C. and Arndt N. (2011) Highly siderophile element systematics of the 3.3 Ga Weltevreden komatiites, South Africa: implications for early Earth history. Earth Planet. Sci. Lett. 311, 253-263.

Dale C. W., Burton K. W., Pearson D. G., Gannoun A., Alard O., Argles T. W. and Parkinson I. J. (2009) Highly siderophile element behaviour accompanying subduction of oceanic crust: Whole rock and mineral-scale insights from a high-pressure terrain. Geochim. Cosmochim. Acta 73, 1394-1416.

Dale C. W., Gannoun A., Burton K. W., Argles T. W. and Parkinson I. J. (2007) Rhenium-osmium isotope and elemental behavior during subduction of oceanic crust and the implications for mantle recycling. Earth Planet. Sci. Lett. 177, 211-225.

Dasgupta R. and Hirschmann M. M. (2006) Melting in the Earth's deep upper mantle caused by carbon dioxide. Nature 440, 659-662.

Dasgupta R., Hirschmann M. M. and Smith N. D. (2007) Partial melting experiments of peridotite + CO_2 at 3 GPa and genesis of alkalic ocean island basalts. J. Petrol. 48, 2093-2124.

Dasgupta R., Mallik A. and Tsuno K., et al. (2013) Carbon- dioxide-rich silicate melt in the Earth's upper mantle. Nature 493, 211-215.

Day J. M. D. (2013) Hotspot volcanism and highly siderophile elements. Chem. Geol. 341, 50-74.

Day J. M. D., Pearson D. G., Macpherson C. G., Lowry D. and Carracedo J.-C. (2010) Evidence for distinct proportions of subducted oceanic crust and lithosphere in HIMU-type mantle beneath El Hierro and La Palma, Canary Islands. Geochim. Cosmochim. Acta 74, 6565-6589.

Eisele J., Sharma M., Galer S. J. G., Blichert-Toft J., Devey C. W. and Hofmann A. W. (2002) The role of sediment recycling in EM-1 inferred from Os, Pb, Hf, Nd, Sr isotope and trace element systematics of the Pitcairn hotspot. Earth Planet. Sci. Lett. 196, 197-212.

Farmer G. L. (2014) Continental Basaltic Rocks. Treatise on Geochemistry, 2nd ed.

Fiorentini M. L., LaFlamme C., Denyszyn S., Mole D., Maas R., Locmelis M., Caruso S. and Bui T. H. (2018) Post-collisional alkaline magmatism as gateway for metal and sulfur enrichment of the continental lower crust. Geochim. Cosmochim. Acta 223, 175-197.

Frezzotti M. L., Selverstone J., Sharp Z. D. and Compagnoni R. (2011) Carbonate dissolution during subduction revealed by diamond-bearing rocks from the Alps. Nat. Geosci. 4, 703-706.

Frost D., Liebske C. and Langenhorst F., et al. (2004) Exper- imental evidence for the existence of iron-rich metal in the Earth's lower mantle. Nature 428, 409-412.

Gaillard F., Malki M., Iacono-Marziano G., Pichavant M. and Scaillet B. (2008) Carbonatite melts and electrical conductivity in the asthenosphere. Science 322, 1363-1365.

Gaillard F., Scaillet B., Pichavant M. and Iacono-Marziano G. (2015) The redox geodynamics linking basalts and their mantle sources through space and time. Chem. Geol. 418, 217-233.

Gannoun A., Burton K. W., Day J. M. D., Harvey J., Schiano P. and Parkinson I. (2016) Highly siderophile element and Os isotope systematics of volcanic rocks at divergent and conver- gent plate boundaries and in intraplate settings. Mineral. Soc. Great Br. 81, 651-724.

Gao S., Rudnick R. L., Xu W. L., Yuan H. L., Liu Y. S., Walker R. J., Puchtel I. S., Liu X. M., Huang H., Wang X. R. and Yang J. (2008) Recycling deep cratonic lithosphere and generation of intraplate magmatism in the North China Craton. Earth Planet. Sci. Lett. 270, 41-53.

Gao T., Ke S., Li R. Y., Meng X. N., He Y. S., Liu C. S., Wang Y., Li Z. J. and Zhu J. M. (2019) High-precision magnesium isotope analysis of geological and environmental reference materials by multiple-collector inductively coupled plasma mass spectrometry. Rapid Commun. Mass Spectrom. 33, 767-777.

Graham S. D., Holwell D. A., McDonald I., Jenkin G. R. T., Hill H. J., Boyce A. J., Smith J. and Sangster C. (2017) Magmatic Cu-Ni-PGE-Au sulfide mineralisation in alkaline igneous sys- tems: an example from the Sron Garbh intrusion, Tyndrum, Scotland. Ore Geol. Rev. 80, 961-984.

Gun S. C. and Luth R. W. (2006) Carbonate reduction by Fe-S-O melts at high pressure and high temperature. Am. Mineral. 91, 1110-1116.

Harvey J., Dale C. W., Gannoun A. and Burton K. W. (2011) Osmium mass balance in peridotite and the effects of mantle- derived sulphides on basalt petrogenesis. Geochim. Cosmochim. Acta 75, 5574-5596.

He D. T., Liu Y. S., Moynier Frédéri, Foley S. F. and Chen C. F. (2020) Platinum group element mobilization in the mantle enhanced by recycled sedimentary carbonate. Earth Planet. Sci. Lett. 541, 116262.

He Y. S., Meng X. N., Ke S. and Wu H. J., et al. (2019) A nephelinitic component with unusual δ56 Fe in Cenozoic basalts from eastern China and its implications for deep oxygen cycle. Earth Planet. Sci. Lett. 512, 175-183.

Hong L. B., Xu Y. G., Zhang L., Wang Y. and Ma L. (2020) Recycled carbonate-induced oxidization of the convective mantle beneath Jiaodong, Eastern China. Lithos.

Huang J., Li S.-G., Xiao Y., Ke S., Li W.-Y. and Tian Y. (2015) Origin of low δ^{26} Mg Cenozoic basalts from South China Block and their geodynamic implications. Geochim. Cosmochim. Acta 164, 298-317.

Huang J. L. and Zhao D. P. (2006) High-resolution mantle tomography of China and surrounding regions. J. Geophys. Res. 111, B09305.

Jackson M. G. and Shirey S. B. (2011) Re-Os isotope systematics in Samoan shield lavas and the use of Os-isotopes in olivine phenocrysts to determine primary magmatic compositions. Earth Planet. Sci. Lett. 312, 91-101.

Jenner F. E. (2017) Cumulate causes for the low contents of sulfide- loving elements in the continental crust. Nat. Geosci. 10, 524-529.

Jin Q. Z., Huang J., Liu S. C. and Huang F. (2020) Magnesium and zinc isotope evidence for recycled sediments and oceanic crust in the mantle sources of continental basalts from eastern China. Lithos 370-371 105627.

Jugo P. J., Wilke M. and Botcharnikov R. E. (2010) Sulfur K-edge XANES analysis of natural and synthetic basaltic glasses: Implications for S speciation and S content as function of oxygen fugacity. Geochim. Cosmochim. Acta 74, 5926-5938.

Karato S. L. (2011) Water distribution across the mantle transition zone and its implications for global material circulation. Earth Planet Sci. Lett. 301, 413-423.

Kelley K. A. and Cottrell E. (2009) Water and the oxidation state of subduction zone magmas. Science 325, 605-607.

Kiseeva E. S., Vasiukov D. M., Wood B. J., McCammon C., Stachel T., Bykov M., Bykova E., Chumakov A., Cerantola V., Harris J. W. and Dubrovinsky L. (2018) Oxidized iron in garnets from the mantle transition zone. Nat. Geosci. 11, 144-147.

Kogiso T., Hirschmann M. M. and Frost D. J. (2003) High- pressure partial melting of garnet pyroxenite: possible mafic lithologies in the source of ocean island basalts. Earth Planet. Sci. Lett. 216, 603-617.

Lee C.-T. A. et al. (2012) Copper systematics in arc magmas and implications for crust-mantle differentiation. Science 336, 64-68.

Li H. Y., Xu Y. G., Ryan J. G., Huang X. L., Ren Z. Y., Guo H. and Ning Z. G. (2016) Olivine and melt inclusion chemical constraints on the source of intracontinental basalts from the eastern North China Craton: discrimination of contributions from the subducted Pacific slab. Geochim. Cosmochim. Acta 178, 1-19.

Li J., Jiang X.-Y., Xu J.-F., Zhong L.-F., Wang X.-C., Wang G.-Q. and Zhao P.-P. (2014) Determination of platinum-group elements and Re-Os isotopes using ID-ICP-MS and N-TIMS from a single digestion after two-stage column separation. Geostand. Geoanal. Res. 38, 37-50.

Li S. G. and Wang Y. (2018) Formation time of the big mantle wedge beneath eastern China and a new lithospheric thinning mechanism of the North China Craton—Geodynamic effects of deep recycled carbon. Sci. China Earth Sci. 61, 853-868.

Li S. G., Yang W., Ke S., Meng X. N., Tian H. C., Xu L. J., He Y. S., Huang J., Wang X. C., Xia Q. K., Sun W. D., Yang X. Y., Ren Z. Y., Wei H. Q., Liu Y. S., Meng F. C. and Yan J. (2017) Deep carbon cycles constrained by a large-scale mantle Mg isotope anomaly in eastern China. Nat. Sci. Rev. 4, 111-120.

Liu J., Xia Q. K., Deloule E., Chen H. and Feng M. (2015a) Recycled oceanic crust and marine sediment in the source of alkali basalts in Shandong, eastern China: Evidence from magma water content and oxygen isotopes. J. Geophys. Res.- Solid Earth 120, 8281-8303.

Liu J. G., Cai R. H., Pearson D. G. and Scott J. M. (2019) Thinning and destruction of the lithospheric mantle root beneath the North China Craton: a review. Earth-Sci. Rev. 196 102873.

Liu J. G., Rudnick R. L., Walker R. J., Gao S., Wu F. Y. and Piccoli P. M. (2010) Processes controlling highly siderophile element fractionations in xenolithic peridotites and their influence on Os isotopes. Earth Planet. Sci. Lett. 297, 287-297.

Liu J. G., Pearson D. G., Shu Q., Sigurdsson H., Thomassot E. and Alard O. (2020) Dating post-Archean lithospheric mantle: Insights from Re-Os and Lu-Hf isotopic systematics of the Cameroon Volcanic Line peridotites. Geochim. Cosmochim. Acta 278, 177-198.

Liu S.-A. and Li S. G. (2019) Tracing the deep carbon cycle using metal stable isotopes: opportunities and challenges. Engineering 5, 448-457.

Liu S.-A., Wang Z.-Z., Li S.-G., Huang J. and Yang W. (2016) Zinc isotope evidence for a large-scale carbonated mantle beneath eastern China. Earth Planet. Sci. Lett. 444, 169-178.

Liu Y. N. and Brenan J. (2015) Partitioning of platinum-group elements (PGE) and chalcogens (Se, Te, As, Sb, Bi) between monosulfide-solid solution (MSS), intermediate solid solution (ISS) and sulfide liquid at controlled fO_2-fS_2 conditions. Geochim. Cosmochim. Acta 159, 139-161.

Liu Y., He D., Gao C., Foley S., Gao S., Hu Z., Zong K. and Chen H. (2015b) First direct evidence of sedimentary carbonate recycling in subduction-related xenoliths. Sci. Rep. 5, 11547.

Luguet A., Nowell G. M. and Pearson D. G. (2008) $^{186}Os/^{188}Os$ and $^{184}Os/^{188}Os$ measurements by NTIMS: Effects of interfer- ing element and mass fractionation correction on data accuracy and precision. Chem. Geol. 248, 342-362.

Marx P. C. (1972) Pyrrhotine and the origin of terrestrial diamonds. Mineral. Mag. 38, 636-638.

Moussallam Y., Oppenheimer C., Scaillet B., Gaillard F., Kyle P., Peters N., Hartley M. E., Berlo K. and Donovan A. (2014) Tracking the changing oxidation state of Erebus magmas, from mantle to surface, driven by magma ascent and degassing. Earth Planet. Sci. Lett. 393, 200-209.

Mungall J. E. (2002) Roasting the mantle: Slab melting and the genesis of major Au and Au-rich Cu deposits. Geology 30, 915-918.

Mungall J. E., Hanley J. J. and Arndt N. T. (2006) Evidence from meimechites and other low-degree mantle melts for redox controls on mantle-crust fractionation of platinum-group elements. PNAS 103, 12695-12700.

Mungall J. E. and Brenan J. M. (2014) Partitioning of platinum- group elements and Au between sulfide liquid and basalt and the origins of mantle-crust fractionation of the chalcophile elements. Geochim. Cosmochim. Acta 125, 265-289.

Mungall J. E. and Su S. G. (2005) Interfacial tension between magmatic sulfide and silicate liquids: constraints on kinetics of sulfide liquation and sulfide migration through silicate rocks. Earth Planet. Sci. Lett. 234, 135-149.

Palyanov Y. N., Borzdov Y. M., Bataleva Y. V., Sokol A. G., Oalyanova G. A. and Kupriyanov I. N. (2007) Reducing role of sulfides and diamond formation in the Earth's mantle. Earth Planet. Sci. Lett. 260, 242-256.

Pearson D. G. and Nowell G. M. (2004) Re-Os and Lu-Hf isotope constraints on the origin and age of pyroxenites from the Beni Bousera peridotite massif implications for mixed peridotite- pyroxenite mantle sources. J. Petrol. 45, 439-455.

Pearson D. G., Irvine G. J., Ionov D. A., Boyd F. R. and Dreibus G. E. (2004) Re-Os isotope systematics and platinum group element fractionation during mantle melt extraction: a study of massif and xenolith peridotite suites. Chem. Geol. 208, 29-59.

Pearson D. G. and Woodland S. J. (2000) Solvent extraction/anion exchange separation and determination of PGE (Os, Ir, Pt, Pd, Ru) and Re-Os isotopes in geological samples by isotope dilution ICP-MS. Chem. Geol. 165, 87-107.

Pearson D. G., Woodhead J. and Janney P. E. (2019) Kimberlites as geochemical probes of Earth's mantle. Elements 15, 387-392.

Pearson, D. G., Canil, D., Shirey, S. B. (2014). Mantle Samples Included in Volcanic Rocks: Xenoliths and Diamonds: Treatise on Geochemistry 2nd Edition, 2, 171-275.

Reekie C. D. J., Jenner F. E., Smythe D. J., Hauri E. H., Bullock E. S. and Williams H. M. (2019) Sulfide resorption during crustal ascent and degassing of oceanic plateau basalts. Nat. Commun. 10(1), 82.

Regier M. E., Pearson D. G., Stachel T., Luth R. W., Sten R. A. and Harris J. W. (2020) The lithospheric to lower mantle carbon cycle recorded in superdeep diamonds. Nature 585, 234-238.

Reisberg L. C., Allègre C. J. and Luck J.-M. (1991) The Re-Os systematics of the ronda ultramafic complex of southern Spain. Earth Planet. Sci. Lett. 105(1), 196-213.

Rohrbach A. and Schmidt M. W. (2011) Redox freezing and melting in the Earth's deep mantle resulting from carbon-iron redox coupling. Nature 472, 209-212.

Rohrbach A., Ballhaus C. and Golla-Schindler U., et al. (2007) Metal saturation in the upper mantle. Nature 449, 456-458.

Rudashevsky, N., Kretser, Y.L., Rudashevsky, V., Sukharzhevs-kaya, E., 2004. A review and comparison of PGE, noble-metal and sulphide mineralization in phoscorites and carbonatites from Kovdor and Phalaborwa. In: Zaitsev, A., Wall, F. (Eds.), Phoscorites and Carbonatites from Mantle to Mine: the Key Example of the Kola Alkaline Province. In: Mineralogical Society Series, 363-393.

Sakuyama T., Tian W., Kimura J. I., Fukao Y., Hirahara Y., Takahashi T., Senda R., Chang Q., Miyazaki T., Obayashi M., Kawabata H. and Tatsumi Y. (2013) Melting of dehydrated oceanic crust from the stagnant slab and of the hydrated mantle transition zone: Constraints from Cenozoic alkaline basalts in eastern China. Chem. Geol. 359, 32-48.

Sobolev A. V., Hofmann A. W., Kuzmin D. V., Yaxley G. M., Arndt N. T., Chung S. L., Danyushevsky L. V., Elliott T., Frey F. A., Garcia M. O., Gurenko A. A., Kamenetsky V. S., Kerr A. C., Krivolutskaya N. A., Matvienkov V. V., Nikogosian I. K., Rocholl A., Sigurdsson I. A., Sushchevskaya N. M. and Teklay M. (2007) The amount of recycled crust in sources of mantle-derived melts. Science 316, 412-417.

Shinotsuka K. and Suzuki K. (2007) Simultaneous determination of platinum-group elements and rhenium in rock samples using isotope dilution inductively coupled plasma-mass spectrometry after cation exchange separation followed by solvent extraction. Anal. Chim. Acta 603, 129-139.

Tassara S., Gonzá lez-Jiménez J. M., Reich M. and Saunders E. (2018) Highly siderophile elements mobility in the subconti- nental lithospheric mantle beneath southern Patagonia. Lithos 314-315, 579-596.

Teng F.-Z. (2017) Magnesium Isotope Geochemistry. Rev. Min- eral. Geochem. 82, 219-287.

Thomson A. R., Walter M. J., Kohn S. C. and Brooker R. A. (2016) Slab melting as a barrier to deep carbon subduction. Nature 529, 76-79.

Turekian K. K. and Wedepohl K. H. (1961) Distribution of the elements in some major units of the Earth's crust. Geol. Soc. Amer. Bull. 72, 175-192.

van Acken D., Becker H., Walker R. J., McDonough W. F., Wombacher F., Ash R. D. and Piccoli P. M. (2010) Formation of pyroxenite layers in the Totalp ultramafic, massif (Swiss Alps) - insights from highly siderophile elements and Os isotopes. Geochim. Cosmochim. Acta 74, 661-683.

Wang Z. Z., Liu S. A., Chen L. H., Li S. G. and Zeng G. (2018a) Compositional transition in natural alkaline lavas through silica-undersaturated melt-lithosphere interaction. Geology 46, 771-774.

Wang Z. and Becker H. (2015a) Abundances of Ag and Cu in mantle peridotites and the implications for the behavior of chalcophile elements in the mantle. Geochim. Cosmochim. Acta 160, 209-226.

Wang Z. and Becker H. (2015b) Fractionation of highly siderophile and chalcogen elements during magma transport in the mantle: constraints from pyroxenites of the Balmuccia peridotite massif. Geochim. Cosmochim. Acta 159, 244-263.

Wang Z., Becker H. and Gawronski T. (2013) Partial reequilibration of highly siderophile elements and the chalcogens in the mantle: a case study on the Baldissero and Balmuccia peridotite massifs (Ivrea Zone, Italian Alps). Geochim. Cosmochim. Acta 108, 21-44.

Wang Z., Becker H., Liu Y. S., Hoffmann E., Chen C. F., Zou Z. Q. and Li Y. (2018b) Constant Cu/Ag in upper mantle and oceanic crust: implications for the role of cumulates during the forma- tion of continental crust. Earth Planet. Sci. Lett. 493, 25-35.

Wang Z. J., Jin Z. M., Mungall J. E. and Xiao X. H. (2020) Transport of coexisting Ni-Cu sulfide liquid and silicate melt in partially molten peridotite. Earth Planet. Sci. Lett. 536 116162.

Wolf R. (2015) Solubility of Au and Ir in Carbonate Melts: Implications for the Precious Metal Potential of Carbonatites and Mantle Metasomatism Bachelor thesis. University of Toronto.

Woodland A. B., Girnis A. V., Bulatov V. K., Brey G. P. and Höfer H. E. (2019) Experimental study of sulfur solubility in silicate-carbonate melts at 5-10.5 GPa. Chem. Geol. 505, 12-22.

Wu F.-Y., Yang J.-H., Xu Y.-G., Wilde S. A. and Walker R. J. (2019) Destruction of the North China craton in the mesozoic. Annu. Rev. Earth Planet. Sci. 47, 173-195.

Xu C., Qi L., Huang Z., Chen Y., Yu X., Wang L. and Li E. (2008) Abundances and significance of platinum group elements in

carbonatites from China. Lithos 105, 201-207.
Xu Y. G., Li H. Y., Hong L. B., Ma L., Ma Q. and Sun M. D. (2018) Generation of Cenozoic intraplate basalts in the big mantle wedge under eastern Asia. Sci. China Earth Sci. 61, 869-886.
Xu Y., Wang Q. and Tang G., et al. (2020a) The origin of arc basalts: new advances and remaining questions. Sci. China Earth. Sci..
Xu Y., Liu J. G., Xiong Q., Su B. X., Scott J. M., Xu B., Zhu D. C. and Pearson D. G. (2020b) The complex life cycle of oceanic lithosphere: a study of Yarlung-Zangbo ophiolitic peridotites, Tibet. Geochim. Cosmochim. Acta 277, 175-191.
Yoshino T., Laumonier M., Mclsaac E. and Katsura T. (2010) Electrical conductivity of basaltic and carbonatite melt-bearing peridotites at high pressures: implications for melt distribution and melt fraction in the upper mantle. Earth Planet. Sci. Lett. 295, 593-602.
Yang W., Teng F.-Z., Zhang H.-F. and Li S.-G. (2012) Magnesium isotopic systematics of continental basalts from the North China craton: implications for tracing subducted carbonate in the mantle. Chem. Geol. 328, 185-194.
Zeng G., Chen L. H., Xu X. S., Jiang S. Y. and Hofmann A. W. (2010) Carbonated mantle sources for Cenozoic intraplate alkaline basalts in Shandong, North China. Chem. Geol. 273(1-2), 3-45.
Zhang H. F., Goldstein S. L., Zhou X. H., Sun M., Zheng J. P. and Cai Y. (2008) Evolution of subcontinental lithospheric mantle beneath eastern China: Re-Os isotopic evidence from mantle xenoliths in Paleozoic kimberlites and Mesozoic basalts. Contrib. Mineral. Petrol. 155, 271-293.
Zhang H.-F., Sun M., Zhou X.-H., Fan W.-M., Zhai M.-G. and Yin J.-F. (2002) Mesozoic lithosphere destruction beneath the North China Craton: evidence from major-, trace-element and Sr-Nd-Pb isotope studies of Fangcheng basalts. Contrib. Mineral. Petrol. 144, 241-254.
Zhang Z. and Hirschmann M. M. (2016) Experimental constraints on mantle sulfide melting up to 8 GPa. Am. Mineral. 101, 181-192.
Zhao Z. F., Dai L. Q. and Zheng Y. F. (2013) Postcollisional mafic igneous rocks record crust-mantle interaction during continental deep subduction. Sci. Rep. 3, 3413.
Zhao D. P., Lei J. S. and Tang R. Y. (2004) Origin of the Changbai intraplate volcanism in Northeast China: evidence from seismic tomography. Chin. Sci. Bull. 49, 1401-1408.
Zheng Y. F., Xu Z., Chen L., Dai L. Q. and Zhao Z. F. (2019) Chemical geodynamics of mafic magmatism above subduction zones. J. Asian Earth Sci. 194 104185.
Zou H., Zindler A., Xu X. and Qi Q. (2000) Major, trace element, and Nd, Sr and Pb isotope studies of Cenozoic basalts in SE China: Mantle sources, regional variations, and tectonic signif- icance. Chem. Geol. 171, 33-47.

Carbonated big mantle wedge extending to the NE edge of the stagnant Pacific slab: Constraints from Late Mesozoic-Cenozoic basalts from far eastern Russia*

Ronghua Cai [1], Shan Xu[1], Dmitri A. Ionov[2], Jian Huang[3], Sheng-Ao Liu[1], Shuguang Li[1,3] and Jingao Liu[1]

1. State Key Laboratory of Geological Processes and Mineral Resources, China University of Geosciences, Beijing 100083, China
2. Géosciences Montpellier, Université de Montpellier, 34095 Montpellier, France
3. CAS Key Laboratory of Crust-Mantle Materials and Environments, School of Earth and Space Sciences, University of Science and Technology of China, Hefei 230026, China

亮点介绍：本文报道了俄罗斯远东地区的中-新生代板内碱性玄武岩详细的主微量元素，Sr-Nd同位素及Mg-Zn同位素数据。发现这一地区的玄武岩与中国东部玄武岩类似，其源区都存在再循环的碳酸盐。结合地幔过渡带中滞留板片的位置，该研究表明碳酸盐化的大地幔楔延伸到了地幔过渡带滞留俯冲板片的东北边缘。

Abstract It has been suggested that the carbonated mantle reflected by Mg-Zn isotopic anomalies of Cenozoic intraplate basalts from East Asia coincides with the stagnant West Pacific slab in the mantle transition zone. However, the northern boundary of such carbonated domain beneath East Asia is uncertain. Late Mesozoic-Cenozoic intraplate basalts widespread in far eastern Russia provide an opportunity to examine this issue. Here we report major-trace element contents and Sr-Nd-Mg-Zn isotopic compositions for 9 Late Mesozoic-Cenozoic basaltic samples from the Khanka Block and Sikhote-Alin accretionary complex. They are characterized by large variations in SiO_2 contents (41 to 50 wt.%) and CaO/Al_2O_3 (0.50 to 0.97), enrichments of large-ion lithophile elements (LILE), positive Nb-Ta anomalies and strongly negative K, Pb, Zr, Hf, Ti, Y anomalies in primitive mantle-normalized trace element spider diagram. Furthermore, the rocks show good correlations of Ti/Ti* with Hf/Hf*, La/Yb, Fe/Mn and trace element contents (e.g., Nb). In addition, they have lighter Mg and heavier Zn isotope compositions than the BSE estimates, coupled with depleted Sr-Nd isotope compositions. These elemental and isotopic characteristics cannot be explained by alteration, magma differentiation or diffusion, but are consistent with the partial melting of carbonated peridotite. By and large, the Late Mesozoic-Cenozoic basalts from far eastern Russia bear very similar geochemical characteristics as those Na-series Cenozoic basalts from eastern China. The extended region of Mg-Zn isotopic anomalies is roughly coincident with the stagnant west Pacific slab beneath East Asia, and all of

* 本文发表在：Jounal of Earth Science, 2022, 33(1): 121-132

these alkali basalts can be generated from mantle sources hybridized by recycled Mg-carbonates from the Pacific slab stagnant in the mantle transition zone. We infer that ① the carbonated big mantle wedge extends to the NE edge of the West Pacific slab and may have also appeared in the Late Mesozoic due to the effect of the paleo-Pacific slab beneath this region, and ② decarbonation of stagnant slabs in the mantle transition zone is a key mechanism for carbon outgassing from deep mantle to surface via intraplate alkali melts.

Key words Mg-Zn isotopes, deep carbon cycling, far eastern Russia, big mantle wedge

0 Introduction

Seismic tomography reveals that the west Pacific slab became stagnant in the mantle transition zone beneath East Asia forming a big mantle wedge (Huang and Zhao, 2006). The roles of the stagnant slab in the generation of intraplate basalts have attracted much attention recently. The widespread Cenozoic basalts from East Asia have similar compositions to oceanic island basalts and are considered as typical intraplate magmas derived from the asthenosphere (Xu et al., 2018). These mantle melts offer a great opportunity to investigate the interaction between the stagnant Pacific slab and upper mantle (Li et al., 2017). Recent studies have demonstrated that the Mg-Zn isotopes of mantle-derived melts have the potential to trace recycled sedimentary carbonate in their mantle sources (Liu and Li, 2019; Li et al., 2017; Liu et al., 2016; Huang et al., 2015; Yang et al., 2012). The large-scale Mg-Zn isotopic anomalies documented in Cenozoic basalts from eastern China have been attributed to recycled Mg-Zn-rich carbonates released from the stagnant West Pacific slab that infiltrated their mantle source (Li et al., 2017; Liu et al., 2016). This finding indicates that the big mantle wedge beneath eastern China could be a huge carbon reservoir. However, two important issues remain unclear: (1) the range of the Mg-Zn isotopic anomalies in terms of both location and time, and (2) whether it coincides with the stagnant slab in the mantle transition zone (Li et al., 2017).

As shown by seismic tomography, the northeastern edge of the stagnant West Pacific slab is located in far eastern Russia (FER) (Huang and Zhao, 2006), where Late Mesozoic-Cenozoic intraplate basalts are widely outcropped (Okamura et al., 2005; Fig. 1). These basalt samples can be used to trace northeastern boundary of the region with Mg-Zn isotopic anomalies and thus address the unresolved issues. We present a comprehensive study on Late Mesozoic-Cenozoic FER basalts, including major-trace elements, Sr-Nd isotopes and Mg-Zn isotopes. Our results reveal that they have similar elemental and isotopic characteristics as those Na-series Cenozoic basalts from eastern China. We conclude that the Late Mesozoic-Cenozoic intraplate basalts from far eastern Russia were derived from carbonated asthenosphere hybridized by the stagnant West Pacific/paleo-Pacific slabs in the mantle transition zone.

1 Geological setting and samples

The eastern Central Asian Orogenic Belt consists of several microcontinental blocks, including Erguna, Xing'an, Songliao and Jiamusi-Khanka-Bureya, all of which are separated by major faults (Fig. 1; Wilde, 2015). These blocks were assembled during the south-north closure of the Paleo-Asian Ocean in the Neoproterozoic to Mesozoic (Zhou et al., 2018). The major part of crust and lithospheric mantle of these blocks is juvenile, reflecting that the Central Asian Orogenic Belt may represent one of the largest areas of Phanerozoic crustal growth (Wu et al., 2011). The rare Precambrian-aged zircons and peridotite xenoliths with ancient Os model ages imply the existence of Precambrian crystalline basement (Ionov et al., 2020; Zhou et al., 2018). Along the Japan Sea lies the Sikhote-Alin accretionary complex composed of accretionary prisms, turbidite basins and arc igneous rocks (Liu K et al., 2020). This complex is largely located in Russia, with a small portion in China referred to as the Nadanhada Terrane (Fig. 1; Wilde, 2015). The Sikhote-Alin complex is considered as a Late Mesozoic accretionary orogen related to the Paleo-Pacific plate subduction (Zhao et al., 2017).

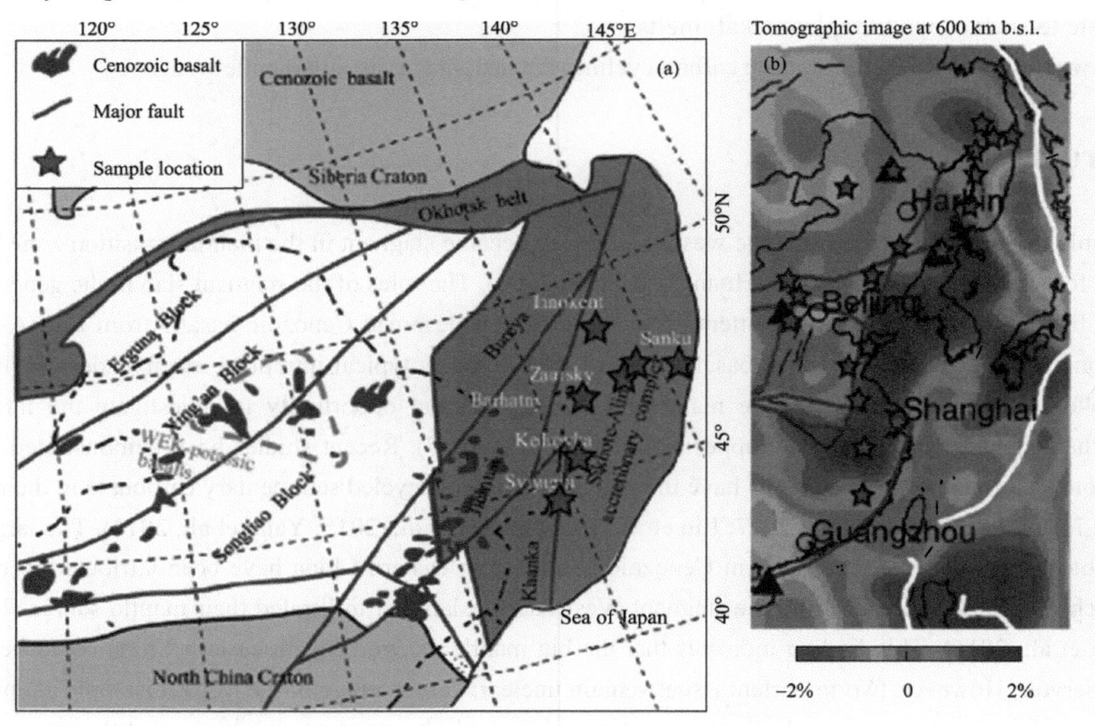

Fig. 1　(a) A geological map of NE China and far eastern Russia which shows the main blocks (modified from Wilde, 2015). (b) P wave velocity image of East Asia at 600 km (modified from Huang and Zhao, 2006). The red stars represent the locations of basalts with Mg-Zn isotopic anomalies. Data sources: Sun et al., (2021), He et al., (2019), Li et al., (2017), Liu et al., 2016, and this study. WEK. Wudalianchi-Erkeshan-Keluo.

The initial subduction of the Paleo-Pacific slab occurred in the late Triassic to early Jurassic, traced by the emplacement of voluminous I-type granitoids during 210−155 Ma and eruption of intermediate-felsic volcanic rocks during 228−202 Ma (Wang F et al., 2015; Guo et al., 2015; Wu et al., 2011). The low-angle subduction of the Paleo-Pacific slab under the Eurasian continent likely took place in the late Early Cretaceous, while the eastward narrowing of the magmatism from the Early Cretaceous to Paleogene reflects the rollback of the Paleo-Pacific slab (Tang et al., 2018). The magmatic gap between 56−46 Ma along the Northeast Asia margin separated two distinct phases of magmatism and may represent the quiescence period prior to the initiation of the Izanagi-Pacific slab subduction (Liu K et al., 2020; Wu and Wu, 2019). With the back-arc spreading of Japan Sea during 25−10 Ma (Chen et al., 2015; Okamura et al., 2005), the tectonic setting in Northeast Asia evolved from the continental margin to the back-arc basin and island arc mode.

Two phases of Cenozoic basaltic magmatism are identified at the North-Eastern Eurasian margin (Okamura et al., 2005). At the pre-opening stage of Japan Sea (55−24 Ma), the basalts from Sikhote-Alin and Sakhalin are characterized by depletion in high field strength elements (HFSE) and have similar geochemical characteristics to arc basalts. Thus they represent the melting products of arc mantle wedge metasomatized by fluids/melts derived from the subducted Pacific slab. By contrast, alkali basalts erupted in far eastern Russia after the opening of Japan Sea (<15 Ma) have OIB-like trace element patterns.

Three out of nine basalt samples in this study (B-A-1, B-A-2 and Kedr1), collected at two localities (NE of Zaursky and Kedrovka) in the northern and central parts of the Sikhote-Alin accretion complex (Fig. 1), are Late Mesozoic in age (~110–120 Ma). The other six samples (An1, Bh1, Bh2, In1, Sa1, Sv1) are from five Mid-Miocene to Pliocene basaltic eruption centers (Zaursky, Barhatny, Innokentiev, Sanku from the Sovgavan lava field, and Sviyagin) located outside the eastern, northern and western margins of the Sikhote-Alin complex; they have $^{40}Ar/^{39}Ar$ ages from 12 to 4 Ma (Okamura et al., 2005; 1998). We did not distinguish the Late Mesozoic and Cenozoic samples because they have similar characteristics.

2 Methods

2.1 Major and trace elements

Fresh basalt fragments were split using wear-resistant wedge into chips <1 cm. The chips were treated with 6 N HCl for 30–60 min, washed with de-ionized water and dried. The leached chips were crushed in a stainless-steel hand mortar, then ground in agate ball mill.

Major element contents were analyzed by wavelength-dispersive X-ray fluorescence (XRF) spectrometry. The rock powders were ignited for 3 h at 1000 ℃, and the loss on ignition (LOI) was calculated. Glass beads, produced by fusing 0.8 g of ignited powders with 4.8 g of dried LiB_4O_7 were analyzed on a Philips PW1404 spectrometer using ultramafic and mafic reference samples as external standards. Trace element concentrations were obtained by a sector-field inductively coupled plasma mass spectrometry following the procedures described in Chen et al. (2017). The analytical results of BCR-2 standard are showed in Table S1.

2.2 Sr-Nd isotopes

Sample preparation and Sr-Nd isotope analyses were done at the Free University of Brussels (ULB). Whole rock powders were dissolved in $HF\text{-}HClO_4\text{-}HNO_3$ mixtures in screw-top Teflon beakers. Strontium and Nd were isolated from the Sr-REE-bearing fraction using Dowex WX50 and HDEHP ion-exchangers. Double-distilled reagents and new resins were used in the dissolution and elution procedures. Isotope ratios of Nd and Sr were measured on a VG54 multiple-collector thermal ionization mass spectrometer (TIMS). Mass fractionation was corrected to $^{86}Sr/^{88}Sr = 0.1194$ and $^{146}Nd/^{144}Nd = 0.7219$, respectively. At the time of this study, the NIST SRM 987 Sr and the Rennes Nd standards yielded $^{87}Sr/^{88}Sr = 0.710273\pm23$ (n=38) and $^{143}Nd/^{144}Nd = 0.511960\pm16$ (n= 32) (2σ), respectively. Total procedural blanks are estimated as 46 pg for Sr and 14 pg for Nd. No corrections for blank contribution or defined values of the standards were applied because they are negligible. Details of the analytical procedures are given in Ionov et al. (2006).

2.3 Mg-Zn isotopes

Magnesium-zinc isotopes were analyzed at the State Key Laboratory of Geological Processes and Mineral Resources, China University of Geosciences (Beijing). For the analysis of Mg isotopes, about 1–5 mg of sample powders were dissolved in a Teflon beaker by a mixture of $HF\text{-}HNO_3$ (3:1 v/v) at 160 ℃. After complete dissolution, the sample solution was evaporated to dryness. The residue was then dissolved in a mixture of $HCl\text{-}HNO_3$ (3:1 v/v) and evaporated to dryness at 80 ℃. Finally, the residue was dissolved in 1 N HNO_3 for Mg separation and purification by AG50W-X8 resin (200–400 mesh chloride form, Bio-Rad, USA). In our routine procedures, the typical total recovery of Mg is more than 99.5% and the procedural blank is <10 ng. Magnesium isotope measurements were conducted on a Thermo-Finnigan Neptune Plus MC-ICP-MS using the sample-standard bracketing method. Each sample was measured four times and the average of those values is reported. More detailed analytical procedures are described in Gao et al. (2019).

For the analysis of Zn isotopes, about 10 mg of sample powders were dissolved respectively in HF-HNO_3 and HCl-HNO_3 (the digestion procedures are the same as those for Mg isotopic analysis), followed by dissolution in 8 N HCl + 0.001% H_2O_2 for column chemistry. Afterwards, 1 ml of sample solution was loaded onto 2 ml anion resin AG-MP-1M (100–200 mesh chloride form, Bio-Rad). Zinc was eluted in 10 ml of 0.5 N HNO_3 after removal of matrix elements, followed by evaporation to dryness and dissolution in 1 ml 3% HNO_3. The typical total recovery of Zn is more than 99.5% and the whole procedural blank less than 6 ng that is negligible compared to >1 μg Zn in samples. The analysis was performed on a Thermo-Finnigan Neptune Plus MC-ICP-MS using the sample-standard bracketing method. More detailed analytical procedures are described in Liu et al. (2016).

The results of Mg and Zn isotope ratios are reported in delta notation against Mg isotope standard DSM-3and Zn isotope standard Lyon JMC, respectively, expressed as follows:

$$\delta^{26}Mg = [(^{26}Mg/^{24}Mg)_{sample} / (^{26}Mg/^{24}Mg)_{DSM3} - 1] \times 1000\ (‰)$$

$$\delta^{66}Zn = [(^{66}Zn/^{64}Zn)_{sample} / (^{66}Zn/^{64}Zn)_{JMC\ 3\text{-}0749L} - 1] \times 1000\ (‰)$$

The reference materials analyzed in this study yielded results (i.e., BHVO-2: $\delta^{26}Mg$=–0.25±0.06‰; $\delta^{66}Zn$= 0.31±0.01‰; BCR-2: $\delta^{26}Mg$=–0.21±0.04‰; $\delta^{66}Zn$=0.26±0.07‰) identical within errors with the literature values (Liu S A et al., 2020; Teng, 2017; Liu et al., 2016).

3 Results

3.1 Major-trace elements and Sr-Nd isotopes

The major-trace element concentrations of the nine FER basalts are provided in Table S1. In the TAS diagram (Fig. 2a), three samples are subdivided into basanites, four samples within the range of alkali basalts and the remaining two tholeiites. These FER basalts exhibit large variations in SiO_2 (41.2–50.0 wt.%), CaO (7.5–10.1 wt.%) and Al_2O_3 (10.2–15.1 wt.%) contents and well-defined correlations between SiO_2 and other major oxides (Fig. 3a-c) including a negative correlation with CaO/Al_2O_3 (0.98–0.51) (Fig. 3d). They have Mg# between 55.4 and 69.9, and MgO contents between 13.6 and 7.0 wt.% that show no robust correlations with other major elements.

The FER basalts are enriched in LILE and show positive Nb-Ta anomalies and negative K, Pb, Zr, Hf, Ti, Y anomalies in the primitive mantle-normalized trace element spider diagram, similar to those in nephelinites and basanites from eastern China (Fig. 4). The FER basalts have high Ce/Pb (45.2–13.4) and Nb/U (38.9–48.1), that are within the range of MORBs & OIBs (Fig. 2d). They display large variations in Ti/Ti* (0.56–1.09) and Hf/Hf* (0.70–0.93) that define good correlations with La/Yb (12.3–45.5), Fe/Mn (48.5–73.8) and trace element contents (e.g., Nb). Similar correlations also appear in Na-series Cenozoic basalts from eastern China. The FER basalts have $^{87}Sr/^{86}Sr$ ratios of 0.70374 to 0.70490 and $\varepsilon_{Nd}(t)$ values of +4.9 to 0 (Table 1).

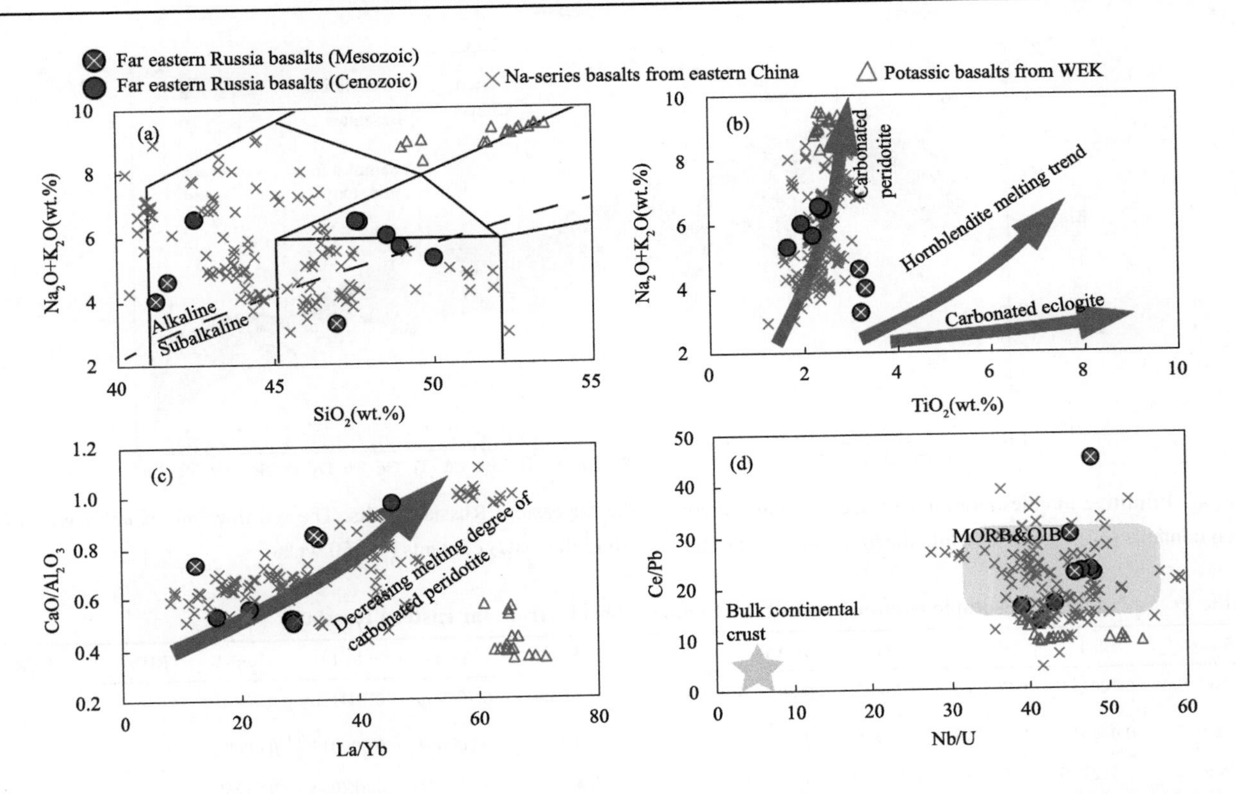

Fig. 2 Total alkali vs. SiO_2 (a), total alkali vs. TiO_2 (b), CaO/Al_2O_3 vs. La/Yb (c) and Ce/Pb vs. Nb/U (d) diagrams of the far eastern Russia basalts. The data of Cenozoic basalts from eastern China are from Huang et al. (2015), Chu et al. (2013) and Zeng et al. (2010). The trends of melting of carbonated peridotite, hornblendite and carbonated eclogite are from Dasgupta et al. (2007), Pilet et al. (2008) and Dasgupta et al. (2006), respectively. Symbols remain the same throughout the figures in text.

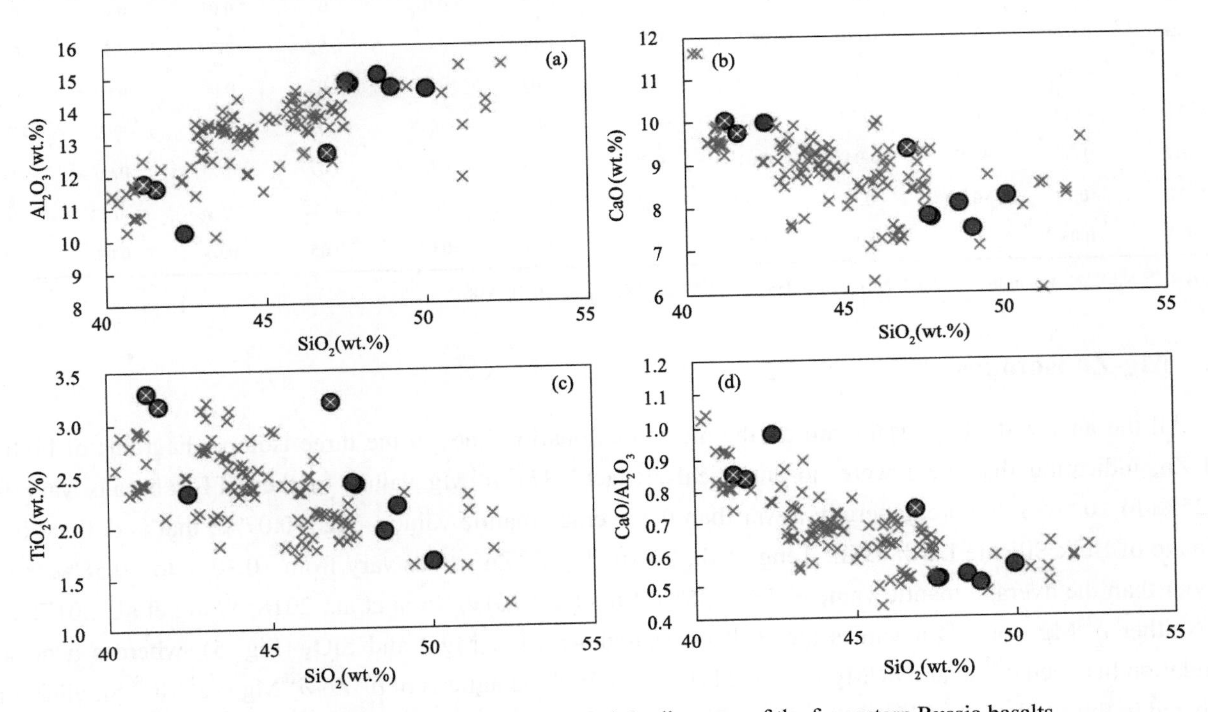

Fig. 3 SiO_2 vs. Al_2O_3, CaO, TiO_2 and CaO/Al_2O_3 diagrams of the far eastern Russia basalts.

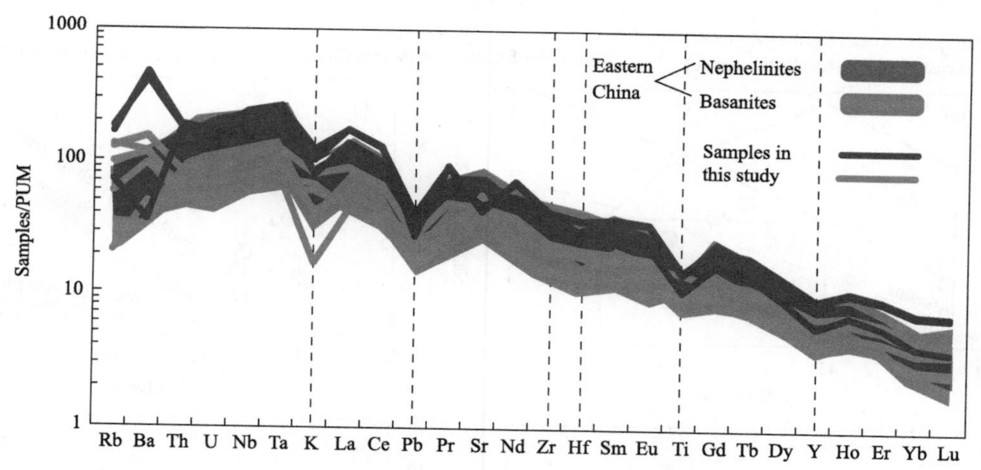

Fig. 4　Primitive mantle-normalized trace element patterns of the far eastern Russia basalts. The red lines are samples with lower SiO_2 contents (42–41 wt.%), while the green lines are samples with higher SiO_2 contents (47–50 wt.%).

Table 1　Sr-Nd-Mg-Zn isotopic compositions of the Cenozoic basalts from far eastern Russia

Sample	Bh-1	Bh-2	B-An-1*	B-An-2*	Sa-1	Sv-9	An-1	In-1	Ked-1	BHVO-2	BCR-2
$^{86}Sr/^{87}Sr$	0.704111	0.704082	0.704033	0.703956	0.703739	0.704081	0.704900	0.704170	0.704850		
2σ	0.000013	0.000013	0.000011	0.000015	0.000013	0.000013	0.000014	0.000012	0.000011		
$^{143}Nd/^{144}Nd$	0.512705	0.512719	0.512769	0.512780	0.512756	0.512729	0.512725	0.512636	0.512891		
2σ	0.000013	0.000013	0.000013	0.000011	0.00001	0.000012	0.000016	0.000010	0.000011		
$\varepsilon_{Nd}(t)$	1.3	1.6	2.6	2.8	2.3	1.8	1.7	0.0	4.9		
$\delta^{66}Zn$	0.38	0.39	0.31	0.45	0.40	0.58	0.36	0.42	0.30	0.31	0.26
2SD	0.07	0.03	0.03	0.06	0.01	0.02	0.01	0	0.01	0.01	0.07
$\delta^{68}Zn$	0.69	0.76	0.60	0.81	0.77	1.11	0.68	0.82	0.59	0.57	0.48
2SD	0.09	0.08	0	0.09	0.04	0.03	0.04	0.02	0.06	0.03	0.12
$\delta^{26}Mg$	−0.44	−0.39	−0.32	−0.25	−0.43	−0.5	−0.48	−0.47	−0.35	−0.25	−0.21
2SD	0.08	0.02	0.04	0.04	0.09	0.05	0.01	0.07	0.05	0.06	0.04
$\delta^{25}Mg$	−0.23	−0.19	−0.16	−0.12	−0.22	−0.25	−0.25	−0.24	−0.18	−0.13	−0.10
2SD	0.04	0.02	0.03	0.02	0.04	0.03	0.03	0.05	0.01	0.03	0.04

* $\varepsilon_{Nd}(t)=[(^{143}Nd/^{144}Nd)_{sample}/(^{143}Nd/^{144}Nd)_{CHUR}-1]\times10000$, $(^{143}Nd/^{144}Nd)_{CHUR}=0.512638$.

3.2　Mg-Zn isotopes

All the analyzed FER basalts fall on the mass fractionation lines in the three-isotope diagrams of both Mg and Zn, indicating that there were no analytical artifacts. The $\delta^{26}Mg$ values of these FER basalts vary from −0.25‰ to −0.50‰ that are generally lower than the average mantle value (−0.25+0.07‰; that is defined by the estimate of Bulk Silicate Earth–BSE; Teng et al., 2010). The $\delta^{66}Zn$ values vary from +0.30‰ to +0.58‰ that are heavier than the average mantle value (0.18+0.05‰; Liu et al., 2019; Sossi et al., 2018; Wang et al., 2017; Table 1). Neither $\delta^{26}Mg$ nor $\delta^{66}Zn$ values are well correlated with Cr, MgO and SiO_2 (Fig. 5), whereas a negative correlation between $\delta^{66}Zn$ and $\delta^{26}Mg$ is defined (Fig. 6c). In the diagrams of $\delta^{66}Zn$-$\delta^{26}Mg$ vs. $^{87}Sr/^{86}Sr$, all samples analyzed in this study fall within the range of Cenozoic basalts from eastern China (Fig. 6).

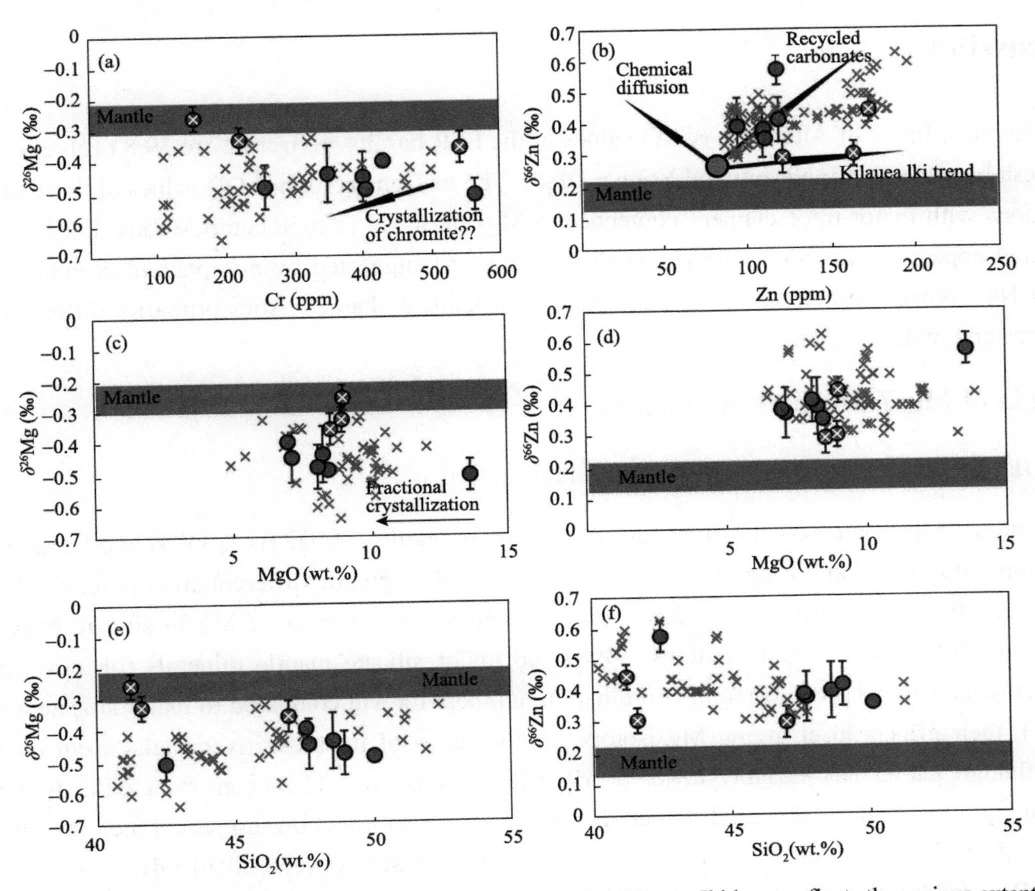

Fig. 5 δ^{66}Zn and δ^{26}Mg vs. Cr, Zn, SiO_2 and MgO diagrams. The trend of Kilauea Iki lavas reflects the various extents of fractional crystallization. The trend of chemical diffusion is from Liu and Li (2019). The δ^{66}Zn and δ^{26}Mg values of the mantle are from Teng (2017), Wang et al. (2017) and Sossi et al. (2018).

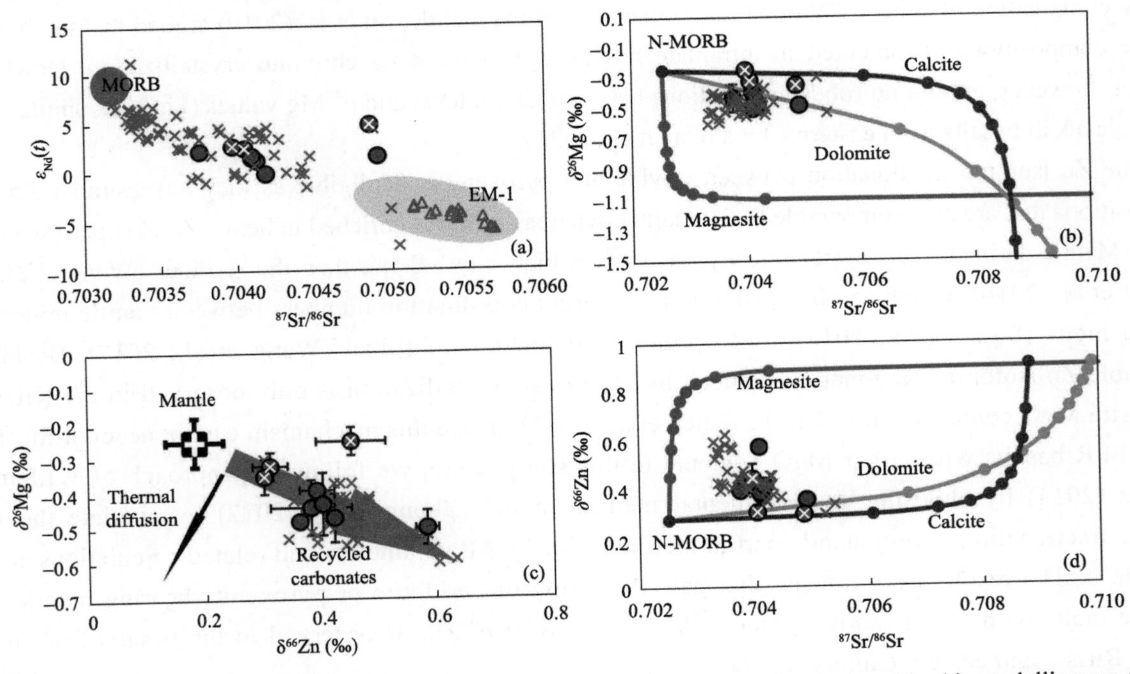

Fig. 6 Sr-Nd-Mg-Zn isotopic compositions of the far eastern Russia basalts. The parameters used in this modelling are from Liu et al. (2016). The trend of thermal diffusion is from Liu and Li (2019). Mixing hyperbolas are marked in 10% increments.

4 Discussion

The Chemical Index of Alteration (CIA) values of the FER basalts are from 27 to 39.8 (Table S1), within the range of fresh basalts (30–45; Nesbitt and Young, 1982). The loss-on-ignition (LOI) values of these samples show no correlations with major-trace element contents and Mg-Zn-Sr-Nd isotope compositions. Furthermore, crustal contamination appears to be negligible in the FER basalts, as suggested by no apparent correlations observed between Sr-Nd isotopes and Nb/U, SiO_2. Thus, their geochemical characteristics primarily reflect those of their mantle source and melting conditions.

4.1 Origin of Mg-Zn isotopic anomalies

4.1.1 Partial melting and magmatic differentiation

The identical Mg isotope compositions among mantle peridotites, MORBs & OIBs and komatiites indicate that the isotopic fractionation during mantle partial melting and magmatic differentiation is limited (Teng, 2017; Teng et al., 2010). This is also supported by similar coordination numbers of Mg in silicate melts (CN=5–6; Shimoda et al., 2007; George and Stebbins, 1998) and major silicate mantle minerals (ol, cpx, opx; CN=6). Although garnet and spinel have different coordination numbers for Mg compared to major silicate minerals (Liu et al., 2011), their effects in changing Mg isotope compositions of mantle-derived melts seem equivocal. For instance, although garnet has variably lower δ^{26}Mg values (down to −1‰), consistent with its higher 8-fold coordination number (Wang et al., 2020; Hu et al., 2016), the garnet effect during partial melting remains poorly known. Zhong et al. (2017) attributed the small Mg isotopic offset between alkali OIBs (−0.31±0.04‰) and tholeiitic OIBs (−0.24±0.02‰) to low-degree melting of garnet-bearing pyroxenite or peridotite source. However, the low-Mg adakitic rocks derived from garnet-bearing lower crust have homogeneous Mg isotopic compositions that are comparable to that of the mantle (Wang et al., 2020). Moreover, in the light of significantly heavier Mg isotope composition of spinel (CN of Mg is 4) than that of the mantle, Su et al. (2019) argued that the lighter Mg isotope compositions documented in intraplate basalts were caused by chromite crystallization. Howbeit, our samples, however, exhibit no robust correlations between Cr contents and δ^{26}Mg values (Fig. 5a), similarly to the Cenozoic alkali basalts from eastern China (Cai et al., 2021).

The Zn isotopic fractionation between olivine and pyroxene is negligible as they have similar Zn isotope compositions that are also comparable to the mantle, whereas spinel is enriched in heavy Zn isotopes (Wang et al., 2017). Mantle-derived melts systematically have δ^{66}Zn higher by ~0.1‰ than the residues (Wang et al., 2017; Doucet et al., 2016), which can be attributed to different coordination numbers between mantle minerals and silicate melts (Sossi et al., 2018), or to preferential melting of spinel (Wang et al., 2017). Besides, the detectable Zn isotopic fractionation induced by fractional crystallization is only observed in the Kilauea Iki lavas with MgO contents below 3 wt.% (Chen et al., 2013), hence this mechanism cannot account for the data of the FER basalts with larger MgO contents in this study. Here, we follow the approach of Williams and Bizimis (2014) by choosing an incremental melting model (Zhong et al., 2017) to simulate the Mg-Zn isotopic fractionation during mantle partial melting (Fig. 7). All parameters and related calculations are listed in Table 2. The results also indicate that partial melting of peridotite or pyroxenite-bearing mantle cannot produce melts with significantly lighter δ^{26}Mg and heavier δ^{66}Zn, as observed in the basalts from both far eastern Russia and eastern China (Fig. 7).

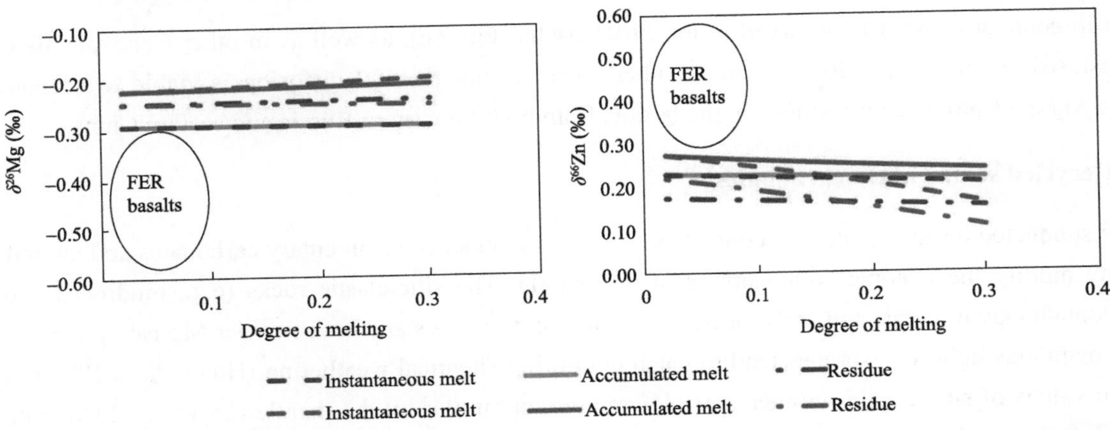

Fig. 7 δ^{66}Zn and δ^{26}Mg values for melt and residue generated by incremental melting of garnet pyroxenite (green) and garnet peridotite (red) against degree of partial melting. See modelling parameters and details in Table 2.

Table 2 The parameters utilized in partial melting modeling

Mineral	Modal		Melting mode		Zn (ppm)	Mg (wt.%)	D_{Zn}	D_{Mg}	$\Delta^{66}Zn_{mineral-Cpx}$	$\Delta^{26}Mg_{mineral-Cpx}$
Ol	0.6	0	0.08	0	50	49	0.96	3.04	0	$-0.17\times10^6/T^2$
Opx	0.07	0	−0.18	0	40	30	0.45	1.96	0	$0.006\times10^6/T^2$
Cpx	0.24	0.75	0.8	0.9	55	22	0.33	1.35	0	0
Grt	0.09	0.25	0.3	0.1	45	22	0.21	1.30	$-0.3\times10^6/T^2$	$-0.94\times10^6/T^2$

The melting mode and mineral abundances for peridotite and pyroxenite are from Davis et al. (2011), Walter (1998), and Pertermann and Hirschmann (2003). Partition coefficients of Zn and Mg are from Davis et al. (2013, 2011). The mineral contents of Mg and Zn are from Wang et al. (2017), Liu S A et al. (2020) and Zhong et al. (2017). The $\Delta^{66}Zn_{mineral-Cpx}$ and $\Delta^{26}Mg_{mineral-Cpx}$ are from McCoy-West et al. (2018) and Huang et al. (2013). The mass melted on each step (F_0) is set to 2% of the initial source mass. Then, the melting degree of the residue of step n is $f_n = \frac{F_n}{1-(n-1)F_n}$. The element abundance in the step n is $C_n^{melt} = \frac{C_{n-1}^{residue}}{D_n + (1-P)f_n}$. The fractionation factor between melt and residue is $\alpha_{melt\text{-}residue} = \alpha_{melt\text{-}cpx} \times \left(\sum_{i=1}^{n} M_i C_i^{Mg/Zn} \Big/ \sum_{i=1}^{n} \alpha_{i\text{-}cpx} M_i C_i^{Mg/Zn} \right)$, where $\alpha_{melt\text{-}cpx}$ is a free variable. Here we choose the $\alpha_{melt\text{-}cpx}$ of Mg is 0.9999 (Stracke et al., 2018), and the $\alpha_{melt\text{-}cpx}$ of Zn is 1.000054 (Sossi et al., 2018).

4.1.2 Chemical and thermal diffusion

Remarkable isotopic fractionation induced by chemical gradient was found in both diffusion couple experiments and natural examples due to faster diffusion of lighter isotopes (Richter et al., 2009; Richter et al., 2008; Pogge von Strandmann et al., 2011). A similar process may also take place between ascending basaltic melts and surrounding mantle (Liu and Li, 2019). Because basaltic melts usually have lower Mg and higher Zn contents than surrounding mantle, Mg tends to diffuse from mantle to melts whereas Zn migrates in the opposite direction (Liu S A et al., 2020). Chemical diffusion can theoretically cause basaltic melts to possess lighter Mg and heavier Zn isotope compositions compared to the host mantle (Fig. 5b). However, if the high δ^{66}Zn values of melts were caused by chemical diffusion of normal mantle melts (e.g., MORBs) to surrounding mantle, a negative correlation between Zn contents and δ^{66}Zn values should be observed and the resultant melts should have relatively low Zn contents (Liu and Li, 2019). This is not consistent with the high Zn contents and the broadly positive correlation between Zn contents and δ^{66}Zn seen in the basalts from East Asia (Fig. 5b).

The inference that thermal gradient can induce isotopic fractionation is implied by thermal diffusion experiments of silicate melts (Huang et al., 2010). Experiments show that the colder end tends to concentrate heavier isotopes than the hotter end (Richter et al., 2008). A similar thermal diffusion may proceed when basaltic melts migrate in the lithosphere, and this process may cause the hot basaltic melts to concentrate lighter Mg-Zn isotopes in contrast to what is observed in the FER basalts (Fig. 6c), as well as in other Cenozoic alkali basalts from East Asia (Liu and Li, 2019). Thus, neither chemical nor thermal diffusion is viable to account for the observed Mg-Zn isotopic compositions in the basalts from both far eastern Russia and eastern China.

4.1.3 Recycled sedimentary carbonates

The subducted oceanic plate that contains siliciclastic sediments, sedimentary carbonates and altered oceanic crust may modify the isotopic compositions of the mantle. The siliciclastic rocks (e.g., mudrocks and shales) mainly contain clay minerals and soils formed as weathering residues and have heavier Mg isotopic compositions than the mantle as light Mg isotopes tend to leach out during chemical weathering (Hu et al., 2017). By contrast, the δ^{66}Zn values of siliciclastic sediments are lower than, or similar to, the mantle (Lv et al., 2020; Little et al., 2016). Altered oceanic crust shows a large variation in δ^{26}Mg from −1.7‰ to 0.21‰ primarily due to precipitation of carbonate and/or saponite (Huang et al., 2018; Huang et al., 2015), whereas some altered oceanic crust materials are slightly enriched in heavy Zn isotopes (~0.5‰) relative to the mantle value (Huang et al., 2016). The subducted mafic crust can preserve primary Mg-Zn isotope compositions up to the ultrahigh-pressure eclogite facies, which implies that the processes of slab subduction and dehydration induce limited isotope fractionation (Inglis et al., 2017; Wang et al., 2014). Hence, the aforementioned recycled materials cannot cause the mantle to have light Mg and heavy Zn isotope compositions.

Previous studies have demonstrated that Mg-Zn-rich carbonate minerals (e.g., dolomite and magnesite) have fairly high Mg-Zn concentrations and exhibit variable isotope compositions relative to the mantle (Liu S A et al., 2020; Liu and Li, 2019; Liu et al., 2017; Pichat et al., 2003). In detail, the Mg isotope compositions of sedimentary carbonates are significantly lighter (δ^{26}Mg = −5.31 to −1.09‰) than the mantle (−0.25±0.07‰; Teng, 2017), and their Zn isotope compositions (δ^{66}Zn=0.99±0.25‰) are heavier than the mantle (+0.18±0.05‰; Wang et al., 2017; Sossi et al., 2018). Hence, the recycled dolomites and magnesites can significantly modify the Mg-Zn isotope compositions of the mantle, and the melts derived from such mantle sources may inherit light Mg and heavy Zn isotopes from recycled Mg-rich carbonates. Similar to the Cenozoic basalts from eastern China, the Mg-Zn isotopic anomalies of the FER basalts also reflect the contribution of recycled Mg-rich carbonates (Fig. 6b, c, d).

Large variation in SiO_2 of the FRE basalts ranging from silica-undersaturated to silica-saturated (Fig. 2a) may be another indication of such an origin. High-pressure experiments reveal that alkaline low-silica melts can be derived from a CO_2-rich mantle source (Dasgupta et al., 2013, 2007). Carbonatite melts produced by low-degree partial melting of carbonated peridotite are characterized by low Al_2O_3, high CaO contents and CaO/Al_2O_3 (Zeng et al., 2010; Foley et al., 2009). Moreover, carbonatite melts are expected to possess negative Zr-Hf-Ti anomalies because the partition coefficients of Zr-Hf-Ti between garnet lherzolite and carbonatite melts are high compared to those of elements with similar general compatibility (Dasgupta et al., 2009). With increasing temperature and/or degree of melting of carbonated peridotite, carbonatite melts will evolve to carbonated silicate melts by incorporating more silicate materials (Dasgupta et al., 2013; Gudfinnsson and Presnall, 2005).

We propose that the systemic correlations defined among Ti/Ti*, Hf/Hf*, La/Yb and trace element contents can be attributed to varying degrees of melting (incongruent melting) of carbonated peridotite (Fig. 8). The endmember (i.e., carbonatite melts) with low Ti/Ti* and Hf/Hf* represents the melts derived from lower degrees of partial melting of carbonated source. Such "carbonatitic signals" will be diluted by silicate melts with increasing melting degrees (Liu et al., 2016; Huang et al., 2015). Given that the melts derived from pyroxenite have higher Fe/Mn than peridotite (Sobolev et al., 2007), the positive correlation between Fe/Mn and Ti/Ti* may imply that the silicate melts were derived from a pyroxenite-bearing mantle and consequently diluted the "carbonatitic signals" (Fig. 8).

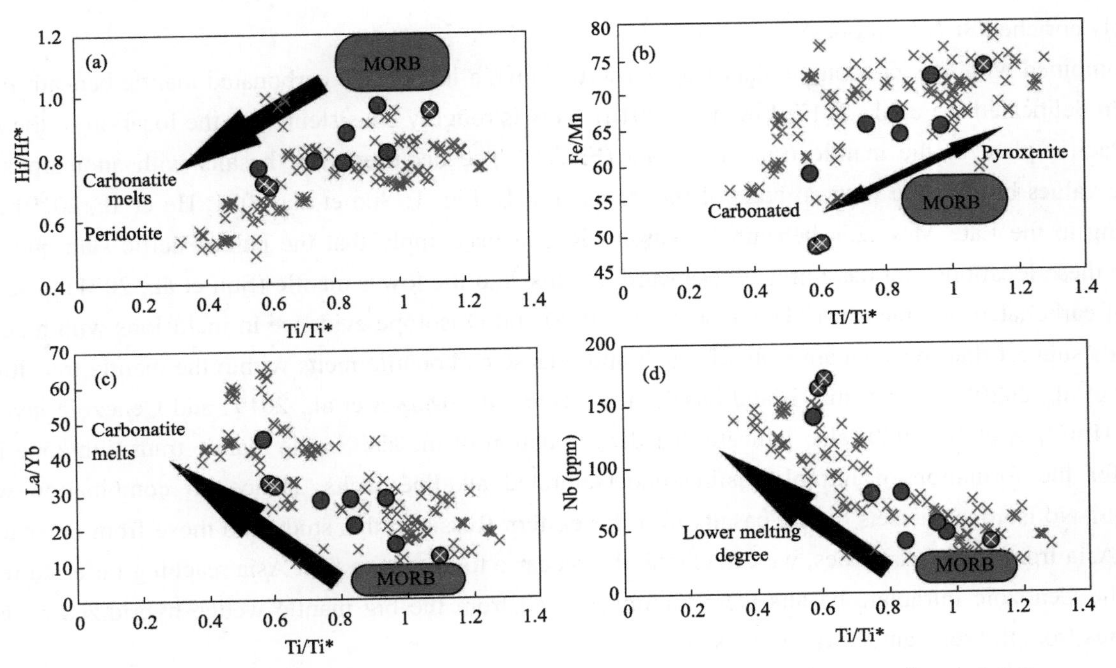

Fig. 8 Ti/Ti* vs. Hf/Hf*, Fe/Mn, La/Yb and Nb contents diagrams of the far eastern Russia basalts.

The FER basalts are characterized by unradiogenic $^{87}Sr/^{86}Sr$ (0.7037–0.7049), in contrast to extremely high Sr contents and radiogenic Sr isotope compositions (Sr: 1300 ppm, $^{87}Sr/^{86}Sr$: 0.709) observed in most sedimentary marine carbonates (Huang and Xiao, 2016). This suggests that the dominant carbonate species in mantle source are dolomite/magnesite with low Sr contents that cannot significantly modify the Sr isotope composition of the mantle (Fig. 6). Although $CaCO_3$ (calcite and aragonite) is the major component in marine carbonate sediments (Plank and Manning, 2019), most of $CaCO_3$ is dissolved into slab-derived fluids at shallow depths and transported into the shallow arc mantle wedge (Frezzotti et al., 2011; Pan et al., 2013). In addition, $CaCO_3$ is transformed to dolomite with increasing pressure to 1–4 GPa (Merlini et al., 2012; Boulard et al., 2011; Isshiki et al., 2004). Experiments reveal that $CaCO_3$ is unstable in the presence of enstatite, and can react with enstatite to form dolomite (Kushiro, 1975). Dolomite can transform further to magnesite in the presence of pyroxene and olivine at 20–50 GPa (Biellmann et al., 1993). Hence, dolomite and magnesite are expected to be the main carbonate species in the upper mantle. This view is supported by the study of ultra-high pressure metamorphic rocks from the Alps where magnesite and dolomite are observed as solid inclusions within garnet while the fluid inclusions within diamond and garnet only contain Ca-rich carbonates (Frezzotti et al., 2011). The Mg-Zn isotope data of most island arc basalts, comparable to mantle values, also indicate that the fluids derived from the shallow slab preferentially dissolve Ca-rich carbonates (Huang J et al., 2018; Li et al., 2017).

4.2 Carbonated big mantle wedge extending to beneath far eastern Russia

The Cenozoic intraplate basalts from East Asia can be divided into Na-series and potassic basalts, with the former occurring widespread (Xu et al., 2018) and the latter only found in the Wudalianchi-Erkeshan-Keluo (WEK) volcanic clusters (Fig. 1; Chu et al., 2013). Based on the data of this study and compiled data from the literature, the Na-series basalts from East Asia have extremely similar characteristics in Mg-Zn-Sr-Nd isotopes and major-trace elements (Fig. 6; Cai et al., 2021; Wang et al., 2018; Xu et al., 2018) that can be essentially explained by mixing between one low-silica endmember (low Hf/Hf*, Ti/Ti* and high La/Yb, CaO/Al_2O_3, depleted Sr-Nd isotopes) and the other high-silica endmember (high Hf/Hf*, Ti/Ti* and low La/Yb, CaO/Al_2O_3, relatively enriched Sr-Nd isotopes).

Combined with Mg-Zn isotopic data from eastern China, a large-scale carbonated mantle beneath East Asia has been delineated (Li et al., 2017; Liu et al., 2016). This is roughly consistent with the location of the stagnant West Pacific plate in the mantle transition zone (Fig. 1). The occurrence of basalts with anomalous Mg-Zn isotopic values beyond the western edge of the stagnant slab (Fig. 1; Sun et al., 2021; He et al., 2019) and also appearing in the Late Mesozoic beneath far eastern Russia may imply that the paleo-Pacific slab once existed beneath these locations, but has retreated eastward or sunk into the lower mantle (Sun et al., 2021). The melting curve of carbonated oceanic crust (Thomson et al., 2016) and O isotope evidence in inclusions within superdeep diamonds suggest that the stagnant slab will melt and release carbonatite melts within the mantle transition zone (Regier et al., 2020). The oceanic island basalts from Bermuda (Mazza et al., 2019) and Cenozoic lavas in SE Xizang (Liu S A et al., 2020) also indicate that decarbonation of the slab in the mantle transition zone is a key factor for the formation of intraplate silica-undersaturated alkaline melts. Hence, in combination with the Mg-Zn-Sr-Nd isotope datasets on the basalts from far eastern Russia in this study and those from other locations in East Asia from previous studies, we can extend the scenario to the entire East Asia reaching far eastern Russia, where the Cenozoic intraplate basalts were mainly derived from the big mantle wedge hybridized by recycled carbonates from the stagnant West Pacific slab.

5 Conclusions

(1) The Late Mesozoic-Cenozoic intraplate basalts from far eastern Russia have positive Nb-Ta anomalies and strongly negative K, Pb, Zr, Hf, Ti, Y anomalies in the trace element spider diagram, coupled with depleted Sr-Nd isotopes and lighter Mg and heavier Zn isotope compositions. These characteristics are very similar as those Na-series Cenozoic basalts from eastern China, reflecting similar features in their mantle sources.

(2) The large-scale Mg-Zn isotopic anomalies are documented in the Cenozoic intraplate basalts from the entire East Asia reaching far eastern Russia. This region is roughly coincident with the stagnant West Pacific slab beneath East Asia that can be attributed to derivation from mantle sources hybridized by recycled Mg-carbonates from the stagnant West Pacific slab in the mantle transition zone. No apparent distinction between the Late-Mesozoic and Cenozoic basalts from far eastern Russia may indicate the effect of the paleo-Pacific slab stagnant beneath this region.

Acknowledgements

This research was financially supported by the National Natural Science Foundation of China (Nos. 41730214, 41822301, and 41790451), the National Key R & D Program of China (Nos. 2019YFA0708400, 2020YFA0714800, and 2019YFC0605403), China “1 000 Youth Talents Program” and the “111” Project (No. B18048), and the pre-research project on Civil Aerospace Technologies (No. D020202) from Chinese National Space Administration. Dmitri A. Ionov thanks V. Prihodko for assistance with sample collection and D. Weiss for access to facilities at the ULB. Dmitri A. Ionov acknowledges Chinese Academy of Sciences President’s International Fellowship Initiative (PIFI) for Visiting Scientists in 2019 (No. 2017VCA0009). Dr. Chunguang Wang and two anonymous reviewers are thanked for providing comments that elevated the quality of this manuscript. This is CUGB petro-geochemical contribution No.PGC-201572 (RIG-No. 11). The final publication is available at Springerviahttps://doi.org/10.1007/s12583-021-1516-x.

Electronic Supplementary material

Supplementary material (Table S1) is available in the online version of this article at https://doi.org/10.1007/s12583-021-1516-x.

References

Biellmann, C., Gillet, P., Guyot, F., et al., 1993. Experimental Evidence for Carbonate Stability in the Earth's Lower Mantle. Earth and Planetary Science Letters, 118(1/2/3/4): 31-41. https://doi.org/10.1016/0012-821x(93)90157-5

Boulard, E., Gloter, A., Corgne, A., et al., 2011. New Host for Carbon in the Deep Earth. PNAS, 108(13): 5184-5187. https://doi.org/10.1073/pnas.1016934108

Cai, R. H., Liu, J. G., Pearson, D. G., et al., 2021. Oxidation of the Deep Big Mantle Wedge by Recycled Carbonates: Constraints from Highly Siderophile Elements and Osmium Isotopes. Geochimica et Cosmochimica Acta, 295: 207-223. https://doi.org/10.1016/j.gca.2020.12.019

Chen, H., Savage, P. S., Teng, F. Z., et al., 2013. Zinc Isotope Fractionation during Magmatic Differentiation and the Isotopic Composition of the Bulk Earth. Earth and Planetary Science Letters, 369/370: 34-42. https://doi.org/10.1016/j.epsl.2013.02.037

Chen, S. S., Liu, J. Q., Chen, S. S., et al., 2015. Variations in the Geochemical Structure of the Mantle Wedge beneath the Northeast Asian Marginal Region from Pre- to Post-Opening of the Japan Sea. Lithos, 224/225: 324-341. https://doi.org/10.1016/j.lithos.2015.03.008

Chu, Z. Y., Harvey, J., Liu, C. Z., et al., 2013. Source of Highly Potassic Basalts in Northeast China: Evidence from Re-Os, Sr-Nd-Hf Isotopes and PGE Geochemistry. Chemical Geology, 357: 52-66. https://doi. org/10.1016/j.chemgeo.2013.08.007

Dasgupta, R., Hirschmann, M. M., McDonough, W. F., et al., 2009. Trace Element Partitioning between Garnet Lherzolite and Carbonatite at 6.6 and 8.6 GPa with Applications to the Geochemistry of the Mantle and of Mantle-Derived Melts. Chemical Geology, 262(1/2): 57-77. https:// doi.org/10.1016/j.chemgeo.2009.02.004

Dasgupta, R., Hirschmann, M. M., Smith, N. D., 2007. Partial Melting Experiments of Peridotite + CO_2 at 3 GPa and Genesis of Alkalic Ocean Island Basalts. Journal of Petrology, 48(11): 2093-2124. https:// doi.org/10.1093/petrology/egm053

Dasgupta, R., Hirschmann, M. M., Stalker, K., 2006. Immiscible Transition from Carbonate-Rich to Silicate-Rich Melts in the 3 GPa Melting Interval of Eclogite + CO_2 and Genesis of Silica-Undersaturated Ocean Island Lavas. Journal of Petrology, 47(4): 647-671. https://doi.org/ 10.1093/petrology/egi088

Dasgupta, R., Mallik, A., Tsuno, K., et al., 2013. Carbon-Dioxide-Rich Silicate Melt in the Earth's Upper Mantle. Nature, 493(7431): 211- 215. https://doi.org/10.1038/nature11731

Davis, F. A., Hirschmann, M. M., Humayun, M., 2011. The Composition of the Incipient Partial Melt of Garnet Peridotite at 3 GPa and the Origin of OIB. Earth and Planetary Science Letters, 308(3/4): 380-390. https://doi.org/10.1016/j.epsl.2011.06.008

Davis, F. A., Humayun, M., Hirschmann, M. M., et al., 2013. Experimentally Determined Mineral/Melt Partitioning of First-Row Transition Elements (FRTE) during Partial Melting of Peridotite at 3 GPa. Geochimica et Cosmochimica Acta, 104: 232-260. https://doi. org/10.1016/j.gca.2012.11.009

Doucet, L. S., Mattielli, N., Ionov, D.A., et al., 2016. Zn Isotopic Heterogeneity in the Mantle:AMelting Control?. Earth and Planetary Science Letters, 451: 232-240. https://doi.org/10.1016/j.epsl.2016.06.040

Foley, S. F., Yaxley, G. M., Rosenthal, A., et al., 2009. The Composition of Near-Solidus Melts of Peridotite in the Presence of CO_2 and H_2O between 40 and 60 kbar. Lithos, 112: 274-283. https://doi.org/10.1016/ j.lithos.2009.03.020

Frezzotti, M. L., Selverstone, J., Sharp, Z. D., et al., 2011. Carbonate Dissolution during Subduction Revealed by Diamond-Bearing Rocks from the Alps. Nature Geoscience, 4(10): 703-706. https://doi.org/ 10.1038/ngeo1246

Gao, T., Ke, S., Li, R. Y., et al., 2019. High-Precision Magnesium Isotope Analysis of Geological and Environmental Reference Materials by Multiple-Collector Inductively Coupled Plasma Mass Spectrometry. Rapid Communications in Mass Spectrometry: RCM, 33(8): 767-777. https://doi.org/10.1002/rcm.8376

George, A. M., Stebbins, J. F., 1998. Structure and Dynamics of Magnesium in Silicate Melts: A High-temperature ^{25}Mg NMR Study. American Mineralogist, 83(9/10): 1022-1029. https://doi.org/10.2138/am-1998-9-1010

Gudfinnsson, G. H., Presnall, D. C., 2005. Continuous Gradations among Primary Carbonatitic, Kimberlitic, Melilititic, Basaltic,

Picritic, and Komatiitic Melts in Equilibrium with Garnet Lherzolite at 3–8 GPa. Journal of Petrology, 46(8): 1645-1659. https://doi.org/10.1093/petrology/egi029

Guo, F., Li, H. X., Fan, W. M., et al., 2015. Early Jurassic Subduction of the Paleo-Pacific Ocean in NE China: Petrologic and Geochemical Evidence from the Tumen Mafic Intrusive Complex. Lithos, 224/225: 46-60. https://doi.org/10.1016/j.lithos.2015.02.014

He, Y., Chen, L. H., Shi, J. H., et al., 2019. Light Mg Isotopic Composition in the Mantle beyond the Big Mantle Wedge beneath Eastern Asia. Journal of Geophysical Research: Solid Earth, 124(8): 8043-8056. https://doi.org/10.1029/2018jb016857

Hu, Y., Teng, F. Z., Plank, T., et al., 2017. Magnesium Isotopic Composition of Subducting Marine Sediments. Chemical Geology, 466: 15-31. https://doi.org/10.1016/j.chemgeo.2017.06.010

Hu, Y., Teng, F. Z., Zhang, H. F., et al., 2016. Metasomatism-Induced Mantle Magnesium Isotopic Heterogeneity: Evidence from Pyroxenites. Geochimica et Cosmochimica Acta, 185: 88-111. https:// doi.org/10.1016/j.gca.2015.11.001

Huang, F., Chakraborty, P., Lundstrom, C. C., et al., 2010. Isotope Fractionation in Silicate Melts by Thermal Diffusion. Nature, 464(7287): 396-400. https://doi.org/10.1038/nature08840

Huang, F., Chen, L. J., Wu, Z. Q., et al., 2013. First-Principles Calculations of Equilibrium Mg Isotope Fractionations between Garnet, Clinopyroxene, Orthopyroxene, and Olivine: Implications for Mg Isotope Thermometry. Earth and Planetary Science Letters, 367: 61- 70. https://doi.org/10.1016/j.epsl.2013.02.025

Huang, J., Li, S. G., Xiao, Y. L., et al., 2015a. Origin of Low δ^{26} Mg Cenozoic Basalts from South China Block and Their Geodynamic Implications. Geochimica et Cosmochimica Acta, 164: 298-317. https: //doi.org/10.1016/j.gca.2015.04.054

Huang, J., Ke, S., Gao, Y. J., et al., 2015b. Magnesium Isotopic Compositions of Altered Oceanic Basalts and Gabbros from IODP Site 1256 at the East Pacific Rise. Lithos, 231: 53-61. https://doi.org/ 10.1016/j.lithos.2015.06.009

Huang, J., Liu, S. -A., Gao, Y. J., et al., 2016. Copper and Zinc Isotope Systematics of Altered Oceanic Crust at IODP Site 1256 in the Eastern Equatorial Pacific. Journal of Geophysical Research: Solid Earth, 121(10): 7086-7100. https://doi.org/10.1002/2016jb013095

Huang, J., Xiao, Y. L., 2016. Mg-Sr Isotopes of Low-δ^{26} Mg Basalts Tracing Recycled Carbonate Species: Implication for the Initial Melting Depth of the Carbonated Mantle in Eastern China. International Geology Review, 58(11): 1350-1362. https://doi.org/10.1080/00206814.2016.1157709

Huang, J., Zhang, X. C., Chen, S., et al., 2018. Zinc Isotopic Systematics of Kamchatka-Aleutian Arc Magmas Controlled by Mantle Melting. Geochimica et Cosmochimica Acta, 238: 85-101. https://doi.org/ 10.1016/j.gca.2018.07.012

Huang, K. J., Teng, F. Z., Plank, T., et al., 2018. Magnesium Isotopic Composition of Altered Oceanic Crust and the Global Mg Cycle. Geochimica et Cosmochimica Acta, 238: 357-373. https://doi.org/ 10.1016/j.gca.2018.07.011

Inglis, E. C., Debret, B., Burton, K. W., et al., 2017. The Behavior of Iron and Zinc Stable Isotopes Accompanying the Subduction of Mafic Oceanic Crust: A Case Study from Western Alpine Ophiolites. Geochemistry, Geophysics, Geosystems, 18(7): 2562-2579. https://doi. org/10.1002/2016gc006735

Ionov, D. A., Guo, P., Nelson, W. R., et al., 2020. Paleoproterozoic Melt- Depleted Lithospheric Mantle in the Khanka Block, far Eastern Russia: Inferences for Mobile Belts Bordering the North China and Siberian Cratons. Geochimica et Cosmochimica Acta, 270: 95-111. https://doi.org/10.1016/j.gca.2019.11.019

Ionov, D. A., Shirey, S. B., Weis, D., et al., 2006. Os-Hf-Sr-Nd Isotope and PGE Systematics of Spinel Peridotite Xenoliths from Tok, SE Siberian Craton: Effects of Pervasive Metasomatism in Shallow Refractory Mantle. Earth and Planetary Science Letters, 241(1/2): 47-64. https://doi.org/10.1016/j.epsl.2005.10.038

Isshiki, M., Irifune, T., Hirose, K., et al., 2004. Stability of Magnesite and Its High-Pressure Form in the Lowermost Mantle. Nature, 427(6969): 60-63. https://doi.org/10.1038/nature02181

Kushiro, I., 1975. Carbonate-Silicate Reactions at High Presures and Possible Presence of Dolomite and Magnesite in the upper Mantle. Earth and Planetary Science Letters, 28(2): 116-120. https://doi. org/ 10.1016/0012-821x(75)90218-6

Li, S. G.,Yang,W., Ke, S., et al., 2017. Deep Carbon Cycles Constrained by a Large-Scale Mantle Mg Isotope Anomaly in Eastern China. National Science Review, 4(1): 111-120. https://doi.org/10.1093/nsr/nww070

Little, S. H., Vance, D., McManus, J., et al., 2016. Key Role of Continental Margin Sediments in the Oceanic Mass Balance of Zn and Zn Isotopes. Geology, 44(3): 207-210. https://doi.org/10.1130/g37493.1

Liu, K., Zhang, J. J., Xiao, W. J., et al., 2020. A Review of Magmatism and Deformation History along the NE Asian Margin from ca. 95 to 30 Ma: Transition from the Izanagi to Pacific Plate Subduction in the Early Cenozoic. Earth-Science Reviews, 209: 103317. https://doi. org/10.1016/j.earscirev.2020.103317

Liu, S. A., Li, S. G., 2019. Tracing the Deep Carbon Cycle Using Metal Stable Isotopes: Opportunities and Challenges. Engineering, 5(3): 448- 457. https://doi.org/10.1016/j.eng.2019.03.007

Liu, S. A., Liu, P. P., Lü, Y. W., et al., 2019. Cu and Zn Isotope Fractionation during Oceanic Alteration: Implications for Oceanic Cu and Zn Cycles. Geochimica et Cosmochimica Acta, 257: 191-205. https://doi.org/10.1016/j.gca.2019.04.026

Liu, S. A., Teng, F. Z., Yang, W., et al., 2011. High-Temperature Inter-Mineral Magnesium Isotope Fractionation in Mantle Xenoliths from the North China Craton. Earth and Planetary Science Letters, 308(1/2): 131-140. https://doi.org/10.1016/j.epsl.2011. 05.047

Liu, S. A., Wang, Z. Z., Li, S. G., et al., 2016. Zinc Isotope Evidence for a Large-Scale Carbonated Mantle beneath Eastern China. Earth and Planetary Science Letters, 444: 169-178. https://doi. org/10.1016/j.epsl.2016.03.051

Liu, S. A., Wang, Z. Z., Yang, C., et al., 2020. Mg and Zn Isotope Evidence for Two Types of Mantle Metasomatism and Deep Recycling of Magnesium Carbonates. Journal of Geophysical Research: Solid Earth, 125(11). https://doi.org/10.1029/2020 jb020684

Liu, S. A., Wu, H. C., Shen, S. Z., et al., 2017. Zinc Isotope Evidence for Intensive Magmatism Immediately before the End-Permian Mass Extinction. Geology, 45(4): 343-346. https://doi.org/10.1130/g38644.1

Lü, Y., Liu, S. A., Teng, F. Z., et al., 2020. Contrasting Zinc Isotopic Fractionation in Two Mafic-Rock Weathering Profiles Induced by Adsorption Onto Fe (Hydr)Oxides. Chemical Geology, 539: 119504. https://doi.org/10.1016/j.chemgeo.2020.119504

Mazza, S. E., Gazel, E., Bizimis, M., et al., 2019. Sampling the Volatile- Rich Transition Zone beneath Bermuda. Nature, 569(7756): 398-403. https://doi.org/10.1038/s41586-019-1183-6

McCoy-West, A. J., Fitton, J. G., Pons, M. L., et al., 2018. The Fe and Zn Isotope Composition of Deep Mantle Source Regions: Insights from Baffin Island Picrites. Geochimica et Cosmochimica Acta, 238: 542-562. https://doi.org/10.1016/j.gca.2018.07.021

Merlini, M., Crichton, W. A., Hanfland, M., et al., 2012. Structures of Dolomite at Ultrahigh Pressure and Their Influence on the Deep Carbon Cycle. Proceedings of the National Academy of Sciences of the United States of America, 109(34): 13509-13514. https://doi. org/ 10.1073/pnas.1201336109

Nesbitt, H. W., Young, G. M., 1982. Early Proterozoic Climates and Plate Motions Inferred from Major Element Chemistry of Lutites. Nature, 299(5885): 715-717. https://doi.org/10.1038/299715a0

Okamura, S., Arculus, R. J., Martynov, Y. A., 2005. Cenozoic Magmatism of the North-Eastern Eurasian Margin: The Role of Lithosphere Versus Asthenosphere. Journal of Petrology, 46(2): 221-253. https://doi.org/10.1093/petrology/egh065

Okamura, S., Martynov, Y. A., Furuyama, K., et al., 1998. K-Ar Ages of the Basaltic Rocks from far East Russia: Constraints on the Tectono- Magmatism Associated with the Japan Sea Opening. The Island Arc, 7(1/2): 271-282. https://doi.org/10.1046/ j.1440-1738.1998.00174.x

Pan, D., Spanu, L., Harrison, B., et al., 2013. Dielectric Properties of Water under Extreme Conditions and Transport of Carbonates in the Deep Earth. Proceedings of the National Academy of Sciences of the United States of America, 110(17): 6646-6650. https://doi.org/10.1073/pnas.1221581110

Pertermann, M., Hirschmann, M. M., 2003. Anhydrous Partial Melting Experiments on MORB-Like Eclogite: Phase Relations, Phase Compositions and Mineral-Melt Partitioning of Major Elements at 2-3 GPa. Journal of Petrology, 44(12): 2173-2201. https:// doi.org/10.1093/petrology/egg074

Pichat, S., Douchet, C., Albarède, F., 2003. Zinc Isotope Variations in Deep-Sea Carbonates from the Eastern Equatorial Pacific over the Last 175 ka. Earth and Planetary Science Letters, 210(1/2): 167-178. https://doi.org/10.1016/s0012-821x(03)00106-7

Pilet, S., Baker, M. B., Stolper, E. M., 2008. Metasomatized Lithosphere and the Origin of Alkaline Lavas. Science, 320(5878): 916-919. https:// doi.org/10.1126/science.1156563

Plank, T., Manning, C. E., 2019. Subducting Carbon. Nature, 574(7778): 343-352. https://doi.org/10.1038/s41586-019-1643-z

Pogge von Strandmann, P. A. E., Elliott, T., Marschall, H. R., et al., 2011. Variations of Li and Mg Isotope Ratios in Bulk Chondrites and Mantle Xenoliths. Geochimica et Cosmochimica Acta, 75(18): 5247-5268. https://doi.org/10.1016/j.gca.2011.06.026

Regier, M. E., Pearson, D. G., Stachel, T., et al., 2020. The Lithospheric-to-Lower-Mantle Carbon Cycle Recorded in Superdeep Diamonds. Nature, 585(7824): 234-238. https://doi.org/10.1038/s41586-020-2676-z

Richter, F. M., Dauphas, N., Teng, F. Z., 2009. Non-Traditional Fractionation of Non-Traditional Isotopes: Evaporation, Chemical Diffusion and Soret Diffusion. Chemical Geology, 258(1/2): 92-103. https://doi.org/10.1016/j.chemgeo.2008.06.011

Richter, F. M., Watson, E. B., Mendybaev, R. A., et al., 2008. Magnesium Isotope Fractionation in Silicate Melts by Chemical and Thermal Diffusion. Geochimica et Cosmochimica Acta, 72(1): 206-220. https://doi.org/10.1016/j.gca.2007.10.016

Shimoda, K., Tobu, Y., Hatakeyama, M., et al., 2007. Structural Investigation of Mg Local Environments in Silicate Glasses by Ultra-High Field ^{25}Mg 3QMAS NMR Spectroscopy. American Mineralogist, 92(4): 695-698. https://doi.org/10.2138/am.2007.2535

Sobolev, A. V., Hofmann, A. W., Kuzmin, D. V., et al., 2007. The Amount of Recycled Crust in Sources of Mantle-Derived Melts. Science, 316(5823): 412-417. https://doi.org/10.1126/science. 1138113
Sossi, P. A., Nebel, O., O'Neill, H. S. C., et al., 2018. Zinc Isotope Composition of the Earth and Its Behaviour during Planetary Accretion. Chemical Geology, 477: 73-84. https://doi.org/10.1016/j.chemgeo.2017.12.006
Stracke, A., Tipper, E. T., Klemme, S., et al., 2018. Mg Isotope Systematics during Magmatic Processes: Inter-Mineral Fractionation in Mafic to Ultramafic Hawaiian Xenoliths. Geochimica et Cosmochimica Acta, 226: 192-205. https://doi.org/10.1016/j.gca. 2018.02.002
Su, B. X., Hu, Y., Teng, F. Z., et al., 2019. Light Mg Isotopes in Mantle-Derived Lavas Caused by Chromite Crystallization, Instead of Carbonatite Metasomatism. Earth and Planetary Science Letters, 522: 79-86. https://doi.org/10.1016/j.epsl.2019.06.016
Sun, Y., Teng, F. Z., Pang, K. N., 2021. The Presence of Paleo-Pacific Slab beneath Northwest North China Craton Hinted by Low-δ^{26}Mg Basalts at Wulanhada. Lithos, 386/387: 106009. https://doi.org/10.1016/j.lithos.2021.106009
Tang, J., Xu, W. L., Wang, F., et al., 2018. Subduction History of the Paleo- Pacific Slab beneath Eurasian Continent: Mesozoic-Paleogene Magmatic Records in Northeast Asia. Science China Earth Sciences, 61(5): 527-559. https://doi.org/ 10.1007/s11430-017-9174-1
Teng, F. Z., 2017. Magnesium Isotope Geochemistry. In: Teng, F. Z.,Watkins, J., Dauphas, N, eds., Non-Traditional Stable Isotopes. De Gruyter, Berlin, Boston. 209-287. https://doi.org/10.1515/9783110545630-008
Teng, F. Z., Li, W. Y., Ke, S., et al., 2010. Magnesium Isotopic Composition of the Earth and Chondrites. Geochimica et Cosmochimica Acta, 74(14): 4150-4166. https://doi.org/10.1016/j.gca.2010.04.019
Thomson, A. R., Walter, M. J., Kohn, S. C., et al., 2016. Slab Melting as a Barrier to Deep Carbon Subduction. Nature, 529(7584): 76-79. https://doi.org/10.1038/nature16174
Walter, M. J., 1998. Melting of Garnet Peridotite and the Origin of Komatiite and Depleted Lithosphere. Journal of Petrology, 39(1): 29-60. https://doi.org/10.1093/petroj/39.1.29
Wang, F., Xu, W. L., Xu, Y. G., et al., 2015. Late Triassic Bimodal Igneous Rocks in Eastern Heilongjiang Province, NE China: Implications for the Initiation of Subduction of the Paleo-Pacific Plate beneath Eurasia. Journal of Asian Earth Sciences, 97: 406-423. https://doi.org/10.1016/j.jseaes.2014.05.025
Wang, S. J., Teng, F. Z., Li, S. G., et al., 2014. Magnesium Isotopic Systematics of Mafic Rocks during Continental Subduction. Geochimica et CosmochimicaActa, 143: 34-48. https://doi.org/10.1016/j.gca.2014.03.029
Wang, X.-C., Wilde, S. A., Li, Q. L., et al., 2015. Continental Flood Basalts Derived from the Hydrous Mantle Transition Zone. Nature Communications, 6: 7700. https://doi.org/10.1038/ncomms8700
Wang, Y., He, Y. S., Ke, S., 2020. Mg Isotope Fractionation during Partial Melting of Garnet-Bearing Sources: an Adakite Perspective. Chemical Geology, 537: 119478. https://doi.org/10.1016/j.chemgeo.2020.119478
Wang, Z. Z., Liu, S.-A., Chen, L. H., et al., 2018. Compositional Transition in Natural Alkaline Lavas through Silica-Undersaturated Melt-Lithosphere Interaction. Geology, 46(9): 771-774. https://doi.org/10.1130/g45145.1
Wang, Z. Z., Liu, S.-A., Liu, J. G., et al., 2017. Zinc Isotope Fractionation during Mantle Melting and Constraints on the Zn Isotope Composition of Earth's upper Mantle. Geochimica et Cosmochimica Acta, 198: 151-167. https://doi.org/10.1016/j.gca.2016. 11.014
Wilde, S. A., 2015. Final Amalgamation of the Central Asian Orogenic Belt in NE China: Paleo-Asian Ocean Closure versus Paleo-Pacific Plate Subduction—A Review of the Evidence. Tectonophysics, 662: 345-362. https://doi.org/10.1016/j.tecto.2015. 05.006
Williams, H. M., Bizimis, M., 2014. Iron Isotope Tracing of Mantle Heterogeneity within the Source Regions of Oceanic Basalts. Earth and Planetary Science Letters, 404: 396-407. https://doi.org/10.1016/j.epsl.2014.07.033
Wu, F. Y., Sun, D. Y., Ge, W. C., et al., 2011. Geochronology of the Phanerozoic Granitoids in Northeastern China. Journal of Asian Earth Sciences, 41(1): 1-30. https://doi.org/10.1016/j.jseaes.2010.11.014
Wu, J. T., Wu, J., 2019. Izanagi-Pacific Ridge Subduction Revealed by a 56 to 46 Ma Magmatic Gap along the Northeast Asian Margin. Geology, 47(10): 953-957. https://doi.org/10.1130/g46778.1
Xu, Y. G., Li, H. Y., Hong, L. B., et al., 2018. Generation of Cenozoic Intraplate Basalts in the Big Mantle Wedge under Eastern Asia. Science China Earth Sciences, 61(7): 869-886. https://doi.org/10.1007/ s11430-017-9192-y
Xu, Z., Zheng, Y. F., 2019. Crust-Mantle Interaction in the Paleo-Pacific Subduction Zone: Geochemical Evidence from Cenozoic Continental Basalts in Eastern China. Earth Science, 44(12): 4135-4143. https://doi.org/10.3799/dqkx.2019.273 (in Chinese with English Abstract)

Yang, W., Teng, F. Z., Zhang, H. F., et al., 2012. Magnesium Isotopic Systematics of Continental Basalts from the North China Craton: Implications for Tracing Subducted Carbonate in the Mantle. Chemical Geology, 328: 185-194. https://doi.org/10.1016/j.chemgeo.2012.05.018

Zeng, G., Chen, L. H., Xu, X. S., et al., 2010. Carbonated Mantle Sources for Cenozoic Intra-Plate Alkaline Basalts in Shandong, North China. Chemical Geology, 273(1/2): 35-45. https://doi.org/10.1016/j.chemgeo.2010.02.009

Zhao, P., Jahn, B. M., Xu, B., 2017. Elemental and Sr-Nd Isotopic Geochemistry of Cretaceous to Early Paleogene Granites and Volcanic Rocks in the Sikhote-Alin Orogenic Belt (Russian Far East): Implications for the Regional Tectonic Evolution. Journal of Asian Earth Sciences, 146: 383-401. https://doi.org/10.1016/j.jseaes.2017.06.017

Zheng, Y. F., Chen, Y. X., 2019. Crust-Mantle Interaction in Continental Subduction Zones. Earth Science, 44(12): 3961-3983. dhttps://doi.org/ 10.3799/dqkx.2019.982 (in Chinese with English Abstract)

Zhong, Y., Chen, L. H., Wang, X. J., et al., 2017. Magnesium Isotopic Variation of Oceanic Island Basalts Generated by Partial Melting and Crustal Recycling. Earth and Planetary Science Letters, 463: 127-135. https://doi.org/10.1016/j.epsl.2017.01.040

Zhou, J. B., Wilde, S. A., Zhao, G. C., et al., 2018. Nature and Assembly of Microcontinental Blocks within the Paleo-Asian Ocean. Earth-Science Reviews, 186: 76-93. https://doi.org/10.1016/j.earscirev.2017.01.012

Tracing deep carbon cycling by metal stable isotopes*

Shu-Guang Li[1,2]

1. State Key Laboratory of Geological Processes and Mineral Resources, China University of Geosciences, Beijing 100083, China
2. CAS Key Laboratory of Crust-Mantle Materials and Environments, School of Earth and Space Sciences, University of Science and Technology of China, Hefei, Anhui 230026, China

亮点介绍：本文是 National Science Review，2022 年第 9 卷（第 6 期）“金属稳定同位素示踪深部碳循环专题”的前言。本文重点介绍了该专题的两个亮点文章：一个是关于板块俯冲过程中碳酸盐-硅酸盐体系的地球化学行为的综述，它回答了西太平洋俯冲板片携带进入地幔过渡带的富镁碳酸盐的来源问题；另一个是一篇研究论文，它首次报道早白垩世华北克拉通岩石圈地幔的破坏和减薄可导致该岩石圈地幔 CO_2 大规模释放的新机制，和 G. Pearson 给予的亮点短评。

Earth is a unique habitable planet in the solar system. One remarkable feature is that Earth's present-day partial pressure of atmospheric carbon dioxide (pCO_2) is 0.001 bar, in sharp contrast to its proto-atmosphere (pCO_2 ~100 bar that was inhabitable, similar to that of the present-day Venus's atmosphere). How Earth's pCO_2 dropped dramatically over geologic history remains an enigma. Deep carbon cycling connects the exospheric and deep reservoirs of carbon via the processes that plate subduction transports surface carbon into Earth's mantle and then deep carbon returns back to the surface via volcanism. Deep carbon cycling has played a critical role in modulating the modern habitable Earth's atmosphere. This Special Topic comprises one Review, one Research Article associated with one Research Highlight, and two Perspectives to address key issues of this research field, such as understanding the source of deeply subducted Mg-rich carbonates and evaluating the outgassing carbon fluxes that are essential to study deep carbon cycling.

Calcium and magnesium are the two most abundant cations in carbonates $(Ca,Mg)CO_3$. Moreover, some carbonates like dolomite and magnesite can have significant amounts of Zn. Given the large or significant isotopic offset of Mg, Zn and Ca between marine carbonates and silicate reservoirs, delivery of surface carbonates into the mantle can significantly modify isotopic compositions of Ca, Mg and Zn in the local mantle domains. As such, Ca, Mg and Zn isotopes have great potential in tracing deep carbon cycling.

Magnesium isotopes have been proved effective in tracing deep carbon cycling. One outstanding progress was the identification of a large-scale light Mg isotopic anomaly in the convecting upper mantle beneath eastern China. This observation has two vital implications: first, during the subduction of the western Pacific plate, marine carbonates were transported into the mantle transition zone and convective upper mantle; second, the carbonates subducted to this depth are dominated by Mg-rich carbonates. How could the subducting slabs carry such massive

* 本文发表在: National Science Review, 2002, 9(6): nwac071

Mg-rich carbonates into the deep mantle, given that carbonates initially entering the subduction zones are mainly Ca-rich? This issue is addressed by Wang and Li, who systematically review the Mg isotopic behaviors of carbonate-silicate systems during subduction [1].

It is a general consensus that the craton lithospheric mantle is a vast carbon reservoir, and its carbon outgassing is facilitated via continental rifting. A Research Article by Wang et al. argues that the destruction of the eastern North China craton was responsible for rapid and massive CO_2 outgassing into the early Cretaceous atmosphere, hence inducing climate change at that time [2]. This new idea is highlighted by Graham Pearson from the University of Alberta [3].

Finally, two Perspectives outlook the future application of metal stable isotopes in a deep C cycling study. Huang and Jacobsen introduce the progresses and difficulties of applying Ca isotopes to tracing deep carbon cycle [4], while Chen et al. discuss whether the subducted carbonates could be delivered to the lower mantle by examining isotopes of plume-related volcanic rocks [5].

References

[1] Wang S-J and Li S-G. Natl Sci Rev 2022; 9: nwac036.
[2] Wang Z-X, Liu S-A and Li S, et al. Natl Sci Rev 2022; 9: nwac001.
[3] Pearson G. Natl Sci Rev 2022; 9: nwac049.
[4] Huang S and Jacobsen S. Natl Sci Rev 2022; 9: nwab173.
[5] Chen L-H, Wang X-J and Liu S.-A., Natl Sci Rev 2022; 9: nwac061.

Contrasting fates of subducting carbon related to different oceanic slabs in East Asia*

Sheng-Ao Liu, Tianhao Wu, Shuguang Li, Zhaoxue Wang and Jingao Liu

State Key Laboratory of Geological Processes and Mineral Resources, China University of Geosciences, Beijing 100083, China

亮点介绍: 中国东部南-北重力梯度线(NSGL)以东玄武岩的 δ^{66}Zn 与其至 NSGL 的距离无关，且分布与地幔过渡带的滞留板片重合，表明了西向俯冲的古太平洋板片携带的碳酸盐较多，交代玄武岩源区的富镁碳酸盐量是类似的；而重力梯度带以西玄武岩的 δ^{66}Zn 随着其至 NSGL 距离的增加而变重，表明古亚洲洋板片东南向俯冲过程中其携带的碳酸盐量在逐步减少。这种差异在于古亚洲洋相比古太平洋较老且扩张速率慢，使得碳主要赋存在蚀变洋壳中，因而随着板片俯冲深度增加其携带的碳酸盐被快速消耗掉。

Abstract Subduction transfers surface carbon into the Earth's interior in a main form of carbonates that influences the global carbon cycles and surface climate through geologic time. Nevertheless, whether the fate of downgoing carbonates significantly varies in past subduction zones is rarely constrained by natural observations. Marine carbonates have remarkably higher zinc isotopic ratios (expressed as $\delta^{66}Zn_{JMC\text{-}Lyon}$) relative to the mantle (0.99±0.24‰ vs. 0.18±0.05‰), making zinc isotopes a sensitive tracer for subducting carbonates. Here we examine this issue through a comparative zinc isotope study on basalts across the North-South Gravity Lineament (NSGL) in East Asia that were genetically related to two different oceanic slabs. Together with existing data, we show that all basalts in the east of the NSGL have high δ^{66}Zn (~0.3–0.6‰; n=134) that do not vary with distances to the trench and are spatially coupled with the horizontally stagnated slab in the transition zone (410–660 km). This indicates that subducting carbonates survived shallow dissolution and were deeply buried during westward subduction of the Paleo-Pacific slab. By contrast, basalts in the west of the NSGL display a gradual decline of δ^{66}Zn from 0.50±0.04‰ to 0.28±0.03‰ (n=35) with increasing distances to the trench. No known magmatic processes (e.g., partial melting, crystal-melt differentiation, melt-peridotite interaction, and degassing) can account for the spatial Zn isotopic variation. The role of slab-derived sulfate rich fluids is also excluded because of the mantle-like Cu isotopic compositions of these basalts. Instead, the gradual decrease of δ^{66}Zn, together with the coupled decline of CaO/Al_2O_3, are best explained as the diminished amounts of dissolved carbonates in their mantle sources. Thus, substantial carbonate dissolution must have occurred during southeastward subduction of the Paleo-Asian slab, which prevents deep burial of subducting carbon. The main differences between the two large slabs include: (i) the Paleo-Asian slab has an extended longevity (~1.1 Ga) and slow spreading rate in

* 本文发表在: Geochimica et Cosmochimica Acta, 2022, 324: 156-173

comparison with the Paleo-Pacific slab, leading to the main incorporation of carbonate minerals into the altered oceanic crust, and (ii) the younger Paleo-Pacific slab contains abundant deep-sea Mg-rich carbonates that were not sufficiently dissolved at shallow depths. These differences demonstrate that subduction of different oceanic slabs can lead to contrasting fates of subducting carbon in ancient subduction zones, depending on the contents and species of carbonate sediments in the oceanic crust.

Keywords Subducting carbonates, Basalts, Zn isotopes, Deep carbon cycling, East Asia

1 Introduction

Since carbon dioxide (CO_2) is the dominant greenhouse gas for most of Earth's history and most of Earth's carbon (>90%) is stored in the crust, mantle and core, the net flux of carbon between the Earth's interior and exterior regulates the atmospheric climate on short to long time scales and is critical for planetary habitability (e.g., Huybers and Langmuir, 2009; Dasgupta and Hirschmann, 2010). Carbonate sediments represent the major form of carbon inputs to the mantle via subduction (~70%), with the remainder (~30%) consisting of organic carbon and carbonate veins in altered oceanic crust and altered peridotites (Plank and Manning, 2019). It has been widely suggested that most of carbonates in subduction zones have undergone fluid-mediated dissolution into the mantle wedge (e.g., Frezzotti et al., 2011; Ague and Nicolescu, 2014; Kelemen and Manning, 2015; Foley and Fischer, 2017). However, some studies proposed that a considerable amount of carbonates may have survived beyond the sub-arc regime and can be delivered to the convective mantle or the mantle transition zone (MTZ) (e.g., Dasgupta et al., 2013; Thomson et al., 2016; Li et al., 2017; Mazza et al., 2019; Liu et al., 2020). Consequently, the efficiency of carbon subduction remains highly uncertain with estimates for carbon entering the convecting mantle ranging from very little (~0) to 60 megaton per year (Mt/yr) (Dasgupta and Hirschmann, 2010; Kelemen and Manning, 2015; Clift, 2017). The carbon input flux in modern subduction zones substantially varies for individual downgoing plates with distinct formation, evolution and sedimentation histories (e.g., Galvez and Pubellier, 2019; Plank and Manning, 2019). This variety is perhaps one of the major reasons for the large uncertainty of carbon input flux estimate in modern subduction zones. Compared with modern subduction zones in which carbon input and output fluxes can be directly observed or estimated (e.g., Marty et al., 1989; Mason et al., 2017), however, the fate of carbonates in paleo subduction zones is rarely known or commonly difficult to observe but can be probed by subduction-related volcanism.

Zinc isotopic systematics is a novel tool of tracking the fates of carbonates in subduction zones and in the mantle sources of basalts (Liu et al., 2016a; Liu and Li, 2019). Zinc is a trace element in the mantle and a huge Zn isotopic difference exists between peridotites and carbonates (Fig. 1). The terrestrial mantle has a $\delta^{66}Zn_{JMC\text{-}Lyon}$ value of 0.18±0.05‰ (Liu et al., 2019; Sossi et al., 2018; Wang et al., 2017), and marine carbonates have the highest $\delta^{66}Zn$ (0.99±0.24‰) among crustal reservoirs (Pichat et al., 2003; Liu et al., 2017; Sweere et al., 2018) due to preferential uptake of isotopically heavy Zn during carbonate precipitation from seawater (Mavromatis et al., 2019). Other crustal reservoirs that can potentially be subducted (e.g., altered oceanic crust, organic and terrigenous sediments) have $\delta^{66}Zn$ values mostly between 0.1‰ and 0.3‰ (Moynier et al., 2017) (Fig. 1). During subduction, partial carbonate dissolution preferentially releases isotopically heavy Zn into fluids (Debret et al., 2018), which further can raise $\delta^{66}Zn$ of the carbonated mantle. Mantle's partial melting and fractional crystallization of basaltic magmas, on the other hand, induce subtle (<0.1‰) Zn isotope fractionation (Chen et al., 2013; Sossi et al., 2018; Wang et al., 2017). All of these characteristics make zinc isotopes a valid proxy for subducting carbonates in the mantle (e.g., Beunon et al., 2020; Liu et al., 2016a, 2020; Liu and Li, 2019). In particular, spatial Zn isotopic variations in basalts that erupted at various distances from the trench may record whether and how carbonates were dissolved during subduction. This in turn determines the fate of downgoing carbonates in past subduction zones.

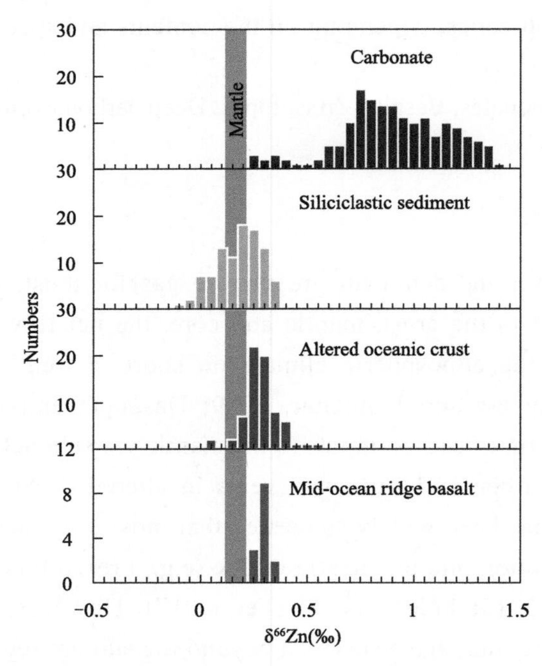

Fig. 1　Summary of Zn isotopic compositions of some components in subducting slabs. Data sources: Carbonates: Pichat et al. (2003), Kunzmann et al. (2013), Liu et al. (2017), and Sweere et al. (2018); Siliciclastic sediments: Maréchal et al. (2000), Sonke et al. (2008), Little et al. (2016) and Vance et al. (2016); Altered oceanic crust (AOC): Huang et al. (2016) and Inglis et al. (2017). Mid-ocean ridge basalt: Wang et al. (2017) and Huang et al. (2018). The grey vertical bar represents the mantle value estimated from mantle peridotites (δ^{66}Zn=0.18±0.05‰, Wang et al., 2017; Sossi et al., 2018; McCoy-West et al., 2018; Liu et al., 2019).

In this study, we examine the fates of subducting carbonates related to different oceanic slabs through a comparative study of zinc isotopes on basaltic lavas across the North-South Gravity Lineament (NSGL) in East Asia. Two large (Paleo-Asian and Paleo-Pacific) oceanic slabs with distinct ages and sedimentation histories have been subducted into the mantle beneath East Asia in the early Paleozoic and late Mesozoic, respectively (Fig. 2a, b; Xiao et al., 2003; Fukao et al., 2009). Basaltic lavas widely distributed in two sides of the NSGL were proposed to record subduction of the two oceanic slabs, respectively, which help ascertain the factors controlling the downgoing fate of carbonates in old subduction zones. Together with existing data, we found that basalts from two sides of the NSGL exhibit strongly contrasting Zn isotopic variations with distances to the trench. We discuss how these spatial Zn isotopic variations have been caused through a combined analysis of Cu isotopes that can help diagnose the effect of slab-derived sulfate rich fluids on the Zn isotopic data. The results demonstrate that subduction of different oceanic slabs can result in strong differences in the final fates of subducting carbonates, which has significance for understanding deep carbon cycling over geologic time.

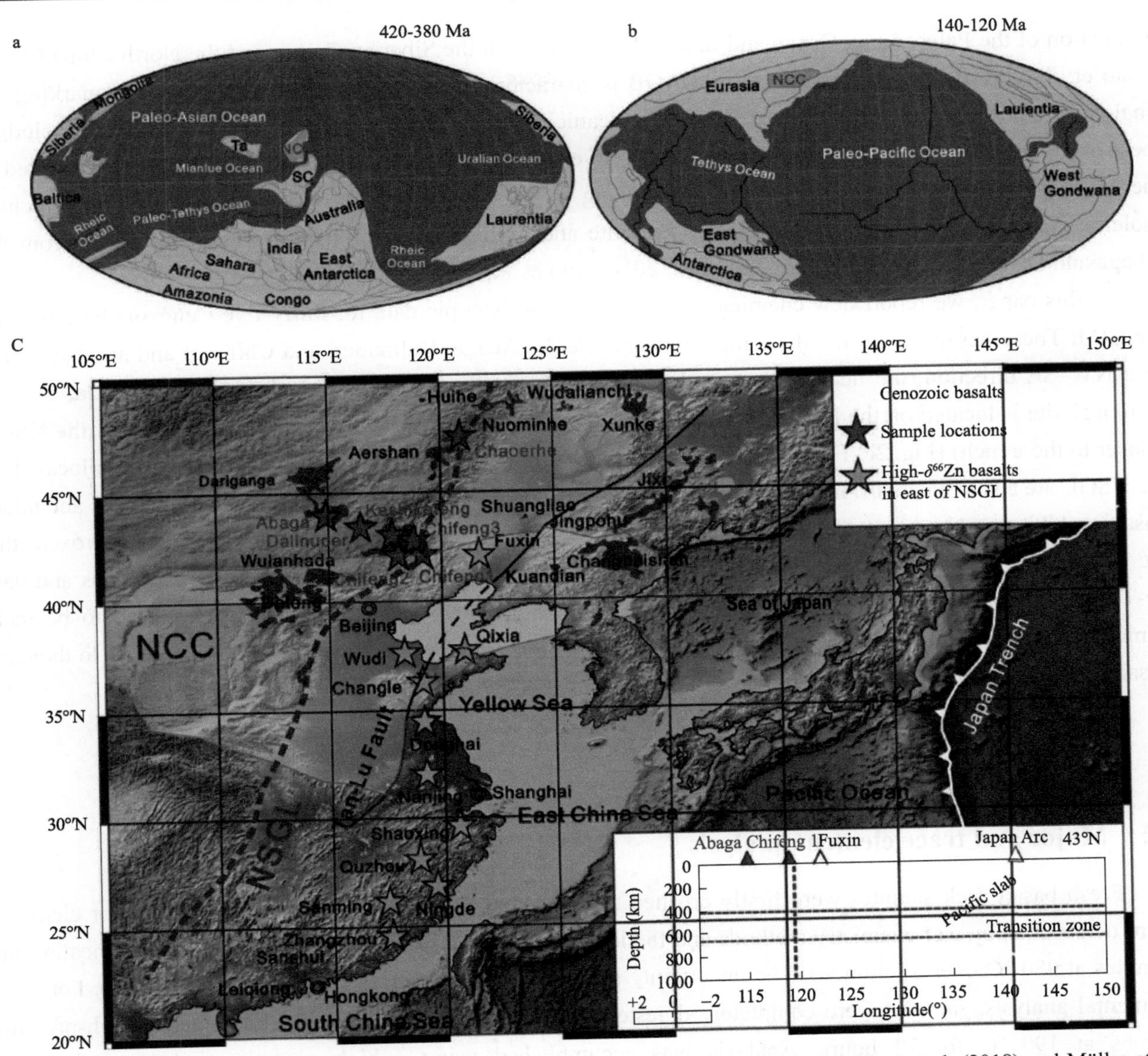

Fig. 2 Global plate reconstructions for (a) 420–380 Ma and (b) 140–120 Ma, modified after Zhao et al. (2018) and Müller et al. (2008b). (c) The distribution of Cenozoic basaltic volcanism in East Asia, highlighting the studied samples from Inner Mongolia (red star). The insert shows the stagnant Paleo-Pacific slab in the MTZ beneath East Asia (Huang and Zhao, 2006). Abbreviations: NCC=North China Craton; SC=South China Craton; Ta=Tarim Craton.

2 Sample distribution and description

Widespread late-Mesozoic to Cenozoic basaltic lavas are separated by the nearly NS-trending NSGL in East Asia (Fig. 2c). The NSGL represents a steep gradient in gravity, elevation, topography, and lithosphere thickness, formed via extensive thinning of the lithosphere in the east of the NSGL related to westward subduction of the Paleo-Pacific slab since the late Mesozoic (e.g., Xu, 2007). The late-Mesozoic to Cenozoic basaltic magmas in the east of the NSGL along the coastal provinces have widely been proposed to be casually a result of this subduction event (e.g., Zeng et al., 2010; Li et al., 2017; Xu et al., 2018). Different from basalts in the east of the NSGL, basalts erupted in central Inner Mongolia (CIM) in the west of the NSGL are far from the coastline. CIM is located within the central part of the Central Asian Orogenic Belt (CAOB), which is mainly composed of micro-continents, island arcs, ophiolitic remnants, and ocean plate stratigraphy and was formed by southward subduction of the Paleo-Asian Ocean and final collision between the Siberian Craton and the North China Craton (Xiao et al., 2003). The termination of the CAOB is characterized by the Solonker suture in CIM, marking the final closure of the Paleo-Asian Ocean. The CIM volcanic province covers a large area of >20,000 km^2, including the Abaga, Dalinuoer and Chifeng volcanic fields (Ho et al., 2008; Fig. S1). Volcanism in this region erupted in the Miocene to Pleistocene (Wang et al., 2015; Pang et al., 2019; Togtokh et al., 2019) and mainly include tholeiite, alkali olivine basalt, and basanite. Peridotite and pyroxenite xenoliths are present in basalts from the Abaga and Dalinuoer volcanic fields (Zou et al., 2014; Guo et al., 2016).

In this paper, we report new chemical and Zn-Cu-Sr-Nd isotopic data for thirty-five Cenozoic basalts from the CIM. These basalts cover the three main volcanic fields (Abaga, Dalinuoer, and Chifeng) and are distributed in a NW–SE direction, the nearly same as the subduction direction of the Paleo-Asian slab (Fig. 2a, c). The Chifeng1 site is located on the NSGL (far away from the trench) and others are situated in the west of the NSGL (closer to the trench) (Fig. 2c; Fig. S1). Cenozoic basalts from Chaoerhe in the north of the CIM that is located on the NSGL are also collected. The samples include basanites, alkali basalts and tholeiites. Most basanites and alkali basalts exhibit a porphyritic texture and contain phenocrysts of predominant olivine and subordinate clinopyroxene that are scattered in a groundmass composed of volcanic glass, acicular plagioclase, tiny clinopyroxene grains and dark Fe-Ti-Cr oxides (Fig. S2). Compared with basanites and alkali basalts, plagioclase phenocrysts tend to be more common in tholeiitic basalts. Overall, phenocryst abundances and sizes decrease gradually from basanites to tholeiitic basalts (Fig. S2). More detailed petrology descriptions can be found in the Supplementary Materials.

3 Methods

3.1 Major and trace element analysis

Fresh basalt rock samples were firstly crushed to powder of 200 mesh in an agate mortar. Major elements were determined by wet-chemistry methods and the loss on ignition (LOI) contents were achieved by heating the samples at 980 °C over 30 minutes. The uncertainty of major element analysis is better than ±1% (1σ). For trace elemental analysis, samples were completely digested in a mixture of HF + HNO_3 +HCl acids in high-pressure bombs at 190 °C for 72 hours. Analysis was accomplished using a high-resolution inductively coupled plasma-mass spectrometry (Thermo Element XR). The USGS basalt standard BHVO-2 was adopted to monitor the whole analytical procedure, with reproducibility better than ±5% (1σ).

3.2 Isotopic analysis

The basalt powder was dissolved independently for Sr, Nd, Zn and Cu isotopic analysis at the China University of Geosciences, Beijing. Samples were firstly transferred to a 1:1 (v/v) mixture of double-distilled HF and HNO_3 in Savillex screw-top beakers and then heated at 160 °C on a hotplate in an exhaust hood (Class 100). The solutions were dried down at 150 °C to expel the fluorine. The dried residues were refluxed with a 1:3 (v/v) mixed HNO_3 and HCl acid, followed by heating and dried down at 80 °C. The samples were refluxed with concentrated HNO_3 until complete dissolution has been achieved. The dissolved samples were passed through columns loaded with pre-cleaned Bio-Rad® cation AG50W-X12 (200–400 mesh) resin for collecting Sr at first, and then Nd was purified using LN resin. The instrumental mass fractionation was corrected by using $^{86}Sr/^{88}Sr$ = 0.1194 and $^{143}Nd/^{144}Nd$ = 0.7219 based on the exponential law.

Dissolved samples were transferred into 1 ml 8 N HCl + 0.001% H_2O_2 in preparation for ion-exchange separation of Zn and Cu, following the procedures outlined by Liu et al. (2014, 2016a). Zinc was separated using

1 ml pre-cleaned Bio-Rad® strong anion resin AG1-X8 (200–400 mesh) and finally collected in 7 ml 0.5 N HNO_3 after matrices were eluted. Bio-Rad AG-MP-1M strong anion exchange resin (100–200 mesh; chloride form) was used for separation of Cu from matrix elements. Copper was collected in 24 ml 8 N HCl + 0.001% H_2O_2 after matrices were leached out. The purified solutions were dried down at 80 °C and then dissolved in 3% HNO_3 (m/m). The recovery of both Zn and Cu after twice column chemistry is >99.5%. Total procedure blank was <6 ng for Zn and <2 ng for Cu, which accounts for negligible portions of the loaded Zn (1–3 μg) and Cu (~0.4 μg).

Copper and zinc isotopic ratios were separately measured on a Thermo Scientific *Neptune Plus* multi-collector inductively coupled plasma mass spectrometry. Sample-standard bracketing method was used in order to correct for instrumental mass fractionation during Cu isotopic analysis. For Zn isotopic analysis, instrumental mass bias correction was achieved through the combination of sample-standard bracketing and Cu-doping. Solutions with concentrations of 200 ng/g Zn and 100 ng/g Cu were introduced into the mass spectrometer using "wet plasma" introduction system combined with a PFA micro concentric nebulizer with an uptake rate of ~50 μL/min. Assuming that the instrumental fractionation is mass-dependent and similar for copper and zinc (Maréchal et al., 1999), an inter-calibration can be achieved by plotting $\ln(^{65}Cu/^{63}Cu)$ against $\ln(^{66,68}Zn/^{64}Zn)$ with excellent correlation factors ($R^2 \geqslant 0.9975$). Using the fitted calibration line of in-house standards and the measured $\ln(^{65}Cu/^{63}Cu)$ ratio in each Cu-added sample, the isotope fractionation between sample and in-house standard can be calculated. Copper and zinc isotope ratios are expressed as the per mil deviation against SRM NIST 976 and JMC 3-0749L (JMC Lyon) in delta notation, respectively:

$$\delta^{65}Cu = ((^{65}Cu/^{63}Cu)_{sample}/(^{65}Cu/^{63}Cu)_{NIST\ 976} - 1) \times 1000$$

$$\delta^{66,68}Zn = ((^{66,68}Zn/^{64}Zn)_{sample}/(^{66,68}Zn/^{64}Zn)_{JMC\ 3\text{-}0749\ L} - 1) \times 1000$$

Since the JMC 3-0749L standard is no longer available, here we use the international reference material IRMM-3702 to bracket the samples, which has been well calibrated against JMC-Lyon with a $\delta^{66}Zn_{IRMM\text{-}3702}$ value of 0.27±0.03‰ (Wang et al., 2017). The long-term external reproducibility (2sd) is±0.05‰ for $\delta^{65}Cu$ (Liu et al., 2014). The reproducibility is±0.04‰ for $\delta^{66}Zn$ and ±0.08‰ for $\delta^{68}Zn$ (Liu et al., 2016a; Wang et al., 2017) obtained from repeated analyses of IRMM 3702 ($\delta^{66}Zn$=0.27±0.04‰; 2sd, n=210) and international rock standards over a period of three years. Two USGS basalt reference materials BHVO-2 and BCR-2, which have chemical compositions analogous to samples in this study (basalts), were analyzed through the whole procedure of this current study. The results ($\delta^{66}Zn$=0.33±0.03‰ and $\delta^{65}Cu$=0.10±0.03‰ for BHVO-2; $\delta^{66}Zn$=0.29±0.04‰ and $\delta^{65}Cu$=0.16±0.04‰ for BCR-2) resemble the literature values (e.g., Sossi et al., 2015; Wang et al., 2017; Moynier et al., 2017).

4 Results

Bulk-rock major and trace elemental data of CIM and Chaoerhe basalts are presented in the Supplementary Table S1, and their Zn-Cu-Sr-Nd isotope data are listed in Table 1. Basalts in the west of the NSGL are mainly basanites or alkali basalts, while those on the NSGL belong to alkali or tholeiitic basalts (Fig. 3a). The overall range of SiO_2 contents of the CIM basalts is from 41.8 wt% to 51.9 wt%. Almost all samples have LOI contents lower than 2.0 wt%, indicating weak post-eruption low-temperature alteration of the studied samples. The overall range of $\delta^{66}Zn$ is from 0.29±0.03‰ to 0.50±0.04‰ (n=35), which are higher than the mantle value of 0.18±0.05‰ and either identical with, or significantly higher than, those of mid-ocean ridge basalts (MORB, 0.28±

0.05‰; Wang et al., 2017; Huang et al., 2018). The CIM basalts (not including Chaoerhe) display covariations of δ^{66}Zn with SiO_2, Nb contents and Hf/Hf* index ($Hf_N/(Sm_N \times Nd_N)^{0.5}$), but not with $^{87}Sr/^{86}Sr$ and Cr contents (Fig. 3). The Chaoerhe basalts display a distinct tendency from the CIM basalts in these plots (Fig. 3). The CIM basalts have fairly homogeneous Cu isotopic compositions with δ^{65}Cu varying between –0.03‰ and +0.09‰, which fall on the range of the terrestrial mantle estimated by peridotites and basalts (0.06 ± 0.10‰, 1sd; Liu et al., 2015).

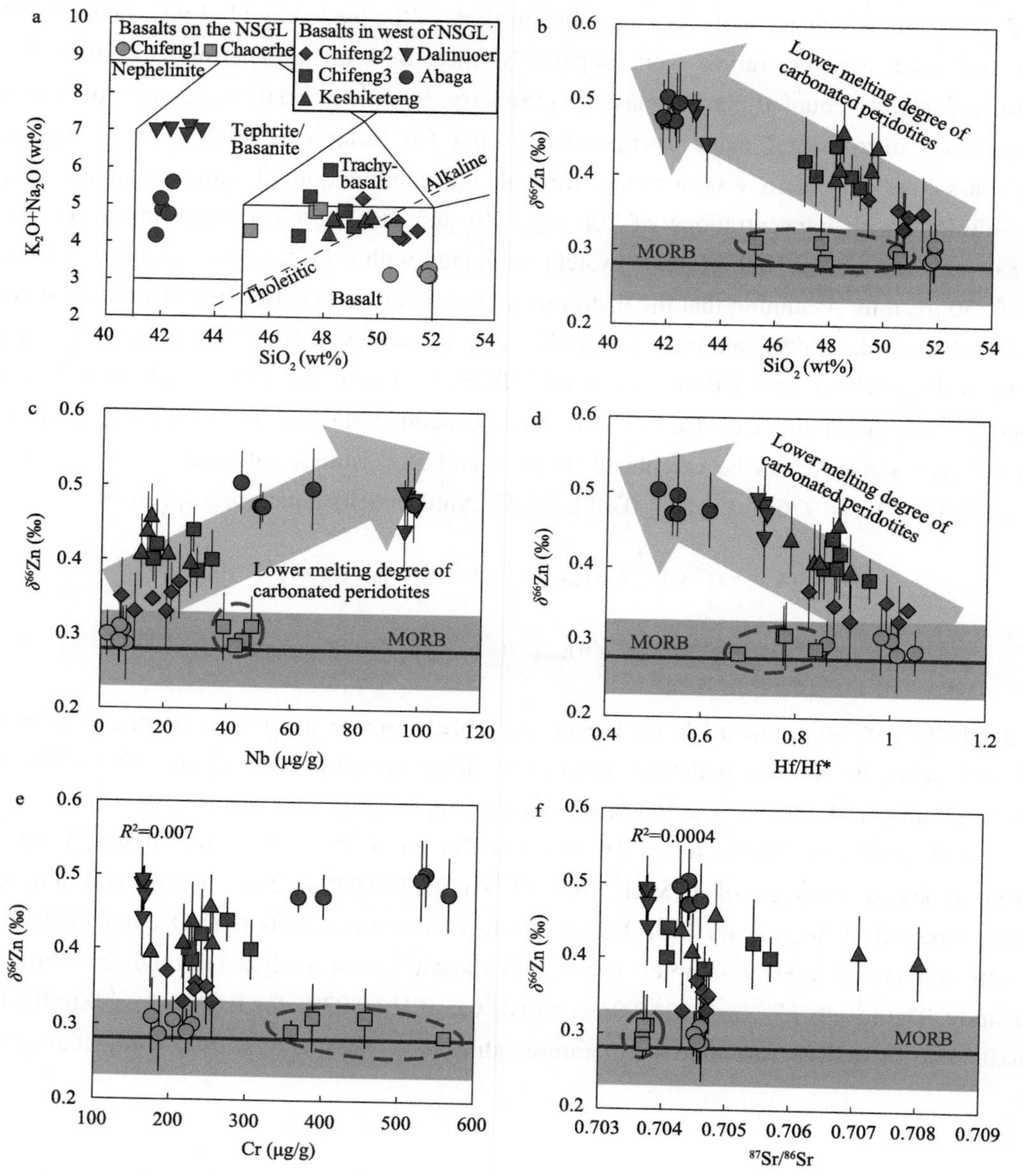

Fig. 3　(a) Total alkali versus SiO_2 diagram for classification of basaltic lavas in this study. (b-f) Plots of δ^{66}Zn values against SiO_2, [Nb], Hf/Hf*, [Cr] and $^{87}Sr/^{86}Sr$ for the CIM lavas reported in this study. δ^{66}Zn of MORB is from Wang et al. (2017).

Spatially, a gradual decline of δ^{66}Zn from higher-than-MORB to MORB-like values toward southeast is observed in the Abaga–Chifeng profile (Fig. 4a). Notably, basalts from Chifeng1 and Chaoerhe that are located on the NSGL have Zn isotopic compositions identical with those of MORB within uncertainty (Fig. 4a). Chemistry of the CIM basalts also varies spatially, with the gradual eastward increase of K/Nb and decrease of CaO/Al_2O_3 and Sm/Yb ratios (Fig. 4b, c, d). $^{87}Sr/^{86}Sr$ and $^{143}Nd/^{144}Nd$ ratios vary from 0.70369 to 0.70802 and from 0.51267 to 0.51291, respectively, which do not systematically vary with space but become more scattered toward southeast.

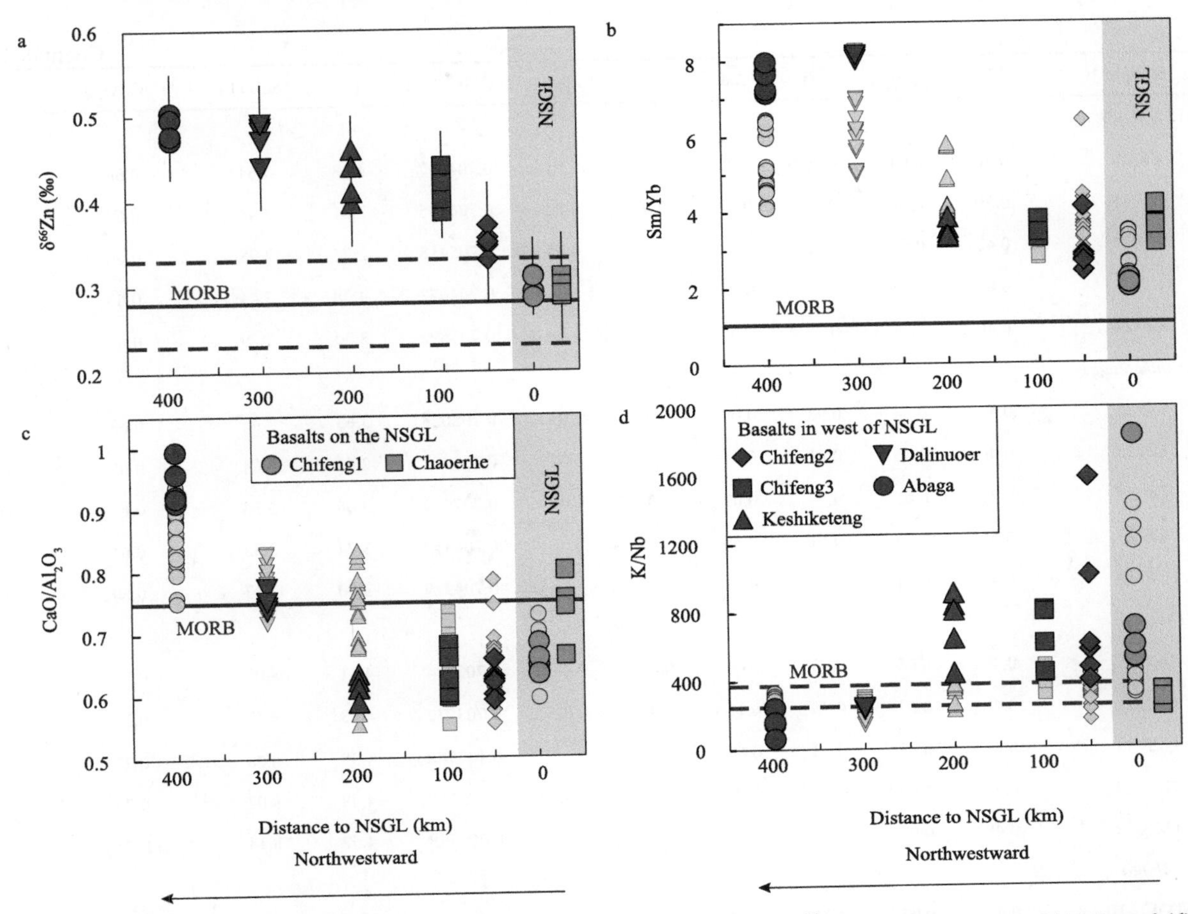

Fig. 4 Spatial variations of chemical and $\delta^{66}Zn$ of basalts in East Asia. Plots of (a) $\delta^{66}Zn$, (b) Sm/Yb, (c) CaO/Al_2O_3 and (d) K/Nb versus distances to the NSGL. Literature data for basalts in CIM are from Ho et al. (2008), Hong et al. (2013), Wang et al. (2015), Zhang and Guo, (2016), Guo et al. (2016), Pang et al. (2019), Togtokh et al. (2019), Guo et al. (2020) and references therein, as shown in smaller symbols with light colors.

Table 1 Bulk-rock Sr, Nd, Zn and Cu isotopic data and selected elemental ratios for basalts from CIM in this study

Sample	$\delta^{66}Zn$	2sd	$\delta^{68}Zn$	2sd	$\delta^{65}Cu$	2sd	$^{87}Sr/^{86}Sr$	ε_{Nd}	Sm/Yb	CaO/Al_2O_3	K/Nb
Chifeng1											
CF-01	0.29	0.05	0.56	0.10	0.01	0.02	0.704623	4.20	2.23	0.65	521
CF-02	0.31	0.05	0.61	0.07			0.704585	3.47	2.10	0.63	616
CF-04	0.31	0.03	0.60	0.05	0.02	0.02	0.704600	3.57	2.09	0.68	2011
CF-06	0.30	0.03	0.61	0.09	0.07	0.05	0.704507	3.92	2.00	0.66	550
CF-08	0.29	0.03	0.59	0.08			0.704554	3.55	2.10	0.63	597
Chifeng2											
CF-09	0.36	0.05	0.66	0.07			0.704639	1.06	2.87	0.63	505
CF-12	0.35	0.05	0.73	0.09	0.04	0.02	0.704693	0.63	4.11	0.60	375
CF-14	0.33	0.04	0.70	0.08			0.704314	2.71	2.73	0.62	367
CF-15	0.37	0.05	0.67	0.04	0.00	0.03	0.704536	3.21	2.81	0.66	1412
CF-17	0.33	0.01	0.69	0.09	−0.01	0.04	0.704688	1.17	2.43	0.62	661
CF-19	0.35	0.05	0.69	0.08			0.704728	0.54	2.68	0.59	1587

Continued

Sample	$\delta^{66}Zn$	2sd	$\delta^{68}Zn$	2sd	$\delta^{65}Cu$	2sd	$^{87}Sr/^{86}Sr$	ε_{Nd}	Sm/Yb	CaO/Al_2O_3	K/Nb
Chifeng3											
CF-20	0.40	0.04	0.80	0.05	0.03	0.01	0.704057	4.48	3.64	0.68	437
CF-21	0.39	0.06	0.82	0.07	0.07	0.02	0.704655	2.81	3.45	0.60	775
CF-22	0.42	0.03	0.83	0.07	0.04	0.05	0.705418	2.73	3.78	0.60	483
CF-23	0.44	0.03	0.79	0.07			0.704072	4.38	3.44	0.66	1090
CF-24	0.40	0.03	0.75	0.08			0.705692	3.35	3.29	0.62	430
Keshiketengqi											
CF-25	0.40	0.04	0.93	0.09	0.03	0.05	0.708023	3.63	3.45	0.63	799
CF-26	0.46	0.05	0.81	0.07	−0.02	0.02	0.704822	2.67	3.84	0.61	620
CF-27	0.41	0.05	0.77	0.09	0.00	0.02	0.707084	3.68	3.56	0.62	475
CF-29	0.41	0.05	0.81	0.08			0.704457	3.74	3.32	0.61	908
CF-33	0.44	0.05	0.87	0.10			0.704279	3.90	3.37	0.59	807
Dalinuoer											
GZS-12	0.49	0.05	0.95	0.06	0.04	0.06	0.703724	4.41	8.09	0.74	241
GZS-20	0.48	0.03	0.95	0.11	−0.01	0.02	0.703732	4.63	8.12	0.74	239
GZS-11	0.47	0.01	0.93	0.01			0.703748	4.38	8.08	0.77	231
GZS-23	0.44	0.05	0.88	0.02	0.08	0.02	0.703750	4.48	8.02	0.74	231
GZS-22	0.49	0.02	0.96	0.00			0.703719	4.52	8.14	0.75	239
Abaga											
HTWL-10	0.47	0.03	0.94	0.07			0.704378	4.14	7.15	0.92	157
HTWL-11	0.47	0.02	0.94	0.05			0.704397	4.02	7.24	0.91	150
HTWL-27	0.50	0.04	0.98	0.04			0.704380	4.02	7.76	0.92	245
HTWL-64	0.50	0.05	0.98	0.07			0.704246	4.12	7.65	0.96	161
HTWL-57	0.48	0.05	0.97	0.04			0.704572	4.34	7.97	0.99	64
Chaoerhe											
CEH-03	0.31	0.04	0.61	0.04			0.703760	5.21	4.13	0.66	273
CEH-02	0.29	0.02	0.59	0.04	0.08	0.02	0.703690	5.00	3.16	0.75	237
CEH-20	0.29	0.01	0.60	0.02	0.10	0.02	0.703701	5.21	4.16	0.74	328
CEH-85	0.31	0.05	0.61	0.06			0.703710	5.29	3.61	0.80	285
Geostandards											
BHVO-2	0.33	0.03	0.67	0.05	0.10	0.03	0.703559				
BCR-2	0.29	0.04	0.58	0.07	0.16	0.04	0.705131				

5 Discussion

In this section, we first discuss the western edge of the subducted Paleo-Pacific slab that is the prerequisite for linking the origins of basalts on two sides of the NSGL in East Asia to different subducting slabs. Then, we investigate the respective fates of subducted carbonates related to the two slabs chiefly based on the different Zn isotope patterns of basalts distributed on the two sides of the NSGL. Finally, we explore possible factors controlling the fates of downgoing carbonates in ancient subduction zones.

5.1 The western edge of stagnant Paleo-Pacific slab in East Asia

The subducted Paleo-Pacific slab is currently horizontally stagnated in the MTZ beneath East Asia, extending over a distance of 800 to 1000 km along the 660 km discontinuity (insert of Fig. 2c; Huang and Zhao, 2006; Togtokh et al., 2019). It has been widely advocated to have influenced mantle dynamics, caused widespread lithospheric reactivation and thinning of the eastern North China craton spanning the late-Mesozoic and Cenozoic eras (e.g., Griffin et al., 1998; Xu, 2007; Zhu et al., 2011). It has been broadly proposed that materials carried by the subducted Paleo-Pacific slab have contributed to chemical and isotopic characteristics of widespread late-Mesozoic to Cenozoic intraplate basalts in East Asia (e.g., Zeng et al., 2010; Li et al., 2017). However, the western edge of the Paleo-Pacific slab remains unknown and largely contentious as yet. Geophysical observations reveal that its present-day western edge is located nearby the NSGL (Huang and Zhao, 2006), while geochemical investigations on the subduction-related basalts suggest it is not. For instance, the chemical and Sr-Nd-Pb isotopic similarity of late-Mesozoic to Cenozoic basalts on two sides of the NSGL leads some studies to propose that the subducted slab once arrived at the west of the NSGL, but has retreated eastward or has sunk into the lower mantle (Togtokh et al., 2019; Sun et al., 2021). In particular, basalts in the west of the NSGL also have lighter-than-mantle magnesium isotopic compositions (Sun et al., 2021), a characteristic that has been widely observed in basalts from the east of the NSGL and interpreted to be related to the subducted Paleo-Pacific slab (e.g., Huang et al., 2015; Li et al., 2017).

Our Zn isotopic results on the Abaga-Chifeng profile in CIM provide novel constraints on this issue. Notably, basalts located on the NSGL have normal (MORB-like) $\delta^{66}Zn$ (Fig. 4a), which are distinct from the late-Mesozoic to Cenozoic basalts in the east of the NSGL that have high $\delta^{66}Zn$ (0.3–0.6‰; Jin et al., 2020; Liu et al., 2016a; Wang et al., 2018). In addition, they are different from basalts in the west of the NSGL that also have higher $\delta^{66}Zn$ than those of MORB (Fig. 4a). If the subducted Paleo-Pacific slab once reached the west of the NSGL (e.g., CIM) and has contributed to origin of basalts therein, it is difficult to coincide with the occurrence of high-$\delta^{66}Zn$ basalts on two sides of the NSGL but the absence in between. For example, basalts from Abaga and Fuxin, which are located on two sides of the NSGL along the nearly same latitude of 43° N (Fig. 2c), have high $\delta^{66}Zn$ (0.47–0.50‰ and 0.41–0.46‰, respectively), but basalts located between them (Chifeng1) have dominantly MORB-like $\delta^{66}Zn$ values of 0.29–0.31‰ (Fig. 4a). Basalts from Chaoerhe that is located in the north of CIM but on the NSGL, similar to basalts in Chifeng1 (both being located on the NSGL), also have MORB-like $\delta^{66}Zn$ (Fig. 4a), although their chemical and Sr isotopic differences from basalts from Chifeng1 (e.g., SiO_2, Nb, Cr, $^{87}Sr/^{86}Sr$; Fig. 3) may reflect initially heterogeneous mantle sources. A straightforward solution to this problem is that the NSGL represents the western edge of the subducted Paleo-Pacific slab and the origin of CIM basalts was related to another subducting slab.

The southeastward subduction of the Paleo-Asian oceanic slab toward the north margin of the North China craton since the early Paleozoic (ca. 500–250 Ma) is the most likely candidate (Fig. 2a). The Paleo-Asian Ocean was formed via separation of the northern margin of the North China craton from southwestern Siberia at ~1.3 Ga and has a prolonged time-scale of ca. 1.1 Ga (e.g., Wan et al., 2018). The trend of Mesoproterozoic to Early Mesozoic ophiolites from the north to the south in the Central Asian Orogenic Belt archives the prolonged suturing history of the archipelagic Paleo-Asian Ocean. The subducted Paleo-Asian slab was widely recorded by a series of Paleozoic ophiolites and retrograded eclogites, Paleozoic to Triassic blueschists, and synchronous igneous rocks (500–250 Ma; Xiao et al., 2003; Zhao et al., 2018). Studies on mantle xenoliths hosted by the CIM basalts also suggest the temporal agreement between the mantle metasomatic event and the Paleo-Asian slab subduction (Zou et al., 2014). In addition, Cenozoic carbonatites in the nearby area had a crustal limestone precursor with the age of recycled limestones between 580 Ma and 360 Ma (Chen et al., 2016). These lines of evidence support a major contribution of the Paleo-Asian oceanic slab to the origin of CIM basalts. Subduction of the two large slabs (Paleo-Asian and Paleo-Pacific) may have widely contributed to metasomatism of mantle domains on two sides of the NSGL, respectively, which served as the sources for the late-Mesozoic–Cenozoic basalts from these regions (see below). The metasomatized mantle domains have not partially molten at subduction stages until large-scale heat upwelling in late-Mesozoic to Cenozoic most likely induced by the rollback of the Paleo-Pacific slab.

5.2 Gradual carbonate dissolution during Paleo-Asian slab subduction

Possible causes for Zn isotopic variation in basaltic magmas include crustal contamination, melt-lithosphere interaction, magmatic processes (melting, differentiation, degassing), and/or source heterogeneity caused by slab-derived fluids. Below we discuss these possible causes for Zn isotopic variation in the CIM basalts in turn.

5.2.1 Crustal contamination or melt-lithospheric mantle interaction

Both crustal contamination and melt–lithospheric mantle interaction may lower Zn isotopic ratios of basaltic melts initially possessing high δ^{66}Zn. Co-variation of δ^{66}Zn with ^{143}Nd/^{144}Nd or ^{87}Sr/^{86}Sr is expected if the primary basaltic melts with high δ^{66}Zn underwent reaction with the overlying enriched lithospheric mantle (Wang et al., 2018; Wang and Liu, 2021) or the overlying ancient continental crust, which is not observed in the CIM basalts (Fig. 3f). All of the CIM basalts have positive ε_{Nd} values (0.54–5.29; Table S1), arguing against significant crustal contamination. Significant crustal contamination during magma descending is also unsupported by the moderate to high MgO contents (mostly between 7 and 12 wt.%) of the CIM basalts (Table S1). Assimilation by crustal carbonates may occur when hot basaltic magmas ascend through the crust (e.g., Mason et al., 2017), which can potentially increase δ^{66}Zn of the reacted melts. Since sedimentary carbonates are characterized by extremely low Nb contents compared to basaltic rocks (Turekian and Wedepohl, 1961), a negative relationship between δ^{66}Zn and Nb contents is expected for crustal carbonate assimilation. This is in contrast to the positive δ^{66}Zn–[Nb] relationship as illustrated in Fig. 3c, which instead reflects higher δ^{66}Zn of the melt generated at lower melting degree (see below).

5.2.2 Magma degassing, crystal-melt differentiation, or “lid effect”

As a moderately volatile element, Zn can become volatilizable at magmatic temperatures with isotopically light Zn preferentially partitioning into the volatile phase, and the volatilization process is commonly accompanied by strong Zn depletion in the residue (e.g., Toutain et al., 2008; Paniello et al., 2012; Kato et al., 2015). The CIM basalts with heavier Zn isotopic compositions generally have higher Zn concentrations (Tables 1 and S1), which is inconsistent with a volatilization effect. In addition, significant loss of isotopically light Zn induced by magmatic degassing contradicts the positive correlation between δ^{66}Zn and the abundance of Nb (Fig. 3c), which is a highly refractory element.

Fractional crystallization of olivine can increase δ^{66}Zn of the residual melt, but the magnitude is not larger than 0.1‰ even during extreme basaltic differentiation (Chen et al., 2013; McCoy-West et al., 2018). This is unlikely to explain the wide δ^{66}Zn range of the CIM basalts (0.29–0.50‰). In particular, the negative correlation between δ^{66}Zn and SiO_2 contents (Fig. 3b) argues against a role of olivine crystallization because it will increase both $\delta^{66}Zn$ and SiO_2 of the melts. A recent study found that Cr-Fe oxide (e.g., chromite), one of the first phases to crystallize during basaltic differentiation (Thy, 1983), is enriched in lighter Zn isotopes relative to silicate mineral/melt and has extremely high Zn concentration (Yang et al., 2021). Fractional crystallization of Cr-Fe oxide can therefore increase $\delta^{66}Zn$ of the residual melt and lower its Cr and Zn contents. There is a lack of correlation between $\delta^{66}Zn$ and Cr contents (R^2=0.007; Fig. 3e) in the CIM basalts, which excludes Cr-Fe oxide crystallization as a cause for the Zn isotopic variation. In addition, the fact that samples with higher $\delta^{66}Zn$ values have higher Zn concentrations is also inconsistent with that is expected by Cr-Fe oxide crystallization.

A recent study proposed that the depth of the lithosphere-asthenosphere boundary layer exerts a primary control on compositional variations of the CIM basalts (so-called "lid effect"; Guo et al., 2020). This model predicts that basaltic melts erupted on thinner lithosphere have chemical characteristics of lower pressure and higher melting degree compared to melts erupted on thicker lithosphere. The southeastward decline of Sm/Yb (Fig. 4b) supports increasing melting degrees toward the southeast for the CIM basalts (Guo et al., 2020). Nevertheless, melting of a pristine peridotitic mantle ($\delta^{66}Zn$=0.18‰) at varying pressures (3–7 GPa) and melt fractions (0.01–0.3) generates melts with $\delta^{66}Zn$ of no greater than 0.30‰ and induces the shift of Zn isotope fractionation by not exceeding 0.08‰ (Doucet et al., 2016; Sossi et al., 2018). This is unlikely to account for the high and widely variable $\delta^{66}Zn$ values of the CIM basalts (0.29–0.50‰). Thereupon, the spatial $\delta^{66}Zn$ variation in these basalts must reflect source inhomogeneity caused by subducted materials. Two potential sources with high $\delta^{66}Zn$ in subduction zones include sulfate-rich fluid and carbonate-rich fluid.

5.2.3 The role of sulfate-rich fluid

Hydrous sulfate and/or carbonate-bearing fluids derived from subducting slabs are the important agents of modifying the oxidation state of the sub-arc mantle (Debret and Sverjensky, 2017). Sulfate is enriched in heavier Zn isotopes relative to sulfide as determined from first-principles calculations (Ducher et al., 2016). Sulfate-rich fluids during dehydration of altered ultramafic rocks in subduction zones have been proposed to dominate slab-derived fluids at depth and to produce high $\delta^{66}Zn$ fluids (Pons et al., 2016). Here we consider the Cu isotope systematic to be an effective tool of identifying the role of sulfate-rich fluid because Cu-sulfate is also strongly enriched in heavier Cu isotope relative to sulfide (Fujii et al., 2013, 2014; Fig. 5). If sulfate-rich fluids play a dominant role in generating high $\delta^{66}Zn$ of a basalt, it should also have high $\delta^{65}Cu$. As shown in Fig. 5, the CIM basalts have $\delta^{65}Cu$ values that fall within uncertainty on the range of MORB and the mantle, and more importantly, they do not vary with space (i.e., the distances to the NSGL). There is a lack of correlation between $\delta^{65}Cu$ and $\delta^{66}Zn$ for CIM basalts (R^2=0.038; Fig. 5), that is, samples with higher-than-MORB $\delta^{66}Zn$ have MORB-like $\delta^{65}Cu$ values. Therefore, the high $\delta^{66}Zn$ of some CIM basalts is unlikely to have been attributed to a dominant role of sulfate-rich fluids in their mantle sources and instead are best explained by a dominant role of recycled carbonates. Although carbonates are also enriched in heavy Cu isotopes relative to sulfides (Fujii et al., 2013), they commonly contain little Cu (mean=~5 μg/g and down to 0.05 μg/g; Fio et al., 2010; Meyer et al., 2012; Zhao and Zheng, 2014) relative to Zn. Here we present a binary mixing model between carbonate or sulfate-rich fluids and the mantle. The results show that carbonate addition remarkably increases Zn isotopic composition of the mantle but does not change its Cu isotopic composition significantly (Fig. 5b). By contrast, metasomatism by sulfate fluids increases both Cu and Zn isotopic compositions of the mantle. This explains why the CIM samples with high $\delta^{66}Zn$ have mantle-like $\delta^{65}Cu$ (Figs. 4, 5) and further supports a dominant role of recycled carbonates in the sources of CIM basalts.

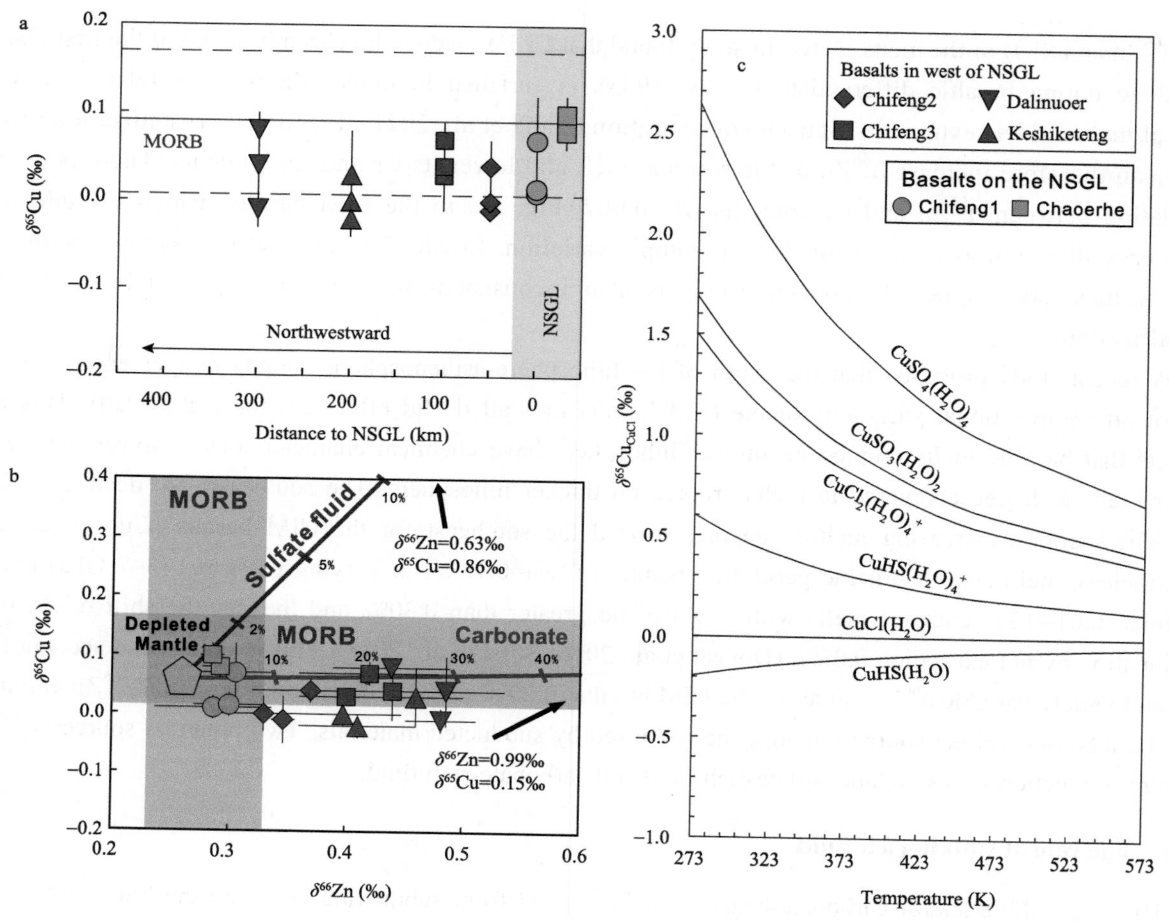

Fig. 5 Plots of Cu isotopic ratios against the distances to the NSGL (a) and δ^{66}Zn (b) for the CIM basalts. (c) Theoretically predicted $\delta^{65}Cu_{CuCl}$ of various Cu complexes calculated by logarithm of the reduced partition function from Fujii et al. (2013, 2014). The red and blue lines in (b) represent the binary mixing between the Depleted Mantle (DM) and recycled carbonate or slab-derived sulfate fluid, respectively. For carbonate, δ^{65}Cu, δ^{66}Zn, [Cu] and [Zn] are set to be 0.15‰ (Zhang et al., 2022), 0.99‰ (Pichat et al., 2003; Sweere et al., 2018), 5 μg/g (e.g., Fio et al., 2010; Meyer et al., 2012; Zhao and Zheng, 2014) and 44 μg/g (Liu et al., 2016b), respectively. For the DM, δ^{65}Cu, δ^{66}Zn, [Cu] and [Zn] are set as 0.06‰, 0.26‰, 20 μg/g and 40 μg/g, respectively, represented by unaltered abyssal peridotites (Liu et al., 2019); Note that an isotopic effect of ~0.1‰ during mantle' partial melting (Sossi et al., 2018) is added to the DM endmember. The element concentrations of sulfate fluid are calculated by a Rayleigh distillation model: $X_{fluids}=(D*X_{protolith})/(1+(1-f)*(D-1))$, where X denotes Cu or Zn, D is the partition coefficient between fluid and protolith, and f is the fraction of Zn or Cu remaining in the rock. The isotopic compositions of sulfate fluid (assumed to be released from dehydration of altered peridotites, e.g., Pons et al., 2016) are calculated by a Rayleigh distillation model: $\delta X_{fluids}=(\delta X_{protolith}+1000)\times\alpha\times f^{(\alpha-1)})-1000$, where X represents ^{65}Cu or ^{66}Zn, and α is the isotope fractionation factor. The Cu-Zn concentrations and isotope ratios of altered peridotites (Cu=25 μg/g, δ^{65}Cu=0.35‰±0.11‰, Zn=56 μg/g, δ^{66}Zn=0.24‰±0.04‰, 2se) are from Liu et al. (2019). α is from Fujii et al. (2013, 2014). D and f are assumed to be 10 and 0.99, respectively (e.g., Keppler and Wyllie, 1991; Zajaca et al., 2008). δ^{65}Cu of MORB is from Liu et al. (2015).

5.2.4 Source heterogeneity caused by various amounts of recycled carbonates

Marine carbonates are typically characterized by higher δ^{66}Zn than that of the peridotitic mantle (Fig. 1). A carbonated mantle source for the CIM basalts is also supported by their major and trace element chemistry. Experimental investigations show that the addition of carbonate components can dramatically lower the solidus of mantle peridotites, and at lower melting degrees, the resultant melts have diminished SiO_2 and enhanced CaO/Al_2O_3 (Dasgupta et al., 2007). At this time, the melt is expected to have higher δ^{66}Zn due to more contributions of carbonate components at lower melt fractions (Liu et al., 2016a). The CIM basalts display elevated δ^{66}Zn values with decreasing SiO_2 and increasing CaO/Al_2O_3 (Figs. 3 and 4), which agrees well with partial melting of carbonated peridotites at various melt fractions. Samples with higher δ^{66}Zn tend to possess higher contents of Nb (a highly incompatible element) and stronger negative anomalies of Hf (i.e., lower Hf/Hf*) (Fig. 3c, d), further supporting a carbonated mantle source. Barium (Ba) is a highly fluid-mobile element in subduction zones (Kessel et al., 2005) and mantle rocks subject to carbonate-rich fluid metasomatism exhibit strong enrichment of Ba (e.g., Ionov et al., 1993). The Ba/Zr and Ba/Y ratios display similar patterns to Zn isotopes in the Abaga-Chifeng profile and are positively correlated with δ^{66}Zn (Fig. 6). This supports the high-δ^{66}Zn endmember to be carbonate-rich fluid released from the subducted slab. Thus, the spatial variation of chemical and Zn isotopic compositions in the CIM basalts (Figs. 4 and 6) reflects a gradual increase in the melting degrees and a gradual decrease in the amounts of recycled carbonates in the sources toward southeast.

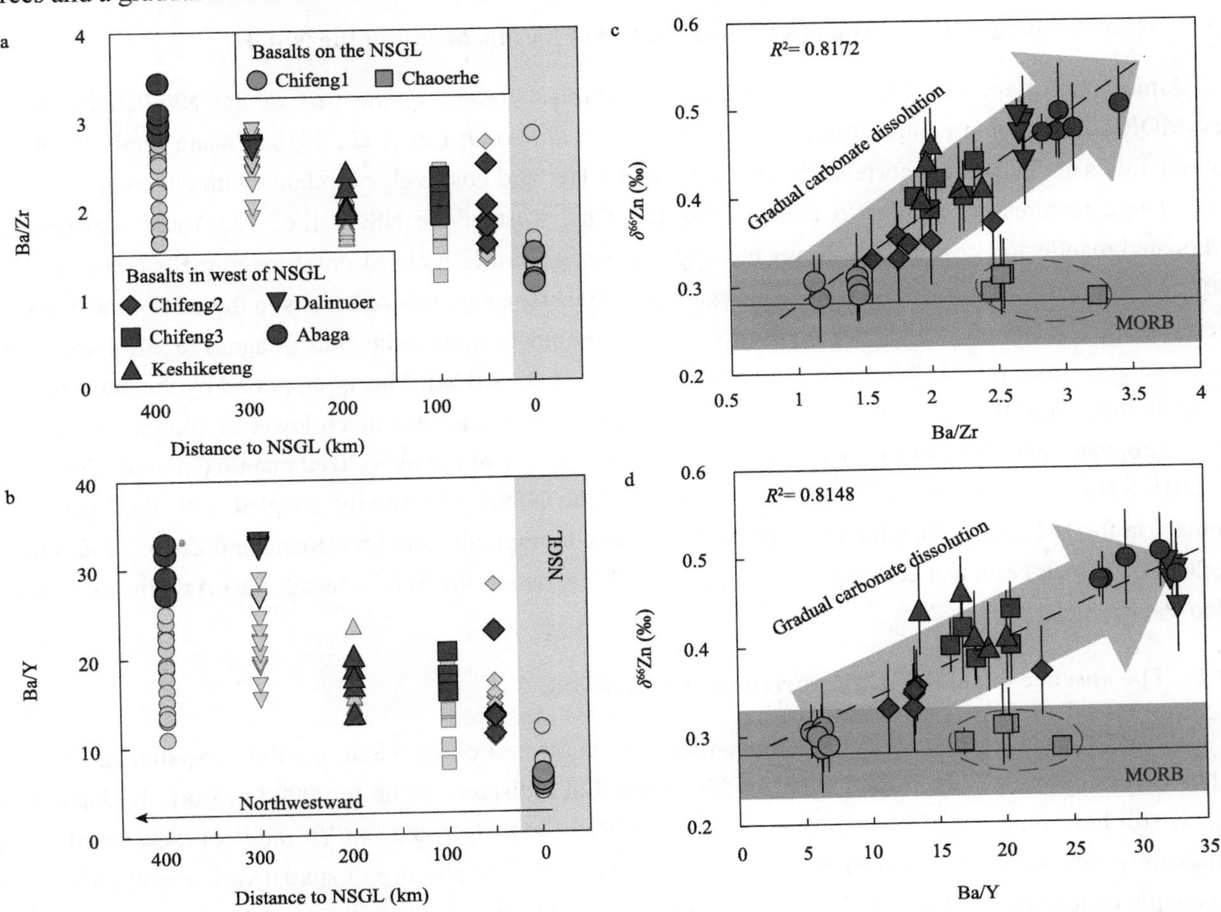

Fig. 6 Spatial variations of Ba/Zr (a) and Ba/Y (b) with the distances to the NSGL and plots of δ^{66}Zn against Ba/Zr (c) and Ba/Y (d) for the CIM basalts. δ^{66}Zn of MORB is from Wang et al. (2017).

Since the solidus of both hot and cold slab surfaces is lower than that of carbonated eclogites at pressures of < ~10 GPa (Thomson et al., 2016), the carbonate-bearing slab may have not undergone melting at subduction stages. Instead, the transfer of downgoing carbonates from the slab to the mantle took place mainly by fluid-mediated dissolution (e.g., Frezzotti et al., 2011; Kelemen and Manning, 2015). During carbonate dissolution, the preferential release of isotopically heavy Zn into the fluids (Debret et al., 2018) lowers the δ^{66}Zn of recycled carbonates as subduction proceeds. This further contributes to the southeastward decline of δ^{66}Zn in the CIM basalts as a result of earlier carbonate dissolution during subduction.

In summary, the above observations suggest that carbonates carried by the Paleo-Asian oceanic slab have gradually been dissolved and released into the mantle as the slab came down in a southeast direction. This further indicates that the CIM basalts were genetically linked to the southeastwardly subducted Paleo-Asian slab but not the northwestward subducted Paleo-Pacific slab as discussed in Section 5.1. Notably, basalts located on the NSGL have δ^{66}Zn reduced to resembling those of MORB (Fig. 4a). To produce basalts with δ^{66}Zn equal to 0.50‰, the required amount of recycled carbonates in the source is 8–10‰, and to produce basalts with δ^{66}Zn not exceeding the range of MORB, this amount is not greater than ~1% (Liu et al., 2020). Therefore, the sources of basalts located on the NSGL contain a very limited amount of recycled carbonates since most of the subducted carbonates have been dissolved and released at earlier subduction stages.

5.3 Deep burial of carbonates during Paleo-Pacific slab subduction

5.3.1 Recycled magnesium carbonates in the sources of basalts in east of the NSGL

Almost all of the late Mesozoic to Cenozoic intraplate basalts in the east of the NSGL have heavier-than-MORB Zn isotopic compositions (~0.3–0.6‰; Jin et al., 2020; Liu et al., 2016a; Wang et al., 2018). Zinc isotopic ratios are negatively correlated with SiO_2 and Hf/Hf* and positively correlated with Nb contents (Fig. 7a, b, c). These features are similar to those of basalts in the west of the NSGL (i.e., CIM) and also point to a carbonated mantle source which, at lower melting degrees, generates more Si-unsaturated melts (Dasgupta et al., 2007). Combined with Mg isotope data, the large-scale Zn isotope anomaly of basalts in the east of the NSGL was interpreted to indicate the widespread presence of recycled magnesium carbonates (magnesite ± dolomite) in the mantle sources since the late Mesozoic (Li et al., 2017; Liu et al., 2016a). This is supported by the low $^{87}Sr/^{86}Sr$ of basalts in the east of the NSGL (Fig. 8b), since magnesium carbonates have much lower Sr contents than those of calcic carbonates and thus have only limited influence on $^{87}Sr/^{86}Sr$ of the hybridized mantle (Li et al., 2017; Liu et al., 2016). Notably, the high-δ^{66}Zn area in the east of the NSGL is spatially coupled with the high-velocity anomaly in the MTZ (Fig. 2), which is deduced by seismic tomography studies (Huang and Zhao, 2006; Fukao et al., 2009). This suggests that carbonates have been deeply carried to the MTZ beneath East Asia by the subducted Paleo-Pacific slab.

5.3.2 The absence of spatial δ^{66}Zn variation and its origin

Unlike the CIM basalts, Zn isotopic ratios of basalts in the east of the NSGL exhibit no spatial variation with eruptive longitude (Fig. 7d). In particular, basalts erupted at a distance of up to 2000 km from the Japan trench (Fig. 2) still have higher δ^{66}Zn values than those of MORB. CaO/Al_2O_3 and Sm/Yb ratios of these basalts do not significantly vary with the eruptive longitude as well (Fig. 7e, f). The absence of spatial variation of chemical and Zn isotopic compositions of basalts in the east of the NSGL indicates that both the amount of recycled carbonates in their mantle sources and the melting degree do not significantly change with various distances to the trench. Thereupon, unlike the subducted Paleo-Asian slab, a considerable amount of carbonates must have survived shallow fluid-mediated dissolution during westward subduction of the Paleo-Pacific slab. Otherwise, a gradual decline of δ^{66}Zn and CaO/Al_2O_3 with increasing distances to the trench should be seen and heavy δ^{66}Zn signals cannot be preserved owing to the preferential release of isotopically heavy Zn during partial carbonate dissolution at shallower depths, as seen in the CIM basalts. In other words, the Paleo-Pacific slab carried a large number of carbonates far away from the trench and into the MTZ, without complete dissolution or decarbonation at shallower depths of <410 km.

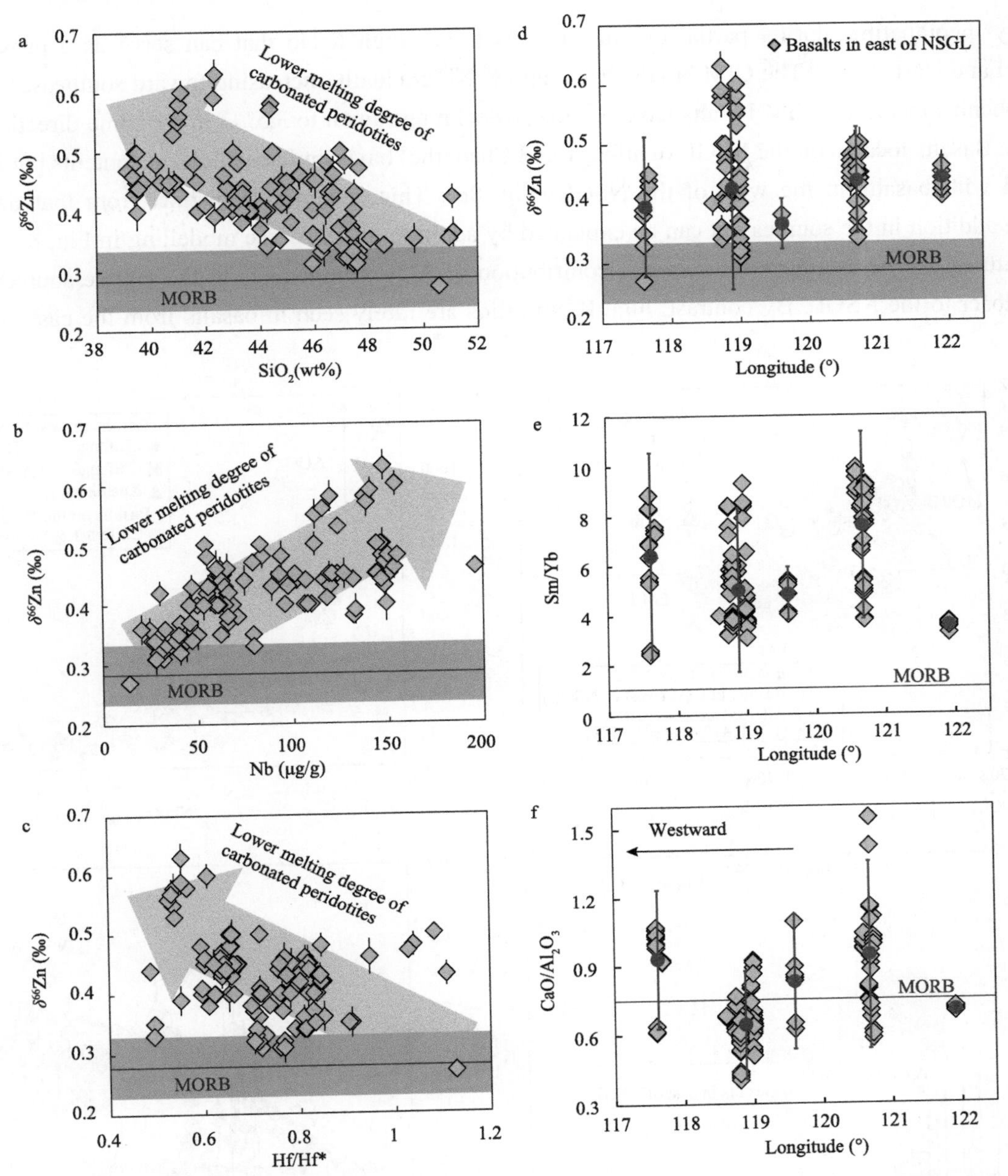

Fig. 7 Plots of (a) SiO_2, (b) Nb, and (c) Hf/Hf* versus δ^{66}Zn. Plots of (d) δ^{66}Zn, (e) Sm/Yb, and (f) CaO/Al_2O_3 versus eruptive longitude of basalts. Data for basalts in the east of the NSGL are from Liu et al. (2016), Wang et al. (2018), and Jin et al. (2020).

5.4 Factors controlling the fates of subducting carbonates

As discussed above, subduction of the two oceanic slabs in East Asia gave rise to strongly contrasting fates of downgoing carbonates, with almost complete dissolution at shallow depths in one subduction zone whereas deep burial in another. Below we explore possible controlling factors by considering ages and sedimentation histories of the two oceanic slabs as well as chemical and isotopic differences of basalts on two sides of the NSGL.

5.4.1 The role of altered oceanic crust

Compared with basalts in the east of the NSGL, the CIM basalts have systematically higher $^{87}Sr/^{86}Sr$ at given $^{143}Nd/^{144}Nd$, which follows the tendency defined by the altered oceanic crust (AOC; Fig. 8a, b). Both K and Nb are highly incompatible during partial melting, but AOC has high K/Nb that can serve as a proxy for AOC (Staudigel and Hart, 1983). The CIM basalts have high K/Nb gradually increasing toward southeast (Fig. 4d, 8c). Olivine phenocrysts in the CIM basalts have $\delta^{18}O_{SMOW}$ declining (down to 3.5‰) in the same direction (Wang et al., 2015). Basalts located on the NSGL (Chifeng1 and Chaoerhe) have similarly high Zn contents but lower $\delta^{66}Zn$ compared with basalts in the west of the NSGL (Fig. 8d). This tendency is distinct from that derived from carbonate addition in the sources but can be explained by addition of AOC (see modelling in Fig. 8). All of these above features can be ascribed to increasing contribution of AOC components in the mantle sources of basalts located closer to the NSGL. By contrast, high K/Nb ratios are rarely seen in basalts from the east of the NSGL (Fig. 8c).

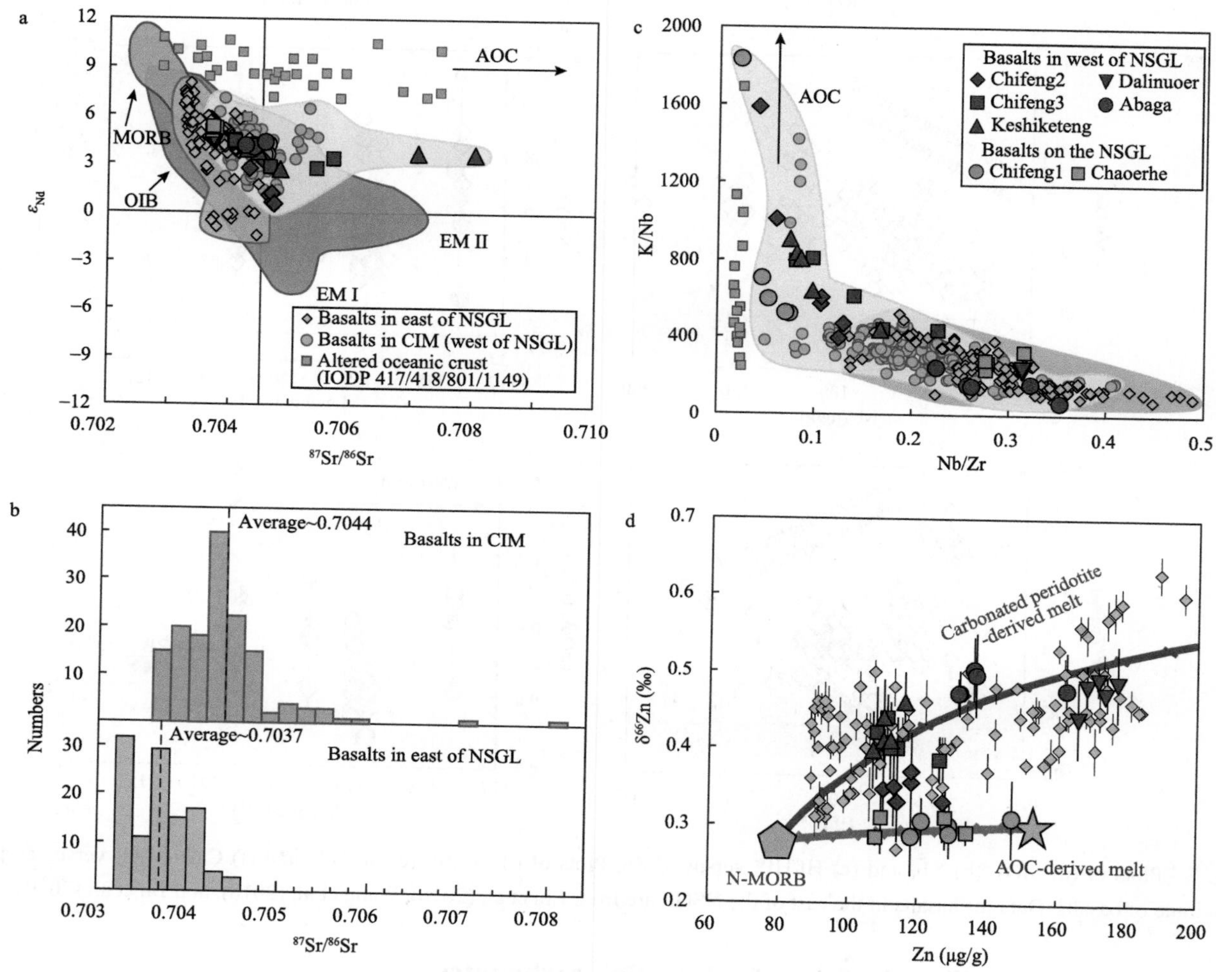

Fig. 8 (a) Plot of ε_{Nd} versus $^{87}Sr/^{86}Sr$. (b) Comparison of $^{87}Sr/^{86}Sr$ between basalts from two sides of the NSGL. (c) Plot of K/Nb versus Nb/Zr. (d) Plot of $\delta^{66}Zn$ versus Zn contents. Data for altered oceanic crust are from Staudigel et al. (1995), Hauff et al. (2003) and Huang et al. (2016). The scope of MORB and OIB and the EMI and EMII arrays are after Hofmann, 1997. Data sources for basalts in CIM and basalts in the east of the NSGL are same as Figs. 4 and 5. Zn content and isotopic data of N-MORB is from Wang et al. (2017). Here we consider a melting-mixing model between MORB-like melts and the altered oceanic crust-derived melts for the Chifeng1 and Chaoerhe basalts. Detailed endmember parameters are provided in the Supplementary Table S2.

The form of carbon in oceanic crust varies with ages and spreading rates of the crust. Crust with older ages and slower spreading rates commonly has higher carbonate contents in AOC (Alt and Teagle, 1999; Gillis and Coogan, 2011). The Paleo-Asian oceanic slab was initially opened at ~1300 Ma and started to be subducted since the Early Paleozoic (Xiao et al., 2003), having an extended longevity of ~1.1 Ga and slower spreading rate in comparison with the western Pacific plate with a shorter span of 100–200 Myr (Müller et al., 2008a) and intermediate- to fast-spreading rate. The long longevity and slow spreading rate would have led to the main incorporation of carbonates into the AOC (e.g., Gillis and Coogan, 2011). Veined carbonates in AOC can be largely dissolved at shallow depths by aqueous fluids released from the AOC, which can enhance carbonate dissolution due to its high carbonate solubility and thus prevent deep sink of subducting carbon (Kelemen and Manning, 2015). The chemical and Zn isotopic variations of the CIM basalts thus reflect almost complete carbonate dissolution at relatively shallow depths, leaving the carbonate-poor AOC to continue subducting to greater depths.

5.4.2 The contents and species of carbonate sediments

The efficiency of carbon subduction is largely controlled by carbonate contents of the sediment column and is partly linked to the latitude of the trench (Clift, 2017). The Paleo-Asian slab subducted toward the North China craton at ~500–250 Ma, whereas the Paleo-Pacific slab started to subduct beneath eastern China later, in the Mesozoic. Notably, the thickness of global deep-sea carbonate sedimentation that can be subducted has substantially increased since ~250 Ma (Ridgwell and Zeebe, 2005) and further increased in the Cretaceous (Dutkiewicz et al., 2019) (Fig. 9). In addition, low latitudes are areas of higher carbonate productivity, while cold ocean waters are more prone to dissolving carbonates produced at high latitudes (Clift, 2017). The Paleo-Asian slab was located on higher latitude than the Paleo-Pacific slab before the Mesozoic (Zhao et al., 2018). Thus, the larger amount of deep-sea carbonates in the Paleo-Pacific slab may allow most of them to have survived shallow dissolution during subduction in comparison with the Paleo-Asian slab.

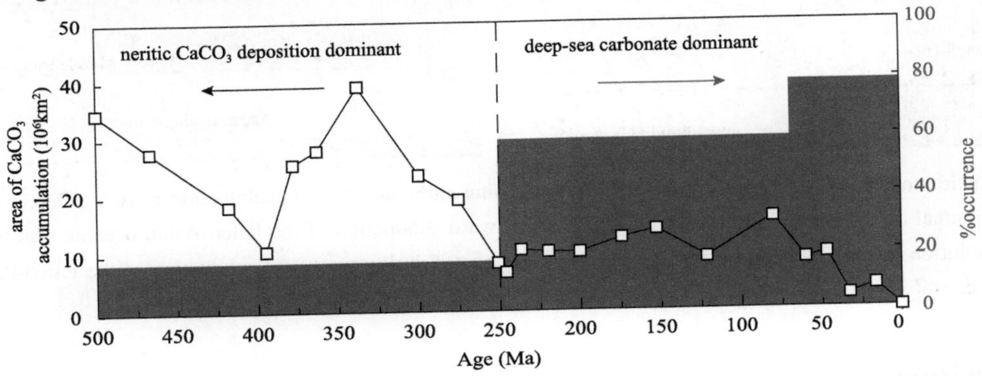

Fig. 9 Percent occurrence of carbonates and reconstructed changes in the total area of platform (shallow water) carbonates since 500 Ma (modified after Ridgwell and Zeebe, 2005). See text for details.

On the other hand, the increased accumulation of deep-sea carbonates since ~250 Ma is suggested to be a result of dramatic decline of neritic $CaCO_3$ deposition and increasing sedimentation of magnesium carbonates (Ridgwell and Zeebe, 2005; Fig. 9). The solubility of carbonate species in hydrous fluids decreases in the order of calcite > dolomite > magnesite at given pressures (Pan et al., 2013). Calcic carbonate is unstable at high pressures and can be transformed into dolomite at 23–45 kbars, or into magnesite at higher pressures (Kushiro, 1975). The Ca–Mg exchange between carbonate and silicate during subduction also favors the formation of magnesian carbonates at higher pressure (Yaxley and Green, 1994). The sedimentary magnesian carbonates in the Paleo-Pacific slab, together with magnesian carbonates formed from Ca–Mg exchange during subduction, have the ability to survive shallow fluid-mediated dissolution.

5.5 Implications for deep carbon cycling

Since carbonate sediments are the major form of carbon inputs to the mantle via subduction, the strongly contrasting fates of subducting carbonates related to different oceanic slabs suggest that global carbon input flux has substantially varied with geologic time in past subduction zones. A cartoon model in Fig. 10 gives a description of the contrasting fates of carbonates during subduction of the two large oceanic slabs in East Asia. The contents and species of carbonate sediments, which are relevant to the ages of oceanic crust and the latitudes of the trench, may play a critical role in the efficiency of carbon subduction in old subduction zones, as have been widely observed in modern subduction zones (Clift, 2017; Plank and Manning, 2019). Hydrous fluids from dehydration of altered oceanic crust could further improve carbonate dissolution at shallow depths. The distinct efficiencies of carbon subduction related to different slabs influence global deep carbon cycles over geologic time. For instance, the drastic increase in both deep-sea carbonate sedimentation and carbon input flux since the late Mesozoic, related to the Paleo-Pacific slab, may have contributed to the diminished atmospheric pCO_2 after the well-known Early Cretaceous Greenhouse.

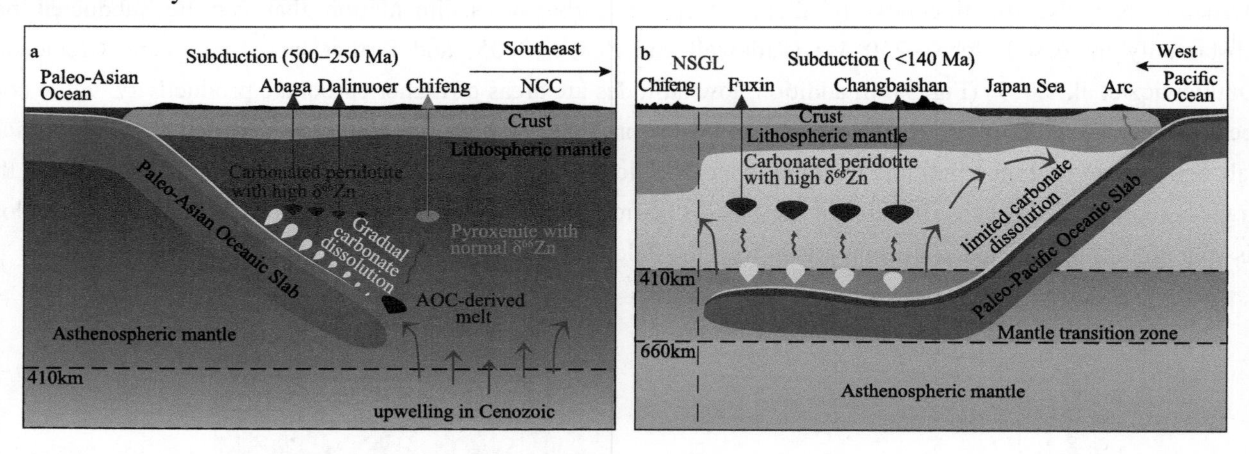

Fig. 10　Schematic models showing contrasting fates of downgoing carbonates during subduction of two different oceanic slabs in East Asia. (a) Gradual dissolution of carbonates during southeastward subduction of the Paleo-Asian oceanic slab in the Paleozoic; (b) Limited dissolution at shallow depths and deep burial of carbonates during westward subduction of the Paleo-Pacific slab since the late Mesozoic.

6 Conclusions

The fates of subducting carbonates related to different oceanic slabs were investigated via a comparative Zn isotopic study on late-Mesozoic to Cenozoic basalts across the NSGL in East Asia. Together with published data, we demonstrate that basalts distributed in two sides of the NSGL were genetically related to different oceanic slabs. Basalts in the west of the NSGL (e.g., CIM) were related to subduction of the Paleo-Asian slab, whereas those in the east were related to the Paleo-Pacific slab. The metasomatized mantle domains beneath the two regions have not partially molten at subduction stages until large-scale heat upwelling in late-Mesozoic to Cenozoic induced by the rollback of the Paleo-Pacific slab.

The higher-than-MORB $\delta^{66}Zn$ of all basalts in the east of the NSGL indicate that carbonates have been deeply carried into the MTZ by the subducted Paleo-Pacific slab since the Mesozoic. By contrast, the gradual decline of $\delta^{66}Zn$ and CaO/Al_2O_3 with increasing distances with the trench in the west of the NSGL indicates gradual carbonate dissolution at shallow depths during subduction of the Paleo-Asian slab. Aqueous fluids released from the AOC could have enhanced carbonate dissolution due to their high carbonate solubility and prevented a deep sink of subducting carbon. The contents and species of carbonate sediments further play a pivotal role in the efficiency of carbon subduction in ancient subduction zones, for which we suggest Zn isotopes to be an effective tool. Our results demonstrate that carbon input flux may significantly vary through deep time given the contrasting fates of downgoing carbonates in different subduction slabs.

Declaration of Competing Interest

The authors declare that they have no known competing financial interests or personal relationships that could have appeared to influence the work reported in this paper.

Acknowledgements

We thank Tian G.C. for help in field work. We appreciate the insightful comments from Stefan Weyer (Associate Editor), Luc Doucet and two anonymous reviewers, which significantly improve the quality of the manuscript. Stefan Weyer and Jeffrey G. Catalano (Executive Editor) are acknowledged for their efficient handling. This work is supported by the National Key R&D Program of China (2019YFA0708400), the National Natural Science Foundation of China (Grant No. 41730214, 41822301), and the "Strategic Priority Research Program" of the Chinese Academy of Sciences (Grant No. XDB18000000).

Appendix A. Supplementary material

Supplementary data to this article can be found online at https://doi.org/10.1016/j.gca.2022.03.009.

References

Alt J. C. and Teagle D. A. H. (1999) The uptake of carbon during alteration of ocean crust. Geochim. Cosmochim. Acta 63, 1527-1535.

Ague J. J. and Nicolescu S. (2014) Carbon dioxide released from subduction zones by fluid-mediated reactions. Nat. Geosci. 7, 355-360.

Beunon H., Mattielli N., Doucet L. S., Moine B. and Debret B. (2020) Mantle heterogeneity through Zn systematics in oceanic basalts: Evidence for a deep carbon cycling. Earth Sci. Rev. 205 103174.

Chen C., Liu Y., Foley S. F., Ducea M. N., He D., Hu Z., Chen W. and Zong K. (2016) Paleo-Asian oceanic slab under the North China craton revealed by carbonatites derived from subducted limestones. Geology 44, 1039-1042.

Chen H., Savage P. S., Teng F.-Z., Helz R. T. and Moynier F. (2013) Zinc isotope fractionation during magmatic differentiation and the isotopic composition of the bulk Earth. Earth Planet. Sci. Lett. 369-370, 34-42.

Clift P. D. (2017) A revised budget for Cenozoic sedimentary carbon subduction. Rev. Geophys. 55, 97-125.

Dasgupta R., Hirschmann M. M. and Smith N. D. (2007) Partial Melting Experiments of Peridotite + CO_2 at 3 GPa and Genesis of Alkalic Ocean Island Basalts. J. Petrol. 48, 2093-2124.

Dasgupta R. and Hirschmann M. M. (2010) The deep carbon cycle and melting in Earth's interior. Earth Planet. Sci. Lett. 298, 1-13.

Dasgupta R., Mallik A., Tsuno K., Withers A. C., Hirth G. and Hirschmann M. M. (2013) Carbon-dioxide-rich silicate melt in the Earth's upper mantle. Nature 493, 211-215.

Debret B. and Sverjensky D. (2017) Highly oxidising fluids generated during serpentinite breakdown in subduction zones. Sci. Rep. 7, 10351.

Debret B., Bouilhol P., Pons M. L. and Williams H. (2018) Carbonate Transfer during the Onset of Slab Devolatilization: New

Insights from Fe and Zn Stable Isotopes. J. Petrol. 59, 1145-1166.
Doucet L. S., Mattielli N., Ionov D. A., Debouge W. and Golovin A. V. (2016) Zn isotopic heterogeneity in the mantle: A melting control? Earth Planet Sci. Lett. 451, 232-240.
Dutkiewicz A., Müller R. D., Cannon J., Vaughan S. and Zahirovic S. (2019) Sequestration and subduction of deep-sea carbonate in the global ocean since the Early Cretaceous. Geology 47, 91-94.
Ducher M., Blanchard M. and Balan E. (2016) Equilibrium zinc isotope fractionation in Zn-bearing minerals from first-principles calculations. Chem. Geol. 443, 83-96.
Fio K., Spangenberg J.E., Vlahović I., Sremac J., Velić I. and Mrinjek E. (2010) Stable isotope and trace element stratigraphy across the Permian-Triassic transition: A redefinition of the boundary in the Velebit Mountain, Croatia. Chem. Geol. 278, 38-57.
Foley S. F. and Fischer T. P. (2017) An essential role for continental rifts and lithosphere in the deep carbon cycle. Nat. Geosci. 10, 897-902.
Frezzotti M. L., Selverstone J., Sharp Z. D. and Compagnoni R. (2011) Carbonate dissolution during subduction revealed by diamond-bearing rocks from the Alps. Nat. Geosci. 4, 703-706.
Fujii T., Moynier F., Abe M., Nemoto K. and Albarède F. (2013) Copper isotope fractionation between aqueous compounds relevant to low temperature geochemistry and biology. Geochim. Cosmochim. Acta 110, 29-44.
Fujii T., Moynier F., Blichert-Toft J. and Albarède F. (2014) Density functional theory estimation of isotope fractionation of Fe, Ni, Cu, and Zn among species relevant to geochemical and biological environments. Geochim. Cosmochim. Acta 140, 553-576.
Fukao Y., Obayashi M. and Nakakuki T. (2009) Stagnant Slab: A Review. Annu. Rev. Earth Planet. Sci. 37, 19-46.
Galvez M. E. and Pubellier M. (2019) How do subduction zones regulate the carbon cycle?. In Deep carbon: Past to present. Cambridge University Press, pp. 276-312.
Gillis K. M. and Coogan L. A. (2011) Secular variation in carbon uptake into the ocean crust. Earth Planet. Sci. Lett. 302, 385-392.
Griffin W. L., Zhang A.-D., O'Reilly S. Y. and Ryan C. G. (1998) In Phanerozoic Evolution of the Lithosphere Beneath the SinoKorean Craton. Amer Geophysical Union, pp. 107-126.
Guo P.-Y., Niu Y.-L, Sun P., Ye L., Liu J.-J, Zhang Y., Feng Y.-X. and Zhao J.-X. (2016) The origin of Cenozoic basalts from central Inner Mongolia, East China: The consequence of recent mantle metasomatism genetically associated with seismically observed paleo-Pacific slab in the mantle transition zone. Lithos 240-243, 104-118.
Guo P., Niu Y., Sun P., Gong H. and Wang X. (2020) Lithosphere thickness controls continental basalt compositions: An illustration using Cenozoic basalts from eastern China. Geology 48, 128-133.
Hauff F., Hoernle K. and Schmidt A. (2003) Sr-Nd-Pb composition of Mesozoic Pacific oceanic crust (Site 1149 and 801, ODP Leg 185): Implications for alteration of ocean crust and the input into the Izu-Bonin-Mariana subduction system. Geochem. Geophys. Geosy. 4, 8913.
Ho K.-S., Liu Y., Chen J.-C. and Yang H.-J. (2008) Elemental and Sr-Nd-Pb isotopic compositions of late Cenozoic Abaga basalts, Inner Mongolia: Implications for petrogenesis and mantle process. Geochem. J. 42, 339-357.
Hofmann A. W. (1997) Mantle geochemistry: The message from oceanic volcanism. Nature 385, 219.
Hong L.-B., Zhang Y.-H., Qian S.-P., Liu J.-Q., Ren Z.-Y. and Xu Y.-G. (2013) Constraints from melt inclusions and their host olivines on the petrogenesis of Oligocene-Early Miocene Xindian basalts, Chifeng area, North China Craton. Contrib. Miner. Petrol. 165, 305-326.
Huang J. and Zhao D. (2006) High-resolution mantle tomography of China and surrounding regions. J. Geophys. Res. 111.
Huang J., Ke S., Gao Y., Xiao Y. and Li S. (2015) Magnesium isotopic compositions of altered oceanic basalts and gabbros from IODP Site 1256 at the East Pacific Rise. Lithos 231, 53-61.
Huang J., Liu S.-A., Gao Y., Xiao Y. and Chen S. (2016) Copper and zinc isotope systematics of altered oceanic crust at IODP Site 1256 in the eastern equatorial Pacific. J. Geophys. Res. Solid Earth 121, 7086-7100.
Huang J., Zhang X.-C., Chen S., Tang L.-M., Wörner G., Yu H.-M. and Huang F. (2018) Zinc isotopic systematics of Kamchatka-Aleutian arc magmas controlled by mantle melting. Geochim. Cosmochim. Acta 238, 85-101.
Huybers P. and Langmuir C. (2009) Feedback between deglaciation, volcanism, and atmospheric CO_2. Earth Planet. Sci. Lett. 286, 479-491.
Ionov D. A., Dupuy C., O'Reilly S. Y., Kopylova M. G. and Genshaft Y. S. (1993) Carbonated peridotite xenoliths from Spitsbergen: implications for trace element signature of mantle carbonate metasomatism. Earth Planet. Sci. Lett. 119, 283-297.
Inglis E. C., Debret B., Burton K. W., Millet M.-A., Pons M.-L., Dale C. W., Bouilhol P., Cooper M., Nowell G. M., McCoy-West A. J. and Williams H. M. (2017) The behavior of iron and zinc stable isotopes accompanying the subduction of mafic oceanic crust:

A case study from Western Alpine ophiolites. Geochem. Geophys. Geosy. 18, 2562-2579.

Jin Q.-Z., Huang J., Liu S.-C. and Huang F. (2020) Magnesium and zinc isotope evidence for recycled sediments and oceanic crust in the mantle sources of continental basalts from eastern China. Lithos 370-371 105627.

Kato C., Moynier F., Valdes M. C., Dhaliwal J. K. and Day J. M.D. (2015) Extensive volatile loss during formation and differentiation of the Moon. Nat. Commun. 6, 7617.

Kelemen P. B. and Manning C. E. (2015) Reevaluating carbon fluxes in subduction zones, what goes down, mostly comes up. Proc. Natl. Acad. Sci. U.S.A. 112, E3997-E4006.

Keppler H. and Wyllie P. J. (1991) Partitioning of Cu, Sn, Mo, W, U, and Th between melt and aqueous fluid in the systems haplogranite-H_2O-HCl and haplogranite-H_2O-HF. Contrib. Mineral. Petrol. 109, 139-150.

Kessel R., Schmidt M. W., Ulmer P. and Pettke T. (2005) Trace element signature of subduction-zone fluids, melts and supercritical liquids at 120-180 km depth. Nature 437, 724-727.

Kunzmann M., Halverson G. P., Sossi P. A., Raub T. D., Payne J. L. and Kirby J. (2013) Zn isotope evidence for immediate resumption of primary productivity after snowball Earth. Geology 41, 27-30.

Kushiro I. (1975) Carbonate-silicate reactions at high pressures and possible presence of dolomite and magnesite in the upper mantle. Earth Planet. Sci. Lett. 28, 116-120.

Li S.-G., Yang W., Ke S., Meng X., Tian H., Xu L., He Y., Huang J., Wang X.-C., Xia Q., Sun W., Yang X., Ren Z.-Y., Wei H., Liu Y., Meng F. and Yan J. (2017) Deep carbon cycles constrained by a large-scale mantle Mg isotope anomaly in eastern China. Natl. Sci. Rev. 4, 111-120.

Liu S.-A., Li D.-D., Li S.-G., Teng F.-Z., Ke S., He Y.-S. and Lu Y.-H. (2014) High-precision copper and iron isotope analysis of igneous rock standards by MC-ICP-MS. J. Anal. At. Spectrom. 29, 122-133.

Liu S.-A., Huang J., Liu J. G., Wörner G., Yang W., Tang Y. J., Chen Y., Tang L.-M., Zheng J.P. and Li S.-G. (2015) Copper isotopic composition of the silicate Earth. Earth Planet. Sci. Lett. 427, 95-103.

Liu S.-A., Liu P.-P., Lv Y., Wang Z.-Z. and Dai J.-G. (2019) Cu and Zn isotope fractionation during oceanic alteration: Implications for Oceanic Cu and Zn cycles. Geochim. Cosmochim. Acta, 257, 191-205.

Liu S.-A., Wang Z.-Z., Li S.-G., Huang J. and Yang W. (2016a) Zinc isotope evidence for a large-scale carbonated mantle beneath eastern China. Earth Planet. Sci. Lett. 444, 169-178.

Liu S.-A., Wu H., Shen S.-Z., Jiang G., Zhang S., Lv Y., Zhang H. and Li S. (2017) Zinc isotope evidence for intensive magmatism immediately before the end-Permian mass extinction. Geology 45, 343-346.

Liu S.-A. and Li S.-G. (2019) Tracing the Deep Carbon Cycle Using Metal Stable Isotopes: Opportunities and Challenges. Engineering 5, 448-457.

Liu S.-A., Wang Z., Yang C., Li S. and Ke S. (2020) Mg and Zn Isotope Evidence for Two Types of Mantle Metasomatism and Deep Recycling of Magnesium Carbonates. J. Geophys. Res. Solid Earth 125.

Liu X. M., Kah L. C., Knoll A. H., Cui H., Kaufman A. J., Shahar A. and Hazen R. M. (2016b) Tracing Earth's O_2 evolution using Zn/Fe ratios in marine carbonates. Geochem. Perspect. Lett. 2, 24-34.

Little S. H., Vance D., Mcmanus J. and Severmann S. (2016) Key role of continental margin sediments in the oceanic mass balance of Zn and Zn isotopes. Geology 44, 207-210.

Maréchal C. N., Télouk P. and Albarède F. (1999) Precise analysis of copper and zinc isotopic compositions by plasma-source mass spectrometry. Chem. Geol. 156, 251-273.

Maréchal C. N., Nicolas E., Douchet C. and Albarède F. (2000) Abundance of zinc isotopes as a marine biogeochemical tracer. Geochem. Geophys. Geosy. 1, 1015.

Marty B., Jambon A. and Sano Y. (1989) Helium isotopes and CO_2 in volcanic gases of Japan. Chem. Geol. 76, 25-40.

Mason E., Edmonds M. and Turchyn A. V. (2017) Remobilization of crustal carbon may dominate volcanic arc emissions. Science 357, 290-294.

Mavromatis V., González A. G., Dietzel M. and Schott J. (2019) Zinc isotope fractionation during the inorganic precipitation of calcite - Towards a new pH proxy. Geochim. Cosmochim. Acta 244, 99-112.

Mazza S.E., Gazel E., Bizimis M., Moucha R., Béguelin P., Johnson E.A., Mcaleer R.J. and Sobolev A.V. (2019) Sampling the volatile-rich transition zone beneath Bermuda. Nature 569, 398-403.

McCoy-West A. J., Fitton J. G., Pons M.-L., Inglis E. C. and Williams H. M. (2018) The Fe and Zn isotope composition of deep mantle source regions: Insights from Baffin Island picrites. Geochim. Cosmochim. Acta 238, 542-562.

Meyer E.E., Quicksall A.N., Landis J.D., Link P.K. and Bostick B.C. (2012) Trace and rare earth elemental investigation of a Sturtian cap carbonate, Pocatello, Idaho: Evidence for ocean redox conditions before and during carbonate deposition. Precambrian Res. 192-195, 89-106.

Moynier F., Vance D., Fujii T. and Savage P. (2017) The Isotope Geochemistry of Zinc and Copper. Rev. Mineral. Geochem. 82, 543-600.

Müller R. D., Sdrolias M., Gaina C. and Roest W. R. (2008a) Age, spreading rates, and spreading asymmetry of the world's ocean crust. Geochem. Geophys. Geosy. 9, Q04006.

Müller R. D., Sdrolias M., Gaina C., Steinberger B. and Heine C. (2008b) Long-Term Sea-Level Fluctuations Driven by Ocean Basin Dynamics. Science 319, 1357-1362.

Pan D., Spanu L., Harrison B., Sverjensky D. A. and Galli G. (2013) Dielectric properties of water under extreme conditions and transport of carbonates in the deep Earth. Proc. Natl. Acad. Sci. U.S.A. 110, 6646-6650.

Pang C.-J., Wang X.-C., Li C.-F., Wilde S. A. and Tian L. (2019) Pyroxenite-derived Cenozoic basaltic magmatism in central Inner Mongolia, eastern China: Potential contributions from the subduction of the Paleo-Pacific and Paleo-Asian oceanic slabs in the Mantle Transition Zone. Lithos 332-333, 39-54.

Paniello R.C., Day J.M.D. and Moynier F. (2012) Zinc isotopic evidence for the origin of the Moon. Nature 490, 376-379.

Pichat S., Douchet C. and Albarède F. (2003) Zinc isotope variations in deep-sea carbonates from the eastern equatorial Pacific over the last 175 ka. Earth Planet. Sci. Lett. 210, 167-178.

Plank T. and Manning C. E. (2019) Subducting carbon. Nature 574, 343-352.

Pons M.L., Debret B., Bouilhol P., Delacour A. and Williams H. (2016) Zinc isotope evidence for sulfate-rich fluid transfer across subduction zones. Nat. Commun. 7, 13794.

Ridgwell A. and Zeebe R. (2005) The role of the global carbonate cycle in the regulation and evolution of the Earth system. Earth Planet. Sci. Lett. 234, 299-315.

Sonke J., Sivry Y., Viers J., Freydier R., Dejonghe L., Andre L., Aggarwal J., Fontan F. and Dupre B. (2008) Historical variations in the isotopic composition of atmospheric zinc deposition from a zinc smelter. Chem. Geol. 252, 145-157.

Sossi P. A., Halverson G. P., Nebel O. and Eggins S. M. (2015) Combined Separation of Cu, Fe and Zn from Rock Matrices and Improved Analytical Protocols for Stable Isotope Determination. Geostand. Geoanal. Res. 39, 129-149.

Sossi P. A., Nebel O., O'Neill H. S. C. and Moynier F. (2018) Zinc isotope composition of the Earth and its behaviour during planetary accretion. Chem. Geol. 477, 73-84.

Staudigel H. and Hart S. R. (1983) Alteration of basaltic glass: Mechanisms and significance for the oceanic crust-seawater budget. Geochim. Cosmochim. Acta 47, 337-350.

Staudigel H., Davies G. R., Hart S. R., Marchant K. M. and Smith B. M. (1995) Large scale isotopic Sr, Nd and O isotopic anatomy of altered oceanic crust: DSDP/ODP sites 417/418. Earth Planet. Sci. Lett. 130, 169-185.

Sweere T. C., Dickson A. J., Jenkyns H. C., Porcelli D., Elrick M., Van Den Boorn S. H. J. M. and Henderson G. M. (2018) Isotopic evidence for changes in the zinc cycle during Oceanic Anoxic Event 2 (Late Cretaceous). Geology 46, 463-466.

Sun Y., Teng F.-Z. and Pang K.-N. (2021) The presence of paleo-Pacific slab beneath northwest North China Craton hinted by low-δ^{26}Mg basalts at Wulanhada. Lithos 386-387 106009.

Thomson A. R., Walter M. J., Kohn S. C. and Brooker R. A. (2016) Slab melting as a barrier to deep carbon subduction. Nature 529, 76-79.

Thy P. (1983) Spinel minerals in transitional and alkali basaltic glasses from Iceland. Contrib. Miner. Petrol. 83, 141-149.

Togtokh K., Miao L., Zhang F., Baatar M., Anaad C. and Bars A. (2019) Major, trace element, and Sr-Nd isotopic geochemistry of Cenozoic basalts in Central-North and East Mongolia: Petrogenesis and tectonic implication. Geol. J. 54, 3660-3680.

Toutain J.-P., Sonke J., Munoz M., Nonell A., Polvé M., Viers J., Freydier R., Sortino F., Joron J.-L. and Sumarti S. (2008) Evidence for Zn isotopic fractionation at Merapi volcano. Chem. Geol. 253, 74-82.

Turekian K. K. and Wedepohl K. H. (1961) Distribution of the Elements in Some Major Units of the Earth's Crust. Geol. Soc. Am. Bull 72, 175-192.

Vance D., Little S. H., Archer C., Cameron V., Andersen M. B., Rijkenberg M. J. A. and Lyons T. W. (2016) The oceanic budgets of nickel and zinc isotopes: the importance of sulfidic environments as illustrated by the Black Sea. Phil. Trans. R. Soc. A 374 (2081), 20150294.

Wan B., Li S.-H., Xiao W.-J. and Windley B.F. (2018) Where and when did the Paleo-Asian ocean form? Precambrian Res. 317, 241-252.

Wang X.-C., Wilde S. A., Li Q.-L., and Yang Y.-N. (2015) Continental flood basalts derived from the hydrous mantle transition zone.

Nat. Commun. 6, 7700.

Wang Z.-Z., Liu S.-A., Liu J., Huang J., Xiao Y., Chu Z.-Y., Zhao X.-M., and Tang L. (2017) Zinc isotope fractionation during mantle melting and constraints on the Zn isotope composition of Earth's upper mantle. Geochim. Cosmochim. Acta 198, 151-167.

Wang Z.-Z., Liu S.-A., Chen L.-H., Li S.-G. and Zeng G. (2018) Compositional transition in natural alkaline lavas through silica-undersaturated melt-lithosphere interaction. Geology 46, 771-774.

Wang Z.-Z. and Liu S.-A. (2021) Evolution of intraplate alkaline to tholeiitic basalts via interaction between carbonated melt and lithospheric mantle. J. Petrol. 62(4), egab025.

Xiao W.-J., Windley B. F., Hao J. and Zhai M. (2003) Accretion leading to collision and the Permian Solonker suture, Inner Mongolia, China: Termination of the central Asian orogenic belt. Tectonics 22, 1069.

Xu Y.-G. (2007) Diachronous lithospheric thinning of the North China Craton and formation of the Daxin'anling-Taihangshan gravity lineament. Lithos 96, 281-298.

Xu Y.-G., Li H., Hong L.-Y., Ma L.-B., Ma Q. and Sun M.-D. (2018) Generation of Cenozoic intraplate basalts in the big mantle wedge under eastern Asia. Sci. China Earth Sci. 61, 869-886.

Yang C., Liu S.-A., Zhang L., Wang Z.-Z., Liu P.-P. and Li S.-G. (2021) Zinc isotope fractionation between Cr-spinel and olivine and its implications for chromite crystallization during magma differentiation. Geochim. Cosmochim. Acta 313, 277-294.

Yaxley G.-M. and Green D.-H. (1994) Experimental demonstration of refractory carbonate-bearing eclogite and siliceous melt in the subduction regime. Earth Planet. Sci. Lett. 128, 313-325.

Zajacz Z., Halter W. E., Pettke T. and Guillong M. (2008) Determination of fluid/melt partition coefficients by LA-ICP-MS analysis of co-existing fluid and silicate melt inclusions: Controls on element partitioning. Geochim. Cosmochim. Acta 72, 2169-2197.

Zeng G., Chen L.-H., Xu X.-S., Jiang S.-Y. and Hofmann A.W. (2010) Carbonated mantle sources for Cenozoic intra-plate alkaline basalts in Shandong, North China. Chem. Geol. 273, 35-45.

Zhang M.-L. and Guo Z.-F. (2016) Origin of Late Cenozoic Abaga-Dalinuoer basalts, eastern China: Implications for a mixed pyroxenite-peridotite source related with deep subduction of the Pacific slab. Gondwana Res. 37, 130-151.

Zhang T., Sun R., Liu Y., Chen L., Zheng W., Liu C.-Q. and Chen J. (2022) Copper and Zinc isotope signatures in scleratinian corals: Implications for Cu and Zn cycling in modern and ancient ocean. Geochim. Cosmochim. Acta 317, 395-408.

Zhao G., Wang Y., Huang B., Dong Y., Li S., Zhang G. and Yu S. (2018) Geological reconstructions of the East Asian blocks: From the breakup of Rodinia to the assembly of Pangea. Earth Sci. Rev. 186, 262-286.

Zhao M.-Y. and Zheng Y.-F. (2014) Marine carbonate records of terrigenous input into Paleotethyan seawater: Geochemical constraints from Carboniferous limestones. Geochim. Cosmochim. Acta 141, 508-531.

Zhu R., Chen L., Wu F. and Liu J. (2011) Timing, scale and mechanism of the destruction of the North China Craton. Sci. China Earth Sci. 54, 789-797.

Zou D., Liu Y., Hu Z., Gao S., Zong K., Xu R., Deng L., He D. and Gao C. (2014) Pyroxenite and peridotite xenoliths from Hexigten, Inner Mongolia: Insights into the Paleo-Asian Ocean subduction-related melt/fluid-peridotite interaction. Geochim. Cosmochim. Acta 140, 435-454.

The fate of subducting carbon tracked by Mg and Zn isotopes: A review and new perspectives *

Sheng-Ao Liu[1], Yuan-Ru Qu[1], Ze-Zhou Wang[1,2], Meng-Lun Li[1], Chun Yang[1,3] and Shu-Guang Li[1,4]

1. State Key Laboratory of Geological Processes and Mineral Resources, China University of Geosciences, Beijing 100083, China
2. Isotope Laboratory, Department of Earth and Space Sciences, University of Washington, Seattle, WA 98195, USA
3. Institut für Planetologie, University of Münster, Wilhelm-Klemm-Straße 10, Münster, 48149, Germany
4. CAS Key Laboratory of Crust-Mantle Materials and Environments, School of Earth and Space Sciences, University of Science and Technology of China, Hefei 230026, China

亮点介绍： 示踪俯冲碳的命运对于理解地球历史上的全球碳循环和气候变化至关重要。镁、锌同位素在沉积碳酸盐与地幔之间的组成存在明显差异，被认为是俯冲碳酸盐的有效示踪剂，这得到了地质样品验证：全球岛弧玄武岩缺乏低δ^{26}Mg和高δ^{66}Zn异常，表明弧下地幔以富钙碳酸盐溶解为主，与实验和理论预测一致；大陆岩石圈地幔的超镁铁质包体和玄武岩普遍具有低δ^{26}Mg和高δ^{66}Zn，表明岩石圈地幔是一个重要的俯冲碳储库；低δ^{26}Mg和高δ^{66}Zn的板内碱性玄武岩分布通常与地幔过渡带滞留板片重合，表明地幔过渡带是另一个以富镁碳酸盐为主的全球性俯冲碳储库。

Abstract Tracking the final fate of subducting carbon is crucial to understanding global carbon cycles and climate changes in the history of the Earth. Available geochemical tracers such as carbon isotopes are apt to identify recycled organic carbon but usually insufficient to discriminate between primordial carbon in the mantle and carbon derived from recycled carbonate sediments. In the past decade, magnesium and zinc isotope systematics have been proposed as novel proxies for subducting carbon owing to the noticeable isotopic offsets between carbonate sediments and the mantle (i.e., $\delta^{26}Mg_{carbonate}<\delta^{26}Mg_{mantle}$; $\delta^{66}Zn_{carbonate}>\delta^{66}Zn_{mantle}$). Nonetheless, isotopic effects induced by subduction-zone processes and crystal-melt differentiation may obscure the information of Mg and Zn isotopic compositions of mantle-derived magmas. In this paper we firstly discuss how these processes modify the Mg and Zn isotopic systematics of mantle-derived magmas. Based on the fact that different carbonate species (calcite, dolomite, and magnesite) possess distinct Mg and Zn contents and their stabilities in subduction zones vary with pressure, we then develop the two isotope systematics as tools to track the final storage depth of subducting carbon. We test this application by collating available Mg and Zn isotopic compositions of ultramafic xenoliths and basaltic lavas sourced from various mantle

* 本文发表在：Earth Science Review, 2022, 228: 104010

depths. The lack of light Mg and heavy Zn isotopic anomalies of global arc lavas supports experimental and theoretical prediction that the dissolved carbonate species in the sub-arc mantle–if any–is dominated by calcium-rich carbonate. The findings of pervasive low-δ^{26}Mg and high-δ^{66}Zn ultramafic xenoliths and basaltic lavas sourced from the sub-continental lithospheric mantle (SCLM) suggest that the SCLM is an important storage of subducting carbon via metasomatism by dolomite that can be substantially dissolved by supercritical fluids at depths of >160 km. Intraplate alkali basalts with low δ^{26}Mg and high δ^{66}Zn are commonly restricted to the regions with stagnant slabs at depths of ~410–660 km, suggesting that the mantle transition zone is another global storage of subducting carbon composed mainly of Mg-rich carbonates. Overall, observations on mantle-derived rocks, with Mg and Zn isotopes as the tracers, indicate that a significant flux of Earth's surface carbon has survived the arc regime and been recycled into the deeper mantle. Future studies that explore a quantitative relationship between Mg-Zn isotopic ratios and the flux of subducting carbon will further promote the application of the paired —isotopic proxies.

Keywords Deep carbon cycles, subducted carbonates, storage depths, Mg isotopes, Zn isotopes

1 Introduction

The presence of carbon and other life-essential volatile elements in the surface environment builds the habitability of Earth, the only life-harboring planet in the inner Solar System. Carbon dioxide (CO_2) is the dominant greenhouse gas for most of Earth's history, and thus understanding long-term climate changes through geologic time requires an understanding of mechanisms that control atmospheric partial pressure of CO_2 (pCO_2) (Kasting and Ackerman, 1986; Marshall et al., 1988). The surface constitutes only ~10% of Earth's carbon with the remaining 90% partitioned between the core, mantle, and crust; thus, the net flux of carbon between the Earth's interior and exterior influences the distribution of carbon at the surface, which in turn affects atmospheric pCO_2 and surface temperature on million-year timescales (Dasgupta and Hirschmann, 2010; Orcutt et al., 2019). Downgoing slabs in subduction zones remove carbon from the surface reservoirs (ocean, atmosphere, and biosphere) through an intermediate step of silicate weathering (Walker et al., 1981; Orcutt et al., 2019). On the other hand, carbon in the solid Earth returns to the atmosphere via volcanic outgassing or diffuse degassing at oceanic ridge, arc, rift, and intraplate areas (Foley and Fischer, 2017; Galvez and Pubellier, 2019; Plank and Manning, 2019). The extraction efficiency of carbon from Earth's interior is linked to its temperature, pressure and oxygen fugacity (fO_2) through time (Rohrbach and Schmidt, 2011; Stagno et al., 2013, 2019), which in turn affects the total budget of carbon outgassing.

The final storage depth of subducting carbon in the mantle is one of the key issues to understanding global deep carbon cycles. It not only bears on the global flux estimate of carbon input via subduction but also is a first-order control on the timescales of return of subducting carbon into the surface primarily due to different migration distances. For example, the circulation duration of carbon between the mantle transition zone (MTZ) and the surface is estimated to be at least tens of millions of years or longer (Li et al., 2017; Liu et al., 2020), which is considerably longer than that in arc systems (10–5 Ma, Kelemen and Manning, 2015) as a consequence of much longer migration distance of carbon in the former circulation system. If carbon is stored in the deeper mantle, the residence time is at least near 1 Ga and perhaps as long as 4.6 Ga (Dasgupta and Hirschmann, 2010). The final fate of global subducting carbon, however, substantially varies among studies adopting different approaches. Experimental studies show that carbonates are difficult to partially melt at the *P-T* conditions of subduction zones, and thus predict more than half of the subducted carbon to have been brought into the convecting mantle (Dasgupta and Hirschmann, 2010). Thermodynamic modellings, however, predict high fluid solubility of $CaCO_3$ under subduction zone conditions and therefore propose the sub-arc mantle to be the main storage of subducting carbon (Caciagli and Manning, 2003; Kelemen and Manning, 2015). The high arc CO_2 fluxes also imply that a substantial amount of subducting carbonates have broken down at sub-arc depths (e.g., Sano and Marty, 1995; De Leeuw et al., 2007). In this respect, fluid-mediated carbonate dissolution acts as an alternative approach to transferring carbon from subducting slabs into the sub-arc mantle (Frezzotti et al., 2011; Manning et al., 2013; Ague and Nicolescu, 2014). More recent estimates, however, suggest that arc outputs represent a lower proportion (~20%) of the inputs (Plank and Manning, 2019), roughly half of that estimated by Kelemen and Manning (2015). Thus, the input flux of subducting carbonates to the sub-arc mantle still remains to be constrained. Minimal amount of carbon was predicted to have entered the lower mantle owing to the abrupt decline of solidus of C-bearing oceanic crust (i.e., carbonated eclogite) at depths of ~400 km (Thomson et al., 2016). However, a deep storage of subducting carbonates at the lower part of the MTZ (>580 km) or even the lower mantle (>670 km) has been suggested on the basis of investigations on mineral inclusions within super-deep diamonds or garnets (e.g., Brenker et al., 2007; Anzolini et al., 2019). In any case, each conceptual model that predicts the final fate of subducting carbon needs to be tested or supported by observations on natural samples.

The carbon that enters subduction zones is mainly stored in the form of carbonate sediments (~70%), with the remainder (~30%) consisting of organic carbon, altered oceanic crust and peridotite that are exposed to alteration by seawater and contain carbonate veins (Clift, 2017; Plank and Manning, 2019). To date, petrological or geochemical proxies that are able to identify recycled surface carbon in the mantle are still scarce. Purely elemental proxies (e.g., CaO/Al_2O_3, Ti/Eu) cannot easily distinguish whether the metasomatic agent is of mantle origin (i.e., primary carbonatitic melt) or derived from recycled crustal carbonate. Carbon isotopes can distinguish organic carbon from inorganic carbon, and recycled organic carbon in the mantle has been distinctly documented by extremely negative carbon isotopic compositions ($\delta^{13}C$; low to −43‰) of diamond inclusions in some deeply derived rocks (e.g., Shirey et al., 2013). Subducted carbonate sediment, however, is commonly difficult to be distinguished by carbon isotopes owing to the modest difference of carbon isotopic composition between carbonate and primordial carbon in the mantle, both of which are chiefly inorganic carbon (Deines, 2002). More importantly, strong carbon isotopic fractionation accompanies magma degassing (e.g., Aubaud et al., 2005). This makes it difficult to obtain the carbon isotopic composition of the mantle sources of basaltic lavas which commonly undergo extensive CO_2 degassing owing to the rather low carbon solubility in basaltic melts (Dixon et al., 1995). Additional geochemical tracers, which are sensitive to subducting carbonates but limitedly affected by magma degassing (e.g., stable isotopes of metals), are required to supplement the application of carbon isotopes as proxies for subducting carbon.

In the past decade, the stable isotope systematics of divalent metals that widely occur as major or trace components in marine carbonates, such as magnesium (expressed as $\delta^{26}Mg$ against the Dead Sea Magnesium, DSM3, in per mil) and zinc (expressed as $\delta^{66}Zn$ against the Zn standard JMC 3-0749L in per mil), have been proposed to be promising tracers for subducting carbon given the remarkable difference in $\delta^{26}Mg$ and $\delta^{66}Zn$ between sedimentary carbonates and the mantle (e.g., for Mg, Yang et al., 2012; Li et al., 2017; Liu and Li, 2019; for Zn, Liu et al., 2016a, 2020; Liu and Li, 2019; Beunon et al., 2020). Yet, more and more studies reveal that some geological processes tend to obscure the information of Mg and Zn isotopic compositions of the subducting carbonates and mantle-derived rocks or magmas, especially the isotopic effects induced by partial carbonate dissolution and carbonate-silicate interaction in subduction zones, fractional crystallization of Fe-Cr-Ti oxides in the magma chamber, and chemical or thermal diffusion in the sources or during magma descending. Besides, some other subduction-zone fluids (e.g., sulfates) can also have high δ^{66}Zn, which are necessarily taken into account before linking the high δ^{66}Zn of basaltic magmas to a major role of recycled carbonates in the sources.

In this review, we firstly discuss the variation of Mg and Zn isotopic compositions of marine carbonates with geologic time and their isotopic differences with the terrestrial mantle. Then, we discuss how and to what extents these geological processes mentioned above affect the application of Mg and Zn isotopes as geochemical probes of subducting carbon. The influence of diffusion-driven isotopic effect on the Mg and Zn isotopic systematics and its bearing on the application to deep carbon cycling have been discussed in the papers of Liu and Li (2019) and Liu et al. (2020). More importantly, here we discuss the stabilities and Mg-Zn abundances of different carbonate species with an emphasis on their tremendous influences on Mg-Zn isotopic composition of the mantle and specifically with a focus on the application of Mg and Zn isotopes as novel proxies of the final storage depths of subducting carbon. To our knowledge, this is the first time to link the Mg and Zn isotopic systematics of mantle-derived rocks to various depths of subducting carbon in the mantle. Finally, we test this application by collating available Mg and Zn isotopic dataset for basaltic lavas and mantle xenoliths worldwide sourced from variable mantle depths. The results demonstrate that the Mg and Zn isotopic systematics of natural samples could provide valuable insights into the final fate of subducting carbon and help examine the conceptual models already proposed for deep carbon cycles.

2 Mg and Zn isotopic systematics of marine carbonates and comparison with the terrestrial mantle

Magnesium isotopic compositions of marine carbonates are highly variable and display a strong mineralogical control, with the median and mean δ^{26}Mg varying from −4‰ for limestones (Wombacher et al., 2011) to −2‰ for dolostones (e.g., Pokrovsky et al., 2011). Other factors such as temperature, precipitation rate, Mg isotopic composition of fluids, kinetic isotope fractionation, vital effects and diagenesis play a subordinate role in the Mg isotopic heterogeneity of marine carbonates (see the review by Teng, 2017). Notably, there is no systematic δ^{26}Mg variation with the formation time for marine carbonates and ancient (e.g., Proterozoic) carbonates also have extremely low δ^{26}Mg values (Fig. 1a). This is essential for magnesium isotopes to be an effective proxy for subducting carbonates in both modern and ancient subduction zones. In this respect, calcium isotopes face substantial challenges because calcium isotopic compositions of marine carbonates greatly varied through time and are either lighter, or heavier, than that of the terrestrial mantle (e.g., Gussone et al., 2020). Although different species of carbonates are distinct in terms of Mg isotopic composition, all of them are isotopically much lighter than the terrestrial mantle (δ^{26}Mg= −0.25±0.04‰) by about −3‰ to −1.5‰ on average (Fig. 1a). The Mg isotopic discrepancy of different carbonate species could result in different isotopic signatures of the mantle metasomatized by recycled carbonates (see below). Notably, siliciclastic sediments also span a broad range of δ^{26}Mg but those with lighter-than-mantle δ^{26}Mg commonly contain a significant portion of carbonates obtained from leaching experiments of bulk sediments (Wang et al., 2015; Qu et al., 2022). Carbonate-free sediments are isotopically similar to or heavier (δ^{26}Mg up to +0.92‰; Li et al., 2010) than the mantle (Fig. 1a). Thus, Mg isotopes can efficiently discriminate between recycled carbonate sediments and recycled carbonate-poor siliciclastic sediments, for which other isotopic systematics such as oxygen isotopes are usually difficult to discriminate.

Marine carbonates also have variable Zn isotopic compositions but exhibit indistinct mineralogical control

(Fig. 1b). Some Neoproterozoic cap carbonates (mainly dolostones) seem to have systematically lower δ^{66}Zn values in comparison with younger carbonates (Fig. 1b), which is most likely attributed to post-depositional diagenesis or resetting by hydrothermal fluids given the positive correlation of δ^{66}Zn with δ^{18}O and negative correlation with ^{87}Sr/^{86}Sr ratios (Lv et al., 2018). Excluding the samples subject to diagenesis, the Zn isotopic compositions of marine carbonates do not correlate with ages (Fig. 1b). The Zn isotopic heterogeneity of marine carbonates is mainly reliant on variation in seawater δ^{66}Zn value (Pichat et al., 2003; Kunzmann et al., 2013; Liu et al., 2017; Sweere et al., 2018), post-depositional diagenesis (Lv et al., 2018), temperature, and/or pH value

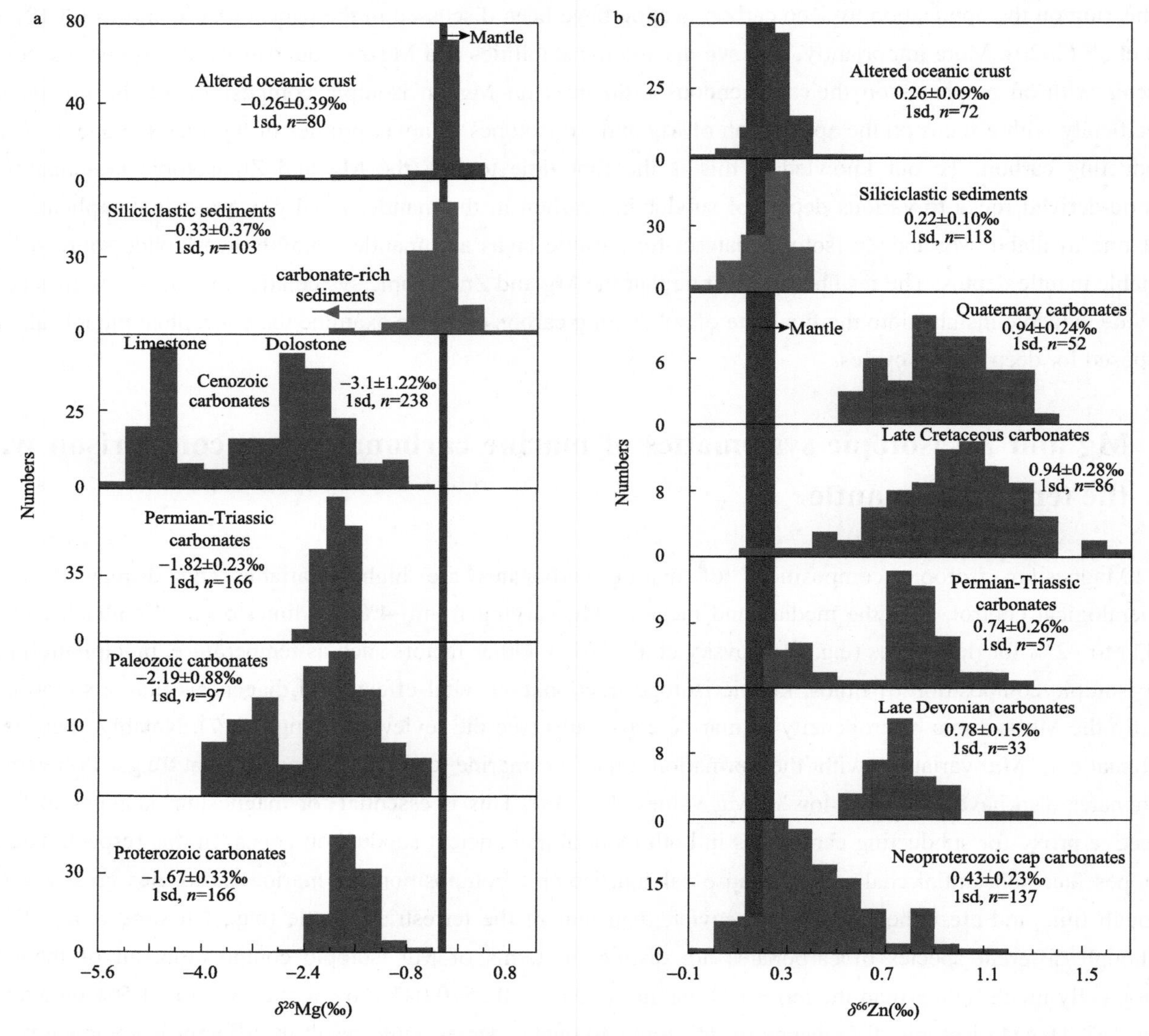

Fig. 1 A compilation of Mg and Zn isotopic compositions of altered oceanic curst (AOC), siliciclastic sediments and carbonates with various ages. Data sources: AOC (δ^{26}Mg: Huang et al., 2015b, 2018c; Zhong et al., 2017; δ^{66}Zn: Huang et al., 2016; Inglis et al., 2017), siliciclastic sediments (δ^{26}Mg: Li et al., 2010; Huang et al., 2013; Wimpenny et al., 2014; Wang et al., 2015; Qu et al., 2022; δ^{66}Zn: Maréchal et al., 2000; Sonke et al., 2008; Little et al., 2016; Vance et al., 2016; Qu et al., 2022), carbonates (δ^{26}Mg: Pokrovsky et al., 2011; Wombacher et al., 2011; Higgins and Schrag, 2012, 2015; Geske et al., 2012; Azmy et al., 2013; Liu et al., 2014; Kasemann et al., 2014; Mavromatis et al., 2014; Fantle and Higgins, 2014; Lavoie et al., 2014; Blättler et al., 2015; Huang et al., 2015c; Peng et al., 2016; Ma et al., 2018; Hu et al., 2021; δ^{66}Zn: Pichat et al., 2003; Kunzmann et al., 2013; Liu et al., 2017; John et al., 2017; Sweere et al., 2018; Wang et al., 2018c; Lv et al., 2018; Chen et al., 2021). The dark bar denotes the mantle's Mg and Zn isotopic compositions (δ^{26}Mg= −0.25±0.04‰, Teng, 2017; δ^{66}Zn=0.18±0.05‰; Wang et al., 2017; Sossi et al., 2018; McCoy-West et al., 2018; Liu et al., 2019).

(Mavromatis et al., 2019) at which the carbonates precipitated. Despite the large range of δ^{66}Zn, the majority of marine carbonates have heavier Zn isotopic composition than that of the mantle (Fig. 1b). For instance, δ^{66}Zn of Cenozoic carbonates (0.94±0.24‰; Pichat et al., 2003) are higher by up to ~0.8‰ on average than that of the mantle (0.18±0.05‰, 2sd; Wang et al., 2017; Sossi et al., 2018; McCoy-West et al., 2018; Liu et al., 2019). Siliciclastic sediments display relatively uniform Zn isotopic compositions (Fig. 1b), which are similar to, or lighter than, the igneous rock average (~0.28‰, Moynier et al., 2017). In particular, organic-rich margin sediments possess δ^{66}Zn values of low to −0.05‰ (Little et al., 2016), which are significantly lower than those of marine carbonates. Overall, given the large isotopic offset between marine carbonate and silicate reservoirs in the Earth (Fig. 1a, b), Mg and Zn isotopes are sensitive to the recycled carbonates in the mantle at various geologic periods. That is, lighter Mg or heavier Zn isotopic compositions of mantle rocks or mantle-derived melts relative to the pristine mantle, if any, may point to the incorporation of marine carbonates into the mantle via subduction, before isotopic effects induced by subduction and magmatic processes can be excluded explicitly.

3 Mg and Zn isotope fractionation in subduction and magmatic processes and its consequences

3.1 Subduction-zone processes

Metamorphic dehydration, carbonate dissolution, and silicate-carbonate reaction occur extensively during subduction (e.g., Kerrick and Connolly, 2001; Frezzotti et al., 2011; Ague and Nicolescu, 2014; Kelemen and Manning, 2015). Because carbon is primarily delivered into the mantle via slab subduction, the behavior of Mg and Zn isotopes during subduction and dehydration processes, particularly during partial carbonate dissolution and carbonate-silicate interaction in the slab, is critical for them to be used as proxies for subducting carbon. The Mg isotopic compositions of subduction-zone rocks experiencing diverse degrees of dehydration and metamorphism are indistinguishable from those of their protoliths within uncertainties, implying that Mg isotopes do not fractionate during mafic crust subduction (Fig. 3a; Wang et al., 2014a). This is chiefly attributed to the fairly low proportion of Mg released during dehydration of mafic rocks. However, carbonate-silicate interaction during subduction could lead to significant Mg isotopic exchange between them, with carbonate components (marbles) gaining heavier Mg relative to their protoliths and carbonated eclogites becoming isotopically lighter (Wang et al., 2014b). Notably, carbonates that interacted with silicates in the subducted crust still have much lighter Mg isotopic composition than that of the mantle (Wang et al., 2014b). In other words, the lighter-than-mantle Mg isotopic signatures of marine carbonates can be preserved after subduction and interaction with silicates in the slabs.

Similar to Mg isotopes, zinc isotope fractionation is also restricted during prograde metamorphism and dehydration of mafic crustal rocks (Fig. 3b; Inglis et al., 2017; Xu et al., 2021). This is attributed to the small variation in Zn contents of subducted rocks indicating limited Zn loss during subduction. However, CO_3^{2-}-rich fluids released during subduction of altered ultramafic rocks preferentially incorporate isotopically heavy Zn, which was documented in some subducted serpentinites with low δ^{66}Zn (Fig. 3b; e.g., Pons et al., 2011; Debret et al., 2018) and Chl-harzburgites with high δ^{66}Zn and Zn contents (Debret et al., 2021). The *ab initio* calculations show that Zn-carbonate is isotopically heavier relative to hydrated Zn^{2+} and Zn sulfide species (Fujii et al., 2011; Ducher et al., 2016). Therefore, potential Zn isotope fractionation during carbonate dissolution in subduction zones must be considered while applying Zn isotopes as a proxy for subducting carbon in the mantle. A recent study carried out leaching experiments on subducting sediments (Qu et al., 2022). Bulk carbonate-bearing and carbonate-free sediments have similar δ^{66}Zn, but the former display lower δ^{26}Mg than the latter, indicating that the recycling of bulk sediments (i.e., no carbonate dissolution) produces low δ^{26}Mg but normal δ^{66}Zn (i.e., decoupling) in mantle-derived magmas. By contrast, dissolved carbonates have higher δ^{66}Zn and lower δ^{26}Mg compared with the silicate residues and the mantle. This means that the mantle-derived magmas will possess a high δ^{66}Zn and low δ^{26}Mg signature (so-called "coupling") if carbonate dissolution occurs during subduction. Therefore, the dissolution of carbonates in subducting sediments can be traced by using a combination of Mg and Zn isotopes (Qu et al., 2022).

In addition to carbonate, distinctive zinc isotopic effect also occurs when zinc is complexed with sulfate ligand. The two components (sulfate- and carbonate-fluids) are commonly considered as the most important oxidizing (high-fO_2) agents to modify the redox state of the mantle wedge (e.g., Evans, 2012). Theoretical calculation predicts that zinc-sulfate complexes preferentially incorporate heavier Zn isotopes relative to sulfides (Black et al., 2011; Ducher et al., 2016). Release of isotopically heavy Zn by sulfate-rich fluids has been observed during dehydration of altered ultramafic rocks (hollow symbols in Fig. 3b; Pons et al., 2016). Because Zn isotopes are fractionated in the same direction during C and S complexation, sulfate-rich fluid is necessary to be considered before attributing the high δ^{66}Zn of subduction-related rocks/magmas to a dominant role of carbonate fluid. In this case, an addition analysis of sulfur contents in the rocks or primary magmas is perhaps required. Alternatively, copper (Cu) isotopes may be considered as an additional tool to identify the role of sulfate fluid (Liu et al., 2022), because Cu-sulfate is also strongly enriched in heavy Cu isotopes relative to sulfides (Fujii et al., 2013). Liu et al. (2022) showed that the incorporation of carbonate to the mantle remarkably increases its Zn isotopic composition but leaves its Cu isotopic composition significantly unchanged due to the usually low Cu contents of marine carbonates; by contrast, metasomatism by sulfate-rich fluids increases both Cu and Zn isotopic compositions of the mantle. Thus, high δ^{66}Zn but mantle-like δ^{65}Cu of basaltic magmas are indicative of a dominant role of recycled carbonates in the sources.

3.2 Partial melting, magmatic differentiation, and degassing

Due to the difficulty of direct sampling of the mantle, ultramafic xenoliths and basaltic lavas provide the unique way of probing the footprint of subducting carbonates in the mantle. Prior to utilizing Mg and Zn isotope systematics of basaltic lavas to characterize the source heterogeneity, isotopic effects induced by magmatic processes such as partial melting, magmatic differentiation and degassing should be considered. Isotopic fractionation during melt extraction and magmatic differentiation is mainly controlled by isotopic partitioning between mineral and melt. Here we collated Mg and Zn isotopic data for major Mg- and Zn-hosting minerals in peridotites and basalts (Fig. 4a). Large isotopic fractionation exists between silicate minerals and oxides or garnet in mantle peridotites and basalts. It is proposed that the preferential melting of isotopically light garnet may cause low-degree partial melts to possess slightly lower δ^{26}Mg relative to source peridotite (Zhong et al., 2017), but the Mg isotopic offset between melt and residual peridotite would become diluted as melt fractions increase. Thus, Mg isotope fractionation during mantle partial melting and basaltic magma differentiation involving olivine and/or pyroxene is typically less than 0.07‰ (Teng et al., 2010; Zhong et al., 2017; Stracke et al., 2018). For Zn isotopes, partial melting of mantle peridotite leads to a slightly heavier isotopic composition for the melt relative to the residue by ~0.04‰ to 0.12‰ (Doucet et al., 2016; Wang et al., 2017; Sossi et al., 2018; McCoy-West et al., 2018). Even if one considers a carbonated source lithology, the Mg and Zn isotopic effects induced by preferential melting of carbonate at high temperatures (e.g., 1573 K) are still limited (Liu et al., 2020). Differentiated melts tend to have heavier Zn isotopic composition by up to ~0.1‰ as a result of olivine crystallization (Chen et al.,

2013; Moynier et al., 2017), given the fact that olivine phenocrysts are isotopically lighter than the hosting melts by ~0.1‰ to 0.4‰ (McCoy-West et al., 2018; Yang and Liu, 2019; Fig. 4b). Overall, the isotopic effect induced by mantle partial melting and magma differentiation involving silicate phases is limited and much smaller in comparison with the isotopic offset between marine carbonates and the mantle (Fig. 1).

Isotopic effect induced by fractional crystallization of oxides in magma systems attracts particular attention in recent years. Both Al-spinels and Cr-spinels are enriched in heavier Mg isotopes relative to silicate minerals by ~0.2‰ to 0.5‰ (Fig. 4a). Titanomagnetite also has heavier Mg isotopic composition than that of the melt in equilibrium (Fig. 4a). Thus, fractional crystallization of these Mg-bearing oxides (e.g., chromite and titanomagnetite) can drive the residual magma toward lighter Mg isotopic composition (Su et al., 2019; Wang et al., 2021a), which must be considered when interpreting Mg isotopic data of differentiated lavas saturated with Fe–Ti–Cr oxides. The effect of oxide crystallization may be evaluated by plotting Mg isotopic ratios against the contents of MgO or Cr + Fe of the magmas. In comparison with Mg, Zn isotopic systematic of oxides such as spinels is more complicated because it heavily depends on the diverse chemical compositions of spinels. Aluminum-spinel is isotopically heavier than coexisting silicates in ultramafic rocks (Wang et al., 2017) whereas chromium-spinel is isotopically lighter (Yang et al., 2021; Fig. 4b). Zinc has a much higher partition coefficient in spinel-group minerals ($D_{\text{spl-melt}}$=5.3; Davis et al., 2013) than that in silicate minerals such as olivine ($D_{\text{ol-melt}}$=~1; Le Roux et al., 2010). As a result, fractional crystallization of Cr-spinel and Al-spinel leads to negative or positive correlation between Zn concentration and δ^{66}Zn in the melt, respectively, which could be utilized to evaluate the effect of oxide crystallization by plotting δ^{66}Zn against Zn concentration (Yang et al., 2021). Overall, Mg and Zn isotope systematics of basaltic magmas could be viable tracers for subducting carbonates in the mantle provided that the influence of oxide crystallization can be excluded.

Zinc is a moderately volatile element with 50% condensation temperature (T_c) of 726 K and lighter Zn isotopes preferentially partition into the volatile phase during volatilization (e.g., Paniello et al., 2012; Kato et al., 2015). Volatile degassing can significantly fractionate Zn isotopes even at magmatic temperatures (Toutain et al., 2008). This isotopic effect must be considered because it results in deviation of Zn isotopic composition of the magma from that of the mantle source, as carbon isotopes do. In particular, the preferential loss of isotopically light Zn during volatilization shifts the melt toward heavier Zn isotopic composition, which is in the same direction expected by the incorporation of recycled carbonates into the sources. Commonly, the volatilization process inducing Zn isotope fractionation is in conjunction with strong Zn depletion in the residue (Toutain et al., 2008; Paniello et al., 2012; Kato et al., 2015). Thus, the isotopic effect induced by magma degassing may be assessed by plotting Zn isotopic ratios against Zn concentration or the abundances of highly refractory elements (e.g., Nb) in the magmas (Liu et al., 2016a, 2022).

4 Stabilities of different carbonate species in subduction zones

The knowledge of the speciation, solubility and stability of carbonate minerals at subduction zone conditions is critical to our understanding of the final fates of downgoing carbonates. The solubility of carbonate species in hydrous fluids is controlled by multiple factors, including thermal parameters of subduction zones, carbonate species, and fluid flux and property (e.g., Connolly and Galvez, 2018; Gorce et al., 2019). Generally, the solubility of carbonate minerals in aqueous fluids decreases in the order of calcite > aragonite > dolomite > magnesite > siderite at given pressures, and for each carbonate species the solubility increases with greater pressure (e.g., Pan et al., 2013). Experimental and thermodynamic studies show that the solubility of calcite and aragonite in H_2O-rich fluids is quite high even at low pressures (e.g., <6 kbar; Caciagli and Manning, 2003; Kelemen and

Manning, 2015). Thus, it is suggested that calcium carbonates can be substantially dissolved at the initial subduction stage (Frezzotti et al., 2011; Kelemen and Manning, 2015). This is supported by observations in the exhumed Cycladic subduction complex from Greece (peak metamorphic condition of ~500–550 °C and 2.0 GPa) where calcite occurs as the dominant daughter mineral in fluid inclusions (Ague and Nicolescu, 2014). Fluid-induced $CaCO_3$ dissolution, coupled with silicate precipitation, was thought to have contributed to up to 60–90% carbon oxide release from subducted slabs in forearc settings (Ague and Nicolescu, 2014). Observations on ultrahigh-pressure metamorphosed (UHPM) rocks from the Alps (600 °C, ⩾3.2 GPa) reveal that Mg-calcite/calcite daughter crystals exist in the fluid inclusions within garnet and diamond but magnesite and dolomite occur as solid inclusions (Frezzotti et al., 2011). This demonstrates that carbon released from the slab at sub-arc mantle depths is chiefly derived from fluid-mediated calcium carbonate dissolution. Although the first evidence for dolomite dissolution was observed in the western Alps calcite-dolomite marbles, the higher abundance of calcite than dolomite in fluid inclusions reveals the dominance of calcium carbonate dissolution at forearc subduction zone conditions (e.g., ~730–750 °C, 4.0–4.5 GPa) (Ferrando et al., 2017).

Magnesium carbonates (dolomite and magnesite) are commonly less soluble than calcic carbonates (calcite and aragonite) in hydrous fluids (Luth, 2001). However, ab initio simulations predict that they can become slightly soluble at the bottom of the upper mantle as the solubility increases with pressure (Pan et al., 2013). At high pressures (>6 GPa), slab-derived supercritical liquids with high solubilities of a large number of trace elements (e.g., Zr, Hf, Nb, Ta, V, Co, Cu, Zn, Sr, REE, Cr, Sb, W, Th) and high Th/U ratios are generated in subduction zones at temperatures of 800–1200 °C (Kessel et al., 2005). Observations on metamorphic zircons from garnet-pyroxenites with high Th/U ratios in an exhumed ultramafic massif in the Dabie orogen, central China, reveal the enrichment of dolomite inclusions, implying that dolomite can be significantly dissolved by supercritical fluids at depths of >160 km (Shen et al., 2018). Thus, Mg-rich carbonates have the potential to be recycled to the deep mantle beyond the sub-arc regime and cause mantle carbonate metasomatism. Many experimental studies show that calcite can be transformed into dolomite at pressures of 23 to 45 kbars or magnesite at higher pressures (e.g., Kushiro, 1975; Luth, 2001; Dasgupta et al., 2004). Metastable phase of $CaCO_3$ (e.g., $MgCO_3$, $ZnCO_3$, $FeCO_3$) can stably exist even at pressures exceeding 62 GPa (e.g., Bouibes and Zaoui, 2014; Boulard et al., 2015). In particular, magnesite is the stable carbonate with respect to dolomite and siderite as a consequence of the high affinity of Ca and Fe for lower mantle silicate minerals (Hammouda 2003; Litasov et al., 2008; Stagno, 2019). Thus, $CaCO_3$ may not be the major host of subducting carbon in the deep mantle and a considerable proportion of Mg-rich carbonates may potentially be retained in subducted slabs and finally carried into the deep mantle. The diverse solubility and stability of different carbonate species in fluids derived from slab dehydration determine their respective fates in subduction zones and ultimately define their storage depths in the mantle.

5 Dependence of Mg and Zn isotopes of the mantle on recycled carbonate species

Apart from different stabilities in subduction zones, various carbonate species have very distinct Mg and Zn contents, which would yield very different impacts on the Mg and Zn isotopic composition of the hybridized mantle if they were recycled into the mantle via subduction.The usually low MgO concentrations of calcium carbonates (average = ~0.8 wt%; Teng, 2017 and reference therein) relative to the peridotitic mantle (37.8 wt%; McDonough and Sun, 1995) imply that Mg isotopic composition of the mantle is very insensitive to fluid metasomatism dominated by calcite or aragonite dissolution. Nevertheless, it is relatively sensitive to metasomatism by magnesium carbonates (dolomite and magnesite) which have Mg concentrations close to that of the mantle. Distinct from Mg, Zn is a subordinate element in carbonate sediments that can replace Mg^{2+} and incorporate into the crystal lattice of carbonate minerals. The Zn concentrations of global marine carbonates have an average of ~44 μg/g despite the large uncertainties (Fig. 2), which is at the same order of Zn concentration of the mantle (~55 μg/g; McDonough and Sun, 1995). This makes Zn systematics of the mantle sensitive to various species of recycled carbonates. This characteristic of Zn is important because the majority of carbon that initially enters into subduction zones including carbon in altered oceanic crust is calcium carbonate (Plank and Manning,

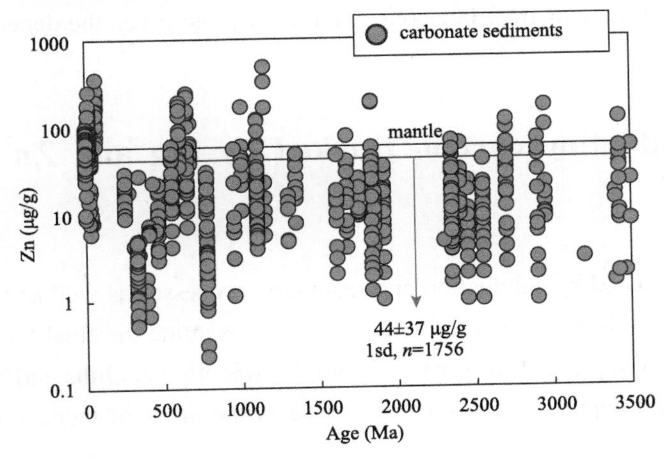

Fig. 2 A compilation of Zn contents for carbonate sediments (Liu et al., 2016b). The black line denotes the Zn concentration of the terrestrial mantle (~55 μg/g; Le Roux et al., 2010). Global marine carbonates have an average Zn concentration of ~44 μg/g, which is close to that of the mantle.

2019), for which Mg isotopes are perhaps difficult to discriminate. Zinc is commonly more abundant in Mg-rich carbonates than in Ca-rich carbonates. For instance, dolomite and magnesite inclusions in deeply subducted slabs have contents of up to 147 μg/g and 449 μg/g respectively, along with high MgO contents of up to 22.0 wt% and 47.6 wt%, respectively (Li et al., 2014; Teng, 2017 and reference therein). Thus, Mg-rich carbonates carried by subducted oceanic slabs can significantly alter Mg and Zn isotopic compositions of the mantle that can be sampled by xenoliths or basaltic lavas.

Since marine carbonates are characterized by heavier Zn and lighter Mg isotopic compositions than those of the pristine mantle (Fig. 1), high δ^{66}Zn and low δ^{26}Mg values of basaltic lavas are potentially indicative of the presence of recycled carbonates in their sources. Here the influence of different species of subducting carbonates on Mg and Zn isotopic composition of the mantle is quantified by binary mixing models. Two kinds of mixing models are presented, including source mixing (carbonate + mantle) and melt-melt mixing (pristine mantle-derived melt + carbonatitic melt). The modelled results are plotted in Fig. 5. A non-model melting model is used for zinc isotope fractionation during partial melting of mantle rocks (Doucet et al., 2016; Sossi et al., 2018; Beunon et al., 2020), and the isotope fractionation values (Δ^{66}Zn) between mantle minerals and basaltic melt are well-obtained in previous studies (Sossi et al., 2018; McCoy-West et al., 2018; Beunon et al., 2020). We choose an incremental melting model to drive the chemical and isotopic variation during mantle melting process (Zhong et al., 2017; Yang and Liu, 2019). The proportions of subducting carbonates required to drive Mg or Zn isotopic composition of the mantle to a specific value are slightly different for each mixing way. However, for each mixing way, addition of different carbonate species consistently results in distinct Mg and Zn isotopic compositions of the mantle mainly owing to their distinct Mg and Zn contents (Fig. 5). For instance, addition of ~5% dolomite into the mantle could produce melts with low δ^{26}Mg and high δ^{66}Zn values exceeding the ranges of pristine mantle-derived melts. Because magnesite has substantially higher Mg and Zn concentrations than those of calcite and dolomite, even a tiny amount of magnesite (<~2%) can induce measurable Mg and Zn isotopic deviation from the normal mantle range. However, Mg isotopic composition of mantle-derived melts is relatively insensitive to calcium carbonate addition so that at least nearly 30% carbonate addition in the sources is required to cause detectable isotopic offset in a basaltic magma (Fig. 5d). Thus, different species of subducting carbonates would impose different influences on the Mg and Zn isotopic compositions of the mantle and mantle-derived magmas. This in turn highlights the two isotopic systematics to be effective tools to identify the species of subducting carbonates in the mantle and thus determining their final fates (e.g., storage depths) given the dependence of their stabilities on pressures or depths.

6 The fate of subducting carbon tracked by Mg and Zn isotopes of natural samples

After isotope effects induced by subduction and magmatic processes as well as by the species of carbonates and the role of sulfate-rich fluids have been discussed, now we examine the final fates of subducting carbonates by using available Mg and Zn isotope data of natural samples (mantle xenoliths and basaltic magmas) formed at different tectonic settings and depths. We discuss this issue in the order of increasing depths, from the sub-arc mantle to the lower mantle.

6.1 Mg and Zn isotopes of arc lavas: Subducting calcium carbonates?

The sub-arc mantle is widely believed to be a vast storage of most subducting carbon (e.g., Ague and Nicolescu, 2014; Caciagli and Manning, 2003; Frezzotti et al., 2011; Manning et al., 2013; Pan et al., 2013; Kelemen and Manning, 2015). The paper of Ague and Nicolescu (2014) suggested that 60–90% of CO_2 can be released by fluid-induced calcium carbonate dissolution based on studies of marble-bearing subduction complexes in the Syros and Tinos islands in Greece. Kelemen and Manning (2015) proposed that hydrous fluids can even dissolve almost all carbonates in subducting slabs to drive the metamorphic decarbonation reaction, which potentially removes almost all carbon from the subducting plates at sub-arc mantle depths. Yet, direct evidence for the substantial involvement of subducting carbonates in the sub-arc mantle sources of arc lavas is rare. Here we examine this issue by collating published Mg and Zn isotopic data for arc basaltic lavas formed above subduction zones, varying from carbonate-rich (e.g., Central America) to carbonate-poor (e.g., Central-Eastern Aleutians) sediment layers (Plank, 2014) and from intermediate (e.g., Central America) to cool (e.g., Central-Eastern Aleutians) thermal regimes (Syracuse et al., 2010). The δ^{26}Mg values of arc lavas vary from −0.38‰ to +0.32‰ (n=78), which are similar to, or significantly higher than, the mantle value of −0.25±0.04‰ (Fig. 6a). The higher-than-mantle δ^{26}Mg values of some arc basalts require the overall composition of the mantle wedge to be buffered and modified by the preferential addition of Mg-rich, high-δ^{26}Mg fluids released from serpentinized oceanic slabs (e.g., Teng et al., 2016; Li et al., 2017; Hu et al., 2020) and/or by assimilation of high-δ^{26}Mg crustal rocks within the crust (e.g., Pang et al., 2020). Notably, a lower-than-mantle Mg isotopic signature that is typically characterized by recycled carbonates have not yet been observed for any arc lavas. Arc basaltic lavas have relatively uniform Zn isotopic compositions (0.16‰–0.32‰; n=42), which are within uncertainties indistinguishable from the δ^{66}Zn values of mid-ocean ridge basalts (MORBs; 0.28±0.05‰; Wang et al., 2017; Huang et al., 2018a) (Fig. 6c). Again, high δ^{66}Zn values as expected by recycling of marine carbonates into the sub-arc mantle sources were not observed yet in any arc lavas. These observations are genuinely surprised because the sub-arc mantle has widely been thought to be the main storage of subducting carbonates.

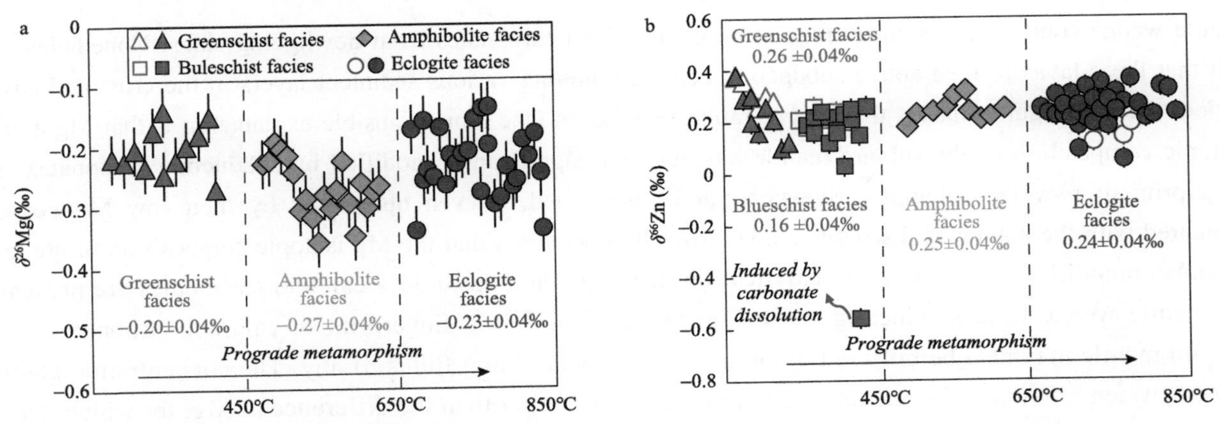

Fig. 3 Evolution of $\delta^{26}Mg$ and $\delta^{66}Zn$ in metamorphic rocks with respect to the prograde metamorphic grade. Data sources: Wang et al., 2014a; Pons et al., 2016; Inglis et al., 2017; Debret et al., 2021; Xu et al., 2021. Metamorphic rocks represented indicated by hollow symbols show a decrease of $\delta^{66}Zn$ with increasing metamorphic grade, which is related to the preferential release of ^{66}Zn-enriched sulfate fluids during subduction (Pons et al., 2016).

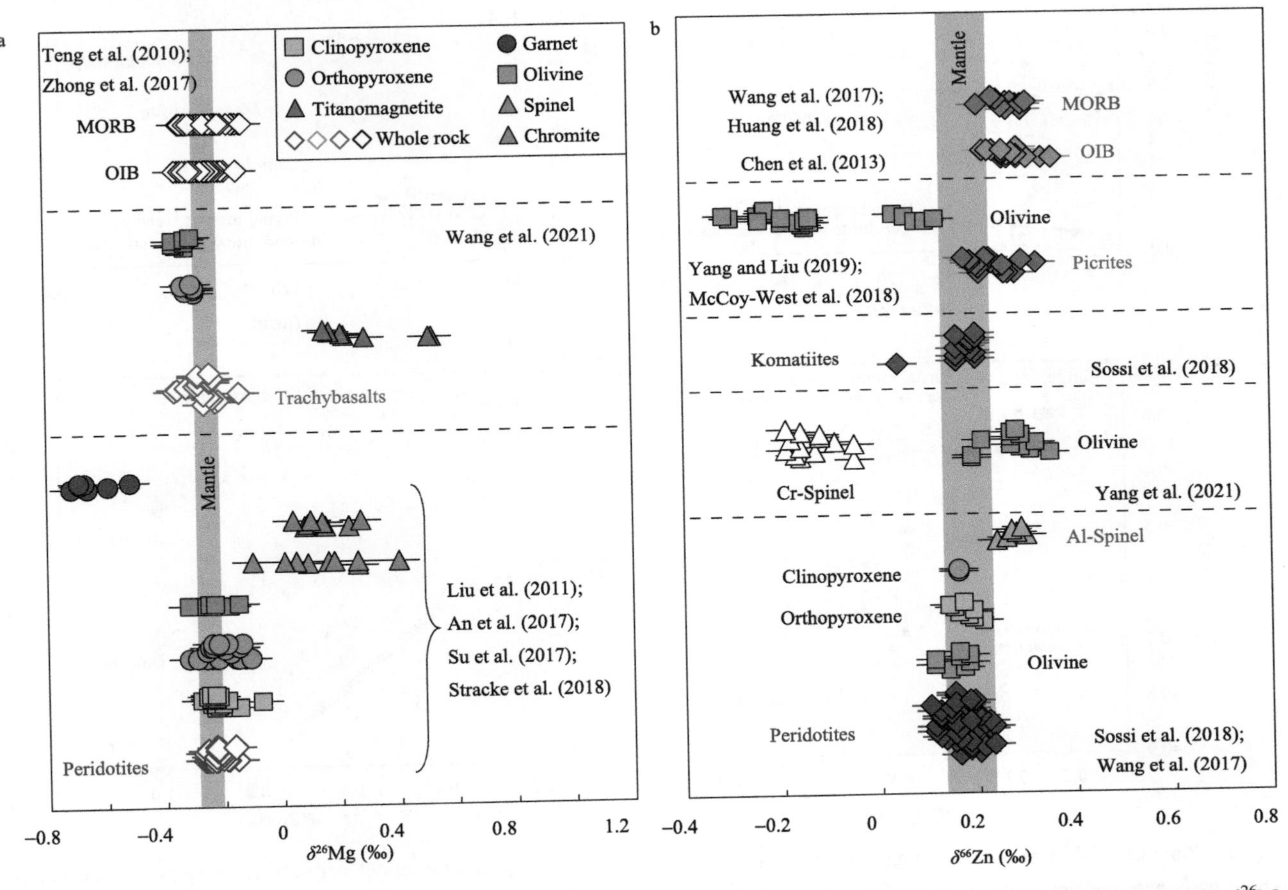

Fig. 4 A compilation of Mg and Zn isotopic compositions of peridotites, basalts, and minerals in peridotites and basalts. The $\delta^{26}Mg$ data are from Teng et al. (2010), Liu et al. (2011), An et al. (2017), Su et al. (2017), Zhong et al. (2017), Stracke et al. (2018) and Wang et al. (2021a). The $\delta^{66}Zn$ data are from Chen et al. (2013), Wang et al. (2017), Sossi et al. (2018), McCoy-West et al. (2018) and Yang and Liu (2019); Yang et al. (2021). The grey band represents the mantle value ($\delta^{26}Mg$=−0.25±0.04‰; Teng, 2017; $\delta^{66}Zn$= 0.18±0.05‰; Wang et al., 2017; Sossi et al., 2018; McCoy-West et al., 2018; Liu et al., 2019).

A possible interpretation on the absence of carbonate-like Mg and Zn isotopic signature of arc lavas (i.e., lower $\delta^{26}Mg$ and higher $\delta^{66}Zn$ than the mantle) is that these lavas investigated have not sampled the sub-arc mantle wedge containing a significant amount of dissolved carbonates from downgoing slabs. Nonetheless, the fact that these lavas formed above subduction zones containing various sediment layers in the crust and having various thermal regimes makes this interpretation less likely. One more plausible explanation is that Mg and Zn isotopic composition of the sub-arc mantle has not been significantly modified by subducting carbonates. The fingerprint of recycled calcium carbonates can be weakened or even obliterated by their low Mg contents compared with the mantle as discussed above. Modelling suggests that the Mg isotopic composition of arc lavas remains unmodified within current analytical uncertainty when up to 30% calcium carbonates are present in the mantle wedge from subducting slabs (Fig. 5). The low-δ^{26}Mg fingerprint of calcium carbonates in the sub-arc mantle may also be buffered by high-δ^{26}Mg serpentinizing fluids, if any. The concentration contrast of Zn between calcium carbonate and the mantle is much smaller than the difference of Mg, for which a lower

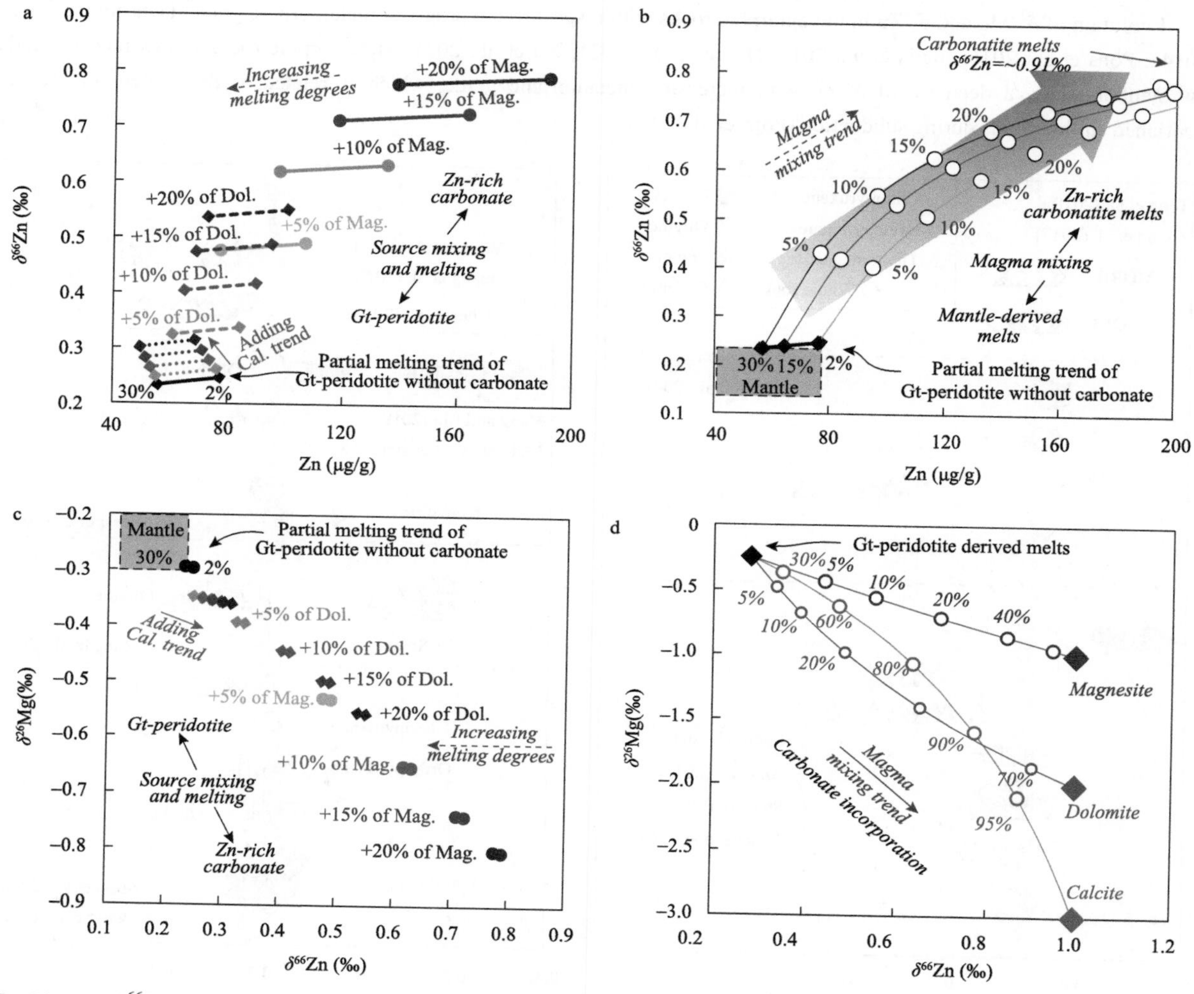

Fig. 5 Plots of δ^{66}Zn against Zn contents (a, b) and δ^{26}Mg (c, d) for a binary mixing model between carbonates and the mantle. Three carbonate species (calcite, dolomite and magnesite) are considered and two ways of mixing (source mixing and melt-melt mixing) are presented. Source mixing model is shown in (a) and (c), and the melt-melt mixing model is shown in (b) and (d). The brown, blue and green lines in (a, c) denote the partial melting trends after adding magnesite, dolomite and calcite to the mantle, respectively. Brown curves in (b) represent the magma mixing lines of Zn-rich carbonatite melt derived from recycled carbonates and peridotite-derived melts formed at various melting degrees (2%, 15%, and 30%). Modelling results for partial melting of normal Gt-peridotite are shown as black curves in these figures, and diamonds denote increasing melting degrees from 2% to 30%. The normal mantle-derived melts are taken to possess δ^{26}Mg of −0.25±0.04‰ (Teng, 2017) and δ^{66}Zn of 0.28±0.05‰ (Wang et al., 2017; Huang et al., 2018a). Modelling parameters are given in Table 1. (For interpretation of the references to colour in this figure legend, the reader is referred to the web version of this article.)

Table 1 Parameters used for mixing model between carbonates and the mantle

Mineral	Modal (wt%)[1]	Melting model	Zn(μg/g)	D_{Zn}	D_{Mg}	Δ^{66}Zn (‰)	Δ^{26}Mg (‰)
Cpx	0.27	0.80		0.33	1.35	$-0.17\times10^6/T^2$	
Ol	0.53	−0.25		0.96	3.04	$-0.17\times10^6/T^2$	
Opx	0.18	0.35		0.45	1.96	$-0.17\times10^6/T^2$	
Grt	0.02	0.10		0.21	1.30	$-0.20\times10^6/T^2$	
Sum	1.00	1.00	55	0.68	2.40	$-0.18\times10^6/T^2$	$-0.12\times10^6/T^2$

End member	Zn (μg/g)	δ^{66}Zn (‰)	MgO (wt%)	δ^{26}Mg (‰)
Mantle	55	0.18±0.05	37.8	−0.25
Normal-mantle derived melts	80	0.28±0.05	7.58	−0.25
Dolomite	132	0.91±0.24	22.00	−2.00
Magnesite	449	0.91±0.24	47.60	−1.00
Calcite/Aragonite	20	0.91±0.24	0.80	−3.00

Sources: normal mantle-derived melts and the mantle (Zn contents and δ^{66}Zn: Le Roux et al., 2010; Wang et al., 2017; Huang et al., 2018a; Sossi et al., 2018; McCoy-West et al., 2018; Liu et al., 2019; MgO contents and δ^{26}Mg: Hofmann, 1988; McDonough and Sun, 1995; Teng, 2017); carbonates (Zn contents and δ^{66}Zn: Turekian and Wedepohl, 1961; Li et al., 2014; Pichat et al., 2003; Sweere et al., 2018; MgO contents and δ^{26}Mg: Teng, 2017 and reference therein). The $\Delta^{66}Zn_{mineral-melt}$ (‰) are taken from Sossi et al. (2018) and McCoy-West et al. (2018), and $\Delta^{26}Mg_{residue-melt}$ (‰) is from the calculation in Wang et al. (2020). Partition coefficients of Zn and Mg are from Le Roux et al. (2015).

proportion of calcium carbonate can change Zn isotopic composition of the mantle wedge (Fig. 5). However, this fingerprint may be buffered by altered oceanic crust-derived fluids or melts (e.g., Zn chlorides enriched in ^{64}Zn; Fujii et al., 2011), and further diluted by high degrees of partial melting of the sub-arc mantle wedge (e.g., Huang et al., 2018a). It is noteworthy that both Mg and Zn isotopic compositions of arc lavas are not correlated with Mg or Zn concentrations and chemical indices of fluid enrichment (e.g., Ba/Th; Fig. 6).This further suggests that the addition of slab-derived fluids has limited influence on the Mg and Zn isotopic composition of the sub-arc mantle wedge.

If the subduction-zone fluid is dominated by calcium carbonate dissolution (i.e., Ca-rich but Mg- and Zn-poor), then the stable isotopic systematics of calcium ($\delta^{44/40}$Ca) is perhaps a more effective tracer for subducting carbonates in the mantle (e.g., Huang et al., 2011). However, several independent studies reported similar, MORB-like $\delta^{44/40}$Ca ratios for modern arc basalts; in particular, no low-$\delta^{44/40}$Ca signature as expected by recycled calcium carbonates has been observed (Zhu et al., 2020; Wang et al., 2021b; Kang et al., 2021). At cold subduction zones where subducting carbonates have relatively low recycling efficiency (<30%) or subducted Ca is mainly in the form of carbonate-bearing altered oceanic crusts (e.g., Tonga and Central America), the buffering effect of the mantle wedge may account for the MORB-like $\delta^{44/40}$Ca of arc magmas (Wang et al., 2021b). At hot subduction zones with high carbonate recycling efficiency (>30%, e.g., Lesser Antilles; Kang et al., 2021), multiple factors may have simultaneously contributed to the MORB-like $\delta^{44/40}$Ca values of arc magmas, including mantle buffering effect, dissolution-reprecipitation, melting-recrystallization, metamorphic decarbonation, and carbonate-eclogite isotope exchange (Kang et al., 2021). Overall, available $\delta^{44/40}$Ca data provide no evidence for a significant flux of calcium carbonates from downgoing slabs to the mantle wedge and/or the data reflect homogenization of variable $\delta^{44/40}$Ca for the slab fluids. The conundrum why substantial calcium carbonate dissolution at sub-arc depths as widely predicted by experimental studies and thermodynamic simulations is not recorded by Mg, Zn and Ca isotopes of arc lavas remains to be explored in the future. Nevertheless, it could be concluded that the Mg and Zn isotopic compositions of arc lavas do not support a significant flux of magnesium carbonates in the sub-arc mantle.

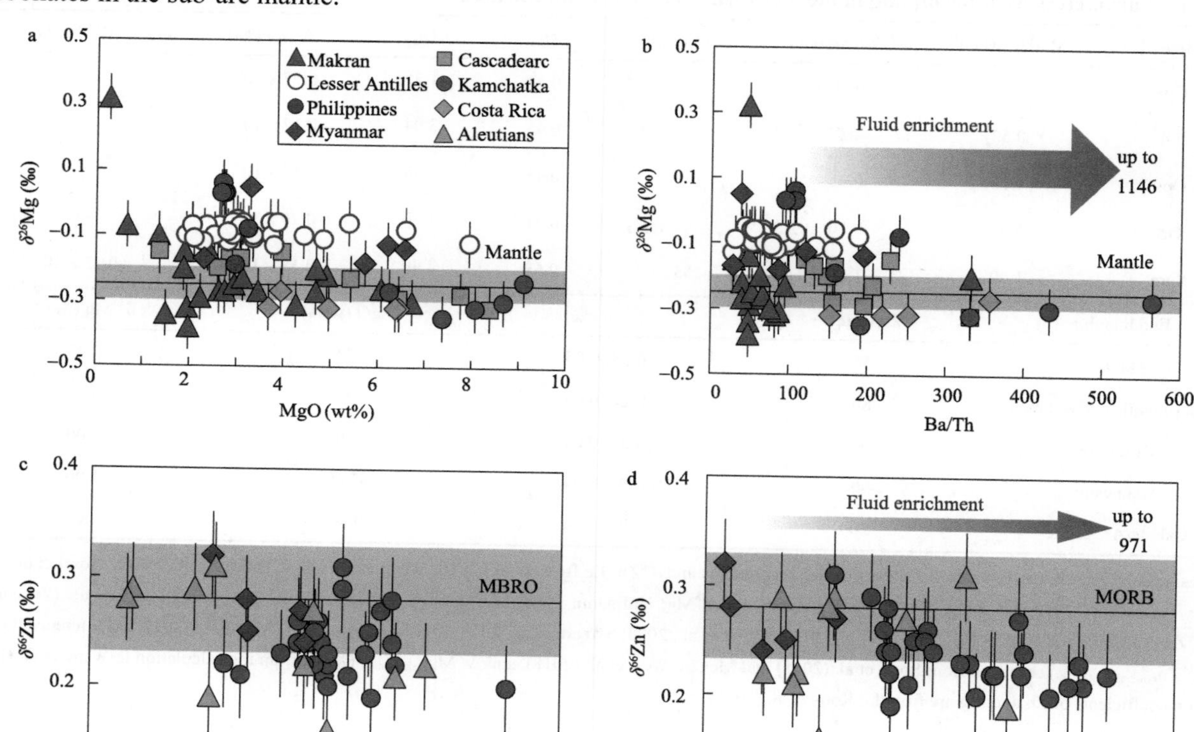

Fig. 6 Published δ^{26}Mg and δ^{66}Zn values for modern arc lavas plotted as a function of MgO or Zn contents and Ba/Th ratios. Data sources: Kamchatka (δ^{26}Mg: Li et al., 2017; δ^{66}Zn: Huang et al., 2018a), Lesser Antilles (Teng et al., 2016), Costa Rica and Philippines (Li et al., 2017), Myanmar (Li et al., 2021), Cascade (Brewer et al., 2018), Makran (Pang et al., 2020) and Aleutians (Huang et al., 2018a). The mantle δ^{26}Mg (− 0.25±0.04‰) is from Teng (2017) and the δ^{66}Zn value of MORBs (0.28±0.05‰) is from Wang et al. (2017) and Huang et al. (2018a).

6.2 Recycled carbonates in the subcontinental lithospheric mantle

Mantle lithosphere, particularly in continental settings, has been hypothesized as a vast and long-term store for carbon, which is incorporated into the sub-arc lithosphere via the accretion of island arcs (Kelemen and Manning, 2015; Foley and Fischer, 2017). This hypothesis is strongly supported by the observations of extensive mantle's CO_2 outgassing along continental rifts or faults such as the East Africa Rift (Lee et al., 2016). The carbon in the subcontinental lithospheric mantle can be further added via melt/fluid metasomatism associated with subduction or mantle plume events over geological timescales (e.g., Foley and Fischer, 2017), which may be recorded in xenoliths or basaltic melts derived from the subcontinental lithospheric mantle. For example, the geochemical similarity of many clinopyroxene-rich peridotites with carbonatites (e.g., high CaO/Al_2O_3 and low Ti/Eu) implies modal carbonate metasomatism of the lithospheric mantle (Ionov et al., 1993; Aulbach et al., 2020). The non-mantle-like δ^{13}C values of mantle xenoliths have often been ascribed to recycling of supracrustal carbon into the lithospheric mantle (Deines, 2002). In recent years, Mg and Zn isotopic analysis on abundant mantle xenoliths has revealed pervasive metasomatism of the lithospheric mantle by subducting carbonates. Compared with non-metasomatized peridotites, metasomatized peridotites and pyroxenites display a much wider range of Mg and Zn isotopic compositions (Yang et al., 2009; Xiao et al., 2013; Hu et al., 2016; Wang et al.,

2016, 2017; Shen et al., 2018; Hu et al., 2019; Su et al., 2019) (Figs. 7, 8). In particular, many of these xenoliths (mainly pyroxenites and wehrlites) have δ^{26}Mg values that are significantly lower than the mantle value and negatively correlated with CaO and CaO/Al_2O_3 (Fig. 7a, b, c). These correlations point to metasomatism of the lithospheric mantle by recycled Mg-rich carbonates (e.g., dolomite) that can be dissolved by supercritical fluids at depths of >160 km as discussed above, providing evidence for a significant flux of subducting carbonates in the sub-continental lithospheric mantle.

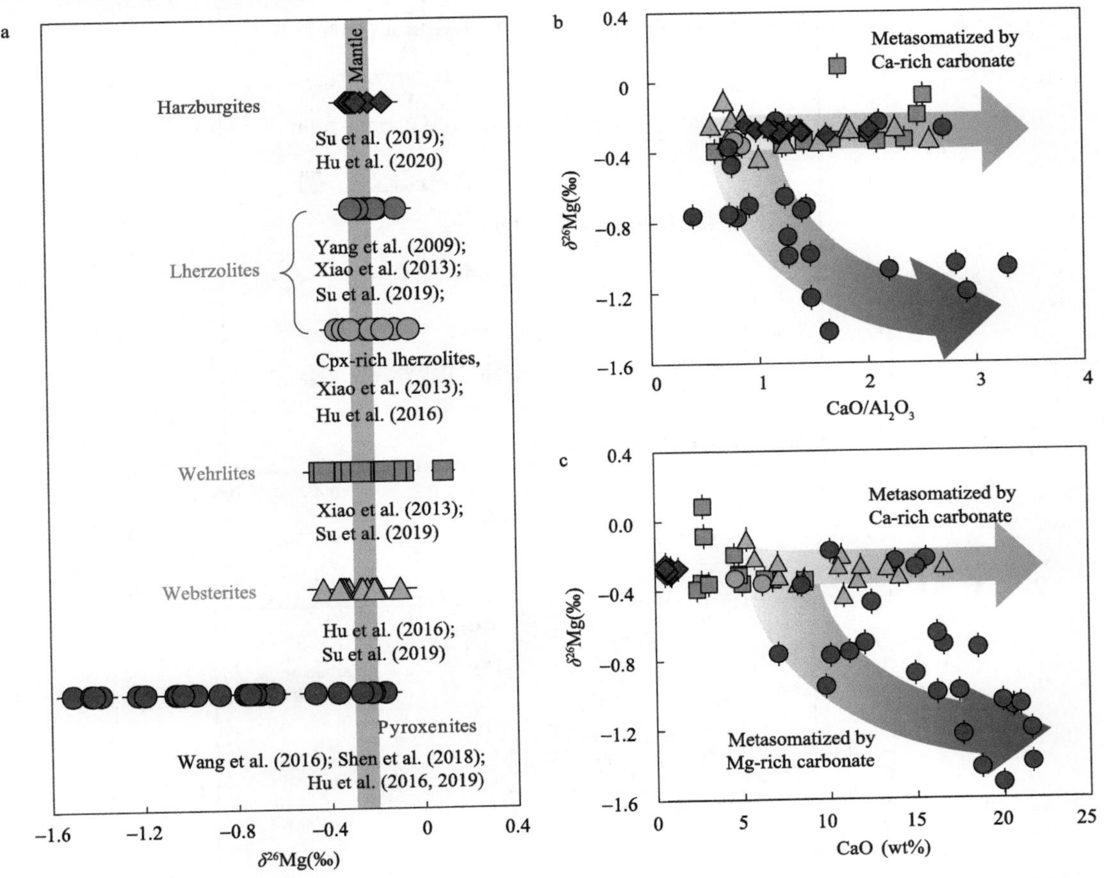

Fig. 7 (a) A compilation of Mg isotopic compositions of mantle xenoliths in the literature. Data sources: harzburgites (Su et al., 2019; Hu et al., 2020), lherzolites (Yang et al., 2009; Xiao et al., 2013; Hu et al., 2016; Su et al., 2019), wehrlites (Xiao et al., 2013; Su et al., 2019), websterites (Hu et al., 2016; Su et al., 2019), and pyroxenites (Hu et al., 2016; Wang et al., 2016; Shen et al., 2018; Hu et al., 2019). The grey band denotes the mantle δ^{26}Mg value (−0.25±0.04‰; Teng, 2017). (b–c) Correlations of δ^{26}Mg values with CaO/Al_2O_3 and CaO contents for mantle xenoliths. Symbols as the same as Fig. 7a.

Notably, some xenoliths also display a prominent enrichment of Ca but have mantle-like Mg isotopic compositions (Fig. 7b, c). This [Ca]–δ^{26}Mg decoupling means that these xenoliths obtained substantial Ca via metasomatism but their Mg isotopic compositions remain unmodified. Since calcium carbonate is commonly too Mg-poor to influence Mg isotopic composition of the mantle (see Section 5), we suggest the mantle-like δ^{26}Mg values of these Ca-rich xenoliths to record calcium carbonate metasomatism at relatively shallower depths where calcium carbonate is not transferred into magnesium carbonate and has been substantially dissolved. Thus, normal δ^{26}Mg values of mantle xenoliths do not necessarily mean the lack of carbonate metasomatism because xenoliths with low δ^{26}Mg only represent a minimum sampling of the carbonated lithospheric mantle. Zinc isotopic compositions of mantle xenoliths worldwide vary from −0.44‰ to +0.46‰ (Fig. 8a), which are either lower or higher than that of the unmodified mantle. The high δ^{66}Zn of some ultramafic xenoliths imply mantle metasomatism by recycled Zn-rich carbonates and those with low δ^{66}Zn are most likely related to kinetic isotope effect during melt-peridotite interaction (e.g., Wang et al., 2017; Huang et al., 2018b) (Fig. 8b). Interaction of carbonated silicate melts with the overlying lithospheric mantle may also increase its Zn isotopic ratios (Wang et al., 2018a; Wang and Liu, 2021).

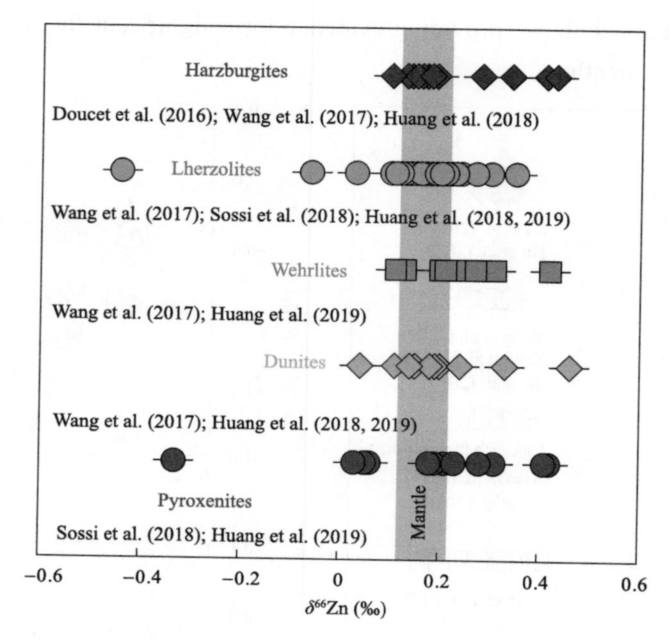

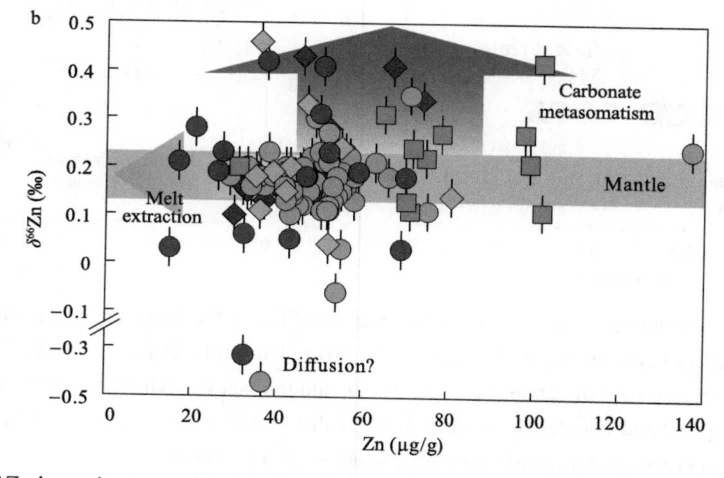

Fig. 8　(a) A compilation of Zn isotopic compositions of mantle xenoliths in the literature. Data sources: harzburgites (Doucet et al., 2016; Wang et al., 2017; Huang et al., 2018b), lherzolites (Wang et al., 2017; Sossi et al., 2018; Huang et al., 2018b, 2019), wehrlites (Wang et al., 2017; Huang et al., 2019), dunites (Wang et al., 2017; Huang et al., 2018b, 2019) and pyroxenites (Sossi et al., 2018; Huang et al., 2019). The grey band denotes the mantle δ^{66}Zn value (0.18±0.05‰; Wang et al., 2017; Sossi et al., 2018; McCoy-West et al., 2018; Liu et al., 2019). (b) Correlations of δ^{66}Zn values with Zn contents for mantle peridotites and pyroxenites.

Mantle eclogite xenoliths from kimberlites in ancient cratons have been considered as remnants of subducted Archean oceanic crust stored in the cratonic lithospheric mantle (e.g., Barth et al., 2001; Jacob, 2004). Studies on many cratonic eclogites reported extremely lower δ^{26}Mg (Wang et al., 2012) and higher δ^{66}Zn (Wang et al., 2022b) than those of the basaltic oceanic crust, which was ascribed to the signatures of carbonated oceanic crustal protoliths. Since studies on Archean greenstones have suggested that hydrothermal carbonatization of oceanic crust dominates the atmospheric and oceanic CO_2 sink in the early and middle Archean (Nakamura and Kato,

2004; Shibuya et al., 2012), the cratonic eclogites of recycled origin might represent the main carbon-bearing components subducted into the Archean mantle.

Magnesium isotopic systematics of lithospheric mantle-derived melts (e.g., K-rich lavas, lamproites, etc.) provide more pervasive evidence for incorporation of subducting carbonates into the lithospheric mantle, in addition to mantle xenoliths. The lithospheric mantle-derived melts with δ^{26}Mg lower than that of the mantle reported so far include Oligocene–Miocene ultrapotassic lavas from the Lhasa terrane, southern Xizang (Liu et al., 2015), Miocene syenites from northwestern Xinjiang (Ke et al., 2016), Eocene high-K trachyandesites from northern Qiangtang block, central Xizang (Tian et al., 2020), Pliocene lamproites from Leucite Hills, Wyoming, USA (Sun et al., 2021), and early Cretaceous lamproites from Shandong in the North China Craton (Wang et al., 2022a). Interestingly, almost all of these magmatic rocks are emplaced in post-collisional or intraplate settings with a thick lithosphere and characterized by enriched Sr and Nd isotopic compositions, which suggests metasomatism of their mantle sources by melts/fluids from silicic sediments (e.g., pelites). The Mg-Sr isotopic mixing model requires a certain amount of magnesium carbonates (dolomites) to have been incorporated into the mantle sources. We suggest that, in a thick lithosphere (>160 km), detectable Mg and Zn isotopic anomalies can be more easily observed in mantle-derived rocks and melts since at such conditions recycled magnesium carbonates are expected to be dissolved by supercritical fluids. An additional example is Tengchong in SE Xizang, where the light Mg and heavy Zn isotopic signatures of basalts have been observed due to dissolution of magnesium carbonates by supercritical fluids (Tian et al., 2018; Qu et al., 2022).

In summary, the light Mg and heavy Zn isotopic compositions of many mantle xenoliths and lithospheric mantle-derived melts demonstrate that the mantle lithosphere is an important carbon reservoir with contribution from multistage and various types of carbonate metasomatism from downgoing slabs. Commonly, the deep part of the thick lithospheric mantle is too reduced to favor stable carbonates (Dasgupta and Hirschmann, 2010). When carbonates are recycled into depths of >120 km, they will be reduced via the following redox reaction: $MgSiO_3 + MgCO_3 = Mg_2SiO_4 + C + O_2$. At depths of 120–170 km, recycled carbonate is transformed to carbon that exists as graphite and at larger depths (>170 km) as diamond (Stagno et al., 2013). During this process, isotopically light Mg and heavy Zn in carbonate components can be incorporated into silicate phases via exchange interaction between carbonate and peridotite. The carbon-bearing mantle source will undergo "redox melting" in response to mantle upwelling or lithospheric extension, generating carbonatitic melts or carbonated silicate melts with light Mg and heavy Zn isotopic signals as widely observed.

6.3 Recycled carbonates in the mantle transition zone

Whether subducting carbonates can be brought into the mantle transition zone and the overlying convecting upper mantle at a global scale is still a debatable issue. The rare occurrences of super-deep diamonds and their containing inclusions may provide evidence for the presence of recycled carbon in the transition zone beneath local areas (Brenker et al., 2007; Anzolini et al., 2019). A recent study found extremely radiogenic ^{206}Pb/^{204}Pb on Bermuda basalts and proposed that the basaltic magmas originated from the transition zone where recycled carbonates are stored (Mazza et al., 2019). Intraplate alkali basaltic magmas, which are away from plate boundaries and spatially coupled with stagnant slabs in the transition zone, may sample the composition and heterogeneity of the transition zone. Horizontally or subhorizontally stagnant slabs in the transition zone have been widely recognized beneath subduction zones in the circum-Pacific, the Mediterranean and the eastern Tibetan areas based on high-velocity anomalies observed at the depths of 410–660 km (Fig. 9a, b; Piromallo and Morelli, 2003; Fukao et al., 2009; Lei et al., 2019). A large-scale low-δ^{26}Mg and high-δ^{66}Zn anomaly is identified in <110 Ma intraplate alkali basalts from East Asia (Yang et al., 2012; Huang et al., 2015a; Liu et al., 2016a, 2022; Li et al., 2017;

Wang et al., 2018a; Jin et al., 2020; Wang and Liu, 2021; Choi and Liu, 2022). The excellent coupling between the spatial distribution of these basaltic lavas characterized by light Mg and heavy Zn isotopic anomalies and the stagnant west Pacific slab at depths of ~440 to 660 km (Fig. 9a, c) suggests a large-scale storage of recycled magnesium carbonates (magnesites ± dolomite) in the transition zone and the overlying convecting upper mantle beneath East Asia since the late Mesozoic (Liu et al., 2016a, 2022; Li et al., 2017). Two recent studies reported low δ^{26}Mg and high δ^{66}Zn on Cenozoic alkali basaltic lavas from SE Xizang and central Myanmar (Liu et al., 2020; Li et al., 2021), where the stagnant Neo-Tethys oceanic slab was also suggested to lie in the transition zone (Fig. 9; Huang and Zhao, 2006). Together with the distinctive chemical signatures of these basalts (e.g., high CaO/Al_2O_3), the Mg and Zn isotopic data provide evidence for recycled carbonates in the transition zone associated with the Neo-Tethys oceanic slab subduction. Collectively, these observations suggest that the transition zone is an important storage of global subducting carbon than previously thought, as has been observed in the west Pacific and Neo-Tethys subduction zones, which are two of the largest oceanic subduction zones during the Phanerozoic.

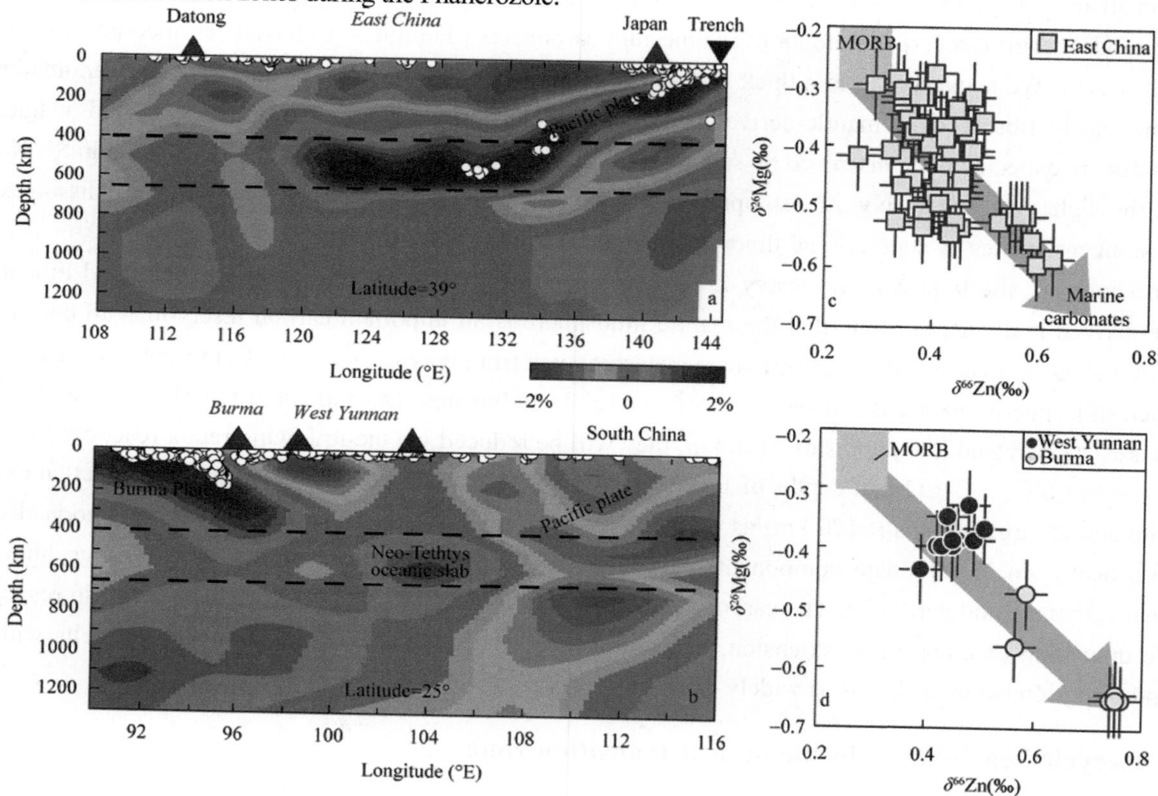

Fig. 9　(a–b) A vertical cross section of P wave velocity perturbations along eastern China, Burma, western Yunnan and the South China Block (after Huang and Zhao, 2006). (c–d) Published δ^{26}Mg and δ^{66}Zn values for alkali basalts. Data sources: East China (Liu et al., 2016a; Li et al., 2017; Jin et al., 2020); West Yunnan in SE Xizang (Liu et al., 2020); Burma (Li et al., 2021). The occurrence of alkali basalts is spatially coincident with the stagnant slabs at depths of ~440 to 660 km, suggesting a carbonate-bearing mantle source in the MTZ.

Whether or not subducting carbonates could pass through the transition zone and enter the Earth's lower mantle is also controversial (e.g., Thomson et al., 2016; Cerantola et al., 2017; Lv et al., 2021). There are some available Mg and Zn isotopic data for oceanic island basalts (OIBs) and picrites and basalts from large igneous provinces (LIPs). Almost all OIBs have mantle-like δ^{26}Mg within current levels of analytical uncertainty (see the review by Teng, 2017), expect the EM1-type OIBs from Pitcairn islands with low δ^{26}Mg (Wang et al., 2018b). Compared with MORBs, some OIBs have slightly higher δ^{66}Zn and were explained as a result of recycling of

C-bearing eclogites into their mantle sources (Beunon et al., 2020). By contrast, there are no any Mg or Zn isotopic anomalies in picrites and basalts from LIPs formed at both continental and oceanic tectonic settings (Tian et al., 2017; McCoy-West et al., 2018; Yang and Liu, 2019). The mantle-like Mg and Zn isotopic signatures of these rocks may reflect the non-ubiquitous presence of recycled carbonates in their sources. Nevertheless, it must be noted that there is no robust evidence for a lower mantle's origin of these rocks and thus the fate of subducting carbon in the lower mantle remains open.

7 Three carbon-cycling models identified by Mg and Zn isotopes

Based on the Mg and Zn isotopic data set of basaltic lavas originating from various mantle depths as discussed above, we develop a schematic model showing the fates of carbonates in subduction zones (Fig. 10). Three carbon-cycling models are presented. At initial subduction stage, calcium carbonate is preferentially dissolved at sub-arc mantle depths due to its high solubility in slab-derived fluids, with subsequent CO_2 released by arc volcanism and diffusive degassing, which constitutes carbon cycles in arc systems (Kelemen and Manning,

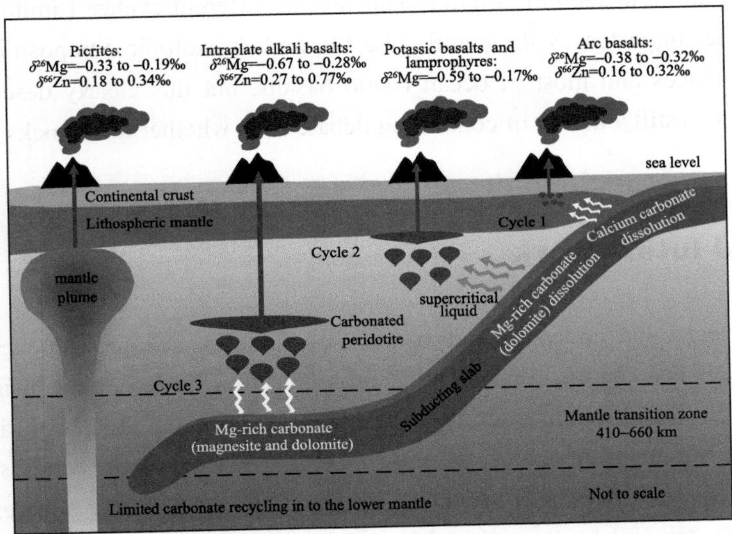

Fig. 10 A schematic model showing the fate of carbonates in subduction zones tracked by Mg and Zn isotopic compositions of mantle-derived rocks. Cycle 1: preferential dissolution of calcium carbonate at sub-arc mantle depths, with CO_2 released by arc volcanism. Cycle 2: Mg-rich carbonate dissolution induced by supercritical liquids and the incorporation into the lithospheric mantle, resulting in potassic lavas with light Mg isotopic compositions. Cycle 3: magnesium carbonates recycled into the transition zone, with carbon released by intraplate volcanism.Data sources: arc basalts (δ^{26}Mg: Teng et al., 2016; Li et al., 2017; Brewer et al., 2018; Pang et al., 2020; Li et al., 2021; δ^{66}Zn: Huang et al., 2018a; Li et al., 2021), potassic lavas and lamprophyres from the lithospheric mantle (δ^{26}Mg: Liu et al., 2015; Ke et al., 2016; Tian et al., 2020; Sun et al., 2021; Wang et al., 2022a), intraplate alkali basalts (δ^{26}Mg: Yang et al., 2012; Huang et al., 2015a; Li et al., 2017; Jin et al., 2020; Wang and Liu, 2021; Li et al., 2021; δ^{66}Zn: Liu et al., 2016a; Wang et al., 2018a; Jin et al., 2020; Li et al., 2021; Wang and Liu, 2021), and picrites (δ^{26}Mg: Tian et al., 2017; δ^{66}Zn: McCoy-West et al., 2018; Yang and Liu, 2019).

2015). The isotopic fingerprint of recycled carbonates in the sub-arc mantle is largely diluted by the low Mg and Zn contents of calcium carbonates and high melting degrees of the mantle wedge as well as buffered by other subduction-zone melts/fluids, consistent with the absence of high-δ^{66}Zn and low-δ^{26}Mg signatures in arc basalts (Cycle 1; Fig. 10). The contribution of Mg-rich carbonates to the sub-arc mantle is likely very limited.

Calcium carbonate is relatively unstable at high pressures and can be transformed into dolomite at 23–45 kbars, or into magnesite at larger pressures/depths (Kushiro et al., 1975; Yaxley and Green, 1994; Dasgupta et al.,

2004). The subducted Mg-rich carbonates (mainly dolomite) can be effectively dissolved by supercritical liquids and released into the lithospheric mantle at depths of >160 km, causing pervasive mantle metasomatism. Recycled carbonates in the lithospheric mantle can be partly or completely reduced into diamond or graphite and subsequently released via redox melting with CO_2 outgassing induced by lifts or intraplate magmatism (Foley and Fischer, 2017; Aulbach et al., 2020; Wang et al., 2022a). In this case, Mg and Zn isotopic compositions of the lithospheric mantle would be significantly modified by recycled Mg-rich carbonates, which is justified by the low-δ^{26}Mg and high-δ^{66}Zn signatures of ultramafic xenoliths and potassic lavas derived from the lithospheric mantle (Cycle 2; Fig. 10).

As subduction proceeds, magnesium carbonates (magnesite ± dolomite) were carried into the depths of mantle transition zone at which the carbonate-bearing slab underwent melting (Thomson et al., 2016). The recycled carbon therein was released by intraplate alkali basaltic volcanism (Liu et al., 2016a; Li et al., 2017). The spatial coupling of large-scale intraplate alkali basalts possessing low δ^{26}Mg and high δ^{66}Zn with the horizontally stagnant slabs at depths of ~440 to 660 km (Fig. 9) supports the pervasive presence of recycled magnesium carbonates in the transition zone (Cycle 3; Fig. 10). The timescales of return of recycled carbon into the surface may be highly variable due to different migration pathways at different cycles. Limited carbonate recycling into the lower mantle may be suggested by the mantle-like Mg and Zn isotopic compositions of basalts and picrites from large igneous provinces and most of ocean island basalts, but this clearly deserves to be investigated in future studies given there is still a matter of continuing debate as to whether these rocks sampled the Earth's lower mantle.

8 Summary and future work

In the past few years our understanding of the Mg and Zn isotopic geochemistry in magmatic and subduction-zone processes has changed dramatically, providing us with important insights into the deep carbon cycling. On the basis of the different stabilities of various carbonate species during subduction and their prominent differences in Mg and Zn contents, we show that Mg and Zn isotopic compositions of basaltic magmas are sensitive to the species and depths of recycled carbonates in their mantle sources. Thus, the final fate of subducting carbon can be tracked by the paired Mg and Zn isotopic systematics and is further verified by the different Mg and Zn isotopic characteristics of basaltic magmas derived from different mantle depths worldwide. Although calcium carbonates have high solubility in aqueous fluids at shallow depths, their fingerprint in the sub-arc mantle is unidentified by the Mg and Zn isotopic composition of arc lavas. More Mg and Zn isotope data for arc lavas produced at hotter subductions or deeper sub-arc mantle depths are still required in the future, given that the solubility of carbonates in hydrous fluids increases with temperature and pressure (e.g., Pan et al., 2013; Manning et al., 2013). Considering the higher geothermal gradient of Archean mantle relative to the present one, magmas generated at Archean subduction zones (e.g., sanukitoids) are also worthy of investigation. Magnesium carbonates have very low solubility at shallow depths but can be substantially dissolved by supercritical fluids at greater depths of >160 km. Given their high Mg and Zn contents, addition of a minute amount (~5%) of magnesium carbonates can cause detectable changes in the Mg and Zn isotopic composition of the mantle. The low-δ^{26}Mg and high-δ^{66}Zn signals widely observed in mafic/ultramafic xenoliths from the lithospheric mantle and alkali basalts from regions with stagnant slabs in the transition zone support that magnesium carbonates are able to survive the arc regime and enter into the deeper mantle. Overall, Mg and Zn isotopic systematics of natural samples shed valuable insights into the final storage depths of subducting carbonates and suggest that the lithospheric mantle and the transition zone are important stores of global subducting carbon.

The recognition of light Mg and heavy Zn isotopic signatures in rocks from different mantle domains represents a tremendous step of qualitatively determining the fate of subducting carbon. To go one step further toward quantifying the global input flux of subducting carbon, however, uncertainties still exist. For instance, future experimental work is required to accurately explore the elemental and isotopic behaviors of Mg and Zn during partial carbonate dissolution and decarbonation at subduction zone conditions. Provided that significant elemental and isotopic fractionation occurs during these processes, the isotopic composition of carbonates remaining behind dissolution and decarbonation will be modified. This could result in inaccurate estimates of the input fluxes of subducting carbon at various mantle depths. Clearly, calibration of these effects with rigorous experiments through a range of parameters (e.g., carbonate proportion in sediments, fluid species, pressure, temperature, and oxygen fugacity) is key to constraining the fates of subducting carbonates and conditions for the formation of isotopically anomalous mantle lithologies and mantle-derived melts. To put further constraints on the carbon storage and fluxes at sub-arc mantle depths, the Mg and Zn isotopic compositions of fluids released from various components in the subducting plates should also be investigated in the future. In addition, although it is possible to estimate the proportion of recycled carbonates in the sources of basalts (e.g., Liu et al., 2020) and the relative fluxes of subducted carbonates related to different subducting slabs (Liu et al., 2022) using Mg and Zn isotopes, which may provide valuable information on atmospheric CO_2 changes through time, the total flux of recycled carbonates in the mantle is difficult to constrain due to incomplete sampling of the carbonate-bearing mantle by basaltic volcanism and limited data on natural samples. Further work is demanded to extend isotopic analyses for mantle xenoliths and basaltic lavas from other cratons, which will help estimate the carbon input flux in the lithospheric mantle.

From the experiences we learnt from the case studies of intraplate magmatism in East Asia and SE Xizang, future studies on magmatism generated in similar tectonic settings from other areas are clearly required to probe recycled carbonates in the transition zone at larger scales. For example, widespread Cenozoic intraplate alkali basalts and ultrapotassic mafic lavas occur in the Circum-Mediterranean area, genetically associated with chemically anomalous mantle sources metasomatized by recycled crustal materials (e.g., Lustrino and Wilson, 2007; Prelević et al., 2008). A large-scale stagnant slab at depths of ~400–600 km was recognized in the Circum-Mediterranean area (Fukao et al., 2009), as seen in SE Xizang and East Asia. Some ultrapotassic mantle-derived lavas are proposed to have been derived from the transition zone containing recycled ancient sediments with K being sequestered by K-hollandite (Murphy et al., 2002). Studies on these rocks and mineral inclusions within them may provide more information on the transition zone as an important storage of global subducting carbon.

Declaration of Competing Interest

All the authors know and concur with the submission of this manuscript to *Earth-Science Reviews* journal and declare no conflict of interests.

Acknowledgements

The critical and thoughtful comments by the anonymous reviewers are gratefully acknowledged and helped to improve the quality of this paper. Special thanks go to editor Christina Yan Wang for the efficient handling and improving the presentation. We thank Tian-Hao Wu, Zi-Tan Shu and Wan-Yu Sun for help with data collection and drawing, and Xiao-Ming Liu for providing the Zn concentration data for global carbonate sediments. This paper is supported by the National Key R&D Program of China (2019YFA0708400), the National Nature Science Foundation of China (Grant No. 41730214), and the "Strategic Priority Research Program" of the Chinese Academy of Sciences (Grant No. XDB18000000) to LSA.

References

Ague, J.J., Nicolescu, S., 2014. Carbon dioxide released from subduction zones by fluid-mediated reactions. Nat. Geosci. 7 (5), 355-360.

An, Y., Huang, J.X., Griffin, W.L., Liu, C., Huang, F., 2017. Isotopic composition of Mg and Fe in garnet peridotites from the Kaapvaal and Siberian cratons. Geochim. Cosmochim. Acta 200, 167-185.

Anzolini, C., Nestola, F., Mazzucchelli, M.L., Alvaro, M., Nimis, P., Gianese, A., Morganti, S., Marone, F., Campione, M., Hutchison, M.T., Harris, J.W., 2019. Depth of diamond formation obtained from single periclase inclusions. Geology 47 (3), 219-222.

Aubaud, C., Pineau, F., Hékinian, R., Javoy, M., 2005. Degassing of CO_2 and H_2O in submarine lavas from the Society hotspot. Earth Planet. Sci. Lett. 235, 511-527.

Aulbach, S., Lin, A.-B., Weiss, Y., Yaxley, G.M., 2020. Wehrlites from continental mantle monitor the passage and degassing of carbonated melts. Geochem. Perspect. Lett. 15, 30-34.

Azmy, K., Lavoie, D., Wang, Z., Brand, U., Al-Aasm, I., Jackson, S., Girard, I., 2013. Magnesium-isotope and REE compositions of lower Ordovician carbonates from eastern Laurentia: Implications for the origin of dolomites and limestones. Chem. Geol. 356, 64-75.

Barth, M.G., Rudnick, R.L., Horn, I., McDonough, W.F., Spicuzza, M.J., Valley, J.W., Hag-gerty, S.E., 2001. Geochemistry of xenolithic eclogites from West Africa, part I: a link between low MgO eclogites and Archean crust formation. Geochim. Cosmochim. Acta 65, 1499-1527.

Beunon, H., Mattielli, N., Doucet, L.S., Moine, B., Debret, B., 2020. Mantle heterogeneity through Zn systematics in oceanic basalts: evidence for a deep carbon cycling. Earth Sci. Rev. 205, 103174.

Black, J.R., Kavner, A., Schauble, E.A., 2011. Calculation of equilibrium stable isotope partition function ratios for aqueous zinc complexes and metallic zinc. Geochim. Cosmochim. Acta 75, 769-783.

Blätler, C.L., Miller, N.R., Higgins, J.A., 2015. Mg and ca isotope signatures of authigenic dolomite in siliceous deep-sea sediments. Earth Planet. Sci. Lett. 419, 32-42.

Bouibes, A., Zaoui, A., 2014. High-pressure polymorphs of $ZnCO_3$: evolutionary crystal structure prediction. Sci. Rep. 4, 1-6.

Boulard, E., Goncharov, A.F., Blanchard, M., Mao, L., W., 2015. Pressure-induced phase transition in $MnCO_3$ and its implications on the deep carbon cycle. J. Geophys. Res. Solid Earth 120 (6), 4069-4079.

Brenker, F.E., Vollmer, C., Vincze, L., Vekemans, B., Szymanski, A., Janssens, K., Szaloki, I., Nasdala, L., Joswig, W., Kaminsky, F., 2007. Carbonates from the lower part of transition zone or even the lower mantle. Earth Planet. Sci. Lett. 260 (1-2), 1-9.

Brewer, A.W., Teng, F.-Z., Mullen, E., 2018. Magnesium Isotopes as a Tracer of Crustal Materials in Volcanic Arc Magmas in the Northern Cascade Arc. Front. Earth Sci. 6.

Caciagli, N.C., Manning, C.E., 2003. The solubility of calcite in water at 6-16 kbar and 500-800℃. Contrib. Mineral. Petrol. 146 (3), 275-285.

Cerantola, V., Bykova, E., Kupenko, I., Merlini, M., Ismailova, L., McCammon, C., Bykov, M., Chumakov, A.I., Petitgirard, S., Kantor, I., Svitlyk, V., Jacobs, J., Hanfland, M., Mezouar, M., Prescher, C., Ruffer, R., Prakapenka, V.B., Dubrovinsky, L., 2017. Stability of iron-bearing carbonates in the deep Earth's interior. Nat. Commun. 8, 15960.

Chen, H., Savage, P.S., Teng, F.-Z., Helz, R.T., Moynier, F., 2013. Zinc isotope fractionation during magmatic differentiation and the isotopic composition of the bulk Earth. Earth Planet. Sci. Lett. 369, 34-42.

Chen, X., Sageman, B.B., Yao, H., Liu, S.A., Han, K., Zou, Y., Wang, C., 2021. Zinc isotope evidence for paleoenvironmental changes during cretaceous Oceanic Anoxic Event 2. Geology 49 (4), 412-416.

Choi, S.H., Liu, S.-A., 2022. Zinc isotopic systematics of the Mt. Baekdu and Jeju Island intraplate basalts in Korea, and implications for mantle source lithologies. Lithos 416-417, 106659.

Clift, P.D., 2017. A revised budget for Cenozoic sedimentary carbon subduction. Rev. Geophy. 55, 97-125.

Connolly, J.A.D., Galvez, M.E., 2018. Electrolytic fluid speciation by Gibbs energy minimization and implications for subduction zone mass transfer. Earth Planet. Sci. Lett. 501, 90-102.

Dasgupta, R., Hirschmann, M.M., 2010. The deep carbon cycle and melting in Earth's interior. Earth Planet. Sci. Lett. 298 (1-2), 1-13.

Dasgupta, R., Hirschmann, M.M., Withers, A.C., 2004. Deep global cycling of carbon constrained by the solidus of anhydrous, carbonated eclogite under upper mantle conditions. Earth Planet. Sci. Lett. 227 (1-2), 73-85.

Davis, F.A., Humayun, M., Hirschmann, M.M., Cooper, R.S., 2013. Experimentally determined mineral/melt partitioning of first-row transition elements (FRTE) during partial melting of peridotite at 3 GPa. Geochim. Cosmochim. Acta 104, 232-260.

De Leeuw, G.A.M., Hilton, D.R., Fischer, T.P., Walker, J.A., 2007. The He-CO_2 isotope and relative abundance characteristics of geothermal fluids in el salvador and honduras: New constraints on volatile mass balance of the central american volcanic arc. Earth Planet. Sci. Lett. 258 (1-2), 132-146.

Debret, B., Bouilhol, P., Pons, M.L., Williams, H., 2018. Carbonate transfer during the onset of slab devolatilization: new Insights from Fe and Zn stable isotopes. J. Petrol. 59, 1145-1166.

Debret, B., Garrido, C.J., Pons, M.-L., Bouilhol, P., Inglis, E., López Sánchez-Vizcaíno, V., Williams, H., 2021. Iron and zinc stable isotope evidence for open-system highpressure dehydration of antigorite serpentinite in subduction zones. Geochim. Cosmochim. Acta 296, 210-225.

Deines, P., 2002. The carbon isotope geochemistry of mantle xenoliths. Earth Sci. Rev. 58, 247-278.

Dixon, J.E., Stolper, E.M., Holloway, J.R., 1995. An experimental study of water and carbon dioxide solubilities in mid-ocean ridge basaltic liquids. Part I: calibration and solubility models. J. Petrol. 36 (6), 1607-1631.

Doucet, L.S., Mattielli, N., Ionov, D.A., Debouge, W., Golovin, A.V., 2016. Zn isotopic heterogeneity in the mantle: a melting control? Earth Planet. Sci. Lett. 451, 232-240.

Ducher, M., Blanchard, M., Balan, E., 2016. Equilibrium zinc isotope fractionation in Znbearing minerals from first-principles calculations. Chem. Geol. 443, 83-96.

Evans, K.A., 2012. The redox budget of subduction zones. Earth-Sci. Rev. 113, 11-32.

Fantle, M.S., Higgins, J., 2014. The effects of diagenesis and dolomitization on ca and Mg isotopes in marine platform carbonates: Implications for the geochemical cycles of ca and Mg. Geochim. Cosmochim. Acta 142, 458-481.

Ferrando, S., Groppo, C., Frezzotti, M.L., Castelli, D., Proyer, A., 2017. Dissolving dolomite in a stable UHP mineral assemblage: evidence from Cal-Dol marbles of the Dora-Maira Massif (Italian Western Alps). Am. Mineral. 102 (1), 42-60.

Foley, S.F., Fischer, T.P., 2017. An essential role for continental rifts and lithosphere in the deep carbon cycle. Nat. Geosci. 10 (12), 897-902.

Frezzotti, M.L., Selverstone, J., Sharp, Z.D., Compagnoni, R., 2011. Carbonate dissolution during subduction revealed by diamond-bearing rocks from the Alps. Nat. Geosci. 4(10), 703-706.

Fujii, T., Moynier, F., Pons, M.-L., Albarède, F., 2011. The origin of Zn isotope fractionation in sulfides. Geochim. Cosmochim. Acta 75, 7632-7643.

Fujii, T., Moynier, F., Abe, M., Nemoto, K., Albarède, F., 2013. Copper isotope fractionation between aqueous compounds relevant to low temperature geochemistry and biology. Geochim. Cosmochim. Acta 110, 29-44.

Fukao, Y., Obayashi, M., Nakakuki, T., 2009. Stagnant Slab: a Review. Annu. Rev. Earth. Pl Sc. 37 (1), 19-46.

Galvez, M.E., Pubellier, M., 2019. How do subduction zones regulate the carbon cycle? In: Orcutt, B.N., Daniel, I., Dasgupta, R. (Eds.), Deep Carbon: Past to Present. Cambridge University Press, Cambridge, pp. 276-312.

Geske, A., Zorlu, J., Richter, D.K., Buhl, D., Niedermayr, A., Immenhauser, A., 2012. Impact of diagenesis and low grade metamorphosis on isotope (δ^{26}Mg, δ^{13}C, δ^{18}O and ^{87}Sr/^{86}Sr) and elemental (Ca, Mg, Mn, Fe and Sr) signatures of Triassic sabkha dolomites. Chem. Geol. 332-333, 45-64.

Gorce, J.S., Caddick, M.J., Bodnar, R.J., 2019. Thermodynamic constraints on carbonate stability and carbon volatility during subduction. Earth Planet. Sci. Lett. 519, 213-222.

Gussone, N., Ahm, A.S.C., Lau, K.V., Bradbury, H.J., 2020. Calcium isotopes in deep time: potential and limitations. Chem. Geol. 544, 119601.

Hammouda, T., 2003. High-pressure melting of carbonated eclogite and experimental constraints on carbon recycling and storage in the mantle. Earth Planet. Sci. Lett. 214, 357-368.

Higgins, J.A., Schrag, D.P., 2012. Records of Neogene seawater chemistry and diagenesis in deep-sea carbonate sediments and pore fluids. Earth Planet. Sci. Lett. 357-358, 386-396.

Higgins, J.A., Schrag, D.P., 2015. The Mg isotopic composition of Cenozoic seawater—evidence for a link between Mg-clays, seawater Mg/Ca, and climate. Earth Planet. Sci. Lett. 416, 73-81.

Hofmann, A.W., 1988. Chemical differentiation of the Earth: the relationship between mantle, continental crust, and oceanic crust. Earth Planet. Sci. Lett. 90 (3), 297-314.

Hu, Y., Teng, F.-Z., Zhang, H.-F., Xiao, Y., Su, B.-X., 2016. Metasomatism-induced mantle magnesium isotopic heterogeneity: evidence from pyroxenites. Geochim. Cosmochim. Acta 185, 88-111.

Hu, J., Jiang, N., Carlson, R.W., Guo, J., Fan, W., Huang, F., Zhang, S., Zong, K., Li, T., Yu, H., 2019. Metasomatism of the crust-mantle boundary by melts derived from subducted sedimentary carbonates and silicates. Geochim. Cosmochim. Acta 260, 311-328.

Hu, Y., Teng, F.-Z., Ionov, D.A., 2020. Magnesium isotopic composition of metasomatized upper sub-arc mantle and its implications to Mg cycling in subduction zones. Geochim. Cosmochim. Acta 278, 219-234.

Hu, Z., Li, W., Zhang, H., Krainer, K., Zheng, Q.-F., Xia, Z., Hu, W., Shen, S.-Z., 2021. Mg isotope evidence for restriction events within the Paleotethys ocean around the Permian-Triassic transition. Earth Planet. Sci. Lett. 556.

Huang, J., Zhao, D., 2006. High-resolution mantle tomography of China and surrounding regions. J. Geophys. Res. Solid Earth 111 (B9), B09305.

Huang, S., Farkaš, J., Jacobsen, S.B., 2011. Stable calcium isotopic compositions of Hawaiian shield lavas: evidence for recycling of ancient marine carbonates into the mantle. Geochim. Cosmochim. Acta 75 (17), 4987-4997.

Huang, K.-J., Teng, F.-Z., Elsenouy, A., Li, W.-Y., Bao, Z.-Y., 2013. Magnesium isotopic variations in loess: Origins and implications. Earth Planet. Sci. Lett. 374, 60-70.

Huang, J., Li, S.-G., Xiao, Y., Ke, S., Li, W.-Y., Tian, Y., 2015a. Origin of low δ^{26}Mg Cenozoic basalts from South China Block and their geodynamic implications. Geochim. Cosmochim. Acta 164, 298-317.

Huang, J., Ke, S., Gao, Y., Xiao, Y., Li, S., 2015b. Magnesium isotopic compositions of altered oceanic basalts and gabbros from IODP site 1256 at the East Pacific rise. Lithos 231, 53-61.

Huang, K.-J., Shen, B., Lang, X.-G., Tang, W.-B., Peng, Y., Ke, S., Kaufman, A.J., Ma, H.-R., Li, F.-B., 2015c. Magnesium isotopic compositions of the Mesoproterozoic dolostones: Implications for Mg isotopic systematics of marine carbonates. Geochim. Cosmochim. Acta 164, 333-351.

Huang, J., Liu, S.-A., Gao, Y., Xiao, Y., Chen, S., 2016. Copper and zinc isotope systematics of altered oceanic crust at IODP Site 1256 in the eastern equatorial Pacific. J. Geophys. Res. Solid Earth 121 (10), 7086-7100.

Huang, J., Zhang, X.-C., Chen, S., Tang, L., Wörner, G., Yu, H., Huang, F., 2018a. Zinc isotopic systematics of Kamchatka-Aleutian arc magmas controlled by mantle melting. Geochim. Cosmochim. Acta 238, 85-101.

Huang, J., Chen, S., Zhang, X., Huang, F., 2018b. Effects of melt percolation on Zn isotope heterogeneity in the mantle: constraints from peridotite massifs in Ivrea-Verbano zone, Italian Alps. J. Geophys. Res. Solid Earth 123, 2706-2722.

Huang, K.-J., Teng, F.-Z., Plank, T., Staudigel, H., Hu, Y., Bao, Z.-Y., 2018c. Magnesium isotopic composition of altered oceanic crust and the global Mg cycle. Geochim. Cosmochim. Acta 238, 357-373.

Huang, J., Ackerman, L., Zhang, X., Huang, F., 2019. Mantle Zn isotopic heterogeneity caused by melt-rock reaction: evidence from Fe-rich peridotites and pyroxenites from the Bohemian massif, Central Europe. J. Geophys. Res. Solid Earth 124, 3588-3604.

Inglis, E.C., Debret, B., Burton, K.W., Millet, M.-A., Pons, M.-L., Dale, C.W., Bouilhol, P., Cooper, M., Nowell, G.M., McCoy-West, A.J., Williams, H.M., 2017. The behavior of iron and zinc stable isotopes accompanying the subduction of mafic oceanic crust: a case study from Western Alpine ophiolites. Geochem. Geophys. Geosyst. 18 (7), 2562-2579.

Ionov, D.A., Dupuy, C., O'Reilly, S.Y., Kopylova, M.G., Genshaft, Y.S., 1993. Carbonated peridotite xenoliths from Spitsbergen: implications for trace element signature of mantle carbonate metasomatism. Earth Planet. Sci. Lett. 119, 283-297.

Jacob, D., 2004. Nature and origin of eclogite xenoliths from kimberlites. Lithos 77(1-4), 295-316.

Jin, Q.-Z., Huang, J., Liu, S.-C., Huang, F., 2020. Magnesium and zinc isotope evidence for recycled sediments and oceanic crust in the mantle sources of continental basalts from eastern China. Lithos 370, 105627.

John, S.G., Kunzmann, M., Townsend, E.J., Rosenberg, A.D., 2017. Zinc and cadmium stable isotopes in the geological record: a case study from the post-snowball Earth Nuccaleena cap dolostone. Palaeogeogr. Palaeoclimatol. Palaeoecol. 466, 202-208.

Kang, J.-T., Qi, Y.-H., Li, K., Bai, J.-H., Yu, H.-M., Zheng, W., Zhang, Z.-F., Huang, F., 2021. Calcium isotope compositions of arc magmas: Implications for Ca and carbonate recycling in subduction zones. Geochim. Cosmochim. Acta 306, 1-19.

Kasemann, S.A., Pogge von Strandmann, P.A.E., Prave, A.R., Fallick, A.E., Elliott, T., Hoffmann, K.-H., 2014. Continental weathering following a Cryogenian glaciation: evidence from calcium and magnesium isotopes. Earth Planet. Sci. Lett. 396, 66-77.

Kasting, J.F., Ackerman, T.P., 1986. Climatic consequences of very high carbon dioxide levels in the Earth's early atmosphere.

Science 234 (4782), 1383-1385.

Kato, C., Moynier, F., Valdes, M.C., Dhaliwal, J.K., Day, J.M.D., 2015. Extensive volatile loss during formation and differentiation of the Moon. Nat. Commun. 6, 7617.

Ke, S., Teng, F.-Z., Li, S.-G., Gao, T., Liu, S.-A., He, Y., Mo, X., 2016. Mg, Sr, and O isotope geochemistry of syenites from Northwest Xinjiang, China: Tracing carbonate recycling during Tethyan oceanic subduction. Chem. Geol. 437, 109-119.

Kelemen, P.B., Manning, C.E., 2015. Reevaluating carbon fluxes in subduction zones, what goes down, mostly comes up. Proc. Natl. Acad. Sci. U. S. A. 112 (30), E3997-E4006.

Kerrick, D.M., Connolly, J.A.D., 2001. Metamorphic devolatilization of subducted marine sediments and the transport of volatiles into the Earth's mantle. Nature 41, 293-296.

Kessel, R., Schmidt, M.W., Ulmer, P., Pettke, T., 2005. Trace element signature of subduction-zone fluids, melts and supercritical liquids at 120-180 km depth. Nature 437 (7059), 724-727.

Kunzmann, M., Halverson, G.P., Sossi, P.A., Raub, T.D., Payne, J.L., Kirby, J., 2013. Zn isotope evidence for immediate resumption of primary productivity after snowball Earth. Geology 41 (1), 27-30.

Kushiro, I., 1975. Carbonate-silicate reactions at high pressures and possible presence of dolomite and magnesite in the upper mantle. Earth Planet. Sci. Lett. 28, 116-120.

Lavoie, D., Jackson, S., Girard, I., 2014. Magnesium isotopes in high-temperature saddle dolomite cements in the lower Paleozoic of Canada. Sediment. Geol. 305, 58-68.

Le Roux, V., Lee, C.T.A., Turner, S.J., 2010. Zn/Fe systematics in mafic and ultramafic systems: Implications for detecting major element heterogeneities in the Earth's mantle. Geochim. Cosmochim. Acta 74 (9), 2779-2796.

Le Roux, V., Dasgupta, R., Lee, C.-T.A., 2015. Recommended mineral-melt partition coefficients for FRTEs (Cu), Ga, and Ge during mantle melting. Am. Mineral. 100, 2533-2544.

Lee, H., Muirhead, J.D., Fischer, T.P., Ebinger, C.J., Kattenhorn, S.A., Sharp, Z.D., Kianji, G., 2016. Massive and prolonged deep carbon emissions associated with continental rifting. Nat. Geosci. 9 (2), 145-149.

Lei, J., Zhao, D., Xu, X., Xu, Y.-G., Du, M., 2019. Is there a big mantle wedge under eastern Tibet? Phys. Earth Planet. Inter. 292, 100-113.

Li, W.-Y., Teng, F.-Z., Ke, S., Rudnick, R.L., Gao, S., Wu, F.-Y., Chappell, B.W., 2010. Heterogeneous magnesium isotopic composition of the upper continental crust. Geochim. Cosmochim. Acta 74 (23), 6867-6884.

Li, J.L., Klemd, R., Gao, J., Meyer, M., 2014. Compositional zoning in dolomite from lawsonite-bearing eclogite (SW Tianshan, China): evidence for prograde metamorphism during subduction of oceanic crust. Am. Mineral. 99 (1), 206-217.

Li, S.-G., Yang, W., Ke, S., Meng, X., Tian, H., Xu, L., He, Y., Huang, J., Wang, X.-C., Xia, Q., Sun, W., Yang, X., Ren, Z.-Y., Wei, H., Liu, Y., Meng, F., Yan, J., 2017. Deep carbon cycles constrained by a large-scale mantle Mg isotope anomaly in eastern China. Natl. Sci. Rev. 4 (1), 111-120.

Li, M.-L., Liu, S.-A., Lee, H.-Y., Yang, C., Wang, Z.-Z., 2021. Magnesium and zinc isotopic anomaly of Cenozoic lavas in Central Myanmar: Origins and implications for deep carbon recycling. Lithos 386-387, 106011.

Litasov, K.D., Fei, Y., Ohtani, E., Kuribayashi, T., Funakoshi, K., 2008. Thermal equation of state of magnesite to 32 GPa and 2073 K. Phys. Earth Planet. Inter. 168, 191-203.

Little, S.H., Vance, D., McManus, J., Severmann, S., 2016. Key role of continental margin sediments in the oceanic mass balance of Zn and Zn isotopes. Geology 44, 207-210. Liu, S.-A., Li, S.-G., 2019. Tracing the Deep Carbon Cycle using Metal Stable Isotopes: Opportunities and challenges. Engineering 5 (3), 448-457.

Liu, S.-A., Teng, F.-Z., Yang, W., Wu, F.-Y., 2011. High-temperature inter-mineral magnesium isotope fractionation in mantle xenoliths from the North China craton. Earth Planet. Sci. Lett. 308 (1-2), 131-140.

Liu, C., Wang, Z., Raub, T.D., Macdonald, F.A., Evans, D.A.D., 2014. Neoproterozoic cap-dolostone deposition in stratified glacial meltwater plume. Earth Planet. Sci. Lett. 404, 22-32.

Liu, D., Zhao, Z., Zhu, D.-C., Niu, Y., Widom, E., Teng, F.-Z., DePaolo, D.J., Ke, S., Xu, J.- F., Wang, Q., Mo, X., 2015. Identifying mantle carbonatite metasomatism through Os-Sr-Mg isotopes in Tibetan ultrapotassic rocks. Earth Planet. Sci. Lett. 430, 458-469.

Liu, S.-A., Wang, Z.-Z., Li, S.-G., Huang, J., Yang, W., 2016a. Zinc isotope evidence for a large-scale carbonated mantle beneath eastern China. Earth Planet. Sci. Lett. 444, 169-178.

Liu, X.M., Kah, L.C., Knoll, A.H., Cui, H., Kaufman, A.J., Shahar, A., Hazen, R.M., 2016b. Tracing Earth's O_2 evolution using Zn/Fe ratios in marine carbonates. Geochem. Perspect. Lett. 2 (1), 24-34.

Liu, S.-A., Wu, H., Shen, S.-Z., Jiang, G., Zhang, S., Lv, Y., Zhang, H., Li, S., 2017. Zinc isotope evidence for intensive magmatism immediately before the end-Permian mass extinction. Geology 45 (4), 343-346.

Liu, S.-A., Liu, P.-P., Lv, Y., Wang, Z.-Z., Dai, J.-G., 2019. Cu and Zn isotope fractionation during oceanic alteration: Implications for Oceanic Cu and Zn cycles. Geochim. Cosmochim. Acta 257, 191-205.

Liu, S.-A., Wang, Z.Z., Yang, C., Li, S.G., Ke, S., 2020. Mg and Zn isotope evidence for two types of mantle metasomatism and deep recycling of magnesium carbonates. J. Geophys. Res. Solid Earth 125 (11), 1-22.

Liu, S.-A., Wu, T., Li, S., Wang, Z., Liu, J., 2022. Contrasting fates of subducting carbon related to different oceanic slabs in East Asia. Geochim. Cosmochim. Acta. https:// doi.org/10.1016/j.gca.2022.03.009.

Lustrino, M., Wilson, M., 2007. The circum-Mediterranean anorogenic Cenozoic igneous province. Earth Sci. Rev. 81 (1-2), 1-65.

Luth, R.W., 2001. Experimental determination of the reaction aragonite + magnesite = dolomite at 5 to 9 GPa. Contrib. Mineral. Petrol. 141 (2), 222-232.

Lv, Y., Liu, S.-A., Wu, H., Hohl, S.V., Chen, S., Li, S., 2018. Zn-Sr isotope records of the Ediacaran Doushantuo Formation in South China: diagenesis assessment and implications. Geochim. Cosmochim. Acta 239, 330-345.

Lv, M., Dorfman, S.M., Badro, J., Borensztajn, S., Greenberg, E., Prakapenka, V.B., 2021. Reversal of carbonate-silicate cation exchange in cold slabs in Earth's lower mantle. Nat. Commun. 12 (1), 1712.

Ma, H., Xu, Y., Huang, K., Sun, Y., Ke, S., Peng, Y., Lang, X., Yan, Z., Shen, B., 2018. Heterogeneous Mg isotopic composition of the early Carboniferous limestone: implications for carbonate as a seawater archive. Acta Geochim. 37 (1), 1-18.

Manning, C.E., Shock, E.L., Sverjensky, D.A., 2013. The Chemistry of Carbon in Aqueous Fluids at Crustal and Upper-Mantle Conditions: Experimental and Theoretical Constraints. Rev. Mineral. Geochem. 75 (1), 109-148.

Maréchal, C.N., Nicolas, E., Douchet, C., Albarède, F., 2000. Abundance of zinc isotopes as a marine biogeochemical tracer. Geochem. Geophys. Geosyst. 1 (5).

Marshall, H.G., Walker, J.C., Kuhn, W.R., 1988. Long-term climate change and the geochemical cycle of carbon. J. Geophys. Res.-Atmos. 93 (D1), 791-801.

Mavromatis, V., Meister, P., Oelkers, E.H., 2014. Using stable Mg isotopes to distinguish dolomite formation mechanisms: a case study from the Peru margin. Chem. Geol. 385, 84-91.

Mavromatis, V., González, A.G., Dietzel, M., Schott, J.M., 2019. Zinc isotope fractionation during the inorganic precipitation of calcite-Towards a new pH proxy. Geochim. Cosmochim. Acta 244, 99-112.

Mazza, S.E., Gazel, E., Bizimis, M., Moucha, R., Beguelin, P., Johnson, E.A., McAleer, R. J., Sobolev, A.V., 2019. Sampling the volatile-rich transition zone beneath Bermuda. Nature 569 (7756), 398-403.

McCoy-West, A.J., Fitton, J.G., Pons, M.-L., Inglis, E.C., Williams, H.M., 2018. The Fe and Zn isotope composition of deep mantle source regions: Insights from Baffin Island picrites. Geochim. Cosmochim. Acta 238, 542-562.

McDonough, W.F., Sun, S.S., 1995. The composition of the Earth. Chem. Geol. 120, 223-253.

Moynier, F., Vance, D., Fujii, T., Savage, P., 2017. The isotope geochemistry of zinc and copper. Rev. Mineral. Geochem. 82, 543-600.

Murphy, D.T., Collerson, K.D., Kamber, B.S., 2002. Lamproites from Gaussberg, Antarctica: possible transition zone melts of Archaean subducted sediments. J. Petrol. 43 (6), 981-1001.

Nakamura, K., Kato, Y., 2004. Carbonatization of oceanic crust by the seafloor hy-drothermal activity and its significance as a CO_2 sink in the early Archean. Geochim. Cosmochim. Acta 68 (22), 4595-4618.

Orcutt, B.N., Daniel, I., Dasgupta, R., Crist, D.T., Edmons, M., 2019. Introduction to deep carbon: Past to present. In: Orcutt, B.N., Daniel, I., Dasgupta, R. (Eds.), Deep Carbon: Past to Present. Cambridge University Press, Cambridge, pp. 1-3.

Pan, D., Spanu, L., Harrison, B., Sverjensky, D.A., Galli, G., 2013. Dielectric properties of water under extreme conditions and transport of carbonates in the deep Earth. Proc. Natl. Acad. Sci. U. S. A. 110 (17), 6646-6650.

Pang, K.-N., Teng, F.-Z., Sun, Y., Chung, S.-L., Zarrinkoub, M.H., 2020. Magnesium isotopic systematics of the Makran arc magmas, Iran: Implications for crust-mantle Mg isotopic balance. Geochim. Cosmochim. Acta 278, 110-121.

Paniello, R.C., Day, J.M.D., Moynier, F., 2012. Zinc isotopic evidence for the origin of the Moon. Nature 490, 376-379.

Peng, Y., Shen, B., Lang, X.-G., Huang, K.-J., Chen, J.-T., Yan, Z., Tang, W.-B., Ke, S., Ma, H.-R., Li, F.-B., 2016. Constraining dolomitization by Mg isotopes: a case study from partially dolomitized limestones of the middle Cambrian Xuzhuang Formation, North China. Geochem. Geophys. Geosyst. 17 (3), 1109-1129.

Pichat, S., Douchet, C., Albarède, F., 2003. Zinc isotope variations in deep-sea carbonates from the eastern equatorial Pacific over the last 175 ka. Earth Planet. Sci. Lett. 210 (1-2), 167-178.

Piromallo, C., Morelli, A., 2003. Pwave tomography of the mantle under the Alpine-Mediterranean area. J. Geophys. Res. Solid Earth 108 (B2).

Plank, T., 2014. The chemical composition of subducting sediments. Treat. Geochem. 607-629.

Plank, T., Manning, C.E., 2019. Subducting carbon. Nature 574 (7778), 343-352.

Pokrovsky, B.G., Mavromatis, V., Pokrovsky, O.S., 2011. Co-variation of Mg and C isotopes in late Precambrian carbonates of the Siberian Platform: a new tool for tracing the change in weathering regime? Chem. Geol. 290 (1-2), 67-74.

Pons, M.L., Quitte, G., Fujii, T., Rosing, M.T., Reynard, B., Moynier, F., Douchet, C., Albarede, F., 2011. Early Archean serpentine mud volcanoes at Isua, Greenland, as a niche for early life. Proc. Natl. Acad. Sci. U. S. A. 108 (43), 17639-17643.

Pons, M.L., Debret, B., Bouilhol, P., Delacour, A., Williams, H., 2016. Zinc isotope evidence for sulfate-rich fluid transfer across subduction zones. Nat. Commun. 7, 13794.

Prelević, D., Foley, S.F., Romer, R., Conticelli, S., 2008. Mediterranean Tertiary lamproites derived from multiple source components in postcollisional geodynamics. Geochim. Cosmochim. Acta 72, 2125-2156.

Qu, Y.R., Liu, S.A., Wu, H., Li, M.L., Tian, H.C., 2022. Tracing carbonate dissolution in subducting sediments by zinc and magnesium isotopes. Geochim. Cosmochim. Acta 319, 56-72.

Rohrbach, A., Schmidt, M.W., 2011. Redox freezing and melting in the Earth's deep mantle resulting from carbon-iron redox coupling. Nature 472, 209-212.

Sano, Y., Marty, B., 1995. Origin of carbon in fumarolic gas from island arcs. Chem. Geol. 119 (1-4), 265-274.

Shen, J., Li, S.-G., Wang, S.-J., Teng, F.-Z., Li, Q.-L., Liu, Y.-S., 2018. Subducted Mg-rich carbonates into the deep mantle wedge. Earth Planet. Sci. Lett. 503, 118-130.

Shibuya, T., Tahata, M., Kitajima, K., Ueno, Y., Komiya, T., Yamamoto, S., Igisu, M., Terabayashi, M., Sawaki, Y., Takai, K., Yoshida, N., Maruyama, S., 2012. Depth variation of carbon and oxygen isotopes of calcites in Archean altered uppero-ceanic crust: implications for the CO_2 flux from ocean to oceanic crust in the Archean. Earth Planet. Sci. Lett. 321-322, 64-73.

Shirey, S.B., Cartigny, P., Frost, D.J., Keshav, S., Nestola, F., Nimis, P., Pearson, D.G., Sobolev, N.V., Walter, M.J., 2013. Diamonds and the Geology of Mantle Carbon. Rev. Mineral. Geochem. 75 (1), 355-421.

Sonke, J., Sivry, Y., Viers, J., Freydier, R., Dejonghe, L., Andre, L., Aggarwal, J., Fontan, F., Dupre, B., 2008. Historical variations in the isotopic composition of atmospheric zinc deposition from a zinc smelter. Chem. Geol. 252 (3-4), 145-157.

Sossi, P.A., Nebel, O., O'Neill, H.S.C., Moynier, F., 2018. Zinc isotope composition of the Earth and its behaviour during planetary accretion. Chem. Geol. 477, 73-84.

Stagno, V., 2019. Carbon, carbides, carbonates and carbonatitic melts in the Earth's interior. J. Geol. Soc. Lond. 176 (2), 375-387.

Stagno, V., Ojwang, D.O., McCammon, C.A., Frost, D.J., 2013. The oxidation state of the mantle and the extraction of carbon from Earth's interior. Nature 493, 84-88.

Stagno, V., Cerantola, V., Aulbach, S., Lobanov, S., McCammon, C.A., Merlini, M., 2019. Carbon-Bearing Phases throughout Earth's Interior: Evolution through Space and Time. Cambridge University Press, pp. 66-88.

Stracke, A., Tipper, E.T., Klemme, S., Bizimis, M., 2018. Mg isotope systematics during magmatic processes: Inter-mineral fractionation in mafic to ultramafic Hawaiian xenoliths. Geochim. Cosmochim. Acta 226, 192-205.

Su, B.-X., Hu, Y., Teng, F.Z., Qin, K.Z., Bai, Y., Sakyi, P.A., Tang, D.M., 2017. Chromite-induced magnesium isotope fractionation during mafic magma differentiation. Sci. Bull. 62, 1538-1546.

Su, B.-X., Hu, Y., Teng, F.-Z., Xiao, Y., Zhang, H.-F., Sun, Y., Bai, Y., Zhu, B., Zhou, X.-H., Ying, J.-F., 2019. Light Mg isotopes in mantle-derived lavas caused by chromite crystallization, instead of carbonatite metasomatism. Earth Planet. Sci. Lett. 522, 79-86.

Sun, Y., Teng, F.-Z., Pang, K.-N., Ying, J.-F., Kuehner, S., 2021. Multistage mantle metasomatism deciphered by Mg- Sr- Nd- Pb isotopes in the Leucite Hills lamproites. Contrib. Mineral. Petrol. 176 (6).

Sweere, T.C., Dickson, A.J., Jenkyns, H.C., Porcelli, D., Elrick, M., van den Boorn, S.H.J. M., Henderson, G.M., 2018. Isotopic evidence for changes in the zinc cycle during Oceanic Anoxic Event 2 (late cretaceous). Geology 46 (5), 463-466.

Syracuse, E.M., van Keken, P.E., Abers, G.A., 2010. The global range of subduction zone thermal models. Phys. Earth Planet. Inter. 183 (1-2), 73-90.

Teng, F.-Z., 2017. Magnesium Isotope Geochemistry. Rev. Mineral. Geochem. 82 (1), 219-287.

Teng, F.-Z., Li, W.Y., Ke, S., Marty, B., Dauphas, N., Huang, S., Wu, F.-Y., Pourmand, A., 2010. Magnesium isotopic composition of the Earth and chondrites. Geochim. Cosmochim. Acta 74 (14), 4150-4166.

Teng, F.Z., Hu, Y., Chauvel, C., 2016. Magnesium isotope geochemistry in arc volcanism. Proc. Natl. Acad. Sci. U. S. A. 113 (26), 7082-7087.

Thomson, A.R., Walter, M.J., Kohn, S.C., Brooker, R.A., 2016. Slab melting as a barrier to deep carbon subduction. Nature 529 (7584), 76-79.

Tian, H.-C., Yang, W., Li, S.-G., Ke, S., 2017. Could sedimentary carbonates be recycled into the lower mantle? Constraints from Mg isotopic composition of Emeishan basalts. Lithos 292-293, 250-261.

Tian, H.-C., Yang, W., Li, S.-G., Ke, S., Duan, X.-Z., 2018. Low δ^{26}Mg volcanic rocks of Tengchong in Southwestern China: a deep carbon cycle induced by supercritical liquids. Geochim. Cosmochim. Acta 240, 191-219.

Tian, H.C., Teng, F.Z., Hou, Z.Q., Tian, S.H., Yang, W., Chen, X.Y., Song, Y.C., 2020. Magnesium and lithium isotopic evidence for a remnant Oceanic Slab beneath Central Tibet. J. Geophys. Res. Solid Earth 125 (1).

Toutain, J.-P., Sonke, J., Munoz, M., Nonell, A., Polvé, M., Viers, J., Freydier, R., Sortino, F., Joron, J.-L., Sumarti, S., 2008. Evidence for Zn isotopic fractionation at Merapi volcano. Chem. Geol. 253, 74-82.

Turekian, K.K., Wedepohl, K.H., 1961. Distribution of the elements in some major units of the earth's crust. Geol. Soc. Am. Bull. 72, 175-192.

Vance, D., Little, S.H., Archer, C., Cameron, V., Andersen, M.B., Rijkenberg, M.J.A., Lyons, T.W., 2016. The oceanic budgets of nickel and zinc isotopes: the importance of sulfidic environments as illustrated by the Black Sea. Phil. Trans. R. Soc. A 374 (2081), 1-26.

Walker, J.C., Hays, P.B., Kasting, J.F., 1981. A negative feedback mechanism for the long-term stabilization of Earth's surface temperature. J. Geophys. Res. Oceans 86 (C10), 9776-9782.

Wang, Z.-Z., Liu, S.-A., 2021. Evolution of Intraplate Alkaline to Tholeiitic Basalts via Interaction between Carbonated Melt and Lithospheric Mantle. J. Petrol. 62 (1), 1-25.

Wang, S.J., Teng, F.Z., Williams, H.M., Li, S.G., 2012. Magnesium isotopic variations in cratonic eclogites: Origins and implications. Earth Planet. Sci. Lett. 359, 219-226.

Wang, S.J., Teng, F.-Z., Li, S.-G., Hong, J.-A., 2014a. Magnesium isotopic systematics of mafic rocks during continental subduction. Geochim. Cosmochim. Acta 143, 34-48.

Wang, S.J., Teng, F.Z., Li, S.G., 2014b. Tracing carbonate-silicate interaction during subduction using magnesium and oxygen isotopes. Nat. Commun. 5, 5328.

Wang, S.J., Teng, F.-Z., Rudnick, R.L., Li, S.-G., 2015. The behavior of magnesium isotopes in low-grade metamorphosed mudrocks. Geochim. Cosmochim. Acta 165, 435-448.

Wang, Z.-Z., Liu, S.-A., Ke, S., Liu, Y.-C., Li, S.-G., 2016. Magnesium isotopic heterogeneity across the cratonic lithosphere in eastern China and its origins. Earth Planet. Sci. Lett. 451, 77-88.

Wang, Z.-Z., Liu, S.-A., Liu, J., Huang, J., Xiao, Y., Chu, Z.-Y., Zhao, X.-M., Tang, L., 2017. Zinc isotope fractionation during mantle melting and constraints on the Zn isotope composition of Earth's upper mantle. Geochim. Cosmochim. Acta 198, 151-167.

Wang, Z.-Z., Liu, S.-A., Chen, L.-H., Li, S.-G., Zeng, G., 2018a. Compositional transition in natural alkaline lavas through silica-undersaturated melt-lithosphere interaction. Geology 46 (9), 771-774.

Wang, X.J., Chen, L.H., Hofmann, A.W., Hanyu, T., Kawabata, H., Zhong, Y., Xie, L.W., Shi, J.H., Miyazaki, T., Hirahara, Y., Takahashi, T., Senda, R., Chang, Q., Vaglarov, B.S., Kimura, J.I., 2018b. Recycled ancient ghost carbonate in the Pitcairn mantle plume. Proc. Natl. Acad. Sci. U. S. A. 115 (35), 8682-8687.

Wang, X., Liu, S.-A., Wang, Z., Chen, D., Zhang, L., 2018c. Zinc and strontium isotope evidence for climate cooling and constraints on the Frasnian-Famennian (~372 Ma) mass extinction. Palaeogeogr. Palaeoclimatol. Palaeoecol. 498, 68-82.

Wang, Y., He, Y., Ke, S., 2020. Mg isotope fractionation during partial melting of garnet-bearing sources: An adakite perspective. Chem. Geol. 537, 119478.

Wang, X.-J., Chen, L.-H., Hanyu, T., Zhong, Y., Shi, J.-H., Liu, X.-W., Kawabata, H., Zeng, G., Xie, L.-W., 2021a. Magnesium isotopic fractionation during basalt differentiation as recorded by evolved magmas. Earth Planet. Sci. Lett. 565, 116954.

Wang, X., Wang, Z., Liu, Y., Park, J.W., Kim, J., Li, M., Zou, Z., 2021b. Calcium stable isotopes of tonga and mariana arc lavas: implications for slab fluid-mediated carbonate transfer in Cold Subduction zones. J. Geophys. Res. Solid Earth 126 (3).

Wang, Z., Liu, S.-A., Li, S., Liu, D., Liu, J., 2022a. Linking deep CO_2 outgassing to cratonic destruction. Natl. Sci. Rev. nwac001 https://doi.org/10.1093/nsr/nwac001.

Wang, Z.-Z., Liu, S.-A., Rudnick, R.L., Teng, F.Z., Wang, S.J., Haggerty, S.E., 2022b. Zinc isotope evidence for carbonate alteration of oceanic crustal protoliths of cratonic eclogites. Earth Planet. Sci. Lett. 580, 117394.

Wimpenny, J., Yin, Q.-Z., Tollstrup, D., Xie, L.-W., Sun, J., 2014. Using Mg isotope ratios to trace Cenozoic weathering changes: a case study from the Chinese Loess Plateau. Chem. Geol. 376, 31-43.

Wombacher, F., Eisenhauer, A., Böhm, F., Gussone, N., Regenberg, M., Dullo, W.C., Rüggeberg, A., 2011. Magnesium stable isotope fractionation in marine biogenic calcite and aragonite. Geochim. Cosmochim. Acta 75 (19), 5797-5818.

Xiao, Y., Teng, F.-Z., Zhang, H.-F., Yang, W., 2013. Large magnesium isotope fractionation in peridotite xenoliths from eastern North China craton: product of melt-rock interaction. Geochim. Cosmochim. Acta 115, 241-261.

Xu, L.-J., Liu, S.-A., Li, S.G., 2021. Zinc isotopic behavior of mafic rocks during continental deep subduction. Geosci. Front. 12 (5).

Yang, C., Liu, S.-A., 2019. Zinc isotope constraints on recycled oceanic crust in the mantle sources of the Emeishan large Igneous Province. J. Geophys. Res. Solid Earth 124 (12), 12537-12555.

Yang, W., Teng, F.-Z., Zhang, H.-F., 2009. Chondritic magnesium isotopic composition of the terrestrial mantle: a case study of peridotite xenoliths from the North China craton. Earth Planet. Sci. Lett. 288 (3-4), 475-482.

Yang, W., Teng, F.-Z., Zhang, H.-F., Li, S.-G., 2012. Magnesium isotopic systematics of continental basalts from the North China craton: Implications for tracing subducted carbonate in the mantle. Chem. Geol. 328, 185-194.

Yang, C., Liu, S.-A., Zhang, L., Wang, Z.-Z., Liu, P.-P., Li, S.-G., 2021. Zinc isotope fractionation between Cr-spinel and olivine and its implications for chromite crystallization during magma differentiation. Geochim. Cosmochim. Acta 313, 277-294.

Yaxley, G.M., Green, D.H., 1994. Experimental demonstration of refractory carbonate-bearing eclogite and siliceous melt in the subduction regime. Earth Planet. Sci. Lett. 128, 313-325.

Zhong, Y., Chen, L.-H., Wang, X.-J., Zhang, G.-L., Xie, L.-W., Zeng, G., 2017. Magnesium isotopic variation of oceanic island basalts generated by partial melting and crustal recycling. Earth Planet. Sci. Lett. 463, 127-135.

Zhu, H., Du, L., Li, X., Zhang, Z., Sun, W., 2020. Calcium isotopic fractionation during plate subduction: Constraints from back-arc basin basalts. Geochim. Cosmochim. Acta 270, 379-393.

Linking deep CO_2 outgassing to cratonic destruction*

Zhao-Xue Wang, Sheng-Ao Liu, Shuguang Li, Di Liu and Jingao Liu

State Key Laboratory of Geological Processes and Mineral Resources, China University of Geosciences, Beijing 100083, China

亮点介绍： 本文发现华北克拉通东部早白垩世高 MgO 煌斑岩具有轻 Mg 同位素组成，结合古生代和中-新生代基性火山岩中幔源包体的低 Mg 同位素组成，表明至少 6.09×10^7 Mt 来自地表碳酸盐的碳俯冲加入至岩石圈地幔。对煌斑岩单斜辉石斑晶中的熔体包裹体研究发现，喷发前的岩浆含有高达 0.5~2.0 wt%的 CO_2，表明克拉通减薄可释放大量地幔 CO_2，并对白垩纪时期地表大气 CO_2 含量上升具有一定影响。

Abstract Outgassing of carbon dioxide from the Earth's interior regulates the surface climate through deep time. Here we examine the role of cratonic destruction in mantle CO_2 outgassing via collating and presenting new data for Paleozoic kimberlites, Mesozoic basaltic rocks and their mantle xenoliths from the eastern North China Craton (NCC), which underwent extensive destruction in the early Cretaceous. High Ca/Al and low Ti/Eu and δ^{26}Mg are widely observed in lamprophyres and mantle xenoliths, which demonstrates that the cratonic lithospheric mantle (CLM) was pervasively metasomatized by recycled carbonates. Raman analysis of bubble-bearing melt inclusions shows that redox melting of the C-rich CLM produced carbonated silicate melts with high CO_2 contents. The enormous quantities of CO_2 in these magmas, together with subtantial CO_2 degassing from the carbonated melt-CLM reaction and crustal heating, indicate that destruction of the eastern NCC resulted in rapid and extensive mantle CO_2 emission, which partly contributed to the early Cretaceous greenhouse climate episode.

Keywords Carbonate metasomatism, lithospheric mantle, CO_2 outgassing, deep carbon cycling, cratonic destruction, North China craton

1 Introduction

Carbon exchange between the Earth's interior and exterior exerts an important influence on the surface climate through geologic time and is critical for planetary habitability. In recent years, it has been increasingly recognized that the cratonic lithospheric mantle (CLM) stores vast amounts of carbon, resulting from gradual enrichment by upward melt infiltration, in addition to the original carbon incorporated during its formation [1,2]. Carbon in the CLM can be extensively remobilized and released via continental rifting [1,3], active island arc volcanism [4,5] and plume-related magmatism [6,7], which represent three main ways proposed for mantle CO_2

* 本文发表在：National Science Review, 2022, 9: nwac001

emission. For example, up to 28 to 34 Mt of carbon per year (expressed as Mt C yr^{-1}) may be released by continental rifting [1]. A quantitative flux estimate for the CO_2 outgassing along with the massive Tan-Lu Fault Belt in eastern China gave 70 ± 58 Mt C yr^{-1} [8]. Extensive CO_2 degassing (71 ± 33 Mt C yr^{-1}) has been estimated through extensional faults along the entire East African Rift [3], which is even comparable to the estimates for CO_2 degassing in island arcs (18–43 Mt C yr^{-1}) and mid-ocean ridges (8–42 Mt C yr^{-1}) [9-11]. Plume-induced CO_2 outgassing has also been proposed to have the ability to have caused abrupt climate changes in Earth's history [12].

Cratons commonly retain tectonic and magmatic quiescence for billions of years [13], but some cratonic regions record extensive crustal deformation and on-craton magmatism that reflect cratonic destruction processes [14,15]. In sharp contrast to many others on Earth, the eastern part of the North China Craton (NCC) is a reactivated craton with the present-day lithosphere being made up of decoupled crust and mantle, i.e., Archaean-Proterozoic crust and Phanerozoic lithospheric mantle [15]. The Archean thick (diamond-bearing) and cold lithospheric keel (>200 km) was partially or even wholly destroyed and removed, and was then replaced by a newly formed thin and hot lithospheric mantle (~75 km), resulting in up to ~120 km of the lithospheric keel being lost [14,16]. Reactivation of the CLM gave rise to a magmatic peak at ~125 Ma including both mafic and felsic magmatism, marking the climax of cratonic lithospheric destruction in the early Cretaceous [16,17]. Since destruction/thinning is confined to the eastern part of the NCC (its western part remains largely intact), the westward subduction of the paleo-Pacific oceanic slab underneath the eastern Asian continent in the early Cretaceous was widely advocated to have resulted in reactivation and destruction of the eastern NCC [18].

In order to examine whether cratonic lithospheric destruction results in massive mantle CO_2 outgassing, we firstly ascertain whether or not the CLM beneath the eastern NCC was initially carbon rich and whether it had been widely subjected to carbonate metasomatism, including carbon from recycled carbonates prior to destruction in the early Cretaceous. For this purpose, we analyzed magnesium (Mg) isotopes for early Cretaceous lamprophyres and collated available chemical and Mg isotopic data for mantle xenoliths in Paleozoic diamond-bearing kimberlites and early Cretaceous mafic igneous rocks as well as orogenic ultramafic massifs in the Dabie orogen located on the south margin of the eastern NCC (Fig. 1). Then, we analyzed the CO_2 components of melt inclusions (MIs) in early Cretaceous lamprophyres that can be used to calculate the CO_2 concentrations in pre-eruptive magmas. Finally, we considered carbonated silicate melt-CLM reaction and crustal heating as additional ways for mantle CO_2 outgassing to occur during cratonic destruction, in addition to mafic magmatism. Our results show that the CLM beneath the eastern NCC had widely interacted with carbonated melts prior to the early Cretaceous and extensive CO_2 emisson had occured as a consequence of cratonic destruction.

2 Primordial carbon in the clm

The terrestial mantle initially contained carbon resulting from accretion and core-mantle differentiation processes [11]. Recent studies provided a rigorous reconstruction of carbon concentration for the MORB source mantle and suggest that the upper mantle contains ~30 ppm C [19]. In addition to the primordial carbon, the continental lithospheric mantle, mainly formed between 2 and 3 Ga contains more carbon (~89 ppm) that is incorporated into the sub-arc lithosphere via the accretion of island arcs [1]. Assuming an area of ~1 000 000 km^2 and lithospheric mantle thickness of ~150 km for the eastern NCC prior to thinning in the Mesozoic, the CLM beneath the eastern NCC initially contains ~4.27×10^7 Mt C. In addition, the model of Foley and Fischer [1] predicts a long-term (>2 Ga) solid storage of carbon in CLM as a result of episodic melt infiltration and redox freezing. If one considers this gradual enrichment from episodic freezing throughout the long evolution history of the ancient NCC (as old as ~3.8 Ga; [20]), carbon concentration in the CLM beneath the eastern NCC may become much higher. Abundant diamonds were discovered in Paleozoic kimberlites from the NCC [21] and are direct evidence for a C-bearing, reduced CLM beneath the eastern NCC prior to the Mesozoic era. For example, the eruption of diamond-bearing kimberlites and high $Mg^{\#}$ (>90; 100×molar $Mg/(Mg+Fe^{2+})$) of olivines in diamond inclusions and xenolith/xenocryst olivines from Mengyin and Fuxian in the eastern NCC (Fig. 1) indicate the existence of a thick, low-density, cold root, which is mainly composed of refractory harzburgite and lherzolite [21].

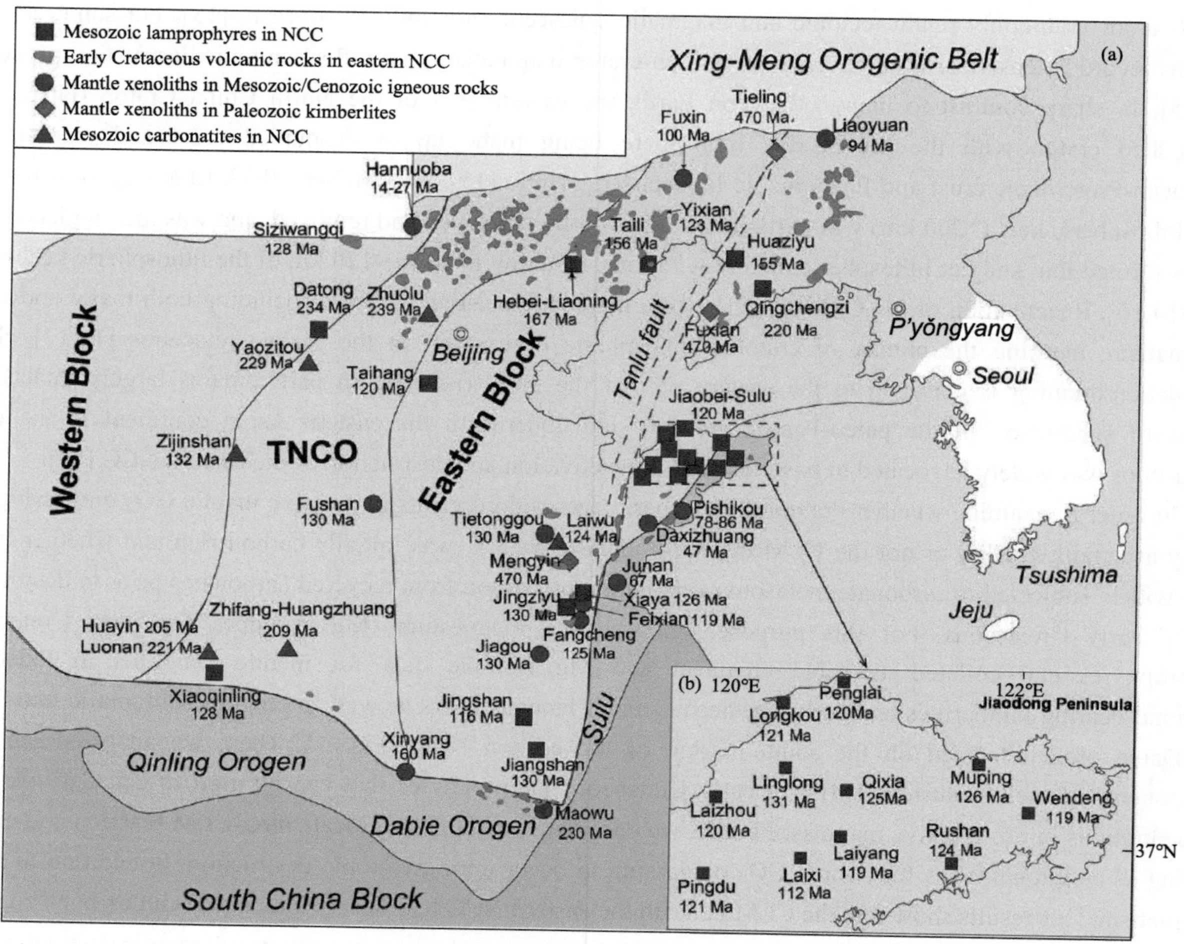

Fig. 1 (a) Spatial distribution of Paleozoic diamond-bearing kimberlites, Mesozoic lamprophyres, carbonatites, and mafic igneous rocks containing mantle xenoliths in the NCC (modified from ref. [36]). (b) The locations of early Cretaceous lamprophyres in the Jiaodong peninsula. More details and data sources are listed in the Supplematry Data.

3 Additional carbon from recycled carbonates

Since the Paleozoic, the NCC further underwent multiple oceanic plate subductions from the south, north and east sides, which potentially added surface carbon into the CLM beneath it. Below we present lines of evidence for more recent carbon addition to the CLM beneath the eastern NCC, from recycled carbonates related to slab subduction.

3.1 Finding of low-δ^{26}Mg lamprophyres

Magnesium isotopes are a novel and efficient tool for identifying recycled carbonates that are isotopically much lighter than the mantle [22,23]. Lamprophyres are typically characterized by a high content of volatiles and commonly record fluid/melt-mantle interaction in their magma sources [24], thereby providing an opportunity to investigate the nature of fluid/melt responsible for CLM metasomatism. Early Cretaceous lamprophyres are widely exposed in the NCC and have a magmatic peak at ~125 Ma (Fig. 1), which is contemporaneous with the climax of cratonic lithospheric destruction of the eastern NCC [25]. Here we present a Mg isotopic dataset for Shandong lamprophyres, and for comparison we collate chemical and Sr-Nd isotopic data for other early Cretaceous lamprophyres widely distributed in the eastern NCC (Fig. 1; Tables S1–S3). Some high-Ti lamprophyres from Shandong have relatively depleted Sr and Nd isotopic compositions (Fig. 2a) and were proposed to have been derived from the asthenospheric mantle [26]. Most of early Cretaceous lamprophyres in the eastern NCC, including those reported in this study, however, have extremely enriched Sr and Nd isotopic compositions ($^{87}Sr/^{86}Sr_{(i)}$ = 0.70520–0.71099, $\varepsilon_{Nd}(t)$ = −18.8 to −8.3) that are in sharp contrast to the High-Ti lamprophyres and Cenozoic alkali basalts in the NCC (Fig. 2a), pointing to an enriched CLM source. According to MgO contents, the Shandong lamprophyres are classified into Low-MgO (MgO<7.5 wt%) and High-MgO (MgO>7.5 wt%) subgroups (Fig. 2). Low-MgO lamprophyres have mantle-like δ^{26}Mg (−0.32‰ to −0.24‰), whereas High-MgO lamprophyres possess significantly lower δ^{26}Mg (−0.59‰ to −0.35‰) than the mantle δ^{26}Mg value of −0.25±0.04‰ (Fig. 2). It has been well demonstrated that Mg isotope fractionation during mantle partial melting and magma differentiation is limited (<0.07‰) [23]. In fact, the negative correlation between δ^{26}Mg and MgO (Fig. 2c) argues against the light δ^{26}Mg of the High-MgO lamprophyres being a result of fractional crystallization of any minerals involving removal of isotopically heavy Mg. More discussions about the influence of magma differentiation, as well as crustal contamination on Mg isotopic systematics of the studied lamprophyres, are provided in the Supplementary Data. Overall, the variation of δ^{26}Mg in Shandong lamprophyres reflects isotopically heterogeneous mantle sources caused by recycled crustal carbonates.

A carbonated mantle source for High-MgO lamprophyres is corroborated by their systematically higher CaO (High-MgO, 9.47±0.84 wt%; Low-MgO, 7.25±1.37 wt%; 1sd; Fig. S1) and CaO/Al_2O_3 ratios (High-MgO, 0.69±0.07; Low-MgO, 0.45±0.1; Fig. 2d) and lower Ni content (High-MgO, 81.96±80.43 ppm; Low-MgO, 154.37±81.89 ppm) in comparison with Low-MgO lamprophyres (Fig. S1), because marine carbonates are commonly Al and Ni poor and carbonate metasomatism can dramatically increase CaO of the mantle [27]. They also have distinct $(Ti/Eu)_N$ ratios of 0.29±0.03 (High-MgO; except for one sample with 0.53) and 0.43±0.07 (Low-MgO), respectively. Mantle carbonate metasomatism also accounts well for the relatively low SiO_2 of the High-MgO lamprophyres (Fig. S1) since partial melting of carbonated peridotites generates more Si-unsaturated melts relative to melting of volatile-poor peridotites [27]. The Mg-Sr isotopic mixing model (details listed in Table S4) suggests that the light-δ^{26}Mg lamprophyres require source metasomatism by at least ~10% recycled Mg-rich carbonates (e.g. dolomites; Fig. 2e). During slab subduction, Ca-rich carbonates can be substantially dissolved by aqueous fluids at initial stages and injected into the sub-arc mantle [9]. At larger depths of >160 km, dolomite dissolution occurs in subducting slabs and can be further enhanced by supercritical fluids [28]. The lithospheric mantle of the eastern NCC had a thickness of >200 km prior to thinning in the early Cretaceous [14,15]. Thus, the finding of low-δ^{26}Mg lamprophyres demonstrates that the CLM beneath the eastern NCC had been metasomatized by dissolved magnesium carbonates from subducting slabs. The Low-MgO rocks are SiO_2 rich and CaO poor and represent partial melts of the CLM metasomatized by recycled siliciclastic sediments, which explains their mantle-like δ^{26}Mg yet highly radiogenic $^{87}Sr/^{86}Sr$ compositions (Fig. 2). We suggest a two-stage

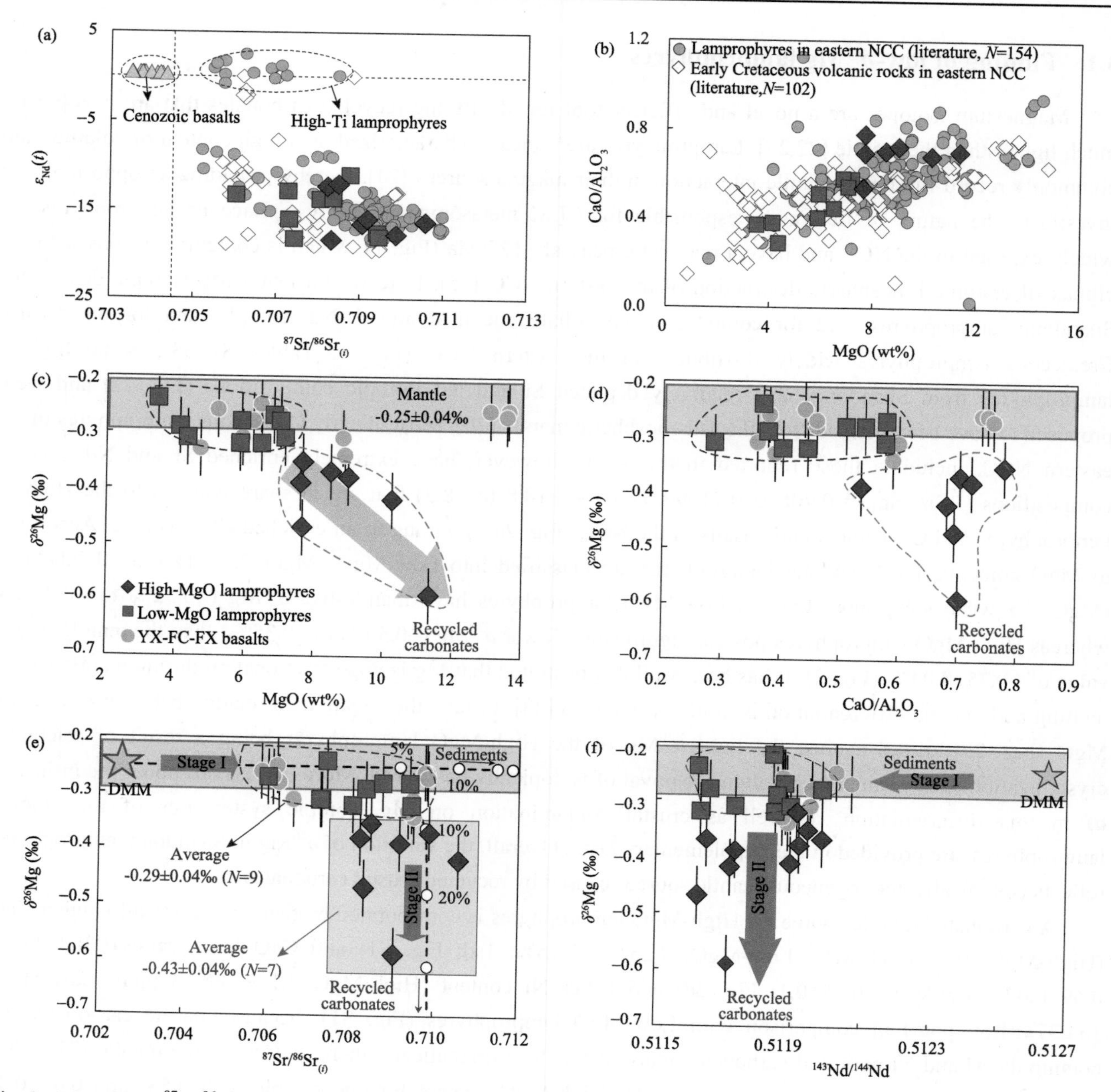

Fig. 2　Plots of (a) $^{87}Sr/^{86}Sr_{(i)}$ versus $\varepsilon_{Nd}(t)$ and (b) MgO versus CaO/Al_2O_3 for the lamprophyres, early Cretaceous volcanic rocks and Cenozoic basalts in the NCC. The gray circles represent the data of lamprophyres in previous studies, the gray triangles represent the data of Cenozoic basalts and the white diamonds represent the data of early Cretaceous volcanic rocks (all data are listed in Table S1-S3). Plots of $\delta^{26}Mg$ versus (c) MgO, (d) CaO/Al_2O_3, (e) $^{87}Sr/^{86}Sr_{(i)}$ and (f) $^{143}Nd/^{144}Nd$ for the studied lamprophyres. Early Cretaceous basalts from Yixian, Feixian and Fangcheng (YX-FX-FC, yellow cycles) in the eastern NCC [29] are shown for comparison. The magnesium isotopic composition of the terrestrial mantle (expressed as $\delta^{26}Mg$ in per mil relative to DSM3) is from ref. [23]. The results of Mg-Sr isotopic modelling show a two-stage source metasomatism: the first stage is associated with siliciclastic sediments, and the second stage is related to carbonate metasomatism. The parameters used for Mg-Sr isotopic modelling are listed in the Supplementary Data (Table S4).

source metasomatism: the first stage is associated with siliciclastic sediments that led to the enriched Sr and Nd isotopic signatures, and the second stage is related to carbonate metasomatism that injected the low $\delta^{26}Mg$ signatures, without significantly affecting Sr-Nd isotopic compositions (Fig. 2e, f). Previous studies found that early Cretaceous basalts from Fangcheng, Yixian, and Feixian in the eastern NCC have mantle-like $\delta^{26}Mg$ [29]

(Fig. 2c and d). It is noteworthy that these basalts have CaO/Al_2O_3 and CaO/TiO_2 ratios similar to those of the Low-MgO lamprophyres with normal $\delta^{26}Mg$. This indicates that the carbonated CLM may be chiefly sampled by lamprophyres that are commonly derived from a volatile-rich source [24]. Indeed, most of the Low-Ti lamprophyres with enriched Sr-Nd isotopic compositions from other regions in the eastern NCC have high CaO/Al_2O_3 and MgO contents, which resembles the Shandong High-MgO lamprophyres with low $\delta^{26}Mg$ (Fig. 2a, b). This implies that the pre-Cenozoic CLM beneath the eastern NCC may have undergone widespread metasomatism by recycled carbonates. Carbonate minerals (e.g. magnesite and calcite) are often observed in coeval lamprophyres in the NCC [30]. Overall, our new Mg isotopic data provide solid evidence for a recycled carbonate component in CLM beneath the eastern NCC at or prior to the early Cretaceous.

3.2 Evidence from xenoliths, ultramafic massifs and carbonatites

Mantle-derived xenoliths in volcanic rocks and orogenic ultramafic massifs sample the lithospheric mantle, serving as a direct window to observe mantle metasomatism. A large number of lherzolite, wehrlite and clinopyroxenite xenoliths carried by Paleozoic diamond-bearing kimberlites, Mesozoic and Cenozoic mafic igneous rocks, and ultramafic massifs in the Dabie orogen derived from the deep mantle wedge (>160 km) on the south margin of the NCC (Fig. 1), record carbonate metasomatism of the CLM beneath the NCC. Wehrlites represent rocks where all, or most, orthopyroxene has been consumed through metasomatic reactions and are considered to be one of the end products of carbonate metasomatism in the CLM [8,31]. Pyroxenites and garnet pyroxenites represent rocks where all olivine and orthopyroxene have been consumed through metasomatic reactions with SiO_2 carried by supercritical fluid or silica-rich melt and are therefore considered to be other end products of carbonate metasomatism in the CLM. Abundant wehrlite xenoliths have been found in Mesozoic basaltic rocks from Tietonggou and Liaoyuan (Fig. 1a) [32,33], which suggests pervasive carbonate metasomatism of the CLM. Pyroxenite xenoliths hosted by the Jiagou intrusion (~130 Ma) in the southeastern NCC (Fig. 1) have extremely low $\delta^{26}Mg$ of –1.23‰ to –0.73‰ (Fig. 3a) [34]. These pyroxenite xenoliths have a metasomatic U-Pb isotopic age of ~400 Ma, suggesting carbonate metasomatism induced by Paleo-Tethys slab subduction. Garnet pyroxenites in the Maowu ultramafic massif have low $\delta^{26}Mg$ of –0.99‰ to –0.65‰ and contain abundant carbonate mineral inclusions and metasomatized zircons with high $\delta^{18}O_{SMOW}$ (up to 12.2‰), suggesting metasomatism of the CLM by recycled carbonates. The age of zircons (457±55 Ma) from the garnet clinopyroxenites also indicates Paleozoic metasomatism by subduction of the Paleo-Tethys oceanic slab [28]. Some pyroxenite and garnet pyroxenite xenoliths hosted by Cenozoic basalts (e.g., Hannuoba) also have low $\delta^{26}Mg$ [35] (Fig. 3a). Because Hannuoba is located to the west of the Daxing'anling-Taihang gravity lineament (DTGL) in the western part of the NCC, in which the CLM has not been affected by the Mesozoic thinning, the presence of low-$\delta^{26}Mg$ xenoliths also indicates mantle carbonate metasomatism of the NCC prior to Cenozoic. It is noted that the Hannuoba xenoliths have $\delta^{26}Mg$ (low to –1.42‰; Fig. 3a) much lower than those of the host basalts and all other Cenozoic basalts in eastern China (–0.6 to –0.3‰) [29], thus their low $\delta^{26}Mg$ is unlikely to have been caused by interaction between low-$\delta^{26}Mg$ basaltic melt and the overlying lithospheric mantle. Generally, there is a negative correlation between $\delta^{26}Mg$ and CaO content for these xenoliths (Fig. 3b), strongly suggesting metasomatism of the CLM by recycled carbonates. Apart from Mg isotopes, Ca/Al, $(La/Yb)_N$ and Ti/Eu ratios of clinopyroxenes are effective indices of mantle carbonatitic metasomatism. As shown in Fig. 3c and d, clinopyroxenes in mantle xenoliths hosted by Paleozoic diamond-bearing kimberlites and Mesozoic mafic igneous rocks have systematically higher Ca/Al and $(La/Yb)_N$ and lower Ti/Eu ratios than those of silicate-metasomatic mantle xenoliths and depleted-MORB-mantle (DMM) peridotites. Along with high $Mg^{\#}$ and low Ti/Eu, these xenoliths are believed to have undergone carbonatitic metasomatism [36].

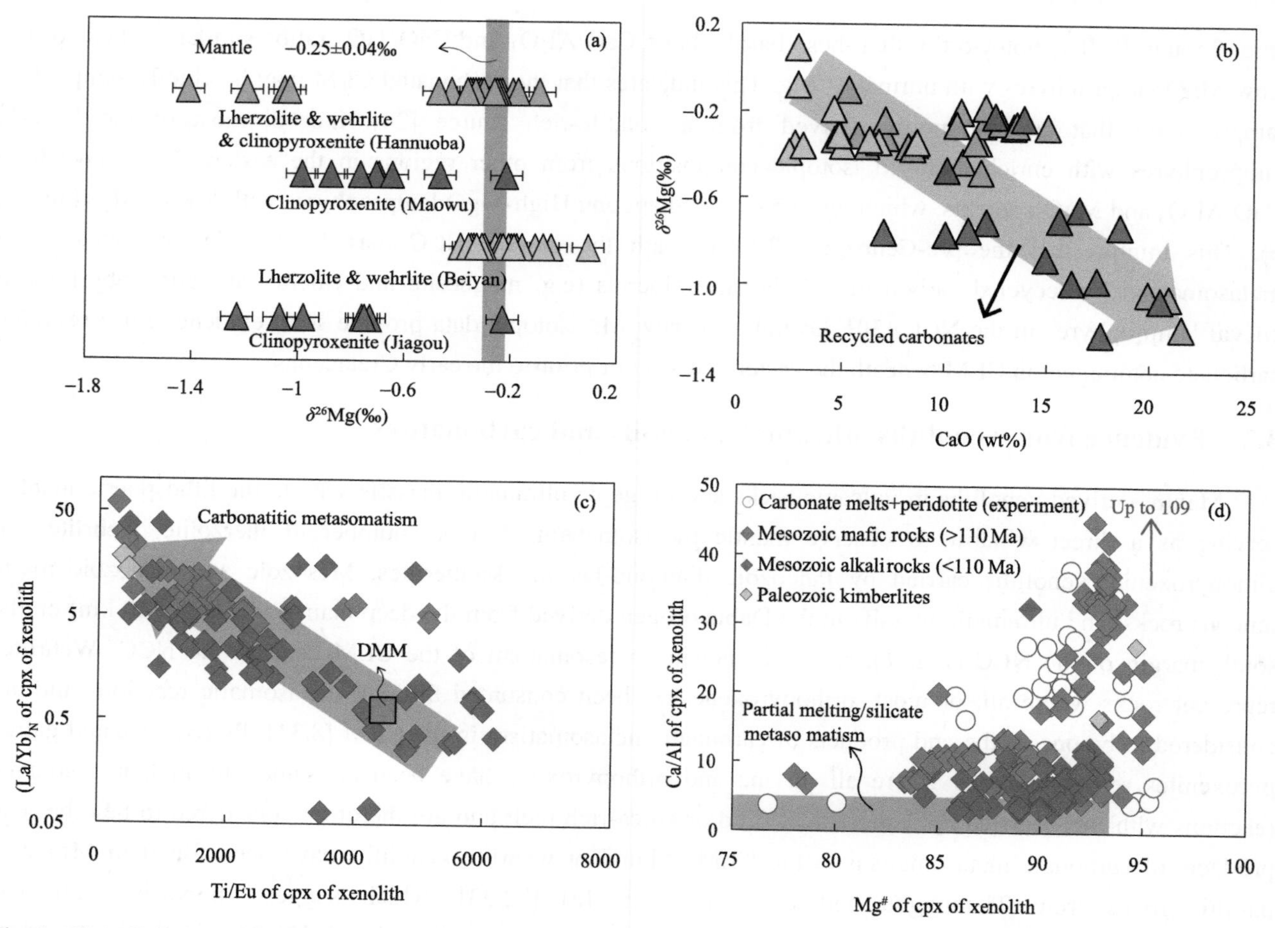

Fig. 3 Compilation of available (a) Mg isotopic and (b, c, d) chemical data for mantle xenoliths hosted by Paleozoic diamond-bearing kimberlites and Mesozoic mafic igneous rocks in the NCC. Data for mantle xenoliths hosted by Cenozoic basalts (Hannuoba, Maowu, Beiyan) and Mesozoic intrusion (Jiagou) [28,34,35,57,58] are also shown (see text for details). Experimental data for carbonate melt-peridotite interaction are from ref. [36] and references therein. The data for depleted MORB source mantle (DMM) and clinopyroxene (CPX) of Paleozoic kimberlites, Mesozoic mafic rocks, and Mesozoic alkaline rocks are summaried in Table S5.

Further evidence for pre-Cenozoic carbonate metasomatism of the CLM comes from carbonatites. The solidus of mantle rocks can be reduced by addition of volatiles such as CO_2 into the mantle and melting of the CO_2-rich mantle would produce alkali-rich and silicon-poor melts, such as carbonatites [27]. Mesozoic carbonatites are exposed at more than 10 locations in the NCC [36-38] (Fig. 1). The enriched Sr and Nd isotopic compositions of these carbonatitic magmas suggest an enriched, carbonated mantle source [37]. Mesozoic carbonatites from Zhuolu and Huairen have high $^{87}Sr/^{86}Sr$ ratios (0.7055–0.7075) and are proposed to have formed by direct melting of recycled sedimentary carbonates in the mantle [38]. Carbonatites intruding on Neogene alkali basalts in Hannuoba on the northern margin of the NCC have high $^{87}Sr/^{86}Sr$ (0.70522–0.70796) and high $\delta^{18}O$ ratios (22.2‰–23.0‰), which are directly linked to the subducted Paleo-Asian oceanic slab beneath the NCC before the Mesozoic era [39].

Collectively, the lines of evidence above strongly suggest that the CLM beneath the eastern NCC has been subject to pervasive carbonate metasomatism since the Paleozoic. The carbonate metasomatism could have been induced by multiple oceanic plate subduction events around the NCC, that is, the Paleo-Asian oceanic slab in the north, Paleo-Tethys oceanic slab in the south, and paleo-Pacific slab in the east (Fig. 4). These subducted slabs carried large amounts of carbonate sediments into the mantle and transformed the CLM into a vast store for carbon. However, the deep part of the mantle is commonly too reduced to favor stable carbonates. That is, when carbonates are recycled into the mantle at depths of >120 km, they will be reduced via the following redox reaction:

$$\underset{\text{enstatite}}{MgSiO_3} + \underset{\text{magnesite}}{MgCO_3} = \underset{\text{olivine}}{Mg_2SiO_4} + \underset{\text{diamond/graphite}}{C} + O_2$$

At depths of 120–170 km, recycled carbonate is transformed into carbon that exists as graphite and at larger depths (>170 km) as diamond [40], although in the CLM diamond is stable to lower pressures at cool conductive geotherms (Fig. 4). It is difficult to quantify the flux of recycled carbon in the mantle of the entire NCC since the Paleozoic, but we can give a rough estimate for this study area. As discussed above, the Mg-Sr isotopic mixing model indicates that the mantle source of low-δ^{26}Mg lamprophyres contains ~10 wt% Mg-rich carbonates (Fig. 2e), which is roughly equivalent to ~1 wt% C. Assuming a density of 3.2 g/cm^3 and a possible 40-km lithosphere depth interval that has been metasomatized, the mass of recycled C in the CLM beneath Shandong peninsula can be calculated. The lithosphere beneath the eastern NCC was >200 km thick before destruction [18,41] and the depth at which Mg-rich carbonates start to dissolve is ~160 km [28,29]; thus we assume an approximately ~40-km interval for carbonate metasomatism. An areal estimate is available for the Shandong peninsula (~73000 km^2), and we assume that about half the area was affected based on the proportion of occurrence of High-MgO lamprophyres with low δ^{26}Mg in the study area (Fig. 2). From this, 6.09×10^7 Mt C is estimated to have been added by recycled carbonates. Together with the primordial carbon (~4.27×10^7 Mt C) in the CLM prior to the Paleozoic,

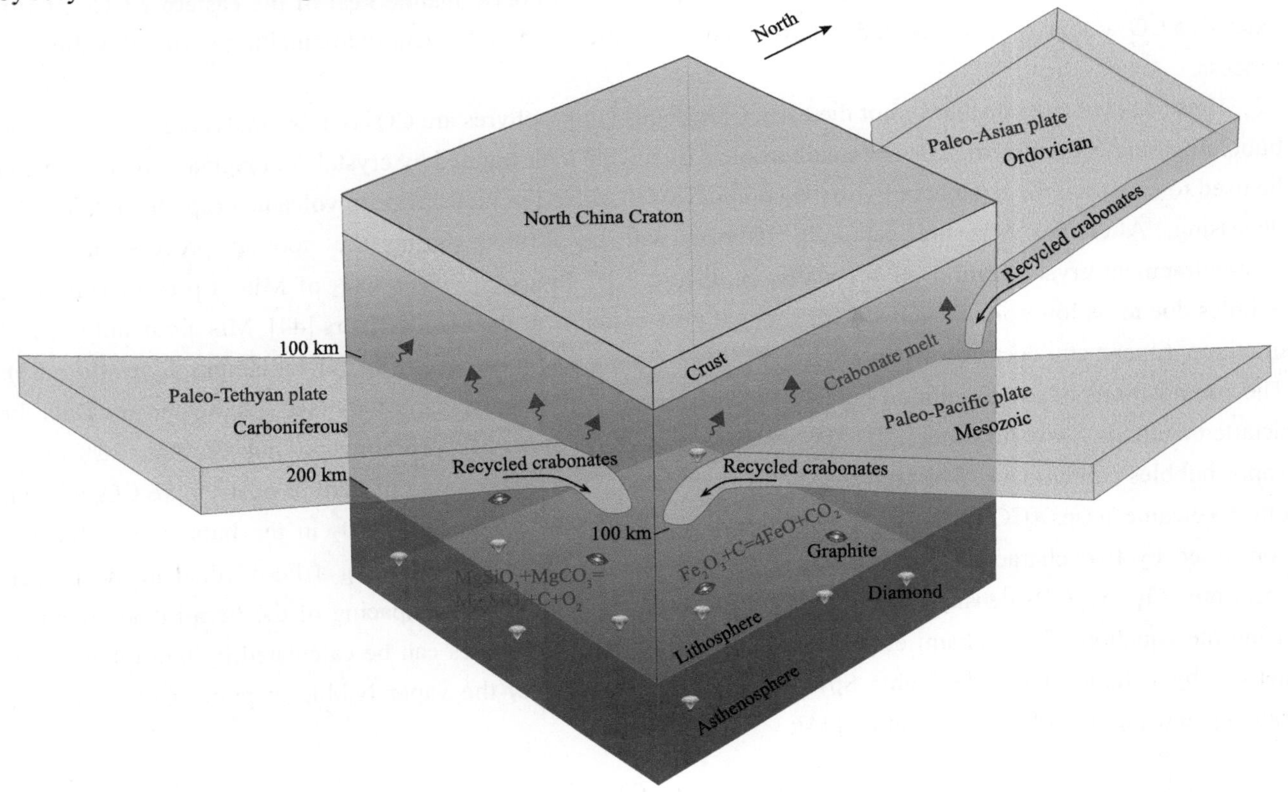

Fig. 4 A cartoon representing multiple subduction events around the NCC at or prior to the early Cretaceous, including the paleo-Asian oceanic slab in the north since Ordovician, paleo-Pacific oceanic slab in the east since Mesozoic, and paleo-Tethyan oceanic slab in the south beginning in the Carboniferous (modified from ref. [59]). Recycled carbonates would be reduced to diamond or graphite at depths of >150 km [40].

the total reservoir of carbon in the CLM beneath the eastern NCC would be 1.04×10^8 Mt C at least, which represents a significant store of carbon in the CLM with important contribution from recycled carbonates. The reduced CLM, with carbon mainly existing as graphite or diamond, has not undergone redox melting and was preserved until the Mesozoic during which it was largely activated and removed.

4 Deep CO_2 outgassing induced by cratonic destruction

Commonly, the deep CLM is primarily reduced as a result of depletion in basaltic melt and the pressure effect on the oxygen fugacity during its formation [42]. A reduced CLM beneath the thick NCC (>200 km) is indicated by the Paleozoic diamond-bearing kimberlites. However, it can become oxidized as the diamondiferous CLM is exhumed to shallower depths due to lithospheric thinning, extension and mantle upwelling [8]. During this process, 'redox melting' would occur and carbon (diamond or graphite) in the CLM would become unstable and be oxidized by the reduction of Fe^{3+} at depths of <170 km [40]. This redox melting would produce carbonatitic melts at depths of ~150 km that could evolve into carbonated silicate melts accompanied by silicate melting at shallower depths. The redox melting of the CLM was probably induced by decompression and rise of the lithosphere-asthenosphere boundary due to slab rollback of the westward subducting paleo-Pacific plate at the early Cretaceous [18]. Carbonatitic melts have much lower viscosity and density relative to silicate melts [43], which could further promote carbonatite metasomatism of the shallower CLM. In the presence of carbonatitic melts, the mantle could be readily fusible, leading to efficient extraction of carbon from the deep interior [11]. Therefore, given the extensive thinning (>120 km) of the lithospheric mantle keel of the eastern NCC [14,16], extensive CO_2 outgassing is expected to have occurred as the CLM underwent redox melting during this thinning process.

Here we evaluate whether or not the early Cretaceous lamprophyres are CO_2 rich by analyzing gas exsolution bubble-bearing MIs in them. MIs are small droplets of silicate melt trapped by crystals in magmatic rocks and can be used to constrain the contents of volatile components dissolved in melts prior to volcanic eruption and, ideally, degassing. After the MIs are captured, bubbles will be formed during the cooling process of melts, post-entrapment crystallization on MI walls, or diffusive H^+ loss [44]. Thus, CO_2 of MIs is present mainly in bubbles due to its low solubility in silicate melts if post-entrapment degassing occurs [44]. MIs are mainly hosted in clinopyroxene and occasionally in olivine and amphibole macrocrysts of the Shandong lamprophyres (Fig. S3). The compositions of gas exsolution bubbles were analyzed by Raman spectroscopy (see Supplementary Data for detailed methods). Among about ~200 MIs we analyzed, >80% of the MIs contain vapor bubbles and ~20% of the vapor bubbles contain CO_2. The analyzed bubbles in most MIs are composed of pure or nearly pure CO_2 without other volcanic gases (CO, CH_4, H_2S, H_2O) being detected. The presence of CO_2 in the bubbles of MIs was confirmed by two characteristic peaks, at ~1285 cm^{-1} and ~1388 cm^{-1}, defining a Fermi diad in the Raman spectrum (Fig. 5). CO_2 density (*d*) of the bubbles can be calculated by the spacing of the Fermi diad (Δ cm^{-1}), using the equation of Kawakami et al. [45]. The mass of CO_2 in bubbles can be calculated by multiplying CO_2 density by volume of bubble (Table S6). Then, the CO_2 content of the vapor bubble in ppm, $[CO_2]_{vb}$, can be calculated using the following equation [44]:

$$[CO_2]_{vb} = (M_{vb}{}^{CO2}/M_{gl}) \times 10^6$$

Where M_{gl} is the mass of glass within the MI, calculated as the glass volume multiplied by a melt density that is assumed to be 2.75 $g\cdot cm^{-3}$[46]. The results show that bubbles in MIs from the lamprophyres contain 323 to

47490 ppm CO_2 (N=29), and >93% bubbles have CO_2 contents of >1000 ppm (0.1 wt%) (Fig. 5f; Table S6). The calculated CO_2 concentrations in MIs of the lamprophyres range from 474 to 47, 641 ppm (N=29), with most (>80%) higher than 5000 ppm. Silicate crystal-hosted MIs, representing melts during various stages of evolving magmatic system, can be analyzed to constrain the CO_2 contents that dissolved in the melt before volcanic eruption and/or degassing [47]. We thus estimate that the measured CO_2 concentrations represent those of the pre-eruptive and possibly evolved lamprophyre magmas, which mostly fall between 0.5 wt% and 2.0 wt%. These contents are similar to or even higher than the CO_2 concentrations (0.5–1.0 wt%) in MIs of the end-Triassic Central Atlantic Magmatic Province basalts, which were estimated by the same method [48]. It should be noted that a high CO_2 concentration of MIs is mainly observed in High-MgO lamprophyres with light δ^{26}Mg values (see Supplementary data), and the number of MIs in High-MgO lamprophyres is much larger than that in Low-MgO lamprophyres. This probably indicates that High-MgO lamprophyres with recycled carbonates in their mantle sources contain more MIs and higher CO_2 concentrations in the pre-eruptive magmas, although low-volume melts could also have extremely high CO_2 content even if the source is not specifically C rich, due to the strong incompatibility of CO_2 in peridotite [49]. Because most of the early Cretaceous lamprophyres in the eastern NCC belong to the High-MgO group with low δ^{26}Mg (Fig. 2), their sources were plausibly most strongly affected by carbonate metasomatism and attendant enrichment in carbon. Melting of this metasomatized CLM then produced primary magmas with high CO_2 content, which may have been further enhanced during pre-eruptive

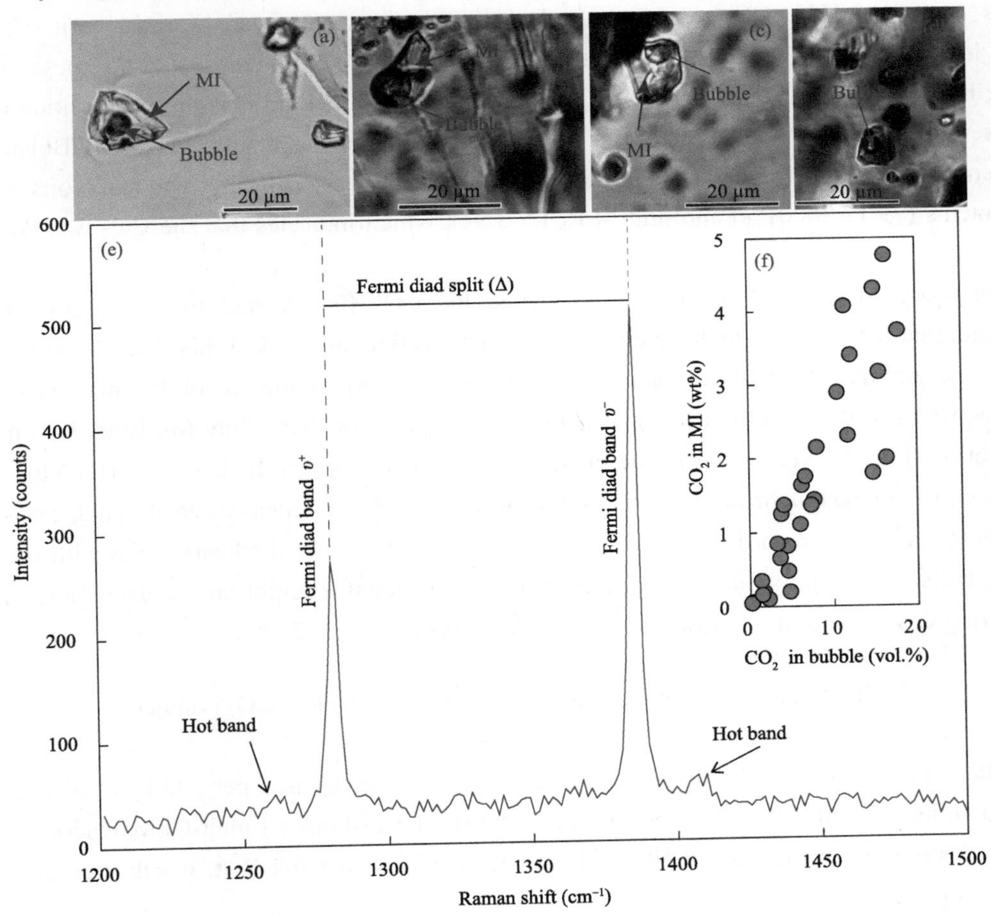

Fig. 5 (a-d) Bubble-bearing MIs in the transmitted light optical microcopy. (e) Raman spectrum of a CO_2-bearing bubble in a clinopyroxene-hosted MI from lamprophyres. The presence of CO_2 is confirmed by the Fermi diad, consisting of two peaks at ~1285 cm^{-1} and ~1388 cm^{-1}, bounded by hot bands, below 1285 cm^{-1} and above1388 cm^{-1}. (f) The relationship between the volume of CO_2 in bubbles and the CO_2 concentrations of MIs.

differentiation. Here we collated geochemical data for early Cretaceous mantle-derived volcanic rocks (see Fig. 1 for locations, Fig. 2a and b and Fig. S4 for chemical compositions) and found that they are widely distributed in the eastern NCC and show similar geochemical characteristics with the lamprophyres. Thus, early Cretaceous mantle-derived magmas in the eastern NCC are much more abundant than those represented by lamprophyres. A larger flux of CO_2 outgassing is thus expected during the period of extensive destruction of NCC, in addition to lamprophyres. Intrusion or eruption of these magmas could have carried a large amount of CO_2 from the mantle into the surface.

Experimental studies show that the solubility of carbon dioxide in melts decreases at lower pressures, and CO_2 can even be directly degassed at mantle depths [50]. Therefore, the thinned lithosphere, as a result of cratonic destruction and extension, can further facilitate CO_2 outgassing via magmatism. There is no evidence for presence of a deep-sourced mantle plume beneath the eastern NCC during the Phanerozoic era. We thus propose that the destruction of the CLM represents another important cause of CO_2 emisson from the mantle, in addition to continental rifting, active island arc volcanism, and mantle plume. During this process, carbon in the CLM can experience gradual oxidation during mantle upwelling, with a change of carbon speciation from a reduced to an oxidized form, and a portion of carbon can be liberated via redox melting from the reduced mantle [11].

Enormous CO_2 reservoirs can be formed by the eruption or intrusion of magmas. There are abundant crustal CO_2 reservoirs in the Songliao and Bohai Bay basins in the eastern NCC, which indicates that the volume of CO_2 degassing is enormous [51]. CO_2 reservoirs in Songliao basin were formed primarily in Cretaceous, and voluminous inorganic CO_2 (mainly mantle-derived and crust-derived) is observed in these reservoirs. For example, the high CO_2 content (>90%) and $\delta^{13}C$ (−4.95‰) and high helium isotopic composition (R/R_a=3.34) of Wanjinta reservoirs indicate that the CO_2 was chiefly sourced from the mantle [51]. The Bohai Bay basin, a Mesozoic-Cenozoic basin, is the central area of destruction of the eastern NCC. The reservoirs there also have high CO_2 contents (79.17–98.61%) and high R/R_a (2–3.34), which indicates that the CO_2 was derived from the mantle [51].

A recent study by Aulbach et al. [8] quantified the CO_2 flux related to the reaction of CLM with silica-undersaturated (carbonated) melt, referred to as wehrlitization, and linked this flux to surficial degassing in rifts and basins. As discussed above, abundant wehrlite and pyroxenite xenoliths are found in Mesozoic mantle-derived rocks in the eastern NCC, and many of these xenoliths have light Mg isotopic compositions and high Ca/Al ratios (Fig. 3). For instance, the characteristics of low Ti/Eu, high Ca/Al, $(La/Yb)_N$ and Zr/Hf of clinopyroxenes in Liaoyuan wehrlites are ascribed to interaction with a silica-undersaturated, carbonated silicate melt [33]. These rocks thus record substantial reaction between carbonated silicate melts with the CLM. In the CLM, these melts are initially out of thermal and compositional equilibrium, causing intensive melt-rock reactions. During this process, the following reaction will happen at ~1.5–2.0 GPa [52]:

$$\text{Enstatite} + \text{dolomite (melt)} = \text{forsterite} + \text{diopside} + CO_2 \text{ (vapor)}.$$

A quantitative estimate suggests that 2.9 to 10.2 kg CO_2 can be released per 100 kg of wehrlite formed [8]. Extensive CO_2 release is suggested to have occurred during the carbonated melt-CLM reaction process in the course of this destruction of the eastern NCC, along with the Tan-Lu Fault Belt, which was most active in the early Cretaceous [33].

Stable continents are long-term storage sites for sedimentary carbonates, and the amount of carbonates stored in continents is thought to be at least 10 times greater than that stored in oceanic crust [53]. Carbonates in crusts can be trapped by plutons that ascend to shallow levels in the arc crust or are transported into the lower crust during later arc stages. Global flare-ups in continental arc volcanism were proposed to have the potential to release CO_2 as a result of magmatic interaction with ancient crustal carbonates stored in the continental crust [54,55]. The eastern NCC is typically characterized by a giant felsic magmatism event at the early Cretaceous with a volume much larger than that of mafic magmatism [17], implying large-scale crustal melting and reworking during the cratonic destruction process. These early Cretaceous felsic magmas (i.e. granites) have high zirconium saturation temperatures and contain an important contribution from the hot upwelling mantle [25]. Decarbonation is expected to widely occur during interaction between the hot felsic magmas and the limestones chronically stored in the continental crust. This process could also contribute to CO_2 release, in addition to mantle CO_2 outgassing via mafic magmatism and carbonated melt-CLM reaction during destruction of the eastern NCC (Fig. 6).

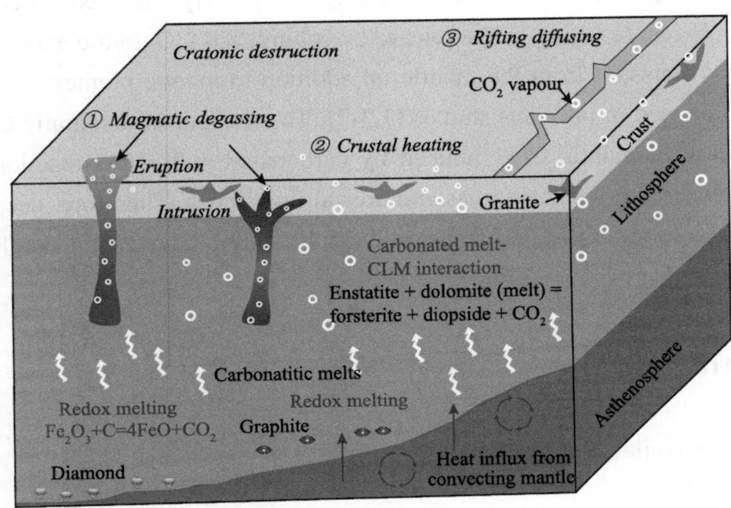

Fig. 6 Schematic cartoon models illustrating mantle CO_2 outgassing in response to cratonic destruction. The carbonated CLM beneath eastern NCC underwent extensive melting and thinning at ~125 Ma as a result of heat upwelling from mantle convection related to rollback of the subducting west Pacific slab. This process produced extensive CO_2-rich magmas with a peak at ~125 Ma, as recorded by lamprophyres, basalts, and carbonatites. The CO_2 vapors were released in three ways: ① magmatic degassing of lamprophyre and eruptive magmas, ② heating of sedimentary carbonates stored in the crust, represented by granites, and ③ the interaction between carbonated melts and CLM, represented by wehrlites and pyroxenites leading to decarbonation and liberation of CO_2 vapor.

5 Possible contributions to the cretaceous greenhouse

The amount of CO_2 outgassing induced by the destruction of the eastern NCC could be significant, particularly if one considers that the removed CLM contained a large amount of recycled carbon from subducted slabs prior to thinning. The fluxes of deep CO_2 outgassing during this destruction process may be large given the short duration of mantle-derived, CO_2-rich magmatism (both intrusions and volcanics; Fig. 1) in the early Cretaceous (~125 Ma) as well as the strong carbonated silicate melt-CLM reaction that resulted in substantial CO_2 release along the massive Tan-Lu Fault Belt [8]. At a larger scale, enormous quantities of CO_2 that were rapidly released into the atmosphere, induced by the destruction, may have perturbed the global climate and partly contribute to the atmospheric CO_2 rise during the Cretaceous, one of the longest Greenhouse periods of Earth's history, with atmospheric CO_2 levels 4 to 10 times higher than those prior to the Industrial Revolution [56].

6 Conclusions

We present the first Mg isotope data for early Cretaceous lamprophyres and collate available chemical and Mg isotopic data for mantle xenoliths in Paleozoic diamond-bearing kimberlites and Mesozoic mafic igneous rocks as well as orogenic ultramafic massifs in the NCC. These results suggest the presence of a widespread, C-rich CLM beneath the NCC at and before lithospheric thinning in the early Cretaceous, with an important contribution from recycled carbonate sediments. Long-term and three-sided–i.e. the south, north and east–oceanic plate subductions underneath the NCC during the Paleozoic and Mesozoic could have contributed a vast amount of carbon to lithospheric mantle of the NCC. Redox melting of the reduced, C-rich CLM as it was exhumed to shallower depths due to lithospheric extension and thinning in the early Cretaceous generated large amounts of basaltic lavas and lamprophyres, resulting in the release of voluminous CO_2 to the exosphere. This may represent an important cause of CO_2 emission from the mantle, in addition to mantle plumes, active island arc volcanism and continental rifts, as proposed in previous studies [1,3-7]. The amount of magmatic CO_2 outgassing is largely supplemented by the release of mantle CO_2 induced by carbonated melt-CLM reaction [8] and decarbonation induced by interaction between hot felsic magmas and crustal limestones. Therefore, deep CO_2 outgassing can be linked to the destruction of a long-term stable craton and can be said to have enhanced global CO_2 inputs into the atmosphere.

Supplementary data

Supplementary data are available at *NSR* online.

Acknowledgements

We are grateful to the editors for editorial handling. Two anonymous reviewers and Sonja Aulbach provided many creative comments that helped improve this manuscript. We thank Wang Z.-Z. for sample collection and help in the lab. We also thank Ke S. for help on Mg isotopic analysis, Liu W.-R. for help during data collection and Liu Y.-H. for help on MIs analysis.

Funding

This work was supported by the National Key Research and Development Program of China (2019YFA0708400), the National Natural Science Foundation of China (41730214 and 41688103), and the 'Strategic Priority Research Program' of the Chinese Academy of Sciences (XDB18000000 to S.-A.L. and S.G.L.)

Author contributions

S.-A.L. and S.G.L. designed the project. Z.-X.W. and D.L. analyzed all data. S.-A.L. and Z.X.W. developed the manuscript with contributions from other co-authors.

Conflict of interest statement. None declared.

References

1. Foley SF and Fischer TP. An essential role for continental rifts and lithosphere in the deep carbon cycle. Nat Geosci 2017; 10: 897-902.
2. Plank T and Manning CE. Subducting carbon. Nature 2019; 574: 343-52.
3. Lee H, Muirhead JD and Fischer TP et al. Massive and prolonged deep carbon emissions associated with continental rifting. Nat Geosci 2016; 9: 145-9.
4. Burton MR, Sawyer GM and Granieri D. Deep carbon emissions from volcanoes. Rev Mineral Geochem 2013; 75: 323-54.
5. Marty B and Tolstikhin IN. CO_2 fluxes from mid-ocean ridges, arcs and plumes. Chem Geol 1998; 145: 233-48.
6. Barnet JSK, Littler K and Kroon D et al. A new high-resolution chronology for the late Maastrichtian warming event: establishing robust temporal links with the onset of Deccan volcanism. Geology 2018; 46: 147-50.
7. Schoene B, Eddy MP and Samperton KM et al. U-Pb constraints on pulsed eruption of the Deccan Traps across the end-Cretaceous mass extinction. Science 2019; 363: 862-6.
8. Aulbach S, Lin AB and Weiss Y et al.Wehrlites from continental mantle monitor the passage and degassing of carbonated melts. Geochem Perspect Lett 2020; 15: 30-4.
9. Kelemen PB and Manning CE. Reevaluating carbon fluxes in subduction zones, what goes down, mostly comes up. Proc Natl Acad Sci USA 2015; 112: E3997-4006.
10. Kagoshima T, Sano Y and Takahata N et al. Sulphur geodynamic cycle. Sci Rep 2015; 5: 8330.
11. Dasgupta R and Hirschmann MM. The deep carbon cycle and melting in Earth's interior. Earth Planet Sci Lett 2010; 298: 1-13.
12. Sobolev SV, Sobolev AV and Kuzmin DV et al. Linking mantle plumes, large igneous provinces and environmental catastrophes. Nature 2011; 477: 312-6.
13. Pearson DG, Scott JM and Liu J et al. Deep continental roots and cratons. Nature 2021; 596: 199-210.
14. Griffin WL, Zhang A and O'Reilly SY et al. Phanerozoic evolution of the lithosphere beneath the Sino-Korean Craton. In: Flower M, Chung S and Luo C et al. (eds.). Mantle Dynamics and Plate Interactions in East Asia. Washington DC: American Geophysical Union, 1998, 107-26.
15. Menzies MA, Fan W and Zhang M. Palaeozoic and Cenozoic lithoprobes and the loss of >120 km of Archaean lithosphere, Sino-Korean craton, China. Geol Soc Spec Publ 1993; 76: 71-81.
16. Menzies M, Xu Y and Zhang H et al. Integration of geology, geophysics and geochemistry: a key to understanding the North China Craton. Lithos 2007; 96: 1-21.
17. Wu F-Y, Yang J-H and Wilde SA et al. Geochronology, petrogenesis and tectonic implications of Jurassic granites in the Liaodong Peninsula, NE China. Chem Geol 2005; 221: 127-56.
18. Zhu R, Xu Y and Zhu G et al. Destruction of the North China Craton. Sci China Earth Sci 2012; 55: 1565-87.
19. Marty B. The origins and concentrations of water, carbon, nitrogen and noble gases on Earth. Earth Planet Sci Lett 2012; 313-14: 56-66.
20. Liu DY, Nutman AP and Compston W et al. Remnants of ⩾ 3800 Ma crust in the Chinese part of the Sino-Korean craton. Geology 1992; 20: 339-42.
21. Zheng J. Comparison of mantle-derived materials from different spatiotemporal settings: implications for destructive and accretional processes of the North China Craton. Sci Bull 2009; 54: 3397-416.
22. Liu S-A and Li S-G. Tracing the deep carbon cycle using metal stable isotopes: opportunities and challenges. Engineering 2019; 5: 448-57.
23. Teng F-Z. Magnesium isotope geochemistry. Rev Mineral Geochem 2017; 82: 219-87.
24. Rock NMS. The nature and origin of lamprophyres: an overview. Geol Soc Spec Publ 1987; 30: 191-226.
25. Wu F-Y, Yang J-H and Xu Y-G et al. Destruction of the North China Craton in the Mesozoic. Annu Rev Earth Planet Sci 2019; 47: 173-95.
26. Ma L, Jiang S-Y and Hofmann AW et al. Lithospheric and asthenospheric sources of lamprophyres in the Jiaodong Peninsula : a consequence of rapid lithospheric thinning beneath the North China Craton? Geochim Cosmochim Acta 2014; 124: 250-71.
27. Dasgupta R, Hirschmann MM and Smith ND. Partial melting experiments of peridotite + CO_2 at 3 GPa and genesis of Alkalic

ocean island basalts. J Petrol 2007; 48:2093-124.
28. Shen J, Li S-G and Wang S-J et al. Subducted Mg-rich carbonates into the deep mantle wedge. Earth Planet Sci Lett 2018; 503: 118-30.
29. Li S-G, YangW and Ke S et al. Deep carbon cycles constrained by a large-scale mantle Mg isotope anomaly in eastern China. Natl Sci Rev 2017; 4: 111-20.
30. Ma L, Jiang S-Y and Hofmann AW et al. Rapid lithospheric thinning of the North China Craton: new evidence from cretaceous mafic dikes in the Jiaodong Peninsula. Chem Geol 2016; 432: 1-15.
31. Neumann ER, Wulff-Pedersen E and Pearson NJ et al. Mantle xenoliths from Tenerife (Canary Islands): evidence for reactions between mantle peridotites and silicic carbonatite melts inducing Ca metasomatism. J Petrol 2002; 43: 825-57.
32. Zhou Q, XuW and Yang D et al. Modification of the lithospheric mantle by melt derived from recycled continental crust evidenced by wehrlite xenoliths in Early Cretaceous high-Mg diorites from western Shandong, China. Sci China Earth Sci 2012; 55: 1972-86.
33. Lin AB, Zheng JP and Aulbach S et al. Causes and consequences of wehrlitization beneath a trans-lithospheric fault: evidence from mesozoic basalt-borne wehrlite xenoliths from the Tan-Lu fault belt, North China Craton. J Geophys Res Solid Earth 2020; 125: e2019JB019084.
34. Wang Z-Z, Liu S-A and Ke S et al. Magnesium isotopic heterogeneity across the cratonic lithosphere in eastern China and its origins. Earth Planet Sci Lett 2016; 451: 77-88.
35. Hu Y, Teng F-Z and Zhang H-F et al. Metasomatism-induced mantle magnesium isotopic heterogeneity: evidence from pyroxenites. Geochim Cosmochim Acta 2016; 185: 88-111.
36. Zong K and Liu Y. Carbonate metasomatism in the lithospheric mantle: implications for cratonic destruction in North China. Sci China Earth Sci 2018; 61: 711-29.
37. Ying J. Geochemical and isotopic investigation of the Laiwu-Zibo carbonatites from western Shandong Province, China, and implications for their petrogenesis and enriched mantle source. Lithos 2004; 75: 413-26.
38. Yan G, Mu B and Zeng Y et al. Igneous carbonatites in North China craton: the temporal and spatial distribution, Sr and Nd isotopic characteristics and their geological significance (in Chinese). Geol J China Univ 2007; 13: 463-73.
39. Chen C, Liu Y and Foley SF et al. Paleo-Asian oceanic slab under the North China craton revealed by carbonatites derived from subducted limestones. Geology 2016; 44: 1039-42.
40. Stagno V, Ojwang DO and McCammon CA et al. The oxidation state of the mantle and the extraction of carbon from Earth's interior. Nature 2013; 493: 84-8.
41. Xu Y. Thermo-tectonic destruction of the Archaean lithospheric keel beneath the Sino-Korean craton in China: evidence, timing and mechanism. Phys Chem Earth Part A 2001; 26: 747-57.
42. Frost DJ and McCammon CA. The redox state of Earth's mantle. Annu Rev Earth Planet Sci 2008; 36: 389-420.
43. Kono Y, Kenney-Benson C and Hummer D et al. Ultralow viscosity of carbonate melts at high pressures. Nat Commun 2014; 5: 5091.
44. Hartley ME, Maclennan J and Edmonds M et al. Reconstructing the deep CO_2 degassing behaviour of large basaltic fissure eruptions. Earth Planet Sci Lett 2014; 393: 120-31.
45. Kawakami Y, Yamamoto J and Kagi H. Micro-Raman densimeter for CO_2 inclusions in mantle-derived minerals. Appl Spectrosc 2003; 57: 1333-9.
46. Passmore E, Maclennan J and Fitton G et al. Mush disaggregation in basaltic magma chambers: evidence from the ad 1783 Laki eruption. J Petrol 2012; 53: 2593-623.
47. Steele-MacInnis M, Esposito R and Moore LR et al. Heterogeneously entrapped, vapor-rich melt inclusions record pre-eruptive magmatic volatile contents. Contrib Mineral Petrol 2017; 172: 18.
48. Capriolo M, Marzoli A and Aradi LE et al. Deep CO_2 in the end-Triassic Central Atlantic Magmatic Province. Nat Commun 2020; 11: 1670.
49. Hirschmann MM. Partial melt in the oceanic low velocity zone. Phys Earth Planet Inter 2010; 179: 60-71.
50. Boudoire G, Rizzo AL and Di Muro A et al. Extensive CO_2 degassing in the upper mantle beneath oceanic basaltic volcanoes: first insights from Piton de la Fournaise volcano (La Réunion Island). Geochim Cosmochim Acta 2018; 235: 376-401.
51. Zhao F, Jiang S and Li S et al. Correlation of inorganic CO_2 reservoirs in East China to subduction of (Paleo-)Pacific Plate (in Chinese). Earth Sci Front 2017; 24: 370-84.
52. Yaxley GM, Green DH and Kamenetsky V. Carbonatite metasomatism in the Southeastern Australian lithosphere. J Petrol 1998;

39: 1917-30.
53. Lee CTA and Lackey JS. Global continental arc flare-ups and their relation to long-term greenhouse conditions. Elements 2015; 11: 125-30.
54. McKenize NR, Horton BK and Loomis SE et al. Continental arc volcanism as the principal driver of icehouse-greenhouse variability. Science 2016; 352: 444-7.
55. Mason E, Edmonds M and Turchyn AV. Remobilization of crustal carbon may dominate volcanic arc emissions. Science 2017; 357: 290-4.
56. Huber BT, Norris RD and MacLeod KG. Deep-sea paleotemperature record of extreme warmth during the Cretaceous. Geology 2002; 30: 123-6.
57. Xiao Y, Teng F-Z and Zhang H-F et al. Large magnesium isotope fractionation in peridotite xenoliths from eastern North China craton: product of melt-rock interaction. Geochim Cosmochim Acta 2013; 115: 241-61.
58. Hu J, Jiang N and Carlson RW et al. Metasomatism of the crust-mantle boundary by melts derived from subducted sedimentary carbonates and silicates. Geochim Cosmochim Acta 2019; 260: 311-28.
59. Xiao Y, Teng F-Z and Su B-X et al. Iron and magnesium isotopic constraints on the origin of chemical heterogeneity in podiform chromitite from the Luobusa ophiolite, Tibet. Geochem Geophys Geosyst 2016; 17: 940-53.

Recycling of carbonates into the deep mantle beneath central Balkan Peninsula: Mg-Zn isotope evidence *

Zi-Tan Shu[1], Sheng-Ao Liu[1], Dejan Prelević[2, 3], Vladica Cvetković[3], Shuguang Li[1]

1. State Key Laboratory of Geological Processes and Mineral Resources, China University of Geosciences, Beijing 100083, China
2. Institute of Geological Sciences, University of Mainz, Becherweg 21, D-55099 Mainz, Germany
3. University of Belgrade, Faculty of Mining and Geology, Đušina 7, 11000 Belgrade, Serbia

亮点介绍： 应用 Mg 和 Zn 同位素示踪深部碳循环的一个重要成就是发现西太平洋俯冲板片滞留在地幔过渡带形成的东亚大地幔楔是一个巨大的俯冲碳酸盐储库。为了检验这一认识是否具有普适性，本文选取了来自东地中海巴尔干半岛的塞尔维亚新生代玄武岩（ESMAR）做了 Mg、Zn 同位素研究。巴尔干半岛陆下的地幔过渡带有大片俯冲板片滞留并形成大地幔楔结构。研究结果表明，ESMAR 具有类似中国东部玄武岩的轻 Mg 同位素和重 Zn 同位素的特征，反映其源区具有再循环的沉积碳酸盐，巴尔干半岛大地幔楔同样是俯冲碳酸盐的大储库。

Abstract Magnesium (Mg) and zinc (Zn) isotopes have been applied to trace whether surface carbonates have been recycled into the deep mantle beneath Balkan Peninsula, where a stagnant slab exists in the mantle transition zone. Here we investigate a suite of Cenozoic sodic alkaline basaltic rocks ($Na_2O/K_2O > 1$), called the East Serbian Mafic Alkaline Rocks (ESMAR) from East Serbia in central Balkan Peninsula, which is an important segment of the Tethyan orogenic belt. The ESMAR have lower $\delta^{26}Mg_{DSM\text{-}3}$ (−0.49‰ to −0.39‰; n=21) and higher $\delta^{66}Zn_{JMC\ 3\text{-}0749L}$ (0.35‰ to 0.45‰; n=21) than those of normal mantle-derived melts (e.g., mid-oceanic ridge basalts; MORBs; $\delta^{26}Mg$ =−0.25±0.06‰ and $\delta^{66}Zn$ = 0.27±0.06‰). After excluding the effects of low-T and magmatic processes, the Mg and Zn isotopic anomalies of the ESMAR are proposed to have been caused by recycled carbonates in their mantle sources. The significantly low Zn contents, Zn/Fe and Fe/Mn ratios of the ESMAR indicate a highly depleted peridotitic mantle source. Their major element compositions (e.g., low Ti_2O and high CaO/Al_2O_3) are consistent with those of experimental partial melts of "peridotite + CO_2". The incremental batch melting and end-member mixing models suggest that ~5–10% of Mg-rich carbonates (dolomite and magnesite) have been recycled into the mantle sources of the ESMAR. Our results provide evidence for deep recycling of carbonates related to subduction of a Tethyan oceanic slab, most likely into the mantle transition zone beneath the Balkan Peninsula. This contributes a significant part of global deep carbon cycling during the Cenozoic.

Keywords Mg and Zn isotopes, deep carbon cycling, Na-rich alkaline basalts, Tethyan orogenic belt, Balkan Peninsula

---

* 本文发表在：Lithos, 2022, 432-433: 106899

1 Introduction

The carbon cycling between the Earth's surface and its interior is crucial to understanding the long-term climate changes because it has a significant impact on the atmospheric CO_2 levels (Marshall et al., 1988). Modern subducting slabs (including sediments, altered oceanic crust and mantle layers) carry 79–88 Megatons per year (Mt/yr) of carbon into the Earth's interior, and 68–79% of the subducting carbon is concentrated in sediments (carbonate and carbonated siliciclastic sediments) (Clift, 2017; Plank and Manning, 2019). Previous studies suggested that most of carbonate-bearing sediments undergo negligible devolatilization and melting under subduction-zone conditions (Kerrick and Connolly, 2001) and are vastly stored in the sub-arc mantle and lithospheric mantle (Foley and Fischer, 2017; Kelemen and Manning, 2015), or can be subducted into the deeper mantle, like the convecting mantle and even the mantle transition zone (MTZ) (e.g., Dasgupta and Hirschmann, 2010; Hirschmann, 2018). Recently, the Mg and Zn isotopes have been developed as tools for tracing recycled carbonates in the deep mantle due to the substantial differences in terms of isotopic composition between sedimentary carbonates and the mantle (see Liu et al., 2022 for a review).

It has been observed that significant Mg and Zn isotopic anomalies exist in Cenozoic intraplate alkali basalts from East Asia, extending >5000 km from Hainan in the south to the Russian border in the north (Li et al., 2017; Liu et al., 2016). Remarkably, the Mg and Zn isotopic anomalies match the "big mantle wedge" system in East Asia with the deeply stagnant Pacific slab in the MTZ (Huang and Zhao, 2006). These alkali basalts have much lower δ^{26}Mg and higher δ^{66}Zn than those of MORBs, suggesting that sedimentary carbonates have been recycled into the MTZ by the stagnant Pacific slab (Huang et al., 2015; Li et al., 2017; Liu et al., 2016; Wang et al., 2018; Yang et al., 2012). In order to get a better grip on the global deep carbon cycling, it is necessary to identify if a similar "coupling of Mg-Zn isotopic anomalies" (i.e., low δ^{26}Mg and high δ^{66}Zn) with stagnant slabs in the MTZ can be observed in other parts of the world (Hofmann, 2007). Besides the Circum-Pacific subduction zone, the Tethyan orogenic belt is another global-scale orogenic belt extending from the Mediterranean region, through Turkey, Iran to Xizang (Tommasini et al., 2011 and references therein), and is related to subduction of Tethyan oceanic slabs. Although a recent study suggested the presence of recycled carbonates in the MTZ beneath SE Xizang (East Tethyan orogenic belt) (Liu et al., 2020), to date, rare is known about the deep carbon cycling within the West Tethyan orogenic belt, like the Circum-Mediterranean region.

The Balkan Peninsula represents a crucial portion of the Tethyan orogenic belt. In the center of Balkans (e.g., East Serbia), a widespread Cenozoic basaltic volcanism occurs with 'anorogenic' affinity, which is generally characterized by the presence of sodic alkaline (Na_2O/K_2O >1) basaltic rocks with typical intraplate geochemical signatures similar to Na-alkaline lavas occurring within oceanic and continental plates worldwide (Lustrino and Wilson, 2007). This volcanism is interpreted to be derived from the asthenospheric mantle (Jovanović et al., 2001), or the deepest part of the lithospheric mantle (Cvetković et al., 2004b), undergoing negligible contamination by upper continental crust material when magma ascended (e.g., carbonate formations; Cvetković et al., 2004b, 2013). Moreover, the upper mantle P wave tomography under the Alpine-Mediterranean area (Piromallo and Morelli, 2003) shows that a stagnant slab probably exists beneath the Balkan Peninsula. Therefore, this anorogenic Na-alkaline basaltic volcanism provides an opportunity to investigate deep carbon cycling related to subduction of Tethyan oceanic slabs.

In this study, we present Mg and Zn isotopic data for a suite of well-characterized basaltic samples from central Balkans, called the East Serbian Mafic Alkaline Rocks (hereafter, ESMAR). Besides, we compare the whole-rock Mg, Zn and Sr isotopic data, as well as major- and trace-element compositions of the ESMAR with these of Cenozoic alkali basalts from East Asia and Southeast (SE) Xizang in an attempt to more thoroughly constrain the source nature in the origin of these alkaline rocks. Our results demonstrate that the Tethyan oceanic slab's subduction has recycled carbonates into the upper mantle beneath the Balkan region.

2 General geology and sample descriptions

The center of Balkan Peninsula (Fig. 1a) has experienced complex geodynamical processes including periods of subduction, collision and extension from the middle Mesozoic to present (Karamata and Krstić, 1996). The ESMAR formed after the convergent/subduction tectonic events related to the final closure of a Mesozoic Tethyan ocean, called the Vardar Ocean (Robertson et al., 2009). The Vardar Ocean separated the northwest Gondwana continent and southeast Eurasian continent, and its oceanic slab began to subduct eastward beneath the Eurasian plate from the Mesozoic and might close mostly between Late Jurassic to Early Cretaceous or even later (e.g., Karamata and Krstić, 1996). The final closure of Vardar Ocean generated a mega-suture named the Vardar Zone, which consisted of ophiolites, accretionary melange and some exotic blocks (Robertson et al., 2009). The subduction of Vardar oceanic slab invoked a great suite of calcalkaline magmatism in Balkan Peninsula, which formed the famous porphyry and epithermal Cu-Au deposits of the Banatite-Timok-Srednjegorje Magmatic and Metallogenetic Belt, starting from northward in Romania, through east Serbia, south Bulgaria and ending up in the northmost part of Turkey (Fig. 1b; Berza et al., 1998). The ESMAR postdated the occurrence of this calcalkaline belt and had a relatively short active time span, from 62 to 39 Ma (Cvetković et al., 2004b). They were distributed in an almost N-S line from Tisa to Pirot, almost paralleling to the Vardar Zone and confined to the eastern tectonic units (i.e., Serbo-Macedonian Massif) (Fig. 1c).

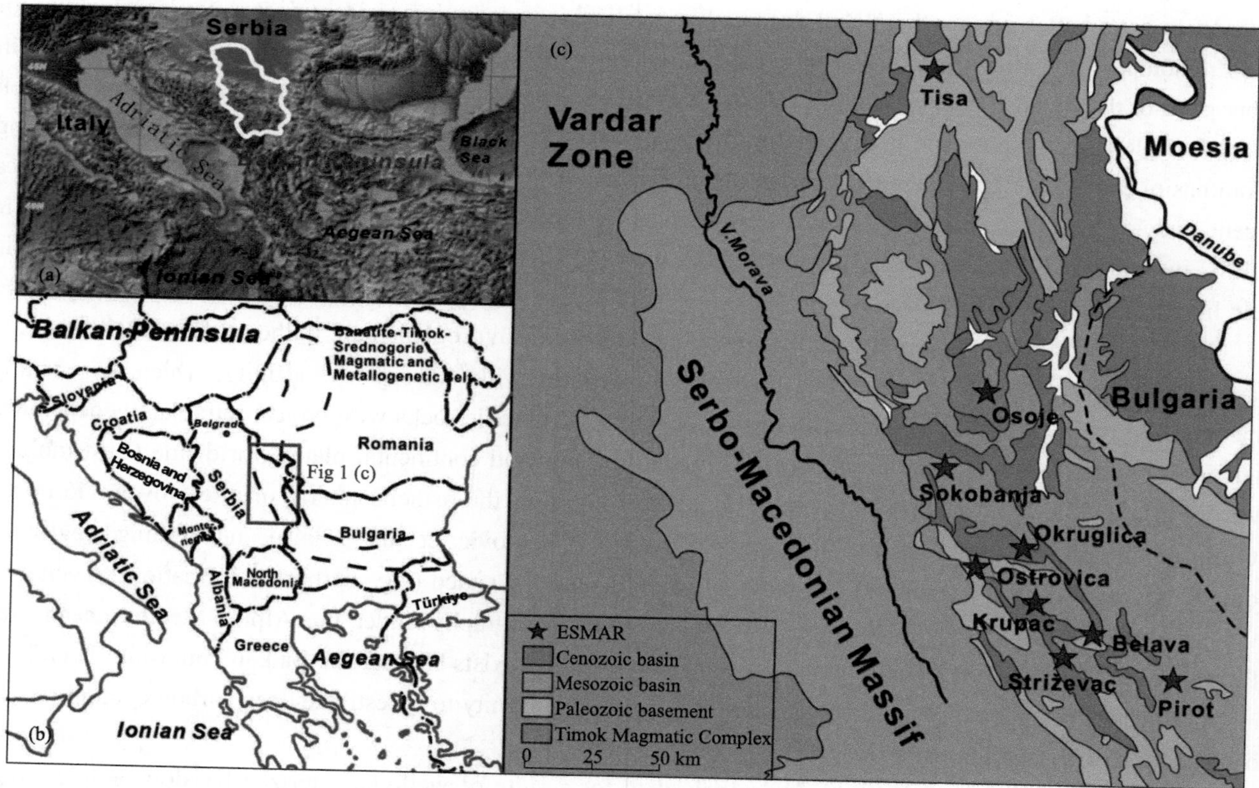

Fig. 1 (a) Digital topographic map of the Balkan Peninsula and its surrounding regions (http://www.ngdc.noaa.gov/mgg/global/global.html). (b) A simplified map of the Balkan Peninsula. (c) A geological sketch of the East Serbia (after Kräutner and Krstić, 2003; Cvetković et al., 2013) along with the ESMAR localities.

Twenty-one ESMAR samples investigated in this study are collected from Tisa to Pirot (Fig. 1c). The ESMAR are almost basanites, with low SiO_2 contents (41.3–43.0 wt%) and high total alkalis (Na_2O+K_2O; 3.08–6.15 wt%), except for the sample Tisa3, with 39.7 wt% of SiO_2, belonging to nephelinite (Fig. 2a). They also have high MgO (8.2–13.5 wt%) and CaO (10.8–16.0 wt%) contents, with typical sodic affinity (Na_2O/K_2O = 1.13–5.42). The ESMAR are characterized by porphyritic texture, comprising olivine (± clinopyroxene ± Fe-Ti oxides) as phenocrysts, and clinopyroxene, plagioclase, nepheline, apatite and opaque minerals in groundmass. The ESMAR are enriched in incompatible elements compared with the MORBs, displaying OIB-like patterns (Fig. 2b), and in contrast with the depleted LILEs (e.g., Rb, K, Pb), they are more enriched in HFSEs (e.g., Nb, Ta) (Fig. 2b). In addition, the ESMAR have depleted Sr and Nd isotopic compositions ($^{87}Sr/^{86}Sr_i$ = 0.7031–0.7048, ε_{Nd} = 2.5–5.9; Cvetković et al., 2013). The ESMAR contain various mantle xenoliths which are mostly composed of spinel-bearing harzburgites and cpx-poor lherzolites with low CaO (<2 wt%), Al_2O_3 (<2 wt%), TiO_2 (<0.1 wt%) and high MgO (almost >43wt%) contents, and they are more depleted than normal European lithosphere (Cvetković et al., 2004a). Four spinel-bearing peridotite xenoliths hosted by the ESMAR from Sokobanja (SB; Fig. 1c) are also analyzed in this study and represent the East Serbian shallow lithospheric mantle. Detailed descriptions and geochemical analysis can be found in Cvetković et al. (2004b, 2013).

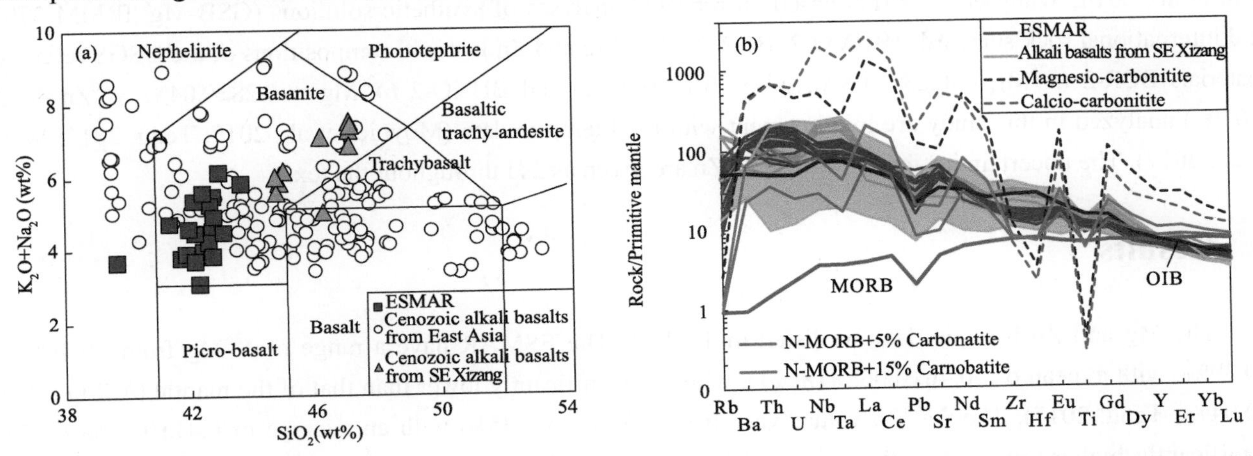

Fig. 2 (a) Geochemical classification of the ESMAR. Total alkali versus SiO_2 adapted from Le Bas et al. (1986). (b) Primitive mantle-normalized trace element patterns. Data source: Cvetković et al. (2013). The grey area represents the range of Cenozoic alkali basalts from East Asia. Data sources: Huang et al. (2015); Jin et al. (2020); Li et al. (2017); Liu et al. (2016); Wang and Liu (2021); Wang et al. (2018); Xu et al. (2022); Yang et al. (2012). Data for Cenozoic alkali basalts from SE Xizang, magnesio-carbonatite and calcio-carbonatite are from Liu et al. (2020); Hoernle et al. (2002) and Bizimis et al. (2003), respectively. The primitive mantle values are from McDonough and Sun (1995). Compositions of average OIB and N-MORB are from Sun and McDonough (1989).

3 Analytical methods

The chemical procedures for purification of Mg and Zn followed methods in Gao et al. (2019) and Liu et al. (2016), and were carried out at the China University of Geosciences, Beijing. After complete dissolution, 100 μl (containing ~20 μg Mg) or 1 ml (containing >1 μg Zn) of each sample solution was loaded on 2 ml Bio-Rad cation exchange resin AG50W-X8 or 1 ml anion exchange resin AG1-X8 for Mg or Zn chromatographic purification, respectively. The Mg was separated from matrix elements (K, Na, Ca, Fe, Al) using 1 N HNO_3. 8 N HCl, 0.4 N HCl and 0.5 N HNO_3 were used in order to achieve Zn purification. To verify that all matrix elements were eliminated, the same column procedures were carried out twice. After thorough drying, samples were dissolved in 1 ml 3% (m/m) HNO_3. More than 99% of Mg and Zn were recovered. Total procedural blank was 12 ng for Mg and 5 ng for Zn, which are neglectable in comparison with the amount of Mg (20 μg) and Zn (>1 μg) in each loaded sample.

Analysis of Mg and Zn isotope ratios was carried out on a Thermo Scientific *Neptune plus* MC-ICP-MS. Purified samples dissolved in 3% (m/m) HNO_3 were introduced into the Ar plasma. For a solution containing 0.4 ppm Mg and 0.2 ppm Zn, 7 V signal of ^{24}Mg and 3 V signal of ^{64}Zn for a standard 10^{11} Ω amplifier were acquired, respectively. For the purpose of correcting instrumental mass bias, the sample-standard bracketing method was applied to Mg isotopes, while a Cu-doping technique following Maréchal et al. (1999) was utilized for Zn isotopes. The Mg and Zn isotope ratios are reported in standard δ-notation in per mil relative to DSM-3 (Galy et al., 2003) and JMC 3-0749L (Maréchal et al., 1999), respectively:

$$\delta^{26,25}\text{Mg}\ (‰) = [(^{26,25}\text{Mg}/^{24}\text{Mg})_{\text{sample}} / (^{26,25}\text{Mg}/^{24}\text{Mg})_{\text{DSM-3}} - 1] \times 1000$$

$$\delta^{68,66}\text{Zn}\ (‰) = [(^{68,66}\text{Zn}/^{64}\text{Zn})_{\text{sample}} / (^{68,66}\text{Zn}/^{64}\text{Zn})_{\text{JMC 3-0749L}} - 1] \times 1000$$

The long-term external reproducibility (2sd) is ±0.06‰ for δ^{26}Mg (Gao et al., 2019) and ± 0.04‰ for δ^{66}Zn (Liu et al., 2016; Wang et al., 2017) based on repeated analyses of synthetic solutions (GSB-Mg, IRMM 3702) and international rock standards (BHVO-2, BCR-2). The Mg and Zn isotopic compositions of the USGS reference materials BCR-2 (δ^{26}Mg =–0.23±0.06‰; δ^{66}Zn=0.27±0.02‰) and BHVO-2 (δ^{26}Mg =–0.28±0.04‰; δ^{66}Zn=0.32± 0.01‰) analyzed in this study are in agreement with the literature data (Moynier et al., 2017; Teng, 2017; Wang et al., 2017). The uncertainties for δ^{26}Mg and δ^{66}Zn are given as 2sd throughout the text.

4 Results

The Mg and Zn isotopic data are listed in Table 1. The ESMAR have a range of δ^{26}Mg from –0.49‰ to –0.39‰, with a mean of –0.46±0.06‰ (n=21), which is significantly lower than that of the mantle (δ^{26}Mg=–0.25 ±0.04‰; Teng, 2017). Their δ^{66}Zn values vary from 0.35‰ to 0.45‰ with an average of 0.41±0.06‰ (n=21), significantly higher than that of the mantle (0.18±0.05‰; Liu et al., 2022 and references therein) and MORBs (0.27±0.06‰; Huang et al., 2018; Wang et al., 2017). Notably, Zn contents of the ESMAR vary from 66.9 to 86.6 μg/g (with a mean of 75.1 μg/g), lower than that of OIBs (~108 μg/g; Beunon et al., 2020 and references therein) and Cenozoic alkali basalts from East Asia (89–196 μg/g; Table S1; Huang et al., 2015; Jin et al., 2020; Li et al., 2017; Liu et al., 2016; Wang and Liu, 2021; Wang et al., 2018; Xu et al., 2022; Yang et al., 2012) and SE Xizang (130–148 μg/g; Table S1; Liu et al., 2020), but similar to those of MORBs (68±12 μg/g; Beunon et al., 2020 and references therein).

Table 1　Mg and Znisotopic ratios (‰), Sr and Nd isotopes, and selected element contents (major elements in wt%; trace elements in μg/g) for all samples

Sample no.	SiO_2	MgO	Cr	Zn	Zn/Fe×10^4	Fe/Mn	δ^{25}Mg	2sd	δ^{26}Mg	2sd	δ^{66}Zn	2sd	δ^{68}Zn	2sd	^{87}Sr/^{86}Sr$_i$	$\varepsilon_{Nd}(t)$
ESMAR																
Tisa3	39.7	8.2	123	67.9	9.0	57.3	–0.24	0.04	–0.46	0.03	0.41	0.01	0.83	0.01	0.70372	4.75
BL-1	42.4	11.1	315	68.5	9.2	60.1	–0.24	0.02	–0.47	0.05	0.35	0.01	0.70	0.04	0.70372	5.02
BL-3	41.7	12.2	322	66.9	8.4	60.3	–0.23	0.03	–0.44	0.02	0.37	0.03	0.73	0.06	0.70367	2.72
OST-140	42.3	11.5	363	73.9	10.0	64.2	–0.24	0.03	–0.47	0.01	0.40	0.04	0.80	0.06	0.70464	3.64
SB61	42.5	11.4	458	73.1	9.8	53.7	–0.24	0.02	–0.46	0.03	0.38	0.04	0.77	0.08	0.70307	5.28

Continued

Sample no.	SiO_2	MgO	Cr	Zn	Zn/Fe×10^4	Fe/Mn	δ^{25}Mg	2sd	δ^{26}Mg	2sd	δ^{66}Zn	2sd	δ^{68}Zn	2sd	^{87}Sr/^{86}Sr$_i$	$\varepsilon_{Nd}(t)$
SB62	42.7	11.1	457	86.6	11.5	54.5	−0.23	0.01	−0.45	0.02	0.44	0.01	0.87	0.08	0.70305	4.01
SB63	42.6	12.0	479	77.2	10.3	54.1	−0.25	0.06	−0.48	0.04	0.42	0.02	0.84	0.02	0.70315	2.55
SB64	41.3	11.5	452	72.3	9.6	57.7	−0.23	0.03	−0.45	0.05	0.40	0.01	0.79	0.00	0.70311	5.92
SB65	43.0	11.1	493	70.2	9.5	56.5	−0.23	0.03	−0.45	0.01	0.37	0.04	0.75	0.03	0.70314	3.92
SB66	41.9	13.2	623	72.9	9.8	56.9	−0.25	0.02	−0.49	0.04	0.39	0.05	0.78	0.08	0.70357	3.48
SB67	42.7	13.5	609	78.8	10.5	54.0	−0.23	0.05	−0.44	0.03	0.42	0.05	0.83	0.03	0.70336	5.25
KR135	42.7	10.4	267	76.7	10.2	57.6	−0.25	0.03	−0.49	0.02	0.45	0.00	0.89	0.08	0.70419	2.92
KR136	43.6	10.3	253	75.3	10.2	60.4	−0.25	0.01	−0.49	0.01	0.45	0.04	0.89	0.04	0.70383	4.52
KR137	42.2	11.4	315	77.5	10.4	61.0	−0.24	0.02	−0.47	0.04	0.43	0.04	0.84	0.06	0.70447	4.27
KR138	42.2	11.9	376	78.6	10.5	61.3	−0.21	0.03	−0.41	0.03	0.39	0.04	0.78	0.04	0.70395	2.5
KR139	42.9	8.3	171	79.0	11.5	52.7	−0.21	0.01	−0.41	0.02	0.44	0.06	0.86	0.05	0.70394	4.7
OKR156	42.6	10.7	301	75.8	10.1	61.0	−0.25	0.01	−0.48	0.01	0.41	0.02	0.81	0.05	0.70382	3.86
OKR157	42.1	10.8	294	77.2	10.2	58.0	−0.23	0.03	−0.44	0.03	0.42	0.04	0.84	0.03	0.70364	3.76
OKR158	42.4	10.7	287	77.9	10.4	61.3	−0.24	0.03	−0.46	0.01	0.45	0.05	0.88	0.08	0.70368	4.04
OKR159	42.7	10.0	253	75.6	10.3	59.7	−0.24	0.02	−0.46	0.02	0.41	0.00	0.82	0.08	0.70387	4.13
PIR161	41.9	9.4	404	75.8	10.1	60.9	−0.20	0.03	−0.39	0.04	0.36	0.01	0.71	0.05	0.70482	3.92
Peridotite xenoliths																
SB01				49.3			−0.09	0.03	−0.21	0.05	0.17	0.03	0.35	0.05		
SB05				50.8			−0.12	0.02	−0.24	0.03	0.18	0.03	0.36	0.05		
SB06				53.8			−0.14	0.02	−0.26	0.04	0.20	0.03	0.41	0.06		
SB07				55.1			−0.13	0.02	−0.25	0.03	0.22	0.04	0.43	0.07		
Rock standards																
BHVO-2				102.5			−0.14	0.02	−0.28	0.04	0.32	0.01	0.64	0.05		
BCR-2				128.7			−0.11	0.03	−0.23	0.06	0.27	0.02	0.53	0.03		

Note: The Mg and Zn isotopic data and Zn contents are from this study. Other data are from Cvetković et al., (2013).

The four peridotite xenoliths from the ESMAR have relatively homogeneous δ^{26}Mg (−0.26‰ to −0.21‰, with an average of −0.24±0.05‰) and δ^{66}Zn (0.17‰ to 0.22‰, with an average of 0.19±0.04‰), in consistence with the values of non-metasomatized refractory peridotites reported in previous studies (Liu et al., 2022; Teng, 2017).

5 Discussion

In this part, we firstly discuss the mantle-source lithology of the ESMAR on the basis of the major-trace elements and Sr-Nd isotopic data. Next, we discuss the origin of their abnormal Mg and Zn isotopic compositions (i.e., lower δ^{26}Mg and higher δ^{66}Zn than those of MORBs), with geodynamic models being proposed for the generation of the ESMAR at last. We also compare the geochemical data of the ESMAR with these of Cenozoic alkali basalts from East Asia and SE Xizang, in order to broaden our insights into the global deep carbon cycling related to different subduction zones.

5.1 Lithology of the ESMAR's mantle source

The ESMAR exhibit a narrow range of ^{87}Sr/^{86}Sr$_i$ (0.70305–0.70482) and $\varepsilon_{Nd}(t)$ values (2.6–5.9; Table 1), indicating a depleted mantle source similar to the European Asthenospheric Reservoir (EAR; Cvetković et al., 2013). The actual source lithology of the ESMAR, however, is still unclear. The Fe, Zn and Mn systematics (e.g.,

Zn/Fe and Fe/Mn) are useful proxies for detecting mantle lithology dominated by olivine (ol) and orthopyroxene (opx) or clinopyroxene (cpx) and garnet in mafic-ultramafic systems (Le Roux et al., 2010; Qin and Humayun, 2008). The inter-mineral exchange coefficients of both $K_{D(ol/melt)}^{Zn/Fe}$ and $K_{D(opx/melt)}^{Zn/Fe}$ are ~0.9–1, much higher than $K_{D(cpx/melt)}^{Zn/Fe}$ and $K_{D(gt/melt)}^{Zn/Fe}$ (~0.56 and 0.2, respectively) (Bedard, 2005; Le Roux et al., 2010). Meanwhile, $K_{D(ol/melt)}^{Fe/Mn}$ (~1.3) and $K_{D(opx/melt)}^{Fe/Mn}$ (~0.85) are much higher than $K_{D(cpx/melt)}^{Fe/Mn}$ and $K_{D(gt/melt)}^{Fe/Mn}$ (~0.69 and 0.6, respectively) (Walter, 1998). By using batch melting calculations, the peridotite-derived melt has obviously lower Zn/Fe ($\times 10^4$) (7.5–12.1) and Fe/Mn (45.9–64.2) compared with pyroxenite-derived melt (12.6–18.9 and 57.4–78.9, respectively) (Fig. 3a). The FC3MS ($FeO^T/CaO-3\times MgO/SiO_2$, all in wt%) is also able to distinguish

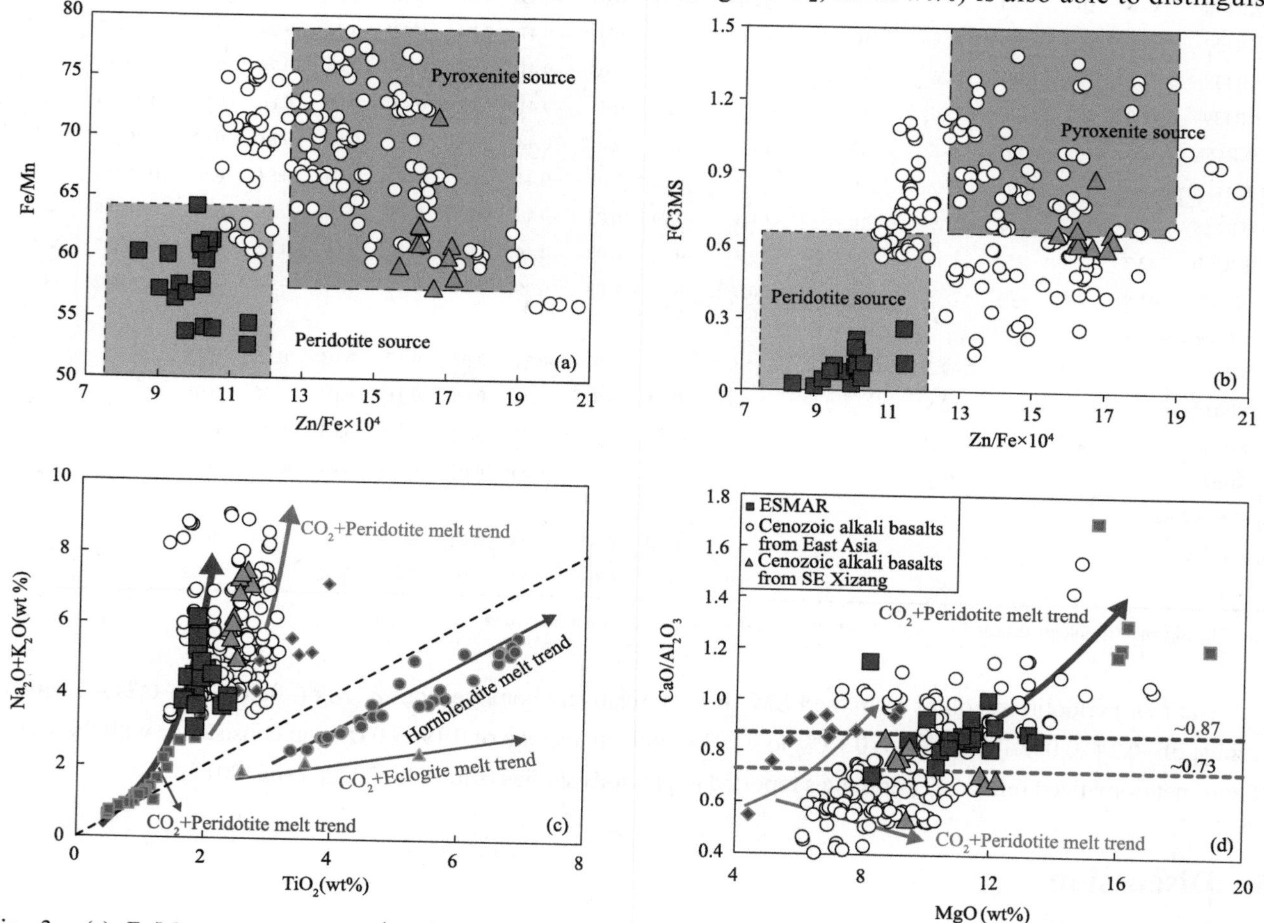

Fig. 3 (a) Fe/Mn versus Zn/Fe×10^4. The green and purple areas denote melts derived from peridotite-dominated and pyroxenite-dominated sources, respectively. Peridotite-melt is calculated using $Zn/Fe_{peridotite}\times 10^4$ = 6.5–10.5, $Fe/Mn_{peridotite}$ = 50–70, $K_{D(peridotite/melt)}^{Zn/Fe}$ =0.85–0.9, and $K_{D(peridotite/melt)}^{Fe/Mn}$ =1.1 (Le Roux et al., 2010; Walter, 1998 and references therein). Pyroxenite-melt is calculated using $K_{D(cpx/melt)}^{Zn/Fe}$ =0.35, $K_{D(gt/melt)}^{Zn/Fe}$ =0.7, $K_{D(cpx/melt)}^{Fe/Mn}$ =0.69 and $K_{D(gt/melt)}^{Fe/Mn}$ =0.6 from a theoretical 70% cpx–30% gt pyroxenite with Zn/Fe×10^4 = 8–12 and Fe/Mn = 40–55 (Bedard, 2005; Le Roux et al., 2010 and references therein). (b) FC3MS versus Zn/Fe×10^4. FC3MS=$FeO^T/CaO-3\times MgO/SiO_2$ (all in wt%) and peridotite-derived melts are distinguished from pyroxenite-derived melts by FC3MS <0.65 (Yang et al., 2016). (c) Total alkali versus TiO_2 and (d) CaO/Al_2O_3 versus MgO. The red and grey dotted lines represent the average CaO/Al_2O_3 ratios of the ESMAR (~0.87) and Cenozoic alkali basalts from East Asia (~0.73), respectively. Experimental melts from partial melting of CO_2 + peridotite (Dasgupta et al., 2010, 2007), CO_2 + pyroxenite (Gerbode and Dasgupta, 2010), hornblendite (Pilet et al., 2008), and CO_2 + eclogite (Dasgupta et al., 2006) are compiled for comparison. Data for Cenozoic alkali basalts from East Asia and SE Xizang are compiled in Table S1. (For interpretation of the references to colour in this figure legend, the reader is referred to the web version of this article.)

peridotite-derived melt from pyroxenite-derived one (FC3MS >0.65) by multi-decadal melting experiments (Yang et al., 2016). The ESMAR have significantly low Zn/Fe ($\times10^4$) (8.4–11.5), Fe/Mn (52.7–64.2) and FC3MS (−0.08–0.25), indicating that they were sourced from peridotitic mantle, in contrast to alkali basalts from the East Asia and SE Xizang which have a mixed "pyroxenite + peridotite" source (Figs. 3a, b). Besides, the high MgO (8.2–13.5 wt%) and CaO/Al_2O_3 (0.7–1.2) and low TiO_2 contents (1.7–2.4 wt%) of the ESMAR display similarities to partial melts of "peridotite + CO_2" from high-pressure melting experiments (Dasgupta and Hirschmann, 2010; Dasgupta et al., 2007; Hirose, 1997; Figs. 3c, d).

Besides the asthenospheric mantle component indicated by OIB-like trace element patterns and depleted Sr-Nd isotopic compositions of the ESMAR, the $^{206}Pb/^{204}Pb$ values (18.1–18.6) of the ESMAR (Cvetković et al., 2013), which are distinctively lower than those of other Cenozoic alkali basalts in Circum-Mediterranean region called CiMACI (Circum-Mediterranean Anorogenic Cenozoic Igneous rocks; $^{206}Pb/^{204}Pb$ ~19.5) (Lustrino and Wilson, 2007), suggest that a highly depleted lithospheric mantle is probably another source component ($^{206}Pb/^{204}Pb$ ~17; Cvetković et al., 2013). Previous studies compiled quantities of mantle xenoliths from the ESMAR, and found that the predominant lithologies are depleted harzburgites and cpx-poor lherzolites with much lower CaO, Al_2O_3, TiO_2 and higher MgO contents than those of the typical European Mantle (Cvetković et al., 2004a), also indicating that the East Serbian lithospheric mantle is highly depleted and maybe a slice of Tethyan oceanic lithosphere accreted during subduction (Cvetković et al., 2007; Cvetković et al., 2010). Therefore, Cvetković et al. (2013) suggested that the primary magma of the ESMAR originated in the lithospheric bottom previously metasomatized by melts from asthenospheric mantle. In addition, the Mautle-like Mg-Zn isotopic data of the four peridotite xenoliths ($\delta^{26}Mg$=−0.24±0.05‰; $\delta^{66}Zn$=0.19±0.04‰) suggest that the generation of ESMAR was not the result of interaction between carbonated melts and shallow lithospheric mantle (Wang et al., 2018; Wang and Liu, 2021). In brief, the lithology of the ESMAR's mantle source is highly depleted peridotite, which is most likely located at the deepest part of the lithospheric mantle.

5.2 Origin of the abnormal δ^{26}Mg and δ^{66}Zn of the ESMAR

5.2.1 Effects of crustal contamination, alteration and weathering

Crustal contamination during magma ascent and post-eruption alteration and weathering could affect the Mg and Zn isotopic ratios of lavas, so these effects must be considered at first. The depleted Sr-Nd isotopic compositions ($^{87}Sr/^{86}Sr_i$=0.70305–0.70482, $\varepsilon_{Nd}(t)$=2.6–5.9; Table 1; Cvetković et al., 2013) and positive Nb-Ta anomalies (Fig. 2b) of the ESMAR suggest that the influence of crustal contamination is negligible. Although some samples have relatively high LOI contents (loss on ignition; 3.1–4.9 wt%), no correlation of δ^{26}Mg and δ^{66}Zn values with LOI can be found (R^2=0.045 and 0.189, respectively). The CIA (chemical index of alteration; $Al_2O_3/(Al_2O_3 + CaO + Na_2O + K_2O)\times100$) of the ESMAR range from 41.4 to 47.7, mostly falling on the range of the un-weathered basalts (30–45; Nesbitt and Young, 1982) and also displaying no correlation with δ^{26}Mg and δ^{66}Zn (both $R^2< 0.001$). In addition, alteration and weathering of silicate rocks on the surface leave behind the residues enriched in heavy Mg and light Zn isotopes (Lv et al., 2020; Teng et al., 2010). Thus, crustal contamination, alteration and weathering have insignificant impact and are not the causes for the low δ^{26}Mg and high δ^{66}Zn of the ESMAR.

5.2.2 Fractional crystallization, partial melting and kinetic isotope fractionation

The ESMAR have high MgO contents (8.2–13.5 wt%; Table 1) but contain phenocrysts of olivine (ol), with few clinopyroxenes (cpx) and Fe-Ti oxides occurring as micro-phenocrysts (Cvetković et al., 2013). Below we discuss the isotopic effect induced by fractional crystallization of ol, cpx and oxides (e.g., Al-spinel and Cr-spinel). Olivine and clinopyroxene commonly have similar Mg isotopic compositions (e.g., Liu et al., 2011), showing no difference with MORBs (−0.25±0.06‰; Teng, 2017). Thus, the fractional crystallization of ol and cpx would result in rapid decrease of MgO contents in basaltic melts but ignorable change of δ^{26}Mg. This is inconsistent with the high MgO contents and low δ^{26}Mg (−0.49‰ to −0.39‰) of the ESMAR (Fig. 4b), excluding the effect of ol or cpx crystallization. Spinels (spl), including Al-spinel and Cr-spinel (chromite), have heavier Mg isotopic compositions than those of silicate minerals ($\delta^{26}Mg_{Al\text{-}spl}$=0.14±0.14‰; $\delta^{26}Mg_{Cr\text{-}spl}$=0.15±0.30‰; Liu et al., 2011; Su et al., 2017), so the crystallization of spl would cause coupled decrease of MgO or Cr contents and δ^{26}Mg. However, no correlation between δ^{26}Mg and MgO (R^2=0.062) or Cr (R^2=0.000) contents of the ESMAR is observed (Figs. 4b, d), precluding the influence of spl crystallization.

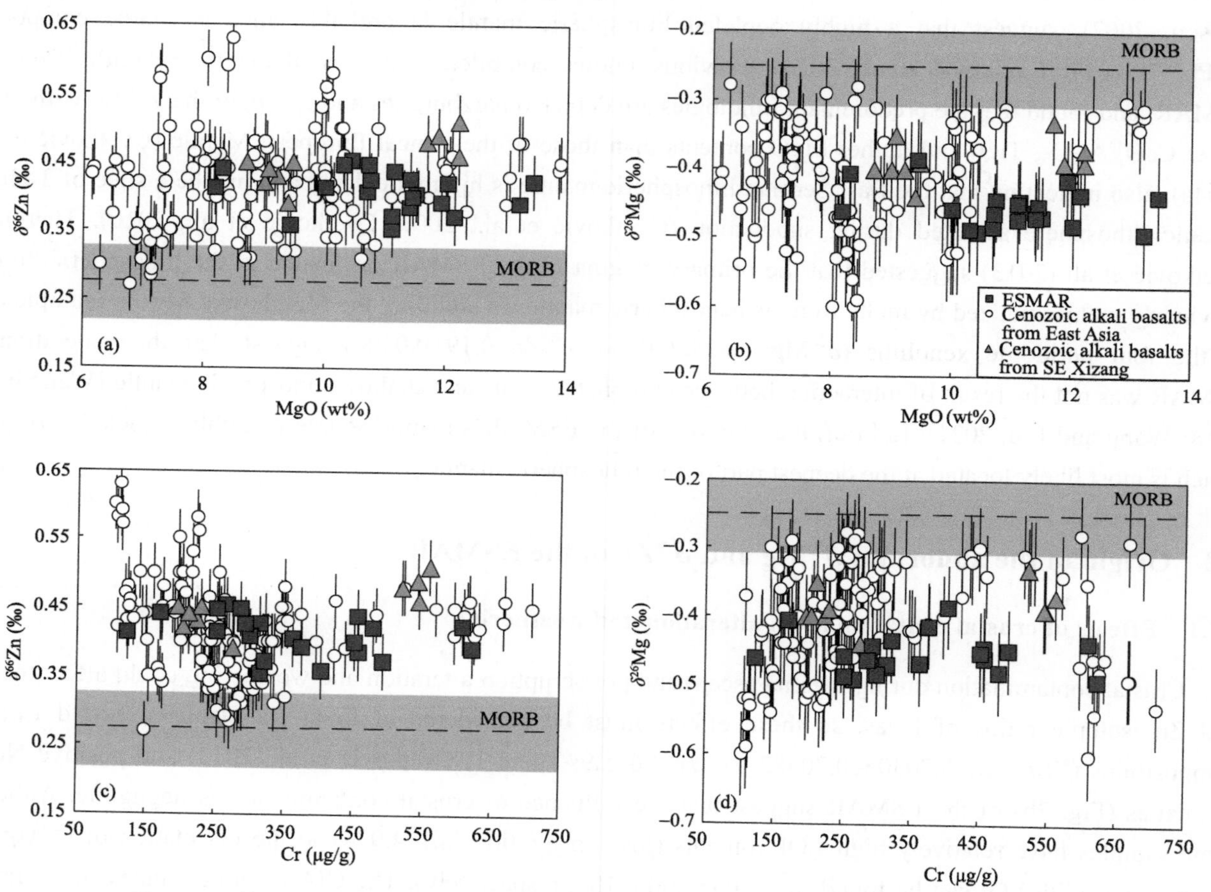

Fig. 4　Plots of δ^{66}Zn and δ^{26}Mg against MgO (a and b) and Cr contents (c and d). Mg and Zn isotopic compositions of MORBs (δ^{26}Mg = −0.25 ± 0.06‰; δ^{66}Zn = 0.27 ± 0.06‰) are from Huang et al. (2018), Teng (2017) and Wang et al. (2017). Data for Cenozoic alkali basalts from East Asia and SE Xizang are compiled in Table S1.

For Zn isotopes, $\delta^{66}Zn_{ol}$ and $\delta^{66}Zn_{cpx}$ (0.15 ± 0.06‰; 0.19±0.09‰; Wang et al., 2017) are lower than MORBs (0.27±0.06‰; Huang et al., 2018; Wang et al., 2017), indicating that ol and cpx crystallization increases δ^{66}Zn with decreasing MgO contents. But δ^{66}Zn values (0.35‰ to 0.45‰) of the ESMAR are not correlated with MgO (R^2=0.037) or Cr (R^2=0.095) contents (Figs. 4a, c). Al-spinels have similar δ^{66}Zn values (0.27‰±0.04‰; Wang et al., 2017) with those of silicate minerals, whereas Cr-spinels possess much lower δ^{66}Zn (−0.13±0.11‰; Yang et al., 2021), thus, crystallization of Cr-spinels can elevate the δ^{66}Zn of melts. Besides, spinel-group minerals are enriched in Zn and have a high partition coefficient of Zn during fractional crystallization ($D_{spl\text{-}melt}$ = 5.3; Davis et al., 2013; Yang et al., 2021). As a result, fractional crystallization of Cr-spinels leads to a negative correlation between Zn contents and δ^{66}Zn in the melts. The ESMAR have similar or higher Zn contents (~75.1 μg/g; Table 1) than that of MORBs (68±12 μg/g; Beunon et al., 2020 and references therein), displaying a slightly positive correlation with δ^{66}Zn (R^2 = 0.4). Therefore, fractional crystallization of ol, cpx and spl is unlikely to cause the high δ^{66}Zn of the ESMAR.

To assess the influence of partial melting of different mantle lithologies on the Mg and Zn isotopic systematics, we used an incremental batch melting model (modified from Sossi et al., 2018; Zhong et al., 2017) for depleted garnet peridotite and garnet pyroxenite, where the Mg and Zn fractionation factors between melts and residue were recalculated at each step (constant mass of melts; 2%), and the fractionation factors ($\alpha_{\text{melt-residue}}$) can be obtained by:

$$\alpha_{\text{melt-residue}} = \alpha_{\text{melt-cpx}} \times \left(\sum_{i=1}^{n} [n_i \text{MgO}_{\text{mineral}}] / \sum_{i=1}^{n} [n_i \text{MgO}_{\text{mineral}} \cdot \alpha_{\text{mineral-cpx}}] \right)$$

$$\alpha_{\text{melt-residue}} = \alpha_{\text{melt-cpx}} \times \left(\sum_{i=1}^{n} [n_i \text{Zn}_{\text{mineral}}] / \sum_{i=1}^{n} [n_i \text{Zn}_{\text{mineral}} \cdot \alpha_{\text{mineral-cpx}}] \right)$$

where $\alpha_{\text{melt-cpx}}$, the fractionation factors between melt and clinopyroxene, are from Zhong et al. (2017) and Sossi et al. (2018). $\alpha_{\text{mineral-cpx}}$ is the fractionation factor between clinopyroxene and its co-existing mineral assemblage (olivine, orthopyroxene and garnet). $\text{MgO}_{\text{mineral}}$ and $\text{Zn}_{\text{mineral}}$ are the MgO and Zn contents of each mineral, respectively. And $\alpha_{\text{mineral-cpx}}$ was estimated by:

$$\Delta_{\text{mineral-cpx}} \approx 10^3 \times \ln \alpha_{\text{mineral-cpx}}$$

The detailed parameters are listed in Table 2. The initial Mg and Zn isotopic compositions of depleted garnet peridotite were set to be $\delta^{26}\text{Mg}_0$=–0.21‰ and $\delta^{66}\text{Zn}_0$=0.19‰ (Table 1). The $\delta^{26}\text{Mg}_0$ and $\delta^{66}\text{Zn}_0$ of garnet pyroxenite were assumed to be –0.20‰ and 0.23‰ (McCoy-West et al., 2018; Zhong et al., 2017). After determining the $\alpha_{\text{melt-residue}}$, we can obtain $\delta^{26}\text{Mg}_{\text{melt}}$ and $\delta^{66}\text{Zn}_{\text{melt}}$ as shown in Fig. 5.

Table 2 Parameters used for incremental batch melting of different mantle lithologies (garnet peridotite and garnet pyroxenite)

Minerals	Ol	Cpx	Opx	Grt
Garnet peridotite				
Initial modal abundance[a]	0.61	0.022	0.332	0.036
Melting mode[a]	0.08	0.8	–0.18	0.3
Kd_{Mg}[b]	3.04	1.35	1.96	1.30
D_{Zn}[c]	0.96	0.40	0.65	0.21
Garnet pyroxenite				
Initial modal abundance[a]		0.3		0.7
Melting mode[a]		0.9		0.1
Kd_{Mg}[b]		2.79		4.49
D_{Zn}[c]		0.40		0.21
$\alpha_{\text{Mg(mineral-cpx)}}$[a]	0.99990	1	0.99995	0.99950
$\alpha_{\text{Zn(mineral-cpx)}}$[d]	0.99993	1	0.99993	0.99992

[a] Simon et al. (2003) and Zhong et al. (2017).

[b] Davis et al. (2011) and Petermann et al. (2004).

[c] Davis et al. (2013) and Le Roux et al. (2011).

[d] McCoy-West et al. (2018) and Sossi et al. (2018).

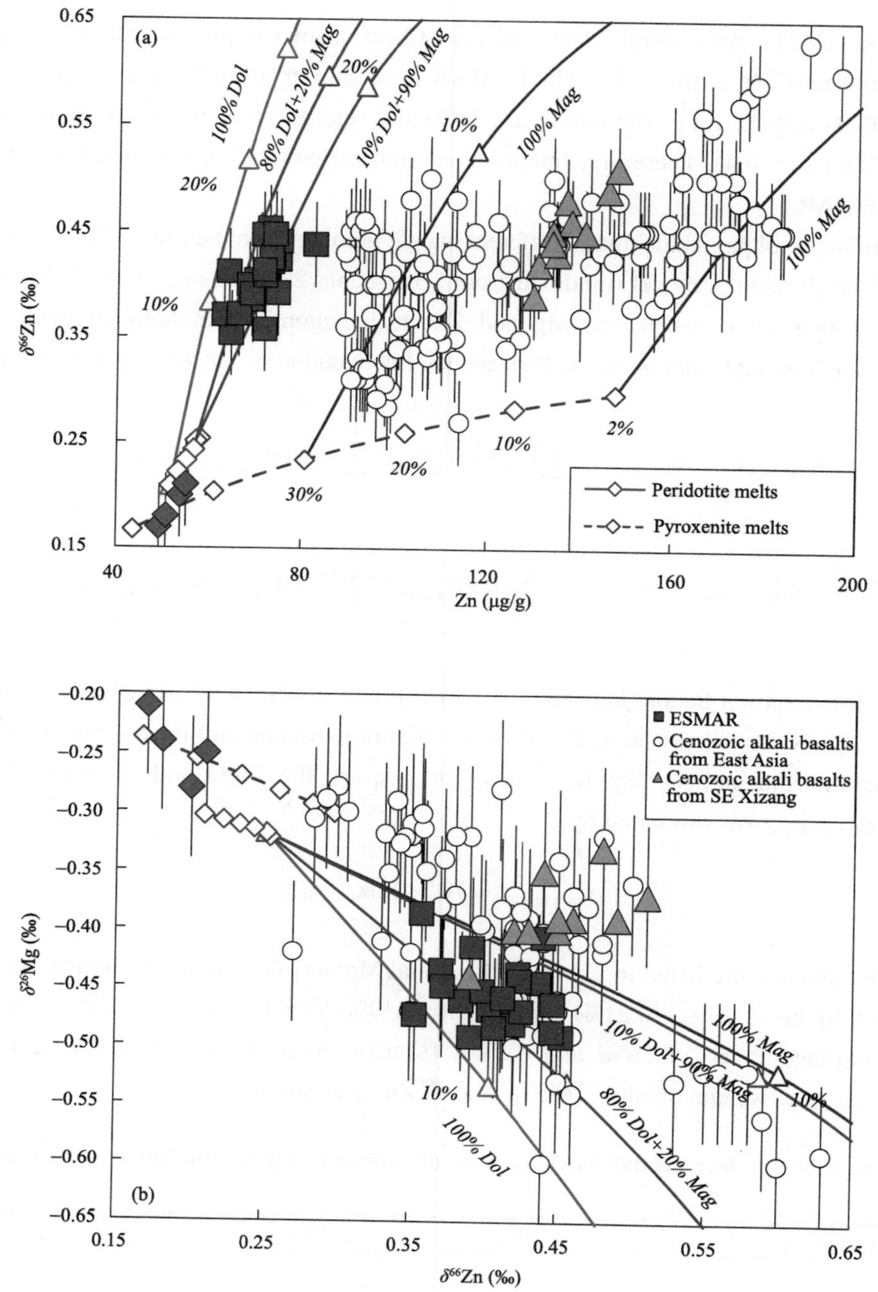

Fig. 5　Plots of δ^{66}Zn against Zn contents (a) and δ^{26}Mg (b). The green rhombus represents peridotite xenoliths hosted by the ESMAR. An incremental batch melting model (the melting degree increases from 2% to 50%) is used for garnet peridotite and garnet pyroxenite, and the results are represented by red open rhombus and blue open rhombus, respectively. The curves denote two end-member mixing between calculated melts and different species of carbonates (Mag = magnesite, Dol = dolomite and Cal = calcite). Data for Cenozoic alkali basalts from East Asia and SE Xizang are compiled in Table S1. Detailed modeling parameters are given in Tables 2 and S2. (For interpretation of the references to colour in this figure legend, the reader is referred to the web version of this article.)

The melting model shows that partial melting of both garnet peridotite and garnet pyroxenite results in limited Mg isotope fractionation (Fig. 5b). As the degree of partial melting (F) increases from 2% to 30%, δ^{26}Mg of the melts from garnet peridotite and garnet pyroxenite range from −0.32‰ to −0.31‰ and −0.30‰ to −0.27‰, respectively, which cannot explain the δ^{26}Mg of the ESMAR as low as −0.49‰. Similarly, δ^{66}Zn of melts from garnet peridotite and garnet pyroxenite range narrowly from 0.26‰ to 0.23‰ and 0.30‰ to 0.23‰ as F increases to 30% (Fig. 5a), significantly lower than δ^{66}Zn of the ESMAR (0.35‰ to 0.45‰). Therefore, partial melting of different source lithologies is not the reason for the abnormally low δ^{26}Mg and high δ^{66}Zn of the ESMAR.

Potential temperature and chemical gradients exist between magma and ambient mantle when magma ascend through the lithosphere, probably inducing kinetic isotope fractionation, i.e., thermal or chemical diffusion. Thermal diffusion drives light isotopes to concentrate in hot places (Richter et al., 2008) and produces both low δ^{26}Mg and δ^{66}Zn in melts (Liu and Li, 2019), which is inconsistent with the low δ^{26}Mg but high δ^{66}Zn of the ESMAR (Fig. 4). Light isotopes diffuse faster than heavier ones when chemical gradient exists (Richter et al., 2008), gradually enriching melts in light Mg and Zn isotopes, with increasing contents of MgO and Zn simultaneously. No such correlation can be found in the ESMAR (Figs. 4b and 5a). Hence, kinetic fractionation is not likely to have caused the ESMAR's abnormal Mg and Zn isotopic compositions.

5.2.3 The key role of recycled carbonates in the ESMAR's mantle source

Collectively, the lower δ^{26}Mg and higher δ^{66}Zn of the ESMAR than those of the mantle should imply source heterogeneity (i.e., mantle metasomatism) induced by recycled crustal materials. Some fertile mantle xenoliths in the ESMAR contain Ti-Al-Cr-rich clinopyroxene (>10% vol.) and Ti-Al-rich spinel, suggesting the presence of source metasomatism by alkaline melts (Cvetković et al., 2010). Sedimentary carbonates typically characterized by extremely low δ^{26}Mg and high δ^{66}Zn (δ^{26}Mg as low as −5.57‰; δ^{66}Zn=0.99±0.25‰; Liu et al., 2022 and references therein) serve as the possible metasomatic agent for the ESMAR mantle source. The ESMAR have strongly negative Rb, Pb, Zr, Hf and Ti anomalies as widely observed in carbonatites (Fig. 2b; Bizimis et al., 2003; Hoernle et al., 2002). And the similar incompatible element patterns between the ESMAR and carbonatites suggest that ~5% to 15% of carbonatitic melts have been added into the primitive melts of the ESMAR (Fig. 2b). Melting experiments of carbonated peridotites reveal that low-degree melting of lherzolites with the addition of CO_2 can produce alkali basaltic melts (Dasgupta et al., 2007; Hirose, 1997). The ESMAR have high $Na_2O + K_2O$ (3.1–6.2 wt%), CaO/Al_2O_3 (0.7–1.2) and MgO (8.2–13.5 wt%), and low TiO_2 contents (1.7–2.4 wt%), indicating a "peridotite + CO_2" mantle source (Figs. 3c, d). High $(La/Yb)_N$ together with low Ti/Eu ratios are regarded as carbonate-metasomatism indicators because carbonate metasomatism can elevate the $(La/Yb)_N$ of mantle source and cause depletion in HFSEs (e.g., Ti/Eu) (e.g., Coltorti et al., 1999). Carbonatitic melts are also typically characterized by extremely low $(Hf/Sm)_N$ (~0.02; Bizimis et al., 2003; Hoernle et al., 2002), much lower that of MORB (~1.1; Sun and McDonough, 1989). As illustrated in Fig. 7, the $(La/Yb)_N$, $(Hf/Sm)_N$ and Ti/Eu ratios of the ESMAR plot between carbonatites and MORB, also implying significant carbonate metasomatism. Therefore, we suggest that the low-δ^{26}Mg and high-δ^{66}Zn signatures of the ESMAR are inherited from recycled sedimentary carbonates.

Different species of sedimentary carbonates have contrasting Mg, Zn and Sr contents and isotopic compositions (Table S2), which allows us to use Mg-Zn-Sr isotopic systematics to specify the species of carbonates in the ESMAR source. Two end-member mixing models (Figs. 5, 6) suggest that about 5–10% of magnesium carbonates (dolomite and magnesite) were recycled into the mantle source, composed of 10–80% dolomite and 20–90% magnesite. Unlike the high solubility of calcite at low pressures, magnesium carbonates are much less soluble than calcite in H_2O-rich fluids and only minimally soluble even at the bottom of the upper mantle (Pan et al., 2013). During subduction, phase changes of carbonate minerals occur from calcite to dolomite and magnesite (e.g., Kushiro, 1975). Besides, magnesium carbonates can also be formed by Ca-Mg exchange between Ca-rich carbonates and silicate at high pressure during subduction (Yaxley and Green, 1994). Rocks from the Alps that have undergone ultrahigh pressure (⩾3.2 Gpa) metamorphosis contain solid inclusions of magnesite and dolomite (Frezzotti et al., 2011). Thus, magnesium carbonates are likely to be recycled into the deep mantle. Furthermore, 5–10% of recycled magnesium carbonates calculated by our models are equivalent to 3–5% CO_2, which is consistent with the high-pressure experiment results (1–3% CO_2 + peridotite; Dasgupta et al., 2007) where compositionally similar alkali basaltic melts were produced.

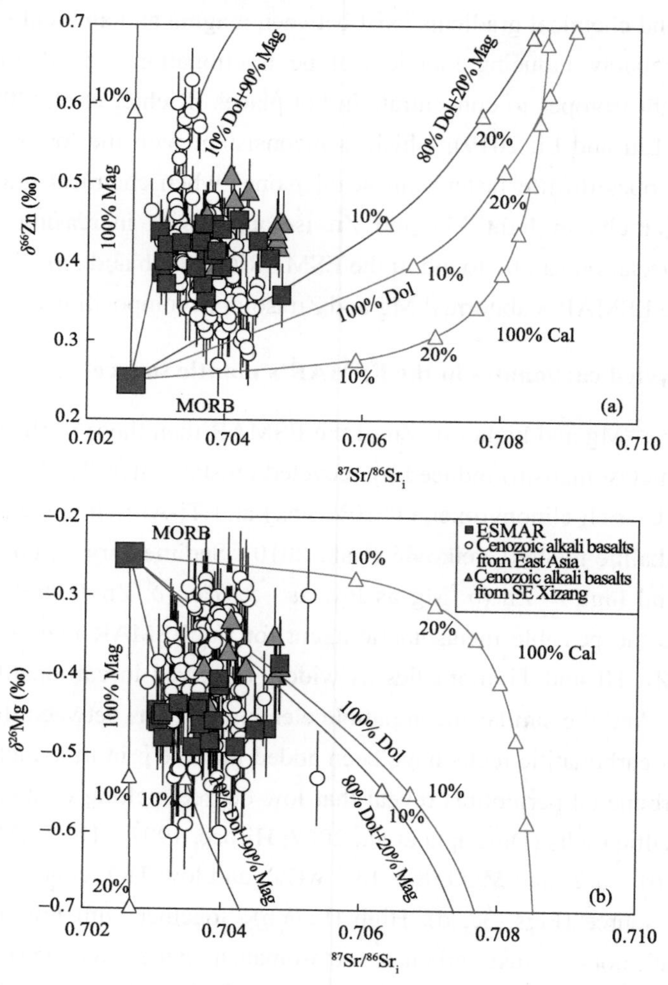

Fig. 6　Plots of δ^{66}Zn (a) and δ^{26}Mg (b) against $^{87}Sr/^{86}Sr_i$. The curves denote two end-member mixing between MORB (blue square) and different species of carbonates (Mag = magnesite, Dol = dolomite and Cal = calcite). Data for Cenozoic alkali basalts from East Asia and SE Xizang are compiled in Table S1. Modeling parameters are given in Table S2. (For interpretation of the references to colour in this figure legend, the reader is referred to the web version of this article.)

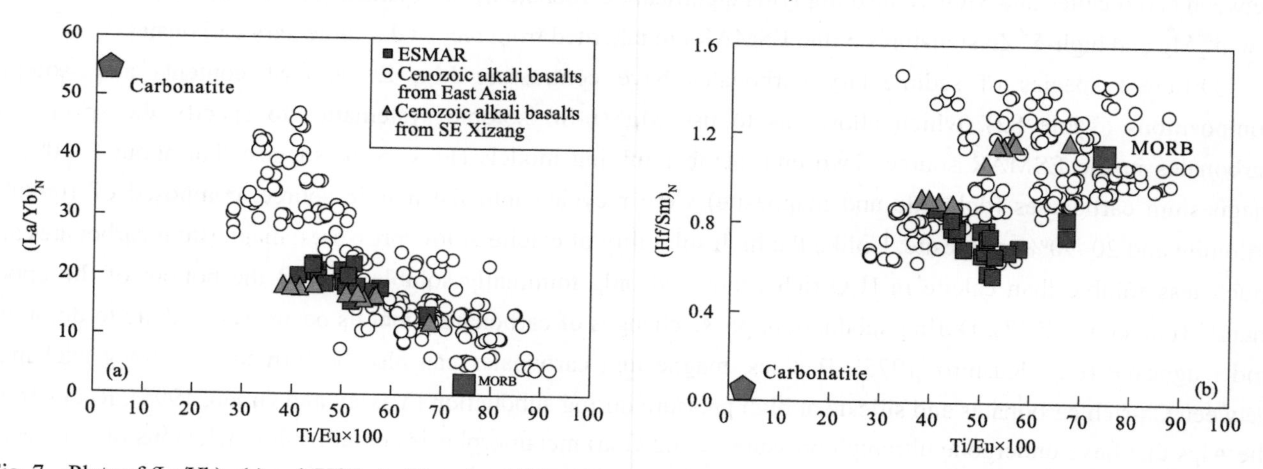

Fig. 7　Plots of $(La/Yb)_N$ (a) and $(Hf/Sm)_N$ (b) against Ti/Eu×100. Data for carbonatite and MORB are from Bizimis et al. (2003), Hoernle et al. (2002) and Sun and McDonough (1989). Data for Cenozoic alkali basalts from East Asia and SE Xizang are compiled in Table S1.

5.3 Geodynamic models for the ESMAR's origin and implications for deep carbon cycling

The slab of a Neo-Tethyan ocean, named the Vardar Ocean, started to subduct eastward beneath the Eurasian plate from the Mesozoic, invoking the widespread calc-alkaline magmatism of the Banatite-Timok-Srednjegorje Magmatic and Metallogenetic Belt in Balkan Peninsula (Fig. 8a). The Vardar Ocean might close completely until the Late Jurassic to Early Cretaceous, generating a suture zone called the Vardar Zone (Karamata and Krstić, 1996). After the cessation of subduction, the eastward subducted Vardar oceanic slab underwent break-off (Cvetković et al., 2004b) and gradually became a stagnant slab in the deep mantle revealed by the high velocity of P wave between ~410–660 km beneath the Balkan region (Figs. 8c, d; Piromallo and Morelli, 2003). Magnesium carbonates carried by the Vardar stagnant slab metasomatized the mantle transition zone and partially melted to produce carbonated silicate melts (e.g., Xu et al., 2020). The carbonated silicate melts ascended to the bottom of the lithospheric mantle (Cvetković et al., 2004b), metasomatizing the highly depleted peridotite there and producing a carbonated mantle source (Fig. 8b). Finally, the extensional processes of the Moesian region during Latest Cretaceous to Paleocene (Cvetković et al., 2004b, 2010) triggered partial melting of this carbonated mantle domain, generating the ESMAR.

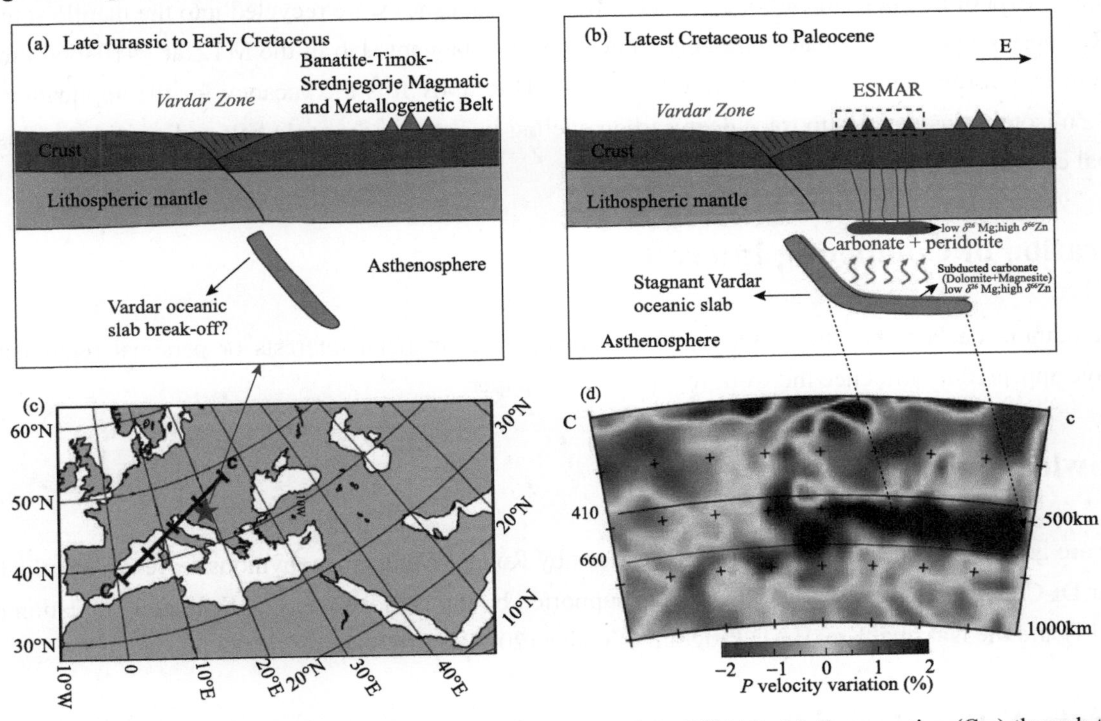

Fig. 8 (a and b) Schematic cartoon models illustrating the generation of the ESMAR. (c) Cross section (C-c) through the Balkan Peninsula, and the red star denotes the locality of East Serbia. (d) P wave velocity structure below the area of the cross section (C-c), modified from Piromallo and Morelli (2003). (For interpretation of the references to colour in this figure legend, the reader is referred to the web version of this article.)

A seismic tomography model of the central Mediterranean reveals an obvious shear-wave low-velocity (Vs) anomaly at depths >180 km, which is interpreted as the rising of carbonate-rich melts with low density and viscosity from the asthenosphere (Frezzotti et al., 2009). The most radiogenic $^{206}Pb/^{204}Pb$ isotopic compositions recorded by Bermuda alkali basalts indicate a volatiles-rich (e.g., carbonate melts) source in the MTZ (Mazza et al., 2019). In addition, the low-δ^{26}Mg and high-δ^{66}Zn anomalies of Cenozoic alkali basalts in East Asia and SE Xizang (Fig. 5) suggest that the MTZ is likely an important carbon reservoir, where the stagnant slabs carried carbonates into depths of ~440 to 660 km (Li et al., 2017; Liu et al., 2022, 2016). This current study reveals the coupling of Mg and Zn isotopic anomalies with presence of recycled carbonates in the MTZ under the Balkan Peninsula, an important segment of the Tethyan orogenic belt, implying that the MTZ is probably an important storage of subducted carbon globally.

6 Conclusions

We present a combined investigation of Mg and Zn isotopes for a collection of well-studied Na-alkaline basaltic rocks (ESMAR) from central Balkan Peninsula (East Serbia). The ESMAR are characterized by abnormally lower δ^{26}Mg (−0.49‰ to −0.39‰) and higher δ^{66}Zn (0.35‰ to 0.45‰) compared with those of MORB, which cannot be explained by any low-*T* or magmatic processes, except source metasomatism by recycled sedimentary carbonates. The distinct chemical compositions of the ESMAR (e.g., low Zn/Fe, Fe/Mn and TiO_2; high MgO and CaO/Al_2O_3) indicate a depleted peridotitic source at the bottom of the lithospheric mantle. Together with Mg-Zn-Sr isotopic data, the incremental batch melting and end-member mixing models suggest that about 5–10% of magnesium carbonates (dolomite and magnesite) were recycled into the mantle source of the ESMAR. These recycled carbonates probably came from the stagnant slab in the MTZ at ~410–660 km beneath the Balkan region revealed by geophysical tomography. This study fills the vacancy for the application of using Mg and Zn isotope systematics to trace deep carbon cycling in Tethyan orogenic belt and provides an insight into the global carbon cycling.

Declaration of Competing Interest

The authors declare that they have no known competing financial interests or personal relationships that could have appeared to influence the work reported in this paper.

Acknowledgements

We are grateful for the comments and suggestions by Rong Xu and an anonymous reviewer and the handling by Editor Di-Cheng Zhu. This work is financially supported by the National Natural Science Foundation of China (41730214) and the National Key R&D Program of China (2019YFA0708400).

Appendix A. Supplementary data

Supplementary data to this article can be found online at https://doi.org/10.1016/j.lithos.2022.106899.

References

Bédard, J.H., 2005. Partitioning coefficients between olivine and silicate melts. Lithos 83, 394-419.

Berza, T., Constantinescu, E., Vlad, S.N., 1998. Upper Cretaceous Magmatic Series and Associated Mineralisation in the Carpathian–Balkan Orogen. Resour. Geol. 48, 291-306.

Beunon, H., Mattielli, N., Doucet, L.S., Moine, B., Debret, B., 2020. Mantle heterogeneity through Zn systematics in oceanic basalts: Evidence for a deep carbon cycling. Earth-Sci. Rev. 205, 103174.

Bizimis, M., Salters, V.J.M., Dawson, J.B., 2003. The brevity of carbonatite sources in the mantle: evidence from Hf isotopes. Contrib. Mineral. Petrol. 145, 281-300.

Clift, P.D., 2017. A revised budget for Cenozoic sedimentary carbon subduction. Rev. Geophys. 55, 97-125.

Coltorti, M., Bonadiman, C., Hinton, R.W., Siena, F., Upton, B.G.J., 1999. Carbonatite Metasomatism of the Oceanic Upper Mantle: Evidence from Clinopyroxenes and Glasses in Ultramafic Xenoliths of Grande Comore, Indian Ocean. J. Petrol. 40, 133-165.

Cvetković, V., Downes, H., Höck, V., Prelević, D., Lazarov, M., 2010. Mafic alkaline metasomatism in the lithosphere underneath East Serbia: Evidence from the study of xenoliths and the host alkali basalts. Geol. Soc. Spec. Publ. 337, 213-239.

Cvetković, V., Downes, H., Prelević, D., Jovanović, M., Lazarov, M., 2004a. Characteristics of the lithospheric mantle beneath East Serbia inferred from ultramafic xenoliths in Palaeogene basanites. Contrib. Mineral. Petrol. 148, 335-357.

Cvetković, V., Downes, H., Prelević, D., Lazarov, M., Resimić-Šarić, K., 2007. Geodynamic significance of ultramafic xenoliths from Eastern Serbia: Relics of sub-arc oceanic mantle? J. Geodyn. 43, 504-527.

Cvetković, V., Prelević, D., Downes, H., Jovanović, M., Vaselli, O., Pécskay, Z., 2004b. Origin and geodynamic significance of Tertiary postcollisional basaltic magmatism in Serbia (central Balkan Peninsula). Lithos 73, 161-186.

Cvetković, V., Šarić, K., Prelević, D., Genser, J., Neubauer, F., Höck, V., von Quadt, A., 2013. An anorogenic pulse in a typical orogenic setting: The geochemical and geochronological record in the East Serbian latest Cretaceous to Palaeocene alkaline rocks. Lithos 180-181, 181-199.

Dasgupta, R., Hirschmann, M.M., 2010. The deep carbon cycle and melting in Earth's interior. Earth Planet. Sci. Lett. 298, 1-13.

Dasgupta, R., Hirschmann, M.M., Smith, N.D., 2007. Partial Melting Experiments of Peridotite + CO_2 at 3 GPa and Genesis of Alkalic Ocean Island Basalts. J. Petrol. 48, 2093-2124.

Dasgupta, R., Hirschmann, M.M., Stalker, K., 2006. Immiscible Transition from Carbonate-rich to Silicate-rich Melts in the 3 GPa Melting Interval of Eclogite + CO_2 and Genesis of Silica-undersaturated Ocean Island Lavas. J. Petrol. 47, 647-671.

Davis, F.A., Hirschmann, M.M., Humayun, M., 2011. The composition of the incipient partial melt of garnet peridotite at 3GPa and the origin of OIB. Earth Planet. Sci. Lett. 308, 380-390.

Davis, F.A., Humayun, M., Hirschmann, M.M., Cooper, R.S., 2013. Experimentally determined mineral/melt partitioning of first-row transition elements (FRTE) during partial melting of peridotite at 3GPa. Geochim. Cosmochim. Acta 104, 232-260.

Foley, S.F., Fischer, T.P., 2017. An essential role for continental rifts and lithosphere in the deep carbon cycle. Nat. Geosci. 10, 897-902.

Frezzotti, M.L., Peccerillo, A., Panza, G., 2009. Carbonate metasomatism and CO_2 lithosphere–asthenosphere degassing beneath the Western Mediterranean: An integrated model arising from petrological and geophysical data. Chem. Geol. 262, 108-120.

Frezzotti, M.L., Selverstone, J., Sharp, Z.D., Compagnoni, R., 2011. Carbonate dissolution during subduction revealed by diamond-bearing rocks from the Alps. Nat. Geosci. 4, 703-706.

Galy, A., Yoffe, O., Janney, P.E., Williams, R.W., Cloquet, C., Alard, O., Halicz, L., Wadhwa, M., Hutcheon, I.D., Ramon, E., Carignan, J., 2003. Magnesium isotope heterogeneity of the isotopic standard SRM980 and new reference materials for magnesium-isotope-ratio measurements. J. Anal. At. Spectrom. 18, 1352.

Gao, T., Ke, S., Li, R., Meng, X. N., He, Y., Liu, C., Wang, Y., Li, Z., Zhu, J.M., 2019. High-precision magnesium isotope analysis of geological and environmental reference materials by multiple-collector inductively coupled plasma mass spectrometry. Rapid Commun. Mass Spectrom. 33, 767-777.

Gerbode, C., Dasgupta, R., 2010. Carbonate-fluxed Melting of MORB-like Pyroxenite at 2.9 GPa and Genesis of HIMU Ocean Island Basalts. J. Petrol. 51, 2067-2088.

Hirose, K., 1997. Partial melt compositions of carbonated peridotite at 3 GPa and role of CO_2 in alkali-basalt magma generation. Geophys. Res. Lett. 24, 2837-2840.

Hirschmann, M.M., 2018. Comparative deep Earth volatile cycles: The case for C recycling from exosphere/mantle fractionation of major (H_2O, C, N) volatiles and from H_2O/Ce, CO_2/Ba, and CO_2/Nb exosphere ratios. Earth Planet. Sci. Lett. 502, 262-273.

Hoernle, K., Tilton, G., Le Bas, M.J., Duggen, S., Garbe-Schönberg, D., 2002. Geochemistry of oceanic carbonatites compared with continental carbonatites: mantle recycling of oceanic crustal carbonate. Contrib. Mineral. Petrol. 142, 520-542.

Hofmann, A.W., 2017. A store of subducted carbon beneath Eastern China. Natl. Sci. Rev. 4, 2-2.

Huang, J., Li, S.-G., Xiao, Y., Ke, S., Li, W.-Y., Tian, Y., 2015. Origin of low δ^{26}Mg Cenozoic basalts from South China Block and their geodynamic implications. Geochim. Cosmochim. Acta 164, 298-317.

Huang, J., Zhang, X.-C., Chen, S., Tang, L., Wörner, G., Yu, H., Huang, F., 2018. Zinc isotopic systematics of Kamchatka-Aleutian arc magmas controlled by mantle melting. Geochim. Cosmochim. Acta 238, 85-101.

Huang, J., Zhao, D., 2006. High-resolution mantle tomography of China and surrounding regions. J. Geophys. Res. Solid Earth 111.

Jin, Q.-Z., Huang, J., Liu, S.-C., Huang, F., 2020. Magnesium and zinc isotope evidence for recycled sediments and oceanic crust in the mantle sources of continental basalts from eastern China. Lithos 370-371.

Jovanović, M., Downes, H., Pecskay, Z., Cvetković, V., Prelević, D., Vaselli, O., 2001. Paleogene mafic alkaline volcanics rocks of East Serbia. Acta Vulcanol. 13, 159-173.

Karamata, S., Krstić, B., 1996. Terranes of Serbia and neighbouring areas. In: Knežević-Djordjević, V., Krstić, B. (Eds.), Terranes of Serbia. Faculty of Mining and Geology, Universityof Belgrade, Belgrade, pp. 25-40.

Kelemen, P.B., Manning, C.E., 2015. Reevaluating carbon fluxes in subduction zones, what goes down, mostly comes up. Proc Natl Acad Sci U S A 112, E3997-4006.

Kerrick, D.M., Connolly, J., 2001. Metamorphic devolatilization of subducted marine sediments and the transport of volatiles into the Earth's mantle. Nature 411, 293.

Kräutner, H.G., Krstić, B.P., 2003. Geological map of the Carpatho-Balkanides between Mehadia. Oravita, Niš and Sofia. Geoinstitut, Belgrade.

Kushiro, I., 1975. Carbonate-silicate reactions at high presures and possible presence of dolomite and magnesite in the upper mantle. Earth Planet. Sci. Lett. 28, 116-120.

Le Bas M. J., Le Maitre R. W., Streckeisen A., Zanettin B. A., 1986. Chemical classification of volcanic rocks based on the total alkali-silica diagram. J. Petrol. 27, 745-750.

Le Roux, V., Lee, C.T.A., Turner, S.J., 2010. Zn/Fe systematics in mafic and ultramafic systems: Implications for detecting major element heterogeneities in the Earth's mantle. Geochim. Cosmochim. Acta 74, 2779-2796.

Li, S.-G., Yang, W., Ke, S., Meng, X., Tian, H., Xu, L., He, Y., Huang, J., Wang, X.-C., Xia, Q., Sun, W., Yang, X., Ren, Z.-Y., Wei, H., Liu, Y., Meng, F., Yan, J., 2017. Deep carbon cycles constrained by a large-scale mantle Mg isotope anomaly in eastern China. Natl. Sci. Rev. 4, 111-120.

Liu, S.-A., Li, S.-G., 2019. Tracing the Deep Carbon Cycle Using Metal Stable Isotopes: Opportunities and Challenges. Engineering 5, 448-457.

Liu, S.-A., Qu, Y.-R., Wang, Z.-Z., Li, M.-L., Yang, C., Li, S.-G., 2022. The fate of subducting carbon tracked by Mg and Zn isotopes: A review and new perspectives. Earth-Sci. Rev. 228.

Liu, S.-A., Teng, F.-Z., Yang, W., Wu, F.-Y., 2011. High-temperature inter-mineral magnesium isotope fractionation in mantle xenoliths from the North China craton. Earth Planet. Sci. Lett. 308, 131-140.

Liu, S.-A., Wang, Z.-Z., Li, S.-G., Huang, J., Yang, W., 2016. Zinc isotope evidence for a large-scale carbonated mantle beneath eastern China. Earth Planet. Sci. Lett. 444, 169-178.

Liu, S.A., Wang, Z.Z., Yang, C., Li, S.G., Ke, S., 2020. Mg and Zn Isotope Evidence for Two Types of Mantle Metasomatism and Deep Recycling of Magnesium Carbonates. J. Geophys. Res. Solid Earth 125.

Lustrino, M., Wilson, M., 2007. The circum-Mediterranean anorogenic Cenozoic igneous province. Earth-Sci. Rev. 81, 1-65.

Lv, Y., Liu, S.-A., Teng, F.-Z., Wei, G.-J., Ma, J.-L., 2020. Contrasting zinc isotopic fractionation in two mafic-rock weathering profiles induced by adsorption onto Fe (hydr)oxides. Chem. Geol. 539.

Maréchal, C.N., Télouk, P., Albarède, F., 1999. Precise analysis of copper and zinc isotopic compositions by plasma-source mass spectrometry. Chem. Geol. 156, 251-273.

Marshall, H.G., Walker, J.C.G., Kuhn, W.R., 1988. Long-term climate change and the geochemical cycle of carbon. J. Geophys. Res. 93, 791-801.

Mazza, S.E., Gazel, E., Bizimis, M., Moucha, R., Beguelin, P., Johnson, E.A., McAleer, R.J., Sobolev, A.V., 2019. Sampling the volatile-rich transition zone beneath Bermuda. Nature 569, 398-403.

McCoy-West, A.J., Fitton, J.G., Pons, M.-L., Inglis, E.C., Williams, H.M., 2018. The Fe and Zn isotope composition of deep mantle source regions: Insights from Baffin Island picrites. Geochim. Cosmochim. Acta 238, 542-562.

McDonough, W., F., Sun, S.S., 1995. The composition of the Earth. Chem. Geol. 120, 223-253.

Moynier, F., Vance, D., Fujii, T., Savage, P., 2017. The Isotope Geochemistry of Zinc and Copper. Rev. Mineral. Geochem. 82, 543-600.

Nesbitt, H.W., Young, G.M., 1982. Early Proterozoic climates and plate motions inferred from major element chemistry of lutites. Nature 299, 715-717.

Pan, D., Spanu, L., Harrison, B., Sverjensky, D.A., Galli, G., 2013. Dielectric properties of water under extreme conditions and transport of carbonates in the deep Earth. Proc Natl Acad Sci U S A 110, 6646-6650.

Pertermann, M., Hirschmann, M.M., Hametner, K., Günther, D., Schmidt, M.W., 2004. Experimental determination of trace element partitioning between garnet and silica-rich liquid during anhydrous partial melting of MORB-like eclogite. Geochem. Geophys. Geosyst. 5.

Pilet, S., Baker, M.B., Stolper, E.M., 2008. Metasomatized lithosphere and the origin of alkaline lavas. Science 320, 916-919.

Piromallo, C., Morelli, A., 2003. Pwave tomography of the mantle under the Alpine-Mediterranean area. J. Geophys. Res. Solid Earth 108.

Plank, T., Manning, C.E., 2019. Subducting carbon. Nature 574, 343-352.

Qin, L., Humayun, M., 2008. The Fe/Mn ratio in MORB and OIB determined by ICP-MS. Geochim. Cosmochim. Acta 72, 1660-1677.

Richter, F.M., Watson, E.B., Mendybaev, R.A., Teng, F.-Z., Janney, P.E., 2008. Magnesium isotope fractionation in silicate melts by chemical and thermal diffusion. Geochim. Cosmochim. Acta 72, 206-220.

Robertson, A., Karamata, S., Šarić, K., 2009. Overview of ophiolites and related units in the Late Palaeozoic–Early Cenozoic magmatic and tectonic development of Tethys in the northern part of the Balkan region. Lithos 108, 1-36.

Simon, N.S.C., Irvine, G.J., Davies, G.R., Pearson, D.G., Carlson, R.W., 2003. The origin of garnet and clinopyroxene in "depleted" Kaapvaal peridotites. Lithos 71, 289-322.

Sossi, P.A., Nebel, O., O'Neill, H.S.C., Moynier, F., 2018. Zinc isotope composition of the Earth and its behaviour during planetary accretion. Chem. Geol. 477, 73-84.

Su, B.-X., Hu, Y., Teng, F.-Z., Qin, K.-Z., Bai, Y., Sakyi, P.A., Tang, D.-M., 2017. Chromite-induced magnesium isotope fractionation during mafic magma differentiation. Sci. Bull. 62, 1538-1546.

Sun, S.S., McDonough, W.F., 1989. Chemical and isotopic systematics of oceanic basalts: implications for mantle composition and processes. Geol. Soc. Spec. Publ. 42, 313-345.

Teng, F.-Z., 2017. Magnesium Isotope Geochemistry. Rev. Mineral. Geochem. 82, 219-287.

Teng, F.-Z., Li, W.-Y., Rudnick, R.L., Gardner, L.R., 2010. Contrasting lithium and magnesium isotope fractionation during continental weathering. Earth Planet. Sci. Lett. 300, 63-71.

Tommasini, S., Avanzinelli, R., Conticelli, S., 2011. The Th/La and Sm/La conundrum of the Tethyan realm lamproites. Earth Planet. Sci. Lett. 301, 469-478.

Walter, M.J., 1998. Melting of Garnet Peridotite and the Origin of Komatiite and Depleted Lithosphere. J. Petrol. 39, 29-60.

Wang, Z.-Z., Liu, S.-A., 2021. Evolution of Intraplate Alkaline to Tholeiitic Basalts via Interaction Between Carbonated Melt and Lithospheric Mantle. J. Petrol. 62, 1-25.

Wang, Z.-Z., Liu, S.-A., Chen, L.-H., Li, S.-G., Zeng, G., 2018. Compositional transition in natural alkaline lavas through silica-undersaturated melt–lithosphere interaction. Geology 46, 771-774.

Wang, Z.-Z., Liu, S.-A., Liu, J., Huang, J., Xiao, Y., Chu, Z.-Y., Zhao, X.-M., Tang, L., 2017. Zinc isotope fractionation during mantle melting and constraints on the Zn isotope composition of Earth's upper mantle. Geochim. Cosmochim. Acta 198, 151-167.

Xu, R., Liu, Y., Lambart, S., Hoernle, K., Zhu, Y., Zou, Z., Zhang, J., Wang, Z., Li, M., Moynier, F., Zong, K., Chen, H., Hu, Z., 2022. Decoupled Zn-Sr-Nd isotopic composition of continental intraplate basalts caused by two-stage melting process. Geochim. Cosmochim. Acta 326, 234-252.

Xu, R., Liu, Y., Wang, X.-C., Foley, S. F., Zhang, Y., Yuan, H., 2020. Generation of continental intraplate alkali basalts and implications for deep carbon cycle. Earth-Sci. Rev. 201.

Yang, C., Liu, S.-A., Zhang, L., Wang, Z.-Z., Liu, P.-P., Li, S.-G., 2021. Zinc isotope fractionation between Cr-spinel and olivine and its implications for chromite crystallization during magma differentiation. Geochim. Cosmochim. Acta 313, 277-294.

Yang, W., Teng, F.-Z., Zhang, H.-F., Li, S.-G., 2012. Magnesium isotopic systematics of continental basalts from the North China craton: Implications for tracing subducted carbonate in the mantle. Chem. Geol. 328, 185-194.

Yang, Z.-F., Li, J., Liang, W.-F., Luo, Z.-H., 2016. On the chemical markers of pyroxenite contributions in continental basalts in Eastern China: Implications for source lithology and the origin of basalts. Earth-Sci. Rev. 157, 18-31.

Yaxley, G. M., Green, D. H., 1994. Experimental demonstration of refractory carbonate-bearing eclogite and siliceous melt in the subduction regime. Earth Planet. Sci. Lett. 128, 313-325.

Zhong, Y., Chen, L.-H., Wang, X.-J., Zhang, G.-L., Xie, L.-W., Zeng, G., 2017. Magnesium isotopic variation of oceanic island basalts generated by partial melting and crustal recycling. Earth Planet. Sci. Lett. 463, 127-135.

Recycled carbonate-bearing silicate sediments in the sources of Circum-Mediterranean K-rich lavas: Evidence from Mg-Zn isotopic decoupling*

Zi-Tan Shu[1], Sheng-Ao Liu[1], Dejan Prelević[2, 3], Yu Wang[4], Stephen F. Foley[5], Vladica Cvetković[3] and Shuguang Li[1]

1. State Key Laboratory of Geological Processes and Mineral Resources, China University of Geosciences, Beijing 100083, China
2. Institute of Geological Sciences, University of Mainz, Becherweg 21, D-55099 Mainz, Germany
3. University of Belgrade, Faculty of Mining and Geology, Đušina 7, 11000 Belgrade, Serbia
4. State Key Laboratory of Isotope Geochemistry, Guangzhou Institute of Geochemistry, Chinese Academy of Sciences, Guangzhou 510640, China
5. School of Natural Sciences, Macquarie University, North Ryde, New South Wales, Australia

亮点介绍：本文对环地中海地区富钾火山岩进行了系统 Mg-Zn 同位素分析，结果显示钾质火山岩的 Zn 同位素组成与 MORB 类似，但其 Mg 同位素组成远远轻于 MORB。与钠质玄武岩轻 Mg 重 Zn 同位素组成的特征不同，我们将钾质岩的这种同位素特征定义为“Mg-Zn 同位素解耦”。通过对环地中海周边地区碳酸盐化硅质沉积物的分析发现，这种类型的沉积物也具有轻 Mg 同位素和与 MORB 类似的 Zn 同位素组成，结合 Sr-Nd-Pb 同位素和微量元素，我们认为钾质岩的地幔源区加入了碳酸盐化硅质沉积物。

Abstract The high flux of subaerial volcanic CO_2 emissions around the circum-Mediterranean region requires the involvement of an unusually carbon-rich reservoir, but the origin of which is still unclear. Here we aim to resolve this problem by analyzing Mg and Zn isotopes for the widely distributed mafic potassic to ultrapotassic lavas in this region. These K-rich lavas have lower δ^{26}Mg but similar δ^{66}Zn compared to mid-ocean ridge basalts (MORB). No known magmatic processes can explain the isotopic data, which must therefore be characteristics of the mantle sources. Recycled carbonate sediments are capable of explaining the low δ^{26}Mg, but they typically also have high δ^{66}Zn. Thus, the low δ^{26}Mg but unfractionated δ^{66}Zn of these K-rich lavas define “Mg-Zn isotopic decoupling” which has not yet been observed for other types of mantle-derived lavas. The carbonate-bearing silicate sediments analyzed here possess low δ^{26}Mg and MORB-like δ^{66}Zn, which can account for the Mg-Zn isotopic decoupling. Therefore, the nature of recycled materials (carbonates vs. carbonate-bearing silicate sediments) in the mantle can be distinguished by the coupling or decoupling of Mg and Zn isotopes of mantle-derived magmas. The input flux of carbon from the sediments to the lithospheric mantle is estimated to be ~8.1

* 本文发表在: Journal of Geophysical. Research-solid. Earth, 2023, 128

Mt/yr, and ~22.4 Mt/yr of CO_2 emissions are predicted, which fit well with the observed output flux of 20.1 ±13.4 Mt/yr. Our results demonstrate that recycled crustal carbon stored in the lithospheric mantle is an important source for the extensive subaerial volcanic CO_2 emissions in the circum-Mediterranean region.

Plain Language Summary Extensive volcanic CO_2 emissions around the circum-Mediterranean region constitute a significant proportion of the global total, but the source of this carbon-rich reservoir is unclear. We analyze Mg and Zn isotopes for the widespread, lithospheric mantle-derived K-rich lavas in this region and find that they have lower Mg but similar Zn isotopic data compared to mid-ocean ridge basalts. This Mg-Zn isotopic feature is defined as "Mg-Zn isotopic decoupling" and explained as resulting from recycled carbonate-bearing sediments in their mantle sources. Thus, the lithospheric mantle beneath the circum-Mediterranean region is carbon-rich, providing an important source for the extensive volcanic CO_2 emissions.

Keywords Mg-Zn isotopic decoupling, K-rich lavas, volcanic CO_2 emissions, lithospheric mantle, circum-Mediterranean region

1 Introduction

The solid Earth contains more carbon than the entire surface reservoir (ocean, atmosphere, and biosphere) in terms of mass: over 90% of carbon is located in the crust, mantle, and core (Dasgupta & Hirschmann, 2010; Wood, 1993). Carbon is released from the Earth's interior mainly in the form of carbon dioxide (CO_2) by degassing from volcanoes (up to 90%) and metamorphic reactions (≈10%; Mason et al., 2017). The flux of CO_2 released from oceanic volcanism (mid-ocean ridges and ocean islands) is ~29–154 Megatons per year (Mt/yr) (Kelemen & Manning, 2015), and the global flux of CO_2 emitted from other volcanic environment, including plume degassing and diffuse degassing, is up to ~108 Mt/yr (Werner et al., 2019), which has a significant time-integrated impact on the concentration of atmospheric CO_2 (161×10^3 Mt; Lee et al., 2019). The circum-Mediterranean region, which experienced a prolonged period of convergence between the African-Arabian and the Eurasian plates from the late Jurassic to Cenozoic (Prelević & Foley, 2007), contributes an anomalously large flux of subaerial volcanic CO_2 emissions (20.1 Mt/yr; Werner et al., 2019), accounting for about 20% of the global total. This enormous CO_2 output flux appears to require an extremely carbon-rich reservoir in the crust or mantle. It has been proposed that CO_2 emissions by arc volcanoes are largely sourced from subducting crustal carbon (>70%; Plank & Manning, 2019). However, there is no explanation so far for the high carbon flux in the circum-Mediterranean region.

Mantle-derived lavas are regarded as an ideal proxy to link volcanic CO_2 outgassing with deep carbon cycling (e.g., Avanzinelli et al., 2018; Bragagni et al., 2022). The circum-Mediterranean region is well-known for having produced widespread subduction-related potassic to ultrapotassic rocks (e.g., lamproites, leucitites, and kamafugites; e.g., Avanzinelli et al., 2009; Conticelli et al., 2002; Lustrino & Wilson, 2007; Prelević et al., 2008). Previous studies have attributed the pronounced enrichments in incompatible elements and characteristic Sr-Nd isotopic compositions (radiogenic $^{87}Sr/^{86}Sr$ but unradiogenic $^{143}Nd/^{144}Nd$) of the Mediterranean K-rich rocks to a strongly enriched lithospheric mantle source to which recycled crustal components have been added (Casalini et al., 2022; Lustrino et al., 2011; Prelević & Foley, 2007; Prelević et al., 2013; Y. Wang et al., 2021). However, whether or not the recycled crustal components are carbonate-rich is not well constrained, and the input flux of crustal carbon to the mantle sources is difficult to estimate using traditional geochemical proxies. Therefore, additional geochemical proxies are required to better constrain the source of carbon in the circum-Mediterranean lithospheric mantle and to provide a robust estimate for the input flux of carbon.

Magnesium and zinc isotopes are effective tools for tracing recycled carbonates in the mantle owing to the contrasting isotopic compositions of marine carbonates (δ^{26}Mg=−3.1±1.22‰; δ^{66}Zn=0.99±0.25‰; e.g., S.-A. Liu et al., 2017; Pichat et al., 2003; Sweere et al., 2018; Teng, 2017) and the mantle (δ^{26}Mg = −0.25 ± 0.04‰ and δ^{66}Zn=0.18±0.05‰; S.-A. Liu et al., 2019; McCoy-West et al., 2018; Sossi et al., 2018; Teng, 2017; Z.-Z. Wang et al., 2017). Importantly, neither Mg nor Zn isotopes fractionate significantly during prograde metamorphic reactions (Inglis et al., 2017; S.-J. Wang et al., 2014) or magmatic processes (see reviews by Moynier et al., 2017; Teng, 2017). Therefore, Mg and Zn isotopic systematics of mantle-derived lavas can be used to identify the presence of recycled carbonates in their mantle sources. In this paper, we report the first Mg and Zn isotopic data for Cenozoic K-rich lavas around the circum-Mediterranean region (Spain, Italy, Serbia, Macedonia, and Turkey) and combine these results with Sr-Nd-Pb isotopes to elucidate the origin of carbonate metasomatism in the lithospheric mantle. In addition, we use Monte Carlo simulations of melting and mixing to estimate the input flux of recycled carbon into the mantle and compare this with the observed volcanic CO_2 outputs in this region. Our results indicate that subducted crustal carbon stored in the lithospheric mantle significantly contributed to the very high levels of subaerial volcanic CO_2 outgassing in the circum-Mediterranean region.

2 Geological Setting and Sample Description

Mantle-derived K-rich lavas occur in all major volcanic provinces along the Alpine–Himalayan orogenic belt (AHOB), which was formed by closure of the Tethyan Ocean in the Mesozoic (Prelević et al., 2013). They are volumetrically minor, but genetically important magmatic rocks as their geochemistry can be used as magmatic proxies to trace many lithospheric processes during orogenesis. The magmatic activity coincides with the major tectonic changes within the region from Cretaceous to Pleistocene (Conticelli et al., 2009), indicating that this episodic activation represents an integral aspect of the broader tectonic development (Lustrino et al., 2011). The circum-Mediterranean region (Fig. 1) forms the western part of the AHOB and is associated with the pre-, syn- and post-collisional effects of anticlockwise convergence of the African-Arabian and Eurasian plates since the Permian–Mesozoic (Lustrino et al., 2011; Prelević & Foley, 2007). Seismic tomographic images reveal that distinct subducted slabs and slab fragments, which transferred crustal materials into the mantle, exist throughout the whole sub-Mediterranean region (Piromallo & Morelli, 2003). Subduction and subsequent post-collisional

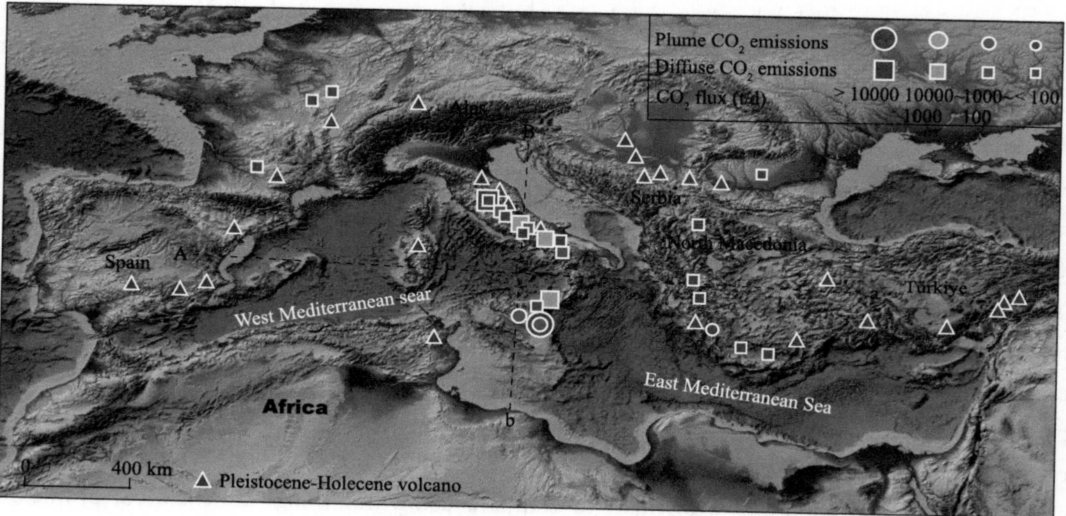

Fig. 1 Digital topographic map of the circum-Mediterranean region (http://www.ngdc.noaa.gov/mgg/global/global.html). Red numbers from 1 to 5 denote localities of K-rich rocks analyzed in this study (Spain, Italy, Serbia, Macedonia, and Turkey in turn). The volcano locations and measured fluxes of CO_2 emissions by plume and diffusing are from Werner et al. (2019) and references therein.

events resulted in the generation of widespread Cenozoic potassic-ultrapotassic lavas around the circum-Mediterranean region that are particularly suitable for investigation of metasomatic agents in the mantle (e.g., Ammannati et al., 2016; Prelević & Foley, 2007).

A total of 39 potassic-ultrapotassic volcanic rock samples from five areas (Fig. 1) were selected for this study: Spain (Murcia-Almeria), Italy (Tuscany), Serbia (NW Vardar zone), Macedonia (Southern Vardar zone) and Turkey (Western Anatolia). These rocks were mainly emplaced in post-collisional extensional tectonic settings from Eocene to Pleistocene (Fritschle et al., 2013). They are mainly characterized by porphyritic textures with olivine (up to 10 vol%), phlogopite (up to 15 vol%) and clinopyroxene phenocrysts and clinopyroxene, sanidine, leucite, phlogopite, apatite and oxides in the groundmass (Fritschle et al., 2013; Prelević et al., 2012). Geochemically, they are extremely enriched in K_2O ($K_2O/Na_2O = 0.1–10.4$; $K_2O+Na_2O = 4.4–13.2$ wt%; Fig. 2a

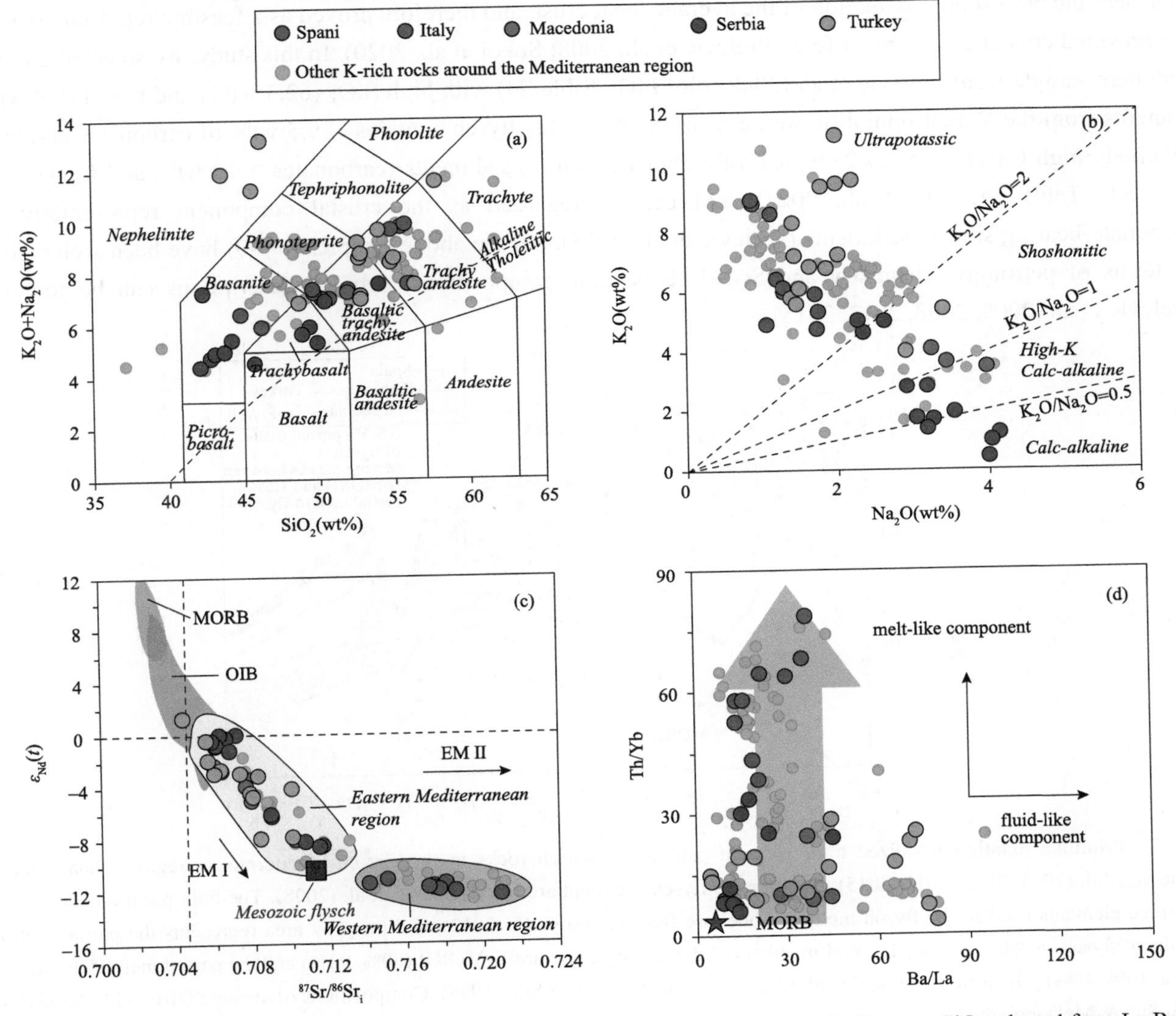

Fig. 2 Geochemical characteristics of the circum-Mediterranean K-rich rocks. (a) Total alkali versus SiO_2 adapted from Le Bas et al. (1986). (b) K_2O versus Na_2O adapted from Foley (1992). (c) $^{87}Sr/^{86}Sr_i$ against $\varepsilon_{Nd}(t)$. The data for MORB and OIB are taken from the GEOROC database (http://georoc.mpchmainz.gwdg.de/georoc/). (d) Th/Yb versus Ba/La, illustrating the control of subduction components on the incompatible element budget of K-rich lavas (Kirchenbaur & Münker, 2015). Red star representing MORB is from Sun & McDonough (1989). Light grey circles represent other K-rich rocks around the Mediterranean region reported in Prelević et al. (2013). The chemical data are collected from Prelević et al. (2005, 2008, 2012, 2015) and listed in Table S4 in Supporting Informations.

and 2b; Table S4) and have broad ranges of SiO_2 (42.1–57.8 wt%) and MgO contents (1.6–16.7 wt%). The Serbian samples have lower K_2O/Na_2O (blue circles in Fig. 2b), but this is due to post-magmatic analcimization of leucites which alters the original ultrapotassic compositions but does not affect most of the other trace elements or isotopic ratios in these samples (Prelević et al., 2004, 2005). The incompatible trace elements, especially the LILEs (e.g., Rb, K, Pb), are extremely enriched in these rocks in comparison with oceanic basalts (Fig.3). The Th peak (high Th/Yb) and Ba trough (low Ba/La) are two key chemical features of Mediterranean K-rich lavas (Fig. 2d), indicating the involvement of sediment-derived melts instead of fluids in their mantle sources (Kirchenbaur & Münker, 2015).

Cretaceous flysch sediments from the Balkans are predominantly derived from the continental margin and are components of the overstep sequence above the ophiolitic mélange. Their geochemistry has been proposed to represent the closest approximation of the average local crust, and therefore proved as a feasible representative for the recycled crustal end-member (e.g., Prelević et al., 2008; Sokol et al., 2020). In this study, we selected a flysch sediment sample from Prelević et al. (2008) (06FL03; Table S4) with high SiO_2 (62.3 wt%) and CaO (10.9 wt%) contents from the Vardar ophiolitic suture zone (Serbia). This flysch contains ~19.5 wt% of carbonates and has a relatively high Ca/Ti ratio (27.2) typical of carbonate-bearing sediments (carbonates > 10 wt% and Ca/Ti > 10; Fig. S1; Table S1; Qu et al., 2022), and can be regarded as the crustal component representative of carbonate-bearing silicate sediments (Prelević et al., 2005). All the above samples (n=40) have been well studied in terms of petrology, chemistry and Sr-Nd-Pb isotopic compositions; detailed descriptions can be found in Prelević et al. (2005, 2008, 2012, 2015).

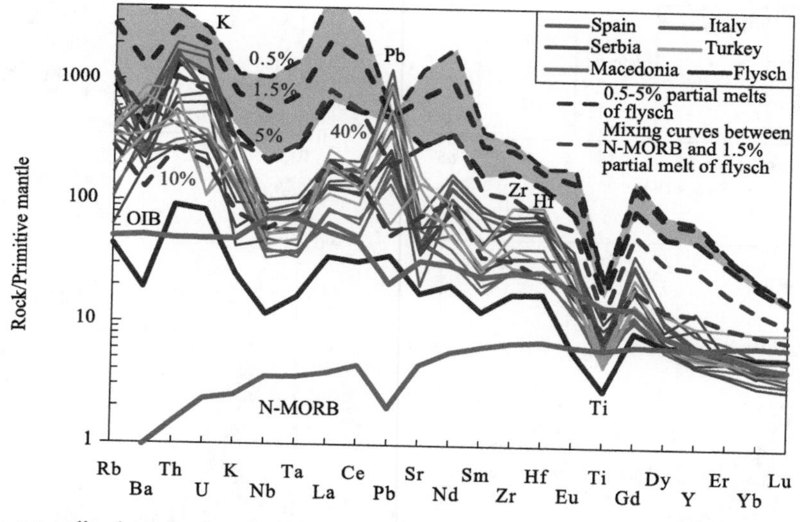

Fig. 3 Primitive mantle-normalized trace element patterns of K-rich rocks in the circum-Mediterranean region. Data sources: Prelević et al. (2005, 2008, 2012, 2015). Data for the flysch sediment are from Prelević et al. (2008). The bulk partition coefficients for trace elements used in the flysch melting model are from Grassi et al. (2012), and the grey area represents the partial melting results of flysch (F = 0.5%–5%). The end-members of the mixing curve are N-MORB (60%–90%) and the partial melt of flysch (F = 1.5%; 10%–40%). The primitive mantle values are from McDonough & Sun (1995). Compositions of average OIB and N-MORB are from Sun & McDonough (1989).

3 Analytical Methods

All chemical procedures for purification of Mg and Zn followed published methods (Gao et al., 2019; S.-A. Liu et al., 2016) and were carried out at the Isotope Geochemistry Laboratory of the China University of

Geosciences, Beijing (CUGB). For Mg isotope analysis, 2 mg whole-rock powder was weighed into 7 ml Teflon vials in a 1:3 (v/v) mixture of concentrated HF-HNO_3 and heated on a hot plate for >12 hours at 130–150°C. For Zn isotope analysis, 20 mg whole-rock powder was weighed and the sample digestion procedure was the same as that of Mg isotope analysis. After complete dissolution, 100 μl (containing ~20 μg Mg) or 1 ml (containing > 1 μg Zn) of each sample solution was loaded on 2 ml Bio-Rad cation exchange resin AG50W-X8 or 1 ml anion exchange resin AG1-X8 for Mg and Zn chromatographic purification, respectively. Magnesium was separated from matrix elements (K, Na, Ca, Fe, Al) by 1 N HNO_3. 8 N HCl, 0.4 N HCl and 0.5 N HNO_3 were used to achieve Zn purification. The same column procedures were repeated in order to assure the complete removal of matrix elements. After thorough drying, samples were dissolved in 1 ml 3% (*m*/*m*) HNO_3. The total recovery of Mg and Zn is more than 99%. Total procedural blank was 12 ng for Mg and 6 ng for Zn, which is very low in comparison with the amount of Mg (~20 μg) and Zn (>1 μg) in each loaded sample.

Analysis of Mg and Zn isotope ratios was carried out on a Thermo Scientific *Neptune plus* MC-ICP-MS. Purified samples dissolved in 3% (*m*/*m*) HNO_3 were introduced into the Ar plasma. Around 7.5 V signal of ^{24}Mg and 3.5 V signal of ^{64}Zn for a standard $10^{11}\Omega$ amplifier were obtained for a solution containing 400 ppb Mg and 200 ppb Zn, respectively. The sample-standard bracketing (SSB) method was used for Mg isotopes to correct instrumental mass bias, while a Cu-doping technique following Maréchal et al. (1999) was utilized for mass bias correction of Zn isotopes. The Mg and Zn isotope ratios are reported in standard δ-notation in per mil relative to DSM-3 (Galy et al., 2003; Equation 1) and JMC 3-0749L (Maréchal et al., 1999; Equation 2), respectively:

$$\delta^{26,25}Mg\ (‰) = [(^{26,25}Mg/^{24}Mg)_{sample}/(^{26,25}Mg/^{24}Mg)_{DSM\text{-}3} - 1] \times 1000 \quad (1)$$

$$\delta^{68,66}Zn\ (‰) = [(^{68,66}Zn/^{64}Zn)_{sample}/(^{68,66}Zn/^{64}Zn)_{JMC\ 3\text{-}0749L} - 1] \times 1000 \quad (2)$$

The long-term external reproducibility (2sd) is±0.06‰ for $\delta^{26}Mg$ (Gao et al., 2019) and±0.04‰ for $\delta^{66}Zn$ (S.-A. Liu et al., 2016; Z.-Z. Wang et al., 2017) based on repeated analyses of synthetic solutions (GSB-Mg, IRMM 3702) and international rock standards (BHVO-2, BCR-2, etc.). The USGS reference materials BCR-2 ($\delta^{26}Mg$=–0.23±0.06‰; $\delta^{66}Zn$=0.27±0.02‰) and BHVO-2 ($\delta^{26}Mg$=–0.28±0.04‰; $\delta^{66}Zn$=0.32±0.01‰) analyzed in this study are consistent with literature data (Moynier et al., 2017; Sossi et al., 2015; Teng, 2017; Z.-Z. Wang et al., 2017). The uncertainties for $\delta^{26}Mg$ and $\delta^{66}Zn$ are given as 2sd throughout the text.

4 Results

The Mg and Zn isotopic data of the circum-Mediterranean K-rich lavas are reported in Table 1. These rocks show a wide range of $\delta^{26}Mg$ values from –0.61‰ to –0.21‰ (*n*=39), which are similar to, or significantly lower than those of MORB and the mantle (–0.25±0.04‰; Teng, 2017) (Fig. 4). Zinc isotopic ratios display a narrow $\delta^{66}Zn$ range of 0.26‰ to 0.35‰ with an average of 0.31±0.04‰ (*n* = 39), which is consistent with that of MORB (0.27±0.06‰; J. Huang et al., 2018; Z.-Z. Wang et al., 2017; Fig. 4) within analytical uncertainty.

Table 1　Mg-Zn isotopic compositions (‰) and Sr-Nd-Pb isotopic data for the Cenozoic K-rich rocks in the circum-Mediterranean region

Sample no.	Rock type	δ^{25}Mg	2sd	δ^{26}Mg	2sd	δ^{66}Zn	2sd	δ^{68}Zn	2sd	^{87}Sr/^{86}Sr$_i$	$\varepsilon_{Nd}(t)$	^{206}Pb/^{204}Pb	^{207}Pb/^{204}Pb	^{208}Pb/^{204}Pb
Spain														
03V16	Lamproite	−0.17	0.03	−0.33	0.02	0.30	0.02	0.59	0.06	0.72092	−12.3	18.78	15.69	38.98
03J09	Lamproite	−0.17	0.03	−0.34	0.05	0.31	0.01	0.62	0.04	0.71401	−11.4	18.8	15.72	39.09
03J10	Lamproite	−0.19	0.01	−0.37	0.01	0.30	0.02	0.59	0.01	0.71497	−11.1	18.84	15.78	39.3
03CX02	Lamproite	−0.14	0.06	−0.27	0.05	0.30	0.03	0.59	0.03	0.71741	−11.6	18.81	15.75	39.18
02CX05	Lamproite	−0.19	0.01	−0.37	0.02	0.31	0.02	0.61	0.03	0.71744	−11.8	18.81	15.74	39.16
03FC01	Lamproite	−0.17	0.04	−0.32	0.03	0.30	0.05	0.60	0.10	0.71803	−11.4	18.79	15.73	39.19
03FC02	Lamproite	−0.18	0.03	−0.34	0.03	0.28	0.04	0.55	0.03	0.71854	−12	18.76	15.7	39.06
Italy														
05RR02	Lamproite	−0.17	0.03	−0.32	0.05	0.35	0.06	0.70	0.06	0.71769	−11.9	18.67	15.71	39.08
05VDA03	Lamproite	−0.20	0.03	−0.37	0.06	0.32	0.03	0.63	0.09	0.71714	−11.7	18.63	15.7	38.99
Serbia														
Bo2-1	Lamproite	−0.12	0.02	−0.24	0.05	0.30	0.05	0.60	0.08	0.70893	−6.22	–	–	–
Bo2-2	Lamproite	−0.20	0.02	−0.39	0.04	0.34	0.03	0.67	0.03	0.70890	−6.07	–	–	–
BK01/3-1	Lamproite	−0.20	0.02	−0.39	0.04	0.31	0.04	0.61	0.02	0.71162	−8.47	18.73	15.68	38.88
BK01/3C	Lamproite	−0.18	0.03	−0.34	0.04	0.29	0.02	0.57	0.06	0.71069	−8.17	18.72	15.67	38.84
BK01/3A-2	Lamproite	−0.16	0.02	−0.32	0.03	0.31	0.01	0.63	0.02	0.71144	−8.55	–	–	–
KR01/1	Ugandite	−0.22	0.02	−0.44	0.04	0.28	0.02	0.55	0.07	0.706	−0.80	18.82	15.66	38.96
NV01/7	Ugandite	−0.16	0.04	−0.31	0.06	0.31	0.03	0.62	0.09	0.70709	0.04	18.75	15.66	38.83
NV01/4	Ugandite	−0.17	0.02	−0.32	0.02	0.29	0.06	0.57	0.03	0.70655	−0.21	18.75	15.67	38.87
NV01/6C	Lct-basanite	−0.11	0.02	−0.22	0.04	0.30	0.05	0.60	0.11	0.70622	0.02	18.76	15.68	38.9
TR01/2	Lct-basanite	−0.21	0.03	−0.41	0.06	0.30	0.02	0.58	0.06	0.70602	−0.74	18.78	15.65	38.88
TR01/1	Lct-basanite	−0.19	0.01	−0.36	0.04	0.32	0.04	0.62	0.04	0.70582	−0.43	18.74	15.64	38.82
DZI/2	Ankaratrite	−0.11	0.03	−0.22	0.04	0.33	0.04	0.66	0.05	0.70674	−1.17	18.76	15.68	38.92
Macedonia														
04CR02	Lamproite	−0.16	0.02	−0.32	0.03	0.31	0.03	0.62	0.05	0.70761	−3.3	18.84	15.72	39
04CR01	Lamproite	−0.18	0.04	−0.35	0.04	0.30	0.02	0.60	0.04	0.70759	−3.9	18.82	15.7	38.92
04SL01	Lamproite	−0.24	0.01	−0.46	0.02	0.35	0.03	0.69	0.02	0.70788	−5	18.81	15.69	38.93
03KK01	Lamproite	−0.12	0.04	−0.24	0.07	0.29	0.05	0.57	0.06	0.71015	−8	18.71	15.71	39.1
Turkey														
05KD012	Lamproite	−0.10	0.01	−0.21	0.01	0.34	0.04	0.68	0.13	0.71001	−4.1	18.91	15.71	38.97
06KD21	Lamproite	−0.13	0.00	−0.24	0.01	0.32	0.06	0.65	0.06	0.70781	−4.4	18.92	15.71	38.98
05GU01	Lamproite	−0.26	0.02	−0.55	0.01	0.27	0.05	0.54	0.14	0.70834	−7.9	18.94	15.75	39.25
05GU02	Lamproite	−0.21	0.02	−0.42	0.03	0.31	0.02	0.61	0.04	0.71002	−7.8	18.98	15.75	39.3
06AF03	Lamproite	−0.26	0.01	−0.51	0.02	0.29	0.08	0.58	0.09	0.70593	−2.3	18.84	15.7	38.88
06KZ01	Lamproite	−0.26	0.02	−0.51	0.03	0.29	0.04	0.57	0.06	0.70824	−3.1	19.07	15.7	38.97
05IL02	Lamproite	−0.12	0.01	−0.23	0.02	0.31	0.06	0.60	0.07	0.70729	−2.9	19.09	15.69	38.98
05BH05	Lamproite	−0.11	0.02	−0.21	0.02	0.30	0.05	0.60	0.09	0.70796	−4.7	18.98	15.7	38.97
31	Shoshonite	−0.12	0.01	−0.28	0.02	0.34	0.03	0.66	0.05	0.70553	−0.4	19.03	15.72	39.16
42	Tephriphonolite	−0.16	0.04	−0.35	0.04	0.31	0.04	0.62	0.00	0.70631	−2.7	18.68	15.7	38.71

Continued

Sample no.	Rock type	δ^{25}Mg	2sd	δ^{26}Mg	2sd	δ^{66}Zn	2sd	δ^{68}Zn	2sd	^{87}Sr/^{86}Sr$_i$	$\varepsilon_{Nd}(t)$	^{206}Pb/^{204}Pb	^{207}Pb/^{204}Pb	^{208}Pb/^{204}Pb
29	Phonotephrite	−0.16	0.01	−0.32	0.02	0.34	0.03	0.66	0.04	0.70432	1.3	–	–	–
23/A	Tephriphonolite	−0.17	0.01	−0.33	0.02	0.32	0.04	0.62	0.10	0.70627	−2.6	18.69	15.7	38.72
19/A	Meliliteleucitite	−0.31	0.01	−0.61	0.01	0.31	0.01	0.61	0.03	0.70594	−2.9	18.85	15.72	38.92
1/A	Meliliteleucitite	−0.24	0.01	−0.46	0.02	0.26	0.04	0.51	0.03	0.70562	−1.9	19.1	15.74	39.24
Flysch														
06FL03		−0.39	0.03	−0.77	0.06	0.27	0.05	0.53	0.06	0.7112	−8.4	18.68	15.67	38.77
Rock standards														
BHVO-2		−0.14	0.02	−0.28	0.04	0.32	0.01	0.64	0.05					
BCR-2		−0.11	0.03	−0.23	0.06	0.27	0.02	0.53	0.03					

Note: Sr-Nd-Pb isotopic data are from Prelević et al. (2005, 2008, 2012, 2015). Sample localities, major and trace element data can be found in Table S4 in Supporting Information S1.

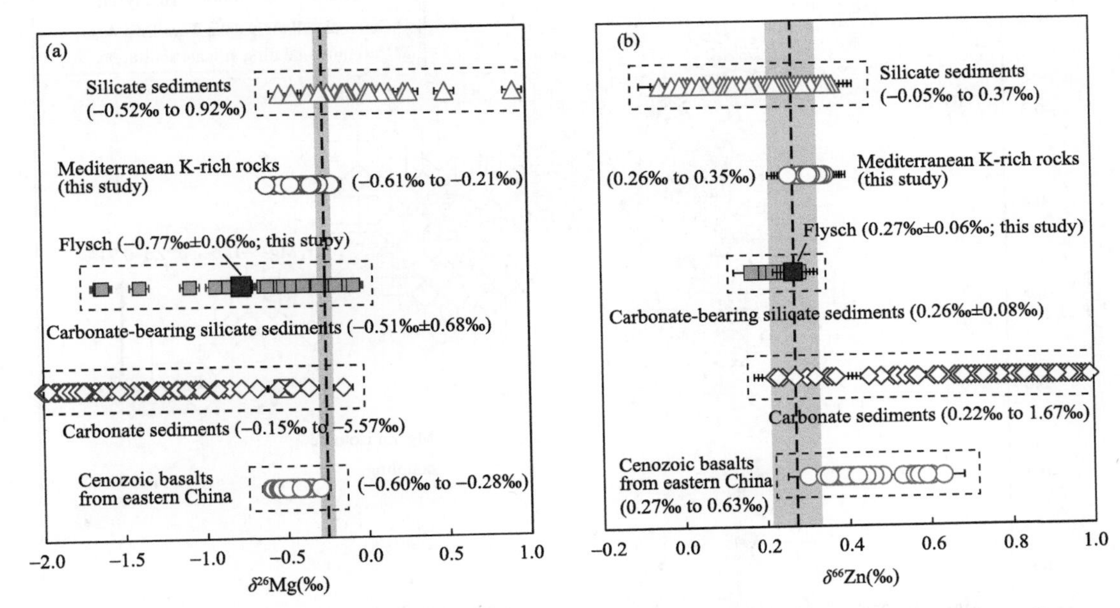

Fig. 4 (a) δ^{26}Mg and (b) δ^{66}Zn of the circum-Mediterranean K-rich rocks, compared to those of silicate sediments, carbonate sediments, carbonate-bearing silicate sediments and <110 Ma intraplate alkali basalts in East Asia. The light blue bands represent the Mg-Zn isotopic data for MORB (δ^{26}Mg =−0.25±0.06‰; Teng, 2017; δ^{66}Zn=0.27±0.06‰; J. Huang et al., 2018; Z.-Z. Wang et al., 2017). δ^{26}Mg data sources: silicate sediments (W.-Y. Li et al., 2010; S.-J. Wang et al., 2015), carbonate sediments (Teng, 2017 and references therein), carbonate-bearing silicate sediments (K.-J. Huang et al., 2013; Qu et al., 2022; S.-J. Wang et al., 2015; Wimpenny et al., 2014) and <110 Ma intraplate alkali basalts in East Asia (compiled in Table S5). δ^{66}Zn data sources: silicate sediments (Little et al., 2016; Maréchal et al., 2000; Vance et al., 2016), carbonate sediments (S.-A. Liu et al., 2017; Pichat et al., 2003; Sweere et al., 2018), carbonate-bearing silicate sediments (Qu et al., 2022) and <110 Ma intraplate alkali basalts in East Asia (Table S5 in Supporting Information S1).

The carbonate-bearing Mesozoic flysch sediment (06FL03) has a δ^{26}Mg value of −0.77±0.06‰, distinctly lower than silicate sediments with δ^{26}Mg ranging from −0.52‰ to 0.92‰ (W.-Y. Li et al., 2010; S.-J. Wang et al., 2015), but falls within the range of carbonate-bearing silicate sediments reported in the literature (−0.51±0.68‰; K.-J. Huang et al., 2013; Qu et al., 2022; S.-J. Wang et al., 2015; Wimpenny et al., 2014; Fig. 4a). The δ^{66}Zn value of this sample (0.27±0.05‰) is much lower than that of marine carbonates (0.99±0.25‰; S.-A. Liu et al., 2017; Pichat et al., 2003; Sweere et al., 2018), but indistinguishable from that of MORB (0.27±0.06‰; J. Huang et al., 2018; Z.-Z. Wang et al., 2017) and carbonate-bearing silicate sediments (0.26 ± 0.08‰; Qu et al., 2022; Fig. 4b).

The most striking Mg and Zn isotopic characteristic of the circum-Mediterranean K-rich lavas is that most of them have δ^{26}Mg lower than mantle values but all samples have δ^{66}Zn values identical to those of MORB (Fig. 4; Table 1). To our knowledge, such a signature has not yet been observed for other types of mantle-derived lavas. For instance, global oceanic island basalts (OIB) have mantle-like δ^{26}Mg (Teng, 2017) and similar or slightly higher δ^{66}Zn than those of MORB (Beunon et al., 2020). Some intraplate alkali basalts (e.g., Cenozoic basalts in East Asia) have lower δ^{26}Mg and higher δ^{66}Zn than these of the mantle (Fig. 5; Table S5), which are labelled as "Mg-Zn isotopic coupling" and have been attributed to the recycling of sedimentary carbonates into the mantle sources (e.g., S.-G. Li et al., 2017; S.-A. Liu et al., 2016, 2022). In contrast to "Mg-Zn isotopic coupling", we consider this specific signature (low δ^{26}Mg but MORB-like δ^{66}Zn) of the circum-Mediterranean K-rich lavas as "Mg-Zn isotopic decoupling" (Fig. 5).

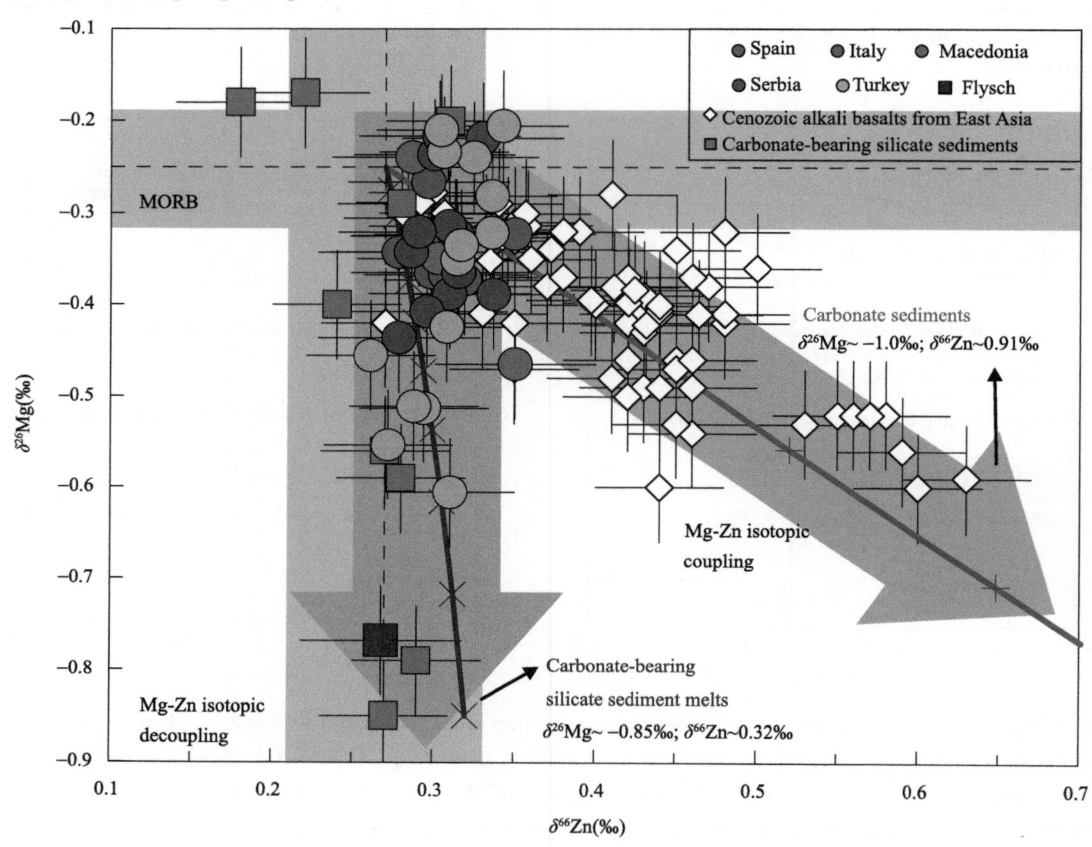

Fig. 5　Plot of δ^{26}Mg versus δ^{66}Zn for the circum-Mediterranean K-rich rocks. The purple curve denotes binary mixing between MORB and carbonate-bearing silicate sediment-derived melts. The light purple squares are Mg-Zn isotopic data for carbonate-bearing silicate sediments in literature (Qu et al., 2022; Table S1). The δ^{66}Zn of carbonate-bearing silicate sediment melts is assumed to be 0.32‰. The orange curve denotes mixing between MORB and carbonate sediments (magnesite). The step sizes are 10%. The grey bands represent the Mg-Zn isotopic data for MORB (δ^{26}Mg=−0.25±0.06‰; Teng, 2017; δ^{66}Zn=0.27±0.06‰; J. Huang et al., 2018; Z.-Z. Wang et al., 2017). The Mg-Zn isotopic data for magnesite and <110 Ma East Asia intraplate alkali basalts are from J. Huang et al. (2015), Jin et al. (2020), S.-G. Li et al. (2017), S.-A. Liu et al. (2016), Z.-Z. Wang et al. (2018), Z.-Z. Wang & S.-A. Liu (2021) and W. Yang et al. (2012) (compiled in Table S5 in Supporting Information S1).

5 Discussion

In this part, we first discuss the origin of Mg-Zn isotopic decoupling in the circum-Mediterranean K-rich lavas, and then estimate the input flux of crustal carbon recycled into the circum-Mediterranean lithospheric mantle. Finally, we simulate the possible CO_2 output flux in this region and compare it with the observed subaerial volcanic CO_2 output flux.

5.1 The origin of Mg-Zn isotopic decoupling in circum-Mediterranean K-rich lavas

5.1.1 Effects of low-temperature and magmatic processes

Before linking the observed Mg and Zn isotopic signatures of the circum-Mediterranean K-rich lavas to mantle source heterogeneity, the potential influence of post-eruption low-temperature alteration, crustal contamination, magmatic differentiation and kinetic isotope fractionation must be evaluated. Although some samples have relatively high loss on ignition (LOI) of 2.1–10.9 wt% (Table S4), their δ^{26}Mg and δ^{66}Zn values display no correlation with LOI (R^2=0.00, 0.08), indicating limited influence of post-magmatic alteration. The incompatible element (e.g., Sr, Nd, and Pb) concentrations of the K-rich rocks are about two times higher than those of continental crust (Prelević et al., 2008), so even considerable crustal contamination cannot explain their high incompatible element concentrations, which instead reflects mantle sources metasomatized by melts of subducted sediments. Moreover, in many cases, the high-MgO clinopyroxene phenocrysts in the K-rich rocks typically have ^{87}Sr/^{86}Sr ratios identical to their host rocks, implying isotopic equilibration between clinopyroxenes and K-rich lavas, which further excludes significant assimilation of crustal materials after clinopyroxene crystallization (Prelević et al., 2013). Fractional crystallization of olivine and/or Fe-Ti-Cr oxides would drive the residual melts to be enriched in ^{66}Zn with increasing SiO_2 because of the low δ^{66}Zn of olivine, while accumulation of olivine would deflect melts towards lower δ^{66}Zn with decreasing SiO_2 (H. Chen et al., 2013; Z.-Z. Wang et al., 2017; C. Yang et al., 2021). Fractional crystallization and accumulation of olivine or pyroxene do not fractionate δ^{26}Mg of melts significantly (Teng et al., 2010). The lack of correlations between δ^{26}Mg or δ^{66}Zn and SiO_2, MgO and Cr contents (R^2=0.01–0.05; Fig. 6a–d) excludes any significant isotopic effects induced by magmatic differentiation. During the mantle partial melting, there is negligible Mg isotope fractionation (Teng, 2017) and only slight Zn isotope fractionation ($\delta^{66}Zn_{melt\text{-}mantle} \leqslant 0.08‰$; Sossi et al., 2018), suggesting that different degrees of partial melting is not a plausible cause for the decoupled Mg-Zn isotopic compositions of these K-rich rocks.

When basaltic melts ascend through the lithospheric mantle, potential temperature and chemical gradients exist between the melts and surrounding mantle peridotite, possibly inducing kinetic isotope fractionation. During thermal diffusion, lighter isotopes tend to concentrate in hot places (Richter et al., 2008) and thus in the ascending K-rich melts. This would result in lower values of both δ^{26}Mg and δ^{66}Zn in melts, which is inconsistent with the low δ^{26}Mg but unfractionated δ^{66}Zn of the K-rich lavas (Fig. 5). During chemical diffusion, lighter isotopes diffuse faster than heavier ones along the chemical gradient (Richter et al., 2008). The K-rich melts have lower MgO contents (1.6–16.6 wt%) and higher Zn concentrations (~85.3 μg/g) than those of the ambient mantle peridotites (MgO ~38 wt% and Zn ~55 μg/g; Le Roux et al., 2010; McDonough & Sun, 1995). If chemical diffusion occurred, δ^{26}Mg and δ^{66}Zn of the melts would become appreciably lower and higher relative to those of MORB, respectively, which is inconsistent with the MORB-like δ^{66}Zn of the K-rich lavas (Fig. 5). Hence, kinetic fractionation during thermal or chemical diffusion is unlikely to lead to the decoupled Mg-Zn isotopic compositions of these K-rich lavas.

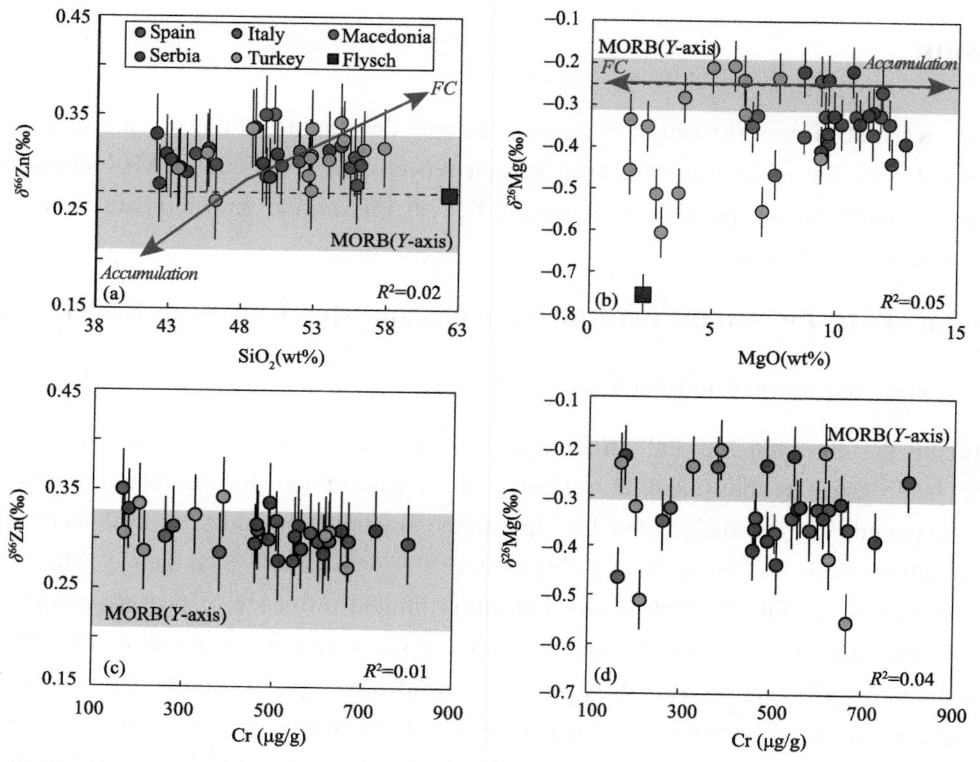

Fig. 6　Plots of δ^{66}Zn versus SiO_2 (a) and δ^{26}Mg versus MgO (b). Grey arrows denote the magma evolution trend of the Kilauea Iki lavas, Hawaii (H. Chen et al., 2013; Teng et al., 2007). The light blue areas represent the Mg-Zn isotopic data for MORB (δ^{26}Mg= −0.25±0.06‰; δ^{66}Zn=0.27±0.06‰; J. Huang et al., 2018; Teng, 2017; Z.-Z. Wang et al., 2017). Samples in this study do not fall on the Kilauea Iki trend, indicating negligible effect of fractional crystallization (FC) and accumulation. (c–d) Plots of δ^{26}Mg and δ^{66}Zn against Cr contents. The lack of correlations indicates that olivine and/or Fe-Ti-Cr oxides crystallization have negligible influence on Mg and Zn isotopic compositions of the circum-Mediterranean K-rich rocks. The chemical data are from Prelević et al. (2005, 2008, 2012, 2015).

5.1.2　The key role of recycled carbonate-bearing silicate sediments in mantle sources

The Mediterranean K-rich rocks are characterized by strong enrichment in incompatible trace elements (especially LILEs; Fig. 3), and high canonical trace element ratios like Pb/Ce (0.03–0.78), Cs/Rb (0.02–0.35) and Th/La (0.14–1.45) (Prelević et al., 2008; Tommasini et al., 2011; Y. Wang et al., 2021). They have highly radiogenic Sr and ^{207}Pb/^{204}Pb and unradiogenic Nd isotopic compositions compared to MORB and OIB (^{87}Sr/^{86}Sr = 0.7043–0.7209, ε_{Nd} = −12.3–1.3, and ^{207}Pb/^{204}Pb = 15.64–15.78; Fig. 7; Table 1). These features require one or more crustal components to have been recycled into their mantle sources (e.g., Conticelli & Peccerillo, 1992; Prelević et al., 2005, 2008). Here, we use Mg and Zn isotopes to decipher the type of recycled crustal components. Marine carbonate sediments typically have lighter Mg and heavier Zn isotopic compositions than those of the mantle (Fig. 4). Thus, recycling of carbonates into the mantle via subduction will drive Mg and Zn isotopic compositions of the hybridized mantle towards lower and higher values, respectively, as widely observed in Cenozoic intraplate alkali basalts from East Asia (Fig. 5; e.g., S.-G. Li et al., 2017; S.-A. Liu et al., 2016). The Mediterranean K-rich rocks have considerably lower δ^{26}Mg (−0.61‰ to −0.21‰) compared to MORB (−0.25± 0.04‰; Teng, 2017), implying that the metasomatic agent most likely contains recycled carbonates given that marine carbonate is the only known crustal reservoir with extremely low δ^{26}Mg. However, their MORB-like δ^{66}Zn (0.26‰ to 0.35‰) are inconsistent with the dominant role of recycled carbonates in the magma sources. The Mg-Zn isotopic decoupling, therefore, suggests a type of carbonate-related metasomatic agent that is distinct from marine carbonate sediments.

The low-δ^{26}Mg but MORB-like δ^{66}Zn signature of the Mediterranean K-rich rocks is in accordance with that of flysch sediment analyzed in this study (δ^{26}Mg=−0.77±0.06‰, δ^{66}Zn=0.27±0.05‰; Fig. 4), as well as carbonate-bearing silicate sediments from East Asia (δ^{26}Mg=−0.85– −0.17‰, δ^{66}Zn = 0.18–0.31‰) reported by Qu et al. (2022). To better understand the range and cause of Mg and Zn isotopic systematics of marine sediments that have the potential to be subducted, we compiled the available Mg-Zn isotopic data for carbonate sediments and silicate sediments (Fig. 4). A large δ^{26}Mg difference of up to 6.49‰ exists between carbonate and silicate sediments (Fig. 4a), indicating that carbonate components have a large impact on the δ^{26}Mg values of bulk sediments since they are the major carrier of light Mg isotopes in sediments. However, the δ^{66}Zn difference between carbonate and silicate sediments (−0.15‰ to 1.72‰; Fig. 4b) is much smaller than that of the δ^{26}Mg difference. Carbonate sediments also have lower Zn concentrations (44±37 μg/g; X. M. Liu et al., 2016 and references therein) than silicate sediments (88±33 μg/g; Plank, 2014 and references therein), although the former are generally similar to the Zn concentration of the mantle (~55 μg/g; Le Roux et al., 2010). Consequently, carbonate components have only a limited impact on the δ^{66}Zn values of bulk sediments, which leads to almost identical Zn isotopic compositions for carbonate-bearing and carbonate-free silicate sediments (Fig. 4b).

To further quantify the distinctive Mg-Zn isotopic compositions of the carbonate-bearing silicate sediments, we applied Monte Carlo models (see details in Table S2 and Data Set S1) in which carbonate components (e.g., $CaCO_3$) account for 5% to 50% (CaO=2.8–28 wt%) of bulk sediments. The results simulate the low δ^{26}Mg (−1.03±1.05‰; n=3000), MORB-like δ^{66}Zn (0.28±0.13‰), limited ^{206}Pb/^{204}Pb (18.74±0.57) and high ^{207}Pb/^{204}Pb (15.69±0.15) characteristics of carbonate-bearing silicate sediments successfully (Fig. 7), implying that Mg-Zn isotopic decoupling with δ^{26}Mg lower than MORB but MORB-like δ^{66}Zn is a common feature of carbonate-bearing silicate sediments. Combined with the trace element patterns of the flysch sediments, namely positive anomalies of Rb, K and Pb and negative anomalies of Nb, Ta and Ti (Fig. 3), we suggest that recycled carbonate-bearing silicate sediments are the most appropriate metasomatic agents operating in the source of the Mediterranean K-rich rocks. In addition, K-rich rocks from the western Mediterranean regions (Spain and Italy) possess significantly higher ^{87}Sr/^{86}Sr (Figs. 7a and 7b) than the eastern ones (Serbia, Macedonia, and Turkey), suggesting that older upper continental crustal materials (UCC; e.g., Jurassic Tethyan sediments from Zermatt-Saas ophiolite with ^{87}Sr/^{86}Sr up to 0.727; Prelević et al., 2008) may have been involved in the western mantle sources.

5.2 Widespread carbonated lithospheric mantle beneath the circum-Mediterranean region

In addition to samples in this study, other K-rich rocks from the central Mediterranean region (kamafugites and plagio-leucitites) have also been demonstrated to have recycled carbonate-rich sediments in their mantle sources, which is clearly constrained by Sr and Nd isotopes and U/Th disequilibria (strong ^{238}U excesses; e.g., Avanzinelli et al., 2007; Conticelli et al., 2002; Conticelli & Peccerillo, 1992). Extremely low Ni (~1000 ppm) and high Ca (up to 4000 ppm) concentrations characterize highly forsteritic olivine phenocrysts of silica-undersaturated leucite-bearing ultrapotassic rocks in Italy, indicating strong mantle metasomatism by melts of carbonate-rich pelites that reacted with orthopyroxene to produce an olivine-rich mantle source (Ammannati et al., 2016; Conticelli et al., 2015, 2013). Among the K-rich rocks analyzed in this study, leucite-free samples (lamproite) also display Mg-Zn isotopic decoupling (δ^{26}Mg=−0.55– −0.21‰, δ^{66}Zn=0.27–0.35‰; n=27; Table 1), which is indistinguishable from that of leucite-bearing rocks (ugandite, leucite-basanite, ankaratrite and melilite-leucitite; δ^{26}Mg=−0.61– −0.22‰, δ^{66}Zn=0.26–0.33‰; n = 9; Table 1). Low δ^{26}Mg values have also been discovered in lamproites from the Leucite Hills (δ^{26}Mg=−0.43– −0.37‰; Sun et al., 2021) and high-MgO lamproites from Eastern China (δ^{26}Mg=−0.59– −0.35‰; Z. X. Wang et al., 2022). Therefore, we suggest that the mantle sources of these circum-Mediterranean K-rich rocks are widely metasomatized by carbonate-bearing silicate sediments.

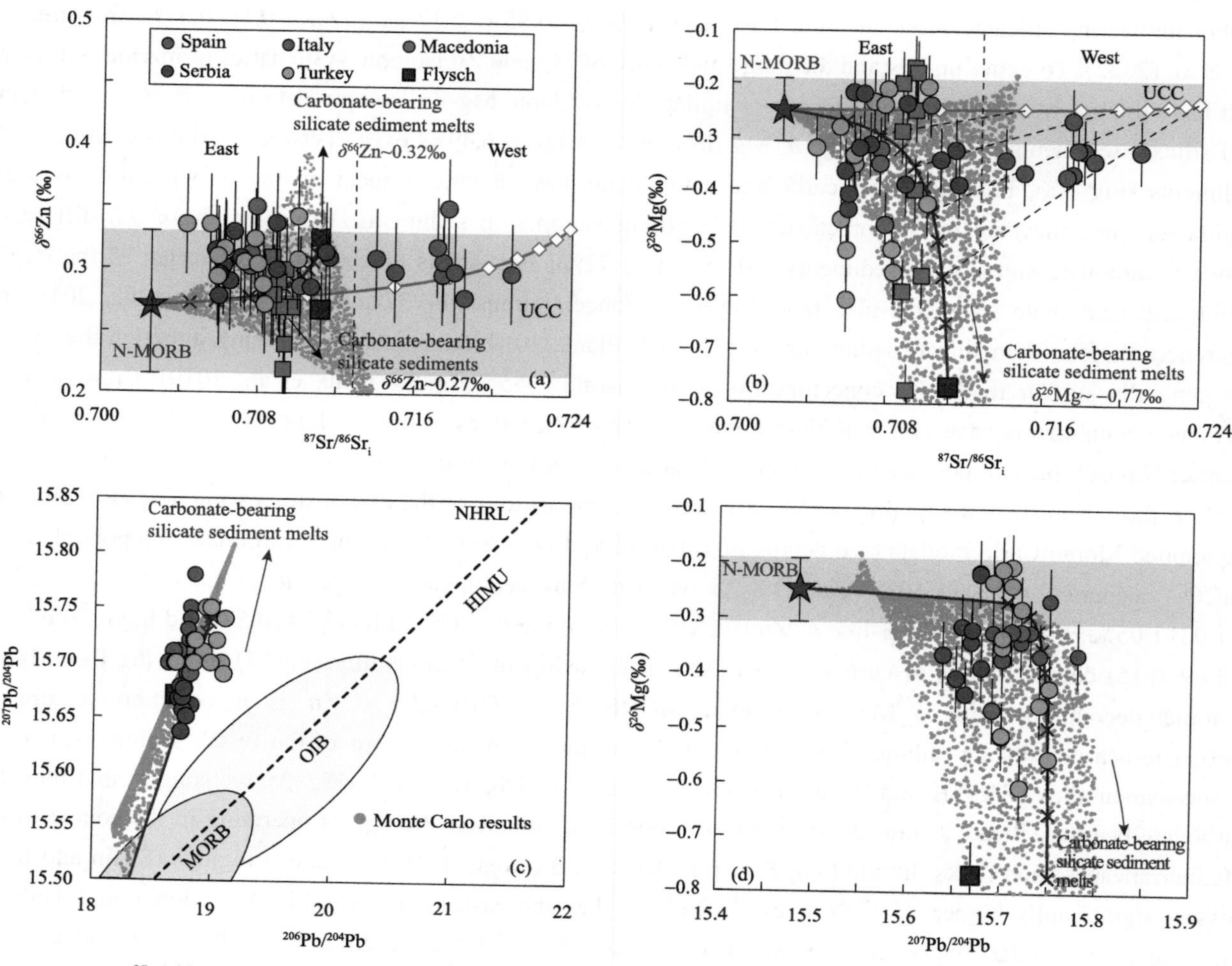

Fig. 7　Plots of ^{87}Sr/^{86}Sr$_i$ against δ^{66}Zn (a) and δ^{26}Mg (b). Plots of ^{207}Pb/^{204}Pb against ^{206}Pb/^{204}Pb (c) and δ^{26}Mg (d). The solid purple curve denotes mixing between N-MORB and carbonate-bearing silicate sediments (flysch). Another crustal component, upper continental crust (UCC), explains the extremely high ^{87}Sr/^{86}Sr of the western Mediterranean K-rich rocks. The grey dots represent the results of the Monte Carlo models, simulating the mixing between N-MORB and melts of carbonate-bearing silicate sediments. The light purple squares represent carbonate-bearing silicate sediment data in literature (Qu et al., 2022; Table S1). The light blue areas denote Mg and Zn isotopic compositions of MORB (δ^{26}Mg=−0.25±0.06‰, Teng, 2017; δ^{66}Zn=0.27±0.06‰; J. Huang et al., 2018; Z.-Z. Wang et al., 2017). The Pb isotopic data for OIB and MORB are taken from the GEOROC database (http://georoc.mpch-mainz.gwdg.de/georoc). NHRL is the Northern Hemisphere Reference Line (Hart, 1984). East refers to Serbia, Macedonia and Turkey; West to Spain and Italy. Details of the end-members, data sources and Monte Carlo models are listed in Table S2 and Data Set S1.

It has previously been suggested that the generation of strongly potassic, mantle-derived melts requires the presence of phlogopite in their mantle sources (Foley, 1992). Recent studies investigating the reaction between continental sediments and depleted peridotite and a mixture of glimmerite and harzburgite demonstrate that potassic to ultrapotassic melts can be formed at 1–3 GPa and 1000–1300 °C, corresponding to shallow lithospheric mantle thermobaric conditions (Förster et al., 2017; Y. Wang et al., 2017). Moreover, high-Mg$^{\#}$ ([atomic Mg/(Mg+Fe)]; up to 0.95) olivine phenocrysts with high-Cr$^{\#}$ spinel inclusions ([atomic Cr/(Cr+Al)]; around 0.95) in Mediterranean K-rich rocks suggest an ultra-depleted lithospheric mantle source, most likely derived from an island-arc oceanic lithosphere accreted during Alpine collisional processes (Prelević & Foley, 2007). Thus, the source region of the circum-Mediterranean K-rich rocks is probably located at lithospheric mantle depths at which carbonate-bearing silicate sediments carried by subducted slabs may be recycled, probably as imbricated blocks (Y. Wang et al., 2021), where they contribute to magma sources. These K-rich rocks plot along the strong curvature of the hyperbolic arrays (Sr-Nd-Pb-Mg isotopes; Figs. 2c and 7b–d), indicating mixing of melts of contrasting compositions in the mantle instead of mixing of source rocks. The eastern Mediterranean samples resulted from binary mixing between carbonate-bearing silicate sediment and MORB-like melts that were most likely derived from the depleted lithospheric mantle upon heating or decompression, while three end-member mixing including melts of carbonate-bearing silicate sediment, the depleted lithospheric mantle and UCC generated the western Mediterranean samples (Fig. 7). Förster et al. (2019) carried out an experiment investigating the role of sediment and peridotite-derived melts in the formation of potassic magmatism. The experimental results show that the partial melts of sediments and dunite form a reaction zone in which the infiltrating melts display major and trace element compositions similar to those of the circum-Mediterranean K-rich lavas (Prelević et al., 2008), implying that mixing between mantle-derived melts (e.g., MORB-like melts) and melts of sediments can generate K-rich lavas. Geophysical tomographic data further support this interpretation: a low S-wave velocity (V_S) layer at a depth of approximately 60–130 km is present beneath the western Mediterranean (Panza, Peccerillo, et al., 2007; Panza, Raykova, et al., 2007) and is attributed to carbon-rich melts originating from melting of continent-derived sediments and/or the subducted African continental crust (Frezzotti et al., 2009). Similar tomographic images are also found for the eastern Mediterranean domain where a high V_S layer indicating a subducting slab is located at 100–200 km and a low V_S layer exists at 50–100 km above the slab (Elgabry et al., 2013).

Geodynamically, the circum-Mediterranean region has experienced long-lasting convergence resulting from subduction of the major Tethyan oceanic slabs. The metasomatic agents (carbonate-bearing silicate sediments) responsible for these K-rich lavas are derived from the subducting slab, and are situated within the lithospheric mantle typically forming veins and dykes dominantly composed of hydrous minerals (e.g., Conceição & Green, 2004; Conticelli & Peccerillo, 1992). Moreover, the closure of small oceanic basins may be facilitated by intra-oceanic subduction which was probably common during Mesozoic Mediterranean Tethyan geodynamics (Robertson, 2002). Supra-subduction-related volcanism would leave behind a strongly depleted mantle area which should be more buoyant relative to surrounding asthenosphere, and the depleted oceanic lithospheric mantle was accreted to the continent during Alpine collisional processes (Prelević & Foley, 2007). The mixing between low-degree melts of these metasomatic veins and melts derived from the depleted lithospheric mantle results in the most extreme ultrapotassic lavas like lamproites and kamafugites, whereas with larger extents of melting, melt-mantle interaction and assimilation of the material from the ambient depleted mantle will result in more voluminous, but less alkaline magmatism (shoshonitic to calc-alkaline series), which is widespread in several major volcanic provinces in the circum-Mediterranean region such as Italy, Spain, Turkey (e.g., Conticelli et al. 2009; Prelević et al., 2008, 2012; Tappe et al., 2006). Therefore, the wide distribution of K-rich lavas (Figure 1) may be an indication of the large-scale presence of carbonated lithospheric mantle beneath the circum-Mediterranean region.

5.3 Estimation of carbon input flux into the lithospheric mantle

To quantify the amount of carbonate-bearing silicate sediment in the sources of the circum-Mediterranean K-rich lavas, we develop a binary mixing simulation between primitive mantle-derived melts (represented by

MORB) and melts derived from partial melting of carbonate-bearing sediments. Different degrees of partial melting (F = 0.5–5%; Fig. 3) are assumed for the carbonate-bearing sediments, following Equation 3 below (Wilson, 1989):

$$C_{X\text{-sediment melts}} = C_{X\text{-sediments}}/[(1-F) \times D_X + F] \tag{3}$$

where X is Mg, Zn, Sr or Pb, $C_{X\text{-sediments}}$ are the element concentrations for flysch sediment listed in Table 1, and D_X is the bulk partition coefficient from Grassi et al. (2012). The binary mixing simulation is calculated with Equations 4 and 5 (Fature & Mensing, 2004):

$$C_X = C_{X\text{-sediment melts}} \times \lambda + C_M \times (1-\lambda) \tag{4}$$

$$R_X = R_{X\text{-sediment melts}} \times \lambda \times (C_{X\text{-sediment melts}}/C_X) + R_M \times (1-\lambda) \times (C_M/C_X) \tag{5}$$

where C_X and R_X are element concentrations and isotopic ratios (δ^{26}Mg, δ^{66}Zn, ^{87}Sr/^{86}Sr, ^{206}Pb/^{204}Pb and ^{207}Pb/^{204}Pb) after mixing, respectively, λ is the proportion of melt derived from carbonate-bearing silicate sediments (flysch), C_M and R_M are element concentrations and isotopic ratios of MORB (data listed in Table S2), and $R_{X\text{-sediment melts}}$ are Mg, Zn, Sr, or Pb isotopic ratios of flysch sediments (Table 1). The δ^{66}Zn of flysch melts is assumed to be 0.32‰.

The results of binary mixing models based on Zn-Mg-Sr-Pb isotopes (Figs. 6 and 7) and trace elements (Fig. 3) indicate that about 10–40% (λ) melts of carbonate-bearing silicate sediment are required to generate these K-rich rocks. To acquire the input flux of recycled carbon (M_{input}), we use a Monte Carlo model as follows:

$$M_{\text{input}}\ (\text{Mt/yr}) = \frac{\text{Aa} \times \text{Bb} \times x \times y \times \lambda \times W_c \times W_{cc} \times \rho_c}{z} \tag{6}$$

where Aa (2300 km) and Bb (900 km) represent the length and width of the region (Fig. 1), x (1–50 km; Panza, Peccerillo, et al., 2007; Panza, Raykova, et al., 2007) is the thickness of the metasomatized lithospheric mantle, y (0–50%) is the percentage by area of the mantle that has been metasomatized, λ (10–40%) is the mass fraction of carbonate-bearing silicate sediment melts calculated above, W_c (5–50%) is the proportion of carbonates in the carbonate-bearing silicate sediments, W_{cc} (13%) is the mass fraction of C in carbonates (assumed to be dolomite), ρ_C (2.4× 10^{15} g/km^3) is the density of carbonate melts in the upper mantle (Ritter et al., 2020), and z (35 Ma) is the initial eruption age of the K-rich magmas (Prelević et al., 2013). The x, y, λ and W_c are taken randomly within the given reasonable range by the Monte Carlo model (see Data Set S2) and an average M_{input} value of about 8.1 Mt/yr is obtained (Fig. S2). A previous study speculated that there might be an unusually high carbon input flux into the Tethyan lithosphere (Plank & Manning, 2019), and our results suggest that a significant flux of carbon (~8.1 Mt/yr) from carbonate-bearing sediments has been recycled and stored in the lithospheric mantle beneath the circum-Mediterranean region, accounting for ~14% of the global carbon input flux from subducting sediments (57–60 Mt/yr; Clift, 2017; Dutkiewicz et al., 2018).

5.4 Implications for subaerial volcanic CO_2 outgassing

Globally, volcanic CO_2 emissions are commonly sourced from the mantle, the downgoing slabs, and/or the overlying crust above subduction zones (Werner et al., 2019 and references therein). Previous studies combined $\delta^{13}C$ with $^3He/^4He$ data to define the origin of arc volcanic CO_2. Carbon isotopic compositions show a striking difference between mantle (−6.0±2.0‰) and surface carbonate (~0‰) (Mason et al., 2017 and references therein). Mantle-derived CO_2 commonly has high $^3He/^4He$ (e.g., MORB=8±1 Ra; Graham, 2002), whereas low $^3He/^4He$ ratios are indicative of a source from the overlying crust. The $\delta^{13}C$-$^3He/^4He$ data of global arc gases reveal that few samples fall in the depleted mantle range, indicating that the mantle carbon is not the major source of volcanic CO_2 (Werner et al., 2019). In the circum-Mediterranean region, studies of volcanic CO_2 have been mostly carried out in Italy, where about 43% of the inorganic carbon in degassing CO_2 is derived from deep sources with heavy $\delta^{13}C$ (−3–1‰), implying mantle sources metasomatized by crustal components (Chiodini et al., 2000, 2004). Compiled gas data with high $^3He/^4He$ ratios (R/Ra up to 4.5; Frezzotti et al., 2009; Minissale, 2004) also imply metasomatized mantle sources rather than crustal assimilation. Therefore, subducted carbon is the main control of the isotopic composition and the amount of CO_2 at arc volcanoes. Combined with the carbon input flux estimated above for the circum-Mediterranean region (~8.1 Mt/yr), subducted carbon recycled into the lithospheric mantle should have a significant impact on volcanic CO_2 emissions.

To form a possible genetic connection between carbon input and output in the study area, we evaluate the flux of subaerial volcanic CO_2 emissions (M_{CO_2}) from the circum-Mediterranean lithospheric mantle using Equation 7:

$$M_{CO_2}\ (\mathrm{Mt/yr}) = M_{\mathrm{input}} \times \varepsilon \times 44/12 \tag{7}$$

Where ε is the rational recycling efficiency of subducted carbon to the atmosphere through subaerial volcanoes (60–90%; Oppenheimer et al., 2014; Plank & Manning, 2019). ε is taken randomly within the given range by the Monte Carlo model (see Data Set S2). The calculated average flux of M_{CO_2} around the circum-Mediterranean region is estimated to be about 22.4 Mt/yr (Fig. S2), which fits the observed value well (20.1±13.4 Mt/yr; 1sd; Table S3; Werner et al., 2019).

In the Mesozoic era, due to the warm climate and high sea levels flooding large expanses of the continents, there were large quantities of sedimentary carbonates when the seafloor of the shallow Tethyan ocean started to subduct (Lee et al., 2012; Plank & Manning, 2019). It is therefore reasonable to infer that abundant carbon was subducted at this time and stored in the lithospheric mantle (8.1 Mt/yr C), later to become the main source of the large subaerial volcanic CO_2 emissions (6.1 Mt/yr C) in the circum-Mediterranean region. Previous studies have suggested that carbonate sediments account for a large part of the subducting flux of carbon (>70%; Kelemen & Manning, 2015; Plank & Manning, 2019), and CO_2-He systematics for arcs worldwide indicate that carbonate sediments contribute about 50–80% CO_2 to volcanic outputs (Oppenheimer et al., 2014). Therefore, subducted carbon and its storage in the lithospheric mantle (Bragagni et al., 2022; Foley & Fischer, 2017) is likely to be an important source for global subaerial volcanic CO_2 emissions, as reflected in the circum-Mediterranean region that accounts for about 20% of the global total.

6 Conclusions

We report the first Mg and Zn isotopic data for the well-studied Cenozoic K-rich lavas around the circum-Mediterranean region (Spain, Italy, Serbia, Macedonia, and Turkey) to investigate the nature of recycled crustal components in their mantle sources, with special emphasis on understanding the source of the high levels of volcanic CO_2 emissions in this area. The K-rich lavas display variable and systematically lower $\delta^{26}Mg$ than that of the mantle, but have δ^{66}Zn values consistent with those of primitive mantle-derived melts. The distinctive Mg-Zn isotopic composition of the circum-Mediterranean K-rich lavas is therefore "decoupled" and can be best explained as a result of source metasomatism by carbonate-bearing silicate sediments. Together with Mg-Zn-Sr-Pb isotopic data, binary mixing simulations suggest that about 10–40% of carbonate-bearing silicate sediments have been recycled to the mantle sources of Mediterranean K-rich lavas. Considering that these K-rich lavas were derived from lithospheric mantle depths, we infer that a large amount of recycled carbon carried by subducted slabs is stored in the lithospheric mantle beneath this region. The carbon input flux from subducted slabs in the circum-Mediterranean region is estimated to be 8.1 Mt/yr. The subaerial volcanic CO_2 output flux is estimated to be 22.4 Mt/yr, which is in accordance with the observed value of 20.1 Mt/yr. Therefore, the recycled carbon stored in lithospheric mantle is likely a critical source for the enormous subaerial volcanic CO_2 emissions around the circum-Mediterranean region.

The combination of Mg and Zn isotopes proves useful to distinguish between the recycling of carbonate-bearing silicate sediments and carbonate-rich sediments, which are characterized by the diverging trends in Fig. 5. This may in turn indicate the tectonic environment in which subduction occurred, with collages of small plates and continental blocks such as in the Mediterranean producing Mg-Zn isotopic decoupling. In contrast, the isotopic coupling may indicate the involvement of carbonate-rich sediments, which may be characteristic for larger-scale convergent margins (C. Chen et al., 2021).

Acknowledgements

The authors are greatly grateful to editor Mark Dekkers and associate editor Philip Janney for their efficient handling and constructive comments. We also thank the anonymous reviewer for the careful comments and suggestions that helped to improve the manuscript. We thank Hai-Bo Ma and Meng-Lun Li for their help in the lab. This work was financially supported by the National Natural Science Foundation of China (41730214) and the National Key R&D Program of China (2019YFA0708400). D. P. and V. C. were supported by the Science Fund of the Republic of Serbia through project RECON TETHYS (7744807). V. C. was supported by the Serbian Academy of Sciences, Projects F9 and F17.

Data Availability Statement

Supporting Information includes Figs. S1–S2, Tables S1–S5 and Data Sets S1–S2. Mg and Zn isotopic data for the Cenozoic K-rich rocks in the circum-Mediterranean region and all supporting information related to this article are available at https://doi.org/10.5281/zenodo.7420592. Cite as: Shu et al. (2022).

References

Ammannati, E., Jacob, D. E., Avanzinelli, R., Foley, S. F., & Conticelli, S. (2016). Low Ni olivine in silica-undersaturated ultrapotassic igneous rocks as evidence for carbonate metasomatism in the mantle. Earth and Planetary Science Letters, 444, 64-74. https://doi.org/10.1016/j.epsl.2016.03.039

Avanzinelli, R., Casalini, M., Elliott, T., & Conticelli, S. (2018). Carbon fluxes from subducted carbonates revealed by uranium excess at Mount Vesuvius, Italy. Geology, 46(3), 259-262. https://doi.org/10.1130/g39766.1

Avanzinelli, R., Elliott, T., Tommasini, S., & Conticelli, S. (2007). Constraints on the Genesis of Potassium-rich Italian Volcanic Rocks from U/Th Disequilibrium. Journal of Petrology, 49(2), 195-223. https://doi.org/10.1093/petrology/egm076

Avanzinelli, R., Lustrino, M., Mattei, M., Melluso, L., & Conticelli, S. (2009). Potassic and ultrapotassic magmatism in the circum-Tyrrhenian region: Significance of carbonated pelitic vs. pelitic sediment recycling at destructive plate margins. Lithos, 113(1-2), 213-227. https://doi.org/10.1016/j.lithos.2009.03.029

Beunon, H., Mattielli, N., Doucet, L. S., Moine, B., & Debret, B. (2020). Mantle heterogeneity through Zn systematics in oceanic basalts: Evidence for a deep carbon cycling. Earth-Science Reviews, 205, 103174. https://doi.org/10.1016/j.earscirev.2020.103174

Bragagni, A., Mastroianni, F., Münker, C., Conticelli, S., & Avanzinelli, R. (2022). A carbon-rich lithospheric mantle as a source for the large CO_2 emissions of Etna volcano (Italy). Geology, 50(4), 486-490. https://doi.org/10.1130/g49510.1

Casalini, M., Avanzinelli, R., Tommasini, S., Natali, C., Bianchini, G., Prelević, D., Mattei, M., & Conticelli, S. (2022). Petrogenesis of Mediterranean lamproites and associated rocks: The role of overprinted metasomatic events in the post-collisional lithospheric upper mantle. Geological Society, London, Special Publications, 513(1), 271-296. https://doi.org/10.1144/sp513-2021-36

Chen, C., Forster, M. W., Foley, S. F., & Liu, Y. (2021). Massive carbon storage in convergent margins initiated by subduction of limestone. Nat Communications, 12(1), 4463. https://doi.org/10.1038/s41467-021-24750-0

Chen, H., Savage, P. S., Teng, F.-Z., Helz, R. T., & Moynier, F. (2013). Zinc isotope fractionation during magmatic differentiation and the isotopic composition of the bulk Earth. Earth and Planetary Science Letters, 369-370, 34-42. https://doi.org/10.1016/j.epsl.2013.02.037

Chiodini, G., Cardellini, C., Amato, A., Boschi, E., Caliro, S., Frondini, F., & Ventura, G. (2004). Carbon dioxide Earth degassing and seismogenesis in central and southern Italy. Geophysical Research Letters, 31(7). https://doi.org/10.1029/2004gl019480

Chiodini, G., Frondini, F., Cardellini, C., Parello, F., & Peruzzi, L. (2000). Rate of diffuse carbon dioxide Earth degassing estimated from carbon balance of regional aquifers: The case of central Apennine, Italy. Journal of Geophysical Research: Solid Earth, 105(B4), 8423-8434. https://doi.org/10.1029/1999jb900355

Clift, P. D. (2017). A revised budget for Cenozoic sedimentary carbon subduction. Reviews of Geophysics, 55(1), 97-125. https://doi.org/10.1002/2016rg000531

Conceição, R., & Green, D. (2004). Derivation of potassic (shoshonitic) magmas by decompression melting of phlogopite+pargasite lherzolite. Lithos, 72(3-4), 209-229. https://doi.org/10.1016/j.lithos.2003.09.003

Conticelli, S., Avanzinelli, R., Ammannati, E., & Casalini, M. (2015). The role of carbon from recycled sediments in the origin of ultrapotassic igneous rocks in the Central Mediterranean. Lithos, 232, 174-196. https://doi.org/10.1016/j.lithos.2015.07.002

Conticelli, S., Avanzinelli, R., Poli, G., Braschi, E., & Giordano, G. (2013). Shift from lamproite-like to leucititic rocks: Sr–Nd–Pb isotope data from the Monte Cimino volcanic complex vs. the Vico stratovolcano, Central Italy. Chemical Geology, 353, 246-266. https://doi.org/10.1016/j.chemgeo.2012.10.018

Conticelli, S., D'Antonio, M., Pinarelli, L., & Civetta, L. (2002). Source contamination and mantle heterogeneity in the genesis of Italian potassic and ultrapotassic volcanic rocks: Sr–Nd–Pb isotope data from Roman Province and Southern Tuscany. Mineralogy and Petrology, 74(2), 189-222. https://doi.org/10.1007/s007100200004

Conticelli, S., Guarnieri, L., Farinelli, A., Mattei, M., Avanzinelli, R., Bianchini, G., Boari, E., Tommasini, S., Tiepolo, M., Prelević, D., & Venturelli, G. (2009). Trace elements and Sr–Nd–Pb isotopes of K-rich, shoshonitic, and calc-alkaline magmatism of the Western Mediterranean Region: Genesis of ultrapotassic to calc-alkaline magmatic associations in a post-collisional geodynamic setting. Lithos, 107(1-2), 68-92. https://doi.org/10.1016/j.lithos.2008.07.016

Conticelli, S., & Peccerillo, A. (1992). Petrology and geochemistry of potassic and ultrapotassic volcanism in central Italy: petrogenesis and inferences on the evolution of the mantle sources. Lithos, 28(3-6), 221-240. https://doi.org/10.1016/0024-4937(92)90008-M.

Dasgupta, R., & Hirschmann, M. M. (2010). The deep carbon cycle and melting in Earth's interior. Earth and Planetary Science Letters, 298(1-2), 1-13. https://doi.org/10.1016/j.epsl.2010.06.039

Dutkiewicz, A., Müller, R. D., Cannon, J., Vaughan, S., & Zahirovic, S. (2018). Sequestration and subduction of deep-sea carbonate in the global ocean since the Early Cretaceous. Geology, 47(1), 91-94. https://doi.org/10.1130/g45424.1

Elgabry, M. N. , Panza, G. F. , Badawy, A. A. , & Korrat, I. M. . (2013). Imaging a relic of complex tectonics: the lithosphere-asthenosphere structure in the eastern mediterranean. Terra Nova, 25(2), 102-109. https://doi.org/10.1111/ter.12011

Faure, G., & Mensing, T. M. (2004). Isotopes: Principles and Applications, third ed. John Wiley, New York

Foley, S. (1992). Petrological characterization of the source components of potassic magmas: geochemical and experimental constraints. Lithos, 28(3-6), 187-204

Foley, S. F., & Fischer, T. P. (2017). An essential role for continental rifts and lithosphere in the deep carbon cycle. Nature

Geoscience, 10(12), 897-902. https://doi.org/10.1038/s41561-017-0002-7

Förster, M. W. , Prelević, D. , Buhre, S. , Mertz-Kraus, R. , & Foley, S. F. (2019). An experimental study of the role of partial melts of sediments versus mantle melts in the sources of potassic magmatism. Journal of Asian Earth Sciences, 177, 76-88. https://doi.org/10.1016/j.jseaes.2019.03.014

Förster, M. W., Prelević, D., Schmück, H. R., Buhre, S., Veter, M., Mertz-Kraus, R., Foley, S. F., & Jacob, D. E. (2017). Melting and dynamic metasomatism of mixed harzburgite + glimmerite mantle source: Implications for the genesis of orogenic potassic magmas. Chemical Geology, 455, 182-191. https://doi.org/10.1016/j.chemgeo.2016.08.037

Frezzotti, M. L., Peccerillo, A., & Panza, G. (2009). Carbonate metasomatism and CO_2 lithosphere–asthenosphere degassing beneath the Western Mediterranean: An integrated model arising from petrological and geophysical data. Chemical Geology, 262(1-2), 108-120. https://doi.org/10.1016/j.chemgeo.2009.02.015

Fritschle, T., Prelević, D., Foley, S. F., & Jacob, D. E. (2013). Petrological characterization of the mantle source of Mediterranean lamproites: Indications from major and trace elements of phlogopite. Chemical Geology, 353, 267-279. https://doi.org/10.1016/j.chemgeo.2012.09.006

Galy, A., Yoffe, O., Janney, P. E., Williams, R. W., Cloquet, C., Alard, O., Halicz, L., Wadhwa, M., Hutcheon, I. D., Ramon, E., & Carignan, J. (2003). Magnesium isotope heterogeneity of the isotopic standard SRM980 and new reference materials for magnesium-isotope-ratio measurements. Journal of Analytical Atomic Spectrometry, 18(11), 1352. https://doi.org/10.1039/b309273a

Gao, T., Ke, S., Li, R., Meng, X. n., He, Y., Liu, C., Wang, Y., Li, Z., & Zhu, J. M. (2019). High-precision magnesium isotope analysis of geological and environmental reference materials by multiple-collector inductively coupled plasma mass spectrometry. Rapid Communications in Mass Spectrometry, 33(8), 767-777. https://doi.org/10.1002/rcm.8376

Graham, D. W. (2002). Noble Gas Isotope Geochemistry of Mid-Ocean Ridge and Ocean Island Basalts: Characterization of Mantle Source Reservoirs. Reviews in Mineralogy and Geochemistry, 47(1), 247-317. https://doi.org/10.2138/rmg.2002.47.8

Grassi, D., Schmidt, M. W., & Günther, D. (2012). Element partitioning during carbonated pelite melting at 8, 13 and 22GPa and the sediment signature in the EM mantle components. Earth and Planetary Science Letters, 327-328, 84-96. https://doi.org/10.1016/j.epsl.2012.01.023

Hart, S. R. (1984). A large-scale isotope anomaly in the Southern Hemisphere mantle. Nature, 309(5971), 753-757. https://doi.org/10.1038/309753a0

Huang, J., Li, S.-G., Xiao, Y., Ke, S., Li, W.-Y., & Tian, Y. (2015). Origin of low δ^{26}Mg Cenozoic basalts from South China Block and their geodynamic implications. Geochimica et Cosmochimica Acta, 164, 298-317. https://doi.org/10.1016/j.gca.2015.04.054

Huang, J., Zhang, X.-C., Chen, S., Tang, L., Wörner, G., Yu, H., & Huang, F. (2018). Zinc isotopic systematics of Kamchatka-Aleutian arc magmas controlled by mantle melting. Geochimica et Cosmochimica Acta, 238, 85-101. https://doi.org/10.1016/j.gca.2018.07.012

Huang, K.-J., Teng, F.-Z., Elsenouy, A., Li, W.-Y., & Bao, Z.-Y. (2013). Magnesium isotopic variations in loess: Origins and implications. Earth and Planetary Science Letters, 374, 60-70. https://doi.org/10.1016/j.epsl.2013.05.010

Inglis, E. C., Debret, B., Burton, K. W., Millet, M.-A., Pons, M.-L., Dale, C. W., Bouilhol, P., Cooper, M., Nowell, G. M., McCoy-West, A. J., & Williams, H. M. (2017). The behavior of iron and zinc stable isotopes accompanying the subduction of mafic oceanic crust: A case study from Western Alpine ophiolites. Geochemistry, Geophysics, Geosystems, 18(7), 2562-2579. https://doi.org/10.1002/2016gc006735

Jin, Q.-Z., Huang, J., Liu, S.-C., & Huang, F. (2020). Magnesium and zinc isotope evidence for recycled sediments and oceanic crust in the mantle sources of continental basalts from eastern China. Lithos, 370-371. https://doi.org/10.1016/j.lithos.2020.105627

Kelemen, P. B., & Manning, C. E. (2015). Reevaluating carbon fluxes in subduction zones, what goes down, mostly comes up. Proc Natl Acad Sci U S A, 112(30), E3997-4006. https://doi.org/10.1073/pnas.1507889112

Kirchenbaur, M., & Münker, C. (2015). The behaviour of the extended HFSE group (Nb, Ta, Zr, Hf, W, Mo) during the petrogenesis of mafic K-rich lavas: The Eastern Mediterranean case. Geochimica et Cosmochimica Acta, 165, 178-199. https://doi.org/10.1016/j.gca.2015.05.030

Le Bas M. J., Le Maitre R. W., Streckeisen A., & Zanettin B. A. (1986). Chemical classification of volcanic rocks based on the total alkali-silica diagram. Journal of Petrology, 27(3), 745-750. https://doi.org/10.1093/petrology/27.3.745

Le Roux, V., Lee, C.-T. A., & Turner, S. J. (2010). Zn/Fe systematics in mafic and ultramafic systems: Implications for detecting major element heterogeneities in the Earth's mantle. Geochimica et Cosmochimica Acta, 74(9), 2779-2796. https://doi.org/10.1016/j.gca.2010.02.004

Lee, C.-T. A., Jiang, H., Dasgupta, R., & Torres, M. (2019). A Framework for Understanding Whole-Earth Carbon Cycling. In B. N. Orcutt, I. Daniel, & R. Dasgupta (Eds.), Deep Carbon: Past to Present (pp. 313-357). Cambridge University Press

Lee, C.-T. A., Shen, B., Slotnick, B. S., Liao, K., Dickens, G. R., Yokoyama, Y., Lenardic, A., Dasgupta, R., Jellinek, M., Lackey, J. S., Schneider, T., & Tice, M. M. (2012). Continental arc-island arc fluctuations, growth of crustal carbonates, and long-term climate change. Geosphere, 9(1), 21-36. https://doi.org/10.1130/ges00822.1

Li, S.-G., Yang, W., Ke, S., Meng, X., Tian, H., Xu, L., He, Y., Huang, J., Wang, X.-C., Xia, Q., Sun, W., Yang, X., Ren, Z.-Y., Wei, H., Liu, Y., Meng, F., & Yan, J. (2017). Deep carbon cycles constrained by a large-scale mantle Mg isotope anomaly in eastern China. National Science Review, 4(1), 111-120. https://doi.org/10.1093/nsr/nww070

Li, W.-Y., Teng, F.-Z., Ke, S., Rudnick, R. L., Gao, S., Wu, F.-Y., & Chappell, B. W. (2010). Heterogeneous magnesium isotopic composition of the upper continental crust. Geochimica et Cosmochimica Acta, 74(23), 6867-6884. https://doi.org/10.1016/j.gca.2010.08.030

Little, S. H., Vance, D., McManus, J., & Severmann, S. (2016). Key role of continental margin sediments in the oceanic mass balance of Zn and Zn isotopes. Geology, 44(3), 207-210. https://doi.org/10.1130/g37493.1

Liu, S.-A., & Li, S.-G. (2019) Tracing the deep carbon cycle using metal stable isotopes: Opportunities and challenges. Engineering, 5, 448-457. https://doi.org/10.1016/j.eng.2019.03.007

Liu, S.-A., Liu, P.-P., Lv, Y., Wang, Z.-Z., & Dai, J.-G. (2019). Cu and Zn isotope fractionation during oceanic alteration: Implications for Oceanic Cu and Zn cycles. Geochimica et Cosmochimica Acta, 257, 191-205. https://doi.org/10.1016/j.gca.2019.04.026

Liu, S.-A., Qu, Y.-R., Wang, Z.-Z., Li, M.-L., Yang, C., & Li, S.-G. (2022). The fate of subducting carbon tracked by Mg and Zn isotopes: A review and new perspectives. Earth-Science Reviews, 228. https://doi.org/10.1016/j.earscirev.2022.104010

Liu, S.-A., Wang, Z.-Z., Li, S.-G., Huang, J., & Yang, W. (2016). Zinc isotope evidence for a large-scale carbonated mantle beneath eastern China. Earth and Planetary Science Letters, 444, 169-178. https://doi.org/10.1016/j.epsl.2016.03.051

Liu, S.-A., Wu, H., Shen, S.-z., Jiang, G., Zhang, S., Lv, Y., Zhang, H., & Li, S. (2017). Zinc isotope evidence for intensive magmatism immediately before the end-Permian mass extinction. Geology, 45(4), 343-346. https://doi.org/10.1130/g38644.1

Liu, X. M., Kah, L. C., Knoll, A. H., Cui, H., Kaufman, A. J., Shahar, A., & Hazen, R. M. (2016). Tracing Earth's O_2 evolution using Zn/Fe ratios in marine carbonates. Geochemical Perspectives Letters, 2(1), 24-34. https://doi.org/10.7185/geochemlet.1603

Lustrino, M., Duggen, S., & Rosenberg, C. L. (2011). The Central-Western Mediterranean: Anomalous igneous activity in an anomalous collisional tectonic setting. Earth-Science Reviews, 104(1-3), 1-40. https://doi.org/10.1016/j.earscirev.2010.08.002

Lustrino, M., & Wilson, M. (2007). The circum-Mediterranean anorogenic Cenozoic igneous province. Earth-Science Reviews, 81(1-2), 1-65. https://doi.org/10.1016/j.earscirev.2006.09.002

Maréchal, C. N., Nicolas, E., Douchet, C., & Albarède, F. (2000). Abundance of zinc isotopes as a marine biogeochemical tracer. Geochemistry, Geophysics, Geosystems, 1(5). https://doi.org/10.1029/1999gc000029

Maréchal, C. N., Télouk, P., & Albarède, F. (1999). Precise analysis of copper and zinc isotopic compositions by plasma-source mass spectrometry. Chemical Geology, 156(1–4), 251-273. https://doi.org/10.1016/S0009-2541(98)00191-0

Mason, E., Edmonds, M., & Turchyn, A. V. (2017). Remobilization of crustal carbon may dominate volcanic arc emissions. Science (New York, N.Y.), 357(6348), 290-294. https://doi.org/10.1126/science.aan5049

McCoy-West, A. J., Fitton, J. G., Pons, M.-L., Inglis, E. C., & Williams, H. M. (2018). The Fe and Zn isotope composition of deep mantle source regions: Insights from Baffin Island picrites. Geochimica et Cosmochimica Acta, 238, 542-562. https://doi.org/10.1016/j.gca.2018.07.021

McDonough, W., F., & Sun, S. s. (1995). The composition of the Earth. Chemical Geology, 120(3-4), 223-253. https://doi.org/10.1016/0009-2541(94)00140-4.

Minissale, A. (2004). Origin, transport and discharge of CO_2 in central Italy. Earth-Science Reviews, 66(1-2), 89-141. https://doi.org/10.1016/j.earscirev.2003.09.001

Moynier, F., Vance, D., Fujii, T., & Savage, P. (2017). The Isotope Geochemistry of Zinc and Copper. Reviews in Mineralogy and Geochemistry, 82(1), 543-600. https://doi.org/10.2138/rmg.2017.82.13

Oppenheimer, C., Fischer, T. P., & Scaillet, B. (2014). Volcanic Degassing: Process and Impact. In H. D. Holland & K. K. Turekian (Eds.), Treatise on Geochemistry (Second Edition) (pp. 111-179). Elsevier. https://doi.org/https://doi.org/10.1016/B978-0-08-095975-7.00304-1

Panza, G. F., Peccerillo, A., Aoudia, A., & Farina, B. (2007). Geophysical and petrological modelling of the structure and composition of the crust and upper mantle in complex geodynamic settings: The Tyrrhenian Sea and surroundings. Earth-Science

Reviews, 80(1-2), 1-46. https://doi.org/10.1016/j.earscirev.2006.08.004

Panza, G. F., Raykova, R. B., Carminati, E., & Doglioni, C. (2007). Upper mantle flow in the western Mediterranean. Earth and Planetary Science Letters, 257(1-2), 200-214. https://doi.org/10.1016/j.epsl.2007.02.032

Pichat, S., Douchet, C., & Albarède, F. (2003). Zinc isotope variations in deep-sea carbonates from the eastern equatorial Pacific over the last 175 ka. Earth and Planetary Science Letters, 210(1-2), 167-178. https://doi.org/10.1016/s0012-821x(03)00106-7

Piromallo, C., & Morelli, A. (2003). Pwave tomography of the mantle under the Alpine-Mediterranean area. Journal of Geophysical Research: Solid Earth, 108(B2). https://doi.org/10.1029/2002jb001757

Plank, T. (2014). The Chemical Composition of Subducting Sediments. In H. D. Holland & K. K. Turekian (Eds.), Treatise on Geochemistry (Second Edition) (pp. 607-629). Elsevier. https://doi.org/https://doi.org/10.1016/B978-0-08-095975-7.00319-3

Plank, T., & Manning, C. E. (2019). Subducting carbon. Nature, 574(7778), 343-352. https://doi.org/10.1038/s41586-019-1643-z

Prelević, D., Akal, C., Foley, S. F., Romer, R. L., Stracke, A., & Van Den Bogaard, P. (2012). Ultrapotassic Mafic Rocks as Geochemical Proxies for Post-collisional Dynamics of Orogenic Lithospheric Mantle: the Case of Southwestern Anatolia, Turkey. Journal of Petrology, 53(5), 1019-1055. https://doi.org/10.1093/petrology/egs008

Prelević, D., Akal, C., Romer, R. L., Mertz-Kraus, R., & Helvac, C. (2015). Magmatic Response to Slab Tearing: Constraints from the Afyon Alkaline Volcanic Complex, Western Turkey. Journal of Petrology, 56(3), 527-562. https://doi.org/10.1093/petrology/egv008

Prelević, D., & Foley, S. F. (2007). Accretion of arc-oceanic lithospheric mantle in the Mediterranean: Evidence from extremely high-Mg olivines and Cr-rich spinel inclusions in lamproites. Earth and Planetary Science Letters, 256(1-2), 120-135. https://doi.org/10.1016/j.epsl.2007.01.018

Prelević, D., Foley, S. F., Cvetković, V., &Romer, R. L. (2004). The analcime problem and its impact on the geochemistry of ultrapotassic roks from Serbia. Mineralogical Magazine 68, 633-648. https://doi.org/10.1180/0026461046840209

Prelević, D., Foley, S. F., Romer, R., & Conticelli, S. (2008). Mediterranean Tertiary lamproites derived from multiple source components in postcollisional geodynamics. Geochimica et Cosmochimica Acta, 72(8), 2125-2156. https://doi.org/10.1016/j.gca.2008.01.029

Prelević, D., Foley, S. F., Romer, R. L., CvetkoviĆ, V., & Downes, H. (2005). Tertiary Ultrapotassic Volcanism in Serbia: Constraints on Petrogenesis and Mantle Source Characteristics. Journal of Petrology, 46(7), 1443-1487. https://doi.org/10.1093/petrology/egi022

Prelević, D., Jacob, D. E., & Foley, S. F. (2013). Recycling plus: A new recipe for the formation of Alpine–Himalayan orogenic mantle lithosphere. Earth and Planetary Science Letters, 362, 187-197. https://doi.org/10.1016/j.epsl.2012.11.035

Qu, Y.-R., Liu, S.-A., Wu, H., Li, M.-L., & Tian, H.-C. (2022). Tracing carbonate dissolution in subducting sediments by zinc and magnesium isotopes. Geochimica et Cosmochimica Acta, 319, 56-72. https://doi.org/10.1016/j.gca.2021.12.020

Richter, F. M., Watson, E. B., Mendybaev, R. A., Teng, F.-Z., & Janney, P. E. (2008). Magnesium isotope fractionation in silicate melts by chemical and thermal diffusion. Geochimica et Cosmochimica Acta, 72(1), 206-220. https://doi.org/10.1016/j.gca.2007.10.016

Ritter, X., Sanchez-Valle, C., Sator, N., Desmaele, E., Guignot, N., King, A., Kupenko, I., Berndt, J., & Guillot, B. (2020). Density of hydrous carbonate melts under pressure, compressibility of volatiles and implications for carbonate melt mobility in the upper mantle. Earth and Planetary Science Letters, 533, 116043. https://doi.org/10.1016/j.epsl.2019.116043

Robertson A. H. F. (2002). Overview of the genesis and emplacement of Mesozoic ophiolites in the Eastern Mediterranean Tethyan region. Lithos, 65, 1-67. https://doi.org/10.1016/S0024-4937(02)00160-3

Shu, Z.-T., Liu, S.-A., Prelević, D., Wang, Y., Foley, S. F., Cvetković, V., & Li, S. (2022). Mg and Zn isotopic data and supplementary information for "Recycled carbonate-bearing silicate sediments in the sources of circum-Mediterranean K-rich lavas: Evidence from Mg-Zn isotopic decoupling". https://doi.org/10.5281/zenodo.7420592

Sokol, K., Prelević, D., Romer, R. L., Božović, M., van den Bogaard, P., Stefanova, E., Kostić, B., & Čokulov, N. (2020). Cretaceous ultrapotassic magmatism from the Sava-Vardar Zone of the Balkans. Lithos, 354-355, 105268. https://doi.org/10.1016/j.lithos.2019.105268

Sossi, P. A., Halverson, G. P., Nebel, O., & Eggins, S. M. (2015). Combined Separation of Cu, Fe and Zn from Rock Matrices and Improved Analytical Protocols for Stable Isotope Determination. Geostandards and Geoanalytical Research, 39(2), 129-149. https://doi.org/10.1111/j.1751-908X.2014.00298.x

Sossi, P. A., Nebel, O., O'Neill, H. S. C., & Moynier, F. (2018). Zinc isotope composition of the Earth and its behaviour during planetary accretion. Chemical Geology, 477, 73-84. https://doi.org/10.1016/j.chemgeo.2017.12.006

Sun, S. S., & McDonough, W. F. (1989). Chemical and isotopic systematics of oceanic basalts: implications for mantle composition and processes. Geological Society, London, Special Publications, 42(1), 313-345. https://doi.org/10.1144/gsl.sp.1989.042.01.19

Sun, Y., Teng, F.-Z., Pang, K.-N., Ying, J.-F., & Kuehner, S. (2021). Multistage mantle metasomatism deciphered by Mg–Sr–Nd–Pb isotopes in the Leucite Hills lamproites. Contributions to Mineralogy and Petrology, 176(6). https://doi.org/10.1007/s00410-021-01801-9

Sweere, T. C., Dickson, A. J., Jenkyns, H. C., Porcelli, D., Elrick, M., van den Boorn, S. H. J. M., & Henderson, G. M. (2018). Isotopic evidence for changes in the zinc cycle during Oceanic Anoxic Event 2 (Late Cretaceous). Geology, 46(5), 463-466. https://doi.org/10.1130/g40226.1

Tappe, S., Foley, S. F., Jenner, G. A., Heaman, L. M., Kjarsgaard, B. A., Romer, R. L., Stracke, A., Joyce, N., & Hoefs, J. (2006). Genesis of Ultramafic Lamprophyres and Carbonatites at Aillik Bay, Labrador: a Consequence of Incipient Lithospheric Thinning beneath the North Atlantic Craton. Journal of Petrology, 47(7), 1261-1315. https://doi.org/10.1093/petrology/egl008

Teng, F.-Z. (2017). Magnesium Isotope Geochemistry. Reviews in Mineralogy and Geochemistry, 82(1), 219-287. https://doi.org/10.2138/rmg.2017.82.7

Teng, F.-Z., Li, W.-Y., Ke, S., Marty, B., Dauphas, N., Huang, S., Wu, F.-Y., & Pourmand, A. (2010). Magnesium isotopic composition of the Earth and chondrites. Geochimica et Cosmochimica Acta, 74(14), 4150-4166. https://doi.org/10.1016/j.gca.2010.04.019

Teng, F.-Z., Wadhwa, M., & Helz, R. T. (2007). Investigation of magnesium isotope fractionation during basalt differentiation: Implications for a chondritic composition of the terrestrial mantle. Earth and Planetary Science Letters, 261(1-2), 84-92. https://doi.org/10.1016/j.epsl.2007.06.004

Tommasini, S., Avanzinelli, R., & Conticelli, S. (2011). The Th/La and Sm/La conundrum of the Tethyan realm lamproites. Earth and Planetary Science Letters, 301(3-4), 469-478. https://doi.org/10.1016/j.epsl.2010.11.023

Vance, D., Little, S. H., Archer, C., Cameron, V., Andersen, M. B., Rijkenberg, M. J. A., & Lyons, T. W. (2016). The oceanic budgets of nickel and zinc isotopes: the importance of sulfidic environments as illustrated by the Black Sea. Philos Trans A Math Phys Eng Sci, 374(2081). https://doi.org/10.1098/rsta.2015.0294

Wang, S.-J., Teng, F.-Z., Li, S.-G., & Hong, J.-A. (2014). Magnesium isotopic systematics of mafic rocks during continental subduction. Geochimica et Cosmochimica Acta, 143, 34-48. https://doi.org/10.1016/j.gca.2014.03.029

Wang, S.-J., Teng, F.-Z., Rudnick, R. L., & Li, S.-G. (2015). The behavior of magnesium isotopes in low-grade metamorphosed mudrocks. Geochimica et Cosmochimica Acta, 165, 435-448. https://doi.org/10.1016/j.gca.2015.06.019

Wang, Y., Foley, S. F., & Prelević, D. (2017). Potassium-rich magmatism from a phlogopite-free source. Geology, 45(5), 467-470. https://doi.org/10.1130/g38691.1

Wang, Y., Foley, S. F., Buhre, S., Soldner, J., & Xu, Y. (2021). Origin of potassic postcollisional volcanic rocks in young, shallow, blueschist-rich lithosphere. Science Advances, 7(29), eabc0291. https://doi.org/10.1126/sciadv.abc0291

Wang, Z. X., Liu, S.-A., Li, S., Liu, D., & Liu, J. (2022). Linking deep CO_2 outgassing to cratonic destruction. Natl Sci Rev, 9(6), nwac001. https://doi.org/10.1093/nsr/nwac001

Wang, Z.-Z., & Liu, S.-A. (2021). Evolution of intraplate alkaline to tholeiitic basalts via interaction between carbonated melt and lithospheric mantle. Journal of Petrology, 62(4). https://doi.org/10.1093/petrology/egab025

Wang, Z.-Z., Liu, S.-A., Chen, L.-H., Li, S.-G., & Zeng, G. (2018). Compositional transition in natural alkaline lavas through silica-undersaturated melt–lithosphere interaction. Geology, 46(9), 771-774. https://doi.org/10.1130/g45145.1

Wang, Z.-Z., Liu, S.-A., Liu, J., Huang, J., Xiao, Y., Chu, Z.-Y., Zhao, X.-M., & Tang, L. (2017). Zinc isotope fractionation during mantle melting and constraints on the Zn isotope composition of Earth's upper mantle. Geochimica et Cosmochimica Acta, 198, 151-167. https://doi.org/10.1016/j.gca.2016.11.014

Werner, C., Fischer, T. P., Aiuppa, A., Edmonds, M., Cardellini, C., Carn, S., Chiodini, G., Cottrell, E., Burton, M., Shinohara, H., & Allard, P. (2019). Carbon Dioxide Emissions from Subaerial Volcanic Regions: Two Decades in Review. In B. N. Orcutt, I. Daniel, & R. Dasgupta (Eds.), Deep Carbon: Past to Present (pp. 188-236). Cambridge University Press.

Wilson, M. (1989). Igneous Petrogenesis: A Global Tectonic Approach. Unwin Hyman, London

Wimpenny, J., Yin, Q.-Z., Tollstrup, D., Xie, L.-W., & Sun, J. (2014). Using Mg isotope ratios to trace Cenozoic weathering changes: A case study from the Chinese Loess Plateau. Chemical Geology, 376, 31-43. https://doi.org/10.1016/j.chemgeo.2014.03.008

Wood, B. J. (1993). Carbon in the core. Earth & Planetary Science Letters, 117(3–4), 593-607. https://doi.org/10.1016/0012-821X(93)90105-I

Yang, C., Liu, S.-A., Zhang, L., Wang, Z.-Z., Liu, P.-P., & Li, S.-G. (2021). Zinc isotope fractionation between Cr-spinel and olivine and its implications for chromite crystallization during magma differentiation. Geochimica et Cosmochimica Acta. https://doi.org/10.1016/j.gca.2021.08.005

Yang, W., Teng, F.-Z., Zhang, H.-F., & Li, S.-G. (2012). Magnesium isotopic systematics of continental basalts from the North China craton: Implications for tracing subducted carbonate in the mantle. Chemical Geology, 328, 185-194. https://doi.org/10.1016/j.chemgeo.2012.05.018

References From the Supporting Information

Hofmann, A. W. (1988). Chemical differentiation of the Earth: The relationship between mantle, continental crust, and oceanic crust. Earth & Planetary Science Letters, 90(3), 297-314. https://doi.org/10.1016/0012-821X(88)90132-X

Huang, J., & Xiao, Y. (2016). Mg-Sr isotopes of low-δ^{26}Mg basalts tracing recycled carbonate species: Implication for the initial melting depth of the carbonated mantle in Eastern China. International Geology Review, 58(11), 1350-1362. https://doi.org/10.1080/00206814.2016.1157709

Ray, J. S., Veizer, J., & Davis, W. J. (2003). C, O, Sr and Pb isotope systematics of carbonate sequences of the Vindhyan Supergroup, India: age, diagenesis, correlations and implications for global events. Precambrian Research, 121(1-2), 103-140. https://doi.org/10.1016/s0301-9268(02)00223-1

Rudnick, R. L., & Gao, S. (2014). Composition of the Continental Crust. in Treatise on Geochemistry (Second Edition), edited by H. D. Holland and K. K. Turekian, pp. 1-51, Elsevier, Oxford, https://doi.org/10.1016/B978-0-08-095975-7.00301-6

Taylor, S. R., & McLennan, S. M. (1985). The continental crust: its composition and evolution. The Journal of Geology, 94(4), 57-72. https://doi.org/10.1086/629067

Workman, R. K., & Hart, S. R. (2005). Major and trace element composition of the depleted MORB mantle (DMM). Earth & Planetary Science Letters, 231(1-2), 53-72. https://doi.org/10.1016/j.epsl.2004.12.005

致　谢

本论文集《李曙光院士论文选集（卷二）》经过半年多的整理、归纳，最终汇编成册。论文集编撰中得到许多老师和同学的支持。大家解决了诸多问题，如论文亮点介绍的概括、图件的重绘、论文的校对和修订等。在此，我们衷心感谢研究生韩颖、耿明婧、王照雪、杨春、李孟伦、舒梓坦、马海波、蔡荣华、刘春阳、孙琬钰、黄子瑄、范雪松、熊芷黎、田晓萍、杨鸿励、汪子腾、李瑞宁、王汝正以及中国科学技术大学和中国科学院地质与地球物理研究所的相关老师和同学的帮助和支持。千言万语汇成一句话：谢谢你们。